CHEMISTRY

Molecules, Matter, and Change

Fourth Edition

CHEMISTRY
Molecules, Matter, and Change

Loretta Jones
University of Northern Colorado

Peter Atkins
Oxford University

W.H. Freeman and Company
New York

Publisher: *Michelle Russel Julet*
Acquisitions Editor: *Jessica Fiorillo*
Marketing Manager: *Kimberly Manzi*
Development Editor: *David Chelton*
Project Editor: *Georgia Lee Hadler*
Copy Editor: *Jodi Simpson*
Cover and Text Designer: *Blake Logan*
Cover Image: *Chapel, John Maher/The Stock Market*
Illustrations: *Peter Atkins with Network Graphics*
Photo Researcher: *Kathy Bendo*
Production and Illustration Coordinator: *Susan Wein*
Composition: *Black Dot Group*
Manufacturing: *RR Donnelley & Sons Company*
Media and Supplements Editors: *Patrick Shriner, Matthew Fitzpatrick*

Library of Congress Cataloging-in-Publication Data

Jones, Loretta.
    Chemistry: molecules, matter, and change.—4th ed./Loretta Jones, Peter Atkins
        p.   cm.
    Includes bibliographical references and index.
    ISBN 0-7167-3254-8
    1. Chemistry.   I. Atkins, P. W. (Peter William).   II. Title.
    QD31.2.A75   1999
    540—dc21

Table of Contents

Why does this compound turn white when heated?

3 Chemical Reactions   89

How do we describe chemical reactions?

4 Chemistry's Accounting: Reaction Stoichiometry   137

5 The Properties of Gases   177

Why is the atmosphere thinner at higher altitudes?

6 Thermochemistry: The Fire Within   219

What causes the colors of fireworks?

9 Molecular Structure   363

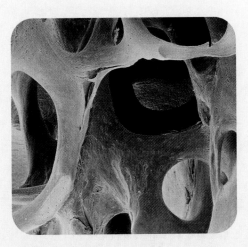

What makes bone tissue strong?

10 Liquids and Solids 421

11 Carbon-Based Materials 471

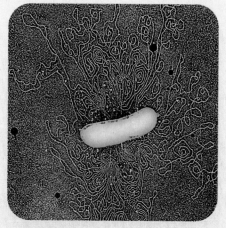

What do we know about the chemistry of genetics?

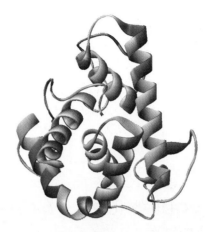

How do enzymes work?

CONNECTION 3: *SPORT DRINKS* 566

12 The Properties of Solutions 523

13 The Rates of Reactions 569

14 Chemical Equilibrium 619

15 Acids and Bases 659

What causes acid rain?

How can spontaneous reactions be reversed?

18 Electrochemistry 791

How can we protect metal objects from corrosion?

19 The Elements: The First Four Main Groups 843

20 The Elements: The Last Four Main Groups 885

What makes some compounds and elements useful as rocket fuels?

21 The *d* Block: Metals in Transition 917

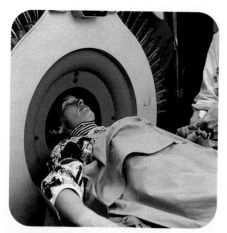

How is nuclear medicine helpful?

22 Nuclear Chemistry   959

About the Authors

Loretta Jones is an award-winning chemical educator who is known for her innovative chemistry curricula. After 13 years teaching general chemistry at the University of Illinois at Urbana-Champaign, she became professor of Chemistry at the University of Northern Colorado. Her work has contributed to an influential body of multimedia instructional materials, including *Exploring Chemistry* (with S. G. Smith and S. D. Gammon). She is active in governance in the American Chemical Society, a Fellow of the American Association for the Advancement of Science, and a recipient of the Scholar of the Year Award from the College of Arts and Sciences at the University of Northern Colorado.

Peter Atkins is professor of chemistry at Oxford University, a fellow of Lincoln College, and the author of more than forty books for students and a general audience. His *Physical Chemistry* broke new ground by stressing understanding over memorization. Now in its sixth edition, it remains the leading text in its field. His other books include *Inorganic Chemistry* (with Duward Shriver), and, for the Scientific American Library, *Molecules, The Second Law,* and *Atoms, Electrons, and Change.* A frequent lecturer in the United States and throughout the world, he has held visiting professorships in Israel, Japan, China, and New Zealand. He serves on the International Union of Pure and Applied Chemistry's committees on physical chemistry and chemical education.

Preface

This book is intended for the full-year general chemistry course. Our underlying philosophy is that a textbook must connect observations that are familiar to students with the concepts of chemistry. We have to connect the world of appearances to the underworld of atoms and their symbolic representations. However, gaining such insight is not enough. Chemistry is a *quantitative* subject, and we know how important it is to guide students through **problem-solving** strategies so that they can compile data, formulate solutions, and arrive at accurate answers. With our textbook, we hope to communicate to students that learning chemistry isn't just about memorizing formulas and getting correct answers. We want students to **visualize** like chemists, **think** like chemists, and **work** like chemists. And we want them to understand and be inspired by the connections between the chemistry encountered in the classroom, practiced in the laboratory, and put to work in the world.

We want students to build insight through applications, problem-solving, and visualization.

Although the themes developed in earlier editions continue to evolve in this edition, we have introduced important changes to make chemistry more meaningful and more comprehensible to students.

Applying Chemical Concepts

General chemistry students often do not understand how they will be using chemical principles throughout their careers. They need to **see the connection between the calculations and concepts they learn in class and the scientific and technological discoveries they read about in magazines and see on television.** In this edition we motivate students by showing them how the material they learn in the course is used in a multitude of applications.

Chapter Opener We have expanded the popular *Chemistry at Work* feature of the third edition to show students at the start of each chapter how the ideas they are about to study are relevant to a wide variety of professions. They will see that chemistry is not just for chemists but also is important for environmentalists, computer designers, corporate purchasing agents, doctors, information engineers—people in careers just beginning to be defined in our fast-moving world.

Applying Chemistry Every chapter contains a case study called *Applying Chemistry*, which shows how the concepts of the chapter apply directly to problems of current interest, such as the greenhouse effect, liquid crystal displays, food and fitness, and novel materials. We intend these case studies not only to be interesting to students, but also to bring together some of the key concepts discussed in the chapter. New exercises that encourage students to relate the *Applying Chemistry* topics to the chapter material are placed at the end of each chapter, in a new section called *Applied Exercises.*

NEW! *Connections* We have added a new type of case study, called *Connections*, which is placed after every group of 3 or 4 chapters. These case studies also deal with current issues, but focus on broader topics that integrate material in several chapters of the book. In this way we hope to show students that what they learn in one chapter has direct bearing on topics in other chapters. There are six *Connections* in the book, showing how chemistry is fundamental to topics as varied as sport drinks and electric cars, and is central to developing new drugs and medicines as well as new sources of energy. Each *Connection* contains suggestions for further reading (sometimes information on related material on the Web) and an *Applying Your Knowledge* problem set that draws on concepts from the preceding group of chapters. The *Connections* present students with challenges to design solutions to real-life problems. These activities, which are suitable for group projects as well as individual assignments, require students to gather information from databases and possibly other resources on the Internet or the CD that accompanies this book.

HARVESTING DRUGS FROM THE SEA

Applying Chemistry: Case Study 2

On one beautiful, sunny day in the Caribbean Sea, off the Isla de Providencia, several boats lay at anchor. Suddenly a group of divers surfaced and climbed aboard the boat shown in the illustration. There they spilled the contents of their collecting pouches onto the deck. They had harvested marine organisms: mostly sponges and tunicates (sea squirts). Like the pleasure-seekers on nearby boats, they gathered eagerly around their catch and selected interesting specimens. However, when the divers entered the large cabin of the boat, any similarity between them and their vacationing neighbors ended. Inside the boat was a fully equipped biochemical laboratory. There was a gas chromatograph connected to a mass spectrometer, a laboratory workbench with pipets and other glassware, and lots of petri dishes containing cultures.

The divers were chemists and chemistry students searching for marine organisms that might be sources of antiviral and antitumor agents. They did not know what

The research vessel *Alpha Helix* is used by chemists from the University of Illinois at Urbana-Champaign to search for ... medicinal value.

... extract were then ... e molar mass of ... pectrometry. The

---

CHEMISTRY IN THE DRUGSTORE

Connection 1

The local pharmacy stocks many different medicines and drugs. Where do all these medicines come from? How was their curative effect discovered? Identifying a vitamin or drug and bringing it to our drugstore shelves involves many of the chemical concepts and procedures that you have been studying in Chapters 1–4.

In antiquity and until very recently, medicines were discovered by trial and error or by accident. There was no way to predict a drug's effect except by testing it on patients. As a result, treatment could create other problems and was sometimes fatal to the patient. For example, arsenic compounds were prescribed for stomach problems, heroin for coughs, and mercury compounds for syphilis. However, some of the medical and herbal remedies of the past were remarkably effective. These medicines now serve as a source of starting material on which to build even more powerful agents that maximize curative effect while reducing side effects. Drugs that target specific infectious agents or tumors while not harming the patient are the "magic bullets" of medicine.

The search for new drugs relies on the skills of synthetic organic chemists and pharmacologists. Because there are so many millions of compounds, it would take too long to start with the elements, combine them in different ways, then test the products. Instead, chemists usually start either by drug *discovery*, the identification of promising medicines that

already exist in nature; or by *rational drug design*, the identification of characteristics of a target bacterium or parasite and the design of new compounds to react with it. Here we look at the process of drug discovery.

In drug discovery, a chemist usually begins by investigating compounds that have already shown medicinal value. A fruitful path is to find a *natural product*, a substance found in nature, that has been shown to have healing characteristics. Nature is the best of all synthetic chemists, with billions of chemicals that fulfill as many different needs. The trick is to find the ones that have curative powers. These substances are found in different ways. In *random* or "blind" collection of samples, a wide variety of plant extracts are collected and tested in the hope that some will be effective. In *guided discovery*, a collection of specific samples is identified with the consultation of native healers who have knowledge of the medicinal value of local herbs and other substances.

Observation of the properties of plants and animals can help to guide a random search. For example, if certain types of fruits are seen to remain fresh while others rot or mold, we might expect them to contain antifungal agents. One example of this type of collection is the gathering of tunicates and sponges in the Caribbean. The antiviral drug didemnin-C and the anticancer drug bryostatin 1 were discovered in marine organisms identified in this way (Applying Chemistry: Case Study 2).

The guided route involves the testing of fewer samples, because the chemist works with a native healer, the ancient lore guiding the modern chemistry. Often an ethnobotanist, a specialist in plants used for native healing, joins the team. This approach saves time for the scientists and can benefit the healers and their nations as well. In 1992, the United Nations Convention on Biological Diversity set guidelines for the conservation of potentially valuable plant and animal species and for the interaction of scientists with native healers. Drugs that have been discovered in this way include a variety of anticancer and antimalarial drugs, blood-clotting agents, antibiotics, and medicines for the heart and digestive system.

Once the empirical and molecular formulas of the active compounds in natural products have been determined (as described in Sections 2.12, 2.13, and 4.5), their structural formulas are deduced. Then chemists look for a way to make them in the laboratory.

(Above) The vitamin supplements on drugstore shelves are the end product of years of discovery, synthesis, and testing. (Right) This field biologist is collecting specimens of plants that may contain compounds with medicinal value.

Analogies

New to this edition are brief **Analogies** that illustrate important concepts by using familiar experiences. Most instructors use analogies to help relate new and familiar ideas, so we have sprinkled many throughout the text.

The unit mole applies to any chemical species, just as 1 dozen means 12 of anything to a grocer, not just the number of eggs in a carton.

An analogy

Problem Solving

Many students think they know the concepts they learn in the course—until they have to apply them to solve problems. We have used our experience in **helping students bridge the gap from qualitative description to quantitative methods** to provide features that help students build confidence in their problem-solving skills.

Toolboxes Students and instructors have responded enthusiastically to the **Toolboxes** introduced in the third edition. These boxes outline strategies students can learn to apply to the major calculations of the course, just as a carpenter uses various tools to build a house, each tool for its own purpose. We have added many more **Toolboxes** in this edition and have revised the basic format to stress when to use these methods, why they work, and how they work. Each **Toolbox** connects the conceptual basis of the procedure to the problem-solving procedure: solving problems, we strongly believe, should be done with understanding, not just with algorithms. **Toolboxes** feature schematic drawings to help students visualize the suggested procedures and are accompanied by related worked **Examples**. The **Toolboxes** summarize the concepts as well as the procedures, and should be an invaluable study guide.

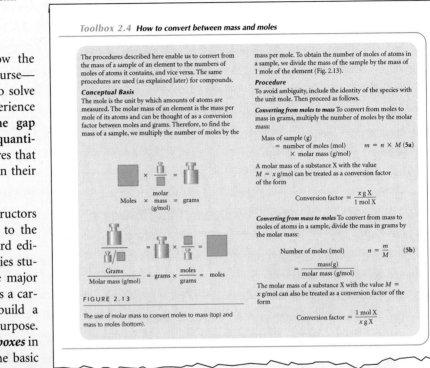

Toolbox 2.4 *How to convert between mass and moles*

The procedures described here enable us to convert from the mass of a sample of an element to the numbers of moles of atoms it contains, and vice versa. The same procedures are used (as explained later) for compounds.

Conceptual Basis

The mole is the unit by which amounts of atoms are measured. The molar mass of an element is the mass per mole of its atoms and can be thought of as a conversion factor between moles and grams. Therefore, to find the mass of a sample, we multiply the number of moles by the mass per mole. To obtain the number of moles of atoms in a sample, we divide the mass of the sample by the mass of 1 mole of the element (Fig. 2.13).

Procedure

To avoid ambiguity, include the identity of the species with the unit mole. Then proceed as follows.

Converting from moles to mass To convert from moles to mass in grams, multiply the number of moles by the molar mass:

Mass of sample (g)
= number of moles (mol) × molar mass (g/mol)

$$m = n \times M \text{ (5a)}$$

A molar mass of a substance X with the value $M = x$ g/mol can be treated as a conversion factor of the form

$$\text{Conversion factor} = \frac{x \text{ g X}}{1 \text{ mol X}}$$

Converting from mass to moles To convert from mass to moles of atoms in a sample, divide the mass in grams by the molar mass:

Number of moles (mol)

$$n = \frac{m}{M} \text{ (5b)}$$

$$= \frac{\text{mass(g)}}{\text{molar mass (g/mol)}}$$

The molar mass of a substance X with the value $M = x$ g/mol can also be treated as a conversion factor of the form

$$\text{Conversion factor} = \frac{1 \text{ mol X}}{x \text{ g X}}$$

Moles × molar mass (g/mol) = grams

$$\frac{\text{Grams}}{\text{Molar mass (g/mol)}} = \text{grams} \times \frac{\text{moles}}{\text{grams}} = \text{moles}$$

FIGURE 2.13

The use of molar mass to convert moles to mass (top) and mass to moles (bottom).

Problem-Solving Representations **Toolboxes** feature visualizations of the steps to be taken to solve a problem. These images are concrete representations of abstract processes.

Worked Examples We have increased the number of worked **Examples** throughout the book, illustrating the major techniques needed to solve problems in chemistry and some of their important variations. We have retained the popular **Strategy** sections that encourage students to reason out the steps involved in solving a problem. In many cases the **Strategies** derive from the preceding **Toolbox**. In others they show students how to break down a problem into solvable steps and show them how to estimate an approximate answer.

Example 1.7 *Naming ionic compounds*

Give the systematic names of (a) a compound that turns from blue to pink in humid weather, $CoCl_2 \cdot 6H_2O$, and (b) a compound used as a disinfectant, $Al(ClO_3)_3$.

Strategy In each case, identify the cation and the anion, by referring to Table 1.3 if necessary. To identify the oxidation number of the cation, note the charge of the anion and the number of anions and cations in the formula; then decide what cation charge is required to cancel the total negative charge. If the cation is a transition metal or an element from Groups 12 through 15 that can have more than one charge, as shown in Fig. 1.26, give its charge as a Roman numeral. If water molecules are included in the formula, then the compound is a hydrate, so add "hydrate" with a prefix corresponding to the number in front of H_2O in the formula.

Solution (a) The cation is a cobalt ion and the anion is a chloride ion, Cl^-. The cation must be Co^{2+} for electrical neutrality; so cobalt (a transition metal) is present as cobalt(II). The $6H_2O$ indicates that the compound is a hexahydrate; so the compound is cobalt(II) chloride hexahydrate. (b) The cation is an aluminum ion, Al^{3+} (the only cation that aluminum forms). The anion is a chlorate ion, ClO_3^-. Hence, the compound is aluminum chlorate.

Self-Test 1.10A Name the compounds (a) $FeCl_2 \cdot 2 H_2O$; (b) $AlBr_3$; (c) $Cr(ClO_3)_2$.
[*Answer:* (a) Iron(II) chloride dihydrate; (b) aluminum bromide; (c) chromium(II) chlorate]

Self-Test 1.10B Name the compounds (a) $AuCl_3$; (b) CaS; (c) Mn_2O_3.

problem solving

Self-Tests Students can sometimes follow the steps of a worked **Example** but have trouble planning these same steps on their own. The pair of **Self-Tests** at the end of every worked **Example** allow students to check their ability to solve an exercise similar to the one worked out in the **Example** and then reinforce their skills with a second exercise. The answer to the first **Self-Test** of each pair is in plain view for instant feedback; the answer to the second **Self-Test** of each pair is given at the back of the book. Occasional **Self-Tests** stand alone in the text, to allow students to check their understanding of a passage.

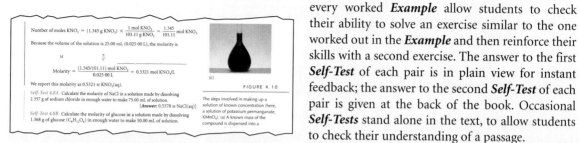

Number of moles KNO_3 = (1.345 g KNO_3) $\times \dfrac{1 \text{ mol } KNO_3}{101.11 \text{ g } KNO_3} = \dfrac{1.345}{101.11}$ mol KNO_3

Because the volume of the solution is 25.00 mL (0.025 00 L), the molarity is

$$M \qquad \dfrac{n}{V}$$

Molarity = $\dfrac{(1.345/101.11) \text{ mol } KNO_3}{0.025\,00 \text{ L}}$ = 0.5321 mol KNO_3/L

We report this molarity as 0.5321 M KNO_3(aq).

Self-Test 4.8A Calculate the molarity of NaCl in a solution made by dissolving 2.357 g of sodium chloride in enough water to make 75.00 mL of solution.
[*Answer:* 0.5378 M NaCl(aq)]

Self-Test 4.8B Calculate the molarity of glucose in a solution made by dissolving 1.368 g of glucose ($C_6H_{12}O_6$) in enough water to make 50.00 mL of solution.

FIGURE 4.10

The steps involved in making up a solution of known concentration (here, a solution of potassium permanganate, $KMnO_4$). (a) A known mass of the compound is dispersed into a

The Bottom Line Each section of the text concludes with a one- or two-sentence summary of its principal points. We have found that this popular feature helps to focus a student's attention and identify the important points of a section.

The partial pressure of a gas is the pressure it would exert if it were alone in the container; it is equal to the mole fraction of the gas times the total pressure. The total pressure of a mixture of gases is the sum of the partial pressures of the components.

Exercises We have included a large number of **Exercises** at the end of each chapter to provide instructors with problems at a wide variety of levels of difficulty, from simple confidence-builders to middle-level exercises that test conceptual understanding to problems that require creative problem-solving skills. Many of these exercises are new to this edition. As before, the first part of each set of exercises is classified by topic, then the **Supplementary Exercises** require students to decide for themselves what topic is being addressed. We have also included two new categories of exercises. **Applied Exercises** include exercises that are related to real-life problems and some that relate to the **Applying Chemistry** case study in that chapter. **Integrated Exercises** require students to use material from previous chapters. **Exercises** classified by topic are paired, and answers to all odd-numbered exercises appear in the back of the book. Solutions to odd-numbered exercises are given in the *Solutions Manual*; solutions to even-numbered exercises appear in the *Instructor's Resource Manual*. Every answer has been independently solved by four different checkers to ensure accuracy.

Applied Exercises

For Exercises 2.95–2.96, see Applying Chemistry: Case Study 2.

2.95 In 1978, scientists extracted a compound with antitumor and antiviral properties from tunicates in the Caribbean Sea. A 2.52-mg sample of the compound, didemnin-C, was analyzed and found to have the following composition: 1.55 mg C, 0.204 mg H, 0.209 mg N, and 0.557 mg O. The mass spectrum of didemnin-C was found to have a parent mass peak at 1014 g/mol that represents the molar mass of the compound. What is the molecular formula of didemnin-C?

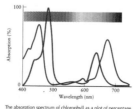

Limiting Reactants

4.21 Complete the drawing below, which represents the reaction $2 H_2(g) + O_2(g) \rightarrow 2 H_2O(g)$. The only reactants available are the numbers of atoms and molecules in the first box. In the second box, draw the system after reaction takes place. Include any unreacted atoms or molecules. In the drawing, red represents an oxygen atom and dark gray, a hydrogen atom.

4.22 Complete the drawing below, which represents the reaction $N_2(g) + 2 O_2(g) \rightarrow N_2O_4(g)$. The only reactants available are the numbers of atoms and molecules in the first box. In the second box, draw the system after reaction takes place. Include any unreacted molecules. In the drawing, red

Investigating Matter Many areas related to the general chemistry course are not essential for understanding other material but are still very useful for students to know. We have included some of these topics in **Investigating Matter** boxes that can be assigned or not as the instructor chooses. Several of these topics include laboratory techniques such as spectroscopy, chromatography, and pH meters; other boxes discuss ideas such as the development of the periodic table and the confirmation of the existence of electron spin. These **Investigating Matter** boxes are designed to give students an introductory idea of the topics, and are not provided with questions or exercises. A few exercises at the end of the chapter draw on these boxes, and are so labeled.

Investigating Matter 9.2: **Ultraviolet and Visible Absorption Spectroscopy**

The measurement of the wavelengths of light emitted by the hot atoms and molecules in stars, flames, or other glowing objects, is called *emission spectroscopy:* the wavelengths tell us the energy given off when an electron falls from a high energy level to a lower one. Electrons also *absorb* wavelengths that correspond to the different energy levels available to them. Chemists use *absorption spectroscopy*, the selective absorption of light, to identify compounds and determine their concentration in samples.

When electromagnetic radiation falls on a molecule, the molecule can be excited to a higher energy state. If the frequency of the radiation is v (nu), it can raise the molecule to a state that differs in energy by ΔE, where

$$\Delta E = hv$$

This is the Bohr frequency condition, (see Section 7.3); h is the Planck constant. For typical ultraviolet wavelengths (300 nm and less, corresponding to a frequency of about 10^{15} Hz), each photon brings enough energy to excite the electrons in a molecule into different energy levels. Provided an unfilled orbital exists at the right energy, the incoming radiation can excite an electron into it, and hence be absorbed. Therefore, the study of visible and ultraviolet absorption gives us information about the electronic energy levels of molecules.

Visible and ultraviolet (UV) absorption spectra are measured in an absorption spectrometer like the one used for infrared spectroscopy (see Investigating Matter 9.1), with the difference that the source gives out visible light or ultraviolet radiation. The wavelengths can be selected with a glass prism for visible light and with a quartz prism or a diffraction grating for ultraviolet radiation (which is absorbed by glass). The

absorption spectrum of chlorophyll is shown in the first illustration. Note that chlorophyll absorbs red and blue light, leaving the green light present in white light to be reflected. That is why most vegetation looks green.

The presence of certain absorption bands in visible and ultraviolet spectra can often be traced to the presence of characteristic groups of atoms in the molecules. These groups of atoms are called *chromophores*, from the Greek words for "color bringer."

The absorption spectrum of chlorophyll as a plot of percentage absorption against wavelength. Chlorophyll *a* is shown in red, chlorophyll *b* in blue.

problem solving

Skills You Should Have Mastered Students often need a way to check that they have learned the key ideas of a chapter and know what they need in order to study the following chapters. We have found that summarizing the chapter for students in the form of **Skills You Should Have Mastered** gives them a straightforward way to check for themselves that they understand the conceptual, problemsolving, and descriptive skills presented. In this edition we have keyed these skills to the **Examples, Toolboxes,** and chapter sections that deal with that skill, so students can review the relevant material.

Visualization

We find that students understand concepts better and remember them better when they can associate them with clear visual images. We are constantly improving the drawings in the text in response to student and reviewer feedback. A large number of figures show accurately scaled models to help students visualize molecules and crystal structures in three dimensions. Many of these illustrations have been prepared with *WebLab ViewerPro* from MSI Inc, and all are included on the CD-ROM that accompanies this text. Other figures combine molecular models with photos of familiar objects that students can relate to, strengthening the link between the macroscopic world of observation and the underlying molecular world that explains these observations.

Net dipole moment

Fig. 9.13

Dynamic Motion Sequences Chemistry is about change as well as structure. Talking to students about molecules often means describing molecules in motion and molecules undergoing modification. We introduced **Motion Sequences** in the third edition, consisting of several linked illustrations that look like stills from an animation. We have improved the clarity and presentation of these sequences so they still present the busy, jostling motion of molecular interactions while staying focused on the pedagogical point under discussion. For complex interactions, we have added insets that focus attention on the main events. Many of these sequences are part of actual animations presented on the CD-ROM that accompanies the book.

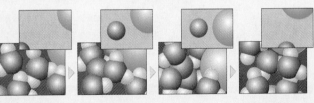

Fig. 3.12

A Chemist's-Eye View
One of the keys to learning and understanding chemistry is the development of the link between the macroscopic world we observe, the microscopic world of atoms and molecules that underlies it, and abstract representations in terms of symbols that chemists use. We have developed multilevel representations that connect these levels, emphasizing what is happening among atoms and molecules that causes the effects we observe. In these diagrams, the different layers represent different ways of viewing and dealing with chemistry. First is the photograph of the observed substance or process. Superimposed on that is the molecular view. On the molecular view may be superimposed a

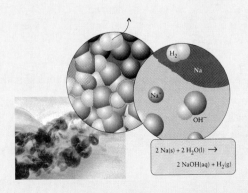

$$2\,Na(s) + 2\,H_2O(l) \rightarrow$$
$$2\,NaOH(aq) + H_2(g)$$

Fig. 3.1

more abstract view that conveys the essence of what is going on. And superimposed on that, at the highest level of abstraction, we often add the symbols that a chemist would commonly use to represent the process. These different layers are overlapped so that the entire representation is before the student in a single place: it is important to acquire the facility to move easily between representations.

Color for Clarity Throughout the text we use color for clarity and as a systematic means of communication. For example:

Orbitals are systematically colored according to subshell: *s*-orbitals and σ-orbitals are blue; *p*-orbitals and π-orbitals are yellow; *d*-orbitals are orange, and *f*-orbitals (on their one appearance) are purple. Hybrid orbitals are green. The same colors are used for corresponding blocks in the periodic table and for the boxes used in atomic and molecular energy-level diagrams.

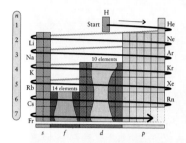

Fig. 7.31

To display trends in physical properties in the periodic table, we use the geographical conventions for depth and altitude, making low values deep blue and high bright red.

Composite processes are depicted by use of pale colored arrows; the overall process, the focus of our attention, with a darker arrow. Energy changes are typically denoted by red arrows, entropy changes by green arrows, and steps in a process by yellow arrows.

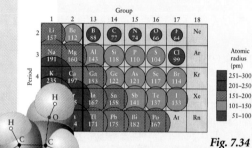

Fig. 7.34

Atoms are colored by common conventions: black for carbon, gray for hydrogen, red for oxygen, blue for nitrogen, and so on. Some atoms are "generic" and are given colors that help to distinguish them and highlight the concept being presented. Pale blue spheres commonly represent solvent molecules, as does a pale blue background when we show less detail.

Fig. 1.23

NEW! *Changes to the Organization of This Edition*

We have rethought and rewritten many portions of the text, from clarifying the wording in a sentence to moving an entire chapter. Some of the more significant changes are:

Chapter 1: Instruction in the nomenclature of inorganic compounds has been enhanced.

Chapter 2: Material on molarity and dilution of solutions now appears in Chapter 4, along with solution stoichiometry.

Chapter 3: Predicting the products of simple reactions is now taught explicitly.

Chapter 5: We have streamlined the presentation of the gas laws and expanded the treatment of mole fraction and molecular speeds.

Chapter 6: This chapter now includes a systematic presentation of the first law of thermodynamics, along with a discussion of potential and kinetic energy, work, and a clarification of the distinction between internal energy and enthalpy.

visualization

Chapter 9: We have added a discussion of molecular orbitals, largely in pictorial form, and include material on infrared and UV-visible spectroscopy. Electrostatic potential diagrams are introduced to help students visualize electron distribution within a molecule.

Chapter 10: Now properties of solids are augmented by a conceptual discussion of band theory. The Clausius-Clapeyron equation is introduced to help students quantify changes in vapor pressure.

Chapter 11: Reactions of hydrocarbons are differentiated and a box on NMR spectroscopy is included.

Chapter 12: Chromatography is included as an ***Investigating Matter*** box and a section on colloids has been added.

Chapter 13: We have placed the chapter on kinetics here, before the chapters on equilibrium. The chapter itself contains more emphasis on practical, experimental data. For those instructors who prefer to do equilibrium before kinetics, this chapter can easily be used out of sequence.

Chapter 15: The relationship between Brønsted acids and Lewis acids is emphasized and the treatment of polyprotic acids expanded.

Chapter 16: Material on selective precipitation and its application to qualitative analysis has been added.

Chapter 19: The first of the four descriptive chemistry chapters begins with a summary of major periodic trends.

Appendix 2: The data tables in Appendix 2A now include molar heat capacities. Appendix 2D now includes more information on ionization energies and the principal oxidation states of the elements.

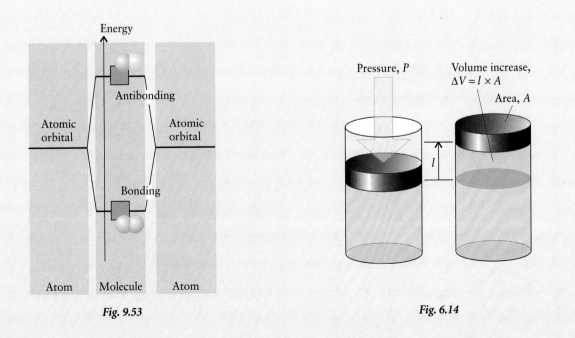

Fig. 9.53

Fig. 6.14

Media and Supplements

A variety of innovative teaching and learning tools have been provided in the supplements and media package for this edition. The student media package consists of two student CDs—one focusing on Visualization, the other on Problem-Solving Skills—as well as a Student Companion Web site.

Media Icons in the Textbook   Throughout the textbook you will notice this Media Icon. Each one indicates that there is an item on the CDs and/or Web site that accompanies that part of the textbook. To find the corresponding item, go to the book's Web site [**www.whfreeman.com/chemistry/**] where you will find a full list of the media offerings by textbook page number.

Visualization CD-ROM   This CD offers a number of exciting and engaging features to help students understand chemistry better and enhance their performance in the course. Responding to demand, we have increased and expanded the number of powerful **animation and simulation movies** from the last edition. Developed by Dr. Roy Tasker, of the University of Western Sydney, Nepean, and CADRE design, these accurate and stunning movies convey a deep understanding of molecular movement. The two popular Simulation Groups from the last edition—**Atomic Orbitals** and **Ions in Solution**—have been revised and expanded. Additionally, three exciting new Simulation Groups have been created for this edition: **Stoichiometry, Equilibrium**, and **Gas Laws**. The animations from the last edition—including *Ice Melting* and *Cesium Chloride Unit Cell*—appear as well.

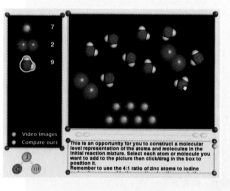

A series of interactive **Nomenclature exercises** to help students learn chemical notation and nomenclature allows students to type in answers—including super- and subscripting. Answers can be checked, and feedback is delivered. CADRE has also fully animated the partial animation sequences represented by two-dimensional diagrams in the textbook.

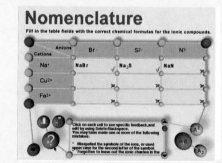

Problem-Solving Skills CD-ROM   This CD contains electronic student study features such as **Q&A** chapter self-quizzes with built-in feedback and **interactive exercises** help students prepare for tests. **Videos** of 22 lab demonstrations offer chemical reactions as they would actually happen. To aid in molecular visualization, a **data base** of all 200+ molecular structures found in the textbook uses **Molecular Simulations Inc. (MSI) WebLab Viewer™** to allow students to rotate the structures as well as change their display style in several ways, including line draw, ball and stick, and space fill. The CD also includes tools that students can use and then take with them as they continue their chemical education: a comprehensive **Periodic Table** visual data base with information not found in the text, such as the isotopic abundance of each element; a **Curve fitter**; a **Plotter**; and six **Calculators** including molar mass, density, temperature conversions, and equilibrium.

Instructor's Resource CD-ROM   Our instructor's CD offers all the images from the textbook, and the animations, videos, and molecular structures in two formats—as part of our **Presentation Manager Pro** software and in **JPEG files**. Presentation Manager lets instructors quickly prepare playlists of images for display during lectures. The JPEG files are for instructors who use commercially available presentation software.

media & supplements

Web Site Our Web site offers an **Online Companion** that serves as an electronic chapter-by-chapter study guide for students. It includes **chapter objectives**, **flash card exercises** to aid in the mastery of the text's key terms, **interactive exercises** and chapter **self-quizzes** to help prepare for tests, and four intensive **practice tests** covering multiple chapters to simulate course exams. For instructors, the site features an **Online Instructor's Guide** with an extensive chapter-by chapter list of specific ideas on using the electronic media for the book, while teaching from this edition.

Additional Web site features include **electronic lab materials**, **expansion modules** examining additional chemistry topics, and a Web version of the CD's **molecular visualization feature.** Web links for each chapter help students explore the world of chemistry on the Internet. Special Web sections offer further ideas for exploration on the topics discussed in each of the textbook's **Applying Chemistry Case Study** and **Connection** features.

www.whfreeman.com/chemistry/

Online Testing Our new **Online Testing** package from Brownstone Research Group uses the Internet to move testing beyond the printed page. Exams and quizzes can be created using our Test Bank questions and instructors can insert multimedia, graphics, movies, sound, or interactive three-dimensional molecular representations into questions. High-level security allows instructors to restrict indi-vidual testing capability to individual computers, and to specify a time block during which a test can be taken. The package also has a suite of grade book and question analysis features.

Traditional Supplements We offer a full complement of print-based supplements to support students and instructors using this edition. The **Student Companion: New Tools and Techniques for Chemistry** by Lynn Geiger, Belia Straushein, and Loretta Jones, of the University of Northern Colorado, features collaborative learning sheets and plans, guided reading strategies, and ideas for teaching with the CD and Web site. Many consider the **Study Guide,** by David Becker of Oakland University and Oakland Community College, the best supplement of its type in this market. It carefully reinforces the key concepts of the text, provides additional practice exercises and answers, and offers a special "pitfalls" section that warns students of the errors most often made in this course. The other student supplement is the indispensable **Student's Solution Manual** by Julie Henderleiter of Grand Valley State University and Charles Trapp of the University of Louisville. It provides the worked-out solutions to all odd-numbered exercises. In addition to the **Instructor's Resource CD,** we offer a number of supplements for those teaching from the book. The **Instructor's Resource Manual with Solutions** by Julie Henderleiter and Charles Trapp provides the solutions for the even-numbered exercises, lecture outlines, chapter overviews, and a number of other teaching hints. A separate **Teaching Assistant Manual** by Dianne Meador of the University of California at Davis addresses specific challenges and concerns of teaching assistants in general chemistry. A set of 200 full-color **Overhead Transparencies** includes key figures from the textbook. Finally, our ever-expanding **Test Bank** (in printed, Windows, and Macintosh formats) by Robert Balahura of the University of Guelph now offers a total of 3 520 questions. Each chapter contains sets of multiple-choice questions and short-answer and drill problems.

media & supplements

Laboratory Manuals

Working with Chemistry Lab Manual
(0-7167-3549-0)

Donald Wink and Sharon Fetzer Gislason,
University of Illinois at Chicago, and
Julie Ellefson Kuehn,
William Rainey Harper College

With this inquiry-based program, students build skills using important chemical concepts and techniques to the point where they are able to design their own solutions to the study of particular problems. Scenarios are drawn from the lives of people who work with chemistry every day, ranging from field ecologists to chemical engineers to health professionals. Instructors can download and class test sample labs from Working with Chemistry at the Freeman Web site,
www.whfreeman.com/chemistry/wwc.

Chemistry in the Laboratory, **Fifth Edition**
(0-7167-3547-4)

James M. Postma,
California State University at Chico, and
Julian L. Roberts, Jr., and J. Leland Hollenberg,
University of Redlands,

This clearly written, class-tested manual has long given students essential hands-on training with key experiments. A new "Consider This" feature gives instructors the opportunity to show students how the experiment is applied to the real world of chemical research and presents opportunities for further inquiry. All experiments are available as lab separates (ask your local representative for details).

Acknowledgments

A project as big as this can come to fruition only with the assistance of many people. First we thank our teachers, students, and colleagues. We try to build on the great store of wisdom and experience that the chemical education community collectively represents—and to push forward the frontiers. We are deeply indebted to all those who shared their time with us: we know that we have benefited greatly from their suggestions.

We especially want to thank those who contributed directly to the quality of this project: Evelyn Erenrich of Rutgers University, for working through all the Self-Tests so conscientiously and for providing us with insights into the way students learn; Carissa Bertin and Jessica Blunt, Grand Valley State University, for their detailed and insightful checking of exercise solutions; Seth Willis, University of Northern Colorado, for his careful, thorough checking of exercise answers; Jetty Duffy-Matzner, for bringing her pedagogical expertise to bear on the entire manuscript. Our supplements authors, particularly Julie Henderleiter, David Becker, and Lynn Geiger, have reached beyond the coasts of their particular responsibilities to give us a great deal of very useful critical advice. We are also grateful to our colleagues, especially Henry Heikkinen, University of Northern Colorado, and Steve Thompson, Colorado State University, who offered many useful suggestions. Our students, of course, have been the unwitting test beds for many of our ideas, and we are grateful to them for the toleration that gradually turned into enjoyment and learning.

The contributions of the staff at W. H. Freeman and Company are summarized on the copyright page, but that simple listing does not convey the deep gratitude that we feel toward them and their assistants. In particular, we would like to acknowledge those who saw the potential of this project and took it on with enthusiasm: Michelle Julet, publisher, who initiated this edition and worked with us throughout; Jessica Fiorillo, chemistry editor, who took us on with grace and competence; David Chelton, development editor, who shaped the project with insights and good cheer; Jodi Simpson, our outstanding coach and trainer who also acted as our copy editor; Georgia Lee Hadler, senior project editor, who oversaw the metamorphosis of the manuscript from its larval phase into the final product; Kathy Bendo, who found the images that will appeal to students; Kimberly Manzi, who enlightened us about the needs of faculty we have not yet met; and Matthew Fitzpatrick and Patrick Shriner, who orchestrated the varied and novel instruments of the supplements package.

We could go on: our thanks could be never ending. But it is time for the words and images to speak for themselves.

Diary Reviewers

Jesse S. Binford, Jr.
University of South Florida

Roger L. Dekock
Calvin College

Max Diem
Hunter College

Jetty L. Duffy-Matzner
Augustana College

Sylvia Esjornson
Southwestern Oklahoma State University

Christopher J. Grayce
University of California at Irvine

Michael Johnson
New Mexico State University

John Krenos
Rutgers University

Alison McCurdy
Harvey Mudd College

Gardiner Myers
University of Florida

J. H. Niewahner
Northern Kentucky University

Steve Ruis
American River College

Jerry Sarquis
Miami University

Jerry P. Suits
McNeese State University

Richard Treptow
Chicago State University

Chapter Reviewers

John E. Adams
University of Missouri at Columbia

Todd L. Austell
University of North Carolina at Chapel Hill

Gabriele Backes
Portland Community College

Robert Balahura
University of Guelph

Danny R. Bedgood, Jr.
Arizona State University

James Byrd
California State University at Stanislaus

Emily Carter
University of California at Los Angeles

Christa Colyer
Wake Forest University

Larry R. Dalton
University of Southern California

Ghislain Deslongchamps
University of New Brunswick

Joseph Franek
University of Minnesota

Ronald S. Friedman
Indiana University, Purdue University at Fort Wayne

Robert M. Hammond
University of Maryland

Dennis Jacobs
University of Notre Dame

Susan A. Jansen
Temple University

Angela G. King
Wake Forest University

Ted Linderman
Colorado College

Glenn L. Millhauser
University of California at Santa Cruz

William H. Myers
University of Richmond

Hans Oesterreicher
University of California at San Diego

Wayne Pearson
United States Naval Academy

Scott S. Perry
University of Houston

Daniel Rabinovich
University of North Carolina at Charlotte

George J. Reilly
University of Delaware

David F. Rieck
Salisbury State University

Silvio Rodriguez
University of the Pacific

Michael Sailor
University of California at San Diego

Barbara Sawrey
University of California at San Diego

Lothar Stahl
University of North Dakota

Uni Susskind
Oakland Community College

John R. Townsend
West Chester University

Gilles Villemure
University of New Brunswick, Fredericton Campus

Barbara J. Weathers
University of Michigan

Gabriela Weaver
University of Colorado at Denver

Gretchen Webb-Kummer
Modesto Junior College

Fourth Edition

CHEMISTRY

Molecules, Matter, and Change

TO THE STUDENT

We wrote this book to help you master chemistry. We remember well the jungle chemistry seemed at first. We remember the trouble we had learning new concepts, the frustration of being unable to solve—and sometimes even begin—the homework problems. However, we are very pleased that we stuck to the course, because we now see the world in a very special way. When we look at a leaf in fall or an opening flower, when we feel the texture of fabrics, when we learn on the evening news about threats to the environment, we appreciate it much more deeply than we did before, because we can see it through chemists' eyes.

As we explain in the preface, one of our principal aims is to impart insight. Insight means being able to bridge what we perceive and what we imagine. A difficulty with learning chemistry is making that connection—between what we observe around us and the explanations that chemists give, in terms of atoms, molecules, and energy. Chemistry is like looking at a city from a thousand miles into space. We know that there is activity down below in the city. However, it is only when we get down to ground level, where we can see all the inhabitants going about their tasks in characteristic ways, that we understand how the city really works. Chemical insight is like knowing those individuals closely. **In a sense, it is about knowing the personalities of atoms and molecules so that we can predict how they are going to behave.**

As we created the study aids in this book, we kept in mind our own difficulties and tried to help you through them. We know that reading a science or technical textbook requires a special kind of concentration and skill. Like physical exercises, mental exercises are useless unless they stretch you. There is usually a lot of information to process and recall later, and some of the concepts are not easy to understand the first time you encounter them. However, this book is organized so that you can make best use of your study time.

A good way to begin a new section is to skim through it first, read the summary at the end, and then return to read the section in detail. Stop to think about what the sentences mean: science can pack a lot of information into a single sentence. Be especially attentive to unfamiliar terms set in bold type, which are collected together in the Glossary at the end of the book. These terms are the basic language of chemistry, so look up words you do not understand. Become familiar with the many study aids built into each chapter, such as the Self-Tests and marginal notes. We describe them in the preface, so be sure to take a look at that.

If you have never studied chemistry before, you may need to spend extra time on the course. Even if you have studied chemistry, perhaps some time ago, it would be a good idea to read through Appendix 1, which reviews the mathematics you will need. If a section has you puzzled, review the sections mentioned in the marginal notes. The accompanying CDROM contains additional hyperlinks, Self-Tests, powerful calculating tools, and other study aids.

Recalling information about the chemical elements is easier if you look for trends. **This textbook emphasizes trends and puts them in the context of the periodic table of the elements.** Beginning in Chapter 1 you will recognize the periodic table as one of the great unifying concepts of chemistry.

Problem-solving skills are best developed through thinking about how to approach a problem, including how to assemble the information you need. The Strategy section in each worked example shows you how to collect your thoughts and assemble the equations you need, and suggests where you can find more information. The Strategy section also often suggests how you can predict the answer by using the chemical insight you have been developing. It is usually sensible to try to make an estimate of the answer before embarking on a calculation so that you can judge whether the result you obtain is plausible. You will get additional help from the Toolboxes, which contain the core calculations and procedures of chemistry—the hammers, screwdrivers, and saws of the calculations and procedures that you use throughout the subject. You may need to come back to these Toolboxes from other chapters, just as in real life you may need a hammer or screwdriver to do many kinds of jobs.

Make good use of the lists of Skills at the end of each chapter to monitor your progress. The skills are divided into three groups, which represent three types of learning goals: concepts, problem solving, and descriptive chemistry. You can use these lists to guide you through the chapter and later for review.

The CD that accompanies the text contains programs specially designed to help you succeed. Interactive problem-solving programs and quizzes will give you feedback on your problem-solving skills. Chemical calculators help guide you through the solving of complex mathematical problems, and a powerful molecular modeling program allows you to create and examine molecular structures in detail. Icons in the text highlight points where the CD is especially useful.

It is very important to keep up with the material in your course. Often topics presented one week are needed to understand topics presented the next week. If you drop behind, you may have trouble catching up again, and the topics will seem unnecessarily difficult. Many general chemistry students have found the following approaches helpful:

- Read the relevant sections of the text before attending class.
- Review and organize class notes daily.
- Form a study group and meet at least twice a week.
- Complete all the homework and check your answers.
- Read related sections of other books in the library to get another slant on a topic.

Above all, stay the course. Even when you feel lost, there are still paths you can take. Go back to read a chapter one more time and work through all the paired Self-Tests. You may check the answers to the second of each pair and to the odd-numbered exercises in the back of the book. Use the learning supplements that your instructor recommends. Use the CD-ROM that accompanies and enriches this text. You can also find additional help on our Web page at http://www.whfreeman.com/chemistry4e

We wish you every success. It may seem that a jungle lies ahead, but you will find the journey one of the great adventures of your mind.

Sincerely,

LORETTA JONES

PETER ATKINS

Cy commencent les croniques
de Sire Jehan Froissart contenās
les nouuelles guerres de France
dangleterre de scece despaigne
Afin que honnora
bles aduenues et
nobles aduentures
fautes en armes
lesquelles sont ad
uenues par les
guerres de France et dangleterre
soient noblement registrees z mises
en memoire perpetuel parquoy
les preux ayent exemple deulx en

dalemaigne et de Bretaigne et
sont diuisees en quatre parties
selon ce quil est contenu en son
prologue
Pourrayez en bien faisant. Je veult
trauuer z recorder histoire z mate
de grant louenge laquele sera di
uisee en quatre parties. Mais ame
que Je le commence Je requier au
sauueur de tout le monde, qui de ne
ant crea toutes choses, que Il vueil
le acer z mettre en moy sens et ente
dement si vertueux que ce liure que
jay commencie Je le puisse cōtinuer

Matter

Chemistry is the study of matter and the changes matter can undergo. **Matter** is anything that has mass and takes up space. Anything you can touch is matter. So are a lot of things you cannot touch, such as the material of flames, rocket exhausts, and stars. Driving a car, cooking a meal, and even taking a breath involve chemistry. A knowledge of chemistry is essential if you want to understand how the world works.

Some terms have specific meanings in science; you will learn these terms and their meanings as you go along. For example, the scientific meaning of the term *substance* is a little different from its everyday meaning. Each different pure kind of matter is called a substance. By *pure*, we mean "the same throughout, even on a microscopic scale." Thus, iron is one substance; water is another. A **substance** in science is a *single, pure* form of matter, not a mixture of several different kinds of matter. According to this definition, flesh, soft drinks, and dirt are not single substances in the scientific sense because they are complex mixtures.

In this chapter you will become acquainted with the major categories of substances and learn how to give names to specific substances and describe their composition. You will also see that an important role of chemistry is the **analysis** of mixtures, the identification of the substances they contain. Chemical analysis is used extensively in fields such as medicine, environmental chemistry and engineering, and *forensic science*, the investigation of criminal evidence (Fig. 1.1). Similar techniques are used by archeologists to determine the diet of early humans by analysis of bones and by forensic laboratories to obtain evidence of crimes such as arson.

This is a page from the illustrated manuscript *Chronicles,* a history of Europe written by Jean Froissart in the 14th century. The picture shows the author presenting his book to the Duchess of Burgundy. The age and origin of the manuscript can be verified by analyzing the paint used. Every material, whether paint, paper, breakfast cereal, or the hull of a spaceship, is made from a set of about 100 elements. Each of these elements has its own distinctive properties that allow it to be identified. This chapter introduces the elements and the types of compounds they form.

FIGURE 1.1

Can you tell that the silver porringer on the left was made by Paul Revere, but that the one on the right is a fake? The fake porringer was detected only by analyzing its elemental composition. Modern chemical instrumentation allows authentic works of art to be distinguished from forgeries without damaging the art itself.

The name *atom* comes from the Greek word for "not cuttable."

THE ELEMENTS

The structure of matter has been a puzzle for thousands of years. The ancient Greeks developed the concept of "elements" as fundamental substances from which all forms of matter can be built. They identified four elements—earth, air, fire, and water—which they believed could produce all other substances when combined in the right proportions. Their concept of "element" is consistent with what we currently believe. However, we now know that there are actually more than 100 elements and that earth, air, fire, and water are not among them. The **chemical elements** are the basic building blocks of matter and in various combinations make up all the matter on Earth (Fig. 1.2). The ancient Greeks identified elements from direct observation. Today, however, we define elements on the basis of differences at the atomic level.

1.1 Atoms

The ancient Greeks asked what would happen if we went on cutting matter into ever smaller pieces. Is there a point at which we would have to stop because the pieces no longer had the same properties as the whole? Or could we go on cutting forever?

We now know that there comes a point at which we have to stop. That is, matter is not continuous but consists of almost unimaginably tiny particles. The smallest particle of an element that can exist is called an **atom.** The first convincing argument for atoms was made in 1807 by the English schoolteacher and chemist John Dalton. Relying on a large number of laborious measurements of the relative masses of elements that combined together, he assembled arguments that strongly indicated the existence of atoms with definite, characteristic masses. For example, in one experiment, he found that 4 g of oxygen combined with 6 g of magnesium and that 8 g of oxygen combined with 12 g of magnesium. The fact that the mass of oxygen that combines is proportional to the mass of magnesium used suggested to him that an atom of oxygen combined with an atom of magnesium. He argued that

FIGURE 1.2

Samples of common elements. Clockwise from the red-brown liquid bromine are the silvery liquid mercury and the solids iodine, cadmium, red phosphorus, and copper.

doubling the mass of magnesium doubles the number of magnesium atoms in the sample, and hence doubles the number of atoms of oxygen that combine (Fig. 1.3).

Today, modern instrumentation provides much more direct evidence of atoms, and we can now make images of individual atoms (Fig. 1.4). We no longer doubt that atoms exist and that they are the units making up the elements:

An **element** is a substance composed of only one kind of atom.

All the atoms in a block of gold are of the same kind. Similarly, all the atoms in a block of lead are of the same kind (but different from those in a block of gold), and so on for all the elements. When scientists make a new element in one of their giant accelerators, they recognize it as new by checking—in ways to be described—whether its atoms are different from those of all known elements. By 1999, 112 elements had been discovered or created, although in some cases in only very small amounts. For instance, only two atoms of the 112th element were made, and they lasted for only a tiny fraction of a second.

To understand chemistry, we need to be able to relate events that we can see and measure to changes at the atomic level. At one level, chemistry is about matter and its transformations. At this level, we can actually *see* the changes, as when fireworks explode, a leaf changes color in the fall, or magnesium burns brightly in air (Fig. 1.5). This level is the **macroscopic** level, where we deal with the properties of large, visible objects. However, these events result from changes in the underworld of atoms, a world that we cannot see directly. This deeper level is called the **microscopic** level. The **symbolic** language of chemistry ties the two levels together by describing macroscopic changes in terms of atoms. We might say that a chemist thinks at the microscopic level, conducts experiments at the macroscopic level, and represents both symbolically. Success in chemistry depends on our ability to move between these viewpoints. The first three chapters in this book will introduce the major symbolic representations of chemistry.

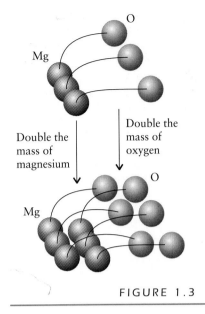

FIGURE 1.3

When the mass of magnesium is doubled, the number of magnesium atoms doubles. As a result, twice the number of oxygen atoms, and therefore twice the mass of oxygen, is needed to react fully with the magnesium.

The atoms of an element are not all *exactly* the same, because they can differ slightly in mass, as discussed in Section 1.4.

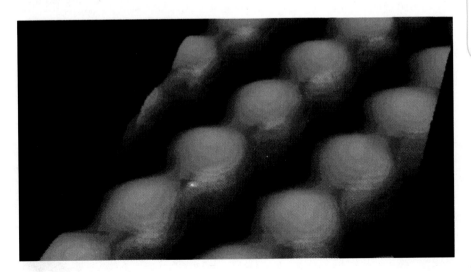

FIGURE 1.4

Individual atoms can be seen as bumps on the surface of a solid by using the technique called scanning tunneling microscopy (STM). This is an image of the surface of gallium arsenide. The gallium atoms are shown as blue and the arsenic atoms as red (these are not their actual colors).

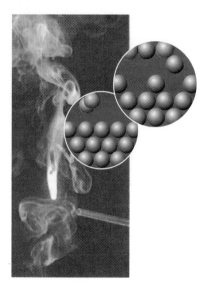

FIGURE 1.5

Magnesium burns brightly in air. In the reaction, magnesium atoms from the metal combine with atoms of oxygen and nitrogen from molecules in the air. No atoms are lost: they simply change their partners.

The names of the elements, together with their symbols, are also listed alphabetically inside the back cover of the book.

All matter is made up of various combinations of the simple forms of matter called the chemical elements. An element is a substance that consists of only one kind of atom.

1.2 Names of the Elements

The names of some elements are very ancient. For example, the name *copper* is derived from Cyprus, where the element was once mined; and the word *gold* is derived from the Old English word meaning "yellow." Some names describe a characteristic property. Chlorine, for instance, is a yellow-green gas, and its name is derived from the Greek word meaning "yellow-green." Vanadium, which forms attractively colored compounds, is named after Vanadis, the Scandinavian goddess of beauty. More recently, elements have been named by their discoverers. Some elements honor people or places; these include americium, europium, einsteinium, meitnerium, and seaborgium.

Chemists have a useful system that saves writing out the full names of the elements. Each element is represented by a **chemical symbol** made up of one or two letters. Many of the symbols are the first one or two letters of the element's name:

hydrogen H	carbon C	nitrogen N	oxygen O
helium He	aluminum Al	nickel Ni	silicon Si

Notice that the first letter of a symbol is always uppercase and the second letter is always lowercase (for example, He, not HE). Some elements have symbols formed from the first letter of the name and a later letter:

magnesium Mg   chlorine Cl   zinc Zn   plutonium Pu

Other symbols are taken from the element's name in Latin, German, or Greek. For example, the symbol for iron is Fe, from its Latin name *ferrum*. Appendix 2D lists the names and chemical symbols of all the elements and gives the origins of their names.

The most abundant element in the universe (in terms of the number of atoms) is hydrogen, followed by helium. The most abundant elements in the crust of the Earth, the top few kilometers of its surface, are oxygen and silicon. If

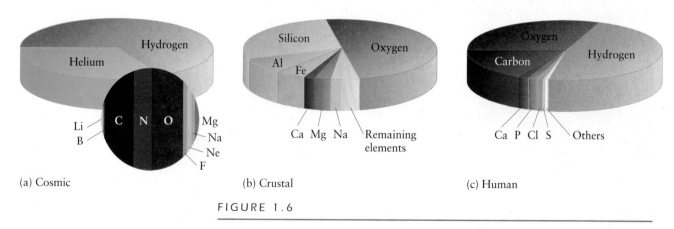

(a) Cosmic

(b) Crustal

(c) Human

FIGURE 1.6

These charts show the relative abundances of the principal elements in (a) the universe as a whole, (b) the crust of the Earth, (c) the human body.

we were to think of a human body as a source of elements, the most abundant are hydrogen, oxygen, and carbon. Figure 1.6 summarizes the distributions of the elements in these three sources.

Each element has a name and a unique chemical symbol.

Self-Test 1.1A Give the symbols of (a) rhenium and (b) boron. Name (c) Hg and (d) Zr.

[*Answer:* (a) Re; (b) B; (c) mercury; (d) zirconium]

Self-Test 1.1B Give the symbols of (a) tin and (b) sodium. Name (c) I and (d) Y.

1.3 The Nuclear Atom

Two hundred years ago, Dalton pictured atoms as featureless, indestructible spheres like billiard balls. Today we know that atoms have an internal structure and are built from even smaller particles. One element differs from another because of the differences in the internal structure of their atoms.

In this section, we see how the development of modern instrumentation and the process of scientific investigation summarized in Investigating Matter 1.1 led to our current model of the atom. This model can be summarized as follows:

1. Atoms are made up of **subatomic particles** called electrons, protons, and neutrons.

2. The protons and neutrons form a compact, central body called the **nucleus** of the atom.

3. The electrons are distributed in space like a cloud around the nucleus.

This model of an atom is called the **nuclear atom** (Fig. 1.7). Table 1.1 summarizes the properties of the three subatomic particles. Protons and neutrons have about the same mass, but protons have one unit of positive electric charge and neutrons (as their name suggests) are electrically neutral. An electron is very much less massive than a proton—nearly 2000 times lighter, in fact—and it has one unit of negative electric charge. The charges of a proton and an electron are exactly equal in magnitude but opposite in sign.

The earliest experimental evidence for the existence of an internal structure of atoms came in 1897 when the British physicist J. J. Thomson was investigating

The sign of electric charge tells us whether particles repel or attract each other. Like charges (+/+ or −/−) repel each other; opposite charges (+/−) attract.

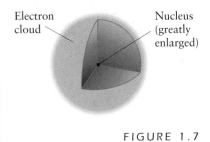

Electron cloud

Nucleus (greatly enlarged)

FIGURE 1.7

The current model of an atom pictures it as a minute central nucleus surrounded by a cloud of electrons. The nucleus is far smaller than drawn here.

Table 1.1 *Properties of subatomic particles*

Particle	Symbol	Charge*	Mass, g
electron	e$^-$	−1	9.109×10^{-28}
proton	p	+1	1.673×10^{-24}
neutron	n	0	1.675×10^{-24}

*Charges are given as multiples of the charge on a proton, which in SI units is 1.602×10^{-19} coulomb.

Scientific inquiry is a human activity with many variations, but its success lies in the use of systematic approaches. There is no one procedure that always leads from a good idea to a scientific discovery. Instead, there are a variety of scientific methods in use. One of the most common methods is a procedure that can be thought of as a series of steps. This procedure is often called the **scientific method.**

The development of the model of the nuclear atom is an example of the scientific method. The first step is the collection of **data** by making observations and measurements on a small **sample** of matter, that is, a representative piece of the material we want to study. When a pattern is observed in a series of data, it is summarized by formulating a scientific **law,** a succinct summary of observations. For example, the law of gravity summarizes measurements of the effect of gravity on objects of different mass.

In the next step, we develop **hypotheses,** possible explanations of the laws in terms of more fundamental concepts. Observation requires careful attention to detail, whereas the development of a hypothesis requires insight, imagination, and creativity. For example, although Dalton could not see individual atoms, he was able to imagine them and formulate his atomic hypothesis. It was a monumental insight that helped others understand the world in a new way.

Once a hypothesis has been developed, we design **experiments,** carefully controlled tests, to verify it. Experiments often require ingenuity and sometimes good luck. If the results of repeated experiments support the hypothesis, then we formulate a **theory,** a formal explanation of a law. A theory is sometimes interpreted through a **model,** a simplified version of the object of study. Theories and models must also be subjected to experiment and revised if they are not supported by experimental results. For example, Rutherford asked his students to perform an experiment he thought would be a simple verification of the atomic model of his time. Instead, they were able to disprove that model and revise it. Our current model of the atom has gone through many formulations and progressive revisions.

Discovering new knowledge requires not only the use of systematic procedures such as collecting data but also creativity in the design of new experiments. These chemistry students are planning and conducting experiments that are based on the same methods used by research scientists.

One outcome of Thomson's work was the invention of the cathode-ray tube used in television. Each time you see a television picture, you are watching a (sometimes) more entertaining version of Thomson's experiment.

the nature of a phenomenon called "cathode rays." These rays are emitted from a metal electrode (electrical contact) when a high potential difference (a high voltage) is applied between that electrode and another one in an evacuated glass tube (Fig. 1.8). Thomson showed that cathode rays are streams of negatively charged particles. They came from inside the atoms that made up one of the electrodes, the one called the cathode.

Thomson found that the charged particles were the same regardless of the metal he used for the cathode and measured the ratio of their charge to their mass. He concluded that they are part of the makeup of *all* atoms. These particles were named **electrons** and denoted e^-. Later workers, most notably the

American Robert Millikan, devised experiments that enabled them to determine the charge of the electron. By combining this information with the ratio of charge to mass found by Thomson, they deduced that the mass of the electron was only 9.109×10^{-28} g.

Although electrons have a negative charge, atoms overall have zero charge. Therefore, each atom must contain enough positive charge to cancel the negative charge. For years, a lively debate took place over the location of the positive charge. Because electrons can be extracted from atoms quite easily, but positive charges cannot, many scientists thought that atoms were blobs of a positively charged jellylike material, with the electrons suspended in it like raisins in pudding.

In 1908, this "plum-pudding" model was overthrown by a simple experiment. The New Zealander Ernest Rutherford was training some of his students in the use of a new piece of apparatus. Rutherford knew that some elements, including radon, emit streams of positively charged particles, which he called **alpha (α) particles.** He asked two students, Hans Geiger and Ernest Marsden, to shoot α particles toward a piece of platinum foil only a few atoms thick (Fig. 1.9). If atoms were indeed like blobs of positively charged jelly, then all the α particles would easily pass through the foil, with only an occasional slight deflection in their paths.

What Geiger and Marsden observed astonished them and everyone around them. Although almost all the α particles did pass through and were deflected only very slightly, about 1 in 20 000 was deflected through more than 90°, and a few α particles bounced straight back in the direction from which they had come.

FIGURE 1.8

A close-up of the glowing path traced by a stream of electrons near the cathode in a simple cathode-ray tube, an apparatus like that used by Thomson. Note the deflection of the cathode ray by a magnetic field.

If you are unfamiliar with scientific notation for numbers (like 9.1×10^{-28}), see Appendix 1C.

α is the first letter in the Greek alphabet.

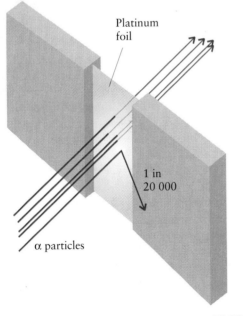

Platinum foil

1 in 20 000

α particles

FIGURE 1.9

Part of the experimental arrangement used by Geiger and Marsden. The α particles came from a sample of the radioactive gas radon. Their deflections were measured by observing the flashes of light (scintillations) produced where they struck a zinc sulfide screen. About 1 in 20 000 α particles was deflected through very large angles; most went through the platinum foil with very little deflection.

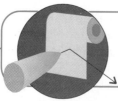

"It was almost as incredible," said Rutherford, "as if you had fired a 15-inch shell at a piece of tissue paper and it had come back and hit you."

The explanation had to be that all the positive charge in an atom is concentrated in one tiny region. Atoms are not blobs of positively charged jelly with electrons suspended in it like raisins. Instead, atoms had to contain dense pointlike centers of positive charge surrounded by a large volume of mostly empty space. Rutherford called the pointlike, positively charged region the **atomic nucleus.** He reasoned that a positively charged α particle that scored a direct hit on one of the platinum nuclei was strongly repelled by its positive charge and deflected through a large angle, like a Ping-Pong ball colliding with a cannonball (Fig. 1.10).

The electrons in an atom are thinly distributed throughout the space around the nucleus. Compared with the size of the nucleus (about 10^{-14} m in diameter), this space is enormous (about 10^{-9} m in diameter; a hundred thousand times greater).

An analogy

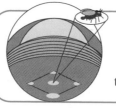

If the nucleus in a hydrogen atom were the size of a fly at the center of a baseball field, then the space occupied by the electron would be about the size of the entire stadium.

In an atom, the positive charge of the nucleus exactly balances the negative charge of the surrounding electrons. So, for each electron outside the nucleus, there must be a matching positively charged particle inside the nucleus. The positively charged particles are the **protons** (denoted p); their properties are given in Table 1.1. As remarked earlier, a proton is nearly 2000 times heavier than an electron and has one unit of positive charge.

The number of protons in an atomic nucleus is called the **atomic number, Z,** of the element. H. G. Moseley, a young British scientist, was the first to determine atomic numbers accurately. Moseley knew that when elements are bombarded with rapidly moving electrons (the cathode rays of Thomson's experiment), they emit x-rays. These rays are like light rays, but they are more energetic and can pass through many substances. He found that the energy of the x-rays emitted by an element depends on its atomic number. By studying the x-rays emitted by many elements, he was able to determine the values of Z for them. Scientists have since determined the atomic numbers of all the known elements. For example, hydrogen has $Z = 1$, so we know that the nucleus of a hydrogen atom contains one proton; helium has $Z = 2$, so its nucleus contains two protons. The most recently discovered elements have atomic numbers of over 100, so each of their nuclei contains more than 100 protons. Because all atoms of a particular element have the same number of protons, the atomic number is the characteristic that distinguishes one element from another.

Because an atom is electrically neutral, the number of electrons surrounding its nucleus must be the same as the number of protons inside the nucleus.

Therefore, Moseley's technique for counting the number of protons in the nucleus is an indirect way of counting the number of electrons too. For hydrogen, $Z = 1$, so we know at once that a hydrogen atom must have one electron. For uranium, $Z = 92$, so we know that every uranium atom has 92 electrons. The atomic numbers of the elements are listed inside the back cover and in Appendix 2D.

> *In the nuclear atom, all the positive charge and almost all the mass is concentrated in the tiny nucleus; the negatively charged electrons surround the nucleus like a cloud. The atomic number is the number of protons in the nucleus; there is an equal number of electrons outside the nucleus.*

Self-Test 1.2A How many electrons are present in an atom of (a) bismuth; (b) calcium?

[*Answer:* (a) 83; (b) 20]

Self-Test 1.2B How many electrons are present in an atom of (a) sulfur; (b) titanium?

This icon indicates a list of media resources at www. whfreeman.com/chemistry

1.4 Isotopes

Technological advances in electronics at the start of the twentieth century led to the invention of the **mass spectrometer,** a device that can be used for determining the mass of a given type of atom (Investigating Matter 1.2). Mass spectrometry was soon applied to all the elements, and we now know, for example, that the masses of all atoms are in the range 1×10^{-24} to 5×10^{-22} g.

As happens so often in science, a new and more precise technique of measurement led to a major discovery. Mass spectrometry showed that not all the atoms of a given element have the same mass. In a sample of perfectly pure neon, for example, most of the atoms have mass 3.32×10^{-23} g, which is about 20 times greater than the mass of a hydrogen atom. Some neon atoms, however, are found to be about 22 times heavier than hydrogen. Others are about 21 times heavier (Fig. 1.11). All three types of atoms have the same atomic number, so they are definitely atoms of neon.

The differences in masses among atoms of a single element suggested to scientists the existence of a third subatomic particle that has no charge and is found in the nucleus: the **neutron** (denoted n). Because neutrons have no electric charge, the number of neutrons in the nucleus affects neither the nuclear charge nor the number of electrons in the atom. However, because neutrons have about the same mass as protons, they do add substantially to the mass of the atom. Neutrons and protons are very similar particles and, because they are found in the nucleus, are jointly known as **nucleons.**

The total number of protons and neutrons in a nucleus is called the **mass number,** A, of the atom. A nucleus of mass number A is about A times heavier than a hydrogen atom, which has a nucleus that consists of a single proton. Conversely, if we know that a particular atom is a certain number of times heavier than a hydrogen atom, then we can deduce the mass number. For example, because the three varieties of neon atoms are about 20, 21, and 22 times heavier than a hydrogen atom, we know that the mass numbers of the three types of

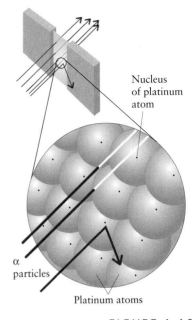

Nucleus of platinum atom

α particles

Platinum atoms

FIGURE 1.10

Rutherford's model of the atom explains why most α particles pass almost straight through, whereas a very few—those scoring a direct hit on the nucleus— undergo very large deflections. The nuclei actually are much smaller relative to the atom than shown here.

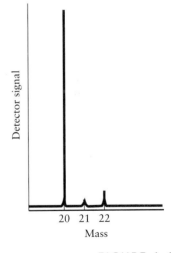

FIGURE 1.11

The mass spectrum of neon (see Investigating Matter 1.2). The locations of the peaks tell us the masses of the atoms, and their heights tell us the relative numbers of atoms with each mass.

Mass spectrometry is used to determine the relative abundance of the different isotopes of an element, but it also has a surprising number of other applications. It is used to study the composition of bone and other body tissues, to analyze the blood of newborns for congenital diseases, to detect vanishingly small concentrations of drugs in urine, to identify fraudulent materials (see Applying Chemistry: Case Study 1) and to determine the structure of the human genome.

To measure the abundance of the isotopes of an element, first we pump all the air out of the spectrometer. Then we feed the sample—either a gaseous element (neon or oxygen, for instance) or the vapor of a liquid or a solid element (such as mercury or zinc)—into the spectrometer's ionization chamber. There the atoms of the sample are struck by a beam of rapidly moving electrons. When one of the accelerated electrons collides with an atom, it knocks another electron out of the atom, thereby creating a positive ion. These ions are accelerated out of the chamber by a strong electric field applied between two metal grids. The speeds reached by the ions depend on their masses, with light ions reaching higher speeds than heavy ones.

As each accelerated ion passes through a magnetic field generated by an electromagnet, its path is bent to an extent that depends on its speed (and hence on its mass). The strength of the magnetic field is slowly changed, and a signal is produced when the magnetic field is just strong enough to bend the beam of ions so that they arrive at the detector. The mass of the type of ion formed is then calculated from the accelerating voltage and the strength of the magnetic field used to produce the signal. The mass spectrum is a plot of the detector signal against the magnetic field. The positions of the peaks are used to calculate the masses of the accelerated ions. The relative

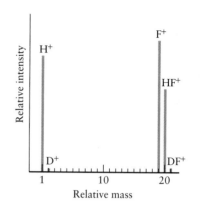

The mass spectrum of HF. The peak produced by the parent ion is that with a mass number of 20.

heights of the peaks indicate the proportions of ions of various masses and thus the relative abundance of each isotope.

Because it is convenient to use simple numerical values in calculations, masses of single atoms are often reported in terms of the **atomic mass unit** (currently denoted u; formerly, amu). One atomic mass unit, 1 u, is defined as exactly one-twelfth the mass of an atom of carbon-12. Because we know experimentally that the mass of an atom of carbon-12 is 1.9926×10^{-23} g, it follows that $1\ u = 1.6605 \times 10^{-24}$ g. The mass of any other type of atom in grams can be expressed in atomic mass units (and vice versa) by using this relation as a conversion factor.

Because a nucleon (a proton or a neutron) has a mass close to 1 u, the mass of any atom in atomic mass units is approximately equal to its mass number (the number of nucleons in its nucleus). Thus, the mass of a magnesium-24 atom is about 24 u and that of a deuterium atom (^2H) is about 2 u.

If the sample being analyzed by mass spectrometry is a compound, the formula and structure of the compound can be determined by studying the fragments. For example, the HF molecule could produce the ions HF^+, H^+, and F^+. Some of the particles that strike the detector are those that result when the molecule simply loses an electron (for example, to produce HF^+ from HF). The mass of this molecular ion, the *parent ion,* is called the *parent mass.* The parent ion has the same mass as the compound itself. For example, the parent mass of HF^+ would be 20 u and the masses of F^+ and H^+ would be 19 u and 1 u, respectively. The mass spectrum of HF would therefore look like that in the illustration, with peaks at 1, 19, and 20 u. The presence of small amounts of ^2H in naturally occurring hydrogen means that very tiny peaks at 2 and 21 are also present at high resolution.

More complex detective work is required to analyze large biomolecules and drugs. However, fragmentation generally follows predictable patterns, and one compound can be identified by comparing its mass spectrum with those of other known compounds with similar structures.

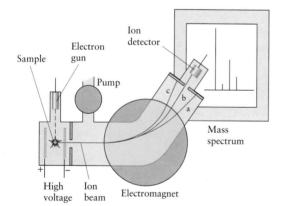

A mass spectrometer is used to measure the mass and abundance of an isotope. As the strength of the magnetic field is changed, the path of the accelerated ions moves from a to c. When the path is at b, the ion detector sends a signal to a recorder. At a fixed magnetic field strength, the three paths represent the trajectories of the ions of isotopes with three different masses, decreasing from a to c.

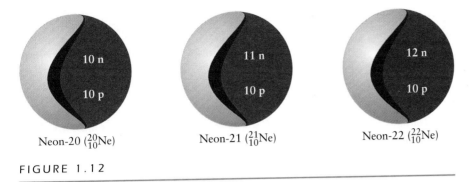

Neon-20 ($^{20}_{10}$Ne)   Neon-21 ($^{21}_{10}$Ne)   Neon-22 ($^{22}_{10}$Ne)

FIGURE 1.12

The nuclei of isotopes have the same numbers of protons but different numbers of neutrons. These three diagrams show the compositions of the nuclei of the three isotopes of neon. On this scale, the atom itself would be about 1 km in diameter. The arrangement of the protons and neutrons inside the nucleus is not shown.

neon atoms are 20, 21, and 22. Because for each of them $Z = 10$, these neon atoms must contain 10, 11, and 12 neutrons, respectively (Fig. 1.12).

Atoms with the same atomic number (belonging to the same element) but with different mass numbers are called **isotopes** of the element. Isotopes of an element have the same number of protons but different numbers of neutrons. An isotope is named by writing its mass number after the name of the element, as in neon-20, neon-21, and neon-22. Its symbol is written by adding the mass number as a superscript to the left of the chemical symbol of the element, as in ^{20}Ne, ^{21}Ne, and ^{22}Ne. You will occasionally see the atomic number included as a subscript on the lower left, as in the symbol $^{22}_{10}$Ne used in Fig. 1.12.

Because isotopes of the same element have the same number of protons and the same number of electrons, they have essentially the same properties. However, for hydrogen, the mass differences between isotopes are relatively large. As a result, the hydrogen isotopes have noticeable differences in some properties. Hydrogen has three isotopes (Table 1.2). The most common (^{1}H) has no neutrons, so its nucleus is a lone proton. The other two isotopes are less common but nevertheless are so important that they are given special names and symbols. One isotope (^{2}H) is called deuterium (D), and the other (^{3}H) is called tritium (T). A

The name *isotope* comes from the Greek words for "equal place."

Table 1.2 *Selected isotopes of some common elements*

Element	Symbol	Atomic number Z	Mass number A	Abundance, %
hydrogen	^{1}H	1	1	99.985
deuterium	^{2}H or D	1	2	0.015
tritium	^{3}H or T	1	3	—*
carbon-12	^{12}C	6	12	98.90
carbon-13	^{13}C	6	13	1.10
oxygen-16	^{16}O	8	16	99.76

*Radioactive, short-lived.

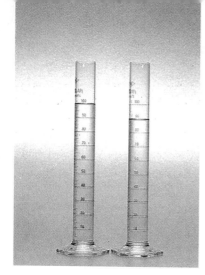

deuterium atom, with a nucleus that consists of one proton and one neutron, has about twice the mass of an ordinary hydrogen atom (more precisely, it is 1.998 times heavier). Ordinary water is composed of hydrogen and oxygen; *heavy water* is a similar combination of deuterium and oxygen. A given mass of heavy water has a volume about 11% less than the same mass of ordinary water (Fig. 1.13).

Isotopes are atoms with the same atomic number but different mass numbers. Their nuclei have the same number of protons but different numbers of neutrons. All the isotopes of a given element have the same number of electrons around the nuclei.

FIGURE 1.13

These two samples, both of which have mass 100 g, illustrate the density difference between ordinary water, H_2O, and heavy water, D_2O. The volume occupied by 100 g of heavy water (right) is 11% less than that occupied by the same mass of ordinary water (left).

Example 1.1 *Using atomic numbers and mass numbers*

A major task in the nuclear industry is the separation of the isotopes uranium-235 and uranium-238. How many protons, neutrons, and electrons are present in an atom of uranium-238?

Strategy The number identifying the isotope gives the mass number A, the total number of protons and neutrons. We find the atomic number Z (and hence the number of protons) by referring to the list of elements inside the back cover. The difference $A - Z$ is the number of neutrons. The number of electrons is the same as the number of protons; hence it is equal to Z.

Solution The mass number of uranium-238 is 238, so the total number of nucleons is 238. Uranium has $Z = 92$, so its nucleus contains 92 protons. The number of neutrons is then $238 - 92 = 146$. The atom is neutral, so it must contain 92 electrons to balance the charge of the protons.

Self-Test 1.3A How many protons, neutrons, and electrons are present in (a) an atom of nitrogen-15; (b) an atom of iron-56?

[***Answer:*** (a) 7, 8, 7; (b) 26, 30, 26]

Self-Test 1.3B How many protons, neutrons, and electrons are present in (a) an atom of oxygen-16; (b) an atom of plutonium-239?

Example 1.2 *Writing the symbols of isotopes*

Write the symbols of the isotopes of (a) argon and (b) calcium that have the same number of neutrons as ^{40}K.

Strategy First, we find the number of neutrons in ^{40}K from the fact that the mass number is the sum of the numbers of protons and neutrons. It follows that the number of neutrons is $A - Z$. Then we find the mass number of the corresponding isotopes of the target elements by adding this number of neutrons to the value of Z for each element. The symbol and value of Z for each element can be found in the list of elements inside the back cover.

Solution The atomic number of potassium is 19; therefore, the isotope ^{40}K has $40 - 19 = 21$ neutrons. (a) The elemental symbol of argon is Ar and $Z = 18$. The mass number of the isotope with 21 neutrons is $A = 21 + 18 = 39$. The symbol is ^{39}Ar. (b) The elemental symbol of calcium is Ca and $Z = 20$. The mass number of the isotope with 21 neutrons is $21 + 20 = 41$. The symbol is ^{41}Ca.

Self-Test 1.4A Write the symbols for isotopes of (a) hafnium and (b) tantalum that have the same mass number as ^{180}W. Identify the number of neutrons in each isotope.

[**Answer:** (a) ^{180}Hf, 108 neutrons; ^{180}Ta, 107 neutrons]

Self-Test 1.4B Write the symbols for the isotopes of germanium that have (a) 41 and (b) 44 neutrons.

1.5 The Origin of the Elements

All the elements except hydrogen and most of the helium were made in the stars. According to our current understanding, seconds after the universe came into being with the Big Bang, the only elements present were the two simplest, hydrogen ($Z = 1$) and helium ($Z = 2$). After millions of years, as the universe cooled, the atoms of hydrogen and helium collected together under the influence of gravity and formed large clouds. These clouds contracted, became hotter and hotter, and in due course burst into incandescence as stars (Fig. 1.14). Within the stars, intense heat causes atoms of hydrogen to smash together, merge, and become atoms of other elements. When two protons and one or two neutrons merge together, the outcome is an atom of helium ($Z = 2, A = 3$ or 4). When a third proton and more neutrons join them, we get an atom of lithium ($Z = 3, A = 6$ or 7), and so on. This merging releases even more heat, which generates starlight. So starlight—and sunlight—are signs that the heavier elements are still being formed.

Many million years after a star is formed, it exhausts its hydrogen fuel. Eventually, it begins to cool, and its outer layers may collapse, like a falling roof, into its core. This mighty starquake produces such great shock waves that the star shrugs off its outer layers and sends them into space in a huge explosion called a supernova. Six such explosions have been detected in our galaxy in the

FIGURE 1.14

Stars are born in immense clouds of molecular hydrogen and stardust such as this one in the Eagle nebula, which is also known as M16. The new stars shining through the dust will emerge as the cloud disperses.

past 1000 years, the most recent in 1987. The shock of the explosion raises the temperature in the star, making it even brighter than before. The Crab nebula (produced by the supernova of 1054) was visible in broad daylight for three weeks. At such high temperatures, even heavy atoms collide violently enough to merge and become still heavier atoms. The very heavy elements now found on Earth, including uranium and gold, were made in this way.

Stardust, the debris of exploding stars, gradually collects together under the influence of gravity and gives rise to a new generation of stars. However, not all the debris collects in a single central body; some collects into smaller bodies that go into orbit around the star. These bodies may form planets, one of which is the Earth. All matter on Earth was formed in this way in long-dead stars. All the elements—other than hydrogen and most of the helium—from which everything is made were formed inside a star. Even our flesh is stardust.

According to current ideas, hydrogen and helium were formed in the Big Bang. The heavier elements were made inside stars and then scattered throughout space.

1.6 The Periodic Table

It may seem overwhelming to realize that there are over 100 elements. How can we possibly be expected to learn all their properties? Fortunately, chemists have discovered that when the elements are listed in order of their atomic number and arranged in a special way, they form families that show regular trends in properties. As a result, we can obtain a reasonable familiarity with an element by knowing to which family it belongs and which elements are its neighbors.

The arrangement of elements that shows their family relationships is called the **periodic table** (it is printed inside the front cover of this book). The vertical columns of the periodic table are called **groups** (Fig. 1.15). The

> The periodic table is so called in recognition of the periodic recurrence of family resemblances as Z increases (Chapter 7).

> In many current versions of the periodic table, you will see a different notation for groups, with the noble gases belonging to Group VIII or Group VIIIA. These alternatives are given in the table printed inside the front cover.

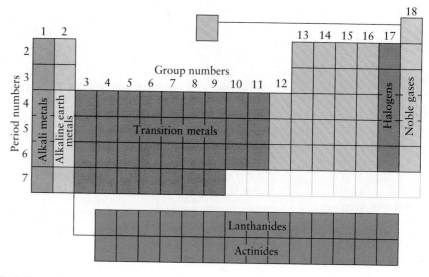

FIGURE 1.15

The structure of the periodic table, showing the names of various regions and groups. The groups are the vertical columns numbered 1 through 18. The periods are the horizontal rows. The main-group elements are hydrogen and those elements in Groups 1, 2, and 13–18.

taller columns (Groups 1, 2, and 13 through 18) are called the **main groups** of the table. The horizontal rows are called **periods** and are numbered, starting at the top of the table. Notice that the first period contains only two elements (hydrogen and helium). Because an element's location in the table is a guide to its properties, it is a good idea to learn the characteristics of the different groups and to note the position of every element we meet for the first time.

The members of each group show similarities with one another and a gradual variation in their properties. The properties of sodium (Na) in Group 1, for example, are a good clue to the properties of the other members of Group 1, namely, lithium (Li), potassium (K), rubidium (Rb), cesium (Cs), and francium (Fr). These elements are called the **alkali metals.** All of them are soft, silvery metals that melt at low temperatures. They all produce hydrogen when they come in contact with water—lithium gently, but with increasing vigor down the group. You can see how vigorously potassium reacts in Fig. 1.16. All the alkali metals must be stored under an oily liquid to keep them from coming in contact with moisture and with the oxygen in the atmosphere.

Next to the alkali metals are the metals of Group 2. These elements resemble the alkali metals in several ways but produce hydrogen less vigorously when they come in contact with water. Calcium (Ca) produces hydrogen from water at room temperature, but magnesium (Mg) does so only if it is heated. Beryllium (Be) does not produce hydrogen from water, even if it is red hot. The elements calcium (Ca), strontium (Sr), and barium (Ba) are called the **alkaline earth metals,** but the name is often extended to all the members of the group.

On the far right of the table, in Group 18, are the elements known as the **noble gases.** They are so called because they combine with very few elements—they are chemically "aloof." In fact, until the 1960s, they were called the *inert gases* because it was thought that they did not combine with any elements at all. All the Group 18 elements are colorless, odorless gases.

Next to the noble gases are the **halogens** of Group 17. Many of the properties of the halogens show regular variations from fluorine (F) through chlorine (Cl) and bromine (Br) to iodine (I). Fluorine, for instance, is a very pale yellow, almost colorless gas; chlorine, a yellow-green gas; bromine, a red-brown liquid; and iodine, a purple-black solid (Fig. 1.17). The trend in physical state down the group is from gas through liquid to solid.

Groups 3 through 11 contain the **transition metals.** They include the important structural metals titanium (Ti) and iron (Fe) and the coinage metals copper (Cu), silver (Ag), and gold (Au). The transition metals take their collective name from their role as a bridge between the chemically active metals of Groups 1 and 2 and the much less active metals of Groups 12, 13, and 14, such as indium and lead.

The long block shown below the main table consists of the **inner transition metals.** It is drawn there simply to save space, for otherwise the periodic table would be inconveniently wide. However, we need to keep in mind where these elements fit into the main part of the table. The elements in the upper row of this block, beginning with lanthanum (element 57) in Period 6, are called the **lanthanides;** and the elements in the lower row, beginning with actinium (element 89) in Period 7, are called the **actinides.**

Right at the head of the periodic table, standing alone, is hydrogen. Some tables place hydrogen in Group 1, others place it in Group 17, and yet others

FIGURE 1.16

The alkali metals react with water, producing gaseous hydrogen and heat. Potassium reacts vigorously, producing so much heat that the hydrogen ignites.

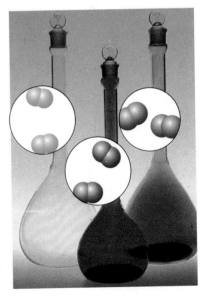

FIGURE 1.17

The halogens are colored elements. From left to right, chlorine is a yellow-green gas, bromine is a red-brown liquid (its vapor fills the flask), and iodine is a blue-black solid (note the small crystals) with a violet vapor. The insets show that the halogens all form molecules consisting of two atoms.

FIGURE 1.18

All metals can be deformed by hammering. Gold can be hammered into a sheet so thin that light can pass through it. Here it is possible to see the light of a candle through the sheet of gold. The inset shows that the atoms of gold lie in a closely packed regular array typical of metals.

Boron is sometimes classified as a nonmetal because of its appearance; here we classify it as a metalloid because of the similarity of its properties to those of silicon.

place it in both groups. We shall treat it as a very special element and place it in none of the groups.

The periodic table is an arrangement of the elements that reflects their family relationships. The members of a group typically show a smooth trend in properties.

1.7 Metals, Nonmetals, and Metalloids

The elements are classified as metals, nonmetals, and metalloids:

A **metal** conducts electricity, has a metallic luster, and is malleable and ductile.

A **nonmetal** does not conduct electricity and is neither malleable nor ductile.

A **metalloid** has the appearance and some physical properties of a metal but behaves chemically like a nonmetal.

A **malleable** substance (from the Latin word for "hammer") is one that can be hammered into thin sheets (Fig. 1.18). A **ductile** substance (from the Latin word for "drawing out") is one that can be drawn out into wires.

Copper, for example, is a metal. It conducts electricity, has a luster when polished, and is malleable. It is so ductile that it is readily drawn out to form electrical wires. Sulfur, on the other hand, is a nonmetal. This brittle yellow solid does not conduct electricity, cannot be hammered into thin sheets, and cannot be drawn out into wires. All elements that are gases at room temperature are nonmetals. The distinctions between metals and metalloids and between metalloids and nonmetals are not very precise (and not always made), but the metalloids are often taken to be the seven elements shown in Fig. 1.19.

A striking feature of the periodic table becomes clear when we mark the positions of the metals, metalloids, and nonmetals: *all the metallic elements occur on the left side and in the middle of the table and all the nonmetallic elements occur on the right.* The metalloids lie in a diagonal band between the metals and nonmetals. Thus, with a glance at the table, we can see whether an element is a metal, a metalloid, or a nonmetal. We may never have heard of indium (In), but its position in the periodic table shows us that it is a metal. Because it is close to the right side of the table, far from the alkali metals, we can even guess that it will not be a very reactive metal. This is only one example of the power of the periodic table: we will encounter many more in the following pages.

Metallic elements lie on the left of the periodic table, nonmetallic elements lie on the right, and the two are separated by a diagonal band of metalloids.

Example 1.3 *Using the periodic table*

Give the symbol and the group and period numbers of each of the following elements. Identify each one as a metal, a nonmetal, or a metalloid: (a) lead; (b) sulfur; (c) arsenic.

Strategy Find the symbol and atomic number of each element in the list of elements inside the back cover. Using the atomic number as a guide, locate the element in the periodic table inside the front cover. The group number is the number at the top of the column containing the element and the period is the number identifying its row. Metalloid elements are identified in Fig. 1.19. Elements to the left of the metalloids are metals and those to the right are nonmetals.

Solution (a) The symbol for lead is Pb. Its atomic number is 82, which places it in Group 14 and Period 6. It is found to the left of the metalloid elements, so it is a metal. (b) The symbol for sulfur is S. Its atomic number is 16, which places it in Group 16 and Period 3. It is found to the right of the metalloid elements, so it is a nonmetal. (c) Arsenic has the symbol As. Its atomic number is 33, which places it in Group 15 and Period 4. It is a metalloid.

Self-Test 1.5A Give the symbol and the group and period numbers of each of the following elements. Identify each one as a metal, a nonmetal, or a metalloid: (a) cobalt; (b) cesium; (c) chlorine.

[***Answer:*** (a) Co, Group 9, Period 4, metal; (b) Cs, Group 1, Period 6, metal; (c) Cl, Group 17, Period 3, nonmetal]

Self-Test 1.5B Give the symbol and the group and period numbers of each of the following elements. Identify each one as a metal, a nonmetal, or a metalloid: (a) selenium; (b) tin; (c) nitrogen.

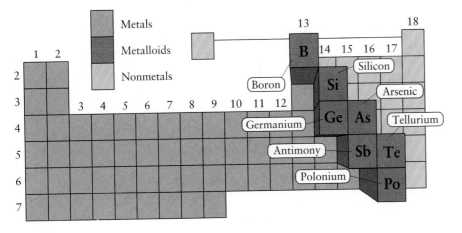

FIGURE 1.19

The location of the seven elements commonly regarded as metalloids: these elements have characteristics of both metals and nonmetals. Other elements, notably beryllium and bismuth, are sometimes included in the classification.

FIGURE 1.20

Sulfur burns with a blue flame and produces the dense gas sulfur dioxide.

COMPOUNDS

Most of the substances around us are combinations of elements rather than single elements. The ability of the elements to combine with one another is responsible for the extraordinary richness of the world, because 100 elements can occur in countless numbers of combinations.

1.8 What Are Compounds?

A **compound** is a substance that consists of two or more different elements with their atoms in a definite, characteristic ratio. Water, for instance, is a compound of hydrogen and oxygen, with two hydrogen atoms for each oxygen atom. Common table salt, sodium chloride, is another compound. This compound invariably contains one sodium atom for each chlorine atom. We would find the same 1:1 ratio in samples of sodium chloride from California, Australia, Siberia, Antarctica, or Mars. This regularity is called the **law of constant composition.** Its discovery was historically important, because it suggested to chemists that compounds consist of specific combinations of atoms.

Compounds are classified as either organic or inorganic. **Organic compounds** are compounds containing the element carbon and usually hydrogen too. They include fuels, such as methane and propane, and sugars, such as glucose. Millions of other substances are also organic compounds. These compounds are called organic because it was once believed that they could be formed only by living organisms. That assumption is now known to be incorrect.

Inorganic compounds are all the other compounds; they include water, calcium sulfate, ammonia, silica, hydrochloric acid, and many, many more. In addition, some very simple carbon compounds, particularly carbon dioxide and the carbonates, which include chalk (calcium carbonate), are treated as inorganic compounds.

The elements in a compound are not just mixed together. Their atoms are actually joined, or *bonded,* to one another in a specific way. For example, when a mixture of hydrogen and oxygen is ignited, an explosion takes place. A lot of heat is released and instead of two colorless gases, we end up with a clear, colorless liquid: water. The atoms of hydrogen and oxygen have formed chemical bonds with one another and each oxygen atom is now joined to two hydrogen atoms. When sulfur is ignited in air, sulfur and oxygen atoms join to form sulfur dioxide. Solid yellow sulfur and odorless oxygen gas have produced a colorless, pungent, poisonous gas (Fig. 1.20).

The atoms of a compound can be bonded together to form molecules, or they can be present as ions:

A **molecule** is a definite and distinct electrically neutral group of atoms bonded together.

An **ion** is a positively or negatively charged atom or bonded group of atoms.

We classify a compound as **molecular** if it consists of molecules and as **ionic** if it consists of ions.

The properties of a compound are a good guide to its classification. All compounds that are gases or liquids at room temperature are molecular. Most organic compounds are molecular. Solid compounds are more difficult to classify. One good classification clue is that ionic compounds melt at high temperatures, whereas molecular compounds usually melt at low temperatures.

The name *ion* comes from the Greek word for "go," because charged particles go either toward or away from a charged electrode.

The different types of compounds are discussed in more detail in Chapters 8–10.

Some elements are also found in molecular form. Except for the noble gases, all the gaseous elements are found as **diatomic molecules,** molecules that consist of two atoms. For example, molecules of hydrogen gas contain two hydrogen atoms bonded together. We represent the hydrogen molecule by the chemical formula H_2 (read "H-two"). The **chemical formula** of a *species* is a statement of its composition in which the chemical symbols tell us which elements are present and the subscripts tell us how many atoms of each element the molecule contains. Oxygen exists as O_2 and chlorine as Cl_2. Solid sulfur exists as S_8 molecules and phosphorus as P_4 molecules.

> We shall use the term *species* when we need to include atoms, molecules, and ions.

Compounds are specific combinations of elements. They are classified as either molecular or ionic.

1.9 Molecular Compounds and Formulas

What do molecules look like? A molecule is a specific *arrangement* of linked atoms that we can portray in different ways, depending on what aspect of the structure we want to emphasize. A simple representation is a **structural formula** like (**1**). Here we represent the atoms by their elemental symbols and use lines to show which atoms are joined together. A structural formula gives very little information about the three-dimensional shape of the molecule, but it is compact and very easy to draw.

Chemists commonly use a **ball-and-stick model** to indicate the shape and size of a molecule (Fig. 1.21). The colored balls depict the atoms and the sticks indicate the links between them. We use ball-and-stick models frequently in this book, because they convey important aspects of the structure clearly and all the atoms can be seen. However, if we want to show a highly complex structure, we might use a **tube structure** (Fig. 1.22). This simple representation omits the balls and represents the atoms and the links between them by colored lengths of tube.

Chemists commonly use a **space-filling model,** in which atoms are represented by colored spheres that fit into one another, to portray the shape

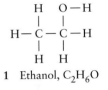

1 Ethanol, C_2H_6O

> Colors are used in molecular graphics to help identify elements; they do not indicate actual colors of atoms.

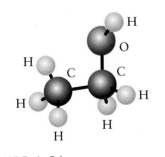

FIGURE 1.21

A ball-and-stick model of a molecule (ethanol, in this case) uses colored balls to depict the atoms and sticks to indicate the links between them. Black denotes carbon, red oxygen, and gray hydrogen. (Other models also use blue for nitrogen and yellow for sulfur.)

FIGURE 1.22

This tube structure of ethanol focuses on the links between atoms by representing the atoms and the links between them by colored lengths of tube. The colored parts of the tubes have the same color coding as the balls in the ball-and-stick model.

FIGURE 1.23

(a)

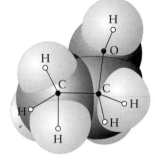

(b)

(a) A space-filling representation of the ethanol molecule, generated by computer graphics. (b) The superimposed lines and labels identify the atoms and show the pattern of bonds between them. The small circles identify the locations of the centers of the atoms.

of a molecule more accurately. Figure 1.23 shows a space-filling model of an ethanol molecule. It is hard, however, to distinguish the atoms in pictures of space-filling models and not easy to draw them freehand. Computer graphics have transformed the ability of chemists to portray molecular structures, but it is still a good idea to use model kits and handle actual molecular models.

In the models of water (**2**), methane (**3**), and carbon dioxide (**4**) molecules pictured, the black spheres represent carbon atoms; the pale gray spheres, hydrogen atoms; and the red spheres, oxygen atoms. We use these color codes throughout this text. As we can see from these diagrams, molecules have characteristic shapes. All molecules of a given substance have the same arrangement of atoms. The spatial arrangement of atoms contributes to the properties of compounds and will play an important role in the remainder of this book.

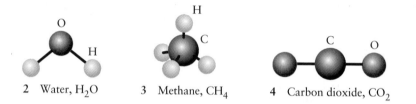

2   Water, H_2O 3   Methane, CH_4 4   Carbon dioxide, CO_2

For molecular compounds, it is common to give the **molecular formula,** a chemical formula that shows how many atoms of each type of element are present in a molecule. For instance, the molecular formula of water is H_2O: the subscript 2 shows that there are two H atoms present in the molecule, and the single O shows that there is one oxygen atom present. Likewise, the molecular formula for methane, CH_4, another molecular compound, shows that each molecule consists of one C atom and four H atoms. The molecular formula for estrone, a female sex hormone, is $C_{18}H_{22}O_2$, showing that a single molecule of estrone consists of 18 C atoms, 22 H atoms, and 2 O atoms. A molecule of a male sex hormone, testosterone, differs by only a few atoms: its molecular formula is $C_{19}H_{28}O_2$.

Molecular models indicate the size and shape of molecules. The molecular formula of a molecular compound indicates the numbers of atoms of each type of element present in each molecule.

1.10 Ions and Ionic Compounds

Ionic compounds consist of positive and negative ions held together by the attraction between their opposite charges. An example is sodium chloride, which consists of sodium ions (denoted Na^+) and chloride ions (denoted Cl^-). Each crystal of the compound is an orderly array of a vast number of alternating Na^+ and Cl^- ions (Fig. 1.24). When you take a pinch of salt, you are picking up crystals, each consisting of huge numbers of ions.

— Na^+

— Cl^-

FIGURE 1.24

An ionic solid consists of an array of cations and anions stacked together. This illustration shows the arrangement of sodium cations (Na^+) and chlorine anions (chloride ions, Cl^-) in a crystal of sodium chloride (common table salt). The faces of the crystal are where the stacks of ions come to an end.

Ions are either positively or negatively charged. A positively charged ion is called a **cation**, and a negatively charged ion is called an **anion**. In sodium chloride, the sodium ions are the cations and the chloride ions are the anions.

Knowing about the nuclear model of the atom helps us understand how **monatomic ions** (single-atom ions) form. Because an atom has exactly the same number of both electrons and protons, an atom itself is electrically neutral. However, the electrons are outside the nucleus and can be removed relatively easily. When an electron is removed, the charge of the remaining electrons no longer balances the positive charge of the nucleus. Because an electron has one unit of negative charge, removing one electron from a neutral atom leaves behind a cation with one unit of positive charge. For example, a sodium cation, Na^+, is a sodium atom that has lost one electron and hence has a single positive charge (Fig. 1.25).

Each electron lost from an atom increases the overall positive charge of the atom by one unit. For instance, when a calcium atom loses two electrons, it becomes the doubly positively charged calcium ion, Ca^{2+} (read "calcium two-plus"). When an aluminum atom loses three electrons, it becomes the triply charged aluminum ion, Al^{3+} ("aluminum three-plus").

Each electron gained by an atom increases the atom's negative charge by one unit. So, when a chlorine atom gains an electron, it becomes the singly negatively charged chloride ion, Cl^-. When an oxygen atom gains two electrons, it becomes the doubly charged oxide ion, O^{2-}. When a nitrogen atom gains three electrons, it becomes the triply charged nitride ion, N^{3-}.

The periodic table can help us decide what charge to expect on cations. *For elements in Groups 1 and 2, the charge of the cations the elements can form is equal to the group number.* Thus, cesium in Group 1 forms Cs^+ ions; barium in Group 2 forms Ba^{2+} ions. *The lighter elements of Groups 12 and 13 tend to form cations with a charge equal to the group number minus 10.* For example, cadmium (Group 12, $12 - 10 = 2$) forms Cd^{2+} ions and gallium (Group 13, $13 - 10 = 3$) forms Ga^{3+} ions. Figure 1.26 lists the cations formed by some of the elements. This illustration also shows that atoms of the transition metals and some of the heavier metals of Groups 12, 13, and 14 can form

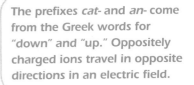

The prefixes *cat-* and *an-* come from the Greek words for "down" and "up." Oppositely charged ions travel in opposite directions in an electric field.

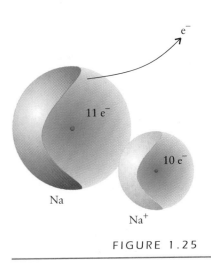

FIGURE 1.25

A neutral sodium atom (left) consists of a nucleus that contains 11 protons and is surrounded by 11 electrons. When 1 electron is lost, the remaining 10 electrons cancel only 10 of the proton charges, and the resulting ion (right) has 1 overall positive charge.

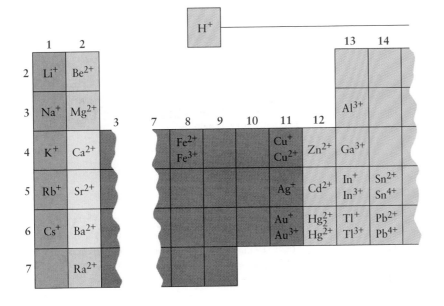

FIGURE 1.26

The typical cations formed by a selection of elements in the periodic table. The transition metals form a wide variety of cations; we have shown only a few.

either of two cations with different charges. An iron atom, for instance, can lose two electrons to become Fe^{2+} or three electrons to become Fe^{3+} (Fig. 1.27). Copper can lose one electron to form Cu^+ or two electrons to become Cu^{2+}.

The periodic table can also help us decide what charge to expect on monatomic anions. Whereas atoms of the metallic elements typically lose electrons and form cations, atoms of nonmetals typically gain electrons and form anions (Fig. 1.28). Some of the anions formed by the elements are listed in Fig. 1.29. Notice that *an element on the right of the table forms an anion with a charge equal to its group number minus 18.* Thus, oxygen (Group 16, $16 - 18 = -2$) forms the oxide ion, O^{2-}. Phosphorus (Group 15, $15 - 18 = -3$) forms the phosphide ion, P^{3-}.

Metallic elements typically form cations, nonmetallic elements typically form anions, and their charges are related to the group to which they belong in the periodic table.

Example 1.4 *Predicting the ion that a main-group element is likely to form*

Selenium is used in some photosensitive devices, including copying machines, and calcium is present in many rigid materials, such as the calcium carbonate of shells and the calcium phosphate of vertebrate skeletons. What ions would you expect (a) selenium and (b) calcium to form?

Strategy To predict whether the element is likely to form a cation or an anion, decide from its location in the periodic table whether it is a metal or a nonmetal. Metals typically form cations; nonmetals typically form anions. To determine the charge of an anion, identify the group to which the element belongs, and then subtract 18. The charge on a cation in Groups 1 and 2 is the same as the group number. The lighter Group 13 elements commonly form +3 cations.

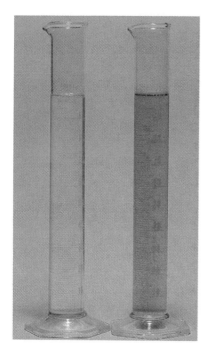

FIGURE 1.27

Iron is an example of an element that can form more than one type of ion. Aqueous solutions containing Fe^{2+} are usually pale green, and solutions containing Fe^{3+} in the form of the complex ion $FeOH(H_2O)_5^{2+}$ are usually yellow-brown.

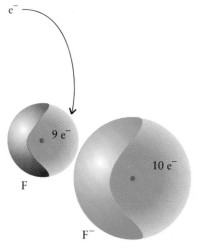

FIGURE 1.28

A neutral fluorine atom (left) consists of a nucleus that contains nine protons and is surrounded by nine electrons. When the atom gains 1 more electron, the 9 proton charges cancel all but one of the 10 electron charges, and the resulting ion (right) has 1 overall negative charge.

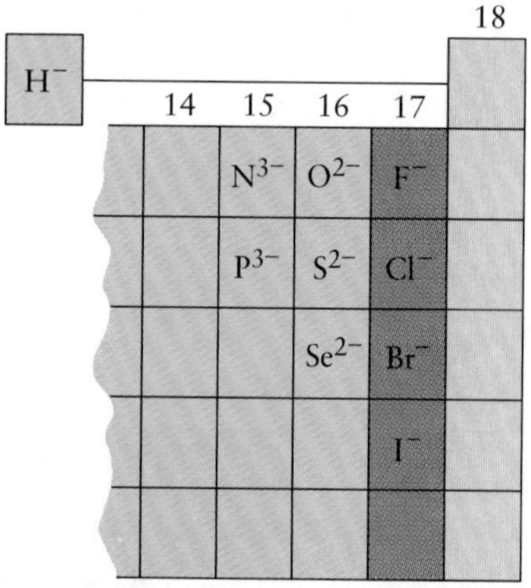

FIGURE 1.29

The typical monatomic anions formed by a selection of elements in the periodic table. Only the nonmetals are shown, for only they form monatomic anions.

Solution (a) Selenium (Se) is a nonmetal, so it can be expected to form an anion. It belongs to Group 16. Because $16 - 18 = -2$, it can be expected to form the anion Se^{2-}. (b) Calcium (Ca) is a metal, so it tends to form cations. It belongs to Group 2 and forms the cation Ca^{2+}.

Self-Test 1.6A What ions are (a) iodine and (b) aluminum likely to form?

[*Answer:* (a) I^-; (b) Al^{3+}]

Self-Test 1.6B What ions are sulfur and potassium likely to form?

1.11 Chemical Formulas of Ionic Compounds

An ionic compound does not consist of individual molecules. Instead, a crystal of an ionic compound (like a crystal of salt) has an arbitrarily large number of ions of each kind (recall Fig. 1.24). We cannot write a molecular formula for an ionic compound. However, because the *ratio* of the number of cations to the number of anions is the same for any sample of a given compound, we can use that ratio to write its formula. The chemical formula of an ionic compound shows the relative numbers of each type of ion in terms of the smallest whole numbers. In sodium chloride, for instance, there is one Na^+ cation for each Cl^- ion, so its formula is NaCl regardless of the size of the crystal. The formula $MgCl_2$ tells us that, for each Mg^{2+} ion in a crystal, there are two Cl^- ions. As we see from these examples, we omit the charges on the ions when writing the formula of a compound. In each case, the ions combine in such a way that the positive and negative charges balance: *all compounds are electrically neutral overall.*

Many ions are **polyatomic,** meaning that they consist of more than one atom bonded together and have an overall positive or negative charge. An example of a polyatomic cation is the ammonium ion, NH_4^+ (**5**). As its structure shows, the ammonium ion consists of one N atom bonded to four H atoms. One N atom and four H atoms together would normally have a total of $7 + (4 \times 1) = 11$ electrons. However, the ammonium ion has only 10, so it is a polyatomic cation with a single positive charge.

The most common polyatomic anions are the **oxoanions,** polyatomic anions that contain oxygen. Examples of polyatomic anions are carbonate (CO_3^{2-}), nitrate (NO_3^-), phosphate (PO_4^{3-}), and sulfate (SO_4^{2-}) anions. A carbonate ion (**6**) has three O atoms bonded to a carbon atom and an additional two electrons. A phosphate ion (**7**) consists of four O atoms bonded to a phosphorus atom, with three additional electrons. Table 1.3 lists the most common polyatomic ions.

In sodium carbonate, there are two Na^+ cations per carbonate ion (CO_3^{2-}), so its formula is Na_2CO_3. When a subscript has to be added to a polyatomic ion, the ion is written within parentheses, as in $(NH_4)_2SO_4$, where $(NH_4)_2$ means that there are two NH_4^+ ions for each SO_4^{2-} ion in ammonium sulfate.

A **formula unit** is a group of ions with a composition given by the formula of an ionic compound. For example, the formula unit of sodium chloride, NaCl, consists of one Na^+ ion and one Cl^- ion (Fig. 1.30a). The formula unit of ammonium sulfate, $(NH_4)_2SO_4$, consists of two NH_4^+ ions and one SO_4^{2-} ion (Fig. 1.30b). The formula unit of an ionic compound can be thought of as the smallest unit of that compound: it is the equivalent of a "molecule" of the compound. The formula unit of a molecular compound is the molecule itself.

Just as a monatomic ion can be thought of as an atom with a charge, a polyatomic ion can be thought of as a molecule with a charge; however, the uncharged parent molecules often do not exist.

5 Ammonium ion, NH_4^+

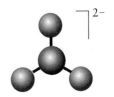

6 Carbonate ion, CO_3^{2-}

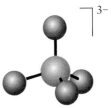

7 Phosphate ion, PO_4^{3-}

(a)

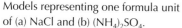

(b)

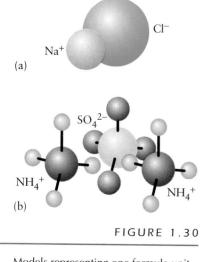

FIGURE 1.30

Models representing one formula unit of (a) NaCl and (b) $(NH_4)_2SO_4$.

Table 1.3 *Common polyatomic ions*

Cations		Anions that end in -ide	
NH_4^+	ammonium	CN^-	cyanide
Hg_2^{2+}	mercury(I)	O_2^{2-}	peroxide
		OH^-	hydroxide

Oxoanions

−1 charge		−2 charge		−3 charge	
$CH_3CO_2^-$	acetate	CO_3^{2-}	carbonate	PO_4^{3-}	phosphate
NO_2^-	nitrite	$C_2O_4^{2-}$	oxalate		
NO_3^-	nitrate	CrO_4^{2-}	chromate		
ClO^-	hypochlorite	$Cr_2O_7^{2-}$	dichromate		
ClO_2^-	chlorite	SO_3^{2-}	sulfite		
ClO_3^-	chlorate	SO_4^{2-}	sulfate		
ClO_4^-	perchlorate				
MnO_4^-	permanganate				

We can often decide whether a substance is an ionic compound, a molecular compound, or an element by examining its formula. **Binary** molecular compounds, those built from two elements, are typically formed from two nonmetals (like hydrogen and oxygen, the elements in water). Ionic compounds are typically formed from the combination of a metallic element with nonmetallic elements (such as potassium, sulfur, and oxygen, the elements in potassium sulfate, K_2SO_4). Ionic compounds typically contain one metallic element; the principal exceptions are compounds containing the ammonium ion, such as ammonium nitrate, which are ionic even though all the elements present are nonmetallic.

> *The chemical formula of an ionic compound shows the ratio of the number of cations to the number of anions present in the compound in terms of the smallest whole numbers. A formula unit of an ionic compound is that collection of ions having the same number of atoms of each element as appears in its formula.*

Example 1.5 *Interpreting chemical formulas*

Identify the type of each substance and give the number of atoms of each element in the formula units or molecules of the compounds (a) N_2O_4; (b) $(NH_4)_3PO_4$; (c) P_4.

Strategy If only one element is present, the substance is an element. If more than one element is present, the substance is a compound. Two nonmetals typically form a molecular compound (but be careful to note whether the ammonium ion is present). A metal and a nonmetal typically form an ionic compound. Each chemical symbol represents an atom of an element. Find the number of atoms of each element from its subscript. If the element is part of a polyatomic ion in parentheses, multiply the subscript of the element by the subscript outside the parentheses to get the total number of atoms.

Solution (a) The compound N_2O_4 contains two nonmetals, so it is a molecular compound. The subscripts 2 on the N and 4 on the O tell us that each N_2O_4 molecule has two N atoms and four O atoms. (b) The compound $(NH_4)_3PO_4$ contains the ammonium ion and thus must be an ionic compound. It consists of two different polyatomic ions, one with a subscript. Each NH_4^+ ion has one N atom and four H atoms. Because there are three NH_4^+ ions in the formula, there are three times as many N and H atoms: 3 N atoms and 12 H atoms. There are also one P atom and four O atoms. The N, H, P, and O atoms are therefore present in the ratios 3:12:1:4 in the formula unit and, in fact, in any sample of the compound. (c) The substance P_4 contains only one element and is a molecular element in which each molecule contains four phosphorus atoms.

Self-Test 1.7A Azurite, $Cu_3(OH)_2(CO_3)_2$, is a blue pigment that was used for centuries. (a) What is the ratio of the numbers of atoms of each element in azurite? (b) What type of substance is it?

[*Answer:* (a) The Cu, O, H, and C atoms are present in azurite in the ratios 3:8:2:2; (b) ionic compound]

Self-Test 1.7B $Ca_3(PO_4)_2$ is the primary component of bone. (a) What is the ratio of the numbers of atoms of each element in $Ca_3(PO_4)_2$? (b) What type of substance is it?

MIXTURES

Each element and compound is a pure substance. However, most materials are neither single elements nor single compounds. Instead, they are **mixtures** of these simpler substances, with one substance mingled with another. We use mixtures every day. Gasoline is a mixture of hydrocarbons and additives blended together to achieve efficient combustion. Many alloys, which are mixtures of metals, are formulated for maximum strength and resistance to corrosion. People and plants are highly complex, highly organized mixtures of mostly water, organic compounds, and calcium phosphate. Typically, a medicine is a mixture of various ingredients chosen to achieve an overall biological effect. Much the same can be said of a perfume.

1.12 Chemical and Physical Properties

The "properties" of matter, its distinguishing characteristics, are used both to identify substances and to separate them from mixtures. A **physical property** of a substance refers to a characteristic that we can observe or measure without changing the identity of the substance. For example, two physical properties of hydrogen are described when we say that hydrogen is a colorless gas. Physical properties include characteristics such as color, melting point (the temperature at which a solid turns into a liquid), hardness, and density. A **chemical property** refers to the ability of a substance to change into another substance. For example, hydrogen is converted into water when it burns in oxygen; this characteristic behavior is a chemical property of hydrogen. Table 1.4 lists various common physical and chemical properties.

An important physical property of a substance is its **physical state**, also called its **state of matter.** The three most common states of matter are solid, liquid, and gas (Fig. 1.31).

Table 1.4 *Common physical and chemical properties**

Physical property	Chemical property
melting point	reaction with acids
boiling point	reaction with bases (alkalis)
vapor pressure	reaction with oxygen (combustion)
color	ability to act as oxidizing agent
state of matter	ability to act as reducing agent
density	reaction with other elements
electrical conductivity	decomposition into simpler substances
solubility	corrosion
adsorption to a surface	
hardness	

*Many of the terms in this table are explained later in the text.

A **solid** is a rigid form of matter.

A **liquid** is a fluid form of matter that occupies the lower part of the container it occupies and has a well-defined surface.

A **gas** is a fluid form of matter that fills any container it occupies.

The term **vapor** is used to denote the gaseous form of a substance that is normally a solid or a liquid. Thus, we speak of ice (the solid form of water), liquid water, and water vapor.

A compound has a fixed composition and therefore has fixed properties; a mixture may have any composition and therefore has variable properties. There are always two H atoms for each O atom in a sample of the compound water, but sugar and sand, for instance, may be mixed in any proportions. Because the components of a mixture are merely mingled with one another rather than

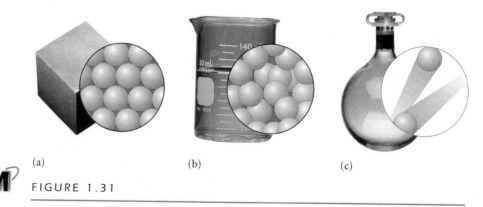

(a)   (b)   (c)

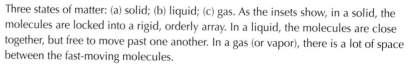

FIGURE 1.31

Three states of matter: (a) solid; (b) liquid; (c) gas. As the insets show, in a solid, the molecules are locked into a rigid, orderly array. In a liquid, the molecules are close together, but free to move past one another. In a gas (or vapor), there is a lot of space between the fast-moving molecules.

Table 1.5 *Differences between mixtures and compounds*

Mixture	Compound
Components can be separated by using physical techniques.	Components cannot be separated by using physical techniques.
Composition is variable.	Compositon is fixed.
Properties are related to those of its components.	Properties are unlike those of its components.

being joined with chemical bonds, they retain their own properties in the mixture. For example, a mixture of sugar and sand is both sweet (from the sugar) and gritty (from the sand). When shaken with water, the sugar dissolves but the sand does not; their different solubilities allow them to be separated easily. In contrast, a compound may have properties very different from those of the elements from which it is built. Water is very different from the gases hydrogen and oxygen from which it is formed, and the hydrogen and oxygen cannot be easily separated from water. Table 1.5 summarizes the differences between compounds and mixtures.

Physical properties are those in which a substance retains its identity.
Chemical properties are those in which the substance changes its identity.

Self-Test 1.8A Determine which of the following properties are chemical and which are physical: (a) the hardness of a metal; (b) the ductility of a metal; (c) the corrosiveness of an acid; (d) the freezing point of a liquid.

[*Answer:* (a) Physical; (b) physical; (c) chemical; (d) physical]

Self-Test 1.8B Identify the following as either a chemical or a physical property: (a) the formation of hydrogen gas when a Group 1 metal is added to water; (b) the formation of tarnish on silver metal; (c) the pressure of air.

1.13 Types of Mixtures

We can identify the different components of some mixtures with the unaided eye or an optical microscope. Such a patchwork of different substances is called a **heterogeneous mixture** (Fig. 1.32). Many of the rocks that form the landscape are heterogeneous mixtures, as are mixtures of sugar and sand, no matter how thoroughly they are mixed. Milk, which looks like a pure substance, is a heterogeneous mixture: through a microscope, we can see the individual globules of butterfat.

In some mixtures, we cannot make out distinct components, even with a very powerful microscope: the molecules or ions of the components are so well mixed that the composition is the same throughout, no matter how large or small the sample. Such a mixture is called a **homogeneous mixture** (Fig. 1.33). For example, syrup is a homogeneous mixture of sugar and water. The molecules of sugar are separated and mixed so thoroughly with the water that no distinctive regions or separate particles can be seen with a microscope.

Homogeneous mixtures are called **solutions.** Vinegar is a solution of acetic acid in water. Seawater is a solution of salt (sodium chloride) and many other

FIGURE 1.32

This piece of granite is a heterogeneous mixture of several substances.

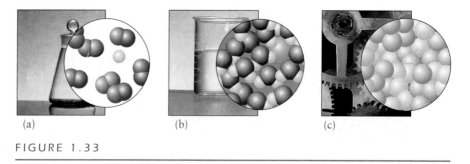

(a)   (b)   (c)

FIGURE 1.33

Three examples of homogeneous mixtures and representations of their nature at the molecular level. (a) Air is a homogeneous mixture of many gases, including the nitrogen, oxygen, and argon depicted here. (b) Table salt dissolved in water consists of sodium ions and chloride ions distributed among water molecules. (c) Brass is a solid mixture of copper and zinc.

substances in water. The component of the solution that is present in the larger amount (water, in these examples) is called the **solvent,** and the dissolved substances are the **solutes** (Fig. 1.34). However, we sometimes identify the less abundant substance as the solvent, especially when it determines the physical state of the solution. For example, in a heavy syrup, the sugar is present in a much greater amount than the water, but water is still considered to be the solvent.

When we use the everyday term *dissolving,* we mean the process of producing a solution. We normally dissolve a solid in a liquid solvent by shaking it up with the solvent or by stirring, as we do when we dissolve sugar in a cup of coffee. The opposite of dissolving occurs when a component comes out of solution. This process is called **crystallization** if the solute slowly comes out of solution as crystals. For example, salt crystals left when water evaporates line the shores of the Great Salt Lake in Utah. In the process of **precipitation,** a solute comes out of solution rapidly as a finely divided powder called a **precipitate.** Precipitation is often almost instantaneous (Fig. 1.35).

Beverages and seawater are examples of **aqueous solutions,** solutions in which the solvent is water. Aqueous solutions are very common both in everyday life and in chemical laboratories; for that reason, most of the solutions discussed in this text are aqueous. **Nonaqueous solutions** are solutions in which the solvent is not water. Although they are less common than aqueous solutions, they have important uses. In "dry cleaning," the grease and dirt on fabrics are dissolved in a solvent such as the compound tetrachloroethene, C_2Cl_4. In **solid solutions** the solvent is a solid. An example is one form of brass, which is a solution of copper in zinc. A sample of dry air could be regarded as a solution of several gases in nitrogen (the major component); however, gaseous solutions are more commonly referred to simply as mixtures.

Mixtures display the properties of their constituents; they differ from compounds, as summarized in Table 1.5. Mixtures are classified as homogeneous or heterogeneous; solutions are homogeneous mixtures.

1.14 Separation Techniques

To analyze the composition of a sample that we suspect is a mixture, we need to separate its components and identify the individual substances present.

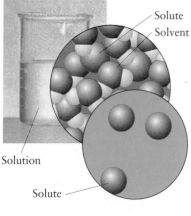

Solute
Solvent

Solution

Solute

FIGURE 1.34

Solutions are homogeneous mixtures. The solvent (in this case, water) is the substance present in the larger amount. The dissolved substance is called the solute.

Methods for separating mixtures all rely on differences in the physical properties of the components. A compound can be separated into its elements only by chemical change (because that separation involves a change in the identity of the substances). However, because the components of a mixture retain their individual physical properties in the mixture, they can be separated by physical means. Therefore, another definition of a mixture is *a form of matter that can be separated into its components by making use of the different physical properties of the components* (Fig. 1.36).

The technique of **filtration** makes use of differences in solubility (the ability to dissolve in a liquid). The sample is shaken with a liquid and then poured through a fine mesh, the filter. The soluble material dissolves in the liquid and passes through the filter. The insoluble material does not dissolve and is captured by the filter. The technique can be used to separate sugar from sand, because sugar is soluble in water but sand is not.

The technique of **distillation** makes use of differences in boiling points. Distillation separates the components of a mixture by vaporizing one or more of the components (Fig. 1.37). It can be used to remove water (which boils at 100°C in the pure state) from salt (which boils at a much higher temperature and is therefore left behind when the water evaporates).

One of the most sensitive techniques available for separating and identifying the components of a mixture is **chromatography.** This technique relies on the different abilities of substances to **adsorb,** or stick, to surfaces. In the simplest version of the technique, the mixture is washed across a strip of filter paper (Fig. 1.38). Substances that adsorb most weakly move farther than others and, if they are colored, give rise to separate patches of color on the paper.

In **gas-liquid chromatography** (GLC), the sample is vaporized and carried in a stream of helium through a long narrow tube. The tube is coated or packed with alumina (aluminum oxide, Al_2O_3) soaked in a liquid that does not vaporize readily. The component of the mixture that sticks to the coated alumina

FIGURE 1.35

Precipitation is the formation of an insoluble substance. Here lead(II) iodide, PbI_2, which is an insoluble yellow solid, precipitates when we mix colorless solutions of lead(II) nitrate, $Pb(NO_3)_2$, and potassium iodide, KI.

The first known distillation apparatus, in the first century AD, is attributed to Mary the Jewess, an alchemist who lived in the Middle East.

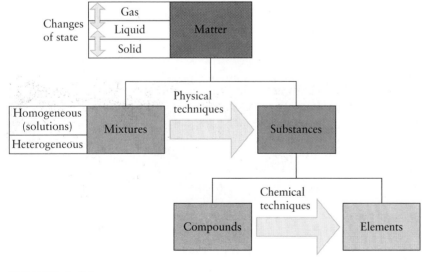

FIGURE 1.36

The hierarchy of materials: matter, mixtures, and substances (which include compounds and elements). Physical techniques of separation—techniques that depend on differences in physical properties—are indicated by the upper horizontal arrow.

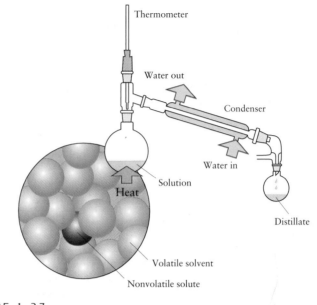

FIGURE 1.37

The technique of distillation, which is used to remove a liquid from a solid. The solution is heated, the liquid boils off, condenses in the water-jacketed tube (the condenser), and is collected.

The colored patterns that appear gave rise to the name chromatography, *which comes from the Greek words for "color writing."*

least strongly emerges first from the other end (Fig. 1.39). As time passes, the other components of the original mixture emerge, and each one is detected electronically and signaled by a peak on a chart called a **chromatogram.** A chromatogram is like a fingerprint of the mixture (Fig. 1.40). Gas-liquid chromatography is used to detect narcotics and explosives in airline baggage; the equipment "sniffs" the baggage and is highly sensitive to very small amounts of substance.

Chromatography is described in more detail in Investigating Matter 12.1.

FIGURE 1.38

In paper chromatography, the components of a mixture are separated by washing them along a paper—the support—with a solvent. A simple form of the technique is shown here. On the left is a dry filter paper to which a drop of food coloring has been applied. Solvent is then added to the center of the filter paper (middle). The filter paper on the right has been allowed to dry after the solvent spread out to the edges of the paper, carrying two components of the coloring matter to different distances as it spread. The dried support showing the separated components is a simple chromatogram.

Chromatography is often used in crime laboratories to separate the substances in a sample used as evidence. Techniques such as mass spectrometry and elemental analysis are then used to identify the composition of each substance (Applying Chemistry: Case Study 1).

Mixtures are separated by making use of the differences in physical properties of the components; separation techniques include filtration, distillation, and chromatography.

THE NOMENCLATURE OF COMPOUNDS

We shall meet many compounds in this text and learn their names as we go along. However, learning some basic rules for forming their names saves a lot of memorization. Many compounds were given **common names** before their compositions were known. Common names include water, salt, sugar, ammonia, and quartz. A **systematic name,** on the other hand, reveals which elements are present and, in some cases, how their atoms are arranged. The systematic name of table salt, for instance, is sodium chloride, which indicates at once that it is a compound of sodium and chlorine. The systematic naming of compounds, which is called **chemical nomenclature,** follows a set of rules. The name of each compound need not be memorized, only the rules.

1.15 Names of Ions

The name of a monatomic cation is the same as the name of the element, with the addition of the word *ion,* as in sodium ion for Na^+. When an element can

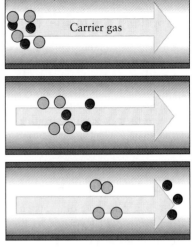

Stationary phase

Carrier gas

FIGURE 1.39

In gas-liquid chromatography (GLC), the mixture of gases or vapors is separated as it travels through a long, coated tube. Some molecules are adsorbed on (stick to) the coating (the stationary phase) more readily than others do and therefore emerge later, but eventually all pass through in the stream of carrier gas (typically, helium). The less readily adsorbed component (red) emerges first, followed by the more readily adsorbed component (yellow).

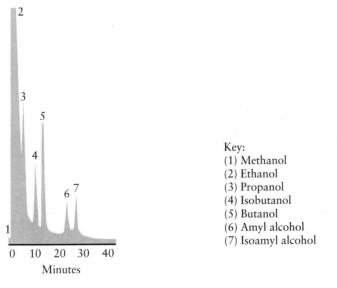

Key:
(1) Methanol
(2) Ethanol
(3) Propanol
(4) Isobutanol
(5) Butanol
(6) Amyl alcohol
(7) Isoamyl alcohol

FIGURE 1.40

A gas chromatogram of bourbon whiskey, showing the components that contribute to the flavor. Mixing the chemicals identified here does not, it is said, recreate the flavor, because the flavor also depends on hundreds of other compounds present in very small amounts.

Applying Chemisty: Case Study 1

When a well-known restaurant burned to the ground, the investigators soon found evidence of arson. How do forensic scientists—scientists who specialize in the study of crime—tell whether a fire has been started deliberately? Separation techniques such as chromatography (Section 1.14) and mass spectrometry (see Investigating Matter 1.2) have greatly enhanced the ability of forensic scientists to detect criminal evidence. These techniques are used to solve mysteries not only by scientists in crime laboratories but also by environmental, agricultural, and wildlife agencies, art museums, airport security services, archeologists, astronomers, and geologists. Only the nose of a trained dog is more sensitive than modern gas chromatography combined with a mass spectrometer (GCMS).

Fast-spreading fires are often started by splashing an *accelerant,* a flammable liquid such as kerosene or gasoline, about a building. Crime laboratory technicians investigating a fire collect charred materials and feed the liquid extracted from them into a gas chromatograph connected to a mass spectrometer to look for the presence of accelerants. Then samples of possible accelerants are exposed to

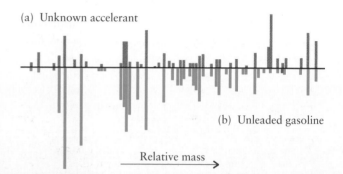

(a) Unknown accelerant

(b) Unleaded gasoline

Relative mass

Two spectra obtained in a GCMS analysis of evidence in a possible arson case. (a) The spectrum of a sample obtained from the scene of the fire. (b) The spectrum of gasoline. In each case, the mass spectrum shows the composition of the compound producing the peak that left the gas chromatograph at the same point in time. The correspondence of peaks shows that the compounds present in gasoline were also present at the crime scene. This finding suggests that gasoline was present and that it is likely that the fire was a result of arson.

This crime lab technician is looking for evidence at a crime scene. The driver of the car had caused a serious accident and then fled.

the same analysis. If a peak in the chromatogram from the crime scene matches a peak from an accelerant, the mass spectra are compared. As shown in the illustration, some of the peaks in the mass spectrum from the crime scene may match peaks in the mass spectrum of an accelerant such as gasoline. It may even be possible to determine the brand of gasoline used from the proportion of chemicals it contains.

The identification of the elements in a substance was a test that was once performed only when necessary, because it meant removing part of the material for testing, a testing technique called *destructive testing*. The techniques were also very time consuming. Fortunately, today there are non-destructive ways to identify the elements in a substance and even find the relative amounts of each present. One method is *pulsed fast-neutron analysis* (PFNA). In this method, a beam of neutrons is fired at an object from different directions. The neutrons are absorbed by the nuclei of the atoms in the object; these nuclei then give off energy as γ rays, which can be thought of as high-energy x-rays. The energy of the γ rays has a characteristic value for each element. These energies can be displayed immediately as different colors that identify the location of each type of element, as shown in the illustration. Such techniques give more information than x-rays and allow bombs and weapons to be detected in airline luggage or cargo before they are loaded onto an airplane.

If a bomb has exploded, the residue of the blast can also be studied to obtain clues to the type of explosive used. Often explosives are identified by testing for certain kinds of ions. This testing is done either by using spectrometry or by dissolving the residue in water and using simple color tests to identify the ions.* Detectives look for traces of nitrates, nitrites, sulfates, sulfites, and sodium and potassium ions, which are given off by common explosive substances.

*You will learn more about some of these color tests in Section 16.16.

X-ray fluorescence analysis is another nondestructive technique. It is used to match paint chips from crime scenes with the paint of automobiles and to authenticate works of art. This technique was used to identify the fake Revere silver dish shown in Fig. 1.1, which was found to contain elements not used in silver pieces until long after Revere's day.

Isotopic analysis conducted with a mass spectrometer is used to authenticate the origins of substances in commercial products such as flavoring agents, maple syrup, and fruit juices. For example, some synthetic flavoring agents are much less expensive than those from natural sources. However, they are usually made from chemicals with isotopic ratios different from that found in the plants from which the naturally occurring flavoring agents are extracted. Thus, the synthetic agents can be detected by mass spectrometry.

A similar technique was used to study a meteorite that is believed to have originated on Mars about 175 million years ago. The investigators found that the isotopic ratios of carbon in the meteorite are those characteristic of a species of Earth's bacteria that generates methane. Such bacteria can live in environments with very low oxygen levels. However, because the history of the meteorite since its impact on Earth is not known, it may have become contaminated with the bacteria after its arrival here.

Key Concepts: elements, isotopes, separation of mixtures, nomenclature of ions

For Further Reading

A. M. Rouhi, Government, industry efforts yield array of tools to combat terrorism, *Chemical and Engineering News*, July 24, 1995, pp. 10–19.

R. Johnson, Will the real Paul Revere please stand up? *Chemical and Engineering News*, February 2, 1998, pp. 28–29.

Related Exercises: 1.92 and 1.93

The red area in this three-dimensional PFNA scan of an airline cargo container shows the presence of materials that may pose a security threat.

form more than one kind of cation, such as Cu^+ and Cu^{2+} from copper, we use the **oxidation number,** a number equal to the charge of the cation. In chemical nomenclature, oxidation numbers are written as Roman numerals. Thus, Cu^+ is a copper(I) ion and Cu^{2+} is a copper(II) ion. Similarly, Fe^{2+} is an iron(II) ion and Fe^{3+} is an iron(III) ion. We always include the oxidation number in the names of the compounds of transition metals and other metals that can occur as ions with more than one charge (recall Fig. 1.26).

An older system of nomenclature is still in use. For example, some cations were once denoted by the endings *-ous* and *-ic* for the ions with lower and higher charges, respectively. In this system, iron(II) ions are called ferrous ions and iron(III) ions are called ferric ions. A list of ions for which these terms are still commonly used can be found in Appendix 3C.

Monatomic anions are named by adding the suffix *-ide* and the word *ion* to the first part of the name of the element (the "stem" of its name). There is no need to give the charge, because most elements that form anions form only one kind of monatomic anion. The ions formed by the halogens are collectively called *halide ions* and include fluoride (F^-), chloride (Cl^-), bromide (Br^-), and iodide (I^-) ions.

Table 1.3 gives the names of the most common polyatomic ions (see also Appendix 3A). The names of most common oxoanions are formed by adding the suffix *-ate* to the stem of the name of the element that is not oxygen, as in the carbonate ion, CO_3^{2-}. However, some elements can form two types of oxoanion, with different numbers of oxygen atoms, so we need names that distinguish them. Nitrogen, for example, forms both NO_2^- and NO_3^-. In such cases, the ion with the larger number of oxygen atoms is given the suffix *-ate* and that with the smaller number of oxygen atoms is given the suffix *-ite*. Thus, NO_3^- is nitrate and NO_2^- is nitrite.

Some elements—particularly the halogens—form more than two oxoanions. The name of the oxoanion with the smallest number of oxygen atoms is formed by adding the prefix *hypo-* to the *-ite* form of the name, as in the *hypo*chlorite ion, ClO^-. The oxoanion with a higher number of oxygen atoms than the *-ate* oxoanion is named with the prefix *per-* added to the *-ate* form of the name. An example is the *per*chlorate ion, ClO_4^- (Table 1.3). Both chlorine and bromine form oxoanions that span the range.

Some anions include hydrogen, such as HS^- and HCO_3^-. We can think of these ions as the result of adding an H^+ ion to the original anion. The charge on a hydrogen-bearing anion is that of the original anion plus one for each additional hydrogen atom. The names of these anions begin with "hydrogen." Thus, HS^- is the hydrogen sulfide ion and HCO_3^- is the hydrogen carbonate ion. These names are sometimes written as a single word, as in hydrogencarbonate ion. In an older system of nomenclature, which is still quite widely used, an anion containing hydrogen is named with the prefix *bi-*, as in bicarbonate ion for HCO_3^-. Anions to which two H^+ ions have been added are named by adding "dihydrogen" to the name of the anion. For example, $H_2PO_4^-$ is the dihydrogen phosphate ion.

The names of monatomic cations are the same as the names of the elements, with the addition of the word "ion." For elements that can form more than one cation, the oxidation number is included. The names of monatomic anions end in -ide. The suffix -ate indicates a greater number of oxygen atoms than the suffix -ite.

Example 1.6 Naming ions

Name the ions (a) N^{3-}; (b) Cr^{3+}; (c) HSO_3^-; (d) Ba^{2+}; (e) NH_4^+.

Strategy Name monatomic cations by giving the name of the element followed by "ion." If the cation comes from an element that exists in only one charge state (as shown in Fig. 1.26), omit the oxidation number. For any other metal, give its charge as an oxidation number, using Roman numerals. Name monatomic anions by replacing the ending of the element name with *-ide* and adding "ion." Identify polyatomic ions by referring to Table 1.3 or Appendix 3A. If an anion contains one hydrogen atom, add "hydrogen" before the name of the anion; if it contains two hydrogen atoms, add "dihydrogen" before the name of the anion.

Solution (a) Monatomic anion: nitride ion; (b) monatomic cation of a transition metal, oxidation number required: chromium(III) ion; (c) polyatomic anion that contains an H atom, named by adding "hydrogen" to the name of the anion; SO_3^{2-} has fewer oxygen atoms than SO_4^{2-}, so it has the ending *-ite* (see Table 1.3): hydrogen sulfite ion; (d) monatomic cation of a Group 2 element: barium ion; (e) polyatomic cation listed in Table 1.3: ammonium ion.

Self-Test 1.9A Name the ions (a) IO_3^-; (b) Mn^{2+}; (c) $H_2PO_3^-$; (d) Cs^+.
[*Answer:* (a) Iodate ion; (b) manganese(II) ion; (c) dihydrogen phosphite ion; (d) cesium ion]

Self-Test 1.9B Name the ions (a) Fe^{3+}; (b) F^-; (c) NO_3^-; (d) Ag^+.

1.16 Names of Ionic Compounds

An ionic compound is named with the cation name first, followed by the name of the anion; the word *ion* is omitted in each case. Potassium chloride, KCl, is a compound containing K^+ and Cl^- ions, and magnesium nitride, Mg_3N_2, contains Mg^{2+} and N^{3-} ions. If the cation is a transition metal or a heavier element from Groups 12 through 15, we can identify its charge from its formula and the charge on the anion. For example, the sulfide ion has a charge of -2, so the total negative charge in the formula unit Cr_2S_3 is $3 \times (-2) = -6$. To balance that charge, each of the two chromium ions must be present as Cr^{3+} to give a total positive charge of $+6$; so the compound is chromium(III) sulfide. The copper chloride that contains Cu^+ ions, CuCl, is called copper(I) chloride; and the chloride that contains Cu^{2+} ions, $CuCl_2$, is called copper(II) chloride.

Some ionic compounds form crystals that incorporate a definite proportion of molecules of water as well as the ions of the compound itself. These compounds are called **hydrates.** For example, copper(II) sulfate normally occurs as blue crystals of composition $CuSO_4 \cdot 5H_2O$ (Fig. 1.41). The centered dot in this formula is used to separate the *water of hydration* from the rest of the formula. This formula indicates that there are five H_2O molecules for each $CuSO_4$ formula unit. Hydrates are named by adding the word *hydrate* and a Greek prefix indicating the number of water molecules in each formula unit (Table 1.6). For example, the name of $CuSO_4 \cdot 5H_2O$ is copper(II) sulfate pentahydrate.

Ionic compounds are named by giving the name of the cation (with the charge indicated by a Roman numeral if more than one charge is possible), followed by the name of the anion; hydrates are named by adding the word "hydrate," preceded by a Greek prefix indicating the number of water molecules.

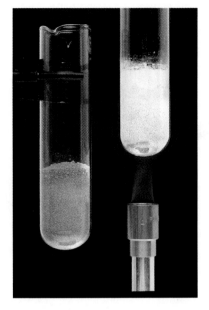

The sample of $CuSO_4 \cdot 5H_2O$ on the left shows the blue color of hydrated copper(II) sulfate. The water molecules can be driven off by heating, resulting in anhydrous ("without water") $CuSO_4$ (right).

The Greek prefixes in the table are used throughout chemistry, so it is a good idea to become familiar with them now.

Table 1.6 Prefixes used for naming compounds

Prefix	Meaning
mono-	1
di-	2
tri-	3
tetra-	4
penta-	5
hexa-	6
hepta-	7
octa-	8
nona-	9
deca-	10
undeca-	11
dodeca-	12

Example 1.7 *Naming ionic compounds*

Give the systematic names of (a) a compound that turns from blue to pink in humid weather, $CoCl_2 \cdot 6H_2O$, and (b) a compound used as a disinfectant, $Al(ClO_3)_3$.

Strategy In each case, identify the cation and the anion, by referring to Table 1.3 if necessary. To identify the oxidation number of the cation, note the charge of the anion and the number of anions and cations in the formula; then decide what cation charge is required to cancel the total negative charge. If the cation is a transition metal or an element from Groups 12 through 15 that can have more than one charge, as shown in Fig. 1.26, give its charge as a Roman numeral. If water molecules are included in the formula, then the compound is a hydrate, so add "hydrate" with a prefix corresponding to the number in front of H_2O in the formula.

Solution (a) The cation is a cobalt ion and the anion is a chloride ion, Cl^-. The cation must be Co^{2+} for electrical neutrality; so cobalt (a transition metal) is present as cobalt(II). The $6H_2O$ indicates that the compound is a hexahydrate; so the compound is cobalt(II) chloride hexahydrate. (b) The cation is an aluminum ion, Al^{3+} (the only cation that aluminum forms). The anion is a chlorate ion, ClO_3^-. Hence, the compound is aluminum chlorate.

Self-Test 1.10A Name the compounds (a) $FeCl_2 \cdot 2 H_2O$; (b) $AlBr_3$; (c) $Cr(ClO_3)_2$.
[*Answer:* (a) Iron(II) chloride dihydrate; (b) aluminum bromide; (c) chromium(II) chlorate]

Self-Test 1.10B Name the compounds (a) $AuCl_3$; (b) CaS; (c) Mn_2O_3.

1.17 Names of Molecular Compounds

Many simple molecular compounds are named by using the Greek prefixes in Table 1.6 to indicate the number of each type of atom present. No prefix is used if only one atom of an element is present; an exception is oxygen, as in carbon monoxide, CO. Most of the common binary molecular compounds—molecular compounds built from two elements—have at least one element from Group 16 or 17. These elements are named second, with their endings changed to *-ide*:

phosphorus trichloride PCl_3 dinitrogen monoxide N_2O
sulfur hexafluoride SF_6 dinitrogen pentoxide N_2O_5

Certain binary molecular compounds have common names that are widely used (Table 1.7). The phosphorus oxides are distinguished by the oxidation number on the P atom, which is calculated by thinking of phosphorus as a metal and the oxygen as present as O^{2-}. Thus, P_4O_6 is thought of as $(P^{3+})_4(O^{2-})_6$ and named phosphorus(III) oxide, and P_4O_{10} is thought of as $(P^{5+})_4(O^{2-})_{10}$ and named phosphorus(V) oxide. These compounds, however, are molecular, so assigning charges to the P and O atoms is only a means to the end of naming them.

Except for organic compounds and a few other cases, both the name and the molecular formula of compounds of hydrogen and nonmetals are written with the H atom first:

hydrogen chloride HCl   hydrogen cyanide HCN

Several simple compounds containing hydrogen are given common names. Thus, NH_3 is ammonia and PH_3 is phosphine.

Table 1.7 *Common names for some simple molecular compounds*

Formula*	Common Name
NH_3	ammonia
N_2H_4	hydrazine
NH_2OH	hydroxylamine
PH_3	phosphine
NO	nitric oxide
N_2O	nitrous oxide
C_2H_4	ethylene
C_2H_2	acetylene

*For historical reasons, the molecular formulas of binary hydrogen compounds of Group 15 elements are written with the Group 15 elements first.

Acids play an enormously important role in chemistry and will be defined formally later. For the time being, they can be defined loosely as compounds that release hydrogen ions in water. Usually, if the hydrogen atoms in a formula appear first, the compound is an acid. Some compounds are named as acids, such as nitric acid, HNO_3, and sulfuric acid, H_2SO_4. Aqueous solutions of the hydrogen halides and some other simple binary compounds of hydrogen are named as acids. Solutions of these compounds are indicated by writing (aq) after the formula, as in HCl(aq), and are named by adding the prefix *hydro-* and replacing the *-ide* ending of the element name with *-ic acid*. For example, HCl(aq), an aqueous solution of the gas hydrogen chloride, is named hydrochloric acid. Likewise, H_2S(aq), an aqueous solution of hydrogen sulfide, is hydrosulfuric acid.

The oxoacids are molecular compounds that can be regarded as the parents of the oxoanions. The formulas of oxoacids are derived from those of the corresponding oxoanions by including enough hydrogen ions to balance the charges. This procedure is only a *formal* way of building the chemical formula, because oxoacids are all molecular compounds. For example, the sulfate ion, SO_4^{2-}, needs two H^+ ions to balance its negative charge, so sulfuric acid is the molecular compound H_2SO_4. Similarly, the phosphate ion, PO_4^{3-}, needs three H^+ ions to achieve charge balance, so its parent acid is the molecular compound H_3PO_4, phosphoric acid. The names of the oxoacids are derived from those of the corresponding oxoanions. In general, *-ic* oxoacids are the parents of *-ate* oxoanions and *-ous* oxoacids are the parents of *-ite* oxoanions. For example, H_2SO_3, the parent acid of the sulfite ion, is sulfurous acid. The more important oxoacids are included in Table 1.8.

Table 1.8 Common anions and their parent acids

Anion	Parent acid	Anion	Parent acid
fluoride ion, F^-	hydrofluoric acid,* HF (hydrogen fluoride)	nitrite ion, NO_2^-	nitrous acid, HNO_2
		nitrate ion, NO_3^-	nitric acid, HNO_3
chloride ion, Cl^-	hydrochloric acid,* HCl (hydrogen chloride)	phosphate ion, PO_4^{3-}	phosphoric acid, H_3PO_4
		hydrogen phosphate ion, HPO_4^{2-}	
bromide ion, Br^-	hydrobromic acid,* HBr (hydrogen bromide)	dihydrogen phosphate ion, $H_2PO_4^-$	
iodide ion, I^-	hydroiodic acid,* HI (hydrogen iodide)	sulfite ion, SO_3^{2-}	sulfurous acid, H_2SO_3
oxide ion, O^{2-}	water, H_2O	sulfate ion, SO_4^{2-}	sulfuric acid, H_2SO_4
hydroxide ion, OH^-		hydrogen sulfate ion, HSO_4^-	
sulfide ion, S^{2-}	hydrosulfuric acid,* H_2S (hydrogen sulfide)	hypochlorite ion, ClO^-	hypochlorous acid, HClO
cyanide ion, CN^-	hydrocyanic acid,* HCN (hydrogen cyanide)	chlorite ion, ClO_2^-	chlorous acid, $HClO_2$
acetate ion, $CH_3CO_2^-$	acetic acid, CH_3COOH	chlorate ion, ClO_3^-	chloric acid, $HClO_3$
carbonate ion, CO_3^{2-}	carbonic acid, H_2CO_3	perchlorate ion, ClO_4^-	perchloric acid, $HClO_4$
hydrogen carbonate (bicarbonate) ion, HCO_3^-			

*The name of the aqueous solution of the compound. The name of the compound itself is in parentheses.

The procedure described in this Toolbox summarizes the rules for naming simple inorganic compounds.

Conceptual Basis

By using a systematic procedure for naming compounds, we can avoid having to memorize lots of names. Ionic and molecular compounds use different procedures, so it is important first to identify which type of compound is present.

Procedure

The rules for naming inorganic compounds are summarized in the procedures for naming ionic compounds and molecular compounds.

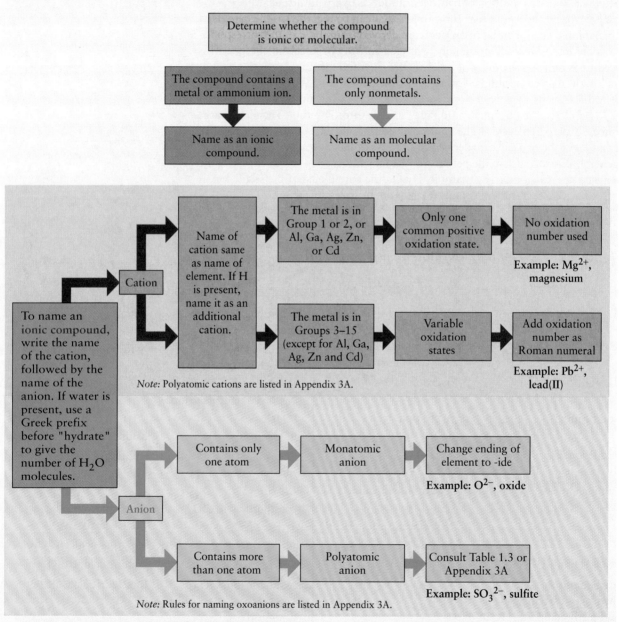

Examples of compound names: $AlCl_3$, aluminum chloride; $Cu(ClO_4)_2 \cdot 6H_2O$, copper(II) perchlorate hexahydrate; $KHCO_3$, potassium hydrogen carbonate

This Toolbox shows how to determine the formula of a compound composed of two elements from its name.

Conceptual Basis

The name of a molecular compound tells us how many atoms of each element are present. However, the name of an ionic compound does not, so we must determine how many atoms of each element are present from the positions of the elements in the periodic table.

Procedure

First determine whether the compound is molecular or ionic, as explained in Toolbox 1.1.

1. Molecular compounds Write the formulas for compounds of two nonmetals by listing the symbols of the elements in the same order as they are in the name and giving them subscripts corresponding to the Greek prefix used.

2. Ionic compounds Compounds are electrically neutral. To write the formula of a binary ionic compound, first we need to know the charge on each ion.

Cations If more than one charge is possible (as shown in Fig. 1.26), give the charge on the cation as a Roman numeral following its name. If only one type of ion is possible, determine the charge on the cation from its group number in the periodic table. Group 1 and 2 metals form ions with a charge equal to the group number, and the lighter elements in Groups 12 and 13 tend to form cations with a charge equal to the group number minus ten

Anions The charge on a monatomic anion is determined by the group number of the element in the periodic table: the charge on a monatomic anion is $G - 18$, where G is the group number.

Write the symbols of the cation and anion. Give them subscripts that will balance their charges.

Strategy Determine whether each compound is molecular or ionic and then follow the strategy set out in Toolbox 1.2.

Solution (a) Each of these compounds consists of a metal and a nonmetal, so both are ionic. Magnesium is a metal in Group 2, so its cation is Mg^{2+}; nitrogen is a nonmetal in Group 15, so the nitride ion has a charge of $15 - 18 = -3$. We need the -6 charge of two N^{3-} ions to balance the $+6$ charge of three Mg^{2+} ions; thus, the formula must be Mg_3N_2 because $(3 \times (+2)) + (2 \times (-3)) = 0$. The II in cobalt(II) signifies that the metal cobalt has a charge of $+2$ in the compound. Chlorine is in Group 17, so the chloride ion has a charge of $17 - 18 = -1$. We can balance the charge of the $+2$ cation with two -1 anions; thus, the formula is $CoCl_2$. Because this compound is a hydrate, we must add the six water molecules indicated by the Greek prefix *hexa-*. The complete formula is therefore $CoCl_2 \cdot 6H_2O$. (b) Each of these compounds consists of two nonmetals, so both are molecular. The subscripts for each atom are taken from the Greek prefixes, thus we have B_2S_3 and $SiCl_4$. Notice that when no prefix is used, as in the case of silicon, only one atom is present in the formula unit.

Self-Test 1.14A Write the formulas for (a) vanadium(V) oxide; (b) beryllium carbide; (c) sulfur tetrafluoride; (d) dinitrogen trioxide.

[*Answer:* (a) V_2O_5; (b) Be_2C; (c) SF_4; (d) N_2O_3]

Self-Test 1.14B Write the formulas for (a) cesium sulfide tetrahydrate; (b) manganese(VII) oxide; (c) hydrogen sulfide (the gas responsible for the smell of rotten eggs); (d) disulfur dichloride.

Skills You Should Have Mastered

Conceptual

☐ **1.** Describe the structure of the nuclear atom, Section 1.3.

☐ **2.** Distinguish by their properties metals, metalloids, and nonmetals and locate them in the periodic table, Example 1.3.

☐ **3.** Identify the type of substance from its chemical formula, Sections 1.8–1.11

☐ **4.** Identify physical and chemical changes, Self-Test 1.8.

☐ **5.** Distinguish heterogeneous mixtures, solutions, compounds, and elements, Section 1.13.

☐ **4.** Interpret chemical formulas in terms of the number of each type of atom present, Example 1.5.

☐ **5.** Name ions, Example 1.6.

☐ **6.** Name binary ionic compounds, compounds with common polyatomic ions, and hydrates, Examples 1.7, 1.10, and 1.11.

☐ **7.** Name molecular compounds, Examples 1.8, 1.9, and 1.10.

☐ **8.** Write the formula of a binary compound from its name, Example 1.11.

Problem-Solving

☐ **1.** Write the symbols for the elements, given their names, and vice versa, Self-Test 1.1.

☐ **2.** State the numbers of neutrons, protons, and electrons in an isotope, given its atomic number and mass number, and write the symbol for an isotope, Examples 1.1 and 1.2.

☐ **3.** Predict the anion or cation that a main-group element is likely to form, Example 1.4.

Descriptive

☐ **1.** Describe the historical development of the model of the nuclear atom, Sections 1.3–1.4.

☐ **2.** Describe the function and operation of a mass spectrometer, Section 1.4.

☐ **3.** Describe how mixtures are separated by each of the techniques described in Section 1.14.

Exercises

The elements are listed in the periodic table inside the front cover and in alphabetical order inside the back cover.

The Nuclear Atom and Isotopes

1.1 (a) What is the charge and mass of the particle in cathode rays? (b) Who first determined the charge of the electron?

1.2 (a) Who proposed the existence of a nucleus in an atom? (b) What experimental evidence motivated the proposal?

1.3 Explain the difference between a law and a theory.

1.4 Rank the following according to increasing credibility among scientists and explain your reasoning: (a) theory; (b) premonition; (c) hypothesis.

1.5 Among other achievements, John Dalton formulated a "law of multiple proportions." This law, which helped to guide his atomic theory, states that when two elements combine to form more than one compound, the ratio of the masses of one element that combines with an equal mass of the second element to form two different compounds is in the form of small integers for each compound. Show that the following data for compounds of sulfur and oxygen support the law of multiple proportions by calculating the ratios of the masses of

sulfur that react to form three different compounds with oxygen:

	Mass of sulfur reacted	Mass of oxygen reacted
(a)	2.00 g	1.00 g
(b)	1.00 g	1.00 g
(c)	0.667 g	1.00 g

1.6 Show that the following data for compounds of carbon and oxygen support the law of multiple proportions. Consult Exercise 1.5 for information.

	Mass of carbon reacted	Mass of oxygen reacted
(a)	0.375 g	1.00 g
(b)	0.750 g	1.00 g
(c)	1.125 g	1.00 g
(d)	1.875 g	1.00 g

1.7 Give the chemical symbol and atomic number of (a) arsenic; (b) sulfur; (c) palladium; (d) gold.

1.8 Give the chemical symbol and atomic number of (a) bromine; (b) barium; (c) bismuth; (d) boron.

1.9 In the following list of element names and their corresponding symbols, mistakes may have been made.

Correct any name or symbol that is in error: (a) krypton, Kr; (b) molybdenum, M; (c) ytrium, Y; (d) strontium, St.

1.10 In the following list of element names and their corresponding symbols, mistakes may have been made. Correct any name or symbol that is in error: (a) radon, Rd; (b) galium, Ga; (c) chromium, Ch; (d) boron, Bo.

1.11 State the number of protons, neutrons, and electrons in an atom of (a) carbon-13; (b) ^{37}Cl; (c) chlorine-35; (d) ^{235}U.

1.12 State the number of protons, neutrons, and electrons in an atom of (a) tritium, ^3H; (b) ^{60}Co; (c) oxygen-16; (d) ^{204}Pb.

1.13 Identify the isotope that has atoms with (a) 63 neutrons, 48 protons, and 48 electrons; (b) 46 neutrons, 36 protons, and 36 electrons; (c) 6 neutrons, 5 protons, and 5 electrons.

1.14 Identify the isotope that has atoms with (a) 78 neutrons, 52 protons, and 52 electrons; (b) 108 neutrons, 73 protons, and 73 electrons; (c) 32 neutrons, 28 protons, and 28 electrons.

1.15 (a) What characteristics do carbon-12, carbon-13, and carbon-14 atoms have in common? (b) In what ways are they different?

1.16 (a) What characteristics do argon-40, potassium-40, and calcium-40 atoms have in common? (b) In what ways are they different?

The Periodic Table

1.17 Name the elements (a) Li; (b) Ga; (c) Xe; (d) K. Give their group numbers in the periodic table and identify each one as a metal, a nonmetal, or a metalloid.

1.18 Name the elements (a) P; (b) Sb; (c) Fe; (d) Ag. Give their group numbers in the periodic table and identify each one as a metal, a nonmetal, or a metalloid.

1.19 Write the chemical symbol of (a) chlorine; (b) cobalt; (c) arsenic. Classify each one as a metal, a nonmetal, or a metalloid.

1.20 Write the chemical symbol of (a) neon; (b) bismuth; (c) tungsten. Classify each one as a metal, a nonmetal, or a metalloid.

1.21 Write the chemical symbol of (a) iodine; (b) chromium; (c) mercury; (d) aluminum. Classify each element as a metal, a nonmetal, or a metalloid.

1.22 Write the chemical symbol of (a) carbon; (b) zinc; (c) barium; (d) germanium. Classify each element as a metal, a nonmetal, or a metalloid.

1.23 List the names, chemical symbols, and atomic numbers of the alkali metals. Describe their reactions with water.

1.24 List the names, chemical symbols, and atomic numbers of the halogens. Identify the normal physical state of each.

Compounds

1.25 Bergamol (also known as linalyl acetate) is an aromatic compound found in bergamot and lavender oils. Bergamol contains atoms of carbon, hydrogen, and oxygen in the ratio 6:10:1 Its molecules each have two oxygen atoms. Write the chemical formula of bergamol.

1.26 The compound melamine is used to make certain kinds of plastics and resins. Melamine contains atoms of carbon, hydrogen, and nitrogen in the ratio 1:2:2. Its molecules each have three carbon atoms. Write the chemical formula of melamine.

1.27 State whether each of the following elements is more likely to form a cation or an anion and write the formula for that ion: (a) sulfur; (b) potassium; (c) strontium; (d) chlorine.

1.28 State whether each of the following elements is more likely form a cation or an anion and write the formula for that ion: (a) zinc; (b) magnesium; (c) nitrogen; (d) oxygen.

1.29 Elements within the same group in the periodic table tend to form compounds with similar formulas. An element ("E") in Period 4 forms the molecular compound H_2E and the ionic compound Na_2E. (a) To which group does the element E belong? (b) Write the name and symbol of element E.

1.30 Elements within the same group tend to form compounds with similar formulas. A main-group element ("E") in Period 3 forms the following ionic compounds: EBr_3 and E_2O_3. (a) To which group does the element E belong? (b) Write the name and symbol of element E.

1.31 How many protons, neutrons, and electrons are present in (a) $^2H^+$; (b) $^9Be^{2+}$; (c) $^{80}Br^-$; (d) $^{32}S^{2-}$?

1.32 How many protons, neutrons, and electrons are present in (a) $^{40}Ca^{2+}$; (b) $^{115}In^{3+}$; (c) $^{127}Te^{2-}$; (d) $^{14}N^{3-}$?

1.33 Write the symbol of the isotopic ion that has (a) 9 protons, 10 neutrons, and 10 electrons; (b) 12 protons, 12 neutrons, and 10 electrons; (c) 52 protons, 76 neutrons, and 54 electrons; (d) 37 protons, 49 neutrons, and 36 electrons.

1.34 Write the symbol of the isotopic ion that has (a) 11 protons, 13 neutrons, and 10 electrons; (b) 13 protons, 14 neutrons, and 10 electrons; (c) 34 protons, 45 neutrons, and 36 electrons; (d) 24 protons, 28 neutrons, and 22 electrons.

Substances and Mixtures

1.35 Identify the following substances as elements or compounds: (a) gold; (b) chlorine gas; (c) sodium chloride (table salt).

1.36 Identify the following substances as mixtures or pure substances: (a) sodium hydrogen carbonate (baking soda); (b) brass; (c) diamond, a crystalline form of carbon.

1.37 Each of the containers pictured below holds either a mixture, a single compound, or a single element. In each case, identify the type of contents.

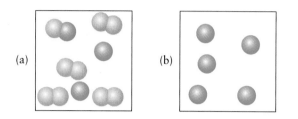

1.38 Each of the containers pictured below holds either a mixture, a single compound, or a single element. In each case, identify the type of contents.

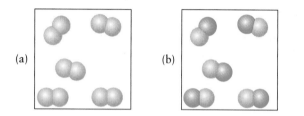

1.39 Classify the following characteristics as either a chemical or a physical property: (a) the color of copper(II) sulfate pentahydrate is blue; (b) the melting point of sodium metal is 98°C; (c) iron rusts in areas with high humidity.

1.40 A chemist investigates the boiling point, melting point, and flammability of a compound called acetone, a component of fingernail polish remover. Which of these properties are physical properties and which are chemical properties?

1.41 Classify the following changes as chemical or physical: (a) the freezing of water; (b) the vaporization of alcohol; (c) the corrosion of aluminum when exposed to air.

1.42 A general chemistry student was asked to decide which of the following changes were physical and which were chemical: melting, rusting, and bending of iron. What is the correct answer?

1.43 Determine which of the following properties are chemical and which are physical: (a) the hardness of a metal; (b) the ductility of a metal; (c) the corrosiveness of an acid; (d) the freezing point of a liquid.

1.44 Identify the following as either a chemical or a physical change: (a) the formation of frost; (b) the tarnishing of silver metal; (c) the production of hydrogen when potassium is in contact with water; (d) the liquefaction of air.

1.45 List all the physical properties and changes in the statement, "The temperature of the land is an important factor for the ripening of oranges, because it affects the evaporation of water and the humidity of the surrounding air."

1.46 List all the chemical properties and changes in the statement, "Copper is a red-brown element obtained from copper sulfide ores by heating in air, a process that forms copper oxide. Heating the copper oxide with carbon produces impure copper metal."

1.47 What physical properties are used for the separation of the components of a mixture by (a) filtration; (b) chromatography; (c) distillation?

1.48 Explain how you would distinguish sugar from salt. Identify the physical and chemical properties used in your explanation.

1.49 Identify the following as homogeneous or heterogeneous mixtures and suggest a technique for separating their components: (a) ethanol and water; (b) chalk and table salt; (c) salt water.

1.50 Identify the following as homogeneous or heterogeneous mixtures and suggest a technique for separating their components: (a) gasoline and motor oil; (b) carbonated water; (c) charcoal and sugar.

Chemical Nomenclature

1.51 Name the ions (a) Cl^-; (b) O^{2-}; (c) C^{4-}; (d) P^{3-}.

1.52 Name the ions (a) Se^{2-}; (b) Br^-; (c) F^-; (d) Mg^{2+}.

1.53 Name the ions (a) PO_4^{3-}; (b) SO_4^{2-}; (c) N^{3-}; (d) SO_3^{2-}; (e) IO_2^-; (f) I^-.

1.54 Name the ions (a) ClO_4^-; (b) NO_2^-; (c) S^{2-}; (d) NO_3^-; (e) ClO_2^-; (f) BrO^-.

1.55 Write the formulas of (a) chlorate ion; (b) nitrate ion; (c) carbonate ion; (d) hypochlorite ion; (e) hydrogen sulfate ion.

1.56 Write the formulas of (a) hydrogen phosphite ion; (b) bromite ion; (c) dihydrogen phosphate ion; (d) periodate ion; (e) hydrogen sulfide ion.

1.57 Write the old and modern names of (a) Pb^{2+}; (b) Fe^{2+}; (c) Co^{3+}; (d) Cu^+.

1.58 Write the names of the following ions, giving both the old and modern names where appropriate: (a) HSO_4^-; (b) Hg_2^{2+}; (c) CN^-; (d) HCO_3^-.

1.59 Write the formulas of (a) copper(II) ion; (b) chlorite ion; (c) phosphide ion; (d) hydride ion.

1.60 Write the formulas of (a) ferric ion; (b) manganese(III) ion; (c) hydrogen ion; (d) sulfate ion.

1.61 Write the formulas of (a) magnesium oxide (magnesia); (b) calcium phosphate (the major inorganic component of bones); (c) aluminum sulfate; (d) calcium nitride.

1.62 Write the formulas of (a) potassium hydroxide; (b) barium sulfate; (c) gold(I) chloride; (d) cupric chloride.

1.63 Name the following ionic compounds. Write both the old and modern names wherever appropriate. (a) K_3PO_4; (b) FeI_2; (c) Nb_2O_5; (d) $CuSO_4$.

1.64 The following ionic compounds are commonly found in laboratories. Write their modern names. (a) $NaHCO_3$ (baking soda); (b) Hg_2Cl_2 (calomel); (c) $NaOH$ (lye); (d) ZnO (calamine).

1.65 Write the names of (a) $Cu(NO_3)_2 \cdot 6H_2O$; (b) $NdCl_3 \cdot 6H_2O$; (c) $NiF_2 \cdot 4H_2O$.

1.66 Write the names of (a) $Na_2CO_3 \cdot 10H_2O$; (b) $Ce(IO_3)_3 \cdot 2H_2O$; (c) $Co(CN)_2 \cdot 3H_2O$.

1.67 Bromine and aluminum react to form a white solid. (a) Classify the solid as ionic or molecular. (b) Write the formula of the solid.

1.68 Magnesium and nitrogen react to form a gray solid. (a) Classify the solid as ionic or molecular. (b) Write the formula of the solid.

1.69 Write the formulas of (a) sodium carbonate monohydrate; (b) indium(III) nitrate pentahydrate; (c) copper(II) perchlorate hexahydrate.

1.70 Write the formulas of (a) lithium nitrite monohydrate; (b) vanadium(III) iodide hexahydrate; (c) chromium(II) sulfate heptahydrate.

1.71 Write the formulas of (a) selenium trioxide; (b) carbon tetrachloride; (c) carbon disulfide; (d) sulfur hexafluoride; (e) diarsenic trisulfide; (f) phosphorus pentachloride; (g) dinitrogen monoxide; (h) chlorine trifluoride.

1.72 Write the formulas of (a) dinitrogen tetroxide; (b) hydrogen sulfide; (c) dichlorine heptoxide; (d) disulfur dichloride; (e) nitrogen triiodide; (f) sulfur dioxide; (g) hydrogen fluoride; (h) diiodine hexachloride.

1.73 Name the molecular compounds (a) SF_4; (b) N_2O_5; (c) NI_3; (d) XeF_4; (e) $AsBr_3$; (f) ClO_2; (g) P_2O_5.

1.74 The following molecular compounds are often found in chemical laboratories. Name each compound: (a) SiO_2 (silica); (b) SiC (carborundum); (c) N_2O (a general anesthetic); (d) P_4O_{10} (a drying agent for organic solvents); (e) CS_2 (a solvent); (f) SO_2 (a bleaching agent); (g) NH_3 (a common reagent).

1.75 The following aqueous solutions are common laboratory acids. What are their names? (a) $HCl(aq)$; (b) $H_2SO_4(aq)$; (c) $HNO_3(aq)$; (d) $CH_3COOH(aq)$; (e) $H_2SO_3(aq)$; (f) $H_3PO_4(aq)$.

1.76 The following acids are also used in chemical laboratories, although they are less common than those in the preceding exercise. Write the formulas of (a) perchloric acid; (b) hypochlorous acid; (c) hypoiodous acid; (d) hydrofluoric acid; (e) phosphorous acid; (f) periodic acid.

1.77 Write formulas for the ionic compounds formed from (a) sodium and oxide ions; (b) potassium and sulfate ions; (c) silver and fluoride ions; (d) zinc and nitrate ions; (e) aluminum and sulfide ions.

1.78 Write formulas for the ionic compounds formed from (a) calcium and chloride ions; (b) iron(III) and sulfate ions; (c) ammonium and iodide ions; (d) lithium and sulfide ions; (e) calcium and phosphide ions.

Supplementary Exercises

1.79 In an industrial reactor, methane is converted to diamonds. However, the result of the reaction includes a large amount of graphite as well as diamonds. Diamonds and graphite are two different forms of carbon. (a) Is the mixture of graphite and diamond a pure substance? (b) How would you classify the result of the reaction, as an element, a compound, a homogeneous mixture, or a heterogeneous mixture?

1.80 The pictures below show either a physical or a chemical change. In each case, identify the type of change.

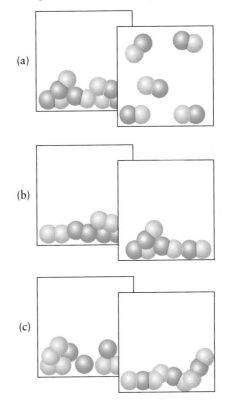

1.81 Sketch the periodic table and mark on it the locations of (a) the alkali metals; (b) the alkaline earth metals; (c) the halogens; (d) the noble gases; (e) the transition metals; (f) the metalloids; (g) the lanthanides; (h) the actinides.

1.82 Identify the following substances as elements, molecular compounds, or ionic compounds: (a) HCl; (b) S_8; (c) CoS; (d) Ar; (e) CS_2; (f) $SrBr_2$.

1.83 What is meant by (a) an ionic compound and (b) a molecular compound? (c) What are the typical properties of the two classes of compounds?

1.84 Name the following compounds, using analogous compounds with phosphorus and sulfur as a guide: (a) H_2TeO_4; (b) Na_3AsO_4; (c) $CaSeO_3$; (d) $Ba_3(SbO_4)_2$; (e) H_3AsO_4; (f) $Co_2(TeO_4)_3$.

1.85 Name the compounds (a) Ag_2S; (b) $ZnCl_2$; (c) ClF_5; (d) $Mg(OH)_2$; (e) $NiSO_4 \cdot 6H_2O$; (f) PCl_5; (g) $Cr(H_2PO_4)_3$; (h) As_2O_3; (i) $MnCl_2$.

1.86 Name the compounds (a) $FeCl_3 \cdot 6H_2O$; (b) $Co(NO_3)_2 \cdot 6H_2O$; (c) $CuCl$; (d) $BrCl$; (e) MnO_2; (f) $Hg(NO_3)_2$; (g) $Ni(NO_3)_2$; (h) N_2O_4; (i) V_2O_5.

1.87 Write formulas for (a) ammonium sulfite; (b) ferric oxide; (c) copper(II) bromate; (d) phosphine; (e) calcium bicarbonate; (f) hydrogen cyanide; (g) lithium bisulfate; (h) selenium tetrafluoride; (i) iron(II) sulfate heptahydrate.

1.88 Write formulas for (a) aluminum phosphate; (b) barium nitrate dihydrate; (c) silicon disulfide; (d) sodium phosphide; (e) perchloric acid; (f) copper(II) oxide; (g) hydroiodic acid; (h) silver(I) sulfate.

1.89 (a) Determine the total number of protons, neutrons, and electrons in one water molecule, H_2O, assuming that only the most common isotopes, ^1H and ^{16}O, are present. (b) What are the total masses of protons, neutrons, and electrons in this water molecule? (c) What fraction of your own mass is due to the neutrons in your body, assuming you consist primarily of water made from this type of molecule?

1.90 Determine the fraction of the total mass of an ^{56}Fe atom that is due to (a) neutrons; (b) protons; (c) electrons. (d) What is the mass of protons in a 1000-kg automobile? Assume the total mass of the vehicle is due to ^{56}Fe.

Applied Exercises

1.91 An antiques dealer has been offered an ingot of metal that is supposed to be platinum. Describe at least two tests that you might conduct to determine whether the claim is true. Use any information in this chapter or in Appendix 2D and assume you have access to any of the instrumentation described. See Applying Chemistry: Case Study 1.

1.92 An investigator studying the residue from an explosion is attempting to identify the explosive and is looking for traces of nitrates, nitrites, sulfates, sulfites, sodium ions, and potassium ions. The laboratory report indicates the presence of the ions below. Name each ion and indicate which, if any, are among the ions being sought: (a) NH_4^+; (b) NO_2^-; (c) S^{2-}; (d) P^{3-}; (e) Na^+. See Applying Chemistry: Case Study 1.

1.93 The mass spectrum data in the following spectrum were obtained for an unknown compound. The mass spectrum of a compound is more complicated than that of an element. The

molecule is blasted apart by the electron beam and the formula of the compound is determined by studying the fragments that result. Some of the particles that strike the detector are those that result when the molecule simply loses an electron, so they have essentially the same mass as the molecule itself. The mass of this ion is called the parent mass. However, the bombardment breaks apart some of the molecules into smaller fragments. The smallest fragments are the individual atoms. The data in the figure give the relative amounts of each fragment present in detectable quantities. (a) Using this information, identify the substance that generated the mass spectrum as either HCl, HBr, H_2S, or Cl_2. (b) Identify the fragments generating each peak. Each element may be present as more than one isotope. See Investigating Matter 1.2. Commonly occurring isotopes are listed in the periodic table database on the CD that accompanies this book.

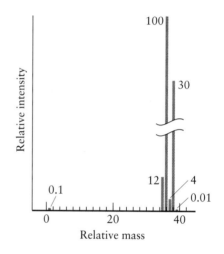

1.94 The mass spectrum data in the following spectrum were obtained for an unknown compound. (a) Using this information and that in the previous exercise, identify the substance that generated the mass spectrum as either H_2O, O_2, CH_4, or CO. (b) Identify the fragments generating each peak. Each element may be present as more than one isotope.

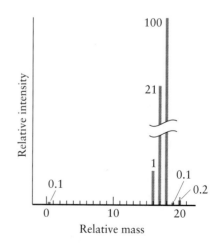

1.95 Water purification, in which impure water is made drinkable, can be achieved by several methods. On the basis of the information in this chapter, suggest a physical process for converting seawater to pure water.

1.96 The British chemist Michael Faraday gave Christmas lectures to children in the mid-nineteenth century. One series of lectures involved investigating what he called the "chemical history of a candle." He would light a candle, then examine the colors and the smoke given off. He weighed the candle before and after burning to see if it lost mass, and it did. He would hold a mirror over it to catch condensation of water and soot. Then he would place a glass tube into the flame and show that thick, white fumes came out. He suggested that the fumes might be candle wax that had been vaporized by the heat. To check for the validity of that suggestion, he lit the fumes and they did indeed burn. Categorize each of Faraday's suggestions and actions as observation, measurement, hypothesis, experiment, or theory.

1.97 Suppose you are a student living in a shared room and that you have been experiencing headaches every evening after dinner when you sit down at your desk to begin your homework. Consider how you might determine the cause of your headaches by using the principles of the scientific method. Discuss what data you could collect, and suggest at least three hypotheses and experiments to test them.

Measurements and Moles

S ome of the luckiest chemists in the world have jobs that require them to visit tropical seas or mountain meadows. There they collect **natural products,** which are materials obtained from the environment. Oceans, mountains, and meadows are a rich source of natural products, some of which have important medical applications. For example, extracts of marine animals such as tunicates (sea squirts) are highly active against tumors and viruses. Recognizing a useful compound and discovering its properties constitute the **qualitative** part of the investigation, the part that involves describing the compound by its properties, such as color, smell, toxicity, and antiviral activity.

If we were traveling with a natural products discovery team, we would soon see that the discovery of a natural product with medicinal properties is only the beginning of the procedure needed to make it useful. The active compound is present in only very small quantities in the plant or animal. **Synthesizing** the compound—making it from simpler components— would make it more widely available. However, before we can start thinking about synthesizing such a chemical, we need to know its composition and, in particular, its chemical formula. The determination of chemical composition is the **quantitative** part of the investigation, the part that involves measuring and using numerical data.

We can illustrate the use of numerical data by considering one particular natural product, vitamin C. Before vitamin C was identified as a vital nutrient, sailors on long voyages had to keep a large supply of citrus fruit on hand to avoid diseases such as scurvy. In 1928, Albert Szent-Györgi, a Hungarian-American chemist, isolated the compound from plants and the adrenal glands of animals and found a way to produce it in bulk. He based his synthesis on data he had collected by measurements on natural samples of vitamin C. Measurements such as those make chemistry a member of the **physical sciences,** the branches of science that make extensive use of quantitative measurements.

This marine biologist is collecting specimens on the Great Barrier Reef off the coast of Australia. Organisms of all kinds produce a multitude of chemicals, some of which have the ability to overcome diseases. However, to study them and to make them in usable quantities, chemists need to determine their composition. Such an investigation involves using the principles in this chapter.

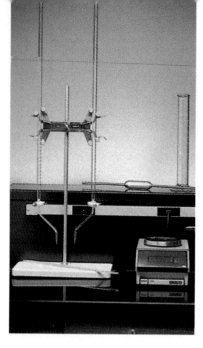

FIGURE 2.1

Equipment commonly used to make measurements in laboratories. Clockwise from the upper left: two burets and a pipet for transferring specific volumes of liquids; a graduated cylinder for measuring volume; a balance for determining mass; and a thermometer for measuring temperature.

MEASUREMENTS AND UNITS

We are already familiar with the way that measurements are reported in everyday life. We report the distance along a highway in miles or kilometers, the volume of gasoline in gallons or liters, and the passage of time in seconds, minutes, and hours. Scientists have made these procedures more systematic and more precise through the use of international standards. By defining standard units of mass, length, and time (and of other quantitites too), we know that a measurement reported from one laboratory means the same thing to chemists in laboratories anywhere in the world (Fig. 2.1).

2.1 The Metric System

The metric system was devised in the 1790s when, with revolutionary fervor, the French swept away their old units as well as their politicians and sought to replace both by more rational systems. The metric system has been developed over the past 200 years and is the basis of the **Système International d'Unités** (SI), which is used for almost all modern scientific work.

When we examine an everyday report of a measurement (such as 5 miles or 10 gallons), we see that it has the form

$$\text{Result of a measurement} = \text{number} \times \text{unit}$$

The same is true in the metric system. For example, the mass of a package might be reported as 5.0 kg—or 5.0 times the unit of mass, which is 1 kilogram (1 kg). If we were to measure the length of a pole, then we might report it as 10 m, which means 10 times the unit of length, 1 meter (1 m). There are seven fundamental units, or SI **base units;** they are listed in Table 2.1. All other units are expressed in terms of these base units. The meter and the second are defined in terms of the properties of light. The kilogram is defined as the mass of an actual physical object, a platinum-iridium cylinder kept in France (Fig. 2.2). The unit called a mole is enormously important in chemistry; it is introduced later in the chapter. The candela is hardly ever used in chemistry and will not appear again in this book.

In science we distinguish between *mass* and *weight*, and the kilogram is a unit of mass, not of weight. The mass of an object is a measure of the *quantity of matter* it contains. The weight of an object is a measure of the *gravitational pull* it experiences. Mass and weight are proportional to each other, but they are not identical. An astronaut would have the same mass (contain the same quantity of matter) on Earth and on Mars. But an astronaut's weight on Mars is less than on Earth because Mars has less gravitational pull than Earth does. Astronauts might occasionally be weightless (experience no gravitational pull), but they are never massless.

The physical properties determined by measurements, such as the mass of an object or the distance between two points, are symbolized by *italic* letters. Three examples are

mass, m

length, l

time, t

Table 2.1 SI base units

Quantity	Base unit
length	1 meter (1 m)
mass	1 kilogram (1 kg)
time	1 second (1 s)
electric current	1 ampere (1 A)
temperature	1 kelvin (1 K)
chemical amount	1 mole (1 mol)
luminous intensity	1 candela (1 cd)

For example, we could write $m = 10$ g if the mass of an object is 10 g. Note that physical properties are symbolized by italic letters (for example, m for mass) and the corresponding units by Roman (upright) letters (for example, g for grams).

A measurement is reported as the numerical multiple of a standard unit.

2.2 Prefixes for Units

The Système International also uses combinations of base units and various prefixes. These prefixes denote *multiples* of powers of 10 of the SI units themselves. The centimeter (cm), for instance, is 1/100 of a meter. In common with all the prefixes, the prefix c can be used with any unit and always means 1/100, or a multiple of 10^{-2}, of the unit:

$$1 \text{ cm} = 10^{-2} \times (1 \text{ m}), \quad \text{or, more simply, } 1 \text{ cm} = 10^{-2} \text{ m}$$

In this book, we shall use the nine prefixes given in Table 2.2. The prefixes in Table 2.2 can be used with any of the units to give expressions like $1 \text{ pm} = 10^{-12}$ m for picometer (pm) and $1 \text{ ms} = 10^{-3}$ s for millisecond (ms). The kilogram is a little out of line: it is a base unit, but it already has a prefix (k). When we use prefixes with the unit of mass, we apply them to the gram instead and note that $1 \text{ kg} = 10^{3}$ g. The mass of this book is about 1 kg; the mass of a 1¢ coin is about 3 g.

We commonly choose the prefix to avoid awkward powers of 10 in a measurement and to suit the scale of the object. For example, centimeters are useful for reporting the size of small objects (the diameter of a small coin is about 2 cm), and meters are useful for measurements of larger objects (a table top is about 1 m above the floor; the dimensions of a room are typically several meters), but picometers are better for describing the size of an atom (the diameter of a nickel atom is 125 pm).

When doing calculations, it is best to replace prefixes by the powers of 10 they represent and then proceed numerically. For example, we replace $1 \text{ } \mu\text{m}$ by 10^{-6} m in calculations. A second point to remember is that an expression like cm^3 (cubic centimeter) means $(\text{cm})^3$:

$$\begin{aligned}
2 \text{ cm}^3 &= 2 \times (1 \text{ cm}) \times (1 \text{ cm}) \times (1 \text{ cm}) \\
&= 2 \times (10^{-2} \text{ m}) \times (10^{-2} \text{ m}) \times (10^{-2} \text{ m}) \\
&= 2 \times 10^{-6} \text{ m}^3
\end{aligned}$$

A volume of 1 cm^3 is about the volume of a small cube about $\frac{1}{2}$ in. on a side, so 2 cm^3 is the volume of two such blocks. Note that when the number is also raised to a power, which would be the case if we wanted the volume of a cube 2 cm on a side, we write

$$\begin{aligned}
(2 \text{ cm})^3 &= (2 \times 10^{-2} \text{ m}) \times (2 \times 10^{-2} \text{ m}) \times (2 \times 10^{-2} \text{ m}) \\
&= 8 \times 10^{-6} \text{ m}^3
\end{aligned}$$

To avoid errors, it is best to show the units at each stage of a calculation, just as though they were numbers, and not to attach the units only in the last line.

Multiples of units that are powers of 10 are represented by prefixes attached to the symbol for the unit.

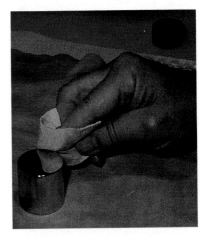

Table 2.2 Common SI prefixes

Prefix	Name	Meaning
G	giga	10^{9}
M	mega	10^{6}
k	kilo	10^{3}
d	deci	10^{-1}
c	centi	10^{-2}
m	milli	10^{-3}
μ	micro	10^{-6}
n	nano	10^{-9}
p	pico	10^{-12}

2.3 Derived Units

In the previous section, we multiplied the symbols for units just as though they were ordinary numbers. We can cancel units too. For example, the result of dividing 6 g by 3 g is a pure number with no units, because the units (grams) cancel:

$$\frac{6\,\cancel{g}}{3\,\cancel{g}} = \frac{6}{3} = 2$$

Another example of the multiplication of units is the calculation we do when we want to find the volume, V, of a box with sides 1.0 m, 2.0 m, and 3.0 m:

$$V = (1.0\ \text{m}) \times (2.0\ \text{m}) \times (3.0\ \text{m}) = 6.0\ \text{m}^3$$

The result is shorthand for $6.0 \times (1\ \text{m}^3)$, where $1\ \text{m}^3$ is the SI unit of volume.

Units built up from base units are called **derived units.** It is common in chemistry to report volumes of liquids and gases in liters (L), where 1 L is defined as exactly $1\ \text{dm}^3$ (corresponding to $10^3\ \text{cm}^3$). By this definition, 1 mL is exactly the same as $1\ \text{cm}^3$.

In some instances, a derived unit includes a unit raised to a negative power. An example is the unit for **density,** d, which is the mass of a sample divided by its volume:

$$\text{Density} = \frac{\text{mass of sample}}{\text{volume of sample}} \qquad d = \frac{m}{V} \tag{1}$$

The SI unit of density is formed by dividing the SI unit of mass by the SI unit of volume:

$$\text{Unit of density} = \frac{1\ \text{kg}}{1\ \text{m}^3} = 1\ \text{kg/m}^3$$

Note that $1/\text{m}^3$ can also be written m^{-3}, just as for numbers. The unit $1\ \text{kg/m}^3$ (or $1\ \text{kg} \cdot \text{m}^{-3}$) is read "one kilogram per cubic meter." For example, if a sample of packing foam has a mass of 12 kg and its volume is $4.0\ \text{m}^3$, then its density would be reported as

$$\text{Density} = \frac{\text{mass}}{\text{volume}} = \frac{12\ \text{kg}}{4.0\ \text{m}^3}$$

$$= \frac{12}{4.0} \times \frac{\text{kg}}{\text{m}^3} = 3.0\ \text{kg/m}^3 \quad \text{or} \quad 3.0\ \text{kg} \cdot \text{m}^{-3}$$

Some properties depend on the size of the sample and some do not. For example, suppose we double the size of the sample. How would that affect the mass, volume, and density of the sample? Doubling the sample doubles both the mass and the volume. Properties that depend on the size of the sample, like mass and volume, are called **extensive.** However, the density (the ratio of mass to volume) remains the same because both mass and volume increase by the same amount when we double the size of the sample. Density is therefore an example of an **intensive property,** a property that is independent of the size of the sample. Temperature, which we discuss more fully later, is another example of an intensive property: if we withdraw 10 mL or 1 L of water from a well-stirred tank and measure its temperature, we obtain the same value in each case.

The unit kilogram per cubic meter for density is often too large to be convenient. So, in most cases, we use gram per cubic centimeter instead, with $1 \text{ g/cm}^3 = 10^3 \text{ kg/m}^3$. The density of water is very close to 1 g/cm^3 (and 1000 kg/m^3). This value implies that the mass of 1 cm^3 of water is very close to 1 g.

Units are multiplied and divided like numbers; derived units are built up from base units.

FIGURE 2.3

Crystals of beryl contaminated by chromium ions have a green color. We know them as emeralds.

Example 2.1 *Determining and using the density*

Beryl is a precious stone with a density of 2.66 g/cm^3 (Fig. 2.3). A rock of mass 35.6 g that appears to be beryl is found in a forest preserve. When it is placed in a cylinder of water that has been filled to the 60.0-mL mark, the level of the water rises to 78.4 mL. (a) Find the density of the rock and decide whether it could be beryl. (b) How many cubic centimeters would 200. g of rock of the same substance occupy?

Strategy (a) Because density is mass divided by volume, we first need to find the volume of the rock. The volume of the rock is equal to the volume of water it displaces. We then divide its mass by its volume. Remember that $1 \text{ mL} = 1 \text{ cm}^3$. Compare the density with that of beryl. If the density is different, the rock cannot be beryl. If the density is the same, the rock could be beryl. (b) Notice that the definition of density in Eq. 1 can be rearranged to give the unknown volume on the left and the density and mass of the sample on the right:

$$\text{Volume} = \frac{\text{mass}}{\text{density}} \qquad V = \frac{m}{d}$$

To find the volume of the sample, we insert the data into this expression.

Solution (a) The volume of water displaced is $78.4 \text{ mL} - 60.0 \text{ mL} = 18.4 \text{ mL}$, or 18.4 cm^3. The density of the rock is therefore

$$\text{Density} = \frac{35.6 \text{ g}}{18.4 \text{ cm}^3} = 1.93 \text{ g/cm}^3$$

This density is very different from that of beryl, so the rock is not beryl. (b) The volume of a sample of the same substance of mass 200. g is

$$\text{Volume} = \frac{200. \text{ g}}{1.93 \text{ g/cm}^3} = \frac{200.}{1.93} \times \frac{\cancel{g} \cdot \text{cm}^3}{\cancel{g}} = 104 \text{ cm}^3$$

Self-Test 2.1A (a) Find the density of a metal if the water in a graduated cylinder rises from 50.0 mL to 61.5 mL when a piece of the metal of mass 35.55 g is placed in it. (b) A door hinge for use on a locker inside the space shuttle was made from the metal and was known to have a volume of 5.00 cm^3. The designers specified a mass of less than 15.0 g. Is the hinge acceptable?

[*Answer:* (a) 3.09 g/cm^3; (b) no (15.4 g)]

Self-Test 2.1B (a) Find the density of a polymeric material that causes the water in a graduated cylinder to rise from 50.0 mL to 83.3 mL when a piece of the polymer of mass 35.0 g is placed in it. (b) The polymer was to be used to make an artificial kneecap, which the designers required to have a mass of no more than 30.0 g. The volume of the polymeric kneecap turned out to be 28 cm^3. Is the kneecap acceptable?

Table 2.3 Relations between units*

Common unit	SI equivalent
1 pound (lb)	453.6 grams (g)
1 inch (in.)	**2.54** centimeters (cm)
1 foot (ft)	**30.48** cm
1 quart (qt)†	0.946 liter (L)
1 minute (min)	**60** s

*A longer list is given inside the back cover of the text. The numbers in bold type are exact equivalents.

†The European and Canadian quart is 1.201 times larger than the American quart used here.

2.4 Unit Conversions

The units in common use in the United States and a few other countries are called either *English units* or *conventional units*. Until the metric system is adopted everywhere, it will be helpful to know how the magnitudes of the SI units compare with those of the familiar English units. Thus, it is helpful to know that 1 m is approximately 3 inches longer than 1 yard, that 1 cm is slightly shorter than half an inch, and that 1 kg is approximately 2.2 lb. Some relations of this kind are summarized in Table 2.3.

For scientific work, we need to convert English units into SI units. For example, in Table 2.3, we see that

$$1 \text{ inch} = 2.54 \text{ cm}$$

Because 1 inch (1 in.) and 2.54 cm are different ways of expressing the same length, it follows that any object that is $x \times$ (1 inch) long is also $x \times$ (2.54 cm) long. Therefore, to convert inches to centimeters, we simply multiply the number of inches by 2.54. However, we also have to cancel the unit *inch* and replace it by the unit *centimeter*. We can carry out both the numerical calculation and the replacement of units in a single step if we multiply our measurement by the following factor:

$$\frac{2.54 \text{ cm}}{1 \text{ in.}}$$

The "in." in the denominator cancels any inches this factor multiplies and the cm in the numerator multiplies the resulting numerical factor. For instance, if the length of a piece of glass tubing is 2.00 inches, then its length in centimeters is

$$\text{Length (cm)} = (2.00 \text{ in.}) \times \left(\frac{2.54 \text{ cm}}{1 \text{ in.}} \right) = 2.00 \times 2.54 \frac{\text{in.} \cdot \text{cm}}{\text{in.}}$$

$$= 5.08 \text{ cm}$$

For the general form of this expression, we write

$$\text{Information required} = \text{information given} \times \text{conversion factor}$$

and use the **conversion factor** in the form

$$\text{Conversion factor} = \frac{\text{units required}}{\text{units given}}$$

One way to remember the form of this factor is to note that the units in the denominator must cancel the units in the data and the units in the numerator must replace the canceled units with the units required. Toolbox 2.1 at the end of this section summarizes this procedure.

We often have to convert a unit that is raised to a power. For example, we might wish to convert a surface area expressed as 256 cm² into square meters (m²). The same procedure is employed each time the unit appears, so in this case the conversion factor appears as its square. We note first that $1 \text{ m} = 10^2 \text{ cm}$ and express this relation as the conversion factor

$$\text{Conversion factor} = \frac{\text{units required}}{\text{units given}} = \frac{1 \text{ m}}{10^2 \text{ cm}}$$

Then we use the conversion factor squared:

$$\text{Area (m}^2) = (256 \text{ cm}^2) \times \left(\frac{1 \text{ m}}{10^2 \text{ cm}}\right)^2$$

$$= \frac{256}{10^4} \times \frac{\text{cm}^2 \cdot \text{m}^2}{\text{cm}^2} = 2.56 \times 10^{-2} \text{ m}^2$$

Note how the cm^2 of the data cancels the cm^2 that comes from squaring the conversion factor. The original units always cancel like this when the conversion factor has been set up correctly, so we know immediately if we are carrying out the conversion properly.

In some cases, we have to convert units that appear in the denominator, such as the units of time in a measurement of speed (in kilometers per second, km/s, for instance, being converted to kilometers per hour, km/h). The conversion is carried out exactly as we have described, but with the conversion factor inverted to give the desired units. For example, to land equipment on Mars, we may need to convert the speed at which Mars orbits the Sun, 25 km/s (25 kilometers per second), into kilometers per hour (km/h). We use the relation 1 h = 3600 s to write the conversion factor

$$\text{Conversion factor} = \frac{\text{units required}}{\text{units given}} = \frac{1 \text{ h}}{3600 \text{ s}}$$

and then invert this factor so that the seconds in the denominator of the speed are canceled by the seconds in the conversion factor:

$$\text{Speed (km/h)} = \left(25 \frac{\text{km}}{\text{s}}\right) \times \left(\frac{3600 \text{ s}}{1 \text{ h}}\right) = 9.0 \times 10^4 \text{ km/h}$$

These conversion procedures are summarized in Toolbox 2.1.

Toolbox 2.1 *How to use conversion factors*

This Toolbox describes how to use conversion factors to convert a measurement made in one set of units into another set of units.

Conceptual Basis
The physical property (the length, for instance) remains the same: the converted value is simply a different numerical multiple of another choice of unit (Fig. 2.4).

Procedure
When information is given in units other than the ones required, convert to the required units by using the general relation

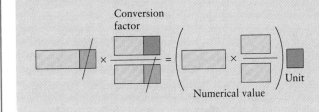

Conversion factor

Numerical value

Information required =
    information given × conversion factor

The conversion factor to use has the form

$$\text{Conversion factor} = \frac{\text{units required}}{\text{units given}}$$

If the unit is raised to a power, use the same procedure for each time that the unit appears. Therefore, if the unit to be converted appears as its nth power, use the nth power of the conversion factor.

If the unit is in the denominator, invert the conversion factor and use it as before.

FIGURE 2.4

When we use a conversion factor, we cancel the old units, replace them with the new units, and make the appropriate numerical conversion.

Example 2.2 *Using a conversion factor*

A food chemist is studying the nutritional content of a new fruit drink and needs the mass of the drink in grams. What is the mass in grams of a 12.0-ounce serving of the fruit drink?

Strategy Find the conversion factor in the listing inside the back cover. Then use the procedure set out in Toolbox 2.1 for converting units.

Solution The mass is given in ounces, but grams are required. The conversion factor inside the back cover reveals that 1.000 oz = 28.35 g. Therefore,

$$\text{Mass (g)} = (12.0 \ \cancel{oz}) \times \left(\frac{28.35 \ \text{g}}{1.000 \ \cancel{oz}}\right) = 340. \ \text{g}$$

Self-Test 2.2A A supplier packages hydrochloric acid in liters, but we need 1.85 quarts (qt). What volume in liters corresponds to 1.85 qt of the acid?

[*Answer:* 1.75 L]

Self-Test 2.2B Express the mass in ounces (oz) of a 250.-g package of breakfast cereal.

Example 2.3 *Converting units that appear in the denominator*

Mercury is used in a number of types of laboratory apparatus because it has the high density typical of many metals but is liquid. Express the density of mercury, 13.5 g/cm³, in kilograms per cubic meter (kg/m³).

Strategy Use the procedures set out in Toolbox 2.1 for converting units in the denominator and units raised to a power. Two conversion factors are involved. To convert from grams (units given) to kilograms (units required) by using 1 kg = 10^3 g, use the form of the factor that has kilograms in the numerator, namely, 1 kg /10^3 g. To convert from centimeters (units given) to meters (units required), use 1 m = 10^2 cm. Because centimeters are in the denominator, use the inverted conversion factor 10^2 cm /1 m.

Solution Each conversion factor is raised to the same power as the unit it is converting, so the conversion factor for length must be inverted and then raised to the third power to provide a conversion factor for volume:

$$\begin{aligned}
\text{Density (kg/m}^3) &= \left(13.5 \ \frac{\text{g}}{\text{cm}^3}\right) \times \left(\frac{1 \ \text{kg}}{10^3 \ \text{g}}\right) \times \left(\frac{10^2 \ \text{cm}}{1 \ \text{m}}\right)^3 \\
&= \left(13.5 \times \frac{\cancel{\text{g}}}{\cancel{\text{cm}^3}}\right) \times \left(\frac{1 \ \text{kg}}{10^3 \ \cancel{\text{g}}}\right) \times \left(\frac{10^6 \ \cancel{\text{cm}^3}}{1 \ \text{m}^3}\right) \\
&= 1.35 \times 10^4 \ \text{kg/m}^3
\end{aligned}$$

Each conversion could be done separately in a series of steps, but it is normally more efficient to carry them out all at the same time.

Self-Test 2.3A Express a density reported as 6.5 g/mm³ in micrograms per cubic nanometer (μg/nm³).

[*Answer:* $6.5 \times 10^{-12} \ \mu\text{g/nm}^3$]

Self-Test 2.3B Express a density of 1.100×10^3 kg/m³ in grams per cubic centimeter.

2.5 Temperature

There are three common units for reporting temperature. The United States is one of the few countries that still uses the **Fahrenheit scale,** in which water freezes at 32°F and boils at 212°F. This familiar but awkward scale is very rarely used in science. Much more common is the Celsius scale devised by Anders Celsius, an eighteenth-century Swedish astronomer. On the **Celsius scale,** water freezes at 0°C (zero degrees Celsius) and boils at 100°C (Fig. 2.5). In a thermometer, the change in length of a mercury column between the freezing and boiling points of water is the same regardless of the scale written beside the column. On the Celsius scale, this difference in length begins at 0 and is divided into 100 degrees; on the Fahrenheit scale, it begins at 32 and is divided into 180 degrees. Because $\frac{180}{100} = \frac{9}{5}$, it follows that a temperature on the Celsius scale is converted to a temperature on the Fahrenheit scale, and vice versa, by using the relation

$$\text{Temperature (°F)} = \left[\frac{9}{5} \times \text{temperature (°C)}\right] + 32 \qquad (2)$$

The Celsius scale (like the Fahrenheit scale) has an arbitrary zero point. However, in 1848, the British scientist Lord Kelvin showed that there is an **absolute zero** of temperature, a temperature below which it is impossible to cool anything. It is natural to set the 0 of the temperature scale to this lowest possible temperature. The absolute zero of temperature lies at −273.15°C. Scientists have cooled objects *almost* to that temperature; in 1995, a research group managed to cool a sample to within 0.000 000 002°C of absolute zero, possibly making the sample colder than anywhere else in the universe.

The **Kelvin scale** has the value 0 at absolute zero, and the size of the degree, which is called the *kelvin* and denoted K, is the same size as the Celsius degree. Therefore, on the Kelvin scale, water freezes at 273.15 K (which corresponds to 273.15°C above absolute zero) and boils 100 K higher, at 373.15 K (Fig. 2.6).

> Daniel Fahrenheit, the German scientist who invented the scale, set body temperature at 100°F (he must have had a slight fever that day) and set the lowest temperature he could reach with a mixture of salt and water at 0°F.

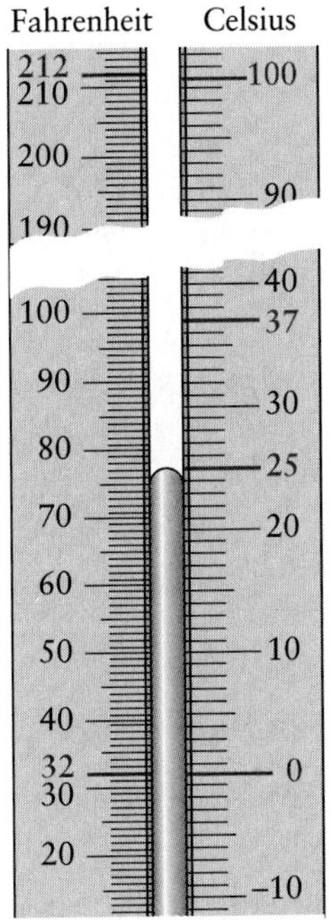

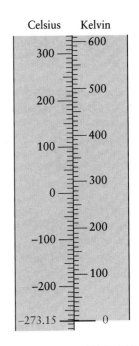

FIGURE 2.5

The Fahrenheit and Celsius temperature scales. The boiling and freezing points of water are marked in red. Two other common temperatures are marked in blue: many data are reported at 25°C, and body temperature is 37°C.

FIGURE 2.6

The Celsius and Kelvin temperature scales. Note that the Kelvin scale does not extend below 0, corresponding to −273.15°C. The Celsius scale is now defined in terms of the Kelvin scale.

Neither the word *degree* nor the degree symbol is used for the kelvin. To express a Celsius temperature on the Kelvin scale, we simply add 273.15. More formally:

$$\text{Temperature (K)} = \text{temperature (°C)} + 273.15 \qquad (3)$$

For example, if a weather report predicted a temperature of 25.00°C for tomorrow, the temperature on the Kelvin scale would be

$$\text{Temperature (K)} = 25.00 + 273.15 = 298.15$$

That is, the Kelvin temperature is 298.15 K. The Kelvin scale is awkward for everyday use. However, its use greatly simplifies many calculations in science.

> *In science, temperatures are measured on the Celsius temperature scale; calculations involving temperature use the Kelvin scale, which starts at 0 for the absolute zero of temperature.*

Because the size of each Celsius degree is the same as the kelvin, a temperature *difference* has the same value on each scale.

Toolbox 2.2 *How to convert between temperature scales*

This Toolbox summarizes the procedures used to convert between temperature scales.

Conceptual Basis
A temperature difference corresponds to a certain length of a column of liquid (or some other physical property) in a thermometer, but different scales divide the length into different numbers of divisions and set the origin of the scale at different temperatures (Fig. 2.7).

Procedure
Celsius and Fahrenheit scales have different units and different zero points. To convert from Celsius to Fahrenheit, use

$$\text{Temperature (°F)} = \left[\tfrac{9}{5} \times \text{temperature (°C)}\right] + 32$$

To convert from Fahrenheit to Celsius, use

$$\text{Temperature (°C)} = \frac{\text{temperature (°F)} - 32}{9/5}$$

Celsius degrees and kelvins are the same size, but the two scales have different zero points. To convert from the Celsius scale to the Kelvin scale, change the zero point by using

$$\text{Temperature (K)} = \text{temperature (°C)} + 273.15$$

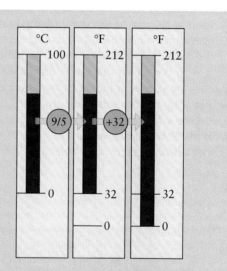

FIGURE 2.7

The conversion between the Celsius and Fahrenheit temperature scales adjusts for the size of the unit and the different zero points of the two scales.

Example 2.4 *Converting between temperature scales*

Express body temperature, about 99°F, on (a) the Celsius scale and (b) the Kelvin scale.

Strategy (a) First, estimate the required temperature by referring to Fig. 2.5. For the precise value, we rearrange Eq. 2 to give the temperature in degrees Celsius in terms

58   CHAPTER 2   MEASUREMENTS AND MOLES

of the temperature in degrees Fahrenheit, as shown in Toolbox 2.2. (b) Convert the Celsius temperature to Kelvin units by adding 273.15 to the numerical value, as explained in Toolbox 2.2.

Solution (a) In Fig. 2.5, we see that a reading of 99 on the Fahrenheit scale should be about the same temperature as 37 on the Celsius scale; so body temperature is about 37°C. To obtain the precise temperature, we use Eq. 2 as shown in Toolbox 2.2:

$$\text{Temperature (°C)} = \frac{99 - 32}{9/5} = 37$$

That is, 99°F corresponds to 37°C, which is consistent with our estimate. (b) To convert 37°C to the Kelvin scale, we use Eq. 3:

$$\text{Temperature (K)} = 37 + 273.15 = 310$$

The temperature 99°F corresponds to 37°C and therefore to 310 K. Note that we have rounded off the sum.

Self-Test 2.4A Convert (a) −223°C to the Fahrenheit scale and (b) −210.°F to the Celsius scale.

[***Answer:*** (a) −369°F; (b) −134°C]

Self-Test 2.4B What is the melting point range of the plant growth regulator gibberellic acid, 233–235°C, on the Kelvin scale?

2.6 The Uncertainty of Measurements

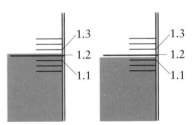

One reason for the high credibility of scientists is the honesty with which they must report measurements. Every measurement is limited by the reliability of the measuring instrument and by the skill of the operator, and the uncertainty must be reported correctly. Although the actual uncertainty of a measurement depends on the precision of the instrument used to record it, we shall use the convention that, unless otherwise specified, *the last digit—the least significant figure, the digit on the right—in the data is imprecise to the extent of ±1.* A volume measurement reported as 1.2 cm³ will be taken to mean that the volume lies between 1.1 and 1.3 cm³. A mass reported as 1.47 g will be taken to mean that it lies between 1.46 and 1.48 g (Fig. 2.8).

The uncertainty of the data determines the uncertainty of the results of calculations based on the data. The digits in a reported measurement are called the **significant figures.** There are two significant figures (written 2 sf) in 1.2 cm³ and 3 sf in 1.78 g. To find the number of significant figures in a measurement:

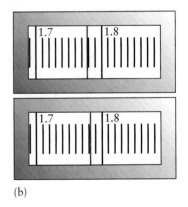

(b)

FIGURE 2.8

Two sets of measurements. (a) The volume of the liquid is reported as 1.23 ± 0.01 and 1.17 ± 0.01 mL. (b) The mass (the reading on this balance scale) is reported as 1.778 ± 0.001 g and 1.781 ± 0.001 g. The precision of the balance reading is greater than that of the volume.

Step 1. Express the data in scientific notation (Appendix 1B), with one nonzero digit in front of the decimal point.

Step 2. Count the number of digits multiplying the factor of ten.

For instance, 0.0025 g is written as 2.5×10^{-3} g, a value with 2 sf.

Some zeros are legitimately measured digits, but other zeros serve only to mark the place of the decimal point (Table 2.4). Trailing zeros (the last ones after a decimal point), as in 22.0 mL, are significant, because they were measured. Thus, 22.0 mL has 3 sf; this number signifies that the volume of the sample lies between 21.9 and 22.1 mL. The "captive" zero in 80.1 kg is also a measured digit, so 80.1 kg has 3 sf. However, the leading digits in 0.0025 g are not significant; they are only placeholders used to indicate powers of 10, not measured numbers. We can see that they are only placeholders by reporting the mass as 2.5×10^{-3} g, which has 2 sf.

What about a length reported as 400 m: does it have 3 sf (4.00×10^2), 2 sf (4.0×10^2), or only 1 sf (4×10^2)? In such cases, the use of scientific notation removes any ambiguity. If it is not convenient to use scientific notation, a final decimal point can be used to indicate that every digit to the left of the decimal is significant. Thus, 400 m is ambiguous and cannot be taken to have more than 1 sf unless other information is given. However, 400. m unambiguously has 3 sf. We shall use this convention throughout the textbook.

Example 2.5 *Counting the number of significant figures*

Report the number of significant figures in (a) 50.00 g; (b) 0.005 01 m; (c) 0.0100 mm; (d) 340 mL.

Strategy In each case, write the information in scientific notation, with one nonzero digit in front of the decimal point and all trailing zeros preserved, then count the total number of digits.

Solution (a) All the zeros in a mass reported as 50.00 g are significant because the value can be written 5.000×10^2 g, with four significant figures (4 sf). (b) A length of 0.005 01 m is the same as 5.01×10^{-3} m, a quantity with 3 sf. (c) A length of 0.0100 mm is the same as 1.00×10^{-2} mm, a quantity with 3 sf. (d) The significance of the final 0 is not given, so we must assume that it is a placeholder; 3.4×10^2 has 2 sf.

Self-Test 2.5A Determine the number of significant figures in (a) 2.1010 kg; (b) 100.000°C; (c) 0.000 000 1 mm.

[***Answer:*** (a) 5 sf; (b) 6 sf; (c) 1 sf]

Self-Test 2.5B Determine the number of significant figures in (a) 5.110 cm; (b) 0.005 00 g; (c) 5.000 505 m.

Table 2.4 *Examples of significant figures*

Decimal notation	Scientific notation	Number of sf
0.751	7.51×10^{-1}	3
0.007 51	7.51×10^{-3}	3
0.070 51	7.051×10^{-2}	4
0.750 100	$7.501\ 00 \times 10^{-1}$	6
7.5010	7.5010	5
7501	7.501×10^2	4
7500	7.5×10^3	2*
7500.	7.500×10^3	4

*In this text, treat trailing zeros as insignificant unless they are followed by a decimal point or other information is given. sf, significant figure.

We need to distinguish between the results of measurements, which are always uncertain, and the results of *counting*, which are exact. For example, the report "12 eggs" means that there are *exactly* 12 eggs present, not a number somewhere between 11 and 13.

Some unit conversion factors are defined exactly, even though they are not whole numbers. For example, 1 in. is defined as *exactly* 2.54 cm. The 273.15 in the conversion between Celsius and Kelvin temperatures is also exact; so 100.000°C converts to 373.150 K. All the bold numbers in Table 2.3 are exact. When they are used in a calculation, the number of significant figures in the answer is the same as the number of significant figures in the data.

In science, we do a lot of calculations on data from measurements. So we need to make sure that the number of significant figures that we use to report the result of a calculation corresponds to the number of significant figures in the data. For instance, if the mass of a sample of plastic is reported as 1.78 g and its volume as 1.2 cm^3, it is wrong to report its density as

$$\text{Density} = \frac{1.78 \text{ g}}{1.2 \text{ cm}^3} = 1.483\,333 \text{ g/cm}^3$$

which may be the display on our calculator. The correct procedure is described in Toolbox 2.3.

The uncertainty of the data determines the uncertainty of the results of calculations based on the data.

Example 2.6 *Using significant figures when adding and subtracting*

Find (a) the total volume of three blocks of copper with measured volumes 11.12, 1.2, and 3.107 cm^3; (b) the total length of three rods with lengths 1550, 233, and 600 mm.

Strategy We need to add the volumes in (a) and the lengths in (b), so we should follow the rule for addition in Toolbox 2.3.

Solution (a) The smallest number of digits after the decimal point is 1, in 1.2 cm^3. Thus, only one digit should follow the decimal point in the total volume:

$$
\begin{aligned}
&11.12 \text{ cm}^3 \\
&1.2 \text{ cm}^3 \\
&\underline{3.107 \text{ cm}^3} \\
\text{Total:} \quad &15.427 \text{ cm}^3; \quad \text{round to } 15.4 \text{ cm}^3
\end{aligned}
$$

(b) All the lengths have the same number of decimal places—none. The significance is determined by the number followed by the largest number of zeros, 600 mm. Thus, no digits smaller than 100 are significant:

$$
\begin{aligned}
&1550 \text{ mm} \\
&233 \text{ mm} \\
&\underline{600 \text{ mm}} \\
\text{Total:} \quad &2383 \text{ mm}; \quad \text{round to } 2400 \text{ mm and report as } 2.4 \times 10^3 \text{ mm}
\end{aligned}
$$

Toolbox 2.3 *How to use significant figures in calculations*

This Toolbox serves as a guide for judging how many significant figures should be retained in the result of a calculation.

Conceptual Basis

The result of a calculation performed on a measurement is only as certain as the original measurement. The uncertainty of the measurement must then be reflected in the result of the calculation.

Procedure

Different rules apply to addition (and its reverse, subtraction) and multiplication (and its reverse, division). Both procedures require us to round off the answers to the correct number of significant figures.

Rounding off In all calculations, round *up* if the last digit is above 5 and round *down* if it is below 5. For numbers ending in the digit 5, always round to the nearest even number. For example, 2.35 rounds to 2.4 and 2.65 rounds to 2.6. This strategy prevents compounding round-off errors in the final answer. The quantity 14.348 cm^3 rounds to 14.3 cm^3 if the answer should have 3 sf, but it would have rounded to 14.35 cm^3 if the data had justified 4 sf. Rounding must be carried out in a single step: 14.348 cm^3 should not first be rounded to 14.35 cm^3 and then to 14.4 cm^3. The correct procedure is to round off only at the final stage of the calculation and to carry all digits in the memory of the calculator until that stage. In this text, we often need to display intermediate rounded results, so the final answer might differ slightly from the answer obtained by rounding only at the end of the calculation.

Addition and subtraction When adding or subtracting, the number of decimal places in the result should be the same as the *smallest number of decimal places* in the data (Fig. 2.9). If there are no decimal points, significance is determined by the least precise measurement.

FIGURE 2.9

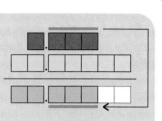

The number of significant figures that result when adding or subtracting.

Multiplication and division When multiplying or dividing, the number of significant figures in the result should be the same as *the smallest number of significant figures* in the data (Fig. 2.10).

Integers and exact numbers When multiplying or dividing by an integer or an exact number, the uncertainty of the result is determined by the measured value.

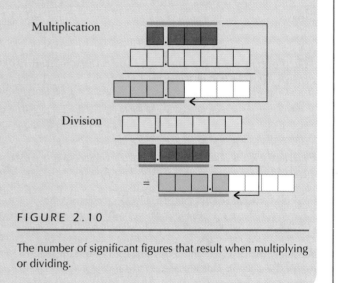

Multiplication

Division

FIGURE 2.10

The number of significant figures that result when multiplying or dividing.

If the lengths had been reported as 1550., 233, and 600. mm, the correct answer would have been 2383 mm.

Self-Test 2.6A What is (a) the total volume of a sample of water prepared by adding 25.6 mL to 50. mL; (b) the temperature in kelvins corresponding to the triple point of oxygen, $-218.789°C$?

[*Answer:* (a) 76 mL; (b) 54.361K]

Self-Test 2.6B Report (a) the total mass of a sample prepared by mixing 1.001 g of sugar, 2.05 g of salt, and 5.0 g of flour; (b) the Celsius temperature corresponding to the melting point of iron, 1813 K.

Example 2.7 *Using significant figures when multiplying or dividing*

The density of the sample of plastic described earlier in this section was calculated from a mass of 1.78 g and a volume of 1.2 cm^3. The experimenter's calculator gives the density as 1.483 333 g/cm^3. How should the experimenter report this density?

Strategy Density is mass divided by volume, so we use the rule for division in Toolbox 2.3.

Solution The smallest number of figures in the data is 2 (in the measurement of volume, 1.2 cm^3), so the experimenter needs to round 1.483 333 g/cm^3 to 2 sf, giving 1.5 g/cm^3.

Self-Test 2.7A (a) Calculate the mass of 250. mL of carbon dioxide that has a density of 2.095 g/L. (b) What would the density of the gas be (in grams per liter) if the same sample were in a container of volume 150 mL?

[*Answer:* (a) 0.524 g; (b) 3.5 g/L]

Self-Test 2.7B (a) Calculate the volume occupied by 1.04 mg of oxygen gas, given that its density is 1.5 g/L. (b) What would its density be (in grams per liter) if the same sample occupied 0.755 mL?

2.7 Accuracy and Precision

To make sure of their data, scientists usually repeat their measurements several times. They never expect to get *exactly* the same result. More often than not, their results are accompanied by two kinds of errors. The **precision** of a measurement refers to how close to one another these repeated measurements are. The **accuracy** of a series of measurements is the closeness of their average value to the true value. Even precise measurements can give inaccurate values. For instance, if there is an unnoticed speck of dust on the pan of a chemical balance that we are using to measure the mass of a sample of vitamin C, then, even though we might be justified in reporting our measurements to 5 sf, the reported mass of the sample will be inaccurate.

A **systematic error** is an error present in every one of a series of repeated measurements. An example is the effect of a speck of dust on a pan, which distorts the mass of each sample in the same direction (the speck makes each sample appear heavier than it is). An **accurate measurement** is one free of systematic error. A **random error** is an error that varies at random and can average to 0 over a series of observations. An example is the effect of drafts of air from an open window. The drafts may move a balance pan either up or down a bit, thereby decreasing or increasing the mass measurements randomly. A **precise measurement** is one free of random error. Scientists attempt to improve the accuracy of their measurements by making many observations and taking the average of the results. This procedure minimizes random errors. Systematic errors are much harder to identify and eliminate. Their elimination sometimes depends on the comparison of results obtained independently in different laboratories.

Measurements with a large number of significant figures are often called *precise measurements*.

The precision of measurements indicates how close to one another the measurements are; the accuracy of a measurement is its closeness to the true value.

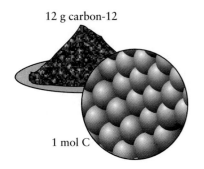

12 g carbon-12

1 mol C

FIGURE 2.11

The definition of the mole. If we measure out exactly 12 g of carbon-12, then we have exactly 1 mol of carbon-12 atoms. There will be exactly an Avogadro number of atoms in the pile.

Self-Test 2.8A Two students determined the mass of a sample of metal by taking three mass measurements. Student A's results were 7.84 g, 7.85 g, and 7.83 g. Student B's results were 7.83 g, 7.92 g, and 7.93 g. The instructor wrote, "One of you has results with high precision, but low accuracy. The other has low precision, but high accuracy." The actual mass of the sample was 7.89 g. (a) Which student achieved greater accuracy? (b) Which student achieved greater precision?

[*Answer:* (a) Student B; (b) Student A]

Self-Test 2.8B A scientist determined the density of a metal by measuring the mass and volume of three pieces of the metal. The masses were 11.24 g, 11.26 g, and 11.25 g. The volumes were 4.10 cm^3, 4.17 cm^3, and 4.15 cm^3. Which set of measurements had the greater precision?

CHEMICAL AMOUNTS

We can now start to apply these ideas to one of the first steps in the discovery of new drugs and other beneficial natural products: the determination of the chemical formulas of compounds (see Applying Chemistry: Case Study 2). A chemical formula tells us the relative numbers of atoms of each element present in a compound. For instance, the scientists who determined the formula of vitamin C first had to find out how many atoms of each element were present in a sample of the compound. However, we saw in Chapter 1 that the mass of an individual atom is so small that astronomical numbers, often as high as 10^{23} or even more, are present in typical samples of matter. Just as it is convenient to use centimeters and kilometers for reporting the lengths of small and large objects, respectively, we need a compact way of reporting the huge numbers of atoms that chemists spoon and pour from one container to another.

2.8 The Mole

The name *mole* comes from the Latin word for "massive heap." The animal of the same name makes massive heaps of soil on lawns.

Chemists keep track of large numbers of atoms, ions, and molecules by using the unit **mole** (abbreviated mol). The unit was invented to provide a simple way of reporting the huge numbers—the "massive heaps"—of atoms and molecules in visible samples. It would be inconvenient to refer to large numbers like 2.0×10^{25} atoms, just as wholesalers would find it inconvenient to count individual items instead of dozens (12) or gross (144). The definition of the unit is as follows (Fig. 2.11):

1 **mole** is the number of atoms in exactly 12 g of carbon-12.

We saw in Investigating Matter 1.2 that mass spectrometry can be used to determine the masses of individual atoms of specified isotopes. The mass of a carbon-12 atom has been found to be $1.992\ 65 \times 10^{-23}$ g, so the number of atoms in exactly 12 g of carbon-12 is

We are being cautious with the number of significant figures; more precise measurements of the mass of a single carbon-12 atom give the number as $6.022\ 14 \times 10^{23}$.

$$\text{Number of carbon-12 atoms} = \frac{12\ \text{g}}{1.992\ 65 \times 10^{-23}\ \text{g}} = 6.0221 \times 10^{23}$$

It follows from the definition of the mole that 1.0000 mol of atoms (of any element) is 6.0221×10^{23} atoms of the element. The same is true of 1 mol of anything—atoms, ions, or molecules:

1.0000 mol of objects always means 6.0221×10^{23} of those objects.

The number of objects per mole, $6.0221 \times 10^{23}/mol$, is called the **Avogadro constant,** N_A. The constant is named in honor of the nineteenth-century Italian scientist Amedeo Avogadro (Fig. 2.12), who helped to establish the existence of atoms.

 The unit mole applies to any chemical species, just as 1 dozen means 12 of anything to a grocer, not just the number of eggs in a carton.

An analogy

Some awe-inspiring illustrations can help you visualize a mole. For instance, 1 mol of chemistry textbooks would cover the surface of the Earth to a height of about 300 km. If you won 1 mol of dollars in a lottery the day you were born and spent a billion dollars a second for the rest of your life, you would still have more than 99.999% of the prize money left when you died at 90.

The Avogadro constant in the form $1 \text{ mol} = 6.0221 \times 10^{23}$ is used as a conversion factor between the number of moles and the number of atoms, ions, or molecules. To convert from numbers of particles of species X (atoms, molecules, ions, or formula units) to number of moles we use

$$\text{Conversion factor} = \frac{1 \text{ mol X}}{6.0221 \times 10^{23} \text{ particles}}$$

The number 6.0221×10^{23}, the number of particles in 1 mol of substance, is widely called *Avogadro's number*. Avogadro's number is a pure number; the Avogadro constant has units (per mole).

To convert from the stated number of moles of a species X to the number of X particles, we invert the conversion factor, as illustrated in Example 2.8.

The formal name for the quantity for which the units are moles is the **chemical amount,** n. Formally, therefore, we would say that the chemical amount of copper atoms in a certain sample is 1.5 mol, and we would write $n = 1.5 \text{ mol Cu}$. However, almost universally, chemists talk informally in terms of the "number of moles" of atoms, molecules, or ions. Moreover, the more we study chemistry, the more we see how rarely chemists refer to the actual *numbers* of atoms, ions, or molecules in a sample. Almost always, it is easier to refer to the numbers of moles of particles and to do calculations by using the mole as a unit. Because mol is a symbol for a unit, it can be used with a prefix. For example, if you were dealing with a tiny quantity of a precious pharmaceutical compound, you might express the amount in millimoles, where $1 \text{ mmol} = 10^{-3} \text{ mol}$.

The amounts of atoms, ions, or molecules in a sample are reported in moles. The Avogadro constant, N_A, is used to convert between numbers of these particles and the numbers of moles.

Example 2.8 *Converting between numbers of atoms and moles*

The chemists who unraveled the structure of vitamin C expressed the number of atoms in a sample in terms of moles. (a) Suppose a sample of vitamin C is known to contain 1.29×10^{24} hydrogen atoms (as well as other kinds of atoms). How many moles of hydrogen atoms does the sample contain? (b) How many oxygen atoms are present in a sample of vitamin C containing 0.500 mol O atoms?

FIGURE 2.12

Lorenzo Romano Amedeo Carlo Avogadro, count of Quaregna and Cerreto (1776–1856).

HARVESTING DRUGS FROM THE SEA

Applying Chemistry: *Case Study 2*

On one beautiful, sunny day in the Caribbean Sea, off the Isla de Providencia, several boats lay at anchor. Suddenly a group of divers surfaced and climbed aboard the boat shown in the illustration. There they spilled the contents of their collecting pouches onto the deck. They had harvested marine organisms: mostly sponges and tunicates (sea squirts). Like the pleasure-seekers on nearby boats, they gathered eagerly around their catch and selected interesting specimens. However, when the divers entered the large cabin of the boat, any similarity between them and their vacationing neighbors ended. Inside the boat was a fully equipped biochemical laboratory. There was a gas chromatograph connected to a mass spectrometer, a laboratory workbench with pipets and other glassware, and lots of petri dishes containing cultures.

The divers were chemists and chemistry students searching for marine organisms that might be sources of antiviral and antitumor agents. They did not know what compounds they were looking for, but they knew that Caribbean tunicates and sponges might be a promising source. Instead of testing every type of creature in the sea, they decided to concentrate on these species.

In the boat's laboratory, the first step was to add drops of fluid extracted from the tunicates to the petri dishes. Some of these dishes contained viruses and others contained leukemia cells. If an extract contained antiviral agents, the viruses in that dish would die. Extracts containing antitumor agents would kill leukemia cells. Soon dead viruses and tumor cells were discovered in the dishes containing extracts from a particular species of tunicate. Once the species was identified, more samples of that species

The research vessel *Alpha Helix* is used by chemists from the University of Illinois at Urbana-Champaign to search for marine organisms that contain compounds of medicinal value.

were collected. The components of its extract were then separated by gas chromatography, and the molar mass of each component determined with mass spectrometry. The individual compounds separated by chromatography were also tested for antiviral and antitumor properties, until the specific group of compounds active against viruses and tumors was found.

On returning to their land-based research laboratory, the chemists analyzed the compounds to determine their molecular formulas and then their structures. The first step

Strategy A good strategy in chemistry is to estimate an approximate answer before going through the calculation, because we can then detect major errors quickly. (a) Because the number of atoms in the sample is greater than 6×10^{23}, we anticipate that *more* than 1 mol of atoms is present. (b) We expect that half a mole of atoms will have approximately half 6×10^{23} atoms. For both calculations, we use the Avogadro constant as a conversion factor.

Solution (a) To convert number of atoms to moles, we write

$$\text{Moles of H atoms (mol)} = (1.29 \times 10^{24} \text{ atoms}) \times \left(\frac{1 \text{ mol H}}{6.0221 \times 10^{23} \text{ atoms}} \right)$$
$$= 2.14 \text{ mol H}$$

We can see how much simpler it is to refer to 2.14 mol H atoms than to the actual number of atoms, 1.29×10^{24} atoms. (b) To convert moles to number of atoms, we write

$$\text{Number of O atoms} = (0.500 \text{ mol O}) \times \left(\frac{6.0221 \times 10^{23} \text{ atoms}}{1 \text{ mol O}} \right)$$
$$= 3.01 \times 10^{23} \text{ atoms}$$

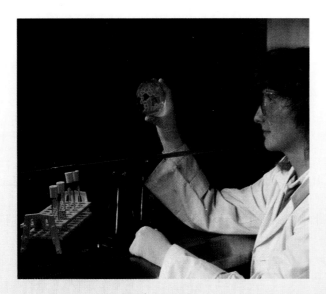

Extracts of marine organisms are spotted on petri dishes containing cancerous cells or viruses. These tests are done in the field to guide further searches for antitumor and antiviral agents.

was to carry out a combustion analysis (see Investigating Matter 2.1) to determine the mass percentage of each element. The molar mass of each compound had already been determined by mass spectrometry, so the molecular formulas could be calculated.

Each compound was further studied to determine its structure. A group of compounds called didemnins, after the tunicate in which they were found, has now been characterized by this process and synthesized. In fact, your physician might prescribe them if you are unfortunate enough to require this type of treatment.

Finding drugs in nature is a practice that has survived from antiquity. However, the healers of old used fresh herbs and remedies and so could serve only a small number of people. Today the chemicals in the remedies can be identified; then they are synthesized so that they can be made available to large numbers of people. Determining the formulas of these substances is the first step to making them widely available.

Key Concepts: mass percentage, molecular formula

For Further Reading

K. Rinehart et al., Marine natural products as sources of antiviral, antimicrobial, and antineoplastic agents, *Pure and Applied Chemistry*, **53**, 795–817, 1981.

Related Exercises: 2.95 and 2.96

Self-Test 2.9A A sample of vitamin C is known to contain 2.58×10^{24} oxygen atoms. How many moles of O atoms are present in the sample?

[*Answer:* 4.28 mol O atoms]

Self-Test 2.9B A small cup of coffee contains 3.14 mol of water molecules. How many H_2O molecules are present?

When reporting the number of moles of atoms, molecules, or ions, we always state explicitly which species (that is, which atoms, molecules, or ions) we mean. For example, hydrogen occurs naturally as a gas, with each gas molecule built from two atoms. Hence, it is denoted H_2. We write 1 mol H if we mean 1 mole of hydrogen atoms, and we write 1 mol H_2 if we mean 1 mole of hydrogen molecules. We do not refer simply to 1 mol of hydrogen, because it would not be clear whether we mean hydrogen atoms or hydrogen molecules. Alternatively, we could write that the number of moles of H atoms (or H_2 molecules) is 1 mol. *The essential point is to be unambiguous.* Recall from Section 1.8 that other elements that occur as *diatomic molecules,* (molecules composed of two atoms) are nitrogen (N_2), oxygen (O_2), and the halogens (Cl_2, for example). Phosphorus exists as P_4 molecules and sulfur as S_8.

Always state explicitly (in some unambiguous way) the identity of the particles to which the term moles refers.

2.9 Molar Mass

Our next task is to see how to determine the number of moles of particles in a sample. We can readily determine the mass of a sample in grams or kilograms, but how do we convert that mass into the number of moles of atoms, molecules, or ions in the sample?

The key concept that acts as a bridge between mass and moles is the **molar mass,** M, the mass per mole of particles:

> The *molar mass of an element* is the mass of the element per mole of its atoms.
>
> The *molar mass of a molecular compound* is the mass of the compound per mole of its molecules.
>
> The *molar mass of an ionic compound* is the mass of the compound per mole of its formula units.

You will still see these properties called atomic weight, molecular weight, and formula weight. However, the properties to which these terms refer are masses per mole, not weights.

For example, the molar mass of the element carbon is the mass per mole of carbon atoms. The molar mass of vitamin C, a molecular compound, is the mass per mole of vitamin C molecules. The molar mass of the ionic compound sodium chloride is the mass per mole of NaCl formula units. The units of molar mass in each case are grams per mole (g/mol).

To convert the mass of a sample into the number of moles in the sample, we need to know the values of molar masses. We already know that the masses of the individual atoms of an element can be measured by using mass spectrometry. To convert the mass of one atom to a mass per mole of atoms, all we have to do is multiply the individual mass by the number of atoms per mole (that is, by the Avogadro constant):

$$\begin{array}{ll} \text{Mass per mole of atoms} \\ \quad = \text{mass of one atom} & M = m_{\text{atom}} \times N_A \qquad \textbf{(4)} \\ \qquad \times \text{ number of atoms per mole} \end{array}$$

For example, the mass of a fluorine atom is 3.155×10^{-23} g, so the molar mass of fluorine is

$$\begin{array}{ll} \text{Mass per mole of F atoms} & = (3.155 \times 10^{-23}\,\text{g}) \times (6.0221 \times 10^{23}/\text{mol}) \\ & = 19.00\,\text{g/mol} \end{array}$$

Once we know the molar mass of an element, we can use it as a conversion factor to convert the mass of a sample to number of moles.

Unlike the molar mass of fluorine, molar masses of most elements cannot be calculated from the known mass of a single isotope: a naturally occurring sample of each of these elements is a mixture of different isotopes. For most elements, we need to calculate the *average* mass of an atom in a sample. Then we multiply this average mass by the number of atoms per mole (the Avogadro constant) to obtain the *average* molar mass of a natural sample. *All molar masses quoted in this text refer to these average values.* Their values are given in the table inside the back cover and in Appendix 2D. They are also included in the periodic table inside the front cover, so we do not have to work them out individually each time.

Calculating average molar masses is very much like calculating the average score on a test: we multiply each score by the number of students who earned that score, add all the results together, and then divide by the total number of students. For example, if in a class of 100 students, 75 students scored 70 and 25 students scored 50, the average score would be

An analogy

$$\text{Average score} = \frac{(75 \times 70) + (25 \times 50)}{100} = 65$$

Notice that the average score is closer to the higher score, which was earned by the larger number of students. Another way of expressing this calculation is to use the fractions of the class ($\frac{75}{100} = 0.75$ and $\frac{25}{100} = 0.25$, respectively) that received each grade:

$$\text{Average grade} = (0.75 \times 70) + (0.25 \times 50) = 65$$

Example 2.9 *Evaluating an average molar mass*

There are two naturally occurring isotopes of chlorine, chlorine-35 and chlorine-37. The mass of an atom of chlorine-35 is 5.807×10^{-23} g and that of an atom of chlorine-37 is 6.139×10^{-23} g. In a typical natural sample of chlorine, 75.77% of the sample is chlorine-35 and 24.23% is chlorine-37. What is the average molar mass of chlorine?

Strategy First, calculate the average mass of the isotopes by adding together the individual masses, each multiplied by the fraction that represents its abundance (as in the Analogy above). Then obtain the average molar mass (the mass per mole of atoms) by multiplying the average atomic mass by the Avogadro constant.

Solution A percentage of 75.77% means a fraction of $\frac{75.77}{100} = 0.7577$; similarly, a percentage given as 24.23% means a fraction of 0.2423. The average mass of an atom of chlorine in a natural sample is

Average mass of a Cl atom
$$= [0.7577 \times (5.807 \times 10^{-23}\,\text{g})] + [0.2423 \times (6.139 \times 10^{-23}\,\text{g})]$$
$$= (4.400 \times 10^{-23}\,\text{g}) + (1.487 \times 10^{-23}\,\text{g})$$
$$= 5.887 \times 10^{-23}\,\text{g}$$

It follows that the molar mass of a typical sample of chlorine atoms is

Molar mass of chlorine
$$= \text{average mass of a Cl atom} \times \text{number of Cl atoms per mole}$$
$$= (5.887 \times 10^{-23}\,\text{g}) \times (6.0221 \times 10^{23}/\text{mol}) = 35.45\,\text{g/mol}$$

Self-Test 2.10A In a typical sample of magnesium, 78.99% of the atoms are magnesium-24 (3.983×10^{-23} g), 10.00% are magnesium-25 (4.149×10^{-23} g), and 11.01% are magnesium-26 (4.315×10^{-23} g). Calculate the average molar mass of a sample of magnesium, given the atomic masses (in parentheses).

[***Answer:*** 24.31 g/mol]

Self-Test 2.10B Calculate the average molar mass of copper, given that a natural sample typically consists of 69.17% copper-63, which has a molar mass of 62.94 g/mol, and 30.83% copper-65, which has a molar mass of 64.93 g/mol.

Toolbox 2.4 How to convert between mass and moles

The procedures described here enable us to convert from the mass of a sample of an element to the numbers of moles of atoms it contains, and vice versa. The same procedures are used (as explained later) for compounds.

Conceptual Basis
The mole is the unit by which amounts of atoms are measured. The molar mass of an element is the mass per mole of its atoms and can be thought of as a conversion factor between moles and grams. Therefore, to find the mass of a sample, we multiply the number of moles by the

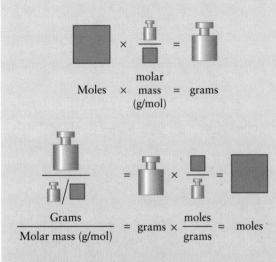

FIGURE 2.13

The use of molar mass to convert moles to mass (top) and mass to moles (bottom).

mass per mole. To obtain the number of moles of atoms in a sample, we divide the mass of the sample by the mass of 1 mole of the element (Fig. 2.13).

Procedure
To avoid ambiguity, include the identity of the species with the unit mole. Then proceed as follows.

Converting from moles to mass To convert from moles to mass in grams, multiply the number of moles by the molar mass:

$$\text{Mass of sample (g)} = \text{number of moles (mol)} \times \text{molar mass (g/mol)} \qquad m = n \times M \ \textbf{(5a)}$$

A molar mass of a substance X with the value $M = x$ g/mol can be treated as a conversion factor of the form

$$\text{Conversion factor} = \frac{x \text{ g X}}{1 \text{ mol X}}$$

Converting from mass to moles To convert from mass to moles of atoms in a sample, divide the mass in grams by the molar mass:

$$\text{Number of moles (mol)} \qquad n = \frac{m}{M} \qquad \textbf{(5b)}$$

$$= \frac{\text{mass(g)}}{\text{molar mass (g/mol)}}$$

The molar mass of a substance X with the value $M = x$ g/mol can also be treated as a conversion factor of the form

$$\text{Conversion factor} = \frac{1 \text{ mol X}}{x \text{ g X}}$$

Once we know the molar mass of an element, we can find the number of moles of atoms present in a sample by measuring its mass. If desired, we can also find the actual number of atoms by multiplying the number of moles by the Avogadro constant. The procedure is summarized in Toolbox 2.4.

Example 2.10 *Converting from moles to mass*

The Denver Mint needs 10.0 mol Cu atoms to make 10 000 pennies (which are 2.5% copper). What mass of copper will be required?

Strategy Find the molar mass of copper from the periodic table in the front of the book and use it as a conversion factor, as shown in Toolbox 2.4, to find the mass of Cu that corresponds to 10.0 mol of Cu atoms.

Solution Use the molar mass of copper given in the periodic table (63.54 g/mol) to write

$$\text{Mass of copper} = \overbrace{(10.0 \ \text{mol Cu})}^{m} \times \overbrace{\left(\frac{63.54 \ \text{g Cu}}{1 \ \text{mol Cu}}\right)}^{n \times M} = 635 \ \text{g Cu}$$

This result means that if we wanted a sample that contained 10.0 mol Cu atoms, then we would need to measure out 635 g of copper.

Self-Test 2.11A What mass of iron contains 1.23 mol Fe atoms?

[*Answer:* 68.7 g Fe]

Self-Test 2.11B In an experiment on the refinement of nuclear fuel, you need a sample of uranium that contains 0.26 mol U atoms. What mass of uranium do you need?

Example 2.11 *Converting from mass to moles*

The city of Toronto uses chlorine to purify its water. In a water research lab in Toronto, a scientist needs 0.300 mol Cl atoms (as chlorine molecules, Cl_2). A tank containing 15.0 g chlorine is available. Is it enough? Calculate the number of moles of Cl atoms in 15.0 g of chlorine.

Strategy Follow the procedure for converting from the mass of a sample to the number of moles set out in Toolbox 2.4.

Solution The molar mass of chlorine is 35.45 g, so

$$\text{Moles of Cl atoms} = \overbrace{(15.0 \ \text{g Cl})}^{n} \times \overbrace{\left(\frac{1 \ \text{mol Cl}}{35.45 \ \text{g Cl}}\right)}^{m/M}$$
$$= 0.423 \ \text{mol Cl}$$

The tank contains 0.423 mol chlorine atoms, more than is needed.

Self-Test 2.12A In an experimental solar-powered decomposition of water, the reaction produced 2.54 g of hydrogen gas for use as a fuel. How many moles of (a) H atoms; (b) H_2 molecules were produced?

[*Answer:* (a) 2.52 mol H atoms; (b) 1.26 mol H_2]

Self-Test 2.12B The same experiment produced 20.16 g of oxygen gas as a by-product. How many moles of O atoms were produced in the experiment?

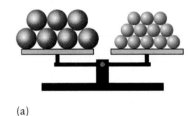
(a)

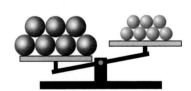
(b)

FIGURE 2.14

(a) The two samples have the same mass, but because the atoms on the right are lighter, the sample on the right consists of a greater number of atoms. (b) The two samples contain the same number of moles of atoms, but because the atoms on the right are lighter, the mass of that sample is smaller.

The molar mass makes it easy to measure out samples of elements that contain 1 mol of atoms: *to measure out 1 mol of atoms, we measure out the numerical value of the molar mass of the element in grams.* So, to obtain 1.000 mol Cu atoms, of molar mass 63.54 g/mol, we measure out 63.54 g of copper; for 1.0000 mol Hg atoms, of molar mass 200.59 g/mol, we measure out 200.59 g of mercury, and so on (Fig. 2.14). This procedure has been used to measure out the "massive heaps" of elements shown in Fig. 2.15. Each of these samples contains about the same number of atoms of the element (approximately 6.022×10^{23} in each case), but the masses vary because the masses of the atoms are different.

The molar mass of an element is the mass per mole of its atoms. It is used to convert between the mass of a sample and the moles of atoms it contains.

FIGURE 2.15

Each sample consists of approximately 1 mol of atoms of the element. Clockwise from the upper left are 12 g of carbon, 32 g of sulfur, 201 g of mercury, 207 g of lead, and 64 g of copper.

2.10 Measuring Out Compounds

We have seen how to measure out a given number of moles of atoms of an element. Now let's see how to measure out a given number of moles of molecules of a molecular compound or formula units of an ionic compound. This is the kind of measurement done by scientists throughout the world, because chemical reactions take place between specific *numbers* of molecules (or formula units), and hence between specific numbers of *moles* of molecules.

Just as for elements, the bridge between the mass of a sample of a compound and the number of moles of molecules or formula units it contains is the molar mass of the compound. Our first task, therefore, is to calculate the molar mass of the compound. Because a molecule or formula unit is made up of a specific number of atoms or ions, *the molar mass of a compound is the sum of the appropriate multiples of molar masses of the elements that make up the compound.* Note that we must take into account the number of times each type of atom appears in the formula. For example, the molar mass of water, which is composed of H_2O molecules, is

$$\text{Molar mass of } H_2O = [2 \times (\text{molar mass of H})] + (\text{molar mass of O})$$
$$= 2 \times 1.0079 + 16.00 \text{ g/mol}$$
$$= 18.02 \text{ g/mol}$$

Similarly, the molar mass of the ionic compound calcium chloride is the mass per mole of $CaCl_2$ formula units, which is given by the following sum:

$$\text{Molar mass of } CaCl_2 = (\text{molar mass of Ca}) + [2 \times (\text{molar mass of Cl})]$$
$$= 40.08 + 2 \times 35.45 \text{ g/mol}$$
$$= 110.98 \text{ g/mol}$$

Example 2.12 *Determining the molar mass of a compound*

Determine the molar mass of (a) glucose, $C_6H_{12}O_6$, a sugar important in human metabolism; (b) calcium chlorate, $Ca(ClO_3)_2$, a disinfectant used to preserve seeds for research.

Strategy To obtain the molar mass of a molecular or an ionic compound, add together the molar masses of the elements, with each molar mass multiplied by the number of times an atom of the element appears in the formula. Remember to multiply the subscripts inside parentheses by the subscript outside. The molar masses of the elements are given in the periodic table inside the front cover and the alphabetical list inside the back cover.

Solution (a) Because the molecular formula of glucose is $C_6H_{12}O_6$, its molar mass is

$$
\begin{array}{rll}
6 \text{ C atoms} \times 12.01 & = & 72.06 \text{ g/mol} \\
+ \ 12 \text{ H atoms} \times 1.0079 & = & 12.0958 \text{ g/mol} \\
+ \ 6 \text{ O atoms} \times 16.00 & = & \underline{96.00 \text{ g/mol}} \\
& & 180.16 \text{ g/mol}
\end{array}
$$

(b) For the ionic compound, the molar mass of $Ca(ClO_3)_2$ is

$$
\begin{array}{rll}
1 \text{ Ca atom} \times 40.08 & = & 40.08 \text{ g/mol} \\
+ \ 2 \text{ Cl atoms} \times 35.45 & = & 70.90 \text{ g/mol} \\
+ \ 6 \text{ O atoms} \times 16.00 & = & \underline{96.00 \text{ g/mol}} \\
& & 206.98 \text{ g/mol}
\end{array}
$$

Self-Test 2.13A Calculate the molar mass of (a) sulfuric acid, H_2SO_4; (b) aluminum oxide, Al_2O_3.

[***Answer:*** (a) 98.08 g/mol; (b) 101.96 g/mol]

Self-Test 2.13B Calculate the molar mass of (a) nitric acid, HNO_3; (b) aluminum sulfate, $Al_2(SO_4)_3$.

The same technique outlined for elements in Toolbox 2.4 allows us to calculate how many moles of molecules or formula units are in a sample of given mass. It also allows us to calculate the mass of sample we need to measure out to get a given number of moles of molecules or formula units. For example, if we measure out 110.98 g of calcium chloride, then we have a sample that contains 1.0000 mol $CaCl_2$ formula units. Similarly, if we want 1.000 mol $C_6H_{12}O_6$ molecules, then we measure out 180.2 g of glucose. In general, to measure out 1 mol of molecules or formula units, we measure out the numerical value of the molar mass of the compound in grams. Some examples of the "massive heaps" that correspond to 1 mol of molecules or formula units of a variety of compounds are illustrated in Figs. 2.16 and 2.17.

FIGURE 2.16

Each sample contains approximately 1 mol of molecules of a molecular compound. From left to right are 18 g of water (H_2O), 46 g of ethanol (C_2H_6O), 180 g of glucose ($C_6H_{12}O_6$), and 342 g of sucrose ($C_{12}H_{22}O_{11}$).

FIGURE 2.17

Each sample contains approximately 1 mol of formula units of an ionic compound. From left to right are 58 g of sodium chloride (NaCl), 100 g of calcium carbonate ($CaCO_3$), 278 g of iron(II) sulfate heptahydrate ($FeSO_4 \cdot 7H_2O$), and 78 g of sodium peroxide (Na_2O_2).

> *The molar mass of a compound is determined by adding the molar masses of its constituent elements, with each molar mass multiplied by the subscript of the element in the formula.*

Example 2.13 *Converting between mass and amount of a compound*

Calculate (a) the number of moles of AgBr formula units in 56.4 g of silver bromide, which is used in photographic emulsions; (b) the mass of the rocket fuel component dinitrogen tetroxide, N_2O_4, in 1.25 mol N_2O_4.

Strategy Use the same procedure set out in Toolbox 2.4 for converting between the mass and amount of an element. The molar mass of the compound is used for the conversion.

Solution (a) The molar mass of AgBr is

$$M = 107.87 + 79.91 \text{ g/mol} = 187.78 \text{ g/mol}$$

The conversion from mass in grams to amount of AgBr in moles is therefore

$$\text{Moles of AgBr} = \overbrace{(56.4 \text{ g AgBr})}^{n} \times \overbrace{\left(\frac{1 \text{ mol AgBr}}{187.78 \text{ g AgBr}}\right)}^{m/M} = 0.300 \text{ mol AgBr}$$

(b) The molar mass of N_2O_4 is

$$M = 2 \times 14.01 + 4 \times 16.00 \text{ g/mol} = 92.02 \text{ g/mol}$$

The conversion from moles to mass in grams is therefore

$$\text{Mass of } N_2O_4 = \overbrace{(1.25 \text{ mol } N_2O_4)}^{m} \times \overbrace{\left(\frac{92.02 \text{ } N_2O_4}{1 \text{ mol } N_2O_4}\right)}^{n \times M} = 115 \text{ g } N_2O_4$$

Self-Test 2.14A (a) Calculate the number of moles of urea molecules, $(NH_2)_2CO$, in 2.3×10^5 kg of urea. Urea is used in facial creams and, on a somewhat bigger scale, as an agricultural fertilizer. (b) Calculate the mass of aluminum oxide corresponding to 6.3 mol Al_2O_3.

[*Answer:* (a) 3.8×10^6 mol $(NH_2)_2CO$; (b) 0.64 kg]

Self-Test 2.14B (a) Calculate the number of moles of $Ca(OH)_2$ formula units in 1.00 kg of slaked lime (calcium hydroxide). Calcium hydroxide is used to adjust the acidity of soils. (b) Calculate the mass of sucrose (cane sugar) corresponding to 1.5 mmol $C_{12}H_{22}O_{11}$.

DETERMINATION OF CHEMICAL FORMULAS

One of the first steps taken to determine the molecular formula of vitamin C was to find out the relative amounts of each element in it; from this information, it was possible to determine the empirical formula. The second step was to determine the molecular formula. Recall from Section 1.8 that

> The **chemical formula** of a compound is the general term for the statement of the composition of the compound in terms of the chemical symbols of the elements present.

When dealing with molecular compounds or ionic compounds containing polyatomic ions, we distinguish between the empirical formula and the molecular formula:

> The **empirical formula** of a compound is a chemical formula that shows the *relative* numbers of atoms of each element, using the smallest whole numbers of atoms.

> The **molecular formula** tells us the *actual* numbers of atoms of each element in a molecule.

For example, the empirical formula of glucose, which is CH_2O, tells us that carbon, hydrogen, and oxygen atoms are present in the ratio 1:2:1. The elements are present in these proportions regardless of the size of the sample. The molecular formula for glucose, which is $C_6H_{12}O_6$, tells us that each glucose molecule consists of 6 carbon atoms, 12 hydrogen atoms, and 6 oxygen atoms.

Vitamin C is a molecular compound. So, to report its chemical formula, we must first determine its empirical formula and then go on to discover its molecular formula.

2.11 Mass Percentage Composition

To determine the empirical formula of a compound, we begin by measuring the *mass* of each element present in a sample. This composition is usually reported as the **mass percentage composition**, the mass of each element, $m_{element}$, expressed as a percentage of the total mass, m_{total}:

$$\text{Mass percentage of element} = \frac{\text{mass of element in sample}}{\text{total mass of sample}} \times 100\% \qquad \text{Mass\%} = \frac{m_{element}}{m_{total}} \times 100\% \qquad (6)$$

In many cases, we already know the formula of a compound and can calculate the mass percentage composition. The procedure is illustrated in the following example.

Example 2.14 *Calculating the mass percentage of an element in a compound on the basis of the compound's formula*

Hydrogen can be generated from water to use as a fuel. What is the mass percentage of hydrogen in water?

Strategy We consider 1 mol of molecules. The molar mass of the compound is the mass of that amount. Next, we find the number of moles of the element of interest that are present in 1 mol of molecules, and multiply that number by the molar mass of the element to calculate the mass of the element in 1 mol of molecules. We then calculate the mass percentage of the element from the definition in Eq. 6, using the masses of the element and the compound present in 1 mol of the compound. When calculating the mass percentage, the units specifying the species are dropped.

Solution The molar mass of water is found from the molar masses of the elements:

$$M = 2 \times 1.0079 + 16.00 \text{ g/mol} = 18.02 \text{ g/mol}$$

There are 2 mol H in 1 mol H_2O. To find the mass percentage of H, divide the mass of hydrogen in 1 mol H_2O by the mass of 1 mol H_2O and multiply by 100%:

$$\text{Mass percentage of H} = \frac{\text{total mass of H atoms}}{\text{mass of } H_2O \text{ molecules}} \times 100\%$$

$$\text{Total mass of H atoms} = (2 \text{ mol H}) \times \left(\frac{1.0079 \text{ g H}}{1 \text{ mol H}}\right) = 2 \times 1.0079 \text{ g}$$

$$\text{Mass of } H_2O \text{ molecules} = (1 \text{ mol } H_2O) \times \left(\frac{18.02 \text{ g } H_2O}{1 \text{ mol } H_2O}\right) = 18.02 \text{ g}$$

$$\text{Mass percentage of H} = \frac{2 \times 1.0079 \text{ g}}{18.02 \text{ g}} \times 100\% = 11.19\%$$

Self-Test 2.15A Calculate the mass percentage of Cl in NaCl.

[*Answer:* 60.66%]

Self-Test 2.15B Calculate the mass percentage of Ag in $AgNO_3$.

The mass percentage of an element in a compound of unknown composition must be determined experimentally. For instance, if we analyzed a sample of vitamin C of total mass 8.00 g, we would obtain the following data:

Carbon	3.27 g
Hydrogen	0.366 g
Oxygen	4.36 g

So, the mass percentage of carbon in vitamin C is

$$\text{Mass percentage of C} = \frac{\text{mass of C in sample}}{\text{total mass of sample}} \times 100\%$$

$$= \frac{3.27 \text{ g}}{8.00 \text{ g}} \times 100\% = 40.9\%$$

If we use the same procedure to find the mass percentages of hydrogen and oxygen, we find the following mass percentage composition of vitamin C:

Carbon   3.27 g   40.9%

Hydrogen   0.366 g   4.58%

Oxygen   4.36 g   54.5%

We now know that any sample of vitamin C has mass percentage composition 40.9% C, 4.58% H, and 54.5% O.

Mass percentage composition is found by calculating the fraction of the total mass contributed by each element present in a compound and expressing the fraction as a percentage.

FIGURE 2.18

Leaves from the eucalyptus tree contain compounds such as eucalyptol, which has important medical uses.

Example 2.15 *Determining mass percentage composition*

For centuries, the Australian aborigines have used the leaves of the eucalyptus tree to alleviate sore throats and other pains (Fig. 2.18). The primary active ingredient has been identified and named eucalyptol. The analysis of a sample of eucalyptol of total mass 3.16 g gave its composition as 2.46 g carbon, 0.373 g hydrogen, and 0.329 g oxygen. Determine the mass percentage of carbon in eucalyptol.

Strategy The mass percentage composition is defined in Eq. 6 as

$$\text{Mass\%} = \frac{m_{\text{element}}}{m_{\text{total}}} \times 100\%$$

To use this definition, we simply substitute the data for each element present in the compound.

Solution The mass percentage of carbon in eucalyptol is

$$\text{Mass percentage of C} = \frac{\text{mass of C in sample}}{\text{total mass of sample}} \times 100\%$$

$$= \frac{2.46 \text{ g}}{3.16 \text{ g}} \times 100\% = 77.8\%$$

Self-Test 2.16A What are the mass percentages of hydrogen and oxygen in eucalyptol?

[*Answer:* 11.8% H and 10.4% O]

Self-Test 2.16B The compound α-pinene, a natural antiseptic found in the resin of the piñon tree, has been used since ancient times by Zuni healers. A 7.50-g sample of α-pinene contains 6.61 g carbon and 0.89 g hydrogen. What are the mass percentages of carbon and hydrogen in α-pinene?

2.12 Determining Empirical Formulas

To determine the empirical formula of a compound, we need the *relative number of atoms* of each element in the sample. Put another way, we need the *relative number of moles* of each type of atom. We can get this information from the mass percentage composition, which is often obtained by a *combustion analysis*

Investigating Matter 2.1: *Combustion Analysis*

Combustion analysis is widely used for the determination of the empirical formulas of organic compounds containing carbon, hydrogen, and oxygen. In this method, a known mass of an organic compound is burned in a plentiful supply of oxygen. The products of the combustion are collected separately and weighed. From the masses of the products, the masses of carbon and hydrogen in the sample can be determined. The mass of oxygen is determined from the difference between the total mass and the masses of carbon and hydrogen. The result is reported as a mass percentage composition.

The illustration below shows a schematic view of the apparatus. The sample is ignited in a tube through which there is a flow of oxygen. All the hydrogen in the compound is converted to water and all the carbon is converted to carbon dioxide. The product gases flow through two more tubes that trap them. The water produced is absorbed in the first tube, so the increase in mass of this tube is equal to the mass of water absorbed. The carbon dioxide is trapped in the second tube, so the increase in mass of this tube is equal to the mass of carbon dioxide produced in the combustion.

In the presence of excess oxygen, each carbon atom in the compound ends up in one molecule of carbon dioxide, as shown in the illustration. Therefore, each *mole* of C atoms in the sample gives 1 mol of CO_2 molecules as product:

1 mol C in sample gives 1 mol CO_2 as product

Hence, by measuring the mass of carbon dioxide produced and then converting that mass to moles of CO_2, we know the number of moles of C atoms in the original sample. We can then use the molar mass of carbon to convert that number of C atoms to the mass of carbon present in the sample.

We find the moles of H atoms in the sample from the mass of water produced. Each H atom in a compound con-

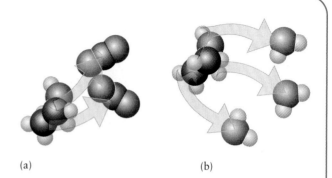

(a)   (b)

When an organic compound burns in a plentiful supply of oxygen, (a) each C atom produces one CO_2 molecule and (b) each pair of H atoms ends up as one H_2O molecule (not necessarily in the same molecule).

tributes to a water molecule when the compound burns. Because a water molecule contains two H atoms, we know that if we obtain 1 mol H_2O as product, then there must have been 2 mol H atoms present in the sample:

2 mol H in sample gives 1 mol H_2O as product

Therefore, if we measure the mass of water produced, we can find the number of moles of H_2O molecules. Multiplying that amount by 2 gives us the number of moles of H atoms in the original sample. Multiplying by the molar mass of hydrogen then gives the mass of hydrogen in the original sample.

Additional information on combustion analysis can be found in Section 4.5.

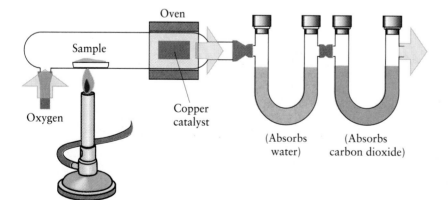

Oven

Sample

Oxygen

Copper catalyst

(Absorbs water)

(Absorbs carbon dioxide)

The apparatus used for a combustion analysis. The masses of carbon dioxide and water produced are obtained from the differences of the masses of the collecting tubes before and after the experiment. The catalyst ensures that any CO produced is oxidized to CO_2.

(Investigating Matter 2.1). To convert the mass percentage of each element into a relative number of moles, imagine that we have a sample of mass 100 g exactly. That way, the mass percentage composition tells us the mass in grams of each element. Then we can use the molar masses of the elements to convert these masses into moles.

We saw in the previous section that the mass of carbon in exactly 100 g of vitamin C is 40.9 g. We find the amount of C atoms in 40.9 g of carbon from the molar mass of carbon, which the periodic table gives as 12.01 g/mol:

$$\text{Moles of C atoms} = (40.9 \text{ g C}) \times \left(\frac{1 \text{ mol C}}{12.01 \text{ g C}} \right) = 3.41 \text{ mol C}$$

In the same way, we find the numbers of moles of H atoms and O atoms in exactly 100 g of vitamin C:

Carbon	40.9 g	3.41 mol C
Hydrogen	4.58 g	4.54 mol H
Oxygen	54.5 g	3.41 mol O

Because number of moles is proportional to number of atoms, we now know that the atoms of each element are present in the ratio (3.41 C):(4.54 H):(3.41 O).

At this point, we might be tempted to write the chemical formula of vitamin C as $C_{3.41}H_{4.54}O_{3.41}$. However, an empirical formula is expressed in terms of the *simplest* whole-number ratios of atoms. Therefore, we divide each number by the smallest value (3.41), which gives a ratio of 1.00:1.33:1.00. One number is still not a whole number, so we must multiply each number by a factor until all numbers are whole numbers or can be rounded off to whole numbers. Because 1.33 is $\frac{4}{3}$ (within experimental error), we multiply through by 3 to obtain 3.00:3.99:3.00, or approximately 3:4:3. Now we know that the empirical formula of vitamin C is $C_3H_4O_3$.

Ionic compounds do not have molecules, so their chemical formulas represent the composition of their formula units. They show the identities of the ions present and the relative numbers of each type of ion.

The empirical formula of a compound is determined from the mass percentage composition and the molar masses of the elements present.

Example 2.16 *Determining an empirical formula from a mass percentage*

A compound that assists in the coagulation of blood has the mass percentage composition 76.71% C, 7.02% H, and 16.27% N. Determine the empirical formula of the compound.

Strategy Mass percentages tell us the mass of each element in exactly 100 g of the compound. Convert each mass percentage to the number of moles found in 100 g of the compound by dividing by the molar mass of the element. Divide the number of moles of each element by the smallest number. If fractional numbers result, then multiply by a factor that will give the smallest whole numbers of moles.

Solution We assume that the sample has a mass of exactly 100 g so that each mass percentage will represent the mass in grams of the element present. The masses are

Carbon	76.71 g
Hydrogen	7.02 g
Nitrogen	16.27 g

Convert these masses to number of moles by using the molar mass of the element:

$$\text{Moles of C atoms} = (76.71 \text{ g C}) \times \left(\frac{1 \text{ mol C}}{12.01 \text{ g C}}\right) = 6.387 \text{ mol C}$$

$$\text{Moles in H atoms} = (7.02 \text{ g H}) \times \left(\frac{1 \text{ mol H}}{1.0079 \text{ g H}}\right) = 6.96 \text{ mol H}$$

$$\text{Moles of N atoms} = (16.27 \text{ g N}) \times \left(\frac{1 \text{ mol N}}{14.01 \text{ g N}}\right) = 1.161 \text{ mol N}$$

At this stage, the composition is known to be

Carbon	76.71 g	6.387 mol
Hydrogen	7.02 g	6.96 mol
Nitrogen	16.27 g	1.161 mol

Divide each number by the smallest number of moles (1.161 mol):

Carbon	76.71 g	6.387 mol	5.501
Hydrogen	7.02 g	6.96 mol	5.99
Nitrogen	16.27 g	1.161 mol	1.000

Recognizing that 5.501 is approximately $\frac{11}{2}$ we multiply all the numbers by 2 to get the ratios 11.002:12.0:2.000. The empirical formula is therefore $C_{11}H_{12}N_2$.

Self-Test 2.17A Use the mass percentage composition of eucalyptol calculated in Example 2.15 and Self-Test 2.16A to determine its empirical formula.

[*Answer:* $C_{10}H_{18}O$]

Self-Test 2.17B The mass percentage composition of the compound thionyl difluoride is 18.59% O, 37.25% S, and 44.16% F. Calculate its empirical formula.

2.13 Determining Molecular Formulas

So far, we know that the empirical formula of vitamin C is $C_3H_4O_3$. However, all this formula tells us is that the C, H, and O atoms are present in the sample in the ratio 3:4:3. We do not yet know the number of atoms in an individual molecule. The same empirical formula would be obtained for $C_3H_4O_3$, $C_6H_8O_6$, $C_9H_{12}O_9$, or any other whole-number multiple of the empirical formula.

To find the molecular formula of a compound, one more piece of information is needed—its molar mass. Once we know the molar mass, we can calculate how many empirical formula units are needed to account for it. For example, the molar mass of vitamin C is found by mass spectrometry to be 176.14 g/mol. The molar mass of a $C_3H_4O_3$ formula unit is

$$\text{Molar mass of } C_3H_4O_3 = (3 \times 12.01) + (4 \times 1.008) + (3 \times 16.00) \text{ g/mol}$$
$$= 88.06 \text{ g/mol}$$

We need two $C_3H_4O_3$ formula units to account for the observed molar mass of vitamin C:

$$\text{Number of empirical formula units per molecule} = \frac{176.14 \text{ g/mol}}{88.06 \text{ g/mol}} = 2.000$$

It follows that the molecular formula of vitamin C is $2 \times (C_3H_4O_3)$, or $C_6H_8O_6$ (1).

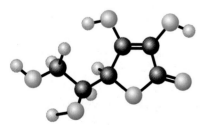

1 Vitamin C, $C_6H_8O_6$

The molecular formula of a molecular compound is found by determining how many empirical formula units are needed to account for the measured molar mass of the compound.

Example 2.17 *Determining a molecular formula from an empirical formula*

The molar mass of ethyl butanoate, a compound that contributes to the flavor of pineapple, is 116 g/mol (Fig. 2.19). Its empirical formula, determined from its mass percentage composition, is C_3H_6O. What is its molecular formula?

Strategy We have to decide how many empirical formula units are required to account for the measured molar mass. Therefore, calculate the molar mass of the empirical formula unit and compare it with the measured molar mass.

Solution The molar mass of the formula unit C_3H_6O is

$$\text{Molar mass of } C_3H_6O = (3 \times 12.01) + (6 \times 1.0079) + 16.00 \text{ g/mol}$$
$$= 58.08 \text{ g/mol}$$

Because the measured molar mass is 116 g/mol, it follows that

$$\text{Number of empirical formula units per molecule} = \frac{116 \text{ g/mol}}{58.08 \text{ g/mol}} = 2.00$$

We conclude that the molecular formula of ethyl butanoate is $C_6H_{12}O_2$.

Self-Test 2.18A The molar mass of styrene, which is used in the manufacture of the plastic polystyrene, is 104 g/mol, and its empirical formula is CH. Deduce its molecular formula.

[**Answer:** C_8H_8]

Self-Test 2.18B The molar mass of oxalic acid, a toxic substance found in rhubarb leaves, is 90.0 g/mol, and its empirical formula is CHO_2. What is its molecular formula?

FIGURE 2.19

The fruit of the pineapple is a source of the chemical ethyl butanoate. The compound is now synthesized from other chemicals and used to flavor candies, cake mixes, and gelatin desserts.

Skills You Should Have Mastered

Conceptual

1. Classify properties as intensive or extensive.
2. Explain the significance of units for reporting measurements.
3. Distinguish between accuracy and precision, Self-Test 2.8.
4. Give the definition of a mole in your own words.

Problem-Solving

1. Calculate the density of a substance from the mass and volume of a sample, Example 2.1.
2. Convert a measurement from one unit to another, Toolbox 2.1 and Examples 2.2 and 2.3.

3. Convert between Kelvin, Celsius, and Fahrenheit temperatures, Toolbox 2.2 and Example 2.4.
4. Use the correct number of significant figures when reporting measurements and the results of calculations, Toolbox 2.3 and Examples 2.5, 2.6, and 2.7.
5. Use the Avogadro constant to convert between number of moles and the number of atoms, molecules, or ions in a sample, Example 2.8.
6. Calculate the average molar mass of an element, given its isotopic composition, Example 2.9.
7. Determine the molar mass of a compound, Example 2.12.
8. Convert between mass and number of moles by using the molar mass, Toolbox 2.4 and Examples 2.10, 2.11, and 2.13.

9. Calculate the mass percentage of an element in a compound from the formula, Example 2.14.

10. Determine the mass percentage composition of a compound from mass measurements, Example 2.15.

11. Calculate the empirical formula of a compound from its mass percentage composition, Example 2.16.

12. Determine the molecular formula of a compound from its empirical formula and its molar mass, Example 2.17.

Descriptive

1. Describe the steps involved in determining the molecular formula of a compound.

Exercises

International System (SI) of Units

2.1 Express the following quantities in scientific notation: (a) 5 556 000 000 000 dollars, U.S. national debt as of October 1998 (to four significant figures); (b) 1 169 811, number of AIDS cases reported worldwide, 1980–1995; (c) 0.000 006 g, recommended daily allowance of vitamin B_{12} for adults; (d) 0.000 000 1 m, diameter of the smallest living cells.

2.2 Express the following quantities in scientific notation: (a) 56 000 000 km, closest approach of Mars to Earth (to two significant figures); (b) 93 900 000 Hz, frequency of a popular FM radio station (to three significant figures); (c) 0.000 000 000 000 000 000 000 020 g, mass of one carbon atom; (d) 0.000 000 000 096 m, approximate O—H bond length in water.

2.3 Convert the following quantities into standard scientific notation: (a) 0.0043×10^2; (b) 1492×10^{-2}; (c) $0.000 051 \times 10^{-4}$; (d) 237×10^{12}.

2.4 Convert the following quantities into standard scientific notation: (a) $0.000 673 \times 10^9$; (b) 986×10^{-4}; (c) 0.0018×10^{-3}; (d) 5305×10^7.

2.5 Complete the following equalities by using the information in Table 2.2:

(a) 250. g = _____ kg

(b) 25.4 mm = _____ cm (1 in.)

(c) 250. μs = _____ ms

(d) 1.49 cm = _____ dm

(e) 2.48 cg = _____ g

(f) 28.35 g = _____ kg (1 ounce)

2.6 Complete the following equalities by using the information in Table 2.2:

(a) 200. μg = _____ g

(b) 88 nm = _____ pm

(c) 0.0789 mg = _____ μg

(d) 454 g = _____ kg (1 lb)

(e) 25.0 mL = _____ kL

(f) 2000. nm = _____ μm

2.7 Complete the following equalities, using scientific notation:

(a) 1 μm = _____ m

(b) 550. nm = _____ mm

(c) 0.10 g = _____ mg

(d) 105 pm = _____ μm

2.8 Express the following measurements in scientific notation:

(a) 186 000 miles/s (3 sf), the speed of light

(b) 0.000 000 002 K, the lowest-ever recorded temperature

(c) 1/100 000 000 m

(d) 0.000 000 535 m, the approximate wavelength of green light

2.9 When a piece of metal of mass 3.60 g is dropped into a graduated cylinder containing 8.3 mL of water, the water level rises to 9.8 mL. What is the density of the metal in grams per cubic centimeter (g/cm^3)?

2.10 When a piece of metal of mass 5.21 g is dropped into a graduated cylinder containing 16.7 mL of water, the water level rises to 18.2 mL. What is the density of the metal in grams per cubic centimeter (g/cm^3)?

2.11 The density of balsa wood is 0.16 g/cm^3. What is the mass of 1.00 ft^3 of balsa wood?

2.12 A supersonic transport (SST) airplane consumes about 18 000 L of kerosene per hour of flight. Kerosene has a density of 0.965 g/mL. What mass of kerosene is consumed on a flight of duration 3.0 h?

2.13 The density of diamond is 3.51 g/cm^3. The international (but non-SI) unit for reporting the masses of diamonds is the "carat", with 1 carat = 200. mg. What is the volume of a diamond of mass 0.300 carat?

2.14 What volume (in cm^3) of lead (of density 11.3 g/cm^3) has the same mass as 100. cm^3 of a piece of redwood (of density 0.38 g/cm^3)?

Conversion Factors

2.15 Use the conversion factors in Table 2.3 and inside the back cover to express the following measurements in the designated units: (a) 25 L to m^3; (b) 25 g/L^2 to mg/dL^2; (c) 1.54 mm/s to pm/μs; (d) 2.66 g/cm^3 to $\mu g/\mu m^3$; (e) 4.2 L/h^2 to mL/s^2; (f) $1.20/gallon to peso/liter (assume 1 dollar = 780 peso).

2.16 Use the conversion factors in Table 2.3 and inside the back cover to express the following measurements in the designated units: (a) 4.82 nm to pm; (b) 1.83 mL/min to mm^3/s; (c) 1.88 ng to kg; (d) 7.01 cm/s to km/h; (e) 0.044 g/L to mg/cm^3; (f) 4.2°C/s to °C/min.

2.17 Convert the following temperatures as indicated: (a) normal body temperature, 98.6°F to °C; (b) −40.°C to °F; (c) absolute zero, 0 K to °F; (d) the boiling point of helium, −269°C to K.

2.18 Certain temperatures occur frequently in chemistry, and it is helpful to know their values on various scales. For future convenience, express the following temperatures on the scale indicated: (a) conventional temperature for reporting standard properties, 298.15 K to °C; (b) the boiling point of nitrogen, used as a low-temperature coolant, 77 K to °C; (c) "hot" water that is still comfortable to the skin, 45°C to K and °F; (d) the temperature of a dry ice-acetone mixture, used as a "cold bath" in the laboratory, −78°C to K.

2.19 Convert the following quantities as indicated:

(a) 1 cm^3 = _____ m^3

(b) 30. m/s = _____ cm/μs

(c) 22 m^2 = _____ cm^2

(d) 25 cm^3 = _____ mL

2.20 Convert the following quantities to the units designated within the brackets: (a) The density of water at 3.98°C [°F] is 1.0 g/mL [mg/L]. (b) The density of oxygen gas is 1.43 g/L [mg/mL] at 0°C [°F]. (c) The volume of a laboratory test tube is 3.0 mL [dL]. (d) A U.S. penny has a mass of 2.5 g, a diameter of 1.9 cm, and a thickness of 1.6 mm. Given that the volume of a cylinder of radius r and height h is $\pi r^2 h$, determine the density of a penny in units of mg/mm^3.

2.21 Rewrite the following statement, using the units in brackets: "A sample of tin of area 1.0 cm^2 [mm^2] was set on a small block of lead of volume 10.0 cm^3 [m^3]. The two metals were placed in a 100.-mL [L] flask and 25.0 mL [cm^3] of acid was added."

2.22 Make the appropriate conversions in the following statement: "A chemist recovered a minute sample of the metal iridium of volume 0.5 mm^3 [μm^3] from a land area of 1.5 km^2 [m^2]. The chemist analyzed a 25-mL [dL] soil sample to find the amount of iridium in 1.0 cm^3 [m^3] of soil as a part of a research project to determine whether there had been a major comet impact on Earth at the time of the extinction of the dinosaurs."

Uncertainty of Measurements and Calculations

2.23 State the number of significant figures in the following quantities: (a) 2.00 g of silver; (b) 0.0200 s; (c) 2.00 × 10^2 mL of water; (d) six thermometers; (e) 0.0023°C; (f) 12 in./ft.

2.24 State the number of significant figures in the following quantities: (a) 3.001 00 g of sugar; (b) 12.011 g/mol; (c) 2.998 × 10^8 m/s (speed of light); (d) 22 beakers; (e) 0.0001 K; (f) 10^3 m/km.

2.25 Shown top, right are two pieces of laboratory glassware graduated in milliliters. Report the volumes, using the correct number of significant figures.

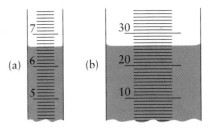

2.26 Record the volumes of solution in milliliters in the two graduated cylinders shown below. Report the volumes, using the correct number of significant figures.

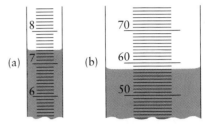

2.27 The masses of copper, zinc, and manganese in a sample of an alloy were measured as 2.011 g, 1.02 g, and 1.4 g, respectively. What is the total mass of the alloy?

2.28 A forensic chemist collected three samples from the scene of a crime. Their masses were 0.220 g, 0.034 76 g, and 0.0001 g. What is the total mass of the collected samples?

2.29 A pharmacist made up a capsule consisting of 0.21 g of one drug, 0.124 g of another, and 1.311 g of a "filler." What is the total mass of the contents of the capsule?

2.30 A diamond of mass 2.001 μg was removed from a sample of soil of mass 1.78 × 10^{-4} g. What is the mass of the remaining sample of soil?

2.31 To how many significant figures should the result of the following calculation be reported?

$$\frac{0.082\,06 \times (273.15 + 1.2)}{1.23 \times 7.004}$$

2.32 To how many significant figures should the result of the following calculation be reported?

$$\frac{534.71 \times 321.83 \times 0.001\,86}{7.529 \times 10^{-3}}$$

2.33 A chemist determined in a set of four experiments that the density of magnesium metal was 1.68 g/cm^3, 1.67 g/cm^3, 1.69 g/cm^3, 1.69 g/cm^3. The accepted value for its density is 1.74 g/cm^3. What can you conclude about the precision and accuracy of the chemist's data?

2.34 The density of a metal was measured by two different methods. In each case, calculate the density. Indicate which set of measurements is more precise. (a) The dimensions of a rectangular block of the metal were measured as 1.10 cm × 0.531 cm × 0.212 cm. Its mass was found to be 0.213 g. (b) The mass of a cylinder of water filled to the

19.65-mL mark was found to be 39.753 g. When a piece of the metal was immersed in the water, the level of the water rose to 20.37 mL and the mass of the cylinder with the metal was found to be 41.003 g.

Fun with Atoms and Moles

2.35 The visible universe is estimated to contain 10^{22} stars. How many moles of stars are there?

2.36 There are approximately 1×10^{11} neurons in your head. How many heads are needed to have 1 mol of neurons?

2.37 (a) The approximate population of Earth is 5.7 billion people. How many moles of people inhabit Earth? (b) If all people were pea pickers and pea counters, then how long would it take for the Earth's population to count out 1 mol of peas at the rate of one pea per second, working 24 hours per day, 365 days per year?

2.38 (a) About 1000 metric tons of sand contain about a trillion (10^{12}) grains of sand. How many metric tons of sand are needed to provide 1 mol of sand? (b) Assuming the volume of a grain of sand is 1 mm^3 and the land area of the continental United States is 3 600 000 mi^2, how deep would the sand pile over the United States be if the surface were evenly covered with 1 mol of grains of sand?

Moles and Molar Masses of Elements

2.39 A 5.00-mol sample of a certain element has a mass of 260. g. Give the name and symbol of the element.

2.40 A 3.00-mol sample of a certain element has a mass of 42.0 g. Give the name and symbol of the element.

2.41 Calculate the average molar mass of carbon in a natural sample, which consists of 98.89% ^{12}C and 1.11% ^{13}C. The mass of an atom of ^{12}C is 1.9926×10^{-23} g and that of ^{13}C is 2.1593×10^{-23} g.

2.42 The nuclear power industry extracts ^{6}Li but not ^{7}Li from natural samples of lithium. As a result, the average molar mass of commercial samples of lithium is increasing. The current abundances of the two isotopes are 7.42% and 92.58%, respectively, and the masses of their atoms are 9.988×10^{-24} g and 1.165×10^{-23} g. (a) What is the current molar mass of a natural sample of lithium? (b) What will the molar mass be when the abundance of ^{6}Li is reduced to 5.67%?

2.43 Calculate the average molar mass of bromine in a natural sample, which consists of 50.54% ^{79}Br (molar mass 78.918 g/mol) and 49.46% ^{81}Br (molar mass 80.916 g/mol).

2.44 Calculate the molar mass of sulfur in a natural sample, which consists of 95.0% ^{32}S (molar mass 31.97 g/mol), 0.8% ^{33}S (molar mass 32.97 g/mol), and 4.2% ^{34}S (molar mass 33.97 g/mol).

2.45 Calculate the amount in moles of (a) 4.82×10^{22} atoms of ^{35}Cl; (b) 2.22 g of copper atoms; (c) 1.11×10^{24} atoms of helium; (d) 8.96 µg of iron atoms.

2.46 Which sample in each of the following pairs contains the greater number of moles of atoms? (a) 25 g of carbon or 35 g of silicon; (b) 1.0 g of Au or 1.0 g of Hg; (c) 2.49×10^{22} atoms of Au or 2.49×10^{22} atoms of Hg.

2.47 Determine the number of atoms in (a) 3.97 mol Xe; (b) 18.3 µg Sc; (c) 12.8 pg Li; (d) 3.78×10^{-4} mol Ar.

2.48 Calculate the mass, in micrograms, of (a) 3.77×10^{18} atoms of Na; (b) 8.22 µmol U; (c) 0.000 006 020 mol K; (d) 6.02×10^{23} boron atoms.

2.49 What mass of nickel contains as many atoms as there are (a) carbon atoms in 12 g of carbon; (b) chromium atoms in 12 g of chromium?

2.50 Determine the mass of aluminum that has the same number of atoms as there are in (a) 6.29 mg of silver; (b) 6.29 mg of gold.

Mass Percentage Composition and Molar Masses of Compounds

2.51 Determine the molar mass of (a) calcium bromide; (b) C_8H_{18}; (c) $NiSO_4 \cdot 6H_2O$; (d) carbon dioxide; (e) CH_4 (methane, the major component of natural gas).

2.52 Determine the molar mass of (a) sulfur tetrafluoride; (b) hydrazine, N_2H_4; (c) sodium cyanide, NaCN; (d) sucrose, $C_{12}H_{22}O_{11}$; (e) copper(II) chloride tetrahydrate.

2.53 Correct each statement, which may be in error or just ambiguous: (a) Two-thirds of the mass of dinitrogen tetroxide, N_2O_4, is due to oxygen. (b) The reaction produced 5.0 moles of oxygen. (c) 1.000 mol Cl_2 contains 6.022×10^{23} atoms.

2.54 Correct each statement, which may be in error or just ambiguous: (a) The molar mass of NaCl is the mass of a mole of NaCl molecules. (b) If the mass percentage of an element in a compound is 25%, then 25% of the atoms in the compound are atoms of that element. (c) One molecule of F_2 has a mass of 38.00 g.

2.55 Calculate the amount (in moles) and the number of molecules (or atoms, if indicated) in (a) 10.0 g of carbon tetrachloride, CCl_4; (b) 1.65 mg of hydrogen iodide, HI; (c) 3.77 mg of hydrazine, N_2H_4; (d) 500. g of sucrose, $C_{12}H_{22}O_{11}$; (e) 2.33 g of oxygen as O atoms and as O_2 molecules.

2.56 Convert the following masses to amounts (in moles) and to number of molecules (or atoms, if indicated). (a) 1.00 kg of H_2O; (b) 1.00 kg of C_2H_5OH (ethanol); (c) 10.0 g of sulfur, as S atoms and as S_8 molecules; (d) 3.0 g of CO_2; (e) 3.0 g of NO_2.

2.57 Calculate the amount (in moles) of (a) Ag^+ ions in 2.00 g of AgCl; (b) UO_3 in 600. g of UO_3; (c) Cl^- ions in 4.19 mg of $FeCl_3$; (d) H_2O in 1.00 g of $AuCl_3 \cdot 2H_2O$.

2.58 Calculate the amount (in moles) of (a) CN^- in 1.00 g of KCN; (b) H atoms in 200. mg of H_2O; (c) $CaCO_3$ in 500. g of $CaCO_3$; (d) H_2O in 5.00 g of $La_2(SO_4)_3 \cdot 9H_2O$.

2.59 (a) Determine the number of formula units in 0.670 mol $AgNO_3$. (b) What is the mass (in mg) of 2.39×10^{20} formula units of Rb_2SO_4? (c) Estimate the number of formula units in 6.66 kg of $NaHCO_2$, sodium formate, which is used in dyeing and printing fabrics.

2.60 (a) How many NaH formula units are present in 3.61 g of NaH? (b) Determine the mass of 5.78×10^{24} formula units of $NaBF_4$, sodium tetrafluoroborate. (c) Calculate the amount (in moles) of 8.52×10^{20} formula units of CeI_3, cerium(III) iodide, a bright yellow, water-soluble solid.

2.61 (a) Calculate the amount (in moles) of molecules in 1.0 mg of testosterone, $C_{19}H_{28}O_2$, a male sex hormone. (b) What is the mass percentage composition of testosterone?

2.62 (a) Aspartame, $C_{14}H_{18}N_2O_5$, is an artificial sweetener sold as NutraSweet. How many molecules are present in 1.0 mg of aspartame? (b) What is the mass percentage composition of aspartame?

2.63 (a) Calculate the mass, in grams, of a water molecule. (b) Determine the number of H_2O molecules in 1.00 g of H_2O.

2.64 Octane, C_8H_{18}, is typical of the molecules found in gasoline. (a) Calculate the mass of one octane molecule. (b) Determine the number of C_8H_{18} molecules in 1.00 mL of C_8H_{18}, the mass of which is 0.82 g.

2.65 A chemist measured out 5.50 g of copper(II) bromide tetrahydrate, $CuBr_2\cdot4H_2O$. (a) How many moles of $CuBr_2\cdot4H_2O$ were measured out? (b) How many moles of Br^- ions are present in the sample? (c) How many water molecules are present in the sample? (d) What fraction of the total mass of the sample was due to copper?

2.66 A chemist wants to extract the gold from 15.0 g of gold(III) chloride dihydrate, $AuCl_3\cdot2H_2O$, by electrolysis of an aqueous solution (this technique is described in Chapter 18). What mass of gold could be obtained from the sample?

Determining Chemical Formulas

2.67 In the following ball-and-stick molecular structures, black indicates carbon, red oxygen, light gray hydrogen, blue nitrogen, and green chlorine. Write the empirical and molecular formulas of each structure. *Hint:* It may be easier to write the molecular formula first.

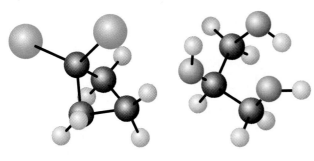

(a)   1,2-Dichlorocyclobutane   (b)   Glycerol

2.68 Write the empirical and molecular formulas of each ball-and-stick structure. See Exercise 2.67 for the color code. *Hint:* It may be easier to write the molecular formula first.

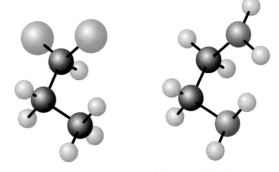

(a)   1,1-Dichloropropane   (b)   1, 2-Diaminoethane

2.69 Determine the empirical formulas from the following analyses. (a) The mass composition of cryolite, a compound used in the production of aluminum, is 32.79% Na, 13.02% Al, and 54.19% F. (b) A compound used to generate O_2 gas in the laboratory has mass composition 31.91% K and 28.93% Cl, the remainder being oxygen. (c) A fertilizer is found to have the following mass composition: 12.2% N, 5.26% H, 26.9% P, and 55.6% O.

2.70 Determine the empirical formula of each compound from the following data. (a) Talc (used in talcum powder) has mass composition 19.2% Mg, 29.6% Si, 42.2% O, and 9.0% H. (b) Saccharin, a sweetening agent, has mass composition 45.89% C, 2.75% H, 7.65% N, 26.20% O, and 17.50% S. (c) Salicylic acid, used in the synthesis of aspirin, has mass composition 60.87% C, 4.38% H, and 34.75% O.

2.71 In an experiment, 4.14 g of the element phosphorus combined with chlorine to produce 27.8 g of a white solid compound. What is the empirical formula of the compound? *Hint:* Determine the mass of chlorine from the mass of the product and that of the phosphorus.

2.72 A chemist found that 4.69 g of sulfur combined with fluorine to produce 15.81 g of a gas. What is the empirical formula of the gas? *Hint:* Determine the mass of fluorine from the mass of the product and that of the sulfur.

2.73 Lindane, used as an insecticide, has mass composition 24.78% C, 2.08% H, and 73.14% Cl and molar mass 290.85 g/mol. What is the molecular formula of lindane?

2.74 Nicotine has mass composition 74.03% C, 8.70% H, and 17.27% N and molar mass 162.23 g/mol. Determine the molecular formula of nicotine.

2.75 Caffeine, a primary stimulant in coffee and tea, has molar mass 194.19 g/mol and mass composition 49.48% C, 5.19% H, 28.85% N, and 16.48% O. What is the molecular formula of caffeine?

2.76 Cacodyl, which has an intolerable garlicky odor and is used in the manufacture of cacodylic acid, a cotton herbicide, has mass composition 22.88% C, 5.76% H, and 71.36% As, and molar mass 209.96 g/mol. What is the molecular formula of cacodyl?

Supplementary Exercises

2.77 Select a convenient SI unit from Tables 2.1 and 2.2 for recording (a) your mass; (b) the diameter of an atom; (c) the mass of a coin; (d) the tolerance (of the order of 0.000 01 m) on a finely machined tool.

2.78 The ångström unit (1 Å $= 10^{-10}$ m) is still widely used to report measurements of the sizes of atoms and molecules. Express the following data in ångströms: (a) the radius of a sodium atom is 180 pm; (b) the wavelength of yellow light is 550 nm. (c) Write a (single) conversion factor between ångströms and nanometers.

2.79 The atomic mass unit (u) is used by nuclear chemists and physicists for masses of individual atoms and nuclei. The atomic mass unit is defined as 1 u $= 1.6605 \times 10^{-24}$ g. (a) How many atoms of carbon-12 are there in 12 u? (b) Name the fundamental constant that is numerically equal to the inverse of the atomic mass unit.

2.80 Express the volume in milliliters of a "1.00-cup" sample of milk given that 2 cups $=$ 1 pint, 2 pints $=$ 1 quart.

2.81 The distance for a marathon is 26 miles and 385 yards. Convert this distance to kilometers, given that 1 mile $=$ 1760 yd.

2.82 Convert (a) the density of bismuth, 8.90 g/cm³, into mg/mm³; (b) the density of gold, 19.3 g/cm³, into kg/m³.

2.83 An international committee of distinguished scientists met to establish the standard snail's pace, P_{sn}. The average snail covered 1 inch in 2 minutes. Express the average P_{sn} in cm/s.

2.84 Rewrite the following statements, using the temperature scales indicated in brackets: (a) The melting point of gold is 1064°C [K]. (b) Sulfur boils at 445°C [°F and K]. (c) Neon boils at −411°F [K] (3 sf). (d) The temperature of outer space is 2.7 K [°C].

2.85 The temperature on the day side of the lunar surface is 127°C and the temperature on the night side of the lunar surface is −183°C. Convert each temperature to K and °F.

2.86 The density of gold is 19.3 g/cm³. What volume of water will a gold nugget of mass 16.7 g displace when placed into a graduated cylinder?

2.87 Copper metal can be extracted from a copper(II) sulfate solution by electrolysis (as described in Chapter 18). If 10.0 g of copper(II) sulfate pentahydrate is dissolved in 100. mL of water and all the copper is electroplated out, what mass of copper would be obtained?

2.88 Epsom salts consists of magnesium sulfate heptahydrate. (a) Write its formula. (b) How many atoms of magnesium are in 2.00 g of Epsom salts? (c) How many formula units of the compound are present in 2.00 g? (d) How many moles of water molecules are in 2.00 g of Epsom salts?

2.89 L-Dopa, a drug used for the treatment of Parkinson's disease, is 54.82% carbon, 5.62% hydrogen, 7.10% nitrogen, and 32.46% oxygen. What is the empirical formula of the compound?

2.90 A chemical analysis of a complex carbohydrate is 40.0% C, 6.71% H, and 53.3% O. Its molar mass is approximately 860 g/mol. (a) What is the empirical formula of the carbohydrate? (b) What is the molecular formula of the carbohydrate?

2.91 The reference points on a newly proposed temperature scale expressed in °X are the freezing and boiling points of water, set equal to 50.°X and 250.°X. (a) Derive a formula for converting temperatures on the Celsius scale to the new scale. (b) Comfortable room temperature is 22°C. What is that temperature in °X?

2.92 A 1.0-cm³ cube of uranium (of density 18.95 g/cm³) is placed next to a 2.0-cm³ cube of niobium (of density 8.57 g/cm³). Which cube contains the greater number of atoms?

2.93 The molar mass of boron atoms in a natural sample is 10.81 g/mol. The sample is known to consist of ¹⁰B (molar mass 10.013 g/mol) and ¹¹B (molar mass 11.093 g/mol). What are the percentage abundances of the two isotopes?

2.94 Certain chemicals used in research are extremely rare and expensive. Suppose you wrote an order for 2.0 g of a chemical at $500.00 per gram. When your sample arrives, you weigh it and observe that the actual mass is only 1.96 g. How much did you overpay for your compound? Considering how you wrote your order, was the chemical company justified in sending the amount they did?

Applied Exercises

For Exercises 2.95–2.96, see Applying Chemistry: Case Study 2.

2.95 In 1978, scientists extracted a compound with antitumor and antiviral properties from tunicates in the Caribbean Sea. A 2.52-mg sample of the compound, didemnin-C, was analyzed and found to have the following composition: 1.55 mg C, 0.204 mg H, 0.209 mg N, and 0.557 mg O. The mass spectrum of didemnin-C was found to have a parent mass peak at 1014 g/mol that represents the molar mass of the compound. What is the molecular formula of didemnin-C?

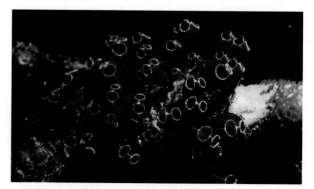

Tunicates, Exercises 2.95–2.96.

2.96 Another compound with medicinal value extracted from Caribbean tunicates is didemnin-A. A 1.78-mg sample of the compound, didemnin-A, was analyzed and found to have the following composition: 1.11 mg C, 0.148 mg H, 0.159 mg N, and 0.363 mg O. The mass spectrum of didemnin-A was found to have a parent mass peak at 942 g/mol that represents the molar mass of the compound. What is the molecular formula of didemnin-A?

2.97 Tu-jin-pi is a root bark used in traditional Chinese medicines for the treatment of athlete's foot. One of the active ingredients in tu-jin-pi is pseudolaric acid A, which is known to contain carbon, hydrogen, and oxygen. A chemist wanting to determine the molecular formula of pseudolaric acid A performed a combustion analysis of 1.000 g of the compound. The products of the combustion were 2.492 g of CO_2 and 0.6495 g of H_2O. (a) Assuming that all the carbon atoms in the CO_2 came from the sample of pseudolaric acid A, what was the mass of carbon in the sample? (b) Assuming that all the hydrogen atoms in the H_2O came from the sample, what was the mass of hydrogen in the sample? (c) Use the information from (a) and (b) to determine how many grams of oxygen were present in the sample and calculate the mass percentage composition of the compound. (d) Determine the empirical formula of the compound. (e) The mass spectrum of pseudolaric acid A showed a parent mass peak at 388.46 g/mol. What is the molecular formula of pseudolaric acid A? See Investigating Matter 1.2 and 2.1 and Section 4.5.

2.98 A folk medicine used in the Anhui province of China to treat acute dysentery is cha-tiao-qi, a preparation of the leaves of *Acer ginnala*. Reacting one of the active ingredients in cha-tiao-qi with water yields gallic acid, a powerful antidysenteric agent. Gallic acid is known to contain carbon, hydrogen, and oxygen. A chemist wanting to determine the molecular formula of gallic acid performed a combustion analysis of 1.000 g of the compound. The products of the combustion were 1.811 g of CO_2 and 0.3172 g of H_2O. (a) Assuming that all the carbon atoms in the CO_2 came from the sample of gallic acid, what was the mass of carbon in the sample? (b) Assuming that all the hydrogen atoms in the H_2O came from the sample, what was the mass of hydrogen in the sample? (c) Use the information from (a) and (b) to determine how many grams of oxygen were present in the sample and calculate the mass percentage composition of the compound. (d) Determine the empirical formula of the compound. (e) The mass spectrum of gallic acid showed a parent mass peak at 170.12 g/mol. What is the molecular formula of gallic acid? See Investigating Matter 1.2 and 2.1 and Section 4.5.

Integrated Exercises

2.99 A metal M forms an oxide with the formula M_2O. The mass percentage of the metal in the oxide is 88.8%. (a) What is the molar mass of the metal? (b) Write the name of the compound.

2.100 A metal M forms an oxide with the formula M_2O_3. The mass percentage of the metal in the oxide is 69.9%. (a) What is the identity of the metal? (b) Write the name of the compound.

2.101 The molar masses of the elements are all relative molar masses because they are determined relative to carbon-12. The isotope silicon-28 has been proposed as a new standard for the molar masses of elements because it can be prepared to a very high degree of purity. The mass of one silicon-28 atom is $4.645\ 67 \times 10^{-23}$ g. If silicon were the standard used for molar mass (instead of carbon-12), one mole would be defined as the number of silicon-28 atoms in 28 g. In that case, what would be (a) the molar mass of carbon-12; (b) the average molar mass of chlorine?

2.102 Suppose it had been decided that the Avogadro constant should be recalculated as the number of gold atoms in 197 g of gold. Gold is a desirable standard because it is not very reactive and because it has only one stable isotope, gold-197. Hence it can be obtained to a very high degree of purity. The mass of one atom of gold-197 is 3.2707×10^{-22} g. (a) Calculate the new value of the Avogadro constant. (b) Using this new value of the Avogadro constant, determine the molar mass of carbon-12 (one atom of carbon-12 has a mass of 1.9926×10^{-23} g).

2.103 (a) Write the formula of copper(II) hydrogen sulfate. (b) Write the formulas of the cation and the anion. (c) How many of each ion is present in each formula unit? (d) In one formula unit of the compound, how many atoms of each element are present? (e) In four *moles* of the compound, how many *moles* of each element are present? (f) What is the molar mass of the compound? (g) How many formula units of the compound are present in 3.45 g? (h) The mass of oxygen is what percentage of the total mass of the compound? (i) What percentage of the atoms in the compound are oxygen atoms? Explain why your answers to (h) and (i) are the same or different.

2.104 (a) Write the formula of cobalt(III) monohydrogen phosphate tetrahydrate. (b) Write the formulas of the cation and the anion. (Ignore the H_2O.) (c) How many of each ion is present in one formula unit? (d) In one formula unit of the compound (include the waters of hydration for the rest of the exercise), how many atoms of each element are present? (e) In two *moles* of the compound, how many *moles* of each element are present? (f) What is the molar mass of the compound? (g) How many formula units of the compound are present in 1.02 g? (h) The mass of oxygen is what percentage of the total mass of the compound? (i) What percentage of the atoms in the compound are oxygen atoms? Explain why your answers to (h) and (i) are the same or different.

C h a p t e r **3**

Chemical Reactions

With every breath we take, we inhale carbon dioxide molecules along with the other gases in air. Each breath connects us physically with history. Some of the carbon atoms in the molecules were once exhaled by our ancestors. In fact, some of them were exhaled by virtually everyone who has ever lived. One atom might have come from Julius Caesar, another from Mohammed, and another from Jesus Christ. Think of *anyone:* when we breathe, we are inhaling a part of them. We are also contributing to forests, to prairies, and to future generations. The air that we exhale contributes to the reservoir of carbon dioxide needed by plants. In the presence of sunlight and water, they use it to produce the compounds called carbohydrates, which include sugars, starches, and cellulose. It is startling to realize that all the vegetation of the world has, in effect, been plucked from the skies.

The complex journey of carbon atoms through the environment is called the **carbon cycle.** A carbon atom in a carbon dioxide molecule in the air may dissolve in seawater, where it is incorporated into the shells of shellfish. After the death of the animal, the shell is compressed to form limestone (calcium carbonate, $CaCO_3$). The carbon atom may remain as limestone or marble (a denser form of limestone) for centuries until, one day, someone converts it into mortar for buildings, uses it to reduce the acidity of soils and to feed nations, or quarries it and carves a statue.

The life story of carbon atoms includes many chemical changes, in which one substance is converted into another. The process that brings about a chemical change is called a **chemical reaction.** All the processes of life involve chemical reactions. To understand how our bodies grow, repair themselves, and reproduce, we need to know about reactions. The chemical industry depends on chemical reactions to make the products essential to modern life. Pollution is a sign of uncontrolled chemical reactions. Clearing up pollution depends on the wiser use of other chemical reactions.

The carbon cycle is evident in fossils like this one, which are found in limestone, a form of calcium carbonate. The carbon atoms in limestone were once part of carbon dioxide molecules in the atmosphere. They were then taken up in the shells of marine organisms. When the organisms died, the shells settled to the bottom of the ocean and became compacted into limestone. Millions of years later, we dig up the limestone and use it to construct buildings. Some of the limestone is also heated to make quicklime in a process that releases the carbon atoms once again to the atmosphere as carbon dioxide.

This chapter, which introduces chemical reactions, is a bridge between Chapters 1 and 2, which focus on individual atoms and elements, and the rest of the book, which deals with the materials that can be made from those atoms and elements. To make those materials, atoms need to change their partners; in this chapter, we see three common ways in which those changes can be achieved.

CHEMICAL EQUATIONS AND CHEMICAL REACTIONS

The combustion of natural gas is a chemical reaction in which the methane of the fuel and the oxygen of the air change into carbon dioxide and water. The starting materials in a chemical reaction are called the **reactants.** In the combustion of methane, the methane and the oxygen are the reactants. The substances formed in a chemical reaction are called the **products.** In the combustion of methane, carbon dioxide and water are the products. Be careful to distinguish *reactants* from *reagents*. Bottles of chemicals kept on hand in a laboratory for use in chemical reactions are called **reagents.** A bottle of hydrochloric acid is a reagent. A reagent is called a reactant only when it is being used in a particular reaction, such as when hydrochloric acid is poured onto zinc metal to make hydrogen gas.

3.1 Symbolizing Chemical Reactions

A chemical reaction is symbolized by an arrow pointing from the reactants to the products:

$$\text{Reactants} \longrightarrow \text{products}$$

You can read the arrow as "go to," "form," or "react to form."

For example, sodium is a soft, shiny metal that reacts vigorously with water. When we drop a small lump of sodium metal into a container of water, hydrogen gas forms rapidly and sodium hydroxide is left in solution (Fig. 3.1). To summarize this reaction *qualitatively*, we could write

$$\text{Sodium} + \text{water} \longrightarrow \text{sodium hydroxide} + \text{hydrogen}$$

In terms of the formulas of the reactants and products, we would write

$$\text{Na} + \text{H}_2\text{O} \longrightarrow \text{NaOH} + \text{H}_2$$

This expression is called a **skeletal equation** because it shows only the bare bones of the reaction (the identities of the reactants and products) in terms of chemical formulas.

The chemical formulas of the species taking part in a reaction are used to provide an efficient way to present important *quantitative* information about the reaction. The use of a chemical equation to convey quantitative information is based on a very old and fundamental observation. In the eighteenth century, chemists noticed that if they carried out a reaction in a sealed container, then there was no change of mass (Fig. 3.2). The preservation of mass during a chemical reaction is called the **law of conservation of mass.** Because each atom of an element has a definite mass, Dalton interpreted the law of conservation of mass as meaning that *atoms are neither created nor destroyed in a chemical reaction:* they simply change their partners. The fact that all the atoms in the reactants are present in the products is symbolized in a chemical equation by

Lavoisier made this observation in France, and Lomonosov made it independently in Russia.

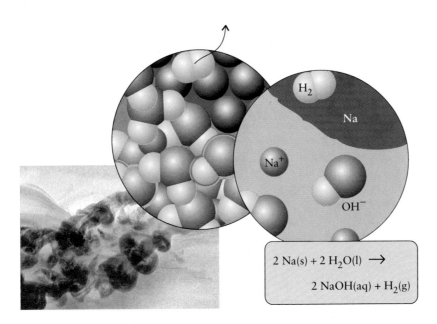

FIGURE 3.1

When a small piece of sodium is dropped into water, a vigorous reaction takes place. Hydrogen gas and sodium hydroxide are formed, and the heat released melts the sodium, which then assumes a rounded shape. A dye that changes color in the presence of sodium hydroxide causes the pink color. Note that two sodium atoms (the dark gray atoms) have given rise to two sodium ions (light purple) and that two water molecules have given rise to one hydrogen molecule (which escapes as a gas) and two hydroxide ions. The chemical equation shows that the reaction is a rearrangement of partners, not a creation or annihilation of atoms.

$$2\,Na(s) + 2\,H_2O(l) \longrightarrow 2\,NaOH(aq) + H_2(g)$$

indicating the same number of atoms of each element on each side of the arrow. There are two H atoms on the left of the skeletal equation above, but three H atoms on the right. So, we change the expression to

$$2\,Na + 2\,H_2O \longrightarrow 2\,NaOH + H_2$$

Now there are four H atoms, two Na atoms, and two O atoms on each side. Section 3.2 provides a more detailed look at balancing chemical equations, and Chapter 4 explores how to use chemical equations to predict how much product a reaction can form.

The numbers multiplying *entire* chemical formulas in chemical equations (for example, the 2 multiplying H_2O) are called the **stoichiometric coefficients** of the substances. A coefficient of 1 (as for H_2) is not written explicitly. When the same number of atoms of each element appear on each side of the arrow, the expression is said to be **balanced**, and it is called a **chemical equation.** We distinguish between the chemical reaction (the actual process) and the chemical equation, which summarizes the reaction in terms of chemical formulas.

The equation for the reaction of sodium with water tells us that

> When two *atoms* of sodium react with two *molecules* of water, they produce two *formula units* of NaOH and one *molecule* of hydrogen.

The word *stoichiometric* comes from the Greek words for "element" and "measure."

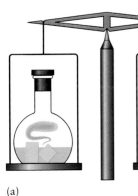
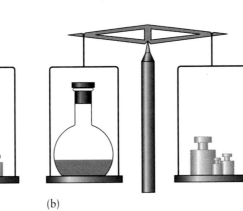

(a)   (b)

FIGURE 3.2

The law of conservation of mass can be verified by experiments in which a chemical reaction is carried out in a sealed flask. The total mass of reactants is measured prior to reaction (a) and is found to be the same as the total mass of products measured after reaction is complete (b).

Because this relationship is valid no matter how large the sample, we can multiply through by the Avogadro number and conclude that

> When 2 *moles* of Na atoms react with 2 *moles* of H_2O molecules, they produce 2 *moles* of NaOH formula units and 1 *mole* of H_2 molecules.

In other words, *the stoichiometric coefficients multiplying the chemical formulas in any balanced chemical equation tell us the relative number of moles of each substance that reacts or is produced in the reaction.*

We can add additional useful information to a chemical equation by indicating the physical state of each reactant and product with a label called a **state symbol:**

Recall from Section 1.13 that an aqueous solution is one in which the solvent is water.

$$\text{(s): solid} \qquad \text{(l): liquid} \qquad \text{(g): gas} \qquad \text{(aq): aqueous solution}$$

For the reaction between solid sodium and water, the complete, balanced chemical equation is therefore

$$2\,Na(s) + 2\,H_2O(l) \longrightarrow 2\,NaOH(aq) + H_2(g)$$

A molecular-level representation of this reaction is shown in Fig. 3.1.

When we want to emphasize that a reaction requires high temperatures, we include the Greek letter Δ (delta) over the arrow. For example, the conversion of limestone to quicklime takes place at about 800°C, and we write

$$CaCO_3(s) \xrightarrow{\Delta} CaO(s) + CO_2(g)$$

The carbon dioxide is released into the atmosphere. Quicklime (calcium oxide, CaO) is widely used in industry, particularly in steelmaking, glassmaking, and the manufacture of cement.

> *A chemical equation shows how atoms are rearranged in a chemical reaction. The stoichiometric coefficients tell us the relative numbers of moles of reactants and products taking part in the reaction.*

3.2 Balancing Chemical Equations

Simple reactions can usually be balanced by inspection; others require more thought. Consider the reaction in which hydrogen and oxygen gases combine to form water. To write the balanced equation, we always start by writing the skeletal equation:

$$H_2 + O_2 \longrightarrow H_2O \qquad \qquad ⚠$$

We use the international *Hazard!* road sign (⚠) to warn that a skeletal equation is not balanced. Then we find the stoichiometric coefficients that balance all the elements. The number of O atoms on each side of the arrow is made the same by multiplying H_2O by 2, to give

$$H_2 + O_2 \longrightarrow 2\,H_2O \qquad \qquad ⚠$$

There are now four H atoms on the right but only two on the left. Therefore, we multiply H_2 by 2 and obtain

$$2\,H_2 + O_2 \longrightarrow 2\,H_2O$$

There are four H atoms and two O atoms on each side of the arrow, so the equation is balanced. At this stage, we add the state symbols:

$$2 H_2(g) + O_2(g) \longrightarrow 2 H_2O(l)$$

Figure 3.3 is a representation of this reaction at the molecular level.

An equation must never be balanced by changing the subscripts in the chemical formulas. That change would suggest that different substances were taking part in the reaction. For example, changing H_2O to H_2O_2 in the skeletal equation and writing

$$H_2 + O_2 \longrightarrow H_2O_2$$

certainly results in a balanced equation. However, it is a summary of a *different* reaction—the formation of hydrogen peroxide, H_2O_2, from its elements. Nor should we write

$$2 H + O \longrightarrow H_2O$$

Although this equation is balanced, it summarizes the reaction between hydrogen and oxygen *atoms,* not the molecules that are the actual starting materials.

Fractional stoichiometric coefficients are perfectly acceptable, as in

$$H_2(g) + \tfrac{1}{2}O_2(g) \longrightarrow H_2O(l)$$

However, because an entire chemical equation can always be multiplied by a numerical factor without affecting its balance, it is common to clear the fractions. This equation can be multiplied by 2 to yield

$$2 H_2(g) + O_2(g) \longrightarrow 2 H_2O(l)$$

Some chemical equations are more difficult to balance. Toolbox 3.1 describes a procedure to use in these cases.

The stoichiometric coefficients in a chemical equation are chosen to show that atoms are neither created nor destroyed in the reaction.

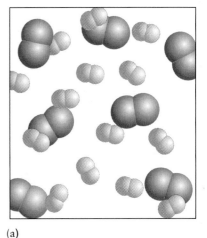
(a)

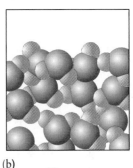
(b)

M

FIGURE 3.3

A molecular representation of the reaction between hydrogen and oxygen that leads to the production of water: (a) reactants; (b) products. No atoms are created or destroyed: they simply change partners. For every two hydrogen molecules that react, one oxygen molecule is consumed and two water molecules are formed.

Example 3.1 *Balancing a chemical equation*

Methane, CH_4, is the main ingredient of natural gas. When methane burns in air, it combines with the oxygen in the air and forms carbon dioxide and water, both produced initially as gases. The skeletal equation for the reaction is

$$CH_4 + O_2 \longrightarrow CO_2 + H_2O \qquad ⚠$$

Balance this equation.

Strategy There are three elements in the skeletal equation: C, H, and O. Balance the atoms of each one of them in turn by using the procedure in Toolbox 3.1.

Solution **Step 1.** Because C and H occur in two formulas and O occurs in three, we begin with C and H. The C atoms are already balanced (one on each side of the arrow). We balance the H atoms by using a stoichiometric coefficient of 2 for H_2O to give four H atoms on each side of the arrow:

$$CH_4 + O_2 \longrightarrow CO_2 + 2 H_2O \qquad ⚠$$

Step 2. Now only O remains to be balanced. Because there is a total of four O atoms

Toolbox 3.1 How to write and balance chemical equations

The procedure in this Toolbox describes how to balance chemical equations that cannot be balanced by inspection.

Conceptual Basis

When we balance a chemical equation, we are symbolizing the fact that all the atoms in the reactants appear in the products. Therefore, we add stoichiometric coefficients to make sure that the same number of atoms of each element appears on each side of the equation. In many cases, when we multiply a chemical formula by a coefficient to balance a particular element, the balance of the other elements in the equation is upset. It is therefore wise to reduce the amount of work by balancing one element at a time (Fig. 3.4).

Procedure

First, make sure that you are using the correct chemical formulas and do not change them during the balancing process: change only the coefficients. Write the skeletal equation for the reaction by writing the formulas of the reactants on the left of the reaction arrow and the formulas of the products on the right. Then, follow these steps:

Step 1. Balance first the element that occurs in the fewest formulas.

Step 2. Balance last the element that is found in the greatest number of formulas.

Step 3. Verify that the coefficients are the smallest whole numbers.

Step 4. Specify the states of each reactant and product.

Groups of atoms such as polyatomic ions (for example, NH_4^+ and PO_4^{3-}) often remain intact during a reaction. When that happens, they can be balanced as single entities. Remember that a subscript outside parentheses shows how many of the polyatomic ions are present.

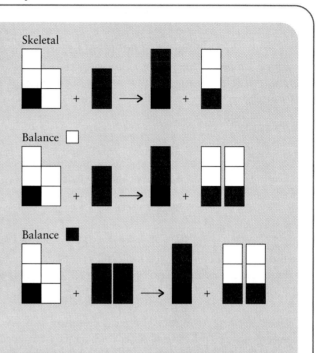

FIGURE 3.4

A representation of a chemical reaction. The differently colored squares represent atoms of different elements. Notice that the skeletal equation (top) lists only the formulas of the reactants and products. The elements are balanced one at a time until the number of atoms of each element is the same on each side of the equation (bottom). The equation is balanced by adding entire molecules involved in the reaction, not separate atoms.

---

on the right but only two on the left, the O_2 needs a stoichiometric coefficient of 2. The result is

$$CH_4 + 2\,O_2 \longrightarrow CO_2 + 2\,H_2O$$

Step 3. Verify that the equation is balanced by counting the number of atoms of each element on each side of the arrow. There are one C atom, four H atoms, and four O atoms on each side of the arrow, so the equation is balanced.

Step 4. Because all reactants and products are gases, we write

$$CH_4(g) + 2\,O_2(g) \longrightarrow CO_2(g) + 2\,H_2O(g)$$

A pictorial representation of this reaction is shown in Fig. 3.5.

Self-Test 3.1A Chlorine gas can be produced in the laboratory by the following reaction:

$$MnO_2(s) + HCl(aq) \longrightarrow Cl_2(g) + MnCl_2(aq) + H_2O(l) \quad \triangle$$

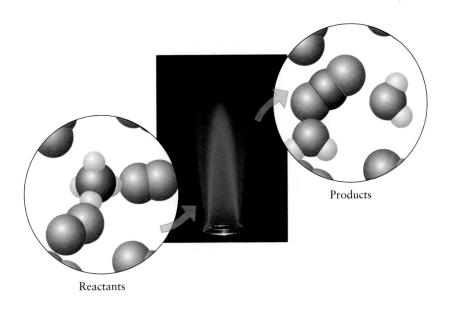

Reactants

Products

FIGURE 3.5

When methane burns, it forms carbon dioxide and water. The blue color is due to the presence of C_2 molecules in the flame. If the oxygen supply is inadequate, these molecules can stick together and form soot, producing a smokey flame. Note that one carbon dioxide molecule and two water molecules are produced for each methane molecule that is consumed. The hydrogen atoms in a water molecule do not necessarily both come from the same methane molecule: the illustration depicts the overall outcome, not the specific outcome of the reaction of one molecule. Because excess oxygen is present, not all the oxygen reacts.

Balance this skeletal equation.

[*Answer:* $MnO_2(s) + 4\,HCl\,(aq) \rightarrow Cl_2(g) + MnCl_2(aq) + 2\,H_2O(l)$]

Self-Test 3.1B Balance the following skeletal equation:

$$Co(NO_3)_3(aq) + (NH_4)_2S(aq) \longrightarrow Co_2S_3(s) + NH_4NO_3(aq)$$ ⚠

Example 3.2 *Writing and balancing a chemical equation*

Write and balance the chemical equation for the combustion of butane, C_4H_{10}, to gaseous carbon dioxide and liquid water.

Strategy First write the skeletal equation. Then balance the equation by using the procedure set out in Toolbox 3.1.

Solution The skeletal equation is

$$C_4H_{10} + O_2 \longrightarrow CO_2 + H_2O$$ ⚠

Step 1. First, balance the carbon and hydrogen atoms:

$$C_4H_{10} + O_2 \longrightarrow 4\,CO_2 + 5\,H_2O$$ ⚠

Step 2. Next, balance the oxygen atoms. In this case, a fractional stoichiometric coefficient is needed:

$$C_4H_{10} + \tfrac{13}{2}O_2 \longrightarrow 4\,CO_2 + 5\,H_2O$$

Step 3. We can clear the fraction by multiplying by 2:

$$2\,C_4H_{10} + 13\,O_2 \longrightarrow 8\,CO_2 + 10\,H_2O$$

Step 4. Finally, we add the physical states:

$$2\,C_4H_{10}(g) + 13\,O_2(g) \longrightarrow 8\,CO_2(g) + 10\,H_2O(l)$$

Self-Test 3.2A When aluminum is melted and heated with solid barium oxide, BaO, a vigorous reaction takes place, and elemental molten barium and solid aluminum oxide, Al_2O_3, are formed. Write the chemical equation for the reaction.

[*Answer:* $2\,Al(l) + 3\,BaO(s) \xrightarrow{\Delta} Al_2O_3(s) + 3\,Ba(l)$]

FIGURE 3.6

When solutions of silver nitrate and potassium chromate are mixed, a precipitate of red silver chromate, Ag_2CrO_4, forms.

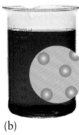

(a)   (b)

FIGURE 3.7

These two beakers contain solutions with different concentrations of the same solute. The lighter color of the solution on the left (a) shows that the solute is less concentrated than in the solution on the right (b). In the molecular-level view, we see that there are more solute particles in a given volume of the more concentrated solution.

Self-Test 3.2B Write the balanced chemical equation for the combustion of propane gas, C_3H_8, to carbon dioxide gas and liquid water.

PRECIPITATION REACTIONS

If we were to let drops of a yellow solution of potassium chromate in water fall into a test tube containing a colorless solution of silver nitrate in water, we would see a dense red cloud form (Fig. 3.6). The cloud is a **precipitate,** a finely divided solid formed from ions in the solution. A familiar example of a precipitate is the mineral deposit called *scale*. Scale, which is primarily calcium carbonate, is found in almost every coffeepot and teakettle and reduces the efficiency of domestic hot-water systems. Reactions that result in the formation of a precipitate are the first of the three important types of reactions that we consider in this chapter. Later sections will introduce the other two types: neutralization reactions and redox reactions.

3.3 Aqueous Solutions

To understand precipitation reactions, we need to know something about the nature of solutions. When we dissolve a substance in water, its molecules disperse evenly throughout the water molecules, forming an aqueous solution (Section 1.13). A **soluble substance** is one that dissolves in a specified solvent. The solvent is often water; and whenever we refer to solubility in this chapter, we mean "soluble in water." An **insoluble substance** is one that does not dissolve significantly in a specified solvent: substances are often regarded as "insoluble" if no more than about 0.1 mol dissolves per liter of solution. Throughout this chapter, the term *insoluble* means "insoluble in water." The **concentration** of the solute refers to how many solute molecules there are in a given volume of solution. For example, the colored solute in the solution on the right in Fig. 3.7 is more concentrated than the one on the left.

Some solutes affect the properties of a solution dramatically. If we make a beaker of pure water a part of an electric circuit, no current flows. If we mix some sugar in the water, still nothing happens. However, if we mix some sodium chloride into the water, a current flows. Because moving ions can conduct electricity, this experiment shows that sodium chloride, but not sugar, is present as ions in water.

An **electrolyte** is a substance that dissolves to give a solution that contains ions; the ions allow the solution to conduct electricity. A **strong electrolyte** is a compound that dissolves to give a solution that contains mainly ions. Aqueous solutions of ionic compounds, such as sodium chloride or potassium nitrate, are solutions of strong electrolytes. These compounds exist as ions in the solid state; when the solid dissolves, the ions are in solution, where they can move and conduct electricity (Fig. 3.8). Aqueous solutions of acids, such as hydrochloric acid or sulfuric acid, which are molecular before they dissolve but form ions in solution, are also electrolyte solutions. A **weak electrolyte** is a compound that dissolves to give a solution that contains mostly molecules and very few ions. We say that weak electrolytes are "partially ionized" in solution. Aqueous acetic acid is a weak electrolyte: in aqueous solution, only a small fraction of CH_3COOH

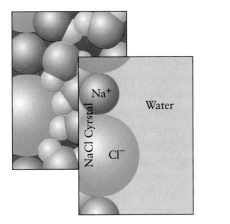
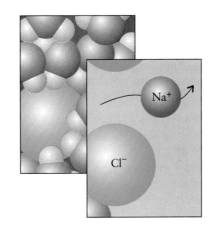

FIGURE 3.8

Sodium chloride consists of sodium ions and chloride ions. When it is added to water (left), the ions separate and spread throughout the solvent (right). The solution consists of water molecules, sodium ions, and chloride ions. There are no NaCl molecules present at any stage. The overlays show only the solute.

molecules ionize to hydrogen ions and acetate ions, $CH_3CO_2^-$. Ammonia, NH_3, is also a weak electrolyte. Figure 3.9 illustrates one consequence of the difference between solutions of strong and weak electrolytes.

A **nonelectrolyte** is a substance that dissolves to give a solution that does not contain ions and therefore does not conduct electricity. Pure water is itself a nonelectrolyte. Aqueous solutions of molecular compounds other than acids, for example, acetone (**1**) and glucose (**2**), are nonelectrolyte solutions—solutions of nonelectrolytes. Sweetened water consists of intact sugar molecules

All weak acids and weak bases, compounds that will be discussed in Sections 3.8 and 3.9, are weak electrolytes.

There are a few ions present in water because H_2O forms some H_3O^+ and OH^- ions. However, these ions are present in very low concentrations. Tap water conducts electricity because it contains dissolved ions, such as Na^+, Ca^{2+}, and HCO_3^-.

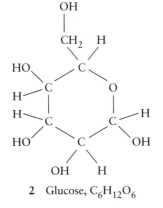
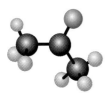

1 Acetone, C_3H_6O **2** Glucose, $C_6H_{12}O_6$

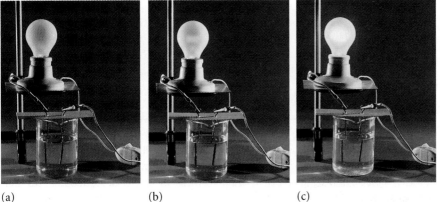

(a)   (b)   (c)

FIGURE 3.9

Pure water is a poor conductor of electricity, as shown by the almost imperceptible glow of the bulb in the circuit (a). However, when ions are present, as in an electrolyte solution, the solution does conduct. The ability of the solution to conduct is low if it is a weak electrolyte (b) but significant if it is a strong electrolyte (c), even if the solute concentration is the same in each case.

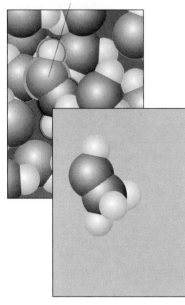

Methanol
molecule

FIGURE 3.10

In a nonelectrolyte solution, the solute remains as molecules and does not break up into ions. Methanol, CH_3OH, is a nonelectrolyte and is present as molecules when it is dissolved in water.

moving among the water molecules. Because there are no ions present, the solution does not conduct electricity. If we could see the individual molecules in a nonelectrolyte solution, we would see the intact solute molecules dispersed among the solvent molecules (Fig. 3.10). The classification of electrolytes appears again in Section 3.9.

In aqueous solution, strong electrolytes form ions that are free to move through the solvent and conduct electricity. Only a small fraction of molecules of weak electrolytes form ions in solution. Nonelectrolytes do not form ions in solution, so they do not conduct electricity.

3.4 Reactions Between Strong Electrolyte Solutions

Sodium chloride, NaCl, dissolves to give a colorless solution consisting of Na^+ cations and Cl^- anions. Similarly, silver nitrate, $AgNO_3$, dissolves to give a colorless solution consisting of Ag^+ cations and NO_3^- anions. However, if we mix these two aqueous solutions, we immediately get a white precipitate. Analysis shows that the precipitate is silver chloride, AgCl, an insoluble white solid (Fig. 3.11). When an insoluble substance is formed in water, it immediately forms a precipitate. The colorless solution remaining above the precipitate in our example contains Na^+ cations and NO_3^- anions. These ions remain in solution because sodium nitrate, $NaNO_3$, is soluble in water.

If we could see the individual ions in a strong electrolyte solution, we would see each ion jostling around throughout the solvent, meeting another ion only occasionally, but sometimes lingering near an ion of opposite charge, then moving off again (Fig. 3.12). Each ion is **hydrated**, which means that it has a number of water molecules closely associated with it (Fig. 3.13). A sodium cation, for example, may have as many as six water molecules stuck to it; and it carries this coating of water molecules through the solution as it moves. A chloride ion

FIGURE 3.11

The formation of a silver chloride precipitate occurs immediately as silver nitrate solution is added to a solution of sodium chloride.

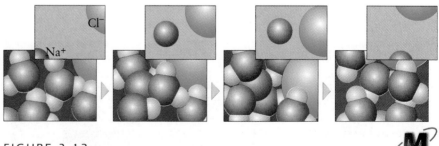

FIGURE 3.12

A series of scenes in a solution of sodium chloride. A sodium ion and a chloride ion move together, linger near each other for a time because of the attraction of their opposite charges, and then move apart. The loose, transient association of oppositely charged ions is called an ion pair. The solution is shown both with solvent molecules, for realism, and without, for clarity.

FIGURE 3.13

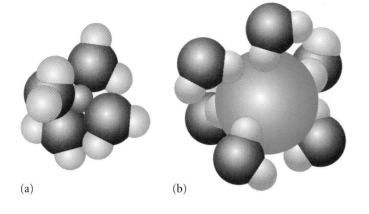

(a)　　　　　　　　(b)

In water, ions are hydrated; that is, they are surrounded by a cluster of water molecules bound loosely to the ion. Note that a hydrated cation (a) is surrounded by water molecules oriented so that the O atom is closest to the ion, whereas a hydrated anion (b) has water molecules attached through their hydrogen atoms. The number of hydrating molecules depends on the size of the ion, but for most ions it is approximately six.

is also hydrated in water. The coating of water molecules on each ion helps to prevent the ions from coming out of solution and forming a solid again.

A **precipitation reaction** is a reaction in which an insoluble solid product is formed when two electrolyte solutions are mixed. We write the chemical equation for the precipitation reaction between sodium chloride and silver nitrate as follows, with (aq) indicating substances that are dissolved in water and (s) a solid that has precipitated:

$$AgNO_3(aq) + NaCl(aq) \longrightarrow AgCl(s) + NaNO_3(aq)$$

Solution　　　　Solution　　　　Precipitate　　　　Solution

Because the ions appear to have exchanged partners, precipitations are also known as *double replacement* or *metathesis* reactions.

A more colorful precipitation reaction is the one that occurs when solutions of lead(II) nitrate, $Pb(NO_3)_2$, and potassium chromate, K_2CrO_4, are mixed. We get an immediate precipitate of insoluble yellow lead(II) chromate, $PbCrO_4$ (Fig. 3.14):

$$Pb(NO_3)_2(aq) + K_2CrO_4(aq) \longrightarrow PbCrO_4(s) + 2 KNO_3(aq)$$

Solution　　　　Solution　　　　Precipitate　　　　Solution

Lead(II) chromate is mixed with lead(II) sulfate, $PbSO_4$, and lead(II) molybdate, $PbMoO_4$, and blended into paint to add a bright, golden yellow color to traffic markings on highways.

A precipitation reaction occurs when solutions of two strong electrolytes are mixed and result in the formation of an insoluble solid.

3.5 Net Ionic Equations

When we write NaCl(aq), we actually mean that we have a solution of Na^+ and Cl^- ions in water. A **complete ionic equation** for a precipitation reaction shows all the ions in solution explicitly. For example, the complete ionic equation for the silver chloride precipitation shown in Fig. 3.11 is

$$Ag^+(aq) + NO_3^-(aq) + Na^+(aq) + Cl^-(aq) \longrightarrow$$
$$AgCl(s) + Na^+(aq) + NO_3^-(aq)$$

In the complete ionic equation, it is easy to see which compounds are dissolved, because they are shown as separate ions.

The complete ionic equation shows that dissolved Ag^+ and Cl^- ions join to form insoluble AgCl. Because the Na^+ and NO_3^- ions appear as both reactants

FIGURE 3.14

In this precipitation reaction, yellow lead(II) chromate is formed when lead(II) nitrate solution is added to a solution of potassium chromate.

FIGURE 3.15

Two depictions of a precipitation reaction that results when the ions in two electrolyte solutions are mixed (left beakers). The top right beakers show the fate of all four types of ions. By imagining the ionic reaction without the spectator ions (bottom right beakers), we can focus on the essential process described by the net ionic equation.

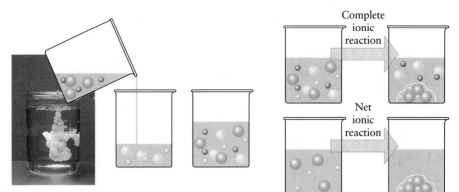

and products, they play no direct role in the reaction. They are **spectator ions,** ions that are present while the reaction takes place but themselves remain unchanged.

An analogy  Spectator ions are like spectators at a sporting event or dance contest. They are present but do not participate in the action.

Because spectator ions remain unchanged we can cancel them on each side of the arrow in the ionic equation:

$$Ag^+(aq) + \cancel{NO_3^-(aq)} + \cancel{Na^+(aq)} + Cl^-(aq) \longrightarrow$$
$$AgCl(s) + \cancel{Na^+(aq)} + \cancel{NO_3^-(aq)}$$

The species remaining after cancellation make up the **net ionic equation** for the reaction, the chemical equation that displays the net change that occurs in the reaction:

$$Ag^+(aq) + Cl^-(aq) \longrightarrow AgCl(s)$$

A net ionic equation shows that the Ag^+ ions supplied by one solution combine with the Cl^- ions supplied by the other solution and that these ions precipitate as solid silver chloride, AgCl. A net ionic equation focuses our attention on the chemical process taking place (Fig. 3.15).

We have just visualized what a solution would look like if we could see the individual molecules and ions. What would we observe if we could see a precipitation reaction such as the formation of silver chloride in progress? In the separate solutions of $AgNO_3$ and NaCl, the hydrated cations and anions each jostle around throughout the solvent. However, when the two solutions are mixed, Ag^+ ions encounter Cl^- ions. They immediately stick together and form clumps of ions that grow larger as more and more ions stick to one another. The resulting compound, silver chloride, is insoluble and falls out of solution. These ionic encounters occur throughout the combined solution as soon as the two initial solutions are mixed, and there is no time for large crystals of silver chloride to grow. Solid particles are formed wherever Ag^+ ions and Cl^- ions meet, so the product is formed as a cloudy, finely divided solid. Nothing happens to the Na^+

Toolbox 3.2 *How to write a net ionic equation*

This Toolbox summarizes how to write the net ionic equation for a reaction.

Conceptual Basis

A net ionic equation shows only the net chemical change (Fig. 3.16). The spectator ions remain unaffected by the reaction and do not appear in the net ionic equation.

Procedure

To write a balanced net ionic equation, follow these steps:

Step 1. Write and balance the chemical equation for the reaction.

Step 2. Write the complete ionic equation, showing all the dissolved ions as they actually exist in solution, as separate, charged ions. Insoluble solids and compounds that remain in molecular form are shown as complete compounds.

Step 3. Identify and cancel the spectator ions, the identical ions in solution that appear on both sides of the arrow. The resulting equation is the net ionic equation.

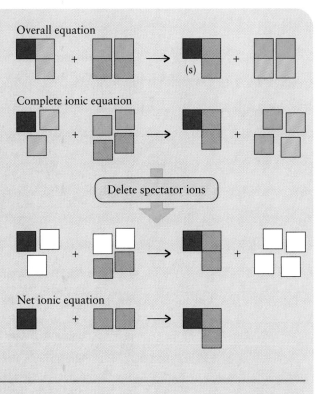

Overall equation

Complete ionic equation

Delete spectator ions

Net ionic equation

FIGURE 3.16

How to write a net ionic equation. Write the balanced overall equation (top), Then show all ionic solutes as separate ions in the complete ionic equation (second line), and delete the spectator ions. The result is the net ionic equation (bottom).

and NO_3^- ions: as spectator ions, they remain in solution. When they encounter one another, they may pause momentarily, but they soon move apart.

A complete ionic equation expresses a reaction in terms of the ions that are present in solution. A net ionic equation is the chemical equation that remains after the cancellation of the spectator ions.

Example 3.3 *Writing a net ionic equation*

When aqueous solutions of barium nitrate, $Ba(NO_3)_2$, and ammonium iodate, NH_4IO_3, are mixed, barium iodate, $Ba(IO_3)_2$, precipitates. Write the net ionic equation for the reaction.

Strategy We expect that the net ionic equation will show the formation of solid barium iodate from barium and iodate ions. Follow the procedure in Toolbox 3.2 to verify this conclusion.

Solution **Step 1.** The chemical equation is

$$Ba(NO_3)_2(aq) + 2\,NH_4IO_3(aq) \longrightarrow Ba(IO_3)_2(s) + 2\,NH_4NO_3(aq)$$

Solution   Solution   Precipitate   Solution

Step 2. The complete ionic equation, with all the dissolved ions written as they exist in the solutions before and after mixing, is

$$Ba^{2+}(aq) + 2\,NO_3^-(aq) + 2\,NH_4^+(aq) + 2\,IO_3^-(aq) \longrightarrow$$
$$Ba(IO_3)_2(s) + 2\,NH_4^+(aq) + 2\,NO_3^-(aq)$$

Step 3. The spectator ions are those that appear as aqueous ions among both the reactants and the products: NH_4^+ and NO_3^-. When they are canceled, we obtain the net ionic equation:

$$Ba^{2+}(aq) + \cancel{2\,NO_3^-(aq)} + \cancel{2\,NH_4^+(aq)} + 2\,IO_3^-(aq) \longrightarrow$$
$$Ba(IO_3)_2(s) + \cancel{2\,NH_4^+(aq)} + \cancel{2\,NO_3^-(aq)}$$

which is the same as

$$Ba^{2+}(aq) + 2\,IO_3^-(aq) \longrightarrow Ba(IO_3)_2(s)$$

Self-Test 3.3A Write the net ionic equation for the reaction in which aqueous solutions of colorless silver nitrate and yellow sodium chromate react to give a precipitate of red silver chromate.

[*Answer:* $2\,Ag^+(aq) + CrO_4^{2-}(aq) \rightarrow Ag_2CrO_4(s)$]

Self-Test 3.3B The mercury(I) ion, Hg_2^{2+}, consists of two Hg^+ ions joined together. Write the net ionic equation for the reaction in which colorless aqueous solutions of mercury(I) nitrate, $Hg_2(NO_3)_2$, and potassium phosphate, K_3PO_4, react to give a white precipitate of mercury(I) phosphate $(Hg_2)_3(PO_4)_2$.

3.6 Making Use of Precipitation

Precipitation reactions are one way to make new compounds. We can make an insoluble compound by choosing starting solutions that, when mixed together, result in the desired precipitate. Then we can separate the solid from the reaction mixture by filtration. This strategy is commonly used in environmental

Table 3.1 Solubility rules for inorganic compounds

Soluble compounds	Insoluble compounds
compounds of Group 1 elements	carbonates (CO_3^{2-}), chromates (CrO_4^{2-}), oxalates ($C_2O_4^{2-}$), and phosphates (PO_4^{3-}), **except** those of the Group 1 elements and NH_4^+
ammonium (NH_4^+) compounds	
chlorides (Cl^-), bromides (Br^-), and iodides (I^-), **except** those of Ag^+, Hg_2^{2+}, and Pb^{2+}*	sulfides (S^{2-}), **except** those of the Group 1 and 2 elements and NH_4^+
nitrates (NO_3^-), acetates ($CH_3CO_2^-$), chlorates (ClO_3^-), and perchlorates (ClO_4^-)	hydroxides (OH^-) and oxides (O^{2-}), **except** those of the Group 1 and 2 elements[†]
sulfates (SO_4^{2-}), **except** those of Ca^{2+}, Sr^{2+}, Ba^{2+}, Pb^{2+}, Hg_2^{2+}, and Ag^+[‡]	

*$PbCl_2$ is slightly soluble.

†$Ca(OH)_2$ and $Sr(OH)_2$ are sparingly (slightly) soluble.

‡Ag_2SO_4 is slightly soluble.

monitoring, where it may be necessary to find out how much lead or mercury is in a sample of water. Precipitating the lead or mercury ions as insoluble compounds allows us to separate them and measure their amounts. The **solubility rules** given in Table 3.1 summarize the observed solubility patterns of common ionic compounds in water. Notice that all nitrates and all compounds of the common Group 1 metals are soluble, so they make useful starting solutions. Reasons for the different solubilities of compounds are explored in Chapter 12.

Any spectator ions can be used, provided they remain in solution and do not react. For example, the table shows that mercury(I) iodide, Hg_2I_2, is insoluble. It is formed as a precipitate whenever solutions containing Hg_2^{2+} ions and I^- ions are mixed:

$$Hg_2^{2+}(aq) + 2I^-(aq) \longrightarrow Hg_2I_2(s)$$

As pointed out in Self-Test 3.3B, the mercury(I) ion is Hg_2^{2+}, with two Hg^+ ions bonded together.

Mercury(I) nitrate, $Hg_2(NO_3)_2$, is soluble (like all nitrates), and potassium iodide, KI, is soluble (like all common compounds of Group 1 metals). We can therefore predict that mixing solutions of mercury(I) nitrate and potassium iodide will result in the precipitation of mercury(I) iodide (Fig. 3.17). The chemical equation for this reaction is

$$Hg_2(NO_3)_2(aq) + 2KI(aq) \longrightarrow Hg_2I_2(s) + 2KNO_3(aq)$$

and the complete ionic equation is

$$Hg_2^{2+}(aq) + 2\,\cancel{NO_3^-(aq)} + 2\,\cancel{K^+(aq)} + 2I^-(aq) \longrightarrow \\ Hg_2I_2(s) + 2\,\cancel{NO_3^-(aq)} + 2\,\cancel{K^+(aq)}$$

The K^+ and NO_3^- ions are spectators and remain in solution. The same precipitate results when we mix any soluble mercury(I) compound with any soluble iodide. For example, we could also form mercury(I) iodide from aqueous solutions of mercury(I) acetate and sodium iodide, both of which are soluble:

$$Hg_2(CH_3CO_2)_2(aq) + 2NaI(aq) \longrightarrow Hg_2I_2(s) + 2NaCH_3CO_2(aq)$$

This chemical equation has the same net ionic equation as the equation for the reaction of $Hg_2(NO_3)_2$ and KI:

$$Hg_2^{2+}(aq) + 2I^-(aq) \longrightarrow Hg_2I_2(s)$$

FIGURE 3.17

Another example of a precipitation reaction. This time, a solution of mercury(I) nitrate is being added to a solution of potassium iodide, and the insoluble product, mercury(I) iodide, is precipitated. Notice that a yellow color forms first. Mercury(I) iodide has two solid forms. The yellow form precipitates first but is soon converted to the more stable orange form.

Example 3.4 *Predicting the outcome of a precipitation reaction*

Predict the likely product when aqueous solutions of sodium phosphate, Na_3PO_4, and lead(II) nitrate, $Pb(NO_3)_2$, are mixed. Write the net ionic equation for the reaction.

Strategy Decide which ions are present in the mixed solutions and consider all possible combinations. Use the solubility rules in Table 3.1 to decide whether any combination of ions would form an insoluble compound. Then write the net ionic equation to match.

Solution The mixed solution will contain Na^+, PO_4^{3-}, Pb^{2+}, and NO_3^- ions. All nitrates and all compounds of Group 1 metals are soluble, but, according to Table 3.1, phosphates of other elements are generally insoluble. Hence, we can predict that Pb^{2+} and PO_4^{3-} ions will form an insoluble compound and that lead(II) phosphate, $Pb_3(PO_4)_2$, will precipitate:

$$3Pb^{2+}(aq) + 2PO_4^{3-}(aq) \longrightarrow Pb_3(PO_4)_2(s)$$

FIGURE 3.18

The shape of this shell is a result of a precipitation reaction in which the shellfish secreted calcium ions at certain points on its surface. The calcium ions reacted with carbonate ions in the surrounding water. The colors of the shell are due to iron impurities that were captured in the solid as it formed.

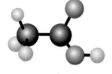

3 Acetic acid, CH_3COOH

Self-Test 3.4A Predict the identity of the precipitate that forms, if any, when aqueous solutions of ammonium sulfide, $(NH_4)_2S$, and copper(II) sulfate, $CuSO_4$, are mixed, and write the net ionic equation for the reaction.

[***Answer:*** Copper(II) sulfide; $Cu^{2+}(aq) + S^{2-}(aq) \rightarrow CuS(s)$]

Self-Test 3.4B Suggest two solutions that can be mixed to prepare strontium sulfate, and write the net ionic equation for the reaction.

Nature makes use of precipitation reactions to form solid structures. For instance, shellfish form their shells from calcium carbonate. The organism secretes calcium ions from cells that are in contact with seawater. Seawater contains dissolved carbon dioxide, some of which is present as carbonate ions, CO_3^{2-}. The ions combine to give a precipitate of calcium carbonate:

$$Ca^{2+}(aq) + CO_3^{2-}(aq) \longrightarrow CaCO_3(s)$$

The organism constructs its shell by secreting the calcium ions at the right spots (Fig. 3.18).

Solubility rules like those in Table 3.1 can be used to predict the products of precipitation reactions.

THE REACTIONS OF ACIDS AND BASES

The second major type of chemical reaction is *neutralization,* the reaction between acids and bases. Reactions between acids and bases play important roles in our daily lives. Acid concentrations in the atmosphere are measured continuously at environmental monitoring sites. In industry, the concentration of acid in a solution is monitored and controlled within strict limits. For example, the *acidity* (the concentration of acid) of a reaction mixture can be critical for the synthesis of a pharmaceutical compound. Even cans of soup must have the correct acidity level or the taste will not be right. Our bodies also function as efficient chemical factories that control the acidity of our blood automatically. If this automatic control were to fail, we would die. Clearly, acids and bases are compounds of great importance to humanity.

Early chemists applied the term *acid* to substances that have a sharp, or sour, taste. Vinegar, for instance, contains acetic acid, CH_3COOH (**3**). **Alkalis,** which are aqueous solutions of substances we now call *bases,* were recognized by their soapy feel. However, tasting and feeling are very dangerous ways to determine classes of compounds: you must *never* taste or touch solutions of chemicals in a laboratory. Fortunately, there are less hazardous ways of recognizing acids and bases. For instance, acids and bases change the color of certain dyes known as indicators (Fig. 3.19). One of the most common indicators is litmus, a vegetable dye obtained from lichen. Aqueous solutions of acids turn litmus red; aqueous solutions of bases (alkalis) turn it blue.

3.7 Acids and Bases in Aqueous Solution

Although many acids are molecular compounds in the pure state, they form ions when they dissolve in water. When a molecule of an acid dissolves in water,

FIGURE 3.19

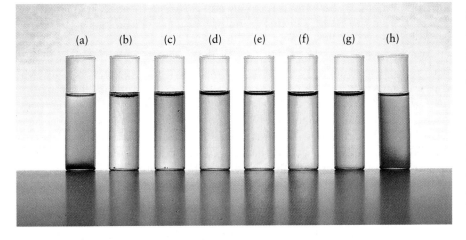

The acidities of various household products can be demonstrated by adding an indicator (an extract of red cabbage in this case) and noting the resulting color. Red indicates acidic, blue basic. From left to right, the household products are (a) lemon juice, (b) soda water, (c) 7-Up, (d) vinegar, (e) ammonia, (f) lye, (g) milk of magnesia, and (h) detergent in water. Ammonia and lye are such strong bases that they destroy the dye, and a yellow color is obtained instead of the expected blue.

it loses a hydrogen ion, H^+, to one of the water molecules. The result is the formation of a *hydronium ion*, H_3O^+ (**4**). For example, when the gas hydrogen chloride, HCl, dissolves in water, it releases a hydrogen ion to a neighboring water molecule. The resulting solution consists of hydronium ions and chloride ions:

$$HCl(g) + H_2O(l) \longrightarrow H_3O^+(aq) + Cl^-(aq)$$

4 Hydronium ion, H_3O^+

The hydronium ion is itself strongly hydrated in solution, and a better representation of the actual ion present in solution may be $H_9O_4^+$ (a hydronium ion, H_3O^+, with one water molecule attached to each H atom). We shall use the simpler formula H_3O^+ because that is enough to emphasize that a proton (an H^+ ion) has been transferred to a water molecule. When we simply wish to show the presence of hydrogen ions in solution and do not need to emphasize the actual transfer of a proton from an acid to water, we shall use the even simpler formula $H^+(aq)$. However, the hydrogen ion, H^+, never actually exists in the free state in water. The (aq) after the formula means that the H^+ ion is attached to at least one other water molecule.

Acids release hydrogen ions (and form hydronium ions) in solution. The Swedish chemist Svante Arrhenius recognized this common feature and classified a compound as an acid if it can act in this way:

An **Arrhenius acid** is a compound that contains hydrogen and releases hydrogen ions (which form hydronium ions) in water.

Whenever we refer to "acids" in this chapter, we mean Arrhenius acids.

We meet a more general definition of acids in Section 15.1.

Hydrogen chloride, HCl, and nitric acid, HNO_3, are acids. Both contain hydrogen that they can release as hydrogen ions (to form hydronium ions) in water, and both turn litmus red. An aqueous solution of hydrogen chloride, HCl(aq), is called hydrochloric acid. Methane, CH_4, is not an acid. Although it contains hydrogen, it does not release hydrogen ions in water and has no effect on litmus. Acetic acid, CH_3COOH, releases one hydrogen ion in water (from the hydrogen atom attached to the oxygen atom), so it is an acid, as its name indicates.

The **acidic hydrogen atom** in a compound is the hydrogen atom that can be released as H^+. It is often written as the first element in a molecular formula, as in HCl, HNO_3, and $HC_2H_3O_2$ (acetic acid). The formulas of carboxylic acids, which are organic compounds that contain the carboxyl group —COOH (**5**), are

5 Carboxyl group, COOH

normally written to show the carboxyl group explicitly. That makes it easier to remember that the H atom in this group of atoms is the acidic one. For example, acetic acid is usually written CH_3COOH.

We can usually recognize which compounds are acids by noting whether their molecular formulas begin with H or—if they are organic compounds—whether they contain a carboxyl group. Thus, we can immediately recognize that HCl, H_2CO_3 (carbonic acid), H_2SO_4 (sulfuric acid), and C_6H_5COOH (benzoic acid) are acids, but that CH_4, NH_3 (ammonia), and $CH_3CO_2^-$ (the acetate ion) are not. The common oxoacids, which are acids containing oxygen, were introduced in Section 1.17 and are listed in Table 1.8.

Arrhenius also recognized that all compounds that turn litmus blue produce hydroxide ions in solution. He classified a species as a base if it can act this way:

An **Arrhenius base** is a compound that produces hydroxide ions in water.

When we refer to "bases" in this chapter, we mean Arrhenius bases.

When the ionic compound sodium hydroxide, NaOH, dissolves in water, its ions separate and the concentration of hydroxide ions is increased; so sodium hydroxide is a base. The molecular compound ammonia, NH_3, does not contain hydroxide ions. However, when it dissolves, it attracts a hydrogen ion from a neighboring water molecule and forms an ammonium ion, NH_4^+, and a hydroxide ion:

$$NH_3(aq) + H_2O(l) \longrightarrow NH_4^+(aq) + OH^-(aq)$$

When a hydrogen ion is transferred from water to ammonia, the concentration of hydroxide ions in the solution is increased, so we classify ammonia as a base. Note the word *compound* in the Arrhenius definition. Sodium reacts with water to produce hydrogen gas and a solution of sodium hydroxide; however, because sodium is not a compound, it is not classified as an Arrhenius base.

An Arrhenius acid contains hydrogen and releases hydrogen ions (that produce hydronium ions) in water. An Arrhenius base is a compound that produces hydroxide ions in aqueous solution.

Self-Test 3.5A Which of the following are acids or bases? (a) HNO_3; (b) C_6H_6; (c) KOH; (d) C_3H_5COOH.

[*Answer:* (a) and (d) are acids; (b) is neither an acid nor a base; (c) is a base.]

Self-Test 3.5B Which of the following are acids or bases? (a) KCl; (b) HClO; (c) HF; (d) $Ca(OH)_2$.

3.8 Strong and Weak Acids

Because acids produce ions in water, they are electrolytes. They can therefore be classified in similar terms:

A **strong acid** is almost completely ionized in aqueous solution.

A **weak acid** is incompletely ionized in aqueous solution.

In this context, "ionized" means that a hydrogen ion has been transferred from the acid molecule to a water molecule to form a hydronium ion. By "completely ionized," we mean that each acid molecule has lost its acidic hydrogen atom by transferring it as a hydrogen ion to a water molecule. By "incompletely ionized," we mean that only a fraction (and usually a very small fraction) of the acid molecules have become ionized.

Hydrochloric acid is a strong acid. An aqueous solution of hydrogen chloride contains hydronium ions, chloride ions, and very few HCl molecules:

$$HCl(g) + H_2O(l) \longrightarrow H_3O^+(aq) + Cl^-(aq)$$

Acetic acid, on the other hand, is a weak acid. In aqueous solution, it undergoes ionization according to the equation

$$CH_3COOH(aq) + H_2O(l) \longrightarrow H_3O^+(aq) + CH_3CO_2^-(aq)$$

However, because it is a weak acid, the solution consists principally of CH_3COOH molecules, with only a small proportion of hydronium ions and acetate ions, $CH_3CO_2^-$, present (Fig. 3.20). The concentration of hydronium ions is much smaller than the concentration of intact acetic acid molecules.

The common strong acids are listed in Table 3.2. They include three acids that are often found as reagents in laboratories, namely, hydrochloric acid, nitric acid, and sulfuric acid. Most acids are weak. All carboxylic acids, HCN, HF, and most acids not listed in Table 3.2 are weak.

Strong acids (the acids listed in Table 3.2) are almost completely ionized in water; weak acids (almost all other acids) are not.

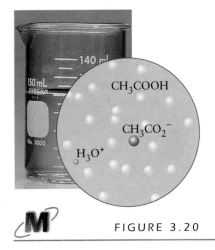

FIGURE 3.20

Acetic acid, like all carboxylic acids, is a weak acid in water. Most of it remains as acetic acid molecules, CH_3COOH (yellow spheres), but a small proportion of these molecules donate a hydrogen ion to a water molecule to give hydronium ions, H_3O^+ (small pink sphere), and acetate ions, $CH_3CO_2^-$ (red sphere).

Self-Test 3.6A What, if anything, is wrong with the following statement: "An aqueous solution that contains 6 mol HCl per liter is a stronger acid than one that contains only 1 mol HCl per liter?"

[**Answer:** The former solution is more concentrated than the latter, but both are solutions of a strong acid. Hydrogen chloride is virtually completely ionized in solution.]

Self-Test 3.6B What, if anything, is wrong with the following statement: "An aqueous solution that contains 0.1 mol of the weak acid HCN per liter contains hydronium ions at a concentration of 0.1 mol per liter?"

Table 3.2 *The strong acids and bases in water*

Strong acids	Strong bases
hydrobromic acid, HBr(aq)	Group 1 hydroxides
hydrochloric acid, HCl(aq)	alkaline earth metal hydroxides*
hydroiodic acid, HI(aq)	
nitric acid, HNO_3	
perchloric acid, $HClO_4$	
chloric acid, $HClO_3$	
sulfuric acid, H_2SO_4 (to HSO_4^-)	

*$Ca(OH)_2$, $Sr(OH)_2$, $Ba(OH)_2$.

Unless other information is given, any acid discussed in this book that is not listed in Table 3.2 is a weak acid.

3.9 Strong and Weak Bases

Bases also produce ions—hydroxide ions—in solution, so they too may be classified like other electrolytes:

A **strong base** is almost completely ionized in aqueous solution.

A **weak base** is not completely ionized in aqueous solution.

In the context of weak bases, ionized means the attachment of a hydrogen ion, as in the formation of NH_4^+ from NH_3.

The common strong bases are the oxides and hydroxides of the alkali metals and alkaline earth metals. They include sodium hydroxide, calcium oxide, and barium hydroxide. Metal hydroxides are bases because they release hydroxide ions when they dissolve in water.

$$NaOH(s) \longrightarrow Na^+(aq) + OH^-(aq)$$

$$Ba(OH)_2(s) \longrightarrow Ba^{2+}(aq) + 2\,OH^-(aq)$$

Metal oxides contain oxide ions, O^{2-}. However, in solution, oxide ions react completely with water to form hydroxide ions:

$$O^{2-}(aq) + H_2O(l) \longrightarrow 2\,OH^-(aq)$$

Calcium oxide is called *quicklime* because it reacts so aggressively with water:

$$CaO(s) + H_2O(l) \longrightarrow Ca^{2+}(aq) + 2\,OH^-(aq)$$

Recall that the true alkaline earth metals are calcium, strontium, and barium.

The reaction is so vigorous and gives out so much heat that the wooden boats once used to transport quicklime sometimes caught fire when water seeped into their holds. Solid calcium hydroxide, $Ca(OH)_2$, is called slaked lime—calcium hydroxide is the product present after the "thirst" of quicklime (CaO) for water has been quenched (slaked), that is, after quicklime has reacted with all the water it can. Slaked lime is widely used in agriculture to adjust the acidity of soil (Fig. 3.21).

Ammonia is a weak base. When it dissolves in water, it gives a solution in which it exists almost entirely as NH_3 molecules (**6**). There is only a small proportion—usually formed by less than one in a hundred molecules—of NH_4^+ cations and OH^- anions. Other common weak bases are the amines, smelly compounds derived from ammonia by replacement of one or more of its hydrogen atoms by an organic group. For example, the replacement of one hydrogen atom in NH_3 by a methyl group, $-CH_3$ (**7**), gives methylamine, CH_3NH_2 (**8**). The replacement of all three hydrogen atoms gives trimethylamine, $(CH_3)_3N$ (**9**), a substance found in decomposing fish and on unwashed dogs.

FIGURE 3.21

Gardeners use slaked lime to make soil more basic.

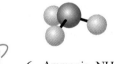

6 Ammonia, NH_3

7 Methyl group, CH_3

8 Methylamine, CH_3NH_2

9 Trimethylamine, $(CH_3)_3N$

Whenever we imagine a solution of a weak base, such as ammonia, we should picture it as consisting almost entirely of the original molecules. Only a very tiny fraction of the molecules are present as cations (such as NH_4^+) and a matching number of OH^- ions. Both weak bases and weak acids are weak electrolytes in water (Table 3.3).

Strong bases (the oxides and hydroxides of the alkali and alkaline earth metals) are completely ionized in solution; oxide ions are converted to hydroxide ions. Weak bases (ammonia and its organic derivatives, the amines) are only partially ionized in solution.

Self-Test 3.7A Identify the following substances as electrolytes or nonelectrolytes and predict which will conduct electricity when dissolved in water: (a) NaOH; (b) HCN.

[*Answer:* (a) Strong base, so strong electrolyte; conducts electricity; (b) weak acid, so weak electrolyte; conducts electricity to a small extent]

Self-Test 3.7B Identify the following substances as electrolytes or nonelectrolytes and predict which will conduct electricity when dissolved in water: (a) HNO_3; (b) NH_3.

3.10 Neutralization

We have seen what acids and bases are, and we are now ready to look at the reaction that takes place between them. The reaction between an acid and a base is called a **neutralization reaction.** The general form of a neutralization reaction is

$$\text{acid } + \text{ base} \longrightarrow \text{salt } + \text{ water or other products}$$

Water is a product when the base is a hydroxide. Other products may include gases (Section 3.13). In chemistry, the term **salt** means any ionic compound that can be formed by the neutralization of an acid with a base. The name is taken from ordinary table salt, sodium chloride, the ionic product of the reaction between hydrochloric acid and sodium hydroxide:

$$\underset{\text{Acid}}{HCl(aq)} + \underset{\text{Base}}{NaOH(aq)} \longrightarrow \underset{\text{Salt}}{NaCl(aq)} + \underset{\text{Water}}{H_2O(l)}$$

The colors of the symbols show that the cation of the salt is provided by the base and the anion is provided by the acid. Another example is the reaction between nitric acid and magnesium hydroxide, which produces magnesium nitrate:

$$\underset{\text{Acid}}{2\,HNO_3(aq)} + \underset{\text{Base}}{Mg(OH)_2(s)} \longrightarrow \underset{\text{Salt}}{Mg(NO_3)_2(aq)} + \underset{\text{Water}}{2\,H_2O(l)}$$

The magnesium nitrate remains in solution as Mg^{2+} and NO_3^- ions. However, if a salt is insoluble, it precipitates as soon as it is formed.

When an acid has two acidic protons, as does sulfuric acid, H_2SO_4, we can write separate equations for each of the two stages of neutralization. The first stage is

$$H_2SO_4(aq) + NaOH(aq) \longrightarrow NaHSO_4(aq) + H_2O(l)$$

Table 3.3	*Classification of electrolytes*

Strong electrolytes

soluble ionic compounds
strong acids
strong bases

Weak electrolytes

weak acids
weak bases

Nonelectrolytes

molecular compounds other than

The second stage is

$$NaHSO_4(aq) + NaOH(aq) \longrightarrow Na_2SO_4(s) + H_2O(l)$$

However, because neutralization is normally not considered complete until all the acidic protons have reacted, we would normally write the equation for the overall reaction obtained by adding together these two equations:

$$H_2SO_4(aq) + 2\,NaOH(aq) \longrightarrow Na_2SO_4(s) + 2\,H_2O(l)$$

In a neutralization reaction, an acid reacts with a base to produce a salt and water.

FIGURE 3.22

Insoluble barium sulfate precipitates when sulfuric acid is added to an aqueous solution of barium hydroxide.

Example 3.5 *Predicting the outcome of a neutralization reaction*

What salt is produced by the reaction between aqueous solutions of barium hydroxide and sulfuric acid? Write the chemical equation for the reaction.

Strategy The base provides the cation and the acid provides the anion. If the salt formed is insoluble in water, it will precipitate. Therefore, check the solubility of the salt in the solubility rules in Table 3.1 before writing the chemical equation. Because the base is a hydroxide, the other product is water.

Solution The base is barium hydroxide, so we expect a barium salt to form. Because the acid is sulfuric acid, we obtain barium sulfate, $BaSO_4$. This salt is insoluble (Table 3.1), so we expect it to precipitate (Fig. 3.22). The chemical equation for the reaction is

$$\underset{\text{Acid}}{H_2SO_4(aq)} + \underset{\text{Base}}{Ba(OH)_2(aq)} \longrightarrow \underset{\text{Salt}}{BaSO_4(s)} + \underset{\text{Water}}{2\,H_2O(l)}$$

Self-Test 3.8A What acid and base solutions could you use to prepare rubidium nitrate? Write the chemical equation for the neutralization.

[*Answer:* $HNO_3(aq) + RbOH(aq) \rightarrow RbNO_3(aq) + H_2O(l)$]

Self-Test 3.8B Write the chemical equation for a neutralization reaction in which calcium phosphate is produced.

3.11 Deeper Insight: Proton Transfer

We have seen how to focus on the essentials of a precipitation reaction by writing the net ionic reaction. Let's see if the same procedure clarifies the nature of a neutralization reaction. The complete ionic equation for the neutralization reaction between nitric acid and magnesium hydroxide described in Section 3.10 is

$$2\,H^+(aq) + 2\,NO_3^-(aq) + Mg^{2+}(aq) + 2\,OH^-(aq) \longrightarrow$$
$$Mg^{2+}(aq) + 2\,NO_3^-(aq) + 2\,H_2O(l)$$

The ions common to both sides cancel, and the net ionic equation of this reaction is therefore

$$2\,H^+(aq) + 2\,OH^-(aq) \longrightarrow 2\,H_2O(l)$$

This equation simplifies (when we multiply through by $\frac{1}{2}$) to

$$H^+(aq) + OH^-(aq) \longrightarrow H_2O(l)$$

We see that *the net outcome of a neutralization reaction between a strong acid and a strong base is the formation of water from hydrogen ions and hydroxide ions.*

Because the hydrogen ion is present in water as a hydronium atom, we can add H_2O to each side of the last equation and obtain

$$H_3O^+(aq) + OH^-(aq) \longrightarrow 2\,H_2O(l)$$

This equation shows that, in a neutralization reaction between a strong acid and a strong base, the essential step is *the transfer of a proton from a hydronium ion to a hydroxide ion.* Later in the text (Chapter 15), we shall see that *the characteristic feature of an acid-base reaction is the transfer of a proton from one species to another.*

The concept of proton transfer helps us understand the neutralization reactions of weak acids and bases. The dominant species in solution is the nonionized form of the weak acid or base (CH_3COOH or NH_3, for instance), so we use these species to write the net ionic equation. For example, the complete ionic equation for the neutralization of the weak acid HCN by the strong base NaOH is

$$HCN(aq) + Na^+(aq) + OH^-(aq) \longrightarrow Na^+(aq) + CN^-(aq) + H_2O(l)$$

After cancellation of the sodium ions, the net ionic equation is

$$HCN(aq) + OH^-(aq) \longrightarrow CN^-(aq) + H_2O(l)$$

Notice the characteristic feature of this reaction: the acid is neutralized when a proton is transferred from the acid molecule to the base (the hydroxide ion). Similarly, because the complete ionic equation for the neutralization of the weak base ammonia by the strong acid HCl is

$$H^+(aq) + Cl^-(aq) + NH_3(aq) \longrightarrow NH_4^+(aq) + Cl^-(aq)$$

the net ionic equation is

$$H^+(aq) + NH_3(aq) \longrightarrow NH_4^+(aq)$$

This reaction, in which the base NH_3 is neutralized by the acid, can also be recognized as a proton transfer reaction if we add H_2O to both sides:

$$H_3O^+(aq) + NH_3(aq) \longrightarrow NH_4^+(aq) + H_2O(l)$$

In reactions between acids and bases, a proton is transferred from one species to another.

Example 3.6 *Writing the net ionic equation for the reaction involving a weak acid or base*

Write the net ionic equation for the following neutralization reactions in aqueous solution. (a) Hydrofluoric acid reacts with calcium hydroxide. (b) The weak base methylamine, CH_3NH_2, reacts with nitric acid.

Strategy Follow the procedure for writing net ionic equations in Toolbox 3.2. The products of a neutralization reaction include a salt and, in the case of a hydroxide, water. Because weak acids and bases are ionized to only a small extent, represent them as intact molecules rather than as ions. Represent an insoluble hydroxide as the solid. If Table 3.1 indicates that the salt formed in the neutralization reaction is insoluble, write its formula as a solid.

Solution (a) *Step 1.* The chemical equation for the reaction is

$$2\,HF(aq) + Ca(OH)_2(aq) \longrightarrow CaF_2(aq) + 2\,H_2O(l)$$
$$\text{Acid} \qquad\qquad \text{Base} \qquad\qquad \text{Salt} \qquad\quad \text{Water}$$

Step 2. The salt formed is soluble, so the complete ionic equation is

$$2\,HF(aq) + Ca^{2+}(aq) + 2\,OH^-(aq) \longrightarrow Ca^{2+}(aq) + 2\,F^-(aq) + 2\,H_2O(l)$$

Step 3. The only spectator ions are the calcium ions, so the net ionic equation is

$$2\,HF(aq) + 2\,OH^-(aq) \longrightarrow 2\,F^-(aq) + 2\,H_2O(l)$$

(b) *Step 1.* The chemical equation for the reaction is

$$HNO_3(aq) + CH_3NH_2(aq) \longrightarrow CH_3NH_3{}^+(aq) + NO_3{}^-(aq)$$

Step 2. Nitric acid is a strong acid and the salt formed, $CH_3NH_3NO_3$, is soluble, so the complete ionic equation is

$$H^+(aq) + NO_3{}^-(aq) + CH_3NH_2(aq) \longrightarrow CH_3NH_3{}^+(aq) + NO_3{}^-(aq)$$

Step 3. The only spectator ions are the nitrate ions, so the net ionic equation is

$$H^+(aq) + CH_3NH_2(aq) \longrightarrow CH_3NH_3{}^+(aq)$$

This reaction can be represented as a proton transfer reaction by adding H_2O to both sides:

$$H_3O^+(aq) + CH_3NH_2(aq) \longrightarrow CH_3NH_3{}^+(aq) + H_2O(l)$$

Self-Test 3.9A Write the net ionic equation for the neutralization of aqueous acetic acid, $CH_3COOH(aq)$, with aqueous potassium hydroxide.

[*Answer:* $CH_3COOH(aq) + OH^-(aq) \rightarrow CH_3CO_2{}^-(aq) + H_2O(l)$]

Self-Test 3.9B Write the net ionic equation for the neutralization reaction of ammonia with the weak acid HCN in aqueous solution.

3.12 Acidic and Basic Character in the Periodic Table

When the oxides of the elements dissolve in water, they commonly form solutions that are either acidic or basic. The position of a main-group element in the periodic table can be used to predict whether its oxide is likely to form acidic or basic solutions.

Most soluble *metal* oxides are strong *bases* in water. We say that most metals have basic oxides. A **basic oxide** is an oxide that reacts with water to form a base (Section 3.9). Barium is a main-group metal, so we expect its oxide, BaO, to be basic. Conversely, many *nonmetal* oxides react with water to give *acids:*

$$CO_2(g) + H_2O(l) \longrightarrow H_2CO_3(aq)$$

$$SO_2(g) + H_2O(l) \longrightarrow H_2SO_3(aq)$$

$$P_4O_{10}(s) + 6\,H_2O(l) \longrightarrow 4\,H_3PO_4(aq)$$

Metals, remember, are found to the left of the table and nonmetals are found to the right.

The transition metal oxides vary in basic character, and some (such as CrO_3) may actually be acidic.

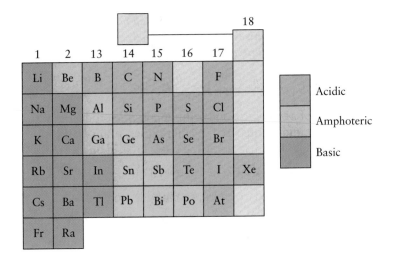

FIGURE 3.23

The locations in the main groups of the periodic table of elements that form acidic, amphoteric, and basic oxides. The diagonal band of amphoteric oxides closely matches the diagonal band of metalloids.

We say that nonmetals have acidic oxides. An **acidic oxide** is an oxide that reacts with water to form an acid. Selenium, for instance, is a nonmetal, so we expect its oxides (SeO_2 and SeO_3) to be acidic. The normal acidity of rain is due to dissolved carbon dioxide. The acidic oxides sulfur dioxide, sulfur trioxide, and various oxides of nitrogen in the atmosphere are responsible for the enhanced acidity of rainwater downwind of industrial centers. These oxides dissolve in water to produce a mixture of acids, sulfuric acid and nitric acid among them, that increases the normal acidity of rain. The strong acids produced are responsible for "acid rain," which eats away at the limestone of buildings, damages plants, and threatens delicate ecosystems.

Acidic oxides react with bases and basic oxides react with acids. For instance, magnesium oxide, a basic oxide, reacts with hydrochloric acid:

$$MgO(s) + 2\,HCl(aq) \longrightarrow MgCl_2(aq) + H_2O(l)$$

Carbon dioxide, an acidic oxide, reacts with sodium hydroxide:

$$2\,NaOH(aq) + CO_2(g) \longrightarrow Na_2CO_3(aq) + H_2O(l)$$

The white crust often seen on pellets of sodium hydroxide is a mixture of sodium carbonate and sodium hydrogen carbonate, $NaHCO_3$, formed in this way.

Recall from Fig. 1.19 that there is a diagonal frontier in the periodic table lying between the metals and nonmetals. On this frontier, metallic character blends into nonmetallic character. The oxides of the elements on this frontier therefore have both acidic and basic character and are classified as **amphoteric,** from the Greek word for "both" (Fig. 3.23). For example, aluminum oxide, Al_2O_3, is amphoteric. It reacts with acids, a reaction showing that it is basic:

$$Al_2O_3(s) + 6\,HCl(aq) \longrightarrow 2\,AlCl_3(aq) + 3\,H_2O(l)$$

Aluminum oxide also reacts with bases, a reaction showing that it is acidic as well:

$$2\,NaOH(aq) + Al_2O_3(s) + 3\,H_2O(l) \longrightarrow 2\,Na[Al(OH)_4](aq)$$

The product in this case is sodium aluminate, a compound that contains the aluminate ion, $Al(OH)_4^-$.

Most metallic elements form basic oxides and all nonmetallic elements form acidic oxides. The elements lying in a diagonal line in the periodic table from beryllium to polonium form amphoteric oxides.

3.13 Gas-Formation Reactions

In some reactions of acids with ionic compounds, one product is a gas. A number of gases are commonly prepared in laboratories in this way. One example is the production of hydrogen halides by the action of acid on metal halides:

$$2\,NaCl(s) \,+\, H_2SO_4(l) \longrightarrow Na_2SO_4(s) \,+\, 2\,HCl(g)$$

This reaction is used to prepare hydrogen chloride from table salt. In this reaction, each Cl^- ion in the solid picks up one hydrogen ion from a H_2SO_4 molecule and escapes as the gas HCl.

Another example is the preparation of hydrogen sulfide, H_2S, by the action of a dilute acid on iron(II) sulfide:

$$FeS(s) \,+\, 2\,HCl(aq) \longrightarrow FeCl_2(aq) \,+\, H_2S(g)$$

Hydrogen sulfide is an intensely poisonous gas and has an odor associated with rotten eggs, because, in fact, as eggs decompose, they give off small amounts of hydrogen sulfide.

The common feature of these reactions is the transfer of a hydrogen ion to the anion of the salt. Therefore these reactions are classified as neutralization reactions or proton transfer reactions. The proton transfer is more easily seen in the net ionic equation. The complete ionic equation for the reaction is

$$FeS(s) \,+\, 2\,H^+(aq) \,+\, 2\,Cl^-(aq) \longrightarrow Fe^{2+}(aq) \,+\, 2\,Cl^-(aq) \,+\, H_2S(g)$$

and the net ionic equation is

$$FeS(s) \,+\, 2\,H^+(aq) \longrightarrow Fe^{2+}(aq) \,+\, H_2S(g)$$

In terms of hydronium ions,

$$FeS(s) \,+\, 2\,H_3O^+(aq) \longrightarrow Fe^{2+}(aq) \,+\, H_2S(g) \,+\, 2\,H_2O(l)$$

We see that each sulfide ion, S^{2-}, in the solid reactant picks up two hydrogen ions from the hydronium ions and escapes as the gas H_2S.

FIGURE 3.24

(a) An acetylene miner's lamp. Impure calcium carbide is a dirty white solid formed from limestone and carbon. When water is poured on it, ethyne (acetylene) gas is formed. (b) Acetylene generates light because it burns with a brilliant white, hot flame in another type of reaction, a redox reaction.

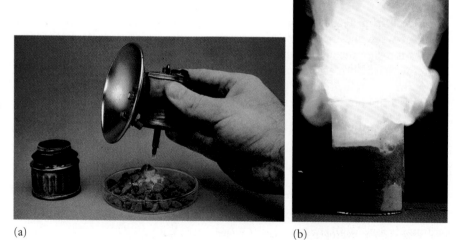

(a)

(b)

Even water can produce a gas by acting as an acid. An example is the reaction of water and calcium carbide, CaC_2, which contains the carbide ion, C_2^{2-}. The reaction produces ethyne, C_2H_2 (**10**), which is commonly called acetylene. It is a highly flammable gas that is used in welding because of the very high temperature (about 3000°C) of its flame. The chemical equation for the formation of acetylene is

$$CaC_2(s) + 2 H_2O(l) \longrightarrow Ca^{2+}(aq) + 2 OH^-(aq) + C_2H_2(g)$$

We see that protons have been transferred from water molecules to the carbide ions. This reaction was once used in automobile and miner's lamps (Fig. 3.24).

In some gas-formation reactions, the hydrogen compound is not itself a gas but decomposes into one as soon as it is formed. One example is the action of dilute sulfuric acid on calcium carbonate, which results in the formation of carbon dioxide (Fig. 3.25):

$$CaCO_3(s) + H_2SO_4(aq) \longrightarrow CaSO_4(s) + H_2O(l) + CO_2(g)$$

The same reaction occurs when acid rain falls on limestone and marble, two forms of calcium carbonate. The action of the acid is, first, to convert the carbonate ions into carbonic acid, H_2CO_3, which then decomposes into carbon dioxide and water:

$$CaCO_3(s) + H_2SO_4(aq) \longrightarrow CaSO_4(s) + H_2CO_3(aq)$$
$$H_2CO_3(aq) \longrightarrow H_2O(l) + CO_2(g)$$

The reaction of acids with salts is a proton transfer reaction that may produce a gas or a compound that decomposes into a gas.

10 Ethyne, C_2H_2

Notice that there are two ions commonly known by the name *carbide*: the monatomic ion C^{4-} and the polyatomic ion C_2^{2-}. When you encounter the name "carbide," check to see which is meant.

Example 3.7 *Devising a means of preparing a gas*

Suggest a method for making the gas phosphine, PH_3. Phosphine is a highly toxic gas that should be made only in a fume hood.

Strategy Because phosphine is a gas, we may be able to devise a reaction in which it is produced by the action of an acid on a suitable salt. To identify a salt we might use as a starting material, we examine the formula of phosphine, remove hydrogen ions (these will be supplied by the acid), and identify a compound that contains the resulting anion.

Solution Removal of three hydrogen ions from phosphine leaves the phosphide ion, P^{3-}. Because any phosphide contains this ion, we can expect to produce phosphine by the action of acid on sodium phosphide:

$$Na_3P(s) + 3 HCl(aq) \longrightarrow 3 NaCl(aq) + PH_3(g)$$

This is, in fact, a method for the production of phosphine.

Self-Test 3.10A Devise a method for producing hydrogen iodide that uses phosphoric acid.

[*Answer:* $3 NaI(s) + H_3PO_4(aq) \rightarrow Na_3PO_4(aq) + 3 HI(g)$]

Self-Test 3.10B Write the chemical equation for the reaction of hydrochloric acid with sodium hydrogen carbonate in aqueous solution.

FIGURE 3.25

Carbon dioxide is formed when a solution of an acid is dropped onto the surface of limestone. This reaction is a convenient test for limestone.

REDOX REACTIONS

This equation summarizes the overall outcome of many individual reactions taking place in a complicated sequence inside green leaves.

One of the most important chemical reactions in the world is photosynthesis, in which carbon dioxide is removed from the atmosphere and converted into carbohydrates:

$$6\,CO_2(g) \,+\, 6\,H_2O(l) \longrightarrow C_6H_{12}O_6(s, glucose) \,+\, 6\,O_2(g)$$

Another very important reaction in technological societies is the combustion of methane, the principal component of natural gas:

$$CH_4(g) \,+\, 2\,O_2(g) \longrightarrow CO_2(g) \,+\, 2\,H_2O(l)$$

At first sight, there seems to be little in common between these two reactions and even less between them and the reaction between magnesium and oxygen:

$$2\,Mg(s) \,+\, O_2(g) \longrightarrow 2\,MgO(s)$$

the reaction of magnesium with chlorine:

$$Mg(s) \,+\, Cl_2(g) \longrightarrow MgCl_2(s)$$

and the reaction that occurs when dilute acid is poured on zinc:

$$Zn(s) \,+\, 2\,HCl(aq) \longrightarrow ZnCl_2(aq) \,+\, H_2(g)$$

Nevertheless, all these reactions *do* belong to the same family. They are all examples of *oxidation-reduction reactions*, which are commonly called *redox reactions*. This is the third and last major class of chemical reactions discussed in this chapter.

3.14 Oxidation and Reduction

To find the common feature of redox reactions, consider the reaction between magnesium and oxygen to produce magnesium oxide (Fig. 3.26). For a long time, chemists have called reaction with oxygen "oxidation" and have referred to this specific reaction as the oxidation of magnesium. Today the term *oxidation* is given a much broader meaning. First, note that, in the course of the reaction, the Mg atoms in the solid magnesium lose electrons to form Mg^{2+} ions and the O atoms in the molecular oxygen, O_2, gain electrons to form O^{2-} ions:

$$2\,Mg(s) \,+\, O_2(g) \longrightarrow 2\,Mg^{2+}(s) \,+\, 2\,O^{2-}(s) \qquad (as\ 2\,MgO(s))$$

Two electrons have been transferred from each magnesium atom to each oxygen atom. *The essential step in the reaction is the transfer of electrons from one species to another.* In the reaction between magnesium and chlorine, the overall outcome is also a transfer of electrons—in this case, from Mg atoms to Cl atoms:

$$Mg(s) \,+\, Cl_2(g) \longrightarrow Mg^{2+}(s) \,+\, 2\,Cl^-(s) \qquad (as\ MgCl_2(s))$$

In the reaction of magnesium with oxygen, the magnesium atoms lose electrons. The same is true of the reaction of magnesium with chlorine. Chemists now define **oxidation** as the loss of electrons, even if no oxygen is involved (Fig. 3.27).

The increase in positive charge of an iron ion from Fe^{2+} to Fe^{3+} shows that a negative charge—an electron—has been removed. The Fe^{2+} ion has been "oxidized." The increase in charge of a species always indicates that it has undergone oxidation, that it has been oxidized. This rule also applies to the conversion of an anion to a neutral species, as in the conversion of chloride ions (Cl^-, charge -1) to chlorine gas (Cl_2, charge 0).

FIGURE 3.26

An example of an oxidation reaction: magnesium burning brightly in air. Magnesium also burns brightly in water and in carbon dioxide; consequently, magnesium fires are very difficult to extinguish.

Now let's consider reduction. The name *reduction* originally referred to extraction of a metal from its oxide, often by reaction with hydrogen, carbon, or carbon monoxide. One example is

$$Fe_2O_3(s) + 3 H_2(g) \xrightarrow{\Delta} 2 Fe(l) + 3 H_2O(g)$$

In the manufacture of steel, the reduction is brought about by using carbon monoxide in place of hydrogen:

$$Fe_2O_3(s) + 3 CO(g) \xrightarrow{\Delta} 2 Fe(l) + 3 CO_2(g)$$

What is the feature common to both reactions?

In both reactions of iron(III) oxide, the Fe^{3+} ions present in Fe_2O_3 are converted into Fe atoms. For this conversion to take place, the Fe^{3+} ions must gain electrons to neutralize their positive charges. The common process in both reactions is electron gain by the Fe^{3+} ions: **reduction** is electron gain.

Another reduction is the conversion of Ni^{2+} to Ni, in which the charge decreases from +2 to 0. There is a gain of electrons (to cancel the positive charge), so the nickel ions have been reduced. The same rule applies when the final charge is negative. For example, when bromine is converted to bromide ions, the charge decreases from 0 (in Br_2) to −1 (in Br^-). The conversion of Br_2 to Br^- is therefore a reduction.

Every reduction reaction must also involve an oxidation and every oxidation reaction must also involve a reduction. Neither reduction nor oxidation can take place alone. For example, in the oxidation of magnesium by chlorine from Mg to Mg^{2+}, chlorine is reduced from Cl_2 to Cl^-.

Oxidation is electron loss; reduction is electron gain.

Self-Test 3.11A Identify the species that have been oxidized or reduced in the reaction $3 Ag^+(aq) + Al(s) \rightarrow 3 Ag(s) + Al^{3+}(aq)$.

[***Answer:*** Al(s) is oxidized, $Ag^+(aq)$ is reduced.]

Self-Test 3.11B Identify the species that have been oxidized or reduced in the reaction $2 Cu^+(aq) + I_2(s) \rightarrow 2 Cu^{2+}(aq) + 2 I^-(aq)$.

3.15 Keeping Track of Electrons: Oxidation Numbers

The change in charge, and the corresponding loss or gain of electrons, is easy to identify when we are dealing with monatomic ions. The conversion of Mg to Mg^{2+} is obviously oxidation (electron loss) and the conversion of Fe^{3+} to Fe is obviously reduction (electron gain). To extend the procedure to molecules and polyatomic ions, chemists have found a way of keeping track of electrons by assigning an **oxidation number** to each element. The oxidation number is defined so that

An *increase* in oxidation number indicates oxidation.

A *decrease* in oxidation number indicates reduction.

The oxidation number of an element in a monatomic ion is the same as its charge. So, the oxidation number of magnesium is +2 when it is present as Mg^{2+} ions, and the oxidation number of chlorine is −1 when it is present as Cl^- ions. The common oxidation numbers of the monatomic ions are shown in Fig. 3.28. *The oxidation number of the elemental form of an element is 0;* so magnesium metal has oxidation number 0 and chlorine in the form of Cl_2 molecules also has oxidation number 0. With these definitions in mind, we can see that the conversion of Mg to Mg^{2+} ions is an oxidation:

FIGURE 3.27

When bromine is poured on red phosphorus, a vigorous reaction takes place. In the reaction phosphorus is oxidized and bromine is reduced.

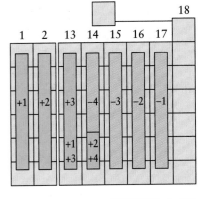

FIGURE 3.28

The common oxidation numbers of main-group elements. Notice the tendency of elements in the same group to assume the same oxidation number.

$$Mg \xrightarrow{\text{remove 2e−}} Mg^{2+}$$

Oxidation number: 0   +2   an increase, signifying oxidation

The conversion of Cl_2 to Cl^- ions is a reduction:

$$Cl_2 \xrightarrow{\text{gain 2e−}} 2\,Cl^-$$

Oxidation number: 0   −1   a decrease, signifying reduction

We use the term **oxidation state** to refer to the actual condition of a species with a specified oxidation number. So, Mg^{2+} is an oxidation state of magnesium with oxidation number $+2$, and Cl^- is an oxidation state of chlorine with oxidation number -1.

When an element is part of a compound, we assign its oxidation number by using the procedure in Toolbox 3.3. According to these rules, an increase in the number of oxygen atoms in an electrically neutral binary compound (for instance, from SO_2 to SO_3) results in an increase in oxidation number of the remaining element and therefore corresponds to an oxidation. However, the overall charge also affects the oxidation number. For example, in the conversion of CO_2 to CO_3^{2-} the oxidation number of carbon does not change, because, although an O atom has been added (indicating an oxidation), the O atom is accompanied by two electrons (indicating a reduction), and the changes cancel.

See Section 8.9 for a deeper understanding of oxidation numbers.

Oxidation increases the oxidation number of an element; reduction decreases the oxidation number.

Example 3.8 *Assigning an oxidation number*

Assign an oxidation number to sulfur in (a) SO_2 and (b) SO_4^{2-}.

Strategy It is not immediately clear whether sulfur is more highly oxidized in sulfur dioxide or in the sulfate ion. Although the ion has more oxygen atoms, it also has a higher negative charge. To find the oxidation numbers, represent the oxidation number of sulfur by x. Then use the rules in Toolbox 3.3 and solve for x.

Solution (a) By rule 2, the sum of oxidation numbers of the atoms in the compound must be 0:

$$[\text{Oxidation number of S}] + [2 \times (\text{oxidation number of O})] = 0$$

That is,

$$x + [2 \times (-2)] = 0$$

S   2 O   Zero charge on neutral molecule

Therefore, the oxidation number of sulfur is $x = +4$. (b) By rule 2, the sum of oxidation numbers of the atoms in the ion is -2, so

$$x + [4 \times (-2)] = -2$$

S   4 O   Total charge on ion

Therefore, $x = +6$. We can conclude that sulfur is more highly oxidized in the sulfate ion than in sulfur dioxide.

Self-Test 3.12A Find the oxidation numbers of sulfur, phosphorus, and nitrogen in (a) H_2S; (b) PO_4^{3-}; (c) NO_3^-.

[*Answer:* (a) -2; (b) $+5$; (c) $+5$]

Self-Test 3.12B Find the oxidation numbers of sulfur, nitrogen, and chlorine in (a) SO_3^{2-}; (b) NO_2^-; (c) $HClO_3$.

Toolbox 3.3 *How to assign oxidation numbers*

This Toolbox explains how to assign oxidation numbers to each atom in a molecule or ion.

Conceptual Basis

If the molecule or ion contains oxygen we imagine it as consisting of oxide ions and cations; the cations are assigned sufficient charge to account for the overall charge of the species. If oxygen is not present, another nonmetallic element such as a halogen is chosen to act as a hypothetical anion and the charges of the other atoms adjusted similarly. The oxidation number of each element is then its "charge" (Fig. 3.29).

Procedure

To assign an oxidation number to an element, we start with two simple rules:

   1. The oxidation number of an element uncombined with other elements is 0.

   2. The sum of the oxidation numbers of all the atoms in a species is equal to its total charge.

The first of these rules tells us that hydrogen, oxygen, iron, and all other elements have oxidation numbers of 0 in their elemental forms. The second rule implies that the oxidation number of K^+ is $+1$, that of Al^{3+} is $+3$, and that of Br^- is -1 but that of F in F_2 is 0.

We assign the oxidation numbers of elements in the compounds included in this part of the text by using these two rules in conjunction with the following specific values:

The oxidation number of hydrogen is $+1$ in combination with nonmetals and -1 in combination with metals.

The oxidation number of oxygen is -2 in most of its compounds.

The oxidation number of all the halogens is -1 in all their compounds, unless the halogen is in combination with oxygen or another halogen higher in Group 17.

The oxidation number of fluorine is -1 in all its compounds.

One common exception for oxygen is the peroxide ion, O_2^{2-}, in which oxygen has an oxidation number of -1.

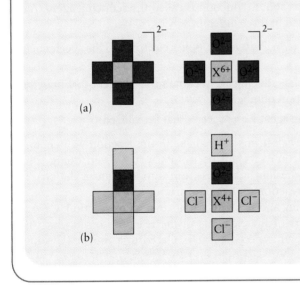

(a)

(b)

FIGURE 3.29

How to determine an oxidation number. Each atom is imagined to be a separate ion. Certain ions are assigned charges by using the rules in Toolbox 3.3, and the charge on the central atom is determined by considering the overall charge on the species. (a) Oxide "ions" in an oxoanion are given the charge of -2; because there are four oxygen atoms and the overall charge on the anion is -2, the charge on the central atom must be $+6$. (b) This molecule has three chlorine atoms with oxidation numbers of -1, an oxygen atom (-2), and a hydrogen atom $(+1)$. The sum of these oxidation numbers is -4 and the overall charge on the molecule is 0. Thus, the central atom must have an oxidation number of $+4$.

3.16 Oxidizing and Reducing Agents

We have seen that the oxidation or reduction of an atom corresponds respectively to an increase or a decrease in its oxidation number. A **redox reaction** is a reaction in which oxidation numbers change. We can verify that the combustion of methane—one of the reactions that we mentioned earlier—is a redox reaction because the oxidation number of oxygen changes from 0 in O_2 to -2 in both CO_2 and H_2O. In fact, all combustion reactions are redox reactions. We can also now see that photosynthesis is a redox reaction, because the oxidation number of some of the oxygen atoms changes from 0 to -2 and that of the carbon atoms changes from $+4$ to 0:

$$6\,CO_2(g) \;+\; 6\,H_2O(l) \longrightarrow C_6H_{12}O_6(s) \;+\; 6\,O_2(g)$$

Oxidation number: $+4\ -2$ $+1\ -2$ $0\ \ +1\ -2$ 0

This analysis of the reaction shows very clearly that oxidation numbers of elements in covalent molecules must not be regarded as true ionic charges: carbon atoms are not really present as C^{4+} ions in carbon dioxide molecules! Oxidation numbers are only a guide to the kind of changes going on in molecules.

The species that causes oxidation in a redox reaction is called the **oxidizing agent.** Oxidizing agents can be elements, ions, or compounds. We can describe the function of oxidizing agents in redox reactions in three different ways:

Electron transfer: The oxidizing agent removes electrons from the species being oxidized.

Oxidation numbers: The oxidizing agent contains an element that undergoes a decrease in oxidation number.

Reaction: The oxidizing agent is the species that is reduced.

For instance, chlorine removes electrons from magnesium in the reaction $Mg(s) + Cl_2(g) \rightarrow MgCl_2(s)$ and forms Mg^{2+} and Cl^- ions. Because chlorine accepts those electrons, its oxidation number decreases from 0 to -1 (a reduction). Chlorine is therefore the oxidizing agent in this reaction.

The species that brings about reduction is called the **reducing agent.** Again the process can be described in three different ways:

Electron transfer: The reducing agent supplies electrons to the species being reduced.

Oxidation numbers: The reducing agent contains an element that undergoes an increase in oxidation number.

Reaction: The reducing agent is the species that is oxidized.

Magnesium metal supplies electrons to chlorine in the reaction written above, thereby causing the reduction of chlorine from Cl_2 to Cl^-. In doing so, the oxidation number of magnesium increases from 0 to $+2$ (an oxidation). It is the reducing agent in the reaction of magnesium and chlorine.

To identify the oxidizing and reducing agents in a redox reaction, we examine the oxidation numbers of the elements in the reaction to see which have changed. For example, when a piece of zinc is placed in a copper(II) solution (Fig. 3.30), the reaction is

$$Zn(s) \;+\; Cu^{2+}(aq) \longrightarrow Zn^{2+}(aq) \;+\; Cu(s)$$

Oxidation number: 0 $+2$ $+2$ 0

We see that the oxidation number of zinc changes from 0 to $+2$ (an oxidation) whereas that of copper changes from $+2$ to 0 (a reduction). Therefore, because zinc is oxidized, zinc metal is the reducing agent in this reaction. Conversely, because copper is reduced, we can regard the copper(II) ions as the oxidizing agent.

The hydrogen ions provided by acids can also act as oxidizing agents. That is their role when hydrochloric acid is poured onto zinc:

$$Zn(s) \;+\; 2\,H^+(aq) \longrightarrow Zn^{2+}(aq) \;+\; H_2(g)$$

Oxidation number: 0 $+1$ $+2$ 0

(a)

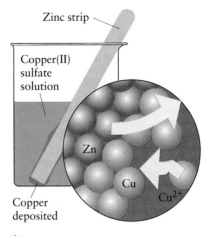

Zinc strip

Copper(II) sulfate solution

Zn

Cu

Cu^{2+}

Copper deposited

(b)

M

FIGURE 3.30

When a strip of zinc is placed in a solution that contains Cu^{2+} ions, the blue solution slowly becomes colorless and copper metal is deposited on the zinc. In this redox reaction, the zinc metal is reducing the Cu^{2+} ions to copper and the Cu^{2+} ions are oxidizing the zinc metal to Zn^{2+} ions. (a) The reaction. (b) A visualization of the process.

We see that the oxidation number of hydrogen changes from $+1$ to 0 (a reduction), a change indicating that hydrogen is the oxidizing agent in this reaction.

Chemists sometimes say that an entire species (an atom or an entire molecule or polyatomic ion) is oxidized if it contains an element that undergoes oxidation (that is, undergoes an increase in oxidation number). Similarly, they say that an entire species is reduced if it contains an element that undergoes reduction (that is, undergoes a decrease in oxidation number). For example, some oxoanions, such as NO_3^- and MnO_4^-, are powerful oxidizing agents, particularly in acidic solution. When copper metal is added to concentrated nitric acid, the product includes nitrogen dioxide (Fig. 3.31):

$$Cu(s) + 4 H^+(aq) + 2 NO_3^-(aq) \longrightarrow Cu^{2+}(aq) + 2 NO_2(g) + 2 H_2O(l)$$

An acidic solution is a solution that contains an acid (usually a strong acid).

(a) 140 mL ±5% 120 100 80 60 40 20

(b) 140 mL ±5% 120 100 80 60 40 20

FIGURE 3.31

(a) Copper reacts slowly with dilute nitric acid to give blue Cu^{2+} ions and the colorless gas nitric oxide, NO. (b) When copper reacts with concentrated nitric acid, nitrogen dioxide, NO_2, is produced instead of NO. The blue solution is turned green by this brown gas.

Because the oxidation number of copper has increased from 0 to $+2$, copper has been oxidized; it is therefore the reducing agent in this reaction. The oxidation number of nitrogen has been reduced from $+5$ in NO_3^- to $+4$ in NO_2, so nitric acid has been reduced: it is therefore the oxidizing agent. Sulfuric acid is also an oxidizing agent. The hot, concentrated acid oxidizes copper, and some of the sulfate ions are reduced to sulfur dioxide:

$$Cu(s) + 2\,H_2SO_4(aq, conc.) \xrightarrow{\Delta} Cu^{2+}(aq) + SO_4^{2-}(aq) + SO_2(g) + 2\,H_2O(l)$$

An oxidizing agent is a species that contains an element that undergoes reduction and in doing so causes oxidation; a reducing agent is a species that contains an element that undergoes oxidation as it causes reduction. Certain oxoanions can act as oxidizing agents, particularly in acidic solution.

Example 3.9 *Identifying oxidizing agents and reducing agents*

Identify the oxidizing agent and the reducing agent in the following net ionic equation:

$$Cr_2O_7^{2-}(aq) + 6\,Fe^{2+}(aq) + 14\,H^+(aq) \longrightarrow 6\,Fe^{3+}(aq) + 2\,Cr^{3+}(aq) + 7\,H_2O(l)$$

Strategy First, determine the oxidation numbers of each of the elements. Oxidation is signaled by an increase in oxidation number; reduction is signaled by a decrease in oxidation number. The oxidizing agent is the species that, in the course of carrying out the oxidation, is itself reduced.

Solution The oxidation numbers of the elements are

$$Cr_2O_7^{2-}(aq) + 6\,Fe^{2+}(aq) + 14\,H^+(aq) \longrightarrow 6\,Fe^{3+}(aq) + 2\,Cr^{3+}(aq) + 7\,H_2O(l)$$
$$\;\;{\scriptstyle +6\;\;-2}\qquad\quad{\scriptstyle +2}\qquad\qquad{\scriptstyle +1}\qquad\qquad{\scriptstyle +3}\qquad\qquad{\scriptstyle +3}\qquad\quad{\scriptstyle +1\;\;-2}$$

The oxidation number of oxygen remains -2 and that of hydrogen remains $+1$. Therefore, we need be concerned only with chromium and iron. The oxidation number, x, of chromium in the $Cr_2O_7^{2-}$ ion is calculated from

$$2x + [7 \times (-2)] = -2$$
$$\underset{\text{2 Cr}}{}\quad\underset{\text{7 O}}{}\quad\underset{\text{Total charge on ion}}{}$$

It follows that $x = +6$. In the reaction, the oxidation number of chromium decreases from $+6$ to $+3$. The dichromate ion therefore undergoes reduction, so it is the oxidizing agent. At the same time, the oxidation number of iron increases from $+2$ to $+3$, so it undergoes oxidation. Fe^{2+} is the reducing agent.

Self-Test 3.13A In the Claus process for the recovery of sulfur from natural gas and petroleum, hydrogen sulfide, H_2S, reacts with sulfur dioxide, SO_2, to form elemental sulfur and water: $2\,H_2S(g) + SO_2(g) \rightarrow 3\,S(s) + 2\,H_2O(l)$. Identify the oxidizing agent and the reducing agent.

[*Answer:* H_2S is the reducing agent; SO_2 is the oxidizing agent.]

Self-Test 3.13B When sulfuric acid reacts with sodium iodide, sodium iodate and sulfur dioxide are produced. Identify the oxidizing and reducing agents in this reaction. *Note:* You need not write the chemical equation, only the formulas for the reactants and products.

3.17 Balancing Simple Redox Equations

Electrons are neither created nor destroyed in chemical reactions; they can only be transferred from one species to another. Therefore, when one species is oxidized, another species taking part in the reaction must be reduced. Indeed, *all* the electrons lost in an oxidation step must be gained by a reduction step. Because electrons are charged, we must write the chemical equation for a redox reaction so that the total charge of the reactants is equal to the total charge of the products.

Consider the equation for the oxidation of copper metal to copper(II) ions by silver ions:

$$Cu(s) + Ag^+(aq) \longrightarrow Cu^{2+}(aq) + Ag(s)$$

At first glance, the equation appears to be balanced, because we see the same number of each kind of atom on each side. However, each copper atom has lost two electrons, whereas each silver atom has gained only one. To balance the electrons, we have to balance the charge by adding a coefficient of 2 for the Ag^+ ion and therefore also for the Ag atom:

$$Cu(s) + 2\,Ag^+(aq) \longrightarrow Cu^{2+}(aq) + 2\,Ag(s)$$

Now all the electrons lost by copper have been gained by silver. Everything is in balance.

Example 3.10 *Writing a net ionic equation for a redox reaction*

Aluminum metal reacts with hydrogen ions in aqueous hydrochloric acid solution to produce hydrogen gas and aluminum ions (Fig. 3.32). Write the net ionic equation for the reaction.

Strategy Write the skeletal equation and balance the elements as described in Toolboxes 3.1 and 3.2. Then balance the charges. In this step, check each side of the equation to ensure that the element balance is not upset.

Solution The skeletal equation (with the physical states included) is

$$Al(s) + H^+(aq) \longrightarrow Al^{3+}(aq) + H_2(g)$$

Now we balance the elements, disregarding charges at this stage:

$$Al(s) + 2\,H^+(aq) \longrightarrow Al^{3+}(aq) + H_2(g)$$

Finally, balance the charge. Currently, there are two positive charges on the left and three on the right. We can obtain six positive charges on each side by increasing the number of H^+ ions to six by multiplying both $2\,H^+$ and H_2 by 3, to give

$$Al(s) + 6\,H^+(aq) \longrightarrow Al^{3+}(aq) + 3\,H_2(g)$$

and then multiplying Al and Al^{3+} by 2, to give

$$2\,Al(s) + 6\,H^+(aq) \longrightarrow 2\,Al^{3+}(aq) + 3\,H_2(g)$$

This is the fully balanced equation.

Self-Test 3.14A When tin is placed in contact with a solution of Fe^{3+} ions, it reduces the iron to iron(II) and is itself oxidized to tin(II) ions. Write the net ionic equation for the reaction.

[*Answer:* $Sn(s) + 2\,Fe^{3+}(aq) \rightarrow Sn^{2+}(aq) + 2\,Fe^{2+}(aq)$]

Self-Test 3.14B Cerium(IV) ions oxidize iodide ions to iodine and are themselves reduced to cerium(III) ions. Write the net ionic equation for the reaction.

FIGURE 3.32

Aluminum reacts vigorously with hydrochloric acid to form soluble aluminum chloride and water.

Applying Chemistry: *Case Study 3*

On Earth, our atmosphere is kept at a relatively constant composition by natural processes, such as photosynthesis and the movement of air. However, a submarine or space ship must use artificial air purifiers and circulators. Otherwise, the crew would be poisoned and suffocated by the carbon dioxide, CO_2, they produce. In such sealed environments, waste gases are removed and oxygen regenerated by a variety of proton transfer and redox chemical reactions.

Because the space shuttle serves only for short flights, its oxygen can be stored on board and need not be replaced, but the CO_2 produced must be removed. To extract carbon dioxide from the air, chemists use the fact that it is an acidic oxide and reacts with bases (see Section 3.12). Canisters on the space shuttle are filled with solid lithium hydroxide, which reacts with the carbon dioxide:

$$CO_2(g) + 2\,LiOH(s) \longrightarrow Li_2CO_3(s) + H_2O(l)$$

Lithium hydroxide is preferred to the hydroxides of the other alkali metals because of its low molar mass. A by-product of this reaction is water, which is a valuable commodity in space.

Astronauts on the space shuttle must change the canisters of lithium hydroxide daily to prevent the accumula-

Astronauts on the space shuttle must change the canisters of lithium hydroxide daily. Here, Sidney Gutierrez carries out the task. Two canisters are used, and one is changed every 12 hours so that the capacity to remove carbon dioxide remains stable.

tion of carbon dioxide in the air. Two canisters are used and one is changed every 12 hours so that the capacity to remove carbon dioxide is never diminished.

A problem with the lithium canisters is that they limit the length of the space flight. Once the LiOH has reacted, the canisters are useless to the crew. A new system that can be regenerated was tested on the space shuttle *Columbia* in April 1998. In the Regenerative Carbon Dioxide Removal System (RCRS), air flows across pellets containing a substance that bonds to the carbon dioxide and removes it from the air. When the pellets have all reacted, the carbon dioxide is released by heating the pellets while they are exposed to the vacuum of space. They can then be used to absorb more carbon dioxide.

During extended missions in a shuttle or on board a space station, a crew might have to regenerate oxygen, O_2, directly from CO_2. Because the oxidation number of oxygen is -2 in CO_2 but 0 in O_2, an oxidizing agent must be used. On the Russian *Salyut* space station, potassium superoxide, KO_2, reacted with the carbon dioxide:

$$4\,KO_2(s) + 2\,CO_2(g) \longrightarrow 2\,K_2CO_3(s) + 3\,O_2(g)$$

In this redox reaction, the oxidation numbers of oxygen have changed from $-\frac{1}{2}$ in O_2^- and -2 in CO_2 to -2 in CO_3^{2-} and 0 in O_2. However, like the LiOH canisters, once all the KO_2 has reacted, it has to be replaced.

Very long space flights need processes that recycle all the oxygen. For example, carbon dioxide can react with hydrogen to form carbon and water in a series of redox reactions for which the net equation is

$$CO_2(g) + 2\,H_2(g) \longrightarrow C(s) + 2\,H_2O(l)$$

Each element can be recovered and reused. The carbon can be used to filter cabin odors, and hydrogen and oxygen gases can be regenerated in another redox reaction by passing an electric current through the water:

$$2\,H_2O(l) \xrightarrow{\text{electric current}} 2\,H_2(g) + O_2(g)$$

Such processes require energy; but the energy can usually be supplied with solar panels, so it is renewable as well.

Key Concepts: proton transfer reactions, redox reactions, acidic oxides

For Further Reading

J. E. Oberg and A. R. Oberg, "Life cycles," in *Pioneering Space*, pp. 78–92, New York: McGraw-Hill, 1986.

Related Exercises: 3.80–3.84

Simple redox reactions involving two elements and the combustion reactions of hydrocarbons can usually be balanced by the procedure in Toolbox 3.1, adjusting coefficients, if necessary, to balance charge. However, some redox reactions have complex equations that require special balancing procedures. These procedures are introduced in Chapter 18.

When balancing the chemical equation for a redox reaction involving ions, the total charge on each side must be balanced.

3.18 Classifying Reactions

We can now look back at the complex life story of carbon dioxide, which introduced this chapter, and see that the carbon cycle illustrates the three types of reactions discussed in this chapter. The combustion reactions that convert organic materials into carbon dioxide are redox reactions. So is the important reaction of photosynthesis, in which carbon dioxide is converted into carbohydrates. Shellfish use precipitation reactions to create their shells. The reaction in which carbon dioxide forms carbonic acid is a proton transfer reaction in which hydrogen ions are transferred. The attack of acid rain on limestone and marble, another result of the transfer of hydrogen ions, results in the formation of carbon dioxide gas. The rich web of carbon's activities in the world are largely the interplay of these three types of reactions—precipitation, proton transfer, and redox.

The same themes will recur throughout the book, because many of the chemical properties of the elements and their compounds can be discussed in terms of the same three fundamental types of reactions. For example, both proton transfer and redox reactions are used to purify air in closed environments (Applying Chemistry: Case Study 3).

One of the skills a chemist needs to develop is the ability to predict the products of a reaction. To make a prediction for a given set of reactants, we first need to know what type of reaction is likely to take place. We can identify a reaction as belonging to one of the three types described in this chapter by looking for the characteristics of each reaction type, as described in Toolbox 3.4. We can then predict its products, also shown in Toolbox 3.4.

Reactions can be classified by identifying the characteristic features of a type of reaction.

Example 3.11 *Classifying a reaction*

Write and balance the equation for each of the following reactions and identify the reaction type: (a) the formation of solid potassium chloride and oxygen gas by heating solid potassium chlorate with a catalyst; (b) the reaction of hydrochloric acid and potassium sulfide in aqueous solution to form hydrogen sulfide gas and aqueous potassium chloride. For reactions in aqueous solution, write the net ionic equation.

Strategy (a) This reaction may be a redox reaction, as a pure element is a product. (b) The reactants are an acid and a sulfide that react to form a gas, so we expect a gas-formation reaction. Verify the predictions by following the procedure in Toolbox 3.4.

Solution (a) The chemical equation for the reaction is

$$KClO_3(s) \xrightarrow{\Delta} KCl(s) + O_2(g) \qquad \triangle$$
$${\scriptstyle +5 \ -2} \phantom{(s) \xrightarrow{\Delta} KCl}{\scriptstyle -1} {\scriptstyle 0}$$

Chlorine and oxygen undergo changes in oxidation number, so this reaction is a redox reaction. Balance oxygen by giving $KClO_3$ a coefficient of 2 and O_2 a coefficient of 3. Finally, give KCl a coefficient of 2 to balance potassium and chlorine. Check to see whether the number of electrons lost equals the number gained.

$$2\,KClO_3(s) \xrightarrow{\Delta} 2\,KCl(s) + 3\,O_2(g)$$

The two chlorine atoms have each gained 6 electrons, for a total of 12. Each of the six oxygen atoms has lost 2 electrons, for a total of 12; therefore, electrons balance. (b) The chemical equation for the reaction is

$$2\,HCl(aq) + K_2S(aq) \longrightarrow H_2S(g) + 2\,KCl(aq)$$

None of the elements undergoes a change in oxidation number, so this reaction is not a redox reaction. There is a transfer of H^+ to S^{2-} that results in a gas, so this is a gas-formation reaction. The net ionic equation is

$$2\,H^+(aq) + S^{2-}(aq) \longrightarrow H_2S(g)$$

Toolbox 3.4 *How to classify reactions and predict their products*

This Toolbox summarizes the characteristic features of three fundamental types of chemical reactions and the products they form.

Conceptual Basis

Each of the three types of reaction described in this chapter has distinguishing characteristics (Fig. 3.33):

A *precipitation reaction* occurs when two soluble salts exchange ions. The product is an insoluble ionic solid.

A *neutralization reaction* occurs when a hydrogen ion (a proton) is transferred from an acid to a base. The products are a salt and water (if the base is a hydroxide). A proton transfer reaction in which one product is a gas is an example of a *gas-formation reaction*.

A *redox reaction* occurs when an electron is transferred from one species to another. The products include species in which the elements have different oxidation numbers from those of the elements in reactant species.

Procedure

To classify a reaction and predict its products (Fig. 3.34):

Step 1. Write the chemical equation for the reaction. Check the oxidation numbers of each element. If there is a change, the reaction is a redox reaction. The product of the reaction of a metal with a nonmetal is typically an ionic solid. The products of combustion reactions of organic compounds are $CO_2(g)$ and $H_2O(l)$. If nitrogen is present in the compound, N_2 is also formed.

Step 2. Check for the transfer of a hydrogen ion, H^+, from an acid to a base. Such a reaction is a neutralization reaction. The products are a salt and (if the base is a hydroxide) water. The acid provides the anion and the base provides the cation of the salt.

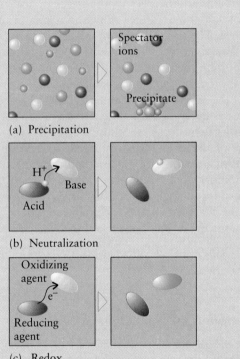

(a) Precipitation

(b) Neutralization

(c) Redox

FIGURE 3.33

The three main types of chemical reactions discussed in this chapter can be distinguished by the type of change taking place. (a) In a precipitation reaction, ions mix and one combination of ions is insoluble. (b) In a neutralization reaction, hydrogen ions are transferred from an acid to a base. (c) In a redox reaction, electrons are transferred from a reducing agent to an oxidizing agent.

Step 3. Look for the transfer of a hydrogen ion, H^+, from an acid to an anion that may form a gas, such as CO_3^{2-} (which forms CO_2) or S^{2-} (which forms H_2S). Such a proton transfer reaction is sometimes called a gas-formation reaction.

Step 4. Check for the formation of a solid ionic compound from two soluble compounds in aqueous solution. Such a reaction is a precipitation reaction. The products are the ionic compounds resulting when ions are exchanged between the two reactants. Check Table 3.1 to see whether either combination is insoluble.

Neutralization reactions that result in the formation of an insoluble salt may also be described as precipitations.

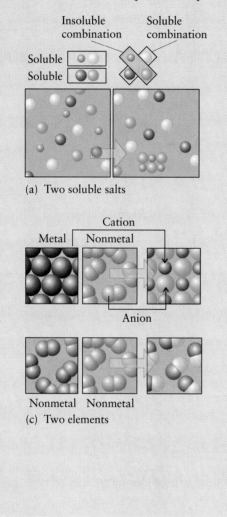

(a) Two soluble salts

(c) Two elements

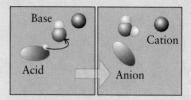

(b) Acid and hydroxide

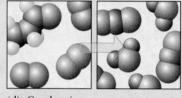

(d) Combustion

FIGURE 3.34

We can predict the products of a reaction by examining the reactants. (a) Two soluble salts may form a precipitate. (b) An acid and a hydroxide react to form a salt and water. (c) When two elements react, a redox reaction generally occurs. A metal and nonmetal react to form an ionic compound and two nonmetals react to form a molecular compound. (d) In combustion reactions, organic compounds react with oxygen to form carbon dioxide and water.

Self-Test 3.15A Write and balance the equation for the reaction in which concentrated aqueous copper(II) sulfate and aqueous silver nitrate solution are mixed and solid silver sulfate is produced, and identify the reaction type.

> [*Answer:* $CuSO_4(aq) + 2 AgNO_3(aq) \rightarrow Ag_2SO_4(s) + Cu(NO_3)_2(aq)$; precipitation]

Self-Test 3.15B Write and balance the equation for the reaction in which ammonia gas reacts with solid copper(II) oxide to form nitrogen gas, copper metal, and liquid water, and identify the reaction type.

Example 3.12 *Predicting the products of a reaction*

Predict the products of the following reactions and write their net ionic equations: (a) $Na_2S(aq) + Cu(NO_3)_2(aq) \rightarrow$ products; (b) $HClO_4(aq) + Co(OH)_3(s) \rightarrow$ products.

Strategy (a) The reactants are two soluble ionic compounds, so we suspect a precipitation reaction (or no reaction if the product is soluble). (b) The reactants include an acid and a base, so we suspect a neutralization reaction. Check the solubility of the salt produced to see whether it is a precipitation as well. Because the base is a hydroxide, water is also produced. Verify the predictions by following the procedure in Toolbox 3.4.

Solution (a) The products will be the compounds obtained by exchanging the ions of $NaNO_3$ and CuS. From Table 3.1, we know that CuS is insoluble, so this is a precipitation reaction. The chemical equation is therefore

$$Na_2S(aq) + Cu(NO_3)_2(aq) \longrightarrow 2\,NaNO_3(aq) + CuS(s)$$

The net ionic equation is

$$Cu^{2+}(aq) + S^{2-}(aq) \longrightarrow CuS(s)$$

(b) The salt produced in the neutralization reaction is cobalt(III) perchlorate, which is soluble, so the products formed are the soluble salt and water:

$$3\,HClO_4(aq) + Co(OH)_3(s) \longrightarrow Co(ClO_4)_3(aq) + 3\,H_2O(l)$$

Perchloric acid is a strong acid and the salt formed is soluble, so the complete ionic equation is

$$3\,H^+(aq) + 3\,ClO_4^{-}(aq) + Co(OH)_3(s) \longrightarrow Co^{3+}(aq) + 3\,ClO_4^{-}(aq) + 3\,H_2O(l)$$

The only spectator ions are the perchlorate ions, so the net ionic equation is

$$3\,H^+(aq) + Co(OH)_3(s) \longrightarrow Co^{3+}(aq) + 3\,H_2O(l)$$

This reaction can also be written as a proton transfer reaction by adding $3\,H_2O$ to both sides:

$$3\,H_3O^+(aq) + Co(OH)_3(s) \longrightarrow Co^{3+}(aq) + 6\,H_2O(l)$$

Self-Test 3.16A (a) Predict the products and write the chemical equation for the reaction $2\,Na(s) + Cl_2(g) \rightarrow$ products. (b) Identify the reaction type.

[***Answer:*** (a) $2\,Na(s) + Cl_2(g) \rightarrow 2\,NaCl(s)$; (b) redox]

Self-Test 3.16B (a) Predict the products and write the net ionic equation for the reaction $Na_2CO_3(aq) + HBr(aq) \rightarrow$ products. (b) Identify the reaction type.

Skills You Should Have Mastered

Conceptual

- [ ] 1. Explain the significance of a balanced chemical equation, Sections 3.1–3.2.
- [ ] 2. Identify substances as electrolytes or nonelectrolytes, Self-Test 3.7.
- [ ] 3. Explain the difference between solutions of strong and weak acids and bases, Sections 3.8–3.9.
- [ ] 4. Define oxidation and reduction in terms of oxidation number and electron transfer, Sections 3.14–3.15.
- [ ] 5. Identify the oxidizing and reducing agents in a reaction, Example 3.9.
- [ ] 6. Identify a reaction as precipitation, neutralization, or redox, Toolbox 3.4 and Example 3.11.

Problem-Solving

- [ ] 1. Write, balance, and label a chemical equation, given the information in a sentence, Toolbox 3.1 and Examples 3.1 and 3.2.
- [ ] 2. Construct the balanced complete ionic and net ionic equations for reactions involving ions, Toolbox 3.2 and Example 3.3.
- [ ] 3. Use the solubility rules to select appropriate solutions that, when mixed, will produce a desired precipitate, Example 3.4.

- [ ] 4. Predict the outcome of a neutralization reaction and write the net ionic equation, Examples 3.5 and 3.6.
- [ ] 5. Determine the oxidation number of an element in an ion or compound, Toolbox 3.3 and Example 3.8.
- [ ] 6. Write the net ionic equation for a redox reaction, Example 3.10.
- [ ] 7. Predict the products of a reaction, Toolbox 3.4 and Example 3.12.

Descriptive

- [ ] 1. Write three balanced chemical equations for reactions that occur in the carbon cycle and describe their roles in the cycle, Section 3.18.
- [ ] 2. Describe the chemical properties of acids and bases, Section 3.7.
- [ ] 3. Recognize common Arrhenius acids and bases from their formulas, Self-Test 3.5.
- [ ] 4. Identify an oxide as acidic or basic from the position of the element in the periodic table, Section 3.12.
- [ ] 5. Describe how a gas is formed by reaction of an acid with an ionic compound and devise a means of preparing a gas in this way, Example 3.7.

Exercises

Balancing Equations

3.1 The shaded squares below represent atoms of different elements. Connected squares represent molecules. Balance the equation represented below by using more of the same symbols.

3.2 The shaded squares below represent atoms of different elements. Connected squares represent molecules. Balance the equation represented below by using more of the same symbols.

3.3 Balance the following skeletal chemical equations:
(a) $P_4O_{10}(s) + H_2O(l) \rightarrow H_3PO_4(l)$
(b) $Cd(NO_3)_2(aq) + Na_2S(aq) \rightarrow CdS(s) + NaNO_3(aq)$
(c) $KClO_3(s) \xrightarrow{\Delta} KClO_4(s) + KCl(s)$
(d) $HCl(aq) + Ca(OH)_2(aq) \rightarrow CaCl_2(aq) + H_2O(l)$

3.4 Balance the following skeletal chemical equations:
(a) $Al(s) + H_2SO_4(aq) \rightarrow Al_2(SO_4)_3(aq) + H_2(g)$
(b) $Pb(NO_3)_2(aq) + Na_3PO_4(aq) \rightarrow Pb_3(PO_4)_2(s) + NaNO_3(aq)$
(c) $KClO_3(s) \xrightarrow{\Delta} KCl(s) + O_2(g)$
(d) $H_3PO_4(aq) + Na_2CO_3(aq) \rightarrow$
$$Na_3PO_4(aq) + CO_2(g) + H_2O(l)$$

3.5 Write balanced chemical equations for the following reactions: (a) Sodium metal reacts with water to produce hydrogen gas and sodium hydroxide. (b) The reaction of sodium oxide, Na_2O, and water produces sodium hydroxide.

(c) Hot lithium metal reacts in a nitrogen atmosphere to produce lithium nitride, Li_3N. (d) The reaction of calcium metal with water leads to the evolution of hydrogen gas and the formation of calcium hydroxide, $Ca(OH)_2$.

3.6 Write balanced chemical equations for the following reactions: (a) One process by which nickel is recovered from nickel(II) sulfide ores is to heat the ore in air. During this "roasting" process, molecular oxygen reacts with the nickel(II) sulfide to produce solid nickel(II) oxide and sulfur dioxide gas. (b) The diamondlike abrasive silicon carbide, SiC, is made by reacting silicon dioxide with elemental carbon at 2000°C to produce silicon carbide and carbon monoxide. (c) The reaction of elemental hydrogen and nitrogen is used for the commercial production of ammonia in the Haber process. (d) Magnesium metal reacts with solid boron oxide, B_2O_3, to form solid elemental boron and solid magnesium oxide.

3.7 In one stage in the commercial production of iron metal in a blast furnace, the iron(III) oxide, Fe_2O_3, reacts with carbon monoxide to form Fe_3O_4 and carbon dioxide. In a second stage, the Fe_3O_4 reacts further with carbon monoxide to produce elemental iron and carbon dioxide. Write the balanced equation for each stage in the process.

3.8 One of the roles of stratospheric ozone, O_3, is to remove damaging ultraviolet radiation from sunlight. One result is the eventual dissociation of ozone into molecular oxygen. Write the balanced equation for the reaction.

3.9 When nitrogen and oxygen react in the cylinder of an automobile engine, nitrogen monoxide is formed. After it escapes into the atmosphere with the other exhaust gases, the nitrogen monoxide reacts with molecular oxygen to produce nitrogen dioxide, one of the precursors of acid rain. Write the two balanced equations for the reactions leading to the formation of nitrogen dioxide.

3.10 The reaction of boron trifluoride, $BF_3(g)$, with sodium borohydride, $NaBH_4(s)$, leads to the formation of sodium tetrafluoroborate, $NaBF_4(s)$, and diborane gas, $B_2H_6(g)$. The diborane reacts with the oxygen in air, forming boron oxide, $B_2O_3(s)$, and water. Write the two balanced equations leading to the formation of boron oxide.

3.11 Hydrofluoric acid is used to etch grooves in glass because it reacts with the silica, $SiO_2(s)$, in glass. The products of the reaction are silicon tetrafluoride and water. Write the balanced equation for the reaction.

3.12 The compound $Sb_4O_5Cl_2(s)$, which has been investigated because of its interesting electrical properties, can be prepared by warming a mixture of antimony(III) oxide and antimony(III) chloride, both of which are solids. Write the balanced equation for the reaction.

Precipitation Reactions

3.13 Explain how a net ionic equation is constructed.

3.14 Explain what is meant by a spectator ion. What are the spectator ions in (a) a precipitation reaction; (b) a neutralization reaction?

3.15 Use the information in Table 3.1 to classify the following ionic compounds as soluble or insoluble in water: (a) lead(II) nitrate, $Pb(NO_3)_2$; (b) lead(II) chloride, $PbCl_2$; (c) silver nitrate, $AgNO_3$; (d) sodium sulfate, Na_2SO_4.

3.16 Use the information in Table 3.1 to classify the following ionic compounds as soluble or insoluble in water: (a) zinc acetate, $Zn(CH_3CO_2)_2$; (b) iron(III) chloride, $FeCl_3$; (c) silver chloride, $AgCl$; (d) copper(II) hydroxide, $Cu(OH)_2$.

3.17 Write the formulas of the principal species present in aqueous solutions of (a) NaI; (b) Ag_2CO_3; (c) $(NH_4)_3PO_4$; (d) $FeSO_4$.

3.18 Write the formulas of the principal species present in aqueous solutions of (a) $PbSO_4$; (b) K_2CO_3; (c) K_2CrO_4; (d) Hg_2Cl_2.

3.19 When the solution in beaker 1 is mixed with the solution in beaker 2, a precipitate forms. Write the net ionic equation describing the formation of the precipitate and then identify the spectator ions.

Beaker 1	Beaker 2
(a) $FeCl_3(aq)$	$NaOH(aq)$
(b) $AgNO_3(aq)$	$KI(aq)$
(c) $Pb(NO_3)_2(aq)$	$K_2SO_4(aq)$
(d) $Na_2CrO_4(aq)$	$Pb(NO_3)_2(aq)$
(e) $Hg_2(NO_3)_2(aq)$	$K_2SO_4(aq)$

3.20 The contents of beaker 1 are mixed with those of beaker 2. If a reaction occurs, write the net ionic equation and indicate the spectator ions.

Beaker 1	Beaker 2
(a) $Fe_2(SO_4)_3(aq)$	$KOH(aq)$
(b) $K_3PO_4(aq)$	$CuCl_2(aq)$
(c) $K_2S(aq)$	$AgNO_3(aq)$
(d) $NiSO_4(aq)$	$(NH_4)_2CO_3(aq)$
(e) Na_2SO_4	$Ca(OH)_2(aq)$

3.21 The five procedures below each result in the formation of a precipitate. For each reaction, write the chemical equations describing the formation of the precipitate: the overall equation, the complete ionic equation, and the net ionic equation. Identify the spectator ions. (a) $(NH_4)_2CrO_4(aq)$ is mixed with $BaCl_2(aq)$. (b) $CuSO_4(aq)$ is mixed with $Na_2S(aq)$. (c) $FeCl_2(aq)$ is mixed with $(NH_4)_3PO_4(aq)$. (d) potassium oxalate, $K_2C_2O_4(aq)$, is mixed with $Ca(NO_3)_2(aq)$. (e) $NiSO_4(aq)$ is mixed with $Ba(NO_3)_2(aq)$.

3.22 The five procedures below each result in the formation of a precipitate. For each reaction, write the chemical equations describing the formation of the precipitate: the overall equation, the complete ionic equation, and the net ionic equation. Identify the spectator ions. (a) $AgNO_3(aq)$ is mixed with $Na_2CO_3(aq)$. (b) $Pb(NO_3)_2(aq)$ is mixed with $KI(aq)$. (c) $Ba(OH)_2(aq)$ is mixed with $K_2SO_4(aq)$. (d) $(NH_4)_2S(aq)$ is mixed with $Cd(NO_3)_2(aq)$. (e) $KOH(aq)$ is mixed with $CuCl_2(aq)$.

3.23 Write the complete ionic equation and the net ionic equation corresponding to each of the following reactions:
(a) $Pb(ClO_4)_2(aq) + 2\,NaBr(aq) \rightarrow PbBr_2(s) + 2\,NaClO_4(aq)$
(b) $AgNO_3(aq) + NH_4Cl(aq) \rightarrow AgCl(s) + NH_4NO_3(aq)$
(c) $2\,NaOH(aq) + Cu(NO_3)_2(aq) \rightarrow$
$$Cu(OH)_2(s) + 2\,NaNO_3(aq)$$

3.24 Write the complete ionic equation and the net ionic equation corresponding to each of the following reactions:
(a) $3\,BaCl_2(aq) + 2\,K_3PO_4(aq) \rightarrow Ba_3(PO_4)_2(s) + 6\,KCl(aq)$
(b) $Hg_2(NO_3)_2(aq) + 2\,KCl(aq) \rightarrow Hg_2Cl_2(s) + 2\,KNO_3(aq)$
(c) $Mg(C_2H_3O_2)_2(aq) + Ba(OH)_2(aq) \rightarrow$
$$Mg(OH)_2(s) + Ba(C_2H_3O_2)_2(aq)$$

3.25 Write the net ionic equation for the formation of each insoluble product in aqueous solution: (a) lead(II) sulfate, $PbSO_4$, a precipitate formed in a lead-acid battery; (b) copper(II) sulfide, CuS; (c) cobalt(II) carbonate, $CoCO_3$. (d) Select two soluble ionic compounds that, when mixed in solution, form each of the above insoluble compounds. Identify the spectator ions.

3.26 Write the net ionic equation for the formation of each of the following insoluble products in aqueous solution: (a) lead(II) carbonate, $PbCO_3$, the white pigment in putty; (b) aluminum hydroxide, $Al(OH)_3$; (c) zinc chromate, $ZnCrO_4$. (d) Select two soluble ionic compounds that, when mixed in solution, form each of the above insoluble compounds. Identify the spectator ions.

3.27 You are given a solution and asked to analyze it for the cations Ag^+, Ca^{2+}, and Zn^{2+}. You add hydrochloric acid and a white precipitate forms. You filter out the solid and add sulfuric acid to the solution. Nothing appears to happen. Then you add hydrogen sulfide. A black precipitate forms. Which ions should you report as present in your solution?

3.28 You are given a solution and asked to analyze it for the cations Ag^+, Ca^{2+}, and Hg^{2+}. You add hydrochloric acid. Nothing appears to happen. You then add sulfuric acid and a white precipitate forms. You filter out the solid and add hydrogen sulfide to the solution. A black precipitate forms. Which ions should you report as present in your solution?

Acids and Bases

3.29 Give the Arrhenius definitions for an acid and a base in aqueous solution.

3.30 Explain how to choose an acid and a base to prepare a specified salt.

3.31 Identify each of the following as either an acid or a base: (a) $NH_3(aq)$; (b) $HCl(aq)$; (c) $NaOH(aq)$; (d) $H_2SO_4(aq)$; (e) $Ba(OH)_2(aq)$.

3.32 Classify each of the following as either an acid or a base: (a) $HNO_3(aq)$; (b) $CH_3NH_2(aq)$, a derivative of ammonia; (c) $CH_3COOH(aq)$; (d) $KOH(aq)$; (e) $HClO_4(aq)$.

3.33 Complete and write the overall equation, the complete ionic equation, and the net ionic equation for the following neutralization reactions. If the substance is a weak acid or base, use its molecular form when writing the equations.
(a) $HCl(aq) + NaOH(aq) \rightarrow$
(b) $NH_3(aq) + HNO_3(aq) \rightarrow$
(c) $CH_3NH_2(aq) + HI(aq) \rightarrow$

3.34 Complete and write the overall equation, the complete ionic equation, and the net ionic equation for the following neutralization reactions. If the substance is a weak acid or base, use its molecular form when writing the equations.
(a) $H_2SO_4(aq) + KOH(aq) \rightarrow$
(b) $Ba(OH)_2(aq) + HCN(aq) \rightarrow$
(c) $NH_3(aq) + HClO_4(aq) \rightarrow$

3.35 Select an acid and a base for a neutralization reaction that results in the formation of (a) potassium bromide, KBr; (b) barium nitrite, $Ba(NO_2)_2$; (c) calcium cyanide, $Ca(CN)_2$; (d) potassium phosphate, K_3PO_4. Write the overall and net ionic equations for each reaction.

3.36 Determine the salt that is produced from the neutralization reaction between (a) potassium hydroxide and acetic acid, CH_3COOH; (b) ammonia, NH_3, and hydroiodic acid; (c) barium hydroxide and sulfuric acid, H_2SO_4 (both H atoms react); (d) sodium hydroxide and hydrocyanic acid, HCN. Write the full ionic equation for each reaction.

3.37 Identify the acid and the base in the following reactions:
(a) $CH_3NH_2(aq) + H_3O^+(aq) \rightarrow CH_3NH_3^+(aq) + H_2O(l)$
(b) $C_2H_5NH_2(aq) + HCl(aq) \rightarrow C_2H_5NH_3^+(aq) + Cl^-(aq)$
(c) $CaO(s) + 2\,HI(aq) \rightarrow CaI_2(aq) + H_2O(l)$

3.38 Identify the acid and the base in the following reactions:
(a) $CH_3COOH(aq) + NH_3(aq) \rightarrow NH_4^+(aq) + CH_3CO_2^-(aq)$
(b) $(CH_3)_3N(aq) + HCl(aq) \rightarrow (CH_3)_3NH^+(aq) + Cl^-(aq)$
(c) $O^{2-}(aq) + H_2O(l) \rightarrow 2\,OH^-(aq)$

3.39 Use the periodic table to determine which oxides form acidic solutions in water and which form basic solutions: (a) CaO; (b) SO_3; (c) N_2O_3; (d) Tl_2O.

3.40 Use the periodic table to determine which oxides form acidic solutions in water and which form basic solutions: (a) P_2O_5; (b) Na_2O; (c) CO_2; (d) MgO.

Redox Reactions

3.41 Define the terms *oxidation* and *reduction* in terms of electron transfer.

3.42 How can we tell from a chemical equation whether a reaction is a redox reaction?

3.43 Write the balanced equations for the following skeletal redox reactions:
(a) $P(s) + Br_2(l) \rightarrow PBr_3(s)$
(b) $Fe^{2+}(aq) + Sn^{4+}(aq) \rightarrow Fe^{3+}(aq) + Sn^{2+}(aq)$
(c) $H_2(g) + S_8(s) \rightarrow H_2S(g)$
(d) $NO(g) + O_2(g) \rightarrow NO_2(g)$

3.44 Write the balanced equations for the following skeletal redox equations:
(a) $Fe(s) + H_2O(l) + O_2(g) \rightarrow Fe(OH)_2(s)$
(b) $KNO_3(s) \rightarrow KNO_2(s) + O_2(g)$
(c) $Al(s) + Cu(NO_3)_2(aq) \rightarrow Cu(s) + Al(NO_3)_3(aq)$
(d) $Na(s) + H_2O(l) \rightarrow NaOH(aq) + H_2(g)$

3.45 Write the balanced equations for the following redox reactions:

(a) displacement of copper(II) ions from solution by magnesium metal:

$$Mg(s) + Cu^{2+}(aq) \longrightarrow Mg^{2+}(aq) + Cu(s)$$

(b) formation of iron(III) ions in the following reaction:

$$Fe^{2+}(aq) + Pb^{4+}(aq) \longrightarrow Fe^{3+}(aq) + Pb^{2+}(aq)$$

(c) synthesis of hydrogen chloride from its elements:

$$H_2(g) + Cl_2(g) \longrightarrow HCl(g)$$

(d) formation of rust (a simplified equation):

$$Fe(s) + O_2(g) \longrightarrow Fe_2O_3(s)$$

3.46 Write the balanced equations for the following redox equations:

(a) displacement of silver ions from solution by copper metal:

$$Ag^+(aq) + Cu(s) \longrightarrow Cu^{2+}(aq) + Ag(s)$$

(b) production of titanium metal by magnesium metal:

$$TiCl_4(g) + Mg(l) \longrightarrow MgCl_2(s) + Ti(s)$$

(c) production of copper metal by the smelting of copper(II) sulfide:

$$CuS(s) + O_2(g) \xrightarrow{\Delta} Cu(s) + SO_2(g)$$

(d) industrial production of elemental bromine from brine:

$$Cl_2(g) + Br^-(aq) \longrightarrow Br_2(l) + Cl^-(aq)$$

Oxidation Numbers

3.47 What is meant by the oxidation number of an element?

3.48 Define oxidation and reduction in terms of oxidation numbers.

3.49 Determine the oxidation number of the italicized element in each compound: (a) XeF_4; (b) $HClO$; (c) NO; (d) HNO_3; (e) SO_2; (f) H_2S.

3.50 Determine the oxidation number of the italicized element in each compound: (a) H_2SO_3; (b) B_2O_3; (c) NH_3; (d) N_2O_3; (e) SO_3; (f) H_3PO_3.

3.51 Identify the oxidation number of the italicized element in each ion: (a) MnO_4^-; (b) $S_2O_3^{2-}$; (c) SO_4^{2-}; (d) MnO_4^{2-}; (e) $Cr_2O_7^{2-}$.

3.52 Identify the oxidation number of the italicized element in each ion: (a) IO_3^-; (b) CrO_4^{2-}; (c) VO^{2+}; (d) BrO_4^-; (e) IO_2^-.

3.53 Indicate which of the following reactions are redox reactions. For the redox reactions, identify the substance oxidized and the substance reduced by the change in oxidation numbers.
(a) $CH_3Br(aq) + OH^-(aq) \rightarrow CH_3OH(aq) + Br^-(aq)$
(b) $BrO_3^-(aq) + 5\,Br^-(aq) + 6\,H^+(aq) \rightarrow$
$$3\,Br_2(l) + 3\,H_2O(l)$$
(c) $2\,F_2(g) + 2\,H_2O(l) \rightarrow 4\,HF(aq) + O_2(g)$

3.54 In each of the following reactions, use oxidation numbers to identify the substance oxidized and the substance reduced.

(a) production of iodine from seawater:

$$Cl_2(g) + 2\,I^-(aq) \longrightarrow I_2(aq) + 2\,Cl^-(aq)$$

(b) reaction to prepare bleach:

$$Cl_2(g) + 2\,NaOH(aq) \longrightarrow NaCl(aq) + NaOCl(aq) + H_2O(l)$$

(c) reaction that destroys ozone in the stratosphere:

$$NO(g) + O_3(g) \longrightarrow NO_2(g) + O_2(g)$$

Oxidizing and Reducing Agents

3.55 Explain why the definitions of oxidizing agent and reducing agent in terms of oxidation numbers are consistent with the definitions in terms of electron transfer.

3.56 Explain how you can identify oxidizing and reducing agents from a chemical equation.

3.57 Which do you expect to be the stronger oxidizing agent? Explain your reasoning. (a) Cl_2 or Cl^-; (b) N_2O or N_2O_5.

3.58 Which do you expect to be the stronger oxidizing agent? Explain your reasoning. (a) $KBrO$ or $KBrO_3$; (b) MnO_4^- or Mn^{2+}.

3.59 Identify the oxidizing agent and the reducing agent in each of the following reactions: (a) $Zn(s) + 2\,HCl(aq) \rightarrow ZnCl_2(aq) + H_2(g)$, a simple means of preparing H_2 gas in the laboratory. (b) $2\,H_2S(g) + SO_2(g) \rightarrow 3\,S(s) + 2\,H_2O(l)$, a reaction used to produce sulfur from hydrogen sulfide, the "sour gas" in natural gas. (c) $B_2O_3(s) + 3\,Mg(s) \rightarrow 2\,B(s) + 3\,MgO(s)$, a preparation of elemental boron.

3.60 Identify the oxidizing agent and the reducing agent in each of the following reactions: (a) $2\,Al(s) + Cr_2O_3(s) \rightarrow Al_2O_3(s) + 2\,Cr(s)$, an example of a thermite reaction used to obtain some metals from their ores. (b) $6\,Li(s) + N_2(g) \rightarrow 2\,Li_3N(s)$, a reaction that shows the similarity of lithium and magnesium. (c) $2\,Ca_3(PO_4)_2(s) + 6\,SiO_2(s) + 10\,C(s) \rightarrow P_4(g) + 6\,CaSiO_3(s) + 10\,CO(g)$, a reaction for the preparation of elemental phosphorus.

3.61 Decide whether to choose an oxidizing agent or a reducing agent to bring about each of the following changes:
(a) $Br^-(aq) \rightarrow BrO_3^-(aq)$
(b) $S_2O_6^{2-}(aq) \rightarrow SO_4^{2-}(aq)$
(c) $NO_3^-(aq) \rightarrow NO(g)$
(d) HCHO (formaldehyde) $\rightarrow CH_3OH$ (methanol)

3.62 Would you choose an oxidizing agent or a reducing agent to make the following conversions?
(a) $ClO_3^-(aq) \rightarrow ClO_2(g)$
(b) $SO_4^{2-}(aq) \rightarrow S^{2-}(aq)$
(c) $Mn^{2+}(aq) \rightarrow MnO_2(s)$
(d) HCHO (formaldehyde) $\rightarrow$ HCOOH (formic acid)

3.63 The industrial production of sodium metal and chlorine gas makes use of the Downs process, in which molten sodium chloride is electrolyzed (Chapter 18). Write the balanced equation for the production of the two elements from molten sodium chloride. Which element is produced by oxidation and which by reduction?

3.64 When water is electrolyzed (Chapter 18), hydrogen is produced at the electrode called the cathode and oxygen is produced at the electrode called the anode. At which electrode does reduction occur, and at which electrode does oxidation occur?

Supplementary Exercises

3.65 Identify each of the following as a strong acid, a weak acid, a base, a soluble ionic compound, or an insoluble ionic compound in water: (a) HNO_3; (b) KOH; (c) NH_3; (d) $CuSO_4$; (e) HCOOH; (f) $Ca_3(PO_4)_2$; (g) ZnS; (h) H_2SO_4.

3.66 Select two soluble ionic compounds that, when mixed in solution, produce the precipitates (a) $MgCO_3$; (b) Ag_2SO_4; (c) $Zn(OH)_2$; (d) $PbCrO_4$.

3.67 Select an acid and a base that, when mixed in solution, produce (a) $Ba(NO_3)_2(aq)$; (b) $Na_2SO_4(aq)$; (c) $KClO_4(aq)$; (d) CsCl(aq).

3.68 Aqueous solutions of potassium chloride conduct electricity. Which picture below best represents KCl in aqueous solution?

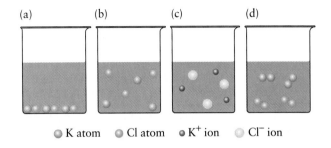

(a)　　(b)　　(c)　　(d)

● K atom　　● Cl atom　　● K^+ ion　　○ Cl^- ion

3.69 Use Table 3.3 to classify an aqueous solution of each of the following compounds as a strong electrolyte, a weak electrolyte, or a nonelectrolyte: (a) CH_3OH; (b) $CaBr_2$; (c) KI.

3.70 Use Table 3.3 to classify an aqueous solution of each of the following compounds as a strong electrolyte, a weak electrolyte, or a nonelectrolyte: (a) HCl; (b) KOH; (c) CH_3COOH.

3.71 Classify the following reactions as precipitation, neutralization or redox. If a precipitation reaction, write a net ionic equation; if a neutralization reaction, identify the acid and the base; if a redox reaction, identify the oxidizing agent and the reducing agent.

(a) Formation of magnesium chloride by the action of hydrochloric acid on magnesium hydroxide:

$$Mg(OH)_2(s) + 2\,HCl(aq) \longrightarrow MgCl_2(aq) + 2\,H_2O(l)$$

(b) Formation of barium sulfate by the action of sulfuric acid on barium hydroxide:

$$Ba(OH)_2(aq) + H_2SO_4(aq) \longrightarrow BaSO_4(s) + 2\,H_2O(l)$$

(c) Production of sulfur trioxide from sulfur dioxide in the manufacture of sulfuric acid:

$$2\,SO_2(g) + O_2(g) \longrightarrow 2\,SO_3(g)$$

3.72 Classify the following reactions as precipitation, neutralization, or redox. If a precipitation reaction, write a net ionic equation; if a neutralization reaction, identify the acid and the base; if a redox reaction, identify the oxidizing agent and the reducing agent.

(a) Formation of nitrogen and water vapor when ammonia burns in air:

$$4\,NH_3(g) + 3\,O_2(g) \longrightarrow 2\,N_2(g) + 6\,H_2O(g)$$

(b) Reaction used to produce elemental phosphorus from its oxide:

$$P_4O_{10}(s) + 10\,C(s) \longrightarrow P_4(s) + 10\,CO(g)$$

(c) Formation of silver chromate, Ag_2CrO_4, from sodium chromate, Na_2CrO_4, and silver nitrate solutions:

$$2\,AgNO_3(aq) + Na_2CrO_4(aq) \longrightarrow$$
$$Ag_2CrO_4(s) + 2\,NaNO_3(aq)$$

3.73 Classify the following reactions as precipitation, neutralization, or redox. If a precipitation reaction, write a net ionic equation; if a neutralization reaction, identify the acid and the base; if a redox reaction, identify the oxidizing agent and the reducing agent.

(a) Reaction used to measure the concentration of carbon monoxide in a gas stream:

$$5\,CO(g) + I_2O_5(s) \longrightarrow I_2(s) + 5\,CO_2(g)$$

(b) Reaction often used to monitor the amount of iodine in a sample:

$$I_2(aq) + 2\,S_2O_3{}^{2-}(aq) \longrightarrow 2\,I^-(aq) + S_4O_6{}^{2-}(aq)$$

(c) Test for bromide ions in solution:

$$AgNO_3(aq) + Br^-(aq) \longrightarrow AgBr(s) + NO_3{}^-(aq)$$

(d) Heating of uranium tetrafluoride with magnesium, one stage in the purification of uranium metal:

$$UF_4(g) + 2\,Mg(s) \longrightarrow U(s) + 2\,MgF_2(s)$$

3.74 Identify the oxidizing agent and the reducing agent for each of the following reactions:

(a) production of tungsten metal from its oxide by the reaction

$$WO_3(s) + 3\,H_2(g) \longrightarrow W(s) + 3\,H_2O(l)$$

(b) generation of hydrogen gas in the laboratory:

$$Mg(s) + 2\,HCl(aq) \longrightarrow H_2(g) + MgCl_2(aq)$$

(c) production of metallic tin from tin(IV) oxide, the mineral cassiterite:

$$SnO_2(s) + 2\,C(s) \xrightarrow{\Delta} Sn(l) + 2\,CO(g)$$

(d) Hydrazine has a low molar mass and releases a significant amount of energy in its reaction with dinitrogen tetroxide. The combination is used as a rocket propellant:

$$2\,N_2H_4(g) + N_2O_4(g) \longrightarrow 3\,N_2(g) + 4\,H_2O(g)$$

3.75 Write a balanced equation for the complete combustion (reaction with oxygen) of octane, C_8H_{18}, the major component of gasoline, to carbon dioxide and water vapor.

3.76 Energy is obtained in the body by a series of reactions which, overall, is equivalent to the combustion of sucrose (cane sugar), $C_{12}H_{22}O_{11}$, to carbon dioxide and liquid water. Write a balanced equation for its combustion.

3.77 The psychoactive drug sold as methamphetamine ("speed"), $C_{10}H_{15}N$, undergoes a series of reactions in the body; the net result of these reactions is the oxidation of solid methamphetamine by oxygen to produce carbon dioxide, liquid water, and nitrogen. Write the balanced equation for this net reaction.

3.78 Write the balanced equation for the combustion of the analgesic (painkiller) sold as Tylenol, $C_8H_9O_2N$, to carbon dioxide, liquid water, and nitrogen.

3.79 Some compounds of hydrogen and oxygen are exceptions to the common observation that H has an oxidation number of $+1$ and O has an oxidation number of -2. Assuming that each metal has the oxidation number of its most common ion, find the oxidation numbers of H and O in the following compounds: (a) KO_2; (b) $LiAlH_4$; (c) Na_2O_2; (d) NaH; (e) KO_3.

Applied Exercises

For Exercises 3.80–3.84, see Applying Chemistry: Case Study 3.

3.80 The Sabatier process has been used to remove CO_2 from artificial atmospheres. An advantage is that it produces methane, CH_4, which can be burned as a fuel, and water, which can be reused. Balance the equation for the process and identify the type of reaction:

$$CO_2(g) + H_2(g) \longrightarrow CH_4(g) + H_2O(l)$$

3.81 Where mass need not be controlled, carbon dioxide can be removed from air by absorption in silicates or silica gel. Balance the equation for the reaction of carbon dioxide and calcium silicate:

$$CO_2(g) + CaSiO_3(s) + H_2O(l) \longrightarrow$$
$$SiO_2(s) + Ca(HCO_3)_2(aq)$$

3.82 Nitrogen in the air of space ships must also be replaced, because it is gradually lost through leakage. One method is to store the nitrogen in the form of hydrazine, $N_2H_4(l)$, from which it is readily obtained in one step by heating. The ammonia produced can be processed further to obtain even more nitrogen:

$$N_2H_4(l) \longrightarrow NH_3(g) + N_2(g)$$

(a) Balance the equation. (b) Give the oxidation number of nitrogen in each compound. (c) Identify the oxidizing and reducing agents.

3.83 A mission to Mars is planned, but lithium is in short supply. Which of the following hydroxides would be best for removing CO_2 from the air supply in the space ship: Na, K, or Cs? Explain your reasoning.

3.84 Once on Mars, you need a way to turn the CO_2 in the atmosphere into oxygen and fuels that can be burned. Two reactions important on Earth could be combined into a two-step process in which some of the compounds are recycled. Balance the following equations and show how they can be combined into an equation in which CO_2 and water are the reactants and the products are CH_4, CO, and O_2:

$$H_2O(g) \longrightarrow H_2(g) + O_2(g)$$

$$CO_2(g) + H_2(g) \longrightarrow CH_4(g) + CO(g) + O_2(g)$$

3.85 Sodium thiosulfate, which as the pentahydrate, $Na_2S_2O_3 \cdot 5H_2O$, forms the large white crystals used as "photographer's hypo," can be prepared by bubbling oxygen through a solution of sodium polysulfide, Na_2S_5, in alcohol and adding water. Sulfur dioxide is a by-product. Sodium polysulfide is made by the action of hydrogen sulfide on a solution of sodium sulfide, Na_2S, in alcohol, which, in turn, is made by the neutralization of hydrogen sulfide, H_2S, with sodium hydroxide. Write the three chemical equations that show how hypo is prepared from hydrogen sulfide and sodium hydroxide.

3.86 The first stage in the production of nitric acid by the Ostwald process is the reaction of ammonia with oxygen, producing nitric oxide, NO, and water. The nitric oxide further reacts with oxygen to produce nitrogen dioxide, which, when dissolved in water, produces nitric acid and nitric oxide. Write the three balanced equations that lead to the production of nitric acid.

Integrated Exercises

3.87 You are asked to identify compound X, which was extracted from a plant seized by customs inspectors. You run a number of tests and accumulate the following information. Compound X is a white, crystalline solid. An aqueous solution of X turns litmus red and conducts electricity weakly, even when present at appreciable concentrations. When you add sodium hydroxide to the solution, a reaction occurs. A solution of the products of the reaction is a good conductor of electricity. An elemental analysis of X shows that the mass percentage composition of the compound is 26.68% C and 2.239% H, with the remainder being oxygen. A mass spectrum of X has a parent molecular ion peak at 90.0 g/mol. See Investigating Matter 1.2. (a) Write the empirical formula of X. (b) Write the molecular formula of X. (c) Write the balanced

chemical equation and the net ionic equation for the reaction of X with sodium hydroxide. (*Hint:* The coefficient of NaOH in the balanced equation is 2.) (d) Name X and the products of the reaction. (e) Classify the reaction by type.

3.88 (a) White phosphorus, which has the formula P_4, burns in air to form compound A, in which the mass percentage composition of phosphorus is 43.64%, with the remainder oxygen. The mass spectrum of A has a parent molecular ion peak at 283.9 g/mol. (b) Compound A reacts with water to form compound B, which has a mass percentage composition of 1.029% H and 31.60% P, with the remainder oxygen. The mass spectrum of B has a parent molecular ion peak at 97.99 g/mol. See Investigating Matter 1.2. (c) Compound B reacts with an aqueous solution of calcium hydroxide to form compound C, a white precipitate. Write balanced chemical and net ionic equations for the reactions in (a), (b), and (c). (d) Name compounds A, B, and C. (e) Identify the type of reaction in (a), (b), and (c).

3.89 Write the names of the ions that are found in aqueous solutions of the following compounds: (a) KCl; (b) $CuCl_2$; (c) $AgNO_3$.

3.90 Write the names of the ions that are found in aqueous solutions of the following compounds: (a) $KMnO_4$; (b) Na_2S; (c) $Co(ClO_3)_2$.

3.91 (a) Determine and tabulate the maximum (most positive) and minimum (most negative) oxidation numbers of the elements in the first seven main groups. *Hint:* Maximum oxidation numbers are often found in the oxoanion with the most oxygen atoms. For minimum oxidation numbers, consult Fig. 3.28. (b) Describe any patterns you see in the data.

3.92 The following redox reactions are important in the refining of certain elements. Balance the equations and, in each case, write the name of the source compound or ore of the element (in boldface) and the oxidation state in that compound of the element that is being extracted in its pure state in the reaction:

(a) $SiCl_4(l) + H_2(g) \rightarrow Si(s) + HCl(g)$

(b) $\mathbf{SnO_2}(s) + C(s) \xrightarrow{1200°C} Sn(l) + CO_2(g)$

(c) $\mathbf{V_2O_5}(s) + Ca(l) \xrightarrow{\Delta} V(s) + CaO(s)$

(d) $\mathbf{B_2O_3}(s) + Mg(s) \rightarrow B(s) + MgO(s)$

Chemistry's Accounting: Reaction Stoichiometry

One day you might ride in the space shuttle. If you do, your journey and your life will depend on the reaction between two elements, hydrogen and oxygen. This reaction powers the main rocket that lifts the shuttle into orbit and the fuel cells that run its life-support systems. But how much hydrogen do you need to take with you? The mission designers need to know how much hydrogen combines with a given amount of oxygen: they cannot afford to include excess fuel because so much energy is needed to get into orbit that every kilogram must be counted and justified. Hydrogen and oxygen are also used as the fuel in the "fuel cells" that provide electricity while the shuttle is in orbit. The amounts of hydrogen and oxygen available for this purpose must be in balance. To answer the kinds of questions that arise when planning a space mission—and in many other areas of life—we need *quantitative* information about the reaction, such as exactly how much of one reactant combines with a given amount of another.

In this chapter, we combine two important concepts from the two preceding chapters. One is the use of moles to measure the amount of a substance. The other is the use of chemical equations to express the con-

Life as an astronaut depends on knowing how much fuel to load for reaching orbit and for survival in the hostile environment of space. These matters of life and death depend on calculations like those in this chapter. We see how to predict how much oxygen needs to be carried to react with a given amount of hydrogen to power the space shuttle and to generate electricity. Back on Earth, much of industry depends on similar calculations.

servation of atoms in a reaction. By interpreting chemical equations in terms of moles we can make very useful quantitative predictions about reactions. For instance, we shall be able to calculate the mass of carbon dioxide a liter of gasoline contributes to the atmosphere when it burns. We shall be able to calculate the maximum amount of ammonia that we can expect to make from a given supply of nitrogen (Fig. 4.1) and the mass of urea fertilizer that we can then make from that ammonia. We can also calculate how much hydrogen we need to react with a given mass of oxygen in a rocket engine or a fuel cell. The methods introduced in this chapter will play an important role throughout the rest of the text.

HOW TO USE REACTION STOICHIOMETRY

We saw in Section 3.1 that the stoichiometric coefficients multiplying the chemical formulas in a balanced chemical equation reflect the fact that no atoms are created or destroyed in the reaction. These coefficients can also be interpreted in terms of numbers of moles. For instance, from the chemical equation

$$2\,H_2(g) + O_2(g) \longrightarrow 2\,H_2O(l)$$

we know that if 1 mol O_2 is consumed in this reaction, then 2 mol H_2O is formed. This interpretation of the coefficients as numbers of moles is the basis of all the calculations described in this chapter.

FIGURE 4.1

A modern plant for producing ammonia by the Haber process. The quantities of nitrogen and hydrogen that must be supplied to produce a given amount of ammonia are determined by applying the concepts described in this chapter.

4.1 Mole-to-Mole Calculations

The numbers of moles of reactants actually used in a reaction are usually different from the stoichiometric coefficients in the equation. For example, we might want to know how much water is formed when 0.25 mol O_2 reacts with hydrogen in a fuel cell. How do we take different starting amounts into account?

The calculation is based on a conversion factor just like the ones we used to convert units in Chapter 2. First, we summarize the information in the chemical equation—that 1 mol O_2 reacts to form 2 mol H_2O—by writing

$$1\,\text{mol}\,O_2 \,\hat{=}\, 2\,\text{mol}\,H_2O$$

The sign $\hat{=}$ is read "is chemically equivalent to," and the expression is called a **stoichiometric relation.** In calculations, the chemical equivalence sign is treated just like an equal sign.

Self-Test 4.1A Write the stoichiometric relation between (a) F_2 and I_2 and (b) F_2 and IF_3 in the formation of iodine trifluoride, IF_3: $I_2(s) + 3\,F_2(g) \rightarrow 2\,IF_3(g)$.

[*Answer:* (a) 3 mol $F_2 \,\hat{=}\, 1$ mol I_2; (b) 3 mol $F_2 \,\hat{=}\, 2$ mol IF_3]

Self-Test 4.1B Write the stoichiometric relation between (a) N_2 and O_2 and (b) N_2 and N_2O_3 in the formation of dinitrogen trioxide, N_2O_3: $2\,N_2(g) + 3\,O_2(g) \rightarrow 2\,N_2O_3(g)$.

Next, the stoichiometric relation is used to set up a conversion factor relating the substance for which the amount is required (water) to the substance for which the amount has been given (oxygen, in our example):

$$\frac{\text{Substance required}}{\text{Substance given}} = \frac{2 \text{ mol } H_2O}{1 \text{ mol } O_2}$$

This factor, which is commonly called the **mole ratio** for the reaction, allows us to relate the moles of water molecules produced to the moles of oxygen molecules consumed. We use it in the same way that we use a conversion factor when converting units:

$$\text{Moles of } H_2O = (0.25 \text{ mol } O_2) \times \frac{2 \text{ mol } H_2O}{1 \text{ mol } O_2}$$
$$= 0.50 \text{ mol } H_2O$$

Note that the unit mol and the species (in this case, O_2 molecules) cancel. This procedure is summarized in Toolbox 4.1.

Toolbox 4.1 *How to carry out mole-to-mole calculations for a chemical reaction*

We use this type of calculation when we need to know how many moles of a substance are consumed or produced in a chemical reaction, given the number of moles of another reactant or product.

Conceptual basis
The procedure is based on two principles: (1) the number of atoms of each element is conserved in a chemical reaction and (2) the coefficients in the balanced chemical equation give the relative numbers of moles of each reactant and product. These numbers of moles are used to construct conversion factors.

Procedure
The procedure is summarized in Fig. 4.2. The calculation is done in three steps:

Step 1. Write the balanced chemical equation for the reaction of interest. The chemical equation tells us the relative numbers of moles of each substance taking part in the reaction and, hence, the stoichiometric relation between the different substances.

Step 2. Find the stoichiometric relation between the two substances of interest, substance given ⇌ substance required, and use it to write the mole ratio in the form

$$\frac{\text{Substance required}}{\text{Substance given}}$$

The numbers in the mole ratios are exact, so they have no effect on the number of significant figures in the result.

Step 3. Multiply the given number of moles by this conversion factor to obtain the number of moles of the required substance. Make sure that the conversion factor is set up with the substance required on top.

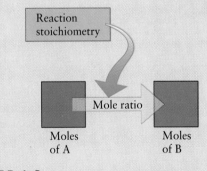

FIGURE 4.2

The schematic procedure for mole-to-mole conversions in a chemical reaction. It is based on the stoichiometric relations in the balanced chemical equation.

Example 4.1 *Determining the number of moles of reactant required or product formed*

Find the number of moles of N_2 needed to produce 5.0 mol NH_3 by reaction with H_2.

Strategy Write down the chemical equation for the reaction, identify the mole ratio between the two substances, and use the mole ratio as a conversion factor as described in Toolbox 4.1.

Solution **Step 1.** The balanced chemical equation is

$$N_2(g) + 3 H_2(g) \longrightarrow 2 NH_3(g)$$

Step 2. It follows from the equation that 1 mol $N_2 \triangleq 2$ mol NH_3, so the mole ratio is

$$\frac{\text{Substance required}}{\text{Substance given}} = \frac{1 \text{ mol } N_2}{2 \text{ mol } NH_3}$$

Step 3. We now apply this conversion factor to the information given and obtain the information required:

$$\text{Moles of } N_2 = (5.0 \text{ mol } NH_3) \times \overbrace{\frac{1 \text{ mol } N_2}{2 \text{ mol } NH_3}}^{\text{Mole ratio}} = 2.5 \text{ mol } N_2$$

Self-Test 4.2A How many moles of NH_3 molecules can be produced from 2.0 mol H_2 in the same reaction as in Example 4.1 if all the hydrogen reacts?

[*Answer:* 1.3 mol NH_3]

Self-Test 4.2B Aluminum metal reacts with molecular oxygen gas to form solid aluminum oxide. How many moles of aluminum are required to produce 25 mol Al_2O_3?

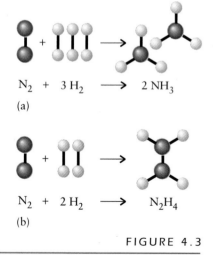

$$N_2 + 3 H_2 \longrightarrow 2 NH_3$$

(a)

$$N_2 + 2 H_2 \longrightarrow N_2H_4$$

(b)

FIGURE 4.3

In the formation of ammonia (a), 1 mol N_2 requires 3 mol H_2, but in the formation of hydrazine (b), 1 mol N_2 requires only 2 mol H_2.

We use mole ratios to convert from moles of a *reactant* to moles of a product or another reactant. We also use them to convert from moles of a *product* to moles of a reactant or another product. For the synthesis of ammonia, for instance, we can write the stoichiometric relation between the two reactants as

$$1 \text{ mol } N_2 \triangleq 3 \text{ mol } H_2$$

Although it is used in the same way, a stoichiometric relation is a little different from the relation between pairs of units. A relation between units, such as 1 in. = 2.54 cm, is universally true. A stoichiometric relation, such as 1 mol $N_2 \triangleq 3$ mol H_2, applies *only to the reaction we are considering* (the formation of ammonia). A different reaction may have a different stoichiometric relation between the same elements (Fig. 4.3). For example, consider the reaction

$$N_2(g) + 2 H_2(g) \longrightarrow N_2H_4(l)$$

where N_2H_4 is hydrazine, a compound similar to the fuel used to maneuver the space shuttle in orbit. For this reaction,

$$1 \text{ mol } N_2 \triangleq 2 \text{ mol } H_2$$

The balanced chemical equation for a reaction is used to set up the conversion factor from one substance to another. That conversion factor, called the mole ratio for the reaction, is applied to the moles given to calculate the moles required.

4.2 Mass-to-Mass Calculations

In some cases, we may need to predict the mass of a product that we can make from a given mass of a reactant. In others, we may need to predict the mass of one reactant that reacts with a given mass of another reactant. We have seen how to relate the moles of reactants to the moles of products. We know from Section 2.9 how to use molar masses to convert between moles and masses. So, by combining the two calculations, we can predict the masses of substances that react. Space engineers do these kinds of calculations to ensure the optimal amounts of hydrogen and oxygen for the space shuttle engines.

Suppose we want to know how many kilograms of iron can be obtained from 10.0 kg of iron(III) oxide, Fe_2O_3, present in iron ore, by the reaction

$$Fe_2O_3(s) + 3\,CO(g) \longrightarrow 2\,Fe(s) + 3\,CO_2(g)$$

This redox reaction takes place inside a blast furnace. We already know how to convert moles of Fe_2O_3 to moles of Fe: we use the stoichiometric relation

$$1\text{ mol }Fe_2O_3 \triangleq 2\text{ mol Fe}$$

in the form of the mole ratio

$$\frac{2\text{ mol Fe}}{1\text{ mol }Fe_2O_3}$$

However, before we can use this ratio as a conversion factor, we have to convert the given mass of iron(III) oxide from kilograms to grams, then to moles of Fe_2O_3. The molar mass of iron(III) oxide is

$$\text{Molar mass of }Fe_2O_3 = 2(55.85) + 3(16.00)\text{ g/mol} = 159.70\text{ g/mol}$$

Therefore,

$$\text{Moles of }Fe_2O_3 = (10.0\text{ kg }Fe_2O_3) \times \frac{10^3\text{ g}}{1\text{ kg}} \times \frac{1\text{ mol }Fe_2O_3}{159.70\text{ g }Fe_2O_3}$$

At this stage, we apply the conversion factor to calculate the moles of Fe atoms produced in the reaction:

$$\text{Moles of Fe} = (10.0\text{ kg }Fe_2O_3) \times \frac{10^3\text{ g}}{1\text{ kg}} \times \frac{1\text{ mol }Fe_2O_3}{159.70\text{ g }Fe_2O_3} \times \frac{2\text{ mol Fe}}{1\text{ mol }Fe_2O_3}$$

Finally, we convert moles of Fe atoms into mass of iron by using the molar mass of iron, which is 55.85 g/mol:

$$\text{Mass of Fe} = (10.0\text{ kg }Fe_2O_3) \times \frac{10^3\text{ g}}{1\text{ kg}} \times \frac{1\text{ mol }Fe_2O_3}{159.70\text{ g }Fe_2O_3}$$
$$\times \frac{2\text{ mol Fe}}{1\text{ mol }Fe_2O_3} \times \frac{55.85\text{ g Fe}}{1\text{ mol Fe}}$$
$$= 6.99 \times 10^3\text{ g Fe}$$

We can therefore expect to extract 6.99 kg of iron from 10.0 kg of the ore. This type of calculation is summarized in Toolbox 4.2.

Toolbox 4.2 *How to carry out mass-to-mass calculations*

The procedure summarized in this Toolbox is used to calculate the mass of a substance consumed or produced in a chemical reaction, given the mass of a reactant or product.

Conceptual basis

The stoichiometric coefficients in a chemical equation tell us how many moles of one substance are needed to produce or react with a given number of moles of another substance. By using molar masses, we can also find the mass of one substance that a given mass of another substance can produce. We first convert the given mass to the corresponding number of moles by using the molar mass of that substance. We then convert the number of moles of the required substance to mass by using its molar mass.

Procedure

The general procedure for mass-to-mass calculations is summarized in Fig. 4.4.

Step 1. Convert the given mass of one substance (in grams) to number of moles by using its molar mass.

Number of moles of substance A =
$$\frac{\text{mass of A (grams)}}{\text{molar mass of A (grams per mole)}} \qquad n_A = \frac{m_A}{M_A}$$

Another approach is to use the molar mass as a conversion factor, as in Example 4.2.

Step 2. Write the balanced chemical equation for the reaction. As in Toolbox 4.1, use the stoichiometric coefficients to construct the mole ratio between the two substances.

Step 3. Use the mole ratio to convert from the given number of moles of one substance to the number of moles of the other substance:

$$\frac{\text{Substance required}}{\text{Substance given}}$$

Step 4. Convert from number of moles of the second substance to mass (in grams) by using the molar mass of the substance.

Mass of B required (grams) =
number of moles of B $\times$ molar mass of B (grams per mole)

$$m_B = n_B \times M_B$$

Again we can use the molar mass as a conversion factor, as in Example 4.2.

If the masses are in units other than grams, first convert to grams.

For the numerical part of the calculation, we can write down the string of conversions in a single line. This approach is used in Example 4.2.

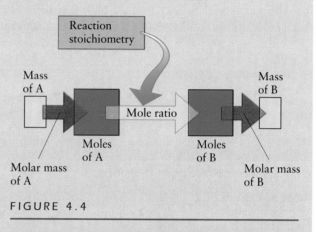

FIGURE 4.4

Mass-to-mass conversions in a chemical reaction depend on the stoichiometric relations from the balanced chemical equation.

Example 4.2 *Calculating the mass of reactant required or product formed*

Propane is a common camping fuel. Calculate the mass of carbon dioxide produced when 100. g of propane (C_3H_8) is burned in a camp stove.

Strategy We anticipate a mass larger than 100. g, because three molecules of carbon dioxide are produced for each molecule of propane that burns, and the molar masses of CO_2 and C_3H_8 are almost the same. Use the mole ratio from the balanced equation to find the mass of product, as described in Toolbox 4.2.

Solution **Step 1.** Convert from mass of propane to moles of C_3H_8 molecules by using the molar mass of propane, which is 44.09 g/mol:

$$\text{Number of moles of } C_3H_8 = (100. \text{ g } C_3H_8) \times \frac{1 \text{ mol } C_3H_8}{44.09 \text{ g } C_3H_8}$$

Step 2. Write the balanced equation for the combustion reaction,

$$C_3H_8(g) + 5 O_2(g) \longrightarrow 3 CO_2(g) + 4 H_2O(l)$$

and find the stoichiometric relation

$$\frac{\text{Substance required}}{\text{Substance given}} = \frac{3 \text{ mol } CO_2}{1 \text{ mol } C_3H_8}$$

Use this mole ratio to convert from mol C_3H_8 to mol CO_2:

$$\text{Number of moles of } CO_2 = (100. \text{ g } C_3H_8) \times \overbrace{\frac{1 \text{ mol } C_3H_8}{44.09 \text{ g } C_3H_8}}^{\text{Convert mass to mol } C_3H_8} \times \overbrace{\frac{3 \text{ mol } CO_2}{1 \text{ mol } C_3H_8}}^{\substack{\text{Convert mol } C_3H_8 \\ \text{to mol } CO_2}}$$

Step 3. Now convert this amount of CO_2 into mass in grams by using the molar mass of carbon dioxide (44.01 g/mol):

$$\text{Mass of } CO_2 = (100. \text{ g } C_3H_8) \times \frac{1 \text{ mol } C_3H_8}{44.09 \text{ g } C_3H_8} \times \frac{3 \text{ mol } CO_2}{1 \text{ mol } C_3H_8} \times \frac{44.01 \text{ g } CO_2}{1 \text{ mol } CO_2}$$

$$= 299 \text{ g } CO_2$$

The combustion of 100. g of propane produces 299 g of carbon dioxide. As anticipated, the mass of carbon dioxide is greater than the initial mass of propane.

Self-Test 4.3A Calculate the mass of oxygen needed to react with 0.450 g of hydrogen gas in the reaction $2 H_2(g) + O_2(g) \rightarrow 2 H_2O(l)$.

[*Answer:* 3.57 g]

Self-Test 4.3B Carbon dioxide can be removed from power plant exhaust gases by combining it with an aqueous slurry of calcium silicate:

$$2 CO_2(g) + H_2O(l) + CaSiO_3(s) \longrightarrow SiO_2(s) + Ca(HCO_3)_2(aq)$$

What mass of $CaSiO_3$ (molar mass 116.17 g/mol) is needed to remove 0.300 kg of carbon dioxide from the gases emitted from a power plant?

FIGURE 4.5

In a gravimetric analysis, the desired compound may be precipitated and filtered from the solution. The filter paper is then dried and weighed. The mass of the filter paper is known and is subtracted from the mass of precipitate and paper to obtain the mass of the dry precipitate.

Mass-to-mass calculations are used in **gravimetric analysis,** which is the use of measurements of mass to determine the amount of substance present. In a typical procedure, an insoluble compound is precipitated from an aqueous solution, the precipitate is filtered and weighed, and from its mass the amount of one of the original substances is calculated (Fig. 4.5). For example, gravimetric analysis is used in environmental monitoring to find out how much lead is present in a sample of water.

> *In a mass-to-mass calculation, convert the given mass to moles, apply the mole-to-mole conversion factor to obtain the moles required, and finally convert moles required to mass.*

Example 4.3 *Using gravimetric analysis to determine the mass of a substance present in a sample*

A 25.4-g sample of solid waste from a photographic developing laboratory must be analyzed to determine its commercial value. The waste is treated with concentrated nitric acid to extract any silver present as Ag^+ ions. Hydrochloric acid is then added to precipitate the silver as silver chloride, which is filtered and dried. The silver chloride is found to have a mass of 16.1 g. Find the mass percentage of silver in the waste.

Strategy First, determine the mass of silver in the waste from the mass of silver chloride by using the procedures in Toolbox 4.2. Then calculate the mass percentage of silver by dividing the mass of silver by the total mass of the sample and multiplying by 100%.

Solution We obtain the number of moles of AgCl in 16.1 g of silver chloride from its molar mass, which is 143.32 g/mol. The net ionic equation for the formation of silver chloride is

$$Ag^+(aq) + Cl^-(aq) \longrightarrow AgCl(s)$$

The stoichiometric relation between the product and silver ions is

$$1 \text{ mol } Ag^+ \simeq 1 \text{ mol AgCl}$$

One Ag^+ ion was obtained from each Ag atom present in the original sample, so we also know that

$$1 \text{ mol } Ag^+ \simeq 1 \text{ mol Ag}$$

Once we have converted to the number of moles of silver, we convert to mass of silver by using the molar mass of silver, which is 107.87 g/mol. The calculation is

$$
\begin{aligned}
\text{Mass of silver} &= (16.1 \text{ g AgCl}) \times \frac{1 \text{ mol AgCl}}{143.32 \text{ g AgCl}} \\
&\quad \times \frac{1 \text{ mol } Ag^+}{1 \text{ mol AgCl}} \times \frac{1 \text{ mol Ag}}{1 \text{ mol } Ag^+} \times \frac{107.87 \text{ g Ag}}{1 \text{ mol Ag}} \\
&= \frac{16.1 \times 107.87}{143.32} \text{ g Ag} = 12.1 \text{ g Ag}
\end{aligned}
$$

The mass percentage silver in the waste is therefore

$$\text{Mass percentage silver} = \frac{12.1 \text{ g}}{25.4 \text{ g}} \times 100\% = 47.7\%$$

Self-Test 4.4A A dietary iron supplement was advertised as containing the RDA (recommended daily allowance) of iron, 18 mg. The tablets were analyzed for iron content by crushing 10 of them (total mass 12.564 g) and dissolving them in nitric acid, which oxidized the iron to Fe^{3+}. The iron was then precipitated as Fe_2O_3 (molar mass 159.70 g/mol) by making the solution basic. After the iron(III) oxide was dried, its mass was found to be 0.261 g. What is the average mass of iron per tablet?

[*Answer:* 18.3 mg]

Self-Test 4.4B A sample of ore of mass 5.324 g from a hillside in Scotland was analyzed for barium by dissolving the sample and adding sulfuric acid to precipitate barium sulfate. After drying, the mass of $BaSO_4$ was found to be 3.752 g. What is the mass percentage of barium in the sample?

THE LIMITS OF REACTION

All the calculations we have done so far have assumed that the reactants are consumed in one chemical reaction and that the reaction has gone to completion. In practice, neither assumption may be correct. Some of the starting materials may be consumed in a **competing reaction,** a reaction occurring at the same time as the one we are interested in and using some of the same reactants, but producing different products. Because of the competing reaction, less reactant is available to produce the product of interest. Even if there is only one reaction going on, some of the starting material might not have reacted at the time we make our measurements. We need a way to assess the efficiency of a reaction, for then we can identify the optimal conditions for the reaction and reduce waste.

In Chapter 14 we shall see that many reactions appear to come to an end well before all the reactants are consumed.

4.3 Reaction Yield

We can see the effect of a competing reaction by considering one of the major contributors to the supply of CO_2 in the atmosphere, the combustion of **fossil fuels**—coal, petroleum, and natural gas. A major source of atmospheric carbon dioxide is the burning of gasoline in automobiles, so we need to know how much CO_2 is likely to be produced from a liter of gasoline (Applying Chemistry: Case Study 4).

Octane, C_8H_{18}, is a representative compound in gasoline. The chemical equation for its combustion is

$$2\,C_8H_{18}(l) + 25\,O_2(g) \longrightarrow 16\,CO_2(g) + 18\,H_2O(l)$$

The **theoretical yield** is the maximum mass of product that can be obtained from a given mass of reactant. The theoretical yield of carbon dioxide is based on the assumptions that this is the *only* reaction taking place and that every molecule of octane is converted into carbon dioxide (and water).

For example, let's calculate the theoretical yield of carbon dioxide when 702 g of octane, the mass of 1.00 L of octane, is burned. We need to convert mass to moles of C_8H_{18} molecules, then moles of C_8H_{18} to moles of CO_2, then moles of CO_2 to mass of carbon dioxide. We can calculate (see Toolbox 4.2) that, when this mass of octane burns in a plentiful supply of oxygen, the mass of carbon dioxide produced is

$$
\begin{aligned}
\text{Mass of CO}_2 &= (702 \text{ g } C_8H_{18}) \times \frac{1 \text{ mol } C_8H_{18}}{114.2 \text{ g } C_8H_{18}} \\
&\quad \times \frac{16 \text{ mol CO}_2}{2 \text{ mol } C_8H_{18}} \times \frac{44.01 \text{ g CO}_2}{1 \text{ mol CO}_2} \\
&= 2.16 \times 10^3 \text{ g CO}_2
\end{aligned}
$$

This 2.16 kg is the theoretical yield of carbon dioxide: the maximum amount that can be produced in the combustion of 1.00 L octane.

When octane burns in a limited supply of oxygen, carbon monoxide is formed as well as carbon dioxide. So, in addition to the reaction written above, the following reaction also takes place:

$$2\,C_8H_{18}(l) + 17\,O_2(g) \longrightarrow 16\,CO(g) + 18\,H_2O(l)$$

This mass corresponds to about 8 kg of CO_2 per gallon of fuel. The average car emits about 1 lb of CO_2 per mile of travel.

Applying Chemistry: *Case Study 4*

A visitor to our solar system is likely to be impressed by the abundance of life on Earth and the mild surface temperatures. In large part, the temperate climate of Earth results from the *greenhouse effect*. About 55% of the radiation received from the sun is reflected away or used in natural processes and most of the remaining 45% escapes as infrared radiation.[*] The greenhouse effect is the trapping of some of this heat by certain gases in the atmosphere. This effect warms the Earth, as if the entire planet were enclosed in a huge greenhouse.

Oxygen and nitrogen, which make up roughly 99% of the atmosphere, do not trap infrared radiation. However, water vapor and carbon dioxide do. Even though these two gases make up only about 1% of the atmosphere, they trap enough radiation to raise the average temperature of the Earth by 33°C. Without this naturally occurring greenhouse effect, the average surface temperature of the Earth would be well below the freezing point of water. However, the greenhouse effect is a dangerous ally, because if not controlled, it could lead to disastrous changes in climate.

The concentration of water vapor in the atmosphere is thought to have remained steady over time, but concentrations of some other greenhouse gases are rising. From the year 1000 or earlier until about 1750, the CO_2 concentration in the atmosphere remained fairly stable at about

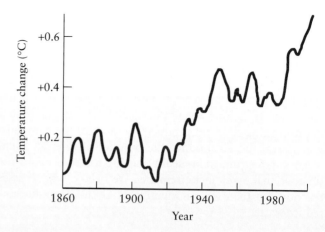

The average surface temperature of the Earth from 1860 to 1990.

280 ± 10 parts per million by volume (ppmv).[†] Since then, the CO_2 concentration has increased by 28% to 360 ppmv. The concentration of methane, CH_4, has more than doubled during this time and is now at its highest level in 160 000 y. Studies of air pockets in ice cores taken from Antarctica show that changes in the concentration of atmospheric CO_2 and methane over the past 160 000 y correlate well with changes in the global surface temperature. The rising concentrations of CO_2 and methane are therefore of some concern.

Human activities are responsible for the additional CO_2. Some is generated when limestone, $CaCO_3$, is heated and decomposed in cement-making (see Section 19.20). Large amounts of CO_2 are also released to the atmosphere by deforestation, which involves burning large areas of brush and trees. However, most of it comes from the burning of fossil fuels, which began on a large scale after 1850. The additional methane is coming mainly from the petroleum industry and from agriculture.

Temperature measurements both on land and on ships at sea, and temperature measurements taken inside boreholes drilled into the Earth, all point to an increase in the surface temperature of about 0.5°C since the late nineteenth century (see graph). If current trends in population growth and energy use continue, then by the middle of the twenty-first century, the concentration of CO_2 in the atmosphere will be about twice its value prior to the industrial revolution. What are the likely consequences of this doubling of the level of CO_2?

The Intergovernmental Panel on Climate Change (IPCC) is an international body of scientists that is studying these changes to the atmosphere and the climate changes likely to result. The IPCC's best estimate is that by

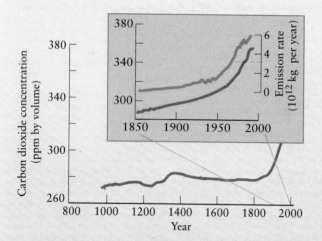

Concentration of carbon dioxide in the atmosphere over the past 1000 years as determined from ice core measurements. The blue line in the inset shows the increase in the rate of emission of CO_2 since 1850.

the year 2100 the Earth will undergo a further increase of 2°C, with a rise in sea level of about 0.5 m. Regional changes in precipitation are also likely to occur. Unless we change the way we use energy, the concentrations of CO_2 and other greenhouse gases will continue to increase even further beyond 2100, causing even more climate changes.

Two degrees may not sound like much. However, the temperature during the last ice age was only 6°C colder than at present. In this context, 2°C is significant. Furthermore, the *rate* of temperature change is likely to be greater than at any time in the last 10,000 y. Ecosystems can often adapt to slow climate changes but not to rapid change.

It will be difficult to stabilize levels of CO_2 below the recommended level of 450 ppmv, but we can take some measures now. For example, we can increase the energy efficiency of transportation, buildings, and appliances. Because of the large savings that would come from using less fuel, many of these changes can be made at no net cost, or even at a profit. Second, we can continue research on various forms of renewable energy, such as solar and wind power. Third, the industrialized countries could make available new, energy-efficient technologies to countries that are undergoing industrialization. Fourth, deforestation contributes both to increased CO_2 emissions and to decreased CO_2 uptake. Therefore, reducing deforestation, or better yet, planting trees to increase the amount of forest land, will help reduce the CO_2 concentration.

Alternatives to fossil fuels, such as hydrogen, are explored in Section 6.15 and Connection 2, which follows Chapter 9. Coal, which is mostly carbon, can also be converted into fuels with a lower proportion of carbon. Its conversion to methane, CH_4, for instance, would reduce CO_2 emissions. We can also work with nature by accelerating the take-up of carbon by the natural processes of the carbon cycle. For example, one proposed solution is to pump CO_2 exhaust deep into the ocean, where it would dissolve to form carbonic acid and bicarbonate ions. Carbon dioxide can also be removed chemically from power plant exhaust gases by passing the exhaust through an aqueous slurry of calcium silicate

$$2\,CO_2(g) + H_2O(l) + CaSiO_3(s) \longrightarrow$$
$$SiO_2(s) + Ca(HCO_3)_2(s)$$

The contribution to the greenhouse effect from methane and the chlorofluorocarbons is not as significant as that from carbon dioxide, but is still important. The use of chlorofluorocarbons is now under strict control (see Applying Chemistry: Case Study 5), but the concentration of methane is rising as a result of the increase in production of rice and livestock.

Key Concept: stoichiometry

For Further Reading
J. Houghton, *Global Warming: The Complete Briefing,* Cambridge: Cambridge University Press, 1997. (A brief summary by the co-chair of the IPCC Scientific Assessment Working Group.)

Related Exercises: 4.76–4.80

---

*Infrared radiation is heat radiation: it lies at longer wavelengths than red light (see Chapter 7).

†Parts per million by volume is the number of milliliters of a substance per 1000 L of sample.

(a)

(b)

The extent to which glaciers and ice caps have receded is shown by these pictures of a boulder in the Andes. The pictures, which were taken in 1978 (a) and 1995 (b), show extensive local warming.

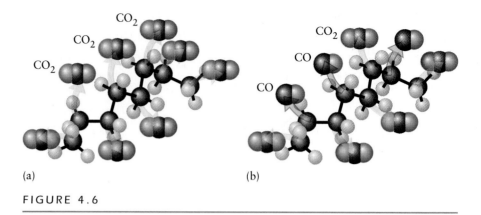

(a)   (b)

FIGURE 4.6

(a) When an octane molecule undergoes complete combustion, it forms carbon dioxide and water: one CO_2 molecule is formed for each carbon atom present (yellow arrows).
(b) However, in a limited supply of oxygen, some of the carbon atoms end up as carbon monoxide molecules, CO, so the yield of carbon dioxide is reduced (blue arrows).

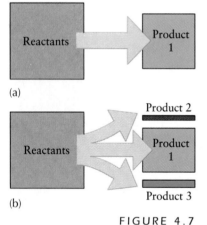

(a)

(b)

FIGURE 4.7

(a) The yield of a product would be 100% if no competing reactions were taking place. (b) However, if a reactant can take part in more than one reaction at the same time, then the yield of a particular product will be less than 100% because other products will also form.

Because this reaction consumes some of the octane, the mass of carbon dioxide formed is less than the theoretically possible mass (Fig. 4.6). To express the extent to which this competing reaction cuts down the mass of carbon dioxide formed, we speak of the "percentage yield." The **percentage yield** of a product is the fraction of the theoretical yield actually obtained, expressed as a percentage (Fig. 4.7):

$$\text{Percentage yield} = \frac{\text{actual yield}}{\text{theoretical yield}} \times 100\%$$

Suppose, for instance, we find that, in a test of an automobile engine to monitor the combustion of 1.00 L of octane under various conditions, only 1.14 kg of carbon dioxide is produced, not the 2.16 kg of the theoretical yield. Then,

$$\text{Percentage yield of } CO_2 = \frac{1.14\text{ kg}}{2.16\text{ kg}} \times 100\% = 52.8\%$$

The percentage yield is by no means fixed: it depends on the conditions of the experiment and the care of the experimenter.

> *The theoretical yield of a product is the maximum mass that can be expected on the basis of the stoichiometry of a chemical equation. The percentage yield is the percentage of the theoretical yield actually achieved.*

Example 4.4 *Calculating the percentage yield of a product*

In a study of old mining lanterns, an excess amount of water was poured on 100. g of calcium carbide and 28.3 g of ethyne (C_2H_2, acetylene) was produced. Calculate the percentage yield of ethyne for the reaction

$$CaC_2(s) + 2\,H_2O(l) \longrightarrow Ca(OH)_2(aq) + C_2H_2(g)$$

Strategy First, we need to know the theoretical yield (in grams) of the product. It is calculated by the technique explained in Toolbox 4.2. We perform the calculation in one step. Then we express the actual yield as a percentage of the theoretical yield.

Solution The molar mass of calcium carbide is 64.10 g/mol and that of ethyne is 26.04 g/mol. The theoretical yield of ethyne is

$$\text{Mass of } C_2H_2 = (100.\text{ g } CaC_2) \times \frac{1 \text{ mol } CaC_2}{64.10 \text{ g } CaC_2} \times \frac{1 \text{mol } C_2H_2}{1 \text{ mol } CaC_2} \times \frac{26.04 \text{ g } C_2H_2}{1 \text{ mol } C_2H_2}$$

$$= \frac{100. \times 26.04}{64.10} \text{ g } C_2H_2 = 40.6 \text{ g } C_2H_2$$

Because the actual yield of ethyne is 28.3 g, its percentage yield is

$$\text{Percentage yield of ethyne} = \frac{28.3 \text{ g}}{40.6 \text{ g}} \times 100\% = 69.7\%$$

Self-Test 4.5A When 24.0 g of potassium nitrate was heated with lead, 13.8 g of potassium nitrite was formed in the reaction $Pb(s) + KNO_3(s) \rightarrow PbO(s) + KNO_2(s)$. Calculate the percentage yield of potassium nitrite.

[***Answer:*** 68.3%]

Self-Test 4.5B Reduction of 15 kg of iron(III) oxide in a blast furnace resulted in the production of 8.8 kg of iron. What is the percentage yield of iron?

4.4 Limiting Reactants

Reactants are not always present in the exact mole ratios required by the chemical equation. For example, one step in the synthesis of the rocket fuel dinitrogen tetroxide, N_2O_4 (a component of the fuel used for orbital maneuvers of the space shuttle), is the formation of nitrogen dioxide from nitrogen monoxide and oxygen:

$$2 NO(g) + O_2(s) \longrightarrow 2 NO_2(g)$$

Every mole of oxygen reacts with two moles of nitrogen monoxide. Therefore, 10 mol O_2 can react with 20 mol NO. However, suppose we had one tank containing 10 mol O_2 and another with only 15 mol NO. This amount of nitrogen monoxide would limit the quantity of product formed. The nitrogen monoxide is an example of a **limiting reactant,** a reactant that governs the maximum yield of product in a given reaction.

> The limiting reactant is sometimes referred to as the *limiting reagent.*

A limiting reactant is like a part in an automobile factory: if there are 200 wheels and 60 car bodies, then the maximum number of automobiles is limited by the number of wheels. Because each body requires four wheels, there are enough wheels for only 50 cars, so the wheels play the role of the limiting reactant. When all the wheels have been used, 10 car bodies remain unused.

An analogy

Once we have identified the limiting reactant as described in Toolbox 4.3, we can calculate the amount of product that can be formed. We can also calculate the amount of excess reactant that remains at the end of the reaction.

__The limiting reactant in a reaction is the species supplied in an amount smaller than that required by the stoichiometric relation between the reactants.__

Toolbox 4.3 *How to identify and use the limiting reactant*

The procedure in this Toolbox allows us to identify the limiting reactant, given the moles or masses of the reactants.

Conceptual Basis

The limiting reactant is the reactant that will be completely used up. All other reactants are in excess. Because the limiting reactant is the one that controls the amounts of products that can be formed, the amount of product predicted is calculated from the amount of the limiting reactant.

Procedure

There are two ways of determining which reactant is the limiting one.

Method 1 In this approach, we use the mole ratio from the chemical equation to determine whether there is enough of one reactant to react with another. This approach is illustrated in Example 4.5.

Step 1. Calculate the amount of each reactant in moles, converting any masses to moles by using the molar mass.

Step 2. Choose one of the substances and use the stoichiometric relation to calculate the number of moles of the second reactant needed for complete reaction (Toolbox 4.1). By using the stoichiometric relation, we calculate the theoretical amount of the second reactant needed, not the actual amount present.

Step 3. Compare the actual amount of the second reactant to that calculated in step 2. If the actual amount is greater than the amount needed, then the second reactant is present in excess and not all of it will react; in this case, the first substance is the limiting reactant. If the actual amount of the second reactant is less than that calculated, then all of it will react, so it is the limiting reactant and the other reactant is in excess (Fig. 4.8).

Method 2 An alternative approach is to calculate the theoretical molar yield of one of the products for each reactant separately, by using the procedure in Toolbox 4.1. This method is a good one to use when there are more than two reactants. The reactant that would produce the smallest amount of product is the limiting reactant. The approach is illustrated in Example 4.5.

Step 1. Calculate the amount of each reactant in moles, if necessary by converting masses to moles by using the molar masses of the substances.

Step 2. Select one of the products. For each reactant, calculate how many moles of the product it can form (Toolbox 4.1).

Step 3. Compare the amount of product each reactant can form. The reactant that can produce the least product is the limiting reactant.

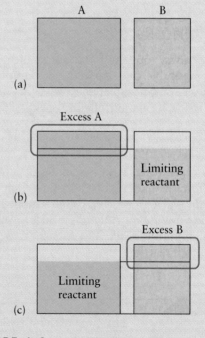

FIGURE 4.8

How to decide which is the limiting reactant. (a) The gold and green boxes depict the relative amounts of each reactant that are required by the stoichiometric relation. (b) If the amount of reactant B is less than that required for all A to react, then B is the limiting reactant. (c) If the amount of A is less than that required for all B to react, then A is the limiting reactant.

Example 4.5 *Identifying the limiting reactant*

As we saw in Example 4.4, calcium carbide reacts with water to form calcium hydroxide and the flammable gas ethyne (acetylene). Which is the limiting reactant when 100. g of water reacts with 100. g of calcium carbide? The chemical equation is

$$CaC_2(s) + 2 H_2O(l) \longrightarrow Ca(OH)_2(aq) + C_2H_2(g)$$

Strategy Follow either procedure in Toolbox 4.3. In this example, we use both to show that they give the same answer.

Solution

Method 1: Step 1. The molar mass of calcium carbide is 64.10 g/mol and that of water is 18.02 g/mol. Therefore, the numbers of moles of the reactants are

$$\text{Number of moles of } CaC_2 = (100. \text{ g } CaC_2) \times \frac{1 \text{ mol } CaC_2}{64.10 \text{ g } CaC_2} = \frac{100.}{64.10} \text{ mol } CaC_2$$

$$\text{Number of moles of } H_2O = (100. \text{ g } H_2O) \times \frac{1 \text{ mol } H_2O}{18.02 \text{ g } H_2O} = \frac{100.}{18.02} \text{ mol } H_2O$$

These two amounts work out as 1.56 mol CaC_2 and 5.55 mol H_2O.
Step 2. Next, we work out from the stoichiometric relation

$$1 \text{ mol } CaC_2 \triangleq 2 \text{ mol } H_2O$$

the amount of H_2O that is needed to react with 100. g of CaC_2:

$$\text{Number of moles of } H_2O = \left(\frac{100.}{64.10} \text{ mol } CaC_2\right) \times \frac{2 \text{ mol } H_2O}{1 \text{ mol } CaC_2}$$

$$= \frac{100. \times 2}{64.10} \text{ mol } H_2O = 3.12 \text{ mol } H_2O$$

Step 3. Because 3.12 mol H_2O is required and 5.55 mol H_2O is supplied, all the calcium carbide can react; so the calcium carbide is the limiting reactant and water is present in excess.

Method 2: In this approach, we take the number of moles of each reactant from step 1 and use the mole ratios from the chemical equation to determine how many moles of C_2H_2 can be formed:

$$\text{Number of moles of } C_2H_2 \text{ from } CaC_2 = \left(\frac{100.}{64.10} \text{ mol } CaC_2\right) \times \frac{1 \text{ mol } C_2H_2}{1 \text{ mol } CaC_2}$$

$$= 1.56 \text{ mol } C_2H_2$$

$$\text{Number of moles of } C_2H_2 \text{ from } H_2O = \left(\frac{100.}{18.02} \text{ mol } H_2O\right) \times \frac{1 \text{ mol } C_2H_2}{2 \text{ mol } H_2O}$$

$$= 2.78 \text{ mol } C_2H_2$$

Because the amount of CaC_2 can form less product than the amount that H_2O can form, CaC_2 is the limiting reactant. The two methods are in agreement.

Self-Test 4.6A In the synthesis of ammonia, which is the limiting reactant when 100. kg of hydrogen reacts with 800. kg of nitrogen?

[*Answer:* Hydrogen]

Self-Test 4.6B Suppose that 28 g of NO_2 and 18 g of water are supplied in the reaction used to produce nitric acid from nitrogen dioxide:

$$3 NO_2(g) + H_2O(l) \longrightarrow 2 HNO_3(l) + NO(g)$$

Which substance is the limiting reactant?

4.5 Combustion Analysis

We first met the technique of combustion analysis in Investigating Matter 2.1. We can now see how the material developed in this chapter is used in combustion analysis to determine the composition of organic compounds.

The sample is burned in a plentiful flow of oxygen, so the sample is the limiting reactant. All the hydrogen in the compound is converted to water and all the carbon is converted to carbon dioxide. The water produced is absorbed by an excess of phosphorus(V) oxide, P_4O_{10}, by the reaction

$$P_4O_{10}(s) + 6\,H_2O(l) \longrightarrow 4\,H_3PO_4(l)$$

Water is the limiting reactant in this reaction, and the increase in mass of this tube is equal to the mass of water produced by the combustion. The carbon dioxide is absorbed by an excess of sodium hydroxide by the reaction

$$NaOH(s) + CO_2(g) \longrightarrow NaHCO_3(s)$$

The carbon dioxide is the limiting reactant in this reaction, and the increase in mass of this tube is equal to the mass of carbon dioxide produced in the combustion.

Each carbon atom in the compound ends up in one molecule of carbon dioxide and each hydrogen atom ends up in a water molecule. For example, in the combustion of vitamin C, $C_6H_8O_6$,

$$C_6H_8O_6(s) + 5\,O_2(g) \longrightarrow 6\,CO_2(g) + 4\,H_2O(g)$$

each C atom present in the compound becomes part of a CO_2 molecule and each H atom part of an H_2O molecule. Therefore, we can write

$$1\text{ mol C in sample} \eqsim 1\text{ mol }CO_2$$
$$2\text{ mol H in sample} \eqsim 1\text{ mol }H_2O$$

These two stoichiometric relations apply to *any* combustion reaction of an organic compound because each C atom in an organic molecule gives rise to one CO_2 molecule, and each two H atoms in the molecule gives rise to one H_2O molecule. Therefore, to find the number of moles of C atoms in the sample, we convert the mass of carbon dioxide produced to moles of CO_2 and then use the first of these two relations to find the number of moles of C atoms in the original sample. We then carry out a similar calculation on the mass of water produced: we convert the mass of water to moles of H_2O and then use the relation above to find the number of moles of H atoms in the original sample.

Some elemental analyzers also analyze samples for the presence of nitrogen. If the compound also contains oxygen, we calculate the mass of oxygen originally present by subtracting the masses of carbon and hydrogen (and nitrogen if it is present) in the sample from the original mass of the sample.

In a combustion analysis, the numbers of moles of C and H atoms in a sample are determined from the masses of carbon dioxide and water produced when the compound burns in excess oxygen.

Example 4.6 *Determining an empirical formula by combustion analysis*

A combustion analysis is carried out on 1.621 g of a newly synthesized compound that is known to contain only C, H, and O. The masses of water and carbon dioxide produced are 1.902 g and 3.095 g, respectively. What is the empirical formula of the compound?

Strategy We need to use the stoichiometric relations given above and the procedures set out in Toolbox 4.2 to find the number of moles of carbon and hydrogen atoms in the sample and to convert those moles to masses. The mass of oxygen in the sample is obtained by subtraction of the total mass of carbon and hydrogen from the mass of the original sample. It is then converted to the number of moles of O atoms. Finally, the relative numbers of atoms are expressed as an empirical formula.

Solution To convert the mass of carbon dioxide to moles of C atoms, we use the molar mass of carbon dioxide (44.01 g/mol) and the stoichiometric relation 1 mol C $\cong$ 1 mol CO_2:

$$\text{Number of moles of C} = (3.095 \text{ g } CO_2) \times \frac{1 \text{ mol } CO_2}{44.01 \text{ g } CO_2} \times \frac{1 \text{ mol C}}{1 \text{ mol } CO_2}$$

$$= \frac{3.095}{44.01} \text{ mol C} = 0.07032 \text{ mol C}$$

The mass of carbon in the sample is therefore

$$\text{Mass of C} = \left(\frac{3.095}{44.01} \text{ mol C}\right) \times \frac{12.01 \text{ g C}}{1 \text{ mol C}} = 0.8446 \text{ g C}$$

To convert mass of water to moles of H atoms, we use the molar mass of water (18.02 g/mol) and the stoichiometric relation 1 mol H_2O $\cong$ 2 mol H to obtain

$$\text{Number of moles of H} = (1.902 \text{ g } H_2O) \times \frac{1 \text{ mol } H_2O}{18.02 \text{ g } H_2O} \times \frac{2 \text{ mol H}}{1 \text{ mol } H_2O}$$

$$= \frac{1.902 \times 2}{18.02} \text{ mol H} = 0.2111 \text{ mol H}$$

This amount of H atoms corresponds to the following mass of hydrogen (of molar mass 1.079 g/mol):

$$\text{Mass of H} = \left(\frac{1.902 \times 2}{18.02} \text{ mol H}\right) \times \frac{1.079 \text{ g H}}{1 \text{ mol H}} = 0.2128 \text{ g H}$$

The total mass of carbon and hydrogen is 0.8446 + 0.2128 g = 1.0574 g, so the mass of oxygen in the sample is

$$\text{Mass of oxygen} = 1.621 - 1.0574 \text{ g} = 0.564 \text{ g}$$

Therefore,

$$\text{Number of moles of O} = (0.564 \text{ g O}) \times \frac{1 \text{ mol O}}{16.00 \text{ g O}} = 0.0352 \text{ mol O}$$

At this stage, we know that the relative numbers of moles, and thus the relative numbers of atoms of each element, are

$$C:H:O = 0.07032 : 0.2111 : 0.0352$$

Division by the smallest number, 0.0352, gives

$$C:H:O = 2.00 : 6.00 : 1.00$$

We conclude that the empirical formula of the new compound is C_2H_6O.

Self-Test 4.7A When 0.528 g of sucrose (a compound of carbon, hydrogen, and oxygen) is burned, 0.306 g of water and 0.815 g of carbon dioxide are formed. Deduce the empirical formula of sucrose.

[***Answer:*** $C_{12}H_{22}O_{11}$]

Self-Test 4.7B When 0.236 g of aspirin is burned in oxygen, 0.519 g of carbon dioxide and 0.0945 g of water are formed. Deduce the empirical formula of aspirin.

SOLUTIONS

Many medications are administered in solution. Chemicals are also stored, measured, and used as reagents in solution. The amount of solute present in a given volume of solution therefore has great practical importance. It is obviously important in medicine and agriculture, because accidental injury or death can result from incorrect dosages or from overexposure to pesticides. It is also very important in chemistry, because we often need to know how much of a substance we are using when we pour one solution into another (Fig. 4.9).

4.6 Molarity

When we have a solution containing a substance we need, how much of the solution should we measure out to obtain a specified amount of the substance? First we need to know the **concentration** of the solute, the amount of solute per given volume of solution or solvent (Section 3.3). The **molar concentration,** *M,* of a solute in solution is commonly referred to informally as the **molarity** of the solute. The molarity is the number of moles of solute molecules or formula units divided by the volume of the solution (in liters).

$$\text{Molarity} = \frac{\text{amount of solute (moles)}}{\text{volume of solution (liters)}} \quad \text{or} \quad M = \frac{n}{V} \quad (1)$$

The units of molarity are moles per liter (mol/L), often denoted M:

$$1\,\text{M} = 1\,\text{mol/L}$$

> You will also see the units of molarity written as *M* or M, but M is the internationally recommended notation.

The symbol M is read "molar." For example, suppose we dissolved 10.0 g of cane sugar in enough water to make 200. mL of solution, which we might do if we were making a glass of lemonade. Cane sugar is sucrose ($C_{12}H_{22}O_{11}$); and from its molar mass (342 g/mol), we can calculate that 10.0 g of cane sugar contains 0.0292 mol of sucrose molecules. The molarity of the solute in the solution is therefore

$$\text{Molarity} = \frac{0.0292 \text{ mol}}{0.200 \text{ L}} = 0.146 \text{ mol/L}$$

We report this molarity as 0.146 M $C_{12}H_{22}O_{11}$(aq). The (aq) indicates an aqueous solution. If we were to dissolve 20.0 g of cane sugar instead of 10.0 g in the same total volume of solution, the sugar in the lemonade would be twice as concentrated (and taste correspondingly sweeter): its molarity would be 0.292 M $C_{12}H_{22}O_{11}$(aq).

Molarity is defined in terms of the volume of *solution,* not the volume of solvent used to prepare the solution. This distinction makes sense, because we need to know what volume of *solution* to measure out to obtain a given amount of solute. The usual way to prepare an aqueous solution of a solid substance is to transfer a known mass of the solid into a volumetric flask, dissolve it in a little water, fill the flask up to the mark with water, and then mix it thoroughly by tipping the flask end over end (Fig. 4.10).

(a)

The molarity of a solute in a solution is the number of moles of solute divided by the volume of the solution in liters.

Example 4.7 *Calculating the molarity*

A student prepared a solution by dissolving 1.345 g of potassium nitrate, KNO_3, in enough water to prepare 25.00 mL of solution. What is the molarity of the solute?

Strategy Because the molarity is the number of moles of formula units divided by the volume of solution, we first convert the mass of solute to moles of formula units by using the molar mass of the solute. Then we calculate the molarity by dividing the moles of solute by the volume of the solution (in liters).

Solution The molar mass of potassium nitrate is 101.11 g/mol. It follows that the number of moles of KNO_3 in 1.345 g of potassium nitrate is

(b)

$$\text{Number of moles KNO}_3 = (1.345 \text{ g KNO}_3) \times \frac{1 \text{ mol KNO}_3}{101.11 \text{ g KNO}_3} = \frac{1.345}{101.11} \text{ mol KNO}_3$$

Because the volume of the solution is 25.00 mL (0.025 00 L), the molarity is

$$\underbrace{\text{Molarity}}_{M} = \underbrace{\frac{(1.345/101.11) \text{ mol KNO}_3}{0.025 \, 00 \text{ L}}}_{\frac{n}{V}} = 0.5321 \text{ mol KNO}_3\text{/L}$$

We report this molarity as 0.5321 M KNO_3(aq).

(c)

Self-Test 4.8A Calculate the molarity of NaCl in a solution made by dissolving 2.357 g of sodium chloride in enough water to make 75.00 mL of solution.

[*Answer:* 0.5378 M NaCl(aq)]

Self-Test 4.8B Calculate the molarity of glucose in a solution made by dissolving 1.368 g of glucose ($C_6H_{12}O_6$) in enough water to make 50.00 mL of solution.

Once we know the molarity of a solution, we can calculate the number of moles of solute in a sample of any volume. We can also find out how much

FIGURE 4.10

The steps involved in making up a solution of known concentration (here, a solution of potassium permanganate, $KMnO_4$). (a) A known mass of the compound is dispensed into a volumetric flask. (b) Some water is added to dissolve it. (c) Finally, water is added up to the mark. The bottom of the solution's meniscus, the curved top surface, should be level with the mark.

solution to use if we want to obtain a given number of moles of solute. These calculations are explained in Toolbox 4.4.

The use of solutions allows us to transfer small amounts of solute readily and precisely; the volume to be transferred is calculated from the molarity.

Toolbox 4.4 *How to use molarity*

The procedure in this Toolbox shows how to calculate the number of moles of solute in a given volume of solution. It also shows how to calculate the volume of solution that contains a given amount of solute.

Conceptual basis
The molarity (molar concentration) of a solute is the number of moles of the solute per liter of solution. So, to determine the number of moles of solute in a given volume, we multiply the volume (in liters) by the molarity (Fig. 4.11). To determine the volume that contains a given number of moles of solute, we divide the number of moles by the molarity.

Procedure
When using molarity, it is best to think about the definition of molarity in Eq. 1:

$$M = \frac{n}{V}$$

We can solve for number of moles or volume (in liters) simply by rearranging the equation.

To find the moles of solute in a given volume of solution The molarity tells us the number of moles of solute per liter of solution. Multiplying the molarity by the stated volume of the solution (in liters) gives the number of moles of solute in that volume:

$$n = VM$$

To find the volume of solution that contains a given amount of solute When the number of moles of solute is given, we calculate the volume of solution, V, that contains that number of moles by rearranging Eq. 1 into

$$V = \frac{n}{M}$$

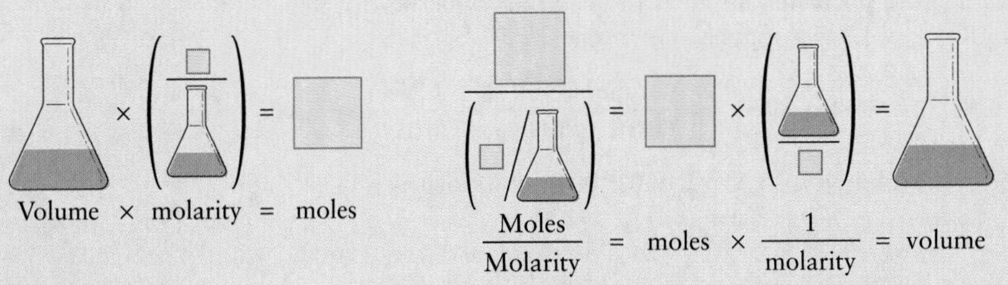

Volume × molarity = moles

$$\frac{\text{Moles}}{\text{Molarity}} = \text{moles} \times \frac{1}{\text{molarity}} = \text{volume}$$

FIGURE 4.11

A schematic summary of how to use molarity to convert volume of solution to moles of solute (left) and the amount of solute to the volume of solution that contains that amount of solute (right).

Example 4.8 *Calculating the number of moles of solute in a given volume*

Find the number of moles of sucrose molecules in 15 mL (0.015 L) of 0.10 M $C_{12}H_{22}O_{11}$(aq) solution.

Strategy We know the volume and concentration. Toolbox 4.4 illustrates how to calculate the number of moles of solute in a given volume of solution by recognizing that the number of moles of solute in a given volume is equal to the volume multiplied by the molarity: $n = VM$.

Solution The molarity of the solute in the solution is 0.10 mol/L. Therefore, the number of moles of solute in 15 mL of solution is

$$\overbrace{\text{Number of moles of } C_{12}H_{22}O_{11}}^{n} = \overbrace{(0.015 \text{ L}) \times (0.10 \text{ mol } C_{12}H_{22}O_{11} / \text{ L})}^{V \times M}$$
$$= 1.5 \times 10^{-3} \text{ mol } C_{12}H_{22}O_{11}$$

Self-Test 4.9A How many moles of NaCl formula units are present in 25.00 mL of 1.85 M NaCl(aq)?

[*Answer:* 4.62×10^{-2} mol NaCl]

Self-Test 4.9B How many moles of urea molecules, $(NH_2)_2CO$, are present in 100.0 mL of 1.45×10^{-2} M $(NH_2)_2CO$(aq)?

Example 4.9 *Calculating the volume of solution that contains a given amount of solute*

Potassium permanganate, $KMnO_4$, is a strong oxidizing agent that is used to purify water. Suppose we have a bottle of 0.0380 M $KMnO_4$(aq) and want to measure out 0.760 mmol $KMnO_4$. What volume of the solution should we use?

Strategy Use the procedure in Toolbox 4.4 for calculating the volume of solution that contains a given amount of solute, which is based on recognizing that we need to divide the number of moles by the molarity ($V = n/M$).

Solution Because the molarity of $KMnO_4$ in the solution is 0.0380 mol/L, the volume that contains 0.760 mmol $KMnO_4$ (that is, 7.60×10^{-4} mol $KMnO_4$) is

$$\overbrace{\text{Volume of solution}}^{V} = \overbrace{\frac{7.60 \times 10^{-4} \text{ mol } KMnO_4}{0.0380 \text{ mol } KMnO_4 / \text{ L}}}^{n/M} = 0.0200 \text{ L}$$

We should therefore transfer 20.0 mL of the potassium permanganate solution by using a buret or pipet. The flask will then contain 0.760 mmol $KMnO_4$.

Self-Test 4.10A What volume of a 1.25×10^{-3} M $C_6H_{12}O_6$(aq) should you transfer to obtain a solution that contains 1.44×10^{-6} mol of glucose molecules?

[*Answer:* 1.15 mL]

Self-Test 4.10B What volume of 0.358 M HCl(aq) should you transfer to have a sample that contains 2.55 mmol HCl?

Example 4.10 *Preparing a solution of specified molarity*

Calculate the mass of potassium permanganate needed to prepare 250. mL of a 0.0380 M $KMnO_4$(aq) solution.

Strategy We need to know the number of moles of solute formula units in the stated volume of solution. For this step, we use the relation $n = VM$ given in Toolbox 4.4.

Then we convert this number of moles to mass of solute. For this step, we use the procedure in Toolbox 4.2, using the molar mass as a conversion factor.

Solution The molarity implies that 1 L contains 0.0380 mol KMnO$_4$. The number of moles of KMnO$_4$ in 250. mL (0.250 L) of solution is therefore

$$\underbrace{\text{Number of moles of KMnO}_4}_{n} = (0.250 \text{ L}) \times \underbrace{\frac{0.0380 \text{ mol KMnO}_4}{1 \text{ L}}}_{V \times M}$$

Because the molar mass of potassium permanganate is 158.04 g/mol, this number of moles corresponds to the following mass:

$$\text{Mass of KMnO}_4 = (0.250 \times 0.0380 \text{ mol KMnO}_4) \times \frac{158.04 \text{ g KMnO}_4}{1 \text{ mol KMnO}_4}$$

$$= 1.50 \text{ g KMnO}_4$$

That is, to make up the solution, dissolve 1.50 g of potassium permanganate in water in a 250-mL volumetric flask, and then add water up to the mark, as in Fig. 4.10.

Self-Test 4.11A Calculate the mass of glucose needed to prepare 150. mL of a 0.442 M C$_6$H$_{12}$O$_6$(aq) solution.

[*Answer:* 11.9 g]

Self-Test 4.11B Calculate the mass of oxalic acid needed to prepare 50.00 mL of a 0.125 M C$_2$H$_2$O$_4$(aq) solution.

4.7 Dilution

Often chemists have only a very tiny amount of the substance they are studying. Here solutions play another very useful role, because we can use them to transfer very small amounts of a substance from one container to another. This procedure is important when only a little sample is available, perhaps because of its rarity or its cost. It can also be important when a sample being examined con-

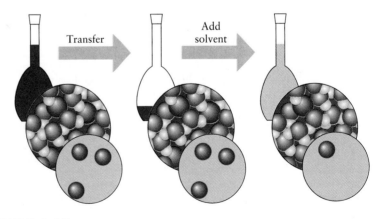

FIGURE 4.12

The steps involved in dilution. A small sample of the original solution is transferred to a volumetric flask, and then water is added up to the mark. The diluted solution (right) has fewer solute molecules per given volume than the concentrated solution.

tains very little of the substance of interest, as in the analysis of an ore for a rare metal or of a blood sample for traces of poison.

Suppose we need to add 0.010 mmol $KMnO_4$ to a solution of the compound we are investigating. However, on hand we have only the 0.0380 M $KMnO_4(aq)$ solution prepared in Example 4.10. First, we need to know the volume of solution that contains 0.010 mmol $KMnO_4$. For this step, we use $V = n/M$:

$$\text{Volume of solution} = \frac{1.0 \times 10^{-5} \text{ mol } KMnO_4}{0.0380 \text{ mol } KMnO_4 / L}$$
$$= 2.6 \times 10^{-4} \text{ L}$$

We could measure out 0.010 mmol $KMnO_4$ by measuring out 0.26 mL of the solution. An instrument called a micropipet can be used to measure out very tiny volumes of liquid accurately, perhaps as little as 10^{-3} mL (1 microliter, 1 μL)—about the volume of a pinhead. A much simpler approach, however, does not require such expensive equipment. We simply **dilute,** or reduce the concentration, of the solute by adding more solvent. After a solution has been diluted, the same amount of solute is present but in a greater volume (Fig. 4.12). An ordinary pipet or buret can then be used to transfer the solution accurately. If the original solution of molarity 0.0380 M is diluted a hundredfold, then 0.010 mmol $KMnO_4$ would be contained in 26 mL of solution, which is much easier to measure than 0.26 mL would be.

Reagents are often stored in concentrated solutions called "stock solutions" and diluted as needed. To dilute a solution we need to reduce the molarity of the solute from an initial value to some final target value. First, we use a pipet to transfer the appropriate volume of the more concentrated solution to a volumetric flask. Then we add enough solvent to increase the volume of the solution to its final value. Toolbox 4.5 shows how to calculate the correct initial volume of solution to use.

Before dilution 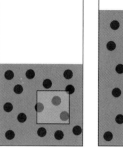 After dilution

FIGURE 4.13

When a solution is diluted, the same number of solute molecules occupy a larger volume, so there is a smaller number of molecules in a given volume (indicated by the small square).

Toolbox 4.5 *How to calculate the volume of solution to dilute*

The procedure in this Toolbox shows how much of an initial concentrated solution should be used to prepare a solution of known concentration.

Conceptual basis

The procedure is based on a simple idea: although we may add more solvent to a given volume of solution, we do not change the number of moles of solute. After dilution, the solute simply occupies a larger volume of solution (Fig. 4.13).

Procedure

The procedure has two steps:

Step 1. Use the final molarity to calculate the number of moles of formula units in the specified volume of the *final* solution (Toolbox 4.4). That is, we calculate

$$n = M_{final}V_{final}$$

where M_{final} is the molarity of the final solution and V_{final} is its volume.

The result of this part of the calculation is the amount of solute to transfer into the volumetric flask.

Step 2. Use the molarity of the initial solution to calculate the volume of the *initial* undiluted solution that contains this amount of solute:

$$V_{initial} = \frac{n}{M_{initial}}$$

A useful way to remember this procedure is to note that because the number of moles of solute in the final solution is the same as the number of moles of solute in the initial volume of solution we can write

$$n_{initial} = n_{final}$$

It then follows that

$$M_{initial}V_{initial} = M_{final}V_{final} \qquad (2)$$

We can rearrange this equation to calculate in a single step any of the four quantities given the other three.

Example 4.11 *Calculating the volume of stock solution to dilute*

Calculate the volume of a 0.0380 M $KMnO_4$(aq) stock solution that we should use to prepare 250. mL of 1.50×10^{-3} M $KMnO_4$(aq).

Strategy The calculation is based on the fact that the number of moles in the diluted solution will be the same as the number contained in the volume of the original solution used in the dilution. Proceed as described in Toolbox 4.5.

Solution **Step 1.** The molarity of solute in the final solution is 1.50×10^{-3} M. Therefore, the amount of $KMnO_4$ in 250. mL (0.250 L) of solution is

$$\underbrace{\text{Number of moles of } KMnO_4}_{n} = \overbrace{(0.250 \text{ L}) \times (1.50 \times 10^{-3} \text{ mol } KMnO_4/L)}^{V_{final} \times M_{final}}$$
$$= 0.250 \times 1.50 \times 10^{-3} \text{ mol } KMnO_4$$

Step 2. The molarity of solute in the original solution is 0.0380 mol $KMnO_4$/L. The volume of this solution that contains the calculated amount of $KMnO_4$ is therefore

$$\underbrace{\text{Volume of stock solution}}_{V_{initial}} = \overbrace{\frac{0.250 \times 1.50 \times 10^{-3} \text{ mol } KMnO_4}{0.0380 \text{ mol } KMnO_4 / L}}^{n/M_{initial}} = 9.87 \times 10^{-3} \text{ L}$$

We conclude that 9.87 mL of the stock solution should be measured into a volumetric flask (using a buret) and water added up to the mark. Note that we could have used Eq. 2 directly, after rearranging it into

$$V_{initial} = \frac{M_{final} V_{final}}{M_{initial}}$$

Substituting the data results in the same value for the initial volume:

$$\text{Volume of stock solution} = \frac{(1.50 \times 10^{-3} \text{ mol } KMnO_4 / L) \times (0.250 \text{ L})}{0.0380 \text{ mol } KMnO_4 / L}$$
$$= 9.87 \times 10^{-3} \text{ L}$$

Self-Test 4.12A Calculate the volume of 0.0155 M HCl(aq) that we should use to prepare 100. mL of a 5.23×10^{-4} M HCl(aq) solution.

[**Answer:** 3.37 mL]

Self-Test 4.12B Calculate the volume of 0.152 M $C_6H_{12}O_6$(aq) that should be used to prepare 25.00 mL of 1.59×10^{-5} M $C_6H_{12}O_6$(aq).

When a small volume of a solution is diluted to a larger volume, the total number of moles of solute in the solution does not change, but the concentration of solute is reduced.

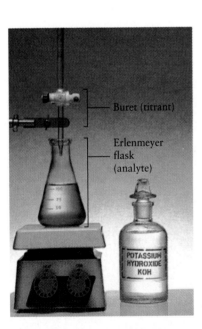

— Buret (titrant)

— Erlenmeyer flask (analyte)

POTASSIUM HYDROXIDE KOH

FIGURE 4.14

The apparatus typically used for a titration: magnetic stirrer; flask containing the analyte; clamp and buret containing the titrant.

4.8 Titrations

A laboratory technique used in virtually all fields for determining the concentration of a solute is called **titration.** Titrations are used daily to monitor water purity and blood composition and for quality control in the food industry.

Titration is a special case of **volumetric analysis,** in which the amount of substance present in a solution is determined from measurements of volume and concentration. One of the most common titration procedures makes use of acid-base neutralization. In an **acid-base titration,** the **analyte,** which is the solution being analyzed, is a base and the **titrant,** which is the solution in the buret, is an acid, or vice versa (Fig. 4.14). In a **redox titration,** one reagent is either reduced by a reducing agent or oxidized by an oxidizing agent.

In a titration, the titrant is added gradually to the analyte, until the reaction is just completed. This point is the **stoichiometric point,** the stage at which the volume of titrant added to the analyte is exactly that required to react with all the analyte. If both reactants are colorless, we need a way to detect the stoichiometric point. Often indicators that change color at this point can be used. In Chapter 3, we saw that acids and bases can be detected by noting the colors of indicators, such as the dyes litmus (red in acidic solution, blue in basic solution) or phenolphthalein (colorless in acidic solution, pink in basic solution). For example, if we titrate hydrochloric acid containing a few drops of the indicator phenolphthalein, the solution is initially colorless. After the stoichiometric point, when excess base is present, the analyte solution is basic and the indicator is pink. The indicator color change is sudden, so it is easy to detect the stoichiometric point. We then note the exact volume of titrant needed to cause the indicator to change color (Fig. 4.15). Indicators are also used for many redox titrations.

To interpret an acid-base or a redox titration, we need to know the stoichiometric relation from the chemical equation for the reaction. This relation is used to write the mole ratio in the usual way. The only new step is to use the molarities of the solutions to convert between the moles of reactants and the volumes of the analyte and titrant. Toolbox 4.6 shows how to combine this information.

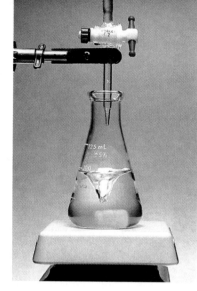

FIGURE 34.15

An acid-base titration in progress. The indicator is phenolphthalein.

> The stoichiometric point is often referred to as the equivalence point.

In a titration, the stoichiometric relation between analyte and titrant species, together with the molarity of the titrant, is used to determine the molarity or amount of the analyte.

Toolbox 4.6 *How to interpret a titration*

The data in a typical titration are the volume of analyte solution, the molarity of the titrant, and the volume of titrant solution needed to reach the stoichiometric point. The goal is usually to determine the molarity of the analyte, but sometimes we only need to determine the number of moles of solute in the analyte solution.

Conceptual basis

The volume of titrant required to reach the stoichiometric point tells us the number of moles of reactant needed to achieve complete reaction (Fig. 4.16). We then use the mole ratio for the reaction to calculate the corresponding number of moles of the analyte. If the molarity of the analyte is needed, we divide by the initial volume of the analyte sample to find the molar concentration (molarity).

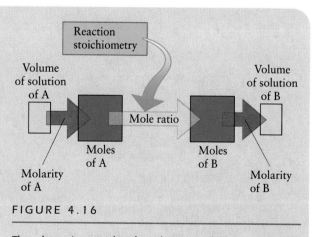

FIGURE 4.16

The schematic procedure for volume-to-volume conversions.

Procedure

The procedure consists of the three steps summarized symbolically in Fig. 4.17:

Step 1. Use the volume of titrant and its molarity to calculate the number of moles of titrant species added from the buret (see Toolbox 4.4):

$$n_{titrant} = M_{titrant}V_{titrant}$$

Because we need to keep track of the substances, we make sure that we specify explicitly the species in the concentration units, for example, by writing 1.0 M HCl(aq) rather than just 1.0 M.

Step 2. Write the stoichiometric relation between the titrant and analyte species from the chemical equation for the reaction and use it to convert the moles of titrant species to moles of analyte species. This step gives us a mole ratio of the form

$$\frac{\text{Substance required}}{\text{Substance given}}$$

which in this case is

$$\frac{\text{Analyte}}{\text{Titrant}}$$

Step 3. Calculate the molarity of the analyte by dividing the number of moles of solute by the initial volume of the solution:

$$\text{Molarity of analyte} = \frac{\text{number of moles of solute}}{\text{volume of solution (liters)}}$$

If the mass of analyte is required, instead of step 3, the molar mass of the analyte is used to convert moles of analyte to grams.

As usual in calculations with multiple steps, it is best to avoid rounding errors by doing the numerical calculation in one step at the end.

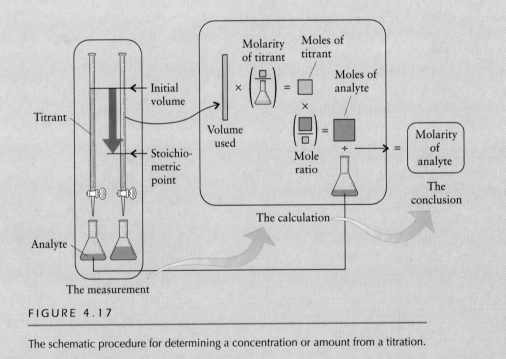

FIGURE 4.17

The schematic procedure for determining a concentration or amount from a titration.

Example 4.12 Interpreting an acid-base titration

Oxalic acid, a toxic compound found in rhubarb leaves, is a useful laboratory reducing agent. Suppose that 25.00 mL of a solution of oxalic acid, $H_2C_2O_4$ (**1**), is titrated with 0.500 M $NaOH(aq)$ and that the stoichiometric point is reached when 38.0 mL of the solution of base is added. What is the molarity of the oxalic acid solution?

Strategy We know the volume and concentration of the titrant. The chemical equation gives the mole ratio needed to convert moles of titrant to moles of analyte. Follow the procedure for determining an unknown concentration in Toolbox 4.6.

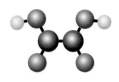

1   Oxalic acid, $(COOH)_2$

Solution **Step 1.** The number of moles of NaOH added is

$$\text{Number of moles of NaOH} = (38.0 \times 10^{-3}\,L) \times (0.500\,\text{mol NaOH/L})$$
$$= 38.0 \times 10^{-3} \times 0.500\,\text{mol NaOH}$$

Step 2. The neutralization reaction is

$$H_2C_2O_4(aq) + 2\,NaOH(aq) \longrightarrow Na_2C_2O_4(aq) + 2\,H_2O(l)$$

It follows that the stoichiometric relation we require is 2 mol NaOH $\triangleq$ 1 mol $H_2C_2O_4$. Therefore, the moles of $H_2C_2O_4$ in the original analyte solution is

$$\text{Number of moles of } H_2C_2O_4 = (38.0 \times 10^{-3} \times 0.500\,\text{mol NaOH}) \times \frac{1\,\text{mol } H_2C_2O_4}{2\,\text{mol NaOH}}$$
$$= \tfrac{1}{2} \times (38.0 \times 10^{-3} \times 0.500)\,\text{mol } H_2C_2O_4$$

Step 3. The molarity of the acid is therefore

$$\text{Molarity of } H_2C_2O_4(aq) = \frac{\tfrac{1}{2} \times (38.0 \times 10^{-3} \times 0.500)\,\text{mol } H_2C_2O_4}{25.00 \times 10^{-3}\,L}$$
$$= 0.380\,(\text{mol } H_2C_2O_4)/L$$

That is, the solution is 0.380 M $H_2C_2O_4(aq)$. We can carry out the calculation in one step by combining the three steps above:

$$\text{Molarity of } H_2C_2O_4(aq) = \frac{\overbrace{(38.0 \times 10^{-3}\,L)}^{\text{Volume of titrant}} \times \overbrace{(0.500\,\text{mol NaOH/L})}^{\text{Concentration of titrant}} \times \overbrace{\dfrac{1\,\text{mol } H_2C_2O_4}{2\,\text{mol NaOH}}}^{(2)\ \text{Convert to moles of analyte}}}{\underbrace{25.00 \times 10^{-3}\,L}_{(3)\ \text{Divide by volume of sample}}}$$

(1) Find moles of titrant

$$= 0.380\,(\text{mol } H_2C_2O_4)/L$$

Self-Test 4.13A A student prepared a sample of hydrochloric acid that contained 0.72 g of hydrogen chloride in 500. mL of solution. This solution was used to titrate 25.0 mL of a solution of calcium hydroxide, and the stoichiometric point was reached when 15.1 mL of acid had been added. What was the molarity of the calcium hydroxide in the solution?

[*Answer:* 0.012 M $Ca(OH)_2(aq)$]

Self-Test 4.13B Many abandoned mines have exposed nearby communities to the problem of acid mine drainage. Certain minerals, such as pyrite (FeS_2), decompose when exposed to air, forming solutions of sulfuric acid. The acidic mine water then drains into lakes and creeks, killing fish and other animals. At a mine in Colorado, a sample of mine water of volume 25.00 mL was neutralized with 16.45 mL of 0.255 M $KOH(aq)$. What is the molar concentration of H_2SO_4 in the water?

Example 4.13 *Interpreting a redox titration*

The iron content of certain ores can be determined by titrating a sample with a solution of potassium permanganate, $KMnO_4$. The ore is dissolved in hydrochloric acid, forming iron(II) ions, and the latter react with MnO_4^-:

$$5\,Fe^{2+}(aq)\,+\,MnO_4^-(aq)\,+\,8\,H^+(aq)\,\longrightarrow\,5\,Fe^{3+}(aq)\,+\,Mn^{2+}(aq)\,+\,4\,H_2O(l)$$

The stoichiometric point is reached when the purple color of the permanganate ion persists (Fig. 4.18). A 0.202-g sample of ore is dissolved in hydrochloric acid and the resulting solution requires 16.7 mL of 0.0108 M $KMnO_4$(aq) to reach the stoichiometric point. (a) What mass of iron(II) ions is present? (b) What is the mass percentage of iron in the ore sample?

Strategy (a) We determine the number of moles of permanganate ion used from its concentration and the volume of solution used. Then we use the mole ratio in the chemical equation to convert moles of permanganate ion to moles of iron(II). These strategies are set out in the first two steps of the procedure in Toolbox 4.6. In this case, we do not need to find the concentration of iron(II) in the solution. We only need the number of moles. Then we convert moles of Fe^{2+} ions to mass by using the molar mass of Fe^{2+}: because the mass of electrons is so small, use the molar mass of elemental iron for the molar mass of iron(II) ions. (b) Divide the mass of iron by the mass of the ore sample and multiply by 100%.

Solution (a) From the chemical equation, the stoichiometric relation between the iron and permanganate ions is

$$5\ mol\ Fe^{2+}\ \backsimeq\ 1\ mol\ MnO_4^-$$

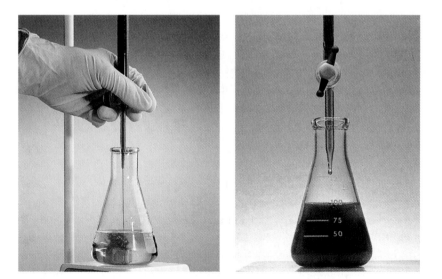

FIGURE 4.18

In this redox titration, the titrant is the oxidizing agent potassium permanganate, $KMnO_4$, and the analyte contains iron(II) ions. The purple color of the permanganate ion disappears as it reacts with iron(II). However, at the stoichiometric point, the purple color persists, showing that all the iron(II) has been oxidized to iron(III).

We can set up the mass calculation in one step:

$$\text{Mass of iron(II)} = \overbrace{(1.67 \times 10^{-2}\,\text{L}) \times (0.0108\,\text{mol MnO}_4^-/\text{L})}^{\text{Number of moles of MnO}_4^-}$$
$$\times \frac{5\,\text{mol Fe}^{2+}}{1\,\text{mol MnO}_4^-} \times \frac{55.85\,\text{g Fe}^{2+}}{1\,\text{mol Fe}^{2+}}$$
$$= 0.0504\,\text{g Fe}^{2+}$$

The sample contains 0.0504 g iron.

(b) The mass percentage of iron in the ore is

$$\text{Mass percentage iron} = \frac{\overbrace{0.0504\,\text{g}}^{\text{Mass of Fe}}}{\underbrace{0.202\,\text{g}}_{\text{Mass of sample}}} \times 100\% = 25.0\%$$

Self-Test 4.14A The iron content of clays that are to be used to make ceramics is critical to some of their properties. A sample of clay of mass 20.750 g was analyzed to determine its iron content. The clay was washed with hydrochloric acid and the iron converted to iron(II) ions. The resulting solution was titrated with cerium(IV) sulfate solution:

$$\text{Fe}^{2+}(\text{aq}) + \text{Ce}^{4+}(\text{aq}) \longrightarrow \text{Fe}^{3+}(\text{aq}) + \text{Ce}^{3+}(\text{aq})$$

In the titration 13.45 mL of 1.340 M $\text{Ce(SO}_4)_2(\text{aq})$ was needed to reach the stoichiometric point. What is the mass percentage of iron in the clay?

[*Answer:* 4.85%]

Self-Test 4.14B The amount of arsenic(III) oxide in a mineral can be determined by dissolving the mineral and titrating it with potassium permanganate:

$$24\,\text{H}^+(\text{aq}) + 5\,\text{As}_4\text{O}_6(\text{s}) + 8\,\text{MnO}_4^-(\text{aq}) + 18\,\text{H}_2\text{O}(\text{l}) \longrightarrow 8\,\text{Mn}^{2+}(\text{aq}) + 20\,\text{H}_3\text{AsO}_4(\text{aq})$$

A sample of industrial waste was analyzed for the presence of arsenic(III) oxide by titration with 0.0100 M KMnO_4. It took 28.15 mL of the titrant to reach the stoichiometric point. How many grams of arsenic(III) oxide did the sample contain?

Skills You Should Have Mastered

Conceptual

☐ 1. Explain the significance of the stoichiometric coefficients in a chemical equation, Section 4.1.

☐ 2. Explain what is meant by a stoichiometric relation and state how it is used, Self-Test 4.1.

☐ 3. Explain how diluting a solution affects the concentration of the solute, Section 4.7.

☐ 4. Explain the significance of the stoichiometric (equivalence) point of a titration, Section 4.8.

Problem-Solving

☐ 1. Carry out mole-to-mole calculations for any two species involved in a chemical reaction, Toolbox 4.1 and Example 4.1.

☐ 2. Carry out mass-to-mole and mass-to-mass calculations for any two species involved in a chemical reaction, Toolbox 4.2 and Example 4.2.

☐ 3. Carry out a gravimetric analysis, Example 4.3.

4. Calculate the theoretical and percentage yields of the products of a reaction, given the mass of starting material, Example 4.4.

5. Identify the limiting reactant of a reaction and calculate the amount of excess reactant present, Toolbox 4.3 and Example 4.5.

6. Determine the empirical formula of an organic compound by combustion analysis, Example 4.6.

7. Calculate the molarity of a solute in a solution, volume of solution, and mass of solute, given the other two quantities, Toolbox 4.4 and Examples 4.7, 4.8, and 4.9.

8. Prepare a solution of given molarity, Example 4.10.

9. Determine the volume of solution needed to prepare a dilute solution of a given molarity, Toolbox 4.5 and Example 4.11.

10. Calculate the molar concentration (molarity) of a solute from titration data, Toolbox 4.6 and Example 4.12.

11. Calculate the mass of a substance that reacts with a known volume of titrant solution, Toolbox 4.6 and Example 4.13.

Descriptive

1. Describe a procedure that can be used to perform a gravimetric analysis, Section 4.2.

2. Describe the purpose and procedure of a titration, Section 4.8.

Exercises

Mole Calculations

4.1 (a) In the commercial manufacture of nitric acid, how many moles of NO_2 produce 7.33 mol HNO_3 in the reaction

$$3 NO_2(g) + H_2O(l) \longrightarrow 2 HNO_3(aq) + NO(g)$$

(b) The concentration of iodide ions in a solution can be determined by its oxidation to iodine with permanganate ion according to the reaction

$$2 MnO_4^-(aq) + 16 H^+(aq) + 10 I^-(aq) \longrightarrow$$
$$2 Mn^{2+}(aq) + 5 I_2(aq) + 8 H_2O(l)$$

How many moles of permanganate ion react with 0.042 mol I^- ions?

4.2 (a) The removal of hydrogen sulfide gas from "sour" natural gas occurs by the reaction

$$16 H_2S(g) + 8 SO_2(g) \longrightarrow 3 S_8(s) + 16 H_2O(l)$$

How many moles of S atoms form from the reaction of 5.0 mol H_2S? (b) The neutralization of phosphoric acid with potassium hydroxide occurs by the reaction

$$H_3PO_4(aq) + 3 KOH(aq) \longrightarrow K_3PO_4(aq) + 3 H_2O(l)$$

Calculate the amount (in moles) of KOH that reacts with 0.22 mol H_3PO_4

4.3 Calculate the number of moles of CO_2 produced when 1.5 mol of hexane molecules, C_6H_{14}, burns in air by the reaction

$$2 C_6H_{14}(l) + 19 O_2(g) \longrightarrow 12 CO_2(g) + 14 H_2O(g)$$

4.4 A method used by the Environmental Protection Agency (EPA) for determining the concentration of ozone in air is to pass the air sample through a bubbler containing iodide ions, which remove the ozone according to the equation

$$O_3(g) + 2 I^-(aq) + H_2O(l) \longrightarrow O_2(g) + I_2(aq) + 2 OH^-(aq)$$

How many moles of iodide ions are needed to remove 3.5×10^{-5} mol O_3?

4.5 Sulfuric acid is a dehydrating agent that extracts water from sucrose, $C_{12}H_{22}O_{11}$, leaving only the carbon atoms. How many moles of water molecules can be extracted in this way from 0.500 mol $C_{12}H_{22}O_{11}$?

4.6 In the industrial synthesis of sulfuric acid, the compound oleum, $H_2S_2O_7$, reacts with water to form sulfuric acid. How many moles of H_2SO_4 can be manufactured from 1.0×10^6 mol $H_2S_2O_7$?

Mass-Mole Relationships

4.7 Ammonia burns in oxygen according to the equation

$$4 NH_3(g) + 3 O_2(g) \longrightarrow 2 N_2(g) + 6 H_2O(g)$$

(a) Calculate the number of moles of H_2O produced in the combustion of 1.0 g of NH_3. (b) How many grams of O_2 are required for the complete reaction of 13.7 mol NH_3?

4.8 Sodium thiosulfate, $Na_2S_2O_3$ (photographer's hypo), reacts with unexposed silver bromide in a film emulsion to form sodium bromide and a soluble compound of formula $Na_3[Ag(S_2O_3)_2]$:

$$2 Na_2S_2O_3(aq) + AgBr(s) \longrightarrow NaBr(aq) + Na_3[Ag(S_2O_3)_2](aq)$$

(a) How many moles of $Na_2S_2O_3$ are needed to dissolve 1.0 mg of AgBr? (b) Calculate the mass of silver bromide that will produce 0.033 mol $Na_3[Ag(S_2O_3)_2]$.

4.9 Small bottles of propane gas are sold in hardware stores as convenient, portable heat sources (for soldering, for example). The combustion reaction of propane is

$$C_3H_8(g) + 5 O_2(g) \longrightarrow 3 CO_2(g) + 4 H_2O(l)$$

(a) What mass of CO_2 is produced from the combustion of 1.55 mol C_3H_8? (b) How many moles of water molecules accompany the production of 4.40 g of CO_2?

4.10 Impure phosphoric acid for use in the manufacture of fertilizers is produced by the action of sulfuric acid on phosphate rock, of which a principal component is $Ca_3(PO_4)_2$. The reaction is

$$Ca_3(PO_4)_2(s) + 3 H_2SO_4(aq) \longrightarrow 3 CaSO_4(s) + 2 H_3PO_4(aq)$$

(a) How many moles of H_3PO_4 are produced from the reaction of 200. kg of H_2SO_4? (b) Determine the mass of calcium sulfate that is produced as a by-product of the reaction of 200. mol $Ca_3(PO_4)_2$.

Mass-Mass Relationships

4.11 The surface atoms of aluminum metal corrode in air to form an impervious aluminum oxide coating that prevents further corrosion of the lower layers of atoms. The oxidation reaction is

$$4\,Al(s) \ + \ 3\,O_2(g) \longrightarrow 2\,Al_2O_3(s)$$

(a) Calculate the mass of aluminum oxide formed from the corrosion of 10.0 g of aluminum. (b) In the reaction of 10.0 g of aluminum, what mass of oxygen is needed?

4.12 A typical problem in the iron industry is to determine the mass of iron that can be obtained from a mixture of iron(III) oxide, the principal component of the ore hematite, and carbon obtained from coal. The iron reduction reaction is

$$Fe_2O_3(s) \ + \ 3\,C(s) \xrightarrow{\ \Delta\ } 2\,Fe(l) \ + \ 3\,CO(g)$$

(a) Calculate the mass of iron that can be produced from 1.0 tonne (1.0 t) of Fe_2O_3 (where 1 t $= 10^3$ kg). (b) What mass of carbon is needed for the reduction of Fe_2O_3 to produce 500. kg of iron?

4.13 The solid fuel in the booster stage of the space shuttle is a mixture of ammonium perchlorate and aluminum powder. Upon ignition, one reaction that occurs is

$$6\,NH_4ClO_4(s) \ + \ 10\,Al(s) \longrightarrow$$
$$5\,Al_2O_3(s) \ + \ 3\,N_2(g) \ + \ 6\,HCl(g) \ + \ 9\,H_2O(g)$$

(a) How many kilograms of aluminum should be mixed with 1.5×10^4 kg of NH_4ClO_4 for this reaction? (b) Determine the mass of Al_2O_3 (alumina, a finely divided white powder that is produced as billows of white smoke) formed in the reaction of 5000. kg of aluminum.

The Space Shuttle booster, Exercise 4.13.

4.14 The compound diborane, B_2H_6, was at one time considered for use as a rocket fuel. The reaction for the combusion of diborane is

$$B_2H_6(g) \ + \ 3\,O_2(l) \longrightarrow 2\,HBO_2(g) \ + \ 2\,H_2O(l)$$

The fact that HBO_2, a reactive compound, was produced rather than the relatively inert B_2O_3 was a factor in the discontinuation of the investigation of diborane as a fuel. (a) What mass of liquid oxygen (LOX) would be needed to burn 50.0 g of B_2H_6? (b) Determine the mass of HBO_2 produced from the combustion of 30.0 g of B_2H_6.

4.15 The camel stores the fat tristearin, $C_{57}H_{110}O_6$, in its hump. As well as being a source of energy, the fat is also a source of water, because, when the fat is oxidized, the reaction

$$2\,C_{57}H_{110}O_6(s) \ + \ 163\,O_2(g) \longrightarrow 114\,CO_2(g) \ + \ 110\,H_2O(l)$$

takes place. (a) What mass of water is available from the oxidation of 2.5 kg of this fat? (b) What mass of oxygen is needed to oxidize 2.5 g of tristearin?

4.16 Potassium superoxide, KO_2, is utilized in closed-system breathing apparatus to remove carbon dioxide and water from exhaled air. The removal of water generates oxygen for breathing by the reaction

$$4\,KO_2(s) \ + \ 2\,H_2O(l) \longrightarrow 3\,O_2(g) \ + \ 4\,KOH(s)$$

The potassium hydroxide removes carbon dioxide from the apparatus by the reaction

$$KOH(s) \ + \ CO_2(g) \longrightarrow KHCO_3(s)$$

(a) What mass of potassium superoxide generates 20.0 g of O_2? (b) What mass of CO_2 can be removed from the apparatus by 100. g of KO_2?

4.17 When a hydrocarbon burns, water is produced as well as carbon dioxide. (For this reason, clouds of condensed water droplets are often seen coming from automobile exhausts, especially on a cold day.) The density of gasoline is 0.79 g/mL. Assume gasoline to be only octane, C_8H_{18}, for which the combustion reaction is

$$2\,C_8H_{18}(l) \ + \ 25\,O_2(g) \longrightarrow 16\,CO_2(g) \ + \ 18\,H_2O(l)$$

Calculate the mass of water produced from the combustion of 1.0 L of gasoline.

4.18 The density of oak wood is 0.72 g/cm^3. Assuming oak wood to have the empirical formula $C_6H_{12}O_6$, calculate the mass of water produced when a log of dimensions 12 cm $\times$ 14 cm $\times$ 25 cm is burned.

Reaction Yield

4.19 When limestone rock, which is principally $CaCO_3$, is heated, carbon dioxide and quicklime, CaO, are produced by the reaction

$$CaCO_3(s) \longrightarrow CaO(s) \ + \ CO_2(g)$$

If 11.7 g of CO_2 is produced from the thermal decomposition of 30.7 g of $CaCO_3$, what is the percentage yield of the reaction?

4.20 Phosphorus trichloride, PCl_3, is produced from the reaction of white phosphorus, P_4, and chlorine:

$$P_4(s) + 6\,Cl_2(g) \longrightarrow 4\,PCl_3(g)$$

A 16.4-g sample of PCl_3 was collected from the reaction of 5.00 g of P_4 with excess chlorine. What is the percentage yield of the reaction?

Limiting Reactants

4.21 Complete the drawing below, which represents the reaction $2\,H_2(s) + O_2(g) \rightarrow 2\,H_2O(g)$. The only reactants available are the numbers of atoms and molecules in the first box. In the second box, draw the system after reaction takes place. Include any unreacted atoms or molecules. In the drawing, red represents an oxygen atom and dark gray, a hydrogen atom.

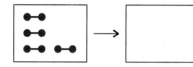

4.22 Complete the drawing below, which represents the reaction $N_2(g) + 2\,O_2(g) \rightarrow N_2O_4(g)$. The only reactants available are the numbers of atoms and molecules in the first box. In the second box, draw the system after reaction takes place. Include any unreacted molecules. In the drawing, red represents an oxygen atom and blue, a nitrogen atom.

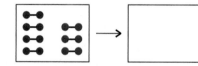

4.23 The souring of wine occurs when ethanol, C_2H_5OH, is converted by oxidation into acetic acid:

$$C_2H_5OH(aq) + O_2(g) \longrightarrow CH_3COOH(aq) + H_2O(l)$$

If 2.00 g of ethanol and 1.00 g of oxygen were sealed in a wine bottle, which would be the limiting reactant for the oxidation?

4.24 Phosphorus trichloride, PCl_3, reacts with water to form phosphorous acid, H_3PO_3, and hydrochloric acid:

$$PCl_3(l) + 3\,H_2O(l) \longrightarrow H_3PO_3(aq) + 3\,HCl(aq)$$

(a) Which is the limiting reactant when 12.4 g of PCl_3 is mixed with 10.0 g of H_2O? (b) What masses of phosphorous acid and hydrochloric acid are formed?

4.25 4.0 L of SO_2 is mixed with 4.0 L of H_2S. Upon heating, a reaction occurs, forming elemental sulfur and water:

$$SO_2(g) + 2\,H_2S(g) \longrightarrow 3\,S(s) + 2\,H_2O(g)$$

Under the conditions of the experiment, the density of SO_2 is 2.86 g/L and that of H_2S is 1.52 g/L. (a) What is the limiting reactant in the formation of sulfur? (b) Calculate the mass of excess reactant in the system. (c) What masses of sulfur and water can form in this experiment? (d) Compare the sum of

the masses of SO_2 and H_2S before reaction with the sum of the masses of the excess reactant, sulfur, and water after reaction.

4.26 Lithium metal is the only member of Group 1 that reacts directly with nitrogen to produce a nitride, Li_3N:

$$6\,Li(s) + N_2(g) \longrightarrow 2\,Li_3N(s)$$

(a) What mass of lithium nitride can form from 48.0 g of lithium and 1.0×10^{24} molecules of N_2? (b) Determine the mass of excess reactant in the system. (c) Confirm that the sum of the masses of lithium and nitrogen before reaction is the same as the sum of the masses of the excess reactant and lithium nitride after reaction.

4.27 The compound aluminum chloride is widely used to control chemical reactions in industry. It is made by the following reaction:

$$Al_2O_3(s) + 3\,C(s) + 3\,Cl_2(g) \xrightarrow{\Delta} 2\,AlCl_3(s) + 3\,CO(g)$$

(a) Which is the limiting reactant if 160. g of Cl_2 is added to a mixture of 100. g of Al_2O_3 and 40. g of C? (b) How many grams of $AlCl_3$ can be produced from these reactants?

4.28 The highly toxic compound hydrogen cyanide is used in the preparation of cyanimid fertilizers and in gas chambers for executions. It is made by the following reaction:

$$2\,CH_4(g) + 2\,NH_3(g) + 3\,O_2(g) \xrightarrow{100°C} 2\,HCN(g) + 6\,H_2O(g)$$

(a) Which is the limiting reactant if 600. g of O_2 is added to a mixture of 400. g of NH_3 and 300. g of CH_4? (b) How many grams of HCN can be produced from these reactants?

Combustion Analysis

4.29 A chemical analysis of a 4.39-g sample of benzoic acid, a common carboxylic acid used as a food preservative, is 3.03 g C, 1.14 g O, and the remainder hydrogen. What is the empirical formula of benzoic acid?

4.30 Oxalic acid occurs in rhubarb leaves. Analysis of a sample of oxalic acid of mass 10.0 g showed that it contained 0.22 g H, 2.7 g C, and the remainder oxygen. What is the empirical formula of oxalic acid?

4.31 The analysis of 0.922 g of aniline, a common organic base used in some varnishes, is 0.714 g C, 0.138 g N, and the remainder hydrogen. What is the empirical formula of aniline?

4.32 Urea is used as a commercial fertilizer because of its nitrogen content. An analysis of 25.0 mg of urea showed that it contained 5.0 mg C, 11.68 mg N, 6.65 mg O, and the remainder hydrogen. What is the empirical formula of urea?

4.33 In a combustion analysis of a 0.152-g sample of the artificial sweetener aspartame, it was found that 0.318 g of carbon dioxide, 0.084 g of water, and 0.0145 g of nitrogen were produced. What is the empirical formula of aspartame? The molar mass of aspartame is 294 g/mol. What is its molecular formula?

4.34 The bitter-tasting compound quinine is a component of tonic water and is used as a protection against malaria. When a sample of mass 0.487 g was burned, 1.321 g of carbon dioxide, 0.325 g of water, and 0.0421 g of nitrogen were produced. The molar mass of quinine is 324 g/mol. Determine the empirical and molecular formulas of quinine.

4.35 The stimulant in coffee and tea is caffeine, a substance of molar mass 194 g/mol. When 0.376 g of caffeine was burned, 0.682 g of carbon dioxide, 0.174 g of water, and 0.110 g of nitrogen were formed. Determine the empirical and molecular formulas of caffeine, and write the equation for its combustion.

4.36 Nicotine, the stimulant in tobacco, causes a very complex set of physiological effects in the body. It is known to have a molar mass of 162 g/mol. When a sample of mass 0.385 g was burned, 1.072 g of carbon dioxide, 0.307 g of water, and 0.068 g of nitrogen were produced. What are the empirical and molecular formulas of nicotine? Write the equation for its combustion.

Solutions

4.37 (a) An aqueous solution was prepared by dissolving 1.567 mol of $AgNO_3$ in enough water to make 250.0 mL of solution. What is the molarity of silver nitrate in the solution? (b) A 2.11-g sample of NaCl is placed into a 1500.-mL volumetric flask. The sample is dissolved and diluted to the mark on the flask with water. What is the molarity of NaCl in the solution?

4.38 (a) A chemist prepared a solution by dissolving 1.230 g of KCl in enough water to make 150.0 mL of solution. What molar concentration of potassium chloride should appear on the label? (b) If the chemist had mistakenly used a 500.0-mL volumetric flask instead of the 150.0-mL flask in (a), what molar concentration of potassium chloride has the chemist actually prepared?

4.39 A chemist studying the properties of photographic emulsions needed to prepare 25.00 mL of 0.155 M $AgNO_3(aq)$. What mass of silver nitrate must be placed into a 25.00-mL volumetric flask and dissolved and diluted to the mark with water?

4.40 What mass of $Na_2CO_3 \cdot 10H_2O$, a compound used in detergents, must be dissolved and diluted to the mark in a 500.0-mL volumetric flask to prepare a 0.10 M $Na_2CO_3(aq)$ solution?

4.41 A student prepared a solution of barium hydroxide by adding 2.577 g of the solid to a 250.0-mL volumetric flask and adding water to the mark. Some of this solution was transferred to a buret. What volume of solution should the student run into a flask to transfer (a) 1.0 mmol $Ba(OH)_2$; (b) 3.5 mmol OH^-; (c) 50.0 mg $Ba(OH)_2$?

4.42 A student investigating the properties of solutions containing carbonate ions prepared a solution containing 7.112 g of Na_2CO_3 in a 250.0-mL volumetric flask. Some of the solution was transferred to a buret. What volume of solution should be dispensed from the buret to provide (a) 5.112×10^{-3} mol Na_2CO_3; (b) 3.451×10^{-3} mol CO_3^{2-}?

4.43 Explain how you would prepare 0.010 M $KMnO_4(aq)$ starting with (a) solid $KMnO_4$; (b) a 0.050 M $KMnO_4(aq)$ stock solution.

4.44 (a) A 12.56-mL sample of 1.345 M $K_2SO_4(aq)$ is diluted to 250.0 mL. What is the molar concentration of K_2SO_4 in the diluted solution? (b) A 25.00-mL sample of 0.366 M HCl(aq) is drawn from a reagent bottle with a pipet. The sample is transferred to a 150.00-mL volumetric flask and diluted to the mark with water. What is the molar concentration of the dilute hydrochloric acid?

4.45 (a) What volume of a 0.778 M $Na_2CO_3(aq)$ solution should be diluted to 150.0 mL with water to reduce its concentration to 0.0234 M $Na_2CO_3(aq)$? (b) An experiment requires the use of 60.0 mL of 0.50 M NaOH(aq). The stockroom assistant can only find a reagent bottle of 2.5 M NaOH(aq). How is the 0.50 M NaOH(aq) solution to be prepared?

4.46 0.094 g of $CuSO_4 \cdot 5H_2O$ is dissolved and diluted to the mark in a 500.0-mL volumetric flask. A 2.00-mL sample of this solution is transferred to a second 500.0-mL volumetric flask and diluted. (a) What is the molarity of $CuSO_4$ in the final solution? (b) To prepare the solution directly, what mass of $CuSO_4 \cdot 5H_2O$ would need to be weighed out?

Titrations

4.47 What is meant by the stoichiometric point of a titration? Give an example.

4.48 Outline the procedure used for an acid-base titration, and describe how the data are interpreted.

4.49 A 15.00-mL sample of sodium hydroxide was titrated to the stoichiometric point with 17.40 mL of 0.234 M HCl(aq). (a) What is the molarity of NaOH in the solution? (b) Calculate the mass of NaOH in the solution.

4.50 A 15.00-mL sample of oxalic acid, $H_2C_2O_4$ (with two acidic protons), was titrated to the stoichiometric point with 17.02 mL of 0.288 M NaOH(aq). (a) What is the molarity of oxalic acid in the solution? (b) Determine the mass of oxalic acid in the sample.

4.51 A 9.670-g sample of barium hydroxide is dissolved and diluted to the mark in a 250.0-mL volumetric flask. It was found that 11.56 mL of this solution was needed to reach the stoichiometric point in a titration of 25.00 mL of a nitric acid solution. (a) Calculate the molarity of HNO_3 in the solution. (b) What mass of HNO_3 was in the initial sample?

4.52 A 10.0-mL sample of 3.0 M KOH(aq) is transferred to a 250.0-mL volumetric flask and diluted to the mark. It was found that 38.5 mL of this diluted solution was needed to reach the stoichiometric point in a titration of 10.0 mL of a phosphoric acid solution, according to the reaction

$$3 KOH(aq) + H_3PO_4(aq) \longrightarrow K_3PO_4(aq) + 3 H_2O(l)$$

(a) Calculate the molarity of H_3PO_4 in the solution. (b) What mass of H_3PO_4 was in the initial sample?

4.53 In a titration, a 3.25-g sample of an acid, HX, requires 68.8 mL of a 0.750 M NaOH(aq) solution for complete reaction. What is the molar mass of the acid?

4.54 In a titration, 16.02 mL of 0.100 M NaOH(aq) was required to titrate 0.2011 g of an unknown acid, HX. What is the molar mass of the acid?

Supplementary Exercises

4.55 Define the terms *theoretical yield* and *percentage yield,* and explain why the latter may be less than 100%.

4.56 Sodium thiosulfate, $Na_2S_2O_3$, is often used in the laboratory to determine the molarity of solutions of iodine (as I_3^-):

$$2 S_2O_3^{2-}(aq) + I_3^-(aq) \longrightarrow 3 I^-(aq) + S_4O_6^{2-}(aq)$$

If 1.134 g of $Na_2S_2O_3$ was needed to react with all the iodine in a sample, how many moles of I_3^- were present?

4.57 A mixture of hydrogen peroxide and hydrazine can be used as a rocket propellant:

$$7 H_2O_2(g) + N_2H_4(l) \longrightarrow 2 HNO_3(aq) + 8 H_2O(l)$$

(a) How many moles of H_2O_2 react with 0.477 mol N_2H_4? (b) How many moles of HNO_3 can be produced in a reaction of 6.77 g of H_2O_2 with excess N_2H_4? (c) What mass of H_2O can be produced in a reaction of 49.6 mg H_2O_2 with excess N_2H_4?

4.58 Nitrogen dioxide contributes to the formation of acid rain as a result of its reaction with water in the air, according to the reaction

$$3 NO_2(g) + H_2O(g) \longrightarrow 2 HNO_3(aq) + NO(g)$$

(a) What mass (in milligrams) of HNO_3 can be produced from the reaction of 2.93 mg of NO_2, assuming an excess of H_2O? (b) If only 1.91 mg of HNO_3 was produced in the reaction, what is the percentage yield?

4.59 Manganese metal can be prepared by the thermite reaction

$$4 Al(s) + 3 MnO_2(s) \longrightarrow 3 Mn(l) + 2 Al_2O_3(s)$$

(a) What mass of manganese metal (in milligrams) can be produced from the reaction of 2.935 mg of aluminum with an excess of MnO_2? (b) If, in the reaction, only 2.386 mg of manganese was produced, what is the percentage yield?

4.60 Consider the reaction

$$10 N_2O(g) + C_3H_8(g) \longrightarrow 10 N_2(g) + 3 CO_2(g) + 4 H_2O(g)$$

What volume of carbon dioxide gas can be produced from the reaction of 8.40 mg of N_2O? Assume an excess of C_3H_8 and take the density of carbon dioxide to be 1.96 g/L under the conditions of the experiment.

4.61 Octane, the principal component of gasoline, burns in excess air by the reaction

$$2 C_8H_{18}(l) + 25 O_2(g) \longrightarrow 16 CO_2(g) + 18 H_2O(l)$$

(a) Calculate the volume of oxygen gas needed to react with 2.27 mg of C_8H_{18}, given that the density of oxygen is 1.43 g/L under the conditions of the experiment. (b) What volume of air is required in (a), given that air is 21% O_2 by volume?

4.62 The equation describing the reaction used for the analysis of ozone in the atmosphere is

$$O_3(g) + 2 KI(aq) + H_2O(l) \longrightarrow O_2(g) + I_2(aq) + 2 KOH(aq)$$

(a) Determine the amount (in moles) of KI needed for the reaction of 0.20 mol of O_3. (b) Calculate the mass of I_2 produced from the reaction of 49.7 mg of O_3. (c) Suppose in (b) 0.118 mg of I_2 was produced. What is the percentage yield in the reaction?

4.63 Nitrogen and hydrogen combine to form ammonia according to this reaction:

$$N_2(g) + 3 H_2(g) \longrightarrow 2 NH_3(g)$$

Assume that you start with the numbers of molecules shown in the accompanying illustration of the reaction container and that the reaction is brought to completion. Draw the flask as it would look at the end of the reaction. (a) Show each molecule of NH_3 that forms and draw all reactant molecules that remain in excess. (b) Which is the limiting reactant?

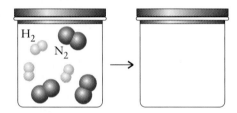

4.64 What mass of the excess reactant remains in a reaction mixture consisting of 1.54 g of $Cr(NO_3)_3$ dissolved in 120. mL of 0.10 M $H_2S(aq)$? The reaction is

$$2 Cr(NO_3)_3(aq) + 3 H_2S(aq) \longrightarrow Cr_2S_3(s) + 6 HNO_3(aq)$$

4.65 The concentration of gold in seawater is 0.011 μg/L. What mass of gold (in kg) is present in the Atlantic Ocean, of volume 3.23×10^{11} km³?

4.66 What mass of $NiSO_4 \cdot 6H_2O$ must be dissolved and diluted to the mark in a 500.0-mL volumetric flask to prepare a 0.15 M $NiSO_4(aq)$ solution?

4.67 (a) Determine the mass of anhydrous copper(II) sulfate that must be used to prepare 250. mL of a 0.20 M $CuSO_4(aq)$ solution. (b) Determine the mass of $CuSO_4 \cdot 5H_2O$ that must be used to prepare 250. mL of a 0.20 M $CuSO_4(aq)$ solution.

4.68 The sulfuric acid solution that is purchased for a stockroom has a molarity of 17.8 M; all sulfuric acid solutions for experiments are prepared by dilution of this stock solution. (a) Determine the volume of 17.8 M $H_2SO_4(aq)$ that must be diluted to 250. mL to prepare a 2.0 M $H_2SO_4(aq)$ solution. (b) An experiment requires a 0.50 M $H_2SO_4(aq)$ solution. The stockroom manager estimates that 6.0 L of the acid is needed. What volume of 17.8 M $H_2SO_4(aq)$ must be used for the preparation?

4.69 A group of students is packing a lunch for a picnic. They are making a special sandwich from two slices of bread, three slices of cheese, one pickle, one slice of tomato, and peanut butter. If they have 16 slices of bread, 21 slices of cheese, 10 pickles, 12 slices of tomato, and an unlimited supply of peanut butter, how many sandwiches can they make?

4.70 A solution of hydrochloric acid was prepared by measuring 10.00 mL of the concentrated acid into a 1.000-L volumetric flask and adding water up to the mark. Another solution was prepared by adding 0.530 g of anhydrous sodium carbonate to a 100.0-mL volumetric flask and adding water up to the mark. Then, 25.00 mL of the latter solution was pipetted into a flask and titrated with the diluted acid. The stoichiometric point was reached after 26.50 mL of acid had been added. (a) Write the balanced equation for the reaction of $HCl(aq)$ with $Na_2CO_3(aq)$. (b) What is the molarity of the original hydrochloric acid?

4.71 A chemist prepared an aqueous solution by mixing 2.50 g of ammonium phosphate trihydrate, $(NH_4)_3PO_4 \cdot 3H_2O$ and 1.50 g of potassium phosphate, K_3PO_4, with 500. g water. (a) Determine the number of moles of formula units of each compound that was measured. (b) How many moles of PO_4^{3-} are present in solution? (c) Calculate the mass of phosphate ions present in the solution. (d) What is the total mass of the water present in the solution?

4.72 To prepare a very dilute solution, it is advisable to perform successive dilutions of a single prepared reagent solution, rather than weighing a very small mass or measuring a very small volume of stock chemical. A solution was prepared by transferring 0.661 g of $K_2Cr_2O_7$ to a 250.0-mL volumetric flask and adding water to the mark. A 1.000-mL sample of this solution was transferred to a 500.0-mL volumetric flask and diluted to the mark with water. Then 10.0 mL of the diluted solution was transferred to a 250.0-mL flask and diluted to the mark with water. (a) What is the final concentration of $K_2Cr_2O_7$ in solution? (b) What mass of $K_2Cr_2O_7$ is in this final solution? (The answer to the last question gives the amount that would have had to have been weighed out if the solution had been prepared directly.)

4.73 A chemist was analyzing the concentration of sulfuric acid in water draining from an abandoned mine. A solution of sodium hydroxide was used to titrate the sample. In the process, the chemist made several mistakes. How would each mistake affect the results of the analysis if it were the only mistake made? In each case, state whether the reported result will show a concentration that is too high or too low, or whether there will be no effect. (a) The flask used to contain the acid was cleaned but not dried, and a small pool of water remained in the flask when the sample was added. (b) The buret used to contain the NaOH was cleaned but not dried. The inside of the buret was still wet when the base was added, and the buret was not rinsed with base first. (c) During the titration, the base was added too rapidly, so that two extra drops of base were added.

Applied Exercises

For Exercise 4.76–4.80, see Applying Chemistry: Case Study 4.

4.74 Ultrapure silicon used in solid-state electronics is produced by the zone refining of impure silicon from the high-temperature reduction of silicon dioxide (SiO_2, the principal component of sand and quartz) with carbon:

$$SiO_2(s) + 2\,C(s) \longrightarrow Si(l) + 2\,CO(g)$$

(a) How many Si atoms are produced in conjunction with 5.00 mg of CO? (b) How many CO molecules can be produced from the reaction of 7.33×10^{22} SiO_2 formula units?

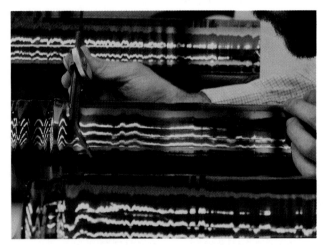

Ultrapure silicon, Exercise 4.74.

4.75 The most heavily manufactured inorganic chemical is sulfuric acid. Concentrated sulfuric acid is sold as an aqueous solution that is 98% by mass H_2SO_4. What is the molarity of this sulfuric acid?

4.76 Methanol, CH_3OH, is a clean-burning liquid fuel being considered as a replacement for gasoline. Calculate the theoretical yield in kilograms of CO_2 produced by the combustion of 1.00 L of methanol (of density 0.791 g/cm^3) and compare it with the 2.16 kg of CO_2 generated by the combustion of 1.00 L of octane. Which fuel contributes more CO_2 per liter to the atmosphere when burned? What other factors would you take into consideration when deciding which of the two fuels to use?

4.77 One of the sources of atmospheric carbon dioxide is the production of quicklime by heating limestone rock:

$$CaCO_3(s) \xrightarrow{\Delta} CaO(s) + CO_2(g)$$

What mass of CO_2 is released during the conversion of 1.00 t (1 t $= 10^3$ kg) of limestone to quicklime?

4.78 In 1992 the concentration of the greenhouse gas methane was estimated to be 1.74 ppm by mass (ppm is parts per million, so this concentration corresponds to 1.74 mg of methane per kilogram of air). Assuming a total atmospheric mass of 5.1×10^{21} g, what was the mass of methane in the atmosphere in 1992?

4.79 What is the mass percentage of carbon in the fuels (a) ethene, C_2H_4; (b) propanol, C_3H_7OH; (c) heptane, C_7H_{16}? (d) Write the chemical equation for the combustion of each compound in oxygen to form carbon dioxide gas and liquid water. (e) Which fuel produces the greatest amount of carbon dioxide per gram of fuel when it burns?

4.80 The reduction of iron(III) oxide to iron metal in a blast furnace is another source of atmospheric carbon dioxide. The reduction takes place in these two steps:

$$2\,C(s) + O_2(g) \longrightarrow 2\,CO(g)$$

$$Fe_2O_3(s) + 3\,CO(g) \longrightarrow 2\,Fe(l) + 3\,CO_2(g)$$

Assume that all the CO generated in the first step reacts in the second. (a) How many C atoms are needed to react with 600. Fe_2O_3 formula units? (b) What is the maximum volume of carbon dioxide (taken to have a density of 1.25 g/L) that can be generated in the production of 1.0 t of iron? (c) Assuming a 67.9% yield, what volume of carbon dioxide is released to the atmosphere in the production of 1.0 t of iron? (d) How many kilograms of O_2 are required for the production of 5.00 kg of Fe? See the illustration below.

4.81 The mass percentage of sulfur in a fuel, such as coal, can be determined by burning the fuel in air and passing the resulting SO_2 and SO_3 gases into dilute H_2O_2, which converts the oxides to sulfuric acid. The amount of the sulfuric acid formed can be determined by titration with a base. In one experiment, 6.43 g of coal was burned in excess air and the resulting H_2SO_4 was titrated to the stoichiometric point with 17.4 mL of 0.100 M NaOH. (a) Determine the amount (in moles) of H_2SO_4 that was produced. (b) What is the mass percentage of sulfur in the coal sample?

4.82 A vitamin C tablet was analyzed to determine whether it did in fact contain, as the manufacturer claimed, 1.0 g of the vitamin. A tablet was dissolved in water to form a 100.00-mL solution, and a 10.0-mL sample was titrated with iodine (as potassium triiodide). It required 10.1 mL of 0.0521 M I_3^-(aq) to reach the stoichiometric point in the titration. Given that 1 mol $I_3^- \simeq 1$ mol vitamin C in the reaction, is the manufacturer's claim correct? The molar mass of vitamin C is 176 g/mol.

4.83 A forensic chemist needed to determine the concentration of HCN in the blood of a suspected homicide victim and decided to titrate a diluted sample of the blood with iodine, using the reaction

$$HCN(aq) + I_3^-(aq) \longrightarrow ICN(aq) + 2\,I^-(aq) + H^+(aq)$$

A diluted blood sample of volume 15.00 mL was titrated to the stoichiometric point with 5.21 mL of an I_3^- solution. The molarity of I_3^- in the solution was determined by titrating it against arsenic(III) oxide, which in solution forms arsenious acid, H_3AsO_3. It was found that 10.42 mL of the triiodide solution was needed to reach the stoichiometric point with a 10.00-mL sample of 0.1235 M H_3AsO_3 in the reaction

$$H_3AsO_3(aq) + I_3^-(aq) + H_2O(l) \longrightarrow$$
$$H_3AsO_4(aq) + 3\,I^-(aq) + 2\,H^+(aq)$$

(a) What is the molarity of triiodide ions in the initial solution? (b) What is the molar concentration of HCN in the blood sample?

A blast furnace, Exercise 4.80.

Integrated Exercises

4.84 When aqueous solutions of calcium nitrate and phosphoric acid are mixed, a white solid precipitates. (a) What is the formula of the solid? (b) How many grams of the solid can be formed from 206 g of calcium nitrate and 150. g of phosphoric acid?

4.85 (a) What is the limiting reactant in a reaction mixture of 12.0 kg of SO_2 and 8.0 kg of H_2S for the reaction

$$8 SO_2(g) + 16 H_2S(g) \longrightarrow 3 S_8(s) + 16 H_2O(l)$$

(b) Identify the oxidizing and reducing agents in this reaction.

4.86 Dinitrogen trioxide reacts with water to form nitrous acid. (a) Write the balanced chemical equation for this reaction. (b) Write the oxidation number of nitrogen in each compound that contains nitrogen. (c) If 2 mol N_2O_3 reacts completely with 1.00 L of water, what will be the molar concentration of HNO_2 in the final solution?

4.87 Antimony reacts with oxygen as follows:

$$4 Sb(s) + 3 O_2(g) \longrightarrow 2 Sb_2O_3(s)$$

(a) What type of reaction is this? (b) What is the limiting reactant when 5.0 mol Sb(s) and 5.0 mol $O_2(g)$ react?

(c) How many moles of the excess reactant remain if reaction is complete? (d) How many moles of product can be formed? (e) If 2.0 mol Sb_2O_3 forms, what is the percentage yield?

4.88 A mixture of 4.94 g of 85.0% pure phosphine, PH_3, and 0.110 kg of $CuSO_4 \cdot 5H_2O$ (of molar mass 249.68 g/mol) is placed in a reaction vessel. (a) Balance the chemical equation for the reaction that takes place, given the skeletal form

$$CuSO_4 \cdot 5H_2O(s) + PH_3(g) \longrightarrow$$
$$Cu_3P_2(s) + H_2SO_4(aq) + H_2O(l)$$

(b) Name each reactant and product. (c) Calculate the mass (in grams) of Cu_3P_2 (of molar mass 252.56 g/mol) formed if a 6.31% yield were produced in the reaction.

4.89 Iodine is a common oxidizing agent, often used as the triiodide ion, I_3^-. Suppose that 25.00 mL of 0.120 M $I_3^-(aq)$ reacts completely with 30.00 mL of a solution containing 19.0 g/L of an ionic compound containing tin and chlorine. The products are iodide ions and another compound of tin and chlorine. The reactant compound is 62.6% tin by mass. (a) Write the formulas of the reactant and product compounds of tin. (b) Write a balanced equation for the reaction. (c) Identify the oxidation state of tin in each compound.

CHEMISTRY IN THE DRUGSTORE

Connection 1

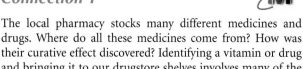

The local pharmacy stocks many different medicines and drugs. Where do all these medicines come from? How was their curative effect discovered? Identifying a vitamin or drug and bringing it to our drugstore shelves involves many of the chemical concepts and procedures that you have been studying in Chapters 1–4.

In antiquity and until very recently, medicines were discovered by trial and error or by accident. There was no way to predict a drug's effect except by testing it on patients. As a result, treatment could create other problems and was sometimes fatal to the patient. For example, arsenic compounds were prescribed for stomach problems, heroin for coughs, and mercury compounds for syphilis. However, some of the medical and herbal remedies of the past were remarkably effective. These medicines now serve as a source of starting material on which to build even more powerful agents that maximize curative effect while reducing side effects. Drugs that target specific infectious agents or tumors while not harming the patient are the "magic bullets" of medicine.

The search for new drugs relies on the skills of synthetic organic chemists and pharmacologists. Because there are so many millions of compounds, it would take too long to start with the elements, combine them in different ways, then test the products. Instead, chemists usually start either by drug *discovery*, the identification of promising medicines that already exist in nature; or by *rational drug design*, the identification of characteristics of a target bacterium or parasite and the design of new compounds to react with it. Here we look at the process of drug discovery.

In drug discovery, a chemist usually begins by investigating compounds that have already shown medicinal value. A fruitful path is to find a *natural product*, a substance found in nature, that has been shown to have healing characteristics. Nature is the best of all synthetic chemists, with billions of chemicals that fulfill as many different needs. The trick is to find the ones that have curative powers. These substances are found in different ways. In random or "blind" collection of samples, a wide variety of plant extracts are collected and tested in the hope that one will be effective. In *guided discovery*, a collection of specific samples is identified with the consultation of native healers who have knowledge of the medical value of local herbs and other substances.

Observation of the properties of plants and animals can help to guide a random search. For example, if certain types of fruits are seen to remain fresh while others rot or mold, we might expect them to contain antifungal agents. One example of this type of collection is the gathering of tunicates and sponges in the Caribbean. The antiviral drug didemnin-C and the anticancer drug bryostatin 1 were discovered in marine organisms identified in this way (Applying Chemistry: Case Study 2).

The guided route involves the testing of fewer samples, because the chemist works with a native healer, the ancient lore guiding the modern chemistry. Often an ethnobotanist, a specialist in plants used for native healing, joins the team. This approach saves time for the scientists and can benefit the healers and their nations as well. In 1992, the United Nations Convention on Biological Diversity set guidelines for the conservation of potentially valuable plant and animal species and for the interaction of scientists with native healers. Drugs that have been discovered in this way include a variety of anticancer and antimalarial drugs, blood-clotting agents, antibiotics, and medicines for the heart and digestive system. Once the empirical and molecular formulas of the active compounds in natural products have been determined (as described in Sections 2.12, 2.13, and 4.5), their structural formulas are deduced. Then chemists look for a way to make them in the laboratory.

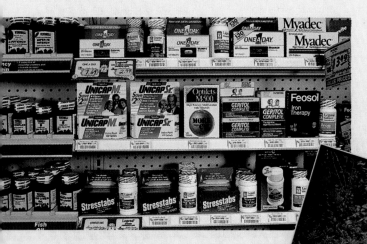

(Above) The vitamin supplements on drugstore shelves are the end product of years of discovery, synthesis, and testing. (Right) This field biologist is collecting specimens of plants that may contain compounds with medicinal value.

Once a good route to a drug has been found, it must be thoroughly tested before being used on patients. Even after a drug is on the market, the pharmaceutical company that manufactures the drug must control quality by monitoring the composition of the product. For example, the ability of an antacid to neutralize acid is measured by titration.

Even everyday medicines like mineral supplements and multivitamin capsules have many years of research behind them. For example, medical research showed that insufficient iron in the diet can lead to a form of weakness called anemia. It also shows that enzymes containing zinc play important roles in metabolism, including the expression of our genes, the digestion of food, and aging.

Many *vitamins,* which are chemicals essential to life in trace quantities, have been discovered by observing that a particular disease can be cured by an alteration of diet. For example, sailors used to be prone to scurvy, a debilitating disease that resulted in many deaths. Then in 1753 the Scottish naval physician James Lind noticed that sailors did not get scurvy when they had fresh citrus fruit to eat. A great deal of research on fruits and vegetables resulted in the identification of vitamin C as an essential nutrient.

The structure of vitamin C was determined in the twentieth century by the Hungarian-American chemist Albert Szent-Györgi. Once the formula and structure of the vitamin were known, it could be synthesized and reproduced, to be made available anywhere in the world, even in our local drugstore.

Vitamin C
(ascorbic acid)

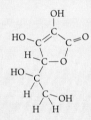

For Further Reading

David Hart, "Designer drugs," *Science Spectra* **8**:52, 1997.

A. M. Rouhi, "Seeking drugs in natural products," *Chemical and Engineering News,* April 7, 1997, p. 14.

Applying Your Knowledge

1. A tablet of antacid is advertised as being able to neutralize 40 times its mass in stomach acid (stomach acid is hydrochloric acid). You are asked to verify the claim, so you weigh one of the tablets and find its mass to be 0.501 g. You crush the tablet and mix it with water, then add an indicator that is blue in base and yellow in acid. Then you titrate the mixture with 0.500 M HCl(aq). Because the tablet does not dissolve well, you make sure to stir thoroughly after each addition of acid. As acid is added, the slurry dissolves. Suddenly the solution turns yellow and you record the volume delivered as 27.56 mL. What mass of HCl was neutralized by the tablet? Was the advertising claim correct?

2. Vitamin C can be analyzed by reacting it with an oxidizing agent such as potassium triiodide. You are asked to analyze a bottle of vitamin-enriched fruit drink to determine how much vitamin C it contains. The manufacturer claims that an 8-oz can of the drink provides 0.50 g of the vitamin. You pipet a 10.00-mL sample of the juice into a flask and titrate it with a solution of iodine (as potassium triiodide). It required 10.1 mL of 0.0117 M I_3^-(aq) to reach the stoichiometric point in the titration. Given that 1 mol I_3^- ≏ 1 mol vitamin C in the reaction, is the manufacturer's claim correct? The molar mass of vitamin C is 176 g/mol.

3. Baptisin is a natural product found in baptisia, or false indigo, a prairie plant native to the central United States. The mass percentage composition of baptisin is 54.93% C and 5.97% H, with the remainder oxygen. The mass spectrum of anhydrous baptisin shows a parent molecular ion at 568.54 g/mol. (a) Write the empirical and molecular formulas of baptisin. (b) If baptisin is dissolved in water and then crystallized, its molar mass increases to 730.66 g/mol and the mass percentage composition is now 42.74% C and 5.52% H, with the remainder O. Explain what has happened and write the new formula for the compound.

4. Some nations recommend the amounts of nutrients considered beneficial in the daily diet. In the United States, the RDA (recommended daily allowance) is used, and in Canada, the RNI (recommended nutrient intake). Design a mineral supplement to meet one of these criteria for your own diet. Your supplement should include iron, calcium, and iodine at the lowest possible cost and without side effects due to toxicity. Determine which of the following chemicals you would use and their costs, using the prices below. The RDA and RNI values are listed in standard reference books and on the CD that accompanies this book. Summarize your answer by creating an imaginative label for your supplement, giving the mass of each compound, the mass of the mineral it provides, and the percentage of the RDA or RNI per tablet for each mineral. For the price of the supplement, multiply the total price of the chemicals by four, to allow for manufacturing, transportation, and marketing costs.

Compound	Price per gram, $/g
$FeCl_2 \cdot 4H_2O$	0.23
$FeSO_4 \cdot 7H_2O$	0.12
$Fe(CN)_2$	0.11
PbI_2	0.20
KI	0.28
AuI_3	120.00
$CaCO_3$	0.20
CaC_2O_4	0.24
$Ca(NO_3)_2 \cdot 4H_2O$	0.10

5. Often food additives are used to improve the appearance, texture, shelf-life, or nutritional value of food. Write the names of the following food additives: (a) ZnO, dietary zinc supplement; (b) $Ca_3(PO_4)_2$, antacid; (c) $KBrO_3$, used to enhance flour and the rising of bread; (d) $MgSO_4$, laxative; (e) KI, added to iodized salt to prevent goiters; (f) $MgCO_3$, keeps salt from caking.

The Properties of Gases

The atmosphere is a precious layer of gas held by gravity to the surface of the Earth. Although the atmosphere is approximately 300 km thick, it is very thin relative to the size of the Earth. From far out in space, where the Earth looks the size of a basketball, the atmosphere looks like it is only 1 mm thick (Fig. 5.1). Yet this thin, delicate layer is vital to life: it shields us from harmful radiation and supplies essential chemicals, such as oxygen, nitrogen, carbon dioxide, and water. In the past, people thought that the atmosphere was so vast that it could absorb anything released into it without harm, so smoke, noxious gases, and exhaust fumes have been released into it indiscriminately since humans first learned to use fire. Now we know that human activities do have an effect on the atmosphere. The changes are small, but they accumulate over decades. Some changes are so important that they could even threaten our survival.

To begin to understand the atmosphere, we need to study smaller samples of gases. The atmosphere itself is difficult to study, largely because it is not uniform in composition, temperature, and density (Table 5.1). One reason for this nonuniformity is the effect of solar radiation, which causes different chemical reactions at different altitudes. Another reason for nonuniformity is gravity. Half the mass of the atmosphere lies below 5.5 km, and an airplane at a typical cruising altitude of 10 km is flying above 70% of the mass of the atmosphere.

So far in this book, we have looked at the behavior of bulk matter. In this chapter, we begin to develop insight into how the properties of bulk matter arise from the behavior of individual atoms and molecules. We shall also see how to apply our knowledge of stoichiometry to determine how much gas is formed or consumed in a chemical reaction.

Some of the properties of gases were discovered by balloonists trying to improve their flight performance. These balloonists are using their knowledge of these properties to stay aloft.

FIGURE 5.1

The delicate film of the Earth's atmosphere, as seen from space.

Table 5.1 *Composition of dry air at sea level*

Constituent	Molar mass,* g/mol	Composition, %	
		Volume	Mass
nitrogen, N_2	28.02	78.09	75.52
oxygen, O_2	32.00	20.95	23.14
argon, Ar	39.95	0.93	1.29
carbon dioxide, CO_2	44.01	0.03	0.05

*The average molar mass of molecules in the air, allowing for their different abundances, is 28.97 g/mol. The percentage of water vapor in ordinary air varies with the humidity.

THE NATURE OF GASES

Eleven elements are gases under normal conditions (Fig. 5.2). The six noble gases are monatomic (they consist of single atoms); all other gases consist of diatomic or polyatomic molecules. Many compounds also are gases. Most compounds that are gases at room temperature have low molar masses. Organic gases include the methane (CH_4) of natural gas and the propane (C_3H_8) and butane (C_4H_{10}) of camping fuel. Inorganic gases include ammonia (NH_3) and sulfur dioxide (SO_2).

> A vapor is a gas, but it usually means the gaseous state of a substance that is ordinarily a liquid or a solid.

We begin our study of gases, the simplest state of matter, by seeing how they differ from solids and liquids. From these differences, we can build a simple molecular model of gases that helps us understand why gases behave in their characteristic ways. Later, we shall build on this knowledge to understand liquids and solids too.

5.1 The Molecular Character of Gases

We can use the distinctive properties of gases to construct an initial model of the gaseous state of matter. As we develop deeper insight into gas behavior, we shall have to refine our model. However, we can use even our simple model to account for most properties of gases under normal conditions.

To build our model, we first need to collect some observations of gas behavior. Suppose we pump all the air out of a flask and allow a colored gas (such as chlorine) to enter it. The gas immediately spreads throughout the flask. Our first conclusion, therefore, is that a gas is a state of matter that fills any container it occupies. If the flask is fitted with a piston, we could compress the gas with little effort (Fig. 5.3). Our second conclusion, therefore, is that a gas is highly **compressible,** or easily confined to a smaller volume. If we were to try to compress a liquid or a solid, we would need a very high pressure to decrease the volume of the sample by even a tiny amount. Solids and liquids are almost incompressible.

The fact that a gas fills its container suggests that its molecules do not interact with one another very strongly, for otherwise they would clump together.

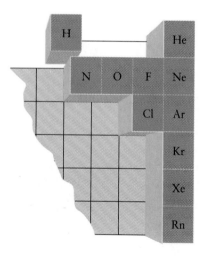

FIGURE 5.2

The 11 elements that are gases under normal conditions. Note how they lie toward the upper right of the periodic table. All the gaseous elements except those in Group 18 are diatomic.

The fact that gases are highly compressible suggests that there is a lot of empty space between the molecules. Moreover, because a gas *immediately* fills its container when released into it, we know that its molecules must be moving very rapidly. These observations suggest that a gas consists of widely separated molecules in ceaseless, rapid, random motion. They zoom from place to place in straight lines, changing direction only when they collide with a wall of the container or another molecule. These collisions change the speed and direction of the molecules. Later, we shall see that molecules do not travel completely independently of one another, so our model is just a starting point for understanding gases.

The molecules of a gas are widely separated and in ceaseless, rapid, random motion.

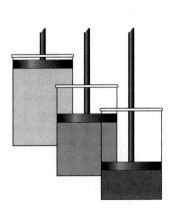

The molecules of a gas act like balls in a three-dimensional cosmic game of pool. At one instant, a molecule may be traveling at the speed of sound. At the next, it might be brought to a standstill in a collision. Then, at the next instant, another molecule may collide with it and send it hurtling off again in a different direction.

An analogy

> The name *gas* comes from the same Greek word as our word *chaos.*

5.2 Pressure

We can specify the state of a sample of gas by reporting four properties: the number of moles of molecules present, the volume of the sample, the temperature, and the pressure. If we change any of these properties, for instance, by adding more molecules, compressing the sample, or changing its temperature, then we change the state of the gas. Pressure is a new property in this chapter, so let's see what it means and how it fits in with our molecular model.

If you have ever pumped up a bicycle tire or squeezed an inflated balloon, you have experienced the opposing force arising from the confined air. The **pressure,** P, of a gas is the force, F, exerted by the gas divided by the area, A, on which the force is exerted:

$$\text{Pressure} = \frac{\text{force}}{\text{area}} \quad \text{or} \quad P = \frac{F}{A}$$

The pressure a gas exerts on the walls of its container results from the collisions of its molecules with the container's surface (Fig. 5.4). A vigorous storm of molecules battering on a surface results in a strong force and hence a high pressure. A less vigorous storm of molecules battering on the same area results in a lower pressure. When we inflate a tire, we pack more and more molecules into the tire with each stroke of the pump (in other words, we change the state of the gas in the tire). The larger number of molecules inside the tire exerts a greater force on the inner surface, and hence a higher pressure. The pressure of the atmosphere arises in the same way. We stand in an invisible storm of molecules that beat on us incessantly and exert a force all over our bodies. At sea level, the storm—and hence the pressure—is great because there are so many molecules in each cubic

FIGURE 5.3

A gas can easily be compressed into a smaller volume by pushing in a piston. This property suggests that there is a lot of space between the molecules. Liquids and solids are almost incompressible, which suggests that their molecules are in contact with one another.

FIGURE 5.4

The pressure of a gas arises from the collisions that its molecules make with the walls of the container. The storm of collisions exerts an almost steady force on the walls.

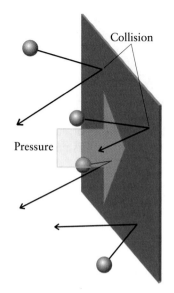

Collision

Pressure

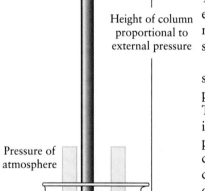

Height of column proportional to external pressure

Pressure of atmosphere

FIGURE 5.5

A barometer is used to measure the pressure of the atmosphere. The pressure of the atmosphere is balanced by the pressure exerted by the column of mercury. The height of the column is proportional to the pressure.

centimeter of air. At high altitudes, where the air is less dense, the battering is less vigorous and so the pressure there is lower.

The pressure exerted by the atmosphere is measured with a **barometer.** This instrument was invented in the seventeenth century by Evangelista Torricelli, a student of Galileo. Torricelli (whose name, coincidentally, means "little tower") formed a little tower of mercury, a dense liquid metal. He sealed a long glass tube at one end, filled it with mercury, and inverted it into a beaker (Fig. 5.5). The height of the column of mercury fell in the tube, leaving a vacuum (an empty space) above it, until the pressure exerted by the weight of the liquid mercury matched the pressure exerted by the atmosphere. The greater the pressure, the taller the column needed to match it.

Because the height of the column is directly proportional to the atmospheric pressure, monitoring the height is a way of monitoring atmospheric pressure. To show this relation, we suppose that the area of the column is A. Then the volume of mercury in the column is the height of the cylinder, h, times its cross-sectional area, or $h \times A$. The mass, m, of mercury in the column is the product of the density, d, and the volume, or $m = d \times hA$. This mass of mercury is pulled down by the force of gravity, and the total force at the base of the column is the product of the mass and the acceleration of free fall, g (a measure of the pull of gravity at the surface of the Earth; its value is close to 9.81 m/s^2), or $F = m \times g$, with $m = dhA$. Therefore, the pressure at the base of the column, the force divided by the area of the column, is

$$P = \frac{\text{force}}{\text{area}} = \frac{F}{A} = \frac{mg}{A} = \frac{dhAg}{A} = dhg \qquad (1)$$

This equation shows that the pressure exerted by the column of mercury, P, is proportional to the height of the column. Because this pressure balances the atmospheric pressure, we see that the height of the column is also proportional to the atmospheric pressure, as we set out to show.

Equation 1 provides a way of measuring the pressure. Suppose the height of the column of mercury in a barometer on a particular day is 760 mm (corresponding to 0.760 m). Given that the density of mercury at 20°C is

13.546 g/cm^3 (corresponding to 13 546 kg/m^3) and the acceleration of free fall is 9.806 65 m/s^2, we can conclude that the atmospheric pressure is

$$P = dgh = (13\ 546 \text{ kg/m}^3) \times (0.760 \text{ m}) \times (9.806\ 65 \text{ m/s}^2)$$
$$= 1.01 \times 10^5 \text{ kg/m·s}^2$$

The SI unit of pressure, the **pascal** (Pa), is defined as

$$1 \text{ Pa} = 1 \text{ kg/m·s}^2$$

The atmospheric pressure in this example is therefore 1.01×10^5 Pa, or 101 kPa (where 1 kPa = 10^3 Pa).

When we measure the pressure of the air in a tire by using a pressure gauge, we are actually measuring the "gauge pressure," the *difference* between the pressure inside the tire and the atmospheric pressure. A flat tire registers a gauge pressure of 0, because the pressure inside the tire is the same as the pressure of the atmosphere.

> *The pressure of a gas arises from the impacts of its molecules with a surface. The pressure is the force of the impacts divided by the area subjected to the force.*

5.3 Units of Pressure

Several units of pressure other than the pascal are in common use (Table 5.2). Because we often deal with substances at close to atmospheric pressure, it is convenient to use the unit called the **atmosphere** (atm), which is defined as

$$1 \text{ atm} = 101.325 \text{ kPa (exactly)}$$

An area of low pressure typically has a pressure of about 0.98 atm at sea level (Fig. 5.6). The pressure in the eye of a hurricane might be as low as 0.90 atm. A typical region of high pressure is about 1.03 atm.

The use of a mercury barometer to measure atmospheric pressure has given rise to two other commonly used units. The height of the mercury column in a

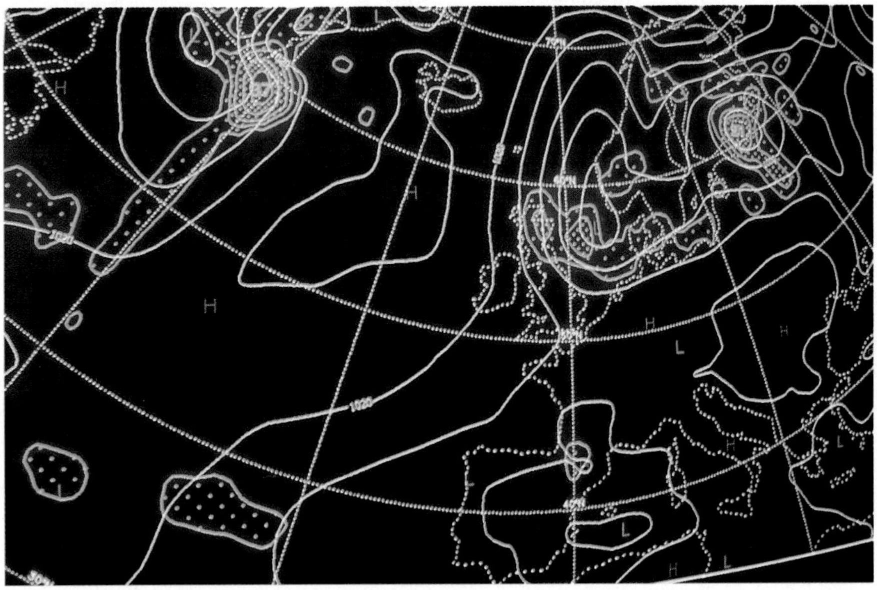

Table 5.2 *Pressure units**

SI unit: Pascal (Pa)

$1 \text{ Pa} = 1 \text{ kg/m·s}^2 = 1 \text{ N/m}^2$

Conventional units

1 bar = 10^5 Pa = **100** kPa
1 atm = **1.013 25 $\times$ 10^5** Pa
        = **101.325** kPa
1 atm = **760** Torr
1 atm = **14.7** lb/inch^2

*Figures in bold are exact. See inside the back cover for more relations. N, newtons.

The modern tendency is to avoid the awkward numerical factor by using the unit *bar*, where 1 bar = 10^5 Pa (that is, 1 bar = 100 kPa exactly).

FIGURE 5.6

A typical weather map showing Europe and the North Atlantic. The closed curves are called isobars and are contours of constant atmospheric pressure. A red H indicates a region of high pressure; a blue L, a region of low pressure.

barometer on a typical day at sea level is about 760 mm (corresponding to the 30 inches referred to in weather forecasts). Hence, a pressure of 760 millimeters of mercury (written 760 mmHg) corresponds to normal atmospheric pressure at sea level.

The unit 1 mmHg has been largely replaced by the unit 1 Torr (named for Torricelli), with

$$760 \text{ Torr} = 1 \text{ atm (exactly)}$$

In all but the most precise calculations, we can use Torr and mmHg interchangeably, with

$$1 \text{ Torr} = 1 \text{ mmHg}$$

The pressure of the atmosphere at the cruising height of a commercial jetliner (10 km) is only about 200 Torr, so airplane cabins must be pressurized. A gas at low pressure, such as a gas sample being used to study reactions high in the atmosphere, might have a pressure of only a few Torr.

The common units for reporting pressure are Torr and atmospheres; the SI unit is the pascal. Pressure units can be interconverted by using the values in Table 5.2.

> Strictly speaking, the name of the unit is torr and its symbol is Torr.

Example 5.1 *Converting between pressure units*

A meteorological map shows the pressure as 99.8 kPa. What is the pressure in Torr?

Strategy Atmospheric pressure is close to 760 Torr, so we expect an answer close to that value. We need to carry out three conversions: from kilopascals to pascals ($1 \text{ kPa} = 10^3 \text{ Pa}$), from pascals to atmospheres ($1.013 \, 25 \times 10^5 \text{ Pa} = 1 \text{ atm}$), and from atmospheres to Torr ($1 \text{ atm} = 760 \text{ Torr}$).

Solution The string of conversion factors we require is

$$\text{Pressure (Torr)} = (99.8 \text{ kPa}) \times \left(\frac{10^3 \text{ Pa}}{1 \text{ kPa}} \right) \times \left(\frac{1 \text{ atm}}{1.013 \, 25 \times 10^5 \text{ Pa}} \right) \times \left(\frac{760 \text{ Torr}}{1 \text{ atm}} \right)$$
$$= 749 \text{ Torr}$$

Self-Test 5.1A Express a pressure of 500. Torr in atmospheres.

[*Answer:* 0.658 atm]

Self-Test 5.1B The atmospheric pressure in Denver, Colorado, on a certain day was 630. Torr. Express this pressure in pascals.

THE GAS LAWS

The first reliable measurements of the properties of gases were made by the Anglo-Irish scientist Robert Boyle in the seventeenth century. More than a century later, two French scientists, Jacques Charles and Joseph-Louis Gay-Lussac, became interested in gases when they took up the exciting new sport of the day, ballooning. To improve their performance, they measured how the temperature of a gas affected its pressure, volume, and density. In the process, they discovered additional gas laws.

Gay-Lussac used his knowledge to establish a world balloon altitude record in 1804 of about 23 000 feet. Charles built the first hydrogen balloon, which became known as the *charlière*.

5.4 Boyle's Law

Boyle observed that when he compressed a fixed amount of gas at constant temperature, the pressure of the gas increased in a certain way. A graph of the dependence is shown in Fig. 5.7. The curve shown there is an **isotherm,** which is a plot of a property at constant temperature. Scientists often look for ways of plotting data in a manner that gives straight lines, because such graphs are easier to analyze and interpret (see Appendix 1E). The data used to construct Fig. 5.7 give a straight line when we plot the pressure against 1/volume (Fig. 5.8). The linear plot implies that, *for a fixed amount of gas at constant temperature, pressure is inversely proportional to volume, V:*

$$\text{Pressure} \propto \frac{1}{\text{volume}} \quad \text{or} \quad P = \frac{\text{constant}}{V}$$

This relation is known as **Boyle's law.** The law implies that, if we compress a sample of a gas into half its initial volume—from 1 L to 0.5 L or from 10 mL to 5 mL, for instance—then the pressure of the gas will double.

The mathematical form of Boyle's law shows that *the product of the pressure and volume of a sample of gas at constant temperature is a constant.* To reach this conclusion, we multiply both sides of the last equation by *V* and obtain

$$PV = \text{constant}$$

Our molecular model of gases is consistent with Boyle's law. As a gas is compressed, its molecules are confined to a smaller volume. The interior of the container becomes more crowded (Fig. 5.9) and molecules collide with the walls more frequently. As a result, they exert a higher total force on the walls, which gives rise to a higher pressure.

> *Boyle's law states that the pressure of a fixed amount of gas at constant temperature is inversely proportional to the volume.*

Example 5.2 *Using Boyle's law*

Suppose a sample of air occupies 120. L at 1.00 atm. What pressure is needed to compress it to 40.0 L at the same temperature?

Strategy The volume of the final state is smaller, so we expect the final pressure to be greater than the initial pressure. Because the volume of a fixed amount of gas is being changed at constant temperature, we can use Boyle's law ($PV = \text{constant}$):

$$P_2V_2 = P_1V_1$$

where 1 stands for the initial state and 2 for the final state. To obtain the final pressure, divide both sides by V_2:

$$P_2 = P_1 \times \frac{V_1}{V_2}$$

In calculations on gases, it is a good idea to collect the data and the unknown in a table.

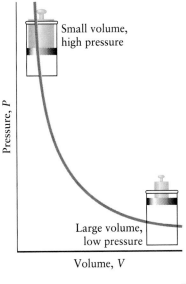

FIGURE 5.7

Boyle's law summarizes the effect of pressure on the volume of a fixed amount of gas at constant temperature.

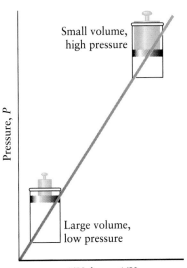

FIGURE 5.8

When the pressure is plotted against 1/volume, a straight line is obtained. For real gases, Boyle's law breaks down at high pressures beyond the scope of this graph, and a straight line is not obtained in these regions.

FIGURE 5.9

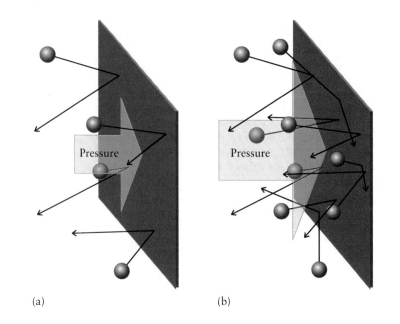

(a) As we have seen, the pressure of a gas arises from the impacts of its molecules on the walls of the container. (b) When the volume of a sample is decreased, there are more molecules in a given volume, so there are more collisions with the walls. Because the total impact on the walls is now greater, so is the pressure. In other words, decreasing the volume occupied by a gas without changing its temperature increases its pressure.

(a)　　　　　　　(b)

Solution The data are as follows:

State	Pressure, P	Volume, V	Amount, n	Temperature, T
1. Initial	1.00 atm	120. L	—	—
2. Final	?	40.0 L	same	same

The final pressure will be

$$P_2 = (1.00 \text{ atm}) \times \left(\frac{120.\text{ L}}{40.0\text{ L}}\right) = 3.00 \text{ atm}$$

We can use a similar approach to calculate the final volume when the pressure of a sample is changed at constant temperature. Notice that we do not need to know the values of properties that do not change.

Self-Test 5.2A What pressure would a sample of argon exert when compressed from 500. mL to 300. mL at constant temperature, given that the initial pressure is 750. Torr?

[***Answer:*** 1.25 kTorr]

Self-Test 5.2B A small cylinder of helium gas has a volume of 2.0 L and a pressure of 10.0 atm at 26°C. If all the gas in the cylinder is used to fill a single large balloon at 2.0 atm and the same temperature, to what volume does the balloon inflate?

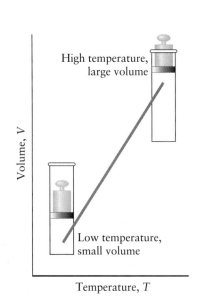

High temperature, large volume

Low temperature, small volume

Temperature, T

Volume, V

FIGURE 5.10

When the temperature of a gas is increased and it is free to change its volume at constant pressure (as depicted by the constant weight acting on the piston), the volume increases. A graph of volume against temperature is a straight line.

5.5 Charles's Law

Charles and Gay-Lussac found that the volume of a gas in a container fitted with a movable wall (to keep the pressure constant) increases as its temperature is raised. If we plot the volume of a gas against the temperature at constant pressure, we obtain a straight line, in agreement with their results (Fig. 5.10). We can conclude that, for a fixed amount of gas under constant pressure, *the volume varies linearly with the temperature.* This relation is called **Charles's law.**

Experiments like those done by Charles and Gay-Lussac reveal a remarkable feature of gases. When graphs like those in Fig. 5.10 are plotted for measurements made on different gases at different pressures, the straight lines extend to the same point (Fig. 5.11). The extension of a graph outside the region where data have been obtained is called **extrapolation.** Therefore, we say that the straight lines extrapolate to the same point. This unique point corresponds to zero volume and −273.15°C. Because the volume cannot be negative, this temperature must be the lowest possible temperature. Indeed, this lowest possible temperature defines 0 on the Kelvin scale, as we saw in Section 2.5. It follows that if we use the absolute temperature, T, then we can write Charles's law as

$$\text{Volume} \propto \text{absolute temperature} \quad \text{or} \quad V = \text{constant} \times T$$

This expression tells us that, if we double the absolute temperature of a fixed amount of gas at constant pressure, then the volume doubles. For example, if we heat a gas from 300 K to 600 K in a container fitted with a piston, then its volume doubles.

A similar expression summarizes the results of measuring the variation in pressure of a sample of gas in a container of fixed volume as the temperature is changed. The pressure decreases linearly as the temperature is reduced and extrapolates to 0 at $T = 0$ (Fig. 5.12). Therefore, we can write

$$\text{Pressure} \propto \text{absolute temperature} \quad \text{or} \quad P = \text{constant} \times T$$

It follows that doubling the absolute temperature doubles the pressure of a gas, provided the volume is constant.

Charles's law suggests another refinement of our molecular model of a gas. We can explain the effect of temperature on the pressure by supposing that as the temperature of a gas is raised, the average speed of the molecules increases. As a result, the molecules strike the walls more often and with greater average force. Therefore, the gas exerts a greater pressure. If the temperature is raised in a container with movable walls, the molecules beat harder and harder against the walls. These impacts drive the walls out, increasing the volume of the container, until the pressure of the gas is equal to the external pressure.

The volume of a fixed amount of gas at constant pressure is directly proportional to the absolute temperature. The pressure of a fixed amount of gas at constant volume is proportional to the absolute temperature.

Example 5.3 *Using Charles's law*

A weather balloon was filled with helium gas to a volume of 262 L on an unseasonably warm winter day in Sweden, when the temperature reached 18°C. However, the balloon could not be released when scheduled, and just before dawn the next morning, the temperature was −10.°C at the same atmospheric pressure. What is the new volume of the helium balloon?

Strategy The temperature has fallen, so we expect the volume of the helium to be smaller. To use Charles's law, we note that because V/T is a constant for a gas sample at constant pressure, the initial and final volumes and temperatures are related by

$$\frac{V_2}{T_2} = \frac{V_1}{T_1}$$

To find the final volume (V_2), we multiply both sides by T_2 and obtain

$$V_2 = V_1 \times \frac{T_2}{T_1}$$

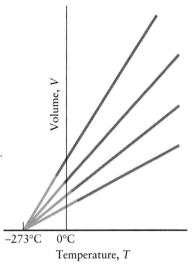

Volume, V

−273°C   0°C

Temperature, T

FIGURE 5.11

The extrapolation of data like that in Fig. 5.10 for a number of gases suggests that the volume of all gases should become 0 at −273°C ($T = 0$ on the Kelvin scale). All gases condense to liquids well before that temperature is reached.

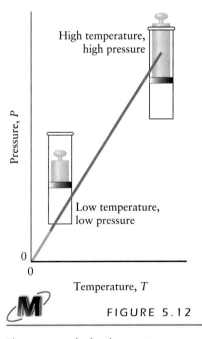

High temperature, high pressure

Pressure, P

Low temperature, low pressure

0

0

Temperature, T

FIGURE 5.12

The pressure of a fixed amount of gas in a vessel of constant volume is proportional to the absolute temperature. The pressure extrapolates to 0 at $T = 0$ on the Kelvin scale.

As usual, it is helpful to collect the data in a table. Remember to convert temperatures to kelvins.

Solution The data and the unknown are as follows:

State	Pressure, P	Volume, V	Amount, n	Temperature, T
1. Initial	—	262 L	—	$(18 + 273.15)$ K $=$ 291 K
2. Final	same	?	same	$(-10. + 273.15)$ K $=$ 263 K

The final volume will be

$$V_2 = (262 \text{ L}) \times \left(\frac{263 \text{ K}}{291 \text{ K}}\right) = 237 \text{ L}$$

A similar procedure can be used to calculate the effect of a change in temperature on the pressure of a gas that is restricted to a constant volume.

Self-Test 5.3A In a study of an internal combustion engine, nitrogen gas in a 125-mL cylinder at 18°C is heated to 322°C. What is the final volume of the gas?

[*Answer:* 256 mL]

Self-Test 5.3B The vapor inside a nearly empty aerosol can has a pressure of 1.01 atm at 25°C. What would the pressure in the can be if it were tossed into an incinerator for disposal and its temperature rose to 1300.°C? Why is the incineration of aerosol cans not recommended?

5.6 Avogadro's Principle

We know from blowing up balloons that adding more gas to a flexible container increases its volume. The Italian scientist Amedeo Avogadro discovered that, at constant temperature and pressure, the volume of a gas is proportional to the amount of gas. We can verify his discovery by comparing the molar volumes of gases. The **molar volume**, V_m, of a substance—any substance, not only a gas—is the volume it occupies divided by the number of moles of atoms, molecules, or formula units, n:

$$\text{Molar volume} = \frac{\text{volume occupied}}{\text{number of moles}} \quad \text{or} \quad V_m = \frac{V}{n}$$

Figure 5.13 shows measurements of the molar volumes of five gases at the same pressure and temperature. We see that, regardless of the identity of the gas, the molar volumes are all very similar. The differences are small under normal conditions and become smaller as the pressure of the gas is reduced. Results like

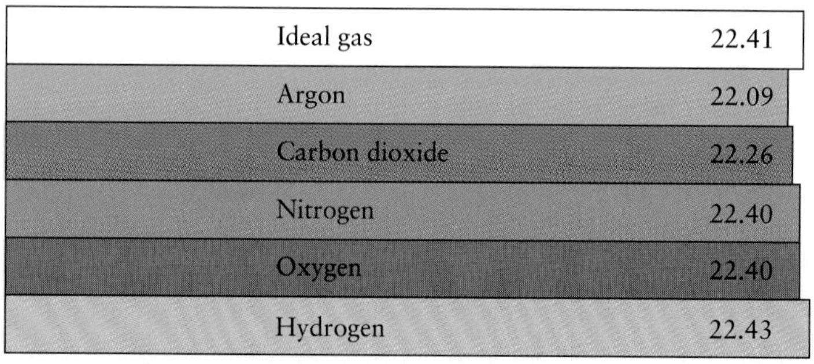

FIGURE 5.13

The molar volumes (in liters per mole) of various gases at 0°C and 1 atm: all are very similar. An ideal gas is introduced in Section 5.7.

Ideal gas	22.41
Argon	22.09
Carbon dioxide	22.26
Nitrogen	22.40
Oxygen	22.40
Hydrogen	22.43

these suggest that, *under the same conditions of temperature and pressure, a given number of molecules occupy the same volume, regardless of their chemical identity.* This observation is now known as **Avogadro's principle.**

It follows from Avogadro's principle that

Volume occupied $\propto$ number of moles   or $V = \text{constant} \times n$

According to this expression, provided the pressure is not too high, doubling the number of moles of molecules at constant temperature and pressure doubles the volume occupied by a gas. This behavior is consistent with our model. In this model, gas molecules are in constant motion and beat against the walls of their container. To keep the pressure constant, the force of the impacts on a given area of the wall must remain constant. Therefore, as more molecules are added, the size of the container must increase.

The volume occupied by a sample of gas at constant pressure and temperature is proportional to the number of moles of molecules present.

Example 5.4 *Using Avogadro's principle*

A sealed flask contains 25 g of a chlorofluorocarbon gas that is being used in atmospheric research. At 20.°C, the pressure of the gas is 278 Torr. The scientists doing the investigation decide they need 35 g of the gas under the same conditions of temperature and pressure. To ensure that the pressure is unchanged, by what factor should they increase the volume of the container after adding 10. g of the chlorofluorocarbon gas?

Strategy With more gas present, but at the same temperature and pressure, we know that a larger volume is required. Because the value of V/n is a constant at constant temperature and pressure, the final and initial volumes and amounts are related by

$$\frac{V_2}{n_2} = \frac{V_1}{n_1}$$

To find V_2, we multiply by n_2 and obtain

$$V_2 = V_1 \times \frac{n_2}{n_1}$$

Because the number of moles of molecules is proportional to the mass of the sample, we can replace the ratio of amounts by the ratio of masses.

Solution The new volume is related to the old by

$$V_2 = V_1 \times \left(\frac{35 \text{ g}}{25 \text{ g}}\right) = V_1 \times 1.4$$

That is, the volume of the flask must be increased by a factor of 1.4.

Self-Test 5.4A A chemist studying ozone finds that at 298 K and 1.00 atm, 0.100 mol $O_3(g)$ occupies 2.45 L. What volume will 0.0250 mol $O_3(g)$ occupy at the same temperature and pressure?

[*Answer:* 0.612 L]

Self-Test 5.4B A 5.00-L tank of carbon dioxide gas contains 0.360 mol $CO_2(g)$. A chemist needs 0.0072 mol $CO_2(g)$ at the same temperature and pressure as the tank. What volume container will the chemist need for the sample?

5.7 The Ideal Gas Law

Charles's law and Avogadro's principle have told us that the volume of a gas is directly proportional to its absolute temperature ($V \propto T$) and the number of moles ($V \propto n$), respectively. Boyle's law has told us that the volume is inversely proportional to the pressure ($V \propto 1/P$). When we combine these three relations, we obtain

$$V \propto \frac{\overbrace{n}^{\text{From Avogadro}} \times \overbrace{T}^{\text{From Charles}}}{\underbrace{P}_{\text{From Boyle}}} \qquad \text{or} \qquad PV \propto nT$$

The constant of proportionality in $PV \propto nT$ is denoted R and called the **gas constant:**

$$PV = nRT \qquad\qquad (2)$$

This expression is called the **ideal gas law.** A gas that obeys this law under all conditions is called an **ideal gas.**

We can find the value of R by measuring P, V, n, and T for a sample of gas at low pressure (when it behaves ideally) and substituting their values into

$$R = \frac{PV}{nT}$$

For pressure in atmospheres and volume in liters, it is found that

$$R = 0.082\ 057\ 8 \ \text{L·atm/K·mol}$$

Table 5.3 gives the values of R for different units of pressure. When doing calculations that use R, either select from the table the value that has units matching the ones you need or convert the units to ones that match the value of R you want to use.

Real gases, or actual gases, such as the nitrogen and oxygen of the atmosphere and those we use in laboratories, behave like ideal gases, provided their pressures are low (below about 2 atm). The ideal gas law is found to be increasingly accurate as the pressure of the gas is reduced. In the limit of zero pressure, every real gas behaves like an ideal gas and the law describes a gas exactly. We say that the ideal gas law is a **limiting law:** it is exactly valid at a certain limit—in this case, the limit of zero pressure.

Our molecular model suggests a reason for this behavior. At moderate pressures, when many molecules are present in a small container, the molecules are close to one another for a lot of the time, and their interactions are not completely negligible. As the pressure is reduced and fewer molecules occupy the container, the interactions become negligible because the molecules rarely meet.

> *The ideal gas law, a limiting law, contains all the relations describing the response of ideal gases to changes in pressure, volume, temperature, and moles of molecules.*

> Unless you are informed otherwise, there is no need to try to remember this value; it is usually supplied in tests and examinations.

Table 5.3 *The gas constant, R*

Units required		R
L	atm	0.082 0578 L·atm/K·mol
L	bar	0.083 1451 L·bar/K·mol
L	Torr	62.364 L·Torr/K·mol
m³	Pa*	8.314 51 L·kPa/K·mol
J		8.314 51 J/K·mol
SI units		8.314 51 J/K·mol

*Note that $1 \text{ Pa} = 1 \text{ N/m}^2$ and $1 \text{ Pa·m}^3 = 1 \text{ J} = 1 \text{ kg·m}^2/\text{s}^2 = 1 \text{ L·kPa}$.

Example 5.5 *Calculating the pressure of a given sample*

Estimate the pressure (in atmospheres) inside a television picture tube, given that its volume is 5.0 L, its temperature is 23°C, and it contains 0.010 mg of nitrogen gas.

This Toolbox shows how to calculate the pressure (or other property) of a given sample of gas under stated conditions and how to predict the effect on a gas of changing the conditions.

Conceptual Basis

The ideal gas law, Eq. 2, summarizes the relations between the properties of an ideal gas. It is reliable for real gases at low pressures. To predict the effect of a change in conditions, it is helpful to visualize the physical significance of the changes:

- At constant volume, an increase in temperature increases the pressure (Fig. 5.14a).
- At constant temperature, an increase in the applied pressure decreases the volume (Fig. 5.14b).
- At constant pressure, an increase in temperature increases the volume (Fig. 5.14c).
- At constant temperature and pressure, an increase in the number of moles increases the volume (Fig. 5.14d).

Procedure

The pressure (or other property) of a given sample

Step 1. To use Eq. 2 directly, rearrange the equation to give the desired quantity on the left and all other quantities on the right.

Step 2. Substitute the data; if necessary, convert from mass of gas to the number of moles. Note that the temperature must be in kelvins.

Step 3. Choose from Table 5.3 the value of R that matches the units of pressure and volume you are using. Alternatively, convert the pressure units to match the value of R you prefer to use.

The procedure is illustrated in Example 5.5.

The response of a gas to a change in conditions
The initial and final conditions are denoted by the subscripts 1 and 2, respectively. The two states may differ in pressure, volume, amount, and temperature, or any combination of the four. In each state, the gas must satisfy

$$\frac{P_1 V_1}{n_1 T_1} = R \qquad \frac{P_2 V_2}{n_2 T_2} = R$$

The constant R is the same, so the two expressions can be equated:

$$\frac{P_1 V_1}{n_1 T_1} = \frac{P_2 V_2}{n_2 T_2}$$

This relation is sometimes called the **combined gas law.** To use it,

Step 1. Draw up a table showing the data, which may include changes in any of the four variables pressure, volume, amount, and temperature. Express the temperatures in kelvins.

Step 2. Rearrange the relation above so that the quantity required is on the left and all other quantities are on the right. Cancel any quantities that are unchanged.

Step 3. Substitute the data and check to see whether the answer agrees with your predictions.

The procedure is illustrated in Example 5.6.

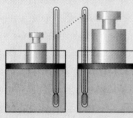
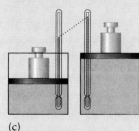

(a)

(b)

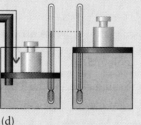

(c)

(d)

FIGURE 5.14

A visualization of the effect of various changes on an ideal gas: (a) increasing temperature at constant volume; (b) increasing pressure at constant temperature; (c) increasing temperature at constant pressure; and (d) increasing the number of moles at constant temperature and pressure.

Strategy We expect a low pressure, because there is very little gas in the tube. Proceed as set out in Toolbox 5.1. Leave the numerical part of the calculation until the end, for that reduces rounding errors.

Solution Equation 2 gives the following expression for the pressure:

$$P = \frac{nRT}{V}$$

The molar mass of N_2 is 28.02 g/mol, so the number of moles of N_2 in the sample is

$$\text{Moles of } N_2(\text{mol}) = (0.010 \text{ mg } N_2) \times \left(\frac{10^{-3} \text{ g}}{1 \text{ mg}}\right) \times \left(\frac{1 \text{ mol } N_2}{28.02 \text{ g } N_2}\right)$$

$$= \frac{1.0 \times 10^{-5}}{28.02} \text{ mol } N_2$$

The temperature, 23°C, corresponds to $T = (273.15 + 23) \text{ K} = 296 \text{ K}$. We want the pressure in atmospheres, so we use the value of R expressed in atmospheres (Table 5.3). The pressure is

$$P = \frac{\overbrace{(1.0 \times 10^{-5}/28.02 \text{ mol})}^{n} \times \overbrace{(0.082\ 06 \text{ L·atm/K·mol})}^{R} \times \overbrace{(296 \text{ K})}^{T}}{\underbrace{5.0 \text{ L}}_{V}}$$

$$= \frac{(1.0 \times 10^{-5}) \times 0.082\ 06 \times 296}{28.02 \times 5.0} \times \frac{\text{mol·L·atm·K}}{\text{K·mol·L}} = 1.7 \times 10^{-6} \text{ atm}$$

A very low pressure is necessary to minimize the collisions between the electrons in the beam and the gas molecules. Deflections of the beam caused by collisions with molecules would give a blurred, dim picture.

Self-Test 5.5A Calculate the pressure in kilopascals (kPa) exerted by 1.0 g of carbon dioxide in a flask of volume 1.0 L at 300.°C.

[***Answer:*** 1.1×10^2 kPa]

Self-Test 5.5B An idling, badly tuned automobile engine releases 1.00 mol CO per minute into the atmosphere. At 27°C and 1.00 atm, what volume of CO, adjusted to 1.00 atm, is emitted per minute?

Example 5.6 *Using the ideal gas law to calculate the effect of a change in conditions*

In an experiment to investigate the properties of the coolant gas used in an air-conditioning system, a sample of volume 500. mL at 28.0°C was found to exert a pressure of 92.0 kPa. What pressure will the sample exert when it is compressed to 300. mL and cooled to −5.0°C?

Strategy Compression increases the pressure, but cooling lowers the pressure; so it is not easy to predict the outcome without doing a calculation. Use the strategy in the second procedure in Toolbox 5.1. Convert temperatures to the Kelvin scale.

Solution The data are summarized in the following table:

State	Pressure, P	Volume, V	Temperature, T	Amount, n
1. Initial	92.0 kPa	500. mL	(28.0 + 273.15) K = 301.2 K	—
2. Final	?	300. mL	(−5.0 + 273.15) K = 268.2 K	same

The final pressure is

$$P_2 = P_1 \times \frac{n_2}{n_1} \times \frac{V_1}{V_2} \times \frac{T_2}{T_1}$$

$$= (92.0 \text{ kPa}) \times 1 \times \frac{500. \text{ mL}}{300. \text{ mL}} \times \frac{268.2 \text{ K}}{301.2 \text{ K}} = 137 \text{ kPa}$$

The net outcome is an increase in pressure, so, in this instance, the compression into a smaller volume has a greater effect than the decrease in temperature.

Self-Test 5.6A Meteorologists refer to localized packets of air that are homogeneous in temperature and pressure as "parcels." A parcel of air of volume 1.00×10^3 L at 20°C and 1.00 atm rises up the side of a mountain range. At the summit, where the pressure is 0.750 atm, the parcel of air has cooled to $-10.$°C. What is the volume of the parcel at that point?

[***Answer:*** 1.20×10^3 L]

Self-Test 5.6B What would be the effect on the pressure exerted by 1 mol of gas molecules if the volume were halved and the absolute temperature were doubled?

5.8 Molar Volume

The ideal gas law can be used to predict the molar volume of a gas under any conditions of temperature and pressure (provided it is behaving ideally):

$$V_m = \frac{V}{n} = \frac{nRT/P}{n} = \frac{RT}{P} \qquad (3)$$

At **standard temperature and pressure** (STP), which means 0°C (273.15 K) and 1.000 atm, the molar volume is

$$V_m = \frac{(0.082\ 058 \text{ L·atm/K·mol}) \times (273.15 \text{ K})}{1.000 \text{ atm}}$$

$$= \frac{0.082\ 058 \times 273.15}{1.000} \times \frac{\text{L·atm·K}}{\text{K·mol·atm}} = 22.41 \text{ L/mol}$$

That is, the volume occupied by 1 mol of ideal gas molecules at STP is close to 22 L, which is about the volume of a cube 1 ft on a side (Fig. 5.15). At 25°C (the temperature for which data are commonly reported in chemistry) and 1.000 atm, the same calculation gives the molar volume of an ideal gas as 24.47 L/mol.

The molar volume is a conversion factor between the volume and the amount of gas at a specified temperature and pressure. When we want to know the volume occupied by a given *mass* of gas, we first convert the mass to moles by using the molar mass and then use $V = n \times V_m$.

The molar volume of an ideal gas at STP (0°C and 1 atm) is 22.41 L/mol.

FIGURE 5.15

The blue cube is the volume occupied by 1 mol of ideal gas molecules at 0°C and 1 atm (22.4 L).

Example 5.7 *Calculating the volume of a given mass of gas*

Calculate the volume occupied by 10. g of carbon dioxide at STP by using the molar volume of an ideal gas.

Strategy We need to carry out two conversions. One is from mass of gas to moles of gas molecules. For this step, we use the molar mass of the gas. The second conversion is from moles of molecules to the volume they occupy under the stated conditions.

For this step, we use the molar volume at the conditions stated and $V = nV_m$. (An alternative procedure is to use the ideal gas law directly in the form $V = nRT/P$.)

Solution The molar mass of carbon dioxide is 44.01 g/mol and the molar volume at STP is 22.41 L/mol. Therefore, the total volume of the sample is

$$V = \underbrace{(10. \text{ g CO}_2)}_{\substack{\text{Mass of} \\ \text{sample}}} \times \underbrace{\left(\frac{1 \text{ mol CO}_2}{44.01 \text{ g CO}_2}\right)}_{\substack{\text{Convert mass} \\ \text{to moles}}} \times \underbrace{\left(\frac{22.41 \text{ L CO}_2}{1 \text{ mol CO}_2}\right)}_{\substack{\text{Convert moles} \\ \text{to volume}}}$$

$$= 5.1 \text{ L CO}_2$$

Self-Test 5.7A Calculate the volume occupied by 1.0 kg of hydrogen gas at STP.

[*Answer:* 1.1×10^4 L]

Self-Test 5.7B Calculate the volume occupied by 2.0 g of helium at 25°C and 1.0 atm.

5.9 The Stoichiometry of Reacting Gases

Sometimes it is important to know the volume of a gas that will be produced in a reaction. One of the most common causes of explosions in industry is the formation of gases in a closed container and the resulting increase in pressure. The gas laws can be used to predict the volume of gas involved in a chemical reaction and help us to anticipate and therefore prevent explosions. Gay-Lussac (Section 5.5) was the first to summarize the relation between volumes of reacting gases, in his **law of combining volumes:**

> At the same temperature and pressure, the volumes of gases react with one another in the ratios of small whole numbers.

We can see that this law is a direct consequence of the stoichiometry of reactions and Avogadro's principle: at constant temperature and pressure, the volume of a gas is proportional to the number of moles present. For example, the stoichiometry of the reaction

$$N_2(g) + 2O_2(g) \longrightarrow 2NO_2(g)$$

in which 1 mol N_2 reacts with 2 mol O_2 to form 2 mol NO_2 can also be expressed in the form (Fig. 5.16)

> 1 L of $N_2(g)$ reacts with 2 L of $O_2(g)$ to form 2 L of $NO_2(g)$

To calculate the volume of gas consumed or produced in a reaction, we combine the material we learned in Chapter 4 with the material presented in

FIGURE 5.16

When measured at the same temperature and pressure, equal volumes of gases contain the same numbers of moles. Therefore, the ratios of the volumes of reactant and product gases in a chemical reaction are the same as the mole ratios.

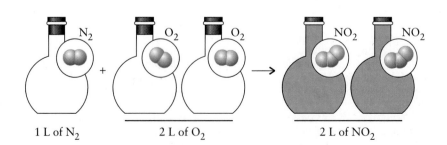

1 L of N_2 2 L of O_2 2 L of NO_2

this chapter. If we need to know what volume of gas a reaction will produce, we first need to calculate how many moles of gas will be produced in a mole-to-mole calculation of the type described in Section 4.1. Then we convert moles of gas to volume by using the molar volume for the given temperature and pressure. The procedures are summarized in Toolbox 5.2.

Example 5.8 *Calculating the volume of gas produced in a reaction*

Calculate the volume of sulfur dioxide produced at 25.00°C and 1.00 atm by the combustion of 10. g of sulfur, according to the reaction $S_8(s) + 8 O_2(g) \rightarrow 8 SO_2(g)$. Sulfur dioxide is a pungent gas that, when formed by the combustion of fuels containing sulfur, contributes to the formation of acid rain.

Strategy From the chemical equation, we know that 1 mol S_8 produces 8 mol SO_2. Because 10. g is much less than the mass of 1 mol S_8, we can predict that the volume of SO_2 produced will be less than the volume occupied by 8 mol (about 200 L). For the quantitative value, follow the procedure in Toolbox 5.2.

Toolbox 5.2 *How to calculate the volume of gas produced or consumed in a reaction*

This Toolbox shows how to calculate the volume of gas produced from a given mass of reactant.

Conceptual Basis
At constant temperature and pressure, $V \propto n$, and any reaction involving gases can be expressed in terms of volumes in place of moles.

Procedure
We calculate the volume of gas produced in a reaction as described in Toolbox 4.2, with one change: instead of using the molar mass to convert from moles of product molecules to mass of product, use the ideal gas law to convert from moles of gas molecules to liters of gas, as described in Toolbox 5.1. The new arrow in the stoichiometry diagram represents this step (Fig. 5.17).

Step 1. Use the molar mass of the reactant to convert the mass of reactant (in grams) to moles of reactant molecules. See Toolbox 4.2.

Step 2. Identify the stoichiometric relation between the reactant and the gaseous product from the chemical equation and use it to convert from moles of reactant to moles of gaseous product. See Toolbox 4.1.

Step 3. Use the molar volume of an ideal gas (at the specified temperature and pressure) to obtain the volume of gas produced ($V = nV_m$, with $V_m = RT/P$).

It is good practice to write these conversions in a string and to perform all the calculations in one step, because that reduces rounding errors. At STP, the molar volume of an ideal gas is 22.41 L/mol; at 25.00°C and 1.000 atm, the molar volume is 24.47 L/mol.

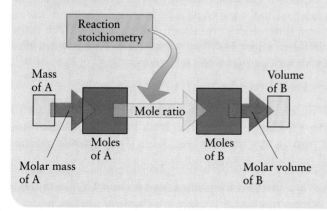

FIGURE 5.17

The volumes of gases produced in a chemical reaction can be determined by using the molar volume of an ideal gas at the temperature and pressure of the reaction to convert moles to volume.

FIGURE 5.18

An explosion caused by the ignition of coal dust. A shock wave is created by the tremendous expansion of volume as large numbers of gas molecules form.

Solution **Step 1.** The molar mass of sulfur molecules is $8 \times (32.06 \text{ g/mol}) = 256.48$ g/mol.

Step 2. For the mole-to-mole step, we use the stoichiometric relation 1 mol $S_8 \simeq$ 8 mol SO_2.

Step 3. The molar volume of sulfur dioxide (treated as an ideal gas) under the stated conditions is 24.47 L/mol. It follows that

$$
\text{Volume of gas (L)} = \overbrace{(10. \text{ g } S_8)}^{(1)\ n_{\text{reactant}}} \times \overbrace{\left(\frac{1 \text{ mol } S_8}{256.48 \text{ g } S_8}\right)}^{} \times \overbrace{\left(\frac{8 \text{ mol } SO_2}{1 \text{ mol } S_8}\right)}^{(2)\ \text{mole ratio}} \times \overbrace{\left(\frac{24.47 \text{ L } SO_2}{1 \text{ mol } SO_2}\right)}^{(3)\ V_m}
$$

$$
= \frac{10. \times 8 \times 24.47}{256.48} \text{ L } SO_2 = 7.6 \text{ L } SO_2
$$

Self-Test 5.8A Calculate the volume of ethyne (acetylene), C_2H_2, produced at 25°C and 1.00 atm when 10. g of calcium carbide reacts completely with water in the reaction $CaC_2(s) + 2 H_2O(l) \rightarrow Ca(OH)_2(s) + C_2H_2(g)$.

[*Answer:* 3.8 L]

Self-Test 5.8B Methanol, CH_3OH, is added to some automobile fuels to improve burning and reduce pollution. What volume of O_2 at STP is consumed in the combustion of 1.00 L of methanol? The density of methanol is 0.791 g/mL. The combustion reaction is $2 CH_3OH(l) + 3 O_2(g) \rightarrow 2 CO_2(g) + 4 H_2O(l)$.

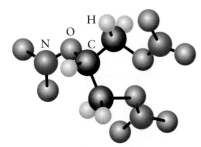

1 Nitroglycerin, $C_3H_5O_9N_3$

The molar volumes of gases are close to 25 L/mol under normal conditions, whereas liquids and solids occupy only about a few tens of milliliters per mole. The molar volume of liquid water, for instance, is only 18 mL/mol. Therefore, when liquids or solids react to form a gas, there may be a thousandfold increase in volume. In other words, 1 mol of gas molecules occupies about a thousand times the volume of 1 mol of molecules in a liquid or solid.

The increase in volume as gaseous products are formed is even larger if several gas molecules are produced from each reactant molecule (Fig. 5.18). The explosive action of liquid nitroglycerin, $C_3H_5(NO_3)_3$ (1), is an example. When subjected to a shock wave from a detonator, nitroglycerin decomposes into many small gaseous molecules:

$$4 C_3H_5(NO_3)_3(l) \longrightarrow 6 N_2(g) + O_2(g) + 12 CO_2(g) + 10 H_2O(g)$$

In this reaction, 4 mol $C_3H_5(NO_3)_3$, which corresponds to a little over 570 mL of the liquid, produces 29 mol of gas molecules of various kinds, giving a total of about 710 L of gas under normal conditions. The pressure wave from the sudden 1300-fold expansion is the destructive shock of the explosion. The detonator, which is typically lead azide, $Pb(N_3)_2$, works on a similar principle; it releases a large volume of nitrogen gas when it is struck:

$$Pb(N_3)_2(s) \longrightarrow Pb(s) + 3 N_2(g)$$

The sudden shock of expansion stimulates the explosive (dynamite, for instance) to react. An explosion of the same kind, but using sodium azide, NaN_3, is used in air bags in automobiles (Fig. 5.19). The explosive release of nitrogen is detonated electrically in a collision.

The molar volume (at a specified temperature and pressure) is used to convert the amount of a gaseous reactant or product in a chemical reaction into a volume.

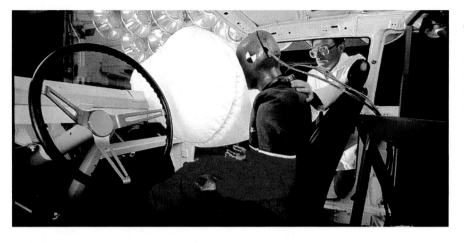

FIGURE 5.19

The rapid decomposition of sodium azide, NaN_3, results in the formation of a large volume of nitrogen gas. The reaction is triggered electrically in this air bag.

5.10 Gas Density

The density of a gas is the mass, m, of the sample divided by its volume, V:

$$\text{Density} = \frac{\text{mass}}{\text{volume}} \quad \text{or} \quad d = \frac{m}{V}$$

We can see from this expression that the density of a gas increases when we compress it because the same number of molecules are confined into a smaller volume. Similarly, heating a gas that is free to expand *reduces* the density of a gas because fewer molecules will be found in a given volume. The lower density of a warmer gas is one reason air rises over warm land; glider pilots ride these currents of warm air, called "thermals," to reach high altitudes. The lower density of the hot air inside a hot-air balloon allows it to rise and even carry a load (Fig. 5.20).

Because all gases have essentially the same molar volume at a given pressure and temperature, a gas with a high molar mass has a higher density than one with a low molar mass. The volume occupied by 1 mol of gas molecules is the same for all gases (by Avogadro's principle), but the mass of the sample is given by the molar mass. Therefore, the ratio of the mass to the volume increases as molar mass increases. For instance, because the mass of 1 mol Cl_2 molecules is 70.90 g and the volume they occupy at 298.15 K and 1.000 atm is 24.47 L, the density of the gas is

$$\text{Density} = \frac{70.90 \text{ g}}{24.47 \text{ L}} = 2.897 \text{ g/L}$$

In contrast, the mass of 1 mol H_2 molecules is only 2.016 g, but the volume they occupy is the same (Fig. 5.21). Therefore,

$$\text{Density} = \frac{2.016 \text{ g}}{24.47 \text{ L}} = 0.082\ 39 \text{ g/L}$$

As expected, chlorine is denser than hydrogen at the same temperature and pressure.

The densities of gases increase with increasing pressure and decreasing temperature, and are proportional to the molar mass.

FIGURE 5.20

The *Virgin Global Challenger* is tested near Reno, Nevada, before beginning an attempt to circumnavigate the globe.

FIGURE 5.21

At the same temperature and pressure, a molecule occupies essentially the same volume in any gas. Hence, the greater the mass of each molecule, the greater the density of the gas. The illustration shows samples of hydrogen and chlorine at the same pressure, volume, and temperature. The numbers are the molar masses in grams per mole.

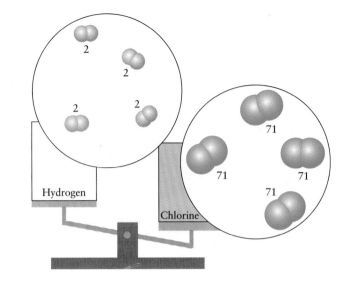

Hydrogen

Chlorine

Example 5.9 *Determining molar mass from gas density*

The volatile organic compound geraniol, a component of oil of roses, is used in perfumes. The density of the vapor at 260.°C is 0.480 g/L when the pressure is 103 Torr. What is the molar mass of geraniol?

Strategy We need a relation between the density and the pressure. To find it, suppose that the sample consists of n moles of molecules of molar mass M. Then the mass of the sample is nM. The volume of the sample is given by the ideal gas law as $V = nRT/P$. Use these two expressions to find an expression for the density d in terms of P and M, and insert the data to find M. If we use R in units that include atmospheres, we need to convert the pressure into atmospheres by using 1 atm = 760 Torr.

Solution Because the mass of the sample is nM and its volume is $V = nRT/P$, the density, d, of the sample is

$$d = \frac{m}{V} = \frac{nM}{nRT/P} = \frac{MP}{RT}$$

This expression rearranges into

$$M = \frac{dRT}{P}$$

The data are

$$d = 0.480 \text{ g/L} \qquad P = 103 \text{ Torr} \qquad T = (273.15 + 260.) \text{ K} = 533 \text{ K}$$

A pressure of 103 Torr corresponds to

$$P = 103 \text{ Torr} \times \frac{1 \text{ atm}}{760 \text{ Torr}} = \frac{103}{760} \text{ atm}$$

When we substitute these values into the equation for M, we get

$$M = \frac{(0.480 \text{ g/L}) \times (0.082\ 057\ 8 \text{ L·atm/K·mol}) \times (533 \text{ K})}{(103/760) \text{ atm}}$$

$$= \frac{0.480 \times 0.082\ 057\ 8 \times 533 \times 760}{103} \times \frac{\text{L·atm·K·g}}{\text{K·mol·atm·L}}$$

$$= 155 \text{ g/mol}$$

Self-Test 5.9A The oil produced from eucalyptus leaves contains the volatile organic compound eucalyptol. At 190°C and 60.0 Torr, a sample of eucalyptol vapor had a density of 0.320 g/L. Calculate the molar mass of eucalyptol.

[*Answer:* 154 g/mol]

Self-Test 5.9B The *Codex Ebers*, an Egyptian medical papyrus, describes the use of garlic to treat many ailments, including the cleansing of wounds. Chemists today have verified that the oxide of diallyl disulfide (the volatile compound responsible for garlic odor) is a powerful antibacterial agent. At 177°C and 200. Torr, a sample of diallyl disulfide vapor had a density of 1.04 g/L. What is the molar mass of diallyl disulfide?

5.11 Mixtures of Gases

So far, we have been considering the behavior of single, pure gases, but the most common gas of all, air, is a mixture. We can treat gas mixtures just like pure gases, because, according to the gas laws, all gases respond in the same way to changes in pressure, volume, and temperature. For the type of calculations in this chapter, it does not matter whether all the molecules in a sample are the same. A mixture of gases that do not react with one another, such as air, behaves like a single pure gas.

John Dalton, whose contribution to atomic theory is described in Section 1.1, showed how to calculate the pressure of a mixture of gases. His reasoning was something like this. Imagine that we introduce a certain amount of oxygen into a container and measure the resulting pressure as 0.60 atm. Then we evacuate the container so that it is empty of all gas. Now introduce into it enough nitrogen to give a pressure of 0.40 atm at the same temperature. Dalton wondered what the total pressure would be if the same amounts of the two gases were present in the container simultaneously. From some fairly crude measurements, he concluded that the total pressure resulting from the pressure of both gases in the same container would be 1.00 atm, the sum of the individual pressures.

Dalton summarized his observations in terms of what he called the **partial pressure** of each gas, which is the pressure the gas would exert if it occupied the container alone. In our example, the partial pressures of oxygen and nitrogen in the mixture are 0.60 atm and 0.40 atm, respectively, because those are the pressures the gases exert when each one is in the container alone. Dalton then described the behavior of gaseous mixtures by his **law of partial pressures:**

> The total pressure of a mixture of gases is the sum of the partial pressures of its components.

If we write the partial pressures of the gases A, B, . . . as P_A, P_B, . . . and the total pressure of the mixture as P, then Dalton's law can be written

$$P = P_A + P_B + \cdots \qquad (4)$$

The law is illustrated in Fig. 5.22. It is exactly true only for gases that behave ideally, but it is a good approximation for nearly all gases under normal conditions.

We can calculate the partial pressure of a gas in a mixture if we know what fraction of the total molecules are molecules of the gas. The **mole fraction,** x_A, is the number of moles of A molecules expressed as a fraction of the total number of moles of gas molecules in the mixture:

$$\text{Mole fraction of A} = \frac{\text{moles of A}}{\text{total number of moles}} \quad \text{or} \quad x_A = \frac{n_A}{n_A + n_B + \cdots} \qquad (5)$$

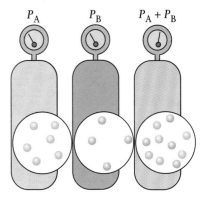

FIGURE 5.22

According to Dalton's law, the total pressure P of a mixture of gases is the sum of the partial pressures P_A and P_B of the components. These partial pressures are the pressures the gases would exert if they were alone in the container (at the same temperature).

$x_{RED} = 0.1$

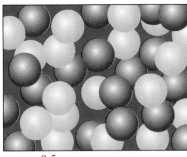

$x_{RED} = 0.5$

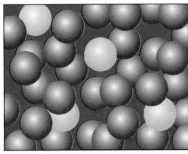

$x_{RED} = 0.9$

FIGURE 5.23

The mole fraction, x, tells us the fraction of molecules of a particular kind in a mixture of two or more kinds of molecules.

The pressure of the vapor of a liquid is called its *vapor pressure;* the topic is treated in more detail in Section 10.14.

For example, if the mixture consists of 2.0 mol N_2 and 3.0 mol O_2, the mole fractions of the two components are

$$x_{N_2} = \frac{2.0 \text{ mol}}{(2.0 + 3.0) \text{ mol}} = \frac{2.0}{5.0} = 0.40$$

$$x_{O_2} = \frac{3.0 \text{ mol}}{(2.0 + 3.0) \text{ mol}} = \frac{3.0}{5.0} = 0.60$$

Note that the sum of the two mole fractions is 1.00. A mole fraction varies from $x_A = 0$ (no A in the mixture) to $x_A = 1$ (the mixture is entirely A), as illustrated by the three different mixtures in Fig. 5.23.

The partial pressure of a component in a gaseous mixture is proportional to the mole fraction of the component. To find the exact relation, we first note that the pressure exerted by a gas A in a container of volume V at a temperature T is given by the ideal gas law as $P_A = n_A RT/V$, where n_A is the number of moles of A. The pressure exerted by the gas B is $P_B = n_B RT/V$. When both gases are present in the same container, their total pressure is determined by the total number of moles of molecules and is $P = (n_A + n_B)RT/V$. Therefore,

$$\frac{P_A}{P} = \frac{n_A RT/V}{(n_A + n_B)RT/V} = \frac{n_A}{n_A + n_B} = x_A$$

That is, the partial pressure of A is related to the mole fraction of A by

$$P_A = x_A P \qquad (6)$$

where P is the total pressure of the gases. A similar expression applies to any other gas in the mixture. It follows that *to calculate the partial pressure of a gas, multiply the total pressure by the mole fraction of that gas.* For instance, if the total pressure of the mixture of 2.0 mol N_2 and 3.0 mol O_2 described earlier is 3.0 atm, then the partial pressures of the two gases would be

$$P_{N_2} = 0.40 \times (3.0 \text{ atm}) = 1.2 \text{ atm} \qquad P_{O_2} = 0.60 \times (3.0 \text{ atm}) = 1.8 \text{ atm}$$

We can use partial pressures to help describe the composition of a humid gas, such as the air in our lungs or a gas collected over water. The total pressure of humid air is the sum of the partial pressure of the air itself and the partial pressure of the water vapor:

$$P = P_{\text{dry air}} + P_{\text{water vapor}}$$

In a closed container, water vaporizes until its partial pressure has reached a certain value, called its "vapor pressure." When there is another gas, such as air, in the container, water still vaporizes in the same way, until its partial pressure equals its vapor pressure. At this point, the gas holds as much water vapor as it can and is said to be **saturated** with water vapor. The vapor pressure of water varies with temperature, and some values are given in Table 5.4. At blood temperature, the temperature in our lungs, the vapor pressure of water is 47 Torr. The partial pressure of water vapor in our lungs is about 47 Torr, so the partial pressure of the air itself is

$$P_{\text{dry air}} = P - 47 \text{ Torr}$$

On a typical day, the total pressure at sea level is 760. Torr, so the pressure due to all the molecules other than H_2O in our lungs is 760. − 47 Torr = 713 Torr.

The partial pressure of a gas is the pressure it would exert if it were alone in the container; it is equal to the mole fraction of the gas times the total pressure. The total pressure of a mixture of gases is the sum of the partial pressures of the components.

Example 5.10 *Calculating partial pressures*

A sample of dry air of total mass 1.00 g consists of 0.76 g of nitrogen and 0.24 g of oxygen. (a) Calculate the partial pressures of these gases and the total pressure (in atmospheres) when this sample is in a flask of volume 1.00 L at 20°C. (b) The sample of air is heated until the total pressure is 1.00 atm. What are the partial pressures of oxygen and nitrogen in the heated sample?

Strategy (a) The partial pressures of the gases are given by the ideal gas law (Eq. 2) applied to each gas in turn, so we need to convert mass to number of moles by using the molar masses of the gases. We use the value of R for pressure expressed in atmospheres ($R = 0.082\,06$ L·atm/K·mol). The total pressure is obtained by adding the two partial pressures together. (b) The total pressure and the mass of each gas are known, so we calculate the mole fraction of each gas and use them in Eq. 6 to determine the partial pressures. Because the mole fractions remain the same as the mixture is heated, but the total pressure rises, we can expect that the partial pressures will rise as the temperature is raised.

Solution (a) The molar masses of N_2 and O_2 are 28.02 g/mol and 32.00 g/mol, respectively. It then follows from Eq. 2 that

$$P_{N_2} = \overbrace{(0.76\text{ g}) \times \left(\frac{1\text{ mol N}_2}{28.02\text{ g N}_2}\right)}^{n_{N_2}} \times \overbrace{\frac{(0.082\,06\text{ L·atm/K·mol}) \times (293\text{ K})}{1.00\text{L}}}^{RT/V}$$

$$= \frac{0.76 \times 0.082\,06 \times 293}{28.02 \times 1.00} \times \frac{\text{g·mol·L·atm·K}}{\text{g·K·mol·L}} = 0.65\text{ atm}$$

$$P_{O_2} = (0.24\text{ g}) \times \left(\frac{1\text{ mol O}_2}{32.00\text{ g O}_2}\right) \times \frac{(0.082\,06\text{ L·atm/K·mol}) \times (293\text{ K})}{1.00\text{ L}}$$

$$= \frac{0.24 \times 0.082\,06 \times 293}{32.00 \times 1.00} \times \frac{\text{g·mol·L·atm·K}}{\text{g·K·mol·L}} = 0.18\text{ atm}$$

The total pressure is the sum of these two partial pressures:

$$P = P_{N_2} + P_{O_2} = (0.65 + 0.18)\text{ atm} = 0.83\text{ atm}$$

(b) The numbers of moles corresponding to the given masses are

$$n_{N_2} = (0.76\text{ g N}_2) \times \left(\frac{1\text{ mol N}_2}{28.02\text{ g N}_2}\right) = 0.027\text{ mol N}_2$$

$$n_{O_2} = (0.24\text{ g O}_2) \times \left(\frac{1\text{ mol O}_2}{32.00\text{ g O}_2}\right) = 0.0075\text{ mol O}_2$$

The mole fractions of the two gases are therefore

$$x_{N_2} = \frac{0.027\text{ mol}}{(0.027 + 0.0075)\text{ mol}} = 0.78 \qquad x_{O_2} = \frac{0.0075\text{ mol}}{(0.027 + 0.0075)\text{ mol}} = 0.22$$

Table 5.4 *Vapor pressure of water*

Temperature, °C	Vapor pressure, Torr
0	4.58
10	9.21
20	17.54
21	18.65
22	19.83
23	21.07
24	22.38
25	23.76
30	31.83
37*	47.08
40	55.34
60	149.44
80	355.26
100	760.00

*Body temperature.

To find the partial pressures, we multiply each mole fraction by the total pressure, 1.00 atm, and obtain

$$P_{N_2} = 0.78 \times (1.00 \text{ atm}) = 0.78 \text{ atm} \qquad P_{O_2} = 0.22 \times (1.00 \text{ atm}) = 0.22 \text{ atm}$$

Note that the total pressure, the sum of the partial pressures, is 1.00 atm, as in the data.

Self-Test 5.10A The composition of the gas in a neon advertising sign of volume 0.75 L is 0.10 g of neon and 0.20 g of xenon. Calculate their partial pressures and the total pressure (in atmospheres) when the tube is operating at 40°C.

[*Answer:* $P_{Ne} = 0.17$ atm, $P_{Xe} = 0.052$ atm, $P = 0.22$ atm]

Self-Test 5.10B A sample of oxygen gas of volume 1.00 L was collected over water at 25°C and a total pressure of 1.00 atm. The partial pressure of water is 23.8 Torr. How many moles of oxygen molecules were collected?

MOLECULAR MOTION IN GASES

So far, we have pictured a gas as a collection of molecules in ceaseless random motion and have seen that the gas laws are consistent with this model. But questions remain. Do all molecules move at the same speed, or do some molecules move faster than others? How does the speed of the molecules vary with the temperature? We can refine our model of a gas and answer these questions by collecting more experimental data.

5.12 Diffusion and Effusion

The experimental data we need for questions like these come from the study of the rate at which a gas spreads away from its original location. The gradual dispersal of one substance through another substance, such as the odor of a skunk spreading through air, is called **diffusion.** The escape of one substance (particularly a gas) through a small hole into a vacuum, like the escape of air through a small hole in a spacecraft, is called **effusion** (Fig. 5.24).

The diffusion of one gas into another explains the spread of pheromones (chemical signals between animals) and perfumes (which may have a similar effect) through air. Diffusion helps to keep the composition of the atmosphere moderately constant, because abnormally high concentrations of a gas disperse by diffusion. At low altitudes, the large-scale motion of air that we call **convection** and experience as wind is a bigger factor, but diffusion is more important at high altitudes (Applying Chemistry: Case Study 5). Chemicals that diffuse into the stratosphere can stay there for years.

A tiny puncture in a tire results in a kind of effusion: in this case, from a high-pressure region to a low-pressure region.

Effusion occurs whenever a gas is separated from a vacuum by a porous barrier (a barrier that contains microscopic holes) or a single pinhole. A gas escapes through a pinhole because there are more collisions with the hole on the high-pressure side than on the low-pressure side. As a result, more molecules pass from the high-pressure region into the low-pressure region than in the opposite direction.

Thomas Graham, a nineteenth-century Scottish chemist, did a series of experiments on the rate of effusion of gases, which is defined as the number of moles of gas molecules that escape from a container in a given interval. He

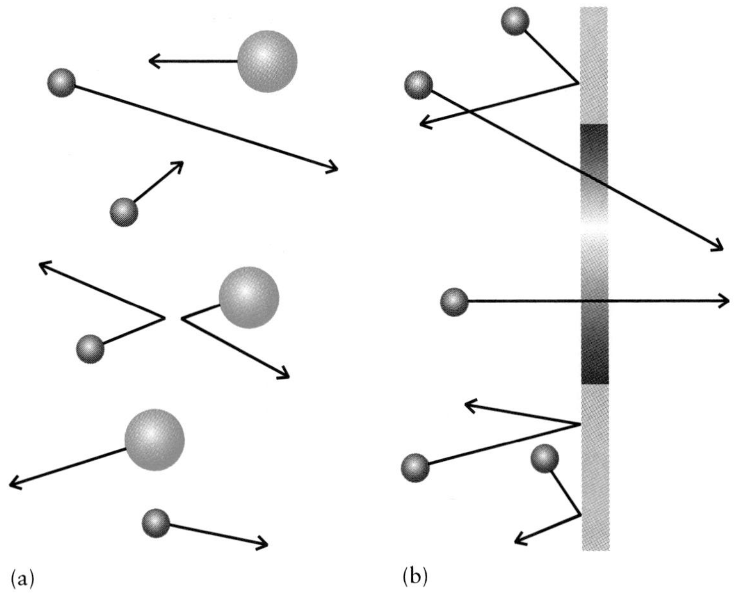

FIGURE 5.24

(a)

(b)

(a) In diffusion, the molecules of one substance spread into the region occupied by molecules of another in a series of random steps, undergoing collisions as they move. (b) In effusion, the molecules of one substance escape through a small hole in a barrier. In both cases, the rate increases with increasing temperature and decreases with increasing molar mass.

found that *the rate of effusion of a gas is inversely proportional to the square root of its molar mass.* This observation is now known as **Graham's law:**

$$\text{Rate of effusion} \propto \sqrt{\frac{1}{\text{molar mass}}} \qquad \text{or} \qquad \text{rate} \propto \sqrt{\frac{1}{M}}$$

The rate of diffusion of a gas has also been found to follow approximately the same inverse square-root dependence, with heavy molecules diffusing more slowly than light molecules.

The *time* a given amount of gas takes to escape is inversely proportional to the rate of its effusion, with rapidly effusing gases taking less time to escape than slowly effusing gases. Graham's law therefore implies that the time, t_{effuse}, required for the effusion of a given number of moles of gas molecules is *directly* proportional to the square root of the molar mass:

$$\text{Effusion time} \propto \sqrt{\text{molar mass}} \qquad \text{or} \qquad t_{\text{effuse}} \propto \sqrt{M}$$

Heavy molecules take longer to escape through a hole, because they move more slowly than light molecules. The ratio of the times it takes the same numbers of moles of two gases, A and B, to effuse under the same conditions is therefore

$$\frac{\text{Effusion time of A}}{\text{Effusion time of B}} = \sqrt{\frac{\text{molar mass of A}}{\text{molar mass of B}}} \qquad \text{or} \qquad \frac{t_{\text{effuse}}(A)}{t_{\text{effuse}}(B)} = \sqrt{\frac{M_A}{M_B}} \qquad (7)$$

The time it takes the molecules of a gas to effuse through an opening or diffuse through another gas is directly proportional to the square root of its molar mass.

Self-Test 5.11A If it takes a certain amount of helium atoms 10. s to effuse through a porous barrier, how long does it take the same amount of methane molecules, CH_4, under the same conditions?

[*Answer:* 20. s]

Self-Test 5.11B Which gas effuses more slowly through a hole of given size, nitrogen dioxide or ozone? Why?

Applying Chemistry: Case Study 5

The abundance of life on Earth is unique in our solar system. Life as we know it requires sunlight, a temperate climate, a plentiful supply of water, and wide availability of certain elements, particularly carbon, hydrogen, oxygen, nitrogen, phosphorus, and sulfur. These elements must be recycled through a complex web of natural processes in the atmosphere. Oxygen and nitrogen are good examples of how this circulation functions.

Millions of years ago, most atmospheric oxygen was in the form of carbon dioxide and water. Some molecular oxygen, O_2, came from the outgassing of rocks and volcanoes

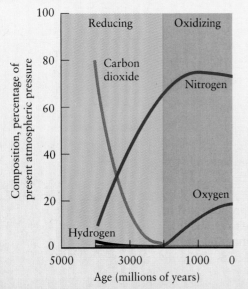

This speculative history of the atmosphere shows a change from a predominantly reducing to a predominantly oxidizing atmosphere (rich in O_2) about 2 billion years ago when photosynthetic plants started to appear. Hydrogen has been lost to compound formation and to outer space. Earth is the only planet in the solar system with an oxidizing atmosphere.

and the effect of solar radiation on water. However, as green plants appeared—initially in the sea—they produced much more O_2 by photosynthesis. Today, the mole fraction of molecular oxygen in dry air is 0.21 and both plants and animals participate in the oxygen cycle.

During electrical storms, some O_2 forms ozone, O_3, a reactive form of oxygen. Ozone in the troposphere (see Investigating Matter 5.1) is regarded as a pollutant, for it reacts with organic components of engine exhausts to produce the eye irritants in smog. However, the ozone in the stratosphere is vital to our existence.

Ozone is formed in the stratosphere in two steps. First, O_2 molecules or other oxygen-containing molecules are broken apart into atoms by sunlight. Then the O atoms react with other O_2 molecules to form ozone:

$$O_2(g) \xrightarrow{\text{sunlight}} 2\,O(g)$$

$$O(g) + O_2(g) \longrightarrow O_3(g)$$

The second of these two reactions releases heat, which raises the temperature of the stratosphere. Some of the ozone is decomposed by ultraviolet radiation:

$$O_3(g) \xrightarrow{\text{uv radiation}} O(g) + O_2(g)$$

Because this reaction absorbs ultraviolet radiation, it helps to shield the Earth from radiation damage. The atmospheric history chart shows that land plants did not appear until there was sufficient molecular oxygen in the atmosphere to produce the protective ozone shield.

The mole fraction of nitrogen gas in dry air is 0.78, but molecular nitrogen is very unreactive. For it to play its part in the global metabolism, nitrogen gas must be converted into compounds that plants and animals can use. This process is called the *fixation* of nitrogen. Atmospheric nitrogen is fixed by lightning, which oxidizes it to oxides such as NO and NO_2. It is also fixed in the soil by bacteria, which oxidize it to the nitrate ion (Section 20.1). Fertilizer factories that use the Haber process (Section 14.12) fix nitrogen industrially by reducing it to ammonia with hydrogen gas.

5.13 The Kinetic Model of Gases

The kinetic model of a gas is also known as the kinetic molecular theory.

Effusion and diffusion have provided us with additional experimental information about gases, and we can now incorporate them into our model. The **kinetic model** of a gas is an elaboration of the simple picture developed in Section 5.1 to account for the gas laws. It is based on four assumptions:

1. A gas consists of a collection of molecules in ceaseless random motion.

2. Gas molecules are negligibly small.

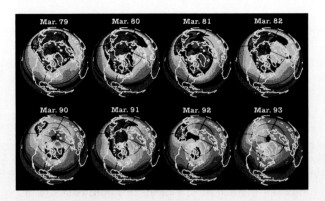

Mar. 79　Mar. 80　Mar. 81　Mar. 82

Mar. 90　Mar. 91　Mar. 92　Mar. 93

These maps of stratospheric ozone concentration over the North Pole show how the ozone has been depleted since 1979. Red areas represent ozone concentrations greater than 500 Dobson units (DU); the concentrations decrease through green, yellow, and blue, to purple at less than 270 DU. Normal ozone concentration at temperate latitudes is about 350 DU. A Dobson unit is equivalent to a layer of pure ozone of thickness 1 μm and a pressure of 1 atm.

Nitrogen monoxide, NO (nitric oxide), is also formed in automobile engines and then oxidized by the oxygen in air to nitrogen dioxide, NO_2. The NO_2 dissolves in water to form nitric acid, which delivers nitrogen to the soil in the form of acid rain (see Applying Chemistry: Case Study 15):

$$3 NO_2(g) + H_2O(g) \longrightarrow 2 HNO_3(aq) + NO(g)$$

Nitrogen oxides that diffuse up to the stratosphere can threaten the ozone layer, because they are decomposed by solar radiation to species that react with ozone.

The normal concentration of gases in the stratosphere is such that ozone formation is balanced by ozone destruction. In the heart of the ozone layer, the concentration of O_3 molecules is about 0.1 mmol/L. However, this balance is in danger of being upset by human activities. Large amounts of nitrogen oxides are generated by jet engines, by automobiles, and by the degradation of nitrogen-based fertilizers. There is also evidence that chlorofluorocarbons (CFCs) may represent the most serious threat to stratospheric ozone. Use of CFCs is now restricted, but because it takes years for CFC molecules to diffuse into the stratosphere, much of the CFCs used during the last decade may still be on the way up. Currently, we are losing about 2% of the ozone in the stratosphere every 10 years.

Key Concepts: diffusion, gaseous mixtures, reactions of gases

For Further Reading

P. J. Crutzen, What is happening to our precious air? *Science Spectra*, **14**: 22–31, 1998.

R. Monastersky, Drop in ozone killers means global gain, *Science News*, Mar. 9, 1996, p. 151.

Related Exercises: 5.109–5.112

An atmospheric scientist prepares an experimental pod for its ascent to the stratosphere in a research balloon from a site in Sweden. The balloon carries equipment that can measure the concentration of 15 pollutants suspected of reducing atmospheric ozone levels.

3. The pointlike particles move in straight lines until they collide.

4. The molecules do not influence one another except during collisions.

The second assumption implies that gas molecules must be very far apart. The fourth assumption means that there are no attractive or repulsive forces between ideal gas molecules except during collisions (Fig. 5.25). The kinetic model is one of the most remarkable models in science, for from a handful of simple assumptions, we can calculate many properties of gases.

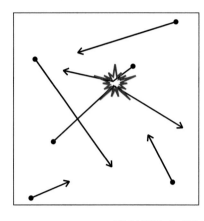

FIGURE 5.25

In the kinetic model of gases, the molecules are regarded as infinitesimal points that travel in straight lines until they undergo instantaneous collisions.

The fact that the ideal gas law is only a limiting law is consistent with the kinetic model. The kinetic model assumes that the molecules are infinitely small and do not interact with one another (except when they collide). Deviations from ideal behavior then arise because real molecules have definite volumes and do attract and repel one another. The attractions and repulsions between molecules are called **intermolecular forces** (Fig. 5.26). They become important at high pressures, when the average separation of the molecules is small. At very low pressures, the average separation of the molecules is so great that intermolecular forces are unimportant. Under these conditions, a real gas acts like an ideal gas (Section 5.15).

Calculations based on the kinetic model help us picture the behavior of an ideal gas in more detail. For instance, the model can be used to calculate the **mean free path** of the molecules, the average distance that a molecule travels between collisions. For a gas like air at 1 atm, the mean free path is about 70 nm, or about 200 molecular diameters. The mean free path decreases as the pressure is increased, because more molecules are packed into a smaller volume. The mean free path is independent of temperature, however, because at constant pressure the distance between collisions is independent of how fast the molecules are traveling.

An analogy

If a molecule were the size of a tennis ball, the mean free path would be about 15 m (about two-thirds as long as a tennis court).

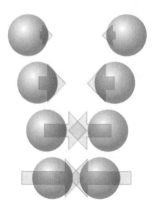

FIGURE 5.26

Molecules attract one another over several molecular diameters. The strength of the attraction (as depicted by the lengths of the arrows) increases as they approach. This attraction is responsible for the condensation of a gas to a liquid.

The kinetic model also leads to the conclusion that the average speed of molecules in a gas is related to the temperature and the molar mass by

$$\text{Average speed} \propto \sqrt{\frac{\text{temperature}}{\text{molar mass}}} \tag{8}$$

We know (from the increase in pressure with temperature) that the average speed of gas molecules increases as the temperature is raised. The kinetic model expresses this dependence quantitatively. It predicts that *the average speed of the molecules of a gas is proportional to the square root of the absolute temperature.* Doubling the temperature of any gas (from 200 K to 400 K, for example) increases the average speed of its molecules by a factor of $\sqrt{2} = 1.4$. Equation 8 also shows that, at a given temperature, the average speed of molecules in a gas is inversely proportional to the square root of the molar mass. This conclusion is consistent with Graham's law, which is based on experiment. Therefore, the kinetic model is supported by experiment.

A more detailed calculation based on the kinetic model shows that a property important in kinetic theory, the **root mean square speed** (rms speed, v) of molecules in a gas, is given by the expression

$$v = \sqrt{\frac{3RT}{M}} \tag{9}$$

The rms speed of molecules is defined as the square root of the average of the *squares* of the speeds of the molecules in the sample:

$$v = \sqrt{\frac{v_1^2 + v_2^2 + \cdots + v_N^2}{N}} \qquad (10)$$

(N is the number of molecules.) For example, if the sample had three molecules ($N = 3$), with speeds 200 m/s, 300 m/s, and 400 m/s, the rms speed would be

$$v = \sqrt{\frac{(200 \text{ m/s})^2 + (300 \text{ m/s})^2 + (400 \text{ m/s})^2}{3}}$$

$$= \sqrt{\frac{2.90 \times 10^5 \text{ m}^2/\text{s}^2}{3}} = 311 \text{ m/s}$$

Note that the rms speed is slightly larger than the mean speed (which is 300 m/s in this example).

We can use Eq. 9 to refine our picture of the atmosphere. At any given temperature, CO_2 molecules have an rms speed that is only 64% that of the lighter H_2O molecules. A sample of humid air at 25°C can therefore be pictured as a storm of molecules, with the CO_2 molecules, the heaviest molecules present, lumbering along at an average speed of 410 m/s and H_2O molecules, the lightest molecules present, zipping along at about 640 m/s (Fig. 5.27). The temperature of the atmosphere varies with altitude (Investigating Matter 5.1), so the rms speeds (and the average molecular speeds) also vary.

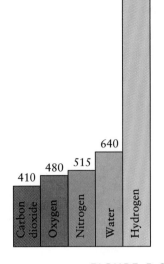

FIGURE 5.27

The average speeds of gas molecules at 25°C in meters per second. The gases are some of the components of air; hydrogen is included to show that light molecules travel much more rapidly on average than heavy molecules.

Example 5.11 *Calculating the root mean square speed of molecules in a gas*

High above the stratosphere, the temperature of the atmosphere may reach 1000°C. However, spacecraft and astronauts do not burn up because the concentration of molecules is very low and the impacts transfer very little energy. What is the rms speed of nitrogen molecules at that temperature?

Strategy We expect a speed of well over 300 m/s (a typical rms speed at room temperature) because of the greater temperature. To calculate the actual value, use Eq. 9 with R in SI units ($R = 8.314\,51$ J/K·mol with 1 J = 1 kg·m²/s²); convert the molar mass to kilograms per mole to ensure that all the units are consistent. The temperature must be expressed in kelvins.

Solution The molar mass of nitrogen is 28.02 g/mol, which corresponds to 2.802×10^{-2} kg/mol. The rms speed is therefore

$$v = \sqrt{\frac{3 \times (8.314\,51 \text{ J/K·mol}) \times (1273 \text{ K})}{2.802 \times 10^{-2} \text{ kg/mol}}}$$

$$= \sqrt{\frac{3 \times 8.314\,51 \times 1273}{2.802 \times 10^{-2}}} \times \sqrt{\frac{\text{J·K·mol}}{\text{K·mol·kg}}}$$

$$= 1064\sqrt{\frac{\text{J}}{\text{kg}}} = 1064\sqrt{\frac{\text{kg·m}^2/\text{s}^2}{\text{kg}}} = 1064 \text{ m/s}$$

Self-Test 5.12A At 25°C, the rms speed of the molecules of an unknown gas is half that of the average speed of helium atoms. What is the molar mass of the unknown gas?

[*Answer:* 16.0 g/mol]

Investigating Matter 5.1: *The Layers of the Atmosphere*

The atmosphere is divided into several regions that vary in temperature according to the chemical reactions taking place in them. These chemical reactions are stimulated by solar radiation, which they convert into heat.

The lowest region of the atmosphere is called the *troposphere*. In this region, the temperature decreases with increasing altitude. Between about 11 and 16 km lies the *tropo-pause*, a region where the temperature remains constant at about −55°C. Above about 16 km, the *stratosphere* begins. Here the temperature rises, until it reaches about 0°C in the *stratopause*, a region lying at about 45 km above sea level. Above that altitude, in the *mesosphere*, the temperature falls again, until the mesopause is reached. After the *mesopause*, in the *thermosphere*, the temperature rises, and at very high altitudes, it can be over 1000°C.

Local variations in the density, composition, motion, and temperature of the atmosphere cause the winds and precipitation that we call "weather." Weather takes place in the troposphere (the name troposphere means "sphere of change"). Packets of less dense warm air rise above cooler, denser air in a process called *convection*. The air cools as it rises into the colder regions of the atmosphere, and the water molecules aggregate into droplets that we see as clouds. These droplets themselves may aggregate and fall to the ground as rain. The cooling also lowers the pressure, so precipitation is often associated with regions of low atmospheric pressure. Adjacent air masses at higher pressures flow into the low-pressure region. This flow of air is what we call "wind."

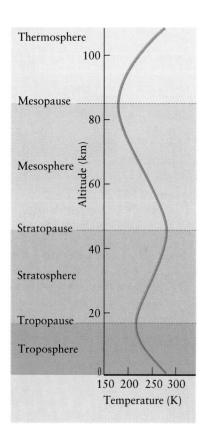

The variation of temperature with altitude in the atmosphere and the various zones into which the atmosphere is divided. The temperature profile represents the consequences of the differing interactions of molecules with solar radiation.

The winds associated with storms result from differences in air pressure. High winds such as the hurricane-generated winds pictured here can cause extensive damage.

Self-Test 5.12B In which gas do the molecules have the higher rms speed, CH_4 or Cl_2, at the same temperature?

From Equation 9, we can deduce an important relation between temperature and the average kinetic energy, E_K, of the gas molecules. The **kinetic energy** of a molecule is the energy it possesses as a result of its motion. For a molecule of mass m moving at a speed v, the kinetic energy is

$$E_K = \tfrac{1}{2}mv^2$$

This expression tells us that a heavy molecule moving rapidly has a higher kinetic energy than a light molecule moving slowly. When we substitute Eq. 9 for v, we obtain

$$E_K = \tfrac{1}{2}m \times \frac{3RT}{M} = \tfrac{1}{2}m \times \frac{3RT}{mN_A} = \frac{3RT}{2N_A}$$

In this calculation, we note that the molar mass M is equal to the mass of one molecule, m, multiplied by the Avogadro constant, N_A.

It follows that *the average kinetic energy of a molecule is proportional to T and independent of the molecular mass.* At a given temperature, the *average speed* of molecules depends on molar mass, but their *average kinetic energy* does not.

The quantity R/N_A is called the Boltzmann constant, and written k.

> **The average speed of the molecules in a gas sample is directly proportional to the square root of the temperature and inversely proportional to the square root of the molar mass.**

5.14 The Maxwell Distribution of Speeds

We can deepen our insight into the nature of an ideal gas even further by noting that Eq. 8 gives only the *average* speed of gas molecules. According to the kinetic model, however, an individual molecule undergoes several billion changes of speed and direction each second. Can we say anything about the *range* of speeds that the molecules of a sample of a gas possess?

Like cars in traffic, individual molecules have different speeds. Moreover, as in a collision between two cars, a molecule might be brought almost to a standstill when it collides with another. Then, in the next instant (but now unlike a collision between cars), it might be struck to one side by another and move off at the speed of sound.

An analogy

The fraction of gas molecules moving at each speed at any instant is called the **distribution of molecular speeds.** The nineteenth-century Scottish scientist James Maxwell used the kinetic model to calculate this distribution, and Fig. 5.28 summarizes his conclusions. The graphs show that heavy molecules (CO_2) travel with speeds close to their average. Light molecules (H_2) not only have a higher average speed, but the speeds of many of them are very different from their average speed. This wide range of speeds implies that light molecules are more likely than heavy molecules to have such high speeds that they escape from the gravitational pull of small planets and go off into space. Consequently, hydrogen molecules and helium atoms, which are both very light, are very rare in the Earth's atmosphere, although they are abundant on massive planets like Jupiter.

Hydrogen is also very reactive, so it is trapped on Earth in compounds.

The curves in the illustration also show that the spread of speeds widens as the temperature increases. At low temperatures, most molecules have speeds

FIGURE 5.28

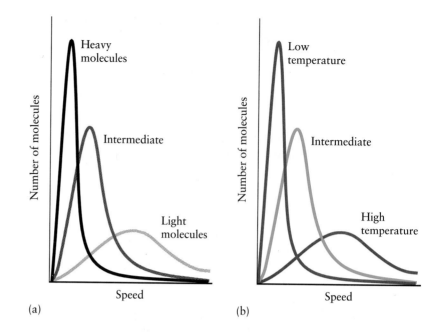

(a) The range of molecular speeds for several gases, as given by the Maxwell distribution. All the curves correspond to the same temperature. The greater the molar mass, the narrower the spread of speeds. (b) The Maxwell distribution again, but now the curves correspond to the speeds of a single substance at different temperatures. The higher the temperature, the broader the spread of speeds.

close to their average speed. At high temperatures, a high proportion have speeds widely different from their average speed. Hot gases have a high proportion of very fast molecules. This feature will prove to be the key to understanding the rates at which reactions take place, and we shall meet it again in Section 13.9.

The molecules of all gases have a wide range of speeds. As the temperature increases, the average speed and the range of speeds increase.

5.15 Real Gases

Real gases—oxygen, carbon dioxide, water vapor, and so on—are only *nearly* ideal. Gases can be condensed to liquids and solids, like the chlorine gas in Fig. 5.29, whereas an ideal gas never condenses to a liquid, however low the temperature or high the pressure. When an ideal gas is compressed or cooled, its volume decreases without limit and may approach 0.

The difference between ideal and real gases is expressed strikingly by plotting the "compression factor" $Z = PV/nRT$ as a function of P (Fig. 5.30). For an ideal gas, the graph is a horizontal line at 1 (because $PV/nRT = 1$). However, if we construct that same plot for several real gases, we get different curves. We have to conclude that PV/nRT is not constant for real gases.

We can explain this behavior if we suppose that molecules interact with one another. As the gas is compressed, the average separation of the molecules decreases and the attractions become more important. Because the molecules move less freely, the pressure is less than predicted by the ideal gas law. However, when the sample is greatly compressed, repulsions between the individual molecules become more important, because the molecules are now very close together; as a result, the pressure is *greater* than the ideal gas law would predict.

Many scientists have suggested ways of changing the ideal gas law so that it takes into account the effect of intermolecular forces. One of the earliest and

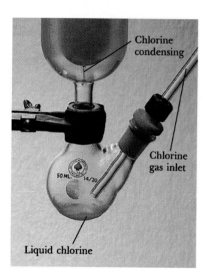

FIGURE 5.29

Chlorine can be condensed to a liquid at atmospheric pressure by cooling it to −35°C or lower. The upper tube contains a cold finger, a smaller tube filled with a very cold mixture of dry ice and acetone.

most useful of these improved equations was proposed by Johannes van der Waals, a nineteenth-century Dutch scientist:

$$\left(P + \frac{an^2}{V^2}\right)(V - nb) = nRT$$

Effect of attractions: $\frac{an^2}{V^2}$

Effect of repulsions: nb

This expression is called the **van der Waals equation.** The constant a represents the effect of attractions between molecules, and the constant b represents the effect of repulsions.

The values of a and b are determined experimentally for individual gases. Table 5.5 gives some values. We can interpret b as a measure of the volume taken up by the gas molecules themselves. Both constants are 0 in an ideal gas, in which the pointlike molecules have zero volume ($b = 0$) and do not attract one another ($a = 0$). The two constants are determined for each gas individually by adjusting their values until the van der Waals equation fits the observed dependence of the pressure on the volume, temperature, and amount of gas being studied. The values of a and b differ from gas to gas because the sizes of molecules differ (which affects b); and their ability to attract each other differs too (which affects a). Unlike the ideal gas equation, the van der Waals equation is different for every gas.

At low pressures, a gas occupies a very large volume. Because V (the volume occupied by the gas) is then very much bigger than nb (the actual volume taken up by the molecules themselves), the term $V - nb$ is then approximately equal to V itself. Moreover, when the volume occupied by the gas is large, an^2/V^2 is very small and we can neglect it relative to P. Now the molecules are so far apart that they do not attract one another significantly. Therefore, at low pressures, an approximate form of the van der Waals equation is

$$PV \approx nRT$$

This expression is the ideal gas law. We see that the equation for a real gas becomes the ideal gas law when the pressure is very low and intermolecular forces can be neglected.

The van der Waals equation describes the behavior of real gases by taking into account the effects of molecular attractions and repulsions.

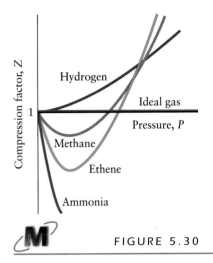

FIGURE 5.30

A plot of the compression factor $Z = PV/nRT$ as a function of pressure for a variety of gases. For an ideal gas, this ratio is equal to 1 for all pressures. For a few real gases with very weak intermolecular attractions, PV/nRT is always greater than 1. For most gases, at low pressures the attractive forces are dominant and $PV/nRT < 1$. At high pressures, repulsive forces become dominant and $PV/nRT > 1$.

Table 5.5 *Van der Waals constants for common gases*

	$a,$ $L^2 \cdot atm/mol^2$	$b,$ L/mol
air	1.4	0.039
ammonia	4.17	0.037
argon	1.35	0.032
carbon dioxide	3.59	0.043
ethene (ethylene)	4.47	0.057
helium	0.034	0.024
hydrogen	0.244	0.027
nitrogen	1.39	0.039
oxygen	1.36	0.032

Skills You Should Have Mastered

Conceptual

☐ 1. Explain the origin of pressure in molecular terms, Sections 5.1–5.2.

☐ 2. List and explain the assumptions of the kinetic theory of gases and show how they can be used to interpret the ideal gas law, Section 5.13.

☐ 3. Describe the effect of molar mass and temperature on the distribution of molecular speeds, Section 5.14.

☐ 4. Explain how real gases differ from ideal gases, Section 5.15.

Problem-Solving

☐ 1. Convert between pressure units, Example 5.1.

☐ 2. Use Boyle's law to calculate the change in pressure due to compression, Example 5.2.

☐ 3. Use Charles's and Gay-Lussac's laws to calculate the change in volume or pressure due to a temperature change, Example 5.3.

☐ 4. Use Avogadro's principle to predict the change in volume due to a change in the number of moles of gas, Example 5.4.

☐ 5. Use the ideal gas law to calculate P, V, T, and n for given conditions or after a change in conditions, Toolbox 5.1 and Examples 5.5 and 5.6.

☐ 6. Calculate the volume of a gas from its mass, Example 5.7.

☐ 7. Calculate the volume of a gas produced or consumed in a reaction, Toolbox 5.2 and Example 5.8.

☐ 8. Determine molar mass from gas density and vice versa, Example 5.9.

☐ 9. Calculate partial pressures of gases and the total pressure of a gas mixture, Example 5.10.

☐ 10. Use Graham's law to account for relative rates of effusion, Self-Test 5.11.

☐ 11. Calculate the root mean square speed of the molecules in a gas, Example 5.11.

Descriptive

☐ 1. Describe the construction and use of a barometer, Section 5.2.

☐ 2. Describe the structure of the atmosphere, Investigating Matter 5.1.

Exercises

Pressure

5.1 Explain how a barometer measures atmospheric pressure.

5.2 Is the space above the mercury in a barometer truly a vacuum? Would the height of the mercury column change if air were introduced into that space?

5.3 Gas pressures are expressed in different units, depending on the application. Make the following conversions: (a) 1.00 bar to atmosphere; (b) 1.00 mmHg to pascal; (c) 1.00 kPa to atmosphere.

5.4 Several different units of pressure will be used in these exercises. Complete the following conversions: (a) 725 Torr to atmosphere; (b) 10.7 lb/inch^2 to Torr; (c) 600. mmHg to atmosphere.

5.5 The pressure needed to make synthetic diamonds from graphite is 80 000 atm. Express this pressure in (a) kilobars; (b) pascals.

5.6 An argon gas cylinder measures a pressure of 22.5 lb/inch^2. Convert this pressure to (a) kilopascal; (b) Torr.

5.7 Suppose you were marooned on a tropical island and had to make a primitive barometer, using seawater (density 1.10 g/cm^3). What height would the water reach in your barometer when a mercury barometer would reach 77.5 cm? The density of mercury is 13.5 g/cm^3.

5.8 Assume that the width of your body (across your shoulders) is 20. inches and the depth of your body (chest to back) is 10. inches. If atmospheric pressure is 14.7 lb/inch^2, what mass of air does your body support when you are in an upright position?

The Gas Laws

5.9 Calculate the final volume for the following constant-temperature processes: (a) the pressure on 1.00 L of H$_2$ changes from 2.20 kPa to 3.00 atm; (b) the pressure on 25.0 mL of CO$_2$ changes from 200. Torr to 0.500 atm.

5.10 (a) A helium balloon has a volume of 22 L at sea level, where the atmospheric pressure is 0.951 atm. When the balloon has risen to a height at which the atmospheric pressure is 550. Torr, the balloon bursts. What is the volume of the balloon just before it bursts? (b) A 1.0-mm^3 bubble at the bottom of a lake where the pressure is 3.5 atm rises to the surface. What is the volume of the bubble when it hits the top where the atmospheric pressure is 780. Torr? Assume a constant temperature in each case.

5.11 Determine the final pressure when (a) 1.0 mL of krypton at 105 kPa is transferred to a 1.0-L vessel; (b) 30.0 cm^3 of O$_2$ at 600. Torr is compressed to 5.0 cm^3. Assume constant temperature.

5.12 (a) A 2.4-L sample of methane at a pressure of 1220 Torr is transferred to a 4.1-L vessel. What is the final pressure of methane if the change occurs at constant temperature? (b) A fluorinated organic gas in a cylinder is compressed from an initial volume of 500. mL at 300. Pa to 175 mL at the same temperature. What is the final pressure?

5.13 When Robert Boyle conducted his experiments, he measured pressure in inches of mercury (inHg). He trapped some air in the tip of a J-tube (**1**) and measured the difference in height of the mercury in the two arms of the tube (h). When h = 12.0 inches, the height of the gas in the tip of the tube was 32.0 inches. Boyle then added additional mercury, and the level rose so that h = 30.0 inches (**2**). (a) If atmospheric pressure that day was 30.0 inHg, what was the pressure of the gas in the tube in (**1**) and in (**2**) in inHg? (b) What was the height of the air space (in inches) in the tip of the tube in (**2**)?

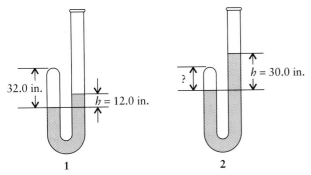

5.14 Boyle continued to add mercury to the apparatus in Exercise 5.13 until the height of the trapped air had been reduced to 10.7 inches. What was the pressure of the trapped air at that point (in inHg)?

5.15 A chemist had prepared a sample of hydrogen bromide and found that it occupied 255 mL at 85°C and 600. Torr. What volume would it occupy at 0.°C at the same pressure?

5.16 A syringe containing 12.0 mL of dry air at 25°C was placed in a hospital sterilizer, where it was heated to 100.°C. The syringe was sealed, but the plunger was free to move, so the volume could change. What was the volume of the syringe when the sterilizer had reached 100.°C? Assume that the process occurs at constant pressure.

5.17 The gas thermometer was an early type of thermometer that measured temperature according to the expansion or contraction of a gas. A certain gas thermometer was calibrated to contain 1.00 L of a gas at 20.°C. (a) What was the temperature when the volume reached 1.03 L? (b) What was the temperature when the volume decreased to 0.95 L?

5.18 One day in Fairbanks, Alaska, the temperature dropped to −35°C. On that day, a couple left a party with a helium balloon that had a volume of 5.00 L at the 22°C temperature of the party. During the 10-min walk home, the temperature of the balloon fell to −35°C. What was the volume of the balloon at that temperature?

5.19 An outdoor storage vessel for hydrogen gas with a volume of 300. m^3 is at 1.5 atm and 10.°C at 2:00 AM. By 2:00 PM, the temperature has risen to 30.°C. What is the new pressure of the hydrogen gas in the vessel?

5.20 (a) A 250-mL aerosol can at 25°C and 1.10 atm was thrown into an incinerator. When the temperature in the can reached 600.°C, it exploded. What was the pressure in the can before it exploded? (b) A helium balloon has a volume of 22 L at 0.951 atm and 18°C. The balloon is cooled at constant pressure until the temperature is −15°C. What is the volume of the balloon at this temperature?

5.21 The pressure in an automobile tire is 30. $lb/inch^2$ at 20.°C, and the atmospheric pressure is 14.7 $lb/inch^2$. The automobile is driven for several hundred kilometers, and the pressure in the tire then reads 34 $lb/inch^2$. Assuming a constant volume and no leaks in the tire, calculate the new temperature of the air in the tire.

5.22 A relief valve on an industrial storage tank operates whenever the pressure of the enclosed carbon dioxide gas exceeds 115 atm. In December, the tank was filled with carbon dioxide at 100. atm when the ambient temperature was −10.°C. On a hot summer day, the temperature rose to 35°C. Should the relief valve have operated?

5.23 A chemist prepares a sample of helium gas at a certain pressure, temperature, and volume, and then removes half the gas molecules. How must the temperature be changed to keep the pressure and volume the same?

5.24 A chemist prepares 0.100 mol Ne(g) at a certain pressure and temperature in an expandable container. Another 0.010 mol of neon atoms is then added to the same container. How must the volume be changed to keep the pressure and temperature the same?

The Ideal Gas Law

5.25 Show how the ideal gas law summarizes Boyle's law, Charles's law, and Avogadro's principle.

5.26 Which of the following plots of functions of an ideal gas will not be linear? (a) P against T; (b) V against T; (c) P against V; (d) V against n. Assume all other variables are constant.

5.27 (a) A 100.-mL flask contains argon at 1.3 atm and 77°C. What amount of Ar is present (in moles)? (b) A 120.-mL flask contains 2.7 μg of O_2 at 17°C. What is the pressure (in Torr)? (c) A 16.7-g sample of krypton exerts a pressure of 100. mTorr at 44°C. What is the volume of the container (in liters)?

5.28 (a) A 350.0-mL flask contains 0.1500 mol Ar at 25°C. What is the pressure of the gas (in kilopascals)? (b) A 23.9-mg sample of bromine trifluoride exerts a pressure of 10. Torr at 100.°C. What is the volume of the container (in milliliters)? (c) A 6000.-m^3 storage tank contains methane at 129 kPa and 15°C. What amount of CH_4 is present (in moles)?

5.29 (a) A 20.-L flask at 200. K and 20. Torr contains nitrogen. What mass of nitrogen is present (in grams)? (b) A 2.6-μL ampoule of xenon has a pressure of 2.00 Torr at 15°C. How many Xe atoms are present?

5.30 (a) A 100.-mL flask contains sulfur dioxide at 0.77 atm and 30.°C. What mass of gas is present? (b) A 1.0-mL ampule of helium has a pressure of 2.00 kPa at −115°C. How many He atoms are present?

5.31 What mass of ammonia will exert the same pressure as 12 mg of hydrogen sulfide, H_2S, in the same container under the same conditions?

5.32 A 2.00-mg sample of argon is confined to a 50.0-mL vial at 20.°C; a 2.00-mg sample of krypton is confined to another 50.0-mL vial. What must the temperature of the krypton be if it is to have the same pressure as the argon?

5.33 (a) Calculate the mass of air (which has an average molar mass of 28.97 g/mol in the troposphere) that is needed to fill a 250.-mL flask to a pressure of 1.22 atm at 25°C. (b) Calculate the mass of nitrogen needed to fill the same 250.-mL flask to a pressure of 1.22 atm at 25°C.

5.34 (a) The mass of ammonia in a 500.-mL flask is 14.6 mg at 28°C. What is the pressure of the ammonia? (b) What mass of ammonia would need to be added to change the pressure to 50. Torr?

5.35 A sample of carbon monoxide has a volume of 150. mL at 10.°C and 750. Torr. What pressure will be exerted by the gas if the temperature is increased to 522°C and the volume changed to 500. mL?

5.36 In a certain chemical process, 2500. L of sulfur dioxide gas is fed from a storage tank at 2000. kPa and 20.°C into a 6000.-L chemical reactor at 150.°C. What is the pressure of the SO_2 in the reactor prior to reaction?

5.37 A measured volume of air (350. cm^3) is exhaled into a machine that measures lung capacity. If the air is exhaled from the lungs at a pressure of 1.08 atm at 37°C but the machine is at ambient conditions of 0.958 atm and 23°C, what is the volume of air measured by the machine?

5.38 A 20.-mL sample of xenon exerts a pressure of 0.48 atm at −15°C. (a) What volume does the sample occupy at 1.00 atm and 298 K? (b) What pressure would it exert if it were transferred to a 12-mL flask at 20.°C? (c) Calculate the temperature needed for the xenon to exert a pressure of 500. Torr in the 12-mL flask.

Molar Volume

5.39 Assume ideal gas behavior and calculate the volume at 1.00 atm and 298 K occupied by (a) 1.00 mol H_2; (b) 27.0 g of Cl_2; (c) 3.00 mol O_2; (d) 14.8 mg of SO_2.

5.40 Assume ideal gas behavior and calculate the volume at 1.00 atm and 298 K occupied by (a) 33.9 μg of SiF_4;
(b) 0.572 mg of Xe; (c) 0.822 mol of BrF_3; (d) 3.55 g of air (of average molar mass 28.97 g/mol).

5.41 Calculate the mass of gas in (a) 2.45 L of O_2; (b) 1.94 mL of SO_3; (c) 6000. L of CH_4; (d) 1.44 mL of CO_2 (all measured at 1.00 atm and 298 K).

5.42 Calculate the number of molecules in (a) 6.99 nL of nitrogen; (b) a spherical flask of ammonia with a diameter of 10. cm ($V = \frac{4}{3}\pi r^3$); (c) an ampoule of krypton with a diameter of 1.4 mm and a length of 16 mm ($V = \pi r^2 l$). (All measurements are at 1.00 atm and 298 K.)

Stoichiometry of Reacting Gases

5.43 Oxygen gas is generated in the laboratory by the thermal decomposition of potassium chlorate. What volume of oxygen (at 1.00 atm and 298 K) is generated from 1.00 g of potassium chlorate in the reaction $2\ KClO_3(s) \rightarrow 2\ KCl(s) + 3\ O_2(g)$?

5.44 Calculate the mass of ammonium nitrate that should be heated to obtain 100. mL of dinitrogen oxide, N_2O, at 1.00 atm and 298 K in the reaction $NH_4NO_3(s) \rightarrow N_2O(g) + 2\ H_2O(g)$.

5.45 Small quantities of hydrogen gas can be generated in the laboratory by the action of dilute hydrochloric acid on zinc metal. When 0.40 g of impure zinc was reacted with an excess of hydrochloric acid, 127 mL of hydrogen was collected over water at 17°C. The external pressure was 737.7 Torr. (a) What volume would the dry hydrogen occupy at 1.00 atm and 298 K? (b) What amount (in moles) of H_2 was collected? (c) What is the percentage purity of the zinc?

Zinc metal in hydrochloric acid, Exercises 5.45.

5.46 One industrial process for the removal of hydrogen sulfide from natural gas is its reaction with sulfur dioxide: $2\ H_2S(g) + SO_2(g) \rightarrow 3\ S(s) + 2\ H_2O(g)$. (a) What volume of sulfur dioxide at 1.00 atm and 298 K is needed to prepare 1.00 kg of sulfur? (b) What volume of sulfur dioxide is needed in (a) if it is supplied at 5.0 atm and 250°C?

5.47 The Haber process for the synthesis of ammonia is one of the most significant industrial processes for the well-being of humanity. (a) What volume of hydrogen at 1.00 atm and 298 K must be supplied to produce 1.0 metric ton (1 t $= 10^3$ kg) of NH_3 by direct reaction with nitrogen? (b) What volume of hydrogen is needed in (a) if it is supplied at 200. atm and 400.°C?

5.48 Nitroglycerin is a shock-sensitive liquid that detonates by the reaction

$$4\,C_3H_5(NO_3)_3(l) \longrightarrow 6\,N_2(g)\,+\,10\,H_2O(g)\,+\,12\,CO_2(g)\,+\,O_2(g)$$

Calculate the total volume of product gases at 150. kPa and 100.°C from the detonation of 1.0 g of nitroglycerin.

5.49 Urea, $CO(NH_2)_2$, is used as a fertilizer and is made by the reaction of carbon dioxide and ammonia:

$$CO_2(g)\,+\,2\,NH_3(g) \longrightarrow CO(NH_2)_2(s)\,+\,H_2O(g)$$

What volumes of CO_2 and NH_3 at 200. atm and 450.°C are needed to produce 2.50 kg of urea?

5.50 Xenon and fluorine react at 350.°C to produce a mixture of XeF_2, XeF_4, and XeF_6. What volumes of xenon and fluorine at 200. kPa and 350.°C are needed to produce 1.00 mg of XeF_4, assuming a 100% yield in the reaction $Xe(g)\,+\,2\,F_2(g) \rightarrow XeF_4(g)$?

5.51 Which of the following starting conditions would produce the larger volume of carbon dioxide by combustion of $CH_4(g)$ with an excess of oxygen gas to produce carbon dioxide and water? Justify your answer. The system is maintained at a temperature of 75°C and 1.00 atm. (a) 2.00 L of $CH_4(g)$; (b) 2.00 g of $CH_4(g)$.

5.52 Which of the following starting conditions would produce the larger volume of carbon dioxide by combustion of $C_2H_4(g)$ with an excess of oxygen gas to produce carbon dioxide and water? Justify your answer. The system is maintained at 45°C and 2.00 atm. (a) 1.00 L of $C_2H_4(g)$; (b) 1.20 g of $C_2H_4(g)$.

Density

5.53 Which gas is the most dense at 1.00 atm and 298 K? (a) N_2; (b) NH_3; (c) NO_2.

5.54 Which gas is the most dense at 1.00 atm and 298 K? (a) CO; (b) CO_2; (c) H_2O.

5.55 What is the density (in g/L) of chloroform ($CHCl_3$) vapor at (a) 1.00 atm and 298 K? (b) 100.0°C and 1.00 atm?

5.56 What is the density (in g/L) of hydrogen sulfide (H_2S) at (a) 1.00 atm and 298 K? (b) 25°C and 0.962 atm?

5.57 A sample of a gas with a mass of 21.3 g is confined to a vessel of volume 7.73 L at 0.880 atm and 30°C. (a) What is the density of the gas at 1.00 atm and 298 K? (b) What is the molar mass of the gas?

5.58 A sample of halogen gas has a mass of 0.239 g and exerts a pressure of 600. Torr at 14°C in a 100.-mL flask. (a) Calculate the density of the gas at 1.00 atm and 298 K. (b) What is the molar mass of the halogen? (c) Identify the halogen.

5.59 The density of a gaseous compound of phosphorus is 3.60 g/L at 420 K when its pressure is 727 Torr. (a) What is the molar mass of the compound? (b) What is the density of the gas at 1.00 atm and 298 K?

5.60 A gaseous fluorinated methane compound has a density of 8.0 g/L at 2.81 atm and 300 K. (a) What is the molar mass of the compound? (b) What is the density of the gas at 1.00 atm and 298 K?

Mixtures of Gases

5.61 A sample of hydrogen chloride gas, HCl, is being collected by bubbling it through liquid benzene into a graduated cylinder. Assume that the molecules pictured as spheres show a representative sample of the mixture of HCl and benzene vapor (● represents an HCl molecule and ○ a benzene molecule). (a) Use the figure to determine the mole fraction of HCl and benzene in the gas inside the container. (b) What are the partial pressures of HCl and benzene in the container when the total pressure inside the container is 0.80 atm?

5.62 In the course of a redox reaction, 220. mL of hydrogen was collected over water at 20.°C when the external pressure was 756.7 Torr. See Table 5.4. (a) What is the partial pressure of hydrogen? (b) What mass of oxygen was also produced in the reaction?

5.63 Calculate the partial pressure of each gas and the total pressure of the following mixtures, each of which occupies a 500.-mL vessel at 0°C: (a) 0.020 mol N_2 and 2.33 g of O_2; (b) 0.015 mol H_2, 4.22 mg of He, and 0.030 mol NH_3.

5.64 The following gases were placed in a 3500.-mL flask at 25°C: (a) 0.0195 g of CH_4 and 0.0195 g of CO_2; (b) 1.00 mg of Ar, 2.00 mg of Kr, and 3.00 mg of Xe. Calculate the partial pressure of each gas and the total pressure.

5.65 A sample of damp air in a 1.00-L container exerts a pressure of 762.0 Torr at 48.5°C; but when it is cooled to −10.0°C, the pressure falls to 607.1 Torr as the water condenses. What mass of water was present? Assume the vapor pressure of the ice at −10.°C to be negligible.

5.66 A 2.00-L flask contains carbon dioxide at 20.°C and 606.0 Torr. After 1.0 g of nitrogen is added to the flask and heated to 200.°C, what is the expected pressure in the flask?

5.67 An apparatus consists of a 4.0-L flask containing nitrogen gas at 25°C and 803 kPa, joined by a tube to a 10.0-L flask containing argon gas at 25°C and 47.2 kPa. The stopcock on the connecting tube is opened and the gases mix. (a) What is the partial pressure of each gas in the combined volume of the two flasks after mixing? (b) What is the total pressure of the gas mixture?

5.68 The total pressure of a mixture of sulfur dioxide and nitrogen gases at 25°C in a 500.-mL vessel is 1.09 atm. The mixture is passed over warm calcium oxide powder, which removes the sulfur dioxide by the reaction $CaO(s) + SO_2(g) \rightarrow CaSO_3(s)$, and is then transferred to a 150.-mL vessel, where the pressure is 1.09 atm at 50°C. (a) What was the partial pressure of the SO_2 in the initial mixture? (b) What is the mass of SO_2 in the initial mixture?

5.69 When a potassium chlorate sample was heated in the presence of MnO_2 (a catalyst for this reaction), 25.7 mL of $O_2(g)$ was collected over water at 14°C and 97.6 kPa. See Table 5.4. What was the mass of $KClO_3$ in the sample? The decomposition reaction is

$$2\ KClO_3(s) \xrightarrow{MnO_2} 2\ KCl(s) + 3\ O_2(g)$$

5.70 An adult takes about 15 breaths per minute, with each breath having a volume of 500. mL. If the air that is inhaled is "dry" but the exhaled air at 1.00 atm is saturated with water vapor at 37°C (body temperature), what mass of water is lost from the body in 1.00 h? The vapor pressure of water at 37°C is 47.1 Torr.

Molecular Motion in Gases

5.71 Which gas effuses more rapidly: (a) argon or sulfur dioxide? (b) D_2 (of molar mass 4.018 g/mol) or H_2?

5.72 Which gas effuses more rapidly: (a) HF or H_2O? (b) CO_2 or CO?

5.73 What is the molar mass of a compound that takes 2.7 times longer to effuse through a porous plug than it did for the same amount of XeF_2 at the same temperature and pressure?

5.74 What is the molecular formula of a compound of empirical formula CH that diffuses 3.22 times more slowly than krypton at the same temperature and pressure?

5.75 Describe Boyle's law in terms of the kinetic model of gases.

5.76 Use the kinetic model to explain why the pressure of a gas is proportional to the temperature.

5.77 Calculate the rms speeds of (a) H_2 molecules at 0°C; (b) Xe atoms at 25°C; (c) He atoms at 25°C.

5.78 (a) Calculate the rms speeds of carbon dioxide at 263 K, 273 K, 298 K, 323 K, 348 K, and 373 K. (b) Plot the rms speed against the temperature. (c) Plot the same information in a manner that results in a straight line.

5.79 (a) By what factor does the rms speed of UF_6 molecules increase when the temperature is increased from 25°C to 100.°C? (b) Compare the relative speed of air molecules at room temperature (at 26°C) with their average speed in a

freezer (at −20.°C). (The average molar mass of air in the troposphere is 28.97 g/mol.)

5.80 (a) Determine the factor by which the speed of helium atoms increases when the temperature increases from 10. K to 20. K. (b) Calculate the ratio of the rms speed of helium atoms on the Earth's surface (25°C) to their speed on the surface of the Sun (at 6500°C).

Real Gases

5.81 Identify the conditions of temperature and pressure under which real gases are most likely to behave ideally. Explain your conclusions.

5.82 Under what circumstances would you expect a real gas to be (a) more compressible than an ideal gas? (b) less compressible than an ideal gas?

5.83 Describe how intermolecular forces vary with distance.

5.84 The pressure of a sample of hydrogen fluoride is lower than expected and rises more quickly as the temperature is increased than the ideal gas law predicts. Suggest an explanation.

5.85 Imagine that a 1.0-L vessel contains 1.0 mol of ideal gas molecules at 273 K and that another 1.0-L vessel contains 1.0 mol C_2H_4, a real gas, at 273 K. (a) Which vessel has the greater pressure? (b) Which vessel has more free space between molecules?

5.86 Imagine that a 1.0-L vessel contains 1.0 mol of ideal gas molecules at 273 K and that another 1.0-L vessel contains 1.0 mol SO_2, a real gas, at 273 K. (a) As the flasks are cooled in the same temperature bath, for which gas does the graph of pressure vs. temperature have the greater slope? Assume that for SO_2, attractive forces dominate repulsive forces. (b) Which substance will have the greater volume at the absolute zero of temperature?

5.87 Use the van der Waals equation and Table 5.5 to calculate the pressure of (a) 10.0 g of carbon dioxide in a container of volume 500. mL at 10°C; (b) 10.0 g of helium in a container of volume 250. mL at 100°C. (c) Repeat (a) and (b), using the ideal gas law equation, and (d) determine the percentage error in assuming that carbon dioxide and helium behave as ideal gases under the same conditions as in (a) and (b).

5.88 Use the van der Waals equation and Table 5.5 to calculate the pressure of (a) 10.0 g of ammonia in a container of volume 500. mL at 25°C; (b) 10.0 g of hydrogen in a container of volume 250. mL vessel at 100°C. (c) Repeat (a) and (b), using the ideal gas law equation, and (d) determine the percentage error in assuming that ammonia and hydrogen behave as ideal gases under the same conditions as in (a) and (b).

Supplementary Exercises

5.89 Explain how you would convert the volume of a gas sample at 0°C and 1 atm to the volume of the same sample at another temperature and pressure.

5.90 Which of the following statements are true? Explain your answer in each case. (a) Real gases act more like ideal gases as the temperature is raised. (b) If n and T are held constant, an increase in P will result in an increase in V. (c) At 1.00 atm and 298 K, every molecule in a sample of a gas has exactly the same speed. (d) At the same temperature and pressure, N_2 gas is denser than NH_3 gas. (e) At constant P and T, a decrease in n will result in a decrease in V.

5.91 A sample of He in a 1.000-L flask and a sample of Cl_2 in a 1.000-L flask, both at 1.00 atm and 298 K differ in which of these properties? (a) number of atoms; (b) temperature; (c) mass; (d) average molecular speed; (e) density; (f) kinetic energy.

5.92 At what temperature will 1.0 mol of ideal gas molecules fill a 1.0-L container at 1.0 atm?

5.93 Low-pressure gauges in research laboratories are occasionally calibrated in inches of water ("inH_2O"). Considering that the density of mercury at 15°C is 13.5 g/cm^3 and the density of water at 5°C is 1.0 g/cm^3, what is the pressure (in Torr) inside a gas cylinder that reads 1.5 inH_2O at 15°C?

5.94 At the cruising altitude of an airplane (10 km), the atmospheric pressure is about 25 kPa. Disregarding the effect of temperature, calculate the volume of a sample of air at that altitude, which at the surface of the Earth would occupy 1.0 L.

5.95 A natural gas well with an approximate volume of 2.0×10^9 L has a reservoir pressure of 1.20 atm at 40.°C. If all the gas is to be transferred to a steel tank where the pressure is not to exceed 6.00 atm on a very cold day (-25°C), what must the volume of the tank be?

5.96 (a) Calculate the final volume of a 100.0-L sample of air that, originally at 0.°C and 2.00 atm, is heated to 100.°C, compressed to 1.55 atm, and then heated again to 175°C at constant pressure. (b) What is the mass of air in the sample? (The average molar mass of the air is 28.97 g/mol.)

5.97 At the top of the troposphere, the temperature is about $-50.$°C and the pressure is about 0.25 atm. If a balloon that is filled with helium to a volume of 10.0 L at 2.00 atm pressure and 27.0°C is released at ground level, what will the volume of the balloon be at the top of the troposphere?

5.98 A 2.10-mL bubble of methane gas forms at the bottom of a deep lake, where the temperature is 8.1°C and the pressure is 6.4 atm, and rises to the top, where the temperature is 25°C and the pressure is 1.00 atm. What is the volume of the methane bubble before it bursts?

5.99 What is the partial pressure of O_2 in the air at a ski resort 3000 m above sea level, where the atmospheric pressure is 520. Torr and the mole fraction of O_2 in air is 0.21?

5.100 A 0.297-g sample of an unidentified gas occupies 250. mL at 300. K and 670. Torr. (a) What is the density of the gas at 1.00 atm and 298 K? (b) What is the density of the gas if it is transferred to a vessel at 170. kPa and 70.0°C? (c) What is the molar mass of the gas?

5.101 Carbon monoxide gas is purchased in a 425-mL bottle at a pressure of 5.00 atm and 23°C. (a) What mass (in grams) of the gas has been purchased? (b) What is the density of the gas in the bottle? (c) If the temperature of the bottle changed to 35.0°C (on a hot day), would the density of the gas change? Explain your answer.

5.102 When titanium tetrachloride is mixed with water, voluminous clouds of a white smoke of titanium dioxide particles are produced by the reaction

$$TiCl_4(g) + 2 H_2O(l) \longrightarrow TiO_2(s) + 4 HCl(g)$$

(a) What mass of TiO_2 is produced from the reaction of 200. L of $TiCl_4$, measured at 500. kPa and 30.°C? (b) Determine the volume of hydrogen chloride gas produced at 1.00 atm and 298 K.

5.103 A sample of argon gas effuses through a porous plug in 147 s. Calculate the time required for the same number of moles of (a) CO_2; (b) C_2H_4; (c) H_2; (d) SO_2 to effuse under the same conditions of pressure and temperature.

5.104 Two identical flasks are each filled with a gas at 0°C. One flask contains 1 mol CO_2 and the other 1 mol Ne. In which flask do the particles have more collisions per second with the walls of the container?

5.105 Compute the rms speed of air molecules at the top of the troposphere (at $-50.$°C) and at ground level (at 20.°C). (The average molar mass of air in the troposphere is 28.97 g/mol.)

5.106 An artificial atmosphere for a commercial fruit storage facility is being designed. If the mole fraction of CO_2 is to be 0.040 and the total atmospheric pressure is to be 640. Torr, what will the partial pressure of CO_2 in the storage facility be?

5.107 (a) Suggest reasons why there is so little hydrogen and helium gas in the Earth's atmosphere. (b) Venus and Earth are similar planets, but the atmosphere of Venus is mainly carbon dioxide. Where has the CO_2 of the Earth's atmosphere gone (see Chapter 3)?

5.108 A group of chemistry students has injected a 32.5-g sample of a gas into an evacuated, constant-volume container at 22°C and atmospheric pressure. Now the students will heat the gas at constant pressure by allowing some of the gas to escape during the heating. What mass of gas must be released if the temperature is raised to 212°C?

Applied Exercises

For Exercises 5.109–5.112, see Applying Chemistry: Case Study 5.

5.109 Molecular speed is an important factor in the diffusion of gases through the atmosphere. Rank O_2, O_3, NO_2, and NO in order of increasing average molecular speed.

5.110 One of the components of acid rain is HNO_3, which forms when NO_2 gas dissolves in water droplets suspended in the air. How many moles of $HNO_3(aq)$ can be produced from 1.00×10^3 L of air at 25.0°C containing NO_2 at a partial pressure of 12.5 Pa?

5.111 The concentrations of pollutants and trace gases in the atmosphere are measured in "parts per million," ppm. A measurement of 4 ppm SO_2 means that, of every 10^6 molecules, four will be molecules of SO_2. The following are alternative ways of expressing this same measurement. Fill in the blanks with either the missing value or the missing unit. Use the SI prefixes for units: (a) _____ mol SO_2 molecules in 10^6 mol air molecules; (b) 4 _____ SO_2 in 1 L of air; (c) 4 _____ SO_2 in 1 m^3 of air; (d) a partial pressure of _____ Torr SO_2 when the total pressure is 760. Torr.

5.112 Ozone, O_3, is a pollutant in the troposphere and its concentration must be monitored. (a) One method to test for ozone is to bubble it through an aqueous solution of iodide ion. The ozone reacts with iodide ion and water to form the triiodide ion, I_3^-, hydroxide ion, and O_2 gas; the triiodide ion is detected with a spectrometer. Write a balanced equation for this reaction. (b) On a smoggy day in Riverside, California, air at 302 K and 0.90 atm was collected by pumping it through 50.0 mL of the solution in (a) at a flow rate of 2.8 L/h for 2.00 h. What volume of air was passed through the solution? (c) Analysis of the solution showed that the concentration of triiodide ion produced was 4.2×10^{-8} mol/L. How many moles of O_3 had been collected? (d) What was the ozone concentration in ppm (see Exercise 5.111) in the air sample?

5.113 An initial volume of 250. L of butane gas at 150.°C and 2.33 atm is burned with an excess of oxygen:

$$2\, C_4H_{10}(g) + 13\, O_2(g) \longrightarrow 8\, CO_2(g) + 10\, H_2O(g)$$

(a) Calculate the mass of carbon dioxide released to the atmosphere. (b) What pressure would the carbon dioxide exert if it were collected and stored in a 4000.-L vessel at 16°C?

5.114 A patient with a severe bronchial infection is having trouble breathing. The physician administers "heliox," a mixture of oxygen and helium with 92.3% by mass O_2. What is the partial pressure of oxygen being administered to the patient if atmospheric pressure is 740. Torr?

Integrated Exercises

5.115 A 200.-mL sample of hydrogen chloride at 690. Torr and 20.°C is dissolved in 100. mL of water. The solution was titrated to the stoichiometric point with 15.7 mL of a sodium hydroxide solution. What is the molar concentration of the NaOH in solution?

5.116 A 2.00-L sample of ethene gas, C_2H_4, at 1.00 atm and 298 K is burned in 2.00 L of oxygen gas at the same pressure and temperature to form carbon dioxide gas and liquid water. Ignoring the volume of water, what is the final volume of the reaction mixture (including products and excess reactant) at 1.00 atm and 298 K if the reaction goes to completion?

5.117 The analysis of a hydrocarbon revealed that it was 85.7% C and 14.3% H by mass. When 1.77 g of the gas was stored in a 1.500-L flask at 17°C, it exerted a pressure of 508 Torr. What is the molecular formula of the hydrocarbon?

5.118 A compound used in the manufacture of Saran is 24.7% C, 2.1% H, and 73.2% Cl by mass. The storage of 3.557 g of the compound in a 750-mL vessel at 0°C results in a pressure of 1.10 atm. What is the molecular formula of the compound?

5.119 A hydrocarbon of empirical formula C_2H_3 takes 349 s to effuse through a porous plug. Under the same conditions of temperature and pressure, it took 210. s for the same amount of argon to effuse. What is the molar mass and molecular formula of the hydrocarbon?

5.120 When 2.36 g of phosphorus was burned in chlorine, 10.5 g of a phosphorus chloride was produced. Its vapor took 1.8 times longer to diffuse than the same amount of CO_2 under the same conditions of temperature and pressure. What is the molar mass and molecular formula of the phosphorus chloride?

5.121 A compound that contains sulfur and has been used in chemical warfare has composition 23.76% S, 23.71% O, and 52.54% Cl by mass. If it takes 46.0 s for this compound to effuse through the same porous plug through which the same amount of argon passed in 25.0 s, what is the molecular formula of the compound?

5.122 At 180.°C, 27.20 mg of the vapor of the antimalarial drug quinine in a 250.-mL container gave rise to a pressure of 9.48 Torr. In a combustion experiment, 17.2 mg of quinine produced 47.0 mg of carbon dioxide, 11.5 mg of water, and 1.51 mg of nitrogen. What is the molecular formula of quinine?

5.123 A 0.473-g sample of a gas that occupies 200. mL at 1.81 atm and 25°C was analyzed and found to contain 0.414 g of nitrogen and 0.0591 g of hydrogen. What is the molecular formula of the compound?

Administering an oxygenated gas mixture to a patient, Exercise 5.114.

5.124 A 1.509-g sample of an osmium oxide (which melts at 40.°C and boils at 130.°C) is placed into a cylinder with a movable piston that can expand against the atmospheric pressure of 745 Torr. When the sample is heated to 200.°C, it is completely vaporized and the volume of the cylinder expands by 235 mL. What is the molar mass of the oxide? Assuming that the oxide is OsO_x, what is the value of x?

5.125 Iron pyrite, FeS_2, is the form in which much of the sulfur occurs in coal. In the combustion of the coal, oxygen reacts with iron pyrite to produce iron(III) oxide and sulfur dioxide:

$$4\ FeS_2(s)\ +\ 11\ O_2(g)\ \longrightarrow\ 2\ Fe_2O_3(s)\ +\ 8\ SO_2(g)$$

(a) Calculate the mass of Fe_2O_3 that is produced from the reaction of 75.0 L of oxygen at 2.33 atm and 150.°C with an excess of iron pyrite. (b) If the sulfur dioxide that is generated in (a) is dissolved to form 5.00 L of aqueous solution, what is the molar concentration of the resulting sulfurous acid, H_2SO_3, solution?

5.126 A 15-mL sample of ammonia at 100. Torr and 30.°C is mixed with 25 mL of hydrogen chloride at 150. Torr and 25°C, and the reaction $NH_3(g)\ +\ HCl(g) \rightarrow NH_4Cl(s)$ occurs. (a) Calculate the mass of NH_4Cl that forms. (b) Identify the gas in excess and determine the partial pressure (in the combined volume of the original two flasks) of the excess gas at 27°C after the reaction is complete.

C h a p t e r **6**

Thermochemistry:
The Fire Within

People lived simply in the Stone Age, without all our modern conveniences. Their energy needs were supplied by the Sun and by the *biomass,* plants and trees that could be burned. Because the biomass is continuously renewed and populations were small, Stone Age energy sources were plentiful. However, this simple lifestyle has almost completely vanished, and modern civilization cannot survive without abundant fuel.

Most of the fuels used by modern civilization are *fossil fuels,* which consist of natural gas, coal, and products derived from petroleum, such as gasoline, diesel fuel, and kerosene. Fossil fuels are not renewable. They are extracted from organic matter that decayed and was deposited in the Earth millions of years ago. One day, they will run out. To manage these energy resources and develop new fuels, we need to understand how energy is released or used in chemical reactions, including the reactions in laboratories, kitchens, factories, plants, and our own bodies.

Energy is the very essence of chemistry as well as of civilization. In chemistry, it determines which reactions can occur and which compounds can exist. We have already met a number of chemical reactions. All these reactions either absorb or release energy, and the information we meet in this chapter is essential for understanding how to measure and report these energy changes. In the following four chapters, we investigate the structures of atoms and how atoms link together to form compounds. The energy changes introduced in this chapter will be important for understanding the formation of those links.

This canopy walkway and net in the Peruvian rainforest allow biomass research to be conducted high above the ground. The source of most of our energy is the Sun. Through the process of photosynthesis, solar radiation causes chemical reactions in green plants that store the energy for future use. We make use of the energy captured by plants when we burn fuels. Fossil fuels contain energy that has been stored for thousands of years, but research into alternative fuels is finding ways to make efficient use of plants for fuel. This chapter presents the basic concepts used in research into the energy changes that accompany all chemical reactions.

ENERGY, HEAT, AND ENTHALPY

The reactions by which we obtain energy from fuels are usually combustions in which a hydrocarbon fuel burns in oxygen to form carbon dioxide and water. A typical reaction is the combustion of natural gas, which is largely methane, CH_4:

$$CH_4(g) + 2\,O_2(g) \longrightarrow CO_2(g) + 2\,H_2O(l) + \text{energy}$$

Energy is released in this reaction, so we have included it as a type of product in the chemical equation. This new "energy" term is going to be the focus of the rest of the chapter. Living organisms make use of a very sophisticated form of combustion to obtain the energy they require to move, to grow, and perhaps to think. Glucose, a component of energy-giving carbohydrates, is the fuel that powers all our thoughts and actions:

$$C_6H_{12}O_6(aq) + 6\,O_2(g) \longrightarrow 6\,CO_2(g) + 6\,H_2O(l) + \text{energy}$$

This reaction does not proceed like a familiar combustion: the glucose does not burn with a flame inside us. Instead, a series of reactions takes place at a more controlled rate. Nevertheless, the overall outcome of the reaction is like that of a combustion, and the process can be discussed in a similar way. The field of **bioenergetics** is the study of the uses of energy by organisms.

6.1 The Conservation of Energy

We have been using a term that we often encounter in daily life: *energy*. Because we use the term in conversation, it is familiar. But what does it really mean?

Energy, *E*, comes in a variety of forms. One contribution to the total energy is *kinetic energy*, E_K, the energy due to motion. As we saw in Section 5.13, the kinetic energy of an object is related to its mass, *m*, and speed, *v*, by $E_K = \frac{1}{2}mv^2$. This equation tells us that a heavy object moving fast has a higher kinetic energy than a light object moving slowly. Another contribution is the **potential energy,** E_P, the energy due to position. No single formula can be given for the potential energy of an object because this contribution to the energy depends on the nature of the forces acting on the object. However, simple expressions can be given in specific cases. For instance, a body of mass *m* at a height *h* above the surface of the Earth has a potential energy

$$E_P = mgh \tag{1}$$

relative to its potential energy on the surface itself (Fig. 6.1). The symbol *g* denotes the **acceleration of free fall** on Earth, the acceleration that an object has in free fall in a vacuum ($g = 9.81$ m/s²). This formula shows that the greater the altitude of an object, the greater its potential energy. As we take an elevator ride up a tall building, our potential energy increases. It decreases again as we ride back down.

The total energy of an object such as a planet, a ball, or an atom is the sum of its kinetic and potential energies: $E = E_K + E_P$. Kinetic energy can be converted into potential energy, and potential energy can be converted into kinetic energy. A clear sign of that interchangeability is the way that a ball thrown upward slows (and loses kinetic energy) as it rises above the ground (and gains

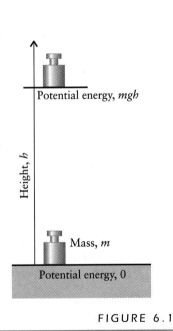

FIGURE 6.1

The potential energy of a mass, *m*, is proportional to its height, *h*, above a surface that is taken to correspond to zero potential energy.

potential energy). However, a fundamental principle of physics, the **law of conservation of energy,** states that

The total energy of an isolated body is constant.

An *isolated body* is an object, like a planet traveling through the vacuum of space (or, to a good approximation, a ball traveling through the air), that cannot exchange energy with its surroundings. At each stage of a ball's flight (provided we neglect air resistance), the *total* energy of the ball remains constant (Fig. 6.2).

In this chapter, we deal with the enormous numbers of atoms and molecules that make up typical samples of matter. To understand energy changes in such collections, we need to learn about three new concepts: *work, heat,* and *internal energy.* Work and heat—which we define more fully later—are two ways in which energy can be transferred from one location to another.

The **internal energy,** U, is the sum of all the kinetic and potential energies of all the atoms and molecules in a sample. The internal energy of an object is a measure of its capacity to do work and supply heat. A wound-up spring—a piece of metal in which the atoms have been pressed together by the act of winding—releases internal energy as it unwinds. The atoms of the metal move apart from one another and thereby reduce their high potential energy. The energy released can be used to do work, such as raising a weight against the pull of gravity. Doing work also includes generating electric currents. It even includes building complicated molecules like proteins inside living cells. A fuel releases internal energy as it burns. This energy can be used to heat the surroundings or to propel a vehicle along a road.

Thermodynamics is the study of the transformations of energy in assemblies of huge numbers of particles, such as the large numbers of atoms in a steel spring or a liter of fuel. We can use thermodynamic principles to study and understand simple, everyday events, like the freezing of water in an ice tray. We can also use them to understand more complicated events, like using a battery to operate a radio or finding alternative fuels to replace our dwindling petroleum reserves. Thermodynamics also helps us to understand a great deal about the chemical properties of matter, and how chemical reactions contribute to the processes of being alive. Bioenergetics is a branch of thermodynamics. So is **thermochemistry,** the study of the heat released or required by chemical reactions. Most of this chapter is concerned with thermochemistry.

> *The total energy is the sum of the kinetic and potential energies. The law of conservation of energy states that the total energy of an isolated body is constant. The internal energy of a sample of matter is the total energy of all its atoms and molecules.*

6.2 Systems and Surroundings

We need to know how to refer to the different locations—a flask or a water bath, for instance—where energy can be stored. A **system** is the part of the world we want to study. In chemistry, the system is usually a reaction mixture in a flask. The **surroundings** consist of everything else outside the system. In principle, the surroundings include the laboratory, the building, the country, and the whole planet. However, it is not practical to measure the changes in such a vast

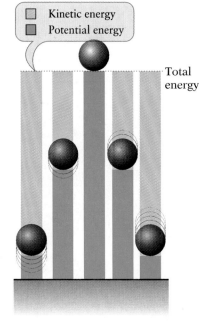

FIGURE 6.2

Kinetic energy (represented by the height of the dark green bar) and potential energy (the light green bar) can be converted into one another. However, their sum (the total height of the bar) is a constant in the absence of external influences, such as air resistance. A ball thrown up from the ground loses kinetic energy as it slows but gains potential energy. The reverse happens as it falls back to Earth.

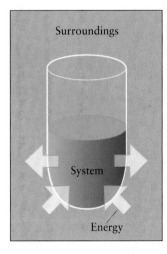

FIGURE 6.3

In thermodynamics, the world is divided into a system (the object of interest) and the surroundings (everything else). In practice, the surroundings may be a constant-temperature water bath. The arrows represent the energy being transferred between the system and its surroundings.

region, so in practice the surroundings may consist of only the water bath used to keep the reaction flask at constant temperature (Fig. 6.3).

A system can be open, closed, or isolated (Fig. 6.4). An **open system** can exchange both matter and energy with the surroundings. Examples are an automobile engine, the human body, and an open flask. A **closed system** has a fixed amount of matter, but it can exchange energy with the surroundings. Examples are electric batteries, the cold packs used for athletic injuries, and a sealed flask. An **isolated system** can exchange neither matter nor energy with its surroundings. We can think of an isolated system as sealed inside rigid, thermally insulated walls. A sealed vacuum flask is a good approximation to an isolated system.

In thermodynamics, the world is divided into a system and its surroundings. An open system can exchange both matter and energy with the surroundings, a closed system can exchange only energy, and an isolated system can exchange neither matter nor energy.

Self-Test 6.1A Identify the following as open, closed, or isolated systems: (a) a space shuttle booster rocket (Fig. 6.5); (b) hot coffee in a perfectly insulated pot.

[***Answer:*** (a) Open; (b) isolated]

Self-Test 6.1B Identify the following as open, closed, or isolated systems: (a) propane gas in a sealed bottle; (b) a burning candle.

6.3 Heat and Work

When a system does work, such as an engine raising a weight, energy is transferred from the system to the surroundings. Energy is also transferred when a system releases heat, as in a combustion. A system with a high internal energy stores a lot of energy and can do a lot of work or provide a lot of heat to the surroundings. Such systems include a tightly coiled spring, a fully charged battery, and the hot steam of a turbine. A system with a low internal energy can do less work or produce only a small amount of heat. Such systems include an uncoiled spring, a nearly exhausted battery, and cold water.

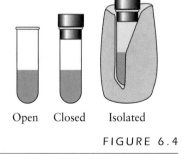

Open   Closed   Isolated

FIGURE 6.4

We classify systems into one of three kinds, according to their interactions with their surroundings. An open system can exchange matter and energy with its surroundings. A closed system can exchange energy but not matter. An isolated system can exchange neither matter nor energy.

FIGURE 6.5

The booster rockets on the space shuttle form an open system. The stream of gases produced by the chemical reaction pours out of the engines and moves the rocket. (See Applying Chemistry: Case Study 20.)

One way to change the internal energy of an open system is to add or to remove matter. An automobile is an open system, and we increase its internal energy by putting gasoline in its tank. However, in most of the systems we consider in this text, the internal energy cannot be changed in this way because they are closed. A closed system is prepared by adding reagents and then sealing the container. Chemical reactions can occur inside the system, but all the reactants and products remain inside.

A second way to change the internal energy of a system is by heating or cooling it. **Heat** *is a transfer of energy that occurs as a result of a temperature difference.* Energy flows as heat from a hot region into a cooler region. For example, suppose the surroundings of the water in a kettle include a hot flame just beneath the kettle. The flame has a higher temperature than the water in a kettle, so energy flows as heat from the flame to the water (Fig. 6.6). On a molecular scale, when a system is heated, some of the chaotic motion of the molecules in the surroundings is transferred to the system. These rapidly moving molecules collide with the walls of the system and pass on their extra energy during the collisions, spreading the energy throughout the system.

The reverse process occurs when a hot object is placed in cool surroundings. Energy flows out of the system and stirs up chaotic motion of the molecules in the surroundings (Fig. 6.7). Random molecular motion is called **thermal motion,** so we can say that *heat stirs up the thermal motion of molecules in the surroundings.* When we say that heat flows from a high-temperature region to a low-temperature region, we mean that energy flows from a region where the thermal motion is vigorous to a region where the thermal motion is less vigorous.

Heating can raise the temperature of a system, or it can cause a change of physical state, such as from liquid to vapor, without affecting the temperature. In each case, the internal energy of the system is increased. For example, if the water in a kettle is boiling, then continued heating simply evaporates the water; the temperature remains the same until all the water is gone. Although the temperature is constant, the internal energy increases as the liquid evaporates. Water vapor has a greater internal energy than the same mass of liquid water at the same temperature because the molecules are far apart (which increases their potential energy).

The third way to change the internal energy of a system is by doing work on the system or by letting the system do work on the surroundings. **Work** *is a transfer of energy that takes place when an object is moved against an opposing force.* A gas inside a container fitted with a piston does work when it pushes the piston out against the opposing pressure of the atmosphere (Fig. 6.8). The gases produced by burning gasoline push out the pistons, which then transfer energy

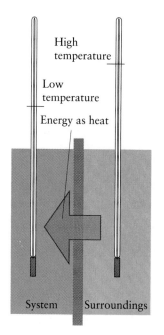

FIGURE 6.6

When we heat a system, we make use of a difference in temperature between it and the surroundings to induce energy to flow through the walls of the system. Heat flows from high temperature to low.

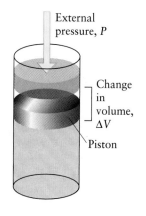

FIGURE 6.8

A system does work when it expands against an external pressure. Here we see a gas that pushes a piston out against a pressure, P. We shall see shortly that the work done is proportional to both the pressure and the change in volume that the system undergoes.

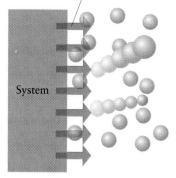

FIGURE 6.7

When energy leaves a system as a result of a temperature difference between the system and the surroundings, we say that the system has lost energy as heat. This transfer of energy stimulates the thermal motion of molecules in the surroundings.

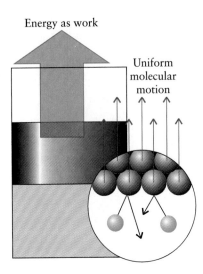

Energy as work

Uniform molecular motion

FIGURE 6.9

When a system expands, it performs work on its surroundings by forcing all the molecules in another object in the surroundings to move in the same direction. Here we see the expansion of a gas raise a weight. The expansion of gases in the cylinders of automobiles do work by pushing on the piston, which turns the gears that move the vehicle.

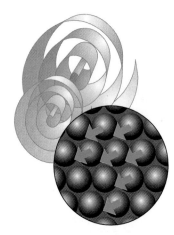

FIGURE 6.10

When we wind a spring, the potential energy of the atoms changes because they are squashed together and repel one another. The internal energy of the spring rises as a result of this increase in potential energy.

to the wheels of our cars. In power plants, steam produced by boiling water is used to rotate the turbines used to generate electricity.

On a molecular scale, work causes *uniform* molecular motion in the surroundings. When the gas in the cylinder of an automobile engine expands, it forces all the atoms in the piston to move in the same direction. If the piston is connected to a weight, all the atoms in the weight move uniformly in the same direction: we see the weight rise (Fig. 6.9). We could extract the same energy as heat, but then the atoms in the piston would only move more vigorously in random directions, and there would be no net motion of the piston or the weight.

There are three ways of changing the internal energy of a system: by adding matter to it, by supplying energy as heat, or by doing work. Heat stimulates thermal motion; work stimulates uniform motion.

6.4 The First Law

As we have seen, the internal energy of a system is the sum of the kinetic energies of all the particles in the system plus the potential energy arising from their interactions with one another. If the molecules are moving rapidly, as they do in a hot gas, then their kinetic energy is high and the internal energy of the sample is high, too. If the particles are close together, as they are in a solid—for example, the atoms in a metal spring—then squashing them even closer together will increase their potential energy and hence increase the internal energy of the solid (Fig. 6.10).

The internal energy of a system depends on its size. For example, a block of iron of mass 2 kg has twice as much internal energy as a block of iron of mass 1 kg at the same temperature and pressure. Therefore, internal energy is an extensive property (one that depends on the size of the sample, Section 2.3). Doubling the size of a system doubles the internal energy of the system. Increasing its size tenfold increases the internal energy tenfold.

The formula for kinetic energy ($E_K = \frac{1}{2}mv^2$) gives us a clue to the units of energy. If mass is measured in kilograms (kg) and speed in meters per second (m/s), the units of E_K turn out to be kg $\times$ (m/s)^2, or kg·m^2/s^2. This combination of base units is the SI unit of energy called the **joule**, J:

$$1\ J = 1\ kg·m^2/s^2$$

Heat and work—which are just energy in transit—are also measured in joules. On the scale of everyday events, the joule is a small unit of energy. For instance, each beat of the human heart uses about 1 J of energy. To lift this book from the floor to a table, you need to use about 15 J of energy. Most energy transfers in chemistry are expressed in kilojoules:

$$1\ kJ = 10^3\ J$$

A unit of energy still widely used in biochemistry and related fields is the **calorie** (cal): 1 cal is the energy needed to raise the temperature of 1 g of water by 1°C (and specifically from 14.5°C to 15.5°C). The modern definition of the calorie is in terms of the joule:

$$1 \text{ cal} = 4.184 \text{ J} \quad \text{(exactly)}$$

It follows that 1.0 J = 0.24 cal; so 1.0 kJ of heat can raise the temperature of 240 g of water (about the amount in a cup of coffee) by 1.0°C.

The internal energy changes when energy enters or leaves a system. It will turn out to be very important to keep track of *changes* in internal energy. We denote a change in internal energy as ΔU, with

$$\Delta U = U_{\text{final}} - U_{\text{initial}} \qquad (2)$$

The Greek uppercase letter Δ (delta) is commonly used to indicate a change in a quantity from an initial value. A positive value of ΔU means that the internal energy of the final state of the system is higher than that of the initial state (Fig. 6.11). This is the case when water is heated or a spring is wound. A negative value of ΔU means that the final state of the system has a lower energy than the initial state. This is the case when water cools or a spring unwinds.

Internal energy is an example of a **state function** or **state property,** a property with a value that depends on the current state of the system (as defined by its pressure, temperature, and composition) and is independent of how the system arrived in that state. A 45-g sample of water at 25°C and 1 atm is in the same state and therefore has exactly the same internal energy as a 45-g sample of water that has previously been heated to 85°C, then cooled, frozen, melted again, and heated back to 25°C. Density is another example of a state property; we shall meet several others in this and later chapters.

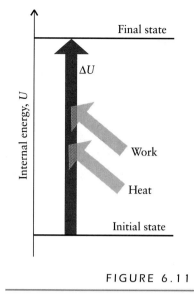

FIGURE 6.11

The internal energy of a system can be changed either by doing work or by heating. The diagram shows that the change in internal energy is positive (*U* increases) when energy is supplied in either way. When energy leaves the system as heat or work, the internal energy falls and ΔU is negative.

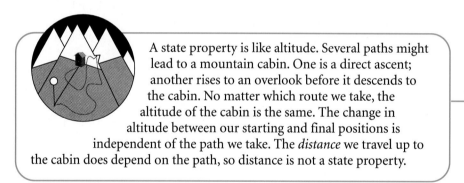

A state property is like altitude. Several paths might lead to a mountain cabin. One is a direct ascent; another rises to an overlook before it descends to the cabin. No matter which route we take, the altitude of the cabin is the same. The change in altitude between our starting and final positions is independent of the path we take. The *distance* we travel up to the cabin does depend on the path, so distance is not a state property.

An analogy

When we do work on a system—for instance, by winding a spring inside it—its internal energy rises. Suppose we do 100 kJ of work; then the system increases its store of energy by 100 kJ and we write $\Delta U = +100$ kJ. If we use w to denote the energy supplied *to* a system as work, we can write

Change in internal energy = energy supplied to system as work
$$\Delta U = w$$

When we supply heat to the system, its internal energy increases by the corresponding amount. Suppose we transfer 40 kJ of energy to a large beaker of water by standing it on a hot surface; then the internal energy of the water increases by 40 kJ and we write $\Delta U = +40$ kJ. If we use q to denote the energy supplied *to* a system as heat, we can write

Change in internal energy = energy supplied to system as heat
$$\Delta U = q$$

Always include the plus sign when reporting a positive change: do not write simply $\Delta U = 100$ kJ.

When we supply energy to the system both by doing work and by heating, the total change in internal energy is

Change in internal energy
= energy supplied to system as heat
+ energy transferred to system as work

$$\Delta U = q + w \qquad (3)$$

For example, if we supply 40. kJ of heat ($q = +40.$ kJ) and do 25 kJ of work on the system ($w = +25$ kJ), then the total change in internal energy is

$$\Delta U = q + w = 40. + 25 \text{ kJ} = +65 \text{ kJ}$$

If, in another process, 40. kJ of heat leaks out of the system while we are doing 25 kJ of work on it, we would write $q = -40.$ kJ, the minus sign indicating that heat has left the system (thereby decreasing the internal energy by 40. kJ), and $w = +25$ kJ (thereby increasing the internal energy by 25 kJ). In this case, the overall change in internal energy is

$$\Delta U = q + w = (-40.) + 25 \text{ kJ} = -15 \text{ kJ}$$

Now there is a net decrease of 15 kJ in the internal energy of the system.

Example 6.1 *Calculating the change in internal energy*

Suppose that a battery drives an electric motor in an aquarium pump, and we consider the battery plus the motor as the system. During a certain period, the system does 555 kJ of work on the pump and releases 124 kJ of heat into the surroundings. What is the change in internal energy of the system?

Strategy All losses of energy as heat or work reduce the internal energy of the system, so they appear in Eq. 3 with negative signs. All gains of energy increase the internal energy, so they appear in Eq. 3 with positive signs.

Solution Both energy transfers are losses, so we write $q = -124$ kJ and $w = -555$ kJ. Therefore, from Eq. 3,

$$\Delta U = q + w = -124 - 555 \text{ kJ} = -679 \text{ kJ}$$

That is, the internal energy of the system decreases by 679 kJ.

Self-Test 6.2A A certain system gains 260. kJ of energy as heat while it is doing 500. kJ of work. What is the change in internal energy?

[*Answer:* $-240.$ kJ]

Self-Test 6.2B An electric motor and its battery together do 500. kJ of work. The battery also releases 250. kJ of energy as heat and the motor releases 50. kJ as heat due to friction. What is the change in internal energy of the system, with the system regarded as (a) the battery alone; (b) the motor alone; (c) the battery and the motor together? For (a) and (b), assume that the battery does 500. kJ of work on the motor, which then does the same amount of work on the surroundings.

A change in internal energy can be brought about either by heat or by work. We can increase the internal energy by 75 kJ by doing 75 kJ of work on it or by supplying 75 kJ of energy as heat. Any combination of the two will do, provided the net transfer of energy is 75 kJ. It was a major achievement of nineteenth-century science (and by Joule, in particular) to recognize that *heat and work are equivalent ways of changing the energy of a system.*

The internal energy is like the reserves of a bank: the bank accepts deposits and withdrawals in either of two currencies (heat or work) but stores them as a common fund, the internal energy.

Now consider an isolated system. Matter cannot be added to or removed from such a system. No work can be done on or by it, because we cannot compress an isolated system or get any mechanical instruments through its walls. We cannot heat the system, because the walls insulate the interior from the surroundings. Because $q = 0$ and $w = 0$, it follows that $\Delta U = 0$. If we could completely isolate a system from its surroundings, then it would have exactly the same energy forever. This observation is stated as the **first law of thermodynamics:**

The internal energy of an isolated system is constant.

In other words, we cannot create or destroy internal energy. We can change the internal energy of a system only by transferring energy through its walls.

A lot of people have tried to become rich by designing systems intended to disprove the first law of thermodynamics. If their devices could create internal energy out of nothing, then they would have an inexhaustible supply of energy, with no fuel required. Many of these so-called *perpetual motion machines* have been announced, but every case has proved to be either a deliberate hoax or technologically faulty (Fig. 6.12).

Internal energy is a state property, a property that depends only on the current state of a system and is independent of how the state was prepared. Heat and work are two equivalent ways of changing the internal energy of a system. The first law of thermodynamics states that the internal energy of an isolated system is constant.

FIGURE 6.12

The inventor of this elaborate device, the Keely motor, claimed that it could generate high pressures and energies from a small amount of water. However, like other perpetual motion machines, the Keely motor was found to be a fraud. The work was accomplished by compressed air from a hidden source.

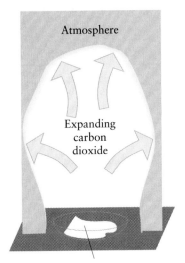

Atmosphere

Expanding
carbon
dioxide

Heated calcium carbonate

FIGURE 6.13

A reaction does work when it generates a gas. For example, carbon dioxide is formed from the thermal decomposition of calcium carbonate. As the gas is formed, it drives back the surrounding atmosphere.

6.5 Transferring Energy as Work

How can we measure changes in the internal energy brought about by doing work? One of the most important forms of work is **expansion work,** the work done when a system changes size and pushes against an external force. Expansion work is done by the hot gases in an internal combustion engine as they push back the pistons. Expansion work is done whenever a system expands and pushes back the atmosphere, because the atmosphere exerts a force on the system that opposes its expansion (Fig. 6.13). A chemical reaction that generates gas, like the action of hydrochloric acid on zinc, does work by pushing back the atmosphere to make room for the gas.

The work done when a system expands is proportional to the change in volume and to the opposing pressure. The greater the expansion, the greater the work the system must do as it expands, and therefore the greater the decrease in the internal energy of the system. The greater the opposing pressure, the greater the work that must be done to expand through a given volume.

We can find the precise relation between work, pressure, and change of volume by considering the arrangement in Fig. 6.14 and using the definition of work in physics. To move an object against an opposing force,

$$\text{Work done} = \text{distance moved} \times \text{opposing force}$$

We saw in Section 5.2 that pressure is force divided by area. The opposing force arises from the external pressure, so

$$\text{Opposing force} = \text{area} \times \text{external pressure}$$

When we combine these two expressions, we get

$$\text{Work done} = \text{distance moved} \times \text{area} \times \text{external pressure}$$

The distance moved multiplied by the area is the change in volume that occurs during the expansion, so

$$\text{Work done} = \text{change in volume} \times \text{external pressure}$$

We can express this relation in symbols. The change in volume (final − initial) is ΔV, and the external pressure is P_{external}. Therefore,

$$w = -P_{\text{external}}\Delta V \tag{4}$$

The negative sign has been included so that w is negative (that is, energy leaves the system and there is a decrease in internal energy) when the system expands (the change in volume, ΔV, is positive).

The units of work (which is energy in transit) are the same as those of energy. We can verify that Eq. 4 gives the correct units: suppose that the external pressure is expressed in pascals ($1\ \text{Pa} = 1\ \text{kg/m·s}^2$, the SI unit of pressure) and the change in volume in cubic meters (m^3). The product of 1 Pa and 1 m^3 is

$$1\ \text{Pa·m}^3 = 1\ \text{kg/m·s}^2 \times 1\ \text{m}^3 = 1\ \text{kg·m}^2/\text{s}^2 = 1\ \text{J}$$

It is easy to keep track of signs by noting whether energy leaves or enters the system: − indicates that energy leaves, + indicates that energy enters.

Therefore, if we use units of pascals and cubic meters, the work is obtained in joules. If we express the pressure in atmospheres and the volume in liters, the answer is in liter-atmospheres. To convert liter-atmospheres into joules, we note that $1\ \text{L} = 10^{-3}\ \text{m}^3$ and $1\ \text{atm} = 101\ 325\ \text{Pa}$ (exactly); therefore,

$$1\ \text{L·atm} = 10^{-3}\ \text{m}^3 \times 101\ 325\ \text{Pa} = 101.325\ \text{Pa·m}^3 = 101.325\ \text{J (exactly)}$$

The work done when a system expands is proportional to the change in volume and to the opposing pressure (see Eq. 4).

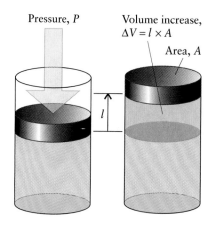

Area, A

FIGURE 6.14

The gas in the piston expands a distance, l, against a pressure, P. The volume increase is $l \times A$, where A is the cross-sectional area of the piston.

Example 6.2 *Calculating the work of expansion*

Calculate the work done when a gas expands by 12.0 L against a pressure of 2.00 atm.

Strategy Work is *done* as the gas expands: work being done implies that the internal energy of the gas (the system) must fall, and therefore that w is negative. To find the numerical value, substitute the data into Eq. 4 and convert liter-atmospheres to joules.

Solution The work done is

$$w = -P_{ex}\Delta V = -(12.0 \text{ L}) \times (2.00 \text{ atm}) = -12.0 \times 2.00 \text{ L·atm}$$

and the work in joules is

$$w = -(12.0 \times 2.00 \text{ L·atm}) \times \left(\frac{101.325 \text{ J}}{1 \text{ L·atm}} \right) = -2.43 \times 10^3 \text{ J}$$

The negative sign means that the internal energy decreases by 2.43×10^3 J, or 2.43 kJ, when the gas expands.

Self-Test 6.3A The hot gaseous products of combustion in an internal combustion engine expand by 2.2 L against an external pressure of 1.0 atm. How much work is done by the expansion?

[*Answer:* 2.2×10^2 J]

Self-Test 6.3B What work is done when 1.00 mol $H_2O(l)$ evaporates against an external pressure of 1.00 atm? Assume that the vapor behaves like an ideal gas and that its temperature is 373 K.

6.6 Transferring Energy as Heat

Now let's see how to measure changes in internal energy by monitoring transfers of energy as heat. Suppose we arrange for a reaction to take place inside a sturdy sealed container. The system cannot expand and therefore it cannot do any expansion work (Fig. 6.15). Because $w = 0$, it follows that

At constant volume: $\Delta U = q$ (5)

Therefore, if we measure the heat released by a reaction taking place in a constant-volume container, we can identify the value with ΔU. For example, if a reaction takes place inside a sealed, rigid flask and releases 75 kJ through its walls (so $q = -75$ kJ), then we know that $\Delta U = -75$ kJ. Conversely, if 75 kJ of heat flows into the flask (so $q = +75$ kJ), then we know that $\Delta U = +75$ kJ.

Now consider the change in internal energy when the system is free to change its size and stay at the same pressure. For example, let's think about what happens when we supply 75 kJ of heat to a container full of gas. If the volume could not change, then all the heat supplied to it would be used to raise the internal energy of the system, and $\Delta U = +75$ kJ. However, when the gas is free to change its volume and stay at the same pressure, some of the 75 kJ supplied is used to drive back the atmosphere as the system expands. Therefore, at constant pressure, only some of the energy supplied as heat is used to increase the internal energy of the system—the rest leaks back into the surroundings as

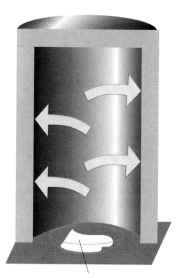

Heated calcium carbonate

FIGURE 6.15

When a reaction (such as the thermal decomposition of calcium carbonate) takes place in a closed, constant-volume container, the gas fills the container but cannot expand against the surrounding atmosphere. As a result, it does no work on the surroundings.

Some energy
leaves as work

Energy enters
as heat

FIGURE 6.16

When a system is free to expand against an external pressure, some of the energy supplied to it as heat escapes back into the surroundings as work. As a result, the change in internal energy is less than the energy supplied.

The name *enthalpy* comes from the Greek words for "heat inside."

work (Fig. 6.16). Suppose 10. kJ is used to expand the system ($w = -10.$ kJ). In this case, although $q = +75$ kJ, the change in internal energy is only

$$\Delta U = q + w = 75 - 10.\ kJ = +65\ kJ$$

Let's generalize this discussion. If the volume of the system changes by ΔV in the course of the reaction, then from Eq. 4 we know that $w = -P_{external}\Delta V$. However, because the system is open to the atmosphere, the pressure of the system, P, is the same as the pressure of the surroundings, $P_{external}$, so we can drop the subscript "external" and write

$$\text{At constant pressure: } \Delta U = q + w = q - P\Delta V \qquad (6)$$

In principle, to determine the change in internal energy, ΔU, we could monitor the heat absorbed or released at constant pressure (the term q in Eq. 6) and combine the result with the work done on the system (as the term $-P\Delta V$). Chemists, however, have found a much more convenient procedure that takes into account the expansion work automatically. They introduce a quantity called the **enthalpy,** H. A change in enthalpy, ΔH, at constant pressure is defined as

$$\Delta H = \Delta U + P\Delta V \qquad (7)$$

When we insert Eq. 6 into this expression, the $P\Delta V$ terms cancel

$$\Delta H = q - \cancel{P\Delta V} + \cancel{P\Delta V}$$

and we are left with

$$\text{At constant pressure: } \Delta H = q \qquad (8)$$

Equation 8 tells us that *the change in enthalpy of a system is equal to the heat transferred to it at constant pressure.* For example, in a combustion reaction taking place in an open vessel, 80.0 kJ of energy might escape into the surroundings as heat ($q = -80.0$ kJ). We conclude that $\Delta H = -80.0$ kJ, and therefore that the enthalpy of the system has decreased by 80.0 kJ. If the system also did 20.0 kJ of work ($w = -20.0$ kJ), we could conclude that $\Delta U = q + w = -80.0 + (-20.0)\ kJ = -100.0$ kJ.

Enthalpy, H, will be the main focus of the rest of the chapter. Most reactions we do in chemistry take place at constant pressure, so we can equate the heat transfers that take place with changes in enthalpy of the system.

Like the internal energy, the enthalpy is a state property: the enthalpy of a system depends only on its current state, such as its current temperature and pressure. It follows that the change in enthalpy, ΔH, is independent of the path between the initial and final states. This conclusion follows from the definition of ΔH. It is equal to the sum of ΔU and $P\Delta V$, both of which are independent of path. The fact that enthalpy is a state property means, for instance, that the change in enthalpy of 100.0 g of water heated from 25°C to 75°C is the same as when the water is heated from 25°C to 250°C and then cooled back down to 75°C, or any more complicated sequence of changes between the given initial and final states.

A change in internal energy can be identified with the heat supplied at constant volume. A change in enthalpy is equal to the heat supplied at constant pressure.

Self-Test 6.4A In a certain reaction at constant pressure, 50 kJ of heat left the system and 30 kJ of energy left the system as expansion work to make room for the products. What are the values of (a) ΔH; (b) ΔU for the reaction?

[*Answer:* (a) -50 kJ; (b) -80 kJ]

Self-Test 6.4B In a certain reaction at constant pressure, 50 kJ of heat entered the system. The products took up less volume than the reactants, and 20 kJ of energy entered the system as work as the outside atmosphere compressed. What are the values of (a) ΔH; (b) ΔU for the reaction?

6.7 Exothermic and Endothermic Processes

An **exothermic process** is a change, such as a chemical reaction, that releases heat (Fig. 6.17). Because, at constant pressure, the release of heat corresponds to a decrease in enthalpy, *an exothermic process taking place at constant pressure is accompanied by a decrease in enthalpy*:

At constant pressure, $\Delta H < 0$ signifies an exothermic process.

All combustions are exothermic. One striking example of an exothermic reaction is the thermite reaction, the reduction of a metal oxide by aluminum:

$$2\,Al(s) + Fe_2O_3(s) \longrightarrow Al_2O_3(s) + 2\,Fe(s) + energy$$

The reaction gives off so much heat that it is used to weld iron rails together.

An **endothermic process** is a change, such as a chemical reaction, that absorbs heat. At constant pressure, the absorption of heat results in an increase in enthalpy, so *an endothermic process taking place at constant pressure is accompanied by an increase in enthalpy*:

At constant pressure, $\Delta H > 0$ signifies an endothermic process.

Most endothermic chemical reactions absorb only a relatively small amount of heat, so there is probably not much future for chemical refrigerators. However, one moderately endothermic reaction takes place when barium hydroxide octahydrate, $Ba(OH)_2 \cdot 8H_2O$, and ammonium thiocyanate, NH_4SCN, are ground together:

$$Ba(OH)_2 \cdot 8H_2O(s) + 2\,NH_4SCN(s) + energy \longrightarrow$$
$$Ba(SCN)_2(aq) + 2\,NH_3(g) + 10\,H_2O(l)$$

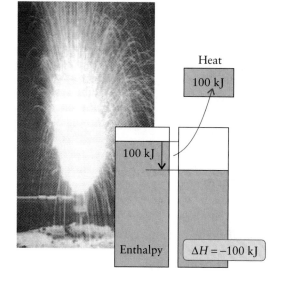

Heat

100 kJ

100 kJ

Enthalpy $\Delta H = -100$ kJ

FIGURE 6.17

The thermite reaction is another highly exothermic reaction—one that can melt the metal it produces. In this reaction, aluminum metal is reacting with iron(III) oxide, Fe_2O_3, causing a shower of molten iron sparks. In an exothermic reaction, energy is lost as heat, the amount lost depending on the amount of reactants available.

FIGURE 6.18

The reaction between ammonium thiocyanate, NH_4SCN, and barium hydroxide octahydrate, $Ba(OH)_2 \cdot 8H_2O$, absorbs a lot of heat and can cause water vapor in the air to freeze on the outside of the beaker. In an endothermic reaction, energy is absorbed as heat.

Heat

100 kJ

100 kJ

Enthalpy $\Delta H = +100$ kJ

When diluting sulfuric acid, always add the acid to the water. Then if the mixture does boil and splash, the spatters will be mainly water.

The reaction mixture becomes so cold that moisture from the air forms a layer of frost on the outside of the beaker (Fig. 6.18).

Dissolving may be either exothermic or endothermic and may involve only a little or a lot of heat, depending on the solute and the solvent. For example, instant cold packs contain one packet of water and another of ammonium nitrate. When the barrier between the salt and the water is broken, the ammonium nitrate dissolves, absorbing a lot of heat. The dissolution of ammonium nitrate in a cold pack is an example of an endothermic process. However, when sulfuric acid is mixed with water, it generates a lot of heat, enough to boil the solution, so the dissolution of sulfuric acid in water is classified as an exothermic process.

Exothermic processes release heat; endothermic processes absorb heat.

6.8 Measuring Heat

The heat released or absorbed in a reaction can be monitored quite easily. We put the reaction vessel in contact with a water bath, let the reaction take place, and measure the rise or fall in temperature of the water in the bath. When heat flows from a reaction flask into the bath (for an exothermic reaction), the temperature of the bath increases in proportion to the heat released. When heat flows from the bath into the reaction vessel (for an endothermic reaction), the temperature of the bath falls in proportion to the heat absorbed.

We need to be able to use the change in temperature (in degrees Celsius) to find the energy transferred (in joules). To make this conversion, we need to know the **heat capacity,** C, which is the heat required to raise the temperature of an object by 1 K:

$$\text{Heat capacity} = \frac{\text{heat supplied}}{\text{temperature rise}} \qquad \text{or } C = \frac{q}{\Delta T} \qquad \textbf{(9)}$$

If we use ΔT in kelvins, the heat capacity has the units joules per kelvin (J/K). However, because a *difference* in temperature has the same value on the Kelvin and Celsius scales, we could also use ΔT in degrees Celsius. In that case, the heat capacity has the units joules per degree Celsius (J/°C). Suppose, for example, that we measure a temperature rise of 2.0°C when we supply 98 kJ of heat to a sample of ethanol. We would report that

$$\text{Heat capacity of ethanol} = \frac{98 \text{ kJ}}{2.0°C} = 49 \text{ J/°C}$$

More heat is needed to raise the temperature of a large sample of a substance by 1°C than is needed for a small sample of the same substance. That is, the larger the sample, the greater its heat capacity (Fig. 6.19). It is therefore common to report the **specific heat capacity,** C_s, (often called just *specific heat*), which is the heat capacity divided by the mass of the sample in grams. Note that heat capacity is an extensive property, but specific heat capacity is intensive. Table 6.1 gives the specific heat capacities of some common substances.

We can calculate the heat capacity of a substance from its mass and its specific heat capacity by using the expression

$$\text{Heat capacity} = \text{mass} \times \text{specific heat capacity}$$
$$C = m \times C_s \tag{10}$$

If we know the mass of a substance, its specific heat capacity, and the temperature rise it undergoes during an experiment, then the heat transferred to it is obtained by combining Eqs. 9 and 10:

$$\text{Heat supplied} = \text{mass (g)} \times \text{specific heat capacity (J/°C·g)}$$
$$\times \text{change in temperature (°C)}$$
$$q = m \times C_s \times \Delta T \tag{11}$$

The specific heat capacity of a dilute solution is normally taken to be the same as that of the pure solvent (which is commonly water).

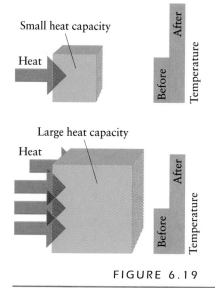

FIGURE 6.19

Heat capacity is an extensive property, so a large object (bottom) has a larger heat capacity than a small object (top) made of the same material.

Example 6.3 *Using specific heat capacity*

An engineer testing a 5.10-g sample of an alloy for use in a sprinkling system heated the sample from 24.2°C to 138.5°C. The alloy has a specific heat capacity of 0.124 J/°C·g. How much energy in joules was supplied to the sample?

Strategy The specific heat capacity tells us how much energy is required to raise the temperature of 1 g of the sample by 1°C. We know the mass of the sample and the temperature rise, so we can calculate the energy required from Eq. 11.

Solution We substitute the data into the equation and solve for heat supplied:

$$\text{Heat supplied} = \overbrace{(5.10 \text{ g alloy})}^{m} \times \overbrace{\left(0.124 \frac{\text{J}}{°C·\text{g alloy}}\right)}^{C_s} \times \overbrace{(138.5 - 24.2°C)}^{\Delta T}$$

$$= \left(5.10 \times 0.124 \frac{\text{J}}{°C}\right) \times (114.3°C)$$

$$= 72.3 \text{ J}$$

Self-Test 6.5A A swimming pool contains 2.0×10^4 L of water. How much energy in joules is required to raise the temperature of the water from 20.°C to 25°C? The specific heat capacity of water is 4.184 J/°C·g.

[**Answer:** 4.2×10^5 kJ]

Table 6.1 *Specific heat capacities of common materials*

Material	Specific heat capacity, J/°C·g
air	1.01
benzene	1.05
brass	0.37
copper	0.38
ethanol	2.42
glass (Pyrex)	0.78
granite	0.80
marble	0.84
polyethylene	2.3
stainless steel	0.51
water: solid	2.03
liquid	4.18
vapor	2.01

FIGURE 6.20

A pyrotechnics expert attaches fuses containing potassium chlorate to fireworks set up in mortars. The highly exothermic reactions of fireworks do not begin until they are intiated by a fuse. The fuse ignites a mixture of carbon and other reducing agents that react with an oxidizer such as potassium perchlorate.

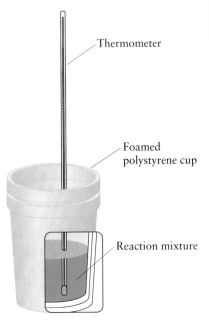

Thermometer

Foamed polystyrene cup

Reaction mixture

FIGURE 6.21

The quantity of heat released or absorbed by a reaction can be measured in this primitive version of a calorimeter. The outer polystyrene cup acts as an extra layer of insulation to ensure that no heat enters or leaves the inner cup. The quantity of heat released or absorbed is proportional to the change in temperature of the calorimeter.

Self-Test 6.5B Potassium perchlorate, $KClO_4$, is used as an oxidizer in fireworks. How much energy in joules is required to raise the temperature of 10.0 g of $KClO_4$ from 25°C to an ignition temperature of 900.°C? The specific heat capacity of $KClO_4$ is 0.8111 J/°C·g (Fig. 6.20).

Heat transfers are measured with a **calorimeter,** which is an insulated container fitted with a thermometer. A calorimeter can be as simple as a Styrofoam cup (Fig. 6.21) or as sophisticated as a bomb calorimeter (Fig. 6.22). The latter consists of a heavy-walled steel container. The sample is placed inside the container, which is then sealed and placed in a water bath. An electric current is passed through the sample to ignite it, and the heat given off by the reaction raises the temperature of the entire apparatus. We measure the increase in temperature and then use the heat capacity of the calorimeter (the entire assembly, including the sample) to convert the change in temperature into heat released by the reaction.

The simple Styrofoam cup calorimeter is open to the atmosphere, so it measures heat transfers at constant pressure. Therefore, measurements made by using this type of calorimeter give us the enthalpy change for the reaction. The bomb calorimeter has a constant volume, so measurements with this type of calorimeter give us the internal energy change for the reaction. The values of ΔH and ΔU are essentially the same for reactions that do not involve gases because their volume changes are so small.

Example 6.4 *Measuring specific heat capacity with a simple calorimeter*

Suppose we have 50. g of water at 20.0°C in a Styrofoam cup calorimeter like that in Fig. 6.21. Then we place 21 g of iron at 90.2°C in the calorimeter and measure a final temperature of 23.2°C. What is the specific heat capacity of the iron?

Strategy All the heat lost by the iron is transferred to the water and the rest of the calorimeter. The heat lost or gained by each substance equals its heat capacity multiplied by its temperature change, and the heat capacity is its specific heat capacity multiplied by its mass. The heat capacity of the Styrofoam is so small that we can safely neglect it. Therefore, we can write

$$\overbrace{(mass)_{metal} \times (specific\ heat)_{metal} \times (fall\ in\ temperature)_{metal}}^{\text{Heat lost by metal}} =$$

$$\underbrace{(mass)_{water} \times (specific\ heat)_{water} \times (rise\ in\ temperature)_{water}}_{\text{Heat gained by water}}$$

There is enough information to find the one unknown, the specific heat capacity of the metal.

Solution We substitute the data into the expression derived in the Strategy and solve for the specific heat capacity of the metal. Note that the temperature of the iron falls from 90.2°C to 23.2°C, or 67.0°C, and the temperature of the water rises from 20.0°C to 23.2°C, or 3.2°C. Therefore,

$$(21\ g) \times (specific\ heat)_{metal} \times (67.0°C) = (50.\ g) \times (4.184\ J/°C·g) \times (3.2°C)$$

That is,

$$(specific\ heat)_{metal} = \frac{(50.\ g) \times (4.184\ J/°C·g) \times (3.2°C)}{(21\ g) \times (67.0°C)}$$
$$= 0.48\ J/°C·g$$

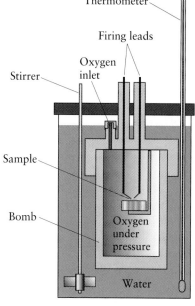

FIGURE 6.22

A bomb calorimeter. The combustion is initiated with an electrical fuse. Once the reaction has begun, energy is released as heat that spreads through the walls of the bomb into the water. The heat released is proportional to the temperature change of the entire calorimeter assembly.

Self-Test 6.6A A piece of copper of mass 19.0 g was heated to 87.4°C and then placed in a calorimeter that contained 55.5 g of water at 18.3°C. The temperature of the water rose to 20.4°C. What is the specific heat capacity of copper?

[*Answer:* 0.38 J/°C·g]

Self-Test 6.6B An alloy of mass 25.0 g was heated to 88.6°C and then placed in a calorimeter that contained 61.2 g of water at 19.6°C. The temperature of the water rose to 21.3°C. What is the specific heat capacity of the alloy?

In practice, a calorimeter is usually calibrated by measuring the temperature rise brought about by a reaction with a known heat output. Then we use that value to determine the heat capacity of the calorimeter. When using a calibrated calorimeter to measure the heat output of a reaction in solution, we must add the same total volume of solution used in the calibration, to ensure that the heat capacity is the same in the two measurements.

> **Heat transfers are measured by using a calibrated calorimeter and are reported in joules or kilojoules. The heat capacity of an object is the ratio of the heat supplied to the temperature rise produced.**

Example 6.5 *Determining the heat output of a reaction*

A reaction known to release 1.78 kJ of heat takes place in a Styrofoam cup calorimeter like that shown in Fig. 6.21 and containing 0.100 L of aqueous sodium chloride solution. The temperature rises by 3.65°C. Next, 50.0 mL of hydrochloric acid and 50.0 mL of aqueous sodium hydroxide are mixed in the same calorimeter and the temperature rises by 1.26°C. What is the heat output of the neutralization reaction?

Strategy First, we calibrate the calorimeter. To do so, we calculate the heat capacity of the calorimeter with 0.0100 L of aqueous solution from the information on the first reaction. For this step, we use Eq. 9. Then we use the value of that heat capacity to convert the temperature rise caused by the neutralization reaction into a heat output. For this step, we use Eq. 9 rearranged to

$$\text{Heat supplied} = \text{heat capacity} \times \text{temperature rise} \quad \text{or} \quad q = C \times \Delta T$$

Note that the calorimeter contains the same volume of liquid in both cases, so its heat capacity is the same. It is good practice to leave all numerical work until the final stage of the calculation, to reduce rounding errors. Because the temperature rises, we know the reaction is exothermic and therefore that ΔH must be negative.

Solution From the data and Eq. 9, the heat capacity of the calorimeter is

$$\text{Heat capacity} = \frac{1.78 \text{ kJ}}{3.65°C} = \frac{1.78}{3.65} \text{ kJ/°C}$$

It follows that the heat released by the neutralization is

$$\text{Heat supplied} = \underbrace{\left(\frac{1.78}{3.65} \text{ kJ/°C}\right)}_{\text{Heat capacity}} \times \overbrace{(1.26°C)}^{\substack{\text{Temperature} \\ \text{rise}}} = 0.614 \text{ kJ}$$

Because the reaction takes place at constant pressure, we can equate this value with the enthalpy change. Because heat is released, the sign of the enthalpy change is negative: $\Delta H = -0.614$ kJ.

Self-Test 6.7A A small piece of calcium carbonate was placed in the same calorimeter, and 100.0 mL of dilute hydrochloric acid was poured over it. The temperature of the calorimeter rose by 3.57°C. How much heat was released by the reaction?

[***Answer:*** 1.74 kJ]

Self-Test 6.7B A calorimeter was calibrated by mixing two 0.100 L aqueous solutions together. The reaction that occurred was known to release 4.16 kJ of heat, and the temperature of the calorimeter rose by 3.24°C. Calculate the heat capacity of this calorimeter when it contains 0.200 L of water.

THE THERMOCHEMISTRY OF PHYSICAL CHANGE

Now we are ready to apply these ideas to actual processes. We start with changes of physical state, such as the conversion of a solid to a liquid or a liquid to a vapor. Most changes of state release or absorb energy as heat; and by monitoring the heat, we can discover valuable information about the process. Later in the chapter, we shall see that many of the same ideas can be used to discuss chemical changes, too.

6.9 Vaporization

Vaporization is an endothermic process, because energy has to be supplied to overcome the forces that hold molecules together as a liquid. For example, when perspiration evaporates from our skin, the water absorbs heat. This is one of the body's strategies for keeping cool. The cold we feel when we step out of a shower is another sign that vaporization is endothermic. Cosmetic cold creams feel cool because they contain water, which evaporates after the mixture is applied to our skin. Because vaporization is endothermic, water vapor at 100°C has a higher

Table 6.2 Standard enthalpies of physical change*

Substance	Formula	Freezing point, K	$\Delta H_{fus}°$, kJ/mol	Boiling point, K	$\Delta H_{vap}°$, kJ/mol
acetone	CH_3COCH_3	177.8	5.72	329.4	29.1
ammonia	NH_3	195.4	5.65	239.7	23.4
argon	Ar	84.2	1.2	87.2	6.5
benzene	C_6H_6	278.6	10.59	353.2	30.8
ethanol	C_2H_5OH	158.7	4.60	351.5	43.5
helium	He	—	0.021	4.22	0.084
mercury	Hg	234.2	2.292	630.2	59.3
methane	CH_4	90.7	0.94	111.7	8.2
methanol	CH_3OH	175.2	3.16	337.2	35.3
water	H_2O	273.2	6.01	373.2	40.7 (44.0 at 25°C)

*Values correspond to the temperature of the phase change. The superscript ° signifies that the change takes place at 1 atm and that the substance is pure.

enthalpy—stores more energy—than the same mass of liquid water at 100°C. We experience this stored energy painfully when scalded by steam: as the steam condenses, it releases energy as heat as it changes back to liquid water.

The enthalpy change per mole of molecules when a liquid vaporizes is called the **enthalpy of vaporization, ΔH_{vap}**. For water at 100.0°C,

$$\Delta H_{vap} = H_{vapor} - H_{liquid} = +40.7 \text{ kJ/mol}$$

This value means that to vaporize 1.00 mol $H_2O(l)$ (18.0 g of water) at 100.0°C, we have to supply 40.7 kJ of energy as heat. To vaporize twice that amount of water, we need to supply twice the energy. Values for other substances are given in Table 6.2. All enthalpies of vaporization are positive, so it is conventional to omit their sign.

Vaporization is endothermic because molecules have a higher energy when they are widely separated than when they are close together.

Example 6.6 *Measuring the enthalpy of vaporization*

Suppose we use an electric heater to raise the temperature of a sample of water to its boiling point in a constant-pressure calorimeter (like the simple Styrofoam cup). We bring the water to its boiling point and then continue heating until 35.0 g of water has vaporized. We calculate from the power rating of the heater and the time taken that the vaporization alone requires 79 kJ of heat. What is the enthalpy of vaporization of water at 100.0°C?

Strategy Because vaporization occurs at constant atmospheric pressure, the heat absorbed by the system (the water) is equal to the change in its enthalpy. We convert this enthalpy change to an enthalpy change per mole of molecules by dividing the observed enthalpy change by the number of moles of molecules vaporized.

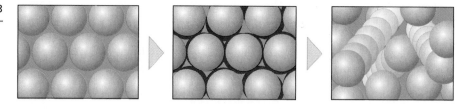

FIGURE 6.23

Melting (fusion) is an endothermic process. As molecules acquire energy, they begin to struggle past their neighbors. Finally the sample changes from a solid with ordered molecules (left) to a liquid with disordered, mobile molecules (right).

Solution We are told that the heat supplied is 79 kJ, so the change in enthalpy when 35.0 g of water vaporizes is +79 kJ. The number of moles of H_2O (of molar mass 18.02 g/mol) in 35.0 g of water is

$$\text{Moles of } H_2O = (35.0 \text{ g } H_2O) \times \left(\frac{1 \text{ mol } H_2O}{18.02 \text{ g } H_2O} \right) = \frac{35.0}{18.02} \text{ mol } H_2O$$

The enthalpy change per mole of H_2O is therefore

$$\Delta H_{vap} = \frac{79 \text{ kJ}}{(35.0/18.02) \text{ mol}} = 41 \text{ kJ/mol}$$

Self-Test 6.8A The same heater was used to heat a sample of benzene, C_6H_6 (molar mass 78.11 g/mol), to 80.1°C, its boiling point. It was found that 28 kJ was required to vaporize 71 g of boiling benzene. What is the enthalpy of vaporization of benzene at its boiling point?

[*Answer:* 31 kJ/mol]

Self-Test 6.8B The same heater was used to heat a sample of ethanol, C_2H_5OH (molar mass 46.07 g/mol), of mass 23 g to its boiling point. It was found that 22 kJ was required to vaporize all the ethanol. What is the enthalpy of vaporization of ethanol at its boiling point?

6.10 Melting and Sublimation

A solid melts when heat is supplied at its melting point. The added energy causes the molecules to oscillate more and more vigorously until they can move past one another and flow as a liquid (Fig. 6.23). The enthalpy change (per mole of molecules) that accompanies melting is called the **enthalpy of fusion, ΔH_{fus}**, of the substance:

$$\Delta H_{fus} = H_{liquid} - H_{solid}$$

Melting is endothermic because the molecules of the solid must be raised to a higher energy state to form the mobile liquid. Therefore, all enthalpies of fusion are positive, and it is conventional to omit their sign. The enthalpy of fusion of water at 0°C is 6.0 kJ/mol: this value means that, to melt 1.0 mol $H_2O(s)$ (18 g of ice) at 0°C, we have to supply 6.0 kJ of energy as heat. Vaporizing the same amount of water takes much more energy (over 40 kJ, in fact), because, when water is vaporized to a gas, its molecules must be separated completely. The enthalpies of fusion of various other substances are included in Table 6.2.

The **enthalpy of freezing** is the change in enthalpy per mole when a liquid turns into a solid. For water at 0°C, the enthalpy of freezing is −6.0 kJ/mol, because 6.0 kJ of heat is released when 1 mol $H_2O(l)$ freezes. This is the heat we remove when making ice in a freezer. All enthalpies of freezing are negative because freezing is an exothermic process: the molecules of the liquid settle into a lower energy state as the solid forms and the excess energy is released as heat.

The enthalpy of freezing of a substance is the negative of its enthalpy of fusion. This relation follows from the fact that enthalpy is a state property. The

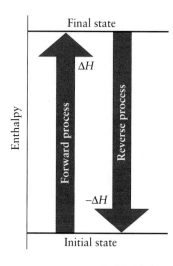

FIGURE 6.24

The enthalpy change for the reverse of a process is the negative of the enthalpy change for the forward process at the same temperature.

enthalpy of water at a given temperature must be the same after being frozen, melted, and then returned to the same temperature as it was initially. Therefore, the same quantity of heat is released in freezing as is absorbed in melting. In general, to obtain the enthalpy change for the reverse of any process, we just take the negative of the enthalpy change for the original process:

$$\Delta H(\text{reverse process}) = -\Delta H(\text{forward process}) \qquad (12)$$

This relation is illustrated in Fig. 6.24. If we find, for example, that the enthalpy of vaporization of mercury is $+59$ kJ/mol at its boiling point, then we immediately know that the enthalpy change occurring when mercury vapor condenses at that temperature is -59 kJ/mol. This value tells us that 59 kJ of heat is released when 1 mol Hg(g) condenses to a liquid.

Sublimation is the direct conversion of a solid into its vapor. Frost disappears on a cold, dry morning as the ice sublimes directly into water vapor. Solid carbon dioxide also sublimes, which is why it is called *dry ice*. Each winter on Mars, solid carbon dioxide is deposited as frost in the polar region; this frost sublimes when the feeble summer arrives (Fig. 6.25). The **enthalpy of sublimation**, ΔH_{sub}, is the enthalpy change per mole when a solid sublimes:

$$\Delta H_{sub} = H_{vapor} - H_{solid}$$

We can use the fact that enthalpy is a state property to calculate the change in enthalpy for the direct conversion of a solid to vapor. We simply add the enthalpy change for melting and vaporization measured at the same temperature (Fig. 6.26):

$$\Delta H_{sub} = \Delta H_{fus} + \Delta H_{vap}$$

For example, at 25°C, the enthalpy of fusion of sodium metal is 2.6 kJ/mol and the enthalpy of vaporization of liquid sodium is 98 kJ/mol. Therefore, the enthalpy of sublimation of solid sodium at 25°C is

$$\Delta H_{sub} = 2.6 + 98 \text{ kJ/mol} = 101 \text{ kJ/mol}$$

We can add enthalpy changes like this only if they correspond to the same temperature.

Enthalpies of changes of state are reported in kilojoules per mole of molecules. The enthalpy change for a reverse process is the negative of the enthalpy change for the forward process. Enthalpy changes can be added to obtain the value for an overall process.

6.11 Heating Curves

The enthalpies of fusion and vaporization affect the appearance of the *heating curve* of a substance. A **heating curve** is the graph showing the variation of the temperature of a sample as it is heated (that is, gains energy as heat) at a constant rate. A **cooling curve** is the graph showing the variation of the temperature as a sample cools (that is, loses energy as heat) at a constant rate. The cooling curve of a sample of a substance mirrors its heating curve.

Consider what happens when we heat a sample of very cold ice. As we see in Fig. 6.27, its temperature rises steadily. Although the molecules are still locked in a solid mass, they are oscillating around their average locations more vigorously. At the melting point, the molecules have enough energy to move past one another. At this temperature, all the added energy is used to overcome the

FIGURE 6.25

The polar ice caps on Mars extend and recede with the seasons. They are solid carbon dioxide and form by direct conversion of the gas to a solid. They disappear by sublimation. Although some water ice is also present in the polar caps, the temperature on Mars never becomes high enough to melt it. On Mars, ice is just another rock.

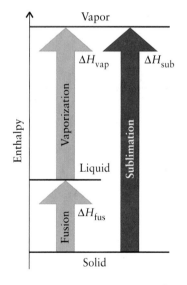

FIGURE 6.26

Because enthalpy is a state property, the enthalpy of sublimation at a given temperature can be expressed as the sum of the enthalpies of fusion and vaporization measured at the same temperature.

FIGURE 6.27

The heating curve of water. The temperature of the solid rises as heat is supplied. At the melting point, the temperature remains constant and the heat is used to melt the sample. When enough heat has been supplied to melt all the solid, the temperature of the liquid begins to rise again. A similar pause in the temperature rise occurs at the boiling point.

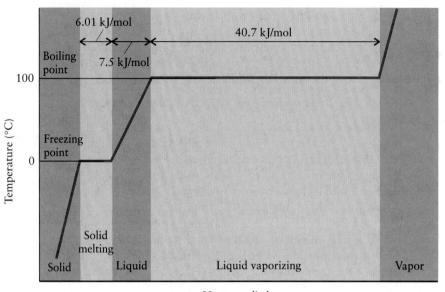

attractive forces between molecules and enable them to move past one another. The temperature remains constant at the melting point as heat continues to be supplied, until all the ice has melted. Only when melting is complete does the temperature rise again, and the rise continues right up to the boiling point. At the boiling point, the temperature rise again comes to a halt. Now the water molecules have enough energy to escape into the vapor, and all the heat supplied is used to form the vapor. From the graph, we can see that boiling requires more energy than melting. Melting releases the molecules from fixed positions, but they are still in contact. When a liquid vaporizes, the molecules separate completely from one another. After all the sample has evaporated, the temperature of the vapor rises again as heating continues and the water molecules move faster and faster.

The temperature of a sample is constant at its melting and boiling points, even though heat is being supplied.

THE ENTHALPY OF CHEMICAL CHANGE

All chemical reactions either release or absorb heat. Exothermic reactions, which are more common, have the form

$$\text{Reactants} \longrightarrow \text{products} + \text{energy as heat}$$

Endothermic reactions, which are less common, have the form

$$\text{Reactants} + \text{energy as heat} \longrightarrow \text{products}$$

We can apply the same principles used to study the energy changes of physical processes to the study of chemical reactions. The concepts and calculations introduced here will recur throughout the text whenever we are discussing energy changes.

6.12 Reaction Enthalpies

Consider the combustion of methane taking place in a vessel immersed in a constant-temperature water bath at constant pressure. The system is kept at constant temperature so that any change in enthalpy is due to the chemical reaction, not to heating or cooling. The chemical reaction is

$$CH_4(g) + 2 O_2(g) \longrightarrow CO_2(g) + 2 H_2O(l) + \text{energy}$$

We need to know how to report the "energy" product for the amounts of reactants and products shown in the chemical equation. Calorimetry shows that burning 1 mol $CH_4(g)$ produces 890. kJ of heat at 298 K and 1 atm. That is, when 1 mol $CH_4(g)$ and 2 mol $O_2(g)$ at 298 K react completely and the products have given off the excess heat and cooled back down to 298 K, the enthalpy of the system has decreased by 890. kJ for each mole of CH_4 consumed. To report this value, we write

$$CH_4(g) + 2 O_2(g) \longrightarrow CO_2(g) + 2 H_2O(l) \qquad \Delta H = -890. \text{ kJ}$$

This entire expression, the chemical equation together with the enthalpy change, is called a **thermochemical equation.** The stoichiometric coefficients in a thermochemical equation tell us the number of moles that react to give the reported change in enthalpy. In this case, the stoichiometric coefficient of CH_4 is 1 and that of O_2 is 2, so the enthalpy change is that resulting from the complete reaction of 1 mol CH_4 and 2 mol O_2.

An alternative method of reporting the same data is to give the **reaction enthalpy,** ΔH_r, the enthalpy change per mole of the species in the chemical equation. For example, we would report that $\Delta H_r = -890.$ kJ/mol for the combustion of methane. In this case, the "per mole" means per mole of CH_4 or per 2 mol O_2 consumed, or per mole of CO_2 or per 2 mol H_2O produced. We normally report ΔH_r, except for thermochemical equations, which use ΔH. In each case, we must know how the chemical equation is written.

The reaction enthalpy refers to the reaction of the number of moles given by the stoichiometric coefficients in the chemical equation. For example, if we write the same reaction with the coefficients all multiplied by 2, then we must also multiply the enthalpy change by 2:

$$2 CH_4(g) + 4 O_2(g) \longrightarrow 2 CO_2(g) + 4 H_2O(l) \qquad \Delta H = -1780. \text{ kJ}$$

Reporting that $\Delta H_r = -1780.$ kJ/mol tells us that the enthalpy change is $-1780.$ kJ per 2 mol CH_4 consumed, and so on. This doubling makes sense, because the equation now represents the burning of twice as much methane.

Just as we saw in Section 6.10 for physical changes, the enthalpy change for the reverse of a chemical reaction is the negative of the enthalpy change of the forward reaction:

$$CO_2(g) + 2 H_2O(l) \longrightarrow CH_4(g) + 2 O_2(g) \qquad \Delta H = +890. \text{ kJ}$$

This reversal of sign also makes sense. If methane and oxygen give off heat when they burn, they must have a higher enthalpy than the products. To go back from products to reactants, we would have to restore that additional enthalpy.

A thermochemical equation is a combination of a chemical equation with the enthalpy change for the reaction as written. The reaction enthalpy is the change in enthalpy per mole for the stoichiometric numbers of moles of reactants in the chemical equation.

Example 6.7 *Calculating the reaction enthalpy*

Research on alternative fuels has shown that methanol produced from coal is a promising alternative to gasoline, for its combustion generates little pollution. When 0.515 g of methanol, CH_3OH, burns in excess oxygen in a calibrated constant-pressure calorimeter with a heat capacity of 551 J/°C, the temperature of the calorimeter rises by 10.6°C. Calculate the reaction enthalpy for

$$2 CH_3OH(l) + 3 O_2(g) \longrightarrow 2 CO_2(g) + 4 H_2O(l)$$

and write the corresponding thermochemical equation.

Strategy First, we note whether the temperature rises or falls. If it rises, then heat is given off, the reaction is exothermic, and ΔH is negative. If the temperature falls, then heat is absorbed, the reaction is endothermic, and ΔH is positive. The enthalpy change is calculated from the temperature change multiplied by the heat capacity. Once we know the enthalpy change for the masses of reactants used, we convert the given mass of reactant to moles by using the molar mass. Then we convert the enthalpy change to the reaction enthalpy for the number of moles of reactants equal to the stoichiometric coefficients. Finally, we express the heat output as the enthalpy change per mole of reactant molecules.

Solution The temperature rises, so we know that the reaction is exothermic. The calorimeter has been calibrated and is known to have a heat capacity of 551 J/°C, so

$$\text{Heat supplied} = (10.6°C) \times (551 \text{ J/°C}) = 10.6 \times 551 \text{ J}$$

Because the molar mass of methanol is 32.04 g/mol, the number of moles of CH_3OH that reacts is

$$\text{Moles of } CH_3OH = (0.515 \text{ g } CH_3OH) \times \left(\frac{1 \text{ mol } CH_3OH}{32.04 \text{ g } CH_3OH}\right) = \frac{0.515}{32.04} \text{ mol } CH_3OH$$

The reaction enthalpy per mole of CH_3OH is therefore

$$\Delta H_r = -\frac{10.6 \times 551 \text{ J}}{(0.515/32.04) \text{ mol } CH_3OH}$$
$$= -3.63 \times 10^5 \text{ J/mol } CH_3OH = -363 \text{ kJ/mol } CH_3OH$$

The negative sign has been included because the reaction is exothermic. We have found that $\Delta H = -363$ kJ for the combustion of 1 mol $CH_3OH(l)$. However, in the chemical equation for the combustion of methanol, CH_3OH has a coefficient of 2. Therefore, we must multiply the enthalpy change by 2 to write the thermochemical equation:

$$2 CH_3OH(l) + 3 O_2(g) \longrightarrow 2 CO_2(g) + 4 H_2O(l) \qquad \Delta H = -726 \text{ kJ}$$

The reaction enthalpy is -726 kJ/mol.

Self-Test 6.9A When 2.31 g of phosphorus reacts with chlorine to form phosphorus trichloride, PCl_3, in a calorimeter of heat capacity 2.16 kJ/°C, the temperature of the calorimeter rises by 11.06°C. Write the thermochemical equation for the reaction.
[*Answer:* $2 P(s) + 3 Cl_2(g) \rightarrow 2 PCl_3(l), \Delta H = -641$ kJ]

Self-Test 6.9B In research on alternative fuels, ethanol, C_2H_5OH (molar mass 46.07 g/mol) is produced from waste cornstalks (Fig. 6.28). When a sample of ethanol of mass 0.461 g was burned in pure oxygen in the same calorimeter used in Self-Test 6.9A, the temperature rose by 6.333°C. The products were gaseous CO_2 and liquid H_2O. Write the thermochemical equation for the reaction.

FIGURE 6.28

A biophysicist monitors an experimental fermentation chamber in which fuel ethanol is being produced from waste biomass by a genetically engineered strain of bacteria.

6.13 Standard Reaction Enthalpies

The heat released or absorbed by a reaction depends on the physical states of the reactants and products (that is, whether they are solid, liquid, or gas) and on the conditions (the temperature and pressure). Let's consider once again the combustion of methane at 25°C and 1 atm. The thermochemical equations for two slightly different reactions are

$$CH_4(g) + 2O_2(g) \longrightarrow CO_2(g) + 2H_2O(g) \qquad \Delta H = -802 \text{ kJ}$$

$$CH_4(g) + 2O_2(g) \longrightarrow CO_2(g) + 2H_2O(l) \qquad \Delta H = -890. \text{ kJ}$$

In the first reaction, all the water is produced as a vapor; in the second, it is produced as a liquid. The heat produced is different in each case. The enthalpy of water vapor is 44 kJ/mol higher than that of liquid water at 25°C (see Table 6.2). As a result, an additional 88 kJ (for 2 mol H_2O) remains stored in the system if water vapor is formed rather than the liquid (Fig. 6.29). If the 2 mol $H_2O(g)$ condenses, 88 kJ is given off as heat.

Enthalpy changes depend on pressure. All the tables in this book give data for reactions in which each of the reactants and products is in its **standard state,** its pure form at 1 atm pressure. The standard state of liquid water is pure water at 1 atm. The standard state of ice is pure ice at 1 atm. A solute in a liquid solution is in its standard state when its concentration is 1 mol/L.

A reaction enthalpy for a reaction in which all the reactants and products are in their standard states is called a **standard reaction enthalpy, $\Delta H_r°$.** For example, we can write

$$CH_4(g) + 2O_2(g) \longrightarrow CO_2(g) + 2H_2O(l) \qquad \Delta H° = -890. \text{ kJ}$$

or, alternatively, report that $\Delta H_r° = -890.$ kJ/mol. Here, pure methane gas at 1 atm is allowed to react with pure oxygen gas at 1 atm, giving pure carbon dioxide gas and pure liquid water, both at 1 atm (Fig. 6.30). Reaction enthalpies do not change very much with pressure, so the standard value gives a good indication of the change in enthalpy for pressures near 1 atm.

Most thermochemical data are reported for 25°C (more precisely, for 298.15 K). Temperature is not part of the definition of standard state. We can have a standard state at any temperature; 298.15 K is simply the most common temperature used in tables of data. All reaction enthalpies used in this text are for 298.15 K unless another temperature is indicated. Professional chemists, however, often need data for other temperatures. For example, a chemical engineer designing a reaction vessel operating at 500°C would need to know the standard reaction enthalpy at that temperature.

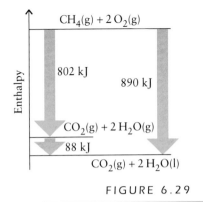

FIGURE 6.29

This diagram shows how the value of the reaction enthalpy depends on the physical states of a product. When water is produced as a vapor rather than as a liquid in the combustion of methane, 88 kJ remains stored in the system for every 2 mol H_2O produced.

The modern definition of standard state replaces 1 atm by 1 bar, where 1 bar = 10^5 Pa. The values of thermochemical quantities are changed only very slightly by this definition.

A superscript circle is commonly used to denote a standard value.

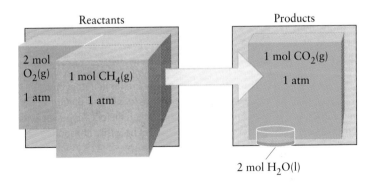

FIGURE 6.30

The standard reaction enthalpy is the difference in enthalpy between the pure products, each at 1 atm, and the pure reactants at the same pressure and the specified temperature (which is commonly but not necessarily 25°C). The scheme here is for the combustion of methane.

Standard reaction enthalpies refer to reactions in which the reactants and products are in their standard states; they are usually reported for 298.15 K.

6.14 Combining Reaction Enthalpies: Hess's Law

We know from comparing enthalpies of fusion and freezing (Section 6.10) that the enthalpy change of any reverse process is the negative of the change for the forward process. We can therefore obtain the standard reaction enthalpy of a reverse reaction from that of the forward reaction simply by changing the sign:

$$P_4(s) + 6\,Cl_2(g) \longrightarrow 4\,PCl_3(l) \qquad \Delta H° = -1279\ kJ$$

$$4\,PCl_3(l) \longrightarrow P_4(s) + 6\,Cl_2(g) \qquad \Delta H° = +1279\ kJ$$

It follows that a reaction that is exothermic in one direction is endothermic in the reverse direction. For example, the combustion of glucose is exothermic:

$$C_6H_{12}O_6(aq) + 6\,O_2(g) \longrightarrow 6\,CO_2(g) + 6\,H_2O(l) \qquad \Delta H° = -2808\ kJ$$

Investigating Matter 6.1: *The World's Energy Resources*

Why is there so much concern about conserving energy when the first law assures us that energy is conserved? Moreover, we are bombarded daily by enormous quantities of energy from the Sun, which must add to our energy resources. Thermochemical principles suggest some reasons to be concerned about energy usage. The fundamental problem is that solar energy is not always available in a useful manner. First, most of the energy received from the Sun is radiated back into space. Much of the remainder is used up evaporating water from the oceans. Less than 1% remains available for energy production.

Solar energy is stored for later use primarily through photosynthesis, in which carbon dioxide and water are converted into glucose. The energy of all the solar radiation absorbed by the vegetation on Earth is enough to build about 6×10^{14} kg of glucose a year. Most of this glucose is converted into starch and cellulose. Dead vegetation remains a store of energy as long as its carbohydrates are not oxidized back to carbon dioxide and water. If losses through forest fires area ignored, this store, the *biomass*, increases by about 10^{19} kJ each year which is about 30 times the current annual global industrial demand for energy.

The energy being stored as biomass seems more than adequate to meet our needs, but it is widely dispersed and hence energy is required to harvest energy from biomass. We need high concentrations of energy-rich substances such as deposits of *fossil fuels*, fuels that are the remains of decayed organic matter laid down millions of years ago. Favorable geological formations have kept these deposits out of contact with the atmosphere, so they have been protected from oxidation. The natural gas that heats our homes, the gasoline that powers our automobiles, and the coal that provides much of our electrical power are fossil fuels that, until we burn them, have escaped oxidation. Vast reserves of coal and petroleum, the source of liquid hydrocarbon fuels such as gasoline, exist in many areas of the world.

Because fossil fuel reserves are limited, they must be extracted wherever they are found. This platform is used to pump petroleum from beneath the ocean; however, the natural gas accompanying it cannot be easily transported and so is burned off.

We immediately know, therefore, that the reverse of this reaction, the formation of glucose and oxygen from carbon dioxide and water, is endothermic:

$$6\,CO_2(g) \;+\; 6\,H_2O(l) \longrightarrow C_6H_{12}O_6(aq) \;+\; 6\,O_2(g) \qquad \Delta H^\circ \;=\; +2808\,kJ$$

This is the process that takes place in photosynthesis, and its energy requirement is supplied by the Sun (Investigating Matter 6.1). It would be very difficult to measure that energy directly in living plants, but it is easy to measure the enthalpy change for the combustion of glucose and then reverse the sign. The reaction enthalpy shows how glucose acts as a store of energy. Forming glucose by photosynthesis is like filling a reservoir with energy, and burning glucose releases the energy it has stored.

We saw in Section 6.10 that the enthalpy change for an overall physical process can be expressed as the sum of the enthalpy changes for a series of individual steps. The same rule applies to chemical reactions. In this context, the rule is known as **Hess's law:**

> The overall reaction enthalpy is the sum of the reaction enthalpies of the steps into which the reaction can be divided.

Unfortunately, of 10^{19} kJ stored annually as biomass by photosynthesis, less than 0.01% survives without being oxidized. Moreover, only a tiny fraction of that (estimated as 0.07% of 0.01%, or 0.0007% of the whole) forms deposits large enough to be economical sources of fuels. The rate at which energy is stored in a useful form is therefore only about 10^{11} kJ per year. We are using it at the rate of 3×10^{17} kJ per year, which is 30 million times more rapidly than it is being stored. In other words, we are not able to replace the fossil fuels at the current rate of use.*

One solution is to make better use of renewable fuels. *Renewable energy resources* are those produced and replenished by the Sun, for example, biomass, hydroelectric plants, and wind turbines. Converting waste biomass readily available from agriculture could also be helpful. Ethanol, CH_3CH_2OH, is produced now from the biological fermentation of the starches in grains, mainly corn. It currently makes up about 10% by volume of gasoline in the United States. Not enough ethanol can be made from grains, but the cellulose in straw and cornstalks left behind when grains are harvested is a promising source (see Connection 2, following Chapter 9).

Reducing our consumption of energy can also help ease the dilemma, as can alternative energy sources such as nuclear power (see Chapter 22). Solar radiation can be harnessed directly to generate electricity by using *photovoltaic cells,* in which silicon crystals convert solar radiation into an electric current. Because an energy resource is practical only if it generates more energy than required to produce it, much energy research goes into increasing the efficiency of both producing fuels and burning them.

An agricultural researcher assesses the growth rate of a seedling. Plant photosynthesis is only about 3% efficient, and conditions that increase this efficiency are actively being sought.

---

*Many experts suspect that our petroleum reserves will last only another 50 years at the present rate of consumption.

The intermediate reactions need not be ones that can actually be carried out. So long as they balance and add up to the equation for the reaction of interest, a reaction enthalpy can be calculated from any convenient sequence of reactions.

Let's take, as an example, the oxidation of carbon to carbon dioxide:

$$C(s) + O_2(g) \longrightarrow CO_2(g)$$

This reaction can be thought of as the outcome of two steps. One step is the oxidation of carbon to carbon monoxide:

$$C(s) + \tfrac{1}{2}O_2(g) \longrightarrow CO(g) \qquad \Delta H° = -110.5 \text{ kJ}$$

The second step is the oxidation of carbon monoxide to carbon dioxide:

$$CO(g) + \tfrac{1}{2}O_2(g) \longrightarrow CO_2(g) \qquad \Delta H° = -283.0 \text{ kJ}$$

This two-step process is an example of a **reaction sequence,** a series of reactions in which the products of one reaction take part as reactants in another reaction. The equation for the overall reaction, the net outcome of the sequence, is the sum of the equations for the intermediate steps:

$$
\begin{array}{lll}
C(s) + \tfrac{1}{2}O_2(g) \longrightarrow CO(g) & \Delta H° = -110.5 \text{ kJ} & (a) \\
CO(g) + \tfrac{1}{2}O_2(g) \longrightarrow CO_2(g) & \Delta H° = -283.0 \text{ kJ} & (b) \\
\hline
C(s) + O_2(g) \longrightarrow CO_2(g) & \Delta H° = -393.5 \text{ kJ} & (a + b)
\end{array}
$$

The same procedure is used to predict the enthalpies of reactions that we cannot measure directly in the laboratory: see Toolbox 6.1.

According to Hess's law, thermochemical equations for the individual steps of a reaction sequence may be combined to obtain the thermochemical equation for the overall reaction.

Example 6.8 *Using Hess's law*

Enthalpies of combustion are readily available, so combustion reactions are often useful for obtaining the enthalpies of reactions involving organic compounds. Use the information given to calculate the standard enthalpy for the synthesis of propane, $3\,C(s) + 4\,H_2(g) \rightarrow C_3H_8(g)$, a gas used as camping fuel.

$$
\begin{array}{lll}
C_3H_8(g) + 5\,O_2(g) \longrightarrow 3\,CO_2(g) + 4\,H_2O(l) & \Delta H° = -2220. \text{ kJ} & (a) \\
C(s) + O_2(g) \longrightarrow CO_2(g) & \Delta H° = -394 \text{ kJ} & (b) \\
H_2(g) + \tfrac{1}{2}O_2(g) \longrightarrow H_2O(l) & \Delta H° = -286 \text{ kJ} & (c)
\end{array}
$$

Strategy Identify equations in which reactants or products appear, then work through the steps in Toolbox 6.1.

Solution **Step 1.** Only (b) and (c) have at least one of the reactants. In both cases, they are on the correct side of the arrow in the overall chemical equation. We select (b) and multiply it through by 3 to give carbon the coefficient it will have in the final equation:

$$3\,C(s) + 3\,O_2(g) \longrightarrow 3\,CO_2(g) \qquad \Delta H° = 3 \times (-394 \text{ kJ}) = -1182 \text{ kJ}$$

Step 2. To obtain C_3H_8 on the right, we reverse equation (a), changing the sign of its reaction enthalpy, and add it to the equation we have just derived:

$$3\,C(s) + 3\,O_2(g) \longrightarrow 3\,CO_2(g) \qquad \Delta H° = -1182 \text{ kJ}$$

$$3\,CO_2(g) + 4\,H_2O(l) \longrightarrow C_3H_8(g) + 5\,O_2(g) \qquad \Delta H° = +2220. \text{ kJ}$$

This Toolbox shows how to use Hess's law to calculate the reaction enthalpy from the enthalpies of other reactions.

Conceptual Basis

Hess's law is based on the fact that enthalpy is a state function. Because the enthalpy change of a system depends only on its initial and final states, we can carry out a reaction in one step or visualize it as the sum of several steps. The enthalpy change for the reaction is the same in each case (Fig. 6.31).

Procedure

The aim is to find a sequence of reactions, the sum of which is the reaction of interest. In many cases, we can see at a glance what reactions to use. When the answer is not obvious, the following procedure is helpful:

Step 1. Select one of the reactants in the overall reaction and write down a chemical equation in which it also appears as a reactant.

Step 2. Select one of the products in the overall reaction and write down a chemical equation in which it also appears as a product. Add this equation to the equation written in Step 1.

Step 3. Cancel unwanted species in the sum obtained in Step 2 by adding an equation that has the same substance or substances on the opposite side of the arrow.

Step 4. Once the sequence is complete, combine the standard reaction enthalpies. In each step, we may need to reverse the equation or multiply it by a factor. Recall that if we want to reverse a chemical equation, then we have to change the sign of the reaction enthalpy. If we multiply the stoichiometric coefficients by a factor, then we must multiply the reaction enthalpy by the same factor.

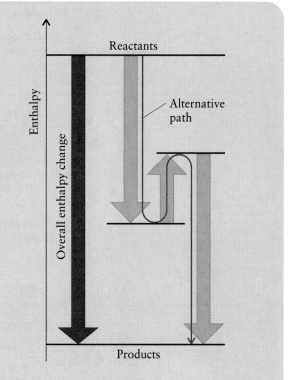

FIGURE 6.31

If the overall reaction can be broken down into a series of steps, then the corresponding overall reaction enthalpy is the sum of the reaction enthalpies of the steps on the alternative path. None of the steps need be a reaction that can actually be carried out in the laboratory.

The sum of these two equations is

$$3\,C(s) + 3\,O_2(g) + 3\,CO_2(g) \longrightarrow C_3H_8(g) + 5\,O_2(g) + 3\,CO_2(g)$$
$$\Delta H^\circ = +1038\ \text{kJ}$$

This equation simplifies to

$$3\,C(s) + 4\,H_2O(l) \longrightarrow C_3H_8(g) + 2\,O_2(g) \qquad \Delta H^\circ = +1038\ \text{kJ}$$

Step 3. To cancel the unwanted reactant H_2O and product O_2, we add equation (c) after multiplying it by 4:

$$3\,C(s) + 4\,H_2O(l) \longrightarrow C_3H_8(g) + 2\,O_2(g) \qquad\qquad \Delta H^\circ = +1038\ \text{kJ}$$

$$4\,H_2(g) + 2\,O_2(g) \longrightarrow 4\,H_2O(l) \qquad \Delta H^\circ = 4 \times (-286\ \text{kJ}) = -1144\ \text{kJ}$$

The sum of these two reactions is

$$3\,C(s) + 4\,H_2(g) + 4\,H_2O(l) + 2\,O_2(g) \longrightarrow C_3H_8(g) + 2\,O_2(g) + 4\,H_2O(l)$$
$$\Delta H^\circ = -106\ \text{kJ}$$

which simplifies to

$$3\,C(s) + 4\,H_2(g) \longrightarrow C_3H_8(g) \qquad \Delta H° = -106\ kJ$$

giving us the quantity we require.

Self-Test 6.10A Gasoline, which contains octane as one component, may burn to carbon monoxide if the air supply is restricted. Derive the standard reaction enthalpy for the incomplete combustion of octane:

$$2\,C_8H_{18}(l) + 17\,O_2(g) \longrightarrow 16\,CO(g) + 18\,H_2O(l)$$

from the standard reaction enthalpies for the combustions of octane and carbon monoxide:

$$2\,C_8H_{18}(l) + 25\,O_2(g) \longrightarrow 16\,CO_2(g) + 18\,H_2O(l) \qquad \Delta H° = -10\,942\ kJ$$

$$2\,CO(g) + O_2(g) \longrightarrow 2\,CO_2(g) \qquad \Delta H° = -566.0\ kJ$$

[**Answer:** −6414 kJ]

Self-Test 6.10B Methanol is a clean-burning liquid fuel being considered as a replacement for gasoline. It can be made from the methane in natural gas:

$$2\,CH_4(g) + O_2(g) \longrightarrow 2\,CH_3OH(l)$$

Find the standard reaction enthalpy for the formation of methanol from methane, given the following three equations:

$$CH_4(g) + H_2O(g) \longrightarrow CO(g) + 3\,H_2(g) \qquad \Delta H° = +206.10\ kJ$$

$$2\,H_2(g) + CO(g) \longrightarrow CH_3OH(l) \qquad \Delta H° = -128.33\ kJ$$

$$2\,H_2(g) + O_2(g) \longrightarrow 2\,H_2O(g) \qquad \Delta H° = -483.64\ kJ$$

THE HEAT OUTPUT OF REACTIONS

Once we know the reaction enthalpy, we can calculate the enthalpy change for any amount, mass, or volume of reactant consumed or product formed. To do this kind of calculation, we carry out a stoichiometry calculation like those described in Chapter 4. We proceed as if heat were a reactant or a product (Fig. 6.32).

6.15 Enthalpies of Combustion

The enthalpy change in an exothermic reaction is an important measure of the value of a fuel. For example, suppose we want to know the heat output at con-

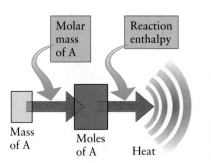

FIGURE 6.32

The amount of heat produced or absorbed in a chemical reaction can be determined from the reaction stoichiometry.

stant pressure from the combustion of 150. g of methane. The thermochemical equation

$$CH_4(g) + 2\,O_2(g) \longrightarrow CO_2(g) + 2\,H_2O(l) \qquad \Delta H° = -890.\,kJ$$

allows us to write the relation

$$1\ mol\ CH_4 \mathrel{\widehat{=}} -890.\,kJ$$

and to use it as a conversion factor in the usual way. We already know how to use the molar mass of methane (16.04 g/mol) to work out the number of moles of CH_4 corresponding to 150. g. From these relations, we can write

$$q = (150.\text{ g CH}_4) \times \overbrace{\left(\frac{1\ mol\ CH_4}{16.04\ g\ CH_4}\right)}^{\substack{\text{Convert from}\\\text{grams to moles}}} \times \overbrace{\left(\frac{-890.\text{kJ}}{1\ mol\ CH_4}\right)}^{\substack{\text{Convert from}\\\text{moles to kilojoules}}}$$

$$= -8.32 \times 10^3\ kJ$$

The negative sign tells us that 8.32 MJ of heat (1 MJ $= 10^6$ J) leaves the system.

The heat absorbed or given off by a reaction can be treated like a reactant or product in a stoichiometric relation.

Example 6.9 *Calculating the heat output of a fuel*

How much propane should a backpacker carry: do we really need to carry a kilogram of gas? Calculate the mass of propane that you would need to burn to obtain 350. kJ of heat, which is just enough energy to heat 1 L of water from 27°C to boiling at sea level (if we ignore all heat losses). The thermochemical equation is

$$C_3H_8(g) + 5\,O_2(g) \longrightarrow 3\,CO_2(g) + 4\,H_2O(l) \qquad \Delta H° = -2220.\,kJ$$

Strategy Because we are given the heat output, the first step is to convert the desired heat output to moles of fuel molecules by using the thermochemical equation. Then we convert from moles of fuel molecules to grams by using the molar mass of the fuel.

Solution The thermochemical equation tells us that

$$1\ mol\ C_3H_8 \mathrel{\widehat{=}} -2220.\,kJ$$

The molar mass of propane is 44.09 g/mol. It follows that

$$\text{Mass of } C_3H_8 \text{ required} = (-350.\,kJ) \times \left(\frac{1\ mol\ C_3H_8}{-2220.\,kJ}\right) \times \left(\frac{44.09\ g\ C_3H_8}{1\ mol\ C_3H_8}\right)$$

$$= 6.95\ g\ C_3H_8$$

That is, just under 7 g of propane is needed to boil the water (and more if we allow for heat losses).

Self-Test 6.11A The thermochemical equation for the combustion of butane is

$$2\,C_4H_{10}(g) + 13\,O_2(g) \longrightarrow 8\,CO_2(g) + 10\,H_2O(l) \qquad \Delta H° = -5756\ kJ$$

What mass of butane must be burned to supply 350. kJ of heat? Would it be easier to pack butane rather than propane?

[*Answer:* 7.07 g. Propane would be very slightly lighter to carry.]

Self-Test 6.11B Ethanol trapped in a gel is another common camping fuel. What mass of ethanol must be burned to supply 350. kJ of heat? The thermochemical equation for the combustion of ethanol is

$$C_2H_5OH(l) + 3\,O_2(g) \longrightarrow 2\,CO_2(g) + 3\,H_2O(l) \qquad \Delta H° = -1368\text{ kJ}$$

Because combustion reactions are so important, their standard enthalpies are given a special symbol, $\Delta H_c°$. This symbol stands for the **standard enthalpy of combustion,** the change in enthalpy per mole of a substance when the substance reacts completely with oxygen under standard conditions (that is, pure substances at 1 atm). In a combustion of an organic compound, carbon forms carbon dioxide and hydrogen forms water; any nitrogen present is released as N_2, unless we specify that nitrogen oxides are formed. For example, from Self-Test 6.11A, we know that the reaction enthalpy for the combustion of 2 mol C_4H_{10} is -5756 kJ. Therefore, the enthalpy of combustion of butane is

$$\Delta H_c° = \frac{-5756\text{ kJ}}{2\text{ mol } C_4H_{10}} = -2878\text{ kJ/mol } C_4H_{10}$$

Standard enthalpies of combustion are listed in Table 6.3 and in the listing of organic compounds in Appendix 2A. We have seen how to use enthalpies of combustion to obtain the standard enthalpies of reactions (see Toolbox 6.1). Here we shall consider another practical application—the choice of a fuel.

One practical measure of a fuel's value is its **specific enthalpy,** the enthalpy of combustion per gram. Fuels with a high specific enthalpy release a lot of heat per gram when they burn. Hence, specific enthalpy is an important criterion when mass is important, as it is in airplanes. When the *volume* occupied by a fuel is important, as it is in automobiles, we assess the value of a fuel in terms of its **enthalpy density,** the enthalpy of combustion per liter. A fuel with a low enthalpy density releases very little heat per liter when it burns (Fig. 6.33).

Three readily available fuels are hydrogen, methane, and octane. Methane is the major component of natural gas, and octane is a typical component of gasoline. The data in Table 6.4 illustrate one advantage of using gasoline: its high enthalpy density of 38 MJ/L (where 1 MJ $= 10^6$ J). That high value means that a gasoline fuel tank need not be large to store a lot of energy.

> When reporting specific enthalpies and enthalpy densities, it is conventional to omit the negative sign.

> Like octane, most of the compounds in gasoline are hydrocarbons with about eight carbon atoms.

Table 6.3 Standard enthalpies of combustion at 25°C*

Substance	Formula	$\Delta H_c°$, kJ/mol
benzene	$C_6H_6(l)$	-3268
carbon	$C(s, \text{graphite})$	-394
ethanol	$C_2H_5OH(l)$	-1398
ethyne (acetylene)	$C_2H_2(g)$	$-1300.$
glucose	$C_6H_{12}O_6(s)$	-2808
hydrogen	$H_2(g)$	-286
methane	$CH_4(g)$	$-890.$
octane	$C_8H_{18}(l)$	-5471
propane	$C_3H_8(g)$	$-2220.$
urea	$CO(NH_2)_2(s)$	-632

*In a combustion, carbon is converted to carbon dioxide, hydrogen to liquid water, and nitrogen to nitrogen gas. More values are given in Appendix 2A.

FIGURE 6.33

During World War II, fuel was in short supply and all manner of ingenious solutions were sought. However, as we can see from this photograph of a vehicle powered by coal gas (a mixture of carbon monoxide and hydrogen) in London, the low enthalpy density of gases creates storage problems. A modern approach to using gases to power a vehicle can be seen in Applying Chemistry: Case Study 18.

A great advantage of hydrogen over hydrocarbon fuels is that it produces no carbon dioxide when it burns. Carbon dioxide is potentially harmful, because it contributes to the greenhouse effect (see Applying Chemistry: Case Study 4). However, although hydrogen forms only water when it burns, its use is limited by its low enthalpy density. Chemists are working to synthesize solid compounds that concentrate hydrogen and release it as needed (Fig. 6.34). Candidates include the hydrides formed when titanium, copper, and other metals are heated in hydrogen. These compounds occupy a much smaller volume than the equivalent amount of hydrogen gas and release hydrogen when heated or treated with acid. One example is an iron titanium hydride of approximate formula $FeTiH_2$. Its enthalpy density is high, but its iron and titanium content make the compound dense, so its specific enthalpy is low. Another promising

Table 6.4 Thermochemical properties of four fuels

Fuel	Combustion equation	ΔH_c°, kJ/mol	Specific enthalpy, kJ/g	Enthalpy density,* kJ/L
hydrogen	$2\,H_2(g) + O_2(g) \longrightarrow$ $2\,H_2O(l)$	-286	142	13
methane	$CH_4(g) + 2\,O_2(g) \longrightarrow$ $CO_2(g) + 2\,H_2O(l)$	$-890.$	55	40.
octane	$2\,C_8H_{18}(l) + 25\,O_2(g) \longrightarrow$ $16\,CO_2(g) + 18\,H_2O(l)$	-5471	48	3.8×10^4
methanol	$2\,CH_3OH(l) + 3\,O_2(g) \longrightarrow$ $2\,CO_2(g) + 4\,H_2O(l)$	-726	23	1.8×10^4

*At atmospheric pressure and room temperature.

FIGURE 6.34

The range and speed of this electric-powered car depend on the type of battery it uses. For example, metal-hydride devices have a longer range than lead-acid storage batteries. As this driver would agree, recharging is generally a slow process. The cable attached to this car may look like a gasoline hose, but it is actually delivering electricity while its owner waits. The use of hydrogen as fuel could reduce the number of refueling stops.

low-density storage medium for hydrogen is offered by carbon nanotubes (see Section 19.15).

Food is just another kind of fuel. Applying Chemistry: Case Study 6 looks at the fuel value of foods such as hamburgers, potatoes, and bread.

The specific enthalpy tells how much heat can be obtained per gram of fuel. The enthalpy density indicates how much heat can be obtained per liter of fuel.

Self-Test 6.12A From the data in Table 6.4, determine whether methane or methanol would make a better fuel for a life-support system on a space ship and explain why.

[*Answer:* Methane. Its specific enthalpy is higher than methanol's, so less mass would be required for the same heat output.]

Self-Test 6.12B From the data in Table 6.4, determine whether methanol or hydrogen gas would be more useful in a small automobile.

6.16 Standard Enthalpies of Formation

Chemists have devised an ingenious way of calculating and tabulating enthalpies of reactions. They report the **standard enthalpy of formation** of each individual substance, ΔH_f°. This quantity is the standard reaction enthalpy for the formation of the substance from its elements in their most stable form; it is expressed in kilojoules per mole of the substance (kJ/mol). To obtain the standard enthalpy of any reaction, we just subtract the standard enthalpies of formation of the reactants from those of the products.

We obtain ΔH_f° for ethanol, for instance, from the thermochemical equation for its formation starting from graphite (the most stable form of carbon) and gaseous hydrogen and oxygen:

$$4\,C(s) + 6\,H_2(g) + O_2(g) \longrightarrow 2\,C_2H_5OH(l) \qquad \Delta H^\circ = -555.38\ kJ$$

Because this thermochemical equation refers to the formation of 2 mol $C_2H_5OH(l)$, the standard reaction enthalpy per mole of ethanol molecules is

$$\Delta H_f^\circ(C_2H_5OH, l) = \frac{-555.38\ kJ}{2\ mol\ C_2H_5OH} = -277.69\ kJ/mol\ C_2H_5OH$$

Note how the substance and its physical state are used to label the enthalpy change, so we know which species and which form of that species we are talking about.

The standard enthalpy of formation of an element in its most stable form is 0. For example, in the reaction

$$C(s, graphite) \longrightarrow C(s, graphite) \qquad \Delta H^\circ = 0$$

nothing happens, and there is no change in enthalpy. However, the enthalpy of formation of an element in a form other than its most stable one is not 0. For example, the conversion of carbon from graphite to diamond is endothermic:

$$C(s, graphite) \longrightarrow C(s, diamond) \qquad \Delta H^\circ = +1.9\ kJ$$

It follows that the standard enthalpy of formation of diamond is $+1.9$ kJ/mol.

When a compound cannot be synthesized from its elements directly (or the reaction is too difficult to study), its enthalpy of formation can still be found from its enthalpy of combustion. Toolbox 6.1 explained how this is done. Some values are listed in Table 6.5, and a longer list can be found in Appendix 2A.

Now let's see how standard enthalpies of formation are combined to give a reaction enthalpy. First, we calculate the total standard enthalpy of formation of all the products of a reaction. Then we calculate the total standard enthalpy of formation of all the reactants. The difference between these two totals is the standard reaction enthalpy:

$$\Delta H_r^\circ = H^\circ_{final} - H^\circ_{initial}$$
$$= \sum n\Delta H_f^\circ(products) - \sum n\Delta H_f^\circ(reactants) \qquad (13)$$

Here Σ (sigma) means a sum, and n stands for the stoichiometric coefficient of each substance. The first sum is the total enthalpy of formation of the products multiplied by their stoichiometric coefficients. The second sum is the similar total for the reactants. Toolbox 6.2 shows how to use this expression.

Table 6.5 Standard enthalpies of formation at 25°C*

Substance	Formula	ΔH_f°, kJ/mol	Substance	Formula	ΔH_f°, kJ/mol
Inorganic compounds			sodium chloride	NaCl(s)	−411.15
ammonia	$NH_3(g)$	−46.11	water	$H_2O(l)$	−285.83
carbon dioxide	$CO_2(g)$	−393.51		$H_2O(g)$	−241.82
carbon monoxide	CO(g)	−110.53	**Organic compounds**		
dinitrogen tetroxide	$N_2O_4(g)$	+9.16	benzene	$C_6H_6(l)$	+49.0
hydrogen chloride	HCl(g)	−92.31	ethanol	$C_2H_5OH(l)$	−277.69
hydrogen fluoride	HF(g)	−271.1	ethyne (acetylene)	$C_2H_2(g)$	+226.73
nitrogen dioxide	$NO_2(g)$	+33.18	glucose	$C_6H_{12}O_6(s)$	−1268
nitric oxide	NO(g)	+90.25	methane	$CH_4(g)$	−74.81

*A much longer list is given in Appendix 2A.

THE THERMOCHEMISTRY OF FITNESS

Applying Chemistry: *Case Study 6*

On a long journey, we often stop to refuel both our vehicle and ourselves. Although the two types of fuel will both be burned to release energy, there are some differences. For example, we usually give vehicles the same kind of fuel each time, but we eat a wide variety of foods; we cannot overfill the gas tank, but we sometimes overeat; and the food in the diner is usually tastier than gasoline.

Gasoline engines are powered by hydrocarbons with formulas similar to that of octane, C_8H_{18}, but we get our energy primarily from carbohydrates, proteins, and fats (see Chapter 11 for a discussion of these classes of compounds). Over 50% by mass of the food in a typical diet is taken in the form of carbohydrates. But not all carbohydrates are digestible. Cellulose, for instance, which is the structural material of plants, is a primary "fiber" in our diet. As dietary fiber, it helps move material through the intestines. The digestible carbohydrates are the starches and sugars. Our bodies break them down into glucose, which is soluble in the bloodstream and can be transported into cells. In animal cells, glucose is used as fuel:

$$C_6H_{12}O_6(aq) + 6O_2(g) \longrightarrow 6CO_2(g) + 6H_2O(l)$$

The standard enthalpy of combustion of glucose is 22.8 MJ/mol (1 MJ = 10^6 J), which corresponds to a specific enthalpy of 16 kJ/g. Therefore, the oxidation of 1.0 g of glucose to carbon dioxide and water produces 16 kJ. That is enough energy to heat 1 L of water by about 4°C. We burn 1 g of glucose in about 1 min when bicycling, or in about 2 min when studying chemistry.

The average specific enthalpy of digestible carbohydrates, including starches, is actually a little higher, about 17 kJ/g. We often see specific enthalpies of food products

Regular exercise not only is good for the metabolism, it can be fun, too, when we make it a part of daily life.

reported in calories. The average specific enthalpy of carbohydrates is therefore about 4 kcal/g. In nutrition, food Calories are denoted Cal, with 1 Cal = 1 kcal. Be very careful with this distinction, or you might overeat a thousandfold! The average specific enthalpy of carbohydrates in food Calories is about 4 Cal/g.

The second major type of nutritional fuel is proteins. These complex compounds of carbon, hydrogen, oxygen, and nitrogen are made up of long chains of small units called amino acids. Proteins are large molecules, with molar masses that sometimes reach 10^6 g/mol. In the body, they are broken down into the individual amino acids and reassembled into new types of proteins the body needs to carry out the functions of a living cell and build muscle. However, proteins are also oxidized to form urea, $CO(NH_2)_2$. This oxidation corresponds to a specific

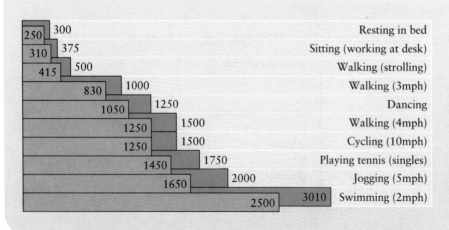

Energy (kJ/h)	Activity
250 | 300	Resting in bed
310 | 375	Sitting (working at desk)
415 | 500	Walking (strolling)
830 | 1000	Walking (3mph)
1050 | 1250	Dancing
1250 | 1500	Walking (4mph)
1250 | 1500	Cycling (10mph)
1450 | 1750	Playing tennis (singles)
1650 | 2000	Jogging (5mph)
2500 | 3010	Swimming (2mph)

Energy consumed (in kilojoules per hour) in typical activities: blue for a 70-kg male and pink for a 58-kg female.

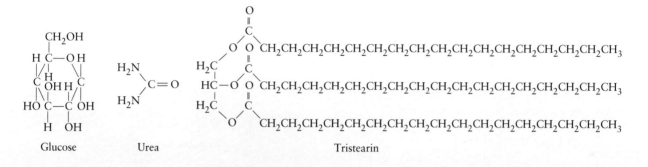

Glucose   Urea   Tristearin

enthalpy of about 17 kJ/g, which is about the same as that of carbohydrates.

Fats are the third major energy source. These compounds have long $-CH_2-CH_2-CH_2-$ chains, a characteristic they share with the hydrocarbons in gasoline. As a result, they behave in some ways like hydrocarbon fuels. One example is tristearin, $C_{57}H_{110}O_6$, a component of beef fat. Tristearin consists of three long hydrocarbon chains attached to a partially oxidized group. The specific enthalpy of combustion of tristearin is about 38 kJ/g, or 9 kcal/g. This value is about twice the specific enthalpy for carbohydrates and, in fact, is closer to the value for gasoline (48 kJ/g). Because fats can be regarded as only very slightly oxidized hydrocarbons, they release more heat during burning than do proteins and carbohydrates, which already contain many oxygen atoms.

Suppose we take in more fuel than our bodies can use. An automobile would simply not accept more than its tank can hold; but our bodies recognize that fuel may not always be available, so they store the extra fuel in the form of fat for future use. Animals, automobiles, and airplanes all need to maximize their efficiency by storing energy in the smallest mass possible. Body fat is the human equivalent of petroleum—energy stored compactly until needed.

The enthalpy values of typical foods are given in the table. The recommended daily consumption for 18- to 20-year-old males is 12 MJ, or about 2800 kcal; for females of the same age, it is 9 MJ, or about 2100 kcal. However, the amount of fuel required by an individual depends on fitness and activity levels. The more active we are, the more fuel we require. In fact, a very active person may need up to twice the enthalpy input that a sedentary person needs. The chart shows the average energy requirements for some common activities.

Beyond simply expending energy, physical activity tends to build muscle and decrease the amount of body fat. Because muscle tissue requires more energy for maintenance than fat tissue does, a physically fit and active person burns more fuel while just breathing than a sedentary person does, and is less likely to gain unwanted weight.

Key Concepts: enthalpy of combustion, specific enthalpy, energy unit conversions

For Further Reading

T. Smith, Chemistry, exercise, and weight control, *Today's Chemist*, February 1990, pp. 10–11, 21.

Related Exercises: 6.95–6.99

Thermochemical properties of some foods

| Food | Percentage composition | | | | Specific enthalpy, kJ/g |
	Water	Protein	Fat	Carbohydrate	
apples	84.3	0.3	0	11.9	2.5
beef	54.3	23.6	21.1	0	13.1
bread	39.0	7.8	1.7	49.7	12.6
cheese	37.0	26.0	33.5	0	17.0
cod	76.6	21.4	1.2	0	3.1
hamburger	40.9	15.8	14.2	29.1	17.3
milk	87.6	3.3	3.8	4.7	2.6
potatoes	80.5	1.4	0.1	19.7	3.5

Standard enthalpies of formation can be combined to obtain the standard enthalpy of any reaction.

Toolbox 6.2 *How to use standard enthalpies of formation*

We use this Toolbox to calculate the standard enthalpy for a reaction from tabulated standard enthalpies of formation.

Conceptual Basis

We imagine that we convert all the reactants into the elements in their standard states, then take those elements and form products. That is the hypothetical path followed when we use enthalpies of formation to calculate an enthalpy of reaction (Fig. 6.35). According to Hess's law (which is based on the first law of thermodynamics), the overall change in enthalpy for the desired reaction is the sum of the reaction enthalpies for the (possibly hypothetical) path.

Procedure

The procedure in this Toolbox is based on the expression

$$\Delta H_r^\circ = \sum n \Delta H_f^\circ(\text{products}) - \sum n \Delta H_f^\circ(\text{reactants})$$

for the standard reaction enthalpy. To use this expression,

Step 1. Write down the chemical equation for the reaction of interest.

Step 2. Add up the individual enthalpies of formation of the products. Each value must be multiplied by the corresponding stoichiometric coefficient. Remember that the standard enthalpy of formation of an element in its most stable form is 0.

Step 3. Calculate the total enthalpy of formation of the reactants in the same way.

Step 4. Subtract the "reactants" sum from the "products" sum.

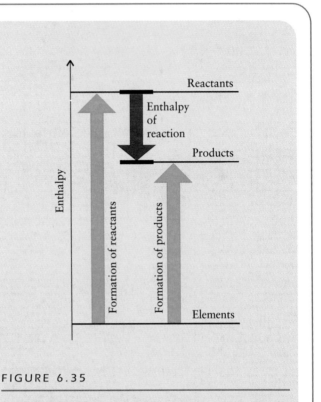

FIGURE 6.35

The reaction enthalpy can be constructed from enthalpies of formation by imagining the formation of both the reactants and the products from their respective elements. The reaction enthalpy is the difference between the two.

Example 6.10 *Using standard enthalpies of formation*

Use the information in Table 6.5 to predict the standard enthalpy of combustion of ethyne (acetylene), which is burned in welding torches with pure oxygen to generate a very hot flame. The products are carbon dioxide gas and liquid water.

Strategy All enthalpies of combustion are negative, because heat is released when a substance burns. To find the numerical value, work through the steps in Toolbox 6.2: calculate the standard reaction enthalpy for the thermochemical equation for the combustion and then convert that enthalpy change to the enthalpy of combustion of ethyne. Finally, report the standard enthalpy of combustion per mole of ethyne molecules.

Solution **Step 1.** The combustion reaction to consider is

$$2\,C_2H_2(g)\ +\ 5\,O_2(g) \longrightarrow 4\,CO_2(g)\ +\ 2\,H_2O(l)$$

Step 2. The total enthalpy of formation of the products is

$$\sum n\Delta H_f^\circ(\text{products})\ =\ 4\Delta H_f^\circ(CO_2, g)\ +\ 2\Delta H_f^\circ(H_2O, l)$$
$$=\ \{4\ \times\ (-393.51)\}\ +\ \{2\ \times\ (-285.83)\}\ \text{kJ/mol}$$
$$=\ -1574.04\ -\ 571.66\ \text{kJ/mol}\ =\ -2145.70\ \text{kJ/mol}$$

Step 3. The total enthalpy of formation of the reactants is

$$\sum n\Delta H_f^\circ(\text{reactants})\ =\ 2\Delta H_f^\circ(C_2H_2, g)\ +\ 5\Delta H_f^\circ(O_2, g)$$
$$=\ \{2\ \times\ 226.7\}3\ +\ \{5\ \times\ (0)\}\ \text{kJ/mol}\ =\ +453.46\ \text{kJ/mol}$$

Step 4. The difference between the two totals is

$$\Delta H_r^\circ\ =\ (-2145.70)\ -\ 453.46\ \text{kJ/mol}\ =\ -2599.16\ \text{kJ/mol}$$

The thermochemical equation for the combustion of ethyne is therefore

$$2\,C_2H_2(g)\ +\ 5\,O_2(g) \longrightarrow 4\,CO_2(g)\ +\ 2\,H_2O(l) \qquad \Delta H^\circ\ =\ -2599.16\ \text{kJ}$$

At this point, we divide by 2 mol to find the enthalpy of combustion per mole of ethyne molecules:

$$\Delta H_c^\circ\ =\ \frac{-2599.16\ \text{kJ}}{2\ \text{mol}\ C_2H_2}\ =\ -1299.58\ \text{kJ/mol}\ C_2H_2$$

Self-Test 6.13A Calculate the standard enthalpy of combustion of glucose from information in Table 6.5 and Appendix 2A. This result is discussed further in Applying Chemistry: Case Study 6.

[***Answer:*** -2808 kJ/mol]

Self-Test 6.13B You have an inspiration: maybe diamonds would make a great fuel! Calculate the standard enthalpy of combustion of diamonds from information in Appendix 2A.

Skills You Should Have Mastered

Conceptual

☐ 1. Distinguish the three types of thermodynamic systems, Self-Test 6.1.

☐ 2. Distinguish between heat and work, Section 6.3.

☐ 3. State, and explain the implications of, the first law of thermodynamics, Section 6.4.

☐ 4. Distinguish between ΔU and ΔH and show how they are related, Self-Test 6.4.

☐ 5. Distinguish between exothermic and endothermic reactions by the direction of heat flow and by the sign of ΔH, Section 6.7.

☐ 6. Outline the various parts of a heating curve and explain its features, Section 6.11.

☐ 7. Explain how Hess's law depends on the fact that enthalpy is a state property, Section 6.14.

Problem-Solving

☐ 1. Calculate the change in internal energy due to heat and work, Example 6.1.

☐ 2. Calculate the work of expansion, Example 6.2.

3. Use specific heat capacity to determine the energy transferred as heat, Example 6.3.

4. Calculate enthalpy changes from calorimetry data, Examples 6.4 and 6.6.

5. Determine the heat output of a reaction, given the temperature change of a calibrated calorimeter, Examples 6.5 and 6.7.

6. Calculate an overall reaction enthalpy from the enthalpies of the reactions in a sequence, Toolbox 6.1 and Example 6.8.

7. Calculate the heat output of a fuel from its standard enthalpy of combustion, Example 6.9.

8. Use standard enthalpies of formation to calculate the standard enthalpy of reaction, Toolbox 6.2 and Example 6.10.

Descriptive

1. Describe the function of calorimeters and how they are calibrated, Section 6.8.

2. Define the standard state of a substance and recognize when a substance is in its standard state, Section 6.13.

3. Define and use the specific enthalpy and the enthalpy density of fuels, Self-Test 6.12.

Exercises

Internal Energy, Heat, and Work

6.1 Identify the following systems as open, closed, or isolated: (a) coffee in a perfectly insulated thermos; (b) coolant in a refrigerator coil; (c) benzene burning in a bomb calorimeter.

6.2 Identify the following systems as open, closed, or isolated: (a) gasoline burning in an automobile engine; (b) mercury in a thermometer; (c) a living plant.

6.3 One tablespoon of a low-fat salad dressing is reported to have an "energy content" of 16 kcal. What is this value in kilojoules?

6.4 It requires about 100. kJ of energy to heat a cup of water from room temperature to 100°C. Express this energy requirement in kilocalories.

6.5 Each of the following pictures shows a molecular visualization of a system undergoing a change. In each case, indicate whether heat is absorbed or given off by the system, whether work is done on or by the system, and predict the signs of q and w for the process.

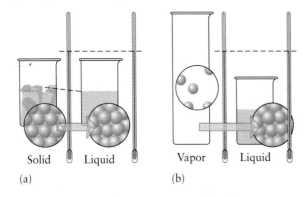

Solid   Liquid   Vapor   Liquid

(a)   (b)

6.6 Each of the following pictures shows a molecular visualization of a system undergoing a change. In each case, indicate whether heat is absorbed or given off by the system, whether expansion work is done on or by the system, and predict the signs of q and w for the process.

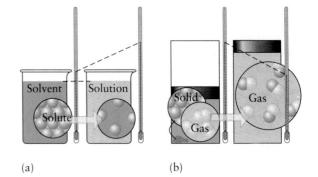

Solvent   Solution   Solid   Gas

Solute   Gas

(a)   (b)

6.7 A gas sample is heated in a cylinder, using 550. kJ of heat. A piston compresses the gas, doing 700. kJ of work. What is the change in internal energy of the gas during this process?

6.8 Calculate the work for a system that absorbs 150. kJ of heat in a process for which the increase in internal energy is 120. kJ. Is work done on or by the system during this process?

6.9 A 100.-W electric heater (1 W = 1 J/s) operates for 20. min to heat the gas in a cylinder. The gas expands from 2.0 L to 2.5 L against an atmospheric pressure of 1.0 atm. What is the change in internal energy of the gas? (1 L·atm = 101.325 J.)

6.10 A gas in a cylinder was placed in a heater and gained 7000. J of heat. If the cylinder increased in volume from 700. mL to 1450. mL against an atmospheric pressure of 750. Torr during this process, what is the change in internal energy of the gas in the cylinder? (1 L·atm = 101.325 J.)

6.11 The change in internal energy for the combustion of 1 mol CH_4 in a cylinder according to the reaction $CH_4(g) + 2 O_2(g) \rightarrow CO_2(g) + 2 H_2O(g)$ is -892.4 kJ. If a piston connected to the cylinder performs 492 kJ of expansion work as a result of the combustion, how much heat is lost from the system (the reaction mixture)?

6.12 In a combustion cylinder, the total internal energy change produced from the burning of a fuel is $-1740.$ kJ. The cooling system that surrounds the cylinder absorbs 470. kJ as heat. How much work can be done by the fuel in the cylinder?

6.13 How does an increase in temperature affect the motion of the molecules in a system?

6.14 If energy is absorbed by a system as heat, what happens to the molecular motion of the surroundings?

Enthalpy Change

6.15 For a certain reaction at constant pressure, $\Delta U = -65$ kJ and 28 kJ of expansion work is done by the system. What is ΔH for this process?

6.16 For a certain reaction at constant pressure, $\Delta H = -15$ kJ and 22 kJ of expansion work is done on the system. What is ΔU for this process?

6.17 Classify the following processes as exothermic or endothermic: (a) the formation of ethyne, which is used for oxyacetylene welding:

$$2\, C(s) + H_2(g) \longrightarrow C_2H_2(g) \quad \Delta H° = +227 \text{ kJ}$$

(b) the freezing of water:

$$H_2O(l) \longrightarrow H_2O(s) \quad \Delta H° = -6.0 \text{ kJ}$$

6.18 State whether the temperature will rise or fall when the following reactions are carried out in an insulated calorimeter: (a) the reaction used in industry to produce carbon disulfide from natural gas:

$$CH_4(g) + 4\, S(s) \longrightarrow CS_2(l) + 2\, H_2S(g) \quad \Delta H° = -106 \text{ kJ}$$

(b) the dissolution of table salt in water:

$$NaCl(s) \longrightarrow NaCl(aq) \quad \Delta H° = +3.9 \text{ kJ}$$

6.19 What is the enthalpy change when 20.0 g of water is cooled from $20.0°C$ to $4.00°C$? Is the process endothermic or exothermic?

6.20 What is the enthalpy change when 50.3 g of copper is heated from $10.°C$ to $150.°C$? Is the process endothermic or exothermic?

Measuring Heat

6.21 (a) Near room temperature, the specific heat capacity of benzene is 1.05 J/°C·g. Calculate the heat needed to raise the temperature of 50.0 g of benzene from $25.3°C$ to $37.2°C$. (b) A 1.0-kg block of aluminum is supplied with 490. kJ of heat. What is the temperature change of the aluminum? The specific heat capacity of aluminum is 0.90 J/°C·g.

6.22 (a) Near room temperature, the specific heat capacity of ethanol is 2.42 J/°C·g. Calculate the heat that must be removed to reduce the temperature of 150.0 g of C_2H_5OH from $50.0°C$ to $16.6°C$. (b) What mass of copper can be heated from $10.°C$ to $200.°C$ when 400. kJ of energy is available?

6.23 The specific heat capacity of water is 4.18 J/°C·g and that of stainless steel is 0.51 J/°C·g. Calculate the heat that must be supplied to a 500.0-g stainless steel vessel containing 450.0 g of water to raise its temperature from $25°C$ to the boiling point of water, $100.°C$. What percentage of the heat is used to raise the temperature of the water?

6.24 The specific heat capacity of water is 4.18 J/°C·g and that of copper is 0.38 J/°C·g. Calculate the heat that must be supplied to a 500.0-g copper kettle containing 450.0 g of water to raise its temperature from $25°C$ to the boiling point of water, $100.°C$. What percentage of the heat is used to raise the temperature of the water? Compare these answers with those of Exercise 6.23.

6.25 (a) Calculate the energy needed to raise the temperature of 10.0 g of a certain type of steel (of specific heat capacity 0.45 J/°C·g) from $25°C$ to $500.°C$. (b) What mass of gold (of specific heat capacity 0.13 J/°C·g) can be heated through the same temperature difference when supplied with the same amount of energy as in (a)?

6.26 (a) How much energy must be removed to lower the temperature of a 25.0-g block of ice (of specific heat capacity 2.03 J/°C·g) from $-12°C$ to $-28°C$? (b) If the same energy were removed from 25.0 g of steam (of specific heat capacity 2.01 J/°C·g), what would be the temperature change?

6.27 A piece of metal of mass 20.0 g at $100.0°C$ is placed in a Styrofoam cup calorimeter containing 50.7 g of water at $22.0°C$. The final temperature of the mixture is $25.7°C$. What is the specific heat capacity of the metal? Assume that all the energy lost by the metal is gained by the water.

6.28 A piece of copper of mass 20.0 g at $100.°C$ is placed in a Styrofoam cup calorimeter containing 50.7 g of water at $22.0°C$. Calculate the final temperature of the water. Assume that all the energy lost by the copper is gained by the water. The specific heat capacity of copper is 0.38 J/°C·g.

6.29 The enthalpy of combustion of benzoic acid, C_6H_5COOH, which is often used to calibrate calorimeters, is -3227 kJ/mol. When 1.236 g of benzoic acid was burned in a calorimeter, the temperature increased by $2.345°C$. What is the heat capacity of the calorimeter?

6.30 A bomb calorimeter was calibrated with an electric heater. It was used to supply 22.5 kJ of energy to the calorimeter, which increased the temperature of the calorimeter, including its water bath, from $22.45°C$ to $23.97°C$. What is the heat capacity of the calorimeter?

6.31 The heat capacity of a calorimeter was measured as 5.24 kJ/°C. With the combustion of a piece of asparagus, the temperature rose from $22.45°C$ to $23.17°C$. What is the enthalpy change for the combustion?

6.32 A calorimeter has a measured heat capacity of 6.27 kJ/°C. The combustion of 1.84 g of magnesium caused the temperature to rise from $21.30°C$ to $28.56°C$. (a) Calculate the enthalpy change of the reaction. (b) What is the enthalpy change for the combustion in kilojoules per mole of magnesium atoms?

6.33 50.0 mL of 0.500 M NaOH(aq) and 50.0 mL of 0.500 M HNO$_3$(aq), both initially at 18.6°C, were mixed and stirred in a calorimeter having a heat capacity equal to 525.0 J/°C when containing 100.0 mL water. The temperature of the mixture rose to 21.3°C. (a) What is the change in enthalpy for the neutralization reaction? (b) What is the change in enthalpy for the neutralization in kilojoules per mole of HNO$_3$?

6.34 The heat capacity of a certain calorimeter is 488.1 J/°C when containing 50.0 mL water. When 25.0 mL of 0.700 M NaOH(aq) was mixed in that calorimeter with 25.0 mL of 0.700 M HCl(aq), both initially at 20.0°C, the temperature increased to 22.1°C. Calculate the enthalpy of neutralization in kilojoules per mole of HCl.

Enthalpy of Physical Change

6.35 (a) The vaporization of 0.235 mol CH$_4$(l) requires 1.93 kJ of heat. What is the enthalpy of vaporization of methane? (b) An electric heater was immersed in a flask of ethanol, C$_2$H$_5$OH, and 22.45 g of ethanol was vaporized when 21.2 kJ of energy was supplied. What is the enthalpy of vaporization of ethanol?

6.36 (a) When 25.23 g of methanol, CH$_3$OH, froze, 4.01 kJ of heat was released. What is the enthalpy of fusion of methanol? (b) A sample of benzene was vaporized at a reduced pressure at 25°C. When 37.5 kJ of heat was supplied, 95 g of the liquid benzene vaporized. What is the enthalpy of vaporization of benzene at 25°C?

6.37 Use the information in Table 6.2 to calculate the enthalpy change for (a) the vaporization of 100.0 g of water at 373.2 K; (b) the melting of 600. g of solid ammonia at its freezing point (195.4 K).

6.38 Use the information in Table 6.2 to calculate the enthalpy change that occurs when (a) 200. g of methanol condenses at its boiling point (337.2 K); (b) 17.7 g of acetone, CH$_3$COCH$_3$, freezes at its freezing point (177.8 K).

6.39 How much heat is needed to melt 50.0 g of ice at 0°C and then heat the water to 25°C? See Tables 6.1 and 6.2.

6.40 If we start with 150. g of water at 30.°C, how much heat must we add to convert all the liquid to steam at 100.°C? See Tables 6.1 and 6.2.

6.41 Use the following information to construct a heating curve for bromine from −7.2°C to 70.0°C. The molar heat capacity of liquid bromine is 75.69 J/K·mol and that of bromine vapor is 36.02 J/K·mol. The enthalpy of vaporization of liquid bromine is 30.91 kJ/mol. Bromine melts at −7.2°C and boils at 58.78°C.

6.42 A 20.0-g ice cube at −14°C was heated until it became steam at 110.°C. (a) Using values from Table 6.2 and the specific heat capacities of water, ice (2.03 J/°C·g), and water vapor (2.01 J/°C·g), determine the total heat that must have been supplied. (b) Draw a heating curve for the process.

Enthalpy of Chemical Change

6.43 What is meant by the *standard state* of a substance?

6.44 Define the term *standard reaction enthalpy*. Identify two methods by which it can be determined without actually carrying out the reaction.

6.45 Carbon disulfide can be prepared from coke (an impure form of carbon) and elemental sulfur:

$$4\,C(s) + S_8(s) \longrightarrow 4\,CS_2(l) \quad \Delta H° = +358.8 \text{ kJ}$$

(a) How much heat is absorbed in the reaction of 0.20 mol S$_8$? (b) Calculate the heat absorbed in the reaction of 20.0 g of carbon with an excess of sulfur. (c) If the heat absorbed in the reaction was 217 kJ, how much CS$_2$ was produced?

6.46 The oxidation of nitrogen in the hot exhaust of jet engines and automobiles occurs by the reaction

$$N_2(g) + O_2(g) \longrightarrow 2\,NO(g) \quad \Delta H° = +180.5 \text{ kJ}$$

(a) How much heat is absorbed in the formation of 0.70 mol NO? (b) How much heat is absorbed in the oxidation of 17.4 L of nitrogen measured at 1.00 atm and 273 K? (c) When the oxidation of N$_2$ to NO was completed in a bomb calorimeter, the heat absorbed was measured as 790. J. What mass of nitrogen gas was oxidized?

6.47 The combustion of octane is expressed by the thermochemical equation

$$2\,C_8H_{18}(l) + 25\,O_2(g) \longrightarrow 16\,CO_2(g) + 18\,H_2O(l)$$
$$\Delta H° = -10\,942 \text{ kJ}$$

(a) Calculate the mass of octane that must be burned to produce 12 MJ of heat. (b) How much heat will be evolved from the combustion of 1.0 gal of gasoline (assumed to be exclusively octane)? The density of octane is 0.70 g/mL.

6.48 Suppose that coal, of density 1.5 g/cm^3, is carbon (it is, in fact, a much more complex substance, but this is a reasonable first approximation). The combustion of carbon is described by the equation

$$C(s) + O_2(g) \longrightarrow CO_2(g) \quad \Delta H° = -394 \text{ kJ}$$

(a) Calculate the heat produced when a lump of coal of size 7.0 cm × 6.0 cm × 5.0 cm is burned. (b) Estimate the mass of water that can be heated from 15°C to 100.°C with this piece of coal.

Hess's Law

6.49 Two successive stages in the industrial manufacture of sulfuric acid are the combustion of sulfur and the oxidation of sulfur dioxide to sulfur trioxide. From the standard reaction enthalpies

$$S(s) + O_2(g) \longrightarrow SO_2(g) \quad \Delta H° = -296.83 \text{ kJ}$$

$$2 S(s) + 3 O_2(g) \longrightarrow 2 SO_3(g) \quad \Delta H° = -791.44 \text{ kJ}$$

calculate the reaction enthalpy for the oxidation of sulfur dioxide to sulfur trioxide in the reaction

$$2 SO_2(g) + O_2(g) \longrightarrow 2 SO_3(g)$$

6.50 In the manufacture of nitric acid by the oxidation of ammonia, the first product is nitric oxide, which is then oxidized to nitrogen dioxide. From the standard reaction enthalpies

$$N_2(g) + O_2(g) \longrightarrow 2 NO(g) \quad \Delta H° = +180.5 \text{ kJ}$$

$$N_2(g) + 2 O_2(g) \longrightarrow 2 NO_2(g) \quad \Delta H° = +66.4 \text{ kJ}$$

calculate the standard reaction enthalpy for the oxidation of nitric oxide to nitrogen dioxide:

$$2 NO(g) + O_2(g) \longrightarrow 2 NO_2(g)$$

6.51 Calculate the standard enthalpy of the reaction

$$P_4(s) + 10 Cl_2(g) \longrightarrow 4 PCl_5(s)$$

from the reactions

$$P_4(s) + 6 Cl_2(g) \longrightarrow 4 PCl_3(l) \quad \Delta H° = -1278.8 \text{ kJ}$$

$$PCl_3(l) + Cl_2(g) \longrightarrow PCl_5(s) \quad \Delta H° = -124 \text{ kJ}$$

6.52 Calculate the standard reaction enthalpy for the reduction of hydrazine to ammonia

$$N_2H_4(l) + H_2(g) \longrightarrow 2 NH_3(g)$$

from the following data:

$$N_2(g) + 2 H_2(g) \longrightarrow N_2H_4(l) \quad \Delta H° = +50.63 \text{ kJ}$$

$$N_2(g) + 3 H_2(g) \longrightarrow 2 NH_3(g) \quad \Delta H° = -92.22 \text{ kJ}$$

6.53 Determine the standard reaction enthalpy for the hydrogenation of ethyne to ethane,

$$C_2H_2(g) + 2 H_2(g) \longrightarrow C_2H_6(g)$$

from the following data:

$$2 C_2H_2(g) + 5 O_2(g) \longrightarrow 4 CO_2(g) + 2 H_2O(l)$$
$$\Delta H° = -2600. \text{ kJ}$$

$$2 C_2H_6(g) + 7 O_2(g) \longrightarrow 4 CO_2(g) + 6 H_2O(l)$$
$$\Delta H° = -3120. \text{ kJ}$$

$$H_2(g) + \tfrac{1}{2} O_2(g) \longrightarrow H_2O(l) \quad \Delta H° = -286 \text{kJ}$$

6.54 Determine the standard reaction enthalpy for the partial combustion of methane to carbon monoxide:

$$2 CH_4(g) + 3 O_2(g) \longrightarrow 2 CO(g) + 4 H_2O(l)$$

Use the following data:

$$CH_4(g) + 2 O_2(g) \longrightarrow CO_2(g) + 2 H_2O(l) \quad \Delta H° = -890. \text{ kJ}$$

$$2 CO(g) + O_2(g) \longrightarrow 2 CO_2(g) \quad \Delta H° = -566.0 \text{ kJ}$$

6.55 Calculate the standard reaction enthalpy for the synthesis of hydrogen chloride gas

$$H_2(g) + Cl_2(g) \longrightarrow 2 HCl(g)$$

from the following data:

$$NH_3(g) + HCl(g) \longrightarrow NH_4Cl(s) \quad \Delta H° = -176.0 \text{ kJ}$$

$$N_2(g) + 3 H_2(g) \longrightarrow 2 NH_3(g) \quad \Delta H° = -92.22 \text{ kJ}$$

$$N_2(g) + 4 H_2(g) + Cl_2(g) \longrightarrow 2 NH_4Cl(s)$$
$$\Delta H° = -628.86 \text{ kJ}$$

6.56 Calculate the standard reaction enthalpy for the formation of anhydrous aluminum chloride

$$2 Al(s) + 3 Cl_2(g) \longrightarrow 2 AlCl_3(s)$$

from the following data:

$$2 Al(s) + 6 HCl(aq) \longrightarrow 2 AlCl_3(aq) + 3 H_2(g)$$
$$\Delta H° = -1049 \text{ kJ}$$

$$HCl(g) \longrightarrow HCl(aq) \quad \Delta H° = -74.8 \text{ kJ}$$

$$H_2(g) + Cl_2(g) \longrightarrow 2 HCl(g) \quad \Delta H° = -1845 \text{ kJ}$$

$$AlCl_3(s) \longrightarrow AlCl_3(aq) \quad \Delta H° = -323 \text{ kJ}$$

6.57 The standard enthalpies of combustion of graphite and diamond are -393.51 and -395.41 kJ/mol, respectively. Calculate the change in enthalpy for the graphite $\rightarrow$ diamond transition.

6.58 Elemental sulfur occurs in several forms, with rhombic sulfur the most stable under normal conditions and monoclinic sulfur slightly less stable. The standard enthalpies of combustion of the two forms to sulfur dioxide are -296.83 and -297.16 kJ/mol, respectively. Calculate the change in enthalpy for the rhombic $\rightarrow$ monoclinic transition.

Heat Output

6.59 A minor component of gasoline is heptane (C_7H_{16}), which has a standard enthalpy of combustion of -4854 kJ/mol and a density of 0.68 g/mL. Calculate the specific enthalpy of heptane and its enthalpy density.

6.60 Another minor component of gasoline is toluene (C_7H_8), with a standard enthalpy of combustion of -3910 kJ/mol and a density of 0.867 g/mL. Calculate the specific enthalpy of toluene and its enthalpy density.

6.61 Calculate the specific enthalpy of magnesium from its enthalpy of combustion to magnesium oxide. Use the standard enthalpy of formation in Appendix 2A. Would aluminum, which burns to aluminum oxide, be a better fuel if specific enthalpy were the only consideration?

6.62 Calculate the specific enthalpy of phosphorus from its enthalpy of combustion in oxygen to P_4O_{10}. Use the standard enthalpy of formation in Appendix 2A. Would sulfur burned to either SO_2 or SO_3 be a more efficient fuel than phosphorus if specific enthalpy were the only consideration?

Enthalpy of Formation

6.63 Write the thermochemical equations that give the values of the standard enthalpies of formation of (a) $KClO_3(s)$; (b) $H_2NCH_2COOH(s)$, glycine; (c) $Al_2O_3(s)$, alumina.

6.64 Write the thermochemical equations that give the values of the standard enthalpies of formation of (a) $CH_3COOH(l)$; (b) $SO_3(g)$; (c) $CO_2(g)$.

6.65 Use the standard enthalpies of formation in Appendix 2A to calculate the enthalpy of reaction of (a) the oxidation of 10.0 g of sulfur dioxide: $2\,SO_2(g) + O_2(g) \rightarrow 2\,SO_3(g)$; (b) the reduction of 1.00 mol $CuO(s)$ with hydrogen: $CuO(s) + H_2(g) \rightarrow Cu(s) + H_2O(l)$.

6.66 Use the standard enthalpies of formation from Appendix 2A to determine the enthalpy of reaction of (a) the hydrogenation of 50.0 g of benzene to cyclohexane: $C_6H_6(l) + 3\,H_2(g) \rightarrow C_6H_{12}(l)$; (b) the hydrogenation of 50.0 g of ethene to ethane: $C_2H_4(g) + H_2(g) \rightarrow C_2H_6(g)$.

6.67 Use the standard enthalpies of formation in Appendix 2A to calculate the standard enthalpy of the following reactions: (a) the replacement of deuterium by ordinary hydrogen in heavy water: $H_2(g) + D_2O(l) \rightarrow H_2O(l) + D_2(g)$; (b) the removal of hydrogen sulfide from natural gas: $2\,H_2S(g) + SO_2(g) \rightarrow 3\,S(s) + 2\,H_2O(l)$; (c) the oxidation of ammonia: $4\,NH_3(g) + 5\,O_2(g) \rightarrow 4\,NO(g) + 6\,H_2O(g)$.

6.68 Using standard enthalpies of formation from Appendix 2A, calculate the standard reaction enthalpy for each reaction: (a) the final stage in the production of nitric acid, when nitrogen dioxide dissolves in and reacts with water: $3\,NO_2(g) + H_2O(l) \rightarrow 2\,HNO_3(aq) + NO(g)$; (b) the formation of boron trifluoride, which is widely used in the chemical industry: $B_2O_3(s) + 3\,CaF_2(s) \rightarrow 2\,BF_3(g) + 3\,CaO(s)$; (c) the formation of a sulfide by the action of hydrogen sulfide on an aqueous solution of a base: $H_2S(aq) + 2\,KOH(aq) \rightarrow K_2S(aq) + 2\,H_2O(l)$.

6.69 Calculate the standard enthalpy of formation of dinitrogen pentoxide from the data

$$2\,NO(g) + O_2(g) \longrightarrow 2\,NO_2(g) \quad \Delta H° = -114.1 \text{ kJ}$$

$$4\,NO_2(g) + O_2(g) \longrightarrow 2\,N_2O_5(g) \quad \Delta H° = -110.2 \text{ kJ}$$

and the standard enthalpy of formation of nitric oxide.

6.70 An important reaction that occurs in the atmosphere is $NO_2(g) \rightarrow NO(g) + O(g)$, which is brought about by sunlight. Calculate the standard enthalpy of the reaction from the data

$$O_2(g) \longrightarrow 2\,O(g) \quad \Delta H° = +498.4 \text{ kJ}$$

$$NO(g) + O_3(g) \longrightarrow NO_2(g) + O_2(g) \quad \Delta H° = -200 \text{ kJ}$$

and additional information from Appendix 2A.

6.71 Calculate the standard enthalpy of formation of $PCl_5(s)$ from the enthalpy of formation of $PCl_3(l)$ (Appendix 2A) and

$$PCl_3(l) + Cl_2(g) \longrightarrow PCl_5(s) \quad \Delta H° = -124 \text{ kJ}$$

6.72 When 1.92 g of magnesium reacts with nitrogen to form magnesium nitride, the heat evolved is 12.2 kJ. Calculate the standard enthalpy of formation of Mg_3N_2.

Supplementary Exercises

6.73 (a) Describe three ways in which you could increase the internal energy of an open system. (b) Which of these methods could you use to increase the internal energy of a closed system? (c) Which, if any, of these methods could you use to increase the internal energy of an isolated system?

6.74 The internal energy of a system increased by 400. J when it absorbed 600. J of heat. (a) Was work done by or on the system? (b) How much work was done?

6.75 (a) Distinguish between ΔU and ΔH for a reaction. (b) Under what circumstances are ΔH and ΔU equal?

6.76 Classify the following properties as intensive or extensive: (a) heat capacity; (b) specific heat capacity; (c) enthalpy; (d) standard enthalpy of formation.

6.77 A process in which no heat is exchanged between the system and surroundings is called an *adiabatic process*. In an adiabatic process, the system can still exchange energy with the surroundings, but only by means of work. Indicate whether each statement about an adiabatic process in a closed system is always true, always false, or true in certain conditions (specify the conditions): (a) $\Delta U = 0$; (b) $q = 0$; (c) q is negative; (d) $\Delta U = q$; (e) $\Delta U = w$.

6.78 A system undergoes a two-step process. In Step 1, it absorbs 50. J of heat at constant volume. In Step 2, it gives off 5 J of heat at a constant pressure of 1 atm as it is returned to its original state. Using the relation 1 L·atm = 101.325 J, find the change in volume of the system during the second step and identify it as an expansion or compression.

6.79 (a) Using the relation 1 L·atm = 101.325 J, calculate the work that must be done at standard temperature and pressure against the atmosphere for the expansion of the gaseous products in the combustion of 2.00 mol $C_6H_6(l)$ in the reaction $2\,C_6H_6(l) + 15\,O_2(g) \rightarrow 12\,CO_2(g) + 6\,H_2O(g)$. (b) Using data in Appendix 2A, calculate the standard enthalpy of the reaction. (c) Calculate the change in internal energy, $\Delta U°$, of the system.

6.80 Five students in a chemistry class determined the specific heat capacity of a metal by the method in Example 6.4. In the process, each made an error. Predict whether the value of the specific heat capacity calculated by each student will be unaffected, too high, too low, or indeterminate (could be either too high or too low). (a) One student used a thermometer that consistently read a temperature 2°C too high to measure the initial temperature of the metal. (b) One added 45 g of water instead of 50 g. (c) One recorded the mass of the metal as greater than it actually was. (d) One used a piece of the wrong metal. (e) The Styrofoam cups were in short supply, so one student used a glass beaker, from which some heat was lost to the surroundings.

6.81 The heat capacity of a certain calorimeter is 8.92 kJ/°C. The combustion of 30. g of cheese produces 460. kJ. What mass of cheese will produce a temperature change in the calorimeter of 2.37°C?

6.82 A 100.-g serving of shrimp provides 91 kcal. If a 20.0-g sample is burned in a calorimeter with a heat capacity of 4.66 kJ/°C, what is the expected temperature change?

6.83 A 50.0-g ice cube at 0°C is added to a glass containing 400. g of water at 45°C. What is the final temperature of the system? Assume that no heat is lost to the surroundings.

6.84 The following reaction can be used for the production of manganese: $3 MnO_2(s) + 4 Al(s) \rightarrow 2 Al_2O_3(s) + 3 Mn(s)$. (a) Use the information in Appendix 2A and the enthalpy of formation of $MnO_2(s)$, which is -521 kJ/mol, to calculate the standard enthalpy of reaction. (b) What is the enthalpy change in the production of 10.0 g of manganese?

6.85 When 3.245 g of lead(IV) oxide is formed from lead metal and oxygen, 3.76 kJ of heat is released. What is the standard enthalpy of formation of $PbO_2(s)$?

6.86 The standard enthalpy of combustion of sulfur to sulfur dioxide is -2374.6 kJ/(mol S_8). What is the standard enthalpy of formation of $SO_2(g)$?

6.87 Calculate the standard reaction enthalpy of the reduction of iron(II) oxide by carbon monoxide, a step in the production of iron, $FeO(s) + CO(g) \rightarrow Fe(s) + CO_2(g)$, given the following thermochemical equations:

$$3 Fe_2O_3(s) + CO(g) \longrightarrow 2 Fe_3O_4(s) + CO_2(g)$$
$$\Delta H° = -47.2 \text{ kJ}$$

$$Fe_2O_3(s) + 3 CO(g) \longrightarrow 2 Fe(s) + 3 CO_2(g)$$
$$\Delta H° = -24.7 \text{ kJ}$$

$$Fe_3O_4(s) + CO(g) \longrightarrow 3 FeO(s) + CO_2(g)$$
$$\Delta H° = +35.9 \text{ kJ}$$

6.88 Calculate the enthalpy of formation of ethyne, $2 C(s) + H_2(g) \rightarrow C_2H_2(g)$, from the following information:

$$2 C_2H_2(g) + 5 O_2(g) \longrightarrow 4 CO_2(g) + 2 H_2O(l)$$
$$\Delta H° = -2600. \text{ kJ}$$

$$C(s) + O_2(g) \longrightarrow CO_2(g) \quad \Delta H° = -394 \text{ kJ}$$

$$2 H_2(g) + O_2(g) \longrightarrow 2 H_2O(g) \quad \Delta H° = -483.6 \text{ kJ}$$

$$H_2O(l) \longrightarrow H_2O(g) \quad \Delta H° = +44 \text{ kJ}$$

6.89 Calculate the standard reaction enthalpy for the hydrogenation of ethyne to ethene: $C_2H_2(g) + H_2(g) \rightarrow C_2H_4(g)$. This reaction cannot be performed easily in the laboratory because of the formation of many by-products in the reaction. Use the following data:

$$2 C_2H_2(g) + 5 O_2(g) \longrightarrow 4 CO_2(g) + 2 H_2O(l)$$
$$\Delta H° = -2600. \text{ kJ}$$

$$2 C_2H_4(g) + 6 O_2(g) \longrightarrow 4 CO_2(g) + 4 H_2O(l)$$
$$\Delta H° = -2822 \text{ kJ}$$

$$2 H_2(g) + O_2(g) \longrightarrow 2 H_2O(l) \quad \Delta H° = -572 \text{ kJ}$$

6.90 Why is the heat of formation of gaseous water less negative than that of liquid water?

6.91 Use the information in Appendix 2A to determine how much heat is evolved in the dissolution of 20.0 g of sodium hydroxide. The process is $NaOH(s) \rightarrow Na^+(aq) + OH^-(aq)$.

6.92 Consider the reaction of iron(II) sulfide with hydrochloric acid, $FeS(s) + 2 H^+(aq) \rightarrow Fe^{2+}(aq) + H_2S(g)$. Use the information in Appendix 2A to calculate the enthalpy change for the production of 30.0 L of hydrogen sulfide at 1.00 atm and 298 K.

6.93 Suppose that, of the heat produced by the combustion of 100. mg of octane, C_8H_{18}, only 70.0% is useful in heating. What will be the resulting temperature change if the fuel is used to heat 250.0 g of ethanol?

6.94 During a heavy rainstorm, 2.13×10^9 L of rain fell. If the density of the rainwater is 1.00 g/cm^3, how much heat is released when this quantity of water condenses from water vapor to rain?

Applied Exercises

For Exercises 6.95–6.99, see Applying Chemistry: Case Study 6.

6.95 A person of average mass burns about 30 kJ per minute playing tennis. Find the time you would have to spend playing tennis to burn up the energy in a 2-oz serving of cheese. Use the table in Applying Chemistry: Case Study 6.

6.96 A premium ice cream contains 16 g of fat and 20. g of sugar per serving (4 oz). A brand of frozen yogurt contains 4 g of fat and 32 g of sugar per serving. Calculate the fuel value of a serving of each item. Both contain 5 g of protein.

6.97 A 60.-kg female student rides a bicycle to class every day, a 10.-mile round trip that takes 30. min in each direction. The same round trip in an automobile would require 0.40 gallons of gasoline. Assume that the student goes to class 150 days per year and that the enthalpy of combustion of gasoline can be approximated by that of octane, which has a density of 0.702 g/cm^3 (3.785 L = 1.000 gal). What is the yearly energy requirement of this journey by (a) bicycle or (b) automobile? Use the information in the table in Applying Chemistry: Case Study 6.

6.98 Some of the energy released by the metabolism of food in our body is stored in the form of combustible fats. Suppose that a "miracle" diet drug is created that causes the body to use all the energy not required to do work to heat the body instead of storing the energy as fat. Suppose a 70.-kg person taking this drug were to eat a meal that provides 4000. kJ but used only 3000. kJ of that energy to do work in the following few hours. By how much would the person's temperature rise in (a) degrees Celsius? (b) degrees Fahrenheit? Assume that the specific heat capacity of the person is 4.2 J/K·g and that none of this additional energy is lost as heat to the surroundings.

6.99 The ABC cereal company is developing a new type of breakfast cereal to compete with a rival product (Brand X). You are asked to compare the energy content of the two cereals to see if the new ABC product is lower in calories, so you burn 1.00-g samples of each of the two cereals in oxygen in a bomb calorimeter with a heat capacity of 600. J/°C. When the Brand X cereal sample burned, the temperature rose from 300.2 K to 309.0 K. When the ABC cereal sample burned, the temperature rose from 299.0 K to 307.5 K. (a) What is the heat output of each sample? (b) One serving of each cereal is 30.0 g. How would you label the packages of the two cereals to indicate the fuel value per 100.-g serving in kilojoules? (c) in nutritional Calories (kilocalories)?

6.100 A solar heat storage system consists of 60 sealed pipes. Each pipe is loaded with 45.6 kg $CaCl_2 \cdot 6H_2O$. Heat from the Sun is absorbed by the salt and stored when the salt melts at 30°C to form an aqueous solution of $CaCl_2$. At night, the heat is released when the salt recrystallizes. How many liters of water can be heated to 25°C by the recrystallization in this heat storage system if the water temperature is initially at 15°C? Assume the density of water to be 1.00 g/mL and $\Delta H_{fus}(CaCl_2 \cdot 6H_2O)$ to be 27 kJ/mol.

6.101 The enthalpy of formation of trinitrotoluene (TNT) is −67 kJ/mol, and the density of TNT is 1.65 g/cm³. In principle, it could be used as a rocket fuel, with the gases resulting from its decomposition streaming out of the rocket to give the required thrust. In practice, of course, it would be extremely dangerous as a fuel because it is sensitive to shock. Explore its potential as a rocket fuel by calculating its enthalpy density for the reaction

$$4\,C_7H_5N_3O_6(s) + 21\,O_2(g) \longrightarrow$$
$$28\,CO_2(g) + 10\,H_2O(g) + 6\,N_2(g)$$

6.102 One problem with fuels containing carbon is that they produce carbon dioxide when they burn, so one consideration governing the selection of a fuel could be the heat generated per mole of CO_2 produced. (a) Calculate this quantity for methane and octane. (b) Calculate the heat produced per mole of CO_2 from the combustion of glucose. (c) Which process produces more carbon dioxide in the environment for each kilojoule generated, eating (resulting in the combustion of glucose) or burning octane?

Integrated Exercises

6.103 An experimental automobile burns hydrogen for fuel. At the beginning of a test drive, the 30.0-L tank was filled with hydrogen at 16.0 atm and 298 K. At the end of the drive, the temperature of the tank was still 298 K, but its pressure was 4.0 atm. (a) How many moles of H_2 were burned during the drive? (b) How much heat, in kilojoules, was given off by the combustion of that amount of hydrogen?

6.104 Calculate the heat evolved from a reaction mixture of 13.4 L of sulfur dioxide at 1.00 atm and 273 K and 15.0 g oxygen in the reaction

$$2\,SO_2(g) + O_2(g) \longrightarrow 2\,SO_3(g) \quad \Delta H° = -198 \text{ kJ}$$

6.105 A "silver tree" can be made in the laboratory by cutting a piece of copper metal in the shape of a tree and placing it into a silver nitrate solution, when the reaction deposits sparkling silver crystals on the tree and forms copper(II) ions in solution. (a) Write the balanced chemical equation for the reaction and identify the type of reaction. (b) Use the information in Appendix 2A to find the change in enthalpy for the formation of 1.88 g of silver. (c) Is the reaction an endothermic or exothermic process?

Fractal-like "branches" of silver being deposited on copper.

6.106 45.00 mL of 0.010 M HCl(aq) was mixed with 65.00 mL of 0.010 M $Ba(OH)_2$(aq) in a Styrofoam cup calorimeter at 26.21°C and allowed to react. (a) Write a balanced chemical equation for the reaction and identify the reaction type. (b) Which reactant is in excess? (c) Assuming that the solutions have the same specific heat capacity as water, calculate the final temperature of the mixture after reaction has gone to completion. The reaction enthalpy is −55.84 kJ/mol H_3O^+.

6.107 A reaction being studied to regenerate oxygen from carbon dioxide on long spaceflights (see Applying Chemistry: Case Study 3) takes place in the following two steps: $CO_2(g) + 2\,H_2(g) \rightarrow C(s) + 2\,H_2O(l)$, followed by $2\,H_2O(l) \rightarrow 2\,H_2(g) + O_2(g)$. (a) Write the thermochemical equation for the overall reaction. (b) What enthalpy change would be associated with the production of 45.0 L of oxygen at 0.80 atm and 300. K? (c) Would the reactor need to be cooled or heated to maintain a constant temperature? (d) If the air purification system is working properly, which will be the limiting reactant in the first step?

6.108 A technician needs to carry out the reaction $2\,SO_2(g) + O_2(g) \rightarrow 2\,SO_3(g)$ at 25°C and 1.00 atm in a constant-pressure cylinder with the ability to expand. Initially, 0.030 mol SO_2 and 0.030 mol O_2 are present in the cylinder. The technician then adds the catalyst that will initiate the reaction. (a) Calculate the volume of the cylinder containing the reactant gases before reaction begins. (b) Which is the limiting reactant? (c) Assuming the reaction goes to completion and the temperature and pressure of the reaction remain constant, what is the final volume of the cylinder? (Be sure to include any excess reactant.) (d) How much work takes place, and is it done by the system or on the system? (e) How much heat is exchanged, and does it leave or enter the system? (f) From your answers to (d) and (e), calculate the change in internal energy for the reaction.

Atomic Structure and the Periodic Table

Colors often fill the night sky. On Earth, their source may be a fireworks display. Far above, and harder to identify unless you have a telescope, the colors are those of stars, luminous interstellar gas, or a rare supernova. In 1835, the philosopher Auguste Comte remarked in reference to the Sun, stars, and planets, "we understand the possibility of determining their shapes, their distances, their sizes and motions, whereas never by any means will we be able to study their chemical composition." However, scientists have come a long way since then. They have discovered how to analyze the chemical compositions of stars and planets by studying their colors with instruments called spectrometers.

The colors of all glowing objects have the same origin: they come from atoms and molecules that have been excited to states of high energy. Atoms in burning fireworks and stars become excited by absorbing energy as heat; they then throw off their excess energy as light. The colors emitted by an atom depend on how its electrons are arranged. So, by investigating the colors an atom emits, we can determine its internal structure.

In Section 1.3, we learned about the nuclear atom. We saw that an atom of atomic number *Z* consists of a central, tiny, dense, positively charged nucleus surrounded by *Z* electrons. This chapter looks at how the electrons are arranged in an atom, the **electronic structure** of an atom. A knowledge of this structure is essential for understanding chemistry and the arrangement of the elements in the periodic table. It will also prove to be important in Chapters 8 and 9 for understanding the types and numbers of bonds that an atom can form.

When Rutherford proposed the nuclear atom, he expected to be able to use *classical mechanics*, the laws of motion proposed by Newton in the seventeenth century, to describe its electronic structure. However, it soon

The Trifid nebula, in the constellation Sagittarius, is 5200 light years from Earth, but we can study its composition from the colors of light emitted. The red color arises from glowing hydrogen atoms excited by hot young stars at the center. The blue glow nearby is light scattered by dust clouds surrounding a star that is not hot enough to excite the surrounding hydrogen atoms.

became clear that classical mechanics fails when it is applied to electrons in atoms. New laws, which came to be known as *quantum mechanics,* were developed in the early part of the twentieth century. Their introduction caused an intellectual earthquake that shook science to its foundations. We shall see a little of that earthquake in this chapter.

OBSERVING ATOMS

The name *spectroscopy* has Latin and Greek origins. It means "image viewer."

The colors given off by the sun, stars, fireworks, and heated samples of matter can be studied by a technique called **spectroscopy,** which is the analysis of the light emitted or absorbed by substances. Spectroscopy is one of the principal methods available for exploring the structure of matter, and all modern chemical laboratories use spectroscopy in one form or another.

7.1 The Characteristics of Light

The instrument used for spectroscopy is called a **spectrometer.** In a spectrometer, the light emitted by a sample first passes through a slit that reduces it to a narrow beam. (Fig. 7.1). That beam is then separated into its different colors by using a device such as a prism. Finally, the separated colors are recorded, either as data on a computer or as a photographic image. The image is a picture of the slit the rays of light passed through, with a separate image for each color of light emitted by the sample. In other words, the individual colors are recorded as **spectral lines.** This set of lines, the **spectrum,** is unique for the atoms of each

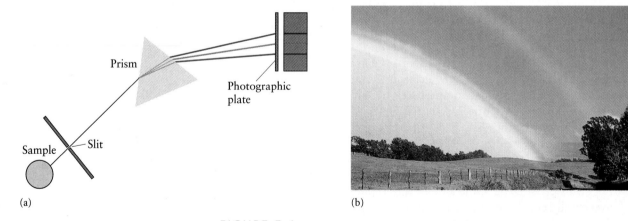

(a)

(b)

FIGURE 7.1

(a) In a spectrometer, the light emitted by an energetically excited sample of an element is passed through a slit, to give a narrow ray, and then through a prism. The prism separates the ray into different colors, which are recorded photographically. The spectral lines on the photograph are the separate images of the slit. (b) A rainbow is formed when white light from the Sun is split into its component colors by raindrops that act as tiny prisms. The light enters the front of the raindrop, reflects from the back, and emerges from the front. Double rainbows are formed when the light reflects a second time inside the drop.

element and is like a fingerprint of the element. The spectrum of colors emitted by atoms that have been heated to high temperatures is called an **atomic emission spectrum.** Astronomers use these spectra to identify the elements present in distant stars.

Light is **electromagnetic radiation,** a wave of electric and magnetic fields. All electromagnetic radiation travels through empty space at 3.00×10^8 m/s, or at just over 670 million miles per hour. This speed is denoted c and is called the **speed of light.**

An **electric field** is an influence that pushes on charged particles, such as electrons. If we were to monitor the electric field of light as it passed by us, we would find that it pushed in one direction, then the opposite direction, over and over again (Fig. 7.2). That is, the field *oscillates* in direction and strength. The number of cycles—complete reversals of direction—per second is called the **frequency,** ν (the Greek letter nu), of the radiation. The unit of frequency is the **hertz** (Hz), which is defined as 1 cycle per second:

$$1 \text{ Hz} = 1/\text{s}$$

The electric field in electromagnetic radiation of frequency 1 Hz would push in one direction, then the opposite direction, and return to the original direction in 1 s. The frequency of electromagnetic radiation we see as light is close to 5×10^{15} Hz, so its electric field changes direction more than a thousand trillion (10^{15}) times a second as it travels past us.

The frequency of light determines its color (see Table 7.1). Our eyes detect different colors because they respond in different ways to light of different frequencies. When the field oscillates at about 6.4×10^{14} Hz, for example, we see blue light. The light from a traffic signal changes frequency from about 5.7×10^{14} Hz to 5.2×10^{14} Hz and then to 4.3×10^{14} Hz as it changes from green to yellow and then to red (Fig. 7.3).

The light waves from a glowing lamp spread through space in all directions. However, suppose we enclose the lamp in a box with only a tiny pinhole and look at the light that escapes. We call a light wave traveling in a single direction a

We can think of a field as a region of influence, like the gravitational field near the Earth.

The unit Hertz honors Heinrich Hertz (1857–1894), one of the pioneers of the study of electromagnetic radiation.

Table 7.1 *Color, frequency, and wavelength of electromagnetic radiation*

Radiation type	Frequency, 10^{14} Hz	Wavelength, nm (2 sf)	Energy per photon, 10^{-19} J
x-rays and γ rays	$\geq 10^3$	≤ 3	$\geq 10^3$
ultraviolet	8.6	350	5.7
visible light			
violet	7.1	420	4.7
blue	6.4	470	4.2
green	5.7	530	3.8
yellow	5.2	580	3.4
orange	4.8	620	3.2
red	4.3	700	2.8
infrared	3.0	1000	2.0
microwaves and radio waves	$\leq 10^{-3}$	$\geq 3 \times 10^6$	$\leq 10^{-3}$

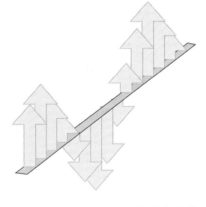

FIGURE 7.2

The electric field of electromagnetic radiation oscillates in space and time. The length of an arrow at any point at a given instant represents the strength of the force that the field exerts on a charged particle at that point. Note the wavelike distribution of the field.

FIGURE 7.3

The color of electromagnetic radiation is determined by its wavelength (and frequency). (a) The wavelengths of the three rays shown here are drawn to scale and you can see that the wavelengths of green, yellow, and red light increase in that order. The perception of color arises from the effect of the radiation on our eyes and the response of our brain. (b) Each lamp in a traffic signal generates white light, a mixture of all colors, but the tinted glass screens allow only certain wavelengths to pass through.

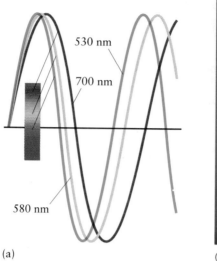

530 nm

700 nm

580 nm

(a)

(b)

ray of light. If we could see an instantaneous snapshot of the ray, it would look like Fig. 7.4. The heights of the peaks above and below the center give the **amplitude,** which is related to the **intensity** of the light, its brightness. The distance between adjacent peaks gives its **wavelength,** λ (the Greek letter lambda). The wavelengths of visible light are close to 500 nm, where 1 nm = 10^{-9} m. Although 500 nm is only half of one-thousandth of a millimeter, it is much longer than the diameters of atoms, which are typically close to 0.2 nm.

> The intensity of the light is proportional to the square of the amplitude of the wave.

An analogy

The effect of the electric field of an electromagnetic wave on an electron is like the effect of ocean waves on a boat. As the waves pass by, the boat is pushed up by the rising water, and then drops back down as the level of the water falls. The greater the frequency of the wave, the faster the boat bobs up and down. The greater the amplitude of the wave, the more violent the boat's motion.

The light wave in Fig. 7.4 is zooming along at the speed *c.* At any given point, this motion results in an oscillation of the field—a push one way, then a push the other way as the peaks and troughs of the wave brush past. Suppose we vary the wavelength of the light. Because all light waves travel at the same speed, the number of peaks that pass a point in one second depends on the distance between the peaks, the wavelength of the light (Fig. 7.5). If the wavelength of the light is short, many oscillations pass a point in a second. If the wavelength is long, fewer oscillations pass the point in a second. That is, short wavelength corresponds to high frequency; long wavelength corresponds to low frequency. The precise relation is

$$\text{Wavelength} \times \text{frequency} = \text{speed} \qquad \text{or} \qquad \lambda \times \nu = c \qquad (1)$$

For example, the wavelength of blue light is

$$\lambda = \frac{c}{\nu} = \frac{3.00 \times 10^8 \text{ m/s}}{6.4 \times 10^{14}/\text{s}} = 4.7 \times 10^{-7} \text{ m}$$

or about 470 nm. Blue light, which has a relatively high frequency, has a shorter wavelength than low-frequency red light (700 nm). The light from the three lamps of a traffic signal changes from about 530 nm to 580 nm and then to 700 nm as they change from green through yellow to red (see Fig. 7.3).

Our eyes detect electromagnetic radiation with wavelengths from 700 nm (red light) to 400 nm (violet light). The electromagnetic radiation in this range is called **visible light.** White light, which includes sunlight, is a mixture of all frequencies of visible light. Electromagnetic radiation includes wavelengths ranging from less than a picometer to more than several kilometers (Fig. 7.6). **Ultraviolet radiation** has a frequency higher than that of violet light; its wavelength is less than about 400 nm. **Infrared radiation,** the radiation we experience as heat, has a frequency lower than that of red light; its wavelength is greater than about 800 nm. **Microwaves,** which are used in radar and microwave ovens, have wavelengths in the millimeter to centimeter range.

The color of light depends on its frequency or wavelength; long-wavelength radiation has a lower frequency than short-wavelength radiation.

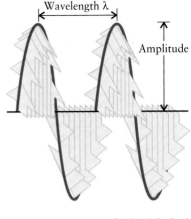

FIGURE 7.4

This diagram represents a "snapshot" of an electromagnetic wave at a given instant. The distance between the peaks is the wavelength of the radiation. The amplitude (height) of the peaks depends on the intensity of the radiation.

Example 7.1 *Finding the frequency of light from the wavelength*

What is the frequency in hertz of infrared radiation used as a range finder in a camera, taking its wavelength to be 1.00 μm?

Strategy The frequency and wavelength are inversely proportional and are related by Eq. 1. Rearrange Eq. 1 to solve for frequency: $\nu = c/\lambda$. Convert the units of wavelength to meters, so that consistent units are used for c and λ.

Solution The wavelength is

$$\lambda = 1.00 \text{ μm} \times \frac{10^{-6} \text{ m}}{1 \text{ μm}} = 1.00 \times 10^{-6} \text{ m}$$

FIGURE 7.5

Wavelength and frequency are inversely related. The two parts of this illustration show the electric field at three instants that you might experience as a single wave flashed by a single point at the speed of light from left to right. The position of the wave at the first instant is the light gray wave and its position at the third instant is the blue wave. (a) Short-wavelength radiation: note how the electric field changes markedly at the three successive instants. (b) For the same three instants, the electric field of the long-wavelength radiation changes much less. Short-wavelength radiation has a high frequency, whereas long-wavelength radiation has a low frequency.

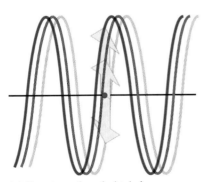

(a) Short wavelength, high frequency

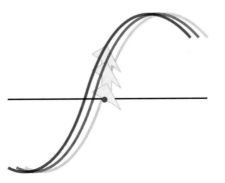

(b) Long wavelength, low frequency

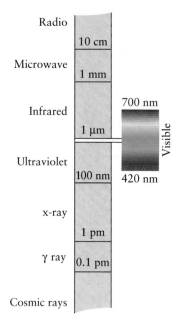

Radio

10 cm

Microwave

1 mm

Infrared

700 nm

1 μm

Ultraviolet

100 nm

420 nm

Visible

x-ray

1 pm

γ ray

0.1 pm

Cosmic rays

FIGURE 7.6

The electromagnetic spectrum and the names of its regions. Note that the region we call "visible light" occupies a very narrow range of wavelengths. The regions are not shown to scale.

We rearrange Eq. 1 and substitute $c = 3.00 \times 10^8$ m/s and $\lambda = 1.00 \times 10^{-6}$ m:

$$\nu = \frac{3.00 \times 10^8 \text{ m/s}}{1.00 \times 10^{-6} \text{ m}} = 3.00 \times 10^{14}/\text{s}$$

That is, the frequency of the light is 3.00×10^{14} Hz.

Self-Test 7.1A What is the wavelength of orange light of frequency 4.8×10^{14} Hz?

[*Answer:* 620 nm]

Self-Test 7.1B Which color has the higher frequency, orange or yellow?

7.2 Quanta and Photons

According to classical mechanics, radiation and matter can have any energy—high, low, and anything in between. For example, a pendulum seems to lose energy continuously by friction, and we seem to be able to give it any energy simply by pushing it harder. The energy output of a lamp apparently can be varied continuously simply by increasing the current through it. However, very precise experiments have shown that classical mechanics is wrong: energy can be transferred only in discrete amounts, or **quanta.** We say that energy is **quantized,** or restricted to discrete values. Classical mechanics gives excellent agreement between theory and observation for large objects, such as baseballs, buildings, jet planes, and planets. However, only **quantum mechanics,** the description of matter and radiation that takes into account quantization, can explain the behavior of objects as small as electrons.

The word *quantum* (plural, quanta) comes from the Latin for "amount"—literally, "How much?".

An analogy

Quantization is like pouring water into a bucket. Water seems to be a continuous fluid, and it seems that any amount can be transferred. However, the smallest amount of water that we can transfer is one H_2O molecule. Likewise, energy seems to be continuously variable, but, in fact, it can be transferred only in discrete amounts.

Electromagnetic radiation carries energy through space. According to quantum mechanics, a ray of electromagnetic radiation is a stream of many packets of electromagnetic energy. These packets are called **photons.** The more intense the light, the greater the number of photons passing a given point each second. A dim source of light emits relatively few photons; a bright source of light emits a dense stream of photons.

Theoretical considerations have predicted and experiments have verified that the energy of a photon is proportional to the frequency of its associated radiation. Photons of ultraviolet radiation are therefore more energetic, and hence more damaging, than photons of visible light. This damaging radiation is responsible for sunburn and tanning and would destroy life on the surface of the Earth if it were not largely absorbed by the ozone layer (see Applying Chemistry: Case Study 5). We write the relation between the energy, E, of a photon and the frequency, ν, of the radiation as

$$E = h\nu \tag{2}$$

Here h is the **Planck constant,** a fundamental constant with the value 6.63×10^{-34} J·s. For example, the energy of each photon of blue light of frequency 6.4×10^{14} Hz is

$$E = (6.63 \times 10^{-34}\,\text{J·s}) \times (6.4 \times 10^{14}\,\text{Hz}) = 4.2 \times 10^{-19}\,\text{J}$$

To derive this value, we have used

$$1\,\text{J·s} \times \text{Hz} = 1\,\text{J·s} \times 1/\text{s} = 1\,\text{J}$$

The constant is named for Max Planck (1858–1947), the German scientist who introduced the idea that energy is transferred in packets.

A ray of red light also consists of a stream of photons, but because the frequency of the light is lower, each of its photons has less energy (2.8×10^{-19} J) than a photon of blue light (4.2×10^{-19} J). We see different colors because photons of radiation with different frequencies and therefore different energies cause different effects in our eyes. The photon energies of light of various colors are included in Table 7.1. Ultraviolet radiation is damaging because its photons have enough energy to break chemical bonds when they strike matter. X-ray photons are even more energetic and are highly damaging to living tissues, which is why we have to take special precautions when using them.

Experimental evidence for the existence of photons and the dependence of their energy on frequency comes from the **photoelectric effect,** the ejection of electrons from a metal when its surface is exposed to electromagnetic radiation (Fig. 7.7). No electrons are ejected unless the radiation has a frequency above a value characteristic of the metal. For example, for sodium, this frequency is 6.7×10^{14} Hz, in the blue region of the spectrum. Light of lower frequencies, such as red or yellow light, cannot eject electrons from sodium, no matter how long the light shines. However, if the frequency is increased to at least the "threshold" level for the metal, electrons are immediately ejected. These observations suggest that electromagnetic energy comes in packets—photons—and that when a photon collides with sufficient vigor with an electron in the metal, the electron will be ejected by the collision. The electron is expelled only if the energy of the photon is great enough. Because $\nu = E/h$, the threshold frequency for each metal is an indication of the minimum energy needed to expel an electron.

A typical lamp emits huge numbers of photons each second. For example, suppose a lamp emits 25 J of yellow light (580 nm) in 1 s. The energy of one photon of yellow light is 3.4×10^{-19} J, so the number of photons that accounts for 25 J of energy is

$$
\begin{aligned}
\text{Number of photons} &= \frac{\text{total energy}}{\text{energy of one photon}} \\
&= \frac{25\,\text{J}}{3.4 \times 10^{-19}\,\text{J}} = 7.4 \times 10^{19}
\end{aligned}
$$

In other words, when we turn on an electric lamp, it generates about 10^{20} photons of yellow light each second. That is why we normally don't see the presence of individual photons: huge numbers are generated by the sources of light we use in everyday life.

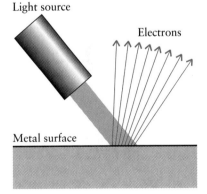

Light source

Electrons

Metal surface

FIGURE 7.7

When a metal is illuminated with light, electrons are ejected, provided the frequency is above a threshold frequency that is characteristic of the metal. Radiation with a lower frequency will not cause electrons to be ejected, no matter how intense it is.

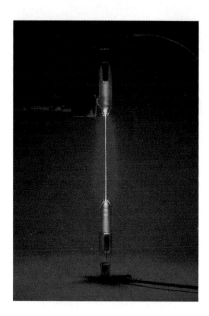

<figure>FIGURE 7.8</figure>

The red glow from this hydrogen discharge lamp comes from excited hydrogen atoms that are returning to a lower energy state and emitting the excess energy as visible radiation.

Electromagnetic radiation can be regarded as consisting of discrete packets of energy called photons. The energy of a photon is proportional to the frequency of the radiation.

Example 7.2 *Determining the energy of photons*

What is the energy in kilojoules per mole of photons of amber light of frequency 5.2×10^{14} Hz?

Strategy Each photon has an energy that corresponds to the frequency of the light by Eq. 2. Use Eq. 2 to find that energy, and multiply by the Avogadro constant to find the energy per mole of photons. Convert to kilojoules per mole by using $1\ kJ = 10^3$ J. Remember that $1\ Hz = 1/s$.

Solution To find the energy per photon, substitute the frequency and the Planck constant into Eq. 2:

$$E = h\nu = (6.63 \times 10^{-34}\ J{\cdot}s) \times (5.2 \times 10^{14}/s) = 6.63 \times 5.2 \times 10^{-20}\ J$$

Now find the energy per mole of photons by multiplying this energy by the Avogadro constant, and convert to kilojoules:

$$\underbrace{(6.63 \times 5.2 \times 10^{-20}\ J)}_{h\nu} \times \underbrace{\left(\frac{6.022 \times 10^{23}}{mol}\right)}_{N_A} \times \left(\frac{1\ kJ}{10^3\ J}\right) = 2.1 \times 10^2\ kJ/mol$$

The photons have an energy of 210 kJ/mol.

Self-Test 7.2A A student's favorite radio station broadcasts a signal at a frequency of 91.5 MHz. What is the energy (in millijoules per mole) of the photons broadcast?

[*Answer:* 36.5 mJ/mol]

Self-Test 7.2B A certain lamp produces 25 J of energy per second in a certain region of the spectrum. In 1.0 s, it emits 5.5×10^{19} photons of light in that region. What is the energy of *one* photon?

7.3 Atomic Spectra and Energy Levels

Now let's see how these ideas help us to interpret atomic spectra. If we pass an electric current through a low-pressure sample of hydrogen gas, many frequencies of electromagnetic radiation are emitted (Fig. 7.8). The current, which is a vigorously moving stream of electrons, breaks up the H_2 molecules and excites the resulting H atoms to higher energies. These atoms discard their excess

<figure>FIGURE 7.9</figure>

The spectrum of atomic hydrogen. The spectral lines have been assigned to various groups of similar wavelength called series; the Balmer and Lyman series are shown here.

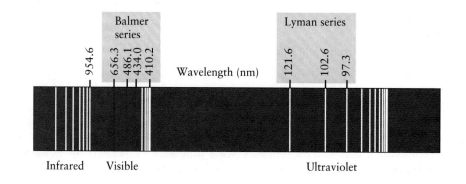

energy by giving off electromagnetic radiation with a wide variety of frequencies; then they combine to form H_2 molecules again.

The spectrum of atomic hydrogen was astonishing to the scientists who first investigated it, because it consists of a series of discrete lines (Fig. 7.9). Why is it that a glowing atom can emit only particular frequencies of radiation and not all possible frequencies? A further puzzle, but also a clue, was the discovery that the lines showed patterns that could be summarized by a simple formula. First, a Swiss schoolteacher, Johann Balmer, noticed that the lines in the visible region of the spectrum all fit the expression

$$\nu = (3.29 \times 10^{15} \text{ Hz}) \times \left(\frac{1}{4} - \frac{1}{n^2}\right) \qquad n = 3, 4, \ldots$$

Then, as more frequencies were detected in the spectrum, the Swedish spectroscopist Johannes Rydberg found that *all* the lines in the spectrum were given by the expression

$$\nu = R_{\text{H}} \times \left(\frac{1}{n_1^2} - \frac{1}{n_2^2}\right) \qquad (3)$$

with $n_1 = 1, 2, \ldots$, and $n_2 = n_1 + 1, n_1 + 2, \ldots$ The constant R_{H} is now called the **Rydberg constant** for hydrogen and has the value 3.29×10^{15} Hz. The **Balmer series** is the set of lines with $n_1 = 2$ (and $n_2 = 3, 4, \ldots$). The **Lyman series,** a set of lines in the ultraviolet region of the spectrum, has $n_1 = 1$ (and $n_2 = 2, 3, \ldots$). The fact that the entire spectrum is described by such a simple expression is an enormously important clue to the structure of the atom; but at the time it left scientists completely baffled.

The existence of photons can be used to explain Rydberg's formula. A photon is emitted from an atom when an electron falls from one energy state to a lower energy state. The observation of discrete frequencies suggests that an electron in a hydrogen atom can exist only in a series of discrete energy states, called **energy levels.** When an electron makes a **transition** from one energy level to another, the difference in energy, ΔE, is carried away as a photon. Because the energy of a photon is $h\nu$, where ν is the frequency of the radiation to which the photon contributes, it follows that

$$\Delta E = h\nu \qquad (4)$$

This relation is called the **Bohr frequency condition.** Each spectral line arises from a specific transition, with different excited hydrogen atoms undergoing different transitions and hence contributing to different spectral lines (Fig. 7.10). The spectrum tells us that, because only certain frequencies are present in the spectrum, only certain energy levels exist in the atom (Fig. 7.11). But why can atoms have only certain energy levels? As we shall see, the explanation had to await the development of quantum mechanics.

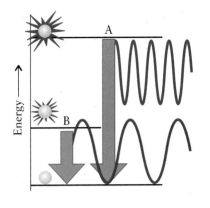

FIGURE 7.10

When an atom undergoes a transition from a state of higher energy to one of lower energy, it loses energy that is carried away as a photon. The greater the energy loss, the higher the frequency (and the shorter the wavelength) of the radiation emitted. Thus, transition A generates light with a higher frequency and shorter wavelength than transition B.

Niels Bohr (1885–1962) was a Danish physicist. He devised the first model of atomic structure that could explain spectral lines.

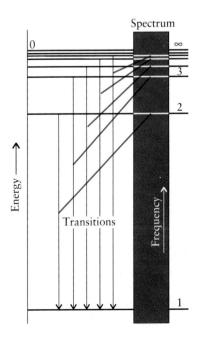

FIGURE 7.11

The spectrum of atomic hydrogen (reproduced on the right) tells us the arrangement of the energy levels of the atom because the frequency of the radiation emitted in a transition is proportional to the energy difference between the two energy levels involved. The 0 on the energy scale corresponds to the completely separated proton and electron. The numbers on the right label the energy levels: they are examples of quantum numbers (see Section 7.7).

The observation of discrete spectral lines suggests that an electron in an atom can have only certain energies. Transitions between these energy levels generate photons of radiation with a frequency in accord with the Bohr frequency condition.

7.4 The Wavelike Properties of Electrons

Our next step is to devise an explanation of the energy levels and Rydberg's amazing formula. The fact that electrons can have only certain discrete energies in atoms may remind us of a guitar string, which can vibrate with only certain discrete frequencies when it is plucked. Like a string of fixed length in a guitar, an electron is confined within an atom. Can its confinement somehow result in its having only certain energies, like the frequencies of a guitar string (Fig. 7.12)?

A major step forward was taken in 1924, when the French scientist Louis de Broglie made the revolutionary suggestion that all matter—not just electrons—has wavelike properties. The **de Broglie relation** states that the wavelength, λ, of *any* particle is related to its mass, m, and velocity, v, by

$$\text{Wavelength} = \frac{h}{\text{mass} \times \text{velocity}} \quad \text{or} \quad \lambda = \frac{h}{mv} \qquad (5)$$

According to this formula, a heavy particle traveling rapidly has a short wavelength and a slow-moving particle of small mass has a relatively long wavelength. A ball bearing with a mass of 1 g traveling at 1 cm/s has a wavelength of only 7×10^{-29} m, which shows that the wave character of everyday particles is undetectably small and can be ignored. However, an electron traveling at high speed in an atom has a wavelength that is comparable to the diameter of the atom, about 10^{-10} m, so its wave character must be taken into account when describing atomic structure.

Example 7.3 *Calculating the wavelength of an object*

Suppose an electron in an atom is traveling at 2.2×10^6 m/s. What is the de Broglie wavelength of the electron?

Strategy Equation 5 gives the relation between wavelength and the mass and speed of an object. To use it, we need the mass of the electron and the value of the Planck constant, which are found inside the back cover of the book. Be sure to express the mass and speed in terms of kilograms, meters, and seconds.

Solution The mass of an electron is sometimes listed as 9.109×10^{-28} g, which corresponds to

$$(9.109 \times 10^{-28}\,\text{g}) \times \frac{10^{-3}\,\text{kg}}{1\,\text{g}} = 9.109 \times 10^{-31}\,\text{kg}$$

The value of h is 6.63×10^{-34} J·s, so the wavelength of the electron is

$$\lambda = \frac{h}{mv} = \frac{6.63 \times 10^{-34}\,\text{J·s}}{(9.109 \times 10^{-31}\,\text{kg}) \times (2.2 \times 10^6\,\text{m/s})} = 3.3 \times 10^{-10}\,\text{m}$$

FIGURE 7.12

When a guitar string vibrates, only certain wavelengths can be sustained over time—those for which the amplitude of the wave goes to 0 at each end. One or more complete half-wavelengths must fit exactly between the end points. (a) A guitar string at rest. (b) One-half wavelength. (c) One full wavelength.

(a) (b) (c)

FIGURE 7.13

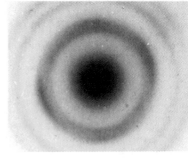

Davisson and Germer showed that electrons give a diffraction pattern when reflected from a crystal. G. P. Thomson working in Aberdeen, Scotland showed that they also give a diffraction pattern when they pass through a very thin gold foil. The latter is shown here. G. P. Thomson was the son of J. J. Thomson, who identified the electron (Section 1.3). Both received Nobel prizes, J. J. for showing that the electron is a particle and G. P. for showing that it is a wave.

Here we have used $1\ J = 1\ kg \cdot m^2/s^2$. This wavelength is 330 pm, which is comparable to the diameter of an atom (about 100–200 pm).

Self-Test 7.3A Calculate the wavelength of a marble of mass 5.00 g traveling at 1.00 m/s.

[***Answer:*** 1.33×10^{-31} m]

Self-Test 7.3B Calculate the wavelength of a rifle bullet of mass 5.00 g traveling at twice the speed of sound (the speed of sound at sea level is 310. m/s).

But is de Broglie's hypothesis valid? One of the first experiments to confirm the wavelike character of electrons was carried out by the American scientists Clinton Davisson and Lester Germer in 1927. They knew that when light waves pass through a grid with a spacing comparable to the wavelength, characteristic patterns of light and dark intensity, called **diffraction patterns,** are obtained. Davisson and Germer showed that electrons reflected from a crystal also give a diffraction pattern on a photographic plate. In this case, the layers of atoms act as the grid. They also found that the pattern corresponds exactly to that expected for electrons with a wavelength given by the de Broglie relation (Fig. 7.13).

We have to conclude that electrons are both particles and waves. The revolutionary aspect of matter that de Broglie identified is called **wave-particle duality,** the possession by matter of both wavelike and particlelike properties. This duality is the basis of our current understanding of the atom.

The wave-particle duality of matter has profound implications for our understanding of the world. We can see how it undermines conventional views about matter by thinking about the location of a particle with a definite velocity. According to the de Broglie relation, a particle with definite velocity is a wave with a well-defined wavelength (as given by Eq. 5). But where is a wave located? A wave spreads through space, and we cannot point to a location and say that the wave is *there*. A wave is everywhere. An implication of the wave-particle duality of matter, therefore, is that *if we know the velocity of a particle, then we cannot know where it is.* Conversely, if we know where a particle is, then it cannot be described by a wave of definite wavelength. That means that *if we know where a particle is, then we cannot know anything about its velocity.* This extraordinary restriction on what we can know is called the **Heisenberg uncertainty principle** (Fig. 7.14). The principle does away with the classical notion that a particle travels along a definite path. However, it has significant implications only for subatomic particles where the wave character of matter is important. In the macroscopic world, where the wave character of matter can be ignored, the uncertainty principle can be disregarded.

> ***Electrons have both wavelike and particlelike properties; their wavelike properties must be taken into account when describing the structure of atoms. The Heisenberg uncertainty principle implies that we cannot know the position and velocity of a particle simultaneously.***

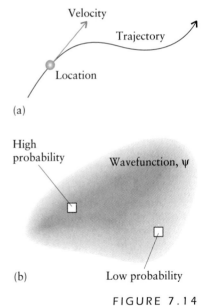

(a)

(b)

FIGURE 7.14

(a) In classical mechanics, a particle follows a path, or trajectory, and its position can be predicted at any instant. (b) In quantum mechanics, the particle is distributed like a wave, so its location cannot be predicted exactly. Where the wavefunction has a high amplitude, there is a high probability of finding the particle; where the amplitude is low, there is only a small probability of finding the particle.

$$-\frac{\hbar}{2m}\nabla^2\psi + V\psi = E\psi$$

FIGURE 7.15

Erwin Schrödinger (1887–1961). The Schrödinger equation is shown superimposed on Schrödinger's head. The constant $\hbar$ stands for $h/2\pi$.

The names of the orbitals, *s*, *p*, *d*, and *f*, come from the fact that spectroscopic lines were once classified as *sharp*, *principal*, *diffuse*, and *fundamental*.

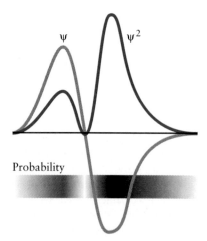

ψ ψ^2

Probability

MODELS OF ATOMS

How do electrons, with their wavelike properties, fit into the nuclear model of the atom? Can this wavelike character explain Rydberg's formula and the spectrum of hydrogen? Our current model of a hydrogen atom was proposed in 1926 by the Austrian scientist Erwin Schrödinger (Fig. 7.15). A simpler model had been proposed earlier by Niels Bohr, but his model—which supposed that the electron travels in a definite orbit around the nucleus—did not take into account the uncertainty principle and had to be rejected.

7.5 Atomic Orbitals

Schrödinger's first step in developing a model of the hydrogen atom was to devise an equation, which is now known as the **Schrödinger equation,** that enabled him to calculate the shape of the wave associated with any particle. Then he went on to solve the equation for the specific case of an electron in a hydrogen atom. First, he found that waves would fit into the atom only for certain energy values. That is exactly what is needed to account for the existence of energy levels. In other words, the quantization of energy follows naturally from the Schrödinger equation, because only certain waves (and their corresponding energies) are allowed. Then he found the mathematical expressions for the shapes of the allowed waves. These expressions are now called **wavefunctions** and denoted ψ (psi).

One of the fundamental principles of quantum mechanics tells us how to interpret wavefunctions. According to the **Born interpretation,** which was proposed by the German physicist Max Born, *the probability of finding the electron at a given point in space is proportional to the square of the wavefunction at that point.* Where the wavefunction has a large amplitude, there is a high chance of finding the electron it describes. Where the wavefunction is small, the electron will only rarely be found. Where the wavefunction is 0, the electron will not be found at all (Fig. 7.16). Another very important difference between classical mechanics and quantum mechanics now becomes clear. In classical mechanics, we can predict the location of a particle, such as the location of a planet (a big particle!) in its orbit around the Sun. In quantum mechanics, we can predict only the *probability* that a particle will be found at a given location.

The wavefunction for an electron in an atom is so important that it is given a special name: an **atomic orbital.** To visualize an atomic orbital, we think of a cloud surrounding the nucleus, with the density of the cloud representing the probability of finding an electron at each point. Denser regions of the cloud represent locations where the electron is more likely to be found. We cannot state precisely *where* an electron in an atomic orbital is; we can speak only of the *probability* of its being in a particular place.

Atomic orbitals have characteristic energies and shapes. The different shapes are identified by different letters. An *s*-orbital is a spherical cloud that

FIGURE 7.16

The Born interpretation of a wavefunction. The probability of the electron being found at a point is proportional to the square of the wavefunction (indicated by ψ^2), as depicted by the density of shading in the band below. Note that the probability density is 0 at a node. A node is a point where the wavefunction passes *through* 0, not merely approaches 0.

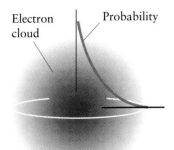

Electron cloud   Probability

FIGURE 7.17

The three-dimensional electron cloud corresponding to an electron in the lowest energy state of hydrogen. This wavefunction is called the 1s-orbital (Section 7.7). The density of shading represents the probability of finding the electron at any point. The superimposed graph shows how the probability varies with the distance from the nucleus along any radius.

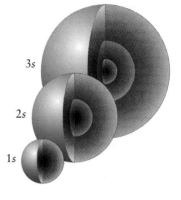

3s

2s

1s

FIGURE 7.18

becomes less dense as the distance from the nucleus increases (Fig. 7.17). In principle, the cloud never thins to exactly 0, so we could think of an atom as being bigger than the Earth! However, there is virtually no chance of finding an electron farther from the nucleus than about 100 pm, so atoms are, in fact, very small. As we can see from the high density of the cloud at the nucleus, an electron in an *s*-orbital, which we call an **s-electron,** has a high probability of being found right at the nucleus itself.

Instead of drawing the *s*-orbital as a cloud, we usually draw its **boundary surface,** the surface that encloses the densest regions of the cloud. The electron is likely to be found only inside the boundary surface of the orbital. An *s*-orbital has a spherical boundary surface, because the electron cloud is spherical (Fig. 7.18). *s*-Orbitals with higher energies have spherical boundary surfaces of bigger diameters. They also have a more complex variation of the density of the electron cloud inside the boundary surfaces.

A **p-orbital** is a cloud with two lobes on opposite sides of the nucleus (Fig. 7.19). The two lobes are separated by a planar region called a **nodal plane,** which cuts through the nucleus. A **p-electron,** an electron in a *p*-orbital, will never be found on this plane, so it is never found at the nucleus. We shall see later that this difference from *s*-orbitals is of major importance for understanding the structure of the periodic table. A *p*-orbital can lie in any of three perpendicular orientations, so there are three *p*-orbitals of a given energy.

The lobes on either side of the nodal plane in Fig. 7.19 have been tinted different colors: these tints represent the different *signs* of the wavefunction (that is, whether the mathematical function for the wave has positive or negative values). Just as the displacement of a wave can be positive or negative, so the wavefunction of an electron can have positive or negative regions too. However, *these signs have no direct physical significance:* they do *not* represent charge, nor do

The simplest way of drawing an atomic orbital is as a boundary surface, a surface within which there is a high probability (typically, 90%) of finding the electron. The spheres here represent the boundary surfaces of the *s*-orbitals in the first three energy levels. Note that *s*-orbitals with *n* > 1 have internal spherical nodes and that the size of the orbital increases with *n*. We shall use blue to denote *s*-orbitals, but that color is only an aid to their identification.

FIGURE 7.19

The boundary surface of a *p*-orbital has two lobes; the nucleus lies on the plane that divides the two lobes, and an electron will, in fact, never be found at the nucleus itself if it is in a *p*-orbital. There are three *p*-orbitals of a given energy, and they lie along three perpendicular axes. We shall use yellow to indicate *p*-orbitals. Note that the orbital has opposite signs (as depicted by the depth of color) on each side of the nodal plane.

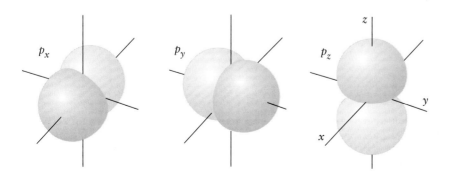

FIGURE 7.20

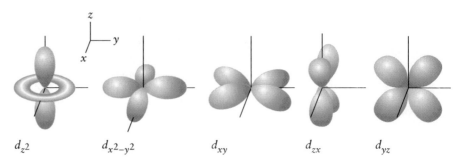

d_{z^2} $d_{x^2-y^2}$ d_{xy} d_{zx} d_{yz}

The boundary surface of a *d*-orbital is more complicated than that of an *s*- or a *p*-orbital. There are five *d*-orbitals of a given energy; four of them have four lobes; one is slightly different. In each case, an electron that occupies a *d*-orbital will not be found at the nucleus. We shall use orange to indicate *d*-orbitals, with different depths of color to indicate different signs.

they represent higher and lower concentrations of the electron. The Born interpretation tells us that to find the probable location of an electron in an atomic orbital, we need to take the square of the wavefunction. Taking the square converts negative regions into positive regions. Therefore, regardless of the sign of the wavefunction, all probabilities are positive. The signs will not play a role in the rest of this chapter but will turn out to be of crucial importance when two atoms are brought together and form bonds (see Section 9.15).

The **d-orbitals** and **f-orbitals** have more complicated shapes. The shapes of the five *d*-orbitals of a given energy are shown in Fig. 7.20. As for *p*-orbitals, the two shades of color denote opposite signs of the wave. The shapes of the seven *f*-orbitals of a given energy are rarely needed to explain chemical properties, so they are not discussed here, but you can see from the one shown in Fig. 7.21 that they have more lobes than *d*-orbitals. Like *p*-orbitals, *d*- and *f*-orbitals have zero electron density at the nucleus.

> *The location of an electron in an atom is best described as a cloud of probable locations. In chemistry, the most important shapes of the clouds are the spherical s-orbitals, the two-lobed p-orbitals, and the more complicated d-orbitals.*

7.6 Energy Levels

Each orbital has a characteristic energy. When he solved his equation, Schrödinger found that the allowed energies of the electron in a hydrogen atom are given by the expression

$$E = -\frac{hR_H}{n^2} \qquad n = 1, 2, \ldots \qquad (6)$$

where R_H, the Rydberg constant for hydrogen, is given by a particular combination of fundamental constants. When Schrödinger evaluated the expression, the constant agreed exactly with the experimental value, 3.29×10^{15} Hz. The negative sign in Eq. 6 simply means that, when an electron is in a hydrogen atom, its energy is lower than when the nucleus and electron are widely separated. The *n* in the equation is an integer (whole number) that labels these energy levels,

Niels Bohr had derived the same expression earlier from his model of the atom.

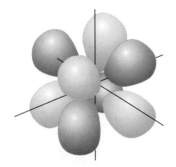

FIGURE 7.21

The boundary surface of one of the seven *f*-orbitals of a shell (with $n \geq 4$). The *f*-orbitals have a complex appearance. Their shapes will not be needed again in this text. However, their existence is important for understanding the periodic table, the presence of the lanthanides and actinides, and the properties of the later *d*-block elements.

from $n = 1$ for the first (lowest) level, $n = 2$ for the second (next higher) level, continuing up to infinity. It is called the *principal quantum number* (quantum numbers are explored in Section 7.7). Figure 7.22 shows the energy levels calculated from Eq. 6.

It must have been a breathtaking moment for Schrödinger when, like Bohr before him, he arrived at Eq. 6 and realized that it was exactly the expression required to account for the spectrum of atomic hydrogen (recall Fig. 7.11). As shown in the following example, all he needed to do was calculate the energy difference between any pair of levels by taking the difference of two expressions like Eq. 6 and setting the result equal to the energy of the photon that flies away from the atom.

The allowed energies of an electron in a hydrogen atom are given by Eq. 6. Whenever the atom emits a photon of radiation, the energy emitted is equal to the difference in two of the allowed energy levels. Each transition corresponds to a line in the spectrum of atomic hydrogen.

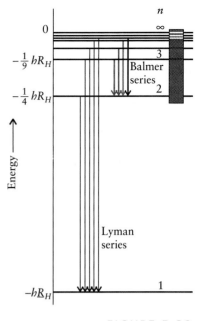

FIGURE 7.22

The permitted energy levels of a hydrogen atom as calculated from Eq. 6. The 0 on the energy scale corresponds to the completely separated proton and electron; the lowest energy state lies at hR_H below the 0 of energy. The levels are labeled with the principal quantum number, n, which ranges from 1 (for the lowest state) to infinity (for the separated proton and electron). Compare this diagram with the experimentally determined array of levels shown in Fig. 7.11.

Example 7.4 *Identifying a line in the hydrogen spectrum*

Calculate the wavelength of a photon emitted by a hydrogen atom when an electron makes a transition from an orbital with $n = 3$ to one with $n = 2$. Identify in Fig. 7.9 the spectral line produced by this transition.

Strategy Calculate the difference between the two energy levels by using Eq. 6, and then calculate the wavelength corresponding to that energy difference. The energy difference between a level with quantum number n_2 and energy $-hR_H/n_2{}^2$ and another level with quantum number n_1 and energy $-hR_H/n_1{}^2$ is

$$\Delta E = -\frac{hR_H}{n_2{}^2} - \left(-\frac{hR_H}{n_1{}^2}\right) = \left(\frac{1}{n_1{}^2} - \frac{1}{n_2{}^2}\right)hR_H$$

To use this expression to find the wavelength of light emitted by an atom, substitute the principal quantum number with the greater value for n_2 and that with the lower value for n_1, then substitute the value of the Rydberg constant for R_H. It is best to leave the expression as a multiple of h, because that constant will cancel in the next step, in which this energy is expressed as a frequency by using Eq. 4. Finally, convert that frequency to a wavelength by using Eq. 1. Because the final state has $n = 2$ (as in the Balmer series), expect a wavelength in the visible region.

Solution The energy difference between the two states is

$$\Delta E = \left(\frac{1}{2^2} - \frac{1}{3^2}\right) \times h \times (3.29 \times 10^{15}\ \text{Hz})$$

Then, from Eq. 4, the frequency of the emitted light is

$$\nu = \frac{\Delta E}{h} = \left(\frac{1}{2^2} - \frac{1}{3^2}\right) \times (3.29 \times 10^{15}\ \text{Hz})$$

Therefore, the wavelength of the radiation is

$$\lambda = \frac{c}{\nu} = \frac{3.00 \times 10^8\ \text{m/s}}{\left(\dfrac{1}{2^2} - \dfrac{1}{3^2}\right) \times (3.29 \times 10^{15}\ \text{Hz})}$$

$$= \frac{3.00 \times 10^8}{\left(\frac{1}{4} - \frac{1}{9}\right) \times (3.29 \times 10^{15})} \times \frac{\text{m}}{\text{Hz·s}} = 6.57 \times 10^{-7}\ \text{m}$$

or 657 nm (note that we have used 1 Hz = 1/s). The transition gives the red line in the spectrum.

Self-Test 7.4A Repeat the calculation for the transition from the state with $n = 4$ to $n = 2$, and identify it as one of the spectral lines in Fig. 7.9.

[***Answer:*** 486 nm; blue line]

Self-Test 7.4B Repeat the calculation for the transition from the state with $n = 5$ to $n = 2$, and identify it as one of the spectral lines in Fig. 7.9.

7.7 Quantum Numbers

Schrödinger found that each atomic orbital is identified by three numbers called *quantum numbers*. In general, a **quantum number** is an integer (in certain cases, a half-integer) that labels a wavefunction and can be used to calculate the value of a property (such as the energy of an electron in an atom). The three quantum numbers discovered by Schrödinger are denoted n, l, and m_l.

The quantum number n is called the **principal quantum number.** The value of n tells us the energy of the electron (by using Eq. 6): the higher the value of n, the greater the energy of the orbital, and hence the less tightly bound the electron in the atom. This quantum number also indicates the average distance of the electron from the nucleus: the greater the value of n, the greater the average distance of the electron from the nucleus. The **orbital angular momentum quantum number,** l, specifies the shape of the orbital (for instance, whether it is an *s*- or *p*-orbital). The **magnetic quantum number,** m_l, specifies the individual orbital of a particular shape (for instance, which of the three *p*-orbitals is meant). To identify the orbital unambiguously, we have to give the values of all three quantum numbers (Table 7.2).

> The orbital angular momentum quantum number is still widely called the *azimuthal quantum number*.

The lowest (most negative) energy corresponds to $n = 1$. This lowest energy state is called the **ground state** of the atom. A hydrogen atom is normally found in its ground state, with its electron in the level with $n = 1$ and energy $-hR_H$. An electron in the ground state of a hydrogen atom is likely to be found close to the nucleus. When the electron absorbs energy, its average distance from the nucleus increases as it occupies orbitals with higher values of n. Its energy climbs up the ladder of levels as n increases and reaches the top of the ladder, corresponding to $E = 0$ and freedom, only when n has reached infinity. At that point, the electron is so far from the nucleus that it has effectively left the atom.

The clouds representing the orbitals get bigger as n increases, so the average distance of an electron from the nucleus also increases as n increases. Therefore, we think of the orbitals as defining a series of concentric **shells.** All the orbitals with a given value of n belong to the same shell of the atom.

Table 7.2 Quantum numbers for electrons in atoms

Name	Symbol	Values	Meaning	Indicates
principal	n	$1, 2, \ldots$	labels shell, specifies energy	size
orbital angular momentum*	l	$0, 1, \ldots, n - 1$	labels subshell	shape
magnetic	m_l	$l, l - 1, \ldots, -l$	labels orbitals of subshell	direction
spin magnetic	m_s	$+\frac{1}{2}, -\frac{1}{2}$	labels spin state	spin direction

*Also called the *azimuthal quantum number*.

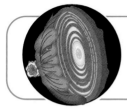

The shells of an atom are something like thick layers of a fuzzy onion. Shells of higher n surround the shells of lower n.

The orbital angular momentum quantum number, l, tells us the general shape of the orbital. For the orbitals belonging to a given shell with quantum number n, the quantum number l can have the following n values:

$$l = 0, 1, 2, \ldots, n - 1$$

Each value of l corresponds to one of the orbital shapes (s, p, d, or f), as shown in Table 7.3. All the orbitals belonging to the same value of l in a given shell are said to belong to the same **subshell** of the shell. Because $n = 1$ for the innermost, lowest energy shell, the only value of l allowed for that shell is 0, corresponding to an s-subshell. When $n = 2$, corresponding to the next higher energy level and the next outer shell, l can have two values, 0 or 1; so there are two subshells. One subshell (with $l = 0$) consists of an s-orbital and the other (with $l = 1$) consists of p-orbitals. The shell with $n = 3$ consists of three subshells composed of s-, p-, and d-orbitals (with $l = 0$, 1, and 2, respectively). When $n = 4$, the f-orbitals ($l = 3$) make their first appearance.

The third quantum number, m_l, labels the specific orbitals of a given subshell and tells us their orientations. The allowed values of m_l are

$$m_l = l, l - 1, l - 2, \ldots, -l$$

For example, for the subshell with $l = 1$, which consists of p-orbitals, m_l can have the values $+1$, 0, and -1, so there are three p-orbitals in this subshell. These three orbitals are normally denoted p_x, p_y, and p_z. Each label corresponds to a possible orientation of the lobes of the orbitals (recall Fig. 7.19). In general, a subshell with quantum number l consists of $2l + 1$ individual orbitals of that type (Fig. 7.23):

When $l = 0, m_l = 0$ (There is one s-orbital in the subshell with $l = 0$.)

When $l = 1, m_l = +1, 0, -1$ (There are three p-orbitals in a subshell with $l = 1$.)

When $l = 2, m_l = +2, +1, 0, -1, -2$ (There are five d-orbitals in a subshell with $l = 2$.)

All three quantum numbers are needed to identify the precise orbital occupied by an electron in an atom. The type of orbital is usually identified by giving the value of n and the letter identifying the subshell. For example, an orbital with $n = 3$ and $l = 2$ is a $3d$-orbital, and one with $n = 2$ and $l = 0$ is a $2s$-orbital.

Table 7.3 *Correspondence of the orbital angular momentum quantum number and subshell*

l	Subshell label
0	s
1	p
2	d
3	f
4	g
5	h

If we think of the atom as a city and the nucleus as the city center, we can think of quantum numbers as addresses for the orbitals. The district in which the orbital is located is specified by n. The lower the number for n, the closer is the district to the city center. The value of l identifies the street within that district, and m_l gives the number of the house itself.

FIGURE 7.23

A summary of the arrangement of shells, subshells, and orbitals in an atom and the corresponding quantum numbers. Note that the quantum number m_l is an alternative label for the individual orbitals: in chemistry, it is more common to use x, y, and z as labels, as shown in Figs. 7.19 and 7.20. There is no direct correspondence between axis designation and the numerical value of m_l.

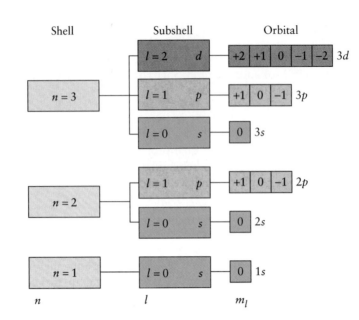

An orbital is specified by three quantum numbers; orbitals are organized into shells and subshells. The relation between shells, subshells, and orbitals is summarized in Fig. 7.23.

Example 7.5 *Identifying the number of orbitals in a shell*

How many orbitals are there in the shell with $n = 4$?

Strategy Decide which subshells are present in the shell (Fig. 7.23), write the number of orbitals in each one, and then add these numbers together. As mentioned earlier, l has whole-number values from 0 up to $n - 1$, and the number of orbitals in a subshell for a given value of l is $2l + 1$.

Solution For $n = 4$, there are four subshells with $l = 0, 1, 2, 3$, consisting of one s-orbital, three p-orbitals, five d-orbitals, and seven f-orbitals, respectively. There are therefore $1 + 3 + 5 + 7 = 16$ orbitals in the shell with $n = 4$ (Fig. 7.24).

Self-Test 7.5A Calculate the total number of orbitals in a shell with $n = 6$.

[*Answer:* 36]

Self-Test 7.5B Consider the number of orbitals in each shell from $n = 1$ to $n = 6$. Deduce from this information a general formula for the number of orbitals in a principal quantum level.

7.8 Electron Spin

Schrödinger's calculation of the orbitals and energies of the hydrogen atom was a milestone in the development of modern atomic theory. Nevertheless, the spectral lines did not have exactly the frequencies he predicted. Two Dutch-American physicists, Samuel Goudsmit and George Uhlenbeck, proposed an explanation for these tiny deviations. They suggested that an electron behaves in some respects like a spinning sphere, something like a planet rotating on its axis, and that the energy of the atom depends on the direction of the electron's rotation.

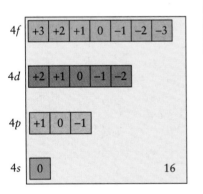

FIGURE 7.24

The solution to Example 7.5. The boxes represent the individual orbitals in each subshell. There are 16 orbitals in the shell with $n = 4$, each of which can hold two electrons.

Electron spin is a very strange property, particularly in that a spinning electron can have only two orientations. Experimental confirmation of this property was obtained by two German scientists in 1920. Otto Stern and Walter Gerlach shot a stream of silver atoms through a highly nonuniform magnetic field. A silver atom has one unpaired electron (all the rest of its 47 electrons are paired), so it behaves like a single unpaired electron riding on a heavy platform (the rest of the atom). As a result of its spin, an electron behaves like a tiny bar magnet, so the atom as a whole also behaves like a tiny magnet.

If a spinning electron behaved like a spinning ball, we would expect it to be able to spin with its axis in any orientation. In the magnetic field they used, Stern and Gerlach might have expected to observe a broad band of silver atoms arriving at their detector. The field would push the silver

atoms by different amounts according to the orientation of the spin. Indeed, that is what they observed when they first carried out the experiment. However, it is a difficult experiment to do properly, because the atoms collide with one another in the beam, and these collisions affect the paths randomly. When Stern and Gerlach repeated their experiment, they used a much less dense beam of atoms, thereby reducing the number of collisions between the atoms. In this experiment, they observed, not a single wide blurred band of atoms, but two narrow bands (see the illustration). One band corresponds to all the atoms flying through the magnetic field with one orientation of their spin; the other band corresponds to the atoms with opposite spin. The fact that only two bands are observed confirms that an electron spin can have only two orientations.

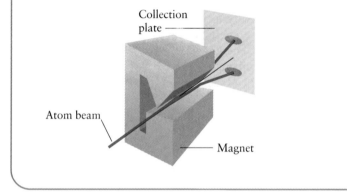

The quantization of electron spin is confirmed by the Stern-Gerlach experiment, in which a stream of atoms splits into two as it passes between the poles of a magnet. The atoms in one stream have an odd ↑ electron, and those in the other an odd ↓ electron.

The rotational motion of an electron is called **spin.** According to quantum mechanics, an electron has two spin states, represented by the arrows ↑ (up) and ↓ (down). For our purposes, we can think of an electron as being able to spin clockwise as viewed from above at a certain rate (the ↓ state) or counterclockwise at exactly the same rate (the ↑ state). Because a spinning electrical charge generates a magnetic field, electrons in these two spin states can be distinguished by their response to a magnetic field (see Investigating Matter 7.1). They are identified by a fourth quantum number, the **spin magnetic quantum number,** m_s. This quantum number can have only two values: $+\frac{1}{2}$ indicates an ↑ electron and $-\frac{1}{2}$ indicates a ↓ electron (Fig. 7.25).

> *An electron has the property of spin; the spin is described by the quantum number m_s, which may have one of two values.*

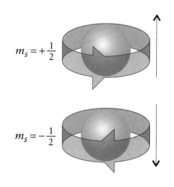

FIGURE 7.25

The two spin states of an electron can be represented as clockwise or counterclockwise rotation around an axis passing through the electron. The two states are labeled by the quantum number m_s and depicted by the arrows shown on the right.

7.9 The Electronic Structure of Hydrogen

Let's review what we now know about the hydrogen atom by imagining what happens to its single electron as the atom acquires energy. Initially, the electron is in the lowest energy level, the ground state of the atom, with $n = 1$. The only orbital in the shell with $n = 1$ is a $1s$-orbital. The electron in the ground state of a hydrogen atom is described by the following values of the four quantum numbers:

$$n = 1 \quad l = 0 \quad m_l = 0 \quad m_s = +\tfrac{1}{2} \text{ or } -\tfrac{1}{2}$$

Either spin state is allowed.

When the hydrogen atom acquires enough energy for its electron to reach the $n = 2$ shell, it can occupy any of the four orbitals in that shell. There is one $2s$-orbital ($l = 0$) and three $2p$-orbitals ($l = 1$) in this shell, and they all have the same energy.

When the atom acquires even more energy, the electron moves into the $n = 3$ shell. There it can occupy any of nine orbitals (one $3s$-, three $3p$-, and five $3d$-orbitals). Still more energy raises the electron further from the nucleus to the $n = 4$ shell, where 16 orbitals are available (one $4s$-, three $4p$-, five $4d$-, and seven $4f$-orbitals).

The electron is ejected from the atom when it absorbs enough energy to overcome the attraction of the nucleus. It then escapes from the atom, and we say that the atom has been **ionized.** To ionize a hydrogen atom from its ground state, we have to supply enough energy to move the electron from the $n = 1$ shell (which lies at $-hR_H$) up to the energy of the separated proton and electron (which is defined as the 0 of energy). Therefore, we have to supply an energy hR_H, or 2.18×10^{-18} J. Ionizing 1 mol H atoms therefore requires 6.02×10^{23} times this energy, or 1.31×10^3 kJ.

> *The state of a hydrogen atom depends on the energy it possesses. The energy hR_H must be provided to ionize a ground-state hydrogen atom.*

Self-Test 7.6A The three quantum numbers for an electron in a hydrogen atom in a certain state are $n = 4, l = 2,$ and $m_l = +1$. In what type of orbital is the electron located?

[*Answer:* $4d$]

Self-Test 7.6B The three quantum numbers for an electron in a hydrogen atom in a certain state are $n = 3, l = 1,$ and $m_l = -1$. In what type of orbital is the electron located?

THE STRUCTURES OF MANY-ELECTRON ATOMS

We have still not explained the different colors that atoms of different elements emit when they are energetically excited. Why, for example, does lithium emit red light and sodium emit yellow light (Fig. 7.26)? The answer lies in the fact that all neutral atoms other than hydrogen have more than one electron and these electrons affect one another. The helium atom ($Z = 2$) has two electrons, the lithium atom ($Z = 3$) has three electrons, and a neutral atom of an element with atomic number Z has Z electrons. All these atoms are examples of **many-electron atoms,** or atoms with more than one electron. We can extend the concept of atomic orbitals developed for hydrogen to these atoms, too, provided we take into account the interactions between the electrons. The resulting description of electronic structure is the basis of the periodic properties of the elements and, ultimately, of the formation of chemical bonds between atoms.

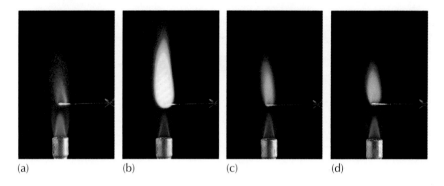

(a)   (b)   (c)   (d)

FIGURE 7.26

Flame tests are used to identify the elements in a compound. They provide an easy way of distinguishing the alkali metals: (a) lithium; (b) sodium; (c) potassium; (d) rubidium. In each case except lithium, the colors come from energetically excited atoms. In lithium's case, LiOH molecules are responsible for the color.

7.10 Orbital Energies

The electrons in a many-electron atom occupy orbitals like those of hydrogen. However, the energies of these orbitals are different. The nucleus of a many-electron atom is more highly charged than the hydrogen nucleus, so it attracts electrons more strongly and hence lowers their energy. The electrons also repel one another, which raises their energy (Fig. 7.27). The combination of these two factors affects the electronic structures and properties of the elements.

In the hydrogen atom, in which there are no electron-electron repulsions, all the orbitals of a given shell have the same energy. For instance, the 2s-orbital and all three 2p-orbitals have the same energy. The same is true of the 3s-orbital, the three 3p-orbitals, and the five 3d-orbitals of the shell with $n = 3$, which all have the same energy in a hydrogen atom. In many-electron atoms, however, electron-electron repulsions cause the energy of an electron in a 2p-orbital to be greater than that of an electron in a 2s-orbital. Similarly, in the $n = 3$ shell, the three 3p-orbitals lie higher than the 3s-orbital, and the five 3d-orbitals lie higher still (Fig. 7.28). The orbitals within a given subshell, though, continue to have the same energy. For example, all three 2p-orbitals have the same energy.

The differences in energy of orbitals in different subshells of the same shell are due to the combined effects of the attraction between the electrons and the nucleus and of the repulsions between the electrons. In other words, in addition to being attracted by the nucleus, each electron is repelled by all the other electrons in the atom. As a result, it is less tightly bound to the nucleus than it would be if those other electrons were absent. We say that each electron is **shielded** from the full attraction of the nucleus by the other electrons in the atom. The shielding effectively reduces the pull of the nucleus on an electron. The **effective nuclear charge,** $Z_{eff}e$, experienced by the electron is always less than the actual nuclear charge, Ze, because the electron-electron repulsions work against the pull of the nucleus. For example, the effect of the nuclear charge experienced by an electron in a helium atom is only about 1.3 times that experienced by an electron in a hydrogen atom, rather than 2 times (Fig. 7.29).

Electrons in different types of orbitals are shielded to different extents. An s-electron of any shell can be found very close to the nucleus, so we say that

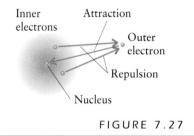

FIGURE 7.27

An electron in an atom is simultaneously attracted to the nucleus and repelled by the other electrons. The repulsion from the other electrons partially reduces the positive charge of the nucleus on the electron.

FIGURE 7.28

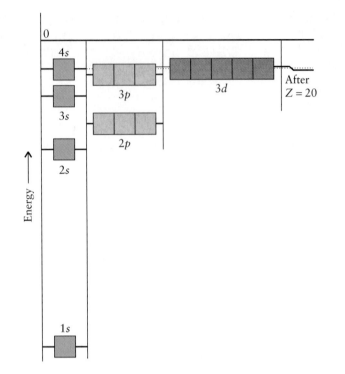

The relative energies of the shells, subshells, and orbitals in a many-electron atom. Each of the boxes can hold up to two electrons. The energies of the $3d$-orbitals lie below that of the $4s$-orbital after $Z = 20$.

it can **penetrate** through the inner shells. As a result of this penetration, an s-electron is not well shielded, is bound strongly to the nucleus, and therefore has a low energy. A p-electron of the same shell penetrates much less: we have seen that its orbital has a nodal plane passing through the nucleus, so there is zero probability of finding a p-electron at the nucleus. Because a p-electron penetrates through the inner shells less than an s-electron does, it is less strongly attracted to the nucleus and therefore has a slightly higher energy. d-Electrons penetrate even less and are bound less tightly than p-electrons of the same shell.

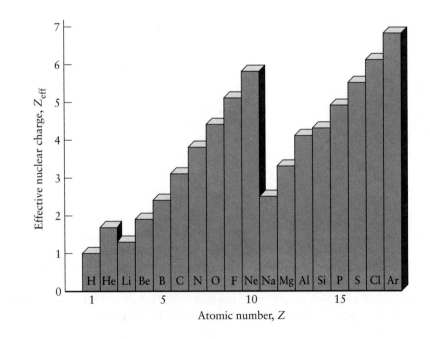

FIGURE 7.29

The variation of the effective nuclear charge on the outermost valence electron as a function of atomic number. Notice that the effective nuclear charge increases from left to right across a period, but drops when electrons enter a higher principal quantum level.

The effects of penetration and shielding can be large. A 4s-electron penetrates through inner shells so effectively that it may be much lower in energy than a 4p- or 4d-electron; it may even be lower in energy than a 3d-electron of the same atom (see Fig. 7.28). The differences in energy levels in atoms of different elements are responsible for the different energies they emit when excited and hence account for the different colors we see in fireworks (Applying Chemistry: Case Study 7).

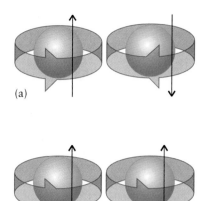

(a)

(b)

FIGURE 7.30

(a) Two electrons are said to be paired if they have opposite spins (one clockwise, the other counterclockwise). (b) Two electrons are classified as parallel if their spins are in the same direction; in this case, both ↑.

In a many-electron atom, because of the effects of penetration and shielding, s-electrons have a lower energy than p-electrons of the same shell; the order of increasing orbital energies within a given shell is $s < p < d < f$.

7.11 The Building-Up Principle

In the ground state of a many-electron atom, the electrons occupy atomic orbitals in such a way that the total energy of the atom is a minimum. We might expect an atom to have its lowest energy when all its electrons are in the lowest energy orbital (the 1s-orbital); but, except for hydrogen and helium, that can never happen. In 1925, the Austrian Wolfgang Pauli discovered why. The **Pauli exclusion principle** states that

No more than two electrons may occupy any given orbital. When two electrons do occupy one orbital, their spins must be paired.

The spins of two electrons are said to be **paired** if one is ↑ and the other is ↓ (Fig. 7.30). Paired spins are denoted ↑↓ and have spin magnetic quantum numbers of opposite signs: $+\frac{1}{2}$ and $-\frac{1}{2}$. Because an atomic orbital is designated by three quantum numbers (n, l, and m_l) and the two spin states are specified by a fourth quantum number, m_s, another way of expressing the exclusion principle for an atom is

No two electrons in an atom can have the same set of four quantum numbers.

The exclusion principle implies that each orbital in the energy level diagram in Fig. 7.28 can hold no more than two electrons. For atoms with $Z > 2$, only two electrons can enter the 1s-orbital; the rest must occupy the orbitals in outer shells, with no more than two electrons in any one orbital.

All orbitals in the same subshell have the same energy. Therefore, when more than one orbital within a subshell is available, an electron will enter an empty orbital rather than pair with an electron already present. In that way, the atom will have a lower energy because the newly arrived electron experiences less repulsion from the electrons already present. This pattern of orbital occupation is summarized by **Hund's rule:**

If more than one orbital in a subshell is available, electrons will fill empty orbitals before pairing in one of them.

The rule is named for the German spectroscopist Friedrich Hund, who first proposed it.

We report the electronic structure of an atom by writing its **electron configuration,** a list of all its occupied orbitals and the number of electrons that each orbital contains. The hydrogen atom in its ground state has one electron in a 1s-orbital. To show this structure in detail, we use a single arrow in a box representing the 1s-orbital (see diagram (**1**), which is a fragment of Fig. 7.28) and report its configuration as $1s^1$ (read "one s one"). The box in this "box diagram" represents one orbital and can hold up to two electrons.

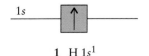

1s

1   H $1s^1$

Applying Chemistry: *Case Study 7**

The brilliant colors observed when a fireworks canister explodes high in the night sky are the outcome of a series of high-temperature chemical processes. The use of fireworks dates back over 1000 years to China and the discovery of *black powder,* a mixture of 75% potassium nitrate, 15% charcoal, and 10% sulfur. A charge of this powder is used to launch the fireworks canister high into the air. A second charge of black powder bursts the cardboard casing and ignites dozens of marble-sized pellets called *stars* that ignite as they spray out to create the displays evocatively called *chrysanthemums, peonies,* and *weeping willows.* Each star contains a mixture of chemicals that burns to produce specific flame colors. The exact formulas for these mixtures are closely guarded by the fireworks manufacturers, but certain general features are known.

The reactions that take place during the combustion of each star are redox reactions. Potassium perchlorate, $KClO_4$, and potassium chlorate, $KClO_3$, are typical pyrotechnic oxidizing agents. The fuels—the compounds oxidized in the reaction—are generally organic compounds. They include natural products, such as charcoal, tree resins, or cornstarch derivatives, and synthetic products, such as polyvinylchloride and chlorinated rubber. The pyrotechnics mixture also contains compounds of elements such as strontium, barium, and sodium, which emit light of different colors when heated to the high temperatures produced during the combustion.

Metallic fuels, such as aluminum, magnesium, and titanium, burn at very high temperatures and are used to produce brilliant white flames. For example, aluminum is used in "sparklers." These mixtures are less frequently used to generate specific flame colors because their high flame temperatures (close to 3500°C) tend to destroy the other more fragile molecules that produce specific colors.

Let's look at how a starburst emits light. First, a species such as SrCl (think of this "molecule" as an ion pair, consisting of a single Sr^+ cation and a Cl^- ion in contact) in an energetically excited state is generated in the combustion reaction. The electrons of the molecule are excited to a high energy level by the high temperature of the exothermic redox reaction. The excited molecule then returns to a lower energy level. In so doing, it emits a photon of red light of specific wavelength corresponding to the energy difference between the excited and ground states.

$$SrCl^* \longrightarrow SrCl + h\nu \qquad (\lambda = 620\text{–}750 \text{ nm})$$

Modern instrumentation has enabled scientists to identify many of the principal species responsible for the other colors produced. Green light (light with wavelengths

This fireworks display in Italy is the result of the simultaneous ignition of several mortars. The colors are determined by the elements used in the individual stars. The brightness of the light depends on the type of fuel used.

In the ground state of a helium atom ($Z = 2$), both electrons are in a $1s$-orbital, which is reported as $1s^2$ ("one s two"). As we see in (**2**), the two electrons are paired. At this point, the $1s$-orbital holds its maximum number of electrons. The shell with $n = 1$ is now complete and no more electrons can occupy it. We say that the helium atom has a **closed shell,** which is a shell containing the maximum number of electrons allowed by the exclusion principle.

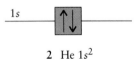

2 He $1s^2$

of 490–560 nm) is emitted by BaCl molecules, and CuCl molecules are the best blue emitters (435–440 nm). Atomic sodium, formed by the vaporization of a sodium compound, produces the bright yellow-orange light (589 nm) seen in some displays.

The chemical mixtures in fireworks produce beautiful red, white, green, and orange flames. However, a deep blue color has remained elusive. The best blue emitter, CuCl, is unstable at the high flame temperatures (above 1500°C) needed to produce intense emissions. In addition, blue light is less efficiently perceived by humans than other colors are. So there is still a need—even after 1000 years of experimentation—for a chemical mixture that burns with a bright blue flame. Perhaps someone is developing the right mixture at this moment. Look for a deep blue color next time you see a fireworks display.

Key Concepts: atomic spectra, electronic transitions

For Further Reading
G. W. Weingart, *Pyrotechnics,* Boulder, CO: Paladin Press, 1983.

Related Exercises: 7.108–7.110

*This case study is an adaptation of a contribution by John A. Conkling, Washington College, Executive Director, American Pyrotechnics Association.

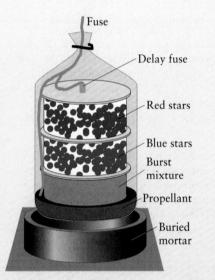

The structure of an aerial shell used in firework displays. The mortar must be buried in the ground, to maintain its upward trajectory (see Fig. 6.20). The black powder on the bottom is set off to launch the shell. A second charge bursts the shell in the air and ignites the stars, which form the graceful arcs of light in the sky.

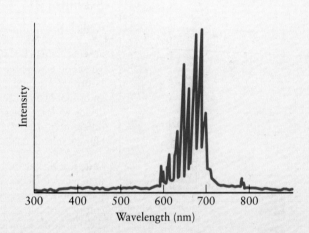

Spectroscopic methods are used to study the light emission from pyrotechnics mixtures. This diagram shows the intensity of light emitted from a red flame produced by burning a mixture of ammonium perchlorate (NH_4ClO_4, an oxidizer), strontium carbonate (a source of strontium atoms), and chlorinated rubber (a fuel and a source of chlorine atoms). In the flame, strontium and chlorine atoms combine to generate SrCl, which emits light in the red region (620–750 nm) of the visible spectrum when exposed to high temperature.

Lithium ($Z = 3$) has three electrons. Two electrons can occupy the 1s-orbital and complete the $n = 1$ shell. The third electron must occupy the next available orbital up the ladder of energy levels, which according to Fig. 7.28 is the 2s-orbital. The ground state of a lithium atom is therefore $1s^2 2s^1$ (**3**). We can think of this atom as consisting of a core made up of the inner $n = 1$ closed shell surrounded by an outer $n = 2$ shell containing a higher energy electron.

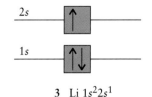

3 Li $1s^2 2s^1$

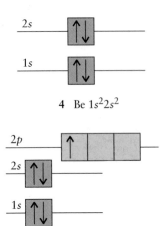

4 Be $1s^2 2s^2$

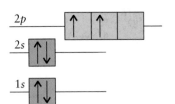

5 B $1s^2 2s^2 2p^1$

Electrons in the outermost shell are called **valence electrons.** In general, only valence electrons can be lost in chemical reactions, because core electrons are too tightly bound. Therefore, lithium will lose only one electron when it forms compounds; hence it will form Li^+ ions, rather than Li^{2+} or Li^{3+} ions.

The element with $Z = 4$ is beryllium, Be, with four electrons. The first three electrons form $1s^2 2s^1$, like Li. The fourth electron pairs with the $2s$-electron, giving $1s^2 2s^2$ (**4**). A Be atom therefore has a heliumlike core surrounded by a valence shell of two paired electrons. Just as lithium has only one valence electron to lose and hence forms Li^+ ions, a Be atom can lose its two valence electrons to form Be^{2+} ions.

The atomic number of boron is 5, so a boron atom has five electrons. Two enter the $1s$-orbital and complete the $n = 1$ shell. The $2s$-orbital can accommodate two of the remaining electrons, so the fifth electron must occupy an orbital of the next available subshell, which Fig. 7.28 shows is a $2p$-subshell. This arrangement of electrons is reported as $1s^2 2s^2 2p^1$ (**5**).

Example 7.6 *Writing the electron configurations of atoms*

Predict the ground-state electron configuration of a carbon atom.

Strategy First, decide how many electrons are present (from the atomic number). Then add arrows representing electrons to the boxes in Fig. 7.28. Start at the $1s$-orbital, and add a pair of electrons to each orbital in a subshell before moving to an orbital in a subshell of higher energy. Carbon has six electrons, so we need to decide whether the sixth electron will join the one already in the $2p$-orbital or enter a different $2p$-orbital. (Remember, there are three p-orbitals in the subshell.) Use Hund's rule to answer this question.

Solution Electrons that occupy different p-orbitals are farther from one another than when they occupy the same orbital. Therefore, they repel each other less than they would if they occupied the same orbital. So, the sixth electron goes into an empty p-orbital, and the ground state of carbon is $1s^2 2s^2 2p_x^1 2p_y^1$ (**6**). We write out the individual subshells like this only when we need to emphasize that electrons occupy different orbitals in a subshell. In most cases, we can write the shorter form, such as $1s^2 2s^2 2p^2$.

Self-Test 7.7A Predict the ground-state configuration of a nitrogen atom.
[***Answer:*** $1s^2 2s^2 2p_x^1 2p_y^1 2p_z^1$ (**7**), or $1s^2 2s^2 2p^3$]

Self-Test 7.7B Predict the ground-state configuration of an oxygen atom.

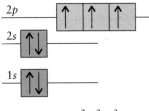

6 C $1s^2 2s^2 2p^2$

7 N $1s^2 2s^2 2p^3$

Note that in the box diagram for the electron configuration of carbon we have drawn the two $2p$-electrons with **parallel spins** ($\uparrow\uparrow$). The notation indicates that they have the same spin magnetic quantum numbers. This arrangement turns out to have slightly lower energy than a paired arrangement. However, it is allowed only when the electrons occupy different orbitals.

We can simplify the description of electron configurations by realizing that once the orbitals of a shell are filled, the elements in the following period have the same core. For example, the configuration of lithium is $1s^2 2s^1$. Because it consists of a single $2s$-electron outside a heliumlike $1s^2$ core, its configuration can be written $[He]2s^1$. Similarly, the configuration of carbon can be written $[He]2s^2 2p^2$. We can think of an atom as having a noble-gas core surrounded by electrons in the **valence shell,** the outermost occupied shell. The valence shell is the occupied shell with the largest value of n.

The procedure we have been using to construct the ground-state electron configurations of atoms is called the **building-up principle.** Let's use the building-up principle to complete the electron configurations across Period 2. From Self-Test 7.7A, we see that the electron configuration of nitrogen can be written as $[He]2s^22p^3$. Each p-electron occupies a different orbital, and the three have parallel spins. Oxygen has $Z = 8$ and one more electron than nitrogen; therefore its configuration is $[He]2s^22p^4$ (**8**). Similarly, fluorine, with $Z = 9$ and one more electron than oxygen, has the configuration $[He]2s^22p^5$ (**9**). The fluorine atom can be pictured as having a heliumlike core surrounded by a valence shell that is complete except for one p-electron. Neon, with $Z = 10$, has one more electron than fluorine. This electron completes the $2p$-subshell, giving $[He]2s^22p^6$ (**10**). According to Fig. 7.28, the next electron enters the $3s$-orbital, the lowest energy orbital of the next shell. The configuration of sodium is therefore $[He]2s^22p^63s^1$, or, more briefly, $[Ne]3s^1$.

Some people call the building-up principle by its equivalent German name, the *Aufbau* principle.

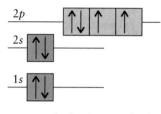

8   O $1s^22s^22p^4$, $[He]2s^22p^4$

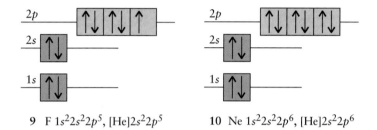

9   F $1s^22s^22p^5$, $[He]2s^22p^5$ 10   Ne $1s^22s^22p^6$, $[He]2s^22p^6$

How to use the building-up principle to write electron configurations is summarized in Toolbox 7.1 and illustrated in Examples 7.7 and 7.8.

The ground-state electron configuration of an atom of an element with atomic number Z is predicted by adding Z electrons to available orbitals so as to obtain the lowest total energy.

Toolbox 7.1   *How to predict the ground-state electron configuration of an atom*

This Toolbox describes how to predict the electron configuration of an atom in its state of lowest energy.

Conceptual Basis
Electrons occupy orbitals in such a way as to minimize the total energy of an atom (Fig. 7.31). The lowest total energy is achieved by occupying the orbitals of lowest energy subject to the requirements of the Pauli exclusion principle. All orbitals in a given subshell have the same energy. When more than one orbital of the same energy is available, electrons take positions as far apart as possible.

Procedure
To assign a ground-state configuration to an element with atomic number Z,

Step 1. Add Z electrons, one after the other, to the orbitals in the order shown in Figs. 7.28 and 7.31, but with

no more than two electrons in any one orbital (the Pauli exclusion principle).

Step 2. If more than one orbital in a subshell is available, add electrons to different orbitals of the subshell before doubly occupying any of them (Hund's rule).

Step 3. Write the orbitals in order of increasing energy. The configuration of a filled shell is represented by the symbol of the noble gas having that configuration, as in [He] for $1s^2$.

Step 4. When drawing a box diagram, show the electrons in different orbitals of the same subshell with parallel spins; electrons sharing an orbital should be shown with paired spins.

In most cases, this procedure gives the configuration of the atom that corresponds to its lowest total energy, taking into account the attraction of the electrons to the nucleus and their repulsion from one another. The few exceptions arise from the stability of half-filled and filled shells and subshells. Any other arrangement is an excited state of the atom. You can check the electron configurations you write by referring to Appendix 2C or the periodic table inside the front cover of this book.

FIGURE 7.31

The order in which atomic orbitals are occupied according to the building-up principle. Each time an electron is added, we move one place to the right. At the end of a period, we move to the start of the next period. The names of the blocks of the periodic table are based on the last subshell occupied in an atom of an element according to the building-up principle.

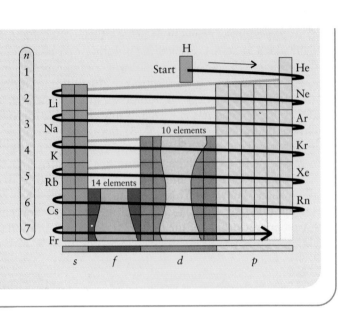

Example 7.7 *Using the building-up principle*

Predict the ground-state electron configuration of a sulfur atom and draw the box diagram.

Strategy Because sulfur and oxygen are in the same group of the periodic table (Group 16), expect S to have a ground-state electron configuration analogous to that of O shown in (**8**). Use the procedure set out in Toolbox 7.1.

Solution **Step 1.** Sulfur has $Z = 16$, so there are 16 electrons to contribute. The 16 electrons enter the orbitals in the order shown in Fig. 7.31, resulting in the configuration $1s^22s^22p^63s^23p^4$.

Step 2. In more detail, three of the four $3p$-electrons enter separate $3p$-orbitals, and the fourth pairs with one of the electrons, giving the configuration $1s^22s^22p^63s^23p_x^23p_y^13p_z^1$.

Step 3. We can simplify the configuration to $[Ne]3s^23p^4$.

Step 4. The two unpaired $3p$-electrons have parallel spins. The box diagram is shown in (**11**).

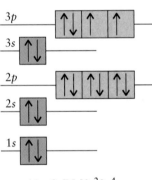

11 S $[Ne]3s^23p^4$

Self-Test 7.8A Predict the ground-state electron configuration of a chlorine atom.
[**Answer:** $1s^22s^22p^63s^23p^5$, or $[Ne]3s^23p^5$]

Self-Test 7.8B Predict the ground-state electron configuration of a magnesium atom.

7.12 The Electron Configurations of Atoms

Once we know the electron configurations of the atoms, the organization of the periodic table falls into place. The principal quantum number, *n*, of an element's valence shell is equal to its period number. For example, the valence shell of elements in Period 2 (lithium to neon) is the shell with $n = 2$. All the atoms in the same period have the same noble-gas core. Thus the atoms of Period 2 elements all have a heliumlike $1s^2$ core, and those of Period 3 elements have a neonlike $1s^2 2s^2 2p^6$ core. Elements belonging to the same group have analogous valence electron configurations but different cores and different values of the principal quantum number of the valence shell. For example, in Group 13, boron has the electron configuration $[\text{He}]2s^2 2p^1$ and aluminum the configuration $[\text{Ne}]3s^2 3p^1$.

In the last section, we built up the electron configurations for the Period 2 elements. The configurations of Period 3 elements are built up in the same way, but what happens after that? The 3s- and 3p-orbitals are full at argon, $[\text{Ne}]3s^2 3p^6$, which is a colorless, odorless, unreactive gas resembling neon. At this point, we might expect electrons to enter the 3d-orbitals, but the fourth period begins with the filling of 4s-orbitals, which are slightly lower in energy than the 3d-orbitals, and the next two configurations are $[\text{Ar}]4s^1$ for potassium and $[\text{Ar}]4s^2$ for calcium. At this point, the 3d-orbitals are the next to be filled, and there is a change in the rhythm of the periodic table.

According to Toolbox 7.1, the next 10 electrons (for scandium, with $Z = 21$, through zinc, with $Z = 30$) enter the 3d-orbitals. The ground-state electron configuration of scandium, for example, is $[\text{Ar}]3d^1 4s^2$, and that of its neighbor titanium is $[\text{Ar}]3d^2 4s^2$. Note that, beginning at scandium, we write the 4s-electrons after the 3d-electrons: the 3d-orbitals now have a lower energy than the 4s-orbital (recall Fig. 7.28).

For transition metals, the electrons are added to the d-orbitals as Z increases. However, there are two exceptions: the half-complete subshell configuration d^5 and the complete subshell configuration d^{10} turn out experimentally to have a lower energy than simple theory suggests. In some cases, the neutral atom has a lower total energy if the 3d-subshell is half-full (d^5) or full (d^{10}) as a result of transferring a 4s-electron into it. For example, the experimental configuration of chromium is $[\text{Ar}]3d^5 4s^1$ and that of copper is $[\text{Ar}]3d^{10}4s^1$. You can find other exceptions to the building-up principle in Appendix 2C.

Because transition metals in the same period differ mainly in the number of d-electrons, their properties are very similar. Most of them form ions in more than one oxidation state, because the d-electrons have similar energies and a variable number can be lost.

Electrons occupy 4p-orbitals once the 3d-orbitals are full. The configuration of germanium, $[\text{Ar}]3d^{10}4s^2 4p^2$, for example, is obtained by adding two electrons to the 4p-orbitals outside the completed 3d-subshell. Period 4 is the first **long period** of the periodic table.

Next in line for occupation at the beginning of Period 5 is the 5s-orbital, followed by the 4d-orbitals. As in Period 4, the energy of the 4d-orbitals falls below that of the 5s-orbital after two electrons have been accommodated in the 5s-orbital. A similar effect is seen in Period 6, but now another set of orbitals, the 4f-orbitals, begins to be occupied. Cerium, for example, has the configuration $[\text{Xe}]4f^1 5d^1 6s^2$. Electrons then continue to occupy the seven 4f-orbitals, which are complete after 14 electrons have been added, at ytterbium, $[\text{Xe}]4f^{14}6s^2$. Next, the 5d-orbitals are occupied. The 6p-orbitals are occupied only after the 6s- and 5d-orbitals are filled at mercury; thallium, for example, has the configuration $[\text{Xe}]4f^{14}5d^{10}6s^2 6p^1$.

We write the ground-state electron configuration of an atom by using the building-up principle in conjunction with Fig. 7.31, the Pauli exclusion principle, and Hund's rule.

Example 7.8 *Predicting a ground-state configuration of a heavy atom*

Predict the ground-state configuration of a lead atom.

Strategy To proceed systematically, use the rules set out in Toolbox 7.1. However, there is a quicker method. First, note the group number, which is a clue to the ground-state valence electron configuration of the atom. Then note the period number, which tells us the value of the principal quantum number of the valence shell. The core consists of the preceding noble-gas configuration together with any completed *d*- and *f*-subshells. Because lead is in the same group as carbon (Group 14), we can anticipate that it will have an analogous valence configuration of the form ns^2np^2.

Solution Lead belongs to Group 14 and Period 6. It therefore has four electrons in its valence shell: two in a 6*s*-orbital and two in different 6*p*-orbitals. The period has complete 5*d*- and 4*f*-subshells, and the preceding noble gas is xenon. The electron configuration of lead is therefore $[Xe]4f^{14}5d^{10}6s^26p^2$.

Self-Test 7.9A Write the ground-state configuration of a bismuth atom.

[*Answer:* $[Xe]4f^{14}5d^{10}6s^26p^3$]

Self-Test 7.9B Write the ground-state configuration of a zirconium atom.

7.13 The Electron Configurations of Ions

A neutral atom becomes a cation when it loses one or more electrons. If the principal quantum number of the valence shell is *n*, any *np*-electrons are lost first, because electrons in that subshell are least tightly bound. Then *ns*-electrons are lost, and finally the *d*-electrons of the previous shell, if any, until the appropriate number of electrons have been removed. Thus, for the Fe^{3+} ion, we work out the configuration of the Fe atom, which is $[Ar]3d^64s^2$, and remove the three least tightly bound electrons from it. There are no 4*p*-electrons, so the first two electrons removed are 4*s*-electrons. The third electron comes from the 3*d*-subshell, giving $[Ar]3d^5$.

Example 7.9 *Writing the electron configurations of cations*

Write the electron configurations of the In^+ and In^{3+} ions.

Strategy Determine the configuration of the neutral atom, as described in Toolbox 7.1. Remove one electron to obtain a singly charged ion, two for a doubly charged ion, and so on. Remove electrons from the valence-shell *p*-orbitals first, then from the *s*-orbitals, and finally, if necessary, from the *d*-orbitals in the next lower shell.

Solution Indium, in Group 13, has three valence electrons. It is in Period 5, so its valence shell has $n = 5$. It is preceded by the Period 4 noble gas krypton, so its noble-gas core is [Kr]. Because indium is in the *p* block, it has filled 4*d*-orbitals. Its ground-state configuration is therefore $[Kr]4d^{10}5s^25p^1$ (**12**). One electron can be lost

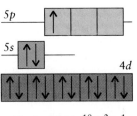

12 In $[Kr]4d^{10}5s^25p^1$

from the 5p-orbital, so In$^+$ has the configuration [Kr]$4d^{10}5s^2$ (**13**). The next two electrons are lost from the 5s-orbital: In^{3+} has the configuration [Kr]$4d^{10}$ (**14**).

Self-Test 7.10A Write the electron configurations of the copper(I) and copper(II) ions.

[***Answer:*** [Ar]$3d^{10}$, [Ar]$3d^9$]

Self-Test 7.10B Write the electron configurations of a manganese(II) ion and manganese in oxidation state +5.

To form monatomic anions, we add enough electrons to complete the valence shell. For example, nitrogen has five valence electrons, so three more electrons are needed to reach a noble-gas configuration, that of neon. Therefore, the ion will be N^{3-}. The configuration of the nitrogen atom is [He]$2s^2 2p^3$, with room for three more electrons in the 2p-subshell. Thus, the N^{3-} ion has the configuration [He]$2s^2 2p^6$ (**15**), the same as that of neon.

> ***To predict the electron configuration of a cation, remove outermost electrons in the order np, ns, and (n − 1)d; for an anion, add electrons until the next noble-gas configuration has been reached.***

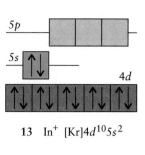

13 In$^+$ [Kr]$4d^{10}5s^2$

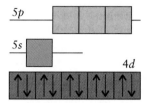

14 In^{3+} [Kr]$4d^{10}$

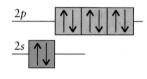

15 N^{3-} [He]$2s^2 2p^6$

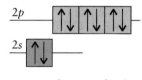

16 O^{2-} [He]$2s^2 2p^6$

Example 7.10 *Writing the electron configurations of anions*

Write the formula and electron configuration of the oxide ion.

Strategy First, establish the ground-state configuration of the neutral oxygen atom by using the building-up principle. Then add enough electrons to complete the vacant orbitals and achieve the closed-shell configuration of the next noble-gas atom (Ne).

Solution We have already seen that the electron configuration of an O atom is [He]$2s^2 2p^4$. The 2p-subshell can accommodate two more electrons, so we expect the oxide ion to be O^{2-}, with configuration [He]$2s^2 2p^6$ (**16**), the same as that of neon.

Self-Test 7.11A Write the chemical formula and electron configuration of the phosphide ion.

[***Answer:*** P^{3-}, [Ne]$3s^2 3p^6$]

Self-Test 7.11B Write the chemical formula and electron configuration of the iodide ion.

7.14 Electronic Structure and the Periodic Table

The periodic table was developed from a consideration of the physical and chemical properties of the elements (Investigating Matter 7.2). However, we can explain the basis for the organization of the periodic table by considering electron configurations. The table is divided into *s, p, d,* and *f* blocks, named for the last subshell that is occupied according to the building-up principle (see Fig. 7.31). Two elements are exceptions. The electron configuration of helium would put it in the *s* block, but its properties put it in the *p* block. It is a gas with properties matching those of the noble gases in Group 18, rather than the reactive metals in Group 2. Its place in Group 18 is justified by the fact that, like all the other Group 18 elements, it has a filled valence shell. Hydrogen occupies a unique position in the periodic table. It has one *s*-electron, so it belongs in

Some periodic tables display hydrogen in Group 1, some in Group 17, and some in both groups.

The periodic table is one of the most important concepts in chemistry. Its development is an example of how scientific discoveries can come from using insight to organize data collected by a large number of people over a period of many years. There are often simultaneous discoveries in science, because discoveries tend to be made when enough of the right kind of data has been collected.

Until the middle of the nineteenth century, the separation and identification of elements and the determination of their relative masses was a primary occupation of chemists. As the number of known elements grew, scientists observed curious similarities in certain groups of elements. The German chemist Johann Döbereiner described *triads* of elements with similar properties. He noticed that the relative atomic mass (which in those days was defined as the atomic mass relative to that of hydrogen*) of the central member of the triad was approximately the mean of the relative atomic masses of the two other members. For example, the average of the relative atomic masses of chlorine and iodine used at that time, $\frac{1}{2}(35.47 + 126.47)$, was 80.97, close to the value of the relative atomic mass of bromine (which at the time was known as 78.38). The English chemist John Newlands responded with his *law of octaves,* in which he organized elements in groups of eight by their properties. Observations like these suggested an intrinsic organization in matter, but no one could find the basis for it. The 63 known elements were simply displayed in long lists.

In 1860, the Congress of Karlsruhe brought together many prominent chemists in an attempt to come to some

Dmitri Ivanovich Mendeleev (1834–1907).

agreement on issues such as the existence of atoms, correct relative atomic masses, and how the elements are related to one another. No agreement was reached, but new experimental evidence, conclusions, and proposals were presented in an atmosphere of stimulating debate. Two scientists attending the congress, Lothar Meyer from Germany and Dmitri Mendeleev from Russia, left with copies of papers and new ideas. They discovered independently in 1869 that a regular repeating pattern of properties could be observed when the elements were arranged in order of increasing relative atomic mass. Mendeleev called this observation the *periodic law.*

Mendeleev, who taught general chemistry and wrote chemistry textbooks, was looking for a way to help students organize information about the elements. He made up cards

The regions of the periodic table were first introduced in Section 1.6.

Group 1. But it is also one electron short of a noble-gas configuration, so it can also act like a member of Group 17.

The *s* and *p* blocks comprise the *main groups.* The atoms of elements in the same main group have analogous valence-shell electron configurations, which differ only in the value of *n.* The similar electron configurations for the elements in a main group are the reason for their similar properties. The group number tells us how many valence-shell electrons are present. In the *s* block, the group number (1 or 2) is the same as the number of valence electrons. In the *p* block, we have to subtract 10 from the group number to find the number of valence electrons. For example, fluorine in Group 17 has seven valence electrons.

Each new period corresponds to the occupation of a shell with a new principal quantum number. This feature explains the different lengths of the periods. Period 1 consists of only two elements, H and He, in which the single 1*s*-orbital of the $n = 1$ shell is being filled with its two electrons. Period 2 consists of the eight elements Li through Ne, in which the one 2*s*- and three 2*p*-orbitals are being filled with eight more electrons. In Period 3 (Na through Ar), the 3*s*- and 3*p*-orbitals are being occupied by eight additional electrons. In Period 4, not only are the 8 electrons of the 4*s*- and 4*p*-orbitals being added, but so are the 10 electrons of the 3*d*-orbitals. Hence there are 18 elements in Period 4.

with the names and properties of the elements and arranged them in different ways, looking for relationships among elements. According to legend, not finding success, he fell asleep discouraged. He then awoke with a new plan, to arrange the elements in rows by increasing relative atomic mass, beginning a new row when the cycle of properties repeated itself. This arrangement put elements with similar properties into columns, forming a table of repeating, or periodic, properties. Meyer found a similar organization, but it was Mendeleev who realized that the arrangement was the key to the fundamental organization of matter.

Mendeleev's chemical insight led him to leave gaps for elements that would be needed to complete the pattern but were unknown at the time. When they were discovered later, he turned out to be strikingly correct. For example, his pattern required an element he named eka-silicon below silicon and between gallium and arsenic. He predicted that the element would have a relative atomic mass of 72 and properties similar to those of silicon. This prediction spurred the German chemist Clemens Winkler in 1886 to discover eka-silicon, which he named germanium. It has a relative atomic mass of 72.6 and properties similar to those of silicon, as shown in the table.

One problem with Mendeleev's table was that some elements seemed to be out of place. For example, when argon was isolated, it did not seem to have the correct mass for its location. Its relative atomic mass of 40 is the same as calcium's, but argon is an inert gas and calcium a reactive metal. Such anomalies led scientists to question the use of relative

Mendeleev's predictions for eka-silicon (germanium)

Property	Eka-silicon (E)	Germanium
molar mass, g/mol	72	72.59
density, g/cm^3	5.5	5.32
melting point, °C	high	937
appearance	dark gray	gray-white
oxide	EO_2; white solid; amphoteric; density 4.7 g/cm^3	GeO_2; white solid; amphoteric; density 4.23 g/cm^3
chloride	ECl_4; boils below 100°C; density 1.9 g/cm^3	$GeCl_4$; boils at 84°C; density 1.84 g/cm^3

atomic mass as the basis for organizing the elements. When Henry Moseley examined x-ray spectra of the elements in the early twentieth century (as described in Section 1.3), he realized that all atoms of the same element had identical nuclear charge and, therefore, the same atomic number. It was soon discovered that elements fall into the uniformly repeating pattern of the periodic table if they are organized according to atomic number, rather than atomic mass.

*Now the relative atomic mass (RAM) is defined relative to the mass of a carbon-12 atom taken as exactly 12.

Period 5 elements add another 18 electrons as the 5s-, 4d-, and 5p-orbitals are filled. In Period 6, a total of 32 electrons are added, because 14 electrons are also being added to the 4f-orbitals. The f-block elements in Period 6 are called the *lanthanides,* or rare earths. They have very similar properties, because their electron configurations differ only in the population of inner f-orbitals, and these electrons do not participate much in bond formation. The same is true of the f-block elements in Period 7, the *actinides,* which are all radioactive metals and have very similar properties.

The blocks of the periodic table are named for the last orbital to be occupied according to the building-up principle. The periods are numbered according to the principal quantum number of the valence shell.

Self-Test 7.12A Write (a) the number of valence electrons and (b) the valence electron configuration of the atoms of the group that contains tin.

[*Answer:* (a) 4; (b) ns^2np^2]

Self-Test 7.12B Write (a) the number of valence electrons and (b) the valence electron configuration of the atoms of the group that contains arsenic.

FIGURE 7.32

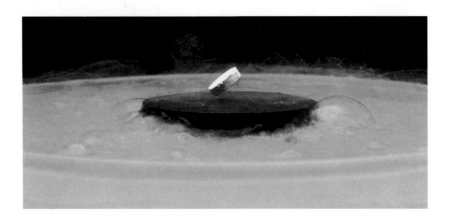

A sample of a high-temperature superconductor, a type of material first produced in 1987. The small magnet hovers over the superconductor, which is cooled by liquid nitrogen. When the material becomes warm, superconductivity is lost and the magnet falls.

THE PERIODICITY OF ATOMIC PROPERTIES

Because the periodic table is a representation of the regularities of atomic structure, we can use it to predict the properties of elements. In particular, we may be able to predict the existence of compounds suitable for making new materials and applying known materials to new purposes. For example, new materials like high-temperature superconductors promise to change many aspects of our lives. Superconductors conduct electricity without resistance and, in a magnetic field, can even be used to levitate matter against the force of gravity (Fig. 7.32). To find the best composition for high-temperature superconductors, materials scientists start with mixtures and compounds known to be effective and prepare variations of them (Fig. 7.33). In looking for the most suitable choices of elements, they select elements on the basis of their location in the periodic table.

In Section 1.6, we used the periodic table to predict the physical and chemical properties of elements. Now we shall see how scientists use the table to predict properties we cannot see directly, such as trends in the radii of atoms and ions. These properties will turn out to be crucial for understanding bonding (Chapters 8 and 9) and the properties of materials (Chapter 10).

FIGURE 7.33

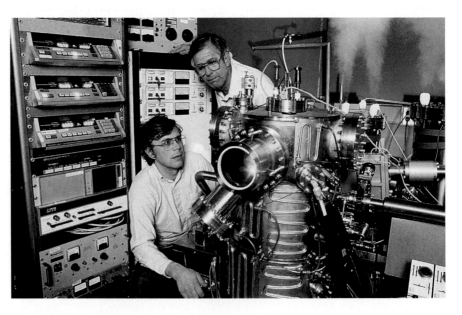

Two materials scientists test a superconductor they have just prepared with a new formulation. The periodic table is the "keyboard" of the materials scientist, who must select elements on the basis of their properties.

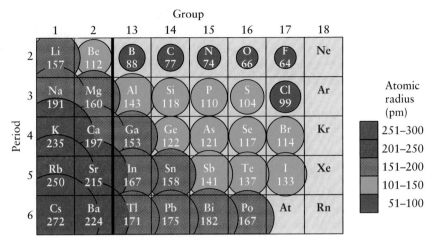

The atomic radii (in picometers, 1 pm = 10^{-12} m) of the main-group elements. The radii decrease from left to right within a period and increase down a group. The colors used here and in subsequent charts represent the general magnitude of the property, as indicated by the key on the right.

7.15 Atomic Radius

Electron clouds do not have distinct boundaries. However, when atoms pack together in solids and molecules, their centers are found at definite, measurable, distances from one another. The **atomic radius** of an element is defined as half the distance between the centers of neighboring atoms (**17**). If the element is a metal, the distance is between the centers of neighboring atoms in a solid sample. Because the distance between neighboring nuclei in solid copper is 256 pm, the atomic radius of copper is 128 pm (where 1 pm = 10^{-12} m). If the element is a nonmetal, the appropriate distance is that between the centers of atoms joined by a chemical bond. Half this distance is also called the **covalent radius** of the element. The distance between the nuclei in a Cl_2 molecule is 198 pm, so the atomic (covalent) radius of chlorine is 99 pm.

Figure 7.34 shows some atomic radii, and Fig. 7.35 shows their variation with atomic number. Note the periodic, sawtooth pattern. Atomic radii generally decrease from left to right across a period and increase down a group.

The increase down a group, like that from Li to Cs, makes sense: with each new period, the outermost electrons occupy shells that lie farther from the nucleus. But how can we account for the decrease across a period, like that from

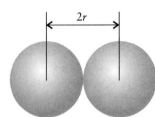

17 Atomic radius

To see how the values are measured, see Investigating Matter 10.1.

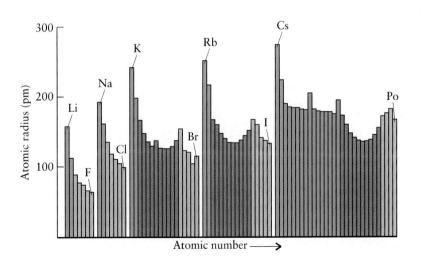

The periodic variation of the atomic radii of the elements. This variation can be explained in terms of the effect of increasing effective nuclear charge as atomic number increases across a period and the occupation of shells with increasing principal quantum number down a group. Numerical values are given in Appendix 2D.

Li to Ne, despite the fact that the number of electrons is increasing? The explanation is that the new electrons are entering the same shell of the atom and are about as close to the nucleus as other electrons in the shell. However, the increasing effective nuclear charge from left to right across a period draws the electrons in and, as a result, the atom is more compact.

Atomic radii generally increase down a group and decrease from left to right across a period.

7.16 Ionic Radius

The **ionic radius** of an element is its share of the distance between neighboring ions in an ionic solid (**18**). The distance between the nuclei of a cation and a neighboring anion is the sum of the two ionic radii. In practice, we use the radius of the oxide ion, 140 pm, to calculate the radii of other ions. For example, because the distance between Mg and O nuclei in magnesium oxide is 212 pm, the radius of the Mg^{2+} ion is calculated as $212 - 140$ pm $= 72$ pm.

Figure 7.36 illustrates some ionic radii, and Fig. 7.37 shows the relative sizes of some ions and their parent atoms. *All cations are smaller than their parent atoms.* The neutral atom loses one or more electrons, and usually all its valence electrons, to form the cation and exposes its smaller core. For example, the atomic radius of Li, with the configuration $1s^2 2s^1$, is 157 pm, but the ionic radius of Li^+, the bare heliumlike $1s^2$ core of the parent atom, is only 58 pm. This size difference is comparable to that between a cherry and its pit. Like atomic radii, cation radii increase down each group because electrons are occupying shells with higher principal quantum numbers.

Figure 7.37 shows that *anions are larger than their parent atoms.* The reason can be traced to the increased number of electrons in the valence shell of the anion and the repulsive effects the electrons exert on one another. The variation in the radii of anions is the same as that for atoms and cations, with the smallest at the upper right of the periodic table, close to fluorine.

Atoms and ions with the same number of electrons are called **isoelectronic.** For example, Na^+, F^-, and Mg^{2+} are isoelectronic. All three ions have the same number of electrons (10) and the same electron configuration, $[He]2s^2 2p^6$, but their radii differ because they have different nuclear charges. The Mg^{2+} ion has the largest nuclear charge of these three species ($Z = 12$), so it has the strongest

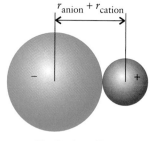

18 Ionic radius

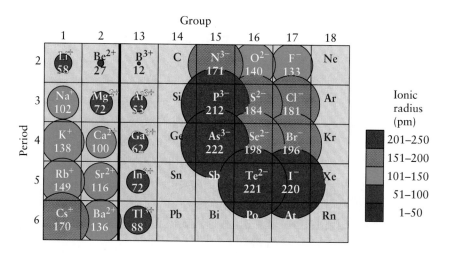

FIGURE 7.36

The ionic radii (in picometers) of the main-group elements. Note that cations (on the left) are typically smaller than anions (on the right), and in some cases very much smaller.

attraction for the 10 electrons and therefore the smallest radius. The F^- ion has the smallest nuclear charge ($Z = 9$), so the smaller pull of its nucleus on the 10 electrons results in a larger radius.

Ionic radii generally increase down a group and decrease from left to right across a period. Cations are smaller than their parent atoms and anions are larger.

Example 7.11 *Predicting the relative sizes of ions*

Arrange each pair of ions in order of increasing ionic radius: (a) Mg^{2+} and Ca^{2+}; (b) O^{2-} and F^-.

Strategy The smaller member of a pair with the same number of electrons will be an ion of an element that lies further to the right in a period, because that element will have the greater nuclear charge. If the two ions are in the same group, the smaller ion will be the one that lies higher in the group, because its electrons are in a shell with a lower principal quantum number and closer to the nucleus.

Solution (a) Because Mg lies above Ca in Group 2, it will have the smaller ionic radius; the actual values are 72 and 100 pm, respectively. (b) Because F lies to the right of O in Period 2, it will have the smaller ionic radius. The actual values are 133 and 140 pm, respectively.

Self-Test 7.13A Arrange each pair of ions in order of increasing ionic radius: (a) Mg^{2+} and Al^{3+}; (b) O^{2-} and S^{2-}.

[***Answer:*** (a) $r(Al^{3+}) < r(Mg^{2+})$; (b) $r(O^{2-}) < r(S^{2-})$]

Self-Test 7.13B Arrange each pair of ions in order of increasing ionic radius: (a) Ca^{2+} and K^+; (b) S^{2-} and Cl^-.

7.17 Ionization Energy

Copper wire is the most common material used to conduct electricity. But why does it conduct at all? An electric current is a flow of electrons, so the reason must lie in the ease with which copper atoms release electrons and let them migrate through the solid. Many chemical properties—particularly the types and numbers of bonds that atoms form—depend on the ease with which atoms give up some of their electrons.

The **ionization energy** of an element is the energy needed to remove an electron from an atom of the element in the gas phase. For the *first* ionization energy, I_1, we start with the neutral atom. For example, for copper,

$$Cu(g) \longrightarrow Cu^+(g) + e^-(g)$$
$$I_1 = (\text{energy of } Cu^+ + e^-) - (\text{energy of } Cu)$$

The experimental value for copper is 785 kJ/mol. The *second* ionization energy, I_2, is the energy needed to remove an electron from a singly charged gas-phase cation. For copper,

$$Cu^+(g) \longrightarrow Cu^{2+}(g) + e^-(g)$$
$$I_2 = (\text{energy of } Cu^{2+} + e^-) - (\text{energy of } Cu^+)$$

For copper, $I_2 = 1958$ kJ/mol. All ionization energies are positive: removing an electron from an atom or a cation always requires energy.

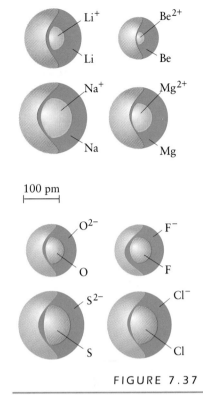

FIGURE 7.37

The relative sizes of cations, anions, and their parent atoms for selected elements. Note that cations are smaller than their parent atoms, whereas anions are larger.

FIGURE 7.38

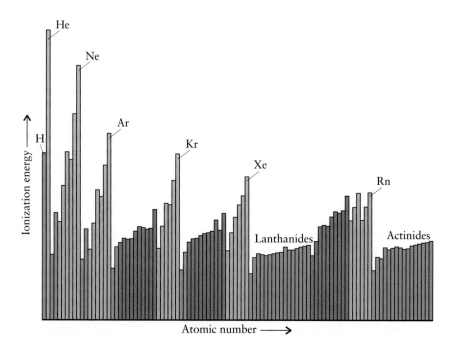

	Group						
	H 1310						18
							He 2370
Period	1	2	13	14	15	16	17
2	**Li** 519	**Be** 900	**B** 799	**C** 1090	**N** 1400	**O** 1310	**F** 1680	**Ne** 2080
3	**Na** 494	**Mg** 736	**Al** 577	**Si** 786	**P** 1011	**S** 1000	**Cl** 1255	**Ar** 1520
4	**K** 418	**Ca** 590	**Ga** 577	**Ge** 784	**As** 947	**Se** 941	**Br** 1140	**Kr** 1350
5	**Rb** 402	**Sr** 548	**In** 556	**Sn** 707	**Sb** 834	**Te** 870	**I** 1008	**Xe** 1170
6	**Cs** 376	**Ba** 502	**Tl** 590	**Pb** 716	**Bi** 703	**Po** 812	**At** 1037	**Rn** 1036

Ionization energy (kJ/mol)

- 2001–2500
- 1501–2000
- 1001–1500
- 501–1000
- 1–500

Figure 7.38 shows that ionization energies generally decrease down a group. The decrease means that it takes less energy to remove an electron from a cesium atom, for instance, than from a sodium atom. Figure 7.39 shows the variation of first ionization energy with atomic number, and we see a periodic sawtooth pattern like that shown by atomic radii. With few exceptions, first ionization energies rise from left to right across a period. Then they fall back to a lower value at the start of the following period. The lowest values occur at the bottom left of the periodic table (near cesium) and the highest at the upper right (near helium). In other words, less energy is needed to remove an electron from atoms of elements near cesium and more energy is needed to remove an electron from atoms of elements near helium. Elements with low ionization energies can be expected to form cations readily and to conduct electricity in

FIGURE 7.39

their solid forms. Elements with high ionization energies are unlikely to form cations and are unlikely to conduct electricity.

The first ionization energy decreases down a group because with each new period the outermost electron of the atom occupies a shell that is farther from the nucleus and therefore less tightly bound than the corresponding electron in the period above. The decrease in ionization energy from left to right across a period results from the increasing effective nuclear charge (recall Fig. 7.29). As a result, the outermost electron is gripped more tightly and the ionization energies generally increase. The small departures from these trends can usually be traced to repulsions between electrons, particularly electrons occupying the same orbital.

Figure 7.40 shows that the second ionization energy of an element is always greater than its first ionization energy and the third ionization energy greater than the second. It takes more energy to remove an electron from a positively charged ion than from a neutral atom (Fig. 7.41). For the Group 1 elements, the second ionization energy is considerably larger than the first, but in Group 2, the first and second ionization energies are similar. This difference makes sense, because the Group 1 elements have an ns^1 valence-shell electron configuration. Although the removal of the first electron requires only a small energy, a second electron must come from the noble-gas core. The core electrons have lower principal quantum numbers and are much closer to the nucleus. They are strongly attracted to the nucleus and a lot of energy is needed to remove them. The trend in ionization energies across a period clearly shows the great increase in energy required to remove a core electron (see Fig. 7.40). The third ionization energies of Group 2 elements are very high for this reason.

The first ionization energy is highest for elements close to helium and is lowest for elements close to cesium. Second ionization energies are higher than first ionization energies (of the same element). An ionization energy is very high if the electron is to be expelled from a closed shell.

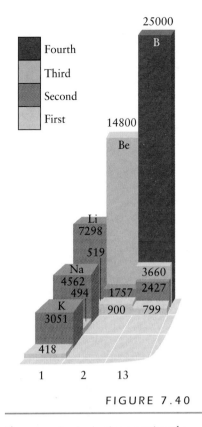

FIGURE 7.40

The successive ionization energies of a selection of main-group elements. Note the great increase in energy required to remove an electron from an inner shell. More values are given in Appendix 2D.

Example 7.12 *Accounting for the variation of ionization energies within a period*

Suggest a reason for the small decrease in ionization energy between nitrogen (1400 kJ/mol) and oxygen (1310 kJ/mol).

Strategy Because the outermost electron is closer to the nucleus in the O atom than in the larger N atom, and the O nucleus is more highly charged, we expect oxygen to have the larger ionization energy. When faced with a conflict between a prediction and an observation, we should look for an influence that has been ignored. In many-electron atoms, the first consideration should be given to the effects of electron-electron repulsion.

Solution The outermost three electrons in N occupy three different 2p-orbitals (recall 7), but one of the 2p-orbitals in O is full (recall 8). The electrons in the doubly occupied 2p-orbital repel each other, and less energy is needed to remove one of them. Hence, the ionization energy of O is less than that of N. (A more detailed analysis would take into account the lowering of the energy of a subshell when it becomes half full.)

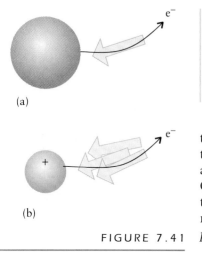

(a)

(b)

FIGURE 7.41

(a) To remove an electron from a neutral atom, energy has to be supplied to drag the electron away from the attraction (represented by the arrow) of the partially shielded nucleus. (b) Much more energy has to be supplied to drag a second electron away from a cation, because now the nucleus is less shielded and exerts a greater attraction (represented by a larger number of arrows).

In some books, you will see electron affinity defined with an opposite sign convention. Those are actually the electron-gain enthalpies.

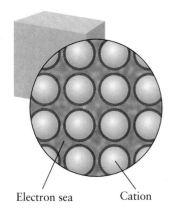

Self-Test 7.14A Account for the decrease in ionization energy between beryllium and boron.

[*Answer:* Boron loses an electron from a higher energy subshell than beryllium does.]

Self-Test 7.14B Account for the fact that the two Group 13 elements aluminum and gallium have approximately the same ionization energies.

The low ionization energies of elements at the lower left of the periodic table account for their metallic character. A block of metal consists of a collection of cations of the element surrounded by a sea of valence electrons that the atoms have lost (Fig. 7.42). For example, a piece of copper consists of a stack of Cu^+ ions held together by a sea of electrons, each of which comes from one of the atoms in the sample. Only elements with low ionization energies—the members of the *s* block, the *d* block, the *f* block, and the lower left of the *p* block—can form metallic solids, because only they can lose electrons easily.

The elements at the upper right of the periodic table have high ionization energies, so they do not readily lose electrons and, hence, are not metals. Our knowledge of electronic structure has helped us to understand a major feature of the periodic table, in this case, why the metals are found at the lower left and the nonmetals are found toward the upper right.

Metals are found toward the lower left of the periodic table because these elements have low ionization energies and can readily lose their electrons.

7.18 Electron Affinity

The **electron affinity,** E_{ea}, of an element is the energy released when an electron is added to a gas-phase atom. For example, the electron affinity of chlorine is the energy *released* in the process

$$Cl(g) + e^-(g) \longrightarrow Cl^-(g) \qquad E_{ea} = \text{(energy of Cl} + e^-) - \text{(energy of Cl}^-)$$

A positive electron affinity means that the energy of the anion is *lower* than that of the neutral atom and a distant electron. This convention matches the everyday meaning of the term *affinity.* Notice that, in this convention, the electron affinity is measured by subtracting the energy of the *product* from that of the *reactants.* Many electron affinities are positive, indicating that the attachment of an electron is generally energetically favorable. A high positive electron affinity means that a lot of energy is released when an electron attaches to an atom. A negative electron affinity means that the energy of the anion is higher than that of the neutral atom and implies that energy must be supplied to push an electron on to the atom. The noble gases have negative electron affinities because any electron that is pushed on to them must occupy an orbital outside a closed shell and far from the nucleus.

Figure 7.43 shows the variation in electron affinity through the periodic table. It is much less periodic than variations in radius and ionization energy.

FIGURE 7.42

A block of metal consists of an array of cations (the spheres) surrounded by a sea of electrons. The charge of the electron sea cancels the charges of the cations. The electrons of the sea are mobile and can move past the cations quite easily and hence conduct an electric current.

Period	1	2	13	14	15	16	17	18
1	H +73							He <0
2	Li +60	Be ≤0	B +27	C +122	N −7	O +141 −844	F +328	Ne <0
3	Na +53	Mg ≤0	Al +43	Si +134	P +72	S +200 −532	Cl +349	Ar <0
4	K +48	Ca +2	Ga +29	Ge +116	As +78	Se +195	Br +325	Kr <0
5	Rb +47	Sr +5	In +29	Sn +116	Sb +103	Te +190	I +295	Xe <0
6	Cs +46	Ba +14	Tl +19	Pb +35	Bi +91	Po +174	At +270	Rn <0

Electron affinity ($kJ{\cdot}mol^{-1}$): >300; 200–300; 100–200; 0–100; <0

FIGURE 7.43

The variation of electron affinity in kilojoules per mole (kJ/mol) of the main-group elements. A positive value indicates a strong affinity. Where two values are given, the first refers to the formation of a singly charged anion and the second refers to the formation of a doubly charged anion from a singly charged anion. The variation is less systematic than for atomic radius and ionization energy, but high values tend to be found close to fluorine (except for the noble gases).

However, one broad trend is clearly visible: electron affinities are highest toward the upper right of the periodic table close to the triangle of oxygen, fluorine, and chlorine. In these atoms, the incoming electron occupies a *p*-orbital close to a highly charged nucleus and is attracted to it strongly.

An atom can gain more than one electron. The energy released when the first electron is attached is called the *first electron affinity* and the additional energy released when the second electron enters is called the *second electron affinity.* An element in Group 17 has a very low second electron affinity. Once an electron has entered the single vacancy in the valence shell of a Group 17 atom, the shell is complete and any additional electron has to begin a new shell. In that new shell, not only is the electron farther from the nucleus, but it is also repelled by the negative charge already present. As a result, the second electron affinity of fluorine is strongly negative, meaning that a lot of energy has to be *supplied* to form F^{2-} from F^-. As a consequence, the ionic compounds of the halogens are built from singly charged ions such as F^- and never from doubly charged ions such as F^{2-}.

A Group 16 atom, such as O or S, has two vacancies in its valence shell *p*-orbitals and can accommodate two additional electrons. The first electron affinity is positive, so energy is released. However, attachment of the second electron requires energy because of the repulsion by the negative charge already present in O^- or S^-. The difference from the halogen case, however, is that there is room to accommodate the additional electron in the valence shell, which still has a single vacancy in the singly charged ion (compare diagrams **8** and **9**). In fact, 141 kJ/mol is released when the first electron adds to the neutral atom to form O^-, but 844 kJ/mol must be supplied to add a second electron to form O^{2-}, so the total energy required to make O^{2-} from O is 703 kJ/mol. As we shall see in Chapter 8, this energy can be achieved in chemical reactions, and O^{2-} ions are typical of metal oxides.

Elements with the highest electron affinities are those close to oxygen, fluorine, and chlorine. Group 17 atoms can acquire one electron with a release of energy and Group 16 atoms can accept two electrons with chemically attainable energies. Consequently, the halogens typically form X^- ions and Group 16 elements typically form X^{2-} ions.

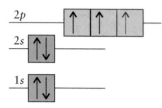

19 C⁻

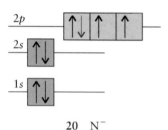

20 N⁻

Example 7.13 *Accounting for the variation in electron affinity*

Suggest a reason for the decrease in electron affinity between carbon and nitrogen.

Strategy We might expect more energy to be released (and the electron affinity to be higher) when an electron attaches to an N atom because it is smaller than a C atom and its nucleus is more highly charged. That the opposite is observed suggests that we may have ignored the effects of electron-electron repulsion in the resulting anions. We should therefore inspect the configurations of the ions.

Solution On forming C⁻ from C, the additional electron occupies an empty 2p-orbital (**19**). On forming N⁻ from N, the additional electron must occupy a 2p-orbital that is already half full (**20**). Although the nuclear charge of nitrogen is greater than that of carbon, the effective nuclear charge experienced by the electron gained by the nitrogen atom is reduced by repulsion from the electron already in the 2p-orbital. Therefore, less energy is released when N⁻ is formed and the electron affinity of nitrogen is lower than that of carbon.

Self-Test 7.15A Account for the large decrease in electron affinity between lithium and beryllium.
> [*Answer:* The additional electron enters a 2s-orbital in Li, but a 2p-orbital in Be; and a 2s-electron has a lower energy than a 2p-electron.]

Self-Test 7.15B Account for the large decrease in electron affinity between fluorine and neon.

7.19 Predicting Periodic Properties

Because the periodic table is a summary of electron configurations, and the electron configurations of the elements govern their properties, we can use it to predict the properties of elements. As we examine the major features of the blocks of the periodic table, we see that we can predict many of the properties of the elements without having to memorize specific data about each one. The general trends are summarized in Toolbox 7.2 at the end of this section.

An atom of an *s*-block element has a low ionization energy, which means its outermost electrons can easily be lost (Fig. 7.44). An atom of a Group 1 element has only one valence electron and is likely to form only +1 ions, such as Li⁺, Na⁺, and K⁺. Group 2 elements similarly form +2 ions, such as Mg²⁺, Ca²⁺, and Ba²⁺. An *s*-block element is likely to be a reactive metal with all the characteristics that the name metal implies (Table 7.4). Because ionization energies are lowest at the bottom of each group, and the elements there lose their valence electrons most easily, the heavy elements cesium and barium react most vigorously of all elements in their respective groups. Francium is not usually included in a discussion of periodic properties because it is radioactive, little is available, and its properties have not been well characterized.

Beryllium, at the head of Group 2, has the highest ionization energy of the block. It therefore loses its valence electrons less readily than other Group 2 elements and has the least pronounced metallic character. The compounds of all the *s*-block elements (with the exception of beryllium) are ionic.

The metalloids, the elements that lie along the frontier between *p*-block metals and nonmetals, run diagonally across the table (Fig. 7.45). Elements on the left of the *p* block, especially the heavier members of the group, have ionization energies that are low enough for the elements to behave as metals.

FIGURE 7.44

All the alkali metals are soft, silvery metals. Sodium is so reactive that it is kept under paraffin oil to protect it from air, and a freshly cut surface soon becomes covered with oxides.

Table 7.4 Characteristics of metals and nonmetals

Metals	Nonmetals
Physical properties	
good conductors of electricity	poor conductors of electricity
ductile	not ductile
malleable, lustrous	not malleable
typically: solid	solid, liquid, or gas
high melting point	low melting point
good conductors of heat	poor conductors of heat
Chemical properties	
react with acids	do not react with acids
form basic oxides	form acidic oxides
(which react with acids)	(which react with bases)
form cations	form anions
form ionic halides	form covalent halides

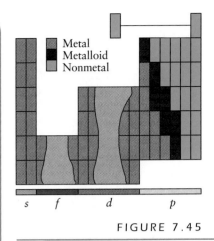

Metal
Metalloid
Nonmetal

s   *f*   *d*   *p*

FIGURE 7.45

The metalloids (red) fall in a diagonal band across the periodic table, between the metals (brown) and the nonmetals (green). The location of the metalloids reflects the diagonal relationship between elements.

However, the ionization energies of the *p*-block metals are quite high, and as a result, they are less reactive than those in the *s* block.

In Group 14, lead and tin are metals; but, although they are malleable and conduct electricity, they are not nearly as reactive as the *s*-block elements and many *d*-block elements (Fig. 7.46). That is why steel cans are tin-plated to protect them against corrosion and why tin has been used for centuries in alloys such as bronze and pewter. Lead, too, was used for about 2000 years to make pipes for domestic water supplies, for it is very resistant to corrosion. It is no longer used because lead compounds are now known to be toxic.

Elements at the right of the *p* block have characteristically high electron affinities: they tend to gain electrons to complete closed shells. Except for the metalloids tellurium and polonium, the members of Groups 16 and 17 are nonmetals (Fig. 7.47). They typically form molecular compounds with each other and react

Romans of the ruling class used lead pots for cooking. Some believe that this practice may have contributed to the fall of the Roman Empire.

FIGURE 7.46

The Group 14 elements. Back row (left to right): silicon, tin. Front row: carbon (as graphite), germanium, lead.

FIGURE 7.47

The Group 16 elements. From left to right: oxygen, sulfur, selenium, tellurium. Note the trend from nonmetal to metalloid.

FIGURE 7.48

Boron (top) and silicon (bottom) have a diagonal relationship. They are both brittle solids with high melting points. They also show a number of chemical similarities.

Zinc, cadmium, and mercury, in Group 12, are not usually considered to be transition metals.

with metals to form the anions in ionic compounds. Fluorine forms ionic or molecular compounds with every element except helium, neon, and argon.

If we compare the properties of elements in adjacent main groups, we notice a **diagonal relationship,** a similarity in properties between diagonal neighbors in the main groups of the periodic table. A part of the reason for this similarity can be seen in Figs. 7.34 and 7.38. These figures show the general trends in atomic radius and ionization energy, both of which affect chemical properties. The colored bands of similar values lie in diagonal stripes across the table. Because these properties lie in a diagonal pattern, it is not surprising to find that the elements within a diagonal band show similar chemical properties (Fig. 7.48).

The diagonal band of metalloids dividing the metals from the nonmetals is one example of a diagonal relationship (see Fig. 7.45). So is the chemical similarity of lithium and magnesium and that of beryllium and aluminum (Fig. 7.49). For example, both lithium and magnesium react directly with nitrogen to form nitrides, specifically Li_3N and Mg_3N_2, although none of the other Group 1 metals reacts directly with nitrogen. Similarly, like aluminum, beryllium reacts with both acids and bases. We shall see more examples of this diagonal similarity when we look at the elements in detail (in Chapter 19).

All *d*-block elements are metals (Fig. 7.50). Their properties are transitional between the *s*- and *p*-block metals, which accounts for their alternative name, the *transition metals*. One characteristic feature of the *d*-block elements is that many of them form compounds with a variety of oxidation states. Iron, we have seen, forms iron(II) and iron(III) ions, Fe^{2+} and Fe^{3+}. Copper forms copper(I) and copper(II) ions, Cu^+ and Cu^{2+}. Potassium, an *s*-block metal, has a $4s^1$ valence configuration like copper but forms only one type of ion, K^+. The reason for the difference between copper and potassium can be seen by comparing their second ionization energies, which are 1958 and 3051 kJ/mol, respectively. To form Cu^{2+}, an electron is removed from the *d*-subshell of $[Ar]3d^{10}$; but to

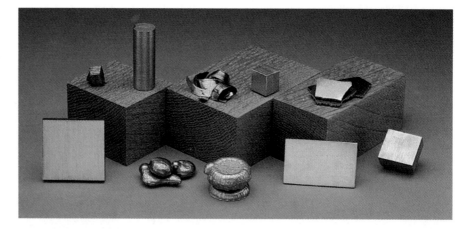

FIGURE 7.49   FIGURE 7.50

The pairs of elements represented by boxes of the same color show a strong diagonal relationship to each other.

The elements in the first row of the *d* block. Top row (left to right): scandium, titanium, vanadium, chromium, manganese. Bottom row: iron, cobalt, nickel, copper, zinc.

form K^{2+}, the electron would have to be removed from potassium's argonlike core. Because such huge amounts of energy are not readily available in chemical reactions, potassium can lose only its $4s$-electron.

The elements in the s block are all reactive metals. The p-block elements tend to gain electrons to complete closed shells; they range from metals through metalloids to nonmetals. Diagonally related pairs of elements often show similar chemical properties. All d-block elements are metals with properties between those of s-block and p-block metals. Many d-block elements form cations with a range of oxidation numbers.

Toolbox 7.2 *How to interpret the periodic table*

The general trends listed in this Toolbox can be used to predict chemical and physical properties of the elements.

Conceptual Basis

The chemical properties of the elements are determined by their electron configurations. The primary factors determining periodic properties are the value of the principal quantum number of the valence electrons and the effective nuclear charge they experience (Fig. 7.51). The greater the value of n, the further the average distance of the valence electrons from the nucleus and the lower the effective nuclear charge. The further to the right in a period, the greater the effective nuclear charge on the valence electrons.

Procedure

Use these general characteristics of the periodic table to predict periodic trends:

- The number of valence electrons of an atom can be inferred from the group number of the element.

- Atoms rarely lose more than their valence electrons in chemical reactions.

- Atoms rarely gain more electrons than are required to reach the next noble-gas configuration.

- Each new period involves the addition of an outer valence shell further from the nucleus than the previous period.

- The effective nuclear charge felt by a valence electron increases from left to right across a period and decreases down a group.

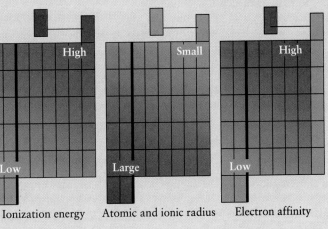

FIGURE 7.51

Periodic trends of several atomic properties.

Example 7.14 *Predicting periodic trends*

Atoms of which element in each pair have the larger radius: (a) Ca or Ba; (b) P or S? Explain your reasoning.

Strategy The characteristics listed in Toolbox 7.2 suggest that atomic radius increases down a group on account of the increasing distance of the valence shell from the nucleus and decreases from left to right across a period on account of the increasing effective nuclear charge pulling in the valence electrons.

Solution (a) Calcium and barium are in the same group, and Ba is lower down. Therefore, Ba has the larger radius. (b) Phosphorus and sulfur are in the same period; but S is further to the right, so its valence electrons have a greater effective nuclear charge. Therefore, P has the larger radius.

Self-Test 7.16A Which species in each pair has the larger radius: (a) Cs or Cs^+; (b) Br or Br^-? Explain your reasoning.

> [*Answer:* (a) Cs, because Cs^+ has lost its outer electron; (b) Br^-, because the additional electron increases the repulsion between the valence electrons.]

Self-Test 7.16B Which of the following elements is most chemically similar (shows a diagonal relationship) to aluminum: (a) beryllium; (b) carbon? Explain your reasoning.

Skills You Should Have Mastered

Conceptual

☐ 1. Describe the nature of light as electromagnetic radiation and as a stream of photons, Sections 7.1 and 7.2.

☐ 2. Explain the significance of the photoelectric effect, Section 7.2.

☐ 3. Use the Bohr frequency condition to explain the origin of the lines in the spectrum of an element, Section 7.3.

☐ 4. Describe the interpretation of atomic orbitals in terms of probabilities, Section 7.5.

☐ 5. Distinguish and sketch the boundary surfaces of *s*-, *p*-, and *d*-orbitals, Section 7.5.

☐ 6. Name and explain the relation of each of the four quantum numbers to the properties of electrons in orbitals, Sections 7.7 and 7.8.

☐ 7. List the allowed energy levels of a bound electron in terms of the quantum numbers *n*, *l*, and m_l, Section 7.7.

☐ 8. Describe the factors affecting the energy of an electron in a many-electron atom, Section 7.10.

Problem-Solving

☐ 1. Calculate the wavelength or frequency of light from the relation $\lambda \times \nu = c$, Example 7.1.

☐ 2. Use the relation $E = h\nu$ to calculate the energy, frequency, or number of photons emitted from a light source, Example 7.2.

☐ 3. Calculate the wavelength of an object, Example 7.3.

☐ 4. Use the expression for the energy levels of a hydrogen atom to correlate the lines in a spectrum with specific energy transitions, Example 7.4.

☐ 5. State how many orbitals of each type are contained in the shell corresponding to a given principal quantum number and how many electrons can be accommodated in those orbitals, Example 7.5.

☐ 6. Write the ground-state electron configuration for an atom, Toolbox 7.1 and Examples 7.6, 7.7, and 7.8.

☐ 7. Write the ground-state electron configuration for an ion, Examples 7.9 and 7.10.

☐ 8. Use the periodic table to predict and explain trends in physical and chemical properties, Toolbox 7.2 and Examples 7.11, 7.12, 7.13, and 7.14.

Descriptive

☐ 1. Describe the general characteristics of elements in the *s*, *p*, and *d* blocks, Section 7.19.

☐ 2. Describe and explain the significance of diagonal relationships in the periodic table, Section 7.19.

Exercises

The Characteristics of Light

7.1 (a) The approximate wavelength range of the visible spectrum is 400 to 700 nm. Express this range in hertz (Hz). (b) The wavelength of an AM radio band is about 250 m. What is the frequency of this band?

7.2 (a) Infrared radiation has wavelengths greater than about 800 nm. What is the frequency of 800-nm infrared radiation? (b) Microwaves, such as those used for radar and to heat food in a microwave oven, have wavelengths greater than 3 mm. What is the frequency of 3.0-mm radiation?

7.3 (a) Violet light has a frequency of 7.1×10^{14} Hz. What is the wavelength (in nanometers) of violet light? (b) When an electron beam strikes a block of copper, x-rays with a frequency of 2.0×10^{18} Hz are emitted. What is the wavelength (in picometers) of these x-rays?

7.4 (a) Radio waves for the FM station "Rock 99" at "99.3" on the FM dial are generated at 99.3 MHz. What is the wavelength of this station? (b) Radioastronomers use 1420.-MHz waves to look at interstellar clouds of hydrogen atoms. What is the wavelength of this radiation?

Quanta and Photons

7.5 Sodium vapor lamps, used for public lighting, emit 589-nm yellow light. How much energy is emitted by (a) an excited sodium atom when it generates a photon? (b) 1.00 mol of excited sodium atoms at this wavelength?

The yellow glow of sodium vapor lamps lighting Chicago streets (Exercise 7.5).

7.6 When an electron beam strikes a block of copper, x-rays with a frequency of 2.0×10^{18} Hz are emitted. How much energy is emitted by (a) an excited copper atom when it generates an x-ray photon? (b) 1.0 mol of excited copper atoms at this wavelength?

7.7 (a) Gamma radiation emitted by the nucleus of an iron-57 atom has a wavelength of 86 pm. Calculate the energy of one photon of gamma radiation. (b) A mixture of argon and mercury vapor used in blue advertising signs emits 470.-nm light. Calculate the energy change resulting from the emission of 1.0 mol of photons at this wavelength.

7.8 (a) The frequency for the FM radio station "Z-95" is 95.5 MHz. Calculate the energy produced in the transmission of 1.00 mol of photons at this frequency. (b) Ultraviolet radiation has wavelengths less than 350. nm. What is the energy accompanying the emission of 1.00 mol of photons at this wavelength?

7.9 In the photoelectric effect, light rays cause electrons to be ejected from metal surfaces. The diagram below shows the dependence of the kinetic energy of the ejected electrons on the frequency of the radiation used. (a) From which of these metals are electrons most easily ejected? (b) Explain why the kinetic energy drops to 0 at a different frequency for each metal. (c) When irradiated with light at the same wavelength, which metal produces the fastest electrons?

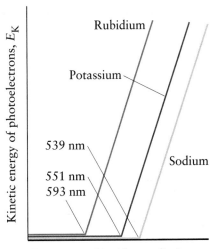

7.10 The minimum energy needed to remove an electron from the surface of a sample of silver is 6.9×10^{-19} J. Could a photoelectric detector for ultraviolet 250-nm light be constructed by using silver as the light-sensitive detector? What is the maximum wavelength of light that could be detected photoelectrically by silver?

Atomic Spectra

7.11 (a) Use the Rydberg formula for atomic hydrogen to calculate the wavelength for the transition from $n = 6$ to $n = 2$. (b) Use Table 7.1 to determine the color in the visible spectrum to which this transition corresponds.

7.12 (a) What is the energy of a photon emitted from a hydrogen atom when an electron makes a transition from $n = 7$ to $n = 2$? (b) What is the wavelength of this photon?

7.13 (a) What is the highest frequency photon that can be emitted from the hydrogen atom? (b) What is the wavelength of this photon? (c) In what part of the electromagnetic spectrum does this photon appear?

7.14 The violet line in the hydrogen spectrum is at 434.0 nm. What is the principal quantum number of the orbital corresponding to the upper energy state of the electron that produces a photon of this wavelength? (*Hint:* For the visible spectrum of hydrogen, the principal quantum number of the lower state is $n = 2$.)

Particles and Waves

7.15 Calculate the wavelengths of (a) an electron and (b) a neutron, both with a velocity of 1.5×10^7 m/s (one-twentieth the speed of light). (Remember to calculate the mass of one particle in kilograms, and that $1 \ J = 1 \ kg \cdot m^2/s^2$.)

7.16 (a) A baseball must weigh between 5.0 and 5.25 ounces (1 oz $= 28.35$ g). What is the wavelength of a 5.0-oz baseball thrown at 92 mph? (b) The velocity of an electron that is emitted from a metallic surface by a photon is 2.2×10^3 km/s. What is the wavelength of the electron? (Remember to calculate the mass of the particle or object in kilograms, and that $1 \ J = 1 \ kg \cdot m^2/s^2$.)

7.17 Who has the shorter wavelength when running at the same speed: a person weighing 60 kg or a person weighing 80 kg? Explain your reasoning.

7.18 (a) Calculate the wavelength of a hydrogen atom traveling at 10. m/s. (b) What would cause the wavelength of the atom to decrease, accelerating the atom to higher speeds, or slowing it down? Explain your reasoning.

Atomic Orbitals and Quantum Numbers

7.19 In the early nineteenth century, John Dalton described atoms as hard spheres. Explain how and why his description should be revised.

7.20 (a) Explain what is meant by an atomic orbital. (b) Sketch the boundary surface of an *s*-orbital, a *p*-orbital, and a *d*-orbital.

7.21 How many subshells are in the shell with n equal to (a) 2? (b) 3? (c) What are the permitted values of l when $n = 3$?

7.22 (a) How many subshells are there in the shell with $n = 5$? (b) Identify the subshells in the form 5s, ... (c) How many orbitals are there in the shell with $n = 5$?

7.23 How many orbitals are in a subshell with l equal to (a) 0? (b) 2? (c) 1? (d) 3?

7.24 How many electrons can occupy a subshell with l equal to (a) 0? (b) 2? (c) 1? (d) 3?

7.25 State the number of (a) subshells with $n = 2$; (b) orbitals with $n = 4$ and $l = 2$; (c) orbitals with $n = 2$; (d) orbitals in a 3*d*-subshell.

7.26 State the number of (a) orbitals in a 4*s*-subshell; (b) subshells with $n = 1$; (c) orbitals in a 5*p*-subshell; (d) orbitals in a 6*d*-subshell.

7.27 Write the subshell notation (in the form 3*d*, for instance) and the number of orbitals having these quantum numbers in an atom: (a) $n = 3, l = 2$; (b) $n = 1, l = 0$; (c) $n = 6, l = 3$; (d) $n = 2, l = 1$.

7.28 Identify the values of the principal quantum number, n, and the orbital angular momentum quantum number, l, for the following subshells: (a) 2*p*; (b) 5*f*; (c) 3*s*; (d) 4*d*.

7.29 How many electrons can have the following quantum numbers in an atom: (a) $n = 2, l = 1$; (b) $n = 4, l = 2$, $m_l = -2$; (c) $n = 2$; (d) $n = 3, l = 2, m_l = +1$?

7.30 How many electrons can occupy the (a) 4*p*-; (b) 3*d*-; (c) 1*s*-; (d) 4*f*-subshells?

7.31 Which of the following subshells cannot exist in an atom: (a) 2*d*; (b) 4*d*; (c) 4*g*; (d) 6*f*?

7.32 Which of the following subshells cannot exist in an atom: (a) 1*p*; (b) 3*f*; (c) 7*p*; (d) 5*d*?

7.33 Identify which of the following sets of four quantum numbers $\{n, l, m_l, m_s\}$ cannot exist for an electron in an atom, and explain why not: (a) $\{4, 2, -1, +\frac{1}{2}\}$; (b) $\{5, 0, -1, +\frac{1}{2}\}$; (c) $\{4, 4, -1, +\frac{1}{2}\}$.

7.34 Identify which of the following sets of four quantum numbers $\{n, l, m_l, m_s\}$ cannot exist for an electron in an atom, and explain why not: (a) $\{2, 2, -1, +\frac{1}{2}\}$; (b) $\{6, 0, 0, +\frac{1}{2}\}$; (c) $\{5, 4, +5, +\frac{1}{2}\}$.

7.35 Identify the possible sets of four quantum numbers $\{n, l, m_l, m_s\}$ for the starred electron in the following diagrams. Select the values of m_l by numbering from $-l$ to $+l$ from left to right.

(a) 2*p* [↑↓] [↑↓*] [↑]

(b) 5*d* [↑↓] [↑↓] [↑↓] [↑*] [↑]

7.36 Identify the possible sets of four quantum numbers $\{n, l, m_l, m_s\}$ for the starred electron in the following diagrams. Select the values of m_l by numbering from $-l$ to $+l$ from left to right.

(a) 6*p* [↑*↓] [↑↓] [↑]

(b) 5*f* [↑↓] [↑↓] [↑↓*] [↑↓] [↑↓] [↑↓]

Orbital Energies

7.37 Explain why, in a many-electron atom, a $2s$-electron is bound more strongly than a $3s$-electron.

7.38 Explain why, in a many-electron atom, a $3s$-electron is bound more strongly than a $3p$-electron.

7.39 (a) List the shells with n ranging from 1 to 4 in order of increasing energy. (b) For the shell with $n = 4$, list the subshells with l ranging from 0 to $n - 1$ in order of increasing energy.

7.40 For electrons in the s-, p-, d-, and f-subshells, list them in order of (a) the probability of being found close to the nucleus; (b) the effectiveness with which they can shield electrons in higher shells.

Electron Configurations and the Building-Up Principle

7.41 Refer to the periodic table and list the subshells $3s$, $5d$, $1s$, $2p$, $3d$ for a many-electron atom in order of increasing energy when empty.

7.42 Refer to the periodic table and list the subshells $2s$, $4d$, $2p$, $1s$ for a many-electron atom in order of increasing energy when empty.

7.43 Predict the ground-state electron configuration of an atom of (a) calcium; (b) nitrogen; (c) bromine; (d) uranium.

7.44 Predict the ground-state electron configuration of an atom of (a) iron; (b) gallium; (c) zirconium; (d) sodium.

7.45 Starting with the previous noble gas, write the ground-state electron configuration of an atom of (a) nickel; (b) cadmium; (c) lead; (d) silver.

7.46 Starting with the previous noble gas, write the ground-state electron configuration of an atom of (a) vanadium; (b) osmium; (c) tellurium; (d) mercury.

7.47 Starting with the previous noble gas, write the electron configurations of the following ions: (a) Fe^{2+}; (b) Cl^-; (c) Tl^+.

7.48 Starting with the previous noble gas, write the electron configurations of the following ions: (a) Ni^{2+}; (b) Pb^{2+}; (c) N^{3-}.

7.49 Identify the element having the following description and then write its electron configuration, starting with the previous noble gas: (a) Group 14, Period 3; (b) Group 18, Period 2; (c) Group 1, Period 6; (d) Group 16, Period 3.

7.50 Identify the element having the following description and then write its electron configuration, starting with the previous noble gas: (a) Group 2, Period 5; (b) Group 18, Period 4; (c) Group 13, Period 3; (d) Group 17, Period 3.

7.51 The valence electron configuration for Group 13 elements is ns^2np^1. Write the valence electron configuration for the elements of (a) Group 2; (b) Group 18.

7.52 The valence electron configuration for Group 13 elements is ns^2np^1. Write the valence electron configuration for the elements of (a) Group 1; (b) the titanium group.

Atomic and Ionic Radius

7.53 Explain why atomic radii increase down a group in the periodic table.

7.54 Explain why atomic radii decrease from left to right across a period.

7.55 (a) From Fig. 7.34, decide which group of elements has the largest atomic radii. (b) From Figs. 7.34 and 7.36, what general statement can be made regarding the ionic radius of a cation relative to its neutral atom? (c) What general statement can be made regarding the ionic radii of cations having the same number of electrons but a different atomic number (for example, Al^{3+} versus Na^+)?

7.56 (a) From Fig. 7.34, decide which group of elements shown have the smallest atomic radii. (b) Compare the relative sizes of the ionic radii of an anion and its parent neutral atom. (c) Compare the relative ionic radii of anions having the same number of electrons but a different atomic number (for example, N^{3-} versus F^-).

7.57 Identify the species with the larger radius in each of the following pairs: (a) Cl or S; (b) Cl^- or S^{2-}; (c) Na or **Mg**; (d) Mg^{2+} or Al^{3+}.

7.58 Which species has the smaller radius: (a) Li^+ or Na^+; (b) Cl or Cl^-; (c) Al or Al^{3+}; (d) N^{3-} or O^{2-}?

Ionization Energy

7.59 Explain why ionization energies decrease down a group and increase across a period.

7.60 (a) Which group of elements has the highest ionization energies? (b) Which group of elements has the lowest ionization energies?

7.61 Predict which atom of each pair has the greater first ionization energy (be alert for any exceptions to the general trend): (a) Na or Mg; (b) C or N; (c) P or S.

7.62 Predict which atom of each pair has the greater first ionization energy (be alert for any exceptions to the general trend): (a) Ba or Ca; (b) Be or B; (c) Ar or Xe.

7.63 In terms of electron configurations obtained from the building-up principle, explain why the ionization energies of Group 16 elements are smaller than those of Group 15 elements.

7.64 In terms of electron configurations obtained from the building-up principle, explain why the ionization energies of Group 13 elements are smaller than those of Group 2 elements.

7.65 Explain why the second ionization energy of sodium (4562 kJ/mol) is significantly higher than the second ionization energy of magnesium (1451 kJ/mol), even though the first ionization energy of sodium is less than that of magnesium (see Appendix 2D).

7.66 Explain why the second ionization energy of sodium (4562 kJ/mol) is larger than the third ionization energy of aluminum (2744 kJ/mol), even though the first ionization energy of sodium is lower than that of aluminum (see Appendix 2D).

Electron Affinity

7.67 (a) Which group of elements is likely to have high electron affinities? (b) What is the general trend in electron affinities across a period?

7.68 (a) Explain why the electron affinity of chlorine is greater than that of bromine. (b) Explain why the electron affinity of the S atom is greater than that of the S^{2-} ion.

7.69 Select the atom that has the greater electron affinity in the following pairs (be alert for any exceptions to the general trend): (a) S or Cl; (b) C or O; (c) Cl or Br.

7.70 Select the atom that has the greater electron affinity in the following pairs (be alert for any exceptions to the general trend): (a) S or Se; (b) Si or P; (c) N or O.

7.71 Describe the correlation in the periodic trends of atomic radii and ionization energies, both across a period and down a group.

7.72 Describe the correlation in the periodic trends of electron affinities and ionization energies, both across a period and down a group.

Trends in Chemical Properties

7.73 What is a diagonal relationship? Give two examples to illustrate the concept.

7.74 State three chemical properties that are typical of (a) metals; (b) nonmetals.

7.75 Describe the general trend in chemical reactivity of metals (a) across a period; (b) down a group.

7.76 Why are the s-block metals more reactive than the p-block metals?

7.77 Identify the following elements as metals, nonmetals, or metalloids: (a) aluminum; (b) carbon; (c) germanium; (d) arsenic.

7.78 Identify the following elements as metals, nonmetals, or metalloids: (a) lead; (b) sulfur; (c) zinc; (d) silicon.

7.79 Which of the following pairs of elements do not exhibit the diagonal relationship: (a) B and Si; (b) Be and Al; (c) Al and Ge; (d) Na and Ca?

7.80 Which of the following pairs of elements do not exhibit the diagonal relationship: (a) O and Cl; (b) Li and Mg; (c) F and Ar; (d) H and Be?

Supplementary Exercises

7.81 The wavelength of radar is about 3 cm. What is the frequency of radar waves?

7.82 X-rays of wavelength 149 pm are generated from a particle accelerator called a synchrotron. What is the frequency of these x-rays?

7.83 The average speed of a hydrogen molecule at 20°C is 1930 m/s. What is the wavelength of the H_2 molecule at this temperature? (Remember to calculate the mass of one H_2 molecule in kilograms, and that $1\ J = 1\ kg \cdot m^2/s^2$.)

7.84 (a) How many values of the quantum number l are possible when $n = 6$? (b) How many values of m_l are allowed for an electron in a 4d-subshell? (c) How many values of m_l are allowed for a 3p-subshell? (d) How many subshells are there in the shell with $n = 4$?

7.85 (a) How many orbitals are there in an h-subshell? (b) How many 5f-orbitals are there? (c) What is the maximum number of electrons that can have $n = 5$ and $l = 2$? (d) How many 5p-electrons are there in an antimony atom?

7.86 Suppose in a certain experiment that the electron in a hydrogen atom could be excited no higher than the shell with $n = 5$. (a) How many different lines could appear in the spectrum as the excited atom falls back to lower states? (b) What would be the range of wavelengths emitted? (To answer this question, find the wavelength of the highest energy transition and that of the lowest energy transition.)

7.87 Which of the following orbitals is a p_y-orbital?

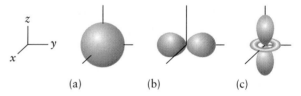

        (a)   (b)   (c)

7.88 (a) How many unpaired electrons are there in the ground state of an iron atom? (b) How many p-electrons (allowing for all shells) are there in the ground state of a phosphorus atom? (c) What is the maximum number of electrons that can be accommodated in a shell with $n = 3$? (d) Which group of elements has a [noble gas]ns^2 electron configuration?

7.89 Which sets of quantum numbers are unacceptable in a ground-state atom?
(a) $n = 3, l = -2, m_l = 0, m_s = +\frac{1}{2}$
(b) $n = 2, l = 2, m_l = -1, m_s = -\frac{1}{2}$
(c) $n = 6, l = 2, m_l = +2, m_s = +\frac{1}{2}$
(d) $n = 4, l = 0, m_l = 0, m_s = -\frac{1}{2}$

7.90 State the following:
(a) The number of orbitals with the quantum numbers $n = 3, l = 2$, and $m_l = 0$.

(b) The number of valence electrons in the outermost *p*-subshell of an S atom.

(c) The number of unpaired electrons in a Mn^{2+} ion.

(d) The subshell with the quantum numbers $n = 4, l = 2$.

(e) The m_l values allowed for a *d*-orbital.

(f) The allowed values of *l* for the shell with $n = 4$.

(g) The number of unpaired electrons in the cobalt atom.

(h) The number of orbitals in a shell with $n = 3$.

(i) The number of orbitals with $n = 3$ and $l = 1$.

(j) The maximum number of electrons with quantum numbers $n = 3$ and $l = 2$.

(k) The allowed values of *l* when $n = 2$.

(l) The possible values of m_l when $l = 2$.

(m) The number of electrons with $n = 4$ and $l = 1$.

(n) The quantum number that characterizes the shape of an atomic orbital.

(o) The designation of the subshell with $n = 3$ and $l = 1$.

(p) The lowest value of *n* for which a *d*-subshell can occur.

7.91 Identify the groups in the periodic table for which the members have the following ground-state valence electron configurations: (a) ns^2np^3; (b) $ns^2(n - 1)d^{10}$; (c) ns^1; (d) ns^2np^6; (e) $ns^2(n - 1)d^6$; (f) ns^2.

7.92 Write the valence electron configurations for the groups in the periodic table with the (a) lowest ionization energies; (b) highest electron affinities; (c) most metallic character.

7.93 Write the ground-state electron configurations (starting with the previous noble gas) for (a) Zr; (b) Se; (c) Rb; (d) Cl; (e) Sb; (f) Pu; (g) Si; (h) Ar.

7.94 Write the ground-state electron configuration (starting with the previous noble gas) for (a) Zn^{2+}; (b) Se^{2-}; (c) I^-; (d) Y; (e) P; (f) In; (g) As; (h) Ir.

7.95 Write the values for the quantum numbers for the starred electron, as in Exercise 7.35:

(a) 3p ↑↓ ↑↓ ↑*

(b) 3s ↑↓*

7.96 Starting with the previous noble gas, write the electron configurations of (a) V; (b) Cl^-; (c) Sn^{2+}; (d) Ni; (e) N^{3-}.

7.97 Select the best of the three choices:

(a) highest first ionization energy	Se	S	Te
(b) smallest radius	Cl^-	Br^-	F^-
(c) lowest electron affinity	K	Rb	Cs
(d) highest first ionization energy	O	S	F
(e) lowest second ionization energy	Ar	K	Ca
(f) greatest number of unpaired electrons	Fe	Co	Ni
(g) largest ionic radius	Ca^{2+}	Mg^{2+}	Ba^{2+}
(h) largest radius	S^{2-}	Cl^-	Cl
(i) highest first ionization energy	C	N	O
(j) highest electron affinity	P	S	Cl
(k) smallest atomic radius	Sn	I	Bi
(l) lowest first ionization energy	K	Na	Ca
(m) impossible subshell designation	4g	5d	4p

(n) number of orbitals with $n = 2$	2	4	8
(o) number of 5*f*-orbitals	14	7	9

7.98 The spheres below show the relative radii of the following species: (a) K^+; (b) Cl^-; (c) Ca^{2+}; (d) O. Rank the radii of these species by matching each species to the correct illustration.

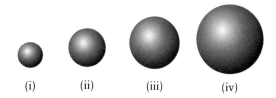

(i)   (ii)   (iii)   (iv)

7.99 Three of the best electrical conductors are silver, copper, and gold. (a) Use electron configurations to show why these elements make such good conductors. (b) Elements in Group 1 have similar valence electron configurations but are not used as conductors. Why not?

7.100 Explain why the elements zinc, cadmium, and mercury have higher ionization energies than the coinage metals (copper, silver, and gold).

7.101 Identify the blocks to which the following elements belong: (a) Zr; (b) Sr; (c) Kr; (d) Au; (e) Pb; (f) Fe.

7.102 In July of 1994, when the Shoemaker-Levy comet struck Jupiter, astronomers focused telescopes fitted with diffraction gratings on Jupiter to collect spectra for future analysis. What would astronomers look for in the spectra, and what did they hope to learn about Jupiter by this means?

Jupiter being struck by fragments of a comet and the glow of the resulting fireball (lower left)

7.103 In some forms of the periodic table, zinc, cadmium, and mercury are classified as belonging to "Group IIB", with the alkaline earth metals in "Group IIA." (a) Justify this classification in terms of electronic structure. (b) How could you explain any differences between "Group IIA" and "Group IIB"?

7.104 The radius of the ammonium ion is 137 pm. (a) If NH_4^+ were actually the cation of the "element" NH_4, where would NH_4 be located in the periodic table, considering its total number of protons? (b) As a result of its placement, describe the properties of the "element" ammonium. (c) Compare its ionic radius to that of the element actually found in the predicted location and explain any differences.

7.105 The German physicist Lothar Meyer observed a periodicity in the physical properties of the elements at about the same time Mendeleev was working on their chemical properties. Some of Meyer's observations can be reproduced by plotting the molar volume, the volume occupied per mole of atoms, for the solid forms of the elements against atomic number. Do this for Periods 2 and 3, given the following densities of the elements in their solid forms:

Element	Density, g/cm³	Element	Density, g/cm³
Li	0.53	Na	0.97
Be	1.85	Mg	1.74
B	2.47	Al	2.70
C	2.27	Si	2.33
N	0.88	P	1.82
O	1.14	S	2.09
F	1.11	Cl	2.03
Ne	1.21	Ar	1.66

Suggest a reason for the sharp change in molar volume between the *s*-block and *p*-block elements.

7.106 Suppose that in some other universe a rule corresponding to our Pauli exclusion principle reads "as many as two electrons in the same atom may have the same set of four quantum numbers." Suppose further that all other factors affecting electron configurations are unchanged. (a) Give the electron configuration of the element in the other universe that has five protons. (b) What is the most likely charge on the ion of this element? (c) Give the value of Z for the second inert gas in the other universe. Explain your reasoning.

7.107 The spectroscopic technique known as "photoelectron spectroscopy" (PES) takes advantage of the photoelectric effect. In PES, ultraviolet light is directed at an atom or molecule. Electrons are ejected from the valence shell, and their kinetic energies are measured. The ionization energy, I, can be deduced from the fact that the total energy is conserved. (a) Show that the speed (v) of the ejected electron and the frequency (v) of the incoming radiation are related by $hv = I + \frac{1}{2}m_e v^2$. (b) Use this relation to calculate the ionization energy of a rubidium atom, given that light of wavelength 58.4 nm produces electrons with a speed of 2450 km/s; recall that $1 J = 1 kg \cdot m^2/s^2$.

Applied Exercises

For Exercises 7.108–7.110, see Applying Chemistry: Case Study 7.

7.108 You decide to make a gathering of friends more festive by using your knowledge of chemistry to make logs in your fireplace that burn with colored flames. You obtain chlorides of sodium, strontium, barium, lithium, and copper and make up solutions in which you will soak the logs. Which of these compounds should you use to produce each color: (a) green; (b) yellow; (c) red; (d) blue?

7.109 Many fireworks mixtures depend on the highly exothermic combustion of magnesium to magnesium oxide, in which the heat causes the oxide to become incandescent and to give out a bright white light. The color of the light can be changed by including nitrates and chlorides of elements that have spectra in the visible region (Fig. 7.26). Barium nitrate is often added to produce a yellow-green color. The excited barium ions generate 487-nm, 524-nm, 543-nm, 553-nm, and 578-nm light. Calculate the change in energy of the atom in each case, and the change in energy per mole of atoms.

7.110 You need to determine whether magnesium produces too much heat to be used in a pyrotechnics show. (a) Calculate the reaction enthalpy for the following reaction, which might occur in fireworks:

$$4\,Mg(s)\ +\ KClO_4(s)\ \longrightarrow\ 4\,MgO(s)\ +\ KCl(s)$$

(b) Calculate the specific enthalpy of the reaction for magnesium (the reaction enthalpy per gram of magnesium). (c) Compare this value with the specific enthalpy of combustion of various fuels cited in Table 6.4. Which fuel has the highest specific enthalpy? Which has the lowest? (d) Why is magnesium not used more as a fuel?

7.111 Gigantic clouds of hot hydrogen gas fill certain regions of outer space. Spectra from these luminous clouds reveal that the hydrogen atoms are very hot and are excited to the point that they ionize. In some hydrogen atoms, electrons may be excited to quantum levels with $n = 100$ or more. (a) Calculate the wavelength that would be detected on Earth if electrons fell from the level with $n = 100$ to $n = 2$. (b) In what series would this transition be found? (c) Some of these high-energy electrons do not fall back to low energy levels within the hydrogen atoms. For instance, transitions between states such as $n = 110$ to $n = 100$ are observed. Would the wavelengths of these transitions be longer or shorter than those in the Balmer series?

7.112 Geiger counters can detect radioactivity because nuclear radiation consists of such energetic particles and electromagnetic radiation that the radiation can ionize atoms. As a result, it is called "ionizing radiation." What is the longest wavelength of radiation that can be detected by a Geiger counter using argon gas as the ionizing medium?

7.113 The photocells that keep some elevator doors from closing on us and that are used in some security systems use the photoelectric effect to detect the presence of an object. A light continually shines on a target made of an element like tungsten or cesium, so electrons are continually being ejected and captured by a positive electrode. When we step into the beam, the flow of electrons stops and a switch is activated. (a) Explain why cesium is a desirable metal for this purpose.

(b) Other elements besides cesium are desired for safety reasons. What safety concern does cesium present?

Integrated Exercises

7.114 Write the symbol for each of the following ions, the formula for the compound that ion would form with oxygen, and the name of the compound: (a) a cobalt ion, $[Ar]3d^6$; (b) a molybdenum ion, $[Kr]4d^3$; (c) a thallium ion, $[Xe]5d^{10}6s^2$.

7.115 On the basis of your knowledge of the gas laws, rank in order of *increasing* density the 11 elements that are gases at 298 K and 1 atm. Assume that all the gases are at the same temperature, pressure, and volume.

7.116 The regions of the periodic table in which elements that form acidic oxides and those that form basic oxides are identified in Fig. 3.23. Use your knowledge of periodic trends to suggest a reason why metal oxides are generally ionic, but nonmetal oxides are not.

7.117 Like the Group 1 metals, the Group 2 metals barium and calcium react with cold water to produce hydrogen and the hydroxide of the metal. Write balanced chemical equations for the reactions of barium and calcium with water.

7.118 Use Hess's law to calculate the enthalpy of vaporization of solid sodium chloride to a gas of ions, $NaCl(s) \rightarrow Na^+(g) + Cl^-(g)$, from the following information and the enthalpy of formation of $NaCl(s)$, which is -411.15 kJ/mol.

Atomization of sodium:
$$Na(s) \longrightarrow Na(g) \qquad \Delta H° = +108.4 \text{ kJ}$$

Ionization of sodium:
$$Na(g) \longrightarrow Na^+(g) + e^-(g) \qquad \Delta H° = +495.8 \text{ kJ}$$

Dissociation of chlorine:
$$Cl_2(g) \longrightarrow 2\,Cl(g) \qquad \Delta H° = +242 \text{ kJ}$$

Electron attachment to chlorine:
$$Cl(g) + e^-(g) \longrightarrow Cl^-(g) \qquad \Delta H° = -348.6 \text{ kJ}$$

7.119 Use Hess's law to calculate the enthalpy of vaporization of solid potassium bromide to a gas of ions, the process $KBr(s) \rightarrow K^+(g) + Br^-(g)$, from the following information and the enthalpy of formation of $KBr(s)$, which is -393.80 kJ/mol.

Atomization of potassium:
$$K(s) \longrightarrow K(g) \qquad \Delta H° = +89.2 \text{ kJ}$$

Ionization of potassium:
$$K(g) \longrightarrow K^+(g) + e^-(g) \qquad \Delta H° = +425.0 \text{ kJ}$$

Vaporization of bromine:
$$Br_2(l) \longrightarrow Br_2(g) \qquad \Delta H° = +30.9 \text{ kJ}$$

Dissociation of bromine:
$$Br_2(g) \longrightarrow 2\,Br(g) \qquad \Delta H° = +192.9 \text{ kJ}$$

Electron attachment to bromine:
$$Br(g) + e^-(g) \longrightarrow Br^-(g) \qquad \Delta H° = -331.0 \text{ kJ}$$

Chemical Bonds

A quiet revolution has swept across the face of the Earth in the past half-century. Materials that were only imagined a few decades ago have now become reality. Polyethylene, polystyrene, and polyvinylchloride (PVC) have made possible portable, lightweight electronic goods. Composite materials have made possible fast, lightweight airplanes and automobiles. Chemists are providing new materials to meet the needs of technicians, designers, engineers, physicians, farmers, and anyone else who contributes to the quality of our lives.

The production of new materials has become possible largely because chemists now understand how atoms bond together in certain definite ways. They also understand how the properties of substances are related to the types of bonds that hold atoms together. Why, for example, is calcium phosphate so rigid that nature has adopted it for the formation of bones? Can we make better bones? Why is it so difficult to make compounds from the nitrogen in air? Can we find an easy way? How can we explain the ability of hemoglobin to form a loosely bonded compound with oxygen, transport it to another part of the body, and then release it in response to a metabolic need?

A **chemical bond** is a link between atoms. A bond forms if the resulting arrangement of atoms has a lower energy than the sum of the energies of the separate atoms. Sodium and chlorine atoms bond together as ions because solid sodium chloride has a lower energy than a gas of widely separated sodium and chlorine atoms. Hydrogen and nitrogen atoms bond together to form ammonia, NH_3, because a gas consisting of NH_3 molecules has a lower energy than a gas consisting of the same number of separate nitrogen and hydrogen atoms. If the lowest energy can be achieved by the *complete transfer* of one or more electrons from one atom to another,

The block being supported by soap bubbles is a piece of SEAgel, a solid foam made from agar. Aerogels like SEAgel have a porous structure that encapsulates gases. It is claimed that the density of an aerogel can be equal to or less than that of air. This material, which was developed for the space program, is a good insulator and can be used in packaging. The properties of SEAgel and other materials are related to the types of bonds between their atoms. In this chapter we examine the formation of bonds.

ions form and the compound is held together by the attraction between these ions. If the lowest energy can be achieved by *sharing* electrons, then the atoms link together to form discrete molecules.

Chemical bonds result from changes in the locations of the *valence electrons* of atoms, the electrons in the outermost shells of atoms. Because the electronic structures of atoms are related to their locations in the periodic table, we can expect to be able to predict the bond-forming ability of an atom on the basis of its group and period (Sections 7.14–7.19).

The American chemist, G. N. Lewis (Fig. 8.1), one of the greatest of all chemists, laid the foundations of our current understanding of bond formation. Although his work was done before the electronic structures of atoms were known, his insights into chemical bonding in large part remain valid today.

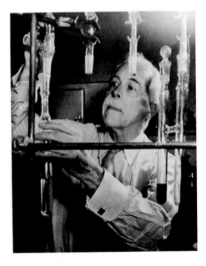

FIGURE 8.1

Gilbert Newton Lewis (1875–1946).

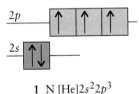

1 N [He]$2s^2 2p^3$

IONIC BONDS

An **ionic bond** is the electrical attraction between the opposite charges of cations and anions. For example, the attraction between Na^+ ions and Cl^- ions results in the ionic compound sodium chloride. No bond is purely ionic, but the **ionic model,** the description of bonding in terms of the attraction between ions, works well for describing compounds of metallic elements, especially those of the *s*-block metals, with nonmetals. Metals have a low ionization energy, so they can give up their valence electrons relatively easily to form cations.

8.1 Lewis Symbols for Atoms and Ions

Lewis devised a very simple way of keeping track of valence electrons when atoms form ionic bonds. The procedure is also useful for predicting the chemical formula of an ionic compound. As we work through this chapter, we shall see how this simple idea can grow into a technique of great usefulness and can be used to explain a wide variety of chemical properties.

A **Lewis symbol** consists of the chemical symbol of an element and a dot for each of its valence electrons. Some examples are

$$H\cdot \quad He: \quad :\overset{\cdot}{\underset{\cdot}{N}}\cdot \quad \cdot\overset{\cdot\cdot}{\underset{\cdot\cdot}{O}}\cdot \quad :\overset{\cdot\cdot}{\underset{\cdot\cdot}{Cl}}\cdot \quad K\cdot \quad Mg:$$

A single dot represents an electron alone in an orbital. A pair of dots represents two paired electrons in the same orbital. The Lewis symbol for nitrogen, for example, represents the valence electron configuration $2s^2 2p_x^1 2p_y^1 2p_z^1$ (**1**), with two electrons paired in an *s*-orbital and three unpaired electrons in separate *p*-orbitals. The Lewis symbol is a visual summary of the valence-shell electron configuration of an atom. It allows us to keep track of the electrons when an ion forms.

To work out the formula of an ionic compound, we first remove the dots from the Lewis symbol for the metal atom. Then we transfer the dots to the Lewis symbol for the nonmetal atom and complete its valence shell. Next, we adjust the numbers of atoms of each kind so that all the dots removed from the metal atom symbols are accommodated by the nonmetal atom symbols. Finally, we add the charges of the ions. A simple example is the formation of the ionic solid potassium chloride:

$$K\cdot + :\overset{\cdot\cdot}{\underset{\cdot\cdot}{Cl}}\cdot \longrightarrow K^+ \left[:\overset{\cdot\cdot}{\underset{\cdot\cdot}{Cl}}:\right]^-$$

The collection of symbols $K^+\left[:\overset{..}{\underset{..}{Cl}}:\right]^-$ is the **Lewis formula** for potassium chloride. The electron supplied by one metal atom is accommodated in the single vacancy of one Cl atom, so the chemical formula of potassium chloride is KCl. We have to combine ions in such a way as to achieve a neutral formula. For example, for calcium chloride, we need two Cl atoms to accommodate the two electrons supplied by one Ca atom:

$$:\overset{..}{\underset{..}{Cl}}\cdot\ +\ Ca:\ +\ :\overset{..}{\underset{..}{Cl}}\cdot\ \longrightarrow\ \left[:\overset{..}{\underset{..}{Cl}}:\right]^-\ Ca^{2+}\ \left[:\overset{..}{\underset{..}{Cl}}:\right]^-$$

It follows that the chemical formula for calcium chloride is $CaCl_2$.

In each case, the transfer of electrons results in the formation of an **octet** of electrons, an s^2p^6 electron configuration, on each of the atoms. Hydrogen, lithium, and beryllium are exceptions, because the $n = 1$ shell is complete with only two electrons. Hydrogen, $\cdot H$, either loses an electron to form a bare proton, H^+, or gains one to form the hydride ion, $:H^-$, with a heliumlike $1s^2$ configuration, a **duplet**. Similarly, lithium ($[He]2s^1$) and beryllium ($[He]2s^2$) lose electrons to form a heliumlike duplet when they become Li^+ and Be^{2+}.

When an s-block metal atom forms a cation, it loses all its valence electrons and is left with only its core. That core has the octet configuration of the *preceding* noble-gas atom (Fig. 8.2). The calcium cation, for instance, has an argonlike configuration. The metals on the left of the p block may also lose their s- and p-electrons. However, when atoms of the p-block metals in Period 4 and below lose their outermost electrons, they are left with a noble-gas core surrounded by an additional, complete subshell of d-electrons. For instance, gallium forms the ion Ga^{3+} with the configuration $[Ar]3d^{10}$. The d-electrons of the p-block atoms are gripped tightly by the nucleus and, in most cases, cannot be lost, so they are not shown in the Lewis symbol of these atoms or their ions.

The valence shell of an anion has an octet of electrons in an ns^2np^6 configuration (or a duplet of electrons in a $1s^2$ configuration in the case of the hydride ion, $:H^-$). The configuration of an anion is that of an atom of the *following* noble gas (Fig. 8.3). The oxide ion, for instance, has a neonlike configuration, and the sulfide ion (S^{2-}) has an argonlike configuration.

> *The formation of ionic bonds is represented in terms of Lewis symbols by the loss or gain of electrons (represented by dots) until both species have reached an octet (or duplet) of electrons.*

Example 8.1 *Predicting the chemical formula of a binary ionic compound from Lewis symbols*

Predict the chemical formula of aluminum oxide.

Strategy We write the Lewis symbol of an atom by arranging the valence electrons (represented as dots) around the symbol of the atom. Note that the main-group elements in Groups 13 through 18 have a number of valence electrons equal to their group numbers minus 10. To write the Lewis formulas of ions, we transfer electrons by moving dots from the metal atoms to the nonmetal atoms. We ensure that all the dots are accommodated by checking that the metal atoms have lost all their valence electrons and that all the valence orbitals of the nonmetal atoms are filled.

Solution Aluminum belongs to Group 13 and has three valence electrons ($:Al\cdot$). Oxygen belongs to Group 16 and has six valence electrons ($\cdot\overset{..}{\underset{.}{O}}\cdot$). Each aluminum

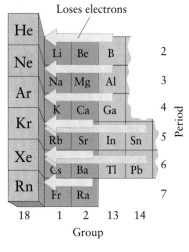

FIGURE 8.2

When a main-group metal atom forms a cation, it loses its valence s- and p-electrons and acquires the electron configuration of the preceding noble-gas atom. The heavier atoms in Groups 13 and 14 behave similarly, but the resulting core consists of the noble-gas configuration and an additional complete subshell of d-electrons.

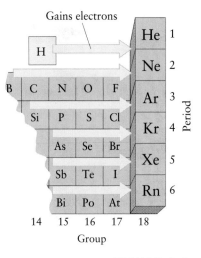

FIGURE 8.3

When p-block elements acquire electrons and form anions, they do so until they have reached the electron configuration of the following noble gas.

FIGURE 8.4

When tin(II) oxide is heated in air, it becomes incandescent as it reacts to form tin(IV) oxide. Even without being heated, it smolders and can ignite.

atom can provide three electrons, and each oxygen atom can accommodate two electrons. Because we need to fill all the oxygen orbitals, we need an even number of electrons from aluminum; so at least two Al atoms must contribute electrons. The $2 \times 3 = 6$ electrons released by two Al atoms can be accommodated by three O atoms, which have $3 \times 2 = 6$ gaps in their valence shells.

$$\cdot\ddot{\text{O}}\cdot \; + \; :\text{Al}\cdot \; + \; \cdot\ddot{\text{O}}\cdot \; + \; :\text{Al}\cdot \; + \; \cdot\ddot{\text{O}}\cdot \; \longrightarrow \; \left[:\ddot{\text{O}}:\right]^{2-} \text{Al}^{3+} \left[:\ddot{\text{O}}:\right]^{2-} \text{Al}^{3+} \left[:\ddot{\text{O}}:\right]^{2-}$$

Hence, the chemical formula of aluminum oxide is Al_2O_3.

Self-Test 8.1A Derive the chemical formula of (a) calcium nitride; (b) sodium telluride by using Lewis symbols.

[***Answer:*** (a) Ca_3N_2; (b) Na_2Te]

Self-Test 8.1B Derive the chemical formula of (a) aluminum sulfide; (b) strontium phosphide by using Lewis symbols.

8.2 Variable Valence

Some elements can form more than one type of cation. For example, consider aluminum and indium in Group 13. An important difference between these two elements is that aluminum forms only Al^{3+} ions, but indium forms both In^{3+} and In^+ ions. The tendency to form cations two units lower in charge than expected from the group number is called the **inert-pair effect.** Group 14 provides another example of the inert-pair effect: tin forms tin(IV) oxide when heated in air, but the heavier lead atom loses only its two *p*-electrons and forms lead(II) oxide. Tin(II) oxide can be prepared, but it is readily oxidized to tin(IV) oxide (Fig. 8.4).

The inert-pair effect is due in part to the different energies of the valence *p*- and *s*-electrons. In the later periods of the periodic table, valence *s*-electrons are very low in energy because they are poorly shielded from the nuclear charge. They may therefore remain attached to the atom. The inert-pair effect is most pronounced among the heaviest members of a group, where the difference in energy between *s*- and *p*-electrons is greatest (Fig. 8.5).

The inert-pair effect implies that the elements highlighted in Fig. 8.5 can lose either their valence *p*-electrons or all their valence *p*- and *s*-electrons. Similarly, we saw in Section 7.19 that many members of the *d* block can lose a variable number of *d*-electrons. In each case, different ionic compounds can be obtained, such as CuCl and $CuCl_2$ for copper. We call the ability to form compounds with different numbers of bonds **variable valence.** To write a chemical formula, we need to know the oxidation number of the metal, which gives the charge on the cation. For example, the copper ion in copper(I) chloride has charge $+1$; and in copper(II) chloride, it has charge $+2$. Although *d*-block metals may lose *d*-electrons when forming compounds, the *d*-electrons are in an inner shell and are not normally considered to be valence electrons. For this reason, Lewis formulas are not usually written for transition metals.

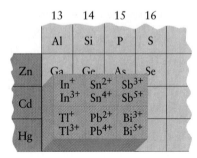

FIGURE 8.5

The typical ions formed by the heavy elements in Groups 13–15 show the influence of the inert-pair effect. These elements have the tendency to form compounds in which the oxidation numbers differ by 2.

The inert-pair effect is the tendency for p-block elements to form cations two units lower in charge than expected from the group number; it is most pronounced for heavy elements.

Self-Test 8.2A Use Lewis symbols to write the formulas of the two chlorides that indium can form.

[***Answer:*** $InCl_3$ (In^{3+} $[:\ddot{\text{Cl}}:]^-$ $[:\ddot{\text{Cl}}:]^-$ $[:\ddot{\text{Cl}}:]^-$), $InCl$ (In^+ $[:\ddot{\text{Cl}}:]^-$)]

8.3 Lattice Enthalpies

The completion of an octet is not an end in itself: atoms do not "want" to acquire a noble-gas configuration. Even though atoms of metallic elements lose electrons down to their core, a considerable investment of ionization energy is needed for them to do so. For example, to form K^+ ions from K atoms requires an input of 418 kJ/mol. The K^+ ions cannot lose more electrons in a chemical reaction because the ionization energies of core electrons are too high. Nonmetals acquire electrons until they have completed their valence octet. They cannot gain more, because that would involve accepting electrons into a higher energy shell. Even the completion of a noble-gas configuration may require considerable energy. For instance, to form an O^{2-} ion from an O atom requires 703 kJ/mol (see Fig. 7.43).

The driving force for the formation of ionic bonds is the considerable lowering of energy that takes place when ions pack together as a solid and attract one another strongly. Ion formation requires energy, but even more energy is released in the strong attraction between opposite charges when the solid forms (Fig. 8.6). The net effect is an overall lowering of energy. A measure of the attraction between ions is the **lattice enthalpy,** the enthalpy change per mole of formula units when a solid is broken up into a gas of widely separated ions. For example, the lattice enthalpy of calcium chloride is the change in enthalpy for

$$CaCl_2(s) \longrightarrow Ca^{2+}(g) + 2\,Cl^-(g)$$

and is 2260. kJ/mol (all lattice enthalpies are positive). Heat equal to the lattice enthalpy is released when the solid lattice forms from gaseous ions (at constant pressure). Therefore, at constant pressure, 2260. kJ is released as heat when widely separated calcium and chloride ions condense to form 1 mol $CaCl_2(s)$.

The energy released when an ionic solid forms from a gas of ions is due to the decrease in potential energy of the ions as they come close together. The potential energy for the interaction between two ions can be expressed as

$$\text{Coulomb potential energy} \propto \frac{\text{charge}_1 \times \text{charge}_2}{\text{separation}} \quad \text{or} \quad V \propto \frac{q_1 q_2}{r_{12}}$$

That is, the potential energy of two charges q_1 and q_2 (such as a cation and an anion, two anions, or two cations) is proportional to their charges and inversely proportional to their separation, r_{12}. These relations imply that the closer the centers of charge (the smaller the radii of the ions) and the greater the charges, the stronger will be the interaction.

Each cation in the solid is surrounded by anions, and the attractions between the cations and their surrounding anions lower the energy of the solid. Beyond those nearest neighbors, there are cations that repel the central cation and thus raise the energy of the solid. However, this repulsion is weaker than the attractions of anions, because the repelling cations are further away than the attracting anions (Fig. 8.7). Each ion in a solid is attracted to all the other oppositely charged ions and repelled by all the other like-charged ions. These repulsions and attractions become progressively weaker as the distance from the central ion increases, and the net outcome is a lowering of energy.

According to Coulomb's law, lattice enthalpies are large when the ions are small and highly charged. To see this correlation, compare, for example, the lattice enthalpies of MgO (3850. kJ/mol) and KCl (717 kJ/mol) in Table 8.1. The

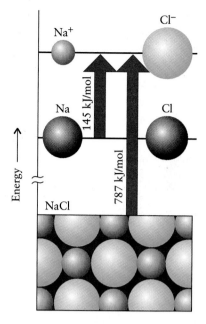

FIGURE 8.6

Considerable energy is needed to produce cations and anions from neutral gas-phase atoms: the ionization energy of the metal atoms must be supplied, and it is only partly recovered from the electron affinity of the nonmetal atoms. The overall lowering of energy that drags the ionic solid into existence is due to the strong attraction between cations and anions in the solid. It takes 145 kJ/mol to produce the ions from the elements, and the solid compound is 787 kJ/mol lower in energy than the separated

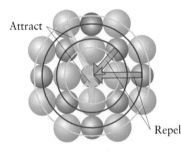

FIGURE 8.7

A two-dimensional slice of an ionic crystal. The greater the distance between two ions, the weaker their attractions. Therefore, the strongest attractions to an ion are those of the adjacent ions of opposite charge. The blue circles denote the distances over which the two closest repulsions occur, and the yellow circles denote the two closest attractions.

Attract

Repel

attractions between the doubly charged ions Mg^{2+} and O^{2-} are stronger than the attractions between the singly charged ions K^+ and Cl^-. Moreover, the attractions are stronger and the potential energy lower if the ions can get closer together, as the smaller radii of the Mg^{2+} and O^{2-} ions allow (see Fig. 7.36).

We cannot measure the lattice enthalpy of a compound directly, but we can obtain it indirectly by combining other measurements. The procedure uses a **Born-Haber cycle,** a thermodynamic cycle that consists of a series of steps starting at the elements and ending at the elements again. In a Born-Haber cycle, we imagine that we break apart the elements into atoms, ionize the atoms, combine the gaseous ions to form the ionic solid, then form the elements again from the ionic solid. Only the lattice enthalpy, the enthalpy of the step in which the ionic solid is formed from the gaseous ions, is unknown. The sum of the enthalpy changes for a complete Born-Haber cycle is 0, because the enthalpy of the system must be the same at both the start and the finish. Therefore, we can use a knowledge of all the other enthalpy changes in the cycle to find the unknown lattice enthalpy. The procedure is described in the following Toolbox and illustrated in Example 8.2. Table 8.1 lists lattice enthalpies obtained in this way.

> The overall enthalpy change around a Born-Haber cycle must be 0 because enthalpy is a state function (Section 6.6).

> We see in Section 12.8 how a knowledge of these values can help us to understand why some compounds are soluble but others are not.

The interactions between ions account for the formation of ionic compounds; the strengths of those interactions are measured by the lattice enthalpy. The lattice enthalpy is large for compounds formed from small, highly charged ions.

Table 8.1 Lattice enthalpies at 25°C in kilojoules per mole

Halides

LiF	1046	LiCl	861	LiBr	818	LiI	759
NaF	929	NaCl	787	NaBr	751	NaI	700.
KF	826	KCl	717	KBr	689	KI	645
AgF	971	AgCl	916	AgBr	903	AgI	887
BeCl$_2$	3017	MgCl$_2$	2524	CaCl$_2$	2260.	SrCl$_2$	2153
		MgF$_2$	2961	CaBr$_2$	1984		

Oxides

MgO	3850.	CaO	3461	SrO	3283	BaO	3114

Sulfides

MgS	3406	CaS	3119	SrS	2974	BaS	2832

Toolbox 8.1 *How to use a thermochemical cycle*

This Toolbox shows how to use a Born-Haber cycle to determine the lattice enthalpy of a compound. A similar procedure can be used to determine any unknown quantity that appears in a thermodynamic cycle.

Conceptual Basis

Because enthalpy is a state property, it must have the same value at the beginning and the end of a complete cycle of changes. Therefore, the sum of enthalpy changes around a complete cycle must be 0 (Fig. 8.8).

Procedure

To use a Born-Haber cycle, we start with the elements—typically a metal and a nonmetal—in the appropriate amounts to form the compound, and then make the changes described in the following steps. We use an arrow to describe each step. For an endothermic process, we draw the arrow pointing up; for an exothermic process, we draw the arrow pointing down.

Step 1. Start with the elements in the proportions in which they appear in the compound and atomize them. Write the corresponding enthalpies of formation of the gas-phase atoms (Appendix 2A) next to the upward pointing arrows in the diagram.

Step 2. Form gaseous cations from the metal atoms. This step requires the ionization energy of the metal and possibly the sum of the first and higher ionization energies (Appendix 2D). The corresponding arrow points upward.

Step 3. Form gaseous anions from the nonmetal atoms. The enthalpy change of this step is called the *electron-gain enthalpy, $\Delta H_{eg}°$*. It is the negative of the electron affinity. If the electron affinity is positive, the electron-gain enthalpy is negative and the corresponding arrow points downward because energy is released. If the electron affinity is negative, the electron-gain enthalpy is positive and the arrow points upward.

Step 4. Let the gas of ions form the solid compound. This step is the reverse of the formation of the ions from the solid, so its enthalpy change is the negative of the lattice enthalpy, $-\Delta H_L$. Denote it by an arrow pointing downward, because the formation of the solid is exothermic. This is the unknown value in the cycle.

Step 5. Complete the cycle with an arrow from the compound to the elements: the enthalpy change for this step is the negative of the enthalpy of formation of the compound from its elements, $\Delta H_f°$. The arrow points up if $\Delta H_f°$ is negative, down if it is positive.

Step 6. Finally, calculate ΔH_L from the fact that the sum of all the enthalpy changes for the complete cycle is 0. The procedure is illustrated in Example 8.2. Lattice enthalpies of other compounds obtained in this way are listed in Table 8.1.

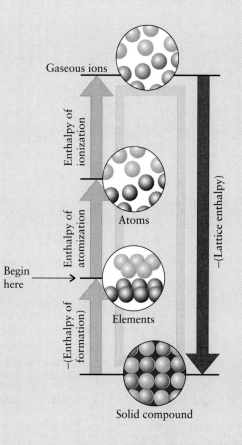

FIGURE 8.8

In a Born-Haber cycle, we select a sequence of steps that starts and ends at the same point (the elements, for instance). The lattice enthalpy is the enthalpy change of the step in which the solid is formed from a gas of ions (the dark red arrow). The sum of enthalpy changes for the complete cycle (yellow line) is 0 because enthalpy is a state property. The "Enthalpy of ionization" step includes the energies required to produce both cations and anions from their respective atoms.

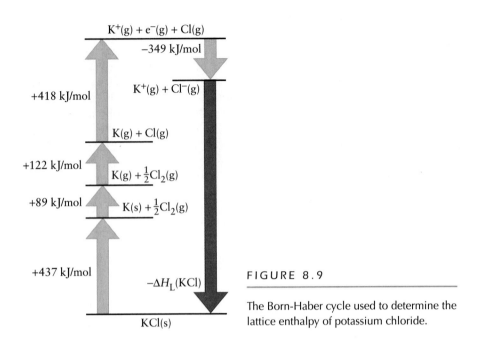

K$^+$(g) + e$^-$(g) + Cl(g)

−349 kJ/mol

+418 kJ/mol

K$^+$(g) + Cl$^-$(g)

K(g) + Cl(g)

+122 kJ/mol

K(g) + $\frac{1}{2}$Cl$_2$(g)

+89 kJ/mol

K(s) + $\frac{1}{2}$Cl$_2$(g)

+437 kJ/mol

−ΔH_L(KCl)

KCl(s)

FIGURE 8.9

The Born-Haber cycle used to determine the lattice enthalpy of potassium chloride.

Example 8.2 *Using a Born-Haber cycle to calculate a lattice enthalpy*

Use a Born-Haber cycle to calculate the lattice enthalpy of potassium chloride.

Strategy We follow the steps set out in Toolbox 8.1. Enthalpies of formation are given in Appendix 2A. Ionization energies and electron affinities are listed in Appendix 2D and some are also given in Figs. 7.38 and 7.43, respectively. We equate the ionization energy with the enthalpy of ionization of the atoms ($\Delta H_{ion} = I$) and the electron affinity with the *negative* of the electron-gain enthalpy ($\Delta H_{eg} = -E_{ea}$). All lattice enthalpies are positive.

Solution The Born-Haber cycle for KCl is shown in Fig. 8.9. The sum of the enthalpy changes for the complete cycle is 0, so we can write

$$\Delta H_f^\circ(K, g) + \Delta H_f^\circ(Cl, g) + I(K) - E_{ea}(Cl) - \Delta H_f(KCl, s) - \Delta H_L = 0$$

On substituting the data, we find

$$\{89 + 122 + 418 - 349 - (-437)\} \text{ kJ/mol} - \Delta H_L = 0$$

and hence

$$\Delta H_L = (89 + 122 + 418 - 349 + 437) \text{ kJ/mol} = +717 \text{ kJ/mol}$$

Therefore, the lattice enthalpy of potassium chloride is 717 kJ/mol.

Self-Test 8.3A Calculate the lattice enthalpy of calcium chloride, CaCl$_2$, by using the data in Appendix 2A and Appendix 2D. Note that 1 mol CaCl$_2$ produces 2 mol Cl$^-$ ions.

[*Answer:* 2259 kJ/mol]

Self-Test 8.3B Calculate the lattice enthalpy of magnesium bromide, MgBr$_2$. Note that 1 mol MgBr$_2$ produces 2 mol Br$^-$ ions.

8.4 The Properties of Ionic Compounds

Ionic solids are assemblies of cations and anions stacked together in a regular array. In sodium chloride, positive sodium ions alternate with negative chloride ions, and large numbers of oppositely charged ions are lined up in all three dimensions (Fig. 8.10). Ionic solids are examples of **crystalline solids,** or solids that consist of atoms, molecules, or ions stacked together in a regular array.

The strong attraction between oppositely charged ions in ionic solids accounts for their typical properties, such as high melting points, high boiling points, and brittleness. It requires a high temperature to get the ions to move past one another to form a liquid, and an even higher temperature to drive them apart into a gas of ions. Ionic solids are brittle because of those strong attractions and repulsions. We cannot just push a block of ions past another block; instead, when we strike an ionic solid, it tends to shatter into fragments (Fig. 8.11).

One way to separate the ions from an ionic solid without heating it is to dissolve it in water. Many ionic solids are soluble in water. As we saw in Section 3.3, when an ionic solid dissolves in water, the ions are separated by the water molecules and give an electrolyte solution. In the solid, the ions are held in place by the coulombic forces; but in aqueous solution, the ions are mobile and conduct electricity as they move through the solution. However, many other ionic solids, such as MgO and AgCl, are insoluble in water: the attractions between their cations and anions are too strong to allow them to be separated from one another.

We can now see why nature has adopted the ionic solid calcium phosphate for our skeletons. The doubly charged, small Ca^{2+} ions and the triply charged PO_4^{3-} ions attract one another very strongly and clamp together to form a rigid, insoluble solid (Fig. 8.12). It is fortunate for us that calcium phosphate does not dissolve in water: if it were soluble, our skeletons would dissolve in our own body fluids! Nevertheless, calcium phosphate is not completely insoluble, and each day a small amount of the calcium in our skeleton dissolves and must be

FIGURE 8.10

An ionic solid—here we see a fragment of sodium chloride, with the sodium ions represented by pink spheres and the chloride ions by green spheres—consists of an almost infinite array of cations and anions stacked together to give the lowest energy arrangement. The pattern shown here is repeated throughout the crystal.

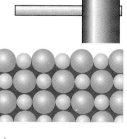
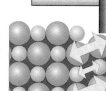
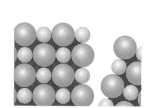

(a)   (b)   (c)

(d)   (e)

FIGURE 8.11

This sequence of images illustrates why ionic solids are brittle. (a) The original solid consists of an orderly array of cations and anions. (b) A hammer blow can push the ions into positions where cations are next to cations and anions are next to anions; there are now strong repulsive forces acting (as depicted by the double-headed arrows). (c) As a result of these repulsive forces, the solid springs apart in fragments. (d) This chunk of calcite consists of several large crystals joined together. (e) The blow of a hammer has shattered the crystal, leaving flat, regular surfaces consisting of planes of ions.

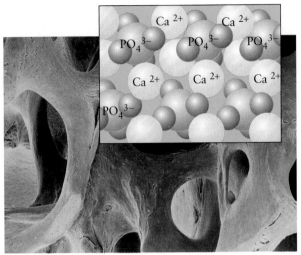

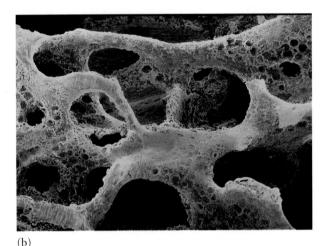

(a)

(b)

FIGURE 8.12

These micrographs show the porous structure of bone. The calcium in bone is extracted by the body if the level of calcium in the diet is low. (a) Healthy bone tissue. (b) Bone that has suffered calcium loss through osteoporosis. The overlay shows the regular arrangement of the calcium and phosphate ions in healthy bone.

replaced or osteoporosis (the disintegration of bone) results. The replacement of bone calcium is one of the reasons why we need calcium in our daily diet.

Ions stack together in regular crystalline structures. Ionic solids typically have high melting and boiling points, are brittle, and form electrolyte solutions if they dissolve in water.

COVALENT BONDS

Nonmetallic elements cannot form monatomic cations, because their ionization energies are too high. However, we know that nonmetals do combine with one another: the existence of millions of different compounds of carbon, hydrogen, and oxygen is evidence enough! The nature of the bond between nonmetals puzzled scientists until 1916, when Lewis found an explanation. With brilliant insight, and before anyone knew about quantum mechanics or electron orbitals, Lewis proposed that a **covalent bond** is a pair of electrons shared between two atoms.

8.5 From Atoms to Molecules

Lewis knew that the negative charges of the two electrons of a shared pair can attract the positively charged nuclei they lie between. Attractions between the electron pair and the nuclei pull the two atoms together (**2**). Lewis had no way of knowing why it had to be a *pair* of electrons and not some other number. The explanation came only with the development of quantum mechanics in the 1920s.

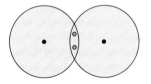

2 Shared electron pair

The simplest example of a bond between two nonmetal atoms is the H_2 molecule (Fig. 8.13). Initially, the two hydrogen atoms each have one valence electron in a $1s$-orbital. As the atoms approach, their orbitals merge and the two electrons that occupy the merged orbitals pair with each other as they would in an atomic orbital. Once the two half-full orbitals have merged into one full orbital, the two $1s$-electrons are equally attracted to and shared by the two nuclei. The electron density *between* the two nuclei is greater than it is on either side. Neither atom has to release an electron totally, so neither has to be supplied with its full ionization energy. Sharing corresponds to a *partial* release of an electron, so less energy is needed. In covalent bonds, each bonded atom may contribute one electron to the shared electron pair, or one atom may contribute both electrons. Either way, the shared pair of electrons lies between the two neighboring atoms and binds them together.

Nonmetals form covalent bonds to one another by sharing pairs of electrons.

8.6 The Octet Rule and Lewis Structures

How many covalent bonds can an atom form? When an ionic bond forms, one atom loses electrons and the other gains them until both atoms have reached a noble-gas configuration—a duplet for the elements close to helium and an octet for all other elements. In covalent bonds, atoms *share* electrons to reach a noble-gas configuration. Lewis called this the **octet rule:**

In covalent bond formation, atoms go as far as possible toward completing their octets by sharing electron pairs.

Nitrogen ($:\!\overset{\cdot}{\text{N}}\!\cdot$) has five valence electrons and needs three more electrons to complete its octet. Chlorine ($:\!\overset{\cdot\cdot}{\underset{\cdot\cdot}{\text{Cl}}}\!\cdot$) has seven valence electrons and needs one more electron to complete its octet. Argon ($:\!\overset{\cdot\cdot}{\underset{\cdot\cdot}{\text{Ar}}}\!:$) already has a complete octet and has no tendency to share any more electrons. Hydrogen (H·) needs one more electron to reach its heliumlike duplet.

The **valence** of an element is the number of covalent bonds an atom of the element forms. This number is most easily found by using the same Lewis symbols that we introduced for the discussion of ionic bonds. Consider molecular hydrogen, H_2. Each atom completes its heliumlike duplet by sharing its electron with the other:

$$\text{H·} + \text{·H} \longrightarrow \text{H:H} \qquad \text{or} \qquad \text{H—H}$$

The symbol H—H, in which the line represents the shared electron pair (the covalent bond between the two atoms), is the simplest example of a **Lewis structure,** a diagram showing how electron pairs are shared between atoms in a molecule. Because hydrogen completes its duplet by sharing one pair of electrons, it has a valence of 1 in all its compounds.

A fluorine atom has seven valence electrons and needs one more to complete its octet. It can achieve an octet by accepting a share in an electron supplied by another atom, such as another fluorine atom:

$$:\!\overset{\cdot\cdot}{\underset{\cdot\cdot}{\text{F}}}\!\cdot + \cdot\!\overset{\cdot\cdot}{\underset{\cdot\cdot}{\text{F}}}\!: \longrightarrow \left(\!:\!\overset{\cdot\cdot}{\underset{\cdot\cdot}{\text{F}}}\!\!\bigcirc\!\!\overset{\cdot\cdot}{\underset{\cdot\cdot}{\text{F}}}\!:\!\right) \qquad \text{or} \qquad :\!\overset{\cdot\cdot}{\underset{\cdot\cdot}{\text{F}}}\!\!-\!\!\overset{\cdot\cdot}{\underset{\cdot\cdot}{\text{F}}}\!:$$

The circles have been drawn around each F atom to show how each one gets an octet by sharing one electron pair. The valence of fluorine is therefore 1, the same as that of hydrogen.

From its Lewis structure, we see that F_2 possesses **lone pairs** of electrons, pairs of valence electrons not involved in bonding. The lone pairs on neighboring F atoms repel one another, and this repulsion is almost enough to overcome the favorable attractions of the bonding pair that holds the F_2 molecule together. This repulsion is one of the reasons why fluorine gas is so reactive: the atoms are bound together as F_2 molecules only very weakly. Among the common diatomic molecules, only H_2 does not have any lone pairs.

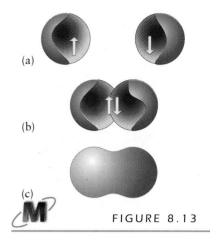

(a)

(b)

(c)

FIGURE 8.13

The formation of a covalent bond between two hydrogen atoms. (a) Two separate hydrogen atoms, each with one electron. (b) The electron cloud that forms when the spins pair and the orbitals merge is most dense between the two nuclei. (c) The boundary surface that we shall use to depict a covalent bond.

Nonmetal atoms share electrons with one another to complete their octets (or duplets). A Lewis structure shows the arrangement of electrons as shared pairs (lines) and lone pairs (pairs of dots).

Self-Test 8.4A Write the Lewis structure for the "interhalogen" compound chlorine monofluoride, ClF, and state how many lone pairs each atom possesses in the compound.

[*Answer:* $:\ddot{\mathrm{Cl}}—\ddot{\mathrm{F}}:$; three on each atom]

Self-Test 8.4B Write the Lewis structure for the compound HBr and state how many lone pairs each atom in the compound possesses.

THE STRUCTURES OF POLYATOMIC SPECIES

We have seen how pairs of atoms form bonds. The atoms in polyatomic molecules and ions are also bound together by covalent bonds. Each atom completes its octet (or duplet for hydrogen) by sharing pairs of electrons with its immediate neighbors. Each shared pair counts as one covalent bond and is represented by a line between the two atoms.

8.7 Lewis Structures

The simplest organic molecule is methane, CH_4. To write its Lewis structure, we count the valence electrons available from all the atoms in the molecule. For methane, the Lewis symbols are

$$:\dot{\mathrm{C}} \quad \mathrm{H}\cdot \quad \mathrm{H}\cdot \quad \mathrm{H}\cdot \quad \mathrm{H}\cdot$$

so there are $4 + (4 \times 1) = 8$ valence electrons. The next step is to arrange the dots representing the electrons so that the C atom has an octet and each H atom has a duplet. We draw the arrangement shown in (**3a**). The Lewis structure of methane is then redrawn as shown in (**3b**). Because the carbon atom is linked by four bonds to other atoms, we say that it is *tetravalent*: it has a valence of 4. Almost all compounds of carbon are tetravalent.

A single shared pair of electrons is called a **single bond.** However, atoms can share two or three electron pairs. Two shared electron pairs form a **double bond,** and three shared electron pairs form a **triple bond.** For instance, a double bond between a carbon atom and an oxygen atom, C::O, appears as C=O in a Lewis structure. Similarly, a triple bond, such as C:::C, appears as C≡C. Double and triple bonds are collectively called **multiple bonds.** As before, each line represents a pair of electrons. A double bond involves a total of four electrons; a triple bond involves six electrons. The **bond order** is the number of electron-pair bonds that link two atoms. The bond order of a single bond is 1, that of a double bond is 2, and that of a triple bond is 3.

Sometimes we may be uncertain about the arrangement of atoms in a polyatomic ion. Just which is the central atom? A good rule of thumb for this situation is to *choose the atom with the lowest ionization energy* for the central atom. This arrangement often leads to the lowest energy because the designated central atom requires the smallest amount of energy to give up its electrons and share them with its neighbors. The atoms with higher ionization energies can hold on to their electrons as lone pairs. Hydrogen is never central, because (except in some peculiar compounds of boron we meet in Chapter 9) it can form only one bond.

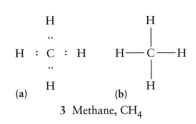

(a)   (b)

3 Methane, CH_4

The only common compound that is an exception to the tetravalence of carbon is CO.

After we have met the concept of electronegativity (Section 8.14), we shall be able to express this rule in a different manner: choose the atom with the lowest electronegativity to be the central atom.

Another rule of thumb for predicting the structure of a molecule is *to arrange the atoms symmetrically around the central atom*. For instance, SO_2 is OSO, not SOO. An exception is dinitrogen monoxide, N_2O (nitrous oxide), which has atoms in the unsymmetrical arrangement NNO. A third rule is that in simple chemical formulas, the central atom is often written first, followed by the atoms attached to it. For example, in the compound with the chemical formula OF_2 (**4**), the arrangement of the atoms is actually FOF, not OFF; and in SF_6, the S atom is surrounded by six F atoms. If the compound is an oxoacid, then the acidic hydrogen atoms are attached to oxygen atoms, which in turn are attached to the central atom. For example, sulfuric acid, H_2SO_4, has the structure $(HO)_2SO_2$.

Recognizing other characteristic patterns helps to simplify writing a Lewis structure. For example, a "terminal" halogen atom, that is, a halogen atom linked to just one other atom, always has a single bond and three lone pairs:

$$—\ddot{\underset{\cdot\cdot}{F}}: \qquad —\ddot{\underset{\cdot\cdot}{Cl}}: \qquad —\ddot{\underset{\cdot\cdot}{Br}}: \qquad —\ddot{\underset{\cdot\cdot}{I}}:$$

Except for CO, terminal oxygen and sulfur atoms form either a single bond with three lone pairs or a double bond with two lone pairs:

$$—\ddot{\underset{\cdot\cdot}{O}}: \text{ or } =\ddot{\underset{\cdot\cdot}{O}} \qquad —\ddot{\underset{\cdot\cdot}{S}}: \text{ or } =\ddot{\underset{\cdot\cdot}{S}}$$

The marginal structures throughout this chapter contain examples of these patterns.

The same general procedure is also used for polyatomic ions, such as the ammonium ion, NH_4^+, or the sulfate ion, SO_4^{2-}. We arrange the atoms in the correct spatial positions, count the dots available for bonds and lone pairs, and then construct the Lewis structure. Each atom provides the number of electrons (dots) equal to the number of electrons in its valence shell, but we have to adjust the total number of dots to represent the overall charge. For a cation, subtract one dot for each positive charge; for an anion, add one dot for each negative charge. The cation and the anion in a compound must be treated separately: they are individual ions and are not linked by shared pairs. The Lewis structure of ammonium sulfate, $(NH_4)_2SO_4$, for instance, is written as three bracketed ions (**5**). The sign at the top right of each bracket shows the charge of each ion: it belongs to the whole ion, not to any particular atom.

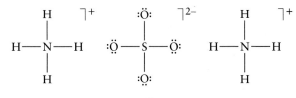

5 Ammonium sulfate, $(NH_4)_2SO_4$

As an example, let's construct the Lewis structure of the HCN molecule. The atoms have $1 + 4 + 5 = 10$ valence electrons, so the molecule has five electron pairs. Carbon has a lower ionization energy than nitrogen, so the HCN molecule is written (with the electron pairs parked on the right):

$$HCN \qquad ::::: $$

We use two pairs to form bonds between the atoms:

$$H:C:N \qquad :::$$

At this point, three of the five electron pairs remain unused. We could try to put all three remaining pairs on the N atom:

$$H\!:\!C\!:\!\ddot{\underset{\cdot\cdot}{N}}\!:$$

However, this arrangement does not complete carbon's octet. If we use the electrons to complete the octet on the C atom, then we do not complete nitrogen's octet:

$$H\!:\!\ddot{\underset{\cdot\cdot}{C}}\!:\!N\!:$$

Therefore, we rearrange the electron pairs to form a triple bond between carbon and nitrogen, which completes octets on both carbon and nitrogen:

$$H\!:\!C\!:\!:\!:\!N\!: \qquad or \qquad H\!-\!C\!\equiv\!N\!:$$

This approach is summarized in the following Toolbox.

Toolbox 8.2 *How to write the Lewis structure of a polyatomic species*

This Toolbox shows how to construct the Lewis structure of a molecule or ion in which every atom other than H and Be observes the octet rule. For molecules that have too many or too few valence electrons for the octet rule to be observed, use strategies such as those in Toolbox 8.3.

Conceptual Basis

Atoms of nonmetals are held together in molecules by covalent bonds that consist of shared pairs of electrons. Valence electrons not used in bonding remain on the atoms as lone pairs. Only valence electrons are shown in Lewis structures.

Procedure

When writing Lewis structures for polyatomic species, it is helpful to break down the procedure into the following steps (Fig. 8.14):

Step 1. Count the total number of valence electrons on each atom and divide by 2 to obtain the number of electron pairs. If the species is a polyatomic ion, add one more electron for each negative charge or subtract one electron for every positive charge.

Step 2. Write the chemical symbols of the atoms to show their layout in the molecule. We can predict the most likely arrangements of atoms by using common patterns and the rules of thumb given earlier.

Step 3. Place one electron pair between each pair of bonded atoms.

Step 4. Complete the octet (or duplet, in the case of H) of each atom by placing any remaining electron pairs as lone pairs around the atoms. If there are not enough

electron pairs to form octets, form multiple bonds. Represent each shared electron pair by a line.

The procedure may be simplified by recalling that a terminal halogen atom has three lone pairs. A terminal oxygen or sulfur atom has either a double bond and two lone pairs or a single bond and three lone pairs.

To check the validity of a Lewis structure, verify that each atom has an octet or duplet.

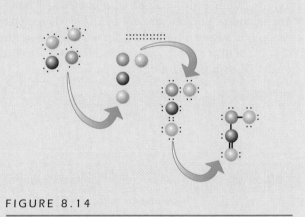

FIGURE 8.14

A diagram of the steps used to construct a Lewis structure. Valence electrons are counted and the atoms set out in the order in which they are bonded. The electrons are then arranged around the atoms in pairs so that each atom has an octet; or, in the case of hydrogen, a duplet.

Example 8.3 *Writing Lewis structures for polyatomic molecules I*

Write the Lewis structure for ammonia, NH_3.

Strategy Identify the group number of each element and thus its number of valence electrons; then write the Lewis structure by working through the four-step procedure outlined in Toolbox 8.2.

Solution **Step 1.** The total number of valence electrons is

$$
\begin{array}{ll}
N & 1 \times 5 = 5 \\
H & 3 \times 1 = \underline{3} \\
& \text{Total} = 8
\end{array}
$$

Because $8 \div 2 = 4$, the molecule has 4 valence electron pairs. **Step 2.** The central atom in ammonia must be nitrogen, because hydrogen atoms can form only one bond. **Step 3.** We use three electron pairs to link neighboring atoms, arranged as shown by the pale gray rectangles in (**6a**). One pair remains. **Step 4.** Each hydrogen atom already has a duplet, so we add the remaining pair to the nitrogen atom as a lone pair. The final Lewis structure is shown in (**6b**).

Self-Test 8.5A Write a Lewis structure for HBrO (the atoms are arranged as HOBr). [***Answer:*** See (7).]

Self-Test 8.5B Suggest a reason why, on the basis of Lewis structures, the C atom in CO can bind to hemoglobin and cause carbon monoxide poisoning, but the C atom in CO_2 cannot.

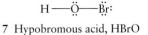

(a)   (b)

6 Ammonia, NH_3

7 Hypobromous acid, HBrO

Example 8.4 *Writing Lewis structures for polyatomic molecules II*

Write the Lewis structure for formic acid, HCOOH. In the —COOH group, both O atoms are attached to the same C atom and one of them is bonded to the final H atom.

Strategy Formic acid has two "central" atoms: the C atom, which is attached to an H atom and both O atoms, and one O atom, which is bonded to the final H atom. The full Lewis structure is obtained by working through the four-step procedure outlined in Toolbox 8.2.

Solution **Step 1.** The total number of valence electrons is

$$
\begin{array}{ll}
C & 1 \times 4 = 4 \\
H & 2 \times 1 = 2 \\
O & 2 \times 6 = \underline{12} \\
& \text{Total} = 18
\end{array}
$$

so the molecule has 9 valence electron pairs. **Step 2.** The atomic arrangement in the molecule, which is suggested by the way the molecular formula is written, is shown in (**8a**). **Step 3.** Use four electron pairs to link neighboring atoms, as shown in (**8b**). Five pairs remain. **Step 4.** Give each atom an octet of electrons by adding two lone pairs to each oxygen atom and allowing the terminal oxygen atom to form a double bond to the carbon atom, as depicted in (**8c**). The final Lewis structure is shown in (**8d**).

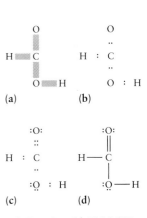

8 Formic acid, HCOOH

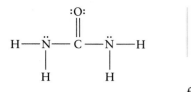

9 Urea, $(NH_2)_2CO$

Self-Test 8.6A Write a Lewis structure for the urea molecule, $(NH_2)_2CO$.

[**Answer:** See (**9**).]

Self-Test 8.6B Write the Lewis structure for ethene, H_2CCH_2.

For molecules with more than one central atom, it is often helpful to look for fragments of molecules with Lewis structures that we already know and to build up the complete structure by combining them. For example, the formula for acetic acid is CH_3COOH, which appears very complex. However, both the acetic acid and formic acid molecules have a —COOH group, with the initial H atom in formic acid replaced by a —CH_3 group in acetic acid. The atoms of acetic acid are arranged as in (**10a**), and its Lewis structure is shown in (**10b**).

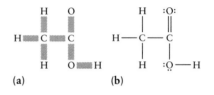

(a)   (b)

10 Acetic acid, CH_3COOH

The Lewis structure of a polyatomic species is obtained by using all the valence electrons to complete the octets (or duplets) of the atoms present, forming single or multiple bonds and leaving some electrons as lone pairs.

Example 8.5 *Writing Lewis structures for polyatomic ions*

Write the Lewis structure for the carbide ion, C_2^{2-}.

Strategy Write the Lewis structure by working through the four-step procedure outlined in Toolbox 8.2. Include an additional two electrons to account for the charge of −2.

Solution **Step 1.** The total number of valence electrons is

$$
\begin{array}{rl}
C \quad 2 \times 4 = & 8 \\
\text{Charge of } -2 = & \underline{2} \\
\text{Total} = & 10
\end{array}
$$

The charge of −2 has contributed two additional electrons for a total of 10 electrons, or 5 electron pairs. **Step 2** is obvious: the atomic arrangement is CC. **Step 3.** We use one electron pair to link neighboring atoms:

$$C\text{—}C \qquad ::::$$

Step 4. If we add the remaining pairs to the carbon atoms as lone pairs, there are not enough for both carbon atoms to have an octet:

$$[:\ddot{C}\text{—}\ddot{C}:]^{2-}$$

Therefore, we need to form a multiple bond between the carbon atoms. A double bond gives one atom an octet, but the other has only six electrons:

$$[:\ddot{C}\text{=}C:]^{2-}$$

However, a triple bond gives both carbon atoms an octet:

$$[:C\equiv C:]^{2-}$$

Self-Test 8.7A Write a Lewis structure for the hydronium ion, H_3O^+.

[*Answer:* See (**11**).]

Self-Test 8.7B Write a Lewis structure for the amide ion, NH_2^-.

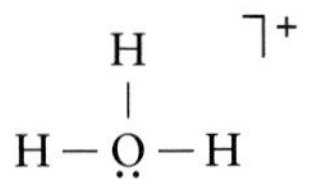

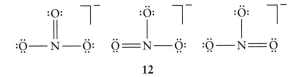

11 Hydronium ion, H_3O^+

8.8 Resonance

In some Lewis structures, the multiple bonds can be written in several equivalent locations. Consider the nitrate ion, NO_3^-. The three Lewis structures shown in (**12**) differ only in the position of the double bond. All are valid structures,

12

and all have exactly the same energy. So which one is correct? The answer is that none *alone* is the correct structure. If one of the pictured structures were correct, we would expect two long single bonds and one short double bond. However, the experimental evidence is that the bonds in a nitrate ion all have the same length (at 124 pm). They are longer than a typical $N=O$ double bond (120 pm), but shorter than a typical $N—O$ single bond (140 pm). The bonds in the nitrate ion have a character intermediate between a pure single bond and a pure double bond. Because all three bonds are identical, a better model of the nitrate ion is a blend of all three Lewis structures. This blending of structures, which is called **resonance,** is depicted in (**13**) by double-headed arrows. The blended structure is a **resonance hybrid** of the contributing Lewis structures.

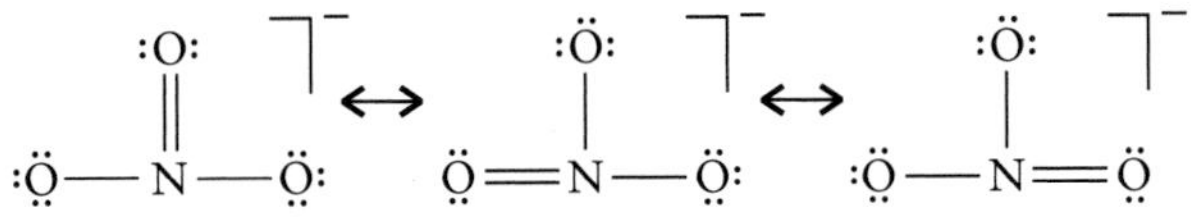

13 Nitrate ion, NO_3^-

Just as a mule is a blend of a horse and a donkey, not a creature that flickers between the two, resonance should be thought of as a *blend* of the individual Lewis structures rather than as the flickering of a molecule between different structures. Note, however, that all analogies have limitations: the horse and donkey actually exist independently, whereas the individual contributors to a resonance hybrid do not exist on their own.

An analogy

Electrons involved in resonance structures are said to be **delocalized.** Delocalization means that the additional electron density due to the second pair of electrons in a double bond is not shared by two particular atoms but is spread out over several atoms.

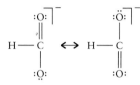

14 Formate ion, HCO_2^-

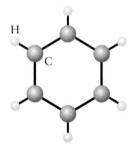

15 Benzene, C_6H_6

The H atoms in (17) are not shown, but there is one attached to each corner (that is, each C atom).

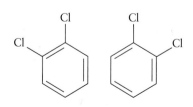

18 Dichlorobenzene

Example 8.6 *Writing a resonance structure*

Suggest two Lewis structures that contribute to the resonance structure for the O_3 molecule. Experimental data show that the two bond lengths are the same.

Strategy Write a Lewis structure for the molecule, using the method outlined in Toolbox 8.2. Decide whether there is another equivalent structure that results from the interchange of a single bond and a double bond. Write the actual structure as a resonance hybrid of these Lewis structures.

Solution Oxygen is a member of Group 16, so each atom has six valence electrons. The total number of valence electrons in the molecule is $3 \times 6 = 18$. One structure is therefore $\ddot{O}=\ddot{O}-\ddot{O}:$. Interchanging the bonds gives $:\ddot{O}-\ddot{O}=\ddot{O}$. The overall structure is the following resonance hybrid:

$$\ddot{O}=\ddot{O}-\ddot{O}: \longleftrightarrow :\ddot{O}-\ddot{O}=\ddot{O}$$

Self-Test 8.8A Write Lewis structures contributing to the resonance hybrid for the formate ion, HCO_2^-. Recall that the structure of HCOOH is described in Example 8.4.
                                                                [*Answer:* See (14).]

Self-Test 8.8B Write Lewis structures contributing to the resonance hybrid for the nitrite ion, NO_2^-.

Benzene, C_6H_6, is another molecule best described as a resonance hybrid. It consists of a hexagonal ring of six carbon atoms, with a hydrogen atom attached to each one (15). A Lewis structure that contributes to the resonance hybrid is shown in (16). It is called a **Kekulé structure** for the German chemist Friedrich Kekulé, who first proposed (in 1865) that benzene had a cyclic structure with alternating single and double bonds. The structure is normally abbreviated by a simple hexagon showing the bonding pattern (17).

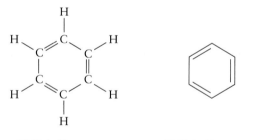

16 Kekulé structure

17 Kekulé structure, shortened form

A single Kekulé structure, however, does not fit all the evidence that chemists have since collected. For one thing, benzene does not undergo reactions typical of compounds with double bonds. Also, a Kekulé structure suggests that benzene should have two different bond lengths: three longer single bonds (154 pm) and three shorter double bonds (134 pm). Instead, all the bonds are found experimentally to have the same length (139 pm). Moreover, all six locations on the ring are the same; if the Kekulé structures were correct, there would be two distinct dichlorobenzenes in which the chlorine atoms are adjacent to each other, as shown in (18). However, only one such dichlorobenzene is known.

These characteristics of the benzene molecule can be explained if we note that there are in fact *two* Kekulé structures: they differ only in the positions of the double bonds (**19**). These two structures have exactly the same energy and blend together as a resonance hybrid. As a result of this resonance, the electron density of the C=C double bonds is spread evenly around the ring, giving each bond a length and an electron density intermediate between that of a single and a double bond. All six C—C bonds are equivalent. This equivalence is implied by the circle inside the hexagon in (**20**). We can now see from (**21**) why there can be only one dichlorobenzene with Cl atoms on adjacent C atoms.

Resonance stabilizes a molecule by lowering its total energy. This stabilization, which can be explained only by quantum mechanics, makes benzene less reactive than expected for a molecule with three carbon-carbon double bonds. Resonance results in the greatest lowering of energy when the contributing structures have equal energies. However, a molecule is a blend of *all reasonable* Lewis structures, including those with different energies. As we shall see, even structures having atoms with more than eight electrons in a valence shell can contribute to resonance. The lowest energy structures contribute most strongly to the overall structure in the sense that, if we think of a resonance hybrid as a blend of Lewis structures, then the structures with lowest energy contribute most to the mixture.

Resonance occurs only between structures with the same arrangement of atoms but with different arrangements of electron pairs. For example, although we might be able to write two hypothetical structures for the dinitrogen oxide molecule, NNO and NON, there is no resonance between them, because the atoms lie in different locations.

Resonance is a blending of structures with the same arrangement of atoms but different arrangements of electrons. It spreads multiple bond character over a molecule and also lowers its energy.

8.9 Formal Charge

The overall electric charge on a polyatomic ion belongs to the ion as a whole. It is possible, however, to divide up the overall charge artificially and to use the resulting charges to help decide which contributions to a resonance hybrid are most important. The same division of charge can be made for molecules with zero charge: for all neutral molecules, the sum of any charges on individual atoms is 0.

To assign a **formal charge** to an atom, we decide how many electrons each atom "owns." We suppose that each atom owns one electron of each bonding pair attached to it. We also imagine that it owns its lone pairs completely (Fig. 8.15). We count the number of electrons assigned to an atom this way and then compare this number with the number of electrons on the free atom. If the atom has more electrons in the molecule than when it is a free, neutral atom, then the atom has a negative formal charge, like a monatomic anion. If the assignment of electrons leaves the atom with fewer electrons than when it is free, then the atom has a positive formal charge, as if it were a monatomic cation.

$$\text{Formal charge} = \text{number of valence electrons in the free atom}$$
$$- \ (\text{number of electrons present as lone pairs})$$
$$- \tfrac{1}{2}(\text{number of electrons shared in bonds})$$

$$\text{FC} = V - \left(L + \tfrac{1}{2}S\right)$$

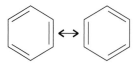

19 Benzene resonance structure

20 Benzene, C_6H_6

21 1,2-Dichlorobenzene, $C_6H_4Cl_2$

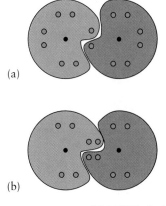

FIGURE 8.15

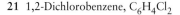

The formal charge on an atom is calculated by dividing the electrons up in a special way, as shown by the colored boundaries. Each atom owns all the electrons of its own lone pairs and owns one electron of any bonding pair it shares. The boundaries show the ownership of electrons for (a) a single bond and (b) a double bond in diatomic molecules.

In the definition, FC is the formal charge, V is the number of valence electrons in the free atom, L is the number of electrons present as lone pairs, and S is the number of shared electrons.

We can think of the formal charge as the charge an atom would have if we exaggerated the purely covalent character of bonds and the perfect sharing of electrons that model implies. Conversely, oxidation number (Section 3.15) is an exaggeration of the *ionic* character of bonds. It is based on a representation in which the atoms are pictured as ions. Formal charges depend on the particular Lewis structure we write; oxidation numbers do not.

Typically, the most stable Lewis structures are those in which the formal charges of the individual nonmetal atoms are closest to 0. The atoms in the preferred structures have undergone the least redistribution of electrons relative to the free atoms. For example, the formal charge rule suggests that the structure OCO is more likely for carbon dioxide than COO, as shown in (**22**). Similarly, it also suggests that the structure NNO is more likely for dinitrogen oxide than NON, as shown in (**23**).

> *The formal charge gives an indication of the extent to which atoms have gained or lost electrons in the process of covalent bond formation. Structures with lowest formal charges are likely to have the lowest energy.*

In Section 8.14, we shall see that the more favored of two Lewis structures with the same total formal charge is the one with the negative charge on the more electronegative atom.

$$\overset{0}{\ddot{\text{O}}}=\overset{0}{\text{C}}=\overset{0}{\ddot{\text{O}}} \qquad \overset{0}{\ddot{\text{O}}}=\overset{+2}{\text{O}}=\overset{-2}{\ddot{\text{C}}}$$

22

$$\overset{-1}{\ddot{\text{N}}}=\overset{+1}{\text{N}}=\overset{0}{\ddot{\text{O}}} \qquad \overset{-1}{\ddot{\text{N}}}=\overset{+2}{\text{O}}=\overset{-1}{\ddot{\text{N}}}$$

23

Example 8.7 *Judging the plausibility of a structure*

Write three Lewis structures with different atomic arrangements for the cyanate ion, NCO^-, and suggest which one is likely to be the most plausible structure. Consider only structures with two double bonds.

Strategy We need to calculate the formal charges on the three possible arrangements of atoms and select the one with formal charges closest to 0. To find the formal charge of each individual atom, draw the Lewis structures showing each electron pair as dots (not as a line). Then draw a closed curve around an atom that captures all its lone pairs of electrons, plus one electron from each of its bonding pairs. Find the number of valence electrons (V) of each free atom from its group in the periodic table. Then subtract the number of electrons assigned to the atom in the Lewis structure $(L + \frac{1}{2}S)$ from the number of valence electrons in the free atom. That is, calculate $FC = V - (L + \frac{1}{2}S)$. The sum of all the formal charges must be equal to the overall charge of the molecule or ion.

Solution Write the Lewis structures for the molecules as shown in Toolbox 8.2. The closed curves that capture the electrons are shown in (**24**).

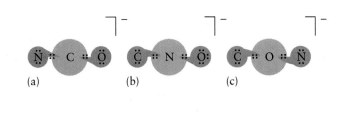

(a)   (b)   (c)

24 Cyanate ion , NCO^-

To calculate the formal charge on each type of atom that is shown in these diagrams, we draw up the following table:

Lewis structure	Number of electrons		Formal charge (FC)
	On free atom (V)	Assigned to bonded atom ($L + \frac{1}{2}S$)	
Central C (**24a**)	4	4	0
Central N (**24b**)	5	4	+1
Central O (**24c**)	6	4	+2
=C	4	6	−2
=N	5	6	−1
=O	6	6	0

The formal charges of the Lewis structures corresponding to the three possible arrangements are shown in (**25**). The individual formal charges are closest to 0 in the first. That structure is therefore the most likely one.

$$\overset{-1}{\underset{\cdot\cdot}{\ddot{N}}}=C=\overset{0}{\underset{\cdot\cdot}{\ddot{O}}} \qquad \overset{-2}{\underset{\cdot\cdot}{\ddot{C}}}=\overset{+1}{N}=\overset{0}{\underset{\cdot\cdot}{\ddot{O}}} \qquad \overset{-2}{\underset{\cdot\cdot}{\ddot{C}}}=\overset{+2}{O}=\overset{-1}{\underset{\cdot\cdot}{\ddot{N}}}$$

25 Cyanate ion, NCO^-

Self-Test 8.9A Suggest a plausible structure for the poisonous gas phosgene, $COCl_2$. Write its Lewis structure and formal charges.

[*Answer:* See (**26**).]

Self-Test 8.9B Calculate the formal charges of the three oxygen atoms in one of the Lewis structures of the ozone resonance hybrid (Example 8.6).

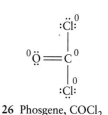

26 Phosgene, $COCl_2$

EXCEPTIONS TO THE OCTET RULE

The octet rule predicts the valence of the elements and the structures of many compounds. Boron, carbon, nitrogen, oxygen, and fluorine atoms obey the octet rule rigorously when it is arithmetically possible for them to do so; that is, when there are enough electrons to go round. Phosphorus, sulfur, and chlorine atoms, however, can accommodate more than eight electrons. In this section, we shall learn to recognize exceptions to the octet rule.

8.10 Radicals and Biradicals

Some species have an odd number of valence electrons, so at least one of their atoms cannot have an octet. Odd-electron species are called **radicals.** They are generally highly reactive. One example is the methyl radical, $\cdot CH_3$, which occurs in the flames of burning hydrocarbon fuels. Under enough stress, a carbon-carbon bond may break, as happens when ethane burns:

$$H_3C-CH_3 \longrightarrow H_3C\cdot + \cdot CH_3$$

The electron pair that formed the carbon-carbon bond has split into two separate electrons, as indicated by the dot on the C atom in each $\cdot CH_3$.

The older name *free radical* is still widely used.

Most radicals are very reactive because they can use their unpaired electron to form a new bond.

Applying Chemistry: *Case Study 8*

Every time we start an internal combustion engine, we contribute to the pollution of the air. The combustion of fossil fuels, especially in vehicles, releases incompletely burned chemicals and oxidized species known as *primary pollutants* into the atmosphere. The dangers they pose range from eye and throat irritation to global warming. Many of the primary pollutants undergo further reaction under the influence of sunlight. The products of these photochemical reactions are called *secondary pollutants*.

The equipment in the illustration monitors air quality from a rooftop in Los Angeles. The concentration of NO_2 in the air due to automobile traffic increases during the day, contributing to the typical brown color of the afternoon sky.

Primary and secondary pollutants—along with *aerosols*, which are suspended fine particles such as water droplets, dust, and soot—contribute to the brown haze we call *smog*.

One crucial primary pollutant is nitric oxide, NO, which is produced in the high-temperature combustion cylinders of automobiles and jet engines:

$$N_2(g) + O_2(g) \xrightarrow{\Delta} 2\,NO(g)$$

In the atmosphere, nitric oxide is oxidized to the brown gas nitrogen dioxide, NO_2, a major constituent of smog:

$$2\,NO(g) + O_2(g) \longrightarrow 2\,NO_2(g)$$

Smog appears brown partly because the nitrogen dioxide absorbs sunlight at wavelengths less than 400 nm. Another contribution is the scattering of light from tiny particles: blue light is scattered more than red light, so we tend to see the red light that is not scattered so much (recall the red Sun at sunset). The short wavelength light and ultraviolet radiation absorbed by nitrogen dioxide molecules dissociate them into NO molecules and O atoms:

$$\cdot NO_2 \xrightarrow{h\nu,\,\lambda\,<\,400\,nm} \cdot NO + \cdot O\cdot$$

Lachrymators, the eye irritants that cause tears, are oxidation products of primary pollutants. The oxygen atom produced by the dissociation of NO_2 is highly reactive. It is a strong oxidizing agent that can oxidize many of the primary pollutants and even reacts with O_2. The major products of its action are ozone and the peroxyacetylnitrates (PAN). The latter compounds are all highly toxic substances; and exposure to them at concentrations of 0.5 ppm for a few minutes can cause eye irritation.

Many reaction paths lead to the formation of a PAN. In one common path, an oxygen atom first removes a

A second example is the hydroxyl radical, $\cdot$OH. This radical forms briefly when a mixture of hydrogen and oxygen is ignited by a spark. It is also present in the upper atmosphere as a result of the action of the Sun's radiation on water molecules. Hydroxyl radicals are found in smoggy areas of the troposphere, where they can react with many substances, including fuel molecules that were only partially oxidized in an engine (Applying Chemistry: Case Study 8). Like most radicals, $\cdot CH_3$ and $\cdot$OH are highly reactive and survive only for very short times under normal conditions.

Radicals control the chemistry of the upper atmosphere, where they contribute to the formation and decomposition of ozone (see Applying Chemistry: Case Study 5). Radicals are also responsible for the rancidity of foods and the degradation of plastics in sunlight. Damage from radicals can be delayed by an additive called an **antioxidant.** Molecules of the antioxidant react rapidly with

The gas NO (left) is colorless. When it is exposed to air (right), it is rapidly oxidized to brown NO_2.

hydrogen atom from a molecule of unburned hydrocarbon fuel to produce two radicals:*

$$CH_4 + \cdot O \cdot \longrightarrow \cdot CH_3 + \cdot OH$$

The hydroxyl radical, $\cdot OH$, is a common oxidizing agent in smog. The methyl radical, $\cdot CH_3$, is also highly reactive and goes on to cause further havoc with the atmosphere.

Hydroxyl radicals react with many substances, including fuel molecules that were only partially oxidized in an

---

*All these reactions take place in the gas phase; we are omitting the state symbols to show the reaction step more clearly.

engine. For example, one oxidation product, acetaldehyde, CH_3CHO, reacts with the highly reactive hydroxyl radical to produce yet another radical:

$$CH_3CHO + \cdot OH \longrightarrow CH_3CO\cdot + H_2O$$

The $CH_3CO\cdot$ radical is also highly reactive and combines with molecular oxygen to form a peroxyl radical:

$$CH_3CO + O_2 \longrightarrow CH_3-\overset{\overset{\displaystyle O}{\|}}{C}-O-O\cdot$$

The peroxyl radical combines with NO_2 (also a radical) to form peroxyacetylnitrate:

$$CH_3-\overset{\overset{\displaystyle O}{\|}}{C}-O-O\cdot + \cdot NO_2 \longrightarrow CH_3-\overset{\overset{\displaystyle O}{\|}}{C}-O-O-NO_2$$

Catalytic converters in automobiles help to control smog by converting the nitrogen oxides back to nitrogen and oxygen. They also complete the oxidation of unburned and partially burned hydrocarbons. Even so, the automobile remains the primary contributor to smog.

Key Concepts: radicals, atmospheric chemistry

For Further Reading

C. Baird, *Environmental Chemistry,* second edition, New York: W. H. Freeman and Company, 1999, pp. 85–171.

Related Exercises: 8.95–8.98

---

radicals to form species in which all electrons are paired, before the radicals have a chance to do their damage. It is believed that human aging is partly due to the action of radicals. Antioxidants such as vitamins C and E may delay the process.

Nitrogen monoxide (nitric oxide, $:\overset{.}{N}=\overset{..}{O}$), which has $5 + 6 = 11$ valence electrons, is a radical. It is formed by the direct reaction of nitrogen and oxygen in the hot exhaust gases of automobile and jet engines:

$$N_2(g) + O_2(g) \longrightarrow 2\,NO(g)$$

Nitric oxide is one of the *neurotransmitters,* small molecules that transmit signals between neurons. It plays a role in blood flow and sexual arousal.

A **biradical** is a molecule with *two* unpaired electrons. The unpaired electrons are usually on different atoms, as depicted in (**27**). One unpaired electron

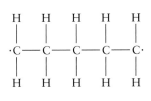

27 A biradical

28 Hydrogenperoxyl, $HO_2\cdot$

is on one carbon atom of the chain, and the second is on another carbon atom several bonds away. In some cases, though, both electrons are on the same atom. One of the most important examples is the oxygen atom itself, $\cdot\ddot{O}\cdot$.

Although not obvious from its Lewis structure, molecular oxygen, O_2, is also a biradical. In fact, experiments have shown that the most plausible Lewis structure, $\ddot{O}=\ddot{O}$, gives a false impression of the arrangement of electrons. In molecular oxygen, two of the electrons that the Lewis structure implies are responsible for the bonds do not in fact pair with one another. This feature cannot be explained by Lewis's model of bonding, but it can be explained by more modern theories (Section 9.16).

A radical is a species with an unpaired electron; a biradical has two unpaired electrons on the same or different atoms.

Self-Test 8.10A Write a Lewis structure for the hydrogenperoxyl radical, $HOO\cdot$, which plays an important role in atmospheric chemistry and which in the body has been implicated in the degeneration of neurons.

[***Answer:*** See (28).]

Self-Test 8.10B Write the Lewis structure of the radical nitrogen dioxide, $\cdot NO_2$.

8.11 Expanded Valence Shells

When the octet rule is obeyed, eight electrons fill a valence shell to give a noble-gas ns^2np^6 configuration. However, when the central atom in a molecule has empty d-orbitals, it may be able to accommodate 10, 12, or even more electrons. The electrons in such an **expanded valence shell** or "expanded octet" may be present as lone pairs or may be used by the central atom to form additional bonds.

Because there must be enough valence orbitals to accommodate the additional electrons, only nonmetal atoms in Period 3 or higher can have expanded octets. For these elements, there are empty d-orbitals in the valence shell of the atom. Another factor—possibly the main factor—in determining whether more atoms than are allowed by the octet rule can bond to a central atom is the size of that atom. A P atom is big enough for up to six Cl atoms to fit comfortably around it, and PCl_5 is a common laboratory chemical. An N atom, however, is too small, and NCl_5 is unknown (Fig. 8.16).

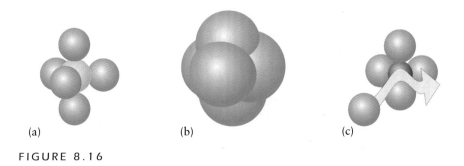

(a) (b) (c)

FIGURE 8.16

(a) A model using small spheres to represent atoms and (b) a space-filling model of PCl_5, showing how closely the chlorine atoms must pack around the central phosphorus atom. (c) A nitrogen atom is significantly smaller than a phosphorus atom, and five chlorine atoms cannot pack around it.

FIGURE 8.17

Phosphorus trichloride is a colorless liquid. When it reacts with chlorine (the pale yellow-green gas in the flask), it forms the very pale yellow solid phosphorus pentachloride (at the bottom of the flask).

Elements that can expand their octets show **variable covalence,** the ability to form different numbers of covalent bonds. Variable covalence means that an element can form one number of bonds in some compounds and a different number in others. Phosphorus is one example. It reacts directly with a limited supply of chlorine to form the toxic, colorless liquid phosphorus trichloride:

$$P_4(g) + 6\,Cl_2(g) \longrightarrow 4\,PCl_3(l)$$

The Lewis structure of the PCl_3 molecule is shown in (**29**), and we see that it obeys the octet rule. However, when excess chlorine is present or when phosphorus trichloride reacts with more chlorine (Fig. 8.17), phosphorus pentachloride, a pale yellow crystalline solid, is produced:

$$P_4(g) + 10\,Cl_2(g) \longrightarrow 4\,PCl_5(s)$$

$$PCl_3(l) + Cl_2(g) \longrightarrow PCl_5(s)$$

Phosphorus pentachloride is an ionic solid consisting of PCl_4^+ cations and PCl_6^- anions. It sublimes at 160°C to a gas of PCl_5 molecules. This peculiar behavior underlines the subtlety of the energy balances in chemical bonding. The Lewis structures of the polyatomic ions and the molecule are shown in (**30**). Although the cation is a polyatomic ion in which the P atom does not need to expand its octet, in the anion the P atom has expanded its octet to 12 by using two of its 3d-orbitals. In PCl_5, the P atom has expanded its octet to 10 by using one 3d-orbital.

29 Phosphorus trichloride, PCl_3

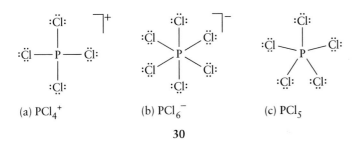

(a) PCl_4^+ (b) PCl_6^- (c) PCl_5

30

Toolbox 8.3 How to write a Lewis structure for a molecule with an expanded valence shell

We use the procedure in this Toolbox to construct the Lewis structure of a molecule in which the central atom does not observe the octet rule.

Conceptual Basis

There are two circumstances that lead us to expect an expanded valence shell on a central atom. One is when there are too many electrons to be accommodated as octets (perhaps because the central atom is attached to more than four atoms). The other is when formal charge arguments favor a resonance structure with an expanded octet (Fig. 8.18).

Procedure

When there are too many electrons Follow the steps in Toolbox 8.2. If there are too many electrons to be accommodated in octets and duplets in Step 4, even when all the bonds are single bonds, place the extra electrons around the central atom. This procedure is illustrated in Example 8.8.

When resonance structures with expanded octets can be written Follow the steps in Toolbox 8.2. If resonance structures are possible, draw all possible resonance structures and determine the formal charges on the atoms. The structure with the lowest formal charges is likely to have the lowest energy. This procedure is illustrated in Example 8.9.

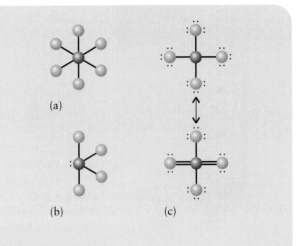

(a)

(b)   (c)

FIGURE 8.18

Two circumstances in which a central atom assumes an expanded octet. First, there are too many electrons to be accommodated in octets because either (a) there are too many atoms attached, or (b) the central atom must accommodate additional lone pairs. (c) Second, a resonance structure with multiple bonds has a favorable energy.

Example 8.8 Writing a Lewis structure for a molecule with an expanded valence shell

A fluoride of composition SF_4 is formed when fluorine diluted with nitrogen is passed over a film of sulfur at $-75°C$ in the absence of oxygen and moisture. Write the Lewis structure of sulfur tetrafluoride.

Strategy A fluorine atom forms only single bonds, so we anticipate that the Lewis structure consists of a shared pair between the central S atom and each of the four surrounding F atoms. However, each F atom has three lone pairs and supplies one bonding electron, and the S atom already has six electrons in its valence shell; so there are two extra electrons. Because sulfur is in Period 3 and has empty $3d$-orbitals available, it can expand its octet.

Solution Sulfur $\cdot\ddot{S}\cdot$ supplies six valence electrons, and each fluorine atom $:\ddot{F}\cdot$ supplies seven. Hence there are $6 + (4 \times 7) = 34$ electrons, or 17 electron pairs, to accommodate. We write each F atom with three lone pairs and a bonding pair shared with the central S atom, then place the two extra electrons on the S atom, as shown in (**31**). All 17 electron pairs are now accommodated. Because the S atom has 10 electrons in its valence shell, which require at least five orbitals, it needs to use one $3d$-orbital in addition to the four $3s$- and $3p$-orbitals.

31 Sulfur tetrafluoride, SF_4

Self-Test 8.11A Write the Lewis structure for xenon tetrafluoride, XeF_4, and give the number of electrons in the expanded octet.

[***Answer:*** See (**32**); 12 electrons.]

Self-Test 8.11B Write the Lewis structure for the I_3^- ion.

32 Xenon tetrafluoride, XeF_4

Example 8.9 *Selecting the favored resonance structure for a molecule*

Determine whether an expanded octet is plausible for the S atom in a sulfate ion by calculating the formal charges on the atoms in the resonance structures shown in (**33**).

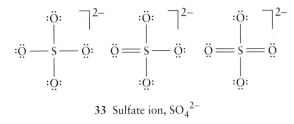

33 Sulfate ion, SO_4^{2-}

Strategy Only the first structure gives every atom an octet. Find the formal charges on the atoms of all three structures by using the strategy in Example 8.7. The structure with formal charges closest to 0 will be the most plausible.

Solution Both O and S belong to Group 16, so the neutral atoms each have six valence electrons and the ion has two additional electrons. The closed curves that capture the electrons are shown in (**34***).

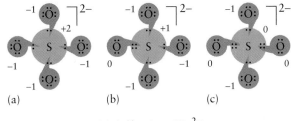

34 Sulfate ion, SO_4^{2-}

To calculate the formal charges that are shown in these diagrams, we draw up the following table:

| Lewis structure | Number of electrons | | Formal charge (FC) |
	On free atom (V)	Assigned to bonded atom $(L + \frac{1}{2}S)$	
S in (**34a**)	6	4	+2
S in (**34b**)	6	5	+1
S in (**34c**)	6	6	0
—Ö:	6	7	−1
=Ö	6	6	0

The individual formal charges are closest to 0 in (**34c**), so the third structure in (**33**) is the most plausible, even though it involves a valence shell on the S atom expanded to 12 electrons.

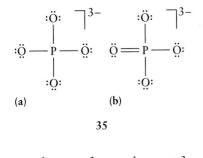

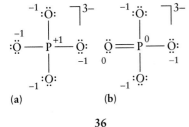

(a)   (b)

35

(a)   (b)

36

$$\ddot{O}=\ddot{S}=\ddot{O}$$

$$\ddot{O}=\ddot{S}-\ddot{O}:$$

$$:\ddot{O}-\ddot{S}=\ddot{O}$$

37   Sulfur dioxide, SO_2

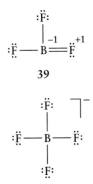

38   Boron trifluoride, BF_3

39

40   Tetrafluoroborate, BF_4^-

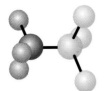

41   NH_3BF_3

Self-Test 8.12A   Calculate the formal charges for the two Lewis structures of the phosphate ion shown in (**35**). Which is more plausible?

[**Answer:** The formal charges are shown in (**36**); (**36b**) is more plausible.]

Self-Test 8.12B   Suggest a plausible structure for the arsenate ion, AsO_4^{3-}. Write its Lewis structure and formal charges.

The sulfate ion, SO_4^{2-}, is a resonance hybrid of octet and expanded-octet Lewis structures. The structures shown in (**33**) are just three of the many possible resonance structures. These and the other contributions to the resonance structure have the same arrangement of atoms, so we can expect the true structure to be a resonance hybrid of them all. Computer calculations show that the expanded-octet structures have a lower energy than the octet structure, so they make the greatest contribution to the hybrid. This conclusion is consistent with the formal charge arguments from Example 8.9.

Sulfur dioxide, SO_2, is another molecule in which expanded-octet Lewis structures are important (**37**). Both experimental data and formal charge calculations suggest that the top Lewis structure makes a bigger contribution to the resonance hybrid than the other two, in which there is no octet expansion.

Octet expansion can occur in elements of Period 3 and later and enables them to form variable numbers of covalent bonds.

LEWIS ACIDS AND BASES

Boron and aluminum atoms need five electrons to complete their octets. In certain circumstances, it may be energetically favorable for them to form fewer bonds than the octet rule implies. As a result, many of the compounds that boron and aluminum form have unusual chemical characteristics. These compounds are of special interest because they introduce a new class of reactions called *Lewis acid-base reactions*. Although we use boron and aluminum to introduce this class of reactions, we shall see that the same type of reaction is shown by many other elements and compounds.

8.12 The Unusual Structures of Group 13 Halides

To introduce this new class of reactions, let's investigate the molecular structure of the colorless gas boron trifluoride, BF_3. The Lewis structure (**38**) indicates that the boron atom has an incomplete octet: its valence shell consists of only six electrons. The molecule could complete its octet by sharing more electrons with fluorine, as depicted in (**39**), but fluorine has such a high ionization energy that this arrangement is unlikely.

The boron octet can be completed if *another* atom or ion with a lone pair of electrons forms a bond by providing the needed pair of electrons. For example, the tetrafluoroborate anion, BF_4^- (**40**), forms when boron trifluoride is passed over a metal fluoride. Now, all the fluorine atoms have their normal valence of 1 and the boron atom has an octet. Another example is the compound formed when boron trifluoride reacts with ammonia:

$$BF_3(g) + NH_3(g) \longrightarrow NH_3BF_3(s)$$

A molecular model of the product, a white molecular solid, is shown in (**41**).

The lone pair on the nitrogen atom of ammonia can be regarded as completing boron's octet by forming a covalent bond to give the Lewis structure shown in (**42**). A bond in which both electrons come from only one of the atoms is called a **coordinate covalent bond.**

Boron trichloride, a colorless, reactive gas of BCl_3 molecules, behaves chemically like BF_3. However, the trichloride of aluminum, which is in the same group as boron, is a volatile white solid that sublimes at 180°C to a gas of Al_2Cl_6 molecules. These molecules survive in the gas up to about 200°C, and only then fall apart into $AlCl_3$ molecules. The Al_2Cl_6 molecule exists because a Cl atom of one $AlCl_3$ molecule uses one of its lone pairs to form a bond to the Al atom of a neighboring $AlCl_3$ molecule. The resulting molecule has the structure shown in (**43**).

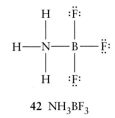

42 NH_3BF_3

Coordinate covalent bonds are described in more detail in Section 21.5.

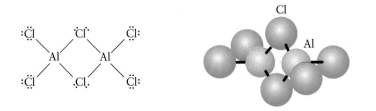

43 Aluminum chloride, Al_2Cl_6

Compounds of boron and aluminum may have unusual Lewis structures in which boron and aluminum have incomplete octets or in which halogen atoms act as bridges.

8.13 Lewis Acid-Base Complexes

When a coordinate covalent bond forms, one species provides a lone pair and the other species accepts it. The species that provides the lone pair is called a *Lewis base,* and the species that accepts it is called a *Lewis acid.* In other words, a **Lewis acid** is an *electron pair acceptor,* and a **Lewis base** is an *electron pair donor.* The product of the reaction between a Lewis acid and a Lewis base, BF_4^- in our first example, is called a **complex.** The form of the reaction is therefore

$$\text{Acid} + \text{:base} \longrightarrow \text{complex}$$

You will sometimes see a Lewis acid-base complex called an *adduct.*

The BF_3 molecule is the Lewis acid, the F^- ion is the Lewis base, and the BF_4^- ion is the Lewis acid-base complex. All bonds formed in a Lewis acid-base reaction are coordinate covalent bonds.

Why do we use the terms *acid* and *base* in this context? Recall from Section 3.7 that acids and bases are normally discussed in terms of the behavior of a hydrogen ion, H^+ (a proton), and that in water an acid such as HCl donates its proton to a neighboring water molecule. We could imagine this reaction as taking place in two steps. First, an HCl molecule releases a hydrogen ion as soon as it dissolves in water. Then that hydrogen ion immediately bonds to a neighboring water molecule. The second step is

We discuss this point again, in more detail, in Section 15.1.

$$H^+ + \overset{\overset{\displaystyle H}{|}}{:\!\underset{..}{O}}\!-\!H \longrightarrow \left[H-\overset{\overset{\displaystyle H}{|}}{\underset{..}{O}}\!-\!H \right]^+$$

This step in the reaction has exactly the same form as the Lewis acid-base reaction, with H^+ playing the role of the Lewis acid, H_2O the role of the Lewis base, and H_3O^+ the resulting complex. It is this analogy to the behavior of H^+ (the species typical of acids) that led to the use of the term *acid* to describe an electron pair acceptor like BF_3. Whenever a species has a vacant orbital, we can suspect that it is likely to act as a Lewis acid.

Similarly, just as the hydroxide ion, OH^-, is a typical base in the conventional sense of being a proton acceptor, it acts as an electron pair donor in the reaction

$$H^+ + [:\ddot{O}-H]^- \longrightarrow \overset{\displaystyle H}{\underset{\displaystyle |}{}} :\ddot{O}-H$$

In the Lewis system, the term *base* is used for any species that can act as an electron pair donor. An ammonia molecule, $:NH_3$, is another example of a typical conventional base that has a lone pair to donate (for example, to BF_3, as described earlier), so it can also be regarded as a Lewis base. Whenever a species has a lone pair of electrons, it may be able to act as a Lewis base.

Example 8.10 *Identifying a Lewis acid and a Lewis base*

When water is added to calcium oxide (quicklime), a vigorous reaction takes place and calcium hydroxide (slaked lime) is formed. Describe the reaction in terms of the formation of a Lewis acid-base complex.

Strategy Look for the species that has a lone pair of electrons it can donate to form a covalent bond: such a species is a Lewis base. The species to which the lone pair is donated is the Lewis acid. Bear in mind (as explained earlier) that the water molecule can act as a source of the Lewis acid H^+.

Solution The oxide ion, $[:\ddot{O}:]^{2-}$, is a Lewis base. A water molecule, H_2O, can be regarded as a Lewis acid-base complex in which H^+ is the Lewis acid and OH^- is the Lewis base. When water is poured on calcium oxide, the Lewis base O^{2-} pulls the Lewis acid H^+ out of the molecule H_2O:

$$[:\ddot{O}:]^{2-} + \overset{\displaystyle H}{\underset{\displaystyle |}{}} H-\ddot{O}: \longrightarrow [:\ddot{O}-H]^- + [:\ddot{O}-H]^-$$

An oxide ion is such a strong Lewis base—a strong electron pair donor—that it never exists as such in water but always reacts to form hydroxide ions.

Self-Test 8.13A Account, in terms of Lewis acid-base behavior, for the formation of ammonium and hydroxide ions when ammonia dissolves in water.

$$[\textbf{\textit{Answer:}}\ H_3N: + \overset{\displaystyle H}{\underset{\displaystyle |}{}} H-\ddot{O}: \longrightarrow NH_4^+ + [:\ddot{O}-H]^-$$
(Lewis acid: H^+, Lewis base: NH_3)]

Self-Test 8.13B Account for the formation of Al_2Cl_6 from $AlCl_3$ molecules in terms of Lewis acids and bases.

Example 8.10 describes the reaction between a basic oxide and water. The reaction between acidic oxides and water and between acidic oxides and bases are also Lewis acid-base reactions. For example, the reaction of carbon dioxide with the hydroxide ion is a Lewis acid-base reaction:

$$\ddot{O}=C=\ddot{O} + :\ddot{O}H^- \longrightarrow \left[H\ddot{O}-\overset{\displaystyle :O:}{\underset{\displaystyle ||}{C}}-\ddot{O}: \right]^-$$

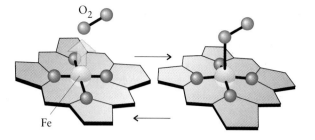

FIGURE 8.19

Oxygen molecules are transported through our bloodstream in the form of a complex with the iron atoms in hemoglobin molecules and then released where they are needed.

In this reaction, the hydroxide ion is the Lewis base. When the oxygen atom shares one of its lone pairs of electrons with the carbon atom in carbon dioxide, one of the carbon-oxygen double bonds becomes a single bond.

Some substances are stronger Lewis acids or bases than others. For example, the Lewis acid-base complex that the oxygen molecule forms with the iron atom in hemoglobin allows oxygen to be carried through the bloodstream (Fig. 8.19). However, carbon monoxide forms a stronger complex with the iron in hemoglobin than oxygen does, which means that the O_2 molecule cannot readily displace the CO molecule from the complex. As a result, exposure to carbon monoxide causes rapid oxygen starvation and death. Factors that enhance the strength of a Lewis acid include positive charge and the availability of empty valence orbitals. The positive charge attracts the electrons of the lone pair and the empty orbitals accommodate them. As a result, cations of d-block metals are often good Lewis acids. The strength of a Lewis base is enhanced by the possession of a negative charge.

Lewis acid-base reactions occur widely in chemistry and biology. They complete our collection of fundamental reaction types that we began to assemble in Chapter 3. Most chemical reactions can be described in terms of one of these four fundamental types of reactions—precipitation, neutralization (proton transfer), oxidation-reduction (redox), and Lewis acid-base reactions. These reactions are often part of the procedures that chemists use to make the new materials of the modern world.

A Lewis acid is an electron pair acceptor; a Lewis base is an electron pair donor. They react to form a Lewis acid-base complex.

IONIC VERSUS COVALENT BONDS

Ionic and *covalent* are terms used to describe two extreme types of chemical bonds. In reality, bonds lie somewhere between purely ionic and purely covalent. When describing bonds between nonmetals, covalent bonding is a good model. When a metal is present, ionic bonding is a good model for most simple compounds. But just how good are these initial descriptions, and how can they be improved?

8.14 Correcting the Covalent Model

A covalent bond acquires some ionic character if one atom has a greater electron-withdrawing power than the other atom. The electron-withdrawing power of an atom when it is part of a bond is called **electronegativity.** When a chemical bond forms between two atoms, the atom of the element with the higher

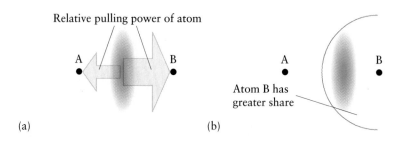

Relative pulling power of atom

(a)

(b)

Atom B has greater share

Electronegativity was first defined by the American chemist Linus Pauling (1901–1994). The interpretation described here was developed by another American chemist, Robert Mulliken.

electronegativity tends to pull the bonding electrons away from the atom of the element with lower electronegativity (Fig. 8.20).

The simplest way to think of the numerical value of the electronegativity is as the average of the ionization energy and electron affinity of the element. If the ionization energy is high, then electrons are given up reluctantly. If the electron affinity is high, then it is energetically favorable to attach electrons to an atom. Elements with both these properties are reluctant to lose their electrons and tend to gain them; hence they are classified as highly electronegative. Conversely, if the ionization energy and the electron affinity are both low, then it takes very little energy for the element to give up its electrons and it has little tendency to gain more; hence the electronegativity is low.

An analogy

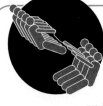

We can think of the electron pair in a bond as the target in a tug-of-war between the two atoms that share the electron pair. If the electronegativities of the two elements in a bond are the same, the atoms have equal pulling power on the electron pair they share and neither wins the tug-of-war. If the electronegativities are very different, the more electronegative atom can acquire the lion's share of the electron pair.

Electronegativities are most useful for nonmetals and are hardly ever used for *d*-block metals.

Figure 8.21 shows the variation of electronegativity for the main-group elements of the periodic table. Because ionization energies and electron affinities are highest at the top right of the periodic table (close to fluorine, but excluding the noble gases), it is not surprising to find that nitrogen, oxygen, bromine, chlorine, and fluorine are the elements with the highest electronegativities. Whenever these elements are present in compounds, we can expect their atoms to pull strongly on electrons shared with their neighbors. Electrons will still be shared with the less electronegative atom, but the sharing is unequal, and the electron cloud will be denser on the atom of the more electronegative element.

If the electronegativities of the two elements in a bond are the same, the atoms share the electrons equally. The covalent model is then a good description of the bonding. If the electronegativities are very different, then the electrons are more likely to be found on the more electronegative atom. Because it has a much larger share of the electrons than the less electronegative atom, the more electronegative atom begins to resemble an anion and the other atom begins to resemble a cation. We say such a bond has a lot of ionic character (Fig. 8.22).

There is no hard-and-fast dividing line between ionic and covalent bonding. However, it is a good rule of thumb that an electronegativity difference of about 2 means that there is enough ionic character for the bond to be regarded

FIGURE 8.21

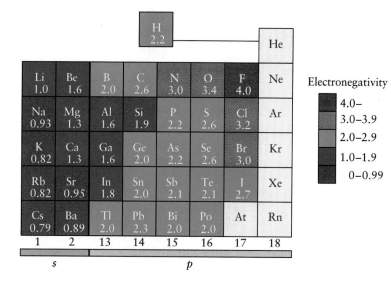

as ionic. For electronegativity differences smaller than about 1.5, a covalent description of the bond is reasonably reliable.

Electronegativity is a measure of the electron-pulling power of an atom on an electron pair in a molecule. Compounds composed of elements with a large difference in electronegativity ($\geq$ 2) tend to have significant ionic character in their bonding.

Example 8.11 *Estimating the relative ionic character of a bond*

(a) In which of the following compounds do the bonds have greater ionic character, ClO_2 or SCl_2? (b) Indicate which atom in each compound has the partial negative charge.

Strategy (a) The greater the difference in electronegativities of the two elements, the greater the extent of ionic character. Find the electronegativities in Fig. 8.21 and compare the differences in the two compounds. (b) The atom with the greater electronegativity will be the one with the partial negative charge.

Solution (a) The electronegativity of S is 2.6, that of O is 3.4, and that of Cl is 3.2. The difference in electronegativities for the O—Cl bond is $3.4 - 3.2 = 0.2$. The difference in electronegativities for the S—Cl bond is $3.2 - 2.6 = 0.6$. The difference is greater for the S—Cl bond, so SCl_2 has bonds with the greater ionic character. (b) In ClO_2, O has the greater electronegativity, so it bears the partial negative charge. In SCl_2, Cl has the greater electronegativity, so it bears the partial negative charge.

Self-Test 8.14A (a) In which of the following compounds do the bonds have greater ionic character: P_2O_5 or PCl_3? (b) Indicate which atom in each compound has the partial negative charge.

[*Answer:* (a) P_2O_5; (b) O in P_2O_5 and Cl in PCl_3]

Self-Test 8.14B (a) In which of the following compounds do the bonds have greater ionic character: NH_3 or NO_2? (b) Indicate which atom in each compound has the partial negative charge.

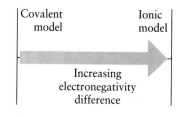

FIGURE 8.22

The electronegativity difference between two elements can be used to predict the most appropriate bonding model for a chemical bond between the elements. When the difference is small (less than about 1.5), the covalent model is good. When the difference is large (greater than about 2), it is most accurate to express the bonding in terms of the ionic model. Note that the numbers we have quoted are only a guide.

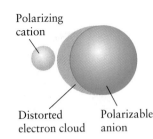

Polarizing cation

Distorted electron cloud

Polarizable anion

FIGURE 8.23

When a small, highly charged cation is close to a large anion, the electron cloud of the latter is distorted in the process we call polarization. Small, highly charged cations are highly polarizing. Large, electron-rich anions are highly polarizable.

8.15 Correcting the Ionic Model

In practice, all ionic bonds have some covalent character. Consider a monatomic anion (such as a chloride ion) next to a cation (such as a sodium cation). As the cation pulls on the anion's electrons, the spherical electron cloud of the anion becomes distorted. We can think of this distortion as the tendency of an electron pair to move into the bonding region between the two nuclei and to form a covalent bond (Fig. 8.23). We can expect ionic bonds to have more covalent character as the distortion of the electron cloud on an ion increases.

Atoms and ions that readily undergo a large distortion are said to be highly **polarizable.** Ions that can *cause* large distortions are said to have a high **polarizing power.** An anion can be expected to be highly polarizable if it is large, like an iodide ion, I^-. In such a large, highly polarizable ion, the nucleus exerts only relatively weak control over its outermost electrons, because they are so far away. As a result, the electron cloud of the large anion is easily distorted. Because the electrons of a cation are so closely held, they are not easily polarized. However, the positive charge on a cation can distort the electron clouds of anions. A cation can be expected to have a strong polarizing power if it is small and highly charged, like an Al^{3+} cation. A small radius means that the center of charge of a highly charged cation can get very close to the anion, where it can exert a strong pull on the anion's electrons. Compounds composed of a small, highly charged cation and a large, polarizable anion tend to have bonds with significant covalent character (Fig. 8.24).

For example, the very small radius of a beryllium atom gives the element properties strikingly different from those of the other Group 2 elements. A Be—Cl bond is significantly covalent; and $BeCl_2$ consists of long chains of cova-

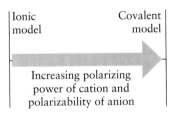

Ionic model

Covalent model

Increasing polarizing power of cation and polarizability of anion

FIGURE 8.24

The polarizability of an anion and the polarizing power of a cation can be used as a guide to judge whether an ionic model of a bond is likely to be valid. As the polarizing power and polarizability increase, the distortion of the anion becomes so great that a covalent description of the bond becomes more reasonable.

FIGURE 8.25

The silver halides, which were formed here by precipitation from silver nitrate and sodium halide solutions, become increasingly insoluble down the group from AgCl to AgI, as the polarizability of the anion increases. The figure shows, from left to right, AgCl, AgBr, and AgI.

lently bonded $BeCl_2$ units in the solid, as shown in (**44**), and individual $BeCl_2$ molecules in the vapor. Another observation we can now explain is the decreasing solubility in water of the silver halides as we go from AgCl to AgI (Fig. 8.25). The covalent character of the silver halides increases from AgCl to AgI as the anion becomes larger and more polarizable. Silver fluoride, which contains the small and almost unpolarizable fluoride ion, is freely soluble in water because it is predominantly ionic.

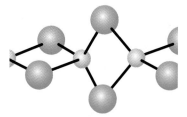

44 Beryllium chloride

Compounds composed of highly polarizing cations and highly polarizable anions have a significant covalent character in their bonding.

Self-Test 8.15A In which of the following compounds do the bonds have greater covalent character: (a) NaBr or (b) $MgBr_2$? Explain your answer.
[*Answer:* (b) because Mg^{2+} is smaller and more highly charged than Na^+]

Self-Test 8.15B In which of the following compounds do the bonds have greater covalent character: (a) CaS or (b) CaO?

Skills You Should Have Mastered

Conceptual

☐ 1. Distinguish between ionic and covalent bonds, Sections 8.1 and 8.5.

☐ 2. Define resonance hybrid and explain its relation to the individual Lewis structures that contribute, Section 8.8.

☐ 3. Explain the characteristics of Lewis acids and bases and how they form bonds, Section 8.13.

☐ 4. Explain the significance of electronegativity and what is meant by the ionic or covalent character of a bond, Section 8.14.

☐ 5. Predict and explain periodic trends in the polarizability of anions and the polarizing power of cations, Section 8.15.

Problem-Solving

☐ 1. Predict the chemical formula of a binary ionic compound and write its formula as a Lewis structure, Example 8.1.

☐ 2. Use a Born-Haber cycle to calculate a lattice enthalpy, Toolbox 8.1 and Example 8.2.

☐ 3. Write the Lewis structure of a molecule or ion, Toolbox 8.2 and Examples 8.3, 8.4, and 8.5.

☐ 4. Write the resonance structure for a molecule, Example 8.6.

☐ 5. Write the Lewis structure of a molecule or ion with an expanded valence shell, Toolbox 8.3 and Example 8.8.

☐ 6. Use formal charges to evaluate alternative Lewis structures, Examples 8.7 and 8.9.

7. Use a table of electronegativities to predict which of two bonds has greater ionic or covalent character, Example 8.11.

Descriptive

1. Describe some of the properties of ionic compounds, Section 8.4.

2. Describe the general characteristics of radical species and how some radicals affect our lives, Section 8.10.

3. Predict which atoms are likely to form molecules in which they have an expanded valence shell, Section 8.11.

4. Identify Lewis acids and Lewis bases, Example 8.10.

Exercises

Ionic Bonds

8.1 Write the most likely charge for the ions formed by each element: (a) Li; (b) S; (c) Ca; (d) Al.

8.2 Write the most likely charge for the ions formed by each element: (a) F; (b) Ba; (c) Se; (d) O.

8.3 Explain why sodium occurs as Na^+ and not Na^{2+} in ionic compounds.

8.4 Explain why chlorine occurs as Cl^- and not Cl^{2-} in ionic compounds.

8.5 Write the Lewis symbols of (a) calcium; (b) sulfur; (c) oxide ion; (d) nitride ion.

8.6 Write the Lewis symbols of (a) chlorine; (b) arsenic; (c) carbon; (d) chloride ion.

8.7 Write the Lewis formulas for the ionic compounds (a) potassium fluoride; (b) aluminum sulfide; (c) calcium nitride.

8.8 Write the Lewis formulas for the ionic compounds (a) cesium carbide; (b) sodium oxide; (c) calcium phosphide.

8.9 Give an example of the manner in which the inert-pair effect shows up in Group 13.

8.10 Explain why the inert-pair effect is seen in Groups 13, 14, and 15, but not in Groups 16 and 17.

Lattice Enthalpies

8.11 Explain why the lattice enthalpy of magnesium oxide (3850. kJ/mol) is greater than that of barium oxide (3114 kJ/mol).

8.12 Explain why the lattice enthalpy of magnesium oxide (3850. kJ/mol) is greater than that of magnesium sulfide (3406 kJ/mol).

8.13 Calculate the lattice enthalpy of silver fluoride from the data in Appendix 2D and the following information:

Enthalpy of formation of Ag(g)	+284 kJ/mol
Ionization energy of Ag(g)	+731 kJ/mol
Enthalpy of formation of F(g)	+79 kJ/mol
Enthalpy of formation of AgF(s)	−205 kJ/mol

8.14 Calculate the lattice enthalpy of calcium sulfide from the data in Appendix 2D and the following information:

Enthalpy of formation of Ca(g)	+178 kJ/mol
Enthalpy of formation of S(g)	+279 kJ/mol
Enthalpy of formation of CaS(s)	−482 kJ/mol
Second ionization energy of Ca(g)	+1150 kJ/mol

8.15 Identify each of the following as endothermic or exothermic processes: (a) ionization of copper; (b) electron attachment to oxygen; (c) sublimation of copper; (d) formation of copper(II) oxide from its ions; (e) formation of copper(II) oxide from the elements.

8.16 Identify each of the following as endothermic or exothermic processes: (a) ionization of silver; (b) electron attachment to chlorine; (c) dissociation of Cl_2 molecules; (d) formation of silver chloride from its ions; (e) formation of silver chloride from its elements.

Lewis Structures

8.17 Write the Lewis structures of (a) hydrogen fluoride; (b) oxygen dichloride; (c) carbon tetrafluoride.

8.18 Write the Lewis structures of (a) hydrogen sulfide; (b) nitrogen trichloride; (c) iodine chloride.

8.19 Write the Lewis structures of (a) ammonium ion, NH_4^+; (b) hypochlorite ion, ClO^-; (c) tetrafluoroborate ion, BF_4^-.

8.20 Write the Lewis structures of (a) nitronium ion, ONO^+; (b) chlorite ion, ClO_2^-; (c) peroxide ion, O_2^{2-}.

8.21 Write the complete Lewis structures for each of the following compounds: (a) ammonium chloride; (b) potassium phosphate; (c) sodium hypochlorite.

8.22 Write the complete Lewis structures for each of the following compounds: (a) zinc cyanide; (b) potassium tetrafluoroborate; (c) barium peroxide (the peroxide ion is O_2^{2-}).

8.23 Molecules that have the same number of valence electrons are called isoelectronic. In the following pairs of isoelectronic molecules and ions, the X and Y represent unknown elements. Identify these elements from the Lewis structures alone.

(a)
$$\ddot{O}=X-\ddot{O}{:}^{2-} \quad \text{and} \quad \ddot{O}=Y-\ddot{O}{:}$$
$$\qquad | \qquad\qquad\qquad |$$
$$\quad :O: \qquad\qquad\qquad :O:$$

(X and Y are in Period 3.)

(b)
$$:\ddot{F}: \qquad\qquad :\ddot{F}: \quad ^{-}$$
$$\quad | \qquad\qquad\qquad |$$
$$:\ddot{F}-X-\ddot{F}: \quad \text{and} \quad :\ddot{F}-Y-\ddot{F}:$$
$$\quad | \qquad\qquad\qquad |$$
$$:\ddot{F}: \qquad\qquad :\ddot{F}:$$

(X and Y are in Period 2.)

8.24 In the following pairs of isoelectronic molecules and ions (see Exercise 8.23), the X and Y represent unknown Period 2 elements. Identify these elements from the Lewis structures alone.
(a) $\ddot{O}=\ddot{X}-\ddot{O}{:}$ and $[\ddot{O}=\ddot{Y}-\ddot{O}{:}]^{-}$
(b) $[{:}X{\equiv}N{:}]^{-}$ and ${:}Y{\equiv}N{:}$

8.25 Write the Lewis structures of the following organic compounds: (a) formaldehyde, H_2CO, which, as its aqueous solution "formalin," is used to preserve biological specimens; (b) methanol, CH_3OH, the toxic compound also called wood alcohol.

8.26 Write the Lewis structures of the following organic compounds: (a) ethanol, CH_3CH_2OH, which is also called grain alcohol (see Fig. 1.21); (b) methylamine, CH_3NH_2, a putrid-smelling substance formed when flesh decays.

Resonance

8.27 Use Lewis structures to describe what is meant by resonance and name two consequences of resonance.

8.28 Do $H-C{\equiv}N$ and $H-N{\equiv}C$ form a pair of resonance structures? Explain your answer.

8.29 Write the Lewis structures that contribute to the resonance hybrid for (a) the nitrite ion, NO_2^{-}; (b) nitryl chloride, $ClNO_2$ (N is the central atom).

8.30 Write the Lewis structures that contribute to the resonance hybrid of (a) ozone, O_3; (b) the formate ion, HCO_2^{-}.

8.31 Write Lewis structures, including typical contributions to the resonance structure (where appropriate, allow for the possibility of octet expansion), for (a) dihydrogen phosphate ion; (b) sulfite ion; (c) chlorate ion. In oxoanions, the O atoms are all attached to the central atom. Each H atom present is attached to one of the O atoms.

8.32 Write Lewis structures, including typical contributions to the resonance structure (where appropriate, allow for the possibility of octet expansion as well as double bonds in different locations), for (a) hydrogen sulfite ion; (b) perchlorate ion; (c) nitrite ion. (See Exercise 8.31.)

Formal Charge

8.33 Justify the definition of formal charge and summarize how the concept is used.

8.34 Why are formal charge and oxidation number artificial measures of the distributions of electrons in molecules and ions?

8.35 Determine the formal charge on each atom in the following Lewis structures:

(a) Hydrazine   (b)   Azide ion

8.36 Determine the formal charge on each atom in the following Lewis structures:

$$\quad H \; H$$
$$\quad | \;\; |$$
$$:\ddot{N}-\ddot{O}{:}$$
$$\quad |$$
$$\quad H$$

(a) Hydroxylamine   (b) Kekulé structure of benzene

8.37 Two contributions to the resonance structure are shown for each species. Determine the formal charge on each atom and then, if possible, identify the Lewis structure of lower energy for each species.

(a) ${:}\ddot{O}-\ddot{S}=\ddot{O} \qquad \ddot{O}=\ddot{S}=\ddot{O}$
$$\qquad\qquad :O: \qquad\qquad :\ddot{O}:$$
(b) $\ddot{O}=\overset{\|}{S}=\ddot{O} \qquad \ddot{O}=\overset{|}{S}=\ddot{O}$

8.38 Two contributions to the resonance structure are shown for each species. Determine the formal charge on each atom and then, if possible, identify the Lewis structure of lower energy for each species.

(a) $\ddot{O}=\overset{:\ddot{O}: \; H}{\underset{}{S}}-\ddot{O}{:}^{-} \qquad {:}\ddot{O}-\overset{:\ddot{O}: \; H}{\underset{}{S}}-\ddot{O}{:}^{-}$

(b) $\ddot{O}=\overset{:\ddot{O}: \; H}{\underset{:\ddot{O}:}{S}}-\ddot{O}{:}^{-} \qquad \ddot{O}=\overset{:O: \; H}{\underset{:\ddot{O}:}{S}}-\ddot{O}{:}^{-}$

8.39 Use formal charge arguments to identify the lower energy contribution to the resonance structure of sulfurous acid.

(a) $:\ddot{O}-\overset{H \;\; :\ddot{O}: \;\; H}{\underset{}{S}}-\ddot{O}{:}$ (b) $:\ddot{O}-\overset{H \;\; :O: \;\; H}{\underset{}{S}}-\ddot{O}{:}$

8.40 Use formal charge arguments to identify the lower energy contribution to the resonance structure of sulfuric acid.

(a) $:\ddot{O}-\overset{H \;\; :\ddot{O}: \;\; H}{\underset{:\ddot{O}:}{S}}-\ddot{O}{:}$ (b) $:\ddot{O}-\overset{H \;\; :O: \;\; H}{\underset{:\ddot{O}:}{S}}-\ddot{O}{:}$

8.41 Select from each of the following pairs of Lewis structures the one that is likely to make the dominant contribution to a resonance hybrid.

(a) $:\ddot{F}-\ddot{\underset{..}{X}}e-\ddot{F}:$ $:\ddot{F}-\ddot{\underset{..}{X}}e=\ddot{F}:$

(b) $\ddot{O}=C=\ddot{O}$ $:\ddot{O}-C\equiv O:$

8.42 Select from each of the following pairs of Lewis structures the one that is likely to make the dominant contribution to a resonance hybrid.

(a) $\ddot{\underset{.}{N}}=N=\ddot{O}$ $:N\equiv N-\ddot{O}:$

(b) $\ddot{O}=\overset{\displaystyle :\ddot{O}:}{\underset{\displaystyle :\ddot{O}:}{P}}-\ddot{O}:$ $\Big]^{3-}$ $\ddot{O}=\overset{\displaystyle :\ddot{O}:}{\underset{\displaystyle :\ddot{O}:}{P}}=\ddot{O}$ $\Big]^{3-}$

Exceptions to the Octet Rule

8.43 What is meant by the term *radical*? Give three examples of radicals, showing their Lewis structures.

8.44 Why are radicals generally highly reactive?

8.45 Write the Lewis structures and note the number of lone pairs on the central atom for (a) sulfur tetrachloride; (b) iodine trichloride; (c) IF_4^-.

8.46 Write the Lewis structures and note the number of lone pairs on the central atom for (a) PF_4^-; (b) ICl_4^+; (c) XeF_4.

8.47 Determine the numbers of electron pairs (both bonding pairs and lone pairs) on the iodine atom in (a) ICl_2^+; (b) ICl_4^-; (c) ICl_3; (d) ICl_5.

8.48 Determine the numbers of electron pairs (both bonding pairs and lone pairs) on the phosphorus atom in (a) PCl_3; (b) PCl_5; (c) PCl_4^+; (d) PCl_6^-.

8.49 Write the Lewis structures of the following species and indicate which are radicals: (a) the superoxide ion, O_2^-; (b) nitrogen difluoride, NF_2; (c) nitrosyl chloride, $NOCl$.

8.50 Write the Lewis structures of the following species and indicate which are radicals: (a) the bleaching agent chlorine dioxide, ClO_2; (b) a substance that is a threat to the ozone layer, ClO; (c) XeO_4.

8.51 Write Lewis structures and state the number of lone pairs on xenon, the central atom of the following species: (a) $XeOF_2$; (b) XeF_2; (c) $HXeO_4^-$.

8.52 Write Lewis structures and state the number of lone pairs on the central atom of the following compounds: (a) ClF_3; (b) AsF_5; (c) SF_6.

Lewis Acids and Bases

8.53 Explain the terms *Lewis acid* and *Lewis base* and give three examples of each type of species.

8.54 Express the formation of magnesium carbonate from carbon dioxide and magnesium oxide as a reaction between a Lewis acid and a Lewis base.

8.55 Explain how an acid-base neutralization reaction in water can be expressed in terms of Lewis acids and bases.

8.56 Explain why redox reactions are not Lewis acid-base reactions.

8.57 Identify the following species as Lewis acids or Lewis bases: (a) NH_3; (b) BF_3; (c) Ag^+; (d) F^-.

8.58 Identify the following species as Lewis acids or Lewis bases: (a) H^+; (b) Al^{3+}; (c) CN^-; (d) NO_2^-.

8.59 Write the Lewis structure of the methoxide ion, CH_3O^- (the C atom is the central atom). Would you expect it to be a Lewis acid or a Lewis base? Explain your answer.

8.60 Write the Lewis structure of the hydride ion. Would you expect it to be a Lewis acid or a Lewis base? Explain your answer.

8.61 Write the Lewis structures of each reactant, identify the Lewis acid and the Lewis base, and then write the product (a complex) for the following Lewis acid-base reactions:
(a) $PF_5 + F^- \longrightarrow$
(b) $SO_2 + Cl^- \longrightarrow$
(c) $Cu^{2+} + 4\,NH_3 \longrightarrow$

8.62 Write the Lewis structures of each reactant, identify the Lewis acid and the Lewis base, and then write the product (a complex) for the following Lewis acid-base reactions:
(a) $GaCl_3 + Cl^- \longrightarrow$
(b) $SF_4 + F^- \longrightarrow$
(c) $Ag^+ + 2\,CN^- \longrightarrow$

Ionic and Covalent Bonding

8.63 Decide which of the following compounds are likely to be ionic and justify your choice: (a) magnesium oxide; (b) nitrogen triiodide; (c) iron(II) oxide.

8.64 Decide which of the following compounds are likely to be ionic and justify your choice: (a) sulfur tetrachloride; (b) cobalt(III) oxide; (c) oxygen difluoride.

8.65 Use the information in Fig. 8.21 to label the following bonds as ionic, covalent, or a significant mix of covalent and ionic character: (a) $Ba-Cl$; (b) $Bi-I$; (c) $Si-H$.

8.66 Use the information in Fig. 8.21 to label the following bonds as ionic, covalent, or a significant mix of covalent and ionic character: (a) $N-H$; (b) $C-S$; (c) $P-F$.

8.67 Arrange the cations Rb^+, Be^{2+}, Sr^{2+} in order of increasing polarizing power. Explain the reasons for your arrangement.

8.68 Arrange the cations K^+, Mg^{2+}, Al^{3+}, Cs^+ in order of increasing polarizing power. Explain the reasons for your arrangement.

8.69 Arrange the anions Cl^-, Br^-, N^{3-}, O^{2-} in order of increasing polarizability, giving reasons for your decisions.

8.70 Arrange the anions N^{3-}, P^{3-}, I^-, At^- in order of increasing polarizability, giving reasons for your decisions.

8.71 In each pair, determine which compound has bonds with greater ionic character: (a) HCl or HI; (b) CH_4 or CF_4; (c) CO_2 or CS_2.

8.72 In each pair, determine which compound has bonds with greater ionic character: (a) PH_3 or NH_3; (b) SO_2 or NO_2; (c) SF_6 or IF_5.

8.73 Classify the bonding in the following compounds as mainly ionic or significantly covalent: (a) AgF; (b) AgI; (c) $AlCl_3$; (d) AlF_3. See Appendix 2D.

8.74 Classify the bonding in the following compounds as mainly ionic or significantly covalent: (a) $BeCl_2$; (b) $CaCl_2$; (c) TlCl; (d) $InCl_3$. See Appendix 2D.

Supplementary Exercises

8.75 Suggest why it is reasonable to write Lewis structures for boron compounds in which boron does not have a complete octet of electrons.

8.76 What is the relation between the electronegativities of two atoms and the type of bond (covalent, polar, or ionic) they are likely to form?

8.77 List the following bonds in order of increasing degree of ionic character: C—Cl, Na—Cl, Al—Cl, Br—Cl.

8.78 Write the Lewis symbols of (a) N^{3-}; (b) S^{2-}; (c) Sn^{2+}.

8.79 Write the Lewis formulas for the following compounds (assuming an ionic model of their bonding): (a) lithium hydride; (b) copper(II) chloride; (c) barium nitride; (d) gallium oxide.

8.80 Complete the following table (all values are in kilojoules per mole):

Compound (MX)	ΔH_f° (M, g)	I (M)	ΔH_f° (X, g)	E_{ea} (X)	ΔH_L° (MX)	ΔH_f° (MX, s)
(a) NaCl	108	494	122	349	787	?
(b) KBr	89	418	97	325	?	−394
(c) RbF	?	402	79	328	774	−558

8.81 Use Appendix 2A, Appendix 2D, and the following data to calculate the lattice enthalpy of (a) Na_2O; (b) $AlCl_3$. $\Delta H_f^\circ(Na_2O) = -409$ kJ/mol; $\Delta H_f^\circ(O, g) = +249$ kJ/mol, $\Delta H_f^\circ(Al, g) = +326$ kJ/mol.

8.82 Draw the requested number of Lewis structures that contribute to the resonance structures for the following compounds:(a) CN_2^{2-} (two); (b) N_2O_3 (three, for the arrangement $ONNO_2$); (c) OCN^- (three); (d) C_3O_2 (two, for the arrangement OCCCO).

8.83 Identify each of the following as a Lewis acid or a Lewis base (or a species that can act as either): (a) BCl_3; (b) Fe^{3+}; (c) GaI_3.

8.84 Write the Lewis structure for each reactant, identify the Lewis acid and Lewis base, and then write the formula of the product (a complex) for the following reactions:
(a) $AlCl_3 + Cl^- \longrightarrow$
(b) $I_2 + I^- \longrightarrow$
(c) $Cr^{3+} + 6 NH_3 \longrightarrow$
(d) $SnCl_4 + 2 Cl^- \longrightarrow$

8.85 In the reaction of sulfur dioxide with hydroxide ion, SO_2 functions as a Lewis acid. Write the balanced chemical equation for the reaction, using Lewis structures.

8.86 Which of the following is a Lewis acid, BF_3 or ClF_3? Explain your reasoning.

8.87 Which of the following is a Lewis base, CH_3^- or CH_4? Explain your reasoning.

8.88 Ionic compounds typically have higher boiling points and lower vapor pressures than covalent compounds. Predict which compound in the following pairs has the lower vapor pressure at room temperature: (a) Cl_2O or Na_2O; (b) $InCl_3$ or $SbCl_3$; (c) LiH or HCl; (d) $MgCl_2$ or PCl_3.

8.89 Determine the formal charge of each atom in the following species. Where two or more Lewis structures are given, identify the structure of lowest energy.

(a) $\ddot{O}=\overset{\overset{\displaystyle :O:}{\|}}{\underset{}{Cl}}-\ddot{O}:$ $:\ddot{O}-\overset{\overset{\displaystyle :\ddot{O}:}{|}}{\underset{}{Cl}}-\ddot{O}:$

(b) $\ddot{O}=C=\ddot{S}$

(c) $H-C\equiv N:$

(d) $\dot{N}=C=\ddot{\overline{N}}:^{2-}$ $:N\equiv C-\ddot{\overline{N}}:^{2-}$

(e) $:\ddot{O}-\overset{\overset{\displaystyle :\ddot{O}:}{|}}{\underset{\underset{\displaystyle :\ddot{O}:}{|}}{As}}-\ddot{O}:^{3-}$ $:\ddot{O}-\overset{\overset{\displaystyle :O:}{\|}}{\underset{\underset{\displaystyle :\ddot{O}:}{|}}{As}}-\ddot{O}:^{3-}$

8.90 Determine the maximum oxidation number that each of the following elements can have in compounds: (a) lead; (b) vanadium; (c) sulfur; (d) chlorine.

8.91 Could you support the view that there is only one type of chemical bond? Explain how covalent bonds, ionic bonds, complex formation, and the octet rule would fit into this view.

8.92 By assuming that the lattice enthalpy of $NaCl_2$ would be the same as that of $MgCl_2$, use enthalpy arguments based on data in Appendix 2A and Appendix 2D to explain why $NaCl_2$ is an unlikely compound.

8.93 By assuming that the lattice enthalpy of MgCl would be the same as that of KCl, use enthalpy arguments based on data in Appendix 2A and Appendix 2D to explain why MgCl is an unlikely compound.

8.94 Show how resonance can occur in the following organic ions: (a) the acetate ion, $CH_3CO_2^-$; (b) an enolate ion, $CH_2COCH_3^-$, which has one resonance structure with a $C=C$ double bond and an $-O^-$ group on the central carbon atom; (c) an allyl cation, $CH_2CHCH_2^+$.

Applied Exercises

For Exercises 8.95–8.98, see Applying Chemistry: Case Study 8.

8.95 Write the Lewis structures for the following components of smog: (a) NO; (b) OH (hydroxyl radical); (c) CH_3CO_2 (acetyl radical); (d) NO_2.

Smog over Frankfurt, Germany, Exercises 8.95 and 8.96.

8.96 (a) The NO in smog can react with NO_2. Write the Lewis structure of the most likely product. (b) The NO_2 in smog also reacts with NO_3 molecules. Write the Lewis structure of the most likely product. (c) Write the balanced chemical equation for the reaction of the product from (b) (an acidic oxide) with water. Name the acid formed.

8.97 Write the Lewis structure for the pollutant acetaldehyde, CH_3CHO. Is resonance likely to be important for this molecule?

8.98 Write the Lewis structures of the following reactive species found to contribute to the destruction of the ozone layer and indicate which are radicals: (a) chlorine monoxide, ClO; (b) dichloroperoxide, $Cl-O-O-Cl$; (c) chlorine nitrate, $ClONO_2$ (the central O atom is attached to the Cl atom and to the N atom of the $-NO_2$ group); (d) chlorine peroxide, $Cl-O-O$.

8.99 Limestone (calcium carbonate) is used to remove impurities from iron ore in a blast furnace. When limestone is heated, it decomposes into calcium oxide and carbon dioxide, and the calcium oxide combines with impurities such as silica, SiO_2, to form calcium silicate, $CaSiO_3$. The latter is a component of slag. Write the chemical equation for the Lewis acid-base decomposition of carbonate ions (the complex) and a

Lewis acid-base reaction for the formation of silicate ions. Identify the Lewis acid and Lewis base in each reaction.

Integrated Exercises

8.100 A chemical analysis of a highly toxic gas that has been used in chemical warfare is 12.1% carbon, 16.2% oxygen, and 71.7% chlorine by mass, and its molar mass is 98.9 g/mol. Write the Lewis structure of the substance.

8.101 Table 3.1 lists common solubility rules. Look up the solubilities of chlorides, carbonates, chromates, phosphates, and nitrates. Judge the trend in polarizability of these ions, considering the polyatomic ions as large spheres with different numbers of electrons. Write a general rule for the effect of covalent character on the solubility of a salt. Justify the differences in solubility for the ions listed above on the basis of covalent character in the ionic bonds in these compounds.

8.102 Thallium and oxygen form two compounds with the following characteristics:

Compound	Mass% Tl	Melting point, °C
I	89.49	717
II	96.23	300.

(a) Determine the chemical formulas of the two compounds. (b) Write the balanced equation for the formation of each compound from its elements. (c) Determine the oxidation number of thallium in each compound. (d) Assuming that the compounds are ionic, write the electron configuration for each thallium ion. (e) Use the melting points to decide which compound has more covalent character in its bonds. Is your finding consistent with what you would predict from the polarizing abilities of the two cations?

8.103 Use your knowledge of electron configurations to explain the following observations: (a) The difference in electronegativity between magnesium and iodine is 1.4, yet magnesium iodide exhibits the properties of ionic bonding. (b) The difference in electronegativity between silicon and fluorine is 2.1, yet silicon tetrafluoride exhibits the properties of covalent bonding.

8.104 The following species have the same number of electrons: Ca, Ti^{2+}, and V^{3+}. (a) Write the electron configurations for each species. Are they the same or different? Explain. (b) How many unpaired electrons, if any, are present in each species? (c) What neutral atom, if any, has the same electron configuration as Ti^{3+}?

8.105 The following species have the same number of electrons: Cd, In^+, and Sn^{2+}. (a) Write the electron configuration for each species. Are they the same or different? Explain. (b) How many unpaired electrons, if any, are present in each species? (c) What neutral atom, if any, has the same electron configuration as In^{3+}?

8.106 The perchlorate ion, ClO_4^-, is described by resonance structures. (a) Draw the Lewis structures that contribute to the resonance hybrid. (b) Identify the most plausible Lewis structures by using formal charge arguments. (c) The average length of a single Cl—O bond is 172 pm and that of a double Cl=O bond can be estimated at 140 pm. The Cl—O bond length in the perchlorate ion is found experimentally to be 144 pm for all four bonds. Identify the most plausible Lewis structures of the perchlorate ion from the experimental data. (d) What is the oxidation number of chlorine in the perchlorate ion? Identify the most plausible Lewis structure by using the oxidation number and by assuming that lone pairs belong to the atom to which they are attached but that all electrons shared in a bond belong to the atom of the more negative element. (e) Are these three approaches consistent? Explain why or why not.

Molecular Structure

How do we see these words? Perception, thinking, and learning depend on the shapes of molecules and how they change. Seeing depends on changes in the shapes of certain molecules in the retinas of our eyes. So does the signal processing that turns those images—in a most amazing, complicated, and almost completely unknown way—into thoughts, memories, actions, and deeds. If we are ever to know at a molecular level what it means to be conscious, then we have to understand why molecules have certain shapes.

The shapes of molecules govern the odors and tastes of compounds and their actions as drugs. They control the reactions that occur throughout our bodies and contribute to the process of being alive. Molecular shapes also govern the properties of the materials around us, including their colors and solubilities and whether a substance is a solid, a liquid, or a gas.

The preceding two chapters introduced the electron configurations of atoms and how the valence electron configurations of atoms are used to construct the Lewis structures of molecules. In this chapter, we see that the arrangements of electrons in atoms also control the three-dimensional shapes of molecules. In the next chapter, we explore how the shapes of molecules affect how they pack together to form liquids and solids. This chapter is therefore a crucial step in forging the connection between the behavior of individual atoms and the properties of bulk matter.

THE SHAPES OF MOLECULES AND IONS

The Lewis structures we encountered in Chapter 8 are only two-dimensional diagrams of the links between atoms. Except in simple cases, a Lewis structure does not show the arrangement of atoms in space. In this section, we see how Lewis's ideas can be extended to account for the three-dimensional shapes of molecules.

New materials, such as the lightweight gear used by these wall climbers on Baffin Island, allow us to explore our world further than thought possible before. The development of new materials like these relies on the principles of molecular structure introduced in this chapter.

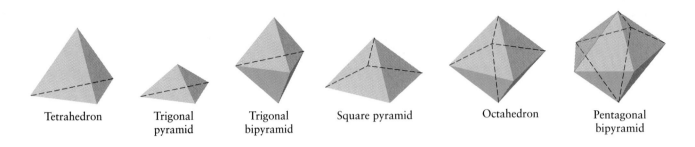

Tetrahedron   Trigonal pyramid   Trigonal bipyramid   Square pyramid   Octahedron   Pentagonal bipyramid

FIGURE 9.1

Some of the basic geometrical shapes that are used to describe the shapes of simple molecules.

9.1 The VSEPR Model

The shape of a molecule is described by reporting the locations of its atoms. Many simple molecules (molecules that have one central atom to which all the other atoms are bonded) have shapes in which the attached atoms lie at the corners of one of the geometrical figures shown in Fig. 9.1. The names of these shapes are given to the molecules. For instance, CH_4 (**1**) is tetrahedral and SF_6 (**2**) is octahedral (Fig. 9.2).

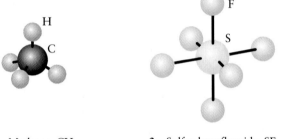

1 Methane, CH_4 **2** Sulfur hexafluoride, SF_6

To report shapes more precisely, we give the **bond angles,** which are the angles between bonds visualized as straight lines joining the atom centers (Fig. 9.3). The bond angle in CH_4, for instance, is the angle (109.5°) between two of the C—H bonds.

The shapes in Fig. 9.2 make sense when we realize that the *regions of high electron concentration found in bonds and lone pairs repel one another.* Bonding

FIGURE 9.2

The names of the shapes of simple molecules and the bond angles certain shapes imply. To name the shape of a molecule, compare only the locations of atoms, not the locations of lone pairs, with the positions shown here.

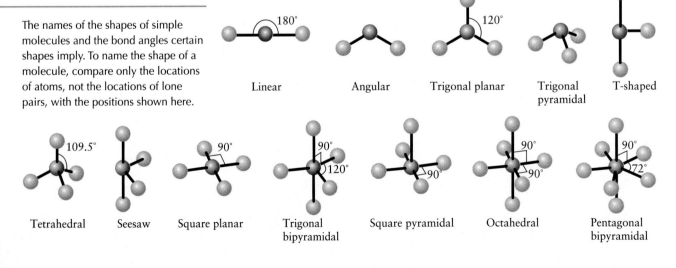

Linear   Angular   Trigonal planar   Trigonal pyramidal   T-shaped

Tetrahedral   Seesaw   Square planar   Trigonal bipyramidal   Square pyramidal   Octahedral   Pentagonal bipyramidal

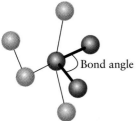

Bond angle

FIGURE 9.3

The bond angle in a molecule is the angle between the two straight lines joining the centers of the atoms concerned.

electrons and lone pairs take up positions as far from one another as possible, for then they repel each other the least. This idea is the central feature of the **valence-shell electron-pair repulsion model,** or VSEPR model, of molecular shape. The following paragraphs explain the model; it is summarized in Toolbox 9.1 at the end of Section 9.3.

In the VSEPR model, we focus on one atom at a time, examining the bonds it forms and any lone pairs it may possess. In a simple molecule, we need to look only at the central atom, such as the B atom in BF_3 or the C atom in CO_2. We then imagine that all the regions of high electron concentration (bonding pairs and lone pairs) on the central atom lie on the surface of an invisible sphere centered on the atom (Fig. 9.4). These electron-rich regions minimize their repulsions by moving as far apart as possible on the surface of the sphere. The regions adopt a characteristic arrangement that depends on their number: two regions take positions on opposite sides of the central atom and so form a linear arrangement, three take positions at the corners of an equilateral triangle, and so on (Fig. 9.5).

Let's see how to predict the shapes of molecules and polyatomic ions in which up to six atoms, but no lone pairs, are attached to one central atom. Beryllium chloride, $BeCl_2$, is a molecule in which two atoms are attached to the central atom. The Lewis structure is $:\ddot{C}l—Be—\ddot{C}l:$. To be as far apart as possible, the bonding pairs, and therefore the Cl atoms, lie on opposite sides of the Be atom. The $BeCl_2$ molecule is therefore expected to be *linear,* with a bond angle of 180° (**3**). That shape is confirmed by experiment.

A boron trifluoride molecule, BF_3, has the Lewis structure shown in (**4**). There are three bonding pairs on the central atom and no lone pairs. According to the VSEPR model, the three bonding pairs, and the fluorine atoms they link, lie at the corners of an equilateral triangle with the boron atom at its center. The molecule is predicted to be *trigonal planar* with all three F—B—F angles 120° (**5**). That is the shape found experimentally.

Methane, CH_4, has four bonding pairs on the central atom, each connecting a hydrogen atom to the carbon atom. The four bonding pairs of electrons are farthest apart if they take up positions at the corners of a tetrahedron, with the C atom at the center (see Fig. 9.5). A hydrogen atom is attached to each electron pair, so we expect the molecule to be *tetrahedral* (see **1**), with a bond angle of 109.5°. This structure is also found experimentally.

This idea was first explored by the British chemists Nevil Sidgwick and Herbert Powell, and has been developed by the Canadian chemist Ronald Gillespie.

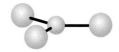

3   Beryllium chloride, $BeCl_2$

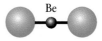

4   Boron trifluoride, BF_3

5   Boron trifluoride, BF_3

FIGURE 9.4

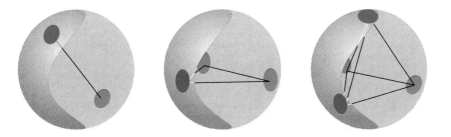

The positions taken up by regions of high electron concentration (shown in green) around a central atom, for three simple cases. Electron pairs or bonds in these positions are as far away from one another as possible and so experience the minimum repulsion from other electrons.

:Ċl:

:Cl—P—Cl:

:Ċl   Ċl:

6   Phosphorus pentachloride, PCl_5

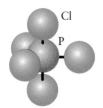

Cl

P

7   Phosphorus pentachloride, PCl_5

There are five bonding electron pairs on the central atom in a phosphorus pentachloride molecule, PCl_5, and no lone pairs (**6**). According to the VSEPR model, the five electron pairs and the atoms they carry are farthest apart when they lie at the corners of a trigonal bipyramid (see Fig. 9.5). In a *trigonal bipyramidal* arrangement, three atoms lie at the corners of an equilateral triangle, with the phosphorus atom at its center, and the other two atoms lie above and below the central atom, on an axis perpendicular to the plane of the triangle (**7**). This structure has two different bond angles. Around the "equator" formed by the triangle, the bond angles are 120°. The angle between the axial and equatorial P—Cl bonds is 90°. Once again, this structure is confirmed experimentally.

A sulfur hexafluoride molecule, SF_6, has six atoms attached to the central S atom and no lone pairs on that atom. According to the VSEPR model, the bonding pairs of electrons (and hence the fluorine atoms) are farthest apart when four pairs lie in a square around the equator, with the S atom at the center, and the remaining two pairs lie above and below the plane of the square (see Fig. 9.5). By referring to Fig. 9.2, we see that the atoms lie at the corners of an octahedron, so the molecule is classified as *octahedral*. All its bond angles are either 90° or 180° (see **2**), and all the bonds and fluorine atoms are equivalent to one another.

The same rules are used to predict the shapes of polyatomic ions. For example, the PCl_6^- ion (six bonds and no lone pairs) is octahedral, and the NH_4^+ ion (four bonds and no lone pairs) is tetrahedral.

> *According to the VSEPR model, bonding pairs and lone pairs, to reduce repulsions, take up positions around an atom that maximize their separations. The shape of the molecule is determined by the locations of the atoms attached to the central atom.*

Self-Test 9.1A   Predict the shape of an AsF_5 molecule, which has no lone pairs on the central atom.

[*Answer:* Trigonal bipyramidal]

Self-Test 9.1B   Predict the shape of a $SiCl_4$ molecule, which has no lone pairs on the central atom.

9.2 Molecules with Multiple Bonds

The VSEPR model does not distinguish between single and multiple bonds. A multiple bond is treated as just another region of high electron concentration. The two electron pairs in a double bond stay together and repel other bonds or lone pairs as a unit. The three electron pairs in a triple bond also stay together and act like a single region of high electron concentration.

For instance, a carbon dioxide molecule has a structure similar to that of $BeCl_2$, except for the presence of double bonds, as in $\ddot{O}{=}C{=}\ddot{O}$. The two pairs

FIGURE 9.5

A summary of the positions taken up by regions of high electron concentration (other atoms and lone pairs) around a central atom. The locations of these regions are given by the straight lines sticking out of the central atom. Use this chart to identify the arrangement of a given number of atoms and lone pairs, and then use Fig. 9.2 to identify the shape of the molecule from the location of only its atoms.

180°

Linear

120°

Trigonal planar

109.5°

Tetrahedral

90°

120°

Trigonal bipyramidal

90°

90°

Octahedral

90°

72°

Pentagonal bipyramidal

of electrons that make up one double bond are treated as a unit. They lie on opposite sides of the C atom and bind the O atoms to it. Therefore, the molecule is linear (**8**). One of the Lewis structures of a carbonate ion, CO_3^{2-}, is shown in (**9**). The two pairs of electrons in the double bond are treated as a unit, and the resulting shape (**10**) is trigonal planar, the same as that of BF_3 (see **5**).

Now let's predict the shape of an ethene (ethylene) molecule, $CH_2{=}CH_2$. There are two "central" atoms in ethene, the two carbon atoms. The first step is to write the Lewis structure (**11**). Each carbon atom has three regions of electron concentration: two single bonds and one double bond. There are no lone pairs. The arrangement around each carbon atom is therefore trigonal planar. We predict that the HCH and HCC angles will both be 120° (**12**), and our prediction is confirmed experimentally.

8   Carbon dioxide, CO_2

9   Carbonate ion, CO_3^{2-}

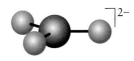

10   Carbonate ion, CO_3^{2-}

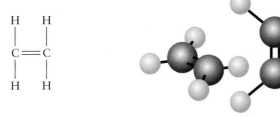

11   Ethene, C_2H_4 12   Ethene, C_2H_4

Example 9.1 *Predicting the shape of a molecule with a multiple bond*

Suggest a shape for the ethyne (acetylene) molecule, HC≡CH.

Strategy Write down the Lewis structure and decide how the electron pairs can be arranged around each "central" atom (two C atoms, in this case) to minimize repulsions. Treat each multiple bond as a single unit. Then identify the overall shape of the molecule (refer to Fig. 9.2 if necessary).

Solution The Lewis structure of the molecule is H—C≡C—H. Each C atom is attached to two other atoms (one H atom and one C atom), and there are no lone pairs. The arrangement of atoms around each C atom is therefore predicted to be linear (**13**). That shape is found experimentally.

Self-Test 9.2A Predict the shape of a formaldehyde molecule, CH_2O.
[*Answer:* Trigonal planar]

Self-Test 9.2B Predict the shape of a hydrogen cyanide molecule, HCN.

13   Ethyne, C_2H_2

Because the VSEPR model treats single bonds and multiple bonds as equivalent, it does not matter which of the Lewis structures in a resonance hybrid we consider, provided the number of lone pairs on the central atom does not change. For example, although we can write several different Lewis structures for a sulfate ion, all of them have four regions of electron concentration around the central S atom (recall Example 8.9); and in each case, we expect a tetrahedral shape (**14**).

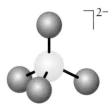

14   Sulfate ion, SO_4^{2-}

> *Electron pairs in multiple bonds are treated as a single unit equivalent to one region of high electron concentration.*

9.3 Molecules with Lone Pairs on the Central Atom

To help us predict the shapes of molecules, we use a special **VSEPR formula** to identify the different combinations of atoms and lone pairs attached to the central atom. In this formula,

> A represents a central atom
>
> X represents an atom bonded to the central atom
>
> E represents a lone pair of electrons on the central atom

For example, in a BF_3 molecule, the central boron atom (A) is attached to three fluorine atoms (X) but no lone pairs (E); it is therefore an example of an AX_3 species. The S atom in the sulfite ion, SO_3^{2-}, has three attached atoms and one lone pair, so the sulfite ion is an example of an AX_3E species (**15**). Some radicals, such as NO_2, have a single nonbonding electron on the central atom. The single electron is treated like a lone pair when predicting molecular shape, so NO_2 is an example of an AX_2E molecule.

The **electron arrangement** of a molecule is the three-dimensional arrangement of all regions of high electron concentration (bonds and lone pairs of electrons) about the central atom. When identifying the electron arrangement around the central atom, we do not distinguish between bonds and lone pairs: single bonds, multiple bonds, and lone pairs are all regions where electrons are concentrated. For instance, the sulfite ion is an example of an AX_3E molecule with four regions of high electron density on the central atom: three bonded atoms and one lone pair. It therefore has a tetrahedral electron arrangement and a trigonal pyramidal shape (**16**). All AX_3E molecules and ions have the same general shape (trigonal pyramidal), but their exact bond angles depend on the identity of the species.

In the VSEPR model, we have to keep in mind the distinction between the electron arrangement and the molecular shape. All electron pairs, whether bonded or not, are included in a description of the electron arrangement. However, only the positions of atoms are considered when describing the shape of a molecule; lone pairs are ignored. Species with an AX_4 formula, such as CH_4, have a tetrahedral electron arrangement and a tetrahedral shape (recall **1**), whereas AX_3E species such as SO_3^{2-} have a tetrahedral electron arrangement, but a trigonal pyramidal shape.

Because the electron arrangement of an SO_3^{2-} ion is tetrahedral, we predict OSO angles of 109.5°, the tetrahedral angle. However, although experiments have shown that the sulfite ion is indeed trigonal pyramidal, its bond angle is only 106° (**17**). Experimental information like this tells us that the VSEPR model as described so far is incomplete.

What do we need to change? We have treated lone pairs the same way we treated bonding pairs. It is likely, however, that lone pairs have a greater repelling effect than bonding pairs because a bonding pair is pinned down by two atoms, not just one. As a result, the electron cloud of a bonding pair cannot spread over as large a volume as a lone pair can, and a bonding pair repels other electrons to a lesser extent than a lone pair does (Fig. 9.6). The VSEPR model therefore uses the following rule:

> The strengths of repulsions are in the order lone pair-lone pair > lone pair-bonding pair > bonding pair-bonding pair

> Notice that lone pairs on the attached atoms are not included in the VSEPR formula, as they have little effect on the shape of the molecule.

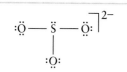

15 Sulfite ion, SO_3^{2-}

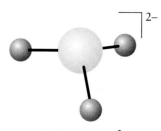

16 Sulfite ion, SO_3^{2-}

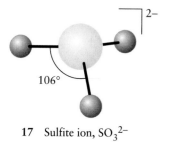

106°

17 Sulfite ion, SO_3^{2-}

This rule implies that it is best for lone pairs to be as far from one another as possible. It is also best for atoms to be far from lone pairs, even though that might bring them close to other atoms.

Our modified model helps to account for the bond angle of the AX₃E sulfite ion (see **17**). The electron pairs adopt a tetrahedral arrangement around the S atom. However, the lone pair exerts a strong repulsion on the bonding electrons, forcing the O atoms together. As a result, the OSO angle is reduced from the 109.5° of the regular tetrahedron to the 106° observed experimentally.

Note that, although we can use the VSEPR model to predict the *direction* of the distortion, we cannot use it to predict its extent. We can predict that in any AX₃E species the angle will be less than 109.5°, but we cannot predict its actual value.

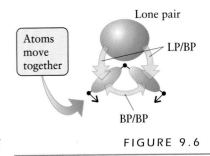

FIGURE 9.6

One possible explanation of why a lone pair has a greater repelling effect than a bonding pair. A lone pair (LP) is more extensive than the bonding pairs (BP) that hold the atoms (black dots) in place, so the peripheral atoms move away from the lone pair in an attempt to lower the repulsion they experience.

Example 9.2 *Using the VSEPR model to predict the electron arrangement and shape of a molecule*

Predict (a) the VSEPR formula of a water molecule; (b) its electron arrangement; and (c) its shape.

Strategy (a) Write the Lewis structure and use it to write the VSEPR formula as AX$_n$E$_m$, where *n* is the number of atoms bonded to the central atom and *m* the number of lone pairs on the central atom. (b) Count the total number of bonds and lone pairs on the central atom and identify the arrangement they adopt, referring to Fig. 9.5 if necessary. (c) Then locate the atoms and identify the shape of the molecule, referring to Fig. 9.2 if necessary. Finally, allow the lone pairs to repel the bonding pairs so that the bonding pairs move together slightly.

Solution (a) The central atom (O) has four electron pairs (two single bonds and two lone pairs, H—Ö—H), so the VSEPR formula is AX₂E₂. (b) These four pairs adopt a tetrahedral arrangement, as illustrated in Fig. 9.7. (c) Two of these tetrahedral locations are occupied by atoms; so, according to Fig. 9.2, the molecule is classified as angular or "bent." The lone pairs push away from each other, which forces the bonding pairs closer together; so we predict an HOH bond angle of less than 109.5°. This prediction is in agreement with the experimental value of 104.5° (**18**).

18 Water, H₂O

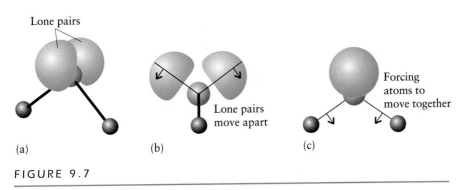

(a)　　　　(b)　　　　(c)

FIGURE 9.7

The electron arrangement of the water molecule results in an angular shape. (a) The basic tetrahedral arrangement of two lone pairs and two bonding pairs. (b) A view parallel to the molecular plane, showing how the two lone pairs move apart. (c) A view perpendicular to the molecular plane, showing how the two bonding pairs move together slightly.

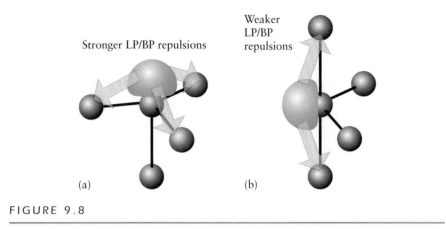

Stronger LP/BP repulsions

Weaker LP/BP repulsions

(a)   (b)

FIGURE 9.8

(a) A lone pair in an axial position is close to the bonding pairs of three equatorial atoms, whereas in an equatorial position (b), a lone pair is close to only two atoms. The latter arrangement is therefore more favorable.

19   Ammonia, NH₃

Self-Test 9.3A Predict (a) the VSEPR formula, (b) the electron arrangement; and (c) the shape of an NH_3 molecule.

[*Answer:* (a) AX_3E; (b) tetrahedral; (c) trigonal pyramidal (**19**), HNH angle less than 109.5°]

Self-Test 9.3B Predict (a) the VSEPR formula; (b) the electron arrangement; and (c) the shape of an NO_2^- ion.

Larger numbers of lone pairs have a greater effect on the bond angle. For example, NH_3 has one lone pair and a bond angle of 107°, whereas H_2O has two lone pairs and a bond angle of 104.5°.

We have another decision to make when we consider an AX_4E molecule or ion. The electron arrangement is trigonal bipyramidal. However, the lone pair can lie in one of two different locations. An **axial lone pair** lies on the axis of the molecule. An **equatorial lone pair** lies on the molecule's "equator." Which position is it likely to adopt? As always, the most likely location is the one that results in the lowest possible energy. Because an axial lone pair repels three electron pairs strongly but an equatorial lone pair repels only two strongly (Fig. 9.8), we can conclude that it is best for a lone pair to be equatorial, so producing a seesaw-shaped molecule.

Now consider an AX_4E_2 molecule, which has two lone pairs. It is best for the two lone pairs to lie opposite each other. Then the four attached X atoms will lie in a plane at the corners of a square, so giving a square planar shape (Fig. 9.9).

Lone pairs on the central atom contribute to the shape of the molecule but are ignored when we name the shape. The molecule adjusts its shape to reduce lone pair–lone pair and lone pair–bonding pair repulsions.

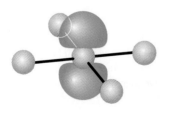

FIGURE 9.9

The square planar arrangement of atoms adopted by AX_4E_2 molecules. The two lone pairs are farthest apart when they are on opposite sides of the central atom.

We use this Toolbox to predict the electron arrangement and shape of a molecule from its molecular formula.

Conceptual Basis

Electron pairs attached to a central atom in a molecule arrange themselves in such a way that repulsions between regions of high electron concentration (bonds and lone pairs) are minimized. Lone pairs repel other electron pairs more strongly than bonding pairs do.

Procedure

The general procedure for predicting the shape of a molecule is as follows (Fig. 9.10):

Step 1. Decide how many electron pairs are present on the central atom by writing a Lewis structure for the molecule. Treat a multiple bond as the equivalent of a single electron pair.

Step 2. Arrange the regions of high electron concentration as far apart as possible to minimize their repulsions (see Fig. 9.5).

Step 3. Decide which positions will be occupied by lone pairs and which by bonds. When lone pairs can be placed in more than one nonequivalent position, they take the positions that reduce repulsions as much as possible.

Step 4. Name the molecular shape from the positions of the atoms (see Fig. 9.2).

Step 5. Allow the molecule to distort so that lone pairs are as far from one another and from bonding pairs as possible. The repulsions are in the order

Lone pair-lone pair $>$ lone pair-bonding pair
$>$ bonding pair-bonding pair

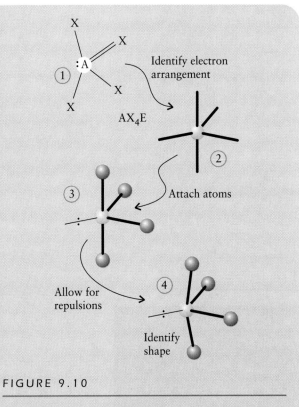

FIGURE 9.10

A schematic summary of the steps in the procedure for using the VSEPR model to predict molecular shape.

Example 9.3 *Predicting the shape of a molecule*

Predict the shape of a sulfur tetrafluoride molecule, SF_4.

Strategy We first write the Lewis structure of the molecule (**20**) (recall Example 8.8) and then apply the procedure outlined in Toolbox 9.1.

20 Sulfur tetrafluoride, SF_4

Solution The molecule is an example of an AX_4E species, with four atoms bonded to the central S atom and one lone pair. The five electron pairs adopt a trigonal bipyramidal arrangement. As shown in (**21**), the lone pair can be either (a) axial or (b) equatorial, but repulsion from the other electrons is lower in the equatorial position. The energy is lowered still further if the atoms move away from the lone pair to give the structure shown in (c). This shape (which resembles a slightly bent seesaw when viewed from the side) is the one found experimentally.

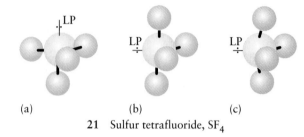

(a) (b) (c)

21 Sulfur tetrafluoride, SF_4

Self-Test 9.4A Predict the shape of a chlorine trifluoride molecule, ClF_3.

[***Answer:*** T-shaped]

Self-Test 9.4B Predict the shape of a xenon tetrafluoride molecule, XeF_4.

CHARGE DISTRIBUTION IN MOLECULES

If we want to understand the properties of a compound, such as its solubility and boiling point, we need to know not only the shape of the molecule but also where its electrons are most likely to be found. Even the activities of life depend on the locations of electrons, because electrons control the shape of the DNA helix and the way it unwinds in the course of reproduction (Fig. 9.11). Electron distribution also controls the shapes of our individual proteins and enzymes, and their shape is crucial to their function. In fact, when proteins lose their shape—for instance, when we suffer burns—they cease to function and we suffer tissue damage, disease, and possibly death.

9.4 Polar Bonds

So far, we have assumed that an electron pair in a covalent bond is shared equally between the atoms. However, the pair may be shared unequally. A covalent bond in which the electron pair is shared unequally has partial ionic character and is called a **polar covalent bond.** The extent to which an atom has a greater or lesser share of electrons in a bond is determined by the difference in electronegativities of the two bonded atoms (see Section 8.14).

An O—H bond is polar because oxygen is more electronegative than hydrogen and gains a greater share in the bonding electron pair. Its greater share of electrons means that oxygen has a **partial negative charge,** which we denote $\delta-$. Because the electron pair has been pulled away from the hydrogen atom, that atom has a **partial positive charge,** denoted $\delta+$. We show the partial charges on the atoms by writing $^{\delta+}H—O^{\delta-}$.

The more electronegative element in a polar covalent bond almost always has a negative partial charge. The partial charges are very small when the two atoms have only a small electronegativity difference, but they increase as the

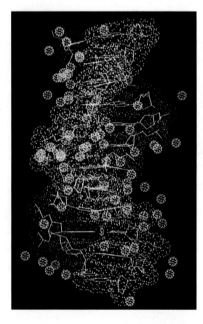

FIGURE 9.11

A DNA molecule consists of two very long chains of four units that repeat in a specific order and determine our genetic makeup. The shapes of the units and their charge distributions cause the chains to wind around each other to form a double helix. The dots show how the electrons are distributed.

difference in electronegativities increases. Because the electronegativities of carbon and hydrogen are very similar (2.6 and 2.2, respectively; see Fig. 8.21), C—H bonds are only slightly polar.

The two atoms in a polar covalent bond are said to possess an **electric dipole**, a positive charge next to an equal but opposite negative charge. A dipole is represented by an arrow that points from the positive charge toward the negative charge (**22**). The size of an electric dipole is reported as the **electric dipole moment**, μ (the Greek letter mu). The dipole moment is the magnitude of the partial charge ($+Q$ or $-Q$) multiplied by r, the distance between the charges: $\mu = Qr$. The greater the partial charges, the greater the dipole moment. The unit of dipole moment is called the **debye** (D). The debye is defined so that a single negative charge (an electron) separated by 100. pm from a single positive charge (a proton) has a dipole moment of 4.80 D. The dipole moment associated with an O—H bond is about 1.2 D. We can think of this dipole as arising from a partial charge of about 25% of an electron's charge on the O atom and an equivalent positive charge on the H atom.

A polar covalent bond is a bond between two atoms that have partial electric charges arising from their difference in electronegativity. Partial charges give rise to an electric dipole moment.

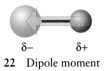

$\delta-$ $\delta+$
22   Dipole moment

In a new notation not used in this text, the arrow head points toward the *positive* end of the dipole.

The SI unit of dipole moment is the coulomb-meter (1 C·m):
1 D = 3.336 × 10^{-30} C·m.

9.5 Polar Molecules

A **polar molecule** is a molecule with a nonzero dipole moment. An HCl molecule, with its polar covalent bond ($^{\delta+}$H—Cl$^{\delta-}$), is a polar molecule. Its dipole moment is 1.1 D, a typical value for a polar diatomic molecule (Table 9.1). All diatomic molecules composed of atoms of different elements are at least slightly polar. A **nonpolar molecule** is a molecule that has zero electric dipole moment. All homonuclear diatomic molecules, such as Cl_2 and H_2, are nonpolar because there are no partial charges on their atoms.

For polyatomic molecules, it is very important to distinguish between a polar *molecule* and a polar *bond*. Although each bond in a polyatomic molecule

Table 9.1 *Dipole moments of selected molecules*

Molecule	Dipole moment, D	Molecule	Dipole moment, D
HF	1.91	PH_3	0.58
HCl	1.08	AsH_3	0.20
HBr	0.80	SbH_3	0.12
HI	0.42	O_3	0.53
CO	0.12	CO_2	0
ClF	0.88	BF_3	0
NaCl*	9.00	CH_4	0
CsCl*	10.42	*cis*-CHCl=CHCl	1.90
H_2O	1.85	*trans*-CHCl=CHCl	0
NH_3	1.47		

*These two species consist of pairs of ions in the gas phase.

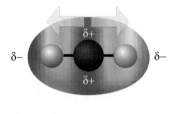

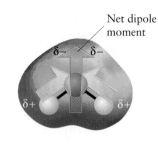

Net dipole moment

FIGURE 9.12

The electrostatic potential diagram of a carbon dioxide molecule shows that electrons are concentrated at the oxygen atoms (the red regions). Electron density at the carbon atom (the blue region) is relatively low. Because the regions of high electron concentration are directly opposite one another, the bond dipoles cancel and carbon dioxide is a nonpolar molecule.

FIGURE 9.13

The electrostatic potential diagram of a water molecule illustrates that the bond dipoles in water do not cancel, because, instead of pointing at one another, as in CO_2, they point to the side. As a result, there is a nonzero dipole moment lying at the midpoint of the HOH angle, and the water molecule is polar.

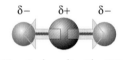

23 Carbon dioxide, CO_2

may be polar, the molecule *as a whole* is nonpolar if the dipoles of the individual bonds cancel one another. For example, the two $^{\delta+}C-O^{\delta-}$ dipoles in carbon dioxide point in opposite directions, so they cancel each other (**23**). As a result, CO_2 is a nonpolar molecule even though its bonds are polar. Modern computational techniques allow us to visualize electron distributions in molecules and to depict them by different colors, as in Fig. 9.12. The illustration is an example of an **electrostatic potential diagram:** red indicates a region of relatively high electron density and blue a region of relatively low electron density. We can see from Fig. 9.12 how the bond dipoles cancel in CO_2. In contrast, the two $^{\delta-}O-H^{\delta+}$ dipoles in H_2O lie at an angle and do not cancel each other, so H_2O is a polar molecule (Fig. 9.13).

The shape of a molecule governs whether or not it is polar. The atoms and bonds are all the same in *cis*-dichloroethene (**24**) and *trans*-dichloroethene (**25**);

The terms *cis* ("this side") and *trans* ("across") refer to the positions of the chlorine atoms relative to the double bond. For more information, see Section 11.12.

24 *cis*-Dichloroethene, CHCl═CHCl   **25** *trans*-Dichloroethene, CHCl═CHCl

but in the trans molecule, the C—Cl bonds point in opposite directions and the dipoles (represented by the arrows) cancel. Therefore, whereas *cis*-dichloroethene is polar, *trans*-dichloroethene is nonpolar. A simple test for the polarity of molecules of a liquid is to see whether a stream of the liquid is deflected by an electrically charged rod (Fig. 9.14).

Tetrahedral molecules with the same atom at each corner, such as tetra-chloromethane (carbon tetrachloride), CCl_4 (**26**), are nonpolar because the

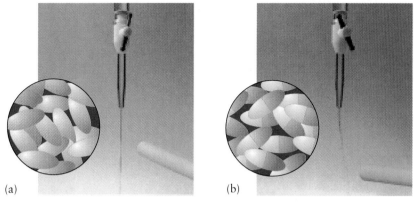

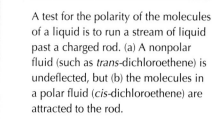

FIGURE 9.14

A test for the polarity of the molecules of a liquid is to run a stream of liquid past a charged rod. (a) A nonpolar fluid (such as *trans*-dichloroethene) is undeflected, but (b) the molecules in a polar fluid (*cis*-dichloroethene) are attracted to the rod.

(a) (b)

dipoles of the four bonds cancel in three dimensions. Highly symmetrical arrangements of polar bonds, such as in AX_2 (linear), AX_3 (trigonal planar), AX_4 (tetrahedral), AX_5 (trigonal bipyramidal), and AX_6 (octahedral) species (in VSEPR notation), where the X represents atoms of the *same element*, result in molecules that are nonpolar, because in these molecules the bond dipoles cancel (Fig. 9.15). However, if one or more atoms are different, as in trichloromethane (chloroform), $CHCl_3$ (**27**), or there are lone pairs, as in NH_3, then the bond dipoles do not necessarily cancel and the molecule may be polar. Examples of such polar molecules are HCN, PH_3, SO_2, and the other structures in the "polar" column in Fig. 9.15.

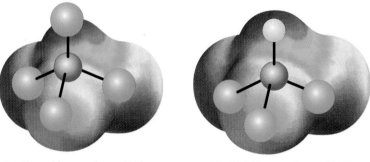

26 Tetrachloromethane, CCl_4 27 Trichloromethane, $CHCl_3$

The importance of shape in determining molecular polarity is demonstrated by the fact that the ozone molecule is polar despite being built of three atoms of the same element. Two positions around the central O atom are linked to O atoms, but the third position is occupied by a lone pair (**28**). Because one position is occupied by a lone pair, making it an AX_2E molecule, the dipoles do not cancel; as a result, the molecule is polar (**29**). This example shows that a molecule can be polar even if all the atoms belong to the same element.

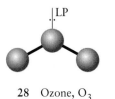

28 Ozone, O_3

29 Ozone, O_3

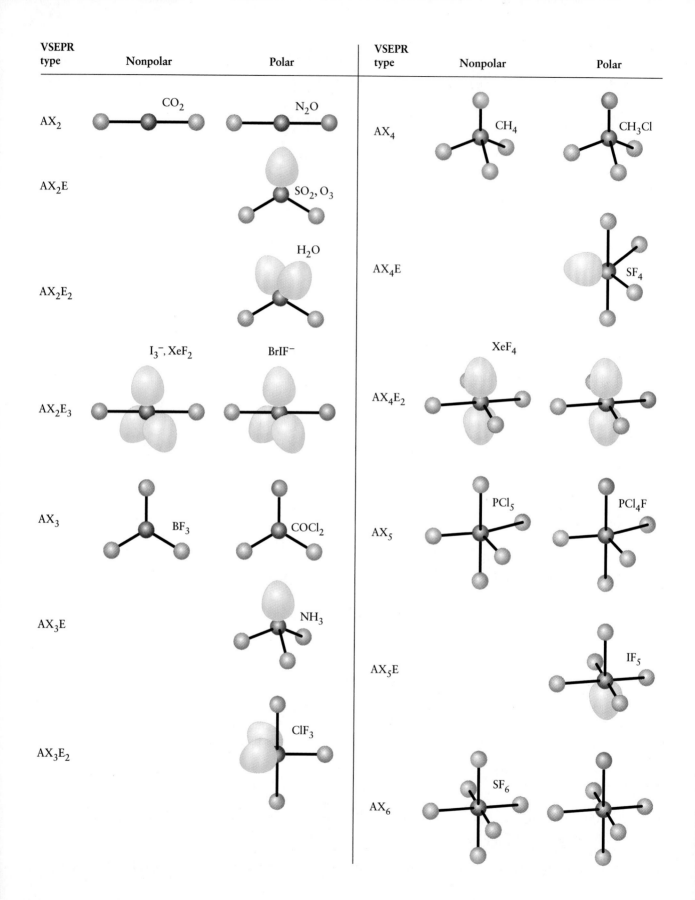

VSEPR type	Nonpolar	Polar
AX_2	CO_2	N_2O
AX_2E		SO_2, O_3
AX_2E_2		H_2O
AX_2E_3	I_3^-, XeF_2	$BrIF^-$
AX_3	BF_3	$COCl_2$
AX_3E		NH_3
AX_3E_2		ClF_3

VSEPR type	Nonpolar	Polar
AX_4	CH_4	CH_3Cl
AX_4E		SF_4
AX_4E_2	XeF_4	
AX_5	PCl_5	PCl_4F
AX_5E		IF_5
AX_6	SF_6	

FIGURE 9.15 *(facing page)*

The arrangements of atoms that give rise to polar and nonpolar molecules, and some examples. In the notation, A stands for a central atom, X for an attached atom, and E for a lone pair. Atoms of the same color are identical; X atoms colored differently represent different elements.

A diatomic molecule is polar if its bond is polar. A polyatomic molecule is polar if it has polar bonds arranged in space in such a way that their dipoles do not cancel.

Example 9.4 *Predicting the polarity of a molecule*

(a) Predict whether a boron trifluoride molecule, BF_3, is polar. (b) Suggest the appearance of its electrostatic potential diagram by sketching the locations of the partially negative (red) and partially positive (blue) areas.

Strategy (a) We use the VSEPR model to decide on the shape of the molecule. Decide by comparing the electronegativities of the atoms whether the bonds are polar; then decide whether the symmetry of the molecule allows the bond dipoles to cancel. If symmetry does not result in cancellation, the molecule is polar. Check the result by referring to Fig. 9.15. (b) Electron density is usually higher on the atoms of the element with the higher electronegativity.

Solution (a) According to Fig. 8.21 or Appendix 2D, we see that boron and fluorine have different electronegativities, so B—F bonds are polar. According to the VSEPR model, BF_3 is a trigonal planar AX_3 molecule. The symmetry results in the cancellation of the bond dipoles, so the molecule is nonpolar (see Fig. 9.15 and **5**). (b) The electronegativity of fluorine is greater than that of boron, so draw the electrostatic potential diagram (**30**) with the red color that indicates high electron density on the fluorine atoms. The partially positive boron atom is shown as a blue area. Despite the polar bonds, the molecule itself is nonpolar, because the three bond dipoles cancel one another.

Self-Test 9.5A Predict whether (a) SF_4 and (b) SF_6 are polar or nonpolar and sketch their electrostatic potential diagrams.

[*Answer:* (a) Polar (**31**); (b) nonpolar (**32**)]

Self-Test 9.5B Predict whether (a) PCl_3 and (b) PCl_5 are polar or nonpolar and sketch their electrostatic potential diagrams.

30 Boron trifluoride, BF_3

31 Sulfur tetrafluoride, SF_4

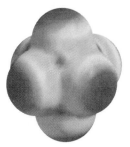

32 Sulfur hexafluoride, SF_6

THE STRENGTHS AND LENGTHS OF BONDS

In addition to its polarity, we can identify other characteristics of the bond between any two atoms A and B. For instance, the length of an A—B bond (of a given order; that is, single, double, or triple) is always approximately the same, regardless of the actual molecule in which the atoms occur. Similarly, the strength of an A—B bond (of a given order) does not vary greatly with the identity of the molecule that contains it. This approximate constancy of bond properties allows us to identify the presence of specific types of bonds in complex molecules (Investigating Matter 9.1). It also enables us to understand the properties of complex molecules by identifying the bonds they contain. For example, we can explain the resistance of polytetrafluoroethylene (Teflon) to chemical

When we feel the warmth of the Sun on our faces, we are responding to infrared radiation that has traveled nearly 150 million kilometers through space. That radiation stimulates the molecules in our skin to vibrate, and specially adapted nerve cells detect the vibration and interpret it as "warmth." The same type of radiation is used to identify molecules in samples and to determine the stiffness of bonds.

Infrared radiation is electromagnetic radiation lying at longer wavelengths (lower frequencies) than red light; a typical wavelength is about 1000 nm. That corresponds to one-thousandth of a millimeter, so it is just possible to imagine its scale. A wavelength of 1000 nm corresponds to a frequency of about 3×10^{14} Hz, which is comparable to the frequency at which molecules vibrate. A molecule can absorb radiation if the dipole moment of the molecule changes as its bonds stretch or bend. Even nonpolar molecules can absorb infrared radiation if the vibration results in a dipole moment, such as the bending motion of carbon dioxide shown in the first illustration.

The frequency at which a molecule vibrates depends on the masses of its atoms and the stiffness of its bonds, their resistance to distortion. A molecule made up of light atoms joined by stiff bonds has a higher vibrational frequency than one made up of heavy atoms joined by loose bonds. The former molecule will therefore absorb higher frequency, shorter wavelength radiation than the latter.

The vibrational absorption spectrum of a molecule is measured by using an *infrared spectrometer*. The source of infrared radiation is a hot filament. The wavelength is selected by diffracting the radiation from a grating. Constructive interference (see Investigating Matter 10.1) results in intense radiation being obtained in a given direction for a given angle of the grating; as that angle is changed, radiation of varying wavelength is passed through the sample. The beam is divided, one beam passing through the sample and the other beam passing through a blank; the intensities of the two beams are then compared at the detector and the reduction in intensity monitored.

A linear molecule consisting of two atoms can vibrate in only one way (a stretching and compressing motion), whereas a molecule with more atoms can vibrate in different ways. That number of ways increases rapidly: a benzene molecule can vibrate in 30 different ways. Some of these ways correspond to expansion and contraction of the ring, some to its distortion into a lozenge shape, and others to flexing and bending. Each way in which a molecule can vibrate is called a *normal mode;* so we say that a benzene molecule has 30 normal modes of vibration. In the illustration, we can see that a

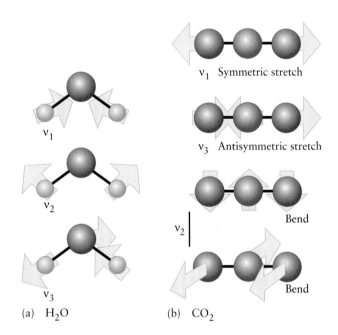

(a) H$_2$O   (b) CO$_2$

(a) The three normal vibrational modes of H$_2$O. Two of these modes involve mainly stretching motions, but mode v_2 is primarily a bending motion. (b) The four normal vibrational modes of CO$_2$. The first two are symmetric and antisymmetric stretching motions, and the last two are perpendicular bending motions.

attack by studying the character of C—F bonds in a much simpler compound, such as tetrafluoromethane, CF$_4$. We can be confident that the C—F bonds in the polymer are similar.

9.6 Bond Strengths

The strength of a chemical bond is measured by its **bond enthalpy,** ΔH_B, the change in enthalpy when the bond in a gaseous molecule is broken (expressed water molecule has three normal modes of vibration and carbon dioxide has four. Each normal mode has a frequency that depends in a complicated way on the masses of the atoms that move during the vibration and the stiffness associated with whatever bending or stretching motions are involved in the mode.

Except in simple cases, it is very difficult to predict the infrared absorption spectrum of a polyatomic molecule, because each of the modes has its characteristic absorption frequency rather than just the single frequency of a diatomic molecule. However, certain groups, such as a benzene ring or a carbonyl group ($\text{C}=\text{O}$), have characteristic frequencies, and their presence can often be detected in a spectrum. An infrared spectrum can therefore be used to identify a compound present in a sample by looking for the characteristic absorption bands associated with various groups of atoms. An example of a spectrum and its analysis are shown. The compound is an amino acid like the ones used to build proteins, but note that, in solution, the spectrum shows that the amino acid exists in an ionic form in which proton transfer has occurred from the carboxyl group to the amino group.

Many compounds have a complex series of absorptions spanning a range of wavelengths. This *fingerprint region* of the spectrum may be too difficult to analyze in detail, but because it is unique to a particular compound, we can identify an unknown compound by comparing its infrared spectrum to a large number of known spectra. This analysis can be done by computer, allowing thousands of comparisons to be made.

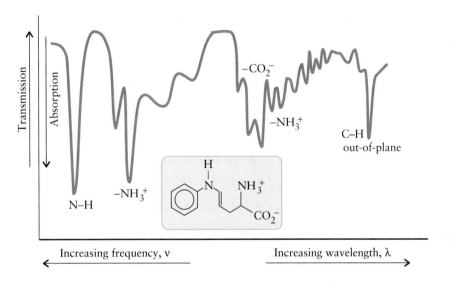

The infrared spectrum of an amino acid, with identification of the groups contributing to some of the peaks. Notice that the spectrum shows absorption; the "peaks" actually appear as valleys.

in kilojoules per mole of molecules). In all the cases we consider, we shall suppose that, when a covalent bond is broken, each atom carries away *one* of the electrons in the bonding pair. Bond breaking of this kind is called **dissociation** of the bond. Bond breaking always requires energy, so all bond enthalpies are positive. For example, the bond enthalpy of H_2 is obtained from the change in enthalpy for the dissociation

$$H_2(g) \longrightarrow 2\,H(g) \qquad \Delta H^\circ = +436 \text{ kJ}$$

More precisely, the process in which each atom carries off one electron from the bonding pair is called *homolytic dissociation.*

FIGURE 9.16

A Teflon coating is being applied to a gasket that will be used in an industrial chemical reactor. Because of its strong bonds, Teflon is very resistant to chemical attack.

We write $\Delta H_B(H—H) = 436$ kJ/mol to report this value (the positive sign is not usually shown because all bond enthalpies are positive). A large bond enthalpy means that the enthalpy of the separated atoms is much higher than that of the original molecule—in other words, a lot of energy has to be supplied to break the bond.

Bonds between different pairs of atoms have different strengths. For example, single bonds between carbon and fluorine are stronger than those between carbon and hydrogen: $\Delta H_B(F—CF_3) = 484$ kJ/mol and $\Delta H_B(H—CH_3) = 412$ kJ/mol. The greater the bond strength, the harder it is to break the bond. Typically, a high bond enthalpy suggests low reactivity. The high bond enthalpy of a carbon-fluorine bond helps to explain why polymers of fluorocarbons are very resistant to chemical attack. They are used to construct valves for corrosive gases and to line the interiors of chemical reactors (Fig. 9.16).

Table 9.2 lists bond enthalpies for a number of diatomic molecules. We see that values range from 151 kJ/mol (for I_2) up to 1074 kJ/mol (for CO, the strongest bond known between two nonmetallic elements). Some of the trends in Table 9.2 can be explained by considering the Lewis structures for the molecules. Consider, for example, the diatomic molecules of nitrogen, oxygen, and fluorine (Fig. 9.17). The bond enthalpy declines as the number of bonds between the atoms decreases—from three in N_2 to one in F_2. A multiple bond is always stronger than a single bond between the same two atoms because more electrons bind the atoms together.

In chemistry, there is rarely a *single* reason for a trend in properties. The Lewis structure of a nitrogen molecule is :N≡N:, with one lone pair on each atom. The Lewis structure of fluorine, :F̈—F̈:, shows that it has three lone pairs on each atom. Lone pairs repel one another, even across the bond, and this repulsion is another contribution to the weakening of the bond in fluorine relative to that in nitrogen. Repulsion between lone pairs also helps to explain why

Table 9.2 Bond enthalpies of diatomic molecules, kJ/mol

Molecule	ΔH_B
H_2	436
N_2	944
O_2	496
CO	1074
F_2	158
Cl_2	242
Br_2	193
I_2	151
HF	565
HCl	431
HBr	366
HI	299

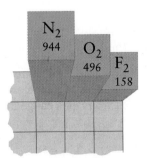

FIGURE 9.17

The bond enthalpies, in kilojoules per mole (kJ/mol), of diatomic nitrogen, oxygen, and fluorine molecules. Note how the bonds weaken as they change from a triple bond in N_2 to a single bond in F_2.

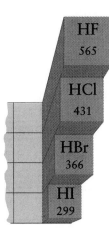

HF	565
HCl	431
HBr	366
HI	299

FIGURE 9.18

The bond enthalpies of the hydrogen halide molecules in kilojoules per mole (kJ/mol). Note how the bonds weaken as the halogen atom becomes larger.

the bond in F_2 is weaker than the bond in H_2, for the latter molecule has no lone pairs.

Trends in atomic radii also contribute to trends in bond strengths. If the nuclei of the bonded atoms cannot get very close to the electron pair lying between them, then the two atoms will be only weakly bonded together. For example, the bond enthalpies of the hydrogen halides decrease from HF to HI, as the radius of the halogen atom increases (Fig. 9.18).

The strength of a bond between two atoms is measured by the bond enthalpy. The bond enthalpy typically increases as the order of the bond increases, decreases as the number of lone pairs on neighboring atoms increases, and decreases as the atomic radius increases.

Assume that ΔH_B is the average bond enthalpy unless it is specified as the enthalpy of a specific bond.

9.7 Bond Strengths in Polyatomic Molecules

The first step in predicting the strength of a bond in a polyatomic molecule is to identify the atoms and the bond order. However, all the other atoms in the molecule exert a pull—through their electronegativities—on all the electrons in the molecule (Fig. 9.19). As a result, although the strength of the bond between a given pair of atoms is largely independent of the molecule, it does vary slightly from one compound to another. Because these variations in strength are not very great, the **average bond enthalpy,** which we also denote ΔH_B, is a guide to the strength of a bond in any molecule (Table 9.3).

As we can see from these average values, a triple bond between two atoms is always stronger than a double bond between the same two atoms. A double bond is always stronger than a single bond between the same two atoms. However, a double bond between two carbon atoms is not twice as strong as a single bond, and a triple bond is a lot less than three times as strong:

$$\Delta H_B(C\!-\!C) = 348 \text{ kJ/mol}$$

$$\Delta H_B(C\!=\!C) = 612 \text{ kJ/mol} \qquad 2 \times \Delta H_B(C\!-\!C) = 696 \text{ kJ/mol}$$

$$\Delta H_B(C\!\equiv\!C) = 837 \text{ kJ/mol} \qquad 3 \times \Delta H_B(C\!-\!C) = 1044 \text{ kJ/mol}$$

Electronegative atom

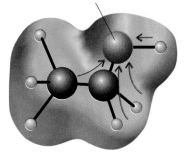

FIGURE 9.19

An electronegative atom can pull electrons toward itself from more distant parts of a molecule. Hence it can influence the strengths of bonds even between atoms to which it is not directly attached.

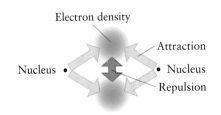

Electron density

Attraction

Nucleus • • Nucleus

Repulsion

FIGURE 9.20

The pairs of electrons in a multiple bond repel each other and can weaken the bond. As a result, a double bond between carbon atoms is not twice as strong as two single bonds would be.

(These are all average values.) The electron pairs in a double bond repel each other slightly, so each pair is not quite as effective at bonding as a pair of electrons in a single bond (Fig. 9.20). Stronger bonds result in lower molecular energy and greater molecular stability, because more energy is required to break strong bonds than weak bonds. Because it takes more energy to break two C—C bonds than one C=C bond, it can be energetically advantageous to replace multiple bonds by more single bonds.

As an example, consider the reaction

$$CH_2{=}CH_2(g) + HCl(g) \longrightarrow CH_3{-}CH_2Cl(g) \qquad \Delta H° = -72 \text{ kJ}$$

The negative value of $\Delta H°$ tells us that the reaction is exothermic and therefore that the energy of the products is lower than that of the reactants. The reaction releases energy because, although the H—Cl bond and the C=C double bond have to be broken before the C—C, C—H, and C—Cl bonds can form, a total of three bonds has replaced two bonds in the reactants (see Example 9.5).

The values in Table 9.3 also show how resonance lowers the energy of a molecule. For example, the strength of the carbon-carbon bonds in benzene is

Table 9.3 Average bond enthalpies, kJ/mol

Bond	Average bond enthalpy	Bond	Average bond enthalpy
C—H	412	C—I	238
C—C	348	N—H	388
C=C	612	N—N	163
C⋯C*	518	N=N	409
C≡C	837	N—O	210.
C—O	360.	N=O	630.
C=O	743	N—F	195
C—N	305	N—Cl	381
C—F	484	O—H	463
C—Cl	338	O—O	157
C—Br	276		

*In benzene.

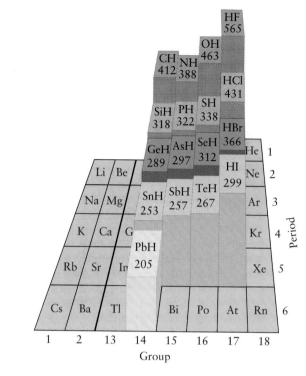

FIGURE 9.21

The bond enthalpies for bonds between hydrogen and the *p*-block elements. The bond strengths decrease from top to bottom of each group as the atoms increase in size.

intermediate between that of a single bond and a double bond. Benzene has a resonance structure, and because the strength of the resonance hybrid bond is closer to that of a double bond, a molecule with six resonance hybrid bonds has a lower energy and is more stable than one with three single and three double bonds.

Average bond enthalpies also help explain the relative stabilities of families of compounds. For example, the enthalpy of the X—H bond, where X is a Group 14 element, decreases down the group, from carbon (412 kJ/mol) to lead (205 kJ/mol), as shown in Fig. 9.21. This weakening of the bond matches the decrease in the stability of the hydrides down the group. Methane (CH_4) can be kept indefinitely in air at room temperature. Silane (SiH_4) bursts into flame on contact with air. Stannane (SnH_4) decomposes into tin and hydrogen. Plumbane (PbH_4) has never been prepared, except perhaps in trace amounts.

When accurate bond enthalpies are not available, enthalpies of reactions in which all reactants and products are gases can be estimated as described in Toolbox 9.2 by using average bond enthalpies compiled from measurements on a series of related compounds. However, reaction enthalpies calculated by this means are not as accurate as those calculated by using standard enthalpies of formation, because the average values may not match the actual values.

Average bond enthalpies can be used to estimate reaction enthalpies and to predict the stability of a molecule.

Toolbox 9.2 *How to use average bond enthalpies*

This Toolbox shows how to estimate a standard reaction enthalpy from tables of bond enthalpies.

Conceptual Basis

A bond absorbs energy when it is broken and releases energy when it is formed. Therefore, by finding the enthalpy required to break the reactant bonds and the enthalpy released when the product bonds are formed, we can calculate the net change in enthalpy (Fig. 9.22).

Procedure

Step 1. Decide which bonds are broken and which bonds are formed (all species must be in the gas phase). Then calculate the change in enthalpy when the bonds are broken in the reactants.

For diatomic molecules, use the data in Table 9.2 for the specific molecule. For other bonds, use the average bond enthalpies in Table 9.3.

Step 2. Calculate in the same way the bond enthalpies for the new product bonds.

Step 3. Calculate the bond *formation* enthalpies for the product bonds from the bond *dissociation* enthalpies obtained in step 2 by reversing the sign.

Step 4. Finally, add the enthalpy change required to break the reactant bonds (a positive value) to the enthalpy change that occurs when the product bonds form (a negative value).

If the reaction involves liquids, the overall enthalpy change must include the enthalpies of vaporization of liquid reactants and the enthalpies of condensation of liquid products.

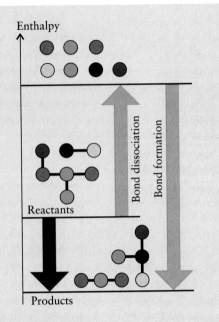

FIGURE 9.22

When we use bond enthalpies to estimate a reaction enthalpy, we are imagining that the reaction occurs in two steps: (1) bond dissociation, the breaking of reactant bonds to form atoms from molecules (which absorbs energy), and then (2) bond formation, the joining of free atoms into new molecules (which releases energy). The dark red arrow shows the resulting reaction enthalpy.

Example 9.5 *Using average bond enthalpies to estimate a reaction enthalpy*

Decide whether the reaction

$$CH_3CH_2I(g) + H_2O(g) \longrightarrow CH_3CH_2OH(g) + HI(g)$$

is exothermic or endothermic.

Strategy We calculate the enthalpy increase needed to break bonds in the reactants and the enthalpy decrease when the new product bonds form. Follow the procedure in Toolbox 9.2.

Solution **Step 1.** We need to break a C—I bond (238 kJ/mol) in CH_3CH_2I and an O—H bond (463 kJ/mol) in H_2O, so the total increase in enthalpy for the reaction of 1 mol CH_3CH_2I is

$$\Delta H^\circ = 238 + 463 \text{ kJ} = +701 \text{ kJ}$$

Step 2. To break 1 mol C—O bonds (360 kJ/mol) in CH_3CH_2OH and 1 mol H—I bonds (299 kJ/mol) results in

$$\Delta H° = 360 + 299 \text{ kJ} = +659 \text{ kJ}$$

Step 3. The enthalpy change when the product bonds are formed is the negative of the bond enthalpy, -659 kJ. **Step 4.** The overall enthalpy change is the sum of the enthalpy changes when the reactant bonds break and when the product bonds form:

$$\Delta H° = 701 + (-659) \text{ kJ} = +42 \text{ kJ}$$

Therefore, the reaction is endothermic, primarily because a relatively large amount of energy is needed to break the O—H bond in water.

Self-Test 9.6A Which of the following reactions is exothermic?
(a) $CCl_3CHCl_2(g) + HF(g) \rightarrow CCl_3CHClF(g) + HCl(g)$;
(b) $CCl_3CHCl_2(g) + HF(g) \rightarrow CCl_3CCl_2F(g) + H_2(g)$.

[*Answer:* (a)]

Self-Test 9.6B Estimate the standard enthalpy of the reaction
$CH_4(g) + 2 F_2(g) \rightarrow CH_2F_2(g) + 2 HF(g)$.

9.8 Bond Lengths

How big are molecules? One way to answer this question is to calculate the lengths of the bonds within a molecule. A **bond length** is the distance between the centers of two atoms joined by a chemical bond. Bond lengths affect the overall size of a molecule. The transmission of hereditary information in DNA, for instance, depends on bond lengths because the two strands of the double helix must fit together like pieces of a jigsaw puzzle (see Fig. 9.11). Bond lengths are also crucial to the action of enzymes, because only if a molecule is the right size—which includes having the right bond lengths—will it fit into the active site of the enzyme molecule (see Section 13.11). As we see from Table 9.4, the lengths of bonds between the elements of Period 2 typically lie in the range 100 to 150 pm. Bond lengths are determined experimentally by using spectroscopy and x-ray diffraction (see Investigating Matter 10.1).

Table 9.4 *Average and actual bond lengths*

Bond	Average bond length, pm	Molecule	Bond length, pm
C—H	109	H_2	74
C—C	154	N_2	110.
C=C	134	O_2	121
C⋯C*	139	F_2	142
C≡C	120.	Cl_2	199
C—O	143	Br_2	228
C=O	112	I_2	268
O—H	96		
N—H	101		

*In benzene.

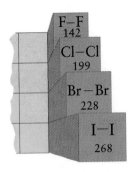

FIGURE 9.23

Bond lengths (in picometers) of the diatomic halogen molecules. Notice how the bond lengths increase from top to bottom of the group as the atomic radii increase.

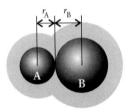

33   Covalent radii

Bonds between heavy atoms tend to be longer than those between light atoms because heavier atoms have larger radii than lighter ones (Fig. 9.23): fat atoms form long bonds. Multiple bonds are shorter than single bonds between the same pair of atoms because the additional bonding electrons pull the nuclei of neighboring atoms closer together: compare the lengths of the various carbon-carbon bonds in Table 9.4. We can also see the averaging effect of resonance. The length of the bond in benzene is intermediate between the lengths of the single and double bonds of a Kekulé structure (but closer to that of a double bond). For bonds between the same pair of atoms, *the shorter the bond, the stronger it is.* Thus, a $C=O$ double bond is both shorter and stronger than a $C-O$ single bond. Similarly, a $C\equiv C$ triple bond is both shorter and stronger than a $C=C$ double bond.

Each atom makes a characteristic contribution, called its *covalent radius*, to the length of a bond (Fig. 9.24). A bond length is approximately the sum of the covalent radii of the two atoms (33). The $O-H$ bond length in ethanol, for example, is the sum of the covalent radii of H and O, $37 + 74$ pm $= 111$ pm. We see from Fig. 9.24 that the covalent radius of an atom taking part in a multiple bond is smaller than that for an atom of the same element involved in a single bond.

Covalent single bond radii typically decrease from left to right across a period, just as atomic radii do (Section 7.15). The reason is the same: the increasing nuclear charge draws in the electrons and makes the atom more compact. Because the atom is smaller, it can get closer to any atom to which it bonds. Again like atomic radii, covalent radii increase down a group because, in successive periods, the valence electrons occupy shells that are more distant from the nucleus and better shielded by the greater number of electrons in the

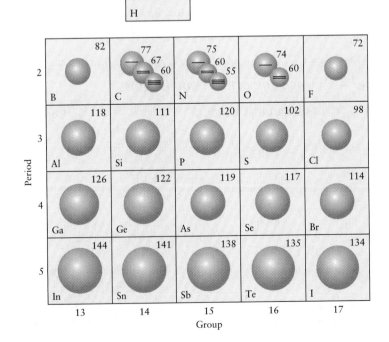

FIGURE 9.24

Covalent radii of hydrogen and the *p*-block elements (in picometers). Where more than one value is given, they refer to single, double, and triple bonds. Covalent radii tend to become smaller toward the top and right sides of the table. A bond length is approximately the sum of the covalent radii of the two atoms involved.

inner core. Such bulkier atoms cannot approach their neighbors very closely; hence they form long, weak bonds: fat atoms form weak bonds.

The covalent radius of an atom is the contribution it makes to the length of a covalent bond. Covalent radii are added together to estimate the lengths of bonds in molecules.

Example 9.6 *Estimating the bond lengths in a molecule*

Estimate all the bond lengths in a molecule of formic acid, HCOOH.

Strategy The first step is to write down the Lewis structure of the molecule and identify the single and multiple bonds. Then we express each bond length as a sum of the appropriate covalent radii taken from Fig. 9.24. Be careful to use the radius corresponding to the correct bond multiplicity (single, double, or triple).

Solution The Lewis structure of formic acid is shown in (**34**). Construct the bond lengths by taking the sums of the appropriate covalent radii given in Fig. 9.24 (**35a**). The resulting bond lengths are shown in (**35b**). Because covalent radii are based on the average values of bond lengths for a variety of compounds, the bond lengths they predict are only estimates.

34 Formic acid, HCOOH

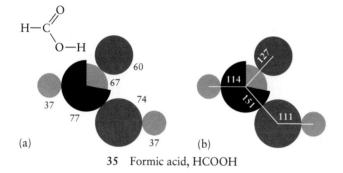

35 Formic acid, HCOOH

Self-Test 9.7A Estimate the bond lengths in the hydrogen cyanide molecule, HCN.

[**Answer:** C—H, 114 pm; C≡N, 115 pm]

Self-Test 9.7B Estimate the bond lengths in the HOCl molecule (the atoms are attached in the order given in the formula).

VALENCE-BOND THEORY

The Lewis model of the chemical bond assumes that each bonding electron pair is located between the two bonded atoms. However, we know from quantum mechanics that, because an electron has wavelike properties (Section 7.4), it cannot have a precise position. The best we can do, according to quantum mechanics, is to note the form of the orbital it occupies and express the probability of finding it in a region of space. Two theories of bonding have arisen from quantum mechanics. In **molecular orbital theory** (MO theory), which is discussed in Sections 9.14–9.17, electrons are described as occupying orbitals that are like atomic orbitals, but spread over entire molecules. An earlier theory, called **valence-bond theory** (VB theory), describes bond formation in terms of the pairing of electrons in atomic orbitals on neighboring atoms. Valence-bond

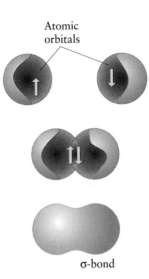

Atomic orbitals

FIGURE 9.25

When the electrons (depicted ↑ and ↓) in two hydrogen 1s-orbitals pair and the s-orbitals overlap, they form a σ-bond, depicted here by the boundary surface of the electron cloud. The cloud forms a cylinder around the internuclear axis and spreads over both nuclei. That is, it has cylindrical symmetry.

σ-bond

The Greek letter sigma, σ, is the equivalent of our letter s. It reminds us that, looking along the internuclear axis, the electron distribution resembles that of an s-orbital. All σ-bonds will be colored blue.

theory has contributed to many terms widely used in chemistry and we discuss it first.

9.9 Sigma-Bonds and Pi-Bonds

We begin with the simplest molecule of all, H_2. A single hydrogen atom consists of an electron in a spherical 1s-orbital that surrounds the nucleus. According to valence-bond theory, when two H atoms come together to form H_2, their 1s-electrons pair (denoted ↑↓, as in the discussion of atomic structure, Section 7.8), and the atomic orbitals they occupy merge together (Fig. 9.25). The resulting sausage-shaped distribution of electrons between the two nuclei is called a **σ-bond.** A hydrogen molecule is held together by one such σ-bond. The merging together of the 1s-orbitals is called the **overlap** of atomic orbitals.

Much the same kind of σ-bond formation ("σ-bonding") occurs in the hydrogen halides. The electron configurations of hydrogen and fluorine are shown in (**36**): the notation is the same as that used in Section 7.11. The unpaired electron on the fluorine atom occupies a $2p_z$-orbital, and the unpaired electron on the hydrogen atom occupies a 1s-orbital. These two electrons are the ones that pair to form a bond. They pair as the orbitals they occupy overlap and merge into a cloud that spreads over both atoms (Fig. 9.26).

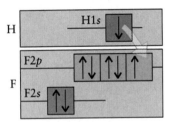

36 Hydrogen fluoride, HF

We identify the bonding orbital on the F atom as a p_z-orbital only by convention. We usually define the z-axis as the one lying along the internuclear axis.

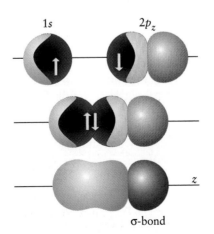

σ-bond

FIGURE 9.26

A σ-bond can also be formed when electrons in 1s- and $2p_z$-orbitals pair (z is the direction along the internuclear axis). The two electrons in the bond are concentrated within the region of space surrounded by the boundary surface.

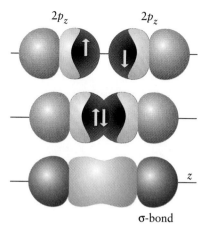

$2p_z$ $2p_z$

σ-bond

FIGURE 9.27

A σ-bond is formed by the pairing of electron spins in two $2p_z$-orbitals on neighboring atoms. At this stage, we are ignoring the effect of the $2p_x$- (and $2p_y$-) orbitals that may also contain unpaired electrons. The electron pair is concentrated within the boundary surface shown in the bottom diagram.

Although the resulting bond has a more complicated shape than the σ-bond in H_2 when viewed from the side, it looks much the same when viewed along the internuclear (z) axis; hence it is also designated a σ-bond.

All single covalent bonds consist of a σ-bond in which two paired electrons have a high probability of being found between the two bonded atoms. A σ-bond can be formed by the pairing of electrons in any two orbitals that overlap end-to-end, such as s-orbitals (as in H_2), an s-orbital and a p-orbital (as in the hydrogen halides), or two p-orbitals (as in a diatomic halogen molecule).

We meet a different type of bond when we consider the structure of a nitrogen molecule, N_2. Suppose we follow the same procedure as before. First, we write the atomic configurations, as shown in (37). There is an electron in each of the three 2p-orbitals on each atom. However, when we try to pair them and form three bonds, only one of the three orbitals can overlap end-to-end to form a σ-bond (Fig. 9.27). The remaining two 2p-orbitals on each atom ($2p_x$ and $2p_y$) are perpendicular to the internuclear axis, and each one contains an unpaired electron (Fig. 9.28, top). When the electrons in

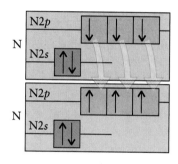

37   Nitrogen, N_2

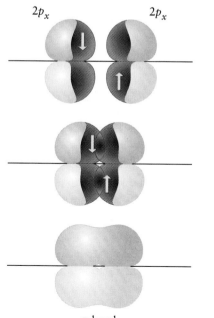

$2p_x$ $2p_x$

π-bond

FIGURE 9.28

A π-bond is formed when electrons in two 2p-orbitals (top) pair and overlap side by side. The middle diagram shows the overlap of the orbitals, and the bottom diagram shows the corresponding boundary surface. Even though the bond has a complicated shape, with two lobes, it is occupied by one pair of electrons and counts as one bond.

FIGURE 9.29

The bonding pattern in a nitrogen molecule, N_2. (a) The two atoms are bonded together by one σ-bond (blue) and two perpendicular π-bonds (yellow). (b) When the three orbitals are put together, the two π-bonds merge to form a long donut-shaped cloud surrounding the σ-bond cloud; the overall structure resembles a cylindrical hot dog.

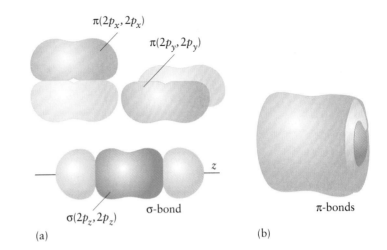

$\pi(2p_x, 2p_x)$

$\pi(2p_y, 2p_y)$

$\sigma(2p_z, 2p_z)$

z

σ-bond

π-bonds

(a)

(b)

The Greek letter pi, π, is the equivalent of our letter *p*. When we imagine looking along the internuclear axis, a π-bond resembles a pair of electrons in a *p*-orbital. All π-bonds will be colored yellow.

two of these orbitals pair, their orbitals overlap in a side-by-side arrangement and form a **π-bond,** a bond in which the two electrons lie in two lobes, one on each side of the internuclear axis (Fig. 9.28, bottom). Although a π-bond has electron density on each side of the internuclear axis, it is only *one* bond with the electron cloud in the form of *two* lobes, just as a *p*-orbital is one orbital with two lobes.

Now we can describe bonding in the nitrogen molecule. A σ-bond (of two electrons) is formed from the end-to-end overlap of two $2p_z$-orbitals directed along the internuclear axis. Two π-bonds (of two electrons each) are formed from the side-by-side overlap of the remaining $2p$-orbitals on the two atoms. There are three bonds in all: one σ-bond and two π-bonds (Fig. 9.29), in agreement with the Lewis structure.

We can generalize from these examples to the description of a multiply bonded species according to valence-bond theory:

A *single bond* is a σ-bond.

A *double bond* is a σ-bond plus one π-bond.

A *triple bond* is a σ-bond plus two π-bonds.

In valence-bond theory, a bond forms when unpaired electrons in valence-shell atomic orbitals on two atoms pair. The atomic orbitals they occupy overlap end to end to form σ-bonds or side by side to form π-bonds.

Self-Test 9.8A How many σ-bonds and how many π-bonds are there in (a) CO_2? (b) CO?

[*Answer:* (a) Two σ, two π; (b) one σ, two π]

Self-Test 9.8B How many σ-bonds and how many π-bonds are there in (a) NH_3? (b) CH_2O?

9.10 Promotion and Hybridization

Valence-bond theory as described so far cannot account for bonding in polyatomic molecules like methane, CH_4. For example, if we tried to apply the theory to methane, we would write down the electron configuration of the carbon atom and look for half-filled orbitals that we can use for bond formation (**38**).

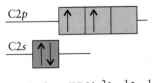

C2p

C2s

38 Carbon, $[He]2s^2 2p_x^1 2p_y^1$

However, two valence electrons are already paired in the 2s-orbital, so there are only two half-filled 2p-orbitals on the carbon atom. It looks as though carbon can form only two bonds and should have a valence of 2. However, we know that carbon almost always has a valence of 4, so we need to revise our model of bonding.

Notice that carbon has an empty 2p-orbital. We can get more half-filled orbitals in carbon if we invest enough energy to **promote** an electron, that is, excite it to a higher energy orbital. When we promote one of the paired 2s-electrons into the empty 2p-orbital, we get the configuration shown in (39). Without promotion, a carbon atom has only two unpaired electrons and so can form only two bonds; after promotion, it has four unpaired electrons and can form four bonds. Each bond releases energy as it forms. Therefore, despite the energy cost of promoting the electron, the overall energy of the CH_4 molecule is lower than it would be if carbon formed only two C—H bonds. Promotion of an electron is possible if the overall change, taking into account the greater number of bonds that can thereby be formed, is toward lower energy.

Despite being able to explain carbon's valence, we still cannot explain methane's tetrahedral shape and its four identical bonds! It looks as though CH_4 should have two different types of bonds: one from the overlap of a hydrogen 1s-orbital and a carbon 2s-orbital, and three more bonds from the overlap of hydrogen 1s-orbitals with each of the three carbon 2p-orbitals. The overlap with the 2p-orbitals should result in three σ-bonds at 90° to one another.

To complete the valence-bond description of the structure of methane, we can introduce a new concept. First, recall from Section 7.5 that s- and p-orbitals are waves centered on the nucleus of an atom. These four wavelike orbitals produce four new patterns where they intersect. These new patterns are called **hybrid orbitals**. The four hybrid orbitals are identical to one another and point toward the corners of a tetrahedron (Fig. 9.30). Each orbital has a node close to the nucleus and, on the other side of the node, a small "tail," in which the s- and p-orbitals do not completely cancel. The four hybrid orbitals are called sp^3 **hybrids** because they are formed from one s-orbital and three p-orbitals. The hybrids are colored green to remind us that they are a blend of (blue) s-orbitals and (yellow) p-orbitals.

Using a box diagram, we represent sp^3 hybridization as the formation of four orbitals of the same energy lying between the energies of the s- and p-orbitals from which they are constructed (40).

We can now account for the bonding in a methane molecule. An unpaired electron occupies each of carbon's sp^3 hybrid orbitals. Each of these four electrons can pair with an electron in a hydrogen 1s-orbital. Their overlapping orbitals form σ-bonds (Fig. 9.31). Because the four hybrid orbitals point toward

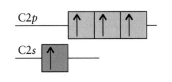

39 Carbon, $[He]2s^1 2p_x^1 2p_y^1 2p_z^1$

A hybrid orbital is a blend of atomic orbitals on a single atom in a molecule. Do not confuse this use of the term *hybrid* with resonance hybrid, which refers to a blend of Lewis structures of entire molecules.

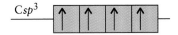

40 sp^3 hybridized carbon

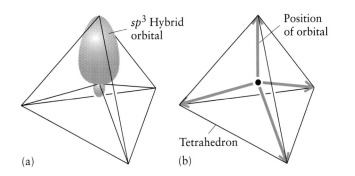

(a)

(b)

FIGURE 9.30

The hybrid orbitals of a carbon atom in methane. (a) One s-orbital and three p-orbitals blend into four sp^3 hybrid orbitals that each point toward one apex of a tetrahedron. (b) The directions of the four orbitals.

FIGURE 9.31

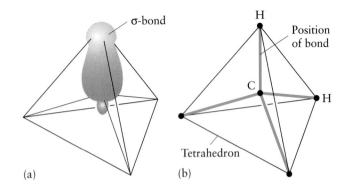

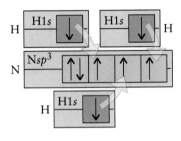

41   Ammonia, NH_3

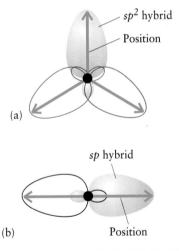

FIGURE 9.32

Two common hybridization schemes. (a) An s- and two p-orbitals can blend together to give three sp^2 hybrid orbitals that point toward the corners of an equilateral triangle. (b) An s-orbital and a p-orbital hybridize into two sp hybrid orbitals that point in opposite directions. Only one of the orbitals is shown in each case.

the corners of a tetrahedron, so do the σ-bonds. All four bonds are identical because they are formed from the same blend of atomic orbitals. The valence-bond description is now consistent with experiment.

Hybridization can also be used to describe bonding in molecules with lone pairs. For example, the four electron pairs in NH_3 take up a tetrahedral arrangement (according to the VSEPR model), so we describe the nitrogen atom in terms of four sp^3 hybrid orbitals. However, because nitrogen has five valence electrons, one of these hybrid orbitals is already doubly occupied (**41**). The two electrons in this orbital are the lone pair in the NH_3 molecule. The lone pair on the nitrogen atom occupies a **nonbonding orbital,** an orbital containing electrons that do not participate in bonding. The 1s-electrons of the three hydrogen atoms pair with the three unpaired electrons in the remaining sp^3 hybrid orbitals. This pairing and overlap results in the formation of three σ-bonds.

It is important to realize that a molecule is not tetrahedral *because* it has sp^3 hybrid orbitals. Hybridization is only a theoretical way of describing the bonds in a given molecular structure. *Hybridization is an interpretation of molecular shape; shape is not a consequence of hybridization.*

> *Hybrid orbitals are constructed on an atom to reproduce the electron arrangement characteristic of the experimentally determined shape of a molecule.*

9.11 Other Hybridization Schemes

The hybridization scheme we use to describe a molecule depends on the number of bonds and lone pairs on the central atom, and therefore on the electron arrangement that determines the molecular shape. For example, methane, CH_4, has a tetrahedral electron arrangement, but formaldehyde, CH_2O, has a trigonal planar electron arrangement about the carbon atom. To explain this structure, we mix one s-orbital with two p-orbitals and obtain three sp^2 **hybrid orbitals.** These hybrid orbitals lie in the same plane and point toward the corners of an equilateral triangle (Fig. 9.32a). One of the sp^2 hybrids forms a σ-bond to the oxygen atom and the other two form σ-bonds to the hydrogen atoms (Fig. 9.33). There is one unhybridized p-orbital on the carbon atom, perpendicular to the plane of the hybrid orbitals. This orbital forms a π-bond by overlapping side-by-side with an orbital on the oxygen atom (Fig. 9.34). For consistency, we can also regard the oxygen atom (and any terminal atom) as hybridized. For instance, when an oxygen atom forms a double bond to a neighbor and has two

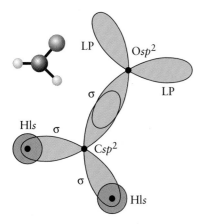

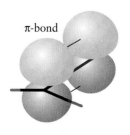

FIGURE 9.33

The atomic orbitals that overlap to form the three σ-bonds in formaldehyde, CH_2O. LP, lone pair.

FIGURE 9.34

Unhybridized p-orbitals on the C and O atoms overlap to form the π-bond in formaldehyde.

lone pairs (as in formaldehyde), we suppose that it is sp^2 hybridized, as depicted in the illustration, with the lone pairs in two of the sp^2 hybrid orbitals.

Carbon dioxide, CO_2, is a linear molecule with two bonds at 180° to each other and no lone pairs on the carbon atom. We describe a linear structure as consisting of two **sp hybrid orbitals,** the result of mixing one s-orbital and one p-orbital (Fig. 9.32b). These hybrid orbitals form σ-bonds to the two oxygen atoms (Fig. 9.35). The carbon atom also has two unhybridized p-orbitals that are perpendicular to each other and to the hybrid orbitals. Each p-orbital overlaps side by side with a p-orbital on an oxygen atom to form a π-bond (Fig. 9.36).

The three types of hybridization found in the Period 2 elements are

Tetrahedral hybridization: four sp^3 hybridized orbitals, angle 109.5°

Trigonal planar hybridization: three sp^2 hybridized orbitals, angle 120°

Linear hybridization: two sp hybridized orbitals, angle 180°

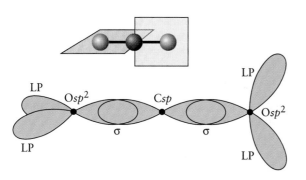

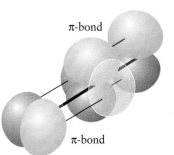

FIGURE 9.35

The atomic orbitals that overlap to form the two σ-bonds in carbon dioxide. The lone pairs are also shown. Note that the planes of σ-bonds around each C atom are perpendicular to each other. LP, lone pair.

FIGURE 9.36

Each of the two unhybridized p-orbitals on the C atom in CO_2 overlaps with an unhybridized p-orbital on an O atom to form a π-bond between each pair of atoms. The two π-bonds are perpendicular to each other.

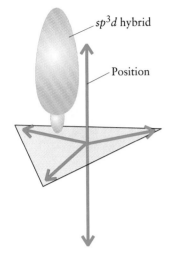

FIGURE 9.37

One of the five sp^3d hybrid orbitals, and their five directions, that may be formed when d-orbitals are available. They form a trigonal bipyramidal arrangement of electron pairs. Each arrow represents the location of a bond or electron pair.

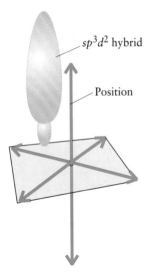

FIGURE 9.38

One of the six sp^3d^2 hybrid orbitals, and their six directions, that may be formed when d-orbitals are available. They form an octahedral arrangement of electron pairs. Each arrow represents the location of a bond or electron pair.

Table 9.5 *Hybridization and molecular shape* *

Electron arrangement	Number of atomic orbitals	Hybridization of the central atom	Number of hybrid orbitals
linear	2	sp	2
trigonal planar	3	sp^2	3
tetrahedral	4	sp^3	4
trigonal bipyramidal	5	sp^3d	5
octahedral	6	sp^3d^2	6

*Other combinations of s-, p-, and d-orbitals can give rise to the same or different shapes, but these combinations are the most common.

No matter how many atomic orbitals we mix together, the number of hybrid orbitals we obtain is always the same as the number of atomic orbitals we started with. The notation sp^3 tells us that each hybrid orbital is built from one s-orbital and three p-orbitals, so we expect four equivalent orbitals of this type. The notation sp^2 tells us that each hybrid orbital is built from one s-orbital and two p-orbitals, giving three of these orbitals and leaving one unhybridized p-orbital. An atom with sp hybridization has two hybrid orbitals, each built from one s- and one p-orbital, leaving two unhybridized p-orbitals. Other characteristic hybridization schemes are listed in Table 9.5. Notice that hybridization is used only to explain the formation of σ-bonds.

In valence-bond theory, bonds are built by pairing electrons in orbitals on neighboring atoms. The number of hybrid orbitals is the same as the number of atomic orbitals hybridized.

Self-Test 9.9A Suggest a structure in terms of hybrid orbitals for BF_3.

[**Answer:** Three σ-bonds are formed from boron sp^2 hybrids and fluorine $2p_z$-orbitals in a trigonal planar arrangement.]

Self-Test 9.9B Suggest a structure in terms of hybrid orbitals for dichloromethane, CH_2Cl_2.

9.12 Hybrids Including d-Orbitals

When the central atom of a molecule is an element in Period 3 or later, it can use its d-orbitals to expand its valence shell and accommodate more than four electron pairs (Section 8.11). When five pairs of valence electrons are present on a central atom in a trigonal bipyramidal arrangement, we need five valence-shell orbitals. In this case, we think of the atom as using one d-orbital in addition to the four s- and p-orbitals of the valence shell (see Table 9.5). The resulting orbitals are called **sp^3d hybrid orbitals** (Fig. 9.37). We use this hybridization when we want to describe a trigonal-bipyramidal molecule, such as PCl_5.

To accommodate six electron pairs around an atom pointing toward the corners of a regular octahedron (Fig. 9.38), we need six orbitals, so we use two d-orbitals in addition to the *s*- and *p*-orbitals (see Table 9.5). We use sp^3d^2 hybridization when we want to describe a molecule based on an octahedral arrangement of electron pairs, as in SF_6 or XeF_4, which has two lone pairs on the xenon atom.

Valence-shell expansion corresponds to the involvement of d-orbitals, which mix with s- and p-orbitals to give hybridization schemes that are consistent with both the VSEPR model and experiment.

Toolbox 9.3 *How to identify the hybridization scheme of a molecule*

This Toolbox shows how to identify a hybridization scheme from the shape of a molecule.

Conceptual Basis

Hybridization is a means of using atomic orbitals to describe the geometry of the electron arrangement about a central atom. We choose the hybridization scheme to match the electron arrangement predicted by the VSEPR model.

Procedure

Carry out the following steps to determine the hybridization scheme of a molecule, using *N* atomic orbitals to form *N* hybrid orbitals (Fig. 9.39).

Step 1. Draw the Lewis structure and determine the electron arrangement about the central atom. The number of σ-bonds and lone pairs required for the electron arrangement is the number of orbitals used by the central atom.

Step 2. Construct the hybrid orbitals from atomic orbitals, using the same number of atomic orbitals as hybrid orbitals required. Start with the *s*-orbital, then add *p*- and *d*-orbitals as needed to create the patterns listed in Table 9.5.

Step 3. Use any remaining *p*-orbitals to form π-bonds with *p*-orbitals on adjacent atoms.

Step 4. Check your solution with Table 9.5.

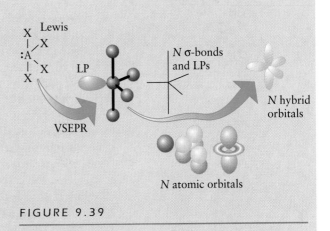

FIGURE 9.39

How the electron arrangement of a molecule is matched to a hybridization scheme.

Example 9.7 *Identifying hybridization schemes*

Describe the electron arrangement, the molecular shape, and the hybridization of the central atom in water, H_2O.

Strategy Follow the procedure in Toolbox 9.2, being alert to the presence of lone pairs.

Solution **Step 1.** The Lewis structure of water (**42**) shows that it has two bonds and two lone pairs. Therefore, H_2O has a tetrahedral electron arrangement and an angular shape (recall Example 9.2). **Step 2.** We can construct four equivalent hybrid

H
|
:O—H
¨

42 Water, H_2O

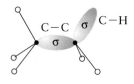

FIGURE 9.40

The valence-bond description of the bonding in an ethane molecule, C_2H_6. The boundary surfaces of only two of the bonds are shown. Each pair of neighboring atoms is linked by a σ-bond formed by the pairing of electrons in either H1s-orbitals or Csp³ hybrid orbitals. All the bond angles are close to 109.5° (the tetrahedral angle).

orbitals by mixing an s-orbital with three p-orbitals, to give sp^3 hybridization.

Step 3. There are no unhybridized p-orbitals remaining after sp^3 hybridization.

Step 4. The result is four sp^3 orbitals, as listed in Table 9.5.

Self-Test 9.10A Describe the electron arrangement, the molecular shape, and the hybridization of the central atom in a phosphorus pentachloride molecule, PCl_5.

[*Answer:* Trigonal bipyramidal; trigonal bipyramidal; sp^3d hybridization]

Self-Test 9.10B Describe the electron arrangement, the molecular shape, and the hybridization of the central atom in sulfur hexafluoride, SF_6.

9.13 Multiple Carbon-Carbon Bonds

We can use hybridization to describe bonds in molecules that have more than one central atom if we concentrate on the atoms one at a time. Let's use this technique to compare the hybridization of three molecules containing two carbon atoms, but with different bond orders. The Lewis structure of ethane is shown in (**43**). According to the VSEPR model, the four electron pairs around each carbon atom take up a tetrahedral arrangement. This arrangement suggests sp^3 hybridization of the carbon atoms. Each C atom has one unpaired electron in each of its four sp^3 hybrid orbitals. As a result, the atom can form four σ-bonds pointing toward the corners of a regular tetrahedron. The pattern of bonds is illustrated in Fig. 9.40.

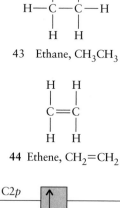

To describe carbon-carbon double bonds, we use the pattern provided by ethene, $CH_2=CH_2$ (**44**). We know from experimental data that all six atoms in ethene lie in the same plane, with H—C—H and C—C—H bond angles of 120°. This angle suggests that we should use sp^2 hybridization. There is one electron in each of the three hybrid orbitals on each C atom. The fourth valence electron of each atom must therefore occupy the unhybridized 2p-orbital to give the arrangement in (**45**). The two carbon atoms form a σ-bond by overlap of an sp^2 hybrid orbital on each atom. The H atoms form σ-bonds with the remaining lobes of the sp^2 hybrids (Fig. 9.41). The unhybridized 2p-orbitals lie perpendicular to the plane formed by the hybrids and therefore to the σ-bonds (Fig. 9.42). The electrons in the two unhybridized 2p-orbitals are free to pair and form a π-bond by side-to-side overlap (Fig. 9.43).

In Section 8.8, we saw that the structure of the benzene molecule is described in the Lewis model as a resonance of two Kekulé structures. Resonance is also used in the valence-bond description of the molecule. In the valence-bond theory, we identify hybrid orbitals that match the 120° bond angles of the hexagonal ring. Therefore, we take each carbon atom to be sp^2 hybridized, as in ethene (Fig. 9.44). There is one electron in each hybrid orbital and one electron in an unhybridized 2p-orbital perpendicular to the plane of the hybrids. We can reproduce each Kekulé structure by representing the π component of each double bond by the overlap of the appropriate pair of

43 Ethane, CH_3CH_3

44 Ethene, $CH_2=CH_2$

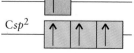

45 sp^2 hybridized carbon

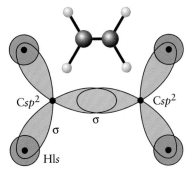

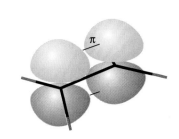

FIGURE 9.41

The atomic orbitals that overlap to form the five σ-bonds in ethene (ethylene).

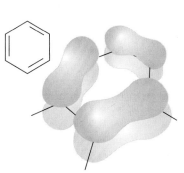

FIGURE 9.42

The formation of a π-bond from the side-by-side overlap of the unhybridized p-orbitals on the C atoms in ethene.

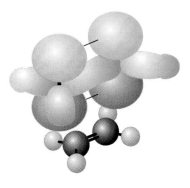

FIGURE 9.43

The bonding pattern in ethene (ethylene), showing the framework of σ-bonds and the single π-bond (represented by two lines above and below the σ-bond) formed by side-to-side overlap of unhybridized C2p-orbitals. The double bond is resistant to twisting because twisting would reduce the overlap between the two C2p-orbitals and weaken the π-bond.

unhybridized 2p-orbitals (Fig. 9.45). Then we allow resonance between the two Kekulé structures, just as in the Lewis description. As a result of this resonance, the π-bond character is spread around the ring as a cloud of electron density above and below the ring (Fig. 9.46).

The Lewis structure of the linear molecule ethyne (acetylene) is H—C≡C—H. To describe the bonding in a linear molecule with no lone pairs on the central atom, we use *sp* hybridization, because that produces two equivalent orbitals at 180° from each other. Each C atom has one electron in each of its two *sp* hybrid orbitals and one electron in each of its two perpendicular unhybridized 2p-orbitals (**46**). The electrons in an *sp* hybrid orbital on each C atom

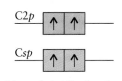

46 *sp* hybridized carbon

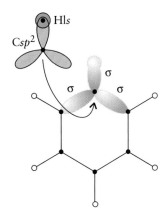

FIGURE 9.44

The framework of σ-bonds in benzene: each carbon atom is *sp²* hybridized, with bond angles of 120° in the hexagonal molecule. Only bonding around one carbon atom is shown explicitly; all the others are the same.

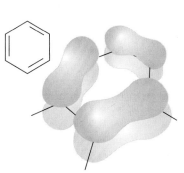

FIGURE 9.45

The unhybridized p-orbital on each C atom in benzene can form a π-bond with either of its immediate neighbors. Two arrangements are possible, each one corresponding to one Kekulé structure. One Kekulé structure and the corresponding π-bonds are shown here.

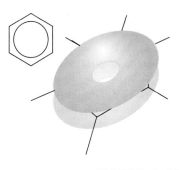

FIGURE 9.46

As a result of resonance between two structures like the one shown in the preceding illustration (corresponding to resonance of the two Kekulé structures), the π-electrons form double donut-shaped clouds, one above and one below the plane of the ring.

FIGURE 9.47

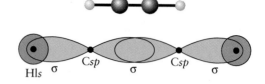

The atomic orbitals that overlap to form the three σ-bonds in ethyne (acetylene).

pair and form a carbon-carbon σ-bond. The electrons in the remaining *sp* hybrid orbitals pair with the hydrogen 1*s*-electrons to form two C—H σ-bonds (Fig. 9.47). The electrons in the perpendicular 2*p*-orbitals pair with a side-by-side overlap, forming two π-bonds at 90° to each other. The resulting bonding pattern is shown in Fig. 9.48. Notice the merging of the two π-clouds, as in N_2, to give a cylindrical distribution of electrons.

Example 9.8 Accounting for the structure of a multiply bonded molecule

Describe the structure of a formic acid molecule, HCOOH, in terms of hybrid orbitals, bond angles, and σ- and π-bonds.

Strategy First, draw the Lewis structure of the molecule and use the VSEPR model to determine the electron arrangement around each atom that is bonded to more than one other atom. Next, identify the hybridization of the atom that reproduces that arrangement as outlined in Toolbox 9.3. Then describe the bonding in the molecule with a σ-bond between each pair of atoms and a π-bond for each additional bond in a multiple bond. Use the hybrid atomic orbitals to form the σ-bonds and the unhybridized *p*-orbitals to form the π-bonds.

Solution The Lewis structure of formic acid is shown in (**47**). According to the VSEPR model, the three electron pairs in σ-bonds around the carbon atom take up a trigonal planar arrangement at 120°, which suggests sp^2 hybridization. The C atom has one unpaired electron in each of its three sp^2 hybrid orbitals and, as a result, can form three σ-bonds pointing toward the corners of an equilateral triangle. Carbon's fourth valence electron is in a *p*-orbital perpendicular to the triangle.

The O atom forming a single bond to the C atom and a single bond to an H atom has two bonds and two lone pairs of electrons, a total of four regions of high electron concentration. In the VSEPR model, these electron pairs lie in a tetrahedral arrangement. We therefore expect a bond angle of about 109.5° and sp^3

47 Formic acid, HCOOH

FIGURE 9.48

The pattern of bonding in ethyne (acetylene). (a) The carbon atoms are *sp* hybridized, and the two remaining *p*-orbitals on each ring form two perpendicular π-bonds. (b) The resulting pattern is very similar to that for nitrogen (see Fig. 9.29), but two C—H groups replace the N atoms.

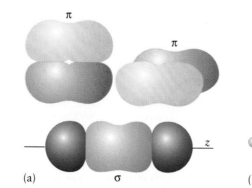

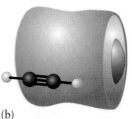

(a)

(b)

hybridization. The terminal O atom has two lone pairs and a double bond to the C atom, formed by the side-by-side overlap of *p*-orbitals on the two atoms.

On the basis of this analysis, we can describe the structure of formic acid as follows: the carbon atom is sp^2 hybridized, and the oxygen atom in the OH group is sp^3 hybridized. There are four σ-bonds and one π-bond in the molecule. All the atoms except the H of the OH group are in the same plane. The pattern of σ-bonds is illustrated in Fig. 9.49.

Self-Test 9.11A Describe the structure of the hydrogen cyanide molecule, HCN, in terms of hybrid orbitals, bond angles, and σ- and π-bonds.

[*Answer:* Linear; bond angle 180°; C is *sp* hybridized; it forms one σ-bond to the H atom and one to the N atom, and two π-bonds to the N atom.]

Self-Test 9.11B Describe the structure of the propene molecule, $CH_3-CH=CH_2$, in terms of hybrid orbitals, bond angles, and σ- and π-bonds.

Carbon-carbon π-bonds are about 84 kJ/mol weaker than carbon-carbon σ-bonds. Valence-bond theory provides an explanation for the relative weakness of carbon-carbon multiple bonds. Whereas a single bond is a σ-bond, the additional π-bonds are weaker because the side-by-side overlap puts their electrons in less favorable locations for holding the atoms together.

Although a carbon-carbon π-bond is weaker than a carbon-carbon σ-bond, it plays a highly important role in determining the shape of a molecule. Two parts of a molecule joined only by a σ-bond can rotate relative to each other. As a result, a singly bonded chain of carbon atoms in a hydrocarbon such as octane is flexible: the chain can roll up into a ball (**48a**) and unroll into a rod (**48b**). A π-bond, however, prevents one part of a molecule from rotating relative to another part. The double bond of ethene, for example, holds the entire molecule flat, as shown in Fig. 9.42. This illustration shows that the two $2p$-orbitals overlap best if the two CH_2 groups lie in the same plane. The π-bond could not form if the CH_2 groups were twisted about the C—C bond.

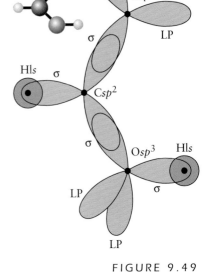

FIGURE 9.49

The pattern of σ-bonds in formic acid (methanoic acid). LP, lone pair.

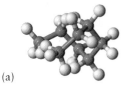

(a)

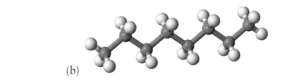

(b)

48   Octane, C_8H_{18}

Double bonds help you read these words. Vision depends on the shape of the molecule *retinal* in the retina of the eye. Retinal is held rigid by its double bonds (**49**). When light enters the eye, it excites an electron out of the π-bond marked by the arrow. The double bond is now weaker, and the molecule is free to rotate around the remaining σ-bond. When the excited electron falls back, the molecule is trapped in its new shape (**50**). This change of shape triggers a signal along the optic nerve and is interpreted by the brain as the sensation of vision.

Structures (49) and (50) are "line structures." Each straight line represents a bond between two C atoms. See Section 11.1.

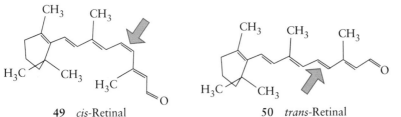

49   *cis*-Retinal

50   *trans*-Retinal

Multiple bonds are formed when an atom forms a σ-bond by using a hybrid orbital and one or more π-bonds by using unhybridized p-orbitals. The side-by-side overlap that forms a π-bond makes a molecule resistant to twisting.

MOLECULAR ORBITAL THEORY

The theory of the chemical bond proposed by G. N. Lewis was brilliant; nevertheless, it was little more than inspired guesswork. Lewis had no way of knowing why the electron pair, the essential feature of his approach, was so important. Another problem was that the Lewis approach failed for some molecules. Valence-bond theory was a major improvement because it showed how electron pairing leads to bond formation. It could even cope with the failures of the Lewis approach, but only in a very cumbersome way. Molecular orbital theory was introduced at about the same time as valence-bond theory and has proved to be the most successful theory of the chemical bond: it overcomes all the deficiencies of Lewis's theory and is easier to use in calculations than valence-bond theory.

9.14 The Limitations of Lewis's Theory

The Lewis approach fails when we try to describe compounds such as diborane (B_2H_6, **51**), a colorless gas that bursts into flame on contact with air. The problem is that diborane has only 12 valence electrons (three from each B atom, one from each H atom); but in the Lewis approach, it needs at least 7 bonds, and therefore 14 electrons, to bind the 8 atoms together! Diborane is an example of an **electron-deficient compound,** a compound with too few valence electrons for it to be assigned a valid Lewis structure.

Another failure is the Lewis description of O_2 as $\ddot{O}{=}\ddot{O}$. Oxygen is a **paramagnetic** substance, which means that it is attracted to a magnetic field (Fig. 9.50). However, paramagnetism is a property of *unpaired* electrons, which act as tiny magnets. The paramagnetism of O_2 contradicts both the Lewis structure and the valence-bond description of the molecule, which require all the electrons to be paired. Most substances contain fully paired electrons and are repelled by a magnetic field: such substances are called **diamagnetic.** Lewis's theory easily accounts for the existence of a diamagnetic substance, but it is completely incapable of accounting for paramagnetism in a molecule like oxygen.

The introduction of molecular orbital theory in the late 1920s overcame these difficulties. In molecular orbital theory, valence electrons occupy precisely

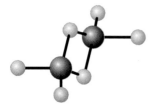

51   Diborane, B_2H_6

Paramagnetism is described in more detail in Section 21.1.

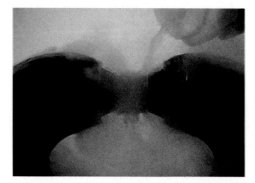

FIGURE 9.50

The paramagnetic properties of oxygen are evident when liquid oxygen is poured between the poles of a magnet. The liquid sticks to the magnet instead of flowing past it.

defined orbitals, called **molecular orbitals,** that spread throughout the entire molecule. As we shall see, the theory accommodates electron-deficient compounds just as naturally as it deals with methane and water. It also predicts that oxygen is paramagnetic.

Molecular orbital theory explains why electron *pairs* are central to the formation of chemical bonds. We saw in Section 7.11 that the Pauli exclusion principle restricts to two the number of electrons that can occupy an atomic orbital, and that the two electrons must have paired spins. Exactly the same is true of electrons in molecular orbitals. *A covalent bond consists of two paired electrons, because that is the greatest number that can occupy a molecular orbital.* Lewis had no idea why an electron pair was so important. Today, we see it is a direct consequence of the Pauli exclusion principle.

Lewis theory cannot account for electron-deficient compounds or the paramagnetism of oxygen. In molecular orbital theory, electrons occupy orbitals that spread throughout the molecule. The Pauli exclusion principle allows no more than two electrons to occupy each molecular orbital.

9.15 Molecular Orbitals in Hydrogen

In molecular orbital theory, we initially disregard the presence of the valence electrons in the atoms and consider the overlap of *empty* valence orbitals. When two atoms approach each other, two *s*-orbitals, or two *p*-orbitals lying head to head, can overlap to form a **σ-orbital,** a sausage-shaped orbital between two atoms. When two electrons occupy this orbital, they form a σ-bond that is like the σ-bond of valence-bond theory (see Fig. 9.25). Two *p*-orbitals lying side by side can overlap to give a **π-orbital.** When a π-orbital is occupied by two electrons, it becomes a π-bond similar to the π-bond of valence-bond theory (see Fig. 9.28).

The important distinction between molecular orbital theory and valence-bond theory is that *a molecular orbital defines a region of space that may be, but is not necessarily, occupied by up to two electrons.* A molecular orbital is formed by the overlap of atomic orbitals, and electrons are added later. In valence-bond theory, the electrons are involved in the merging of the orbitals, and we never think of an "empty" bonding region. A molecular orbital built up from atomic orbitals on atoms is called a **linear combination of atomic orbitals** (LCAO).

Let's consider a hydrogen molecule in terms of molecular orbital theory. Each atom provides a 1*s*-orbital (at this stage, we ignore the presence of the electrons). As the distance between the atoms approaches the bonding distance, the orbitals of the two atoms act like waves and interfere with each other. If the waves interfere *constructively,* in the sense that their amplitudes add together, they form a **bonding orbital,** which we denote σ_{1s} (Fig. 9.51). This notation shows both the identity of the parent atomic orbitals (1*s*) and the type of molecular orbital they form (σ). We write the wavefunction of a bonding orbital as

$$\psi = \psi_A + \psi_B \tag{1}$$

where ψ_A is the atomic orbital on atom A and ψ_B is the atomic orbital on atom B. The plus sign means that the two waves add together where they overlap. Because the wavefunction is enhanced between the nuclei, where the atomic orbitals interfere constructively, there is a greater probability of finding the electron there. As a result, any electron that occupies this molecular orbital can

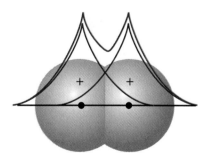

FIGURE 9.51

When two 1*s*-orbitals overlap in such a way that they have the same signs in the same regions of space, their wavefunctions (red lines) interfere constructively and give rise to a region of enhanced amplitude between the two nuclei (blue line).

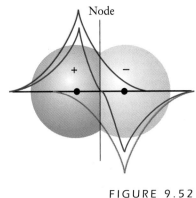

FIGURE 9.52

When two 1s-orbitals overlap in the same region of space with opposite signs, their wavefunctions (red lines) interfere destructively and give rise to a region of diminished amplitude and a node between the two nuclei (blue line).

interact strongly with both nuclei, instead of just one, as is the case in the separated atoms. Consequently, a bonding orbital has a lower energy than the atomic orbitals from which it is made.

Alternatively, the two atomic orbitals may interfere *destructively* and form a molecular orbital with a reduced amplitude between the nuclei (Fig. 9.52). This molecular orbital is written

$$\psi = \psi_A - \psi_B \qquad (2)$$

The minus sign means that one atomic orbital subtracts from the other where they overlap. In fact, on a plane halfway between the nuclei, the cancellation is complete; so the orbital has a *nodal plane* (a plane on which the wavefunction passes through 0) between the two atoms. This type of combination is called an **antibonding orbital** because any electron that occupies it is excluded from the internuclear region and therefore has a higher energy than in the isolated atom. We denote this orbital σ_{1s}^*, where an asterisk denotes an antibonding orbital.

The relative energies of the original 1s-orbitals on the H atoms and the bonding and antibonding molecular orbitals in the H_2 molecule are shown in a **molecular orbital energy-level diagram** like that in Fig. 9.53. The increase in energy resulting when an antibonding orbital is occupied is a little greater than the lowering in energy resulting when the corresponding bonding orbital is occupied.

At this point, we accommodate the electrons provided by the two hydrogen atoms. In molecular orbital theory, electrons fill the molecular orbitals according to the same building-up principle that we used to derive the electron configurations of atoms (Section 7.11). That is,

1. Electrons are accommodated first in the lowest energy molecular orbital, then in orbitals of increasingly higher energy.

2. According to the Pauli exclusion principle, each orbital can accommodate up to two paired electrons.

3. If more than one orbital of the same energy is available, the electrons enter them singly and adopt parallel spins.

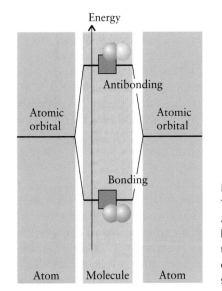

FIGURE 9.53

A molecular orbital energy-level diagram for the bonding and antibonding molecular orbitals that can be built from two s-orbitals. The signs of the s-orbitals are depicted by the different shades of blue.

Two electrons are available in H_2, one from each atom. Both occupy the bonding orbital (the lower energy orbital), resulting in the configuration σ_{1s}^2 (Fig. 9.54). Because only the bonding orbital is occupied, the energy of the molecule is lower than that of the separate atoms, and hydrogen exists as H_2 molecules.

Why are the noble gases monatomic? Molecular orbital theory provides a simple answer. Let's construct the energy-level diagram for the hypothetical molecule He_2. Because both the He_2 and H_2 molecules are built from $1s$-orbitals, we can use the diagram in Fig. 9.54 to discuss He_2, but with four electrons instead of two. The first two electrons enter σ_{1s}, which is then full. The remaining two must enter σ_{1s}^*. The configuration of He_2 is therefore $\sigma_{1s}^2\sigma_{1s}^{*2}$ (Fig. 9.55), and the bonding effect of σ_{1s}^2 is canceled by the antibonding effect of σ_{1s}^{*2}. Hence, there is no net bonding in He_2 and it is not a stable species. A similar argument applies to all the noble gases.

> **When two atoms are part of a molecule, their valence-shell atomic orbitals overlap to form molecular orbitals. The overlap of two atomic orbitals gives rise to a bonding molecular orbital and an antibonding molecular orbital. Each molecular orbital can hold up to two electrons.**

Self-Test 9.12A Is He_2^{2+} a possible species?

[*Answer:* Yes]

Self-Test 9.12B Does H_2^+ have a higher or a lower energy than an H atom a long distance away from a proton? [*Hint:* Think about which orbital the electron occupies in each case.]

9.16 Molecular Orbitals in Period 2 Diatomic Molecules

Let's see how to use molecular orbital theory to account for the bonding in homonuclear diatomic molecules, such as O_2 and N_2, formed between atoms of Period 2 elements.

The first step is to build up the molecular orbital energy-level diagram by constructing the LCAOs from the valence-shell atomic orbitals of the two atoms in the molecule. According to molecular orbital theory, in this first step we consider all the valence atomic orbitals, disregarding any electrons present. Because Period 2 atoms have one $2s$- and three $2p$-orbitals in their valence shells, we form molecular orbitals from the overlap of these four atomic orbitals. The rule to remember is that *from N atomic orbitals, we can build N molecular orbitals.* Because each atom contributes four valence orbitals, we can expect to form eight molecular orbitals from the two atoms.

Two $2s$-orbitals overlap to form two sausage-shaped molecular orbitals, one bonding (the σ_{2s}-orbital) and the other antibonding (the σ_{2s}-orbital). The three $2p$-orbitals on each neighboring atom can overlap in two distinct ways. The two $2p$-orbitals that are directed toward each other along the internuclear axis form a bonding σ-orbital (denoted σ_{2p}) and an antibonding σ^*-orbital (denoted σ_{2p}^*) where they overlap. The two $2p$-orbitals that are perpendicular to the internuclear axis overlap side by side to form bonding and antibonding π-orbitals. Because there are two perpendicular $2p$-orbitals on each atom, two bonding π_{2p}-orbitals and two antibonding π_{2p}^*-orbitals are formed by their

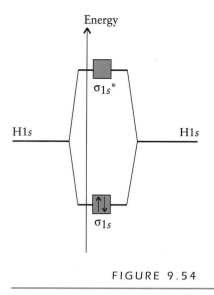

FIGURE 9.54

The two electrons in an H_2 molecule occupy the lower (bonding) molecular orbital and result in a stable molecule.

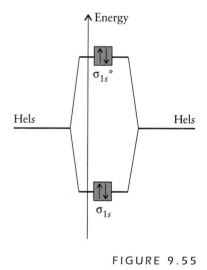

FIGURE 9.55

Two of the four electrons in a hypothetical He_2 molecule occupy the bonding orbital, but the Pauli principle forces the remaining two electrons to occupy the antibonding orbital. As a result, the He_2 molecule does not have a lower energy than two separate He atoms and does not exist as a stable species.

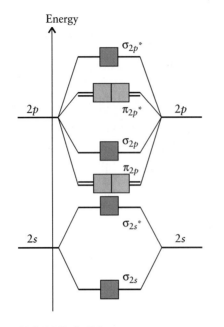

FIGURE 9.56

A typical molecular orbital energy-level diagram for the homonuclear diatomic molecules Li_2 through N_2. Each box represents one molecular orbital and can accommodate up to two electrons.

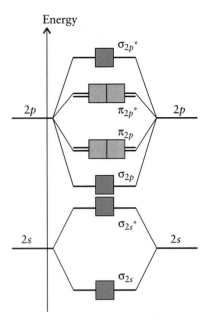

FIGURE 9.57

The molecular orbital energy-level diagram for the homonuclear diatomic molecules to the right of Period 2, specifically O_2 and F_2.

The differences in order in different molecules stem from the differences in the relative energies of the 2s- and 2p-orbitals that are used to form the molecular orbitals.

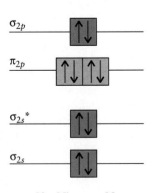

52   Nitrogen, N_2

overlap. Detailed calculations for Li_2 through N_2 result in the energy-level diagram shown in Fig. 9.56. There are some small differences in the order of energy levels as the number of electrons on the atoms increases, and in O_2 and F_2 molecules, the energy of the σ_{2p}-orbital drops below that of the π_{2p}-orbitals (Fig. 9.57). The order of the energy levels in O_2 and F_2 is easy to understand, because the end-to-end overlap of $2p$-orbitals to give σ_{2p}- and $\sigma_{2p}{}^*$-orbitals is more effective than the side-by-side overlap to give π_{2p}- and $\pi_{2p}{}^*$-orbitals.

In the ground state, all the valence electrons will be found in the lowest energy orbitals available, just as electrons in atomic orbitals are. Excitation of electrons to higher energy levels and the release of energy when they return to the ground state can be detected by spectroscopy (Investigating Matter 9.2).

Now that we know what molecular orbitals are available, we can construct the ground-state electron configurations of the molecules by using the building-up principle. Let's consider the nitrogen molecule, N_2. Because nitrogen belongs to Group 15, each atom supplies five valence electrons. We must therefore assign a total of 10 electrons to the molecular orbitals shown in Fig. 9.56. Two fill the σ_{2s}-orbital. The next two fill the $\sigma_{2s}{}^*$-orbital. Next in line for occupation are the two π_{2p}-orbitals, which can hold four electrons. The last two electrons then enter the σ_{2p}-orbital. This configuration is shown as (**52**), where the boxes now represent *molecular* orbitals. It is written:

$$N_2: \quad \sigma_{2s}{}^2 \, \sigma_{2s}{}^{*2} \, \pi_{2p}{}^4 \, \sigma_{2p}{}^2$$

At first sight, the molecular orbital description of N_2 looks quite different from the Lewis description ($:N\equiv N:$). However, it is in fact very closely related. We can see this by calculating the *bond order* (BO), which in molecular orbital theory is the net number of bonds after allowing for the cancellation of bonds by antibonds:

Bond order $= \frac{1}{2} \times$ (number of electrons in bonding orbitals
$\qquad\qquad\qquad$ $-$ number of electrons in antibonding orbitals)

$$\text{BO} = \frac{1}{2} \times (B - A) \qquad\qquad\qquad (3)$$

Here B is the number of electrons in bonding molecular orbitals and A is the number of electrons in antibonding molecular orbitals. This definition of bond order is a generalization of the one in Section 8.7, where we simply counted shared pairs. In N_2, there are eight electrons in bonding orbitals and two in antibonding orbitals, so

$$\text{BO} = \frac{1}{2} \times (8 - 2) = 3$$

Because its bond order is 3, N_2 effectively has three bonds between the N atoms, just as the Lewis structure suggests.

Toolbox 9.4 *How to determine the ground-state electron configuration and bond order of a diatomic molecule or ion*

Use this Toolbox to build up the ground-state electron configuration of a homonuclear diatomic molecule from its valence atomic orbitals and identify its bond order.

Conceptual Basis

The valence atomic orbitals of the two atoms combine to form an equal number of molecular orbitals. All the available valence electrons are accommodated in the orbitals that result in the lowest energy. The bond order is the net number of bonds that hold the molecule together.

Procedure

Step 1. Identify the atomic orbitals in the valence shells of the two atoms by noting the period of the element. Use *all* the orbitals in the valence shells, ignoring how many electrons they contain. Atoms of Period 1 elements each have one atomic orbital and atoms of Period 2 elements each have four atomic orbitals (one $2s$ and three $2p$).

Step 2. Use each pair of valence-shell atomic orbitals to build a bonding and an antibonding molecular orbital (still ignoring the electrons). The molecular orbital energy-level diagrams for the Period 2 elements are summarized in Figs. 9.56 and 9.57.

Step 3. Note how many electrons are present in the valence shells and accommodate them in the molecular orbitals according to the building-up principle. (Add the electrons to the lowest energy available orbital, adding no more than two electrons in one orbital. If more than one orbital has the same energy, add the electrons to separate orbitals with parallel spins before pairing electrons in any one of the orbitals.)

Step 4. To determine the bond order, subtract the number of electrons in antibonding orbitals from the number in bonding orbitals and divide the result by two (Eq. 3).

If the species is negatively charged, include an additional electron for each negative charge. If the species is positively charged, remove an electron for each positive charge.

The measurement of the wavelengths of light emitted by the hot atoms and molecules in stars, flames, or other glowing objects, is called *emission spectroscopy.* The wavelengths tell us the energy given off when an electron falls from a high energy level to a lower one. Electrons also *absorb* wavelengths that correspond to the different energy levels available to them. Chemists use *absorption spectroscopy,* the selective absorption of light, to identify compounds and determine their concentration in samples.

When electromagnetic radiation falls on a molecule, the molecule can be excited to a higher energy state. If the frequency of the radiation is ν (nu), it can raise the molecule to a state that differs in energy by ΔE, where

$$\Delta E = h\nu$$

This is the Bohr frequency condition, (see Section 7.3); h is the Planck constant. For typical ultraviolet wavelengths (300 nm and less, corresponding to a frequency of about 10^{15} Hz), each photon brings enough energy to excite the electrons in a molecule into different energy levels. Provided an unfilled orbital exists at the right energy, the incoming radiation can excite an electron into it, and hence be absorbed. Therefore, the study of visible and ultraviolet absorption gives us information about the electronic energy levels of molecules.

Visible and ultraviolet (UV) absorption spectra are measured in an absorption spectrometer like the one used for infrared spectroscopy (see Investigating Matter 9.1), with the difference that the source gives out visible light or ultraviolet radiation. The wavelengths can be selected with a glass prism for visible light and with a quartz prism or a diffraction grating for ultraviolet radiation (which is absorbed by glass). The

absorption spectrum of chlorophyll is shown in the first illustration. Note that chlorophyll absorbs red and blue light, leaving the green light present in white light to be reflected. That is why most vegetation looks green.

The presence of certain absorption bands in visible and ultraviolet spectra can often be traced to the presence of characteristic groups of atoms in the molecules. These groups of atoms are called *chromophores,* from the Greek words for "color bringer."

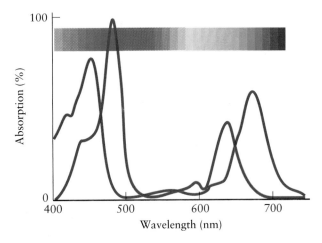

The absorption spectrum of chlorophyll as a plot of percentage absorption against wavelength. Chlorophyll *a* is shown in red, chlorophyll *b* in blue.

Example 9.9 *Deducing the electron configurations and bond orders of Period 2 diatomic molecules*

Deduce the electron configuration of the fluorine molecule and calculate its bond order.

Strategy Because the Lewis structure of F_2 is $\ddot{\text{F}}—\ddot{\text{F}}$, we anticipate that the bond order is 1. To calculate it formally, we need to know the numbers of electrons in bonding and antibonding orbitals in the molecule. Therefore, set up the molecular orbital energy-level diagram and use the building-up principle to accommodate the valence electrons as described in Toolbox 9.4. Then calculate the bond order from the resulting configuration.

Solution Each atom provides one 2s-orbital and three 2p-orbitals, so from the eight atomic orbitals we can build eight molecular orbitals. The molecular orbital energy-level diagram to use is the one shown in Fig. 9.57. There are $2 \times 7 = 14$ electrons to accommodate. The first 10 electrons repeat the N_2 configuration (apart from the

One important chromophore is a carbon-carbon double bond. In terms of the language of molecular orbital theory, the electronic transition involved in the absorption is the excitation of an electron from a bonding π orbital to the corresponding antibonding orbital, π^* (see Section 9.16); this transition (see the illustration) is therefore known as a "π-to-π^*" (pi to pi star) transition; it occurs at about 160 nm, which is in the ultraviolet. The transition occurs in the visible region for compounds that have many possible energy levels, a condition that results in a small energy separation between bonding and antibonding orbitals. Molecules with a chain of alternating single and double bonds, which are called "conjugated" double bonds, have many possible energy levels and so absorb visible light. We can see conjugated double bonding in the structure of the compound carotene, which is partly responsible for the color of carrots, mangoes and persimmons. Related compounds account for the colors of shrimps and flamingoes. This transition is also responsible for the primary act of vision, as explained in Section 9.13.

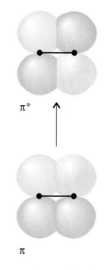

In a π-to-π^* transition, an electron in a bonding π-orbital is excited into an empty antibonding π^*-orbital.

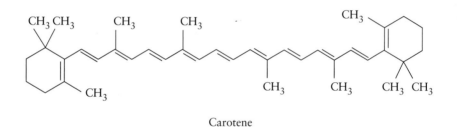

Carotene

---

change in order of the σ_{2p}- and π_{2p}-orbitals). The remaining four can enter the two antibonding π_{2p}^*-orbitals. This configuration is illustrated in (53) and is written:

$$F_2: \quad \sigma_{2s}^{2} \sigma_{2s}^{*2} \sigma_{2p}^{2} \pi_{2p}^{4} \pi_{2p}^{*4}$$

The bond order is

$$BO = \tfrac{1}{2} \times [(2 + 2 + 2 + 2) - (2 + 2 + 2)] = 1$$

Hence, F_2 is a singly bonded molecule, in agreement with the Lewis structure $:\ddot{F}-\ddot{F}:$. Because the π-bond and π-antibond cancel, the surviving bond in F_2 is a σ-bond.

Self-Test 9.13A Deduce the electronic configuration and bond order of the carbide ion (C_2^{2-}).

[*Answer:* $\sigma_{2s}^{2} \sigma_{2s}^{*2} \pi_{2p}^{4} \sigma_{2p}^{2}$, BO = 3]

Self-Test 9.13B Suggest a configuration for the nitrogen monoxide molecule, NO, and state its bond order.

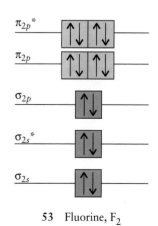

53   Fluorine, F_2

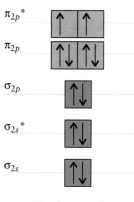

π_{2p}^*

π_{2p}

σ_{2p}

σ_{2s}^*

σ_{2s}

54 Oxygen, O_2

In Section 9.14, we remarked that the Lewis approach cannot account for the paramagnetism of O_2. However, it is easily explained in molecular orbital theory. Oxygen has 12 valence electrons (6 from each atom). We build its electron configuration by feeding these valence electrons into the molecular orbitals shown in Fig. 9.57. The first 10 repeat the N_2 configuration (apart from the change in order of the σ_{2p}- and π_{2p}-orbitals). According to Hund's rule, the last two electrons occupy the two separate π_{2p}^*-orbitals and do so with parallel spins. The configuration is shown in (**54**). This conclusion is a minor triumph for molecular orbital theory. Because the last two spins are not paired, their magnetic fields do not cancel, and the molecule is predicted to be paramagnetic—exactly as observed. The bond order of O_2 is

$$BO = \tfrac{1}{2} \times [(2 + 2 + 2 + 2) - (2 + 1 + 1)] = 2$$

This bond order is consistent with the Lewis structure $\ddot{O}=\ddot{O}$. However, the Lewis structure conceals the fact that the double bond is actually a σ-bond plus two "*half* π-bonds"—each one consisting of a *single* electron in a π-orbital—so the molecule has two unpaired electrons.

ELECTRONIC SUNSCREENS

Applying Chemistry: Case Study 9

Every time you apply a sunscreen, you are spreading a nearly invisible shield of molecular orbitals over your skin. Sunscreens contain chemicals that have delocalized electrons that can absorb ultraviolet (UV) radiation from the Sun.

Solar radiation consists of a very wide range of wavelengths of electromagnetic radiation. This radiation is essential to our lives, for it stimulates photosynthesis and the formation of food. However, the high-energy UV radiation can be harmful, because it is energetic enough to break bonds. Solar UV radiation is divided into three wavelength ranges, or "bands": UVA (315–400 nm), UVB (290–315 nm), and UVC (220–290 nm). The ozone in the stratosphere absorbs nearly all radiation with wavelengths shorter than 220 nm and most of the UVC band. The intensity of solar radiation over these three ranges is shown by the violet line in the diagram. Superimposed on that line is a green line that represents the sensitivity of living things to electromagnetic radiation. It is clear that radiation in the UVB band can cause serious harm.

Although organisms on Earth have evolved to live in this environment, overexposure of unprotected skin to sunlight can result in burning and contributes to skin cancer. However, many people favor the skin tones achieved by exposure to the Sun. It is primarily the UVA band that results in tanning, by promoting the production of the chemical pigment melanin. The UVB band destroys the skin cells instead of producing a tan. Sunscreen research

Sunscreens protect our skin from ultraviolet radiation, which can cause burns and even skin cancer.

has therefore focused on developing products that allow radiation that promotes tanning to penetrate but stops UVB radiation.

Zinc oxide creams contain a powder that protects skin by blocking all radiation, so no tanning occurs under

The electron configurations of diatomic molecules are constructed by forming molecular orbitals from all the valence-shell atomic orbitals of the two atoms and adding the valence electrons to the molecular orbitals in order of increasing energy, in accord with the building-up principle.

9.17 Orbitals in Polyatomic Molecules

The molecular orbital theory of polyatomic molecules follows the same principles as for diatomic molecules. However, remember that molecular orbitals spread over *all* the atoms in the molecule. Therefore, an electron pair that occupies a bonding orbital helps to bind together the *whole* molecule, not just an individual pair of atoms (Applying Chemistry: Case Study 9).

This picture of **delocalized electrons,** which are electrons in molecular orbitals that spread over the entire molecule, accounts very neatly not only for the existence of conventional molecules but also for the existence of electron-deficient compounds. Electron delocalization implies that the bonding influence of an electron pair is spread over all the atoms in the molecule. Therefore,

regions covered by such creams. Zinc oxide creams are favored by high-altitude skiers to protect sensitive regions of the face. Tanning creams contain compounds such as *p*-aminobenzoic acid (PABA). These compounds contain systems of conjugated (alternating) double bonds.

For Further Reading

T. E. Graedel and P. J. Crutzen, *Atmosphere, Climate, and Change,* New York: Scientific American Library, 1995, pp. 108–109.

Related Exercises: 9.75–9.76

p-Aminobenzoic acid

Electrons in these extended π-networks absorb energy in the UVB region and are promoted to antibonding orbitals (see Investigating Matter 9.2). The absorbed energy is lost as thermal motion of the surrounding molecules. As a result, this energy is far less damaging than a direct hit by a single UVB photon. The next time you apply a sunscreen, think of the electrons that are protecting your skin by becoming excited into orbitals of higher energy and thereby converting harmful UV radiation into heat.

Key Concepts: molecular orbital theory, electron delocalization

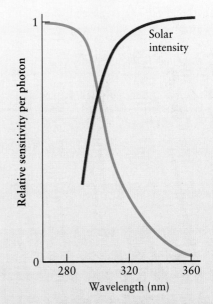

The sensitivity of living things to electromagnetic radiation in the ultraviolet region (green) superimposed over the ultraviolet spectrum of solar radiation (violet).

there is no need to provide one pair of electrons for each pair of atoms that are bonded together: a smaller number of pairs of electrons spread throughout the molecule may be able to bind all the atoms together. This is the case in B_2H_6, where six electron pairs can hold the eight nuclei together (**55**).

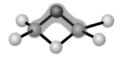

55 Three-center bond

According to molecular orbital theory, the delocalization of electrons in a polyatomic molecule spreads the bonding effects of electrons over the entire molecule.

Skills You Should Have Mastered

Conceptual

☐ **1.** Explain why electron pairs are more likely to be found in certain locations around a central atom and how and why they affect the bond angles in a molecule, Sections 9.1–9.3.

☐ **2.** Define the electric dipole moment of a bond and explain how it depends on the electronegativities of the two atoms in a bond, Section 9.4.

☐ **3.** Explain how bond enthalpy is related to bond multiplicity, atomic radius, and the presence of unpaired electrons, Sections 9.6–9.7.

☐ **4.** List the factors affecting bond length and explain the effect of each, Section 9.8.

☐ **5.** Distinguish σ- and π-bonds by their shapes, properties, and component orbitals, Section 9.9.

☐ **6.** Explain how hybridization arises from atomic orbitals and how it is used to describe molecular shape, Sections 9.10–9.13.

☐ **7.** Explain how molecular orbitals are constructed for diatomic molecules, Sections 9.14–9.16.

Problem-Solving

☐ **1.** Predict the shape of a molecule or polyatomic ion from its formula, giving each bond angle approximately, Toolbox 9.1 and Examples 9.1–9.3.

☐ **2.** Predict the polarity of a molecule, Example 9.4.

☐ **3.** Use average bond enthalpies to estimate a reaction enthalpy, Toolbox 9.2 and Example 9.5.

☐ **4.** Estimate the bond lengths in a molecule, Example 9.6.

☐ **5.** Account for the structure of a molecule in terms of hybrid orbitals and σ- and π-bonds, Toolbox 9.3 and Examples 9.7 and 9.8.

☐ **6.** Deduce the ground-state electron configurations and bond orders of Period 2 diatomic molecules, Toolbox 9.4 and Example 9.9.

Descriptive

☐ **1.** Describe the bonding in molecules of ethane, ethene, and ethyne, Section 9.13.

☐ **2.** Describe π-bonding in benzene, according to valence-bond theory, Section 9.13.

Exercises

The Shapes of Molecules and Ions

9.1 What is the expected shape and bond angle or angles in a molecule of VSEPR type (a) AX_5; (b) AX_2; (c) AX_3E; (d) AX_2E_2; (e) AX_4?

9.2 What is the expected shape and bond angle or angles in a molecule of VSEPR type (a) AX_4E; (b) AX_3; (c) AX_6; (d) AX_3E_2; (e) AX_2E?

9.3 Below are ball-and-stick models of two molecules, with arbitrary shading assigned to the atoms. In each case, indicate whether or not there must be, may be, or cannot be one or more lone pairs of electrons on the central atom:

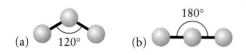

(a) 120°   (b) 180°

9.4 Below are ball-and-stick models of two molecules, with arbitrary shading assigned to atoms. In each case, indicate whether or not there must be, may be, or cannot be one or more lone pairs of electrons on the central atom:

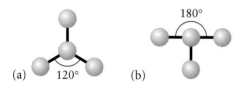

(a) 120°   (b) 180°

9.5 Using Lewis structures and the VSEPR model, predict the shape of each of the following molecules: (a) HClO (for which the structure is HOCl); (b) PF_3; (c) N_2O; (d) O_3.

9.6 Using Lewis structures and the VSEPR model, predict the shape of each of the following molecules: (a) H_2S; (b) CS_2; (c) NO_2; (d) Cl_2O.

9.7 Using Lewis structures and the VSEPR model, predict the shape of each of the following ions: (a) H_3O^+; (b) SO_4^{2-}; (c) IF_4^+; (d) NO_3^-.

9.8 Using Lewis structures and the VSEPR model, predict the shape of each of the following ions: (a) $CH_3NH_3^+$; (b) ClO_3^-; (c) ClO_4^-; (d) PF_4^+.

9.9 Using Lewis structures and the VSEPR model, predict the shape of each of the following species: (a) sulfur tetrachloride; (b) iodine trichloride; (c) IF_4^-; (d) xenon trioxide.

9.10 Using Lewis structures and the VSEPR model, predict the shape of each of the following species: (a) PF_4^-; (b) ICl_4^+; (c) phosphorus pentachloride; (d) xenon tetrafluoride.

9.11 Predict the molecular shape and bond angles in each of the following: (a) I_3^-; (b) IF_3; (c) IO_4^-; (d) TeF_6.

9.12 Predict the molecular shape and bond angles in each of the following: (a) I_3^+; (b) PCl_3; (c) SeO_3^{2-}; (d) GeH_4.

9.13 Write the Lewis structures and the VSEPR formula, name the shape, and predict the approximate bond angles in (a) CF_3Cl; (b) GaI_3; (c) CH_3^-; (d) SF_6; (e) $XeOF_4$.

9.14 Write the Lewis structures and the VSEPR formula, name the shape, and predict the approximate bond angles in (a) PCl_3F_2; (b) SnF_4; (c) SnF_6^{2-}; (d) IF_5; (e) XeO_4.

9.15 When BF_3 reacts with NH_3, a white solid of molecular formula $F_3B—NH_3$ is formed. Write Lewis structures of the reactants and the product, and estimate the bond angles.

9.16 Hydrogen fluoride reacts with antimony pentafluoride to give a salt in which the cation is H_2F^+ and the anion is SbF_6^-. Write Lewis structures of the reactants and the product, and estimate the bond angles.

Charge Distribution in Molecules

9.17 What do the terms *electric dipole* and *electric dipole moment* mean?

9.18 Explain why a molecule may be nonpolar even though it has polar bonds.

9.19 Indicate the direction of the electric dipole moment in the following bonds: (a) O—H; (b) O—F; (c) F—Cl; (d) O—S.

9.20 Indicate the direction of the electric dipole moment in the following bonds: (a) N—O; (b) C—O; (c) C—N; (d) N—H.

9.21 Identify in the following molecules the bonds that are polar and nonpolar: (a) Br_2; (b) H_2NNH_2; (c) CH_4; (d) O_3.

9.22 Identify in the following molecules the bonds that are polar and nonpolar: (a) I_2; (b) S_8; (c) CH_3CH_3; (d) H_2O_2.

9.23 Classify the following molecules as polar or nonpolar. Use the VSEPR model to determine their shapes; assume that all Xs are atoms of the same element. (a) AX_3; (b) AX_4E_2; (c) AX_3E; (d) AX_2E_2.

9.24 Classify the following molecules as polar or nonpolar. Use the VSEPR model to determine their shapes; assume that all Xs are atoms of the same element. (a) AX_3E_2; (b) AX_4E; (c) AX_5; (d) AX_4.

9.25 Write the Lewis structures and predict whether each of the following molecules is polar or nonpolar: (a) CCl_4; (b) CS_2; (c) PCl_5; (d) XeF_4.

9.26 Write the Lewis structures and predict whether each of the following molecules is polar or nonpolar: (a) CH_2Cl_2; (b) H_2S; (c) PCl_3; (d) SF_4.

9.27 There are three different dichlorobenzenes, $C_6H_4Cl_2$, which differ in the relative positions of the chlorine atoms in the benzene ring. (a) Which of the three forms are polar and which are nonpolar? (b) Which has the largest dipole moment?

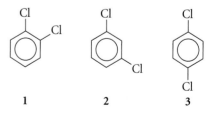

1   2   3

9.28 There are three different dichloroethenes, $C_2H_2Cl_2$, which differ in the locations of the chlorine atoms. (a) Which of the forms are polar and which are nonpolar? (b) Which has the largest dipole moment?

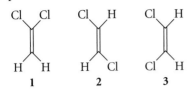

Bond Strengths and Bond Lengths

Refer to Tables 9.2, 9.3, and 9.4 for the following exercises.

9.29 Use bond enthalpies to estimate the enthalpy change that occurs when each molecule is dissociated into its atoms: (a) H_2O; (b) CO_2; (c) CH_3COOH; (d) CH_3NH_2 (methylamine).

9.30 Use bond enthalpies to estimate the enthalpy change that occurs when each molecule is dissociated into its atoms: (a) NH_3; (b) C_2H_4 (ethene); (c) C_2H_5OH (ethanol); (d) C_6H_6 (benzene).

9.31 Use bond enthalpies to estimate the standard enthalpy of formation, ΔH_f°, of the following molecules in the gas phase: (a) HCl; (b) H_2O_2; (c) CCl_4; (d) NH_3.

9.32 Use bond enthalpies to estimate the standard enthalpy of formation, ΔH_f°, of the following molecules in the gas phase: (a) HF; (b) CH_2Cl_2; (c) N_2O (NNO); (d) NH_2OH.

9.33 Use bond enthalpies to estimate the reaction enthalpy for
(a) $HCl(g) + F_2(g) \rightarrow HF(g) + ClF(g)$, given that
$\Delta H_B(Cl—F) = 256$ kJ/mol
(b) $C_2H_4(g) + HCl(g) \rightarrow CH_3CH_2Cl(g)$
(c) $C_2H_2(g) + 2 H_2(g) \rightarrow CH_3CH_3(g)$

9.34 Use bond enthalpies to estimate the enthalpy of reaction for
(a) $N_2(g) + 3 F_2(g) \rightarrow 2 NF_3(g)$
(b) $CH_3CH=CH_2(g) + H_2O(g) \rightarrow CH_3CH(OH)CH_3(g)$
(c) $CH_4(g) + Cl_2(g) \rightarrow CH_3Cl(g) + HCl(g)$

9.35 List the carbon-oxygen bonds in the following compounds in order of increasing length: (a) CH_3CH_2OH; (b) H_2CO; (c) CO.

9.36 List the nitrogen-nitrogen bonds in the following compounds in order of increasing length: (a) H_2NNH_2 (hydrazine); (b) N_2; (c) HNNH (diazene).

9.37 Use the information in Fig. 9.24 to estimate bond lengths in (a) H_2NNH_2 (hydrazine); (b) CO_2; (c) $OC(NH_2)_2$ (urea); (d) HNNH (diazene).

9.38 Use the information in Fig. 9.24 to estimate bond lengths in (a) CH_2O (formaldehyde); (b) CH_3OCH_3 (dimethyl ether); (c) CH_3OH; (d) CH_3SH.

Orbitals and Bonds

9.39 What atomic orbitals overlap to form a bond when the Cl_2 molecule forms? Classify the bond as σ or π.

9.40 What atomic orbitals overlap to form a bond when the HBr molecule forms? Classify the bond as σ or π.

9.41 Name the electron arrangements corresponding to the following hybrid orbitals: (a) sp^3; (b) sp; (c) sp^3d^2; (d) sp^2.

9.42 What is the hybridization of the orbitals used by the central atom of a molecule with the following electron arrangements: (a) tetrahedral; (b) trigonal bipyramidal; (c) octahedral; (d) linear?

9.43 State the hybridization of the atom in boldface in the following molecules: (a) $\mathbf{S}F_4$; (b) $\mathbf{B}Cl_3$; (c) $\mathbf{N}H_3$; (d) $(CH_3)_2\mathbf{Be}$.

9.44 State the hybridization of the atom in boldface in the following molecules: (a) $\mathbf{S}F_6$; (b) $Cl_2\mathbf{O}$; (c) $H_2\mathbf{N}—CH_3$ (methylamine); (d) $OC\mathbf{C}l_2$ (phosgene).

9.45 Use an orbital box diagram to identify the hybrid orbitals used by the central atoms for bonding in the following species: (a) CH_3^+; (b) $AlCl_4^-$; (c) ClO_2^+; (d) BI_3.

9.46 Use an orbital box diagram to identify the hybrid orbitals used by the central atoms for bonding in the following species: (a) CH_3^-; (b) BiI_4^-; (c) XeF_5^+; (d) NH_2^-.

9.47 Identify hybrid orbitals used by C, O, and P in the following molecules: (a) C_2H_6; (b) C_2H_2; (c) PCl_5; (d) HOCl.

9.48 Identify the hybrid orbitals used by C, N, O, S, and I in the following species: (a) N_2H_4; (b) CH_3OH; (c) SF_4; (d) IF_4^+.

Molecular Orbitals

9.49 Write the valence-shell electron configurations and bond orders of (a) Be_2; (b) B_2; (c) Ne_2.

9.50 Write the valence-shell electron configurations and bond orders of (a) O_2^{2-} (peroxide ion); (b) C_2^{2+}; (c) O_2^{2+}.

9.51 Which of the following species are paramagnetic: (a) O_2; (b) O_2^- (superoxide ion); (c) O_2^+?

9.52 Which of the following species are paramagnetic: (a) N_2^-; (b) O_2^{2-}; (c) F_2^+?

9.53 Determine the bond orders and use them to predict which species of each pair has the stronger bond: (a) F_2 or F_2^{2-}; (b) B_2 or B_2^+.

9.54 Determine the bond orders and use them to predict which species of each pair has the stronger bond: (a) C_2^{2+} or C_2; (b) O_2 or O_2^{2+}.

9.55 Would you expect the HeH^- ion to exist? What would be its bond order? Compare the stability of HeH^- with that of HeH^+.

9.56 Molecular orbital theory predicts that O_2 is paramagnetic. What other Period 2 homonuclear diatomic molecules are also paramagnetic?

Supplementary Exercises

9.57 Write the Lewis structure of each of the following compounds, predict the shape about each central atom, and state whether it is polar or nonpolar: (a) O_2SCl_2 (S is the central atom); (b) $AsCl_5$; (c) SiF_4; (d) SF_4; (e) HCN; (f) ICl_3.

9.58 Predict the types of hybrid orbitals used in bonding by the central atom, the shape, and the bond angles in (a) $In(CH_3)_3$; (b) PCl_3; (c) ICl_2^-; (d) SiF_6^{2-}; (e) CH_3^+; (f) H_3O^+; (g) SO_3; (h) SiH_4.

9.59 Arrange the carbon-nitrogen bonds in the following compounds in order of increasing length: (a) CH_3NH_2; (b) HCN; (c) CH_2NH.

9.60 For each molecule or ion, write the Lewis structure, list the number of lone pairs on the central atom, identify the shape, and estimate the bond angle: (a) XeF_5^+; (b) $XeOF_2$; (c) SF_5^+; (d) TeH_2; (e) BrO_3^-.

9.61 Write the Lewis structures and predict the shapes of (a) TeF_4; (b) NH_2^-; (c) CS_2; (d) NH_4^+; (e) GeH_4; (f) OCS.

9.62 Write the Lewis structures and predict the shapes of (a) $OCCl_2$; (b) $OSCl_2$; (c) $OSbCl_3$. In each case, the second element in the formula is the central atom.

9.63 Write the Lewis structures and give the approximate bond angles of (a) C_2H_4; (b) ClCN; (c) $OPCl_3$; (d) N_2H_4.

9.64 Investigate whether the replacement of a carbon-carbon double bond by single bonds is energetically favored by using Tables 9.2 and 9.3 to calculate the reaction enthalpy for the conversion of ethene, C_2H_4, to ethane, C_2H_6. The reaction is $H_2C=CH_2 + H_2 \rightarrow CH_3-CH_3$.

9.65 Use the covalent radii in Fig. 9.24 to calculate the bond lengths in (a) CF_4; (b) SiF_4; (c) SnF_4. Account for the trend in the values you calculate.

9.66 The halogens form compounds among themselves. These compounds, called the interhalogens, have the formulas XX', XX'_3, and XX'_5, where X is the heavier halogen. Predict their structures and bond angles. Which of them are polar?

9.67 Oxygen difluoride is almost the only compound in which oxygen is assigned a positive oxidation number. Write its Lewis structure and then determine the composition of the hybrid orbitals it uses to form bonds to the fluorine atoms. Predict the bond angle.

9.68 The bond enthalpy in NO is 632 kJ/mol and that of each N—O bond in NO_2 is 469 kJ/mol. Using Lewis structures and the average bond enthalpies in Table 9.3, explain (a) the difference in bond enthalpies between the two molecules; (b) the fact that the bond enthalpies of the two bonds in NO_2 are the same.

9.69 Benzene is more stable and less reactive than would be predicted from its Kekulé structures. Use the average bond enthalpies in Table 9.3 to calculate the lowering in energy per mole that occurs when resonance is allowed between the Kekulé structures of benzene.

9.70 The following molecules are bases that are part of the nucleic acids involved in the genetic code. Identify (a) the hybridization of each C and N atom; (b) the number of σ- and π-bonds; (c) the number of lone pairs of electrons in each molecule.

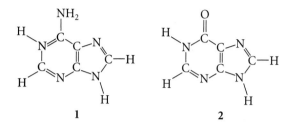

9.71 Borazine, $B_3N_3H_6$, a compound that has been called "inorganic benzene" on account of its similar planar hexagonal structure (but with alternating B and N atoms in place of C atoms), is the basis of a large class of boron-nitrogen compounds. Write its Lewis structure and predict the composition of the hybrid orbitals used by each B and N atom.

9.72 Iodine heptafluoride, IF_7, has a pentagonal bipyramidal structure.

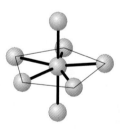

(a) How many lone pairs, if any, are on the I atom in this molecule? (b) What are the bond angles about the I atom? (c) What are the hybrid orbitals that I uses in this compound? (d) Is it likely that ClF_7 exists? Why or why not?

9.73 Write the valence-shell electron configurations and bond orders of (a) NO^+; (b) N_2^+; (c) O_2^-. (Figure 9.56 is approximately valid for diatomic molecules composed of dissimilar atoms.)

9.74 Predict the valence-shell electron configurations and bond orders of each of the following species. On that basis, predict which member of each pair will have the greater bond enthalpy: (a) CO or CO^+; (b) CN or CN^-; (c) O_2 or O_2^+. (Figure 9.56 is approximately valid for diatomic molecules composed of dissimilar atoms.)

Applied Exercises

For Exercises 9.75–9.76, see Applying Chemistry: Case Study 9.

9.75 Phenolphthalein is an indicator used to detect the stoichiometric point in acid-base titrations because it is colorless in acid and bright pink in base. The structures of the two forms of phenolphthalein are shown below. (a) Redraw them as Lewis resonance structures and show each atom, each bond, and each lone pair. (b) Give the hybridization of the carbon atom marked with an *. (c) Suggest an explanation for why the base form is colored but the acid form is not.

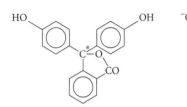
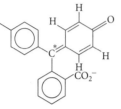

Phenolphthalein (acid form)   Phenolphthalein (base form)

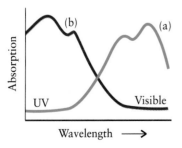

Phenolphthalein changing color as sodium reacts with water to produce a basic solution, Exercise 9.75.

9.76 The absorption spectra of two compounds are shown below. Which of the compounds could probably act as a sunscreen?

9.77 Three biologically important diatomic species are (a) CO, which binds to hemoglobin and prevents the transfer of oxygen in the body; (b) NO, a neurotransmitter; and (c) CN^-, which is a potent poison that works by interrupting the electron-transfer chain. The biological action of these species is related to their molecular orbital structure. Write their ground-state electron configurations.

9.78 The N_2O_3 molecule is formed in polluted air by the reaction of the radicals NO and NO_2. (a) Describe the molecular shape of N_2O_3 by giving the bond angles about each N atom. (b) Are all the atoms in the same plane? Explain your answer.

9.79 The structure of acrolein, an eye irritant in smoke with a pungent smell, is given below. (a) Estimate the bond angles marked with arcs and lowercase letters. (b) State the hybridization of each C atom. (c) State the number of σ- and π-bonds in the molecule.

9.80 The structure of peroxyacetylnitrate, an eye irritant in smog, is given below. (a) Estimate the bond angles marked with arcs and lowercase letters. (b) State the hybridization of each C and N atom. (c) State the number of σ- and π-bonds in the molecule.

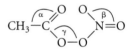

Integrated Exercises

9.81 An organic compound distilled from wood was found to have a molar mass of 32.04 g/mol and the following composition by mass: 37.5% C, 12.6% H, and 49.9% O. (a) Write the Lewis structure of the compound and estimate the bond angles about the carbon and oxygen atoms. (b) Give the hybridization of the carbon and oxygen atoms. (c) Predict whether the molecule is polar or not.

9.82 Knowing that carbon has a valence of four in nearly all its compounds and can form chains and rings of C atoms, (a) write two possible Lewis structures for C_3H_4; (b) determine all bond angles in each structure; (c) determine the hybridization of each carbon atom in the two structures; (d) ascertain whether the two structures are resonance structures or not and explain your reasoning.

9.83 Xenon forms XeO_3, XeO_4^{2-}, and XeO_6^{4-}, all of which are powerful oxidizing agents. Give (a) their Lewis structures; (b) their bond angles; (c) the hybridization of the xenon atom; (d) the oxidation number of Xe in each species.

9.84 1.00 L of chlorine gas at 1.00 atm and 298 K reacts completely with 1.00 L of nitrogen gas and 2.00 L of oxygen gas at the same temperature and pressure. When reaction is complete, there is a single gaseous product, which fills a 2.00-L flask at 1.00 atm and 298 K. Use this information to determine the following characteristics of the product: (a) empirical formula; (b) molecular formula; (c) Lewis structure (the central atom is an N atom); (d) molecular shape.

9.85 Resonance can strengthen the bonds in a molecule. (a) Calculate the standard enthalpies of formation of $N_2O(g)$ and $NO_2(g)$ from bond enthalpies and compare your results with the standard enthalpies of formation in Appendix 2A. (b) Assuming that any differences between the calculated and tabulated values result from resonance stabilization, calculate the resonance stabilization energy of each compound.

(c) Explain why one compound may experience greater stabilization than the other.

9.86 Use the information in Tables 6.2, 9.2, and 9.3 to estimate the enthalpy of formation of benzene, C_6H_6, in the liquid state, given that the standard enthalpy of sublimation of carbon is 717 kJ/mol: (a) assume no resonance; (b) assume resonance.

9.87 The electron configurations of molecules derived by using Figs. 9.56 and 9.57 depend on the Pauli exclusion principle. Suppose that we lived in a universe in which *three* electrons could occupy one orbital. (a) What would be the complete electronic structure of O_2? (b) Could Ar_2 exist? (c) Suggest what rules for writing Lewis structures you would propose if you were the G. N. Lewis living in that universe.

Connection 2

Our modern way of life has been made possible by the use of *fossil fuels,* fuels that are the remains of decayed organic matter deposited millions of years ago. The natural gas that heats our homes, the gasoline that powers our vehicles, and the coal that provides much of our electrical power are fossil fuels. Coal and vast reserves of petroleum, the source of liquid hydrocarbon fuels such as gasoline, exist in many areas of the world. However, although large, these reserves are limited, and we are using them up at a much faster rate than they can be replaced (Investigating Matter 6.1). In addition, the burning of fossil fuels harms our environment, and the carbon dioxide produced as a combustion product contributes to global warming (Applying Chemistry: Case Study 4). Alternative power generation methods, such as hydroelectric power, wind power, solar power, and alternative fuels are being sought to reduce the demand for fossil fuels. However, as you can see from the table, they make up only a small part of our energy resources.

In the combustion of a fuel, the fuel is the reducing agent and the oxygen of the air is the oxidizing agent. A good fuel has relatively weak bonds and contains elements that can form strong bonds with oxygen. The combustion products should also be nontoxic and easy to dispose of. Hydrocarbons make good fuels because they form strong $C=O$ and $O-H$ bonds when they burn and produce carbon dioxide gas and water vapor (Section 9.7). Some fuels already contain some oxygen, such as methanol, CH_3OH, and ethanol, CH_3CH_2OH. These fuels do not give off as much heat per gram as hydrocarbons (see Table 6.4). However, they may have other advantages, such as high enthalpy density, the enthalpy of combustion per liter (Section 6.15), or the ability to burn cleanly, which reduces pollution.

Three of the most promising alternative fuels are hydrogen, ethanol, and methane. Hydrogen is obtained from ocean water by using an electric current. Ethanol is obtained by fermenting *biomass,* the name given to plant materials that can be burned or reacted to produce fuels. Methane is generated by bacterial digestion of wastes such as sewage and agricultural wastes. In each case, the fuel is *renewable,* which means that the source of the fuel is replenished every year by the Sun. Methane is also obtained from the *gasification* of coal, but here we look only at renewable sources of fuels.

Ethanol, CH_3CH_2OH, is produced from the biological fermentation of the starches in grains, mainly corn. It currently makes up about 10% by volume of gasoline in the United States, which reduces pollution as well as usage of petroleum. The oxygen atom in the ethanol molecule reduces emissions of carbon monoxide and hydrocarbons by helping to ensure complete combustion. One bushel of corn (about

U.S. Energy consumption, 1995

Energy source	Energy consumed, 10^{18} J
Fossil fuels	
petroleum	36.5
natural gas	23.4
coal	20.7
nuclear power	7.6
Total fossil fuels	88.2
Renewable energy sources	
hydroelectric power	3.7
biomass	3.1
geothermal energy	0.3
solar energy	0.7
wind energy	0.3
Total renewable fuels	8.1

Source: U.S. Department of Energy

30 L) can produce nearly 10 L of ethanol. Ethanol passed a test as a reliable fuel when Brazil adopted it as its primary transportation fuel for light-duty vehicles in 1975.

A problem with ethanol as a fuel has been that the sugars and starches fermented to produce it are rather expensive. However, the cellulose in straw and cornstalks left behind when grains are harvested are now attracting attention. Cellulose is the structural material in plants. It is made up of simple sugars, just as starches are, but the bacteria that ferment starches cannot digest cellulose. Research is now being conducted on enzymes that break down cellulose into sugars that can be digested. Some species of fungi and bacteria contain certain enzymes that have the ability to break down cellulose into simple sugars that can be fermented into ethanol. These enzymes, called *cellulases,* are very complex molecules that would be difficult to synthesize. Instead the researchers grow fungi and bacteria, and extract the enzymes from them. Enzymes *catalyze* the reactions—make them take place at an acceptable rate—but are not themselves consumed in the process. A small amount of enzyme can catalyze the production of a relatively large amount of ethanol.

The production of ethanol from waste biomass would greatly increase the amount of biomass available for fuel production, because straw, wood, grass, and in fact, nearly any plant materials could be used for fuel. The amount of ethanol that could be obtained from biomass in this way would be sufficient to replace all the gasoline now used. The disadvantage is that the biomass would then not be available for soil enrichment.

Methane, CH_4, is also obtained from biological materials, but the digestion is *anaerobic,* which means that it takes place in the absence of oxygen. Currently, many sewage treatment plants have anaerobic digesters that produce sufficient methane to operate the plants. To generate methane by anaerobic digestion of waste on a large-scale basis, additional materials, such as sugars from the enzymatic breakdown of biomass, would need to be used. Methane would be less use-

A biotechnologist investigating new alternative fuels tests a sample of oils produced by microalgae. Some of these oils can be converted into a "biodiesel" fuel that burns in diesel engines with minimal pollution.

Ethanol can now be produced not only from the corn in this field, but from the stalks and leaves of the plants as well.

ful than ethanol as a transportation fuel, because of its low enthalpy density. However, it can be used wherever natural gas is used.

Methane and ethanol produce carbon dioxide when burned and therefore contribute to the greenhouse effect and possibly global warming. However, because the carbon dioxide released through burning the fuel replaces carbon dioxide absorbed by plants in the previous year, they make little net contribution to the greenhouse effect. In addition, renewable fuels generate less carbon dioxide per gram than gasoline and will be available for as long as the Sun shines and produces green plants. Other renewable fuels include methyl-*tert*-butyl ether, $C_5H_{12}O$ (MTBE), and ethyl-*tert*-butyl ether, $C_6H_{14}O$ (ETBE), which are made from ethanol; they burn very well in automobiles, but are more toxic than ethanol. "Biodiesel" fuels made from vegetable oils are also being studied.

The most promising fuel is hydrogen. It is so abundant in the form of water that it could meet all our energy needs. When it burns, the only product is water, so there is no pollution other than the nitrogen oxides formed from air in the heat of the combustion (Applying Chemistry: Case Study 8). The two disadvantages of hydrogen are the energy required to extract it from water and the difficulty of transporting it safely and compactly. The process that has been studied most intensively for the generation of hydrogen is electrolysis (see Section 19.3); metal hydrides and the new technologies stemming from the discovery of carbon nanotubes are beginning to show promise as means of storing and transporting hydrogen.

For Further Reading

J. Houghton, Energy and transport for the future, *Global Warming: The Complete Briefing*, 2nd Ed., Cambridge: Cambridge University Press, 1997, pp. 189–228.

C. Wu, Fill'er up.... with veggie oil, *Science News*, **154** (December 5, 1998), 364–366.

EIA, Renewable energy annual, http://www.eia.doe.gov/solar.renewables/renewable.energy.annual/contents.html, Washington, DC: Department of Energy, 1997.

Applying Your Knowledge

You may need to consult Chapters 5–9 and occasionally earlier chapters to answer these questions.

1. Find the enthalpy densities of hydrogen, methane, methanol, and octane in Table 6.4. Assume that the enthalpy densities of octane and gasoline are the same. (a) Calculate the enthalpy densities for the combustion of ethanol and ETBE. The density of ethanol at 298 K is 0.79 g/cm^3 and that of ETBE is 0.75 g/cm^3. (b) What is the volume of the tank, in liters, that would be required to hold enough fuel to travel as far as can be achieved with a 50-L tank of gasoline for each of the fuels listed in Table 6.4? Assume that the gases hydrogen and methane are carried under an average pressure of 10. atm and that the standard enthalpy of formation of ETBE is -661 kJ/mol.

2. For each fuel listed in question 1, determine the volume of carbon dioxide gas that would be produced at 298 K and 1 atm for each 3.8×10^4 kJ produced. This is about the energy required to drive a car 5–10 miles, depending on the efficiency of the vehicle. Which fuel contributes most to global warming per kilojoule of heat released? Which contributes least?

3. Three compounds that could be produced biologically and used as fuels are methylamine, H_2N-CH_3, a gas that occurs in herring brines and animal urine, dimethyl ether, $H_3C-O-CH_3$, a gas that can be produced from methanol, and acetic acid, H_3CCOOH, a liquid ($\Delta H_{vap} = 23.70$ kJ/mol) obtained from the fermentation of sugars . (a) Write the Lewis structures of each of these compounds (acetic acid has a structure similar to that of formic acid, Example 8.4). (b) Use bond enthalpies to calculate the specific enthalpy of each fuel, assuming that they burn to produce gaseous CO_2, liquid H_2O, and, in the case of glycine, N_2. (c) Look up these compounds and their toxicity in reference books such as *The Merck Index* or the *Handbook of Chemistry and Physics*. Would any make a useful fuel?

4. Hydrides of various elements have been studied as possible means of storing hydrogen. (a) Use your knowledge of the periodic table to write the formulas of lithium hydride, calcium hydride, and gallium hydride. Often bonds between elements with very different electronegativies are stronger than those between elements with similar electronegativities. (b) Use relative electronegativities to predict which of these three hydrides will have the strongest bonds (be most stable).

5. Greenhouse gases contribute to global warming by absorbing infrared radiation. Only molecules having dipole moments or undergoing bending and stretching motions that create momentary dipoles (such as CO_2, see Investigating Matter 9.1) can absorb infrared radiation. Which of the following gases, all of which occur naturally in air, can function as greenhouse gases? (a) CO; (b) O_2; (c) O_3; (d) SO_2; (e) N_2O; (f) Ar.

6. On the basis of your answers to the above questions and on any literature or Internet research that you may have done, recommend at least two alternative fuels for use in automobiles. Consider features such as enthalpy density, specific enthalpy (Section 6.15), safety, and contribution to global warming. Describe the advantages and disadvantages of each fuel.

7. Burning fossil fuels to light our homes is a very inefficient process. If we disregard losses from transmission and other causes, we find that only 33% of the change in enthalpy accompanying the burning of coal in a power plant is converted to electrical energy. An incandescent light bulb is only 5% efficient, further reducing the available energy. (a) Ignoring any additional energy loss during transmission and assuming that you study with a 60.-W incandescent light bulb, what mass of coal must be burned to light your lamp for 1.0 h? Use 1 W = 1 J/s and assume that coal is pure carbon in the form of graphite that is burned completely to form carbon dioxide gas. (b) Repeat the calculation for a fluorescent desk lamp, which requires only 15 W to provide the same illumination and is 20.% efficient.

Liquids and Solids

Water is a familiar substance, but it has a number of remarkable properties that are critical for the survival of life. For instance, unlike most substances, the solid form of water (ice) is less dense than the liquid form. As a result, when lakes and rivers freeze, ice floats on the surface rather than sinking to the deeps. The surface layer helps to insulate the water below, so fish and other aquatic life can survive harsh winters. Water also has an unusually high boiling point, which enables it to exist as a liquid on Earth, and a high molar heat capacity, which allows it to moderate the climate.

The unusual properties of water raise all kinds of questions. For instance, why is water a liquid? Indeed, why do *any* liquids form? Why do liquids freeze to solids as the temperature is lowered? The answer must lie in the ability of molecules to bind to one another, but how do they do that? How do the physical properties of solids and liquids depend on the nature of the atoms, ions, and molecules that compose them?

The answers to these questions illustrate a very important step in chemistry: accounting for bulk properties in terms of individual atoms, molecules, and ions. We took the first step along this path in the preceding chapter when we saw that the shapes of molecules and the distribution of their electrons determine whether a particular molecule is polar. In this chapter, we take the next step and see—among other things—how molecular polarity helps bind molecules together to form bulk materials.

Some unusual properties of water, such as its high boiling point and high molar heat capacity, allow it to exist as a liquid on Earth, to moderate the climate, and to support life. The theme of this chapter is the effect of the nature of molecules (or the individual atoms or ions) on the structure and properties of the liquids and solids they form.

Table 10.1 *Interionic and intermolecular forces**

Type of interaction	Typical energy, kJ/mol	Interacting species
ion-ion	250	ions only
dipole-dipole	2	stationary polar molecules
	0.3	rotating polar molecules
London (dispersion)	2	all types of molecules
hydrogen bonding	20	N, O, F; the link is a shared H atom

*The total force experienced by a species is the sum of all the forces in which it can participate.

INTERMOLECULAR FORCES

The attractive and repulsive forces between molecules are called *intermolecular forces*. These forces are also called **van der Waals forces,** after Johannes van der Waals, the nineteenth-century Dutch scientist who studied the effects of intermolecular forces on the behavior of real gases. There are three main types of intermolecular forces; their names and strengths are summarized in Table 10.1. The total intermolecular force acting between two molecules is the sum of all the forces they exert on each other.

10.1 London Forces

Every substance, even the noble gases, can be condensed to a liquid. We have seen that the atoms of noble gases do not form bonds to one another, yet some kind of attractive force must act between them if the gases are to condense. The same is true of molecules in which the atoms have formed all the bonds they can.

It turns out that all molecules interact by the attractive **London force.** For noble gas atoms and nonpolar molecules, this is the only attractive force that acts between them. The London force arises because the electron clouds of atoms and molecules are not static but more like swirling fogs (Fig. 10.1). If we could take a snapshot of the electron cloud in a molecule at a given instant, it would look slightly deformed. As the electrons move about in the molecule, they may pile up in one region of a molecule, leaving the nucleus of an atom in another region partially exposed. As a result, one region of the molecule will have a momentary partial negative charge and another region will have a momentary partial positive charge. This uneven charge distribution produces a transient **instantaneous dipole moment,** even in noble gas atoms. The instantaneous partial charges on different molecules attract one another, and as a result, the molecules stick together and at low temperatures condense to a liquid.

Fritz London (1900–1954) was the German-American physicist who first explained the force. The London force is also called the "dispersion force."

Only a movie filmed at the impossible rate of about 10^{16} frames a second would show the swirling motion.

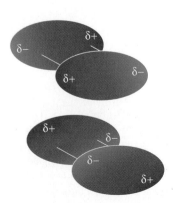

FIGURE 10.1

London forces arise from the attraction between two instantaneous dipoles. The dipoles are due to fluctuations in electron locations in the molecules. Although the instantaneous dipoles on two molecules are continuously changing direction, they remain in step long enough to attract each other.

The strength of the London force increases with the number of electrons in the molecule and therefore with molar mass (Fig. 10.2). Heavier molecules have more electrons and, therefore, have bigger fluctuations in their partial charges as the electrons flicker between different positions. Because an H_2 molecule has only two electrons, the fluctuations in its electron cloud are very small and its molecules interact so weakly that hydrogen gas condenses to a liquid only if the temperature is lowered to 20 K ($-253°C$) at 1 atm. The variation of strength of London forces with number of electrons explains the different states of the halogens at room temperature: fluorine and chlorine are gases, bromine a liquid, and iodine a solid (Table 10.2). As we descend the group, the halogen molecules have more electrons and hence exert stronger London forces, which means that they do not have to be cooled to such a low temperature to be condensed.

The strength of the London force is also influenced by the shape of the molecule. Both pentane (**1**) and 2,2-dimethylpropane (**2**), for instance, have the molecular formula C_5H_{12}, so they have the same numbers of electrons.

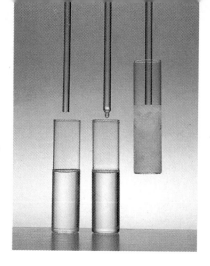

FIGURE 10.2

Hydrocarbons show how London forces increase in strength with molar mass. Pentane, C_5H_{12}, is a mobile fluid (left); pentadecane, $C_{15}H_{32}$, a viscous liquid (middle); and octadecane, $C_{18}H_{38}$, a waxy solid (right). The effect of increasing intermolecular forces is enhanced by the ability of long-chain molecules to become entangled with one another.

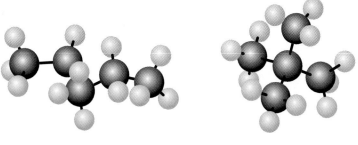

1   Pentane, C_5H_{12} 2   2,2-Dimethylpropane

We might therefore expect them to have the same boiling points. However, pentane boils at 36°C, whereas 2,2-dimethylpropane boils at 10°C. The difference can be traced to the different shapes of the molecules. Pentane molecules are rod-shaped and can lie together compactly. The instantaneous partial charges on adjacent rod-shaped molecules can interact strongly. In contrast, the instantaneous partial charges on more spherical molecules like 2,2-dimethylpropane cannot get so close to one another because only a small region of each molecule can be in contact (Fig. 10.3). As a result, the boiling point of pentane is greater than that of 2,2-dimethylpropane. In general, the London forces between rod-shaped molecules are stronger than those between spherical molecules of the same mass.

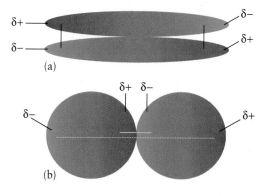

FIGURE 10.3

(a) The instantaneous dipole moments (represented by the color gradient) in two neighboring rod-shaped molecules tend to be close together and to interact strongly. (b) Those on neighboring spherical molecules tend to be far apart and to interact weakly.

Table 10.2 Normal melting and boiling points of substances*

Substance	Melting point, °C	Boiling point, °C	Substance	Melting point, °C	Boiling point, °C
Noble gases			**Small inorganic species**		
He	—†	−269 (4.2 K)	H_2	−259	−253
Ne	−249	−246	N_2	−210	−196
Ar	−189	−186	O_2	−218	−183
Kr	−157	−153	H_2O	0	100
Xe	−112	−108	H_2S	−86	−60
Halogens			NH_3	−78	−33
			CO_2	—	−78s‡
F_2	−220	−188	SO_2	−76	−10
Cl_2	−101	−34	**Organic compounds**		
Br_2	−7	59			
I_2	114	184	CH_4	−182	−162
Hydrogen halides			CF_4	−150	−129
			CCl_4	−23	77
HF	−93	20	C_6H_6	6	80
HCl	−114	−85	CH_3OH	−94	65
HBr	−89	−67	glucose	142	d‡
HI	−51	−35	sucrose	184d‡	—

*The normal melting and boiling points of a substance are the temperatures at which it melts and boils, respectively, when the atmospheric pressure is 1 atm.

†Solid helium does not form at 1 atm, however low the temperature.

‡s, solid sublimes; d, solid decomposes.

The London force arises from the attraction between instantaneous electric dipoles on neighboring molecules and acts between all types of molecules; its strength increases with increasing number of electrons and hence with molar mass. If two molecules have the same formula, the London forces are stronger for the molecule with the more linear shape.

10.2 Dipole-Dipole Interactions

Polar molecules have permanent partial charges in addition to the fleeting, instantaneous partial charges that arise from fluctuations in their electron clouds. The permanent partial charges of one polar molecule can interact with the permanent partial charges of a neighboring molecule and give rise to a **dipole-dipole interaction** (Fig. 10.4). This interaction is in addition to the London force, which is common to all molecules. However, the London force is often more important, particularly when the molecules are rotating freely, as they do in the gas phase, because rotation weakens the dipole-dipole force but not the London force.

To judge the relative strength of a dipole-dipole interaction, let's compare the enthalpy of vaporization of a solid composed of polar molecules with the lattice enthalpy of an ionic solid. In both cases, the enthalpy change is a measure

Lattice enthalpies were introduced in Section 8.3; they measure the strength of the coulombic (electrostatic) interactions between ions.

of the energy needed to separate the molecules or ions. The enthalpy of vaporization of solid hydrogen chloride is 18 kJ/mol; the lattice enthalpy of sodium chloride is 787 kJ/mol, nearly 50 times bigger. The dipole-dipole interaction is weaker than the ion-ion interactions in ionic solids because polar molecules have only partial charges.

Because the dipole moment of a diatomic molecule increases with the electronegativity difference between the two atoms, the strength of the dipole-dipole interaction between diatomic molecules also increases with the difference in electronegativity. For polyatomic molecules, their dipolar character and, therefore, the strength of their dipole-dipole forces also depend on the shape of the molecule. A highly symmetrical molecule in which bond dipoles cancel has no net dipole moment even if each bond is polar (Section 9.5). Only London forces are important between nonpolar molecules.

The stronger the intermolecular forces in a liquid, the greater is the energy required to separate the molecules. Therefore, liquids with strong intermolecular forces have high enthalpies of vaporization and high boiling points, and we can use boiling points as indicators of the relative strength of intermolecular forces.

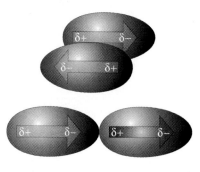

FIGURE 10.4

Polar molecules attract each other by the interaction between the permanent partial charges of their electric dipoles (represented by the arrows). Both the relative orientations shown (end-to-end and side-by-side) are energetically favorable.

Polar molecules form liquids and solids partly as a result of dipole-dipole interactions, the attraction between the partial charges of their molecules.

Boiling points are discussed more fully in Section 10.15. For now, we take a boiling point to be the temperature above which a liquid changes into a vapor when the pressure is 1 atm.

Example 10.1 *Predicting relative boiling points*

Which would you expect to have the higher boiling point, *p*-dichlorobenzene (**3**) or *o*-dichlorobenzene (**4**)?

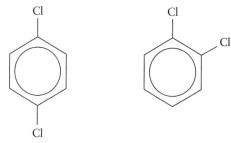

3   *p*-Dichlorobenzene   4   *o*-Dichlorobenzene

Strategy Stronger intermolecular forces result in higher boiling points. When two molecules have different dipole moments but are otherwise very similar, we expect the molecules with the bigger electric dipole moment to interact more strongly. Therefore, assign the higher boiling point to the more strongly polar compound. To decide whether a molecule is polar, decide whether or not the dipole moments of the bonds cancel one another, as explained in Section 9.5.

Solution The two C—Cl bonds in *p*-dichlorobenzene lie directly across the ring and their dipole moments cancel, so the molecule is nonpolar. An *o*-dichlorobenzene molecule is polar, because the two C—Cl dipole moments do not cancel. We therefore predict that *o*-dichlorobenzene should have a higher boiling point than *p*-dichlorobenzene. The experimental values are 180°C for *o*-dichlorobenzene and 174°C for *p*-dichlorobenzene.

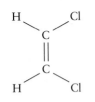

H—C—Cl, C=C, H—C—Cl

5 *cis*-Dichloroethene

H—C—Cl, C=C, Cl—C—H

6 *trans*-Dichloroethene

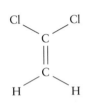

Cl—C—Cl, C=C, H—C—H

7 1,1-Dichloroethene

Self-Test 10.1A Which will have the higher boiling point, *cis*-dichloroethene (**5**) or *trans*-dichloroethene (**6**)?

[**Answer:** *cis*-Dichloroethene]

Self-Test 10.1B Which will have the higher boiling point, 1,1-dichloroethene (**7**) or *trans*-dichloroethene (**6**)?

Example 10.2 *Accounting for a trend in boiling points*

Explain the trend in the boiling points of the hydrogen halides: HCl, $-85°C$; HBr, $-67°C$; HI, $-35°C$. Electronegativity differences between hydrogen and the halogen decrease from HCl to HI.

Strategy We can use trends in boiling points to decide whether London forces or dipole-dipole forces are dominant. Dipole moment and, therefore, the strength of the dipole-dipole interaction increase with the difference in electronegativity between the hydrogen and halogen atoms. The strength of the London force increases with the number of electrons. Use the data to decide which is the dominant effect.

Solution Because electronegativity differences decrease from HCl to HI, the dipole moments decrease in that order. Therefore, dipole-dipole forces decrease too, and we can expect boiling points to decrease from HCl to HI. This prediction conflicts with the data, so we examine the London forces. The number of electrons in a molecule increases from HCl to HI, so the London interaction increases too. Hence, the boiling points should increase from HCl to HI, in accord with the data. This analysis suggests that London forces dominate dipole-dipole interactions for these molecules.

Self-Test 10.2A Account for the trend in boiling points of the noble gases, which increase from helium to radon.

[**Answer:** London interactions increase as the number of electrons increases.]

Self-Test 10.2B Suggest a reason why trifluoromethane, CHF_3, has a higher boiling point than tetrafluoromethane, CF_4.

10.3 Hydrogen Bonding

The existence of a third and stronger type of intermolecular interaction becomes apparent when we plot the boiling points of the hydrides of the elements in Groups 14 to 17 (Fig. 10.5). The trend in Group 14 is what we expect for similar compounds that differ in their number of electrons—the boiling points increase with molar mass because London forces increase as the number of electrons increases. However, the hydrides of nitrogen, oxygen, and fluorine show anomalous behavior. Water boils at a much higher temperature (100°C) than hydrogen sulfide ($-60°C$). In fact, hydrogen sulfide is a gas at room temperature, even though an H_2S molecule has many more electrons—and hence stronger London forces—than an H_2O molecule. Ammonia and hydrogen fluoride have higher boiling points than phosphine, PH_3, and hydrogen chloride respectively. The exceptionally high boiling points of water, ammonia, and hydrogen fluoride are evidence for unusually strong attractive interactions between their molecules. These interactions are known as *hydrogen bonds*.

A **hydrogen bond** consists of a hydrogen atom lying between two small, strongly electronegative atoms with lone pairs of electrons—specifically, N, O, or F. A hydrogen bond is denoted by dots to distinguish it from a true covalent bond, as in $O—H\cdots N$, for example. To understand the formation of the bond,

FIGURE 10.5

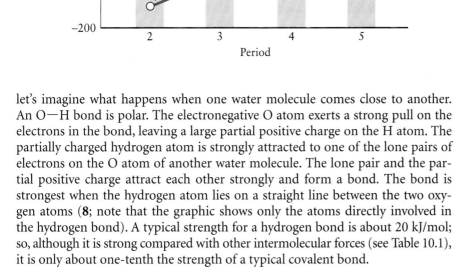

let's imagine what happens when one water molecule comes close to another. An O—H bond is polar. The electronegative O atom exerts a strong pull on the electrons in the bond, leaving a large partial positive charge on the H atom. The partially charged hydrogen atom is strongly attracted to one of the lone pairs of electrons on the O atom of another water molecule. The lone pair and the partial positive charge attract each other strongly and form a bond. The bond is strongest when the hydrogen atom lies on a straight line between the two oxygen atoms (**8**; note that the graphic shows only the atoms directly involved in the hydrogen bond). A typical strength for a hydrogen bond is about 20 kJ/mol; so, although it is strong compared with other intermolecular forces (see Table 10.1), it is only about one-tenth the strength of a typical covalent bond.

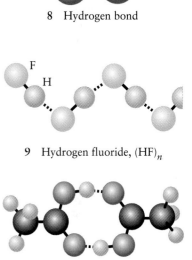

8 Hydrogen bond

Self-Test 10.3A Which of the following intermolecular links can be made by hydrogen bonds: (a) CH_3NH_2 to CH_3NH_2; (b) CH_3OCH_3 to CH_3OCH_3; (c) HBr to HBr?
[***Answer:*** Only (a) has H bonded directly to N, O, or F.]

Self-Test 10.3B Which of the following molecules can take part in hydrogen bonding in the pure state: (a) CH_3OH; (b) PH_3; (c) HClO (which has the structure Cl—O—H)?

9 Hydrogen fluoride, $(HF)_n$

When it occurs, a hydrogen bond is strong enough to dominate all the other types of intermolecular interactions. In fact, hydrogen bonding is so strong that, in some cases, it survives even in a vapor. Liquid hydrogen fluoride, for instance, contains zigzag chains of HF molecules (**9**), and the vapor contains short fragments of the chains and $(HF)_6$ rings. Acetic acid vapor contains *dimers,* or pairs of molecules, linked by two hydrogen bonds (**10**).

The shape of a protein molecule is governed partly by hydrogen bonds. A protein molecule contains long chains of atoms that are held in specific shapes by hydrogen bonds. Once the bonds are broken, the delicately organized protein molecule loses its function. When we cook an egg, the clear albumen becomes milky white because thermal motion breaks the hydrogen bonds, which allows the chains to unravel and collapse into a random jumble. Trees stand upright on account of hydrogen bonds. Cellulose molecules (which have many —OH groups) can form many hydrogen bonds with one another, and the strength of wood is due in large part to the strength of the hydrogen bonds between neighboring ribbonlike cellulose molecules (see Section 11.17). Many wood and paper glues are substances that form hydrogen bonds to the two surfaces they join. Hydrogen bonding also binds the two chains of a DNA molecule

10 Acetic acid dimer

The role of hydrogen bonding in protein molecules is described in Section 11.16.

together, so hydrogen bonding is a key to understanding reproduction (see Section 11.18).

Hydrogen bonding is responsible for the peculiar properties of water—peculiarities that make life possible. Hydrogen bonds bind water molecules together strongly, so water has a much higher boiling point than expected on the basis of its molar mass. Hydrogen bonds hold water molecules apart in an open structure when water freezes to ice, with the result that ice is less dense than liquid water and hence floats on its surface. Because water has so many hydrogen bonds, and these bonds can store energy supplied as heat, water has a high molar heat capacity. The high heat capacity of liquid water helps to moderate Earth's climate, because large masses of water are slow to heat and to cool.

Hydrogen bonding, which occurs between oxygen, nitrogen, and fluorine atoms bonded to hydrogen atoms, is the strongest type of intermolecular force.

LIQUID STRUCTURE

The molecules in a liquid are in contact with their neighbors, but they can move about and tumble over one another. When we think of a liquid, we can imagine a jostling crowd of molecules, each one constantly changing places with its neighbors.

An analogy

A liquid at rest is like a crowd of people milling about in a large room; a flowing liquid is like a crowd of people leaving a room.

In the next two sections, we look at two of the most characteristic properties of liquids: their ability to flow and their possession of a sharply defined surface. In Section 10.14, we shall also examine their ability to vaporize and exert a vapor pressure. These properties are all related to the strengths of the intermolecular forces within the liquid.

10.4 Viscosity

The **viscosity** of a liquid is its resistance to flow: the higher the viscosity, the more sluggish the flow. A liquid with a high viscosity (like molasses at room temperature, or molten glass) is said to be *viscous*. Figure 10.6 shows the relative viscosities of several liquids. Viscosity arises from the forces between molecules: strong intermolecular forces hold molecules together and do not let them move past one another easily. Therefore, we should be able to explain the trends shown in Fig.10.6 in terms of intermolecular forces.

FIGURE 10.6

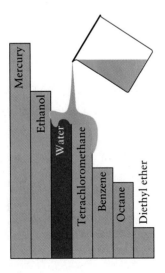

The relative viscosities of several liquids, compared with water. Liquids composed of molecules that cannot form hydrogen bonds are generally less viscous than those that can form hydrogen bonds. Mercury is an exception: its atoms stick together by a kind of metallic bonding, and its viscosity is relatively high.

Because hydrogen bonding is so strong, water has a greater viscosity than benzene. Phosphoric acid, H_3PO_4, and glycerol, $C_3H_8O_3$ (**11**), are very viscous at room temperature because of the numerous hydrogen bonds their molecules can form. However, London forces between large nonpolar molecules, such as hydrocarbon waxes, can also be strong enough to cause high viscosity.

Viscosity usually decreases as the temperature rises. Molecules have more energy at high temperatures and can wriggle past their neighbors more readily. The viscosity of water at 100°C, for instance, is only one-sixth of its value at 0°C, so six times the volume of water would flow through a section of tube at 100°C than would flow at 0°C in the same time interval.

The greater the viscosity of a liquid, the more slowly it flows; hydrogen-bonded liquids typically have high viscosities. Viscosity usually decreases with increasing temperature.

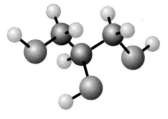

11 Glycerol, $C_3H_8O_3$

10.5 Surface Tension

The surface of a liquid is smooth because intermolecular forces tend to pull the molecules together and inward (Fig. 10.7). The **surface tension** of a liquid is the net inward pull. The surface tension of water is about three times higher than that of most other common liquids, as a result of its strong hydrogen bonds. The surface tension of mercury is higher still—more than six times that of water. This difference suggests that there are very strong bonds between mercury atoms.

A droplet of liquid suspended in air is spherical because the surface tension pulls the molecules into the most compact shape, a sphere. The droplet is approximately spherical even when it rests on a waxed surface (Fig. 10.8). The attractive forces between water molecules are greater than those between water and wax, which is largely hydrocarbon.

Water has strong interactions with paper, wood, and cloth because strong hydrogen bonds form between their molecules, which allow water to spread over these materials; in other words, water wets them. The attraction between water and materials such as glass accounts for **capillary action,** the rise of liquids up narrow tubes. The liquid rises because there are favorable attractions between its molecules and the tube's inner surface. These are forces of **adhesion,** forces that bind a substance to a surface, as distinct from the forces of **cohesion,** the forces that bind the molecules of a substance together to form a bulk material. Narrow tubes and high adhesive forces result in tall columns of liquid. Wide tubes and low adhesive forces result in short columns of liquid. Capillary action is partly responsible for the ability of a paper towel to mop up a water spill, because the narrow channels between the fibers of the paper act like capillaries.

The **meniscus** of a liquid is the curved surface it forms in a narrow tube (Fig. 10.9). The meniscus of water in a glass capillary is curved upward at the edges (forming a concave shape) because the adhesive forces between water molecules and the oxygen atoms and —OH groups of the glass surface are stronger than the cohesive forces between water molecules. The water therefore

FIGURE 10.7

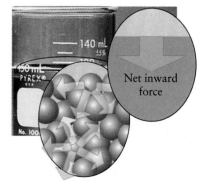

Surface tension arises from the attractive forces acting on the molecules at the surface, as shown in the inset. A molecule within the liquid experiences forces from all directions, but a molecule at the surface experiences a net inward force.

FIGURE 10.8

The nearly spherical shape of these beads of water on the waxy surface of a leaf arises from the effect of surface tension. The beads are flattened slightly by the effect of the Earth's gravity.

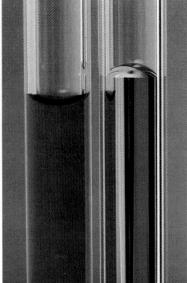

FIGURE 10.9

When the adhesive forces between a liquid and glass are stronger than the cohesive forces within the liquid, the liquid forms a concave meniscus, as shown here for water in glass (left). When the cohesive forces are stronger than the adhesive forces (as they are for mercury in glass), the surface is convex, curved downward (right).

tends to spread over the greatest possible area of glass. The meniscus of mercury curves downward in glass (forming a convex shape). This shape is a sign that cohesive forces between mercury atoms are stronger than their adhesion to glass, for in this case the liquid tends to reduce its contact with the glass.

Surface tension arises from the imbalance of intermolecular forces at the surface of a liquid. It is responsible for the tendency of liquids to form droplets and for capillary action.

Self-Test 10.4A How do you expect surface tension to be affected by temperature?
[*Answer:* It decreases as temperature increases.]

Self-Test 10.4B Which of the following liquids do you expect to have the greater surface tension: (a) ethanol, CH_3CH_2OH; (b) benzene, C_6H_6?

SOLID STRUCTURES

Almost all liquids freeze to solids when the temperature is low enough. In the solid state, the atoms, ions, or molecules are bound together so tightly that they cannot move past one another; as a result, they form a rigid mass. The nature and properties of the solid depend on the types of forces that hold the atoms, ions, or molecules together in a tightly packed array.

10.6 Classification of Solids

Solids are classified as crystalline or amorphous. A **crystalline solid** is a solid in which the atoms, ions, or molecules lie in an orderly array (Fig. 10.10). An **amorphous solid** is a solid in which the atoms, ions, or molecules lie in a random jumble, as in butter, rubber, and glass (Fig. 10.11). Crystalline solids typically have flat, well-defined surfaces called **faces,** which have definite angles at their edges. These faces are formed by orderly stacks of atoms. Amorphous

FIGURE 10.10

Crystalline solids have well-defined faces and an orderly internal structure. Each face is the edge of a stack of atoms, molecules, or ions. The crystal in the photograph is galena, PbS.

solids do not have well-defined faces unless they have been molded or cut. When some solids melt, they become **liquid crystals,** a form of matter with some properties of both liquids and crystalline solids (Applying Chemistry: Case Study 10).

Crystalline solids include metallic elements (such as copper and iron) and alloys (such as brass), in which the atoms of the elements are stacked in regular arrays, like that in Fig. 10.10. The solid forms of the nonmetallic elements, such as sulfur, phosphorus, iodine, and solid argon are also crystalline. Ionic compounds, such as sodium chloride and potassium nitrate, crystallize into structures that allow cations and anions to stack together in electrostatically favorable arrangements, sometimes with hydrating water molecules (such as $CuSO_4 \cdot 5H_2O$). In many substances, the atoms, ions, or molecules can adopt more than one arrangement, depending on the conditions. These different arrangements result in the different solid forms of the substance. Forms of an element that differ in the pattern of bonds between atoms are called **allotropes.** For example, diamond and graphite are two different allotropes of carbon, with the carbon atoms arranged differently in each form. Different allotropes of an element have different physical and chemical properties, such as melting point, density and reactivity. The specific arrangement of atoms, ions, and molecules within a crystal can be determined by x-ray diffraction (Investigating Matter 10.1).

We classify crystalline solids according to the bonds that hold their atoms, ions, or molecules in place (Table 10.3):

Metallic solids, commonly called *metals,* consist of cations held together by a sea of valence electrons.

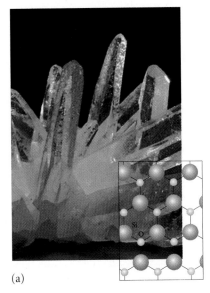

(a)

(b)

FIGURE 10.11

(a) Quartz is a crystalline form of silica, SiO_2; and the atoms are arranged in an orderly network, represented in two dimensions in the overlay. (b) When molten silica solidifies rapidly, it becomes glass. Now the atoms form a disorderly network.

X-ray diffraction is one of the most powerful techniques for determining the structures of solids and the molecules they contain. We use it not only to determine the separations of layers of atoms but also to determine the locations of individual atoms in enzymes containing thousands of atoms. The technique is primarily responsible for the enormous advances in molecular biology over the past few decades. With it, we can understand the processes of life more deeply than we have ever done before.

To understand the principles involved, we need to know that waves may *interfere* with one another. Imagine two waves of electromagnetic radiation in the same region of space. Where the peaks of one wave coincide with the peaks of the other wave, they add together to give a stronger wave, as shown in the illustration. This strengthening is called *constructive interference*. When the combined wave is detected photographically or electronically, the spot obtained is brighter than would be obtained by either ray alone. However, the waves partially cancel each other when the peaks of one coincide with the troughs of the other; the result is a weaker wave. This cancellation is called *destructive interference*. When the combined wave is detected, we see a dimmer spot than would be obtained from either ray alone. No spot at all is detected when the peaks and troughs match exactly and cancellation is complete.

Diffraction is the interference between waves caused by an object in their path (recall from Section 7.4 that the wave nature of electrons was discovered from their ability to be diffracted). The resulting pattern of bright spots against a dark background is called a *diffraction pattern*. A grid of lines can cause light to diffract. The regular array of atoms in a crystal

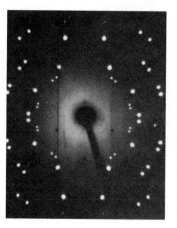

The diffraction pattern formed when x-rays pass through a single crystal of sodium chloride. The bright spots are the photographic record of the pattern formed where x-rays interfere constructively.

acts like an extremely fine grid. Consequently, a crystal can cause diffraction in a beam of x-rays, and a bright spot of constructive interference is obtained when the crystal is held at a certain angle to the beam. The angle θ (the Greek letter theta) at which constructive interference occurs is related to the wavelength λ and the distance between the atoms (see the illustration below) by

$$2d \sin \theta = \lambda$$

This formula, which is called the *Bragg equation*, enables us to measure the spacing between layers of atoms. For example, if we find a spot at 17.5° when we use x-rays of wavelength 154 pm, we can conclude that there are layers of atoms separated by a distance

$$d = \frac{\lambda}{2 \sin \theta} = \frac{154 \text{ pm}}{2 \sin 17.5°} = 256 \text{ pm}$$

Because $\sin \theta$ cannot be greater than 1, the smallest distance d that can be measured in this way is $\frac{1}{2}\lambda$. That is why x-rays need to be used: only their wavelengths are short enough to measure distances comparable to the separations of atoms in crystals.

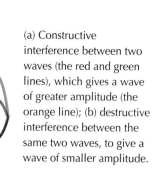

(a) Constructive interference between two waves (the red and green lines), which gives a wave of greater amplitude (the orange line); (b) destructive interference between the same two waves, to give a wave of smaller amplitude.

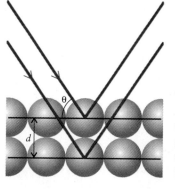

The distance *d* between layers of atoms determines the angle θ at which the x-rays reflected from the two layers are in phase and interfere constructively.

Table 10.3 *Typical characteristics of solids*

Bonding type	Examples	Characteristics
metallic	*s*- and *d*-block elements	malleable, ductile, lustrous, electrically and thermally conducting
ionic	$NaCl$, KNO_3, $CuSO_4 \cdot 5H_2O$	hard, rigid, brittle; high melting and boiling points; those soluble in water give conducting solutions
network	B, C, black P, BN, SiO_2	hard, rigid, brittle; very high melting points; insoluble in water
molecular	$BeCl_2$, S_8, P_4, I_2, ice, glucose, naphthalene	relatively low melting and boiling points; brittle if pure

Ionic solids consist of ions held together by the mutual attractions of cations and anions.

Network solids consist of atoms bonded to their neighbors covalently throughout the extent of the solid.

Molecular solids are collections of discrete molecules held in place by intermolecular forces.

The following sections describe the structure and properties of each type of solid.

Crystalline solids have a regular internal arrangement of atoms; solids are classified as metallic, ionic, network, or molecular.

10.7 Metallic Crystals

To describe the arrangements adopted by atoms in a metallic element, we can use identical spheres to represent the atoms and then think about how these spheres can be packed together. In many metallic elements, the spheres pack together to give a **close-packed structure,** an arrangement in which the spheres representing the atoms are stacked together with the least waste of space.

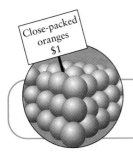

The spheres in a close-packed structure are stacked together like oranges in a grocery display.

An analogy

To see how to stack identical spheres together to give a close-packed structure, look at Fig. 10.12. The spheres in the first layer (which we call A) each touch their six neighbors. The spheres of the second (upper) layer (which we call B) each lie in a dip of the first layer. The third layer of spheres may lie in the dips of the second layer that are directly above the atoms of the first layer (Fig. 10.13). The third layer then duplicates layer A, the next layer duplicates B,

LIQUID CRYSTALS

Applying Chemistry: Case Study 10

A thin, flat television or computer display was once considered impossible, but now they are part of our everyday lives. Ultrathin, flexible computer and television displays and fast-reacting thermometers have been made possible because of a special kind of material that seems to be neither solid nor liquid. *Liquid crystals* are substances that flow like viscous liquids, but their molecules lie in a moderately orderly array, like those in a crystal. They are substances in an intermediate state of matter with the fluidity of a liquid and some of the molecular order of a solid.

A typical liquid crystal molecule is long and rodlike; *p*-azoxyanisole is an example:

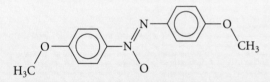

The rodlike shape causes the molecules to stack together like uncooked spaghetti: they lie parallel but are free to slide past one another along their long axes. Liquid crystals are anisotropic because of this ordering. *Anisotropic* materials have properties that depend on the direction of measurement. The viscosity of liquid crystals is least in the direction parallel to the molecules. It is easier for the long rod-shaped molecules to slip past one another along their axes than to move sideways.

Isotropic materials have properties that do not depend on the direction of measurement. Ordinary liquids are isotropic: their viscosities, for instance, are the same in every direction. Liquid crystals can become isotropic liquids when they are heated above a characteristic temperature because then the molecules have enough energy to overcome the attractions that restrict their movement.

There are three classes of liquid crystals, which differ in the arrangement of their molecules. In the *nematic* phase, the molecules lie together, all in the same direction, but staggered, like cars on a busy multilane highway. In the

A scanning tunneling microscope image of liquid crystal molecules aligned in a smectic phase.

smectic phase, the molecules line up like soldiers on parade and form layers. Cell membranes are composed mainly of smectic liquid crystals. In the *cholesteric* phase, the molecules form nematiclike layers, but neighboring layers have molecules at slightly different angles, always turning in the same direction going up the stack of layers, so the liquid crystal has a helical arrangement of molecules.*

We can also classify liquid crystals by their mode of preparation. *Thermotropic* liquid crystals are made by melting the solid phase. The liquid crystal phase exists over a short temperature range between the solid and liquid states. *p*-Azoxyanisole is a thermotropic liquid crystal. Thermotropic liquid crystals are highly viscous and can be either translucent or opaque. They are used in applications such as watches, computer screens, and thermometers. *Lyotropic* liquid crystals are layered structures that result from the action of a solvent on a solid. Examples are aqueous solutions of detergents, lipids (fats), and cell membranes. These molecules, like the detergent sodium lauryl

---

*Nematic comes from the Greek word for "thread"; smectic comes from the Greek words for "soapy"; cholesteric is related to the word cholesterol, which comes from the Greek for "bile solid."

and so on. This arrangement results in an ABABAB... pattern of layers called the **hexagonal close-packed** (hcp) structure. This structure is so called because we can make out a hexagonal pattern in the arrangement of atoms (Fig. 10.14). Magnesium and zinc are examples of metals that crystallize in this way.

As can be seen from the illustration, in an hcp structure each sphere has three nearest neighbors in the plane below, six in its own plane, and three in the

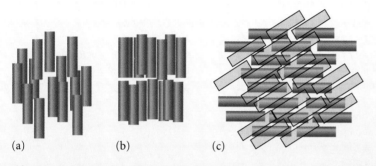

(a)   (b)   (c)

(a) In the nematic phase of a liquid crystal, the long molecules lie parallel to one another. (b) In the smectic phase of a liquid crystal, not only do the molecules lie parallel to one another, they also form sheets. (c) In the cholesteric phase of a liquid crystal, layers of parallel molecules are rotated relative to their neighbors and form a helical structure.

sulfate, have long, nonpolar hydrocarbon chains attached to polar heads:

$$\text{\Large\textbackslash/\textbackslash/\textbackslash/\textbackslash/}\,SO_3^-\quad Na^+$$

Dilute solutions of detergents—more formally, *surfactant* molecules (surface-active molecules)—tend to form layers on the surface of water (see Section 12.4). For example, detergent molecules line up next to one another on the water surface, with their polar heads in the water and their nonpolar tails in the air. In high concentrations, detergents form *micelles*, small clusters in which the polar heads point out toward the water and the hydrocarbon tails point inward. The hydrocarbon tails of the detergent molecules are attracted to greasy dirt and form micellelike structures around it. The polar heads surrounding the micelle keep it suspended in water, and it can be washed away. The lipids that form cell membranes have structures similar to those of micelles. When mixed with water, they spontaneously form sheets in which the molecules are aligned in rows, forming a double layer, with their polar heads facing outward on each side of the sheet. These sheets form the protective membranes of our cells.

Electronic displays make use of the fact that the orientation of the molecules in liquid crystals changes in the presence of an electric field. This reorientation may cause a change in their optical properties, making them opaque or transparent, and hence forming a pattern on a screen. Cholesteric liquid crystals are also of interest because the helical structure unwinds slightly as the temperature is

changed. Because the twist of the helical structure affects the optical properties of the liquid crystal, these properties change with temperature. The effect is utilized in liquid crystal thermometers.

Key Concepts: intermolecular forces, crystalline structures

For Further Reading
P. Ball, A soft and sticky world, *Designing the Molecular World*, Princeton: Princeton University Press, 1994, pp. 216–255.

Related Exercises: 10.93–10.96

A cross section through one of the two layers of a cell membrane. The long, narrow molecules are aligned with their polar heads toward the surfaces of the membrane.

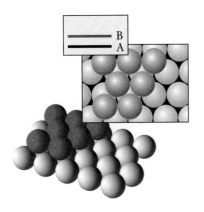

FIGURE 10.12

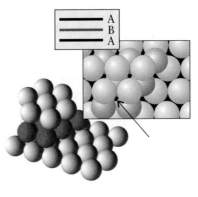

FIGURE 10.13

FIGURE 10.14

A close-packed structure can be built up in stages. The first layer (A) is laid down with minimum waste of space, and the second layer (B) lies in one-half the dips—the depressions—of the first. The inset shows a top-down view.

The third layer of spheres can be deposited so that the atoms lie directly above the atoms of the first layer to give an ABABAB . . . structure. Notice in the top-down view (inset) that the holes in each layer are lined up so that in certain places (indicated by the arrow) you can see all the way through the crystal. You can also see this effect when you stack pennies in an ABAB structure.

A fragment of the structure formed as described in Fig. 10.13 shows the hexagonal symmetry of the arrangement, and the origin of its name "hexagonal close-packed."

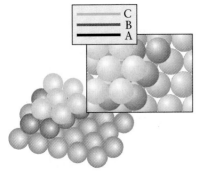

FIGURE 10.15

As an alternative to the scheme shown in Fig. 10.13, the spheres of the third layer can lie in the dips above the dips in the first layer to give an ABCABC . . . arrangement of layers. In this case, you cannot see through the layers because each hole has an atom either on top of it or below it.

one above, giving 12 in all. The **coordination number,** the number of nearest neighbors of each atom, in the solid is 12. It is impossible to pack identical spheres together with a coordination number greater than 12.

In an alternative arrangement, the spheres of the third layer lie in the dips of the second layer that do *not* lie directly over the atoms of the first layer (Fig. 10.15). If we call this third layer C, then the resulting structure has an ABCABC. . . pattern of layers and a **cubic close-packed** (ccp) structure. When we view a ccp structure from an angle, we can see that the spheres form a cubic pattern, with one sphere at each corner and one at the center of each face (Fig. 10.16). The coordination number is also 12: each sphere has three nearest neighbors in the layer below, six in its own layer, and three in the layer above. Aluminum, copper, silver, and gold are examples of metals that crystallize in this way.

Which close-packed structure—if either—a metal adopts depends on which structure corresponds to the lower energy, and that depends on details of the electronic structure of the atoms. In fact, some atoms achieve a lower energy by adopting a different arrangement entirely. In a **body-centered cubic** (bcc) structure, for instance, a single atom lies at the center of a cube formed by eight other atoms (Fig. 10.17). This structure is not close packed. Iron, sodium, and potassium are examples of metals that crystallize in this way.

Many metals have close-packed structures, with the atoms stacked in either a hexagonal or a cubic arrangement; close-packed atoms have a coordination number of 12.

FIGURE 10.16

A fragment of the structure formed as described in Fig. 10.15 shows the origin of the names "cubic close-packed" or "face-centered cubic" for this arrangement. The layers A, B, C can be seen along the diagonals and are indicated by the different colors of atoms.

FIGURE 10.17

The body-centered cubic (bcc) structure. This structure is not packed as closely as the others we have illustrated. It is less common among metals than the close-packed structures, but some ionic structures are based on it.

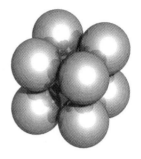

FIGURE 10.18

A primitive cubic structure (which later we see is a unit cell) has an atom at each of the eight corners.

Self-Test 10.5A What is the coordination number of an atom in a body-centered cubic (bcc) structure?

[*Answer:* 8]

Self-Test 10.5B What is the coordination number of an atom in a primitive cubic structure (Fig. 10.18), in which there is one atom at each corner of a cube (**12**)?

Each of the small units shown in Figs. 10.14, 10.16, 10.17, and 10.18 is a **unit cell,** that is, the smallest hypothetical unit that, when stacked together repeatedly without any gaps, can reproduce the entire crystal (Fig. 10.19).

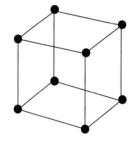

12 Primitive cubic cell

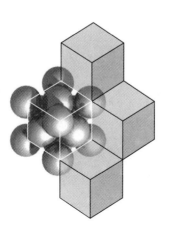

FIGURE 10.19

The entire crystal structure is constructed from a unit cell by stacking the cells together without any gaps in between. In this cubic close-packed (face-centered cubic) cell, each corner atom is shared by eight cubes that touch at the corner. Each atom on a face is shared by two adjacent cubes.

FIGURE 10.20

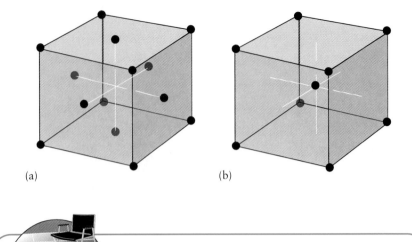

(a)   (b)

An analogy

A unit cell is to a crystal in three dimensions as a tile is to a tiled floor in two dimensions. We can use tiles to cover, without gaps, a floor that stretches virtually endlessly in all directions. Likewise, we can build an indefinitely large three-dimensional crystal by stacking together unit cells without any gaps.

Unit cells are sometimes drawn by representing each atom by a dot that marks the location of its center (Fig. 10.20). A cubic close-packed unit cell has a dot at the center of each face; for this reason, it is also called a **face-centered cubic** (fcc) structure.

All cubic close-packed structures are fcc, but not all fcc structures are close packed (see Section 10.11).

The number of atoms in a unit cell is counted by noting how they are shared between neighboring cells. For example, an atom at the center of a cell belongs entirely to that cell, but one on a face is shared between two cells, so only half the atom belongs to that cell. For an fcc structure, each of the eight corner atoms is shared by eight cells, so overall they contribute $8 \times \frac{1}{8} = 1$ atom to the cell. Each atom at the center of each of the six faces contributes half an atom, so jointly they contribute $6 \times \frac{1}{2} = 3$ atoms (Fig. 10.21). The total number of atoms in an fcc unit cell is therefore $1 + 3 = 4$, and the mass of the unit cell is four times the mass of one atom.

FIGURE 10.21

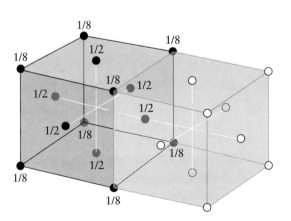

Example 10.3 Counting the number of atoms in a unit cell

How many atoms are there in a bcc cell (see Fig. 10.20b)?

Strategy As explained above, count 1 for each atom that is not shared by other cells, and $\frac{1}{2}$ for each atom on a face (where it is shared by two cells). When an atom is on an edge, it counts as $\frac{1}{4}$ (because it is shared by four cells). Each atom at a corner (where it is shared by eight cells) counts as $\frac{1}{8}$.

Solution Inspection of Fig. 10.20b shows that there is one central atom (count 1) and eight corner atoms (count $8 \times \frac{1}{8} = 1$). Therefore, each cell has two atoms.

Self-Test 10.6A How many atoms are there in a primitive cubic cell (see Fig. 10.18)?
[*Answer:* 1]

Self-Test 10.6B How many atoms are there in the structure made up of unit cells like the one shown in (**13**), which has an atom at each corner, two on opposite faces, and two inside the cell on a diagonal?

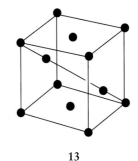

13

Because many metals are close packed, they are generally quite dense: the close packing means that a lot of matter is present in a small volume. Metals that are not close packed, such as those having a body-centered cubic structure, can often be forced under pressure into a close-packed form. In some cases, the type of unit cell in a metal can be deduced from its density, as described in Toolbox 10.1 and Example 10.4.

> *The atoms in a unit cell are counted by determining what fraction of each atom resides within the cell.*

Example 10.4 Using density to identify structure

The atomic radius of copper is 128 pm, and the density of copper is 8.93 g/cm^3. Is copper metal close packed?

Strategy Calculate the density of a cubic close-packed (fcc) unit cell as described in Toolbox 10.1 and compare the calculated value with the actual density. Because the density is given in grams per cubic centimeter, convert lengths to centimeters.

Solution From Fig. 10.22c and the Pythagorean theorem, with

$$r = 128 \text{ pm} = 1.28 \times 10^{-10} \text{ m} = 1.28 \times 10^{-8} \text{ cm}$$

the volume of the unit cell of side a is

$$a^3 = (8^{1/2}r)^3 = (8^{1/2} \times 1.28 \times 10^{-8} \text{ cm})^3 = (8^{1/2} \times 1.28 \times 10^{-8})^3 \text{ cm}^3$$

The mass, m, of one atom of copper is

$$m = \frac{63.54 \text{ g/mol}}{6.022 \times 10^{23}/\text{mol}} = \frac{63.54}{6.022 \times 10^{23}} \text{ g}$$

Each unit cell contains $(8 \times \frac{1}{8}) + (6 \times \frac{1}{2}) = 4$ copper atoms. The total mass of a close-packed unit cell is four times the mass of one atom. Therefore,

$$\text{Density} = \frac{m}{a^3} = \frac{4 \times 63.54/(6.022 \times 10^{23}) \text{ g}}{(8^{1/2} \times 1.28 \times 10^{-8})^3 \text{ cm}^3} = 8.89 \text{ g/cm}^3$$

The value is close to the experimental value for a close-packed structure.

Toolbox 10.1 How to deduce the structure of a crystalline solid from its density

We use this Toolbox to decide which type of unit cell is the most likely structure for a metal with a known density.

Conceptual Basis

Density is an intensive property, so the density of the bulk material is the same as that of one unit cell. We can calculate the density of a unit cell from the mass of the atoms it contains and the volume of the cell, and then compare the result with the measured density to determine the type of unit cell that makes up the crystal. Figure 10.22 summarizes the relation between unit cell geometry and atomic mass and radius for three types of unit cells.

Procedure

The mass of a unit cell is the sum of the masses of the atoms it contains. The mass of each atom is equal to the molar mass of the element divided by the Avogadro constant. The volume of a cubic unit cell is the cube of the length of one of its sides. That length can be obtained from the radius of the metal atom, the Pythagorean theorem, and the geometry of the cell. The Pythagorean theorem states that the square of the hypotenuse of a right-angled triangle is equal to the sum of the squares of the other two sides. That is, if the hypotenuse is c and the other two sides are a and b, then $a^2 + b^2 = c^2$.

Primitive cubic unit cell A primitive cubic unit cell contains one atom at each of the eight corners (Fig. 10.22a). The number of atoms in the cell is $\frac{1}{8} \times 8 = 1$, and the mass of each atom is $m = M/N_A$. The length of a side of the cell, a, is twice the radius of an atom: $a = 2r$, so the volume of a unit cell is $a^3 = (2r)^3$. The density of the cell is therefore

$$d = \frac{m}{a^3} = \frac{M/N_A}{(2r)^3} = \frac{M}{8N_A r^3}$$

Body-centered cubic unit cell To calculate the density of a bcc unit cell, we let the length of the diagonal of a face of the cell be f and the length of the diagonal through the body of the cell be b. Then, from Fig. 10.22b and the Pythagorean theorem,

$$a^2 + f^2 = b^2 = (4r)^2$$

The Pythagorean theorem also tells us that $f^2 = 2a^2$, so

$$a^2 + f^2 = a^2 + 2a^2 = 3a^2$$

It follows that $3a^2 = (4r)^2$ and therefore that

$$a = \frac{4r}{3^{1/2}}$$

Each unit cell contains $(8 \times \frac{1}{8}) + 1 = 2$ spheres, so the total mass of a body-centered cubic unit cell is $2M/N_A$. Therefore,

$$d = \frac{m}{a^3} = \frac{2M/N_A}{(4r/3^{1/2})^3} = \frac{3^{3/2}M}{32N_A r^3}$$

Close-packed unit cell A cubic close-packed structure has the same density as a hexagonal close-packed structure, but the former is easier to treat. A cubic close-packed structure has a face-centered cubic unit cell (Fig. 10.22c). The length of the diagonal of the face of the cube is $4r$, where r is the

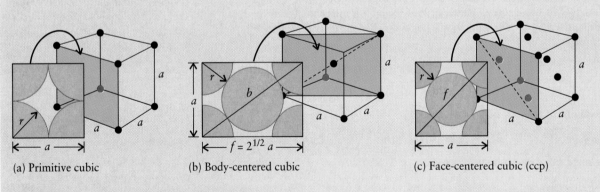

(a) Primitive cubic (b) Body-centered cubic (c) Face-centered cubic (ccp)

FIGURE 10.22

The geometries of three cubic unit cells, showing the relation of the dimensions of each cell to the radius of a sphere representing an atom or ion, r. The side of a cell is a; the diagonal of the body of a cell, b; and the diagonal of a face, f. (a) Primitive cubic cell; (b) body-centered cubic cell; (c) face-centered cubic cell.

radius of each sphere. Therefore, from the Pythagorean theorem,

$$a^2 + a^2 = (4r)^2 = 16r^2$$

It follows that

$$a^2 = \tfrac{1}{2}(16r^2) = 8r^2$$

Taking the square root of both sides gives $a = 8^{1/2}r$. There are 8 spheres at the corners, but only $\frac{1}{8}$ of each sphere projects into the cube, so these spheres contribute $\frac{1}{8} \times 8 = 1$ sphere to the cube. There is half a sphere on each of the 6 faces, so these spheres contribute $\frac{1}{2} \times 6 = 3$ spheres, giving 4 in all. The density is therefore

$$d = \frac{m}{a^3} = \frac{4M/N_A}{(8^{1/2}r)^3} = \frac{4M}{8^{3/2}N_Ar^3}$$

This procedure is illustrated in Example 10.4.

Self-Test 10.7A The atomic radius of silver is 144 pm and its density is 10.5 g/cm³. Is the structure close packed?

[***Answer:*** Yes]

Self-Test 10.7B The atomic radius of iron is 124 pm and its density is 7.87 g/cm³. Is this density consistent with a body-centered cubic structure?

10.8 Properties of Metals

In a metallic solid, cations lie in a regular array and are surrounded by a sea of electrons. This structure gives metals their unique properties. Their characteristic luster is due to the mobility of their electrons. An incident light wave is an oscillating electric field (Section 7.1). When it strikes the surface of a metal, the electric field pushes the mobile electrons backward and forward. These oscillating electrons radiate light, and we see it as a luster—essentially a reflection of the incident light (Fig. 10.23). The electrons oscillate in step with the incident light, so they give out light having the same frequency as the incident light. In other words, red light reflected from a metallic surface is red and blue light is reflected as blue. That is why an image in a mirror—a thin coating of a metal, usually silver or aluminum, on glass—is a faithful portrayal of the reflected object. When we look at ourselves in a mirror, what we see are the oscillations of

FIGURE 10.23

(a) When light of a particular color shines on the surface of a metal, the electrons at the surface oscillate in step. This oscillating motion gives rise to an electromagnetic wave that we perceive as the reflection of the source. (b) Each of these solar mirrors in California is positioned at the best angle to reflect sunlight of all wavelengths into a collector that uses it to generate electricity.

Incident waves   Reflected waves

Oscillating electrons

(a)   (b)

the mobile electrons in the metal film, with different parts of the film oscillating at different frequencies according to the color being reflected there.

The malleability and ductility of metals are also due to the mobility of their electrons. Because the cations are surrounded by an electron sea, there is very little directional character in the bonding. As a result, a cation can be pushed past its neighbors without too much effort. A blow from a hammer can drive large numbers of cations past their neighbors. The electron sea immediately adjusts to ensure that the atoms are not simply broken off but remain attached in their new positions (Fig. 10.24).

One of the most striking properties of a metal is its ability to conduct an **electric current,** which is the flow of electric charge. In **electronic conduction,** the charge is carried by electrons. Electronic conduction is the mechanism of conduction in metals. In **ionic conduction,** the charge is carried by ions. This is the mechanism of electrical conduction—the transport of charge—in a molten salt or an electrolyte solution. Because ions are too bulky to travel easily through most solids, the flow of charge through solids is almost always a result of electronic conduction.

The ability of a substance to conduct electricity is measured by its **resistance:** the lower the resistance, the better it conducts. We classify substances according to their resistance and how it varies with temperature (Fig. 10.25):

An **insulator** is a substance that has such a high resistance that it does not conduct electricity.

A **metallic conductor** is an electronic conductor with a resistance that increases as the temperature is raised.

A **semiconductor** is an electronic conductor with a resistance that decreases as the temperature is raised.

A **superconductor** is an electronic conductor that conducts electricity with zero resistance, usually at very low temperatures.

FIGURE 10.24

(a) A metal is malleable because, when cations are displaced by a blow from a hammer, the mobile electrons can immediately respond and follow the cations to their new positions. (b) This piece of lead has been flattened by hammering, whereas crystals of ionic compounds, such as the orange compound lead(II) oxide shown here, shatter when struck.

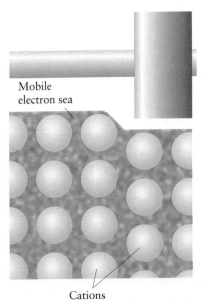

Mobile electron sea

Cations

(a)

(b)

FIGURE 10.25

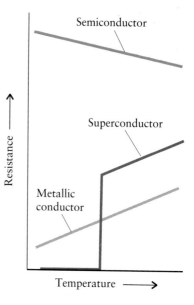

A metallic conductor is a substance with a resistance that increases with temperature. A semiconductor is a substance with a resistance that decreases with increasing temperature. A superconductor is a substance that has zero resistance below a certain temperature. An insulator behaves like a semiconductor with a very high resistance.

Insulators include gases, most ionic solids, most network solids, almost all organic compounds, and all molecular and covalent liquids and solids. Metallic conductors include all metals and certain other solids, such as graphite. An example of a semiconductor is a crystal of silicon.

Electronic conduction in metals results from the motion of electrons past the stationary cations. The resistance to the motion of the electrons increases as the temperature is raised because the cations vibrate more violently and hinder the passage of the electrons.

Semiconductors are solids in which the valence electrons, although not initially free, can be removed from their parent atoms by a modest rise in temperature. As the temperature is raised, more electrons are liberated, so the ability of the solid to carry an electric current increases. In other words, its resistance decreases as temperature increases.

Until 1987, the known superconductors (mostly metals, such as lead) had to be cooled to close to absolute zero. However, in 1987, the first *high-temperature superconductors* were reported. They can be used at about 100 K, which means that liquid nitrogen (which boils at 77 K) can be used to cool them rather than the very much more expensive liquid helium (see Fig. 7.32). Some high-temperature superconductors are **ceramics,** inorganic solids that have undergone heat treatment (see Applying Chemistry: Case Study 19). Superconducting ceramics are complex ionic oxides, such as $YBa_2Cu_3O_7$.

> *The mobility of the valence electrons in a metal accounts for its luster, malleability, ductility, and electrical conductivity. The resistance of conductors increases with temperature, but the resistance of semiconductors decreases with temperature.*

10.9 The Band Theory of Solids

Electrical conduction in metals can be explained in terms of delocalized molecular orbitals that spread throughout the solid. We saw in Section 9.16 that when N atomic orbitals merge together in a molecule, they form N molecular orbitals. The same is true of a metal, but now N is enormous (around 10^{23} for 10 g of copper, for example). Instead of the few molecular orbitals with widely spaced energies typical of small molecules, the huge number of molecular orbitals in a metal are so close together in energy that they form a nearly continuous band (Fig. 10.26). The electrons that occupy these orbitals are delocalized throughout the complete solid—even if the solid is a cable many miles long.

Consider a metal such as sodium. Each atom contributes one valence orbital (the 3s-orbital in this case) and one valence electron. The N 3s-orbitals in the sample merge to give a band of N molecular orbitals. Half these orbitals have a net bonding character and half have a net antibonding character. Because each of the N atoms supplies one valence electron, a total of N electrons must be accommodated in the orbitals. These N electrons occupy the orbitals according

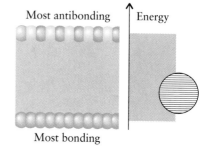

FIGURE 10.26

A metallic solid contains so many bonded atoms that their molecular orbitals form an almost continuous band. At the lower edge of the band, the molecular orbital is fully bonding; at the upper edge, the molecular orbital is fully antibonding. The inset shows that, although the band of allowed energies appears to be continuous, it is in fact composed of discrete, closely spaced levels.

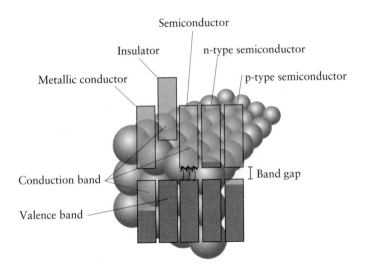

FIGURE 10.27

The band structure typical of a variety of solids. The occupied regions of the bands are shaded tan. Note that the distinction between a solid insulator and a semiconductor is the width of the gap between the valence and conduction bands.

to the building-up principle, filling the lower $\frac{1}{2}N$ bonding orbitals (because two electrons can occupy each orbital). The orbitals in the band lie so close together in energy that it takes very little additional energy to excite the electrons from the topmost filled molecular orbitals to the empty orbitals just above. An incompletely filled band of molecular orbitals such as this is called a **conduction band** (Fig. 10.27). The electrons move freely through the conduction band and can carry an electric current from one end of the solid to the other.

In an insulator, the valence electrons fill all the orbitals of a band and give rise to a **valence band,** a full band of orbitals. There is a substantial **band gap,** a range of energies for which there are no orbitals, before the next band of orbitals begins (see Fig. 10.27). The electrons in the valence band can be excited into the upper empty band only by a large injection of energy. Because the valence band is full, the electrons are not mobile unless they are excited into the empty orbitals.

In a semiconductor, a conduction band lies close in energy to a valence band. As a result, as the solid is warmed, electrons can be excited from the valence band into the conduction band, where they can travel through the solid. The resistance of a semiconductor falls as it is heated because more electrons are excited into the conduction band. The ability of a semiconductor to carry an electric current can also be enhanced by adding electrons to the conduction band or removing some from the valence band. This modification is carried out chemically by **doping** the solid, that is, by spreading small amounts of impurities through it.

In an **n-type semiconductor,** a minute amount of an element from a group with a higher number is added to the semiconductor. For example, an n-type semiconductor is formed when a Group 15 element such as arsenic is added to very pure silicon. The arsenic increases the number of valence electrons in the solid: each Si atom (Group 14) has four valence electrons, whereas each As atom (Group 15) has five. The additional electrons enter the upper, normally empty conduction band of silicon and allow the solid to conduct (see Fig. 10.27). This type of material is called an n-type semiconductor because it contains excess *n*egatively charged electrons. When silicon (Group 14) is doped with indium (Group 13) instead of arsenic, the solid has fewer valence electrons than pure silicon, so the valence band is no longer completely full. We say that the band now contains "holes." Because the valence band is no longer full, the electrons in

FIGURE 10.28

This solar-powered vehicle makes use of silicon-based semiconductors to convert sunlight into electrical energy.

it can carry an electric current (see Fig. 10.27). This type of semiconductor is called a **p-type semiconductor,** because the absence of electrons is equivalent to the presence of *p*ositively charged holes.

Solid-state electronic devices such as diodes, transistors, and integrated circuits contain **p-n junctions** in which a p-type semiconductor is in contact with an n-type semiconductor. Moreover, because solar radiation can excite electrons into a conduction band, semiconductors can be used to generate an electrical current by solar radiation. There is a hope that one day they will be used to power pollution-free transportation (Fig. 10.28).

Bonding in solids may be described in terms of bands of molecular orbitals. In metals, the conduction bands are incompletely filled orbitals that allow electrons to flow. In insulators, the valence bands are full and the large band gap prevents promotion of electrons to empty orbitals. In semiconductors, empty levels are close in energy to filled levels.

10.10 Alloys

Alloys are mixtures of two or more metals. In a **homogeneous alloy,** atoms of the different elements are distributed uniformly. Examples are brass, bronze, and the coinage alloys. A **heterogeneous alloy,** like tin-lead solder and the mercury amalgam sometimes used to fill teeth, consists of a mixture of crystalline regions with different compositions. Examples of common alloys are listed in Table 10.4.

The structures of alloys are more complicated than those of pure metals, because the two or more types of metal atoms have different radii. However, because the metallic radii of the *d*-block elements are all quite similar, for them the stacking problem is easily solved. They form an extensive range of alloys with each other because one type of atom can take the place of another with little distortion of the crystal structure. One example is the copper-zinc alloy that was used in the United States prior to 1982 for "copper" pennies and was 5% zinc. Copper atoms crystallize in a cubic close-packed structure. Because zinc

An amalgam is an alloy of a metal in mercury. Dental amalgam is silver in mercury together with small amounts of copper, tin, and zinc.

Table 10.4 *Compositions of typical alloys*

Alloy	Mass percentage composition
brass	up to 40% zinc in copper
bronze	a metal other than zinc or nickel in copper (casting bronze: 10% Sn and 5% Pb)
cupronickel	nickel in copper (coinage cupronickel: 25% Ni)
pewter	6% antimony and 1.5% copper in tin
solder	tin and lead
stainless steel*	over 12% chromium in iron

*For more detailed information on steels, see Tables 21.2 and 21.3.

atoms are nearly the same size as copper atoms and have similar electronic properties, they can take the place of some of the copper atoms in the crystal lattice. An alloy in which atoms of one metal are substituted for atoms of another metal is called a **substitutional alloy** (Fig. 10.29). Elements that can form substitutional alloys have atoms with atomic radii that differ by no more than about 15%.

Because there are slight differences in size and electronic structure, the solute atoms in a substitutional alloy distort the shape of the lattice and hinder the flow of electrons. Moreover, because the lattice is distorted, it is harder for one plane of atoms to slip past another. Although a substitutional alloy has lower electrical and thermal conductivity than the pure element, it is harder and stronger.

Steel is an alloy of generally less than 2% (by mass) carbon in iron. Carbon atoms are much smaller than iron atoms, so they cannot substitute for iron atoms in the crystal lattice. They are so small that they can fit into the gaps, or

> Although carbon is not a metal, it is conventional to regard steel as an alloy because the iron is modified by its presence.

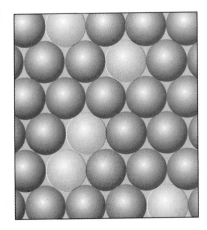

FIGURE 10.29

In a substitutional alloy, the positions of some of the atoms of one metal are taken by atoms of another metal. The two elements must have similar atomic radii.

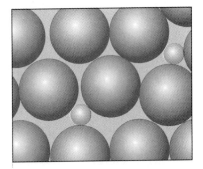

FIGURE 10.30

In an interstitial alloy, the atoms of one metal lie in the gaps between the atoms of another metal. The two elements must have markedly different atomic radii.

FIGURE 10.31

Many racing bicycle frames like this one are built from high-strength, low-density steels formed by alloying iron with metals such as manganese, molybdenum, and titanium.

interstices, in the iron lattice. The resulting material is called an **interstitial alloy** (Fig. 10.30). For two elements to form an interstitial alloy, the atomic radius of the solute element must be less than 60% of the atomic radius of the host metal. The atoms in the interstices interfere with electrical conductivity and with the movement of the atoms forming the lattice. This restricted motion makes an interstitial alloy harder and stronger than the pure host metal.

Stainless steel is an alloy of iron with metals such as chromium and nickel, which help it resist corrosion. Adding other metals to stainless steel produces a range of desired properties. For example, steel that contains manganese and molybdenum has high resistance to mechanical shock and is used in racing cycles (Fig. 10.31).

Some alloys are softer or have lower melting points than the component metals. The presence of big bismuth atoms helps to soften a metal and lower its melting point, much as grapefruit destabilize a stack of oranges because they do not fit together well. A low-melting-point alloy of lead, tin, cadmium, and bismuth called "Wood's alloy" is employed to control water sprinklers used in certain fire-extinguishing systems. The heat of the fire melts the alloy, which activates the sprinklers before the fire can spread.

Alloys of metals tend to be harder and have lower electrical conductivity than pure metals. In substitutional alloys, atoms of the solute metal take the place of some atoms of a metal of similar atomic radius. In interstitial alloys, atoms of the solute element fit into the gaps in the lattice formed by atoms of a metal with a larger atomic radius.

10.11 Ionic Structures

Many ionic solids are hard, brittle solids with high melting points. Most do not conduct electricity in the solid state, but they do act as ionic conductors when molten. These properties are related to the structures of ionic solids. They are formed from oppositely charged ions that are bound tightly together by their electrostatic attractions. However, the ions have different radii and must pack together in the arrangement that maximizes the attractions of unlike charges and minimizes the repulsion of like charges. The crystal is electrically neutral

FIGURE 10.32

The arrangement of ions in the rock-salt structure. (a) The unit cell, showing the packing of the individual ions. (b) A representation of the structure in terms of dots that identify the centers of the ions. The pink spheres are cations and the green spheres are anions.

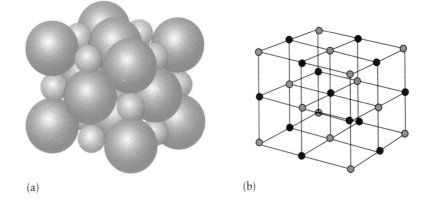

(a) (b)

overall, so each unit cell must have the same stoichiometry as the compound and be electrically neutral. No wonder ionic solids can have quite complicated structures! Two simple structures, however, are quite common.

The **rock-salt structure** takes its name from the mineral form of sodium chloride. In it, the Cl^- anions lie at the corners and in the centers of the faces of a cube, forming a face-centered cubic unit cell. The smaller Na^+ cations fit snugly between the anions. If we look carefully at Fig. 10.32, we can see that each cation is surrounded by six anions and each anion is surrounded by six cations. The pattern repeats over and over, and each ion is surrounded by six oppositely charged ions. The three-dimensional array of a vast number of these little cubes gives rise to a crystal of sodium chloride (Fig. 10.33).

In an ionic solid, the **coordination number** means the number of ions of opposite charge immediately surrounding each type of ion. In the rock-salt structure, the coordination numbers of the cations and the anions are both 6, and the structure overall is described as having *(6,6)-coordination*. In this notation, the first number is the cation coordination number (the number of anions around each cation) and the second is that of the anion (the number of cations around each anion). The rock-salt structure is found for a number of other minerals with ions of the same charge type, including KBr, RbI, MgO, CaO, and AgCl. It is common whenever the cations and anions have very different radii. Specifically, a rock-salt structure is expected when the **radius ratio,** the radius of the smaller ion (normally the cation) divided by that of the larger ion (normally

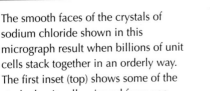

The ions are not completely close packed.

FIGURE 10.33

The smooth faces of the crystals of sodium chloride shown in this micrograph result when billions of unit cells stack together in an orderly way. The first inset (top) shows some of the stacked unit cells, viewed from one side of the crystal. The middle inset identifies the individual ions. The third inset (lower right) illustrates the coordination of an anion to its six cation neighbors.

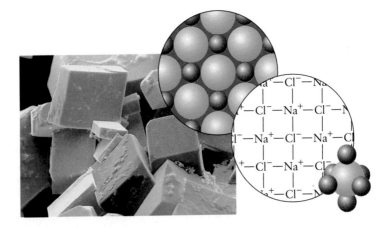

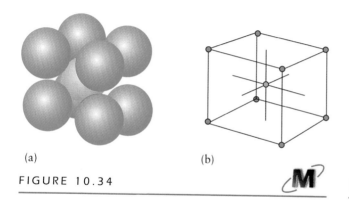

(a)   (b)

FIGURE 10.34

The cesium-chloride structure: (a) the unit cell and (b) the locations of the centers of the ions. The pink spheres are cations and the green spheres are anions.

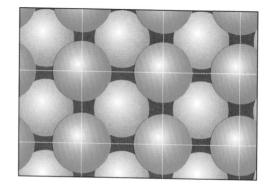

FIGURE 10.35

The repetition of the cesium-chloride unit cell recreates the entire crystal. This view from one side of the crystal shows six unit cells.

the anion), is in the range 0.4 to 0.7. For sodium chloride,

$$\text{Radius ratio} = \frac{r_{\text{cation}}}{r_{\text{anion}}} = \frac{102 \text{ pm}}{181 \text{ pm}} = 0.564$$

When the radii of the cations and anions are similar and the radius ratio is greater than 0.7, the cations cannot fit into the spaces between the anions in the rock-salt structure. Instead, the ions adopt the **cesium-chloride structure,** which has a body-centered cubic unit cell. As illustrated in Fig. 10.34, the Cs^+ cation is at the center of the cell, with eight Cl^- ions at its corners. Equivalently, each chloride ion can be regarded as being at the center of a body-centered unit cell, with eight cesium ions at its corners. The coordination number of each type of ion is 8, and overall the structure has (8,8)-coordination. This pattern repeats throughout the crystal (Fig. 10.35). The cesium-chloride structure is much less common than the rock-salt structure, but it is found for CsCl, CsI, and (with some distortion) brass of composition CuZn.

In these and many other ionic structures, we treat ions as spheres of the appropriate radii and stack them together in the arrangement that has the lowest total energy. However, the simple sphere-packing arrangement may break down if the bonding is not purely ionic. If the bonding has partial covalent character, the ions may have to lie in specific positions around one another. An example is nickel arsenide, Ni_3As_2. In this solid, the small Ni^{2+} cations polarize the big As^{3-} anions, and the bonds have some covalent character. The ions pack together in an arrangement that is quite different from a purely ionic, sphere-packing model (Fig. 10.36).

> *Ions stack together in the lowest energy regular crystalline structure; two common arrangements are the rock-salt structure and the cesium-chloride structure. Covalent character in an ionic bond imposes a directional character on the bonding.*

Self-Test 10.8A How many Na^+ ions are present in the NaCl unit cell? Account for each one.

[***Answer:*** 4; 1 center atom and $\frac{1}{4} \times 1$ on each of 12 edges]

Self-Test 10.8B How many Cl^- ions are there in one unit cell of CsCl?

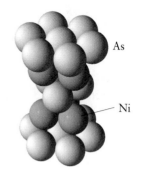

FIGURE 10.36

The structure of nickel(II) arsenide. Structures like this are found when covalent bonding is starting to become important and the ions have to take up specific positions relative to one another to maximize their bonding.

10.12 Molecular Solids

Molecular solids—solids composed of individual molecules—often have low melting points: only a small amount of thermal motion is needed to overcome the relatively weak intermolecular forces that hold the molecules in place. The physical properties of molecular solids depend on the strengths of the intermolecular forces holding them together. Some molecular solids are as soft as paraffin wax, which is a mixture of long-chain hydrocarbons. These molecules lie together in a disorderly way, and the forces between them are so weak that they can be pushed past one another very easily. Other molecular solids are brittle and hard. For example, sucrose molecules, $C_{12}H_{22}O_{11}$, are held together by hydrogen bonds between their numerous —OH groups. Hydrogen bonding between sucrose molecules is so strong that by the time the melting point has been reached (at 184°C), the molecules themselves have started to decompose. The partly decomposed mixture of products, called caramel, is used to add flavor and color to food.

Because molecules of different compounds have such widely varying shapes, they stack together in a variety of different ways. In ice, each O atom is surrounded by four H atoms at the corners of a tetrahedron. Two of these H atoms are linked to the O atom through short σ-bonds. The other two belong to neighboring H_2O molecules and are linked to the O atom by relatively long hydrogen bonds. As a result, the structure of ice is an open network of H_2O molecules held in place by hydrogen bonds (Fig. 10.37). Some of the hydrogen bonds break when ice melts. As the orderly arrangement collapses, the molecules pack less uniformly but more densely. The openness of the network in ice compared with that in the liquid explains why it has a lower density than liquid water (0.92 g/cm³ and 1.00 g/cm³, respectively, at 0°C). Solid benzene and solid carbon dioxide, in contrast, have higher densities than their liquids. Their molecules are held in place by London forces, which are much less directional than hydrogen bonds, so they can pack together more closely in the solid than in the liquid (Fig. 10.38).

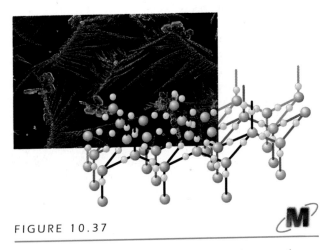

FIGURE 10.37

As the crystalline structure of ice forms, the regular array of molecules creates intricate designs. The water molecules in ice are held together by hydrogen bonds in a relatively open structure. Each O atom (red) is surrounded tetrahedrally by four hydrogen atoms, two of which are σ-bonded to it and two of which are hydrogen bonded to it.

FIGURE 10.38

As a result of an open structure, ice is less dense than water and floats in it (left). Benzene molecules can pack more tightly than water molecules in the solid state and, as a result, solid benzene is denser than liquid benzene; "benzenebergs" sink in liquid benzene (right).

Molecular solids are typically less hard and less brittle than ionic solids and melt at lower temperatures.

10.13 Network Solids

The atoms in network solids are joined to their neighbors by covalent bonds. These bonds form a network that extends throughout the crystal. Network solids exhibit the strength of the covalent bonds that bind them by being very hard, rigid materials with high melting and boiling points. They are generally insoluble in water or other solvents, and most are electrical insulators.

The carbon allotropes diamond and graphite are network solids. Each C atom in diamond is covalently bonded to four neighbors through sp^3 hybrid σ-bonds (Fig. 10.39). The tetrahedral framework extends throughout the solid like the steel framework of a large building. This structure accounts for the great hardness of the solid. Diamond is so hard it is used to protect drill bits and as a long-lasting abrasive. Diamond is also one of the best conductors of heat and is used as a base for some integrated circuits so that they do not overheat. Its high thermal conductivity results from the rigid network structure of the crystal, because the vigorous vibration of an atom in a hot part of the crystal is rapidly transmitted to distant, cooler parts through the covalent bonds, rather like the effect of slamming a door in a steel-framed building.

Graphite, the "lead" of pencils, is a black, lustrous, electrically conducting, slippery solid that sublimes at about 3700°C. It consists of flat sheets of sp^2 hybridized carbon atoms bonded covalently into hexagons like chicken wire (Fig. 10.40). There are also weak bonds between the sheets. In the commercially available forms of graphite, there are many impurity atoms trapped between the sheets. These atoms weaken the already weak intersheet bonds and let the sheets of atoms slide over one another. So, in contrast to the hardness of diamonds, graphite is soft and slippery. When we write with a pencil, the mark left on the

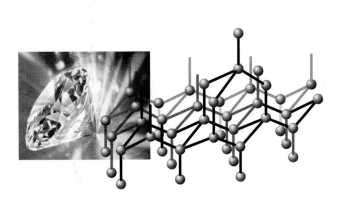

FIGURE 10.39

Part of the structure of a diamond. Each sphere represents the location of the center of a carbon atom. Each atom forms an sp^3 hybrid covalent bond to each of its four neighbors. The highly regular structure results in the smooth crystal faces seen in the photograph.

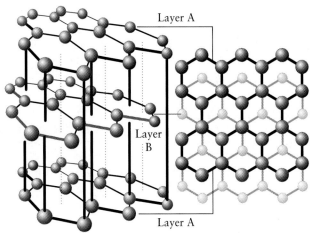

FIGURE 10.40

Graphite consists of staggered layers of hexagonal rings of sp^2 hybridized carbon atoms. On the left is a view along the layers; on the right, a view perpendicular to them. The slipperiness of graphite is due to the ease with which the layers can slide over one another when impurity atoms are present.

paper consists of rubbed-off layers of graphite. Electrons can move easily within the sheets but much less readily from one sheet to another. Hence, graphite conducts electricity better parallel to the sheets than perpendicular to them.

In nature, diamond is found embedded in a soft rock called *kimberlite*. This rock rises in columns from deep in the Earth, where the diamonds are formed under intense pressure. To make synthetic industrial diamonds, we need to recreate the geological conditions that produce natural diamonds by compressing graphite at pressures of more than 80 000 atm and temperatures above 1500°C. Small amounts of metals such as chromium and iron are included. It is thought that the molten metals dissolve the graphite and then, as they cool, deposit crystals of diamond, which are less soluble than graphite in the molten metal. Graphite is produced naturally as a result of changes in ancient organic remains. Most commercial graphite is produced by heating carbon rods to a high temperature in an electric furnace for several days.

Network solids are typically hard and rigid, and have high melting and boiling points.

PHASE CHANGES

We have seen that a substance can have more than one solid form. The term **phase** denotes the physical form of a substance and refers, not only to the three states of matter, but also to different varieties of these states. A phase may be solid, liquid, or gas, or one of several different solid forms of a substance. Thus, diamond and graphite are two solid phases of carbon. Ice, liquid water, and water vapor are three phases of water. The conversion of one phase to another, such as the melting of ice, the vaporization of water, or the conversion of graphite into diamond, is called a **phase transition.**

10.14 Vapor Pressure

Although molecules are bound together in a liquid, their energies can have a wide range of values, just as they do in the gas phase (recall Section 5.14). Some molecules have enough energy to escape from their neighbors and to become vapor. We know this from personal experience: water evaporates from open containers and the odors of perfumes drift through a room. The conversion of a liquid into a vapor is called **vaporization** or, when vaporization continues to dryness, **evaporation.** The ease with which a liquid vaporizes depends on the temperature and on the strength of the intermolecular forces within the liquid.

We can measure the tendency of a liquid to vaporize by using an apparatus like that shown in Fig. 10.41. When the stopcock between the flask and the tube is opened, vapor rushes into the vacuum above the mercury and the pressure of the vapor pushes the surface of the mercury down a few centimeters. At a fixed temperature, and provided some liquid remains in the flask, the vapor exerts a characteristic pressure called the **vapor pressure** of the liquid (Table 10.5). For example, at 40°C, water vapor presses the mercury down by 55 mm, so the pressure of water at that temperature is 55 Torr. The vapor pressure is the same whether there is 1 mL or 100 mL of liquid water present in the flask.

Liquids with high vapor pressures are said to be **volatile.** If a volatile liquid is left in an open flask, a lot of the vapor soon spreads throughout the room. For example, we can see from Fig. 10.42 that ethanol is more volatile than water.

Table 10.5 Vapor pressures at 25°C

Substance	Vapor pressure, Torr
benzene	94.6
ethanol	58.9
mercury	0.0017
methanol	122.7
water*	23.8

*For values at other temperatures, see Table 5.4.

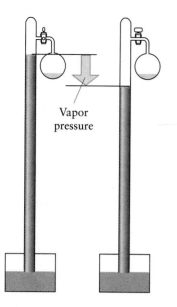

FIGURE 10.41

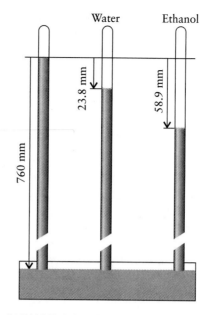

FIGURE 10.42

The apparatus on the left shows a mercury barometer, with a vacuum over the mercury column and a flask containing a liquid and its vapor. The stopcock between the top of the barometer and flask is closed. In the apparatus on the right, the stopcock has been opened and the vapor pressure exerted by the liquid in the flask (in Torr) can be measured by recording the distance in millimeters by which the column of mercury has been lowered. The vapor pressure is the same, however much liquid is present.

Three tubes of mercury; the tube in the middle is attached to a flask of water like that in Fig. 10.41, and the one on the right is similarly attached to a flask of ethanol. Neither flask is shown. The fact that ethanol has a higher vapor pressure than water is shown by the fact that the level of mercury in the tube connected to the ethanol flask is lower than the level in the tube connected to the flask with water. The total pressure at the base of each column is the sum of the pressure due to the weight of the mercury and the vapor pressure of the liquid.

Solids also exert a vapor pressure, but it is usually much lower than the vapor pressures of liquids because the molecules in a solid are gripped more tightly than they are in a liquid. The very low volatility of most solids is one reason why solids do not commonly have pronounced odors, and why our buildings and vehicles don't evaporate.

Let's consider vaporization in a closed container. A vapor is formed as molecules leave the surface of the liquid. As the number of molecules in the vapor phase increases, more of them strike the surface of the liquid. Eventually the number of molecules returning to the liquid each second exactly matches the number escaping. The vapor is now condensing as fast as the liquid is vaporizing. We say that the liquid and vapor are in **dynamic equilibrium,** a condition in which a forward process and its reverse take place at equal rates (Fig. 10.43). We write

Rate of vaporization = rate of condensation

The word *dynamic* is important: it implies continuous activity, unlike the static equilibrium of a ball resting at the foot of a hill.

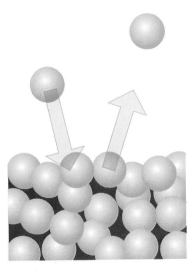

FIGURE 10.43

When a liquid and its vapor are in dynamic equilibrium inside a closed container, the rate at which the molecules leave the liquid is equal to the rate at which they return.

An analogy

A liquid in dynamic equilibrium with its vapor is like a busy store that has customers continually arriving and leaving, with the number of customers inside the store remaining constant.

The dynamic equilibrium between liquid water and its vapor is denoted

$$H_2O(l) \rightleftharpoons H_2O(g)$$

Whenever we see the symbol $\rightleftharpoons$ (which is read "in equilibrium with"), it means that the species on both sides are in dynamic equilibrium. Although products are being formed from reactants, the products are changing back into reactants at a matching rate. We can now see that *the vapor pressure of a liquid (or a solid) is the pressure exerted by its vapor when the vapor and the liquid (or the solid) are in dynamic equilibrium.* Most vaporization takes place from the surface of the liquid or solid because the molecules there can escape to the vapor more easily than those in the bulk.

Suppose another gas is present above the liquid. That would be the case if we had a stoppered flask containing water and air. Now the vapor pressure of the liquid is the *partial* pressure of its vapor at that temperature.

Vapor pressure increases with temperature. Molecules move more energetically in a hot liquid than in a cool liquid and can escape more readily from their neighbors (Fig. 10.44). We saw in Section 5.14 that average molecular speeds in gases increase with temperature. Much the same is true of molecules in liquids. As the temperature increases, the molecules of the liquid move more vigorously and those at the surface can break away from their neighbors more easily and escape as vapor.

The vapor pressure of a liquid also depends on the strength of the intermolecular forces that bind it together. Strong intermolecular forces hinder the escape of molecules and result in a low vapor pressure. We therefore expect liquids composed of molecules capable of forming hydrogen bonds to be less volatile than liquids that cannot form such bonds (see Fig. 10.44). We can see

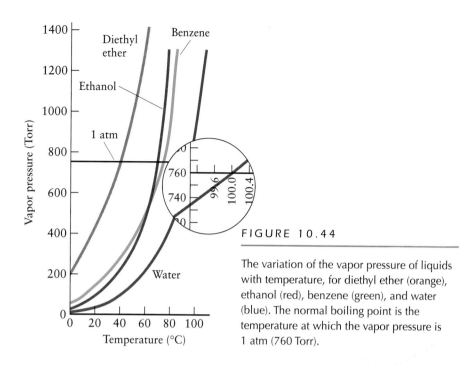

FIGURE 10.44

The variation of the vapor pressure of liquids with temperature, for diethyl ether (orange), ethanol (red), benzene (green), and water (blue). The normal boiling point is the temperature at which the vapor pressure is 1 atm (760 Torr).

the effect of hydrogen bonding clearly by comparing dimethyl ether (**14**) and ethanol (**15**), both of which have the molecular formula C_2H_6O and so might be expected to have similar vapor pressures. However, ethanol molecules each have an —OH group and can form hydrogen bonds to one another. The ether molecules cannot form hydrogen bonds to one another because all their hydrogen atoms are attached to carbon atoms. As a result, ethanol is a liquid at room temperature, whereas dimethyl ether is a gas. Even diethyl ether, $(C_2H_5)_2O$, which has a much higher molar mass than ethanol, has a higher vapor pressure than ethanol (see Fig. 10.44). However, it is important to be cautious when judging relative volatilities. Hydrocarbons with long chains, for instance, have lower vapor pressures than ethanol because their molecules have many more electrons than those of ethanol and their London forces are stronger.

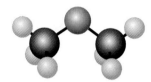

14   Dimethyl ether, CH_3OCH_3

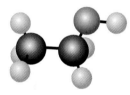

15   Ethanol, CH_3CH_2OH

16   *cis*-Dibromoethene

17   *trans*-Dibromoethene

Example 10.5 *Predicting the relative vapor pressures of substances*

Which liquid do you expect to have the higher vapor pressure at room temperature, *cis*-dibromoethene (**16**) or *trans*-dibromoethene (**17**)?

Strategy The two molecules have the same molar masses and differ only in the arrangement of their atoms. Decide which intermolecular interactions they can have and then decide which has the stronger interactions. That substance is likely to have the lower vapor pressure.

Solution Both substances can interact by London forces; the strengths will be about the same in each case because they are composed of the same atoms. However, only the cis compound is polar, so only it can also have dipole-dipole interactions. Therefore, we can expect *cis*-dibromoethene to have stronger intermolecular forces than *trans*-dibromoethene. It follows that we should expect *trans*-dibromoethene to be the more volatile.

Self-Test 10.9A Which liquid do you expect to have the higher vapor pressure at 175°C, tetrabromomethane, CBr_4, or tetraiodomethane, CI_4? Give reasons.

[*Answer:* CBr_4; weaker London forces]

Self-Test 10.9B Which liquid do you expect to have the higher vapor pressure at −50°C, CH_3CHO or $CH_3CH_2CH_3$?

There is a simple relation between the vapor pressure of a liquid and the temperature that shows how intermolecular forces affect vapor pressure. According to the **Clausius-Clapeyron equation,** which can be derived from thermodynamics, the vapor pressure P of a liquid at a temperature T is given by

$$\ln P = \text{constant} - \frac{\Delta H_{vap}}{RT} \qquad (1)$$

where ΔH_{vap} is the enthalpy of vaporization of the liquid, R is the gas constant, and the constant in the equation depends on the substance. Figure 10.44 shows the variation of the vapor pressure with temperature. The higher the enthalpy of vaporization, the more sharply the vapor pressure increases with temperature. For example, as shown in Fig. 10.44, the vapor pressure of ethanol (for which $\Delta H_{vap}° = 43.5$ kJ/mol) rises more steeply than that of benzene, which has a lower enthalpy of vaporization ($\Delta H_{vap}° = 30.8$ kJ/mol). For substances with high enthalpies of vaporization, only a few molecules have enough energy to escape from the liquid, and even a small rise in temperature can make a big difference in the fraction that can escape.

In practice, the Clausius-Clapeyron equation is useful for predicting the vapor pressure at one temperature when it is known at another. To do this, we write the Clausius-Clapeyron equation for the two temperatures:

$$\ln P_2 = \text{constant} - \frac{\Delta H_{vap}}{RT_2} \qquad \ln P_1 = \text{constant} - \frac{\Delta H_{vap}}{RT_1}$$

Then we subtract the second equation from the first. The constants cancel, and we are left with

$$\ln\left(\frac{P_2}{P_1}\right) = \frac{\Delta H_{vap}}{R}\left(\frac{1}{T_1} - \frac{1}{T_2}\right) \qquad (2)$$

(To get this equation, we have used $\ln a - \ln b = \ln(a/b)$.) Example 10.6, in Section 10.15, shows one way to use this equation.

The vapor pressure of a liquid is the pressure of the vapor in dynamic equilibrium with the liquid. Vapor pressure is low for substances with strong intermolecular forces and increases with increasing temperature.

10.15 Boiling

Suppose a liquid is left in an open container. The vapor formed spreads away from the liquid. Little, if any, is recaptured at the surface, so the rate of condensation never increases to the point at which it matches the rate of vaporization. As a result, dynamic equilibrium is never reached and, given enough time, the liquid evaporates completely. We can expect volatile compounds to evaporate more rapidly than less volatile compounds, because their molecules escape more readily from their neighbors.

Now consider what happens as we heat a liquid in an open container. As the temperature is raised, the molecules can escape more readily from the surface of the liquid and the vapor pressure rises. Eventually, at a high enough temperature, the vapor pressure becomes equal to the atmospheric pressure. At this point, bubbles of vapor can form in the body of the liquid because the vapor has a pressure great enough to drive back the atmosphere and make room for itself. We call this condition "boiling." The **normal boiling point,** T_b, of a liquid is the temperature at which a liquid boils when the atmospheric pressure is 1 atm.

Example 10.6 *Estimating the boiling point of a liquid*

Experiments show that the vapor pressure of benzene is 75 Torr at 20°C. Estimate the normal boiling point of benzene, given that its enthalpy of vaporization is 30.8 kJ/mol.

Strategy The normal boiling point of a liquid is the temperature at which its vapor pressure is 1 atm (760 Torr exactly). Use Eq. 2 to find an expression for the temperature T_2 of a liquid at which it has the pressure $P_2 = 760$ Torr, given its vapor pressure P_1 at a temperature T_1. Remember to express all temperatures in kelvins and to use R in joules per kelvin per mole (to match the units of ΔH_{vap}).

Solution To find T_2, we rearrange Eq. 2 into

$$\frac{1}{T_2} = \frac{1}{T_1} - \frac{R}{\Delta H_{vap}} \ln\left(\frac{P_2}{P_1}\right)$$

At this point, we insert the data ($T_1 = 293$ K, $P_1 = 75$ Torr, $P_2 = 760$ Torr, and $\Delta H_{vap} = 30.8 \times 10^3$ J/mol):

$$\frac{1}{T_2} = \frac{1}{293 \text{ K}} - \frac{8.3145 \text{ J/K·mol}}{30.8 \times 10^3 \text{ J/mol}} \times \ln\left(\frac{760 \text{ Torr}}{75 \text{ Torr}}\right)$$

$$= \frac{1}{293 \text{ K}} - \frac{8.3145}{30.8 \times 10^3 \text{ K}} \times \ln\left(\frac{760}{75}\right) = \frac{1}{359 \text{ K}}$$

Therefore, $T_2 = 359$ K, corresponding to 86°C. The experimental value is 80.1°C. The discrepancy arises in large part from the assumption that the enthalpy of vaporization is independent of the temperature.

Self-Test 10.10A The vapor pressure of acetone, C_3H_6O, at 7.7°C is 100. Torr, and its enthalpy of vaporization is 29.1 kJ·mol^{-1}. Estimate the normal boiling point of acetone.
[*Answer:* 62.3°C (actual: 56.2°C)]

Self-Test 10.10B The vapor pressure of methanol, CH_3OH, at 49.9°C is 400. Torr and its enthalpy of vaporization is 35.3 kJ·mol^{-1}. Estimate the normal boiling point of methanol.

Boiling occurs at a temperature higher than the normal boiling point when the pressure is greater than 1 atm, as it is in a pressure cooker. As a result of the higher temperature of the boiling water, food cooks more quickly in a pressure cooker. Boiling occurs at a temperature lower than the normal boiling point when the pressure is lower than 1 atm. At the summit of Mt. Everest—where the pressure is about 240 Torr—water boils at only 71°C, which is nearly cool enough to hold your hand in.

> *Boiling occurs when the vapor pressure of a liquid is equal to the atmospheric pressure. Strong intermolecular forces usually lead to high normal boiling points.*

10.16 Freezing and Melting

A liquid solidifies when its molecules have such low energies that they are unable to wriggle past their neighbors. In the solid, the molecules vibrate about their average positions but rarely move from place to place. The freezing temperature varies slightly as the pressure is changed, and the **normal freezing point,** T_f, of a liquid is the temperature at which it freezes at 1 atm. In practice, a liquid sometimes does not freeze until the temperature is below its freezing point. A substance that remains a liquid below its freezing point is said to be **supercooled.**

Most liquids freeze at higher temperatures when subjected to pressure because the pressure helps to hold the molecules together. However, except at very high pressures, the effect of pressure is usually quite small. Iron, for example, melts at 1813 K under 1 atm, and only a few degrees higher when the pressure is a thousand times greater. At the center of the Earth, the pressure is so high that iron is thought to be solid despite the high temperatures there.

Water is an exception to the general rule: it freezes at a *lower* temperature under pressure. Liquid water has a higher density than ice because the collapse of many of the hydrogen bonds in ice (which hold the molecules apart in an open structure) allows the molecules to occupy a smaller volume. This process is favored by the application of pressure. The melting of ice under pressure is thought to contribute to the advance of glaciers. The weight of ice pressing on the edges of rocks deep under the glacier results in very high local pressures. The liquid forms despite the low temperature, and the glacier slides slowly downhill on a film of water.

Most liquids freeze at a higher temperature under pressure. Water's hydrogen bonds make it anomalous: it freezes at a lower temperature under pressure.

10.17 Phase Diagrams

A **phase diagram** is a summary of the temperatures and pressures at which the various solid, liquid, and vapor phases of a pure substance exist. Figure 10.45 shows the phase diagram for water. In the region marked "Liquid," the liquid phase is the most stable. We should expect to find water as a liquid at the pressures and temperatures corresponding to any points on the graph lying in this region. If we raise the temperature at constant pressure, we cross into the region marked "Vapor." At any point in this region, the vapor phase is most stable. For example, if the pressure is 760. Torr, we cross the frontier at 100.°C, the boiling point of water.

Self-Test 10.11A A sample of water is kept at $-10.$°C and 320. Torr. Which phase is the most stable?

[*Answer:* Solid]

Self-Test 10.11B A sample of water is kept at 150.°C and 750. Torr. Which phase is the most stable?

The lines separating the areas in the diagram, the frontiers between the regions, are called **phase boundaries.** The points on a phase boundary show the conditions under which two phases coexist in dynamic equilibrium. The solid and liquid are in equilibrium along the line showing the dependence of melting point on pressure. For example, ice and liquid water are in equilibrium at 0°C

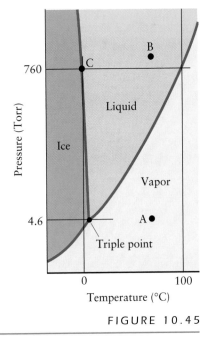

FIGURE 10.45

Phase diagram for water (not to scale). The solid lines define the boundaries of the regions of pressure and temperature at which each phase is the most stable. Note that the freezing point decreases with increasing pressure. The letters A and B are referred to in Example 10.7.

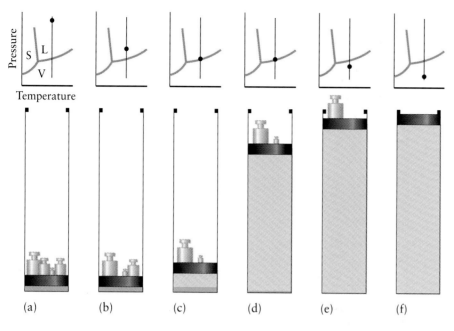

(a) (b) (c) (d) (e) (f)

and 760. Torr (point C in Fig. 10.45). The phase boundary between liquid and vapor is the line showing the pressures at which vapor and liquid are in equilibrium at a given temperature. It is therefore a plot of the vapor pressure of the liquid against temperature. The solid-vapor phase boundary is the plot of the vapor pressure of the solid, because it shows the pressures at which the solid is in equilibrium with its vapor at any temperature.

To understand the significance of a phase diagram, imagine that we have a sample of water in a cylinder fitted with a piston, that the temperature is 50.°C, and that the weights on the piston exert a pressure of 1.0 atm (Fig. 10.46). The piston presses on the surface of the liquid, as in (a). Now we gradually reduce the pressure by removing some of the weights (b). At first, nothing seems to happen. The high pressure is keeping all the water molecules in the liquid state, and the volume of a liquid changes very little with pressure. However, when so many weights have been removed that the pressure has fallen to 0.12 atm (93 Torr, the vapor pressure of water at that temperature), vapor begins to appear (c). We can now pull the piston up by an arbitrary extent (d) until enough water evaporates to maintain the pressure of 0.12 atm. We are still at the vapor-liquid boundary line on the phase diagram. If a tiny mass is loaded on the piston, the vapor is squashed back into the liquid and we move into the "liquid" region of the diagram again (b). If a tiny mass is removed from a piston that is exerting 0.12 atm, we move into the "vapor" region (e). Now the liquid evaporates completely and drives the piston out as far as needed to accommodate its vapor (f).

The slope of a phase boundary allows us to predict how the phase equilibrium it represents responds to pressure. For example, the solid-liquid phase boundary of water leans over slightly to the left. This slope signifies that the melting point of ice—the temperature at which the solid and liquid phases are in dynamic equilibrium—decreases as the pressure is raised. Therefore, if we increase the pressure on a sample of ice just below 0°C, it will melt, just as we predicted from the relative densities of ice and water in Section 10.16.

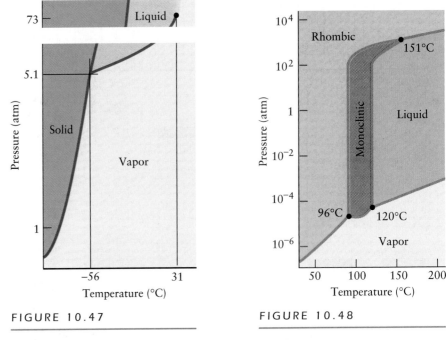

FIGURE 10.47

Phase diagram for carbon dioxide. Note that carbon dioxide sublimes to a vapor at 1 atm. Liquid carbon dioxide can exist only at pressures above 5.1 atm.

FIGURE 10.48

Phase diagram for sulfur. Notice that there are three triple points and that the pressure scale covers a very wide range of values.

Self-Test 10.12A From the phase diagram for carbon dioxide (Fig. 10.47), predict which is more dense, the solid or liquid phase. Explain your conclusion.

[*Answer:* The solid, because higher pressures favor its formation from the liquid.]

Self-Test 10.12B From the phase diagram for sulfur (Fig. 10.48), predict which solid phase is more dense. Explain your conclusion.

A **triple point** is a point on a phase diagram where three phase boundaries meet. The triple point of water occurs at 4.6 Torr and 0.010°C (see Fig. 10.45). Under these conditions (and only these), all three phases (ice, liquid, and vapor) coexist in dynamic equilibrium. Under these conditions, water molecules leave ice to become liquid and return to form ice at the same rate; liquid vaporizes and vapor condenses at the same rate; and ice sublimes and vapor condenses directly to ice again at the same rate. Unlike the freezing or boiling points of a substance, which depend on the applied pressure, the triple point of water is a fixed property. It is, in fact, used to define the size of the kelvin: by definition, there are exactly 273.16 K between absolute zero and the triple point of water. The normal freezing point of water is found to lie 0.01 K below the triple point.

A phase diagram summarizes the pressures and temperatures at which each phase is most stable (see Toolbox 10.2). The phase boundaries show the conditions under which two phases can coexist in equilibrium with each other. Three phases coexist at equilibrium at the triple point.

Toolbox 10.2 *How to interpret and use a phase diagram*

This Toolbox shows how to use a phase diagram to predict the stable phase and the phase transitions of a substance.

Conceptual Basis

The regions in a phase diagram indicate the most stable phase of a substance at each temperature and pressure (Fig. 10.49). The lines are phase boundaries; they indicate that two phases are in equilibrium. At a triple point, three phases are in equilibrium.

Procedure

Find on the diagram the point corresponding to the pressure and temperature of the substance.

1. If the point lies in a region, then that phase is the most stable and the substance can be expected to be in that phase alone.

2. If the point lies on a phase boundary, both neighboring phases are present and in dynamic equilibrium.

3. If the point corresponds to the intersection of three phase boundaries (at the triple point), then those three phases are in mutual dynamic equilibrium.

The solid-liquid phase boundary slopes up to the right (the melting point increases with pressure) if the liquid is less dense than the solid; it slopes up to the left if the liquid is more dense than the solid.

A liquid phase does not exist above the critical temperature (Section 10.18), the terminal point of a liquid-vapor phase boundary.

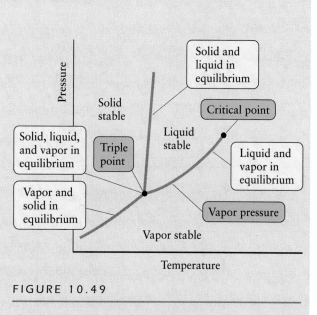

FIGURE 10.49

The phase diagram for a one-component system and the interpretation of each region, line, and point.

Example 10.7 *Interpreting a phase diagram*

Use the phase diagram in Fig. 10.45 to describe the physical states and phase changes of water as the pressure increases from 5 Torr to 800 Torr at 70°C.

Strategy Locate the points corresponding to the initial and final conditions on the phase diagram and determine the state of the substance, as described in Toolbox 10.2. If the path crosses one of the lines, then the phases meeting at the line remain at equilibrium until the phase transition is complete.

Solution Although the phase diagram is not to scale, we can find the approximate locations of the points we need. Point A is at 5 Torr and 70°C, so it lies in the vapor region. Increasing the pressure takes the system to the liquid-vapor phase boundary, at which point liquid begins to form. At this pressure, liquid and vapor are in equilibrium. The pressure is increased further to 800 Torr, which takes it to point B, in the liquid region. The sample is a liquid in this region because the applied pressure is greater than the vapor pressure and all the molecules have been pushed back into the condensed phase.

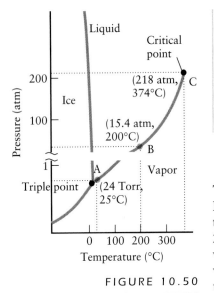

FIGURE 10.50

The phase diagram for water shown with more detail. The triple point and critical point are labeled.

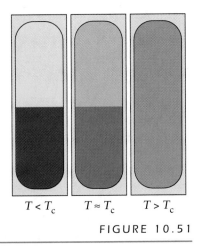

FIGURE 10.51

When the temperature of a liquid in a sealed, constant-volume container is raised, the density of the liquid decreases and the density of the vapor increases as the liquid evaporates. At the critical temperature, T_c, the density of the vapor becomes the same as the density of the liquid; above that temperature, a single uniform phase fills the container.

Self-Test 10.13A The phase diagram for carbon dioxide is shown in Fig. 10.47. Describe the physical states and phase changes of carbon dioxide as it is heated at 2 atm from $-155°C$ to $25°C$.

[*Answer:* Solid CO_2 is heated until it begins to sublime at the solid-vapor boundary. The temperature remains constant until all the CO_2 has vaporized. The vapor is then heated to $25°C$.]

Self-Test 10.13B Describe what happens when liquid carbon dioxide in a container at 60 atm and $25°C$ is released into a room at 1 atm and the same temperature.

10.18 Critical Properties

The liquid-vapor boundary terminates at a point (the point labeled C in Fig. 10.50). To see what happens at that point, suppose that the sealed, rigid tube shown in Fig. 10.51 contains liquid water and water vapor at $25°C$ and 24 Torr (the vapor pressure of water at $25°C$). This system lies on the liquid-vapor curve (point A in Fig. 10.50), so the two phases are in equilibrium, with water molecules leaving the liquid as rapidly as other water molecules return. As the water and vapor are heated, the vapor pressure increases, with the system continuing to lie higher and higher on the liquid-vapor curve. However, the water never boils, even at $100°C$, because the vapor cannot escape. Instead, the pressure continues to rise as the water vaporizes. At $200°C$, the vapor pressure has risen to 15.4 atm, point B; liquid and vapor are still in dynamic equilibrium, but now the vapor is very dense because it is at such a high pressure.

At the end of the boundary line (at point C), the system has reached $374°C$ and a vapor pressure of 218 atm—the container must be very strong! The density of the vapor is now so great that it is equal to that of the remaining liquid. Suddenly the surface separating liquid and vapor vanishes. A liquid phase can no longer be identified because there is no longer a dividing surface. Instead, a single uniform substance fills the container (as shown on the right in Fig. 10.51). Because a substance that fills any container it occupies is by definition a gas, we conclude that we have reached a temperature above which the liquid state does not exist. We have reached the **critical temperature**, T_c, of water, the temperature at and above which it cannot be condensed to a liquid. The corresponding pressure is called the **critical pressure**, P_c, of the substance. The critical pressure of water is 218 atm; that of carbon dioxide is 73 atm.

A substance can be liquefied by applying pressure only if the temperature is below the critical temperature of the substance. For example, the critical temperature of carbon dioxide is $31°C$, and we can liquefy a sample only if its temperature is below $31°C$. The critical temperatures and pressures of several substances are listed in Table 10.6. Their values are important in practice because it is useless to try to liquefy a gas by applying pressure if it is above its critical temperature. For example, the critical temperature of oxygen is $-118°C$, so we know that it cannot be compressed to a liquid at room temperature. Oxygen tanks contain only gaseous oxygen. Because the critical temperature of helium is 5.2 K, the gas must be cooled almost to absolute zero before it can be liquefied by applying pressure. Critical temperatures typically increase with the strength of intermolecular forces. Light molecules (such as oxygen and nitrogen), with their weak intermolecular forces, have low critical temperatures. Heavier molecules (and molecules that can form hydrogen bonds, such as water), with their stronger intermolecular forces, have higher critical temperatures.

A substance just above its critical temperature and pressure is called a **supercritical fluid.** It may be so dense that, although it is a gas, it can act as a solvent for liquids and solids. Supercritical carbon dioxide can dissolve organic compounds. It is used to remove caffeine from coffee beans and to extract fragrances from flowers without contaminating the extracts with potentially harmful solvents. The carbon dioxide is easily separated by allowing it to evaporate at a lower pressure. Supercritical hydrocarbons are used to dissolve coal and separate it from ash, and they have been proposed for extracting oil from oil-rich tar sands.

A substance cannot be converted to a liquid by the application of pressure if the temperature is above the critical temperature of the substance.

Self-Test 10.14A Look for trends in the data in Table 10.6 that show the effect of London forces on the critical temperature.

[*Answer:* Noble gases have higher critical temperatures as atomic number increases, so critical temperature increases with the strength of London forces.]

Self-Test 10.14B Look for trends in the data in Table 10.6 that indicate the effect of hydrogen bonding on the critical temperature.

Table 10.6 Critical temperatures and pressures of substances

Substance	Critical temperature, °C	Critical pressure, atm
He	−268 (5.2 K)	2.3
Ne	−229	27
Ar	−123	48
Kr	−64	54
Xe	17	58
H_2	−240	13
O_2	−118	50
H_2O	374	218
N_2	−147	34
NH_3	132	111
CO_2	31	73
CH_4	−83	46
C_6H_6	289	49

Skills You Should Have Mastered

Conceptual

☐ 1. Explain how London forces arise and how they vary with the number of electrons in an atom or molecule, Section 10.1.

☐ 2. Describe dipole-dipole interactions and hydrogen bonds and explain why hydrogen bonds are the strongest type of intermolecular force, Sections 10.2 and 10.3.

☐ 3. Explain why viscosity and surface tension vary with temperature and depend on the strengths of intermolecular forces, Sections 10. 4 and 10.5.

☐ 4. Explain how an electrical current is carried in metallic conductors and semiconductors, Section 10.9.

☐ 5. Explain the meaning of dynamic equilibrium, Section 10.14.

☐ 6. Interpret the major features of a phase diagram, Toolbox 10.2.

Problem-Solving

☐ 1. Predict the relative boiling points, vapor pressures, and enthalpies of vaporization of substances from the strengths of their intermolecular forces, Section 10.2 and Examples 10.1 and 10.2.

☐ 2. Identify molecules that can take part in hydrogen bonding, Self-Test 10.3.

☐ 3. Determine the coordination number of an atom or ion in a given crystal lattice, Self-Test 10.5.

☐ 4. Deduce the number of atoms or ions in a given unit cell, Example 10.3.

☐ 5. Deduce the structure of a solid from its density, Toolbox 10.1 and Example 10.4.

☐ 6. Predict the relative vapor pressures of two substances, Example 10.5.

☐ 7. Predict the boiling temperature of a substance, Example 10.6.

☐ 8. Use a phase diagram to identify the stable phase of a substance at a given temperature and pressure and predict phase transitions, Toolbox 10.2 and Example 10.7.

Descriptive

☐ 1. List properties of water that are anomalous as a result of hydrogen bonding, Sections 10.3 and 10.16.

☐ 2. Distinguish substitutional and interstitial alloys and predict which combinations of elements produce which type of alloy, Section 10.10.

☐ 3. Distinguish metals, ionic solids, network solids, and molecular solids by their structures and by their properties, Sections 10.6–10.13.

☐ 4. Describe cubic and hexagonal close-packing and primitive and body-centered cubic structures, Section 10.7.

☐ 5. Describe the rock-salt and cesium-chloride structures, Section 10.11.

☐ 6. Describe the structures of graphite and diamond, and explain how their structures account for their properties, Section 10.13.

Exercises

Intermolecular Forces

10.1 Identify the kinds of intermolecular forces that might arise between molecules of the following substances: (a) Cl_2; (b) HCl; (c) C_6H_6; (d) C_6H_5Cl.

10.2 Identify the kinds of intermolecular forces that might arise between molecules of the following substances: (a) N_2; (b) H_2O; (c) $CH_2=CH_2$; (d) CHCl=CHCl (consider both the cis and trans isomers).

10.3 Write the Lewis structure and indicate the dominant type of intermolecular forces for (a) CO; (b) H_2O; (c) NF_3; (d) CH_4.

10.4 Write the Lewis structure and indicate the dominant type of intermolecular forces for (a) CH_3Cl; (b) OF_2; (c) Xe; (d) SO_2Cl_2 (sulfur is the central atom).

10.5 Explain the structure of a hydrogen bond.

10.6 List four consequences of hydrogen bonding and explain why they occur.

10.7 Which of the following molecules are able to form hydrogen bonds: (a) HF; (b) CH_4; (c) NH_3; (d) CH_3OH?

10.8 Which of the following molecules are able to form hydrogen bonds: (a) D_2O; (b) CH_3COOH; (c) CH_3CH_2OH; (d) H_3PO_4?

10.9 Write the Lewis structures and give the shapes of each of the following molecules. Predict which substance of each pair has the higher boiling point: (a) SF_4 or SF_6; (b) BF_3 or ClF_3; (c) SF_4 or CF_4; (d) cis-CHCl=CHCl or trans-CHCl=CHCl.

10.10 Write the Lewis structures and give the shapes of each of the following molecules. Predict which substance of each pair has the higher boiling point: (a) PF_3 or PCl_3; (b) SO_2 or CO_2; (c) BF_3 or BCl_3; (d) $AsCl_3$ or $AsCl_5$.

Liquid Structure

10.11 Water "beads" on a waxed surface. What physical property accounts for this observation?

10.12 Explain why water forms a concave meniscus in a glass graduated cylinder but a convex meniscus in a plastic graduated cylinder.

10.13 Predict which substance has the greater viscosity in its liquid form at 0°C: (a) ethanol, CH_3CH_2OH, or dimethyl ether, CH_3-O-CH_3; (b) butane, C_4H_{10}, or propanone, CH_3COCH_3.

10.14 Predict which liquid has the greater surface tension: (a) cis-dichloroethene or trans-dichloroethene (refer to structures (**5**) and (**6**)); (b) benzene at 20°C or benzene at 60°C.

Classification of Solids

10.15 Use Table 10.3 to classify each of the following solids according to its bonding type: (a) sodium chloride; (b) solid nitrogen; (c) sugar (sucrose); (d) copper.

10.16 Use Table 10.3 to classify each of the following solids according to its bonding type: (a) quartz; (b) sulfur; (c) iron(II) sulfate; (d) brass.

10.17 Three unknown substances were tested in order to classify them. The following table shows the results of the tests. Use Table 10.3 to classify substances A, B, and C as metallic solids, ionic solids, network solids, or molecular solids.

Substance	Appearance	Melting point, °C	Electrical conductivity	Solubility in water
A	hard, colorless	800	only if melted or dissolved in water	soluble
B	lustrous, malleable	1500	high	insoluble
C	soft, yellow	113	none	insoluble

10.18 Three unknown substances were tested in order to classify them. The following table shows the results of the tests. Use Table 10.3 to classify substances X, Y, and Z as metallic, ionic, network, or molecular solids.

Substance	Appearance	Melting point, °C	Electrical conductivity	Solubility in water
X	hard, colorless	146	none	soluble
Y	very hard, colorless	1600	none	insoluble
Z	hard, orange	398	only if melted or dissolved in water	soluble

Metallic Crystals

10.19 Determine (a) the number of atoms per unit cell and (b) the coordination number of an atom in a body-centered cubic (bcc) structure.

10.20 Determine (a) the number of atoms per unit cell and (b) the coordination number of an atom in a face-centered cubic (fcc) structure.

10.21 Above 500°C, cobalt crystallizes in a face-centered cubic structure. Its metallic radius is 125 pm. (a) What is the length of the side of the unit cell? (b) How many unit cells are there in 1.00 cm^3 of cobalt?

10.22 Calcium crystallizes in a face-centered cubic structure. Its metallic radius is 180. pm. (a) What is the length of the side of the unit cell? (b) How many unit cells are there in 1.00 cm^3 of calcium?

10.23 The metal polonium (which was named by Marie Curie after her homeland, Poland) crystallizes in a primitive cubic structure, with an atom at each corner of a cubic unit cell. The atomic radius of polonium is 167 pm. Sketch the unit cell and determine (a) the number of atoms per unit cell; (b) the coordination number; (c) the length of the side of the unit cell.

10.24 Potassium crystallizes in a bcc structure. The atomic radius of potassium is 235 pm. Sketch the unit cell and determine (a) the number of atoms per unit cell; (b) the coordination number of the lattice; (c) the length of the side of the unit cell.

10.25 Determine the density of each of the following metals from the data: (a) nickel (fcc structure), atomic radius 125 pm; (b) rubidium (bcc structure), atomic radius 250. pm.

10.26 Calculate the density of each of the following metals from the data: (a) platinum (fcc structure), atomic radius 139 pm; (b) cesium (bcc structure), atomic radius 272 pm.

10.27 Calculate the atomic radius of the following elements from the data: (a) gold (fcc structure), density 19.3 g/cm³; (b) vanadium (bcc structure), density 6.11 g/cm³.

10.28 Calculate the atomic radius of each of the following elements from the data: (a) iridium (fcc structure), density 22.6 g/cm³, the densest of all known elements; (b) molybdenum (bcc structure), density 10.2 g/cm³.

Metals and Alloys

10.29 Explain the difference between electronic and ionic conduction.

10.30 Explain the difference between a superconductor and a semiconductor.

10.31 An alloy of tin in lead is used as solder in electrical circuits. (a) Is the alloy interstitial or substitutional? Justify your answer. (b) How do you expect the melting point of the alloy to differ from that of pure lead?

10.32 When iron surfaces are exposed to ammonia at high temperatures, "nitriding," the incorporation of nitrogen into the iron lattice, occurs. The atomic radius of iron is 124 pm. (a) Is the alloy interstitial or substitutional? Justify your answer. (b) How do you expect nitriding to change the properties of the iron?

The Band Theory of Solids

10.33 State whether each of the following materials is an n- or a p-type semiconductor: (a) Si doped with P; (b) Si doped with In; (c) Ge doped with Sb.

10.34 State whether each of the following materials is an n- or a p-type conductor: (a) Si doped with Al; (b) Ge doped with Ga; (c) Ge doped with As.

10.35 Distinguish between an electrical conductor and an electrical insulator in terms of molecular orbitals.

10.36 Silicon is a semiconductor even without being doped (diamond is too). Account for this property in terms of bands of molecular orbitals, using as a clue the fact that at absolute zero ($T = 0$) the conductivities of silicon and carbon are 0.

10.37 Zinc oxide is a semiconductor that loses oxygen atoms in a vacuum but gains additional oxygen atoms when heated in oxygen. Its conductivity increases when it is heated in a vacuum but decreases when it is heated in oxygen. Account for these observations.

10.38 The electrical conductivity of graphite parallel to its planes is different from that perpendicular to them. Parallel to the planes, the conductivity decreases as the temperature is raised; but perpendicular to them, it rises. In what sense is graphite a metallic conductor or a semiconductor?

Ionic Solids

10.39 Calculate the number of cations, anions, and formula units per unit cell in the following solids: (a) the rock-salt unit cell shown in Fig. 10.32; (b) the fluorite, CaF_2, unit cell shown below. What are the coordination numbers of the ions in fluorite?

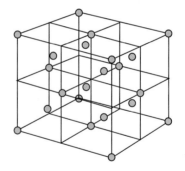

○ Ca (at opposite corners of small cubes)
◐ F (at centers of small cubes)

10.40 Calculate the number of cations, anions, and formula units per unit cell in the following solids: (a) the cesium-chloride unit cell shown in Fig. 10.34; (b) the rutile, TiO_2, unit cell shown below. What are the coordination numbers of the ions in rutile?

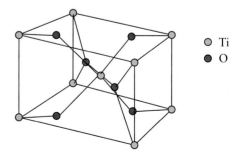

○ Ti
● O

10.41 A unit cell of the mineral perovskite is drawn below. What is its formula?

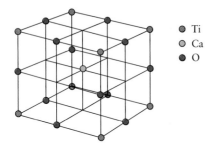

Ti
Ca
O

10.42 A unit cell of one of the new high-temperature superconductors is shown below. What is its formula?

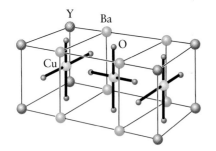

10.43 How many unit cells of the kind shown in Fig. 10.32 are there in a 1.00-mm³ grain of table salt? (The density of sodium chloride is 2.17 g/cm³.)

10.44 How many unit cells of the kind shown in Fig. 10.32 are there in a 1.00-mm³ grain of potassium bromide? (The density of potassium bromide is 2.75 g/cm³.)

10.45 Use radius ratios to predict the coordination number of the cation in (a) KBr; (b) LiBr; (c) BaO. (See Fig. 7.36.)

10.46 Use radius ratios to predict the coordination number of the cation in (a) RbF; (b) MgO; (c) NaBr. (See Fig. 7.36.)

Network Solids

10.47 Name three elements that, in at least one allotropic form, are network solids.

10.48 Name two compounds that exist as network solids at room temperature and pressure.

10.49 The enthalpy of sublimation of diamond is 713 kJ/mol. What is the C—C bond enthalpy in the solid?

10.50 The enthalpy of formation of BN(s) is −254 kJ/mol, and that of BN(g) is +647 kJ/mol. Given that the enthalpies of formation of gaseous B and N atoms are +563 kJ/mol and +473 kJ/mol, respectively, calculate the B—N bond enthalpy in the solid.

Molecular Solids

10.51 Identify the types of intermolecular forces that are responsible for the existence of each of the following molecular solids: (a) solid argon; (b) ice; (c) solid HCl.

10.52 Identify the types of intermolecular forces that are responsible for the existence of each of the following molecular solids: (a) iodine; (b) oxalic acid, $H_2C_2O_4$; (c) solid benzene.

10.53 The figure below shows a unit cell of a two-dimensional molecular compound of elements A, B, and C. What is its molecular formula?

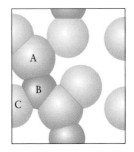

A
B
C

10.54 State the molecular formula of the imaginary element E with the two-dimensional unit cell shown below.

Vapor Pressure

10.55 What is meant by "dynamic equilibrium"? How does it differ from static equilibrium?

10.56 How, by using isotopes, could you show that an equilibrium is dynamic and not static?

10.57 Hydrogen peroxide, H_2O_2, is a syrupy liquid with a vapor pressure lower than that of water and a boiling point of 152°C. Account for the differences between these properties and those of water.

10.58 Explain how the vapor pressure of a liquid is affected by each of the following changes in conditions: (a) an increase in temperature; (b) an increase in surface area of the liquid; (c) an increase in volume above the liquid; (d) the addition of air to the volume above the liquid.

10.59 Suppose you were to collect 1.0 L of air by slowly passing it through water at 20.°C and into a container. Estimate the mass of water vapor in the collected air, assuming that the air is saturated with it. At 20.°C, the vapor pressure of water is 17.5 Torr.

10.60 Suppose you were to collect 500. mL of nitrogen gas by slowly passing it through ethanol at 25°C, at which temperature ethanol's vapor pressure is 58.9 Torr. Estimate the mass of ethanol vapor in the collected gas, assuming that the nitrogen is saturated with it.

10.61 Use the vapor pressure curve in Fig. 10.44 to estimate the boiling point of water when the atmospheric pressure is (a) 735 Torr; (b) 750. Torr.

10.62 Use the vapor pressure curve in Fig. 10.44 to estimate the boiling point of ethanol when the atmospheric pressure is (a) 600. Torr; (b) 400. Torr.

Phase Changes

10.63 The vapor pressure of ethanol at 34.9°C is 13.3 kPa. Use the information in Table 6.2 to estimate the normal boiling point of ethanol.

10.64 Tetrachloromethane, CCl_4, which is now known to be carcinogenic, was once used as a dry cleaning solvent. The enthalpy of vaporization of CCl_4 is 33.05 kJ/mol, and its vapor pressure at 57.8°C is 405 Torr. Estimate the normal boiling point of tetrachloromethane.

10.65 Use Fig. 10.50 (the phase diagram for water) to predict the state of a sample of water under the following sets of conditions: (a) 1 atm, 200.°C; (b) 100. atm, 200.°C; (c) 3 Torr, 25°C; (d) 218 atm, 375°C.

10.66 Using Fig. 10.47 (the phase diagram for carbon dioxide) to predict the state of a sample of CO_2 under the following sets of conditions: (a) 6 atm, −80.°C; (b) 1 atm, −56°C; (c) 80. atm, 25°C; (d) 5.1 atm, −56°C.

10.67 A vacuum pump is attached to a flask of water at 0°C and 2 atm, and the pressure on the liquid is decreased to 5 Torr at constant temperature. (a) Explain what would be observed, based on the phase diagram of water, Fig. 10.50. (b) Explain what would be observed if the water were at 50°C instead of 0°C.

10.68 Use the phase diagram for carbon dioxide (Fig. 10.47) to predict what would happen to a sample of carbon dioxide gas at −50.°C and 1 atm if its pressure were suddenly increased to 73 atm. What would be the final physical state of the carbon dioxide?

10.69 The phase diagram for helium is shown on the right. Use it to answer the following questions: (a) What is the maximum temperature at which superfluid helium-II can exist? (b) What is the minimum pressure at which solid helium can exist? (c) What is the normal boiling point of helium-I? (d) Can solid helium sublime?

10.70 The phase diagram for carbon is shown top, right: it indicates the extreme conditions that are needed to form diamonds from graphite. Use the diagram to answer the following questions: (a) At 2000 K, what is the minimum pressure needed before graphite changes to diamond?

(b) What is the minimum temperature at which liquid carbon can exist at pressures below 10 000 atm? (c) At what pressure does graphite melt at 3000 K? (d) Are diamonds stable under normal conditions? If not, why is it that people can wear them without the diamonds having to be compressed and heated?

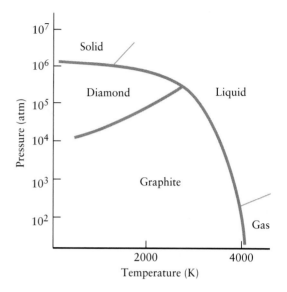

10.71 Use the phase diagram for helium (below) (a) to describe the phases in equilibrium at each of helium's two triple points; (b) to state which liquid phase is more dense, helium-I or helium-II.

10.72 Use the phase diagram for carbon (above) (a) to describe the phase transitions that carbon would experience if compressed at a constant temperature of 2000 K from 100 atm to 1×10^6 atm; (b) to rank the diamond, graphite, and liquid phases of carbon in order of increasing density.

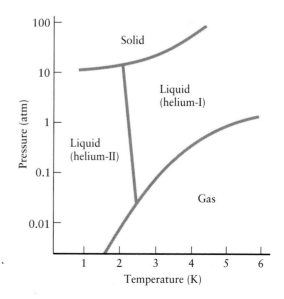

Supplementary Exercises

10.73 Explain in terms of intermolecular forces why the skin feels cooler when wet with rubbing alcohol (isopropanol) than when wet with water.

10.74 Explain the effect an increase in temperature has on each of the following properties: (a) viscosity; (b) surface tension; (c) vapor pressure; (d) evaporation rate.

10.75 Account for the following observations in terms of the type and strength of intermolecular forces. (a) The melting point of xenon is $-112°C$ and that of argon is $-189°C$. (b) The critical temperature of HI is $151°C$ and that of HCl is $52°C$. (c) The vapor pressure of diethyl ether ($C_2H_5OC_2H_5$) is greater than that of water.

10.76 Account for the following observations in terms of the type and strength of intermolecular forces. (a) The boiling point of NH_3 is $-33°C$ and that of PH_3 is $-87°C$. (b) The critical temperature of H_2O is $374°C$ and that of HF is $188°C$. (c) The vapor pressure of CH_3OH is less than that of CH_3SH at $25°C$.

10.77 Suggest, giving reasons, which substance in each pair is likely to have the higher normal boiling point (Lewis structures may help your arguments): (a) HCl or NaCl; (b) CH_4 or SiH_4; (c) HF or HCl; (d) H_2O (to compare, write as HOH) or CH_3OH.

10.78 Complete the following statements about the effect of intermolecular forces on the physical properties of a substance: (a) The higher the boiling point of a liquid, the (stronger, weaker) its intermolecular forces. (b) Substances with strong intermolecular forces have (high, low) vapor pressures. (c) Substances with strong intermolecular forces typically have (high, low) surface tensions. (d) The higher the vapor pressure of a liquid, the (stronger, weaker) its intermolecular forces. (e) Because nitrogen, N_2, has (strong, weak) intermolecular forces, it has a (high, low) critical temperature. (f) Substances with high vapor pressures have correspondingly (high, low) boiling points. (g) Because water has a relatively high boiling point, it must have (strong, weak) intermolecular forces and a correspondingly (high, low) enthalpy of vaporization.

10.79 Select the one substance that has the corresponding property. Justify your answer in each case: (a) the strongest hydrogen bonding: H_2O, H_2S, CH_3OH (as liquids); (b) the greatest surface tension: CH_3OH, C_2H_5OH, C_3H_7OH (as liquids); (c) the highest vapor pressure: CO_2, SO_2, SiO_2 (as solids); (d) the lowest viscosity: HCl, HBr, HI (as liquids); (e) the lowest enthalpy of vaporization: H_2O, H_2S, H_2Te (as liquids); (f) the lowest critical temperature: O_2, N_2, H_2 (as gases); (g) the highest boiling point: AsH_3, PH_3, NH_3 (as liquids); (h) the highest enthalpy of fusion: Na_2O, H_2O, Cl_2O (as solids); (i) the strongest dipole-dipole forces: H_2S, SCl_2, SF_2; (j) the strongest London forces: CH_4, SiF_4, GeF_4.

10.80 The normal boiling temperature of water is $100.°C$. Suppose a cyclonic region (a region of low pressure) moves into the area. State and explain what happens to the boiling point of the water.

10.81 Relative humidity at a particular temperature is defined as

$$\text{Relative humidity} = \frac{\text{partial pressure of water}}{\text{vapor pressure of water}} \times 100\%$$

The vapor pressure of water at various temperatures is given in Table 5.4. (a) What is the relative humidity at $30.°C$ when the partial pressure of water is 25.0 Torr? (b) Explain what would be observed if the temperature of the air fell to $25°C$.

10.82 Krypton crystallizes with a face-centered cubic unit cell of edge 559 pm. (a) What is the density of solid krypton? (b) What is the atomic radius of krypton? (c) What is the volume of one krypton atom? (The volume of a sphere of radius r is $\frac{4}{3}\pi r^3$.) (d) What percentage of the unit cell is empty space if each atom is treated as a hard sphere?

10.83 "Graphite bisulfates" are formed by heating graphite with a mixture of sulfuric and nitric acids. In the reaction, the graphite planes are partially oxidized (so that there is one positive charge shared between about 24 carbon atoms) and the HSO_4^- anions are distributed between the planes. What effects should this have on the electrical conductivity?

10.84 Aluminum metal has a density of 2.70 g/cm^3 and crystallizes in a lattice with a cubic unit cell. (a) What type of cubic unit cell is formed by aluminum? (b) What is the coordination number of aluminum?

10.85 All the noble gases except helium crystallize with cubic close-packed structures. Find an equation relating the atomic radius to the density of a cubic close-packed solid of given molar mass and apply it to deduce the atomic radii of the noble gases, given the following densities (in g/cm^3): Ne, 1.21; Ar, 1.66; Kr, 2.82; Xe, 3.56; Rn, 4.4.

10.86 All the alkali metals crystallize into bcc structures. (a) Find an equation relating the metallic radius to the density of a bcc solid of an element in terms of its molar mass and use it to deduce the atomic radii of the alkali metals, given the following densities (in g/cm^3): Li, 0.53; Na, 0.97; K, 0.86; Rb, 1.53; Cs, 1.87. (b) Find a factor for converting a bcc density to a cubic close-packed density of the same element (see Exercise 10.85). (c) Calculate what the densities of the alkali metals would be if they were cubic close-packed. (d) Which, if any, would float on water?

10.87 Metals with bcc structures, like tungsten, are not close packed. Therefore, their densities would be greater if they were to change to a cubic close-packed structure (under pressure, for instance). What would the density of tungsten be if it had a cubic close-packed structure rather than bcc? The density of the bcc form of tungsten is 19.3 g/cm^3.

10.88 To create a special effect for a magic show, a group of chemistry students put chunks of dry ice into a tub of water. Vigorous bubbling began as soon as the dry ice had been added, and white clouds poured from the flask and created a fog along the floor. (a) What was the maximum temperature of the dry ice in the tub? (b) What was the gas in the bubbles? (c) What was the composition of the white fog? (d) Why did the fog stay along the floor instead of rising?

10.89 The phase diagram below belongs to which Period 2 element? Explain how you came to your conclusion.

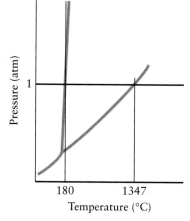

10.90 A new substance developed in a laboratory has the following properties: normal melting point, 83.7°C; normal boiling point, 177°C; triple point, 200. Torr and 38.6°C. Sketch the approximate phase diagram and label the solid, liquid, and vapor phases and the solid-liquid, liquid-vapor, and solid-vapor phase boundaries.

10.91 A reflection from a cubic crystal found in a moon rock was observed at a glancing angle of 12.1° when x-rays of wavelength 152 pm were used. What is d, the separation of the layers causing this reflection? See Investigating Matter 10.1.

10.92 At what glancing angle would you detect the reflection arising from the same crystal as in the previous exercise, but reflected from the second layer away (so d is twice as large)? See Investigating Matter 10.1.

Applied Exercises

For Exercises 10.93–10.96, see Applying Chemistry: Case Study 10.

10.93 Identify the following substances as isotropic or anisotropic: (a) a crystal of NaCl; (b) an aqueous solution of NaCl; (c) a smectic liquid crystal; (d) a snowflake; (e) a cell membrane.

10.94 Explain how a nematic liquid crystal differs from (a) a solid glass material; (b) an isotropic liquid.

10.95 Nonpolar molecules with shapes similar to that of *p*-azoxyanisole do not form liquid crystals. Explain how polar groups contribute to the properties of a liquid crystal.

10.96 1-Pentanol, an organic compound with the formula $CH_3CH_2CH_2CH_2CH_2OH$, has molecules with a nonpolar hydrocarbon chain of medium length attached to a polar —OH group. It is insoluble in water. However, if detergent is added, a lyotropic liquid crystal forms and suspends the pentanol in the water. Propose a structure for the lyotropic liquid crystal, describing the arrangement of the water, pentanol, and detergent molecules.

10.97 (a) Suggest, in terms of energy bands, why the electrical conductivity of some substances increases when they are exposed to light. (b) The band gap in amorphous selenium, which is used in the xerographic photocopying process, is 2.9×10^{-19} J. What is the longest wavelength of light that can make it conduct? (Recall from Chapter 7 that $E_{photon} = h\nu$.) (c) Can amorphous selenium be used in infrared burglar alarms?

10.98 Paper towels are effective in cleaning up water spills. Towels made from a synthetic fiber (such as a polyamide or polyester) may be more economical to produce. Explain why paper towels are effective. What properties should the synthetic fiber exhibit to be more effective than paper towels?

Integrated Exercises

10.99 Liquid ammonia is a polar solvent in which polar organic compounds are soluble. However, because the dipole moment of ammonia is lower than that of water, ammonia does not solvate ions as well as water does. Consequently, the only ionic compounds that are soluble in ammonia are those with appreciable covalent character. Predict which compound in each of the following pairs will be more soluble in ammonia: (a) CH_4, CH_3OH; (b) KCl, KI; (c) LiBr, CsBr.

10.100 The vapor pressure of ethanol at 25°C is 58.9 Torr. A sample of ethanol vapor at 25°C and 58.9 Torr partial pressure is in equilibrium with a very small amount of liquid ethanol in a 10-L container also containing dry air, at a total pressure of 750. Torr. The volume of the container is then reduced to 5.0 L. (a) What is the partial pressure of ethanol in the smaller volume? (b) What is the total pressure of the mixture?

10.101 Some compounds of *d*-metal ions form nonstoichiometric solids, for which the formula may vary. These compounds are actually mixtures that include the metal ion in more than one oxidation state. Iron forms a nonstoichiometric compound with oxygen that has the formula $Fe_{1.7}O_2$. What percentage of the iron atoms is in the +2 oxidation state and what percentage is in the +3 oxidation state?

10.102 Nickel forms a nonstoichiometric compound with oxygen that has the formula $Ni_{0.97}O$. What percentage of the nickel is in the +2 oxidation state and what percentage is in the +3 oxidation state? (See Exercise 10.101.)

10.103 The van der Waals constant *a* for NH_3 is greater than that for CO_2, which in turn is greater than that for O_2. Use your knowledge of intermolecular forces to explain why this is so.

Chapter 11

Carbon-Based Materials

Carbon is a most versatile element. Almost every piece of clothing we wear is made of compounds of carbon. So is most of the food we eat and the fuel that keeps us warm. We put its compounds to use in our automobiles, our computers, and our calculators. Polymer chemists are learning how to create custom materials based on carbon for medicine, industry, and research. Although computer chips manufactured today are based on silicon, even that may change, and in the future our computers may be made entirely of carbon compounds. The unique feature that makes carbon so adaptable is the ability of its atoms to link to one another through covalent bonds to form chains and rings of endless variety. In this chapter, we see how chemists use the information in the previous three chapters to form bonds between carbon atoms and build these important and fascinating materials.

Despite the amazing variety of molecules that carbon forms and the many reactions they undergo, the properties of organic compounds can be understood in terms of small groups of atoms. **Organic chemistry,** the chemistry of the compounds of carbon, is therefore the chemistry of *families* of compounds and of patterns of reaction shown by the groups of atoms these compounds contain.

Whenever we cook food, we are acting as organic chemists. The proteins, fats, and carbohydrates that make up our food are organic compounds, and cooking the food causes chemical reactions to occur that decompose some of the compounds into tastier ones. In this chapter we see a little of the amazing variety of organic compounds and study some of their reactions.

HYDROCARBONS

Toolbox 11.1 in Section 11.4 explains how to name hydrocarbons.

The simplest organic compounds are the **hydrocarbons,** which are compounds that contain only carbon and hydrogen. Even though hydrocarbons consist of only two elements, there are enormous numbers of them because of the many ways in which carbon atoms can bond with one another.

11.1 Types of Hydrocarbons

The primary sources of hydrocarbons are the fossil fuels petroleum and coal (see Investigating Matter 6.1). Petroleum is a mixture of many different hydrocarbons, which are used as fuels and raw materials for synthetic fabrics, medicines, and construction materials.

Because there are so many hydrocarbons, some of which are quite complicated, we need a simple way to represent their structures. To show the spatial arrangements of atoms, we use a **structural formula,** which shows how the atoms are connected in detail. Alternatively, we can use a **line structure,** which represents a chain of carbon atoms as a zigzag line: each angle or unlabeled end of a line in the zigzag represents a carbon atom. Because carbon nearly always has a valence of four in organic compounds, we do not need to show any C—H bonds explicitly. We just fill in the correct number of hydrogen atoms mentally.

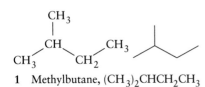

1 Methylbutane, $(CH_3)_2CHCH_2CH_3$

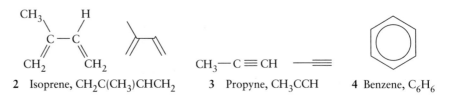

2 Isoprene, $CH_2C(CH_3)CHCH_2$ 3 Propyne, CH_3CCH 4 Benzene, C_6H_6

Examples of structural and line formulas are methylbutane (**1**), isoprene (**2**), and propyne (**3**). A benzene ring is represented in both types of formulas by a circle inside a hexagon, and we need to remember that one hydrogen atom is attached to each carbon atom (**4**).

There are several ways of writing the molecular formulas of hydrocarbons. Often it is sufficient to show only how the atoms are grouped together. For instance, we write $CH_3CH_2CH_2CH_3$ for butane and $CH_3CH(CH_3)CH_3$ for methylpropane. The parentheses around the CH_3 group show that it is attached to the carbon atom on its left. In general, when writing a formula, we look for a short, unambiguous way to describe the compound. For instance, when several CH_2 groups are repeated, we collect them together. Thus, we can write butane more compactly as $CH_3(CH_2)_2CH_3$ and methylpropane as $CH(CH_3)_3$. These expressions are examples of a **condensed structural formula,** a compact representation of a structural formula.

You can tell whether a group in parentheses is a side chain or part of the main chain by recalling that carbon has a valence of four: $(CH_2)_2$ must be part of the main chain, but $(CH_3)_2$ must be side chains or groups at the end of a chain.

Aromatic hydrocarbons have a benzene ring as a part of their molecular structure; **aliphatic hydrocarbons** do not. The compound shown as (**5**) is aromatic; the compound in (**6**) is aliphatic. In complex molecules, such as (**7**), we speak of aliphatic parts and aromatic parts, rather than trying to classify the molecule as a whole. A **saturated hydrocarbon** is an aliphatic hydrocarbon with no multiple bonds. An **unsaturated hydrocarbon** has one or more double or triple bonds. More hydrogen can be added to compounds in which there are multiple bonds, but compounds with only single bonds are

A saturated hydrocarbon is one in which the carbon atoms are bonded to as many hydrogen atoms as possible. It can be thought of as "saturated" with hydrogen.

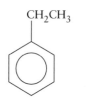

5   Ethylbenzene, $C_6H_5CH_2CH_3$

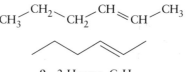

6   Pentane, C_5H_{12}

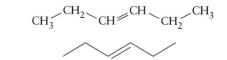

7   3,4-Dimethyl-1-phenylpentane

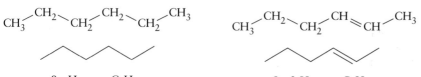

8   Hexane, C_6H_{14}

9   2-Hexene, C_6H_{12}

10   3-Hexene, C_6H_{12}

"saturated" with hydrogen. Compound (**8**) is saturated; compounds (**9**) and (**10**) are unsaturated.

Aromatic hydrocarbons contain benzene rings; aliphatic hydrocarbons do not. Saturated hydrocarbons have only single bonds.

Example 11.1 *Drawing structural formulas from condensed structural formulas and line formulas*

Draw the structural formulas of (a) 2,2-dimethylbutane, $CH_3C(CH_3)_2CH_2CH_3$, which is obtained from petroleum; (b) cyclohexene (**11**), which is found in coal tar, a sticky substance distilled from coal.

Strategy (a) Insert single bonds between all atoms, unless a multiple bond is indicated by the presence of a C atom attached to fewer than four atoms. Such a C atom must have its valence of four satisfied by the formation of a multiple bond. A group of atoms enclosed in parentheses is attached to the preceding atom through the first atom in the group, unless the group is at the beginning of the chain, in which case it is attached to the following atom. (b) Draw a C atom at the end of each line segment that is not occupied by another atom. Add enough H atoms to provide each C with four bonds.

Solution (a) Draw a chain of four C atoms, with each H atom and group in parentheses attached to the preceding C atom (**12**). (b) Draw six C atoms in a hexagon, with a double line between two C atoms and single lines between the others. Add two H atoms to each of the four C atoms with only single bonds and one H atom to each of the two C atoms connected with a double bond (**13**).

11   Cyclohexene, C_6H_{10}

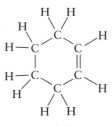

13   Cyclohexene

12 2,2-Dimethylbutane, C_6H_{10}

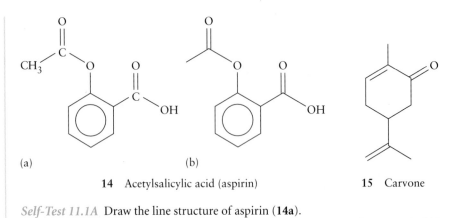

(a)   (b)

14   Acetylsalicylic acid (aspirin)   **15   Carvone**

Self-Test 11.1A Draw the line structure of aspirin (**14a**).

[***Answer:*** (**14b**)]

Self-Test 11.1B Write the molecular structure of carvone (**15**).

11.2 Alkanes

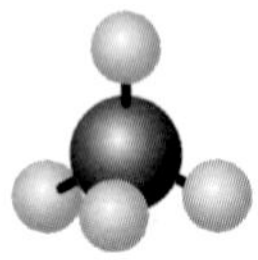

16   Methane, CH$_4$

Saturated hydrocarbons are called **alkanes.** The simplest alkane is methane, CH$_4$ (**16**), the main component of natural gas. Methane is also given off by decaying organic matter and is collected for use as a fuel from animal wastes on farms and in municipal waste-treatment facilities (see Connection 2, following Chapter 9).

The alkanes form a family in which the formula of each succeeding member is derived from CH$_4$ by inserting CH$_2$ groups between pairs of atoms. Thus, we have ethane, CH$_3$—CH$_2$—H, which is also found in natural gas, and propane, CH$_3$CH$_2$—CH$_2$—H, which is also a gas. Because the critical temperature of propane is above room temperature, it can be liquefied by the application of pressure and is sold in liquid form as "LP gas." Butane, CH$_3$CH$_2$CH$_2$CH$_3$, which boils at −0.5°C, is used in cigarette lighters. Alkanes with a longer chain of carbon atoms are named by combining a Greek prefix for the number of carbon atoms with the suffix *-ane* (Table 11.1). Alkanes that have ringlike structures are

Table 11.1 *Alkane nomenclature*

Number of carbon atoms	Formula	Name of alkane	Name of alkyl group
1	CH$_4$	methane	methyl
2	CH$_3$CH$_3$	ethane	ethyl
3	CH$_3$CH$_2$CH$_3$	propane	propyl
4	CH$_3$(CH$_2$)$_2$CH$_3$	butane	butyl
5	CH$_3$(CH$_2$)$_3$CH$_3$	pentane	pentyl
6	CH$_3$(CH$_2$)$_4$CH$_3$	hexane	hexyl
7	CH$_3$(CH$_2$)$_5$CH$_3$	heptane	heptyl
8	CH$_3$(CH$_2$)$_6$CH$_3$	octane	octyl
9	CH$_3$(CH$_2$)$_7$CH$_3$	nonane	nonyl
10	CH$_3$(CH$_2$)$_8$CH$_3$	decane	decyl
11	CH$_3$(CH$_2$)$_9$CH$_3$	undecane	undecyl
12	CH$_3$(CH$_2$)$_{10}$CH$_3$	dodecane	dodecyl

called **cycloalkanes.** Cyclopropane, C_3H_6 (**17**), and cyclohexane, C_6H_{12} (**18**), are cycloalkanes. Side chains on hydrocarbons are named as **alkyl groups,** with the ending *-ane* replaced by *-yl.* For example, if one of the H atoms is removed from a methane molecule, we are left with $CH_3—$, the methyl group. Similarly, the $CH_3CH_2—$ group is the ethyl group, and so on. The names of the hydrocarbons are important because they form a basis for naming all organic molecules.

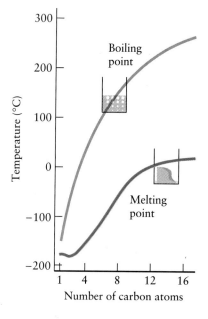

FIGURE 11.1

The melting and boiling points of the unbranched alkanes from CH_4 to $C_{16}H_{34}$.

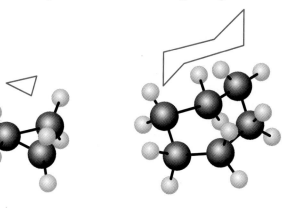

17 Cyclopropane, C_3H_6 **18** Cyclohexane, C_6H_{12}

As noted in Section 9.10, the bonds to each carbon atom in alkanes lie in a tetrahedral arrangement, with sp^3 hybridization. Because all the carbon-carbon bonds are single, one part of a molecule can easily rotate relative to the rest of the molecule around any C—C bond. As a result, in liquids and gases, the chains are normally twisting and turning in constant motion.

The dominant force between alkane molecules is the London force. Because this force increases with increasing number of electrons, the alkanes become less volatile with increasing molar mass (Fig. 11.1). Differences in volatility are used to separate the hydrocarbons in petroleum by fractional distillation. The petroleum is boiled in tall columns in which the temperature decreases with height. The different fractions are condensed from the vapor at different heights, according to their boiling points. As we have seen, the lightest alkanes, methane through butane, are gases at room temperature, and so rise to the top of the column. However, alkanes have increasingly higher boiling points as carbon atoms are added. Pentane (C_5H_{12}) is a volatile liquid, and hexane through undecane $(C_{11}H_{24})$ are moderately volatile liquids that are present in gasoline (Table 11.2). Kerosene, a fuel used in jet and diesel engines, contains a number of alkanes in the range C_{10} to C_{16}. Lubricating oils are mixtures in the range C_{17} to C_{22}. The heavier members of the series include the paraffin waxes and asphalt.

Table 11.2 Hydrocarbon constituents of petroleum

Hydrocarbons	Boiling range, °C	Name
C_1 to C_4	-160 to 0	gas
C_5 to C_{11}	30 to 200	gasoline
C_{10} to C_{16}	180 to 400	kerosene, fuel oil
C_{17} to C_{22}	350 and above	lubricants
C_{23} to C_{34}	low-melting-point solids	paraffin wax
C_{35} upward	soft solids	asphalt

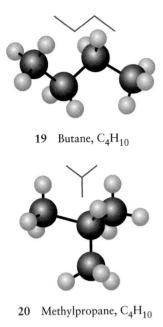

19 Butane, C_4H_{10}

20 Methylpropane, C_4H_{10}

The name *isomer* comes from the Greek words for "equal parts." We can think of isomeric molecules as being built from the same kit of parts.

As soon as there are four carbon atoms in an alkane molecule, at C_4H_{10}, we start to see another reason for the variety of compounds that carbon can form. The carbon atoms of C_4H_{10} can link together in two different ways: in a chain to form butane (**19**) or in a Y-shape to form the alkane called methylpropane (**20**).

Compounds with the same molecular formula but with their atoms in different arrangements are called **isomers.** *Straight-chain* isomers are those in which all the C atoms are linked in a linear chain, like butane. *Branched* isomers have one or more **side chains,** shorter chains attached to the longest chain. For example, methylpropane has a side chain consisting of a methyl group. Although they have the same molecular formula, isomers are distinct compounds with different physical and chemical properties. As we have seen, the normal boiling point of butane is $-0.5°C$, but methylpropane is less volatile, with a normal boiling point of $-11.6°C$. In general, alkanes with long, unbranched chains have higher melting points, boiling points, and enthalpies of vaporization than their branched isomers, because the atoms of neighboring branched molecules cannot get as close together as their unbranched isomers can; hence the attractive forces between branched molecules are weaker (Fig. 11.2).

Self-Test 11.2A Which compound has the higher boiling point: (a) $CH_3CH_2CH_2CH_2CH_2CH_3$ or (b) $(CH_3)_3CCH_2CH_3$? Why?

[*Answer:* (a), because it is not branched.]

Self-Test 11.2B Which compound has the higher boiling point: (a) $CH_3CH_2CH_2CH_2CH_3$ or (b) $CH_3CH_2CH_2CH_3$? Why?

Despite their wide use as fuels, the alkanes were once called the *paraffins*, from the Latin words for "little affinity." As their name suggests, they are not very reactive with common reagents. They are unaffected by concentrated sulfuric acid, by boiling nitric acid, by strong oxidizing agents like potassium permanganate, and by boiling aqueous sodium hydroxide. One reason for their resistance to this chemical onslaught is thermodynamic: the C—C and C—H bonds are strong (see Table 9.3), and, in most cases, there is little energy advantage in replacing them with other bonds.

The most familiar reaction of alkanes is the one used in our furnaces and Bunsen burners, their combustion to carbon dioxide and water:

$$CH_4(g) + 2\,O_2(g) \longrightarrow CO_2(g) + 2\,H_2O(g)$$

FIGURE 11.2

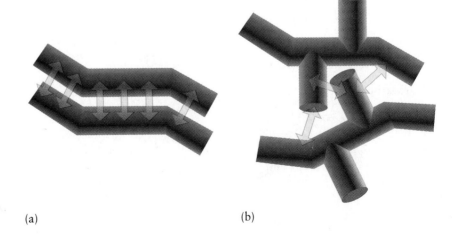

(a) The atoms in neighboring straight-chain alkanes, represented by the tubelike structures, can lie close together. (b) Fewer of the atoms of neighboring branched alkane molecules can get so close together, so the London forces (represented by double-headed arrows) are weaker and branched alkanes are more volatile.

(a)

(b)

In this reaction, the strong carbon-hydrogen bonds (which have bond enthalpy 412 kJ/mol) are replaced by the even stronger hydrogen-oxygen bonds (463 kJ/mol), and the oxygen-oxygen bond (496 kJ/mol) is replaced by two very strong carbon-oxygen double bonds (743 kJ/mol each). The energy released is given off as heat (Fig. 11.3). Similar exothermic reactions make the heavier alkanes our most valuable fuels (Section 6.15).

The second common type of reaction characteristic of alkanes is substitution. In an alkane **substitution reaction,** an atom or group of atoms replaces a hydrogen atom in the original molecule (Fig. 11.4). An example is the reaction between methane and chlorine. A mixture of these two gases survives unchanged in the dark; but when heated or exposed to ultraviolet radiation, they react:

$$CH_4(g) + Cl_2(g) \xrightarrow{\text{UV radiation or heat}} CH_3Cl(g) + HCl(g)$$

Chloromethane, CH_3Cl, is only one of the products of this reaction. Dichloromethane, CH_2Cl_2, trichloromethane, $CHCl_3$, and tetrachloromethane, CCl_4, may also be formed, depending on the relative amounts of methane and chlorine present initially. Trichloromethane is better known as *chloroform* and was one of the early anesthetics. Tetrachloromethane, which is commonly called carbon tetrachloride, is nonflammable; it was used in fire extinguishers and as a solvent until it was identified as a carcinogen.

The strength of the London forces between alkane molecules increases as the molar mass of the molecules increases. Alkanes are not very reactive but do undergo oxidation and substitution.

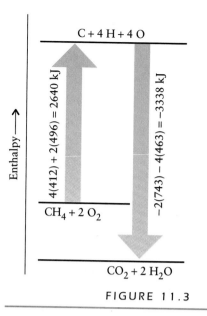

FIGURE 11.3

The enthalpy changes accompanying the combustion of methane. Although the bonds in the reactants are strong, they are even stronger in the products; and the overall process is exothermic.

11.3 Alkenes and Alkynes

The simplest unsaturated hydrocarbon is ethene, C_2H_4, commonly called ethylene (**21**). Ethene is the parent of the **alkenes,** a series of compounds with one carbon-carbon double bond. Their formulas are derived from ethene by adding $-CH_2-$ groups. For example, the next member of the family is propene, $CH_3CH{=}CH_2$. The names of the alkenes are the same as the names of the corresponding alkanes, except that they end in *-ene.* The **alkynes** are hydrocarbons that have at least one carbon-carbon triple bond. The simplest is ethyne,

See Toolbox 11.1 for more information on the nomenclature of alkenes and alkynes.

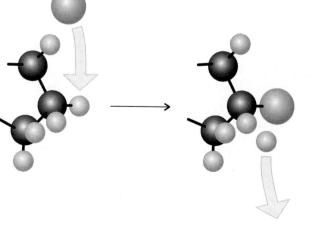

FIGURE 11.4

In an alkane substitution reaction, an incoming atom or group of atoms (represented by the orange sphere) replaces a hydrogen atom in the alkane molecule.

HC≡CH, which is commonly called acetylene (**22**). Alkynes are named like alkanes, but with the suffix *-yne*.

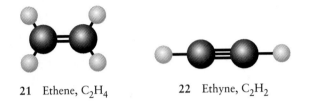

21 Ethene, C_2H_4 22 Ethyne, C_2H_2

How does the presence of a double bond affect the properties of a hydrocarbon? As we saw in Section 9.13, the C=C double bond of alkenes consists of a σ-bond and a π-bond. The double bond makes alkene molecules more rigid than alkanes and is responsible for differences in the physical properties of the two types of hydrocarbons. All four atoms attached to the C=C group lie in the same plane and are locked into that arrangement by the π-bond (Fig. 11.5). Because alkene molecules cannot roll up into a compact ball, they cannot pack together as closely as alkanes can; so their London forces are less effective and they have lower melting points (Fig. 11.6).

Alkenes are important feedstocks for the production of polymers. In fact, by the end of the twentieth century, ethene had become the most heavily manufactured chemical in the United States. One of the first steps in the petrochemical industry is to convert some of the abundant alkanes into alkenes. This conversion is achieved by removing hydrogen atoms from neighboring carbon atoms:

$$CH_3CH_3(g) \xrightarrow{Cr_2O_3} CH_2{=}CH_2(g) + H_2(g)$$

This reaction is an example of an **elimination reaction,** a reaction in which two groups or two atoms on neighboring carbon atoms are removed, or eliminated, from a molecule, leaving a multiple bond between the carbon atoms (Fig. 11.7). The formula for chromium(III) oxide written above the arrow indicates that it is a *catalyst,* a substance that facilitates a reaction but remains unchanged itself.

The most characteristic chemical reaction of an alkene is an **addition reaction,** in which atoms supplied by the other reactant are attached to the two

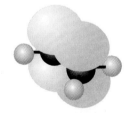

FIGURE 11.5

The π-bond (yellow electron clouds) in an alkene molecule makes the molecule resistant to twisting around a double bond, so all six atoms lie in the same plane.

Catalysts are treated in more detail in Section 13.10.

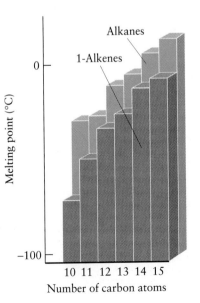

FIGURE 11.6

The melting point of an alkene is usually lower than that of the alkane with the same number of carbon atoms. The values shown are for unbranched alkanes and 1-alkenes (that is, alkenes in which the double bond is at the end of the carbon chain).

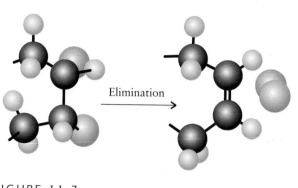

Elimination

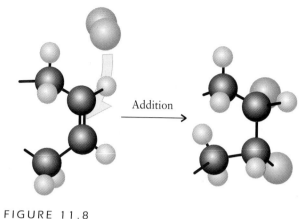

Addition

FIGURE 11.7

In an elimination reaction, two atoms (the orange and purple spheres) attached to neighboring carbon atoms are removed from the molecule, leaving a double bond in their place.

FIGURE 11.8

In an addition reaction, the atoms provided by an incoming molecule are attached to the carbon atoms originally joined by a multiple bond. Addition is the reverse of elimination.

atoms joined by the double bond (Fig. 11.8). An addition reaction is the reverse of an elimination reaction. An example is the addition of a halogen (Fig. 11.9). The two atoms of the halogen molecule are added to the alkene molecule, as in the formation of 1,2-dichloroethane:

$$CH_2{=}CH_2 \; + \; Cl_2 \; \longrightarrow \; CH_2Cl{-}CH_2Cl$$

Hydrocarbons containing a double bond are called alkenes; those with a triple bond are called alkynes. Alkenes and alkynes undergo addition reactions at the multiple bond.

11.4 Aromatic Compounds

Aromatic hydrocarbons as a group are called **arenes.** The parent compound of aromatic compounds is benzene, C_6H_6. Aromatic compounds also include fused-ring compounds, such as naphthalene, $C_{10}H_8$ (**23**), and anthracene, $C_{14}H_{10}$ (**24**). Both compounds can be obtained by the distillation of coal. Indeed, coal itself, which is a very complex mixture, consists of vast networks of atoms and

FIGURE 11.9

When bromine dissolved in a solvent (the brown liquid) is mixed with an alkene (the colorless liquid), the bromine atoms add to the molecule at the double bond, a reaction giving a colorless product.

has regions in which aromatic rings can be identified (Fig. 11.10). When coal is destructively distilled—heated in the absence of oxygen so that it decomposes and vaporizes—the sheetlike molecules break up, and the fragments include the aromatic hydrocarbons and their derivatives. The complicated liquid mixture that results is called *coal tar*.

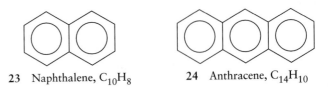

23   Naphthalene, $C_{10}H_8$ 24   Anthracene, $C_{14}H_{10}$

Unlike alkenes, arenes predominantly undergo *substitution* reactions, not addition reactions. Because the aromatic ring is stabilized by resonance, it is very stable and its π-bonds are usually left intact. As an example, bromine immediately adds to a double bond of an alkene; however, it reacts with benzene only in the presence of a catalyst—typically iron(III) bromide—and it does not affect the bonding in the ring. Instead, one of the bromine atoms substitutes for a hydrogen atom:

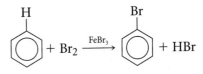

Aromatic rings are much less reactive than their double bond character would suggest; they commonly undergo substitution rather than addition.

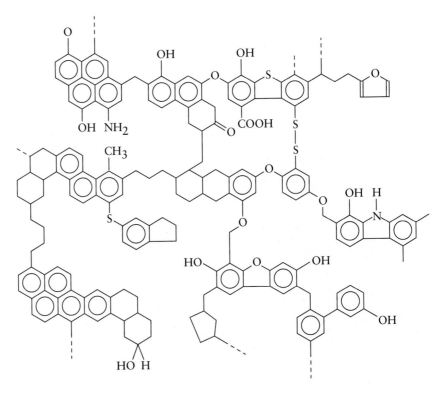

FIGURE 11.10

A highly schematic representation of a portion of the structure of coal. When coal is heated in the absence of oxygen, the structure breaks up and a complex mixture of products—many of them aromatic—is obtained as coal tar.

Toolbox 11.1 *How to name hydrocarbons*

This Toolbox describes how to name hydrocarbons from their formulas and how to identify the structure of a hydrocarbon from its name.

Conceptual Basis

Because the kinds of carbon-carbon bonds present tend to dominate the properties of the molecule, a hydrocarbon is first classified according to the carbon-carbon bonds present (alkane, alkene, alkyne, or arene). Then the longest chain of carbon atoms is made the "root" of the name. Other hydrocarbon groups attached to the longest chain are named as side chains.

Procedure

Identify the type of hydrocarbon by checking for the presence of a multiple bond or benzene ring, then use the corresponding procedure below.

Alkanes The names of straight-chain alkanes are given in Table 11.1: they all end in *-ane*. To name a branched alkane, treat the side chains as substituent groups.

Step 1. Identify the longest unbranched chain of carbon atoms and give it the name of the corresponding alkane.

Example:

$$CH_3CH_2CH_2\underset{\underset{\displaystyle CH_3}{|}}{C}HCH_3 \text{ is a substituted pentane}$$

Step 2. Name alkyl substituent groups by changing the *-ane* suffix to *-yl* (see Table 11.1). Use Greek prefixes to indicate how many of each substituent are in the molecule. When different groups are present, list them in alphabetical order (disregarding the Greek prefixes) and attach them to the root name.

Step 3. Indicate the locations of the substituents by numbering the backbone C atoms from whichever end of the molecule results in the lower numbers of locations for the substituents. The locations are then written before each substituent, separated by commas.

Examples:

$$\overset{5}{C}H_3\overset{4}{C}H_2\overset{3}{C}H_2\overset{2}{\underset{\underset{\displaystyle CH_3}{|}}{C}}H\overset{1}{C}H_3 \qquad \overset{1}{C}H_3\overset{2}{C}(CH_3)_2\overset{3}{C}H_2\overset{4}{C}H_3$$

2-Methylpentane   2,2-Dimethylbutane

Alkenes and alkynes Proceed as for alkanes, but change the suffix *-ane* to either *-ene* (for alkenes) or *-yne* (for alkynes).

In all cases except ethene (ethylene) and propene (propylene) and the corresponding alkynes, it is necessary to specify the location of the multiple bond. Therefore,

Step 4. Number the C atoms in the backbone in the order that gives the lower numbers to the two atoms joined by the multiple bond. The multiple bond has priority over the numbering of substituents.

Example:

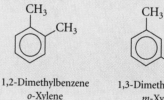

$$\overset{1}{C}H_3\overset{2}{C}H=\overset{3}{C}H\overset{4}{C}H_2\overset{5}{\underset{\underset{\displaystyle CH_3}{|}}{C}}H\overset{6}{C}H_3$$

5-Methyl-2-hexene

Arenes When the benzene ring, as C_6H_5—, is treated as a substituent, it is called a phenyl group. In general, aromatic groups are called *aryl* groups. For derivatives of benzene, give the locations of substituents on the ring as numbers that run clockwise or counterclockwise. Select the direction that corresponds to the lower substituent numbers. In an older but still widely used system of nomenclature, when there is a substituent in location 1, locations 2, 3, or 4 are denoted *ortho-* (abbreviated *o-*), *meta-* (*m-*), and *para-* (*p-*), respectively.

Examples:

1,2-Dimethylbenzene   1,3-Dimethylbenzene
o-Xylene   *m*-Xylene

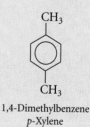

1,4-Dimethylbenzene
p-Xylene

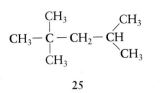

25

Example 11.2 Naming aliphatic hydrocarbons

(a) Name the compound shown as (25) and (b) write the structural formula of 2,3-dimethyl-4-ethylcyclohexene.

Strategy Identify each compound as alkane, alkene, alkyne, or aromatic. Work through the corresponding procedure in Toolbox 11.1.

Solution **Step 1.** (a) The compound has only single bonds, so it is an alkane. It is a substituted pentane because its longest carbon chain (in bold type) has five C atoms. **Step 2.** We identify the substituent groups: there are three methyl groups (CH$_3$—). The molecule is therefore a trimethylpentane. **Step 3.** We number the backbone C atoms from whichever end of the molecule results in the lower numbers for the substituents:

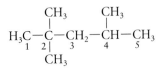

The molecule is 2,2,4-trimethylpentane. (b) Draw the longest chain of carbon atoms first. The last part of the name, cyclohexene, tells us that the molecule has a ring of six carbon atoms with a double bond:

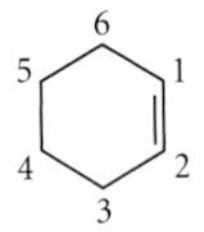

The double bond has priority in numbering, so it is between C atoms 1 and 2. The prefix "2,3-dimethyl" tells us there are two methyl groups, one on C atom 2 and one on C atom 3. We then add an ethyl group (from the "4-ethyl" part of the name) to C atom four. Add hydrogen atoms as needed to give each carbon atom a valence of four:

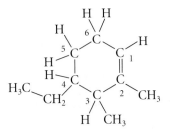

Self-Test 11.3A (a) Name compound (26) and (b) write the condensed structural formula of 5-ethyl-2,2-dimethyloctane.

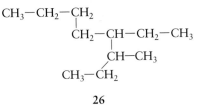

26

[**Answer:** (a) 4-Ethyl-3-methyloctane;
(b) CH$_3$C(CH$_3$)$_2$CH$_2$CH$_2$CH(CH$_2$CH$_3$)CH$_2$CH$_2$CH$_3$]

Self-Test 11.3B (a) Name the compound (CH$_3$)$_2$CHCH$_2$CH(CH$_2$CH$_3$)$_2$ and (b) give the condensed structural formula of 3,3,5-triethylheptane.

Example 11.3 *Naming aromatic hydrocarbons*

Name the compound

Strategy Use the rule for naming arenes in Toolbox 11.1; note that hydrocarbon substituents are named in alphabetical order.

Solution An ethyl group is located at position 1 and a methyl group is located at position 3, so the compound is 1-ethyl-3-methylbenzene.

Self-Test 11.4A Name the compound

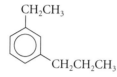

[*Answer:* 1-Ethyl-3-propylbenzene]

Self-Test 11.4B Name the compound

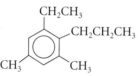

Example 11.4 *Predicting the products of an organic reaction*

Write the condensed structural formula of the compound formed by the addition of hydrogen bromide to ethene.

Strategy As set out in Toolbox 11.2, the reaction typical of alkenes is addition, so there will be only one product. The double bond will be replaced by a single bond and new single bonds will form between the carbon atoms previously involved in the double bond and atoms of the reactant being added.

Solution The H atom of HBr adds to one carbon atom and the Br atom adds to the other. The addition reaction is

$$CH_2{=}CH_2 \ + \ HBr \ \longrightarrow \ CH_3{-}CHBr$$

The product is bromoethane, CH_3CHBr.

Self-Test 11.5A Write the condensed structural formula of the compound that is produced when chlorine is added to (a) a flask of propene, $CH_3CH{=}CH_2$, in the dark; (b) a flask of benzene to which a catalyst has been added.

[*Answer:* (a) $CH_3CHClCH_2Cl$; (b) C_6H_5Cl]

Toolbox 11.2 *How to predict the characteristic reactions of hydrocarbons*

This Toolbox summarizes the types of reactions undergone by the different types of hydrocarbons: alkanes, alkenes, alkynes, and arenes. It can be used to predict the reactions a hydrocarbon might undergo and the structural formulas of their products.

Conceptual Basis

The types of bonds present in a hydrocarbon molecule determine its chemical properties. Alkanes contain strong carbon-carbon σ-bonds, so they are not very reactive, except to substitution of hydrogen atoms and, when ignited, to combustion with oxygen. Alkenes and alkynes contain relatively weak and reactive π-bonds, to which other molecules can add. Arenes contain multiple bonds; but the aromatic ring undergoes resonance stabilization, so it is very stable. Thus, an arene primarily undergoes reactions involving the substitution of another atom or group for one of its hydrogen atoms (Fig. 11.11).

Procedure

We use the following properties of hydrocarbons to predict their typical reactions:

1. All hydrocarbons burn in oxygen to form carbon dioxide and water.

2. The typical reaction of alkanes is substitution in the presence of heat or ultraviolet radiation. In a substitution reaction, an atom or group replaces a hydrogen atom. A second product contains the hydrogen atom that is displaced in the reaction. In the presence of a catalyst, other reactions may occur. For example, in the presence of Cr_2O_3, alkanes undergo elimination reactions. Two H atoms are removed and an alkene is formed.

3. The typical reaction of alkenes and alkynes is addition. The hydrocarbon retains all its atoms and gains the atoms of the compound being added. The addition compound is the sole product.

4. The typical reaction of arenes is substitution, with preservation of the benzene ring. The arene loses an H atom for each new atom or group added. A second product contains the hydrogen atoms displaced in the reaction.

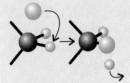
(a) Substitution of alkanes

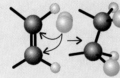
(b) Addition to alkenes

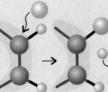
(c) Substitution of arenes

FIGURE 11.11
_____

A summary of the reactions typical of the different classes of hydrocarbons.

Self-Test 11.5B Write the condensed structural formula of the principal product formed when 1 mol $Cl_2(g)$ is added to a flask containing 2 mol $CH_3CH_3(g)$ (a) in the dark; (b) after a flash of light.

FUNCTIONAL GROUPS

Certain groups of atoms dominate the properties and reactions (the "functions") of many organic compounds. These **functional groups** are attached to the carbon chains and rings. Examples are the chlorine atom in chloroethane and the $>C=O$ group in acetone (propanone), CH_3COCH_3.

Table 11.3 lists the more common functional groups. These are the groups that chemists use to build new pharmaceuticals, plastics, and fuels. Living cells use them to form proteins and participate in the biochemical processes of life. Functional groups are like a kit of parts from which the whole range of organic molecules, and ultimately living beings, are constructed. They are also the key to the synthesis of many polymers. Functional groups can be identified by their reactions, by spectroscopy (see Investigating Matter 9.1), and by nuclear magnetic resonance (Investigating Matter 11.1).

11.5 Alcohols

The **hydroxyl group,** $-OH$, is an $-O-H$ group covalently bonded to a carbon atom. It can be thought of as a water molecule, HOH, from which one of the H atoms has been removed. The hydroxyl group must be distinguished from the hydroxide ion (OH^-) of inorganic hydroxides, which is a diatomic ion.

An **alcohol** is an organic compound that contains a hydroxyl group not connected directly to a benzene ring or to a $>C=O$ group. An alcohol can be thought of as a water molecule in which one H atom has been replaced by an organic group, R, to give the molecule $R-OH$. The simplest alcohol is methanol, CH_3OH, which is a liquid fuel sometimes used in vehicles (see Connection 2, following Chapter 9). One of the best-known organic compounds is ethanol, CH_3CH_2OH, which is also called *ethyl alcohol* and *grain alcohol*. Ethanol is produced traditionally in large amounts from the bacterial fermentation of grains and sugars; commercial ethanol is also obtained from petrochemicals.

Alcohols are divided into three classes on the basis of the number of other organic groups attached to the carbon atom that is linked to the $-OH$ group. A **primary alcohol** has the form RCH_2-OH, in which R represents an alkyl group; a **secondary alcohol,** the form R_2CH-OH; and a **tertiary alcohol,** the form R_3C-OH. The R groups need not all be the same. CH_3CH_2OH is a primary alcohol and $CH_3CHOHCH_2CH_3$ is a secondary alcohol.

A further aspect of carbon's variety is that molecules of its compounds can contain more than one functional group. Ethylene glycol, or 1,2-ethanediol, $HOCH_2CH_2OH$ (**27**), is an example of a **diol,** a compound with two hydroxyl groups. It is used as a component of antifreeze and in the manufacture of some synthetic fibers.

Alcohols with low molar masses are liquids. This property is a sign of the importance of hydrogen bonding, for alcohols have much lower vapor pressures than do hydrocarbons with approximately the same molar mass. For example, ethanol is a liquid at room temperature, but butane, which has a higher molar mass than ethanol, is a gas. Because of the hydrogen bonding ability of the $-OH$ group, alcohols with low molar masses are soluble in water. As the number of carbon atoms increases, the presence of the functional group has less effect on the physical properties of the compound as a whole, so alcohols with high molar masses, such as decanol, $C_{10}H_{21}OH$, are not soluble in water.

The formula of an alcohol is derived from H_2O by replacing one of the hydrogen atoms with an organic group. Like water, alcohols form intermolecular hydrogen bonds.

Table 11.3 *Common functional groups*

Group	Class of compound
$-OH$	alcohol, phenol
$-N<$	amine
$-CHO$	aldehyde
$-CO-$	ketone
$-COOH$	carboxylic acid
$-COOR$	ester
$-CO-N<$	amide
$-O-$	ether
$-X$	halide (X = F, Cl, Br, or I)

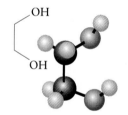

27 1,2-Ethanediol, $HOCH_2CH_2OH$

Nuclear magnetic resonance (NMR) is the premier technique for the identification of organic compounds and is among the leading techniques for the determination of structure. NMR has also been developed, as *magnetic resonance imaging* (MRI), for use as an investigative technique in medicine. The basis of the technique is the fact that many atomic nuclei behave like small bar magnets and have energies that depend on their orientation in a magnetic field. An NMR spectrometer detects transitions between these energy levels. The most important nucleus with magnetic properties is the proton, and we shall concentrate on that.

Recall from Section 7.8 that electrons possess the property of spin, which can be thought of as a spinning motion. Protons also have a spin. This spin always occurs at the same rate, but it can be oriented in either of two directions, which are denoted ↑ (or α) and ↓ (or β). Because a proton has an electric charge and because a moving charge generates a magnetic field, the proton acts as a tiny bar magnet that can have either of two orientations.

Suppose an external magnetic field is applied; now these two spin orientations correspond to different energies, and the separation between them is proportional to the strength of the applied field. If the sample is also exposed to electromagnetic radiation, the nuclei are flipped from one orientation to the other when the energy of the incoming photons matches the energy separation. This is the "resonance" condi-

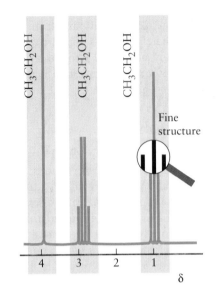

The NMR spectrum of ethanol. The red letters denote the protons that give rise to the associated peaks.

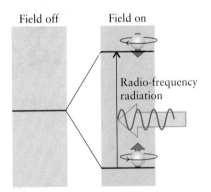

The two orientations of a nuclear spin have the same energy in the absence of a magnetic field. When a field is applied, the energy of the α spin falls and that of the β spin increases. When the separation between the two energy levels is equal to the energy of a radio-frequency photon, there is a strong absorption of radiation, giving a peak in the NMR spectrum.

tion of the technique. The energy of the photons in the radiation that causes resonance is equal to $h\nu$, where ν (nu) is the frequency of the radiation.

In modern spectrometers, superconducting magnets are used to generate very high, uniform fields, and resonance requires radio-frequency radiation at about 400 MHz. After a sample is inserted, the strength of the magnetic field is gradually increased. The resonance condition is detected by a strong absorption of the radiation, and a sharp peak is observed in the detector output.

Every organic compound has a characteristic NMR fingerprint, and many compounds can be recognized by comparing the observed pattern with a library of patterns from known substances, or by calculating the expected pattern of lines. The illustration shows the NMR spectrum of ethanol, and we see that there are three groups of peaks and a characteristic pattern of splitting within the groups. If we see this pattern in an NMR spectrum, we know immediately that the sample contains ethanol.

The separation of the absorption into groups of lines stems from the presence of protons in different environments within the molecule. In ethanol, CH_3CH_2OH, three protons are present in the methyl group (CH_3), two are present in the methylene group (CH_2), and one is present in the hydroxyl group (OH). In an applied magnetic field, the protons in each group experience slightly different local magnetic fields. As a result, slightly different magnetic fields are needed to bring them into resonance and three groups of absorption lines are observed in the NMR spectrum.

We say that each group of protons has a characteristic *chemical shift* from a standard sample. The measurement of the chemical shift helps to identify the type of group responsible for the absorption, so we can start to identify the components of the molecule. The chemical shift of a group of lines is expressed in terms of the δ scale (delta scale), which is a measure of the difference between the resonance frequency of a peak and that of a standard compound.

Another aid to identification is the relative areas under the peaks, which are proportional to the numbers of protons they contain. The three peaks in the ethanol spectrum, for example, have overall areas in the ratio 3:2:1, which is what we would expect for the three methyl, two methylene, and one hydroxyl protons. Each peak is further split into individual lines, called the *fine structure* of the peak. The fine structure tells us how many protons are on adjacent carbon atoms.

Magnetic Resonance Imaging

Magnetic resonance imaging (MRI) is a noninvasive structural technique for complex systems of molecules. In its simplest form, MRI is a portrayal of the concentration of protons in a sample. If the sample—which may be a living human body—is exposed to a uniform magnetic field in an NMR spectrometer, and we work at a resolution that does not show any chemical shifts or fine structure, the protons give rise to a single resonance line. However, if the magnetic field varies across the sample, the protons will resonate at different frequencies according to their location in the field. Moreover, the intensity of the resonance at a given field will be proportional to the number of protons at the location corresponding to that particular field value.

If the field gradient is rotated into different orientations, another portrayal of the concentration of protons through the sample is obtained. After many such measurements of the absorption intensity, each at a slightly different angle of rotation, the data can be analyzed on a computer, and a two-dimensional image of any section through the sample can be constructed. The image is the distribution of protons—in large measure, the distribution of water in a body. A bucket of water would give a completely uniform MRI scan. However, in the body, water is distributed in different ways in different tissues and individual organs can be identified.

A major advantage of MRI over x-rays is that the patient is exposed only to radio-frequency radiation, so the damage caused by x-rays is avoided. Another advantage is that with MRI a "slice" of the body can be viewed without interference from structures in the front or back of it. If a series of "slices" are obtained, they can even be assembled into a three-dimensional image that yields a much more accurate image than is possible with x-rays.

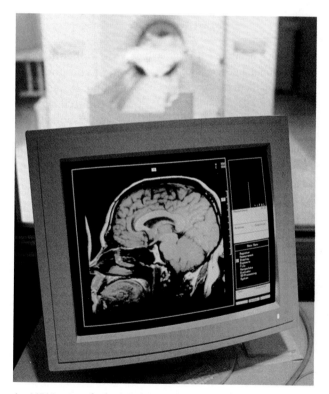

An MRI image of a human brain. The patient must lie within the strong magnetic field (background) and the detectors can be rotated around the patient's head, allowing many different views to be recorded.

11.6 Ethers

An **ether** is a compound with formula R—O—R′, where the substituents R and R′ can be either alkyl or aryl groups and need not be the same. Just as we can think of an alcohol as being derived from HOH by the replacement of one H atom with an alkyl group, so we can think of an ether as an HOH molecule in which both H atoms have been replaced by alkyl or aryl groups to give a compound of the form R—O—R′:

H—O—H CH_3CH_2—O—H CH_3CH_2—O—CH_2CH_3

Water   Ethanol   Diethyl ether

Ethers are more volatile than the alcohols of the same molar mass because they have no hydrogen atom bonded to the oxygen atom and hence do not form hydrogen bonds (Fig. 11.12).

Because ethers are not very reactive, they are useful solvents for other organic compounds. However, ethers are flammable; diethyl ether is easily ignited and must be used with great care.

> *Ethers are more volatile and much less reactive than the corresponding alcohols.*

11.7 Phenols

In a **phenol**, a hydroxyl group is attached directly to an aromatic ring. The parent compound, phenol itself, C_6H_5OH (**28**), is a white, crystalline, molecular solid. It was once obtained from the distillation of coal tar, but now it is mainly produced synthetically from benzene. Many phenols occur naturally, and some are responsible for the fragrances of plants. They are often components of essential oils, the oils that can be distilled from flowers and leaves. Thymol (**29**), for instance, is the active ingredient of oil of thyme, and eugenol (**30**) provides most of the scent and flavor of oil of cloves.

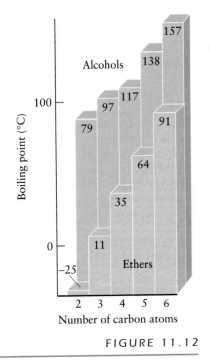

FIGURE 11.12

The boiling points of ethers (given on each column, in degrees Celsius) are lower than those of isomeric alcohols, because hydrogen bonding occurs in alcohols but not in ethers. All the molecules referred to here are unbranched.

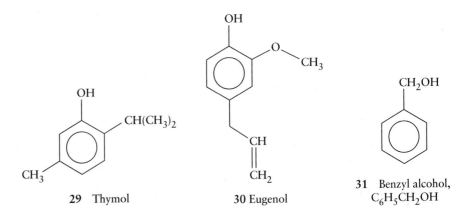

28   Phenol, C_6H_5OH 29   Thymol   30 Eugenol   31   Benzyl alcohol, $C_6H_5CH_2OH$

Phenols differ from alcohols in that they are weak acids. Benzyl alcohol (**31**) is so weak a proton donor that it is best regarded as a neutral compound; phenol, though, is definitely acidic and is, in fact, also known as *carbolic acid*. Phenols tend to be insoluble in water; however, because they react with bases to form anions, they do dissolve in basic aqueous solutions. This behavior can be used to separate and identify phenols in a mixture of organic compounds.

> *Phenols are weak acids.*

11.8 Aldehydes and Ketones

The **carbonyl group**, $\mathrm{\small >\!C\!=\!O}$, occurs in two closely related families of compounds:

Aldehydes are compounds of the form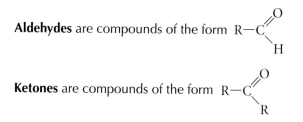

Ketones are compounds of the form $\mathrm{R\!-\!C}\overset{\displaystyle O}{\underset{\displaystyle R}{\big\|}}$

Notice that, in an aldehyde, the carbonyl group is found at the end of the carbon chain; whereas, in a ketone, it is at an intermediate position. The R groups may be either aliphatic or aromatic. The group characteristic of the ketones is written $-CO-$, as in CH_3COCH_3, propanone (acetone), a common laboratory solvent. The group characteristic of the aldehydes is normally written $-CHO$, as in HCHO (formaldehyde), the simplest aldehyde. The liquid *formalin,* which is used to preserve biological specimens, is an aqueous solution of formaldehyde. Wood smoke contains formaldehyde, and formaldehyde's destructive effect on bacteria is one of the reasons why smoking food helps to preserve it.

Aldehydes occur naturally in essential oils and are often responsible for the flavors of fruits and the odors of plants. Benzaldehyde, C_6H_5CHO (**32**), contributes to the characteristic aroma of cherries and almonds. Cinnamaldehyde (**33**) occurs in oil of cinnamon, and vanillin (**34**) in oil of vanilla. Ketones can also be fragrant; carvone (**15**), for example, the essential oil of spearmint, is used to flavor chewing gum.

Formaldehyde is prepared industrially (for the manufacture of phenol-formaldehyde resins) by the catalytic oxidation of methanol:

$$2\,CH_3OH(g) \;+\; O_2(g) \xrightarrow{\;600°C,\;Ag\;} 2\,HCHO(g) \;+\; 2\,H_2O(g)$$

When oxidizing a primary alcohol to an aldehyde, we can avoid further oxidation of the product to a carboxylic acid (see Section 11.9) by using a mild oxidizing agent. There is less risk of further oxidation for ketones than for aldehydes, because a C—C bond would have to be broken for a carboxylic acid to form from a ketone. Dichromate oxidation of secondary alcohols produces the ketone in good yield, with little additional oxidation. Schematically,

$$CH_3CH_2CH(OH)CH_3 \xrightarrow{\;Na_2Cr_2O_7(aq),\;H_2SO_4(aq)\;} CH_3CH_2COCH_3$$

One indication of the difference between the ease of oxidation of aldehydes and that of ketones is the ability of aldehydes but not ketones to produce a silver mirror—a coating of silver on the inside walls of a test tube or flask—with *Tollens reagent,* a solution of Ag^+ ions in aqueous ammonia (Fig. 11.13). In summary,

Aldehydes:
$$CH_3CH_2CHO + Ag^+(\text{from Tollens reagent}) \longrightarrow CH_3CH_2COOH + Ag(s)$$

Ketones:
$$CH_3COCH_3 + Ag^+(\text{from Tollens reagent}) \longrightarrow \text{no reaction}$$

Aldehydes and ketones can be prepared by the oxidation of alcohols. Aldehydes are reducing agents; ketones are not.

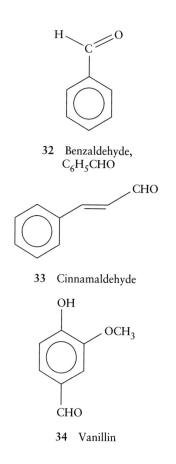

32 Benzaldehyde, C_6H_5CHO

33 Cinnamaldehyde

34 Vanillin

FIGURE 11.13

An aldehyde (left) produces a silver mirror with Tollens reagent, but a ketone (right) does not.

11.9 Carboxylic Acids and Esters

The **carboxyl group** is $-C\!\!\underset{O-H}{\overset{O}{\diagdown}}$. It is the functional group found in **carboxylic acids,** which are weak acids of the form R—COOH. Although the carboxyl group is normally abbreviated to —COOH, it must be remembered that the group does not contain an O—O bond. The simplest carboxylic acid is formic acid, HCOOH, the acid in ant venom; one of the most common carboxylic acids is acetic acid, CH_3COOH, the acid of vinegar. Acetic acid is formed when the ethanol in wine is oxidized by air:

$$CH_3-CH_2-OH(aq) + O_2(g) \longrightarrow H_3C-C\!\!\underset{OH}{\overset{O}{\diagup}}(aq) + H_2O(l)$$

Carboxylic acids can be prepared by oxidizing primary alcohols and aldehydes with a strong oxidizing agent such as acidified aqueous potassium permanganate solution. In some cases, alkyl groups can be oxidized directly to carboxyl groups. This process is very important industrially and, among other applications, is used for the oxidation of *p*-xylene:

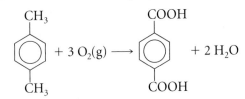

The product, terephthalic acid, is used for the production of artificial fibers.

The product of the reaction between a carboxylic acid and an alcohol is called an **ester** (**35**). Acetic acid and ethanol, for example, react when heated to about 100°C in the presence of a small amount of a strong acid. The products of this esterification are ethyl acetate, a fragrant liquid, and water:

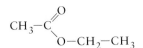

35 Ethyl acetate, $CH_3COOCH_2CH_3$

$$H_3C-C\!\!\underset{OH}{\overset{O}{\diagup}} + H-OCH_2CH_3 \xrightarrow{H^+,\Delta} H_3C-C\!\!\underset{OCH_2CH_3}{\overset{O}{\diagup}} + H_2O$$

Many esters have pleasant odors and contribute to the flavors of fruits. Other naturally occurring esters include fats and oils. For example, the animal fat tristearin (**36**), which is a component of beef fat, is an ester formed from glycerol and stearic acid. Esters have the general formula RCOOR′.

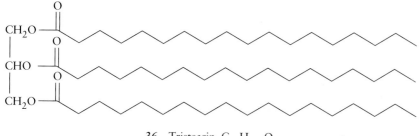

36 Tristearin, $C_{57}H_{110}O_6$

Ester formation is an example of a **condensation reaction.** In this type of reaction, two molecules combine to form a larger one and a small molecule is eliminated (Fig. 11.14); the two reactant molecules are "condensed" into one. In an esterification, the eliminated molecule is H_2O. Notice that, although an alcohol and carboxylic acid can each form hydrogen bonds to themselves, an ester cannot because it does not have a hydrogen atom bonded to an oxygen atom.

> *Carboxylic acids have an* $-OH$ *group attached to a carbonyl group. Alcohols condense with carboxylic acids to form esters.*

Self-Test 11.6A (a) Write the structural formula of the ester formed from the reaction between propanoic acid, CH_3CH_2COOH, and methanol, CH_3OH. (b) Write the structural formulas of the acid and alcohol that react to form 1-pentyl ethanoate, $CH_3COOC_5H_{11}$, a contributor to the flavor of bananas.
[*Answer:* (a) $CH_3CH_2COOCH_3$; (b) CH_3COOH and $CH_3(CH_2)_3CH_2OH$]

Self-Test 11.6B (a) Write the structural formula of the ester formed from the reaction between formic acid, $HCOOH$, and ethanol, CH_3CH_2OH. (b) Write the structural formulas of the acid and alcohol that react to form methyl butanoate, $CH_3(CH_2)_2COOCH_3$, a contributor to the flavor of apples.

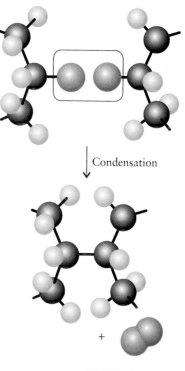

Condensation

FIGURE 11.14

In a condensation reaction, two molecules are linked as a result of removing two atoms or groups of atoms (the orange and purple spheres) as a small molecule (typically, water).

11.10 Amines and Amides

An **amine** is a compound derived from ammonia by replacing various numbers of H atoms by organic groups:

$$H-\underset{\underset{H}{|}}{N}-H \qquad CH_3-\underset{\underset{H}{|}}{N}-H \qquad CH_3-\underset{\underset{H}{|}}{N}-CH_3 \qquad CH_3-\underset{\underset{CH_3}{|}}{N}-CH_3$$

Ammonia   Methylamine   Dimethylamine   Trimethylamine

In each case, the N atom is sp^3 hybridized, with one lone pair of electrons and three σ-bonds. Like ammonia, amines are also weak bases (Section 3.9). An amine can form a **quaternary ammonium ion,** a tetrahedral ion of the form R_4N^+, such as the tetramethylammonium ion, $(CH_3)_4N^+$. Quaternary ammonium ions have little effect on pH unless one of the groups is an H atom.

Amines are widespread naturally. Many of them have a pungent, often unpleasant odor. Because proteins are organic polymers containing nitrogen, amines are present in the decomposing remains of living matter and, together with sulfur compounds, are responsible for the stench of decaying flesh. The common names of two diamines—putrescine, $NH_2(CH_2)_4NH_2$, and cadaverine, $NH_2(CH_2)_5NH_2$—speak for themselves.

An **amino acid** is a carboxylic acid that contains an **amino group** $(-NH_2)$ or one of its derivatives $(-NHR,$ or $-NR_2)$ as well as a carboxyl group. The

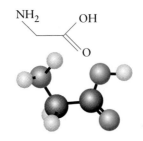

37 Glycine, NH_2CH_2COOH

simplest example is glycine, NH_2CH_2COOH (**37**). Notice that an amino acid has both a basic group ($-NH_2$) and an acidic group ($-COOH$) in the same molecule. We can predict that the presence of a basic group and an acidic group in the same molecule leads to some remarkable properties. This is, in fact, the case, as we shall see in Section 11.16.

Like alcohols, amines also condense with carboxylic acids:

$$H_3C-C\overset{O}{\underset{OH}{\diagdown}} \;+\; \underset{H-N-CH_3}{\overset{H}{|}} \longrightarrow H_3C-C\overset{O}{\underset{NHCH_3}{\diagdown}} \;+\; H_2O$$

The product is an **amide.** One example of an amide is the pain-relieving drug sold as Tylenol (**38**). Many amides have N—H bonds that can take part in hydrogen bonding, so the intermolecular forces between their molecules are relatively strong.

Amines are derived from ammonia by replacing hydrogen atoms with organic groups. Amides result from the condensation of amines with carboxylic acids. Amines and amides with an N—H bond take part in hydrogen bonding.

Self-Test 11.7A Predict whether an ester or an amine of the same molar mass would have the higher boiling point and explain why.

[*Answer:* The amine, if there is an H atom attached to the N atom; because it can form hydrogen bonds with the $-NH_2$ group, but the RCOOR′ group cannot form hydrogen bonds in the pure state.]

Self-Test 11.7B What are the hybridization schemes of the C and N atoms in formamide, $HCONH_2$?

> In practice, amides are best prepared by the reaction between an amine and an acid chloride, RCOCl, a compound in which the —OH of the carboxylic acid group has been replaced by a —Cl group.

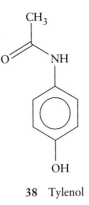

38 Tylenol

Toolbox 11.3 *How to name compounds with functional groups*

This Toolbox outlines the systematic nomenclature of organic compounds with functional groups.

Conceptual Basis
Because a functional group tends to dominate the properties of a compound, the compound is named as a member of the series featuring that functional group. The common functional groups are shown in Table 11.3.

Procedure
Names of compounds containing functional groups follow the same conventions and numbering system as do names of hydrocarbons (Toolbox 11.1). However, the ending of the name is changed to indicate the functional group.

Alcohols Identify the parent alkane and replace the ending *-e* by *-ol.* Indicate the location of the hydroxyl group by numbering the C atoms of the backbone, as described in Toolbox 11.1, starting at the end of the chain that results in the lower number for the —OH group. Thus, the systematic name of the alcohol $CH_3CH_2CHOHCH_3$ is 2-butanol. When —OH is named as a substituent, it is called *hydroxy-*.

Ethers Name each of the hydrocarbon groups attached to the O atom separately and alphabetically. Thus, $CH_3OCH_2CH_3$ is ethyl methyl ether.

Aldehydes and ketones For aldehydes, identify the parent alkane: include the C of —CHO in the count of carbon atoms. Then change the final -*e* of the alkane name to -*al*. Thus, CH_3CH_2CHO is propanal. The —CHO group can occur only at the end of a carbon chain and is given the number 1 only if other substituents need to be located. For ketones, change the -*e* of the parent alkane to -*one*. The location of the C=O group is indicated by selecting a numbering order that gives it the lower number. Thus, $CH_3CH_2CH_2COCH_3$ is 2-pentanone.

Carboxylic acids Change the -*e* of the parent alkane to -*oic acid*. To identify the parent acid, include the C atom of the —COOH group when counting carbon atoms. Thus, $CH_3CH_2CH_2COOH$ is butanoic acid.

Esters Change the -*ol* of the alcohol to -*yl* and the -*oic acid* of the parent acid to -*oate*. Thus, $CH_3CH_2COOCH_3$ is methyl propanoate.

Amines Specify the groups attached to the nitrogen atom in alphabetical order, followed by the suffix -*amine*. Thus, $(CH_3CH_2)_2NCH_3$ is diethylmethylamine. Amines with two amino groups are called diamines. The —NH_2 group is called *amino-* when it is a substituent.

Halides Name the halogen atom as a substituent by changing the -*ine* part of its name to -*o*. Thus, CH_3Br is bromomethane.

A functional group has priority in numbering over an alkyl group. In other words, we number the carbon chain to give functional groups the lower numbers, regardless of the location of alkyl groups.

Example 11.5 *Naming compounds with functional groups*

Name (a) $CH_3CH(CH_3)CHOHCH_3$; (b) $CH_3CH_2CH_2COCH_3$; (c) $(CH_3CH_2)_2NCH_2CH_2CH_3$.

Strategy First identify the functional group present, using Table 11.3 as a guide if necessary. Then follow the procedures in Toolbox 11.3 for the functional group present in the molecule. It can be helpful to draw the structural formula.

Solution (a) The —OH group identifies the compound as an alcohol, so its name ends in -*ol*. The longest chain has four carbon atoms, so the compound is a butanol. Because functional groups have priority in numbering and the —OH group is on the second carbon atom from one end, the compound is a 2-butanol. Finally, note the methyl group on carbon atom 3 and name the molecule 3-methyl-2-butanol. (b) Because the compound has a carbonyl group (CO) as part of its chain, it is a ketone, and its name will end in -*one*. The chain has five carbon atoms, so it is based on pentane. The carbonyl group is built on the second carbon from one end, so the name of the compound is 2-pentanone. (c) The compound has two ethyl groups and one propyl group attached to a nitrogen atom. Attach the prefix *di*- to the ethyl groups to indicate the number of each and list the groups in alphabetical order, disregarding prefixes. The compound is diethylpropylamine.

Self-Test 11.8A Name the compounds (a) $CH_3CH(CH_2CH_2OH)CH_3$; (b) $CH_3CH(CHO)CH_2CH_3$; (c) $(C_6H_5)_3N$.

[**Answer:** (a) 3-Methyl-1-butanol; (b) 2-methylbutanal; (c) triphenylamine]

Self-Test 11.8B Name the compounds (a) $CH_3CH_2CHOHCH_2CH_3$; (b) $CH_3CH_2COCH_2CH_3$; (c) $CH_3CH_2NHCH_3$.

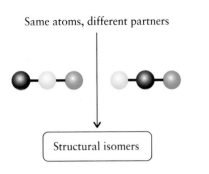

Same atoms, different partners

Structural isomers

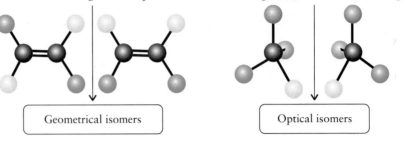

Same atoms, same partners, different arrangement in space

Geometrical isomers

Same atoms, same partners, same neighbors, different mirror image

Optical isomers

FIGURE 11.15

A summary of the various types of isomerism that occur in molecular compounds. Geometrical and optical isomers are both types of stereoisomers.

ISOMERS

We have seen that isomers are different compounds built from the same kit of atoms. Hydrocarbons in general have large numbers of isomers, and that number is immense when we allow for the presence of functional groups. Figure 11.15 summarizes the various types of isomerism.

11.11 Structural Isomers

Molecules built from the same atoms but connected differently are called **structural isomers.** For example, we can insert the C atom and two H atoms of a —CH_2— group into the C_3H_8 molecule in two different ways to give two different compounds with the formula C_4H_{10}:

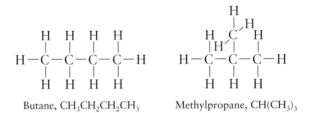

Butane, $CH_3CH_2CH_2CH_3$ Methylpropane, $CH(CH_3)_3$

Although the —CH_2— group could be inserted in other places, the resulting molecules can all be twisted into one or the other of these two isomers. It is important to remember that structural formulas are two-dimensional representations of three-dimensional structures and that one part of the molecule can rotate relative to another, producing a different appearance. Singly bonded structures are continuously writhing and twisting, and —CH_3 groups rotate like tiny propellers. Two structures that look like distinct isomers on paper may actually be only different **conformations,** different orientations of the same molecule produced by rotating about bonds. Example 11.6 shows how to identify structures as different isomers or merely different conformations of the same isomer.

Structural isomers have identical molecular formulas, but their atoms are linked to different neighbors. Different conformations of the same molecule can be twisted into one another.

Example 11.6 *Writing the formulas of isomeric molecules*

Draw two-dimensional molecular structures for all the alkanes of formula C_5H_{12} and write the condensed structural formulas for each.

Strategy Decide which isomers can be constructed by inserting one $-CH_2-$ group between different pairs of atoms of the formulas of the two C_4H_{10} molecules given earlier. Different isomers cannot be changed into one another simply by rotating either the entire formula or parts of the formula on the page, but different conformations of the same isomer can be changed into one another in that way. Discard formulas that repeat those already obtained.

Solution From butane, we can form

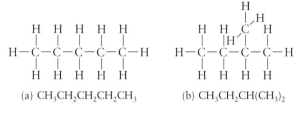

(a) $CH_3CH_2CH_2CH_2CH_3$ (b) $CH_3CH_2CH(CH_3)_2$

From methylpropane, we can form

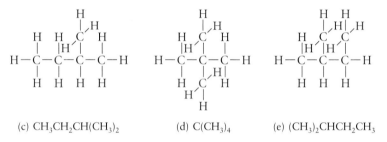

(c) $CH_3CH_2CH(CH_3)_2$ (d) $C(CH_3)_4$ (e) $(CH_3)_2CHCH_2CH_3$

Molecules (b) and (c) are the same. The atoms of molecule (e) are joined together in the same arrangement as (b) and (c) even though they look different as drawn on the page; so (b), (c), and (e) are different conformations of the same isomer. There are therefore only three distinct isomers with formula C_5H_{12}: (a), (b), and (d).

Self-Test 11.9A Write the structural formulas for the five isomeric alkanes of formula C_6H_{14}.

[***Answer:*** $CH_3(CH_2)_4CH_3$ (**39a**); $CH_3(CH_2)_2CH(CH_3)_2$ (**39b**); $CH_3CH_2CH(CH_3)CH_2CH_3$ (**39c**); $CH_3CH_2C(CH_3)_3$ (**39d**); $(CH_3)_2CHCH(CH_3)_2$ (**39e**)]

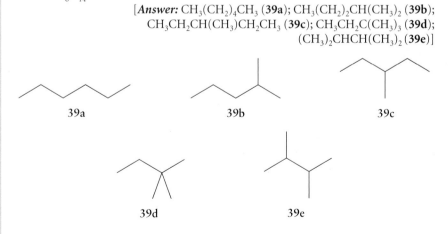

Self-Test 11.9B Write the structural formulas for the four isomers with the molecular formula C_4H_9Br.

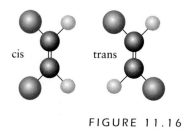

FIGURE 11.16

A pair of geometrical isomers in which two groups are either both on the same side of a double bond (cis) or on opposite sides (trans). Notice that the bonded neighbors of each atom are the same in both cases, but nevertheless the arrangements of the atoms in space are different.

11.12 Geometrical and Optical Isomers

Stereoisomers are isomers that have atoms linked to the same neighbors but arranged differently in space. We distinguish two types of stereoisomerism, geometrical and optical.

Geometrical isomers are stereoisomers that are not mirror images of each other (Fig. 11.16). For example, there are two different compounds with the name 2-butene. They have the same molecular formula, the same structural formula, and one is not the mirror image of the other. They are therefore geometrical isomers. Geometrical isomers of molecules with double bonds or rings are distinguished by the prefixes *cis* and *trans*. A cis isomer is one in which the substituents are on the same side of the double bond or ring; in a trans isomer, the substituents are on opposite sides of the double bond or ring (Fig. 11.17).

Self-Test 11.10A Identify (**40a**) and (**40b**) as cis or trans.

> [*Answer:* (**40a**) is *trans*-2-pentene; (**40b**) is *cis*-2-pentene.]

Self-Test 11.10B Identify (**41a**) and (**41b**) as cis or trans.

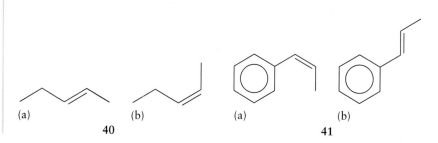

(a)   (b)   (a)   (b)

40   41

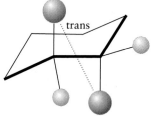

FIGURE 11.17

Compounds with rings can also exhibit geometrical isomerism. Groups attached to carbon atoms in a ring can be both on the same face of the ring (cis) or across the plane of the ring from each other (trans).

Optical isomers are stereoisomers that are non-superimposable mirror images of each other (Fig. 11.18). If we look at the illustration carefully (or, better, build models), then we see that, no matter how we twist and turn the molecules, we cannot superimpose the mirror image molecule on the original molecule. It is like trying to superimpose your right hand on your left hand. The two molecules are, in fact, two distinct compounds. Optical isomerism occurs whenever four different groups are attached to a carbon atom, as in the amino acid alanine, $NH_2CH(CH_3)COOH$.

A **chiral molecule** is one that is not identical to its mirror image and typically contains an atom with four different groups attached. We say that alanine is chiral, and that a chiral molecule and its mirror image form a pair of **enantiomers**. Glycine, NH_2CH_2COOH (**37**), does not have four different groups attached to either central carbon atom. Its mirror image can therefore be rotated and superimposed on the original molecule. We say that glycine is an **achiral molecule.**

Enantiomers have identical chemical properties, except when they react with other chiral compounds. One consequence of this difference in reactivity is that enantiomers may have different odors and pharmacological activities. The molecule has to fit into a cavity, or slot, of a certain shape, either in an odor receptor in the nose or in an enzyme. Only one member of the enantiomeric pair may be able to fit.

Enantiomers differ in one physical property: chiral molecules display **optical activity,** the ability to rotate the plane of polarization of light. In ordinary light, the wave motion occurs in random orientations around the direction of

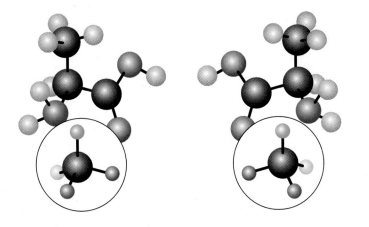

FIGURE 11.18

The molecule on the right is the mirror image of the molecule on the left, as can be seen more clearly by inspecting the simplified representations in the circles. Because the two molecules cannot be superimposed, they are distinct optical isomers.

travel. In plane-polarized light, the wave lies in a single plane (Fig. 11.19). Plane-polarized light can be prepared by passing ordinary light through a special filter, like the material used to make polarized sunglasses. One type of chiral molecule rotates the plane of polarization clockwise; its mirror image partner rotates it by the same amount the other way.

Optical activity is detected through the use of a *polarimeter* (Fig. 11.20). The polarized light is passed through a sample cell about 10 cm long. The cell contains a solution of the sample at a known concentration. When the light emerges from the far end of the cell, its plane of polarization may be different from its original value. To determine the angle of rotation, the light is passed through an *analyzer* that contains another polarizing filter. The analyzer is rotated until the intensity of the light that has passed through the polarizer, sample, and analyzer reaches its maximum. The angle of the plane of polarization is determined from this maximum setting. If the sample is not optically active, the analyzer gives the maximum intensity at an angle of 0. The sample is optically active if the angle of rotation is different from 0. The actual value depends on the identity of the compound, its concentration, and the length of the sample cell.

Amino acids synthesized in the laboratory are **racemic mixtures,** or mixtures of enantiomers in equal proportions. Because molecules that form a pair of enantiomers rotate the plane of polarization of light in opposite directions, a

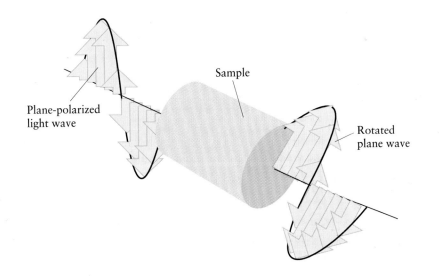

Sample

Plane-polarized light wave

Rotated plane wave

FIGURE 11.19

Plane-polarized light consists of radiation in which all the wave motion lies in one plane (as represented by the orange arrows on the left). When such light passes through a solution of an optically active substance, the plane of the polarization is rotated through a characteristic angle that depends on the concentration of the solution and the length of the path through it.

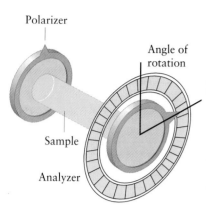

Polarizer

Angle of rotation

Sample

Analyzer

FIGURE 11.20

This polarimeter measures the optical activity of compounds in solution. Light is plane polarized by passage through a polarizer and is then sent through a sample. An analyzer on the right of the sample is rotated until the angle at which the light is brightest is found. That angle is the angle of rotation for the sample.

racemic sample is not optically active. In contrast, reactions in living cells that produce amino acids lead to only one enantiomer. It is a remarkable feature of nature (which is not yet fully understood) that all naturally occurring amino acids in animals have the same handedness.

> *Stereoisomers have the same molecular and structural formulas but the atoms have different arrangements in space. Molecules in which two different groups can be positioned on either side of a double bond or ring are geometrical isomers. Molecules with four different groups attached to a single carbon atom are chiral and are not superimposable on their mirror images. A chiral molecule and its mirror image are optical isomers.*

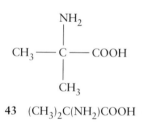

42   $CH_3CH_2CH(NH_2)COOH$

43   $(CH_3)_2C(NH_2)COOH$

Example 11.7 *Predicting whether a compound is chiral*

Predict whether the isomeric amino acids (a) $CH_3CH_2CH(NH_2)COOH$ and (b) $(CH_3)_2C(NH_2)COOH$ are chiral.

Strategy To decide whether a compound is chiral, draw the structural formula and check for a carbon atom with four different groups attached.

Solution The two amino acids are (**42**) and (**43**). We see that (**42**) contains a carbon atom that has four different groups attached and hence is chiral. However, (**43**) does not have such a carbon atom, it is identical to its mirror image, and therefore it is not chiral.

Self-Test 11.11A Which of the following chlorofluorocarbons is chiral: (a) CH_3CF_2Cl; (b) CH_3CHFCl; (c) CH_2FCl?

[*Answer:* (b)]

Self-Test 11.11B Which of the following alcohols is chiral: (a) CH_3CH_2OH; (b) $CH_3CH(OH)CH_3$; (c) $CH_3CH(OH)CH_2CH_3$?

POLYMERS

Many organic materials are **polymers,** giant molecules consisting of chains or networks made by linking together small organic molecules. Some, like polypropylene and Teflon, are synthetic. Others, like silk and cellulose, occur naturally. *Plastics* are polymers that can be molded into shape.

The key to understanding polymers is their repetitive structures. They consist of the same **repeating unit,** a small group of atoms that is repeated over and over again to create the long polymer chains. Big as polymer molecules may be, they are prepared by using reactions like the ones we have been discussing so far. Furthermore, although a polymer molecule might have thousands of functional groups, the properties we have already met apply to them too.

Table 11.4 Addition polymers

Monomer name	Formula	Polymer formula	Common name
ethene*	$CH_2{=}CH_2$	$+CH_2{-}CH_2\frac{}{n}$	polyethylene
vinyl chloride	$CHCl{=}CH_2$	$+CHCl{-}CH_2\frac{}{n}$	polyvinyl chloride
styrene	$CH(C_6H_5){=}CH_2$	$+CH(C_6H_5)CH_2\frac{}{n}$	polystyrene
acrylonitrile	$CH(CN){=}CH_2$	$+CH(CN){-}CH_2\frac{}{n}$	Orlon, Acrilan
propene*	$CH(CH_3){=}CH_2$	$+CH(CH_3){-}CH_2\frac{}{n}$	polypropylene
methyl methacrylate	$CH_3O\overset{\overset{O}{\|}}{C}C(CH_3){=}CH_2$	$\left(\!\!\begin{array}{c} CH_3 \\ \| \\ C{-}CH_2 \\ \| \\ COOCH_3 \end{array}\!\!\right)_{\!n}$	Plexiglas, Lucite
tetrafluoroethene*	$CF_2{=}CF_2$	$+CF_2{-}CF_2\frac{}{n}$	Teflon, PTFE[†]

*The suffix *-ene* is replaced by *-ylene* in the common names of these compounds, hence the names of the corresponding polymers.

[†]PTFE, polytetrafluoroethylene.

11.13 Addition Polymerization

Alkenes react with themselves in a process called **addition polymerization.** For example, an ethene molecule may form a bond to another ethene molecule, another ethene molecule may add to that, and so on, until a long hydrocarbon chain has grown. The original alkene, such as ethene (commonly called ethylene), is called the **monomer.** The product, the chain of covalently linked monomers, is a polymer. The simplest addition polymer is polyethylene, $+CH_2CH_2\frac{}{n}$, which consists of long chains of CH_2CH_2 units. Many polymer molecules also have a number of branches, which are side chains that sprout from intermediate points in the original chain.

The plastics industry has developed addition polymers from a number of monomers with the formula $CHX{=}CH_2$, where X is a single atom (such as the Cl in vinyl chloride, $CHCl{=}CH_2$) or a group of atoms (such as the CH_3 in propene). These substituted alkenes give polymers of formula $+CHXCH_2\frac{}{n}$ and include polyvinyl chloride (PVC), $+CHClCH_2\frac{}{n}$, and polypropylene, $+CH(CH_3)CH_2\frac{}{n}$. Some of these polymers are listed in Table 11.4. They differ in appearance, rigidity, transparency, and resistance to weathering. Many plastic materials made of addition polymers can be recycled by melting and reprocessing. The recycling code on the bottom of a plastic container indicates the polymer used to make the container (Table 11.5).

A widely used polymerization procedure is **radical polymerization,** in which the reaction depends on the presence of radicals (Section 8.10). In a typical procedure, a monomer (such as ethene) is compressed to about 1000 atm and heated to 100°C in the presence of a small amount of an organic peroxide (a compound of formula $R{-}O{-}O{-}R$, where R is an organic group). The reaction is initiated by dissociation of the $O{-}O$ bond, giving two radicals:

$$R{-}O{-}O{-}R \longrightarrow R{-}O\cdot + \cdot O{-}R$$

These radicals attack monomer molecules $CHX{=}CH_2$ (with X = H for ethene itself) and form a new, highly reactive radical:

> In general, the formula for the repeating unit of an addition polymer is the same as that of the monomer.

Table 11.5 Recycling codes

Recycling code		Polymer
1	PET	polyethylene terephthalate
2	HDPE	high-density polyethylene
3	PVC or VC	polyvinyl chloride
4	LDPE	low-density polyethylene
5	PP	polypropylene
6	PS	polystyrene

$$R{-}O{\cdot} + H_2C{=}C\overset{\displaystyle H}{\underset{\displaystyle X}{\Big|}} \longrightarrow R{-}O{-}CH_2{-}\overset{\displaystyle H}{\underset{\displaystyle X}{\overset{|}{C\cdot}}}$$

This radical attaches to another monomer molecule, which attaches to another, and the chain grows longer:

$$R{-}O{-}CH_2{-}\overset{\displaystyle H}{\underset{\displaystyle X}{\overset{|}{C\cdot}}} + H_2C{=}C\overset{\displaystyle H}{\underset{\displaystyle X}{\Big|}} \longrightarrow R{-}O{-}CH_2{-}\overset{\displaystyle H}{\underset{\displaystyle X}{\overset{|}{C}}}{-}CH_2{-}\overset{\displaystyle H}{\underset{\displaystyle X}{\overset{|}{C\cdot}}}$$

The reaction continues until all the monomer has been used up. The product consists of long chains of formula $-(CH_2CX_2)_n-$ in which n can reach many thousands.

The backbone carbon atoms of some polymers are chiral, with structures like those in (**44**). This chirality was an early problem with polypropylene, because the arrangements of H and CH_3 groups on each chiral C atom in the chain were random and the resulting material was amorphous, sticky, and nearly useless. Now, however, the chirality of the atoms can be controlled by using a special catalyst known as a **Ziegler-Natta catalyst** in the polymerization. A Ziegler-Natta catalyst consists of titanium tetrachloride and an aluminum alkyl compound, such as triethylaluminum.

A polymer in which the units all take the same handedness or lie in the same direction along the chain is called **stereoregular.** In **isotactic** polypropylene, the carbon atoms forming the chain all have the same handedness, so all the methyl groups are on the same side of the chain, as in Fig. 11.21a. In **syndiotactic** polypropylene, the carbon atoms in the chain alternate in handedness, so the methyl groups alternate in direction, as in Fig. 11.21b. **Atactic** polypropylene is not stereoregular, because the methyl groups point randomly to either side of the chain (Fig. 11.21c).

The high density of polymers produced by Ziegler-Natta catalysts results from the regular arrangement of groups along the polymer chain. As a result of this regular arrangement, the chains can pack together well and form a highly crystalline, dense material that can be spun into fibers; polypropylene fibers are now used in sportswear. Catalysts are also used to control the structure of conducting polymers (Applying Chemistry: Case Study 11), to enhance their electrical characteristics.

Rubber is a natural polymer with isoprene monomers (**2**). Natural rubber is obtained from the bark of the rubber tree as a milky white liquid, called latex (Fig. 11.22), that consists of a suspension of rubber particles in water. The rubber itself is a soft white solid that becomes even softer when warm. It is used for pencil erasers and was once used as crepe for the soles of shoes.

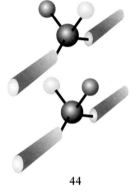

44

FIGURE 11.21

The stereoregular polymers produced by using Ziegler-Natta catalysts may be (a) isotactic (all on one side) or (b) syndiotactic (alternating). (c) In an atactic polymer, the substituents lie on random sides of the chain.

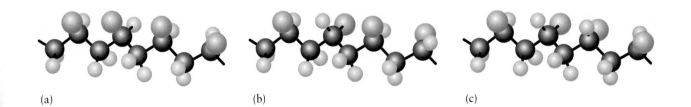

(a)

(b)

(c)

Applying Chemistry: *Case Study 11*

One day, if you want to check your email, you might simply pull a pen-shaped object from your pocket and unroll it to form a shiny plastic sheet. When you activate the tiny microprocessor in the sheet, your messages will appear and you will reply by writing on the screen or speaking to it.

The remarkable materials required to build this pocket computer are not made of the metals and metalloids found in today's computers. Metals conduct electricity because of their very low ionization energies (Section 7.17), but some corrode easily and many have high densities. *Conducting polymers* are molecular compounds made of nonmetals. They conduct electricity, but do not corrode; they have low densities, and they are flexible. These materials have the potential to change our lives in interesting ways. To understand how they operate, we need to know something about the kinds of bonds carbon atoms form.

Conducting polymers were discovered by accident in the early 1970s when a chemist who was polymerizing ethyne (acetylene) added a thousand times too much catalyst. Instead of a synthetic rubber, he made a thin, flexible film. It looked like metal tinted pink (see the photograph), and—very much like a metal—it conducted electricity. An electric current is the flow of electrons. Metals conduct electricity because their valence electrons move easily from atom to atom. Most covalently bonded solids do not conduct electricity and are classified as insulators. Conducting polymers provide a new and exciting alternative. They can be molded or drawn into light-weight rust-free shells, fibers, or thin plastic sheets that conduct electricity as well as many metals do. They can be made to glow with almost any color and to change conductivity with conditions. Imagine cases of food labeled with polymer tags that change in conductivity or appearance when the cases are left unrefrigerated for too long.

How can polymer molecules conduct electricity? To answer this question, we need to think about how electricity is conducted through a solid. In metals, there is a conduction band consisting of very closely spaced energy levels. Can such a band exist in a molecular compound? All conducting polymers have a common feature: a long chain of sp^2 hybridized carbon atoms, often with nitrogen or sulfur atoms included in the chains. Polyacetylene, the first conducting polymer to be made, is also the simplest: it consists of thousands of $+CH{=}CH)_n$ units:

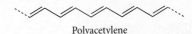

Polyacetylene

The double bonds alternate, so each C atom has an unhybridized *p*-orbital that can overlap with the *p*-orbital on either side. This arrangement allows electrons to be delocalized along the entire chain like a very long, one-dimensional version of benzene. Because the number of carbon *p*-orbitals contributing to the π-orbitals is very large, there is a huge number of energy levels available. As a result, these energy levels are very closely spaced, forming a nearly continuous band. Because the band is nearly continuous, electrons can easily move from one level to another and conduct electricity through the polymer.

Polyaniline is a conducting polymer used in flexible coaxial cable, rechargeable, buttonlike batteries, and laminated films that could be used as flexible computer or television displays.

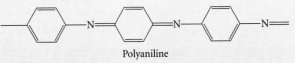

Polyaniline

Polyaniline could one day serve both as a nonmetallic solder for the printed circuit boards inside a computer and as electrical shielding for the case. All-plastic transistors are also being studied, raising the possibility of an all-plastic computer that could survive highly corrosive conditions, such as those at marine research sites.

Key Concepts: polymers, hybridization of carbon atoms, π-orbitals, electron delocalization

For Further Reading

M. G. Kanatzidis, Conductive polymers, *Chemical and Engineering News,* Dec. 3, 1990, pp. 36–54.

P. Yam, Plastics get wired, *Scientific American,* July, 1995, pp. 83–89.

Related Exercises: 11.109–11.112

This flexible polyacetylene sheet was peeled from the walls of the reaction flask in which it was made from acetylene.

FIGURE 11.22

Collecting latex from a rubber tree in Malaysia, one of its principal producers.

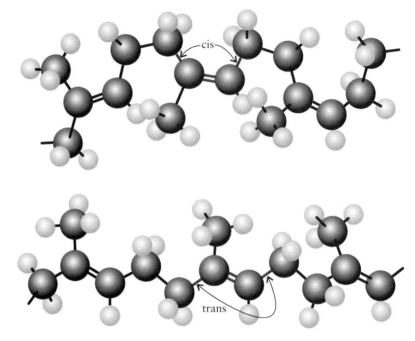

FIGURE 11.23

In natural rubber, the isoprene units are polymerized to be all cis. The harder material, gutta-percha, is the all-trans polymer.

Chemists were unable to synthesize rubber for a long time, even though they knew it was a polymer of isoprene. The enzymes in the rubber tree produce a polymer in which all the links between monomers are in a cis arrangement, so natural rubber is *cis*-polyisoprene (Fig. 11.23). Straightforward radical polymerization, however, produced a random mixture of cis and trans links and a sticky, useless product. Now a Ziegler-Natta catalyst is used to produce almost pure, rubbery *cis*-polyisoprene. *trans*-Polyisoprene, in which all the links are trans, is also produced naturally: it is the hard material *gutta-percha,* once used inside golf balls and still used to fill root canals in teeth.

> **Alkenes undergo addition polymerization. If a Ziegler-Natta catalyst is used, the polymer is stereoregular and has a relatively high density.**

11.14 Condensation Polymerization

Although hydrocarbon polymers are highly important commercially, synthetic fibers are primarily produced by another kind of polymerization reaction. Polymers formed by linking together monomers that have carboxylic acid groups with those that have alcohol groups are called **polyesters.** Polyesters are widely used for making artificial fibers and are examples of **condensation polymers,** which are polymers formed by a series of condensation reactions. Polyesters have many uses besides fibers. For example, they can be molded and used in surgical implants, such as artificial hearts, or made into thin films for cassette tapes.

The most common polyester is Dacron, or Terylene, a polymer produced from the esterification of terephthalic acid with ethylene glycol; its technical name is polyethylene terephthalate. The first condensation is

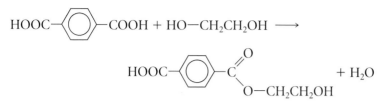

A new ethylene glycol molecule can condense with the carboxyl group on the left of the product, and another terephthalic acid molecule can condense with the hydroxyl group on the right. As a result, the polymer grows at each end and in due course becomes

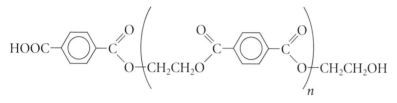

Because condensation polymerization occurs only at the functional group, condensation polymers are unbranched and make good fibers. In the radical addition polymerization of alkenes, side chains can start to grow out from the main chain. However, in condensation polymerization, growth can occur only at the functional groups, so chain branching is much less likely. As a result, polyester molecules can be made to lie side by side by stretching the heated product and forcing it through a small hole (Fig. 11.24). The fibers produced can then be spun into smooth yarns (Fig. 11.25).

Condensation polymerization of amines with carboxylic acids leads to the **polyamides,** substances more commonly known as *nylons.* A common polyamide is nylon-66, which is a polymer of 1,6-diaminohexane, $H_2N(CH_2)_6NH_2$, and adipic acid, $HOOC(CH_2)_4COOH$. The 66 in the name indicates the numbers of C atoms in the two monomers.

For condensation polymerization, it is necessary to have two functional groups on each monomer and to mix stoichiometric amounts of the reactants.

FIGURE 11.24

Synthetic fibers are made by extruding liquid polymer from small holes in an industrial version of the spider's spinneret.

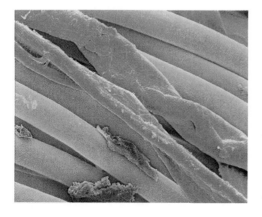

FIGURE 11.25

A scanning electron micrograph of Dacron polyester and cotton fibers in a blended shirt fabric. The cotton fibers have been colored green. Compare the smooth cylinders of the polyester with the irregular surface of cotton. The smooth polyester fibers resist wrinkles, and the irregular cotton fibers produce a more comfortable and absorbent texture.

In the case of polyamide formation, the starting materials form nylon salt by proton transfer, as in

$$HOOC(CH_2)_4COOH + H_2N(CH_2)_6NH_2 \longrightarrow$$
$$^-O_2C(CH_2)_4CO_2^- + {}^+H_3N(CH_2)_6NH_3^+$$

At this point, the excess acid or amine can be removed. Then, when the nylon salt—an amine salt of a carboxylic acid—is heated, the condensation begins. The first step is

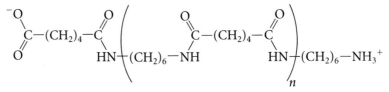

The amide grows at both ends by further condensations (Fig. 11.26), and the final product is

The long polyamide (nylon) chains can be spun into fibers (like polyesters) or molded. Some of the strength of nylon fibers arises from the $\diagup$N—H$\cdots$O=C$\diagdown$ hydrogen bonding that can occur between neighboring chains (Fig. 11.27). This

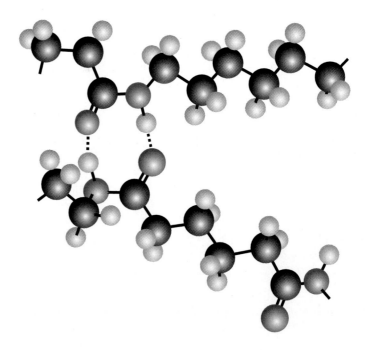

FIGURE 11.26

A rather crude nylon fiber can be made by dissolving the salt of the amine in water and dissolving the acid in a layer of hexane, which floats on the water. The polymer forms at the interface of the two layers, and a long string can be slowly pulled out.

FIGURE 11.27

The strength of nylon fibers is yet another sign of the presence of hydrogen bonds, this time between neighboring polyamide chains.

CHAPTER 11   CARBON-BASED MATERIALS

ability to form hydrogen bonds also accounts for nylon's tendency to absorb moisture, because H_2O molecules can form hydrogen bonds to the chains and worm their way in among the chains. The ability of an —NH— group to become charged, forming —NH_2^+—, for example, accounts for the buildup of electrostatic charge on nylon fabrics and carpets when they are subjected to the friction of a footstep.

Condensation polymers are formed by condensing a carboxylic acid with an alcohol to form a polyester or with an amine to form a polyamide.

Example 11.8 *Determining the formulas of polymers and monomers*

Write the formulas of (a) the monomers of Kevlar, a strong fiber used to make bulletproof vests:

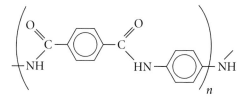

(b) two repeating units of the polymer that is formed when peroxides are added to $CH_3CH_2CH{=}CH_2$.

Strategy (a) Look at the backbone of the polymer, the long chain to which the other groups are attached. If the atoms are all carbon atoms, it is an addition polymer. If ester groups are present in the backbone, the polymer is a polyester and the monomers will be an acid and an alcohol. If the backbone contains amide groups, the polymer is a polyamide and the monomers will be an acid and an amine. If the polyester or polyamide groups all face in the same direction, both groups are on the same molecule and there is only one monomer; otherwise, there are two different monomers. Monomers of condensation polymers have two functional groups in each molecule. (b) If the monomer is an alkene or alkyne, the monomers will add to one another; a π-bond will be replaced by new σ-bonds between the monomers. If the monomers are an acid and an alcohol or amine, a condensation polymer forms. Draw the ester or amide that would result from the loss of a molecule of water.

Solution (a) Amide groups are present in the backbone, so the polymer is a polyamide. The amide groups face in opposite directions, so there are two different monomers, one with two acid groups and one with two amine groups. As we split each amide group apart, we add a molecule of water:

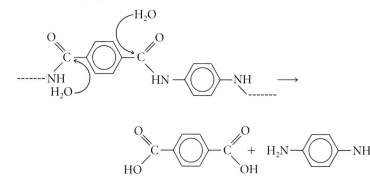

The two compounds in the product are terephthalic acid (more formally, 1,4-benzenedicarboxylic acid) and 1,4-diaminobenzene. (b) The monomer is an alkene (1-butene), so it forms an addition polymer. Replace the π-bond by two additional σ-bonds, one to each adjacent monomer:

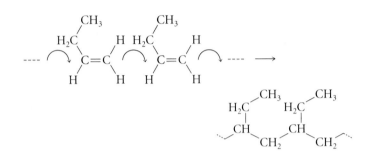

The product is called polybutene (or polybutylene).

Self-Test 11.12A (a) Write the formula for the monomer of Teflon, $\text{-}(CF_2\text{--}CF_2)_n$. (b) The polymer of lactic acid (**45**) is used in surgical sutures that dissolve in the body; write the formula for the repeating unit of this polymer.

45 Lactic acid, $CH_3CH(OH)COOH$

[**Answer:** (a) $CF_2\text{=}CF_2$; (b) $\text{-}(O\text{--}CH(CH_3)\text{--}CO)_n$]

Self-Test 11.12B Write the formula for (a) the monomer of polymethylmethacrylate,

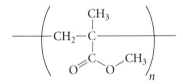

used in contact lenses; (b) two repeating units of polyalanine, the polymer of the amino acid alanine, $CH_3CH(NH_2)COOH$.

11.15 Physical Properties

Synthetic polymers do not have definite molar masses because they consist of molecules with different lengths. Two polymer molecules with the same repeating unit, but different lengths, are still considered molecules of the same substance. So, we speak of the *average* molar mass and the *average* chain length of a polymer. The presence of a mixture of molecules means that polymers do not have definite melting points but soften gradually as the temperature is raised. The viscosity of a polymer, its ability to flow when molten (Section 10.4), depends on its chain length. The longer the chains, the more entangled they may become, and hence the slower their flow.

FIGURE 11.28

The two samples of polyethylene in the test tube were produced by different processes. The floating, low-density polymer was produced by high-pressure polymerization. The high-density polymer at the bottom was produced with a Ziegler-Natta catalyst. As the insets show, the higher density results from the greater linearity of the chains, allowing them to pack together better.

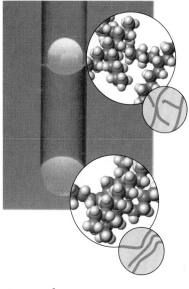

Chain length affects mechanical strength. Tearing a tangled long-chain polymer apart means breaking backbone covalent bonds, which are much stronger than intermolecular forces, the only forces holding very short chains together. Longer chains thus produce higher mechanical strength.

We can think of polymer chains as strands of cooked spaghetti. Spaghetti that has been chopped into small pieces represents short chains. It can be torn apart easily. However, long strands of spaghetti, which represent long-chain polymers, are very difficult to separate (and even to ladle out of a bowl), because the chains become entangled.

An analogy

The way the chains of a polymer pack together affects polymer strength. Packing arrangements that maximize intermolecular contact also maximize intermolecular forces and increase the strength of the material. Long, unbranched chains can line up next to one another like raw spaghetti and form crystalline regions that allow maximum interaction between chains. The closer the contact, the denser the material and the stronger the forces. Branched polymer chains cannot fit together as closely (Fig. 11.28).

The polarity of the functional groups in a polymer affects the strength of intermolecular forces, which in turn affects mechanical strength. For example, nylon is a polyamide, which can take part in hydrogen bonding. Polyethylene has only London forces holding its molecules together. For chains of the same average length, stronger intermolecular forces produce greater mechanical strength.

The **elasticity** of a polymer refers to its ability to return to its original shape when stretched, as a rubber band snaps back after being stretched. Natural rubber has low elasticity and is easily softened by heating. These properties limited the use of rubber until the early nineteenth century, when Charles Goodyear was searching for a means to improve the elasticity of rubber. In 1839, he discovered, partly by accident, that vulcanizing rubber produces the desired properties.

In **vulcanization,** rubber is heated with sulfur. The sulfur atoms form cross-links between the polyisoprene chains and produce a three-dimensional network of atoms (Fig. 11.29). Because separate chains are now linked together, vulcanized rubber does not soften as much as natural rubber does when the temperature is raised. It is also much more resistant to deformation when stretched, because the cross-links pull it back. Materials that return to their

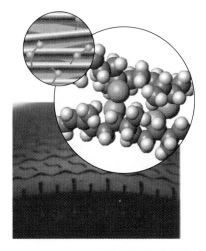

FIGURE 11.29

Automobile tires are made of vulcanized rubber and a number of additives, including carbon. The gray cylinders in the small inset represent polyisoprene molecules, and the beaded yellow strings represent disulfide ($-S-S-$) links that are introduced when the rubber is vulcanized, that is, heated with sulfur. These cross-links increase the resilience of the treated rubber and make it more useful than natural rubber.

FIGURE 11.30

This high-performance race car is made of a composite material that is stronger than steel and can withstand great stress.

original shapes after stretching are called **elastomers.** However, extensive cross-linking can produce a rigid network that resists stretching. For example, high concentrations of sulfur result in very extensive cross-linking and the hard material called *ebonite.*

One way to modify the properties of a polymer is to mix it with another substance. **Composite materials** contain two or more separate materials that have been solidified together. Composite materials like fiberglass have inorganic solids mixed into a polymer matrix. The result is a material that has great strength yet remains flexible. Some lightweight composites—such as the graphite composite used for tennis rackets, the body of the space shuttle, and high performance aircraft—can have three times the strength-to-density ratio of steel (Fig. 11.30).

Polymers melt over a range of temperatures, and polymers consisting of long chains tend to have high viscosities. Polymer strength increases with increasing chain length, polarity of functional groups in the polymer, and the extent of crystallization. Elastomers contain cross-links that increase their strength and elasticity. Composite materials blend the properties of more than one component material.

BIOPOLYMERS

We would be surrounded by polymers even if there were no polymer industry. Many natural materials are polymers, including the cellulose of wood, natural fibers like cotton and silk, the proteins and carbohydrates in our food, and the nucleic acids of our genes.

11.16 Proteins

Amino acids are the building blocks of proteins. In a sense, proteins are a very elaborate form of nylon, because they result from condensation reactions that produce amide links. However, whereas nylon is a monotonous alternation of the same two monomers, the polypeptide chains of proteins are made by linking up to 20 different naturally occurring amino acids distinguished only by their side chains (Table 11.6), so there are billions of possible combinations. Every cell in our body contains more than 5000 different kinds of proteins, each kind with a specific function to perform. Our bodies can synthesize 11 amino acids in sufficient amounts for our needs. The other nine, which are known as the **essential amino acids,** must be included in our diet. The uniqueness of every individual stems from the small differences in body proteins that result from the way these 20 amino acid building blocks are arranged.

A molecule formed from two or more amino acids is called a **peptide.** An example is the combination of glycine and alanine, denoted Gly-Ala:

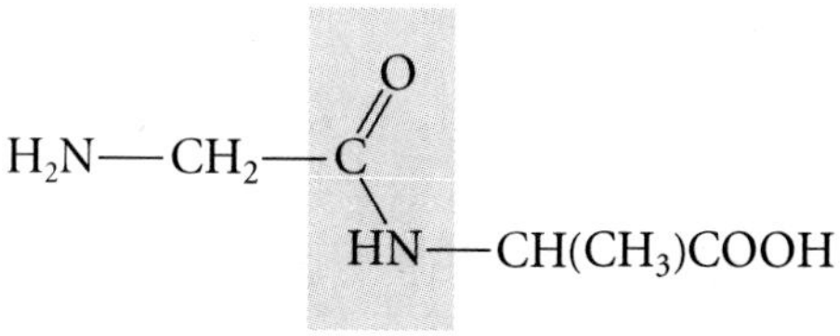

$$H_2N-CH_2-\overset{\overset{O}{\|}}{C}$$
$$HN-CH(CH_3)COOH$$

The $-CO-NH-$ link shown in the box is called a **peptide bond.** Each amino acid unit in a peptide is called a **residue,** because H_2O has been lost in the formation of the peptide bond. A typical protein is a polypeptide chain of more than a hundred residues joined through peptide bonds and arranged in a strict order. When only a few amino acid residues are present, we call the molecule an **oligopeptide.** The artificial sweetening agent aspartame is a type of oligopeptide called a **dipeptide** because it has two residues.

The sequence of residues in the peptide chain is the protein's **primary structure.** The primary structure of the peptide shown above is Gly-Ala. Three fragments of the primary structure of human hemoglobin are

Leu–Ser–Pro–Ala–Asp–Lys–Thr–Asn–Val–Lys–...

...–Val–Lys–Gly–Trp–Ala–Ala–...

...–Ser–Thr–Val–Leu–Thr–Ser–Lys–Ser–Lys–Tyr–Arg

Table 11.6 The naturally occurring amino acids, $X-CH(NH_2)COOH$

$-X$	Name	Abbreviation	$-X$	Name	Abbreviation
$-H$	glycine	Gly	$-CH_2(CH_2)_3NH_2$	lysine*	Lys
$-CH_3$	alanine	Ala	$-CH_2(CH_2)_2NH-\underset{\underset{NH}{\|\|}}{C}-NH_2$	arginine	Arg
$-CH_2$⬡	phenylalanine*	Phe			
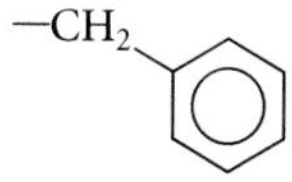			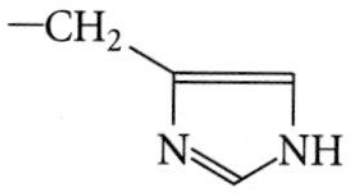 $-CH_2$ (imidazole ring, N NH)	histidine*	His
$-CH(CH_3)_2$	valine*	Val	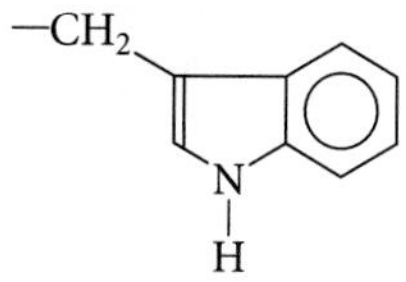		
$-CH_2CH(CH_3)_2$	leucine*	Leu	$-CH_2$ (indole ring, N H)	tryptophan*	Trp
$-CH(CH_3)CH_2CH_3$	isoleucine*	Ile			
$-CH_2OH$	serine	Ser			
$-CH(OH)CH_3$	threonine*	Thr			
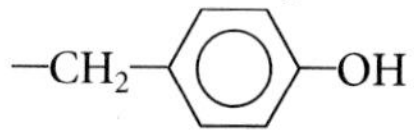 $-CH_2$⬡$-OH$	tyrosine	Tyr			
$-CH_2COOH$	aspartic acid	Asp	$-CH_2CONH_2$	asparagine	Asn
$-CH_2CH_2COOH$	glutamic acid	Glu	$-CH_2CH_2CONH_2$	glutamine	Gln
$-CH_2SH$	cysteine	Cys	(pyrrolidine ring, NH COOH)	proline†	Pro
$-CH_2CH_2SCH_3$	methionine*	Met			

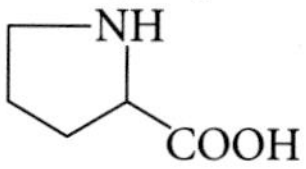

*Essential amino acids for humans.

†The entire amino acid is shown.

FIGURE 11.31

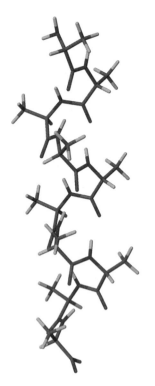

A representation of part of an α helix, one of the secondary structures adopted by polypeptide chains. The tubes represent the atoms and their bonds, with colors that correspond to the colors commonly used to represent different atoms. The narrow lines indicate hydrogen bonds. The methyl group side chains show that this molecule is polyalanine.

The way a polypeptide chain coils together or arranges itself is called the **secondary structure** of the molecule. The most common secondary structure in animal proteins is the **α helix,** a specific conformation of a polypeptide chain held in place by hydrogen bonds between residues (Fig. 11.31). An alternative secondary structure is the **β-pleated sheet** form of the protein we know as silk. In this structure, the protein molecules lie side by side to form nearly flat sheets. The molecules of many proteins consist of alternating regions of α helix and β-pleated sheet (Fig. 11.32).

Protein molecules also adopt a specific **tertiary structure,** the shape into which the α helix, β-pleated sheet, and other regions are folded as a result of interactions between residues lying in different parts of the primary structure. The globular form of each chain in hemoglobin is an example. The links that create the tertiary structure are primarily intermolecular forces. However, one important type of link responsible for tertiary structure is the covalent **disulfide link,** —S—S—, between amino acids containing sulfur.

Proteins may also have a **quaternary structure,** in which neighboring polypeptide chains stack together in a specific arrangement. The hemoglobin molecule, for example, has a quaternary structure of four polypeptide units like the one shown in Fig. 11.32.

Any modification of the primary structure of a protein—the replacement of one amino acid residue by another—may lead to the malfunction we call con-

FIGURE 11.32

One of the four polypeptide chains that make up the human hemoglobin molecule. Each chain consists of alternating regions of α helix (represented by red ribbons) and β-pleated sheet. The oxygen molecules we inhale attach to the iron atom (blue sphere) and are carried through the bloodstream to be released where they are needed.

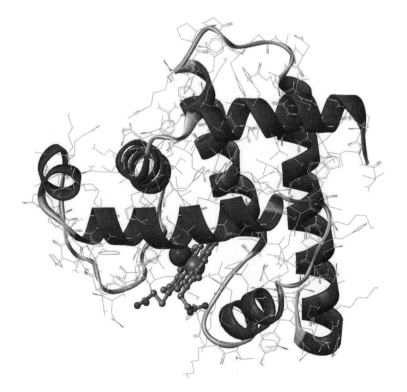

genital disease. Even one different amino acid in the chain can disrupt the normal function of the molecule (Fig. 11.33).

The loss of structure of a protein is called **denaturation.** This structural change may be a loss of quaternary, tertiary, or secondary structure, or even the degradation of the primary structure by cleavage of the peptide bonds. Mild heating can cause irreversible denaturation of some proteins, as happens when we cook an egg and the albumen denatures into a white mass. The permanent waving of hair—which consists primarily of long α helices of the protein keratin—is a result of partial denaturation. Mild reducing agents are applied to the hair to sever the disulfide links between the protein strands that make up hair. The links are then reformed by applying a mild oxidizing agent while the hair is stretched and twisted into the desired arrangement.

We wear proteins as well as eat them. Natural protein-based materials include fibers such as silk and wool and some adhesives. Some animals, such as spiders, also use proteins as a structural material of remarkable strength (Fig. 11.34). Chemists are duplicating nature by making artificial spider silk (Fig. 11.35), which is one of the strongest fibers known. It can be spun into thread or wound into cables strong enough to support suspension bridges.

> *The primary structure of a polypeptide is the sequence of amino acid residues; the secondary structure is the geometrical arrangement of the chains in space; the tertiary structure is the folding of helices and sheets; the quaternary structure is the packing of individual polypeptide chains together.*

11.17 Carbohydrates

The **carbohydrates** are so called because they often have the empirical formula CH_2O, which suggests a hydrate of carbon. They include starches, cellulose, and

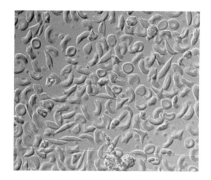

FIGURE 11.33

The sickle-shaped red blood cells that form when a certain glutamic acid residue in hemoglobin (see Fig. 11.32) is replaced by valine.

FIGURE 11.34

The protein made by spiders to produce a web is a form of silk that can be exceptionally strong.

FIGURE 11.35

The thread on these spools is synthetic spider silk, one of the strongest fibers known. It can be used as the thin, tough thread shown here or wound into cables strong enough to support suspension bridges.

FIGURE 11.36

The amylose molecule, a component of starch, is a polysaccharide. A polymer of glucose, it consists of glucose units linked together to give a structure like this but with a moderate degree of branching.

sugars like glucose, which is an aldehyde, $C_6H_{12}O_6$ (**46**), and fructose (fruit sugar), a structural isomer of glucose that is a ketone (**47**). Carbohydrates have many —OH groups and form numerous hydrogen bonds with one another and with water.

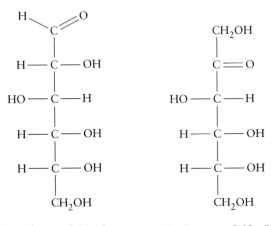

    46 Glucose, $C_6H_{12}O_6$ **47** Fructose, $C_6H_{12}O_6$

Many **polysaccharides** are polymers of glucose. They include starch, which we can digest, and cellulose, which we cannot. Starch is made up of two components, amylose and amylopectin. Amylose, which constitutes 20–25% of most starches, is composed of chains made up of several thousand glucose units linked together (Fig. 11.36). Amylopectin is also made up of glucose chains (Fig. 11.37), but its chains are linked into a branched structure and its molecules are much larger. Each molecule consists of about a million glucose units.

Cellulose is the structural material of plants. Like starch, it is a polymer of glucose, but the units link differently, forming flat, ribbonlike strands (Fig. 11.38). These strands can lock together through hydrogen bonds into a rigid structure that for us (but not for termites) is indigestible. The difference between cellulose and starch shows nature at its most economical and elegant, for only a small modification of the linking between glucose units results, on the one hand, in an important foodstuff and, on the other, in a versatile construction material. Cellulose is the most abundant organic chemical in the world, and billions of tons of it are produced annually by photosynthesis.

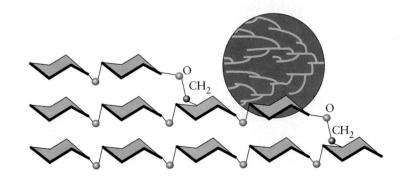

FIGURE 11.37

The amylopectin molecule is another component of starch. It has a more highly branched structure than amylose.

(a)

FIGURE 11.38

(a) Cellulose is yet another polysaccharide constructed from glucose units. The linking between the units in cellulose results in long, flat ribbons that can produce a fibrous material through hydrogen bonding. (b) These long tubes of cellulose formed the structural material of an aspen tree.

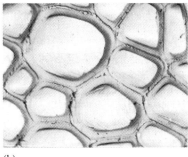

(b)

Glucose is an alcohol and an aldehyde that polymerizes to form starch and cellulose.

11.18 DNA and RNA

People have used materials to store information since the first cave paintings. Today, we can speculate about the possibility of storing information in single molecules. Arrays of such molecules could serve as data storage devices of enormous capacity. Nature, however, has already used this technique for millions of years. It uses the molecule called *deoxyribonucleic acid* (DNA) to store the genetic information that enables life forms to reproduce (Fig. 11.39).

Every living cell contains at least one DNA molecule to control the production of proteins and carry genetic information from one generation of cells to the next. Human DNA molecules are immense: if one could be extracted without damage from a cell nucleus and drawn out straight from its highly coiled natural shape, it would be about 2 m long (Fig. 11.40). The *ribonucleic acid* (RNA) molecule is closely related to DNA. One of its functions is to carry information stored by DNA to a region of the cell where that information can be used in protein synthesis.

The best way to understand the structure of DNA is to see how it gets its name. It is a polymer built up of repeating units derived from the sugar ribose (**48**). For DNA, the ribose molecule has been modified by removing the oxygen atom at carbon atom 2, the second carbon atom from the ether oxygen atom in the five-membered ring. Therefore, the repeating unit—the monomer—is called deoxyribose (**49**).

48   Ribose, $C_5H_{10}O_5$

49   Deoxyribose, $C_5H_{10}O_4$

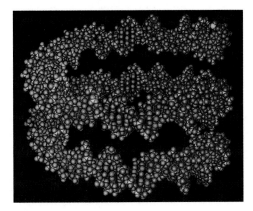

FIGURE 11.39

A computer graphics image of a short section of a DNA molecule, which consists of two entwined helices. In this illustration, the double helix is also coiled around itself in a shape called a superhelix.

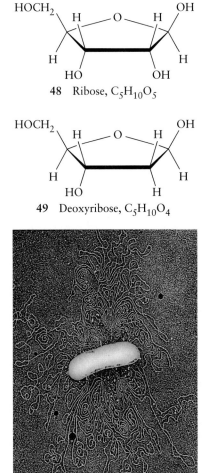

FIGURE 11.40

A DNA molecule is very large, even in bacteria. In this micrograph, a DNA molecule has spilled out through the damaged cell wall of a bacterium.

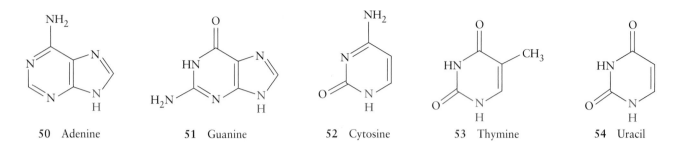

50 Adenine   51 Guanine   52 Cytosine   53 Thymine   54 Uracil

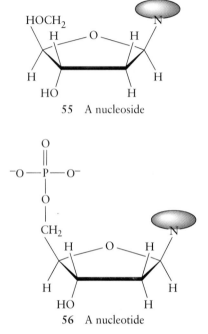

55 A nucleoside

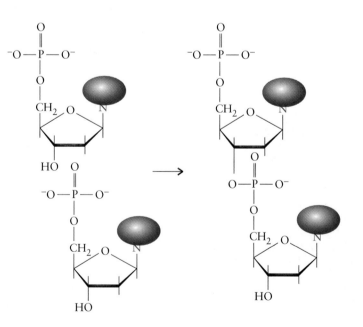

56 A nucleotide

Attached by a covalent bond to carbon atom 1 of the deoxyribose ring is an amine (and therefore a base), which may be adenine, A (**50**), guanine, G (**51**), cytosine, C (**52**), or thymine, T (**53**). In RNA, uracil, U (**54**), occurs in place of thymine. The base bonds to carbon atom 1 of deoxyribose through the nitrogen of the $\diagdown$N H group (printed in red); and the compound so formed is called a **nucleoside.** All nucleosides have a similar structure, which we can summarize as the shape shown in (**55**). The lens-shaped object represents the attached amine.

At this stage, the DNA monomers are each completed by a phosphate group, $-O-PO_3^{2-}$, covalently bonded to carbon atom 5 to give a compound called a **nucleotide** (**56**). Because there are four possible nucleoside monomers (one for each base), there are four possible nucleotides in each type of nucleic acid.

Molecules of DNA and RNA are *polynucleotides,* polymeric species built from nucleotide units. Polymerization occurs when the phosphate group of one nucleotide condenses with the —OH group on carbon atom 3 of another nucleotide, thereby releasing a molecule of water. As this condensation continues, it results in a structure like that shown in Fig. 11.41, a compound known as a *nucleic acid.* The DNA molecule itself is a double helix in which two long nucleic acid strands are wound around each other and held in place by hydrogen bonds between the nucleotides.

The ability of DNA to replicate lies in its double helical structure. There is a precise correspondence between the bases in the two strands. Adenine in one

FIGURE 11.41

The condensation of nucleotides that leads to the formation of a nucleic acid—a polynucleotide. The lens-shaped object is an attached amine.

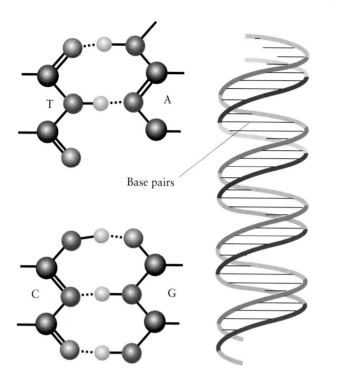

Base pairs

FIGURE 11.42

The bases in the DNA double helix fit together by virtue of the hydrogen bonds that they can form as shown on the left. Once formed, the AT and GC pairs are almost identical in size and shape. As a result, the turns of the helix shown on the right are regular and consistent.

strand always forms hydrogen bonds to thymine in the other, and guanine always forms hydrogen bonds to cytosine, so the base pairs are always AT and GC (Fig. 11.42).

As well as replication—the production of copies of itself for reproduction and cell division—DNA governs the production of proteins by generating molecules of RNA. These molecules, with U in place of T, carry information about segments of the genetic message out of the animal or plant cell nucleus to the locations where protein synthesis takes place.

Nucleic acids are copolymers of four nucleotides linked with phosphate groups. The nucleotide sequence stores all genetic information.

Skills You Should Have Mastered

Conceptual

☐ 1. Distinguish alkanes, alkenes, alkynes, and arenes by differences in bonding, structure, and reactivity, Sections 11.1–11.4.

☐ 2. Explain how intermolecular forces and hydrogen bonds affect the physical properties of compounds, Section 11.6 and Self-Tests 11.2 and 11.7.

☐ 3. Distinguish between addition and condensation polymers, Sections 11.13–11.14.

☐ 4. Explain the role of chain length, crystallinity, network formation, cross-linking, and intermolecular forces in determining the physical properties of polymers, Section 11.15.

Problem-Solving

☐ 1. Draw a structural formula from a condensed structural formula or a line formula, Example 11.1.

☐ 2. Name hydrocarbons and write their structural formulas from their names, Toolbox 11.1 and Examples 11.2 and 11.3.

☐ 3. Predict the products of given elimination, addition, and substitution reactions, Toolbox 11.2 and Example 11.4.

☐ 4. Write the structural formula of an ester or an amide formed from the condensation reaction of a given carboxylic acid with a given alcohol or amine, Self-Test 11.6.

☐ 5. Name compounds with functional groups, Toolbox 11.3 and Example 11.5.

6. Identify two molecules, given their structural formulas, as structural, geometrical, or optical isomers, Example 11.6 and Self-Test 11.10.

7. Judge whether a compound is chiral, Example 11.7.

8. Predict the type of polymer a given monomer can form and identify monomers, given the repeating unit of a polymer, Example 11.8.

Descriptive

1. Describe general trends in the physical properties of alkanes, Section 11.2.

2. Recognize a simple haloalkane, alcohol, ether, phenol, aldehyde, ketone, carboxylic acid, amine, amide, or ester, given a molecular structure, Sections 11.5–11.10.

3. Describe the composition of proteins and distinguish their primary, secondary, tertiary, and quaternary structures, Section 11.16.

4. Describe the composition of carbohydrates, Section 11.17.

5. Describe the structures and functions of DNA and RNA, Section 11.18.

Exercises

Structures and Reactions of Aliphatic Compounds

11.1 Why do branched-chain alkanes have lower melting points and boiling points than straight-chain alkanes with the same number of carbon atoms?

11.2 Why do alkenes have lower melting points and boiling points than alkanes with the same number of carbon atoms?

11.3 Identify the type and number of bonds on carbon atom 2 in (a) pentane; (b) 2-pentene; (c) 2-pentyne.

11.4 Predict the geometry and hybridization of the orbitals used in bonding on carbon atom 2 in (a) pentane; (b) 2-pentene; (c) 2-pentyne.

11.5 Classify each of the following reactions as addition or substitution and write its chemical equation: (a) chlorine reacts with methane when exposed to light; (b) bromine reacts with ethene in the absence of light.

11.6 Classify each of the following reactions as addition or substitution and write its chemical equation: (a) hydrogen reacts with 2-pentene in the presence of a nickel catalyst; (b) hydrogen chloride reacts with propene.

11.7 Classify as addition or substitution and write the chemical equations for the reaction that occurs (if any) when chlorine gas is mixed with (a) ethane and exposed to sunlight; (b) ethene; (c) ethyne. Assume that each molecule of the organic product contains two chlorine atoms.

11.8 Classify as addition or substitution and write the chemical equations for the reactions that occur (if any) when an excess of hydrogen is mixed in the presence of a platinum catalyst with (a) 2-methylpropene; (b) propyne; (c) propane.

Nomenclature of Aliphatic Compounds

11.9 Name each of the following as an unbranched alkane: (a) C_2H_6; (b) C_6H_{14}; (c) C_8H_{18}; (d) C_7H_{16}.

11.10 Name each of the following as an unbranched alkane: (a) $C_{10}H_{22}$; (b) C_4H_{10}; (c) C_9H_{20}; (d) C_5H_{12}.

11.11 Name the following groups: (a) $CH_3—$; (b) $CH_3(CH_2)_3CH_2—$; (c) $CH_3CH_2CH_2—$.

11.12 Name the following groups: (a) $Cl—$, (b) $CH_3CH_2—$; (c) $CH_3CH_2CH_2CH_2—$.

11.13 Give the systematic name of (a) $CH_3CH_2CH_3$; (b) CH_3CH_3; (c) $CH_3(CH_2)_3CH_3$.

11.14 Give the systematic name of (a) $CH_3CH_2CH_2CH_2CH_2CH_3$; (b) $CH_3CH(CH_2CH_3)C(CH_3)_2CH_3$; (c) $(CH_3)_3CCH_2CH(CH_3)CH_3$.

11.15 Name the compounds (a) $CH_3CH{=}CHCH(CH_3)_2$; (b) $CH_3CH_2CH(CH_3)C(C_6H_5)(CH_3)_2$.

11.16 Name the compounds (a) $CH_2{=}CHCH(C_6H_5)(CH_2)_4CH_3$; (b) $(CH_3)_2CHCH(CH_3)CHClC{\equiv}CH$.

11.17 Give the systematic name of the following compounds; if geometrical isomers are possible, write the names of each one: (a) $CH_2{=}CHCH_3$; (b) $CH_3CH{=}CH(CH_2)_2CH_3$; (c) $CH{\equiv}CCH_2CH_3$; (d) $CH_3C{\equiv}CCH_3$.

11.18 Give the systematic name of the following compounds; if geometrical isomers are possible, write the names of each one: (a) $CH_3CH_2CH{=}CHCH_3$; (b) $CH_3CH_2CH{=}CH_2$; (c) $(CH_3)_2C{=}CHCH_2CH_3$; (d) $CH_3C{\equiv}CCH(CH_3)CH_2CH_3$.

11.19 Write the shortened (condensed) structural formula of (a) 3-methyl-1-pentene; (b) 4-ethyl-3,3-dimethylheptane; (c) 5,5-dimethyl-1-hexyne; (d) 3-ethyl-2,4-dimethylpentane.

11.20 Write the shortened (condensed) structural formula of (a) 4-ethyl-2-methylhexane; (b) 3,3-dimethyl-1-pentene; (c) *cis*-4-methyl-2-hexene; (d) *trans*-2-butene.

11.21 Write the structural formula of (a) 4,4-dimethylnonane; (b) 4-propyl-5,5-diethyl-1-decyne; (c) 2,2,4-trimethylpentane; (d) *trans*-3-hexene.

11.22 Write the structural formula of
(a) 2,7,8-trimethyldecane; (b) 2,3,5-trimethyl-4-propylheptane; (c) 1,3-dimethylcyclohexane; (d) 5-ethyl-3-heptene.

Aromatic Compounds

11.23 Name the following compounds:

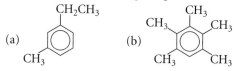

(a)   (b)

11.24 Name the following compounds:

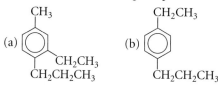

(a)   (b)

11.25 The common name of methylbenzene is toluene. Write the structural formulas of (a) toluene; (b) *p*-chlorotoluene; (c) 1,3-dimethylbenzene; (d) 2-chloromethylbenzene.

11.26 Write the structural formulas of (a) *p*-xylene; (b) 1,2-dichlorobenzene; (c) 3-phenylpropene; (d) 2-ethyl-1,4-dimethylbenzene.

11.27 (a) Draw all the isomeric dichlorobenzenes. (b) Name each one and indicate which are polar and which are nonpolar.

11.28 (a) Draw all the isomeric trichlorobenzenes. (b) Name each one and indicate which are polar and which are nonpolar.

Identifying and Naming Functional Groups

11.29 Write the general formula of each of the following types of compounds, using R to denote an organic group: (a) amine; (b) alcohol; (c) carboxylic acid; (d) aldehyde.

11.30 Write the general formula of each of the following types of compounds, using R to denote an organic group: (a) ether; (b) ketone; (c) ester; (d) amide.

11.31 Identify each functional group: (a) R—O—R; (b) R—CO—R; (c) R—NH$_2$; (d) R—COOR.

11.32 Identify each functional group: (a) R—CHO; (b) R—COOH; (c) R—CONHR; (d) R—OH.

11.33 Write the formulas of the following compounds and state whether each one is a primary, secondary, or tertiary alcohol or a phenol: (a) 1-butanol; (b) 2-butanol; (c) *p*-hydroxymethylbenzene.

11.34 Write the formulas of the following compounds and state whether each one is a primary, secondary, or tertiary alcohol or a phenol: (a) 2-methyl-2-propanol; (b) 1-propanol; (c) *o*-hydroxytoluene (see Exercise 11.25).

11.35 (a) Name the compound CH$_3$CH$_2$OCH$_2$CH$_3$. (b) Write the formula of methyl ethyl ether.

11.36 (a) Name the compound CH$_3$CH$_2$CH$_2$OCH$_2$CH$_3$. (b) Write the formula of methyl propyl ether.

11.37 Identify each compound as an aldehyde or ketone and give its systematic name: (a) CH$_3$CHO; (b) CH$_3$COCH$_3$; (c) (CH$_3$CH$_2$)$_2$CO.

11.38 Identify each compound as an aldehyde or ketone and give its systematic name:
(a) CH$_3$CH$_2$CHO; (b) CH$_3$—⟨◯⟩—CHO;
(c) (CH$_3$CH$_2$)$_2$CHCH$_2$COCH$_3$.

11.39 Give the systematic names of (a) CH$_3$COOH; (b) CH$_3$CH$_2$CH$_2$COOH; (c) CH$_2$(NH$_2$)COOH.

11.40 Give the systematic names of (a) CH$_3$CH$_2$COOH; (b) CHCl$_2$COOH; (c) CH$_3$(CH$_2$)$_7$COOH.

11.41 Give the systematic names of: (a) CH$_3$NH$_2$; (b) (CH$_3$CH$_2$)$_2$NH; (c) *o*-CH$_3$C$_6$H$_4$NH$_2$.

11.42 Give the systematic names of: (a) CH$_3$CH$_2$CH$_2$NH$_2$; (b) (CH$_3$CH$_2$CH$_2$)$_3$N; (c) *p*-ClC$_6$H$_4$NH$_2$.

11.43 Name the following compounds:
(a) CH$_3$CH(OH)CH$_3$; (b) CH$_3$OCH$_3$; (c) HCHO; (d) (CH$_3$CH$_2$)$_2$CO; (e) CH$_3$NH$_2$.

11.44 Name the following compounds: (a) CH$_3$COOCH$_3$; (b) CH$_3$COCH$_3$; (c) HCOOH; (d) CH$_3$CH$_2$CH$_2$CH$_2$OH; (e) (CH$_3$CH$_2$)$_3$N.

11.45 Identify all the functional groups in the following compounds: (a) vanillin, the compound responsible for vanilla flavor

(b) carvone, the compound responsible for spearmint flavor

(c) caffeine, the stimulant in coffee, tea, and cola drinks

11.46 Identify all the functional groups in the following compounds: (a) zingerone, the pungent, hot component of ginger

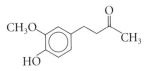

(b) capsaicin, the biting, hot component of chili peppers

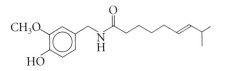

(c) procaine, a local anesthetic

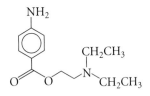

Structures and Reactions of Functional Groups

11.47 Write the structural formulas of (a) methanal; (b) propanone; (c) 2-heptanone.

11.48 Write the structural formulas of (a) 2-ethyl-2-methylpentanal; (b) 3,5-dihydroxy-4-octanone; (c) octanal.

11.49 Suggest an alcohol that could be used for the preparation of each of the following compounds and indicate how the reaction would be carried out: (a) ethanal; (b) 2-octanone; (c) 5-methyloctanal.

11.50 Suggest an alcohol that could be used for the preparation of each of the following compounds and indicate how the reaction would be carried out: (a) methanal; (b) propanone; (c) 6-methyl-5-decanone.

11.51 Write the structural formulas of (a) benzoic acid, C_6H_5COOH; (b) 2-methylbutanoic acid; (c) propanoic acid.

11.52 Write the structural formulas of (a) 2-methylpropanoic acid; (b) 2,2-dichlorobutanoic acid; (c) 2,2,2-trifluoroethanoic acid.

11.53 Write the structural formulas of (a) o-methylphenylamine; (b) triethylamine; (c) tetramethylammonium ion.

11.54 Write the structural formulas of (a) methylpropylamine; (b) dimethylamine; (c) m-methylphenylamine.

11.55 Write the structural formulas of the principal products formed from the following condensation reactions: (a) butanoic acid with 2-propanol; (b) ethanoic acid with 1-pentanol; (c) hexanoic acid with methylethylamine; (d) ethanoic acid with propylamine.

11.56 Write the structural formulas of the principal products formed from the following condensation reactions: (a) octanoic acid with methanol; (b) propanoic acid with ethanol; (c) propanoic acid with methylamine; (d) methanoic acid with diethylamine.

11.57 You are given samples of propanal, 2-propanone, and ethanoic acid. Describe how you would use chemical tests, such as acid-base indicators and oxidizing agents, to distinguish between the three compounds.

11.58 You are given samples of 1-propanol, pentane, and ethanoic acid. Describe how you would use chemical tests, such as aqueous solubility and acid-base indicators, to distinguish between the three compounds.

Isomerism

11.59 Write the structural formulas and name three alkene isomers having the formula C_4H_8.

11.60 Write the structural formulas and name two cycloalkane structural isomers having the formula C_4H_8.

11.61 Write the structural formulas for and name all structural and geometrical isomers of the alkene C_4H_8.

11.62 Write the structural formulas for and name all structural and geometrical isomers of the alkene C_5H_{10}.

11.63 Identify each of the following pairs as structural isomers, geometrical isomers, or not isomers: (a) butane and cyclobutane; (b) pentane and 2,2-dimethylpropane; (c) cyclopentane and pentene;

(d)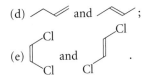

(e) ◁ and ═/ .

11.64 Identify each of the following pairs as structural isomers, geometrical isomers, or not isomers: (a) ethanol and dimethyl ether; (b) 1-hydroxy-2-pentene and 3-methylbutanal; (c) 1-chlorohexane and chlorocyclohexane;

(d) ⌇ and ⌇ ;

(e) [Cl structures] and [Cl structure] .

11.65 The branched hydrocarbon C_4H_{10} reacts with chlorine in the presence of light to give two branched structural isomers with the formula C_4H_9Cl. Write the structural formulas of (a) the hydrocarbon; (b) the isomeric products.

11.66 The branched hydrocarbon C_6H_{14} reacts with chlorine in the presence of light to give only two structural isomers with the formula $C_6H_{13}Cl$. Write the structural formulas of (a) the hydrocarbon; (b) the two isomeric products.

11.67 Indicate which molecules display optical activity and identify the chiral carbons in those that do: (a) $CH_3CHBrCH_2CH_3$; (b) $CH_3CH_2CHCl_2$; (c) 1-bromo-2-chloropropane; (d) 1,2-dichloropentane.

11.68 Indicate which molecules display optical activity and identify the chiral carbons in those that do: (a) $CH_3CHOHCOOH$; (b) $CH_3CH_2CHClCH_3$; (c) 2-bromo-2-chloropropane; (d) 1,2-dibromobutane.

11.69 Identify each chiral carbon in camphor, used in cooling salves

11.70 Identify each chiral carbon in menthol, the flavor of peppermint

Polymers

11.71 Sketch three repeating units of the polymer formed from (a) $CH_2=C(CH_3)_2$; (b) $CH_2=CHC\equiv N$ (only the double bond polymerizes); (c) isoprene

11.72 Sketch three repeating units of each of the polymers formed from (a) tetrafluoroethene; (b) phenylethene; (c) $CH_3CH=CHCH_3$.

11.73 Write the structural formulas of the monomers of the following polymers, for which one repeating unit is shown: (a) polyvinylchloride (PVC), $+CHClCH_2\overline{)_n}$; (b) Kel-F, $+CFClCF_2\overline{)_n}$.

11.74 Write the structural formulas of the monomers of the following polymers, for which one repeating unit is shown: (a) a polymer used to make carpets, $+OC(CH_3)_2-CO\overline{)_n}$; (b) $+CH(CH_3)CH_2\overline{)_n}$; (c) the polypeptide $+NHCH_2CO\overline{)_n}$.

11.75 Write the structural formula of two units of the polymer formed from (a) the reaction of oxalic acid (ethanedioic acid, HOOCCOOH) with 1,4-diaminobutane, $H_2NCH_2CH_2CH_2CH_2NH_2$; (b) the polymerization of the amino acid alanine (2-aminopropanoic acid).

11.76 Write the structural formula of two units of the polymer formed from (a) the reaction of terephthalic acid (see top of next column) with 1,2-diaminoethane, $H_2NCH_2CH_2NH_2$;

(b) the polymerization of 4-hydroxybenzoic acid (see below).

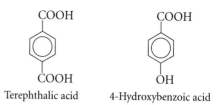

Terephthalic acid   4-Hydroxybenzoic acid

11.77 Distinguish between an isotactic polymer, a syndiotactic polymer, and an atactic polymer.

11.78 Explain why a syndiotactic polymer is stronger than an atactic polymer.

11.79 How does average chain length affect the following polymer characteristics: (a) softening point; (b) viscosity; (c) strength?

11.80 How does the polarity of the functional groups in a polymer affect the following characteristics of the polymer: (a) softening point; (b) viscosity; (c) strength?

11.81 Describe how the linearity of the polymer chain affects polymer strength.

11.82 Describe how cross-linking affects the elasticity and rigidity of a polymer.

Biopolymers

11.83 (a) Draw the structure of the peptide bond that links the amino acids in proteins. (b) Identify the functional group formed. (c) Identify the type of polymer formed (addition or condensation).

11.84 (a) Draw the structure of the link between glucose units that creates amylose. (b) Identify the functional group formed. (c) Identify the type of polymer formed (addition or condensation).

11.85 Draw the line structure of the amino acid lysine (see Table 11.6).

11.86 Draw the line structure of the amino acid threonine (see Table 11.6).

11.87 Name the amino acids in Table 11.6 that contain side groups capable of hydrogen bonding. This interaction contributes to the tertiary structures of proteins.

11.88 Name the amino acids in Table 11.6 that contain nonpolar side groups. These groups can contribute to the tertiary structure of a protein by avoiding contact with water.

11.89 Write the structural formula of the peptide formed from the reaction of the acid group of tyrosine with the amine group of glycine.

11.90 Write the structural formula of the peptide formed from the reaction of the acid group of glycine with the amine group of tyrosine.

11.91 Identify the functional groups in the glucose molecule shown in (**46**).

11.92 Identify the chiral carbon atoms in the glucose molecule shown in (**46**).

11.93 Write the complementary nucleic acid sequence that would pair with the following DNA sequences: (a) CATGAGTTA; (b) TGAATTGCA.

11.94 Write the complementary nucleic acid sequence that would pair with the following DNA sequences: (a) GGATCTCAG; (b) CTAGCCTGT.

Supplementary Exercises

11.95 Explain why there is a large number and variety of organic compounds.

11.96 Explain the difference in chemical bonding in arenes and alkenes.

11.97 Acrylonitrile, CH_2CHCN, is used in the synthesis of acrylic fibers (polyacrylonitriles), such as Orlon. Write the Lewis structure of acrylonitrile and describe the hybrid orbitals on each carbon atom. What is the approximate value of the CCC bond angle?

11.98 Write the structural formulas of (a) methylcyclopropane; (b) 2,4,6-trimethyltoluene (toluene is the common name of methylbenzene); (c) 2-phenyl-2-methylpentane; (d) cyclohexene; (e) 2-methyl-2-butene; (f) *m*-bromomethylbenzene.

11.99 Write a balanced equation for the production of 2,3-dibromobutane from 2-butyne.

11.100 Estimate the bond angles and determine the hybrid orbitals used by the carbon atoms in the following compounds: (a) acetaldehyde (ethanal), CH_3CHO; (b) acetic acid, CH_3COOH.

11.101 Pheromones are commonly called sex attractants, although they have more complex signaling functions. The structure of a pheromone in the queen bee is *trans-*$CH_3CO(CH_2)_5CH{=}CHCOOH$. (a) Write the structural formula of the pheromone. (b) Identify and name the functional groups in the molecule. (c) State the number of σ- and π-bonds in one molecule. (d) Indicate the hybridization of each carbon and oxygen atom (omit terminal oxygen atoms).

11.102 Write the condensed structural formulas of the principal products of the reaction that occurs when (a) ethylene glycol (1,2-ethanediol) is heated with stearic acid, $CH_3(CH_2)_{16}COOH$; (b) ethanol is heated with oxalic acid, $HOOCCOOH$; (c) 1-butanol is heated with propanoic acid.

11.103 Two structural isomers can result when hydrogen bromide reacts with 2-pentene. Write their structural formulas.

11.104 Why do polymers not have definite molar masses? How does the fact that polymers have average molar masses affect their melting points?

11.105 The condensed structural formula of the repeating unit of a certain polyester is
$\{OCH_2{-}C_6H_4{-}CH_2O{-}CO{-}C_6H_4{-}CO\}_n$.
Identify the monomers of the polyester.

11.106 (a) Explain the differences among the primary, secondary, tertiary, and quaternary structures of a protein. (b) Identify the forces holding each structure together as covalent bonds or primarily intermolecular forces.

11.107 Concentrated sulfuric acid is a strong dehydrating agent, in the sense that it can remove the elements of water, HOH, from a compound. What principal product results from the reaction of $CH_3(CH_2)_2CH(OH)CH_3$ with concentrated sulfuric acid at 120°C?

11.108 Dopamine is a neurotransmitter in the central nervous system, and dopamine deficiency is related to Parkinson's disease. Two systematic names for dopamine are 4-(2-aminoethyl)-1,2-benzenediol and 3,4-dihydroxyphenylethylamine. Write the structural formula of dopamine.

Applied Exercises

For Exercises 11.109–11.112, see Applying Chemistry: Case Study 11.

11.109 (a) Write the Lewis structure of ethyne and sketch its valence bond structure, showing σ- and π-bonds. What is the hybridization of the carbon atoms in ethyne? (b) Draw two resonance structures for a three-carbon fragment of polyacetylene, $\{CH{=}CH\}_n$. (c) What is the hybridization of the carbon atoms in polyacetylene? (d) Are polyacetylene molecules likely to be fairly stiff, or do the molecules rotate and coil freely about their carbon-carbon bonds? Explain your reasoning.

11.110 Draw three of the repeating units of the polyacetylene formed from $CH_3C{\equiv}CH$.

11.111 The monomer of the conducting polymer polyaniline is the compound aniline (aminobenzene). (a) Draw the structural formula of the aniline monomer. What is the hybridization of the N atom in (b) aniline? (c) polyaniline? (d) Indicate the locations of lone pairs, if any, in polyaniline. (e) Do the N atoms help to carry the current? Explain your reasoning.

11.112 Fibers of the conducting polymer polypyrrole are woven into radar camouflage cloth. Because it absorbs microwaves, rather than reflecting them back to their source, the cloth appears to be a patch of empty space on radar. The structure of polypyrrole is shown on the next page. (a) What is the hybridization of the N atom in polypyrrole? (b) Redraw the structure and trace the path of electron flow in polypyrrole. (c) Explain why polypyrrole can absorb long-wavelength radiation such as microwave radiation, but small organic molecules cannot. See Investigating Matter 9.2.

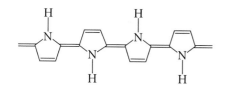

Integrated Exercises

11.113 In a combustion experiment, a 2.14-g sample of a hydrocarbon formed 3.32 g of water and 6.48 g of carbon dioxide. Deduce its empirical formula and state whether it is likely to be an alkane, an alkene, or an alkyne. Explain your reasoning.

11.114 In a combustion experiment, 3.69 g of a hydrocarbon formed 4.50 g of water and 11.7 g of carbon dioxide. Deduce its empirical formula and state whether it is likely to be an alkane, an alkene, or an alkyne. Explain your reasoning.

11.115 (a) Write the structural formulas of diethyl ether and 1-butanol (note that they are isomers). (b) Whereas the solubility of both compounds in water is about 8 g per 100 mL, the boiling point of 1-butanol is 117°C, but that of diethyl ether is 35°C. Account for these observations in terms of intermolecular forces.

11.116 A hydrocarbon has the mass percentage composition 90.% carbon and 10.% hydrogen and has molar mass 40. g/mol. It reacts with bromine in the absence of light and 0.73 g of the hydrocarbon reacts with 800. mL of hydrogen at 1.0 atm and 273 K in the presence of a nickel catalyst. Write the molecular formula of the hydrocarbon and the structural formulas of two possible isomers.

11.117 How many liters of hydrogen at 1.00 atm and 298 K are needed to hydrogenate completely 1.00 mol of (a) cyclohexene? (b) benzene? (c) Estimate the enthalpy change of each reaction from the average bond enthalpies in Tables 9.2 and 9.3. (d) The Lewis structures imply that the enthalpy of hydrogenation of benzene is three times that of cyclohexene. Do your calculations support this? Explain any differences.

11.118 What is the oxidation number of the carbon atom in (a) methane; (b) methanol; (c) methanal; (d) methanoic acid?

11.119 A 0.400-L sample of propene gas at 10.0 atm and 298 K is allowed to react with 500.0 mL of a 0.220 M Br_2 solution in hexane. (a) Write the equation for the reaction that occurs. (b) Which is the limiting reactant? (c) What mass of product can be produced?

11.120 Many of the phenols are insoluble in water, but soluble in base, because the proton of the —OH group is acidic. Write the balanced chemical equation for the reaction of 3-methylphenol with aqueous sodium hydroxide.

The Properties of Solutions

Solutions are of universal importance in biology, industry, the environment, and the laboratory. All our bodily fluids are solutions. Every time we take a breath, oxygen is taken up by our red blood cells and carried through the bloodstream to the cells that need it. Red blood cells and white blood cells are washed through our arteries, capillaries, and veins in plasma—an aqueous solution of salts, sugars, and other nutrients and waste products. The same processes by which nutrients are carried into our cells when they are required are used by trees to carry water to leaves. A healthy organism automatically maintains optimal concentrations of nutrients in the solutions that make up its various fluids. However, disease or strenuous exercise can upset that balance and upset the transport of molecules and ions across membranes.

Other solutions occur on a vast scale. The largest liquid solutions we encounter on Earth are the oceans, which account for 1.4×10^{12} kg of the Earth's surface water. That amounts to nearly 300 000 tons of water for each inhabitant, more than enough to supply all the drinking water needed for a thirsty world. However, seawater cannot be used directly as drinking water, because of the high concentrations of dissolved salts. The main solutes in seawater are Na^+ and Cl^- ions (Table 12.1), but it also contains huge amounts of other substances and at least a trace of every naturally occurring element. Transforming seawater into potable water is a major challenge to water engineers. A number of techniques have been proposed, all of which apply the properties of solutions we meet in this chapter.

A medical research scientist inspects sheets of collagen that will be used to make samples of artificial bone. Materials like artificial bone are used in joint replacement and reconstructive surgery, when real tissues are not available. To be successful, biomimetic materials (materials that mimic living tissues) must have chemical and physical properties like those of the natural tissue and be completely inert in the body. This chapter describes some of the chemistry of artificial tissues and the fluids in our bodies.

Table 12.1 Principal ions found in seawater

Element	Principal form	Concentration, g/L
Cl	Cl^-	19.0
Na	Na^+	10.5
Mg	Mg^{2+}	1.35
S	SO_4^{2-}	0.89
Ca	Ca^{2+}	0.40
K	K^+	0.38
Br	Br^-	0.065
C	HCO_3^-, H_2CO_3, CO_2	0.028

In the five previous chapters, we examined the properties of pure substances. In this chapter, we see how those properties determine the solubility of one substance in another. We also see how the presence of a solute affects the properties of the solution.

SOLUTES AND SOLVENTS

Solutions—homogeneous mixtures—can be solid, liquid, or gaseous (Section 1.13). Solid solutions of metalloids and nonmetals, such as silicon with a tiny amount of phosphorus as solute, are the primary materials of the electronics industry. Gaseous solutions—which are more commonly referred to as gaseous mixtures—provide special environments for deep-sea divers and food storage facilities. We shall concentrate mainly on liquid solutions.

Chemists carry out many reactions in liquid solutions, partly because molecules and ions of solid substances are more mobile in liquids than they would be in the solid state and so can react with one another more quickly. Chemists use both aqueous solutions, which are solutions in water, and nonaqueous solutions, which are solutions in liquids other than water, such as hydrocarbons, alcohols, and ethers. We shall focus on aqueous solutions because they are so important, but much of this material also applies to nonaqueous solutions.

Throughout the chapter, we shall need to refer to the concentrations of solutes in solution. Table 12.2 summarizes the measures of concentration commonly used. The most common measure, the molarity, was introduced in Section 4.6. A measure often used in environmental chemistry is parts per million, ppm, which can refer to mass or volume. Thus, the number of **parts per million by mass** is the mass of a solute in milligrams divided by the mass of solvent in kilograms. The number of **parts per million by volume** is the volume of a solute in microliters divided by the volume of solvent in liters. For example, 0.1 ppm by mass of nitrate ion in drinking water is $(0.1 \text{ mg } NO_3^-)/(1 \text{ kg water})$, and 0.1 ppm by volume of hydrogen sulfide in air is $(0.1 \text{ μL } H_2S)/(1 \text{ L air})$. The

Table 12.2 Concentration units

Measure	Units	Note
molarity (molar concentration)	moles per liter, mol/L, M	moles of solute per liter of solution
molality	moles per kilogram, mol/kg, m	moles of solute per kilogram of solvent
volume percentage	%	volume of component expressed as a percentage of the total volume, both measured prior to mixing
mass percentage	%	mass of component expressed as a percentage of the total mass
mole fraction (x)	—	number of moles of component expressed as a fraction of the total number of moles
parts per million by volume	ppm (by volume)	volume in microliters per liter of sample
parts per million by mass	ppm (by mass)	mass in milligrams per kilogram of sample

molality is a measure of concentration useful in certain specific applications and is described in Section 12.11.

12.1 The Molecular Nature of Dissolving

Let's imagine what happens to the molecules of a solute as it dissolves in a liquid. Suppose we add a crystal of glucose to some water. At the surface of the crystal, the glucose molecules are in contact with water molecules and hydrogen bonds begin to form between the two types of molecules. The surface glucose molecules are pulled into solution by water molecules but are held back by other glucose molecules. When the water molecules are successful, the surface glucose molecules break away from the crystal and drift off into the solvent, surrounded by water molecules.

A similar process occurs when an ionic solid dissolves. The oxygen atoms in H_2O molecules have partial negative charges. These partially charged atoms surround the cations and pry them away from the crystal lattice. At the same time, hydrogen atoms in the water molecules surround anions at the crystal surface and begin to attract the anions away from their cation partners (Fig. 12.1). Stirring or shaking speeds the process because it brings more free water molecules to the surface of the solid and sweeps the hydrated ions away.

Solids dissolve in water when individual molecules or ions are attracted to water molecules and break away from the solid.

12.2 Solubility

A given quantity of solvent can dissolve only so much solute. If we add 20 g of glucose to 100 mL of water at room temperature, then all the glucose dissolves. As we add more glucose, it continues to go into solution, until no more can dissolve. If we add so much glucose that some remains undissolved, then we say that the solution is "saturated" with glucose (Fig 12.2). A solution is said to be **saturated** when the solvent has dissolved all the solute it can and some undissolved solute remains.

The concentration of a solute has reached its greatest value in a saturated solution. In other words, a saturated solution represents the limit of a solute's ability to dissolve in a given quantity of solvent. The **molar solubility,** *S*, of a substance is the molar concentration of the substance in a saturated solution. For instance, if we measure the concentration of glucose in its saturated solution at 20°C, we find it to be 6 mol/L; so the molar solubility of glucose in water at 20°C is reported as 6 mol/L.

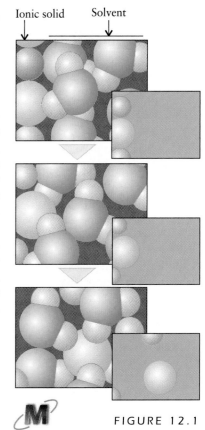

Ionic solid   Solvent

FIGURE 12.1

The events that take place at the interface of a solid solute and a solvent (water). The ions at the surface of the solid become hydrated; and as they do, they move off into the solution. The insets show only the solute ions.

FIGURE 12.2

When a little glucose is stirred into 100 mL of water, it all dissolves (left). However, when a large amount is added, the solution becomes saturated and some undissolved glucose remains (right).

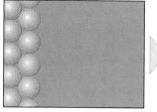 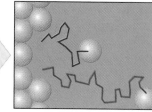 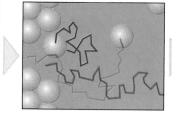 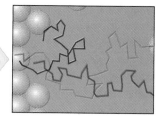

FIGURE 12.3

The solute in a saturated solution is in dynamic equilibrium with the undissolved solute. If we could follow the history of a solute particle (yellow spheres), we should sometimes find it in solution and at other times forming the solid. Red, green, and blue lines represent the paths of individual solute particles. The solvent molecules are not shown.

The solid solute in contact with a saturated solution continues to dissolve, but the rate at which it dissolves exactly matches the rate at which other molecules or ions of the solute return to the solid (Fig. 12.3). Therefore, there is no further *net* change in concentration, because solute continues to enter and leave the solution at matching rates. In other words, *in a saturated solution, the dissolved and undissolved solute are in dynamic equilibrium.* For example, in a saturated solution of calcium hydroxide, the dynamic equilibrium is

$$Ca(OH)_2(s) \rightleftharpoons Ca^{2+}(aq) + 2\,OH^-(aq)$$

with Ca^{2+} and OH^- ions leaving and returning to the surface of the solid at equal rates.

Impure solids are sometimes purified by dissolving them in a solvent in which the impurities and the desired solid have different solubilities. If the impurities are soluble but the desired solid is not, the mixture can be filtered. The impurities pass through the filter paper, whereas the pure solid remains on the paper. If the desired solid is more soluble, filtration can still be used, but now the impurities remain on the filter paper. The pure solid can then be allowed to crystallize from the filtered solution as it cools.

Some solutes, such as sodium acetate, are difficult to crystallize because they can form **supersaturated solutions,** which are solutions in which the concentration is higher than indicated by the value of that solute's molar solubility. A supersaturated solution may last for a long time, but it is unstable: the addition of a single crystal of the solute will cause the rapid crystallization of excess solute (Fig. 12.4). The crystal acts as a "seed" upon which other ions can deposit.

The molar solubility of a substance is its molar concentration in a saturated solution. A saturated solution is one in which the dissolved and undissolved solute are in dynamic equilibrium with each other.

FIGURE 12.4

(a) A highly supersaturated solution of sodium acetate can remain liquid for a long period of time, until a small "seed" crystal of sodium acetate is added. (b) Crystals begin to form immediately, until (c) the solution is no longer supersaturated.

FACTORS AFFECTING SOLUBILITY

The solubility of a substance depends on the identities of the substance and solvent. The most important molecular property that affects solubility is the strength and type of the interactions between solute and solvent molecules (recall Sections 10.1–10.3). Solubilities of all substances also depend on the temperature, and the solubility of a gas depends on its partial pressure.

12.3 Solubilities of Ionic Compounds

We know from our everyday experience that some substances, like sugar and sodium chloride, are soluble in water and some, like chalk, are not. We have also seen that we can predict whether a given ionic solid is likely to be soluble in water by using the rules in Table 3.1. These rules help us to understand many features of the world around us. For example, nitrates are rarely found in mineral deposits because they are soluble in groundwater, the water that trickles through the ground and washes away soluble substances. There are exceptions, such as the large deposits of impure sodium nitrate in the arid coastal region of Chile, where groundwater is absent (Fig. 12.5).

What explains the differing solubilities of ionic compounds? There are many factors, some of which we discuss in Sections 12.6–12.10; but we can get a lot of information just by analyzing simple trends. For example, if we compare the solubilities of ionic compounds with different oxoanions, we notice that *ionic compounds of large oxoanions with low charges are often soluble, but ionic compounds of oxoanions with high charges tend to be insoluble.* A low charge means that the anion is attracted to cations weakly, so the crystal lattice is easily broken up by the solvent. If the anion is large, then the separation between the centers of neighboring cations and anions is large and, as a result, the interactions between the cations and anions may be weak even if the anion has a high charge. For example, nitrates are soluble because the NO_3^- ion is quite bulky but has only a single negative charge. As a result, it is easy for water to detach the ions from solid nitrates.

Phosphate ions are more bulky than nitrate ions, but they are triply charged. As a result, they are more strongly attracted to cations than the nitrate ions are, and it is much harder for water to break them out of solids. The low solubility of most phosphates is an advantage for animals with skeletons, because bone consists largely of calcium phosphate (much of the rest is the protein collagen). However, this insolubility is inconvenient for agriculture, because it means that phosphorus is slow to circulate through the ecosystem. One of chemistry's achievements has been the development of manufacturing processes to turn phosphates into soluble fertilizers. Phosphate fertilizers are obtained from phosphate rocks, principally the apatites—hydroxyapatite, $Ca_5(PO_4)_3OH$, and fluorapatite, $Ca_5(PO_4)_3F$—by treating them with concentrated sulfuric acid:

$$2\,Ca_5(PO_4)_3OH(s) + 7\,H_2SO_4(aq) \longrightarrow$$
$$7\,CaSO_4(s) + 3\,Ca(H_2PO_4)_2(aq) + 2\,H_2O(l)$$

The phosphate rocks themselves were once alive, for they are the crushed and compressed remains of the skeletons of prehistoric animals.

The attachment of a hydrogen ion to an oxoanion—a process that occurs in the formation of a hydrogen phosphate ion, HPO_4^{2-}, from a phosphate ion, PO_4^{3-}, for instance—reduces the charge of the ion without substantially

FIGURE 12.5

This saltpeter (impure sodium nitrate) is highly soluble in water. It has survived in the arid region of Chile because there is too little groundwater and rainwater to dissolve it and wash it away.

changing its size. We can therefore expect weaker cation-anion interactions with the protonated anion and therefore higher solubilities of the compounds that contain them. As predicted, calcium dihydrogen phosphate, $Ca(H_2PO_4)_2$, is more soluble than calcium phosphate because the $H_2PO_4^-$ ion has only a single charge. It is used in commercial phosphate fertilizers (see Section 20.4). Similarly, hydrogen carbonates (bicarbonates, HCO_3^-) are more soluble than carbonates.

The solubility difference between carbonates and bicarbonates is responsible for the behavior of "hard water," which is water that contains dissolved calcium and magnesium salts. Hard water originates as rainwater, which dissolves carbon dioxide from the air and forms carbonic acid:

$$CO_2(g) + H_2O(l) \longrightarrow H_2CO_3(aq)$$

As the water runs over and through the ground, the carbonic acid reacts with the calcium carbonate of limestone or chalk and forms the more soluble hydrogen carbonate:

$$CaCO_3(s) + H_2CO_3(aq) \longrightarrow Ca^{2+}(aq) + 2\,HCO_3^-(aq)$$

These two reactions are reversed when water containing $Ca(HCO_3)_2$ is heated in a kettle or furnace:

$$Ca^{2+}(aq) + 2\,HCO_3^-(aq) \longrightarrow CaCO_3(s) + H_2O(l) + CO_2(g)$$

The calcium carbonate is deposited as a solid called *scale*.

Ionic compounds formed from large oxoanions with low charges are often soluble.

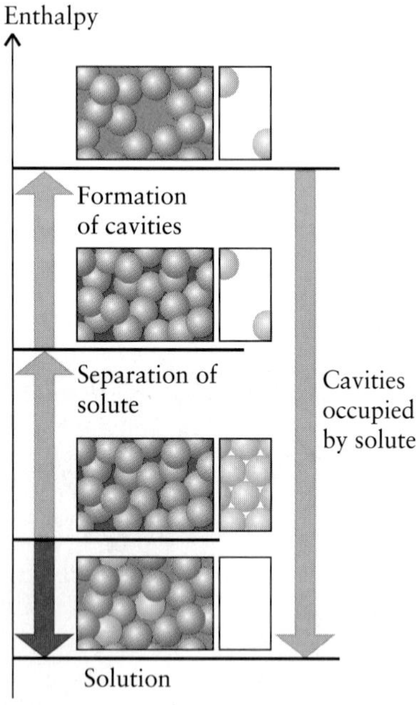

12.4 The Like-Dissolves-Like Rule

The solubility of a substance depends on the solvent as well as on the substance itself. For instance, sodium chloride is very soluble in water but insoluble in benzene. Conversely, grease (a mixture of long-chain hydrocarbonlike molecules) dissolves in benzene but not in water. The technique of chromatography makes use of the different solubilities of one substance in another to separate and identify solutes (Investigating Matter 12.1).

Suppose we need to remove some spilled wax. How can we know what solvent to use? A good guide is the rule that *like dissolves like*. A polar liquid, such as water, is generally the best solvent for ionic and polar compounds. Conversely, nonpolar liquids, including benzene and tetrachloroethene, $Cl_2C{=}CCl_2$, which is used for dry cleaning, are often better solvents for other nonpolar compounds like wax, which are held together by London forces.

The like-dissolves-like rule reflects the fact that, to form a solution, attractions between solute molecules and some attractions between solvent molecules must be replaced by solute-solvent attractions. We can think of the act of solution as taking place in three hypothetical steps (Fig. 12.6). First, the solute molecules or ions separate, next some solvent molecules move apart to form a cavity in the solution, and finally the widely separated solute particles slip into the cavities in the solvent. If the new attractions are similar to those replaced, very little energy is required for the solution to form. For example, when the main cohesive forces in a solute are hydrogen bonds, the solute is more likely to dissolve in a hydrogen-bonded solvent than in other solvents. The molecules can go into

FIGURE 12.6

A molecular representation of the energy changes associated with the formation of a dilute solution. First, solute molecules are separated from one another. Then some of the solvent molecules move apart and leave cavities in the solvent. Finally, the solute molecules occupy the cavities in the solvent, and release energy. The overall energy change for the process is shown by the dark red arrow. In the actual solution process, these steps do not occur independently.

solution only if they can replace solute-solute hydrogen bonds with solute-solvent hydrogen bonds. Glucose, for example, has —OH groups that can form hydrogen bonds; it dissolves readily in water but not in benzene.

If the principal cohesive forces between solute molecules are London forces, then the best solvent is likely to be one held together by the same kind of forces. For example, a good solvent for nonpolar substances is nonpolar carbon disulfide, CS_2. Carbon disulfide is a much better solvent than water for sulfur, which consists of S_8 molecules held together in the solid by London forces (Fig. 12.7). The sulfur molecules slip easily between the nonpolar carbon disulfide molecules. They cannot penetrate into the strongly hydrogen-bonded structure of water because they cannot replace those bonds with interactions of similar strength.

Soaps and detergents are a practical application of the like-dissolves-like rule. Soaps are the sodium salts of long-chain carboxylic acids, including sodium stearate (**1**). The anions of these acids have a polar carboxylate ($-CO_2^-$) group, called the *head group*, at one end of a nonpolar hydrocarbon chain. The head group is **hydrophilic,** or water-attracting, whereas the nonpolar hydrocarbon tails are **hydrophobic,** or water-repelling. The like-dissolves-like rule suggests that the hydrophilic head group of the anion ought to dissolve in water and the hydrophobic hydrocarbon tails ought to dissolve in grease. This dual ability makes soap very effective at removing grease. The hydrocarbon tails sink into the blob of grease up to the head groups. These hydrophilic groups remain on the surface of the blob (Fig. 12.8) and enable the entire unit to dissolve in water.

Soaps are made by heating a fat such as beef fat with sodium hydroxide. Beef fat contains an ester formed between glycerol and stearic acid. The sodium hydroxide attacks the ester and releases the stearic acid as sodium stearate. A problem with soaps, however, is that they form a scum in hard water. The scum is a precipitate of calcium stearate, which forms because calcium salts are less soluble than sodium salts. One way to avoid the formation of scum is to precipitate the Ca^{2+} ions from the solution before adding the soap. For example, sodium carbonate (washing soda) can be added to hard water to precipitate Ca^{2+} ions as calcium carbonate, which is easily rinsed away:

$$Ca(HCO_3)_2(aq) + Na_2CO_3(aq) \longrightarrow CaCO_3(s) + 2\,NaHCO_3(aq)$$

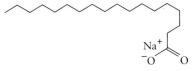

1 Sodium stearate, $NaCH_3(CH_2)_{16}CO_2$

Esters are described in Section 11.9.

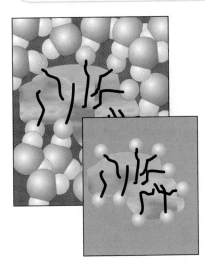

FIGURE 12.8

The hydrocarbon tails of soap or surfactant molecules dissolve in grease, but the water-attracting head groups remain on the surface, where they can interact favorably with water. As a result, the whole blob can dissolve in water and be washed away.

FIGURE 12.7

Sulfur is a nonpolar molecular solid that does not dissolve in water (left), but it does dissolve in carbon disulfide (right), with which the S_8 molecules have favorable London interactions.

Chromatography is used in hospitals to diagnose illness, in forensic laboratories to search for drugs or other evidence, by food manufacturers to test product quality, and on Mars to search for evidence of life. All these techniques depend on subtle differences in intermolecular forces to separate compounds.

If an aqueous solution of a compound in water is placed into a container with another liquid that is *immiscible* (mutually insoluble) with water, some of the compound may dissolve in the other solvent. For example, molecular iodine, I_2, is very slightly soluble in water, but highly soluble in carbon tetrachloride, CCl_4, which is immiscible with water. When carbon tetrachloride is added to water containing iodine, most of the iodine dissolves in the CCl_4. The solute is said to *partition* itself between the two solvents. When the CCl_4 is separated from the water, it takes most of the iodine with it. This separation technique, which is called *solvent extraction*, is used to separate plant flavors and aromas from aqueous slurries of plants.

Liquid chromatography

A problem with solvent extraction of plants is that it extracts the many compounds in plants as a mixture. However, in the early part of the twentieth century, the Russian botanist M. S. Tsvet found a way to separate the pigments in flowers and leaves. He ground up the plant materials and dissolved the pigments, then poured the solution over the top of a vertical tube of ground chalk. The different pigments trickled through the chalk at different rates, producing colored bands in the tube, which inspired the name *chromatography* ("color writing"). The separation occurred because the different pigments were absorbed to different extents by the chalk.

Chromatography is one of the most widely used and powerful means for separating mixtures, because it is often inexpensive and can be used to provide quantitative as well as qualitative information (Section 1.14). The simplest method is *paper chromatography*, in which a drop of solution is placed near the bottom edge of an absorbent support, such as a strip of paper, which is called the *stationary phase*. A solvent, called the *mobile phase*, is added below the spot and the solvent is adsorbed by the support. As the level of the solvent rises in the support, the materials in the spot also begin to travel upward, the rate depending on how strongly they adhere to the support. In a sense, there is a continuous extraction going on between the support and the solvent, which is also called the *carrier solvent*. Those solutes attracted more strongly to the solvent than to the support will move the fastest.

The same concepts apply to *column chromatography*, where the stationary phase is normally small particles of silica, SiO_2, or alumina, Al_2O_3. These substances are not very reactive and have specially prepared surfaces to increase their ability to adsorb solvents. The column is saturated with solvent and a small volume of solution containing the solutes is poured on to the top. As soon as it has soaked in, more solvent is added. The solutes travel slowly down the column and are *eluted* (removed as fractions) at the bottom. If the mobile phase is less polar than the stationary phase, the less polar solutes will be eluted first and the more polar ones last.

A difficulty with column chromatography is that, for a good separation, the column needs to be long, so elution may

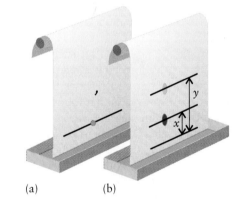

(a)   (b)

Two stages in a paper chromatography separation of a mixture of two components. (a) Before separation; (b) after separation. The relative values of the distances *x* and *y* are used to identify the components.

Modern commercial detergents are mixtures. Their most important component is a **surfactant,** or surface active agent, which takes the place of the soap. Surfactant molecules are organic compounds that resemble the one shown as (**2**). Like the stearate ion, they have a hydrophilic head group and a hydrophobic

2   A typical surfactant ion

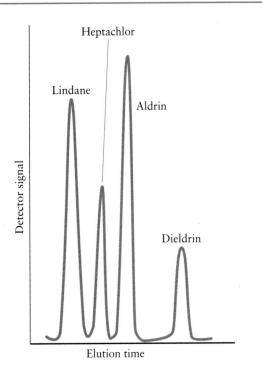

A gas chromatogram of a mixture of pesticides from farmland. The areas under the peaks indicate the relative abundances of the compounds.

those for liquid chromatography, but the output is more often a *chromatogram*, like that in the illustration, rather than a series of eluted samples. The chromatogram shows when each solute was eluted, and the peak areas tell how much of each component is present. The identity of the solute producing each peak can be determined by comparing its location against a database of known compounds. The amount of each solute present is determined from the area under the peak.

The stationary phase is either a liquid or a solid that coats the particles in the tube or the walls of the tube. Often the tube itself is very narrow and long, perhaps 100 m, and has to be coiled. The different compounds travel at different rates through the tube, depending on their relative polarities. The more polar the compound, the more strongly it is adsorbed onto the stationary phase. Because tiny amounts of solutes can be detected by gas chromatography, many environmental monitoring and forensic applications (recall Applying Chemistry: Case Study 1) have been developed.

take a very long time: days may be required. An improvement is *high-performance liquid chromatography* (HPLC), in which the mobile phase is forced under pressure through a long, narrow column. This technique yields an excellent separation in a relatively short time. HPLC has become the primary means of monitoring the use of therapeutic drugs.

Gas chromatography

Volatile compounds can be separated in a *gas chromatograph*, in which the mobile phase is a relatively inert gas such as helium, nitrogen, or hydrogen. The principles are the same as

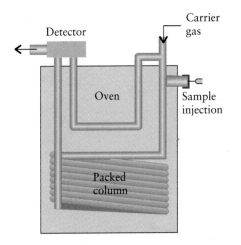

A schematic diagram of the arrangement in a gas chromatograph. The coiled column may be as much as 100 m in length.

tail and act similarly. Most commercial detergents also contain additives such as fillers (to generate bulk), foam-reducing agents, and bleach. Often, additives that fluoresce (absorb ultraviolet radiation and then give out visible light) are added to detergents to enhance brightness and give the impression of greater cleanliness.

A general guide to the suitability of a solvent is the rule that like dissolves like. Soaps and detergents contain surfactant molecules that have both hydrophobic and hydrophilic regions.

Example 12.1 *Predicting solubility in a given solvent*

Which of the following molecular substances is likely to be the most soluble in water: (a) carbon tetrachloride, CCl_4; (b) bromine, Br_2; (c) methanol, CH_3OH?

Strategy Substances tend to dissolve in solvents having molecules with similar polarity and similar intermolecular forces. Water is highly polar and is capable of hydrogen bonding. Determine whether the molecule is polar and whether it can participate in hydrogen bonding.

Solution The molecules in (a) and (b) are nonpolar and cannot participate in hydrogen bonding. The molecule in (c) is polar and the —OH group can take part in hydrogen bonding. Therefore, methanol, CH_3OH, is likely to be the most soluble in water, as is found by experiment.

Self-Test 12.1A Which of the following molecular substances is likely to be the most soluble in carbon disulfide, CS_2: (a) NH_3; (b) Cl_2; (c) HCl?

[***Answer:*** Cl_2]

Self-Test 12.1B Which of the following molecular substances is likely to be the most soluble in methanol, CH_3OH: (a) methane, CH_4; (b) Cl_2; (c) formaldehyde, H_2CO?

12.5 Colloids

The molecules of some solutes clump together and form a **suspension,** a widely dispersed mist of tiny particles. For example, when a soap or surface active agent dissolves in clean water, the molecules cluster together in groups of hundreds to form **micelles.** These nearly spherical particlelike collections of molecules are dispersed throughout the solvent. Together, the micelles and solvent are an example of a **colloid,** which is the general term for a form of matter consisting of a huge number of particles of diameter 1 nm to 10 μm dispersed throughout a liquid, solid, or gas. Many foods are colloids, as are clay particles, smoke, foams, and many of the fluids in living cells. Colloidal particles are much larger than most molecules but are too small to be seen with an optical microscope, so they can be thought of as having characteristics between those of solutions and those of heterogeneous mixtures. Because the particles are so small, colloids have a homogeneous appearance; but colloids can be distinguished from solutions by the fact that the suspended particles are large enough to scatter light (Fig. 12.9).

Colloids are classified according to the phases of the substances involved (Table 12.3). A colloid composed of tiny solid particles dispersed throughout a liquid is called a **sol.** Muddy water is a sol in which tiny flakes of clay are dispersed in water. A colloid composed of tiny droplets of liquid dispersed throughout another liquid is called an **emulsion.** Mayonnaise is an emulsion in which small droplets of water are dispersed in vegetable oil. A **solid emulsion** consists of small particles of a liquid or solid phase dispersed throughout a solid. Butter is a solid emulsion of milk in butterfat. Gelatin desserts are a type of solid emulsion called a **gel,** which is soft but holds its shape. Many liquid crystalline arrays can be considered colloids.

Aqueous colloids are either hydrophilic or hydrophobic. Emulsions of fat in water, such as milk, are hydrophobic. The micelles formed by soap molecules or ions, and the droplets of grease coated in detergent molecules, are hydrophilic. The macromolecules of the proteins in the colloid we call gelatin have many

FIGURE 12.9

Laser beams are invisible. However, they can be traced when they pass through smoke or clouds because the light scatters from the colloidal particles suspended in air. Light from the laser beams in this photograph is being scattered by colloidal water droplets in the air.

Table 12.3 *The classification of colloids**

Dispersed phase	Dispersion medium	Technical name	Examples
solid	gas	aerosol	smoke
liquid	gas	aerosol	hairspray, mist, fog
solid	liquid	sol or gel	printing ink, paint
liquid	liquid	emulsion	milk, mayonnaise
gas	liquid	foam	fire-extinguisher foam
solid	solid	solid dispersion	ruby glass (Au in glass); some alloys
liquid	solid	solid emulsion	bituminous road paving; ice cream
gas	solid	solid foam	insulating foam

*Based on R. J. Hunter, *Foundations of Colloid Science*, Vol. 1 (Oxford: Oxford University Press, 1987).

hydrophilic groups that attract water: they uncoil in hot water and their numerous polar groups form hydrogen bonds with the water. When the mixture cools, the protein chains link together again, but now they enclose many water molecules within themselves, as well as molecules of sugar, dye, and flavoring agents. The result is a gel: an open network of protein chains that hold the water in a flexible solid structure. Materials based on gels and other colloids provide a means of delivering medications to targeted locations in the body (Applying Chemistry: Case Study 12).

The tiny particles of a colloid display **Brownian motion,** the motion resulting from constant bombardment by solvent molecules. This motion helps to stop the particles from coagulating and separating from the solvent. Colloids are also stabilized by the adsorption of ions on the surfaces of the particles. The ions attract a layer of water molecules that prevents the particles from adhering to one another. However, if the concentration of ions in solution becomes too high, the opposite effect occurs and the colloid particles start to attract one another and may coalesce. This effect is responsible for the formation of silt in estuaries, when the clay colloids washed down in river water meet the salt water of the sea and coagulate.

Colloids consist of particles of one substance dispersed throughout another medium. The particles are too small to be seen with a microscope but are large enough to scatter light.

12.6 Pressure and Gas Solubility

Knowing about gas solubility could save your life. A deep-sea diver, for instance, must know the concepts in this section, because they are essential for the diver's survival.

We have already seen (in Section 5.2) that the pressure of a gas arises from the impact of its molecules. When the gas is in contact with a liquid, its molecules can burrow into a liquid like meteorites plunging into the ocean. Because

Applying Chemistry: *Case Study 12*

The complexity and versatility of materials made by nature are the envy of human scientists. We are only beginning to be able to create materials that have the strong, yet porous structure of bone or that have the strength and flexibility of spider silk (Section 11.16). *Biomimetic materials* are materials that are modeled after naturally occurring materials. Gels or flexible polymers modeled after natural membranes and tissues are biomimetic materials with remarkable properties. Some can be made to crawl on their own like tiny nanometer worms, others pulsate to an internally generated rhythm, and still others respond in seemingly lifelike ways to stimuli.

You might actually be wearing one of these new materials the next time you use in-line skates. One brand of these skates contains a "smart gel" that is a liquid at normal temperatures but sets to a firm, rubbery gel when exposed to body temperature. The gel fills the space between the lining of the skate boot and the sides, so once the skate has been secured, it sets to conform to the exact shape of the

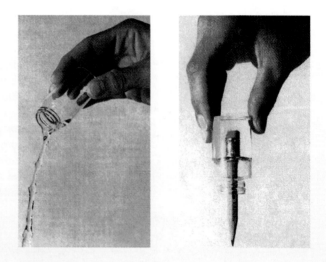

This sample of a smart gel expands in response to an increase in temperature.

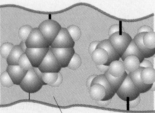

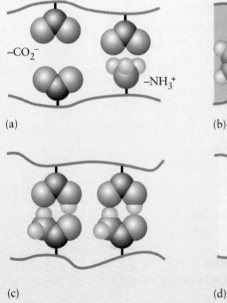

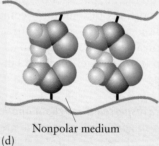

(a)

(b) Water

(c)

(d) Nonpolar medium

Four types of forces responsible for the adaptive behavior of smart gels. The different forces become active when the network of polymer chains composing the gel is disturbed. (a) Charged ionic regions can attract or repel one another. (b) Nonpolar regions can exclude water. (c) Hydrogen bonds may form from one chain to another. (d) Dipole-dipole interactions can attract or repel chains.

the number of impacts increases as the pressure of the gas increases, we predict that the solubility of the gas should increase with pressure. This is, in fact, the case, as shown by Fig. 12.10. If the gas above the liquid is a mixture (like air), then the solubility of each component depends on the number of collisions that foot. The gel goes through the same phase transition every time you put on the skate.

A gel like this generally consists of an aqueous suspension of organic molecules with long chains containing different regions, some of which have strong intermolecular forces and some of which have weak intermolecular forces. The temperature determines how the chains are coiled. At low temperatures, the regions with strong intermolecular forces are turned inward, so the molecules remain in liquid solution. However, at higher temperatures, those regions are turned outward, where they can attract other chains, and a flexible, but firm, network is created.

Another type of gel expands and contracts according to electrical signals. It is being investigated for use in artificial limbs that would respond and feel like real ones. However, humanoid robots are still in the future; natural muscle contracts within 100 ms of receiving the instruction from the brain, which is faster than most gels can respond.

Because the molecules that make up these gels are similar to biological molecules, their use in medicine is promising. For example, delivering insulin to diabetic patients involves some uncertainty. In a nondiabetic, insulin is delivered by the body only when blood sugar rises. However, a diabetic must take insulin at specified times of the day and the same amount each time. If the blood sugar is already low, a hypoglycemic reaction, and possibly coma, could result. A possible insulin-delivery system that would be responsive to blood sugar levels is now being investigated. The system would make use of a smart gel incorporating an enzyme that converts glucose (blood sugar) into an acid. The gel fills the pores of a sac that contains insulin and could be implanted into the body. When the level of glucose rises above a certain concentration, the acid formed in the vicinity of the gel would break down the gel network, allowing insulin to be released. Controlling the amount released, however, might be difficult.

Controlled-release drug delivery systems can also be fashioned from liquid crystals (Applying Chemistry: Case Study 10). Compounds called *lipids* are found in fats and form the membranes of living cells. These molecules are liquid crystals similar to detergent molecules in structure: they have a polar head and a nonpolar hydrocarbon tail

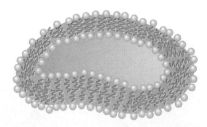

The structure of a liposome, in which a bilayer membrane is formed from liquid crystalline lipid molecules.

(Section 12.4). Some of these lipids form structures spontaneously in water, assembling into sheets consisting of double layers of these molecules lined up next to one another with their polar heads toward the water on each side. The sheets can be coaxed into forming hollow spheres called *liposomes* (see the illustration). They are similar to micelles, except liposomes are formed from a double layer of molecules, like the membrane of a living cell.

When a drug is present in the aqueous solution in which liposomes are being formed, some of the drug is encapsulated inside each liposome, forming tiny containers for the drug. The liposomes can then be injected into the body, where they stick only to certain types of cells—cancer cells, for instance. By this means, the drug is delivered to target cells and less of it is required. Liposome drug delivery has already been used to treat patients with leukemia. A smaller dosage is required and side effects are greatly reduced.

Key Concepts: solubility, effect of relative polarity on solubility

For Further Reading

R. Dagani, Intelligent gels, *Chemical and Engineering News,* June 9, 1997.

P. Ball, Chapter 7: A soft and sticky world: The self-organizing magic of colloid chemistry, *Designing the Molecular World,* Princeton: Princeton University Press, 1994, pp. 216–255.

Related Exercises: 12.89–12.90

the molecules of that component makes with the liquid. Therefore, the solubility of a gas in a mixture is proportional to the *partial* pressure of that gas.

Deep-sea divers breathe compressed air, which contains nitrogen that is normally not very soluble in blood. However, at great depths, a diver's body is

> Partial pressure (Section 5.11) is the contribution that the gas makes to the total pressure of the sample.

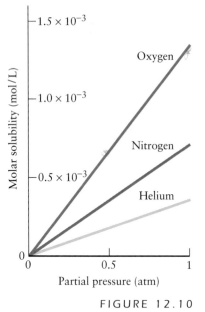

FIGURE 12.10

The variation of the molar solubilities of oxygen, nitrogen, and helium with partial pressure. Note that the solubility of each gas is doubled when its partial pressure is doubled.

In champagne, the carbon dioxide results from fermentation in the sealed bottle. In soft drinks, the liquid is carbonated by dissolving carbon dioxide in it under pressure.

exposed to very high pressures and the nitrogen becomes more soluble because its partial pressure is higher. The dissolved nitrogen comes out of solution rapidly when the divers return to the surface, and numerous small bubbles form in the bloodstream. These bubbles can burst the capillaries—the narrow vessels that distribute the blood—or block them and starve the tissues of oxygen. This condition is called "the bends" (Fig. 12.11). In some cases, the bends can be fatal. The risk is reduced if helium is used instead of nitrogen to dilute the diver's oxygen supply, because helium is less soluble than nitrogen in blood plasma. Moreover, because helium atoms are so small, unlike nitrogen molecules they can pass through cell walls without damaging them.

The straight lines in Fig. 12.10 show that the molar solubility, S, of a gas is directly proportional to its partial pressure, P. This observation was first made in 1801 by the English chemist William Henry and is now known as **Henry's law:**

$$\text{Molar solubility} = \text{constant} \times \text{partial pressure} \qquad S = k_H \times P \qquad (1)$$

The constant k_H, which is called the **Henry's law constant,** depends on the gas, the solvent, and the temperature (Table 12.4). The law implies that, at constant temperature, doubling the partial pressure of a gas doubles its solubility.

The production of fizzy soft drinks and champagne is a familiar illustration of Henry's law in action. Each of these drinks contains carbon dioxide dissolved under pressure. When a bottle of champagne is opened, the partial pressure of carbon dioxide above the solution is suddenly reduced and the gas **effervesces,** or bubbles out of solution, with a festive pop.

The solubility of a gas is proportional to its partial pressure, because an increase in pressure corresponds to an increase in the rate at which gas molecules strike the surface of the solvent.

Example 12.2 *Using Henry's law to predict solubilities*

The lowest concentration of oxygen that can support aquatic life is about 1.3×10^{-4} mol/L. The partial pressure of oxygen is 0.21 atm at sea level. (a) Is the concentration of oxygen in water at sea level adequate to maintain aquatic life? (b) What is the lowest partial pressure of oxygen that can support life?

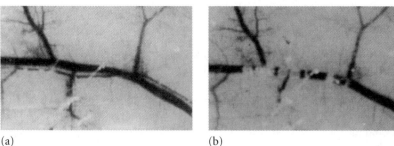

(a)   (b)

FIGURE 12.11

Small bubbles of air that form during decompression are responsible for the bends. (a) Normal blood vessels; (b) catastrophic collapse as bubbles escape from solution in the blood plasma.

CHAPTER 12   THE PROPERTIES OF SOLUTIONS

Strategy Because the solubility of a gas is proportional to its partial pressure, use Henry's law and the information from Table 12.4. To find the minimum partial pressure, solve Eq. 1 for the partial pressure corresponding to the stated minimum concentration of O_2.

Solution (a) The molar solubility of oxygen is

$$S = \{1.3 \times 10^{-3} \text{ mol/(L·atm)}\} \times (0.21 \text{ atm}) = 2.7 \times 10^{-4} \text{ mol/L}$$

This molar concentration is more than adequate to sustain life (it corresponds to about 8.6 mg of oxygen per liter of water). (b) The minimum partial pressure is

$$P = \frac{S}{k_H} = \frac{1.3 \times 10^{-4} \text{ mol/L}}{1.3 \times 10^{-3} \text{ mol/(L·atm)}} = 0.10 \text{ atm}$$

Self-Test 12.2A At the elevation of Bear Lake in Rocky Mountain National Park, 2900 m, the partial pressure of oxygen is 0.14 atm. What is the solubility of oxygen in Bear Lake at 20.°C?

[*Answer:* 1.8×10^{-4} mol/L]

Self-Test 12.2B Use the information in Table 12.4 to calculate the number of moles of CO_2 that will dissolve in 900. mL of water at 20.°C if the partial pressure of CO_2 is 1.00 atm.

Table 12.4 *Henry's law constants for gases in water at 20°C*

Gas	k_H, mol/L·atm
air	7.9×10^{-4}
argon	1.5×10^{-3}
carbon dioxide	2.3×10^{-2}
helium	3.7×10^{-4}
hydrogen	8.5×10^{-4}
neon	5.0×10^{-4}
nitrogen	7.0×10^{-4}
oxygen	1.3×10^{-3}

12.7 Temperature and Solubility

Most substances dissolve more quickly at higher temperatures. However, faster dissolving does not necessarily mean more extensive dissolving. In a number of cases, the solubility turns out to be lower at higher temperatures. We must always distinguish the effect of temperature on the rate of a process from its effect on the final outcome.

Most gases become less soluble as the temperature is raised. The lower solubility of gases in warm water is responsible for the tiny bubbles that appear when cool water from the faucet is left to stand in a warm room. The bubbles consist of air that dissolved when the water was cooler; it comes out of solution as the temperature rises.

Thermal pollution is the damage caused to the environment by the waste heat of an industrial process. Oxygen is less soluble in the warm water discharged from power stations. Moreover, the lower density of warm water causes it to rise to the surface of rivers, where it acts like a suffocating blanket that prevents absorption of oxygen and its penetration to the cooler water below. As a result, marine life (principally, fish) may die.

Most ionic and molecular solids are more soluble in hot water than in cold (Fig. 12.12). We make use of this characteristic in the laboratory to dissolve a substance and to grow crystals by letting a saturated solution cool slowly. However, a few solids, such as lithium sulfate, are less soluble at high temperatures than at low. A smaller number of compounds show a mixed behavior. For example, the solubility of sodium sulfate increases up to 32°C but then decreases as the temperature is raised further.

The variation of solubility with temperature is difficult to predict, except in the case of gases, where raising the temperature nearly always lowers the solubility.

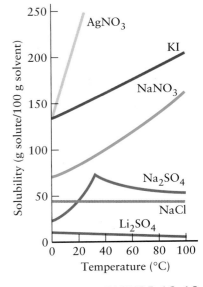

FIGURE 12.12

The variation with temperature of the solubilities of six substances in water.

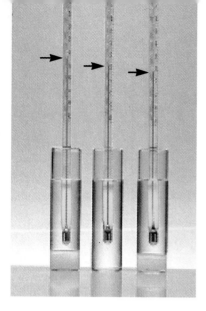

FIGURE 12.13

The exothermic dissolution of lithium chloride (left) is shown by the rise in temperature above that of the original water (center). In contrast, ammonium nitrate (right) dissolves endothermically.

WHY DOES ANYTHING DISSOLVE?

We have seen that like dissolves like because good solvents mimic the surroundings of the solute molecules: partial charges on polar solvent molecules play the role of ionic charges, hydrogen bonds replace hydrogen bonds, and London forces exerted by a solvent mimic those of the solute molecules themselves. We can discover the relative strengths of these forces by measuring how much heat is released or absorbed when a solute dissolves and then identifying that heat with the change in enthalpy. The information we obtain will help us understand why some substances dissolve but others do not.

12.8 The Enthalpy of Solution

The change in enthalpy per mole when a substance dissolves to form a very dilute solution is called the **enthalpy of solution.** It can be measured by calorimetry (Section 6.8). The experimental values in Table 12.5 show that some solids dissolve exothermically ($\Delta H < 0$) and others dissolve endothermically ($\Delta H > 0$). Lithium chloride has a negative enthalpy of solution, indicating an exothermic process: when the salt is added to water, the solution becomes noticeably warm (Fig. 12.13). In contrast, when ammonium nitrate is added to water, the temperature falls, thereby indicating that ammonium nitrate dissolves endothermically. The puzzle we have to answer is why some substances dissolve even though they require energy to do so.

In Section 12.4, we imagined dissolving as a three-step process. Now we see how to calculate the contributions of these steps to the enthalpy of solution. First, we imagine the solid breaking up to form a gas of molecules or ions (the first step in Fig. 12.6). Then we imagine the separated molecules or ions entering cavities forming in the solvent (Fig. 12.14). The enthalpy change of the first process is the change of enthalpy when an ionic solid becomes a gas of ions. We saw in Section 8.3 that this value is the lattice enthalpy, ΔH_L, of the solid

---

Table 12.5 Enthalpies of solution, ΔH_{sol}, at 25°C for very dilute aqueous solutions, in kilojoules per mole*

Cation	Anion							
	fluoride	chloride	bromide	iodide	hydroxide	carbonate	nitrate	sulfate
lithium	+4.9	−37.0	−48.8	−63.3	−23.6	−18.2	−2.7	−29.8
sodium	+1.9	+3.9	−0.6	−7.5	−44.5	−26.7	+20.4	−2.4
potassium	−17.7	+17.2	+19.9	+20.3	−57.1	−30.9	+34.9	+23.8
ammonium	−1.2	+14.8	+16.0	+13.7	—	—	+25.7	+6.6
silver	−22.5	+65.5	+84.4	+112.2	—	+41.8	+22.6	+17.8
magnesium	−12.6	−160.0	−185.6	−213.2	+2.3	−25.3	−90.9	−91.2
calcium	+11.5	−81.3	−103.1	−119.7	−16.7	−13.1	−19.2	−18.0
aluminum	−27	−329	−368	−385	—	—	—	−350.

*The value for silver iodide, for example, is the entry found where the row labeled "silver" intersects the column labeled "iodide." A positive value of ΔH_{sol} indicates an endothermic process.

Table 12.6 Enthalpies of hydration, ΔH_{hyd}, at 25°C, of some halides, in kilojoules per mole*

Cation	Anion			
	F^-	Cl^-	Br^-	I^-
H^+	−1613	−1470.	−1439	−1426
Li^+	−1041	−898	−867	−854
Na^+	−927	−784	−753	−740.
K^+	−844	−701	−670.	−657
Ag^+	−993	−850.	−819	−806
Ca^{2+}	—	−2337	—	—

*The entry where the row labeled Na^+ intersects the column labeled Cl^-, for instance, is the enthalpy change, −784 kJ/mol, for the process $Na^+(g) + Cl^-(g) \rightarrow Na^+(aq) + Cl^-(aq)$; the values here apply only when the resulting solution is very dilute.

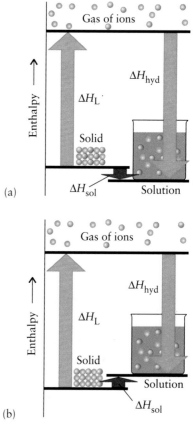

(a)

(b)

FIGURE 12.14

The enthalpy of solution, ΔH_{sol}, is the sum of the enthalpy changes required to form the solution: the enthalpy required to separate the molecules or ions of the solute (the lattice enthalpy, ΔH_L) and the enthalpy change accompanying their hydration, ΔH_{hyd}. The outcome is finely balanced: (a) in some cases, it is exothermic, (b) in others, it is endothermic. For gaseous solutes, the lattice enthalpy is 0, because the molecules are already widely separated.

(Table 8.1). For sodium chloride, for instance, the lattice enthalpy is the molar enthalpy change for the process

$$NaCl(s) \longrightarrow Na^+(g) + Cl^-(g)$$

Compounds formed from small, highly charged ions (such as Mg^{2+} and O^{2-}) have high lattice enthalpies: a lot of energy is needed to break up the solid. Because it is hard for solvent molecules to pull apart such strongly attached ions, these compounds are usually insoluble.

In the second process, heat is released as cavities form between solvent molecules and the ions enter them, becoming surrounded by solvent molecules. When the solvent is water, this process is called **hydration**; for other solvents, the process is called **solvation**. The molar enthalpy of the process

$$Na^+(g) + Cl^-(g) \longrightarrow Na^+(aq) + Cl^-(aq)$$

is called the **enthalpy of hydration, ΔH_{hyd}**, of the compound (Table 12.6). The hydration of ions is always exothermic on account of the favorable interaction between their charges and the partial charges of the polar solvent molecules. This **ion-dipole interaction** is much larger than the energy needed to form cavities in the solvent to accommodate the incoming ion. Hydration is also exothermic for molecules that can form hydrogen bonds, such as glucose and sucrose.

The enthalpy of solution is the sum of the enthalpy changes for the two processes: the lattice enthalpy and the enthalpy of hydration of the compound. For sodium chloride, we can write

$$NaCl(s) \longrightarrow Na^+(g) + Cl^-(g) \qquad \Delta H = +787 \text{ kJ}$$

$$Na^+(g) + Cl^-(g) \longrightarrow Na^+(aq) + Cl^-(aq) \qquad \Delta H = -784 \text{ kJ}$$

The sum of these two thermochemical equations is

$$NaCl(s) \longrightarrow Na^+(aq) + Cl^-(aq) \qquad \Delta H = +3 \text{ kJ}$$

Because, in this case, the enthalpy of solution is positive, there is a net inflow of energy as heat. Sodium chloride therefore dissolves endothermically, but only to

Because of experimental uncertainties, the calculated value of +3 kJ/mol differs slightly from the measured value, +3.9 kJ/mol, given in Table 12.5.

the extent of 3 kJ/mol. As this example shows, the overall change in enthalpy depends on a very delicate balance between the lattice enthalpy and the enthalpy of hydration. If sodium chloride had only a 0.5% smaller lattice enthalpy ($+783$ instead of $+787$ kJ/mol), it would dissolve exothermically instead of endothermically.

Predictions of enthalpies of solution based on tabulated values of lattice enthalpies and enthalpies of hydration are not very reliable. A small inaccuracy in either quantity can lead to an answer with the wrong sign. It is like trying to calculate the mass of the captain of a ship by measuring the mass of the ship before and after the captain comes aboard. Moreover, the enthalpy of hydration changes as the concentration of the solution changes, and the values calculated from tables apply only to the formation of very dilute solutions. However, it is reasonably safe to expect a positive enthalpy of solution if the lattice enthalpy is very high. Dissolving is then endothermic. Similarly, we can expect a negative enthalpy of solution if the lattice enthalpy is low and the enthalpy of hydration is large. Dissolving is then exothermic.

Enthalpies of solution in dilute solution can be expressed as the sum of the lattice enthalpy and the enthalpy of hydration of the compound.

12.9 Individual Ion Hydration Enthalpies

The enthalpies of hydration in Table 12.6 are the enthalpy changes that occur when widely separated cations and anions in a gas form a very dilute solution. It would be useful to know the values for the cations and anions separately, for then we could predict the enthalpy of hydration of a gas of sodium and chloride ions, for instance, by adding together the values for $Na^+(g)$ and $Cl^-(g)$. Unfortunately, in general we cannot measure enthalpies of hydration of cations without anions being present too. The value for the hydrogen ion, however, has been measured by a special technique and has the following value:

$$H^+(g) \longrightarrow H^+(aq) \qquad \Delta H = -1130.\,kJ$$

That is, when 1 mol $H^+(g)$ ions dissolves in a very large volume of water, 1130. kJ of energy is released as heat. This large value is about the same as the energy obtained by burning about 50 mL of gasoline.

Once we know the enthalpy of hydration for hydrogen ions, we can combine it with the experimentally determined values in Table 12.6 to obtain the individual hydration enthalpies for other ions. For example, the combined enthalpy of hydration of H^+ and Cl^- ions is $-1470.$ kJ/mol. The ion hydration enthalpy of Cl^- itself is therefore $-1470. - (-1130.)$ kJ/mol $= -340.$ kJ/mol. Table 12.7 gives values for a selection of ions: all the values are negative because the hydration of any ion releases energy.

An important procedure in science is to find patterns in data, for recognizing patterns deepens our understanding of nature and helps us find an explanation of an observation. The individual ion hydration enthalpies in Table 12.7 show two patterns:

> 1. The transfer of an ion from gas to water is more exothermic the more highly charged the ion.

Table 12.7 Individual ion hydration enthalpies

Ion	Hydration enthalpy, kJ/mol
Cations	
H^+	$-1130.$
Li^+	-558
Na^+	-444
K^+	-361
Ag^+	$-510.$
Be^{2+}	-2533
Mg^{2+}	-2003
Ca^{2+}	-1657
Sr^{2+}	-1524
Al^{3+}	-4704
Anions	
F^-	-483
Cl^-	$-340.$
Br^-	-309
I^-	-296

As an illustration of this trend, we can compare Li^+ (-558 kJ/mol), Be^{2+} (-2533 kJ/mol), and Al^{3+} (-4704 kJ/mol). The increasingly large negative values reflect the much greater strength of the interaction between a highly charged ion and the polar water molecules.

2. The transfer of an ion from gas to water is more exothermic the smaller the ion.

To see this trend, compare the ion hydration enthalpies for Li^+ (-558 kJ/mol), Na^+ (-444 kJ/mol), and K^+ (-361 kJ/mol). The trend arises because the partial charges of a water molecule can approach the center of charge of a small ion such as Li^+ more closely and hence interact with it more strongly.

Patterns of behavior are important because a deviation of an individual value from the general trend can give further insight—providing we can interpret it! As an example, Ag^+ is bigger than Na^+, but its hydration is more exothermic. The explanation of this anomaly may be that the Ag^+ ion can form covalent bonds with the hydrating water molecules.

Small, highly charged ions have high enthalpies of hydration.

Example 12.3 *Predicting relative ion hydration enthalpies*

Which ion is likely to have a more negative hydration enthalpy, Ca^{2+} or Sr^{2+}?

Strategy We need to consider two factors, the charges of the ions (the higher the charge, the more negative the hydration enthalpy), and their radii (the smaller the ion, the more negative the hydration enthalpy). Ionic radii of main group elements are given in Fig. 7.36.

Solution The ions have the same charge. Strontium is lower down Group 2 than calcium, so we expect it to be larger. Therefore, we predict that Ca^{2+} will be hydrated more strongly and have a more negative hydration enthalpy. This prediction is confirmed by the data in Table 12.7.

Self-Test 12.3A Why does Cl^- have a more negative hydration enthalpy than I^-?

[**Answer:** Cl^- has the smaller radius.]

Self-Test 12.3B Would you expect Rb^+ to have a more negative or less negative hydration enthalpy than Sr^{2+}? Explain your answer.

12.10 Solubility and Disorder

Exothermic dissolving tells us that the energy released during hydration is more than enough to compensate for breaking up the solid solute. We might expect such substances to be soluble because the ions or molecules have a lower energy in the solution than in the solid. However, not all substances with a negative enthalpy of solution are soluble, so there must be another factor at work. Moreover, some very soluble substances like ammonium nitrate have *positive* enthalpies of solution in water, so the solution has a higher enthalpy than separate containers of pure water and solid ammonium nitrate; nevertheless, the substance dissolves.

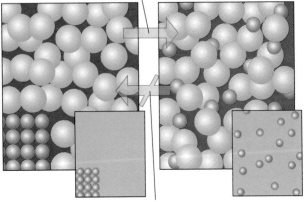

Natural

Unnatural

FIGURE 12.15

Matter has a natural tendency to disperse in a disorderly way. The reverse process is unnatural and has no tendency to occur. The dispersal of matter accounts for the expansion of gases and the spreading of matter through the environment, and contributes to the tendency of solids to dissolve. We shall see later that it also contributes to the tendency of chemical reactions to occur. The insets show the solute particles alone.

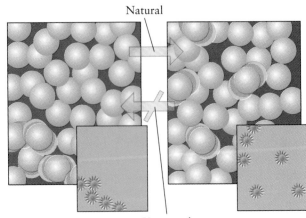

Natural

Unnatural

FIGURE 12.16

Energy has a natural tendency to disperse in a disorderly way. The reverse process is unnatural and has no tendency to occur. The molecules with the greater energy are shown in the drawing by suggesting a shaking motion; think of the vigorous shaking motion as being passed on from molecule to molecule as they bump into one another. The insets show only the higher energy molecules.

> The tendency to disorder is described by the second law of thermodynamics and the property of entropy. See Chapter 17.

The real reason why substances dissolve can be found in a very simple idea: *energy and matter tend to disperse.* By disperse, we mean spread out in a disorderly way (Figs. 12.15 and 12.16). The tendency of energy and matter to disperse is a fundamental scientific principle. As we shall see again and again in the following chapters, it explains why any change occurs.

Let's apply this idea to dissolving. We need to consider the total disorder produced when a solution forms, taking into account both the solution and the surroundings. Suppose a solute dissolves exothermically. Disorder normally increases in a solution when a substance dissolves because the solute molecules are more disordered in the solution than in the original solid. If the dissolving is exothermic, energy spreads into the surroundings, stirring the molecules there into disorder (Fig. 12.17). Because disorder increases in both the system (the solution) and the surroundings, exothermic dissolving is natural.

A solution and its surroundings undergo opposing shifts in disorder when a substance dissolves endothermically. As usual, the solution becomes more disordered as the solute dissolves. However, the disorder of the surroundings decreases as they lose energy to the solution (Fig. 12.18). If the dissolution is only slightly endothermic (as it is for sodium chloride), the disorder of the solution increases more than the disorder of the surroundings decreases. In this

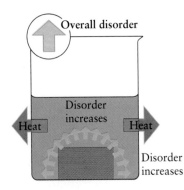

Overall disorder

Disorder increases

Heat   Heat

Disorder increases

FIGURE 12.17

Many substances dissolve to give a system with more disorder than was present initially. If their dissolving is also exothermic, we can be confident that the dissolving is a natural process.

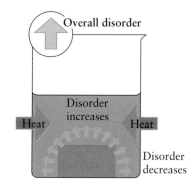

FIGURE 12.18

If dissolution is endothermic, it will occur only if the disorder caused by the solute dispersing is great enough to produce an overall increase in disorder.

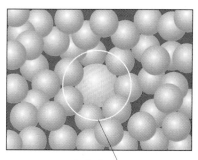

Orderly arrangement

FIGURE 12.19

When a nonpolar compound (the yellow sphere) dissolves in water, the water molecules may become organized around it. As a result, the disorder of the solvent is reduced when the solution forms.

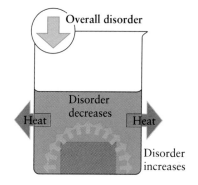

FIGURE 12.20

Even though dissolving may be weakly exothermic, and hence increase the disorder of the surroundings, the reduced disorder of the solvent illustrated in Fig. 12.19 might cancel the increase. As a result, dissolution does not take place.

case, the substance dissolves. In contrast, the dissolution of MgO would be highly endothermic (because its lattice enthalpy is so great). As a result, the decrease in disorder of the surroundings would be so great that the small increase in disorder of the solution would not compensate for it. Hence, a substance with a strongly endothermic enthalpy of solution, such as MgO, is likely to be insoluble.

In a few cases, the presence of a solute causes the solvent molecules to take up a more organized arrangement than they had in the pure liquid (Fig. 12.19). Now the disorder of the solution is *lower* than that in the pure liquid. Even though energy is released into the surroundings, the increase in the disorder of the surroundings might not be enough to overcome the decrease in disorder of the solution (Fig. 12.20). As a result, the solution does not form. This situation is illustrated by the inability of some hydrocarbons to dissolve in water, even though they have weakly negative enthalpies of solution.

> ***Dissolving depends on the balance between the change in disorder of the solution and the change in disorder of the surroundings: overall there must be an increase in disorder.***

COLLIGATIVE PROPERTIES

Now we shall see how the presence of a solute affects the physical properties of the solvent. Experiments have shown that some physical properties are affected in the same way by solutes, regardless of the chemical identity of the solute. Properties that depend on the relative number of solute molecules and not on their chemical identity are called **colligative properties.** The four colligative properties we consider are the lowering of the vapor pressure of the solvent, the raising of its boiling point, the lowering of its freezing point, and the tendency of a solvent to flow through a membrane into a solution. The last property is of enormous importance, because it contributes to the flow of nutrients through biological cell walls and the movement of solutions through plants.

12.11 Measures of Concentration

Only the concentration of the solute, not its identity, is important for predicting the magnitude of a colligative property. For example, the water in an aqueous glucose solution in which 1 out of every 100 molecules is a glucose molecule has the same vapor pressure, boiling point, and freezing point as the water in an aqueous sucrose solution in which 1 out of every 100 molecules is a sucrose molecule. Therefore, to predict the magnitude of a colligative property, we need a measure of the relative numbers of solute and solvent molecules.

Two measures of concentration important for the study of colligative properties are mole fraction and molality. We first met the mole fraction, x, in Section 5.11, where we saw that it is the ratio of the number of moles of a species to the total number of moles of all the species present in a mixture. For solute molecules in a nonelectrolyte solution,

$$x_{solute} = \frac{\text{moles of solute molecules}}{\text{total moles of solute and solvent molecules}} \qquad x_{solute} = \frac{n_{solute}}{n_{solute} + n_{solvent}} \qquad (2)$$

where n_{solute} and $n_{solvent}$ are the numbers of moles of solute and solvent molecules, respectively. A similar expression defines the mole fraction of the solvent. The mole fractions of a solvent and all the solute species always sum to 1:

$$x_{solute} + x_{solvent} = 1$$

This relation follows directly from the definition of x. For example, if there is a single solute; then

$$x_{solute} + x_{solvent} = \frac{n_{solute}}{n_{solute} + n_{solvent}} + \frac{n_{solvent}}{n_{solute} + n_{solvent}}$$
$$= \frac{n_{solute} + n_{solvent}}{n_{solute} + n_{solvent}} = 1$$

Example 12.4 *Calculating the mole fractions of species in a solution*

Calculate the mole fractions of the nonelectrolyte sucrose, $C_{12}H_{22}O_{11}$, and water in a solution prepared by dissolving 5.00 g of sucrose in 100.0 g of water.

Strategy The procedure is essentially the same as that for gases (Example 5.10b). First, find the numbers of moles of solute and solvent molecules in the solution by using molar mass to convert mass to moles. The total number of moles is the sum of the moles of solute and solvent. Then use Eq. 2 to calculate the mole fractions as ratios of the moles of the species divided by the total number of moles of species present.

Solution The molar masses of sucrose and water are 342.3 and 18.02 g/mol, respectively. Therefore, the numbers of moles of each species are

$$\text{Moles of } C_{12}H_{22}O_{11} = (5.00 \text{ g}) \times \left(\frac{1 \text{ mol } C_{12}H_{22}O_{11}}{342.3 \text{ g}}\right) = \frac{5.00}{342.3} \text{ mol } C_{12}H_{22}O_{11}$$

$$\text{Moles of } H_2O = (100.0 \text{ g}) \times \left(\frac{1 \text{ mol } H_2O}{18.02 \text{ g}}\right) = \frac{100.0}{18.02} \text{ mol } H_2O$$

(These two values are 0.0146 and 5.549 mol, respectively, but we avoid rounding errors by delaying the calculation.) The total number of moles of molecules in the solution is the sum,

$$\text{Total number of moles} = \frac{5.00}{342.3} \text{mol} + \frac{100.0}{18.02} \text{mol}$$

$$= \left(\frac{5.00}{342.3} + \frac{100.0}{18.02}\right) \text{mol}$$

(This expression works out to 5.564 mol.) The two mole fractions are therefore

$$x_{\text{sucrose}} = \frac{(5.00/342.3) \text{ mol}}{\{(5.00/342.3) + (100.0/18.02)\} \text{ mol}} = 0.002\,62$$

$$x_{\text{water}} = \frac{(100.0/18.02) \text{ mol}}{\{(5.00/342.3) + (100.0/18.02)\} \text{ mol}} = 0.997$$

Note that $0.002\,62 + 0.997 = 1.000$.

Self-Test 12.4A Calculate the mole fractions of H_2O and CH_3CH_2OH in a mixture of equal masses of water and ethanol.

[***Answer:*** $x_{\text{water}} = 0.719$; $x_{\text{ethanol}} = 0.281$]

Self-Test 12.4B Calculate the mole fractions of the cations, anions, and water molecules present in a solution prepared by adding 5.00 g of sodium chloride to 100. g of water. (*Hint:* Treat the three species as individual entities.)

The **molality** of a solution is the number of moles of solute in a solution divided by the mass of the solvent in kilograms:

$$\text{Molality} = \frac{\text{moles of solute (mol)}}{\text{mass of solvent (kg)}} \qquad (3)$$

There is a key difference between molarity and molality. Whereas the molarity of a solute is defined in terms of the volume of the *solution* (not the volume of the solvent), the molality is defined in terms of the mass of the *solvent* used to prepare the solution (not the mass of the solution). The unit of molality is 1 mole per kilogram of solvent (1 mol/kg); this unit is often denoted *m*. For example, a solution prepared by dissolving 1.0 mol $NiSO_4$ in 1.0 kg of water would be reported as 1.0 *m* $NiSO_4$(aq) (Fig. 12.21).

FIGURE 12.21

(a) A solution of given molarity is prepared by measuring the required mass of solute and then adding it to a flask of known volume. Solvent is added up to the mark. (b) The steps taken to prepare a solution of given molality. First (left), the required masses of solute and solvent are measured out. Then (right), the solute is dissolved in the solvent.

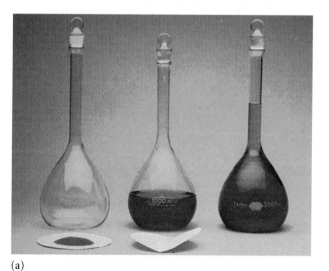

(a)

(b)

Example 12.5 *Preparing a solution of given molality*

What is the molality of an NaCl solution prepared by dissolving 10.5 g of sodium chloride in 250. g of water?

Strategy The molality of the solution is the number of moles of solute per kilogram of solvent. Convert the mass of NaCl to moles (by dividing the mass of solute by the molar mass) and divide it by the mass of water in kilograms. Convert grams to kilograms by using $1 \text{ kg} = 10^3 \text{ g}$.

Solution First, we find the number of moles of solute:

$$\text{Moles of NaCl} = (10.5 \text{ g NaCl}) \times \left(\frac{1 \text{ mol NaCl}}{58.44 \text{ g NaCl}}\right) = \frac{10.5}{58.44} \text{ mol NaCl}$$

Next, we convert grams of water to kilograms of water:

$$\text{Mass of water (kg)} = 250. \text{ g} \times \frac{1 \text{ kg}}{10^3 \text{ g}} = 0.250 \text{ kg}$$

Finally, we calculate the molality:

$$\text{Molality of NaCl} = \frac{(10.5/58.44) \text{ mol}}{0.250 \text{ kg}} = 0.719 \text{ mol/kg}$$

We report this molality as 0.719 *m* NaCl(aq).

Self-Test 12.5A Calculate the molality of $ZnCl_2$ in a solution prepared by dissolving 4.11 g of zinc chloride in 150. g of water.

[*Answer:* 0.201 mol/kg]

Self-Test 12.5B Calculate the molality of a $KClO_3$ solution prepared by dissolving 7.36 g of potassium chlorate in 20.0 g of water.

Toolbox 12.1 shows how to use molality. One advantage of using molality is that it does not change with temperature; the molarity does depend on temperature because the volume of the solution changes as the temperature is changed.

To emphasize the relative numbers of solute and solvent particles in a solution, express the composition in terms of either mole fraction or molality.

Toolbox 12.1 *How to use molality*

This Toolbox shows how to calculate the moles of solute present in a given mass of solvent from the molality, how to convert between molality and mole fraction, and how to convert between molarity and molality.

Conceptual Basis
The molality is the concentration of solute in moles per kilogram of solvent (Fig. 12.22a). The mole fraction is the amount of a species expressed as a fraction of the total number of moles of ions and molecules in the solution (Fig. 12.22b). These quantities are calculated from measurements made before mixing. The molarity is the number of moles of solute per liter of solution (Fig. 12.22c). It is calculated by measuring the mass of solute before mixing and the volume of solution *after* mixing.

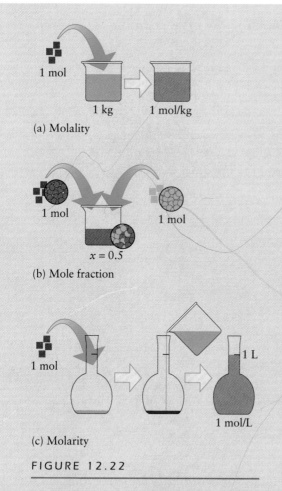

(a) Molality

$x = 0.5$

(b) Mole fraction

1 L

1 mol/L

(c) Molarity

FIGURE 12.22

Visualizations of three common measures of concentration: (a) molality; (b) mole fraction; (c) molarity.

Procedure

Calculating the mass of solute from molality and mass of solvent

Step 1. Calculate the moles of solute molecules present in a given mass of solvent by rearranging the equation defining molality (Eq. 3) into

Moles of solute $=$ molality $\times$ mass of solvent (kg)

Step 2. Use the molar mass of the solute to find the mass of solute present.

See Example 12.6 for an illustration.

Calculating the molality from a mole fraction

The mole fraction tells us that in a total of 1 mol of molecules in a sample, there is x_{solute} mol of solute molecules and $(1 - x_{\text{solute}})$ mol of solvent molecules.

Step 1. Use the molar mass to convert the number of moles of solvent molecules to mass of solvent.

At this point, we know both the moles of solute molecules (x_{solute} mol) and the mass of solvent in a sample that contains a total of 1 mol of molecules.

Step 2. Calculate the molality by dividing the moles of solute molecules by the mass of solvent (in kilograms):

$$\text{Molality} = \frac{n_{\text{solute}} \ (\text{mol})}{m_{\text{solvent}} \ (\text{kg})}$$

See Example 12.7 for an illustration.

Calculating the mole fraction from the molality

We consider a solution in which there is exactly 1 kg of solvent present.

Step 1. Convert that mass of solvent to moles of solvent molecules by using the molar mass of the solvent. We already know the number of moles of solute molecules in the solution (from the molality).

Step 2. Calculate the mole fractions from the numbers of moles of molecules by using Eq. 2:

$$x_{\text{solute}} = \frac{n_{\text{solute}}}{n_{\text{solute}} + n_{\text{solvent}}}$$

Calculating the molality from the molarity

Molality is defined as moles of solute divided by the mass of *solvent*, whereas molarity is defined as moles of solute divided by the volume of the *solution*. To find the mass of solvent in the solution of given molarity, we need its density, which must either be provided as data or measured.

Step 1. Calculate the total mass of exactly 1 L of solution by using the density of the solution and converting to kilograms.

Step 2. Use the molarity to calculate the mass of solute in 1 L of solution and convert to kilograms.

Step 3. Subtract the mass of solute from the total mass to find the mass of solvent in 1 L of solution.

Step 4. Divide the moles of solute by the mass of the solvent in kilograms to obtain the molality.

See Example 12.8 for an illustration.

Example 12.6 Estimating the mass required for a given molality

When a chemist makes up a solution, the first step is to estimate the mass of solute needed, then to spoon out *approximately* this mass of solute and weigh it precisely. The precise mass is used to calculate the actual molality of the solution. Estimate the mass of potassium nitrate to add to 250. g of water to prepare 0.200 m $KNO_3(aq)$.

Strategy To find the mass of solute required, we determine the number of moles required and convert to mass, as outlined in the first procedure in Toolbox 12.1.

Solution **Step 1.** The moles of KNO_3 in 250. g (0.250 kg) of solvent is

$$\text{Moles of } KNO_3 = \overbrace{\{0.200 \text{ (mol } KNO_3)/\text{kg}\}}^{\text{Molality}} \times \overbrace{(0.250 \text{ kg})}^{\text{Mass of solvent}} = 0.200 \times 0.250 \text{ mol } KNO_3$$

Step 2. The molar mass of KNO_3 is 101.1 g/mol, therefore the mass of potassium nitrate required is

$$\text{Mass of } KNO_3 \text{ (g)} = (0.200 \times 0.250 \text{ mol } KNO_3) \times \left(\frac{101.1 \text{ g } KNO_3}{1 \text{ mol } KNO_3}\right)$$
$$= 5.06 \text{ g } KNO_3$$

Self-Test 12.6A What mass of potassium permanganate is needed to prepare 0.150 m $KMnO_4(aq)$ with 500. g of water?

[*Answer:* 11.9 g of $KMnO_4$]

Self-Test 12.6B What mass of glucose, $C_6H_{12}O_6$, is needed to prepare 0.255 m $C_6H_{12}O_6(aq)$ with 250. g of water?

Example 12.7 Calculating a molality from a mole fraction

What is the molality of a solution of benzene, C_6H_6, in toluene, $CH_3C_6H_5$, in which the mole fraction of benzene is 0.150?

Strategy Convert the moles of solvent to kilograms, as described in the second procedure in Toolbox 12.1.

Solution **Step 1.** From the definition of mole fraction, 1.000 mol of solution molecules contains 0.150 mol C_6H_6 and $1.000 - 0.150$ mol $= 0.850$ mol $CH_3C_6H_5$. The mass of this amount of toluene (of molar mass 92.13 g/mol) is

$$\text{Mass of toluene (kg)} = (0.850 \text{ mol } CH_3C_6H_5) \times \left(\frac{92.13 \text{ g } CH_3C_6H_5}{1 \text{ mol } CH_3C_6H_5}\right) \times \left(\frac{1 \text{ kg}}{10^3 \text{ g}}\right)$$
$$= 0.850 \times 92.13 \times 10^{-3} \text{ kg } CH_3C_6H_5$$

Step 2. The molality of benzene in the solution is therefore

$$\text{Molality of } C_6H_6 = \frac{0.150 \text{ mol}}{0.850 \times 92.13 \times 10^{-3} \text{ kg}} = 1.92 \text{ mol/kg}$$

Self-Test 12.7A Calculate the molality of a solution of toluene in benzene, given that the mole fraction of toluene is 0.150.

[*Answer:* 2.26 mol/kg]

Self-Test 12.7B Calculate the molality of a solution of methanol in water, given that the mole fraction of methanol is 0.250.

Example 12.8 *Calculating the molality from the molarity*

What is the molality of 1.06 M $C_{12}H_{22}O_{11}$(aq), which is known to have density 1.14 g/cm^3 (which is the same as 1.14 g/mL)?

Strategy First, find the mass of solvent in 1 L of solution, then convert that mass to kilograms, as described in the last procedure in Toolbox 12.1.

Solution **Step 1.** The mass of exactly 1 L (1×10^3 mL) of solution is

$$\text{Mass of solution} = \text{density} \times \text{volume}$$
$$= (1.14 \text{ g/mL}) \times (1 \times 10^3 \text{ mL}) = 1.14 \times 10^3 \text{ g}$$
$$= 1.14 \text{ kg}$$

Step 2. The number of moles of solute molecules in exactly 1 L of the 1.06 M solution is 1.06 mol, and the molar mass of sucrose is 342.3 g/mol. The mass of sucrose in exactly 1 L of solution is therefore

$$\text{Mass of sucrose (g)} = (1.06 \text{ mol } C_{12}H_{22}O_{11}) \times \left(\frac{342.3 \text{ g } C_{12}H_{22}O_{11}}{1 \text{ mol } C_{12}H_{22}O_{11}} \right)$$
$$= 1.06 \times 342.3 \text{ g } C_{12}H_{22}O_{11} = 363 \text{ g } C_{12}H_{22}O_{11}$$

or 0.363 kg $C_{12}H_{22}O_{11}$.

Step 3. The mass of water present in 1 L of solution is therefore

$$\text{Mass of water} = 1.14 \text{ kg} - 0.363 \text{ kg} = 0.78 \text{ kg}$$

Step 4. Because 1.06 mol $C_{12}H_{22}O_{11}$ is dissolved in 0.78 kg of water, we conclude that the molality of the solution is

$$\text{Molality of } C_{12}H_{22}O_{11} = \frac{1.06 \text{ mol}}{0.78 \text{ kg}} = 1.4 \text{ mol/kg}$$

Self-Test 12.8A Battery acid is 4.27 M H_2SO_4(aq) and has a density of 1.25 g/cm^3. What is the molality of H_2SO_4 in the solution?

[***Answer:*** 5.14 *m* H_2SO_4(aq)]

Self-Test 12.8B The density of 1.83 M NaCl(aq) is 1.07 g/cm^3. What is the molality of NaCl in the solution?

12.12 Vapor-Pressure Lowering

Salt lakes evaporate more slowly than freshwater lakes. Small salt lakes scattered throughout the desert regions of the southwestern United States remain when all other water has dried up, tempting unwary animals to drink their poisonous waters. The lower rate of evaporation in salt lakes implies that the presence of the salt has reduced the vapor pressure. We see the same phenomenon in the laboratory. The vapor pressure of a solvent is found to be lower when a nonvolatile solute is present (Fig. 12.23). For example, the vapor pressure of pure water at 40°C is 55 Torr but that of 1.0 *m* NaCl(aq) at the same temperature is only 54 Torr.

The nineteenth-century French scientist François-Marie Raoult spent much of his life measuring vapor pressures. He found that the effect of a solute could be summarized in what we now know as **Raoult's law:**

The vapor pressure of a solvent in the presence of a nonvolatile solute is proportional to the mole fraction of the solvent.

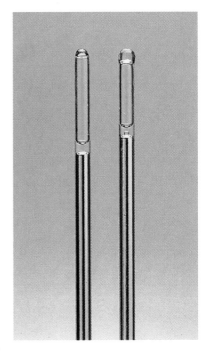

FIGURE 12.23

The vapor pressure of a solvent is lowered by a nonvolatile solute. The barometer tube on the left has a small volume of pure water floating on the mercury. That on the right has a small volume of 10 *m* NaCl(aq), and a lower vapor pressure. Note that the column on the right is depressed less by the vapor in the space above the mercury than the one on the left, showing that the vapor pressure is lower when the solute is present.

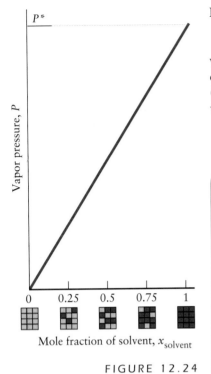

Mole fraction of solvent, $x_{solvent}$

FIGURE 12.24

Raoult's law predicts that the vapor pressure of a solvent in a solution should be proportional to the mole fraction of the solvent molecules. The vapor pressure of the pure solvent is P^*.

Raoult's law is normally written

$$P = x_{solvent} \times P_{pure} \qquad (4)$$

where P_{pure} is the vapor pressure of the pure solvent, $x_{solvent}$ is the mole fraction of the solvent, and P is the vapor pressure of the solvent in the solution (Fig. 12.24). Notice that the vapor pressure of the solvent is directly proportional to the mole fraction of the solvent in the solution.

Example 12.9 Using Raoult's law

Calculate the vapor pressure of water at 100.°C in a solution prepared by dissolving 5.00 g of sucrose in 100. g of water.

Strategy Once we know the mole fraction of the solvent (water) in the solution, the calculation is a straightforward application of Raoult's law. That mole fraction was calculated in Example 12.4. To use Raoult's law, we need the vapor pressure of the pure solvent. We could get the value from Table 5.4, but because the normal boiling point of water is 100.°C, we know that its vapor pressure at that temperature is 760. Torr. Expect a lower vapor pressure when the solute is present.

Solution From Example 12.4, we know that $x_{water} = 0.997$. Therefore, because $P_{pure} = 760.$ Torr, the vapor pressure of the water in the solution is

$$P = 0.997 \times (760.\ \text{Torr}) = 758\ \text{Torr}$$

Self-Test 12.9A Calculate the vapor pressure of water at 90.°C for a solution prepared by dissolving 5.00 g of glucose ($C_6H_{12}O_6$) in 100. g of water. The vapor pressure of pure water at 90.°C is 524 Torr.

[*Answer:* 521 Torr]

Self-Test 12.9B Calculate the vapor pressure of ethanol in kilopascals (kPa) at 19°C for a solution prepared by dissolving 2.00 g of cinnamaldehyde, C_9H_8O, in 50.0 g of ethanol, C_2H_5OH. The vapor pressure of pure ethanol at that temperature is 5.3 kPa.

To understand why a nonvolatile solute lowers the vapor pressure of the solvent, we can think about the molecules of solvent and solute. The vapor pressure of the pure solvent represents the tendency of the liquid to become more disordered by forming a gas. When a solute is present, the solution is more disordered than the pure solvent. Because it is already more disordered, the solvent has a lower tendency to become more disordered by forming a vapor. That is, it has a lower vapor pressure than the pure solvent (Fig. 12.25). The higher the concentration of the solute, the lower the vapor pressure of the solvent.

A hypothetical solution that obeys Raoult's law exactly at all concentrations is called an **ideal solution**. In an ideal solution, the forces between solute and solvent molecules are the same as the forces between solvent molecules, so the solute molecules mingle freely and invisibly with the solvent molecules. Solutes that form nearly ideal solutions are often similar in composition and structure to the solvent molecules. For instance, methylbenzene (toluene) forms nearly ideal solutions with benzene.

Real solutions do not obey Raoult's law at all concentrations; but the lower the solute concentration, the more they act like ideal solutions. A solution that does not obey Raoult's law at a particular solute concentration is called a **nonideal solution**. Real solutions are approximately ideal at solute concentrations

FIGURE 12.25

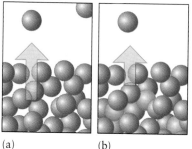

(a) A pure solvent is a largely random collection of molecules, and it has a tendency to form an even more random vapor. Vaporization continues until the liquid and vapor phase molecules are in dynamic equilibrium, with the rate of vaporization equal to the rate of condensation. (b) When a nonvolatile solute is present, the liquid phase has a greater randomness, and therefore a smaller tendency to form the vapor. As a result, the vapor pressure over the solution is lower than that over the pure liquid. The relative length of the arrow represents the relative tendency of the solvent to vaporize.

(a)   (b)

below about 0.1 M for nonelectrolyte solutions and 0.01 M for electrolyte solutions. The problem with electrolyte solutions is that interactions between ions occur over a long distance and hence have a pronounced effect. We can safely assume that all the solutions we meet are ideal, because we shall treat only dilute solutions.

> *The vapor pressure of a solvent is reduced by the presence of a nonvolatile solute: in an ideal solution, the vapor pressure of the solvent is proportional to the mole fraction of solvent.*

12.13 Boiling-Point Elevation and Freezing-Point Depression

The lowering of the vapor pressure of a solvent explains another colligative property: the raising of the boiling point by a nonvolatile solute. This increase is called the **boiling-point elevation.** Because the vapor pressure of a solvent in a solution is lower than that of the pure solvent at the same temperature, a higher temperature is needed to raise the vapor pressure up to atmospheric pressure, when boiling begins (Fig. 12.26). However, the increase is usually quite small and is of little practical importance. The normal boiling point of a 0.1 *m* aqueous sucrose solution, for instance, is 100.05°C.

More important is the **freezing-point depression,** the lowering of the freezing point of the solvent. For example, seawater, an aqueous solution rich in Na^+ and Cl^- ions, freezes about 1°C lower than fresh water. In regions with cold

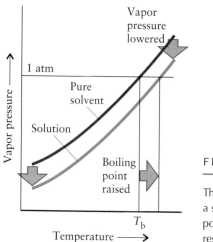

FIGURE 12.26

The lowering of vapor pressure of the solvent in a solution results in an increase in its boiling point because a higher temperature is needed to restore the vapor pressure to 1 atm.

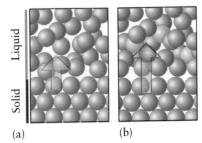

(a)   (b)

FIGURE 12.27

(a) A pure solid (lower part of diagram) has a high degree of order, whereas a liquid (upper part) is disordered. The solid has a tendency to form the liquid and does so when the temperature reaches a certain value (the melting point). (b) When a solute is present in the liquid, the latter has a higher degree of disorder. As a result, the solid has a stronger tendency to form the liquid, and so freezes at a temperature lower than that at which the pure liquid freezes. Note that the solid consists only of pure solvent, even if a solute is present in the liquid phase.

winters, salt is spread on highways to lower the freezing point of water and delay ice formation. Chemists make use of the depression of the freezing point to judge the purity of a solid compound: a melting point lower than the accepted value is a sign that impurities are present.

The depression of freezing point has the same origin as the depression of vapor pressure: the disorder introduced into the solution by the solute. The tendency to melt reflects the tendency of the solid to become disordered. When a solute is available to mingle with the molten solvent, the tendency to melt is enhanced because the resulting solution is more disordered than the pure solvent would be. The enhanced tendency to melt means that melting can occur at a lower temperature than it does in the absence of the solute (Fig. 12.27). In other words, the melting point—and hence the freezing point too—is lowered by the solute.

Calculations show that the freezing-point depression, ΔT, in an ideal solution is proportional to the molality of the solute. For a nonelectrolyte solution,

$$\text{Freezing-point depression} = \text{constant} \times \text{molality} \qquad \Delta T = k_f \times \text{molality} \quad (5)$$

The constant k_f is called the **freezing-point constant** of the solvent; it has the units kelvin-kilogram per mole (K·kg/mol). It is different for each solvent and must be determined experimentally (Table 12.8). A 0.1 m aqueous sucrose solution, for instance, has

$$\text{Freezing-point depression} = (1.86 \text{ K·kg/mol}) \times (0.1 \text{ mol/kg}) = 0.2 \text{ K}$$

Hence the solution freezes at $-0.2°C$.

In an electrolyte solution, each formula unit contributes two or more ions. Sodium chloride, for instance, dissolves to give Na^+ and Cl^- ions, and both kinds of ions contribute to the depression of the freezing point. The cations and anions affect colligative properties nearly independently in very dilute solutions, so the total solute molality is twice the molality in terms of NaCl formula units. In more concentrated solutions, the ions do not move independently. For instance, some stick together to form ion pairs and other aggregates of small numbers of ions. The effect of the solute on the freezing point in these solutions is very difficult to predict. We can write

$$\Delta T = ik_f \times \text{molality} \qquad (6)$$

where i is known as the **van't Hoff i factor**; it is determined experimentally. In a very dilute solution, in which all ions act independently, $i = 2$ for MX salts and $i = 3$ for MX_2 salts such as $CaCl_2$, and so on. For dilute nonelectrolyte solutions, $i = 1$. The i factor is so unreliable that it is best to confine quantitative treatments of freezing-point depression to nonelectrolyte solutions. Even these solutions must be dilute enough to be approximately ideal.

The i factor is named for Jacobus van't Hoff (1852–1911), a Dutch chemist who in 1901 was awarded the first Nobel prize for chemistry (but not for introducing this simple factor).

Table 12.8 Boiling-point and freezing-point constants

Solvent	Freezing point, °C	k_f, K·kg/mol	Boiling point, °C	k_b, K·kg/mol
acetone	−95.35	2.40	56.2	1.71
benzene	5.5	5.12	80.1	2.53
camphor	179.8	39.7	204	5.61
carbon tetrachloride	−23	29.8	76.5	4.95
cyclohexane	6.5	20.1	80.7	2.79
naphthalene	80.5	6.94	217.7	5.80
phenol	43	7.27	182	3.04
water	0	1.86	100.0	0.51

The i factor can be used to help determine the extent to which an acid forms ions in solution. For example, in dilute solution, HCl has an i factor of 1 in toluene and 2 in water. These values suggest that HCl retains its molecular form in toluene but is present as ions in water. If 5% of acid molecules form ions in water, the i factor would be $(0.05 \times 2) + 0.95 = 1.05$, because each acid molecule produces two ions.

Self-Test 12.10A How many moles of ions are present in a solution containing 0.10 mol Na_2SO_4, assuming complete dissociation into Na^+ and SO_4^{2-} ions? Estimate the i factor.

[*Answer:* 0.30 mol (0.20 mol Na^+ ions and 0.10 mol SO_4^{2-} ions); $i = 3$]

Self-Test 12.10B How many moles of ions are present in a solution containing 0.25 mol $CoCl_3$, assuming complete dissociation into Co^{3+} and Cl^- ions? Estimate the i factor.

Freezing-point depression is the basis of the technique called **cryoscopy,** the determination of the molar mass of a solute from the freezing-point depression it causes. Camphor is often used as the solvent for organic compounds because it has a large freezing-point constant, so solutes significantly depress its freezing point.

> Note that the percentage deprotonation of an acid is a mole percentage, not a mass percentage.

Example 12.10 *Determining molar mass from a freezing-point depression*

The addition of 0.24 g of sulfur to 100. g of carbon tetrachloride lowers the freezing point of carbon tetrachloride by 0.28°C. What is the molar mass and molecular formula of sulfur?

Strategy First, convert the observed freezing-point depression into solute molality by rearranging Eq. 5 into

$$\text{Molality} = \frac{\text{freezing-point depression}}{k_f}$$

Take the freezing-point constant from Table 12.8. Calculate the moles of solute in the sample by multiplying this molality by the mass of solvent in kilograms. Determine the molar mass of the solute by dividing the given mass of solute by the calculated number of moles of solute. For the molecular formula, decide how many atoms of sulfur are needed in each molecule to account for the molar mass.

Solution From the data,

$$\text{Molality of } S_x = \frac{\overbrace{0.28 \text{ K}}^{\Delta T}}{\underbrace{29.8 \text{ K·kg/mol}}_{k_f}} = \frac{0.28}{29.8 \text{ kg/mol}} = \frac{0.28}{29.8} \text{ mol/kg}$$

Because the mass of solvent is 100. g (that is, 0.100 kg), we find

$$\text{Moles of } S_x = (0.100 \text{ kg}) \times \left(\frac{0.28}{29.8} \text{ mol/kg}\right) = \frac{0.100 \times 0.28}{29.8} \text{ mol}$$

It then follows that

$$\text{Molar mass of } S_x = \frac{0.24 \text{ g}}{(0.100 \times 0.28/29.8) \text{ mol}} = \frac{0.24 \times 29.8}{0.100 \times 0.28} \text{ g/mol}$$

This expression evaluates to 260 g/mol (2 sf). Because the molar mass of atomic sulfur is 32.06 g/mol, we can find the value of x in the molecular formula S_x from

$$x = \frac{\overbrace{(0.24 \times 29.8)/(0.100 \times 0.28) \text{ g/mol}}^{\text{Molar mass of } S_x}}{\underbrace{32.06 \text{ g/mol}}_{\text{Molar mass of S}}} = \frac{0.24 \times 29.8}{0.100 \times 0.28 \times 32.06} = 8.0$$

Elemental sulfur is therefore composed of S_8 molecules.

Self-Test 12.11A When 250.0 mg of eugenol, the compound responsible for the odor of oil of cloves, was added to 100.0 g of camphor, it lowered the freezing point of camphor by 0.62°C. Calculate the molar mass of eugenol.

[*Answer:* 1.6×10^2 g/mol (actual: 164.2 g/mol)]

Self-Test 12.11B When 200.0 mg of linalool, a fragrant compound extracted from Ceylon cinnamon oil, was added to 100.0 g of camphor, it lowered the freezing point of camphor by 0.51°C. What is the molar mass of linalool?

FIGURE 12.28

A mixture of water and a commercial antifreeze at −13.3°C. The mixture is still a liquid; pure water would be frozen solid at this temperature.

An aqueous solution of antifreeze, typically the organic compound ethylene glycol (1,2-ethanediol, **3**), has a much lower freezing point than that of pure water (Fig. 12.28). However, antifreeze is used in such high concentrations that the solutions cannot be considered ideal, and the freezing-point depression is not really an example of a colligative property. A better explanation is that ethylene glycol and water molecules pack together so badly that it is difficult for them to form a rigid structure.

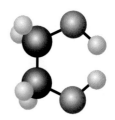

3 1,2 - Ethanediol, $HOCH_2CH_2OH$

The presence of a solute lowers the freezing point of a solvent; if the solute is nonvolatile, the boiling point is also raised. The freezing-point depression can be used to calculate the molar mass of the solute. If the solute is an electrolyte, the extent of its dissociation and ionization must also be taken into account.

12.14 Osmosis

The experiment shown in Fig. 12.29 demonstrates a colligative property of great importance in our lives. A solution in the tube is separated from the pure solvent in the flask by a thin sheet of cellulose acetate. Initially, the heights of the solution and the pure solvent are the same. However, the level of the solution inside the tube begins to rise as pure solvent passes through the membrane into the solution. At equilibrium, the pressure exerted by the rising column of solution forces solvent molecules back through the membrane, and the rate at which molecules enter the column matches the rate at which they leave.

The flow of solvent through a membrane into a more concentrated solution is called **osmosis.** The membrane is **semipermeable,** which means that only certain types of molecules or ions can pass through it. Cellulose acetate allows water molecules to pass through it, but not solute molecules or ions with their bulky coating of hydrating water molecules. The pressure needed to stop the flow of solvent is called the **osmotic pressure,** Π (the Greek uppercase letter pi). The greater the osmotic pressure, the greater the height of the solution needed

Cellulose acetate is widely used as a transparent wrapper on candy boxes, so this experiment is easy to repeat.

The name *osmosis* comes from the Greek word for "push."

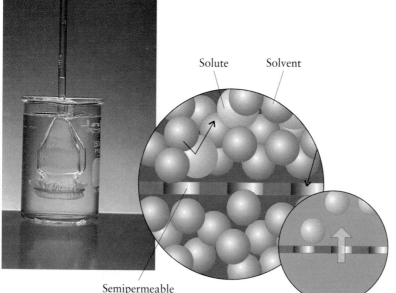

Solute   Solvent

Semipermeable membrane

FIGURE 12.29

An experiment to illustrate osmosis. The tube contains a sucrose solution and the beaker contains pure water. The initial heights of the two liquids were the same. At the stage shown here, water has passed into the solution through the membrane by osmosis and its level has risen above that of the pure water. The large inset shows the solvent molecules (below the membrane) tending to join those in the solution (above the membrane) because there the disorder of the molecules is greater (on account of the presence of the solute molecules). The small inset shows just the solute molecules; the arrow shows the direction of flow of solvent molecules.

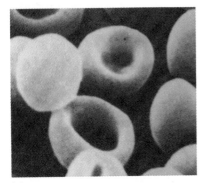

(a)

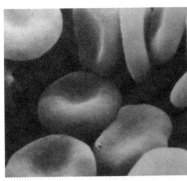

(b)

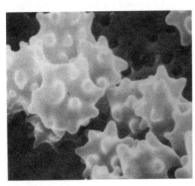

(c)

Note that the molarity, not the molality, occurs in this expression.

FIGURE 12.30

(a) Red blood cells need to be in an isotonic solution if they are to function properly.
(b) When the solution is too dilute (hypotonic), water passes into them and they burst.
(c) When the solution is too concentrated (hypertonic), water flows out of them and they shrivel up.

to stop the net flow. The pressure exerted by a column of fluid is called the **hydrostatic pressure.**

Osmosis plays an important role in maintaining life (Connection 3, which follows this chapter). Biological cell walls act as semipermeable membranes that allow water, small molecules, and hydrated ions to pass through. However, they block the passage of the enzymes and proteins that have been synthesized within the cell. The fluids outside the cell are lower in total solute molality than the fluid inside the cell; they therefore have a lower osmotic pressure and are called **hypotonic.** The difference in total concentrations of solute inside and outside a cell gives rise to an osmotic pressure, and water passes into the more concentrated solution in the interior of the cell, carrying small nutrient molecules with it. This influx of water also keeps the cell *turgid* (swollen). When the fluid outside the cell is **hypertonic** (meaning that it has a higher osmotic pressure than the fluids inside the cell), water flows out of the cell, the cell becomes dehydrated, the turgidity is lost, and metabolic activity ceases. In a plant, this dehydration results in wilting. Salted meat is preserved from bacterial attack by osmosis. In this case, the concentrated salt solution dehydrates—and kills—the bacteria by causing water to flow out of them. Solutions administered intravenously must be **isotonic** with blood cells (have the same osmotic pressure) (Fig. 12.30).

Kidney dialysis machines use osmosis to purify blood. The blood to be cleansed is brought into contact with a sterile solution on the other side of a semipermeable membrane. That membrane allows the passage of small dissolved species, such as inorganic salts and urea, but not the blood cells and proteins in plasma. The unwanted migration of amino acids, glucose, and other desirable compounds is prevented by adding them to the sterile solution: they then migrate through the membrane as fast as their counterparts migrate in the opposite direction.

To understand the origin of osmosis, we can think once again about the disorder introduced into the solvent when a solute is added. Because the disorder is greater on the solution side of the membrane than on the pure solvent side, the solvent has a tendency to flow into the solution, for that action increases the disorder (recall the large inset in Fig. 12.29). This tendency is opposed by the increasing hydrostatic pressure of the solution as the height of the column of solution increases. At equilibrium, the hydrostatic pressure exerted by the column of solution just balances the tendency of the solvent to acquire greater disorder.

The same van't Hoff responsible for the *i* factor showed that the osmotic pressure of a nonelectrolyte solution is related to the molarity of the solute:

$$\Pi = iRT \times \text{molarity} \qquad (7)$$

where *R* is the gas constant and *T* is the absolute temperature. This expression is now known as the **van't Hoff equation.** It follows that the osmotic pressure of a solution of any 0.010 M nonelectrolyte solution at 25°C (298 K) is

$$\Pi = 1 \times (0.0821 \text{ L·atm/(K·mol)}) \times (298 \text{ K}) \times (0.010 \text{ mol/L}) = 0.24 \text{ atm}$$

This pressure is enough to push a column of water to a height of over 2 m (the height needed to exert a hydrostatic pressure of 0.24 atm). The van't Hoff equation is also used to determine the molar mass of a solute from osmotic pressure measurements. This technique, which is called **osmometry,** is very sensitive and is commonly used to determine large molar masses, such as those of polymer molecules.

Osmosis is the flow of solvent through a semipermeable membrane into a solution. The osmotic pressure is proportional to the molar concentration of the solute.

Toolbox 12.2 *How to use osmometry*

This Toolbox shows how to use osmometry to determine molar mass.

Conceptual Basis

In osmometry, a solution containing a known mass of a compound is placed in a tall tube with a semipermeable membrane at the bottom. The tube is then placed into a container of the pure solvent (Fig. 12.31). Because only solvent molecules can pass through the semipermeable membrane, the solvent flows through the membrane into the solution, thereby causing its level to rise in the tube to a height that depends on the molarity of the solute. Because the mass of solute is known, its molar mass can be calculated from the molarity.

Procedure

Prepare a solution by dissolving a known mass of solute in sufficient solvent to make a solution of known volume. Determine the height of the column of solution at its maximum.

Step 1. From the measured osmotic pressure (or height of solution), use the van't Hoff equation (with $i = 1$) to calculate the molarity of the solute:

$$\text{Molarity} = \frac{\text{osmotic pressure}}{RT} = \frac{\Pi}{RT}$$

Use the gas constant in liters and the same pressure units as those used for the osmotic pressure (Table 5.3).

Step 2. Use the molarity as a conversion factor to calculate the number of moles of solute in the solution.

$$\text{Moles of solute} = \text{volume} \times \text{molarity} \qquad n = V \times M$$

Step 3. We now know the mass of solute and the number of moles of solute present. Therefore, to find the molar mass of the solute, divide the mass of solute by the number of moles of solute.

This procedure is illustrated in Example 12.11.

To calculate the osmotic pressure from the height of the column of liquid, use the hydrostatic pressure exerted by a column of height, h, of a solution of density d, like that in Fig. 12.31:

$$\text{Osmotic pressure} = g \times \text{height} \times \text{density} \qquad \Pi = ghd$$

where g is the acceleration of free fall (inside the back cover).

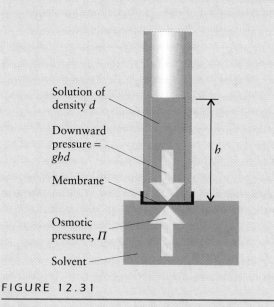

Solution of density d

Downward pressure $= ghd$

h

Membrane

Osmotic pressure, Π

Solvent

FIGURE 12.31

The height of a column of liquid, h, and its density, d, along with the acceleration of free fall, g, are used to measure the osmotic pressure a solution exerts, Π.

Example 12.11 Using osmometry to determine molar mass

The osmotic pressure due to 2.20 g of polyethylene (PE), a molecular compound, dissolved in enough benzene to produce 100. mL of solution was 1.10×10^{-2} atm at 25°C. Calculate the average molar mass of the polymer.

Strategy Use the procedure in Toolbox 12.2. The osmotic pressure is given in atmospheres, so use R in liter-atmospheres per kelvin-mole. The value of i is 1, because the polymer is a molecular compound.

Solution **Step 1.** The molarity of the solution is

$$\text{Molarity} = \frac{\overbrace{1.10 \times 10^{-2} \text{ atm}}^{\Pi}}{\underbrace{1}_{i} \times \underbrace{(0.082\ 06 \text{ L} \cdot \text{atm/K} \cdot \text{mol})}_{R} \times \underbrace{(298 \text{ K})}_{T}}$$

$$= \frac{1.10 \times 10^{-2}}{0.082\ 06 \times 298} \text{ L/mol} = \frac{1.10 \times 10^{-2}}{0.082\ 06 \times 298} \text{ mol/L}$$

Step 2. The number of moles of PE in 100. mL (0.100 L) of solution is

$$\text{Moles of PE} = \overbrace{(0.100 \text{ L})}^{V} \times \left(\overbrace{\frac{1.10 \times 10^{-2}}{0.082\ 06 \times 298} \text{ mol/L}}^{M} \right)$$

$$= \frac{0.100 \times 1.10 \times 10^{-2}}{0.082\ 06 \times 298} \text{ mol}$$

(This expression evaluates to 4.5×10^{-5} mol PE.)
Step 3. Because the mass of this amount of PE is 2.20 g, we conclude that

$$\text{Average molar mass of PE} = \frac{\overbrace{2.20 \text{ g}}^{m}}{\underbrace{\{(0.100 \times 1.10 \times 10^{-2})/(0.082\ 06 \times 298)\} \text{ mol}}_{n}}$$

$$= \frac{2.20 \times 0.082\ 06 \times 298}{0.100 \times 1.00 \times 10^{-2}} \text{ g/mol} = 4.9 \times 10^4 \text{ g/mol}$$

The molar mass of the polymer is therefore 49 kg/mol.

Self-Test 12.12A The osmotic pressure of 3.0 g of polystyrene dissolved in enough benzene to produce 150. mL of solution was 1.21 kPa at 25°C. Calculate the average molar mass of the sample of polystyrene.

[*Answer:* 41 kg/mol]

Self-Test 12.12B The osmotic pressure of 1.50 g of polymethyl methacrylate dissolved in enough methylbenzene to produce 175 mL of solution was 2.11 kPa at 20°C. Calculate the average molar mass of the sample of polymethyl methacrylate.

In **reverse osmosis,** a pressure greater than the osmotic pressure is applied to the solution side of the semipermeable membrane. This application of pressure increases the rate at which solvent molecules leave the solution and consequently reverses the flow of solvent, forcing it to flow from the solution to pure solvent. Reverse osmosis is used to remove salts from seawater to produce fresh water for drinking and irrigation. The water is almost literally squeezed out of the salt solution through the membrane. The technological challenge is to fabricate new membranes that are strong enough to withstand high pressures and that do not easily become clogged. Commercial plants use cellulose acetate at pressures as high as 70 atm. Each cubic meter of membrane can produce about 250 000 L of pure water a day.

Osmometry is used to determine the molar masses of polymers and natural macromolecules; osmosis helps to transport nutrients in plants; reverse osmosis is used in water purification.

Skills You Should Have Mastered

Conceptual

☐ 1. Explain the basis for the like-dissolves-like rule, Section 12.4.

☐ 2. Explain how solutions result from the tendency of energy and matter to disperse, Section 12.10.

☐ 3. Explain how a nonvolatile solute lowers the vapor pressure, raises the boiling point, and lowers the freezing point of a solvent, Sections 12.12 and 12.13.

☐ 4. Interpret a measured value of the van't Hoff *i* factor, Section 12.13.

☐ 5. Calculate the amount of solute present in a given mass of solvent from the molality, Toolbox 12.1 and Example 12.6.

☐ 6. Convert between molality and mole fraction or molarity, Toolbox 12.1 and Examples 12.7 and 12.8.

☐ 7. Calculate the vapor pressure of a solvent in a solution by using Raoult's law, Example 12.9.

☐ 8. Find the molar mass of a substance from a freezing-point depression, Example 12.10.

☐ 9. Find the molar mass of a substance from its osmotic pressure in a solution, Toolbox 12.2 and Example 12.11.

Problem-Solving

☐ 1. Predict the relative solubility of solutes in a given solvent, Example 12.1.

☐ 2. Calculate the solubility of a gas at a given pressure, given its Henry's law constant, Example 12.2.

☐ 3. Predict the relative hydration enthalpies of two ions, Example 12.3.

☐ 4. Calculate the mole fraction and molality of a solute, given the masses of solute and solvent, Examples 12.4 and 12.5.

Descriptive

☐ 1. Describe the formation of a solution at the molecular level, Sections 12.1 and 12.2.

☐ 2. Describe how surfactants allow grease to be suspended in water, Section 12.4.

☐ 3. Identify the different types of colloids and describe the role of Brownian motion in stabilizing colloids, Section 12.5.

☐ 4. Describe the processes of osmosis and reverse osmosis, Section 12.14.

Exercises

Solubility

12.1 Explain why CH_3OH is miscible (soluble in all proportions) in water but not completely miscible in toluene, $C_6H_5CH_3$.

12.2 Explain why calcium hydrogen phosphate is soluble in water, whereas calcium phosphate is not.

12.3 Which would be the better solvent, water or benzene, for each of the following: (a) KCl; (b) CCl_4; (c) CH_3COOH?

12.4 Which would be the better solvent, H_2O or CCl_4, for each of the following: (a) NH_3; (b) HCl; (c) I_2?

12.5 Which of the following compounds will have the higher solubility in water? (a) SF_6 or SF_4; (b) IF_5 or AsF_5.

12.6 Which solute is likely to have the greater solubility in benzene? (a) CCl_4 or CH_3OH; (b) H_2CO (formaldehyde) or C_8H_{18} (octane).

12.7 The following groups are found in some organic molecules. Which are hydrophilic and which are hydrophobic: (a) $-NH_2$; (b) $-CH_3$; (c) $-COOH$?

12.8 The following groups are found in some organic molecules. Which are hydrophilic and which are hydrophobic: (a) $-OH$; (b) $-CH_2CH_3$; (c) $-CONH_2$?

12.9 (a) Why is grease soluble in gasoline but not in water? (b) How can grease be suspended in water by the addition of a soap? Draw a molecular picture to show how the detergent suspends the grease; identify the hydrophilic and hydrophobic regions of the detergent molecules.

12.10 Water is insoluble in oil, but when detergent is added to the oil, water becomes suspended in the oil. (a) Explain the role of the detergent in this process. (b) The micelles that form in the suspension of water and detergent in oil are called inverted micelles. Draw a picture of one of these inverted micelles, showing the locations of the detergent, water, and oil molecules, and identifying the hydrophilic and hydrophobic regions of the detergent molecules.

12.11 Classify these foods by type of colloid: (a) yogurt; (b) marshmallows.

12.12 Classify these foods by type of colloid: (a) orange juice; (b) ice cream.

Ice cream is a type of colloid.

Gas Solubility

12.13 State the molar solubility (in moles per liter) in water of (a) O_2 at 50. kPa; (b) CO_2 at 500. Torr; (c) CO_2 at 0.10 atm. The temperature in each case is 20.°C and the pressures are partial pressures of the gas. Use the information in Table 12.4.

12.14 Calculate the aqueous solubility (in milligrams per liter) of (a) air at 1.0 atm; (b) helium at 1.0 atm; (c) helium at 25 kPa. The temperature is 20.°C in each case and the pressures are partial pressures of the gas. Use the information in Table 12.4.

12.15 A soft drink is made by dissolving CO_2 at 3.00 atm in a flavored solution and sealing the solution in an aluminum can at 20.°C. What volume of CO_2 is released when a 355-mL can

is opened to 1.00 atm at 20.°C and all the CO_2 is allowed to escape?

12.16 The minimum mass concentration of oxygen required for fish life is 4 mg/L. (a) Assuming the density of the solution to be 1.00 g/mL, express this concentration in parts per million (mg/kg; see Table 12.2). (b) What is the minimum partial pressure of O_2 that would supply this concentration in water at 20.°C? (c) What is the minimum atmospheric pressure that would give this partial pressure, assuming oxygen exerts about 21% of the atmospheric pressure?

12.17 The carbon dioxide gas dissolved in a sample of water in a partially filled, sealed container has reached equilibrium with its partial pressure in the air above the solution. Explain what happens to the solubility of the CO_2 if (a) the partial pressure of the CO_2 gas is doubled by the addition of more CO_2; (b) the total pressure of the gas above the liquid is doubled by the addition of nitrogen.

12.18 Explain what happens to the solubility of the CO_2 in Exercise 12.17 if (a) the partial pressure of the CO_2 gas is increased by compressing the air to a third of its original volume; (b) the temperature is raised.

Enthalpy of Solution

12.19 Describe and explain the trend in the ion hydration enthalpies for the halide ions. See Table 12.7.

12.20 Describe and explain the trend in ion hydration enthalpies for the isoelectronic cations Na^+, Mg^{2+}, and Al^{3+}. See Table 12.7.

12.21 The enthalpy of solution of lithium sulfate in water is negative. Which is larger for lithium sulfate, the lattice enthalpy or the enthalpy of hydration?

12.22 Ammonium nitrate dissolves endothermically in water. Which is larger for NH_4NO_3, the lattice enthalpy or the enthalpy of hydration?

12.23 Calculate the heat evolved or absorbed when 10.0 g (a) NaCl; (b) NaBr; (c) $AlCl_3$; (d) NH_4NO_3 is dissolved in 100.0 g of water. Assume that the enthalpies of solution in Table 12.5 are applicable.

12.24 Determine the temperature change when 10.0 g of (a) KCl; (b) $MgBr_2$; (c) KNO_3; (d) NaOH is dissolved in 100.0 g of water. Assume that the specific heat capacity of the solution is 4.18 J/K·g and that the enthalpies of solution in Table 12.5 are applicable.

12.25 Estimate the enthalpy of solution of $SrCl_2$ from the data for the ion enthalpies of hydration and the lattice enthalpy in Tables 12.7 and 8.1.

12.26 (a) Calculate the enthalpy of hydration of Br^- by using the appropriate data from Table 12.6 and the enthalpy of hydration of H^+ (−1130. kJ/mol). (b) Use the value obtained in (a) to determine the enthalpy of hydration of Rb^+, given that the enthalpy of solution of RbBr is $+22$ kJ/mol and its lattice enthalpy is 651 kJ/mol.

12.27 Considering the relative ability of ions to become hydrated in aqueous solution, which solution do you expect to be more disordered, 0.010 M LiCl(aq) or 0.010 M CsCl(aq)?

12.28 Considering the relative ability of ions to become hydrated in aqueous solution, which solution do you expect to be more disordered, 0.010 M $NaNO_3$(aq) or 0.010 M $AgNO_3$(aq)?

12.29 Describe the changes that take place in the energy and disorder of (a) the system and (b) the surroundings when a solute dissolves exothermically.

12.30 Describe the changes that take place in the energy and disorder of (a) the system and (b) the surroundings when a solute dissolves endothermically.

Measures of Concentration

See Table 12.2 for definitions of concentration units.

12.31 Determine the mass percentage of each solute in the following solutions: (a) 4.0 g of NaCl dissolved in enough water to form a total of 100. g of aqueous solution; (b) 4.0 g of NaCl dissolved in 100. g of water; (c) 1.66 g of $C_{12}H_{22}O_{11}$ dissolved in 200. g of water.

12.32 What mass (in grams) of each solute should be added to make the following solutions: (a) to 200. g water to form a 3.0% by mass KBr solution; (b) to 25.0 g water to form a 6.0% by mass $AgNO_3$ solution; (c) to 500. g water to form a 10.5% by mass C_2H_5OH solution?

12.33 Calculate the mole fraction of each component in the following solutions: (a) 25.0 g of water and 50.0 g of ethanol, C_2H_5OH; (b) 25.0 g of water and 50.0 g of methanol, CH_3OH.

12.34 Calculate the mole fraction of each component in the following solutions: (a) 25.0 g of benzene, C_6H_6, and 50.0 g of toluene, $C_6H_5CH_3$; (b) 25.0 g of benzene, 10.0 g of carbon tetrachloride, CCl_4, and 50.0 g of naphthalene, $C_{10}H_8$.

12.35 Calculate the molality of the solute in each of the following solutions: (a) 10.0 g of NaCl dissolved in 250. g water; (b) 0.48 mol of KOH dissolved in 50.0 g water; (c) 1.94 g of urea, $CO(NH_2)_2$, dissolved in 200. g water.

12.36 (a) What mass (in grams) of NaOH must be mixed with 250. g of water to prepare a 0.22 m NaOH solution? (b) Calculate the amount (in moles) of ethylene glycol, HOC_2H_4OH, that should be added to 2.0 kg of water to prepare 0.44 m HOC_2H_4OH(aq).

12.37 Calculate the mass of the solute $KClO_3$ required (a) to dissolve in 20.0 g of water to prepare a 3.0% by mass

$KClO_3$(aq) solution; (b) to dissolve in 20.0 g of water to prepare 3.0 m $KClO_3$(aq).

12.38 Calculate the mass of the solute $Na_2Cr_2O_7$ required (a) to dissolve in 200. g of water to prepare a 1.5% by mass $Na_2Cr_2O_7$(aq) solution; (b) to dissolve in 200. g water to prepare 1.5 m $Na_2Cr_2O_7$(aq).

12.39 In a laboratory exercise, a student mixes 25.0 g of ethanol, C_2H_5OH, with 150. g of water. (a) What is the mole fraction of ethanol in the solution? (b) What is the molality of ethanol in the solution?

12.40 In the laboratories of an oil company, a research chemist dissolved 1.0 g of octane, C_8H_{18}, in 50.0 g of benzene, C_6H_6. (a) What is the mole fraction of octane in the solution? (b) What is the molality of octane in the solution?

12.41 Calculate the mole fractions of the cations, anions, and water in (a) 0.10 m NaCl(aq); (b) 0.20 m Na_2CO_3(aq). Assume complete dissociation of the ionic solute.

12.42 What are the mole fractions of the cations and anions in (a) 0.10 m $MgSO_4$(aq); (b) 0.55% by mass $MgBr_2$? Assume complete dissociation of the ionic solute.

12.43 The density of 0.35 M $(NH_4)_2SO_4$(aq) is 1.027 g/mL. Determine (a) the molality; (b) the mole fraction of ammonium sulfate in the solution.

12.44 A 5.00% K_3PO_4 aqueous solution has a density of 1.043 g/mL. Determine (a) the molality; (b) the molarity of potassium sulfate in solution.

Vapor-Pressure Lowering

12.45 Calculate the vapor pressure of the solvent in each solution listed below. Use the data in Table 5.4 for the vapor pressure of water at various temperatures. (a) An aqueous solution at 100.0°C in which the mole fraction of sucrose is 0.100. (b) An aqueous solution at 100.0°C in which the molality of sucrose is 0.100 mol/kg.

12.46 What is the vapor pressure of the solvent in each of the solutions listed below? Use the data in Table 5.4 for the vapor pressure of water at various temperatures. (a) The mole fraction of glucose is 0.050 in an aqueous solution at 80.0°C. (b) 0.10 m urea, $CO(NH_2)_2$, at 25°C, where $CO(NH_2)_2$ is a nonelectrolyte.

12.47 (a) Calculate the vapor pressure of a 1.0% by mass aqueous ethylene glycol, $C_2H_4(OH)_2$, solution at 0°C. (b) What is the vapor pressure of 0.10 m NaOH(aq) at 80.0°C? (c) Predict the change in the vapor pressure of water when 6.6 g of $CO(NH_2)_2$ is dissolved in 100. g water at 10.°C. For data, see Tables 5.4 and 12.2.

12.48 (a) Calculate the change in vapor pressure from that of pure water for an aqueous solution at 40.0°C in which the mole fraction of fructose is 0.22. (b) Predict the vapor pressure of a saturated magnesium fluoride solution at 20.0°C. The solubility of MgF_2 at 20.°C is 8 mg/100 g water. (c) What is the vapor pressure of 0.010 *m* $Fe(NO_3)_3$(aq) at 0°C? For data, see Tables 5.4 and 12.2.

12.49 Two beakers, one containing 0.010 *m* NaCl(aq) and the other containing pure water, are placed inside a bell jar and sealed. The beakers are left until the water vapor has come to equilibrium with any liquid in the beakers. The levels of the liquids at the beginning of the experiment are the same, as pictured below. Draw the level of the liquid in each beaker after equilibrium has been reached. Explain your reasoning.

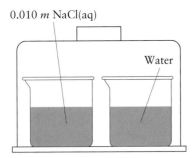

0.010 *m* NaCl(aq)

Water

12.50 Two beakers, one containing 0.010 *m* NaCl(aq) and the other containing 0.010 *m* $AlCl_3$(aq), are placed inside a bell jar and sealed. The beakers are left until the water vapor has come to equilibrium with any liquid in the beakers. The levels of the liquids at the beginning of the experiment are the same, as pictured below. Draw the level of the liquid in each beaker after equilibrium has been reached. Explain your reasoning.

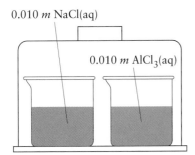

0.010 *m* NaCl(aq)

0.010 *m* $AlCl_3$(aq)

12.51 When 8.05 g of an unknown compound X was dissolved in 100. g of benzene, C_6H_6, the vapor pressure of the benzene decreased from 100. Torr to 94.8 Torr at 26°C. What is (a) the mole fraction and (b) the molar mass of X?

12.52 The normal boiling point of ethanol, C_2H_5OH, is 78.4°C. When 9.15 g of a soluble nonelectrolyte was dissolved in 100. g of ethanol, the vapor pressure of the solution at that temperature was 740. Torr. (a) What are the

mole fractions of ethanol and solute? (b) What is the molar mass of the solute?

Boiling-Point Elevation and Freezing-Point Depression

12.53 Estimate the boiling-point elevation and the normal boiling points of (a) 0.10 *m* $C_{12}H_{22}O_{11}$(aq); (b) 0.22 *m* NaCl(aq); (c) a saturated solution of LiF, of solubility 230. mg/100.0 g water at 100.0°C. Assume complete dissociation of ionic solutes.

12.54 Estimate the boiling-point elevation and the normal boiling points of (a) 0.22 *m* $CaCl_2$(aq); (b) a saturated solution of Li_2CO_3, of solubility 0.72 mg/100.0 g water at 100.0°C; (c) 1.7% by mass $CO(NH_2)_2$(aq). Assume complete dissociation of ionic solutes.

12.55 A 1.05-g sample of a molecular compound is dissolved in 100.0 g of carbon tetrachloride, CCl_4. The normal boiling point of the solution is 61.51°C; the normal boiling point of CCl_4 is 61.20°C. What is the molar mass of the compound?

12.56 When 2.25 g of an unknown compound is dissolved in 150.0 g of cyclohexane, the boiling point increased by 0.481°C. Determine the molar mass of the compound.

12.57 Estimate the freezing-point depression and the freezing point of each of the following aqueous solutions: (a) 0.10 *m* $C_{12}H_{22}O_{11}$(aq); (b) 0.22 *m* NaCl(aq); (c) a saturated solution of LiF, of solubility 120. mg/100.0 g water at 0°C. Assume complete dissociation of ionic solutes.

12.58 Estimate the freezing point of the following aqueous solutions: (a) 0.10 *m* $C_6H_{12}O_6$(aq); (b) 0.22 *m* $CaCl_2$(aq); (c) a saturated solution of Li_2CO_3, of solubility 1.54 mg/100.0 g water at 0°C. Assume complete dissociation of ionic solutes.

12.59 A 1.14-g sample of a molecular substance dissolved in 100.0 g camphor (freezing point 179.8°C) freezes at 177.3°C. What is the molar mass of the substance?

12.60 When 2.11 g of a nonpolar solute was dissolved in 50.0 g phenol, the freezing point of the phenol was lowered by 1.753°C. Calculate the molar mass of the solute.

12.61 (a) What would be the freezing point of a benzene solution that boils at 82.0°C, rather than 80.1°C, the normal boiling point of pure benzene? The freezing point of benzene is 5.5°C. (b) An aqueous solution freezes at −3.04°C. What is the molality of the solution?

12.62 (a) A benzene solution has a vapor pressure of 740 Torr at 80.1°C, the normal boiling point of pure benzene. What is the expected freezing point of the solution? The freezing point of benzene is 5.5°C. (b) How many moles of $CO(NH_2)_2$ are present in 1.20 kg water, given that the freezing point of the solution is −4.02°C?

12.63 An aqueous solution that is 1.00% NaCl by mass has a freezing point of −0.593°C. (a) Estimate the van't Hoff *i* factor from the data. (b) Calculate the percentage dissociation of NaCl in this solution. (*Hint:* The molality calculated from the freezing-point depression is the sum of the molalities of the undissociated ion pairs, the cations, and the anions.)

12.64 A 1.00% by mass $MgSO_4(aq)$ solution has a freezing point of $-0.192°C$. (a) Estimate the van't Hoff i factor for the solution. (b) Calculate the percentage dissociation of the $MgSO_4$ in this solution. (See the hint in Exercise 12.63.)

12.65 Determine the freezing point of a 0.10 mol/kg aqueous solution of a weak electrolyte that is 7.5% dissociated into two ions. (See the hint in Exercise 12.63.)

12.66 A 0.124 m $CCl_3COOH(aq)$ solution has a freezing point of $-0.423°C$. What is the percentage deprotonation of the acid? (See the hint in Exercise 12.63.)

Osmosis and Osmometry

12.67 What is the osmotic pressure of (a) 0.010 M $C_{12}H_{22}O_{11}(aq)$? (b) 1.0 M $HCl(aq)$? (c) 0.010 M $CaCl_2(aq)$ at 20.°C?

12.68 Which of the following solutions has the highest osmotic pressure at 50.0°C: (a) 0.10. M $KCl(aq)$; (b) 0.60 M $CO(NH_2)_2(aq)$; (c) 0.30 M $K_2SO_4(aq)$? Justify your answer by calculating the osmotic pressure of each solution.

12.69 A 0.40-g sample of a polypeptide dissolved in 1.0 L of an aqueous solution at 27°C has an osmotic pressure of 3.74 Torr. What is the molar mass of the polypeptide?

12.70 When 0.10 g of insulin is dissolved in 200.0 mL of water, the osmotic pressure is 2.30 Torr at 20.0°C. What is the molar mass of insulin?

12.71 A 0.10-g sample of a polymer dissolved in 100.0 mL of toluene has an osmotic pressure of 5.4 Torr at 20.°C. What is the molar mass of the polymer?

12.72 A solution prepared by adding 0.50 g of a polymer to 200.0 mL of toluene (an organic solvent) showed an osmotic pressure of 0.582 Torr at 20.0°C. What is the molar mass of the polymer?

12.73 Calculate the osmotic pressure at 20.0°C of each of the solutions listed below; assume complete dissociation for any ionic solutes. (a) 0.050 M $C_{12}H_{22}O_{11}(aq)$; (b) 0.0010 M $NaCl(aq)$; (c) a saturated solution of AgCN of solubility 2.3×10^{-5} g/100.0 g water.

12.74 Calculate the osmotic pressure at 20.0°C of the solutions specified below. Assume complete dissociation of ionic compounds. (a) 3.0×10^{-3} M $C_6H_{12}O_6(aq)$; (b) 2.0×10^{-3} M $CaCl_2(aq)$; (c) 0.010 M $K_2SO_4(aq)$.

Supplementary Exercises

12.75 A saturated solution of copper(II) sulfate was prepared by adding an excess of small crystals of copper(II) sulfate to a

flask of water. The flask, which contained the saturated solution and many small crystals of copper(II) sulfate, was sealed and placed in a cabinet for several months. When it was removed, most of the small crystals were gone and several large crystals had appeared. However, the concentration of copper(II) sulfate in the solution was the same. Explain, describing the events that occur at the molecular level, how it is possible for the crystals to change without a change in concentration of the solution.

12.76 Consider the enthalpy of solution of (a) NaF; (b) NaCl; (c) NaBr; (d) NaI in Table 12.5. Identify and explain the trend in the enthalpy of solution on proceeding down the group of halides.

12.77 A 10.0% $H_2SO_4(aq)$ solution has a density of 1.07 g/cm³. (a) How many milliliters of solution contain 6.32 g of H_2SO_4? (b) What is the molality of the solution? (c) What mass (in grams) of H_2SO_4 is in 300.0 mL of solution?

12.78 The density of a 16.0% by mass $C_{12}H_{22}O_{11}(aq)$ solution is 1.0635 g/cm³ at 20.0°C. (a) What is the molarity of $C_{12}H_{22}O_{11}(aq)$? (b) What is the vapor pressure of the solution at 20.0°C? (c) Predict the boiling point of the solution. See Table 12.2.

12.79 Nitric acid is purchased from chemical suppliers as a solution that is 70.% HNO_3 by mass. What mass (in grams) of a 70.% $HNO_3(aq)$ solution is needed to prepare 250. g of 2.0 m $HNO_3(aq)$? The density of 70.% $HNO_3(aq)$ is 1.42 g/cm³. See Table 12.2.

12.80 Organic chemists once used freezing-point and boiling-point measurements to determine the molar mass of compounds they had synthesized. When 0.30 g of a nonvolatile solute is dissolved in 30.0 g of CCl_4, the boiling point increases from 76.54°C (the normal boiling point of CCl_4) to 77.19°C. What is the molar mass of the compound?

12.81 When determining a molar mass from freezing-point depression, it is possible to make each of the following errors (among others). In each case, predict whether the error would cause the reported molar mass to be greater or less than the actual molar mass: (a) There was dust on the balance, causing the mass of solute to appear greater than it actually was. (b) The solvent was measured by volume, assuming a density of 1.00 g/cm³, but the solvent was warmer and less dense than assumed. (c) The thermometer was not calibrated accurately, so the temperature of the freezing point was actually 0.5°C higher than recorded. (d) The solution was not stirred sufficiently, so that not all the solute dissolved.

12.82 Intravenous medications are often administered in 5.0% glucose, $C_6H_{12}O_6$, aqueous solutions by mass. What is the osmotic pressure of such solutions at 37°C (body temperature)?

12.83 Catalase, a liver enzyme, dissolves in water. A 10.0-mL solution containing 0.166 g of catalase exhibits an osmotic pressure of 1.2 Torr at 20.°C. What is the molar mass of catalase?

12.84 Interpret the following verse from Coleridge's *The Rime of the Ancient Mariner:*

> Water, water, every where,
> And all the boards did shrink;
> Water, water every where,
> Nor any drop to drink.

12.85 Suggest a reason why saltwater fish die when they are suddenly transferred to a freshwater aquarium.

12.86 (a) Determine the mass of $Na_2CO_3 \cdot 10H_2O$ needed to prepare 1.0 m $Na_2CO_3(aq)$, using 250. g water. (b) Find a general expression for the mass of solute needed to prepare a solution of the solute B of molality m when the hydrated solute has the formula $B \cdot xH_2O$.

12.87 The height of a column of liquid that can be supported by a given pressure can be calculated from the expression at the end of Toolbox 12.2. An aqueous solution of 0.010 g of a protein in 10.0 mL of water at 20.°C shows a 5.22-cm rise in the apparatus shown in Fig. 12.29 (see also Fig. 12.31). Assume the density of the solution to be 0.998 g/cm³. (a) What is the molar mass of the protein? (b) What is the freezing point of the solution? (c) Which colligative property is best for measuring the molar mass of these large molecules? Give reasons for your answer.

12.88 A 0.020 M $C_6H_{12}O_6(aq)$ solution is separated from a 0.050 M $CO(NH_2)_2(aq)$ solution by a semipermeable membrane at 25°C. (a) Which solution has the higher osmotic pressure ($i = 1$ for both solutions)? (b) Which solution becomes more dilute with the passage of H_2O molecules through the membrane? (c) To which solution should an external pressure be applied in order to maintain an equilibrium flow of H_2O molecules across the membrane? (d) What external pressure (in atm) should be applied in (c)?

Applied Exercises

For Exercises 12.89–12.90, see Applying Chemistry: Case Study 12.

12.89 (a) Identify the functional groups shown in panels (a), (c), and (d) in the molecular illustration in Applying Chemistry: Case Study 12. (b) In which panel of the four will the interactions be the strongest? (c) In which will they be the weakest?

12.90 Pentanol, an organic compound with the formula $CH_3CH_2CH_2CH_2CH_2OH$, has molecules with a nonpolar hydrocarbon chain of medium length attached to a polar —OH group. It is insoluble in water. However, if detergent is added, the molecules of the three compounds form a layered structure that suspends the pentanol in the water. Propose a description of the layered structure, giving the arrangement of the water, pentanol, and detergent molecules.

12.91 The volume of blood in the body of a certain deep-sea diver is about 6.00 L. Blood cells make up about 55% of the blood volume, and the remaining 45% is the aqueous solution called plasma. What is the maximum volume of nitrogen measured at 1.00 atm and 37°C that could dissolve in the diver's blood plasma at a depth of 93 m, where the pressure is 10.0 atm? (This is the volume that could come out of solution suddenly, causing the bends if the diver ascended too quickly.) Assume that the Henry's law constant for nitrogen at 37°C (body temperature) is 5.8×10^{-4} mol/L·atm.

12.92 The legal limit for mercury in natural waters and fish sold for food is 0.5 ppm. Mercury pollution in Lake Saint Clair, which lies between the United States and Canada, rose to 6 ppm in the 1970s but has since dropped below 1 ppm. How many grams of mercury would you ingest if you ate 8.0 ounces of a fish containing 1 ppm mercury?

12.93 During a forensic analysis, a pair of amino acids is separated in a column in which the stationary phase is saturated with water and the carrier solvent is methanol, CH_3OH. The more polar the acid, the more strongly it is absorbed by the stationary phase. The amino acids that were separated in this column are (a) $HOOC—CHNH_2—CH_2—COOH$ and (b) $HOOC—CHNH_2—CH(CH_3)_2$. Which amino acid would you expect to travel farther? Explain your reasoning. See Investigating Matter 12.1.

12.94 Which pesticide represented in the chromatogram in Investigating Matter 12.1 is present in the greatest amount?

Integrated Exercises

12.95 The average molar mass of a sample of polypropylene was determined by measuring the osmotic pressure of a solution of 3.16 g of the polypropylene in 1.00 L of benzene. A pressure of 0.0112 atm was observed at 25°C. (a) What is the average molar mass of the polymer? (b) What is the average number of propene monomer units in each chain? (c) Assuming that this sample consists only of linear chains and that the carbon-carbon bond lengths in the polymer are close to their average value, what is the average chain length, in nanometers?

12.96 The combustion analysis of L-carnitine, an organic compound thought to build muscle strength, yielded 52.15% C, 9.38% H, 8.69% N, and 29.78% O. The osmotic pressure of a 100.00-mL aqueous solution of 0.322 g of L-carnitine was found to be 0.501 atm at 32°C. Assuming that L-carnitine does not ionize in methanol, determine (a) the molar mass of L-carnitine; (b) the molecular formula of L-carnitine.

12.97 A 0.010 M aqueous solution of acetic acid has an osmotic pressure of 0.24 atm. However, a 0.010 M solution of acetic acid in acetone has an osmotic pressure only about half as great. Use drawings of the molecular structure of two acetic acid molecules to show how their interaction could reduce the apparent number of solute species.

12.98 What volume of 0.010 M NaOH(aq) is required to react completely with 30.0 g of an aqueous acetic acid solution in which the mole fraction of acetic acid is 0.15?

12.99 A 10.0-g sample of *p*-dichlorobenzene, a component of mothballs, is dissolved in 80.0 g benzene, C_6H_6. The freezing point of the solution is 1.20°C. The freezing point of pure benzene is 5.48°C. (a) What is an approximate molar mass of *p*-dichlorobenzene? (b) An elemental analysis of *p*-dichlorobenzene indicated that the empirical formula is C_3H_2Cl. What is the molecular formula of *p*-dichlorobenzene? (c) Using the atomic molar masses from the periodic table, calculate a more accurate molar mass of *p*-dichlorobenzene.

12.100 An elemental analysis of adrenaline (epinephrine) is 59.0% carbon, 26.2% oxygen, 7.15% hydrogen, and 7.65% nitrogen by mass. When 0.64 g of adrenaline was dissolved in 36.0 g benzene, the freezing point decreased by 0.50°C. (a) Determine the empirical formula of adrenaline. (b) What is the molar mass of adrenaline? (c) Deduce the molecular formula of adrenaline.

Connection 3

Playing football can be exhausting in any climate, but it is particularly tough in a hot climate such as that of Florida, where water is lost at a rapid rate. The coach of the University of Florida football team noticed that his players regularly became fatigued by the second half of the game. However, if they drank fruit juices, which contain both sugar for energy and water for rehydration, the players became nauseated. The coach turned to two campus physicians for help, and out of that collaboration came a drink they named Gatorade® after their team, the Florida Gators. From that drink arose the entire industry of sport drinks.

Our bodies are largely aqueous solutions. Fluids fill the blood vessels and the millions of cells in our tissues. Extracellular fluid circulates between the cells. Intravenous solutions must have the same concentration of total dissolved solute as blood, in order to maintain the normal osmotic pressure of blood. Such solutions are called *isotonic* (Section 12.14). A *hypotonic* solution would have a concentration of dissolved solute that is too dilute. A hypotonic solution has a low osmotic pressure and would cause blood cells to swell up and possibly burst. A *hypertonic* solution is too concentrated and would cause blood cells to shrivel.

When we perspire, extracellular fluid—mainly water—is lost, but some of the electrolytes—ions such as sodium, potassium, and chloride ions—are lost as well. Extensive perspiring reduces the volume of fluid in the body but leaves some of the dissolved salts in the body. As a result, the concentration of salts in the extracellular fluid is increased and the increased osmotic pressure causes water to flow out of the blood vessels into the extracellular fluids. The lowered blood volume leads to reduced blood pressure and impaired athletic performance. To ensure that tissues are kept hydrated during exercise, athletes need to drink fluids frequently. Because water in an isotonic solution is absorbed more rapidly by the small intestine than pure water is, an isotonic solution is recommended for a drink to be used during or immediately after vigorous exercise. However, exercising athletes do not always feel thirsty. Therefore, a sport drink needs to taste good as well as provide the needed nutrients.

In addition to water, electrolytes, and flavor, a sport drink must also provide fuel for the body. During vigorous exercise, we may burn about 2000 kJ/h, depending on our mass and the level of exertion (see the table in Applying Chemistry: Case Study 6). The body has only a small amount of fuel in readily available form, so it needs continued input to maintain a high level of activity. Fats and proteins require too much time to digest, so carbohydrates must be used. In addition, the carbohydrates need to be in the form of simple sugars, the food that the body can digest most easily and rapidly. Complex carbohydrates such as starches (Section 11.17) must first be broken into the individual sugar monomers before they can be used by the body to generate energy. However, simple sugars such as glucose and fructose can be easily and quickly oxidized by cells.

Athletes involved in strenuous exercise must be careful to avoid dehydration. Sport drinks were invented to provide a beverage that would be quickly absorbed and that would be sufficiently appetizing to encourage the intake of fluids.

Sport drinks are formulated to hydrate tissues during vigorous exercise.

Sport drinks contain sugars for energy and sodium and potassium salts for electrolyte balance. However, if the concentration of sugars or salts is too high, the solution is hypertonic and has an osmotic pressure so high that water flows in the wrong direction, from the bloodstream into the small intestine. The result is temporarily increased dehydration and, often, nausea. Therefore, the concentration of carbohydrates needs to be kept below 7% by mass, but high enough to provide energy and a pleasant taste. An effective sport drink provides about 1 kJ/mL. Flavorings such as citric acid, the acid found in citrus fruit, and other natural or artificial flavors are used to make the drink tasty, so that enough will be ingested. A small amount of vegetable gum is usually added to create a colloidal dispersion (Section 12.5), to simulate the appearance of fruit juice. The large, hydrophilic molecules of vegetable gums contain many — OH groups, which can form hydrogen bonds with water that help them to stay suspended (Section 10.3). However, sport drinks usually contain no juice, only sugars and salts.

Applying Your Knowledge

You may need to consult Chapters 10–12 and occasionally earlier chapters, such as Chapters 6 and 9, to answer these questions.

1. Any solution used for intravenous administration or administered to the eyes must be isotonic with body fluids. *Normal saline solution* is an isotonic solution containing 0.9% NaCl by mass. (a) What is the total molarity of the solution, assuming complete dissociation of the NaCl? (Assume that the density of the solution is 1.0 g/mL.) (b) What mass of glucose would you need to make up 500.0 mL of an isotonic solution of glucose in water for administration to a patient?

2. A paramedic treating injuries in a remote area has 300.0 g of a 1% by mass solution of boric acid, $B(OH)_3$, that needs to be

made isotonic (assume $i = 1$ in this solution). What mass of NaCl should be added? Assume that the NaCl will be completely dissociated in the solution and that the density of the solution remains at 1.0 g/mL.

3. One of the main flavoring agents in sport drinks is citric acid, shown below. (a) Identify each functional group in citric acid. (b) Indicate the hybridization of each C atom. (c) Can citric acid undergo hydrogen bonding? (d) Predict from a consideration of intermolecular forces whether citric acid is a gas, liquid, or solid at 25°C and whether it is soluble in water.

$$HOOC \overset{\overset{\displaystyle CH_2}{|}}{\underset{\underset{\displaystyle HO \quad COOH}{|}}{C}} \overset{\displaystyle CH_2}{} COOH$$

Citric acid

4. Find a bottle of a sport drink in a store and copy down the name of each ingredient and the nutritional analysis. Identify the role of the ingredient and, if possible, find its formula and write it. Determine whether the drink is isotonic, hypotonic, or hypertonic (see Question 1).

5. Design your own sport drink. (a) List the ingredients and the mass of each that you would need for a 1-L bottle of the drink. For flavorings and other trace ingredients, simply indicate "trace amount" for the mass. Indicate the reason for adding each ingredient. (b) Give your drink an appealing name. (c) Calculate the total molality of all species in solution (ignoring substances present in trace amounts) and show that your drink is isotonic. Draw up a label for your drink with all the pertinent information, including the caloric value of an 8-oz serving of the drink and its name.

The Rates of Reactions

Good health depends on the complex interplay of large numbers of reactions in the cells of the body. In a healthy body, these reactions take place at the optimal rate, in the right place, and at the right time. They are controlled by enzymes, the biological equivalent of catalysts (substances that make chemical reactions go faster). The part of chemistry concerned with the rates of chemical reactions is called **chemical kinetics.** The term *kinetics* implies motion, and chemical kinetics looks at how reactions take place as well as at how fast they occur. We can use the information obtained from chemical kinetics to study the body's metabolism and to model the effects of pollutants on our atmosphere. The development of new catalysts is another branch of chemical kinetics, one that is crucial to the success of the chemical industry.

Chemical kinetics is like a microscope through which we get insight into how chemical reactions occur. We know how atoms bond together to form molecules. But how do atoms change their partners? What goes on between molecules when a hydrogen molecule meets an iodine molecule in the reaction that produces hydrogen iodide? What goes on when a biologically important molecule contributes to the processes that keep us alive? How is the chemical industry able to get exactly the right product from a reaction that has hundreds of possible outcomes?

St. John's wort is an herb that is thought to create a sense of tranquility. Herbs and other medicines have been used through the ages to cure disease and to relieve pain. In many cases, the medicine is effective because it controls the rates of reactions within the body. In this chapter, we examine the rates of chemical reactions and the mechanisms by which they take place.

CONCENTRATION AND RATE

Rates in chemistry are defined like rates in other fields: as the change in a property divided by the time it takes for that change to occur. For example, the rate at which an automobile travels (its speed) is the distance traveled divided by the time taken (in miles per hour, for instance). In chemistry, we are concerned with how quickly reactants are used up or products formed, so the rate of a reaction is expressed in terms of changes in concentration. In a slow reaction, there is little change in the concentration of reactant during a given period of time, whereas in a fast reaction, there is a large change. Because concentration is easily measured by spectroscopy (see Investigating Matter 9.2), we can follow the rate of a reaction by measuring the light absorbed by a reactant or a product. From these data, we can determine how the concentration of the species changes over time.

13.1 The Definition of Reaction Rate

Suppose we heat a flask containing hydrogen iodide gas at a concentration of 4.00 mmol/L, where 1 mmol = 10^{-3} mol. As the HI decomposes by the reaction

$$2\,HI(g) \longrightarrow H_2(g) + I_2(g)$$

its concentration in the flask decreases. If we represent the molar concentration of HI as [HI], then the **reaction rate** is defined as the change in concentration of HI, $\Delta[HI]$, divided by the time interval, Δt, in which that change occurs:

$$\text{Rate of reaction} = -\frac{\text{change in concentration of reactant}}{\text{time interval}} = -\frac{\Delta[HI]}{\Delta t}$$

Because the concentration of a reactant decreases during a reaction ($\Delta[HI]$ is negative), we include a minus sign in the rate expression to ensure that the rate is positive, which is the normal convention in chemical kinetics. If we follow the concentration of one of the products, such as the iodine, then we would express the rate as

$$\text{Rate of reaction} = \frac{\text{change in concentration of product}}{\text{time interval}} = \frac{\Delta[I_2]}{\Delta t}$$

Because the concentration of a product increases during a reaction ($\Delta[I_2]$ is positive), the rate is already positive.

To determine the rate of the decomposition of hydrogen iodide to hydrogen and iodine gases, we can measure the concentration of HI at a series of times after the start of the reaction and report the data as shown in Table 13.1. Because the data show that the molar concentration of HI decreases from 100. mmol/L to 17 mmol/L in 500.0 s, we would report the reaction rate as

$$\text{Rate of reaction} = -\frac{(17 - 100.)\,(\text{mmol HI})/L}{500.0\ s}$$
$$= 0.17\,(\text{mmol HI})/L \cdot s$$

The rate of a reaction is the change in concentration of a species divided by the time it takes the change to occur. All reaction rates are positive.

Table 13.1 *The concentration of hydrogen iodide at 800 K*

Time, s	Molar concentration, mmol/L
0	100.
100.0	51
200.0	34
300.0	26
400.0	20.
500.0	17
600.0	15
700.0	13
800.0	11
900.0	10.
1000.0	6

The units of this rate would be read as "moles of HI per liter per second."

Example 13.1 *Expressing the reaction rate for different substances*

If the rate of the reaction above is reported as 5.0×10^{-3} (mol HI)/L·s, what is the reaction rate in terms of the concentration of hydrogen?

Strategy The stoichiometric relation between species lets us convert from the rate with respect to one species to the rate with respect to another. Write the chemical equation for the reaction and use the stoichiometric coefficients to write the mole ratio between the two species of interest. Apply that ratio as a conversion factor to the given rate.

Solution The chemical equation is

$$2 \, HI(g) \longrightarrow H_2(g) + I_2(g)$$

so we use 2 mol HI $\rightleftharpoons$ 1 mol H_2 to write a conversion factor. Because H_2 is a product, the rate is positive as calculated:

$$\text{Rate of reaction} = \{5.0 \times 10^{-3} \, (\text{mol HI})/ \text{L·s}\} \times \frac{1 \, \text{mol} \, H_2}{2 \, \text{mol HI}}$$

$$= 2.5 \times 10^{-3} \, (\text{mol} \, H_2)/\text{L·s}$$

We could report this rate more simply as 2.5 (mmol H_2)/L·s (because 1 mmol = 10^{-3} mol).

Self-Test 13.1A The rate of formation of ammonia in the reaction $N_2(g) + 3 H_2(g) \rightarrow 2 NH_3(g)$ was reported as 1.15 (mol NH_3)/L·h. What is the rate of disappearance of H_2?

[**Answer:** 1.72 (mol H_2)/L·h]

Self-Test 13.1B What is the rate of disappearance of N_2 in the reaction in Self-Test 13.1A?

13.2 The Instantaneous Rate of Reaction

Most reactions slow down as the reactants are used up. Consider, for instance, the decrease in concentration of penicillin stored at 25°C (Table 13.2). Figure 13.1 shows the data graphically, and we see at a glance that the concentration of the reactant changes more rapidly at the beginning of the period than toward the end. At the end of the reaction, all the reactant has disappeared. Similar remarks apply to the concentrations of products. Now, however, their concentrations start off at 0, and rise ever more slowly toward their final value: the rate at which they are produced decreases as the reaction proceeds.

Suppose we want to calculate the reaction rate at 5.0 weeks after the start of the penicillin experiment. The rate changes with time; so, if we used the concentrations at 0 and 10. weeks to calculate the slope, we would get only an *average* value, 5.6 mmol/L per week (Fig. 13.2). We would get a slightly different average value if we used the concentrations at 2.5 and 7.5 weeks to calculate the slope, 6.2 mmol/L per week. To find the rate of reaction at *a given instant* (for instance, at 5.0 weeks), we draw a *tangent* to the graph at the time of interest. For example, the rate at 5.0 weeks is 6.8 mmol/L per week. The slope of this straight line is called the **instantaneous rate** of the reaction at a given time.

Table 13.2 *The concentration of penicillin at 25°C*

Time, weeks	Molar concentration, mmol/L
0	0.076
5.0	0.046
10.0	0.020
15.0	0.010
20.0	0.0050
25.0	0.0020
30.0	0.0010

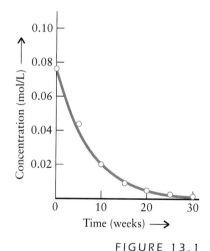

FIGURE 13.1

The activity of penicillin declines over several weeks when it is stored at room temperature in the absence of stabilizers. The shape of this graph of concentration of penicillin as a function of time is typical of the behavior of chemical reactions, although the time span may vary from fractions of a second to years.

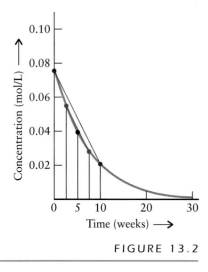

FIGURE 13.2

This graph shows two examples of how the rate of consumption of penicillin can be monitored while it is being stored. The red line shows the average rate calculated from measurements at 0 and 10 weeks, and the blue line shows the average rate calculated from measurements at 2.5 and 7.5 weeks. The instantaneous rate at 5 weeks is the tangent to the curve at that time (not shown).

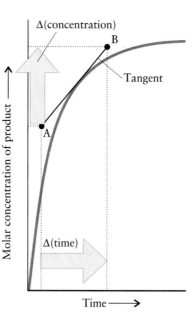

FIGURE 13.3

To calculate the instantaneous reaction rate, we draw the tangent to the curve at the time of interest and then calculate the slope of this tangent. To calculate the slope, we identify any two points, A and B, on the straight line and identify the molar concentrations and times to which they correspond. The slope is then worked out by dividing the difference in concentrations by the difference in times. Notice that this graph shows the concentration of a product.

From now on, whenever we speak of a reaction rate, we shall always mean an instantaneous rate (Fig. 13.3). The instantaneous rate is always reported as a positive quantity, so if the tangent slopes down from left to right (as in Fig. 13.2, where the slope gives the rate of disappearance of reactant), we remove its negative sign.

> *The instantaneous reaction rate is the (positive) slope of a tangent drawn to the graph of concentration as a function of time; it varies as the reaction proceeds.*

13.3 Rate Laws

One way to understand how reactions occur is to identify patterns in the **initial rate of reaction**, the instantaneous rate at the very beginning of the reaction (the tangent at the start). At the start of the reaction, there are no products present and patterns are easier to find.

Suppose we were to measure different masses of dinitrogen pentoxide, N_2O_5, into different flasks and immerse them all in a water bath at 65°C to vaporize the solid. At regular time intervals, we would measure the concentrations of reactants and products in the reaction

$$2 N_2O_5(g) \longrightarrow 4 NO_2(g) + O_2(g)$$

The data might be like that plotted in Fig. 13.4. We can calculate the initial rate of disappearance of N_2O_5 in each flask by drawing the tangents to the curves at $t = 0$ (the black lines in Fig. 13.4). This procedure gives the instantaneous rate at $t = 0$. As the slopes of the black lines show, the initial rate of decomposition

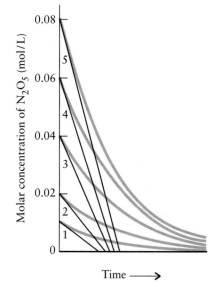

FIGURE 13.4

The initial rate of reaction for the decomposition of N_2O_5 in five experiments. The initial rate of disappearance of a reactant is determined by drawing a tangent to the curve at the start of the reaction.

FIGURE 13.5

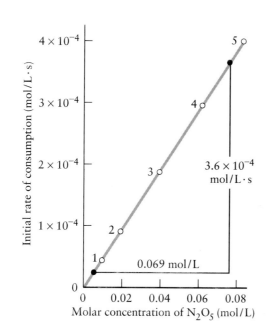

of the vapor is higher in the flasks with higher initial concentrations of N_2O_5. Now we come to the important point. When we plot the initial rates against the corresponding initial concentrations of N_2O_5 (Fig. 13.5), we get a straight line. That is, *the initial rates are directly proportional to the initial concentration of the reactant.* Doubling the initial concentration doubles the initial rate, tripling the concentration triples the rate, and so on. For this reaction, therefore, we can write

$$\text{Initial rate} = k \times \text{initial concentration}$$

Here k is a constant, called the **rate constant** for the reaction. The rate constant must be determined by experiment. For the experiments plotted in Fig. 13.5, the value of k is the slope of the straight line, therefore, $k = 5.2 \times 10^{-3}/s$.

What about the rate later in the reaction? Does it follow the same pattern of initial rates? We take the data for a single experiment, any one of the curves in Fig. 13.4, and calculate the instantaneous rates at a series of times, noting the concentration of N_2O_5 at each of those times. Then we plot the instantaneous rates against the corresponding concentrations. Once again, the result is a straight line (Fig. 13.6). The straight line tells us that the rate is directly proportional to the concentration of N_2O_5 at all stages of the reaction. We therefore conclude that at any stage of the reaction

$$\text{Rate} = k \times \text{concentration}$$

This equation is an example of a **rate law,** an equation expressing the dependence of the instantaneous reaction rate on the concentrations of the species appearing in the chemical equation for the reaction. If we denote the molar concentration of N_2O_5 as $[N_2O_5]$, the explicit form of the rate law for the decomposition of N_2O_5 is

$$\text{Rate of disappearance of } N_2O_5 = k[N_2O_5]$$

Each reaction has its own characteristic rate law and rate constant k. The rate constant is independent of the concentrations of the reactants but depends on the temperature. For the decomposition of N_2O_5 at 65°C, $k = 5.2 \times 10^{-3}/s$.

FIGURE 13.6

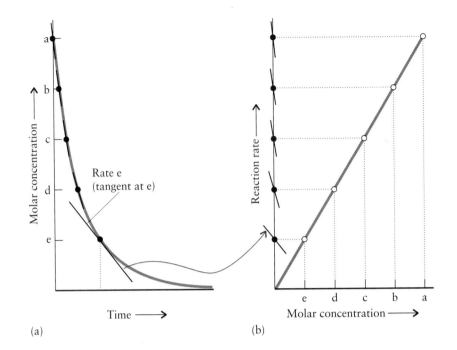

(a) The instantaneous reaction rates for the decomposition of N_2O_5 at five different times during a single experiment are obtained from the slopes of the tangents to the line at each of the five points. (b) When these rates (the slopes) are plotted as a function of the concentration of N_2O_5 remaining, the result is a straight line with a slope equal to the rate constant. In (b), we have indicated the rates by redrawing the tangents.

(a) (b)

When similar measurements are made on the reaction

$$2\,NO_2(g) \longrightarrow 2\,NO(g) + O_2(g)$$

we find that plotting the rate against concentration of NO_2 does not give a straight line. However, plotting the rate as a function of the *square* of the concentration of NO_2 does give a straight line, so the rate is proportional to the square of the concentration:

$$\text{Rate} = k \times (\text{concentration})^2$$

and explicitly for this reaction

$$\text{Rate of disappearance of } NO_2 = k[NO_2]^2$$

where $[NO_2]$ is the molar concentration of NO_2. It is found experimentally that $k = 0.54\ \text{L}/(\text{mol } NO_2)\cdot\text{s}$ at 300°C.

The rates of the decompositions of N_2O_5 and NO_2 have different patterns, one being proportional to the concentration and the other to the square of the concentration. Both rate laws, however, have the form

$$\text{Rate} = k \times (\text{concentration})^a$$

with $a = 1$ for the N_2O_5 reaction and $a = 2$ for the NO_2 reaction. This observation suggests that we can classify reactions according to the form of their rate laws and that we can expect some similarities of behavior for reactions with the same type of rate law. The N_2O_5 reaction is an example of a **first-order reaction,** a reaction for which the rate is proportional to the first power of the concentration (that is, $a = 1$). Doubling the concentration of a reactant in a first-order reaction doubles the reaction rate. The NO_2 reaction is an example of a **second-order reaction,** a reaction for which the rate is proportional to the second power (the square) of the concentration of a substance (that is, $a = 2$). Doubling the concentration of the reactant in any second-order reaction increases the reaction rate by a factor of $2^2 = 4$.

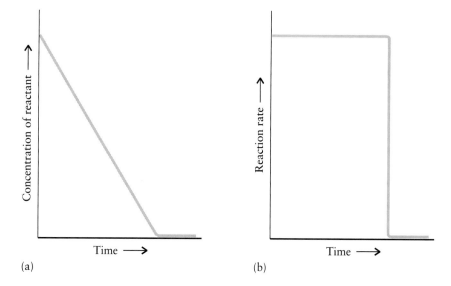

FIGURE 13.7

(a) The concentration of the reactant in a zero-order reaction falls at a constant rate until the reactant is exhausted. (b) The rate of a zero-order reaction is independent of the concentration of the reactant and remains constant until all the reactant has been consumed, when it falls abruptly to 0.

Notice that, although the reactant has a stoichiometric coefficient of 2 in each case, one reaction is first order and the other is second order. In general, the order of a reaction with respect to a reactant is not related to the reactant's stoichiometric coefficient in the balanced equation. *Rate laws are determined by experiment:* except in the special cases we detail later, *they cannot be written from the coefficients in the chemical equation for the reaction.*

Most of the reactions we shall discuss will be either first or second order. However, other orders may occur. For example, ammonia decomposes on a hot platinum wire:

$$2\,NH_3(g) \longrightarrow N_2(g) + 3\,H_2(g) \qquad Rate = k$$

> The order of a reaction with respect to a reactant is an indication of how the reaction takes place, not its stoichiometry (see Section 13.13).

The ammonia decomposes at a constant rate k until it has all disappeared. This reaction is an example of a **zero-order reaction,** a reaction with a rate that is independent of concentration. Zero-order reactions continue at the same rate until all the reactants have been consumed. Then they stop abruptly (Fig. 13.7). They are called zero-order reactions because

$$Rate = k \times (concentration)^0 = k$$

> $x^0 = 1$, whatever the value of x. Try it on your calculator.

Example 13.2 *Identifying the order of a reaction*

A group of students carried out a series of experiments on the reaction

$$S_2O_8{}^{2-}(aq) + 3\,I^-(aq) \longrightarrow 2\,SO_4{}^{2-}(aq) + I_3{}^-(aq)$$

in which iodide ions are oxidized by persulfate ions, $S_2O_8{}^{2-}$. They found that the rate of the reaction doubles when the concentration of iodide ions is doubled while the concentration of persulfate ions is held constant. What is the order with respect to I^- in the rate law?

Strategy In a zero-order reaction, the rate is independent of the concentration; in a first-order reaction, the rate is proportional to the concentration; and in a second-order reaction, the rate is proportional to the square of the concentration. Decide which one of these three possibilities fits the data. Denote the molar concentration of a species X by [X].

Solution When [I$^-$] doubles, the rate doubles. Therefore, the reaction rate is proportional to [I$^-$]. We can conclude that the reaction is first order with respect to the iodide ion.

Self-Test 13.2A The rate of decomposition of dinitrogen oxide, N_2O, in the reaction $2 N_2O(g) \rightarrow 2 N_2(g) + O_2(g)$ falls by a factor of 3 when the concentration of dinitrogen oxide is reduced from 6×10^{-3} mol/L to 2×10^{-3} mol/L. What is the reaction order with respect to N_2O?

[*Answer:* First order]

Self-Test 13.2B The rate of reaction of nitrogen monoxide, NO, with bromine gas in the reaction $2 NO(g) + Br_2(g) \rightarrow 2 NOBr(g)$ falls by a factor of 4 when the concentration of NO is reduced from 2×10^{-2} mol/L to 1×10^{-2} mol/L and the concentration of bromine is held constant. What is the reaction order with respect to NO?

Some reactions have rate laws that depend on the concentrations of more than one substance. An example is the redox reaction between persulfate and iodide ions in Example 13.2. Experiments show that the rate of this reaction is proportional to both the I$^-$ concentration and the concentration of $S_2O_8^{2-}$ ions. If we write their molar concentrations as [I$^-$] and [$S_2O_8^{2-}$], the rate law is

$$\text{Rate} = k[\text{I}^-][\text{S}_2\text{O}_8^{2-}]$$

We say that the reaction is first order with respect to (or "in") $S_2O_8^{2-}$ and first order in I$^-$, because the concentration of each reactant is raised to the power 1. Doubling either the $S_2O_8^{2-}$ ion concentration or the I$^-$ ion concentration doubles the reaction rate. We say that its *overall* order is 2. In general, if

$$\text{Rate} = k[\text{A}]^a[\text{B}]^b \ldots$$

then the **overall order** is the sum of all the powers: $a + b + \ldots$.

The units of k depend on the order of the reaction: the rate of a reaction is always expressed in moles per liter per second (mol/L·s) or related units (such as mol/L·min), but k multiplies various powers of concentration. Therefore, we must assign to k units that cancel the powers of the units of concentration and replace them by mol/L·s. For a first-order reaction, the single concentration on the right of the rate law is in mol/L, and we get the rate units mol/L·s by multiplying that concentration by 1/s; so the units of k are 1/s. For a second-order reaction, the rate constant is multiplied by (mol/L)^2; so, to obtain a rate in mol/L·s, the rate constant k must have units of L/mol·s. For a zero-order reaction, the rate is equal to the rate constant, so k must have the units mol/L·s. The rate laws and rate constants for some reactions are listed in Table 13.3.

Self-Test 13.3A (a) The rate law for the reduction of bromate ions, BrO_3^-, by bromide ions in acidic solution is rate $= k[\text{BrO}_3^-][\text{Br}^-][\text{H}^+]^2$. What are the orders with respect to the reactants and the overall order? (b) What are the units of the rate constant?

[*Answer:* (a) First order in BrO_3^- and Br^-; second order in H^+; fourth order overall; (b) L^3/mol^3·s]

Self-Test 13.3B The rate law for the reaction between bromomethane, CH_3Br, and hydroxide ion in aqueous solution is rate $= k[\text{CH}_3\text{Br}][\text{OH}^-]$. (a) What are the orders with respect to the reactants and the overall order? (b) What are the units of the rate constant?

When more than one reactant is present, we can separate the effects of increasing their concentrations by using the **method of initial rates.** In this method, several rate experiments are conducted, but the concentration of only

Table 13.3 *Rate laws and rate constants*

Reaction	Rate law*	Temperature, K	Rate constant
Gas phase			
$H_2 + I_2 \longrightarrow 2\,\mathbf{HI}$	$k[H_2][I_2]$	500.	4.3×10^{-7} L/mol·s
		600.	4.4×10^{-24}
		700.	6.3×10^{-22}
		800.	2.6
$2\,\mathbf{HI} \longrightarrow H_2 + I_2$	$k[HI]^2$	500.	6.4×10^{-9} L/mol·s
		600.	9.7×10^{-6}
		700.	1.8×10^{-3}
		800.	9.7×10^{-2}
$2\,\mathbf{N_2O_5} \longrightarrow 4\,NO_2 + O_2$	$k[N_2O_5]$	298	3.7×10^{-5}/s
		318	5.1×10^{-4}
		328	1.7×10^{-3}
		338	5.2×10^{-3}
$2\,\mathbf{N_2O} \longrightarrow 2\,N_2 + O_2$	$k[N_2O]$	1000.	0.76/s
		1050.	3.4
$2\,\mathbf{NO_2} \longrightarrow 2\,NO + O_2$	$k[NO_2]^2$	573	0.54 L/mol·s
$\mathbf{C_2H_6} \longrightarrow 2\,CH_3$	$k[C_2H_6]$	973	5.5×10^{-4}/s
cyclopropane $\longrightarrow$ propene	$k[\text{cyclopropane}]$	773	6.7×10^{-4}/s
Aqueous solution			
$\mathbf{H^+} + OH^- \longrightarrow H_2O$	$k[H^+][OH^-]$	298	1.5×10^{11} L/mol·s
$\mathbf{CH_3CH_2Br} + OH^- \longrightarrow CH_3CH_2OH + Br^-$	$k[CH_3CH_2Br][OH^-]$	298	2.8×10^{-4} L/mol·s
$\mathbf{C_{12}H_{22}O_{11}} + H_2O \longrightarrow 2\,C_6H_{12}O_6$	$k[C_{12}H_{22}O_{11}][H^+]$	298	1.8×10^{-4} L/mol·s

*For the rate of consumption or formation of the substance in bold type in the reaction column.

one reactant at a time is changed. For example, we can write the general rate law:

$$\text{Rate} = k[A]^a[B]^b$$

and carry out a series of experiments, holding the concentration of reactant B constant. Then we note that if the initial rate doubles when [A] doubles, then $a = 1$. If the initial rate increases by a factor of 4 when [A] doubles, then $a = 2$. We can use a similar strategy to determine b.

Many reactions can be classified according to their order in a particular species. The order is the power to which the concentration of a species is raised in the rate law. The overall order is the sum of all the individual orders.

Example 13.3 *Determining reaction order by using the method of initial rates*

Four experiments were conducted to discover how the initial rate of disappearance of BrO_3^- ions in the reaction

$$BrO_3^-(aq) + 5\,Br^-(aq) + 6\,H^+(aq) \longrightarrow 3\,Br_2(aq) + 3\,H_2O(l)$$

varies as the concentrations of the reactants are changed. Use the experimental data in the table below to determine the order of the reaction with respect to each reactant and the overall order. Write the rate law for the reaction and determine the value of k.

Experiment	Initial concentration, mol/L			Initial rate, $(mol\ BrO_3^-)/L \cdot s$
	BrO_3^-	Br^-	H^+	
1	0.10	0.10	0.10	1.2×10^{-3}
2	0.20	0.10	0.10	2.4×10^{-3}
3	0.10	0.30	0.10	3.5×10^{-3}
4	0.20	0.10	0.15	5.4×10^{-3}

Strategy Suppose the concentration of a substance is increased by a certain factor f, but all other concentrations are held constant. From the rate law, rate $= k(\text{concentration})^a$, we know that if the rate increases by f^a, then the reaction has order a in that substance. Therefore, inspect the data to see how the rate changes when the concentration of each substance is changed. To isolate the effect of each substance, compare experiments that differ in the concentration of only one substance at a time.

Solution Comparing experiment 1 with experiment 2, we see that when the concentration of BrO_3^- is doubled, the rate also doubles. Therefore, the reaction is first order in BrO_3^-. Comparing experiment 1 with experiment 3, we see that when the concentration of Br^- is changed by a factor of 3.0, the rate changes by a factor of $3.5/1.2 = 2.9$. Allowing for experimental error, we can deduce that the reaction is also first order in Br^-. When the concentration of hydrogen ions is increased from experiment 2 to experiment 4, by a factor of 1.5, the rate increases by a factor of $5.4/2.4 = 2.2$. Thus we need to solve $1.5^a = 2.2$ for a. Trial and error shows by setting $a = 0$, then 1, then 2, and so on, that $(1.5)^2 = 2.2$, so the reaction is second order in H^+. The rate law is therefore

Rate of disappearance of BrO_3^- ions $= k[BrO_3^-][Br^-][H^+]^2$

Because the sum of the individual orders is $1 + 1 + 2 = 4$, the reaction is fourth order overall. We find k by substituting the values from one of the experiments into the rate law and solving for k. For example, from experiment 4,

$$5.4 \times 10^{-3}\ \text{mol/L} \cdot s = k \times (0.20\ \text{mol/L}) \times (0.10\ \text{mol/L}) \times (0.15\ \text{mol/L})^2$$

so

$$k = \frac{5.4 \times 10^{-3}\ \text{mol/L} \cdot s}{(0.20\ \text{mol/L}) \times (0.10\ \text{mol/L}) \times (0.15\ \text{mol/L})^2} = 12\ L^3/mol^3 \cdot s$$

Note that, as we have done here, it is common to omit the species from the units, especially in high-order reactions, for otherwise they would be too cumbersome.

Self-Test 13.4A Write the rate law for the disappearance of persulfate ions in the reaction

$$S_2O_8^{2-}(aq) + 3\,I^-(aq) \longrightarrow 2\,SO_4^{2-}(aq) + I_3^-(aq)$$

with respect to each reactant and determine the value of k, given the following data:

Experiment	Initial concentration, mol/L		Initial rate, $(mol\ S_2O_8^{2-})/L \cdot s$
	$S_2O_8^{2-}$	I^-	
1	0.15	0.21	1.14
2	0.22	0.21	1.70
3	0.22	0.12	0.98

[*Answer:* $k[S_2O_8^{2-}][I^-]$, $k = 36$ L/mol$\cdot$s]

For a more systematic strategy for finding a by using logarithms, see Appendix 1C.

In practice, the values found for k from several experiments are averaged for greater accuracy.

Self-Test 13.4B Write the rate law and determine the value of k for the reaction $2\,NO(g) + O_2(g) \rightarrow 2\,NO_2(g)$, which contributes to smog formation, given the following data collected at a certain temperature:

Experiment	Initial concentration, mol/L		Initial rate, (mol NO)/L·s
	NO	O_2	
1	0.012	0.020	0.102
2	0.024	0.020	0.408
3	0.024	0.040	0.816

13.4 More Complicated Rate Laws

A few reactions have negative orders, as in (concentration)$^{-1}$. An example is one of the reactions that takes place in the upper atmosphere, the decomposition of ozone (O_3):

$$2\,O_3(g) \longrightarrow 3\,O_2(g)$$

The rate law for this reaction is

$$\text{Rate} = k\frac{[O_3]^2}{[O_2]} = k[O_3]^2[O_2]^{-1}$$

A negative order signifies that the reaction rate decreases as the concentration of a substance—in this case, oxygen—increases. Often, as here, that species is a product, and we cannot use initial rates alone to determine the rate law.

Some reactions have fractional orders, as in (concentration)$^{1/2}$. For example, the oxidation of sulfur dioxide to sulfur trioxide in the presence of platinum (Fig. 13.8) in the reaction

$$2\,SO_2(g) + O_2(g) \xrightarrow{\text{Pt}} 2\,SO_3(g)$$

is found to have the rate law

$$\text{Rate} = k\frac{[SO_2]}{[SO_3]^{1/2}} = k[SO_2][SO_3]^{-1/2}$$

The reaction is first order in SO_2 and minus one-half order in SO_3. Because $1 + \left(-\frac{1}{2}\right) = \frac{1}{2}$, the reaction is one-half order overall. Again, the negative order implies that the rate of the reaction decreases as the concentration of SO_3 increases. A knowledge of this rate law is important to the economics of sulfuric acid manufacture (for which this reaction is an intermediate step). The rate law shows that we can increase the reaction rate by keeping the concentration of SO_2 high and removing the SO_3 as it is formed.

The order of a reaction with respect to a reactant can be fractional or negative.

13.5 First-Order Integrated Rate Laws

How much ozone will be present in a region of the atmosphere in April? How much sulfur trioxide can be produced in an hour? How much penicillin will be left after six months? These questions can be answered by using an **integrated**

In the language of calculus, a rate law is a "differential equation." Differential equations are solved by the technique called "integration," hence the term "integrated rate law."

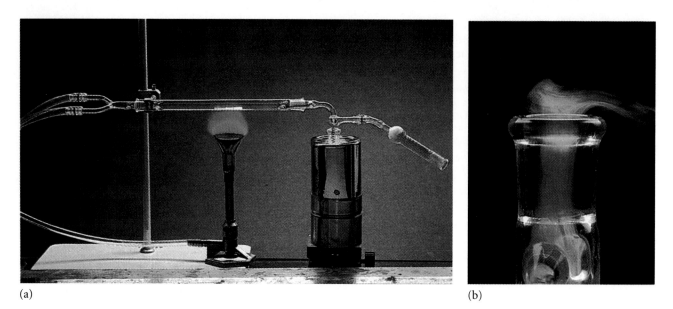

(a)　　　　　　　　　　　　　　　　　　　　　　　　　(b)

FIGURE 13.8

(a) When sulfur dioxide and oxygen are passed over hot platinum foil, they combine to form sulfur trioxide. (b) The sulfur trioxide forms dense white fumes of sulfuric acid when it comes into contact with moisture in the atmosphere. The rate law for the formation of sulfur trioxide shows that its rate of formation decreases as its concentration increases, so the sulfur trioxide must be removed as it is formed if the reaction is to proceed rapidly.

rate law, a formula that gives the concentration of reactants or products at any time after the start of the reaction, if the initial concentrations of reactants are known.

An analogy　　Finding the integrated rate law from a rate law is like calculating the distance that a car will have traveled by knowing its speed at each moment of the journey. A car that travels at high speed for much of its journey (a reaction that is fast) will cover a long distance (will consume a lot of reactant and form a lot of product) in a given time. The opposite is true if the car is slow.

The integrated rate law for a first-order reaction allows us to predict the concentration of reactant remaining after a given time. If we write the molar concentration of a reactant A as $[A]_t$ at time t and denote its initial concentration $[A]_0$, then the integrated rate law for a reaction first order in A is

e = 2.718 28 . . .; see Appendix 1C.

$$[A]_t = [A]_0 e^{-kt} \tag{1}$$

where k is the rate constant and t is the elapsed time; the e in this expression is the "exponential e," the base of natural logarithms. The variation of concentra-

FIGURE 13.9

The characteristic shape of the graph showing the time dependence of the concentration of a reactant in a first-order reaction is an exponential decay, as shown here. The larger the rate constant, the faster the decay from the same initial concentration.

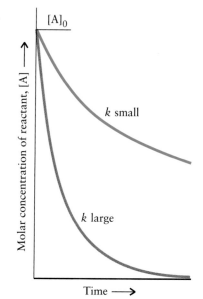

tion with time predicted by this equation is shown in Fig. 13.9. Notice that the bigger the rate constant, the more rapid the initial decay. The behavior shown in the illustration is called an **exponential decay.** The change in concentration is initially rapid, but the concentration changes more slowly as time goes on and the reactant is used up.

Example 13.4 *Calculating a concentration from a first-order integrated rate law*

Calculate the concentration of N_2O_5 remaining 600. s (10.0 min) after the start of its decomposition at 65°C when its concentration was 0.040 mol/L. The reaction and its rate law are

$$2\,N_2O_5(g) \longrightarrow 4\,NO_2(g) + O_2(g)$$
$$\text{Rate of disappearance of } N_2O_5 = k[N_2O_5]$$

with $k = 5.2 \times 10^{-3}/s$.

Strategy The concentration will be smaller than 0.040 mol/L because some of the reactant decomposes as the reaction proceeds. Because the reaction is first order, use Eq. 1 to predict the concentration at any time after the reaction begins. To work out e^x on a calculator, enter x and press the e^x or "inv ln" key.

Solution Substituting the data in Eq. 1 gives

$$
\begin{aligned}
[N_2O_5]_t &= [N_2O_5]_0 \times e^{-kt} \\
&= \{0.040\ (\text{mol } N_2O_5)/L\} \times e^{-(0.0052/s) \times (600.\,s)} \\
&= 1.8 \times 10^{-3}\ (\text{mol } N_2O_5)/L
\end{aligned}
$$

That is, after 600. s, the concentration of N_2O_5 will have fallen from 40. mmol/L to 1.8 mmol/L.

Self-Test 13.5A Calculate the concentration of N_2O remaining after the first-order decomposition

$$2\,N_2O(g) \longrightarrow 2\,N_2(g) + O_2(g) \qquad \text{Rate of decomposition of } N_2O = k[N_2O]$$

has continued at 780°C for 100. ms, if the initial concentration of N_2O was 0.20 mol/L and $k = 3.4/s$.

[***Answer:*** 0.14 (mol N_2O)/L]

Self-Test 13.5B Calculate the concentration of cyclopropane, C_3H_6, remaining after the first-order isomerization

$$C_3H_6(g) \longrightarrow CH_3CH{=}CH_2(g)$$
$$\text{Rate of isomerization of cyclopropane} = k[C_3H_6]$$

has continued at 773 K for 200. s, if the initial concentration of C_3H_6 was 0.100 mol/L and $k = 6.7 \times 10^{-4}/s$.

1 Cyclopropane, C_3H_6

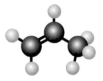

2 Propene, C_3H_6

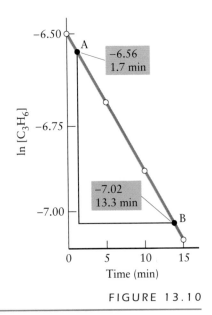

FIGURE 13.10

We test for a first-order reaction by plotting the natural logarithm of the reactant concentration as a function of time. The slope of the line, which is calculated as indicated in the illustration by selecting two points A and B, is equal to the negative of the rate constant.

Integrated rate laws are also used to determine whether a reaction is first order and, if it is, to measure its rate constant. If we take logarithms of both sides of Eq. 1, we obtain

$$\ln[A]_t = \ln[A]_0 - kt \tag{2}$$

This expression is the equation of a straight line:

$$y = \text{intercept} + \text{slope} \times x$$

That is, if we plot $\ln[A]_t$ as a function of t, then we should get a straight line with slope $-k$ (see Appendix 1E). If we get a straight line, the reaction is first order and we can determine its rate constant. If we do not get a straight line, the reaction is not first order and we have to explore whether it is second or some other order.

In a first-order reaction, the concentration of reactant decays exponentially with time. To verify that a reaction is first order, plot the natural logarithm of its concentration as a function of time and expect a straight line; the slope of the straight line is $-k$.

Example 13.5 *Measuring the rate constant of a first-order reaction*

When cyclopropane (**1**) is heated to 500.°C (773 K), it changes into propene (**2**). The following data were obtained in one experiment:

t, min	Concentration of cyclopropane, mol/L
0	1.50×10^{-3}
5.0	1.24×10^{-3}
10.0	1.00×10^{-3}
15.0	0.83×10^{-3}

Confirm that the reaction is first order in cyclopropane and calculate the rate constant.

Strategy Plot $\ln[\text{cyclopropane}]_t$ as a function of t and see whether a straight line is obtained. If so, then the reaction is first order and the slope of the graph has the value $-k$.

Solution For the graphical procedure, begin by setting up the following table:

t, min	$\ln[\text{cyclopropane}]_t$
0	-6.50
5.0	-6.69
10.0	-6.91
15.0	-7.09

The points are plotted in Fig. 13.10. The graph is a straight line, confirming that the reaction is first order in cyclopropane. The slope of the line is

$$\text{Slope} = \frac{(-7.02) - (-6.56)}{(13.3 - 1.7)\ \text{min}} = -0.040/\text{min}$$

Therefore, because $k = -$slope, $k = 0.040$/min. This value is equivalent to $k = 6.7 \times 10^{-4}$/s, the value in Table 13.3.

Self-Test 13.6A Data on the decomposition of N_2O_5 at 25°C are

t, min	Concentration of N_2O_5, mol/L
0	1.5×10^{-2}
200.0	9.6×10^{-3}
400.0	6.2×10^{-3}
600.0	4.0×10^{-3}
800.0	2.5×10^{-3}
1000.0	1.6×10^{-3}

Confirm that the reaction is first order and find the value of k.

[**Answer:** 2.2×10^{-3}/min]

Self-Test 13.6B Use the data on the decomposition of penicillin in Table 13.2 to confirm that the reaction is first order in penicillin, and find the rate constant.

13.6 Half-Lives for First-Order Reactions

The **half-life,** $t_{1/2}$, of a substance is the time needed for its concentration to fall to one-half its initial value. For instance, if the half-life of an environmental pollutant is 18 months, then its concentration will fall to half its initial value in 18 months. The half-lives of CFCs and other atmospheric species are important for assessing their environmental impact. If their half-lives are short, they may not survive long enough to reach the stratosphere, where they can destroy ozone. The half-life of mercury in human blood, in the form of Hg^{2+} ions, is about 6 days; in this case, half-life refers to the interval between ingestion and excretion.

The half-life of a first-order reaction is related to the rate constant, because the higher the value of k, the more rapid the disappearance of the reactant and hence the shorter the half-life (Fig. 13.11). We can find the precise relation by solving Eq. 2 for the time needed for the concentration of the reactant A to fall to half its initial value. First, we rearrange the equation into

$$\ln\left(\frac{[A]_0}{[A]_t}\right) = kt$$

Then we set t equal to $t_{1/2}$ and $[A]_t = \frac{1}{2}[A]_0$ (that is, half the initial concentration of A), and obtain

$$\ln\left(\frac{[A]_0}{\frac{1}{2}[A]_0}\right) = kt_{1/2}$$

Now we can cancel the initial concentrations and obtain

$$\ln 2 = kt_{1/2}$$

It follows that

$$t_{1/2} = \frac{\ln 2}{k} \approx \frac{0.693}{k}$$

(3)

Although ln 2 ≈ 0.693, it is more accurate to use ln 2 on your calculator.

As we expected, the greater the value of the rate constant k, the shorter the half-life for the reaction. Note that the concentration does not appear: *for a first-order reaction, the half-life depends only on the rate constant.*

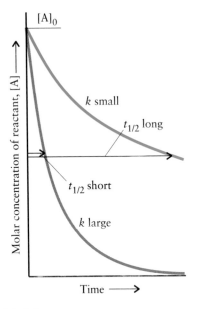

FIGURE 13.11

The half-life of a reactant is short if the first-order rate constant is large, because the exponential decay of the concentration of the reactant is then faster.

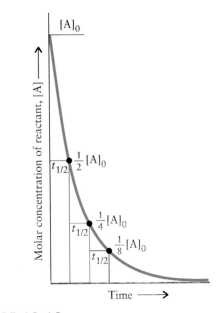

FIGURE 13.12

The half-life of a substance is the time needed for it to fall to one-half its initial concentration. For first-order reactions, the half-life is the same whatever the concentration at the start of the chosen period. Therefore, it takes one half-life for the concentration to fall to half the initial value, two half-lives for the concentration to fall to one-fourth the initial value, three half-lives to fall to one-eighth, and so on.

Figure 13.12 shows the significance of the half-life of a first-order reaction. We see that the concentration falls to half its initial value in the time $t_{1/2}$. The concentration falls to half its new value in the same time, and then to half that new value in another half-life. For each time interval equal to the half-life, the concentration falls to half its value at the start of the interval.

The half-life of a first-order reaction is characteristic of the reaction and independent of the initial concentration. A reaction with a large rate constant has a short half-life.

13.7 Second-Order Integrated Rate Laws

For second-order reactions with rate law

$$\text{Rate} = k[A]^2$$

the integrated rate law is

$$[A]_t = \frac{[A]_0}{1 + [A]_0 kt} \quad \text{or} \quad \frac{1}{[A]_t} = \frac{1}{[A]_0} + kt \tag{4}$$

For a second-order reaction, the plot of $1/[A]$ against t is linear with slope k. Figure 13.13 shows that the concentration of the reactant decreases rapidly at

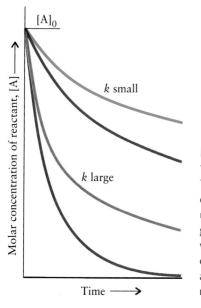

Molar concentration of reactant, [A] →

$[A]_0$

k small

k large

Time →

FIGURE 13.13

The characteristic shapes (orange and green lines) of the time dependence of the concentration of a reactant during two second-order reactions. The gray lines are the curves for first-order reactions with the same initial rates. Note how the concentrations for second-order reactions fall away much less rapidly than those for first-order reactions do.

first but then changes more slowly than a first-order reaction with the same initial rate. Because many pollutants disappear by second-order reactions, they remain at low concentration in the environment for long periods.

Toolbox 13.1 shows how to use integrated rate laws to predict the change in concentration of a reactant or product for first- and second-order reactions.

A second-order reaction has a long tail of low concentration at long reaction times.

Toolbox 13.1 *How to use integrated rate laws*

This Toolbox summarizes the principal applications of the rate laws of first- and second-order reactions.

Conceptual Basis

A rate law, which is determined from experimental data, gives the relation between the rate of a reaction and the concentrations of reactants or products. The integrated rate law is a version of the rate law that tells us the concentration of the reactant or product as a function of time. Each rate law can be rearranged into a form that suggests a plot that will give a straight line if the rate law has a particular order (Fig. 13.14). The rate constant can then be obtained from the slope of the plot that gives a straight line.

Procedure

To find the amount of reactant remaining after a given period of time, use the integrated rate law. Test for a particular order by plotting the integrated rate law for

that order. If the plot is linear, then the reaction is that order in A.

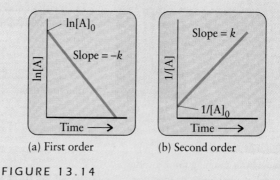

(a) First order   (b) Second order

FIGURE 13.14

A summary of the plots that give straight lines for first- and second-order reactions.

Example 13.6 Using half-lives

A pollutant escapes into a local picnic site. Studies had shown that the pollutant decays by a first-order reaction with rate constant 3.8×10^{-3}/h. Calculate the time needed for the concentration to fall to (a) one-half and (b) one-fourth its initial value.

Strategy Because the reaction is first order, the time it takes for the concentration to fall to half its initial value is $t_{1/2} = (\ln 2)/k$. The total time required for the concentration to fall to one-fourth its initial value is the sum of two successive half-lives. For a first-order reaction, the two half-lives are equal, so the time required is $2t_{1/2}$.

Solution Because $k = 3.8 \times 10^{-3}$/h,

$$t_{1/2} = \frac{\ln 2}{3.8 \times 10^{-3}/\text{h}} = 1.8 \times 10^2\ \text{h}$$

(a) The time required for half the original amount to react is therefore 1.8×10^2 h, or just over a week. (b) The time required for three-fourths of the original amount to react is twice this interval, or about 15 days.

Self-Test 13.7A Calculate the time needed for the concentration of N_2O to fall to (a) one-half and (b) one-eighth of its initial value as it decomposes in a first-order reaction at 1000 K. Consult Table 13.3 for the rate constant.

[*Answer:* (a) 0.91 s; (b) 2.7 s]

Self-Test 13.7B Calculate the time needed for the concentration of C_2H_6 to fall to (a) one-half and (b) one-sixteenth of its initial value as it decomposes to CH_3 radicals in a first-order reaction at 973 K. Consult Table 13.3 for the rate constant.

FIGURE 13.15

A solid iron rod can be heated in a flame without catching fire. However, a powder of very finely divided iron oxidizes rapidly in air to form Fe_2O_3, because the powder presents a much greater surface area for reaction with oxygen.

CONTROLLING REACTION RATES

Thus far we have seen that the rate of a reaction usually depends on the concentration of one or more reactants, so one way to modify the reaction rate is to change the concentrations of the reactants. When a solid reacts, the rate depends on the surface area of the solid (Fig. 13.15), so rates of reactions can also be changed by changing the surface area of any solid reactant. However,

FIGURE 13.16

The *Hindenburg,* a hydrogen-filled airship, caught fire as it came in to land at Lakehurst, New Jersey in 1937 after its inaugural flight across the Atlantic. The hydrogen was ignited when the airship collided with an electrical tower.

some reactions do not seem to take place, however much we increase the concentrations or surface areas. For example, a mixture of hydrogen and oxygen can exist for years without reacting, and dynamite can be safely stored and transported. Yet these materials react in an instant when subjected to a spark or jolt (Fig. 13.16). It is also true that the rates of most reactions increase as the temperature increases. To understand these properties, we need to know how reactions occur and how energy affects their rates.

13.8 The Effect of Temperature

Most reactions go faster as the temperature is raised. An increase of 10°C from room temperature typically doubles the rate of reaction of organic compounds in solution—more precisely, the rate usually increases by a factor of between 1.5 and 4 for a 10°C rise in temperature. An everyday application of this temperature dependence is cooking: we cook foods to accelerate the breakdown of cell walls and proteins to give desirable flavors and aromas and as an aid to digestion. In contrast, we refrigerate foods to slow down the natural chemical reactions that lead to bacterial growth and decomposition.

In a series of experiments, the Swedish chemist Svante Arrhenius got a straight line when he plotted the logarithm of the rate constant of a reaction against $1/T$, where T is the absolute temperature (Fig. 13.17). That is, he established that many rate constants obey an expression of the form

$$\ln k = \text{intercept} + \text{slope} \times \frac{1}{T}$$

The intercept is normally denoted $\ln A$, where A is called the **pre-exponential factor.** For reasons that will become clear, the slope is equal to $-E_a/R$, where R is the gas constant and E_a is called the **activation energy.** When we introduce these two quantities into the equation, it becomes

$$\ln k = \ln A - \frac{E_a}{RT} \qquad (5)$$

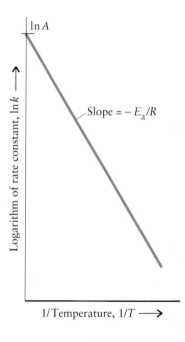

FIGURE 13.17

An Arrhenius plot is a graph of $\ln k$ against $1/T$. If, as here, the line is straight, then the reaction is said to show Arrhenius behavior in the temperature range studied. The activation energy for the reaction is obtained by setting the slope of the line equal to $-E_a/R$.

A reaction that obeys this equation—and many do—is said to show **Arrhenius behavior,** and the equation itself is called the **Arrhenius equation.** The values given in Table 13.4 were obtained from graphs of the kind shown in Fig. 13.17.

Example 13.7 *Measuring an activation energy*

The rate constant for the second-order reaction between bromoethane and hydroxide ions in water,

$$C_2H_5Br(aq) + OH^-(aq) \longrightarrow C_2H_5OH(aq) + Br^-(aq)$$

was measured at several temperatures, with the following results:

Temperature, °C	k, L/mol·s
25	8.8×10^{-5}
30.	1.6×10^{-4}
35	2.8×10^{-4}
40.	5.0×10^{-4}
45	8.5×10^{-4}
50.	1.4×10^{-3}

Find the activation energy of the reaction.

Strategy In the graphical procedure, activation energies are found by plotting $\ln k$ against $1/T$ (as in Eq. 5) and measuring the slope of the line. Begin by converting the temperatures to kelvins and drawing up a table of $1/T$ and $\ln k$.

Solution The table we require for drawing the graph is

Temperature, °C	T, K	$1/T$, 1/K	$\ln k$
25	298	3.35×10^{-3}	-9.34
30.	303	3.30×10^{-3}	-8.74
35	308	3.25×10^{-3}	-8.18
40.	313	3.19×10^{-3}	-7.60
45	318	3.14×10^{-3}	-7.07
50.	323	3.10×10^{-3}	-6.57

The points are plotted in Fig. 13.18. We calculate the slope from two points, such as those marked A and B:

$$\text{Slope} = -\frac{3.22}{0.30 \times 10^{-3}/\text{K}} = -\frac{3.22}{0.30} \times 10^3 \text{ K}$$

Because the slope is equal to $-E_a/R$, we have

$$E_a = -R \times \text{slope} = -(8.3145 \text{ J/K·mol}) \times \left(-\frac{3.22}{0.30} \times 10^3 \text{ K} \right)$$
$$= 8.9 \times 10^4 \text{ J/mol}$$

We see that the activation energy is 89 kJ/mol.

> Although the slope could be calculated from any two of the data points, it is better scientific practice to use the plot, for that minimizes the effect of errors in the data.

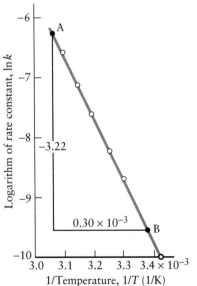

FIGURE 13.18

An Arrhenius plot of the data in Example 13.7. The slope of the line has been worked out by using points A and B.

Table 13.4 Arrhenius parameters

Reaction	A	E_a, kJ/mol
First order, gas phase		
cyclopropane $\longrightarrow$ propene	1.6×10^{15}/s	272
$CH_3NC \longrightarrow CH_3CN$	4.0×10^{13}/s	160.
$C_2H_6 \longrightarrow 2\,CH_3$	2.5×10^{17}/s	384
$N_2O \longrightarrow N_2 + O$	8.0×10^{11}/s	250.
$2\,N_2O_5 \longrightarrow 4\,NO_2 + O_2$	4.0×10^{13}/s	103
Second order, gas phase		
$O + N_2 \longrightarrow NO + N$	1×10^{11} L/mol·s	315
$OH + H_2 \longrightarrow H_2O + H$	8×10^{10} L/mol·s	42
$2\,CH_3 \longrightarrow C_2H_6$	2×10^{10} L/mol·s	0
Second order, in aqueous solution		
$CH_3CH_2Br + OH^- \longrightarrow CH_3CH_2OH + Br^-$	4.3×10^{11} L/mol·s	90.
$CO_2 + OH^- \longrightarrow HCO_3^-$	1.5×10^{10} L/mol·s	38
$C_{12}H_{22}O_{11} + H_2O \longrightarrow 2\,C_6H_{12}O_6$	1.5×10^{15} L/mol·s	108

Self-Test 13.8A The rate constant for the second-order gas-phase reaction $HO(g) + H_2(g) \rightarrow H_2O(g) + H(g)$ varies with the temperature as follows:

Temperature, °C	k, L/mol·s
100.	1.1×10^{-9}
200.	1.8×10^{-8}
300.	1.2×10^{-7}
400.	4.4×10^{-7}

Calculate the activation energy.

[*Answer:* 41 kJ/mol]

Self-Test 13.8B The rate constant for a first-order reaction varies with the temperature as follows:

Temperature, °C	k, /s
18	3.00
30.	4.35
58	9.14
97	20.6

What is the activation energy of the reaction?

One reason why it is useful to know the activation energy of a reaction is that *the higher the activation energy E_a, the stronger the temperature dependence of the reaction rate.* Reactions with low activation energies (about 10 kJ/mol, with not very steep slopes for their Arrhenius plots) have rates that increase only slightly with temperature. Reactions with high activation energies (above about 60 kJ/mol, with steep slopes) have rates that depend strongly on the temperature (Fig. 13.19).

FIGURE 13.19

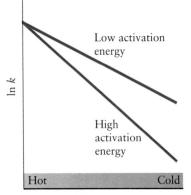

The variation of the rate constant with temperature for different values of the activation energy. Note that the higher the activation energy, the more strongly the rate constant varies with temperature.

Another important use of the Arrhenius equation is to express the rate constant at one temperature in terms of its value at another temperature. First, we write the equation for the two temperatures, which we call T and T':

$$\text{At temperature } T': \ln k' = \ln A - \frac{E_a}{RT'}$$

$$\text{At temperature } T: \ln k = \ln A - \frac{E_a}{RT}$$

Then we eliminate $\ln A$ by subtracting the second equation from the first:

$$\ln k' - \ln k = -\frac{E_a}{RT'} - \left(-\frac{E_a}{RT}\right)$$

That is,

$$\ln\left(\frac{k'}{k}\right) = \frac{E_a}{R}\left(\frac{1}{T} - \frac{1}{T'}\right) \qquad (6)$$

Toolbox 13.2 and Example 13.8 summarize how this equation is used.

An Arrhenius plot of ln k against 1/T is used to determine the Arrhenius parameters of a reaction; a large activation energy signifies a high sensitivity of the rate to changes in temperature.

Toolbox 13.2 *How to describe the temperature dependence of reaction rates*

This Toolbox describes how to determine the Arrhenius parameters for a reaction and how to use them to predict the temperature dependence of rate constants.

Conceptual Basis
The activation energy is obtained from experimental data, using an Arrhenius plot of $\ln k$ against $1/T$ (see Fig. 13.17). The greater the activation energy of a reaction, the more strongly the rate depends on the temperature and the greater the temperature dependence of the rate constant (as shown by Fig. 13.19).

Procedure
The temperature dependence of a rate constant is summarized by the Arrhenius equation:

$$\ln k = \ln A - \frac{E_a}{RT}$$

To determine the Arrhenius parameters Plot the natural logarithm of the rate constant against the inverse absolute temperature. The slope of the line is equal to $-E_a/R$, and the intercept at $1/T = 0$ is the value of $\ln A$. (If common logarithms are used, the slope is unchanged but the intercept is $\log A$.)

To predict a rate constant at one temperature, given its value at another temperature Use the following expression derived from the Arrhenius equation written for the two temperatures T and T':

$$\ln\left(\frac{k'}{k}\right) = \frac{E_a}{R}\left(\frac{1}{T} - \frac{1}{T'}\right)$$

Example 13.8 *Using the activation energy to predict a rate constant*

One of the processes taking place inside us as a part of the digestive process is the hydrolysis of sucrose, in which a sucrose molecule is broken down into a glucose molecule and a fructose molecule. How strongly does the rate of this reaction depend on our body temperature? Calculate the rate constant for the hydrolysis of sucrose at 35°C, given that it is equal to 1.0×10^{-3} L/mol·s at 37°C (normal body temperature) and that the activation energy of the reaction is 108 kJ/mol.

Strategy We expect a lower rate constant at the lower temperature. Base the quantitative calculation on Eq. 6 and substitute the data into this expression. Remember to express temperature in kelvins. Use $R = 8.3145$ J/K·mol.

Solution We take $T = 310$ K and $T' = 308$ K. Then

$$\ln\left(\frac{k'}{k}\right) = \frac{1.08 \times 10^5 \text{ J/mol}}{8.3145 \text{ J/K·mol}}\left(\frac{1}{310 \text{ K}} - \frac{1}{308 \text{ K}}\right) = -0.27$$

When we take the natural antilogarithm (e^x or "inv ln") of this number, we obtain

$$\frac{k'}{k} = 0.76$$

Finally, because $k = 1.0 \times 10^{-3}$ L/mol·s,

$$k' = 0.76 \times (1.0 \times 10^{-3} \text{ L/mol·s}) = 7.6 \times 10^{-4} \text{ L/mol·s}$$

at 35°C. The high activation energy of the reaction means that its rate is very sensitive to temperature.

Self-Test 13.9A The rate constant of the second-order reaction between CH_3CH_2Br and OH^- ions is 2.8×10^{-4} L/mol·s at 25°C; what is it at 50.°C? See Table 13.4 for data.

[*Answer:* 4.7×10^{-3} L/mol·s]

Self-Test 13.9B The rate constant of the first-order isomerization of cyclopropane, C_3H_6, to propene, $CH_3CH=CH_2$, is 6.7×10^{-4}/s at 500.°C; what is it at 300.°C? See Table 13.4 for data.

13.9 The Origin of the Temperature Dependence

The Arrhenius equation is an empirical expression that applies to a wide variety of reactions. What does it reveal about the way reactions take place and what is the significance of the Arrhenius parameters? First, we look at gas-phase reactions, then at reactions that take place in solution.

According to the **collision theory** of gas-phase reactions, molecules react when they collide. We can picture molecules as behaving like defective billiard balls that bounce apart when they collide at low speed but might smash into pieces when the impact is really energetic (Fig. 13.20). In collision theory, the activation energy is a measure of the energy needed for reaction to occur. It is represented as an energy barrier between the reactants and the products, with a height equal to the activation energy. The plot in Fig. 13.21, which shows the energies of reactants and products as well as the energy barrier, is called a **reaction profile.** At great separations, the potential energy of interaction of the reactant molecules is 0, but the molecules have a certain amount of kinetic energy.

FIGURE 13.20

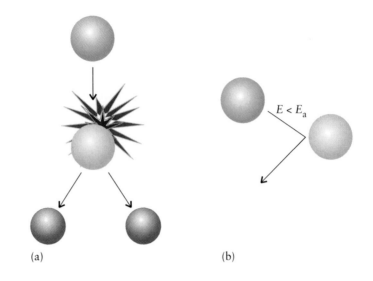

$E < E_a$

(a)   (b)

As the reactant molecules approach each other, they lose kinetic energy but gain potential energy. If initially their kinetic energy is less than E_a, they cannot pass over the barrier, so they bounce apart and separate again. If initially their kinetic energy is at least E_a, they can pass over the barrier and form products.

An analogy

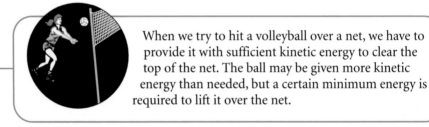

When we try to hit a volleyball over a net, we have to provide it with sufficient kinetic energy to clear the top of the net. The ball may be given more kinetic energy than needed, but a certain minimum energy is required to lift it over the net.

The fraction of molecules that collide with a kinetic energy equal to or greater than the activation energy, E_a, can be deduced from the Maxwell distribution of speeds (Section 5.14). As shown by the shaded area under the blue curve in Fig.13.22, at room temperature, very few molecules have enough kinetic energy to cross the barrier. At higher temperatures, a much larger fraction of molecules can react, as represented by the shaded area under the red

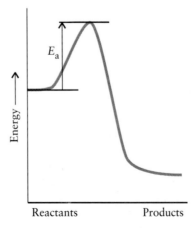

E_a

Energy ⟶

Reactants   Products

FIGURE 13.21

A reaction profile for an exothermic reaction. In the collision theory of reaction rates, the potential energy (the energy due to position) increases as the reactant molecules approach each other, reaches a maximum as the molecules distort, and then decreases as the atoms rearrange into the bonding pattern characteristic of the products and these products separate. Only molecules with sufficient energy can cross the barrier and react to form products.

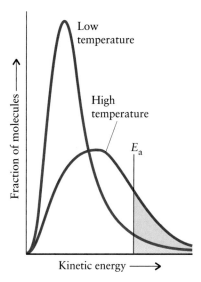

FIGURE 13.22

The fraction of molecules that collide with a kinetic energy that is at least equal to the activation energy, E_a, is denoted by the shaded areas under each curve. The fraction increases rapidly as the temperature is raised.

curve. It turns out that, at a temperature T, the fraction of collisions with at least the energy E_a is proportional to $e^{-E_a/RT}$, where R is the gas constant. Therefore, the rate constant—which is proportional to the number of molecules with kinetic energy equal to or greater than the activation energy—is proportional to the same factor, and we can write

$$k = Ae^{-E_a/RT} \tag{7}$$

This expression takes into account the rate of collisions (through the term A). The factor A also takes into account the possibility that the success of a collision depends on the relative direction of approach. For example, in the gas-phase reaction of chlorine atoms with HI molecules, $HI + Cl \rightarrow HCl + I$, the Cl atom reacts with the HI molecule only if it approaches HI from the correct direction (Fig. 13.23). This direction-dependence is called the **steric requirement** of the reaction.

To confirm that the collision theory is in agreement with Arrhenius behavior, we take logarithms of both sides of Eq. 7 and obtain

$$\ln k = \ln(Ae^{-E_a/RT}) = \ln A + \ln(e^{-E_a/RT}) = \ln A - \frac{E_a}{RT}$$

exactly as the Arrhenius equation requires.

How do we know that? When beams of molecules oriented a certain way are fired at one another, we look at how the outcome varies with the relative orientations of the molecules.

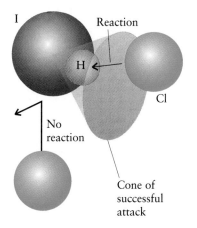

FIGURE 13.23

Whether or not a reaction occurs when two species collide in the gas phase depends on their relative orientation as well as on their energies. In the reaction between a Cl atom and an HI molecule, for example, only those collisions in which the Cl atom approaches the HI molecule along a path that lies inside a cone centered on the H atom lead to reaction, even though the energies of collisions in other orientations may exceed the activation energy.

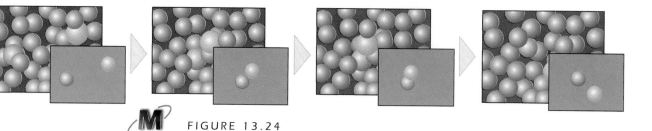

FIGURE 13.24

This sequence of images illustrates the motion of two reactant molecules (red and green) in solution. The blue spheres represent solvent molecules. We see the reactants drifting together, lingering near each other for some time, and then drifting apart again. Reaction may occur during the relatively long period of encounter. To highlight the positions of the reactant molecules, the insets show the solvent molecules as a solid blue background.

Now consider reactions in solution, most of which also show Arrhenius behavior. In **activated complex theory,** two molecules are considered to approach and to distort as they meet. In solution, the approach is a zigzag walk among solvent molecules (Fig. 13.24), and the distortion might not take place until after the two reactant molecules have met and receive a particularly vigorous kick from the solvent molecules around them (Fig. 13.25). The kick results in the formation of an **activated complex,** an unstable combination of the two molecules that can either go on to form products or fall apart into the unchanged reactants. In the activated complex, the original bonds have lengthened and weakened, whereas the new bonds are only partially formed.

The activated complex corresponds to the peak of the energy barrier (Fig. 13.26). As in collision theory, the rate of the reaction depends on the rate at which reactants can climb to the top of the barrier and form the complex. The resulting expression for the rate is very similar to the one given in Eq. 7, so this more general theory also accounts for the form of the Arrhenius equation and the observed dependence of the reaction rate on temperature.

> *According to the collision theory of gas-phase reactions, a reaction occurs only if the reactant molecules collide with a kinetic energy of at least the activation energy. In activated complex theory, a reaction occurs only if two molecules acquire enough energy, perhaps from the surrounding solvent, to form an activated complex and cross an energy barrier.*

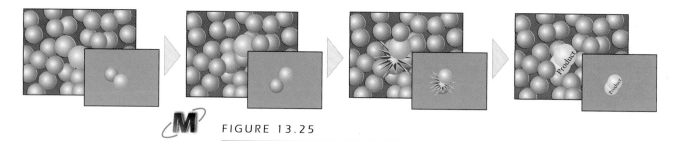

FIGURE 13.25

We join this sequence of images at the moment when the reactant molecules are in the middle of their encounter. They may acquire enough energy by impacts from the solvent molecules to form an activated complex, which may go on to form products. Once again, the insets highlight the reactants by showing the solvent molecules as a solid blue background.

FIGURE 13.26

The reaction profile for the activated complex theory of reactions in solution, in which the reactants form an activated complex, provided they encounter each other with at least the activation energy.

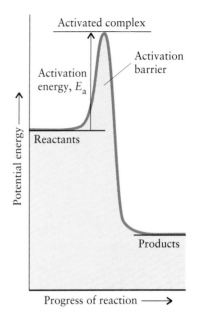

13.10 Catalysis

In some cases, we may want to increase the rate of a reaction but do not want to raise the temperature: high temperatures make reactions difficult to control and may result in undesirable products. One solution is to use a **catalyst,** a substance that increases the rate of a reaction without being consumed in the reaction. In many cases, only a small amount of catalyst is necessary, because it acts over and over again. However, there is no such thing as a "universal catalyst." Not all reactions can be accelerated with a catalyst, and the choice of catalyst depends on the reaction.

Although there is no universal catalyst, there is a common principle behind the action of all catalysts. We shall see in Section 13.12 that a chemical reaction takes place by a sequence of steps—a reaction pathway—in which atoms change their partners. A catalyst speeds up a reaction by providing an alternative pathway—an alternative sequence of steps—from reactants to products. This new pathway has a lower activation energy than the original pathway (Fig. 13.27), so a greater fraction of reactant molecules have enough energy to form products and consequently the reaction is faster.

A catalyst is classified as **homogeneous** if it is present in the same phase as that of the reactants. For reactants that are gases, a homogeneous catalyst is also a gas. If the reactants are liquids, a homogeneous catalyst is a dissolved liquid or solid (Fig. 13.28). Dissolved bromine is a homogeneous liquid-phase catalyst for the decomposition of aqueous hydrogen peroxide:

$$2\ H_2O_2(aq) \xrightarrow{\ Br_2\ } 2\ H_2O(l)\ +\ O_2(g)$$

The name *catalyst* comes from the Greek words meaning "breaking down by coming together." The Chinese characters for catalyst, which translate as "marriage broker," capture the sense quite well.

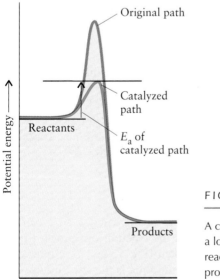

FIGURE 13.27

A catalyst provides a new reaction pathway with a lower activation energy, thereby allowing more reactant molecules to cross the barrier and form products. Notice that E_a for the reverse reaction is also lowered on the catalyzed path.

(a)   (b)

FIGURE 13.28

A small amount of catalyst—in this case, potassium iodide in aqueous solution—can accelerate the decomposition of hydrogen peroxide to water and oxygen. This effect is shown (a) by the slow inflation of the balloon when no catalyst is present and (b) by its rapid inflation when a catalyst is present.

In the absence of bromine or another catalyst, a solution of hydrogen peroxide can be stored for a long time at room temperature; however, bubbles of oxygen form as soon as a drop of bromine is added. Bromine's role in this reaction is believed to be its reduction to Br^- in one step, followed by the oxidation of Br^- back to Br_2 in a second step:

$$Br_2(aq) + H_2O_2(aq) \longrightarrow 2\,Br^-(aq) + 2\,H^+(aq) + O_2(g)$$

$$2\,Br^-(aq) + H_2O_2(aq) + 2\,H^+(aq) \longrightarrow Br_2(aq) + 2\,H_2O(l)$$

When we add the two steps, the H^+, Br_2, and Br^- each cancel, leaving the overall equation as the one on the previous page. Therefore, even though Br_2 molecules have taken part in the reaction, they are not consumed and can be used over and over again.

A catalyst is classified as **heterogeneous** if it is in a phase different from that of the reactants. The most common heterogeneous catalysts are finely divided or porous solids used in gas-phase or liquid-phase reactions. They are finely divided or porous in order to provide a large surface area for reaction. One example is the iron catalyst used in the Haber process for ammonia. Another is finely divided vanadium pentoxide (V_2O_5), which is used in the contact process for the production of sulfuric acid:

$$2\,SO_2(g) + O_2(g) \xrightarrow{V_2O_5} 2\,SO_3(g)$$

The reactant **adsorbs,** or attaches, to the surface of the catalyst. As a reactant molecule adsorbs, its bonds are weakened and the reaction can proceed more quickly because the bonds are more easily broken (Fig. 13.29). One important step in the Haber process for the synthesis of ammonia is the adsorption of N_2 molecules on iron and the weakening of the strong $N\equiv N$ triple bond.

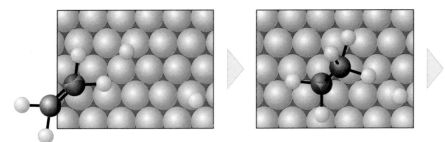

FIGURE 13.29

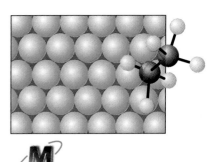

The reaction between ethene, CH$_2$=CH$_2$, and hydrogen on a catalytic metal surface. In this sequence of images, we see the ethene molecule approaching the metal surface to which hydrogen molecules have already adsorbed: when they adsorb, they dissociate and stick to the surface as hydrogen atoms. After sticking to the surface, the ethene molecule meets a hydrogen atom and forms a bond. At this stage (center), a ·CH$_2$CH$_3$ radical is attached to the surface by one of its carbon atoms. Finally, the radical and another hydrogen atom meet, ethane is formed, and the product escapes from the surface.

One of the biggest advances in industrial chemistry in recent years has been the development of catalysts based on zeolites (Fig. 13.30). These materials are aluminosilicates (see Section 19.18) with an open network structure of tunnels and cavities. The network structure gives zeolites an enormous surface area, which is largely on the "inside" of the solid. Moreover, because molecules of only a certain size can worm their way through the tunnels, the catalysts are highly selective in the reactions they promote.

Catalysts are used in the catalytic converters of automobiles to bring about the complete and rapid combustion of unburned fuel (Fig. 13.31). The mixture of gases leaving an engine includes carbon monoxide, unburned hydrocarbons, and the nitrogen oxides collectively referred to as NO$_x$. Air pollution is decreased if the carbon compounds are oxidized to carbon dioxide and the NO$_x$ reduced, by another catalyst, to nitrogen.

Catalysts can be **poisoned,** or inactivated. A common cause of such poisoning is the adsorption of a molecule so tightly to the catalyst that it seals its surface against further reaction. Some heavy metals, especially lead, are very potent poisons for heterogeneous catalysts, which is why lead-free gasoline must be used for engines fitted with catalytic converters. The elimination of lead has the further benefit of decreasing the amount of poisonous lead in the environment. The challenges for the designers of new automobile catalysts include catalyzing both the oxidation of some pollutants (carbon monoxide and unburned fuel) and the reduction of others (nitrogen oxides to nitrogen).

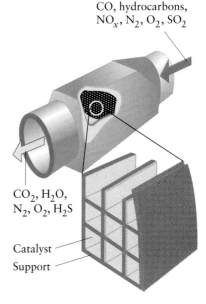

CO, hydrocarbons,
NO$_x$, N$_2$, O$_2$, SO$_2$

CO$_2$, H$_2$O,
N$_2$, O$_2$, H$_2$S

Catalyst

Support

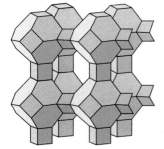

FIGURE 13.30

The internal structure of a zeolite is like a honeycomb of passages and cavities. As a result, a zeolite presents a huge surface area. It can also permit the entry and exit of molecules of a certain size into the active regions within the holes. This zeolite is ZSM-5.

FIGURE 13.31

The structure of a typical catalytic converter for an automobile exhaust. The gases flow through a honeycomb-like porous ceramic support covered with catalyst.

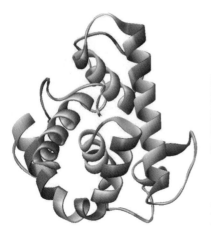

FIGURE 13.32

The lysozyme molecule shown here is a typical enzyme molecule. Lysozyme occurs in a number of places, including tears and the mucus in the nose. One of its functions is to attack the cell walls of bacteria and destroy them. In this illustration the winding ribbon represents the long chain that makes up the molecule.

Catalysts participate in reactions but are not themselves used up; they provide a reaction pathway with a lower activation energy. Catalysts are classified as either homogeneous or heterogeneous.

Self-Test 13.10A How does a catalyst affect (a) the rate of the reverse reaction; (b) ΔH for the reaction?

[*Answer:* (a) Increases it; (b) has no effect]

Self-Test 13.10B How does a homogeneous catalyst affect (a) the rate law; (b) the path by which a reaction occurs?

13.11 Living Catalysts: Enzymes

Living cells contain thousands of different kinds of large proteins called **enzymes** that act as catalysts of biochemical reactions (Fig. 13.32). Enzymes have a slot-like **active site,** where reaction takes place. The **substrate,** the molecule on which the enzyme acts, fits into the slot as a key fits into a lock (Fig. 13.33). However, unlike an ordinary lock, a protein molecule distorts slightly as the substrate molecule approaches, and its ability to undergo the correct distortion also determines whether the "key" will fit. This refinement of the original lock-and-key model is known as the **induced-fit mechanism** of enzyme action.

Once in the slot, the substrate reacts (for instance, by accepting a proton) and the modified substrate molecule is released for use in the next stage, which is controlled by another enzyme. The original enzyme molecule is then free to receive the next substrate molecule (see Applying Chemistry: Case Study 13). One example is the enzyme amylase present in your mouth. The amylase in saliva helps to break down starches in food into the more easily digested glucose. If you chew a cracker long enough, you can notice the increased sweetness.

One form of biological poisoning is similar to the effect of lead on a catalytic converter. The activity of enzymes is destroyed if an alien substrate attaches too strongly to the reactive site, for then the site is blocked and made unavailable to the true substrate (Fig. 13.34). As a result, the chain of biochemical reactions in the cell stops, and the cell dies. The action of nerve gases is believed to stem from their ability to block the enzyme-controlled reactions that allow impulses to travel through nerves. Arsenic, that favorite of fictional poisoners, acts in a similar way. After ingestion as As(V) in the form of arsenate ions (AsO_4^{3-}), it is reduced to As(III), which binds to enzymes and inhibits their action.

Enzymes are biological catalysts that function by modifying substrate molecules to promote reaction.

FIGURE 13.33

In the lock-and-key model of enzyme action, the correct substrate is recognized by its ability to fit into the active site like a key into a lock. In a refinement of this model, the enzyme changes its shape slightly as the key enters.

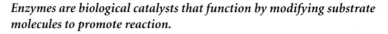
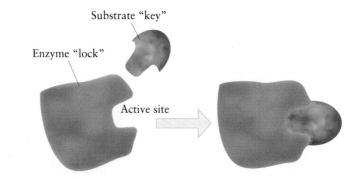

Substrate "key"

Enzyme "lock"

Active site

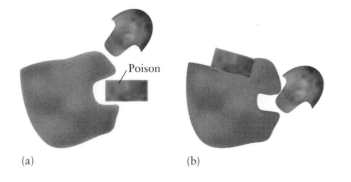

FIGURE 13.34

(a) An enzyme poison (represented by the mottled green rectangle) can act by attaching so strongly to the active site that it blocks the site, thereby taking the enzyme out of action.
(b) Alternatively, the poison molecule may attach elsewhere, so distorting the enzyme molecule and its active site that the substrate no longer fits.

(a)   (b)

REACTION MECHANISMS

How does a molecule of ozone change into a molecule of oxygen? How do chlorine atoms punch holes in the ozone layer? How do molecules react in the hot engine of an automobile? Kinetic data provide clues about the processes that occur when chemical reactions take place. They reveal information about the approach of one molecule to another, the reorganization of bonds that takes place during a collision, and the possibility that the molecules break up in the collision to produce fragments that go on to attack other molecules.

13.12 Elementary Reactions

The reason we cannot predict a rate law from the chemical equation for a reaction is that all but the simplest reactions are the outcome of several, and sometimes many, steps called **elementary reactions.** To describe *how* a reaction occurs, we propose a **reaction mechanism,** a sequence of elementary reactions that describes the atomic changes we believe take place as reactants are transformed into products.

For example, in the decomposition of ozone, $2 O_3(g) \rightarrow 3 O_2(g)$, we could imagine the reaction occurring in one step, when two O_3 molecules collide and rearrange their atoms to form three O_2 molecules. Alternatively, we could propose a mechanism involving two elementary reactions: in the first step, an O_3 molecule is energized by solar radiation and dissociates into an O atom and an O_2 molecule. Then, in a second step, the O atom attacks another O_3 molecule to produce two more O_2 molecules. The O atom is a **reaction intermediate,** a species that plays a role in a reaction but does not appear in the chemical equation for the overall reaction. Because they are neither reactants nor products, intermediates do not appear in rate laws for reactions.

Which, if either, of these two mechanisms is correct? Although we cannot prove that a particular mechanism is the one that actually occurs, we can determine the rate laws for the two proposed mechanisms and compare each of them with the experimental rate law for the reaction. If there is a match, then that mechanism is considered acceptable. To make this comparison, we begin by writing the equation for each step in the two proposed mechanisms.

Elementary reactions are summarized by chemical equations written without state symbols. These equations show how *individual* atoms and molecules take part in the reaction (Fig. 13.35). In general, the **molecularity** of an elementary reaction is the number of reactant molecules involved in it. For example,

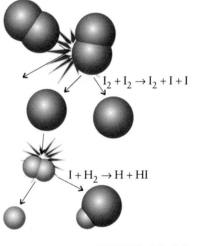

$$I_2 + I_2 \rightarrow I_2 + I + I$$

$$I + H_2 \rightarrow H + HI$$

FIGURE 13.35

The chemical equations for elementary reactions show the individual events that take place when atoms and molecules encounter one another. This illustration shows two of the steps believed to occur during the formation of hydrogen iodide from hydrogen and iodine vapor. In one, $I_2 + I_2 \rightarrow I_2 + I + I$, a collision between two iodine molecules results in the dissociation of one of them. In the second step, $I + H_2 \rightarrow H + HI$, one of the I atoms produced in the first step attacks a hydrogen molecule and forms a hydrogen iodide molecule and a hydrogen atom.

Applying Chemistry: *Case Study 13*

Seeking pleasure and avoiding pain are two of the primary motivators of human activity. Some of the first healing herbs used in prehistoric times were painkillers. In medieval times, alchemists sought an elixir of happiness that would be free of the sensory numbing and addictive nature of opiates. Today, science fiction writers speculate about a future in which we would be able to control our moods at will.

Can chemicals make us happy? The answer is complex. Our moods are regulated through our body chemistry, which normally keeps our moods in balance. However, when body mechanisms go awry, chemistry can help restore the balance. For example, the symptoms of disorders such as depression and schizophrenia can be treated with daily medication.

Many medications work by controlling the action of an enzyme, a natural biological catalyst (see Section 13.11).

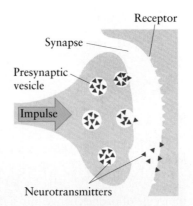

A diagram of a synapse. The triangles represent neurotransmitters that travel from the neuron on the left to the receptors in the neuron on the right. The concentration of neurotransmitters in the synapse is controlled by enzymes.

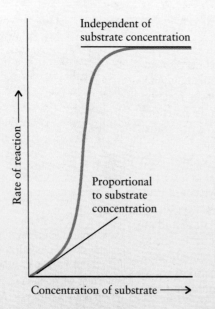

A plot of Michaelis-Menten enzyme kinetics. At low substrate concentrations, the rate of reaction is directly proportional to substrate concentration. However, at high substrate concentrations, the rate is constant, as the enzyme molecules are "saturated" with substrate.

The kinetics of enzyme reactions were first studied by Leonor Michaelis and Maud Menten in the early part of the twentieth century. They found that the rate of an enzyme-catalyzed reaction increases with the concentration of the substrate, as shown in the plot; but when the concentration of substrate is high, the reaction rate depends only on the concentration of the enzyme.

The receptors in our central nervous system are analogous to heterogeneous catalysts. Chemicals called *neurotransmitters* are released from neurons on one side of a space called a *synapse* and attach to receptors in another neuron on the other side. When the receptor is finished with the neurotransmitter, it releases the molecule, which is then reabsorbed into the original neuron and decomposed or modified by a type of enzyme called a monoamine oxidase. The neurotransmitter must be destroyed to prevent a continued buildup of its concentration within the synapse.

When the levels of neurotransmitters or *endorphins* such as dopamine and norepinephrine are optimal, a feeling of well-being is produced through their action on the receptors. Depression can result if too little norepinephrine or dopamine is available. Conversely, if so much dopamine is present that the receptors are flooded with it, then the nervous system becomes overstimulated, and hallucinations and other symptoms of schizophrenia result.

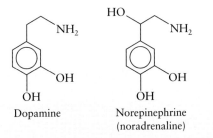

Dopamine

Norepinephrine
(noradrenaline)

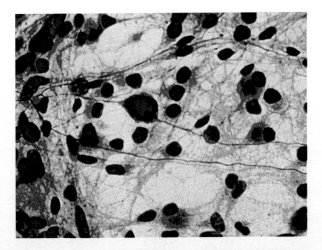

The neurons in the human brain affect how we think and feel and how we perceive reality, including chemistry books.

Depression can be treated by inhibiting the action of the monoamine oxidases, so that the neurotransmitter is not destroyed and its concentration can build up to an acceptable level in the synapse. Some antidepressant drugs such as iproniazid inhibit the monoamine oxidases directly. They bind to them more strongly than the neurotransmitters do, thereby blocking the active sites of the enzymes. Other antidepressants, such as amitriptyline hydrochloride (Elavil), increase the level of norepinephrine by inhibiting its reabsorption into the neuron.

Schizophrenia is treated by drugs that inhibit the action of the receptors by binding to them, blocking the attachment of the dopamine molecules. Chlorpromazine hydrochloride (Thorazine) and haloperidol (Haldol) are commonly used for this purpose.

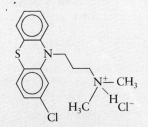

Chlorpromazine hydrochloride (Thorazine)

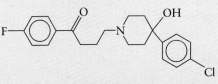

Haloperidol (Haldol)

With fewer receptors available, the excess dopamine goes unused and is reabsorbed. A comparable drug treatment for alcoholism uses naltrexone, which has a molecular shape similar to that of the endorphins. It fits into the receptor sites without creating the sensation of euphoria; therefore, it inhibits the action of the endorphins and reduces the dependency of the individual on alcohol. A similar treatment for cocaine addiction is being studied.

Stimulants such as amphetamines cause the overproduction of dopamine and, when abused, can generate the symptoms of schizophrenia. Self-medication for mood regulation is extremely dangerous because of the delicate balance of enzymes and receptors needed for proper functioning of our nervous systems. Even medically prescribed drug treatments for mood regulation involve side effects that must be carefully monitored.

Key Concepts: catalysis, enzyme kinetics

For Further Reading

M. Balter, New clues to brain dopamine control, cocaine addiction, *Science,* 271:909, 1996.

J. G. Cannon, The chemical roots of pharmacology, *Today's Chemist at Work,* February, 1996, pp. 47–53

D. Dressler and H. Potter, *Discovering Enzymes,* New York: Scientific American Library, 1991.

Related Exercises: 13.99–13.101

one step proposed for the second mechanism for the decomposition of ozone would be written

$$O_3 \longrightarrow O_2 + O$$

This equation tells us that an *individual* O_3 molecule breaks apart on its own, if its energy is high enough. It is an example of a **unimolecular reaction,** an elementary reaction in which only one reactant species is involved. We can picture an ozone molecule acquiring energy from sunlight and vibrating so vigorously that it shakes itself apart.

In the second step of that mechanism, the O atom produced by the dissociation of O_3 goes on to attack another O_3 molecule:

$$O + O_3 \longrightarrow O_2 + O_2$$

This elementary reaction is an example of a **bimolecular reaction,** an elementary reaction in which two reactant species must come together for reaction to occur. Note that all species are shown explicitly in the equation for an elementary reaction: we do not write $O + O_3 \to 2\,O_2$, for instance.

One criterion for the acceptability of a proposed mechanism is that the sum of the equations for the elementary reactions must be the chemical equation for the overall reaction. For our proposed two-step mechanism, we have

$$O_3 \longrightarrow O_2 + O$$
$$\underline{O + O_3 \longrightarrow O_2 + O_2}$$

Overall reaction: $\quad 2\,O_3 \longrightarrow 3\,O_2$

In principle, the reverse of any forward reaction also occurs, although some reverse reactions are so slow that they can be ignored. Therefore, the unimolecular decay of O_3 is accompanied by the formation reaction

$$O_2 + O \longrightarrow O_3$$

This step is bimolecular. Similarly, the bimolecular attack of O on O_3 is accompanied, in principle at least, by its reverse reaction

$$O_2 + O_2 \longrightarrow O + O_3$$

This step is also bimolecular. These two reactions add up to give the reverse of the overall reaction: $3\,O_2 \to 2\,O_3$.

The proposed one-step mechanism also yields a chemical equation that matches the overall reaction:

$$O_3 + O_3 \longrightarrow O_2 + O_2 + O_2$$

The reverse of this elementary reaction is a **termolecular reaction,** an elementary reaction requiring the simultaneous collision of three molecules:

$$O_2 + O_2 + O_2 \longrightarrow O_3 + O_3$$

Termolecular reactions are uncommon: it is very unlikely that three molecules will collide simultaneously with one another under normal conditions.

The term *molecularity* applies only to individual elementary reactions; it is not used for the overall reaction.

Many reactions occur by a series of elementary reactions. The molecularity of an elementary reaction indicates how many reactant species are involved in that step.

Self-Test 13.11A What is the molecularity of the elementary reaction
(a) $C_2N_2 \rightarrow CN + CN$? (b) $NO_2 + NO_2 \rightarrow NO + NO_3$?

[***Answer:*** (a) Unimolecular; (b) bimolecular]

Self-Test 13.11B What is the molecularity of the elementary reaction
(a) $C_2H_5Br + OH^- \rightarrow C_2H_5OH + Br^-$? (b) $Br_2 \rightarrow Br + Br$?

13.13 The Rate Laws of Elementary Reactions

To determine whether a proposed mechanism is plausible, we must construct its overall rate law and check to see whether it is consistent with the experimentally determined rate law. To write the rate law for a mechanism, we first write the rate laws for each of the elementary reactions we have proposed. Then we combine these rate laws to arrive at the predicted rate law for the overall reaction. It is important to realize, however, that although the rate law obtained in this way may agree with the experimental rate law, the proposed mechanism may still be incorrect: a different mechanism may lead to the same rate law. Kinetic information can only support a proposed mechanism; it can never prove that a mechanism is correct. The acceptance of a suggested mechanism is more like the process of proof in an ideal court of law than in mathematics, with evidence being assembled to give a convincing, consistent picture.

We use the following two rules for writing down the rate laws of elementary reactions:

1. A unimolecular elementary reaction depends on a single energized species shaking itself apart and always has a first-order rate law.

$$X \longrightarrow products \qquad \text{Rate of disappearance of } X = k[X]$$

2. A bimolecular reaction depends on collisions between two species, so its rate law is second order. The two reactant species may be the same or different:

$$X + X \longrightarrow products \qquad \text{Rate of disappearance of } X = k[X]^2$$

$$X + Y \longrightarrow products \qquad \text{Rate of disappearance of } X = k[X][Y]$$

Writing rate laws from chemical equations is permissible *only* for elementary reactions. Two examples of this procedure are

$$O_3 \longrightarrow O_2 + O \qquad \text{Rate of disappearance of } O_3 = k_1[O_3]$$

$$O + O_3 \longrightarrow O_2 + O_2 \qquad \text{Rate of disappearance of } O_3 = k_2[O][O_3]$$

Knitting together a collection of individual rate laws into an overall rate law can be very tricky. However, there are some cases where it is quite simple, and we will consider only them. The simplest case is when a reaction mechanism includes a **rate-determining step,** an elementary reaction that is much slower than the rest and governs the rate of the overall reaction. The rate of the overall reaction is then equal to the rate of this step. Notice that a rate-determining step must be slow *and* govern the overall rate of formation of products. A very slow elementary reaction that appears in the mechanism does not control the rate if it can be side-stepped by a faster one (Fig. 13.36).

A rate-determining step in a mechanism is like a slow ferry between two cities. The rate at which the traffic arrives on the other side of the river is governed by the rate at which it is ferried across, because this part of the journey is much slower than any of the others. It is the "bottleneck" in the path. However, if a bridge is built beside the ferry—representing a new reaction pathway—the rate of arrival on the other side is much faster and is independent of the slow ferry crossing.

Let's see how to propose a mechanism and how to construct its rate law. The reaction

$$NO_2(g) + CO(g) \longrightarrow NO(g) + CO_2(g)$$

is one of the very complex set of reactions occurring inside hot internal combustion engines. The experimental rate law for this reaction is

$$\text{Rate of disappearance of } NO_2 = k[NO_2]^2$$

Note that CO does not appear in the rate law. The reaction is second order in NO_2 and zero order in CO. We could propose two mechanisms. One is the single-step mechanism

$$NO_2 + CO \longrightarrow NO + CO_2$$

in which a CO molecule collides with an NO_2 molecule and plucks out an O atom. The rate law for this bimolecular elementary reaction is

$$\text{Rate of disappearance of } NO_2 = k[NO_2][CO]$$

This rate law is different from the one observed experimentally, so we discard it. Next, we might propose a two-step mechanism:

Step 1. $NO_2 + NO_2 \longrightarrow NO_3 + NO$ Rate of disappearance of $NO_2 = k_1[NO_2]^2$

Step 2. $NO_3 + CO \longrightarrow NO_2 + CO_2$ Rate of disappearance of $CO = k_2[NO_3][CO]$

Overall: $\overline{NO_2 + CO \longrightarrow NO + CO_2}$

The sum of these two chemical equations is the overall reaction, but which step is the rate-determining step? If the second step is the slow one, then the rate

Reactants   Intermediates   Products

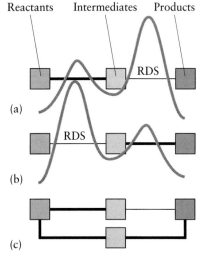

(a)

(b)

(c)

RDS

FIGURE 13.36

The rate of a reaction is controlled by the rate-determining step (RDS). (a) If the rate-determining step is the second step, then the rate law for that step determines the rate law for the overall reaction. The orange curve shows the reaction profile for such a mechanism, with a high activation energy for the slow step. The concentrations of intermediates can usually be expressed in terms of reactants and products by taking into account the steps preceding the RDS. (b) If the rate-determining step is the first step, then the rate law for that step must match the rate law for the overall reaction. Later steps do not affect the rate or the rate law. (c) If two parallel paths lead to products, the faster one (in this case, the lower one) determines the rate of the reaction.

would depend on the concentration of CO and CO would appear in the rate law. Because it does not, the first step must be the rate-determining step (the "ferry") and the second step must be fast (the highway following the ferry). The rate law for both steps combined is the same as that for the first step, because fast steps following the rate-determining step do not affect the reaction rate. We then equate the rate law of the overall reaction with the rate law for the rate-determining step: the rate law is second-order in NO_2, the same as the experimental rate law. We can conclude that the mechanism and its rate law are

Step 1. $NO_2 + NO_2 \longrightarrow NO_3 + NO$ (slow)

Step 2. $NO_3 + CO \longrightarrow NO_2 + CO_2$ (fast)

$$\text{Rate of disappearance of } NO_2 = k_1[NO_2]^2$$

A different approach is needed when the rate-determining step occurs later in the reaction sequence. For example, nitrogen monoxide, NO, reacts with bromine vapor to form gaseous NOBr:

$$2\,NO(g) + Br_2(g) \longrightarrow 2\,NOBr(g)$$

The rate law determined from experiment is

$$\text{Rate} = k[NO]^2[Br_2]$$

Because a termolecular collision is unlikely, we reject a one-step mechanism in which three reactant molecules collide. Therefore, we expect two or more steps. One mechanism that has been proposed is

Step 1, forward. $NO + Br_2 \longrightarrow NOBr_2$ $\qquad$ $\text{Rate} = k_1[NO][Br_2]$

Step 1, reverse. $NOBr_2 \longrightarrow NO + Br_2$ $\qquad$ $\text{Rate} = k_1{'}[NOBr_2]$

Step 2. $NOBr_2 + NO \longrightarrow NOBr + NOBr$ $\quad$ $\text{Rate} = k_2[NOBr][NO]$

To find the rate-determining step, we can make two assumptions and test them.

Let's assume that the forward reaction in the first step is slow and the second step is fast. In that case, the rate law would be the rate law of the slow first step:

$$\text{Rate} = k_1[NO][Br_2]$$

However, this rate law does not match the experimental one.

Now assume that the second step is the slow step. The rate law for this mechanism is equal to that of the second step:

$$\text{Rate} = k_2[NOBr_2][NO]$$

However, $NOBr_2$ is an intermediate. Because the concentration of an intermediate cannot appear in the rate law for the overall reaction, we have to find a way to express the concentration of $NOBr_2$ in terms of the reactants.

Let's suppose that the forward and reverse reactions in step 1 are both much faster than step 2. At the start of the reaction, the forward step results in the formation of $NOBr_2$; but as the concentration of the intermediate grows, the reverse reaction becomes more and more important. After a short time, the rate of the reverse reaction matches that of the forward reaction. From now on, there is no further change in concentration of $NOBr_2$, for any $NOBr_2$ that is formed is immediately removed by the reverse reaction. At this stage of the reaction, therefore, the rate of production of the intermediate is equal to the rate of its disappearance, and we can write

$$k_1[NO][Br_2] = k_1'[NOBr_2]$$

Now we can solve for the concentration of $NOBr_2$:

$$[NOBr_2] = \frac{k_1}{k_1'}[NO][Br_2]$$

and substitute this concentration into the rate law for the second step:

$$\text{Rate} = \frac{k_2 k_1}{k_1'}[NO]^2[Br_2]$$

We recognize this rate law as the same as the experimental rate law if we set $k = k_2 k_1 / k_1'$.

> *Elementary unimolecular reactions have first-order rate laws; elementary bimolecular reactions have second-order rate laws. A rate law can be derived from a proposed mechanism. Steps following the rate-determining step do not affect the rate.*

Example 13.9 *Setting up an overall rate law from a proposed mechanism*

Hydrogen reacts with iodine chloride gas to produce hydrogen chloride and iodine gas. The proposed mechanism for the reaction is $ICl + H_2 \rightarrow HI + HCl$ and its reverse, both of which are fast, followed by $HI + ICl \rightarrow HCl + I_2$. Find the rate law on the assumption that the second step is slow.

Strategy Write down the chemical equation for the overall reaction as the sum of the equations for the elementary forward reactions. Identify any reaction intermediates as species that are formed and consumed in the mechanism but do not appear in the overall equation. Identify the rate-determining step, and set the rate of the overall reaction equal to the rate of that step. If the resulting rate law includes a reaction intermediate, express its concentration in terms of the reactants (and possibly products) by assuming that the rates of the fast forward and reverse reactions are the same.

Solution The overall reaction is

$$2\,ICl(g) + H_2(g) \longrightarrow 2\,HCl(g) + I_2(g)$$

and the mechanism can be summarized as

Step 1, forward. $ICl + H_2 \longrightarrow HI + HCl$ (fast)   Rate $= k_1[ICl][H_2]$

Step 1, reverse. $HI + HCl \longrightarrow ICl + H_2$ (fast)   Rate $= k_1'[HI][HCl]$

Step 2. $HI + ICl \longrightarrow HCl + I_2$ (slow)   Rate $= k_2[HI][ICl]$

The hydrogen iodide does not appear in the overall equation, so it is a reaction intermediate. The second step is slow, so it is the rate-determining step. The rate of this step, and therefore of the overall reaction, is

$$\text{Rate} = k_2[HI][ICl]$$

This rate law depends on the concentration of the intermediate HI. We find its concentration in terms of the concentrations of the reactants by equating the rates of the fast forward and reverse reaction in step 1:

$$k_1[ICl][H_2] = k_1'[HI][HCl]$$

From this equality, we find

$$[HI] = \frac{k_1[ICl][H_2]}{k_1'[HCl]}$$

It then follows that

$$\text{Rate} = k_2[HI][ICl] = \frac{k_1k_2[ICl]^2[H_2]}{k_1'[HCl]}$$

The rate constant for the overall reaction is equal to k_1k_2/k_1'.

Self-Test 13.12A The following two-step mechanism has been proposed for the reaction $2\,NO_2(g) + F_2(g) \rightarrow 2\,NO_2F(g)$:

Step 1. $NO_2 + F_2 \longrightarrow NO_2F + F$ (slow)
Step 2. $F + NO_2 \longrightarrow NO_2F$ (fast)

Find the rate law and the relation between the rate constant k for the overall reaction and the rate constants for the elementary reactions.

[***Answer:*** Rate $= k_1[NO_2][F_2]$; $k = k_1$]

Self-Test 13.12B The following two-step mechanism has been proposed for the reaction $2\,NO(g) + O_2(g) \rightarrow 2\,NO_2(g)$, by which atmospheric NO_2 is generated from NO in automobile exhaust:

Step 1, forward. $NO + NO \longrightarrow N_2O_2$ (fast)

Step 1, reverse. $N_2O_2 \longrightarrow NO + NO$ (fast)

Step 2. $N_2O_2 + O_2 \longrightarrow NO_2 + NO_2$ (slow)

Find the rate law and the relation between the rate constant k for the overall reaction and the rate constants for the elementary reactions.

13.14 Chain Reactions

In a **chain reaction,** a highly reactive intermediate is produced and reacts to produce another highly reactive intermediate, which reacts to produce another, and so on (Fig. 13.37). In many cases, the reaction intermediate—which in this context is called a **chain carrier**—is a radical, and the reaction is called a **radical chain reaction.** In a radical chain reaction, one radical reacts with a molecule to produce another radical, which goes on to attack another molecule, and so on.

The synthesis of HBr from the elements is an example of a chain reaction. The chain carriers are hydrogen atoms (H·) and bromine atoms (Br·). The first step in any chain reaction is **initiation,** the formation of chain carriers from a reactant. In this reaction, the initiation step is

$$Br_2 \xrightarrow{\text{heat or radiation}} Br\cdot + Br\cdot$$

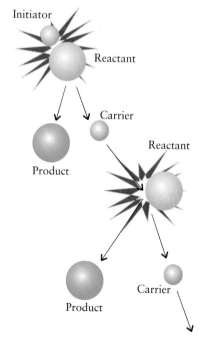

Initiator

Reactant

Product

Carrier

Reactant

Product

Carrier

FIGURE 13.37

In a chain reaction, the product of one reaction step is a reactant in a subsequent step, which in turn produces species that can take part in subsequent reaction steps.

FIGURE 13.38

This flame front was caught during the rapid combustion that occurs inside an internal combustion engine every time a spark plug ignites gasoline vapor. This radical chain reaction occurs in automobile engines. The products, which are hot gases, push a piston out, initiating a chain of events that ultimately moves the vehicle.

Once chain carriers have been formed, the chain can **propagate** in a series of reactions in which a carrier reacts with a reactant molecule to produce another carrier. The elementary reactions for the propagation of the chain are

$$Br\cdot + H_2 \longrightarrow HBr + H\cdot$$

$$H\cdot + Br_2 \longrightarrow HBr + Br\cdot$$

The chain carriers—radicals here—produced in these reactions can go on to attack other reactant (H_2 and Br_2) molecules, thereby allowing the chain to continue. The elementary reactions that end the chain, a process called **termination**, occur when chain carriers combine to form products that do not act as carriers:

$$Br\cdot + Br\cdot \longrightarrow Br_2$$

In some cases, a chain can propagate explosively. Explosions can be expected when **chain branching** occurs, when more than one chain carrier is formed in a propagation step. The characteristic pop that occurs when a mixture of hydrogen and oxygen is ignited is a consequence of chain branching. The two gases combine in a radical chain reaction in which the initiation step may be the formation of hydrogen atoms:

$$\text{Initiation:} \quad H_2 \xrightarrow{\text{spark}} H\cdot + H\cdot$$

Two new radicals form when one hydrogen atom attacks an oxygen molecule:

$$\text{Branching:} \quad H\cdot + O_2 \longrightarrow HO\cdot + \cdot O\cdot$$

The oxygen atom, with valence electron configuration $2s^2 2p_x^2 2p_y^1 2p_z^1$, has two electrons with unpaired spins. Two radicals are also produced when the oxygen atom attacks a hydrogen molecule:

$$\text{Branching:} \quad \cdot O\cdot + H_2 \longrightarrow HO\cdot + H\cdot$$

As a result of these branching processes, the chain produces a large number of radicals that can take part in even more branching steps. The reaction rate increases rapidly, and an explosion typical of many combustion reactions may occur (Fig. 13.38).

Chain reactions begin with the initiation of a reactive intermediate that propagates the chain and concludes with termination when radicals combine. Branching chain reactions can be explosively fast.

Conceptual

☐ 1. Show how the instantaneous rate is obtained by drawing a tangent to the graph of concentration against time, Section 13.2.

☐ 2. Explain how collision theory and activated complex theory account for the temperature dependence of reactions, Section 13.9.

☐ 3. Interpret a reaction profile, Section 13.9.

Problem-Solving

☐ 1. Show how the rate of change of one species in a reaction is related to that of another species, Example 13.1.

☐ 2. Determine the order of a reaction, its rate law, and its rate constant from experimental data, Examples 13.2 and 13.3.

☐ 3. Calculate a concentration or rate constant by using an integrated rate law, Toolbox 13.1 and Examples 13.4 and 13.5.

☐ 4. For a first-order process, calculate the half-life, given the rate constant, and vice versa, Toolbox 13.1 and Example 13.6.

☐ 5. Use the Arrhenius equation and rate constants measured at different temperatures to determine an activation energy, Toolbox 13.2 and Example 13.7.

☐ 6. Use the Arrhenius equation and the activation energy to find the rate constant at a given temperature, Toolbox 13.2 and Example 13.8.

☐ 7. Determine the molecularity of an elementary reaction and write its rate law, Self-Test 13.11 and Section 13.13.

☐ 8. Deduce a rate law from a mechanism, Example 13.9.

Descriptive

☐ 1. Explain the action of enzymes and catalysts, Sections 13.10–13.11.

☐ 2. Describe the various steps in a chain reaction and explain how each affects the progress of the reaction, Section 13.14.

Exercises

Reaction Rates

13.1 Complete the following statements relating to the production of ammonia by the Haber process, for which the overall reaction is $N_2(g) + 3 H_2(g) \rightarrow 2 NH_3(g)$.
(a) The rate of decomposition of N_2 is _____ times the rate of decomposition of H_2.
(b) The rate of formation of NH_3 is _____ times the rate of decomposition of H_2.
(c) The rate of formation of NH_3 is _____ times the rate of decomposition of N_2.

13.2 (a) In the reaction $3 ClO^-(aq) \rightarrow 2 Cl^-(aq) + ClO_3^-(aq)$, the rate of formation of Cl^- is 3.6 mol/L·min. What is the rate of reaction of ClO^-? (b) In the Haber process for the industrial production of ammonia, $N_2(g) + 3 H_2(g) \rightarrow 2 NH_3(g)$, the rate of ammonia production is 2.7×10^{-3} mol/L·s. What is the rate of reaction of H_2?

13.3 (a) The rate of formation of O_2 is 1.5 mmol/L·s in the reaction $2 O_3(g) \rightarrow 3 O_2(g)$. What is the rate of decomposition of ozone? (b) The rate of formation of dichromate ions is 0.14 mol/L·s in the reaction $2 CrO_4^{2-}(aq) + 2 H^+(aq) \rightarrow Cr_2O_7^{2-}(aq) + H_2O(l)$. What is the rate of reaction of chromate ion in the reaction?

13.4 (a) NO_2 decomposes at the rate of 6.5 mmol/L·s by the reaction $2 NO_2(g) \rightarrow 2 NO(g) + O_2(g)$. Determine the rate of formation of O_2. (b) Manganate ions, MnO_4^{2-}, form permanganate ions and manganese(IV) oxide in an acidic solution at a rate of 2.0 mol/L·min: $3 MnO_4^{2-}(aq) + 4 H^+(aq) \rightarrow 2 MnO_4^-(aq) + MnO_2(s) + 2 H_2O(l)$. What is the rate of formation of permanganate ions? What is the rate of reaction of $H^+(aq)$?

Rate Laws

13.5 The decomposition of gaseous dinitrogen pentoxide in the reaction $2 N_2O_5(g) \rightarrow 4 NO_2(g) + O_2(g)$ gives the following data at 298 K:

Time, h	$[N_2O_5]$, mmol/L
0	2.15
1.11	1.88
2.22	1.64
3.33	1.43
4.44	1.25

(a) Plot the concentration of N_2O_5 as a function of time.
(b) Estimate the rate of decomposition of N_2O_5 at each time.
(c) Plot the concentrations of NO_2 and O_2 against the time on the same graph.

13.6 The decomposition of gaseous hydrogen iodide, $2 HI(g) \rightarrow H_2(g) + I_2(g)$, gives the following data at 700 K:

Time, s	$[HI]$, mmol/L
0	10.0
1000.	4.4
2000.	2.8
3000.	2.1
4000.	1.6
5000.	1.3

(a) Plot the concentration of HI against the time. (b) Estimate the rate of decomposition of HI at each time. (c) Plot the concentrations of H_2 and I_2 as a function of time on the same graph.

13.7 Write the rate laws for the following reactions. (a) The reaction $X + 2Y \rightarrow Z$ is second order in X and one-half order in Y. (b) The reaction $2A + B \rightarrow C$ is found to be first order in A, first order in B, and three-halves order overall.

13.8 Write the rate laws for the following reactions. (a) The reaction $A + 2B + C \rightarrow D + 4E$ is found to be first order in A, first order in B, and zero order in C. (b) The reaction $2A + B \rightarrow C + D$ is found to be first-order in A, one-half order in B, three-halves order in D, and zero order overall.

13.9 Express the units for rate constants when the concentrations are in moles per liter and time is in seconds for (a) zero-order reactions; (b) first-order reactions; (c) second-order reactions.

13.10 Because partial pressures are proportional to concentrations, rate laws for gas-phase reactions can also be expressed in terms of partial pressures, for instance, as rate $= kP_X$ for a first-order reaction of a gas X. What are the units for the rate constants when partial pressures are expressed in Torr and time is expressed in seconds for (a) zero-order reactions; (b) first-order reactions; (c) second-order reactions?

13.11 If $k = 1.35 \text{ L}^2/\text{mol}^2 \cdot \text{h}$, what is the order of the reaction?

13.12 If $k = 10/\text{Torr} \cdot \text{s}$ at 298 K, what is the order of the reaction?

13.13 Dinitrogen pentoxide, N_2O_5, decomposes by a first-order reaction. What is the initial rate for the decomposition of N_2O_5 when 2.0 g of N_2O_5 is confined in a 1.0-L container and heated to 65°C? From Table 13.3, $k = 5.2 \times 10^{-3}/\text{s}$ at 65°C.

13.14 Ethane, C_2H_6, decomposes to methyl radicals at 700.°C by a first-order reaction. If a 100.-mg sample of ethane is confined to a 250.-mL reaction vessel and heated to 700.°C, what is the initial rate of ethane decomposition? From Table 13.3, $k = 5.5 \times 10^{-4}/\text{s}$ at 700.°C.

13.15 A 0.15-g sample of H_2 and a 0.32-g sample of I_2 are confined to a 500.-mL reaction vessel and heated to 700. K, when they react by a second-order process, with $k = 0.063 \text{ L/mol} \cdot \text{s}$. (a) What is the initial reaction rate? (b) By what factor does the reaction rate increase if the mass of H_2 present in the mixture is doubled?

13.16 A 100.-mg sample of NO_2, confined to a 200.-mL reaction vessel, is heated to 300.°C, when it decomposes by a second-order process, with $k = 0.54 \text{ L/mol} \cdot \text{s}$. (a) What is the initial reaction rate? (b) How does the reaction rate change (and by what factor) if the mass of NO_2 present in the container is increased to 200. mg?

13.17 State the order of the reaction with respect to each species and the overall order of each of the following reactions:
(a) $H_2(g) + I_2(g) \rightarrow 2HI(g)$, rate $= k[H_2][I_2]$
(b) $2SO_2(g) + O_2(g) \xrightarrow{Pt} 2SO_3(g)$, rate $= k[SO_2][SO_3]^{-1/2}$
(c) $A(g) + 2B(g) + C(g) \rightarrow$ products, rate $= k[A]^2[C]^{-2}$

13.18 State the order of the reaction with respect to each species and the overall order of each of the following reactions:
(a) $S_2O_8{}^{2-}(aq) + 3I^-(aq) \rightarrow 2SO_4{}^{2-}(aq) + I_3{}^-(aq)$, rate $= k[S_2O_8{}^{2-}][I^-]$
(b) $2NO(g) + Cl_2(g) \rightarrow 2NOCl(g)$, rate $= k[NO]^2[Cl_2]$
(c) $A(g) + 3B(g) \rightarrow$ products, rate $= k[A]^2[B]^{3/2}$

13.19 In the reaction $CH_3Br(aq) + OH^-(aq) \rightarrow CH_3OH(aq) + Br^-(aq)$, when the OH^- concentration alone was doubled, the rate doubled. When the CH_3Br concentration alone was increased by a factor of 1.2, the rate increased by 1.2. Write the rate law for the reaction.

13.20 In the reaction $2NO(g) + O_2(g) \rightarrow 2NO_2(g)$, when the NO concentration alone was doubled, the rate increased by a factor of 4. When both the NO and O_2 concentrations were increased by a factor of 2, the rate increased by a factor of 8. What is the rate law for the reaction?

13.21 The following kinetic data were obtained for the reaction $A(g) + 2B(g) \rightarrow$ product:

Experiment	Initial concentration, mol/L		Initial rate, mol/L·s
	$[A]_0$	$[B]_0$	
1	0.60	0.30	12.6
2	0.20	0.30	1.4
3	0.60	0.10	4.2
4	0.17	0.25	?

(a) What is the order with respect to each reactant, and the overall order of the reaction? (b) Write the rate law for the reaction. (c) From the data, determine the value of the rate constant. (d) Use the data to predict the reaction rate for experiment 4.

13.22 The following kinetic data were obtained for the reaction $3A_2B(g) + CX_3(g) \rightarrow$ product:

Experiment	Initial concentration, mol/L		Initial rate, mol/L·s
	$[A_2B]_0$	$[CX_3]_0$	
1	1.72	2.44	0.68
2	3.44	2.44	5.44
3	1.72	0.10	2.8×10^{-2}
4	2.91	1.33	?

(a) What is the order with respect to each reactant, and the overall order of the reaction? (b) Write the rate law for the reaction. (c) From the data, determine the value of the rate constant. (d) Use the data to predict the reaction rate for experiment 4.

13.23 The following kinetic data were obtained for the reaction $2 ICl(g) + H_2(g) \rightarrow I_2(g) + 2 HCl(g)$:

Experiment	Initial concentration, mmol/L		Initial rate, mmol/L·s
	$[ICl]_0$	$[H_2]_0$	
1	1.5	1.5	0.37
2	3.0	1.5	0.74
3	3.0	4.5	2.2
4	4.7	2.7	?

(a) Write the rate law for the reaction. (b) From the data, determine the value of the rate constant. (c) Use the data to predict the initial rate for experiment 4.

13.24 The following kinetic data were obtained for the reaction $NO_2(g) + O_3(g) \rightarrow NO_3(g) + O_2(g)$:

Experiment	Initial concentration, mmol/L		Initial rate, mol/L·s
	$[NO_2]_0$	$[O_3]_0$	
1	0.21	0.70	6.3
2	0.21	1.39	12.5
3	0.38	0.70	11.4
4	0.66	0.18	?

(a) Write the rate law for the reaction. (b) What is the order of the reaction? (c) From the data, determine the value of the rate constant. (d) Use the data to predict the initial rate for experiment 4.

13.25 The following data were obtained for the reaction $A + B + C \rightarrow$ products:

Experiment	Initial concentration, mmol/L			Initial rate, mmol/L·s
	$[A]_0$	$[B]_0$	$[C]_0$	
1	1.25	1.25	1.25	8.7
2	2.50	1.25	1.25	17.4
3	1.25	3.02	1.25	50.8
4	1.25	3.02	3.75	457
5	3.01	1.00	1.15	?

(a) Write the rate law for the reaction. (b) What is the order of the reaction? (c) Determine the value of the rate constant. (d) Use the data to predict the reaction rate for experiment 5.

13.26 For the reaction $A + 2 B + C \rightarrow$ products, the following data were collected:

Experiment	Initial concentration, mmol/L			Initial rate, mmol/L·s
	$[A]_0$	$[B]_0$	$[C]_0$	
1	2.06	3.05	4.00	3.7
2	0.87	3.05	4.00	0.66
3	0.50	0.50	0.50	0.013
4	1.00	0.50	1.00	0.072
5	1.60	2.00	3.00	?

(a) Write the rate law for the reaction. (b) What is the order of the reaction? (c) Determine the value of the rate constant.

Integrated Rate Laws

13.27 Determine the rate constants for the following first-order reactions: (a) $A \rightarrow B$, given that the concentration of A decreases to one-half its initial value in 1000. s. (b) $A \rightarrow B$, given that the concentration of A decreases from 0.33 mol/L to 0.14 mol/L in 47 s. (c) $2 A \rightarrow B + C$, given that $[A]_0 = 0.050$ mol/L and that after 120. s the concentration of B rises to 0.015 mol/L.

13.28 Determine the rate constants for the following first-order reactions: (a) $2 A \rightarrow B + C$, given that the concentration of A decreases to one-third its initial value in 25 min. (b) $2 A \rightarrow B + C$, given that $[A]_0 = 0.020$ mol/L and that after 1.3 h the concentration of B increases to 0.0060 mol/L. (c) $2 A \rightarrow 3 B + C$, given that $[A]_0 = 0.050$ mol/L and that after 7.7 min the concentration of B rises to 0.050 mol/L.

13.29 The half-life for the first-order decomposition of A is 200. s. How much time must elapse for the concentration of A to decrease to (a) one-half; (b) one-sixteenth; (c) one-ninth of its initial concentration?

13.30 The first-order rate constant for the photodissociation of A is 0.173/s. Calculate the time needed for the concentration of A to decrease to (a) one-fourth; (b) one-thirty-second; (c) one-fifth of its initial concentration.

13.31 Dinitrogen pentoxide, N_2O_5, decomposes by first-order kinetics with a rate constant of 3.7×10^{-5}/s at 298 K. (a) What is the half-life (in hours) for the decomposition of N_2O_5 at 298 K? (b) If $[N_2O_5]_0 = 2.33 \times 10^{-2}$ mol/L, what will be the concentration of N_2O_5 after 2.0 h? (c) How much time (in minutes) will elapse before the N_2O_5 concentration decreases from 23.3 mmol/L to 17.6 mmol/L?

13.32 Dinitrogen pentoxide, N_2O_5, decomposes by first-order kinetics with a rate constant of 0.15/s at 353 K. (a) What is the half-life (in seconds) for the decomposition of N_2O_5 at 353 K? (b) If $[N_2O_5]_0 = 2.33 \times 10^{-2}$ mol/L, what will be the concentration of N_2O_5 after 2.0 s? (c) How much time in minutes will elapse before the N_2O_5 concentration decreases from 23.3 mmol/L to 17.6 mmol/L?

13.33 Sulfuryl chloride, SO_2Cl_2, decomposes by first-order kinetics, and $k = 2.81 \times 10^{-3}$/min at a certain temperature. (a) Determine the half-life for the reaction. (b) Determine the time needed for the concentration of a SO_2Cl_2 sample to decrease to 10.% of its initial concentration. (c) If a 14.0-g sample of SO_2Cl_2 is sealed in a 2500.-L reaction vessel and heated to the specified temperature, what mass will remain after 1.5 h?

13.34 Ethane, C_2H_6, forms $\cdot CH_3$ radicals at 700.°C in a first-order reaction, for which $k = 1.98/h$. (a) What is the half-life for the reaction? (b) Calculate the time needed for the amount of ethane to fall from 1.15 mmol to 0.235 mmol in a 500.-mL reaction vessel at 700.°C. (c) How much of a 6.88-mg sample of ethane in a 500.-mL reaction vessel at 700.°C will remain after 45 min?

13.35 Substance A forms B in the first-order reaction $A \rightarrow 2B$, in which the concentration of A falls to 20.% of its original concentration in 120. s. (a) What is the rate constant for the reaction? (b) Determine the time required for the concentration of A to fall to 10.% of its original concentration.

13.36 In the first-order reaction $2A \rightarrow B + C$, it was observed that the initial concentration of A decreased to 80.% of its original concentration in 175 min. How much time would be needed for the concentration to fall to 20.% of its original value?

13.37 From a set of kinetic data, a plot of the logarithm of the concentration of a reactant as a function of time yielded a straight line with a negative slope. What is the order of the reaction?

13.38 From a set of kinetic data, a plot of the reciprocal concentration of a reactant as a function of time yielded a straight line with a positive slope. What is the order of the reaction?

13.39 The following data were collected for the reaction $2N_2O_5(g) \rightarrow 4NO_2(g) + O_2(g)$ at 25°C:

Time, s	$[N_2O_5]$, mmol/L
0	2.15
4000.	1.88
8000.	1.64
12 000.	1.43
16 000.	1.25

(a) Plot the data to confirm that the reaction is first order.
(b) From the graph, determine the rate constant.

13.40 The following data were collected for the reaction $C_2H_6(g) \rightarrow 2\cdot CH_3(g)$ at 700°C:

Time, s	$[C_2H_6]$, mmol/L
0	1.59
1000.	0.92
2000.	0.53
3000.	0.31
4000.	0.18
5000.	1.10

(a) Plot the data to confirm that the reaction is first order.
(b) From the graph, determine the rate constant.

13.41 The following data were collected for the reaction $2HI(g) \rightarrow H_2(g) + I_2(g)$ at 580 K:

Time, s	$[HI]$, mmol/L
0	1000.
1000.	112
2000.	61
3000.	41
4000.	31

(a) Plot the data to confirm that the rate law is rate $= k[HI]^2$.
(b) From the graph, determine the rate constant.

13.42 The following data were collected for the reaction $H_2(g) + I_2(g) \rightarrow 2HI(g)$ at 780 K:

Time, s	$[I_2]$, mmol/L
0	1.00
1	0.43
2	0.27
3	0.20
4	0.16

(a) Plot the data to confirm that the reaction is second order.
(b) From the graph, determine the rate constant.

Collision Theory and Arrhenius Behavior

13.43 Explain why the rate constant increases with temperature, and justify the form of the Arrhenius equation.

13.44 Which reaction has a rate that is more sensitive to temperature, one with a high activation energy or one with a low activation energy?

13.45 Draw a reaction profile for an endothermic reaction. Label the activation energy, E_a, for the forward reaction. Mark the location of the activated complex. What is the relation of the activation energy of the reverse reaction, E_a', to that of the forward reaction? For this exercise, do not distinguish between energy and enthalpy.

13.46 The activation energy for a certain reaction is 125 kJ/mol and the activation energy for the reverse of that reaction is 155 kJ/mol. Draw a reaction profile for the reaction. Indicate the location of the activated complex.

13.47 (a) Calculate the activation energy for the conversion of cyclopropane to propene from an Arrhenius plot of the following data:

T, K	k, 1/s
750.	1.8×10^{-4}
800.	2.7×10^{-3}
850.	3.0×10^{-2}
900.	0.26

(b) What is the value of the rate constant at 600.°C?

13.48 (a) Determine the activation energy for $C_2H_5I(g) \rightarrow C_2H_4(g) + HI(g)$ from an Arrhenius plot of the following data:

T, K	k, 1/s
660.	7.2×10^{-4}
680.	2.2×10^{-3}
720.	1.7×10^{-2}
760.	0.11

(b) What is the value of the rate constant at 400.°C?

13.49 The rate constant of the first-order reaction $2 N_2O(g) \rightarrow 2 N_2(g) + O_2(g)$ is 0.38/s at 1000. K and 0.87/s at 1030. K. Calculate the activation energy of the reaction.

13.50 The rate constant of the second-order reaction $2 HI(g) \rightarrow H_2(g) + I_2(g)$ is 2.4×10^{-6} L/mol·s at 575 K and 6.0×10^{-5} L/mol·s at 630. K. Calculate the activation energy of the reaction.

13.51 The rate constant of the reaction $O(g) + N_2(g) \rightarrow NO(g) + N(g)$, which occurs in the stratosphere, is 9.7×10^{10} L/mol·s at 800.°C. The activation energy of the reaction is 315 kJ/mol. Determine the rate constant at 700.°C.

13.52 The rate constant of the reaction between CO_2 and OH^- in aqueous solution, to give the HCO_3^- ion is 1.5×10^{10} L/mol·s at 25°C. Determine the rate constant at blood temperature (37°C), given that the activation energy for the reaction is 38 kJ/mol.

13.53 The rate constant for the decomposition of N_2O_5 at 45°C is $k = 5.1 \times 10^{-4}$/s. From Table 13.4, the activation energy for the reaction is 103 kJ/mol. Determine the value of the rate constant at 50.°C.

13.54 Ethane, C_2H_6, dissociates into methyl radicals at 700.°C with a rate constant, $k = 5.5 \times 10^{-4}$/s. Determine the rate constant at 800.°C, given that the activation energy of the reaction is 384 kJ/mol.

Catalysis

13.55 The presence of a catalyst provides a reaction pathway in which the activation energy of a certain reaction is reduced from 100. kJ/mol to 50. kJ/mol. By what factor does the rate constant of the reaction increase at 400. K, all other factors being equal?

13.56 The presence of a catalyst provides a reaction pathway in which the activation energy of a certain reaction is reduced from 88 kJ/mol to 62 kJ/mol. By what factor does the rate constant of the reaction increase at 300. K, all other factors being equal?

13.57 The rate constant of a reaction increases by a factor of 1000. in the presence of a catalyst at 25°C. The activation energy of the original pathway is 98 kJ/mol. What is the activation energy of the new pathway, all other factors being equal?

13.58 The rate constant of a reaction increases by a factor of 500. in the presence of a catalyst at 37°C. The activation energy of the original pathway is 106 kJ/mol. What is the activation energy of the new pathway, all other factors being equal?

Reaction Mechanisms

13.59 Each of the following steps is an elementary reaction. Write its rate law and indicate its molecularity: (a) $NO + NO \rightarrow N_2O_2$; (b) $Cl_2 \rightarrow Cl + Cl$; (c) $NO_2 + NO_2 \rightarrow NO + NO_3$. (d) Which of these reactions might be radical chain initiating (produce radicals)?

13.60 Each of the following is an elementary reaction. Write its rate law and indicate its molecularity. (a) $O + CF_2Cl_2 \rightarrow ClO + CF_2Cl$; (b) $OH + NO_2 + N_2 \rightarrow HNO_3 + N_2$ (the N_2 takes part in a three-body collision, but is left unchanged chemically); (c) $ClO^- + H_2O \rightarrow HClO + OH^-$. (d) Which of these reactions might be radical chain propagating (produce radicals)?

13.61 The contribution to the destruction of the ozone layer caused by high-flying aircraft has been attributed to the following mechanism:

$$\textit{Step 1. } O_3 + NO \longrightarrow NO_2 + O_2$$
$$\textit{Step 2. } NO_2 + O \longrightarrow NO + O_2$$

(a) Write the overall reaction. (b) Write the rate law for each step and indicate its molecularity. (c) What is the catalyst in the reaction? (d) What is the reaction intermediate?

13.62 A reaction was believed to occur by the following mechanism:

$$\textit{Step 1. } A_2 \longrightarrow A + A, \text{ and its reverse, } A + A \longrightarrow A_2$$
$$\textit{Step 2. } A + A + B \longrightarrow A_2B$$
$$\textit{Step 3. } A_2B + C \longrightarrow A_2C + B$$

(a) Write the overall reaction. (b) Write the rate law for the forward reaction of each step and indicate its molecularity. (c) What is the catalyst in the reaction? (d) Which species are the reaction intermediates?

13.63 Write the overall reaction for the mechanism proposed below and identify any reaction intermediates:

$$\textit{Step 1. } HA + B^{2-} \longrightarrow HB^- + A^-$$
$$\textit{Step 2. } HA + HB^- \longrightarrow H_2B + A^-$$

13.64 Write the overall reaction for the mechanism proposed below and identify any reaction intermediates:

> Step 1. $Cl_2 \longrightarrow Cl + Cl$
> Step 2. $Cl + CO \longrightarrow COCl$
> Step 3. $COCl + Cl \longrightarrow COCl_2$

13.65 The following mechanism has been proposed for the reaction between nitric oxide and chlorine:

> Step 1. $NO + Cl_2 \longrightarrow NOCl_2$ (slow)
> Step 2. $NOCl_2 + NO \longrightarrow NOCl + NOCl$ (fast)

Write the rate law implied by this mechanism.

13.66 The following mechanism has been proposed for the reaction between chlorine and chloroform, $CHCl_3$:

> Step 1. $Cl_2 \longrightarrow Cl + Cl$, and its reverse,
> $Cl + Cl \longrightarrow Cl_2$ (both fast)
> Step 2. $CHCl_3 + Cl \longrightarrow CCl_3 + HCl$ (slow)
> Step 3. $CCl_3 + Cl \longrightarrow CCl_4$ (fast)

Write the rate law implied by this mechanism.

13.67 The production of phosgene, $COCl_2$, from carbon monoxide and chlorine is believed to take place by the following mechanism:

> Step 1. $Cl_2 \longrightarrow Cl + Cl$, and its reverse
> $Cl + Cl \longrightarrow Cl_2$ (both fast)
> Step 2. $Cl + CO \longrightarrow COCl$, and its reverse
> $COCl \longrightarrow CO + Cl$ (both fast)
> Step 3. $COCl + Cl_2 \longrightarrow COCl_2 + Cl$ (slow)

Write the rate law implied by this mechanism.

13.68 The mechanism proposed for the oxidation of iodide ion by the hypochlorite ion in aqueous solution is as follows:

> Step 1. $OCl^- + H_2O \longrightarrow HOCl + OH^-$, and its reverse
> $HOCl + OH^- \longrightarrow OCl^- + H_2O$ (both fast)
> Step 2. $I^- + HOCl \longrightarrow HOI + Cl^-$ (slow)
> Step 3. $HOI + OH^- \longrightarrow IO^- + H_2O$ (fast)

Write the rate law implied by this mechanism.

13.69 Three mechanisms for the reaction $2 O_3(g) \rightarrow 3 O_2(g)$ have been proposed:
(a) A one-step mechanism: $O_3 + O_3 \rightarrow O_2 + O_2 + O_2$

(b)
> Step 1. $O_3 \longrightarrow O_2 + O$ (slow)
> Step 2. $O + O_3 \longrightarrow O_2 + O_2$ (fast)

(c)
> Step 1. $O_3 \longrightarrow O_2 + O$, and its reverse
> $O_2 + O \longrightarrow O_3$ (both fast)
> Step 2. $O + O_3 \longrightarrow O_2 + O_2$ (slow)

Decide which mechanism agrees with the experimental rate law and explain your reasoning:

$$\text{Rate} = k\frac{[O_3]^2}{[O_2]} = k[O_3]^2[O_2]^{-1}$$

13.70 When the rate of the reaction $2 NO(g) + O_2(g) \rightarrow 2 NO_2(g)$ was studied, it was found that the rate doubled when the O_2 concentration alone was doubled, but quadrupled when the NO concentration alone was doubled. Which of the following mechanisms accounts for these observations?

(a)
> Step 1. $NO + O_2 \longrightarrow NO_3$, and its reverse (both fast)
> Step 2. $NO + NO_3 \longrightarrow NO_2 + NO_2$ (slow)

(b)
> Step 1. $NO + NO \longrightarrow N_2O_2$ (slow)
> Step 2. $O_2 + N_2O_2 \longrightarrow N_2O_4$ (fast)
> Step 3. $N_2O_4 \longrightarrow NO_2 + NO_2$ (fast)

Supplementary Exercises

13.71 A reaction of the form $3 A + B \rightarrow$ products was determined to be first order in A and one-half order in B. (a) What is the rate law for the reaction? (b) What are the units of the rate constant? Assume that the concentrations are expressed in moles per liter and time in minutes.

13.72 When the concentration of 2-bromo-2-methyl-propane, C_4H_9Br, is doubled, the rate of the reaction $C_4H_9Br(aq) + OH^-(aq) \rightarrow C_4H_9OH(aq) + Br^-(aq)$ increases by a factor of 2. When the C_4H_9Br and OH^- concentrations are both doubled, the rate also increases by a factor of 2. What are the reactant orders and overall order of the reaction?

13.73 The first-order decomposition of compound X, a gas, is carried out and the data are represented below. The green spheres represent the molecules of the compound; the decomposition products are not shown. The times at which the images were taken are shown below each flask. (a) Determine the half-life of the reaction. (b) Draw the appearance of the molecular image at 8 s.

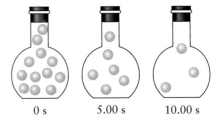

0 s   5.00 s   10.00 s

13.74 The following rate data were collected for the reaction $2 A(g) + 2 B(g) + C(g) \rightarrow 3 G(g) + 4 F(g)$:

	Initial concentration, mmol/L			Initial rate,
Experiment	$[A]_0$	$[B]_0$	$[C]_0$	(mmol G)/L·s
1	10.0	100.0	700.0	2.0
2	20.0	100.0	300.0	4.0
3	20.0	200.0	200.0	16.0
4	10.0	100.0	400.0	2.0
5	4.62	0.177	12.4	?

(a) What is the order for each reactant and the overall order of the reaction? (b) Write the rate law for the reaction.

(c) Determine the reaction rate constant. (d) Predict the initial rate for experiment 5.

13.75 A reaction has an activation energy of 100. kJ/mol and the reverse reaction has an activation energy of 300. kJ/mol. (a) Is the reaction exothermic or endothermic? (b) Which rate increases more when the temperature is raised, that of the forward reaction or that of the reverse reaction? Explain.

13.76 (a) The rate law for the thermal decomposition of acetaldehyde, $CH_3CHO(g) \rightarrow CH_4(g) + CO(g)$, under certain conditions is rate $= k[CH_3CHO]^{3/2}$. What is the order of the reaction? (b) The rate law for the "hot wire" decomposition of ammonia, $2\,NH_3(g) \rightarrow N_2(g) + 3\,H_2(g)$, is rate $= k$. What is the order of the reaction?

13.77 All radioactive decay processes follow first-order kinetics. The half-life of the radioactive isotope tritium (^3H or T) is 12.3 years. How much of a 1.0-μg sample of tritium would remain after 5.2 years?

13.78 The concentration of a species A was initially 0.20 mol/L and decreased by the reaction $2\,A \rightarrow B + C$ to 0.10 mol/L in 100. s by first-order kinetics. Calculate the time needed for the concentration of A to fall to (a) one-eighth; (b) one-thirty-second of its initial concentration.

13.79 Why might a bimolecular collision with more than the minimum energy required for reaction not produce products?

13.80 Explain why a finely divided solid catalyst is more effective in increasing a reaction rate than the same mass of the catalyst in large pellets.

13.81 Which of the following plots will be linear? (a) [A] as a function of time for a reaction first order in A; (b) [A] as a function of time for a reaction zero order in A; (c) ln[A] as a function of time for a reaction first order in A; (d) 1/[A] against time for a reaction second order in A; (e) ln k against temperature; (f) initial rate against [A] for a reaction first order in A.

13.82 Determine the molecularity of the following elementary reactions:
(a) ○ + ●—● ⟶ ○—● + ●
(b) ○—○ ⟶ ○ + ○
(c) ○ + ○ + ● ⟶ ● + ○—○ (the blue atom takes part in the collision, but is itself unchanged.)

13.83 The rate law of the reaction $2\,NO(g) + 2\,H_2(g) \rightarrow N_2(g) + 2\,H_2O(g)$ is rate $= k[NO]^2[H_2]$, and the mechanism that has been proposed is

> **Step 1.** $NO + NO \longrightarrow N_2O_2$ and its reverse, $N_2O_2 \longrightarrow NO + NO$
>
> **Step 2.** $N_2O_2 + H_2 \longrightarrow N_2O + H_2O$
>
> **Step 3.** $N_2O + H_2 \longrightarrow N_2 + H_2O$

(a) Which step in the mechanism is likely to be rate determining? Explain your answer. (b) Sketch a reaction profile

for the (exothermic) overall reaction. Label the activation energies of each step.

13.84 The mechanism of the reaction A $\rightarrow$ B, consists of two steps, involving the formation of a reaction intermediate. Overall, the reaction is exothermic. (a) Sketch the reaction profile, labeling the activation energies for each step. (b) Indicate on the same diagram the effect of a catalyst on the first step of the reaction.

13.85 The activation energy of the decomposition of ammonia to its elements is reduced from 350. kJ/mol to 162 kJ/mol in the presence of a tungsten catalyst. By what factor is the rate constant increased at 700.°C (all other factors being equal)?

13.86 The activation energy of the reaction $H_2(g) + I_2(g) \rightarrow 2\,HI(g)$ is reduced from 184 kJ/mol to 59 kJ/mol in the presence of a platinum catalyst. By what factor is the rate constant increased by the platinum at 600. K, all other factors being equal?

13.87 (a) Calculate the activation energy for the reaction between methyl bromide, CH_3Br, and hydroxide ions in water from an Arrhenius plot of the following data:

Temperature, °C	k, mL/mol·s
24	1.3
28	2.0
32	3.0
36	4.4
40.	6.4

(b) Calculate the rate constant at 25°C.

13.88 (a) Calculate the activation energy for the acid hydrolysis of sucrose from an Arrhenius plot of the following data:

Temperature, °C	k, mL/mol·s
24	4.8
28	7.8
32	13
36	20.
40.	32

(b) Calculate the rate constant at 37°C (body temperature).

13.89 Determine the rate constants for the following second-order reactions: (a) $2\,A \rightarrow B + 2\,C$, given that the concentration of A decreases from 2.5 mmol/L to 1.25 mmol/L in 100. s. (b) $3\,A \rightarrow 2\,D + C$, given that $[A]_0 = 0.30$ mol/L, and that the concentration of C increases to 0.010 mol/L in 200. s.

13.90 When a 1:1 mole ratio mixture of hydrogen and chlorine is exposed to sunlight, the reaction $H_2(g) + Cl_2(g) \rightarrow 2\,HCl(g)$ occurs explosively. The proposed chain reaction mechanism is thought to be

Step 1. $Cl_2 \longrightarrow Cl + Cl$, an initiation step caused by light

Step 2. A propagation step that results in the formation of a hydrogen atom

Step 3. A propagation step in which the hydrogen atom reacts with Cl_2

Steps 4, 5, and 6. three termination steps that occur between the two radicals

Write the equations for the elementary reactions in steps 2 to 6.

13.91 The half-life for the first-order reaction of a substance A is 50.5 s when $[A]_0 = 0.84$ mol/L. Calculate the time needed for the concentration of A to decrease to (a) one-sixteenth; (b) one-fourth; (c) one-fifth of its original value.

13.92 The first-order rate constant for the decomposition of N_2O (to N_2 and O_2) at 1000 K is 0.76 L/mol·s. Calculate the time for an initial NO_2 concentration of 0.20 mol/L to decrease to (a) one-half; (b) one-sixteenth; (c) one-ninth of its initial concentration.

13.93 The following data were obtained for the reaction $2 A \rightarrow B$:

Time, s	[A], mmol/L
0	100.
5	14.1
10.	7.8
15	5.3
20.	4.0

(a) Plot the data to determine the order of the reaction. (*Hint:* The integrated rate law that gives a linear plot is the one that corresponds to the reaction order.) (b) Determine the rate constant.

13.94 The following data were obtained on the reaction $2 A \rightarrow B$:

Time, s	[A], mmol/L
0	250.
100.	143
200.	81
300.	45
400.	25

(a) Plot the data to determine the order of the reaction. (*Hint:* The integrated rate law that gives a linear plot is the one that corresponds to the reaction order.) (b) Determine the rate constant.

13.95 The half-life of a substance taking part in a second-order reaction A $\rightarrow$ products is inversely proportional to the initial concentration of A. How may this half-life be used to predict the time needed for the concentration to fall to

(a) one-half; (b) one-fourth; (c) one-sixteenth of its initial value?

13.96 Under certain conditions, the reaction $H_2(g) + Br_2(g) \rightarrow 2 HBr(g)$ obeys the rate law rate $= k[H_2][Br_2]^{1/2}$. However, the reaction hardly proceeds at all if another substance is added that rapidly removes hydrogen and bromine atoms. Suggest a mechanism for the reaction.

13.97 Determine the time required for each of the following second-order reactions to occur:
(a) $2 A \rightarrow B + C$, for the concentration of A to decrease from 0.10 mol/L to 0.080 mol/L, given that $k = 0.010$ L/mol·min.
(b) $A \rightarrow 2 B + C$, when $[A]_0 = 0.45$ mol/L, for the concentration of B to increase to 0.45 mol/L, given that $k = 0.0045$ L/mol·min.

13.98 The reaction of methane, CH_4, with chlorine gas occurs by a mechanism similar to that described in Exercise 13.90. (a) Write the mechanism. (b) Identify the initiation, propagation, and termination steps. (c) What products are predicted from the mechanism?

Applied Exercises

For Exercises 13.99–13.101, see Applying Chemistry: Case Study 13.

13.99 The rates of many enzyme-catalyzed reactions exhibit the dependence on substrate concentration shown in the plot in Applying Chemistry: Case Study 13. The catalyzed reaction is first order in both enzyme and substrate at low concentrations of substrate. What is the order of the reaction in the substrate at high concentrations of substrate? Explain your reasoning.

13.100 The following mechanism for enzyme catalysis, in which E represents the enzyme, S the substrate, and P the product formed by S, has been proposed for low substrate concentrations:

$$E + S \rightleftharpoons ES \text{ (fast)}$$

$$ES \longrightarrow E + P \text{ (slow)}$$

(a) Write the rate law consistent with this mechanism.
(b) Would this mechanism be valid for high concentrations of substrate? Explain your conclusion.

13.101 How would the presence of an enzyme inhibitor affect the slope of the plot in Applying Chemistry: Case Study 13?

13.102 Raw milk sours in about 4 h at 28°C, but in about 48 h in a refrigerator at 5°C. What is the activation energy for the souring of milk?

13.103 The following mechanism has been suggested to explain the contribution of chlorofluorocarbons to the destruction of the ozone layer:

$$\textbf{Step 1.} \ O_3 + Cl \longrightarrow ClO + O_2$$

$$\textbf{Step 2.} \ ClO + O \longrightarrow Cl + O_2$$

(a) What is the catalyst in the reaction? (b) What is the reaction intermediate? (c) Identify the radicals in the mechanism. (d) Identify the steps as initiating, propagating, or terminating. (e) Write a chain-terminating step for the reaction.

13.104 The radioactive decay of carbon-14 is first order, and the half-life is 5730 years. While a plant or animal is living, it has a constant proportion of carbon-14 (relative to carbon-12) in its composition. When the organism dies, the proportion of carbon-14 decreases as a result of radioactive decay, and the age of the organism can be determined if the proportion of carbon-14 in its remains is measured. If the proportion of carbon-14 in an ancient piece of wood is found to be one-fourth that in living trees, how old is the sample?

Integrated Exercises

13.105 The rate law for the reaction $H_2SeO_3(aq) + 6 I^-(aq) + 4 H^+(aq) \rightarrow Se(s) + 2 I_3^-(aq) + 3 H_2O(l)$ is rate $= k[H_2SeO_3][I^-]^3[H^+]^2$ with $k = 5.0 \times 10^5$ $L^5/mol^5 \cdot s$. (a) What is the initial reaction rate when $[H_2SeO_3]_0 = [I^-]_0 = 0.020$ mol/L and $[H^+]_0 = 0.010$ mol/L? (b) Which is the limiting reactant? (c) If the reaction mixture has a total initial volume of 1.00 L, what mass of selenium is produced?

13.106 At 328 K, the *total* pressure of the dinitrogen pentoxide decomposition to NO_2 and O_2 varied with time as shown by the data that follow. Use the data to find the rate in moles per liter per minute (mol/L·min) at each time.

Time, min	Pressure, kPa
0	27.3
5.0	43.7
10.0	53.6
15.0	59.4
20.0	63.0
30.0	66.3

13.107 The half-life for the (first-order) decomposition of azomethane, $CH_3N=NCH_3$, in the reaction $CH_3N=NCH_3(g) \rightarrow N_2(g) + C_2H_6(g)$ is 1.02 s at 300.°C. A 45.0-mg sample of azomethane is placed in a 300.-mL reaction vessel and heated to 300.°C. (a) What mass (in milligrams) of azomethane remains after 10.0 s? (b) Determine the partial pressure exerted by the $N_2(g)$ in the reaction vessel after 3.0 s.

13.108 The second-order gas-phase reactions in Table 13.4 show wide variations in activation energy. In an activated complex, the reactant bonds are lengthened while the product bonds are beginning to form. Consider what bonds need to be stretched to form the activated complex and use bond enthalpies (Sections 9.6–9.7) to explain the differences in activation energies.

Chemical Equilibrium

Whenever chemical engineers design a factory to make a new chemical—a drug, a new kind of paint or fabric, or a fuel—they have a problem to overcome. They want the best possible yield from the reactants, but they find that most reactions do not go to completion. Reactions continue at changing rates until a certain mixture of reactants and products has formed. Then, gradually the concentration of product stops increasing. It looks as though the reaction has come to a halt despite the presence of plenty of reactants. Similarly, in reactions carried out in a laboratory, it often seems impossible to get a yield of 100%. What is going on? Why do reactions seem to stop before they reach completion? Can we modify the conditions to improve the yield?

To answer these questions, we need to know about **chemical equilibrium,** the stage of a reaction at which there is no further tendency for the composition to change. We have already considered several physical processes that reach dynamic equilibrium. For example, a liquid in a sealed container reaches equilibrium with its vapor (Section 10.14), and a solute in a saturated solution is at equilibrium with undissolved solute (Section 12.2). This chapter shows how to apply the same ideas to chemical reactions. Chemical equilibria are dynamic equilibria, with the forward and reverse reactions occurring at the same rate. The key idea is that *dynamic equilibria are responsive to changes in the conditions.*

All reactions have a tendency to approach equilibrium, whether they seem to go to completion or whether they seem to form no products at all. In many cases (combustion is a good example), the equilibrium composition corresponds to virtually complete conversion of reactants into products (subject to the presence of any limiting reactant). Reactions that seem not to go at all, like the corrosion of gold or the decomposition of water into hydrogen and oxygen, also tend to approach equilibrium; but in these

This chemical engineer is testing a process for the formation of new liquid fuels from coal and petroleum. Such methods may transform the generation of energy worldwide.

FIGURE 14.1

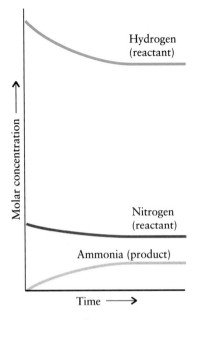

FIGURE 14.1

In the synthesis of ammonia, the molar concentrations of N_2, H_2, and NH_3 change with time until there is no further net change and the concentrations settle into values corresponding to a mixture in which all three substances are present.

reactions the equilibrium composition corresponds to virtually unchanged reactants. Many reactions, however, approach an equilibrium corresponding to a composition in which there are significant concentrations of both reactants and products. Finding and controlling the position of equilibrium in such reactions is a primary focus of this chapter.

EQUILIBRIUM AND COMPOSITION

The world was in desperate need of nitrogen-based fertilizers at the opening of the twentieth century. Almost all the nitrates used for fertilizers and explosives were quarried from deposits in Chile, and the limited supply could not keep up with the demand in a world preparing for war. The problem was solved by the German chemist Fritz Haber, who, through determination, persistence, and a modicum of luck, found an economical way to capture the nitrogen of the air for agriculture. His success, which won him a Nobel prize, was due to his application of the principles of equilibria and kinetics introduced in this and the previous chapter.

Haber was looking for a way to "fix" nitrogen—that is, a way to turn atmospheric nitrogen into compounds. The problem is that nitrogen has a strong triple bond, and a correspondingly high bond enthalpy (944 kJ/mol). As a result, it is very unreactive. Ammonia is much more reactive than nitrogen (the average N—H bond enthalpy in ammonia is only 388 kJ/mol), so Haber investigated ways to convert nitrogen into ammonia by using the reaction

$$N_2(g) + 3 H_2(g) \longrightarrow 2 NH_3(g)$$

Microorganisms in the roots of certain plants fix nitrogen, but they use complicated enzymes that chemists are still trying to replicate. Haber was looking for a simple, efficient, and economical way to produce ammonia, using the technological resources of the early twentieth century. To do that, he had to understand what happens when a reaction reaches equilibrium.

14.1 The Reversibility of Chemical Reactions

In one series of experiments, Haber started with known amounts of nitrogen and hydrogen maintained at high temperature and pressure and, at regular intervals, determined the amount of ammonia present. He worked out from the reaction stoichiometry how much of the nitrogen and hydrogen remained (Fig. 14.1). As the graph shows, after a certain time, the composition of the mixture remains the same, even though most of the reactants are still present. The reaction has reached equilibrium.

To imagine what is going on, we have to consider both the forward reaction, which is given above, and the reverse reaction:

$$2 NH_3(g) \longrightarrow N_2(g) + 3 H_2(g)$$

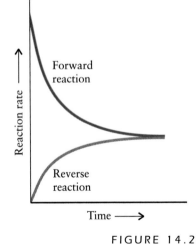

FIGURE 14.2

When we plot the rates of the forward and reverse reactions for the formation of ammonia on one graph, we can see that as the forward rate decreases, the reverse rate increases, until they are equal. At this point, the reaction is at equilibrium and the rates remain constant.

An NH$_3$ molecule forms by a complicated mechanism in which N$_2$ and H$_2$ molecules collide and exchange atoms. The NH$_3$ molecules that form also collide with one another and with any N$_2$ and H$_2$ molecules still present. As a result of these collisions, the NH$_3$ molecules decompose back into nitrogen and hydrogen. As the concentration of the product grows, collisions involving product molecules take place more frequently and the reverse reaction goes faster. At the same time, reactants are being depleted and the forward reaction slows down. The system is at equilibrium when the forward and reverse reactions take place at the same rate (Fig. 14.2). To symbolize the condition in which the rates of the forward and reverse reaction are equal, we write

$$N_2(g) \; + \; 3\,H_2(g) \; \rightleftharpoons \; 2\,NH_3(g) \qquad \textbf{(A)}$$

The symbol $\rightleftharpoons$ always signifies a condition of chemical equilibrium in which the forward and reverse reactions still continue, but at equal rates.

At the molecular level, chemical reactions at equilibrium never actually stop, even though the concentrations remain constant. We could perform an experiment to show that the forward and reverse reactions still continue at equilibrium. For example, we could carry out two ammonia syntheses with exactly the same starting conditions, but with D$_2$ (deuterium) in place of H$_2$ in one of them (Fig. 14.3). The two reaction mixtures reach equilibrium with the same composition, except that N$_2$, D$_2$, and ND$_3$ are present in one system and N$_2$, H$_2$, and NH$_3$ in the other. Suppose we now combine the two equilibrium mixtures and leave them for a while. Later, we find that the concentration of ammonia is just the same as before. However, when we analyze the sample with a mass spectrometer, we find that all isotopic forms of ammonia (NH$_3$, NH$_2$D, NHD$_2$, and ND$_3$) and all isotopic forms of hydrogen (H$_2$, HD, and D$_2$) are present. The presence of both H and D atoms in the same molecules must result from a continuation of the forward and reverse reactions in the mixture. If the reactions had simply stopped when they reached equilibrium, there would have been no mixing of isotopes in this way.

> *Chemical reactions reach a state of dynamic equilibrium in which the rates of the forward and reverse reactions are equal and there is no net change in composition.*

14.2 The Equilibrium Constant

In 1864, long before Haber began his work, the Norwegians Cato Guldberg (a mathematician) and Peter Waage (a chemist) had discovered the mathematical relation that describes quantitatively the composition of a reaction mixture at equilibrium. For example, look at the data in Table 14.1 for the reaction between SO$_2$ and O$_2$:

$$2\,SO_2(g) \; + \; O_2(g) \; \rightleftharpoons \; 2\,SO_3(g)$$

In the experiments reported in Table 14.1, several different mixtures with different initial compositions were prepared at 1000 K and allowed to reach equilibrium. The concentration of each species at equilibrium is listed in the table. At first sight, there seems to be no pattern in the data. However, Guldberg and Waage found that the value of the following quantity was the same for every experiment, no matter what the initial compositions were:

$$K_c = \frac{[SO_3]^2}{[SO_2]^2[O_2]}$$

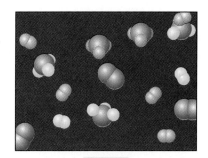

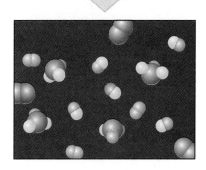

FIGURE 14.3

In an experiment showing that equilibrium is dynamic, a reaction mixture in which N$_2$ (pairs of blue spheres), D$_2$ (pairs of yellow spheres), and ND$_3$ have reached equilibrium is mixed with one with the same concentrations of N$_2$, H$_2$ (pairs of gray spheres), and NH$_3$. After some time, the concentrations of nitrogen, hydrogen, and ammonia are found to be the same, but the D atoms are distributed among the hydrogen and ammonia molecules.

This reaction is used in another important industrial process, the production of sulfuric acid.

Table 14.1 *Equilibrium data and the equilibrium constant for the reaction* $2\,SO_2(g) + O_2(g) \rightleftharpoons 2\,SO_3(g)$ *at 1000.K*

[SO$_2$], mol/L	[O$_2$], mol/L	[SO$_3$], mol/L	K_c
0.660	0.390	0.0840	0.0415
0.0380	0.220	0.00360	0.0409
0.110	0.110	0.00750	0.0423
0.950	0.880	0.180	0.0408
1.44	1.98	0.410	0.0409

Average: 0.0413

The subscript c on K_c tells us that K is defined in terms of molar concentrations, and [X] is the molar concentration of X, with the units struck out. For example, if the molar concentration of SO$_2$ happens to be 2.3×10^{-3} mol/L, then we write [SO$_2$] $= 2.3 \times 10^{-3}$. Within experimental error, the same value of K_c is obtained in each case. This remarkable result shows that K_c is characteristic of the composition of the reaction at equilibrium. It is known as the **equilibrium constant** for the reaction.

The equilibrium composition of any reaction can be expressed in a similar way. The relationship is called the **law of mass action:**

The composition of the equilibrium mixture of the reaction

$$a\,A + b\,B \rightleftharpoons c\,C + d\,D$$

is characterized by an equilibrium constant calculated from the expression

$$K_c = \frac{[C]^c[D]^d}{[A]^a[B]^b} \tag{1}$$

with [X] representing the molar concentration (with units omitted) of X at equilibrium.

The products (C and D) occur in the numerator, and the reactants (A and B) in the denominator. Each [X] is raised to a power equal to its stoichiometric coefficient in the balanced chemical equation for the reaction. The same expression applies to reactions in solution, for which [X] is the numerical value of the molarity of the species X. The rules for writing equilibrium constants presented here and in the next two sections are summarized in Toolbox 14.1 at the end of Section 14.4.

> Note that although [X] retains the units mol/L in kinetics calculations, it is unitless in equilibrium calculations.

Example 14.1 *Writing an equilibrium constant*

One of the steps in the commercial production of nitric acid is the oxidation of ammonia to nitrogen monoxide (oxidation state $+2$). Write the equilibrium constant K_c for the reaction

$$4\,NH_3(g) + 5\,O_2(g) \rightleftharpoons 4\,NO(g) + 6\,H_2O(g)$$

Strategy Identify the stoichiometric coefficients in the chemical equation and write the equilibrium constant in the form

$$K_c = \frac{\text{molar concentrations of products}}{\text{molar concentrations of reactants}}$$

with each concentration raised to the power of the corresponding stoichiometric coefficient.

Solution The equilibrium constant is

$$K_c = \frac{[NO]^4[H_2O]^6}{[NH_3]^4[O_2]^5}$$

with the molar concentrations measured at equilibrium.

Self-Test 14.1A Write the equilibrium constant for the reaction $3\,ClO_2^-(aq) \rightleftharpoons 2\,ClO_3^-(aq) + Cl^-(aq)$.

[**Answer:** $K_c = [ClO_3^-]^2[Cl^-]/[ClO_2^-]^3$]

Self-Test 14.1B Write the equilibrium constant for the reaction $2\,O_3(g) \rightleftharpoons 3\,O_2(g)$.

Equilibrium constants are normally given for chemical equations that have been written with the smallest possible whole numbers for the stoichiometric coefficients. However, if we change the stoichiometric coefficients in a chemical equation (for instance, by multiplying through by a factor), then we must make sure that the equilibrium constant reflects that change. The law of mass action is still valid, but the form of the expression and the calculated value of K_c reflect the way we write the chemical equation. For example, at 700 K,

$$H_2(g) + I_2(g) \rightleftharpoons 2\,HI(g) \qquad K_{c1} = \frac{[HI]^2}{[I_2][H_2]} = 54$$

If we write the equation as

$$2\,H_2(g) + 2\,I_2(g) \rightleftharpoons 4\,HI(g)$$

then the equilibrium constant becomes

$$K_{c2} = \frac{[HI]^4}{[I_2]^2[H_2]^2} = \left(\frac{[HI]^2}{[I_2][H_2]}\right)^2$$
$$= (K_{c1})^2 = (54)^2 = 2.9 \times 10^3$$

In general, if we multiply a chemical equation by a factor n, then we raise K_c to the nth power.

Now suppose we reverse the original equation for the reaction:

$$2\,HI(g) \rightleftharpoons H_2(g) + I_2(g)$$

This equation still describes the same equilibrium, but how is the equilibrium constant related to the one we wrote earlier? To find out, we simply write the equilibrium constant from this equation and obtain

$$K_{c3} = \frac{[H_2][I_2]}{[HI]^2} = \frac{1}{K_{c1}} = \frac{1}{54} = 1.8 \times 10^{-2}$$

In general, the equilibrium constant for an equilibrium written in one direction is the inverse of the equilibrium constant for the equilibrium written in the opposite direction.

If a chemical equation can be expressed as the sum of two or more chemical equations, then the equilibrium constant for the overall reaction is the *product* of the equilibrium constants for the component reactions. For example, consider the three gas-phase reactions

$$2\,P(g)\,+\,3\,Cl_2(g) \rightleftharpoons 2\,PCl_3(g) \qquad K_c' = \frac{[PCl_3]^2}{[P]^2[Cl_2]^3}$$

$$PCl_3(g)\,+\,Cl_2(g) \rightleftharpoons PCl_5(g) \qquad K_c'' = \frac{[PCl_5]}{[PCl_3][Cl_2]}$$

$$2\,P(g)\,+\,5\,Cl_2(g) \rightleftharpoons 2\,PCl_5(g) \qquad K_c = \frac{[PCl_5]^2}{[P]^2[Cl_2]^5}$$

The third reaction is the sum of the first reaction and twice the second reaction:

$$
\begin{aligned}
2\,P(g)\,+\,3\,Cl_2(g) &\rightleftharpoons 2\,PCl_3(g)\\
PCl_3(g)\,+\,Cl_2(g) &\rightleftharpoons PCl_5(g)\\
PCl_3(g)\,+\,Cl_2(g) &\rightleftharpoons PCl_5(g)\\
\hline
2\,P(g)\,+\,5\,Cl_2(g) &\rightleftharpoons 2\,PCl_5(g)
\end{aligned}
$$

and its equilibrium constant, K_c, can be written

$$
\begin{aligned}
K_c &= \frac{[PCl_5]^2}{[P]^2[Cl_2]^5} = \frac{[PCl_3]^2}{[P]^2[Cl_2]^3} \times \frac{[PCl_5]}{[PCl_3][Cl_2]} \times \frac{[PCl_5]}{[PCl_3][Cl_2]}\\
&= K_c' \times K_c'' \times K_c''
\end{aligned}
$$

Notice that because we used the second chemical equation twice in the sum, the equilibrium constant of that reaction appears twice in the product.

Self-Test 14.2A At 500 K, K_c for $H_2(g)\,+\,D_2(g) \rightleftharpoons 2\,HD(g)$ is 3.6. What is the value of K_c for $2\,HD(g) \rightleftharpoons H_2(g)\,+\,D_2(g)$?

[*Answer:* 0.28]

Self-Test 14.2B At 500 K, K_c for $F_2(g) \rightleftharpoons 2\,F(g)$ is 7.3×10^{-13}. What is the value of K_c for $\frac{1}{2}\,F_2(g) \rightleftharpoons F(g)$?

Each reaction has its own characteristic equilibrium constant, with a value that can be changed only by varying the temperature (Table 14.2). Whatever the initial composition of the reaction mixture, at a given temperature its equilibrium composition will always correspond (in practice, within about 5%) to the value of K_c for that reaction. Therefore, to find the numerical value of an equilibrium constant at a certain temperature, we can take any initial mixture of reagents for the reaction and allow the reaction to reach equilibrium at the temperature of interest. Then we measure the molar concentrations of the reactants and products, and substitute them into the expression for K_c.

As we see in more detail shortly, the physical significance of the value of the equilibrium constant is that when K_c is large, the equilibrium composition is rich in products; when K_c is small, we can expect the equilibrium composition to be rich in reactants. In other words, a very large equilibrium constant corresponds to a reaction that "goes"; a very small equilibrium constant corresponds to a reaction that "does not go."

Equilibrium constants are written by dividing the product concentrations (raised to powers equal to their stoichiometric coefficients) by the reactant concentrations (raised to powers equal to their coefficients).

Table 14.2 Equilibrium constants, K_c, for various reactions

Reaction	Temperature, K*	K_c
$H_2(g) + Cl_2(g) \rightleftharpoons 2\,HCl(g)$	300	4.0×10^{31}
	500	4.0×10^{18}
	1000	5.1×10^8
$H_2(g) + Br_2(g) \rightleftharpoons 2\,HBr(g)$	300	1.9×10^{17}
	500	1.3×10^{10}
	1000	3.8×10^4
$H_2(g) + I_2(g) \rightleftharpoons 2\,HI(g)$	298	794
	500	160
	700	54
$2\,BrCl(g) \rightleftharpoons Br_2(g) + Cl_2(g)$	300	377
	500	32
	1000	5
$2\,HD(g) \rightleftharpoons H_2(g) + D_2(g)$	100	0.52
	500	0.28
	1000	0.26
$F_2(g) \rightleftharpoons 2\,F(g)$	500	7.3×10^{-13}
	1000	1.2×10^{-4}
	1200	2.7×10^{-3}
$Cl_2(g) \rightleftharpoons 2\,Cl(g)$	1000	1.2×10^{-7}
	1200	1.7×10^{-5}
$Br_2(g) \rightleftharpoons 2\,Br(g)$	1000	4.1×10^{-7}
	1200	1.7×10^{-5}
$I_2(g) \rightleftharpoons 2\,I(g)$	800	3.1×10^{-5}
	1000	3.1×10^{-3}
	1200	6.8×10^{-2}

*All temperatures except 298 K are correct to ± 10 K.

14.3 Rates and Equilibrium

At equilibrium, the rates of the forward and reverse reactions must be equal. Because the rates depend on rate constants and concentrations, we can expect to find a relation between rate constants for elementary reactions and equilibrium constants for the overall reaction.

The equilibrium constant for a chemical reaction that has the form $2\,A \rightleftharpoons C + D$ is

$$K_c = \frac{[C][D]}{[A]^2}$$

Suppose that experiments show that both the forward reaction and the reverse reaction are elementary second-order reactions, with the following rate laws:

$$A + A \longrightarrow C + D \qquad Rate = k[A]^2$$
$$C + D \longrightarrow A + A \qquad Rate = k'[C][D]$$

FIGURE 14.4

The equilibrium constant for a reaction is equal to the ratio of the rate constants for the forward and reverse elementary reactions that continue in a state of dynamic equilibrium. (a) A relatively large forward rate constant means that the forward rate can match the reverse rate even though only a small amount of reactants is present. (b) Conversely, if the reverse rate constant is relatively large, then the forward and reverse rates are equal when only small amounts of products are present.

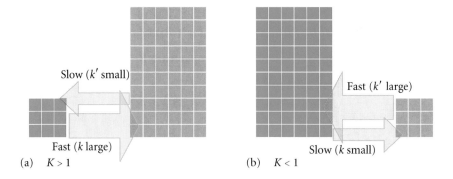

Slow (k' small)

Fast (k' large)

Fast (k large)

Slow (k small)

(a) $K > 1$

(b) $K < 1$

At equilibrium, these two rates are equal, so we can write

$$k[A]^2 = k'[C][D]$$

It follows that, at equilibrium,

$$\frac{[C][D]}{[A]^2} = \frac{k}{k'}$$

Because this expression has the same form as the expression for the equilibrium constant, we see that

$$K_c = \frac{k}{k'} \qquad (2)$$

That is, *the equilibrium constant for a reaction is equal to the ratio of the rate constants for the forward and reverse elementary reactions that contribute to the overall reaction.*

The relation we have just derived helps us understand why different reactions have different values for the equilibrium constant. The equilibrium constant is much larger than 1 (and products are favored) when the rate constant for the forward direction (k) is much larger than the rate constant for the reverse direction (k'). In this case, the fast forward reaction builds up a high concentration of products before reaching equilibrium (Fig. 14.4). In contrast, the equilibrium constant is very small (and reactants are favored) when $k \ll k'$. Now the reverse reaction removes the products rapidly, and at equilibrium only a tiny amount is present.

We derived the relation between the equilibrium constant and the rate constant for a single-step reaction. However, suppose that a reaction has a complex mechanism in which the elementary reactions have rate constants k_1, k_2, and so on, and the reverse elementary reactions have rate constants k_1', k_2', and so on. Then, by an argument similar to that for the single-step reaction, the overall equilibrium constant is related to the rate constants as follows:

$$K_c = \frac{k_1}{k_1'} \times \frac{k_2}{k_2'} \times \dots \qquad (3)$$

The equilibrium constant for an elementary reaction is equal to the ratio of the forward and reverse rate constants of the reaction.

14.4 Heterogeneous Equilibria

Chemical equilibria in which all reactants and products are in the same phase are called **homogeneous equilibria.** All the equilibria described so far in this chapter are homogeneous. Equilibria in systems having more than one phase are called **heterogeneous equilibria.** The equilibrium between water vapor and liquid water in a closed system is heterogeneous:

$$H_2O(l) \rightleftharpoons H_2O(g)$$

In this reaction, there is a gas phase and a liquid phase. Likewise, the equilibrium between a solid and its saturated solution,

$$Ca(OH)_2(s) \rightleftharpoons Ca^{2+}(aq) + 2\,OH^-(aq)$$

is a heterogeneous equilibrium. Here the phases are liquid solution and pure solid.

Like the equilibria in these examples, heterogeneous equilibria often involve a pure solid or liquid. We can simplify the equilibrium expression for heterogeneous equilibria involving a pure liquid or a pure solid by realizing that *the molar concentration of a pure solid or liquid is a constant, independent of the amount present.* To see why, we note that the molar concentration of a substance is measured by dividing the number of moles of the substance by the volume it occupies. However, moles divided by volume is proportional to mass divided by volume, or the density of the substance. Because density is an intensive property, it follows that the molar concentration must be intensive too. Therefore, the molar concentration of a pure solid or liquid always has the same value, no matter how much or how little of the substance is present.

Because the molar concentration of a pure solid or a pure liquid does not change as a reaction approaches equilibrium, we can ignore it in all equilibrium calculations. Therefore, when we write an equilibrium constant for a reaction involving pure solids or liquids, we ignore substances in separate solid or liquid phases. For example, we write

$$Ca(OH)_2(s) \rightleftharpoons Ca^{2+}(aq) + 2\,OH^-(aq) \qquad K_c = [Ca^{2+}][OH^-]^2$$

Note that, for an electrolyte, we use the ionic equation for the reaction, with each species written out explicitly and making its own contribution to the law of mass action. Similarly, in the equilibrium between nickel, carbon monoxide, and nickel carbonyl that is used in the purification of nickel,

$$Ni(s) + 4\,CO(g) \rightleftharpoons Ni(CO)_4(g) \qquad K_c = \frac{[Ni(CO)_4]}{[CO]^4}$$

The pure substances must be present for the equilibrium to exist, but they do not appear in the expression for the equilibrium constant. The concentrations of the gases appear because their concentrations change during the reaction.

Some reactions in solution involve the solvent as a reactant or a product. When the solution is very dilute, the change in solvent concentration due to the reaction is insignificant. In such cases, the solvent is treated as a pure substance and ignored when writing K_c.

> *Pure liquids and solids are ignored when writing expressions for equilibrium constants.*

We do not divide by the total volume of the system; we divide by the volume the phase actually occupies.

Recall from Section 2.3 that an intensive property is a property with a value that is independent of the size of the sample.

Self-Test 14.3A Write the equilibrium constant for $Ag_2S(s) \rightleftharpoons 2\,Ag^+(aq) + S^{2-}(aq)$.
[*Answer:* $K_c = [Ag^+]^2[S^{2-}]$]

Self-Test 14.3B Write the equilibrium constant for $P_4(s) + 5\,O_2(g) \rightleftharpoons P_4O_{10}(s)$.

14.5 Gaseous Equilibria

An example of an equilibrium that involves gases is the thermal decomposition of calcium carbonate:

$$CaCO_3(s) \rightleftharpoons CaO(s) + CO_2(g) \qquad K_c = [CO_2]$$

There are two pure solids ($CaCO_3$ and CaO) in the chemical equation, and neither appears in the expression for K_c. Next, we note that the molar concentration of a gas, n_X/V for the gas X, is proportional to its partial pressure, P_X, as we can see by rearranging the ideal gas law $PV = nRT$ into

$$\text{concentration} = \frac{n_X}{V} = \frac{P_X}{RT}$$

Partial pressure was introduced in Section 5.11.

Therefore, if we express the equilibrium constant in terms of partial pressures instead of molar concentrations, the result is still a constant. The equilibrium constant in terms of partial pressures is denoted K_p (Table 14.3). To keep K_p a pure number, we write all pressures in atmospheres, and then strike out the units. For example, for the calcium carbonate decomposition, we write

$$K_p = P_{CO_2}$$

It follows from this expression that, for the decomposition of calcium carbonate, the equilibrium partial pressure of the CO_2 in the system is equal to K_p. This relation gives us a very easy way to measure K_p: we simply measure the pressure of carbon dioxide in equilibrium with calcium carbonate and calcium oxide at the temperature of interest. At 800°C, for instance, that pressure is 0.22 atm, so $K_p = 0.22$ at that temperature.

Self-Test 14.4A Write the expression for K_p for the reaction $N_2(g) + 3 H_2(g) \rightleftharpoons 2 NH_3(g)$.

[*Answer:* $K_p = P_{NH_3}{}^2/P_{N_2}P_{H_2}{}^3$]

Self-Test 14.4B Write the expression for K_p for the reaction $2 S(s) + 3 O_2(g) \rightleftharpoons 2 SO_3(g)$.

Table 14.3 *Equilibrium constants, K_p, for various reactions*

Reaction	Temperature, K*	K_p
$N_2(g) + 3 H_2(g) \rightleftharpoons 2 NH_3(g)$	298	6.8×10^5
	400	41
	500	3.6×10^{-2}
$H_2(g) + I_2(g) \rightleftharpoons 2 HI(g)$	298	794
	500	160.
	700	54
$2 SO_2(g) + O_2(g) \rightleftharpoons 2 SO_3(g)$	298	4.0×10^{24}
	500	2.5×10^{10}
	700	3.0×10^4
$N_2O_4(g) \rightleftharpoons 2 NO_2(g)$	298	0.98
	400	47.9
	500	1700.

*All temperatures except 298 K are correct to ± 10 K.

The numerical values of K_p and K_c may be different, so it is important to specify which one is being used. Provided the gases are ideal, the two quantities are related as follows:

$$K_p = (RT)^{\Delta n} K_c = (0.082\ 06 \times T)^{\Delta n} K_c \qquad (4)$$

where Δn is the difference in the number of gas-phase molecules (products − reactants) in the chemical equation and, for this equation only, R is the gas constant in liter-atmospheres per kelvin-mole and T is the temperature in kelvins, with their units struck out. For example, $\Delta n = -2$ for the reaction $N_2(g) + 3 H_2(g) \rightleftharpoons 2 NH_3(g)$ because 4 reactant molecules become 2 product molecules, so for that reaction at 298 K,

$$K_p = \frac{K_c}{(0.082\ 06 \times 298)^2} = \frac{K_c}{598}$$

> K_p can be derived from K_c by using the ideal gas law to write $[X] = \dfrac{RT}{P_X}$; see the Web site for this book.

Self-Test 14.5A Find the relation between K_p and K_c for the reaction $2 NO_2(g) \rightleftharpoons N_2O_4(g)$ at 400. K.

[***Answer:*** $K_p = K_c/32.8$]

Self-Test 14.5B Find the relation between K_p and K_c for the reaction $Cl_2(g) + 3 F_2(g) \rightleftharpoons 2 ClF_3(g)$ at 74°C.

When should we use K_p and when K_c? When data are given as molar concentrations, we use K_c. When gas pressures are reported, we use K_p. Many relations apply to both K_c and K_p, and in such cases, we shall normally write simply K. We also use K to describe the equilibrium composition when a reaction involves both solutes in liquid solution, given as molar concentrations, and gases, given as partial pressures.

Equilibrium constants for gaseous reactions can be written by using either molar concentrations or partial pressures.

Example 14.2 *Writing the equilibrium constant from a chemical equation*

Write the equilibrium constants (a) K for $SO_2(g) + H_2O(l) \rightleftharpoons H_2SO_3(aq)$; (b) K_c for $H_2SO_3(aq) + H_2O(l) \rightleftharpoons H_3O^+(aq) + HSO_3^-(aq)$; (c) K for $SO_2(g) + 2 H_2O(l) \rightleftharpoons H_3O^+(aq) + HSO_3^-(aq)$.

Strategy Follow the procedure set out in Toolbox 14.1; note that the chemical equation in (c) is the sum of the chemical equations in (a) and (b).

Solution (a) Because H_2O is a pure liquid, the equilibrium constant is

$$K = \frac{[H_2SO_3]}{P_{SO_2}}$$

(b) For this reaction,

$$K_c = \frac{[H_3O^+][HSO_3^-]}{[H_2SO_3]}$$

Toolbox 14.1 *How to write equilibrium constants*

This Toolbox summarizes the rules for writing equilibrium constants and shows the relation between the different forms of the equilibrium constant.

Conceptual Basis

The composition of a reaction mixture at equilibrium must be consistent with the equilibrium constant. An equilibrium constant is the ratio of the concentrations (or partial pressures) of the products to those of the reactants, each raised to a power equal to its stoichiometric coefficient in the balanced chemical equation (Fig. 14.5). All equilibrium constants are dimensionless (without units).

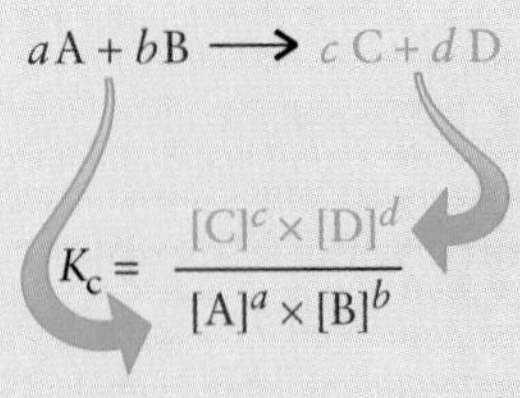

Procedure

The equilibrium constant K_p for a gaseous reaction has the following form:

$$a\,A(g) + b\,B(g) \rightleftharpoons c\,C(g) + d\,D(g) \qquad K_p = \frac{P_C^c P_D^d}{P_A^a P_B^b}$$

where P_X is the partial pressure at equilibrium, in atmospheres (with the units deleted).

The equilibrium constant K_c for a reaction that may be either gaseous or in solution has the following form:

$$a\,A + b\,B \rightleftharpoons c\,C + d\,D \qquad K_c = \frac{[C]^c[D]^d}{[A]^a[B]^b}$$

where [X] is the molar concentration of X at equilibrium (with the units mol/L deleted). Pure solids, pure liquids, and solvents in very dilute solutions are ignored when writing an equilibrium constant.

In some cases, both solutes and gas-phase species occur in the reaction: then write the equilibrium constant in terms of molar concentrations of the solutes and partial pressures of the gases (if pressure data are supplied): denote this equilibrium constant K.

The value of the equilibrium constant

Substitute the equilibrium concentrations (or partial pressures) of each substance into the expression for K_c (or K_p). Be sure to raise each concentration (or partial pressure) to a power equal to the stoichiometric coefficient of that species in the chemical equation.

$$a\,A + b\,B \longrightarrow c\,C + d\,D$$

$$K_c = \frac{[C]^c \times [D]^d}{[A]^a \times [B]^b}$$

FIGURE 14.5

The equilibrium constant is the ratio of the concentrations or partial pressures of the products to those of the reactants, each concentration raised to a power equal to its stoichiometric coefficient in the balanced equation.

Relations between equilibrium constants

Consider the equilibrium

$$a\,A + b\,B \rightleftharpoons c\,C + d\,D \qquad K_1$$

If the chemical equation is multiplied through by n, the equilibrium constant is raised to the nth power.

$$na\,A + nb\,B \rightleftharpoons nc\,C + nd\,D \qquad K_2 = K_1^{\,n}$$

If the chemical equation is written in reverse, the equilibrium constant is the inverse of that written above:

$$c\,C + d\,D \rightleftharpoons a\,A + b\,B \qquad K_3 = 1/K_1$$

The relation between K_p and K_c for a gas-phase equilibrium is

$$K_p = (0.082\,06 \times T)^{\Delta n} K_c$$

where T is the numerical value of the temperature in kelvins and Δn is the change in the number of gas-phase molecules in the chemical equation. For a reaction in which the number of gas-phase reactant and product molecules is the same, $K_c = K_p$.

Combinations of equilibrium constants

If a chemical equation for an equilibrium can be expressed as the sum of chemical equations with equilibrium constants $K_1, K_2, \ldots$, the equilibrium constant K for the overall reaction is the product of the individual equilibrium constants:

$$K = K_1 \times K_2 \times \cdots$$

(c) We note that, because the chemical equation is the sum of the equations in (a) and (b), the equilibrium constant is the product of the two equilibrium constants found above:

$$K = \frac{[H_2SO_3]}{P_{SO_2}} \times \frac{[H_3O^+][HSO_3^-]}{[H_2SO_3]} = \frac{[H_3O^+][HSO_3^-]}{P_{SO_2}}$$

Self-Test 14.6A Write the equilibrium constant K for $3\,NO_2(g) + H_2O(l) \rightleftharpoons 2\,HNO_3(aq) + NO(g)$. The strong acid HNO_3 is present as $H^+(aq)$ and $NO_3^-(aq)$ in solution.

$$[\textbf{\textit{Answer:}}\ K = \frac{[H^+]^2[NO_3^-]^2 \times P_{NO}}{P_{NO_2}{}^3}]$$

Self-Test 14.6B Write (a) the equilibrium constant K for $CO_2(g) + H_2O(l) \rightleftharpoons H_2CO_3(aq)$; (b) K_c for $H_2CO_3(aq) + H_2O(l) \rightleftharpoons H_3O^+(aq) + HCO_3^-(aq)$.

Example 14.3 *Determining the equilibrium constant, given equilibrium concentrations*

Haber mixed some nitrogen and hydrogen and allowed it to react at 500. K until the mixture reached equilibrium with the product, ammonia. When he analyzed the equilibrium mixture, he found it to consist of 0.796 mol/L NH_3, 0.305 mol/L N_2, and 0.324 mol/L H_2. What is the equilibrium constant for reaction A in Section 14.1?

Strategy First, write the chemical equation and the expression for the equilibrium constant. Then, substitute the equilibrium concentrations of each substance into the expression for K_c, as described in the first procedure in Toolbox 14.1.

Solution The chemical equation (reaction A) and the expression for the equilibrium constant are

$$N_2(g) + 3\,H_2(g) \rightleftharpoons 2\,NH_3(g) \qquad K_c = \frac{[NH_3]^2}{[N_2][H_2]^3}$$

Now, substitute the equilibrium molar concentrations (without their units) into the expression for K_c:

$$K_c = \frac{(0.796)^2}{0.305 \times (0.324)^3} = 61.1$$

Self-Test 14.7A A reaction important in the gasification of coal is $2\,CO(g) + 2\,H_2(g) \rightleftharpoons CH_4(g) + CO_2(g)$. Determine the equilibrium constant for this reaction at 298 K, given the following equilibrium concentrations: 4.30×10^{-9} mol/L CO; 1.15×10^{-8} mol/L H_2; 5.14×10^{-2} mol/L CH_4; and 4.12×10^{-2} mol/L CO_2.
$$[\textbf{\textit{Answer:}}\ 8.66 \times 10^{29}]$$

Self-Test 14.7B Determine the equilibrium constant for $2\,BrCl(g) \rightleftharpoons Br_2(g) + Cl_2(g)$ at 500 K, given the following equilibrium concentrations: 0.131 mmol/L BrCl; 3.51 mmol/L Br_2; and 0.156 mmol/L Cl_2.

USING EQUILIBRIUM CONSTANTS

The numerical value of an equilibrium constant is important because it tells us whether we can expect a reaction mixture at equilibrium to contain a high or a low concentration of product. It also allows us to predict the direction in which the reaction will tend to proceed. This information helps us to predict, for instance, whether a proposed industrial process is likely to make enough product to be worthwhile.

14.6 The Extent of Reaction

The product concentrations or partial pressures appear in the numerator of K, and the reactant concentrations or partial pressures appear in the denominator. If the products are relatively abundant at equilibrium, the numerator will be large and the denominator small. Therefore, K is large when the equilibrium mixture consists mostly of products. In contrast, when the numerator is small and the denominator large, K is small and equilibrium is reached after very little reaction has occurred (Fig. 14.6). Some general guidelines are:

Large values of K (larger than about 10^3): products dominate at equilibrium.

Intermediate values of K (approximately in the range 10^{-3} to 10^3): neither reactants nor products dominate at equilibrium.

Small values of K (smaller than about 10^{-3}): reactants dominate at equilibrium.

For instance, consider the reaction

$$H_2(g) + 2\,Cl_2(g) \rightleftharpoons 2\,HCl(g) \qquad K_p = \frac{P_{HCl}^{\,2}}{P_{H_2}P_{Cl_2}} \qquad \textbf{(B)}$$

Experiment shows that $K_p = 4.0 \times 10^{31}$ at 300 K. Such a large value for K_p tells us that the system does not reach equilibrium until most of the reactants have been converted to HCl. We say that the equilibrium *favors* HCl. In fact, this reaction is fast as well as favorable, because it goes explosively as soon as the reactants are mixed and exposed to sunlight.

Now consider the equilibrium

$$N_2(g) + O_2(g) \rightleftharpoons 2\,NO(g) \qquad K_p = \frac{P_{NO}^{\,2}}{P_{N_2}P_{O_2}} \qquad \textbf{(C)}$$

It is found that $K_p = 4.8 \times 10^{-31}$ at 298 K. The very small value of K_p implies that the reactants N_2 and O_2 will be the dominant species in the system at equilibrium at 298 K.

Example 14.4 *Calculating an equilibrium concentration*

Suppose that in the equilibrium between H_2, Cl_2, and HCl in reaction B, $H_2(g) + Cl_2(g) \rightleftharpoons 2\,HCl(g)$, the equilibrium molar concentration of H_2 is 1.0×10^{-17} mol/L and that of Cl_2 is 2.0×10^{-16} mol/L. What is the equilibrium molar concentration of HCl at 300 K, given $K_c = 4.0 \times 10^{31}$ for the reaction?

Strategy At equilibrium, the molar concentrations of the reactants and products satisfy the expression for K_c (note that we are working with molar concentrations). Write the expression for K_c and rearrange it to give the one unknown concentration. Then substitute the data.

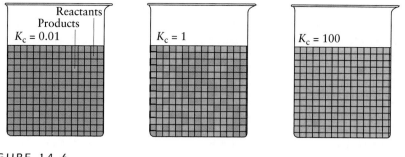

FIGURE 14.6

The size of the equilibrium constant indicates whether the reactants (blue squares) or the products (yellow squares) are favored. Note that reactants are favored when K_c is small (left), products are favored when K_c is large (right), and reactants and products are in almost equal abundance when K_c is close to 1 (middle). Here, for simplicity, we compare reactions with $K_c = 10^{-2}$ and $K_c = 10^2$.

Solution The expression for K_c is

$$K_c = \frac{[HCl]^2}{[H_2][Cl_2]}$$

To find [HCl], we rearrange this expression into

$$[HCl] = \sqrt{K_c[H_2][Cl_2]}$$

and substitute the data:

$$[HCl] = \sqrt{(4.0 \times 10^{31}) \times (1.0 \times 10^{-17}) \times (2.0 \times 10^{-16})} = 0.28$$

That is, the molar concentration of HCl at equilibrium is 0.28 mol/L. This result means that, at equilibrium, the amount of product in the system is overwhelming compared with the amounts of reactants.

Self-Test 14.8A Suppose that the equilibrium molar concentrations of H_2 and Cl_2 at 300 K are both 1.0×10^{-16} mol/L. What is the equilibrium molar concentration of HCl, given $K_c = 4.0 \times 10^{31}$?

[***Answer:*** 0.63 mol/L, about 10^{16} times greater than the concentrations of H_2 and Cl_2.]

Self-Test 14.8B Suppose that the equilibrium molar concentrations of N_2 and O_2 at 298 K are 0.0010 mol/L. What is the equilibrium molar concentration of NO in the reaction $N_2(g) + O_2(g) \rightleftharpoons 2\,NO(g)$?

An example of a reaction with an equilibrium constant of intermediate value is

$$H_2(g) + I_2(g) \rightleftharpoons 2\,HI(g) \qquad K_c = \frac{[HI]^2}{[H_2][I_2]} \qquad \textbf{(D)}$$

for which $K_c = 46$ at 783 K. The equilibrium mixture consists of hydrogen and hydrogen iodide gases and iodine vapor. If the equilibrium concentrations of two of the species are known, we can use the same type of calculation used in Example 14.4 to calculate the third. For example, if the concentrations of H_2 and I_2 are each 1.0×10^{-3} mol/L, then the equilibrium concentration of HI is 6.8×10^{-3} mol/L. In this case, the equilibrium concentrations of the reactants and products are all approximately the same.

If K is large, then products are favored at equilibrium; if K is small, then reactants are favored.

Example 14.5 *Using K to determine a partial pressure*

The equilibrium constant for the gas-phase reaction $PCl_5(g) \rightleftharpoons PCl_3(g) + Cl_2(g)$ is $K_p = 25$ at 298 K. The partial pressures of PCl_5 and Cl_2 at equilibrium are 0.0021 and 0.48 atm, respectively. What is the equilibrium partial pressure of PCl_3?

Strategy The equilibrium constant has an intermediate value, so we suspect that the equilibrium partial pressures of all the components will be comparable to one another. To calculate the value of the one unknown, from the chemical equation write the expression for the equilibrium constant in terms of the partial pressures. Rearrange the expression to give the one unknown on the left, and then substitute the data.

Solution The equilibrium is

$$PCl_5(g) \rightleftharpoons PCl_3(g) + Cl_2(g) \qquad K_p = \frac{P_{PCl_3}P_{Cl_2}}{P_{PCl_5}} \qquad \text{(E)}$$

The unknown quantity is the partial pressure of PCl_3, so we rearrange the expression for K into

$$P_{PCl_3} = K_p \times \frac{P_{PCl_5}}{P_{Cl_2}}$$

Substitution of the data gives

$$P_{PCl_3} = \frac{25 \times 0.0021}{0.48} = 0.11$$

That is, the partial pressure of PCl_3 at equilibrium in the mixture is 0.11 atm.

Self-Test 14.9A Nitrosyl chloride, NOCl, decomposes into NO and Cl_2 according to the equation $2\,NOCl(g) \rightleftharpoons 2\,NO(g) + Cl_2(g)$, with $K_p = 0.018$ at 500 K. An analysis of a reaction mixture at equilibrium indicates that the partial pressures of NO and Cl_2 are 0.11 and 0.84 atm, respectively. What is the equilibrium partial pressure of NOCl?

[*Answer:* 0.75 atm]

Self-Test 14.9B Carbon monoxide reacts with oxygen to produce carbon dioxide: $2\,CO(g) + O_2(g) \rightleftharpoons 2\,CO_2(g)$. At 1000 K, $K_p = 2.8 \times 10^{20}$ for this reaction. An analysis of a reaction mixture at equilibrium indicates that the partial pressures of O_2 and CO_2 are 1.4×10^{-9} and 75 atm, respectively. What is the equilibrium partial pressure of CO?

14.7 The Direction of Reaction

Very often in industry, reaction mixtures are prepared with different initial conditions. If we need to know whether a certain reaction mixture will tend to form more products or decompose into reactants, we can compare the actual concentrations with the equilibrium concentrations. We first calculate the **reaction quotient,** Q. This quantity is defined in exactly the same way as the equilibrium constant, but with the molar concentrations (for solutes) or partial pressures (for gases) at *any* stage of the reaction. For a homogeneous gas-phase reaction with data in the form of molar concentrations, we denote the reaction quotient Q_c; for gaseous reactions with the data in the form of partial pressures, we use Q_p. For example, if at a certain stage in the reaction $H_2(g) + I_2(g) \rightleftharpoons 2\,HI(g)$, we know the partial pressures of H_2, I_2, and HI are 0.1, 0.2, and 0.4 atm, respectively, then the reaction quotient at that stage of the reaction is

$$Q_p = \frac{P_{HI}^2}{P_{H_2}P_{I_2}} = \frac{(0.4)^2}{0.1 \times 0.2} = 0.8$$

Note that, like the equilibrium constant, a reaction quotient is a pure (unitless) number.

To predict whether a particular mixture of reactants and products will tend to produce more products or more reactants, we compare Q with K. If $Q > K$, then the concentrations or partial pressures of the products are too high (or the concentrations or partial pressures of the reactants too low) for equilibrium. Hence, the reaction has a tendency to proceed in the reverse direction, toward reactants. If $Q < K$, then the reaction tends to go forward and form products. If $Q = K$, then the mixture has its equilibrium composition and has no tendency to change in either direction. This pattern is summarized in Fig. 14.7.

Notice that we say there is a *tendency* toward either reactants or products. The reaction may be so slow that it never reaches equilibrium in the time we are prepared to wait. A mixture of hydrogen and oxygen has a tendency to form water; nevertheless, at room temperature and in the absence of an initiating spark, the reaction is almost infinitely slow. A bottle of benzene has a tendency to decompose into diamonds and hydrogen gas, but that change will never happen, even if we waited a million years.

A reaction has a tendency to form products if $Q < K$ and to form reactants if $Q > K$.

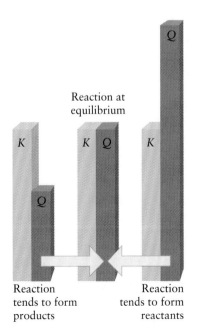

Reaction at equilibrium

K K Q K

Q

Reaction tends to form products

Reaction tends to form reactants

FIGURE 14.7

The relative sizes of the reaction quotient Q and the equilibrium constant K indicate the direction in which a reaction mixture tends to change. The arrows point from reactants to products when $Q < K$ (left) or from products to reactants when $Q > K$ (right). There is no tendency to change once the reaction quotient has become equal to the equilibrium constant.

Example 14.6 *Predicting the direction of reaction*

A chemist introduced a mixture of hydrogen, iodine, and hydrogen iodide, each at 0.0020 mol/L, into a container heated to 783 K. At this temperature, $K_c = 46$ for the reaction $H_2(g) + I_2(g) \rightleftharpoons 2 HI(g)$. Predict whether or not more HI has a tendency to form.

Strategy Because the equilibrium constant is given in terms of molar concentrations, calculate Q_c and compare it with K_c. If $Q_c > K_c$, the products need to decompose until their concentrations match K_c. The opposite is true if $Q_c < K_c$: in that case, more products need to form.

Solution The reaction quotient is

$$Q_c = \frac{[HI]^2}{[H_2][I_2]} = \frac{(0.0020)^2}{0.0020 \times 0.0020} = 1.0$$

Because $Q_c < K_c$, we conclude that the reaction will tend to form more product and consume reactants. That is, more HI will tend to form.

Self-Test 14.10A A mixture of H_2, N_2, and NH_3 with molar concentrations 3.0×10^{-3} mol/L, 1.0×10^{-3} mol/L, and 2.0×10^{-3} mol/L, respectively, was prepared and heated to 500 K, at which temperature $K_c = 62$ for the reaction $N_2(g) + 3 H_2(g) \rightleftharpoons 2 NH_3(g)$. Decide whether ammonia tends to form or to decompose.

[*Answer:* $Q_c = 1.5 \times 10^5 > K_c$; tends to decompose]

Self-Test 14.10B For the reaction $N_2O_4(g) \rightleftharpoons 2 NO_2(g)$ at 298 K, $K_p = 0.98$. A mixture of N_2O_4 and NO_2 with initial partial pressures of 2.4 and 1.2 atm, respectively, was prepared at 298 K. Which compound will tend to increase its partial pressure?

14.8 Equilibrium Tables

We can use equilibrium constants to calculate the equilibrium composition of a reaction mixture for any initial composition. The key idea is that the changes in concentration or partial pressure of each component are linked by the reaction stoichiometry. The easiest way to proceed is to draw up an **equilibrium table** that shows the initial composition, the changes needed to reach equilibrium, and the final equilibrium composition.

Example 14.7 *Calculating the equilibrium constant for a reaction*

Haber started one experiment with a mixture consisting of 0.500 mol/L N_2 and 0.800 mol/L H_2 and allowed it to reach equilibrium with the product, ammonia, in a constant-volume vessel. He found that, at equilibrium at a certain temperature, the molar concentration of NH_3 is 0.150 mol/L. Calculate the equilibrium constant for the reaction at that temperature.

Strategy Always begin equilibrium calculations by writing the equilibrium constant expression from the chemical equation. Identify the change in molar concentration of one substance, and use the reaction stoichiometry to calculate the changes in the concentrations of the other substances. The reaction vessel has constant volume, so the molar concentration of a species is proportional to the number of moles of the species. To keep track of changes, draw up a table with columns headed by each substance as it appears in the chemical equation and rows giving the initial concentration, the change in concentration, and the final, equilibrium concentration. Finally, substitute the equilibrium concentrations into the expression for K_c.

Solution The equation for the reaction is

$$N_2(g) + 3 H_2(g) \rightleftharpoons 2 NH_3(g) \qquad K_c = \frac{[NH_3]^2}{[N_2][H_2]^3}$$

The equation implies that 1 mol $N_2 \simeq$ 2 mol NH_3 and 3 mol $H_2 \simeq$ 2 mol NH_3. Therefore, because the molar concentration of NH_3 increases by 0.150 mol/L to reach equilibrium, the concentration of N_2 decreases by half that much, 0.075 mol/L, and the concentration of H_2 by 1.5 times that amount, or 0.225 mol/L. The equilibrium table, with all molar concentrations in moles per liter, is therefore

	Species		
	N_2	H_2	NH_3
1. Initial molar concentration	0.500	0.800	0
2. Change in molar concentration	−0.075	−0.225	+0.150
3. Equilibrium molar concentration	0.425	0.575	0.150

It follows that

$$K_c = \frac{[NH_3]^2}{[N_2][H_2]^3} = \frac{(0.150)^2}{0.425 \times (0.575)^3} = 0.278$$

Self-Test 14.11A A mixture of 0.0020 mol/L hydrogen gas and 0.0020 mol/L iodine gas is allowed to form hydrogen iodide at 773°C by the reaction $H_2(g) + I_2(g) \rightarrow$ 2 HI(g). At equilibrium, the concentration of HI is 0.0020 mol/L. Calculate the equilibrium constant for reaction at 773°C.

[*Answer:* 4.0]

Self-Test 14.11B In one experiment in the synthesis of ammonia, nitrogen at a concentration of 0.40 mol/L and hydrogen at 0.90 mol/L were allowed to react and to reach equilibrium with the product, ammonia, in a constant-volume vessel. At equilibrium at a certain temperature, the molar concentration of NH_3 is 0.20 mol/L. Calculate the equilibrium constant for reaction A at that temperature.

Much more commonly, we know the equilibrium constant and need to calculate the equilibrium composition. The procedure is exactly the same, but we write one of the unknown changes as x and use the reaction stoichiometry to find the other changes. The procedure is summarized in Toolbox 14.2 and illustrated in the examples that follow.

To calculate the equilibrium composition of a reaction mixture, set up an equilibrium table in terms of a change x in one of the species, express the equilibrium constant in terms of x, and solve the resulting equation for x.

Toolbox 14.2 *How to set up and use an equilibrium table*

This Toolbox shows how to set up an equilibrium table for a reaction taking place in a constant-volume container.

Conceptual Basis

Reaction proceeds until the concentrations of reactants and products satisfy the equilibrium constant. Any change in one species is related to changes in the others by the stoichiometry of the reaction. When K is large, we can expect a high proportion of products at equilibrium. When K is small, we expect a high proportion of reactants at equilibrium.

Procedure

Begin by writing the expression for the equilibrium constant from the balanced chemical equation for the reaction. Then draw up an equilibrium table as follows:

Step 1. Set up a table with columns labeled by the reactants and products in the same order as in the balanced equation. In the first row, show the initial molar concentrations or partial pressures of each species.

This step shows how the reaction system is prepared. As always, strike out the units from molar concentrations (in moles per liter) and partial pressures (in atmospheres). Because the molar concentrations of pure solids and liquids are unchanged by reaction, there is no need to include them in the table.

Step 2. Write the changes in the molar concentrations or partial pressures that are needed for the reaction to reach equilibrium.

It is often the case that we do not know the changes, so write one of them as x, usually multiplied by its stoichiometric coefficient in the chemical equation, and then use the reaction stoichiometry to express the other changes as multiples of that x.

Step 3. Write expressions for the equilibrium molar concentrations or partial pressures by adding the change in molar concentration or partial pressure (from step 2) to the initial value for each substance (from step 1).

Although a change may be positive (an increase) or negative (a decrease), the value of the concentration or partial pressure itself must always be positive.

Step 4. Use the equilibrium constant to determine the value of x, the unknown molar concentration or partial pressure at equilibrium.

Use the expressions in the last line of the equilibrium table to calculate the equilibrium concentrations.

Quadratic equations In some cases, the equation for x is a quadratic equation of the form

$$ax^2 + bx + c = 0$$

The two solutions of this equation are

$$x = \frac{-b \pm \sqrt{b^2 - 4ac}}{2a}$$

Decide which of the two solutions given by this expression is valid (the one with the $+$ sign or the one with the $-$ sign in front of the square root) by seeing which solution is physically possible. See Example 14.8 for an illustration of this procedure.

In other cases, the equation for x in terms of K may be quite complicated. There are then two possible ways forward. One is to use a computer and the chemical equilibrium option in the calculator tool provided on the CD-ROM that accompanies this text. The alternative is to look for an approximate solution.

Solution by approximation An approximation technique can greatly simplify calculations when the change in molar concentration (x) is less than about 5% of the initial concentrations. To use it, assume that x is so small that, when it is added to or subtracted from a concentration, the change in concentration is negligible. Thus, we can replace all expressions in which x is added to or subtracted from a concentration, like $A + x$ or $A - 2x$, for example, by A. When x occurs on its own (not added to or subtracted from another number), it is left unchanged. So, an expression like $(0.1 - x)x^2$ simplifies to $0.1x^2$.

At the end of the calculation, it is important to verify that the calculated value of x is indeed smaller than about 5% of the initial concentrations. If it is, the approximation is valid. If not, we must solve the equation without making an approximation. The approximation procedure is illustrated in Example 14.9.

Example 14.8 *Calculating the equilibrium composition*

Suppose that 0.150 mol PCl_5 is placed in a reaction vessel of volume 500. mL and allowed to reach equilibrium with its decomposition products phosphorus trichloride and chlorine at 250°C, when $K_c = 1.80$ for $PCl_5(g) \rightleftharpoons PCl_3(g) + Cl_2(g)$.

What is the composition of the equilibrium mixture? All three substances are gases at 250°C.

Strategy The general procedure is like that set out in Toolbox 14.2: write the expression for the equilibrium constant, then set up an equilibrium table with x used to denote the change in the molar concentration of the reactant. Calculate the initial molar concentration of the reactant by dividing the number of moles by the volume of the container. Use the reaction stoichiometry to express the molar concentrations of the products in terms of x. Substitute these equilibrium values into the expression for K_c to verify the answer.

Solution The chemical equation is

$$PCl_5(g) \rightleftharpoons PCl_3(g) + Cl_2(g) \qquad K_c = \frac{[PCl_5]}{[PCl_3][Cl_2]}$$

The initial molar concentration of PCl_5 is

$$\text{Molar concentration of } PCl_5 = \frac{0.150 \text{ mol}}{0.500 \text{ L}} = 0.300 \text{ mol/L}$$

We suppose that the change in the molar concentration of PCl_5 is $-x$ mol/L. Using the chemical equation above, we draw up the following table, with all molar concentrations in moles per liter.

	Species		
	PCl$_5$	PCl$_3$	Cl$_2$
Step 1. Initial molar concentration	0.300	0	0
Step 2. Change in molar concentration	$-x$	$+x$	$+x$

The stoichiometry of the reaction implies that if the molar concentration of PCl_5 decreases by x, then the molar concentrations of PCl_3 and Cl_2 both increase by x.

Step 3. Equilibrium molar concentration	$0.300 - x$	x	x

The values in step 3 are the sums of the initial molar concentrations (step 1) and the changes in their concentrations brought about by reaction (step 2).

Step 4. Substitution of these equilibrium values into the expression for the equilibrium constant gives

$$K_c = \frac{[PCl_3][Cl_2]}{[PCl_5]} = \frac{x \times x}{0.300 - x}$$

Because we are told that $K_c = 1.80$, the equation we have to solve is

$$1.80 = \frac{x^2}{0.300 - x}$$

We can rearrange this expression into the quadratic equation

$$x^2 + 1.80x - 0.540 = 0$$

The solutions of this equation (obtained by using the quadratic formula in Toolbox 14.2) are

$$x = \frac{-1.80 \pm \sqrt{(1.80)^2 - 4(1)(-0.540)}}{2} = 0.262 \text{ and } -2.06$$

Because the concentrations must be positive and because (from step 3) x is the molar concentration of each of the products, we select 0.262 as the solution. It follows that, at equilibrium,

$$[PCl_5] = 0.300 - x = 0.300 - 0.262 = 0.038$$

$$[PCl_3] = x = 0.262$$

$$[Cl_2] = x = 0.262$$

That is, the equilibrium concentrations of PCl_5, PCl_3, and Cl_2 are 0.038, 0.262, and 0.262 mol/L, respectively. To check the solution, we substitute these values into the expression for K_c:

$$K_c = \frac{0.262 \times 0.262}{0.038} = 1.81$$

Which is very close to the actual value.

Self-Test 14.12A Bromine monochloride, BrCl, decomposes into bromine and chlorine and reaches the equilibrium $2\,BrCl(g) \rightleftharpoons Br_2(g) + Cl_2(g)$, for which $K_c = 32$ at 500 K. If initially pure BrCl is present at a concentration of 3.30×10^{-3} mol/L, what is its molar concentration in the mixture at equilibrium?

[*Answer:* 3×10^{-4} mol/L]

Self-Test 14.12B Chlorine and fluorine react at a certain temperature to produce ClF. The reaction is $Cl_2(g) + F_2(g) \rightleftharpoons 2\,ClF(g)$, with $K_c = 20$. If 0.200 mol Cl_2 and 0.100 mol F_2 are added to a 1.00-L container and allowed to come to equilibrium at that temperature, what is the molar concentration of ClF in the equilibrium mixture?

Example 14.9 *Calculating the equilibrium composition by approximation*

Under certain conditions, nitrogen and oxygen react to form nitrous oxide (dinitrogen monoxide, N_2O). Suppose that a mixture of 0.482 mol N_2 and 0.933 mol O_2 is placed in a reaction vessel of volume 10.0 L and allowed to form N_2O at a temperature for which $K_c = 2.0 \times 10^{-37}$. What will the composition of the equilibrium mixture be?

Strategy Proceed exactly as set out in Toolbox 14.2. However, because K_c is so small, we can expect that very little N_2O is produced; hence, terms like $A - x$ can be approximated by A itself.

Solution The equilibrium is

$$2\,N_2(g) + O_2(g) \rightleftharpoons 2\,N_2O(g) \qquad K_c = \frac{[N_2O]^2}{[N_2]^2[O_2]}$$

The initial molar concentrations are

$$\text{Molar concentration of } N_2 = \frac{0.482 \text{ mol}}{10.0 \text{ L}} = 0.0482 \text{ mol/L}$$

$$\text{Molar concentration of } O_2 = \frac{0.933 \text{ mol}}{10.0 \text{ L}} = 0.0933 \text{ mol/L}$$

Because there is no N_2O present initially, the equilibrium table, with all molar concentrations in moles per liter, is

	Species		
	N_2	O_2	N_2O
Step 1. Initial molar concentration	0.0482	0.0933	0
Step 2. Change in molar concentration	$-2x$	$-x$	$+2x$
Step 3. Equilibrium molar concentration	$0.0482 - 2x$	$0.0933 - x$	$2x$

Step 4. Substitution of the values in the last line of the table into the expression for K_c gives

$$K_c = \frac{[N_2O]^2}{[N_2]^2[O_2]} = \frac{(2x)^2}{(0.0482 - 2x)^2(0.0933 - x)}$$

When rearranged, this equation is a cubic equation (an equation in x^3). It is difficult to solve cubic equations exactly. However, because K_c is very small, we anticipate that x will turn out to be so small that we can replace $0.0482 - 2x$ by 0.0482 and $0.0933 - x$ by 0.0933. These approximations simplify the equation above to

$$K_c \approx \frac{4x^2}{(0.0482)^2(0.0933)}$$

This equation is easily solved by rearranging it to

$$x \approx \sqrt{\frac{(0.0482)^2 \times 0.0933 \times K_c}{4}}$$

Because $K_c = 2.0 \times 10^{-37}$, we obtain

$$x \approx \sqrt{\frac{(0.0482)^2 \times 0.0933 \times (2.0 \times 10^{-37})}{4}} = 3.3 \times 10^{-21}$$

The value of x is very small compared with 0.0482 (far smaller than 5% of it), so the approximation is valid. We conclude that at equilibrium

$$[N_2] = 0.0482 - 2x \approx 0.0482$$

$$[O_2] = 0.0933 - x \approx 0.0933$$

$$[N_2O] = 2x \approx 6.6 \times 10^{-21}$$

Therefore, the equilibrium concentrations of N_2, O_2, and N_2O are reported as 0.0482, 0.0933, and 6.6×10^{-21} mol/L, respectively. To verify the answer, we calculate K_c from the final equilibrium concentrations:

$$K_c = \frac{(6.6 \times 10^{-21})^2}{(0.0482)^2 \times 0.0933} = 2.0 \times 10^{-37}$$

Self-Test 14.13A The initial concentrations of nitrogen and hydrogen are 0.010 and 0.020 mol/L, respectively. The mixture is heated to a temperature at which $K_c = 1.1 \times 10^{-3}$ for the reaction $N_2(g) + 3 H_2(g) \rightleftharpoons 2 NH_3(g)$. What is the equilibrium composition of the mixture?

[*Answer:* 0.010 mol/L N_2; 0.020 mol/L H_2; 3.0×10^{-6} mol/L NH_3]

Self-Test 14.13B Hydrogen chloride gas is added to a reaction vessel containing solid iodine until its concentration reaches 0.012 mol/L. At the temperature of the experiment, $K_c = 3.5 \times 10^{-32}$ for the reaction $2 HCl(g) + I_2(s) \rightleftharpoons 2 HI(g) + Cl_2(g)$. What is the equilibrium composition of the mixture?

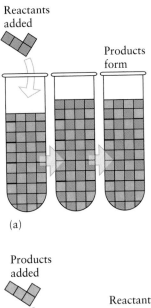

(a)

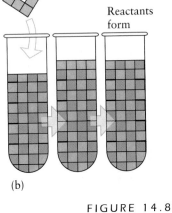

(b)

FIGURE 14.8

(a) When a reactant (blue) is added to a reaction mixture at equilibrium, products have a tendency to form. (b) When a product (yellow) is added instead, reactants tend to be formed. For this reaction, we have used $K_c = 1$.

THE RESPONSE OF EQUILIBRIA TO CHANGES IN THE CONDITIONS

Chemical equilibria are dynamic and therefore respond to changes in the conditions. When we add or remove a reactant, the equilibrium composition shifts to compensate. The composition also might change in response to a change in pressure. In the following sections, we shall see how Haber used this responsiveness to manipulate the equilibrium between ammonia, nitrogen, and hydrogen to increase the production of ammonia.

14.9 Adding and Removing Reagents

The French chemist Henri Le Chatelier discovered a general principle that enables us to predict how the composition of a reaction mixture at equilibrium tends to change when the conditions—the concentrations, the pressure, or the temperature—are changed. **Le Chatelier's principle** states that

> When a stress is applied to a system in dynamic equilibrium, the equilibrium tends to adjust to minimize the effect of the stress.

Le Chatelier's principle only suggests an outcome; it does not provide an explanation or lead to a quantitative prediction. It is more a rule of thumb than a fundamental principle; and to understand it, we need to think about what is happening to the molecules in the reaction mixture.

Let's suppose that the reaction $N_2(g) + 3H_2(g) \rightleftharpoons 2NH_3(g)$ has reached equilibrium and we pump in more hydrogen gas. According to Le Chatelier's principle, the reaction will tend to minimize the increase in the number of hydrogen molecules. Hydrogen will tend to react with nitrogen and, as a result, additional ammonia will be formed. If, instead of hydrogen, we were to add some ammonia, then the reaction would tend instead to form reactants at the expense of the added ammonia. Figure 14.8 shows the effect at the molecular level of adding a reactant or a product to a reaction mixture at equilibrium.

We can find the explanation of this application of Le Chatelier's principle in the discussion of the relative sizes of Q and K as summarized in Fig. 14.7. When reactants are added, the value of the reaction quotient Q falls below that of K momentarily, because the reactants appear in the denominator of Q. As we have seen, when $Q < K$, the reaction mixture responds by forming products. Likewise, when products are added, Q rises above K momentarily, because products appear in the numerator. Then, because $Q > K$, the reaction mixture responds by forming reactants at the expense of the products until $Q = K$ again.

Example 14.10 *Predicting the effect on a chemical equilibrium of adding or removing reactants and products*

Consider the equilibrium $4NH_3(g) + 3O_2(g) \rightleftharpoons 2N_2(g) + 6H_2O(g)$. Predict the effect on each equilibrium concentration of (a) the addition of N_2; (b) the removal of NH_3; (c) the removal of H_2O.

Strategy Le Chatelier's principle implies that an equilibrium mixture tends to minimize the effect of a change. Therefore, when a substance is added, the reaction tends to reduce the concentration of that substance. When a substance is removed, the reaction tends to replace it.

Solution (a) The addition of N_2 to the equilibrium mixture causes the reaction to shift in the direction that minimizes the increase in N_2 concentration. Therefore, the reaction shifts toward reactant formation. Forming more reactants increases the amounts of NH_3 and O_2 while decreasing the amount of H_2O. The concentration of N_2 remains slightly higher than its original equilibrium value, but lower than its concentration immediately after the additional N_2 was supplied. (b) When NH_3 is removed from the system at equilibrium, the reaction shifts to minimize its loss. The reaction tends to increase the amount of O_2 and decrease the amounts of N_2 and H_2O. The concentration of NH_3 will be somewhat lower than its original equilibrium value, but not as low as it was immediately after the removal of NH_3. (c) The removal of H_2O causes the equilibrium to shift in favor of products to restore (partially) the amount removed. This shift increases the amount of N_2 while decreasing the amounts of NH_3 and O_2. The concentration of H_2O is somewhat lower than its original value.

Self-Test 14.14A Consider the equilibrium $SO_3(g) + NO(g) \rightleftharpoons SO_2(g) + NO_2(g)$. Predict the effect on the equilibrium of (a) the addition of NO; (b) the addition of SO_2; (c) the removal of NO_2.

> [*Answer:* The equilibrium tends to shift toward
> (a) products; (b) reactants; (c) products.]

Self-Test 14.14B Consider the equilibrium $CO(g) + 2H_2(g) \rightleftharpoons CH_3OH(g)$. Predict the effect on the equilibrium of (a) the addition of H_2; (b) the removal of CH_3OH; (c) the removal of CO.

A good way of ensuring that a reaction goes on generating a substance is to remove products as they are formed. Industrial processes rarely reach equilibrium. Instead, the product is removed as soon as it is formed: then, in its continuing hunt for equilibrium, the reaction generates more product. Removal of the product is an important feature of the Haber process. Haber's ammonia synthesis never reaches equilibrium; instead, ammonia is continually removed to encourage its further production. Reactions in living systems also tend to be far from equilibrium, for the same reason: there is a ceaseless exchange of products and reactants with the surroundings (Applying Chemistry: Case Study 14).

> ***When a reactant is added to a reaction mixture at equilibrium or a product removed from it, the reaction tends to form products; when a reactant is removed or a product added, more reactant tends to form.***

14.10 Compressing a Reaction Mixture

A gas-phase equilibrium is affected by the compression—a reduction in volume—of the reaction vessel. According to Le Chatelier's principle, the composition will tend to change in a way that minimizes the increase in pressure resulting from the compression. A pressure increase is reduced when reaction occurs in the direction that results in fewer moles of gas molecules. For example, Fig. 14.9 illustrates the effect of compression on the dissociation of a diatomic molecule. In the formation of NH_3 from N_2 and H_2, the number of gas-phase molecules in the container decreases because 4 mol of reactant molecules produce 2 mol of product molecules. The forward reaction therefore decreases the pressure the mixture exerts. When the mixture is compressed, the equilibrium composition tends to shift in favor of product, for that minimizes the increase

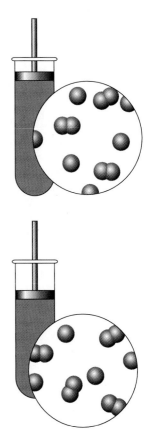

FIGURE 14.9

Le Chatelier's principle predicts that when a reaction at equilibrium is compressed, the number of molecules in the gas phase will tend to decrease. For example, in the dissociation of a diatomic molecule, the number of diatomic molecules in the system increases as it is compressed. Compression reverses the dissociation.

Applying Chemistry: *Case Study 14*

How is it that we can stay alive for more than a few minutes, if all the reactions inside us are proceeding toward equilibrium? The methods our bodies use to avoid equilibrium are similar to those industrial chemists and chemical engineers use to achieve a different result: the conversion of as much reactant as possible into product. All reactions have a tendency to move toward equilibrium, but if the equilibrium constant is small, very little product will form before equilibrium is reached. However, we can obtain more product even when the equilibrium constant is small by removing product as soon as it is formed, thereby forcing the reaction to go on forming product in a fruitless chase for equilibrium.

The Haber process, for instance, would be commercially useless if the manufacturers used a "batch process," in which a mixture of reactants—hydrogen and nitrogen, in this case—is simply allowed to reach equilibrium. Even a very effective catalyst will not produce more product. Therefore, industrial chemists and chemical engineers have devised methods to "cheat" equilibrium. In the Haber process, the reaction is carried out under high pressure and low temperature to achieve the most favorable equilibrium mixture and a catalyst is used to accelerate the process. To produce even more ammonia, the equilibrium mixture is cycled through a refrigeration unit, in which only the ammonia condenses. The nitrogen and hydrogen are then

Installing a platinum/rhodium gauze catalyst for the production of nitric acid by the oxidation of ammonia. The catalyst will accelerate the rate of the reaction without affecting the equilibrium composition.

returned to the reaction chamber to react further. The equilibrium constant is a *guide* to the amount of conversion expected, but the commercial process never reaches equilibrium.

Similar methods are used to make economical use of the ammonia in further reactions. For instance, ammonia is used to make nitric acid. Here, the first step is the oxidation of ammonia in the presence of a catalyst, usually an alloy of platinum with 10% rhodium (see the illustration):

in pressure. The opposite response, a tendency for the product to decompose, occurs in an expansion. Haber realized that in order to increase the yield of ammonia, he needed to carry out the synthesis with highly compressed gases. The actual industrial process uses pressures of 250 atm and more (Fig. 14.10).

Example 14.11 *Predicting the effect of compression on an equilibrium*

Predict the effect of compression on the equilibrium composition of the reaction mixture in which the equilibrium (a) $N_2O_4(g) \rightleftharpoons 2 NO_2(g)$; (b) $H_2(g) + I_2(g) \rightleftharpoons 2 HI(g)$ has been established.

Strategy A glance at the chemical equation shows which direction corresponds to a decrease in the number of gas-phase molecules. The composition of the equilibrium mixture tends to shift in that direction when the reaction mixture is compressed.

Solution (a) In the reverse reaction, two NO_2 molecules combine to form one N_2O_4 molecule. Hence, compression favors the formation of N_2O_4. (b) Because neither

$$4\,NH_3(g) \;+\; 5\,O_2(g) \xrightarrow{\;Pt/Rh\;} 6\,H_2O(g) \;+\; 4\,NO(g)$$

Excess oxygen is used, in part to increase a reactant concentration, which shifts the reaction forward, and also to remove a product by oxidizing the NO further to NO_2:

$$2\,NO(g) \;+\; O_2(g) \longrightarrow 2\,NO_2(g)$$

Continuous removal of the NO from the first reaction keeps the oxidation of ammonia from reaching equilibrium. Nor is the oxidation of NO allowed to reach equilibrium. The NO_2 is removed as it is formed by spraying the reaction mixture with water, which reacts with the NO_2:

$$3\,NO_2(g) \;+\; 3\,H_2O(l) \longrightarrow$$
$$2\,H_3O^+(aq) \;+\; 2\,NO_3^-(aq) \;+\; NO(g)$$

The NO produced in the last step reacts with more oxygen to produce more NO_2. Notice how each step is encouraged to proceed by removing a product. The efficiency of industry depends on cheating equilibrium, for equilibrium limits production.

Living systems are constantly struggling to find ways to cheat equilibrium, for equilibrium is death. Equilibrium is avoided in living systems by a continual supplying and discarding of matter and energy. Our very existence is the result of cheating equilibrium, temporarily at least.

Key Concepts: reaction quotient, equilibrium constant, Le Chatelier's principle

For Further Reading

V. Smil, Global population and the nitrogen cycle, *Scientific American,* 277 (July): 76–81, 1997.

Related Exercises: 14.87–14.90

Living beings are open systems and require frequent input of energy in the form of food and heat to maintain life.

FIGURE 14.10

One of the high-pressure vessels inside which the commercial synthesis of ammonia takes place.

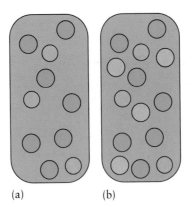

FIGURE 14.11

When an inert gas (orange spheres) is added to a system at equilibrium, the relative concentrations of reactants (pink spheres) and products (green spheres) are unaffected, because their partial pressures remain the same. (a) A system at equilibrium. (b) The same system after the addition of an inert gas.

(a)   (b)

direction corresponds to a reduction of gas-phase molecules, compressing the mixture should have little effect on the composition of the equilibrium mixture.

Self-Test 14.15A Predict the effect of compression on the equilibrium composition for the reaction $CH_4(g) + H_2O(g) \rightleftharpoons CO(g) + 3 H_2(g)$.

[*Answer:* Reactants favored]

Self-Test 14.15B Predict the effect of compression on the equilibrium composition of the reaction $2 NO(g) + Cl_2(g) \rightleftharpoons 2 NOCl(g)$.

We can explain the response of a gaseous equilibrium to compression as soon as we realize that compressing a system changes the concentrations and hence affects the value of Q. Consequently, the reaction tends to adjust in the direction that restores Q to K. For example, suppose we want to discover the effect of compression on the equilibrium

$$N_2O_4(g) \rightleftharpoons 2 NO_2(g) \qquad K_c = \frac{[NO_2]^2}{[N_2O_4]}$$

First, we express K_c in terms of the volume of the system. The molar concentration of each gas is the amount, n, of that gas divided by the volume, V, of the reaction vessel:

$$\text{Molar concentration of } NO_2 = \frac{n_{NO_2}}{V}$$

$$\text{Molar concentration of } N_2O_4 = \frac{n_{N_2O_4}}{V}$$

Hence, the expression for the equilibrium constant has the form

$$K_c = \frac{(n_{NO_2}/V)^2}{(n_{N_2O_4}/V)} = \frac{n_{NO_2}{}^2}{n_{N_2O_4}} \times \frac{1}{V}$$

For this expression to remain constant when the volume of the system is reduced, the ratio $n_{NO_2}{}^2/n_{N_2O_4}$ must decrease. That is, the amount of NO_2 must decrease and the amount of N_2O_4 must increase. Therefore, as the volume of the system is decreased, the equilibrium shifts to a smaller total number of gas molecules, just as Le Chatelier's principle predicts.

Suppose that, instead of decreasing the volume, we were to increase the total pressure inside a reaction vessel by pumping in argon or some other inert gas. In this case, the equilibrium composition is unaffected. The reacting gases continue to occupy the same volume, so their individual molar concentrations and partial pressures remain unchanged despite the presence of an inert gas (Fig. 14.11).

> *Compression of a reaction mixture at equilibrium tends to drive the reaction in the direction that reduces the number of gas-phase molecules; increasing the pressure by introducing an inert gas has no effect on the equilibrium composition.*

14.11 Temperature and Equilibrium

We can see from Tables 14.2 and 14.3 that the equilibrium constant depends on the temperature. It is found experimentally that, for an exothermic reaction, the formation of products is favored by lowering the temperature. For an endothermic reaction, the products are favored by raising the temperature. In terms of the equilibrium constant, the experimental observations tell us that

K increases with decreasing temperature if the forward reaction is exothermic.

K increases with increasing temperature if the forward reaction is endothermic.

Le Chatelier's principle is consistent with these observations because we can imagine the heat released in an exothermic reaction as helping to offset the lowering of temperature. Similarly, we can imagine the heat absorbed in an endothermic reaction as helping to offset the increase in temperature.

We can find the *explanation* of the effect of temperature by referring to the relation between the equilibrium constant and the forward and reverse rate constants, Eq. 2. If the forward reaction is endothermic, then the activation energy is higher for the forward direction than for the reverse direction (Fig. 14.12). The higher activation energy means that the rate constant of the forward reaction depends more strongly on temperature than does the rate constant of the reverse reaction. Therefore, when the temperature is raised, the rate constant for

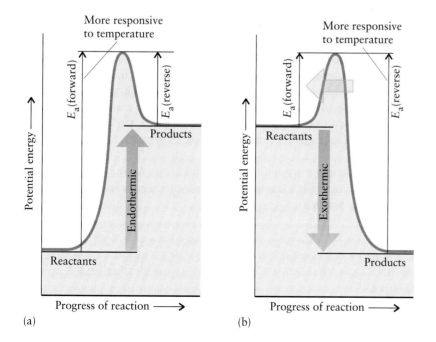

(a)

(b)

FIGURE 14.12

(a) The activation energy for an endothermic reaction is larger in the forward direction than in the reverse direction, so the rate of the forward reaction is more sensitive to temperature, and the equilibrium shifts toward products as the temperature is raised. (b) The opposite is true for an exothermic reaction: the reverse reaction is more sensitive to temperature. In this case, the equilibrium shifts toward reactants as the temperature is raised.

FIGURE 14.13

A radar image of the surface of Venus. Although the rocks are very hot, the partial pressure of carbon dioxide in the atmosphere is so great that carbonates may be abundant.

the forward reaction increases more than that of the reverse reaction. As a result, K will increase and the products will become more favored, just as Le Chatelier's principle predicts.

An example is the decomposition of carbonates. A reaction such as

$$CaCO_3(s) \rightleftharpoons CaO(s) + CO_2(g)$$

is strongly endothermic in the forward direction, and an appreciable partial pressure of carbon dioxide is present at equilibrium only if the temperature is high. For instance, at 800°C, the partial pressure is 0.22 atm. If the heating takes place in an open container, this partial pressure is never reached because equilibrium is never reached. The gas drifts away, and the calcium carbonate decomposes completely, leaving a solid residue of CaO. However, if the system is already so rich in carbon dioxide that the partial pressure of CO_2 exceeds 0.22 atm, then virtually no decomposition occurs: for every CO_2 molecule that is formed, one is converted back to carbonate. This is probably what happens on the hot surface of Venus (Fig. 14.13), where the partial pressure of carbon dioxide is about 87 atm. This high value of the pressure has led to speculation that the planet's surface is rich in carbonates, in spite of its high temperature (about 500°C).

Raising the temperature of an exothermic reaction favors the formation of reactants; raising the temperature of an endothermic reaction favors the formation of products.

Example 14.12 *Predicting the effect of temperature on an equilibrium*

One stage in the manufacture of sulfuric acid is the formation of sulfur trioxide by the reaction of SO_2 with O_2 in the presence of a vanadium(V) oxide catalyst. Predict how the equilibrium composition for the sulfur trioxide synthesis will tend to change when the temperature is raised.

Strategy Heating an equilibrium mixture will tend to shift its composition in the endothermic direction of the reaction. A positive reaction enthalpy indicates that the reaction is endothermic in the forward direction. A negative reaction enthalpy indicates that the reaction is endothermic in the reverse direction. To find the standard reaction enthalpy, use the standard enthalpies of formation of the species given in Appendix 2A.

Solution The chemical equation is

$$2\,SO_2(g) + O_2(g) \rightleftharpoons 2\,SO_3(g)$$

The standard reaction enthalpy is therefore

$$\begin{aligned}
\Delta H_r^\circ &= \{2\Delta H_f^\circ(SO_3, g)\} - \{2\Delta H_f^\circ(SO_2, g)\} \\
&= \{2 \times (-395.72\ \text{kJ/mol})\} - \{2 \times (-296.83\ \text{kJ/mol})\} \\
&= -197.78\ \text{kJ/mol}
\end{aligned}$$

Because the formation of SO_3 is exothermic, the reverse reaction is endothermic. Hence, raising the temperature of the equilibrium mixture favors the reverse reaction, the decomposition of SO_3.

Self-Test 14.16A Predict the effect of raising the temperature on the equilibrium composition of the reaction $N_2O_4(g) \rightleftharpoons 2\,NO_2(g)$. See Appendix 2A for data.

[***Answer:*** NO_2 favored]

Self-Test 14.16B Predict the effect of lowering the temperature on the equilibrium composition of the reaction $2\,CO(g) + O_2(g) \rightleftharpoons 2\,CO_2(g)$. See Appendix 2A for data.

14.12 Catalysts and Haber's Achievement

We have already met the term *catalyst,* a substance that increases the rate of a chemical reaction without being consumed itself (Section 13.10). A catalyst increases the rate at which a reaction reaches equilibrium, but it does not affect the composition at equilibrium. A catalyst acts by providing a faster route to the same destination. At a molecular level, we can think of a catalyst as speeding up both the forward and reverse reactions by the same amount. Therefore, the dynamic equilibrium is unaffected. That is, *a catalyst has no effect on the equilibrium composition of a reaction mixture.*

The use of a catalyst was the final step in Haber's achievement. Haber realized that he had to compress the gases and remove the ammonia as it was formed. As we have seen, compression shifts the equilibrium composition in favor of ammonia, which increases the yield of product. Removing the ammonia also encourages more to be formed. In addition, Haber wanted to work at low temperatures: the synthesis reaction is exothermic, and low temperature favors the formation of product. However, nitrogen and hydrogen combine too slowly at low temperatures, so Haber had to find a way to increase the rate of the reaction. He solved this part of his problem by discovering an appropriate catalyst. The process Haber devised in collaboration with the chemical engineer Carl Bosch is still in use throughout the world. In the United States alone, it accounts for almost the entire annual production of over 1.6×10^{10} kg of ammonia (Fig. 14.14).

A catalyst does not affect the equilibrium composition of a reaction mixture.

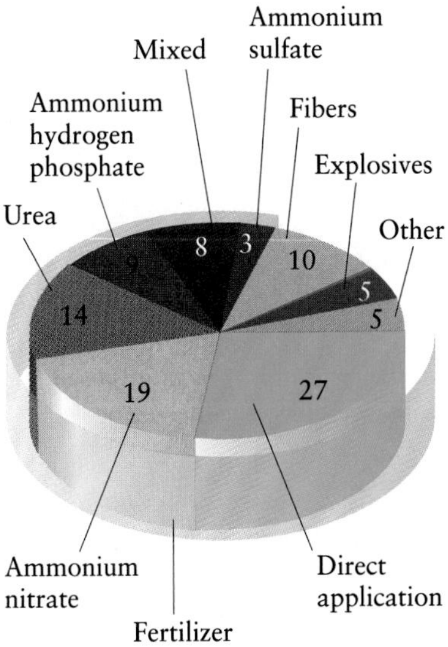

FIGURE 14.14

The Haber process is still used to produce almost all the ammonia manufactured in the world. This pie chart shows how the ammonia is used. The figures are percentages. Note that 80 percent—as shown by the green band—is used as fertilizer, either directly or after conversion to another compound.

Skills You Should Have Mastered

Conceptual

☐ 1. Interpret chemical equilibrium as a dynamic process involving change at the molecular level, Section 14.1.

☐ 2. Show how the equilibrium constant is related to the forward and reverse rate constants of the elementary reactions contributing to an overall reaction, Section 14.3.

☐ 3. Explain how the value of the reaction quotient, Q, allows the direction of reaction to be predicted, Section 14.7.

☐ 4. Explain the basis of Le Chatelier's principle, Sections 14.9–14.11.

Problem-Solving

☐ 1. Write the equilibrium constant for a chemical reaction, given its balanced equation, Toolbox 14.1 and Examples 14.1 and 14.2.

☐ 2. Calculate the effect on K of reversing a reaction or multiplying the chemical equation by a factor, Toolbox 14.1 and Self-Test 14.2.

☐ 3. Determine the equilibrium constant given equilibrium concentrations, Example 14.3.

☐ 4. Determine an equilibrium concentration or partial pressure, given K and all other equilibrium concentrations, Examples 14.4 and 14.5.

☐ 5. Determine the direction of a reaction, given K and the concentrations of all reactants and products, Example 14.6.

☐ 6. Use an equilibrium table to calculate K and equilibrium composition, Toolbox 14.2 and Examples 14.7, 14.8, and 14.9.

☐ 7. Use Le Chatelier's principle to predict how the equilibrium composition of a reaction mixture is affected by adding or removing reagents, Example 14.10.

☐ 8. Determine the direction in which a pressure or temperature change will shift an equilibrium, Examples 14.11 and 14.12.

Descriptive

☐ 1. Distinguish homogeneous and heterogeneous equilibria and write equilibrium constants for both types, Section 14.4.

☐ 2. Describe how the Haber process is designed to maximize the yield of product at a low temperature, Section 14.12 and Applying Chemistry: Case Study 14.

Exercises

Equilibrium and Equilibrium Constants

14.1 Explain what is wrong with the following statements. (a) Once a reaction has reached equilibrium, all reaction stops. (b) If more reactant is used, the equilibrium constant will have a larger value.

14.2 Explain what is wrong with the following statements. (a) If we can make a reaction go faster, we can increase the amount of product at equilibrium. (b) The reverse reaction does not begin until all the reactants have formed products.

14.3 Figure 14.1 shows how the concentrations of reactants and product change as a function of time for the synthesis of ammonia. Why do the concentrations of the reactants and product stop changing?

14.4 Copy Fig. 14.1 onto another piece of paper and on the same plot draw a second set of curves for the same reaction at the same temperature, but starting with pure NH_3. In what ways do your plots show the reversibility of reactions?

14.5 A 0.10-mol sample of pure ozone, O_3, is placed in a sealed 1.0-L container, and the reaction $2 O_3(g) \rightleftharpoons 3 O_2(g)$ is allowed to reach equilibrium. A 0.50-mol sample of pure

$O_3(g)$ is placed in a second 1.0-L container at the same temperature and allowed to reach equilibrium with oxygen. Without doing any calculations, predict which of the following will be different in the two containers at equilibrium. Which will be the same? (a) Amount of O_2; (b) concentration of O_2; (c) the ratio $[O_2]/[O_3]$; (d) the ratio $[O_2]^3/[O_3]^2$; (e) the time it takes for equilibrium to be established. Explain each of your answers.

14.6 A 0.10-mol sample of H_2 gas and a 0.10-mol sample of Br_2 gas are placed into a 2.0-L container. The reaction $H_2(g) + Br_2(g) \rightleftharpoons 2 HBr(g)$ is then allowed to come to equilibrium. A 0.20-mol sample of HBr is placed into a second 2.0-L sealed container at the same temperature and allowed to reach equilibrium with H_2 and Br_2. Which of the following will be different in the two containers at equilibrium? Which will be the same? Explain each of your answers. (a) Amount of Br_2; (b) concentration of H_2; (c) the ratio $[HBr]/[H_2][Br_2]$; (d) the ratio $[HBr]^2/[H_2][Br_2]$; (e) the total pressure in the container. Explain each of your answers.

14.7 Write the equilibrium constants (in terms of molar concentrations) for the following reactions:
(a) $CO(g) + Cl_2(g) \rightleftharpoons COCl(g) + Cl(g)$
(b) $H_2(g) + Br_2(g) \rightleftharpoons 2 HBr(g)$
(c) $2 H_2S(g) + 3 O_2(g) \rightleftharpoons 2 SO_2(g) + 2 H_2O(g)$

14.8 Write the equilibrium constants (in terms of partial pressures) for the following reactions:
(a) $2 NO(g) + O_2(g) \rightleftharpoons 2 NO_2(g)$
(b) $SbCl_5(g) \rightleftharpoons SbCl_3(g) + Cl_2(g)$
(c) $N_2(g) + 2 H_2(g) \rightleftharpoons N_2H_4(g)$

14.9 For the reaction $N_2(g) + 3 H_2(g) \rightleftharpoons 2 NH_3(g)$ at 400 K, $K_p = 41$. Find the value of K_p for each of the following reactions at the same temperature:
(a) $2 NH_3(g) \rightleftharpoons N_2(g) + 3 H_2(g)$
(b) $\frac{1}{2} N_2(g) + \frac{3}{2} H_2(g) \rightleftharpoons NH_3(g)$
(c) $2 N_2(g) + 6 H_2(g) \rightleftharpoons 4 NH_3(g)$

14.10 The equilibrium constant for the reaction $2 SO_2(g) + O_2(g) \rightleftharpoons 2 SO_3(g)$ has the value $K_p = 2.5 \times 10^{10}$ at 500 K. Find the value of K_p for each of the following reactions at the same temperature:
(a) $SO_2(g) + \frac{1}{2} O_2(g) \rightleftharpoons SO_3(g)$
(b) $SO_3(g) \rightleftharpoons SO_2(g) + \frac{1}{2} O_2(g)$
(c) $3 SO_2(g) + \frac{3}{2} O_2(g) \rightleftharpoons 3 SO_3(g)$

Rate Constants and Equilibrium

14.11 Explain why the following statements are wrong. (a) At equilibrium, the rate constants of the forward and reverse reactions are equal. (b) For a reaction with a very large K_c, at equilibrium the rate of the reverse reaction is much larger than the rate of the forward reaction.

14.12 Explain why the following statements are wrong. (a) The equilibrium constant for a reaction equals the ratio of the forward and reverse rates. (b) A reaction with a large equilibrium constant is faster than a reaction with a small equilibrium constant.

Equilibrium Constants from Equilibrium Amounts

14.13 Use the following data, which were collected at 460°C and are equilibrium molar concentrations, to determine K_c for the reaction $H_2(g) + I_2(g) \rightleftharpoons 2 HI(g)$:

$[H_2]$, mol/L	$[I_2]$, mol/L	$[HI]$, mol/L
6.47×10^{-3}	0.594×10^{-3}	0.0137
3.84×10^{-3}	1.52×10^{-3}	0.0169
1.43×10^{-3}	1.43×10^{-3}	0.0100

14.14 Determine K_p from the following equilibrium data collected at 24°C for the reaction $NH_4HS(s) \rightleftharpoons NH_3(g) + H_2S(g)$:

P_{NH_3}, atm	P_{H_2S}, atm
0.309	0.309
0.364	0.258
0.539	0.174

14.15 Determine K_c for each of the following equilibria from the value of K_p:

(a) $2 NOCl(g) \rightleftharpoons 2 NO(g) + Cl_2(g), K_p = 1.8 \times 10^{-2}$ at 500. K
(b) $CaCO_3(s) \rightleftharpoons CaO(s) + CO_2(g), K_p = 167$ at 1073 K

14.16 Determine K_c for each of the following equilibria from the value of K_p:
(a) $2 SO_2(g) + O_2(g) \rightleftharpoons 2 SO_3(g), K_p = 3.4$ at 1000. K;
(b) $NH_4HS(s) \rightleftharpoons NH_3(g) + H_2S(g), K_p = 9.4 \times 10^{-2}$ at 24°C.

Heterogeneous Equilibria

14.17 Explain why the following equilibria are heterogeneous, and write the expression for K_p for each one.
(a) $NH_4HS(s) \rightleftharpoons NH_3(g) + H_2S(g)$
(b) $NH_4(NH_2CO_2)(s) \rightleftharpoons 2 NH_3(g) + CO_2(g)$
(c) $2 KNO_3(s) \rightleftharpoons 2 KNO_2(s) + O_2(g)$

14.18 Explain why the following equilibria are heterogeneous, and write the expression for K_p for each one.
(a) $NH_4Cl(s) \rightleftharpoons NH_3(g) + HCl(g)$
(b) $Na_2CO_3 \cdot 10H_2O(s) \rightleftharpoons Na_2CO_3(s) + 10 H_2O(g)$
(c) $2 KClO_3(s) \rightleftharpoons 2 KCl(s) + 3 O_2(g)$

14.19 Write the expression for K_c for
(a) $Cu(s) + Cl_2(g) \rightleftharpoons CuCl_2(s)$
(b) $NH_4NO_3(s) \rightleftharpoons N_2O(g) + 2 H_2O(g)$
(c) $MgCO_3(s) \rightleftharpoons MgO(s) + CO_2(g)$

14.20 Write the expression for K_c for
(a) $NH_4HS(s) \rightleftharpoons NH_3(g) + H_2S(g)$
(b) $CuSO_4(s) + 5 H_2O(l) \rightleftharpoons CuSO_4 \cdot 5H_2O(s)$
(c) $ZnO(s) + CO(g) \rightleftharpoons Zn(s) + CO_2(g)$

The Extent and Direction of Reactions

14.21 In a gas-phase equilibrium mixture of H_2, I_2, and HI at 500. K, $[HI] = 2.21$ mmol/L and $[I_2] = 1.46$ mmol/L. Given the value of the equilibrium constant in Table 14.2, calculate the concentration of H_2.

14.22 In a gas-phase equilibrium mixture of H_2, Cl_2, and HCl at 1000. K, $[HCl] = 1.45$ mmol/L and $[Cl_2] = 2.45$ mmol/L. Use the information in Table 14.2 to calculate the concentration of H_2.

14.23 In a gas-phase equilibrium mixture of PCl_5, PCl_3, and Cl_2 at 500 K, $P_{PCl_5} = 0.15$ atm and $P_{Cl_2} = 0.20$ atm. What is the partial pressure of PCl_3, given that $K_p = 25$ for the reaction $PCl_5(g) \rightleftharpoons PCl_3(g) + Cl_2(g)$?

14.24 In a gas-phase equilibrium mixture of $SbCl_5$, $SbCl_3$, and Cl_2 at 500 K, $P_{SbCl_5} = 0.15$ atm and $P_{SbCl_3} = 0.20$ atm. Calculate the equilibrium partial pressure of Cl_2, given that $K_p = 3.5 \times 10^{-4}$ for the reaction $SbCl_5(g) \rightleftharpoons SbCl_3(g) + Cl_2(g)$.

14.25 For the reaction $H_2(g) + I_2(g) \rightleftharpoons 2 HI(g), K_c = 160.$ at 500. K. An analysis of a reaction mixture at 500. K showed that it had the composition 4.8 mmol/L H_2, 2.4 mmol/L I_2, and 2.4 mmol/L HI. (a) Calculate the reaction quotient. (b) Is the reaction mixture at equilibrium? (c) If not, is there a tendency to form more reactants or more products?

14.26 Analysis of a reaction mixture showed that it had the composition 0.417 mol/L N_2, 0.524 mol/L H_2, and 0.122 mol/L NH_3 at 800. K, at which temperature $K_c = 0.278$ for $N_2(g) + 3 H_2(g) \rightleftharpoons 2 NH_3(g)$. (a) Calculate the reaction quotient. (b) Is the reaction mixture at equilibrium? (c) If not, is there a tendency to form more reactants or more products?

14.27 A 500.-mL reaction vessel at 700. K contains 1.20×10^{-3} mol $SO_2(g)$, 5.0×10^{-4} mol $O_2(g)$, and 1.0×10^{-4} mol $SO_3(g)$. At 700. K, $K_c = 1.7 \times 10^6$ for the equilibrium $2 SO_2(g) + O_2(g) \rightleftharpoons 2 SO_3(g)$. (a) Calculate the reaction quotient Q_c. (b) Will more $SO_3(g)$ tend to form?

14.28 Given that $K_c = 61$ for the reaction $N_2(g) + 3 H_2(g) \rightleftharpoons 2 NH_3(g)$ at 500. K, calculate whether more ammonia will tend to form when a mixture of composition 2.23×10^{-3} mol/L N_2, 1.24×10^{-3} mol/L H_2, and 1.12×10^{-4} mol/L NH_3 is present in a container at 500. K.

Equilibrium Constants from Initial Amounts

14.29 When 0.0172 mol HI(g) is heated to 500. K in a 2.00-L sealed container, the resulting equilibrium mixture contains 1.90 g of HI. Calculate K_c for the decomposition reaction $2 HI(g) \rightleftharpoons H_2(g) + I_2(g)$.

14.30 When 1.00 g of gaseous I_2 is heated to 1000. K in a 1.00-L sealed container, the resulting equilibrium mixture contains 0.830 g of I_2. Calculate K_c for the equilibrium $I_2(g) \rightleftharpoons 2 I(g)$.

14.31 A 25.0-g sample of ammonium carbamate, $NH_4(NH_2CO_2)$, was placed in an evacuated 250.-mL flask and kept at 25°C. At equilibrium, 17.4 mg of CO_2 was present. What is the value of K_c for the decomposition of ammonium carbamate into ammonia and carbon dioxide? The reaction is $NH_4(NH_2CO_2)(s) \rightleftharpoons 2 NH_3(g) + CO_2(g)$.

14.32 Carbon monoxide and water vapor, each at 200. Torr, were introduced into a 250.-mL container. When the mixture had reached equilibrium at 700°C, the partial pressure of CO_2 was 88 Torr. Calculate the value of K_p for the equilibrium $CO(g) + H_2O(g) \rightleftharpoons CO_2(g) + H_2(g)$.

14.33 Data for the equilibrium $CH_3COOH + C_2H_5OH \rightleftharpoons CH_3COOC_2H_5 + H_2O$ at 100°C were obtained for reaction in an organic solvent. The initial molar concentrations of the reactants are reported in columns 1 and 2 in the following table and the equilibrium concentration of $CH_3COOC_2H_5$ is reported in column 3. Calculate $[H_2O]$ and the value of K_c for each experiment.

Initial concentration		Equilibrium concentration
$[CH_3COOH]$, mol/L	$[C_2H_5OH]$, mol/L	$[CH_3COOC_2H_5]$, mol/L
1.00	0.180	0.171
1.00	1.00	0.667
1.00	8.00	0.966

14.34 A 1.0-L reaction vessel was filled to the initial partial pressures of N_2O_4 given in column 1 in the following table, and the gas was allowed to decompose into NO_2. At 25°C, the total pressure ($P = P_{N_2O_4} + P_{NO_2}$) of the system at equilibrium is listed in column 2. Write the balanced chemical equation for the reaction, using the smallest possible whole-number coefficients, and calculate K_p for each set of data.

$P_{N_2O_4}$, atm	P, atm
0.154	0.212
0.333	0.425

Equilibrium Table Calculations

14.35 (a) A sample consisting of 2.0 mmol Cl_2 was sealed into a 2.0-L reaction vessel and heated to 1000. K to study its dissociation into Cl atoms. Use the information in Table 14.2 to calculate the equilibrium composition of the mixture. What is the percentage decomposition of the Cl_2? (b) If 2.0 mmol F_2 were placed into the reaction vessel instead of the chlorine, what would be its equilibrium composition at 1000. K? (c) Use your results from (a) and (b) to determine which is more stable at 1000. K, Cl_2 or F_2.

14.36 (a) A sample consisting of 5.0 mmol Cl_2 was sealed into a 2.0-L reaction vessel and heated to 1200. K, and the dissociation equilibrium was established. What is the equilibrium composition of the mixture? What is the percentage decomposition of the Cl_2? Use the information in Table 14.2. (b) If 5.0 mmol of Br_2 were placed into the reaction vessel instead of the chlorine, what would be the equilibrium composition at 1200. K? (c) Use your results from (a) and (b) to determine which is more stable at 1200. K, Cl_2 or Br_2.

14.37 The initial concentration of HBr in a reaction vessel is 1.2 mmol/L. If the vessel is heated to 500. K, what is the percentage decomposition of the HBr and the equilibrium composition of the mixture? For $2 HBr(g) \rightleftharpoons H_2(g) + Br_2(g)$ at 500 K, $K_c = 7.7 \times 10^{-11}$.

14.38 The initial concentration of BrCl in a reaction vessel is 1.4 mmol/L. If the vessel is heated to 500. K, what is the percentage decomposition of the BrCl and what is the equilibrium composition of the mixture? See Table 14.2 for data on the reaction.

14.39 The equilibrium constant $K_c = 1.1 \times 10^{-2}$ for the reaction $PCl_5(g) \rightleftharpoons PCl_3(g) + Cl_2(g)$ at 400. K. (a) Given that 1.0 g of PCl_5 was initially placed in a 250.-mL reaction vessel, determine the molar concentrations in the mixture at equilibrium. (b) What percentage of the PCl_5 is decomposed at 400. K?

14.40 For the reaction $PCl_5(g) \rightleftharpoons PCl_3(g) + Cl_2(g)$, $K_c = 0.61$ at 500. K. (a) Calculate the equilibrium molar concentrations of the components in the mixture when 2.0 g of PCl_5 is placed in a 300.-mL reaction vessel and allowed to come to equilibrium. (b) What percentage of PCl_5 is decomposed at 500. K?

14.41 When solid NH_4HS and 0.400 mol of gaseous NH_3 were placed into a 2.0-L vessel at 24°C, the equilibrium $NH_4HS(s) \rightleftharpoons NH_3(g) + H_2S(g)$, for which $K_c = 1.6 \times 10^{-4}$, was reached. What are the equilibrium concentrations of NH_3 and H_2S?

14.42 When solid NH_4HS and 0.200 mol $NH_3(g)$ were placed into a 2.0-L vessel at 24°C, the equilibrium $NH_4HS(s) \rightleftharpoons NH_3(g) + H_2S(g)$, for which $K_c = 1.6 \times 10^{-4}$, was reached. What are the equilibrium concentrations of NH_3 and H_2S?

14.43 At 760.°C, $K_c = 33.3$ for the reaction $PCl_5(g) \rightleftharpoons PCl_3(g) + Cl_2(g)$. If a mixture that consists of 0.200 mol PCl_5 and 0.600 mol PCl_3 is placed in a 4.00-L reaction vessel and heated to 760.°C, what is the equilibrium composition of the system?

14.44 At 760.°C, $K_c = 33.3$ for the reaction $PCl_5(g) \rightleftharpoons PCl_3(g) + Cl_2(g)$. If a mixture that consists of 0.200 mol PCl_3 and 0.600 mol Cl_2 is placed in a 8.00-L reaction vessel and heated to 760.°C, what is the equilibrium composition of the system?

14.45 The equilibrium constant K_c for the reaction $N_2(g) + O_2(g) \rightleftharpoons 2 NO(g)$ at 1200.°C is 1.00×10^{-5}. Calculate the equilibrium molar concentrations of NO, N_2, and O_2 in a 1.00-L reaction vessel that initially held 0.114 mol N_2 and 0.114 mol O_2.

14.46 The equilibrium constant K_c for the reaction $N_2(g) + O_2(g) \rightleftharpoons 2 NO(g)$ at 1200.°C is 1.00×10^{-5}. Calculate the equilibrium concentrations of NO, N_2, and O_2 in a 10.0-L reaction vessel that initially held 0.014 mol N_2 and 0.214 mol O_2.

14.47 A reaction mixture that consisted of 0.400 mol H_2 and 1.60 mol I_2 was prepared in a 3.00-L flask and heated. At equilibrium, 60.0% of the hydrogen gas had reacted. What is the equilibrium constant for the reaction $H_2(g) + I_2(g) \rightleftharpoons 2 HI(g)$ at this temperature?

14.48 A reaction mixture that consisted of 0.20 mol N_2 and 0.20 mol H_2 was placed into a 25.0-L reactor and heated. At equilibrium, 5.0% of the nitrogen gas had reacted. What is the value of the equilibrium constant K_c for the reaction $N_2(g) + 3 H_2(g) \rightleftharpoons 2 NH_3(g)$ at this temperature?

Effect of Added Reagents

14.49 Consider the equilibrium $CO(g) + H_2O(g) \rightleftharpoons CO_2(g) + H_2(g)$. (a) If the partial pressure of CO_2 is increased, what happens to the partial pressure of H_2? (b) If the partial pressure of CO is decreased, what happens to the partial pressure of CO_2? (c) If the concentration of CO is increased, what happens to the concentration of H_2? (d) If the concentration of H_2O is decreased, what happens to the equilibrium constant for the reaction?

14.50 Consider the equilibrium $CH_4(g) + 2 O_2(g) \rightleftharpoons CO_2(g) + 2 H_2O(g)$. (a) If the partial pressure of CO_2 is increased, what happens to the partial pressure of CH_4? (b) If the partial pressure of CH_4 is decreased, what happens to the partial pressure of CO_2? (c) If the concentration of CH_4 is increased, what happens to the equilibrium constant for the reaction? (d) If the concentration of H_2O is decreased, what happens to the concentration of CO_2?

14.51 The four gases NH_3, O_2, NO, and H_2O are mixed in a reaction vessel and allowed to reach equilibrium in the reaction $4 NH_3(g) + 5 O_2(g) \rightleftharpoons 4 NO(g) + 6 H_2O(g)$. Certain changes (see the following table) are then made to this mixture. Considering each change separately, state the effect (increase, i; decrease, d; or no change, nc) that the change has on the original equilibrium value of the quantity in the second column (or K_c, if that is specified). The temperature and volume are constant unless otherwise noted.

Change	Quantity	Effect		
add NO	amount of H_2O	i	d	nc
add NO	amount of O_2	i	d	nc
remove H_2O	amount of NO	i	d	nc
remove O_2	amount of NH_3	i	d	nc
add NH_3	K_c	i	d	nc
remove NO	amount of NH_3	i	d	nc
add NH_3	amount of O_2	i	d	nc

14.52 The four substances HCl, I_2, HI, and Cl_2 are mixed in a reaction vessel and allowed to reach equilibrium in the reaction $2 HCl(g) + I_2(s) \rightleftharpoons 2 HI(g) + Cl_2(g)$. Certain changes (which are specified in the first column in the following table) are then made to this mixture. Considering each change separately, state the effect (increase, i; decrease, d; or no change, nc) that the change has on the original equilibrium value of the quantity in the second column (or K_c, if that is specified). The temperature and volume are constant unless otherwise noted.

Change	Quantity	Effect		
add HCl	amount of HI	i	d	nc
add I_2	amount of Cl_2	i	d	nc
remove HI	amount of Cl_2	i	d	nc
remove Cl_2	amount of HCl	i	d	nc
add HCl	K_c	i	d	nc
remove HCl	amount of I_2	i	d	nc
add I_2	K_c	i	d	nc

Response to Pressure

14.53 State whether reactants or products will be favored by compression of each of the following equilibria. If no change occurs, explain why that is so.
(a) $2 O_3(g) \rightleftharpoons 3 O_2(g)$
(b) $H_2O(g) + C(s) \rightleftharpoons H_2(g) + CO(g)$
(c) $4 NH_3(g) + 5 O_2(g) \rightleftharpoons 4 NO(g) + 6 H_2O(g)$
(d) $2 HD(g) \rightleftharpoons H_2(g) + D_2(g)$
(e) $Cl_2(g) \rightleftharpoons 2 Cl(g)$

14.54 State what happens to the concentration of the indicated substance when each of the following equilibria is compressed:
(a) $NO_2(g)$ in
$2 Pb(NO_3)_2(s) \rightleftharpoons 2 PbO(s) + 4 NO_2(g) + O_2(g)$
(b) $NO(g)$ in $3 NO_2(g) + H_2O(l) \rightleftharpoons 2 HNO_3(aq) + NO(g)$
(c) $HI(g)$ in $2 HCl(g) + I_2(s) \rightleftharpoons 2 HI(g) + Cl_2(g)$
(d) $SO_2(g)$ in $2 SO_2(g) + O_2(g) \rightleftharpoons 2 SO_3(g)$
(e) $NO_2(g)$ in $2 NO(g) + O_2(g) \rightleftharpoons 2 NO_2(g)$

14.55 Consider the equilibrium $4 NH_3(g) + 5 O_2(g) \rightleftharpoons 4 NO(g) + 6 H_2O(g)$. (a) What happens to the partial pressure of NH_3 when the partial pressure of NO is increased?
(b) Does the partial pressure of O_2 increase or decrease when the partial pressure of NH_3 decreases?

14.56 Consider the equilibrium $2 SO_2(g) + O_2(g) \rightleftharpoons 2 SO_3(g)$. (a) What happens to the partial pressure of SO_3 when the partial pressure of SO_2 is decreased? (b) If the partial pressure of SO_2 increases, what happens to the partial pressure of O_2?

Response to Temperature

14.57 Predict whether each of the following equilibria will shift toward products or reactants with a temperature increase:
(a) $N_2O_4(g) \rightleftharpoons 2 NO_2(g)$, $\Delta H° = +57$ kJ
(b) $Ni(s) + 4 CO(g) \rightleftharpoons Ni(CO)_4(g)$, $\Delta H° = -161$ kJ
(c) $CO_2(g) + 2 NH_3(g) \rightleftharpoons CO(NH_2)_2(s) + H_2O(g)$, $\Delta H° = -90.$ kJ

14.58 Predict whether each of the following equilibria will shift toward products or reactants with a temperature increase:
(a) $CH_4(g) + H_2O(g) \rightleftharpoons CO(g) + 3 H_2(g)$, $\Delta H° = +206$ kJ
(b) $CO(g) + H_2O(g) \rightleftharpoons CO_2(g) + H_2(g)$, $\Delta H° = -41$ kJ
(c) $2 SO_2(g) + O_2(g) \rightleftharpoons 2 SO_3(g)$, $\Delta H° = -198$ kJ

14.59 A mixture consisting of 2.23 mmol N_2 and 6.69 mmol H_2 in a 500.-mL container was heated to 600. K and allowed to reach equilibrium. Will more ammonia be formed if that equilibrium mixture is then heated to 700. K? For $N_2(g) + 3 H_2(g) \rightleftharpoons 2 NH_3(g)$, $K_p = 1.7 \times 10^{-3}$ at 600. K and 7.8×10^{-5} at 700. K.

14.60 A mixture consisting of 1.1 mmol SO_2 and 2.2 mmol O_2 in a 250.-mL container was heated to 500. K and allowed to reach equilibrium. Will more sulfur trioxide be formed if that equilibrium mixture is cooled to 25°C? For the reaction $2 SO_2(g) + O_2(g) \rightleftharpoons 2 SO_3(g)$, $K_p = 2.5 \times 10^{10}$ at 500. K and 4.0×10^{24} at 25°C.

Supplementary Exercises

14.61 The endothermic dissociation of a diatomic molecule, $X_2(g) \rightleftharpoons 2 X(g)$ occurs at 500 K. Picture (1) shows the equilibrium state of the dissociation, and picture (2) shows the equilibrium state in the container after a change. Which of the following changes will produce the composition in (2)?

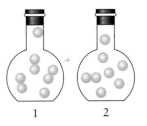

1   2

(a) Increasing the temperature; (b) adding X atoms; (c) decreasing the volume; (d) adding a catalyst.

14.62 The value of the equilibrium constant K_c for the reaction $F_2(g) \rightleftharpoons 2 F(g)$ is 2.7×10^{-3} at 1200. K. Determine the value of K_p for the reactions (a) $F_2(g) \rightleftharpoons 2 F(g)$; (b) $2 F(g) \rightleftharpoons F_2(g)$.

14.63 The value of the equilibrium constant K_p for the reaction $2 SO_2(g) + O_2(g) \rightleftharpoons 2 SO_3(g)$ is 3.0×10^4 at 700. K. Determine the value of K_c for the reactions (a) $2 SO_2(g) + O_2(g) \rightleftharpoons 2 SO_3(g)$; (b) $SO_3(g) \rightleftharpoons SO_2(g) + \frac{1}{2} O_2(g)$.

14.64 At 500.°C, $K_c = 0.061$ for $N_2(g) + 3 H_2(g) \rightleftharpoons 2 NH_3(g)$. Calculate the value of K_c at 500.°C for the reactions
(a) $\frac{1}{6} N_2(g) + \frac{1}{2} H_2(g) \rightleftharpoons \frac{1}{3} NH_3(g)$
(b) $NH_3(g) \rightleftharpoons \frac{1}{2} N_2(g) + \frac{3}{2} H_2(g)$
(c) $4 NH_3(g) \rightleftharpoons 2 N_2(g) + 6 H_2(g)$

14.65 At 500. K, $K_c = 0.061$ for $N_2(g) + 3 H_2(g) \rightleftharpoons 2 NH_3(g)$. If analysis shows that the composition is 3.00 mol/L N_2, 2.00 mol/L H_2, and 0.500 mol/L NH_3, is the reaction at equilibrium? If not, in which direction does the reaction tend to proceed to reach equilibrium?

14.66 At 2500. K, the equilibrium constant is $K_c = 20.$ for the reaction $Cl_2(g) + F_2(g) \rightleftharpoons 2 ClF(g)$. An analysis of a reaction vessel at 2500. K revealed the presence of 0.18 mol/L Cl_2, 0.31 mol/L F_2, and 0.92 mol/L ClF. Will ClF tend to form or to decompose as the reaction proceeds toward equilibrium?

14.67 At 500. K, the equilibrium constant is $K_c = 0.031$ for the reaction $Cl_2(g) + Br_2(g) \rightleftharpoons 2 BrCl(g)$. If the equilibrium composition is 0.22 mol/L Cl_2 and 0.097 mol/L BrCl, what is the equilibrium concentration of Br_2?

14.68 The equilibrium constant is $K_p = 3.5 \times 10^4$ for reaction $PCl_3(g) + Cl_2(g) \rightleftharpoons PCl_5(g)$ at 760.°C. At equilibrium, the partial pressure of PCl_5 was 2.2×10^{-4} atm and that of PCl_3 was 1.33 atm. What was the equilibrium partial pressure of Cl_2?

14.69 A reaction mixture consisting of 2.00 mol CO and 3.00 mol H_2 is placed into a 10.0-L reaction vessel and heated to 1200. K. At equilibrium, 0.478 mol CH_4 was present in the system. Determine the value of K_c for the reaction $CO(g) + 3 H_2(g) \rightleftharpoons CH_4(g) + H_2O(g)$.

14.70 A mixture consisting of 1.0 mol $H_2O(g)$ and 1.0 mol $CO(g)$ is placed in a 10.0-L reaction vessel at 800. K. At equilibrium, 0.665 mol $CO_2(g)$ is present as a result of the reaction $CO(g) + H_2O(g) \rightleftharpoons CO_2(g) + H_2(g)$. What are (a) the equilibrium concentrations for all substances and (b) the value of K_c?

14.71 A reaction mixture was prepared by mixing 0.100 mol SO_2, 0.200 mol NO_2, 0.100 mol NO, and 0.150 mol SO_3 in a 5.00-L reaction vessel. The reaction $SO_2(g) + NO_2(g) \rightleftharpoons NO(g) + SO_3(g)$ is allowed to reach equilibrium at 460°C, when $K_c = 85.0$. What is the equilibrium concentration of each substance?

14.72 A 0.100-mol sample of H_2S is placed in a 10.0-L reaction vessel and heated to 1132°C. At equilibrium, 0.0285 mol H_2 is present. Calculate the value of K_c for the reaction $2 H_2S(g) \rightleftharpoons 2 H_2(g) + S_2(g)$.

14.73 A mixture of 0.0560 mol O_2 and 0.0200 mol N_2O is placed in a 1.00-L reaction vessel at 25°C. When the reaction $2 N_2O(g) + 3 O_2(g) \rightleftharpoons 4 NO_2(g)$ is at equilibrium, 0.0200 mol NO_2 is present. (a) What are the equilibrium concentrations of O_2 and N_2O? (b) What is the value of K_c?

14.74 At 500. K, 1.0 mol NOCl is 9.0% dissociated in a 1.0-L vessel. Calculate the value of K_c for the reaction $2 NOCl(g) \rightleftharpoons 2 NO(g) + Cl_2(g)$.

14.75 At 25°C, $K_p = 3.2 \times 10^{-34}$ for the reaction $2 HCl(g) \rightleftharpoons H_2(g) + Cl_2(g)$. If a 1.0-L reaction vessel is charged with HCl at a pressure of 0.22 atm, what are the equilibrium partial pressures of HCl, H_2, and Cl_2?

14.76 If 4.00 L of HCl(g) at 1.00 atm and 273 K and 26.0 g of $I_2(s)$ are transferred to a 12.00-L reaction vessel and heated to 25°C, what will the equilibrium concentrations of HCl, HI, and Cl_2 be? $K_c = 1.6 \times 10^{-34}$ at 25°C for $2 HCl(g) + I_2(s) \rightleftharpoons 2 HI(g) + Cl_2(g)$.

14.77 A 30.1-g sample of NOCl is placed into a 200.-mL reaction vessel and heated to 500. K. The value of K_p for the decomposition of NOCl at 500. K in the reaction $2 NOCl(g) \rightleftharpoons 2 NO(g) + Cl_2(g)$ is 1.13×10^{-3}. (a) What are the equilibrium partial pressures of NOCl, NO, and Cl_2? (b) What is the percentage decomposition of NOCl? (You will need to solve a cubic equation.)

14.78 At 25°C, $K_c = 4.01 \times 10^{-2}$ for $N_2O_4(g) \rightleftharpoons 2 NO_2(g)$. If 2.50 g of N_2O_4 and 0.33 g of NO_2 are placed in a 2.0-L reaction vessel, what are the equilibrium concentrations of N_2O_4 and NO_2?

14.79 The equilibrium constant K_c of the reaction $2 CO(g) + O_2(g) \rightleftharpoons 2 CO_2(g)$ is 0.66 at 2000°C. If 0.28 g of CO and 0.032 g of O_2 are placed in a 2.0-L reaction vessel and heated to 2000°C, what will the equilibrium composition of the system be? (You will need a graphing program or calculator to solve the cubic equation in x.)

14.80 Use Le Chatelier's principle to predict the effect (increase, i; decrease, d; or no change, nc) that the change given in the first column of the table below has on the quantity in the second column for the following equilibrium system:

$$5 CO(g) + I_2O_5(s) \rightleftharpoons I_2(g) + 5 CO_2(g), \Delta H° = -1175 \text{ kJ}$$

Each change is applied separately to the system.

Change	Quantity	Effect		
decrease volume	K_c	i	d	nc
increase volume	amount of CO	i	d	nc
raise temperature	K_c	i	d	nc
add I_2	amount of CO_2	i	d	nc
add I_2O_5	amount of I_2	i	d	nc
remove CO_2	amount of I_2	i	d	nc
compress	amount of CO	i	d	nc
reduce temperature	amount of CO_2	i	d	nc
add CO_2	amount of I_2O_5	i	d	nc
add CO_2	amount of CO_2	i	d	nc

14.81 At 25°C, $K_c = 4.66 \times 10^{-3}$ for the reaction $N_2O_4(g) \rightleftharpoons 2 NO_2(g)$. If a 2.50-g N_2O_4 sample is placed in a 2.00-L reaction flask, what will be the composition of the equilibrium mixture?

14.82 Let the equilibrium constants for the reactions $2 H_2O(g) \rightleftharpoons 2 H_2(g) + O_2(g)$ and $2 CO_2(g) \rightleftharpoons 2 CO(g) + O_2(g)$ be K_{p1} and K_{p2}, respectively. Show that the equilibrium constant for the reaction $CO_2(g) + H_2(g) \rightleftharpoons H_2O(g) + CO(g)$ is $K_{p3} = (K_{p2}/K_{p1})^{1/2}$, and evaluate it at 1565 K, at which temperature $K_{p1} = 1.6 \times 10^{-11}$ and $K_{p2} = 1.3 \times 10^{-10}$.

14.83 Let α be the fraction of PCl_5 molecules that have decomposed to PCl_3 and Cl_2 in the reaction $PCl_5(g) \rightleftharpoons PCl_3(g) + Cl_2(g)$ in a constant-volume container, so that the amount of PCl_5 at equilibrium is $n(1 - \alpha)$, where n is the amount present initially. Derive an equation for K_p in terms of α and the total pressure P, and solve it for α in terms of P. Calculate the fraction decomposed at 556 K, at which temperature $K_p = 4.96$, and the total pressure is (a) 0.50 atm; (b) 1.0 atm.

14.84 The reaction $N_2O_4 \rightleftharpoons 2\,NO_2$ is allowed to reach equilibrium in chloroform solution at 25°C. The equilibrium concentrations are 0.405 mol/L N_2O_4 and 2.13 mol/L NO_2. (a) Calculate K_c for the reaction. (b) An additional 1.00 mol NO_2 is added to 1.00 L of the solution and the system is allowed to reach equilibrium again at the same temperature. Use Le Chatelier's principle to predict the direction of change (increase, decrease, or no change) for N_2O_4, NO_2, and K_c after the addition of NO_2. (c) Calculate the final equilibrium concentrations after the addition of NO_2, and confirm that your predictions in (b) were valid. If they do not agree, check your procedure and repeat it if necessary.

14.85 The van't Hoff equation relates the equilibrium constant K_p' at a temperature T' to its value K_p at T:

$$\ln\!\left(\frac{K_p'}{K_p}\right) = \frac{\Delta H_r^\circ}{R}\left(\frac{1}{T} - \frac{1}{T'}\right)$$

where ΔH_r° is the reaction enthalpy of the forward process. The temperature dependence of the equilibrium constant of the reaction $N_2(g) + O_2(g) \rightleftharpoons 2\,NO(g)$, which makes an important contribution to atmospheric nitrogen oxides, can be expressed as $\ln K_p' = 2.5 - (21\,700\ \text{K})/T'$. What is the standard enthalpy of the forward reaction?

14.86 The dissociation vapor pressure of solid ammonium hydrogen sulfide (NH_4HS) is 501 Torr at 298.3 K and 919 Torr at 308.8 K. Using the information in Exercise 14.85, estimate its enthalpy of dissociation into $NH_3(g)$ and $H_2S(g)$ and the temperature at which it would sublime into these products when the external pressure is 1 atm. *Hint:* The dissociation vapor pressure is the sum of the partial pressures of NH_3 and H_2S at equilibrium.

Applied Exercises

For Exercises 14.87–14.90, see Applying Chemistry: Case Study 14.

14.87 In the Haber process for ammonia synthesis, $K_p = 0.036$ for $N_2(g) + 3\,H_2(g) \rightleftharpoons 2\,NH_3(g)$ at 500. K. (a) If a 2.0-L reactor is charged with 100.0 atm of N_2 and 100.0 atm of H_2, what will the equilibrium partial pressures in the mixture be? (b) When the mixture is chilled, the ammonia produced condenses and is removed. If the mixture of N_2 and H_2 is then heated to 500. K, what will the new equilibrium concentrations be?

14.88 Which of the following steps might you take to increase the yield of nitrate ion in the endothermic reaction of NO_2 gas with liquid water, $3\,NO_2(g) + 3\,H_2O(l) \rightarrow 2\,H_3O^+(aq) + 2\,NO_3^-(aq) + NO(g)$? (a) Decrease the volume. (b) Add sodium hydroxide to the solution. (c) Reduce the temperature. (d) Dilute the solution. Explain your answers.

14.89 The photosynthesis reaction is $6\,CO_2(g) + 6\,H_2O(l) \rightarrow C_6H_{12}O_6(s) + 6\,O_2(g)$, and $\Delta H_r^\circ = +2802$ kJ/mol. In an experiment on artificial photosynthesis, the reaction is at equilibrium, but the experimenters want to increase the amount of glucose produced. Which of the following changes will shift the reaction toward the formation of products? (a) The partial pressure of O_2 is increased. (b) The system is compressed. (c) The amount of CO_2 is increased. (d) The temperature is increased. (e) Some of the $C_6H_{12}O_6$ is removed. (f) The partial pressure of CO_2 is decreased.

14.90 A reaction important in the body is the conversion of carbon dioxide gas to the hydrogen carbonate ion in the bloodstream, $CO_2(g) + 2\,H_2O(l) \rightleftharpoons H_3O^+(aq) + HCO_3^-(aq)$. Shock or stress can cause hyperventilation (breathing very hard and fast), which eliminates carbon dioxide from the bloodstream. (a) Explain how hyperventilation would affect the concentration of hydrogen carbonate ion in the bloodstream. (b) What first aid remedy could you use to help someone who has been hyperventilating? Explain your answers.

Integrated Exercises

14.91 A pharmacologist is preparing a new drug. In one step, acetic acid, CH_3COOH, reacts with ethanol, C_2H_5OH, to form the ester ethyl acetate, $CH_3COOC_2H_5$, and water:

$$CH_3COOH + C_2H_5OH \rightleftharpoons CH_3COOC_2H_5 + H_2O$$
$$K_c = 4.0 \text{ at } 100°C$$

(a) If the initial concentrations of CH_3COOH and C_2H_5OH are 0.32 and 6.3 mol/L, respectively, and no products are present initially, what will the equilibrium concentration of the ester be? *Hint:* Water is a product, not a solvent. (b) Next the solution must be cooled to 37°C. Use enthalpies of formation from Appendix 2A to predict in which direction the reaction will shift when the temperature is decreased. The enthalpy of formation of ethyl acetate is −479.3 kJ/mol.

A pharmacologist adjusting the temperature of a reaction mixture for use in the preparation of a new drug.

14.92 Use your knowledge of bond enthalpies to predict whether the following reactions will shift toward products or reactants with a temperature increase:
(a) $Cl_2(g) \rightleftharpoons 2\,Cl(g)$
(b) $O(g) + N_2(g) \rightleftharpoons NO(g) + N(g)$
(c) $OH(g) + H_2(g) \rightleftharpoons H_2O(g) + H(g)$

14.93 Write the equilibrium expression for each of the following dissolution reactions, then use the data in Table 3.1 to determine whether the equilibrium constant will be large or small:
(a) $PbCrO_4(s) \rightleftharpoons Pb^{2+}(aq) + CrO_4^{2-}(aq)$

(b) $Cs_2SO_4(s) \rightleftharpoons 2\,Cs^+(aq) + SO_4^{2-}(aq)$
(c) $Fe(OH)_3(s) \rightleftharpoons Fe^{3+}(aq) + 3\,OH^-(aq)$

14.94 Use Le Chatelier's principle to predict the effect of temperature on the proportion of product in a mixture in which the forward and reverse reactions (both of which are first order) have reached dynamic equilibrium and (a) the forward reaction is exothermic; (b) the forward reaction is endothermic. (c) Explain this effect in terms of the temperature dependence of the rate constants of the forward and reverse reactions.

Acids and Bases

A cid rain is a problem that we are just beginning to learn how to control. Automobile engines and coal-fired power plants generate oxides of sulfur and nitrogen in great abundance. These oxides react with the water in raindrops to form acids that are much stronger than the acids occurring naturally in rain; as a result, acid rain can damage forests, lakes, buildings, and human health. But what does it mean to say that some acids are "stronger" than others? What difference does it make if rain contains strong or weak acids? Indeed, why do the oxides of many nonmetals form acidic solutions in water?

Acid rain is less of a problem in the western and southwestern United States than in the northeastern parts of the country. In the west and southwest, the acids in rain are often neutralized before they fall. Mineral dust from the dry soils of the region contains metal oxides and hydroxides and is blown up into the clouds. These compounds—which are bases—react with the acids. But how do metal oxides and hydroxides react with acids? What do we really mean by a base?

Acids and bases play important roles in our lives. They are fundamentally important in industry, and life depends on the complex networks of acid-base reactions going on inside organisms. To understand life—and death—we need to understand acids and bases and the reactions they undergo.

In this chapter, we build on the information we have gathered since we first discussed acids and bases in Chapter 3. In particular, we see how to use equilibrium calculations to discuss their properties quantitatively. We shall also see how the strengths of acids are related to their molecular structures.

This forester in Poland is investigating the growth of insects in spruce trees damaged by acid rain. Even remote forests, lakes, and mountain tundras are affected by acid rain if they are downwind of cities. The acidity of rain and natural waters is monitored by using techniques described in this chapter.

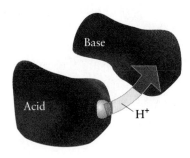

FIGURE 15.1

A Brønsted acid is a proton donor, and a Brønsted base is a proton acceptor. A proton is an H^+ ion.

WHAT ARE ACIDS AND BASES?

In Chapter 3, we saw that an Arrhenius acid is a molecular compound, such as HCl or CH_3COOH, that contains hydrogen and produces hydronium ions (H_3O^+) in water. We also saw that an Arrhenius base is a compound, such as NaOH, CaO, or NH_3, that produces hydroxide ions (OH^-) in water. However, early in the twentieth century, new definitions were developed that are much broader in scope and more helpful for understanding acids and bases.

15.1 Brønsted-Lowry Acids and Bases

In 1923, the Danish chemist Johannes Brønsted proposed the following definitions:

An **acid** is a proton donor.

A **base** is a proton acceptor.

The same definitions were proposed independently by the English chemist Thomas Lowry, and the theory based on them is called the **Brønsted-Lowry theory** of acids and bases (Fig. 15.1).

Brønsted acids (proton donors) include all the Arrhenius acids (Section 3.7). For example, when a molecule of the Arrhenius acid hydrogen chloride dissolves in water, it loses (donates) a proton to a neighboring H_2O molecule (Fig. 15.2). The loss of a proton is called **deprotonation**:

$$HCl(aq) + H_2O(l) \longrightarrow H_3O^+(aq) + Cl^-(aq)$$

The Brønsted definition, however, also includes acids that are not Arrhenius acids. For example, an ion can be a Brønsted acid. The hydrogen carbonate ion, HCO_3^-, one of the species present in natural waters, is not an Arrhenius acid (it is an ion, not an electrically neutral compound), but it can act as a proton donor:

$$HCO_3^-(aq) + H_2O(l) \longrightarrow H_3O^+(aq) + CO_3^{2-}(aq)$$

The action of HCO_3^- as a Brønsted acid contributes to the formation of stalactites and stalagmites in caves (Fig. 15.3).

In the Brønsted-Lowry theory, the strength of an acid depends on the extent to which it donates protons to the solvent. Therefore, when the solvent is water, the definition of a strong Brønsted acid is the same as that of a strong Arrhenius acid (Section 3.8):

A strong acid is fully ionized in water.

A weak acid is only partially ionized in water.

The "ionization" of an acid in water means the donation of a proton to a water molecule. "Partially ionized" means that only a small fraction of the acid species

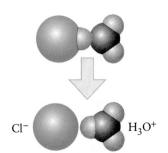

Cl^- H_3O^+

FIGURE 15.2

When an HCl molecule dissolves in water, a hydrogen bond forms between the H atom of HCl and the O atom of a neighboring H_2O molecule, and the nucleus of the hydrogen atom (H^+) is pulled out of the HCl molecule.

FIGURE 15.3

Stalactites hang from the roof of a cave and stalagmites grow from the floor. They grow when soluble hydrogen carbonates carried into the cave by groundwater lose their hydrogen ions; the carbonates that result are insoluble.

have lost a proton. Hydrogen chloride is a strong acid in water because almost every HCl molecule donates its proton to a water molecule. However, very few HCO_3^- ions give up their protons, so the hydrogen carbonate ion is an example of a weak acid in water. The seven common strong acids were listed in Table 3.2; for convenience, that information is summarized here as Table 15.1. All other acids can be treated as weak acids unless additional information is given.

In this chapter, we begin to see the importance of the presence of hydronium ions in aqueous solutions of acids. Therefore, because in each case "ionization" results in the formation of a hydronium ion, an equivalent way to express the distinction between strong and weak acids is as follows:

A strong acid reacts completely with water to produce hydronium ions.

A weak acid reacts incompletely with water to produce hydronium ions.

Because proton transfer is the central idea of the Brønsted-Lowry theory, we can express the distinction between strong and weak acids in yet another, but equivalent, way:

A strong acid is fully deprotonated in water.

A weak acid is only partially deprotonated in water.

"Partially deprotonated" does not mean that the proton is not completely transferred. It means that only a small fraction of the acid species has donated a proton.

Brønsted bases (proton acceptors) include all the Arrhenius bases. For instance, when a metal oxide (such as CaO) dissolves in water, the high negative charge of the small O^{2-} ion pulls a proton out of a neighboring H_2O molecule (Fig. 15.4). The process of accepting a proton is called **protonation.** The oxide ion forms a coordinate covalent bond to the proton, by providing both the electrons in the bond, and becomes a hydroxide ion. Lewis structures help us to understand how that occurs. The curved arrows show the direction in which the electrons move:

Table 15.1 *Common strong acids in water*
HBr
HCl
HI
HNO_3
$HClO_4$
$HClO_3$
H_2SO_4 (to HSO_4^-)

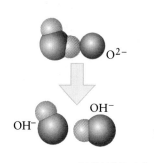

FIGURE 15.4

When an oxide ion is added to water, it exerts such a strong attraction on the nucleus of a hydrogen atom in a neighboring water molecule that that nucleus is pulled out of the molecule. As a result, the oxide ion and water form two hydroxide ions.

$$:\ddot{O}:^{2-} \quad H \qquad\qquad :\ddot{O}-H^- $$
$$:\ddot{O}-H \longrightarrow \qquad :\ddot{O}-H^-$$

Because the oxide ion accepts a proton, it is classified as a base. Many molecular compounds that contain nitrogen are also bases because the lone pair of electrons on the nitrogen atom can attract a proton. For example, when ammonia, NH_3, dissolves in water, some of the molecules undergo the following reaction:

The lone pair of electrons on the N atom in NH_3 has much less proton-pulling power than the full negative charge of an ion, and only a very small proportion of the NH_3 molecules are converted into NH_4^+ ions (Fig. 15.5). Therefore, in water, ammonia is an example of a weak Brønsted base. All amines (Section 11.10) are weak bases in water. Every oxide ion accepts a proton in water, so O^{2-} is an example of a strong base. In aqueous solution,

A strong base reacts completely with water to produce hydroxide ions.

A weak base reacts incompletely with water to produce hydroxide ions.

s-Block hydroxides already contain hydroxide ions, so they do not need to react to produce OH^- ions.

As in the discussion of acids, because we focus on the transfer of protons in Brønsted-Lowry theory, we can also distinguish strong and weak bases as follows:

A strong base is completely protonated in water.

A weak base is only partially protonated in water.

"Partially protonated" means that only a small fraction of the base species have accepted a proton. For example, the amide ion, NH_2^-, is fully protonated in solution (to form NH_3), so it is a strong base in water; the ammonia molecule is only partially protonated (to form NH_4^+), so it is a weak base. The hydroxide ion appears to be excluded from this definition, because it survives in water, but it is also a strong base. When an OH^- ion is protonated, the H_2O molecule acting as donor becomes another OH^- ion, so there is no net change:

$$OH^-(aq) + H_2O(l) \longrightarrow H_2O(l) + OH^-(aq)$$

That is, OH^- ions effectively survive in water even if every OH^- ion added accepts a proton! A solution of sodium hydroxide, for instance, is a solution of Na^+ and OH^- ions, but the OH^- ions in solution are not necessarily those provided by the solute.

The three different definitions of acids and bases (Arrhenius, Brønsted, and Lewis) are all closely related. Recall from Section 8.13 that a Lewis acid is defined as an electron pair acceptor and a Lewis base as an electron pair donor. A proton (H^+) is an electron pair acceptor, and therefore a Lewis acid, because it

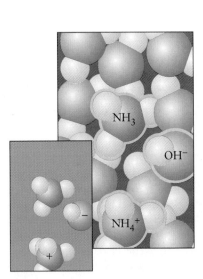

FIGURE 15.5

In this molecular-level representation of a solution of ammonia in water, we see that there are NH_3 molecules still present, because only a few of them have accepted hydrogen ions from water. In practice, only about 1 in 100 NH_3 molecules is protonated in a typical solution. The overlay shows only the solute species.

can attach to a lone pair of electrons on, for instance, H_2O. The chemical equations above show that *a Brønsted acid is a supplier of one particular Lewis acid, a proton.* Likewise, *a Brønsted base is a special kind of Lewis base, one that can use a lone pair to bond to a proton.* Notice how the definitions of acids and bases have became broader and capture a wider range of species as "acids" and "bases":

An Arrhenius acid or base is defined according to the ability of a compound to produce hydronium ions or hydroxide ions in water.

A Brønsted acid or base is defined according to the ability of a species to donate or accept a proton. Water need not be involved.

A Lewis acid or base is defined according to the ability of a species to donate or accept a pair of electrons and to form a coordinate covalent bond. A proton need not be involved.

We can see from our earlier discussion of the reaction of oxide ions with water that oxide ions are bases by all three definitions: an O^{2-} ion gives rise to OH^- ions in solution (Arrhenius), it accepts a proton (Brønsted), and it can use its lone pairs to attach to other species (Lewis). The acidic character of non-metal oxides, such as carbon dioxide, is more subtle. It stems from their ability to act as Lewis acids:

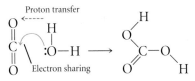

For clarity, we are showing only one of the lone pairs.

In this reaction, the C atom of CO_2, the Lewis acid, accepts an electron pair from the O atom of a water molecule, the Lewis base, and a proton migrates from one oxygen atom to another. The product, an H_2CO_3 molecule, then acts as a Brønsted acid.

The transfer of protons is very important in chemistry. Therefore, from now on, unless we specify "Arrhenius" or "Lewis," whenever we refer to acids and bases in this book, we mean the Brønsted-Lowry definitions.

An acid is a proton donor and a base is a proton acceptor.

15.2 Conjugate Acids and Bases

Recall from Section 3.11 that a proton is transferred when an Arrhenius base neutralizes an Arrhenius acid. This transfer of a proton is central to the reactions of Brønsted acids and bases. When CH_3COOH dissolves in water, it donates a proton to water and forms hydronium ions and acetate ions:

$$CH_3COOH(aq) + H_2O(l) \rightleftharpoons H_3O^+(aq) + CH_3CO_2^-(aq)$$

The water molecule is acting as a Brønsted base and acetic acid is a Brønsted acid. Because the acetate ion, a base, is formed from acetic acid by proton loss, it is called the **conjugate base** of acetic acid. In general,

$$acid \xrightarrow{\text{donates } H^+} conjugate\ base$$

According to this definition, Cl^- is the conjugate base of HCl and OH^- is the conjugate base of H_2O. An acid is converted to its conjugate base by the loss of one proton.

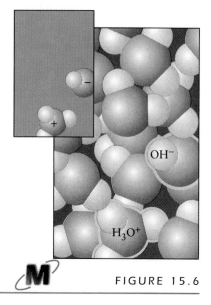

When we dissolve sodium acetate in water, an acetate ion can accept a proton from a water molecule and be converted into an acetic acid molecule:

$$H_2O(l) + CH_3CO_2^-(aq) \rightleftharpoons CH_3COOH(aq) + OH^-(aq)$$

In this reaction, water is a Brønsted acid and the acetate ion is a Brønsted base. Because a CH_3COOH molecule is an acid formed by attaching a proton to an acetate ion, it is the **conjugate acid** of the base $CH_3CO_2^-$. In general,

$$\text{base} \xrightarrow{\text{accepts } H^+} \text{conjugate acid}$$

Likewise, HCl is the conjugate acid of Cl^- and H_2O is the conjugate acid of OH^-. A base is converted to its conjugate acid by the gain of one proton. Even the OH^- ion is a conjugate acid, because it has one more proton than the O^{2-} ion, its conjugate base.

The conjugate base of an acid is formed when the acid has donated a proton. The conjugate acid of a base is formed when the base has accepted a proton.

Self-Test 15.1A Write chemical formulas for (a) the conjugate acids of CH_3NH_2 and CN^-; (b) the conjugate bases of H_3PO_3 and HI.

[*Answer:* (a) $CH_3NH_3^+$ and HCN; (b) $H_2PO_3^-$ and I^-]

Self-Test 15.1B Write chemical formulas for (a) the conjugate acids of NH_3 and CO_3^{2-}; (b) the conjugate bases of NH_3 and HNO_3.

FIGURE 15.6

As a result of autoprotolysis, pure water consists of hydronium ions and hydroxide ions as well as water molecules. The concentration of ions that results from autoprotolysis is only about 10^{-7} mol/L, so only about 1 molecule in 200 million is ionized. The overlay shows only the ions.

15.3 Proton Exchange Between Water Molecules

Is water an acid, a base, or both? We have seen that a water molecule accepts a proton from an acid molecule to form an H_3O^+ ion. So, water is a base. However, a water molecule can donate a proton to a base and become an OH^- ion. So, water is also an acid. We say that water is **amphiprotic,** meaning that it can act as both a proton donor and a proton acceptor.

The prefix *amphi-* comes from the Greek word for "both."

Protons migrate between water molecules, even in the absence of another acid or base:

$$2\,H_2O(l) \longrightarrow H_3O^+(aq) + OH^-(aq)$$

Because H_3O^+ is an acid and OH^- is a base, the reverse reaction

$$H_3O^+(aq) + OH^-(aq) \longrightarrow 2\,H_2O(l)$$

also occurs. The transfer of protons is very rapid, and the equilibrium

$$2\,H_2O(l) \rightleftharpoons H_3O^+(aq) + OH^-(aq)$$

is always present in water and aqueous solutions. This type of reaction, in which one molecule transfers a proton to another molecule of the same kind, is called **autoprotolysis** (Fig. 15.6).

The equilibrium constant for the autoprotolysis of water is

$$K_c = \frac{[H_3O^+][OH^-]}{[H_2O]^2}$$

The general term for the process in which the molecules of a single substance form ions by reacting with one another is *autoionization.*

where, as in Chapter 14, [X] means the numerical value of the molarity of the species X (the molarity with the units moles per liter deleted). In water itself and in dilute aqueous solutions (the only ones we consider), the solvent, water, is very nearly pure, so the molar concentration of water can be treated as a constant and combined with K_c. The resulting expression is called the **autoprotolysis constant** of water and is written K_w:

$$K_w = K_c[H_2O]^2 = [H_3O^+][OH^-] \tag{1}$$

The older, and still widely used, names for this quantity are the *autoionization constant of water* and *ion-product constant for water*.

The molarities of H_3O^+ and OH^- in pure water at 25°C are known by experiment to be 1.0×10^{-7} mol/L, so

$$K_w = (1.0 \times 10^{-7}) \times (1.0 \times 10^{-7}) = 1.0 \times 10^{-14}$$

The concentrations of H_3O^+ and OH^- are very low in pure water, which explains why pure water is such a poor conductor of electricity.

 To imagine the very tiny extent of autoprotolysis, think of each letter in this book as a water molecule. We would need to search through more than 50 books to find one H_3O^+ ion.

An analogy

Because K_w is an equilibrium constant, and the autoprotolysis equilibrium is reached very rapidly, *the product of the molarities of H_3O^+ and OH^- ions is equal to K_w in every aqueous solution.* We can increase the concentration of H_3O^+ ions by adding acid, but then the concentration of OH^- ions must decrease to preserve the value of K_w. Alternatively, we can increase the concentration of OH^- ions by adding base, but then the concentration of H_3O^+ ions must decrease. We cannot increase the molarities of *both* hydronium ions *and* hydroxide ions in the same solution. The equilibrium linking the molarities of H_3O^+ and OH^- ions is like a seesaw: when one goes up, the other goes down (Fig. 15.7). Therefore, if we know the concentration of one ion, we can calculate the concentration of the other:

$$[OH^-] = \frac{K_w}{[H_3O^+]} \quad \text{and} \quad [H_3O^+] = \frac{K_w}{[OH^-]}$$

In an aqueous solution of a strong acid, virtually all the acid molecules have donated their protons to water molecules. The concentration of hydronium ions is therefore equal to the **initial concentration** of acid, the concentration of the acid as initially prepared, before it lost its protons. For example, for hydrochloric acid, $[H_3O^+] = [HCl]_{initial}$. In an aqueous solution of the hydroxide of a Group 1 or 2 metal, the concentration of hydroxide ions in solution is equal to the concentration of hydroxide ions added. For example, for NaOH, $[OH^-] = [NaOH]_{initial}$. Therefore, we can find the concentrations of H_3O^+ and OH^- in a solution of a strong acid or base from the initial concentration of the

FIGURE 15.7

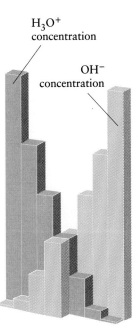

H_3O^+ concentration

OH^- concentration

The product of the concentrations of hydronium and hydroxide ions in water (pure water and aqueous solutions) is a constant. If the concentration of one of these ions increases, the concentration of the other must decrease to keep the product of concentrations constant.

Example 15.10 shows how to calculate the concentration of H_3O^+ ions in sulfuric acid.

acid or base. The strong acid H_2SO_4 is a special case: it has two acidic protons, but only one of them is lost completely in water.

The autoprotolysis of water also contributes H_3O^+ and OH^- ions to aqueous solutions. However, except for extremely dilute solutions, the amount contributed by autoprotolysis is insignificant compared with that resulting from the added acid or base and can be ignored.

In aqueous solutions, the molarities of H_3O^+ and OH^- ions are related by the autoprotolysis equilibrium; if one concentration is increased, then the other must decrease to maintain the value of K_w.

Example 15.1 *Calculating the molarities of ions in a solution of a strong acid*

What are the molarities of H_3O^+ and OH^- ions in 0.020 M HCl(aq) at 25°C?

Strategy First, decide whether the acid is strong. For a strong acid, which is almost completely deprotonated in water, the concentration of H_3O^+ ions in the solution is equal to the initial concentration of the acid. To find the molarity of OH^- ions, use $K_w = [H_3O^+][OH^-]$.

Solution Table 15.1 lists hydrochloric acid as a strong acid. The molarity of H_3O^+ ions is equal to the initial concentration of HCl, which is 0.020 mol/L. We find the molarity of OH^- from

$$[OH^-] = \frac{K_w}{[H_3O^+]} = \frac{1.0 \times 10^{-14}}{0.020} = 5.0 \times 10^{-13}$$

That is, the molarity of OH^- ions in the solution is only 5.0×10^{-13} mol/L.

Self-Test 15.2A Estimate the molarities of H_3O^+ and OH^- ions in 6.0×10^{-5} M HI(aq).

[*Answer:* 6.0×10^{-5} mol/L H_3O^+; 1.7×10^{-10} mol/L OH^-]

Self-Test 15.2B Estimate the molarities of H_3O^+ and OH^- ions in 0.010 M HNO_3(aq).

Example 15.2 *Calculating the molarities of ions in a solution of a strong base*

What are the molarities of H_3O^+ and OH^- ions in 0.0030 M $Ba(OH)_2$(aq) at 25°C?

Strategy Strong bases are virtually completely protonated in water; for the special case of OH^- ions, "complete protonation" implies that the number of OH^- ions present is equal to the number added. First, decide whether the base is strong. If it is, decide from the chemical formula how many OH^- ions are provided by each formula unit. Then calculate the molarity of these ions in the solution. To find the molarity of H_3O^+ ions, use $K_w = [H_3O^+][OH^-]$.

Solution Barium hydroxide is a hydroxide of an alkaline earth metal, so it is a strong base (see Section 3.9). The equation

$$Ba(OH)_2(s) \longrightarrow Ba^{2+}(aq) + 2\,OH^-(aq)$$

tells us that 1 mol $Ba(OH)_2 \cong 2$ mol OH^-. Because the molarity of $Ba(OH)_2$ in the solution is 0.0030 mol/L, it follows that the molarity of OH^- is twice that value, or 0.0060 mol/L. Then, for the molarity of H_3O^+ ions, we write

$$[H_3O^+] = \frac{K_w}{[OH^-]} = \frac{1.0 \times 10^{-14}}{0.0060} = 1.7 \times 10^{-12}$$

That is, the molarity of H_3O^+ ions in the solution is only 1.7×10^{-12} mol/L.

Self-Test 15.3A Calculate the molarities of H_3O^+ and OH^- ions in 0.052 M KOH(aq).
[**Answer:** 1.9×10^{-13} mol/L H_3O^+; 0.052 mol/L OH^-]

Self-Test 15.3B Calculate the molarities of H_3O^+ and OH^- ions in 2.2×10^{-3} M NaOH(aq).

15.4 The pH Scale

In a hospital emergency room or in a manufacturing plant, it may be necessary to report concentrations of H_3O^+ and OH^- ions and make comparisons quickly and easily, in a manner that reduces error. However, the molarities of these ions vary over many orders of magnitude (powers of 10). In some solutions, they may be as high as 1 mol/L and in others as low as 10^{-14} mol/L. We can avoid the awkwardness of dealing with such a wide range of values by using logarithms, which condense these values into a much smaller and more convenient range (Fig. 15.8). For example, the logarithm of 10^{-7} is -7. Chemists usually report hydronium ion molarities in aqueous solution by giving the **pH** of the solution, the negative of the common logarithm of the molarity:

The pH scale was introduced by the Danish chemist Søren Sørenson in 1909 for his quality control work at a brewery.

A common logarithm is a logarithm to the base 10. See Appendix 1C.

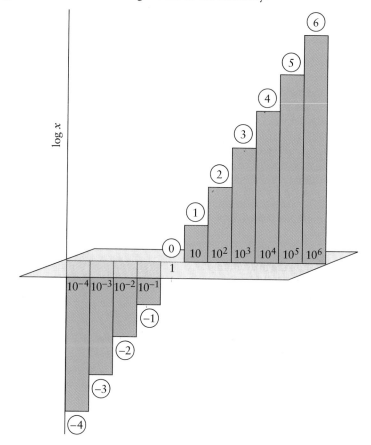

FIGURE 15.8

Very large ranges of numbers are difficult to represent graphically. However, their logarithms span a much smaller range and can be represented easily. Note how the numbers shown here range over 10 orders of magnitude (from 10^{-4} to 10^6), but their logarithms range over 10 units (from -4 to 6). Negative values of logarithms correspond to numbers between 0 and 1; positive values correspond to numbers greater than 1.

$$pH = -\log[H_3O^+] \tag{2}$$

As usual, $[H_3O^+]$ is the molarity of H_3O^+ ions, with the units of moles per liter struck out. For example, the pH of pure water, in which the molarity of H_3O^+ ions is 1.0×10^{-7} mol/L at 25°C, is

$$pH = -\log(1.0 \times 10^{-7}) = -(-7.00) = 7.00$$

Most aqueous solutions have a pH in the range 0 to 14, but values outside this range are possible. The pH scale is used across all scientific, medical, engineering, and technical fields, but only for aqueous solutions.

The negative sign in the definition of pH means that *the higher the H_3O^+ molarity, the lower the pH*. Therefore,

The pH of a basic solution is greater than 7.

The pH of pure water is 7.

The pH of an acidic solution is less than 7.

Because pH is the negative of the common logarithm of the concentration, a change of one pH unit means the molarity of H_3O^+ ions has changed by a factor of 10. For example, when the pH decreases from 5 to 4, the H_3O^+ molarity increases by a factor of 10, from 10^{-5} mol/L to 10^{-4} mol/L.

Example 15.3 *Calculating a pH*

What is the pH of (a) human blood, in which the molarity of H_3O^+ ions is 4.0×10^{-8} mol/L; (b) 0.020 M HCl(aq); (c) 0.040 M KOH(aq)?

Strategy Use the definition of pH given earlier: take the logarithm of the H_3O^+ molarity (make sure the logarithm is to the base 10), and change the sign. For strong acids, the molarity of H_3O^+ is equal to the molarity of the acid; for acids, expect pH < 7. For strong bases, carry out a calculation like that in Example 15.2: first find the molarity of OH^-, then convert that molarity to $[H_3O^+]$ by using $[H_3O^+] = K_w/[OH^-]$; for bases, expect pH > 7. The number of digits *following* the decimal point in a pH value is equal to the number of significant figures in the corresponding molarity. This rule follows from the fact that the digits preceding the decimal point simply report the power of 10 in the data (as in log $10^5 = 5$ exactly).

Solution (a) For a solution in which the molarity of H_3O^+ ions is 4.0×10^{-8} mol/L, we write $[H_3O^+] = 4.0 \times 10^{-8}$ and obtain

$$pH = -\log(4.0 \times 10^{-8}) = 7.40$$

Notice that because the hydronium ion concentration is between 10^{-7} and 10^{-8}, the pH is between 7 and 8. (b) Because HCl is a strong acid, the molarity of H_3O^+ is 0.020 mol/L. Hence,

$$pH = -\log 0.020 = 1.70$$

(c) Each formula unit of KOH (a strong base) provides one OH^- ion; therefore the molarity of OH^- is 0.040 mol/L and

$$[H_3O^+] = \frac{K_w}{[OH^-]} = \frac{1.0 \times 10^{-14}}{0.040} = 2.5 \times 10^{-13}$$

Hence

$$pH = -\log(2.5 \times 10^{-13}) = 12.60$$

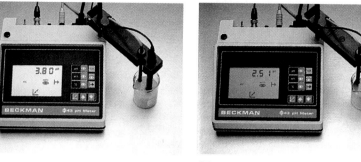

(a)   (b)

Self-Test 15.4A Calculate the pH of (a) household ammonia, for which the OH^- molarity is about 3×10^{-3} mol/L; (b) 6.0×10^{-5} M $HClO_4$(aq).

[**Answer:** (a) 11.5; (b) 4.22]

Self-Test 15.4B Calculate the pH of 0.077 M NaOH(aq).

We can measure the pH of an aqueous solution with a strip of universal indicator paper, which turns different colors at different pH values. More precise measurements are made with a pH meter (Fig. 15.9), which reports the pH directly (the operation of this instrument is explained in Investigating Matter 18.1). In the United States, the Environmental Protection Agency (EPA) defines waste as corrosive if its pH is either lower than 3.0 (highly acidic) or higher than 12.5 (highly basic). The results of measuring the pH of a selection of liquids and beverages to assess their relative acidities are shown in Fig. 15.10. Fresh lemon juice has an average pH of 2.5, corresponding to an H_3O^+ molarity of 3×10^{-3} mol/L, more than 20 times more acidic than orange juice (see Fig. 15.9). Natural rainwater, with an acidity largely due to dissolved carbon dioxide, has a pH of about 5.7. This pH corresponds to an H_3O^+ ion molarity of 2×10^{-6} mol/L. The pH of some of the rainfall experienced by Scandinavian countries has been as low as 4, which corresponds to an H_3O^+ molarity of 1×10^{-4} mol/L, or about 50 times more concentrated (see Applying Chemistry: Case Study 15).

The pH scale is used to report H_3O^+ molarity: high pH denotes a basic solution, low pH an acidic solution. A neutral solution has pH = 7.

15.5 The pOH of Solutions

Many expressions involving acids and bases are greatly simplified by writing quantities in terms of their logarithms, and the values are easier to remember. The quantity "pX" is a generalization of pH:

$$pX = -\log X$$

For instance, **pOH** is defined as

$$pOH = -\log[OH^-]$$

The pOH is a convenient scale for reporting the molarities of OH^- ions in solution. For example, in pure water, where the molarity of OH^- ions is 1.0×10^{-7} mol/L, the pOH is 7.00. Similarly, by pK_w, we mean

$$pK_w = -\log K_w = -\log(1.0 \times 10^{-14}) = 14.00$$

pH

pH	
1	
2	Corrosive
3	Lemon juice / Vinegar
4	Soda Wine
5	Beer Acid rain
6	Tomato juice
7	Milk Tap water Urine
8	Blood Saliva
9	
10	Detergents
11	
12	Household ammonia
13	
14	Corrosive

The pX notation results in simple, useful relations. For example, let's take the logarithms of both sides of the expression $K_w = [H_3O^+][OH^-]$. Because $\log ab = \log a + \log b$,

$$\log[H_3O^+] + \log[OH^-] = \log K_w$$

Multiplication of both sides of the equation by -1 gives

$$-\log[H_3O^+] - \log[OH^-] = -\log K_w$$

which is the same as

$$pH + pOH = pK_w \qquad (3)$$

Because $pK_w = 14.00$ at 25°C, at that temperature

$$pH + pOH = 14.00$$

In fact, it is simpler to use this relation to calculate the pH of a solution of a strong base than the procedure used in Example 15.3c. Because $[OH^-] = 0.040$ for the solution treated there, $pOH = -\log(0.040) = 1.40$; hence $pH = 14.00 - 1.40 = 12.60$.

The pH and pOH of a solution are related by $pH + pOH = pK_w$.

Toolbox 15.1 *How to use the pH and pOH*

This Toolbox shows how to calculate the pH and pOH from the concentrations of H_3O^+ or OH^- ions, and how to calculate the concentrations of H_3O^+ and OH^- ions from the pH or pOH.

Conceptual Basis

The pH is the negative of the common logarithm of the concentration of H_3O^+ ions (Fig. 15.11). Similarly, the pOH is the negative of the common logarithm of the concentration of OH^- ions. These concentrations are related, because the ions are in equilibrium with water molecules. At 25°C, $pH + pOH = 14.00$, the value of pK_w, in every aqueous solution. The pH increases with *decreasing* H_3O^+ concentration (decreasing acidity). A high pH indicates a basic solution:

Acidic solution: $[H_3O^+] > [OH^-]$ $pH < 7$

Neutral solution: $[H_3O^+] = [OH^-]$ $pH = 7$

Basic solution: $[H_3O^+] < [OH^-]$ $pH > 7$

Procedure

To convert from molarity to pH Because the pH is the negative logarithm of the H_3O^+ concentration, we take the

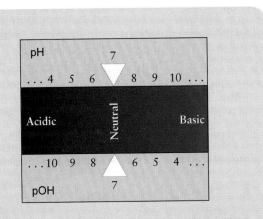

FIGURE 15.11

The numbers along the top of the rectangle are values of pH and those on the bottom are the values of pOH that correspond to the pH directly above. Note that neutrality corresponds to a pH of 7; smaller pH values correspond to acidic solutions, larger values to basic solutions. Most values of pH and pOH lie in the range 1 to 14, but in principle pH and pOH values can lie outside this range and even be negative.

logarithm to the base 10 of the H_3O^+ concentration and change its sign. If we know only the OH^- concentration, we take its logarithm to the base 10 and change its sign to find the pOH. Then, provided the temperature is 25°C, we find the pH by subtracting the pOH from 14.00.

To convert from pH or pOH to molarity We need to find the antilogarithm of the *negative* of the pH or pOH:

$$[H_3O^+] = 10^{-pH} \quad \text{and} \quad [OH^-] = 10^{-pOH}$$

To carry out this procedure with a calculator, first enter the pH or pOH, press the $+/-$ key, then the 10^x key (or the inverse of the "log" key, depending on the calculator).

If we need the H_3O^+ concentration, but are given the pOH, a good procedure is to calculate the pH from the pOH first and then calculate the corresponding concentration. If we need the OH^- concentration and are given the pH, it is best to convert pH to pOH and then use the equation above.

Example 15.4 *Calculating the H_3O^+ and OH^- concentrations from the pH*

The pH of a solution is 4.83. What are the concentrations of H_3O^+ and OH^- in the solution?

Strategy Because pH < 7, the solution is acidic and $[H_3O^+] > [OH^-]$. Find the H_3O^+ and OH^- concentrations by taking the antilogarithms of the negative of the pH and pOH, as in the second procedure in Toolbox 15.1.

Solution First change the sign of the pH, then take its antilogarithm:

$$[H_3O^+] = 10^{-4.83} = 1.5 \times 10^{-5}$$

The concentration of hydronium ions is 1.5×10^{-5} mol/L. The pOH is $14.00 - 4.83 = 9.17$, so $[OH^-] = 10^{-9.17} = 6.8 \times 10^{-10}$. The concentration of hydroxide ions is therefore 6.8×10^{-10} mol/L.

Self-Test 15.5A The pH of stomach fluids is about 1.7. What is the H_3O^+ molarity in the stomach?

[*Answer:* 2×10^{-2} mol/L]

Self-Test 15.5B The pH of pancreatic fluids, which help to digest food once it has left the stomach, is about 8.2. What is the approximate H_3O^+ molarity of pancreatic fluids?

WEAK ACIDS AND BASES

If we were to measure the pH of 0.10 M $CH_3COOH(aq)$, we would find that it has a higher pH, and hence a lower H_3O^+ molarity, than 0.10 M HCl(aq). Similarly, we would find that 0.10 M $NH_3(aq)$ has a lower pH than 0.10 M NaOH(aq). The explanation must be that CH_3COOH is not fully deprotonated and NH_3 is not fully protonated in water. That is, the two compounds are examples of a weak acid and a weak base. Incomplete deprotonation is part of the reason why the carbonic acid in rain is not as harmful as strong acids like nitric acid. It also explains why solutions of HCl and CH_3COOH react at different rates even though they have the same molarities (Fig. 15.12). Although the initial concentrations of the two acids are the same, the concentration of hydronium ions is greater in the solution of strong acid.

15.6 Proton Transfer Equilibria

Proton transfer is one of the fastest reactions known in aqueous solution. Because it is so fast, we can be confident that conjugate acids and bases, such as CH_3COOH and $CH_3CO_2^-$ or NH_4^+ and NH_3, are always in equilibrium with each other in water. Therefore, forward and reverse reactions like

$$CH_3COOH(aq) + H_2O(l) \longrightarrow H_3O^+(aq) + CH_3CO_2^-(aq)$$

and

$$H_3O^+(aq) + CH_3CO_2^-(aq) \longrightarrow CH_3COOH(aq) + H_2O(l)$$

can be combined into

$$\underset{\text{Acid}}{CH_3COOH(aq)} + \underset{\text{Base}}{H_2O(l)} \rightleftharpoons \underset{\substack{\text{Conjugate}\\\text{acid}}}{H_3O^+(aq)} + \underset{\substack{\text{Conjugate}\\\text{base}}}{CH_3CO_2^-(aq)}$$

This equilibrium is a **proton transfer equilibrium.** It is common to all acids and bases in water, and we met it earlier in the autoprotolysis equilibrium of water. Another example is

$$\underset{\text{Acid}}{H_2O(l)} + \underset{\text{Base}}{NH_3(aq)} \rightleftharpoons \underset{\substack{\text{Conjugate}\\\text{acid}}}{NH_4^+(aq)} + \underset{\substack{\text{Conjugate}\\\text{base}}}{OH^-(aq)}$$

In each proton transfer equilibrium, a proton is transferred from an acid to a base. The products are the conjugate base and conjugate acid.

Self-Test 15.6A Identify (a) the Brønsted acid and base in the reaction $HNO_3(aq) + HPO_4^{2-}(aq) \rightarrow NO_3^-(aq) + H_2PO_4^-(aq)$; (b) the conjugate base and acid formed.

[*Answer:* (a) Acid, HNO_3; base, HPO_4^{2-}; (b) conjugate base, NO_3^-; conjugate acid, $H_2PO_4^-$]

Self-Test 15.6B Identify (a) the Brønsted acid and base in the reaction $HCO_3^-(aq) + NH_4^+(aq) \rightarrow H_2CO_3(aq) + NH_3(aq)$; (b) the conjugate base and acid formed.

Proton transfer equilibria are described by equilibrium constants. For example, for acetic acid in water,

$$CH_3COOH(aq) + H_2O(l) \rightleftharpoons H_3O^+(aq) + CH_3CO_2^-(aq)$$

$$K_c = \frac{[H_3O^+][CH_3CO_2^-]}{[CH_3COOH][H_2O]}$$

Because the solutions we consider are dilute and the solvent is almost pure water, $[H_2O]$ can be treated as a constant and combined with the equilibrium constant K_c. The resulting expression is called an **acidity constant,** denoted K_a:

$$K_a = \frac{[H_3O^+][CH_3CO_2^-]}{[CH_3COOH]}$$

The experimental value of K_a for acetic acid at 25°C is 1.8×10^{-5}. This small value tells us that only a small proportion of CH_3COOH molecules lose their protons when dissolved in water. In Toolbox 15.4, we shall see how to estimate the fraction of molecules deprotonated, and in Example 15.8, we shall see that, depending on the concentration, about 99 out of 100 CH_3COOH molecules remain intact. This value is typical of weak acids in water (Fig. 15.13).

FIGURE 15.12

Equal masses of magnesium metal have been added to solutions of HCl, a strong acid (top) and CH_3COOH, a weak acid (bottom). Although the acids have the same concentrations, the rate of hydrogen evolution, which depends on the concentration of hydronium ions, is much greater in the strong acid.

The acidity constant is also widely called the *ionization constant* and sometimes the *dissociation constant* of the acid. The constants K_a and K_c are related by $K_a = K_c[H_2O]$.

CHAPTER 15   ACIDS AND BASES

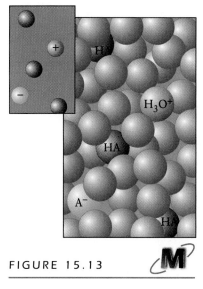

FIGURE 15.13

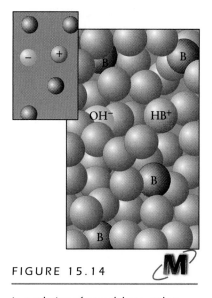

FIGURE 15.14

In a solution of a weak acid, only some of the acidic hydrogen atoms are present as hydronium ions (the red spheres), and the solution contains a high proportion of the original acid molecules (HA, gray spheres). The green spheres represent the conjugate base of the acid. The overlay shows only the solute species.

In a solution of a weak base, only a small proportion of the base molecules (B, represented here by the gray spheres) have accepted protons from water (the blue spheres) to form HB^+ ions (the red spheres) and OH^- ions (green spheres). The overlay shows only the solute species.

An equilibrium constant can also be written for the proton transfer equilibrium of a base in water. For aqueous ammonia, for instance,

$$NH_3(aq) + H_2O(l) \rightleftharpoons NH_4^+(aq) + OH^-(aq) \qquad K_b = \frac{[NH_4^+][OH^-]}{[H_2O][NH_3]}$$

In dilute solutions, the water is almost pure, and its nearly constant molar concentration can be combined with K_c to form the **basicity constant,** K_b:

$$K_b = \frac{[NH_4^+][OH^-]}{[NH_3]}$$

The name *base ionization constant* is still widely used.

The experimental value of K_b for ammonia in water at 25°C is 1.8×10^{-5}. This small value tells us that only a small proportion of NH_3 molecules are present as NH_4^+ under normal conditions. Calculations of the type described in Section 15.12 suggest that only about 1 in 100 molecules is protonated in a typical solution (Fig. 15.14).

The fact that K_b for ammonia has the same numerical value as K_a for acetic acid is a coincidence.

Acidity and basicity constants are commonly reported as their negative logarithms, by defining

$$pK_a = -\log K_a \qquad pK_b = -\log K_b$$

The lower the value of K_a, the higher the value of pK_a. Hence, we can conclude that the higher the value of pK_a, the weaker the acid. Thus, the pK_a of acetic acid is $-\log(1.8 \times 10^{-5}) = 4.75$ and that of boric acid, a much weaker acid, is 9.14.

The larger the acidity constant, the greater the proton-donating strength of an acid; the larger the basicity constant, the greater the proton-accepting strength of a base.

15.7 The Conjugate Seesaw

How are the strengths of an acid and its conjugate base related? The stronger the acid, the more easily its acidic proton is lost. Hydrochloric acid is a strong acid because its conjugate base, Cl^-, is unable to remove protons from H_3O^+. As a result, HCl is fully deprotonated in water. Conversely, acetic acid is a weak acid because its conjugate base, the acetate ion, $CH_3CO_2^-$, is a relatively good proton acceptor and forms CH_3COOH molecules in water.

The stronger the base, the more tightly it holds on to any proton it accepts. Because a base is converted into its conjugate acid when it accepts a proton, *the stronger the base, the weaker its conjugate acid.* For example, methylamine, CH_3NH_2, is a stronger base in water than ammonia (Table 15.2). The conjugate acid of methylamine (the methylammonium ion, $CH_3NH_3^+$) must therefore be less able than NH_4^+, the conjugate acid of ammonia, to donate a proton to H_2O. We can conclude that

The stronger the acid, the weaker its conjugate base.

The stronger the base, the weaker its conjugate acid.

Because the strengths of conjugate acids and bases have a seesaw relation, we expect the K_b of a base (such as NH_3) to be inversely related to the K_a of its conjugate acid (here NH_4^+). To find the relation, consider the proton transfer equilibrium of the base NH_3 in water:

$$NH_3(aq) + H_2O(l) \rightleftharpoons NH_4^+(aq) + OH^-(aq) \qquad K_b = \frac{[NH_4^+][OH^-]}{[NH_3]}$$

and the proton transfer equilibrium of ammonia's conjugate acid, NH_4^+, in water,

$$NH_4^+(aq) + H_2O(l) \rightleftharpoons H_3O^+(aq) + NH_3(aq) \qquad K_a = \frac{[H_3O^+][NH_3]}{[NH_4^+]}$$

When we multiply these two equilibrium constants together, we obtain

$$K_a \times K_b = \frac{[H_3O^+][NH_3]}{[NH_4^+]} \times \frac{[NH_4^+][OH^-]}{[NH_3]} = [H_3O^+][OH^-]$$

That is,

$$K_a \times K_b = K_w \qquad (4)$$

This important relation applies to all conjugate acid-base pairs. It is the quantitative version of the seesaw relation between conjugate acid and base strengths. If K_b of a base is large (that is, the base is relatively strong), then K_a of its conjugate acid must be small (that is, the conjugate acid is relatively weak) to keep the product $K_a \times K_b$ equal to K_w.

If we take logarithms of both sides of the relation $K_a \times K_b = K_w$, we obtain

$$\log K_a + \log K_b = \log K_w$$

Table 15.2 *Conjugate acid-base pairs arranged by strength*

Acid name	Formula	Base formula	Name
Strong acid		**Very weak base**	
hydroiodic acid	HI	I^-	iodide ion
perchloric acid	$HClO_4$	ClO_4^-	perchlorate ion
hydrobromic acid	HBr	Br^-	bromide ion
hydrochloric acid	HCl	Cl^-	chloride ion
sulfuric acid	H_2SO_4	HSO_4^-	hydrogen sulfate ion
chloric acid	$HClO_3$	ClO_3^-	chlorate ion
nitric acid	HNO_3	NO_3^-	nitrate ion
hydronium ion	H_3O^+	H_2O	*water*
hydrogen sulfate ion	HSO_4^-	SO_4^{2-}	sulfate ion
hydrofluoric acid	HF	F^-	fluoride ion
nitrous acid	HNO_2	NO_2^-	nitrite ion
acetic acid	CH_3COOH	$CH_3CO_2^-$	acetate ion
carbonic acid	H_2CO_3	HCO_3^-	hydrogen carbonate ion
hydrosulfuric acid	H_2S	HS^-	hydrogen sulfide ion
ammonium ion	NH_4^+	NH_3	ammonia
hydrocyanic acid	HCN	CN^-	cyanide ion
hydrogen carbonate ion	HCO_3^-	CO_3^{2-}	carbonate ion
methylammonium ion	$CH_3NH_3^+$	CH_3NH_2	methylamine
water	H_2O	OH^-	*hydroxide ion*
ammonia	NH_3	NH_2^-	amide ion
hydrogen	H_2	H^-	hydride ion
methane	CH_4	CH_3^-	methide ion
hydroxide ion	OH^-	O^{2-}	oxide ion
Very weak acid		**Strong base**	

When we multiply through by -1, this expression becomes

$$pK_a + pK_b = pK_w \qquad (5)$$

For example, because the pK_b of NH_3 is 4.75, the pK_a of NH_4^+ at 25°C is

$$pK_a = pK_w - pK_b = 14.00 - 4.75 = 9.25$$

This value shows that NH_4^+ ($pK_a = 9.25$) is a weaker proton donor than acetic acid ($pK_a = 4.75$) but stronger than hypoiodous acid (HIO, $pK_a = 10.64$), Fig. 15.15.

> *The lower the value of K_a for an acid, the higher the value of K_b for its conjugate base. Similarly, the lower the value of K_b for a base, the higher the value of K_a for its conjugate acid.*

15.8 The Special Role of Water

Imagine what happens when a strong acid dissolves in water. We can picture the process as a tug-of-war between the proton-accepting powers of H_2O and A^-,

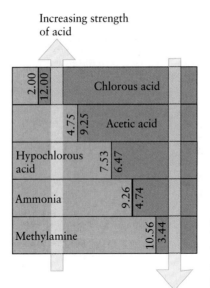

Increasing strength of acid

2.00	12.00	Chlorous acid
	4.75 / 9.25	Acetic acid
Hypochlorous acid	7.53 / 6.47	
Ammonia	9.26 / 4.74	
Methylamine	10.56 / 3.44	

Increasing strength of base

FIGURE 15.15

The sum of the pK_a of an acid (pink) and the pK_b of its conjugate base (blue) is constant, and equal to pK_w, which is 14.00 at 25°C. The values of the constants enable us to arrange all acids and bases in order of strength on a single chart.

Note that a specific base can be either electrically neutral or negatively charged. The symbol A^- may be used to represent species such as NH_3, NH_2^-, or PO_4^{3-}.

the conjugate base of the strong acid HA. For instance, if the acid is HCl, then A^- is Cl^-; and if the acid is HNO_3, then A^- is NO_3^-. Because the conjugate base of a strong acid, A^-, is a weaker proton acceptor than water, the battle is resolved in favor of H_2O. As a result, the solute species consist only of H_3O^+ ions and A^- ions. In other words, *the proton donor in an aqueous solution of a strong acid is the H_3O^+ ion.* Because all strong acids behave as though they were solutions of the acid H_3O^+, we say that strong acids are **leveled** to the strength of the acid H_3O^+ in water.

This distinction applies to any solvent: an acid is strong if it is a stronger proton donor than the conjugate acid of the solvent. In such a case, the tug-of-war for the proton is resolved in favor of the solvent, so the acid is completely deprotonated. Acetic acid, CH_3COOH, for instance, is a stronger proton donor than NH_4^+ (the conjugate acid of NH_3), so acetic acid, although weak in water, is strong in liquid ammonia.

Now, suppose that the base A^- is a stronger proton acceptor than H_2O. In this case, the conjugate acid HA will be the dominant species in aqueous solution. For example, if the base is CN^-, then the acid is HCN, and a high proportion of HCN molecules survive in solution. Such an acid is weak, for it is only slightly deprotonated in aqueous solution.

We can now see that *the strength of H_2O as a proton acceptor marks the frontier between strong and weak acids in water.* Any acid that is a stronger proton donor than H_3O^+ is a strong acid in aqueous solution. These acids are the ones that lie above H_3O^+ in Table 15.2. They are listed in Table 15.1. Any acid that lies below H_3O^+ in Table 15.2, such as CH_3COOH, is a weak acid because it is a weaker proton donor than water.

Base strength in water is determined by a similar competition, but this time between the proton-*accepting* powers of OH^- and the base. A base is a strong base in water if it lies below OH^- in the listing of bases in Table 15.2. Such a base is virtually completely protonated in water, so the base that survives in water is the OH^- ion. Once again, water determines the location of the frontier between weak and strong in aqueous solution.

Self-Test 15.7A Label each of the following species as a strong or a weak acid in water: (a) $HClO_4$; (b) NH_4^+; (c) HNO_2.

[*Answer:* (a) Strong acid; (b) weak acid; (c) weak acid]

Self-Test 15.7B Label each of the following species as a strong or a weak base in water: (a) SO_4^{2-}; (b) NH_2^-; (c) $CH_3CO_2^-$.

A proton donor stronger than H_3O^+ is a strong acid in water. A proton donor weaker than H_3O^+ is a weak acid in water. A proton acceptor weaker than OH^- is a weak base in water. A proton acceptor stronger than OH^- is a strong base in water.

Toolbox 15.2 *How to predict the relative strengths of conjugate acids and bases*

This Toolbox shows how to predict, from an inspection of pK_a or pK_b (Table 15.3) or from a comparison of strengths of conjugate base or acid (Table 15.2), which of two species is the stronger acid or the stronger base. It also summarizes the relation between K_a and K_b, and shows how one can be calculated from the other.

Conceptual Basis

The strength of an acid in water depends on the ability of the conjugate base of the acid to accept protons from water molecules (Fig. 15.16). The conjugate acid of a strong base is very weak, because its conjugate base holds tightly to its protons. On the other hand, if an acid is strong, its conjugate base must be very weak, as it gives up essentially all its protons to water. The pK_a and pK_b of a conjugate acid-base pair are related in a seesaw fashion: if one is large, the other must be small (recall Fig. 15.15).

Procedure

Qualitative comparison To determine the relative strengths of acids and bases, remember the following points:

The larger the K_a of a weak acid, the stronger the acid and the weaker its conjugate base.

The larger the K_b of a weak base, the stronger the base and the weaker its conjugate acid.

An acid stronger than H_3O^+ is a strong acid in water; acids weaker than H_3O^+ are weak acids in water.

A base stronger than OH^- is a strong base in water; bases weaker than OH^- are weak bases in water.

Calculating K_a and K_b The seesaw dependence of K_a and K_b is expressed by Eqs. 4 and 5:

$$K_a \times K_b = K_w \qquad pK_a + pK_b = pK_w$$

If one is known, the other can be calculated from these expressions.

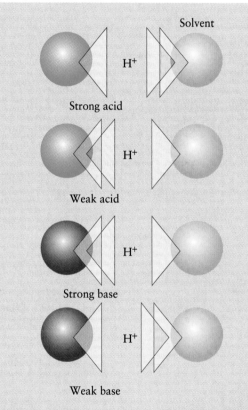

FIGURE 15.16

The ability of a molecule or ion to act as a strong or a weak acid or base depends on its proton-attracting power (represented by the number of triangles) relative to that of the solvent.

Table 15.3 Acidity and basicity constants at 25°C*

Acid	K_a	pK_a	Base	K_b	pK_b
trichloroacetic acid, CCl_3COOH	3.0×10^{-1}	0.52	urea, $CO(NH_2)_2$	1.3×10^{-14}	13.90
benzene sulfonic acid, $C_6H_5SO_3H$	2.0×10^{-1}	0.70	aniline, $C_6H_5NH_2$	4.3×10^{-10}	9.37
iodic acid, HIO_3	1.7×10^{-1}	0.77	pyridine, C_5H_5N	1.8×10^{-9}	8.75
sulfurous acid, H_2SO_3	1.5×10^{-2}	1.81	hydroxylamine, NH_2OH	1.1×10^{-8}	7.97
chlorous acid, $HClO_2$	1.0×10^{-2}	2.00	nicotine, $C_{10}H_{14}N_2$	1.0×10^{-6}	5.98
phosphoric acid, H_3PO_4	7.6×10^{-3}	2.12	morphine, $C_{17}H_{19}O_3N$	1.6×10^{-6}	5.79
chloroacetic acid, $CH_2ClCOOH$	1.4×10^{-3}	2.85	hydrazine, NH_2NH_2	1.7×10^{-6}	5.77
lactic acid, $CH_3CH(OH)COOH$	8.4×10^{-4}	3.08	ammonia, NH_3	1.8×10^{-5}	4.75
nitrous acid, HNO_2	4.3×10^{-4}	3.37	trimethylamine, $(CH_3)_3N$	6.5×10^{-5}	4.19
hydrofluoric acid, HF	3.5×10^{-4}	3.45	methylamine, CH_3NH_2	3.6×10^{-4}	3.44
formic acid, $HCOOH$	1.8×10^{-4}	3.75	dimethylamine, $(CH_3)_2NH$	5.4×10^{-4}	3.27
benzoic acid, C_6H_5COOH	6.5×10^{-5}	4.19	ethylamine, $C_2H_5NH_2$	6.5×10^{-4}	3.19
acetic acid, CH_3COOH	1.8×10^{-5}	4.75	triethylamine, $(C_2H_5)_3N$	1.0×10^{-3}	2.99
carbonic acid, H_2CO_3	4.3×10^{-7}	6.37			
hypochlorous acid, $HClO$	3.0×10^{-8}	7.53			
hypobromous acid, $HBrO$	2.0×10^{-9}	8.69			
boric acid, $B(OH)_3^†$	7.2×10^{-10}	9.14			
hydrocyanic acid, HCN	4.9×10^{-10}	9.31			
phenol, C_6H_5OH	1.3×10^{-10}	9.89			
hypoiodous acid, HIO	2.3×10^{-11}	10.64			

*The K_a and K_b listed here have been calculated from pK_a and pK_b values with more significant figures than shown so as to minimize rounding errors. Values for polyprotic acids—those capable of donating more than one proton—refer to the first deprotonation.

†The proton transfer equilibrium is $B(OH)_3(aq) + 2 H_2O(l) \rightleftharpoons H_3O^+(aq) + B(OH)_4^-(aq)$.

Example 15.5 Predicting relative strengths of acids and bases

Use the information in Table 15.3 to decide which member of each of the following pairs of species is the stronger acid or base in water: (a) acid: HF or HIO_3; (b) base: NO_2^- or CN^-.

Strategy Find the K_a or K_b of each substance in Table 15.3. As described in Toolbox 15.2, the acid with the larger K_a (smaller pK_a) is the stronger acid. Similarly, the base with the larger K_b (smaller pK_b) is the stronger base.

Solution (a) Because $K_a(HIO_3) > K_a(HF)$, it follows that HIO_3 is the stronger acid. (b) Because $K_a(HNO_2) > K_a(HCN)$, and because the stronger acid has the weaker conjugate base, it follows that NO_2^- is a weaker base than CN^-. Hence, CN^- is the stronger base.

Self-Test 15.8A Use Table 15.3 to decide which species of each pair is the stronger acid or base: (a) acid: HF or HIO; (b) base: $C_6H_5CO_2^-$ or $CH_2ClCO_2^-$; (c) base: $C_6H_5NH_2$ or $(CH_3)_3N$; (d) acid: $C_6H_5NH_3^+$ or $(CH_3)_3NH^+$.

[*Answer:* Stronger acids: (a) HF; (d) $C_6H_5NH_3^+$.
Stronger bases: (b) $C_6H_5CO_2^-$; (c) $(CH_3)_3N$]

Self-Test 15.8B Use Table 15.3 to decide which species of each pair is the stronger acid or base: (a) base: C_5H_5N or NH_2NH_2; (b) acid: $C_5H_5NH^+$ or $NH_2NH_3^+$; (c) acid: HIO_3 or $HClO_2$; (d) base: ClO_2^- or HSO_3^-.

15.9 Why Are Some Acids Weak and Others Strong?

The actual values of K_a and K_b are difficult to predict because they depend on a number of factors. These factors include the strength of the H—A bond, the strength of the O—H bond in H_3O^+, and the extent to which the conjugate base, A^-, of the acid is hydrated in water. However, we can identify dominant factors by looking for trends among series of compounds with similar structures.

The ease with which an acid molecule, HA, donates a proton to a water molecule depends in part on the strength of the hydrogen bond it forms with the O atom of the H_2O molecule:

$$H_2O{\cdots}H{-}A \longrightarrow H_2O{-}H^+ + A^-$$

The stronger the hydrogen bond, the easier it is for the water molecule to pull the proton from the acid and form an H_3O^+ ion. We know that the more polar the H—A bond, the greater the partial positive charge on H and the stronger the $O{\cdots}H{-}A$ hydrogen bond. Therefore, we expect an acid HA with a very polar H—A bond to be a stronger acid than one with a less polar H—A bond. Because the polarity of the H—A bond increases with the electronegativity of A, we can also predict that *the greater the electronegativity of A, the stronger the acid HA.*

For example, the electronegativity difference of the atoms is 0.8 in the N—H bond and 1.8 in F—H. Therefore, the H—F bond is markedly more polar than the N—H bond. Experimentally, HF is acidic and NH_3 is basic in aqueous solutions. For these two compounds at least, bond polarity determines the relative acid strengths. In general, bond polarity dominates the trend of acid strengths for binary acids of elements of the same period.

Another factor affecting acid strength is the strength of the H—A bond. *The weaker the H—A bond, the easier it is for the proton to leave and the stronger the acid HA.* An example is the anomalous position of HF among the hydrohalic acids. Even though the H—F bond is the most polar of the group, HF is a weak acid in water, whereas all other hydrohalic acids are strong. A part of the reason is the great strength of the H—F bond, which means that the proton is lost from HF only with difficulty.

The relative strengths of the hydrohalic acids can be measured in a solvent that is a poorer proton acceptor than water (such as pure acetic acid), because they are not leveled in this solvent. The acid strengths so found lie in the order HF < HCl < HBr < HI. This order is consistent with the weakening of the H—A bond down the group. The same trend is also found for the Group 16 acids in aqueous solution; the acid strengths lie in the order H_2O < H_2S < H_2Se < H_2Te. The bond strengths and bond polarities both decrease down the group. In general, for binary acids of elements in the same group, bond strength dominates the trend in acid strengths.

> The bond strengths of binary hydrides decrease down a group of the periodic table (recall Fig. 9.21).

The more polar the H—A bond in a given period, the stronger the acid. The weaker the H—A bond in a given group, the stronger the acid.

15.10 The Strengths of Oxoacids

The proton of an —OH group in an oxoacid molecule is acidic because of the high polarity of the O—H bond. For example, phosphorous acid, H_3PO_3, has

Table 15.4 *Correlation of acid strength and oxidation number**

Acid	Structure[†]	Oxidation number of chlorine atom	pK_a
perchloric acid, $HClO_4$	:O: ‖ Ö=Cl—Ö—H ‖ :O:	+7	strong
chloric acid, $HClO_3$	:O: ‖ :Cl—Ö—H ‖ :O:	+5	strong
chlorous acid, $HClO_2$	:O: ‖ :Cl—Ö—H	+3	2.00
hypochlorous acid, $HClO$	:Cl—Ö—H	+1	7.53

*The structures drawn are those with the most favorable formal charges. The actual structures of $HClO_3$ and $HClO_4$ have less electron density on the Cl atom.

[†]The red arrows indicate the direction of the shift of electron density away from the O—H bond.

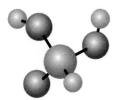

1 Phosphorous acid, H_3PO_3

the structure $(HO)_2PHO$ (**1**): it can donate the protons from its two —OH groups but not the proton attached directly to the phosphorus atom.

To identify features of molecular structure that affect the strength of oxoacids, we can study variations in acidity among families of oxoacids. First, let's consider a family of oxoacids in which the number of oxygen atoms varies, as in the chlorine oxoacids $HClO$, $HClO_2$, $HClO_3$, and $HClO_4$, or the sulfur oxoacids H_2SO_3 and H_2SO_4. *The greater the number of oxygen atoms attached to the central atom, the stronger the acid.* Because the oxidation number of the central atom increases as the number of O atoms increases, we can also conclude that *the greater the oxidation number of the central atom, the stronger the acid* (Table 15.4).

Now let's consider a family of oxoacids in which the number of O atoms is constant, as in the hypohalous acids $HClO$, $HBrO$, and HIO. It is found that the greater the electronegativity of the central atom (the halogen atoms in these examples), the stronger the oxoacid (Table 15.5). We can understand this trend

Table 15.5 *Correlation of acid strength and electronegativity*

Acid, HXO	Structure*	Electronegativity of atom X	pK_a
hypochlorous acid, $HClO$	:Cl—Ö—H	3.2	7.53
hypobromous acid, $HBrO$	:Br—Ö—H	3.0	8.69
hypoiodous acid, HIO	:I—Ö—H	2.7	10.64

*The red arrows indicate the direction and magnitude of the shift of electron density away from the O—H bond.

as a withdrawing of electrons from the O—H bond that increases as the electronegativity of the central atom increases. As these bonding electrons are pulled away, the O—H bond becomes more polar and the proton can be donated to H_2O more readily. A central atom with high electronegativity also makes the conjugate base weaker, because electronegative atoms do not readily share their electrons. We can conclude that *the more electronegative the central atom of an oxoacid, the stronger the acid.*

We can see the effect of the number of O atoms on the strengths of organic acids by comparing alcohols and carboxylic acids. Alcohols are organic compounds in which an —OH group is attached to a carbon atom, as in ethanol (**2**). Carboxylic acids have another O atom bonded to the carbon atom to which the —OH group is attached, as in acetic acid (**3**). Although carboxylic acids are weak acids, they are much stronger acids than alcohols, partly on account of the electron-withdrawing power of the second O atom.

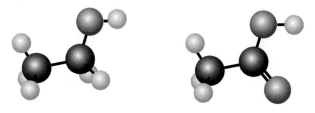

2   Ethanol, CH_3CH_2OH 3   Acetic acid, CH_3COOH

The strength of a carboxylic acid is also increased relative to that of an alcohol by electron delocalization in the conjugate base. The second O atom of the carboxyl group provides an additional electronegative atom over which the negative charge of the conjugate base can spread: this electron delocalization stabilizes the anion by resonance (**4**). Moreover, because the charge is spread over several atoms, it is less effective at attracting a proton. A carboxylate ion is therefore a weaker base than the conjugate base of an alcohol (for example, the ethoxide ion, $CH_3CH_2O^-$, **5**).

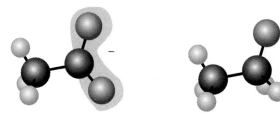

4   Acetate ion, $CH_3CO_2^-$ 5   Ethoxide ion, $CH_3CH_2O^-$

The strengths of carboxylic acids also vary with total electron-withdrawing power of the atoms bonded to the carboxyl group. Because hydrogen is less electronegative than chlorine, the —CH_3 group bonded to —COOH in acetic acid is less electron withdrawing than the —CCl_3 group in trichloroacetic acid. Therefore, we expect CCl_3COOH to be a stronger acid than CH_3COOH. In agreement with this prediction, the pK_a of acetic acid is 4.75, whereas that of trichloroacetic acid is 0.52. The rules for predicting the relative strengths of acids are summarized in Toolbox 15.3.

The greater the number of oxygen atoms and the more electronegative the atoms present in a molecule, the stronger the acid.

Toolbox 15.3 How to predict the relative strengths of acids

This Toolbox shows how to predict which of two acids is the stronger acid on the basis of their molecular structures.

Conceptual Basis

The strength of an acid depends on how readily an acidic proton can be lost. The more polar the bond, the more easily a base can remove the proton. The weaker the bond to the acidic proton, the more easily the bond can be broken. Groups that withdraw electrons from the atom attached to the acidic proton stabilize the conjugate base and hence contribute to increased acid strength (Fig. 15.17).

Procedure

Determine the types of acids being compared, then consider the following rules.

Binary acids

1. The more polar the H—A bond, the stronger the acid. (This effect is dominant for acids of the same period.)

2. The weaker the H—A bond, the stronger the acid. (This effect is dominant for acids of the same group.)

Oxoacids

1. The greater the number of O atoms attached to the central atom (the greater the oxidation number of the central atom), the stronger the acid.

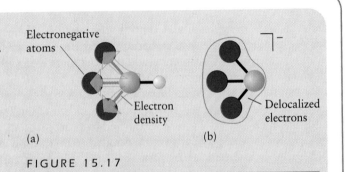

FIGURE 15.17

(a) Electron-withdrawing groups attached to an atom bonded to an acidic proton enhance the acidity of the proton by allowing it to have a larger partial positive charge.
(b) Electron-withdrawing groups also stabilize the conjugate base by delocalizing the negative charge over a larger part of the ion.

2. When the same number of O atoms is attached to the central atom, then the greater the electronegativity of the central atom, the stronger the acid.

Carboxylic acids

The greater the electronegativities of the atoms in the group attached to the carboxyl group, the stronger the acid.

Example 15.6 Predicting the relative strengths of inorganic acids

In the following pairs, predict which acid is stronger and explain why: (a) H_2S and H_2Se; (b) H_2SO_4 and H_2SO_3; (c) H_2SO_4 and H_3PO_4.

Strategy First determine the types of acids being compared, then use the corresponding guidelines in Toolbox 15.3. (a) Use the procedure for binary acids; (b) and (c) use the procedure for oxoacids.

Solution (a) Sulfur and selenium are in the same group, and we expect the H—Se bond to be weaker than the H—S bond. Thus, H_2Se can be expected to be the stronger acid. (b) H_2SO_4 has the greater number of O atoms bonded to the S atom and the oxidation number of sulfur in this acid is +6; whereas, in H_2SO_3, the sulfur has an oxidation number of only +4. Therefore, H_2SO_4 is expected to be the stronger acid. (c) Both acids have four O atoms bonded to the central atom; because the electronegativity of sulfur is greater than that of phosphorus, H_2SO_4 is expected to be the stronger acid.

Self-Test 15.9A In the following pairs, predict which acid is stronger and give the reason for your choice: (a) H_2S and HCl; (b) HNO_2 and HNO_3; (c) H_2SO_3 and $HClO_3$.

[***Answer:*** (a) HCl, binary acid, rule 1; (b) HNO_3, oxoacid, rule 1; (c) $HClO_3$, oxoacid, rule 2]

Self-Test 15.9B In the following pairs, predict which acid is stronger and give the reason for your choice: (a) $HClO$ and $HClO_2$; (b) HBr and HI.

Example 15.7 *Predicting the trend in strengths of carboxylic acids*

List the following carboxylic acids in order of increasing strength: CH_3COOH, CH_2FCOOH, CHF_2COOH, and CF_3COOH.

Strategy The greater the electron-withdrawing power of the group attached to the carboxyl group, the stronger the acid. Therefore, judge the electron-withdrawing power of that group by comparing the electronegativities of the atoms.

Solution Because fluorine is more electronegative than hydrogen, the groups attached to —COOH increase in electron-attracting power in the order $CH_3 < CH_2F < CHF_2 < CF_3$. Therefore the acid strengths increase in the order $CH_3COOH < CH_2FCOOH < CHF_2COOH < CF_3COOH$.

Self-Test 15.10A List the following carboxylic acids in order of increasing strength: $CH_2ClCOOH$, $CH_2BrCOOH$, and CH_2FCOOH.

[***Answer:*** $CH_2BrCOOH < CH_2ClCOOH < CH_2FCOOH$]

Self-Test 15.10B List the following carboxylic acids in order of increasing strength: $CHCl_2COOH$, CH_3COOH, and $CH_2ClCOOH$.

THE pH OF SOLUTIONS OF WEAK ACIDS AND BASES

A weak acid in solution is in equilibrium with hydronium ions and its conjugate base. Therefore, a solution of a weak acid in water consists of H_3O^+ ions, the conjugate base of the acid, and acid molecules that have not donated their protons to water molecules. To find the H_3O^+ molarity, and hence the pH of the solution, we need the equilibrium constant K_a. Similarly, to calculate the pH of a solution of a weak base, we have to use K_b.

15.11 Solutions of Weak Acids

Weak acids produce a lower concentration of H_3O^+ ions in aqueous solution than do strong acids of the same initial concentration. For example, 0.01 M $HCl(aq)$ has a pH of 2; however, 0.01 M $CH_3COOH(aq)$ has a much lower concentration of H_3O^+ ions, and its pH is 3. To find the H_3O^+ molarity in a solution of a weak acid, we have to take into account the equilibrium between the acid HA and its conjugate base A^-:

Remember that an increase in pH by 1 corresponds to a tenfold decrease in concentration of hydronium ions.

$$HA(aq) + H_2O(l) \rightleftharpoons H_3O^+(aq) + A^-(aq) \qquad K_a = \frac{[H_3O^+][A^-]}{[HA]}$$

Once we know the H_3O^+ concentration, we can calculate the percentage of HA molecules that are deprotonated. From the stoichiometry of the equilibrium, we see that the concentration of deprotonated molecules, $[A^-]$, is equal to the concentration of hydronium ions, $[H_3O^+]$. Therefore,

$$\text{Percentage deprotonated} = \frac{\text{molarity of deprotonated HA}}{\text{initial molarity of HA}} \times 100\%$$

$$= \frac{[H_3O^+]}{[HA]_{\text{initial}}} \times 100\%$$

A small percentage of deprotonated molecules indicates that the solute consists primarily of the acid HA. The concentrations of H_3O^+ ions and conjugate base ions must then be very low.

Because a weak acid and its conjugate base are always in equilibrium, we can find the H_3O^+ molarity (and thus the pH) in a solution of a weak acid by using the methods discussed in Chapter 14 to calculate the equilibrium composition of the solution. The procedure is described in Toolbox 15.4.

Toolbox 15.4 *How to calculate the pH of a solution of a weak acid*

M

This Toolbox describes how to calculate the pH of an aqueous solution of a weak acid.

Conceptual Basis

When a weak acid is added to water, the molecules of the acid take part in a proton transfer equilibrium with water molecules. Because the equilibrium concentrations of acid, hydronium ion, and conjugate base of the acid satisfy the acidity constant for the acid, the hydronium ion concentration and thus the pH can be calculated by using an equilibrium table.

Procedure

Set up an equilibrium table as described in Toolbox 14.2 and follow the steps outlined there.

Step 1. Set up a table with columns labeled by the acid HA, its conjugate base A^-, and H_3O^+. In the first row, show the initial molarities of each species.

For the initial values, assume that no acid molecules have donated a proton, so $[H_3O^+] = [A^-] = 0$ and $[HA]$ is its initial concentration.

Step 2. Write the changes in the molarities that are needed for the reaction to reach equilibrium. The change in molarity of the conjugate base is equal to the change in molarity of H_3O^+.

Assume that the molarity of the acid decreases by

x mol/L as a result of deprotonation. Use the reaction stoichiometry to express the other changes in terms of x.

Step 3. Write the equilibrium molarities by adding the change in molarities (step 2) to the initial values for each substance (step 1).

Although a change in concentration may be positive (an increase) or negative (a decrease), the value of the concentration must be positive.

Step 4. Use the value of K_a to calculate the value of x.

If the value of x is less than 5% of the initial molarity, we can simplify the calculation by ignoring changes in the initial molarity, as shown in Toolbox 14.2. At the end of the calculation, we must check that this approximation is valid by calculating the percentage of acid deprotonated. If this percentage is greater than 5%, we must go back and solve the exact expression for x; this calculation often involves solving a quadratic equation, as explained in Toolbox 14.2. The calculator tool on the CD that accompanies this book also has a special option for equilibrium calculations that eliminates the need for approximations.

If the solution is so dilute that the pH is between 6 and 8, the autoprotolysis of water must be taken into account. Procedures for coping with this type of calculation are included on the Web site for this book.

Example 15.8 *Calculating the pH of a weak acid solution*

Calculate the pH and percentage deprotonation of CH_3COOH in 0.10 M $CH_3COOH(aq)$, given that K_a for acetic acid is 1.8×10^{-5}.

Strategy Acetic acid is a weak acid; consequently, we expect the molarity of H_3O^+ ions to be less than 0.10 mol/L and, therefore, its pH to be greater than 1.0 (because $-\log 0.10 = 1.0$). To find the actual value, we set up an equilibrium table as described in Toolbox 15.4. Although we suspect that the percentage deprotonation will be low for such a weak acid, we should be aware that even a weak acid may be extensively deprotonated in *very* dilute solutions.

Solution The proton transfer equilibrium we consider is

$$CH_3COOH(aq) + H_2O(l) \rightleftharpoons H_3O^+(aq) + CH_3CO_2^-(aq)$$

$$K_a = \frac{[H_3O^+][CH_3CO_2^-]}{[CH_3COOH]}$$

The equilibrium table, with the concentrations in moles per liter, is

	Species		
	CH₃COOH	**H₃O⁺**	**CH₃CO₂⁻**
Step 1. Initial molarity	0.10	0	0
Step 2. Change in molarity	$-x$	$+x$	$+x$
Step 3. Equilibrium molarity	$0.10 - x$	x	x

Step 4. Substitute these equilibrium molarities into the expression for the acidity constant:

$$K_a = 1.8 \times 10^{-5} = \frac{x \times x}{0.10 - x}$$

If we anticipate that $x \ll 0.1$, we can approximate this expression to

$$K_a = 1.8 \times 10^{-5} \approx \frac{x^2}{0.10}$$

Solving for x gives

$$x \approx \sqrt{(0.10) \times (1.8 \times 10^{-5})} = 1.3 \times 10^{-3}$$

From step 3, $x = [H_3O^+] = 1.3 \times 10^{-3}$, so

$$pH \approx -\log(1.3 \times 10^{-3}) = 2.89$$

We have assumed that x is less than about 5% of 0.10. The percentage deprotonated is

$$\text{Percentage deprotonated} = \frac{[H_3O^+]}{[CH_3COOH]_{initial}} \times 100\%$$

$$= \frac{1.3 \times 10^{-3}}{0.10} \times 100\% = 1.3\%$$

and the approximation is valid.

Self-Test 15.11A Calculate the pH of 0.20 M lactic acid. See Table 15.3 for K_a. Be sure to check any approximation to see whether it is valid.

[*Answer:* 1.90 (must use exact solution)]

Self-Test 15.11B Calculate the pH of 0.22 M chloroacetic acid.

The procedure in Toolbox 15.4 ignores the H_3O^+ ions that result from the autoprotolysis of water. Suppose the acid is so dilute or weak that the calculation predicts an H_3O^+ molarity of less than 10^{-7} mol/L. In this case, we must not report its pH as more than 7 because the autoprotolysis already provides H_3O^+ ions at a molarity of 10^{-7} mol/L! We can ignore the contribution of the autoprotolysis of water when the calculated H_3O^+ molarity is substantially (about 10 times) higher than 10^{-7} mol/L (that is, for solutions in which pH is less than about 6 or, in the case of a base, greater than 8).

> *To calculate the pH of a solution of a weak acid, set up an equilibrium table and determine the H_3O^+ molarity by using the acidity constant.*

15.12 Solutions of Weak Bases

A weak base has a lower pH than a strong base of the same initial molarity: a weak base establishes a dynamic equilibrium with its conjugate acid and therefore produces fewer OH^- ions than a strong base of comparable initial concentration. A 0.01 M NaOH(aq) solution has a pH of 12, but the pH of 0.01 M NH_3(aq) is 11. We calculate the pH of solutions of weak bases in the same way that we calculate the pH of solutions of weak acids—by using an equilibrium table. To calculate the pH of the solution, we first calculate the molarity of OH^- ions at equilibrium, express that as pOH, and then calculate pH from the relation pH + pOH = 14.00 at 25°C.

In some applications, we need to know the fraction of base molecules that have been protonated. We report this fraction by calculating the **percentage protonated.** If we use the symbol B to represent a specific base, whether it is neutral or negatively charged (NH_3 or CO_3^{2-}, for instance), and HB^+ to represent the conjugate acid (NH_4^+ or HCO_3^-, for instance), we can write

$$\text{Percentage protonated} = \frac{\text{molarity of protonated base}}{\text{initial molarity of base}} \times 100\%$$

$$= \frac{[HB^+]}{[B]_{\text{initial}}} \times 100\%$$

The percentage protonated varies with concentration and depends on the strength of the base, as illustrated in Example 15.9.

> *To calculate the pH of a solution of a weak base, use the equilibrium table to calculate pOH from the value of K_b, and convert that pOH to pH by subtracting it from 14.*

Example 15.9 *Calculating the pH of a solution of a weak base*

Calculate the pH and percentage protonation of a 0.20 M aqueous solution of methylamine, CH_3NH_2. The K_b for CH_3NH_2 is 3.6×10^{-4}.

Strategy Expect pH > 7, because the solution is basic. Calculate the molarity of OH^- ions from the reaction of the base with water by using the equilibrium table as explained in Toolbox 15.4, but use K_b instead of K_a. Calculate $[OH^-]$, convert it to pOH, and then convert that pOH to pH by using the relation pH + pOH = 14.00.

Solution The proton transfer equilibrium is

$$H_2O(l) + CH_3NH_2(aq) \rightleftharpoons CH_3NH_3^+(aq) + OH^-(aq)$$

$$K_b = \frac{[CH_3NH_3^+][OH^-]}{[CH_3NH_2]}$$

The equilibrium table, with all concentrations in moles per liter, is

	Species		
	CH_3NH_2	$CH_3NH_3^+$	OH^-
Step 1. Initial molarity	0.20	0	0
Step 2. Change in molarity	$-x$	$+x$	$+x$
Step 3. Equilibrium molarity	$0.20 - x$	x	x

Step 4. Substituting the equilibrium molarities into the expression for the basicity constant and using $K_b = 3.6 \times 10^{-4}$ yields

$$3.6 \times 10^{-4} = \frac{x \times x}{0.20 - x}$$

We now anticipate that x is less than 5% of 0.20 and approximate this expression by

$$3.6 \times 10^{-4} \approx \frac{x^2}{0.20}$$

Therefore,

$$x \approx \sqrt{(0.20) \times (3.6 \times 10^{-4})} = 8.5 \times 10^{-3}$$

According to step 3, $[OH^-] = x = 8.5 \times 10^{-3}$, so

$$pOH \approx -\log(8.5 \times 10^{-3}) = 2.07$$

and hence

$$pH \approx 14.00 - 2.07 = 11.93$$

The percentage of base molecules protonated is

$$\text{Percentage protonated} = \frac{[CH_3NH_3^+]}{[CH_3NH_2]} \times 100\%$$
$$= \frac{8.5 \times 10^{-3}}{0.20} \times 100\% = 4.2\%$$

That is, 4.2% of the methylamine is present as the protonated form, $CH_3NH_3^+$, and the approximation is valid. Because the pH is greater than 8, the assumption that the equilibrium here dominates the pH is valid.

Self-Test 15.12A Estimate the pH and percentage of protonated base in 0.15 M $NH_2OH(aq)$ (aqueous hydroxylamine).

[*Answer:* 9.61; 0.027%]

Self-Test 15.12B Estimate the pH and percentage of protonated base in 0.012 M $C_{10}H_{14}N_2(aq)$ (nicotine).

15.13 Polyprotic Acids and Bases

Brønsted acids that can donate more than one proton are called **polyprotic acids.** Common examples of polyprotic acids are sulfuric acid, H_2SO_4, and carbonic acid, H_2CO_3, each of which can donate two protons (and hence are called "diprotic" acids), and phosphoric acid, H_3PO_4, which can donate three protons (and is called a "triprotic" acid). A **polyprotic base** is a species that can accept more than one proton. Examples include the CO_3^{2-} anion and the oxalate

Table 15.6 *Acidity constants of polyprotic acids*

Acid	K_{a1}	pK_{a1}	K_{a2}	pK_{a2}	K_{a3}	pK_{a3}
sulfuric acid, H_2SO_4	strong		1.2×10^{-2}	1.92		
oxalic acid, $(COOH)_2$	5.9×10^{-2}	1.23	6.5×10^{-5}	4.19		
sulfurous acid, H_2SO_3	1.5×10^{-2}	1.81	1.2×10^{-7}	6.91		
phosphorous acid, H_3PO_3	1.0×10^{-2}	2.00	2.6×10^{-7}	6.59		
phosphoric acid, H_3PO_4	7.6×10^{-3}	2.12	6.2×10^{-8}	7.21	2.1×10^{-13}	12.68
tartaric acid, $C_2H_4O_2(COOH)_2$	6.0×10^{-4}	3.22	1.5×10^{-5}	4.82		
carbonic acid, H_2CO_3	4.3×10^{-7}	6.37	5.6×10^{-11}	10.25		
hydrosulfuric acid, H_2S	1.3×10^{-7}	6.89	7.1×10^{-15}	14.15		

anion, $C_2O_4^{2-}$, both of which can accept two protons, and the PO_4^{3-} anion, which can accept three protons.

Carbonic acid has the following deprotonation equilibria:

$$H_2CO_3(aq) + H_2O(l) \rightleftharpoons H_3O^+(aq) + HCO_3^-(aq) \qquad K_{a1} = 4.3 \times 10^{-7}$$

$$HCO_3^-(aq) + H_2O(l) \rightleftharpoons H_3O^+(aq) + CO_3^{2-}(aq) \qquad K_{a2} = 5.6 \times 10^{-11}$$

The conjugate base of H_2CO_3 in the first equilibrium, HCO_3^-, acts as an acid in the second equilibrium, producing in turn its own conjugate base, CO_3^{2-}. The equation for the overall deprotonation,

$$H_2CO_3(aq) + 2\,H_2O(l) \rightleftharpoons 2\,H_3O^+(aq) + CO_3^{2-}(aq)$$

is the sum of the two individual equations. We saw in Section 14.2 that an overall equilibrium constant is the product of the equilibrium constants for each contributing reaction. The same is true of acidity constants, so we can write

$$K_a = K_{a1} \times K_{a2}$$

Table 15.6 gives the acidity constants of some polyprotic acids. The data show that the strengths of polyprotic acids decrease as protons are lost:

$$K_{a1} > K_{a2} > \dots$$

The decrease is reasonable: it is harder to lose a positively charged proton from a negatively charged ion (such as HCO_3^-) than from the original uncharged molecule (H_2CO_3). Sulfuric acid, for example, is a strong acid, and H_2SO_4 loses its first proton to give its conjugate base, the hydrogen sulfate ion, HSO_4^-, which is a weak acid. However, some proteins can donate dozens of protons in reactions with little decrease in acid strength. Because the protons are donated from widely separated sites, the loss of one proton has little influence on the loss of the next (Fig. 15.18).

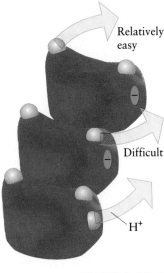

Relatively easy

Difficult

H^+

FIGURE 15.18

A proton is removed from an enzyme molecule (lower left). The loss of a second proton from a location close to the site of one that has been lost already (green area with negative charge) is difficult because of the attraction between opposite charges (middle). However, the loss of the first proton has relatively little effect on the ease with which a more distant proton can be lost (upper right).

Example 15.10 *Calculating the pH of a solution of sulfuric acid*

Calculate the pH of 0.010 M $H_2SO_4(aq)$ at 25°C. Use the appropriate information from Table 15.6.

Strategy The pH depends on the total H_3O^+ molarity, so both deprotonation steps must be taken into account. Sulfuric acid is the only common polyprotic acid for which the first deprotonation is complete. The second deprotonation adds to the H_3O^+ molarity slightly, so the overall pH will be slightly less than that due to the first deprotonation alone. To find its contribution, set up the equilibrium table in the usual way but use as the initial H_3O^+ and HSO_4^- molarities the values obtained by assuming that the first proton is lost completely. Because K_{a2} is relatively large (0.012), there are no shortcuts: it will be necessary to solve a quadratic equation (see Toolbox 14.2).

Solution The first deprotonation step,

$$H_2SO_4(aq) + H_2O(l) \longrightarrow H_3O^+(aq) + HSO_4^-(aq)$$

results in an H_3O^+ molarity equal to the original molarity of the acid before deprotonation, 0.010 mol/L. This value corresponds to pH = 2.0. The second deprotonation equilibrium is

$$HSO_4^-(aq) + H_2O(l) \rightleftharpoons H_3O^+(aq) + SO_4^{2-}(aq) \qquad K_{a2} = 1.2 \times 10^{-2}$$

and the equilibrium table, with all concentrations in moles per liter, is

	Species		
	HSO_4^-	H_3O^+	SO_4^{2-}
Step 1. Initial molarity	0.010	0.010	0
Step 2. Change in molarity	$-x$	$+x$	$+x$
Step 3. Equilibrium molarity	$0.010 - x$	$0.010 + x$	x

Step 4. Now substitute the molarities in step 3 into the expression for the second acidity constant, and use $K_{a2} = 0.012$, which gives

$$0.012 = \frac{(0.010 + x) \times x}{0.010 - x}$$

To find x, rearrange this expression into a quadratic equation:

$$x^2 + 0.022x - (1.2 \times 10^{-4}) = 0$$

Solving this equation by using the quadratic formula (Toolbox 14.2) yields $x = 4.5 \times 10^{-3}$. The total H_3O^+ molarity is

$$[H_3O^+] = 0.010 + x = 0.010 + (4.5 \times 10^{-3}) = 1.4 \times 10^{-2}$$

so

$$pH \approx -\log(1.4 \times 10^{-2}) = 1.85$$

The pH of the solution is less than 2.0, as was predicted.

Self-Test 15.13A Estimate the pH of 0.050 M H_2SO_4(aq).

[*Answer:* 1.23]

Self-Test 15.13B Estimate the pH of 0.10 M H_2SO_4(aq).

Almost all polyprotic acids other than sulfuric are weak for all their deprotonation stages. To calculate their pH, we take only the first deprotonation into account: subsequent deprotonations are much less important than the first and can be ignored. For carbonic acid, H_2CO_3, for instance, $K_{a2} = 5.6 \times 10^{-11}$, which is much smaller than $K_{a1} = 4.3 \times 10^{-7}$. We just use K_{a1} and treat it as a monoprotic acid, as illustrated in Toolbox 15.4. However, to find the concentrations of the anions in the solution, we would need to use both K_as.

This type of calculation is illustrated on the Web site for this book.

ACID RAIN

Applying Chemistry: Case Study 15

Scientists in the Netherlands noticed in 1989 that the great tit, a forest songbird, was producing eggs with thin, porous shells. The insecticide DDT had caused a similar problem in the 1960s and 1970s, but no evidence of toxic pesticides could be found. Scientists therefore investigated the birds' supply of calcium, which is needed for strong shells. The birds normally got sufficient calcium from the snails that make up most of their diet. However, the snails had virtually vanished from the forests. Dry forest soil normally contains 5–10 g calcium per kilogram; the calcium content of this soil had fallen to 0.3 g/kg, which is too low for snails to survive. Without snails to eat, the birds had begun raiding chicken houses and picnic grounds, seeking discarded eggshells to supplement their diets.

The fall in the calcium content of soil in Europe and the United States has been traced to acid rain, in particular to rain containing sulfuric acid. As the first illustration shows, acid rain is a regional phenomenon. The lines on

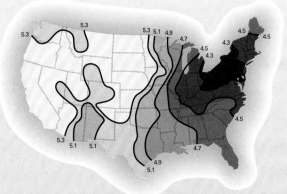

The curves on this map of the United States, measured in 1990, represent pH isopleths, regions in which the precipitation has the same pH. Notice that the average pH of precipitation decreases from west to east.

the map are contours showing the pH of rain, which decreases downwind of heavily populated areas. The low pH in heavily industrialized and populated areas is thought to be caused by the acidic oxides sulfur dioxide, SO_2, and the nitrogen oxides NO and NO_2.

Rain unaffected by human activity contains mostly weak acids and has a pH of 5.7. The primary acid present is carbonic acid, H_2CO_3, which results from the dissolving of atmospheric carbon dioxide, an acidic oxide, in water. The serious pollutants in acid rain are strong acids. Atmospheric nitrogen and oxygen can react to form NO at the high temperatures of automobile internal combustion engines and electrical power stations:

$$N_2(g) + O_2(g) \longrightarrow 2\,NO(g)$$

Nitric oxide, NO, is not very soluble in water, but it is oxidized further in air to form nitrogen dioxide:

$$2\,NO(g) + O_2(g) \longrightarrow 2\,NO_2(g)$$

The NO_2 reacts with water, forming nitric acid and nitric oxide:

$$3\,NO_2(g) + 3\,H_2O(l) \longrightarrow$$
$$2\,H_3O^+(aq) + 2\,NO_3^-(aq) + NO(g)$$

Catalytic converters in automobiles reduce the nitrogen in NO to N_2 and are required in the United States for all new cars and trucks.

Sulfur dioxide is produced as a by-product of the burning of fossil fuels. It may combine with water directly, to form sulfurous acid, a weak acid:

$$SO_2(g) + H_2O(l) \longrightarrow H_2SO_3(aq)$$

Alternatively, in the presence of particulate matter and aerosols, sulfur dioxide may react with atmospheric oxygen to form sulfur trioxide, which forms sulfuric acid in water:

$$2\,SO_2(g) + O_2(g) \longrightarrow 2\,SO_3(g)$$
$$SO_3(g) + 2\,H_2O(l) \longrightarrow H_3O^+(aq) + HSO_4^-(aq)$$

Example 15.11 Calculating the pH of a solution of a polyprotic acid

Use the information in Table 15.6 to calculate the pH of 0.020 M $H_2S(aq)$.

Strategy The primary solute species in a solution of a polyprotic acid is the acid itself, in this case, H_2S. Find the pH of the solution by assuming that the acid loses only one proton and treating it as a monoprotic weak acid. Set up an equilibrium

One consequence of acid rain is vividly illustrated by these two photographs of a forest site in Germany. The one on the left was taken in 1970, the one on the right in 1983. Acid rain in this region has now been reduced, as a result of better control of power plant emissions.

Sulfuric acid is a strong acid that is especially damaging to soil because it causes the leaching of calcium ions. Most soil contains clay particles, which are surrounded by layers of ions, including Ca^{2+}. However, calcium ions on the clay particles can be replaced by hydrogen ions from sulfuric acid. Because calcium sulfate is insoluble in water, it can no longer circulate through the soil or be taken up by plants. If the calcium leached from soil is not replaced, plants suffer and entire forests can be affected.

Since 1989, important steps have been taken to reduce the acidity of rain. In both Europe and the United States, the emissions of sulfur have dropped significantly, partly due to the increased use of scrubbers. In a scrubber, exhaust gases from power plants are passed through an aqueous slurry of solid bases, such as calcium carbonate, that react with acidic oxides. Although it may take many years, we must continue to reduce acidic oxide emissions if we are to maintain our quality of life without losing our precious natural heritage.

Key Concepts: pH, acidic oxides

For Further Reading

T. Adler, Acid soil blamed for thinning eggshells, *Science News*, 145:212, 1994.

R. A. Kerr, Acid rain control: Success on the cheap, *Science*, 282:1024–1027, 1998.

S. E. Schwartz, Acid deposition: Unraveling a regional phenomenon, *Science*, 243:753–763, 1989.

Related Exercises: 15.82–15.84

table like the one in Toolbox 15.4 and determine the H_3O^+ molarity by using the first acidity constant, K_{a1}.

Solution The primary proton transfer equilibrium is

$$H_2S(aq) + H_2O(l) \rightleftharpoons H_3O^+(aq) + HS^-(aq)$$

and the equilibrium table, with the concentrations in moles per liter, is

	Species		
	H_2S	H_3O^+	HS^-
Step 1. Initial molarity	0.020	0	0
Step 2. Change in molarity	$-x$	$+x$	$+x$
Step 3. Equilibrium molarity	$0.020 - x$	x	x

Step 4. Substitute the equilibrium molarities into the acidity constant expression:

$$K_{a1} = 1.3 \times 10^{-7} = \frac{x^2}{0.020 - x}$$

Because of the small value of the acidity constant, we anticipate that $x \ll 0.020$ and approximate the expression to

$$1.3 \times 10^{-7} \approx \frac{x^2}{0.020}$$

Solving for x gives

$$x \approx \sqrt{(0.020) \times (1.3 \times 10^{-7})} = 5.1 \times 10^{-5}$$

so

$$pH \approx -\log(5.1 \times 10^{-5}) = 4.29$$

Self-Test 15.14A Estimate the pH of 0.10 M H_3PO_3(aq). Refer to Table 15.6 for K_a values.

[**Answer:** 1.62]

Self-Test 15.14B Estimate the pH of 0.10 M H_2SO_3(aq), using the data in Table 15.6.

Proton transfer explains the effervescence of carbon dioxide in antacid tablets (Fig. 15.19). When an acid is added to a carbonate or a hydrogen carbonate, the equilibrium shifts in accord with Le Chatelier's principle. At first, the added H_3O^+ ions shift the equilibrium composition to the left, favoring the formation of H_2CO_3:

$$H_2CO_3(aq) + H_2O(l) \rightleftharpoons H_3O^+(aq) + HCO_3^-(aq)$$

However, carbonic acid in solution is also in equilibrium with dissolved CO_2 molecules:

$$H_2CO_3(aq) \rightleftharpoons H_2O(l) + CO_2(aq)$$

Therefore, when the H_2CO_3 concentration increases as a result of the addition of acid, the CO_2 concentration increases in response. The increase in concentration of CO_2 is so great that the CO_2 bubbles out of solution with the familiar fizz. As in this everyday example, chemical effects are often transmitted along chains of equilibria, the disturbance of one equilibrium affecting another. This effect helps to explain the impact of acid rain. The increased acidity of natural waters affects other reactions in lakes, rivers, and soil (see Applying Chemistry: Case Study 15).

To estimate the pH of a polyprotic acid for which all deprotonations are weak, use only the first deprotonation equilibrium and assume that further deprotonation is insignificant. An exception is sulfuric acid, the only common polyprotic acid that is a strong acid in its first deprotonation.

FIGURE 15.19

The effervescence that occurs when an antacid tablet containing hydrogen carbonate ions is dissolved in acid is the result of a chain of effects that links several equilibria.

Skills You Should Have Mastered

Conceptual

☐ 1. Distinguish Arrhenius, Brønsted, and Lewis acids and bases, Section 15.1.

☐ 2. Explain how the pH of a solution is related to its hydronium ion and hydroxide ion concentrations, Section 15.4.

☐ 3. Explain why solutions of weak acids have higher pH values than solutions of strong acids at the same concentration, Sections, 15.6–15.8.

☐ 4. Show how the acidity constant of an acid is related to the basicity constant of its conjugate base, Section 15.7.

☐ 5. Use K_a values to predict the relative strengths of two acids or two bases, Toolbox 15.2 and Example 15.5.

Problem-Solving

☐ 1. Write the formulas for conjugate acids and bases, Self-Test 15.1.

☐ 2. Calculate the molar concentrations of hydronium and hydroxide ions in a solution of strong acid or base, Examples 15.1 and 15.2.

☐ 3. Calculate the pH and pOH of a solution of a strong acid or base, Toolbox 15.1 and Example 15.3.

☐ 4. Calculate the molar concentrations of hydronium and hydroxide ions from the pH or pOH of a solution of a strong acid or base, Toolbox 15.1 and Example 15.4.

☐ 5. Calculate the pH of a weak acid or base, Toolbox 15.4 and Examples 15.8 and 15.9.

☐ 6. Calculate the pH of a solution of a polyprotic acid, Examples 15.10 and 15.11.

Descriptive

☐ 1. Identify the common strong acids and bases, Section 15.1.

☐ 2. Describe and predict trends in acid strength in series of binary acids, oxoacids, and carboxylic acids, Toolbox 15.3 and Examples 15.6 and 15.7.

Exercises

Unless stated otherwise, assume that all solutions are aqueous and that the temperature is 25 °C.

Brønsted Acids and Bases

15.1 Classify each of the following species as an acid, a base, or amphiprotic: (a) H_2O; (b) CH_3NH_2; (c) PO_4^{3-}; (d) $C_6H_5NH_3^+$.

15.2 Classify each of the following species as an acid, a base, or amphiprotic: (a) HSO_4^-; (b) HI; (c) IO^-; (d) OH^-.

15.3 Write the formulas for the conjugate acids of (a) CH_3NH_2 (methylamine); (b) NH_2NH_2; (c) HCO_3^-; and the conjugate bases of (d) HCO_3^-; (e) C_6H_5OH (phenol); (f) CH_3COOH.

15.4 Write the formulas for the conjugate acids of (a) H_2O; (b) OH^-; (c) $C_6H_5NH_2$ (aniline); and the conjugate bases of (d) H_2S; (e) HPO_4^{2-} (hydrogen phosphate ion); (f) $HClO_4$ (perchloric acid).

15.5 Below are the molecular models of two oxoacids. Write the name of each acid and then draw the model of its conjugate base. (Red = O, light gray = H, green = Cl, and blue = N.)

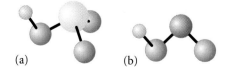

(a)   (b)

15.6 Below are the molecular models of two oxoacids. Write the name of each acid and then draw the model of its conjugate base. (Red = O, light gray = H, green = Cl, and blue = N.)

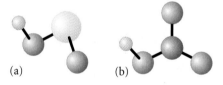

(a)   (b)

15.7 Write the proton transfer equilibria for the following acids in aqueous solution and identify the conjugate acid-base pairs in each one: (a) H_2SO_4; (b) $C_6H_5NH_3^+$ (anilinium ion); (c) $H_2PO_4^-$ (dihydrogen phosphate ion); (d) HCOOH (formic acid); (e) $NH_2NH_3^+$ (hydrazinium ion).

15.8 Write the proton transfer equilibria for the following bases in aqueous solution and identify the conjugate acid-base pairs in each one: (a) CN^-; (b) NH_2NH_2 (hydrazine); (c) CO_3^{2-}; (d) HPO_4^{2-}; (e) NH_2CONH_2 (urea).

15.9 Write the two proton transfer equilibria that demonstrate the amphiprotic character of (a) HCO_3^-; (b) HPO_4^{2-}, and identify the conjugate acid-base pairs in each equilibrium.

15.10 Write the two proton transfer equilibria that show the amphiprotic character of (a) $H_2PO_4^-$; (b) $HC_2O_4^-$ (hydrogen oxalate ion); and identify the conjugate acid-base pairs in each equilibrium.

15.11 Identify (a) the Brønsted acid and base for the reactants in the following reaction, and (b) the conjugate base and acid formed: $HNO_3(aq) + HPO_4^{2-}(aq) \rightarrow NO_3^-(aq) + H_2PO_4^-(aq)$.

15.12 Identify (a) the Brønsted acid and base for the reactants in the following reaction, and (b) the conjugate base and acid formed: $HSO_3^-(aq) + NH_4^+(aq) \rightarrow NH_3(aq) + H_2SO_3(aq)$.

Autoprotolysis of Water

15.13 Calculate the molarity of OH^- in solutions with the following H_3O^+ molarities: (a) 3.1×10^{-2} mol/L; (b) 1.0×10^{-4} mol/L; (c) 0.20 mol/L.

15.14 Estimate the molarity of H_3O^+ in solutions with the following OH^- molarities: (a) 5.6×10^{-3} mol/L; (b) 8.5×10^{-5} mol/L; (c) 0.12 mol/L.

15.15 The value of K_w for water at body temperature (37°C) is 2.5×10^{-14}. (a) What is the molarity of H_3O^+ ions and the pH of neutral water at 37°C? (b) What is the molarity of OH^- in neutral water at 37°C?

15.16 The molarity of H_3O^+ ions at the freezing point of water is 3.9×10^{-8} mol/L. (a) Calculate K_w and pK_w at 0°C. (b) What is the pH of neutral water at 0°C?

pH and pOH of Strong Acids and Bases

15.17 Calculate the initial molarity of HCl and the molarities of H_3O^+, Cl^-, and OH^- in an aqueous solution that contains 0.48 mol HCl in 500.0 mL of solution.

15.18 Calculate the initial molarity of HNO_3 and the molarities of H_3O^+, NO_3^-, and OH^- in an aqueous solution that contains 0.062 mol HNO_3 in 250.0 mL of solution.

15.19 Calculate the initial molarity of $Ba(OH)_2$ and the molarities of Ba^{2+}, OH^-, and H_3O^+ in an aqueous solution that contains 0.50 g of $Ba(OH)_2$ in 100.0 mL of solution.

15.20 Calculate the initial molarity of KNH_2 and the molarities of K^+, NH_2^-, OH^-, and H_3O^+ in an aqueous solution that contains 1.0 g of KNH_2 in 250.0 mL of solution.

15.21 The pH of several solutions was measured in the research laboratories of a food company; convert each of the following pH values to the molarity of H_3O^+ ions: (a) 3.3 (the pH of sour orange juice); (b) 6.7 (the pH of a saliva sample); (c) 4.4 (the pH of beer); (d) 5.3 (the pH of a coffee sample). (e) List the samples in order of increasing acidity.

15.22 The pH of several solutions was measured in a hospital laboratory; convert each of the following pH values to the molarity of H_3O^+ ions: (a) 5.0 (the pH of a urine sample); (b) 2.3 (the pH of a sample of lemon juice); (c) 7.4 (the pH of blood); (d) 10.5 (the pH of milk of magnesia). (e) List the samples in order of increasing acidity.

15.23 The molarity of H_3O^+ ions in the following solutions was measured at 25°C. Calculate the pH and pOH of the solutions: (a) 2.0×10^{-5} mol/L (sample of rainwater); (b) 1.0 mol/L; (c) 5.0×10^{-14} mol/L; (d) 5.02×10^{-5} mol/L.

15.24 The molarity of OH^- ions in the following solutions was measured at 25°C. Calculate the pH and pOH of the solutions: (a) 1.5×10^{-7} mol/L (a sample of milk); (b) 2.2 mmol/L; (c) 1.00×10^{-6} mol/L (a sample of tap water); (d) 7.09×10^{-4} mol/L.

15.25 Calculate the pH and pOH of each of the following aqueous solutions of strong acid or base: (a) 0.010 M $HNO_3(aq)$; (b) 0.22 M $HCl(aq)$; (c) 1.0×10^{-3} M $Ba(OH)_2(aq)$; (d) 14.0 mg of NaOH in 250.0 mL of solution.

15.26 Calculate the pH and pOH of each of the following solutions of strong acid or base: (a) 0.0149 M $HI(aq)$; (b) 0.0602 M $HCl(aq)$; (c) 1.73×10^{-3} M $Ba(OH)_2(aq)$; (d) 4.4 mg of KOH in 10.0 mL of aqueous solution.

Acidity and Basicity Constants

15.27 Refer to Table 15.3. Name the following acids, write their K_a and pK_a values, and list the acids in order of increasing strength: (a) HCOOH; (b) CH_3COOH; (c) CCl_3COOH; (d) C_6H_5COOH.

15.28 Refer to Table 15.3. Name the following acids, write their K_a and pK_a values, and list the acids in order of increasing strength: (a) HCN; (b) HIO_3; (c) HNO_2; (d) HF.

15.29 Give the K_a values and list the following acids in order of increasing strength: (a) phosphoric acid, H_3PO_4, $pK_{a1} = 2.12$; (b) phosphorous acid, H_3PO_3, $pK_{a1} = 2.00$; (c) arsenic acid, H_3AsO_4, $pK_{a1} = 2.26$; (d) arsenious acid, H_3AsO_3, $pK_{a1} = 9.29$.

15.30 Give the pK_b values for the following bases and list them in order of increasing strength: (a) ammonia, NH_3, $K_b = 1.8 \times 10^{-5}$; (b) deuterated ammonia, ND_3, $K_b = 1.1 \times 10^{-5}$; (c) hydrazine, NH_2NH_2, $K_b = 1.7 \times 10^{-6}$; (d) hydroxylamine, NH_2OH, $K_b = 1.1 \times 10^{-8}$.

15.31 (a) Write the names and formulas for the strongest and weakest conjugate bases of the acids listed in Exercise 15.29. (b) What are the K_b values for the two bases? (c) Which base, dissolved in water to a given concentration, would produce a solution with the highest pH?

15.32 (a) Write the names and formulas for the strongest and weakest conjugate acids of the bases listed in Exercise 15.30. (b) What are the K_a values for the two acids? (c) Which acid, dissolved in water to a given concentration, would produce a solution with the highest pH?

Structures and Strengths of Acids

15.33 The pK_a for HIO (hypoiodous acid) is 10.64 and that for HIO_3 (iodic acid) is 0.77. Account for the differences in acid strength.

15.34 The pK_a for HClO (hypochlorous acid) is 7.53 and that for HBrO (hypobromous acid) is 8.69. Account for the differences in acid strength.

15.35 Determine which acid is stronger and explain why: (a) HF or HCl; (b) HClO or HClO$_2$; (c) HBrO$_2$ or HClO$_2$; (d) HClO$_4$ or H$_3$PO$_4$; (e) HNO$_3$ or HNO$_2$; (f) H$_2$CO$_3$ or H$_2$GeO$_3$.

15.36 Determine which acid is stronger and explain why: (a) H$_3$AsO$_4$ or H$_3$PO$_4$; (b) HBrO$_3$ or HBrO; (c) H$_3$AsO$_4$ or H$_3$AsO$_3$; (d) H$_2$Te or H$_2$Se; (e) H$_2$S or HCl; (f) HClO or HIO.

15.37 Suggest an explanation for the difference between the strengths of (a) acetic acid and trichloroacetic acid and (b) acetic acid and formic acid.

15.38 Suggest an explanation of the difference in base strengths between (a) ammonia and methylamine and (b) hydrazine and hydroxylamine.

15.39 The values of K_a for phenol and 2,4,6-trichlorophenol (see following structures) are 1.3×10^{-10} and 1.0×10^{-6}, respectively. Which is the stronger acid? Account for the differences in acid strength.

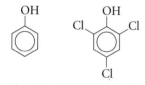

Phenol   2,4,6-Trichlorophenol

15.40 The value of pK_b for aniline is 9.37 and that for 4-chloroaniline is 9.85 (see following structures). Which is the stronger base? Account for the differences in base strength.

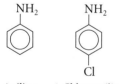

Aniline   4-Chloroaniline

15.41 Arrange the following bases in order of increasing strength on the basis of the pK_a values of their conjugate acids, which are given in parentheses: (a) ammonia (9.26); (b) methylamine (10.56); (c) ethylamine (10.81); (d) aniline (4.63). Is there a simple pattern of strengths?

15.42 Arrange the following bases in order of increasing strength on the basis of the pK_a values of their conjugate acids, which are given in parentheses: (a) aniline (4.63); (b) 2-hydroxyaniline (4.72); (c) 3-hydroxyaniline (4.17); (d) 4-hydroxyaniline (5.47). Is there a simple pattern of strengths?

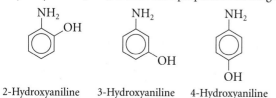

2-Hydroxyaniline   3-Hydroxyaniline   4-Hydroxyaniline

Weak Acid and Weak Base Calculations

Refer to Table 15.3 for the appropriate K_a and K_b values for the following exercises.

15.43 Determine the concentration of H$_3$O$^+$ and OH$^-$ in (a) 0.20 M C$_6$H$_5$COOH(aq); (b) 0.20 M NH$_2$NH$_2$(aq), hydrazine; (c) 0.20 M (CH$_3$)$_3$N(aq), trimethylamine.

15.44 Determine the concentration of H$_3$O$^+$ and OH$^-$ in (a) 0.15 M HCOOH(aq); (b) 0.10 M C$_6$H$_5$SO$_3$H(aq), benzenesulfonic acid; (c) 0.10 M NH$_3$(aq).

15.45 Calculate the pH of the following solutions: (a) 0.20 M HCOOH(aq); (b) 0.12 M NH$_2$NH$_2$(aq) (hydrazine); (c) 0.15 M C$_6$H$_5$COOH(aq) (benzoic acid); (d) 0.0034 M C$_{10}$H$_{14}$N$_2$(aq) (nicotine, a base).

15.46 Calculate the pH of the following solutions: (a) 0.0477 M HCN(aq) (hydrocyanic acid); (b) 1.5×10^{-5} M CH$_3$COOH(aq); (c) 0.023 M HBrO(aq).

15.47 Calculate the pH and pOH of the following aqueous solutions: (a) 0.15 M CH$_3$COOH(aq); (b) 0.15 M CCl$_3$COOH(aq); (c) 0.15 M HCOOH(aq).

15.48 Calculate the pH and pOH of the following aqueous solutions: (a) 0.20 M CH$_3$CH(OH)COOH(aq) (lactic acid); (b) 1.0×10^{-5} M CH$_3$CH(OH)COOH(aq); (c) 0.10 M C$_6$H$_5$SO$_3$H(aq) (benzenesulfonic acid).

15.49 Calculate the pH, pOH, and percentage protonation of solute in the following aqueous solutions: (a) 0.10 M NH$_3$(aq); (b) 0.017 M NH$_2$OH(aq); (c) 0.20 M (CH$_3$)$_3$N(aq); (d) 0.020 M codeine, given that the pK_a of its conjugate acid is 8.21. Codeine, a cough suppressant, is extracted from opium.

15.50 Calculate the pOH, pH, and percentage protonation of solute in the following aqueous solutions: (a) 0.11 M C$_5$H$_5$N(aq), pyridine; (b) 0.0058 M C$_{10}$H$_{14}$N$_2$(aq), nicotine; (c) 0.020 M quinine, given that the pK_a of its conjugate acid is 8.52; (d) 0.011 M strychnine, given that the K_a of its conjugate acid is 5.49×10^{-9}.

15.51 (a) When the pH of 0.10 M HClO$_2$(aq) was measured, it was found to be 1.2. What are the values of K_a and pK_a of chlorous acid? (b) The pH of 0.10 M propylamine, C$_3$H$_7$NH$_2$(aq) was measured as 11.86. What are the values of K_b and pK_b of propylamine?

15.52 (a) The pH of 0.015 M HNO$_2$(aq) was measured as 2.63. What are the values of K_a and pK_a of nitrous acid? (b) The pH of 0.10 M butylamine, C$_4$H$_9$NH$_2$(aq) was measured as 12.04. What are the percentage protonation and the values of K_b and pK_b of butylamine?

15.53 Use the information in Table 15.3 to find the initial concentration of the weak acid or base in each of the following aqueous solutions: (a) a solution of HClO with pH = 4.6; (b) a solution of hydrazine, NH$_2$NH$_2$, with pH = 10.2.

15.54 Use the information in Table 15.3 to find the initial concentration of the weak acid or base in each of the following aqueous solutions: (a) a solution of HCN with pH $= 5.3$; (b) a solution of pyridine, C_5H_5N, with pH $= 8.8$.

15.55 The percentage deprotonation of benzoic acid in a 0.110 M solution is 2.4%. What is the pH of the solution and the K_a of benzoic acid?

15.56 The percentage deprotonation of veronal (diethylbarbituric acid) in a 0.0200 M aqueous solution is 0.14%. What is the pH of the solution and the K_a of veronal?

15.57 The percentage protonation of octylamine (an organic base) in a 0.100 M aqueous solution is 6.7%. What is the pH of the solution and the K_b of octylamine?

15.58 Cacodylic acid is used as a cotton defoliant. Cacodylic acid is 0.77% deprotonated in a 0.011 M aqueous solution. What is the pH of the solution and the K_a of cacodylic acid?

Polyprotic Acids

Refer to Table 15.6 for the K_a values needed for the following exercises.

15.59 Write the stepwise proton transfer equilibria for the deprotonation of (a) sulfuric acid, H_2SO_4; (b) arsenic acid, H_3AsO_4; (c) phthalic acid, $C_6H_4(COOH)_2$.

15.60 Write the stepwise proton transfer equilibria for the deprotonation of (a) phosphoric acid, H_3PO_4; (b) adipic acid, $(CH_2)_4(COOH)_2$; (c) succinic acid, $(CH_2)_2(COOH)_2$.

15.61 Calculate the pH of 0.15 M $H_2SO_4(aq)$ at 25°C.

15.62 Calculate the pH of 0.010 M $H_2SeO_4(aq)$, given that K_{a1} is very large and $K_{a2} = 1.2 \times 10^{-2}$.

15.63 Calculate the pH of the following diprotic acid solutions at 25°C, ignoring second deprotonations: (a) 0.0010 M $H_2CO_3(aq)$; (b) 0.20 M $H_2S(aq)$.

15.64 Calculate the pH of the following diprotic acid solutions at 25°C; ignore second deprotonations: (a) 0.10 M $H_2S(aq)$; (b) 0.10 M $(COOH)_2(aq)$.

Supplementary Exercises

15.65 What condition is required for an aqueous solution to be considered neutral?

15.66 Write the formula for the conjugate acid of (a) the carbonate ion, CO_3^{2-}; (b) the acetate ion, $CH_3CO_2^-$, and the conjugate base of (c) the dihydrogen phosphate ion, $H_2PO_4^-$; (d) the hydroxide ion, OH^-.

15.67 Identify the conjugate acid-base pairs and write the proton transfer equilibria for (a) propionic acid, C_2H_5COOH; (b) chloric acid, $HClO_3$; (c) acetylsalicylic acid (aspirin), $C_8H_7O_2COOH$; (d) caffeine (a base), $C_8H_{10}N_4O_2$; (e) pyridine (a base), C_5H_5N.

15.68 The value of K_w at 40°C is 3.8×10^{-14}. What is the pH of pure water at 40°C?

15.69 Under what conditions can the pH of solutions be (a) negative; (b) greater than 14?

15.70 Calculate the pH and pOH of (a) 0.026 M $HNO_3(aq)$; (b) 0.012 M $Ba(OH)_2(aq)$; (c) 1.47×10^{-4} M $HBr(aq)$; (d) 3.19×10^{-3} M $KOH(aq)$.

15.71 Calculate the molarity of H_3O^+ ions in a solution having a pH of (a) 9.33; (b) 7.95; (c) 0.01; (d) 4.33; (e) 1.99; (f) 11.95.

15.72 Use the information in Table 15.2 to decide whether carbonic acid is a strong or a weak acid in liquid ammonia solvent. Explain your answer.

15.73 The K_a of phenol (C_6H_5OH, which is also called carbolic acid) is 1.3×10^{-10} and that for the ammonium ion is 5.6×10^{-10}. (a) Write the proton transfer equilibria for each in aqueous solution. (b) Calculate the pK_a for each. (c) Which acid is the stronger acid?

15.74 Determine the value of K_b for the following anions (all of which are the conjugate bases of acids) and list them in order of increasing base strength (refer to Tables 15.3 and 15.6 for information): F^-, $CH_2ClCO_2^-$, CO_3^{2-}, IO_3^-, Cl^-.

15.75 Identify the stronger acid in each of the following pairs, and give reasons for your choice: (a) $HBrO_4$ or HIO_4; (b) HF or HI; (c) HIO_2 or HIO_3; (d) H_3AsO_4 or H_2SeO_4.

15.76 Calculate the pOH and pH of each of the following solutions: (a) 0.029 M $C_6H_5NH_2(aq)$; (b) 0.10 M $C_6H_{11}NH_2(aq)$, cyclohexylamine, for which $pK_b = 3.36$; (c) 0.0194 M $C_{17}H_{19}O_3N(aq)$, morphine, a base; (d) 0.015 M $NH_2CH_2CH_2NH_2(aq)$, ethylenediamine, given that the K_a of its conjugate acid is 1.9×10^{-11}.

15.77 When 0.150 g of an organic base of molar mass 31.06 g/mol is dissolved in 50.0 mL of water, the pH is found to be 10.05. What is the percentage protonation of the base? Calculate the pK_b of the base and the pK_a of its conjugate acid.

15.78 Write the stepwise proton transfer equilibria for the deprotonation in water of (a) hydrosulfuric acid, H_2S; (b) tartaric acid, $C_2H_4O_2(COOH)_2$; (c) malonic acid, $CH_2(COOH)_2$.

15.79 Although many chemical reactions take place in water, it is often necessary to use other solvents instead, and liquid ammonia (b.p. $-33°C$) has been used extensively. Many of the reactions that occur in water have analogous reactions in liquid ammonia. (a) Write the chemical equation for the autoprotolysis of NH_3. (b) What are the formulas of the acid and base species that result from the autoprotolysis of liquid ammonia? (c) The autoprotolysis constant, K_{am}, of liquid ammonia has the value 1×10^{-33} at $-35°C$. What is the value of pK_{am} at that temperature? (d) What is the molarity of NH_4^+ ions in neutral liquid ammonia? (e) Evaluate pNH_4 and pNH_2 in neutral liquid ammonia at $-35°C$. (f) Determine the relation between pNH_4, pNH_2, and pK_{am}.

15.80 Calculate the pH of the following acid solutions at 25°C; ignore second deprotonations only when that approximation is justified. (a) 0.015 M H_3PO_4(aq); (b) 0.10 M H_2SO_3(aq).

15.81 Heavy water, D_2O, is used in some nuclear reactors (see Chapter 22). The K_w for heavy water at 25°C is 1.35×10^{-15}. (a) Write the chemical equation for the autoprotolysis of D_2O. (b) Evaluate pK_w for D_2O at 25°C. (c) Calculate the molarities of D_3O^+ and OD^- in neutral heavy water at 25°C. (d) Evaluate the pD and pOD of neutral heavy water at 25°C. (e) Find the relation between pD, pOD, and pK_w.

Applied Exercises

For Exercises 15.82–15.84, see Applying Chemistry: Case Study 15.

15.82 U.S. Forest Service scientists found that the pH of a remote wilderness lake near Aspen, Colorado was 3.39. (a) Calculate the hydronium ion and hydroxide ion concentrations in the lake. (b) If unaffected by human activity, the lake has a pH of 5.79. By what factor has the concentration of hydronium ion in the lake been increased?

Scientists measuring the pH of a wilderness lake.

15.83 1.00 metric ton (1 t $= 10^3$ kg) of coal that contains 2.5% sulfur by mass is burned in a coal-fired plant. (a) What mass of SO_2 is produced? (b) What is the pH of rainwater when the SO_2 dissolves in a volume of water equivalent to

2.0 cm of rainfall over 2.6 km²? (The pK_{a1} of sulfurous acid is 1.81. Consider the water to be initially pure and at a pH of 7.) (c) If the SO_2 is first oxidized to SO_3 before the rainfall occurs, what would the pH of the same rainwater be?

15.84 One process used to clean SO_2 from the emissions of coal-fired plants is to pass the stack gases along with air through a wet calcium carbonate slurry, where the following reaction occurs: $CaCO_3(s) + SO_2(g) + \frac{1}{2}O_2(g) \rightarrow CaSO_4(s) + CO_2(g)$. What mass of limestone ($CaCO_3$) is needed to remove 50.0 kg sulfur dioxide from stack gases if the removal process is 90% efficient?

Integrated Exercises

15.85 The partial pressure of CO_2 in air saturated with water vapor at 25°C and 1.00 atm is 3.04×10^{-4} atm. Henry's constant for CO_2 in water is 2.3×10^{-2} mol/L·atm, and for carbonic acid, $pK_{a1} = 6.37$. Verify by calculation that the pH of "normal" rainwater is about 5.7.

15.86 Use Table 15.3 to determine the percentage deprotonation of 1.00 M lactic acid in water (assume that the density of the solution is 1.00 g/cm³). At what temperature will the solution freeze?

15.87 Calculate the pH and pOH of each of the following aqueous solutions of strong acid or base: (a) 10.0 mL of 0.022 M KOH(aq) after dilution to 250.0 mL; (b) 50.0 mL of 0.000 43 M HBr(aq) after dilution to 250.0 mL.

15.88 Calculate the pH and pOH of each of the following aqueous solutions of strong acid or base: (a) 10.0 mL of 0.0022 M NaOH(aq) after dilution to 500.0 mL; (b) 5.0 mL of 0.000 43 M $HClO_4$(aq) after dilution to 50.0 mL.

15.89 Estimate the enthalpy of the deprotonation of formic acid at 25°C given that K_a is 1.765×10^{-4} at 20°C and 1.768×10^{-4} at 30°C. (*Hint:* Use the van't Hoff equation, Exercise 14.85.)

15.90 Convert the van't Hoff equation (see Exercise 14.85) for the temperature dependence of an equilibrium constant to an expression for the temperature dependence of pK_w. Estimate the value of pK_w at the normal boiling point of water from the enthalpy of the water autoprotolysis reaction, which is $+57$ kJ for the reaction $2 H_2O(l) \rightarrow H_3O^+(aq) + OH^-(aq)$. What is the pH of pure water at that temperature?

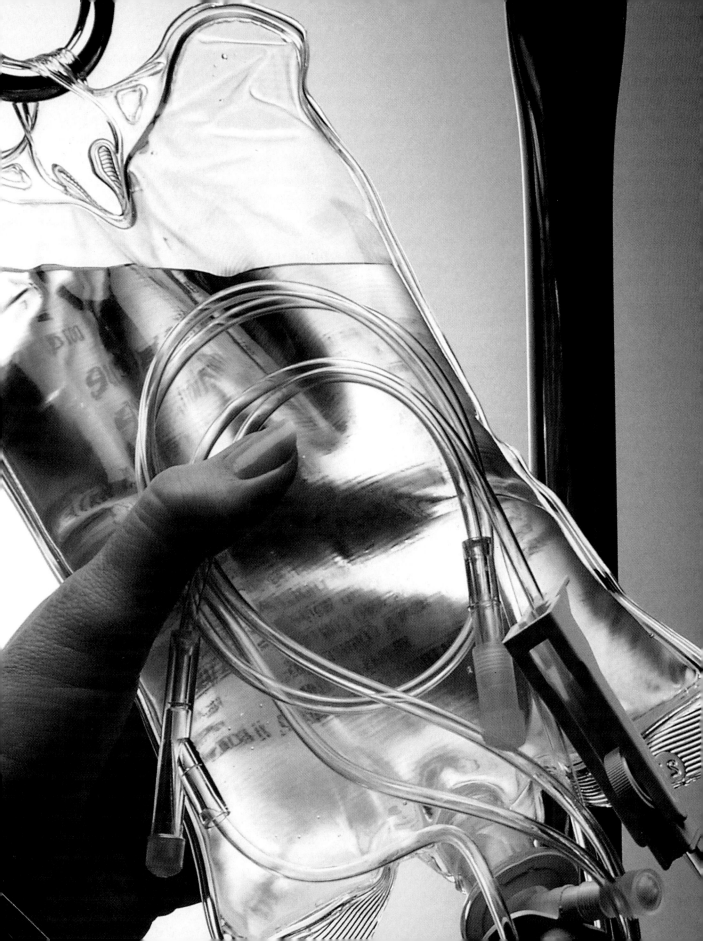

Aqueous Equilibria

You are likely to die if the pH of your blood plasma falls by more than 0.4 from its normal value of 7.4. The pH can fall as a result of disease or shock, both of which would generate acidic conditions in your body. You are also likely to die if the pH of your blood plasma rises to 7.8, as could happen during the early stages of recovery from severe burns. To survive, your body must control its own pH; and if your control systems fail, then medical professionals must intervene rapidly. Intravenous administration of electrolyte solutions is often the first treatment in emergency rooms.

In this chapter, we see how different ions affect pH and how they can be used to control it. We also look at the factors that determine the solubility of salts in water and see how to identify ions from their solubility behavior. This chapter builds on and extends Chapter 15, because many of the properties of aqueous solutions involve proton transfer equilibria between Brønsted acids and bases. Reactions between ions in solution reach equilibrium very rapidly, and the values of the equilibrium constants of these reactions—the acidity constant, the basicity constant, and the autoprotolysis constant of water—have to be maintained. If salts are present in solution, the equilibrium constants for their dissolution in water must also be satisfied.

SALTS IN WATER

The pH of aqueous solutions—not only blood plasma, but seawater, detergents, sap, and reaction mixtures—is controlled by the transfer of protons between ions and water molecules. First, let's analyze what happens to the pH when we dissolve a salt in water (Fig. 16.1). The pH rises above 7 if the salt provides ions that are bases. For example, sodium

The intravenous solution in this bag might save a life. The concentration of each solute in the solution is carefully selected to keep the total solute concentration within an optimal range and to maintain the pH of the blood. Because even slight deviations in blood pH can be fatal, the first treatment administered to an injured person is usually an intravenous solution.

FIGURE 16.1

Solutions of salts in water give rise to acidic, neutral, and basic solutions, as shown here by the color of the indicator bromothymol blue (see Table 16.3). From left to right are solutions of ammonium chloride, sodium chloride, and sodium acetate. Throughout this chapter, we use the Brønsted definitions of acids and bases.

acetate, $NaCH_3CO_2$, provides the acetate ion, the conjugate base of acetic acid, so we expect a solution of sodium acetate to have pH > 7. On the other hand, we expect pH < 7 if the salt provides ions that act as acids. An example is NH_4Cl, which provides the NH_4^+ ion, the conjugate acid of ammonia. The ions provided by a salt such as sodium chloride, NaCl, are such weak acids and bases that we can consider them "neutral," and the pH of the solution is close to 7.

16.1 Ions as Acids and Bases

Any cation that is the conjugate acid of a weak base functions as an acid and lowers the pH of the solution. For example, the ammonium ion, NH_4^+, the conjugate acid of the weak base NH_3, is an acid:

$$NH_4^+(aq) + H_2O(l) \rightleftharpoons H_3O^+(aq) + NH_3(aq)$$

Although K_a for NH_4^+ is small (5.6×10^{-10}), enough of the salt's NH_4^+ ions donate protons to lower the pH appreciably. Table 16.1 lists some cations that are acidic in water.

Small, highly charged metal cations that can act as Lewis acids in water, such as Al^{3+} and Ti^{3+}, also produce acidic solutions, even though the cations themselves have no hydrogen ions to donate (Fig. 16.2). The protons come from the water molecules that hydrate the ions in solution (Fig. 16.3). These molecules act as Lewis bases and share electrons with the metal cation. This partial loss of electrons weakens the O—H bonds and allows one or more hydrogen ions to be lost. Small, highly charged ions exert the greatest pull on the electrons. Some cations that act as acids in this way are included in Table 16.1.

Table 16.1 Acidic character and K_a values of common cations in water*

Character	Examples	K_a	pK_a
Acidic			
conjugate acids of weak bases	anilinium ion, $C_6H_5NH_3^+$	2.3×10^{-5}	4.64
	pyridinium ion, $C_5H_5NH^+$	5.6×10^{-6}	5.24
	ammonium ion, NH_4^+	5.6×10^{-10}	9.25
	methylammonium ion, $CH_3NH_3^+$	2.8×10^{-11}	10.56
small, highly charged metal cations	$Fe^{3+}(aq)$	3.5×10^{-3}	2.46
	$Cr^{3+}(aq)$	1.3×10^{-4}	3.89
	$Al^{3+}(aq)$	1.4×10^{-5}	4.85
	$Fe^{2+}(aq)$	1.3×10^{-6}	5.89
	$Cu^{2+}(aq)$	3.2×10^{-8}	7.49
	$Ni^{2+}(aq)$	9.3×10^{-10}	9.03
Neutral			
Group 1 and 2 cations; metal cations with charge +1	Li^+, Na^+, K^+ Mg^{2+}, Ca^{2+}, Ag^+		
Basic	none		

*As in Table 15.3, the experimental pK_a values have more significant figures than shown here, and the K_a values have been calculated from these better data.

Table 16.2 *Acidic and basic character of common anions in water*

Character	Examples
Acidic	
very few	$HSO_4^-, H_2PO_4^-$
Neutral	
conjugate bases of strong acids	$Cl^-, Br^-, I^-, NO_3^-, ClO_4^-$
Basic	
conjugate bases of weak acids	$F^-, O^{2-}, OH^-, S^{2-}, HS^-, CN^-, CO_3^{2-},$ $PO_4^{3-}, NO_2^-, CH_3CO_2^-,$ other carboxylate ions

FIGURE 16.2

A solution of titanium(III) sulfate is so acidic that it can release H_2S from some sulfides.

The cations of Group 1 and 2 metals, and those of charge $+1$ from other groups, are such weak Lewis acids that they do not act as acids when hydrated. These metal cations are too large or have too low a charge to have an appreciable polarizing effect on the water molecules that surround them. As a result, the hydrating water molecules do not readily release their protons. Only a few anions are acids (Table 16.2); it is difficult for the positively charged proton to leave an anion.

All anions that are the conjugate bases of weak acids act as proton acceptors, so we can expect them to give basic solutions. For example, formic acid, HCOOH, the acid in ant venom, is a weak acid; hence the formate ion acts as a base in water:

$$H_2O(l) + HCO_2^-(aq) \rightleftharpoons HCOOH(aq) + OH^-(aq)$$

Acetate ions and the other ions listed in Table 16.2 also act as bases in water.

We saw in Chapter 15 that the stronger the acid, the weaker its conjugate base (Fig. 16.4). It follows that the anions of strong acids—which include Cl^-, Br^-, I^-, NO_3^-, and ClO_4^-—are *very* weak bases. They have no significant effect on the pH of a solution.

Salts that contain the conjugate acids of weak bases produce acidic aqueous solutions; so do salts that contain small, highly charged metal cations. Salts that contain the conjugate bases of weak acids produce basic aqueous solutions.

Self-Test 16.1A Use Tables 16.1 and 16.2 to decide whether aqueous solutions of the salts (a) $Ba(NO_2)_2$; (b) $CrCl_3$; (c) NH_4NO_3 are acidic, neutral, or basic.

[*Answer:* (a) Basic; (b) acidic; (c) acidic]

Self-Test 16.1B Decide whether aqueous solutions of (a) Na_2CO_3; (b) $AlCl_3$; (c) KNO_3 are acidic, neutral, or basic.

FIGURE 16.3

In water, Al^{3+} cations exist as hydrated ions that can act as Brønsted acids. Although, for clarity, only four water molecules are shown here, metal cations typically have six H_2O molecules attached to them.

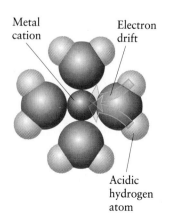

Metal cation

Electron drift

Acidic hydrogen atom

FIGURE 16.4

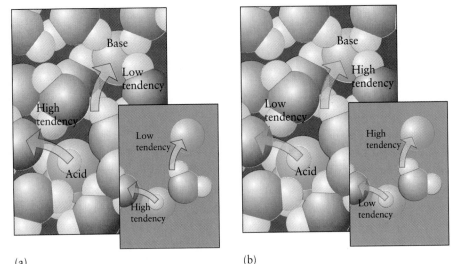

The relative strengths of conjugate acids and bases have a reciprocal relation. (a) When a species has a high tendency to donate a proton, the resulting conjugate base has a low tendency to accept one. (b) When a species has a low tendency to donate a proton, the resulting conjugate base has a high tendency to accept one. In the insets, a solid blue color represents the water molecules.

(a)   (b)

16.2 The pH of a Salt Solution

Applying Chemistry: Case Study 16 contains additional information on how salt solutions are used in emergency medicine.

We sometimes need to know the pH of a salt solution. For example, in hospitals, nurses administer salt solutions intravenously to regulate the pH of bodily fluids. The pH of the water in swimming pools and aquaria is also regulated by adding salts.

To calculate the pH of a salt, we treat an acidic ion as a weak acid and a basic ion as a weak base. Then, as in Chapter 15, we set up an equilibrium table. Initially, the instant the solution is prepared, we suppose that it consists of only water molecules and the cations and anions of the salt. Then proton transfer between the cations, anions, and water immediately adjusts the composition to equilibrium. The procedure is summarized in Toolbox 16.1.

Aqueous solutions of salts with acidic cations have a pH lower than 7; salts with basic anions produce a pH higher than 7 in aqueous solution.

Toolbox 16.1 *How to calculate the pH of an electrolyte solution*

This Toolbox shows how to calculate the pH of a solution of a salt that contains an acidic or basic ion.

Conceptual Basis

Because ions can act as Brønsted acids or bases, they can donate or accept protons and participate in proton transfer equilibria with water. If a salt contains a cation that is the conjugate acid of a weak base, then the pH of the solution will be less than 7; if the salt contains an anion that is the conjugate base of a weak acid, then the pH of the solution will be greater than 7 (Fig. 16.5).

Procedure

The procedure is very similar to the one set out in Toolbox 15.4. The only difference is that the acid or base is now the cation or anion of a salt. Although we shall calculate pH and pK values to the number of significant figures appropriate to the data, the answers are often considerably less reliable than that. For instance, we might calculate the pH of a solution as 8.82; but, in practice, the answer is unlikely to be reliable to more than one decimal place (pH = 8.8). One reason is that we are ignoring interactions between the ions in solution. Because

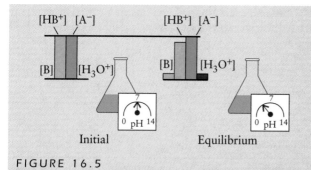

FIGURE 16.5

The initial (left) and equilibrium (right) composition of a solution of a salt composed of the cation HB^+ and the anion A^-, where HB^+ is a weak acid and A^- is neutral. (left) The hypothetical initial situation that we imagine before deprotonation. (right) At equilibrium, the acidic cation is partially deprotonated, and the solution is acidic.

of this unreliability, it is best to reduce by one the number of significant figures in a calculated pH. The acidic and basic characters of ions are summarized in Tables 16.1 and 16.2.

A salt with an acidic ion First, write the chemical equation for proton transfer to water and the corresponding expression for K_a. The products are the hydronium ion and the conjugate base of the acidic ion. Then set up an equilibrium table:

Step 1. The initial molarity of the acidic ion is the molarity of the initially completely protonated ion. We assume that the initial molarities of its conjugate base and H_3O^+ are 0.

Step 2. Write the increase in molarity of H_3O^+ as x mol/L and use the reaction stoichiometry to write the corresponding changes for the acidic ion and its conjugate base. Ignore H_3O^+ from the autoprotolysis of water; this approximation is valid if $[H_3O^+]$ is substantially (about 10 times) greater than 1×10^{-7}. Check the validity of this approximation at the end of the calculation.

Step 3. Write the equilibrium molarities of the species in terms of x.

Step 4. Express the acidity constant for the ion in terms of x and solve the equation for x; assume that x is small, but check the validity of this assumption at the end of the calculation. If K_a is not available, obtain it from the value of K_b for the conjugate base by using $K_a = K_w/K_b$ (Section 15.7). Because x mol/L is the H_3O^+ molarity, the pH of the solution is $-\log x$.

This procedure is illustrated in Example 16.1.

A salt with a basic ion In this case, we expect pH > 7. We follow the same procedure as that used for an acidic ion, except that now proton transfer from water to the ion results in the formation of the OH^- ion and the conjugate acid of the ion. We therefore use K_b, and the equilibrium table leads to a value for pOH. At the end of the calculation, convert pOH to pH by using pH + pOH = 14.00. This procedure is illustrated in Example 16.2.

An amphiprotic anion An amphiprotic anion can act as both an acid and a base ($H_2PO_4^-$ is an example). Special techniques, which are discussed on the web site for this book, lead to the result that, at any concentration of the salt,

$$pH = \tfrac{1}{2}(pK_a + pK_b)$$

Example 16.1 *Calculating the pH of a salt solution with an acidic cation*

A student prepared 0.15 M $NH_4Cl(aq)$, dipped in a piece of litmus paper, and found that the paper turned red. Should the student have been surprised? Estimate the pH of the solution.

Strategy Think about the acidic or basic character of the ions. For the calculation, use the procedure for a salt with an acidic cation in Toolbox 16.1. The equilibrium to consider is

$$NH_4^+(aq) + H_2O(l) \rightleftharpoons H_3O^+(aq) + NH_3(aq) \qquad K_a = \frac{[H_3O^+][NH_3]}{[NH_4^+]}$$

Make the initial assumption that the extent of deprotonation of the weakly acidic cation is very small, but verify and correct that assumption if necessary.

Solution Because NH_4^+ cations are acids and Cl^- anions are neutral, we expect an acidic solution with pH < 7. We construct the following equilibrium table, with all concentrations in moles per liter:

	Species		
	NH_4^+	H_3O^+	NH_3
Step 1. Initial molarity	0.15	0	0
Step 2. Change in molarity	$-x$	$+x$	$+x$
Step 3. Equilibrium molarity	$0.15 - x$	x	x

Step 4. The acidity constant K_a for NH_4^+ is obtained from the value of K_b for NH_3 in Table 15.3:

$$K_a = \frac{K_w}{K_b} = \frac{1.0 \times 10^{-14}}{1.8 \times 10^{-5}} = 5.6 \times 10^{-10}$$

Substitution of this value and the information from step 3 into the expression for K_a gives

$$5.6 \times 10^{-10} = \frac{x \times x}{0.15 - x}$$

Because K_a is very small, we now suppose that x is less than 5% of 0.15 and simplify this expression to

$$\frac{x^2}{0.15} \approx 5.6 \times 10^{-10}$$

The solution of this equation is

$$x \approx \sqrt{0.15 \times (5.6 \times 10^{-10})} = 9.2 \times 10^{-6}$$

The approximation that x is less than 5% of 0.15 is valid. Moreover, the H_3O^+ molarity (9.2×10^{-6} mol/L) is much larger than that generated by the autoprotolysis of water (1.0×10^{-7} mol/L), so the neglect of the latter contribution is also valid. Because the H_3O^+ molarity is 9.2×10^{-6} mol/L, the pH of the solution is

$$pH = -\log(9.2 \times 10^{-6}) = 5.04$$

or about 5.0.

Self-Test 16.2A Estimate the pH of 0.10 M $CH_3NH_3Cl(aq)$, an aqueous solution of methylammonium chloride; the cation is $CH_3NH_3^+$.

[*Answer:* 5.78, or about 5.8]

Self-Test 16.2B Estimate the pH of 0.10 M $NH_4NO_3(aq)$.

Example 16.2 *Calculating the pH of a salt solution with a basic anion*

Estimate the pH of 0.15 M $Ca(CH_3CO_2)_2(aq)$.

Strategy Calcium ions are neutral, but acetate ions are basic, so we expect pH > 7. Use the procedure for a salt with a basic anion described in Toolbox 16.1. The proton transfer equilibrium is

$$H_2O(l) + CH_3CO_2^-(aq) \rightleftharpoons CH_3COOH(aq) + OH^-(aq)$$

$$K_b = \frac{[CH_3COOH][OH^-]}{[CH_3CO_2^-]}$$

Make the initial assumption that the extent of protonation of the weakly basic anion is very small, but verify and correct this assumption if necessary.

Solution The initial molarity of $CH_3CO_2^-$ is 2×0.15 mol/L $= 0.30$ mol/L, because each formula unit of salt provides two $CH_3CO_2^-$ ions.

	Species		
	$CH_3CO_2^-$	CH_3COOH	OH^-
Step 1. Initial molarity	0.30	0	0
Step 2. Change in molarity	$-x$	$+x$	$+x$
Step 3. Equilibrium molarity	$0.30 - x$	x	x

Step 4. Because Table 15.3 gives the K_a of CH_3COOH as 1.8×10^{-5}, the K_b of its conjugate base, the $CH_3CO_2^-$ ion, is

$$K_b = \frac{K_w}{K_a} = \frac{1.0 \times 10^{-14}}{1.8 \times 10^{-5}} = 5.6 \times 10^{-10}$$

Next, we insert this value and the information from step 3 into the expression for K_b and obtain

$$5.6 \times 10^{-10} = \frac{x \times x}{0.30 - x}$$

Because K_b is so small, we assume that x is less than 5% of 0.30 and simplify this expression to

$$\frac{x^2}{0.30} \approx 5.6 \times 10^{-10}$$

It follows that

$$x \approx \sqrt{0.30 \times (5.6 \times 10^{-10})} = 1.3 \times 10^{-5}$$

which is far less than 5% of 0.30, so our assumption is valid. The OH^- molarity (1.3×10^{-5} mol/L, from step 3) arising from the proton transfer equilibrium is much larger than that arising from the autoprotolysis of water (1.0×10^{-7} mol/L), so the neglect of autoprotolysis is also valid. Because the molarity of OH^- ions is 1.3×10^{-5} mol/L,

$$pOH = -\log(1.3 \times 10^{-5}) = 4.89$$

and therefore

$$pH = 14.00 - 4.89 = 9.11$$

or about 9.1. The solution is basic, as expected.

Self-Test 16.3A Estimate the pH of 0.10 M $KC_6H_5CO_2$(aq), potassium benzoate. (The conjugate acid of the benzoate ion is benzoic acid, C_6H_5COOH, $K_a = 6.5 \times 10^{-5}$.)

[*Answer:* 8.59, or about 8.6]

Self-Test 16.3B Estimate the pH of 0.020 M KF(aq); see Table 15.3.

16.3 The pH of Mixed Solutions

We saw in Section 15.11 how to estimate the pH of a solution of a weak acid when it is the only solute. But suppose a salt containing the conjugate base of the acid were also added: how would that affect the pH? For example, we might want to find the pH of a solution that contains equal amounts of sodium acetate and acetic acid. We can use our knowledge of proton transfer equilibria to predict what happens: because acetate ions are bases, increasing their concentration increases the pH of the solution. Similarly, suppose we have a solution of a base and add a salt containing the conjugate acid of the base to it (for instance, NH_4Cl added to aqueous ammonia). Because the salt contains an acidic cation, we can expect the pH to fall.

> The effect on equilibrium of increasing the concentration of a product is discussed in Section 14.9.

To estimate the pH of mixed solutions, we set up an equilibrium table and use the acidity or basicity constant to calculate the unknown concentrations as described in Toolboxes 15.4 and 16.1. The only difference is that the initial concentrations of the conjugate acid and base are both specified in the problem. These are the values we use in step 1 of the equilibrium table. For example, in a solution that is 0.010 M HCN(aq) and 0.020 M NaCN(aq), the initial concentration of HCN is 0.010 mol/L and that of the CN^- ion is 0.020 mol/L.

The pH of a solution of a weak acid increases when a salt containing its conjugate base is added. The pH of a solution of a weak base decreases when a salt containing its conjugate acid is added.

Example 16.3 *Calculating the pH of a solution of a weak acid and its salt*

Calculate the pH of a solution that is 0.500 M HNO_2(aq) and 0.100 M KNO_2(aq). From Table 15.3, $K_a = 4.3 \times 10^{-4}$ for HNO_2.

Strategy The solution contains NO_2^-, which is a weak base (the conjugate base of the acid HNO_2), so we expect the pH to be higher than that of nitrous acid alone. The K^+ ion supplied by the salt has no effect on the pH of the solution. To calculate the pH of the mixed solution, set up an equilibrium table and consider the initial molarity of HNO_2 to be 0.500 mol/L. Because nitrite ions have also been added to the solution, set their initial molarity equal to the molarity of the salt (each KNO_2 formula unit supplies one NO_2^- anion). Then proceed as described in Toolbox 15.4.

Solution The proton transfer equilibrium to consider is

$$HNO_2(aq) + H_2O(l) \rightleftharpoons H_3O^+(aq) + NO_2^-(aq) \qquad K_a = \frac{[H_3O^+][NO_2^-]}{[HNO_2]}$$

The equilibrium table, with all concentrations in moles per liter, is

	Species		
	HNO_2	H_3O^+	NO_2^-
Step 1. Initial molarity	0.500	0	0.100
Step 2. Change in molarity	$-x$	$+x$	$+x$
Step 3. Equilibrium molarity	$0.500 - x$	x	$0.100 + x$

Step 4. Substitution of the information from step 3 and $K_a = 4.3 \times 10^{-4}$ into the expression for K_a gives

$$4.3 \times 10^{-4} = \frac{x \times (0.100 + x)}{0.500 - x}$$

Assuming that x is less than 5% of 0.100 (and therefore also less than 5% of 0.500), we write

$$4.3 \times 10^{-4} \approx \frac{x \times 0.100}{0.500}$$

The solution of this equation is $x \approx 2.2 \times 10^{-3}$. To check the validity of the assumption, we evaluate

$$\frac{2.2 \times 10^{-3}}{0.100} \times 100\% = 2.2\%$$

Because the value of x is 2.2% of 0.100, it is only 0.44% of 0.500, so the assumptions are valid. It follows that the equilibrium molarity of H_3O^+ ions is 2.2×10^{-3} mol/L. This is much larger than the molarity arising from the autoprotolysis of water, so the neglect of autoprotolysis is also valid. The pH of the solution is

$$pH = -\log(2.2 \times 10^{-3}) = 2.66$$

or about 2.7. The pH of 0.500 M $HNO_2(aq)$ is 1.8; so, as anticipated, the pH of the mixed solution is higher.

Self-Test 16.4A Calculate the pH of a solution that is 0.300 M $CH_3NH_2(aq)$ and 0.146 M $CH_3NH_3Cl(aq)$. From Table 15.3, the K_b of CH_3NH_2 is 3.6×10^{-4}.

[*Answer:* 10.87]

Self-Test 16.4B Calculate the pH of a solution that is 0.010 M $HClO(aq)$ and 2.0×10^{-4} M $NaClO(aq)$.

TITRATIONS

Sometimes we need to know not just the pH, but how much acid or base is present in a sample. For example, an environmental chemist studying a lake in which fish are dying must know exactly how much acid is present in a sample of the water. We saw in Section 4.8 that titration involves adding a solution called the *titrant* from a buret to a flask containing the sample, called the *analyte*. The success of an acid-base titration depends on how accurately we can detect the stoichiometric point. At that point, the number of moles of H_3O^+ (or OH^-) added as titrant is equal to the number of moles of OH^- (or H_3O^+) initially present in the analyte. At the stoichiometric point, the solution consists of a salt in water. The solution is acidic (pH < 7) if acidic ions are present, basic (pH > 7) if basic ions are present, and neutral (pH = 7) if the ions are neither acidic nor basic.

16.4 Strong Acid-Strong Base Titrations

As a strong acid is gradually added to a strong base, it reacts with the base and the pH gradually falls. A **pH curve** is a plot of the pH of the analyte solution as titrant is added (Fig. 16.6). We can calculate the points on the curve as illustrated in Example 16.4. Initially, the pH falls slowly. When the stoichiometric point is passed, there is a sudden fall through pH = 7 as the H_3O^+ molarity increases sharply. At this point, an indicator changes color or an automatic titrator detects the stoichiometric point by responding electronically to this sudden fall in pH. The pH then falls slowly toward the value of the acid itself as the dilution caused by the original analyte solution becomes less and less important.

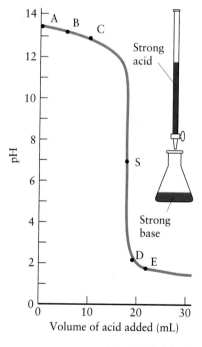

FIGURE 16.6

The variation of pH during the titration of a strong base with a strong acid. This curve is for 25.00 mL of 0.250 M NaOH(aq) titrated with 0.340 M HCl(aq). The stoichiometric point occurs at pH = 7 (point S). The other points are discussed in the text.

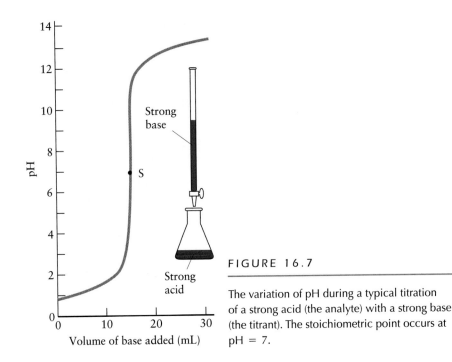

FIGURE 16.7

The variation of pH during a typical titration of a strong acid (the analyte) with a strong base (the titrant). The stoichiometric point occurs at pH = 7.

The pH curve for the titration of a strong acid (the analyte) with a strong base (the titrant) mirrors the titration of a strong base with a strong acid (Fig. 16.7). The initial pH is low, it increases only slightly until just before the stoichiometric point, and then it increases sharply. The pH continues to increase after the stoichiometric point, but then levels off as a result of the presence of excess strong base in the solution. We can calculate the shape of the curve by using the procedure in Toolbox 16.2. We also use these procedures to calculate the pH of any solution that results from the reaction between a strong acid and a strong base.

Toolbox 16.2   *How to calculate the pH during a strong acid-strong base titration*

This Toolbox illustrates how to calculate the pH during the titration of a strong acid with a strong base and vice versa.

Conceptual Basis
The pH during the titration of a strong acid with a strong base is determined by the major species in solution (Fig. 16.8). Because the conjugate base of the strong acid has little effect on the pH, the pH is determined by whichever is in excess, the strong acid or the strong base. We calculate the amount of acid or base in excess, then determine the concentration of the excess reagent from the reaction stoichiometry and the total volume of solution. Finally, the pH can be calculated from that concentration.

Procedure
The procedure combines three concepts we have studied so far: reaction stoichiometry, molar concentration and pH.

Use the reaction stoichiometry to find the number of moles of excess acid or base (Toolboxes 4.1 and 4.4)

Step 1. Calculate the moles of H_3O^+ ions (if the analyte is a strong acid) or OH^- ions (if the analyte is a strong base) in the original analyte solution from its molarity and its volume.

Step 2. Calculate the moles of OH^- ions (if the titrant is a strong base) or H_3O^+ ions (if the titrant is a strong acid) in the volume of titrant added.

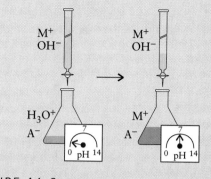

FIGURE 16.8

The composition of the solutions initially (left) and at the stoichiometric point (right) in the titration of a strong acid (the analyte) with a strong base (the titrant).

Step 3. Write the chemical equation for the neutralization reaction and use the reaction stoichiometry to find the moles of H_3O^+ ions (or OH^- ions if the analyte is a strong base) that remain in the analyte solution. Each mole of H_3O^+ ions reacts with 1 mol OH^- ions; therefore, subtract the number of moles of H_3O^+ or OH^- ions that have reacted from the initial number of moles of H_3O^+ or OH^- ions.

Determine the concentration (Toolbox 4.4)

Step 4. Divide the remaining number of moles of H_3O^+ (or OH^- if the analyte is a strong base) by the total volume of the combined solutions (analyte plus added titrant), which gives the molarity of the H_3O^+ (or OH^-) ions in the solution.

Calculate the pH (Toolbox 15.1)

Step 5. Take the negative logarithm of the relevant molarity to find the pH (or the pOH if the analyte is a strong base). Convert pOH to pH by using the relation pH + pOH = 14.00.

Example 16.4 *Calculating points on the pH curve for a strong acid-strong base titration*

Suppose we carry out a titration in which 0.340 M HCl(aq) is the titrant and the analyte initially consists of 25.0 mL of 0.250 M NaOH(aq). What is the pH of the analyte solution after the addition of 5.00 mL of titrant?

Strategy After the addition of 5.00 mL of the acid titrant, we can expect the pH to have decreased from its initial value. Follow the procedure in Toolbox 16.2.

Solution **Step 1.** Initially, the pOH of the analyte is pOH = $-\log 0.250$ = 0.602, so the pH of the solution is pH = 14.00 − 0.602 = 13.40. This is point A in Fig. 16.6. The amount of OH^- ions initially present is

$$\text{Moles of } OH^- \text{ (from the base)} = (25.00 \times 10^{-3} \text{ L}) \times (0.250 \text{ mol/L})$$
$$= 6.25 \times 10^{-3} \text{ mol}$$

or 6.25 mmol (because 1 mmol = 10^{-3} mol).
Step 2. The amount of H_3O^+ ions supplied by the titrant is

$$\text{Moles of } H_3O^+ \text{ (from the acid)} = (5.00 \times 10^{-3} \text{ L}) \times (0.340 \text{ mol/L})$$
$$= 1.70 \times 10^{-3} \text{ mol}$$

or 1.70 mmol.
Step 3. After reaction of all the H_3O^+ ions added, the amount of OH^- remaining is

$$(6.25 - 1.70) \text{ mmol} = 4.55 \text{ mmol}$$

Step 4. Because the total volume of the solution is now 30.00 mL, or 0.030 00 L, the molarity of OH^- is

$$\text{Molarity of } OH^- = \frac{4.55 \times 10^{-3} \text{ mol}}{0.030\ 00 \text{ L}} = 0.152 \text{ mol/L}$$

Step 5. Because pOH $= -\log 0.152 = 0.82$, we know that pH $= 13.18$, point B in Fig. 16.6. Note that the pH has fallen, as expected, but only by a very small amount. This small change is consistent with the shallow slope of the pH curve at the start of the titration.

Self-Test 16.5A What is the pH of the solution that results from the addition of a further 5.00 mL of the HCl(aq) titrant to the analyte?

[*Answer:* 12.911, or 12.91, point C]

Self-Test 16.5B What is the pH of the solution that results from the addition of another 2.00 mL of titrant to the analyte?

The technique illustrated in Example 16.4 can be applied at any stage of the calculation. For example, to demonstrate the sharp change in pH close to the stoichiometric point, we could calculate the pH of the solution at the stoichiometric point itself (that is easy: because the salt produced by the reaction of a strong acid and a strong base is neutral, the solution is neutral, so pH $= 7$) and at a nearby point. The volume of titrant needed to reach the stoichiometric point is calculated from the stoichiometry of the reaction by using the techniques described in Toolbox 4.6. We know that 6.25 mmol NaOH react completely with 6.25 mmol HCl. Therefore,

$$\text{Volume of acid} = \frac{\text{moles of HCl}}{\text{molarity of HCl}} = \frac{6.25 \times 10^{-3} \text{ mol}}{0.340 \text{ mol/L}}$$
$$= 18.4 \times 10^{-3} \text{ L}$$

The stoichiometric point occurs when 18.4 mL of titrant has been added. In the following example, we calculate the pH for a solution to which 19.4 mL has been added, which takes it 1.00 mL beyond the stoichiometric point.

> *In the titration of a strong acid with a strong base (or a strong base with a strong acid), the pH increases (or decreases) slowly initially, increases (or decreases) rapidly through pH $= 7$ at the stoichiometric point, and then increases (or decreases) slowly again.*

Example 16.5 *Calculating the pH after the stoichiometric point of a strong base-strong acid titration*

Calculate the pH of the solution in Example 16.4 after 19.4 mL of the acid has been added.

Strategy After the stoichiometric point, all the base in the analyte has reacted with acid. Because the amount of acid added now exceeds the amount of base in the analyte, we expect a pH of less than 7. We can use the reaction stoichiometry to determine how much of the acid added remains after neutralization. Then we use the total volume of solution to find the molar concentration of H_3O^+ and convert it to pH.

Solution The moles of H_3O^+ ions supplied by the solution is

$$\text{Moles of } H_3O^+ \text{ supplied} = (19.4 \times 10^{-3} \text{ L}) \times (0.340 \text{ mol/L})$$
$$= 6.60 \times 10^{-3} \text{ mol, or } 6.60 \text{ mmol}$$

The number of moles of OH^- ions initially present (from Example 16.4) is 6.25 mmol. More moles of acid have been supplied than there were moles of base originally, so we know that we are past the stoichiometric point. After all the OH^- ions have reacted,

$$\text{Moles of } H_3O^+ \text{ remaining} = (6.60 - 6.25) \text{ mmol} = 0.35 \text{ mmol}$$

Because the total volume of the solution is 25.0 + 19.4 mL, or 44.4 mL, which is 0.0444 L, the molar concentration of H_3O^+ is

$$\text{Molarity of } H_3O^+ = \frac{3.5 \times 10^{-4} \text{ mol}}{0.0444 \text{ L}} = 7.9 \times 10^{-3} \text{ mol/L}$$

Hence,

$$pH = -\log(7.9 \times 10^{-3}) = 2.10$$

A pH of 2.10, or about 2.1 (point D), is well below the pH at the stoichiometric point, although only 1 mL more acid has been added.

Self-Test 16.6A Calculate the pH of the solution after the addition of 20.4 mL of titrant.

[***Answer:*** 1.82, or about 1.8, point E]

Self-Test 16.6B Calculate the pH of the solution after the addition of 25.0 mL of titrant.

16.5 Weak Acid-Strong Base and Strong Acid-Weak Base Titrations

At the stoichiometric point of the titration of formic acid, HCOOH (**1**), with sodium hydroxide, the solution consists of sodium formate, $NaHCO_2$, and water. Formate ion, HCO_2^- (**2**), is a base, and the Na^+ ions have virtually no effect on pH; so overall the solution is basic, even though the acid has been completely neutralized. Because the solute is a salt with a basic anion, we can therefore expect the pH to be greater than 7 at the stoichiometric point. A good rule of thumb is that a strong base dominates a weak acid, so the solution is basic.

At the stoichiometric point of the titration of aqueous ammonia with hydrochloric acid, the solute is ammonium chloride. Because NH_4^+ is an acid, we expect the solution to be acidic with pH less than 7. The same is true for the titration of any weak base and strong acid. The rule of thumb is that a strong acid dominates a weak base, so the solution is acidic at the stoichiometric point.

1   Formic acid, HCOOH

2   Formate ion, HCO_2^-

Toolbox 16.3 *How to calculate the pH during a titration of a weak acid or weak base*

This Toolbox shows how to calculate the pH of a solution of a weak acid or weak base at any point in a titration.

Conceptual Basis

As strong base is added to a solution of a weak acid, a salt of the conjugate base of the weak acid is formed. This salt

also affects the pH and needs to be taken into account. The pH is determined by the major solute species present in solution. If the weak acid is HA and the base a hydroxide, MOH, where M is a metal ion, then the major species are (Fig. 16.9)

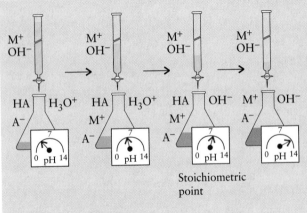

FIGURE 16.9

The composition of the solutions in the course of the titration of a weak acid with a strong base. From left to right: at the start of the titration, just before the stoichiometric point, at the stoichiometric point, and well after the stoichiometric point. At the stoichiometric point, the OH^- ions are mainly those arising from proton transfer from H_2O to A^-.

Initial solution: HA, H_3O^+, A^-. Find the pH from the concentration of the weak acid.

Before the stoichiometric point: HA, A^-, H_3O^+, M^+. The amount of titrant is less than required to react with the amount of analyte present. Find the pH from the concentrations of the excess weak acid and the conjugate base formed.

At the stoichiometric point: M^+, A^-. The amount of titrant is exactly the amount required to react with the amount of analyte present. The system now consists of a solution of the salt formed in the neutralization reaction. The concentration of this salt determines the pH, as A^- and H_2O form HA and OH^-.

After the stoichiometric point: M^+, OH^-, A^-. The amount of titrant is greater than required to react with the amount of analyte present, so the stoichiometric point has been passed and the pH is determined by the excess titrant present.

For the titration of a weak base by a strong acid, the procedure is essentially the same, except that the pH of the solution before the stoichiometric point is determined from the concentration of the excess weak base and any conjugate acid formed.

Procedure

Use the reaction stoichiometry to find the number of moles of excess acid or base (Toolboxes 4.1 and 4.4)

Step 1. Calculate the moles of weak acid (or weak base) in the original analyte solution from its molarity and its volume.

Step 2. Calculate the moles of OH^- ions (or H_3O^+ ions if the analyte is a weak base) in the volume of titrant added.

Step 3. Write the chemical equation for the reaction between acid and base and use the reaction stoichiometry to calculate the following:

Weak acid with a strong base: the number of moles of conjugate base formed by the reaction of the acid with added base, and the moles of weak acid remaining.

Weak base with a strong acid: the number of moles of conjugate acid formed by the reaction of the base with added acid, and the moles of weak base remaining.

Determine the concentration (Toolbox 4.4)

Step 4. Find the molarities of the conjugate acid and base in solution by dividing the number of moles of each species by the total volume of the solution.

Set up an equilibrium table (Toolbox 16.1 and Example 16.3)

Step 5. For a weak acid, use an equilibrium table to find the H_3O^+ ion concentration and, in each case, assume that the contribution of the autoprotolysis of water to the pH is insignificant if the pH is less than 6 or greater than 8. If necessary, convert between K_a and K_b by using $K_w = K_a \times K_b$.

Calculate the pH (Toolbox 15.1) Convert the hydronium ion concentration to pH. For a weak base, use an equilibrium table to find the OH^- ion concentration and convert first to pOH, then to pH.

Example 16.6 *Estimating the pH at the stoichiometric point of the titration of a weak acid with a strong base*

Estimate the pH at the stoichiometric point of the titration of 25.00 mL of 0.100 M HCOOH(aq) with 0.150 M NaOH(aq).

Strategy The salt present at the stoichiometric point is sodium formate, $NaHCO_2$. The pH of the salt solution is calculated as described in Toolbox 16.1 and Example 16.2. The K_b of HCO_2^- is related to the K_a of its conjugate acid HCOOH by $K_a \times K_b = K_w$; K_a is listed in Table 15.3.

Solution **Steps 1 and 2.** The number of moles of HCOOH in the initial analyte solution of volume 25.00 mL (2.500×10^{-2} L) is

$$\text{Moles of HCOOH} = (2.500 \times 10^{-2}\,\text{L}) \times (0.100\,\text{mol/L})$$
$$= 2.50 \times 10^{-3}\,\text{mol}$$

or 2.50 mmol. Because all the HCOOH has reacted to form HCO_2^- ions, the number of moles of HCO_2^- at the stoichiometric point is also 2.50 mmol.

Step 3. The stoichiometry of the reaction requires 1 mol OH^- for 1 mol HCOOH:

$$HCOOH(aq) + NaOH(aq) \longrightarrow NaHCO_2(aq) + H_2O(l)$$

Step 4. We need the volume of 0.150 M NaOH(aq) containing that number of moles of OH^- ions:

$$\text{Volume} = \frac{2.50 \times 10^{-3}\,\text{mol}}{0.150\,\text{mol/L}} = 16.7 \times 10^{-3}\,\text{L}$$

or 16.7 mL. Therefore, the total volume of the combined solutions at the stoichiometric point is $25.00 + 16.7\,\text{mL} = 41.7\,\text{mL}$. It follows that, at the stoichiometric point, we have a solution of sodium formate:

$$\text{Molarity of } HCO_2^- = \frac{2.50 \times 10^{-3}\,\text{mol}}{41.7 \times 10^{-3}\,\text{L}} = 0.0600\,\text{mol/L}$$

Because the formate ion is a weak base, the equilibrium to consider is

$$HCO_2^-(aq) + H_2O(l) \rightleftharpoons HCOOH(aq) + OH^-(aq)$$

From Table 15.3, $K_a = 1.8 \times 10^{-4}$ for formic acid; therefore, $K_b = K_w/K_a = 5.6 \times 10^{-11}$. Now set up the equilibrium table, with all concentrations in moles per liter:

	Species		
	HCO_2^-	HCOOH	OH^-
Step 1. Initial molarity	0.0600	0	0
Step 2. Change in molarity	$-x$	$+x$	$+x$
Step 3. Equilibrium molarity	$0.0600 - x$	x	x

Step 4. Substitution of the values from step 3 into the expression for K_b gives

$$5.6 \times 10^{-11} = \frac{[HCOOH][OH^-]}{[HCO_2^-]} = \frac{x \times x}{0.0600 - x}$$

If we suppose that x is less than 5% of 0.0600, we can write

$$5.6 \times 10^{-11} \approx \frac{x^2}{0.0600}$$

and obtain

$$x \approx \sqrt{0.0600 \times K_b} = \sqrt{0.0600 \times (5.6 \times 10^{-11})} = 1.8 \times 10^{-6}$$

It follows from step 3 in the equilibrium table that the molarity of OH^- is 1.8×10^{-6} mol/L, which is about 18 times greater than the molarity of OH^- ions that come from the autoprotolysis of water (1.0×10^{-7} mol/L), so the neglect of the latter is reasonable. At this stage, we can write

$$pOH = -\log(1.8 \times 10^{-6}) = 5.74$$

and therefore

$$pH = 14.00 - 5.74 = 8.26$$

or about 8.3.

Notice that the pH at the stoichiometric point is greater than 7, because the solution at that point consists of the salt formed in the neutralization reaction. The salt is basic, because it contains a basic anion (the formate ion) and a neutral cation.

Self-Test 16.7A Calculate the pH at the stoichiometric point of the titration of 25.00 mL of 0.020 M $NH_3(aq)$ with 0.015 M HCl(aq). (For NH_4^+, $K_a = 5.6 \times 10^{-10}$.)

[*Answer:* 5.66, or about 5.7]

Self-Test 16.7B Calculate the pH at the stoichiometric point of the titration of 25.00 mL of 0.010 M HClO(aq) with 0.020 M KOH(aq). See Table 15.3 for K_a.

The complete pH curves for two typical types of titrations are shown in Figs. 16.10 and 16.11. As for strong acid–strong base titrations, there is a sharp change in pH close to the stoichiometric point. The rapid change acts like a switch and gives a clear signal that the titration has passed through the stoichiometric point. The slow change in pH about halfway to the stoichiometric point is another characteristic feature. This feature is also of great practical importance, because it accounts for the stabilization of the pH of solutions, such as that of oceans, lakes, cell fluids, and blood plasma.

We have already seen how to estimate both the pH of the initial analyte when only weak acid or weak base is present (point A in Fig. 16.10, for instance) and the pH at the stoichiometric point (point S). Other points correspond to a mixed solution of some weak acid (or base) and some salt. We can therefore use the techniques described in Toolbox 16.2 and Examples 16.1–16.3 to predict the shape of the entire curve.

Example 16.7 *Calculating the pH before the stoichiometric point in a weak acid-strong base titration*

Calculate the pH of the solution resulting when 5.00 mL of 0.150 M NaOH(aq) is added to 25.00 mL of 0.100 M HCOOH(aq). Use $K_a = 1.8 \times 10^{-4}$ for HCOOH.

Strategy First, write the chemical equation for the reaction and use the reaction stoichiometry to calculate the number of moles of HCO_2^- ions formed by the reaction of the acid with added base and the moles of HCOOH remaining. Then find the molarities of HCOOH and HCO_2^- in the solution by dividing the number of moles by the total volume of the solution. Finally, use an equilibrium table to find the pH as in Example 16.3.

Solution **Step 1.** The initial amount of HCOOH in the analyte is (25.00 mL) × (0.100 mol/L) = 2.50 mmol HCOOH.

Step 2. The amount of OH^- in 5.00 mL of the titrant is (5.00 mL) × (0.150 mol/L) = 0.750 mmol OH^-.

Step 3. The chemical equation

$$HCOOH(aq) + OH^-(aq) \longrightarrow HCO_2^-(aq) + H_2O(l)$$

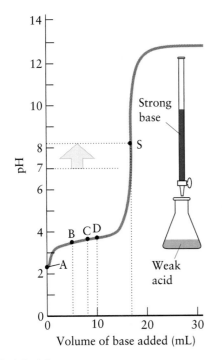

FIGURE 16.10

The pH curve for the titration of a weak acid with a strong base. This curve is for the titration of 25.00 mL of 0.100 M HCOOH(aq) with 0.150 M NaOH(aq). The stoichiometric point (S) occurs on the basic side of pH = 7 because the anion HCO_2^- is a base.

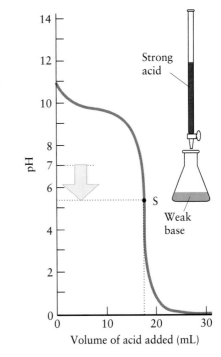

FIGURE 16.11

A typical pH curve for the titration of a weak base with a strong acid. The stoichiometric point (S) occurs on the acidic side of pH = 7 because the salt formed by the neutralization reaction has an acidic cation.

shows that 1 mol OH^- ≏ 1 mol HCOOH and 1 mol OH^- ≏ 1 mol HCO_2^-. So, 0.750 mmol OH^- produces 0.750 mmol HCO_2^- and leaves 2.50 − 0.750 mmol = 1.75 mmol HCOOH.

Step 4. The total volume of the solution at this stage is (25.00 + 5.00) mL = 30.00 mL, so the molarities of acid and conjugate base are

$$\text{Molarity of HCOOH} = \frac{1.75 \times 10^{-3}\,\text{mol}}{30.00 \times 10^{-3}\,\text{L}} = 0.0583\,\text{mol/L}$$

$$\text{Molarity of } HCO_2^- = \frac{7.50 \times 10^{-4}\,\text{mol}}{30.00 \times 10^{-3}\,\text{L}} = 0.0250\,\text{mol/L}$$

Step 5. The proton transfer equilibrium for HCOOH in water is

$$HCOOH(aq) + H_2O(l) \rightleftharpoons H_3O^+(aq) + HCO_2^-(aq) \qquad K_a = \frac{[H_3O^+][HCO_2^-]}{[HCOOH]}$$

The equilibrium table, with concentrations in moles per liter, is

	Species		
	HCOOH	**H_3O^+**	**HCO_2^-**
Step 1. Initial molarity	0.0583	0	0.0250
Step 2. Change in molarity	−x	+x	+x
Step 3. Equilibrium molarity	0.0583 − x	x	0.0250 + x

Step 4. Substitution of the values from step 3 and $K_a = 1.8 \times 10^{-4}$ into the expression for K_a gives

$$1.8 \times 10^{-4} = \frac{x \times (0.0250 + x)}{0.0583 - x}$$

Assuming that x is less than 5% of 0.0250 (and therefore also less than 5% of 0.0583), we write

$$1.8 \times 10^{-4} \approx \frac{x \times 0.0250}{0.0583}$$

The solution is $x \approx 4.2 \times 10^{-4}$. This value is less than 5% of 0.0250, so the assumptions are valid. It follows that the equilibrium H_3O^+ molarity is 4.2×10^{-4} mol/L. Therefore, the pH of the solution is

$$pH = -\log(4.2 \times 10^{-4}) = 3.38$$

or about 3.4. This is point B in Fig. 16.10. As anticipated, the pH of the mixed solution is higher than that of the original acid.

Self-Test 16.8A Calculate the pH of the solution after the addition of another 5.00 mL of 0.150 M NaOH(aq).

[*Answer:* 3.92, or about 3.9, point D]

Self-Test 16.8B Calculate the pH of the solution after the addition of yet another 5.00 mL of 0.150 M NaOH(aq).

In the region of the pH curve between points B and D in Fig. 16.10, the pH changes very slowly as base is added. Both HCOOH molecules and HCO_2^- ions are present in similar concentrations in solutions corresponding to the slowly changing region of the pH curve. When base is added from the buret, the HCOOH molecules supply protons to react with the OH^- ions being added, and the pH of the solution barely changes.

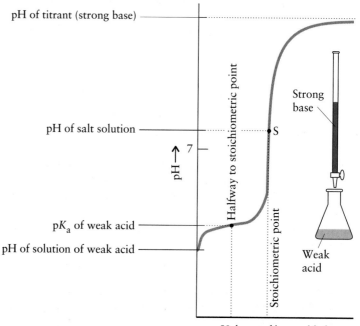

FIGURE 16.12

The pK_a of an acid can be determined by carrying out a titration of the weak acid with a strong base (or vice versa) and locating the pH of the solution after the addition of half the volume of titrant needed to reach the stoichiometric point.

Halfway to the stoichiometric point, half the acid molecules have been converted to their conjugate base and $[HCO_2^-] = [HCOOH]$. The expression for the acidity constant then becomes

$$K_a = \frac{[H_3O^+][HCO_2^-]}{[HCOOH]} = [H_3O^+]$$

because $[HCO_2^-]$ and $[HCOOH]$ cancel. Therefore, by taking the negative logarithm of both sides, we see that, *at the halfway point of a weak acid-strong base titration,*

$$pH = pK_a$$

For the formic acid titration, $pK_a = 3.75$, so halfway to the stoichiometric point, $pH = 3.75$ (point C in Fig. 16.10). The same expression applies to the titration of a weak base with a strong acid, but now K_a refers to the conjugate acid of the weak base. For example, if we were to titrate NH_3 with HCl, halfway to the stoichiometric point, $pH = pK_a$ of NH_4^+.

It is now simple to determine pK_a: we plot the pH curve during a titration and then identify the pH halfway to the stoichiometric point (Fig. 16.12). To obtain the pK_b of a weak base, we find pK_a the same way but go on to use $pK_a + pK_b = pK_w$. The values recorded in Table 15.3 were obtained by this method.

Well after the stoichiometric point in the titration of a weak acid with a strong base, the pH depends only on the concentration of excess strong base. For example, suppose we went on to add several liters of strong base from a giant buret. The presence of salt—the product of the neutralization reaction—would be negligible relative to the concentration of excess base. The pH would be that of the nearly pure titrant (the original base solution).

Figure 16.12 summarizes the changes in pH of a solution during a titration of a weak acid. Halfway to the stoichiometric point, the pH is equal to the pK_a of the acid present in the solution. The pH is greater than 7 at the stoichiometric point of the titration of a weak acid and strong base. The pH is less than 7 at the stoichiometric point of the titration of a weak base and strong acid (Fig. 16.11).

16.6 Titrating a Polyprotic Acid

The titration of a polyprotic acid follows the same rules used for monoprotic acid titrations, with one exception: if the values of the successive K_as are sufficiently different in magnitude, we can detect as many stoichiometric points in the titration as there are acidic hydrogen atoms in a polyprotic acid molecule. For example, suppose we are titrating the triprotic acid H_3PO_4 with a solution of NaOH. The experimentally determined pH curve is shown in Fig. 16.13. Notice that there are three stoichiometric points (B, D, and F). As we add the hydroxide solution, we assume that initially NaOH reacts completely with the acid to form the diprotic conjugate base $H_2PO_4^-$:

$$H_3PO_4(aq) + OH^-(aq) \longrightarrow H_2PO_4^-(aq) + H_2O(l)$$

Once all the H_3PO_4 molecules have lost one acidic proton, the system is at point B in Fig. 16.13, and the primary species in solution are the diprotic conjugate base and sodium ion—we have a solution of $NaH_2PO_4(aq)$. Point B is the first

FIGURE 16.13

The variation of the pH of the analyte solution during the titration of a triprotic acid (phosphoric acid) and the major species present in solution. SP1, SP2, and SP3 indicate the stoichiometric points. Points A, C and E are the points at which pH = pK_a for each deprotonation. Points B, D, and F are discussed in the text.

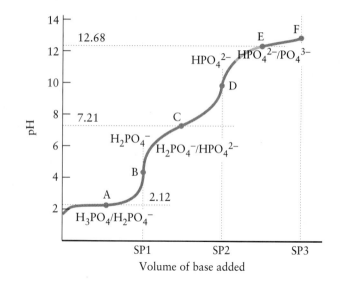

stoichiometric point, and to reach it, we need to supply one mole of base for each mole of acid.

As we continue to add base, it reacts with the dihydrogen phosphate ion to form the latter's conjugate base, HPO_4^{2-}. Enough base takes us to the second stoichiometric point, Point D. The primary species in solution at this point are the monoprotic HPO_4^{2-} anions and the Na^+ cations, which form a solution of $Na_2HPO_4(aq)$. To reach the second stoichiometric point, an additional mole of base is required for each mole of acid originally present. A total of 2 mol of base for each mole of acid has now been added:

$$H_2PO_4^-(aq) + OH^-(aq) \longrightarrow HPO_4^{2-}(aq) + H_2O(l)$$

Additional base reacts with HPO_4^{2-} to produce phosphate ion, PO_4^{3-}. When this reaction is complete, the primary species in solution are phosphate ions and sodium ions, which form a solution of $Na_3PO_4(aq)$. To reach this stoichiometric point, we have to add another mole of base for each mole of acid present (point F in the plot).

$$HPO_4^{2-}(aq) + OH^-(aq) \longrightarrow PO_4^{3-}(aq) + H_2O(l)$$

At this point, a total of 3 mol OH^- has been added for each mole of H_3PO_4. Notice that the third stoichiometric point is so indistinct that it cannot be detected, largely because K_{a3} is comparable to K_w.

The titration of a polyprotic acid has a stoichiometric point corresponding to the removal of each proton. The product at each stoichiometric point is the salt of a conjugate base of the acid or one of its hydrogen-bearing anions.

Self-Test 16.9A What volume of 0.010 M NaOH(aq) is required to reach the (a) first; (b) second stoichiometric point in a titration of 25.00 mL of 0.010 M $H_2SO_3(aq)$?

[*Answer:* (a) 25.00 mL; (b) 50.00 mL]

Self-Test 16.9B What volume of 0.020 M NaOH(aq) is required to reach the (a) first; (b) second; (c) third stoichiometric point in a titration of 30.00 mL of 0.010 M $H_3PO_4(aq)$?

An automatic titrator monitors the pH of the analyte solution continuously and detects the stoichiometric point by the characteristic rapid change in pH (Fig. 16.14). When automatic equipment is not available, the stoichiometric point can be detected with a pH meter or by observing the change in color of an indicator. We first met indicators in Section 3.7. We can now explain their color changes in the context of Brønsted-Lowry theory and proton transfer equilibria.

An **acid-base indicator** is a weak acid that has one color in its acid form (HIn, where In stands for indicator) and another color in its conjugate base form (In^-). The color change occurs because the acidic proton in the acid form, HIn, modifies the structure of the molecule in such a way that the light absorption characteristics of HIn are different from those of In^-. When the concentration of HIn is much greater than that of In^-, the solution has the color of the acid form of the indicator. When the concentration of HIn is much less than that of In^-, the solution has the color of the base form of the indicator. Because the indicator is itself an acid, only a few drops are used, so as not to upset the accuracy of the titration.

Because it is a weak acid, an indicator in solution takes part in a proton transfer equilibrium:

$$HIn(aq) + H_2O(l) \rightleftharpoons H_3O^+(aq) + In^-(aq) \qquad K_{In} = \frac{[H_3O^+][In^-]}{[HIn]}$$

The **end point** of a titration is defined as the point at which the concentrations of the acid and base forms of the indicator are equal: $[HIn] = [In^-]$. At this point, the color of the indicator is halfway between the colors of its acid and base forms. If we substitute this equality into the preceding expression, we see that $[H_3O^+] = K_{In}$ at the end point. That is, the color change occurs when

$$pH = pK_{In}$$

> Be careful to distinguish the end point of a titration from the stoichiometric point. The former refers to the color change of the indicator; the latter refers to the stage at which reaction is complete.

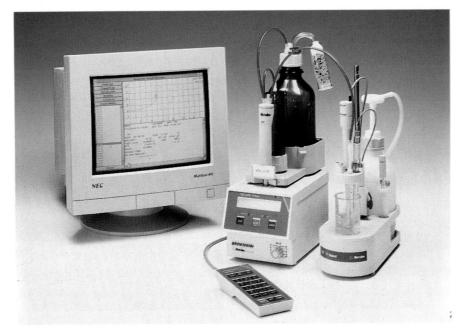

FIGURE 16.14

A commercially available automatic titrator. The stoichiometric point of the titration is detected by a sudden change in pH; the pH is monitored electrically, using a technique described in Investigating Chemistry 18.1. The pH can be plotted as the reaction proceeds, as shown on the left.

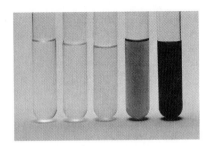

FIGURE 16.15

The stoichiometric point of an acid-base titration may be detected by the color change of an indicator. Here we see the colors of solutions containing a few drops of phenolphthalein at (from left to right) pHs of 7.0, 8.5, 9.4 (its end point), 9.8, and 12.0. At the end point, equal amounts of the conjugate acid and base forms of the indicator are present.

One common indicator is phenolphthalein. The acid form of this large organic acid molecule (**3**) is colorless; its conjugate base form (**4**) is pink (Fig. 16.15). The pK_{In} of phenolphthalein is 9.4, so the end point occurs in slightly basic solution. The change from colorless to pink starts at pH = 8.2 and is complete by pH = 10. Litmus, another well-known indicator, has pK_{In} = 6.5; it is red for pH < 5 and blue for pH > 8.

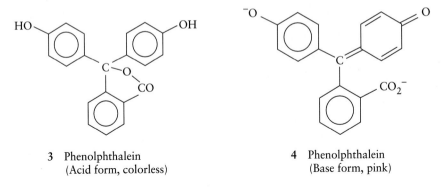

3 Phenolphthalein
(Acid form, colorless)

4 Phenolphthalein
(Base form, pink)

There are many naturally occurring indicators, such as extract of red cabbage (see Fig. 3.19) and the colors of many flowers. For example, a single compound is responsible for the colors of red poppies and blue cornflowers (Fig. 16.16): the pH of the sap is different in the two plants.

It is important when planning a titration to select an indicator with an end point close to the stoichiometric point (Fig. 16.17). In practice, the pK_{In} of the indicator should be within about 1 pH unit of the stoichiometric point of the titration:

$$pK_{In} \approx pH(\text{stoichiometric point}) \pm 1$$

Phenolphthalein can be used for titrations with a stoichiometric point near pH = 9, such as a titration of a weak acid with a strong base (Fig. 16.18). Methyl orange changes color between pH = 3.2 and pH = 4.4 and can be used in the titration of a weak base with a strong acid (Fig. 16.19). Ideally, indicators for strong acid-strong base titrations should have end points close to pH = 7; however, in strong acid-strong base titrations, the pH changes rapidly over several pH units, and even phenolphthalein can be used. The properties of several common indicators are listed in Table 16.3.

Acid-base indicators are weak acids that change color close to pH = pK_{In}; an indicator should be chosen so that its end point is close to the stoichiometric point of the titration.

FIGURE 16.16

The same dye is responsible for the red of poppies (a) and the blue of corn-flowers (b). The color difference is a consequence of the more acidic sap of poppies.

(a)   (b)

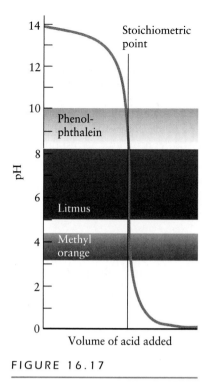

FIGURE 16.17

Ideally, an indicator should have a sharp color change close to the stoichiometric point of the titration (at pH = 7 for a strong acid-strong base titration). However, the change in pH is so abrupt that phenolphthalein can also be used. The color change of methyl orange, however, would give a less accurate result.

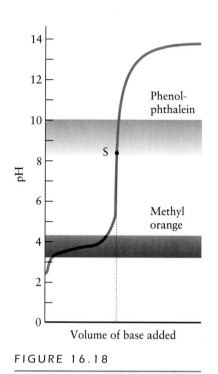

FIGURE 16.18

Phenolphthalein can be used to detect the stoichiometric point of a weak acid-strong base titration, but methyl orange would give a very inaccurate indication of the stoichiometric point. The pH curves are superimposed on approximations to the colors of the indicators in the neighborhoods of their end points.

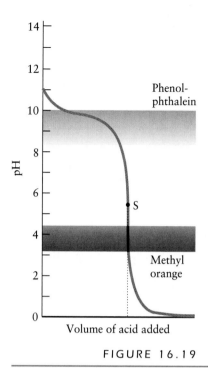

FIGURE 16.19

Methyl orange can be used for a weak base-strong acid titration. Phenolphthalein would be inappropriate because its color change occurs well away from the stoichiometric point. The pH curves are superimposed on approximations to the colors of the indicators in the neighborhoods of their end points.

Table 16.3 *Indicator color changes*

Indicator	Color of acid form	pH range of color change	pK_{In}	Color of base form
thymol blue	red	1.2 to 2.8	1.7	yellow
	yellow	8.0 to 9.6		blue
methyl orange	red	3.2 to 4.4	3.4	yellow
bromophenol blue	yellow	3.0 to 4.6	3.9	blue
bromocresol green	yellow	3.8 to 5.4	4.7	blue
methyl red	red	4.8 to 6.0	5.0	yellow
bromothymol blue	yellow	6.0 to 7.6	7.1	blue
litmus	red	5.0 to 8.0	6.5	blue
phenol red	yellow	6.6 to 8.0	7.9	red
thymol blue	yellow	8.0 to 9.6	8.9	blue
phenolphthalein	colorless	8.2 to 10.0	9.4	pink
alizarin yellow R	yellow	10.1 to 12.0	11.2	red
alizarin	red	11.0 to 12.4	11.7	purple

BUFFER SOLUTIONS

Figure 16.12 shows how slowly the pH changes halfway to the stoichiometric point, close to pH $=$ pK_a. When a small amount of strong acid or base is added to a solution corresponding to this region, the pH changes much less than when the same amount is added to pure water. Solutions that resist changes in pH when small amounts of strong acid or base are added are called **buffers.** Buffers are a central component of the mechanisms of *homeostasis,* the self-regulating process by which organisms maintain stability but continue to be responsive to changes in their environment.

16.8 The Action of Buffers

An **acid buffer** solution consists of a weak acid and its conjugate base; it has pH $<$ 7. A **base buffer** solution consists of a weak base and its conjugate acid; it has pH $>$ 7. Buffer solutions are particularly important in biological systems. For example, blood and other cell fluids are buffered at pH $=$ 7.4. The oceans are maintained at about pH $=$ 8.4 by a complex buffering process that depends on the presence of hydrogen carbonates and silicates.

We can understand the action of buffers by considering how the acetic acid equilibrium

$$CH_3COOH(aq) + H_2O(l) \rightleftharpoons H_3O^+(aq) + CH_3CO_2^-(aq)$$

responds to the addition of a strong acid or base (Fig. 16.20). Suppose a strong acid is added to a solution that contains $CH_3CO_2^-$ ions and CH_3COOH molecules in about equal concentrations. As we see from Fig. 16.20, the newly arrived H_3O^+ ions transfer protons to the $CH_3CO_2^-$ ions to form CH_3COOH molecules. Only a small fraction of these newly formed CH_3COOH molecules deprotonate, so the pH is left nearly unchanged. If a small amount of strong base is added instead, the OH^- ions remove protons from the CH_3COOH molecules. As a result, the concentration of $CH_3CO_2^-$ increases, but the OH^- concentration remains nearly unchanged. Consequently, the H_3O^+ concentration (and the pH) is also left nearly unchanged. We can summarize this action as follows:

> *Acid buffer action*
>
> The weak acid transfers protons to the OH^- ions from added strong base.
>
> The conjugate base of the weak acid accepts protons from the H_3O^+ ions supplied by a strong acid.

Now consider the equilibrium

$$NH_3(aq) + H_2O(l) \rightleftharpoons NH_4^+(aq) + OH^-(aq)$$

in a solution that contains similar concentrations of $NH_3(aq)$ and $NH_4^+(aq)$ and which has been prepared by adding ammonium chloride to aqueous ammonia. When a strong base is added, the incoming OH^- ions remove protons from NH_4^+ ions to make NH_3 and H_2O molecules. When a strong acid is added, the incoming protons attach to NH_3 molecules to make NH_4^+ ions, and hence are removed from the solution. In each case, the pH is left almost the same. We can summarize this action as follows:

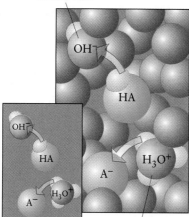
Added base

Added acid

FIGURE 16.20

Buffer action depends on the donation of protons by the weak acid molecules, HA, when a strong base is added and the acceptance of protons by the conjugate base ions, A^-, when a strong acid is added. In the inset, a solid blue color represents water.

Base buffer action

The weak base accepts protons from the H_3O^+ ions supplied by a strong acid.

The conjugate acid of the weak base transfers protons to the OH^- ions from added strong base.

The flattest part of the pH curve of a solution of a weak acid and its conjugate base occurs where their molarities are equal. The pH is on the acid side of neutral and equal to the pK_a of the weak acid. Such a mixture is an acid buffer that stabilizes the solution at pH $\approx pK_a$. Similarly, a mixture of a weak base and its conjugate acid is a base buffer that stabilizes the pH on the basic side of neutrality, at pH $\approx pK_a$, where K_a is the acidity constant of the conjugate acid of the base. Mixtures in which the salt and acid (or base) have unequal molarities are also buffers but are usually less effective than those in which the molarities are nearly equal. Table 16.4 lists several typical buffer systems.

> **A buffer is a mixture of weak conjugate acids and bases that stabilizes the pH of a solution by both accepting and providing protons for reaction with added acid or base.**

16.9 Selecting a Buffer

Buffer solutions are used to calibrate pH meters, to culture bacteria, and to control the pH of solutions in which chemical reactions are taking place. They are also administered intravenously to critically ill patients. The question we face is how does the pH depend on the composition of the buffer? Equivalently, how can we design a buffer with a specified pH?

Table 16.4 *Typical buffer systems*

Composition	pK_a
Acid buffers	
$CH_3COOH/CH_3CO_2^-$	4.74
HNO_2/NO_2^-	3.37
$HClO_2/ClO_2^-$	2.00
Base buffers	
NH_4^+/NH_3	9.25
$(CH_3)_3NH^+/(CH_3)_3N$	9.81
$H_2PO_4^-/HPO_4^{2-}$	7.21

Because acid buffer solutions are mixed solutions in the sense that they contain both an acid and the salt of its conjugate base, we can calculate their pH as illustrated in Example 16.3. For a solution containing a weak acid HA and a salt that provides the conjugate base anion, A^-, the proton transfer equilibrium is

$$HA(aq) + H_2O(l) \rightleftharpoons H_3O^+(aq) + A^-(aq) \qquad K_a = \frac{[H_3O^+][A^-]}{[HA]}$$

We now rearrange the expression for K_a into

$$[H_3O^+] = K_a \times \frac{[HA]}{[A^-]}$$

and take the negative logarithm of both sides:

$$-\log[H_3O^+] = -\log K_a - \log\left(\frac{[HA]}{[A^-]}\right)$$

That is,

$$pH = pK_a - \log\left(\frac{[HA]}{[A^-]}\right) = pK_a + \log\left(\frac{[A^-]}{[HA]}\right)$$

Because a weak acid HA typically loses only a tiny fraction of its protons and because only a tiny fraction of the anions of the weak base A^- accepts protons, we can approximate [HA] and $[A^-]$ by the molarities of the acid and the salt as added initially. Then

$$pH \approx pK_a + \log\left(\frac{[\text{base}]_{\text{initial}}}{[\text{acid}]_{\text{initial}}}\right)$$

This relation is called the **Henderson-Hasselbalch equation.** Because so many chemical reactions in our bodies occur in buffered environments, biochemists commonly use the equation for quick estimates of pH. In practice, we use the equation to estimate the concentrations of acid and conjugate base required to make a desired buffer. Then we adjust the pH to the value we require by adding more acid or base and monitoring the effect with a pH meter.

Toolbox 16.4 *How to calculate the pH of a buffer solution*

This Toolbox describes how to calculate the pH of a buffer solution and how to determine the effect of added acid or base on the pH of the solution.

Conceptual Basis
We can estimate the pH of a buffer solution from the initial concentrations of weak acid and conjugate base (or weak base and conjugate acid) because their relatively large concentrations usually dominate the hydronium ion concentration. A buffer solution maintains a relatively constant pH by providing both a weak acid and a weak base that act as a source and a sink, respectively, for protons (Fig. 16.21).

Procedure

Original buffer solution There are two equivalent methods for calculating the pH of a buffer. Either set up an equilibrium table to calculate the pH as in Example 16.3 or

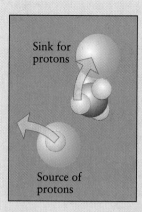

FIGURE 16.21

A buffer solution contains a sink for protons (a weak base) that are supplied when a strong acid is added and a source of protons (a weak acid) to supply to a strong base that is added. The joint action of the source and the sink keeps the pH constant when a small amount of either strong acid or strong base is added.

work through the calculation described in the text that led to the Henderson-Hasselbalch equation:

$$pH \approx pK_a + \log\left(\frac{[\text{base}]_{\text{initial}}}{[\text{acid}]_{\text{initial}}}\right)$$

If you use the Henderson-Hasselbalch equation, it is advisable to work through the derivation each time rather than to use this expression blindly; then approximations can be identified. Calculate the pH of the solution by substituting the initial concentrations of the conjugate acid and base that form the buffer solution.

Effect of added acid or base

Step 1. Determine the number of moles of conjugate acid and base in the original buffer solution by multiplying their molarities by the volume of the solution.

Step 2. Allow reaction between added strong base and the weak acid to occur to completion. The product is additional weak base. Find the amount of weak acid remaining by subtracting the amount reacted from the original amount. Find the new amount of weak base by adding the additional amount of weak base to the original amount.

If strong acid is added, it reacts with the weak base to produce additional weak acid. Find the amount of weak base remaining by subtracting the amount reacted from the original amount. Find the new amount of weak acid by adding the additional amount of weak acid to the original amount.

Step 3. Calculate the final molarities of the weak acid and weak base by dividing by the total volume of the solution.

Step 4. Write the proton transfer equilibrium and calculate the new pH either by substituting the final molarities in the equilibrium expression or the Henderson-Hasselbalch equation or by using the procedure in Example 16.3.

Example 16.8 *Calculating the pH of a buffer solution*

Calculate the pH of a buffer solution that is 0.040 M $NaCH_3CO_2(aq)$ and 0.080 M $CH_3COOH(aq)$ at 25°C.

Strategy Begin by identifying the weak acid and its conjugate base (the acid has one more H atom than the base). Then write the proton transfer equilibrium between them, rearrange the expression for K_a to give $[H_3O^+]$, and find the pH, as in Toolbox 16.4.

Solution **Step 1.** The acid is CH_3COOH and its conjugate base is $CH_3CO_2^-$. The equilibrium to consider is

$$CH_3COOH(aq) + H_2O(l) \rightleftharpoons H_3O^+(aq) + CH_3CO_2^-(aq)$$

$$K_a = \frac{[H_3O^+][CH_3CO_2^-]}{[CH_3COOH]}$$

Step 2. From Table 15.3, $pK_a = 4.75$. Rearranging the expression for K_a, we obtain first

$$[H_3O^+] = K_a \times \frac{[CH_3COOH]}{[CH_3CO_2^-]}$$

Step 3. Then we write

$$pH = pK_a + \log\left(\frac{[CH_3CO_2^-]}{[CH_3COOH]}\right)$$

Step 4. We now approximate the concentrations of acid and base by their initial values:

$$pH \approx 4.75 + \log\left(\frac{0.040}{0.080}\right) = 4.45$$

That is, the solution acts as a buffer close to pH $= 4.4$.

Self-Test 16.10A Calculate the pH of a buffer solution that is 0.040 M $NH_4Cl(aq)$ and 0.030 M $NH_3(aq)$.

[*Answer:* 9.13, or about 9.1]

Self-Test 16.10B Calculate the pH of a buffer solution that is 0.15 M $HNO_2(aq)$ and 0.20 M $NaNO_2(aq)$.

Example 16.9 *Calculating the pH change when acid or base is added to a buffer solution*

Suppose that 1.2 g of sodium hydroxide (0.030 mol NaOH) is dissolved in 500.0 mL of the buffer solution described in Example 16.8. Calculate the pH of the resulting solution and the change in pH. Assume that the volume is constant.

Strategy The 0.030 mol OH^- added to the buffer solution reacts with the acid of the buffer system, CH_3COOH, thereby decreasing its amount by 0.030 mol and increasing the amount of its conjugate base, $CH_3CO_2^-$, by 0.030 mol. There is no significant change in the volume of solution. Follow the procedure described in Toolbox 16.4 to find the pH of the final solution.

Solution **Step 1.** The number of moles of the acid CH_3COOH in the solution initially (from the data in Example 16.8) is

$$\text{Moles of } CH_3COOH = (0.5000 \text{ L}) \times (0.080 \text{ mol/L}) = 0.040 \text{ mol}$$

Step 2. The added sodium hydroxide reacts with the acetic acid:

$$CH_3COOH(aq) + OH^-(aq) \rightleftharpoons CH_3CO_2^-(aq) + H_2O(l)$$

The amount of CH_3COOH that reacts is

$$0.030 \text{ mol } OH^- \times \frac{1 \text{ mol } CH_3COOH}{1 \text{ mol } OH^-} = 0.030 \text{ mol } CH_3COOH$$

Therefore, the amount of CH_3COOH remaining is $0.040 - 0.030$ mol $= 0.010$ mol.
Step 3. The resulting molarity of CH_3COOH is now

$$\text{Molarity of } CH_3COOH = \frac{0.010 \text{ mol}}{0.5000 \text{ L}} = 0.020 \text{ mol/L}$$

Similarly, the initial number of moles of the base $CH_3CO_2^-$ in the 500.0 mL of solution (using data from Example 16.8) is

$$\text{Moles of } CH_3CO_2^- = (0.5000 \text{ L}) \times (0.040 \text{ mol/L}) = 0.020 \text{ mol}$$

The addition of 0.030 mol OH^- ions increases this amount to $0.020 + 0.030 \text{ mol} = 0.050 \text{ mol}$ through the reaction with CH_3COOH. Hence the molarity of the base $CH_3CO_2^-$ is increased to

$$\text{Molarity of } CH_3CO_2^- = \frac{0.050 \text{ mol}}{0.5000 \text{ L}} = 0.10 \text{ mol/L}$$

Step 4. The proton transfer equilibrium is

$$CH_3COOH(aq) + H_2O(l) \rightleftharpoons H_3O^+(aq) + CH_3CO_2^-(aq)$$

We assume that x is very small relative to 0.10 and 0.020; we obtain

$$pH \approx pK_a + \log\left(\frac{[CH_3CO_2^-]_{initial}}{[CH_3COOH]_{initial}}\right) = 4.75 + \log\left(\frac{0.10}{0.020}\right) = 5.45$$

or about 5.4. That is, the pH of the solution changes from about 4.4 to about 5.4. The small change is the result of the buffering action. If the solution had contained HCl at a pH of 4.4, the addition of the same amount of NaOH would have raised the pH to 12.8.

Self-Test 16.11A Suppose that 0.0100 mol HCl(g) is dissolved in 500.0 mL of the buffer solution of Example 16.8. Calculate the pH of the resulting solution and the change in pH.

[**Answer:** 4.1, a decrease of 0.3]

Self-Test 16.11B Suppose that 0.0200 mol NaOH(s) is dissolved in 300.0 mL of the buffer solution of Example 16.8. Calculate the pH of the resulting solution and the change in pH.

Commercially available buffer solutions can be purchased for virtually any desired pH. For example, pH meters are often calibrated by using a mixed solution of 0.025 M $Na_2HPO_4(aq)$ and 0.025 M $KH_2PO_4(aq)$, which has pH = 6.87 at 25°C. The method demonstrated in Example 16.8 predicts pH = 7.2 for this solution; however, as noted at the start of the chapter, these calculations ignore ion-ion interactions, which modify the pH slightly.

> **The pH of a buffer solution is close to the pK_a of the weak acid component when the conjugate acid and base have similar concentrations.**

Self-Test 16.12A Which of the buffer systems listed in Table 16.4 would be a good choice for preparing a buffer with pH = 5 ± 1?

[**Answer:** $CH_3COOH/CH_3CO_2^-$]

Self-Test 16.12B Which of the buffer systems listed in Table 16.4 would be a good choice for preparing a buffer with pH = 10 ± 1?

16.10 Buffer Capacity

The **buffer capacity** is an indication of the amount of acid or base that can be added before the buffer loses its ability to resist the change in pH. A buffer with a high capacity can maintain its buffering action after the addition of more strong acid or base than can one with a small capacity. The buffer becomes

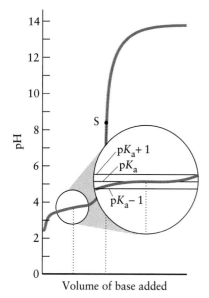

FIGURE 16.22

When conjugate acid and base are present at similar concentrations, the pH changes very little as more strong base (or strong acid) is added. As the inset shows, the pH lies between $pK_a \pm 1$ for a wide range of concentrations.

exhausted when most of the weak base has been converted to acid or when most of the weak acid has been converted to base.

A buffer has a high capacity when the amount of base present is at least 10% or more of the amount of acid, for otherwise the base gets used up quickly. Similarly, the amount of acid present should be at least 10% or more of the amount of base, for otherwise the acid gets used up quickly. The Henderson-Hasselbalch equation then shows us that the corresponding pH range of the buffer is from

$$pH \approx pK_a + \log\left(\frac{[A^-]_{initial}}{10[A^-]_{initial}}\right) = pK_a + \log\left(\frac{1}{10}\right) = pK_a - 1$$

when the acid is 10 times more abundant than the base, to

$$pH \approx pK_a + \log\left(\frac{10[HA]_{initial}}{[HA]_{initial}}\right) = pK_a + \log\left(\frac{10}{1}\right) = pK_a + 1$$

when the base is 10 times more abundant than the acid. The buffer can act effectively within this range (Fig. 16.22). Buffer capacity also depends on the absolute amount of conjugate acid and base available. A dilute buffer solution will have a lower buffer capacity than the same volume of a more concentrated solution of the same buffer.

The composition of blood plasma, with the concentration of HCO_3^- ions about 20 times that of H_2CO_3, seems to be outside the range for optimum buffering. However, the principal waste products of living cells are carboxylic acids, such as lactic acid. Plasma, with its relatively high concentration of HCO_3^- ions, can absorb a significant surge of hydrogen ions from the carboxylic acids. The high proportion of HCO_3^- also helps us to withstand disturbances that lead to excess acid, such as disease and shock due to burns (see Applying Chemistry: Case Study 16).

A buffer is most effective in the range $pK_a \pm 1$.

Example 16.10 Determining the ratio of molarities for a buffer solution with a given pH

Calculate the ratio of the molarities of CO_3^{2-} and HCO_3^- ions required to achieve buffering at pH = 9.50. The pK_{a2} of H_2CO_3 is 10.25.

Strategy The pH of an HA/A^- buffer is calculated from the Henderson-Hasselbalch equation:

$$pH \approx pK_a + \log\left(\frac{[A^-]_{initial}}{[HA]_{initial}}\right)$$

Identify the acid and base, and rearrange this expression to give the desired concentration ratio.

Solution The base is CO_3^{2-} and the acid is HCO_3^-; hence,

$$\log\left(\frac{[CO_3^{2-}]}{[HCO_3^-]}\right) \approx pH - pK_a = 9.50 - 10.25 = -0.75$$

The antilogarithm of -0.75 is 0.18, so the ratio of molarities is

$$\frac{[CO_3^{2-}]}{[HCO_3^-]} = 0.18$$

Therefore, the solution will act as a buffer with a pH close to 9.50 if it is prepared by mixing the solutes in the ratio 0.18 mol CO_3^{2-} to 1.0 mol HCO_3^-.

Self-Test 16.13A Calculate the ratio of the molarities of acetate ions and acetic acid needed to buffer a solution at pH = 5.25. The pK_a of CH_3COOH is 4.75.

[*Answer:* 3.2:1]

Self-Test 16.13B Calculate the ratio of the molarities of benzoate ions and benzoic acid (C_6H_5COOH) needed to buffer a solution at pH = 3.50. The pK_a of C_6H_5COOH is 4.19.

SOLUBILITY EQUILIBRIA

The pH is only one property of natural waters that must be monitored to ensure safe drinking water (see Connection 4, following this chapter). Natural waterways are inspected regularly to detect contamination by toxic substances, and tap water is treated with chemicals that disinfect and soften it. Some important treatment and monitoring procedures make use of the limited solubilities of certain salts (Fig. 16.23). By learning to predict just how soluble specific salts are, we can control when precipitates form and use precipitation in the laboratory to separate mixtures. The calculations involved are applications of equilibria: no matter how insoluble a salt, provided some solid salt is present, it is always in equilibrium with its dissolved ions in a saturated solution.

16.11 The Solubility Product

K_{sp} is also called the *solubility product constant,* the *solubility constant,* or simply the *ion product.*

The equilibrium constant for a solubility equilibrium is called the **solubility product,** K_{sp}, of the solute. For example, the solubility product for bismuth sulfide, Bi_2S_3, is

$$Bi_2S_3(s) \rightleftharpoons 2\,Bi^{3+}(aq) + 3\,S^{2-}(aq) \qquad K_{sp} = [Bi^{3+}]^2[S^{2-}]^3$$

In this expression, $[Bi^{3+}]$ and $[S^{2-}]$ are the numerical values of the molar concentrations of the ions in the saturated solution. Solid Bi_2S_3 does not appear in the expression for K_{sp} because it is a pure solid (Section 14.4).

A solubility product is no different from any other equilibrium constant and is used in the same way. However, because ion-ion interactions in concentrated electrolyte solutions can complicate its interpretation, a solubility product is generally used only for sparingly soluble salts. One of the simplest ways to

FIGURE 16.23

A pollution control officer collects samples of river water in England. The water will be tested for the presence of heavy metal ions by adding a solution containing anions such as sulfide ions. If a precipitate appears, further tests will be conducted to identify the particular ions present.

Applying Chemistry: Case Study 16

Buffer systems are so vital to the existence of living organisms that the most immediate threat to the survival of a person with severe injury or burns is a change in blood pH. One of a paramedic's first steps is therefore to administer intravenous fluids.

Metabolic processes normally maintain the pH of human blood within a narrow range (7.35–7.45). To control blood pH, the body uses primarily the carbonic acid-hydrogen carbonate (bicarbonate) ion system. The normal ratio of HCO_3^- to H_2CO_3 in the blood is 20:1, with most of the carbonic acid in the form of dissolved CO_2. When the

Three solutions commonly given intravenously: lactated Ringer's solution, 0.9% sodium chloride, and 5% dextrose (glucose). The first two help to control electrolyte levels, the third maintains the blood sugar level, and all three help to maintain blood volume.

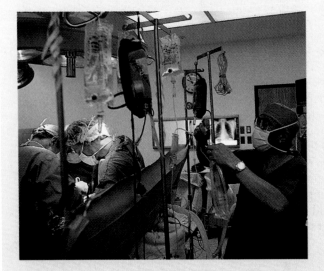

Patients undergoing surgery often need to be supplied with several kinds of intravenous solutions. These doctors are operating on a patient's liver, a procedure that requires the administration of three different intravenous solutions and a blood transfusion.

concentration of HCO_3^- increases further relative to that of H_2CO_3, blood pH rises. If the pH rises above the normal range, the condition is called *alkalosis*. Conversely, blood pH decreases when the ratio decreases; and when blood pH falls below the normal range, the condition is called *acidosis*. These conditions are life threatening and death can result within minutes.

Blood carbonic acid and hydrogen carbonate concentrations are controlled by independent mechanisms. Carbonic acid concentration is controlled by respiration: as we exhale, we deplete our system of CO_2, and hence deplete it of H_2CO_3, too. This decrease in acid concentration raises the blood pH. Breathing faster and more deeply increases the amount of CO_2 exhaled and hence decreases the carbonic acid concentration in the blood, which in turn raises

Molar solubility was introduced in Section 12.2.

determine K_{sp} is to measure the *molar solubility, S,* of the compound (the molarity of the compound in a saturated solution). In calculations of equilibrium constants, we use *s*, the numerical value of S (the value of S with the units moles per liter struck out). For example, if the molar solubility of a compound is 1.5×10^{-6} mol/L, we write $s = 1.5 \times 10^{-6}$. More advanced and accurate methods of determining K_{sp} are also available. Table 16.5 gives some experimental values.

> **The solubility product is the equilibrium constant for the equilibrium between an undissolved salt and its ions in a saturated solution.**

the blood pH. Hydrogen carbonate ion concentration is controlled mainly by its rate of excretion in urine.

Respiratory acidosis is the fall in pH that results when decreased respiration raises the concentration of CO_2 in the blood. Asthma, pneumonia, emphysema, or inhaling smoke can all cause respiratory acidosis. So can any condition that reduces a person's ability to breathe. Respiratory acidosis is usually treated with a mechanical ventilator, to assist the victim's breathing. The improved exhalation increases the excretion of CO_2 and raises blood pH. In many cases of asthma, chemicals can also facilitate respiration by opening constricted bronchial passages.

Metabolic acidosis is caused by the release into the bloodstream of excessive amounts of lactic acid and other acidic by-products of metabolism. These acids enter the bloodstream, react with hydrogen carbonate ion to produce H_2CO_3, and shift the ratio of HCO_3^- to H_2CO_3 to a lower value. Heavy exercise, diabetes, or fasting can produce metabolic acidosis. The normal response of the body is to increase the rate of breathing in order to get rid of some of the CO_2. Thus, we pant heavily when running uphill.

Metabolic acidosis can also result when a person is severely burned. Blood plasma leaks from the circulatory system into the injured area, producing swelling (edema) and reducing the blood volume. If the burned area is large, this loss of blood volume may be sufficient to reduce blood flow and oxygen supply to all the body's tissues. Lack of oxygen, in turn, causes the tissues to produce an excessive amount of lactic acid, causing metabolic acidosis. To minimize the decrease in pH, the injured individual breathes harder to eliminate the excess CO_2. However, if blood volume drops below levels for which the body can compensate, a vicious circle ensues in which blood flow decreases still further, blood pressure falls, CO_2 excretion diminishes, and acidosis becomes more severe. Individuals in this state are said to be in *shock* and will die if not treated promptly.

The dangers of shock are avoided or treated by intravenous infusion of large volumes of a salt-containing solution, usually one known as *lactated Ringer's solution*. The added liquid increases blood volume and blood flow, which improves oxygen delivery; the $[HCO_3^-]/[H_2CO_3]$ ratio then increases toward normal, thereby allowing the severely injured person to survive.

Respiratory alkalosis is the rise in pH associated with excessive respiration. Hyperventilation, which can result from anxiety or high fever, is a common cause. The body may control blood pH in a hyperventilating individual by fainting, which results in slower respiration. An intervention that may avoid fainting is to have a hyperventilating person breathe into a paper bag, which allows much of the CO_2 to be taken up again.

Metabolic alkalosis is the increase in pH resulting from illness or chemical ingestion. Repeated vomiting or the overuse of diuretics can cause metabolic alkalosis. Once again the body compensates, this time by decreasing the rate of respiration.

Key Concepts: buffers, buffer capacity

For Further Reading

I. Asimov, Life with air, in *Life and Energy,* New York: Avon Books, 1962

R. J. Beynon and J. S. Easterby, *Buffer Solutions: The Basics,* Oxford: BIOS Scientific Publishers, 1996

http://www.madsci.com/manu/indexgas.htm

Related Exercises: 16.111–16.114

*This box is an adaptation of a contribution by B. A. Pruitt, M. D., and A. D. Mason, M. D., U. S. Army Institute of Surgical Research.

Example 16.11 *Determining the solubility product*

The molar solubility of silver chromate, Ag_2CrO_4, is 6.5×10^{-5} mol/L. Determine the value of K_{sp}.

Strategy Write the chemical equation for the solubility equilibrium and the expression for the equilibrium constant. The molarity of the pure solid does not appear in the expression for K_{sp}. To evaluate K_{sp}, we need to know the molarities of each type of ion formed by the salt. Determine them from the relation between the molarity of each type of ion and s, the numerical value of the molar solubility of the salt.

Table 16.5 Solubility products at 25°C

Compound	Formula	K_{sp}	Compound	Formula	K_{sp}
aluminum hydroxide	$Al(OH)_3$	1.0×10^{-33}	fluoride	PbF_2	3.7×10^{-8}
antimony sulfide	Sb_2S_3	1.7×10^{-93}	iodate	$Pb(IO_3)_2$	2.6×10^{-13}
barium carbonate	$BaCO_3$	8.1×10^{-9}	iodide	PbI_2	1.4×10^{-8}
fluoride	BaF_2	1.7×10^{-6}	sulfate	$PbSO_4$	1.6×10^{-8}
sulfate	$BaSO_4$	1.1×10^{-10}	sulfide	PbS	8.8×10^{-29}
bismuth sulfide	Bi_2S_3	1.0×10^{-97}	magnesium		
calcium carbonate	$CaCO_3$	8.7×10^{-9}	ammonium phosphate	$MgNH_4PO_4$	2.5×10^{-13}
fluoride	CaF_2	4.0×10^{-11}	carbonate	$MgCO_3$	1.0×10^{-5}
hydroxide	$Ca(OH)_2$	5.5×10^{-6}	fluoride	MgF_2	6.4×10^{-9}
sulfate	$CaSO_4$	2.4×10^{-5}	hydroxide	$Mg(OH)_2$	1.1×10^{-11}
copper(I) bromide	$CuBr$	4.2×10^{-8}	mercury(I) chloride	Hg_2Cl_2	1.3×10^{-18}
chloride	$CuCl$	1.0×10^{-6}	iodide	Hg_2I_2	1.2×10^{-28}
iodide	CuI	5.1×10^{-12}	mercury(II) sulfide, black	HgS	1.6×10^{-52}
sulfide	Cu_2S	2.0×10^{-47}	sulfide, red	HgS	1.4×10^{-53}
copper(II) iodate	$Cu(IO_3)_2$	1.4×10^{-7}	nickel(II) hydroxide	$Ni(OH)_2$	6.5×10^{-18}
oxalate	$Cu(C_2O_4)$	2.9×10^{-8}	silver bromide	$AgBr$	7.7×10^{-13}
sulfide	CuS	1.3×10^{-36}	carbonate	Ag_2CO_3	6.2×10^{-12}
iron(II) hydroxide	$Fe(OH)_2$	1.6×10^{-14}	chloride	$AgCl$	1.6×10^{-10}
sulfide	FeS	6.3×10^{-18}	hydroxide	$AgOH$	1.5×10^{-8}
iron(III) hydroxide	$Fe(OH)_3$	2.0×10^{-39}	iodide	AgI	8.0×10^{-17}
lead(II) bromide	$PbBr_2$	7.9×10^{-5}	sulfide	Ag_2S	6.3×10^{-51}
chloride	$PbCl_2$	1.6×10^{-5}	zinc hydroxide	$Zn(OH)_2$	2.0×10^{-17}
			sulfide	ZnS	1.6×10^{-24}

Solution The chemical equation and the expression for the solubility product are

$$Ag_2CrO_4(s) \rightleftharpoons 2\,Ag^+(aq) + CrO_4^{2-}(aq) \qquad K_{sp} = [Ag^+]^2[CrO_4^{2-}]$$

Because 2 mol $Ag^+ \mathrel{\hat{=}} 1$ mol Ag_2CrO_4 and 1 mol $CrO_4^{2-} \mathrel{\hat{=}} 1$ mol Ag_2CrO_4,

$$[Ag^+] = 2s = 2 \times (6.5 \times 10^{-5}) = 1.30 \times 10^{-4}$$

$$[CrO_4^{2-}] = s = 6.5 \times 10^{-5}$$

Therefore

$$K_{sp} = [Ag^+]^2[CrO_4^{2-}] = (2s)^2(s)$$
$$= (1.30 \times 10^{-4})^2 \times (6.5 \times 10^{-5}) = 1.1 \times 10^{-12}$$

Self-Test 16.14A The molar solubility of lead(II) iodate, $Pb(IO_3)_2$, at 26°C is 4.0×10^{-5} mol/L. What is the value of K_{sp} for lead(II) iodate?

[*Answer:* 2.6×10^{-13}]

Self-Test 16.14B The molar solubility of calcium iodate, $Ca(IO_3)_2$, at 10°C is 3.8×10^{-3} mol/L. What is the value of K_{sp} for calcium iodate?

Example 16.12 *Calculating the molar solubility from the solubility product*

According to Table 16.5, $K_{sp} = 1.4 \times 10^{-8}$ for lead(II) iodide in water. Estimate its molar solubility.

Strategy Write the chemical equation for the solubility equilibrium and the expression for K_{sp}. The molar solubility, s, is the numerical value of the molarity of formula units in the saturated solution. Because each formula unit produces a known number of cations and anions in solution, express the molarities of the cations and anions in terms of s. Then express K_{sp} in terms of s and solve for s.

Solution The solubility equilibrium is

$$PbI_2(s) \rightleftharpoons Pb^{2+}(aq) + 2\,I^-(aq) \qquad K_{sp} = [Pb^{2+}][I^-]^2$$

Because 1 mol $PbI_2 \triangleq 1$ mol Pb^{2+} and 1 mol $PbI_2 \triangleq 2$ mol I^-,

$$[Pb^{2+}] = s \qquad [I^-] = 2s$$

It follows that

$$K_{sp} = [Pb^{2+}][I^-]^2 = s \times (2s)^2 = 4s^3$$

This expression is easily solved for s:

$$s = \left(\tfrac{1}{4}K_{sp}\right)^{1/3}$$

Now we can substitute the numerical value of K_{sp}:

$$s = \{\tfrac{1}{4} \times (1.4 \times 10^{-8})\}^{1/3} = 1.5 \times 10^{-3}$$

The molar solubility of PbI_2 is therefore 1.5×10^{-3} mol/L, or 1.5 mmol/L.

Self-Test 16.15A The solubility product of silver bromide is 7.7×10^{-13}. Calculate the molar solubility of the salt.

[*Answer:* 8.8×10^{-7} mol/L]

Self-Test 16.15B The solubility product of silver sulfate is 1.4×10^{-5}. Calculate the molar solubility of the salt.

> The expression $x^{1/3}$ means the cube root of x: it is easily found by using the y^x command on a calculator, with $x = \tfrac{1}{3}$.

16.12 The Common-Ion Effect

Suppose we have a saturated solution of silver chloride in water:

$$AgCl(s) \rightleftharpoons Ag^+(aq) + Cl^-(aq) \qquad K_{sp} = [Ag^+][Cl^-]$$

Experimentally, $K_{sp} = 1.6 \times 10^{-10}$ at 25°C, and the molar solubility of AgCl in water is 1.3×10^{-5} mol/L. If we add sodium chloride to the solution, the concentration of Cl^- ions increases. The system responds by tending to minimize the effect of adding more Cl^- ions, so some of the Cl^- ions form solid AgCl, taking some Ag^+ ions out of solution (Fig. 16.24). Because there is now less Ag^+ in solution, the solubility of AgCl is lower in a solution of NaCl than it is in pure water. A similar decrease in solubility occurs whenever two salts have an ion in common and is called the **common-ion effect** (Fig. 16.25).

Predicting the common-ion effect quantitatively is more difficult. Because ions interact with one another strongly, simple equilibrium calculations are rarely valid. However, we can still get an idea of the size of the common-ion effect by considering the solubility product.

FIGURE 16.24

If the concentration of one of the ions in solution is increased, then the concentration of the other is decreased to maintain a constant value of K_{sp}. (a) The cations (pink) and anions (green) in solution. (b) When more anions are added (together with their accompanying spectator ions, which are not shown), the concentration of cations decreases. In other words, the solubility of the original compound is reduced by the presence of a common ion. In the insets, a solid blue color represents the water molecules.

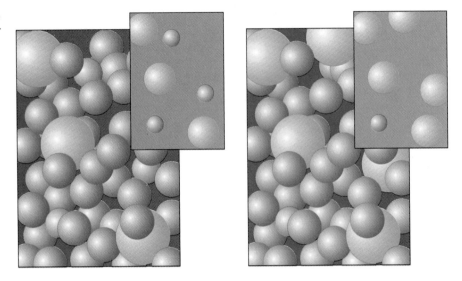

The common-ion effect is the reduction in the solubility of a sparingly soluble salt by the addition of a soluble salt that has an ion in common with it.

Example 16.13 *Estimating the effect of a common ion*

What is the approximate molar solubility of silver chloride in 0.10 M NaCl(aq)?

Strategy In this solution, $[Cl^-] = 0.10$. Silver chloride will dissolve until the product of the concentrations of Ag^+ and Cl^- ions is equal to the solubility product of AgCl, 1.6×10^{-10}.

Solution The silver chloride dissolves until the concentration of Ag^+ ions is given by

$$[Ag^+] = \frac{K_{sp}}{[Cl^-]} = \frac{1.6 \times 10^{-10}}{0.10} = 1.6 \times 10^{-9}$$

FIGURE 16.25

(a) A saturated solution of zinc acetate in water. (b) When acetate ions are added (as solid sodium acetate in the spatula shown in (a)), the solubility of the zinc acetate is significantly reduced, and additional zinc acetate precipitates.

(a)

(b)

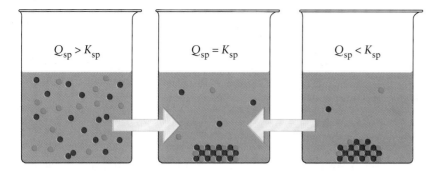

FIGURE 16.26

The solution in the middle is at equilibrium. The solution on the left is supersaturated, so it forms a precipitate until solid and solute are in equilibrium. In the solution on the right, solid has been added to pure water. As the solid begins to dissolve, the solute concentration is lower than the equilibrium concentration. In this case, the solid dissolves until the equilibrium concentration is reached.

The molarity of Ag^+ ions, and hence the solubility of AgCl formula units, is only 1.6×10^{-9} mol/L. That is a 10 000-fold decrease from its solubility in pure water.

Self-Test 16.16A What is the molar solubility of calcium carbonate in 0.20 M $CaCl_2(aq)$?

[*Answer:* 4.4×10^{-8} mol/L]

Self-Test 16.16B What is the molar solubility of silver bromide in 0.10 M $CaBr_2(aq)$?

16.13 Predicting Precipitation

In Section 14.7, we saw how to predict the direction in which a reaction would occur by using Q, the reaction quotient. We can use the techniques introduced in Example 14.6 to decide whether a precipitate is likely to form from the ions in two electrolyte solutions that are mixed. In this case, the equilibrium constant is the solubility product, K_{sp}, and the solubility quotient is denoted Q_{sp}. Precipitation occurs when $Q_{sp} > K_{sp}$ (Fig. 16.26).

For example, we might want to know whether a precipitate of PbI_2 will form when equal volumes of 0.2 M solutions of lead(II) nitrate and potassium iodide are mixed. The overall equation for the reaction is

$$Pb(NO_3)_2(aq) + 2 KI(aq) \longrightarrow 2 KNO_3(aq) + PbI_2(s)$$

and the net ionic equation is

$$Pb^{2+}(aq) + 2 I^-(aq) \rightleftharpoons PbI_2(s)$$

The reverse of this equation is the dissolution equilibrium of PbI_2:

$$PbI_2(s) \rightleftharpoons Pb^{2+}(aq) + 2 I^-(aq) \qquad K_{sp} = [Pb^{2+}][I^-]^2$$

We know from Table 16.5 that $K_{sp} = 1.4 \times 10^{-8}$ at 25°C. The instant the solutions are mixed, the molarities of the Pb^{2+} and I^- ions are very high. Because equal volumes were mixed together, the new volume is twice as large, so the new molarities are half (0.1 mol/L $Pb^{2+}(aq)$ and 0.1 mol/L $I^-(aq)$) their original values. Therefore,

$$Q_{sp} = [Pb^{2+}][I^-]^2 = 0.1 \times (0.1)^2 = 1 \times 10^{-3}$$

This value is considerably higher than K_{sp}, so a precipitate will form. Precipitations occur immediately: we never have to wait more than a millisecond or so, because ions in electrolyte solutions are very mobile and clump together rapidly (Fig. 16.27).

A salt precipitates if Q_{sp} is greater than or equal to K_{sp}.

FIGURE 16.27

When a lead(II) nitrate solution is added to a solution of potassium iodide, yellow lead(II) iodide immediately precipitates.

Self-Test 16.17A Does a precipitate of silver chloride form when we mix 200.0 mL of 1.0×10^{-4} M $AgNO_3(aq)$ and 900.0 mL of 1.0×10^{-6} M $KCl(aq)$?

[*Answer:* No ($Q_{sp} = 1.5 \times 10^{-11} < K_{sp}$)]

Self-Test 16.17B Does a precipitate of barium fluoride form when we mix 100.0 mL of 1.0×10^{-3} M $Ba(NO_3)_2(aq)$ with 200.0 mL of 1.0×10^{-3} M $KF(aq)$? Ignore possible protonation of F^-.

16.14 Selective Precipitation

In some cases, we can use **selective precipitation** to separate different cations in a solution by adding a soluble salt containing an anion with which the cations form insoluble salts. For a good separation, the salts formed must differ greatly in their solubility. For example, seawater is a mixture of many different ions. It is possible to precipitate magnesium ions from seawater by adding hydroxide ions, which form insoluble magnesium hydroxide. However, other cations are also present in seawater. Their individual concentrations and the relative solubilities of their hydroxides determine which will precipitate first if a certain amount of hydroxide is added. Example 16.14 illustrates a strategy for predicting the order of precipitation.

> *Ions in solution can be separated by adding an anion with which they form salts having very different solubilities.*

Example 16.14 *Assessing the order of precipitation*

One of the major sources of magnesium is seawater, from which it is precipitated as the hydroxide. However, seawater contains calcium ions as well. Suppose we add sodium hydroxide to seawater to precipitate magnesium hydroxide. Will calcium hydroxide also precipitate? Seawater contains the following concentrations of the ions: 0.050 mol/L $Mg^{2+}(aq)$ and 0.010 mol/L $Ca^{2+}(aq)$. Use the information in Table 16.5 to determine the order in which each ion precipitates as solid NaOH is added, and give the concentration of OH^- when precipitation of each begins. Assume no volume change on addition of the NaOH.

Strategy A salt precipitates when the concentrations of its ions are such that $Q_{sp} \geq K_{sp}$. Calculate the value of $[OH^-]$ required for each salt to precipitate by writing the expression for K_{sp} for each salt and then use the data provided to determine $[OH^-]$. The values of K_{sp} are found in Table 16.5.

Solution For $Ca(OH)_2$, $K_{sp} = 5.5 \times 10^{-6} = [Ca^{2+}][OH^-]^2$. Therefore,

$$[OH^-] = \sqrt{\frac{K_{sp}}{[Ca^{2+}]}} = \sqrt{\frac{5.5 \times 10^{-6}}{0.010}} = 0.023$$

That is, a hydroxide concentration of 0.023 mol/L is required for calcium hydroxide to precipitate. For $Mg(OH)_2$, $K_{sp} = 1.1 \times 10^{-11} = [Mg^{2+}][OH^-]^2$. Therefore,

$$[OH^-] = \sqrt{\frac{K_{sp}}{[Mg^{2+}]}} = \sqrt{\frac{1.1 \times 10^{-11}}{0.050}} = 1.5 \times 10^{-5}$$

Magnesium hydroxide will begin to precipitate at a hydroxide concentration of 1.5×10^{-5} mol/L. We can conclude that the hydroxides precipitate in the order $Mg(OH)_2$ at 1.5×10^{-5} mol/L $OH^-(aq)$ and $Ca(OH)_2$ at 0.023 mol/L $OH^-(aq)$. The solubility of $Ca(OH)_2$ is so much greater than that of $Mg(OH)_2$ that calcium hydroxide is the base normally used to precipitate magnesium as the hydroxide from seawater.

Self-Test 16.18A Potassium sulfate is added to a solution containing the following concentrations of soluble cations: 0.030 mol/L $Ba^{2+}(aq)$ and 0.0010 mol/L $Ca^{2+}(aq)$. Use the information in Table 16.5 to determine the order in which each ion precipitates as the concentration of K_2SO_4 is increased and give the concentration of SO_4^{2-} when precipitation of each begins.

[*Answer:* $CaSO_4$ precipitates first, at 2.4×10^{-2} mol/L SO_4^{2-}, then SO_4 at 3.7×10^{-9} mol/L SO_4^{2-}]

Self-Test 16.18B Bromide ion is added to a solution containing the following concentrations of soluble cations: 0.020 mol/L $Pb^{2+}(aq)$ and 0.0010 mol/L $Ag^+(aq)$. Use the information in Table 16.5 to determine the order in which each ion precipitates as the concentration of bromide ion is increased and give the concentration of Br^- when precipitation of each begins.

16.15 Dissolving Precipitates

It may be necessary to dissolve a precipitate if we want to identify it or to use it in another reaction. For example, once a silver salt precipitates, it may be necessary to dissolve it again so that it can be electroplated as metallic silver. One strategy is to remove one of the ions from the solubility equilibrium, so that the precipitate continues to dissolve.

Suppose, for example, a solid hydroxide is in equilibrium with its ions in solution:

$$Fe(OH)_3(s) \rightleftharpoons Fe^{3+}(aq) + 3\,OH^-(aq)$$

To dissolve more of the solid, we can add acid. As the H_3O^+ ions from the acid convert the OH^- into water, the equilibrium shifts to the right, producing more $Fe^{3+}(aq)$ and $OH^-(aq)$ in an attempt to reach the value of K_{sp}; consequently, more solid dissolves.

Many carbonate, sulfite, and sulfide precipitates also can be dissolved by the addition of acid, because the anions react with the acid to form a gas that bubbles out of solution. For example, in a saturated solution of zinc carbonate, solid $ZnCO_3$ is in equilibrium with its ions:

$$ZnCO_3(s) \rightleftharpoons Zn^{2+}(aq) + CO_3^{2-}(aq)$$

The CO_3^{2-} ions react with acid to form CO_2:

$$CO_3^{2-}(aq) + 2\,HCl(aq) \rightleftharpoons H_2CO_3(aq) + 2\,Cl^-(aq)$$
$$H_2CO_3(aq) \rightleftharpoons H_2O(l) + CO_2(g)$$

The loss of carbonate ions as carbon dioxide gas results in the dissociation equilibrium of $ZnCO_3$ shifting to the right, so more zinc carbonate dissolves.

An alternative procedure to remove an ion from solution is to change its oxidation number. The metal ions in very insoluble heavy metal sulfide precipitates can be dissolved by oxidizing the sulfide ion to elemental sulfur. For example, copper(II) sulfide, CuS, takes part in the equilibrium

$$CuS(s) \rightleftharpoons Cu^{2+}(aq) + S^{2-}(aq)$$

However, when nitric acid is added, the sulfide ions are oxidized to elemental sulfur:

$$3\,S^{2-}(aq) + 8\,HNO_3(aq) \longrightarrow 3\,S(s) + 2\,NO(g) + 4\,H_2O(l) + 6\,NO_3^-(aq)$$

This complicated oxidation removes the sulfide ions from the equilibrium, and Cu^{2+} ions dissolve as $Cu(NO_3)_2$.

> *The solubility of an ionic solid can be increased by removing one of its ions from solution; acid can be used to dissolve a hydroxide or carbonate precipitate, and nitric acid can be used to oxidize metal sulfides to sulfur and a soluble salt.*

16.16 Complex Ions and Solubilities

The formation of a complex can also remove an ion, which disturbs the solubility equilibrium until more solid dissolves. We first met complexes in Section 8.13, where we saw that they were species formed by the reaction of a Lewis acid and a Lewis base. In this section, we consider complexes in which the Lewis acid is a metal cation, such as Ag^+. An example is the formation of $Ag(NH_3)_2{}^+$ when aqueous ammonia (a Lewis base) is added to a solution of silver chloride:

$$Ag^+(aq) + 2\,NH_3(aq) \rightleftharpoons Ag(NH_3)_2{}^+(aq)$$

Complex formation removes some of the Ag^+ ions from solution. As a result, to preserve the value of K_{sp}, more silver chloride will dissolve. If enough ammonia is present, all the precipitate dissolves. We can conclude that the qualitative effect of complex formation is to *increase* the solubility of a sparingly soluble compound.

To treat complex formation quantitatively, we note that complex formation and dissolving are at equilibrium; so we can write

$$AgCl(s) \rightleftharpoons Ag^+(aq) + Cl^-(aq) \qquad K_{sp} = [Ag^+][Cl^-] \qquad \textbf{(A)}$$

$$Ag^+(aq) + 2\,NH_3(aq) \rightleftharpoons Ag(NH_3)_2{}^+(aq) \qquad K_f = \frac{[Ag(NH_3)_2{}^+]}{[Ag^+][NH_3]^2} \quad \textbf{(B)}$$

The equilibrium constant for the formation of the complex ion is called the **formation constant**, K_f. At 25°C, $K_f = 1.6 \times 10^7$ for this reaction. Values for other equilibria are given in Table 16.6 and the example below shows how to use them.

> *Salts are more soluble if a complex ion can be formed.*

Table 16.6 *Formation constants in water at 25°C*

Equilibrium	K_f
$Ag^+(aq) + 2\,CN^-(aq) \rightleftharpoons Ag(CN)_2{}^-(aq)$	5.6×10^8
$Ag^+(aq) + 2\,NH_3(aq) \rightleftharpoons Ag(NH_3)_2{}^+(aq)$	1.6×10^7
$Au^+(aq) + 2\,CN^-(aq) \rightleftharpoons Au(CN)_2{}^-(aq)$	2.0×10^{38}
$Cu^{2+}(aq) + 4\,NH_3(aq) \rightleftharpoons Cu(NH_3)_4{}^{2+}(aq)$	1.2×10^{13}
$Hg^{2+}(aq) + 4\,Cl^-(aq) \rightleftharpoons HgCl_4{}^{2-}(aq)$	1.2×10^5
$Fe^{2+}(aq) + 6\,CN^-(aq) \rightleftharpoons Fe(CN)_6{}^{4-}(aq)$	7.7×10^{36}
$Ni^{2+}(aq) + 6\,NH_3(aq) \rightleftharpoons Ni(NH_3)_6{}^{2+}(aq)$	5.6×10^8

Example 16.15 Calculating molar solubility in the presence of complex formation

Calculate the molar solubility of silver chloride in 0.10 M $NH_3(aq)$, given that $K_{sp} = 1.6 \times 10^{-10}$ for silver chloride and $K_f = 1.6 \times 10^7$ for the ammonia complex of Ag^+ ions, $Ag(NH_3)_2^+$.

Strategy Write the chemical equation for the equilibrium between the solid solute and the complex in solution and express this equilibrium as the sum of the solubility and complex formation equilibria. The equilibrium constant for the overall equilibrium is the product of the equilibrium constants for the two contributing processes. Then set up an equilibrium table, express the equilibrium constant in terms of the unknown concentrations, and solve for the equilibrium concentrations of ions in solution.

Solution The overall equilibrium is

$$AgCl(s) + 2 NH_3(aq) \rightleftharpoons Ag(NH_3)_2^+(aq) + Cl^-(aq)$$

This equation is the sum of the equations for reactions A and B in the text. The equilibrium constant for this equilibrium,

$$K = \frac{[Ag(NH_3)_2^+][Cl^-]}{[NH_3]^2}$$

is therefore the product of the equilibrium constants for reactions A and B:

$$K = K_{sp} \times K_f$$

Because 1 mol AgCl $\simeq$ 1 mol $Ag(NH_3)_2^+$, the molar solubility of AgCl is equal to the molarity of $Ag(NH_3)_2^+$ in the saturated solution. The equilibrium table, with all concentrations in moles per liter, is

	Species		
	NH_3	$Ag(NH_3)_2^+$	Cl^-
Step 1. Initial molarity	0.10	0	0
Step 2. Change in molarity	$-2x$	$+x$	$+x$
Step 3. Equilibrium molarity	$0.10 - 2x$	x	x

Step 4. We find K and substitute the information from step 3 for the equilibrium concentrations:

$$K = K_{sp} \times K_f = (1.6 \times 10^{-10}) \times (1.6 \times 10^7) = 2.6 \times 10^{-3}$$

$$K = \frac{[Ag(NH_3)_2^+][Cl^-]}{[NH_3]^2} = \frac{x \times x}{(0.10 - 2x)^2} = 2.6 \times 10^{-3}$$

Take the square root of each side and obtain

$$\frac{x}{0.10 - 2x} = 5.1 \times 10^{-2}$$

Rearranging and solving for x gives $x = 4.6 \times 10^{-3}$. Therefore, from step 3, $[Ag(NH_3)_2^+] = 4.6 \times 10^{-3}$ mol/L. Hence, the molar solubility of silver chloride in 0.10 M $NH_3(aq)$ is 4.6×10^{-3} mol/L, greater by a factor of over 100 from the molar solubility of silver chloride in pure water, 1.3×10^{-5} mol/L.

Self-Test 16.19A Calculate the molar solubility of silver bromide in 1.0 M $NH_3(aq)$.

[*Answer:* 3.5 mmol/L]

Self-Test 16.19B Calculate the molar solubility of copper(II) iodate in 1.2 M $NH_3(aq)$.

16.17 Qualitative Analysis

Complex formation, selective precipitation, and control of the pH of a solution all play important roles in qualitative analysis. There are many different schemes of analysis, but they have common features and follow the same general principles. In this section, we look at a typical scheme for the identification of several cations in the laboratory.

Suppose we have a solution that contains lead(II), mercury(I), silver, copper(II), and zinc ions. The scheme is set out in Table 16.7. Most chlorides are soluble; so when hydrochloric acid is added to a mixture of salts, only certain chlorides precipitate (see Table 16.7). Silver and mercury(I) chlorides have such small values of K_{sp} that even low concentrations of Cl^- ions raise Q_{sp} above K_{sp} in solutions of those cations, and the chlorides precipitate. Lead(II) chloride, which is slightly soluble, will also precipitate if the chloride ion concentration is high enough (Fig. 16.28). The hydronium ions provided by the acid play no role in this step; they simply accompany the chloride ions. At this point, the precipitate can be separated from the solution by using a centrifuge to compact the solid, then **decanting** (pouring off) the solution. The solution now contains zinc and copper(II) ions, whereas the solid consists of $PbCl_2$, Hg_2Cl_2, and $AgCl$.

Because $PbCl_2$ is slightly soluble, if the precipitate is rinsed with hot water, the lead(II) chloride dissolves and the solution is separated from the precipitate. If sodium chromate is then added to that solution, any lead(II) present will precipitate as lead(II) chromate (see Fig. 3.14):

$$Pb^{2+}(aq) + CrO_4^{2-}(aq) \longrightarrow PbCrO_4(s)$$

At this point, the silver(I) and mercury(I) chlorides remain as precipitates. When ammonia is added to the solid mixture, the silver precipitate dissolves as a result of the formation of the complex ion $Ag(NH_3)_2^+$ (Fig. 16.29):

$$Ag^+(aq) + 2 NH_3(aq) \longrightarrow Ag(NH_3)_2^+(aq)$$

> A large excess of concentrated hydrochloric acid must be avoided to prevent the formation of the soluble complex ions $AgCl_2^-$ and $PbCl_4^{2-}$.

Table 16.7 Part of a qualitative analysis scheme

Step	Possible precipitate	K_{sp}
Add HCl(aq)	AgCl	1.6×10^{-10}
	Hg_2Cl_2	1.3×10^{-18}
	$PbCl_2$	1.6×10^{-5}
Add H_2S(aq) (in acid solution, low S^{2-} concentration)	Bi_2S_3	1.0×10^{-97}
	CdS	4.0×10^{-29}
	CuS	7.9×10^{-45}
	HgS	1.6×10^{-52}
	Sb_2S_3	1.6×10^{-93}
Add base to H_2S(aq) (in basic solution, higher S^{2-} concentration)	FeS	6.3×10^{-18}
	MnS	1.3×10^{-15}
	NiS	1.3×10^{-24}
	ZnS	1.6×10^{-24}

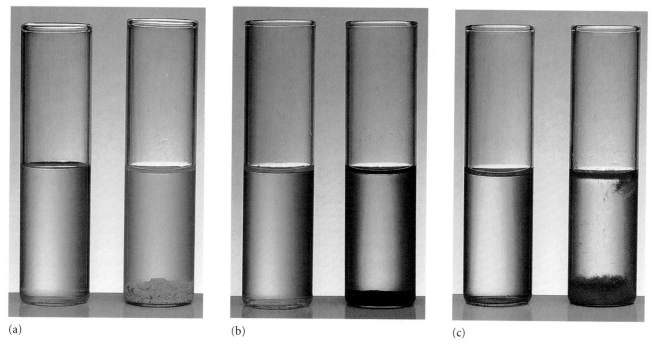

FIGURE 16.28

The sequence for the analysis of cations by selective precipitation. (a) The original solution contains Pb^{2+}, Hg_2^{2+}, Ag^+, Cu^{2+}, and Zn^{2+} ions. Addition of HCl(aq) precipitates AgCl, Hg_2Cl_2, and $PbCl_2$, which can be removed by decanting and filtration. (b) Addition of H_2S to the remaining solution precipitates CuS, which can also be removed. (c) Making the resulting solution basic by adding ammonia precipitates ZnS.

and mercury(I) reacts to form a gray mixture of mercury(II) ions precipitated as white $HgNH_2Cl(s)$ and metallic mercury, which looks black:

$$Hg_2Cl_2(s) + 2\,NH_3(aq) \longrightarrow Hg(l) + HgNH_2Cl(s) + NH_4^+(aq) + Cl^-(aq)$$

The solution can be separated from the solid and the presence of silver ion in the solution verified by addition of nitric acid. This acid reacts with the ammonia, pulling the ammonia molecules away from the silver ions and allowing silver chloride to precipitate:

$$Ag(NH_3)_2^+(aq) + Cl^-(aq) + 2\,H_3O^+(aq) \longrightarrow$$
$$AgCl(s) + 2\,NH_4^+(aq) + 2\,H_2O(l)$$

At a later stage, certain sulfides are precipitated by the addition of S^{2-} ions (see Table 16.7). The goal at this stage is selective precipitation. To achieve it, we make use of the widely different solubilities (and solubility products) of different sulfides. Some metal sulfides (such as CuS, HgS, and Sb_2S_3) have very small solubility products and are precipitated if there is the merest trace of S^{2-} ions in the solution. Such a very low concentration of S^{2-} ions is achieved by adding hydrogen sulfide, H_2S, to an acidified solution. A higher hydronium ion concentration shifts the equilibrium

$$H_2S(aq) + H_2O(l) \rightleftharpoons 2\,H_3O^+(aq) + S^{2-}(aq)$$

FIGURE 16.29

When an aqueous solution of ammonia is added to a silver chloride precipitate, the precipitate dissolves. However, when ammonia is added to a precipitate of mercury(I) chloride, mercury metal and mercury(II) ions are formed by disproportionation and the mass turns gray. Left to right: silver chloride in water, silver chloride in aqueous ammonia, mercury(I) chloride in water, mercury(I) chloride in aqueous ammonia. The different responses of AgCl and Hg_2Cl_2 to the addition of ammonia allow them to be distinguished in a mixture.

to the left, so almost all the H_2S is in its fully protonated form. As a result, very little S^{2-} is present. Nevertheless, that little will result in the precipitation of highly insoluble solids if the appropriate cations are present.

After any precipitates have been removed, ammonia is added to the solution. The base removes the hydronium ion from the H_2S equilibrium, which shifts the equilibrium in favor of S^{2-} ions. The higher concentration of S^{2-} ion increases the Q_{sp} values of any remaining metal sulfides above their K_{sp} values, and they precipitate. This step detects the presence of metal sulfides that have larger solubility constants than those in the preceding step (Table 16.7), including MnS and ZnS.

Qualitative analysis involves the separation and identification of ions by selective precipitation, complex formation, and the control of pH.

Skills You Should Have Mastered

Conceptual

1. Explain why salts of weak bases produce acidic solutions and salts of weak acids produce basic solutions, Section 16.1.

2. Interpret the features of the pH curve for the titration of a strong or weak acid with a strong base, or a strong or weak base with a strong acid, Sections 16.4–16.6.

3. Explain how buffer solutions resist changes in pH, Section 16.8.

4. Select an appropriate buffer for a given pH, Section 16.9.

5. Explain what is meant by the buffer capacity of a solution, Section 16.10.

Problem-Solving

1. Calculate the pH of a salt solution, Toolbox 16.1 and Examples 16.1 and 16.2.

2. Calculate the pH of a solution of a weak acid and its conjugate base, Example 16.3.

3. Calculate the pH at any point in a strong base-strong acid, strong base-weak acid, or weak base-strong acid titration, Toolboxes 16.2 and 16.3 and Examples 16.4–16.7.

4. Calculate the pH of a buffer solution, Toolbox 16.4 and Example 16.8.

5. Calculate the pH change when acid or base is added to a buffer solution, Toolbox 16.4 and Example 16.9.

6. Determine the relative concentrations of conjugate acid and base needed to prepare a buffer solution with a given pH, Example 16.10.

7. Determine a solubility constant from molar solubility and vice versa, Examples 16.11 and 16.12.

8. Estimate the effect of a common ion on solubility, Example 16.13.

9. Assess the order of precipitation of a series of salts, Example 16.14.

10. Calculate molar solubility in the presence of complex ion formation, Example 16.15.

Descriptive

1. Show how small, highly charged hydrated metal cations produce acidic solutions, Section 16.1.

2. Explain how indicators change color with pH and select an appropriate indicator for a titration, Section 16.7.

3. Show how ions can be separated and detected by a qualitative analysis scheme, Section 16.17.

Exercises

The values for the acidity and basicity constants are listed in Tables 15.3, 15.6, and 16.1. Unless stated otherwise, assume the solutions are aqueous and at 25°C.

Ions as Acids and Bases

16.1 Determine whether aqueous solutions of the following salts have a pH equal to, greater than, or less than 7. If pH > 7 or pH < 7, write a chemical equation to justify your answer. (a) NH_4Br; (b) Na_2CO_3; (c) KF; (d) KBr; (e) $AlCl_3$; (f) $Cu(NO_3)_2$.

16.2 Determine whether aqueous solutions of the following salts have a pH equal to, greater than, or less than 7. If pH > 7 or pH < 7, write a chemical equation to justify your answer. (a) $K_2C_2O_4$ (potassium oxalate); (b) $Ca(NO_3)_2$; (c) CH_3NH_3Cl; (d) K_3PO_4; (e) $FeCl_3$; (f) C_5H_5NHCl (pyridinium chloride).

16.3 Determine the value of the acidity or basicity constants of the following ions: (a) NH_4^+; (b) CO_3^{2-}; (c) F^-; (d) ClO^-; (e) HCO_3^-; (f) $(CH_3)_3NH^+$ (trimethylammonium ion).

16.4 Determine the value of the acidity or basicity constants of the following ions: (a) $N_2H_5^+$; (b) PO_4^{3-}; (c) NO_2^-; (d) NO_3^-; (e) HPO_4^{2-}; (f) $C_5H_5NH^+$.

16.5 Calculate the pH of the following solutions: (a) 0.20 M $NaCH_3CO_2(aq)$; (b) 0.10 M $NH_4Cl(aq)$; (c) 0.10 M $AlCl_3(aq)$; (d) 0.15 M KCN(aq).

16.6 Calculate the pH of the following solutions: (a) 0.15 M $CH_3NH_3Cl(aq)$; (b) 0.20 M $Na_2SO_3(aq)$; (c) 0.30 M $FeCl_3(aq)$; (d) 0.10 M $KClO_2(aq)$.

16.7 (a) A 10.0-g sample of potassium acetate, KCH_3CO_2, is dissolved in 250.0 mL of solution. What is the pH of the solution? (b) What is the pH of a solution resulting from the dissolution of 5.75 g of ammonium bromide, NH_4Br, in 100.0 mL of solution?

16.8 (a) A 1.00-g sample of sodium hydrogen sulfite, $NaHSO_3$, is dissolved in 50.0 mL of solution. What is the pH of the solution? (b) What is the pH of a solution resulting from the dissolution of 100. mg of silver nitrate, $AgNO_3$, in 10.0 mL of solution?

16.9 (a) A 200.0-mL sample of 0.200 M $NaCH_3CO_2(aq)$ is diluted to 500.0 mL. What is the concentration of acetic acid at equilibrium? (b) What is the pH of a solution resulting from the dissolution of 5.75 g of ammonium bromide, NH_4Br, in 400.0 mL of solution?

16.10 (a) A 50.0-mL sample of 0.630 M KCN(aq) is diluted to 125.0 mL. What is the concentration of hydrocyanic acid present at equilibrium? (b) A 1.00-g sample of sodium hydrogen carbonate, $NaHCO_3$, is dissolved in 150.0 mL of solution. What is the pH of the solution?

Mixed Solutions

16.11 Explain what happens to (a) the concentration of H_3O^+ ions in an acetic acid solution when solid sodium acetate is added; (b) the percentage deprotonation of benzoic acid in a benzoic acid solution when hydrochloric acid is added; (c) the pH of the solution when solid ammonium chloride is added to an ammonia solution.

16.12 Explain what happens to (a) the pH of a phosphoric acid solution after the addition of solid sodium dihydrogen phosphate; (b) the percentage deprotonation of HCN in a hydrocyanic acid solution after the addition of hydrobromic acid; (c) the concentration of H_3O^+ ions when pyridinium chloride is added to a pyridine solution.

16.13 A solution of equal concentrations of lactic acid and sodium lactate was found to have pH = 3.08. (a) What are the values of pK_a and K_a of lactic acid? (b) What would the pH be if the acid had twice the concentration of the salt?

16.14 A solution containing equal concentrations of saccharin and its sodium salt was found to have pH = 11.68. (a) What are the values of pK_a and K_a of saccharin? (b) What would the pH be if the salt had twice the concentration of the acid?

16.15 Calculate the concentration of hydronium ions in (a) a solution that is 0.20 M HBrO(aq) and 0.10 M KBrO(aq); (b) a solution that is 0.010 M $(CH_3)_2NH(aq)$ and 0.150 M $(CH_3)_2NH_2Cl(aq)$; (c) a solution that is 0.10 M HBrO(aq) and 0.20 M KBrO(aq); (d) a solution that is 0.020 M $(CH_3)_2NH(aq)$ and 0.030 M $(CH_3)_2NH_2Cl(aq)$.

16.16 What is the concentration of hydronium ions in (a) a solution that is 0.050 M HCN(aq) and 0.030 M NaCN(aq); (b) a solution that is 0.10 M $NH_2NH_2(aq)$ and 0.50 M NaCl(aq); (c) a solution that is 0.030 M HCN(aq) and 0.050 M NaCN(aq); (d) a solution that is 0.15 M $NH_2NH_2(aq)$ and 0.15 M $NH_2NH_3Br(aq)$?

16.17 Determine the pH and pOH of (a) a solution that is 0.40 M $NaHSO_4(aq)$ and 0.80 M $Na_2SO_4(aq)$; (b) a solution that is 0.40 M $NaHSO_4(aq)$ and 0.20 M $Na_2SO_4(aq)$; (c) a solution that is 0.40 M $NaHSO_4(aq)$ and 0.40 M $Na_2SO_4(aq)$.

16.18 Calculate the pH and pOH of (a) a solution that is 0.17 M $Na_2HPO_4(aq)$ and 0.25 M $Na_3PO_4(aq)$; (b) a solution that is 0.66 M $Na_2HPO_4(aq)$ and 0.42 M $Na_3PO_4(aq)$; (c) a solution that is 0.12 M $Na_2HPO_4(aq)$ and 0.12 M $Na_3PO_4(aq)$.

16.19 Calculate the pH of the solution that results from mixing (a) 20.0 mL of 0.050 M HCN(aq) with 80.0 mL of 0.030 M NaCN(aq); (b) 80.0 mL of 0.030 M HCN(aq) with 20.0 mL of 0.050 M NaCN(aq); (c) 25.0 mL of 0.105 M HCN(aq) with 25.0 mL of 0.105 M NaCN(aq).

16.20 Calculate the pH of the solution that results from mixing (a) 100.0 mL of 0.020 M $(CH_3)_2NH(aq)$ with 300.0 mL of 0.030 M $(CH_3)_2NH_2Cl(aq)$; (b) 65.0 mL of 0.010 M $(CH_3)_2NH(aq)$ with 10.0 mL of 0.150 M $(CH_3)_2NH_2Cl(aq)$; (c) 50.0 mL of 0.015 M $(CH_3)_2NH(aq)$ with 125.0 mL of 0.015 M $(CH_3)_2NH_2Cl(aq)$.

Strong Acid-Strong Base Titrations

16.21 Calculate the pH of the following solutions: (a) 25.0 mL of 0.30 M HCl(aq) was added to 25.0 mL of 0.20 M NaOH(aq); (b) 25.0 mL of 0.15 M HCl(aq) was added to 50.0 mL of 0.15 M KOH(aq); (c) 21.7 mL of 0.27 M $HNO_3(aq)$ was added to 10.0 mL of 0.30 M NaOH(aq).

16.22 Calculate the pH of the following solutions: (a) 17.3 mL of 0.25 M HCl(aq) was added to 15.0 mL of 0.33 M NaOH(aq); (b) 21.8 mL of 0.15 M NaOH(aq) was added to 50.0 mL of 0.073 M HCl(aq); (c) 15.94 mL of 0.101 M NaOH(aq) was added to 25.0 mL of 0.094 M $HNO_3(aq)$.

16.23 A 14.0-g sample of NaOH was dissolved in 250.0 mL of solution, and then 25.0 mL was pipetted into 50.0 mL of 0.20 M HBr(aq). What is the pH of the resulting solution?

16.24 A 0.150-g sample of $Ba(OH)_2$ was dissolved in 50.0 mL of solution, and then 25.0 mL was pipetted into 100.0 mL of 0.0010 M HCl(aq). What is the pH of the resulting solution?

16.25 What volume of 0.0631 M HCl(aq) is required to neutralize 25.0 mL of 0.0497 M KOH(aq)?

16.26 It was found that 24.7 mL of 0.184 M HI(aq) was required to neutralize 20.0 mL of $Ba(OH)_2(aq)$. What is the molarity of the $Ba(OH)_2$ in the solution?

16.27 Sketch reasonably accurately the pH curve for the titration of 20.0 mL of 0.10 M HCl (aq) with 0.20 M KOH(aq). Mark on the curve (a) the initial pH; (b) the pH at the stoichiometric point.

16.28 Sketch reasonably accurately the pH curve for the titration of 20.0 mL of 0.10 M $Ba(OH)_2(aq)$ with 0.20 M HCl(aq). Mark on the curve (a) the initial pH; (b) the pH at the stoichiometric point.

16.29 Calculate the volume of 0.150 M HCl(aq) required to neutralize (a) one-half; (b) all the hydroxide ions in 25.0 mL of 0.110 M NaOH(aq). (c) What is the molarity of Na^+ ions at the stoichiometric point? (d) Calculate the pH of the solution after the addition of 20.0 mL of 0.150 M HCl(aq) to 25.0 mL of 0.110 M NaOH(aq).

16.30 Calculate the volume of 0.116 M HCl(aq) required to neutralize (a) one-half; (b) all the hydroxide ions in 25.0 mL of 0.215 M KOH(aq). (c) What is the molarity of Cl^- ions at the stoichiometric point? (d) Calculate the pH of the solution after the addition of 40.0 mL of 0.116 M HCl(aq) to 25.0 mL of 0.215 M KOH(aq).

16.31 (a) A 2.54-g sample of NaOH was dissolved in 25.0 mL of solution and titrated with HCl(aq). What volume (in milliliters) of 0.150 M HCl(aq) is required to reach the stoichiometric point? (b) What is the molarity of Cl^- ions at the stoichiometric point?

16.32 (a) A 2.88-g sample of KOH was dissolved in 25.0 mL of solution and titrated with $HNO_3(aq)$. What volume of 0.200 M $HNO_3(aq)$ is required to reach the stoichiometric point? (b) What is the molarity of NO_3^- ions at the stoichiometric point?

16.33 Calculate the pH at each stage in the titration of 25.0 mL of 0.110 M NaOH(aq) by 0.150 M HCl(aq) (a) initially; (b) after the addition of 5.0 mL of acid; (c) after the addition of a further 5.0 mL; (d) at the stoichiometric point; (e) after the addition of 5.0 mL of acid beyond the stoichiometric point; (f) after the addition of 10.0 mL of acid beyond the stoichiometric point.

16.34 Calculate the pH at each stage in the titration of 25.0 mL of 0.215 M KOH(aq) by 0.116 M HCl(aq) (a) initially; (b) after the addition of 5.0 mL of acid; (c) after the addition of a further 5.0 mL; (d) at the stoichiometric point; (e) after the addition of 5.0 mL of acid beyond the stoichiometric point; (f) after the addition of 10.0 mL of acid beyond the stoichiometric point.

Weak Acid-Strong Base and Weak Base-Strong Acid Titrations

16.35 A 25.0-mL sample of 0.10 M $CH_3COOH(aq)$ is titrated with 0.10 M NaOH(aq). The K_a for CH_3COOH is 1.8×10^{-5}. (a) What is the initial pH of the 0.10 M $CH_3COOH(aq)$? (b) What is the pH after the addition of 10.0 mL of 0.10 M NaOH(aq)? (c) What volume of 0.10 M NaOH(aq) is required to reach halfway to the stoichiometric point? (d) Calculate the pH at that halfway point. (e) What volume of 0.10 M NaOH(aq) is required to reach the stoichiometric point? (f) Calculate the pH at the stoichiometric point. (g) Suggest a suitable indicator.

16.36 A 30.0-mL sample of 0.20 M $C_6H_5COOH(aq)$ solution is titrated with 0.30 M KOH(aq). The K_a for C_6H_5COOH is 6.5×10^{-5}. (a) What is the initial pH of the 0.20 M $C_6H_5COOH(aq)$? (b) What is the pH after the addition of 15.0 mL of 0.30 M KOH(aq)? (c) What volume of 0.30 M KOH(aq) is required to reach halfway to the stoichiometric point? (d) Calculate the pH at the halfway point. (e) What volume of 0.30 M KOH(aq) is required to reach the stoichiometric point? (f) Calculate the pH at the stoichiometric point. (g) Suggest a suitable indicator.

16.37 A 15.0-mL sample of 0.15 M $NH_3(aq)$ solution is titrated with 0.10 M HCl(aq). The K_b for NH_3 is 1.8×10^{-5}.

(a) What is the initial pH of the 0.15 M NH$_3$(aq)? (b) What is the pH after the addition of 15.0 mL of 0.10 M HCl(aq)? (c) What volume of 0.10 M HCl(aq) is required to reach halfway to the stoichiometric point? (d) Calculate the pH at the halfway point. (e) What volume of 0.10 M HCl(aq) is required to reach the stoichiometric point? (f) Calculate the pH at the stoichiometric point. (g) Suggest a suitable indicator.

16.38 A 50.0-mL sample of 0.25 M CH$_3$NH$_2$(aq) solution is titrated with 0.35 M HCl(aq). The K_b for CH$_3$NH$_2$ is 3.6 × 10^{-4}. (a) What is the initial pH of the 0.25 M CH$_3$NH$_2$(aq)? (b) What is the pH after the addition of 15.0 mL of 0.35 M HCl(aq)? (c) What volume of 0.35 M HCl(aq) is required to reach halfway to the stoichiometric point? (d) Calculate the pH at the halfway point. (e) What volume of 0.35 M HCl(aq) is required to reach the stoichiometric point? (f) Calculate the pH at the stoichiometric point. (g) Suggest a suitable indicator.

16.39 Calculate the pH of 25.0 mL of 0.110 M aqueous lactic acid being titrated with 0.150 M NaOH(aq) (a) initially; (b) after the addition of 5.0 mL of base; (c) after the addition of a further 5.0 mL of base; (d) at the stoichiometric point; (e) after the addition of 5.0 mL of base beyond the stoichiometric point; (f) after the addition of 10.0 mL of base beyond the stoichiometric point. (g) Suggest a suitable indicator.

16.40 Calculate the pH of 25.0 mL of 0.215 M chloroacetic acid being titrated with 0.116 M NaOH(aq) (a) initially; (b) after the addition of 5.0 mL of base; (c) after the addition of a further 5.0 mL of base; (d) at the stoichiometric point; (e) after the addition of 5.0 mL of base beyond the stoichiometric point; (f) after the addition of 10.0 mL of base beyond the stoichiometric point. (g) Suggest a suitable indicator.

Indicators

For the pH ranges over which common indicators change color, see Table 16.3.

16.41 Over what pH range can each of the following indicators be used for detecting the stoichiometric point in a titration: (a) methyl orange; (b) litmus; (c) methyl red; (d) phenolphthalein?

16.42 Over what pH range can each of the following indicators be used for detecting the stoichiometric point in a titration: (a) thymol blue; (b) phenol red; (c) bromophenol blue; (d) alizarin.

16.43 Which indicators could you use for a titration of 0.20 M CH$_3$COOH(aq) with 0.20 M NaOH(aq): (a) methyl orange; (b) litmus; (c) thymol blue; (d) phenolphthalein? Explain your selections.

16.44 Which indicators could you use for a titration of 0.20 M NH$_3$(aq) with 0.20 M HCl(aq): (a) bromocresol green;

(b) methyl red; (c) phenol red; (d) thymol blue? Explain your selections.

Buffers

16.45 Identify which of the following mixed systems can function as a buffer solution and write an equilibrium equation for each buffer system: (a) equal volumes of 0.10 M HCl(aq) and 0.10 M NaCl(aq); (b) a solution that is 0.10 M HClO(aq) and 0.10 M NaClO(aq); (c) a solution that is 0.10 M (CH$_3$)$_3$N(aq) and 0.10 M (CH$_3$)$_3$NHCl(aq); (d) equal volumes of 0.20 M CH$_3$COOH(aq) and 0.10 M NaOH(aq); (e) equal volumes of 0.20 M HNO$_3$(aq) and 0.20 M NaOH(aq).

16.46 Identify which of the following mixed systems can function as a buffer solution. Write an equilibrium equation for each buffer system: (a) equal volumes of 0.10 M C$_6$H$_5$COOH(aq) and 0.10 M NaC$_6$H$_5$CO$_2$(aq); (b) a solution that is 0.10 M HNO$_3$(aq) and 0.10 M NaNO$_3$(aq); (c) a solution that is 0.10 M C$_5$H$_5$N(aq) and 0.10 M C$_5$H$_5$NHCl(aq); (d) equal volumes of 0.10 M NH$_3$(aq) and 0.10 M HCl(aq); (e) a solution that is 0.10 M HNO$_2$(aq) and 0.10 M NaNO$_2$(aq).

16.47 Sodium hypochlorite, NaClO, is the active ingredient in many bleaches. Calculate the ratio of the concentrations of ClO$^-$ and HClO in a bleach solution having a pH adjusted to 6.50 by using strong acid or strong base.

16.48 Aspirin (shown below) is a derivative of salicylic acid, which has K_a = 1.1 × 10^{-3}. Calculate the ratio of the concentrations of the salicylate ion (its conjugate base) to salicylic acid in a solution that has a pH adjusted to 2.50 by using strong acid or strong base.

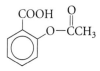

16.49 Predict the pH region in which each of the following buffers will be effective, assuming equal molarities of the acid and its conjugate base: (a) sodium lactate and lactic acid; (b) sodium benzoate and benzoic acid; (c) potassium hydrogen phosphate and potassium phosphate; (d) potassium hydrogen phosphate and potassium dihydrogen phosphate; (e) hydroxylamine and hydroxylammonium chloride.

16.50 Predict the pH region in which each of the following buffers will be effective, assuming equal molarities of the acid and its conjugate base: (a) sodium nitrite and nitrous acid; (b) sodium formate and formic acid; (c) sodium carbonate and sodium hydrogen carbonate; (d) ammonia and ammonium chloride; (e) pyridine and pyridinium chloride.

16.51 Use Tables 15.3 and 15.6 to suggest a conjugate acid-base system that would be an effective buffer at a pH close to (a) 2; (b) 7; (c) 3; (d) 12.

16.52 Use Tables 15.3 and 15.6 to suggest a conjugate acid-base system that would be an effective buffer at a pH close to (a) 4; (b) 9; (c) 5; (b) 11.

16.53 (a) What must be the ratio of the concentrations of CO_3^{2-} and HCO_3^- ions in a buffer solution having a pH of 11.00? (b) What mass of K_2CO_3 must be added to 1.00 L of 0.100 M $KHCO_3(aq)$ to prepare a buffer solution with a pH of 11.00? (c) What mass of $KHCO_3$ must be added to 1.00 L of 0.100 M $K_2CO_3(aq)$ to prepare a buffer solution with a pH of 11.00? (d) What volume of 0.200 M $K_2CO_3(aq)$ must be added to 100.0 mL of 0.100 M $KHCO_3(aq)$ to prepare a buffer solution with a pH of 11.00?

16.54 (a) What must be the ratio of the molarities of PO_4^{3-} and HPO_4^{2-} ions in a buffer solution having a pH of 12.00? (b) What mass of K_3PO_4 must be added to 1.00 L of 0.100 M $K_2HPO_4(aq)$ to prepare a buffer solution with a pH of 12.00? (c) What mass of K_2HPO_4 must be added to 1.00 L of 0.100 M $K_3PO_4(aq)$ to prepare a buffer solution with a pH of 12.00? (d) What volume of 0.150 M $K_3PO_4(aq)$ must be added to 50.0 mL of 0.100 M $K_2HPO_4(aq)$ to prepare a buffer solution with a pH of 12.00?

16.55 A 100.0-mL buffer solution is 0.10 M $CH_3COOH(aq)$ and 0.10 M $NaCH_3CO_2(aq)$. (a) What is the pH of the buffer solution? (b) What are the pH and the pH change resulting from the addition of 3.0 mmol NaOH to the buffer solution? (c) What are the pH and the pH change resulting from the addition of 6.0 mmol HNO_3 to the initial buffer solution?

16.56 A 100.0-mL buffer solution is 0.15 M $Na_2HPO_4(aq)$ and 0.10 M $KH_2PO_4(aq)$. In Table 15.6, the K_{a2} for phosphoric acid is 6.2×10^{-8}. (a) What is the pH of the buffer solution? (b) What are the pH and the pH change resulting from the addition of 8.0 mmol NaOH to the buffer solution? (c) What are the pH and the pH change resulting from the addition of 10.0 mmol HNO_3 to the initial buffer solution?

16.57 A 100.0-mL buffer solution is 0.10 M $CH_3COOH(aq)$ and 0.10 M $NaCH_3CO_2(aq)$. (a) What are the pH and the pH change resulting from the addition of 10.0 mL of 0.950 M $NaOH(aq)$ to the buffer solution? (b) What are the pH and the pH change resulting from the addition of 20.0 mL of 0.10 M $HNO_3(aq)$ to the initial buffer solution? (*Hint:* Consider the dilution stemming from the addition of strong base or acid.)

16.58 A 100.0-mL buffer solution is 0.15 M $Na_2HPO_4(aq)$ and 0.10 M $KH_2PO_4(aq)$. (a) What are the pH and the pH change resulting from the addition of 80.0 mL of 0.010 M $NaOH(aq)$ to the buffer solution? (b) What are the pH and the pH change resulting from the addition of 10.0 mL of 1.0 M $HNO_3(aq)$ to the initial buffer solution? (*Hint:*

Consider the dilution stemming from the addition of strong base or acid.)

Solubility Products

The values for the solubility products of some sparingly soluble salts are listed in Table 16.5.

16.59 Write the expression for the solubility products of the following substances: (a) AgBr; (b) Ag_2S; (c) $Ca(OH)_2$; (d) Ag_2CrO_4.

16.60 Write the expressions for the solubility products of the following substances: (a) AgI; (b) AgSCN; (c) Sb_2S_3; (d) $Mg_3(PO_4)_2$.

16.61 Determine the K_{sp} for the following sparingly soluble substances, given their molar solubilities: (a) AgBr, 8.8×10^{-7} mol/L; (b) $PbCrO_4$, 1.3×10^{-7} mol/L; (c) $Ba(OH)_2$, 0.11 mol/L; (d) MgF_2, 1.2×10^{-3} mol/L.

16.62 Determine the K_{sp} for the following sparingly soluble compounds, given their molar solubilities: (a) AgI, 8.9×10^{-9} mol/L; (b) $Ca(OH)_2$, 0.011 mol/L; (c) $BaCrO_4$, 9.0×10^{-5} mol/L; (d) Hg_2Cl_2, 5.2×10^{-7} mol/L.

16.63 Use the data in Table 16.5 to calculate the molar solubility of (a) MgF_2; (b) $BaSO_4$; (c) CuI.

16.64 Use the data in Table 16.5 to determine the molar solubility of (a) $PbSO_4$; (b) Hg_2I_2; (c) $Fe(OH)_2$.

16.65 The molarity of CrO_4^{2-} in a saturated Tl_2CrO_4 solution is 6.3×10^{-5} mol/L. What is the K_{sp} of Tl_2CrO_4?

16.66 The molar solubility of cerium(III) hydroxide, $Ce(OH)_3$, is 5.2×10^{-6} mol/L. What is the K_{sp} of cerium(III) hydroxide?

Common-Ion Effect

16.67 Use the data in Table 16.5 to calculate the molar solubility of each sparingly soluble substance in its respective solution: (a) silver chloride in 0.20 M $NaCl(aq)$; (b) mercury(I) chloride in 0.10 M $NaCl(aq)$; (c) lead(II) chloride in 0.10 M $CaCl_2(aq)$; (d) iron(II) hydroxide in 1.0×10^{-4} M $FeCl_2(aq)$.

16.68 Use the data in Table 16.5 to calculate the solubility of each sparingly soluble substance in its respective solution: (a) silver bromide in 1.0×10^{-3} M $NaBr(aq)$; (b) magnesium carbonate in 4.2×10^{-5} M $Na_2CO_3(aq)$; (c) lead(II) sulfate in 0.10 M $Na_2SO_4(aq)$; (d) nickel hydroxide in 3.7×10^{-5} M $NiSO_4(aq)$.

16.69 (a) What molar concentration of Ag^+ ions is required for the formation of a precipitate when added to 1.0×10^{-5} M $NaCl(aq)$? (b) What mass (in micrograms) of $AgNO_3$ needs to be added for the onset of precipitation in 100.0 mL of the solution in (a)?

16.70 It is necessary to add iodide ions to precipitate lead(II) ion from 0.0020 M $Pb(NO_3)_2(aq)$. (a) What (minimum) iodide ion concentration is required for the onset of PbI_2 precipitation? (b) What mass (in grams) of KI must be added for PbI_2 to form?

16.71 Determine the pH required for the onset of precipitation of $Ni(OH)_2$ from 0.010 M $NiSO_4(aq)$.

16.72 Limestone is composed primarily of calcium carbonate. A 1.0-mm³ chip of limestone was accidentally dropped into a water-filled swimming pool, measuring 10. m × 7 m × 2 m. Assuming that the carbonate ion does not function as a Brønsted base and that the pH of the water is 7.0, will the pebble dissolve entirely? The density of calcium carbonate is 2.71 g/cm³.

Predicting Precipitation Reactions

16.73 Decide whether a precipitate will form when the following solutions are mixed: (a) 27.0 mL of 0.0010 M NaCl(aq) and 73.0 mL of 0.0040 M $AgNO_3(aq)$; (b) 1.0 mL of 1.0 M $K_2SO_4(aq)$, 10.0 mL of 0.0030 M $CaCl_2(aq)$, and enough water to dilute the solution to 100.0 mL.

16.74 Decide whether a precipitate will form when the following solutions are mixed: (a) 5.0 mL of 0.10 M KI(aq) and 1.00 L of 0.010 M $AgNO_3(aq)$; (b) 3.3 mL of 1.0 M HCl(aq), 4.9 mL of 0.0030 M $AgNO_3(aq)$, and enough water to dilute the solution to 50.0 mL.

16.75 Suppose that there are typically 20 average-sized drops in 1 mL of an aqueous solution. Will a precipitate form when 1 drop of 0.010 M NaCl(aq) is added to 10.0 mL of (a) 0.0040 M $AgNO_3(aq)$; (b) 0.0040 M $Pb(NO_3)_2(aq)$?

16.76 Assuming 20 drops per milliliter, will a precipitate form if (a) 7 drops of 0.0029 M KOH(aq) are added to 25.0 mL of 0.0018 M $CaCl_2(aq)$; (b) 10 drops of 0.010 M NaOH(aq) are added to 10.0 mL of 0.0040 M $AgNO_3(aq)$?

16.77 Which sulfide precipitates first when sulfide ions are added to a solution containing equal amounts of Fe^{2+}, Cu^{2+}, and Zn^{2+}? Explain your conclusion.

16.78 In the process of separating Cu^{2+} ions from Pb^{2+} ions as sparingly soluble iodates, what is the Cu^{2+} concentration when $Pb(IO_3)_2$ just begins to precipitate from a solution that is 0.0010 M $PbCl_2(aq)$ and 0.0010 M $CuCl_2(aq)$?

16.79 The concentrations of magnesium, calcium, and nickel(II) ions in an aqueous solution are 0.0010 mol/L. (a) In what order do they precipitate when a KOH solution is added? (b) Determine the pH at which each salt precipitates.

16.80 Suppose that two hydroxides MOH and $M'(OH)_2$ both have $K_{sp} = 1.0 \times 10^{-12}$ and that initially both cations are present in a solution at concentrations of 0.0010 mol/L. Which hydroxide precipitates first, and at what pH, when NaOH is added?

Dissolving Precipitates and Qualitative Analysis

16.81 Calculate the solubility of silver bromide in 0.10 M KCN(aq). Refer to Tables 16.5 and 16.6.

16.82 Precipitated silver chloride dissolves in ammonia solutions as a result of the formation of the $Ag(NH_3)_2^+$ ion. What is the solubility of silver chloride in 1.0 M $NH_3(aq)$?

16.83 Use the data in Table 16.5 to calculate the solubility of each sparingly soluble substance in its respective solution: aluminum hydroxide at (a) pH = 7.0; (b) pH = 4.5; zinc hydroxide at (c) pH = 7.0; (d) pH = 6.0.

16.84 Use the data in Table 16.5 to calculate the solubility of each sparingly soluble compound in its respective solution: iron(III) hydroxide at (a) pH = 11.0; (b) pH = 3.0; iron(II) hydroxide at (c) pH = 8.0; (d) pH = 6.0.

16.85 Consider the two equilibria

$$CaF_2(s) \rightleftharpoons Ca^{2+}(aq) + 2 F^-(aq), K_{sp} = 4.0 \times 10^{-11}$$

$$F^-(aq) + H_2O(l) \rightleftharpoons HF(aq) + OH^-(aq),$$
$$K_b = 2.9 \times 10^{-11}$$

(a) Write the chemical equation for the overall equilibrium and determine the corresponding equilibrium constant. (b) Determine the solubility of CaF_2 at pH = 7.0. (c) Determine the solubility of CaF_2 at pH = 5.0.

16.86 Consider the two equilibria

$$BaF_2(s) \rightleftharpoons Ba^{2+}(aq) + 2 F^-(aq), K_{sp} = 1.7 \times 10^{-6}$$

$$F^-(aq) + H_2O(l) \rightleftharpoons HF(aq) + OH^-(aq),$$
$$K_b = 2.9 \times 10^{-11}$$

(a) Write the chemical equation for the overall equilibrium and determine the corresponding equilibrium constant. (b) Determine the solubility of BaF_2 at pH = 7.0. (c) Determine the solubility of BaF_2 at pH = 4.0.

Supplementary Exercises

16.87 Predict whether aqueous solutions of the following salts will be acidic, basic, or neutral (and justify your prediction): (a) KI; (b) CsF; (c) CrI_3; (d) $C_6H_5NH_3Cl$; (e) Na_2CO_3; (f) $Cu(NO_3)_2$.

16.88 Write an equilibrium that shows that (a) a $CrCl_3$ solution is acidic; (b) a $(CH_3)_3NHCl$ (trimethylammonium chloride) solution is acidic; (c) a $NaC_2H_5CO_2$ (sodium propionate) solution is basic; (d) a Na_3PO_4 solution is basic.

16.89 A solution is prepared by mixing 200.0 mL of 0.27 M $Na_3PO_4(aq)$ and 150.0 mL of 0.62 M KCl(aq). What is the pH of the mixed solution?

16.90 Describe the principal features that distinguish the pH curve of a strong acid-strong base titration from that of a weak acid-strong base titration.

16.91 Distinguish between the end point and the stoichiometric point of an acid-base titration.

16.92 The pH at the stoichiometric point in the titration of a weak base with a strong acid occurs near pH = 4.0. Which indicators in Table 16.3 would have a suitable end point?

16.93 In the determination of the heat evolved in the neutralization of a strong acid by a strong base, a total of 21.0 mL of 3.0 M $HNO_3(aq)$ was mixed with 25.2 mL of 2.50 M $NaOH(aq)$. What is the pH of the resulting solution?

16.94 A 25.0-mL sample of 6.0 M $HCl(aq)$ is diluted to 1.0 L, and 15.7 mL of this diluted solution was required to reach the stoichiometric point in the titration of 25.0 mL of $KOH(aq)$. What is the molarity of the KOH solution?

16.95 A 20.0-mL sample of 0.020 M $HCl(aq)$ solution was titrated with 0.035 M $KOH(aq)$. Calculate the pH at the following points in the titration and sketch the pH curve: (a) no KOH added; (b) 5.00 mL of KOH(aq) added; (c) an additional 5.00 mL of KOH(aq) (for a total of 10.0 mL) added; (d) another 5.00 mL of KOH(aq) added; (e) another 5.00 mL of KOH(aq) added. (f) Determine the volume of KOH(aq) required to reach the stoichiometric point.

16.96 An old bottle labeled "Standardized 6.0 M $NaOH(aq)$" was found on the back of a shelf in the stockroom. Over time, some of the NaOH had reacted with the glass and the solution was no longer 6.0 M. To determine its concentration, 5.0 mL of the solution was diluted to 100.0 mL and titrated to the stoichiometric point with 11.8 mL of 2.05 M $HCl(aq)$. What is the molarity of the sodium hydroxide solution?

16.97 (a) What volume of 0.0400 M $NaOH(aq)$ is required to reach the stoichiometric point in the titration of 10.00 mL of 0.0633 M $HBrO(aq)$? (b) What is the pH at the stoichiometric point? (c) Use Table 16.3 to suggest a suitable indicator.

16.98 The narcotic cocaine is a weak base with pK_b = 5.59. Calculate the ratio of the concentration of cocaine and its conjugate acid in a solution of pH = 8.00.

16.99 Novocaine, which is used by dentists as a local anesthetic, is a weak base with pK_b = 5.05. Blood has a pH of 7.4. What is the ratio of the novocaine concentration to that of its conjugate acid in the bloodstream?

16.100 A buffer solution is prepared by mixing 50.0 mL of 0.022 M $C_6H_5COOH(aq)$ and 20.0 mL of 0.032 M $NaC_6H_5CO_2(aq)$. (a) What is the pH of the buffer solution? (b) What are the pH and the change in pH after the addition of 0.054 mmol HCl to the buffer solution? (c) What would be the pH change if the 0.054 mmol HCl had been added to pure water instead of the buffer solution? (d) What are the pH and the change in pH after the addition of 10.0 mL of 0.054 M $HCl(aq)$ to the original buffer solution?

16.101 Describe, with accompanying calculations, the procedure for preparing a buffer solution for pH = 10.0, starting with solid Na_2CO_3 and solid $NaHCO_3$.

16.102 What is the ideal pH range for a buffer solution that uses HBrO and NaBrO as the acid-base pair?

16.103 (a) What is the molar solubility of barium sulfate, which is used to enhance x-rays of the gastrointestinal tract? (b) What is its molar solubility in 2.0 × 10^{-4} M $Ba(NO_3)_2(aq)$? (c) What mass of $BaSO_4$ will dissolve in 10.0 L of the solution in (b)?

16.104 Limewater is a saturated aqueous calcium hydroxide solution. (a) What is the pH of limewater? (b) What volume of 0.010 M $HCl(aq)$ is required to titrate 25.0 mL of limewater to the phenolphthalein end point?

16.105 Will Ag_2CO_3 precipitate from a solution formed from a mixture of 100.0 mL of 1.0 × 10^{-4} M $AgNO_3(aq)$ and 100.0 mL of a solution containing 1.0 × 10^{-4} mol/L $CO_3^{2-}(aq)$?

16.106 A 25.0-mL sample of 0.20 M $(COOH)_2(aq)$, oxalic acid, is titrated with 0.20 M $NaOH(aq)$. For oxalic acid, K_{a1} = 5.9 × 10^{-2} and K_{a2} = 6.5 × 10^{-5} (see Table 15.6). (a) What volume of 0.20 M $NaOH(aq)$ is required to reach the first stoichiometric point? (b) What is the salt present at that point? (c) Calculate the pH at the first stoichiometric point. (d) What (total) volume of 0.20 M $NaOH(aq)$ is required to reach the second stoichiometric point? What is the salt present at that point? (e) Calculate the pH at the second stoichiometric point. (f) Suggest a suitable indicator to detect the first stoichiometric point and a second indicator for the second stoichiometric point.

16.107 What is the pH at each stoichiometric point in the titration of 0.20 M $H_2SO_4(aq)$ with 0.20 M $NaOH(aq)$?

16.108 What volume (in liters) of a saturated aqueous solution of mercury(I) iodide, Hg_2I_2, contains an average of one mercury(I) ion, Hg_2^{2+}?

16.109 Barium ions can be separated from lead(II) ions by precipitating barium sulfate from the solution. What sulfate ion concentrations are required for the precipitation of (a) $BaSO_4$ and (b) $PbSO_4$ from a solution containing 0.010 mol/L Ba^{2+} and 0.010 mol/L Pb^{2+}? (c) What is the concentration of barium ions when lead(II) sulfate begins to precipitate?

16.110 It is often useful to know whether two ions can be separated by selective precipitation from a solution. Generally, a 99% separation is considered "separated." A solution is 0.010 mol/L Pb^{2+} and 0.010 mol/L Ag^+. Chloride ions are added from a sodium chloride solution. (a) Determine the chloride ion concentration required for the precipitation of each cation. (b) Which cation precipitates first? (c) What is the molarity of the first cation that precipitates when the second cation begins to precipitate?

Applied Exercises

For Exercises 16.111–16.114, see Applying Chemistry: Case Study 16.

16.111 Very rapid, deep breathing called hyperventilation can result from anxiety. It causes dizziness and disturbs the

HCO$_3^-$/H$_2$CO$_3$ equilibrium. (a) Predict in which direction the equilibrium is driven by hyperventilation and explain your answer. (b) Should paramedics treating an accident victim experiencing hyperventilation anticipate acidosis or alkalosis?

16.112 Blood pH is controlled by a complex mixture of acids and bases. What would be the pH of a buffer with a 20:1 ratio of HCO$_3^-$ to H$_2$CO$_3$ if no other acids and bases were present? Take pK_a = 6.1 for H$_2$CO$_3$ in blood.

16.113 What is the effective pH range of the [HCO$_3^-$]/[H$_2$CO$_3$] buffer in water?

16.114 To simulate blood conditions, a phosphate buffer system with a pH = 7.40 must be prepared. (a) What must be the ratio of the concentrations of HPO$_4^{2-}$ to H$_2$PO$_4^-$ ions? (b) What mass of Na$_2$HPO$_4$ must be added to 500.0 mL of 0.10 M NaH$_2$PO$_4$(aq) in the preparation of the buffer?

16.115 Fluoridation of city water supplies produces a fluoride ion concentration close to 5 × 10^{-5} mol/L. Will CaF$_2$ precipitate in hard water in which the Ca^{2+} ion concentration is 2 × 10^{-4} mol/L?

16.116 The fluoride ions in drinking water convert the hydroxyapatite, Ca$_5$(PO$_4$)$_3$OH, of teeth into fluorapatite, Ca$_5$(PO$_4$)$_3$F. The K_{sp} of the two compounds are 1.0 × 10^{-36} and 1.0 × 10^{-60}, respectively. What are the molar solubilities of each substance? The solubility equilibria to consider are

$$Ca_5(PO_4)_3OH(s) \rightleftharpoons 5\ Ca^{2+}(aq) + 3\ PO_4^{3-}(aq) + OH^-(aq)$$

$$Ca_5(PO_4)_3F(s) \rightleftharpoons 5\ Ca^{2+}(aq) + 3\ PO_4^{3-}(aq) + F^-(aq)$$

Integrated Exercises

16.117 A 25.00-mL sample of 0.010 M H$_2$SO$_4$(aq) is titrated with 0.010 M Ba(OH)$_2$(aq). The leads of a device like that in Fig. 3.9 are inserted into the analyte. (a) Write the equation for the reaction that occurs during the titration. (b) At the beginning of the titration, the bulb is glowing brightly. As the titrant is added at a steady rate, the bulb begins to dim and then goes out. As more titrant is added the bulb begins to glow again and is eventually as bright as in the beginning. How many milliliters of the titrant have been added when the bulb goes out? (c) Explain the changes in the intensity of the light

from the bulb in the apparatus and indicate which ions are carrying the current at the beginning and end of the titration.

16.118 Predict the effect of each of the following on the solubility of Ag$_2$CO$_3$: (a) adding nitric acid; (b) heating the solution; (c) adding dry ice (solid carbon dioxide); (d) adding ammonia.

16.119 When 10.0 mg of sodium barbituate are dissolved in 250.0 mL of solution, the resulting pH is 7.71. The molar mass of sodium barbituate is 150 g/mol. Determine (a) the percentage protonation of barbituate ions; (b) the K_a of barbituric acid.

16.120 A 1.331-g sample of impure barium hydroxide was dissolved in 250.0 mL of aqueous solution. A 35.0-mL portion of this solution was titrated to the stoichiometric point with 17.6 mL of 0.0935 M HCl(aq). What is the percentage purity of the original sample?

16.121 In an attempt to determine the amount of sulfur dioxide in the air near a power plant, two students set up a bubbler that passes air through 50.00 mL of 1.00 × 10^{-4} M NaOH(aq). The temperature is 22°C, and the atmospheric pressure 753 Torr. The air is pumped for 2.5 h at a flow rate of 3.0 L/h. The students then returned to the laboratory and titrated the solution with 1.50 × 10^{-4} M HCl(aq), using phenolphthalein indicator, to see how much NaOH was left unreacted. They found that 30.2 mL of HCl(aq) was required to reach the stoichiometric point. (a) Write the balanced chemical equation for the reaction of sulfur dioxide and aqueous sodium hydroxide. (b) How much NaOH had reacted with SO$_2$? (c) What was the concentration of sulfur dioxide in the air, in parts per million?

16.122 Consider the two equilibria

$$ZnS(s) \rightleftharpoons Zn^{2+}(aq) + S^{2-}(aq)\ K_{sp} = 1.6 \times 10^{-24}$$

$$S^{2-}(aq) + 2\ H_2O(l) \rightleftharpoons H_2S(aq) + 2\ OH^-(aq),$$
$$K_a = K_{a1}K_{a2} = 9.3 \times 10^{-22}$$

(a) Write the chemical equation for the overall equilibrium and determine the corresponding equilibrium constant. (b) Determine the solubility of ZnS in a saturated H$_2$S (0.1 M H$_2$S(aq)) solution adjusted to pH = 7.0. (c) Determine the solubility of ZnS in a saturated H$_2$S (0.1 M H$_2$S(aq)) solution adjusted to pH = 10.0.

Connection 4

We can buy bottled water from a store, but we can also drink tap water nearly free. Is bottled water really necessary? How pure is our water? When we boil tap water, we may notice that a hard scale has been deposited in the pot. Clearly, tap water contains more than just water. What substances are commonly found in tap water and how is it made safe to drink?

Water is available on a huge scale worldwide, but in various states of purity; consequently, it is treated by nearly every city and town. Snowmelt and glacier water are relatively free of contaminants; and water flowing through underground aquifers (porous rock layers) is relatively pure, although still subject to underground pollution. River water and lake water usually need the heaviest treatment, as they may contain industrial effluent, agricultural runoff, and other waste products from cities upstream.

Assuring a potable (consumable) drinking water supply is the responsibility of water chemists and sanitary engineers. Domestic water must be free of color, odor, suspended solids, toxic compounds, and bacteria. Water treatment makes use of some of the physical separation techniques discussed in Section 1.14 as well as catalysts (Section 13.10) and the neutralization and precipitation reactions discussed in Chapters 15 and 16. All these reactions rely on the principles of chemical equilibrium introduced in Chapter 14.

The extent of municipal water treatment depends on the purity of the raw (untreated) water but generally includes the

Some bottled water is distilled water, which contains only ions, such as hydrogen carbonate ions, that are formed when gases from the atmosphere dissolve in water. Other bottled water comes from wells and underground springs. Naturally carbonated water is produced when water trickles through limestone rock.

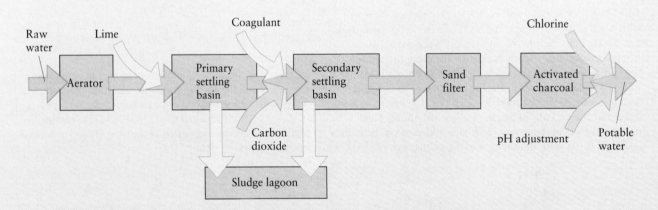

The steps commonly used in water purification. Following an initial sedimentation and filtration (not shown), the water is aerated, lime may be added to soften it, and solids are precipitated, flocculated, allowed to settle, and removed by filtration. Then the water may be decolorized and deodorized, the pH is adjusted, and chlorine is added as a disinfectant.

steps shown in the illustration. Raw water may contain any or all of the following:

- Large debris, such as sand, plants, fish, and insects
- Small suspended (colloidal) particles, such as clay and microbes
- Acids or bases
- Undesirable organic compounds
- Toxic heavy metal ions
- Calcium and magnesium ions, usually in the form of hydrogen carbonates
- Dissolved carbon dioxide and harmless salts

Additional materials are added to the water during treatment. Let's follow what happens to raw water as it makes its way through the steps in a typical treatment plant.

Sedimentation   First, the water is allowed to settle in sedimentation tanks, which removes much of the large debris; then it is filtered through gravel and sand to remove smaller debris such as insects and leaves.

Aeration   The next step for highly polluted water is aeration. In this step, air is bubbled through the water to remove foul-smelling dissolved gases such as H_2S, to oxidize some organic compounds to CO_2, and to add oxygen and nitrogen. Nonaerated water, such as well water or distilled water, tastes flat. Aeration also oxidizes any Fe^{2+} ions to Fe^{3+} ions.

Softening   Hard water contains relatively high concentrations of Ca^{2+} and Mg^{2+} ions, often in the form of hydrogen carbonates. Unless they are removed, these cations will precipitate as scale in furnace boilers and coffee pots, and react with soap to form a scum that is hard to remove from laundry. Magnesium and calcium hydrogen carbonates can be precipitated from hard water by adding calcium ions in the form of slaked lime:

$$Mg(HCO_3)_2(aq) + Ca(OH)_2(aq) \longrightarrow$$
$$Mg(OH)_2(s) + Ca(HCO_3)_2(aq)$$

$$Ca(HCO_3)_2(aq) + Ca(OH)_2(aq) \longrightarrow$$
$$2\,CaCO_3(s) + 2\,H_2O(l)$$

It seems ironic at first that calcium ions can be removed by adding more calcium ions. This softening method works because the strong base $Ca(OH)_2$ converts the soluble hydrogen carbonates into insoluble carbonates, which can be removed from the water by filtration.

Precipitation   If toxic heavy metal ions are present, additional precipitation steps may be required. Usually either

The pools of water at this wastewater treatment plant are settling basins.

sodium or potassium hydroxide is used to precipitate the cations as insoluble hydroxides.

Flocculation   After the precipitates form, the water is pumped into a primary settling basin. The precipitates tend to form as a colloidal suspension (Section 12.5), so either $Fe_2(SO_4)_3$ or an alum such as $Al_2(SO_4)_3 \cdot 18H_2O$ is added to coagulate and flocculate the precipitates so that they can be filtered. *Coagulation* involves pulling the ions together to form larger particles. *Flocculation* involves joining particles together in a fluffy gel that can be filtered. Aluminum is amphoteric and forms soluble $Al(OH)_4^-$ ions in basic solution. Carbon dioxide is often added to raise the acidity of the water, which precipitates the aluminum as $Al(OH)_3$ so that it can be removed.

Filtration   As the precipitates settle slowly in a secondary basin, the particles *adsorb* (attract to their surfaces) any remaining suspended $CaCO_3$, bacteria, and other suspended particles, such as dirt and algae. Precipitates from the primary and secondary basins are combined in a sludge lagoon for disposal. The clear water is then decanted and passed through a sand filter to remove any remaining suspended particles. The *turbidity*, or cloudiness, of the water is reduced in this step.

Decolorizing and deodorizing   Often river water contains organic compounds, which come both from natural sources such as soil and from industrial wastes. An activated charcoal filter is used to remove the organic compounds. Activated charcoal is made of finely divided carbon with a highly porous surface that adsorbs organic compounds as the water passes through it.

pH adjustment   The water is made slightly basic to reduce acid corrosion of the pipes. Typically, municipal water has a pH of about 8 when it leaves the treatment plant.

Disinfection   At this point, a disinfectant, usually chlorine, is added. In the United States, the chlorine level is required by law to be greater than 1 g of Cl_2 per 10^3 kg of water at the point of consumption. Even when the raw water is pure, chlorine is needed to prevent the growth of microbes, molds, and algae in the water pipes. In water, chlorine forms hypochlorous acid, which is highly toxic to bacteria:

$$Cl_2(g) + 2 H_2O(l) \longrightarrow H_3O^+(aq) + Cl^-(aq) + HClO(aq)$$

Fluoridation   Many municipalities also add a very small concentration of ionic fluorides to the water, as it is believed that a low concentration of fluorides in the diet strengthens teeth.

High-purity water for special applications is obtained by distillation or by *ion exchange*, the exchange of one type of ion in a solution by another. In ion exchange, water passes through a *zeolite*, an aluminosilicate with a very open structure that can capture ions such as Mg^{2+} and Ca^{2+} and exchange them for H^+ ions. Depending on the source and condition of the water, additional water purification steps, such as reverse osmosis (Section 12.14) may be required.

How is bottled water treated? Some bottled water is distilled, so it contains very few ions. Other types of bottled water may be obtained from groundwater sources such as natural springs or wells. Some groundwater sources are very pure, but usually bottled water is filtered and sterilized before bottling. Generally, bottled water is low in fluorides, and a recent increase in cavities after years of a much lower incidence of tooth decay is being attributed by some to increased drinking of nonfluoridated bottled water instead of tap water.

For Further Reading

C. Baird, The purification of polluted water, *Environmental Chemistry*, 2nd ed., New York: W. H. Freeman and Company, 1999, pp. 461–501.

The Water-Wastewater Web, http://www.w-ww.com/plants/index.html

Applying Your Knowledge

You may need to consult Chapters 13–16 and occasionally earlier chapters to answer these questions.

1. (a) Slaked lime is used to precipitate heavy metal ions. Write the complete chemical equations and the net ionic equations for the precipitation of Pb^{2+} and Fe^{3+} ions by slaked lime. (b) Write the balanced chemical equation for the oxidation of ethanol, C_2H_5OH, to CO_2 and H_2O by O_2 during aeration.

2. Some of the anions that may be found in raw water are Cl^-, SO_4^{2-}, NO_3^-, HCO_3^-, CN^- (in mine runoff), and HPO_4^{2-}. (a) How will each anion affect the pH of the water? (b) Nontoxic ions are left in the water, but toxic ions must be removed. Which of these anions are toxic? You may want to consult a reference book such as *The Merck Index*.

3. Explain why the high surface area of zeolites and activated charcoal is important for purifying running water.

4. (a) Find out the source of your municipal water supply and identify the primary pollutants. (b) What is the pH of the raw water and what is the pH of the water after it leaves the treatment plant? (c) Identify the reactions that cause the change in pH. *Note: Some cities post water treatment information on the World-Wide Web.*

5. The water source for Boulder, Colorado is the pure water that melts from the ice of the Arapahoe Glacier, high in the mountains. The characteristics of the water change as it runs down a rocky canyon and through the city, which has both a municipal water treatment plant and a wastewater treatment plant. One day the following data were recorded. Suggest explanations for each of the values:

	At the glacier	Entering the city	Leaving the city
pH	3.9	7.2	8.4
Hardness (mg/L)	15	87	140
Total dissolved salts (mg/L)	9	49	380

6. Design a water treatment plant for river water, which may contain calcium and magnesium hydrogen carbonates, iron(III) salts, organic pollutants, and various kinds of large and small debris and has a foul odor. Decide if you want to fluoridate your water supply. Draw a schematic diagram of your water treatment facility. For each treatment step, indicate the purpose of the step, list the chemicals that would be needed, and, if a chemical reaction is involved, write a balanced chemical equation for the reaction.

Chapter

The Direction of Chemical Change

Have you ever wondered why water evaporates? Why hot objects cool? Why hydrogen combines with oxygen? Why leaves turn red in the fall? Why *anything* happens? A part of the answer is related to the availability of energy. We need energy to think, to move, and to live. Landscapes emerge and change as a result of the energy associated with volcanos, hurricanes, earthquakes, and continental drift. Agriculture depends on energy, not only to run machinery, but also because plants and the animals that feed on plants use the energy of sunlight to grow. On a smaller scale, every chemical reaction, whether in a test tube or part of a metabolic process, makes use of energy to rearrange the bonds between atoms.

Thermodynamics, the branch of chemistry we first encountered in Chapter 6, deals with questions like these. That chapter introduced us to the first law of thermodynamics, which describes the conservation of energy. There we were concerned with the *quantity* of energy a reaction produced or absorbed. We saw how the internal energy, *U*, is used to keep track of energy transfers in the form of heat and work. We also saw how to use the enthalpy, *H*, to keep track of transfers of energy as heat at constant pressure. In this chapter, we change the focus of the discussion and meet another law of thermodynamics that governs the *direction* of natural change. This "second law" enables us to predict whether or not a reaction has a tendency to occur.

The second law of thermodynamics is of fundamental importance in chemistry. It provides a basis for discussing, explaining, and predicting equilibrium, the subject of the previous three chapters. It is also the foundation of the whole field of electrochemistry, the subject of the following chapter.

Life thrives on Earth because there is a constant supply of energy from the Sun. Some of this energy is stored in plants through photosynthesis. Although photosynthesis is only about 3% efficient, it supports nearly all plants on Earth and the animals that feed on them. The concepts in this chapter are critical for understanding the conversion of energy from one form to another and for research into the availability and deployment of energy.

THE DIRECTION OF SPONTANEOUS CHANGE

We have seen how to account for the exchange of atoms and energy in chemical reactions, but we still cannot say what determines the *direction* of a chemical reaction. Why does methane burn in oxygen to form carbon dioxide and water, but carbon dioxide and water not react to form methane? Why do reactions tend to run only in the direction that leads to equilibrium? To answer these questions, we need to learn about another property of energy beyond the fact that it is conserved.

17.1 Spontaneous Change

The technical term for a natural change is a *spontaneous* change: a **spontaneous change** is a change that tends to occur without needing to be driven by an external influence. One simple example is the cooling of a block of hot metal to the temperature of its surroundings (Fig. 17.1). The reverse change, a block of metal spontaneously growing hotter than its surroundings, has never been observed. A **nonspontaneous change** is a change that occurs only if it is driven—for example, by forcing an electric current through the metal to heat it. The expansion of a gas into a vacuum is spontaneous (Fig. 17.2). A gas has no tendency to contract spontaneously into one part of a container. However, we can drive a gas into a smaller volume and bring about this nonspontaneous change by doing work—in this case, by pushing in a piston.

A spontaneous change need not be fast. Molasses has a spontaneous tendency to flow out of an overturned can, but at low temperatures that flow may be very slow. The reaction in which water is formed from a mixture of hydrogen and oxygen gases is spontaneous, but the mixture can be kept safely for centuries, provided we do not ignite it with a spark. A spontaneous process is a change that has a natural *tendency* to occur; it does not necessarily take place at a significant rate. Spontaneity is a *tendency* to change; that tendency may not be realized in practice.

> *A spontaneous change has a tendency to occur without being driven by an external influence; spontaneous changes need not be fast.*

17.2 Entropy and Disorder

The key idea that accounts for spontaneous change is that *energy and matter tend to become more disordered.* A hot block of metal tends to cool because the energy stored as thermal motion of its atoms tends to spread into the surroundings (Fig. 17.3). The reverse change is very unlikely to be observed, because it is very unlikely that energy will leave the very extensive surroundings and collect inside a small block of metal. Similarly, it is natural for randomly moving gas molecules to spread throughout their container, so it is natural for a gas to expand and fill the entire container. It is very unlikely that the random motion of gas molecules will bring them all simultaneously back into one corner (Fig. 17.4).

The thermodynamic property that measures the disorder of a system is the **entropy,** S. The internal energy, U, is a measure of the *quantity* of energy stored in a system; the entropy is a measure of the *quality* of that stored energy, the orderliness with which the energy is stored. Low entropy means little disorder;

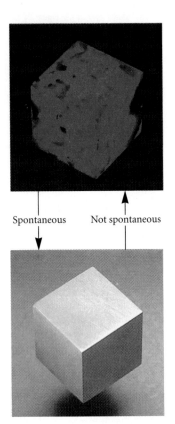

Spontaneous   Not spontaneous

FIGURE 17.1

The direction of spontaneous change is for a hot block of metal (top) to cool to the temperature of its surroundings (bottom). A block at the same temperature as its surroundings does not spontaneously become hotter.

FIGURE 17.2

The direction of spontaneous change for a gas is toward filling its container. A gas that already fills its container does not collect spontaneously in a small region of the container. A glass cylinder containing a brown gas (upper piece of glassware in the top illustration) is attached to an empty flask. When the stopcock between them is opened, the brown gas fills both upper and lower vessels (bottom illustration). The brown gas is nitrogen dioxide.

Spontaneous   Not spontaneous

high entropy means great disorder. Like internal energy, entropy is a state property (Section 6.4); so its value depends only on the current state of the system, and any change in entropy is independent of the path between the initial and final states.

These two ideas—that energy and matter tend to disperse and that entropy is a measure of disorder—are combined in the **second law of thermodynamics,** which says that

The entropy of an isolated system tends to increase.

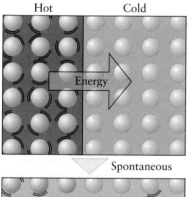

Hot   Cold

Energy

Spontaneous

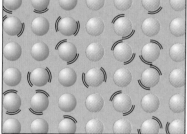

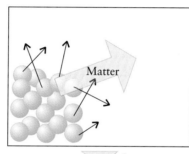

Matter

Spontaneous

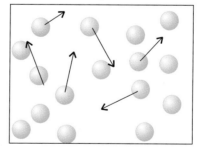

FIGURE 17.3

We can understand the natural direction of the migration of heat from a hot region to a cold region by thinking about the jostling between the vigorously moving atoms in the hot region. Molecules jostle their neighbors, and the thermal motion spreads.

FIGURE 17.4

We can understand the natural direction of the migration of matter by visualizing how the random motion of molecules results in their spreading throughout the available space.

FIGURE 17.5

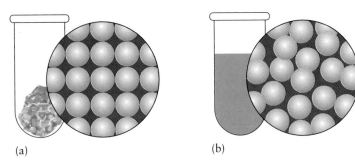

(a)   (b)

A representation of the arrangement of molecules in (a) a solid and (b) a liquid. When the solid melts, there is an increase in the disorder of the system and hence a rise in entrdopy.

In short, things get worse.

This statement summarizes the tendency of energy and matter to spread and become more disordered. In some cases, the system of interest is truly isolated. In others—where the system of interest can influence its immediate surroundings—we get an "isolated system" only by imagining that the boundary has been expanded to include all the parts of the immediate surroundings that our system can affect. In yet others, the "isolated system" may be the entire universe!

Before we explore the consequences of the second law, we need to develop our insight into entropy. We need a sense of when to expect high entropy—much disorder, with a lot of energy stored as thermal motion and atoms or molecules widely dispersed—and when to expect low entropy—an orderly arrangement, with little energy in the system and atoms or molecules highly localized.

The entropy of a substance can be increased in two ways. Entropy is increased by heating, which stimulates the thermal motion of the molecules, or by increasing the number of locations into which the molecules can spread. Heating increases the **thermal disorder,** the disorder arising from the thermal motion of the molecules. Increasing the volume of a substance or mixing it with another substance (as we saw in Section 12.10) spreads the molecules of the substance into additional locations and increases the **positional disorder,** the disorder related to the locations of the molecules.

The relative entropies of different physical states of the same substance are also easy to predict. When a solid melts, its molecules have more energy (melting is endothermic) and have more freedom to move (Fig. 17.5). Because the liquid form of a substance has a higher positional disorder than its solid form at the same temperature, its entropy is higher, too (Fig. 17.6). Conversely, the entropy of a liquid decreases when it freezes to a solid and the molecules settle into orderly arrays. An even bigger increase in entropy is expected when a substance vaporizes. When a substance turns into a vapor, an endothermic process, its molecules gain additional energy and occupy a much greater volume. We can expect gases to have much higher molar entropies than solids and liquids.

Example 17.1 *Predicting the relative entropies of two samples*

Which has the greater entropy: (a) 1 g of pure solid NaCl or 1 g of NaCl dissolved in 100 mL of water; (b) 1 g of water at 25°C or 1 g of water at 50°C?

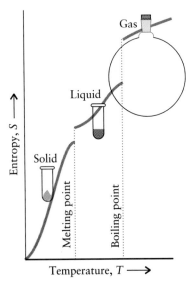

Entropy, S →

Gas

Liquid

Solid

Melting point

Boiling point

Temperature, T →

FIGURE 17.6

The entropy of a solid increases as its temperature is raised. The entropy increases sharply when the solid melts to form the more disordered liquid and then gradually increases again up to the boiling point. A second, larger jump in entropy occurs when the liquid turns into a vapor.

Strategy All other things being equal, the sample with the higher temperature has the greater thermal disorder and the sample spread out over the greater volume has the greater positional disorder.

Solution (a) Pure solid NaCl is concentrated in a much smaller space than the same amount of NaCl dissolved in water, so the NaCl in solution has the greater positional disorder and therefore the greater entropy. (b) One gram of water at 50°C has greater thermal disorder than the same mass of water at 25°C, so the sample at the higher temperature has the greater entropy.

Self-Test 17.1A Which has the greater entropy, 1 mol $CO_2(s)$ or 1 mol $CO_2(g)$ at the same temperature?

[***Answer:*** Gaseous carbon dioxide]

Self-Test 17.1B Which has the greater entropy, a sample of liquid mercury at $-15°C$ or the same sample at $0°C$?

Before we can calculate the entropy change of a process, we need to define entropy more precisely. The Austrian Ludwig Boltzmann defined entropy in terms of the properties and arrangements of molecules (Investigating Matter 17.1). The French engineer Sadi Carnot and the German physicist Rudolph Clausius defined entropy in terms of thermodynamic quantities such as heat and temperature. It turns out that the two approaches are equivalent. We use the thermodynamic approach here but draw on the molecular approach to devise explanations where appropriate, because it gives such deep insight into thermodynamics.

The thermodynamic definition of a *change* in entropy, ΔS, when a system undergoes a change at a constant temperature (such as vaporization) is

$$\text{Change in entropy} = \frac{\text{heat supplied reversibly}}{\text{temperature at which the transfer takes place}} \qquad \Delta S = \frac{q_{rev}}{T} \qquad (1)$$

> Remember that wherever T appears, it means the *absolute* temperature (in kelvins).

where q_{rev} is the heat required to bring about the change reversibly. This expression shows that, with heat expressed in joules and the temperature in kelvins, entropy is expressed in joules per kelvin (J/K). The units of molar entropy, S_m, the entropy divided by the number of moles of molecules in the sample, are joules per kelvin-mole (J/K·mol).

What is the significance of the term *reversibly* in the definition? In everyday language, a reversible process is one that can take place in either direction, like heating and cooling. This common usage is refined in science. In thermodynamics, a **reversible process** is one that can be reversed by an *infinitesimal* change in a variable. For example, if two blocks of metal have the same temperature, then when they are put together, there is no flow of heat between them. If the temperature of one block is increased infinitesimally, then heat will flow from that block into the other. If, instead, the temperature of the first block is decreased infinitesimally, then heat flows in the other direction. Because we can reverse the flow of heat with an infinitesimal change in temperature, the flow of heat is reversible. On the other hand, if one block is 5°C warmer than the other, then an infinitesimal change in temperature of either block does not reverse the flow of heat. In this case, the transfer of energy as heat is irreversible. The term *reversibly* in the definition in Eq. 1 therefore means that when we transfer heat to the system, the source must have the same temperature as the system itself.

Let's consider the meaning of each term in Eq. 1, to reconcile it with our qualitative understanding of entropy. First, we see that *the greater the energy*

In the mid-nineteenth century, there was an explosion of knowledge about how matter and energy interact. This knowledge provided the foundations of thermodynamics. However, although thermodynamic properties could be predicted, they could not be explained. The Austrian physicist Ludwig Boltzmann finally established the link between measurable thermodynamic properties and the behavior of the vast collection of atoms, molecules, and ions that make up matter. Unfortunately, the existence of atoms was not yet universally accepted, and his work was actively discredited. He eventually took his own life in despair.

Boltzmann proposed that the reason systems move spontaneously toward increased entropy is simply that disordered states are more numerous and thus more probable. He proposed that entropy is a measure of the probability that a system is in a certain state and found a formula for relating the entropy of a substance to the number of arrangements of atoms that correspond to the same state:

$$S = k \ln W$$

The fundamental constant k, which is called the *Boltzmann constant*, is 1.38066×10^{-23} J/K. The term W is the number of ways that the atoms or molecules in the sample can be arranged and yet have the same total energy.*

Let's use the Boltzmann formula to find W for a very simple system, a tiny solid made up of 20 diatomic molecules

---

*The term $\ln W$ is the natural logarithm of W (see Appendix 1C).

Ludwig Boltzmann (1844–1906). His formula for entropy (using an earlier notation for natural logarithms) became his epitaph.

of a binary compound such as carbon monoxide, CO. Suppose that the 20 molecules have formed a perfectly ordered crystal and that, because $T = 0$, all motion has

---

transferred to the system as heat, the greater the increase in entropy. Next, we see that for a given transfer of energy, the increase in entropy is inversely proportional to the temperature: *if the transfer is made to a hot system, the increase in entropy is smaller than when the same amount of energy is transferred to a cool system.* The entropy increase is small in a hot system because there is already a lot of thermal motion, and the transfer of more energy has little additional effect. However, because a cool system has little thermal motion, the transfer of energy to a cool system can cause a lot of additional disorder.

An analogy

The transfer of energy to a hot system compared with the transfer of energy to a cool system is like the effect of a sneeze in a busy street compared with the effect of the same sneeze in a quiet library. In a busy street (the hot system), there is already a lot of disorder, and the sneeze has only a slight additional effect. In contrast, the same sneeze in a quiet library (the cool system) will cause a lot of disorder.

ceased (see left panel of figure). We expect the sample to have zero entropy, because there is no disorder in either location or energy. This expectation is confirmed by the Boltzmann formula: because there is only one way of arranging the molecules in the perfect crystal, $W = 1$ and (because $\ln 1 = 0$)

$$S = k \ln 1 = 0$$

Now suppose that the compound is frozen in a disordered state, such that each molecule can point in either of two directions in the solid, yet still have the same energy (see right panel of figure). Because each of the 20 molecules can be in one of two orientations, the total number of ways of arranging the molecules is

$$W = (2 \times 2 \times 2 \ldots)_{\text{twenty factors}} = 2^{20}$$

or over 1 million different arrangements. The entropy of this tiny disorderly solid is therefore

$$\begin{aligned} S &= k \ln 2^{20} = (1.38 \times 10^{-23} \text{ J/K}) \times (20 \ln 2) \\ &= 1.9 \times 10^{-22} \text{ J/K} \end{aligned}$$

The entropy of the disordered solid is higher than that of the perfectly ordered solid. For a solid that contained 1.00 mol CO, corresponding to 6.02×10^{23} CO molecules, each of which could point in either of two directions, the entropy would be

$$\begin{aligned} S &= (1.38 \times 10^{-23} \text{ J/K}) \times (6.02 \times 10^{23}) \ln 2 \\ &= 5.76 \text{ J/K} \end{aligned}$$

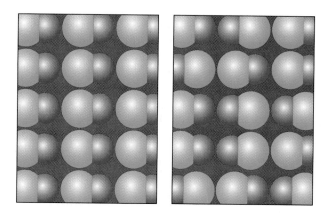

Some of the 20 heteronuclear diatomic molecules in a perfectly ordered arrangement at $T = 0$ (left). The sample has zero spatial and thermal disorder, and hence zero entropy ($S = 0$). This sample represents a perfect crystal at absolute zero. Each of the molecules in the sample on the right (15 of which are shown here) can take up either of two orientations without affecting the energy. There are $2 \times 2 \times \ldots = 2^{20}$, or 1 048 576, different possible arrangements, and this illustration shows a portion of just one of them.

Chemists now use calculations like this one to determine the entropies of more complicated substances and get very good agreement with experimental values. In some cases, experimental values of S are not available, and the Boltzmann formula has to be used to obtain their values.

Example 17.2 *Calculating the change in entropy of a system arising from the transfer of heat*

Calculate the change in entropy of a large tank of water when a total of 100. J of energy is transferred to it reversibly as heat at 20.°C.

Strategy Remember that q is positive when heat enters a system. Because heat is transferred to the water, we expect the entropy of the water to increase. The fact that the sample is large means that its temperature barely changes when the heat is supplied, so we can use Eq. 1 to calculate the change in entropy. The temperature in the denominator is an absolute temperature, in kelvins.

Solution The transfer takes place at 293 K, and $q_{\text{rev}} = +100.$ J. Therefore,

$$\Delta S = \frac{q_{\text{rev}}}{T} = \frac{(+100. \text{ J})}{293 \text{ K}} = +0.341 \text{ J/K}$$

As expected, the entropy of the water increases.

Self-Test 17.2A Calculate the change in entropy of a large iron block at 24.0°C when 1500. J of energy escapes as heat from the block to the surroundings.

[***Answer:*** -5.05 J/K]

Self-Test 17.2B Calculate the change in entropy of a large swimming pool at 28.0°C when 240. J of energy escapes from the pool as heat to the surroundings.

The **entropy of vaporization** of a substance, ΔS_{vap}, is the difference in molar entropies of the vapor and liquid phases of a substance at a stated temperature. Similarly, the **entropy of fusion**, ΔS_{fus}, is the difference in molar entropies of the liquid and solid phases of a substance at a stated temperature.

To calculate the entropy change for a substance undergoing a transition from one phase to another, we need to note two facts. First, at the temperature of the phase transition (such as the boiling point if the transition is vaporization), the temperature of the substance remains constant as heat is supplied. All the energy supplied goes into driving the transition, such as converting liquid into vapor, rather than into raising the temperature. The T in the denominator of Eq. 1 is therefore equal to the transition temperature. The second fact is that, at the temperature of a phase transition, the transfer of heat is reversible: provided the external pressure is fixed (at 1 atm, for instance), raising the temperature of the surroundings an infinitesimal amount results in complete vaporization, and lowering it causes complete condensation.

> *The second law states that the entropy of an isolated system tends to increase. The entropy change of a process is calculated from Eq. 1, which assumes that heat is transferred reversibly. Melting at the melting point and boiling at the boiling point are examples of reversible processes.*

Example 17.3 *Calculating the entropy of vaporization*

Calculate the change in molar entropy when water vaporizes at its boiling point.

Strategy We expect a positive change in entropy because the compact liquid phase is converted into the highly disordered vapor phase. The energy transferred as heat is the enthalpy of vaporization of the liquid, because vaporization takes place at constant pressure. Even though energy is absorbed, there is no change in temperature at the boiling point, because all the energy goes into the separation of the molecules. Therefore, we can use Eq.1 with the numerator replaced by the enthalpy of vaporization (Table 6.2) and the denominator set equal to the boiling point (in kelvins).

Solution The change in molar entropy is

$$\Delta S_{vap} = \frac{q_{rev}}{T_b} = \frac{\Delta H_{vap}{}^{\circ}}{T_b} = \frac{(+40.7 \text{ kJ/mol})}{373 \text{ K}}$$

$$= +\frac{40.7 \times 10^3 \text{ J/mol}}{373 \text{ K}} = +109 \text{ J/K·mol}$$

As expected, the entropy change is positive (and large).

Self-Test 17.3A Calculate the entropy of fusion of ice at its melting point (use the data in Table 6.2).

[*Answer:* +22.0 J/K·mol]

Self-Test 17.3B Calculate the entropy of vaporization of ammonia at its boiling point (use the data in Table 6.2).

17.3 Absolute Entropies

Like the internal energy, entropy is defined in Eq. 1 in terms of a *change* in its value. However, unlike the internal energy, the *absolute* entropy of a substance can be calculated. In a perfect crystal at room temperature, all the atoms are in a perfectly ordered array, and we can expect the entropy to be very low. The entropy is not 0, because the atoms have thermal disorder due to their jiggling motion around their mean positions. However, as the crystal is cooled, the jiggling motion lessens, and thermal disorder vanishes as T approaches 0. This behavior is summarized by the **third law of thermodynamics:**

> The entropy of a perfect crystal approaches 0 as the absolute temperature approaches 0.

All substances have some degree of thermal disorder at temperatures above $T = 0$, so it follows that the entropy of any substance at room temperature is greater than 0. In other words, *all absolute entropies are positive.*

To obtain the absolute entropy of a sample, we determine the change in its entropy by using Eq. 1 at each value of T as a sample is heated from $T = 0$ up to the temperature that interests us (typically 298 K). We do not go into the details of the technique here, but the technique allows us to measure the **standard molar entropy,** $S_m°$, of a substance, the entropy per mole of the pure substance at 1 atm. Table 17.1 lists a selection of values, and more can be found in Appendix 2A.

Some of the values in Table 17.1 are quite easy to rationalize. For example, compare the molar entropy of diamond, 2.4 J/K·mol, with the much higher value for lead, 64.8 J/K·mol. The low entropy of diamond is what we should expect for a solid that has rigid bonds (recall the discussion of diamond in Section 10.13) and, for quantum mechanical reasons, lighter atoms. At room temperature, its atoms are not able to jiggle around as much as the loosely bonded atoms of lead can, so diamond has less thermal disorder than lead does at 298 K. Complex compounds commonly have higher entropies than simpler compounds do (compare $CaCO_3$ with CaO or NH_3 with H_2). There are more ways of storing the energy in molecular rotations (for gases) and vibrations (for gases, liquids, and solids) when the formula unit of a compound has many atoms (see Investigating Matter 9.1).

Molar entropies increase as the temperature is raised, as we see from the data for water in Table 17.2. As the temperature is raised, the thermal disorder of the molecules increases. The large increase in molar entropy at the boiling point of water, from 87 J/K·mol for the liquid to 197 J/K·mol for the vapor, shows the increase in positional disorder that occurs when a liquid changes to a much more chaotic gas. A smaller increase occurs when solids melt, because a liquid is only slightly more disordered than a solid (recall Fig. 17.6).

The entropies of ions in solution present a problem. There is no way that we can study cations without anions being present, so the entropies of ions in solution cannot be given absolute values. Chemists adopt the convention that the standard entropy of an ion in solution is its entropy *relative* to the entropy of hydrogen ions in solution. That is why some values are positive and others negative. For example, the entropy of Ca^{2+} ions in water is reported as −53.1 J/K·mol, which may be puzzling at first sight. However, all it means is that the entropy of Ca^{2+}(aq) is 53.1 J/K·mol lower than the entropy of H^+(aq). The entropy of Pb^{2+}(aq), which is +10.5 J/K·mol, is 10.5 J/K·mol higher than that of H^+(aq).

Table 17.1 *Standard molar entropies at 25°C**

Substance	$S_m°$, J/K·mol
Gases	
ammonia, NH_3	192.4
carbon dioxide, CO_2	213.7
hydrogen, H_2	130.7
nitrogen, N_2	191.6
oxygen, O_2	205.1
Liquids	
benzene, C_6H_6	173.3
ethanol, C_2H_5OH	160.7
water, H_2O	69.9
Solids	
calcium oxide, CaO	39.8
calcium carbonate, $CaCO_3^†$	92.9
diamond, C	2.4
graphite, C	5.7
lead, Pb	64.8

*Additional values are given in Appendix 2A.
†Calcite.

Table 17.2 *Standard molar entropy of water at various temperatures*

Phase	Temperature, °C	$S_m°$, J/K·mol
solid	−273 (0 K)	3.4*
	0	43.2
liquid	0	65.2
	20	69.6
	50	75.3
	100	86.8
vapor	100	196.9
	200	204.1

*This value is not 0 because the hydrogen bonding results in disorder even at $T = 0$.

The entropy of a perfect crystal tends to 0 as the temperature tends to 0. The standard molar entropies of gases are higher than those of comparable solids and liquids at the same temperature. The entropy of a substance increases when it melts, when it vaporizes, and as its temperature is raised.

Toolbox 17.1 *How to predict the change in entropy*

This Toolbox shows how to predict the sign of an entropy change and how to calculate an entropy change due to the transfer of heat or a phase transition.

Conceptual Basis

Calculations of the entropy from thermochemical data are based on the definition

$$\Delta S = \frac{q_{rev}}{T}$$

and the third law of thermodynamics, that S approaches 0 as T approaches 0. The use of this expression is illustrated in Examples 17.2 and 17.3. Expect a large increase in entropy when a lot of energy is transferred to a system at a low temperature (Fig. 17.7).

Procedure

The sign of an entropy change To predict the sign of the entropy change for a process, note that entropy is higher for

- Higher temperature
- Larger volume
- More complex structures
- Larger sample size
- Heavier atoms
- Vapor relative to liquid or solid
- Liquid relative to solid

Entropy of phase transition For the molar entropy of phase transition *at the transition temperature*, divide the enthalpy

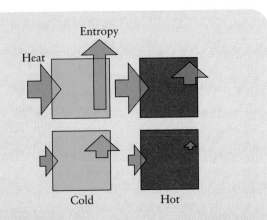

Entropy

Heat

Cold   Hot

FIGURE 17.7

The entropy change due to heat transfer depends on both the amount of heat transferred and the temperature of the system. A lot of heat transferred to a cold system (upper left) results in a large increase in the entropy of the system. A small quantity of heat transferred to a hot system (lower right) results in a small increase in entropy of the system.

of phase transition at the transition temperature by the transition temperature (in kelvins):

$$\Delta S_{fus}^{\circ} = \frac{\Delta H_{fus}^{\circ}}{T_f} \qquad \Delta S_{vap}^{\circ} = \frac{\Delta H_{vap}^{\circ}}{T_b}$$

where T_f is the normal melting (fusion) point and T_b is the normal boiling point. These expressions are illustrated in Example 17.3. Expect a large increase in entropy at the boiling point and a smaller increase at the melting point.

Example 17.4 *Estimating the relative value of the molar entropy*

Which substance in each pair has the higher molar entropy: (a) CO_2 at 25°C and 1 atm or CO_2 at 25°C and 3 atm; (b) $Br_2(l)$ or $Br_2(g)$ at the same temperature and pressure; (c) methane gas, CH_4, or propane gas, $CH_3CH_2CH_3$, at the same temperature and pressure?

Strategy Decide which state of a substance is likely to have the greater molecular disorder. For a given substance and temperature, the disorder is greater in the gas phase than in the liquid, and greater in the liquid than in the solid. Disorder increases with available space (for a gas) and with increasing temperature. The molar entropy of a substance also depends on its molecular complexity.

Solution (a) The fact that the same amount of CO_2 exerts a greater pressure at the same temperature tells us that the sample at 3 atm must be compressed more than the one at 1 atm. Therefore, the sample at a pressure of 1 atm occupies a greater volume and hence has the higher molar entropy. (b) Bromine vapor has more molecular disorder than liquid bromine at the same temperature, so bromine vapor has the higher molar entropy. (c) Propane has more atoms than methane, so it has more ways of storing energy in the gas phase. We can expect its molar entropy to be greater when both gases are at the same temperature and pressure.

Self-Test 17.4A Which substance in each pair has the higher molar entropy: (a) He at 25°C or He at 100°C in a container of the same volume; (b) Br(g) or $Br_2(g)$ at the same temperature and pressure?

[*Answer:* (a) He at 100°C; (b) $Br_2(g)$]

Self-Test 17.4B Which substance in each pair has the higher molar entropy at the same temperature and pressure: (a) Pb(s) or Pb(l); (b) $SbCl_3(g)$ or $SbCl_5(g)$?

17.4 Reaction Entropy

One of the principal uses of the data in Table 17.1 is to calculate the entropy change of a chemical reaction. The **standard reaction entropy,** ΔS_r°, is the difference between the standard molar entropies of the products and the reactants:

$$\Delta S_r^\circ = \sum n S_m^\circ (\text{products}) - \sum n S_m^\circ (\text{reactants}) \qquad (2)$$

The first term on the right is the total standard molar entropy of the products and the second term is that of the reactants; n denotes the various stoichiometric coefficients in the chemical equation.

The following example shows how to use Eq. 2. However, even without such a detailed calculation, we can anticipate when the reaction entropy is positive or negative. Because the molar entropy of a gas is so much greater than that of solids and liquids, a change in the amount of gas normally dominates any other entropy change in a reaction. A net increase in the amount of gas therefore usually results in a positive reaction entropy. A net consumption of gas usually results in a negative reaction entropy. Entropy changes are much more finely balanced for reactions in which there is no net change in the amount of gas; for these reactions, we have to use numerical data to predict the sign of a reaction entropy.

The standard reaction entropy is usually positive (an increase in entropy) if there is a net production of gas in a reaction; it is usually negative (a decrease) if there is a net consumption of gas.

Example 17.5 *Calculating the standard reaction entropy*

Calculate the standard reaction entropy for $N_2(g) + 3 H_2(g) \rightarrow 2 NH_3(g)$.

Strategy We expect a decrease in entropy because there is a net reduction in the amount of gas. To find the numerical value, use the chemical equation to write an expression for ΔS_r°, and then substitute values from Table 17.1.

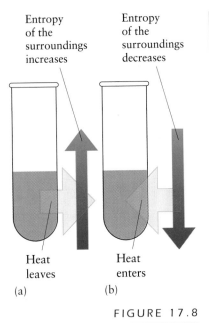

Entropy of the surroundings increases

Entropy of the surroundings decreases

Heat leaves

Heat enters

(a)

(b)

FIGURE 17.8

(a) In an exothermic process, heat escapes into the surroundings and increases their entropy. (b) In an endothermic process, the entropy of the surroundings decreases. The blue-green arrows indicate the direction of entropy change in the surroundings.

Whenever a symbol appears without a subscript, such as ΔS, it refers to a property of the system.

Solution From the chemical equation, we can write

$$
\begin{aligned}
\Delta S_r^\circ &= 2S_m^\circ(NH_3, g) - \{S_m^\circ(N_2, g) + 3S_m^\circ(H_2, g)\} \\
&= (2 \times 192.4) - \{191.6 + (3 \times 130.7)\} \text{ J/K·mol} \\
&= -198.9 \text{ J/K·mol}
\end{aligned}
$$

Because ΔS_r° is negative, the product is less disordered than the reactants, as we expected.

Self-Test 17.5A Use data from Appendix 2A to calculate the standard reaction entropy of $N_2O_4(g) \rightarrow 2\,NO_2(g)$ at 25°C.

[***Answer:*** +175.83 J/K·mol]

Self-Test 17.5B Use data from Appendix 2A to calculate the standard reaction entropy of $C_2H_4(g) + H_2(g) \rightarrow C_2H_6(g)$ at 25°C.

17.5 The Entropy of the Surroundings

We can see from Table 17.2 that at 0°C the entropy of liquid water is 22.0 J/K·mol higher than the entropy of ice. This difference is expected, because liquid water is more disordered than ice. Therefore, during freezing (the reverse of melting), water becomes less disordered as it forms ice, and its entropy decreases. However, water freezes spontaneously below 0°C. This conclusion may seem puzzling at first, because an entropy *increase* is a signal of spontaneous change. What have we overlooked?

We have overlooked the rest of the universe. The second law implies that the *total* entropy of a system and its surroundings has a tendency to increase. Whenever we use arguments based on entropy, we must always take into account the sum of the changes in the system, ΔS, and the surroundings, ΔS_{surr}:

$$\text{Total entropy change} = \text{entropy change of system}$$
$$+ \text{ entropy change of surroundings}$$
$$\Delta S_{tot} = \Delta S + \Delta S_{surr} \quad \quad (3)$$

The entropy of the substance that is freezing (the system) decreases. However, because freezing is an exothermic process, heat passes into the surroundings. This heat stirs up the thermal motion of the atoms in the surroundings, which increases their disorder and hence their entropy (Fig. 17.8). In general, there will be an overall increase in the total disorder whenever the increase in disorder of the surroundings is greater than the decrease in the disorder of the system. When the *total* entropy increases, the change is spontaneous.

To see how to calculate the change in entropy of the surroundings, suppose 1 mol H_2O freezes in the system. The change in enthalpy is -6 kJ; so 6 kJ of heat flows into the surroundings and stirs up thermal motion there. In general, if the enthalpy change of the system is ΔH (in this case, -6 kJ), then the heat that flows into the surroundings is $-\Delta H$ (that is, $+6$ kJ). The sign changes because heat that leaves the system enters the surroundings. Now we use this heat transfer in the definition of entropy change, Eq. 1, and obtain

$$
\begin{aligned}
\text{Entropy change of surroundings} &= \frac{\text{heat transferred to surroundings}}{\text{temperature of surroundings}} \\
&= -\frac{\text{enthalpy change of system}}{\text{temperature of surroundings}} \quad (4) \\
\Delta S_{surr} &= -\frac{\Delta H}{T}
\end{aligned}
$$

FIGURE 17.9

In an exothermic reaction, (a) the overall entropy change is certainly positive when the entropy of the system increases. (b) The overall entropy change is positive even when the entropy of the system decreases, provided that the entropy increase in the surroundings is greater. The reaction is spontaneous in both cases.

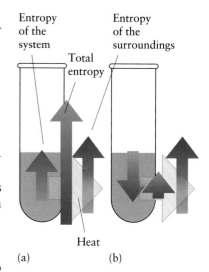

(a)   (b)

The formula applies only at constant pressure, because only then can we identify ΔH with the heat transferred.

The same formula applies to the change in the entropy of the surroundings that accompanies a chemical reaction. For example, for the synthesis of NH_3 in the reaction

$$N_2(g) + 3 H_2(g) \longrightarrow 2 NH_3(g)$$

the standard reaction enthalpy is -92.22 kJ/mol. The reaction is exothermic, so we expect the entropy of the surroundings to increase as heat spreads into them. In fact, if the reaction were to go to completion,

$$\Delta S_{surr} = -\frac{(-92.22 \times 10^3 \text{ J/mol})}{298 \text{ K}} = +309 \text{ J/K·mol}$$

Self-Test 17.6A An exothermic reaction releases 35.7 kJ of heat to the surroundings at 300. K. What is the change in entropy of the surroundings for this reaction?

[***Answer:*** $+119$ J/K]

Self-Test 17.6B An endothermic reaction absorbs 71.5 kJ of heat from the surroundings at 150. K. What is the change in entropy of the surroundings for this reaction?

Provided ΔH_r° is negative, with a reasonably high numerical value, a reaction with ΔS_r° either positive or negative may be spontaneous (Fig. 17.9). In fact, many exothermic reactions are spontaneous because entropy changes in the system are quite small relative to the large increase in entropy of the surroundings.

Endothermic reactions were a puzzle for nineteenth-century chemists, who believed that reactions ran in the direction of decreasing energy of the system. An endothermic reaction is one in which the products have a higher enthalpy than the reactants, so it seemed that in endothermic reactions reactants were spontaneously rising to higher energies, like a weight suddenly leaping up from the floor to a table. However, *whenever we consider the tendency to change, we have to consider the entropy, not the energy.* Because heat flows from the surroundings into the system, the entropy of the surroundings decreases in the course of an endothermic reaction. Nevertheless, there can still be an overall increase in entropy, provided the disorder of the system increases enough. Every endothermic reaction must be accompanied by increased disorder within the system if it is to be spontaneous (Fig. 17.10).

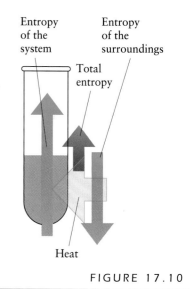

FIGURE 17.10

An endothermic reaction is spontaneous only when the entropy of the system increases enough to overcome the decrease in entropy of the surroundings, as it does here.

A chemical reaction is spontaneous if it is accompanied by an increase in the total entropy of the system and the surroundings. Spontaneous exothermic reactions are common because they release heat that increases the entropy of the surroundings. Endothermic reactions are spontaneous only if the reaction mixture undergoes a large increase in entropy.

Example 17.6 *Judging whether a process is spontaneous*

Is the dissolution of ammonium nitrate to form a dilute aqueous solution spontaneous at 25°C?

Strategy Begin by writing the chemical equation for the process. To assess the spontaneity quantitatively, calculate the enthalpy and entropy of solution from the data in Appendix 2A, and convert those values into a total entropy change. The dissolution is spontaneous if the total entropy change is positive.

Solution The chemical equation for the dissolving of ammonium nitrate in water is

$$NH_4NO_3(s) \longrightarrow NH_4^+(aq) + NO_3^-(aq)$$

The standard enthalpy of solution is

$$\Delta H_{sol}° = \{\Delta H_f°(NH_4^+, aq) + \Delta H_f°(NO_3^-, aq)\} - \Delta H_f°(NH_4NO_3, s)$$
$$= (-132.51 - 205.0) - (-365.56) \text{ kJ/mol} = +28.0 \text{ kJ/mol}$$

Note that the dissolution is endothermic. The entropy change in the surroundings is therefore

$$\Delta S_{surr} = -\frac{(+28.0 \times 10^3 \text{ J/mol})}{298 \text{ K}} = -94.0 \text{ J/K·mol}$$

The standard entropy of dissolution is

$$\Delta S_{sol}° = \{S_m°(NH_4^+, aq) + S_m°(NO_3^-, aq)\} - S_m°(NH_4NO_3, s)$$
$$= (113.4 + 146.4) - 151.08 \text{ J/K·mol} = +108.7 \text{ J/K·mol}$$

The overall change in entropy is therefore

$$\Delta S_{tot}° = 108.7 - 94.0 \text{ J/K·mol} = +14.7 \text{ J/K·mol}$$

There is an increase in total entropy, so ammonium nitrate does dissolve spontaneously. It is, in fact, a highly soluble substance, despite its positive enthalpy of solution.

Self-Test 17.7A The combustion of wood is a typical combustion reaction. Show that it is spontaneous at 25°C by treating wood as glucose and the products as carbon dioxide gas and water vapor. Glucose, $C_6H_{12}O_6$, is a model for wood, which is largely cellulose, a polymer of glucose. (Write the equation with the smallest whole number coefficients.)

[***Answer:*** +7.53 kJ/K·mol]

Self-Test 17.7A Is the reaction of aqueous bromide ions with solid elemental iodine to produce liquid elemental bromine and aqueous iodide ions spontaneous at 25°C? (Write the equation with the smallest whole number coefficients.)

FREE ENERGY

In Chapter 14, we saw that chemical reactions proceed spontaneously in the direction of equilibrium. We therefore suspect that the position of equilibrium is closely related to the entropy changes in the system and surroundings. When a reaction reaches equilibrium, it has no tendency to form more products, nor do the products have any tendency to decompose into reactants. The forward and reverse reactions are still continuing, but there is no *net* change of composition. In terms of the second law, the absence of any tendency to form products or reactants means that, at equilibrium, the total entropy change is 0 both for

are nonspontaneous and must be driven by an external source (Applying Chemistry: Case Study 17).

To find the temperature at which a nonspontaneous reaction first becomes spontaneous, we need to find the temperature at which a positive ΔG falls to 0, for that value marks the frontier between spontaneous and nonspontaneous reactions. Because ΔH and ΔS change only very slightly with temperature, we can consider them approximately constant over small temperature ranges. Therefore, we can use Eq. 5 to calculate the temperature at which $\Delta G = 0$. If $\Delta S > 0$, an *endothermic* reaction will be spontaneous *above* that temperature. If $\Delta S < 0$, an *exothermic* reaction will be spontaneous *below* that temperature.

For a reaction that is at equilibrium at constant pressure and temperature, $\Delta G = 0$.

Example 17.7 *Predicting the boiling point of a substance*

Liquid metals, such as mixtures of sodium and potassium, are used as coolants in some nuclear reactors. Predict the normal boiling point of liquid sodium, given that the standard entropy of vaporization of liquid sodium is 84.8 J/K·mol and that its standard enthalpy of vaporization is 98.0 kJ/mol.

Strategy At the normal boiling point, a liquid and its vapor are in equilibrium at 1 atm. In other words, there is no change in free energy when one phase changes into the other ($\Delta G = 0$). We are given the entropy and enthalpy changes of the system at 1 atm (their standard values). Use these values to find the temperature at which $\Delta G = \Delta H - T\Delta S$ is 0.

Solution We want the temperature at which $\Delta G = 0$:

$$\Delta H_{vap} - T_b\Delta S_{vap} = 0$$

This equation can be solved for T:

$$T_b = \frac{\Delta H_{vap}}{\Delta S_{vap}}$$

from which we find that

$$T_b = \frac{98.0 \times 10^3 \text{ J/mol}}{84.8 \text{ J/K·mol}} = 1.16 \times 10^3 \text{ K}$$

or about 890°C. The experimental value is 883°C.

Self-Test 17.9A Predict the melting point of solid chlorine, given that its enthalpy of fusion is 6.41 kJ/mol and its entropy of fusion is 37.3 J/K·mol.

[*Answer:* 172 K (experimental: 172 K)]

Self-Test 17.9B Predict the boiling point of methanol, given that its enthalpy of vaporization is 35.3 kJ/mol and its entropy of vaporization is 104.7 J/K·mol.

17.8 Standard Reaction Free Energies

We can calculate ΔH_r° and ΔS_r° for a reaction and, from them, find the **standard free energy of reaction, ΔG_r°**, where

$$\Delta G_r^\circ = \Delta H_r^\circ - T\Delta S_r^\circ$$

Applying Chemistry: *Case Study 17*

How can we explain life? Why do the molecules in our body form a highly organized, complex structure, rather than slime, ooze, or gas? Every cell of a living being is organized to an extraordinary extent. Thousands of different compounds, each one having a specific function to perform, move in the intricately choreographed dance we call life. We are examples of systems with very low entropy. In fact, our existence seems at first thought a contradiction of the second law of thermodynamics.

The ultimate example of low entropy is in our genes. Every one of the billions of atoms in the long double spiral of a DNA molecule has an appointed place, and that location contributes to the blueprint for our bodies. These molecules carry all the information needed to create a human being. The protein molecules that act as enzymes or structural molecules (as in our hair and nails) are also highly organized. If a protein loses its highly organized shape, then it ceases to function. Heightened entropy—a loss of order—means disease and perhaps even death.

One of the most important processes that supports life and ultimately maintains the structure of DNA is photosynthesis, the series of reactions that uses sunlight to convert carbon dioxide and water into carbohydrates and oxygen. Without photosynthesis, the Earth would be a warm, wet rock with no green plants or animals. The photosynthesis reaction is itself nonspontaneous:

$$6\,CO_2(g) + 6\,H_2O(l) \longrightarrow C_6H_{12}O_6(s) + 6\,O_2(g)$$

It is accompanied by a decrease in entropy of the system because many small molecules must be assembled into the larger glucose molecule. It is also accompanied by a decrease in entropy of the surroundings because it is endothermic. So how can it occur at all, let alone on a megaton scale worldwide? The answer lies in sunlight.

The bubbles on the leaves of this underwater plant are oxygen produced by photosynthesis. Molecules such as the chlorophyll that colors the leaves green capture sunlight to begin the transformation of carbon dioxide and water to glucose and oxygen.

Solar radiation floods a leaf with energy, and some of that energy is captured by chlorophyll. This captured energy is used by subsequent chemical reactions to generate even more entropy than is lost in the glucose-construction reaction itself.

The Sun's energy also accounts for nonspontaneous reactions in our bodies; we get our energy, not directly from the Sun, but from chemicals—food—that have stored the energy of sunlight. In all living things, one biochemical process, which generates a lot of entropy as it runs in its spontaneous direction, drives other reactions in a nonspontaneous direction, perhaps to build a protein or contribute to the construction of a DNA molecule. In other words, biochemical processes are coupled: one may be

The value of ΔG_r° is that for reactants and products in their standard states (pure, at 1 atm). For species in solution, the standard state is a concentration of 1 mol/L.

There is a simpler way to find ΔG_r° for many reactions. In Section 6.16, we saw that standard enthalpies of formation are handy means of tabulating thermochemical data and that they can be combined to obtain the standard reaction enthalpy of any reaction. The same approach can be used to calculate standard reaction free energies.

The **standard free energy of formation**, ΔG_f°, of a compound is the standard reaction free energy per mole for the formation of a compound from its elements in their most stable form. The most stable form of an element is the driven uphill in free energy by another reaction that rolls downhill. Staying alive is very much like the effect of a heavy weight tied to another weight by a string that passes over a pulley. The lighter weight could never fly up into the air on its own. However, when it is connected to a heavier weight falling downward on the other side of a pulley, it can soar upward.

When we eat food containing glucose, we consume a fuel. Like all fuels, its combustion is spontaneous. If we were simply to burn glucose, it would give off a great deal of heat and light. However, in our body, the combustion is slower and represents a highly controlled and sophisticated version of burning. Burning glucose in an open container does no work other than pushing back the atmosphere, so it gives off a lot of heat. The slow oxidation of glucose in our cells, under the control of enzymes and the intricate biochemical mechanisms of the body, does much more work and, in fact, makes us very efficient at utilizing the energy in our food.

To harness as much free energy as possible, a process must proceed reversibly, which in practice commonly means very slowly and carefully. In such a controlled reaction, the work the process can do approaches ΔG_r. Human cells are therefore like small, efficient power plants. We copy nature when we design practical fuel cells (see Applying Chemistry: Case Study 18).

When we die, we no longer ingest the second-hand sunlight stored in molecules of carbohydrate, protein, and fat. Now the natural direction of change becomes dominant, and our intricate molecules start to decompose to the ooze and slime that we managed to avoid becoming during our lives. Life is a constant battle to generate enough entropy in our surroundings to go on building and maintaining our intricate interiors. As soon as we stop the battle, we stop generating that external entropy, and our bodies decay.

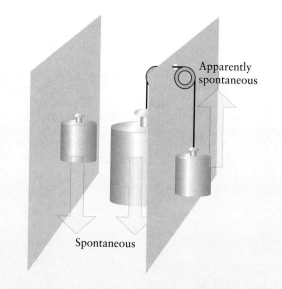

A weight with a small mass can be lifted into the air by another weight of the same or greater mass. What would appear unnatural if we saw it by itself (a weight rising) is actually part of a spontaneous event overall. The "natural" fall of the heavier weight causes the "unnatural" rise of the smaller weight.

Key Concepts: entropy, reaction free energy

For Further Reading

P. W. Atkins, *The Second Law*, New York: Scientific American Books, 1984, pp. 165–175.

Related Exercises: 17.79–17.81

---

state with the lowest free energy (Table 17.4). For example, the standard free energy of formation of hydrogen iodide gas at 25°C is $\Delta G_f^\circ = +1.70$ kJ/mol. It is the standard reaction free energy for

$$\tfrac{1}{2} H_2(g) + \tfrac{1}{2} I_2(s) \longrightarrow HI(g)$$

Standard free energies of formation can be determined in a variety of ways. One straightforward way is to combine the enthalpy and entropy data from Tables 6.5 and 17.1. A list of some of the resulting values is given in Table 17.5, and a more extensive one appears in Appendix 2A. The standard free energies of formation of elements in their most stable forms are 0.

Table 17.4 The most stable forms of some elements at 25°C and 1 atm

Element	Most stable form
H_2, O_2, Cl_2, Xe	gas
Br_2, Hg	liquid
C	graphite
Na, Fe, I_2	solid

Example 17.8 Calculating a standard free energy of formation

Calculate the standard free energy of formation of HI(g) at 25°C from its standard entropy and standard enthalpy of formation.

Strategy First, we write the chemical equation, with the compound of interest having a stoichiometric coefficient of 1. Then we calculate the standard reaction free energy from $\Delta G_r^\circ = \Delta H_r^\circ - T\Delta S_r^\circ$. The standard reaction enthalpy is found from the standard enthalpies of formation by using data from Appendix 2A. The standard reaction entropy is found as in Example 17.5, by using the data from Table 17.1 or Appendix 2A.

Solution The chemical equation is

$$\tfrac{1}{2} H_2(g) + \tfrac{1}{2} I_2(s) \longrightarrow HI(g)$$

From data in Appendix 2A, and $\Delta H_f^\circ = 0$ for each of the elements,

$$\Delta H_r^\circ = \Delta H_f^\circ(HI, g) - \{\tfrac{1}{2}\Delta H_f^\circ(H_2, g) + \tfrac{1}{2}\Delta H_f^\circ(I_2, s)\}$$
$$= \Delta H_f^\circ(HI, g) = +26.48 \text{ kJ/mol}$$

From data in Appendix 2A, noting that the standard entropies of elements are not 0,

$$\Delta S_r^\circ = S_m^\circ(HI, g) - \{\tfrac{1}{2}S_m^\circ(H_2, g) + \tfrac{1}{2}S_m^\circ(I_2, s)\}$$
$$= 206.59 - \{(\tfrac{1}{2} \times 130.68) + (\tfrac{1}{2} \times 116.14)\} \text{ J/K·mol}$$
$$= +83.18 \text{ J/K·mol}$$

We now convert the value of ΔS from joules to kilojoules and use $T = 298$ K (corresponding to 25°C),

$$\Delta G_r^\circ = \Delta H_r^\circ - T\Delta S_r^\circ$$
$$= 26.48 \text{ kJ/mol} - (298 \text{ K}) \times (83.18 \times 10^{-3} \text{ kJ/K·mol})$$
$$= +1.69 \text{ kJ/mol}$$

This value is the standard free energy of formation of HI(g), which we write $\Delta G_f^\circ(HI, g)$.

Self-Test 17.10A Calculate the standard free energy of formation of $NH_3(g)$ at 25°C.

[*Answer:* −16.5 kJ/mol]

Self-Test 17.10B Calculate the standard free energy of formation of $C_3H_6(g)$, cyclopropane, at 25°C.

The standard free energy of formation of a compound is a measure of its stability relative to its elements. If $\Delta G_f^\circ < 0$ at a certain temperature, then the elements have a spontaneous tendency to form the compound at that temperature. Under standard conditions, the compound is more stable than its elements at that temperature (Fig. 17.12). If $\Delta G_f^\circ > 0$, the reverse reaction, the decomposition of the compound, is spontaneous. Under standard conditions, the compound has a tendency to decompose into its elements. If $\Delta G_f^\circ > 0$, there is no point in trying to synthesize the compound from its elements at that temperature under standard conditions.

For example, the standard free energy of formation of benzene is +124 kJ/mol at 25°C. Therefore, the reaction

$$6 C(s) + 3 H_2(g) \longrightarrow C_6H_6(l)$$

(with all the species in their standard states) is not spontaneous. There is no point in trying to make benzene by exposing carbon to hydrogen gas at 25°C

Table 17.5 Standard free energies of formation at 25°C*

Substance	ΔG_f°, kJ/mol
Gases	
ammonia, NH_3	−16.45
carbon dioxide, CO_2	−394.4
nitrogen dioxide, NO_2	+51.3
water, H_2O	−228.6
Liquids	
benzene, C_6H_6	+124.3
ethanol, C_2H_5OH	−174.8
water, H_2O	−237.1
Solids	
calcium carbonate, $CaCO_3$†	−1128.8
silver chloride, AgCl	−109.8

*Additional values are given in Appendix 2A.
†Calcite.

and 1 atm, because pure benzene has a higher free energy than graphite and hydrogen at 1 atm. However, the reverse reaction

$$C_6H_6(l) \longrightarrow 6\,C(s) + 3\,H_2(g)$$

has $\Delta G_f^\circ = -124$ kJ/mol and is spontaneous at 25°C. The system and its surroundings become more disordered if pure benzene decomposes into graphite and pure hydrogen gas at 1 atm. We say, therefore, that benzene is thermodynamically unstable with respect to its elements under standard conditions.

A **thermodynamically stable compound** is a compound with a negative free energy of formation. Such a compound has no tendency to decompose into its elements. A **thermodynamically unstable compound** is a compound with a positive standard free energy of formation. Such a compound has a thermodynamic tendency to decompose into its elements. However, that tendency may not be realized in practice because the decomposition may be very slow. Benzene can, in fact, be kept indefinitely without decomposing at all. Because the activation energy for its decomposition is so high, benzene is said to be "nonlabile" or "kinetically stable."

The standard free energy of formation of a substance is the standard free energy of reaction per mole of compound when it is prepared from its elements in their most stable forms. The sign of ΔG_f° tells us whether a compound is stable or unstable with respect to its elements.

Self-Test 17.11A Is glucose stable relative to its elements at 25°C and under standard conditions? (See the listing of organic compounds at the end of Appendix 2A.) Explain your answer.

[**Answer:** Yes; for glucose, $\Delta G_f^\circ = -910$ kJ/mol]

Self-Test 17.11B Is methylamine, CH_3NH_2, stable relative to its elements at 25°C and under standard conditions? (See the listing of organic compounds at the end of Appendix 2A.) Explain your answer.

17.9 Using Free Energies of Formation

We combine standard free energies of formation to obtain standard reaction free energies, just as we combine standard enthalpies of formation to obtain standard enthalpies of reaction (Toolbox 6.2):

$$\Delta G_r^\circ = \sum n\Delta G_f^\circ(\text{products}) - \sum n\Delta G_f^\circ(\text{reactants}) \qquad (6)$$

Here, as usual, n denotes the stoichiometric coefficients in the chemical equation. For example, for the oxidation of ammonia

$$4\,NH_3(g) + 5\,O_2(g) \longrightarrow 4\,NO(g) + 6\,H_2O(g)$$

we have (from Appendix 2A, for 298 K)

$$\begin{aligned}\Delta G_r^\circ &= \{4\Delta G_f^\circ(NO, g) + 6\Delta G_f^\circ(H_2O, g)\} - \{4\Delta G_f^\circ(NH_3, g) + 5\Delta G_f^\circ(O_2, g)\} \\ &= \{4 \times 86.55 + 6 \times (-228.57)\} - \{4 \times (-16.45) + 0\}\ \text{kJ/mol} \\ &= -959.42\ \text{kJ/mol}\end{aligned}$$

The negative sign shows that the reaction is spontaneous under standard conditions; that is, pure ammonia and oxygen each at 1 atm have a strong tendency to form pure nitric oxide and water vapor, each at 1 atm. However, because the

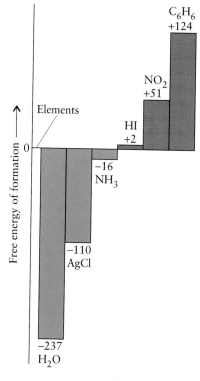

FIGURE 17.12

The standard free energies of formation of compounds are defined as the standard reaction free energy for their formation from the elements. They represent a thermodynamic "altitude" with respect to the elements at "sea level." The numerical values are in kilojoules per mole for 298K.

activation energy is high, the reaction is so slow that it does not proceed without a catalyst.

Standard free energies of formation are combined to calculate the standard free energy of a reaction.

Example 17.9 *Calculating the standard reaction free energy from free energies of formation*

Calculate the standard free energy of the reaction $2\,CO(g) + O_2(g) \rightarrow 2\,CO_2(g)$ from standard free energies of formation at 298 K.

Strategy Find the standard free energy of reaction from the free energies of formation in Appendix 2A by using Eq. 6, with the stoichiometric coefficients given in the chemical equation.

Solution We substitute the free energies of formation from Appendix 2A for each of the reactants and products:

$$\Delta G_r^\circ = 2\Delta G_f^\circ(CO_2, g) - \{2\Delta G_f^\circ(CO, g) + \Delta G_f^\circ(O_2, g)\}$$
$$= \{2 \times (-394.36)\} - \{2 \times (-137.17) + 0\}\ kJ/mol$$
$$= -514.38\ kJ/mol$$

This strongly negative value tells us that the oxidation of carbon monoxide is spontaneous under standard conditions at 298 K.

Self-Test 17.12A Calculate the standard free energy for the reaction $2\,SO_2(g) + O_2(g) \rightarrow 2\,SO_3(g)$ from free energies of formation.

[**Answer:** $-141.74\ kJ/mol$]

Self-Test 17.12B Calculate the standard free energy for the reaction $6\,CO_2(g) + 6\,H_2O(l) \rightarrow C_6H_{12}O_6(s) + 6\,O_2(g)$ for glucose formation from free energies of formation.

17.10 Free Energy and Composition

The free energy of a reaction mixture varies with the composition of the mixture. When we plot the free energy against its changing composition, we get a lopsided ∪-shaped curve (Fig. 17.13). The reaction tends to proceed toward the composition at the lowest point of the curve, because that is the direction of decreasing free energy. *The composition at the lowest point of the curve, the point of minimum free energy, corresponds to equilibrium.* For a system at equilibrium, a change in composition in either direction results in an increase in free energy, so neither change is spontaneous. When the free-energy minimum lies very close to the pure products, the equilibrium composition strongly favors products, and the reaction goes nearly to completion ($K \gg 1$, Fig. 17.14). When the free-energy minimum lies very close to the pure reactants, the equilibrium composition strongly favors the reactants, and the reaction forms only a very small

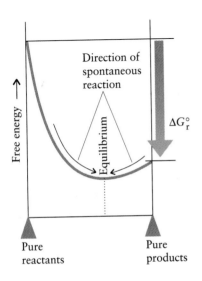

Direction of spontaneous reaction

Equilibrium

ΔG_r°

Free energy →

Pure reactants

Pure products

FIGURE 17.13

At constant temperature and pressure, the direction of spontaneous change is toward lower free energy. The equilibrium composition of a reaction mixture corresponds to the lowest point on the curve. In this example, substantial quantities of both reactants and products are present at equilibrium, and K is close to 1.

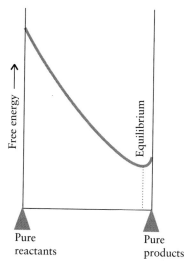

FIGURE 17.14

In this reaction, the free energy is a minimum when products are much more abundant than reactants. The equilibrium lies in favor of the products, and $K \gg 1$. This reaction goes almost to completion.

amount of product ($K \ll 1$, Fig. 17.15). When the minimum lies approximately halfway between pure products and pure reactants, all the species are present in similar concentrations at equilibrium ($K \approx 1$, as in Fig. 17.13).

To make these ideas quantitative, we need to distinguish between the standard reaction free energy, ΔG_r°, and the reaction free energy, ΔG_r:

ΔG_r° is the difference in standard free energies of formation of the pure products and the pure reactants.

ΔG_r is the reaction free energy at a definite, fixed composition of the reaction mixture.

The value of ΔG_r° is the difference in molar free energy between the *pure* products and the *pure* reactants in their standard states. Pressures of gases are all taken to be 1 atm and concentrations of substances in solution are all taken to be 1 mol/L. On the other hand, the value of ΔG_r refers to any chosen, fixed composition of the reaction mixture and represents the difference in molar free energy between the products and reactants at the concentrations and partial pressures present in the mixture.

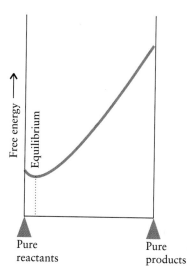

FIGURE 17.15

In this reaction, the free energy is a minimum when the reactants are much more abundant than the products. The equilibrium lies in favor of the reactants, and $K \ll 1$. This reaction "does not go."

To fix the meaning of ΔG_r, and to see how a change can refer to a system at fixed composition, think of a very large reaction mixture—a swimming pool full, for instance—and consider the change in free energy when 1 mol of reactant molecules changes into products. The extent of reaction is so small relative to the huge amount of material present that the change has a negligible effect on the composition of the reaction mixture. The concentrations of reactants and products are essentially the same after the reaction as before. For example, suppose nitrogen and hydrogen react to form ammonia in a very large container, with 10 000 mol of each substance present. The free energy of reaction is the change in free energy when 1 mol N_2 and 3 mol H_2 form 2 mol NH_3 in such an environment.

The value of ΔG_r changes as the composition of the reaction mixture changes. If the composition of the swimming pool of reaction mixture contains relatively more reactants than it would have at equilibrium, then ΔG_r at that composition will be negative, and the reaction will have a tendency to form more products. If the reaction mixture happens to contain a higher proportion of products than at equilibrium, then ΔG_r will be positive, and the reverse reaction, the formation of reactants, will be spontaneous. However, if the reaction mixture happens to have its equilibrium composition, then $\Delta G_r = 0$ and there is no tendency for the reaction to run in either direction.

Thermodynamic arguments can be used to show that the reaction free energy is related to the composition of the reaction mixture by

$$\Delta G_r = \Delta G_r^\circ + RT \ln Q \qquad (7)$$

Here Q is the reaction quotient (Section 14.7) and R is the gas constant. The reaction quotient is expressed in terms of partial pressures for any gases in the reaction and in terms of molar concentrations for any solutes. Because most standard free energies are tabulated for 25°C, most calculations are carried out for that temperature. We can then use

$$RT = (8.314\,51\ \text{J/K·mol}) \times (298.15\ \text{K}) = 2.4790\ \text{kJ/mol at 298.15 K}$$

Equation 7 shows that the spontaneous direction of a reaction depends on the relative concentrations of reactants and products. We can use this relationship to determine the value of ΔG_r at any concentrations of reactants and products at a given temperature.

The reaction free energy, ΔG_r, is the difference in free energy between the products and reactants in a reaction mixture with a specific composition; it is calculated from Eq. 7.

Example 17.10 *Calculating the reaction free energy from the reaction quotient*

The standard reaction free energy for $N_2(g) + 3\,H_2(g) \rightarrow 2\,NH_3(g)$ is -32.90 kJ/mol at 25°C. What is the reaction free energy when the partial pressure of each gas is 100. atm? What is the spontaneous direction of the reaction under these conditions?

Strategy Calculate the reaction quotient and substitute it and the data into Eq. 7. If ΔG_r turns out to be negative, then the formation of ammonia is spontaneous at the given composition; if ΔG_r is positive, then ammonia has a spontaneous tendency to decompose into its elements at the given composition.

Solution The reaction quotient is

$$Q = \frac{P_{NH_3}^2}{P_{N_2}P_{H_2}^3} = \frac{(100.)^2}{(100.) \times (100.)^3} = 1.00 \times 10^{-4}$$

Therefore, the reaction free energy at this composition is

$$\begin{aligned}
\Delta G_r &= \Delta G_r^{\circ} + RT \ln Q \\
&= (-32.90 \text{ kJ/mol}) + (2.4790 \text{ kJ/mol})\ln(1.00 \times 10^{-4}) \\
&= -55.7 \text{ kJ/mol}
\end{aligned}$$

Because the reaction free energy is negative, the formation of products is spontaneous at this composition and temperature.

Self-Test 17.13A The standard reaction free energy for $H_2(g) + I_2(g) \rightarrow 2\,HI(g)$ is $\Delta G_r^{\circ} = -21.1$ kJ/mol at 500. K. What is the value of ΔG_r when the partial pressures of the gases are $P_{H_2} = 1.5$ atm, $P_{I_2} = 0.88$ atm, and $P_{HI} = 0.065$ atm? What is the spontaneous direction of the reaction?

[***Answer:*** -45 kJ/mol; toward products]

Self-Test 17.13B The standard reaction free energy for $N_2O_4(g) \rightarrow 2\,NO_2(g)$ is $\Delta G_r^{\circ} = +4.73$ kJ/mol at 298 K. What is the value of ΔG_r when the partial pressures of the gases are $P_{N_2O_4} = 0.80$ atm and $P_{NO_2} = 2.10$ atm? What is the spontaneous direction of the reaction?

17.11 Equilibrium Constants

At equilibrium, $\Delta G_r = 0$, because the reaction mixture has no tendency to change in either direction. Therefore, at equilibrium, the left-hand side of Eq. 7 is equal to 0. Moreover, as we saw in Section 14.7, at equilibrium, $Q = K$, the equilibrium constant for the reaction. The equation then becomes

$$0 = \Delta G_r^{\circ} + RT \ln K$$

This relation rearranges to

$$\Delta G_r^{\circ} = -RT \ln K \qquad (8)$$

This equation is probably one of the most important in the whole of chemical thermodynamics. It is the link between the equilibrium constant for any change—physical or chemical—and the standard reaction free energy. We can use Eq. 8 to calculate K whenever free energy data are available (see Toolbox 17.2). Note that, when using Eq. 8, K is expressed in terms of the numerical values of the partial pressures of gases and as the numerical values of the molar concentrations of solutes. To obtain K_c for a gas-phase reaction from ΔG_r°, we first calculate K_p, then convert from K_p to K_c by using the procedure in Toolbox 14.1.

An important conclusion from Eq. 8 is that

$$K < 1 \text{ when } \Delta G_r^{\circ} > 0$$

$$K > 1 \text{ when } \Delta G_r^{\circ} < 0$$

These relations highlight the significance of the standard reaction free energy. The value of ΔG_r° does *not* tell us whether a reaction is "spontaneous" or not: the spontaneity of a reaction depends on the value of ΔG_r and the composition of the reaction mixture ($\Delta G_r < 0$, spontaneous at the given composition; $\Delta G_r > 0$, not spontaneous at the given composition). The value of ΔG_r° *does* tell us whether the equilibrium composition of the reaction mixture favors

Toolbox 17.2 *How to calculate equilibrium constants*

This Toolbox shows how to calculate an equilibrium constant from tables of thermodynamic data.

Conceptual Basis

The relation between ΔG_r° for a reaction and the equilibrium constant shows that a reaction with $\Delta G_r^\circ < 0$ has $K > 1$ and a reaction with $\Delta G_r^\circ > 0$ has $K < 1$ (Fig. 17.16). Because $\Delta G_r^\circ = \Delta H_r^\circ - T\Delta S_r^\circ$, it follows that K can be expected to be large for an exothermic reaction ($\Delta H_r^\circ < 0$) and small for an endothermic reaction ($\Delta H_r^\circ > 0$). Only when ΔS_r° is large and positive can we expect $K > 1$ for an endothermic reaction.

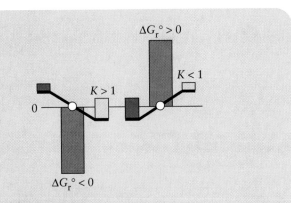

FIGURE 17.16

A negative value of the standard reaction free energy corresponds to an equilibrium constant greater than 1 and to products (yellow) favored over reactants (purple) at equilibrium. A positive value of the standard reaction free energy corresponds to an equilibrium constant of less than 1 and to reactants favored over products at equilibrium.

Procedure

To calculate the equilibrium constant for a reaction we combine Eqs. 6 and 8 as follows.

Step 1. Write down the chemical equation for the reaction and the expression for the equilibrium constant K.

In this procedure, we must work with the equilibrium constant K, written down by using the rules set out in Toolbox 14.1 (that is, partial pressures for gases, molar concentrations for solutes). If the reaction involves only gases, $K = K_p$.

Step 2. Use the data in Appendix 2A to calculate the standard reaction free energy:

$$\Delta G_r^\circ = \sum n\Delta G_f^\circ(\text{products}) - \sum n\Delta G_f^\circ(\text{reactants})$$

The free energies of formation of elements in their most stable states are 0. This step was illustrated in Example 17.9.

Step 3. Use the value of ΔG_r° from step 2 in $\Delta G_r^\circ = -RT \ln K$ to calculate K:

$$K = e^{-\Delta G_r^\circ/RT}$$

Use $R = 8.314\ 51$ J/K·mol and the temperature in kelvins; the units of RT (joules per mole, J/mol) then cancel the units of ΔG_r after kilojoules per mole (kJ/mol) are converted to joules per mole.

Step 4. If desired, convert K_p for a gas-phase reaction (which is equal to K) to K_c by using the relation in Toolbox 14.1, $K_p = (RT)^{\Delta n}K_c$.

---

products ($\Delta G_r^\circ < 0$, corresponding to $K > 1$) or reactants ($\Delta G_r^\circ > 0$, corresponding to $K < 1$).

The equilibrium constant of a reaction is related to the standard free energy of reaction by $\Delta G_r^\circ = -RT \ln K$.

Example 17.11 *Calculating an equilibrium constant from free energies*

Calculate K_p at 25°C for the equilibrium $N_2O_4(g) \rightleftharpoons 2\ NO_2(g)$.

Strategy The species are in the gas phase, so the equilibrium constant obtained from tables of thermodynamic data, K, is equal to K_p. Work through the procedure in Toolbox 17.2.

Solution **Step 1.** The chemical equation and the expression for K_p are

$$N_2O_4(g) \rightleftharpoons 2\,NO_2(g) \qquad K_p = \frac{P_{NO_2}{}^2}{P_{N_2O_4}}$$

Step 2. From the data in Appendix 2A,

$$\begin{aligned}\Delta G_r^\circ &= 2\Delta G_f^\circ(NO_2, g) - \Delta G_f^\circ(N_2O_4, g)\\ &= (2 \times 51.31) - 97.89 \text{ kJ/mol} = +4.73 \text{ kJ/mol}\end{aligned}$$

Step 3. After rearranging Eq. 8 and substituting the value of ΔG_r° (at 25°C) and the temperature, we have

$$\begin{aligned}\ln K_p &= -\frac{\Delta G_r^\circ}{RT} = -\frac{4.73 \times 10^3 \text{ J/mol}}{(8.3145 \text{ J/K·mol}) \times (298 \text{ K})}\\ &= -\frac{4.73 \times 10^3}{8.3145 \times 298}\end{aligned}$$

This expression evaluates to -1.91; but we avoid rounding errors—which can be severe when working with logarithms and antilogarithms—by taking the antilogarithm (e^x) of both sides in this form. The result is $K_p = 0.15$.

Self-Test 17.14A Calculate K for the reaction $N_2(g) + 3\,H_2(g) \rightleftharpoons 2\,NH_3(g)$ at 25°C.
[*Answer:* 5.8×10^5]

Self-Test 17.14B Calculate K for the reaction $2\,NO(g) + O_2(g) \rightleftharpoons 2\,NO_2(g)$ at 25°C.

> Press the e^x button on a calculator to obtain the natural antilogarithm of x. This key is sometimes labeled "inv ln."

17.12 The Effect of Temperature

An exothermic reaction with a negative standard reaction entropy (ΔH_r° and ΔS_r° both negative) has $\Delta G_r^\circ < 0$ at low temperatures. However, as the temperature is raised, the $-T\Delta S_r^\circ$ term in $\Delta G_r^\circ = \Delta H_r^\circ - T\Delta S_r^\circ$, which is positive, becomes more significant. That is, although $K > 1$ at low temperatures, and products are favored, $K < 1$ at high temperatures, and the reactants become favored.

The reverse is true for an endothermic reaction with a positive standard reaction entropy (so ΔH_r° and ΔS_r° are both positive). In this case, ΔG_r° is positive at low temperatures but may become negative if the temperature is raised above the point at which $T\Delta S_r^\circ > \Delta H_r^\circ$. Such a reaction has $K < 1$ at low temperatures but $K > 1$ when the temperature is raised sufficiently high. This relation is illustrated in the following example.

For an endothermic reaction with a negative standard reaction entropy ($\Delta H_r^\circ > 0$ and $\Delta S_r^\circ < 0$), there is no temperature at which $\Delta G_r^\circ < 0$, and therefore no temperature at which $K > 1$. These conclusions are summarized in Table 17.3.

The free energy increases with temperature for reactions with a negative ΔS_r° and decreases with temperature for reactions with a positive ΔS_r°.

Example 17.12 *Estimating the minimum temperature at which K > 1*

Estimate the temperature at which it is thermodynamically possible, in the sense that $K > 1$, for carbon to reduce iron(III) oxide to iron by the endothermic reaction
$$2\,Fe_2O_3(s) + 3\,C(s) \rightarrow 4\,Fe(s) + 3\,CO_2(g).$$

Strategy The reaction becomes feasible in the sense that $K > 1$ when the standard reaction free energy, $\Delta G_r^\circ = \Delta H_r^\circ - T\Delta S_r^\circ$, becomes negative. Therefore, if ΔS_r° is positive, an endothermic reaction becomes spontaneous at temperatures above

$$T = \frac{\Delta H_r^\circ}{\Delta S_r^\circ}$$

Calculate the standard reaction enthalpy and entropy from the data in Appendix 2A, assuming that neither changes appreciably with temperature.

Solution From data in Appendix 2A, and recalling that the standard enthalpies of formation of the elements are 0,

$$\begin{aligned}\Delta H_r^\circ &= 3\Delta H_f^\circ(CO_2, g) - 2\Delta H_f^\circ(Fe_2O_3, s)\\&= 3(-393.5) - 2(-824.2) \text{ kJ/mol} = +467.9 \text{ kJ/mol}\end{aligned}$$

Similarly, but noting that the standard molar entropies of elements are not 0,

$$\begin{aligned}\Delta S_r^\circ &= \{4S_m^\circ(Fe, s) + 3S_m^\circ(CO_2, g)\} - \{2S_m^\circ(Fe_2O_3, s) + 3S_m^\circ(C, s)\}\\&= \{(4 \times 27.3) + (3 \times 213.7)\} - \{(2 \times 87.4) + (3 \times 5.7)\} \text{ J/K·mol}\\&= +558.4 \text{ J/K·mol}\end{aligned}$$

The temperature above which ΔG_r° is negative is

$$\begin{aligned}T &= \frac{\Delta H_r^\circ}{\Delta S_r^\circ} = \frac{467.9 \text{ kJ/mol}}{558.4 \text{ J/K·mol}}\\&= \frac{467.9 \times 10^3}{558.4} \frac{\text{J/mol}}{\text{J/K·mol}} = 838 \text{ K}\end{aligned}$$

The minimum temperature at which reduction occurs is therefore about 565°C.

Self-Test 17.15A What is the minimum temperature at which magnetite, Fe_3O_4, can be reduced to iron by using carbon (to produce CO_2)?

[*Answer:* 943 K]

Self-Test 17.15B Estimate the temperature at which magnesium carbonate can be expected to decompose to magnesium oxide and carbon dioxide.

Skills You Should Have Mastered

Conceptual

☐ 1. State and explain the implications of the second law of thermodynamics, Section 17.2.

☐ 2. Explain how temperature, volume, and state of matter affect the entropy of a substance, Sections 17.2 and 17.3.

☐ 3. Show how ΔS_{surr} is related to ΔH for a change at constant temperature and pressure and justify the relationship, Section 17.5.

☐ 4. Show how the free energy change accompanying a process is related to the direction of spontaneous reaction and the position of equilibrium, Sections 17.7 and 17.10.

Problem-Solving

☐ 1. Predict which of two systems has the greater entropy, given their compositions and conditions, Toolbox 17.1 and Example 17.1.

☐ 2. Calculate the change in entropy of a system due to heat transfer and phase changes, Toolbox 17.1 and Examples 17.2 and 17.3.

☐ 3. Estimate the relative molar entropies of two substances, Toolbox 17.1 and Example 17.4.

☐ 4. Calculate the standard reaction entropy from standard molar entropies, Example 17.5.

Exercises

Entropy

17.1 Which substance in each pair has the higher molar entropy at 298 K: (a) HBr(g), HF(g); (b) NH_3(g), Ne(g); (c) I_2(s), I_2(l); (d) Ar(g) at 1.00 atm or Ar(g) at 2.00 atm?

17.2 Which substance in each pair has the higher molar entropy? (Assume the temperature is 298 K unless otherwise specified.) (a) CH_4(g), C_3H_8(g); (b) KCl(aq), KCl(s); (c) He(g), Kr(g); (d) O_2(g) at 273 K and 1.00 atm, O_2(g) at 450. K and 1.00 atm.

17.3 List the following substances in order of increasing molar entropy at 298 K: H_2O(l), H_2O(g), H_2O(s), C(s). Explain your reasoning.

17.4 List the following substances in order of increasing molar entropy at 298 K: CO_2(g), Ar(g), H_2O(l), Ne(g). Explain your reasoning.

17.5 The following pictures show a molecular visualization of a system undergoing a change. Decide from the picture whether entropy increases or decreases during the process. Explain your reasoning.

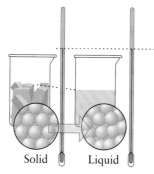

Solid   Liquid

17.6 The following pictures show a molecular visualization of a system undergoing a change. Decide from the picture

whether entropy increases or decreases during the process. Explain your reasoning.

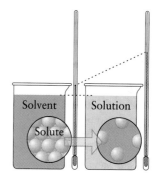

Solvent   Solution

Solute

17.7 Without performing any calculations, state whether the entropy of the system increases or decreases during each of the following processes: (a) the oxidation of nitrogen, N_2(g) + 2 O_2(g) → 2 NO_2(g); (b) the sublimation of dry ice, CO_2(s) → CO_2(g); (c) the cooling of water from 50°C to 4°C. Explain your reasoning.

17.8 Without performing any calculations, state whether the entropy of the system increases or decreases during each of the following processes: (a) the dissolution of table salt, NaCl(s) → NaCl(aq); (b) the photosynthesis of glucose: 6 CO_2(g) + 6 H_2O(l) → $C_6H_{12}O_6$(s) + 6 O_2(g); (c) the evaporation of water from damp clothes. Explain your reasoning.

17.9 Use data from Table 6.2 or Appendix 2A to calculate the entropy change for (a) the freezing of 1 mol H_2O(l) at 0°C; (b) the vaporization of 50.0 g of ethanol, C_2H_5OH(l), at 351.5 K.

17.10 Use data from Table 6.2 or Appendix 2A to calculate the entropy change for (a) the vaporization of 1 mol H_2O(l) at 100.°C and 1 atm; (b) the freezing of 3.33 g of NH_3(l) at 195.3 K.

17.11 Use data from Table 17.1 or Appendix 2A to calculate the standard reaction entropy for each of the following reactions at 25°C. For each reaction, interpret the sign and magnitude of the reaction entropy.
(a) The combustion of hydrogen to water vapor:
$2 H_2(g) + O_2(g) \rightarrow 2 H_2O(g)$
(b) The oxidation of carbon monoxide:
$2 CO(g) + O_2(g) \rightarrow 2 CO_2(g)$
(c) The decomposition of calcite:
$CaCO_3(s) \rightarrow CaO(s) + CO_2(g)$
(d) The following reaction of potassium chlorate:
$4 KClO_3(s) \rightarrow 3 KClO_4(s) + KCl(s)$

17.12 Use data from Table 17.1 or Appendix 2A to calculate the standard reaction entropy for each of the following reactions at 25°C. For each reaction, interpret the sign and magnitude of the reaction entropy.
(a) The synthesis of carbon disulfide from natural gas (methane):
$2 CH_4(g) + S_8(s) \rightarrow 2 CS_2(l) + 4 H_2S(g)$
(b) The production of acetylene from calcium carbide:
$CaC_2(s) + 2 H_2O(l) \rightarrow Ca(OH)_2(s) + C_2H_2(g)$
(c) The oxidation of ammonia, the first step in the production of nitric acid:
$4 NH_3(g) + 5 O_2(g) \rightarrow 4 NO(g) + 6 H_2O(l)$
(d) The industrial synthesis of urea, a common fertilizer:
$CO_2(g) + 2 NH_3(g) \rightarrow CO(NH_2)_2(s) + H_2O(l)$

Entropy Change in the Surroundings

17.13 Account for the spontaneity of the following processes in terms of the entropy changes in the system and the surroundings: (a) hot coffee cools; (b) an organic compound dissolves in water with no change in temperature; (c) gasoline burns; (d) helium gas diffuses through argon gas.

17.14 Account for the spontaneity of the following processes in terms of the entropy changes in the system and the surroundings: (a) sodium reacts with chlorine to produce sodium chloride, giving off heat and light; (b) water vaporizes at 100.1°C at 1 atm; (c) water freezes at −0.01°C; (d) milk spills out of an overturned bottle.

17.15 The following picture shows a molecular visualization of a system undergoing a spontaneous change. Account for the

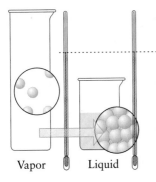

Vapor   Liquid

spontaneity of the process in terms of the entropy changes in the system and the surroundings.

17.16 The following picture shows a molecular visualization of a system undergoing a spontaneous change. Account for the spontaneity of the process in terms of the entropy changes in the system and the surroundings.

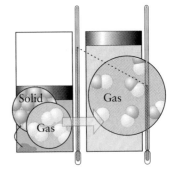

Solid   Gas

Gas

17.17 A human body generates heat at the rate of about 100 W (1 W = 1 J/s). (a) At what rate does your body heat generate entropy in your surroundings, taken to be at 20°C? (b) How much entropy do you generate each day? (c) Would the entropy generated be greater or less if you were in a room kept at 30°C? Explain your answer.

17.18 An electric heater is rated at 2 kW (1 W = 1 J/s). (a) At what rate does it generate entropy in a room maintained at 28°C? (b) How much entropy does it generate in the course of a day? (c) Would the entropy generated be greater or less if the room were maintained at 25°C? Explain your answer.

17.19 (a) Calculate the change in entropy of a block of copper at 25°C that absorbs 5 J of energy from a heater. (b) If the block of copper is at 100.°C and it absorbs 5 J of energy from the heater, what is its entropy change? (c) Explain any difference in entropy change.

17.20 (a) Calculate the change in entropy of 1.0 L of liquid water at 0°C when it absorbs 500. J of energy from a heater. (b) If the 1.0 L of water is at 99°C, what is its entropy change? (c) Explain any difference in entropy change.

17.21 Use the information in Table 6.2 to calculate the change in entropy of the surroundings and the system for (a) the vaporization of 1.0 mol $CH_4(l)$ at its normal boiling point; (b) the melting of 1.0 mol $C_2H_5OH(l)$ at its normal melting point; (c) the freezing of 1.0 mol $C_2H_5OH(l)$ at its normal freezing point.

17.22 Use the information in Table 6.2 to calculate the change in entropy of the surroundings and the system for (a) the melting of 1.0 mol $NH_3(s)$ at its normal melting point; (b) the freezing of 1.0 mol $CH_3OH(l)$ at its normal freezing point; (c) the vaporization of 1.0 mol $H_2O(l)$ at its normal boiling point.

17.23 Consider the reaction for the production of formaldehyde:

$$H_2(g) + CO(g) \longrightarrow HCHO(g),$$
$$\Delta H° = +1.96 \text{ kJ}, \Delta S° = -109.58 \text{ J/K}$$

Calculate the change in entropy of the surroundings and predict whether the reaction is spontaneous at 25°C.

17.24 Consider the reaction in which deuterium exchanges with ordinary hydrogen in water:

$$D_2(g) + H_2O(l) \longrightarrow H_2(g) + D_2O(l),$$
$$\Delta H° = -7.38 \text{ kJ}, \Delta S° = +114.15 \text{ J/K}$$

Calculate the change in entropy of the surroundings and predict whether the reaction is spontaneous at 25°C.

17.25 Why are many exothermic reactions spontaneous?

17.26 Explain how an endothermic reaction can be spontaneous.

Free Energy

17.27 Predict the sign of ΔG for a reaction that is
(a) exothermic and accompanied by an increase in entropy;
(b) endothermic and accompanied by an increase in entropy.
(c) Can a temperature change affect the sign of ΔG in (a) or (b)? If so, how?

17.28 Predict the sign of ΔG for a reaction that is
(a) exothermic and accompanied by a decrease in entropy;
(b) endothermic and accompanied by a decrease in entropy.
(c) Can a temperature change affect the sign of ΔG in (a) or (b)? If so, how?

17.29 The entropy of vaporization of liquid chlorine is 85.4 J/K·mol and its enthalpy of vaporization is 20.4 kJ/mol. What is its boiling point?

17.30 Estimate the boiling point of fluorine by assuming that its enthalpy and entropy of vaporization are similar to the values for argon (see Table 6.2).

17.31 The entropy of vaporization of benzene is approximately 85 J/K·mol. (a) Estimate the enthalpy of vaporization of benzene at its normal boiling point of 80.1°C. (b) What is the entropy change of the surroundings when 10. g of benzene, C_6H_6, vaporizes at its normal boiling point?

17.32 The entropy of vaporization of acetone is approximately 85 J/K·mol. (a) Estimate the enthalpy of vaporization of acetone at its normal boiling point of 56.2°C. (b) What is the entropy change of the surroundings when 10. g of acetone, CH_3COCH_3, condenses at its normal boiling point?

Standard Reaction Free Energies

17.33 Calculate the standard free energy for each reaction, using the equation $\Delta G_r° = \Delta H_r° - T\Delta S_r°$ and data from

Appendix 2A. (Note that $\Delta S_r°$ for each reaction was calculated in Exercise 17.11.) For each reaction, interpret the sign and magnitude of the free energy.
(a) The combustion of hydrogen to water vapor:
$$2 H_2(g) + O_2(g) \rightarrow 2 H_2O(g)$$
(b) The oxidation of carbon monoxide:
$$2 CO(g) + O_2(g) \rightarrow 2 CO_2(g)$$
(c) The decomposition of calcite:
$$CaCO_3(s) \rightarrow CaO(s) + CO_2(g)$$
(d) The following reaction of potassium chlorate:
$$4 KClO_3(s) \rightarrow 3 KClO_4(s) + KCl(s)$$

17.34 Calculate the standard free energy for each reaction, using the equation $\Delta G_r° = \Delta H_r° - T\Delta S_r°$ and data from Appendix 2A. (Note that $\Delta S_r°$ for each reaction was calculated in Exercise 17.12.) For each reaction, interpret the sign and magnitude of the free energy.
(a) The synthesis of carbon disulfide from natural gas (methane):
$$2 CH_4(g) + S_8(s) \rightarrow 2 CS_2(l) + 4 H_2S(g)$$
(b) The production of acetylene from calcium carbide:
$$CaC_2(s) + 2 H_2O(l) \rightarrow Ca(OH)_2(s) + C_2H_2(g)$$
(c) The oxidation of ammonia, the first step in the production of nitric acid:
$$4 NH_3(g) + 5 O_2(g) \rightarrow 4 NO(g) + 6 H_2O(l)$$
(d) The industrial synthesis of urea, a common fertilizer:
$$CO_2(g) + 2 NH_3(g) \rightarrow CO(NH_2)_2(s) + H_2O(l)$$

17.35 Write a chemical equation for the formation reaction and then calculate the standard free energy of formation of each of the following compounds from the enthalpies of formation and the standard molar entropies, using $\Delta G_r° = \Delta H_r° - T\Delta S_r°$: (a) $NH_3(g)$; (b) $H_2O(g)$; (c) $CO(g)$; (d) $NO_2(g)$.

17.36 Write a chemical equation for the formation reaction and then calculate the standard free energy of formation of each of the following compounds from the enthalpies of formation and the standard molar entropies, using $\Delta G_r° = \Delta H_r° - T\Delta S_r°$: (a) $SO_3(g)$; (b) $C_6H_6(l)$; (c) $C_2H_5OH(l)$; (d) $CaCO_3(s)$.

17.37 Determine which of the following compounds are stable with respect to decomposition into their elements at 25°C (see Appendix 2A): (a) $PCl_5(g)$; (b) $HCN(g)$; (c) $NO(g)$; (d) $SO_2(g)$.

17.38 Determine which of the following compounds are stable with respect to decomposition into their elements at 25°C (see Appendix 2A): (a) $CuO(s)$; (b) $C_6H_{12}(l)$, cyclohexane; (c) $SbCl_3(g)$; (d) $N_2H_4(l)$.

17.39 On the basis of the equation $\Delta G_r° = \Delta H_r° - T\Delta S_r°$, which of the following compounds becomes more unstable with respect to its elements as the temperature is raised: (a) $PCl_5(g)$; (b) $HCN(g)$; (c) $NO(g)$; (d) $SO_2(g)$?

17.40 On the basis of the equation $\Delta G_r° = \Delta H_r° - T\Delta S_r°$, which of the following compounds becomes more unstable

with respect to its elements as the temperature is raised:
(a) CuO(s); (b) C_6H_{12}(l), cyclohexane; (c) $SbCl_3$(g); (d) N_2H_4(l)?

17.41 Use the standard free energies of formation in Appendix 2A to calculate the standard free energy of each of the following reactions at 25°C. Comment on the spontaneity of each reaction at 25°C.
(a) $2\,SO_2(g) + O_2(g) \rightarrow 2\,SO_3(g)$
(b) $CaCO_3$(s, calcite) $\rightarrow CaO(s) + CO_2(g)$
(c) $SbCl_5(g) \rightarrow SbCl_3(g) + Cl_2(g)$
(d) $2\,C_8H_{18}(l) + 25\,O_2(g) \rightarrow 16\,CO_2(g) + 18\,H_2O(l)$

17.42 Use the standard free energies of formation in Appendix 2A to calculate the standard free energy of each of the following reactions at 25°C. Comment on the spontaneity of each reaction at 25°C.
(a) $NH_4Cl(s) \rightarrow NH_3(g) + HCl(g)$
(b) $H_2(g) + D_2O(l) \rightarrow D_2(g) + H_2O(l)$
(c) $2\,NO_2(g) \rightarrow N_2O_4(g)$
(d) $2\,CH_3OH(g) + 3\,O_2(g) \rightarrow 2\,CO_2(g) + 4\,H_2O(l)$

Free Energy and Composition

17.43 How does the free energy of a reaction mixture change as it approaches equilibrium?

17.44 Use entropy and free energy arguments to account for the observation at 1 atm that ice melts if the temperature is raised above 0°C and water freezes if the temperature is cooled below 0°C.

17.45 Calculate the equilibrium constant at 25°C for each of the following reactions. (Note that the standard free energy of each reaction was determined in Exercise 17.33.)
(a) The combustion of hydrogen to water vapor:
$2\,H_2(g) + O_2(g) \rightarrow 2\,H_2O(g)$
(b) The oxidation of carbon monoxide:
$2\,CO(g) + O_2(g) \rightarrow 2\,CO_2(g)$
(c) The decomposition of calcite:
$CaCO_3(s) \rightarrow CaO(s) + CO_2(g)$

17.46 Calculate the equilibrium constant at 25°C for each reaction. (Note that the standard free energy of each reaction was determined in Exercise 17.34.)
(a) The synthesis of carbon disulfide from natural gas (methane):
$2\,CH_4(g) + S_8(s) \rightarrow 2\,CS_2(l) + 4\,H_2S(g)$
(b) The production of acetylene from calcium carbide:
$CaC_2(s) + 2\,H_2O(l) \rightarrow Ca(OH)_2(s) + C_2H_2(g)$
(c) The oxidation of ammonia, the first step in the production of nitric acid:
$4\,NH_3(g) + 5\,O_2(g) \rightarrow 4\,NO(g) + 6\,H_2O(l)$
(d) The industrial synthesis of urea, a common fertilizer:
$CO_2(g) + 2\,NH_3(g) \rightarrow CO(NH_2)_2(s) + H_2O(l)$

17.47 Calculate the standard free energy of each of the following reactions:
(a) $N_2(g) + 3\,H_2(g) \rightleftharpoons 2\,NH_3(g)$, $K_p = 41$ at 400. K
(b) $2\,SO_2(g) + O_2(g) \rightleftharpoons 2\,SO_3(g)$, $K_p = 3.0 \times 10^4$ at 700. K

17.48 Calculate the standard free energy for each of the following reactions:
(a) $H_2(g) + I_2(g) \rightleftharpoons 2\,HI(g)$, $K_p = 160.$ at 500. K
(b) $N_2O_4(g) \rightleftharpoons 2\,NO_2(g)$, $K_p = 47.9$ at 400. K

17.49 If $Q_p = 1.0$ for the reaction $N_2(g) + O_2(g) \rightleftharpoons 2\,NO(g)$ at 25°C, will the reaction have a tendency to form products or reactants, or will it be at equilibrium?

17.50 If $Q_p = 1.0 \times 10^{50}$ for the reaction $C(s) + O_2(g) \rightleftharpoons CO_2(g)$ at 25°C, will the reaction have a tendency to form products or reactants, or will it be at equilibrium?

17.51 Calculate the reaction free energy of $2\,SO_2(g) + O_2(g) \rightarrow 2\,SO_3(g)$ at 700. K ($K_p = 3.0 \times 10^4$) when the partial pressures of SO_2 and O_2 are each 0.026 atm and that of SO_3 is 5.00 atm. What is the spontaneous direction of the reaction?

17.52 Calculate the reaction free energy of $SbCl_5(g) \rightleftharpoons SbCl_3(g) + Cl_2(g)$ at 500. K ($K_p = 3.5 \times 10^{-4}$) in a gas-phase equilibrium mixture having the following partial pressures: $P_{SbCl_5} = 0.20$ atm, $P_{SbCl_3} = 0.020$ atm, and $P_{Cl_2} = 0.10$ atm. What is the spontaneous direction of the reaction?

17.53 (a) Calculate the reaction free energy of $N_2(g) + 3\,H_2(g) \rightleftharpoons 2\,NH_3(g)$ when the partial pressures of N_2, H_2, and NH_3 are 1.0, 4.2, and 63 atm, respectively, and the temperature is 400. K. For this reaction, $K_p = 41$ at 400. K. (b) Indicate whether this reaction mixture is likely to form reactants, is likely to form products, or is at equilibrium.

17.54 (a) Calculate the reaction free energy of $H_2(g) + I_2(g) \rightleftharpoons 2\,HI(g)$ when the partial pressures of H_2, I_2, and HI are 0.026, 0.33, and 1.84 atm, respectively, and the temperature is 700. K. For this reaction $K_p = 54$ at 700. K. (b) Indicate whether this reaction mixture is likely to form reactants, is likely to form products, or is at equilibrium.

17.55 For the reaction, $N_2O_4(g) \rightleftharpoons 2\,NO_2(g)$, $\Delta H_r^\circ = +57.2$ kJ/mol and $\Delta S_r^\circ = +175.83$ J/K·mol at 25°C. (a) What is the standard reaction free energy at 25°C? (b) What is the standard reaction free energy at 75°C? (Assume that ΔH_r° and ΔS_r° are unaffected by temperature changes.) (c) Determine the temperature at which $\Delta G_r^\circ = 0$ and the value of K at that temperature.

17.56 For the reaction, $2\,NO(g) + O_2(g) \rightleftharpoons 2\,NO_2(g)$, $\Delta H_r^\circ = -114.1$ kJ/mol and $\Delta S_r^\circ = -146.54$ J/K·mol.
(a) What is the standard reaction free energy at 25°C?
(b) What is the reaction free energy at 700.°C? (Assume that ΔH_r° and ΔS_r° are unaffected by temperature changes.)
(c) Temperatures in an internal combustion cylinder approach 2500°C. In the presence of an ample supply of oxygen, which nitrogen oxide is favored at that temperature? (d) Which nitrogen oxide is favored at the temperature of the exhaust gases, which is approximately 50.°C?

17.57 (a) Do $Cu^+(aq)$ ions have a thermodynamic tendency to form $Cu^{2+}(aq)$ ions and copper metal in water at 25°C? (b) Is the tendency greater at higher temperatures or lower temperatures?

17.58 (a) At what temperature does $NO_2(g)$ have a thermodynamic tendency to form $N_2O_4(g)$? (b) Is the tendency greater at higher temperatures or lower temperatures?

17.59 Determine whether titanium dioxide can be reduced by carbon at 1.00×10^3 K in each of the following reactions
(a) $TiO_2(s) + 2\,C(s) \rightarrow Ti(s) + 2\,CO(g)$
(b) $TiO_2(s) + C(s) \rightarrow Ti(s) + CO_2(g)$
given that, at 1.00×10^3 K, $\Delta G_f°(CO, g) = -200.$ kJ/mol, $\Delta G_f°(CO_2, g) = -396$ kJ/mol, and $\Delta G_f°(TiO_2, s) = -762$ kJ/mol.

17.60 Determine whether manganese(IV) oxide can be reduced by carbon at 1.00×10^3 K in each of the following reactions:
(a) $MnO_2(s) + 2\,C(s) \rightarrow Mn(s) + 2\,CO(g)$
(b) $MnO_2(s) + C(s) \rightarrow Mn(s) + CO_2(g)$
given that, at 1.00×10^3 K, $\Delta G_f°(CO, g) = -200.$ kJ/mol, $\Delta G_f°(CO_2, g) = -396$ kJ/mol, and $\Delta G_f°(MnO_2, s) = -405$ kJ/mol.

Supplementary Exercises

17.61 How does entropy differ from enthalpy? Explain why both the entropy and enthalpy changes of a system are necessary to determine whether a process is spontaneous.

17.62 How would the entropy of water vapor change if it were compressed at 150.°C?

17.63 The standard molar entropy of $Cl(g)$ is 165.2 J/K·mol and that of $Cl_2(g)$ is 223.07 J/K·mol. Suggest a reason why the standard molar entropy of the diatomic molecules is greater than that of the atoms at 25°C.

17.64 Without performing any calculations, state whether an increase or decrease in entropy occurs for each of the following processes:
(a) $Cl_2(g) + H_2O(l) \rightarrow HCl(aq) + HClO(aq)$
(b) $Cu_3(PO_4)_2(s) \rightarrow 3\,Cu^{2+}(aq) + 2\,PO_4^{3-}(aq)$
(c) $SO_2(g) + Br_2(g) + 2\,H_2O(g) \rightarrow H_2SO_4(aq) + 2\,HBr(g)$
(d) $2\,Fe(s) + 3\,Cl_2(g) \rightarrow 2\,FeCl_3(s)$

17.65 Calculate the change in the entropy of the surroundings when (a) 1 mJ is released to the surroundings at 2×10^{-7} K; (b) 1 J, the energy of a single heartbeat, is released to the surroundings at 37°C (normal body temperature).

17.66 Does methanol have a thermodynamic tendency to decompose into carbon monoxide and hydrogen at 25°C and 1 atm? Would the tendency be greater at a higher or lower temperature?

17.67 Use data from Table 17.1 and Appendix 2A to calculate the standard reaction entropy for each of the following reactions and interpret the sign and magnitude of each answer.
(a) The decomposition of hydrogen peroxide:
$2\,H_2O_2(l) \rightarrow 2\,H_2O(l) + O_2(g)$
(b) The preparation of hydrofluoric acid from fluorine and water:
$2\,F_2(g) + 2\,H_2O(l) \rightarrow 4\,HF(aq) + O_2(g)$
(c) The production of synthesis gas, a low-grade industrial fuel:
$CH_4(g) + H_2O(g) \rightarrow CO(g) + 3\,H_2(g)$
(d) The thermal decomposition of ammonium nitrate:
$NH_4NO_3(s) \rightarrow N_2O(g) + 2\,H_2O(g)$

17.68 For the reaction $PCl_3(g) + Cl_2(g) \rightarrow PCl_5(g)$ at 25°C, $\Delta G_r° = -37.2$ kJ/mol and $\Delta H_r° = -87.9$ kJ/mol. (a) Calculate $\Delta S_r°$ for the reaction at 25°C. (b) Does the sign of the entropy change agree with your expectations?

17.69 The reaction for the production of a synthetic fuel from coal is $C(s, graphite) + H_2O(g) \rightleftharpoons CO(g) + H_2(g)$. (a) Calculate $\Delta H_r°$, $\Delta S_r°$, and $\Delta G_r°$ at 25°C. (b) Assuming that $\Delta H_r°$ and $\Delta S_r°$ are unaffected by temperature changes, calculate the temperature at which $\Delta G_r° = 0$ and the value of K at that temperature.

17.70 (a) Calculate $\Delta H_r°$ and $\Delta S_r°$ at 25°C for the reaction $C_2H_2(g) + 2\,H_2(g) \rightarrow C_2H_6(g)$. (b) Calculate the standard reaction free energy from the equation $\Delta G_r° = \Delta H_r° - T\Delta S_r°$. (c) Interpret the calculated values for $\Delta H_r°$ and $\Delta S_r°$. (d) Determine the temperature at which $\Delta G_r° = 0$. What is the significance of that temperature? (Assume that $\Delta H_r°$ and $\Delta S_r°$ are unaffected by temperature changes.)

17.71 For the reaction $CO(g) + Cl_2(g) \rightleftharpoons COCl(g) + Cl(g)$, $K = 9.1 \times 10^{-30}$ at 25°C. Calculate the standard free energy of the reaction.

17.72 For the phase transition $M(s_1) \rightarrow M(s_2)$ of a metal M from solid(1) to solid(2), $\Delta H° = -14.07$ kJ and $\Delta S° = -0.0480$ kJ/K at $-84°C$. (a) What is the value of $\Delta G°$ for the transition? (b) At what temperature does $\Delta G° = 0$ and what is the significance of that temperature? (c) Which phase is favored at 0°C?

17.73 Assuming that $\Delta H°$ and $\Delta S°$ are unaffected by temperature changes, use data from Appendix 2A to estimate (a) the normal boiling point of $PCl_3(l)$; (b) the transition temperature of the phase change $S(s, rhombic) \rightarrow S(s, monoclinic)$; (c) the sublimation temperature of iodine; (d) the normal boiling point of heavy water, D_2O.

17.74 Calculate the equilibrium constant at 25°C for each of the following reactions. (Note that the standard free energy of each reaction was determined in Exercise 17.42.)
(a) $NH_4Cl(s) \rightarrow NH_3(g) + HCl(g)$
(b) $H_2(g) + D_2O(l) \rightarrow D_2(g) + H_2O(l)$

(c) $2 NO_2(g) \rightarrow N_2O_4(g)$
(d) $2 CH_3OH(g) + 3 O_2(g) \rightarrow 2 CO_2(g) + 4 H_2O(l)$

17.75 Calculate the entropy change for (a) the vaporization of 1.0 g of water at its normal boiling point and (b) the melting of 1.0 g of ice at its normal melting point. (c) Why is the entropy change for the vaporization so much larger?

17.76 Determine the equilibrium constant of the reaction $SbCl_5(g) \rightleftharpoons SbCl_3(g) + Cl_2(g)$ at 25°C.

17.77 Explain why each of the following statements is false: (a) Reactions with negative reaction free energies occur spontaneously and rapidly. (b) If $\Delta G_r° = 0$, the reaction is at equilibrium. (c) Every sample of a pure element, regardless of its physical state, is assigned a free energy of formation equal to 0. (d) An exothermic reaction producing more moles of gas than are consumed has a positive standard reaction free energy.

17.78 The enthalpy of vaporization of heavy water is 41.6 kJ/mol at its normal boiling point of 101.4°C. (a) Calculate the entropy change of the surroundings when 10.0 g of heavy water vaporizes at its normal boiling point. (b) Use data in Appendix 2A to determine the vapor pressure of heavy water at 25°C. *Hint:* For a vaporization equilibrium, K is the same as the vapor pressure in atmospheres.

Applied Exercises

For Exercises 17.79–17.81, see Applying Chemistry: Case Study 17.

17.79 (a) Calculate the standard reaction enthalpy, entropy, and free energy for the combustion of glucose (the reverse of the photosynthetic reaction). (b) Does the energy available to do nonexpansion work ($\Delta G_r°$) increase or decrease as the temperature is raised? Explain.

17.80 Calculate the standard reaction free energy of a fuel cell in which methane is oxidized by oxygen to carbon dioxide and water.

17.81 The key metabolic reaction by which free energy is generated in living systems is the hydrolysis of adenosine triphosphate (ATP) to adenosine diphosphate (ADP). If ΔG_r for the hydrolysis of ATP to ADP is -30.5 kJ/mol, what is the minimum number of moles of ATP molecules that would have to be hydrolyzed in a plant to provide the free energy needed for the production of 1 mol of glucose molecules by photosynthesis under standard conditions?

17.82 Octane, C_8H_{18}, a hydrocarbon used in gasoline, does not burn smoothly in automobile engines, causing a noisy condition known as knocking. Isooctane, a branched isomer of octane with the formal name 2,2,4-trimethylpentane, burns more smoothly. The "octane rating" of isooctane is 100, that of octane is close to 0. To improve the octane rating of gasoline, some of the octane is converted into isooctane. (a) Calculate

the change in entropy, the change in free energy and the equilibrium constant at 25°C for the conversion, using the data given below. (b) Which hydrocarbon is favored at equilibrium in the conversion? (c) Is the specific enthalpy of the hydrocarbon increased or decreased in the conversion?

Compound	$\Delta H_f°$, kJ/mol	$\Delta G_f°$, kJ/mol	$S_m°$, J/K·mol
octane	−249.9	+7.6	+358
isooctane	−259.3	−21	+328

17.83 The depletion of ozone in the stratosphere can be summarized by the net equation, $2 O_3(g) \rightarrow 3 O_2(g)$. (a) From values in Appendix 2A, determine the standard reaction free energy and the standard reaction entropy for the reaction. (b) What is the equilibrium constant of the reaction? (c) What is the significance of your answers with regard to ozone depletion?

Integrated Exercises

17.84 Potassium nitrate readily dissolves in water and its enthalpy of solution is $+34.9$ kJ/mol. (a) How does the enthalpy of solution favor or not favor the dissolving process? (b) Is the entropy change of the system positive or negative when the salt dissolves? (c) Is the entropy change of the system primarily a result of changes in positional disorder or thermal disorder? (d) Is the entropy change of the surroundings primarily a result of changes in positional disorder or thermal disorder? (e) What is the driving force for the dissolution of KNO_3?

17.85 (a) What is the free energy of vaporization of pure water at 100.°C and 1 atm? (b) What is the free energy of vaporization of water at 100.°C from an aqueous solution that is 10.0% by mass glucose ($C_6H_{12}O_6$)? (c) Explain why your answers to (a) and (b) are the same or different. (See the hint in Exercise 17.78.)

17.86 The standard reaction free energy change for the deprotonation of a weak monoprotic acid in water is 32.0 kJ/mol. What percentage of the molecules of acid are deprotonated in a 0.100 M aqueous solution of the acid?

17.87 Use data from Appendix 2A to calculate $\Delta G_r°$, $\Delta S_r°$, and K_{sp} for the dissolution of the following sparingly soluble compounds at 25°C. If the sign of $\Delta S_r°$ is positive, explain in terms of the total entropy change why the compound does not dissolve: (a) AgI; (b) $CaCO_3$ (aragonite); (c) $Ca(OH)_2$.

17.88 Use data from Appendix 2A to calculate $\Delta G_r°$, $\Delta S_r°$, and K_{sp} for the dissolution of the following sparingly soluble compounds at 25°C. If the sign of $\Delta S_r°$ is positive, explain in terms of the total entropy change why the compound does not dissolve: (a) AgBr; (b) AgCl; (c) $PbBr_2$.

17.89 (a) Calculate the standard reaction free energy for the autoprotolysis of water, $2 H_2O(l) \rightarrow H_3O^+(aq) + OH^-(aq)$, from K_w at 25°C. (b) Use the data in Appendix 2A to determine whether the concentration of hydronium ions in pure water increases or decreases as the temperature is raised.

17.90 (a) Calculate the standard reaction free energy for the deprotonation of formic acid, $HCOOH(aq) + H_2O(l) \rightarrow H_3O^+(aq) + HCO_2^-(aq)$, from its K_a at 25°C. (b) The percentage deprotonation is greater in dilute solutions of formic acid than in concentrated solutions, although the value of $\Delta G_r°$ is the same, regardless of concentration. Explain this finding.

17.91 Calculate the reaction free energy of $I_2(g) \rightarrow 2 I(g)$ at 1200 K ($K_c = 6.8 \times 10^{-2}$) when the concentrations of I_2 and I are 0.026 and 0.0084 mol/L, respectively. What is the spontaneous direction of the reaction? *Hint:* You will need to convert K_c to K_p and Q_c to Q_p to calculate the reaction free energy. See Toolbox 14.1.

17.92 Calculate the reaction free energy of $PCl_3(g) + Cl_2(g) \rightarrow PCl_5(g)$ at 230.°C when the concentrations of PCl_3, Cl_2, and PCl_5 are 0.22, 0.41, and 1.33 mol/L, respectively. What is the spontaneous direction of reaction, given that $K_c = 49$ at 230.°C? *Hint:* You will need to convert K_c to K_p and Q_c to Q_p to calculate the reaction free energy. See Toolbox 14.1.

Electrochemistry

In Chapter 3, we learned about one of the main types of reactions, redox reactions. Now that we have studied equilibrium and thermodynamics, we can investigate these reactions in more depth. Redox reactions are of fundamental importance. For instance, redox reactions keep us alive. They capture the energy of the Sun by photosynthesis and then use that energy to power our muscles and our minds. But redox reactions can also kill, for they can result in explosions and flames. Redox reactions are used throughout the chemical industry to extract metals from their ores and to convert petroleum into petrochemicals and pharmaceuticals. However, redox reactions also corrode the artifacts that industry produces. What redox reactions achieve, redox reactions can destroy.

In Chapter 17, we saw that some chemical reactions take place spontaneously. **Electrochemistry** is the branch of chemistry that deals with the use of spontaneous chemical reactions to produce electricity and the use of electricity to bring about nonspontaneous chemical change. We shall also see that electrochemical techniques allow us to monitor concentrations electronically and to measure pH and K_a.

TRANSFERRING ELECTRONS

This chapter has another link to earlier chapters. We saw in Chapters 15 and 16 that a great deal of chemistry could be discussed in terms of the transfer of protons. Now we see that another large area of chemistry can be understood in terms of the transfer of another fundamental particle, the electron. Proton transfer is the basis of acid-base reactions; electron

The batteries we use to power portable computers and other electronic devices all rely on redox reactions to generate an electric current. Redox reactions are central to the development of small, light, long-lasting power sources. To a large extent, the future development of technology depends on the capabilities of these power sources. In this chapter, we see what is involved in using chemical reactions to generate electricity.

transfer is the basis of redox reactions. In particular, as we saw in Chapter 3, *oxidation* is electron loss and *reduction* is electron gain. An *oxidizing agent* is a species that oxidizes another species and, as a result, is itself reduced. A *reducing agent* is a species that reduces another species and, as a result, is itself oxidized.

18.1 Half-Reactions

Although oxidation always occurs together with reduction, it is helpful to consider each process separately by writing two **half-reactions,** chemical equations showing the changes involved only in oxidation or reduction. To focus on the oxidation that goes on in a redox reaction, we write the chemical equation for an *oxidation half-reaction.* For example,

$$Mg(s) \longrightarrow Mg^{2+}(aq) + 2\,e^-$$

Notice that the number of electrons lost is equal to the increase in charge on the atom. In other words, we balance charge as well as numbers of atoms of each element.

A half-reaction is only a *conceptual* way of reporting an oxidation: the electrons are not necessarily free. For example, when oxygen is used to oxidize magnesium, the electrons move directly from magnesium atoms to the oxygen atoms. We do not write the state of the electrons in half-reactions: they are "in transit" and do not have a definite physical state. In an oxidation half-reaction, the electrons released always appear on the right of the arrow. The reduced and oxidized species in a half-reaction jointly form a **redox couple.** In this example, the redox couple consists of Mg^{2+} and Mg, and is denoted Mg^{2+}/Mg. A redox couple has the form Ox/Red, where Ox is the oxidized form of the species and Red is the reduced form.

To show the gain of electrons by a species that is being reduced, we write the equation for the corresponding *reduction half-reaction.* For example, for the reduction of iron(III) to elemental iron:

$$Fe^{3+}(aq) + 3\,e^- \longrightarrow Fe(s)$$

In a reduction half-reaction, the electrons gained always appear on the left of the arrow. In this example, the redox couple is Fe^{3+}/Fe.

> *Half-reactions express the two contributions (oxidation and reduction) to an overall redox reaction.*

Self-Test 18.1A Write the equations for the half-reactions for (a) the oxidation of iron(II) ions to iron(III) ions in aqueous solution; (b) the reduction of copper(II) ions in aqueous solution to copper metal.

[*Answer:* (a) $Fe^{2+}(aq) \rightarrow Fe^{3+}(aq) + e^-$; (b) $Cu^{2+}(aq) + 2\,e^- \rightarrow Cu(s)$]

Self-Test 18.1B Write the equations for the half-reactions in which (a) aluminum metal is oxidized to Al^{3+} in aqueous solution; (b) solid sulfur (which exists as S_8 molecules) is reduced to an aqueous solution of sulfide ions (S^{2-}).

18.2 Balancing Redox Equations

Many redox reactions look as though they will be a nightmare to balance. An example is the oxidation of oxalic acid, $H_2C_2O_4$, by permanganate ions, MnO_4^-, to form manganese(II) ions, Mn^{2+}, and carbon dioxide gas, CO_2. From this information, we can write the skeletal equation

$$MnO_4^-(aq) + H_2C_2O_4(aq) \longrightarrow Mn^{2+}(aq) + CO_2(g) \qquad \triangle$$

Can you balance this by inspection? Your authors can't. One problem is that hydrogen appears in a reactant, but not in a product. For many redox reactions in aqueous solution, we may have to include H^+, OH^-, or H_2O to balance the hydrogen and oxygen atoms. The balancing process can be very intricate. Here, for instance, we need to add a product that contains hydrogen atoms.

Fortunately, there is a way to simplify the balancing process. By concentrating on the half-reactions that make up the overall redox reaction, we break a complicated problem down into two simpler problems. That is, we balance the reduction and oxidation half-reactions separately. Then we bring the two balanced half-reactions together by matching the number of electrons released by oxidation with the number used in reduction. The stoichiometric coefficients of the electrons in the two half-reactions must match, because electrons are neither created nor destroyed in chemical reactions. The procedure is outlined in Toolbox 18.1.

The chemical equation of a reduction half-reaction is added to that of an oxidation half-reaction to form the chemical equation for the overall redox reaction.

Example 18.1 *Balancing a redox equation in acidic solution*

Permanganate ions, MnO_4^-, react with oxalic acid, $H_2C_2O_4$, in acidic aqueous solution to produce manganese(II) ions and carbon dioxide gas. The partial skeletal equation is

$$MnO_4^-(aq) + H_2C_2O_4(aq) \longrightarrow Mn^{2+}(aq) + CO_2(g) \qquad \triangle$$

Balance this equation.

Strategy Because the solution is acidic, we use H^+ to balance H atoms. Work through the procedure set out in Toolbox 18.1, using the rule for balancing in acid solution in step 4.

Solution **Step 1.** Because the oxidation number of manganese decreases from $+7$ in MnO_4^- to $+2$ in Mn^{2+}, the MnO_4^- ion is reduced in the reaction. The oxidation number of carbon increases from $+3$ to $+4$, so oxalic acid is oxidized.

Step 2. From the information in step 1, we can write the skeletal equations:

$$\text{Reduction:} \quad MnO_4^- \longrightarrow Mn^{2+}$$

$$\text{Oxidation:} \quad H_2C_2O_4 \longrightarrow CO_2$$

Step 3. Now balance all elements except H and O:

$$\text{Reduction:} \quad MnO_4^- \longrightarrow Mn^{2+}$$

$$\text{Oxidation:} \quad H_2C_2O_4 \longrightarrow 2\,CO_2$$

Step 4. Balance the O atoms by using H_2O:

$$\text{Reduction:} \quad MnO_4^- \longrightarrow Mn^{2+} + 4\,H_2O$$

$$\text{Oxidation:} \quad H_2C_2O_4 \longrightarrow 2\,CO_2$$

Toolbox 18.1 *How to balance redox equations*

This Toolbox describes how to balance redox reactions by the half-reaction method.

Conceptual Basis

In a chemical reaction, both the number of atoms of each type of element and the number of electrons are conserved. In aqueous solution, hydrogen and oxygen are balanced by using H_2O, H^+ ions, and OH^- ions. Note that, although the hydrogen ion exists as hydronium ions in water, it is conventional to write it as $H^+(aq)$ rather than as H_3O^+ when writing equations for redox reactions. Oxidation and reduction half-reactions are balanced separately and are then added together. Any electrons released in oxidation must be used in reduction.

Procedure

The chemical equation for a redox reaction can be balanced as follows (Fig. 18.1):

Step 1. Identify the species being oxidized and the species being reduced from the changes in their oxidation numbers. The species being oxidized contains the element that undergoes an increase in oxidation number. The species being reduced contains the element that undergoes a decrease in oxidation number.

Step 2. Write the two skeletal (unbalanced) equations for the oxidation and reduction half-reactions. Do not include the electrons yet.

Step 3. Balance all elements in the half-reactions except O and H.

Step 4. In *acidic solution,* balance O by adding H_2O to the side of each half-reaction that needs O, and then balance H by adding H^+ to the side that needs H. In *basic solution,* balance O by adding H_2O to the side that needs O. Then balance H by adding H_2O to the side that needs H, and for each H_2O molecule added, add an OH^- ion to the other side.

When we add

$$\ldots OH^- \ldots \longrightarrow \ldots H_2O \ldots$$

we are effectively adding one H atom to the right. When we add

$$\ldots H_2O \ldots \longrightarrow \ldots OH^- \ldots$$

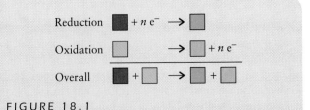

Reduction	■ $+ n\,e^-$	$\longrightarrow$	□
Oxidation	□	$\longrightarrow$	□ $+ n\,e^-$
Overall	■ + □	$\longrightarrow$	□ + □

FIGURE 18.1

A schematic diagram of how to balance a redox equation by balancing the half-reactions separately and then combining them.

we are effectively adding an H atom to the left.

An alternative method for basic solution A simple way to balance half-reactions in basic solution is to balance hydrogen and oxygen by using H^+ ions and H_2O molecules, just as for acidic solution. Then, because the solution is basic, add enough OH^- ions to *both* sides of the half-reaction to "neutralize" all the H^+ ions. In the final step, simplify the equations by canceling the number of water molecules that appear on both sides of the equation. Both methods are demonstrated in Example 18.2.

Step 5. Balance the electric charges by adding electrons (denoted e^-) to the left for reductions and to the right for oxidations to give the same total charge on each side of the half-reaction. If charge cannot be balanced in this way, the species being oxidized and reduced may have been misidentified, so return to step 1.

Step 6. Multiply either one or both half-reactions through by a factor so that the number of electrons in each half-reaction is the same (because the number of electrons released in the oxidation must match that used in the reduction), and then add the half-reactions.

Finally, simplify the appearance of the equation by canceling species that appear on both sides of the equation. It is always a good idea to check that both numbers of atoms and charge are balanced.

Balance the H atoms on the two sides of both half-reactions with H^+:

$$\text{Reduction:} \quad MnO_4^- + 8\,H^+ \longrightarrow Mn^{2+} + 4\,H_2O$$

$$\text{Oxidation:} \quad H_2C_2O_4 \longrightarrow 2\,CO_2 + 2\,H^+$$

Step 5. Now balance the electric charge. Add electrons to the left of the equation in the reduction step and to the right of the equation in the oxidation step:

$$\text{Reduction:} \quad MnO_4^- + 8\,H^+ + 5\,e^- \longrightarrow Mn^{2+} + 4\,H_2O$$

$$\text{Oxidation:} \quad H_2C_2O_4 \longrightarrow 2\,CO_2 + 2\,H^+ + 2\,e^-$$

The reduction half-reaction has a total charge of $+2$ on each side and the oxidation half-reaction has a total charge of 0 on each side, so charge is balanced in each case.
Step 6. Multiply the permanganate half-reaction by 2 and the oxalic acid half-reaction by 5, so that 10 electrons are transferred in each:

$$\text{Reduction:} \quad 2\,MnO_4^- + 16\,H^+ + 10\,e^- \longrightarrow 2\,Mn^{2+} + 8\,H_2O$$

$$\text{Oxidation:} \quad 5\,H_2C_2O_4 \longrightarrow 10\,CO_2 + 10\,H^+ + 10\,e^-$$

Now add the equation for the second half-reaction to the first:

$$2\,MnO_4^- + 5\,H_2C_2O_4 + 16\,H^+ \longrightarrow 2\,Mn^{2+} + 8\,H_2O + 10\,CO_2 + 10\,H^+$$

Ten of the H^+ ions on the left are canceled by the 10 H^+ ions on the right. After adding the physical states of the species, we get

$$2\,MnO_4^-(aq) + 5\,H_2C_2O_4(aq) + 6\,H^+(aq) \longrightarrow 2\,Mn^{2+}(aq) + 8\,H_2O(l) + 10\,CO_2(g)$$

This is the fully balanced net ionic equation.

Self-Test 18.2A When a piece of copper metal is placed in dilute nitric acid, copper(II) nitrate and the gas nitric oxide, NO, are formed. Write the balanced net ionic equation for the reaction.

$$[\textbf{\textit{Answer:}}\ 3\,Cu(s) + 2\,NO_3^-(aq) + 8\,H^+(aq) \rightarrow$$
$$3\,Cu^{2+}(aq) + 2\,NO(g) + 4\,H_2O(l)]$$

Self-Test 18.2B When acidified potassium permanganate is mixed with a solution of sulfurous acid, H_2SO_3, sulfuric acid and manganese(II) ions are produced. Write the balanced net ionic equation for the reaction. In acidic aqueous solution, the weak acid H_2SO_3 is present in molecular form and sulfuric acid is present as HSO_4^- ions.

Example 18.2 *Balancing a redox equation in basic solution*

The products of the reaction of bromide ions with permanganate ions in basic aqueous solution are solid manganese(IV) oxide, MnO_2, and bromate ions, BrO_3^-. Balance the chemical equation for the reaction.

Strategy Work through the procedure set out in Toolbox 18.1, using the rule for balancing in basic solution in step 4.

Solution **Step 1.** The oxidation number of manganese changes from $+7$ in MnO_4^- to $+4$ in MnO_2, so MnO_4^- is reduced. The oxidation number of bromine increases from -1 as Br^- to $+5$ in BrO_3^-, so Br^- is oxidized.
Step 2. The skeletal equations for the oxidation and reduction half-reactions are

$$\text{Reduction:} \quad MnO_4^- \longrightarrow MnO_2$$

$$\text{Oxidation:} \quad Br^- \longrightarrow BrO_3^-$$

Step 3. Both equations are already balanced with respect to Mn and Br.

Step 4. First, balance the O atoms by using H_2O:

$$\text{Reduction:} \quad MnO_4^- \longrightarrow MnO_2 + 2\,H_2O$$

$$\text{Oxidation:} \quad Br^- + 3\,H_2O \longrightarrow BrO_3^-$$

Method I: Next, balance H by adding H_2O to the side of each equation that needs hydrogen (one H_2O molecule for each H atom needed) and an equal number of OH^- ions to the opposite side of each arrow:

$$\text{Reduction:} \quad MnO_4^- + 4\,H_2O \longrightarrow MnO_2 + 2\,H_2O + 4\,OH^-$$

$$\text{Oxidation:} \quad Br^- + 3\,H_2O + 6\,OH^- \longrightarrow BrO_3^- + 6\,H_2O$$

In the final part of this step, simplify each half-reaction by canceling like species on opposite sides of the arrow (in this case, H_2O):

$$\text{Reduction:} \quad MnO_4^- + 2\,H_2O \longrightarrow MnO_2 + 4\,OH^-$$

$$\text{Oxidation:} \quad Br^- + 6\,OH^- \longrightarrow BrO_3^- + 3\,H_2O$$

Method II: To use the alternative method described in Toolbox 18.1, after balancing the O atoms by adding H_2O, add H^+ to balance H:

$$\text{Reduction:} \quad MnO_4^- + 4\,H^+ \longrightarrow MnO_2 + 2\,H_2O$$

$$\text{Oxidation:} \quad Br^- + 3\,H_2O \longrightarrow BrO_3^- + 6\,H^+$$

Because the solution is basic, next add OH^- ions to *both* sides of the equations to "neutralize" the H^+ ions:

$$\text{Reduction:} \quad MnO_4^- + 4\,H_2O \longrightarrow MnO_2 + 2\,H_2O + 4\,OH^-$$

$$\text{Oxidation:} \quad Br^- + 3\,H_2O + 6\,OH^- \longrightarrow BrO_3^- + 6\,H_2O$$

Simplify by canceling two water molecules in the first half-reaction and three in the second:

$$\text{Reduction:} \quad MnO_4^- + 2\,H_2O \longrightarrow MnO_2 + 4\,OH^-$$

$$\text{Oxidation:} \quad Br^- + 6\,OH^- \longrightarrow BrO_3^- + 3\,H_2O$$

Notice that both methods give the same result.

Step 5. Now balance the electric charges by using electrons. Electrons are reactants in the reduction half-reaction and products in the oxidation half-reaction:

$$\text{Reduction:} \quad MnO_4^- + 2\,H_2O + 3\,e^- \longrightarrow MnO_2 + 4\,OH^-$$

$$\text{Oxidation:} \quad Br^- + 6\,OH^- \longrightarrow BrO_3^- + 3\,H_2O + 6\,e^-$$

Step 6. To ensure that the same number of electrons appears in both half-reactions, multiply the reduction half-reaction by 2:

$$\text{Reduction:} \quad 2\,MnO_4^- + 4\,H_2O + 6\,e^- \longrightarrow 2\,MnO_2 + 8\,OH^-$$

$$\text{Oxidation:} \quad Br^- + 6\,OH^- \longrightarrow BrO_3^- + 3\,H_2O + 6\,e^-$$

Six electrons are transferred in this redox reaction. Now add the two half-reactions, simplify the equation by canceling species that appear on both sides (in this case, electrons, H_2O, and OH^-), and attach state symbols. We obtain the following fully balanced net ionic equation:

$$2\,MnO_4^-(aq) + Br^-(aq) + H_2O(l) \longrightarrow 2\,MnO_2(s) + BrO_3^-(aq) + 2\,OH^-(aq)$$

Self-Test 18.3A A basic aqueous solution of hypochlorite ions, ClO^-, reacts with solid chromium(III) hydroxide, $Cr(OH)_3$, to produce chromate ions, CrO_4^{2-}, and chloride ions. Write the balanced net ionic equation for the reaction.

[*Answer:* $2\,Cr(OH)_3(s) + 4\,OH^-(aq) + 3\,ClO^-(aq) \rightarrow$
$2\,CrO_4^{2-}(aq) + 5\,H_2O(l) + 3\,Cl^-(aq)]$

Self-Test 18.3B When iodide ions are added to a solution of iodate ions, IO_3^-, in basic aqueous solution, triiodide ions, I_3^-, are formed. Write the balanced net ionic equation for the reaction. (Note that the same product is obtained in each half-reaction.)

GALVANIC CELLS

An **electrochemical cell** is a device in which an electric current—a flow of electrons through a circuit—is either produced by a spontaneous chemical reaction or is used to bring about a nonspontaneous reaction. Here we concentrate on a **galvanic cell,** an electrochemical cell in which a spontaneous chemical reaction is used to generate an electric current. The battery in a portable CD player is an example of a galvanic cell, because it makes use of a chemical reaction inside the cell to produce an electric current through a circuit attached to it. An electrochemical cell in which an electric current is used to drive a chemical reaction is called an **electrolytic cell.** We deal with them later in the chapter.

18.3 The Structure of a Galvanic Cell

A galvanic cell consists of two **electrodes,** or metallic conductors, that are in electrical contact with an *electrolyte,* an ionically conducting medium, inside the cell. The electrolyte is typically an aqueous solution of an ionic compound. In a galvanic cell, the oxidation and reduction half-reactions take place in different regions of the cell and are commonly called *electrode reactions.* Oxidation takes place at the electrode called the **anode,** and the species being oxidized releases electrons into the anode. Reduction takes place at the other electrode, called the **cathode,** and the species being reduced collects electrons from the cathode (Fig. 18.2). The overall chemical reaction sets up a flow of electrons through an external wire joining the two electrodes. Electrons are deposited into one electrode by the oxidation half-reaction and collected from the other electrode by the reduction half-reaction.

When we complete the circuit—by turning on a CD player or some other piece of equipment—we allow the chemical reaction in the cell to take place. The electrons are forced from the anode of the cell through the electrical circuit, where they turn the electric motor and power the electronics and then return to the cell to complete the reaction at the cathode. The electrons from the external circuit entering the cell at the cathode are used in the reduction half-reaction. If you look at a commercial galvanic cell, you will see one electrode marked with a +. The + marks the cathode, the electrode at which electrons passing through the external circuit enter the cell. The other electrode is marked with a − sign, which indicates where electrons leave the cell. That electrode must therefore be the site of oxidation, the anode.

An early example of a galvanic cell, a **Daniell cell,** was invented by the British chemist John Daniell in 1836 when the growth of telegraphy created an

Electron flow

FIGURE 18.2

In an electrochemical cell, a reaction takes place in two separate regions. Oxidation occurs at one electrode, and the electrons released travel through the external circuit to the other electrode, where they cause reduction. The site of oxidation is called the anode, and the site of reduction is called the cathode.

Notice how the − sign signifies loss of electrons from a galvanic cell and the + sign signifies their entry into the cell.

FIGURE 18.3

(a) (b)

(a) When a bar of zinc is placed in a beaker of copper(II) sulfate solution, copper is deposited on the zinc and the blue copper(II) ions are gradually replaced by colorless zinc ions. (b) The residue in the beaker is copper metal. No more copper ions can be seen in solution.

urgent need for a reliable, steady source of electric current. Daniell knew that the redox reaction

$$Zn(s) + Cu^{2+}(aq) \longrightarrow Zn^{2+}(aq) + Cu(s)$$

is spontaneous, because, when a piece of zinc metal is placed in an aqueous copper(II) sulfate solution, metallic copper is deposited on the surface of the zinc (Fig. 18.3). In atomic terms, as the reaction takes place, electrons are transferred from zinc atoms to Cu^{2+} ions nearby in the solution (Fig. 18.4). These electrons reduce the Cu^{2+} ions to Cu atoms, which stick to the surface of the zinc or form a finely divided solid deposit in the flask. The piece of zinc slowly diminishes as its atoms give up electrons and form Zn^{2+} ions that drift off into the solution.

Daniell realized that if he could separate the reactants in a redox reaction, he could set up an electric current consisting of the electrons being transferred from the zinc metal to the copper ions. He separated the oxidation and reduction half-reactions in his cell by using the arrangement shown in Fig. 18.5. The redox reaction is the same, but the reactants are separated by a porous cup. Now the electrons can travel from Zn atoms to Cu^{2+} ions only by passing through the external circuit (the wire and the light bulb). The Cu^{2+} ions are converted into Cu atoms in one compartment by the reduction half-reaction

$$Cu^{2+}(aq) + 2 e^- \longrightarrow Cu(s)$$

Zinc atoms are converted to Zn^{2+} ions in the other compartment by the oxidation half-reaction

$$Zn(s) \longrightarrow Zn^{2+}(aq) + 2 e^-$$

Ions move between the two compartments (through the porous cup) to preserve electrical neutrality inside the cell and complete the electrical circuit. Anions (for instance, the SO_4^{2-} ions in the solution) travel toward the anode to balance the charges of the Zn^{2+} ions formed by the oxidation of the zinc electrode. Cations travel toward the cathode, to replace the positive charges of the Cu^{2+} ions that have been deposited as copper metal.

The electrodes in the Daniell cell are the metals involved as reactants or products in the reaction. However, not all electrode reactions involve a conducting solid as a reactant or product. For example, the reduction $2 H^+(aq) + 2 e^- \longrightarrow H_2(g)$ produces a gas, so it is necessary to use a chemically inert electrical conductor to supply or remove electrons from the compartment where the reaction takes place. Platinum is most commonly used for the electrode, and hydrogen gas is bubbled over it as it dips into a solution that contains hydronium ions.

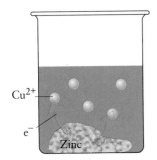

Cu^{2+}

e^-

Zinc

FIGURE 18.4

The reaction shown in Fig. 18.3 takes place all over the surface of the zinc as electrons are transferred to the Cu^{2+} ions in solution.

This arrangement is called a **hydrogen electrode.** Another electrode material used in industry is graphite, which is written C(gr) or C(s, graphite).

In a galvanic cell, a spontaneous chemical reaction tends to draw electrons into the cell through the cathode, the site of reduction, and to release them at the anode, the site of oxidation.

18.4 The Notation for Cells

Chemists use a shorthand notation to specify the structures of electrodes in galvanic cells. The two electrodes in the Daniell cell, for instance, are denoted

$$Zn(s)|Zn^{2+}(aq) \qquad \text{and} \qquad Cu^{2+}(aq)|Cu(s)$$

The ordering is reactant|product for each half-reaction. Each vertical line represents a junction between phases, in this case, between solid metal and ions in solution. A hydrogen electrode is denoted

$$H^+(aq)|H_2(g)|Pt(s)$$

when it acts as a cathode (and H^+ is reduced). When the same electrode acts as an anode (and H_2 is oxidized), we write it

$$Pt(s)|H_2(g)|H^+(aq)$$

An electrode consisting of a platinum wire dipping into a solution of iron(II) and iron(III) ions is denoted

$$Fe^{3+}(aq),Fe^{2+}(aq)|Pt(s)$$

In this case, the oxidized and reduced species are both in the same phase, so a comma is used to separate them.

The structure of an entire cell is reported by writing a **cell diagram,** a combination of the notations for the two electrodes. The cell diagram for the Daniell cell, for instance, is

$$Zn(s)|Zn^{2+}(aq)|Cu^{2+}(aq)|Cu(s)$$

There is one complication with the Daniell cell and others in which electrode compartments consist of two different solutions. The two solutions must be in contact to complete the circuit. However, when different ions mingle together at the interface between their solutions, they affect the voltage in ways that are difficult to predict. In the Daniell cell, zinc sulfate and copper(II) sulfate solutions meet inside the porous barrier. To avoid the complications arising from the mingling of ions, chemists often use a **salt bridge** to join the two electrode compartments and complete the electrical circuit. A salt bridge typically consists of a gel containing a concentrated aqueous salt solution in a bridge-shaped tube (Fig. 18.6). The bridge still allows a flow of ions, so it completes the electrical circuit, but the ions do not affect the cell reaction (often KCl is used). The symbol for a salt bridge is a double vertical line ($\|$), so the arrangement in Fig. 18.6 is denoted

$$Pt(s)|Fe^{2+}(aq),Fe^{3+}(aq)\|Cu^{2+}(aq)|Cu(s)$$

An electrode is designated by representing the interfaces between phases by |. A cell diagram depicts the physical arrangement of species and interfaces, with a salt bridge denoted by $\|$.

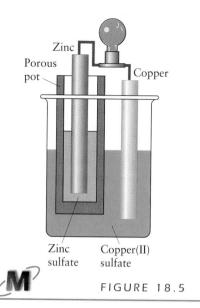

FIGURE 18.5

The Daniell cell consists of copper and zinc electrodes dipping into solutions of copper(II) sulfate and zinc sulfate, respectively. The two solutions make contact through the porous pot, which allows ions to pass through to complete the electrical circuit.

The precise order in which the electrodes are written is important; it is explained in Section 18.5.

FIGURE 18.6

This cell is typical of galvanic cells used in the laboratory. The two electrodes are connected by an external circuit (not shown) and a salt bridge. The latter completes the electrical circuit within the cell.

18.5 Cell Potential

The **cell potential,** *E,* is a measure of the ability of a cell reaction to force electrons through a circuit. The cell potential is sometimes called the *electromotive force* (emf) of the cell or—more colloquially—its *voltage.* A reaction with a strong tendency to push electrons into the anode and pull them from the cathode generates a high cell potential. A reaction with little pulling and pushing power, like a reaction close to equilibrium, generates only a small potential. An exhausted battery is a cell in which the reaction is at equilibrium, so it has lost its power to move electrons and has zero potential.

Cell potentials are measured with electronic voltmeters calibrated in the SI unit of potential, the **volt** (V, Fig. 18.7). A positive reading (such as +1.1 V) means that the + terminal of the meter is connected to the + terminal of the cell, the *cathode.* So, we can determine experimentally which electrode is the cathode by connecting a voltmeter to the cell and seeing which terminal is positive. The other electrode, the − electrode, is the *anode.* Then we write the cell diagram with the anode on the left and the cathode on the right, as in

$$Zn(s)|Zn^{2+}(aq)\|Cu^{2+}(aq)|Cu(s) \qquad E = 1.1\ V$$

The cathode is the site of reduction and, therefore, the electrode at which electrons enter the cell. The anode is the site of oxidation and the electrode at which electrons leave the cell (Fig. 18.8). From the cell diagram for the Daniell cell on the previous page, we know that the copper electrode is the cathode.

FIGURE 18.7

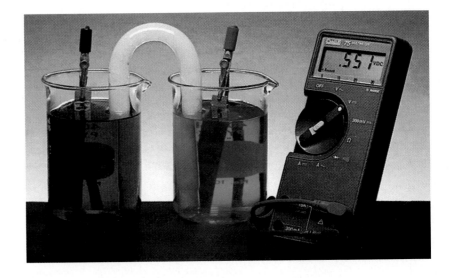

The cell potential is measured with an electronic voltmeter, a device that draws negligible current so that the composition of the cell does not change during the measurement. The display shows a positive value when the + terminal of the meter is connected to the cathode of the galvanic cell.

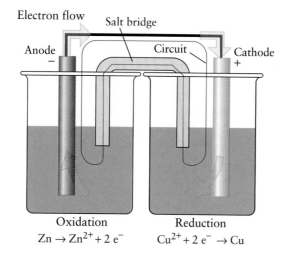

Electron flow

Salt bridge

Anode
−

Circuit

Cathode
+

Oxidation
$Zn \rightarrow Zn^{2+} + 2\,e^-$

Reduction
$Cu^{2+} + 2\,e^- \rightarrow Cu$

FIGURE 18.8

Electrons produced by oxidation leave a galvanic cell at the anode (−), travel through the external circuit, and reenter the cell at the cathode (+), where they cause reduction. The circuit is completed inside the cell by migration of ions through the salt bridge. A salt bridge is unnecessary when the two electrodes share a common electrolyte.

Therefore, the spontaneous reaction in the cell is the reduction of copper(II) ions and the oxidation of zinc:

$$Cu^{2+}(aq) + Zn(s) \longrightarrow Cu(s) + Zn^{2+}(aq)$$

Toolbox 18.2 explains how to write the cell diagram for a given cell.

The cell potential is an indication of the electron-pulling and -pushing power of the cell reaction; cell reactions at equilibrium generate zero potential.

Example 18.3 *Describing a galvanic cell and identifying the cell reaction*

(a) Write the chemical equation for the cell reaction resulting from the following half-reactions:

$$Hg_2^{2+}(aq) + 2\,e^- \longrightarrow 2\,Hg(l)$$

$$2\,Hg(l) + 2\,Cl^-(aq) \longrightarrow Hg_2Cl_2(s) + 2\,e^-$$

(b) Write the cell diagram and draw a schematic picture of the cell and its contents, label the anode and cathode, and indicate the direction of electron flow. Assume that the cell contains a salt bridge filled with a KCl gel.

Strategy Because oxidation takes place at the anode, the mercury metal there is oxidized to mercury(I) ions that immediately react with chloride ions and precipitate as the solid Hg_2Cl_2. At the cathode, mercury(I) ions from the solution are reduced to metallic mercury. Use the procedure set out in Toolbox 18.2 to write the cell reaction and draw a schematic picture of the cell.

Solution **Step 1.** The number of electrons gained matches the number lost, so we add the two half-reactions:

$$2\,Hg(l) + Hg_2^{2+}(aq) + 2\,Cl^-(aq) + 2\,e^- \longrightarrow Hg_2Cl_2(s) + 2\,Hg(l) + 2\,e^-$$

This equation simplifies to

$$Hg_2^{2+}(aq) + 2\,Cl^-(aq) \longrightarrow Hg_2Cl_2(s)$$

Toolbox 18.2 *How to describe a galvanic cell and identify the cell reaction*

This Toolbox explains how to write the chemical equation for the reaction taking place inside a galvanic cell from two half-reactions, how to describe the cell with a cell diagram, how to identify the anode and cathode, and how to trace the flow of electrons in the cell.

Conceptual Basis

The oxidation half-reaction takes place at the anode; the reduction half-reaction at the cathode. We add the two half-reactions together to obtain the cell reaction. In the cell diagram, the anode is written on the left and the cathode on the right. Because electrons are released into the anode, they flow through the external circuit of the cell from anode to cathode (Fig. 18.9). The electric current is carried within the cell (and through the salt bridge, if one is present) by the flow of ions.

Procedure

Step 1. Ensure that the numbers of electrons are the same in each half-reaction (if necessary, multiply through by a suitable factor), and add the two half-reactions together to obtain the cell reaction.

Step 2. Write the cell diagram with the anode on the left and the cathode on the right. Start with the electrode for the anode, then add the other reactants and products, with phases separated by a single line. A double line indicates a salt bridge and usually separates anode from cathode. Repeat the process for the cathode, ending with the electrode material.

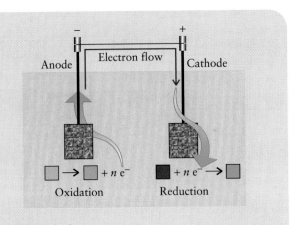

FIGURE 18.9

This schematic picture of a galvanic cell indicates the identities of the anode and cathode, displays the oxidation and reduction half-reactions, and shows the direction of electron flow.

Step 3. Draw a schematic picture of the cell, indicating the contents of each electrode compartment. Label the anode (the site of oxidation) and the cathode (the site of reduction). Draw an arrow on the external wire from anode to cathode to indicate the direction of electron flow.

---

which corresponds to the precipitation of mercury(I) chloride. Notice that a cell reaction does not need to be a redox reaction; all that is necessary is that it can be described as the sum of two half-reactions.

Step 2. Mercury serves as the electrode at both anode and cathode. Write the reactants and product of the oxidation half-reaction first, because oxidation occurs at the anode. Draw two vertical lines to indicate a salt bridge and then write the reactant and product of the reduction half-reaction:

$$Hg(l)|Hg_2Cl_2(s)|HCl(aq)||Hg_2(NO_3)_2(aq)|Hg(l)$$

Step 3. Figure 18.10 shows a schematic picture of the cell and the flow of ions and electrons.

Self-Test 18.5A (a) Write the chemical equation for the cell reaction resulting from the following half-reactions: $H_2(g) \rightarrow 2\,H^+(s) + 2\,e^-$ and $Co^{3+}(aq) + e^- \rightarrow Co^{2+}(aq)$. (b) Assuming that platinum electrodes are used, write the cell diagram

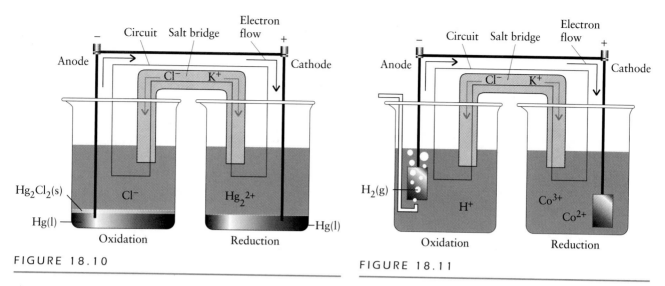

FIGURE 18.10

The cell described in Example 18.3.

FIGURE 18.11

The cell described in Self-Test 18.5A.

and draw a schematic picture of the cell and its contents, label the anode and cathode, and indicate the direction of electron flow.

[**Answer:** (a) $H_2(g) + 2 Co^{3+}(aq) \rightarrow 2 H^+(aq) + 2 Co^{2+}(aq)$;
(b)$Pt(s)|H_2(g)|H^+(aq)||Co^{3+}(aq), Co^{2+}(aq)|Pt(s)$; see Fig. 18.11]

Self-Test 18.5B (a) Write the chemical equation for the cell reaction resulting from the following half-reactions: $Cd(s) + 2 OH^-(aq) \rightarrow Cd(OH)_2(s) + 2 e^-$ and $Cd^{2+}(aq) + 2 e^- \rightarrow Cd(s)$. (b) Write the cell diagram and draw a schematic picture of the cell and its contents, label the anode and cathode, and indicate the direction of electron flow.

18.6 Standard Cell Potentials

The cell potential when all the species are in their standard states is called the **standard cell potential** and is written $E°$. The standard state of a gas is the pure gas at 1 atm, and the standard state of an electrolyte in solution is a concentration of 1 mol/L. There are thousands of possible galvanic cells and therefore thousands of possible standard cell potentials. It is a great simplification to think of each electrode as making a characteristic contribution called its **standard potential.** Standard potentials of electrodes are also denoted $E°$, but it is always clear from the context what kind of potential is meant. Standard potentials are always written for reduction half-reactions, so they are sometimes called *standard reduction potentials* or *standard electrode potentials*.

Each standard potential is a measure of the electron-pulling power of a single electrode. In a galvanic cell, the electrodes pull in opposite directions, so the overall pulling power of the cell, the cell's standard potential, is the difference of the standard potentials of the two electrodes (Fig. 18.12). That difference is always written

$$E° = E°(\text{cathode}) - E°(\text{anode})$$

For example, the standard potential of the cell

$$Fe(s)|Fe^{2+}(aq)||Ag^+(aq)|Ag(s) \qquad E° = 1.24 V$$

> From a list of 50 standard potentials, we can form $(50 \times 49)/2 = 1225$ different combinations; so the list summarizes 1225 different cells.

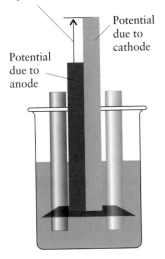

Cell potential

Potential due to cathode

Potential due to anode

FIGURE 18.12

The cell potential can be thought of as being the difference of the two reduction potentials produced by the two electrodes. The cell potential is positive if the cathode has a higher potential than the anode.

is the difference in the potentials of the silver electrode and the iron electrode under standard conditions (all solutes present at 1 mol/L, all gases at 1 atm, and all solid and liquid substances pure):

$$E° = E°(Ag^+ + e^- \rightarrow Ag) - E°(Fe^{2+} + 2\,e^- \rightarrow Fe)$$

It is convenient to abbreviate the half-reactions associated with each standard potential by using the redox couple notation, as in $E°(Ag^+/Ag)$.

A problem with compiling a list of standard potentials for use in this way is that we know only the *overall* cell potential, not the contribution of each electrode. A voltmeter placed between the two electrodes of a galvanic cell measures the *difference* of their potentials, not the individual values. To get around this difficulty, we set the standard potential of one particular electrode, the hydrogen electrode, equal to 0 at all temperatures:

$$2\,H^+(g) + 2\,e^- \longrightarrow H_2(g) \qquad E° = 0$$

In redox couple notation, $E°(H^+/H_2) = 0$. The hydrogen electrode is used to define the standard potential of all other electrodes. For example, to find the standard potential of a zinc electrode, we measure the standard potential of the cell in which the hydrogen electrode is one electrode and a zinc electrode is the other:

$$Zn(s)|Zn^{2+}(aq)\|H^+(aq)|H_2(g)|Pt(s) \qquad E° = 0.76\text{ V}$$

Because, according to convention, the hydrogen electrode contributes 0 to the standard cell potential, the standard potential, 0.76 V, is attributed entirely to the zinc electrode. Notice that a positive voltage is measured when the zinc electrode is the anode. Because the cell potential is the difference of the standard potentials of the two electrodes, $E° = E°(\text{cathode}) - E°(\text{anode})$, the standard potential of the zinc electrode is subtracted from that of hydrogen:

$$E°(H^+/H_2) - E°(Zn^{2+}/Zn) = 0.76\text{ V}$$

It follows that

$$E°(Zn^{2+}/Zn) = -0.76\text{ V}$$

H								He
0								

Li	Be	B	C	N	O	F	Ne
−3.05	−1.85				+1.23	+2.87	

Na	Mg	Al	Si	P	S	Cl	Ar
−2.71	−2.36	−1.66			−0.48	+1.36	

K	Ca	Ga	Ge	As	Se	Br	Kr
−2.93	−2.87	−0.49			−0.67	+1.09	

Rb	Sr	In	Sn	Sb	Te	I	Xe
−2.93	−2.89	−0.34	−0.14		−0.84	+0.54	

Cs	Ba	Tl	Pb	Bi	Po	At	Rn
−2.92	−2.91	−0.34	−0.13	+0.20			

Fr	Ra	13	14	15	16	17	18
	−2.92						

1	2

FIGURE 18.13

The variation of standard potentials in the main groups of the periodic table. Note that the most negative values occur in the *s* block and the most positive values occur close to fluorine.

Table 18.1 Standard potentials at 25°C*

Species†	Reduction half-reaction	E°, V
Oxidized form is strongly oxidizing		
F_2/F^-	$F_2(g) + 2e^- \longrightarrow 2F^-(aq)$	+2.87
Au^+/Au	$Au^+(aq) + e^- \longrightarrow Au(s)$	
Ce^{4+}/Ce^{3+}	$Ce^{4+}(aq) + e^- \longrightarrow Ce^{3+}(aq)$	+1.61
$MnO_4^-, H^+/Mn^{2+}$	$MnO_4^-(aq) + 8H^+(aq) + 5e^- \longrightarrow Mn^{2+}(aq) + 4H_2O(l)$	+1.51
Cl_2/Cl^-	$Cl_2(g) + 2e^- \longrightarrow 2Cl^-(aq)$	+1.36
$Cr_2O_7^{2-}, H^+/Cr^{3+}$	$Cr_2O_7^{2-} + 14H^+(aq) + 6e^- \longrightarrow 2Cr^{3+}(aq) + 7H_2O(l)$	+1.33
$O_2, H^+/H_2O$	$O_2(g) + 4H^+(aq) + 4e^- \longrightarrow 2H_2O(l)$	+1.23
		+0.82 at pH = 7
Br_2/Br^-	$Br_2(l) + 2e^- \longrightarrow 2Br^-(aq)$	+1.09
$NO_3^-, H^+/NO$	$NO_3^-(aq) + 4H^+(aq) + 3e^- \longrightarrow NO(g) + 2H_2O(l)$	+0.96
Ag^+/Ag	$Ag^+(aq) + e^- \longrightarrow Ag(s)$	+0.80
Fe^{3+}/Fe^{2+}	$Fe^{3+}(aq) + e^- \longrightarrow Fe^{2+}(aq)$	+0.77
I_2/I^-	$I_2(s) + 2e^- \longrightarrow 2I^-(aq)$	+0.54
O_2/OH^-	$O_2(g) + 2H_2O + 4e^- \longrightarrow 4OH^-(aq)$	+0.40
		+0.82 at pH = 7
Cu^{2+}/Cu	$Cu^{2+}(aq) + 2e^- \longrightarrow Cu(s)$	+0.34
$AgCl/Ag, Cl^-$	$AgCl(s) + e^- \longrightarrow Ag(s) + Cl^-(aq)$	+0.22
H^+/H_2	$2H^+(aq) + 2e^- \longrightarrow H_2(g)$	0, by definition
Fe^{3+}/Fe	$Fe^{3+}(aq) + 3e^- \longrightarrow Fe(s)$	−0.04
$O_2/HO_2^-, OH^-$	$O_2(g) + H_2O(l) + 2e^- \longrightarrow HO_2^-(aq) + OH^-(aq)$	−0.08
Pb^{2+}/Pb	$Pb^{2+}(aq) + 2e^- \longrightarrow Pb(s)$	−0.13
Sn^{2+}/Sn	$Sn^{2+}(aq) + 2e^- \longrightarrow Sn(s)$	−0.14
Fe^{2+}/Fe	$Fe^{2+}(aq) + 2e^- \longrightarrow Fe(s)$	−0.44
Zn^{2+}/Zn	$Zn^{2+}(aq) + 2e^- \longrightarrow Zn(s)$	−0.76
$H_2O/H_2, OH^-$	$2H_2O(l) + 2e^- \longrightarrow H_2(g) + 2OH^-(aq)$	−0.83
		−0.42 at pH = 7
Al^{3+}/Al	$Al^{3+}(aq) + 3e^- \longrightarrow Al(s)$	−1.66
Mg^{2+}/Mg	$Mg^{2+}(aq) + 2e^- \longrightarrow Mg(s)$	−2.36
Na^+/Na	$Na^+(aq) + e^- \longrightarrow Na(s)$	−2.71
K^+/K	$K^+(aq) + e^- \longrightarrow K(s)$	−2.93
Li^+/Li	$Li^+(aq) + e^- \longrightarrow Li(s)$	−3.05
Reduced form is strongly reducing		

*For a more extensive table, see Appendix 2B.
†In the notation X/Y, X is the oxidized species (the reactant, the oxidizing agent) and Y is the reduced species (the product, the reducing agent) in the half-reaction.

We report the standard potential of the zinc electrode as −0.76 V. The signs of standard potentials are always given, for positive values as well as negative values.

Table 18.1 gives a selection of standard potentials (at 25°C); a longer table can be found in Appendix 2B. Standard potentials of elements vary in a complicated way through the periodic table (Fig. 18.13). However, the most negative are usually found toward the left of the periodic table and the most positive are found toward the upper right.

Notice that in Appendix 2B standard potentials are listed both by voltage and alphabetically, to make it easy to find the one you want.

The standard potential of an electrode is the standard potential of a cell in which the other electrode is a hydrogen electrode. The latter is assigned zero potential. If the test electrode is found to be the anode, then it is assigned a negative potential; if it is the cathode, then its standard potential is positive.

Example 18.4 *Deducing the standard potential of an electrode*

The standard potential of a Zn^{2+}/Zn electrode is -0.76 V, and the standard potential of the cell

$$Zn(s)|Zn^{2+}(aq)||Cu^{2+}(aq)|Cu(s)$$

is 1.10 V. What is the standard potential of the Cu^{2+}/Cu electrode?

Strategy The standard potential of one of the electrodes and the overall potential are known; the value for the other electrode can be calculated from them. The cell diagram reveals which electrode is the anode (the site of oxidation, the one on the left) and which is the cathode (the site of reduction, the one on the right). The difference of the standard potentials, $E°(\text{cathode}) - E°(\text{anode})$, is equal to the overall potential of the cell.

Solution The zinc electrode is the anode, so zinc is oxidized and copper is reduced. The standard potential of zinc is -0.76 V. The copper electrode must contribute a potential such that overall the cell potential is 1.10 V:

$$E°(Cu^{2+}/Cu) - E°(Zn^{2+}/Zn) = 1.10\,V$$

Therefore,

$$
\begin{aligned}
E°(Cu^{2+}/Cu) &= 1.10\,V + E°(Zn^{2+}/Zn) \\
&= 1.10\,V - 0.76\,V = +0.34\,V
\end{aligned}
$$

Self-Test 18.6A The standard potential of the Ag^+/Ag electrode is $+0.80$ V, and the standard potential of the cell $Pt(s)|I_2(s)|I^-(aq)||Ag^+(aq)|Ag(s)$ is 0.26 V at the same temperature. What is the standard potential of the iodine electrode?

[***Answer:*** $+0.54$ V]

Self-Test 18.6B The standard potential of the Fe^{2+}/Fe electrode is -0.44 V and the standard potential of the cell $Fe(s)|Fe^{2+}(aq)||Pb^{2+}(aq)|Pb(s)$ is 0.31 V. What is the standard potential of the lead electrode?

18.7 The Significance of Standard Potentials

The sign and magnitude of a standard potential is an indication of the ability of a species to act as a reducing agent. The positive sign of $E°(Cu^{2+}/Cu)$, $+0.34$ V, tells us that under standard conditions in a cell with a hydrogen electrode, copper is the cathode, the site of reduction. That is, the positive sign tells us that under standard conditions, copper ions have a tendency to be reduced by hydrogen gas:

$$Cu^{2+}(aq) + H_2(g) \longrightarrow Cu(s) + 2\,H^+(aq)$$

Copper could, in principle, reduce the hydrogen ions and drive this reaction to the left. However, the positive sign tells us that copper is a weaker reducing agent than hydrogen and that copper does not drive this reaction to the left.

In contrast, the negative sign of $E°(Zn^{2+}/Zn)$, -0.76 V, tells us that under standard conditions in a cell with a hydrogen electrode, zinc is the anode, the site of oxidation. That is, the negative sign indicates that, under standard conditions, zinc has a tendency to reduce hydrogen ions to hydrogen gas:

$$Zn(s) + 2 H^+(aq) \longrightarrow Zn^{2+}(aq) + H_2(g)$$

Hydrogen gas could, in principle, reduce the Zn^{2+} ions and drive the reaction to the left. However, the negative sign tells us that zinc metal is a stronger reducing agent than hydrogen gas and drives the reaction to the right.

We can conclude that because zinc metal (with its negative potential) is a stronger reducing agent than hydrogen, and hydrogen is a stronger reducing agent than copper metal (with its positive potential), *the more negative the standard potential, the stronger the reducing power of the electrode* (Fig. 18.14).

We can use standard potentials to predict the spontaneous direction of reactions. For example, magnesium, iron, indium, tin, lead, and other metals with negative standard potentials have a thermodynamic tendency to reduce hydrogen ions in a 1 M acid solution to hydrogen gas and to be oxidized themselves. On the other hand, because metals with positive standard potentials (those above hydrogen in the table) cannot reduce hydrogen ions, we know that they cannot produce hydrogen gas when acted on by 1 M acid. For example, copper and the noble metals silver, platinum, and gold are not oxidized by hydrogen ions. That is one of the reasons why copper, silver, and gold have been used for coinage for so long.

A thermodynamic tendency is not always realized in practice, often because the reaction is very slow or because a protective oxide is formed. For example, the standard potential of aluminum (-1.66 V) suggests that, like magnesium, it should give hydrogen with acid. Aluminum can be oxidized by hydrochloric acid. However, it does not react with the more strongly oxidizing nitric acid because any Al^{3+} ions that are produced in nitric acid immediately form a layer of oxide on the surface of the metal (Fig. 18.15). This layer prevents further reaction, and we say that the metal has been **passivated**, or protected from further reaction, by a surface film. The passivation of aluminum enables the metal to be used, among other things, for beverage cans, airplanes, and window

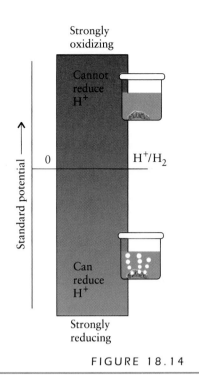

FIGURE 18.14

The significance of standard potentials. Only couples with negative standard potentials (and hence lying below hydrogen) can reduce hydrogen ions to hydrogen gas. The reducing power increases as the standard potential becomes more negative.

FIGURE 18.15

Although aluminum has a negative standard potential, signifying that it can be oxidized by hydrogen ions (as in the hydrochloric acid, left), nitric acid (right) stops reacting with it as soon as an impenetrable layer of aluminum oxide has formed on its surface. This resistance to further reaction is termed passivation of the metal.

frames. Aluminum containers are used to transport nitric acid because, once the surface is passivated, no further reaction occurs.

> *A metal with a negative standard potential has a thermodynamic tendency to reduce hydrogen ions in solution; the ions of a metal with a positive standard potential have a tendency to be reduced by hydrogen gas.*

18.8 The Electrochemical Series

Just as pK_a values allow us to rank acids according to their strengths, standard potentials provide a way to rank oxidizing and reducing agents. *The more positive the standard potential of a species, the greater its oxidizing power* and the more readily it gains electrons. *The more negative the standard potential, the greater its reducing power* and the more readily it releases electrons.

When Table 18.1 is viewed as a table of relative strengths of oxidizing and reducing agents, it is called the **electrochemical series.** The list is arranged with positive values at the top and negative values at the bottom, so oxidizing strength increases going up the list and reducing strength increases going down the list. A reduced member of a couple (on the right of the half-reactions in the list) can function as a reducing agent and has a tendency to reduce the oxidized species in any couple that lies above it. The strongest reducing agents are at the bottom right of the table. For example, lithium metal is the strongest reducing agent in the table. Equivalently, an oxidized member of a couple (on the left of the half-reactions in the list) can function as an oxidizing agent and has a tendency to oxidize the reduced member of any couple that lies below it. The strongest oxidizing agents are at the upper left of the table. For example, F_2 is a strong oxidizing agent, whereas Li^+ is a very, very poor oxidizing agent. We can use Table 18.1 to predict the spontaneous direction under standard conditions of any redox reaction that can be constructed from the half-reactions in the list.

An element in an oxidation state that is intermediate in its range of possible values can undergo either oxidation or reduction. Some such species (such as Fe^{2+}, an intermediate oxidation state of iron, between 0, for iron metal, and +3, for Fe^{3+}) are unstable with respect to both their oxidation and reduction products and spontaneously decompose to them in solution. For example, the reaction

$$3\ Fe^{2+}(aq) \longrightarrow Fe(s) + 2\ Fe^{3+}(aq)$$

occurs slowly in solutions of iron(II) compounds. A reaction in which a single element simultaneously undergoes both oxidation and reduction, as in this example, is called **disproportionation.**

> *The oxidizing and reducing power of a substance can be determined by its position in the electrochemical series. The strongest oxidizing agents are at the top of the table as reactants; the strongest reducing agents are at the bottom of the table as products.*

Example 18.5 *Predicting relative oxidizing strength by using the electrochemical series*

Can aqueous potassium permanganate be used to oxidize iron(II) to iron(III) under standard conditions in acidic solution?

Recall from Section 18.6 that all standard potentials refer to reductions.

Strategy All we need to do is look for the relative positions of the species in the electrochemical series in Table 18.1 or Appendix 2B. The species that has the greater potential (lies higher in the series) is more strongly oxidizing.

Solution We find the half-reactions and their potentials in Table 18.1:

$$MnO_4^-(aq) + 8\,H^+(aq) + 5\,e^- \longrightarrow Mn^{2+}(aq) + 4\,H_2O(l) \qquad E^\circ = +1.51\,V$$

$$Fe^{3+}(aq) + e^- \longrightarrow Fe^{2+}(aq) \qquad E^\circ = +0.77\,V$$

The permanganate half-reaction lies higher in the table, which indicates that it has a greater tendency to occur than the iron half-reaction. Therefore, permanganate ions can oxidize Fe^{2+} to Fe^{3+} in acidic aqueous solution.

Self-Test 18.7A Can mercury produce zinc from aqueous zinc sulfate under standard conditions?

> [*Answer:* No; the mercury potential lies above that of zinc.]

Self-Test 18.7B Can chlorine gas oxidize water to oxygen gas under standard conditions in basic solution?

To find a standard cell potential arising from a spontaneous reaction, we must combine the standard potential of the cathode half-reaction (reduction) with that of the anode half-reaction (oxidation) in such a way as to obtain a positive value. The overall potential must be positive because that corresponds to a spontaneous process, and only a spontaneous process can generate a potential. If the calculation results in a negative potential, it means that the reverse reaction is spontaneous.

To determine a standard cell potential, combine two half-reactions by reversing the equation and the sign of the standard potential of the half-reaction with the lower standard potential and then adding the half-reactions and potentials together.

Toolbox 18.3 **How to calculate a standard cell potential**

This Toolbox shows how to use standard potential data to calculate the standard potential of a cell.

Conceptual Basis

The value of E° for a half-reaction indicates the tendency for reduction to occur in a galvanic cell in which all reactants and products are in their standard states. Because a spontaneous cell reaction generates a positive potential, the cell must be arranged, and the two corresponding half-reactions combined, in a way that results in a positive potential.

Procedure

We can anticipate which direction is likely to be spontaneous by noting that the species corresponding to the lower standard potential will be more strongly reducing, and therefore likely to be oxidized at the anode.

Step 1. Inspect Table 18.1 or Appendix 2B to see which half-reaction lies above the other. Note that Appendix 2B lists potentials both alphabetically and by potential.

Step 2. To construct the chemical equation, reverse the half-reaction lying lower in the table and change the sign of its corresponding standard potential. Then add the two half-reactions, multiplying by factors if necessary to match numbers of electrons. Then add the two potentials (the multiplication has no effect on the values of the standard potentials). If the cell potential is positive, the cell reaction is positive as written. If it is negative, reverse the equation to obtain the spontaneous reaction.

Step 3. Construct the cell from two half-cells, in which all reactants and products in solution are present at 1 mol/L and all gases are at 1 atm. Write the cell diagram, with the anode on the left.

Example 18.6 Predicting relative oxidizing strengths and standard cell potentials

Is an acidified permanganate solution a more powerful oxidizing agent than an acidified dichromate solution under standard conditions? Write the chemical equation for the spontaneous reaction and determine the standard cell potential. Write the cell diagram for an electrochemical cell that could be used to verify your answer.

Strategy We need to find the standard potentials in Appendix 2B and work through the procedure in Toolbox 18.3.

Solution **Step 1.** We find the following two half-reactions in the table:

$$MnO_4^-(aq) + 8\,H^+(aq) + 5\,e^- \longrightarrow Mn^{2+}(aq) + 4\,H_2O(l) \qquad E° = +1.51\,V$$

$$Cr_2O_7^{2-}(aq) + 14\,H^+(aq) + 6\,e^- \longrightarrow 2\,Cr^{3+}(aq) + 7\,H_2O(l) \qquad E° = +1.33\,V$$

Because $E° = +1.51\,V$ lies above $E° = +1.33\,V$, MnO_4^- is the stronger oxidizing agent. Therefore, MnO_4^- ions are more strongly oxidizing than $Cr_2O_7^{2-}$ ions in aqueous solution.

Step 2. To construct the spontaneous reaction, reverse the equation (and the sign of the potential) for the half-reaction with the lower potential and add the result to the first equation, which gives a standard cell potential of

$$E° = 1.51 - 1.33\,V = 0.18\,V$$

(Reversing the first half-reaction would have given a negative cell potential.) The corresponding equation for the second half-reaction is now an oxidation:

$$2\,Cr^{3+}(aq) + 7\,H_2O(l) \longrightarrow Cr_2O_7^{2-}(aq) + 14\,H^+(aq) + 6\,e^-$$

To match electrons, we multiply the first half-reaction by 6 and the second by 5. Their sum is then

$$6\,MnO_4^-(aq) + 11\,H_2O(l) + 10\,Cr^{3+}(aq) \longrightarrow$$
$$6\,Mn^{2+}(aq) + 22\,H^+(aq) + 5\,Cr_2O_7^{2-}(aq)$$

To test the answer, we could measure the potential of the following cell:

$$Pt(s)\big|Cr^{3+}(aq),Cr_2O_7^{2-}(aq),H^+(aq)\big\|H^+(aq),MnO_4^-(aq),Mn^{2+}(aq)\big|Pt(s)$$

Self-Test 18.8A Which metal is the stronger reducing agent, zinc or nickel? Evaluate the standard cell potential and write the chemical equation for the spontaneous reaction.

[*Answer:* Zinc; $+0.53\,V$; $Zn(s) + Ni^{2+}(aq) \rightarrow Zn^{2+}(aq) + Ni(s)$]

Self-Test 18.8B Which is the stronger oxidizing agent, Cu^{2+} or Ag^+? Evaluate the standard cell potential for the reaction and write the equation for the corresponding cell reaction.

18.9 Standard Potentials, Free Energy, and Equilibrium Constants

We saw in Section 18.5 that the cell potential is a measure of the driving force of the cell reaction. A cell with a potential of 2 V has a greater driving force than one with a potential of only 1 V. But what determines these potentials, and what can we deduce from their measured values? To make sense of them, we need to recall from Section 17.8 that the thermodynamic measure of the tendency of a reaction to take place is the reaction free energy.

When the reaction free energy ΔG_r is negative, the cell reaction is spontaneous and the cell generates a positive potential. When ΔG_r is *large* as well as negative, the cell potential is high as well as positive. This relationship suggests that $-\Delta G_r$ and E are proportional to each other. Thermodynamic arguments confirm this proportionality and show that

$$\Delta G_r = -nFE \qquad (1)$$

where n is the number of moles of electrons that are transferred between the electrodes for the cell reaction as written in the chemical equation. The **Faraday constant,** F, is the magnitude of the charge per mole of electrons:

$$F = N_A e = 9.6485 \times 10^4 \text{ C/mol}$$

It follows from the definitions of the SI units coulombs and volts that

$$1 \text{ C·V} = 1 \text{ J}$$

so we can write $F = 9.6485 \times 10^4$ J/V·mol, or $F = 96.485$ kJ/V·mol.

We get the value of n from the balanced chemical equation, either from the half-reactions or more directly from the changes in oxidation numbers. For the reaction in the Daniell cell, $n = 2$ because 2 mol of electrons migrates from Zn to Cu:

$$\text{Zn(s)} + \text{Cu}^{2+}\text{(aq)} \longrightarrow \text{Zn}^{2+}\text{(aq)} + \text{Cu(s)} \qquad E = 1.1 \text{ V}, n = 2$$

It then follows from Eq. 1 that

$$\Delta G_r = -nFE = -2 \times (96.485 \text{ kJ/V·mol}) \times (1.1 \text{ V})$$
$$= -2.1 \times 10^5 \text{ J/mol}$$

For the Daniell cell, the reaction free energy is about -210 kJ/mol.

We usually employ Eq. 1 for the *standard* reaction free energy, ΔG_r°, when it becomes

$$\Delta G_r^\circ = -nFE^\circ \qquad (2)$$

In this expression, E° is the standard cell potential, the cell potential measured when all the species taking part are in their standard states. For example, to measure the standard potential of the Daniell cell, we should use 1 M $CuSO_4$(aq) in the copper electrode compartment and 1 M $ZnSO_4$(aq) in the zinc compartment.

It is important to notice that, although ΔG_r° changes when the chemical equation is multiplied by a factor, the cell potential E° does not: n is changed by the same factor, so E is left unchanged. The standard cell potential is the same no matter how large or small the cell or how the chemical equation is written. The value of E° for a given cell depends only on the temperature.

The value of ΔG_r is very important in electrochemistry, because it is a measure of how much electrical work an electrochemical cell can do. As the reactants are used up, the cell potential drops to 0 (corresponding to $\Delta G_r = 0$), at which point the reaction has reached equilibrium and can no longer provide electrons to do work.

Example 18.7 *Assessing the reaction free energy from a cell potential*

The cell $Cr(s)|Cr^{3+}(aq)\|Cu^{2+}(aq)|Cu(s)$ was found to have $E° = +1.08$ V at 298 K. (a) Write the balanced net ionic equation for the cell reaction; (b) determine n; and (c) calculate the standard reaction free energy at 298 K.

Strategy Because the standard cell potential is positive, expect a large, negative standard reaction free energy (corresponding to $K \gg 1$). Write the cell reaction from the cell diagram. Use the balanced equation with the smallest whole number coefficients to determine the number of electrons transferred (n) and insert the data in Eq. 1. Note that $1 \, C \cdot V = 1$ J.

Solution (a) The electrode on the left of the cell diagram is the anode, so that half-reaction is written as an oxidation:

$$Cr(s) \longrightarrow Cr^{3+}(aq) + 3\,e^-$$

The electrode on the right is the cathode, so that half-reaction is written as a reduction:

$$Cu^{2+}(aq) + 2\,e^- \longrightarrow Cu(s)$$

To write the equation for the cell reaction, we multiply the oxidation half-reaction by 2 and the reduction half-reaction by 3 to cancel electrons, then add the two half-reactions together. The equation is

$$3\,Cu^{2+}(aq) + 2\,Cr(s) \longrightarrow 3\,Cu(s) + 2\,Cr^{3+}(aq)$$

(b) We note from the fact that *three* copper(II) ions change oxidation state from $+2$ to 0 that $n = 6$. It then follows that

$$\Delta G_r° = -nFE° = -6 \times (96.485 \text{ kJ/V·mol}) \times (+1.08 \text{ V})$$
$$= -625 \text{ kJ/mol}$$

As expected, the standard reaction free energy is large and negative.

Self-Test 18.9A The following cell was set up: $Hg(l)|Hg_2Cl_2(s)|HCl(aq)\|$ $Hg_2(NO_3)_2(aq)|Hg(l)$, $E° = +0.52$ V at 298 K. (a) Write the equation for the cell reaction; (b) determine n; and (c) calculate the standard reaction free energy at 298 K.

[*Answer:* (a) $Hg_2^{2+}(aq) + 2\,Cl^-(aq) \rightarrow Hg_2Cl_2(s)$; (b) 2; (c) -1.0×10^2 kJ/mol]

Self-Test 18.9B The reaction between zinc metal and iodine in water generates 1.30 V under standard conditions. Determine (a) n and (b) $\Delta G_r°$ for the cell reaction $Zn(s) + I_2(aq) \rightarrow Zn^{2+}(aq) + 2\,I^-(aq)$.

A major use of tables of standard potentials of electrochemical cells is to calculate equilibrium constants, even for reactions that are not redox reactions. For example, the equilibrium constant for the neutralization of an acid by a base, a precipitation (recall Example 18.3), or any chemical reaction can be calculated from electrochemical data, provided the cell reaction can be written as the sum of two half-reactions.

We saw in Section 17.11 that the standard reaction free energy, $\Delta G_r°$, is related to the equilibrium constant, K, of the reaction by

$$\Delta G_r° = -RT \ln K$$

In this chapter, we have seen that the standard free energy is related to the standard cell potential of a galvanic cell by

$$\Delta G_r^\circ = -nFE^\circ$$

When we combine the two equations, we get

$$-nFE^\circ = -RT \ln K$$

This expression can be reorganized to allow us to calculate the equilibrium constant from the standard cell potential:

$$\ln K = \frac{nFE^\circ}{RT} \tag{3a}$$

The combination RT/F occurs often in electrochemistry; at 25°C (298.15 K) it has the value 2.5693×10^{-2} J/C, or 0.025 693 V; so at that temperature,

$$\ln K = \frac{nE^\circ}{0.025\ 693\ \text{V}} \tag{3b}$$

Because we can calculate E° from standard potentials, we can now also calculate equilibrium constants for any reaction that can be expressed in terms of two half-reactions. For example, the standard cell potential for the reaction

$$\text{Zn(s)} + \text{Cu}^{2+}\text{(aq)} \rightleftharpoons \text{Zn}^{2+}\text{(aq)} + \text{Cu(s)} \qquad K = \frac{[\text{Zn}^{2+}]}{[\text{Cu}^{2+}]}$$

is 1.10 V and $n = 2$ for the reaction as written; therefore,

$$\ln K = \frac{2 \times (1.10\ \text{V})}{0.025\ 693\ \text{V}} = 85.6$$

Taking the natural antilogarithm gives $K = 1.6 \times 10^{37}$. Now we know not only that the reaction is spontaneous as written but also that equilibrium is reached only when the concentration of Zn^{2+} ions is over 10^{37} times greater than that of Cu^{2+} ions. For all practical purposes, the reaction goes to completion.

The magnitude of E° is an indication of the equilibrium composition. A reaction with a large positive E° has $K \gg 1$. A reaction with a large negative calculated E° has $K \ll 1$. The procedure for calculating equilibrium constants is summarized in Toolbox 18.4.

The equilibrium constant of a reaction can be calculated from standard potentials by combining the equations for the half-reactions to give the reaction of interest and determining the standard potential of the corresponding cell.

Example 18.8 *Calculating the equilibrium constant of a reaction*

The solubility product is the equilibrium constant for the dissolution of a salt. Calculate the solubility product of silver chloride:

$$\text{AgCl(s)} \rightleftharpoons \text{Ag}^+\text{(aq)} + \text{Cl}^-\text{(aq)} \qquad K_{sp} = [\text{Ag}^+][\text{Cl}^-]$$

Toolbox 18.4 How to calculate equilibrium constants from electrochemical data

This Toolbox shows how to use tables of standard potentials to calculate the equilibrium constants of a variety of chemical reactions.

Conceptual Basis
A reaction with a large *positive* standard potential has a large *negative* reaction free energy, and therefore can be expected to have a high abundance of products at equilibrium. The opposite is true of a reaction with a large negative standard potential (Fig. 18.16).

Procedure
The procedure for calculating an equilibrium constant is as follows:

Step 1. Write the balanced chemical equation. Then find two half-reactions to combine to give the equation of interest: reverse one, multiplying by appropriate factors, if necessary, to make sure the numbers of electrons transferred in each case is the same. Then add them together.

Step 2. Identify the value of n from the change in oxidation numbers, or by examining the half-reactions (*after* multiplication by appropriate factors) for the number of electrons transferred in the balanced equation.

Step 3. Find the standard potential of the reaction by changing the sign of the standard potential of the half-reaction that was reversed and adding the result to the standard potential of the half-reaction that was left as a reduction.

Step 4. Use the relation $\ln K = nFE^\circ/RT$ to calculate the value of K for the reaction as written in step 1. Often the data are for 25°C, in which case, use $RT/F = 0.025\ 693$ V.

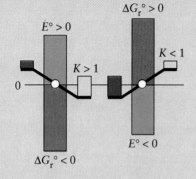

FIGURE 18.16

The relation between the standard potential of a reaction (reactants, purple; products, yellow) and the equilibrium constant.

Strategy Because we know that AgCl is insoluble, we know that K_{sp} is very small, and therefore we expect E° to be negative. Provided we can express the solubility equilibrium (which is not a redox reaction) as a difference of two reduction half-reactions, we can use the procedure in Toolbox 18.4.

Solution **Step 1.** The two reduction half-reactions required are

$$AgCl(s) + e^- \longrightarrow Ag(s) + Cl^-(aq) \qquad E^\circ = +0.22 \text{ V}$$

$$Ag^+(aq) + e^- \longrightarrow Ag(s) \qquad E^\circ = +0.80 \text{ V}$$

To obtain the reaction of interest, we need to reverse the second half-reaction:

$$Ag(s) \longrightarrow Ag^+(aq) + e^- \qquad E^\circ = -0.80 \text{ V}$$

and add it to the first half-reaction to obtain the equation for the equilibrium:

$$AgCl(s) \rightleftharpoons Ag^+(aq) + Cl^-(aq)$$

Step 2. From the half-reactions that were added to obtain the balanced equation, we see that $n = 1$.

Step 3. The standard cell potential is the sum

$$E° = 0.22 + (-0.80) \text{ V} = -0.58 \text{ V}$$

Step 4. It follows that

$$\ln K_{sp} = \frac{nFE°}{RT} = \frac{-0.58 \text{ V}}{0.025\ 693 \text{ V}} = -\frac{0.58}{0.025\ 693}$$

Taking the natural antilogarithm gives

$$K_{sp} = e^{-0.58/0.025\ 693} = 1.6 \times 10^{-10}$$

Many of the solubility products listed in tables, such as those in Table 16.5, have been determined in this way.

Self-Test 18.10A Use the tables in Appendix 2B to calculate the solubility product of mercury(I) chloride, Hg_2Cl_2.

[***Answer:*** 2.6×10^{-18}]

Self-Test 18.10B Use the tables in Appendix 2B to calculate the solubility product of cadmium hydroxide, $Cd(OH)_2$.

18.10 The Nernst Equation

Nearly everyone has experienced the disappointment of a battery that has run down at an inconvenient time. A battery fails because the composition of the cell has changed and the cell reaction has reached equilibrium. Once the reaction reaches equilibrium, it loses its pushing and pulling power. To understand this behavior, we need to know how the cell potential varies with the composition of the cell.

The formula for predicting the variation of cell potential with concentration and pressure is expressed by an equation first derived by the German chemist Walther Nernst. We already know how ΔG_r varies as the composition of a mixture changes in the course of a reaction:

$$\Delta G_r = \Delta G_r° + RT \ln Q$$

where Q is the reaction quotient (this expression is Eq. 7 of Section 17.10). Because $\Delta G_r = -nFE$ and $\Delta G_r° = -nFE°$, it follows that

$$-nFE = -nFE° + RT \ln Q$$

When we divide through by $-nF$, we get the **Nernst equation:**

$$E = E° - \frac{RT}{nF} \ln Q \tag{4}$$

We can use the Nernst equation to calculate the cell potential for any concentration of reactants and products: all we do is substitute their values into the expression for Q and use Eq. 4. We can also use the equation to determine concentrations by rearranging it to give Q in terms of the measured cell potential. This approach is used to measure pH (Investigating Matter 18.1). Another point is that the Nernst equation applies to half reactions as well as to overall reactions. In this application, the "stateless" electrons are ignored when writing down the expression for Q.

The variation of cell potential with composition is expressed by the Nernst equation.

The pH of a solution can be measured electrochemically with a device called a *pH meter*. The technique makes use of a cell in which one electrode is sensitive to the hydrogen ion concentration and the second electrode serves as a reference. One combination is a hydrogen electrode connected through a salt bridge to a calomel electrode (*calomel* is the common name for mercury(I) chloride, Hg_2Cl_2). The reduction half-reaction for the calomel electrode is

$$Hg_2Cl_2(s) + 2\,e^- \longrightarrow 2\,Hg(l) + 2\,Cl^-(aq)$$
$$E° = +0.27\,V$$

The complete reaction and the reaction quotient at 1 atm are

$$Hg_2Cl_2(s) + H_2(g) \longrightarrow 2\,H^+(aq) + 2\,Hg(l) + 2\,Cl^-(aq)$$
$$Q = [H^+]^2[Cl^-]^2$$

Because the pressure of hydrogen is 1 atm, we have omitted it from the reaction quotient. To find the concentration of hydrogen ions in the anode compartment at 25°C, we write the Nernst equation

$$\begin{aligned}
E &= E° - \tfrac{1}{2}(0.0257\,V)\ln[H^+]^2[Cl^-]^2 \\
&= E° - \tfrac{1}{2}(0.0257\,V)\ln[Cl^-]^2 - \tfrac{1}{2}(0.0257\,V)\ln[H^+]^2 \\
&= E° - (0.0257\,V)\ln[Cl^-] - (0.0257\,V)\ln[H^+]
\end{aligned}$$

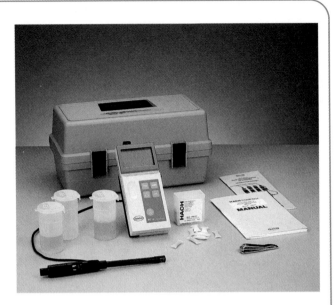

This durable portable pH meter can be used for quick measurements of pH in the field. Its accuracy is not as high as that of a laboratory pH meter.

The Cl^- concentration is fixed for a calomel electrode when it is manufactured, so it is a constant. Therefore, we can combine the first term on the right into a single constant, $E' = -(0.0257\,V)\ln[Cl^-]$. Then, because we can write $\ln x = 2.303\log x$,

$$\begin{aligned}
E &= E' - 2.303 \times (0.0257\,V)\log[H^+] \\
&= E' + (0.0592\,V) \times pH
\end{aligned}$$

Therefore, by measuring the cell potential, E, we can determine the pH. If E is first measured for a solution of known pH, it is not necessary to calculate E'.

A *glass electrode,* a thin-walled glass bulb containing an electrolyte, is much easier to use than a hydrogen electrode and has a potential that is proportional to the pH. Often there is a calomel electrode built into the probe that makes contact with the test solution through a miniature salt bridge. A pH meter therefore usually has only one probe, which contains a complete electrochemical cell. The meter is calibrated with a buffer of known pH, and the measured cell potential is then automatically converted into the pH of the solution, which is displayed. Solid-state electrodes have made possible durable, pocket-size pH meters that can be used to measure pH at remote locations and under extreme conditions.

Commercially available electrodes called *pX meters* are sensitive to other ions, such as Na^+, Ca^{2+}, NH_4^+, CN^-, and S^{2-}. They are used to monitor industrial processes and in pollution control.

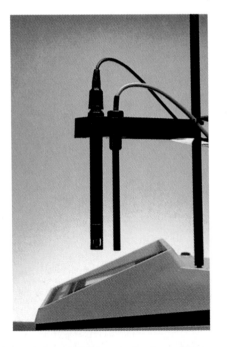

A glass electrode in a protective plastic sleeve (left) is used to measure pH. It is sometimes used in conjunction with a calomel electrode (right) in pH meters such as this one.

Example 18.9 *Using the Nernst equation to predict a cell potential*

Calculate the potential at 25°C of a Daniell cell in which the concentration of Zn^{2+} ions is 0.10 mol/L and that of the Cu^{2+} ions is 0.0010 mol/L.

Strategy First, write the balanced equation. Then determine $E°$ from the standard potentials in Table 18.1 (or Appendix 2B) and note the value of n. Determine the value of Q for the stated conditions. Calculate the cell potential by substituting these values into the Nernst equation. At 25°C, $RT/F = 0.025\ 693$ V.

Solution The reaction in the Daniell cell is

$$Cu^{2+}(aq) + Zn(s) \longrightarrow Zn^{2+}(aq) + Cu(s) \qquad E° = 1.10\ V, n = 2$$

The reaction quotient is

$$Q = \frac{[Zn^{2+}]}{[Cu^{2+}]} = \frac{0.10}{0.0010}$$

The Nernst equation gives

$$E = 1.10\ V - \left(\frac{0.025\ 693\ V}{2}\right) \ln\left(\frac{0.10}{0.0010}\right)$$
$$= 1.10\ V - 0.059\ V = +1.04\ V$$

Self-Test 18.11A Calculate the potential of the galvanic cell $Zn(s)|Zn^{2+}(aq, 1.50\ mol/L)\|Fe^{2+}(aq, 0.10\ mol/L)|Fe(s)$ at 298 K.

[***Answer:*** +0.29 V]

Self-Test 18.11B Calculate the potential of the galvanic cell $Ag(s)|Ag^+(aq, 0.0010\ mol/L)\|Ag^+(aq, 0.010\ mol/L)|Ag(s)$ at 298 K.

18.11 Practical Cells

The commercial galvanic cells we know as *batteries* are familiar items in our daily lives, but they may play an even more important role in the future. Batteries are currently the subject of intense research by scientists who are using them to solve problems associated with the environment, health, communication, and transport.

Table 18.2 summarizes the chemical reactions used to power typical commercial galvanic cells. A **primary cell** produces electricity from chemicals that are sealed into it when it is made. This type of galvanic cell cannot be recharged; once the cell reaction has reached equilibrium, the cell is discarded. The workhorse of primary cells is the *dry cell* (Fig. 18.17), which is widely used to power portable electric equipment. A dry cell produces about 1.5 V initially; but with use, its potential falls to about 0.8 V as reaction products accumulate inside it. A *fuel cell* is like a primary cell, but the reactants (the fuel) are continuously supplied from outside. The cell can produce a current for as long as fuel is supplied.

> Originally, the term *battery* referred to several galvanic cells connected in series. However, it is now commonly used to refer to any portable source of electric current.

FIGURE 18.17

Carbon rod (cathode)

MnO_2 + carbon black + NH_4Cl

Zinc cup (anode)

A commercial dry cell consists of a graphite cathode in a zinc container; the latter acts as the anode. The container is filled with a moist paste of NH_4Cl, MnO_2, finely divided carbon, and an inert filler such as starch. In the cell reaction, manganese(IV) is reduced to manganese(III) and zinc metal is oxidized to Zn^{2+} ions.

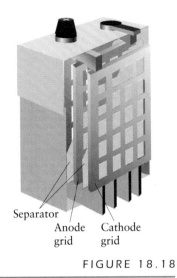

Separator
Anode   Cathode
grid   grid

FIGURE 18.18

One cell of a lead-acid battery like those used in automobiles. A lead-acid battery is an example of a secondary cell. It needs to be charged before it can produce a current. The electrolyte is dilute sulfuric acid.

The *silver cell* is a primary cell that uses zinc as the anode and has a cathode made of Ag_2O. The relatively high potential of a silver cell, with its solid reactants and products, is maintained with great reliability over long periods of time, making it desirable for medical implants such as pacemakers, for hearing aids, and in cameras.

A **secondary cell** must be charged before it can be used and is normally rechargeable. The batteries used in portable computers and automobiles are secondary cells. In the charging process, an external source of electricity temporarily reverses the spontaneous cell reaction and restores a nonequilibrium mixture of reactants. After charging, the cell can again produce electricity as the reaction once more sinks toward equilibrium.

One of the most common secondary cells is the *lead-acid cell* of an automobile battery. Each cell contains several grids that act as electrodes (Fig. 18.18). Because the total surface area of these grids is large, the battery can generate large currents on demand—at least for short periods, like the time needed for starting an engine. The electrodes are initially a hard lead-antimony alloy covered with a paste of lead(II) sulfate. The electrolyte is dilute sulfuric acid. During the first charging, some of the lead(II) sulfate is reduced to lead on one of the electrodes; the same electrode will act as the anode during discharge. Simultaneously, lead(II) sulfate is oxidized to lead(IV) oxide on the electrode that will later act as the cathode. The chemical equations for the cell reactions (Table 18.2) show that sulfuric acid is used up during discharge. When the cell is recharged, the cell reactions are driven in reverse by the external supply, and sulfuric acid is produced. The state of charge of the cell can therefore be judged from the concentration of the sulfuric acid solution, and that concentration in turn can be judged from the density of the electrolyte.

Table 18.2 *Reactions in commercial galvanic cells*

Primary cells

dry cell

$Zn(s)|ZnCl_2(aq), NH_4Cl(aq)|MnO(OH)(s)|MnO_2(s)|graphite, 1.5\ V$

Anode: $Zn(s) \longrightarrow Zn^{2+}(aq) + 2\ e^-$
followed by $Zn^{2+}(aq) + 4\ NH_3(g) \longrightarrow Zn(NH_3)_4^{2+}(aq)$
Cathode: $MnO_2(s) + H_2O(l) + e^- \longrightarrow MnO(OH)(s) + OH^-(aq)$
followed by $NH_4^+(aq) + OH^-(aq) \longrightarrow H_2O(l) + NH_3(g)$

mercury cell

$Zn(s)|ZnO(s)|KOH(aq)|HgO(s)|Hg(l)|steel, 1.3\ V$

Anode: $Zn(s) + 2\ OH^-(aq) \longrightarrow ZnO(s) + H_2O(l) + 2\ e^-$
Cathode: $HgO(s) + H_2O(l) + 2\ e^- \longrightarrow Hg(l) + 2\ OH^-(aq)$

silver cell

$Zn(s)|ZnO(s)|KOH(aq)|Ag_2O(s)|Ag(s)|steel, 1.6\ V$

Anode: $Zn(s) + 2\ OH^-(aq) \longrightarrow ZnO(s) + H_2O(l) + 2\ e^-$
Cathode: $Ag_2O(s) + H_2O(l) + 2\ e^- \longrightarrow 2\ Ag(s) + 2\ OH^-(aq)$

Secondary cells

lead-acid battery

$Pb(s)|PbSO_4(s)|H^+(aq), HSO_4^-(aq)|PbO_2(s)|PbSO_4(s)|Pb(s), 2\ V$

Anode: $Pb(s) + HSO_4^-(aq) \longrightarrow PbSO_4(s) + H^+(aq) + 2\ e^-$
Cathode: $PbO_2(s) + 3\ H^+(aq) + HSO_4^-(aq) + 2\ e^- \longrightarrow PbSO_4(s) + 2\ H_2O(l)$

nicad cell

$Cd(s)|Cd(OH)_2(s)|KOH(aq)|Ni(OH)_3(s)|Ni(OH)_2(s)|Ni(s), 1.25\ V$

Anode: $Cd(s) + 2\ OH^-(aq) \longrightarrow Cd(OH)_2(s) + 2\ e^-$
Cathode: $2\ Ni(OH)_3(s) + 2\ e^- \longrightarrow 2\ Ni(OH)_2(s) + 2\ OH^-(aq)$

FIGURE 18.19

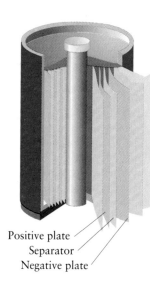

A rechargeable nickel-cadmium (nicad) cell. The electrodes are assembled in a jelly roll arrangement and separated by a layer of paper soaked in moist sodium or potassium hydroxide.

Positive plate
Separator
Negative plate

A *nicad cell* (Fig. 18.19) is a secondary cell widely used to power electronic equipment. The source of electrons in a circuit powered by a nicad cell is oxidation of cadmium to insoluble cadmium hydroxide, which adheres to the grid as the oxidation occurs. The insoluble nickel(II) hydroxide produced from the reduction of nickel(III) hydroxide also adheres to the stainless steel grid and is thus readily available when the cell is charged (when the reactions are reversed). Because no gases are produced in either the charging or the discharging processes, the cells can be sealed, which makes nicad cells useful for portable equipment.

Fuel cells make highly efficient use of resources, because they produce very little waste heat. Therefore, fuel cells that oxidize natural gas or hydrogen are promising alternatives to conventional power plants (see Applying Chemistry: Case Study 18). In a simple version of a fuel cell, hydrogen gas—the fuel—is passed over one electrode, oxygen is passed over the other, and the electrolyte is aqueous potassium hydroxide. A version of this type of cell is used on the space shuttle to power the life-support systems, one advantage being that the crew can drink the product of the cell reaction.

Electric eels are mobile, natural fuel cells (Fig. 18.20). They generate their electric charge in an "electric organ," a battery of biological electrochemical cells, each cell providing about 0.15 V and an overall potential difference of about 700 V. It is an incidental feature of nature that the eel's head is its cathode and its tail the anode. The electric catfish has the opposite polarity.

Practical galvanic cells are classified as primary cells, secondary cells, and fuel cells.

Example 18.10 *Predicting the potential of a fuel cell*

Suppose you were part of a team designing a fuel cell for campers, which would use the oxidation of propane gas to generate electricity. What is the maximum potential you can expect such a cell to generate when all the reactants and products are in their standard states?

Strategy First, write down the chemical equation for the oxidation reaction and identify the number of electrons transferred in the oxidation. Then calculate the standard reaction free energy by using tables of standard free energies of formation (Appendix 2A). Finally, convert the standard reaction free energy to a standard cell potential by using Eq. 2. Note that $1 \text{ J} = 1 \text{ C·V}$.

Solution The chemical equation for the combustion is

$$C_3H_8(g) + 5\,O_2(g) \longrightarrow 3\,CO_2(g) + 4\,H_2O(l)$$

FIGURE 18.20

The electric eel (*Electrophorus electricus*) lives in the Amazon. The average potential difference it produces along its length (1 m) is about 700 V.

Applying Chemistry: *Case Study 18*

In the early years of the space program, life support in space was a critical problem. The cabin needed a source of electrical power and the astronauts needed enough water for drinking and washing. Because the mass of a spacecraft must be kept as low as possible, batteries—which usually provide energy from the oxidation of a metal—would be too heavy. Electricity can be obtained from combustion reactions by burning a fuel to create heat, which runs a generator. However, producing electricity from burning fuels is very inefficient because most of the energy is wasted as heat.

The problem was solved when Francis Bacon, a British scientist and engineer, developed a new type of galvanic cell, based on a device invented many years before by Sir William Grove at the Royal Institution in London. The

One of the three alkali fuel cells used on the space shuttle. Although only one cell is needed to provide life-support, electricity, and drinking water, shuttle flight rules require that all three be functioning.

invention, a *fuel cell*, generates electricity directly from a chemical reaction, as in a battery, but with reactants supplied continuously, as in an automobile engine.

Fuel cells underwent intensive development by the aerospace industry, and eventually a fuel cell that runs on hydrogen and oxygen was produced for the space shuttle. An advantage is that the only product of the cell reaction, water, can be drunk by the crew. No pollutants are generated and the total mass of the three fuel cells carried aboard the shuttle is relatively small.

Most of our engines and power plants that burn fossil fuels such as natural gas or coal to generate electricity are so inefficient that no more than about 30% of the energy released by combustion is actually used to do work. The development of the fuel cell has led to an entirely new kind of technology for making use of redox reactions electrochemically and hence more efficiently. Fuel cells can make efficient use of resources because little waste heat is produced.

Many varieties of fuel cells are possible, and in some the electrolyte is a solid polymer membrane or a ceramic. Three of the most promising fuel cells are the alkali fuel cell and the phosphoric acid fuel cell, both of which involve the formation of water from hydrogen and oxygen, and the molten carbonate fuel cell.

The cell used in the space shuttle is called an *alkali fuel cell*, because it has an alkaline electrolyte. Both electrodes are platinum; and in some versions, a catalyst is used to improve efficiency. Water is produced at the anode in this cell:

Alkali fuel cell

Anode: $2\,H_2(g) + 4\,OH^-(aq) \longrightarrow 4\,H_2O(l) + 4\,e^-$

Electrolyte: $KOH(aq)$

Cathode: $O_2(g) + 4\,e^- + 2\,H_2O(l) \longrightarrow 4\,OH^-(aq)$

Although its expense prohibits its use in many applications, the alkali fuel cell is the primary fuel cell used in the

---

Because ten O atoms undergo a change of oxidation number from 0 (in O_2) to -2 (in the products), there must be 20 electrons transferred, so $n = 20$ for the reaction as written. The standard free energy for the reaction as written is

$$\Delta G_r^\circ = \{3\Delta G_f^\circ(CO_2, g) + 4\Delta G_f^\circ(H_2O, l)\} - \{\Delta G_f^\circ(C_3H_8, g) + 5\Delta G_f^\circ(O_2, g)\}$$
$$= \{3 \times (-394.36) + 4 \times (-237.13)\} - \{(-23.49) + 5 \times (0)\}\,kJ/mol$$
$$= -2108.11\,kJ/mol$$

We now use Eq. 2, with $n = 20$, to convert this free energy to a potential:

$$E^\circ = -\frac{\Delta G_r^\circ}{nF} = -\frac{(-2108.11\,kJ/mol)}{20 \times (96.4853\,kJ/V \cdot mol)}$$
$$= +1.0925\,V$$

We can expect a maximum potential of about 1.09 V.

High-pressure hydrogen tanks run across the top of this bus provided by Ballard Power Systems for testing hydrogen-oxygen fuel cells in Chicago. The bus is pollution free and can go 250 miles before needing to be refueled.

aerospace industry, because it has a very high specific energy (see Connection 5, following Chapter 18).

If an acid electrolyte is used, water is produced at the cathode. An example is the *phosphoric acid fuel cell:*

Phosphoric acid fuel cell

Anode: $2\,H_2(g) \longrightarrow 4\,H^+(aq) + 4\,e^-$

Electrolyte: $H_3PO_4(aq)$

Cathode: $O_2(g) + 4\,e^- + 4\,H^+ \longrightarrow 2\,H_2O(l)$

This fuel cell has shown promise for *combined heat and power* (CHP) *systems.* In such systems, the waste heat is used to heat buildings or to do work. Efficiency in a CHP

plant can reach 80%. These fuel cells could replace heating plants and power sources in colleges and universities, hotels, and apartment buildings. Because fuel cells have few or no moving parts, little maintenance is required. However, hydrogen is still very expensive.

In the *molten carbonate fuel cell,* methane is used as the fuel. This cell uses a molten mixture of lithium and potassium carbonates as electrolyte.

Molten carbonate fuel cell (direct)

Anode: $CH_4(g) + 4\,CO_3^{2-}(l) \longrightarrow$
$$2\,H_2O(g) + 5\,CO_2(g) + 8\,e^-$$

Electrolyte: $K_2CO_3(l)/Li_2CO_3(l)$

Cathode: $O_2(g) + 2\,CO_2(g) + 4\,e^- \longrightarrow 2\,CO_3^{2-}(l)$

Two difficulties with this cell are that it must run at a high temperature and that the electrolyte is highly corrosive.

There are many obstacles that must be overcome before fuel cells reach their potential to provide us with pollution-free energy. The hydrogen fuel cells are the most attractive, because of their use of a renewable fuel. Hydrogen can be obtained from the water in the oceans. The challenge is to extract it from seawater by a process using less energy than that given off in the fuel cell and to find safe means of transportation and storage.

Key Concepts: half-reactions, cell potential, free energy, practical cells

For Further Reading

L. J. Blomen and M. N. Mugerwa, Eds., *Fuel Cell Systems,* New York: Plenum Press, 1992.

Beyond Batteries, http://www.sciam.com/explorations/122396explorations.html

Related Exercises: 18.115–18.117

Self-Test 18.12A One fuel cell being tested for use in power plants uses the reaction of methane gas and oxygen gas to produce carbon dioxide gas and liquid water. What is the maximum potential you can expect such a cell to generate when all the reactants and products are in their standard states at 25°C?

[*Answer:* 1.06 V]

Self-Test 18.12B A promising alternative fuel for automobiles is methanol, CH_3OH. What is the maximum potential you can expect such a cell to generate when all the reactants and products are in their standard states at 25°C?

18.12 Corrosion

Corrosion is the unwanted oxidation of a metal. It cuts short the lifetimes of steel products such as bridges and automobiles; replacing corroded metal parts

FIGURE 18.21

Iron nails stored in oxygen-free water (left) do not rust because the oxidizing power of water itself is weak. When oxygen is present (as a result of air dissolving in the water, right), oxidation is thermodynamically spontaneous and rust soon forms.

costs billions of dollars a year. Corrosion is an electrochemical process, and the electrochemical series gives us insight into why corrosion occurs and how it can be prevented.

The main culprit in corrosion is water. Any metal lower in the electrochemical series than -0.83 V can be oxidized by water under standard conditions as a result of the half-reaction

$$2 H_2O(l) + 2 e^- \longrightarrow H_2(g) + 2 OH^-(aq) \qquad E^\circ = -0.83 \text{ V}$$

This standard potential is for an OH^- concentration of 1 mol/L, which corresponds to pH $= 14$, a strongly basic solution. However, from the Nernst equation, at pH $= 7$, this couple has $E^\circ = -0.42$ V. Because iron has almost the same potential ($E^\circ = -0.44$ V for $Fe^{2+}(aq) + 2 e^- \rightarrow Fe(s)$), iron has only a very slight tendency to be oxidized by pure water. For this reason, iron can be used for making pipes in water supply systems and can be stored in oxygen-free water without rusting (Fig. 18.21). However, when iron is exposed to damp air, with both oxygen and water present, the half-reaction

$$O_2(g) + 4 H^+(aq) + 4 e^- \longrightarrow 2 H_2O(l) \qquad E^\circ = +1.23 \text{ V}$$

must be taken into account. The potential of this couple at pH $= 7$ is $E^\circ = +0.82$ V, which lies well above the value for iron. Hence, oxygen and water can jointly oxidize iron to Fe^{2+}.

A drop of water on the surface of iron acts as the electrolyte in a tiny electrochemical cell (Fig. 18.22). At the edge of the drop, dissolved oxygen oxidizes the iron by the reaction given earlier. However, the electrons withdrawn from the metal by this oxidation can be restored from another part of the conducting metal—in particular, from iron lying beneath the oxygen-poor region in the center of the drop. The iron atoms there give up their electrons, form Fe^{2+} ions, and drift away into the surrounding water. This process results in the formation of tiny pits in the surface. The Fe^{2+} ions are oxidized to Fe^{3+} ions by the dissolved oxygen. These ions then precipitate as hydrated iron(III) oxide, $Fe_2O_3 \cdot H_2O$, the brown, insoluble substance we call *rust*. Water is more highly conducting when it has dissolved ions, and the formation of rust is then accelerated. That is one reason why the salt air of coastal cities and salt used for de-icing highways is very damaging to exposed metal.

The simplest way to prevent corrosion is to protect the surface of the metal from exposure to air and water by painting it. One way to achieve greater protection is to **galvanize** the metal, a process that involves coating the metal with a film of zinc (Fig. 18.23). Zinc lies below iron in the electrochemical series, so if a scratch exposes the metal beneath, the more strongly reducing zinc releases electrons to the iron. Thus, the zinc, not the iron, is oxidized. The zinc itself survives

FIGURE 18.22

The mechanism of rust formation. (a) Oxidation of the iron occurs at a point out of contact with the oxygen of the air, and the surface of the metal acts as an anode in a tiny galvanic cell. (b) Further oxidation of Fe^{2+} to Fe^{3+} results in the deposition of rust on the surface.

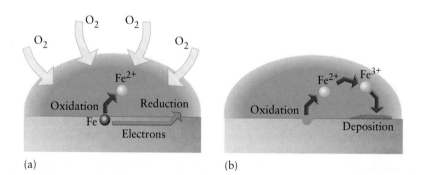

(a)

(b)

FIGURE 18.23

Metal girders are galvanized by immersion in a bath of molten zinc.

exposure on the surface because, like aluminum, it is passivated by a protective oxide.

It is not possible to galvanize large structures—for example, ships, underground pipelines or gasoline storage tanks, and bridges—but **cathodic protection** can be used. A block of a more strongly reducing metal, such as zinc or magnesium, is buried in moist soil and connected to the underground pipeline (Fig. 18.24). The block of zinc or magnesium is preferentially oxidized and supplies electrons to the iron for the reduction of oxygen. The block of metal, which is called a **sacrificial anode,** protects the iron pipeline and is cheaper to replace than the pipeline itself. For similar reasons, automobiles generally have negative ground systems as part of their electrical circuitry, which means that the body of the car is connected to the anode of the battery. The decay of the anode in the battery is the sacrifice that helps preserve the vehicle itself.

The corrosion of iron is accelerated by the presence of oxygen, moisture, and salt. Corrosion can be inhibited by coating the surface with paint or zinc or by cathodic protection.

Self-Test 18.13A Which of the following procedures helps to prevent the corrosion of an iron rod in water: (a) decreasing the concentration of oxygen in the water; (b) painting the rod; (c) increasing the concentration of ions in solution?

> [*Answer:* (a) and (b). (c) will decrease corrosion only at such a high concentration of ions that oxygen cannot dissolve.]

Self-Test 18.13B Which of (a) copper, (b) zinc, or (c) tin can act as a sacrificial metal for iron?

ELECTROLYSIS

Redox reactions that have a positive reaction free energy are not spontaneous but can be made to occur electrochemically. For example, fluorine is so highly reactive that it cannot be isolated as the element by any common chemical reaction. However, in 1886, the French chemist Henri Moissan isolated fluorine by passing an electric current through an anhydrous (water-free) molten mixture

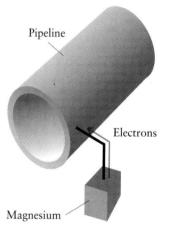

Pipeline

Electrons

Magnesium

FIGURE 18.24

In the cathodic protection of a buried pipeline, or other large metal construction, the artifact is connected to a number of buried blocks of metal, such as magnesium or zinc. The sacrificial anodes (the magnesium block in this illustration) supply electrons to the pipeline (the cathode of the cell), thereby preserving it from oxidation.

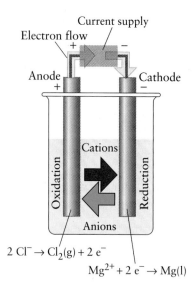

Current supply
Electron flow

Anode
+

Cathode
−

Cations

Oxidation

Reduction

Anions

$2\,Cl^- \rightarrow Cl_2(g) + 2\,e^-$

$Mg^{2+} + 2\,e^- \rightarrow Mg(l)$

FIGURE 18.25

A schematic picture of the electrolytic cell used in the Dow process for magnesium. The electrolyte is molten magnesium chloride. As the current generated by the external source passes through the cell, magnesium metal is produced at the cathode and chlorine gas is produced at the anode.

of potassium fluoride and hydrogen fluoride. Fluorine is still prepared commercially by the same process.

18.13 Electrolytic Cells

Electrolysis is the process of driving a reaction in a nonspontaneous direction by using an electric current. An *electrolytic cell* is an electrochemical cell in which an electric current from an external source is used to drive a *nonspontaneous* chemical reaction. Electrolytic cells are constructed differently from galvanic cells. Specifically, the two electrodes usually share the same compartment, there is usually only one electrolyte, and concentrations and pressures are usually far from standard.

Figure 18.25 shows the layout of an electrolytic cell used for the commercial production of magnesium metal from molten magnesium chloride (the *Dow process*). As in a galvanic cell, oxidation occurs at the anode and reduction occurs at the cathode, electrons travel through the external wire from anode to cathode, cations move through the electrolyte toward the cathode, and anions move toward the anode. Unlike the process in a galvanic cell, however, a current supplied by an external electrical power source drives electrons through the wire in a predetermined direction, forcing oxidation to occur at one electrode and reduction at the other:

$$\text{Anode reaction: } 2\,Cl^-(l) \longrightarrow Cl_2(g) + 2\,e^-$$

$$\text{Cathode reaction: } Mg^{2+}(l) + 2\,e^- \longrightarrow Mg(l)$$

A rechargeable battery functions as a galvanic cell when it is doing work and as an electrolytic cell when it is being recharged.

> *In an electrolytic cell, current supplied by an external source is used to drive a nonspontaneous redox reaction.*

18.14 The Potential Needed for Electrolysis

To drive a reaction in a nonspontaneous direction, the external supply must generate a potential greater than would be produced by the reverse, spontaneous reaction (Fig. 18.26). For example, because

$$2\,H_2(g) + O_2(g) \longrightarrow 2\,H_2O(l) \qquad E = +1.23\,V \text{ at pH} = 7$$

is spontaneous, to produce hydrogen and oxygen gases in the nonspontaneous reaction

$$2\,H_2O(l) \longrightarrow 2\,H_2(g) + O_2(g) \qquad E = -1.23\,V \text{ at pH} = 7$$

we must apply at least 1.23 V from the external source to overcome the reaction's natural "pushing power" in the opposite direction. In practice, the applied potential must usually be greater than the cell potential to reverse a spontaneous cell reaction. The additional potential, which varies with the type of electrode, is

The anode of an electrolytic cell is labeled + and the cathode −, the opposite of a galvanic cell.

FIGURE 18.26

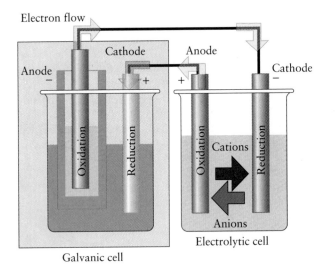

Electron flow

Galvanic cell

Electrolytic cell

In this schematic picture of an electrolysis experiment, the electrons emerge from a galvanic cell at its anode (−) and enter the electrolytic cell at its cathode (−), where they bring about reduction. Electrons are drawn out of the electrolytic cell through its anode (+) and into the galvanic cell at its cathode (+). If the cell reaction in the galvanic cell is more strongly spontaneous than the reaction in the electrolytic cell is nonspontaneous, then the overall process is spontaneous. This experiment is an example of one reaction driving another to which it is coupled.

called the **overpotential.** Much research on electrolytic cells involves attempts to reduce the overpotential and thereby increase efficiency. For platinum electrodes, the overpotential for the production of water from hydrogen and oxygen is about 0.6 V, so about 1.8 V (0.6 + 1.23 V) is actually required to electrolyze water when platinum electrodes are used.

For the electrolysis of water to occur, an electrolyte must be added to carry the current. Consequently, we have to consider the possibility that the added ions can also be oxidized or reduced by the electric current. The potential for the reduction of oxygen in water at pH = 7 is +0.82 V:

$$O_2(g) + 4H^+(aq) + 4e^- \longrightarrow 2H_2O(l) \qquad E = +0.82\ V$$

To reverse this half-reaction and bring about the oxidation of water needs a potential of at least 0.82 V. If any ions present are more readily oxidized than water, then they may be oxidized rather than the water. For example, when Cl^- ions are present at 1 mol/L in water, is it possible that they, and not the water, will be oxidized? From Table 18.1, we see that the standard potential for the reduction of chlorine is +1.36 V:

$$Cl_2(g) + 2e^- \longrightarrow 2Cl^-(aq) \qquad E° = +1.36\ V$$

To reverse this reaction and oxidize chloride ions, we have to supply at least 1.36 V. Because only 0.82 V is needed to force the oxidation of water, but 1.36 V is needed to force the oxidation of Cl^-, it appears that oxygen should be the product at the cathode. However, the overpotential for oxygen production can be very high, and in practice chlorine might also be produced.

Suppose the solution contains I^- ions at 1 mol/L instead of chloride ions. From Table 18.1, we know that at least 0.54 V is needed to oxidize I^-:

$$I_2(s) + 2e^- \longrightarrow 2I^-(aq) \qquad E° = +0.54\ V$$

Because only 0.54 V is needed to reverse this reduction and hence bring about the oxidation of iodide ions, the I^- ion would be oxidized in preference to water.

The potential supplied to an electrolytic cell must be at least as great as that of the cell reaction to be reversed. If there is more than one reducible species in solution, the species with the greater potential for reduction is preferentially reduced. The same principle applies to oxidation.

Example 18.11 *Predicting the likely products of electrolysis*

Predict the products resulting from the electrolysis of 1 M $ZnNO_2(aq)$ at pH = 7.

Strategy Write the half-reactions that are likely to occur for the cation and the anion. Find the standard potentials of these half-reactions in Appendix 2B and compare them with those of water. *Anode:* At least 0.82 V is needed to oxidize water to oxygen at pH = 7. If one of the species is a product in a half-reaction with a standard potential below 0.82 V, that species is more readily oxidized than water and will be oxidized at the anode. *Cathode:* A potential of only −0.42 V is needed to reduce water to molecular hydrogen at pH = 7. If one of the species in solution is a reactant in a half-reaction with a standard potential above −0.42 V, then that species is more easily reduced than water and will be reduced at the cathode.

Solution The two half-reactions are

$$Zn^{2+}(aq) + 2\,e^- \longrightarrow Zn(s) \qquad E° = -0.76\ V$$

$$NO_3^-(aq) + H_2O(l) + 2\,e^- \longrightarrow NO_2^-(aq) + 2\,OH^-(aq) \qquad E° = +0.01\ V$$

Anode: Nitrite ion is a product in the second half-reaction; its potential is below 0.82 V. This half-reaction is reversed during electrolysis, so nitrite ion oxidized to nitrate ion at a lower potential than water can be oxidized to molecular oxygen. Therefore, nitrate ion will be produced at the anode. *Cathode:* Zinc ions are reduced in the first half-reaction, but the standard potential is lower than −0.42 V, so they will not be reduced. Hydrogen will be produced at the cathode.

Self-Test 18.14A Predict the products resulting from the electrolysis of 1 M $AgNO_3(aq)$ at pH = 7.

[*Answer:* Cathode, Ag; anode, O_2, H^+]

Self-Test 18.14B Predict the products resulting from the electrolysis of 1 M NaBr(aq) at pH = 7.

18.15 The Products of Electrolysis

Now we see how to calculate the amount of product formed by a given amount of electricity. The calculation is based on observations made by Michael Faraday (Fig. 18.27) and summarized as follows:

> **Faraday's law of electrolysis:** In electrolysis, the moles of electrons supplied and the moles of product formed are in the mole ratio given by the half-reaction.

Once we know the number of moles of product formed, we can calculate the masses of the products or, if they are gases, their volumes. For example, copper is refined electrolytically by using an impure form of copper metal (called *blister copper*) as the anode in an electrolytic cell (Fig. 18.28). The current causes the oxidation of the blister copper to copper(II) ions,

$$Cu(s, blister) \longrightarrow Cu^{2+}(aq) + 2\,e^-$$

These ions are then reduced at the cathode:

$$Cu^{2+}(aq) + 2\,e^- \longrightarrow Cu(s)$$

We see that 2 mol e⁻ ≏ 1 mol Cu. Therefore, if 4.0 mol e⁻ is supplied, the amount of copper produced is

$$\text{Moles of Cu} = (4.0 \text{ mol e}^-) \times \frac{1 \text{ mol Cu}}{2 \text{ mol e}^-} = 2.0 \text{ mol Cu}$$

The charge passed through an electrolytic cell is the product of the current, I, and the time, t, for which it is supplied. Electric current, the rate of flow of charge, is measured in the SI unit ampere (A), where 1 A = 1 C/s. Therefore, the charge supplied is

$$\text{Charge supplied (C)} = \text{current (A)} \times \text{time (s)}$$

For example, if a current of 3.00 A is passed for 121 s, the charge supplied to the cell is

$$\text{Charge supplied} = (3.00 \text{ A}) \times (121 \text{ s}) = 363 \text{ A·s} = 363 \text{ C}$$

We have used 1 A·s = 1 C. To convert charge to moles of electrons transferred, we use the fact that the Faraday constant, F, is the magnitude of the charge per mole of electrons (recall Section 18.9). Because *charge = moles of electrons × F*, it follows that

$$\text{Moles of e}^- = \frac{\text{Charge supplied}}{F} = \frac{\text{current (A)} \times \text{time (s)}}{9.6485 \times 10^4 \text{ C/mol}} \quad \text{or} \quad n = \frac{It}{F} \quad \textbf{(5)}$$

So, by measuring the current and the time for which it flows, we can determine the number of moles of electrons supplied.

The amount of product in an electrolysis reaction is calculated from the stoichiometry of the half-reaction and the current and time for which the current flows.

> A current of 1 A corresponds to about 6×10^{18} electrons passing a given point each second.

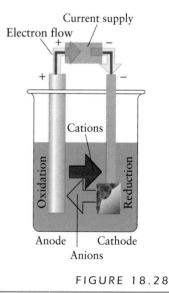

FIGURE 18.28

A schematic picture showing the electrolytic process for refining copper. The anode is impure copper. It undergoes oxidation, and the Cu^{2+} ions so produced migrate to the cathode, where they are reduced to pure copper metal. A similar arrangement is used for electroplating objects.

FIGURE 18.27

Michael Faraday (1791–1867) giving a public lecture on chemistry at the Royal Institution in London.

This Toolbox shows how to determine the amount of product that can be produced from electrolysis or the amount of time a current must be passed to obtain a given amount of product.

Conceptual Basis

The number of moles of electrons required in an electrolysis depends on the amount of product desired and the half-reactions for its production. Once we measure the current and the time for which it flows, we can use Eq. 5 to determine the moles of electrons supplied. By combining the number of moles of electrons supplied with the mole ratio from the stoichiometry of the half-reaction, we can deduce the amount of product obtained (Fig. 18.29).

Procedure

To determine the amount of product produced

Step 1. Write the half-reaction for the formation of the product and find the mole ratio between electrons and product.

Step 2. Determine the charge supplied from the current and the time (1C = 1A·S), and convert charge to number of moles of electrons by using the Faraday constant (inside the back cover), as in Eq. 5:

$$n(e^-) = \frac{It}{F}$$

Use the mole ratio as a conversion factor and then convert from moles of product to mass by using the molar mass of the product. Example 18.12 illustrates this procedure.

To determine the time a current must flow to produce a certain mass of product

Step 1. Write the half-reaction for the formation of the product and find the mole ratio between electrons and product.

Step 2. Convert the mass of product to moles of product by dividing it by the molar mass; then convert to moles of electrons by using the mole ratio from step 1.

Step 3. Multiply by the Faraday constant to convert moles of electrons to charge, then divide by the current to obtain the time required. That is, the time required is found from

$$t = \frac{F \times n(e^-)}{I} \tag{6}$$

Example 18.13 illustrates this procedure.

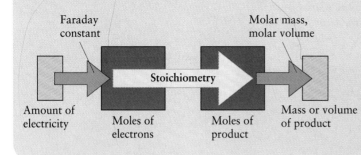

FIGURE 18.29

A schematic representation of the stoichiometric relations that are used to calculate the amount of product formed by electrolysis or the amount of time current must flow to produce a given product.

Example 18.12 *Calculating the amount of product produced from electrolysis*

Aluminum is produced by electrolysis of its oxide dissolved in molten cryolite (Na_3AlF_6). Calculate the mass of aluminum that can be produced in 24.0 h in an electrolytic cell operating continuously at 1.00×10^5 A. The cryolite does not react.

Strategy The number of moles of electrons supplied during the electrolysis is given by Eq. 5. The number of moles of electrons is converted to number of moles of product and then mass, as described in Toolbox 18.5.

Solution **Step 1.** The aluminum metal is produced from the reduction of Al^{3+}, present in the molten Al_2O_3. The balanced cathodic half-reaction is $Al^{3+} + 3\,e^- \rightarrow Al$, from which it follows that 3 mol e⁻ $\rightleftharpoons$ 1 mol Al. **Step 2.** Because the molar mass of aluminum is 26.98 g/mol, the string of conversions we need is

$$\text{Mass of Al} = \frac{(1.00 \times 10^5 \text{ C/s}) \times \overbrace{(24.0 \text{ h} \times 3600 \text{ s/h})}^{\text{Seconds in a day}}}{\underbrace{9.6485 \times 10^4 \text{ C/(mol e}^-)}_{\text{Moles of electrons}}} \times \frac{1 \text{ mol Al}}{3 \text{ mol e}^-} \times \frac{26.98 \text{ g Al}}{1 \text{ mol Al}}$$

$$= \frac{1.00 \times 10^5 \times 24.0 \times 3600 \times 26.98}{9.6485 \times 10^4 \times 3} \text{ g Al}$$

$$= 8.05 \times 10^5 \text{ g Al}$$

or 805 kg. The fact that the production of 1 mol Al requires 3 mol e⁻ accounts for the very high consumption of electricity characteristic of aluminum production plants.

Self-Test 18.15A Determine the mass, in grams, of sodium metal that can be produced from molten sodium chloride in the Downs process, using a current of 0.0125 A for 48.0 h.

[**Answer:** 0.515 g]

Self-Test 18.15B Fluorine is produced by electrolysis of a mixture of fluorides. Calculate the mass of fluorine that can be produced by electrolysis, using a current of 2.50 A for 36.0 h.

Example 18.13 *Calculating the time required to produce a given mass of product*

How many hours are required to plate 25.00 g of copper metal from 1.00 M $CuSO_4(aq)$, using a current of 3.00 A?

Strategy The time required to supply the charge needed for the electrolysis is given by Eq. 6. The mass of product is converted to moles of electrons and then to charge, as described in Toolbox 18.5. We can set up the calculation in one step.

Solution **Step 1.** The half-reaction for the electrolysis is $Cu^{2+}(aq) + 2 e^- \rightarrow Cu(s)$, so we can write 2 mol e⁻ $\rightleftharpoons$ 1 mol Cu.

Step 2. To find the number of moles of electrons, $n(e^-)$, we need to convert mass of copper to moles of copper and moles of copper to moles of electrons:

$$n(e^-) = (25.00 \text{ g Cu}) \times \left(\frac{1 \text{ mol Cu}}{63.54 \text{ g Cu}}\right) \times \left(\frac{2 \text{ mol e}^-}{1 \text{ mol Cu}}\right) = \frac{25.00 \times 2}{63.54} \text{ mol e}^-$$

Step 3. Substitute the number of moles of electrons, the current, and the Faraday constant into Eq. 6:

$$t = \frac{(9.6485 \times 10^4 \text{ C/mol e}^-)}{3.00 \text{ C/s}} \times \left(\frac{25.00 \times 2}{63.54} \text{ mol e}^-\right) \times \left(\frac{1 \text{ h}}{3600 \text{ s}}\right) = 7.03 \text{ h}$$

Self-Test 18.16A Determine the time, in hours, required to electroplate 7.00 g of magnesium metal from molten magnesium chloride, using a current of 7.30 A.

[**Answer:** 2.11 h]

Self-Test 18.16B How many hours are required to plate 12.00 g of chromium metal from a 1 M solution of CrO_3 in dilute sulfuric acid, using a current of 6.20 A?

FIGURE 18.30

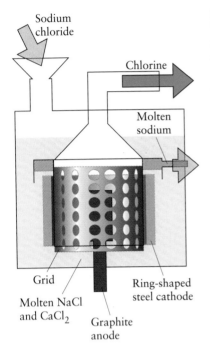

Sodium chloride

Chlorine

Molten sodium

Grid

Molten NaCl and CaCl₂

Graphite anode

Ring-shaped steel cathode

In the Downs process, molten sodium chloride is electrolyzed with a graphite anode (at which the Cl⁻ ions are oxidized to chlorine) and a steel cathode (at which the Na⁺ ions are reduced to sodium). The sodium and chlorine are kept apart by the hoods surrounding the electrodes. Calcium chloride is present to lower the melting point of sodium chloride to a more economical value.

18.16 Applications of Electrolysis

We have already described the electrolytic extraction of aluminum, magnesium, and fluorine and the refining of copper. Another important industrial application of electrolysis is the production of sodium metal by the *Downs process,* the electrolysis of molten rock salt (Fig. 18.30):

$$\text{Cathode reaction: } Na^+(l) + e^- \longrightarrow Na(l)$$

$$\text{Anode reaction: } 2\,Cl^-(l) \longrightarrow Cl_2(g) + 2\,e^-$$

Sodium chloride is plentiful as rock salt, but the solid does not conduct electricity, because the ions are locked into place. Therefore, sodium chloride must be molten for electrolysis to occur. The electrodes in the cell are made of inert materials like carbon, and the cell is designed to keep the sodium and chlorine produced by the electrolysis out of contact with each other and with air. Over 2×10^7 kg of sodium is produced annually by the Downs process, most of it for use in the extraction of other metals. Other active metals, such as lithium, magnesium, and calcium, are also prepared by the electrolysis of their molten chloride salts. In a modification of the Downs process, the electrolyte is an aqueous solution of sodium chloride. The products of this *chloralkali process* are chlorine and aqueous sodium hydroxide.

Electroplating is the electrolytic deposition of a thin film of metal on an object. The object to be electroplated (either metal or plastic coated with graphite) is made the cathode, and the electrolyte is an aqueous solution of a salt of the plating metal. Metal is deposited on the cathode from ions in the electrolyte solution. These cations are either supplied by the added salt or from oxidation of the anode, which is made of the plating metal.

FIGURE 18.31

Chromium plating lends decorative flair as well as protection to the steel of this motorcycle. Large quantities of electricity are needed to plate chromium because six electrons are required to produce each atom of chromium.

For chromium plating (Fig. 18.31), the electrolyte is prepared by dissolving CrO_3 in dilute sulfuric acid. The electrolysis reduces chromium(VI) first to chromium(III) and then to Cr(0):

$$CrO_3(aq) + 6H^+(aq) + 6e^- \longrightarrow Cr(s) + 3H_2O(l)$$

The chromium deposits on the cathode as a hard protective film. Because six electrons must be supplied for each atom of chromium deposited, a large amount of electricity is needed.

Electrolysis is used industrially to produce aluminum and magnesium, to extract metals from their salts, to prepare chlorine, fluorine, and sodium hydroxide, to refine copper, and to carry out electroplating.

Skills You Should Have Mastered

Conceptual

☐ 1. Distinguish galvanic and electrolytic cells and describe their operation; identify anode, cathode, and the direction of current flow for each, Sections 18.3 and 18.13.

☐ 2. Explain the relationship between the standard free energy, standard cell potential, and equilibrium constant of a reaction, Section 18.9.

☐ 3. Predict the effect of changes in concentration of reactants and products on the cell potential, Section 18.10.

Problem-Solving

☐ 1. Balance chemical equations for redox reactions by the half-reaction method, Toolbox 18.1 and Examples 18.1 and 18.2.

☐ 2. Write the chemical equation, cell diagram, and description of the galvanic cell in which two given half-reactions take place, Toolbox 18.2 and Example 18.3.

☐ 3. Deduce the standard potential of an electrode, Example 18.4.

☐ 4. Predict relative oxidizing strength by using the electrochemical series, Example 18.5.

☐ 5. Calculate a standard cell potential, given the standard electrode potentials, and predict the spontaneous direction of the cell reaction, Toolbox 18.3 and Example 18.6.

☐ 6. Assess the reaction free energy from a cell potential, Example 18.7.

☐ 7. Calculate the equilibrium constant for a reaction from electrochemical data, Toolbox 18.4 and Example 18.8.

☐ 8. Use the Nernst equation to calculate a cell potential under nonstandard conditions, Example 18.9.

☐ 9. Predict the standard potential of a fuel cell, Example 18.10.

☐ 10. Predict the products of electrolysis of an aqueous solution from the electrochemical series, Example 18.11.

☐ 11. Calculate the amount of product formed in electrolysis, Toolbox 18.5 and Example 18.12.

☐ 12. Calculate the time required to produce a given mass of product by electrolysis, Toolbox 18.5 and Example 18.13.

Descriptive

☐ 1. Distinguish and give practical examples of primary cells, secondary cells, and fuel cells, Section 18.11.

☐ 2. Describe corrosion and means of protecting iron from corrosion, Section 18.12.

☐ 3. Describe the Downs process for producing sodium metal and chlorine gas, Section 18.16.

☐ 4. Describe electroplating and give an example, Section 18.16.

Exercises

Assume a temperature of 25°C (298 K) for the following exercises unless instructed otherwise.

Balancing Redox Equations

18.1 Balance the following partial skeletal equations for each half-reaction occurring in an acidic solution. Identify each half-reaction as an oxidation or a reduction.
(a) The conversion of vanadyl ions to vanadium(III) ions, $VO^{2+}(aq) \rightarrow V^{3+}(aq)$

(b) One of the half-reactions that occurs when a lead-acid battery is charged, $PbSO_4(s) \rightarrow PbO_2(s) + SO_4^{2-}(aq)$
(c) The decomposition of hydrogen peroxide to oxygen gas, $H_2O_2(aq) \rightarrow O_2(g)$

18.2 Balance the following partial skeletal equations for each half-reaction occurring in an acidic solution. Identify each half-reaction as an oxidation or a reduction.
(a) $Cr_2O_7^{2-}(aq) \rightarrow Cr^{3+}(aq)$
(b) $I^-(aq) \rightarrow IO_3^-(aq)$
(c) $NO(g) \rightarrow NO_3^-(aq)$

18.3 Balance the following partial skeletal equations for each half-reaction occurring in a basic solution. Identify each half-reaction as an oxidation or a reduction.
(a) Conversion of hypochlorite ions to chloride ions (a part of the action of bleach), $ClO^-(aq) \rightarrow Cl^-(aq)$
(b) Conversion of iodate ions to hypoiodite ions, $IO_3^-(aq) \rightarrow IO^-(aq)$
(c) Conversion of sulfite ions to dithionite ions, $SO_3^{2-}(aq) \rightarrow S_2O_4^{2-}(aq)$

18.4 Balance the following partial skeletal equations for each half-reaction occurring in a basic solution. Identify each half-reaction as an oxidation or a reduction.
(a) $NO_3^-(aq) \rightarrow NO_2^-(aq)$
(b) $CrO_4^{2-}(aq) \rightarrow Cr(OH)_3(s)$
(c) $MnO_4^{2-}(aq) \rightarrow MnO_2(s)$

18.5 Balance the following partial skeletal equations by using oxidation and reduction half-reactions. All the reactions occur in acidic solution. Identify the oxidizing agent and reducing agent in each reaction.
(a) Reaction of thiosulfate ion with chlorine gas, $Cl_2(g) + S_2O_3^{2-}(aq) \rightarrow Cl^-(aq) + SO_4^{2-}(aq)$
(b) Action of the permanganate ion on sulfurous acid, $MnO_4^-(aq) + H_2SO_3(aq) \rightarrow Mn^{2+}(aq) + HSO_4^-(aq)$
(c) Reaction of hydrosulfuric acid with chlorine, $H_2S(aq) + Cl_2(g) \rightarrow S(s) + Cl^-(aq)$
(d) Reaction of chlorine in water, $Cl_2(g) \rightarrow HClO(aq) + Cl^-(aq)$

18.6 Balance the following partial skeletal equations by using oxidation and reduction half-reactions. All reactions occur in acidic solution. Identify the oxidizing agent and reducing agent in each reaction.
(a) Conversion of iron(II) to iron(III) by dichromate ion, $Fe^{2+}(aq) + Cr_2O_7^{2-}(aq) \rightarrow Fe^{3+}(aq) + Cr^{3+}(aq)$
(b) Formation of acetic acid from ethanol by the action of permanganate ion, $C_2H_5OH(aq) + MnO_4^-(aq) \rightarrow Mn^{2+}(aq) + CH_3COOH(aq)$
(c) Reaction of iodide ion with nitric acid, $I^-(aq) + NO_3^-(aq) \rightarrow I_2(aq) + NO(g)$
(d) Reaction of arsenic(III) sulfide with nitric acid, $As_2S_3(s) + NO_3^-(aq) \rightarrow H_3AsO_4 + S(s) + NO(g)$

18.7 Balance the following partial skeletal equations by using oxidation and reduction half-reactions. All reactions occur in basic solution. Identify the oxidizing agent and reducing agent in each reaction.
(a) Action of ozone on bromide ions, $O_3(aq) + Br^-(aq) \rightarrow O_2(g) + BrO_3^-(aq)$
(b) Reaction of bromine in water, $Br_2(l) \rightarrow BrO_3^-(aq) + Br^-(aq)$
(c) Formation of chromate ions from chromium (III) ions, $Cr^{3+}(aq) + MnO_2(s) \rightarrow Mn^{2+}(aq) + CrO_4^{2-}(aq)$
(d) Reaction of elemental phosphorus to form phosphine, PH_3, a poisonous gas with the odor of decaying fish, $P_4(s) \rightarrow H_2PO_2^-(aq) + PH_3(g)$

18.8 Balance the following partial skeletal equations by using oxidation and reduction half-reactions. All reactions occur in basic solution. Identify the oxidizing agent and reducing agent in each reaction.
(a) Production of chlorite ions from dichlorine heptoxide, $Cl_2O_7(g) + H_2O_2(aq) \rightarrow ClO_2^-(aq) + O_2(g)$
(b) Action of permanganate ions on sulfide ions, $MnO_4^-(aq) + S^{2-}(aq) \rightarrow S(s) + MnO_2(s)$
(c) Reaction of hydrazine with chlorate ions, $N_2H_4(g) + ClO_3^-(aq) \rightarrow NO(g) + Cl^-(aq)$
(d) Reaction of plumbate ions with hypochlorite ions, $Pb(OH)_4^{2-}(aq) + ClO^-(aq) \rightarrow PbO_2(s) + Cl^-(aq)$

18.9 The hydrogen sulfite ion, HSO_3^-, is a moderately strong reducing agent in acidic solutions and, depending on the conditions, is oxidized to either the hydrogen sulfate ion, HSO_4^-, or the dithionate ion, $S_2O_6^{2-}$. (a) Write the equations for each half-reaction. (b) The reaction of the hydrogen sulfite ion with iodine to form iodide and hydrogen sulfate ions is used to determine its concentration in solution in the laboratory. Write the equation for the reduction half-reaction and the net ionic equation for the overall reaction.

18.10 The thiosulfate ion, $S_2O_3^{2-}$, is a moderately strong reducing agent in acidic solutions. It is used to determine the concentration of elemental iodine in solution, forming iodide ions and tetrathionate ions, $S_4O_6^{2-}$, in the reaction. Write the equation for each half-reaction and the net ionic equation for the overall reaction.

18.11 Potassium permanganate is an excellent oxidizing agent for laboratory use and in sewage treatment. It readily reacts with the organic compounds in sewage to produce carbon dioxide and water. Write the equation for each half-reaction and the overall equation for the oxidation of glucose, $C_6H_{12}O_6$. The partial skeletal form is $MnO_4^-(aq) + C_6H_{12}O_6(aq) \rightarrow Mn^{2+}(aq) + CO_2(g) + H_2O(l)$.

18.12 Nitrogen monoxide gas can be produced from the reaction of nitrite ions with each other in a dilute sulfuric acid solution. The partial skeletal equation for the reaction is $NO_2^-(aq) \rightarrow NO(g) + NO_3^-(aq)$. Write the equation for each half-reaction and the overall equation for the reaction.

Galvanic Cells

18.13 Complete the following statements. (a) In a galvanic cell, oxidation occurs at the (anode, cathode). (b) The cathode is the (positive, negative) electrode.

18.14 Complete the following statement. In a galvanic cell, the anions migrate toward the (anode, cathode) and the electrons flow through an external circuit from the (anode, cathode) to the (anode, cathode).

18.15 Write the reduction half-reaction for (a) zinc metal in contact with Zn^{2+} ions in solution; (b) a solution of iron(II) and iron(III) salts; (c) chlorine gas in contact with Cl^- ions in solution; (d) the calomel electrode (see Investigating Matter 18.1).

18.16 Write the oxidation half-reaction for (a) silver metal in contact with Ag^+ ions in solution; (b) Ce^{3+} and Ce^{4+} ions in solution; (c) oxygen gas in contact with hydroxide ions in water; (d) the silver-silver iodide electrode (silver in contact with $AgI(s)$ and a solution of I^- ions).

18.17 Write cathode and anode half-reactions, the balanced equation for the cell reaction, and the cell diagram for the following skeletal equations:
(a) $Ni^{2+}(aq) + Zn(s) \rightarrow Ni(s) + Zn^{2+}(aq)$
(b) $Ce^{4+}(aq) + I^-(aq) \rightarrow I_2(s) + Ce^{3+}(aq)$
(c) $Cl_2(g) + H_2(g) \rightarrow HCl(aq)$
(d) $Au^+(aq) \rightarrow Au(s) + Au^{3+}(aq)$

18.18 Write cathode and anode half-reactions, the balanced equation for the cell reaction, and the cell diagram for the following skeletal equations:
(a) $Mn(s) + Ti^{2+}(aq) \rightarrow Mn^{2+}(aq) + Ti(s)$
(b) $Fe^{3+}(aq) + H_2(g) \rightarrow Fe^{2+}(aq) + H^+(aq)$
(c) $Cu^+(aq) \rightarrow Cu(s) + Cu^{2+}(aq)$
(d) $MnO_4^-(aq) + H^+(aq) + Cl^-(aq) \rightarrow$
$$Cl_2(g) + Mn^{2+}(aq) + H_2O(l)$$

18.19 Write cathode and anode half-reactions and the balanced equation for the cell reaction for each of the following galvanic cells:
(a) $C(gr)|H_2(g)|H^+(aq)||Cl^-(aq)|Cl_2(g)|Pt(s)$
(b) $U(s)|U^{3+}(aq)||V^{2+}(aq)|V(s)$
(c) $Pt(s)|O_2(g)|H^+(aq)||OH^-(aq)|O_2(g)|Pt(s)$
(d) $Pt(s)|Sn^{4+}(aq),Sn^{2+}(aq)||Cl^-(aq)|Hg_2Cl_2(s)|Hg(l)$

18.20 Write cathode and anode half-reactions and the balanced equation for the cell reaction for each of the following galvanic cells:
(a) $Cu(s)|Cu^{2+}(aq)||Cu^+(aq)|Cu(s)$
(b) $Ag(s)|AgI(s)|I^-(aq)||Cl^-(aq)|AgCl(s)|Ag(s)$
(c) $Hg(l)|Hg_2Cl_2(s)|Cl^-(aq)||Cl^-(aq)|AgCl(s)|Ag(s)$
(d) $Cu(s)|Cu^{2+}(aq)||Pb^{4+}(aq),Pb^{2+}(aq)|C(gr)$

18.21 (a) Write balanced cathode and anode half-reactions for the redox reaction of an acidified solution of potassium permanganate and iron(II) chloride (see Table 18.1). (b) Write the balanced equation for the cell reaction. (c) Write the cell diagram and draw a schematic picture of the cell and its contents, label the anode and cathode, and indicate the direction of electron flow.

18.22 (a) Write balanced cathode and anode half-reactions for the redox reaction between sodium dichromate and mercury(I) nitrate in an acidic solution (see Table 18.1). (b) Write the balanced equation for the cell reaction. (c) Write the cell diagram and draw a schematic picture of the cell and its contents, label the anode and cathode, and indicate the direction of electron flow.

18.23 Write cathode and anode half-reactions and devise a galvanic cell (write a cell diagram) to study each of the following reactions:
(a) $AgBr(s) \rightarrow Ag^+(aq) + Br^-(aq)$, a solubility equilibrium
(b) $H_3O^+(aq) + OH^-(aq) \rightarrow 2 H_2O(l)$, the Brønsted neutralization reaction
(c) $Cd(s) + 2 Ni(OH)_3(s) \rightarrow Cd(OH)_2(s) + 2 Ni(OH)_2(s)$, the reaction in the nickel-cadmium cell

18.24 Write balanced cathode and anode half-reactions and devise a galvanic cell (write a cell diagram) to study each of the following reactions:
(a) $AgNO_3(aq) + KI(aq) \rightarrow AgI(s) + KNO_3(aq)$, a precipitation reaction
(b) $H_3O^+(aq, concentrated) \rightarrow H_3O^+(aq, dilute)$
(c) $Zn(s) + Ag_2O(s) \rightarrow ZnO(s) + 2 Ag(s)$, the reaction in a silver cell

Cell Potential, Free Energy, and the Electrochemical Series

18.25 Find the value of n, the number of moles of electrons transferred, for each of the following chemical equations:
(a) $C_6H_{12}O_6(s) + 6 O_2(g) \rightarrow 6 CO_2(g) + 6 H_2O(l)$
(b) $2 B(s) + 3 Cl_2(g) \rightarrow 2 BCl_3(g)$
(c) $SiO_2(l) + 2 C(s) \rightarrow Si(l) + 2 CO(g)$

18.26 Find the value of n, the number of moles of electrons transferred, for each stage in the Ostwald process for the production of nitric acid from ammonia:
(a) $4 NH_3(g) + 5 O_2(g) \rightarrow 4 NO(g) + 6 H_2O(g)$
(b) $2 NO(g) + O_2(g) \rightarrow 2 NO_2(g)$
(c) $3 NO_2(g) + H_2O(l) \rightarrow 2 HNO_3(aq) + NO(g)$

18.27 Predict the standard potential of the galvanic cells with the following cell reactions:
(a) $2 Cr^{2+}(aq) + Cu^{2+}(aq) \rightarrow 2 Cr^{3+}(aq) + Cu(s)$
(b) $AgCl(s) + I^-(aq) \rightarrow Cl^-(aq) + AgI(s)$
(c) $Hg_2^{2+}(aq) + 2 Cl^-(aq) \rightarrow Hg_2Cl_2(s)$

18.28 Predict the standard potential of the galvanic cells with the following cell reactions:
(a) $Ag^+(aq) + Fe^{2+}(aq) \rightarrow Fe^{3+}(aq) + Ag(s)$
(b) $3 V^{2+}(aq) + 2 U(s) \rightarrow 2 U^{3+}(aq) + 3 V(s)$
(c) $Sn^{2+}(aq) + Pb^{4+}(aq) \rightarrow Sn^{4+}(aq) + Pb^{2+}(aq)$

18.29 Calculate the standard reaction free energy for the cell reactions in Exercise 18.27.

18.30 Predict the standard cell potential and calculate the standard reaction free energy for the following cell reactions:
(a) $Zn(s) + Fe^{2+}(aq) \rightarrow Zn^{2+}(aq) + Fe(s)$
(b) $2 H_2(g) + O_2(g) \rightarrow 2 H_2O(l)$ in an acidic solution
(c) $Ag^+(aq) + Cl^-(aq) \rightarrow AgCl(s)$
(d) $3 Au^+(aq) \rightarrow 2 Au(s) + Au^{3+}(aq)$

18.31 Arrange the following metals in order of increasing strength as reducing agents: (a) Cu, Zn, Cr, Fe; (b) Li, Na, K, Mg; (c) U, V, Ti, Al; (d) Ni, Sn, Au, Ag.

18.32 Arrange the following species in order of increasing strength as oxidizing agents: (a) Co^{2+}, Cl_2, Ce^{4+}, In^{3+}; (b) NO_3^-, ClO_4^-, $HBrO$, $Cr_2O_7^{2-}$, all in acidic solution; (c) H_2O_2, O_2, MnO_4^-, $HClO$, all in acidic solution; (d) Ti^{3+}, Sn^{4+}, Hg_2^{2+}, Fe^{2+}

18.33 The standard potential of a Co^{2+}/Co electrode is -0.28 V and the standard potential of the cell $Pt(s)|Ti^{2+}(aq),Ti^{3+}(aq)||Co^{2+}(aq)|Co(s)$ is 0.09 V. What is the standard potential of the Ti^{3+}/Ti^{2+} electrode?

18.34 The standard potential of a La^{3+}/La electrode is -2.52 V and the standard potential of the cell $La(s)|La^{3+}(aq)||U^{3+}(aq)|U(s)$ is 0.73 V. What is the standard potential of the U^{3+}/U electrode?

18.35 The following half-reactions are joined to form a galvanic cell that generates a current under standard conditions. Identify the oxidizing agent and the reducing agent, write a cell diagram, and calculate the standard cell potential:
(a) $Pt^{2+}(aq) + 2e^- \rightarrow Pt(s)$ and $AgF(s) + e^- \rightarrow Ag(s) + F^-(aq)$
(b) $Cr^{3+}(aq) + e^- \rightarrow Cr^{2+}(aq)$ and $I_3^-(aq) + 2e^- \rightarrow 3I^-(aq)$

18.36 The following half-reactions are joined to form a galvanic cell that generates a current under standard conditions. Identify the oxidizing agent and the reducing agent, write a cell diagram, and calculate the standard cell potential:
(a) $2H^+(aq) + 2e^- \rightarrow H_2(g)$ and $Ni^{2+}(aq) + 2e^- \rightarrow Ni(s)$
(b) $O_3(g) + H_2O(l) + 2e^- \rightarrow O_2(g) + 2OH^-(aq)$ and $O_3(g) + 2H^+(aq) + 2e^- \rightarrow O_2(g) + H_2O(l)$

18.37 Answer the following questions, and for each "yes" response, write a balanced cell reaction and calculate the standard cell potential: (a) Can H_2 reduce Ti^{2+} ions in aqueous solution to titanium metal? (b) Can chromium metal reduce Pb^{2+} ions in aqueous solution to lead metal? (c) Can permanganate ions oxidize copper metal to Cu^{2+} ions in an acidic solution? (d) Can Fe^{3+} ions in aqueous solution oxidize mercury metal to mercury(I)?

18.38 Identify the spontaneous reactions in the following list, and for each spontaneous reaction, identify the oxidizing agent and calculate the standard cell potential:
(a) $Cl_2(g) + 2Br^-(aq) \rightarrow 2Cl^-(aq) + Br_2(l)$
(b) $MnO_4^-(aq) + 8H^+(aq) + 5Ce^{3+}(aq) \rightarrow$ $5Ce^{4+}(aq) + Mn^{2+}(aq) + 4H_2O(l)$
(c) $2Pb^{2+}(aq) \rightarrow Pb(s) + Pb^{4+}(aq)$
(d) $2NO_3^-(aq) + 4H^+(aq) + Zn(s) \rightarrow$ $Zn^{2+}(aq) + 2NO_2(g) + 2H_2O(l)$

18.39 Identify the spontaneous reactions among the following reactions, and for the spontaneous reactions,

write balanced reduction and oxidation half-reactions. Show that the reaction is spontaneous by calculating the standard free energy of the reaction. (a) $I_2(s) + H_2(g) \rightarrow$?; (b) $Mg^{2+}(aq) + Cu(s) \rightarrow$?; (c) $Al(s) + Pb^{2+}(aq) \rightarrow$?.

18.40 Identify the spontaneous reactions among the following reactions, and for the spontaneous reactions, write balanced reduction and oxidation half-reactions. Show that the reaction is spontaneous by calculating the standard free energy of the reaction. (a) $Hg_2^{2+}(aq) + Ce^{3+}(aq) \rightarrow$?; (b) $Zn(s) + Sn^{2+}(aq) \rightarrow$?; (c) $O_2(g) + H^+(aq) + Hg(l) \rightarrow$?.

18.41 Chlorine is used to displace bromine from brine that contains sodium bromide. Could oxygen in an acidified solution be used instead? If so, why is it not used? *Hint:* See Section 18.14.

18.42 A chemist is interested in the compounds formed by the *d*-block element manganese and wants to find a way to prepare an aqueous solution of Mn^{3+} from Mn^{2+}. Would an acidified solution of sodium dichromate be suitable?

Equilibrium Constants

18.43 Write the expression for the equilibrium constants of the following cell reactions:
(a) $Pt(s)|H_2(g)|H^+(aq),Cl^-(aq)|AgCl(s)|Ag(s)$
(b) $Pt(s)|Fe^{3+}(aq),Fe^{2+}(aq),NO_3^-(aq),H^+(aq)|NO(g)|Pt(s)$

18.44 Write the expression for the equilibrium constant of the following cell reactions:
(a) $Cr(s)|Cr^{3+}(aq)||Br^-(aq)|AgBr(s)|Ag(s)$
(b) $Bi(s)|Bi^{3+}(aq)||OH^-(aq)|O_2(g)|Pt(s)$

18.45 Use standard potentials to calculate the equilibrium constants for the following cells and cell reactions at 298 K:
(a) $Mn(s) + Ti^{2+}(aq) \rightarrow Mn^{2+}(aq) + Ti(s)$
(b) A Pb^{2+}/Pb redox couple in combination with a Hg_2^{2+}/Hg redox couple (write the equation with the smallest whole number coefficients)
(c) $In^{3+}(aq) + U^{3+}(aq) \rightarrow In^{2+}(aq) + U^{4+}(aq)$

18.46 Use standard potentials to calculate the equilibrium constants for the following cells and cell reactions at 298 K:
(a) An $AgI/Ag,I^-$ redox couple in combination with a I_2/I^- redox couple (write the equation with the smallest whole number coefficients)
(b) $2Fe^{3+}(aq) + H_2(g) \rightarrow 2Fe^{2+}(aq) + 2H^+(aq)$
(c) $Cr(s) + Zn^{2+}(aq) \rightarrow Cr^{2+}(aq) + Zn(s)$

18.47 A chemist wants to make a range of silver(II) compounds. Could aqueous sodium persulfate be used to oxidize silver(I) to silver(II)? If so, what would be the equilibrium constant for the reaction?

18.48 A chemist suspects that manganese(III) might be involved in an unusual biochemical reaction and wants to prepare some of its compounds. Could aqueous potassium permanganate be used to oxidize manganese(II) to manganese(III) in either basic or acidic solution? If so, what would be the equilibrium constant for the reaction?

The Nernst Equation

18.49 What would be the effect of each of the following changes on the potential of the Daniell cell (Fig. 18.5): (a) increasing the concentration of the zinc(II) ions; (b) adding sodium hydroxide to the cathode compartment; (c) diluting the solution in the anode compartment; (d) using a larger zinc electrode?

18.50 What would be the effect of each of the following changes on the potential of the galvanic cell in Fig. 18.10: (a) decreasing the concentration of mercury(I) ions in the cathode compartment; (b) removing some of the liquid mercury from the cathode compartment; (c) adding silver nitrate to the anode compartment; (d) increasing the concentration of chloride ions?

18.51 Calculate the reaction quotient Q for the cell reaction, given the measured values of the cell potential:
(a) $Pb^{4+}(aq) + Sn^{2+}(aq) \rightarrow Pb^{2+}(aq) + Sn^{4+}(aq)$, $E = 1.33 \text{ V}$
(b) $2 Cr_2O_7^{2-}(aq) + 16 H^+(aq) \rightarrow$ $4 Cr^{3+}(aq) + 8 H_2O(l) + 3 O_2(g)$, $E = 0.10 \text{ V}$

18.52 Calculate the reaction quotient Q for the cell reaction, given the measured values of the cell potential:
(a) $ClO_4^-(aq) + 2 H^+(aq) + 2 Ag(s) \rightarrow$ $ClO_3^-(aq) + H_2O(l) + 2 Ag^+(aq)$, $E = 0.40 \text{ V}$
(b) $2 Au^{3+}(aq) + 6 Cl^-(aq) \rightarrow 2 Au(s) + 3 Cl_2(g)$, $E = 0.00 \text{ V}$

18.53 A *concentration cell* consists of the same redox couples at the anode and the cathode, with different concentrations of the ions in the respective compartments. Calculate $E(\text{cell})$ for the following concentration cells:
(a) $Cu(s)|Cu^{2+}(aq, 0.0010 \text{ mol/L})\|Cu^{2+}(aq, 0.010 \text{ mol/L})|Cu(s)$
(b) $Pt(s)|H_2(g, 1 \text{ atm})|H^+(aq, pH = 4.0)\|H^+(aq, pH = 3.0)|H_2(g, 1 \text{ atm})|Pt(s)$

18.54 A concentration cell consists of the same redox couples at the anode and the cathode with different concentrations of the ions in the respective compartments. Determine the unknown concentration in in the following cells:
(a) $Pb(s)|Pb^{2+}(aq, ?)\|Pb^{2+}(aq, 0.10 \text{ mol/L})|Pb(s)$, $E = 0.050 \text{ V}$
(b) $Pt(s)|Fe^{3+}(aq, 0.10 \text{ mol/L}), Fe^{2+}(aq, 1.0 \text{ mol/L})\|Fe^{3+}(aq, ?), Fe^{2+}(aq, 0.0010 \text{ mol/L})|Pt(s)$, $E = 0.10 \text{ V}$

18.55 Determine the potentials of the galvanic cells with the following cell reactions (or descriptions) and concentrations: (a) $Ni^{2+}(aq) + Zn(s) \rightarrow Ni(s) + Zn^{2+}(aq)$, with 0.10 mol/L $Zn^{2+}(aq)$ and 0.0010 mol/L $Ni^{2+}(aq)$; (b) $2 H^+(aq) + 2 Cl^-(aq) \rightarrow H_2(g) + Cl_2(g)$, with Cl_2 at 100. Torr and 1.0 M HCl(aq) in the chlorine electrode compartment, and H_2 at 450. Torr and 0.010 M HCl(aq) in the hydrogen electrode compartment; (c) a galvanic cell with one half-cell consisting of a tin electrode and 0.020 mol/L $Sn^{2+}(aq)$ and the other half-cell consisting of a platinum electrode, 0.060 mol/L $Sn^{4+}(aq)$, and 1.0 mol/L $Sn^{2+}(aq)$.

18.56 Determine the potentials of the galvanic cells with the following cell reactions and concentrations: (a) $3 Pb^{2+}(aq) + 2 Cr(s) \rightarrow 3 Pb(s) + 2 Cr^{3+}(aq)$, with 0.10 mol/L $Cr^{3+}(aq)$ and 1.00×10^{-5} mol/L $Pb^{2+}(aq)$; (b) a galvanic cell with one half-cell consisting of a graphite electrode, 0.0030 mol/L $Sn^{4+}(aq)$, and 0.10 mol/L $Sn^{2+}(aq)$, and the other half-cell consisting of a platinum electrode, 0.10 mol/L $Fe^{3+}(aq)$, and 1.0×10^{-4} mol/L $Fe^{2+}(aq)$; (c) $AgCl(s) + I^-(aq) \rightarrow AgI(s) + Cl^-(aq)$, with 0.010 mol/L $I^-(aq)$ and 1.0×10^{-6} mol/L $Cl^-(aq)$.

18.57 Determine the unknown in the following cells:
(a) $Pt(s)|H_2(g, 1.0 \text{ atm})|H^+(pH = ?)\|$ $Cl^-(aq, 1.0 \text{ mol/L})|Hg_2Cl_2(s)|Hg(l)$, $E = 0.33 \text{ V}$
(b) $C(gr)|Cl_2(g, 1.0 \text{ atm})|Cl^-; (aq, ?)\|$ $MnO_4^-(aq, 0.010 \text{ mol/L}), H^+(pH = 4.0),$ $Mn^{2+}(aq, 0.10 \text{ mol/L})|Pt(s)$, $E = -0.30 \text{ V}$

18.58 Determine the unknown in the following cells:
(a) $Pt(s)|H_2(g, 1.0 \text{ atm})|H^+(pH = ?)\|$ $Cl^-(aq, 1.0 \text{ mol/L})|AgCl(s)|Ag(s)$, $E = 0.30 \text{ V}$
(b) $Pb(s)|Pb^{2+}(aq, ?)\|Ni^{2+}(aq, 0.10 \text{ mol/L})|Ni(s)$, $E = 0.040 \text{ V}$

Practical Cells

18.59 Explain the difference between a primary cell and a secondary cell.

18.60 Explain the difference between a primary cell and a fuel cell.

18.61 What is (a) the electrolyte; (b) the oxidizing agent in a mercury cell (shown below)? (c) Write the overall cell reaction for a mercury cell (see Table 18.2).

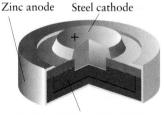

Zinc anode   Steel cathode

HgO in KOH and $Zn(OH)_2$

18.62 What is (a) the electrolyte; (b) the oxidizing agent during discharge in a lead-acid battery? (c) Write the reaction that occurs at the cathode during the charging of the lead-acid battery.

18.63 Explain how a dry cell generates electricity.

18.64 The density of the electrolyte in a lead-acid battery is measured to assess its state of charge. Explain how the density reflects the state of charge of the battery.

18.65 (a) What is the electrolyte in a nickel-cadmium cell? (b) Write the reaction that occurs at the anode when the cell is being charged.

18.66 (a) Suggest a reason why lead-antimony grids are used as electrodes in the lead-acid battery rather than smooth plates. (b) What is the reducing agent in the lead-acid battery? (c) The lead-acid cell potential is about 2 V. How, then, does a car battery produce 12 V for its electrical system?

Corrosion

18.67 A chromium-plated steel bicycle handlebar is scratched. Will rusting of the iron in the steel be encouraged or retarded by the chromium?

18.68 A solution consists of the ions, Cu^{2+}, Ni^{2+}, and Ag^+, each present at 1 mol/L. What will happen if a strip of tin metal is placed in the solution?

18.69 (a) What is the approximate chemical formula of rust? (b) What is the oxidizing agent in the formation of rust? (c) How does the presence of salt accelerate the rusting process?

18.70 (a) What is the electrolyte solution in the formation of rust? (b) How are steel (iron) objects protected by galvanizing and by sacrificial anodes? (c) Suggest two metals that could be used in place of zinc for galvanizing iron.

18.71 (a) Suggest two metals that could be used for the cathodic protection of a titanium pipeline. (b) What factors other than relative positions in the electrochemical series need to be considered in practice? (c) Often copper piping is connected to iron pipes in household plumbing systems. What is a possible effect of the copper on the iron pipes?

18.72 (a) Could aluminum be used for the cathodic protection of an underground storage container? (b) Which of the metals zinc, silver, copper, and magnesium cannot be used as a sacrificial anode in the protection of a buried iron pipeline? Explain your answer. (c) What is the electrolyte solution for the cathodic protection of an underground pipeline by a sacrificial anode?

Electrolysis

For the exercises in this section, base your answers on the potentials listed in Table 18.1 or Appendix 2B, with the exception of the reduction and oxidation of water at pH = 7:

$$2 H_2O(l) + 2 e^- \longrightarrow H_2(g) + 2 OH^-(aq),$$
$$E = -0.42 \text{ V at pH} = 7$$

$$O_2(g) + 4 H^+(aq) + 4 e^- \longrightarrow 2 H_2O(l),$$
$$E = +0.82 \text{ V at pH} = 7$$

Ignore other factors such as passivation or overpotential.

18.73 Complete the following statements. (a) In an electrolytic cell, oxidation occurs at the (anode, cathode). (b) The anode is the (positive, negative) electrode.

18.74 Complete the following statement. In an electrolytic cell, the anions migrate toward the (anode, cathode) and the electrons flow from the (anode, cathode) to the (anode, cathode).

18.75 A 1 M $CoSO_4$(aq) solution was electrolyzed, using inert electrodes. Write (a) the cathode reaction; (b) the anode reaction. (c) Assuming no overpotential or passivity at the electrodes, what is the minimum potential that must be supplied to the cell for the onset of electrolysis?

18.76 A 1 M CsI(aq) solution was electrolyzed, using inert electrodes. Write (a) the cathode reaction; (b) the anode reaction. (c) Assuming no overpotential or passivity at the electrodes, what is the minimum potential that must be supplied to the cell for the onset of electrolysis?

18.77 Write the half-reaction and specify the electrode at which each of the following processes occurs in an electrolytic cell: (a) the deposition of copper from a Cu^{2+} solution; (b) the production of sodium metal in the Downs cell; (c) the production of chlorine gas in the Downs cell; (d) the production of hydrogen gas from water.

18.78 Write the half-reaction and specify the electrode at which each of the following processes occurs in an electrolytic cell: (a) the production of aluminum from molten Al_2O_3; (b) the electroplating of silver onto a spoon; (c) the production of oxygen gas from an acidic aqueous solution; (d) the production of hypochlorite ions from brine (concentrated aqueous sodium chloride solution).

18.79 Aqueous solutions of (a) Mn^{2+}; (b) Al^{3+}; (c) Ni^{2+}; (d) Au^{3+} were electrolyzed. Determine whether the metal ion or water will be reduced at the cathode.

18.80 The anode of an electrolytic cell was constructed from (a) Cr; (b) Pt; (c) Cu; (d) Ni. Determine whether oxidation of the electrode or oxidation of water will occur at the anode.

18.81 Determine the amount (in moles) of electrons needed to produce the indicated substance in an electrolytic cell: (a) 5.12 g of copper from a copper(II) sulfate solution; (b) 200. g of aluminum from molten aluminum oxide dissolved in cryolite; (c) 200. L of oxygen gas at 273 K and 1.00 atm from an aqueous sodium sulfate solution.

18.82 A total charge of 96.5 kC is passed through an electrolytic cell. Determine the quantity of substance produced in each case: (a) the mass (in grams) of silver metal from a silver nitrate solution; (b) the volume (in liters at 273 K and 1.00 atm) of chlorine gas from a brine solution (concentrated aqueous sodium chloride solution); (c) the mass of copper (in grams) from a copper(II) chloride solution.

18.83 (a) How much time is required to electroplate 4.4 mg of silver from a silver nitrate solution, using a current of 0.50 A? (b) When the same current is used for the same length of time, what mass of copper could be electroplated from a copper(II) sulfate solution?

18.84 (a) When a current of 150 mA is used for 8.0 h, what volume (in liters at 273 K and 1.0 atm) of fluorine gas can be produced from a molten mixture of potassium and hydrogen fluorides? (b) With the same current and time period, how many liters of oxygen gas at 273 K and 1.0 atm could be produced from the electrolysis of water?

18.85 (a) What current is required to produce 4.0 g of chromium metal from chromium(VI) oxide in 24 h? (b) What current is required to produce 4.0 g of sodium metal from molten sodium chloride during the same period?

18.86 What current is required to electroplate 6.66 μg of gold in 30.0 min from a gold(III) chloride aqueous solution? (b) How much time is required to electroplate 6.66 μg of chromium from a potassium dichromate solution, using a current of 100. mA?

18.87 When a titanium chloride solution was electrolyzed for 500. s with a 120.-mA current, 15.0 mg of titanium was deposited. What is the oxidation number of the titanium in the titanium chloride?

18.88 A 0.26-g sample of mercury was produced from a mercury nitrate aqueous solution when a current of 210. mA was applied for 1200. s. What is the oxidation number of the mercury in the mercury nitrate?

18.89 Thomas Edison was faced with the problem of measuring the electricity that each of his customers had used. His first solution was to use a zinc "coulometer," an electrolytic cell in which the quantity of electricity is determined by measuring the mass of zinc deposited. Only some of the current used by the customer passed through the coulometer. What mass of zinc would be deposited in one month (of 31 days) if 1.0 mA of current passed through the cell continuously?

18.90 An alternative solution to the problem described in Exercise 18.89 is to collect the hydrogen produced by electrolysis and measure its volume. What volume would be collected at 273 K and 1.00 atm under the same conditions?

Supplementary Exercises

18.91 Balance the skeletal equation for the following half-reactions and state whether each half-reaction is an oxidation or a reduction.
(a) $I^-(aq) \rightarrow I_3^-(aq)$
(b) $SeO_4^{2-}(aq) + H_2O(l) \rightarrow SeO_3^{2-}(aq) + OH^-(aq)$

18.92 A galvanic cell has the cell reaction, $M(s) + 2 Zn^{2+}(aq) \rightarrow 2 Zn(s) + M^{4+}(aq)$. The standard potential of the cell is 0.16 V. What is the standard potential of the M^{4+}/M redox couple?

18.93 Suppose the reference electrode for Table 18.1 were the standard calomel electrode, $Hg_2Cl_2/Hg,Cl^-$, with $E°$ for it

set equal to 0. Under this system, what would be the standard potential for (a) the hydrogen electrode; (b) the Cu^{2+}/Cu redox couple?

18.94 Calculate (a) the standard cell potential; (b) the standard free energy; (c) the equilibrium constant of the reaction $2 MnO_4^-(aq) + 16 H^+(aq) + 5 Sn^{2+}(aq) \rightarrow 5 Sn^{4+}(aq) + 2 Mn^{2+}(aq) + 8 H_2O(l)$.

18.95 Gold metal can be oxidized to gold (III) by permanganate ions but not by dichromate ions in an acidic solution. Explain this observation.

18.96 What is the standard potential for the reduction of oxygen to water in (a) an acidic solution? (b) a basic solution? (c) Is MnO_4^{2-} more stable in an acidic or a basic aerated solution (a solution saturated with oxygen gas at 1 atm)? Explain your conclusion.

18.97 For each reaction that is spontaneous under standard conditions, determine the standard cell potential, and calculate $\Delta G_r°$.
(a) $2 NO_3^-(aq) + 8 H^+(aq) + 6 Hg(l) \rightarrow$
$\qquad 3 Hg_2^{2+}(aq) + 2 NO(g) + 4 H_2O(l)$
(b) $2 Hg^{2+}(aq) + 2 Br^-(aq) \rightarrow Hg_2^{2+}(aq) + Br_2(l)$
(c) $Cr_2O_7^{2-}(aq) + 14 H^+(aq) + 6 Pu^{3+}(aq) \rightarrow$
$\qquad 6 Pu^{4+}(aq) + 2 Cr^{3+}(aq) + 7 H_2O(l)$

18.98 Use standard potentials to determine the equilibrium constant for each of the following reactions:
(a) $PbSO_4(s) \rightarrow Pb^{2+}(aq) + SO_4^{2-}(aq)$
(b) $2 Pb^{2+}(aq) \rightarrow Pb(s) + Pb^{4+}(aq)$
(c) $Hg_2Cl_2(s) \rightarrow Hg_2^{2+}(aq) + 2 Cl^-(aq)$
(d) $2 V^{3+}(aq) + Co(s) \rightarrow Co^{2+}(aq) + 2 V^{2+}(aq)$

18.99 Use standard potential data to calculate the solubility of (a) $Cd(OH)_2(s)$; (b) $Hg_2Cl_2(s)$; (c) $PbSO_4(s)$.

18.100 A technical handbook contains tables of thermodynamic quantities for common reactions. If you want to know whether a certain reaction is spontaneous under standard conditions, which of the following properties would give you that information directly (on inspection)? Which would not? Explain. (a) $\Delta G_r°$; (b) $\Delta H_r°$; (c) $\Delta S_r°$; (d) $\Delta U_r°$; (e) $E°$; (f) K.

18.101 The following reaction is carried out in a galvanic cell: $Fe^{2+}(aq) + Ag^+(aq) \rightarrow Fe^{3+}(aq) + Ag(s)$. (a) Calculate the standard cell potential. (b) Calculate the potential of the cell when the following concentrations are present: 1.0 mol/L $Fe^{3+}(aq)$, 0.0010 mol/L $Fe^{2+}(aq)$, and 0.010 mol/L $Ag^+(aq)$. Compare and comment on your answers.

18.102 The following items are obtained from the stockroom for the construction of a galvanic cell: two 250-mL beakers, a salt bridge, a voltmeter with attached wires and clips, 200. mL of 0.010 M $CrCl_3(aq)$, 200. mL of 0.16 M $CuSO_4(aq)$, a piece of copper wire, and a chrome-plated piece of metal.
(a) Describe the construction of the galvanic cell. (b) Write the anode and cathode half-reactions. (c) Write the cell

reaction. (d) Write the cell diagram for the galvanic cell. (e) What is the expected cell potential?

18.103 The potential of a galvanic cell with the cell reaction $Zn(s) + Pb^{2+}(aq) \rightarrow Zn^{2+}(aq) + Pb(s)$ is 0.66 V when the concentration of $Pb^{2+}(aq)$ ions is 0.10 mol/L. (a) Use standard potentials to calculate the standard cell potential for the reaction. (b) What is the molar concentration of Zn^{2+} ions in the cell?

18.104 What are the charge-carrying species in (a) the external circuit and (b) the internal circuit of a galvanic cell?

18.105 A current of 15.0 A electroplated 50.0 g of hafnium metal from an aqueous solution in 2.00 h. What was the oxidation number of hafnium in the solution?

18.106 A mass loss of 12.57 g occurred in 6.00 h at a titanium anode when a current of 4.70 A was used in an electrolytic cell. What is the oxidation number of the titanium in solution?

18.107 A piece of copper metal is to be electroplated on all sides with silver to a thickness of 1.0 μm. If the metal strip measures 50.0 mm × 10.0 mm × 1.0 mm, how long must the solution, which contains $Ag(CN)_2^-$ ions, be electrolyzed, using a current of 100.0 mA? The density of silver metal is 10.5 g/cm³.

18.108 It was decided to electroplate copper metal from 1 M $CuSO_4(aq)$, using a platinum anode. Assuming no overpotential or passivity at the electrodes, what is the minimum potential that must be supplied to the electrolytic cell for the onset of the electroplating of copper metal?

18.109 A metal forms the salt MCl_3. Electrolysis of the molten salt with a current of 0.70 A for 6.63 h produced 3.00 g of the metal. (a) What is the molar mass of the metal? (b) What is the identity of the metal?

18.110 Consider the galvanic cell $Pt(s)|Sn^{4+}(aq, 0.010 \text{ mol/L})$, $Sn^{2+}(aq, 0.10 \text{ mol/L})\|O_2(g, 1 \text{ atm})|H^+(aq, pH = 4.00)|C(gr)$. (a) What is the standard cell potential? (b) Write the cell reaction. (c) Calculate the actual cell potential. (d) Determine the equilibrium constant of the cell reaction. (e) Calculate the free energy of the cell reaction. (f) Suppose the cell potential measured 0.89 V. Assuming all other concentrations are precise, what is a more accurate pH of the solution?

18.111 Show how a silver-silver chloride electrode (silver in contact with solid AgCl and a solution of Cl^- ions) and a hydrogen electrode can be used to measure (a) pH; (b) pOH. See Investigating Matter 18.1.

18.112 When a pH meter was standardized with a boric acid-borate buffer with a pH of 9.40, the cell potential was 0.060 V. When the buffer was replaced with a solution of unknown hydronium ion concentration, the cell potential was 0.22 V. What is the pH in the solution? See Investigating Matter 18.1.

18.113 (a) How many moles of H_3O^+ ion are produced at a platinum anode in the electrolysis of 200.0 mL of a $CuSO_4$ aqueous solution, using a current of 4.00 A for 30.0 min? (b) If the pH of the solution was initially 7.0, what will the pH of the solution be after the electrolysis? Assume no volume change.

18.114 In the electrolytic refining of copper, blister copper is used as the anode and oxidized. The copper(II) ion that is produced from its oxidation is then reduced at the cathode to give a metal with a much higher purity. The impurities in the blister copper include iron, nickel, silver, gold, cobalt, and trace amounts of other metals. The material that is not oxidized at the anode falls to the bottom of the electrolytic cell and is called "anode mud." What are some of the components of the anode mud? Explain your choices.

Applied Exercises

For Exercises 18.115–18.117, see Applying Chemistry: Case Study 18.

18.115 A fuel cell in which hydrogen reacts with nitrogen instead of oxygen is proposed. (a) Write the chemical equation for the reaction, assuming that aqueous ammonia is the only product. (b) What would be the change in free energy for the consumption of 28.0 kg of nitrogen? (c) Is this type of fuel cell thermodynamically feasible?

18.116 The "aluminum-air fuel cell" is used as a reserve battery in remote locations. In this cell, aluminum reacts with the oxygen in air in basic solution. (a) Write the oxidation and reduction half-reactions for this cell. (b) Calculate the standard cell potential.

18.117 The body functions as a kind of fuel cell that uses oxygen from the air to oxidize glucose:

$$C_6H_{12}O_6(aq) + 6 O_2(g) \longrightarrow 6 CO_2(g) + 6 H_2O(l)$$

During normal activity, a person uses the equivalent of about 10 MJ of energy a day. Assuming this value represents ΔG, estimate the average current through your body in the course of a day, assuming that all the energy we use arises from the reduction of O_2 in the glucose oxidation reaction.

18.118 A photoelectrochemical cell is an electrochemical cell that uses light to carry out a chemical reaction. This type of cell is being considered for the production of hydrogen from water. The silicon electrodes in a photoelectrochemical cell react with water:

$$SiO_2(s) + 4 H^+(aq) + 4 e^- \longrightarrow Si(s) + 2 H_2O(l)$$
$$E° = -0.84 \text{ V}$$

Calculate the standard cell potential for a photoelectrochemical cell that produces hydrogen from water and in which silicon reacts with water.

An experimental photoelectrochemical cell (Exercise 18.118).

Suggest a reason why, when you accidentally bite on a piece of aluminum foil with a tooth containing a silver filling, you may feel pain. Write a balanced chemical equation to support your suggestion.

18.121 What kind of system (open, closed, or isolated) is each of the following cells: (a) dry cell; (b) fuel cell; (c) nicad battery?

18.122 One stage in the extraction of gold from rocks involves dissolving the metal from the rock with a basic solution of sodium cyanide that has been thoroughly aerated. This stage results in the formation of soluble $Au(CN)_2^-$ ions. The next stage is to precipitate the gold by the addition of zinc dust, forming $Zn(CN)_4^{2-}$. Write the balanced equations for the half-reactions and the overall redox equation for both stages.

Integrated Exercises

18.123 Given the data in Appendix 2B and the fact that, for the half-reaction $F_2(g) + 2H^+(aq) + 2e^- \rightarrow 2HF(aq)$, $E° = +3.03$ V, calculate the value of K_a for HF.

18.124 An aqueous solution of Na_2SO_4 was electrolyzed for 30.0 min; 25.0 mL of oxygen was collected at the anode over water at 22°C at a total pressure of 722 Torr. Determine the current that was used to produce the gas. See Table 5.4 for the vapor pressure of water.

18.125 A brine solution is electrolyzed, using a current of 2.0 A. How much time is required to collect 20.0 L of chlorine if the gas is collected over water at 20.0°C and the total pressure is 770. Torr? Assume that the water is already saturated with chlorine, so no more dissolves. See Table 5.4 for the vapor pressure of water.

18.126 Determine the current required to produce 15.0 L of chlorine at 273 K and 1.00 atm from molten sodium chloride in 1.0 h.

18.127 In the Dow process for the production of magnesium, a current of 10. kA is used. (a) What mass (in kilograms) of magnesium can be produced in 24 hours? (b) What volume of chlorine gas at 273 K and 1.00 atm is produced during the same period?

18.119 In a neuron (a nerve cell), the concentration of K^+ ions inside the cell is about 20 to 30 times that outside. What potential difference between the inside and the outside of the cell would you expect to measure if the difference is due only to the imbalance of potassium ions?

18.120 Dental amalgam, a solid solution of silver and tin in mercury, is sometimes used for filling tooth cavities. Two of the reduction half-reactions that the filling can undergo are

$$3Hg_2^{2+}(aq) + 4Ag(s) + 6e^- \longrightarrow 2Ag_2Hg_3(s),$$
$$E° = +0.85 \text{ V}$$

$$Sn^{2+}(aq) + 3Ag(s) + 2e^- \longrightarrow Ag_3Sn(s), E° = -0.05 \text{ V}$$

Connection 5

The automobile you buy in a few years time may have a very different engine from the ones used today. The energy to run your engine may come from a battery; and when the battery runs down, you may take it to a service station and exchange it for a freshly charged one or simply plug it into an electric outlet. Research on electric cars and vans has been growing rapidly, primarily because of the need to reduce air pollution (see Connection 2, following Chapter 9), but also to increase the efficiency of automobile engines. It is alarming to realize that 80% of the energy generated by the gasoline burning in a typical internal combustion engine is wasted. Such engines are only 20% efficient.

The development of electric vehicles and vehicles powered by fuel cells experienced rapid growth when California passed a law requiring that by 2000, 5% of new vehicles must be zero-emission. "Zero-emission" means that they produce no pollution. Vehicles powered by rechargeable batteries or hydrogen fuel cells are considered to be the most promising solutions.

The most important factor in developing an electric vehicle is the battery that provides the power. An ideal battery for an electric vehicle should meet the following criteria:

High specific energy (large energy production per kilogram)

High energy density (large energy production per liter)

Long range of operation per charge

Long cycle life (many cycles of recharging without loss of potential)

Environmentally benign when discarded

Inexpensive to operate

Both the mass and volume of a battery are critical parameters. The electrolyte in a battery uses as little water as possible, both to reduce leakage of the electrolyte and to keep the mass low. Much of the research on batteries is devoted to raising the *specific energy,* the reaction free energy per kilogram (typically expressed in kilowatt-hours per kilogram, $kW \cdot h/kg$) or the *energy density,* the reaction free energy per liter (in kilowatt hours per liter, $kW \cdot h/L$).* The battery in a vehicle consists of a large number of separate cells (see illustration). The potential difference generated by the cells depends on the choice of reactants and their concentrations and also on the number of cells used.

The batteries used in most portable computers and automobiles are secondary cells, which can be recharged (see Section 18.11). In the charging process, an external source of

*$1 \, kW \cdot h = (10^3 \, J/s) \times (3600 \, s) = 3.6 \times 10^6 \, J = 3.6 \, MJ$ exactly.

The EV1, an electric vehicle manufactured by General Motors. The automobile is available with either advanced lead-acid batteries or NiMH batteries. It has an aluminum frame to reduce weight and a range of about 80 miles.

electricity reverses the spontaneous cell reaction and creates a nonequilibrium mixture of reactants. After charging, the cell can again produce electricity as the reaction once more sinks back toward equilibrium.

The *lead-acid cell* of an automobile battery is described in Section 18.11. It is inexpensive, has a long shelf life, and recharges well. However, it releases lead to the environment when discarded and sometimes generates hydrogen gas at the cathode when charging and hence can generate an explosive mixture of gases. It also has a low specific energy due to lead's high mass density.

The *nickel-cadmium (nicad) cell* is a secondary cell that is also discussed in Section 18.11. It has a high energy density, recharges rapidly, and is relatively inexpensive. It is used for portable equipment such as camcorders, cellular phones, and laptop computers. However, it has a low cell potential, short shelf life, and loses its ability to maintain its maximum potential after several recharges. When discarded, it releases cadmium, a toxic heavy metal, into the environment.

The *nickel-metal hydride (NiMH) cell* makes use of newer technology to replace the cadmium in a nicad cell with hydrogen. The hydrogen is stored as a metal hydride, using an alloy of several metals, commonly including titanium, vanadium, chromium, and nickel. It is used in electronic equipment such as laptop computers and cellular phones. This battery has a lower mass than nicad batteries of similar storage capacity, and hence higher energy density. It charges rapidly, has a long cycle life, and is nontoxic. However, it is sensitive to overcharging and is expensive.

The *lithium-ion cell* and *lithium-polymer cell* are displacing NiCd and NiMH cells in some applications, particularly laptop computers, because they have a very long cycle life. The lithium-ion cell makes use of a special electrolyte in which polypropylene oxide or polyethylene oxide is dissolved in low-melting-point mixtures of lithium salts, and then allowed to cool. The resulting rubbery materials serve as good conductors of Li^+ ions at normal temperatures. The low mass density of lithium gives it a very high energy density, and lithium's very negative electrode potential provides a high voltage. The battery holds a charge up to 10 years, charges rapidly, and can be recharged up to 1000 times. However, it is expensive and sensitive to overcharging. Lithium is also highly reactive and could cause injuries if the case is broken.

The *zinc-air cell* is being developed for applications in which a high specific energy is needed. The cathode in this cell makes use of the oxygen in air, greatly reducing the mass of the battery and allowing very thin, flat, "wafer" shapes. The cell can be made in large or small sizes and, as it is so cheap, is being intensively studied for use in electric cars. The anode is zinc, the electrolyte is a solution of potassium hydroxide, and the cathode is a very thin layer of metal with air holes that can be sealed when the battery is not in use, to prevent unwanted reaction with air. The zinc-air battery has a very high energy density and is nontoxic and inexpensive. However, it has a short cycle life, takes a long time to recharge, and is sensitive to overcharging.

A *sodium-sulfur cell* is one of the more startling batteries. It has liquid reactants (sodium and sulfur) and a solid electrolyte (a porous aluminum oxide ceramic); it operates at a temperature of about 320°C; and it is highly dangerous in case of breakage. In its most common application, to power electric vehicles, 2000 sodium-sulfur cells are placed inside an insulated container. Once the vehicle is operating, the heat generated by the battery is sufficient to maintain the temperature and to keep it hot for several hours after the car stops. The battery has a very high specific energy, due to the low mass density of sodium. However, it is expensive, six hours are required to recharge the battery, the cycle life is short, and sodium is highly reactive and could cause injuries if the case is broken.

For Further Reading

D. Sperling, The case for electric vehicles, *Scientific American,* November, 1996, pp. 54–59.

S. L. Wilkinson, Electric vehicles gear up, *Chemical and Engineering News,* October 13, 1997, pp. 18–24.

Applying Your Knowledge

You may need to consult Chapters 17 and 18 and occasionally earlier chapters in order to answer these questions.

1. Suppose that the lithium-polymer battery becomes the major source of power in automobiles and then a shortage of lithium arises. Consider the properties of lithium that make it useful in batteries for electric vehicles and suggest a metal you could substitute for lithium. Explain how you made your decision.

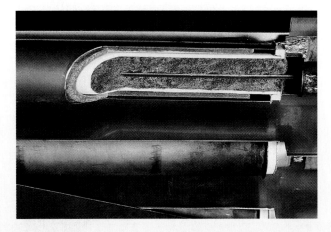

One of the sodium-sulfur cells used to power electric vehicles. The cutaway shows that the sodium is surrounded by a ceramic electrolyte, then by a layer of sulfur. The battery must be heated until both the sodium and sulfur are molten in order to operate.

2. The hydrogen fuel cell is a nonpolluting power source for vehicles. (a) Calculate the specific energy of an alkali fuel cell from its enthalpy of reaction and the mass of hydrogen required to generate that enthalpy (see Applying Chemistry: Case Study 18), using 1 W·h = 3600 J exactly. Assume that the oxygen is obtained from the air. Compare the specific energy of an alkali fuel cell with those of the batteries in the data table on the CD or web site for the book. (b) Calculate the energy density of the alkali fuel cell in watt-hours per liter (W·h/L), assuming that the hydrogen is at 16 atm and 25°C and compare this value to those of the same batteries. (c) In what situations would an alkali fuel cell be a better choice for a vehicle than a rechargeable battery?

3. Which of the batteries described in Connection 5 would be your first choice for each of the following situations: (a) low-cost operation; (b) a small automobile with a very small space for the battery; (c) a long journey; (d) space travel, for which the mass must be kept at a minimum; (e) your own personal automobile? In each case, justify your answer.

4. You have been hired as part of a team to develop an electric automobile using a new kind of rechargeable battery. Use information in the Appendix to design a battery in which each cell has a potential of at least 1 V. You will want to reduce mass as much as possible, so consider the size of the automobile and its construction materials in your design. Indicate whether the automobile would be used primarily as a long-range family car, for short-range commuting, or for sport, and give it a name.

5. Electric vehicles can be used to reduce pollution, but have other environmental impacts. What would you expect to be the major environmental or safety disadvantages of the following batteries: (a) lead-acid; (b) sodium-sulfur; (c) nickel-cadmium?

Chapter 19

The Elements: The First Four Main Groups

There were 113 elements known when this text was written, and there may be more by the time you read it. The elements combine to form millions of compounds, and hundreds of new compounds are made or discovered every year. In the next three chapters, we get to know a small selection of the elements and see how their properties are related to their location in the periodic table. Once we have learned to interpret the periodic table, we shall be in a good position to predict the properties of an element from its location and shall come to see why chemists think of the periodic table as one of the great unifying concepts of chemistry.

Each group in the periodic table has its own unique characteristics. However, the main groups have certain features in common:

1. The members of a group have analogous valence-shell electron configurations.

2. The element at the head of a group (the lightest element in the group) often has a character that is quite distinct from the other members of the group.

3. There are diagonal relationships between elements, especially between Periods 2 and 3.

4. There is a trend toward metallic character on going to the left along a period and on going down a group.

This work of art is being created from a glass formed from a mixture of ordinary sand, sodium carbonate, and other minerals. By varying the composition of the mixture, the glass blower can give the glass different colors and different properties. Selecting the right material, whether for glass or for chemical reactions, requires knowing something about the properties of the elements.

PERIODIC TRENDS

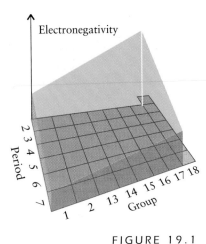

FIGURE 19.1

The electronegativities of the elements tend to increase from left to right across a period and decrease down a group. This diagram is a highly schematic representation of those trends.

> Many of these binary compounds have hydrogen in its +1 oxidation state, so the name "hydride" is not really appropriate. However, it is the conventional term.

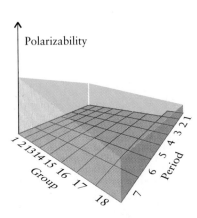

FIGURE 19.2

The polarizabilities of the atoms of the elements tend to decrease from left to right across a period and increase down a group. This diagram is a highly schematic representation of those trends.

In Chapter 7 we saw that periodic trends in atomic properties can be explained by trends in electron configurations. Here we look at periodic trends in chemical properties. The valence-shell electron configuration of the atoms of each element is similar for all members of a group. However, partly because its atoms are so small and have no d-orbitals available, the element at the head of each main group (the lightest element in the group) often has a character that is quite distinct from the other members of the group. We also have to remember (from Section 7.19) that there are diagonal relationships between elements (especially between Periods 2 and 3) and that there is a trend toward metallic character on going to the left along a period and down a group.

Trends in atomic radius, ionization energy, and electron affinity are summarized in Toolbox 7.2. The lowest ionization energies are found toward the lower left of the periodic table, near cesium, so we expect these elements to be metals and to exist as cations in compounds. Trends in the consequences of electronegativity and polarizability for the covalent or ionic character of bonds are summarized in Sections 8.14–8.15. When judging these trends, bear in mind that electronegativities typically increase from left to right across a period and decrease down a group (Fig. 19.1; see also Fig. 8.21). Polarizability typically decreases from left to right along a period and increases down a group (Fig. 19.2).

19.1 Chemical Properties: Hydrides

The periodic trends in main-group elements become apparent when we compare the binary compounds they form with one specific element. All the main-group elements, with the exception of the noble gases and, possibly, indium and thallium, form binary compounds with hydrogen, so these **binary hydrides** can be examined to look for periodic trends.

The formulas of binary hydrides are related directly to the group number and reveal the typical valences of the elements (Fig. 19.3). For instance, carbon (Group 14) forms CH_4, nitrogen (Group 15) forms NH_3, oxygen (Group 16) forms H_2O, and fluorine (Group 17) forms HF.

The nature of a particular binary hydride is related to the characteristics of its parent element (Fig. 19.4). Strongly electropositive metallic elements form ionic compounds with hydrogen in which the latter is present as a hydride ion, H^-, and has oxidation number -1. These ionic compounds are called **saline hydrides** (or "saltlike hydrides"). They are formed by all members of the s block with the exception of beryllium, and most are made by heating the metal in hydrogen:

$$2\,K(s) + H_2(g) \longrightarrow 2\,KH(s)$$

The saline hydrides are white, high-melting-point solids with crystal structures that resemble those of the corresponding halides. The alkali metal hydrides, for instance, have the rock-salt structure (see Fig. 10.32).

The **metallic hydrides** are black, powdery, electrically conducting solids. They are formed by heating certain of the d-block metals in hydrogen (Fig. 19.5):

$$2\,Cu(s) + H_2(g) \xrightarrow{\Delta} 2\,CuH(s)$$

Because the metallic hydrides release their hydrogen (as H_2) when heated or treated with acid, they are being investigated for storing and transporting hydrogen. Both saline and metallic hydrides have the high enthalpy densities desirable in a portable fuel.

Nonmetals form covalent **molecular hydrides,** which consist of discrete molecules. These compounds have low melting points and are volatile; many are Brønsted acids. Most are gases such as ammonia, the hydrogen halides (HF, HCl, HBr, HI), and the lighter hydrocarbons such as methane, ethane, ethene, and ethyne. Liquid compounds include water and hydrocarbons such as octane and benzene.

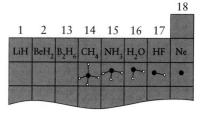

FIGURE 19.3

The binary hydrides show periodic trends in properties. Hydrides of s-block metals are classified as saline, those of the p block are mainly molecular, and those of the d block are metallic.

The chemical formulas of the hydrides of the elements in the main groups display the typical valences of the elements. The formula of the simplest boron hydride, borane, shows it to have the empirical formula BH_3.

19.2 Chemical Properties: Oxides

All the main-group elements except the noble gases react with oxygen. Like hydrides, the resulting oxides also reveal periodic trends in the elements. For instance, we can use their properties to classify elements as metals or nonmetals: oxides of main-group metals are basic, and oxides of nonmetals are acidic. The trend in bonding type is from soluble ionic oxides on the left of the periodic table, through insoluble, high-melting-point oxides on the left of the *p* block, to low-melting-point and often gaseous molecular oxides on the right.

Metallic elements with low ionization energies commonly form ionic oxides. As remarked in Section 15.1, the oxide ion is a strong base, so the oxides of most of these metals form basic solutions in water. Magnesium is an exception because its oxide, MgO, is only very slightly soluble. However, even this oxide reacts with acids, so it is classified as basic. Elements with intermediate ionization energies, such as beryllium, aluminum, and most of the metalloids, form amphoteric oxides. These oxides do not react with water, but many dissolve in acidic or basic solutions.

Xenon forms an oxide, but by an indirect route.

Recall from Section 3.12 that amphoteric substances react with both acids and bases.

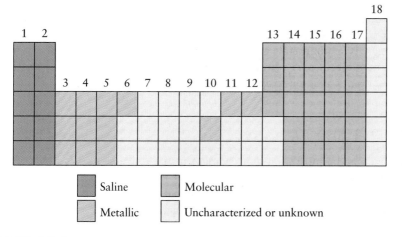

FIGURE 19.4

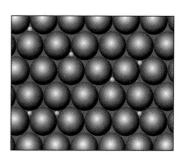

FIGURE 19.5

The different classes of binary hydrogen compounds and their distribution throughout the periodic table.

In a metallic hydride, the small hydrogen atoms (the small light gray spheres) occupy gaps—called interstices—between the larger metal atoms (the large dark gray spheres).

Many oxides of nonmetals are gaseous molecular compounds, such as CO_2, NO, and SO_3. Most can act as Lewis acids, because the electronegative oxygen atoms withdraw electrons from the central atom, enabling it to act as an electron pair acceptor. For instance, carbon dioxide can react with the oxides of metals because the oxide ion is a strong Lewis base:

$$CO_2(g) \; + \; Na_2O(s) \longrightarrow Na_2CO_3(s)$$

Many oxides of nonmetals form acidic solutions in water; the familiar laboratory acids HNO_3 and H_2SO_4, for instance, are derived from acidic binary oxides, which are therefore classified as acid anhydrides. An **acid anhydride** is an oxide that differs from an acid by the removal of the elements of water (H, H, and O) and, when hydrated, forms the acid. Thus, SO_3 is the anhydride of sulfuric acid. A **formal anhydride** of an acid is the molecule obtained by striking out the elements of water from the molecular formula of the acid, but that molecule does not react with water to form the acid. Carbon monoxide, for instance, is the formal anhydride of formic acid, HCOOH, although it does not react with water to form the acid. The acidity of oxoacids depends on the electronegativity of the nonmetal at the center of the O atoms and on its oxidation number: the higher the electronegativity and the higher the oxidation number, the stronger the acid (Section 15.10).

The oxides of most metalloids and some of the less electropositive elements are amphoteric. Aluminum oxide, for instance, reacts with acids and with aqueous solutions of bases. The oxides reveal a strong diagonal relationship between beryllium and aluminum, for beryllium oxide is also amphoteric.

The oxides of main-group elements show periodic trends in properties. Oxides of metals tend to be ionic and to form basic solutions in water. Oxides of nonmetals are molecular and some are the anhydrides of acids.

HYDROGEN

Hydrogen atoms are the simplest of all atoms, for most hydrogen atoms are made up of just two subatomic particles, a proton and an electron. Although it has the same valence electron configuration as the Group 1 elements, ns^1, hydrogen is a nonmetal; and in the elemental state, it is found as gaseous diatomic molecules, H_2. It has few similarities to the alkali metals; and in this text, we do not assign it to any group.

19.3 The Element

Hydrogen is the most abundant element in the universe. It was formed within the first few seconds after the Big Bang that most scientists believe marked the beginning of the universe. However, even though 89% of all atoms in the universe are thought to be hydrogen atoms, there is little free hydrogen on Earth because H_2 molecules are very light. As a result, they move at such high average speeds that, unless trapped underground or sealed in a container, they soon escape from Earth's gravity. It takes heavier atoms like oxygen to anchor hydrogen to the planet in compounds. A lot of Earth's hydrogen is present as water, H_2O, either in the oceans or trapped inside minerals.

The three isotopes of hydrogen were introduced in Section 1.4.

Natural supplies of hydrogen gas are far too small to satisfy the needs of industry, so most commercial hydrogen is obtained as a by-product of petroleum refining in a series of two catalyzed reactions. The first is the *re-forming reaction,* in which a hydrocarbon and steam are converted to carbon monoxide and hydrogen over a nickel catalyst:

$$CH_4(g) + H_2O(g) \xrightarrow{Ni} CO(g) + 3 H_2(g)$$

The mixture of products, which is called *synthesis gas,* is the starting point for the manufacture of numerous compounds, including methanol. The re-forming reaction is followed by the *shift reaction,* in which the carbon monoxide in the synthesis gas reacts with more water:

$$CO(g) + H_2O(g) \xrightarrow{Fe/Cu} CO_2(g) + H_2(g)$$

where Fe/Cu denotes a catalyst made of iron and copper.

Hydrogen is also produced by the electrolysis of water, but that process is economical only where electricity is cheap. Chemists are currently seeking ways of using sunlight to drive the water-splitting reaction, the photochemical decomposition of water into its elements:

$$2 H_2O(l) \longrightarrow 2 H_2(g) + O_2(g)$$

Hydrogen is prepared in the laboratory by reducing hydrogen ions with a metal having a negative standard potential, such as zinc:

$$Zn(s) + 2 HCl(aq) \longrightarrow ZnCl_2(aq) + H_2(g)$$

Hydrogen is a colorless, odorless, tasteless gas (Table 19.1). Because H_2 molecules are nonpolar, they attract each other only by London forces. Each molecule has only two electrons and hence only a very small instantaneous electric dipole, so these forces are very weak. In fact, its intermolecular interactions are so weak that it is almost completely insoluble in polar solvents such as water and does not condense to a liquid until it is cooled to 20 K. One striking physical property of liquid hydrogen is its very low density (0.071 g/cm^3), which is less than one-tenth that of water (Fig. 19.6). This low density makes hydrogen a very lightweight fuel. Hydrogen has the highest specific enthalpy of any known fuel, so liquid hydrogen is used to power the space shuttle's main rocket engines (Fig. 19.7).

Each year, about half the 3×10^8 kg of hydrogen used in industry is converted into ammonia by the Haber process (Section 14.12). Through the reactions of ammonia, hydrogen finds its way into numerous other important

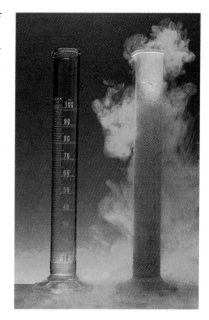

FIGURE 19.6

The two measuring cylinders contain the same mass of liquid. The liquid on the left is water, that on the right is liquid hydrogen, which is much less dense.

The specific enthalpy of rocket fuels is discussed further in Applying Chemistry: Case Study 20.

Table 19.1 Physical Properties of hydrogen

Valence configuration: $1s^1$
Normal form*: colorless, odorless gas

Z	Name	Symbol	Molar mass, g/mol	Abundance,[†] %	Melting point, °C	Boiling point, °C	Density, g/L
1	hydrogen	H	1.0079	99.98	−259 (14 K)	−253 (20. K)	0.089
1	deuterium	^{2}H or D	2.014	0.02	−254 (19 K)	−249 (24 K)	0.18
1	tritium	^{3}H or T	3.016	radioactive	−252 (21 K)	−248 (25 K)	0.27

*Normal form means the state and appearance of the element at 25°C and 1 atm. The density refers to the same conditions.

[†]The abundance is the fraction of atoms of each isotope, expressed as a percentage.

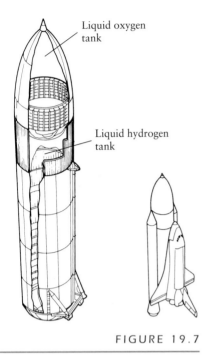

FIGURE 19.7

The arrangement of fuel tanks in the space shuttle and its booster. Note the large size of the hydrogen tank (bottom) relative to that of the oxygen tank (top). The high specific enthalpy of liquid hydrogen makes it a valuable fuel for space missions.

nitrogen compounds such as hydrazine and sodium amide (Section 20.2). Hydrogen also enters the economy through the production of methanol:

$$2\,H_2(g) + CO(g) \longrightarrow CH_3OH(l)$$

About a third of the hydrogen manufactured is used for the hydrometallurgical extraction of copper and other metals, the extraction from their ores by reduction in aqueous solution:

$$Cu^{2+}(aq) + H_2(g) \longrightarrow Cu(s) + 2\,H^+(aq)$$

In this process, ores containing copper(II) oxide and copper(II) sulfide are dissolved in sulfuric acid, and then hydrogen is bubbled through the solution. The reduction is thermodynamically favored, because the standard potential for $Cu^{2+}(aq) + 2\,e^- \rightarrow Cu(s)$ is positive ($E° = +0.34$ V). Metals with negative standard potentials, such as zinc ($E° = -0.76$ V) and nickel ($E° = -0.23$ V), cannot be extracted by hydrogen. In other words, hydrogen cannot reduce ions such as Zn^{2+} or Ni^{2+}.

Hydrogen is used in the food industry to convert vegetable oils into shortening (Fig. 19.8). There it is added to carbon-carbon double bonds in a *hydrogenation reaction*:

$$H_2(g) + \cdots C{=}C \cdots \longrightarrow \cdots CH{-}CH \cdots$$

This reaction converts a C=C double bond into a C—C single bond. Oil and fat molecules both have long hydrocarbon chains, but oils have more double bonds. Because double bonds resist twisting, oil molecules do not pack together well, so the result is a liquid. When the double bonds are replaced by single bonds, the chains become much more flexible, so the molecules pack together better and form a solid.

Hydrogen is produced as a by-product of the refining of fossil fuels and by electrolysis. It has a low density and weak intermolecular forces. Hydrogen is a good reducing agent for species with positive standard potentials.

19.4 Compounds of Hydrogen

Because hydrogen forms compounds with so many other elements (Table 19.2), we shall meet many of its compounds when we study the other elements. Here we consider only its binary compounds. As remarked in Section 19.1, hydrides are classified as saline, molecular, and metallic.

The saline hydrides contain the hydride ion, H^-. The hydride ion is very large, with a radius of 154 pm, between that of the fluoride and chloride ions. The single positive charge of the hydrogen atomic nucleus can barely manage to

FIGURE 19.8

When the runny oil (top) is hydrogenated, it is converted into a solid—a fat (bottom). The hydrogen converts carbon-carbon double bonds into single bonds. The resulting, more flexible molecules can pack together more closely and so form a solid.

Table 19.2 *Chemical properties of hydrogen*

Reactant	Reaction with hydrogen
Group 1 metals (M)	$2\,M(s) + H_2(g) \longrightarrow 2\,MH(s)$
Group 2 metals (M, not Be or Mg)	$M(s) + H_2(g) \longrightarrow MH_2(s)$
some d-block metals (M)	$2\,M(s) + x\,H_2(g) \longrightarrow 2\,MH_x(s)$
oxygen	$O_2(g) + 2\,H_2(g) \longrightarrow 2\,H_2O(l)$
nitrogen	$N_2(g) + 3\,H_2(g) \longrightarrow 2\,NH_3(g)$
halogen (X_2)	$X_2(g, l, s) + H_2(g) \longrightarrow 2\,HX(g)$

keep control over the two electrons in the H$^-$ ion, so they are easily lost. The low energy required for a hydride ion to lose an electron results in saline hydrides being very powerful reducing agents, and $E° = -2.25$ V for $H_2(g) + 2 e^- \rightarrow 2 H^-(aq)$. This value is similar to the standard potential of $Na^+(aq) + e^- \rightarrow Na(s)$, and, like sodium metal, hydride ions reduce water as soon as they come into contact with it:

$$NaH(s) + H_2O(l) \longrightarrow NaOH(aq) + H_2(g)$$

Because this reaction produces hydrogen, saline hydrides are potentially useful as transportable sources of hydrogen.

The metallic hydrides are so called because they are electrically conducting and retain other properties of metals. Many metallic hydrides act like interstitial alloys, in which hydrogen atoms occupy the interstices in the metal. However, the chemical interactions between the hydrogen atoms and metal atoms suggest that metallic hydrides are true compounds.

All the molecular hydrides are covalently bonded compounds. Some release hydrogen ions, so they are considered to be acids. The strength of the acid increases with the polarity of the bond to the acidic hydrogen atom and decreases as the strength of that bond increases. Thus, HF is an acid that can donate hydrogen ions to water molecules, but NH_3 is not (see Section 15.9). The molecular hydrides of nitrogen, oxygen, and fluorine take part in the very strong intermolecular interactions known as hydrogen bonding (see Section 10.3). Hydrogen bonding dominates the bulk physical properties of these compounds.

Some of these molecular hydrides can be prepared by **protonation,** or proton transfer from an acid to a base, such as S^{2-}:

$$FeS(s) + 2 HCl(aq) \longrightarrow FeCl_2(aq) + H_2S(g)$$

Volatile binary acids can be prepared by using a less volatile acid as the proton donor:

$$CaF_2(s) + H_2SO_4(l) \longrightarrow CaSO_4(s) + 2 HF(g)$$

The reaction proceeds to the right, because the volatile product is removed as a gas.

Saline hydrides are strong reducing agents. Some binary molecular compounds of hydrogen are characterized by acidic character.

Self-Test 19.1A What common compounds of hydrogen are more powerful reducing agents than hydrogen itself?

[*Answer:* The saline hydrides]

Self-Test 19.1B What are some of the properties that argue against the classification of hydrogen as a Group 1 element?

GROUP 1: THE ALKALI METALS

The members of Group 1 are called the *alkali metals.* Their valence electron configurations are ns^1, where n is the period number, and their physical and chemical properties are dominated by the ease with which the single valence electron can be removed (Tables 19.3 and 19.4).

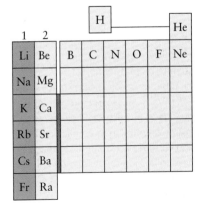

Table 19.3 *The Group 1 elements*

Valence configuration: ns^1
Normal form*: soft, silver-gray metals

Z	Name	Symbol	Molar mass, g/mol	Melting point, °C	Boiling point, °C	Density, g/cm³
3	lithium	Li	6.94	181	1347	0.53
11	sodium	Na	22.99	98	883	0.97
19	potassium	K	39.10	64	774	0.86
37	rubidium	Rb	85.47	39	688	1.53
55	cesium	Cs	132.91	28	678	1.87
87	francium†	Fr	223	27	677	—

*Normal form means the state and appearance of the element at 25°C and 1 atm.
†Radioactive.

19.5 The Elements

The alkali metals are the most violently active of all the metals. They are too easily oxidized to be found in the free state in nature. Common, cheap materials are too feeble as reducing agents to extract them from their compounds. The pure metals are obtained by electrolysis of their molten salts, as in the electrolytic Downs process (Section 18.16) or, in the case of potassium, by exposing

(a)

(b)

FIGURE 19.9

The alkali metals of Group 1:
(a) lithium; (b) sodium; (c) potassium;
(d) rubidium and cesium. Francium
has never been isolated in visible
quantities. The first three elements
corrode rapidly in moist air; rubidium
and cesium are even more reactive and
have to be stored (and photographed)
in sealed, airless containers.

(c)

(d)

Table 19.4 *Chemical properties of the Group 1 metals*

Reactant	Reaction with alkali metal (M)
hydrogen	$2\,M(s) + H_2(g) \longrightarrow 2\,MH(s)$
oxygen	$4\,Li(s) + O_2(g) \longrightarrow 2\,Li_2O(s)$
	$2\,Na(s) + O_2(g) \longrightarrow Na_2O_2(s)$
	$M(s) + O_2(g) \longrightarrow MO_2(s),\quad M = K, Rb, Cs$
nitrogen	$6\,Li(s) + N_2(g) \longrightarrow 2\,Li_3N(s)$
halogen (X_2)	$2\,M(s) + X_2(g, l, s) \longrightarrow 2\,MX(s)$
water	$2\,M(s) + 2\,H_2O(l) \longrightarrow 2\,MOH(aq) + H_2(g)$

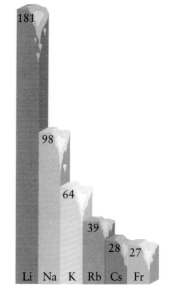

FIGURE 19.10

The melting points of the alkali metals decrease down the group. The numerical values shown here are degrees Celsius.

molten potassium chloride to sodium vapor:

$$KCl(l) + Na(g) \xrightarrow{\Delta} NaCl(s) + K(g)$$

Although the equilibrium constant for this reaction is not particularly favorable, the reaction runs to the right because potassium is more volatile than sodium. The potassium vapor is driven off by the heat and condensed in a cooled collecting vessel.

All the Group 1 elements are soft, silver-gray metals (Fig. 19.9). Lithium is the hardest, but even so it is softer than lead. The melting points decrease down the group (Fig. 19.10). Cesium melts at 28°C, just above room temperature. Some alloys of sodium and potassium are liquid at room temperature, because their atoms pack together poorly and hence produce a fluid structure, like a pile of oranges and grapefruit, which is less stable than a pile of oranges or grapefruit alone (Fig. 19.11). Liquid sodium and potassium mixtures are used as coolants in breeder nuclear reactors. Such a mixture has a high boiling point and conducts heat very well, and the metals are not decomposed by radiation. Lithium metal had few applications until thermonuclear weapons (which use lithium-6) were developed after World War II.

The alkali metals melt at low temperatures; they are the most reactive metals.

19.6 Chemical Properties of the Alkali Metals

Because the first ionization energies of the alkali metals are so low, they are most commonly found as singly charged cations, such as Na^+, and consequently most of their compounds are ionic (Table 19.5). Because of their low ionization energies, the alkali metals are excellent reducing agents. Molten sodium metal is used to produce zirconium and titanium from their chlorides:

$$TiCl_4(g) + 4\,Na(l) \longrightarrow 4\,NaCl(s) + Ti(s)$$

Because their standard potentials are so strongly negative, the alkali metals can even reduce the hydrogen in water:

$$2\,Na(s) + 2\,H_2O(l) \longrightarrow 2\,NaOH(aq) + H_2(g)$$

FIGURE 19.11

Sodium atoms (represented by the pink spheres) and potassium atoms (yellow) pack together poorly, and an alloy of the two metals is a liquid at room temperature.

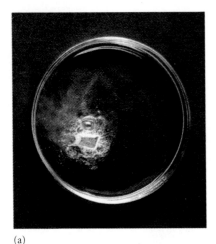
(a)

(b)   (c)

FIGURE 19.12

The alkali metals react with water, producing gaseous hydrogen and a solution of the alkali metal hydroxide. (a) Lithium reacts quietly. (b) Sodium reacts so vigorously that the heat released melts the unreacted metal. (c) Potassium reacts even more vigorously, producing so much heat that the hydrogen produced in the reaction is ignited.

The vigor of this reaction increases uniformly down the group (Fig. 19.12). The reaction is dangerously explosive with rubidium and cesium. Rubidium and cesium are denser than water, so they sink and react beneath the surface. The rapidly evolved hydrogen gas then forms a shock wave that can shatter the vessel.

The alkali metals also release their valence electrons when they dissolve in liquid ammonia, but the outcome is different. The electrons occupy cavities formed by groups of NH_3 molecules and give ink-blue metal-ammonia solutions (Fig. 19.13). These solutions of electrons (and cations of the metal) are often used to reduce organic compounds. As the metal concentration is increased, the blue gives way to a metallic bronze, and the solutions begin to conduct electricity like liquid metals.

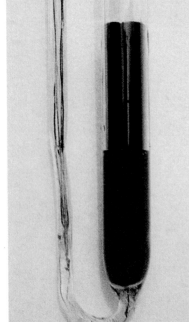

FIGURE 19.13

Sodium dissolves in liquid ammonia to form the deep blue solution in the lower half of the tube. At higher sodium concentrations, the metal-ammonia solution becomes bronze in color, as shown in the top half of the tube.

Table 19.5 Properties of Group 1 compounds

Compound	Formula*	Comment
oxides	M_2O	formed by decomposition of carbonates; strong bases; react with water to form hydroxides
hydroxides	MOH	formed by reduction of water with the metal or from the oxide; strong bases
carbonates	M_2CO_3	soluble in water; most decompose into oxides when heated
hydrogen carbonates	$MHCO_3$	weak bases in water; can be obtained as solids
nitrates	MNO_3	decompose to nitrite and evolve oxygen when strongly heated

*M stands for a Group 1 metal.

FIGURE 19.14

Although the alkali metals give a mixture of products when they react with oxygen, lithium gives mainly the oxide (left), sodium the very pale yellow peroxide (center), and potassium the yellow superoxide (right).

All alkali metals react directly with most nonmetals (other than the noble gases). However, only lithium reacts with nitrogen, which it reduces to the nitride ion:

$$6 \, Li(s) \; + \; N_2(g) \longrightarrow 2 \, Li_3N(s)$$

The principal product of the reaction of the alkali metals with oxygen varies systematically down the group (Fig. 19.14). It is commonly found that ionic compounds formed from cations and anions of similar radius are more stable than those formed from ions with markedly different radii. Lithium forms mainly the oxide, Li_2O. Sodium, which has a larger cation (recall Fig. 7.36), forms predominantly the pale yellow sodium peroxide, Na_2O_2. Potassium, with an even bigger cation, forms mainly the superoxide, KO_2. Potassium superoxide, a yellow-orange solid, is used in closed-system breathing apparatus such as gas masks, submarines, and space vehicles (see Applying Chemistry: Case Study 3).

> *The alkali metals are usually found as singly charged cations. They reduce water with increasing vigor down the group.*

19.7 Compounds of Lithium, Sodium, and Potassium

As head of its group, lithium differs significantly from other Group 1 elements. The differences stem in part from the small size of the Li^+ cation. This smallness makes it so strongly polarizing that the bonds it forms have a pronounced covalent character. Its small size also means that a Li^+ ion has strong ion-dipole interactions, so many lithium salts are hydrated.

The availability of lithium has increased during the past few decades, and so has the variety of its applications. Lithium compounds are used in ceramics, lubricants, and batteries (Fig. 19.15). They are also used in medicine, and small daily doses of lithium carbonate have been found to be an effective treatment for manic-depressive disorder; the mode of action is still not fully understood. Lithium soaps—the lithium salts of long-chain carboxylic acids—are used as

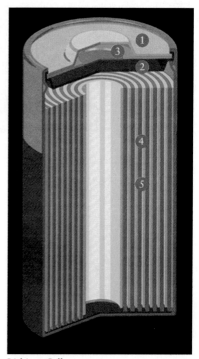

Lithium Cell

1. Positive Cap
Formed nickel-plated steel disk which serves as positive terminal

2. Seal
Molded plastic part which contains cell materials and provides a safety vent

3. PTC Device
Polymeric device which shuts off current in a cell that has been accidentally short-circuited

4. Electrode Jell Roll
Assembly of positive and negative electrodes which are separated by microporous polymer separator

5. Electrolyte
Solution of complex salts which are dissolved in organic solvents

FIGURE 19.15

Lithium ion batteries like these can store more energy than nickel-cadmium batteries of similar size. Hence, lithium ion batteries are used in laptop computers and electric cars.

FIGURE 19.16

An evaporation pond. The blue color is due to a dye added to the brine to increase heat absorption and hence speed up evaporation.

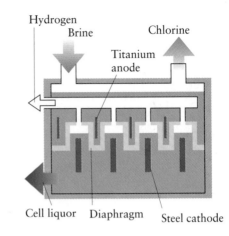

FIGURE 19.17

A diaphragm cell for the electrolytic production of sodium hydroxide from brine (aqueous sodium chloride solution). The diaphragm prevents the chlorine produced at the anode from mixing with the hydrogen and the sodium hydroxide. The liquid (cell liquor) is drawn off and the water is partially evaporated. The unconverted sodium chloride crystallizes, leaving the sodium hydroxide in solution.

thickeners in lubricating greases for high-temperature applications because they have higher melting points than more conventional sodium and potassium soaps.

Sodium compounds are important largely because they are generally soluble in water and inexpensive. It is readily mined as rock salt or obtained from the evaporation of brine (Fig. 19.16). Sodium chloride is used in large quantities in the electrolytic production of chlorine and sodium hydroxide from brine.

Sodium hydroxide, NaOH, is a soft, waxy, white, corrosive solid. It is an important industrial chemical because it is an inexpensive base for the production of other sodium salts. The amount of electricity used to electrolyze brine to produce NaOH is second only to the amount used to extract aluminum from its ores. The process produces chlorine and hydrogen as well as sodium hydroxide:

$$2\,NaCl(aq) + 2\,H_2O(l) \longrightarrow 2\,NaOH(aq) + Cl_2(g) + H_2(g)$$

Most modern production uses a *diaphragm cell,* in which the compartments containing the steel and titanium electrodes are separated by porous diaphragms to isolate the products (Fig. 19.17).

Anhydrous sodium sulfate, Na_2SO_4, which is called *salt cake,* is used to make glass and paper. Much of it comes from natural sources, particularly the sulfate-rich underground brines found in Texas. In countries lacking natural supplies, it is produced by the action of concentrated sulfuric acid on sodium chloride:

$$H_2SO_4(aq, conc) + 2\,NaCl(s) \longrightarrow Na_2SO_4(s) + 2\,HCl(g)$$

This reaction is driven forward by the escape of a volatile product. The same reaction is used to generate hydrogen chloride in the laboratory.

Sodium hydrogen carbonate, $NaHCO_3$ (sodium bicarbonate), is commonly called *bicarbonate of soda* or *baking soda.* The rising action of baking soda in bread and doughs depends on the reaction of a weak acid with the hydrogen carbonate ions:

$$NaHCO_3(aq) + H_3O^+(aq) \longrightarrow Na^+(aq) + 2\,H_2O(l) + CO_2(g)$$

FIGURE 19.18

Double-acting baking powder first forms small cavities in the dough when it is moistened. These are later inflated by a second release of carbon dioxide during baking.

The release of gas causes the dough to rise. The weak acids are provided by the recipe, generally in the form of lactic acid from sour milk or buttermilk, citric acid from lemons, or acetic acid from vinegar. Baking powder contains a solid acid as well as the hydrogen carbonate, and carbon dioxide is released when water is added (Fig. 19.18).

Sodium carbonate decahydrate, $Na_2CO_3 \cdot 10H_2O$, was widely used earlier in the twentieth century as *washing soda*. Anhydrous sodium carbonate, or *soda ash*, is used in large amounts in the glass industry as a source of sodium oxide, into which it decomposes when heated.

The principal mineral sources of potassium are *sylvite*, KCl (Fig. 19.19), and *carnallite*, $KCl \cdot MgCl_2 \cdot 6H_2O$. Potassium chloride is incorporated directly into fertilizers as a source of essential potassium, but potassium nitrate is used instead for crops such as potatoes and tobacco, which cannot tolerate high Cl^- ion concentrations.

Potassium compounds are generally more expensive than the corresponding sodium compounds; but, in some applications, their advantages outweigh their expense. Potassium nitrate, KNO_3, releases oxygen when heated, in the reaction

$$2 \, KNO_3(s) \xrightarrow{\Delta} 2 \, KNO_2(s) + O_2(g)$$

and is used to facilitate the ignition of matches. It is less **hygroscopic** (water absorbing) than the corresponding sodium compounds, because the K^+ cation is larger and is less strongly hydrated by H_2O molecules.

Sodium compounds are soluble in water, plentiful, and inexpensive. Potassium compounds are generally less soluble and less hygroscopic than sodium compounds.

GROUP 2: THE ALKALINE EARTH METALS

Calcium, strontium, and barium are called the alkaline earth metals, because their "earths"—the old name for oxides—are basic (alkaline). The name alkaline earth metals is often extended loosely to all the members of the group (Table 19.6).

The mineral sylvite is a form of potassium chloride. It is found in the beds of ancient lakes and seas. The potassium was probably collected by plants that grew around the lake and extracted it from groundwater.

Table 19.6 The Group 2 elements

Valence configuration: ns^2
Normal form*: soft, silver-gray metals

Z	Name	Symbol	Molar mass, g/mol	Melting point, °C	Boiling point, °C	Density, g/cm³
4	beryllium	Be	9.01	1285	2470	1.85
12	magnesium	Mg	24.31	650	1100	1.74
20	calcium	Ca	40.08	840	1490	1.53
38	strontium	Sr	87.62	770	1380	2.58
56	barium	Ba	137.34	710	1640	3.59
88	radium†	Ra	226.03	700	1500	5.00

*Normal form means the state and appearance of the element at 25°C and 1 atm.

†Radioactive.

19.8 The Elements

It is very unlikely that you will be asked to remember the complicated formulas of minerals like beryl. Your authors certainly can't!

All the Group 2 elements are metals and too reactive to occur in the uncombined state in nature (Fig. 19.20). Instead, they are generally found as doubly charged cations. The element beryllium occurs mainly as *beryl*, $3BeO \cdot Al_2O_3 \cdot 6SiO_2$, sometimes in crystals so big that they weigh several tons. The gemstone *emerald* is a form of beryl; its green color is caused by the Cr^{3+} ions present as impurities (Fig. 19.21). Magnesium occurs in seawater and as the mineral *dolomite*, $CaCO_3 \cdot MgCO_3$. Calcium also occurs as $CaCO_3$ in compressed deposits of the shells of ancient marine organisms and exoskeletons of tiny one-celled organisms; the minerals derived from these deposits are *limestone*, *calcite*, and *chalk* (a softer variety of limestone).

Beryllium's low density makes it important in the construction of missiles and satellites. It is used as windows for x-ray tubes because beryllium atoms have so few electrons that thin sheets of the metal are transparent to x-rays and allow the rays to escape. Much of the metal that is produced is added in small amounts to copper; the small Be atoms pin the Cu atoms together in an interstitial alloy that is more rigid than pure copper but still conducts electricity well. The hard, electrically conducting alloy is formed into nonsparking tools for use in oil refineries and grain elevators, where there is a risk of explosion.

(a)

(b)

(c)

(d)

(e)

FIGURE 19.20

The elements of Group 2: (a) beryllium; (b) magnesium; (c) calcium; (d) strontium; (e) barium. The four central elements of the group (magnesium through barium) were discovered by Humphry Davy in a single year (1808). The two outer elements were discovered later: beryllium in 1828 (by Friedrich Wöhler) and radium (which is not shown here) in 1898 (by Pierre and Marie Curie).

Magnesium is a silver-white metal that is protected from extensive oxidation in air by a film of white oxide; hence, it looks dull gray soon after it has been cut. Its low density is only two-thirds that of aluminum, and it is widely used as a component of alloys in applications where lightness and toughness are needed—in airplanes, for instance.

Metallic magnesium is produced by reduction of its compounds, either electrolytically or with a chemical reducing agent. In the chemical reduction of magnesium oxide obtained from the decomposition of dolomite, ferrosilicon (an alloy of iron and silicon) is used as the reducing agent at about 1200°C. At this temperature, the magnesium produced is immediately vaporized; so even though the equilibrium constant does not favor the reduction, the process continues because the product is removed as soon as it is formed.

The electrolytic method uses seawater as its principal raw material (Fig. 19.22). The first stage is the precipitation of magnesium hydroxide with slaked lime (calcium hydroxide):

$$Mg^{2+}(aq) + Ca(OH)_2(aq) \longrightarrow Mg(OH)_2(s) + Ca^{2+}(aq)$$

The lime is produced by the thermal decomposition of calcium carbonate in shells dredged up from the ocean floor. The precipitated magnesium hydroxide is filtered off and treated with hydrochloric acid:

$$Mg(OH)_2(s) + 2\,HCl(aq) \longrightarrow MgCl_2(aq) + 2\,H_2O(l)$$

Finally, the magnesium chloride is dried and then added to an electrolytic cell. Magnesium is produced from the molten salt at the cathode and chlorine is produced at the anode:

$$MgCl_2(l) \xrightarrow{\Delta,\ \text{electrolysis}} Mg(s) + Cl_2(g)$$

The true alkaline earth metals—calcium, strontium, and barium—are obtained either by electrolysis or by reduction with aluminum in a version of the thermite process (see Section 6.7). For example,

$$3\,BaO(s) + 2\,Al(s) \longrightarrow Al_2O_3(s) + 3\,Ba(s)$$

FIGURE 19.21

An emerald is a crystal of beryl with some Cr^{3+} ions, which are responsible for the color.

FIGURE 19.22

A magnesium extraction plant at Freeport, Texas.

FIGURE 19.23

Calcium reacts gently with water at room temperature to produce hydrogen and calcium hydroxide, $Ca(OH)_2$.

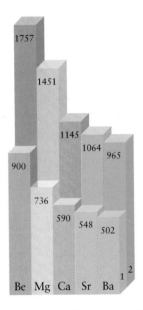

FIGURE 19.24

The first and second ionization energies (in kilojoules per mole) of the Group 2 elements. Although the second ionization energies are larger than the first, they are not enormous, and both valence electrons are lost from each atom in all the compounds of these elements.

Table 19.7 *Chemical properties of the Group 2 metals*

Reactant	Reaction with Group 2 metal (M)
hydrogen	$M(s) + H_2(g) \longrightarrow MH_2(s)$, not Be or Mg
oxygen	$2\,M(s) + O_2(g) \longrightarrow 2\,MO(s)$
nitrogen	$3\,M(s) + N_2(g) \longrightarrow M_3N_2(s)$
halogen (X_2)	$M(s) + X_2(g,l,s) \longrightarrow MX_2(s)$
water	$M(s) + 2\,H_2O(l) \longrightarrow M(OH)_2(aq) + H_2(g)$, not Be

Some important reactions of the Group 2 elements are summarized in Table 19.7.

Magnesium burns vigorously in air with a brilliant white flame, partly because it reacts with the nitrogen and carbon dioxide in air as well as with oxygen—especially when it is sprayed with water. Neither water nor CO_2 fire extinguishers should ever be used on a magnesium fire, because they will make the fire worse! As in Group 1, reactions of the metals with oxygen and water increase in vigor going down the group. Beryllium, magnesium, calcium, and strontium are partially passivated in air by a protective surface layer of oxide. Barium, however, does not form a protective oxide and may ignite in moist air.

All the Group 2 elements with the exception of beryllium reduce water; for example,

$$Ca(s) + 2\,H_2O(l) \longrightarrow Ca(OH)_2(aq) + H_2(g)$$

Beryllium does not react with water, even when red hot: its protective oxide film survives even at high temperatures. Magnesium reacts with hot water, and calcium reacts with cold water (Fig. 19.23). The metals reduce hydrogen ions to hydrogen, but neither beryllium nor magnesium dissolves in nitric acid, because they become passivated by a film of oxide.

The valence electron configuration of the atoms of the Group 2 elements is ns^2, where n is the period number. The second ionization energy is low enough to be recovered from the increased lattice enthalpy (Fig. 19.24). Hence, the Group 2 elements occur with an oxidation number of $+2$, as the cation M^{2+}, in all their compounds (Table 19.8). Apart from a tendency toward nonmetallic character in beryllium, the elements have all the chemical characteristics of metals, such as having basic oxides and hydroxides.

Beryllium shows a hint of nonmetallic character, but the other elements are all typical metals. The vigor of reaction with water and with oxygen increases down the group.

19.9 Compounds of Beryllium, Magnesium, and Calcium

Beryllium compounds are very toxic and must be handled with great caution. Their properties are dominated by the highly polarizing character of the Be^{2+} ion and its small size. The strong polarizing power results in moderately covalent compounds, and its small size limits to four the number of groups that can attach to the ion. These two features together are responsible for the prominence of the tetrahedral BeX_4 unit (**1**), like that in the beryllate ion, $Be(OH)_4{}^{2-}$,

Table 19.8 *Properties of Group 2 compounds*

Compound	Formula*	Comment
oxides	MO	formed by decomposition of carbonates; strong bases (BeO is amphoteric), react with water to from hydroxides; withstand high temperatures
hydroxides	M(OH)$_2$	formed by action of water on oxides or by precipitation from salt solutions; sparingly soluble in water (except Ba); strong bases (Be is amphoteric)
carbonates	MCO$_3$	very slightly soluble in water; most decompose into oxides when heated
hydrogen carbonates	M(HCO$_3$)$_2$	unstable as solids; more soluble than carbonates
nitrates	M(NO$_3$)$_2$	decompose when heated; soluble in water

*M stands for a Group 1 metal.

formed when beryllium reacts with sodium hydroxide solution. A tetrahedral unit is also found in the chloride (**2**) and the hydride (**3**). The chloride is made by the action of chlorine on the oxide in the presence of carbon:

$$BeO(s) + C(s) + Cl_2(g) \xrightarrow{600-800°C} BeCl_2(g) + CO(g)$$

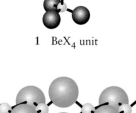

1 BeX$_4$ unit

The Be atoms in BeCl$_2$ act as Lewis acids and accept electron pairs from the Cl atoms of neighboring BeCl$_2$ groups, forming a chain of tetrahedral BeCl$_4$ units.

Magnesium has more pronounced metallic properties than beryllium does, and its compounds are primarily ionic, with some covalent character. Magnesium oxide, MgO, is formed when magnesium burns in air. However, the product is contaminated by magnesium nitride. To prepare the pure oxide, the hydroxide or the carbonate is heated. Magnesium oxide dissolves only very slowly and slightly in water. One of its most striking properties is that it is **refractory,** or able to withstand high temperatures, for it melts only at 2800°C. This high stability can be traced to the small ionic radii of the Mg^{2+} and O^{2-} ions, and hence to their very strong electrostatic interaction with each other. The oxide has two other characteristics that make it useful: it conducts heat very well and it conducts electricity poorly. All three properties lead to its use as an insulator in electric heaters.

2 Beryllium chloride, BeCl$_2$

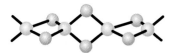

3 Beryllium hydride, BeH$_2$

Magnesium hydroxide, Mg(OH)$_2$, is not very soluble in water but forms instead a white colloidal suspension (a mist of small particles dispersed through a liquid; see Section 12.5), which is known as *milk of magnesia* and is used as a stomach antacid and laxative. Magnesium sulfate, or *Epsom salts,* MgSO$_4$, is also a common purgative. Its action appears to inhibit the absorption of water from the intestine. The resulting increased flow of water into the intestine triggers the mechanism that results in defecation.

Arguably the most important compound of magnesium is chlorophyll (Fig. 19.25). This green organic compound consists of large molecules that capture light from the Sun and channel its energy into photosynthesis.

Calcium carbonate, CaCO$_3$, occurs naturally as chalk and limestone. Marble is a dense form of calcium carbonate that can be given a high polish; it is often colored by impurities, most commonly iron cations (Fig. 19.26). The two most common forms of pure calcium carbonate are calcite and aragonite. All these carbonates are the fossilized remains of marine life. Calcium carbonate decomposes to calcium oxide, CaO, or *quicklime,* when heated:

The name *chlorophyll* comes from the Greek words for "green leaf."

$$CaCO_3(s) \xrightarrow{\Delta} CaO(s) + CO_2(g)$$

FIGURE 19.25

The decomposition of $CaCO_3$ requires a higher temperature (about 800°C) than does that of $MgCO_3$. The difference can be explained by recognizing that the larger Ca^{2+} ion is less effective than Mg^{2+} at removing an O^{2-} ion from a neighboring CO_3^{2-} ion.

Quicklime is produced in enormous quantities throughout the world. About 40% of the quicklime produced is used in metallurgy. In ironmaking (Section 21.3), it is used as a Lewis base; its O^{2-} ion reacts with silica (SiO_2) impurities in the ore to form a liquid slag:

$$CaO(s) + SiO_2(s) \xrightarrow{\Delta} CaSiO_3(l)$$

About 50 kg of quicklime is needed to produce 1 ton of iron. Slaked lime, $Ca(OH)_2$, is also used as an inexpensive base in industry, to adjust soil pH and to soften water (Connection 4, following Chapter 16).

Calcium compounds are often used as structural materials; their rigidity stems from the strength with which the small, highly charged Ca^{2+} cation interacts with its neighbors. *Mortar* consists of about one part lime and three parts sand (largely silica, SiO_2). It sets to a hard mass as the lime reacts with the carbon dioxide of the air to form the carbonate (Fig. 19.27). Calcium is also found in the rigid structural components of living things, either as the calcium carbonate of the shells of shellfish or the calcium phosphate of bone (see Section 8.4). About 1 kg of calcium is present in an adult human body, mostly in the form of insoluble calcium phosphate, but also as Ca^{2+} ions in other fluids inside our cells. The calcium in newly formed bone is in dynamic equilibrium with the calcium ions in the body fluids, so calcium must be part of our daily diet to maintain bone strength.

Tooth enamel is a *hydroxyapatite*, a phosphate mineral of composition $Ca_5(PO_4)_3OH$. Tooth decay begins when acids attack the enamel:

$$Ca_5(PO_4)_3OH(s) + 4 H_3O^+(aq) \longrightarrow 5 Ca^{2+}(aq) + 3 HPO_4^-(aq) + 5 H_2O(l)$$

The principal agents of tooth decay are the carboxylic acids produced when bacteria act on the remains of food. A more resistant coating forms when the OH^- ions in the apatite are replaced by F^- ions. The resulting mineral is called *fluorapatite*:

$$Ca_5(PO_4)_3OH(s) + F^-(aq) \longrightarrow Ca_5(PO_4)_3F(s) + OH^-(aq)$$

The addition of fluoride ions to domestic water supplies (by addition of NaF) is now widespread. Fluoridated toothpastes, containing either tin(II) fluoride or

FIGURE 19.26

Marble is a dense form of calcium carbonate. The red color is due to iron cations. Other ions are responsible for different colors of marble.

sodium monofluorophosphate (MFP, Na_2FPO_3), are also recommended to strengthen tooth enamel.

Beryllium compounds have a pronounced covalent character, and the structural unit is commonly tetrahedral. The small size of the magnesium cation results in a thermally stable oxide with very low solubility in water. Calcium compounds are common in structural materials, because the small, highly charged Ca^{2+} ion results in rigid structures.

Self-Test 19.2A Explain why beryllium compounds have covalent characteristics.

[*Answer:* The small size and high charge on the beryllium ion make it highly polarizing.]

Self-Test 19.2B Explain why MgO has such a high melting point.

GROUP 13: THE BORON FAMILY

Group 13 is the first group of the *p* block. Its members have an ns^2np^1 electron configuration (Table 19.9), so we expect a maximum oxidation number of +3. The oxidation numbers of B and Al are +3 in almost all their compounds (Table 19.10). However, the heavier elements in the group are more likely to keep their *s*-electrons (the inert-pair effect, Section 8.2), so the oxidation number +1 becomes increasingly important on going down the group, and Tl(I) compounds are as common as Tl(III) compounds. We shall concentrate on the two most important members of the group, boron and aluminum.

19.10 The Elements

Boron, at the head of Group 13, is mined as *borax* and *kernite*, $Na_2B_4O_7\cdot xH_2O$, with $x = 10$ and 4, respectively. Large deposits are found in volcanic regions, such as the Mojave Desert region of California (Fig. 19.28). In the extraction process, the minerals are converted into boron oxide with acid and then reduced with magnesium to an impure brown, amorphous form of boron:

$$B_2O_3(s) + 3\,Mg(s) \xrightarrow{\Delta} 2\,B(s) + 3\,MgO(s)$$

FIGURE 19.27

An electron micrograph of the surface of mortar, showing how tiny interlocking crystals grow as carbon dioxide reacts with calcium hydroxide and silica.

								H		
	2	13	14							He
Li	Be	B	C	N	O	F	Ne			
	Mg	Al	Si							
	Ca	Ga	Ge							
	Sr	In	Sn							
	Ba	Tl	Pb							
	Ra									

Table 19.9 *The Group 13 elements*

Valence configuration: ns^2np^1

Z	Name	Symbol	Molar mass, g/mol	Melting point, °C	Boiling point, °C	Density, g/cm^3	Normal form*
5	boron	B	10.81	2030	3700	2.47	brown nonmetallic powder
13	aluminum	Al	26.98	660	2350	2.70	silver-white metal
31	gallium	Ga	69.72	30	2070	5.91	silver metal
49	indium	In	114.82	157	2050	7.29	silver-white metal
81	thallium	Tl	204.37	304	1460	11.87	soft metal

Normal form means the state and appearance of the element at 25°C and 1 atm.

Table 19.10 *Chemical properties of the Group 13 elements*

Reactant	Reaction with Group 13 element (E)
oxygen	$4\,E(s) + 3\,O_2(g) \longrightarrow 2\,E_2O_3(s)$
nitrogen	$2\,E(s) + N_2(g) \longrightarrow 2\,EN(s), \quad E = B,\,Al$
halogen (X_2)	$2\,B(s) + 3\,X_2(g,l,s) \longrightarrow 2\,BX_3(g)$
	$2\,E(s) + 3\,X_2(g,l,s) \longrightarrow E_2X_6(g), \quad E = Al,\,Ga,\,Tl$
	$2\,Tl(s) + X_2(g,l,s) \longrightarrow 2\,TlX(s)$
water	$2\,Tl(s) + 2\,H_2O(l) \longrightarrow 2\,TlOH(aq) + H_2(g)$
acid	$2\,E(s) + 6\,H_3O^+(aq) \longrightarrow 2\,E^{3+}(aq) + 6\,H_2O(l) + 3\,H_2(g), \quad E = Al,\,Ga,\,Tl$
base	$2\,E(s) + 6\,H_2O(l) + 2\,OH^-(aq) \longrightarrow 2\,E(OH)_4^-(aq) + 3\,H_2(g), \quad E = Al,\,Ga,$

Boron production remains quite low despite the element's desirable properties of hardness and lightness.

Elemental boron has several allotropes. It is typically either a gray-black, high-melting-point solid or a dark brown powder with a structure based on clusters of 12 atoms (Fig. 19.29). When boron fibers are incorporated in plastics, the result is a very tough material that is stiffer than steel yet lighter than aluminum and is used in aircraft, missiles, and body armor. The element is very inert and is attacked by only the strongest oxidizing agents.

Boron is often classified as a metalloid, but it is best regarded as a nonmetal in most of its chemical properties. It has acidic oxides and forms an interesting and extensive range of binary molecular hydrides.

Aluminum is the most abundant metallic element in the Earth's crust and, following oxygen and silicon, the third most abundant element. However, the aluminum content in most minerals is low, and the commercial source of aluminum, bauxite, is a hydrated, impure oxide, $Al_2O_3 \cdot xH_2O$, where x can range up to 3. The *bauxite* ore, which is red from the iron oxides it contains, is processed to obtain alumina, Al_2O_3.

Aluminum metal is obtained by the **Hall process.** Charles Hall discovered in 1886 that by mixing the mineral *cryolite*, Na_3AlF_6, with alumina, he got a mixture that melted at a much more economical temperature, 950°C, instead of the 2050°C of pure alumina. The melt is electrolyzed in a cell that uses graphite (or carbonized petroleum) anodes and a carbonized steel-lined vat that serves as the cathode (Fig. 19.30). The electrolysis reactions are

$$\text{Cathode reaction: } Al^{3+}(\text{melt}) + 3\,e^- \longrightarrow Al(l)$$

$$\text{Anode reaction: } 2\,O^{2-}(\text{melt}) + C(s,\,gr) \longrightarrow CO_2(g) + 4\,e^-$$

The overall reaction is

$$4\,Al^{3+}(\text{melt}) + 6\,O^{2-}(\text{melt}) + 3\,C(s,\,gr) \longrightarrow 4\,Al(l) + 3\,CO_2(g)$$

Note that the carbon electrode takes part in the reaction. From the reaction stoichiometry, we can calculate that a current of 1 A must flow for 80 h to produce 1 mol Al (27 g of aluminum, about enough for two soft-drink cans). The very high electricity consumption can be greatly reduced by recycling, which requires less than 5% of the electricity needed to extract aluminum from bauxite.

FIGURE 19.28

Boron, California, a major source of borax and hence of the element boron.

Aluminum has a low density; it is a strong metal and an excellent electrical conductor. Although it is strongly reducing and thus easily oxidized, it is resistant to corrosion because its surface is passivated in air by a stable oxide film. The thickness of the oxide layer can be increased by making aluminum the anode of an electrolytic cell; the result is called *anodized aluminum*. Dyes may be added to the dilute sulfuric acid electrolyte used in the anodizing process, to produce surface layers with different colors. Brown and bronze anodized aluminum parts produced in this way are widely used in modern architecture.

Aluminum's low density, wide availability, and corrosion resistance make it ideal for construction. For use in airplanes, it is usually alloyed with copper and silicon. Its lightness and good electrical conductivity have led to its use for overhead power lines.

Aluminum is sufficiently far to the right in the periodic table to show a hint of nonmetallic character. Thus, aluminum is amphoteric, reacting both with nonoxidizing acids (such as hydrochloric acid) to form aluminum ions:

$$2\,Al(s) + 6\,H^+(aq) \longrightarrow 2\,Al^{3+}(aq) + 3\,H_2(g)$$

and with hot aqueous alkali to form aluminate ions:

$$2\,Al(s) + 2\,OH^-(aq) + 6\,H_2O(l) \longrightarrow 2\,[Al(OH)_4]^-(aq) + 3\,H_2(g)$$

Boron is a hard, largely nonmetallic element. Aluminum is a light, strong, amphoteric, reactive metallic element with a surface that becomes passivated when exposed to air.

19.11 Group 13 Oxides

The oxides of boron and aluminum are important in their own right, as sources of the elements and as the starting point for the manufacture of other compounds.

Boric acid, $B(OH)_3$, is a white solid that melts at 171°C. It is toxic to bacteria and many insects as well as humans and has long been used as a mild antiseptic and pesticide. Because the boron atom in $B(OH)_3$ has an incomplete octet, it can act as a Lewis acid and form a bond by accepting a lone pair of electrons from an H_2O molecule acting as a Lewis base:

$$(OH)_3B + \!:OH_2 \longrightarrow (OH)_3B{-}OH_2$$

The compound so formed is a weak *mono*protic acid:

$$B(OH)_3OH_2(aq) + H_2O(l) \rightleftharpoons H_3O^+(aq) + B(OH)_4{}^-(aq) \qquad pK_a = 9.14$$

Boric acid also retards the spread of flames in cellulosic materials, particularly paper. The scrap paper used to manufacture home insulation contains about 5% boric acid, to reduce the risk of fire. However, the major use of boric acid is as the starting material for its anhydride, boron oxide, B_2O_3. Because it melts (at 450°C) to a liquid that dissolves many metal oxides, boron oxide (often as the acid) is used as a *flux*, a substance that cleans metals before they are soldered or welded. Boron oxide is also used to make fiberglass and borosilicate glass, a glass that expands very little when heated, such as Pyrex.

We can recognize some of the oxides of the *p*-block elements as acid anhydrides. For example, CO_2 forms carbonic acid, H_2CO_3. A number of acid

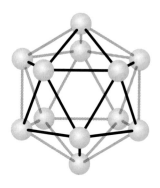

FIGURE 19.29

The structure of boron is based on linked 12-atom units. The unit has 20 faces, so it is called an icosahedron (from the Greek words for "twenty-faced").

Before recycling of aluminum was widespread, the aluminum industry in the United States consumed each day the electricity used by 100 000 people in one year!

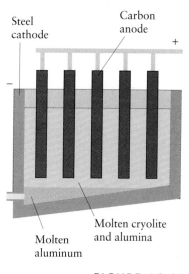

FIGURE 19.30

In the Hall process, aluminum oxide is dissolved in molten cryolite and the mixture is electrolyzed in a cell with carbon anodes and a steel cathode.

FIGURE 19.31

Some of the impure forms of α-alumina are prized as gems. (a) Ruby is alumina with Cr^{3+} replacing some Al^{3+} ions. (b) Sapphire is alumina with Fe^{3+} and Ti^{4+} impurities. (c) Topaz is alumina with Fe^{3+} impurities.

(a) (b) (c)

anhydrides can be formed by simply heating the oxoacid. This is the case with boric acid:

$$2\,B(OH)_3(s) \xrightarrow{\Delta} B_2O_3(s) + 3\,H_2O(g)$$

Aluminum oxide, Al_2O_3, is almost universally known as *alumina*. It exists with a variety of crystal structures. As α-alumina, it is the very hard substance *corundum*; impure microcrystalline corundum is the purple-black abrasive known as *emery*. Some impure forms of alumina are beautiful, rare, and highly prized (Fig. 19.31). A less dense and more reactive form of the oxide is γ-alumina. This form absorbs water and is used as the stationary phase in chromatography (Investigating Matter 12.1).

γ-Alumina is produced by heating aluminum hydroxide. It is quite reactive and is amphoteric, dissolving readily in bases to produce the aluminate ion and in acids to produce the hydrated Al^{3+} ion:

$$Al_2O_3(s) + 2\,OH^-(aq) + 3\,H_2O(l) \longrightarrow 2\,Al(OH)_4^-(aq)$$

$$Al_2O_3(s) + 6\,H_3O^+(aq) + 3\,H_2O(l) \longrightarrow 2\,[Al(H_2O)_6]^{3+}(aq)$$

The strong polarizing effect of the small, highly charged Al^{3+} ion on the water molecules around it results in the $[Al(H_2O)_6]^{3+}$ ion being an acid. Solutions of aluminum salts are therefore acidic.

One of the most important aluminum salts prepared by the action of an acid on alumina is aluminum sulfate, $Al_2(SO_4)_3$:

$$Al_2O_3(s) + 3\,H_2SO_4(aq) \longrightarrow Al_2(SO_4)_3(aq) + 3\,H_2O(l)$$

Aluminum sulfate is called *papermaker's alum* and is used in the paper industry to coagulate cellulose fibers into a hard, nonabsorbent surface. True *alums* (from which aluminum takes its name) are mixed sulfates of formula $M^+M'^{3+}(SO_4^{2-})_2\cdot12H_2O$, and include potassium alum, $KAl(SO_4)_2\cdot12H_2O$ (which is used in water and sewage treatment), and ammonium alum, $NH_4Al(SO_4)_2\cdot12H_2O$ (which is used for pickling cucumbers). Other alums are used for waterproofing fabrics and as mordants, compounds that help dyes adhere to fabrics, in dying and printing textiles.

FIGURE 19.32

Aluminum hydroxide, $Al(OH)_3$, forms as a white, fluffy precipitate. The fluffy form of the solid captures impurities and is used in the purification of water.

Sodium aluminate, $NaAl(OH)_4$, is used along with aluminum sulfate in water purification. When mixed with aluminate ions, the acidic hydrated Al^{3+} cation from the aluminum sulfate produces aluminum hydroxide:

$$Al^{3+}(aq) + 3\,Al(OH)_4^-(aq) \longrightarrow 4\,Al(OH)_3(s)$$

The aluminum hydroxide is formed as a fluffy, gelatinous network that entraps impurities as it settles, and this precipitate can be removed by filtration (Fig. 19.32). Using aluminum for both the cation and the anion gives the greatest possible bulk of impurity-collecting alumina, because the reaction has no by-products.

Boron oxide is an acid anhydride. Aluminum shows some nonmetallic character in that its oxide is amphoteric.

19.12 Carbides, Nitrides, and Halides

When boron is heated to high temperatures with carbon, it forms boron carbide, $B_{12}C_3$, a high-melting-point solid that is almost as hard as diamond. The solid consists of B_{12} groups that are pinned together by C atoms. When boron is heated to white heat in ammonia, boron nitride, BN, is formed as a fluffy, slippery powder:

$$2\,B(s) + 2\,NH_3(g) \xrightarrow{\Delta} 2\,BN(s) + 3\,H_2(g)$$

Its structure resembles that of graphite, with flat planes of hexagons of alternating B and N atoms similar to the flat planes of carbon hexagons in graphite (Fig. 19.33). Unlike graphite, however, it is white and does not conduct electricity. Under high pressure, boron nitride is converted to a very hard, diamondlike crystalline form called Borazon. In recent years, boron nitride nanotubes similar to those formed by carbon (Section 19.15) have been synthesized, and they have been found to be semiconducting. They have the potential for interesting applications in microelectronics.

The boron halides are made either by direct reaction of the elements at high temperature or from the oxide. The most important is boron trifluoride, BF_3, an industrial catalyst produced by the reaction between boric oxide, calcium fluoride, and sulfuric acid:

$$B_2O_3(s) + 3\,CaF_2(s) + 3\,H_2SO_4(l) \xrightarrow{\Delta} 2\,BF_3(g) + 3\,CaSO_4(s) + 3\,H_2O(l)$$

Boron trichloride, BCl_3, which is also widely used as a catalyst, is produced commercially by the action of the halogen on the oxide in the presence of carbon:

$$B_2O_3(s) + 3\,C(s) + 3\,Cl_2(g) \xrightarrow{500°C} 2\,BCl_3(g) + 3\,CO(g)$$

The B atom has an incomplete octet in all its trihalides (Section 8.12). The compounds consist of planar triangular molecules with an empty 2p-orbital perpendicular to the molecular plane. The empty orbital allows the molecules to act as Lewis acids, which accounts for the catalytic action of BF_3 and BCl_3.

Aluminum chloride, $AlCl_3$, another major industrial catalyst, is made by the action of chlorine on aluminum or on alumina in the presence of carbon:

$$2\,Al(s) + 3\,Cl_2(g) \longrightarrow 2\,AlCl_3(s)$$

$$Al_2O_3(s) + 3\,C(s) + 3\,Cl_2(g) \longrightarrow 2\,AlCl_3(s) + 3\,CO(g)$$

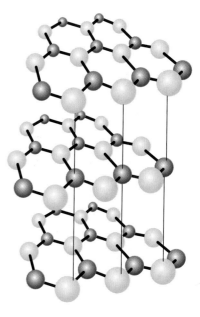

FIGURE 19.33

The structure of boron nitride, BN, resembles that of graphite, consisting of flat planes of hexagons. In boron nitride, however, the hexagons consist of alternating B and N atoms (in place of C atoms) and the layers are stacked differently. (Compare with Fig. 19.37.)

FIGURE 19.34

When anhydrous aluminum chloride is left exposed to moist air, it reacts to form hydrochloric acid. Here white fumes of ammonium chloride form as the hydrochloric acid reacts with ammonia released in the vicinity.

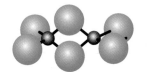

4 Aluminum chloride dimer, Al_2Cl_6

Aluminum chloride is an ionic solid in which each Al^{3+} ion is surrounded by six Cl^- ions. However, it sublimes at 192°C to a vapor of Al_2Cl_6 molecules (**4**). An Al_2Cl_6 molecule is an example of a **dimer,** the union of two identical molecules.

Aluminum halides react with water with a considerable evolution of heat. When the anhydrous chloride is exposed to moist air, it produces fumes of hydrochloric acid (Fig. 19.34). The ionic aluminum chloride hexahydrate, $AlCl_3 \cdot 6H_2O$, is used as a deodorant and antiperspirant, one of its roles being to kill the bacteria that feed on perspiration and produce unpleasant smells.

Boron and aluminum halides act as Lewis acids.

19.13 Boranes and Borohydrides

Sodium borohydride is a white crystalline solid produced from the reaction between sodium hydride and boron trichloride dissolved in a nonaqueous solvent:

$$4\,NaH + BCl_3 \longrightarrow NaBH_4 + 3\,NaCl$$

Sodium borohydride is a very useful reducing agent. At pH = 14 (strongly alkaline conditions), the standard potential of the half-reaction

$$H_2BO_3^-(aq) + 5\,H_2O(l) + 8\,e^- \longrightarrow BH_4^-(aq) + 8\,OH^-(aq)$$

is −1.24 V, well below that of the $Ni^{2+}(aq) + 2\,e^- \rightarrow Ni(s)$ half-reaction (−0.23 V). Therefore, the borohydride ion can reduce Ni^{2+} ions to metallic nickel. This reduction is the basis of the **chemical plating** of nickel (Fig. 19.35). The advantage of chemical plating over electroplating is that the item being plated does not have to be an electrical conductor.

The boranes are an extensive series of binary compounds of boron and hydrogen, somewhat analogous to the hydrocarbons. The starting point for borane production is the reaction (in an organic solvent) of sodium borohydride with boron trifluoride:

$$4\,BF_3 + 3\,BH_4^- \longrightarrow 3\,BF_4^- + 2\,B_2H_6$$

FIGURE 19.35

It is difficult to coat nonconducting objects with a metal surface. One technique is by chemical reduction. Another is by vapor deposition, the technique used to coat this figurine of the Star Trek character Worf, shown in ritual attire.

The product B_2H_6 is diborane (**5**), a colorless gas that bursts into flame in air. On contact with water, it is immediately oxidized to boric acid as it reduces the hydrogen in the water:

$$B_2H_6(g) + 6\,H_2O(l) \longrightarrow 2\,B(OH)_3(aq) + 6\,H_2(g)$$

When diborane is heated to a high temperature, it decomposes into hydrogen and pure boron:

$$B_2H_6(g) \xrightarrow{\Delta} 2\,B(s) + 3\,H_2(g)$$

This sequence of reactions is a useful route to the pure element, but more complex boranes form when the heating is less severe. When diborane is heated to 100°C, for instance, it forms decaborane, $B_{10}H_{14}$, a solid that melts at 100°C. Decaborane is stable in air, is oxidized by water only slowly, and is an example of the general rule that heavier boranes are less flammable than boranes of low molar mass.

The boranes are examples of electron-deficient compounds, compounds for which valid Lewis structures cannot be written because too few electrons are available. For instance, in diborane, there are eight atoms, so we need at least seven bonds; however, there are only 12 electrons, so we can form at most six electron-pair bonds. According to molecular orbital theory (the modern theory of the chemical bond), electron pairs are spread over the entire molecule. Consequently, the bonding power of one electron pair can be shared by several atoms. In the case of diborane, a single electron pair is considered to be delocalized over a B—H—B unit, and it binds all three atoms together. There are two such bridging three-center bonds in the molecule (**6**).

The boranes are an extensive series of highly reactive electron-deficient binary compounds of boron and hydrogen; the boranes have three-center bonds.

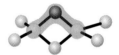

5   Diborane, B_2H_6

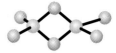

6   Three-center bond

Self-Test 19.3A Write the formulas of the anhydrides of the acids (a) H_2SO_4; (b) $B(OH)_3$.

[***Answer:*** (a) SO_3; (b) B_2O_3]

Self-Test 19.3B Why is aluminum resistant to corrosion, even though it is strongly electropositive?

GROUP 14: THE CARBON FAMILY

Carbon is one of the elements central to life and natural intelligence. Silicon and germanium, in the same group, are the elements central to electronic technology and artificial intelligence (Fig. 19.36, Table 19.11). The half-filled valence shell of these elements gives them special properties that straddle the line between metals and nonmetals. Carbon, at the head of the group, forms so many compounds that it has its own branch of chemistry, organic chemistry (Chapter 11). We carbon-based life forms exploit silicon-based devices to augment our own intelligence, but the intelligence of silicon-based machines is growing and may be converging on ours.

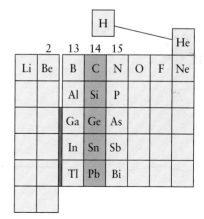

19.14 The Elements

The elements show increasing metallic character down the group (Table 19.12). Carbon has definite nonmetallic properties: it forms covalent compounds with

FIGURE 19.36

The elements of Group 14. Back row, from left to right: silicon, tin. Front row: carbon (graphite), germanium, lead.

nonmetals and ionic compounds with metals. The oxides of carbon and silicon are acidic. Germanium is a typical metalloid in that it exhibits metallic or nonmetallic properties according to the other element present in the compound. Tin and, even more so, lead have definite metallic properties. However, even though tin is classified as a metal, it is not far from the metalloids in the periodic table, and it does have some amphoteric properties. For example, it reacts with both hot concentrated hydrochloric acid and hot alkali:

$$Sn(s) + 2\,HCl(aq) \longrightarrow SnCl_2(aq) + H_2(g)$$

$$Sn(s) + 2\,OH^-(aq) + 2\,H_2O(l) \longrightarrow [Sn(OH)_4]^{2-}(aq) + H_2(g)$$

Because it stands at the head of its group, we expect carbon to be different from the other members of the group. In fact, the differences are more pronounced in Group 14 than anywhere else in the periodic table. For example, some of the differences between carbon and silicon compounds stem from the wide occurrence of C=C and C=O double bonds, compared with the rarity of

Table 19.11 The Group 14 elements

Valence configuration: ns^2np^2

Z	Name	Symbol	Molar mass, g/mol	Melting point, °C	Boiling point, °C	Density, g/cm³	Normal form*
6	carbon	C	12.01	3700s†	—	1.9 to 2.3	black nonmetal (graphite)
						3.2 to 3.5	transparent nonmetal (diamond)
							orange nonmetal (fullerite)
14	silicon	Si	28.09	1410	2620	2.33	gray metalloid
32	germanium	Ge	72.59	937	2830	5.32	gray-white metalloid
50	tin	Sn	118.69	232	2720	7.29	white lustrous metal
82	lead	Pb	207.19	328	1760	11.34	blue-white lustrous metal

*Normal form means the state and appearance of the element at 25°C and 1 atm.

†The symbol s denotes that the element sublimes.

Table 19.12 Chemical properties of the Group 14 elements

Reactant	Reaction with Group 13 element (E)
hydrogen	$C(s) + 2H_2(g) \longrightarrow CH_4(g)$ and other hydrocarbons
oxygen	$E(s) + O_2(g) \longrightarrow EO_2(s),\quad E = C, Si, Ge, Sn$
	$2Pb(s) + O_2(g) \longrightarrow 2PbO(s)$
halogen (X_2)	$E(s) + 2X_2(g, l, s) \longrightarrow EX_4(s, l, g),\quad E = C, Si, Ge, Sn$
	$Pb(s) + X_2(g, l, s) \longrightarrow PbX_2(s)$
water	$C(s) + H_2O(g) \xrightarrow{\Delta} CO(g) + H_2(g)$
	$Si(s) + 2H_2O(g) \xrightarrow{\Delta} SiO_2(s) + 2H_2(g)$
acid	$E(s) + 2H_3O^+(aq) \longrightarrow E^{2+}(aq) + 2H_2O(l) + H_2(g),\quad E = Sn, Pb$
base	$E(s) + 2H_2O(l) + 2OH^-(aq) \longrightarrow 2E(OH)_4^{2-}(aq) + H_2(g),\quad E = Sn, Pb$

$Si=Si$ and $Si=O$ double bonds. Carbon dioxide, which consists of discrete $O=C=O$ molecules, is a gas that we breathe. Silicon dioxide (silica), which consists of networks of $-O-Si-O-$ groups, is a mineral we stand on.

Singly bonded silicon compounds can also act as Lewis acids, whereas carbon compounds can function as Lewis acids only when the carbon is doubly bonded. Because a silicon atom is bigger than a carbon atom and can expand its valence shell by using its d-orbitals, it can accommodate the lone pair of an attacking Lewis base. A carbon atom is smaller and has no low-lying d-orbitals, so it cannot act as a Lewis acid when bonded to four other atoms.

The valence electron configuration is ns^2np^2 for all members of the group. All four electrons are approximately equally available for bonding in the lighter elements, and carbon and silicon are characterized by their ability to form four covalent bonds. However, on descending the group, the energy separation between the s- and p-orbitals increases and the s-electrons become progressively less available for bonding; in fact, the most common oxidation number for lead is $+2$.

Carbon is the only member of Group 14 that commonly forms multiple bonds. Silicon can expand its valence shell.

19.15 The Many Faces of Carbon

Solid carbon exists as graphite, diamond, and—as we now know—solids formed by the fullerenes. Graphite is the thermodynamically most stable of these allotropes under normal conditions. Pure graphite is produced in industry by passing a heavy electric current for several days through rods of coke, the solid left after distillation of the volatile components of coal. Natural sources of diamonds are rare. Synthetic diamonds are made at high pressure and high temperature (Section 10.13) or by thermal decomposition of methane. In the latter technique the carbon atoms settle on a cool surface as graphite and diamond. However, hydrogen atoms produced in the decomposition react more quickly with graphite to form volatile hydrocarbons, so more diamond than graphite survives.

Soot and *carbon black* contain very small crystals of graphite. Carbon black, which is produced by heating gaseous hydrocarbons to nearly 1000°C in the

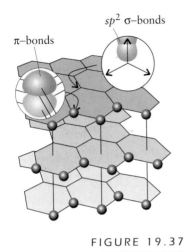

FIGURE 19.37

The structure of graphite consists of flat planes of hexagons lying one above another. When impurities are present, the planes can slide over one another quite easily. Graphite conducts electricity well within the planes, but less well perpendicular to the planes.

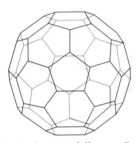

7 Buckminsterfullerene, C_{60}

The 1996 Nobel prize for chemistry was awarded to R. F. Curl, H. W. Kroto, and R. E. Smalley for the discovery of buckminsterfullerene.

absence of air, is used for reinforcing rubber, for pigments, and for printing inks, such as the ink on this page. *Activated carbon,* which is also called activated charcoal, consists of granules of microcrystalline carbon. It is produced by heating waste organic matter in the absence of air and then processing it to increase the porosity. The very high surface area (up to about 2000 m^2/g) of the porous carbon enables it to remove organic impurities from liquids and gases by adsorption. It is used in air purifiers, gas masks, and aquarium water filters. On a larger scale, activated carbon is used in emission-control canisters of automobiles to minimize the release of unburned hydrocarbons and in water purification plants to remove organic compounds from drinking water.

Differences in bonding explain the differences in properties between the carbon allotropes. Graphite consists of planar sheets of sp^2 hybridized carbon atoms in a hexagonal network (Fig. 19.37). Electrons are free to move from one carbon atom to another through a delocalized π-network formed by the overlap of unhybridized p-orbitals on each carbon atom. This network spreads across the entire plane. Because of the electron delocalization, graphite is a black, lustrous, electrically conducting solid; indeed, graphite is used as an electrical conductor in industry. Its slipperiness, which results from the ease with which the flat planes move past one another when impurities are present, leads to its use as a lubricant. In diamond, each carbon atom is sp^3 hybridized and linked tetrahedrally to its four neighbors, with all electrons localized in C—C σ-bonds (see Fig. 19.44b). Diamond is a rigid, transparent, electrically insulating solid. It is the hardest substance known and the best conductor of heat, being about five times better than copper. These last two properties make it an ideal abrasive, for it can scratch all other substances, yet the heat generated by friction is quickly conducted away.

Chemists were greatly surprised when soccer ball-shaped carbon molecules were first identified in 1985, particularly because they might be even more abundant than graphite and diamond! The C_{60} molecule (7) is named buckminsterfullerene after the American architect whose "geodesic domes" they resemble. Within two years, scientists had succeeded in making crystals of them: these solid samples are called *fullerite* (Fig. 19.38). Like the discovery of benzene, these molecules opened up the prospect of a whole new field of chemistry. For instance, the interior of a C_{60} molecule is big enough to hold an atom of another element, and chemists are now busily preparing a whole new periodic table of these "shrink-wrapped" atoms. The *fullerenes* are members of the family of molecules resembling buckminsterfullerene but having more than 60 atoms. The reason they might be very abundant is that they are formed in smoky flames and by red giants (stars with low surface temperatures and large diameters), so the universe might contain huge numbers of them.

Graphite and diamond are network solids that are insoluble in liquid solvents. However, the fullerenes, which are molecular, can be dissolved by suitable solvents (such as benzene); buckminsterfullerene itself gives a red-brown solution. Fullerite currently has few uses, but some of the compounds of the fullerenes have great promise. For example, K_3C_{60} is a superconductor below 18 K, and other compounds appear to be active against cancer and diseases such as AIDS.

Spurred on by the discovery of fullerenes, chemists are also busily—and excitedly—looking into the properties of a new form of fibrous carbon that consists of concentric tubes with walls like sheets of graphite rolled into cylinders (Fig. 19.39). These tiny structures, called *nanotubes,* hold out the promise of forming strong, conducting fibers. They can be thought of as narrow sheets of a million or more carbon atoms linked together in six-membered benzene-like rings connected as in graphite, but rolled into a very long cylinder that is only 1–3 nm in diameter internally. To illustrate how narrow the tubes are, a carbon nanotube long enough to reach from Earth to the Moon could be rolled into a ball the size of a poppy seed.

Carbon nanotubes conduct electricity because, like conducting polymers, they have an extended network of delocalized π-bonds. Electrons are delocalized from one end of the tube to the other. Along the long axis of the tube, the conductivity of a carbon nanotube can be high enough to be considered metallic. The tubes are very strong and their tensile strength parallel to the axis of the tube is the greatest of any material. Because they have a very low density, their strength-to-mass ratio is 40 times that of steel.

The very large surface area provided by the highly porous nature of nanotubes means that atoms of gases are readily adsorbed into the tubes. In addition to their uses in electronics, nanotubes carrying hydrogen molecules could therefore become a high enthalpy density storage medium for hydrogen-powered vehicles (see Applying Chemistry: Case Study 18). Such a material would solve a major obstacle to the use of hydrogen fuel cells, the problem of a safe, compact storage medium for hydrogen. Hydrogen adsorbed into nanotubes can be stored in a volume much smaller than that required to store the gas.

Carbon has an important series of allotropes: diamond, graphite, and the fullerenes.

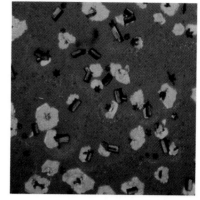

FIGURE 19.38

The small dark crystals are fullerite, in which buckminsterfullerene molecules are packed together in a close-packed structure like those shown in Figs. 10.13–10.16.

19.16 Silicon, Tin, and Lead

Silicon is the second most abundant element in the Earth's crust. It occurs widely in rocks as silicates (compounds containing the silicate ion, SiO_3^{2-}) and as the silica (SiO_2) of sand (Fig. 19.40). Pure silicon is obtained from *quartzite*, a granular form of *quartz*, by reduction with high-purity carbon in an electric arc furnace:

$$SiO_2(s) + 2\,C(s) \xrightarrow{\Delta} Si(s) + 2\,CO(g)$$

The crude product is exposed to chlorine to form silicon tetrachloride, which is then distilled and reduced with hydrogen to a purer form of the element:

$$SiCl_4(l) + 2\,H_2(g) \longrightarrow Si(s) + 4\,HCl(g)$$

"Ultrapure" silicon for use in semiconductors is produced by **zone refining**, in which a hot, molten zone is dragged from one end of a cylindrical sample to the other, collecting impurities as it goes (Fig. 19.41).

Tin and lead are obtained very easily from their ores and have been known since antiquity. Tin occurs chiefly as the mineral *cassiterite*, SnO_2, and is obtained from it by reduction with carbon at 1200°C. The principal lead ore is *galena*, PbS. It is roasted in air, which converts it to PbO, and this oxide is then reduced with coke:

$$2\,PbS(s) + 3\,O_2(g) \xrightarrow{\Delta} 2\,PbO(s) + 2\,SO_2(g)$$
$$PbO(s) + C(s) \xrightarrow{\Delta} Pb(s) + CO(g)$$

Tin is expensive and not very strong, but it is resistant to corrosion. Its main use is in tinplating, which accounts for about 40% of its consumption. Tin is also used in alloys.

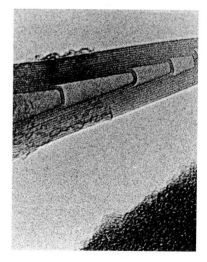

FIGURE 19.39

A slice through a concentric series of carbon nanotubes. Spacing between layers is about 0.34 nm, slightly farther apart than layers of graphite. The end of a carbon nanotube can pucker to form a cap if five-membered rings begin to form. Here we can see caps on the four innermost tubes.

(a)

(b)

(c)

FIGURE 19.40

Three of the common forms of silica, SiO_2: (a) quartz; (b) quartzite; (c) cristobalite. The black parts of the sample of cristobalite are obsidian, a volcanic rock that contains silica.

Lead's durability (its chemical inertness) and malleability make it useful in the construction industry. The inertness of lead under normal conditions can be traced to the passivation of its surface by oxides, chlorides, and sulfates. Passivation allows lead to be used for transporting hot concentrated sulfuric acid but not nitric acid, because lead nitrate is soluble. Another important property of lead is its high density, which makes it useful as a radiation shield, because its numerous electrons absorb high-energy radiation. The main use of lead today is in the electrodes of storage batteries.

Metallic character increases significantly down Group 14.

19.17 Oxides of Carbon

Such is its importance that carbon dioxide, CO_2, has been described at length throughout the text. It is formed when organic matter burns in a plentiful supply of air. Carbon dioxide is the acid anhydride of carbonic acid, H_2CO_3, which forms when the gas dissolves in water. However, not all the dissolved molecules react to form the acid, and a solution of carbon dioxide in water is an equilibrium mixture of CO_2, H_2CO_3, HCO_3^-, and a very small amount of CO_3^{2-}.

Carbon monoxide, CO, is produced when carbon or organic compounds burn in a limited supply of air, as in cigarettes and badly tuned automobile engines. It is produced commercially as synthesis gas by the re-forming reaction (Section 19.3):

$$CH_4(g) + H_2O(g) \longrightarrow CO(g) + 3 H_2(g)$$

Carbon monoxide is the formal anhydride of formic acid, HCOOH, and can be produced in the laboratory by the dehydration of that acid with hot, concentrated sulfuric acid:

$$HCOOH(l) \xrightarrow{150°C,\ H_2SO_4} CO(g) + H_2O(l)$$

Carbon monoxide is a colorless, odorless, flammable, almost insoluble, very toxic gas that condenses to a colorless liquid at $-90°C$. It is not very reactive, largely because its bond enthalpy ($1074\ kJ\cdot mol^{-1}$) is the highest for any molecule. However, it is a Lewis base, and the lone pair on the carbon atom forms

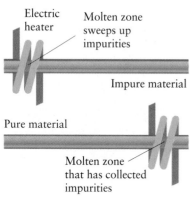

FIGURE 19.41

In the technique of zone refining, a heater that creates a molten zone is passed repeatedly from one end of the solid sample to the other. The impurities collect in the zone and are dragged through the sample.

FIGURE 19.42

covalent bonds with *d*-block atoms and ions. An example of this behavior is its reaction with nickel to give nickel carbonyl, a toxic, volatile liquid:

$$Ni(s) + 4\,CO(g) \xrightarrow{\text{50°C, 1 atm}} Ni(CO)_4(l)$$

Although nickel carbonyl is intensely poisonous, it is used in the *Mond process* for the refinement of nickel (see Section 21.3). Complex formation is also responsible for carbon monoxide's toxicity: it attaches more strongly than oxygen to the iron in hemoglobin and prevents it from accepting oxygen from lungs. As a result, the victim suffocates.

Carbon monoxide is a reducing agent used in the production of a number of metals, most notably iron in blast furnaces (Section 21.3):

$$Fe_2O_3(s) + 3\,CO(g) \xrightarrow{\text{> 800°C}} 2\,Fe(l) + 3\,CO_2(g)$$

Carbon has two important oxides, carbon dioxide and carbon monoxide. Carbon dioxide is the acid anhydride of carbonic acid, the parent acid of the hydrogen carbonates and the carbonates.

19.18 Oxides of Silicon: The Silicates

Silica, SiO_2, is a hard, rigid network solid that is insoluble in water. It occurs naturally as quartz and as sand, which consists of small fragments of quartz, usually colored golden brown by iron oxide impurities. Some precious and semiprecious stones are impure silica (Fig. 19.42). *Flint* is silica colored black by carbon impurities.

Silica gets its strength from its covalently bonded network structure. In silica itself, each silicon atom is at the center of a tetrahedron of oxygen atoms, and each corner O atom is shared by two Si atoms. Hence, each tetrahedron contributes one Si atom and $4 \times \frac{1}{2} = 2$ oxygen atoms to the solid, resulting in the empirical formula SiO_2 (Fig. 19.43). The structure of quartz is complicated, for it is built from helical chains of SiO_4 units wound around one another. When it is heated to about 1500°C, it changes to another arrangement, that of the mineral *cristobalite* (Fig. 19.44). This structure is easier to describe: its Si atoms are arranged like the C atoms in diamond, but in cristobalite an O atom lies between each pair of neighboring Si atoms.

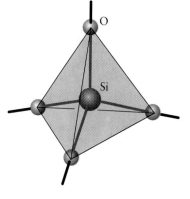

FIGURE 19.43

The structures of silicates are built up from SiO_4 tetrahedra. In different silicates, different numbers of O atoms are shared. In some cases, neighboring tetrahedra share one O atom; in others, they share two O atoms.

FIGURE 19.44

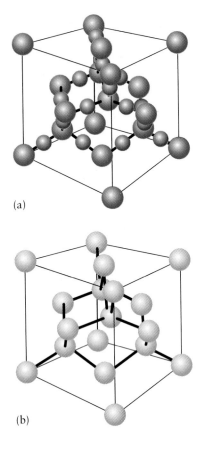

(a)

(b)

In the structure of cristobalite (a) the Si atoms lie in a tetrahedral arrangement similar to that of the C atoms in the diamond structure (b) except that an O atom (red) lies between each Si atom (purple).

Metasilicic acid, H_2SiO_3, is a weak acid. Sodium metasilicate is a basic salt used in detergents, partly as a basic buffer and partly to keep dirt from settling back onto the fabric. The SiO_3^{2-} ions attach to dirt particles, giving the particles a negative charge and thereby preventing them from merging with others into larger, insoluble particles (Fig. 19.45). Orthosilicic acid, H_4SiO_4, is also a weak acid. However, when a solution of sodium orthosilicate is acidified, instead of H_4SiO_4, a gelatinous precipitate of silica is produced:

$$4\,H_3O^+(aq) + SiO_4^{4-}(aq) + x\,H_2O(l) \longrightarrow SiO_2(s) \cdot x H_2O(gel) + 6\,H_2O(l)$$

After it is washed and dried, this *silica gel* has a very high surface area (about 700 m^2/g) and is useful as a drying agent, a support for catalysts, a packing for chromatography columns, and a thermal insulator.

Silicates can be viewed as arrangements of tetrahedral oxoanions of silicon. Each Si—O bond has considerable covalent character. The differences between the various silicates arise from the number of negative charges on each tetrahedron, the number of corner O atoms shared with other tetrahedra, and the manner in which chains and sheets of the linked tetrahedra lie together. Differences in the internal structures of these highly regular network solids lead to a wide array of materials, ranging from gemstones to fibers.

The simplest silicates, the orthosilicates, are built from SiO_4^{4-} ions. They are not very common but include the mineral *zircon*, $ZrSiO_4$, which is used as a substitute for diamond in jewelry. The *pyroxenes* consist of chains of SiO_4 units in which two corner O atoms are shared by neighboring units (Fig. 19.46); the repeating unit is the metasilicate ion, SiO_3^{2-}. Electrical neutrality is provided by cations regularly spaced along the chain. The pyroxenes include *jade*, $NaAl(SiO_3)_2$.

The chains of units can link together to form the ladderlike structures that include *tremolite*, $Ca_2Mg_5(Si_4O_{11})_2(OH)_2$. Tremolite is one of the minerals called *asbestos*, which are characterized by a fibrous structure and an ability to withstand heat (Fig. 19.47). Their fibrous quality reflects the way the ladders of SiO_4 units lie together but can easily be torn apart. Because of their great resistance to fire, asbestos fibers were once widely used for heat insulation in buildings. However, these fibers can lodge in lung tissue, where they cannot be absorbed. Eventually, fibrous scar tissue can form around them, giving rise to the disease asbestosis and creating a susceptibility to lung cancer. In some minerals, the SiO_4 tetrahedra link together to form sheets. An example is *talc*, a hydrated magnesium silicate, $Mg_3(Si_2O_5)_2(OH)_2$. Talc is soft and slippery because the silicate sheets slide over one another.

Repulsion between like charges

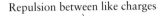

SiO_3^{2-} group

FIGURE 19.45

When SiO_3^{2-} ions (represented by green spheres) attach to dirt particles (the orange and green blobs), they repel one another and the dirt particles are prevented from collecting into larger, insoluble particles. The blue spheres represent water molecules.

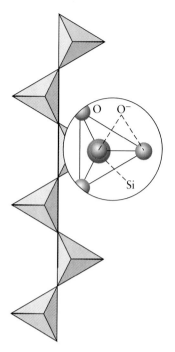

FIGURE 19.46

The basic structural unit of the minerals called pyroxenes. Each tetrahedron is an SiO_4 unit (like that in Fig. 19.43). A shared corner represents a shared O atom (the two red spheres on the left in the inset). The two unshared O atoms each carry a negative charge.

FIGURE 19.47

The minerals commonly called asbestos (from the Greek words for "not burning") are fibrous because they consist of long chains based on SiO_4 tetrahedra linked through shared oxygen atoms.

More complex (and more common) structures result when some of the Si^{4+} ions in silicates are replaced by Al^{3+} ions to form the aluminosilicates. The missing positive charge is made up by extra cations. These cations account for the difference in properties between the silicate talc and the aluminosilicate *mica* (Fig. 19.48). One form of mica is $KMg_3(Si_3AlO_{10})(OH)_2$. In this mineral, the sheets of tetrahedra are held together by extra K^+ ions. Although it cleaves neatly into layers when the sheets are torn apart, mica is not slippery like talc.

The *feldspars* are aluminosilicates in which more than half the Si^{4+} ions have been replaced by Al^{3+} ions. They are the most abundant silicate materials on Earth and are a major component of *granite,* a compressed mixture of mica, quartz, and feldspar (Fig. 19.49). When some of the cations between the crystal layers are washed away as these rocks weather, the structure crumbles to clay, one of the main inorganic components of soil. A typical feldspar has the formula $KAlSi_3O_8$. Its weathering by carbon dioxide and water can be described by the equation

$$2\ KAlSi_3O_8(s) + 2\ H_2O(l) + CO_2(g) \longrightarrow$$
$$K_2CO_3(aq) + Al_2Si_2O_5(OH)_4(s) + 4\ SiO_2(s)$$

The potassium carbonate is soluble and washes away, but the aluminosilicate remains as the clay. Clays were the raw materials for some of the first manufactured containers, ceramic pots, which have been used since prehistoric times (Applying Chemistry: Case Study 19).

Cements are obtained when aluminosilicates are melted and then allowed to solidify. The most widely used is *Portland cement,* which is made by heating a mixture of silica, clay, and limestone to about 1500°C. The cooled mass is then crushed and gypsum ($CaSO_4 \cdot 2H_2O$) is added. The main components of the complex mixture are various calcium silicates and aluminates. When water is added, complex reactions occur and the mass sets to a solid. Concrete is the very strong, durable material that results when sand and gravel are mixed into cement.

FIGURE 19.48

The aluminosilicate mica cleaves into thin transparent sheets. It is used for windows in furnaces.

Applying Chemistry: *Case Study 19*

Just as past ages have been known as the Bronze Age and the Iron Age, according to the new materials that changed lives in important ways, our times may be known one day as the Ceramic Age. Communication networks and automobile engines may soon be made of glass and ceramic materials. The beauty and utility of glasses and ceramics have been valued since antiquity. Today, high technology has found new uses for these ancient materials, including superconductors and the optical fibers connecting computers.

A *glass* is an ionic solid with an amorphous structure resembling that of a liquid. Glass has a network structure based on a nonmetal oxide, usually silica, SiO_2, that has been melted together with metal oxides, which act as "network modifiers." In a glass factory, silica in the form of sand is heated to about 1600°C. Metal oxides, such as MO (where M is a metal cation) are added to the silica. As the mixture melts, many of the Si—O bonds break and the orderly structure of the individual crystals is lost. When the melt cools, the Si—O bonds re-form but are prevented from forming a

The colors of stained glass are produced by mixing selected materials into the molten glass.

Glass fibers such as this one, about the diameter of a human hair, are used in high-performance computing systems and communications networks when large amounts of information must be transmitted in a short period of time.

crystalline lattice because some of the silicon atoms bond with the O^{2-} ions of the metal oxide to give $-Si-O^{-}-M^{2+}$ groups in place of some of the $-Si-O-Si-$ links present in pure silica. Compare the amorphous structure of glass with the long-range order of the crystalline silicates described in Section 19.18 (also recall Fig. 10.11). Silicate glasses are generally transparent and durable, and can be formed into flat sheets, blown into bottles, or molded into desired shapes.

Almost 90% of all manufactured glass combines sodium and calcium oxides with silica in *soda-lime glass*. This glass, which is used for windows and bottles, contains about 12% Na_2O prepared by the action of heat on sodium carbonate (the soda) and 12% CaO (the lime). When the

FIGURE 19.49

The mineral granite is actually a compressed mixture of mica, quartz, and feldspar.

proportions of soda and lime are reduced and 16% B_2O_3 is added, a *borosilicate glass,* such as Pyrex, is produced. Because borosilicate glasses do not expand much when heated, they survive rapid heating and cooling, and are used for ovenware and laboratory beakers.

Colored glass is produced by adding small amounts of other substances; cadmium sulfide and selenide, for instance, give ruby glass, which is red. Ordinary soda-lime glass is usually very pale green as a result of iron impurities in the form of Fe^{2+}. You can see the green color when you view a glass pane edge on. Cobalt blue glass is colored by Co^{2+} ions. Brown beer-bottle glass is colored by iron sulfides. Amber glass, such as that used for medicine bottles, is colored with a mixture of sulfur and iron oxides that give tints from pale yellow to amber. The colored oxides absorb harmful radiation.

Glass is resistant to attack by most chemicals. However, the silica in glass reacts with the strong Lewis base F^- from hydrofluoric acid to form fluorosilicate ions:

$$SiO_2(s) + 6\,HF(aq) \longrightarrow SiF_6^{2-}(aq) + 2\,H_3O^+(aq)$$

It is also attacked by the Lewis base OH^- in hot, molten sodium hydroxide and by O^{2-} in the carbonate anion of hot molten sodium carbonate:

$$SiO_2(s) + Na_2CO_3(l) \xrightarrow{1400°C} Na_2SiO_3(s) + CO_2(g)$$

The process by which silica is dissolved from glass by the ions F^- (from HF), OH^-, and CO_3^{2-} is called *etching.*

A *ceramic* is an inorganic material that has been hardened by heating to a high temperature. Many ceramics are created by heating aluminosilicate clays to drive out the water between the sheets of tetrahedra. This procedure leaves a rigid heterogeneous mass of tiny interlocking crystals bound together by glassy silica. *China clay,* which is used to make porcelain and china, is a form of aluminum aluminosilicate that can be obtained reasonably free of the iron impurities that make many clays look reddish brown. It is used in large amounts to coat paper (such as this page) to give a smooth, nonabsorbent surface.

Ceramics have internal structures that make them extremely hard, but brittle. Their stability at high temperatures has made them useful as furnace liners and has led to interest in ceramic automobile engines, which could endure overheating. High-temperature superconductors are ceramic materials formed from mixtures containing certain metal oxides, usually oxides of barium, copper, and yttrium.

Ceramics and glasses in the form of pots, jewelry, and art work are often the only materials that survive to enlighten archaeologists about early cultures. Perhaps one day we will be known by our glass and ceramic remains; future archaeologists will find electrical insulators, superconductors, and fiberglass skateboards, as well as pots and vases.

Key Concepts: silicates, structure of glass, Lewis acid-base reactions

For Further Reading
Ben Selinger, Chemistry of hardware and software, *Chemistry in the Marketplace,* Sydney: Harcourt Brace Jovanovich, 1989, pp. 193–299.

Related Exercises: 19.85–19.88

Glass is etched by reaction with hydrofluoric acid. The surface of the glass is covered with wax, a design is scratched on the wax, and acid is poured over it. This etched glass bowl was designed by the artist Frederick Carder in the 1920s.

Silicones consist of long $-O-Si-O-$ chains with the remaining silicon bonding positions occupied by organic groups, such as the methyl group, CH_3 (Fig. 19.50). Silicones are used to waterproof fabrics because their oxygen atoms attach to the fabric, leaving the hydrophobic (water-repelling) methyl groups like tiny, inside-out umbrellas sticking up out of the fabric's surface. For similar reasons, these methyl silicones are biologically inert and survive intact when exposed to body fluids. Because they do not coagulate blood and do not usually stick to body tissues, they are used for joint replacement and jaw reconstruction.

Silicate structures are based on SiO_4 tetrahedra with different negative charges and different numbers of shared O atoms.

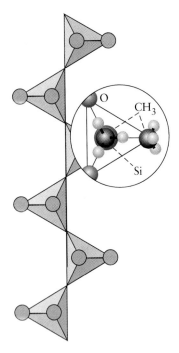

FIGURE 19.50

A typical silicone structure. The hydrocarbon groups give the substance a water-repelling quality. Note the similarity of this structure to that of the purely inorganic pyroxenes in Fig. 19.46.

19.19 Carbides

Carbides are binary compounds of carbon with metals and metalloids. There are three classes of carbides: saline carbides, covalent carbides, and interstitial carbides. The **saline carbides** (or saltlike carbides) are formed most commonly from the Group 1 and 2 metals, aluminum, and a few other metals, and contain either C_2^{2-} or C^{4-} anions. The s-block metals form saline carbides when their oxides are heated with carbon. All the C^{4-} carbides produce methane and the corresponding hydroxide in water:

$$Al_4C_3(s) + 12 H_2O(l) \longrightarrow 4 Al(OH)_3(s) + 3 CH_4(g)$$

The species C_2^{2-} is called the acetylide ion; the carbides that contain it react with water to produce ethyne (acetylene, the conjugate acid of the acetylide ion) and the corresponding hydroxide (see Fig. 3.24):

$$CaC_2(s) + 2 H_2O(l) \longrightarrow Ca(OH)_2(s) + HC\equiv CH(g)$$

Calcium carbide, CaC_2, is the most common saline carbide.

The **covalent carbides** are compounds of carbon and other nonmetals, including silicon carbide, SiC, which is sold as *Carborundum:*

$$SiO_2(s) + 3 C(s) \xrightarrow{2000°C} SiC(s) + 2 CO(g)$$

Pure silicon carbide is colorless, but iron impurities normally impart an almost black color to the crystals. Carborundum is an excellent abrasive because it is very hard, with a diamondlike structure that fractures into pieces with sharp edges (Fig. 19.51).

The **interstitial carbides** are compounds formed by the direct reaction of a d-block metal and carbon at temperatures above 2000°C. In these compounds, the C atoms occupy the gaps between the metal atoms, as do the H atoms in metallic hydrides (Fig. 19.52). Here, however, the C atoms pin the metal atoms together into a rigid structure, resulting in extremely hard substances with melting points often well above 3000°C. Tungsten carbide, WC, is used for the cutting surfaces of drills; and iron carbide, Fe_3C, is an important component of steel.

Carbon forms ionic carbides with Group 1 and Group 2 metals, covalent carbides with nonmetals, and interstitial carbides with d-block metals.

FIGURE 19.51

Carborundum crystals, showing the sharp fractured edges that give the substance its abrasive power.

FIGURE 19.52

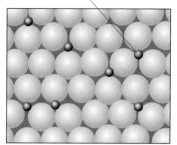

Carbon atom

The structure of an interstitial carbide, in which the carbon atoms (represented by the black spheres) lie between the larger metal atoms (the gray spheres), so producing a rigid structure.

Self-Test 19.4A Explain why graphite can conduct electricity but diamond cannot.
[*Answer:* Graphite has a network of π-bonds, through which electrons can be delocalized. In diamond, electrons are localized in σ-bonds.]

Self-Test 19.4B Explain the fact that SiH_4 reacts with water containing OH^- ions but CH_4 does not.

Skills You Should Have Mastered

Conceptual

☐ 1. Explain hydrogen's special placement in the periodic table.

☐ 2. Explain trends in the properties of binary hydrides and oxides.

☐ 3. Explain why the Period 2 elements in Groups 1, 2, 13, and 14 have properties distinct from those of the other members of their groups.

☐ 4. Distinguish the allotropes of carbon by their structures and show how their structures affect their properties.

Problem-Solving

☐ 1. Predict trends in metallic properties across a period or down a group.

☐ 2. Write the formula for an acid, given that of the anhydride, and vice versa.

Descriptive

☐ 1. Describe the names, properties, and reactions of the principal compounds of hydrogen and the Period 2 and 3 elements in Groups 1, 2, 13, and 14.

☐ 2. Describe and write balanced equations for the principal reactions used to produce hydrogen and the Period 2 and 3 elements in Groups 1, 2, 13, and 14.

☐ 3. Describe the major uses of hydrogen, sodium, potassium, beryllium, magnesium, boron, aluminum, carbon, and silicon.

☐ 4. Describe the reactions of the alkali metals with water and with nonmetals.

☐ 5. Identify amphoteric elements in Groups 2, 13, and 14.

☐ 6. Identify elements in Groups 2, 13, and 14 that are passivated by the formation of protective oxides.

☐ 7. Distinguish the principal silicate structures and describe their properties.

Exercises

Periodic Trends

19.1 Classify each of the following compounds as a saline, molecular, or metallic hydride: (a) KH; (b) NH_3; (c) HBr; (d) TaH.

19.2 Classify each of the following compounds as a saline, molecular, or metallic hydride: (a) B_2H_6; (b) SiH_4; (c) CaH_2; (d) Pd_2H.

19.3 Identify the products and write a balanced equation for the reaction of hydrogen with (a) chlorine; (b) sodium; (c) phosphorus.

19.4 Identify the products and write a balanced equation for the reaction of hydrogen with (a) nitrogen; (b) fluorine; (c) potassium.

19.5 Write the Lewis structures of (a) LiH; (b) SiH_4 (silane); (c) SbH_3.

19.6 Fluorine forms one binary hydride, oxygen forms two binary hydrides, and three of the binary hydrides that nitrogen forms are NH_3, N_2H_4, and HN_3. Write the Lewis structures for all six hydrides.

19.7 Describe the trend in acidity (a) of the binary hydrogen compounds of Period 2 elements and account for the trend in terms of electronegativity; (b) of the oxides of Period 2 elements.

19.8 Write chemical equations that represent the acidic or basic character of (a) two metal oxides and (b) two nonmetal oxides.

Hydrogen

19.9 Describe evidence for the statement that hydrogen can act as both a reducing agent and an oxidizing agent. Give chemical equations to support your evidence.

19.10 Explain why hydrogen differs from the Group 1 elements in its chemical and physical properties. Give two examples to support your explanation.

19.11 Write the chemical equations for (a) the shift reaction and the reactions between (b) lithium and water; (c) magnesium and hot water; (d) potassium and hydrogen.

19.12 Write the chemical equations for (a) the reaction between sodium hydride and water; (b) the formation of synthesis gas; (c) the hydrogenation of ethene, $H_2C=CH_2$; (d) the reaction of magnesium with hydrochloric acid.

19.13 Use Appendix 2A to determine the standard reaction enthalpy at 25°C for (a) the re-forming reaction; (b) the shift reaction; (c) the overall reaction of these two processes.

19.14 Use Appendix 2A to determine the standard free energy at 25°C of (a) the re-forming reaction; (b) the shift reaction; (c) the overall reaction of these two processes.

19.15 About 3×10^8 kg of hydrogen is produced each year in the United States. If 1.5×10^9 kg of ammonia is produced annually by the Haber process, what fraction of the H_2 is used for this purpose?

19.16 Calculate the volume that the annual United States production of hydrogen (3×10^8 kg) would occupy if it were stored as liquid hydrogen. The density of liquid hydrogen is 0.089 g/cm^3.

Group 1: The Alkali Metals

19.17 Refer to Fig. 7.26 and describe the color of the flame test for (a) lithium; (b) potassium; (c) sodium; (d) rubidium.

19.18 (a) Write the valence electron configuration for the alkali metal atoms. (b) Explain why the alkali metals are strong reducing agents in terms of electron configurations and ionization energies.

19.19 Write the chemical equations for the reactions between (a) lithium and oxygen; (b) lithium and nitrogen; (c) sodium and water; (d) potassium superoxide and water.

19.20 Write the chemical equations for the reactions between (a) potassium and oxygen; (b) sodium oxide and water; (c) lithium and hydrochloric acid; (d) cesium and iodine.

19.21 Complete and balance the following equations:
(a) $Ca(s) + H_2(g) \rightarrow$
(b) $NaHCO_3(s) \xrightarrow{\Delta}$

19.22 Complete and balance the following equations:
(a) $Na(s) + H_2O(l) \rightarrow$
(b) $K(s) + F_2(g) \rightarrow$

19.23 Give the chemical names and formulas of the minerals (a) rock salt; (b) sylvite.

19.24 Write the chemical formulas for the compounds (a) washing soda; (b) soda ash.

19.25 Sodium carbonate is often supplied as the decahydrate, $Na_2CO_3 \cdot 10H_2O$. What mass of this solid should be used to prepare 250. mL of 0.100 M $Na_2CO_3(aq)$?

19.26 Sodium metal is produced from the electrolysis of molten sodium chloride in the Downs process (Chapter 18). Determine (a) the standard free energy of the reaction in which 1.000 mol $Cl_2(g)$ is produced (see Appendix 2A), and (b) the current needed to produce 1.00 kg of sodium in 10.0 h.

Group 2: The Alkaline Earth Metals

19.27 Predict and explain the trend in strengths of the Group 2 metals as reducing agents, moving down the group.

19.28 Explain the decreasing lattice enthalpies of the chlorides of the Group 2 metals, moving down the group.

19.29 Name the minerals used as sources of (a) beryllium; (b) calcium; (c) magnesium.

19.30 Distinguish among limestone, lime, quicklime, slaked lime, chalk, marble, and calcite.

19.31 Give the chemical names and write the formulas of (a) Epsom salts; (b) limestone; (c) milk of magnesia. (See Appendix 3B.)

19.32 Give the chemical names and write the formulas of (a) calcite; (b) dolomite; (c) quicklime.

19.33 Bearing in mind that aluminum and beryllium have a diagonal relationship, write the chemical equations for the reaction of (a) aluminum with aqueous sodium hydroxide; (b) beryllium with aqueous sodium hydroxide.

19.34 Write the chemical equations for (a) the industrial preparation of magnesium metal from the magnesium chloride in seawater; (b) the action of water on calcium metal.

19.35 Predict the products of the following reactions and then balance the equations:
(a) $Mg(OH)_2(s) + HCl(aq) \rightarrow$
(b) $Ca(s) + H_2O(l) \rightarrow$
(c) $BaCO_3(s) \xrightarrow{\Delta}$

19.36 Predict the products of the following reactions and then balance the equations:
(a) $Mg(s) + Br_2(l) \rightarrow$
(b) $Ca(NO_3)_2(s) \xrightarrow{\Delta}$
(c) $CaO(s) + SiO_2(s) \xrightarrow{\Delta}$

19.37 (a) Write the Lewis structure for $BeCl_2$ (see Chapter 8) and (b) predict the $Cl—Be—Cl$ bond angle. (c) What hybrid orbitals are used in the bonding?

19.38 (a) Write the Lewis structure of $MgCl_2$ (see Chapter 8). (b) How does its structure differ from that of $BeCl_2$?

19.39 What is the mass percentage of water in the magnesium sulfate hydrate sold as Epsom salts? See Appendix 3B.

19.40 The concentration of magnesium in seawater is about 1.35 g/L. What volume of water must be processed to collect 1.0 kg of magnesium, assuming an 80% removal?

Group 13: The Boron Family

19.41 Write a balanced equation for the industrial preparation of aluminum from its oxide.

19.42 Write a balanced equation for the industrial preparation of impure boron.

19.43 Write the formula of (a) boric acid; (b) alumina; (c) borax.

19.44 Write the formula of (a) potassium alum; (b) corundum; (c) diborane.

19.45 Complete and balance the following equations:
(a) $B_2O_3(s) + Mg(l) \xrightarrow{\Delta}$
(b) $Al(l) + Cl_2(g) \rightarrow$
(c) $Al(s) + O_2(g) \rightarrow$

19.46 Complete and balance the following equations:
(a) $Al_2O_3(s) + OH^-(aq) \rightarrow$
(b) $Al_2O_3(s) + H_3O^+(aq) \rightarrow$
(c) $B(s) + NH_3(g) \xrightarrow{\Delta}$

19.47 Identify a use for (a) $AlCl_3$; (b) α-alumina; (c) $B(OH)_3$.

19.48 Identify a use for (a) BF_3; (b) $NaBH_4$; (c) $Al_2(SO_4)_3$.

19.49 Balance the following skeletal equations:
(a) $B_2H_6 + H_2O \rightarrow B(OH)_3 + H_2$
(b) $B_2H_6 + O_2 \rightarrow B_2O_3 + H_2O$

19.50 Balance the following skeletal equations:
(a) $B_2H_6 + NaBH_4 \rightarrow Na_2B_{12}H_{12} + H_2$
(b) $B_2O_3 + C + Cl_2 \rightarrow BCl_3 + CO$

Group 14: The Carbon Family

19.51 Describe the sources of silicon and write balanced equations for the three steps in the industrial preparation of silicon.

19.52 Describe the sources of carbon and how it is converted to graphite.

19.53 Compare the hybridizations and structures of carbon in diamond and graphite. How do these features explain the physical properties of the two allotropes?

19.54 Explain why there is no silicon analogue of the graphite structure.

19.55 Write formulas for (a) carborundum; (b) silica; (c) zircon.

19.56 Write formulas for (a) nickel carbonyl; (b) tungsten carbide; (c) silica gel.

19.57 Complete and balance the equations for the following reactions (refer to Table 19.12, if necessary):
(a) $SiCl_4(l) + H_2(g) \rightarrow$
(b) $SiO_2(s) + C(s) \rightarrow$
(c) $Ge(s) + F_2(g) \rightarrow$
(d) $CaC_2(s) + H_2O(l) \rightarrow$

19.58 Complete and balance the equations for the following reactions (refer to Table 19.12, if necessary):
(a) $Sn(s) + H_2O(l) + OH^-(aq) \rightarrow$
(b) $C(s) + H_2O(g) \xrightarrow{\Delta}$
(c) $CH_4(g) + H_2O(g) \xrightarrow{\Delta}$
(d) $Al_4C_3(s) + H_2O(l) \rightarrow$

19.59 Write a Lewis structure for the orthosilicate anion, SiO_4^{4-}, and deduce the formal charges and oxidation numbers of the atoms. Use the VSEPR model to predict the shape of the ion.

19.60 Use the VSEPR model to estimate the $Si—O—Si$ bond angle in silica.

19.61 Calculate the mass percentage of silicon in silica.

19.62 Calculate the mass percentage of silicon in feldspar, $KAlSi_3O_8$.

19.63 Determine the mass of HF(aq), hydrofluoric acid, that is required to etch 2.00 mg of SiO_2 from a glass plate in the reaction $SiO_2(s) + 6 HF(aq) \rightarrow 2 H_3O^+(aq) + SiF_6^{2-}(aq)$.

19.64 What mass of coke containing 98% carbon is needed to reduce the silicon in 1.0 kg of 88.5% pure silica?

19.65 Determine the surface area in square meters of 1.00 mol C as activated carbon if the surface area of 1.0 g is $2.0 \times 10^3 \text{ m}^2$.

19.66 Explain why silicon tetrachloride reacts with water to produce SiO_2, but carbon tetrachloride does not react with water. (Consider the Lewis acidity of each compound.)

19.67 Describe the structures of two silicates in which the silicate tetrahedra share (a) one O atom; (b) two O atoms.

19.68 What is the chemical formula of a potassium silicate in which the silicate tetrahedra share (a) one O atom or (b) two O atoms in a long chain? In each case, there are single negative charges on the unshared O atoms.

Supplementary Exercises

19.69 (a) Plot standard potential against atomic number for the elements of Groups 1 and 2. Refer to Appendix 2B for data. (b) What generalizations can be deduced from the graph?

19.70 Use data from Fig. 7.38 and Appendix 2B to plot ionization energy against standard potential for the elements of Groups 1 and 2. What generalizations can be drawn from the graph?

19.71 (a) Write equations for the reactions of hydrogen gas with the halogens, from fluorine to iodine. (b) Comment on the relative vigor of the reactions. (c) Name the products when they are dissolved in aqueous solutions.

19.72 (a) Name the type of reaction that occurs between calcium oxide and silica in a blast furnace. (b) Write the chemical equation for a related reaction, that between calcium oxide and carbon dioxide.

19.73 Write the Lewis structures of (a) BaO_2; (b) BeH_2; (c) Na_2O_2; (d) $Be(OH)_4^{2-}$.

19.74 State a use for the elemental form of (a) boron; (b) aluminum; (c) beryllium; (d) silicon; (e) tin.

19.75 (a) Write Lewis structures for boric acid, $B(OH)_3$, and boron trifluoride. (b) Using the VSEPR model, predict the structure and bond angles of boric acid and boron trifluoride. (c) What type of hybridization can be ascribed to the boron atom in the bonding in boric acid and boron trifluoride?

19.76 Arrange the elements aluminum, gallium, indium, thallium, tin, and germanium in order of increasing electronegativity (see Fig. 8.21) and increasing reducing strength.

19.77 (a) State the trends in first ionization energies and atomic radii down Groups 13 and 14. (b) Account for the trends. (c) How do the trends correlate with the properties of the elements?

19.78 (a) Suggest a reason for the observations that methane is stable in an aqueous alkaline solution, but silane, SiH_4, reacts rapidly in the same solution. (b) Write a balanced chemical equation for the reaction of silane with water in an aqueous alkaline solution.

19.79 Complete and balance the following equations:

(a) $AlCl_3(s) + H_2O(l) \rightarrow$

(b) $B_2H_6 \xrightarrow{\text{high temperatures}}$

(c) $BF_3 + BH_4^- \xrightarrow{\text{organic solvent}}$

19.80 Complete and balance the following equations:
(a) $Na_2CO_3(s) + HCl(aq) \rightarrow$
(b) $SiCl_4 + H_2O \rightarrow$
(c) $SnO_2 + C \rightarrow$

19.81 Is there any chemical support for the view that hydrogen should be classified as a member of Group 1? Would it be better to consider hydrogen a member of Group 17? Give evidence that supports each view.

19.82 What justification is there for regarding the ammonium ion as an analogue of a Group 1 metal cation? Consider properties such as solubility, charge, and size.

19.83 It has been proposed that the ability of a cation to polarize anions is proportional to its charge divided by its radius (see Section 8.15). (a) Use this criterion to arrange the s-block elements in order of increasing polarizing power. (b) Do the resulting values support the diagonal relationships within the block?

19.84 Suppose that the stability of carbonates when heated depends on the ability of the metal cation to polarize the carbonate ion and remove an oxide ion from it, thereby releasing carbon dioxide. (a) Predict the order of thermal stability of the Group 1 and 2 metal carbonates. Comment on the likely stability of aluminum carbonate. (b) When the mineral dolomite, $CaCO_3 \cdot MgCO_3$, is heated, it gives off carbon dioxide and forms a mixture of a metal oxide and a metal carbonate. Which oxide is formed, CaO or MgO? Which carbonate is formed, $CaCO_3$ or $MgCO_3$?

Applied Exercises

For Exercises 19.85–19.88, see Applying Chemistry: Case Study 19.

19.85 Write the chemical equation for a reaction between OH^- and SiO_2. (*Hint:* Silicon can act as a Lewis acid.)

19.86 Water adheres to glass. What kinds of intermolecular processes are involved when water adheres to silicate glass? Sketch the intermolecular interactions between water molecules and SiO_2 units in glass.

19.87 Which of the following would solidify as a glass when cooled: (a) tar (which contains many different long-chain hydrocarbons); (b) sodium chloride; (c) molten granite (see Section 19.18); (d) water; (e) low-density polyethylene; (f) a highly branched polymer (Section 11.15)?

19.88 Select the physical properties that you would expect to differ between the glassy and crystalline phases of the same substance and explain your answers: (a) ability to cleave along a plane; (b) rigidity; (c) sharp melting point; (d) transparency; (e) isotropy (appearing the same in all directions).

19.89 (a) What is "hard water" (see Section 12.3)? (b) Write chemical equations for the softening of hardness due to HCO_3^- ions in water that can be achieved by using slaked lime. Refer to Connection 4, following Chapter 16.

Integrated Exercises

19.90 What mass of calcium oxide can be produced from the thermal decomposition of 200. g of calcium carbonate?

19.91 Magnesium is produced from the electrolysis of molten magnesium chloride. (a) Calculate the mass of magnesium that can be produced in 1.5 h, using a current of 100. A. See Chapter 18. (b) Calculate the volume (at STP) of chlorine gas produced when 1000. kg of magnesium metal is obtained from this process.

19.92 Calcium hydride is used as a portable source of hydrogen on account of its reaction with water:

$$CaH_2(s) + 2 H_2O(l) \longrightarrow Ca(OH)_2(s) + 2 H_2(g)$$

(a) What volume of H_2 gas (at STP) can be produced from 500. g of CaH_2? (b) How many milliliters of water should be supplied for the reaction? Assume the density of water to be 1.0 g/mL.

19.93 What volume of hydrogen gas (at STP) is required for the reduction of 20.0 g of WO_3 in the reaction $WO_3(s) + 3 H_2(g) \rightarrow W(s) + 3 H_2O(l)$?

19.94 Hydrogen burns in an atmosphere of bromine to give hydrogen bromide. If 120. mL of H_2 gas at STP combines with a stoichiometric amount of bromine and the resulting hydrogen bromide dissolves to form 150. mL of an aqueous solution, what is the molar concentration of the resulting hydrobromic acid solution?

19.95 The standard enthalpies of formation of $BH_3(g)$ and diborane are $+100.$ kJ/mol and $+36$ kJ/mol, respectively, and the enthalpies of formation of $B(g)$ and $H(g)$ are $+563$ kJ/mol and $+218$ kJ/mol, respectively. Use these values to calculate the mean bond enthalpies of the B—H bonds in each case and to estimate the bond enthalpies of the terminal B—H and H—B—H bonds. Which bonds would you expect to be the longer?

19.96 Determine the values of $\Delta H°$, $\Delta S°$, and $\Delta G°$ for the production of high-purity silicon by the reaction $SiO_2(s) + 2 C(s, \text{graphite}) \rightarrow Si(s) + 2 CO(g)$ at 25°C and estimate the temperature at which the reaction becomes spontaneous.

19.97 Determine the values of $\Delta H°$, $\Delta S°$, and $\Delta G°$ for the reaction $2 CO(g) + O_2(g) \rightarrow 2 CO_2(g)$ at 25°C and estimate the temperature at which the reaction ceases to be spontaneous.

19.98 Hydrogen can be produced from the electrolysis of water. (a) Write the equation for the half-reaction for the production of hydrogen. (b) Is the hydrogen produced at the anode or cathode? (c) Calculate the volume of hydrogen (at STP) that is produced if a current of 10.0 A is passed through an electrolytic cell for 30. min.

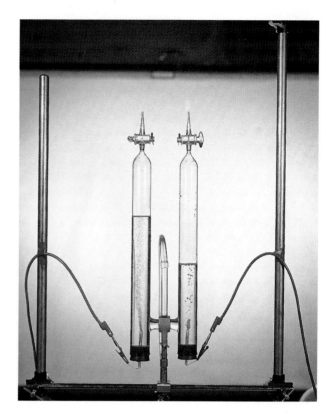

Water undergoing electrolysis

19.99 The reduction of tin(IV) oxide to (white) tin metal by graphite proceeds at moderately low temperatures. However, tin(IV) oxide is normally reduced to the metal at temperatures above 980 K by an excess of carbon monoxide. (a) Write the two chemical equations for the reduction of SnO_2. (b) Determine the value of $\Delta G°$ for each reaction at 25°C. (c) How is the spontaneity affected by temperature in each case?

19.100 The standard reduction potential of the Al^{3+}/Al redox couple is -1.66 V. Calculate the standard free energy of formation of $Al^{3+}(aq)$. Account for any differences between the standard free energy of formation of $Tl^{3+}(aq)$ ($+215$ kJ/mol at 25°C) and that of $Al^{3+}(aq)$.

19.101 What mass of aluminum can be produced by the Hall process in a period of 8.0 h, using a current of 1.0×10^5 A?

19.102 In the production of 1.0×10^3 kg of aluminum by the Hall process, what mass of carbon is lost at the anode?

19.103 Heat is generated in the formation of slaked lime from the reaction $CaO(s) + H_2O(l) \rightarrow Ca(OH)_2(s)$. (a) Calculate the standard enthalpy of reaction. (b) If all this heat could be transferred to 250. g of water, what would be the resulting temperature change?

The Elements: The Last Four Main Groups

We are now in the heart of the *p* block of the periodic table. In Groups 15 to 18, we find nonmetallic elements of great variety. Some are richly colored, most are relatively soft, and some are even gases. Except for carbon and hydrogen, the gases of the air are made up solely of elements from this part of the *p* block, some as elements and some as compounds. Every time lightning strikes, it initiates chemical reactions among these gases and produces other compounds.

Metallic character increases down each group in this region of the periodic table, but the only element considered to be metallic is bismuth, at the foot of Group 15. Arsenic, antimony, tellurium, and polonium are metalloids. All the elements have moderately high ionization energies and form compounds with other nonmetals by sharing electrons in covalent bonds. When they react with metals, these elements form ionic bonds in which they accept electrons and become anions.

The firing of the powerful space shuttle booster rockets is an impressive display of the vigor of the reactions that take place between some of the elements on the right of the periodic table. These elements play an important role in many aspects of space flight, not only as fuel. Their reactions are also important in maintaining life on our planet.

	H		
	14	15	16	He

Li	Be	B	C	N	O	F	Ne
			Si	P	S		
			Ge	As	Se		
			Sn	Sb	Te		
			Pb	Bi	Po		

GROUP 15: THE NITROGEN FAMILY

The Group 15 elements (Table 20.1) all have the valence configurations ns^2np^3. They range in character from the nonmetals nitrogen and phosphorus to the largely metallic bismuth (Fig. 20.1). This range of behavior is reflected in their chemical properties (Table 20.2). For example, all the oxides of nitrogen and phosphorus are acidic, whereas bismuth's oxide is basic.

20.1 The Elements

Nitrogen is the principal component of air (76% by mass) and is obtained by the distillation of liquid air. In this process, air is cooled below $-196°C$ and then warmed. The nitrogen (b.p. $-196°C$) boils off, but most of the oxygen (b.p. $-183°C$) remains as a liquid. Any oxygen that does boil off is removed by passing the gases over hot copper:

$$2\,Cu(s) + O_2(g) \xrightarrow{\Delta} 2\,CuO(s)$$

The strong $N\equiv N$ bond (944 kJ/mol) in N_2 makes nitrogen almost as inert as the noble gases. To be available for organisms, it must first be **fixed,** or combined with other elements. Once fixed, nitrogen can be converted into other compounds for use as fertilizers, explosives, and plastics. Lightning converts some nitrogen to its oxides, which rain then washes into the soil. Some species of bacteria also fix nitrogen in nodules on the roots of clover, beans, peas, alfalfa, and other legumes (Fig. 20.2). At present, the Haber synthesis of ammonia is the main industrial route to fixing nitrogen at high temperatures and pressures (Section 14.12), but the search is on for catalysts that work at normal temperatures.

Unlike the other Group 15 elements, nitrogen is highly electronegative ($\chi = 3.0$, the same as that of bromine). Because its atoms are small, it can form multiple bonds by using its p-orbitals; but its valence shell ($n = 2$) has no d-orbitals. These characteristics account for many of the differences between the chemical and physical properties of nitrogen and those of the other Group 15

Table 20.1 The Group 15 elements

Valence configuration: ns^2np^3

Z	Name	Symbol	Molar mass, g/mol	Melting point, °C	Boiling point, °C	Density, g/cm³ at 25°C	Normal form*
7	nitrogen	N	14.01	−210.	−196	1.04‡	colorless gas
15	phosphorus	P	30.97	44	280.	1.82	white nonmetal
33	arsenic	As	74.92	613s†	—	5.78	gray metalloid
51	antimony	Sb	121.75	631	1750.	6.69	blue-white lustrous metalloid
83	bismuth	Bi	208.98	271	1650.	8.90	white-pink metal

*Normal form means the appearance and state of the element at 25°C and 1 atm.

†The symbol s denotes that the element sublimes.

‡For the liquid at its boiling point.

FIGURE 20.1

The elements of Group 15: (back row, from left to right) nitrogen (cooled to the liquid), red phosphorus, arsenic; (front row, from left to right) antimony and bismuth.

elements. For example, nitrogen can form bonds to no more than four atoms at a time, whereas phosphorus can bind to six. Moreover, the compounds that nitrogen forms with hydrogen are the only ones in Group 15 that take part in hydrogen bonding. Nitrogen has one of the widest ranges of oxidation numbers of any element: nitrogen compounds are known for each whole-number oxidation number from -3 (in NH_3) to $+5$ (in nitric acid and the nitrates). It also occurs with fractional oxidation numbers, such as $-\frac{1}{3}$ in the azide ion, N_3^-.

Phosphorus is obtained from *apatites,* which are mineral forms of calcium phosphate, $Ca_3(PO_4)_2$. The rocks are heated in an electric furnace with carbon and sand:

$$2\,Ca_3(PO_4)_2(s) + 6\,SiO_2(s) + 10\,C(s) \xrightarrow{\Delta} P_4(g) + 6\,CaSiO_3(l) + 10\,CO(g)$$

The phosphorus vapor condenses as *white phosphorus,* a soft, white, poisonous molecular solid consisting of tetrahedral P_4 molecules (**1**). This allotrope is highly reactive, in part because of the strain associated with the acute angles between the bonds. It bursts into flame on contact with air, and is normally stored under water. White phosphorus changes into red phosphorus when heated in the absence of air (see Fig. 20.1). This allotrope is less reactive, but it can be ignited by friction. Red phosphorus is used in the striking surfaces of matchbooks and on the sides of boxes of safety matches. The friction created by rubbing a match across the rough surface ignites the phosphorus sufficiently to light the highly flammable material in the match head. The structure of red phosphorus may consist of chains of linked P_4 tetrahedra.

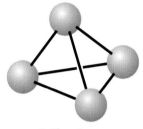

1 Phosphorus

The name *phosphorus* comes from the Greek words for "light bringer."

Table 20.2 Chemical properties of the Group 15 elements

Reactant	Reaction with Group 15 element (E)
hydrogen	$N_2(g) + 3\,H_2(g) \longrightarrow 2\,NH_3(g)$ $P_4(s) + 6\,H_2(g) \longrightarrow 4\,PH_3(g)$
oxygen	$N_2(g) + x\,O_2(g) \longrightarrow 2\,NO_x(g)$ $P_4(s) + 3\ or\ 5\,O_2(g) \longrightarrow P_4O_6(s)\ or\ P_4O_{10}(s)$ $4\,As(s) + 3\,O_2(g) \longrightarrow As_4O_6(s)$ $4\,E(s) + 3\,O_2(g) \longrightarrow 2\,E_2O_3(s),\quad E = Sb,\ Bi$
water	no reaction
halogen	$2\,E(s) + 3\,X_2(s,l,g) \longrightarrow 2\,EX_3(s,l),\quad E = P,\ As,\ Sb,\ Bi$ $2\,E(s) + 5\,X_2(s,l,g) \longrightarrow 2\,EX_5(s,l),\quad E = P,\ As,\ Sb$

FIGURE 20.2

The bacteria that inhabit these nodules on the roots of a pea plant are able to fix atmospheric nitrogen and make it available to the plant.

FIGURE 20.3

The minerals (from left to right) orpiment, As_2S_3; stibnite, Sb_2S_3; and realgar, As_4S_4.

Arsenic and antimony are metalloids. They have been known in the pure state since ancient times because they are easily reduced from their ores (Fig. 20.3). In the elemental state, they are used primarily in the lead alloys employed as electrodes in storage batteries. Bismuth is used in low-melting alloys and in medicines that relieve indigestion. It is also used to make type for the printing industry, although that usage has decreased. Like ice, solid bismuth is less dense than the liquid. As a result, molten bismuth does not shrink when it solidifies in type molds.

Nitrogen is highly unreactive as an element, largely because of its strong triple bond; white phosphorus is highly reactive.

20.2 Compounds with Hydrogen and the Halogens

By far the most important hydrogen compound of a Group 15 element is ammonia, NH_3, which is prepared in huge amounts by the Haber process (Section 14.12). Small quantities of ammonia are present naturally in the atmosphere as a result of the bacterial decomposition of organic matter in the absence of air. This decomposition typically occurs in lake and river beds, in swamps, and in cattle feedlots.

Ammonia is a pungent, toxic gas that condenses to a colorless liquid at $-33°C$. The liquid resembles water in its physical properties, including its ability to act as a solvent for a wide range of substances. Gaseous ammonia is very soluble in water because the NH_3 molecules can form hydrogen bonds to H_2O molecules. Ammonia is a weak Brønsted base in water; it is also a reasonably strong Lewis base, particularly toward *d*-block elements. For example, it reacts with $Cu^{2+}(aq)$ ions to give a deep blue complex (Fig. 20.4):

$$Cu^{2+}(aq) + 4\,NH_3(aq) \longrightarrow [Cu(NH_3)_4]^{2+}(aq)$$

Ammonium salts decompose when heated:

$$(NH_4)_2CO_3(s) \xrightarrow{\Delta} 2\,NH_3(g) + CO_2(g) + H_2O(g)$$

The pungent smell of decomposing ammonium carbonate made it an effective smelling salt, a type of medication used to revive people who had fainted.

The pungency of hot ammonium chloride was known to the Ammonians, the worshippers of the Egyptian god Ammon, and they used it in their ceremonies.

Hydrazine, NH_2NH_2, is an oily, colorless liquid. It is prepared by the gentle oxidation of ammonia with alkaline hypochlorite solution:

$$2\,NH_3(aq) + ClO^-(aq) \longrightarrow N_2H_4(aq) + Cl^-(aq) + H_2O(l)$$

Its physical properties are very similar to those of water; for instance, its melting point is 2.0°C and its boiling point is 113°C. However, it is dangerously explosive and is normally kept in aqueous solution. Hydrazine is used as a rocket fuel (see Applying Chemistry: Case Study 20). It is also added to the water used in high-pressure, high-temperature steam furnaces to eliminate dissolved, corrosive oxygen from the water:

$$N_2H_4(aq) + O_2(g) \longrightarrow N_2(g) + 2\,H_2O(l)$$

Ammonia can be regarded formally as the parent of the nitrides, which contain the nitride ion, N^{3-}. Magnesium nitride, Mg_3N_2, is formed together with the oxide when magnesium is burned in air (Fig. 20.5):

$$3\,Mg(s) + N_2(g) \xrightarrow{\Delta} Mg_3N_2(s)$$

Magnesium nitride, like all nitrides, dissolves in water to produce ammonia and the corresponding hydroxide:

$$Mg_3N_2(s) + 6\,H_2O(l) \longrightarrow 3\,Mg(OH)_2(s) + 2\,NH_3(g)$$

In this reaction, the nitride ion acts as a strong base, accepting protons from water to form ammonia.

Hydrazoic acid, HN_3, is the parent of the azides, which contain the azide ion, N_3^-, a highly reactive polyatomic anion of nitrogen. Its most common salt, sodium azide, NaN_3 (Fig. 20.6), is prepared from dinitrogen oxide and molten sodium amide:

$$N_2O(g) + 2\,NaNH_2(l) \xrightarrow{175°C} NaN_3(l) + NaOH(l) + NH_3(g)$$

Lead azide, $Pb(N_3)_2$, like most azide salts, is shock sensitive and is used as a detonator for some explosives.

The hydrogen compounds of other members of Group 15 are much less stable than ammonia and decrease in stability down the group. Phosphine, PH_3, is a poisonous gas that smells faintly of garlic and bursts into flame in air if the phosphine is slightly impure. It is much less soluble than ammonia in water because PH_3 cannot form hydrogen bonds to water. Aqueous solutions of phosphine are neutral, for PH_3 has only a very weak tendency to accept a proton ($pK_b = 27.4$).

Phosphorus trichloride, PCl_3, and phosphorus pentachloride, PCl_5, are the two most important halides of phosphorus. The trichloride is prepared by direct chlorination of phosphorus. It is a major intermediate for the production of pesticides, oil additives, and flame retardants. Phosphorus pentachloride is made by allowing the trichloride to react with more chlorine (recall Fig. 8.17).

A typical reaction of the nonmetal halides is their reaction with water to give oxoacids, without a change in oxidation number:

$$PCl_3(l) + 3\,H_2O(l) \longrightarrow H_3PO_3(aq) + 3\,HCl(g)$$

This reaction is an example of a **hydrolysis reaction,** a reaction with water in which new element-oxygen bonds are formed. Another example is the reaction of PCl_5 with water to produce phosphoric acid, H_3PO_4:

$$PCl_5(s) + 4\,H_2O(l) \longrightarrow H_3PO_4(aq) + 5\,HCl(g)$$

This reaction is violent and dangerous.

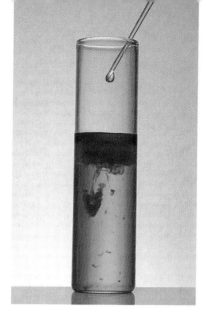

FIGURE 20.4

When aqueous ammonia is added to a copper(II) sulfate solution, first a light-blue precipitate of $Cu(OH)_2$ forms. The precipitate disappears when more ammonia is added to form the dark blue complex $[Cu(NH_3)_4]^{2+}$ by a Lewis acid-base reaction.

Section 5.9 describes the action of detonators and their use in automobile air bags.

FIGURE 20.5

Magnesium nitride is formed when magnesium burns in an atmosphere of nitrogen.

FIGURE 20.6

When sodium azide (left) is heated in a sealed, evacuated tube, it decomposes into sodium metal and nitrogen gas. The sodium condenses on the walls of the tube (right).

The 1998 Nobel prize for chemistry was awarded to Robert Furchgott, Louis Ignarro, and Ferid Murad, who identified the physiological role of nitric oxide.

The important compounds of nitrogen with hydrogen are ammonia, hydrazine, and hydrazoic acid, the parent of the shock-sensitive azides. Phosphine forms neutral solutions in water; hydrolysis of nonmetal halides produces oxoacids with no change in oxidation number.

20.3 Nitrogen Oxides and Oxoacids

The numerous nitrogen oxides may seem confusing at first. Fortunately, their properties can be understood by keeping track of their oxidation numbers. All nitrogen oxides are acidic, and some are the acid anhydrides of the nitrogen oxoacids (Table 20.3). In atmospheric chemistry, where the oxides play an important role in both maintaining and polluting the atmosphere, they are referred to collectively as NO_x (read "nox").

Dinitrogen monoxide, N_2O (oxidation number $+1$), is commonly called nitrous oxide. It is formed by gently heating ammonium nitrate:

$$NH_4NO_3(s) \xrightarrow{250°C} N_2O(g) + 2 H_2O(g)$$

Because it is tasteless, is nontoxic in small amounts, and dissolves readily in fats, N_2O is sometimes used as a foaming agent and propellant for whipped cream.

Nitrogen oxide (or nitrogen monoxide), NO (oxidation number $+2$), is commonly called nitric oxide. It is prepared industrially by the catalytic oxidation of ammonia:

$$4 NH_3(g) + 5 O_2(g) \xrightarrow{1000°C, Pt} 4 NO(g) + 6 H_2O(g)$$

Atmospheric nitrogen is also converted into NO in hot airplane and automobile engines. The NO contributes to the problem of acid rain, the formation of smog, and the destruction of the ozone layer. Nitric oxide is synthesized in minute quantities in our bodies, where it acts as a neurotransmitter and participates in the physiological changes accompanying sexual arousal.

Nitrogen dioxide, NO_2 (oxidation number $+4$), is a choking, poisonous, brown gas that contributes to the color and odor of smog. It is an odd-electron molecule; and in the gas phase, it exists in equilibrium with its colorless dimer N_2O_4. Only the dimer exists in the solid, so the brown gas condenses to a colorless solid. When it dissolves in water, NO_2 *disproportionates;* that is, it reacts in

Table 20.3 The oxides and oxoacids of nitrogen

Oxidation number	Oxide		Oxoacid	
	Formula	Name	Formula	Name
5	N_2O_5	dinitrogen pentoxide	HNO_3	nitric acid
4	NO_2*	nitrogen dioxide	—	
	N_2O_4	dinitrogen tetroxide	—	
3	N_2O_3	dinitrogen trioxide	HNO_2	nitrous acid
2	NO	nitrogen monoxide nitric oxide	—	
1	N_2O	dinitrogen monoxide nitrous oxide	$H_2N_2O_2$	hyponitrous acid

*$2 NO_2 \rightleftharpoons N_2O_4$.

such a way that the oxidation number of nitrogen in some of the molecules increases whereas the oxidation number of nitrogen in other NO_2 molecules decreases. In other words, both an oxidation product and a reduction product are formed from the same reactant. The products in this case are nitric acid and nitric oxide:

$$3\,NO_2(g) + H_2O(l) \longrightarrow 2\,HNO_3(aq) + NO(g)$$

Oxidation number: +4   +5   +2

Nitrogen dioxide in the atmosphere undergoes the same reaction and contributes to the formation of acid rain. It also initiates a complex sequence of smog-forming photochemical reactions in the atmosphere.

Nitrous acid, HNO_2 (oxidation number +3), has not been isolated in pure form but is widely used in aqueous solution. Its acid anhydride is the dark blue liquid dinitrogen trioxide, N_2O_3 (**2**; Fig. 20.7). Nitrites are produced by the reduction of nitrates with hot metal:

$$KNO_3(s) + Pb(s) \xrightarrow{350°C} KNO_2(s) + PbO(s)$$

Most nitrites are soluble in water and mildly toxic. Despite their toxicity, nitrites are used in the processing of meat products because they form a pink complex with hemoglobin and inhibit the oxidation of blood (a reaction that would turn the meat brown if it were allowed to occur). Nitrites are responsible for the pink color of ham, sausages, and other cured meat.

Nitric acid, HNO_3 (oxidation number +5), is used extensively in the production of fertilizers and explosives. It is produced by the three-stage *Ostwald process*:

Step 1. Oxidation of ammonia, from oxidation number −3 to +2:

$$4\,NH_3(g) + 5\,O_2(g) \xrightarrow{850°C,\ 5\ atm,\ Pt/Rh} 4\,NO(g) + 6\,H_2O(g)$$

Step 2. Oxidation of nitrogen monoxide, from oxidation number +2 to +4:

$$2\,NO(g) + O_2(g) \longrightarrow 2\,NO_2(g)$$

Step 3. Disproportionation in water, from oxidation number +4 to +5 and +2:

$$3\,NO_2(g) + H_2O(l) \longrightarrow 2\,HNO_3(aq) + NO(g)$$

Nitric acid is a colorless liquid that boils at 83°C and is normally used in aqueous solution. Concentrated nitric acid is often pale yellow as a result of partial decomposition of the acid to NO_2. Because nitrogen has its highest oxidation number (+5) in HNO_3, nitric acid is an oxidizing agent as well as an acid.

> *Nitrogen forms oxides in each of its integral oxidation states from +1 to +5; the properties of the oxides and oxoacids can be explained by considering the oxidation state of nitrogen.*

20.4 Phosphorus Oxides and Oxoacids

The structures of the phosphorus oxides are based on the tetrahedral PO_4 unit. White phosphorus burns in a limited supply of air to form phosphorus(III) oxide, P_4O_6 (**3**):

$$P_4(s,\ white) + 3\,O_2(g) \longrightarrow P_4O_6(s)$$

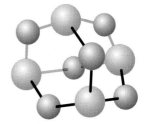

FIGURE 20.7

Dinitrogen trioxide, N_2O_3, condenses to a deep blue liquid that freezes at −100°C to a pale blue solid. On standing, it turns green as a result of partial decomposition into nitrogen dioxide (not shown), a yellow-brown gas.

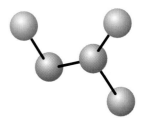

2 Dinitrogen trioxide, N_2O_3

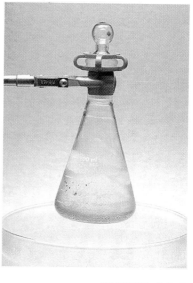

3 Phosphorus(III) oxide, P_4O_6

> Phosphorus(III) oxide is often called phosphorus trioxide because its empirical formula is P_2O_3.

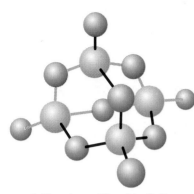

4 Phosphorus(V) oxide, P_4O_{10}

Phosphorus(V) oxide is often called phosphorus pentoxide because its empirical formula is P_2O_5.

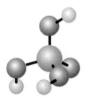

5 Phosphoric acid, H_3PO_4

The molecules are tetrahedral, like P_4, but an O atom lies between each pair of P atoms. Phosphorus(III) oxide is the anhydride of phosphorous acid, H_3PO_3. Although its formula suggests that it should be a triprotic acid, H_3PO_3 is, in fact, diprotic because one of the H atoms is attached directly to the P atom and the P—H bond is nonpolar (see Section 15.10).

When phosphorus burns in an ample supply of air, it forms phosphorus(V) oxide, P_4O_{10} (4). This white solid has such a strong attraction for water that it is widely used in the laboratory as a drying agent. It is also used as a dehydrating agent to remove water from compounds, for example, in the preparation of other acid anhydrides. Phosphorus(V) oxide reacts with water to form phosphoric acid, H_3PO_4 (5), the parent acid of the phosphates. Phosphate rock is mined in huge quantities in Florida and Morocco. After being crushed, it is treated with sulfuric acid to give a mixture of sulfates and phosphates called *superphosphate*, a major fertilizer:

$$Ca_3(PO_4)_2(s) + 2\,H_2SO_4(l) \longrightarrow 2\,CaSO_4(s) + Ca(H_2PO_4)_2(s)$$

The phosphate rock can be treated with phosphoric acid rather than sulfuric acid to produce a mixture with a higher phosphate content:

$$Ca_3(PO_4)_2(s) + 4\,H_3PO_4(l) \longrightarrow 3\,Ca(H_2PO_4)_2(s)$$

The resulting mixture of calcium phosphates is sold as the fertilizer *triple superphosphate*.

When phosphoric acid is heated, it undergoes a condensation reaction (Section 11.9). Two of the molecules combine and release a molecule of water:

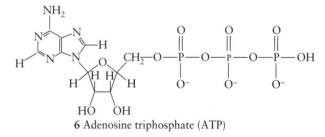

The product, $H_4P_2O_7$, is pyrophosphoric acid. Further heating gives even more complicated products that have chains and rings of PO_4 groups and are called polyphosphoric acids. The most important polyphosphate is adenosine triphosphate, ATP (6), for it is found in every living cell.

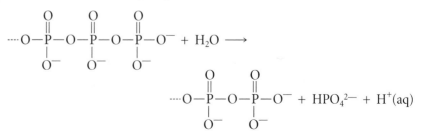

6 Adenosine triphosphate (ATP)

The triphosphate part of this molecule is a chain of three phosphate groups. Its conversion to adenosine diphosphate, ADP, in the reaction

$$\cdots O-\underset{O^-}{\overset{O}{\underset{|}{\overset{||}{P}}}}-O-\underset{O^-}{\overset{O}{\underset{|}{\overset{||}{P}}}}-O-\underset{O^-}{\overset{O}{\underset{|}{\overset{||}{P}}}}-O^- + H_2O \longrightarrow$$

$$\cdots O-\underset{O^-}{\overset{O}{\underset{|}{\overset{||}{P}}}}-O-\underset{O^-}{\overset{O}{\underset{|}{\overset{||}{P}}}}-O^- + HPO_4^{2-} + H^+(aq)$$

FIGURE 20.10

A collection of sulfide ores (from left to right): galena, PbS; cinnabar, HgS; pyrite, FeS$_2$; sphalerite, ZnS. Pyrite has a lustrous golden color and has frequently been mistaken for gold; hence it is also known as fool's gold. Gold and fool's gold are readily distinguished by their densities.

of metallurgical processes, especially the extraction of copper from its sulfide ores, and is removed from sulfur-rich petroleum by the *Claus process.* In this process, some of the H$_2$S that occurs in oil and natural gas wells is first oxidized to sulfur dioxide:

$$2\,H_2S(g) \,+\, 3\,O_2(g) \longrightarrow 2\,SO_2(g) \,+\, 2\,H_2O(l)$$

This SO$_2$ is then used to oxidize the remainder of the hydrogen sulfide:

$$2\,H_2S(g) \,+\, SO_2(g) \xrightarrow{300°C,\,Al_2O_3} 3\,S(s) \,+\, 2\,H_2O(l)$$

Deposits of sulfur are mined by the *Frasch process,* in which superheated water and compressed air are pumped down concentric pipes, which forces the molten sulfur up to the surface. Most of the sulfur produced is used to make sulfuric acid, but an appreciable amount is used to vulcanize rubber (Section 11.16).

Elemental sulfur is a yellow, tasteless, almost odorless, insoluble, nonmetallic molecular solid of crownlike S$_8$ rings (**7**). The most stable crystalline form under normal conditions is *rhombic sulfur* (Fig. 20.11). When sulfur is melted, the sulfur atoms remain in S$_8$ rings but are free to move. As molten sulfur is heated, however, its molecular structure changes and its viscosity increases. For example, just above 113°C, rhombic sulfur is a mobile, straw-colored liquid of S$_8$ molecules. The viscosity increases when the liquid is heated further because the S$_8$ rings break open into chains that become tangled. The viscosity falls again at still higher temperatures and the liquid becomes red-brown, because the S$_8$ chains break up into smaller, more mobile, highly colored, S$_2$ and S$_3$ molecules. Sulfur vapor has a blue tint from the S$_2$ molecules present in it (recall Fig. 1.20). These S$_2$ molecules are paramagnetic, like O$_2$.

Selenium and tellurium occur in sulfide ores; they are also recovered from the anode sludge formed during the electrolytic refining of copper. Both elements have several allotropes, the most stable consisting of long zigzag chains of

7   Sulfur, S$_8$

(a)   (b)

FIGURE 20.11

One of the two most common forms of sulfur is the blocklike rhombic form (a). It differs from the needlelike monoclinic sulfur (b) in the manner in which the S$_8$ rings are stacked together.

FIGURE 20.12

Two of the Group 16 elements: selenium (left) and tellurium (right).

atoms. Although these allotropes look like silver-white metals, they are poor electrical conductors (Fig. 20.12). The conductivity of selenium is increased in the presence of light, so it is used in photoelectric devices and in photocopying machines.

Electronegativities decrease down the group (Fig. 20.13), and atomic and ionic radii increase (Fig. 20.14). The differences between oxygen and sulfur are emphasized by the latter's striking ability to **catenate,** that is, to form chains or rings of atoms. Oxygen's ability to form chains is very limited, with H_2O_2, O_3, and the anions O_2^-, O_2^{2-}, and O_3^- the only examples. Sulfur's ability is much more pronounced, and appears, for instance, in the existence of S_8 rings and their fragments, which can polymerize when heated to about 200°C. The existence of —S—S— links that form cross-links between different parts of the polypeptide chains of proteins is another example of catenation. These **disulfide links** contribute to the shapes of proteins and so help to keep us alive.

Metallic character increases down Group 16 as electronegativity decreases. Oxygen and sulfur occur naturally in the elemental state. Sulfur forms long chains and rings with itself, but oxygen does not.

20.6 Compounds with Hydrogen

By far the most important compound of oxygen and hydrogen is water, H_2O. Water is available on a huge scale worldwide, but in various states of purity. Municipal water supplies normally undergo several stages of purification, as described in Connection 4, following Chapter 16. High-purity water for special applications is obtained by distillation or by **ion exchange,** the replacement of one type of ion in a solution by another. In ion exchange, water passes through a *zeolite*, an aluminosilicate with a very open structure that can capture ions such as Mg^{2+} and Ca^{2+} and exchange them for H^+ ions.

Water is an oxidizing agent:

$$2 H_2O(l) + 2 e^- \longrightarrow 2 OH^-(aq) + H_2(g) \qquad E = -0.42 \text{ V at pH} = 7$$

One example is its reaction with the alkali metals, as in

$$2 Na(s) + 2 H_2O(l) \longrightarrow 2 NaOH(aq) + H_2(g)$$

However, unless the other reactant is a strong reducing agent, water acts as an oxidizing agent only at high temperatures, as in the reforming reaction:

$$CH_4(g) + H_2O(g) \xrightarrow{\Delta} CO(g) + 3 H_2(g)$$

Water is a very mild reducing agent, the half-reaction being

$$2 H_2O(l) \longrightarrow 4 H^+(aq) + O_2(g) + 4 e^- \qquad E = -0.81 \text{ V at pH} = 7$$

However, few substances besides fluorine are strong enough oxidizing agents to remove electrons from water.

Water is also a Lewis base, for an H_2O molecule can donate one of its lone pairs to a Lewis acid and form complexes such as $[Fe(H_2O)_6]^{3+}$. Its ability to act as a Lewis base is also the origin of water's ability to hydrolyze substances. The reaction between water and phosphorus pentachloride mentioned in Section 20.2 is an example.

Hydrogen peroxide, H_2O_2 (**8**), is a very pale blue liquid that is appreciably denser than water (1.44 g/mL at 25°C) but similar in other physical properties: its melting point is −0.4°C, and its boiling point is 152°C. Chemically, though,

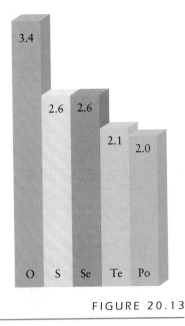

FIGURE 20.13

The electronegativities of the Group 16 elements decrease down the group.

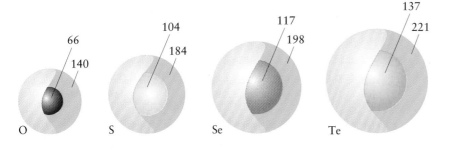

FIGURE 20.14

66
140
O

104
184
S

117
198
Se

137
221
Te

The atomic and ionic radii of the Group 16 elements increase down the group. The values shown are in picometers, and the anion (shown in green in each case) is substantially larger than the neutral parent atom.

hydrogen peroxide and water differ greatly. The presence of the second oxygen atom makes H_2O_2 a very weak acid ($pK_{a1} = 11.75$). Hydrogen peroxide is also a good oxidizing agent in both acidic and basic solution. For example, H_2O_2 oxidizes Fe^{2+} and Mn^{2+} in acidic and basic solutions. It can also act as a reducing agent. For example, hydrogen peroxide reduces permanganate ions and chlorine in acidic and basic solutions. The O_2 formed by the oxidation of H_2O_2 is sometimes produced in an energetically excited state and emits light as it discards its excess energy. This process is an example of **chemiluminescence,** the emission of light by products formed in energetically excited states (Fig. 20.15).

Hydrogen peroxide is normally sold for industrial use as a 30% by mass aqueous solution. When used as a hair bleach (as a 6% solution), it acts by oxidizing the pigments in the hair. Because it oxidizes unpleasant effluents without producing any harmful by-products, H_2O_2 is increasingly widely used as an oxidizing agent in the control of pollution. A 3% H_2O_2 aqueous solution is used as a mild antiseptic in the home. Contact with blood catalyzes the disproportionation of hydrogen peroxide into water and oxygen gas, which cleanses the wound.

All the H_2X compounds of the Group 16 elements other than water are toxic gases with offensive odors. They are insidious poisons because they paralyze the olfactory nerve and, as a result, cannot be smelled soon after exposure. Rotten eggs smell of hydrogen sulfide, H_2S, because egg proteins contain sulfur and give off the gas when they decompose. Another sign of the formation of sulfides in eggs is the pale green discoloration sometimes seen in cooked eggs where the white meets the yolk: this discoloration is a deposit of iron(II) sulfide. Hydrogen sulfide is also responsible for the sour smell of underground natural gas reserves.

Aqueous solutions of hydrogen sulfide slowly become cloudy. This cloudiness is a result of the oxidation of hydrogen sulfide by dissolved air and the formation of a colloidal dispersion of small particles of sulfur. Hydrogen sulfide is a weak diprotic acid and the parent acid of the hydrogen sulfides (which contain the HS^- ion) and the sulfides (which contain the S^{2-} ion). The sulfides of the s-block elements are moderately soluble, whereas the sulfides of the heavy p- and d-block metals are generally very insoluble.

The sulfur analogue of hydrogen peroxide also exists and is an example of a *polysulfane*, a catenated molecular compound of composition $HS-S_n-SH$, where n can take on values from 0 through 6. The polysulfide ions obtained from the polysulfanes include two ions found in lapis lazuli (Fig. 20.16).

Water can act as a Brønsted acid, a Brønsted base, a Lewis base, an oxidizing agent, and a weak reducing agent. Hydrogen peroxide is a strong oxidizing agent. Hydrogen sulfide is a weak acid. Sulfur can catenate to form polysulfanes.

8   Hydrogen peroxide, H_2O_2

FIGURE 20.15

Chemiluminescence, the emission of light as the result of a chemical reaction, occurs when hydrogen peroxide is added to a solution of the organic compound perylene. Although hydrogen peroxide itself can fluoresce, in this case the light is emitted by the perylene.

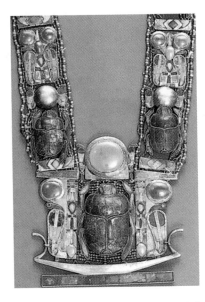

FIGURE 20.16

The blue stones in this ancient Egyptian ornament are lapis lazuli. This semiprecious stone is an aluminosilicate colored by S_2^- and S_3^- impurities. The blue color is due to S_3^-, and its hint of green to S_2^-.

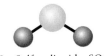

9   Sulfur dioxide, SO_2

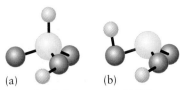

(a)   (b)

10   Sulfurous acid, H_2SO_3

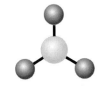

11   Sulfur trioxide, SO_3

Self-Test 20.2A Determine from the Lewis structures of the following molecules which are paramagnetic and explain how you decided: (a) N_2; (b) NO; (c) N_2O.
[*Answer:* (b) is paramagnetic, because it has an unpaired electron.]

Self-Test 20.2B Write the half-reactions and the overall reaction for the oxidation of water by F_2. Determine the standard cell potential for the reaction.

20.7 Sulfur Oxides and Oxoacids

The most important oxides and oxoacids of sulfur are the dioxide and trioxide and the corresponding sulfurous and sulfuric acids.

Sulfur burns in air to form sulfur dioxide, SO_2 (**9**), a colorless, choking, poisonous gas. Sulfur oxides in the atmosphere are referred to as SO_x (read "sox"). About 7×10^{10} kg of the dioxide result from the decomposition of vegetation and from volcanic emissions. In addition, the approximately 1×10^{11} kg of naturally occurring hydrogen sulfide can be oxidized to the dioxide by atmospheric oxygen:

$$2\,H_2S(g) + 3\,O_2(g) \longrightarrow 2\,SO_2(g) + 2\,H_2O(g)$$

Industry and transport contribute another 1.5×10^{11} kg of the dioxide, of which about 70% comes from oil and coal combustion—mainly in electricity-generating plants. The average concentration of SO_x in the atmosphere in rural areas in the northern hemisphere is found to be about 1 μmol/L. However, the concentration is much higher in industrialized areas (see Applying Chemistry: Case Study 15).

Sulfur dioxide is an acidic oxide, the anhydride of sulfurous acid, H_2SO_3, which is the parent acid of the hydrogen sulfites (or bisulfites) and the sulfites:

$$SO_2(g) + H_2O(l) \longrightarrow H_2SO_3(aq)$$

Sulfurous acid is an equilibrium mixture of two molecules, (**10a**) and (**10b**). In the form shown in **10a,** it resembles phosphorous acid, with one of the H atoms attached directly to the S atom. These molecules are also in equilibrium with molecules of SO_2, each of which is surrounded by a cage of water molecules. The evidence for this equilibrium is that crystals of composition $SO_2 \cdot xH_2O$, with x about 7, are obtained when the solution is cooled. Substances like this, in which a molecule sits in a cage of other molecules, are called **clathrates.** Methane, carbon dioxide, and the noble gases also form clathrates with water.

Sulfur dioxide is easily liquified under pressure and can therefore be used as a refrigerant. It is also a preservative for dried fruit and a bleach for textiles and flour. Its most important use is in the production of sulfuric acid.

The oxidation number of sulfur in sulfur dioxide and the sulfites is +4, an intermediate value in sulfur's range from −2 to +6. Hence these compounds can act as either oxidizing agents or reducing agents (Fig. 20.17). By far the most important reaction of sulfur dioxide in the atmosphere is its oxidation to sulfur trioxide, SO_3 (**11**), in which sulfur has the oxidation number +6:

$$2\,SO_2(g) + O_2(g) \longrightarrow 2\,SO_3(g)$$

The direct reaction is very slow, so an SO_2 molecule survives for a few days in the atmosphere before it is oxidized to SO_3. The oxidation is catalyzed by the metal ions in the minerals on the walls of buildings and metal cations like Fe^{3+}, which may be dissolved in droplets of water. Other routes from SO_2 to SO_3

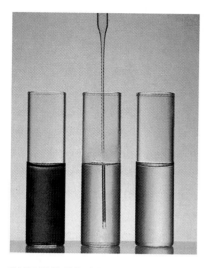

FIGURE 20.17

Sulfur dioxide is a reducing agent. When it is bubbled through an aqueous solution of bromine (left), it reduces the Br_2 to colorless bromide ions (right). The SO_2 is oxidized to H_2SO_4.

FIGURE 20.18

Sulfuric acid is an oxidizing agent. When concentrated acid is poured onto solid sodium bromide, NaBr, the bromide ions are oxidized to bromine, which colors the solution red-brown.

include reaction with ·OH radicals, H_2O_2, and O_3 formed by the effect of sunlight on air and water vapor. The sulfur trioxide produced by these processes reacts with atmospheric water to form dilute sulfuric acid:

$$SO_3(g) + H_2O(l) \longrightarrow H_2SO_4(aq)$$

The product of this reaction, together with the nitric acid formed by dissolved NO_x, falls as acid rain.

Sulfuric acid, H_2SO_4, is a colorless, corrosive, oily liquid that boils (and decomposes) at about 300°C. It has three chemically important properties: it is a strong acid, a dehydrating agent, and an oxidizing agent (Fig. 20.18). It is produced commercially in the *contact process*, in which sulfur is first burned in oxygen and the SO_2 produced is oxidized to SO_3 over a V_2O_5 catalyst:

$$2\,SO_2(g) + O_2(g) \xrightarrow{500°\,C,\,V_2O_5} 2\,SO_3(g)$$

Sulfur trioxide cannot be added directly to water, because it forms a corrosive acid mist with the vapor over the water. Instead, it is absorbed in 98% concentrated sulfuric acid to give the dense, oily liquid called *oleum*:

$$SO_3(g) + H_2SO_4(l) \longrightarrow H_2S_2O_7(l)$$

Oleum is converted to the acid by dilution with water:

$$H_2S_2O_7(l) + H_2O(l) \longrightarrow 2\,H_2SO_4(l)$$

Sulfuric acid is the most heavily produced inorganic chemical worldwide, the annual production being over 4×10^{10} kg in the United States alone. The low cost of sulfuric acid leads to its widespread use in industry, particularly for the production of fertilizers, petrochemicals, dyestuffs, and detergents. About two-thirds is used in the manufacture of ammonium sulfate and phosphate fertilizers. Its

(a)   (b)   (c)

FIGURE 20.19

Sulfuric acid is a dehydrating agent. (a) When concentrated sulfuric acid is poured onto
sucrose, (b) the sucrose, a carbohydrate, is dehydrated, (c) leaving a frothy black mass
of carbon.

reaction with phosphate rock to produce superphosphate fertilizer was discussed in Section 20.4.

Sulfuric acid is a powerful dehydrating agent, as seen when a little concentrated acid is poured on sucrose, $C_{12}H_{22}O_{11}$. A black, frothy mass of carbon forms as a result of the extraction of H_2O (Fig. 20.19):

$$C_{12}H_{22}O_{11}(s) \longrightarrow 12\,C(s) + 11\,H_2O(l)$$

The froth is caused by CO and CO_2 gases formed in side reactions.

> **Sulfur dioxide is the acid anhydride of sulfurous acid, and sulfur trioxide is the anhydride of sulfuric acid. Sulfuric acid is a strong acid, a dehydrating agent, and an oxidizing agent.**

Self-Test 20.3A Concentrated sulfuric acid is 98.0% by mass H_2SO_4. Calculate its molarity, given that its density is 1.84 g/mL.

[*Answer:* 18.4 M]

Self-Test 20.3B When concentrated sulfuric acid and water are mixed, the total volume is less than the sum of the individual volumes before mixing. Explain this volume contraction in terms of intermolecular interactions.

GROUP 17: THE HALOGENS

The unique properties of the halogens (Table 20.5), the members of Group 17, can be traced to their valence configurations, ns^2np^5, which need only one more electron to reach a closed-shell configuration. The elements form a family that shows smooth trends in physical properties, electronegativity (Fig. 20.20), and atomic and ionic radii (Fig. 20.21).

20.8 The Elements

Fluorine occurs widely in many minerals, including *fluorspar*, CaF_2; *cryolite*, Na_3AlF_6; and the *fluorapatites*, $Ca_5F(PO_4)_3$. Fluorine is the most strongly oxidizing element ($E° = +2.87$ V). Only an anode can be made more oxidizing (by

Table 20.5 *The Group 17 elements*

Valence configuration: ns^2np^5

Z	Name	Symbol	Molar mass, g/mol	Melting point, °C	Boiling point, °C	Density, g/cm³ at 25°C	Normal form*
9	fluorine	F	19.00	−220.	−188	1.51†	almost colorless gas (O_2)
17	chlorine	Cl	35.45	−101	−34	1.66†	yellow-green gas
35	bromine	Br	79.91	−7	59	3.12	red-brown liquid
53	iodine	I	126.90	114	184	4.95	purple-black nonmetallic solid
85	astatine‡	At	210	300.	350.		nonmetallic solid

Normal form means the appearance and state of the element at 25°C and 1 atm.

†For the liquid at its boiling point.

‡Radioactive

increasing its positive charge); hence, only in an electrolytic cell can fluorine be driven out of its compounds by removing one of the tightly held electrons from the F⁻ ion. Fluorine is produced by electrolyzing an anhydrous molten mixture of potassium fluoride and hydrogen fluoride at about 75°C with a carbon anode.

Fluorine is a reactive, colorless gas of F_2 molecules. It was little used before the development of the nuclear industry but is now produced on a large scale, at about 5×10^6 kg a year in the United States (Fig. 20.22). Most of the fluorine production is used to make the volatile solid UF_6 used for processing nuclear fuel (Section 22.11). Much of the rest is used in the production of sulfur hexafluoride, an electrically insulating gas used in electrical equipment. Fluorinated

FIGURE 20.20

The electronegativities of the halogens decrease steadily down the group.

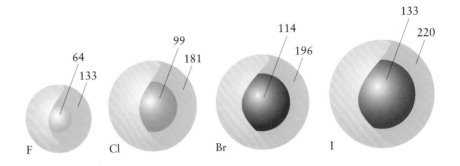

FIGURE 20.21

The atomic and ionic radii of the halogens increase steadily down the group as electrons occupy outer shells of the atoms further from the nuclei. The values shown are in picometers. In all cases, ionic radii (represented by the green spheres) are larger than atomic radii.

hydrocarbons, such as Teflon and Freon, are made by the reaction between hydrocarbons and hydrogen fluoride. They are relatively inert chemically: they are inert to oxidation by air, hot nitric acid, concentrated sulfuric acid, and other strong oxidizing agents.

Chlorine is obtained from sodium chloride by electrolysis, either of molten rock salt or of brine (recall Fig. 19.17). It is a pale yellow-green gas of Cl_2 molecules that condenses at $-34°C$. It reacts directly with nearly all the elements (the exceptions being carbon, nitrogen, oxygen, and the noble gases). It is a strong oxidizing agent and oxidizes metals to high oxidation states; for example, anhydrous iron(III) chloride, not iron(II) chloride, is formed when chlorine reacts with iron (Fig. 20.23):

$$2 \, Fe(s) \; + \; 3 \, Cl_2(g) \longrightarrow 2 \, FeCl_3(s)$$

Chlorine is used in a number of industrial processes, including the manufacture of plastics, solvents, and pesticides. It is also used as a bleach in the paper and

FIGURE 20.22

Fluorine is prepared on a large scale by an adaptation of the electrolytic method that was used to isolate it originally. These electrolytic cells are producing fluorine in a commercial preparation plant.

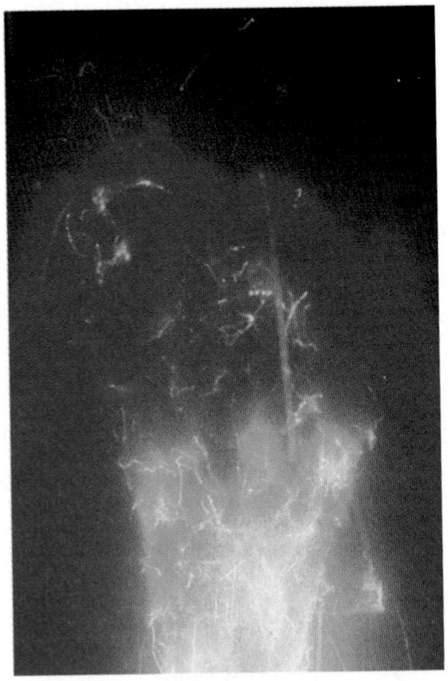

FIGURE 20.23

Iron reacts vigorously and exothermically with chlorine to form anhydrous iron(III) chloride.

textile industries and as a disinfectant in water treatment. The use of chlorine to provide potable water has made our modern lifestyles, especially life in large cities, feasible. The chlorine smell of chlorinated water comes largely from amines (Section 11.10) in which $-NH_2$ groups have become chlorinated (converted to $-NHCl$ groups).

Chlorine is also used to produce bromine from brine wells through the oxidation of Br^- ions (Fig. 20.24):

$$2\,Br^-(aq)\ +\ Cl_2(g)\ \longrightarrow\ Br_2(l)\ +\ 2\,Cl^-(aq)$$

Air is bubbled through the solution to vaporize the bromine and drive it out.

Bromine is a corrosive, red-brown fuming liquid of Br_2 molecules, with a penetrating odor. It is increasingly used in industrial chemistry because of the ease with which it can be added to and removed from the organic chemicals used for complicated syntheses. Organic bromides are incorporated into textiles as fire retardants and are used as pesticides; inorganic bromides, particularly silver bromide, are used as photographic emulsions. Saturated aqueous zinc bromide has a very high density and is used in the oil industry to control the escape of oil from deep wells. The tall column of liquid formed when this solution is poured down a well exerts a very high pressure at its base.

Iodine occurs as iodide ions in brines and as an impurity in Chile saltpeter. It was once obtained from seaweed, which contains high concentrations accumulated from seawater: 2000 kg of seaweed produce about 1 kg of iodine. The best modern source is the brine from oil wells, for the oil itself was produced by the decay of marine organisms that had accumulated the iodine while they were alive. The element is produced by oxidation with chlorine:

$$Cl_2(g)\ +\ 2\,I^-(aq)\ \longrightarrow\ I_2(aq)\ +\ 2\,Cl^-(aq)$$

The blue-black lustrous solid sublimes easily and boils to a purple vapor at 185°C. The I_2 molecules retain their separate identities in the solid, like the other halogens but unlike carbon, silicon, phosphorus, and sulfur, which all form chains and networks of atoms.

Iodine dissolves in organic solvents to give a variety of colors that arise from the different interactions between the I_2 molecules and the solvent (Fig. 20.25). Iodine is only slightly soluble in water, unless I^- ions are present, in which case the soluble, brown triiodide ion, I_3^-, is formed. The element itself has few direct

FIGURE 20.24

Chlorine is an oxidizing agent. When chlorine is bubbled through a colorless solution of bromide ions, it oxidizes them to bromine, which colors the solution red-brown.

FIGURE 20.25

Solutions of iodine in a variety of solvents. From left to right, the solvents are tetrachloromethane (carbon tetrachloride), water, and potassium iodide solution. In the solution at the far right, a little starch has been added to a solution of iodine in potassium iodide solution; starch acts as an indicator for the presence of iodine.

Table 20.6 Known interhalogens

Interhalogen	Normal form*
XF_n	
ClF	colorless gas
ClF_3	colorless gas
ClF_5	colorless gas
BrF	pale brown gas
BrF_3	pale yellow liquid
BrF_5	colorless liquid
IF	unstable
IF_3	yellow solid
IF_5	colorless liquid
IF_7	colorless gas
XCl_n	
BrCl	red-brown gas
ICl	red solid
I_2Cl_6	yellow solid
XBr_n	
IBr	black solid

*Normal form means the appearance and state of the element at 25°C and 1 atm

uses; but, dissolved in alcohol, it is familiar as a mild oxidizing antiseptic. Iodine is an essential trace element for living systems; a deficiency in humans leads to a swelling of the thyroid gland in the neck. Iodides are added to table salt (to produce iodized salt) to prevent this deficiency.

As the first member of the group, fluorine has a number of unique properties that stem from its high electronegativity, small size, and lack of available d-orbitals. In particular, it oxidizes other elements to high oxidation states. (Its smallness helps, for it allows several F atoms to pack around a central atom, as in IF_7.) Because the fluoride ion is so small, the lattice enthalpies of its ionic compounds tend to be high (see Table 8.1). One result is the lower solubilities of most fluorides relative to those of other halides (exceptions include AgF, which is soluble). This difference in solubility is one reason why the oceans are salty with chlorides rather than fluorides, even though fluorine is more abundant than chlorine in the Earth's crust: chlorides are more readily dissolved in groundwater and washed out to sea.

The halogens show smooth trends in chemical properties down the group. Fluorine has some anomalous properties, such as its strength as an oxidizing agent and the lower solubilities of most fluorides.

20.9 Compounds of the Halogens

The halogens form compounds among themselves. These *interhalogens* have the formulas XX', XX'$_3$, XX'$_5$, and XX'$_7$, where X is the heavier (and larger) of the two halogens. Only some of the possible combinations have been prepared (Table 20.6). They are prepared by direct reaction of the two halogens. The product formed depends on the proportions of reactants used. For example,

$$Cl_2(g) + 3 F_2(g) \longrightarrow 2 ClF_3(g)$$

$$Cl_2(g) + 5 F_2(g) \longrightarrow 2 ClF_5(g)$$

The interhalogens have physical properties intermediate between those of their parent halogens. Their chemical properties are dominated by the decreasing X—X' bond enthalpy as X becomes heavier. For example, the fluorides of the heavier halogens are all very reactive. Bromine trifluoride is so reactive a gas that even asbestos burns in it.

The hydrogen halides, HX, can be prepared by the direct reaction of the elements:

$$H_2(g) + X_2(g) \longrightarrow 2 HX(g)$$

Table 20.7 The hydrogen halides

Compound	Molar mass, g/mol	Melting point, °C	Boiling point, °C	pK_a in water	Bond enthalpy, kJ/mol	Bond length, pm
HF	20.01	−83	20.	3.45	565	92
HCl	36.46	−115	−85	strong	421	127
HBr	80.92	−89	−67	strong	366	141
HI	127.91	−51	−35	strong	299	161

Fluorine reacts explosively by a radical chain reaction as soon as the gases are mixed. A mixture of hydrogen and chlorine explodes when exposed to light. Bromine and iodine react with hydrogen much more slowly. A less hazardous laboratory source of the hydrogen halides is the action of a nonvolatile acid on a metal halide, as in

$$CaF_2(s) + 2\,H_2SO_4(aq, conc) \longrightarrow Ca(HSO_4)_2(aq) + 2\,HF(g)$$

Because Br^- and I^- are oxidized by sulfuric acid, phosphoric acid is used in the preparation of HBr and HI:

$$KI(s) + H_3PO_4(aq) \xrightarrow{\Delta} KH_2PO_4(aq) + HI(g)$$

All the hydrogen halides are colorless, pungent gases (Table 20.7), but hydrogen fluoride is a liquid at temperatures below 20°C. Its low volatility is a sign of extensive hydrogen bonding, and short zigzag chains of hydrogen-bonded molecules, up to about $(HF)_5$, survive to some extent in the vapor. All the hydrogen halides dissolve in water to give acidic solutions. Hydrofluoric acid has the distinctive property of attacking glass and silica (see Applying Chemistry: Case Study 19). The interiors of lamp bulbs are frosted by the vapors from a solution of hydrofluoric acid and ammonium fluoride (Fig. 20.26).

The hypohalous acids, HXO (oxidation number +1, Table 20.8), are prepared by direct reaction of the halogen with water. For example, chlorine gas disproportionates in water to produce hypochlorous acid and hydrochloric acid:

$$Cl_2(g) + H_2O(aq) \longrightarrow HClO(aq) + HCl(aq)$$

Hypofluorous acid is so unstable that it survives only below the freezing point of water. Sodium hypochlorite, NaClO, is produced from the electrolysis of brine when the electrolyte is rapidly stirred, and the chlorine gas produced at the anode reacts with the hydroxide ion generated at the cathode. The chlorine gas disproportionates to produce hypochlorite and chloride ions:

$$Cl_2(g) + 2\,OH^-(aq) \longrightarrow ClO^-(aq) + Cl^-(aq) + H_2O(l)$$

(a)

(b)

FIGURE 20.26

When a mixture of hydrofluoric acid and ammonium fluoride is swirled inside a flask (a), the reaction with the silica in the glass frosts the glass surface (b).

Table 20.8 Halogen oxoacids

Oxidation number	General formula	General acid name	Known examples	pK_a in water
7	HXO_4	perhalic acid	$HClO_4$	strong
			$HBrO_4$	strong
			HIO_4	1.64
5	HXO_3	halic acid	$HClO_3$	strong
			$HBrO_3$	strong
			HIO_3	0.77
3	HXO_2	halous acid	$HClO_2$	2.00
			$HBrO_2$	unstable
1	HXO	hypohalous acid	HFO	unstable
			HClO	7.53
			HBrO	8.69
			HIO	10.64

Calcium hypochlorite, which is marketed as a dry bleach and for purifying the water in home swimming pools, is produced by passing chlorine gas over dry calcium oxide (quicklime). Calcium hypochlorite is used in preference to sodium hypochlorite for purifying the water in swimming pools because the Ca^{2+} ions form insoluble calcium carbonate, which is removed by filtration. Sodium would remain in solution and make the water too salty.

Chlorate ions, ClO_3^- (oxidation number +5), form when chlorine reacts with hot concentrated aqueous alkali:

$$3\,Cl_2(g) + 6\,OH^-(aq) \xrightarrow{\Delta} ClO_3^-(aq) + 5\,Cl^-(aq) + 3\,H_2O(l)$$

They decompose when heated, to an extent that depends on whether or not a catalyst is present:

$$4\,KClO_3(s) \xrightarrow{\Delta} 3\,KClO_4(s) + KCl(s)$$

$$2\,KClO_3(s) \xrightarrow{\Delta,\ MnO_2} 2\,KCl(s) + 3\,O_2(g)$$

The catalyzed reaction is a convenient laboratory source of oxygen.

Chlorates are useful oxidizing agents. Potassium chlorate is used as an oxygen supply in fireworks and in safety matches. The heads of safety matches consist of a paste of potassium chlorate, antimony sulfide, sulfur, and powdered glass to create friction when the match is struck; the striking strip contains red phosphorus, which ignites the match head (see Section 20.1). The principal use of sodium chlorate is as a source of chlorine dioxide, ClO_2. The chlorine in ClO_2 has oxidation number +4, so the chlorate must be reduced to form it. Sulfur dioxide is a convenient reducing agent for this reaction, which is carried out in dilute sulfuric acid:

$$2\,NaClO_3(aq) + SO_2(g) + H_2SO_4(aq) \longrightarrow 2\,NaHSO_4(aq) + 2\,ClO_2(g)$$

Chlorine dioxide has an odd number of electrons and is a paramagnetic yellow gas. It is used to bleach paper pulp because it can oxidize the various pigments in the pulp without degrading the wood fibers.

The perchlorates, ClO_4^- (oxidation number +7), are prepared by electrolytic oxidation of aqueous chlorates:

$$ClO_3^-(aq) + H_2O(l) \longrightarrow ClO_4^-(aq) + 2\,H^+(aq) + 2\,e^-$$

Perchloric acid, $HClO_4$, is prepared by the action of concentrated hydrochloric acid on sodium perchlorate, followed by distillation. It is a colorless liquid and the strongest of all common acids. Because chlorine has its highest oxidation number in these compounds, the perchlorates are also powerful oxidizing agents; contact between perchloric acid and even a small amount of organic material can result in a dangerous explosion. One spectacular example of their oxidizing ability under controlled conditions is the use of a mixture of ammonium perchlorate and aluminum powder in the booster rockets of the space shuttle (Applying Chemistry: Case Study 20).

The interhalogens have properties intermediate between those of the constituent halogens. Nonmetals form covalent halides; metals form ionic halides. The oxoacids of chlorine are all oxidizing agents. Acidity and oxidizing strength of oxoacids both increase as the oxidation number of the halogen increases.

Self-Test 20.4A Predict the trend in oxidizing strength of the halogens in aqueous solution.

[***Answer:*** Decreases down the group: F > Cl > Br > I]

Self-Test 20.4B Which halide ion is the strongest reducing agent in aqueous solution?

GROUP 18: THE NOBLE GASES

The elements in Group 18, the noble gases, get their name from their very low reactivity (Table 20.9). Their closed-shell electron configurations (ns^2np^6) prompted the belief that these elements were chemically inert. However, the first noble-gas compound, xenon hexafluoroplatinate, $XePtF_6$, was synthesized in 1962 by the reaction of xenon with platinum hexafluoride; and shortly after that, xenon tetrafluoride, XeF_4, was produced from a high-temperature mixture of xenon and fluorine.

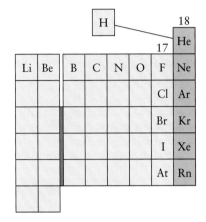

20.10 The Elements

All the Group 18 elements occur in the atmosphere as monatomic gases; the most common is argon, which is the third most abundant gas in the atmosphere (after nitrogen and oxygen and discounting the variable amount of water vapor). All except helium and radon are obtained by the distillation of liquid air. Helium, the second most abundant element in the universe after hydrogen, is rare on Earth because its atoms are so light that a large fraction of them reach high speeds and escape from the atmosphere. However, it is found as a component of natural gases trapped under rock formations in some locations (notably Texas), where it has collected as a result of the emission of α (alpha) particles by radioactive elements. An α particle is a helium nucleus (He^{2+}); and an atom of the element forms when the particle picks up two electrons from its surroundings.

Helium gas is twice as dense as hydrogen under the same conditions. However, because its density is still very low and it is nonflammable, helium is used to

Radioactivity and the changes that nuclei undergo are discussed in Chapter 22.

Table 20.9 *The Group 18 elements (the noble gases)*

Valence configuration: ns^2np^6
Normal form: colorless monatomic gases

Z	Name	Symbol	Molar mass, g/mol	Melting point, °C	Boiling point, °C
2	helium	He	4.00	—	−269 (4.2 K)
10	neon	Ne	20.18	−249	−246
18	argon	Ar	39.95	−189	−186
36	krypton	Kr	83.80	−157	−153
54	xenon	Xe	131.30	−112	−108
86	radon*	Rn	222	−71	−62

*Radioactive

Applying Chemistry: *Case Study 20*

A rocket technician once dramatically demonstrated the high reactivity of a rocket propellant: he sprayed a few drops onto the snow-covered ground and it burst into flame. Fuels used in space travel are very different from fuels we use in our automobiles, because they are designed for very different conditions. The preferred fuel for automobiles is gasoline, which burns in air and has a high enthalpy density (see Section 6.15). That is, it releases a lot of heat per liter as it burns. In space flight, the mass of fuel is more important than the volume it occupies, because the entire assembly has to be thrust up out of the Earth's gravitational field. Hence, a rocket fuel needs a high specific enthalpy (enthalpy of combustion per gram of substance)

Powdered reactants are mixed with a liquid polymer base and hardened inside the space shuttle booster rocket shell.

rather than a high enthalpy density. Moreover, a rocket-powered space ship must carry an oxidizer as well as a fuel, because the fuel must burn in the absence of air.

The volume of gaseous exhaust emitted by an automobile is of little concern, although the nature of the emissions is of interest. A rocket, however, is moved by the thrust created when gases are ejected from its engine. Therefore, a desirable rocket fuel will be a low-density liquid or solid that produces a lot of heat and a large volume of gases when it burns.

Liquid and solid fuels are used in stages to propel the space shuttle into orbit. First, the solid booster rockets are ignited to lift the shuttle from the ground and take it high into the atmosphere, at which point the rocket shells are released and fall into the ocean for recovery. The solid fuel in the booster rockets consists of aluminum powder (the fuel), ammonium perchlorate (the oxidizing agent as well as a fuel), and iron(III) oxide (the catalyst) mixed into a liquid polymer that is then set into a solid inside the rocket shell. A variety of products can be produced. One of the reactions that occurs is

$$3\,NH_4ClO_4(s)\ +\ 3\,Al(s)\ \xrightarrow{Fe_2O_3}$$
$$Al_2O_3(s)\ +\ AlCl_3(s)\ +\ 6\,H_2O(g)\ +\ 3\,NO(g)$$

The solid products form the thick clouds of white powder emitted by the solid rocket boosters during liftoff.

The shuttle is propelled into orbit by the reaction between two gases that are stored as liquids. For a typical shuttle mission, about 1.4×10^6 L (1.4 ML) of liquid hydrogen and 0.54 ML of liquid oxygen are used. The gases are stored as liquids to reduce the volume: huge gas-filled tanks would be a source of drag. The hydrogen and oxygen are vaporized and ignited as they mix. The rapid expansion of the product, water vapor, at the high temperature of the combustion provides the thrust for the main engines. The net reaction for the vaporization and combustion is

$$2\,H_2(l)\ +\ O_2(l)\ \longrightarrow\ 2\,H_2O(g) \qquad \Delta H^\circ\ =\ -475\ kJ$$

provide buoyancy in airships such as blimps. It is also used to dilute the oxygen used in deep-sea diving, to pressurize rocket fuels, as a coolant, and in helium-neon lasers. The element has the lowest boiling point of any substance (4.2 K), and it does not freeze to a solid at any temperature unless pressure is applied to hold the light, mobile atoms together. These properties make helium useful for **cryogenics,** the study of matter at very low temperatures. Below 2 K, liquid helium shows the remarkable property of **superfluidity,** the ability to flow without viscosity. Helium is the only substance known to have more than one liquid phase.

Neon, which emits a red glow when an electric current flows through it, is widely used in advertising signs (Fig. 20.27). Argon is used to provide an inert

As we go to press, reports are coming in of a possible superfluid phase of water.

Another difference between rocket fuels and automobile fuels lies in their relative ease of handling. Gasoline can be handled by any motorist at a self-service filling station. However, rocket fuels are optimized for function, rather than for safe handling, and are often very hazardous. For example, methyl hydrazine is a deadly poison, and N_2O_4 is a highly reactive compound that must be kept in corrosion-resistant containers.

Key Concepts: compounds of nitrogen and oxygen, specific enthalpy, enthalpy density

For Further Reading
G. P. Sutton, *Rocket Propulsion Elements: An Introduction to the Engineering of Rockets,* New York: Wiley, 1992.

Related Exercises: 20.87–20.90

The white smoke emitted by the space shuttle booster rockets consists of powdered aluminum oxide and aluminum chloride.

Different fuel and oxidizer were needed on the Apollo missions for the lunar lander. Liquid hydrogen and oxygen are too volatile, and the enthalpy density of liquid hydrogen is too low. The problem with solid rocket fuel is that the combustion is difficult to extinguish and cannot be restarted once the flame has been quenched. Apollo used a mixture of hydrazine derivatives (such as methyl hydrazine, CH_3NHNH_2) and liquid N_2O_4 when landing on and leaving the moon. These two liquids ignite as soon as they mix, instantly producing a large volume of gas:

$$4\,CH_3NHNH_2(l) + 5\,N_2O_4(l) \longrightarrow$$
$$9\,N_2(g) + 12\,H_2O(g) + 4\,CO_2(g)$$

The Apollo lunar lander was powered by a mixture of hydrazine derivatives and dinitrogen tetroxide.

atmosphere for welding (to prevent oxidation) and to fill some types of light bulbs, where its function is to conduct heat away from the filament. Krypton gives an intense white light when a current is passed through it, so it is used in airport runway lighting. Because krypton is produced by nuclear fission, its atmospheric abundance is one measure of worldwide nuclear activity. Xenon is used in halogen lamps for automobile headlights and in high-speed photographic flash tubes because an electric discharge through it—a miniature lightning flash—gives an intense white light. Xenon is also being investigated as an anesthetic.

Radon, which is a radioactive gas, seeps out of the ground as a product of radioactive processes deep in the Earth. There is now some concern that the

FIGURE 20.27

The colors of fluorescent lighting art by Tom Anthony are due to emissions from noble-gas atoms. Neon is responsible for the red light; when it is mixed with a little argon, the color becomes blue-green. The yellow color is achieved by coating the inside of the glass with substances that give off yellow light when excited.

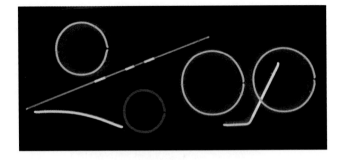

accumulation of radon and its nuclear decay products in buildings can lead to dangerously high levels of radiation.

The noble gases are all found naturally as unreactive monatomic gases. Helium has two liquid phases; the lower temperature phase is a superfluid.

20.11 Compounds of the Noble Gases

The ionization energies of the noble gases are relatively high but decrease down the group (Fig. 20.28); xenon's ionization energy is low enough for electrons to be lost to very electronegative elements, especially fluorine. No compounds of helium, neon, and argon exist, except under very special conditions. Radon is known to react with fluorine; but its radioactivity makes the study of its compounds difficult and dangerous, so little is known about them. Krypton forms only one known stable neutral molecule, KrF_2. In 1988, a compound with a Kr—N bond was reported, but it is stable only below $-50°C$. This leaves xenon as the noble gas with the richest chemistry. It forms several compounds with fluorine and oxygen, and compounds with Xe—N and Xe—C bonds have been reported.

The starting point for the synthesis of xenon compounds is the preparation of xenon difluoride, XeF_2 (**12**), and xenon tetrafluoride, XeF_4 (**13**), by heating a mixture of the elements to $300°C$. At higher pressures, fluorination proceeds as far as xenon hexafluoride, XeF_6 (**14**). All three fluorides are crystalline solids (Fig. 20.29). In the gas phase, all are molecular compounds. Solid xenon hexafluoride, however, is ionic, with a complex structure consisting of XeF_5^+ cations bridged by F^- anions.

The xenon fluorides are used as powerful fluorinating agents (reagents for attaching fluorine atoms to other substances). The tetrafluoride will even fluorinate platinum:

$$Pt(s) + XeF_4(s) \longrightarrow Xe(g) + PtF_4(s)$$

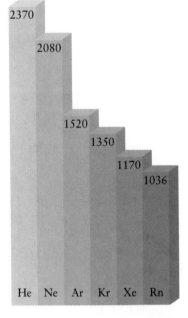

FIGURE 20.28

The ionization energies of the noble gases decrease steadily down the group. The values shown are in kilojoules per mole.

Chemists working with some of the first xenon compounds found that they needed special equipment, because the compounds set fire to stopcock grease.

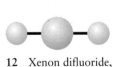

12 Xenon difluoride, XeF_2

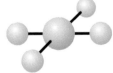

13 Xenon tetrafluoride, XeF_4

14 Xenon hexafluoride, XeF_6

The xenon fluorides are used to prepare a series of xenon oxides and oxoacids and, in a sequence of disproportionations, to bring the oxidation number of xenon up to $+8$. Xenon trioxide, XeO_3, is the anhydride of xenic acid, H_2XeO_4. It reacts with aqueous alkali to form a hydrogen xenate ion, $HXeO_4^-$. This ion slowly disproportionates into xenon and the octahedral perxenate ion, XeO_6^{4-}, in which the oxidation number of xenon is $+8$. Aqueous perxenate solutions are yellow and are very powerful oxidizing agents as a result of the high oxidation number of xenon.

When barium perxenate is treated with sulfuric acid, it is dehydrated to the anhydride of perxenic acid, xenon tetroxide, XeO_4, (15). With this compound, which is an explosively unstable gas, our journey through the main groups of the periodic table comes to an end with a bang.

FIGURE 20.29

Crystals of xenon tetrafluoride, XeF_4. This compound was first prepared in 1962 by the reaction of xenon and fluorine at 6 atm and 400°C.

Only xenon is known to form an extensive series of compounds with fluorine and oxygen. Xenon fluorides are powerful fluorinating agents, and xenon oxides are powerful oxidizing agents.

Self-Test 20.5A Suggest a reason for the increase in the ease of compound formation down Group 18.

[*Answer:* Ionization energy decreases down the group, so it is easier for the heavier elements to share their electrons.]

Self-Test 20.5B Draw the Lewis structure for $XeOF_4$ and predict its molecular shape.

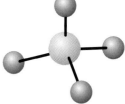

15 Xenon tetroxide, XeO_4

Skills You Should Have Mastered

Conceptual

□ **1.** Explain why nitrogen, oxygen, and fluorine have properties that differ from those of the other members of their groups.

□ **2.** Explain trends in the properties of the elements in Groups 15, 16, 17, and 18.

□ **3.** Rationalize the properties of the oxides and oxoacids of the nonmetals in terms of the oxidation number of the nonmetal and the identification of acid anhydrides.

□ **4.** Explain the low reactivity of the noble gases and the fact that the ease of compound formation increases down the group.

Problem-Solving

□ **1.** Write the formula of an acid, given the formula of its anhydride, and the formula of an anhydride, given that of the acid it forms.

□ **2.** Predict trends in chemical and physical properties across a period and down a group.

Descriptive

□ **1.** Identify the valence electron configurations of Groups 15, 16, 17, and 18.

□ **2.** Describe how nitrogen is obtained from air and converted into useful compounds.

□ **3.** Describe the structures and properties of the nitrogen and phosphorus oxides.

□ **4.** Describe the acidic and basic character of water and hydrogen peroxide and compare their functions as oxidizing and reducing agents.

□ **5.** Describe the names, properties, and reactions of the principal compounds of the members of Groups 15, 16, 17, and 18.

□ **6.** Describe and write balanced equations for the principal reactions used to produce the elements in Groups 15, 16, and 17.

Exercises

Groups 15–18

20.1 Write the ground-state electron configuration of (a) He; (b) O; (c) F; (d) As.

20.2 Write the ground-state electron configuration of (a) Ar; (b) I; (c) P; (d) S.

20.3 Explain why SO_2 can act as a reducing agent, but SO_3 cannot.

20.4 Refer to Appendix 2B and arrange S, O_2 (to H_2O), Cl_2, and H_2O_2 (to H_2O) in order of increasing oxidizing strength in water.

20.5 Arrange the elements P, O, As, S in order of increasing electronegativity.

20.6 Arrange the elements S, Br, O, F in order of increasing atomic radius.

Group 15

20.7 Describe the industrial production of liquid nitrogen.

20.8 Describe, with chemical equations, the industrial production of white and red phosphorus.

20.9 Name the following compounds: (a) HNO_2; (b) NO; (c) H_3PO_4; (d) N_2O_3.

20.10 Name the following compounds: (a) HNO_3; (b) N_2O; (c) H_3PO_3; (d) N_2O_5.

20.11 Write the chemical formula of (a) ammonium nitrate; (b) magnesium nitride; (c) calcium phosphide; (d) hydrazine.

20.12 Write the chemical formula of (a) sodium azide; (b) dinitrogen trioxide; (c) phosphorus(V) oxide; (d) ammonium dihydrogen phosphate.

20.13 Determine the oxidation number of nitrogen in (a) NO; (b) N_2O; (c) HNO_2; (d) N_3^-.

20.14 Determine the oxidation number of phosphorus in (a) P_4; (b) PH_3; (c) H_3PO_4; (d) P_4O_6.

20.15 Urea, $CO(NH_2)_2$, reacts with water to form ammonium carbonate. Write the chemical equation for the reaction and calculate the mass of ammonium carbonate that can be obtained from 5.0 kg of urea.

20.16 Nitrous acid reacts with hydrazine in acidic solution to form hydrazoic acid, HN_3. (a) Write the chemical equation and determine the mass of hydrazoic acid that can be produced from 20.0 g of hydrazine. (b) Suggest a method for preparing sodium azide, NaN_3, from dinitrogen monoxide, N_2O, and sodium amide, $NaNH_2$.

20.17 The common acid anhydrides of nitrogen are N_2O_3 and N_2O_5. Write (a) the formulas of their corresponding acids and (b) the chemical equations for the formation of the acids by the reaction of the anhydrides with water.

20.18 The common acid anhydrides of phosphorus are P_4O_6 and P_4O_{10}. Write (a) the formulas of their corresponding acids and (b) the chemical equations for the formation of the acids by the reaction of the anhydrides with water.

20.19 Solid phosphorus pentachloride exists as $PCl_4^+PCl_6^-$. Write the Lewis structures for the two ions and predict their shapes from VSEPR theory.

20.20 Solid dinitrogen pentoxide exists as $NO_2^+NO_3^-$. Write Lewis structures for the two ions and predict their shapes from VSEPR theory.

20.21 What is the mass percentage of phosphorus in (a) superphosphate? (b) triple superphosphate?

20.22 Calculate the mass percentage of nitrogen in the fertilizers (a) ammonia, NH_3; (b) ammonium nitrate, NH_4NO_3; (c) urea, $CO(NH_2)_2$.

Group 16

20.23 Write the chemical formula of (a) sulfuric acid; (b) calcium sulfite; (c) ozone; (d) barium peroxide.

20.24 Write the chemical formula of (a) sulfurous acid; (b) hydrogen selenide; (c) sodium sulfite; (d) disulfur dichloride.

20.25 Write equations for (a) the burning of lithium in oxygen; (b) the reaction of sodium metal with water; (c) the reaction of fluorine gas with water; (d) the oxidation of water at the anode of an electrolytic cell.

20.26 Write equations for the reaction of (a) sodium oxide and water; (b) sodium peroxide and water; (c) sulfur dioxide and water; (d) sulfur dioxide and oxygen, using a vanadium pentoxide catalyst.

20.27 Complete and balance the following equations:
(a) $H_2S(g) + O_2(g) \rightarrow$
(b) $PCl_5(s) + H_2O(l) \rightarrow$

20.28 Complete and balance the following equations:
(a) $Cl_2(g) + H_2O(l) \rightarrow$
(b) $CaO(s) + H_2O(l) \rightarrow$

20.29 Write the Lewis structure of H_2O_2 and predict the approximate H—O—O bond angle.

20.30 (a) Write the Lewis structure of SO_3 in which the formal charge on each atom is 0. (b) What is the shape of the molecule? (c) What is the hybridization of the sulfur atom?

20.31 Write the Lewis structure and predict the shape of (a) SO_2; (b) SF_4; (c) SO_4^{2-}.

20.32 Write the Lewis structure and predict the shape of (a) SF_6; (b) H_2S; (c) SO_3^{2-}.

Group 17

20.33 List the natural sources of fluorine and chlorine.

20.34 List the natural sources of bromine and iodine.

20.35 Write chemical equations that describe the preparation of fluorine and chlorine.

20.36 Write chemical equations that describe the preparation of bromine and iodine.

20.37 Name the following compounds or solutions: (a) HBr(aq); (b) IBr; (c) ClO_2; (d) $NaIO_3$.

20.38 Name the following compounds or solutions: (a) HI(aq); (b) IF_3; (c) HClO(aq); (d) NaClO.

20.39 Write the chemical formulas of (a) perchloric acid; (b) sodium chlorate; (c) hydroiodic acid; (d) sodium triiodide.

20.40 Write the chemical formulas of (a) potassium periodate; (b) ammonium perchlorate; (c) chloric acid; (d) potassium bromide.

20.41 Identify the oxidation number of the halogen atoms in (a) hypoiodous acid; (b) ClO_2; (c) dichlorine heptoxide; (d) $NaIO_3$.

20.42 Identify the oxidation number of the halogen atoms in (a) IF_7; (b) sodium periodate; (c) hypobromous acid; (d) $HClO_2$.

20.43 Write the Lewis structure for each of the following species, describe its electronic arrangement, and predict its shape: (a) ClO_4^-; (b) IO_3^-; (c) IF_3.

20.44 Write the Lewis structure for each of the following species, describe its electronic arrangement, and predict its shape: (a) BrO_2^-; (b) ClF_5; (c) I_3^-; (d) BrF_4^-.

20.45 Write the balanced chemical equations for (a) the thermal decomposition of potassium chlorate without a catalyst; (b) the reaction of bromine with water; (c) the reaction between sodium chloride and concentrated sulfuric acid. (d) Identify each reaction as a Brønsted acid-base, Lewis acid-base, or redox reaction.

20.46 Write the balanced chemical equations for the reaction of chlorine in (a) neutral aqueous solution; (b) dilute basic solution; (c) hot, concentrated basic solution. (d) Verify that each reaction is a disproportionation reaction.

20.47 (a) Arrange the chlorine oxoacids in order of increasing oxidizing strength. (b) Suggest an interpretation of that order in terms of oxidation numbers.

20.48 (a) Arrange the hypohalous acids in order of increasing acid strength. (b) Suggest an interpretation of that order in terms of electronegativities.

20.49 Write the Lewis structure for Cl_2O. Predict the shape of the Cl_2O molecule and estimate the $Cl-O-Cl$ bond angle.

20.50 Write the Lewis structure for BrF_3. What is the hybridization of the bromine atom in the molecule?

20.51 Plot a graph of the standard free energy of formation of the hydrogen halides against the period number of the halogens. What conclusions can be drawn from the graph?

20.52 Use the data from Table 20.7 to plot the normal boiling points of the hydrogen halides against the period number of the halogens. Account for the trend revealed by the graph.

Group 18

20.53 What are the sources for the production of helium and argon?

20.54 What are the sources for the production of krypton and xenon?

20.55 Determine the oxidation number of the noble gas atom in (a) KrF_2; (b) XeF_6; (c) KrF_4; (d) XeO_4^{2-}.

20.56 Determine the oxidation number of the noble gas atom in (a) XeO_3; (b) XeO_6^{4-}; (c) XeF_2; (d) $HXeO_4^-$.

20.57 Xenon tetrafluoride is a powerful oxidizing agent. In an acidic solution, it is reduced to xenon gas and fluoride ions. Write the corresponding half-reaction.

20.58 Xenon hexafluoride reacts with water to produce xenon trioxide and hydrofluoric acid. Write the chemical equation for the reaction.

20.59 Predict the relative acid strengths of H_2XeO_4 and H_4XeO_6. Explain your conclusions.

20.60 Predict the relative oxidizing strengths of H_2XeO_4 and H_4XeO_6. Explain your conclusions.

20.61 Write the Lewis structure for XeF_4. Estimate the $F-Xe-F$ bond angle.

20.62 Write the Lewis structure for XeO_3. What is the hybridization on the xenon atom in the molecule?

Supplementary Exercises

20.63 List the evidence that shows that metallic character decreases from the lower left of the periodic table to the upper right.

20.64 Summarize, with examples and explanations, the principal differences between the elements in the s and p blocks.

20.65 Arsenic(III) sulfide is oxidized by acidic hydrogen peroxide solution to the arsenate ion, AsO_4^{3-}. Write the chemical equation and the reduction and oxidation half-reactions for the reaction.

20.66 When the enthalpy of vaporization of water is divided by its boiling point (on the Kelvin scale), the result is 110 J/K·mol. For hydrogen sulfide, the same calculation gives 88 J/K·mol. Explain why the value for water is greater than that for hydrogen sulfide.

20.67 Determine the volume (in liters) of concentrated sulfuric acid (18.4 M) that is required for the production of 1000. kg of phosphoric acid from phosphate rock by the reaction

$$Ca_3(PO_4)_2(s) + 3\,H_2SO_4(l) \rightarrow 2\,H_3PO_4(l) + 3\,CaSO_4(s)$$

20.68 Balance each skeletal equation and classify the reaction as Brønsted acid-base, Lewis acid-base, or redox:

(a) $KClO_3(s) \rightarrow KCl(s) + O_2(g)$

(b) $CaF_2(s) + H_2SO_4(conc) \rightarrow Ca(HSO_4)_2(aq) + HF(g)$

(c) $OF_2(g) + OH^-(aq) \rightarrow O_2(g) + F^-(aq) + H_2O(l)$

(d) $H_2S(g) + O_2(g) \rightarrow SO_2(g) + H_2O(l)$

20.69 Balance each skeletal equation and classify the reaction as Brønsted acid-base, Lewis acid-base, or redox:

(a) $SO_2(g) + H_2O(l) \rightarrow H_2SO_3(aq)$

(b) $F_2(g) + NaOH(aq) \rightarrow OF_2(g) + NaF(aq) + H_2O(l)$

(c) $S_2O_3^{2-}(aq) + Cl_2(g) + H_2O(l) \rightarrow$
$$HSO_4^-(aq) + H_3O^+(aq) + Cl^-(aq)$$

(d) $XeF_6(s) + OH^-(aq) \rightarrow$
$$XeO_6^{4-}(aq) + Xe(g) + O_2(g) + F^-(aq) + H_2O(l)$$

20.70 Considering (a) Cl_2O; (b) Cl_2O_7 to be acid anhydrides, write the formulas of the corresponding acids.

20.71 Complete and balance the following skeletal equations:
(a) $I_2(s) + F_2(g) \rightarrow$ (any one of a number of products)
(b) $I_2(aq) + I^-(aq) \rightarrow$
(c) $Cl_2(g) + H_2O(l) \rightarrow$
(d) $F_2(g) + H_2O(l) \rightarrow$

20.72 Explain why iodine is much more soluble in a solution of potassium iodide than in pure water.

20.73 Write equations for reactions used to synthesize gaseous samples of (a) hydrogen chloride; (b) hydrogen bromide; (c) hydrogen iodide.

20.74 Give the names and formulas of three compounds in which (a) nitrogen; (b) phosphorus has oxidation number (i) +3; (ii) +5.

20.75 (a) Summarize and account for the differences between fluorine and the other Group 17 elements. (b) What do we mean when we say that fluorine is a reactive element?

20.76 Discuss the chemistry of a safety match: identify the chemicals that are used and the function of each. See Section 19.7, too.

20.77 Write chemical equations for the preparation of (a) XeF_4; (b) PtF_4. See Section 20.11.

20.78 Xenon hexafluoride exists as the ionic solid $XeF^{5+}F^-$. Write the Lewis structure for XeF^{5+} and, from the VSEPR model, predict its shape.

20.79 Describe the trends in the physical and chemical properties of the noble gases.

20.80 Suggest reasons why the noble gas compounds were unknown until late in the twentieth century.

20.81 Identify one interesting compound of each (a) Period 2; (b) Period 3 element that would leave the world a much poorer place if it did not exist, and give your reasons for each choice.

20.82 Summarize the evidence for or against the statement that metallic character increases down Group 16 but is less pronounced than in Group 15, which, in turn, is less pronounced than in Group 14.

20.83 Chapters 19 and 20 have been organized by group. (a) Discuss whether it would be helpful to organize the elements according to period and give examples of trends that could be displayed helpfully in that way. (b) Would a different organization be useful for the transition metals? Explain your decision.

20.84 The azide ion has radius of 148 pm and forms many ionic and covalent compounds that are similar to those of the halides. (a) Write the Lewis formula for the azide ion and predict the N—N—N bond angle. (b) On the basis of its ionic radius, where in Group 17 would you place the azide ion? (c) Compare the acidity of hydrazoic acid with those of the hydrohalic acids and explain any differences (for HN_3,

$K_a = 1.7 \times 10^{-5}$). (d) Write the formulas of three ionic or covalent azides.

20.85 Account for the observation that melting and boiling points generally decrease from fluoride to iodide for ionic halides, but increase from fluoride to iodide for molecular halides.

20.86 Account for the observation that solubility in water generally increases from chloride to iodide for ionic halides with low covalent character (such as the potassium halides), but decreases from chloride to iodide for ionic halides in which the bonds are significantly covalent (such as the silver halides).

Applied Exercises

For Exercises 20.87–20.90, see Applying Chemistry: Case Study 20.

20.87 In the reaction of the solid-fuel booster rockets, one element is being reduced and two are being oxidized:

$$3\,NH_4ClO_4(s) + 3\,Al(s) \xrightarrow{Fe_2O_3}$$
$$Al_2O_3(s) + AlCl_3(s) + 6\,H_2O(g) + 3\,NO(g)$$

Identify these three elements and write half-reactions for each.

20.88 Using Appendix 2A and the equation in Exercise 20.87, determine the standard enthalpy of reaction for the solid fuel in the booster rocket. Assume that ΔH_f° of $NH_4ClO_4(s)$ is -295 kJ/mol.

20.89 The rocket fuel methyl hydrazine is oxidized by dinitrogen tetroxide:

$$4\,CH_3NHNH_2(l) + 5\,N_2O_4(l) \longrightarrow$$
$$9\,N_2(g) + 12\,H_2O(g) + 4\,CO_2(g)$$

Using Appendix 2A, determine the standard enthalpy of the reaction. Assume that ΔH_f° of $CH_3NHNH_2(l)$ is $+54$ kJ/mol.

20.90 (a) Calculate the specific enthalpy of liquid hydrogen in its reaction with liquid oxygen in the main engine of the space shuttle. (b) Would gaseous hydrogen have a larger, smaller, or the same specific enthalpy as liquid hydrogen? Explain your reasoning.

20.91 (a) Write the chemical equation for the production of chlorine from the electrolysis of an aqueous sodium chloride solution. (b) The annual production of chlorine gas in the United States in 1997 was 1.2×10^{10} kg. How many coulombs of electricity are required for the production of this chlorine? (c) If the chlorine is generated 24 hours a day for 365 days a year, what is the average current (in amperes) that must be used?

20.92 The sodium iodate found as an impurity in Chile saltpeter was once the major source of iodine. The element was obtained by the reduction of an acidic solution of the iodate ion with sodium hydrogen sulfite. (a) Write the chemical equation for the reaction, assuming the oxidized product to be HSO_4^-. (b) Calculate the mass of sodium hydrogen sulfite needed to produce 50.0 g of iodine.

Integrated Exercises

20.93 Lead azide, $Pb(N_3)_2$, is used as a detonator. (a) What volume of nitrogen at STP does 1.0 g of lead azide produce when it decomposes into lead metal and nitrogen gas? (b) Would 1.0 g of mercury(II) azide, $Hg(N_3)_2$, which is also used as a detonator, produce a larger or smaller volume, given that its decomposition products are liquid mercury and nitrogen gas?

20.94 Sodium azide is used to inflate protective air bags in automobiles. What mass of solid sodium azide is needed to provide 100. L of N_2 at 1.5 atm and 20°C?

20.95 Hydrogen peroxide is unstable with respect to decomposition into water and oxygen when exposed to light, heat, or a catalyst. If 500. mL of a 3.00% H_2O_2 aqueous solution (like that sold in drugstores) decomposes, what volume of oxygen at 273 K and 1.00 atm will be produced? Assume the density of the solution to be 1.0 g/mL.

20.96 If 2.00 g of sodium peroxide is dissolved to form 200. mL of an aqueous solution, what would be the pH of the solution? For H_2O_2, $K_{a1} = 1.8 \times 10^{-12}$ and K_{a2} is negligible.

20.97 If you were titrating 0.10 M $H_2S(aq)$ with 0.10 M $NaOH(aq)$, at what pH would you have a 0.050 M $NaHS(aq)$ solution?

20.98 (a) Calculate the standard enthalpy and entropy of reaction for the formation of ozone from oxygen. (b) Is the reaction favored at high temperatures or low temperatures? (c) Does the reaction entropy favor the spontaneous formation of ozone? Explain your conclusions.

20.99 When lead(II) sulfide is treated with hydrogen peroxide, the possible products are either (a) lead(IV) oxide and sulfur dioxide or (b) lead(II) sulfate. In terms of the reaction free energy, which product or products are more likely?

20.100 The standard free energy of formation of $HI(g)$ is $+1.70$ kJ/mol at 25°C. Determine the equilibrium constant for the formation of $HI(g)$ from its elements according to the equation $\frac{1}{2} H_2(g) + \frac{1}{2} I_2(s) \rightleftharpoons HI(g)$.

20.101 The standard free energy of formation of $HCl(g)$ is -95.3 kJ/mol at 25°C. Determine the equilibrium constant for the formation of $HCl(g)$ according to the equation $\frac{1}{2} H_2(g) + \frac{1}{2} Cl_2(g) \rightleftharpoons HCl(g)$.

20.102 The concentration of F^- ions can be measured by adding an excess of lead(II) chloride solution and weighing the lead(II) chlorofluoride (PbClF) precipitate. Calculate the molarity of F^- ions in 25.00 mL of a solution that gave a lead(II) chlorofluoride precipitate of mass 0.765 g.

20.103 Suppose 25.00 mL of an aqueous solution of iodine were titrated with 0.025 M $Na_2S_2O_3(aq)$, using starch as the indicator. The blue color of the starch-iodine complex disappeared when 28.45 mL of the thiosulfate solution had

been added. What was the molar concentration of I_2 in the original solution? The titration reaction is
$$I_2(aq) + 2 S_2O_3^{2-}(aq) \rightarrow 2 I^-(aq) + S_4O_6^{2-}(aq).$$

20.104 (a) Determine the values of $\Delta H_r°$ and $\Delta S_r°$ for each stage of the Ostwald process for the production of nitric acid. (b) Predict the conditions of pressure and temperature that favor the formation of the products in each case.

Stage 1: $4 NH_3(g) + 5 O_2(g) \rightarrow 4 NO(g) + 6 H_2O(g)$
Stage 2: $2 NO(g) + O_2(g) \rightarrow 2 NO_2(g)$
Stage 3: $3 NO_2(g) + H_2O(l) \rightarrow 2 HNO_3(aq) + NO(g)$

20.105 The annual production of sulfuric acid in the United States is approximately 4×10^{10} kg. (a) If the acid were all produced from elemental sulfur, what mass of sulfur would be used for the production of sulfuric acid? (b) What volume of sulfur trioxide (at 25°C and 5.0 atm) is required for the annual production of sulfuric acid?

20.106 The concentration of Cl^- ions can be measured gravimetrically by precipitating silver chloride, with silver nitrate as the precipitating reagent in the presence of dilute nitric acid. The white precipitate is filtered off and its mass is determined. (a) Calculate the Cl^- ion concentration in 25.00 mL of a solution that gave a silver chloride precipitate of mass 3.050 g. (b) Why is the method inappropriate for measuring the concentration of fluoride ions?

20.107 The concentration of nitrate ion in a basic solution can be determined by the following sequence of steps: (1) zinc metal reduces nitrate ions to ammonia in a basic aqueous solution; (2) the ammonia is passed into a solution containing a known, but excess, amount of $HCl(aq)$; (3) the unreacted $HCl(aq)$ is titrated with a standard $NaOH(aq)$ solution. (a) Write balanced chemical equations for the three reactions. (b) A 25.00-mL sample of water from a rural well contaminated with $NO_3^-(aq)$ was treated with an excess of zinc metal. The evolved ammonia gas was passed into 50.00 mL of a 0.250 M $HCl(aq)$ solution. The unreacted $HCl(aq)$ was titrated to the stoichiometric point with 28.22 mL of 0.150 M $NaOH(aq)$. What was the molar concentration of nitrate ion in the well water?

20.108 The concentration of hypochlorite ions in a solution can be determined by adding a sample of known volume to a solution containing excess I^- ions, which are oxidized to iodine:
$$ClO^-(aq) + 2 I^-(aq) + H_2O(l) \longrightarrow$$
$$I_2(aq) + Cl^-(aq) + 2 OH^-(aq)$$

The iodine concentration is then measured by titration with sodium thiosulfate:
$$I_2(aq) + 2 S_2O_3^{2-}(aq) \longrightarrow 2 I^-(aq) + S_4O_6^{2-}(aq)$$

In one experiment, 10.00 mL of ClO^- solution was added to a KI solution, which in turn required 28.34 mL of 0.110 M $Na_2S_2O_3(aq)$ to reach the stoichiometric point. Calculate the molar concentration of ClO^- ions in the original solution.

The *d* Block: Metals in Transition

The *d*-block metals are the workhorse elements of the periodic table. Iron and copper helped civilization rise from the Stone Age and are still industrial metals of the greatest importance. Other members of the block include the metals of newer technologies, such as titanium for the aerospace industry and vanadium for catalysts in the petrochemical industry. The precious metals—silver, platinum, and gold—are prized as much for their appearance, rarity, and durability as for their usefulness. Compounds of *d*-block metals give color to paint, turn sunlight into electricity, serve as powerful oxidizing agents, and form the basis of cancer treatments.

Atoms of the *d*-block elements have electron configurations in which, according to the building-up principle, *d*-orbitals are the last to be occupied. For example, in Period 4, the electron configurations range from $[Ar]3d^14s^2$ for scandium, the first member of the block, to $[Ar]3d^{10}4s^2$ for zinc, the last member. Because there are five *d*-orbitals in a given shell and because each one can accommodate up to two electrons, there are 10 elements in each row of the *d* block. Just after the third row of the *d* block has been started, at lanthanum, the seven 4*f*-orbitals begin to be occupied, and the lanthanides therefore delay the completion of that row. The elements in Period 7 are all radioactive, and most do not occur naturally in measurable quantities on Earth.

Steel girders like these form the skeletons of most of our buildings. Iron is one of the *d*-block elements called transition metals because their properties are transitional between the reactive metals of the *s* block and the relatively inert metals of the *p* block. The transistion metals are also responsible for transitions in human history. Our evolution from Stone Age to Bronze Age to Iron Age to the age of steel and the industrial revolution arose from technological improvements that allowed more and more sophisticated processing of metals.

FIGURE 21.1

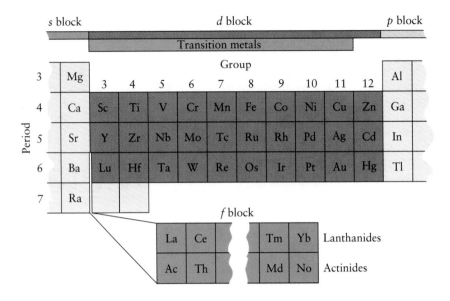

THE *d*-BLOCK ELEMENTS AND THEIR COMPOUNDS

The *d*-block elements lose their valence *s*-electrons when they form compounds. Most of them can also lose a variable number of *d*-electrons. This variable valence makes these elements useful as catalysts and, in partnership with proteins, as enzymes. The only elements of the block that do not use their *d*-electrons in compound formation are the members of Group 12 (zinc, cadmium, and mercury). The elements in Groups 3 through 11 represent a transition from the highly reactive metals of the *s* block to the much less reactive metals of Group 12 and the *p* block; they are called the *transition metals* (Fig. 21.1).

21.1 Trends in Physical Properties

All the *d*-block elements are metals. Most of these "*d*-metals" are good electrical conductors. Most are malleable, ductile, lustrous, and silver-white in color, and they generally have higher melting and boiling points than the main-group elements do. There are a few notable exceptions: mercury has such a low melting point that it is a liquid at room temperature; copper is red-brown and gold is yellow.

Many properties of the *d*-block elements can be traced to the shapes of *d*-orbitals (recall Fig. 7.20). First, the lobes of two *d*-orbitals on the same atom occupy markedly different regions of space. Because they are relatively far apart, electrons in separate *d*-orbitals repel one another weakly. Second, electron density in *d*-orbitals is low near the nucleus. Because *d*-electrons are far from the nucleus, they are not very effective at shielding other electrons from the nuclear charge.

The orbital shapes help account for the trends in atomic radii of the *d*-block metals (Fig. 21.2). Nuclear charge and the number of *d*-electrons both increase from left to right across each row (from scandium to zinc, for instance). Because the repulsion between *d*-electrons is weak, the increasing nuclear charge can

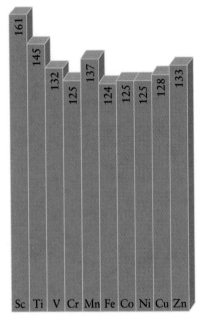

FIGURE 21.2

The atomic radii (in picometers) of the elements of the first row (Period 4) of the *d* block. The radii decrease from Sc to Ni because of the increasing nuclear charge and the poor shielding effect of *d*-electrons.

draw them inward, so the atom becomes smaller. Further across the block, though, the radii begin to increase slightly. For these elements, there are so many *d*-electrons that the electron-electron repulsion increases more than the effective nuclear charge. However, these attractions and repulsions are finely balanced, and the range of *d*-metal atomic radii is not very great. Because the atoms have similar sizes, the atoms of one *d*-metal can replace the atoms of another in a crystal lattice without causing too much strain (Fig. 21.3). The *d*-metals can therefore be mixed together to form a wide range of alloys, including the many varieties of steel.

The atomic radii of *d*-metals in the second row of the block (Period 5) are typically greater than those in the first row (Period 4). However, the radii in the third row (Period 6) are about the same as those in the second row. This similarity can be traced to the presence of the *f*-block elements near the beginning of Period 6 (Fig. 21.4). There is a pronounced decrease in atomic radius along the first row of the *f* block (the lanthanides). When the *d* block resumes, the atomic radius has fallen from 224 pm for barium to 172 pm for lutetium. As a result, the atoms of all the following elements are smaller than expected. This effect is called the **lanthanide contraction.** The contraction is explained by the higher nuclear charge and the poor shielding effect of the *f*-electrons: the atomic radii decrease as the effective nuclear charge increases and pulls the valence electrons inward.

The lanthanide contraction is responsible for the high density of the Period 6 elements (Fig. 21.5). A block of iridium, for example, contains about as many atoms as a block of rhodium of the same volume; but each iridium atom is nearly twice as heavy as each rhodium atom, so the density of the sample is nearly twice as great. Another effect of the contraction is the low reactivity of gold and platinum. Because their valence electrons are relatively close to the nucleus, they are tightly bound and not readily available for chemical reactions.

The presence of unpaired *d*-electrons in the ground states of the *d*-block elements explains why some of these metals—most notably, iron, cobalt, and nickel—are used to make permanent magnets. Recall that *paramagnetism* is the tendency of a substance to move into a magnetic field; it arises when an atom or molecule has at least one unpaired electron (see Section 20.5). Because the spins on neighboring atoms or molecules are aligned almost randomly, paramagnetism is very weak. In some *d*-metals, however, the unpaired electrons of large numbers of neighboring atoms can align with one another, producing the much stronger effect of **ferromagnetism.** The aligned spins form **domains,** or regions of aligned spins (Fig. 21.6), which survive even after the applied field is turned off. Ferromagnetic materials are used in the coatings of cassette tapes and computer disks (see also Section 21.3).

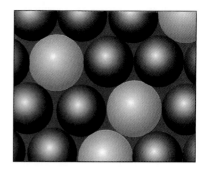

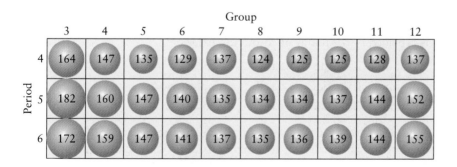

Group

	3	4	5	6	7	8	9	10	11	12
4	164	147	135	129	137	124	125	125	128	137
5	182	160	147	140	135	134	134	137	144	152
6	172	159	147	141	137	135	136	139	144	155

Period

FIGURE 21.4

The atomic radii of the *d*-block elements (in picometers). Notice the similarity of all the values and, in particular, the close similarity between the second and third rows as a result of the lanthanide contraction.

FIGURE 21.5

The densities (in grams per cubic centimeter, g/cm³) of the d-metals at 25°C. The lanthanide contraction has a pronounced effect on the densities of the elements in Period 6 (front row in this illustration), which are among the densest of all the elements.

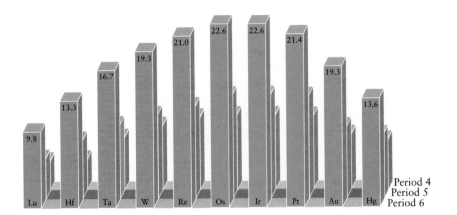

The atomic radii of the d-block elements decrease with increasing atomic number on the left of a row and then increase slightly on the right; the radii of the atoms of Periods 5 and 6 are similar as a result of the lanthanide contraction. Iron, cobalt, and nickel are ferromagnetic.

21.2 Trends in Chemical Properties

Figure 21.7 shows the range of oxidation states of each d-metal. Except for mercury, the elements at the ends of each row of the d block occur in only one oxidation state other than 0. Scandium, for example, is found only in oxidation state +3, and zinc only in +2. All the other elements of each row are found in at least two oxidation states. The most common oxidation states for copper, for example, are +1 (as in CuCl) and +2 (as in $CuCl_2$). Elements close to the center of each row have the widest range of oxidation states. Manganese, at the center of its row, has seven oxidation states. Elements in the second and third rows of the block are more likely to reach higher oxidation states than those in the first row.

It is difficult to predict the likely reaction of an element, but some broad principles are often helpful. For instance, a species in which an element is in a high oxidation state tends to be a good oxidizing agent. In the permanganate ion, MnO_4^-, for example, manganese has oxidation number +7, so permanganate ion is a good oxidizing agent in acidic solution:

$$MnO_4^-(aq) + 8\,H^+(aq) + 5\,e^- \longrightarrow Mn^{2+}(aq) + 4\,H_2O(l) \qquad E° = +1.51\,V$$

The presence of a high concentration of H^+ ions in acidic solution helps to drive this reaction to the right. The H^+ ions assist in the removal of oxygen from the permanganate ion.

Although most d-metal oxides are basic, the oxides of a given element show a shift toward acidic character with increasing oxidation number. A good example is provided by the family of chromium oxides:

chromium(II) oxide	CrO	basic
chromium(III) oxide	Cr_2O_3	amphoteric
chomium(VI) oxide	CrO_3	acidic

Chromium(VI) oxide, CrO_3, is the anhydride of chromic acid, H_2CrO_4, the parent acid of the chromates. In this highly oxidized state, chromium is electron poor and the O atoms attached to it give up their electrons to the Cr atom and

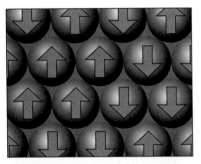

(a)

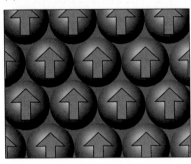

(b)

FIGURE 21.6

Ferromagnetic materials include iron, cobalt, nickel, gadolinium, chromium (IV) oxide, and the iron oxide mineral magnetite. They contain regions of atoms in which electrons of many atoms spin in the same direction and give rise to a strong magnetic field. (a) Before magnetization, the spins are almost randomly aligned. (b) After magnetization the spins are aligned in the same direction. The orange arrows represent the electron spins of each atom.

FIGURE 21.7

8 7 6 5 4 3 2 1

Sc Ti V Cr Mn Fe Co Ni Cu Zn

Y Zr Nb Mo Tc Ru Rh Pd Ag Cd

Lu Hf Ta W Re Os Ir Pt Au Hg

The oxidation numbers of the *d*-block elements. The orange blocks mark the principal oxidation numbers for each element; the green blocks mark the other known states.

FIGURE 21.8

hence are less likely to be able to share electrons with a proton. The conjugate base of chromic acid, $HCrO_4^-$, does not attract protons strongly, despite its overall negative charge, and is therefore a very weak base.

The elements on the left of the *d* block resemble the metals of the *s* block in being much more difficult to extract from their ores than the *d*-block metals on the right. Indeed, if we go to the right side of the *d* block and move back across it from right to left, then we encounter the elements in very roughly the order in which they became available for exploitation. On the far right are copper and zinc, which jointly were responsible for the Bronze Age. That age was succeeded by the Iron Age, as higher processing temperatures became attainable and iron ore reduction feasible. Finally, at the left of the block, we have metals like titanium that require such extreme conditions for their extraction—including the use of other active metals or electrolysis—that they have become widely available only in this century (Fig. 21.8).

These three artifacts represent the progress that has been made in the extraction of *d*-metals. (a) An ancient bronze chariot axle cap from China, made from an alloy of metals that are easy to extract. (b) A nineteenth-century iron steam engine made from a metal that was moderately easy to extract once high temperatures could be achieved. (c) A twentieth-century airplane engine with titanium components that had to await advanced, high-temperature technology before the element became widely available.

(a)

(b)

(c)

The range of oxidation states of a d-block element increases toward the center of the block. High-oxidation-state compounds tend to be oxidizing; low-oxidation-state compounds tend to be reducing. The acidic character of oxides increases with the oxidation state of the element.

Self-Test 21.1A Predict trends in ionization energies of the *d*-block metals.

[*Answer:* Ionization energy increases from left to right across a row and decreases down a group.]

Self-Test 21.1B Predict trends in the densities of the *d*-block metals.

21.3 Scandium Through Nickel

Table 21.1 summarizes the physical properties of the elements in the first row of the *d* block from scandium to nickel. Here we consider only their most important chemical properties.

Scandium, Sc, which was first isolated in 1937, is a reactive metal: it reacts with water about as vigorously as calcium does. It has few uses. The small, highly charged Sc^{3+} ion is strongly hydrated in water (like Al^{3+}), and the resulting hydrated ion is about as strong a Brønsted acid as acetic acid.

Titanium, Ti, is a light, strong metal that is used where these properties are vital—as it is in airplanes (Fig. 21.9). Unlike scandium, it is resistant to corrosion because it is passivated by a protective skin of oxide that forms on its surface. The principal sources of the metal are the ores *ilmenite*, $FeTiO_3$, and *rutile*, TiO_2.

Titanium requires strong reducing agents for extraction from its ores. It was not exploited commercially until quite recently, when demand from the aerospace industry increased. The metal is obtained first by treating the ores with chlorine in the presence of coke to form titanium(IV) chloride. The chloride is then reduced to the metal by passing it through liquid magnesium:

$$TiCl_4(g) + 2\,Mg(l) \xrightarrow{700°C} Ti(s) + 2\,MgCl_2(s)$$

The most stable oxidation state of titanium is +4, with both its 4*s*-electrons and its two 3*d*-electrons lost. Its most important compound is titanium(IV) oxide, TiO_2, which is almost universally known as titanium dioxide. It is a brilliantly white, nontoxic, stable solid used as the white pigment in paints and paper. Titanium dioxide acts as a semiconductor in the presence of light, so it

Table 21.1 *Properties of the d-block elements scandium through nickel*

Z	Name	Symbol	Electron configuration	Melting point, °C	Boiling point, °C	Density, g/cm³
21	scandium	Sc	$3d^1\,4s^2$	1540	2800	2.99
22	titanium	Ti	$3d^2\,4s^2$	1660	3300	4.55
23	vanadium	V	$3d^3\,4s^2$	1920	3400	6.11
24	chromium	Cr	$3d^5\,4s^1$	1860	2600	7.19
25	manganese	Mn	$3d^5\,4s^2$	1250	2120	7.47
26	iron	Fe	$3d^6\,4s^2$	1540	2760	7.87
27	cobalt	Co	$3d^7\,4s^2$	1494	2900	8.80
28	nickel	Ni	$3d^8\,4s^2$	1455	2150	8.91

FIGURE 21.9

This piece of titanium has been formed and then machined for use in a high-performance military jet plane. Titanium has excellent corrosion resistance, low density, and high strength, making it ideal for aerospace applications.

has been studied for use as an electrode in *photoelectrochemical cells*, cells that use light to produce chemicals, principally hydrogen from water.

Titanium forms a series of oxides called *titanates*, which are prepared by heating TiO_2 with a stoichiometric amount of the oxide or carbonate of a second metal. Barium titanate, $BaTiO_3$, is **piezoelectric,** which means that it becomes electrically charged when it is mechanically distorted. This property leads to its use for underwater sound detection, in which a mechanical vibration is converted into an electrical signal.

Vanadium, V, a soft silver-gray metal, is produced by reducing vanadium(V) oxide with calcium:

$$V_2O_5(s) + 5\,Ca(l) \xrightarrow{\Delta} 2\,V(s) + 5\,CaO(s)$$

or reducing vanadium(II) chloride with magnesium:

$$VCl_2(s) + Mg(l) \xrightarrow{\Delta} V(s) + MgCl_2(s)$$

Vanadium metal is also produced commercially by electrolysis of molten vanadium(II) chloride.

Vanadium is used to make tough steels for automobile and truck springs. Because it is not economical to add the pure metal to iron, a *ferroalloy* of the metal—an alloy with iron and carbon that is less expensive to produce—is used instead. *Ferrovanadium,* a mixture of about 86% V, 12% C, and 2% Fe by mass, is prepared by reducing vanadium(V) oxide with aluminum in the presence of iron. The ferrovanadium is then added to a molten mixture of iron and carbon to make the vanadium steel.

Vanadium(V) oxide, V_2O_5, commonly known as vanadium pentoxide, is the most important compound of vanadium. This orange-yellow solid is used as an oxidizing agent and as an oxidizing catalyst in the contact process for the manufacture of sulfuric acid (Section 20.7). The wide range of colors of vanadium compounds, including the blue of the vanadyl ion, VO^{2+} (Fig. 21.10), has led to their use as glazes in the ceramics industry.

Chromium, Cr, is a bright, lustrous, corrosion-resistant metal that gets its name from its colorful compounds. It is obtained from the mineral *chromite*, $FeCr_2O_4$, by reduction with carbon in an electric arc furnace:

$$FeCr_2O_4(s) + 4\,C(s) \xrightarrow{\Delta} Fe(s) + 2\,Cr(s) + 4\,CO(g)$$

Chromium metal is also produced by the thermite process (Section 6.2):

$$Cr_2O_3(s) + 2\,Al(s) \xrightarrow{\Delta} Al_2O_3(s) + 2\,Cr(l)$$

The thermite process, in which aluminum is a vigorous reducing agent, is also used to reduce other metals, such as manganese and tantalum, from their ores.

Most of the chromium metal produced is used in steelmaking and for chromium plating (Section 18.16). Chromium(IV) oxide, CrO_2, is a ferromagnetic material that is used for coating "chrome" recording tapes because they

It is helpful to remember that, by coincidence, V (which is also the Roman numeral for 5) is in Group 5.

The name *chromium* comes from *chroma*, the Greek word for "color."

FIGURE 21.10

Many vanadium compounds form vividly colored aqueous solutions. They are also used to color pottery glazes. The blue colors here are due to the vanadyl ion, VO^{2+}.

respond better to high-frequency magnetic fields than do conventional "ferric" (Fe_2O_3) tapes.

Sodium chromate, Na_2CrO_4, is a yellow solid. Its importance lies in its use as a source of most other chromium compounds. The chromate ion changes into the orange dichromate ion, $Cr_2O_7{}^{2-}$, in the presence of acid (Fig. 21.11):

$$2\,CrO_4{}^{2-}(aq) + 2\,H^+(aq) \longrightarrow Cr_2O_7{}^{2-}(aq) + H_2O(l)$$

In the laboratory, acidified solutions of dichromates, in which the oxidation number of chromium is +6, are useful oxidizing agents:

$$Cr_2O_7{}^{2-}(aq) + 14\,H^+(aq) + 6\,e^- \longrightarrow 2\,Cr^{3+}(aq) + 7\,H_2O(l) \qquad E^\circ = +1.33\,V$$

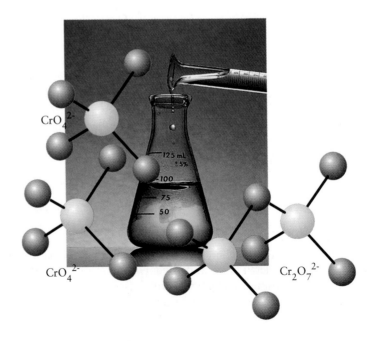

FIGURE 21.11

The chromate ion, $CrO_4{}^{2-}$, is yellow. When acid is added to a chromate solution, the ions form dichromate ions, $Cr_2O_7{}^{2-}$, which are orange.

Table 21.2 *Composition of steels*

Element blended into iron	Typical amount, %	Effect
manganese	0.5 to 1.0	increases strength and hardness but lowers ductility
	13	increases wear resistance
nickel	<5	increases strength and shock resistance
	>5	increases corrosion resistance (stainless) and hardness
chromium	variable	increases hardness and wear resistance
	>12	increases corrosion resistance (stainless)
vanadium	variable	increases hardness
tungsten	<20	increases hardness, especially at high temperatures

Sodium chromate and sodium dichromate are both starting points for the production of a number of pigments, corrosion inhibitors, fungicides, and ceramic glazes.

Manganese, Mn, is a gray metal that resembles iron. Unlike chromium, manganese is rarely used alone, because it is much less resistant to corrosion, but it is an important component of alloys. In steelmaking, it removes sulfur by forming a sulfide. It also increases hardness, toughness, and resistance to abrasion (see Table 21.2). Another useful alloy is manganese bronze (39% Zn, 1% Mn, some iron and aluminum, and the rest copper), which is very resistant to corrosion and is used for the propellers of ships. Manganese is alloyed with aluminum to increase the stiffness of beverage cans.

A rich supply of manganese lies in the manganese nodules that litter the ocean floors (Fig. 21.12). These nodules range in diameter from millimeters to meters and are lumps of the oxides of iron, manganese, and other elements. However, because this source is technically difficult to exploit, manganese is currently obtained by the thermite process from *pyrolusite,* a mineral form of MnO_2:

$$3\,MnO_2(s)\ +\ 4\,Al(s)\ \xrightarrow{\Delta}\ 3\,Mn(l)\ +\ 2\,Al_2O_3(s)$$

Manganese lies near the center of its row (in Group 7) and occurs in a wide variety of oxidation states. The most stable state is +2, but +4, +7, and to some extent +3, are also common in manganese compounds. Its most important compound is manganese(IV) oxide, MnO_2, which is commonly called manganese dioxide. This compound is a brown-black solid used in dry cells, as a decolorizer to conceal the green tint of glass, and as the starting point for the production of other manganese compounds. Potassium permanganate, in which the manganese is in its highest oxidation state (+7), is a widely used strong oxidizing agent in acidic solution. Its usefulness stems not only from its thermodynamic tendency to oxidize other species but also from its ability to act by a variety of mechanisms; hence, it is likely to be able to find a path with low activation energy. Potassium permanganate is used for oxidations in organic chemistry and also as a mild disinfectant.

Iron, Fe, the most widely used of all the *d*-metals, is the most abundant element on Earth as a whole and the second most abundant metal in Earth's crust (aluminum is the most abundant metal in the crust). Its principal ores are the oxides *hematite,* Fe_2O_3, and *magnetite,* Fe_3O_4. The sulfide mineral *pyrite,* FeS_2 (see Fig. 20.10), is also widely available, but it is not used because the sulfur is difficult to remove.

FIGURE 21.12

Manganese nodules litter the ocean floor and are potentially a valuable source of the element.

FIGURE 21.13

The reduction of iron ore takes place in a blast furnace containing a mixture of the ore with coke and limestone. Different reactions occur in different zones when the blast of air and oxygen is admitted. The ore, an oxide, is reduced to the metal by reduction with carbon monoxide produced in the furnace. The different zones are regions of different temperatures.

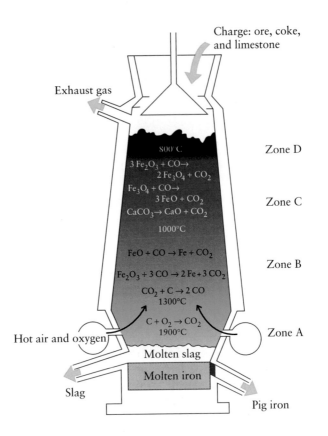

Charge: ore, coke, and limestone

Exhaust gas

Zone D

$800°C$

$3 Fe_2O_3 + CO \rightarrow 2 Fe_3O_4 + CO_2$

$Fe_3O_4 + CO \rightarrow 3 FeO + CO_2$

$CaCO_3 \rightarrow CaO + CO_2$

Zone C

$1000°C$

$FeO + CO \rightarrow Fe + CO_2$

$Fe_2O_3 + 3 CO \rightarrow 2 Fe + 3 CO_2$

$CO_2 + C \rightarrow 2 CO$

$1300°C$

Zone B

$C + O_2 \rightarrow CO_2$

$1900°C$

Zone A

Hot air and oxygen

Molten slag

Molten iron

Slag

Pig iron

Iron ores are primarily used to make steel. The ore is reduced to iron metal in a blast furnace (Fig. 21.13), which is a marvel of efficient use of materials. Even the waste products of the fire that heats it are used as reactants. The furnace, which is approximately 40 m in height, is filled from the top with a mixture of ore, coke, and limestone. Each kilogram of iron produced requires about 1.75 kg of ore, 0.75 kg of coke, and 0.25 kg of limestone. The limestone, which is primarily calcium carbonate, undergoes thermal decomposition to calcium oxide (lime) and carbon dioxide. The calcium oxide, which contains the Lewis base O^{2-}, helps to remove the acidic (nonmetal oxide) anhydride and amphoteric impurities from the ore:

$$CaO(s) + SiO_2(s) \xrightarrow{\Delta} CaSiO_3(l)$$

$$CaO(s) + Al_2O_3(s) \xrightarrow{\Delta} Ca(AlO_2)_2(l)$$

$$6 CaO(s) + P_4O_{10}(s) \xrightarrow{\Delta} 2 Ca_3(PO_4)_2(l)$$

The mixture of products, which is known as *slag*, is molten at the temperatures in the blast furnace and floats on the denser molten iron. It is drawn off and used to make rocklike material for the construction industry.

Molten iron is produced through a series of reactions in the different temperature zones of the blast furnace. The molten iron (of density 7.9 g/cm^3) is run off as *pig iron*, consisting of 90–95% iron, 3–5% carbon, 2% silicon, and trace amounts of other elements found in the original ore. *Cast iron* is similar to pig iron, but some impurities have been removed. Its carbon content is still usually greater than 2%.

Pure iron is relatively flexible and malleable, but the carbon atoms make cast iron very hard and brittle. It is used where there is little mechanical and

Table 21.3 *The effect of carbon on iron*

Type of steel	Carbon content, %	Properties and applications
low-carbon steel	<0.15	ductility and low hardness, iron wire
mild-carbon steel	0.15 to 0.25	cables, nails, chains, and horseshoes
medium-carbon steel	0.20 to 0.60	nails, girders, rails, and structural purposes
high-carbon steel	0.61 to 1.5	knives, razors, cutting tools, drill bits

FIGURE 21.14

In the basic oxygen process, a blast of oxygen and powdered limestone is used to purify molten iron by, respectively, oxidizing and combining with the impurities present in it.

thermal shock, such as for ornamental railings, engine blocks, brake drums, and transmission housings.

Steels have various hardnesses, tensile strengths, and ductilities; and the higher the carbon content, the harder and more brittle the steel (Table 21.3). The corrosion resistance of iron is greatly improved by alloying to form a variety of steels. Stainless steels are highly corrosion resistant; they typically contain about 15% chromium by mass.

Steelmaking begins with pig iron. The first stage is to lower the carbon content of the iron and to remove the remaining impurities, which include silicon, phosphorus, and sulfur. In the **basic oxygen process,** oxygen and powdered limestone are forced through the molten metal (Fig. 21.14). In the second stage, steels are produced by adding the appropriate metals, often as ferroalloys, to the molten iron.

A healthy adult human body contains about 3 g of iron, mostly as hemoglobin. Close to 1 mg is lost daily (in sweat, feces, and hair), so iron must be ingested daily in order to maintain the balance. Women lose about 20 mg during menstruation. Iron deficiency, or anemia, results in reduced transport of oxygen to the brain and muscles, and an early symptom is chronic tiredness.

Cobalt, Co, ores are often found in association with copper(II) sulfide. Cobalt is a silver-gray metal and is used mainly for alloying with iron. *Alnico steel,* an alloy of iron, nickel, cobalt, and aluminum, is used to make permanent magnets like those used in loudspeakers. Cobalt steels are hard enough to be used as surgical steels, drill bits, and lathe tools. Cobalt(II) oxide is heated with silica and alumina to produce a blue pigment used to color glass and ceramic glazes. We need cobalt in our diet, for it is a component of vitamin B_{12}.

Nickel, Ni, is also used to alloy with iron. About 70% of the Western World's supply comes from iron and nickel sulfide ores that were brought close to the surface by the impact of a meteor at Sudbury, Ontario. Impure nickel is refined in the **Mond process,** in which it is first exposed to carbon monoxide to form nickel carbonyl, $Ni(CO)_4$:

$$Ni(s) + 4\,CO(g) \xrightarrow{50°C} Ni(CO)_4(g)$$

Nickel carbonyl is a volatile, poisonous liquid that boils at 43°C and so can be removed from impurities. Nickel metal is then obtained by heating pure nickel tetracarbonyl to about 200°C, which reverses the reaction for its formation.

Nickel is a hard, silver-white metal used mainly for the production of stainless steel and for alloying with copper to produce *cupronickels,* the alloys used

Miners called cobalt minerals that interfered with copper production *Kobold* (which means "evil spirit" in German).

The origin of the name *nickel* is "Old Nick," used for much the same reason as the German name for cobalt.

for nickel coins (which consist of about 25% Ni and 75% Cu). Cupronickels are slightly yellow but are whitened by the addition of small amounts of cobalt. Nickel is also used as a catalyst, especially for the addition of hydrogen to organic compounds, for example, in the manufacture of edible solid fats from vegetable oils.

The d-block elements titanium through nickel are obtained chemically from their ores and have many industrial uses. Scandium is highly reactive. Iron is reduced by the action of coke and limestone in blast furnaces. Steel is an alloy of iron with carbon and other metals, which affect its properties.

21.4 Groups 11 and 12

Group 11, close to the right-hand edge of the *d* block, contains the coinage metals—copper, silver, and gold—which have $(n - 1)d^{10}ns^1$ electron configurations (Table 21.4). Group 12 contains zinc, cadmium, and mercury, with configurations $(n - 1)d^{10}ns^2$. The low reactivity of the coinage metals is partly due to the poor shielding of the *d*-electrons and hence the tight grip that the nucleus can exert on the outermost electrons. The inertness of gold is also enhanced by the lanthanide contraction.

Copper, Cu, is unreactive enough for some to be found native, but most is produced from its sulfides, particularly the ore *chalcopyrite,* $CuFeS_2$ (Fig. 21.15). The crushed and ground ore is separated from excess rock by the process called **froth flotation,** which makes use of the fact that sulfide ores are wetted by oils, but not by water. In this process (Fig. 21.16), the powdered ore is combined with oil, water, and detergents. Then air is blown through the mixture; the oil-coated sulfide mineral floats to the surface with the froth, and the unwanted copper-poor residue, which is called *gangue,* sinks to the bottom.

Processes for extracting metals from their ores are generally classified as **pyrometallurgical,** if high temperatures are used, or **hydrometallurgical,** if aqueous solutions are used. Copper is extracted by both methods. In the pyrometallurgical process for the extraction of copper, the enriched ore is **roasted,** or heated in air:

$$2\,CuFeS_2(s) + 3\,O_2(g) \xrightarrow{\Delta} 2\,CuS(s) + 2\,FeO(s) + 2\,SO_2(g)$$

The CuS is then **smelted,** a process in which metal ions are reduced by melting with another compound (Fig. 21.17). At the same time, the sulfur is oxidized to

Table 21.4 *Properties of elements in Groups 11 and 12*

Z	Name	Symbol	Electron configuration	Melting point, °C	Boiling point, °C	Density, g/cm^3
29	copper	Cu	$3d^{10}4s^1$	1083	2567	8.93
47	silver	Ag	$4d^{10}5s^1$	962	2212	10.50
79	gold	Au	$5d^{10}6s^1$	1064	2807	19.28
30	zinc	Zn	$3d^{10}4s^2$	420.	907	7.14
48	cadmium	Cd	$4d^{10}5s^2$	321	765	8.65
80	mercury	Hg	$5d^{10}6s^2$	−39	357	13.55

FIGURE 21.15

Three important copper ores (from left to right): chalcopyrite, $CuFeS_2$; malachite, $CuCO_3 \cdot Cu(OH)_2$; and chalcocite, Cu_2S.

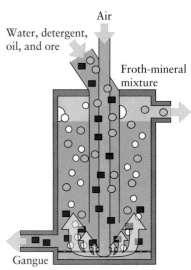

FIGURE 21.16

In the froth flotation process, a stream of bubbles is passed through a mixture of ore (orange circles), rock (brown rectangles), and detergent. The ore is buoyed up by the froth of bubbles and is removed from the top of the chamber. The unwanted gangue is washed away through the bottom of the container.

SO_2. This oxidation is accomplished by blowing compressed air through the mixture:

$$CuS(s) + O_2(g) \xrightarrow{\Delta} Cu(l) + SO_2(g)$$

These processes could contribute a damaging amount of SO_2 to the atmosphere if precautions were not taken to remove it. Limestone and sand, which are added to the mixture, form a molten slag that removes many of the impurities

FIGURE 21.17

In this industrial copper refinery, the molten impure copper produced by smelting is poured into molds. Next, the copper will be purified by electrolysis.

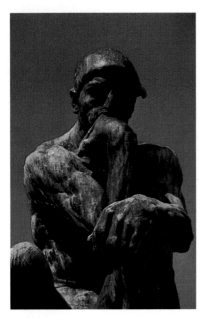

FIGURE 21.18

Copper corrodes in air to form a pale green layer of basic copper carbonate. This patina, or incrustation, passivates the surface.

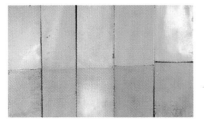

FIGURE 21.19

The color of commercial gold depends on its composition (left to right): 8-carat gold, 14-carat gold, white gold, 18-carat gold, and 24-carat gold. White gold consists of 6 parts Au and 18 parts Ag by mass.

as well as the SO_2. For example, calcium oxide (a basic oxide) from the limestone reacts with the SO_2 (an acidic oxide) to produce calcium sulfite:

$$CaO(s) + SO_2(g) \longrightarrow CaSO_3(s)$$

The copper product is known as *blister copper* because of the appearance of air bubbles in the solidified metal.

The impure copper from either process is refined electrolytically: it is made into anodes and plated onto cathodes of pure copper. The rare metals—most notably, platinum, silver, and gold—obtained from the anode sludge are sold to recover much of the cost of the electricity used in the electrolysis.

Copper is an excellent electrical conductor, but it needs to be very pure for use in the electrical industry. It is also used in alloys: with zinc to form brass, with tin to form bronze, and with nickel to form cupronickel (Table 10.4). Because copper lies above hydrogen in the electrochemical series, it cannot displace hydrogen from acidic solutions; however, it can be oxidized by oxidizing acids, such as nitric acid and sulfuric acid.

Copper corrodes in moist air in the presence of carbon dioxide:

$$2\,Cu(s) + H_2O(l) + O_2(g) + CO_2(g) \longrightarrow Cu_2(OH)_2CO_3(s)$$

The pale green product is called *basic copper carbonate* and is responsible for the green patina of copper and bronze objects (Fig. 21.18). The patina adheres to the surface, protects the metal, and gives a pleasing appearance.

Like all the coinage metals, copper forms compounds with oxidation number +1. However, in water, copper(I) salts disproportionate into metallic copper and copper(II) ions. The latter exist as pale blue $[Cu(H_2O)_6]^{2+}$ ions in water. This pale blue is the color of copper(II) sulfate solutions; the deeper blue of the solid pentahydrate, $CuSO_4 \cdot 5H_2O$, is due to $[Cu(H_2O)_4]^{2+}$ ions (the fifth H_2O links this ion to the sulfate anion). A mixture of copper(II) carbonate with stoichiometric amounts of Y_2O_3 and $BaCO_3$ heated to approximately 1000°C forms a *123 superconductor* of composition $YBa_2Cu_3O_{6.5-7.0}$ (the numbers 1, 2, and 3 denote the proportions in which the metal atoms Y, Ba, and Cu, respectively, are present in the compound).

Silver, Ag, is rarely found native (as the metal); most is obtained as a by-product of the refining of copper and lead, and a considerable amount is recycled through the photographic industry. Silver has a positive standard potential, so it does not reduce $H^+(aq)$ to hydrogen.

Silver(I) does not disproportionate in aqueous solution, and almost all silver compounds have oxidation number +1. Apart from silver nitrate and silver fluoride, silver salts are generally only sparingly soluble in water. Silver nitrate, $AgNO_3$, is the most important compound of silver and the starting point for the manufacture of the silver halides used in photography (Applying Chemistry: Case Study 21).

Gold, Au, is so inert that most of it is found native. Pure gold is classified as 24-carat gold; and its alloys with silver and copper, which differ in hardness and hue, are classified according to the proportion of gold they contain (Fig. 21.19). For example, 10- and 14-carat golds contain, respectively, 10/24 and 14/24 parts by mass of gold. Gold is a highly malleable metal, and 1 g of gold can be worked into a leaf covering about 1 m^2 or a wire over 2 km in length. Gold leaf is used for decoration, like that on dishes and books and in cathedrals and temples (Fig. 21.20).

Gold lies well above hydrogen in the electrochemical series and is too noble to react even with strong oxidizing agents such as nitric acid. Both the following gold couples lie above H^+/H_2 and $NO_3^-,H^+/NO,H_2O$:

$$Au^+(aq) + e^- \longrightarrow Au(s) \qquad E° = +1.69\,V$$

$$Au^{3+}(aq) + 3\,e^- \longrightarrow Au(s) \qquad E° = +1.40\,V$$

$$NO_3^-(aq) + 4\,H^+(aq) + 3\,e^- \longrightarrow NO(g) + 2\,H_2O(l) \qquad E° = +0.96\,V$$

However, gold does react with aqua regia (a mixture of concentrated nitric and hydrochloric acids) because the complex ion $AuCl_4^-$ forms:

$$Au(s) + 6\,H^+(aq) + 3\,NO_3^-(aq) + 4\,Cl^-(aq) \longrightarrow$$
$$AuCl_4^-(aq) + 3\,NO_2(g) + 3\,H_2O(l)$$

Even though the equilibrium constant for the formation of Au^{3+} from gold is very unfavorable, the reaction proceeds because any Au^{3+} ions formed are immediately removed by being hidden away inside a complex.

Zinc, Zn, is found mainly as its sulfide ZnS in sphalerite, often in association with lead ores (see Fig. 20.10). The ore is concentrated by froth flotation, and the metal is extracted by roasting and then smelting with coke:

$$2\,ZnS(s) + 3\,O_2(g) \xrightarrow{\Delta} 2\,ZnO(s) + 2\,SO_2(g)$$

$$ZnO(s) + C(s) \xrightarrow{\Delta} Zn(l) + CO(g)$$

Cadmium, Cd, is obtained in a similar manner. Zinc and cadmium are silvery, reactive metals. Zinc is used mainly for galvanizing iron; like copper, it is protected by a hard film of basic carbonate, $Zn_2(OH)_2CO_3$, which forms on contact with air.

Zinc and cadmium are similar to each other but differ sharply from mercury. Zinc is amphoteric (like its main-group neighbor aluminum). It reacts with acids to form Zn^{2+} ions and with alkalis to form the zincate ion, $[Zn(OH)_4]^{2-}$:

$$Zn(s) + 2\,OH^-(aq) + 2\,H_2O(l) \longrightarrow [Zn(OH)_4]^{2-}(aq) + H_2(g)$$

Galvanized containers should therefore not be used for transporting alkalis. Cadmium, which is lower down the group and is more metallic, has a more basic oxide, but its use is limited by the fact that cadmium salts are toxic. Zinc and cadmium have an oxidation number of $+2$ in all their compounds.

Mercury, Hg, occurs mainly as HgS in the mineral *cinnabar* (see Fig. 20.10), from which it is separated by froth flotation and then roasting:

$$HgS(s) + O_2(g) \xrightarrow{\Delta} Hg(g) + SO_2(g)$$

The volatile metal is distilled off and condensed. Mercury is unique in being the only metallic element that is liquid at room temperature. It has a long liquid range, from its melting point of $-39°C$ to its boiling point of $357°C$, so it is well suited for its use in thermometers, silent electrical switches, and high-vacuum pumps.

Mercury lies above hydrogen in the electrochemical series, so it is not oxidized by hydrogen ions. However, it does react with nitric acid. In compounds, mercury has the oxidation number $+1$ or $+2$. Its compounds with oxidation number $+1$ are unusual in that the mercury(I) cation is the covalently bonded diatomic ion $(Hg—Hg)^{2+}$, written Hg_2^{2+}. Both mercury(I) and silver ions tend

FIGURE 21.20

The thin gold leaf coating these cathedral domes in Russia not only adds long-lasting beauty, it protects the structure beneath from corrosion.

Applying Chemistry: *Case Study 21*

One day, the windows in our homes may be made of *photochromic* glass, which darkens when exposed to light, then becomes colorless and transparent again when the light dims. Photochromic windows would control the interior climate automatically, keeping homes cool in bright sunlight, without reducing transparency in the evening.

Photochemical materials are materials that undergo reactions when exposed to light. Photochromic glass is only one of the newest photochemical materials. Practical applications of photochemistry have their roots in photographic film. In the nineteenth century, the French artist Louis Daguerre plated silver onto sheets of copper, then treated them with iodine vapor. The iodine oxidized the silver:

$$2\,Ag(s)\;+\;I_2(g)\;\longrightarrow\;2\,AgI(s)$$

Daguerre kept the coated "plates" in the dark, then used a lens to focus the image of a person or scene on it. Silver ions are rapidly reduced to silver metal in the presence of light, so everywhere light fell on the coating, a black deposit of finely divided silver formed. The image was called a "daguerreotype."

Modern photography works on essentially the same principle as that used by Daguerre. Black-and-white photographic film consists of a plastic sheet coated with an emulsion such as gelatin. These emulsions contain microscopic crystals of silver bromide, about 500 nm in diameter, called "grains," and small amounts of silver iodide.

The reaction that records the image is an example of a *photochemical reaction*, a reaction caused by light. When the emulsion is exposed to light, an electron is driven out of a bromide ion wherever the light falls. The liberated electron wanders through the grain and reduces a nearby silver cation. The resulting redox reaction,

$$Ag^+(s)\;+\;Br^-(s)\;\xrightarrow{\;light\;}\;Ag(s)\;+\;Br(s)$$

occurs only where light falls, creating small clusters of silver atoms within the grains.

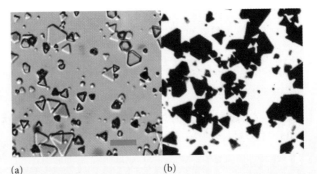

(a)   (b)

Micrographs of the silver halide crystals in a photographic emulsion before (a) and after (b) development. The scale bar represents 10 μm.

The film is developed by reducing the silver ions remaining in the exposed grains, but not those in the unexposed grains. This reaction is carried out with a mild reducing agent, typically the organic compound hydroquinone, HOC_6H_4OH, which has the structure

The developing process is completed by "fixing" the film in a solution that stops the reduction, so the image does not fade. Then the unreacted silver halide is dissolved in aqueous sodium thiosulfate, $Na_2S_2O_3 \cdot 5H_2O$, or photographer's hypo (a shortening of its old name, sodium hyposulfite). The thiosulfate anions coordinate to the silver ions, forming the soluble complex ion $[Ag(S_2O_3)_2]^{3-}$, which is then rinsed away, leaving the metallic silver behind and removing the light sensitivity of the film. The result is a negative, a plastic film that is transparent where it was not exposed to light and dark where it was.

The negative is then placed between a light source and a lens to enlarge the image. Light-sensitive paper coated

to form insoluble salts. The precipitate that forms when a chloride solution is added to a solution containing a soluble mercury(I) or silver salt is a common test for the presence of the ions (see Section 16.17).

Mercury compounds, particularly its organic compounds, are acutely poisonous. The discharge of industrial waste containing mercury compounds into a shallow sea resulted in the death of 52 people in Minimata, Japan, in 1952. In the sea, the mercury was converted into CH_3HgSCH_3, which was ingested by the fish that were a staple in the diet of the local residents. Mercury vapor is also an with AgCl is placed under the lens and exposed to the light traveling through the negative. The paper is then treated to stop further reaction, and the photograph is complete.

Color photographs are printed from negatives that are made of layers of silver halide emulsions. Each layer contains a dye that makes it sensitive to one of the three primary colors: red, green, or blue. The layers are developed separately into colored transparent negatives of the desired images. The colors in the negatives are the complementary colors of the colors in the original image.

Photochromic glass is made by mixing silver nitrate, copper(I) nitrate, and a metal halide into a borosilicate glass mixture and heating it until it melts, at about 1200°C. As the glass cools, small crystallites of the salts form. The crystallites (about 10 nm in diameter) are too small to scatter or absorb visible light, so the glass appears transparent. As in the photographic process, sunlight reduces the Ag^+ ions to silver metal:

$$Ag^+(s) + Cu^+(s) \xrightarrow{\text{light}} Ag(s) + Cu^{2+}(s)$$

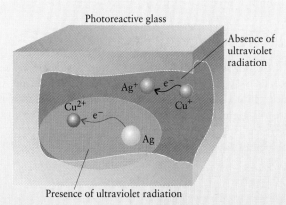

Photoreactive glass

Absence of ultraviolet radiation

Presence of ultraviolet radiation

Light causes the reduction of Ag^+ to Ag metal, darkening photochromic glass. In the absence of radiation, Ag is oxidized by Cu^+ and the glass becomes colorless again.

However, in this case, the reaction can be reversed, because the oxidized reducing agent, Cu^{2+}, cannot escape; and as soon as the light is removed, it oxidizes the silver again. Prescription sunglasses made of photochromic glass allow us to read in dim light, while also protecting our eyes from bright sunlight outdoors. Because photochromic glass darkens when the sun is shining but reverts to a clear and transparent state at dusk, it could reduce the amount of energy required for cooling homes and allow light intensity to be controlled in greenhouses.

Photochromic sunglasses. Here only one of the lenses has been exposed to ultraviolet light.

Key Concepts: redox reactions of transition metals, complex formation

For Further Reading
H. B. Gray, J. D. Simon, and W. C. Trogler, Photochemistry, in *Braving the Elements*, Sausalito, CA: University Science Books, 1995, pp. 287–314.

Related Exercises: 21.93–21.96

insidious poison because its effect is cumulative. Frequent exposure to low levels of mercury vapor can allow mercury to accumulate in the body. The effects include memory loss and other ailments.

> *Metals in Groups 11 and 12 are easily reduced from their compounds and have low reactivity as a result of poor shielding by the d-electrons. Copper is extracted from its ores by either pyrometallurgical or hydrometallurgical processes.*

Self-Test 21.2A Use standard free energies of formation to calculate ΔG_r° at 298 K for the reaction by which 1.000 mol of Cu(s) is obtained by smelting CuS ore with oxygen. (ΔG_f° for CuS is -49.0 kJ/mol at 298 K.)

[***Answer:*** -251.2 kJ/mol]

Self-Test 21.2B Pig iron is much cheaper than steel but is not used in construction. Explain why pig iron is not suitable for erecting buildings and bridges.

COMPLEXES OF THE d-BLOCK ELEMENTS

If you have ever admired an oil painting, used glazed pottery, or taken your vitamins, you have encountered some of the many complexes that *d*-block metals form. The *d*-block elements are excellent Lewis acids (electron pair acceptors, Section 8.13) and form coordinate covalent bonds with molecules or ions that can act as Lewis bases (electron pair donors). Complexes formed in this way participate in many biological reactions. Hemoglobin and vitamin B_{12}, for example, are both complexes—the former of iron and the latter of cobalt. The *d*-metal complexes are often brightly colored and magnetic, and are used in chemical analysis, color science, and catalysis. They are the target of current research in solar-energy conversion, in nitrogen fixation, and in pharmaceuticals. Complexes are used as dyes, to dissolve ions (Section 16.16), in the electroplating of metals, and in photography (see Applying Chemistry: Case Study 21).

21.5 The Structures of Complexes

A *d*-metal complex consists of a central metal atom or ion to which are attached a number of molecules or ions (typically, four or six) known as **ligands.** One example is the hexacyanoferrate(II) ion, $Fe(CN)_6^{4-}$, in which the Lewis acid Fe^{2+} forms bonds by accepting electron pairs provided by the CN^- ions. An example of a neutral complex is $Ni(CO)_4$, in which the Ni atom acts as the Lewis acid and the CO molecules act as Lewis bases.

> The hexacyanoferrate(II) ion was formerly called the ferrocyanide ion, and you will still see that name used.

FIGURE 21.21

When cyanide ions (in the form of potassium cyanide) are added to an aqueous solution of iron(II) sulfate, they replace the H_2O ligands from the $[Fe(H_2O)_6]^{2+}$ complex and produce a new complex, the more strongly colored hexacyanoferrate(II) ion, $[Fe(CN)_6]^{4-}$.

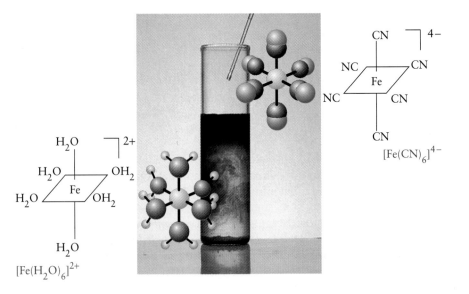

$[Fe(H_2O)_6]^{2+}$

$[Fe(CN)_6]^{4-}$

FIGURE 21.22

Some of the highly colored compounds that result when complexes are formed (left to right): aqueous solutions of the complexes $[Fe(SCN)(H_2O)_5]^{2+}$, $[Co(SCN)_4(H_2O)_2]^{2-}$, $[Cu(NH_3)_4(H_2O)_2]^{2+}$, and $[CuBr_4]^{2-}$.

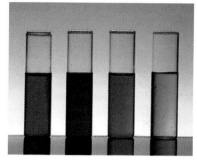

Each ligand in a complex is a Lewis base with at least one lone pair of electrons that forms a coordinate covalent bond to the central atom or ion. Because both electrons in the bond come from the ligand, we say that the ligand **coordinates** to the metal. **Coordination compounds** are compounds with coordinate covalent bonds. They include electrically neutral complexes, such as $Ni(CO)_4$, and ionic compounds in which at least one of the ions is a complex, as in $K_4[Fe(CN)_6]$. The ligands directly attached to the central ion in a complex (conventionally enclosed within brackets) make up the **coordination sphere** of the central ion. The number of points where ligands are attached to the central metal atom is called the **coordination number** of the complex: the coordination number is 4 in $[Ni(CO)_4]$ and 6 in $[Fe(CN)_6]^{4-}$.

Because water is a Lewis base, it forms complexes with most d-block metal ions when they dissolve in it. Aqueous solutions of d-metal ions are usually solutions of their H_2O complexes, such as $[Fe(H_2O)_6]^{2+}$: these complexes are the "hydrated ions," such as $Fe^{2+}(aq)$, that we have met previously. Many other complexes are prepared simply by mixing aqueous solutions of a d-metal ion with the appropriate Lewis base (Fig. 21.21):

$$[Fe(H_2O)_6]^{2+} + 6\,CN^-(aq) \longrightarrow [Fe(CN)_6]^{4-}(aq) + 6\,H_2O$$

This reaction is an example of a **substitution reaction,** in which one Lewis base takes the place of another: in this example, the CN^- ions drive out H_2O molecules from the $[Fe(H_2O)_6]^{2+}$ complex. A less complete replacement occurs when certain other ions, such as Cl^-, are added to an iron(II) solution:

$$[Fe(H_2O)_6]^{2+}(aq) + Cl^-(aq) \longrightarrow [FeCl(H_2O)_5]^+(aq) + H_2O(l)$$

The color of a d-metal complex depends on the identity of the ligands as well as that of the metal, and impressive changes of color accompany many substitution reactions (Fig. 21.22).

In the great majority of cases, complexes with a coordination number of 6, such as $[Fe(CN)_6]^{4-}$, have their ligands at the corners of an octahedron, with the metal ion at the center, and are called **octahedral complexes** (Fig. 21.23). We can represent the structures of these octahedral complexes by a simplified diagram that emphasizes the arrangement of the ligands (**1**). In complexes with a coordination number of 4, the ligands are found either at the corners of a tetrahedron in a **tetrahedral complex,** as in $[Cu(NH_3)_4]^{2+}$, or (most notably for d^8 electron configurations, such as Pt^{2+} and Au^{3+}) at the corners of a square, in **square planar complexes.**

Some ligands can occupy more than one binding site. Ethylenediamine, $NH_2CH_2CH_2NH_2$ (**2**), has a nitrogen lone pair at each end. This ligand is widely used in coordination chemistry and is abbreviated to en, as in $[Co(en)_3]^{3+}$ (**3**). The metal atom in $[Co(en)_3]^{3+}$ lies at the center of the three ligands as though pinched by three molecular claws. It is an example of a **chelate,** a complex containing one or more ligands that form a ring of atoms that includes the central metal atom. Another example is the ethylenediaminetetraacetate ion, edta^{4-} (**4**).

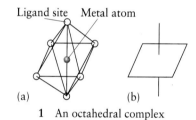

Ligand site   Metal atom

(a)   (b)

1   An octahedral complex

2   Ethylenediamine, $NH_2CH_2CH_2NH_2$

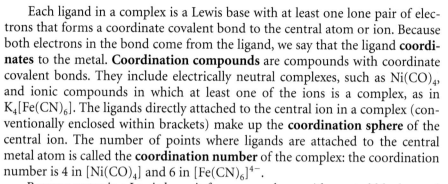

3 $[Co(en)_3]^{3+}$

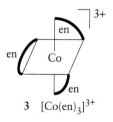

The name *chelate* comes from the Greek word for "claw."

FIGURE 21.23

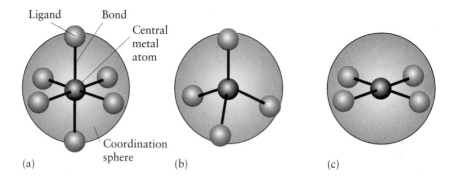

(a)   (b)   (c)

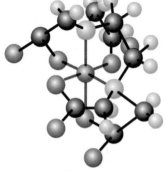

4 edta complex

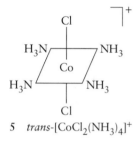

5 *trans*-$[CoCl_2(NH_3)_4]^+$

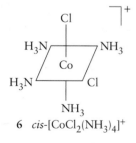

6 *cis*-$[CoCl_2(NH_3)_4]^+$

This ligand forms complexes with many metal ions, including Pb^{2+}, and hence is used as an antidote to lead poisoning.

Chelating ligands are quite common in nature. Mosses and lichens secrete chelating ligands to harvest essential metal ions from the rocks they dwell on. Chelate formation also lies behind the body's strategy of producing a fever when infected by bacteria. The higher temperature kills bacteria by reducing their ability to synthesize a particular iron-chelating ligand.

Table 21.5 gives the names of some common ligands and their abbreviations. The rules for building up the names of complexes are listed in Appendix 3D.

A complex is formed between a Lewis acid (the metal atom or ion) and a number of Lewis bases (the ligands).

21.6 Isomers

Many complexes and coordination compounds exist as isomers. Recall from Section 11.11 that isomers are compounds that contain the same numbers of the same atoms, but in different arrangements. For example, the ions shown in (**5**) and (**6**) differ only in the positions of the Cl^- ligands, but they are distinct species, for they have different physical and chemical properties. Both structural isomers and stereoisomers are found in coordination compounds. The structural isomers can be classified according to four different types: ionization, hydrate, linkage, and coordination. The varieties of isomerism are summarized by the chart in Toolbox 21.1 at the end of this section.

Ionization isomers have formulas like [MX]Y and [MY]X, where X is in the coordination sphere in one compound but outside the coordination sphere in the other compound. For instance, $[CoBr(NH_3)_5]SO_4$ and $[CoSO_4(NH_3)_5]Br$ are ionization isomers because the Br^- ion is a ligand of the cobalt in the former but an accompanying anion in the latter. The two compounds have different physical properties: the bromo complex (the compound with Br within the brackets) is violet and the sulfato complex is red (Fig. 21.24). The chemical properties of the two ionization isomers are also different: the red isomer forms an off-white precipitate of AgBr when Ag^+ ions are added, but no precipitate forms after the addition of Ba^{2+}:

$$[CoSO_4(NH_3)_5]Br(aq, red) + Ag^+(aq) \longrightarrow AgBr(s) + [CoSO_4(NH_3)_5]^+(aq)$$

$$[CoSO_4(NH_3)_5]Br(aq, red) + Ba^{2+}(aq) \longrightarrow \text{no reaction}$$

Table 21.5 Common ligands

Formula	Name
Neutral ligands	
H_2O	aqua
NH_3	ammine
NO	nitrosyl
CO	carbonyl
$NH_2CH_2CH_2NH_2$	ethylenediamine (en)*
$NH_2CH_2CH_2NHCH_2CH_2NH_2$	diethylenetriamine (dien)†
Anionic ligands	
F^-	fluoro
Cl^-	chloro
Br^-	bromo
I^-	iodo
OH^-	hydroxo
O^{2-}	oxo
CN^-	cyano (as M—CN)
NC^-	isocyano (as M—NC)
SCN^-	thiocyanato (as M—SCN)
NCS^-	isothiocyanato (as M—NCS)
NO_2^-	nitrito (as M—ONO)
NO_2^-	nitro (as M—NO_2)
CO_3^{2-}	carbonato
$C_2O_4^{2-}$	oxalato (ox)*
(structure)	ethylenediaminetetraacetato (edta)‡
SO_4^{2-}	sulfato

*Bidentate (attaches to two sites).

†Tridentate (attaches to three sites).

‡Hexadentate (attaches to six sites).

FIGURE 21.24

Solutions of the two coordination compounds [CoBr(NH$_3$)$_5$]SO$_4$ (left) and [CoSO$_4$(NH$_3$)$_5$]Br (right). Although the isomers are built from the same atoms in the same proportions, they are different compounds, with their own characteristic physical and chemical properties.

On the other hand, the violet isomer is unaffected by the addition of low concentrations of Ag^+ ions, but it forms a white precipitate of $BaSO_4$ after the addition of Ba^{2+} ions:

$$[CoBr(NH_3)_5]SO_4(aq, violet) + Ag^+(aq) \longrightarrow \text{no reaction}$$

$$[CoBr(NH_3)_5]SO_4(aq, violet) + Ba^{2+}(aq) \longrightarrow$$
$$BaSO_4(s) + [CoBr(NH_3)_5]^{2+}(aq)$$

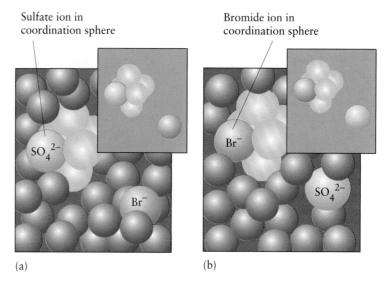

Sulfate ion in coordination sphere

Bromide ion in coordination sphere

SO_4^{2-}

Br^-

Br^-

SO_4^{2-}

(a)

(b)

FIGURE 21.25

(a) When this red coordination complex dissolves, the bromide ion (the orange sphere) is not part of the coordination sphere; it is free to move and is available for reaction. (b) In contrast, when this violet coordination complex dissolves, the bromide ion remains attached to the cobalt ion in the complex and is not available for reaction. In this case, the sulfate ion (purple sphere) is free to react.

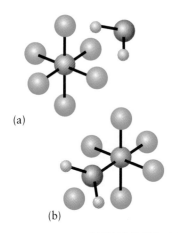

(a)

(b)

FIGURE 21.26

Hydrate isomers. In (a), the water molecule is simply part of the surrounding solvent; in (b), the water molecule is present in the coordination sphere and an ion (or neutral species) is now present in the solution.

These observations show that ions bound in a coordination sphere are not free to react. When the red complex dissolves in water, bromide ions become free to move through the solution (Fig. 21.25a); and when they encounter a silver ion, they form a precipitate. Conversely, when the violet complex dissolves in water, the Br^- ion remains attached to the cobalt ion and is not free to react with Ag^+ ions (Fig. 21.25b). However, the sulfate ion is not a part of the coordination sphere and is free to react.

Hydrate isomers have formulas like $[MX] \cdot H_2O$ and $[M(H_2O)]X$, where X is a ligand and H_2O is water of hydration in the former compound but X is outside the coordination sphere and H_2O is a ligand in the latter compound (Fig. 21.26). (We can think of hydrate isomerism as a special case of ionization isomerism.) For example, the solid hexahydrate of chromium(III) chloride, $CrCl_3 \cdot 6H_2O$, may be any of the three compounds

$[Cr(H_2O)_6]Cl_3$	violet
$[CrCl(H_2O)_5]Cl_2 \cdot H_2O$	blue-green
$[CrCl2(H_2O)_4]Cl \cdot 2 H_2O$	green

The addition of $AgNO_3$ to solutions of the compounds results in the precipitation of different amounts of AgCl from each of them, because only the chloride ions outside the coordination sphere will react with the silver ions.

Linkage isomers have formulas like $[M \cdots (X-Y)]$ and $[M \cdots (Y-X)]$; that is, they differ in the identity of the atom that a given ligand uses to attach to the metal ion (Fig. 21.27). Common ligands that show linkage isomerism are SCN^- versus NCS^-, NO_2^- versus ONO^-, and CN^- versus NC^-, where the coordinating atom is written first in each pair. For example, NO_2^- can form the yellow nitro complex $[CoCl(NO_2)(NH_3)_4]^+$ or the red nitrito complex $[CoCl(ONO)(NH_3)_4]^+$.

Coordination isomers have formulas like [MX][M'Y] and [MY][M'X], where M and M' are different metals. That is, they occur when one or more ligands are exchanged between the coordination spheres of two complex ions that form a coordination compound (Fig. 21.28). An example of a pair of coordination isomers is $[Cr(NH_3)_6][Fe(CN)_6]$ and $[Fe(NH_3)_6][Cr(CN)_6]$.

Self-Test 21.3A Identify the type of isomers represented by the following pairs: (a) $[Cu(NH_3)_4][PtCl_4]$ and $[Pt(NH_3)_4][CuCl_4]$; (b) $[Cr(OH)_2(NH_3)_4]Br$ and $[CrBr(OH)(NH_3)_4]OH$.

[***Answer:*** (a) Coordination; (b) ionization]

Self-Test 21.3B Identify the type of isomers represented by the following pairs: (a) $[Co(NCS)(NH_3)_5]Cl_2$ and $[Co(SCN)(NH_3)_5]Cl_2$; (b) $[CrCl(H_2O)_5]Cl_2 \cdot H_2O$ and $[CrCl_2(H_2O)_4]Cl \cdot 2H_2O$.

Although they are built from the same numbers and kinds of atoms, structural isomers have different chemical formulas, because the formulas show how the atoms are grouped within or outside the coordination sphere. In contrast, *stereoisomers* have the same formulas: they differ only in the spatial arrangement of the ligands within the coordination sphere. As for organic compounds (Section 11.12), we have to distinguish between geometrical isomers and optical isomers. **Geometrical isomers** differ in the spatial arrangement of ligands within the coordination sphere but are not related to each other as mirror images. **Optical isomers** are also complexes that differ in the spatial arrangement of ligands within the coordination sphere, but in such a way that they are nonsuperimposable mirror images of each other (like left and right hands).

The two complexes of formula $[CoCl_2(NH_3)_4]^{2+}$ shown earlier in (**5**) and (**6**) are geometrical isomers. As in organic compounds, the isomer with the Cl^- ligands on opposite sides of the central atom is called the **trans isomer;** the isomer with the ligands on the same side is called the **cis isomer.** Geometrical isomers exist for square planar and octahedral complexes, but they do not exist for tetrahedral complexes.

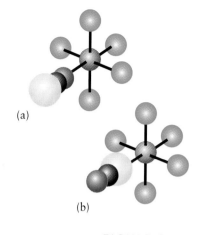

(a)

(b)

FIGURE 21.27

Linkage isomers. In (a), the ligand (here, NCS^-) is attached through its N atom, but in (b), it is attached through its S atom. The different arrangements of the ligand give the two complexes different chemical and physical properties.

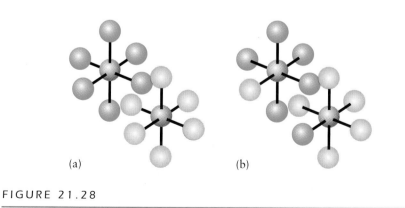

(a)   (b)

FIGURE 21.28

The compounds in (a) and (b) are coordination isomers. In these compounds, a ligand has been exchanged between the cationic and anionic complexes.

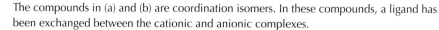

The chemical and physiological properties of geometrical isomers can differ greatly. For example, *cis*-[PtCl$_2$(NH$_3$)$_2$] (**7**) is pale orange-yellow, has a solubility of 0.252 g per 100 g of water, and is used for chemotherapy treatment of cancer patients. Conversely, *trans*-[PtCl$_2$(NH$_3$)$_2$] (**8**) is dark yellow, has a solubility of 0.037 g per 100 g of water, and exhibits no chemotherapeutic effect. Antitumor activity is also found in other cis isomers of platinum(II), such as *cis*-[PtCl$_2$(en)], *cis*-[Pt(ox)(NH$_3$)$_2$], and *cis*-[Pt(NO$_3$)$_2$(C$_6$H$_4$(NH$_2$)$_2$)]. An octahedral complex that exhibits geometrical isomerism is [Fe(CN)$_4$(NH$_3$)$_2$]$^-$ (**9**) and (**10**).

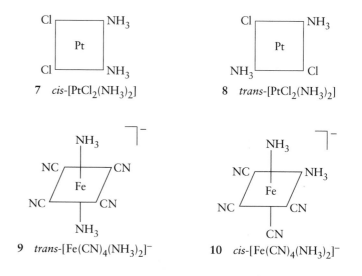

Figure 21.29 shows an example of optical isomerism. Both geometrical and optical isomerism can occur in an octahedral complex, as in [CoCl$_2$(en)$_2$]$^+$: the trans isomer is green (**11**), and the two alternative cis isomers (**12**) and (**13**), which are optical isomers of one another, are violet. Optical isomers have almost identical physical and chemical properties but differ in the direction in which they rotate polarized light.

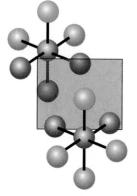

FIGURE 21.29

Optical isomers. The two complexes are each other's mirror image, and no matter how we rotate them, one complex cannot be superimposed on the other.

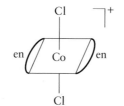

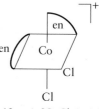

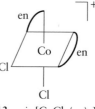

CHAPTER 21   THE *d* BLOCK: METALS IN TRANSITION

The possibility of isomerism increases the number of compounds that can be produced by the same set of atoms; the varieties of isomerism are summarized in Fig. 21.30. Optical isomers rotate the plane of polarization of light in opposite directions.

Toolbox 21.1 *How to classify isomers*

This Toolbox explains how to classify pairs of isomers by type.

Conceptual Basis

Isomers are built from the same kit of atoms but have different "connectivities"; that is, they are joined together in different patterns. In structural isomers, these patterns differ in the way the atoms are positioned in different parts of a compound; in stereoisomers, they differ in the way the ligands are positioned around the central atom of the complex.

Procedure

Identify the type of isomerism relating a pair of complexes, if any, by comparing the two complexes with the different options in the chart in Fig. 21.30.

Ionization isomers A ligand is inside the coordination sphere of one isomer and outside the coordination sphere of the other.

Hydrate isomers A water molecule is inside the coordination sphere of one isomer and outside the coordination sphere of the other.

Linkage isomers A ligand is attached to the central atom by one atom in the first isomer and by a different atom in the second.

Coordination isomers A ligand is in the coordination sphere of one metal ion in the first coordination compound and in the coordination sphere of the other metal ion in the second compound.

Geometrical isomers A ligand is in one position relative to the central atom in one complex, and in a different position in the other.

Optical isomers The ligands are arranged so that two isomers are mirror images but cannot be superimposed.

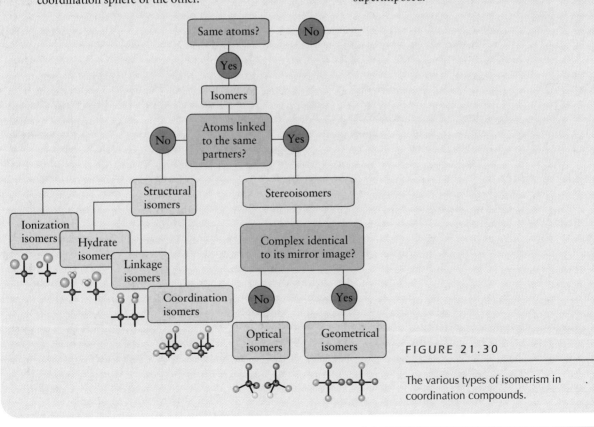

FIGURE 21.30

The various types of isomerism in coordination compounds.

FIGURE 21.31

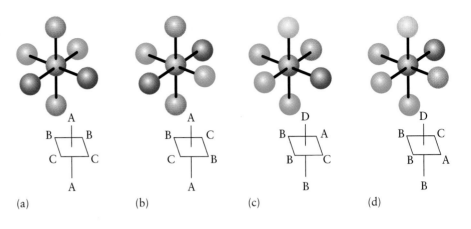

The complexes referred to in Example 21.1. Note that we also display the abbreviated form of each structure.

(a)　　(b)　　(c)　　(d)

Example 21.1 *Identifying optical isomerism*

Which of the complexes in Fig. 21.31 are optically active and which form pairs of isomers?

Strategy Draw the mirror image of each complex and mentally rotate the original complex; judge whether any rotation will cause the original molecule to match its mirror image. If not, then the complex is optically active. Determine which complexes form pairs of isomers by finding pairs that are the nonsuperimposable mirror images of each other.

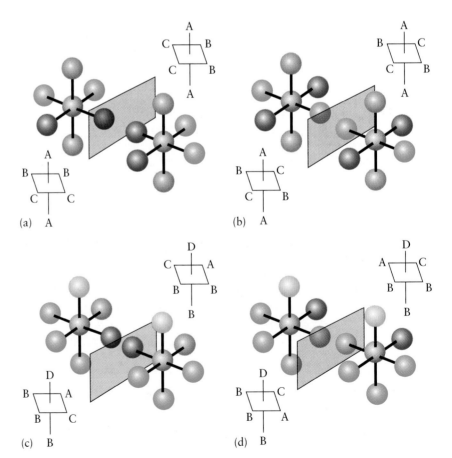

(a)　　(b)　　(c)　　(d)

FIGURE 21.32

The reflections referred to in the solution of Example 21.1.

FIGURE 21.33

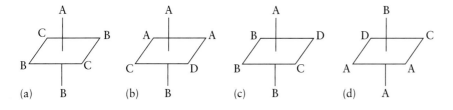

Solution In Fig. 21.32, the original complexes are on the left of each pair, and the mirror image of each complex is on the right. (a, b) Neither is optically active. (a) If we rotate the mirror image about A—A, we obtain a structure identical to the original. (b) The mirror image is identical to the original. (c, d) The original complex is optically active because no rotation can make either match its mirror image. However, when the mirror image of (c) is rotated by 90° around the vertical B–D axis, it becomes the complex (d); hence (c) and (d) form a pair of isomers.

Self-Test 21.4A Repeat the exercise for the complexes shown in Fig. 21.33.

[*Answer:* (a) Not optically active; (c) optically active; (b, d) optically active and a pair of isomers]

Self-Test 21.4B Repeat the exercise for the complexes shown in Fig. 21.34.

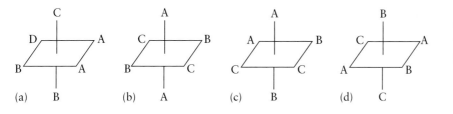

FIGURE 21.34

CRYSTAL FIELD THEORY

Many coordination compounds are colored and many are paramagnetic. These properties can be discussed in terms of **crystal field theory.** This description of bonding in complexes was originally devised to explain the colors of crystalline solids, particularly ruby, which owes its color to Cr^{3+} ions surrounded by an octahedron of O^{2-} ions acting like ligands (Fig. 21.35).

> A more complete version of crystal field theory, which makes use of molecular orbitals, is called *ligand field theory.*

21.7 The Effects of Ligands on *d*-Electrons

In crystal field theory, each ligand is represented by a point charge: these negative charges represent the ligand lone pair directed toward the central metal atom (Fig. 21.36). The electronic structure of the complex is then expressed in terms of the coulombic interactions between these point charges and the electrons of the central metal ion. We begin by considering a complex with a single *d*-electron, such as $[Ti(H_2O)_6]^{3+}$, in which the electron configuration of Ti^{3+} is $[Ar]3d^1$.

FIGURE 21.35

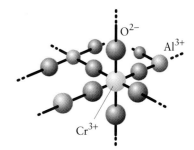

In a crystal of ruby, Cr^{3+} ions take the place of some Al^{3+} ions in the alumina crystal; each Cr^{3+} ion is surrounded by an octahedron of oxide ions.

FIGURE 21.36

In the crystal field theory of complexes, the lone pairs of electrons (the Lewis base sites) on the ligands (a) are treated as equivalent to point negative charges (b).

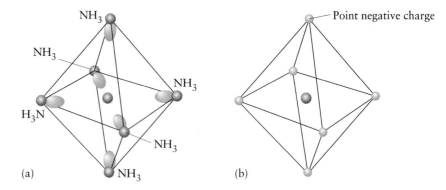

(a)

(b)

Point negative charge

Because the metal atom at the center of a complex is usually positively charged, the negative charges representing the ligands are attracted to it. This attraction results in the formation of the complex. The $[Ti(H_2O)_6]^{3+}$ ion, for instance, is a stable complex because there is a strong attraction between the Ti^{3+} ion and the six negative point charges, each of which represents a lone pair on one of the six H_2O ligands.

However, the single $3d$-electron of the Ti^{3+} ion interacts differently with the point charges, depending on which of the five $3d$-orbitals it occupies. In an octahedral complex such as $[Ti(H_2O)_6]^{3+}$, the six ligands (represented by point charges) lie around the central metal ion and along the x-, y-, and z-axes. From the drawings of the d-orbitals in Fig. 21.37, we can see that three of the orbitals (d_{xy}, d_{yz}, and d_{zx}) have their lobes directed between the point charges. These three d-orbitals are called **t-orbitals** in crystal field theory. The other two d-orbitals (d_{z^2} and $d_{x^2 - y^2}$) are directed straight toward the point charges. These two orbitals are called **e-orbitals.** Because of their different arrangement in space, electrons in t-orbitals are repelled less by the negative point charges of the ligands than electrons in e-orbitals are. As a result, the t-orbitals are of lower energy than the e-orbitals. The difference between the energy of the t-orbitals and that of the e-orbitals is typically about 10% of the total interaction energy between the central ion and its ligands. So, the stability of the complex is largely due to the attraction between the central positive ion and the negative charges

The t- and e-orbitals are normally given the more elaborate designations t_{2g} and e_g, respectively.

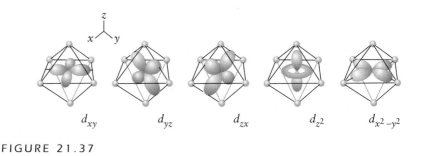

FIGURE 21.37

In an octahedral complex with a central d-metal atom or ion, a d_{xy}-orbital is directed between the ligand sites, and an electron that occupies it has a relatively low energy. The same lowering of energy occurs for d_{yz}- and d_{zx}-orbitals. A d_{z^2}-orbital points directly toward two ligands, and an electron that occupies it has a relatively high energy. The same rise in energy occurs for a $d_{x^2-y^2}$-electron.

FIGURE 21.38

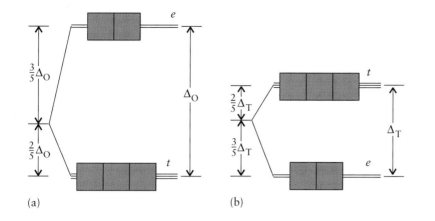

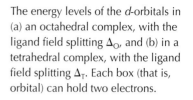

The energy levels of the d-orbitals in (a) an octahedral complex, with the ligand field splitting Δ_O, and (b) in a tetrahedral complex, with the ligand field splitting Δ_T. Each box (that is, orbital) can hold two electrons.

representing the ligands. However, the 10% difference in energy of the t- and e-orbitals is important because it accounts for differences in the colors and magnetic properties of complexes.

The energy-level diagram in Fig. 21.38a shows how the energies of the d-orbitals are affected by the presence of ligands.. The energy separation between the two sets of orbitals is called the **ligand field splitting,** Δ_O (the O denotes octahedral). The three t-orbitals lie at an energy that is $\frac{2}{5}\Delta_O$ below the average d-orbital energy, and the two e-orbitals lie at an energy $\frac{3}{5}\Delta_O$ above the average. Because the t-orbitals have the lower energy, we can predict that, in the ground state of the $[Ti(H_2O)_6]^{3+}$ complex, the electron occupies one of them in preference to an e-orbital and hence that the electron configuration of the complex is t^1. This configuration is represented by the box diagram in (**14**).

In a tetrahedral complex, the three t-orbitals point more directly at the ligands than the two e-orbitals do. As a result, in a tetrahedral complex, the t-orbitals have a higher energy than the e-orbitals (Fig. 21.38b). The ligand field splitting, Δ_T (where the T denotes tetrahedral), is generally smaller than in octahedral complexes (typically, $\Delta_T \approx \frac{4}{9}\Delta_O$) because the d-orbitals do not point so directly at the ligands and there are fewer repelling ligands.

In octahedral complexes, the e-orbitals (d_{z^2} and $d_{x^2-y^2}$) lie higher in energy than the t-orbitals (d_{xy}, d_{yz}, and d_{zx}) do.

21.8 The Effects of Ligands on Color

The t-electron of the octahedral $[Ti(H_2O)_6]^{3+}$ complex can be excited into one of the e-orbitals if it absorbs a photon of energy equal to Δ_O (Fig. 21.39). The wavelength of radiation absorbed by a complex can therefore be used to determine the ligand field splitting.

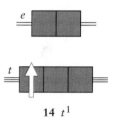

14 t^1

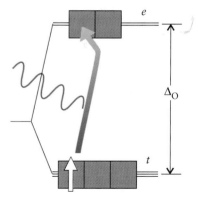

FIGURE 21.39

When a complex is exposed to light of the correct frequency, an electron can be excited to a higher energy orbital (from t to e if the complex is octahedral), and the light is absorbed.

Example 21.2 *Determining the ligand field splitting*

The complex $[Ti(H_2O)_6]^{3+}$ absorbs light of wavelength 510. nm. What is the ligand field splitting in the complex in kilojoules per mole?

Strategy Because a photon carries an energy $h\nu$, where h is the Planck constant and ν is the frequency of the radiation, it can be absorbed if its frequency satisfies

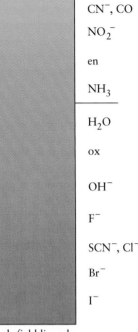

Strong-field ligands

CN⁻, CO

NO₂⁻

en

NH₃

H₂O

ox

OH⁻

F⁻

SCN⁻, Cl⁻

Br⁻

I⁻

Weak-field ligands

FIGURE 21.40

The spectrochemical series. Strong-field ligands give rise to a large splitting between the *t*- and *e*-orbitals, whereas weak-field ligands give rise to only a small splitting. The horizontal line marks the frontier between the two kinds of ligands. The change in color of the bar represents the increasing energy of light absorbed as the field strength increases.

Electromagnetic radiation was first discussed in Section 7.1.

$$\Delta_O = h\nu$$

where h is the Planck constant. The wavelength, λ, of light is related to the frequency by $\lambda = c/\nu$, where c is the speed of light (Section 7.1). Therefore, the wavelength of light absorbed and the ligand field splitting are related by

$$\Delta_O = \frac{hc}{\lambda} \qquad \text{alternatively,} \quad \lambda = \frac{hc}{\Delta_O}$$

That is, the greater the splitting, the shorter the wavelength of the light that is absorbed by the complex. At this point, substitute the data.

Solution Because the wavelength absorbed is 510. nm, which is 5.10×10^{-7} m, it follows that the ligand field splitting is

$$\Delta_O = \frac{(6.626 \times 10^{-34}\text{ J·s}) \times (2.998 \times 10^8\text{ m/s})}{5.10 \times 10^{-7}\text{ m}} = 3.90 \times 10^{-19}\text{ J}$$

Multiplication by the Avogadro constant turns the energy into a molar quantity:

$$\Delta_O = (6.022 \times 10^{23}/\text{mol}) \times (3.90 \times 10^{-19}\text{ J}) = 2.35 \times 10^5\text{ J/mol}$$

or 235 kJ/mol.

Self-Test 21.5A The complex $[Fe(H_2O)_6]^{3+}$ absorbs light of wavelength 700. nm. What is the value (in kilojoules per mole) of the ligand field splitting?

[***Answer:*** 171 kJ/mol]

Self-Test 21.5B The complex $[Fe(CN)_6]^{4-}$ absorbs light of wavelength 305 nm. What is the value (in kilojoules per mole) of the ligand field splitting?

Ligands can be arranged in a **spectrochemical series** according to the magnitude of the Δ_O they produce, as shown in Fig. 21.40. Those below the horizontal line are called **weak-field ligands,** and those above it are called **strong-field ligands.**

A complex of a given metal atom has a smaller Δ_O value if it contains weak-field ligands than if it contains strong-field ligands. Less energy is needed to promote the electron across a small Δ_O than across a large one, hence the complex absorbs longer wavelength light if it has weak-field ligands than if it has strong-field ligands. Because a CN^- ligand produces a stronger field splitting than H_2O, the complex $[Fe(CN)_6]^{4-}$ absorbs shorter wavelength radiation than $[Fe(H_2O)_6]^{2+}$ does.

White light is a mixture of all wavelengths of electromagnetic radiation from about 400 nm (violet) to 800 nm (red). When some of these wavelengths are removed from a beam of white light by passing the light through a sample, the emerging light is no longer white. For example, if red light is absorbed from white light, then the light that remains appears green. Conversely, if green is removed, then the light appears red. We say that red and green are each other's **complementary color**—each is the color that white light becomes when the other is removed.

Complementary colors are shown on the "color wheel" in Fig. 21.41. We can see from the color wheel that, if a substance looks blue (as does a copper(II) sulfate solution, for instance), then it is absorbing orange (580 to 620 nm) light. Conversely, if we know the wavelength (and therefore the color) of the light that a substance absorbs, then we can predict the color of the substance by noting the complementary color on the color wheel.

The prediction of color is, in fact, quite difficult. One problem is that compounds absorb light over a range of wavelengths and may actually absorb in several regions of the spectrum. Chlorophyll, for example, absorbs both red and blue light, leaving only the wavelengths near green to be reflected from vegetation (see Investigating Matter 9.2).

Because weak-field ligands give small splittings, the complexes they form absorb low-energy, long-wavelength radiation. The long wavelengths correspond to red light, so these complexes exhibit colors near green. Because strong-field ligands give large splittings, the complexes they form should absorb high-energy, short-wavelength radiation, corresponding to the violet end of the visible spectrum. Such complexes can therefore be expected to have colors near orange and yellow (Fig. 21.42). For example, when ammonia is added to aqueous copper(II) sulfate, strong-field NH_3 ligands replace four of the weak-field H_2O ligands of the $[Cu(H_2O)_6]^{2+}$ ion. The absorption shifts to higher energies and shorter wavelengths, from orange to yellow, and the perceived color shifts from blue toward violet (see Fig. 20.4).

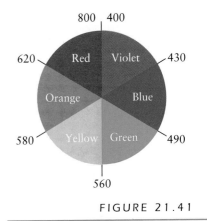

FIGURE 21.41

The perceived color of a complex in white light is the complementary color of the light it absorbs. In this color wheel, complementary colors are opposite each other. The numbers are approximate wavelengths in nanometers.

Transitions between d-orbitals in complexes give rise to color. The spectrochemical series summarizes the relative magnitudes of the ligand field splitting.

Self-Test 21.6A Predict which of the following complexes absorbs light of the shorter wavelength. Explain your reasoning. (a) $[Co(H_2O)_6]^{3+}$ or (b) $[Co(en)_3]^{3+}$.
[**Answer:** (b); en is a stronger-field ligand than H_2O.]

Self-Test 21.6B Predict which of the following complexes absorbs light of the shorter wavelength. Explain your reasoning: (a) $[Fe(CN)_6]^{4-}$ or (b) $[Fe(NH_3)_6]^{2+}$.

21.9 The Electronic Structures of Many-Electron Complexes

The electron configurations of complexes depend on the ligand field splitting. With no ligands present, all five d-orbitals have the same energy. Electrons therefore enter each orbital until each has one electron, and only after that do any additional electrons pair. However, in a complex, we have to take into account the different energies of the t- and e-orbitals. To write the electron configuration, we use the orbital energy-level diagram in Fig. 21.38a for octahedral complexes and the diagram in Fig. 21.38b for tetrahedral complexes.

At this point, we can explore how the presence of the ligands affects the electron configuration of a d-metal atom or ion in an octahedral complex. There are three t-orbitals, and because all three have the same energy, for complexes with up to three d-electrons (that is, d^1, d^2, and d^3 complexes), each electron can occupy a separate t-orbital. According to Hund's rule (Section 7.11), these electrons will have parallel spins, (**15**) and (**16**).

FIGURE 21.42

The effect of changing the ligands in octahedral cobalt(III) complexes in aqueous solution. The ligand field strengths increase from left to right.

15 t^2

16 t^3

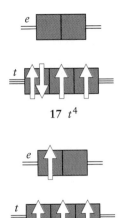

17 t^4

18 t^2e^1

Now consider a d^4 complex, in which there are four d-electrons. The fourth electron could enter a t-orbital, thereby producing a t^4 configuration. However, to do so, it must enter an orbital that is already half full, and hence experience a strong repulsion from the electron already there (**17**). To avoid this repulsion, it could occupy an empty e-orbital to give a t^3e^1 configuration (**18**). Now, though, it experiences a strong repulsion from the ligands. Which configuration has the lower energy depends on the ligands attached. If Δ_O is large (as it is for strong-field ligands), signifying strong ligand repulsion of an e-electron, the energy difference between the t- and e-orbitals will be large and t^4 will give the lower energy. If Δ_O is small (as it is for weak-field ligands), t^3e^1 will be the lower energy configuration and hence the one adopted.

Table 21.6 gives the configurations for d^1 through d^{10} complexes, including the alternative configurations for d^4 through d^7 octahedral complexes. A d^n complex with the maximum number of unpaired spins is called a **high-spin complex.** High-spin complexes are expected for weak-field ligands because the electrons can easily occupy both the t- and e-orbitals, and the greatest number of electrons then have parallel spins. Tetrahedral complexes are almost always high-spin because they do not have enough ligands to give a large ligand field splitting even if the ligands are classified as strong-field ligands for octahedral complexes. A d^n complex with the minimum number of unpaired spins is called a **low-spin complex.** A low-spin complex is expected for strong-field ligands, because the strong repulsions from the ligands raise the energy of the e-orbitals: electrons then enter the t-orbitals until they are completely full, even though they have to pair their spins.

We can predict whether an octahedral complex is likely to be a high-spin or a low-spin complex by noting where the ligands lie in the spectrochemical series. If they are strong-field ligands, we expect a low-spin complex; if they are weak-field ligands, we expect a high-spin complex.

The electron configurations of complexes are obtained by applying the building-up principle to the d-orbitals, taking into account the strength of the ligand field splitting.

Table 21.6 *Electronic configurations of d^n complexes*

Number of d-electrons	Configuration		
	Octahedral complexes		**Tetrahedral complexes**
d^1	t^1		e^1
d^2	t^2		e^2
d^3	t^3		e^2t^1
	Low spin	High spin	
d^4	t^4	t^3e^1	e^2t^2
d^5	t^5	t^3e^2	e^2t^3
d^6	t^6	t^4e^2	e^3t^3
d^7	t^6e^1	t^5e^2	e^4t^3
d^8	t^6e^2		e^4t^4
d^9	t^6e^3		e^4t^5
d^{10}	t^6e^4		e^4t^6

Toolbox 21.2 How to predict the electron configurations of d-metal complexes

This Toolbox explains how to predict the ground-state electron configurations of octahedral and tetrahedral d-metal complexes and summarizes their properties.

Conceptual Basis

In an octahedral complex, the d-orbitals are split into a low-energy set of three t-orbitals and a high-energy set of two e-orbitals. The opposite is true in a tetrahedral complex. The n electrons of a d^n complex occupy the orbitals in such a way as to achieve the lowest total energy subject to the Pauli exclusion principle (Fig. 21.43).

Procedure

To predict the electron configuration

Step 1. Identify the number of electrons to be accommodated (n in a d^n complex).

Step 2. For an octahedral complex, decide from Fig. 21.40 whether the ligands are strong-field or weak-field. For a tetrahedral complex, treat all ligands as weak-field.

Step 3. For strong-field ligands, occupy the lower set of orbitals first, pairing electrons if necessary, before occupying the upper orbitals. For weak-field ligands, occupy the lower set of orbitals with parallel spins, then the upper set with parallel spins, and only after that allow electrons to pair with those already present.

To predict the color of the complex

Step 1. A photon can be absorbed if it has sufficient energy to excite an electron from the lower set of orbitals to the upper set. Estimate the frequency of that photon from $\nu = \Delta/h$ or its wavelength from $\lambda = hc/\Delta$.

Step 2. Use the color wheel in Fig. 21.41 to identify the complementary color of the color absorbed.

Caution: The prediction of color is very difficult; it is more reliable to use this approach to estimate shifts in color between complexes related by ligand substitution reactions.

To predict the magnetism of a complex (see Section 21.10)

Step 1. Count the number of unpaired d-electrons in the ground-state electron configuration of the complex.

Step 2. A complex with no unpaired electrons will be diamagnetic; the paramagnetism increases with the number of unpaired electrons.

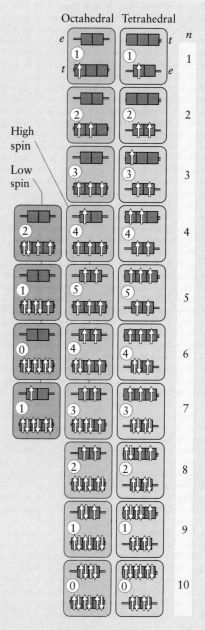

FIGURE 21.43

The electron configurations of d-metal complexes. The circled number gives the number of unpaired electrons.

Example 21.3 *Predicting the electron configuration of a complex*

Predict the electron configuration of an octahedral d^5 complex with (a) strong-field ligands and (b) weak-field ligands. Give the number of unpaired electrons in each case.

Strategy Electrons occupy the orbitals that result in the lowest energy configuration. If the splitting Δ_O is small, electron-electron repulsion is stronger than the weak repulsions between electrons and ligands. In this case, electrons are likely to occupy all the vacant orbitals, even those at the higher energy, before pairing. If the splitting Δ_O is large, the repulsion between electrons and ligands is great. In this case, electrons are likely to pair in the lower energy orbitals and fill them completely before occupying any of the higher energy orbitals. Use the procedure in Toolbox 21.2 to predict the electron configuration from these principles.

Solution (a) In the strong-field case, all five electrons enter the t-orbitals; and to do so, some of them must pair (**19**). There is one unpaired electron in this configuration. (b) In the weak-field case, the five electrons occupy all five orbitals without pairing (**20**). There are now five unpaired electrons.

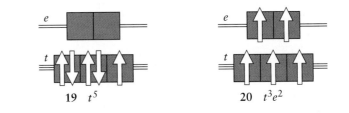

19 t^5

20 $t^3 e^2$

Self-Test 21.7A Predict the electron configurations and the number of unpaired electrons of an octahedral d^6 complex with (a) strong-field ligands and (b) weak-field ligands.

[***Answer:*** (a) t^6 (0); (b) $t^4 e^2$ (4)]

Self-Test 21.7B Predict the electron configurations and the number of unpaired electrons of an octahedral d^7 complex with (a) strong-field ligands and (b) weak-field ligands.

21.10 Magnetic Properties of Complexes

As we saw in Section 9.14, a species with unpaired electrons is *paramagnetic* and is pulled into a magnetic field. A substance with no unpaired electrons is *diamagnetic* and is pushed out of a magnetic field. Paramagnetism and diamagnetism can be distinguished experimentally by using the apparatus shown in Fig. 21.44: a sample is hung from a balance so that it lies between the poles of an electromagnet. When the magnet is turned on, a paramagnetic substance is pulled into the field and appears to weigh more than when the magnet is off. A diamagnetic substance is pushed out of the field and appears to weigh less.

Many d-metal complexes have unpaired d-electrons and are therefore paramagnetic. We have just seen that a high-spin d^n complex has more unpaired electrons than a low-spin d^n complex. The former is therefore more strongly paramagnetic and is drawn more strongly into a magnetic field. Moreover,

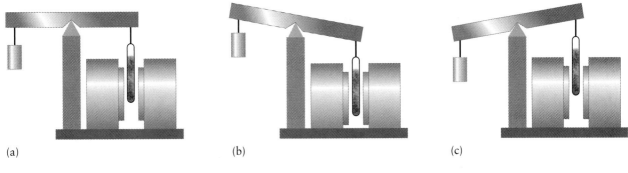

(a)   (b)   (c)

FIGURE 21.44

The magnetic character of a complex can be studied with the apparatus shown here, which is called a Gouy balance. (a) A sample is hung from a balance so that it lies partly between the poles of an electromagnet. (b) When the magnetic field is turned on, a paramagnetic sample is drawn into it, so the sample seems to weigh more. (c) In contrast, a diamagnetic sample is pushed out of the field when the field is turned on, so it seems to weigh less.

whether a complex is high-spin or low-spin depends on the ligands present. Strong-field ligands create a large energy difference between the t- and e-orbitals (Fig. 21.45a). Their d^4–d^7 complexes therefore tend to be low spin and diamagnetic or only weakly paramagnetic. Weak-field ligands create a small energy gap, so electrons fill the higher energy orbitals rather than pairing in the lower energy orbitals (Fig. 21.45b). Their d^4–d^7 complexes therefore tend to be high spin and strongly paramagnetic. This correlation suggests that it should be possible to modify the magnetic properties of an octahedral complex by changing the ligands with which it is coordinated.

> *The magnetic properties of a complex depend on the magnitude of the ligand field splitting. Strong-field ligands tend to form low-spin, weakly paramagnetic complexes; weak-field ligands tend to form high-spin, strongly paramagnetic complexes.*

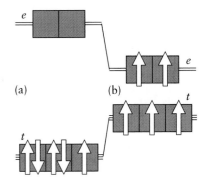

(a)   (b)

FIGURE 21.45

(a) A strong-field ligand is likely to lead to a low-spin complex (in this case, the configuration is that of Fe^{3+}). (b) A weak-field ligand is likely to result in a high-spin complex.

Example 21.4 *Predicting the magnetic properties of a complex*

Compare the magnetic properties of $[Fe(CN)_6]^{4-}$ with those of $[Fe(H_2O)_6]^{2+}$.

Strategy Decide from their positions in the spectrochemical series whether the ligands in the complex are weak-field or strong-field ligands. Then judge whether each complex is high spin or low spin. If one complex is high spin and the other low spin, expect a difference in their magnetic properties.

Solution The Fe^{2+} ion is a d^6 ion. Because H_2O is a weak-field ligand, we predict a high-spin configuration with four unpaired electrons for $[Fe(H_2O)_6]^{2+}$. The ion is therefore predicted (and found) to be paramagnetic. When cyanide ions are added to an aqueous solution of Fe^{2+} ions, they form the $[Fe(CN)_6]^{4-}$ ion. Now the ligands are strong-field ligands, and the resulting complex is a low-spin t^6 complex; it has no unpaired electrons, so it is not paramagnetic. In other words, the ligand substitution reaction has had the effect of quenching the paramagnetism.

Self-Test 21.8A What change in magnetic properties can be expected when NO_2^- ligands in an octahedral complex are replaced by Cl^- ligands in (a) a d^6 complex? (b) a d^3 complex?

[**Answer:** (a) The complex becomes paramagnetic; (b) no change in magnetic properties]

Self-Test 21.8B Compare the magnetic properties of $[Ni(en)_3]^{2+}$ with those of $[Ni(H_2O)_6]^{2+}$.

Skills You Should Have Mastered

Conceptual

☐ 1. Explain trends in chemical and physical properties among the *d*-block elements, Sections 21.1 and 21.2.

☐ 2. Interpret the formation and structure of a *d*-metal complex in terms of Lewis acids and bases, Section 21.5.

☐ 3. Explain how the colors of *d*-metal complexes are related to the ligand field splitting, Section 21.8.

☐ 4. Explain how weak-field and strong-field ligands affect the electron configuration of a *d*-metal ion in a complex, Section 21.9.

Problem-Solving

☐ 1. Predict the effect of oxidation number on the oxidizing or reducing ability of compounds of a *d*-block element and on the acidity of its oxides, Section 21.2.

☐ 2. Identify pairs of ionization, linkage, hydrate, coordination, geometrical, and optical isomers, Toolbox 21.1 and Example 21.1.

☐ 3. Calculate the ligand field splitting from the frequency or wavelength of light absorbed by a *d*-metal complex, Example 21.2.

☐ 4. Use the spectrochemical series to predict the effect of a ligand on the color, electron configuration, and magnetic properties of a *d*-metal complex, Toolbox 21.2 and Examples 21.3 and 21.4.

Descriptive

☐ 1. Distinguish between ferromagnetism, paramagnetism, and diamagnetism, and identify common *d*-metals that are ferromagnetic, Sections 21.1 and 21.10.

☐ 2. Describe and write balanced equations for the principal reactions used to produce the elements in the first row of the *d* block and in Groups 11 and 12, Sections 21.3 and 21.4.

☐ 3. Describe the names, properties, and reactions of some of the principal compounds of the elements in the first row of the *d* block, Sections 21.3 and 21.4.

☐ 4. Describe how iron is refined from the ore and how steel is made by the basic oxygen process, Section 21.3.

☐ 5. Describe the composition of steel and how the properties of steel are affected by the proportion of carbon, Section 21.3.

Exercises

The d-Block Elements and Their Electron Configurations

21.1 Write the ground-state electron configuration of (a) manganese; (b) cadmium; (c) zinc; (d) zirconium.

21.2 Write the ground-state electron configuration of (a) mercury; (b) iron; (c) titanium; (d) gold.

21.3 State the number of unpaired electrons in a ground-state atom of (a) Sc; (b) V; (c) Cu.

21.4 State the number of unpaired electrons in a ground-state atom of (a) nickel; (b) silver; (c) rhenium.

21.5 Identify the elements in the *d* block that are ferromagnetic.

21.6 Which members of the *d* block, those at the left or at the right of the block, tend to have strongly negative standard potentials? Name six elements in the *d* block that have standard potentials greater than 0.

Trends in Properties

21.7 Identify the element with the larger atomic radius in each of the following pairs: (a) scandium and titanium; (b) copper and gold; (c) vanadium and niobium.

21.8 Identify the element with the larger atomic radius in each of the following pairs: (a) cobalt and manganese; (b) copper and zinc; (c) chromium and molybdenum.

21.9 Identify the element with the higher first ionization energy in each of the following pairs: (a) scandium and titanium; (b) nickel and copper; (c) iron and zinc.

21.10 Identify the element with the higher first ionization energy in each of the following pairs: (a) iron and cobalt; (b) manganese and iron; (c) vanadium and chromium.

21.11 Explain what is meant by the "lanthanide contraction" and give examples of its effect.

21.12 Is there evidence for a "d-block contraction" analogous to the lanthanide contraction?

21.13 Explain why the density of mercury ($13.5 \ g/cm^3$) is significantly higher than that of cadmium ($8.65 \ g/cm^3$), whereas the density of cadmium is only slightly greater than that of zinc ($7.14 \ g/cm^3$).

21.14 Assuming the same crystal structure, explain why the density of iron ($7.87 \ g/cm^3$) is significantly less than that of cobalt ($8.80 \ g/cm^3$).

21.15 Describe the trend in the stability of oxidation states moving down a group in the d block (for example, from chromium to molybdenum to tungsten).

21.16 Which oxoanion, MnO_4^- or ReO_4^-, is expected to be the stronger oxidizing agent? Explain your choice.

21.17 Which of the elements vanadium, chromium, and manganese is most likely to form an oxide with the formula MO_3? Explain your answer.

21.18 Which of the elements zirconium, chromium, and iron is most likely to form a chloride with the formula MCl_4? Explain your answer.

Scandium Through Nickel

21.19 Outline a process, using chemical equations where possible, by which (a) titanium; (b) vanadium is prepared.

21.20 Outline a process, using chemical equations where possible, by which (a) nickel; (b) chromium is prepared.

21.21 Predict the major products of each of the following reactions and then balance each skeletal equation:
(a) $TiCl_4(g) \ + \ Mg(l) \xrightarrow{\Delta}$
(b) $CoCO_3(s) \ + \ HNO_3(aq) \rightarrow$
(c) $V_2O_5(s) \ + \ Ca(l) \xrightarrow{\Delta}$

21.22 Predict the products of each of the following reactions and then balance each equation:
(a) $Ti(s) \ + \ F_2(g, \text{in excess}) \rightarrow$ liquid

(b) $CrO_4^{2-}(aq) \ + \ H_3O^+(aq) \rightarrow$
(c) $MnO_2(s) \ + \ Al(s) \xrightarrow{\Delta}$

21.23 Give the systematic name and chemical formula of the principal component of (a) rutile; (b) hematite; (c) pyrolusite.

21.24 Give the systematic name and chemical formula of the principal component of (a) magnetite; (b) pyrite; (c) ilmenite; (d) chromite.

21.25 Use Appendix 2B to predict the products of the following reactions: (a) vanadium with 1 M HCl(aq); (b) mercury with 1 M HCl(aq); (c) cobalt with 1 M HCl(aq).

21.26 Use Appendix 2B to predict products of the following reactions: (a) nickel with 1 M HCl(aq); (b) titanium with 1 M HCl(aq); (c) platinum with 1 M $KMnO_4$(aq) in 1 M HCl(aq).

Groups 11 and 12

21.27 Describe the chemical evidence for treating the coinage metals as a single group.

21.28 Describe the chemical evidence for treating zinc, cadmium, and mercury as a single group.

21.29 Give the systematic name and chemical formula of the principal component of (a) chalcopyrite; (b) sphalerite; (c) cinnabar.

21.30 What are the major sources of (a) silver; (b) gold; (c) cadmium?

21.31 Outline a process, using chemical equations where possible, by which (a) zinc; (b) mercury can be produced.

21.32 Outline a process, using chemical equations where possible, by which copper is extracted and purified from chalcopyrite by the pyrometallurgical process.

d-Metal Complexes

21.33 Determine the oxidation number of the metal atom in each of the following complexes: (a) $[Fe(CN)_6]^{4-}$; (b) $[Co(NH_3)_6]^{3+}$; (c) $[Co(CN)_5(H_2O)]^{2-}$; (d) $[Co(SO_4)(NH_3)_5]^+$.

21.34 Determine the oxidation number of the metal atom in each of the following complexes: (a) $[Fe(CN)_6]^{3-}$; (b) $[Fe(OH)(H_2O)_5]^{2+}$; (c) $[CoCl(NH_3)_4(H_2O)]^{2+}$; (d) $[Ir(en)_3]^{3+}$.

21.35 Use Table 21.5 to determine the coordination number of the metal ion in each of the following complexes: (a) $[NiCl_4]^{2-}$; (b) $[Ag(NH_3)_2]^+$; (c) $[PtCl_2(en)_2]^{2+}$; (d) $[Cr(edta)]^-$.

21.36 Use Table 21.5 to determine the coordination number of the metal ion in each of the following complexes: (a) $[Ir(en)_3]^{3+}$; (b) $[Fe(ox)_3]^{3-}$ (the oxalato ligand attaches at two points); (c) $[PtCl_2(NH_3)_2]$; (d) $[Fe(CO)_5]$.

21.37 Use the information in Table 21.5 and Appendix 3D to name the following complexes: (a) $[Fe(CN)_6]^{4-}$; (b) $[Co(NH_3)_6]^{3+}$; (c) $[Co(CN)_5(H_2O)]^{2-}$; (d) $[Co(SO_4)(NH_3)_5]^+$.

21.38 Use the information in Table 21.5 and Appendix 3D to name the following complexes: (a) $[Fe(CN)_6]^{3-}$; (b) $[Fe(OH)(H_2O)_5]^{2+}$; (c) $[CoCl(NH_3)_4(H_2O)]^{2+}$; (d) $[Ir(en)_3]^{3+}$.

21.39 Use the information in Table 21.5 and Appendix 3D to write the formula for each of the following coordination compounds: (a) potassium hexacyanochromate(III); (b) pentaaminesulfatocobalt(III) chloride; (c) tetraaminediaquacobalt(III) bromide; (d) sodium diaquabis(oxalato)ferrate(III).

21.40 Use the information in Table 21.5 and Appendix 3D to write the formula for each of the following coordination compounds: (a) triamineaquadihydroxochromium(III) chloride; (b) potassium tetrachloroplatinate(II); (c) tetraaquadichloronickel(IV) iodide; (d) potassium tris(oxalato)rhodate(III); (e) sodium chlorohydroxobis(oxalato)rhodate(III) octahydrate.

Isomerism

21.41 Determine the type of structural isomerism that exists in the following pairs of compounds: (a) $[Co(NO_2)(NH_3)_5]Br_2$ and $[Co(ONO)(NH_3)_5]Br_2$; (b) $[Pt(SO_4)(OH)(NH_3)_4]OH$ and $[Pt(OH)_2(NH_3)_4]SO_4$; (c) $[CoCl(SCN)(NH_3)_4]Cl$ and $[CoCl(NCS)(NH_3)_4]Cl$; (d) $[CrCl(NH_3)_5]Br$ and $[CrBr(NH_3)_5]Cl$.

21.42 Determine the type of structural isomerism that exists in the following pairs of compounds or complex ions: (a) $[Cr(en)_3][Co(ox)_3]$ and $[Co(en)_3][Cr(ox)_3]$; (b) $[CoCl_2(NH_3)_4]Cl \cdot H_2O$ and $[CoCl(NH_3)_4(H_2O)]Cl_2$; (c) $[Co(CN)_5(NCS)]^{3-}$ and $[Co(CN)_5(SCN)]^{3-}$; (d) $[Pt(NH_3)_4][PtCl_6]$ and $[PtCl_2(NH_3)_4][PtCl_4]$.

21.43 Write the formulas for the hydrate isomers of a compound having the empirical formula $CoCl_3 \cdot 6H_2O$ and a coordination number of 6.

21.44 Write the formula of a coordination isomer of $[Co(NH_3)_6][Cr(NO_2)_6]$.

21.45 Write the formula of a linkage isomer of $[CoCl(NO_2)(en)_2]Cl$.

21.46 Write the formula of an ionization isomer of $[CoCl(NO_2)(en)_2]Cl$.

21.47 Which of the following coordination compounds can have cis and trans isomers? If such isomerism exists, draw the two structures and name the compound. (a) $[CoCl_2(NH_3)_4]Cl \cdot H_2O$; (b) $[CoCl(NH_3)_5]Br$; (c) $[PtCl_2(NH_3)_2]$, a square planar complex.

21.48 Which of the following complexes can have cis and trans isomers? If such isomerism exists, draw the two structures and name the compound. (a) $[Fe(OH)(H_2O)_5]^{2+}$; (b) $[RuBr_2(NH_3)_4]^+$; (c) $[Co(NH_3)_4(H_2O)_2]^{3+}$.

21.49 Can a tetrahedral complex show (a) stereoisomerism; (b) geometrical isomerism; (c) optical isomerism?

21.50 Draw the structure of the *cis*-diammine-*cis*-diaqua-*cis*-dichlorochromium(III) ion and comment on its isomerism. What kind of isomerism is possible for the *trans*-diammine isomer?

21.51 Draw the geometrical isomers of $[Cr(ox)_2(H_2O)_2]^-$.

21.52 Draw the structures of (a) *cis*-$[CrCl_2(NH_3)_4]^+$ (violet); (b) *trans*-$[CrCl_2(NH_3)_4]^+$ (bright green).

21.53 Is either of the following complexes chiral? If both complexes are chiral, do they form an enantiomeric pair?

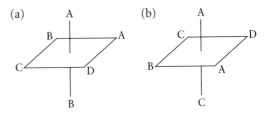

21.54 Is either of the following complexes chiral? If both complexes are chiral, do they form an enantiomeric pair?

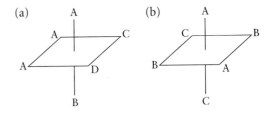

21.55 Draw the structures of the optical isomers of $[CoCl_2(en)_2]^+$.

21.56 Draw the structures of the isomeric forms of $[CrClBrI(NH_3)_3]$. Which isomers are chiral?

Crystal Field Theory

21.57 The valence electron configurations of the *d*-block ions can be summarized as d^n. Summarize the valence electron configuration for each of the following ions: (a) Co^{2+}; (b) Ni^{2+}; (c) Mn^{2+}; (d) Cr^{3+}.

21.58 The valence electron configurations of the *d*-block ions can be summarized as d^n. Summarize the valence electron configuration for each of the following ions: (a) Cr^{2+}; (b) Fe^{2+}; (c) V^{2+}; (d) Co^{3+}.

21.59 Draw an orbital energy-level diagram (like those in Fig. 21.43) showing the configuration of *d*-electrons on the metal ion in each of the following complexes: (a) $[Co(NH_3)_6]^{3+}$; (b) $[NiCl_4]^{2-}$ (tetrahedral); (c) $[Fe(H_2O)_6]^{3+}$; (d) $[Fe(CN)_6]^{3-}$.

21.60 Draw an orbital energy-level diagram (like those in Fig. 21.43) showing the configuration of d-electrons on the metal ion in each of the following complexes: (a) $[Zn(H_2O)_6]^{2+}$; (b) $[CoCl_4]^{2-}$ (tetrahedral); (c) $[Co(CN)_6]^{3-}$; (d) $[CoF_6]^{3-}$.

21.61 The complexes (a) $[Co(en)_3]^{3+}$ and (b) $[Mn(CN)_6]^{3-}$ have low-spin electron configurations. How many unpaired electrons are there in each complex?

21.62 The complexes (a) $[FeF_6]^{3-}$ and (b) $[Co(ox)_3]^{4-}$ have high-spin electron configurations. How many unpaired electrons are there in each complex?

21.63 Explain the difference between a weak-field ligand and a strong-field ligand. What measurements can be used to classify them as such?

21.64 Describe the changes that may occur in a compound's properties when weak-field ligands are replaced by strong-field ligands.

21.65 Of the two complexes (a) $[CoF_6]^{3-}$ and (b) $[Co(en)_3]^{3+}$, one appears yellow in an aqueous solution and the other appears blue. Match the complex to the color and explain your choice.

21.66 A concentrated solution of copper(II) bromide in the presence of potassium bromide is deep violet due to the presence of tetrabromocuprate(II) ions. However, the solution becomes light blue upon dilution with water, as water molecules replace bromide ions in the coordination sphere and form hexaaquacopper(II) ions. (a) Write the formulas of the two complex ions of copper(II). (b) Is the difference in color of the two complexes expected? Explain your reasoning.

21.67 State the color of a sample that absorbs light of wavelength (a) 410. nm; (b) 650. nm; (c) 480. nm; (d) 590. nm.

21.68 State the wavelength range over which the following colors are absorbed from a sample and then predict the color of the complex: (a) blue; (b) red; (c) violet; (d) green.

21.69 Suggest a reason why $Zn^{2+}(aq)$ ions are colorless. Would you expect zinc compounds to be paramagnetic? Explain your answer.

21.70 Suggest a reason why copper(II) compounds are often colored but copper(I) compounds are colorless. Which oxidation number gives paramagnetic compounds?

21.71 Estimate the ligand field splitting for (a) $[CrCl_6]^{3-}$ ($\lambda_{max} = 740.$ nm); (b) $[Cr(NH_3)_6]^{3+}$ ($\lambda_{max} = 460.$ nm); (c) $[Cr(H_2O)_6]^{3+}$ ($\lambda_{max} = 575$ nm), where λ_{max} is the wavelength of the most intensely absorbed light. Arrange the ligands in order of increasing ligand field strength.

21.72 Estimate the ligand field splitting for (a) $[Co(CN)_6]^{3-}$ ($\lambda_{max} = 295$ nm); (b) $[Co(NH_3)_6]^{3+}$ ($\lambda_{max} = 435$ nm);

(c) $[Co(H_2O)_6]^{3+}$ ($\lambda_{max} = 540.$ nm), where λ_{max} is the wavelength of the most intensely absorbed light. Arrange the ligands in order of increasing ligand field strength.

Supplementary Exercises

21.73 How do (a) diamagnetism and (b) ferromagnetism differ from paramagnetism?

21.74 Iron, cobalt, and nickel are often grouped together as the *iron triad.* What similarities do they show that justifies this grouping? How might these similarities be explained?

21.75 Write a chemical equation showing in what respect a Sc^{3+} ion is a Brønsted acid in aqueous solution.

21.76 Explain in terms of electron configurations why the atomic radius of manganese is larger than that of chromium.

21.77 Explain as fully as possible, with diagrams of the structures of the hydrated ions, the changes that occur when (a) anhydrous copper(II) sulfate is moistened; (b) it is dissolved in water; and (c) an excess of aqueous ammonia is added to the resulting solution.

Water being added to anhydrous $CuSO_4$.

21.78 Zinc metal is used to galvanize steel because of the cathodic protection it provides. What would result if copper were used instead? (See Section 18.12.)

21.79 (a) Explain why the dissolution of a chromium(III) salt produces an acidic solution. (b) Explain why the slow addition of hydroxide ions to a solution containing chromium(III) ions first produces a gelatinous precipitate that subsequently dissolves with further addition of hydroxide ions. Write chemical equations showing these aspects of the behavior of chromium(III) ions.

21.80 Use the information in Table 21.5 and Appendix 3D to name each of the following complexes. Determine the coordination number and oxidation number of the d-metal ion: (a) $[Zr(ox)_4]^{4-}$; (b) $[CuCl_4(H_2O)_2]^{2-}$; (c) $[PtCl_3(NH_3)]^-$; (d) $[Mo(O)_2(CN)_4]^{4-}$.

21.81 By considering electron configurations, suggest a reason why iron(III) compounds are readily prepared from iron(II), but the conversions of nickel(II) and cobalt(II) to nickel(III) and cobalt(III) are much more difficult.

21.82 Draw all possible isomers of the square planar complex $[PtBrCl(NH_3)_2]$.

21.83 How can the existence of the isomers in Exercise 21.82 be used to show that the complex is square planar rather than tetrahedral?

21.84 Suggest a chemical test for distinguishing between (a) $[Ni(SO_4)(en)_2]Cl_2$ and $[NiCl_2(en)_2]SO_4$; (b) $[NiI_2(en)_2]Cl_2$ and $[NiCl_2(en)_2]I_2$.

21.85 Two chemists prepared a complex and determined its formula, which they wrote as $CrNH_3Cl_3 \cdot 2H_2O$. When they dissolved 2.11 g of the compound in water and added an excess of silver nitrate, 2.87 g of AgCl precipitated. Write the correct formula of the compound and draw the structure of the complex, including all possible isomers.

21.86 For which of the ions (a) Mn^{2+}; (b) V^{2+}; (c) Ni^{2+}; (d) Cr^{2+}, would there be no difference in the magnetic properties of the octahedral complexes of strong-field and weak-field ligands? Draw orbital energy-level diagrams to support your conclusions.

21.87 What change in magnetic properties can be expected when NH_3 ligands replace the H_2O ligands in $[Co(H_2O)_6]^{3+}$?

21.88 Suggest the form the orbital energy-level diagram would take for a square planar complex and discuss how the building-up principle applies. *Hint:* The d_{z^2}-orbital has more electron density in the x–y plane than the d_{zx}- or d_{yz}-orbitals do, but less than the d_{xy}-orbital.

21.89 Before the structures of complexes had been determined, various means were used to explain the fact that transition metal ions could bind to a greater number of ligands than expected on the basis of their charges. For example, Co^{3+} can bind to six ligands, not just three. An early theory attempted to explain this behavior by postulating that once three ligands had attached to a metal ion of charge $+3$, the others bonded to the attached ligands, forming chains of ligands. Thus, the compound $[Co(NH_3)_6]Cl_3$, which we now know to have the octahedral structure, would have been described as $Co(NH_3-NH_3-Cl)_3$. Show how the chain theory is not consistent with at least two properties of coordination compounds.

21.90 The relative thermodynamic stability of two complexes can be predicted from a comparison of their standard potentials. Determine which complex of the following pair is the more stable and state your conclusions about the relationship between the stability of a complex and (a) the

oxidation number of the central atom; (b) the field strength of the attached ligands. Give your reasoning for each conclusion and suggest an explanation for the relationship.

$$[Co(NH_3)_6]^{3+}(aq) + e^- \longrightarrow$$
$$[Co(NH_3)_6]^{2+}(aq) \quad E° = +0.11 \text{ V}$$

$$[Co(H_2O)_6]^{3+}(aq) + e^- \longrightarrow$$
$$[Co(H_2O)_6]^{2+}(aq) \quad E° = +1.81 \text{ V}$$

Applied Exercises

For Exercises 21.93–21.96, see Applying Chemistry: Case Study 21.

21.91 Identify the element, compound, or mixture that best fits the description: (a) bronze; (b) the green patina on the copper plate of the Statue of Liberty; (c) 24-carat gold; (d) the densest of the d-block elements.

The green patina on the Statue of Liberty.

21.92 Identify the element, compound, or mixture that best fits the description: (a) a substance used as the white pigment in paints and in papermaking; (b) the catalyst used for the production of sulfuric acid; (c) the second most abundant metal in the Earth's crust; (d) pig iron.

21.93 Explain why the images on daguerreotypes fade, the images on photographic film are permanent, and the colors of photochromic sunglasses can be reversed.

21.94 (a) Write the net ionic equation for the reaction of aqueous sodium thiosulfate with solid silver bromide. (b) Predict the magnetic properties of the $[Ag(S_2O_3)_2]^{3-}$ complex.

21.95 When the price of silver rose rapidly in the 1980s, alternatives to silver were sought for use in photography. What other metals might have been reasonably considered?

21.96 Suggest a reason why the dim lights used to illuminate photographic darkrooms are red.

21.97 (a) What reducing agent is used in the production of iron from its ore? (b) Write chemical equations for the production of iron in a blast furnace. (c) What is the major impurity in the product of the blast furnace?

21.98 (a) What is the purpose of adding limestone to a blast furnace? (b) Write chemical equations that show the reactions of lime in a blast furnace.

Integrated Exercises

21.99 Use the information in Appendix 2B to determine whether an acidic sodium dichromate solution can oxidize (a) bromide ions to bromine; (b) silver(I) ions to silver(II) ions under standard conditions.

21.100 Use the information in Appendix 2B to determine whether an acidic potassium permanganate solution can oxidize (a) chloride ions to chlorine; (b) mercury metal to mercury(I) ions under standard conditions.

21.101 Use the information in Appendix 2B to determine the equilibrium constant for the disproportionation of copper(I) ions in aqueous solution at 25°C to copper metal and copper(II) ions.

21.102 Use the information in Appendix 2B to determine the equilibrium constant for the disproportionation of mercury(I) ions in aqueous solution at 25°C to mercury metal and mercury(II) ions.

21.103 Determine the mass of $FeCr_2O_4$ (from chromite ore) that is needed to produce 1.00 kg of sodium chromate by the reaction

$$4\,FeCr_2O_4(s)\ +\ 8\,Na_2CO_3(s)\ +\ 7\,O_2(g)\ \xrightarrow{\Delta}$$
$$8\,Na_2CrO_4(s)\ +\ 2\,Fe_2O_3(s)\ +\ 8\,CO_2(g)$$

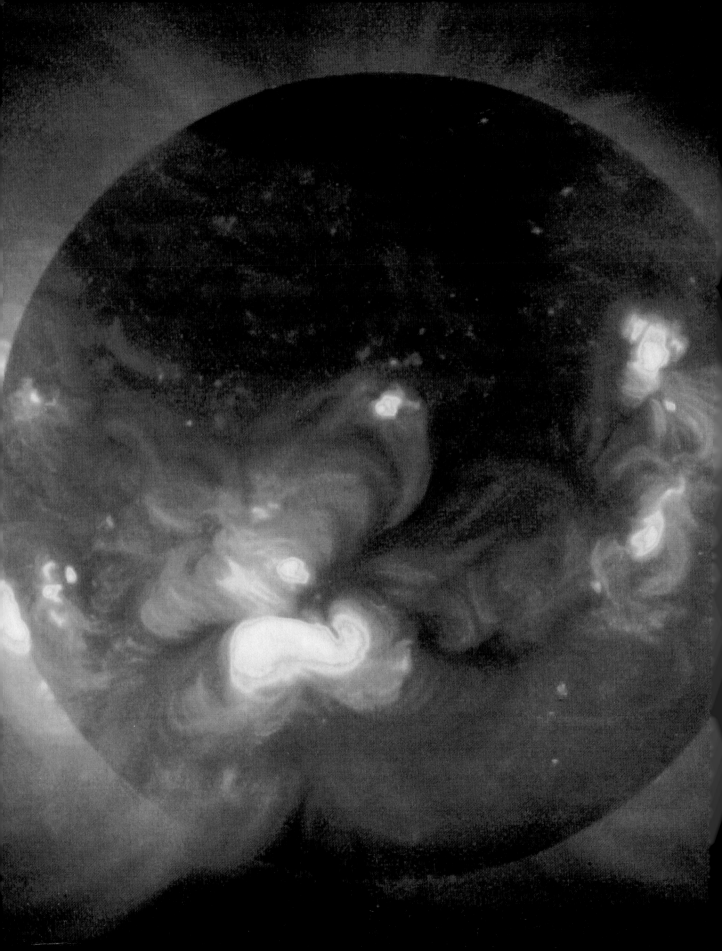

Nuclear Chemistry

Nuclear energy is viewed as both a blessing and a curse, a cure for disease and an instrument of destruction, a source of abundant energy and a waste-disposal nightmare. Like other powerful resources, nuclear energy presents us not only with great promise but also with technical challenges. We face many important decisions about the control and use of nuclear energy that will affect future generations for centuries.

So far, we have viewed the atomic nucleus as an unchanging passenger in chemical reactions. However, nuclei can change, and **nuclear chemistry** is the branch of chemistry that explores the consequences of those changes. Nuclear chemistry is part of our lives in many ways. It is central to the development of nuclear energy, because techniques must be found for the purification and recycling of nuclear fuels and for the disposal of hazardous radioactive waste. It is used in medicine to treat cancer and to produce images of the body's interior, in the rest of chemistry to investigate reaction mechanisms, in archeology to date archeological objects, and by governments as part of the military defense strategies of many nations. However, the use of nuclear energy also presents hazards and waste-disposal problems. In this chapter, we see how chemistry is used to look for solutions to these problems. Currently, there are no clear-cut solutions, but the material presented here may help you assess the risks and benefits of some of the approaches that are being explored.

NUCLEAR STABILITY

At first glance, the existence of nuclei seems impossible. Nuclei consist of positively charged protons packed together into a very tiny volume, with no negative charge to prevent them from flying apart. In fact, some nuclei do fly apart; and when these unstable nuclei break up, they emit radiation.

Nuclear Stability
22.1   Spontaneous nuclear decay
22.2   Nuclear reactions
22.3   The pattern of nuclear stability
22.4   Nucleosynthesis

Radioactivity
22.5   The effects of radiation
22.6   Measuring radioactivity
22.7   The law of radioactive decay

Nuclear Energy
22.8   Mass-energy conversion
22.9   Nuclear fission
22.10   Nuclear fusion
22.11   The chemistry of nuclear power

X-ray emission from the Sun in false color shows the swirling shapes of the high-temperature regions that often lie over sunspots. Nuclear processes taking place in the Sun release immense amounts of energy, a tiny fraction of which is intercepted by our planet and used to support life. Similar nuclear processes have many useful functions on Earth and can be beneficial if properly managed.

959

FIGURE 22.1

Henri Becquerel discovered radioactivity when he noticed that an unexposed photographic plate left near some uranium oxide became fogged. This photograph shows one of his original plates.

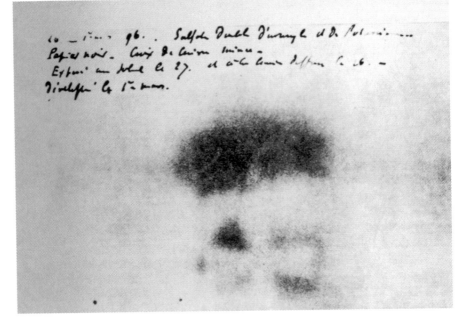

22.1 Spontaneous Nuclear Decay

The radiation emitted by nuclei was detected before the existence of the nucleus was known. The French scientist Henri Becquerel discovered a strange, highly penetrating type of radiation in 1896 when he stored a uranium compound in a drawer with some photographic plates (Fig. 22.1). Shortly after, a young Polish student, Marie Sklodowska Curie (Fig. 22.2), went to Becquerel looking for a topic for her doctoral dissertation research. She found that uranium and certain other elements gave off radiation that was unchanged by their state of chemical combination, so it had to be a property of the atoms themselves. When nuclei were discovered by Ernest Rutherford in 1908 (recall Section 1.3), it become clear that the source of the radiation must be the nuclei of the atoms, for nuclei are unchanged by the state of chemical combination. Becquerel's radiation is *nuclear* radiation. We now define **radioactivity** as high-energy radiation emitted by nuclei.

Becquerel's radioactivity was studied by Ernest Rutherford. He identified three different types of radioactivity by observing the effect of electric fields on radioactive emissions (Fig. 22.3), and he called them α (alpha), β (beta), and γ (gamma) radiation. Rutherford found that α radiation is attracted to a negatively charged electrode. This observation led him to propose that it consists of positively charged particles, which he called **α particles** (alpha particles). From the charge and mass of the particles, he was able to identify them as the nuclei of helium atoms, $^4_2\text{He}^{2+}$. An α particle is denoted $^4_2\alpha$, or simply α. We can think of it as a tightly bound cluster of two protons and two neutrons (Fig. 22.4). When a nucleus emits an α particle, we say the nucleus undergoes **α decay.** The supplies of helium on Earth are formed as a result of α decay deep underground: the α particles collect electrons from their surroundings and form He atoms.

Rutherford also found that β radiation is attracted to a positively charged electrode. This observation suggested to him that β rays consist of a stream of

Marie Curie's work with radioactivity won her two Nobel prizes, one shared with her husband, the French physicist Pierre Curie, and Becquerel.

FIGURE 22.2

Marie Sklodowska Curie (1867–1934).

FIGURE 22.3

The effects of an electric field on nuclear radiation. Deflection toward the negative plate identifies α particles as positively charged, and deflection toward the positive plate identifies β particles as negatively charged. The fact that γ rays are not deflected toward either plate identifies them as uncharged.

FIGURE 22.4

An α particle has two positive charges and a mass number of 4. It consists of two protons and two neutrons, and is the same as the nucleus of a helium-4 atom.

negatively charged particles. Measurement of the charge and mass of these particles showed that they are, in fact, electrons. The rapidly moving electrons emitted by nuclei are called **β particles** (beta particles) and are denoted β. Because a β particle has no protons or neutrons, its mass number is 0. Because its charge is −1, it is convenient (when balancing nuclear equations) to denote a β particle as $_{-1}^{0}e$, but the subscript is not a true atomic number. When a nucleus emits a β particle, we say the nucleus undergoes **β decay**. We can regard β decay as the result of the conversion of a neutron into a proton within the nucleus:

$$_{0}^{1}n \longrightarrow {}_{1}^{1}p + {}_{-1}^{0}e, \quad \text{or more simply} \quad n \longrightarrow p + \beta^{-}$$

γ Radiation (gamma radiation) is electromagnetic radiation like light, but of much higher frequency—greater than about 10^{20} Hz—and with wavelengths less than about 1 pm. γ Radiation can be regarded as a stream of very high energy photons; each photon is emitted by a single nucleus as that nucleus discards energy. Like all photons, γ ray photons are massless, uncharged, and unaffected by electric fields.

Heavy elements are more likely to give off α radiation, whereas β radiation is more typical of lighter elements. α and β Radiation is often accompanied by γ radiation because the ejection of an α or β particle often leaves the nucleons in the product nucleus in a high-energy arrangement (Fig. 22.5); a γ ray photon is then emitted when the nucleons collapse into a state of lower energy. As in the case of radiation emitted by electrons in excited atoms, the emitted photon has a frequency, ν (nu), given by the relation $\Delta E = h\nu$ (see Section 7.3). γ Rays have very high frequencies because the energy difference between the excited and ground *nuclear* states is very large.

Radioactive nuclei commonly emit three types of radiation: α particles (the nuclei of helium atoms), β particles (fast electrons ejected from the nucleus), and γ rays (high-energy electromagnetic radiation).

High-energy arrangement

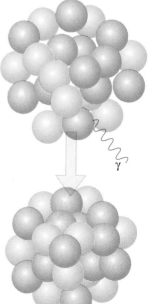

Low-energy arrangement

FIGURE 22.5

After a nucleus decays, the nucleons remaining in the nucleus may be left in a high-energy state, as shown by the loose arrangement in the upper part of the illustration. As the nucleons adjust to a lower energy arrangement (bottom), the excess energy is released as a γ ray photon.

22.2 Nuclear Reactions

We saw in Sections 1.3 and 1.4 that nuclei are composed of protons and neutrons (Fig. 22.6); these subatomic particles are collectively called *nucleons*. A specific nucleus with a given atomic number and mass number is called a **nuclide.** Thus, ^1H, ^2H, and ^{16}O are three different nuclides; the first two are *isotopes* of the same element. Nuclei that change their structure spontaneously and emit radiation are called **radioactive.** Many such unstable nuclei occur naturally: for instance, all nuclei from polonium (Z = 84) onward in the periodic table are radioactive, and all the lighter elements have some unstable isotopes.

Radioactivity is a sign of **nuclear decay,** the partial breakup of a nucleus. The α and β particles originate in the nucleus, so they leave behind a nucleus with a number of protons different from that of the original atom. The product, which is called the **daughter nucleus,** is the nucleus of an atom of a different element. For example, when a radon-222 nucleus emits an α particle, a polonium-218 nucleus is formed. A **nuclear transmutation,** the conversion of one element into another, has taken place. Nuclear transmutation, particularly of lead into gold, was the dream of the alchemists, and the search for it was a root of modern chemistry. However, chemical means cannot change nuclei. Only nuclear processes, which were not recognized and developed until the twentieth century, can change one element into another.

The changes that nuclei undergo are called **nuclear reactions.** They differ in important ways from chemical reactions, which involve changes only in the valence-shell electrons:

1. Isotopes of one element show almost identical chemical properties, but they undergo different nuclear reactions.

2. A nuclear reaction typically results in the formation of an element not initially present, whereas a chemical reaction never does.

3. The energy changes are very much greater for nuclear reactions than for chemical reactions.

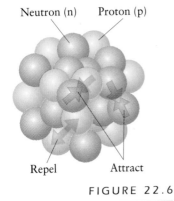

Neutron (n)   Proton (p)

Repel   Attract

FIGURE 22.6

A nucleus can be pictured as a collection of protons (pink) and neutrons (gray). The protons repel one another electrically, but a strong force that acts between all the particles holds the nucleus together.

The combustion of 1 g of methane produces about 50 kJ of energy as heat. In contrast, a nuclear reaction of 1 g of uranium-235 produces about 8×10^7 kJ! Obviously, the "test tubes" used for nuclear reactions must be far more sturdy, complex, and expensive than those used for chemical reactions.

To predict the identity of a daughter nucleus produced in a nuclear reaction, we note how the atomic number and the mass number change when a particle is ejected from the parent nucleus. When an α particle (Z = 2, A = 4) is ejected from a nucleus, it carries away four nucleons, two of which are protons and two neutrons. The loss of two protons reduces the atomic number of the element by 2. The loss of four nucleons reduces the mass number by 4. For example, when a radium-226 nucleus, with Z = 88, undergoes α decay, the fragment remaining is a nucleus of atomic number 86 (radon) and mass number 222, so the daughter nucleus is radon-222:

$$^{226}_{88}Ra \longrightarrow \, ^{222}_{86}Rn + \, ^4_2\alpha$$

This expression is an example of a **nuclear equation,** a summary of the changes that occur in a nuclear reaction.

Table 22.1 Nuclear radiation

Type	Degree of penetration	Speed*	Particle†	Example
α	not penetrating but damaging	10% of c	helium-4 nucleus $^4_2He^{2+}$, $^4_2\alpha$, α	$^{226}_{86}Ra \longrightarrow {}^{222}_{86}Rn + {}^4_2\alpha$
β	moderately penetrating	less than 90% of c	electron $^0_{-1}e$, β^-, β	$^3_1H \longrightarrow {}^3_2He + {}^0_{-1}e$
γ	very penetrating; often accompanies other radiation	c	photon	$^{60}_{27}Co^{*‡} \longrightarrow {}^{60}_{27}Co + \gamma$
β^+	moderately penetrating	less than 90% of c	positron $^0_{+1}e$, β^+	$^{22}_{11}Na \longrightarrow {}^{22}_{10}Ne + {}^0_{+1}e$
p	moderate/low penetration	10% of c	proton $^1_1H^+$, ^1_1p, p	$^{53}_{27}Co \longrightarrow {}^{52}_{26}Fe + {}^1_1p$
n	very penetrating	less than 10% of c	neutron ^1_0n, n	$^{137}_{53}I \longrightarrow {}^{136}_{53}I + {}^1_0n$

*c is the speed of light.
†Alternative symbols are given for the particles; often it is sufficient to use the simplest (the one on the right).
‡An energetically excited state of a nucleus is usually denoted by an asterisk.

In addition to α, β, and γ radiation, other types of nuclear radiation have since been identified; their properties are summarized in Table 22.1, and some of them are illustrated in Figs. 22.7–22.10. In the process called **electron capture,** a nucleus captures one of the surrounding electrons; a proton is turned into a neutron and, although there is no change in mass number, the atomic number is reduced by 1. An example is the nuclear reaction

$$^{44}_{22}Ti + {}^0_{-1}e \longrightarrow {}^{44}_{21}Sc$$

The subatomic particle called a **positron** is denoted $^0_{+1}e$, or more simply β^+; it has the same tiny mass as an electron but a single positive charge. Positron emission can be thought of as the positive charge being shrugged off by a proton as it is converted into a neutron. As a result, the atomic number decreases by 1, but there is no change in mass number. The resulting change in the nucleus is the same as that for electron capture:

$$^{43}_{22}Ti \longrightarrow {}^{43}_{21}Sc + {}^0_{+1}e$$

Proton emission and neutron emission are less common and occur only in special cases. Loss of a proton decreases both mass number and atomic number by 1. Loss of a neutron decreases only the mass number by 1:

$$^{57}_{30}Zn \longrightarrow {}^{56}_{29}Cu + {}^1_1p$$
$$^{91}_{34}Se \longrightarrow {}^{90}_{34}Se + {}^1_0n$$

The general procedure for predicting the identity of daughter nuclei is described in Toolbox 22.1 and illustrated in Examples 22.1 and 22.2.

Nuclear reactions involve large energy changes, are different for different isotopes, and may result in the formation of different elements. The transmutation of a nucleus can be predicted by balancing the atomic numbers and the mass numbers in the nuclear equation for the process.

Toolbox 22.1 How to identify the products of a nuclear reaction

This Toolbox shows how to predict the products of a nuclear reaction by balancing the nuclear equation for the reaction.

Conceptual Basis

The number of nucleons and the total electric charge are conserved in the simple nuclear reactions that we consider. If the mass numbers and atomic numbers of the particles emitted from a nucleus are known, the product nucleus can be identified by balancing mass number and atomic number in the nuclear equation.

Procedure

Step 1. Identify the type of decay and write the nuclear equation, with the mass number and the atomic number of the daughter nucleus written as A and Z, respectively.

Step 2. Find the values of A and Z for the daughter nucleus from the requirement that the sums of the mass numbers and the sums of the nuclear charges remain unchanged in the decay.

Step 3. Once A and Z are known, identify the daughter nucleus by referring to the periodic table.

The masses and charges of the most common particles ejected or captured by nuclei are as follows:

Process	Mass number of particle	Charge	Example	Illustration
α decay	4	+2	$^{211}_{84}\text{Po} \longrightarrow {}^{207}_{82}\text{Pb} + {}^{4}_{2}\alpha$	Fig. 22.7
β decay	0	−1	$^{24}_{11}\text{Na} \longrightarrow {}^{24}_{12}\text{Mg} + {}^{0}_{-1}\text{e}$	Fig. 22.8
electron capture	0	−1	$^{44}_{22}\text{Ti} + {}^{0}_{-1}\text{e} \longrightarrow {}^{44}_{21}\text{Sc}$	Fig. 22.9
positron emission	0	+1	$^{43}_{22}\text{Ti} \longrightarrow {}^{43}_{21}\text{Sc} + {}^{0}_{+1}\text{e}$	Fig. 22.10
proton emission	1	+1	$^{57}_{30}\text{Zn} \longrightarrow {}^{56}_{29}\text{Cu} + {}^{1}_{1}\text{p}$	
neutron emission	1	0	$^{91}_{34}\text{Se} \longrightarrow {}^{90}_{34}\text{Se} + {}^{1}_{0}\text{n}$	

More information on these particles is given in Table 22.1.

Example 22.1 Writing a nuclear equation I

Write the nuclear equation for (a) the positron decay of oxygen-15 and (b) the α decay of thorium-232.

Strategy Use the rule given in Toolbox 22.1—that the total number of nucleons (mass number) and total charge must be the same on both sides of the equation—to determine the identity of the unknown isotope. (a) A positron has a positive charge but a very tiny mass. Loss of a positron decreases the atomic number by 1 but leaves the mass number unchanged. (b) An α particle has a charge of $+2$ and a mass number of 4, so emission of an α particle decreases the atomic number by 2 and the mass number by 4.

Solution (a) Write the equation with $^{A}_{Z}\text{E}$ representing the daughter nucleus:

$$^{15}_{8}\text{O} \longrightarrow {}^{A}_{Z}\text{E} + {}^{0}_{+1}\text{e}$$

Because mass number and charge must be the same on each side of the equation, we find the mass number of the daughter nucleus from $15 = A + 0$, so $A = 15$. We find the nuclear charge (atomic number) from $8 = 1 + Z$, so $Z = 7$ and the daughter nucleus is $^{15}_{7}\text{N}$. The balanced nuclear equation is

$$^{15}_{8}\text{O} \longrightarrow {}^{15}_{7}\text{N} + {}^{0}_{+1}\text{e}$$

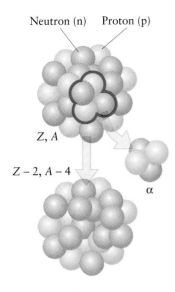

Neutron (n)   Proton (p)

Z, A

Z – 2, A – 4

α

FIGURE 22.7

When a nucleus ejects an α particle, the atomic number of the atom decreases by 2 and the mass number decreases by 4. The nucleons ejected from the upper nucleus have been indicated by the blue boundary.

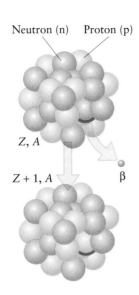

Neutron (n)   Proton (p)

Z, A

Z + 1, A

β

FIGURE 22.8

When a nucleus ejects a β particle, the atomic number of the atom increases by 1 and the mass number remains unchanged. The neutron that we can regard as the source of the electron is indicated by the blue boundary in the upper part of the diagram.

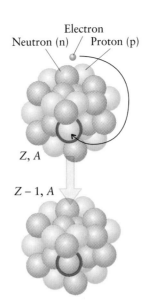

Electron
Neutron (n)   Proton (p)

Z, A

Z – 1, A

FIGURE 22.9

In electron capture, a nucleus captures one of the surrounding electrons. The effect is to convert a proton (outlined in blue) into a neutron. As a result, the atomic number decreases by 1 but the mass number remains the same.

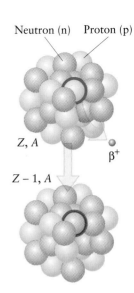

Neutron (n)   Proton (p)

Z, A

Z – 1, A

β+

FIGURE 22.10

In positron (β+) emission, the nucleus ejects a positron. The effect is to convert a proton (outlined in blue) into a neutron. As a result, the atomic number decreases by 1, but the mass number remains the same because the number of nucleons is unchanged.

(b) Write the equation as described in (a):

$$^{232}_{90}\text{Th} \longrightarrow {}^{A}_{Z}\text{E} + {}^{4}_{2}\alpha$$

To balance mass, we find from $232 = A + 4$ that $A = 228$. To balance charge, Z must equal $90 - 2 = 88$, so the daughter nucleus is $^{228}_{88}\text{Ra}$. The balanced nuclear equation is

$$^{232}_{90}\text{Th} \longrightarrow {}^{228}_{88}\text{Ra} + {}^{4}_{2}\alpha$$

Self-Test 22.1A Write the nuclear equation for (a) the α decay of plutonium-242; (b) positron emission by sodium-22.

[*Answer:* (a) $^{242}_{94}\text{Pu} \longrightarrow {}^{238}_{92}\text{U} + {}^{4}_{2}\alpha$; (b) $^{22}_{11}\text{Na} \longrightarrow {}^{22}_{10}\text{Ne} + {}^{0}_{+1}\text{e}$]

Self-Test 22.1B Write the nuclear equation for (a) α decay of uranium-235; (b) positron emission by carbon-11.

Example 22.2 *Writing a nuclear equation II*

Write the nuclear equation for (a) the decay by electron capture of calcium-41; (b) the β decay of lithium-9.

Strategy Use the rule given in Toolbox 22.1—that the total number of nucleons (mass number) and total charge must be the same on both sides of the equation—to determine the identity of the unknown isotope required to balance the nuclear equation. (a) The electron is captured from the electrons surrounding the nucleus. Because it has a negative charge but zero mass number, the electron decreases the atomic number by 1 but leaves the mass number unchanged. (b) A β particle has a charge of −1 and zero mass number, so emission of a β particle increases the atomic number by 1 but leaves the mass number unchanged.

Solution (a) Write the equation with $^A_Z E$ representing the daughter nucleus:

$$^{41}_{20}Ca + {}^{0}_{-1}e \longrightarrow {}^A_Z E$$

Because mass number and charge must be the same on both sides of the equation, we find the mass number of the daughter nucleus from $A = 41 + 0 = 41$; so $A = 41$. We find the atomic number from $Z = 20 - 1 = 19$; so $Z = 19$, and the daughter nucleus is $^{41}_{19}K$. The balanced nuclear equation is

$$^{41}_{20}Ca + {}^{0}_{-1}e \longrightarrow {}^{41}_{19}K$$

(b) Write the equation as described in (a):

$$^{9}_{3}Li \longrightarrow {}^A_Z E + {}^{0}_{-1}e$$

To balance mass, we find from $9 = A + 0$ that $A = 9$. To balance charge, Z must equal $3 - (-1) = 4$, so the daughter nucleus is $^{9}_{4}Be$. The balanced nuclear equation is

$$^{9}_{3}Li \longrightarrow {}^{9}_{4}Be + {}^{0}_{-1}e$$

Self-Test 22.2A Write the nuclear equation for (a) β decay of indium-115; (b) electron capture by astatine-210.

[*Answer:* (a) $^{115}_{49}In \rightarrow {}^{115}_{50}Sn + {}^{0}_{-1}e$; (b) $^{210}_{85}At + {}^{0}_{-1}e \rightarrow {}^{210}_{84}Po$]

Self-Test 22.2B Write the nuclear equation for (a) electron capture by beryllium-7; (b) β decay of radium-228.

22.3 The Pattern of Nuclear Stability

Which nuclides are stable and which are not? Nuclei with an even number of protons and an even number of neutrons are more stable than those with any other combination of nucleons. Conversely, nuclei with odd numbers of both protons and neutrons are the least stable (Fig. 22.11). Nuclei are likely to be more stable if they are built from certain **magic numbers** of either kind of nucleons, namely, 2, 8, 20, 50, 82, and 126. For example, there are 10 stable isotopes of tin ($Z = 50$), the most of any element, but only two stable isotopes of its neighbor antimony ($Z = 51$). The α particle itself is a "doubly magic" nucleus, with two protons and two neutrons. We shall see later that a number of actinides decay through a series of steps until they reach ^{208}Pb, another doubly magic nuclide, with 126 neutrons and 82 protons. The magic number 126 suggests that it may be possible to synthesize an element with $Z = 126$. This element would be more stable than the known transuranium elements.

A similar pattern of stability applies to electrons in atoms, for we have already seen that the noble gas atoms have 2, 10, 18, 36, 54, and 86 electrons. This pattern strongly suggests that a nucleus also has a shell structure in which nucleons occupy a series of shells, just as electrons in atoms do.

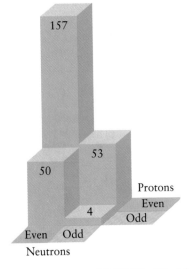

FIGURE 22.11

The numbers of stable nuclides for even and odd numbers of neutrons and protons. By far the greatest number of stable nuclides have even numbers of protons and even numbers of neutrons.

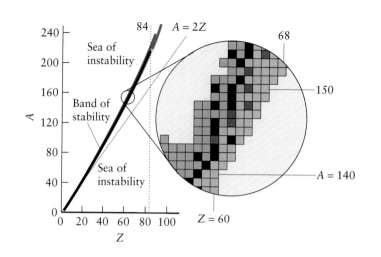

FIGURE 22.12

The manner in which nuclear stability depends on the atomic number and the mass number. Nuclides along the narrow black band (the band of stability) are generally stable. Nuclides in the blue region are likely to emit a β particle, and those in the red region are likely to emit an α particle. Nuclei in the pink region are likely to emit either positrons or to undergo electron capture. The magnified view of the diagram near $Z = 60$ shows the structure of the band of stability.

Figure 22.12 is a plot of mass number against atomic number for known nuclides. Stable nuclei are found in a **band of stability** surrounded by a **sea of instability,** the region of unstable nuclides that decay with the emission of radiation. For atomic numbers up to about 20, the stable nuclides have approximately equal numbers of neutrons and protons, so A is close to $2Z$. For higher atomic numbers, all known nuclides—both stable and unstable—have more neutrons than protons (so $A > 2Z$). We can explain this behavior in terms of charge. Protons have mutually repulsive electrostatic charges, whereas neutrons have none. Therefore, neutrons can contribute to the **strong force,** the short-range force that holds the nucleus together, without increasing repulsion. In a nucleus with many protons, a lot of neutrons are needed to overcome the mutual electric repulsion of the protons and help glue them together.

Figure 22.12 can be used to predict the type of disintegration a nuclide is likely to undergo. Unstable isotopes above the band of stability are **neutron-rich nuclei:** they have a high proportion of neutrons. These nuclei, such as $^{14}_{6}C$, can reach stability by ejecting a β particle:

$$^{14}_{6}C \longrightarrow \, ^{14}_{7}N + \, ^{0}_{-1}e$$

Unstable isotopes of elements that lie below the band of stability have a low proportion of neutrons and are classified as **proton-rich nuclei.** For example, $^{29}_{15}P$ is proton rich and can move toward the band of stability by emitting a positron:

$$^{29}_{15}P \longrightarrow \, ^{29}_{14}Si + \, ^{0}_{+1}e$$

Alternatively, proton-rich nuclides can decrease their proton count by proton emission:

$$^{43}_{21}Sc \longrightarrow \, ^{42}_{20}Ca + \, ^{1}_{1}p + \gamma$$

or by electron capture:

$$^{7}_{4}Be + \, ^{0}_{-1}e \longrightarrow \, ^{7}_{3}Li$$

All nuclides with $Z > 83$ are unstable and radioactive; they decay mainly by α emission. Very few nuclides with $Z < 60$ emit α particles. However, to achieve stability, nuclides of elements with $Z > 83$ discard protons to reduce their atomic number, and they generally need to lose neutrons too. They decay in a stepwise manner and give rise to a **radioactive series,** a specific sequence of

FIGURE 22.13

The uranium-238 decay series. The times are the half-lives of the nuclides (see Section 22.7).

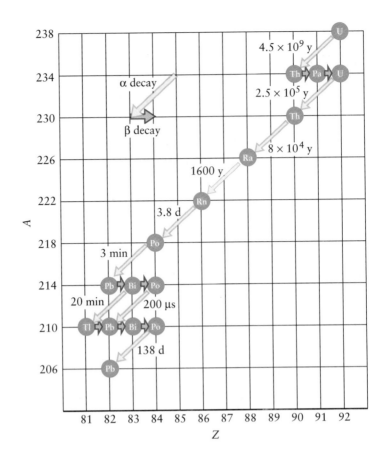

nuclides (Fig. 22.13). First, one α particle is ejected, then another α particle or a β particle is ejected, and so on, until a stable nuclide is reached—usually the final nuclide is an isotope of lead (the element with magic atomic number 82). Three radioactive series begin from naturally occurring nuclides:

Uranium-238 series: starts at uranium-238 and ends at lead-206

Uranium-235 series: starts at uranium-235 and ends at lead-207

Thorium-232 series: starts at thorium-232 and ends at lead-208

Radioactive series were important when radioactivity was first studied, because radioactive materials could be obtained only from the decay of heavier elements. They still help to summarize the behavior of nuclear fuels.

The pattern of nuclear stability can be used to predict the likely mode of radioactive decay: neutron-rich nuclei tend to reduce their neutron count; proton-rich nuclei tend to reduce their proton count. In general, only heavy nuclides emit α particles.

Self-Test 22.3A Which of the processes—(a) electron capture; (b) proton emission; (c) β emission; (d) β^+ emission—might a $^{145}_{64}$Gd nucleus undergo to begin to reach stability? Refer to Fig. 22.12.

[*Answer:* a, b, d]

Self-Test 22.3B Which of the same set of processes listed in Self-Test 22.3A might a $^{148}_{58}$Ce nucleus undergo to begin to reach stability?

22.4 Nucleosynthesis

Nucleosynthesis is the formation of elements. Hydrogen and helium have been present since the beginning of the universe. Nuclear reactions in the interior of stars have produced all the other naturally occurring elements on Earth: in this sense, virtually everything around us is made of stardust. Some elements are not found naturally on Earth. For example, the radioactive elements technetium and promethium were discovered in the spectra of stars; but on Earth, they are found only in minute quantities in uranium ores. No element with an atomic number greater than 92 occurs naturally. To make these elements on Earth, we must overcome the energy barriers to nuclear synthesis by simulating the very high energy conditions found inside a star (Fig. 22.14). There are two ways to do this. One is to heat a substance to the very high temperatures (millions of degrees) found in the interiors of stars to increase the speed of the particles. Another is to bombard nuclei with elementary particles or other nuclei that have been accelerated to high speeds in a particle accelerator (Fig. 22.15). High speed is essential if the projectile particles are positively charged, because, to approach the target nucleus closely, they must overcome its electrostatic repulsion.

Rutherford achieved the first artificial nuclear transmutation in 1919. He bombarded nitrogen-14 nuclei with α particles. These particles travel at high speed because their positive charge receives a hefty "kick" from the positive charge of the nucleus that ejects them. The products of the transmutation are oxygen-17 and a proton:

$$^{14}_{7}\text{N} + {}^{4}_{2}\alpha \longrightarrow {}^{17}_{8}\text{O} + {}^{1}_{1}\text{p}$$

A similar process occurs in stars and results in the formation of oxygen-16 from carbon-12:

$$^{12}_{6}\text{C} + {}^{4}_{2}\alpha \longrightarrow {}^{16}_{8}\text{O} + \gamma$$

A large number of nuclides have been synthesized on Earth by using a similar procedure. For instance, technetium ($Z = 43$) was prepared for the first time

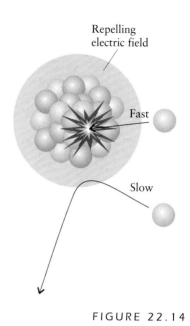

Repelling electric field

Fast

Slow

FIGURE 22.14

When a positively charged particle approaches a nucleus, it is repelled strongly. However, if it is traveling very fast, it can reach the nucleus before the repulsion turns it aside and a nuclear reaction may occur.

Charged particles can be accelerated to high energies by a cyclotron, a synchrocyclotron, or a linear accelerator. Most of the transuranium elements were originally produced in laboratories in California, Germany, and Russia.

FIGURE 22.15

An aerial view of the Fermi National Accelerator Laboratory in Batavia, Illinois. The largest circle is the main accelerator; three experimental lines are tangential to it.

on Earth in 1937 by the reaction between molybdenum ($Z = 42$) and **deuterons** (deuterium nuclei):

$$^{97}_{42}\text{Mo} + {}^{2}_{1}\text{H} \longrightarrow {}^{97}_{43}\text{Tc} + 2\,{}^{1}_{0}\text{n}$$

Moderately abundant supplies of technetium are now available; technetium-99 is used in medical applications, particularly for imaging the heart.

It is easier for a neutron to get close to a target nucleus because it is not repelled by the nuclear charge. An example of neutron-induced transmutation is the three-step formation of cobalt-60, which is used in the radiation treatment of cancer, from iron-58. First, iron-59 is produced from iron-58:

$$^{58}_{26}\text{Fe} + {}^{1}_{0}\text{n} \longrightarrow {}^{59}_{26}\text{Fe}$$

The second step is β decay of the iron-59 to cobalt-59:

$$^{59}_{26}\text{Fe} \longrightarrow {}^{59}_{27}\text{Co} + {}^{0}_{-1}\text{e}$$

In the final step, the cobalt-59 absorbs another neutron from the incident neutron beam and is converted into cobalt-60:

$$^{59}_{27}\text{Co} + {}^{1}_{0}\text{n} \longrightarrow {}^{60}_{27}\text{Co}$$

The overall reaction is

$$^{58}_{26}\text{Fe} + 2\,{}^{1}_{0}\text{n} \longrightarrow {}^{60}_{27}\text{Co} + {}^{0}_{-1}\text{e}$$

Self-Test 22.4A Complete the following nuclear reactions: (a) ? $+ {}^{4}_{2}\alpha \rightarrow$ $^{243}_{96}\text{Cm} + {}^{1}_{0}\text{n}$; (b) $^{242}_{96}\text{Cm} + {}^{4}_{2}\alpha \rightarrow {}^{245}_{98}\text{Cf} + $?.

[*Answer:* (a) $^{240}_{94}\text{Pu}$; (b) $^{1}_{0}\text{n}$]

Self-Test 22.4B Complete the following nuclear reactions: (a) $^{250}_{98}\text{Cf} + $? $\rightarrow$ $^{257}_{103}\text{Lr} + 4\,{}^{1}_{0}\text{n}$; (b) ? $+ {}^{12}_{6}\text{C} \rightarrow {}^{254}_{102}\text{No} + 4\,{}^{1}_{0}\text{n}$.

The **transuranium elements** are the elements following uranium in the periodic table; they have $Z > 92$. All are synthetic and are produced by the bombardment of target nuclei with a smaller projectile. The elements through meitnerium (Mt, $Z = 109$) have been formally named. The elements beyond meitnerium (including hypothetical nuclides that have not yet been made) are named systematically, at least until they have been identified and there is agreement on a permanent name. The systematic nomenclature of these elements uses the terms in Table 22.2. For example, the element with $Z = 111$, one atom of which was first made in 1994, will be called unununium, Uuu, until it is finally named by international agreement.

Meitnerium is named after Lise Meitner, the Austrian-born physicist who discovered protactinium and whose studies laid the foundation for our use of nuclear fission (see Section 22.9).

New elements and isotopes of known elements are made by nucleosynthesis; the repulsive electrical forces of like-charged particles are overcome when very fast particles collide.

RADIOACTIVITY

Nuclear radiation is sometimes called **ionizing radiation,** because it is energetic enough to eject electrons from atoms. These high-energy rays can be used—with care—to cure disease, but they can also harm biological tissues (Applying Chemistry: Case Study 22). The extent of damage depends on the strength of the source, the type of radiation, and the length of exposure. Another important aspect of radioactivity is its persistence: how long do we need to store radioactive material before it can be considered safe?

Table 22.3 *Shielding requirements of α, β, and γ radiation*		
Radiation	**Relative penetrating power**	**Shielding required**
α	1	paper, skin
β	100	3 mm aluminum
γ	10 000	concrete, lead

22.5 The Effects of Radiation

The three main types of nuclear radiation penetrate matter to different extents (Table 22.3). The least penetrating is α radiation, which can be stopped by a sheet of paper. The massive, highly charged α particles interact so strongly with matter that they slow down, capture electrons from surrounding matter, and change into helium atoms before traveling very far. However, even though α particles do not penetrate far into matter, they are very damaging because the energy of their impact can knock atoms out of molecules and displace ions from their sites in crystals. The damage caused by α particles when they strike human tissue can lead to serious illness and even death. If DNA and the enzymes that interpret its protein-building messages are damaged, then the result may be cancer. Most α radiation is absorbed by the surface layer of dead skin, where it can do little harm. However, inhaled and ingested particles can cause serious internal damage. For example, plutonium, considered to be one of the most toxic radioactive materials, is an α emitter and can be handled safely with minimal shielding (Fig. 22.16). However, it is easily oxidized to plutonium(IV), which has chemical properties similar to those of iron(III). It can take the place of iron in the body and be absorbed rather than excreted, remaining to cause radiation sickness and cancer.

Next in penetrating power is β radiation. The fast electrons that make up these rays can penetrate about 1 cm deep into flesh before their electrostatic interactions with the electrons and nuclei of molecules bring them to a standstill.

FIGURE 22.16

Even relatively large pieces of plutonium-239 can be handled with minimum protection, as long as they are isolated from the atmosphere. Here, a glove box is used to provide a sealed environment. Plutonium is warm to the touch because of the constant α decay, and larger pieces can be too hot to touch.

Applying Chemistry: *Case Study 22*

Radioisotopes are widely used in medicine to diagnose, study, and treat illness. For example, a physician can determine how and at what rate the thyroid gland takes up iodine by using iodine-131 as a radioactive tracer; cobalt-60 is used to kill rapidly growing cancer cells. Because the same powerful energies that diagnose illness and cure it are also capable of damaging healthy tissue, a great deal of attention has been paid to reducing the risks associated with the medical use of radioisotopes.

Nuclear chemistry has transformed the field of medical diagnosis. For example, radioactive tracers are used to measure organ function, sodium-24 is used to monitor blood flow, and strontium-87 is used to study bone growth. However, the most dramatic impact of radioisotopes on diagnosis has been in the field of imaging. The better the picture of an organ, the better the diagnosis a physician can make. Technetium-99m is a metastable isotope (the m denotes a "metastable" state, an excited state with an appreciable lifetime) that is the most widely used radioactive nuclide in medicine for studying the functioning of internal organs, including the heart; for instance, as $^{99}TcO_4^-$, it is used to detect and pinpoint brain tumors. Technetium-99m gives off a γ ray and decays to technetium-99, with a half-life of 6 h. Thus, it is highly active but has a very short half-life. The fact that the isotope emits only γ rays adds to its desirability for imaging. The heavier α rays cause more damage in the body and are absorbed by body tissues before they can be detected by instruments.

Positron emission tomography (PET) is a technique that uses a positron emitter such as fluorine-18 to image human tissue in a degree of detail not possible with x-rays. For example, the hormone estrogen can be labeled with fluorine-18 and injected into an individual with a tumor. The fluorine-bearing compound is preferentially absorbed by the tumor. The positrons given off by the fluorine atoms are quickly annihilated, and the γ rays resulting from the annihilations are detected by a scanner that moves slowly over the part of the body with the tumor. The growth of the tumor can be quickly and accurately estimated with this technique. An advantage of this technique is that the positron emitters have very short half-lives; for example, the half-life of fluorine-18 is 110 min. In fact, a PET imaging facility must be located near a cyclotron, so that the positron emitters can be quickly incorporated into the desired compounds as soon as they are created.

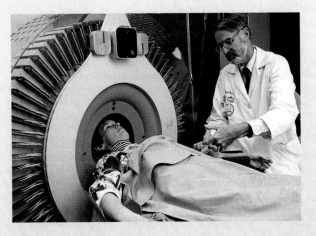

This patient is about to undergo a PET scan of brain function.

Most penetrating of all is γ radiation. The uncharged, high-energy γ ray photons can pass right through the body and most building materials, and can cause damage by ionizing the molecules in their path. Protein molecules and DNA that have been damaged in this way cannot carry out their functions, and the result can be radiation sickness and cancer. Intense sources of γ rays must be surrounded by walls built from lead bricks or thick concrete to shield people from this penetrating radiation.

The **absorbed dose** of radiation is the energy deposited in the sample (in particular, the human body) when it is exposed to radiation. The SI unit of absorbed dose is the **gray** (Gy), which corresponds to an energy deposit of 1 J/kg. If the radiation were absorbed uniformly throughout the body, a dose of 1 Gy would correspond to a 65-kg person absorbing a total of 6.5 mJ. The same energy is transferred as heat when a droplet of boiling water of mass 2 mg touches the skin. However, the energy of each radioactive particle is highly

For a long time, the unit for reporting dose has been the *rad*, which stands for radiation absorbed dose, and is the amount of radiation that deposits 10^{-2} J of energy per kilogram of tissue; so 1 Gy = 100 rad.

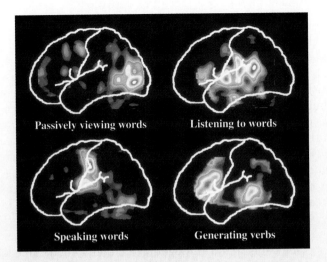

These four PET scans show how blood flow to different parts of the brain is affected by various mental activities. In this case, an oxygen isotope that is taken up by the hemoglobin in blood is used as a source of positrons.

The radioisotopes used to cure cancer are associated with much greater risks than those associated with imaging techniques, because the therapeutic radiation must be fairly strong to destroy the tumor. However, two relatively new techniques can considerably reduce the risk of radiation treatment. The first is the use of *monoclonal antibodies*. Antibodies are compounds that the body produces to fight disease. Monoclonal antibodies are artificial antibodies that biotechnologists create to attack specific types of cancer. When radioisotopes are incorporated into a mono-clonal antibody that is then injected into the body, the antibody attaches to a malignant tumor and the radiation is concentrated in the cancerous tissue. Iodine-131 ($t_{1/2}$ = 8.05 d) and yttrium-90 ($t_{1/2}$ = 64 h) are frequently used in monoclonal antibodies. Their short half-lives ensure that the activity in the body will not last long.

Boron neutron capture therapy is unusual in that boron-10, the isotope injected, is not radioactive. However, when it is bombarded with neutrons, boron-10 gives off highly destructive α particles. In boron neutron capture therapy, the boron-10 is incorporated into a compound that is preferentially absorbed by tumors. The patient is then exposed to neutron bombardment, but only for brief periods of time. As soon as the bombardment ceases, the boron-10 stops generating α particles.

It is unfortunate that Marie Curie was not able to benefit from the results of her work. She died of leukemia induced by her exposure to radiation before the dangers were known; today leukemia can be treated with radioactive isotopes.

Key Concepts: radioisotopes, nuclear stability, half-life

For Further Reading

R. L. Rawls, Bringing boron to bear on cancer, *Chemical and Engineering News*, March 22, 1999, 26–29.

S. E. Peterson, P. T. Fox, M. I. Posner, M. A. Mintum, and M. E. Raichle, *Nature* **331**:585–589, 1988.

Related Exercises: 22.79–22.82

localized, like the impact of a bullet, but on a much smaller scale, and not spread evenly over a region the size of a water droplet. As a result, the incoming particles can break individual bonds as they collide with molecules in their path.

The extent of radiation damage to living tissue depends on the type of radiation and the type of tissue. A dose of γ radiation causes about the same amount of damage as the same dose of β radiation, but α particles are about 20 times more damaging (even though they are the least penetrating). A factor called the **relative biological effectiveness**, *Q*, must therefore be included when assessing the damage that a given dose of each type of radiation may cause. For β and γ radiation, *Q* is set arbitrarily at about 1; but for α radiation, *Q* is close to 20. The precise figures depend on the total dose, the rate at which the dose accumulates, and the type of tissue, but these values are typical.

The **dose equivalent** is the actual dose modified to take into account the different destructive powers of the various types of radiation in combination with various types of tissues. It is obtained by multiplying the actual dose (in gray) by the value of Q for the radiation type. The SI unit of dose equivalent is the **sievert** (Sv):

$$\text{Dose equivalent (Sv)} = Q \times \text{absorbed dose (Gy)}$$

A dose of 0.3 Gy (30 rad) of γ radiation corresponds to a dose equivalent of 0.3 Sv (30 rem), enough to cause a reduction in the number of white blood cells (the cells that fight infection); but 0.3 Gy (30 rad) of α radiation corresponds to 6 Sv (600 rem), which is enough to cause death. A typical average annual dose equivalent that we each receive from natural sources is about 2 mSv/y (where 1 mSv $= 10^{-3}$ Sv), but this figure varies, depending on our lifestyle and where we live. About 20% is radiation from our own bodies. About 30% comes from cosmic rays (a mix of γ rays and high-energy elementary particles from outer space) that continuously bombard the Earth, and 40% comes from radon seeping out of the ground. The remaining 10% is a result largely of medical diagnoses (a typical chest x-ray gives a dose equivalent of about 70 μSv). Emissions from nuclear power plants and other nuclear facilities contribute about 0.1% in countries where they are widely used. The principal natural source of radioactivity in the human body is potassium-40, which occurs in about 0.01% abundance among the potassium ions found throughout the body. About 37 000 potassium-40 nuclei disintegrated in your body in the time it took you to read this sentence (about 10 s).

We are not in danger from this constant bombardment from **background radiation,** the radiation to which we are exposed daily and that has always been present on Earth. It is only when we are exposed to large amounts of radiation that the body's defense mechanisms are in danger of being overwhelmed.

Human health in the presence of radiation is monitored by reporting the absorbed dose and the dose equivalent; the latter takes into account the effects of different types of radiation on tissues.

22.6 Measuring Radioactivity

The ability of nuclear radiation to ionize atoms can be used to measure its intensity. Becquerel first used this ionization to gauge the intensity of radiation by the degree to which it blackened a photographic film. The blackening results from the same redox processes as those occurring in ordinary photography (see Applying Chemistry: Case Study 21), except that the initial oxidation of the halide ions is caused by nuclear radiation instead of light. Becquerel's technique is still used in the film badges that monitor the exposure of workers to radiation (Fig. 22.17).

A **Geiger counter** monitors radiation by detecting the ionization of a low-pressure gas (Fig. 22.18). The radiation ionizes atoms inside a hollow cylinder and allows a brief flow of current between the electrodes. The resulting electrical signal can be recorded directly or converted into an audible click. The frequency of the clicks then indicates the intensity of the radiation. A **scintillation counter** makes use of the fact that certain substances (notably, zinc sulfide) give a flash of light—a scintillation—when exposed to radiation. The intensity of the radiation is measured by counting the scintillations electronically (Fig. 22.19).

Each click of a Geiger counter or flash of a scintillation counter indicates that one nuclear disintegration has occurred. The **activity** of a sample is the number of nuclear disintegrations occurring per second. The more active the

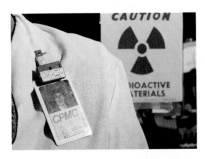

FIGURE 22.17

A film badge used by workers in areas where exposure to radiation is a risk.

(a)

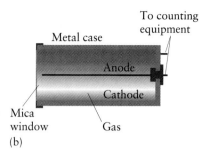

Metal case
To counting equipment
Anode
Cathode
Mica window
Gas
(b)

FIGURE 22.18

(a) A Geiger counter displaying the level of radioactivity detected in a piece of uranium ore.
(b) The detector in a Geiger counter consists of a gas (typically argon and a little ethanol vapor, or neon and some bromine vapor) in a container with a high potential difference (of 500 to 1200 V) between the walls and the central wire. An electric current passes between the electrodes (walls and wire) when radiation ionizes the gas.

FIGURE 22.19

This scintillation counter detects each radioactive decay as a flash of light. The total number of flashes per second is recorded electronically.

source, the greater the number of nuclear disintegrations per second. The SI unit of activity is the **becquerel** (Bq): 1 Bq = 1 nuclear disintegration per second. Another commonly used (non-SI) unit of radioactivity is the **curie** (Ci). It is equal to 3.7×10^{10} nuclear disintegrations per second, the radioactive output of 1 g of radium-226. Because the curie is a very large unit, most activities are expressed in millicuries (1 mCi = 10^{-3} Ci) and microcuries (1 μCi = 10^{-6} Ci). These units and their conversions are summarized in Table 22.4.

Radioactivity is measured by using its ionizing effects; the activity of a source is a measure of the number of nuclear disintegrations per second.

Table 22.4 Radiation units*

Property	Unit name	Symbol	Definition
activity	curie	Ci	3.7×10^{10} disintegrations per second
	becquerel	Bq	1 disintegration per second
			(1 Ci = 3.7×10^{10} Bq)
absorbed dose	radiation absorbed dose	rad	10^{-2} J/kg
	gray	Gy	1 J/kg
			(1 Gy = 100 rad)
dose equivalent	roentgen equivalent man	rem	$Q \times$ absorbed dose[†]
	sievert	Sv	100 rem

*The SI units are becquerel, gray, and sievert.
[†]Q is the relative biological effectiveness of the radiation. Normally, $Q \approx 1$ for γ, β, and most other radiation, but $Q \approx 20$ for α radiation and fast neutrons. A further factor of 5 (that is, 5Q) is used for bone under certain circumstances.

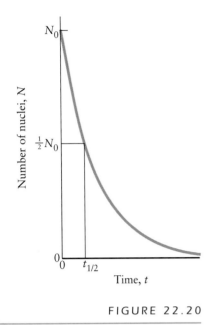

Number of nuclei, N

N_0

$\frac{1}{2}N_0$

$t_{1/2}$

Time, t

FIGURE 22.20

The exponential decay of the number of radioactive nuclei in a sample implies that the activity of the sample also decays exponentially with time. The curve is characterized by the half-life, $t_{1/2}$.

22.7 The Law of Radioactive Decay

We can use Geiger counters and scintillation counters to study the rate at which the nuclei of a radioactive isotope, also called a **radioisotope,** decays. The nuclear equation

$$\text{Parent nucleus} \longrightarrow \text{daughter nucleus} + \text{radiation}$$

has exactly the same form as the equation for a unimolecular elementary reaction, with an unstable nucleus taking the place of an excited molecule. As in the case of a unimolecular chemical reaction (Sections 13.12 and 13.13), the rate law for nuclear decay is first order. That is, the relation between the rate of decay and the number N of radioactive nuclei present is given by the **law of radioactive decay:**

$$\text{Activity} = \text{rate of decay} = k \times N \qquad (1)$$

In this context, k is called the **decay constant.** The law tells us that the more radioactive nuclei there are in the sample, the greater the rate of nuclear decay and hence the more active the sample. Each nuclide has a characteristic value of k, so the rate of decay also depends on the identity of the nuclide.

As explained in Section 13.5, a first-order rate law implies an exponential decay. Therefore, the number of radioactive nuclei left in a sample and its activity fall exponentially toward 0 (Fig. 22.20). As for the concentration of the reactant in a first-order chemical reaction, it follows that the number, N, of nuclei remaining after a time t is given by

$$N = N_0 e^{-kt} \qquad (2)$$

where N_0 is the number of radioactive nuclei present initially (at $t = 0$).

> **Example 22.3** *Determining the amount of an isotope remaining after a specified time*
>
> One of the reasons why thermonuclear weapons have to be serviced is the radioactive decay of the tritium they contain. Suppose a sample of tritium of mass 1.0 g is stored. What mass of that isotope will remain after 5.0 y? The decay constant is 0.0564/y.
>
> *Strategy* The number of tritium nuclei in the sample is proportional to the total mass m of tritium, so Eq. 2 can be expressed as
>
> $$m = m_0 e^{-kt}$$
>
> where m_0 is the initial mass of tritium. The question can now be answered by substituting the data.
>
> *Solution* Substitution of the data gives
>
> $$m = (1.0 \text{ g}) \times e^{-(0.0564/y) \times (5.0 \text{ y})} = 0.75 \text{ g}$$
>
> *Self-Test 22.5A* The decay constant for fermium-254 is 210./s. What mass of the isotope will be present if a sample of mass 1.00 mg is kept for 10. ms?
>
> [*Answer:* 0.12 mg]
>
> *Self-Test 22.5B* The decay constant for neptunium-237 is 3.3×10^{-7}/y. What mass of the isotope will be present if a sample of mass 5.0 μg is stored for 1.0×10^6 y?

Half-lives were first introduced in Section 13.6 in connection with first-order chemical reactions.

Radioactive decay is normally discussed in terms of the **half-life,** $t_{1/2}$, the time needed for half the initial number of nuclei to disintegrate. Just as we did in Section 13.6, we find the relation between $t_{1/2}$ and k by setting $N = \frac{1}{2}N_0$ and $t = t_{1/2}$ in Eq. 2:

$$t_{1/2} = \frac{\ln 2}{k} \qquad (3)$$

The larger the value of k, the shorter the half-life of the nuclide. Nuclides with short half-lives are less stable than nuclei with long half-lives. That means they are more likely to decay in a given period of time and are thus "hotter" than nuclides with long half-lives.

Table 22.5 shows that half-lives span a very wide range. As an example, consider strontium-90, for which the half-life is 28 y. This nuclide occurs in nuclear **fallout,** the fine radioactive dust that settles from clouds of airborne particles after a nuclear bomb test or an accidental release of radioactive materials into the air. Because it is chemically very similar to calcium, strontium may accompany that element through the environment and become incorporated into bones; once there, it continues to emit radiation for many years. Even after three half-lives (84 y), one-eighth of the original strontium-90 still survives. About 10 half-lives (for strontium-90, 280 y) must pass before the activity of a sample has fallen to 1/1000 of its initial value. Plutonium-239, which is present in nuclear waste, has a half-life of 2.4×10^4 y. Consequently, very long term storage facilities are required for plutonium waste, and land that becomes contaminated with plutonium cannot be inhabited again for thousands of years. Serious contamination from radioactive fallout occurred within a 30-km radius of the nuclear reactor that caught fire in Chernobyl, Ukraine. The land is contaminated with plutonium and with cesium-137 and has been sealed off indefinitely from human use (Fig. 22.21).

Table 22.5 Radioactive half-lives*

Nuclide	Half-life, $t_{1/2}$
tritium	12.3 y
carbon-14	5.73×10^3 y
carbon-15	2.4 s
potassium-40	1.26×10^9 y
cobalt-60	5.26 y
strontium-90	28.1 y
iodine-131	8.05 d
cesium-137	30.17 y
radium-226	1.60×10^3 y
uranium-235	7.1×10^8 y
uranium-238	4.5×10^9 y
fermium-244	3.3 ms

*d = day, y = year.

Example 22.4 *Determining the time required for the activity of a radioactive isotope to decay below a certain value*

Use Table 22.5 to calculate the time needed for the activity of a 10.-Ci cobalt-60 source to decay to 8.0 Ci.

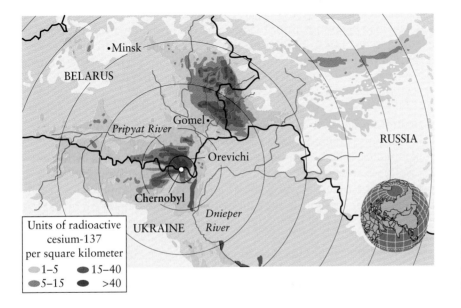

Units of radioactive cesium-137 per square kilometer
- 1–5
- 5–15
- 15–40
- >40

FIGURE 22.21

A map of the region near the Chernobyl nuclear power plant. The colors indicate the relative intensity of radiation remaining in the soil. Areas with the darkest colors are uninhabitable.

Strategy The activity gives the rate of decay. Because the activity at a given instant is proportional to N, as shown in Eq. 1, we can substitute initial activity for N_0 and final activity for N in Eq. 2 and rearrange the equation:

$$\frac{\text{Final activity}}{\text{Initial activity}} = e^{-kt}$$

Then we solve for t:

$$t = -\frac{1}{k}\ln\left(\frac{\text{final activity}}{\text{initial activity}}\right)$$

We can obtain the value of k from the half-life by using

$$k = \frac{\ln 2}{t_{1/2}}$$

Solution The half-life of cobalt-60 in Table 22.5 is 5.26 y, so the decay constant is

$$k = \frac{\ln 2}{5.26 \text{ y}}$$

The time at which the final activity is 8.0 Ci is given by

$$t = -\left(\frac{5.26 \text{ y}}{\ln 2}\right)\ln\left(\frac{8.0 \text{ Ci}}{10. \text{ Ci}}\right) = 1.7 \text{ y}$$

Self-Test 22.6A Use Table 22.5 to calculate the time needed for the activity of a 20.-mCi strontium-90 source to decay to 0.25 mCi.

[*Answer:* 180 y]

Self-Test 22.6B Use Table 22.5 to calculate the time needed for the activity of a 15-µCi cesium-137 source to decay to 1.5 µCi.

The constancy of the half-life of a nuclide is put to practical use in determining the ages of archeological artifacts. In isotopic dating, we measure the activity of the radioactive isotopes the artifacts contain. Isotopes used for dating objects include uranium-238, potassium-40, and tritium. Some applications are described in Investigating Matter 22.1. However, the most important example is radiocarbon dating, which uses the β decay of carbon-14, for which the half-life is 5.73×10^3 y. Carbon-14 is a naturally occurring isotope of carbon and is present in all living things. Prior to the industrial revolution, which released carbon stored in fossil fuels into the atmosphere, the supply of carbon-14 atoms in the environment was nearly constant over archeological time. The atoms are produced when nitrogen nuclei in the atmosphere are bombarded by neutrons:

There are small variations with sunspot activity.

$$^{14}_{7}\text{N} + ^{1}_{0}\text{n} \longrightarrow ^{14}_{6}\text{C} + ^{1}_{1}\text{p}$$

The neutrons originate from the collisions of cosmic rays with other nuclei. The carbon-14 atoms produced in the atmosphere enter living organisms as $^{14}\text{CO}_2$ through photosynthesis and digestion. They leave the organisms by the normal processes of excretion and respiration, and also because the nuclei decay at a steady rate. As a result, all living things have a constant proportion of the isotope among their very much more numerous carbon-12 atoms. In other words, there is a fixed ratio (of about $1/10^{12}$) of carbon-14 atoms to carbon-12 atoms in living tissues.

When the organism dies, it no longer exchanges carbon with its surroundings. However, carbon-14 nuclei already inside the organism continue to disintegrate with a constant half-life. Hence the ratio of carbon-14 to carbon-12 decreases after death, and the ratio observed in a sample of dead tissue can be used to estimate the time since death. In the modern version of the technique, which requires only milligrams of sample, the carbon atoms are converted into C^- ions by bombardment of the sample with cesium atoms. The C^- ions are then accelerated with electric fields, and the carbon isotopes are separated and counted with a mass spectrometer (Fig. 22.22). In a simpler version of the technique, similar to the original method developed by Willard Libby in Chicago in the late 1940s, larger samples are used and β radiation from the sample is measured. The procedure is illustrated in the following example.

The law of radioactive decay implies that the number of radioactive nuclei decay exponentially with time with a characteristic half-life. Radioactive isotopes are used to determine the ages of objects.

Example 22.5 *Interpreting carbon-14 dating*

A sample of carbon of mass 1.00 g from wood found in an archeological site in Arizona gave 7900 carbon-14 disintegrations in a period of 20. h. In the same period, 1.00 g of carbon from a modern source underwent 18 400 disintegrations. Calculate the ages of the sample.

Strategy First, we rearrange Eq. 2 to give an expression for the time:

$$t = -\frac{1}{k}\ln\left(\frac{N}{N_0}\right)$$

Then we replace k by $t_{1/2}$ by using Eq. 3:

$$t = -\frac{t_{1/2}}{\ln 2}\ln\left(\frac{N}{N_0}\right)$$

The number of disintegrations reported in the same time period is proportional to the number of carbon-14 nuclei present in the two samples. If we assume that the

FIGURE 22.22

In the modern version of the carbon-14 dating technique, a mass spectrometer is used to determine the proportion of carbon-14 nuclei in the sample relative to the number of carbon-12 nuclei.

Carbon-14 is used to date organic materials, but to determine the age of very old substances such as rocks, we have to use materials with longer half-lives. Uranium-238 ($t_{1/2} = 4.5 \times 10^9$ y) and potassium-40 ($t_{1/2} = 1.26 \times 10^9$ y) are used to date very old rocks. For dating relatively recent samples, the activity due to tritium ($t_{1/2} = 12.3$ y) is measured.

Uranium-238 decays through a series of α and β emissions to lead-206. To determine the age of a rock that contains uranium, we measure the ratio of ^{238}U to ^{206}Pb present in it. Potassium-40 decays by electron capture to form argon-40, so to use this isotope we need to measure the amount of argon trapped in the rock. To do so, the rock is placed under vacuum, crushed, and a mass spectrometer is used to measure the amount of argon gas that escapes. This technique was used to determine the age of the rocks along the California coastline and on the surface of the Moon.

Radioactive *tracers* are isotopes that are used to trace a path—either a physical path or a reaction pathway. For example, oil companies inject radioactive tracers with a short half-life into pipelines. Radiation detectors are set up at regular distances along the pipeline to determine how fast the tracer is traveling. Because the half-life of the tracer is short, the oil is no longer radioactive when it emerges from the pipe.

Chemists use tracers to identify the mechanisms of reactions. For example, a sample of sugar can be "labeled" with carbon-14: that is, some carbon-12 atoms in sugar molecules are replaced by carbon-14 atoms, which can then be detected by radiation counters. Then the progress of the sugar molecule through the body can be monitored. Fertilizers labeled with radioactive nitrogen, phosphorus, and potassium are used to follow the mechanism of plant growth and the passage of these elements through the environment.

Radioactive nuclides also have important commercial applications. For example, americium-243 is used in smoke detectors, where it ionizes smoke particles, allowing a current to flow and set off the alarm. Carbon-14 is incorporated into steel to test for wear and into plastics to check for uniformity. Exposure to radiation is also used to sterilize food and inhibit the sprouting of potatoes. The radiation kills the bacteria that spoil food but does not produce harmful substances.

This facility at Lawrence Livermore Laboratory is used to study the irradiation of food. A container of fruit is being lowered into a pool containing an accelerator that generates radiation. The water protects the technician from the harmful effects of the radiation.

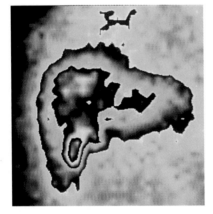

An iodine-123 scan of a normal liver. Notice how clearly the shape of the organ is revealed by the radioisotope. The orange bean-shaped object is the gallbladder.

proportion of carbon-14 in the atmosphere was the same when the ancient sample was alive as it is now, then the original activity of carbon-14 in the ancient sample can be assumed to be the same as the activity of the modern sample. We can therefore set N/N_0 equal to the ratio of the number of disintegrations in the ancient and modern samples.

Solution From the expression above, we obtain

$$t = -\frac{5.73 \times 10^3 \, y}{\ln 2} \times \ln\left(\frac{7900}{18\,400}\right) = 6.99 \times 10^3 \, y$$

We conclude that approximately 7000 y has elapsed since the piece of wood was part of a living tree.

Self-Test 22.7A A sample of carbon of mass 250. mg from wood found in a tomb in Israel underwent 2480 carbon-14 disintegrations in 20. h. Estimate the time since death of the tree, assuming the same activity for a modern sample, as we did in Example 22.5.

[***Answer:*** $5.1 \times 10^3 \, y$]

Self-Test 22.7B A sample of carbon of mass 1.00 g from scrolls found near the Dead Sea underwent 1.4×10^4 carbon-14 disintegrations in 20. h. Estimate the approximate time since the sheepskins composing the scrolls were removed from the sheep.

NUCLEAR ENERGY

Nuclear reactions can release huge amounts of energy, and the large-scale application of nuclear reactions to generate power provides a substantial amount of the world's electricity. However, its use is controversial. The benefits of nuclear energy are balanced by risks and dangers, such as the problem of nuclear waste disposal. The following sections outline some of the principles involved. They introduce nuclear fuels and how they are processed, how nuclear reactions are controlled, and critical issues associated with the use of nuclear power.

22.8 Mass-Energy Conversion

Energy is released when the protons and neutrons in a nucleus adopt a more stable arrangement. The change in energy that accompanies this rearrangement can be found by comparing the masses of the nuclear reactants and the nuclear products. Einstein's theory of relativity tells us that the mass of an object is a measure of its energy content: the greater the mass of an object, the greater its energy. Specifically, the total energy, E, and the mass, m, are related by Einstein's famous equation

$$E = mc^2 \qquad (4)$$

where c is the speed of light ($2.997\,92 \times 10^8$ m/s).

Mass loss accompanies all energy loss, but it is normally far too small to detect. For example, when 100. g of water cools from 100.°C to 20.°C, it loses 33 kJ of energy as heat; this energy loss corresponds to a mass loss of only 3.7×10^{-10} g. Even in a strongly exothermic chemical reaction, such as one that releases 10^3 kJ of energy, the mass of the products is only 10^{-8} g less than that of the reactants; this change in mass is outside the range of all but extremely precise measurements. However, in a nuclear reaction, the energy changes are very large, the corresponding mass loss is measurable, and the observed change in mass can be used to calculate the energy released.

The **nuclear binding energy,** E_{bind}, is the energy *released* when protons and neutrons come together to form a nucleus. A positive value of the binding energy means that the energy of a nucleus is lower than that of its constituent nucleons; the greater the binding energy, the lower the energy of the nuclide. We can use Einstein's equation to calculate the nuclear binding energy from the difference in mass, Δm, between the nucleus and the separated nucleons. For example, iron-56 has 26 protons, each of mass m_p, and 30 neutrons, each of mass m_n. The difference in mass between the nucleus and the separate nucleons is

$$\Delta m = \sum m(\text{products}) - \sum m(\text{reactants})$$
$$= m\left(^{56}_{26}\text{Fe nucleus}\right) - [(26 \times m_p) + (30 \times m_n)]$$

We then use Einstein's formula to write

$$E_{\text{bind}} = |\Delta m| \times c^2 \tag{5}$$

> $|\Delta m|$ means the absolute value of Δm; all binding energies are positive.

Before we see an example of this calculation, we need to note two points. One is that the masses of nuclides are normally reported in **atomic mass units,** u (formerly, amu). An atomic mass unit is defined as exactly $\frac{1}{12}$ the mass of one atom of carbon-12:

$$1\,\text{u} = 1.6605 \times 10^{-27}\,\text{kg}$$

The second point is that it is more convenient to use tables of isotope masses for calculations of binding energies than to use the masses of the bare nuclei. However, an isotope mass is the mass of the neutral atom, not the nucleus alone, so it includes the mass of the electrons and the mass equivalent of their binding energy to the nucleus. We can take into account the mass of the electrons by using the mass of a hydrogen atom instead of the mass of each proton. With this approach, the masses of electrons will be the same on both sides of the equation. The electron-nucleus binding energy is only about 0.000 001 u per proton, so it can be ignored in elementary calculations.

Example 22.6 *Calculating the nuclear binding energy*

Calculate the nuclear binding energy for helium-4 given the following masses: ^4He, 4.0026 u; ^1H, 1.0078 u; n, 1.0087 u.

Strategy The nuclear binding energy is the energy released in the formation of the nucleus from its nucleons. Use H atoms instead of protons to account for the masses of the electrons in the He produced:

$$2\,^1\text{H} + 2\,^1_0\text{n} \longrightarrow\,^4\text{He}$$

Begin by calculating the difference in masses between the products and the reactants; then convert the result from atomic mass units to kilograms. Finally, use the Einstein relation to calculate the energy corresponding to this loss of mass.

Solution The change in mass is

$$\Delta m = 4.0026\,\text{u} - [(2 \times 1.0078) + (2 \times 1.0087)]\,\text{u} = -0.0304\,\text{u}$$

In kilograms, this mass difference is

$$\Delta m = (-0.0304\,\text{u}) \times \frac{1.6605 \times 10^{-27}\,\text{kg}}{1\,\text{u}} = -3.04 \times 1.6605 \times 10^{-29}\,\text{kg}$$

Then, from Eq. 5,

$$E_{\text{bind}} = |-3.04 \times 1.6605 \times 10^{-29}\,\text{kg}| \times (3.00 \times 10^8\,\text{m/s})^2$$
$$= 4.54 \times 10^{-12}\,\text{kg·m}^2/\text{s}^2 = 4.54 \times 10^{-12}\,\text{J}$$

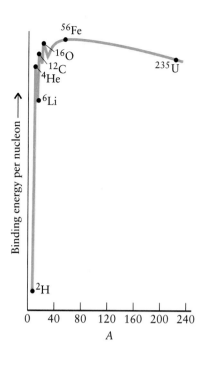

FIGURE 22.23

The variation of the nuclear binding energy per nucleon. The maximum binding energy per nucleon occurs close to iron and nickel. Their nuclei have the lowest energy of all because their nucleons are most tightly bound. (The vertical axis is E_{bind}/A.)

(In the last line, we have used $1\ kg\cdot m^2/s^2\ =\ 1\ J$.) Therefore, 4.54×10^{-12} J of energy is released when one He nucleus forms from its nucleons. Multiplying by the Avogadro constant and converting joules to kilojoules, we find that the molar binding energy of helium-4 is 2.7×10^9 kJ/mol, which is 10 million times larger than the energy of a typical chemical bond.

Self-Test 22.8A Calculate the molar binding energy of carbon-12 nuclei.

[**Answer:** 8.9×10^9 kJ/mol]

Self-Test 22.8B Calculate the molar binding energy of uranium-235 nuclei. The mass of one uranium atom is 235.0439 u.

Figure 22.23 shows the binding energy per nucleon, E_{bind}/A, for the elements. The greater the binding energy per nucleon, the more stable are the nucleons in the nucleus. The graph shows that nucleons are bonded together most strongly in iron and nickel. This high binding energy is one of the reasons why these elements are so abundant on a rocky planet like the Earth. We can infer that nuclei of atoms lighter than iron and nickel become more stable when they "fuse" together and that nuclei of heavier elements become more stable when they undergo "fission" and split into lighter nuclei.

> The actual binding energy of iron-56, for instance, is 56 times the value shown in the graph (because there are 56 nucleons in the nucleus), and that of uranium-235 is 235 times the value shown for that nucleus.

Nuclear binding energies are determined from the mass difference between the nucleus and its components and the use of Einstein's formula. Iron and nickel have the highest binding energy per nucleon.

22.9 Nuclear Fission

Nuclear **fission** is the breaking of a nucleus into two or more smaller nuclei of similar mass. When a uranium atom disintegrates into smaller nuclei, energy is released. Because fission reactions release huge amounts of energy, they are used to generate electricity in nuclear power plants. The energy released in fission can be calculated by subtracting the total mass of the fission products from the mass of the starting materials. The change in mass is then converted to a change in energy by using Einstein's equation.

Toolbox 22.2 *How to calculate the energy released in a nuclear reaction*

This Toolbox shows how to determine the energy change that occurs in a nuclear reaction from a measurement of the mass difference between products and reactants.

Conceptual Basis
The energy of an object and its mass are related by Eq. 4. The energy released in a nuclear reaction is so great that the mass loss is readily detectable. The mass loss is calculated from the masses of the reactants and products of a nuclear reaction, and then Einstein's equation is used to express the mass loss as a change in energy.

Procedure

Calculating the energy change in a nuclear reaction
Step 1. Write the nuclear equation for the reaction, substituting a hydrogen atom for each proton and the mass of an atom of an isotope for the mass of the corresponding nucleus.
Step 2. Subtract the total mass of reactants from the total mass of products. When calculating the energy change accompanying a nuclear reaction, retain the sign of the mass difference.
Step 3. Convert atomic mass units to kilograms by using the conversion factor $1\ u = 1.6605 \times 10^{-27}$ kg.
Step 4. Use the Einstein relation in the form $\Delta E = \Delta m \times c^2$, with mass in kilograms and c in meters per second, to convert the change in mass into a change in energy (in joules).
Step 5. Multiply the change in energy by the Avogadro constant to obtain the molar energy change (in joules per mole). Convert to kilojoules per mole. Use the molar mass of an isotope to convert to the energy change for a given mass of that isotope.

Calculating binding energy
When calculating binding energy, follow the procedure given above but use the absolute value of the mass difference.

Example 22.7 *Calculating the energy released during fission*

When uranium-235 nuclei are bombarded with neutrons, they can split apart in a variety of ways, like glass balls that shatter into pieces of different sizes. In one process, uranium-235 forms barium-142 and krypton-92:

$$^{235}_{92}U + ^{1}_{0}n \longrightarrow ^{142}_{56}Ba + ^{92}_{36}Kr + 2\,^{1}_{0}n$$

Calculate the energy (in joules) released when 1.0 g of uranium-235 undergoes this fission reaction. The masses of the particles are $^{235}_{92}U$, 235.04 u; $^{142}_{56}Ba$, 141.92 u; $^{92}_{36}Kr$, 91.92 u; n, 1.0087 u.

Strategy If we know the mass loss, we can find the energy released by using Einstein's equation. Therefore, we calculate the total mass of the particles on each side of the equation and then substitute the difference into Eq. 4, as described in Toolbox 22.2.

Solution **Steps 1 and 2.** The nuclear equation and isotopic masses are given above.

$$\text{Mass of products} = m(Ba) + m(Kr) + (2 \times m_n)$$
$$= 141.92 + 91.92 + (2 \times 1.0087)\ u = 235.86\ u$$
$$\text{Mass of reactants} = m(U) + m_n$$
$$= 235.04 + 1.0087\ u = 236.05\ u$$

Step 3. The change in mass is the difference in the masses of products and reactants:

$$\Delta m = 235.86 - 236.05\ u = -0.19\ u$$
$$= (-0.19\ u) \times \left(\frac{1.6605 \times 10^{-27}\ \text{kg}}{1\ u}\right) = -1.9 \times 1.6605 \times 10^{-28}\ \text{kg}$$

Because nuclear reactions are elementary processes, we do not cancel species that appear on both sides of a nuclear equation. The equation shows how the process occurs. For instance, in Example 22.7, a neutron is needed to initiate the fission and two neutrons are given off during a fission event.

Step 4. The energy change accompanying the fission of one ^{235}U nucleus to these products is therefore

$$\Delta E = -(1.9 \times 1.6605 \times 10^{-28}\,\mathrm{kg}) \times (3.00 \times 10^8\,\mathrm{m/s})^2$$
$$= -1.9 \times 1.6605 \times (3.00)^2 \times 10^{-12}\,\mathrm{J}$$

Step 5. This energy release accompanies the fission of one nucleus. The number of atoms in 1.0 g (1.0×10^{-3} kg) of uranium is

$$\text{Number of atoms} = \frac{\text{mass of sample}}{\text{mass of 1 atom}}$$
$$= \frac{1.0 \times 10^{-3}\,\mathrm{kg}}{235.04\,\mathrm{u} \times \left(\dfrac{1.6605 \times 10^{-27}\,\mathrm{kg}}{1\,\mathrm{u}}\right)} = \frac{1.0 \times 10^{24}}{235.04 \times 1.6605}$$

The total energy change is therefore

$$\Delta E = \overbrace{\frac{1.0 \times 10^{24}}{235.04 \times 1.6605}}^{\text{Number of atoms}} \times \overbrace{(-1.9 \times 1.6605 \times (3.00)^2 \times 10^{-12}\,\mathrm{J})}^{\text{Energy change for one atom}}$$
$$= -7.4 \times 10^{10}\,\mathrm{J}$$

(Note that the 1.6605 factors cancel.)

Self-Test 22.9A Another mode in which uranium-235 can undergo fission is $^{235}_{92}U + ^{1}_{0}n \rightarrow ^{135}_{52}Te + ^{100}_{40}Zr + ^{1}_{0}n$. Calculate the energy change when 1.0 g of uranium-235 undergoes fission in this way. The masses needed are U, 235.04 u; n, 1.0087 u; Te, 134.92 u; Zr, 99.92 u.

[***Answer:*** -7.7×10^{10} J]

Self-Test 22.9B A nuclear reaction that can cause great destruction is one of many that take place in the ^{235}U atomic bomb: $^{235}_{92}U + ^{1}_{0}n \rightarrow ^{138}_{56}Ba + ^{86}_{36}Kr + 12\,^{1}_{0}n$. How much energy is released when 1.0 g of uranium-235 undergoes fission in this manner? The additional masses needed are Ba, 137.91 u; Kr, 85.91 u.

Spontaneous nuclear fission, nuclear fission that does not need to be initiated, takes place when the natural shaking motion of the nucleons in a heavy nucleus causes it to break into two nuclei of similar mass (Fig. 22.24). An example is the disintegration of americium-244 into iodine and molybdenum:

$$^{244}_{95}Am \longrightarrow ^{134}_{53}I + ^{107}_{42}Mo + 3\,^{1}_{0}n$$

FIGURE 22.24

In spontaneous nuclear fission, the oscillations of the heavy nucleus in effect tear the nucleus apart, thereby forming two or more smaller nuclei of similar mass. Some nucleons may also be released during fission.

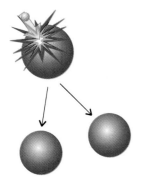

FIGURE 22.25

In induced nuclear fission, the impact of an incoming neutron causes the nucleus to break apart.

Fission does not occur in precisely the same way in every instance: the fission products may include different isotopes of these elements and also isotopes of other elements.

Induced nuclear fission is fission caused by bombarding a heavy nucleus with neutrons (Fig. 22.25). Nuclei that can undergo induced fission are called **fissionable.** For most nuclei, fission is induced only if the impinging neutrons are moving so rapidly that they can smash into the nucleus and drive it apart with the shock of impact; uranium-238 is one such nuclide. Some nuclei, however, can be nudged into breaking apart even if the incoming neutrons are slow. Such nuclides are called **fissile.** They include uranium-235, uranium-233, and plutonium-239, the fuels of nuclear power plants.

Once nuclear fission has been induced, it can continue, even if the supply of neutrons from outside is discontinued, as long as more neutrons are produced by the fission event than are used to induce them initially. Such "self-sustaining fission" occurs with uranium-235, which undergoes numerous fission processes, including

$$\ce{^{235}_{92}U + ^{1}_{0}n -> ^{139}_{56}Ba + ^{94}_{36}Kr + 3 ^{1}_{0}n}$$

If the three product neutrons strike three other fissile nuclei, then, after the next round of fission, there will be nine neutrons, which can induce fission in nine more nuclei. An average of 2.5 neutrons are released per fission event when uranium-235 is bombarded with neutrons. In the language of Section 13.14, neutrons are carriers in a branched chain reaction (Fig. 22.26). When a nuclear branched chain reaction is allowed to run freely, the cascade of released neutrons can result in the fissioning of all the available uranium-235 in only a fraction of a second. The result is a nuclear explosion.

Neutrons produced in a chain reaction are moving very fast, and most escape into the surroundings without colliding with another fissionable nucleus. However, if a large enough number of uranium nuclei are present in the sample, enough neutrons can be captured to sustain the chain reaction. That is, there is a **critical mass,** a mass of fissionable material below which so many neutrons escape from the sample that the fission chain reaction is not sustained. If a sample is **supercritical,** with a mass in excess of the critical value, then the reaction is self-sustaining and may result in an explosion. The critical mass for a solid sphere of pure plutonium of normal density is about 15 kg, a sphere about the size of a grapefruit. The critical mass is smaller if the metal is compressed by detonating a conventional explosive that surrounds it. Then the nuclei are pressed closer together and become more effective at blocking the escape of neutrons. The critical mass can be as low as 5 kg for highly compressed plutonium.

The technically difficult step in designing a nuclear weapon is to convert a given mass of fissile material into a supercritical mass so rapidly that the chain reaction occurs uniformly throughout the metal. This can be done by shooting two blocks toward each other (as was done in the bomb that fell on Hiroshima) or by implosion of a single subcritical mass (the technique used in the bomb that destroyed Nagasaki). A strong neutron emitter, typically polonium, is also included to initiate the chain reaction. The products of the fission process, which consist of dozens of different nuclides, most of them radioactive, are dispersed rapidly in the explosion. They are carried many miles by air currents before they settle to the ground, becoming the fallout of a nuclear explosion.

Explosive fission cannot occur in a nuclear power plant because the fuel used in a **nuclear reactor,** a device for achieving controlled nuclear fission, is much less dense than that used to induce explosive fission (Fig. 22.27). Instead,

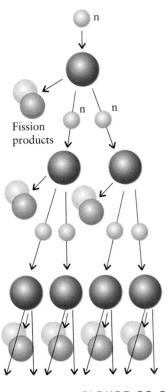

FIGURE 22.26

A self-sustaining chain reaction, in which neutrons are the chain carriers, occurs when induced fission produces more than one neutron per fission event. These newly produced neutrons can go on to stimulate fission in other nuclei.

a much slower, controlled chain reaction is sustained by making efficient use of a limited supply of neutrons by slowing the neutrons down. Nuclear fuel consists of UO_2 pellets, enriched to about 3% uranium-235 and shaped into long rods in a zirconium alloy tube. The fuel rods are inserted into a **moderator,** a material that slows down the neutrons as they pass between fuel rods; the slower neutrons have a greater probability of being captured by a nucleus. Slow neutrons have three significant roles: they do not induce the fission of fissionable (as distinct from fissile) material; they are most effectively absorbed by the fissile uranium-235; and they allow the control rods to act most efficiently. The first moderator used was graphite. **Light-water reactors** (LWRs) use ordinary water as a moderator. They are the most common type of nuclear reactors in the United States (Fig. 22.28).

The rate of the chain reaction must be kept below a certain level, or the reactor will become too hot and begin to melt. Control rods made from neutron-absorbing elements, such as boron or cadmium, are inserted between the fuel rods to control the number of neutrons available for inducing fission and thus control the rate of nuclear reaction. Explosions that have occurred in nuclear power plants have been chemical explosions, generally associated with overheating.

One of the many problems connected with nuclear power is the availability of fuel: uranium-235 reserves are only about 0.7% those of the nonfissile uranium-238. One solution is to synthesize fissile nuclides from other elements. In a **breeder reactor,** a reactor that is used to create nuclear fuel, the neutrons are not moderated. Their high speeds result in the formation of not only uranium-235 but also some fissile plutonium-239, which can be used as fuel (or for warheads). Breeder reactors are more hazardous to operate than nuclear power plants. They run very hot, and the fast reactions require more careful control than a reactor used for nuclear power generation. Their use is still controversial.

Nuclear energy can be extracted by arranging for a nuclear chain reaction to occur, with neutrons as the chain carriers. A moderator is used to reduce the speed of the neutrons in a reactor that uses fissile material.

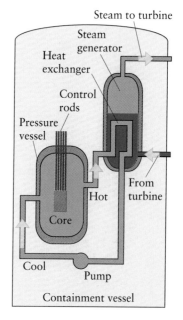

FIGURE 22.27

A schematic diagram of a pressurized water reactor, in which both the coolant and the moderator are water under pressure. The fission reactions produce heat, which boils water in the steam generator.

22.10 Nuclear Fusion

Fission reactors generate radioactive waste products that must be controlled; we shall discuss that topic in Section 22.11. However, another kind of nuclear reaction is essentially free of long-lived hazardous waste products, and its abundant fuel is extracted from seawater. The reaction is nuclear **fusion,** the merging of smaller nuclei into larger ones, specifically hydrogen nuclei into helium nuclei. We see from the plot of nuclear binding energy in Fig. 22.23 that there is a large increase in nuclear binding energy (and hence a lowering of total energy) on going from deuterium to the first few light elements.

The strong electrical repulsion between protons makes it difficult for them to approach each other closely enough to fuse together. The heavier isotopes of

FIGURE 22.28

The reactor core of a typical light-water reactor (LWR) nuclear power plant is immersed in water.

hydrogen are therefore used in fusion reactions; their nuclei fuse together more readily, because the additional neutrons help to overcome the repulsion between the approaching protons. The high kinetic energies needed for successful collisions are achieved by raising the temperature to millions of degrees.

One fusion scheme uses deuterium (D) and tritium (T) in the following sequence of nuclear reactions

$$D + D \longrightarrow {}^3He + n$$
$$D + D \longrightarrow T + p$$
$$D + T \longrightarrow {}^4He + n$$
$$D + {}^3He \longrightarrow {}^4He + p$$

Overall reaction: $\quad 6\,D \longrightarrow 2\,{}^4He + 2\,p + 2\,n$

The method described in Section 22.9 can be used to show that the overall reaction releases 3×10^8 kJ for each gram of deuterium consumed, corresponding to the energy generated when the Hoover Dam operates at full capacity for about an hour. Additional tritium is supplied to the reaction chamber to facilitate the process. Because it has a very low natural abundance and is radioactive, the tritium is generated by bombarding lithium-6 with neutrons in the immediate surroundings of the reaction zone:

$${}^6Li + n \longrightarrow T + {}^4He$$

Nuclear fusion is very difficult to achieve in practice because of the vigor with which the charged nuclei must be hurled at each other. One way of accelerating them to sufficiently high speeds is to heat them with a fission explosion: this method is used to produce a **thermonuclear explosion,** in which a fission bomb (using uranium or plutonium) is used to ignite a lithium-6 fusion bomb. A more constructive and controlled approach is to heat a plasma, or ionized gas, by passing an electric current through it. The very fast ions in the plasma are kept away from the walls of the container with magnetic fields. This method of achieving controlled fusion is the subject of intense research and is beginning to show signs of success (Fig. 22.29).

Nuclear fusion makes use of the energy released when light nuclei fuse together to form heavier nuclei.

FIGURE 22.29

Research into controlled nuclear fusion is being carried out in several countries. Here we see the Tokamak fusion test reactor at the Princeton Plasma Physics Laboratory in the United States.

22.11 The Chemistry of Nuclear Power

Chemistry is the key to the safe use of nuclear power. Chemistry is used in the preparation of the fuel itself, the recovery of important fission products, and the safe disposal or utilization of nuclear waste.

The fuel of nuclear reactors is uranium, which is mined as several minerals. The most important is *pitchblende*, UO_2 (Fig. 22.30). Uranium is refined not only to reduce the ore to the metal but also to **enrich** it, that is, to increase the abundance of a specific isotope—in this case, uranium-235. The natural abundance of uranium-235 is about 0.7%, and the goal of enrichment is to increase this fraction to about 3%.

The enrichment procedure uses the small mass difference between the hexafluorides of uranium-235 and uranium-238 to separate them. The first procedure to be developed converts the uranium into uranium hexafluoride, UF_6, which can be vaporized readily. The different effusion rates of the two isotopic fluorides are then used to separate them. It follows from Graham's law (rate of effusion is inversely proportional to the square root of the molar mass, Section 5.12) that the rates of effusion of $^{235}UF_6$ (molar mass 349.0 g/mol) and $^{238}UF_6$ (molar mass 352.0 g/mol) should be in the ratio

$$\frac{\text{Rate of effusion of } ^{235}UF_6}{\text{Rate of effusion of } ^{238}UF_6} = \sqrt{\frac{352.0}{349.0}} = 1.004$$

Because the ratio is so close to 1, the vapor must be allowed to effuse repeatedly through porous barriers consisting of screens with large numbers of minute holes. In practice, it is allowed to do so thousands of times. Consequently, the plants must be very large. The original plant in Oak Ridge, Tennessee (called, somewhat loosely, the gaseous diffusion plant), uses 4000 stages and covers an area of 43 acres (Fig. 22.31). We can see why such plants, which are vital to nuclear defense and nuclear power generation, are difficult to hide from foreign powers who want to monitor one another's nuclear capabilities.

Because the effusion process is technically demanding and uses a lot of energy, scientists and engineers continue to look for alternative enrichment procedures. One of these approaches uses a centrifuge that rotates samples of uranium hexafluoride vapor at very high speed. This rotation causes the heavier $^{238}UF_6$ molecules to be thrown outward and collected as a solid on the outer parts of the rotor, leaving a higher proportion of $^{235}UF_6$ closer to the axis of the rotor, from which position it can be removed.

The processing of spent nuclear fuel is a much more complex task than the fuel's initial preparation. Any remaining uranium-235 must be recovered, any plutonium produced must be extracted, and the highly radioactive and largely useless fission products must be safely stored (Fig. 22.32).

The storage of nuclear waste is a serious problem. Currently, thousands of tons of nuclear waste are stored in the United States alone, some of the isotopes having half-lives of many thousands of years. The highly radioactive fission (HRF) products from used nuclear fuel rods must be stored for about 10 half-lives before their level of radioactivity is no longer dangerous. Generally they are buried underground, but even burial of radioactive wastes is not without problems. Metal storage drums can corrode and allow liquid radioactive waste to seep into aquifers that supply drinking water (Fig. 22.33). Leakage can be minimized by incorporating the HRF products into a hard ceramic material or a glass—a solid, complex network of silicon and oxygen atoms. Most of the

FIGURE 22.30

Pitchblende is a common uranium ore. It is a variety of uranite, UO_2.

FIGURE 22.31

The individual diffusion stages in the original uranium-235 diffusion plant at Oak Ridge, Tennessee. There are thousands of such stages in the entire plant. Note the size of the components relative to that of the technician.

FIGURE 22.32

Containers of high-level waste products, including cesium-137 and strontium-90, glow under a protective layer of water. The radiation from these canisters is so intense that direct contact with one could cause death within about 4 s.

FIGURE 22.33

This 35-year-old drum of radioactive waste has become corroded and has leaked radioactive materials into the soil. The drum was located in one of the nuclear waste disposal sites at the U. S. Department of Energy's nuclear manufacturing and research facility at Hanford, Washington. Several storage sites at this facility have become seriously contaminated.

fission products are oxides of the type that form one of the components of glass—they are network formers (see Applying Chemistry: Case Study 19); that is, they promote the formation of a relatively disorderly —Si—O— network rather than inducing crystallization into an orderly array of atoms (Fig. 22.34). Crystallization is dangerous because crystalline regions easily form cracks that leave the incorporated radioactive materials exposed to moisture, which might dissolve them and carry them away from the storage area.

A chief attraction of fusion reactors is their virtual freedom from long-lived radioactive waste generation. The promise of fusion reactors for abundant, pollution-free power is great. We are at the point in history when the elements have completed their journey from the stars, have come alive, and have begun to control their own destiny. Now the cycle can be completed. The fusion reactions that power the stars may soon be recreated on Earth. Here they will emulate the processes that produced the elements from which we humans are made.

Uranium is extracted by a series of reactions that lead to uranium hexafluoride, and then the isotopes are separated by a variety of procedures. Some radioactive waste is currently converted to glass for storage in cool, dry caverns.

FIGURE 22.34

Molten glass for storing nuclear waste, being poured from a platinum crucible into a steel bar mold.

CHAPTER 22 **NUCLEAR CHEMISTRY**

Skills You Should Have Mastered

Conceptual

☐ 1. Explain how nuclear reactions differ from chemical reactions, Section 22.2.

☐ 2. Explain why accelerators are required for the transmutation of elements by nucleosynthesis, Section 22.4.

☐ 3. Distinguish nuclear fission from nuclear fusion and predict which nuclides undergo each type of process, Sections 22.9 and 22.10.

Problem-Solving

☐ 1. Write, complete, and balance nuclear equations, Toolbox 22.1 and Examples 22.1 and 22.2.

☐ 2. Use the band of stability to predict the types of decay a certain radioactive nucleus is likely to undergo, Self-Test 22.3.

☐ 3. Predict the amount of a radioactive sample that will remain after a specified time period, given the decay constant or half-life of the sample, Example 22.3.

☐ 4. Use the law of radioactive decay to determine the time required for the activity of a radioactive isotope to decay below a certain value, Example 22.4.

☐ 5. Use the half-life of an isotope to determine the age of a sample, Example 22.5.

☐ 6. Calculate the nuclear binding energy of a given nuclide, Example 22.6.

☐ 7. Calculate the energy released during a nuclear reaction, Toolbox 22.2 and Example 22.7.

Descriptive

☐ 1. Distinguish α, β, and γ radiation by their response to an electric field, penetrating power, and relative biological effectiveness, Sections 22.1 and 22.5.

☐ 2. Describe electron capture and the emission of γ rays, positrons, protons, and neutrons, Section 22.2.

☐ 3. Describe some applications of radioactive isotopes in chemistry, industry, and medicine, Investigating Matter 22.1.

☐ 4. Describe the construction and operation of a nuclear reactor, Section 22.9.

☐ 5. Describe the problems associated with the management of nuclear waste and some solutions to the problems, Section 22.11.

Exercises

Nuclear Structure and Radiation

22.1 Distinguish α particles, β particles, and γ rays by composition, mass, charge, and penetrating power.

22.2 Explain how Rutherford determined the nature of α particles.

22.3 Determine the number of protons, neutrons, and nucleons in (a) ^{2}H; (b) ^{24}Mg; (c) ^{263}Rf; (d) ^{60}Co; (e) ^{238}Pu; (f) ^{258}Md.

22.4 Determine the number of protons, neutrons, and nucleons in (a) ^{12}C; (b) ^{90}Sr; (c) ^{257}Db; (d) ^{99}Mo; (e) ^{3}T; (f) ^{57}Fe.

22.5 Write the nuclear symbol and determine the number of protons, neutrons, and nucleons in (a) bromine-81; (b) krypton-90; (c) curium-244; (d) iodine-128; (e) sulfur-32; (f) americium-241.

22.6 Write the nuclear symbol and determine the number of protons, neutrons, and nucleons in (a) chlorine-37; (b) berkelium-247; (c) dubnium-262; (d) gold-197; (e) californium-249; (f) uranium-233.

Radioactive Decay

22.7 Identify the daughter nucleus in each of the following decays and write the balanced nuclear equation: (a) β decay of tritium; (b) β^+ decay of yttrium-83; (c) β decay of krypton-87; (d) α decay of protactinium-225.

22.8 Identify the daughter nucleus in each of the following decays and write the balanced nuclear equation: (a) β decay of actinium-228; (b) α decay of radon-212; (c) α decay of francium-221; (d) electron capture by protactinium-230.

22.9 Write the balanced nuclear equation for the following radioactive decays: (a) β^+ decay of boron-8; (b) β decay of nickel-63; (c) α decay of gold-185; (d) electron capture by beryllium-7.

22.10 Write the balanced nuclear equation for the following radioactive decays: (a) β decay of uranium-233; (b) proton emission of cobalt-56; (c) β^+ decay of holmium-158; (d) α decay of polonium-212.

22.11 Determine the particle emitted and write the balanced nuclear equation for the following nuclear transitions: (a) sodium-24 to magnesium-24; (b) ^{128}Sn to ^{128}Sb; (c) lanthanum-140 to barium-140; (d) ^{228}Th to ^{224}Ra.

22.12 Determine the particle emitted and write the balanced nuclear equation for the following nuclear transitions: (a) carbon-14 to nitrogen-14; (b) neon-19 to fluorine-19; (c) gold-188 to platinum-188; (d) uranium-229 to thorium-225.

The Pattern of Nuclear Stability

22.13 In each case, predict which nuclide is more stable: (a) ^{40}Ca or ^{39}Ca; (b) ^{208}Pb or ^{208}Bi.

22.14 In each case, predict which nuclide is more stable: (a) ^{4}He or ^{3}He; (b) ^{118}Sn or ^{117}In.

22.15 What condition of the neutron-to-proton ratio favors the emission of (a) α particles? (b) β particles?

22.16 What characteristics of a nucleus favor the emission of a positron from a radioactive nuclide?

22.17 The following nuclides lie outside the band of stability. Predict whether they are most likely to undergo β decay, β^+ decay, or α decay and identify the daughter nucleus: (a) copper-68; (b) cadmium-103; (c) berkelium-243; (d) dubnium-260.

22.18 The following nuclides lie outside the band of stability. Predict whether they are most likely to undergo β decay, β^+ decay, or α decay and identify the daughter nucleus: (a) copper-60; (b) xenon-140; (c) americium-246; (d) neptunium-240.

22.19 Identify the daughter nuclides in each step of the radioactive decay of uranium-235, if the string of particle emissions is $\alpha, \beta, \alpha, \beta, \alpha, \alpha, \alpha, \beta, \alpha, \beta, \alpha$.

22.20 Neptunium-237 undergoes an $\alpha, \beta, \alpha, \alpha, \beta, \alpha, \alpha, \alpha, \beta, \alpha, \beta$ sequence of radioactive decays. Determine the daughter nuclide after each decay and write a balanced nuclear equation for each step.

Nucleosynthesis

22.21 Complete the following nuclear equations:
(a) ^{14}N + ? $\rightarrow$ ^{17}O + $^{1}_{1}$p
(b) ? + $^{1}_{0}$n $\rightarrow$ ^{249}Bk + $^{0}_{-1}$e
(c) ^{243}Am + $^{1}_{0}$n $\rightarrow$ ^{244}Cm + ? + γ
(d) ^{13}C + $^{1}_{0}$n $\rightarrow$? + γ

22.22 Complete the following nuclear equations:
(a) ? + $^{1}_{1}$p $\rightarrow$ ^{21}Na + γ
(b) ^{1}H + $^{1}_{1}$p $\rightarrow$ ^{2}H + ?
(c) ^{15}N + $^{1}_{1}$p $\rightarrow$ ^{12}C + ?
(d) ^{20}Ne + ? $\rightarrow$ ^{24}Mg + γ

22.23 Complete the following nuclear equations for transmutation reactions:
(a) ^{20}Ne + α $\rightarrow$? + ^{16}O
(b) ^{20}Ne + ^{20}Ne $\rightarrow$ ^{16}O + ?
(c) ^{44}Ca + ? $\rightarrow$ γ + ^{48}Ti
(d) ^{27}Al + ^{2}H $\rightarrow$? + ^{28}Al

22.24 Complete the following nuclear equations for transmutation reactions:
(a) ? + γ $\rightarrow$ $^{0}_{-1}$e + ^{20}Ne
(b) ^{44}Ti + $^{0}_{-1}$e $\rightarrow$ $^{0}_{+1}$e + ?
(c) ^{241}Am + ? $\rightarrow$ 4 $^{1}_{0}$n + ^{248}Fm
(d) ? + $^{1}_{0}$n $\rightarrow$ $^{0}_{-1}$e + ^{244}Cm

22.25 Write nuclear equations that represent the following processes: (a) oxygen-17 produced by α particle bombardment of nitrogen-14; (b) americium-240 produced by neutron bombardment of plutonium-239.

22.26 Write the nuclear equations for the following transformations: (a) ^{257}Rf produced by the bombardment of californium-245 with carbon-12 nuclei; (b) the first synthesis of ^{266}Mt by the bombardment of bismuth-209 with iron-58 nuclei. Given that the first decay of meitnerium is by α emission, what is the daughter nucleus?

22.27 Each of the following equations represents a fission reaction. Complete and balance the equations:
(a) ^{244}Am $\rightarrow$ ^{134}I + ^{107}Mo + 3 ?
(b) ^{235}U + $^{1}_{0}$n $\rightarrow$? + ^{138}Te + 2 $^{1}_{0}$n
(c) ^{235}U + $^{1}_{0}$n $\rightarrow$ ^{101}Mo + ^{132}Sn + ?

22.28 Complete each of the following nuclear equations for fission reactions:
(a) ^{239}Pu + $^{1}_{0}$n $\rightarrow$ ^{98}Mo + ^{138}Te + ?
(b) ^{239}Pu + $^{1}_{0}$n $\rightarrow$ ^{100}Tc + ? + 4 $^{1}_{0}$n
(c) ^{239}Pu + $^{1}_{0}$n $\rightarrow$? + ^{133}In + 3 $^{1}_{0}$n

Measuring Radioactivity and Its Effects

22.29 The activity of a certain radioactive source is 3.7×10^6 Bq. Express this activity in curies.

22.30 The activity of a sample containing carbon-14 is 12.9 Bq. Express this activity in microcuries.

22.31 Determine the number of disintegrations per second for radioactive sources of the following activities: (a) 1.0 Ci; (b) 82 mCi; (c) 1.0 μCi.

22.32 A certain Geiger counter is known to respond to only 1 of every 1000 nuclear disintegrations emitted from a sample. Calculate the activity of each radioactive source in curies, given the following data: (a) 370. clicks in 10. s; (b) 1.4×10^5 clicks in 1.0 h; (c) 266 clicks in 1.0 min.

22.33 A 1.0-kg sample absorbs an energy of 1.0 J as a result of its exposure to β radiation. Calculate the dose in grays and the dose equivalent in sieverts.

22.34 A 5.0-g sample of muscle tissue absorbs 2.0 J of energy as a result of its exposure to α radiation. Calculate the dose in grays and the dose equivalent in sieverts.

22.35 Someone is exposed to a source of β radiation that results in a dose rate of 1.0 rad/day. Given that nausea begins after a dose equivalent of about 100 rem, after what period will that symptom of radiation sickness be apparent?

22.36 Someone is exposed to a source of α radiation that results in a dose rate of 2.0 mrad/day. If nausea begins after a dose equivalent of about 100 rem, after what period will nausea become apparent?

Rate of Nuclear Disintegration

22.37 Determine the decay constant for (a) tritium, $t_{1/2} = 12.3$ y; (b) lithium-8, $t_{1/2} = 0.84$ s; (c) nitrogen-13, $t_{1/2} = 10.0$ min.

22.38 Determine the half-life of (a) potassium-40, $k = 5.3 \times 10^{-10}$/y; (b) cobalt-60, $k = 0.132$/y; (c) nobelium-255, $k = 3.85 \times 10^{-3}$/s.

22.39 Use Table 22.5 to calculate the time needed for the activity of each source to change as indicated: (a) a 1.0-Ci radium-226 source to decay to 0.10 Ci; (b) a 1.0-μCi potassium-40 source to decay to 10. nCi.

22.40 Use Table 22.5 to calculate the time needed for the activity of each source to change as indicated: (a) a 0.010-mCi strontium-90 source to decay to 0.0010 mCi; (b) a 1.0-Ci iodine-131 source to decay to 1.0 mCi.

22.41 Estimate the activity of a 4.4-Ci cobalt-60 source, $t_{1/2} = 5.26$ y, after 50. y have passed.

22.42 The activity of a strontium-90 source is 3.0×10^4 Bq. What is its activity after 50. y have passed?

22.43 (a) What percentage of a carbon-14 sample remains after 1000. y? (b) Determine the percentage of a tritium sample that remains after 20.0 y.

22.44 (a) What percentage of a strontium-90 sample remains after 10.0 y? (b) Determine the percentage of an iodine-131 sample that remains after 5.0 d.

22.45 (a) What fraction of the original activity of ^{238}U remains after 9.0 billion years? (b) Potassium-40, which is presumed to exist at the formation of the Earth, is used for dating minerals. If one-half of the original potassium-40 exists in a rock, how old is the rock?

22.46 (a) A sample of krypton-85 ($t_{1/2} = 10.8$ y) is released into the atmosphere. What fraction of the krypton-85 remains after 15.0 y? (b) A piece of wood, found in an archeological dig, has a carbon-14 activity that is 90% of the current carbon-14 activity. How old is the piece of wood?

22.47 A 250.-mg sample of charcoal was found by archeologists at a location in the Middle East thought to be the site of the legendary city of Ubar. The sample undergoes 1500 disintegrations in 10.0 h. If a current 1.00-g sample of carbon shows 920 disintegrations per hour, how old is the charcoal?

22.48 A current 1.00-g sample of carbon shows 920 disintegrations per hour. If a 1.00-g sample of carbon from charcoal in an archeological dig in a limestone cave in Slovenia shows 5500 disintegrations in 24.0 h, what is the age of the charcoal sample?

22.49 Use the law of radioactive decay to determine the activity of (a) a 1.0-mg sample of radium-226 ($t_{1/2} = 1.60 \times 10^3$ y); (b) a 2.0-μg sample of strontium-90 ($t_{1/2} = 28.1$ y); (c) a 0.43-mg sample of promethium-147 ($t_{1/2} = 2.6$ y). The mass of each atom in atomic mass units (u) is equal to its mass number, within two significant figures.

22.50 Use the law of radioactive decay to determine the activity of (a) a 1.0-g sample of ^{235}UO$_2$ ($t_{1/2} = 7.1 \times 10^8$ y); (b) a 1.0-g sample of cobalt containing 1.0% ^{60}Co ($t_{1/2} = 5.26$ y); (c) a 5.0-mg sample of thallium-200 ($t_{1/2} = 26.1$ h). The mass of each atom in atomic mass units (u) is equal to its mass number, within two significant figures.

Nuclear Energy

The unit 1 u is defined in Section 22.8.

22.51 What is the meaning of "critical mass" in nuclear fission?

22.52 Distinguish between nuclear fission and nuclear fusion. Explain why heavy nuclides are most likely to undergo fission, whereas light nuclides are most likely to undergo fusion.

22.53 Calculate the energy in joules that is equivalent to (a) 1.0 g of matter; (b) 1 electron, of mass 9.109×10^{-28} g.

22.54 Calculate the energy in joules that is equivalent to (a) 1.0 pg of matter; (b) 1 proton, of mass 1.673×10^{-24} g.

22.55 The Sun emits radiant energy at the rate of 3.9×10^{26} J/s. What is the rate of mass loss (in kilograms per second) of the Sun?

22.56 For the fusion reaction $6 \, D \rightarrow 2 \, {}^4He + 2 \, {}^1H + 2 \, {}^1_0n$, 3×10^8 kJ of energy is released by a certain sample of deuterium. What is the mass loss (in grams) for the reaction?

Archeologists uncovering remains of an ancient city thought to be Ubar (Exercises 22.47).

22.57 Calculate the binding energy per nucleon (J/nucleon) for (a) ^{4}He, 4.0026 u; (b) ^{239}Pu, 239.0522 u; (c) ^{2}H, 2.0141 u; (d) ^{56}Fe, 55.9349 u. Which nuclide is the most stable?

22.58 Calculate the binding energy per nucleon (J/nucleon) for (a) ^{98}Mo, 97.9055 u; (b) ^{151}Eu, 150.9196 u; (c) ^{10}B, 10.0129 u; (d) ^{232}Th, 232.0382 u. Which nuclide is the most stable?

22.59 Calculate the energy released per gram of starting material in the fusion reaction represented by each of the following equations.
(a) $D + D \rightarrow {}^3He + {}^1_0n$ (D, 2.0141 u; ^{3}He, 3.0160 u)
(b) $^3He + D \rightarrow {}^4He + {}^1H$ (^{4}He, 4.0026 u; ^{1}H, 1.0078 u)
(c) $^7Li + {}^1H \rightarrow 2\,{}^4He$ (^{7}Li, 7.0160 u)
(d) $D + T \rightarrow {}^4He + {}^1H$ (T, 3.0160 u)

22.60 Calculate the energy released per gram of starting material in the nuclear reaction represented by each of the following equations.
(a) $^7Li + {}^1H \rightarrow {}^1_0n + {}^7Be$ (^{1}H, 1.0078 u; ^{7}Li, 7.0160 u; ^{7}Be, 7.0169 u)
(b) $^{59}Co + D \rightarrow {}^1H + {}^{60}Co$ (^{59}Co, 58.9332 u; ^{60}Co, 59.9529 u; D, 2.0141 u)
(c) $^{40}K + e^- \rightarrow {}^{40}Ar$ (^{40}K, 39.9640 u; ^{40}Ar, 39.9624 u)
(d) $^{10}B + {}^1_0n \rightarrow {}^4He + {}^7Li$ (^{10}B, 10.0129 u; ^{4}He, 4.0026 u)

Supplementary Exercises

22.61 Explain how α particles can cause severe biological damage, even though their penetrating power is relatively low.

22.62 Describe the technique of radiocarbon dating and explain how it can be used to date objects.

22.63 Describe the evidence for the shell model of the nucleus.

22.64 Write balanced nuclear equations for the radioactive decay of the following nuclides: (a) ^{74}Kr, β^+ emission; (b) ^{174}Hf, α emission; (b) ^{98}Tc, β emission; (d) ^{41}Ca, electron capture.

22.65 Complete the following equations for nuclear reactions:
(a) $^{11}B + ? \rightarrow 2\,{}^1_0n + {}^{13}N$
(b) $? + D \rightarrow {}^1_0n + {}^{36}Ar$
(c) $^{96}Mo + D \rightarrow ? + {}^{97}Tc$
(d) $^{45}Sc + {}^1_0n \rightarrow \alpha + ?$

22.66 Tritium undergoes β decay, and the emitted β particle has an energy of 0.0186 MeV (1 MeV $= 1.602 \times 10^{-13}$ J). If a 1.0-g sample of tissue absorbs 10.0% of the decay products of 1.0 mg of tritium, what dose equivalent does the tissue absorb?

22.67 (a) How many radon-222 nuclei ($t_{1/2} = 3.82$ d) decay per minute to produce an activity of 4 pCi? (b) A bathroom in the basement of a home measures 2.0 m $\times$ 3.0 m $\times$ 2.5 m. If the activity of radon-222 in the room is 4.0 pCi/L, how many nuclei decay during a shower lasting 5.0 min?

22.68 It is found that 1.0×10^{-5} mol of radon-222 atoms ($t_{1/2} = 3.82$ d) has seeped into a closed basement of volume 2.0×10^3 m^3. (a) What is the initial activity of the radon in

picocuries per liter (pCi/L)? (b) How many atoms of ^{222}Rn will remain after one day (24 h)? (c) How long will it take for the radon to decay to below the EPA-recommended level of 4 pCi/L?

22.69 What mass of a 15.0-mg sample of ^{47}V ($t_{1/2} = 33.0$ min) will remain after 45.0 min?

22.70 A sample of ^{244}Fm ($t_{1/2} = 3.3$ ms) having an activity of 0.10 mCi was produced in a nuclear reactor. What will the activity of the ^{244}Fm be after 1.0 s?

22.71 (a) A sample of phosphorus-32 has an initial activity of 58 counts per second. After 12.3 days, the activity was 32 counts per second. What is the half-life of phosphorus-32? (b) If phosphorus-32 is used in an experiment to monitor the consumption of phosphorus by plants, what fraction of the nuclide will remain after 30. d?

22.72 A sample containing sulfur-35 ($t_{1/2} = 88$ d) that is being used to study the reactions by which sulfur is utilized by bacteria has an activity of 10.0 Ci. What mass of sulfur-35 is present? The mass of a sulfur-35 atom is 35 u.

22.73 What is the activity (in curies) of a 22-μg sample of ^{210}Po ($t_{1/2} = 138.4$ d)? The mass of a polonium-210 atom is about 210 u.

22.74 Sodium-24 (23.990 96 u) decays to magnesium-24 (23.985 04 u). Determine (a) the change in the binding energy per nucleon; (b) the change in energy that accompanies the decay. (See Section 22.8 for the definition of 1 u.)

22.75 How much energy is emitted in each α decay of uranium-234? (^{234}U, 234.0409 u; ^{230}Th, 230.0331 u; see Section 22.8 for the definition of 1 u.)

22.76 Uranium-238 decays through a series of α and β emissions to lead-206, with an overall half-life for the entire process of 4.5×10^9 y. How old is the ore sample if a uranium-bearing ore is found to have a ^{238}U/^{206}Pb ratio of (a) 1.00? (b) 1.25?

22.77 The age of a bottle of wine was determined by monitoring the tritium level in the wine. The activity of the tritium is determined to be 8.3% that of a sample of fresh grape juice from the same region where the wine was bottled. How old is the wine?

22.78 Sodium-24 is used for monitoring blood circulation. (a) If a 2.0-μg sample of sodium-24 has an activity of 17.3 Ci, what is its decay constant and its half-life? (b) What mass of a 2.0-μg sample remains after 2.0 days? The mass of a sodium-24 atom is 24 u.

Applied Exercises

For Exercises 22.79–22.82, see Applying Chemistry: Case Study 22.

22.79 Do nuclei that are positron emitters lie above or below the band of stability? Which of the following isotopes might be suitable for PET scans? Explain your reasoning and write the equation for the decay: (a) ^{18}O; (b) ^{13}N; (c) ^{11}C; (d) ^{20}F; (e) ^{15}O.

22.80 Write the nuclear equation for the reaction of boron-10 and a neutron in which an α particle is produced.

22.81 Technetium-99m is produced by a sequence of reactions in which molybdenum-98 is bombarded with neutrons to form molybdenum-99, which undergoes β decay to technetium-99m. Write the balanced nuclear equations for this sequence.

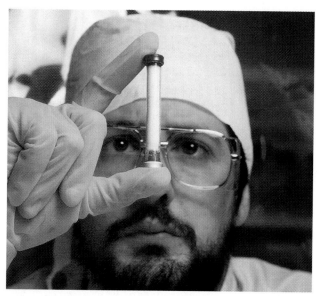

A column that is used to generate technetium-99m from molybdenum-99.

22.82 The mass of an electron is 9.109×10^{-28} g; a positron has the same mass but the opposite charge. When a positron emitted during a PET scan encounters an electron, annihilation occurs in the body: electromagnetic energy is produced and no matter remains. How much energy (in joules) is produced in the annihilation?

22.83 (a) How long will it take for the loss of 99.0% of a sample of radon-222 gas ($t_{1/2} = 3.82$ d)? (b) Comment on radon's lifetime and chances of reaching the surface if formed deep in the Earth's crust. (c) How might a homeowner reduce the rate at which radon enters a home?

22.84 In the mid-1940s, it was proposed that atomic bombs could be used to simplify excavation work, such as that required to dig large canals, and the issue has been raised again. (a) Which type of bomb would be more suitable for such work, fusion or fission? Explain your reasoning. (b) Present arguments for and against this proposal.

22.85 Explain why containers of radioactive waste are more susceptible to corrosion than containers of nonradioactive waste with the same chemical reactivity.

22.86 Explain how uranium ore is enriched.

22.87 What is radioactive "fallout" and why is it used as an argument against nuclear bomb tests?

22.88 Discuss the problems associated with storing waste nuclear fuel.

Integrated Exercises

22.89 Calculate the wavelength and the energy per mole of photons of γ radiation of frequency (a) 9.4×10^{19} Hz; (b) 5.7×10^{21} Hz; (c) 3.7×10^{20} Hz; (d) 7.3×10^{22} Hz.

22.90 Calculate the frequency and wavelength of the γ radiation emitted as a result of a rearrangement of nucleons in a daughter nucleus through an energy (a) 9.5×10^{-14} J; (b) 3.9×10^{-13} J; (c) 2.6×10^{-16} kJ; (d) 4.7×10^{-16} kJ.

22.91 A common energy unit is the electronvolt (eV), the energy an electron gains by passing through a potential difference of 1 V: 1 eV $= 1.602 \times 10^{-19}$ J. In nuclear processes, energies of decay products and transmutation reactions are often expressed in millions of electronvolts (MeV). In the nucleon rearrangement of the following daughter nuclei, the energy changes by the amount shown and a γ ray photon is emitted. Determine the frequency and wavelength of the γ rays in each case: (a) cobalt-60, 1.33 MeV; (b) arsenic-80, 1.64 MeV; (c) iron-59, 1.10 MeV.

22.92 Determine the frequency and wavelength of the γ rays emitted in the decay of the following nuclides, where the energy change of the nucleus is given in MeV (see Exercise 22.91): (a) carbon-15, 5.30 MeV; (b) scandium-50, 0.26 MeV; (c) bromine-87, 5.4 MeV.

22.93 Calculate the mass loss or gain for each of the following processes: (a) a 250-g sample of copper (specific heat capacity, 0.39 J/(°C)·g) is heated from 35°C to 250.°C; (b) a 50.0-g sample of water freezes at 0°C ($\Delta H_{melt}^{\circ} = 6.01$ kJ/mol); (c) 2.0 mol $PCl_5(g)$ is formed from its elements under standard conditions at 25°C.

22.94 Calculate the mass loss or gain for each of the following processes: (a) a 50.0-g block of iron (specific heat capacity, 0.45 J/°C·g) cools from 600.°C to 25°C; (b) a 100.-g sample of ethanol vaporizes at its normal boiling point ($\Delta H_{vap}^{\circ} = 43.5$ kJ/mol); (c) 10.0 g of $SO_2(g)$ is formed from its elements under standard conditions at 25°C.

22.95 Breeder reactors run so hot that a high-boiling-point mixture of molten sodium and potassium is used as the coolant, instead of water. Unfortunately, the steel cooling pipes are susceptible to embrittlement and fracture. It is suspected that the steel becomes brittle because the liquid metals leach carbon from it. Suppose that you had at your disposal the following materials: a fully equipped research laboratory, a small research foundry for making steel, a gas chromatograph, a scintillation counter, and supplies of sodium, potassium, iron, molybdenum, and carbon-14. Design an experiment to test the hypothesis that liquid sodium and potassium mixtures can dissolve carbon from steel.

Connection 6

You have come to the end of your general chemistry course. What can you do with all that hard-won knowledge? How about terraforming Mars? That is, take a whole, inhospitable planet and make it habitable. As far as we know, Mars is currently lifeless, so we would not be destroying life or other civilizations. Far from it: colonization can be regarded as a crucially important step in the fulfillment of the potential of humanity. At a stroke, you can solve many of the pressing problems on Earth and set humans on the path to populating the galaxy.

Terraforming Mars will not be easy. For one thing, no one yet knows how to do it, and no one has even been to the planet. A key will be the availability of energy. But energy alone is not enough: we need to understand the problems and then devise imaginative solutions. General chemistry won't carry us all the way, but if you have followed the material in this text, you will be in a good position to start. Perhaps yours will be the generation that sets Mars on its path to being the home for your descendants.

What do we know about Mars today and what is required to make it habitable? The radius of Mars is 53% that of Earth, its mass is only 11% that of the Earth, and the strength of gravity at its surface is 38% that of Earth. On Mars, a person of mass 65 kg would feel as light as someone of mass 25 kg on Earth. Mars is 1.5 times further from the Sun than Earth is, so the daily input of solar energy is only about half that on Earth. As a result, the surface temperature on Mars averages only 230 K ($-43°C$), with wide variations. When dust storms occur, about once a Martian year, the temperature of the atmosphere may rise by nearly 50°C, because the dust particles absorb radiation. During the winter, the temperature is low enough at the poles for carbon dioxide to freeze (Chapter 10), and the polar ice caps are mainly carbon dioxide. The atmosphere on Mars is mostly carbon dioxide, which is poisonous to humans (Chapter 3). Moreover, its pressure at ground level is only 1% that of Earth. That low pressure is partly due to the small amount of gases present, the low temperature, and the weak pull of gravity (Chapter 5).

There is no surface water on Mars but plenty of evidence that water has run there in the past. There may be a lot of

Mars as it is today, a parched desert world hostile to life.

Your project is showing signs of success. Water is emerging on to the surface from aquifers, but the temperature is so low that it immediately freezes and the atmospheric pressure is still too low for comfort.

water underground in aquifers, water-bearing layers of the *regolith,* the granular material that lies on top of the bedrock. Deep down, as a result of the warming caused by radioactive decay, the water may even be liquid.

Surveys of the Martian surface for mineral deposits are still in their infancy, but we can assume that the composition of the crust is similar to that of Earth, except for the absence of organic deposits like coal and petroleum (Chapter 11). One of your initial missions, therefore, would be to perform a thorough survey of the resources of the planet.

To create conditions on Mars that would support life, you will need to transform the atmosphere and to raise the surface temperature. These two aims are linked, because the greenhouse effect of an atmosphere that is abundant in carbon dioxide would help to raise the temperature, just as it does on Earth (Chapter 4). However, although more abundant carbon dioxide is beneficial in the short term, it would need to be replaced by other gases, specifically nitrogen and oxygen, if plants and animals are to thrive. One judgment you need to make is whether to try to increase the partial pressure of carbon dioxide significantly, to kick-start the warming

process, or to find alternative ways of heating that avoid the later problem of removing the excess carbon dioxide.

Plenty of carbon dioxide is available on the planet, frozen in the polar ice caps and locked into the carbonate rocks of the surface. If you could generate enough energy, you could sublime the ice caps (Chapter 6) or decompose the carbonates into carbon dioxide and oxides by reversing the Lewis acid-base reactions that formed them (Chapter 8). You might consider using a global flood of sulfuric acid to release CO_2 from the carbonate ions, but that might well prove to be more trouble than it's worth in the long term, and you would need to establish a sulfuric acid plant first (Chapter 20). That would be difficult in the initial absence of water. So, you need abundant energy to effect atmospheric change.

Three sources of energy are available to you. One is geothermal energy, the heat from within the planet. Just as on Earth, the high temperature of the interior of the planet is due to radioactive decay. To release that heat to the surface, you will need to bore numerous holes all over the planet, reaching down through the mantle. Unfortunately, boring holes takes energy, so you still need another source of energy

You are now well into your project: the atmosphere has been thickened and the temperature raised to the point that subsurface water has begun to vaporize, condensing on the higher places and pooling as liquid at lower altitudes.

You have completed your project successfully. There is now a full ocean in the northern latitudes, where evidence suggests there may have been an ancient ocean. Now your task is to populate the planet, allow societies to flourish and preserve this hard-won world for posterity.

initially. One approach would be to establish a number of nuclear reactors on the surface (Chapter 22). The first of these reactors would need to be imported, but once it was up and running, it output could be used to build more reactors and to mine more uranium for fuel.

The third source of energy available to you is the Sun. Although the Sun's power is diminished at such great distances, it still pours huge quantities of energy on to the surface, and you could enhance the influx of energy by ringing the planet with huge orbital mirrors. One problem with mirrors is that they tend to be blown out of orbit by radiation pressure and the solar wind (the stream of fast moving particles ejected by the Sun), so you would have to select their locations very carefully. Note, incidentally, that all three sources of energy available to you initially are nuclear: geothermal, uranium reactor, and solar energies are all based on either the fission or fusion of nuclei (Chapter 22). Fossil fuels (Chapter 6) will not be available, so you will probably decide to adopt hydrogen as the primary portable fuel.

As the planet slowly warms, you will need to find sources of water, oxygen, and nitrogen (Chapters 19 and 20). If there is water underground, it could be brought to the surface once global warming has taken hold. Then you could initially flood a few craters and, later, great stretches of the planet. With luck, the rainfall that would accompany the Great Flood (as it might be recalled by future generations of Martian humans) would wash a lot of the carbon dioxide out of the air in the form of mildly acidic rain (Chapter 15). You would have to be very careful that the resulting loss of atmospheric carbon dioxide did not freeze the planet again. One option would be to synthesize fluorocarbons that could be released on a massive scale into the upper atmosphere and act as greenhouse gases. You would need to investigate their spectroscopic absorption properties very carefully, because you want the molecules to be transparent to the beneficial ultraviolet radiation bands but to absorb infrared radiation (Chapter 9). They should be stable in the presence of the intense radiation they experience high in the atmosphere. In your models of different atmospheres, you will need to take into account the rates of reactions that might take place (Chapter 13) and your knowledge of the properties of gases (Chapter 5).

You will need to generate soil and make it fertile. One problem is that much of the mixture that makes up soil on Earth is living matter or its remains. So you will need to establish colonies of bacteria that can thrive at low temperatures and which are highly effective at fixing nitrogen. Almost certainly, you will have to resort to the use of genetic engineering (Chapter 11) to combine the ability of some bacteria to fix nitrogen (Chapter 19) with the ability of other bacteria to survive intense cold and intense radiation.

The first colonies you establish will live in bubbles inside which the atmosphere will be kept suitable for humans. The development of lightweight tenting material that can withstand radiation will be a first priority and will depend on the skills of chemists. The supporting ropes might be made of carbon nanotubes reinforced with strands of artificial diamonds (Chapters 10 and 19). So the bulk production of these materials should be a priority. The strength of carbon-fluorine bonds (Chapter 9) suggests that fluorocarbons might be suitable for the covering material for the tents, so it would be sensible to search for abundant supplies of fluorides on the planet, and to use your nuclear reactors to electrolyze them (Chapter 18). There are no hydrocarbons on the planet, so you will need to synthesize them from inorganic materials. By the time that synthesis is taking place, supplies on Earth may have run out, so the terraforming of Mars may be of use to the sustenance of life on Earth.

All in all, terraforming another planet will be a colossal task. But you can be confident that it will draw on just about every aspect of the chemistry you have learned in this book. Good luck!

For Further Reading

K. S. Robinson, *Red Mars, Green Mars,* and *Blue Mars,* a trilogy, New York: Harper Collins, 1992.

http://www.hawastsoc.org/solar/eng/mars.htm Information about Mars and many pictures.

http://www.users.globalnet.co.uk/~mfogg/index.htm The Terraforming Information Site.

http://www.mars2030.net/ A U. S. national effort to generate ideas for terraforming Mars.

Applying Your Knowledge

You have only one assignment: devise a plan to begin the terraforming of Mars. Your plan should indicate how you will accomplish these things:

(a) Raise the temperature of the surface.

(b) Increase the atmospheric pressure.

(c) Increase the amount of oxygen, nitrogen, and water.

(d) Reduce harmful radiation at the surface.

You may use as raw materials either local Martian materials or materials that can be found on asteroids. For example, it has been proposed that nitrogen could be added to the atmosphere of Mars by redirecting ammonia-rich asteroids into a collision course with Mars. The Web sites listed can be used as resources.

Mathematical Information

1A Algebra Rules

Most of the chemical calculations described in this book can be expressed as algebraic equations. This section of Appendix 1 reviews the basic rules of algebra and arithmetic that you will need in this course.

1. Negative Numbers

Adding a negative number is the same as subtracting a positive number. *Example:* To add -7 to 3, write $3 - 7 = -4$.

Subtracting a negative number is the same as adding a positive number. *Example:* To subtract -7 from 3, write $3 - (-7) = 3 + 7 = 10$.

Multiplying or dividing two numbers gives a negative result if one of the numbers is negative and a positive result if the two numbers are either both positive or both negative. *Example:* $(-7) \times 3 = -21$, but $(-7) \times (-3) = 21$.

2. Fractions and Percentages

A fraction is the ratio of two numbers and may be less than or greater than 1. The number on top is the *numerator* and the number on the bottom is the *denominator*.

A fraction can be simplified by dividing both the numerator and the denominator by the same number; the value of the fraction does not then change because this action is the same as multiplying the fraction by 1. *Example:*

$$\frac{4}{6} = \frac{4/2}{6/2} = \frac{2}{3}$$

When multiplying fractions, multiply the numerators together and multiply the denominators together. *Example:*

$$\frac{2}{3} \times \frac{5}{4} = \frac{2 \times 5}{3 \times 4} = \frac{10}{12} = \frac{5}{6}$$

When dividing fractions, invert the divisor. *Example:*

$$\frac{2}{5} \bigg/ \frac{4}{3} = \frac{2}{5} \times \frac{3}{4} = \frac{6}{20} = \frac{3}{10}$$

Sometimes you will see $\frac{2}{5}\big/\frac{4}{3}$ written

$$\frac{2/5}{4/3}$$

Convert fractions to decimal notation by dividing them. *Examples:* $\frac{5}{4} = 1.25$, $\frac{2}{3} = 0.66 \ldots$. If there is one digit to the right of the decimal point, the corresponding fraction has 10 in the denominator, so 0.3 is $\frac{3}{10}$. If there are two digits, use 100 in the denominator, so 0.33 is $\frac{33}{100}$, and so on.

A percentage is a fraction of 100, so 5% is $\frac{5}{100}$. To find a percentage from a fraction, multiply the fraction by 100%. *Example:* $\frac{2}{3} \times 100\% = 67\%$.

3. Solving a Linear Equation for an Unknown

An algebraic equation expresses an unknown quantity, x, in terms of other numbers. *Example:* $ax + b = c$, where a, b, and c are any positive or negative numbers.

We can manipulate both sides of an algebraic equation in the following ways:

add or subtract the same number

multiply or divide by the same number

invert (express each side as 1 divided by that side)

Examples: Subtract b from both sides of $ax + b = c$, to get $ax = c - b$. Divide both sides of the resulting equation by a to get $x = (c - b)/a$. Thus, if $3x + 1 = 7$, $x = (7 - 1)/3 = 2$.

4. Ratios

A ratio is a fraction formed by dividing one number by another. Ratios are frequently used in chemistry and often appear in algebraic expressions. *Example:* Solve $a/x = b/c$ for x. First, multiply each side by c and then by x, which gives $ca = xb$. Then divide each side by b to give $x = ca/b$.

5. Factors and Sums

When two quantities in a sum or difference can be divided by the same number, it is possible to simplify the expression by factoring. *Example:* $ax + ab = a(x + b)$.

The product of two sums is the sum of the products of each term. *Example:* $(x + a)(y + b) = xy + ay + bx + ab$.

When x appears in each of two terms that are multiplied together, the product is a quadratic equation. *Example:* The expression $(x + a)(x + b) = 0$ is the same as $x^2 + (a + b)x + ab = 0$.

1B Scientific Notation

In **scientific notation,** a number is written as $A \times 10^a$, where A is a decimal number with one nonzero digit in front of the decimal point and a is a whole number. For example, 333 is written 3.33×10^2 in scientific notation, because $10^2 = 10 \times 10 = 100$:

$$333 = 3.33 \times 100 = 3.33 \times 10^2$$

On a scientific calculator, this number is entered as

$$\boxed{3} \quad \boxed{.} \quad \boxed{3} \quad \boxed{3} \quad \boxed{EXP} \quad \boxed{2}$$

(On some calculators, the $\boxed{EXP}$ key is labeled $\boxed{EE}$ or $\boxed{EXX}$.) We use

$$10^1 = 10$$
$$10^2 = 10 \times 10 = 100$$
$$10^3 = 10 \times 10 \times 10 = 1000$$
$$10^4 = 10 \times 10 \times 10 \times 10 = 10\,000$$

and so on. Note that the number of zeros following 1 is equal to the power of 10. Thus, 10^6 is 1 followed by six zeros:

$$10^6 = 10 \times 10 \times 10 \times 10 \times 10 \times 10 = 1\,000\,000$$

Numbers between 0 and 1 are expressed in the same way, but with a negative power of 10; they have the form $A \times 10^{-a}$, with $10^{-1} = \frac{1}{10} = 0.1$, and so on. Thus, 0.0333 in decimal notation is 3.33×10^{-2} because

$$10^{-2} = \frac{1}{10} \times \frac{1}{10} = \frac{1}{100}$$

$$0.0333 = 3.33 \times \frac{1}{100} = 3.33 \times 10^{-2}$$

On a scientific calculator, this number is entered as

$$\boxed{3} \quad \boxed{.} \quad \boxed{3} \quad \boxed{3} \quad \boxed{EXP} \quad \boxed{+/-} \quad \boxed{2}$$

(Be sure to use the $\boxed{+/-}$ key, which is sometimes labeled $\boxed{CHS}$, to enter the negative power of 2 and not the $\boxed{-}$ key.)

We use

$$10^{-2} = 10^{-1} \times 10^{-1} = 0.01$$
$$10^{-3} = 10^{-1} \times 10^{-1} \times 10^{-1} = 0.001$$
$$10^{-4} = 10^{-1} \times 10^{-1} \times 10^{-1} \times 10^{-1} = 0.0001$$

When a negative power of 10 is written out as a decimal number, the number of zeros following the decimal point is one less than the number (disregarding the sign) to which 10 is raised. Thus, 10^{-6} is written as a decimal point followed by $6 - 1 = 5$ zeros and then a 1:

$$10^{-6} = 10^{-1} \times 10^{-1} \times 10^{-1} \times 10^{-1} \times 10^{-1} \times 10^{-1}$$
$$= 0.000\,001$$

To multiply numbers in scientific notation, the decimal parts of the numbers are multiplied and the powers of 10 are added:

$$(A \times 10^a) \times (B \times 10^b) = (A \times B) \times 10^{a+b}$$

An example is

$$(1.23 \times 10^2) \times (4.56 \times 10^3) = 1.23 \times 4.56 \times 10^{2+3}$$
$$= 5.61 \times 10^5$$

This rule holds even if the powers of 10 are negative:

$$(1.23 \times 10^{-2}) \times (4.56 \times 10^{-3})$$
$$= 1.23 \times 4.56 \times 10^{-2-3} = 5.61 \times 10^{-5}$$

The keystrokes for this calculation are

$$\boxed{1} \quad \boxed{.} \quad \boxed{2} \quad \boxed{3} \quad \boxed{EXP} \quad \boxed{+/-} \quad \boxed{2} \quad \boxed{\times}$$
$$\boxed{4} \quad \boxed{.} \quad \boxed{5} \quad \boxed{6} \quad \boxed{EXP} \quad \boxed{+/-} \quad \boxed{3} \quad \boxed{=}$$

The results of such calculations are adjusted so that there is one digit in front of the decimal point:

$$(4.56 \times 10^{-3}) \times (7.65 \times 10^6) = 34.88 \times 10^3$$
$$= 3.488 \times 10^4$$

When dividing two numbers in scientific notation, we divide the decimal parts of the numbers and subtract the powers of 10:

$$\frac{A \times 10^a}{B \times 10^b} = \frac{A}{B} \times 10^{a-b}$$

An example is

$$\frac{4.31 \times 10^5}{9.87 \times 10^{-8}} = \frac{4.31}{9.87} \times 10^{5-(-8)}$$
$$= 0.437 \times 10^{13} = 4.37 \times 10^{12}$$

Before adding and subtracting numbers in scientific notation, we rewrite the numbers as decimal numbers multiplied by the *same* power of 10:

$$1.00 \times 10^3 + 2.00 \times 10^2 = 1.00 \times 10^3 + 0.200 \times 10^3$$
$$= 1.20 \times 10^3$$

When raising a number in scientific notation to a particular power, we raise the decimal part of the number to the power and *multiply* the power of 10 by the power:

$$(A \times 10^a)^b = A^b \times 10^{a \times b}$$

For example, 2.88×10^4 raised to the third power is

$$(2.88 \times 10^4)^3 = 2.88^3 \times (10^4)^3 = 2.88^3 \times 10^{3 \times 4}$$
$$= 23.9 \times 10^{12} = 2.39 \times 10^{13}$$

The key sequence on a scientific calculator for this calculation is

$$2 \quad . \quad 8 \quad 8 \quad \boxed{EXP} \quad 4 \quad \boxed{y^x} \quad 3 \quad \boxed{=}$$

The rule follows from the fact that

$$(10^4)^3 = 10^4 \times 10^4 \times 10^4 = 10^{4+4+4} = 10^{3 \times 4}$$

1C Logarithms

The *common logarithm* of a number x is denoted $\log x$ and is the power to which 10 must be raised to equal x. Thus, the logarithm of 100 is 2, written $\log 100 = 2$, because $10^2 = 100$. The logarithm of 1.5×10^2 is 2.18 because

$$10^{2.18} = 10^{0.18 + 2} = 10^{0.18} \times 10^2 = 1.5 \times 10^2$$

The number in front of the decimal point in the logarithm (the 2 in $\log(1.5 \times 10^2) = 2.18$) is called the *characteristic* of the logarithm; the decimal fraction (the numbers following the decimal point; the 0.18 in the example) is called the *mantissa*. The characteristic is the power of 10 in the original number (the power 2 in 1.5×10^2) and the mantissa is the logarithm of the decimal number written with one nonzero digit in front of the decimal point (the 1.5 in the example).

The distinction between the characteristic and the mantissa is important when we have to decide how many significant figures to retain in a calculation that involves logarithms (as we do in the calculation of pH): *the number of significant figures in the mantissa is equal to the number of significant figures in the decimal number.* Because the decimal number 1.5×10^2 has two significant figures, its mantissa is written 0.18 (two significant figures); so its logarithm is 2.18, as written above. Just as the power of 10 in a decimal number indicates only the location of the decimal point and plays no role in the determination of significant figures, so the characteristic of a logarithm is not included in the count of significant figures in a logarithm.

The *common antilogarithm* of a number x is the number that has x as its common logarithm. In practice, the common antilogarithm of x is simply another name for 10^x, so the common antilogarithm of 2 is $10^2 = 100$ and that of 2.18 is

$$10^{2.18} = 10^{0.18 + 2} = 10^{0.18} \times 10^2 = 1.5 \times 10^2$$

In keeping with the remarks above, the *mantissa* of the logarithm (the 0.18 in 2.18) determines the number of significant figures in the antilogarithm (1.5×10^2, two significant figures).

The logarithm of a number greater than 1 is positive, and that of a number smaller than 1 (but greater than 0) is negative:

If $x > 1, \log x > 0$

If $x = 1, \log x = 0$

If $x < 1, \log x < 0$

On a scientific calculator, the common logarithm of a number x is calculated by entering x and pressing the $\boxed{\log x}$ key on the calculator. For example, the logarithm of 4.33×10^{-5} is determined with the following sequence of keystrokes:

$$4 \quad . \quad 3 \quad 3 \quad \boxed{EXP} \quad \boxed{+/-} \quad 5 \quad \boxed{\log x}$$

(It is important to use the $\boxed{+/-}$ key, not the $\boxed{-}$ key when entering the negative power of 10.) The decimal number has three significant figures in this example, so the mantissa should also be written with three significant figures, and the answer reported as -4.364. Likewise, the common antilogarithm of x is found by entering x and pressing the $\boxed{10x}$ key (or $\boxed{INV}$ and $\boxed{\log}$ keys); so, to calculate the antilogarithm of 11.68, the keystrokes are

$$1 \quad 1 \quad . \quad 6 \quad 8 \quad \boxed{10^x}$$

Because the mantissa (0.68) of the original number has two significant figures, the antilogarithm should be written with two significant figures, and the correct answer is 4.8×10^{11}.

The *natural logarithm* of a number x is denoted $\ln x$ and is the power to which the number e = 2.718 ... must be raised to equal x. Thus, $\ln 10.0 = 2.303$, signifying that $e^{2.303} = 10.0$. The number e may seem a peculiar choice, but it occurs naturally in a number of mathematical expressions, and its use simplifies many formulas. On an electronic calculator, the natural logarithm of x is calculated by entering x and then pressing the $\boxed{\ln x}$ key. There is no simple rule for assessing the correct number of significant figures when natural logarithms are used: one way is to convert natural logarithms to common logarithms (see below) and then to use the rules specified above.

Common and natural logarithms are related by the expression

$$\ln x = \ln 10 \times \log x = 2.303 \times \log x$$

The *natural antilogarithm* of x is normally called the *exponential* of e; it is the value of e raised to the power of x. On a calculator, it is obtained by entering x and pressing the $\boxed{e^x}$ key (or the $\boxed{INV}$ and $\boxed{\ln x}$ keys). Thus, the natural antilogarithm of 2.303 is $e^{2.303} = 10.0$.

The following relationships between logarithms are useful. They are written here mainly for common logarithms, but they apply to natural logarithms as well.

Relation	Example
$\log 10^x = x$	$\log 10^{-7} = -7$
$\ln e^x = x$	$\ln e^{-kt} = -kt$
$\log x + \log y = \log xy$	$\log[\text{Ag}^+] + \log[\text{Cl}^-] =$ $\log[\text{Ag}^+][\text{Cl}^-]$
$\log x - \log y = \log\left(\dfrac{x}{y}\right)$	$\log A_0 - \log A = \log\left(\dfrac{A_0}{A}\right)$
$x \log y = \log y^x$	$2\log[\text{H}^+] = \log[\text{H}^+]^2$
$\log\left(\dfrac{1}{x}\right) = -\log x$	$\log\left(\dfrac{1}{[\text{H}^+]}\right) = -\log[\text{H}^+]$

Examples:

1. pH

The pH of a solution is the negative logarithm of $[\text{H}_3\text{O}^+]$. If the molar concentration of H_3O^+ ions in a solution is 0.0024 mol/L, then the pH of the solution is $-\log 0.0024$. To calculate the pH on a scientific calculator, first convert the concentration to scientific notation and then enter

$$\boxed{2}\ \boxed{.}\ \boxed{4}\ \boxed{\text{EXP}}\ \boxed{+/-}\ \boxed{3}\ \boxed{\log x}\ \boxed{+/-}$$

The mantissa of this pH can have only two significant figures (preceded by the characteristic), so the pH is reported as 2.62. Conversely, if the pH of the solution is measured as 7.4, the hydrogen ion concentration is the value of $10^{-7.4}$, which is evaluated by using the keystrokes

$$\boxed{7}\ \boxed{.}\ \boxed{4}\ \boxed{+/-}\ \boxed{10^x}$$

There is only one significant figure in the mantissa (the 4 in pH = 7.4), so the hydrogen ion concentration is reported as 4×10^{-8} mol/L, with only one significant figure. If your hydronium ion concentration is impossibly large, you may have forgotten to press the $\boxed{+/-}$ key *before* pressing $\boxed{10^x}$.

2. Rate Laws

Logarithms are useful for solving expressions of the form

$$a^x = b$$

for the unknown x. (This type of calculation can arise in the study of chemical kinetics when the order of a reaction is being determined.) We take logarithms of both sides

$$\log a^x = \log b$$

and from a relation given above:

$$x \log a = \log b$$

Therefore,

$$x = \frac{\log b}{\log a}$$

1D Quadratic and Cubic Equations

A *quadratic equation* is an equation of the form

$$ax^2 + bx + c = 0$$

The two *roots* of the equation (the solutions) can be found most easily by using a graphing calculator or plotting program (Fig. A.1). However, they can also be calculated from the expressions:

$$x_1 = \frac{-b + \sqrt{(b^2 - 4ac)}}{2a}$$

$$x_2 = \frac{-b - \sqrt{(b^2 - 4ac)}}{2a}$$

When a quadratic equation arises in connection with a chemical calculation, we accept only the root that leads to a physically plausible result. For example, if x is a concentration, then it must be a positive number, and a negative root can be ignored. However, if x is a *change* in concentration, then it may be either positive or negative. In such a case, we would have to determine which root led to an acceptable (positive) final concentration.

On occasion, an equilibrium table (or some other type of calculation) results in a *cubic equation*:

$$ax^3 + bx^2 + cx + d = 0$$

Cubic equations are very tedious to solve exactly, so it is better to use a graphing calculator or plotter, such as the one on the CD that accompanies this book (Fig. A.2).

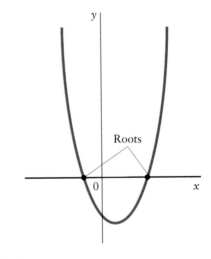

FIGURE A.1

The two roots of a quadratic equation can be found from a plot of the equation.

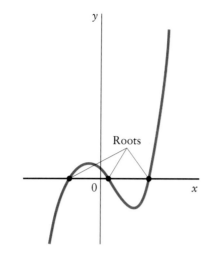

FIGURE A.2

The three roots of a cubic equation can be determined by plotting the equation.

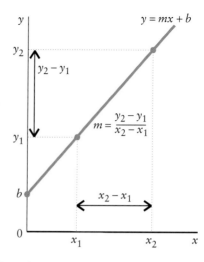

FIGURE A.3

The straight line $y = mx + b$. Its intercept at $x = 0$ is b and its slope is m.

1E Graphs

Experimental data can often be analyzed by plotting a graph. In many cases, the best procedure is to find a way of plotting the data so that a straight line results. This is more useful than plotting a curve, because it is quite easy to tell whether or not the points do in fact fall on a straight line, whereas small deviations from a curve are harder to detect. Moreover, it is also easy to calculate the slope of a straight line, to *extrapolate* (or extend) a straight line beyond the range of the data, and to *interpolate* between the data (that is, find a value between two measured values).

The formula of a straight line graph of y (the vertical axis) plotted against x (the horizontal axis) is

$$y = mx + b$$

Here b is the *intercept* of the graph with the y-axis (Fig. A.3), the value of y where the graph cuts through the vertical axis at $x = 0$. The *slope* of the graph, its gradient, is m. The slope can be calculated by choosing two points, x_1 and x_2, and their corresponding values on the y-axis, y_1 and y_2, and substituting the values into the formula

$$m = \frac{y_2 - y_1}{x_2 - x_1}$$

Because b is the intercept and m is the slope, the equation of the straight line is equivalent to

$$y = \text{slope} \times x + \text{intercept}$$

Many of the equations we meet in the text can be rearranged to give a straight line graph when plotted. These equations include

Application	$y = $ slope $\times$ x	$+$	intercept
Temperature scale conversions	$°C = 1 \times K$	$-$	273.15
	$°F = 1.8 \times °C$	$+$	32
Ideal gas law	$P = nRT \times \dfrac{1}{V}$		
First-order integrated rate law	$\ln[A] = -k \times t$	$+$	$\ln[A]_0$
Second-order integrated rate law	$\dfrac{1}{[A]} = k \times t$	$+$	$\dfrac{1}{[A]_0}$
Arrhenius law	$\ln k = \dfrac{-E_a}{R} \times \dfrac{1}{T}$	$+$	A

We can speak of *the* slope of a straight line because the slope is the same at all points. On a curve, however, the slope changes from point to point. The *tangent* of a curve at a specified point is the straight line that has the same slope as the curve at that point. The tangent can be found by a series of approximations, as shown in Fig. A.4. Approximation 1 is found by drawing a point on the curve on each side of the point of interest (corresponding to equal distances along the x-axis) and joining them by a straight line. A better approxi-

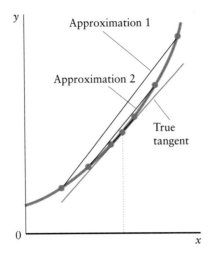

Approximation 1

Approximation 2

True
tangent

mation (approximation 2) is obtained by moving the points an equal distance closer to the point of interest and drawing a new line. The "exact" tangent is obtained when the two points are virtually coincident with the point of interest. Its slope is then equal to the slope of the curve at the point of interest. This technique can be used to measure the rate of a chemical reaction at a specified time (see Section 13.2).

FIGURE A.4

Successive approximations to the tangent are obtained as the two points defining the straight line come closer together and finally coincide.

2A Thermodynamic Data at 25°C

INORGANIC SUBSTANCES

Substance	Molar mass, M, g/mol	Enthalpy of formation, $\Delta H_f°$, kJ/mol	Free energy of formation, $\Delta G_f°$, kJ/mol	Molar heat capacity, $C_{P,m}$, J/K·mol	Molar entropy,* $S_m°$, J/K·mol
Aluminum					
Al(s)	26.98	0	0	24.35	28.33
Al^{3+}(aq)	26.98	−524.7	−481.2	—	−321.7
Al_2O_3(s)	101.96	−1675.7	−1582.3	79.04	50.92
$Al(OH)_3$(s)	78.00	−1276	—	—	—
$AlCl_3$(s)	133.33	−704.2	−628.8	91.84	110.67
Antimony					
Sb(s)	121.75	0	0	25.23	45.69
SbH_3(g)	124.77	+145.11	+147.75	41.05	232.78
$SbCl_3$(g)	228.10	−313.8	−301.2	76.69	337.80
$SbCl_5$(g)	299.00	−394.34	−334.29	121.13	401.94
Arsenic					
As(s), gray	74.92	0	0	24.64	35.1
As_2S_3(s)	246.02	−169.0	−168.6	116.3	163.6
AsO_4^{3-}(aq)	138.92	−888.14	−648.41	—	−162.8
Barium					
Ba(s)	137.34	0	0	28.07	62.8
Ba^{2+}(aq)	137.34	−537.64	−560.77	—	+9.6
BaO(s)	153.34	−553.5	−525.1	47.78	70.42
$BaCO_3$(s)	197.35	−1216.3	−1137.6	85.35	112.1
$BaCO_3$(aq)	197.35	−1214.78	−1088.59	—	−47.3
Boron					
B(s)	10.81	0	0	11.09	5.86
B_2O_3(s)	69.62	−1272.8	−1193.7	62.93	53.97
BF_3(g)	67.81	−1137.0	−1120.3	50.46	254.12
Bromine					
Br_2(l)	159.82	0	0	75.69	152.23
Br_2(g)	159.82	+30.91	+3.11	36.02	245.46
Br(g)	79.91	+111.88	+82.40	20.79	175.02
Br^-(aq)	79.91	−121.55	−103.96	—	+82.4
HBr(g)	80.92	−36.40	−53.45	29.14	198.70
Calcium					
Ca(s)	40.08	0	0	25.31	41.42
Ca(g)	40.08	+178.2	+144.3	20.79	154.88
Ca^{2+}(aq)	40.08	−542.83	−553.58	—	−53.1
CaO(s)	56.08	−635.09	−604.03	42.80	39.75
$Ca(OH)_2$(s)	74.10	−986.09	−898.49	87.49	83.39
$Ca(OH)_2$(aq)	74.10	−1002.82	−868.07	—	−74.5

(continued)

Substance	Molar mass, M, g/mol	Enthalpy of formation, ΔH_f°, kJ/mol	Free energy of formation, ΔG_f°, kJ/mol	Molar heat capacity, $C_{P,m}$, J/K·mol	Molar entropy,* S_m°, J/K·mol
$CaCO_3(s)$, calcite	100.09	−1206.9	−1128.8	81.88	92.9
$CaCO_3(s)$, aragonite	100.09	−1207.1	−1127.8	81.25	88.7
$CaCO_3(aq)$	100.09	−1219.97	−1081.39	—	−110.0
$CaF_2(s)$	78.08	−1219.6	−1167.3	67.03	68.87
$CaF_2(aq)$	78.08	−1208.09	−1111.15	—	−80.8
$CaCl_2(s)$	110.98	−795.8	−748.1	72.59	104.6
$CaCl_2(aq)$	110.98	−877.1	−816.0	—	59.8
$CaBr_2(s)$	199.90	−682.8	−663.6	72.59	130.
$CaC_2(s)$	64.10	−59.8	−64.9	62.72	69.96
$CaSO_4(s)$	136.14	−1434.11	−1321.79	99.66	106.7
$CaSO_4(aq)$	136.14	−1452.10	−1298.10	—	−33.1

Carbon (for organic compounds, see the next table)

Substance	Molar mass, M, g/mol	Enthalpy of formation, ΔH_f°, kJ/mol	Free energy of formation, ΔG_f°, kJ/mol	Molar heat capacity, $C_{P,m}$, J/K·mol	Molar entropy,* S_m°, J/K·mol
$C(s)$, graphite	12.01	0	0	8.53	5.740
$C(s)$, diamond	12.01	+1.895	+2.900	6.11	2.377
$C(g)$	12.01	+716.68	+671.26	20.84	158.10
$CO(g)$	28.01	−110.53	−137.17	29.14	197.67
$CO_2(g)$	44.01	−393.51	−394.36	37.11	213.74
$CO_3^{2-}(aq)$	60.01	−677.14	−527.81	—	−56.9
$CCl_4(l)$	153.81	−135.44	−65.21	131.75	216.40
$CS_2(l)$	76.13	+89.70	+65.27	75.7	151.34
$HCN(g)$	27.03	+135.1	+124.7	35.86	201.78
$HCN(l)$	27.03	+108.87	+124.97	70.63	112.84
$HCN(aq)$	27.03	+107.1	+119.7	—	124.7

Cerium

Substance	Molar mass, M, g/mol	Enthalpy of formation, ΔH_f°, kJ/mol	Free energy of formation, ΔG_f°, kJ/mol	Molar heat capacity, $C_{P,m}$, J/K·mol	Molar entropy,* S_m°, J/K·mol
$Ce(s)$	140.12	0	0	26.94	72.0
$Ce^{3+}(aq)$	140.12	−696.2	−672.0	—	−205
$Ce^{4+}(aq)$	140.12	−537.2	−503.8	—	−301

Chlorine

Substance	Molar mass, M, g/mol	Enthalpy of formation, ΔH_f°, kJ/mol	Free energy of formation, ΔG_f°, kJ/mol	Molar heat capacity, $C_{P,m}$, J/K·mol	Molar entropy,* S_m°, J/K·mol
$Cl_2(g)$	70.90	0	0	33.91	223.07
$Cl(g)$	35.45	+121.68	+105.68	21.84	165.20
$Cl^-(aq)$	35.45	−167.16	−131.23	—	+56.5
$HCl(g)$	36.46	−92.31	−95.30	29.12	186.91
$HCl(aq)$	36.46	−167.16	−131.23	—	56.5

Copper

Substance	Molar mass, M, g/mol	Enthalpy of formation, ΔH_f°, kJ/mol	Free energy of formation, ΔG_f°, kJ/mol	Molar heat capacity, $C_{P,m}$, J/K·mol	Molar entropy,* S_m°, J/K·mol
$Cu(s)$	63.54	0	0	24.44	33.15
$Cu^+(aq)$	63.54	+71.67	+49.98	—	+40.6
$Cu^{2+}(aq)$	63.54	+64.77	+65.49	—	−99.6
$Cu_2O(s)$	143.08	−168.6	−146.0	63.64	93.14
$CuO(s)$	79.54	−157.3	−129.7	42.30	42.63
$CuSO_4(s)$	159.60	−771.36	−661.8	100.0	109
$CuSO_4 \cdot 5H_2O(s)$	249.68	−2279.7	−1879.7	280.	300.4

Deuterium

Substance	Molar mass, M, g/mol	Enthalpy of formation, ΔH_f°, kJ/mol	Free energy of formation, ΔG_f°, kJ/mol	Molar heat capacity, $C_{P,m}$, J/K·mol	Molar entropy,* S_m°, J/K·mol
$D_2(g)$	4.028	0	0	29.20	144.96
$D_2O(g)$	20.028	−249.20	−234.54	34.27	198.34
$D_2O(l)$	20.028	−294.60	−243.44	34.27	75.94

Substance	Molar mass, M, g/mol	Enthalpy of formation, $\Delta H_f°$, kJ/mol	Free energy of formation, $\Delta G_f°$, kJ/mol	Molar heat capacity, $C_{P,m}$, J/K·mol	Molar entropy,* $S_m°$, J/K·mol
Fluorine					
$F_2(g)$	38.00	0	0	31.30	202.78
$F^-(aq)$	19.00	-332.63	-278.79	—	-13.8
$HF(g)$	20.01	-271.1	-273.2	29.13	173.78
$HF(aq)$	20.01	-330.08	-296.82	—	88.7
Hydrogen (see also Deuterium)					
$H_2(g)$	2.0158	0	0	28.82	130.68
$H(g)$	1.0079	$+217.97$	$+203.25$	20.78	114.71
$H^+(aq)$	1.0079	0	0	0	0
$H_2O(l)$	18.02	-285.83	-237.13	75.29	69.91
$H_2O(g)$	18.02	-241.82	-228.57	33.58	188.83
$H_2O_2(l)$	34.02	-187.78	-120.35	89.1	109.6
$H_2O_2(aq)$	34.02	-191.17	-134.03	—	143.9
$H_3O^+(aq)$	19.02	-285.83	-237.13	75.29	$+69.91$
Iodine					
$I_2(s)$	253.80	0	0	54.44	116.14
$I_2(g)$	253.80	$+62.44$	$+19.33$	36.90	260.69
$I^-(aq)$	126.90	-55.19	-51.57	—	$+111.3$
$HI(g)$	127.91	$+26.48$	$+1.70$	29.16	206.59
Iron					
$Fe(s)$	55.85	0	0	25.10	27.28
$Fe^{2+}(aq)$	55.85	-89.1	-78.90	—	-137.7
$Fe^{3+}(aq)$	55.85	-48.5	-4.7	—	-315.9
$Fe_3O_4(s)$, magnetite	231.55	-1118.4	-1015.4	143.43	146.4
$Fe_2O_3(s)$, hematite	159.70	-824.2	-742.2	103.85	87.40
$FeS(s, \alpha)$	87.91	-100.0	-100.4	50.54	60.29
$FeS(aq)$	87.91	—	$+6.9$	—	—
$FeS_2(s)$	119.98	-178.2	-166.9	62.17	52.93
Lead					
$Pb(s)$	207.19	0	0	26.44	64.81
$Pb^{2+}(aq)$	207.19	-1.7	-24.43	—	$+10.5$
$PbO_2(s)$	239.19	-277.4	-217.33	64.64	68.6
$PbSO_4(s)$	303.25	-919.94	-813.14	103.21	148.57
$PbBr_2(s)$	367.01	-278.7	-261.92	80.12	161.5
$PbBr_2(aq)$	367.01	-244.8	-232.34	—	175.3
Magnesium					
$Mg(s)$	24.31	0	0	24.89	32.68
$Mg(g)$	24.31	$+147.70$	$+113.10$	20.79	148.65
$Mg^{2+}(aq)$	24.31	-466.85	-454.8	—	-138.1
$MgO(s)$	40.31	-601.70	-569.43	37.15	26.94
$MgCO_3(s)$	84.32	-1095.8	-1012.1	75.52	65.7
$MgBr_2(s)$	184.13	-524.3	-503.8	—	117.2

(continued)

Substance	Molar mass, M, g/mol	Enthalpy of formation, ΔH_f°, kJ/mol	Free energy of formation, ΔG_f°, kJ/mol	Molar heat capacity, $C_{P,m}$, kJ·mol	Molar entropy,* S_m°, J/K·mol
Mercury					
$Hg(l)$	200.59	0	0	27.98	76.02
$Hg(g)$	200.59	+61.32	+31.82	20.79	174.96
$HgO(s)$	216.59	−90.83	−58.54	44.06	70.29
$Hg_2Cl_2(s)$	472.08	−265.22	−210.75	102	192.5
Nitrogen					
$N_2(g)$	28.02	0	0	19.12	191.61
$NO(g)$	30.01	+90.25	+86.55	29.84	210.76
$N_2O(g)$	44.02	+82.05	+104.20	38.45	219.85
$NO_2(g)$	46.01	+33.18	+51.31	37.20	240.06
$N_2O_4(g)$	92.02	+9.16	+97.89	77.28	304.29
$HNO_3(l)$	63.02	−174.10	−80.71	109.87	155.60
$HNO_3(aq)$	63.02	−207.36	−111.25	—	146.4
$NO_3^-(aq)$	62.02	−205.0	−108.74	—	+146.4
$NH_3(g)$	17.03	−46.11	−16.45	35.06	192.45
$NH_3(aq)$	17.03	−80.29	−26.50	—	111.3
$NH_4^+(aq)$	18.04	−132.51	−79.31	—	+113.4
$NH_2OH(s)$	33.03	−114.2	—	—	—
$HN_3(g)$	43.04	+294.1	+328.1	98.87	238.97
$N_2H_4(l)$	32.05	+50.63	+149.34	139.3	121.21
$NH_4NO_3(s)$	80.05	−365.56	−183.87	84.1	151.08
$NH_4Cl(s)$	53.49	−314.43	−202.87	—	94.6
$NH_4ClO_4(s)$	117.49	−295.31	−88.75	—	186.2
Oxygen					
$O_2(g)$	32.00	0	0	29.36	205.14
$O_3(g)$	48.00	+142.7	+163.2	39.29	238.93
$OH^-(aq)$	17.01	−229.99	−157.24	—	−10.75
Phosphorus					
$P(s)$, white	30.97	0	0	23.84	41.09
$P_4(g)$	123.88	+58.91	+24.44	67.15	279.98
$PH_3(g)$	33.99	+5.4	+13.4	37.11	210.23
$P_4O_{10}(s)$	283.88	−2984.0	−2697.0	—	228.86
$H_3PO_3(aq)$	81.99	−964.8	—	—	—
$H_3PO_4(l)$	97.99	−1266.9	—	—	—
$H_3PO_4(aq)$	97.99	−1288.34	−1142.54	—	158.2
$PCl_3(l)$	137.32	−319.7	−272.3	—	217.18
$PCl_3(g)$	137.32	−287.0	−267.8	71.84	311.78
$PCl_5(g)$	208.22	−374.9	−305.0	112.8	364.6
$PCl_5(s)$	208.22	−443.5	—	—	—
Potassium					
$K(s)$	39.10	0	0	29.58	64.18
$K(g)$	39.10	+89.24	+60.59	20.79	160.34
$K^+(aq)$	39.10	−252.38	−283.27	—	+102.5
$KOH(s)$	56.11	−424.76	−379.08	64.9	78.9

Substance	Molar mass, M, g/mol	Enthalpy of formation, ΔH_f°, kJ/mol	Free energy of formation, ΔG_f°, kJ/mol	Molar heat capacity, $C_{P,m}$, J/K·mol	Molar entropy,* S_m°, J/K·mol
KOH(aq)	56.11	−482.37	−440.50	—	91.6
KF(s)	58.10	−567.27	−537.75	49.04	66.57
KCl(s)	74.55	−436.75	−409.14	51.30	82.59
KBr(s)	119.01	−393.80	−380.66	52.30	95.90
KI(s)	166.00	−327.90	−324.89	52.93	106.32
KClO$_3$(s)	122.55	−397.73	−296.25	100.25	143.1
KClO$_4$(s)	138.55	−432.75	−303.09	112.38	151.0
K$_2$S(s)	110.26	−380.7	−364.0	—	105
K$_2$S(aq)	110.26	−471.5	−480.7	—	190.4
Silicon					
Si(s)	28.09	0	0	20.00	18.83
SiO$_2$(s, α)	60.09	−910.94	−856.64	44.43	41.84
Silver					
Ag(s)	107.87	0	0	25.35	42.55
Ag$^+$(aq)	107.87	+105.58	+77.11	—	+72.68
Ag$_2$O(s)	231.74	−31.05	−11.20	65.86	121.3
AgBr(s)	187.78	−100.37	−96.90	52.38	107.1
AgBr(aq)	187.78	−15.98	−26.86	—	155.2
AgCl(s)	143.32	−127.07	−109.79	50.79	96.2
AgCl(aq)	143.32	−61.58	−54.12	—	129.3
AgI(s)	234.77	−61.84	−66.19	56.82	115.5
AgI(aq)	234.77	+50.38	+25.52	—	184.1
AgNO$_3$(s)	169.88	−124.39	−33.41	93.05	140.92
Sodium					
Na(s)	22.99	0	0	28.24	51.21
Na(g)	22.99	+107.32	+76.76	20.79	153.71
Na$^+$(aq)	22.99	−240.12	−261.91	—	+59.0
NaOH(s)	40.00	−425.61	−379.49	59.54	64.46
NaOH(aq)	40.00	−470.11	−419.15	—	48.1
NaCl(s)	58.44	−411.15	−384.14	50.50	72.13
NaBr(s)	102.90	−361.06	−348.98	51.38	86.82
NaI(s)	149.89	−287.78	−286.06	52.09	98.53
Sulfur					
S(s), rhombic	32.06	0	0	22.64	31.80
S(s), monoclinic	32.06	+0.33	+0.1	23.6	32.6
S^{2-}(aq)	32.06	+33.1	+85.8	—	−14.6
SO$_2$(g)	64.06	−296.83	−300.19	39.87	248.22
SO$_3$(g)	80.06	−395.72	−371.06	50.67	256.76
H$_2$SO$_4$(l)	98.08	−813.99	−690.00	138.9	156.90
SO$_4^{2-}$(aq)	96.06	−909.27	−744.53	—	+20.1
HSO$_4^-$(aq)	97.07	−887.34	−755.91	—	+131.8
H$_2$S(g)	34.08	−20.63	−33.56	34.23	205.79
H$_2$S(aq)	34.08	−39.7	−27.83	—	121
SF$_6$(g)	146.06	−1209	−1105.3	97.28	291.82

(continued)

INORGANIC SUBSTANCES (*continued*)

Substance	Molar mass, M, g/mol	Enthalpy of formation, ΔH_f°, kJ/mol	Free energy of formation, ΔG_f°, kJ/mol	Molar heat capacity, $C_{P,m}$, J/K·mol	Molar entropy,[*] S_m°, J/K·mol
Tin					
Sn(s), white	118.69	0	0	26.99	51.55
Sn(s), gray	118.69	−2.09	+0.13	25.77	44.14
SnO(s)	134.69	−285.8	−256.9	44.31	56.5
SnO$_2$(s)	150.69	−580.7	−519.6	52.59	52.3
Zinc					
Zn(s)	65.37	0	0	25.40	41.63
Zn^{2+}(aq)	65.37	−153.89	−147.06	—	−112.1
ZnO(s)	81.37	−348.28	−318.30	40.25	43.64

[*]The entropies of individual ions in solution are determined by setting the entropy of H$^+$ in water equal to 0 and then defining the entropies of all other ions relative to this value; hence a negative entropy is one that is lower than the entropy of H$^+$ in water. All *absolute* entropies are positive, and no sign need be given; all entropies of ions are relative to that of H$^+$, and are listed here with a sign (either + or −).

ORGANIC COMPOUNDS

Substance	Molar mass, M, g/mol	Enthalpy of combustion, ΔH_c°, kJ/mol	Enthalpy of formation, ΔH_f°, kJ/mol	Free energy of formation, ΔG_f°, kJ/mol	Molar heat capacity, $C_{P,m}$, J/K·mol	Molar entropy, S_m°, J/K·mol
Hydrocarbons						
CH$_4$(g), methane	16.04	−890.	−74.81	−50.72	35.31	186.26
C$_2$H$_2$(g), ethyne (acetylene)	26.04	−1300.	+226.73	+209.20	43.93	200.94
C$_2$H$_4$(g), ethene (ethylene)	28.05	−1411	+52.26	+68.15	43.56	219.56
C$_2$H$_6$(g), ethane	30.07	−1560.	−84.68	−32.82	52.63	229.60
C$_3$H$_6$(g), propene (propylene)	42.08	−2058	+20.42	+62.78	63.89	266.6
C$_3$H$_6$(g), cyclopropane	42.08	−2091	+53.30	+104.45	55.94	237.4
C$_3$H$_8$(g), propane	44.09	−2220.	−103.85	−23.49	73.5	270.2
C$_4$H$_{10}$(g), butane	58.12	−2878	−126.15	−17.03	97.45	310.1
C$_5$H$_{12}$(g), pentane	72.14	−3537	−146.44	−8.20	120.2	349
C$_6$H$_6$(l), benzene	78.11	−3268	+49.0	+124.3	136.1	173.3
C$_6$H$_6$(g)	78.11	−3302	+82.9	+129.72	81.67	269.31

Substance	Molar mass, M, g/mol	Enthalpy of combustion, $\Delta H_c°$, kJ/mol	Enthalpy of formation, $\Delta H_f°$, kJ/mol	Free energy of formation, $\Delta G_f°$, kJ/mol	Molar heat capacity, $C_{P,m}$, J/K·mol	Molar entropy, $S_m°$, J/K·mol
C_7H_8(l), toluene	92.13	−3910.	+12.0	+113.8	—	221.0
C_7H_8(g)	92.13	−3953	+50.0	+122.0	103.6	320.7
C_6H_{12}(l), cyclohexane	84.15	−3920.	−156.4	+26.7	156.5	204.4
C_6H_{12}(g)	84.15	−3953	—	—	—	—
C_8H_{18}(l), octane	114.22	−5471	−249.9	+6.4	—	358
Alcohols and phenols						
CH_3OH(l), methanol	32.04	−726	−238.86	−166.27	81.6	126.8
CH_3OH(g)	32.04	−764	−200.66	−161.96	43.89	239.81
C_2H_5OH(l), ethanol	46.07	−1368	−277.69	−174.78	111.46	160.7
C_2H_5OH(g)	46.07	−1409	−235.10	−168.49	65.44	282.70
C_6H_5OH(s), phenol	94.11	−3054	−164.6	−50.42	—	144.0
Carboxylic acids						
HCOOH(l), formic acid	46.02	−255	−424.72	−361.35	99.04	128.95
CH_3COOH(l), acetic acid	60.05	−875	−484.5	−389.9	124.3	159.8
CH_3COOH(aq)	60.05	—	−485.76	−396.46	—	86.6
$(COOH)_2$(s), oxalic acid	90.04	−254	−827.2	−697.9	117	120.
C_6H_5COOH(s), benzoic acid	122.12	−3227	−385.1	−245.3	146.8	167.6
Aldehydes and ketones						
HCHO(g), methanal (formaldehyde)	30.03	−571	−108.57	−102.53	35.40	218.77
CH_3CHO(l), ethanal (acetaldehyde)	44.05	−1166	−192.30	−128.12	—	160.2
CH_3CHO(g)	44.05	−1192	−166.19	−128.86	57.3	250.3
CH_3COCH_3(l), propanone (acetone)	58.08	−1790.	−248.1	−155.4	124.7	200.
Sugars						
$C_6H_{12}O_6$(s), glucose	180.15	−2808	−1268	−910.	—	212
$C_6H_{12}O_6$(aq)	180.15	—	—	−917	—	—
$C_6H_{12}O_6$(s), fructose	180.15	−2810.	−1266	—	—	—
$C_{12}H_{22}O_{11}$(s), sucrose	342.29	−5645	−2222	−1545	—	360.
Nitrogen compounds						
$CO(NH_2)_2$(s), urea	60.06	−632	−333.51	−197.33	93.14	104.60
$C_6H_5NH_2$(l), aniline	93.13	−3393	+31.6	+149.1	—	191.3
NH_2CH_2COOH(s), glycine	75.07	−969	−532.9	−373.4	99.2	103.51
CH_3NH_2(g), methylamine	31.06	−1085	−22.97	+32.16	53.1	243.41

2B   Standard Potentials at 25°C
POTENTIALS IN ELECTROCHEMICAL ORDER

Reduction half-reaction	$E°$, V	Reduction half-reaction	$E°$, V
Strongly oxidizing		$NO_3^- + H_2O + 2\,e^- \longrightarrow NO_2^- + 2\,OH^-$	+0.01
$H_4XeO_6 + 2\,H^+ + 2\,e^- \longrightarrow XeO_3 + 3\,H_2O$	+3.0	$Ti^{4+} + e^- \longrightarrow Ti^{3+}$	0.00
$F_2 + 2\,e^- \longrightarrow 2\,F^-$	+2.87	$2\,H^+ + 2\,e^- \longrightarrow H_2$	0, by definition
$O_3 + 2\,H^+ + 2\,e^- \longrightarrow O_2 + H_2O$	+2.07	$Fe^{3+} + 3\,e^- \longrightarrow Fe$	−0.04
$S_2O_8^{2-} + 2\,e^- \longrightarrow 2\,SO_4^{2-}$	+2.05	$O_2 + H_2O + 2\,e^- \longrightarrow HO_2^- + OH^-$	−0.08
$Ag^{2+} + e^- \longrightarrow Ag^+$	+1.98	$Pb^{2+} + 2\,e^- \longrightarrow Pb$	−0.13
$Co^{3+} + e^- \longrightarrow Co^{2+}$	+1.81	$In^+ + e^- \longrightarrow In$	−0.14
$H_2O_2 + 2\,H^+ + 2\,e^- \longrightarrow 2\,H_2O$	+1.78	$Sn^{2+} + 2\,e^- \longrightarrow Sn$	−0.14
$Au^+ + e^- \longrightarrow Au$	+1.69	$AgI + e^- \longrightarrow Ag + I^-$	−0.15
$Pb^{4+} + 2\,e^- \longrightarrow Pb^{2+}$	+1.67	$Ni^{2+} + 2\,e^- \longrightarrow Ni$	−0.23
$2\,HClO + 2\,H^+ + 2\,e^- \longrightarrow Cl_2 + 2\,H_2O$	+1.63	$V^{3+} + e^- \longrightarrow V^{2+}$	−0.26
$Ce^{4+} + e^- \longrightarrow Ce^{3+}$	+1.61	$Co^{2+} + 2\,e^- \longrightarrow Co$	−0.28
$2\,HBrO + 2\,H^+ + 2\,e^- \longrightarrow Br_2 + 2\,H_2O$	+1.60	$In^{3+} + 3\,e^- \longrightarrow In$	−0.34
$MnO_4^- + 8\,H^+ + 5\,e^- \longrightarrow Mn^{2+} + 4\,H_2O$	+1.51	$Tl^+ + e^- \longrightarrow Tl$	−0.34
$Mn^{3+} + e^- \longrightarrow Mn^{2+}$	+1.51	$PbSO_4 + 2\,e^- \longrightarrow Pb + SO_4^{2-}$	−0.36
$Au^{3+} + 3\,e^- \longrightarrow Au$	+1.40	$Ti^{3+} + e^- \longrightarrow Ti^{2+}$	−0.37
$Cl_2 + 2\,e^- \longrightarrow 2\,Cl^-$	+1.36	$In^{2+} + e^- \longrightarrow In^+$	−0.40
$Cr_2O_7^{2-} + 14\,H^+ + 6\,e^- \longrightarrow 2\,Cr^{3+} + 7\,H_2O$	+1.33	$Cd^{2+} + 2\,e^- \longrightarrow Cd$	−0.40
$O_3 + H_2O + 2\,e^- \longrightarrow O_2 + 2\,OH^-$	+1.24	$Cr^{3+} + e^- \longrightarrow Cr^{2+}$	−0.41
$O_2 + 4\,H^+ + 4\,e^- \longrightarrow 2\,H_2O$	+1.23	$Fe^{2+} + 2\,e^- \longrightarrow Fe$	−0.44
$MnO_2 + 4\,H^+ + 2\,e^- \longrightarrow Mn^{2+} + 2\,H_2O$	+1.23	$In^{3+} + 2\,e^- \longrightarrow In^+$	−0.44
$ClO_4^- + 2\,H^+ + 2\,e^- \longrightarrow ClO_3^- + H_2O$	+1.23	$S + 2\,e^- \longrightarrow S^{2-}$	−0.48
$Pt^{2+} + 2\,e^- \longrightarrow Pt$	+1.20	$In^{3+} + e^- \longrightarrow In^{2+}$	−0.49
$Br_2 + 2\,e^- \longrightarrow 2\,Br^-$	+1.09	$Ga^+ + e^- \longrightarrow Ga$	−0.53
$Pu^{4+} + e^- \longrightarrow Pu^{3+}$	+0.97	$O_2 + e^- \longrightarrow O_2^-$	−0.56
$NO_3^- + 4\,H^+ + 3\,e^- \longrightarrow NO + 2\,H_2O$	+0.96	$U^{4+} + e^- \longrightarrow U^{3+}$	−0.61
$2\,Hg^{2+} + 2\,e^- \longrightarrow Hg_2^{2+}$	+0.92	$Se + 2\,e^- \longrightarrow Se^{2-}$	−0.67
$ClO^- + H_2O + 2\,e^- \longrightarrow Cl^- + 2\,OH^-$	+0.89	$Cr^{3+} + 3\,e^- \longrightarrow Cr$	−0.74
$Hg^{2+} + 2\,e^- \longrightarrow Hg$	+0.85	$Zn^{2+} + 2\,e^- \longrightarrow Zn$	−0.76
$NO_3^- + 2\,H^+ + e^- \longrightarrow NO_2 + H_2O$	+0.80	$Cd(OH)_2 + 2\,e^- \longrightarrow Cd + 2\,OH^-$	−0.81
$Ag^+ + e^- \longrightarrow Ag$	+0.80	$2\,H_2O + 2\,e^- \longrightarrow H_2 + 2\,OH^-$	−0.83
$Hg_2^{2+} + 2\,e^- \longrightarrow 2\,Hg$	+0.79	$Te + 2\,e^- \longrightarrow Te^{2-}$	−0.84
$AgF + e^- \longrightarrow Ag + F^-$	+0.78	$Cr^{2+} + 2\,e^- \longrightarrow Cr$	−0.91
$Fe^{3+} + e^- \longrightarrow Fe^{2+}$	+0.77	$Mn^{2+} + 2\,e^- \longrightarrow Mn$	−1.18
$BrO^- + H_2O + 2\,e^- \longrightarrow Br^- + 2\,OH^-$	+0.76	$V^{2+} + 2\,e^- \longrightarrow V$	−1.19
$MnO_4^{2-} + 2\,H_2O + 2\,e^- \longrightarrow MnO_2 + 4\,OH^-$	+0.60	$Ti^{2+} + 2\,e^- \longrightarrow Ti$	−1.63
$MnO_4^- + e^- \longrightarrow MnO_4^{2-}$	+0.56	$Al^{3+} + 3\,e^- \longrightarrow Al$	−1.66
$I_2 + 2\,e^- \longrightarrow 2\,I^-$	+0.54	$U^{3+} + 3\,e^- \longrightarrow U$	−1.79
$I_3^- + 2\,e^- \longrightarrow 3\,I^-$	+0.53	$Be^{2+} + 2\,e^- \longrightarrow Be$	−1.85
$Cu^+ + e^- \longrightarrow Cu$	+0.52	$Mg^{2+} + 2\,e^- \longrightarrow Mg$	−2.36
$Ni(OH)_3 + e^- \longrightarrow Ni(OH)_2 + OH^-$	+0.49	$Ce^{3+} + 3\,e^- \longrightarrow Ce$	−2.48
$O_2 + 2\,H_2O + 4\,e^- \longrightarrow 4\,OH^-$	+0.40	$La^{3+} + 3\,e^- \longrightarrow La$	−2.52
$ClO_4^- + H_2O + 2\,e^- \longrightarrow ClO_3^- + 2\,OH^-$	+0.36	$Na^+ + e^- \longrightarrow Na$	−2.71
$Cu^{2+} + 2\,e^- \longrightarrow Cu$	+0.34	$Ca^{2+} + 2\,e^- \longrightarrow Ca$	−2.87
$Hg_2Cl_2 + 2\,e^- \longrightarrow 2\,Hg + 2\,Cl^-$	+0.27	$Sr^{2+} + 2\,e^- \longrightarrow Sr$	−2.89
$AgCl + e^- \longrightarrow Ag + Cl^-$	+0.22	$Ba^{2+} + 2\,e^- \longrightarrow Ba$	−2.91
$Bi^{3+} + 3\,e^- \longrightarrow Bi$	+0.20	$Ra^{2+} + 2\,e^- \longrightarrow Ra$	−2.92
$SO_4^{2-} + 4\,H^+ + 2\,e^- \longrightarrow H_2SO_3 + H_2O$	+0.17	$Cs^+ + e^- \longrightarrow Cs$	−2.92
$Cu^{2+} + e^- \longrightarrow Cu^+$	+0.15	$Rb^+ + e^- \longrightarrow Rb$	−2.93
$Sn^{4+} + 2\,e^- \longrightarrow Sn^{2+}$	+0.15	$K^+ + e^- \longrightarrow K$	−2.93
$AgBr + e^- \longrightarrow Ag + Br^-$	+0.07	$Li^+ + e^- \longrightarrow Li$	−3.05
		Strongly reducing	

POTENTIALS IN ALPHABETICAL ORDER

Reduction half-reaction	$E°$, V	Reduction half-reaction	$E°$, V
$Ag^+ + e^- \longrightarrow Ag$	$+0.80$	$In^{2+} + e^- \longrightarrow In^+$	-0.40
$Ag^{2+} + e^- \longrightarrow Ag^+$	$+1.98$	$In^{3+} + e^- \longrightarrow In^{2+}$	-0.49
$AgBr + e^- \longrightarrow Ag + Br^-$	$+0.07$	$In^{3+} + 2\,e^- \longrightarrow In^+$	-0.44
$AgCl + e^- \longrightarrow Ag + Cl^-$	$+0.22$	$In^{3+} + 3\,e^- \longrightarrow In$	-0.34
$AgF + e^- \longrightarrow Ag + F^-$	$+0.78$	$K^+ + e^- \longrightarrow K$	-2.93
$AgI + e^- \longrightarrow Ag + I^-$	-0.15	$La^{3+} + 3\,e^- \longrightarrow La$	-2.52
$Al^{3+} + 3\,e^- \longrightarrow Al$	-1.66	$Li^+ + e^- \longrightarrow Li$	-3.05
$Au^+ + e^- \longrightarrow Au$	$+1.69$	$Mg^{2+} + 2\,e^- \longrightarrow Mg$	-2.36
$Au^{3+} + 3\,e^- \longrightarrow Au$	$+1.40$	$Mn^{2+} + 2\,e^- \longrightarrow Mn$	-1.18
$Ba^{2+} + 2\,e^- \longrightarrow Ba$	-2.91	$Mn^{3+} + e^- \longrightarrow Mn^{2+}$	$+1.51$
$Be^{2+} + 2\,e^- \longrightarrow Be$	-1.85	$MnO_2 + 4\,H^+ + 2\,e^- \longrightarrow Mn^{2+} + 2\,H_2O$	$+1.23$
$Bi^{3+} + 3\,e^- \longrightarrow Bi$	$+0.20$	$MnO_4^- + e^- \longrightarrow MnO_4^{2-}$	$+0.56$
$Br_2 + 2\,e^- \longrightarrow 2\,Br^-$	$+1.09$	$MnO_4^- + 8\,H^+ + 5\,e^- \longrightarrow Mn^{2+} + 4\,H_2O$	$+1.51$
$BrO^- + H_2O + 2\,e^- \longrightarrow Br^- + 2\,OH^-$	$+0.76$	$MnO_4^{2-} + 2\,H_2O + 2\,e^- \longrightarrow MnO_2 + 4\,OH^-$	$+0.60$
$Ca^{2+} + 2\,e^- \longrightarrow Ca$	-2.87	$NO_3^- + 2\,H^+ + e^- \longrightarrow NO_2 + H_2O$	$+0.80$
$Cd^{2+} + 2\,e^- \longrightarrow Cd$	-0.40	$NO_3^- + 4\,H^+ + 3\,e^- \longrightarrow NO + 2\,H_2O$	$+0.96$
$Cd(OH)_2 + 2\,e^- \longrightarrow Cd + 2\,OH^-$	-0.81	$NO_3^- + H_2O + 2\,e^- \longrightarrow NO_2^- + 2\,OH^-$	$+0.01$
$Ce^{3+} + 3\,e^- \longrightarrow Ce$	-2.48	$Na^+ + e^- \longrightarrow Na$	-2.71
$Ce^{4+} + e^- \longrightarrow Ce^{3+}$	$+1.61$	$Ni^{2+} + 2\,e^- \longrightarrow Ni$	-0.23
$Cl_2 + 2\,e^- \longrightarrow 2\,Cl^-$	$+1.36$	$Ni(OH)_3 + e^- \longrightarrow Ni(OH)_2 + OH^-$	$+0.49$
$ClO^- + H_2O + 2\,e^- \longrightarrow Cl^- + 2\,OH^-$	$+0.89$	$O_2 + e^- \longrightarrow O_2^-$	-0.56
$ClO_4^- + 2\,H^+ + 2\,e^- \longrightarrow ClO_3^- + H_2O$	$+1.23$	$O_2 + 4\,H^+ + 4\,e^- \longrightarrow 2\,H_2O$	$+1.23$
$ClO_4^- + H_2O + 2\,e^- \longrightarrow ClO_3^- + 2\,OH^-$	$+0.36$	$O_2 + H_2O + 2\,e^- \longrightarrow HO_2^- + OH^-$	-0.08
$Co^{2+} + 2\,e^- \longrightarrow Co$	-0.28	$O_2 + 2\,H_2O + 4\,e^- \longrightarrow 4\,OH^-$	$+0.40$
$Co^{3+} + e^- \longrightarrow Co^{2+}$	$+1.81$	$O_3 + 2\,H^+ + 2\,e^- \longrightarrow O_2 + H_2O$	$+2.07$
$Cr^{2+} + 2\,e^- \longrightarrow Cr$	-0.91	$O_3 + H_2O + 2\,e^- \longrightarrow O_2 + 2\,OH^-$	$+1.24$
$Cr_2O_7^{2-} + 14\,H^+ + 6\,e^- \longrightarrow 2\,Cr^{3+} + 7\,H_2O$	$+1.33$	$Pb^{2+} + 2\,e^- \longrightarrow Pb$	-0.13
$Cr^{3+} + 3\,e^- \longrightarrow Cr$	-0.74	$Pb^{4+} + 2\,e^- \longrightarrow Pb^{2+}$	$+1.67$
$Cr^{3+} + e^- \longrightarrow Cr^{2+}$	-0.41	$PbSO_4 + 2\,e^- \longrightarrow Pb + SO_4^{2-}$	-0.36
$Cs^+ + e^- \longrightarrow Cs$	-2.92	$Pt^{2+} + 2\,e^- \longrightarrow Pt$	$+1.20$
$Cu^+ + e^- \longrightarrow Cu$	$+0.52$	$Pu^{4+} + e^- \longrightarrow Pu^{3+}$	$+0.97$
$Cu^{2+} + 2\,e^- \longrightarrow Cu$	$+0.34$	$Ra^{2+} + 2\,e^- \longrightarrow Ra$	-2.92
$Cu^{2+} + e^- \longrightarrow Cu^+$	$+0.15$	$Rb^+ + e^- \longrightarrow Rb$	-2.93
$F_2 + 2\,e^- \longrightarrow 2\,F^-$	$+2.87$	$S + 2\,e^- \longrightarrow S^{2-}$	-0.48
$Fe^{2+} + 2\,e^- \longrightarrow Fe$	-0.44	$SO_4^{2-} + 4\,H^+ + 2\,e^- \longrightarrow H_2SO_3 + H_2O$	$+0.17$
$Fe^{3+} + 3\,e^- \longrightarrow Fe$	-0.04	$S_2O_8^{2-} + 2\,e^- \longrightarrow 2\,SO_4^{2-}$	$+2.05$
$Fe^{3+} + e^- \longrightarrow Fe^{2+}$	$+0.77$	$Se + 2\,e^- \longrightarrow Se^{2-}$	-0.67
$Ga^+ + e^- \longrightarrow Ga$	-0.53	$Sn^{2+} + 2\,e^- \longrightarrow Sn$	-0.14
$2\,H^+ + 2\,e^- \longrightarrow H_2$	0, by definition	$Sn^{4+} + 2\,e^- \longrightarrow Sn^{2+}$	$+0.15$
$2\,HBrO + 2\,H^+ + 2\,e^- \longrightarrow Br_2 + 2\,H_2O$	$+1.60$	$Sr^{2+} + 2\,e^- \longrightarrow Sr$	-2.89
$2\,HClO + 2\,H^+ + 2\,e^- \longrightarrow Cl_2 + 2\,H_2O$	$+1.63$	$Te + 2\,e^- \longrightarrow Te^{2-}$	-0.84
$2\,H_2O + 2\,e^- \longrightarrow H_2 + 2\,OH^-$	-0.83	$Ti^{2+} + 2\,e^- \longrightarrow Ti$	-1.63
$H_2O_2 + 2\,H^+ + 2\,e^- \longrightarrow 2\,H_2O$	$+1.78$	$Ti^{3+} + e^- \longrightarrow Ti^{2+}$	-0.37
$H_4XeO_6 + 2\,H^+ + 2\,e^- \longrightarrow XeO_3 + 3\,H_2O$	$+3.0$	$Ti^{4+} + e^- \longrightarrow Ti^{3+}$	0.00
$Hg_2^{2+} + 2\,e^- \longrightarrow 2\,Hg$	$+0.79$	$Tl^+ + e^- \longrightarrow Tl$	-0.34
$Hg^{2+} + 2\,e^- \longrightarrow Hg$	$+0.85$	$U^{3+} + 3\,e^- \longrightarrow U$	-1.79
$2\,Hg^{2+} + 2\,e^- \longrightarrow Hg_2^{2+}$	$+0.92$	$U^{4+} + e^- \longrightarrow U^{3+}$	-0.61
$Hg_2Cl_2 + 2\,e^- \longrightarrow 2\,Hg + 2\,Cl^-$	$+0.27$	$V^{2+} + 2\,e^- \longrightarrow V$	-1.19
$I_2 + 2\,e^- \longrightarrow 2\,I^-$	$+0.54$	$V^{3+} + e^- \longrightarrow V^{2+}$	-0.26
$I_3^- + 2\,e^- \longrightarrow 3\,I^-$	$+0.53$	$Zn^{2+} + 2\,e^- \longrightarrow Zn$	-0.76
$In^+ + e^- \longrightarrow In$	-0.14		

2C Ground-State Electron Configurations*

Z	Symbol	Configuration	Z	Symbol	Configuration
1	H	$1s^1$	31	Ga	$[Ar]3d^{10}4s^24p^1$
2	He	$1s^2$	32	Ge	$[Ar]3d^{10}4s^24p^2$
			33	As	$[Ar]3d^{10}4s^24p^3$
3	Li	$[He]2s^1$	34	Se	$[Ar]3d^{10}4s^24p^4$
4	Be	$[He]2s^2$	35	Br	$[Ar]3d^{10}4s^24p^5$
			36	Kr	$[Ar]3d^{10}4s^24p^6$
5	B	$[He]2s^22p^1$			
6	C	$[He]2s^22p^2$	37	Rb	$[Kr]5s^1$
7	N	$[He]2s^22p^3$	38	Sr	$[Kr]5s^2$
8	O	$[He]2s^22p^4$			
9	F	$[He]2s^22p^5$	39	Y	$[Kr]4d^15s^2$
10	Ne	$[He]2s^22p^6$	40	Zr	$[Kr]4d^25s^2$
			41	Nb	$[Kr]4d^45s^1$
11	Na	$[Ne]3s^1$	42	Mo	$[Kr]4d^55s^1$
12	Mg	$[Ne]3s^2$	43	Tc	$[Kr]4d^55s^2$
			44	Ru	$[Kr]4d^75s^1$
13	Al	$[Ne]3s^23p^1$	45	Rh	$[Kr]4d^85s^1$
14	Si	$[Ne]3s^23p^2$	46	Pd	$[Kr]4d^{10}$
15	P	$[Ne]3s^23p^3$	47	Ag	$[Kr]4d^{10}5s^1$
16	S	$[Ne]3s^23p^4$	48	Cd	$[Kr]4d^{10}5s^2$
17	Cl	$[Ne]3s^23p^5$			
18	Ar	$[Ne]3s^23p^6$	49	In	$[Kr]4d^{10}5s^25p^1$
			50	Sn	$[Kr]4d^{10}5s^25p^2$
19	K	$[Ar]4s^1$	51	Sb	$[Kr]4d^{10}5s^25p^3$
20	Ca	$[Ar]4s^2$	52	Te	$[Kr]4d^{10}5s^25p^4$
			53	I	$[Kr]4d^{10}5s^25p^5$
21	Sc	$[Ar]3d^14s^2$	54	Xe	$[Kr]4d^{10}5s^25p^6$
22	Ti	$[Ar]3d^24s^2$			
23	V	$[Ar]3d^34s^2$	55	Cs	$[Xe]6s^1$
24	Cr	$[Ar]3d^54s^1$	56	Ba	$[Xe]6s^2$
25	Mn	$[Ar]3d^54s^2$			
26	Fe	$[Ar]3d^64s^2$	57	La	$[Xe]5d^16s^2$
27	Co	$[Ar]3d^74s^2$	58	Ce	$[Xe]4f^15d^16s^2$
28	Ni	$[Ar]3d^84s^2$	59	Pr	$[Xe]4f^36s^2$
29	Cu	$[Ar]3d^{10}4s^1$	60	Nd	$[Xe]4f^46s^2$
30	Zn	$[Ar]3d^{10}4s^2$			

Z	Symbol	Configuration	Z	Symbol	Configuration
61	Pm	$[Xe]4f^5 6s^2$	91	Pa	$[Rn]5f^2 6d^1 7s^2$
62	Sm	$[Xe]4f^6 6s^2$	92	U	$[Rn]5f^3 6d^1 7s^2$
63	Eu	$[Xe]4f^7 6s^2$	93	Np	$[Rn]5f^4 6d^1 7s^2$
64	Gd	$[Xe]4f^7 5d^1 6s^2$	94	Pu	$[Rn]5f^6 7s^2$
65	Tb	$[Xe]4f^9 6s^2$	95	Am	$[Rn]5f^7 7s^2$
66	Dy	$[Xe]4f^{10} 6s^2$	96	Cm	$[Rn]5f^7 6d^1 7s^2$
67	Ho	$[Xe]4f^{11} 6s^2$	97	Bk	$[Rn]5f^9 7s^2$
68	Er	$[Xe]4f^{12} 6s^2$	98	Cf	$[Rn]5f^{10} 7s^2$
69	Tm	$[Xe]4f^{13} 6s^2$	99	Es	$[Rn]5f^{11} 7s^2$
70	Yb	$[Xe]4f^{14} 6s^2$	100	Fm	$[Rn]5f^{12} 7s^2$
71	Lu	$[Xe]4f^{14} 5d^1 6s^2$	101	Md	$[Rn]5f^{13} 7s^2$
72	Hf	$[Xe]4f^{14} 5d^2 6s^2$	102	No	$[Rn]5f^{14} 7s^2$
73	Ta	$[Xe]4f^{14} 5d^3 6s^2$			
74	W	$[Xe]4f^{14} 5d^4 6s^2$			
75	Re	$[Xe]4f^{14} 5d^5 6s^2$	103	Lr	$[Rn]5f^{14} 6d^1 7s^2$
76	Os	$[Xe]4f^{14} 5d^6 6s^2$	104	Rf	$[Rn]5f^{14} 6d^2 7s^2(?)$
77	Ir	$[Xe]4f^{14} 5d^7 6s^2$	105	Db	$[Rn]5f^{14} 6d^3 7s^2(?)$
78	Pt	$[Xe]4f^{14} 5d^9 6s^1$	106	Sg	$[Rn]5f^{14} 6d^4 7s^2(?)$
79	Au	$[Xe]4f^{14} 5d^{10} 6s^1$	107	Bh	$[Rn]5f^{14} 6d^5 7s^2(?)$
80	Hg	$[Xe]4f^{14} 5d^{10} 6s^2$	108	Hs	$[Rn]5f^{14} 6d^6 7s^2(?)$
			109	Mt	$[Rn]5f^{14} 6d^7 7s^2(?)$
81	Tl	$[Xe]4f^{14} 5d^{10} 6s^2 6p^1$			
82	Pb	$[Xe]4f^{14} 5d^{10} 6s^2 6p^2$			
83	Bi	$[Xe]4f^{14} 5d^{10} 6s^2 6p^3$			
84	Po	$[Xe]4f^{14} 5d^{10} 6s^2 6p^4$			
85	At	$[Xe]4f^{14} 5d^{10} 6s^2 6p^5$			
86	Rn	$[Xe]4f^{14} 5d^{10} 6s^2 6p^6$			
87	Fr	$[Rn]7s^1$			
88	Ra	$[Rn]7s^2$			
89	Ac	$[Rn]6d^1 7s^2$			
90	Th	$[Rn]6d^2 7s^2$			

*The electron configurations followed by a question mark are speculative.

2D   The Elements

Element	Symbol	Atomic number	Molar mass, g/mol	Normal state*	Density, g/cm³	Melting point, °C
actinium	Ac	89	227.03	s, m	10.07	1230
(Greek *aktis*, ray)						
aluminum	Al	13	26.98	s, m	2.70	660
(from alum, salts of the form $KAl(SO_4)_2 \cdot 12H_2O$)						
americium	Am	95	241.06	s, m	13.67	990
(the Americas)						
antimony	Sb	51	121.75	s, md	6.69	631
(probably a corruption of an old Arabic word; Latin *stibium*)						
argon	Ar	18	39.95	g, nm	1.66†	−189
(Greek *argos*, inactive)						
arsenic	As	33	74.92	s, md	5.78	613‡
(Greek *arsenikos*, male)						
astatine	At	85	210.	s, nm	—	300
(Greek *astatos*, unstable)						
barium	Ba	56	137.34	s, m	3.59	710
(Greek *barys*, heavy)						
berkelium	Bk	97	249.08	s, m	14.79	986
(Berkeley, California)						
beryllium	Be	4	9.01	s, m	1.85	1285
(from the mineral beryl, $Be_3Al_2SiO_{18}$)						
bismuth	Bi	83	208.98	s, m	8.90	271
(German *weisse Masse*, white mass)						
bohrium	Bh	107	262.12	—	—	—
(Niels Bohr)						
boron	B	5	10.81	s, md	2.47	2030
(Arabic *buraq*, borax, $Na_2B_4O_7 \cdot 10H_2O$; *bor*(ax) + (carb)*on*)						
bromine	Br	35	79.91	l, nm	3.12	−7
(Greek *bromos*, stench)						
cadmium	Cd	48	112.40	s, m	8.65	321
(Greek *Cadmus*, founder of Thebes)						
calcium	Ca	20	40.08	s, m	1.53	840
(Latin *calx*, lime)						
californium	Cf	98	251.08	s, m	—	—
(California)						
carbon	C	6	12.01	s, nm	2.27	3700‡
(Latin *carbo*, coal or charcoal)						

(continued)

Boiling point, °C	Ionization energies, kJ/mol	Electron affinity, kJ/mol	Electronegativity	Principal oxidation states	Atomic radius, pm	Ionic radius,§ pm
3200	499, 1170., 1900.	—	1.1	+3	188	118(3+)
2350	577, 1817, 2744	+43	1.6	+3	143	53(3+)
2600	578	—	1.3	+3	184	107(3+)
1750	834, 1794, 2443	+103	2.1	−3, +3, +5	141	89(3+)
−186	1520.	<0	—	0	174	—
—	947, 1798	+78	2.2	−3, +3, +5	121	222(3−)
350	1037, 1600.	+270.	2.0	−1	—	227(1−)
1640	502, 965	+14	0.89	+2	224	136(2+)
—	601	—	1.3	+3	—	87(4+)
2470	900., 1757	<0	1.6	+2	112	27(2+)
1650	703, 1610., 2466	+91	2.0	+3, +5	182	96(3+)
—	660.	—	—	+5	128	83(5+)
5100‡	799, 2430, 3660., 25 000	+27	2.0	+3	88	12(3+)
59	1140., 2104	+325	3.0	−1, +1, +3, +4, +5, +7	114	196(1−)
765	868, 1631	<0	1.7	+2	152	103(2+)
1490	590, 1145, 4910.	+2	1.3	+2	197	100.(2+)
—	608	—	1.3	+3	—	117(2+)
—	1090., 2352, 4620.	+122	2.6	−4, −1, +2, +4	77	260.(4−)

Element	Symbol	Atomic number	Molar mass, g/mol	Normal state*	Density, g/cm³	Melting point, °C
cerium (the asteroid Ceres, discovered 2 days earlier)	Ce	58	140.12	s, m	6.71	800
cesium (Latin *caesius*, sky blue)	Cs	55	132.91	s, m	1.87	28
chlorine (Greek *chloros*, yellowish green)	Cl	17	35.45	g, nm	1.66†	−101
chromium (Greek *chroma*, color)	Cr	24	52.00	s, m	7.19	1860
cobalt (German *Kobold*, evil spirit; Greek *kobalos*, goblin)	Co	27	58.93	s, m	8.80	1494
copper (Latin *cuprum*, from Cyprus)	Cu	29	63.54	s, m	8.93	1083
curium (Marie Curie)	Cm	96	247.07	s, m	13.30	1340
dubnium (Dubna, a town in Russia)	Db	105	262.11	s, m	29	—
dysprosium (Greek *dysprositos*, hard to get at)	Dy	66	162.50	s, m	8.53	1410
einsteinium (Albert Einstein)	Es	99	254.09	s, m	—	—
erbium (Ytterby, a town in Sweden)	Er	68	167.26	s, m	9.04	1520
europium (Europe)	Eu	63	151.96	s, m	5.25	820
fermium (Enrico Fermi, an Italian physicist)	Fm	100	257.10	s, m	—	—
fluorine (Latin *fluere*, to flow)	F	9	19.00	g, nm	1.51†	−220
francium (France)	Fr	87	223	s, m	—	27
gadolinium (Johann Gadolin, a Finnish chemist)	Gd	64	157.25	s, m	7.87	1310
gallium (Latin *Gallia*, France; also a pun on the Latin form of the discoverer's forename, Le Coq)	Ga	31	69.72	s, m	5.91	30
germanium (Latin *Germania*, Germany)	Ge	32	72.59	s, md	5.32	937
gold (Anglo-Saxon *gold*; Latin *aurum*, gold)	Au	79	196.97	s, m	19.28	1064

(continued)

Boiling point, °C	Ionization energies, kJ/mol	Electron affinity, kJ/mol	Electronegativity	Principal oxidation states	Atomic radius, pm	Ionic radius,§ pm
3000	527, 1047, 1949	<50	1.1	+3, +4	183	107(3+)
678	376, 2420.	+46	0.79	+1	272	170.(1+)
−34	1255, 2297	+349	3.2	−1, +1, +3, +4, +5, +6, +7	99	181(1−)
2600	653, 1592, 2987	+64	1.7	+2, +3	129	84(2+)
2900	760., 1646, 3232	+64	1.9	+3, +6	125	64(3+)
2567	785, 1958, 3554	+118	1.9	+1, +2	128	72(2+)
—	581	—	1.3	+3	—	99(3+)
—	640.	—	—	+5	139	68(5+)
2600	572, 1126, 2200.	—	1.2	+3	177	91(3+)
—	619	<50	1.3	+3	203	98(3+)
2600	589, 1151, 2194	<50	1.2	+3	176	89(3+)
1450	547, 1085, 2404	<50	—	+3	204	98(3+)
—	627	—	1.3	+3	—	91(3+)
−188	1680., 3374	+328	4.0	−1	64	133(1−)
677	400	+44	0.7	+1	270.	180.(1+)
3000	592, 1167, 1990.	<50	1.2	+2, +3	180.	97(3+)
2070	577, 1979, 2963	+29	1.6	+1, +3	153	62(3+)
2830	784, 1557, 3302	+116	2.0	+2, +4	122	90.(2+)
2807	890., 1980.	+223	2.5	+1, +3	144	91(3+)

(continued)

Element	Symbol	Atomic number	Molar mass, g/mol	Normal state*	Density, g/cm^3	Melting point, °C
hafnium (Latin *Hafnia*, Copenhagen)	Hf	72	178.49	s, m	13.28	2230
hassium (Hesse, the German state)	Hs	108	265	—	—	—
helium (Greek *helios*, the sun)	He	2	4.00	g, nm	0.12†	—
holmium (Latin *Holmia*, Stockholm)	Ho	67	164.93	s, m	8.80	1470
hydrogen (Greek *hydro* + *genes*, water-forming)	H	1	1.0079	g, nm	0.089†	−259
indium (from the bright indigo line in its spectrum)	In	49	114.82	s, m	7.29	157
iodine (Greek *ioeidēs*, violet)	I	53	126.90	s, nm	4.95	114
iridium (Greek and Latin *iris*, rainbow)	Ir	77	192.2	s, m	22.56	2447
iron (Anglo-Saxon *iron*; Latin *ferrum*)	Fe	26	55.85	s, m	7.87	1540
krypton (Greek *kryptos*, hidden)	Kr	36	83.80	g, nm	2.82†	−157
lanthanum (Greek *lanthanein*, to lie hidden)	La	57	138.91	s, m	6.17	920
lawrencium (Ernest Lawrence, an American physicist)	Lr	103	262.1	s, m	—	—
lead (Anglo-Saxon *lead*; Latin *plumbum*)	Pb	82	207.19	s, m	11.34	328
lithium (Greek *lithos*, stone)	Li	3	6.94	s, m	0.53	181
lutetium (*Lutetia*, ancient name of Paris)	Lu	71	174.97	s, m	9.84	1700
magnesium (Magnesia, a district in Thessaly, Greece)	Mg	12	24.31	s, m	1.74	650
manganese (Greek and Latin *magnes*, magnet)	Mn	25	54.94	s, m	7.47	1250
meitnerium (Lise Meitner)	Mt	109	266	—	—	—
mendelevium (Dmitri Mendeleev)	Md	101	258.10	—	—	—

(continued)

Boiling point, °C	Ionization energies, kJ/mol	Electron affinity, kJ/mol	Electronegativity	Principal oxidation states	Atomic radius, pm	Ionic radius,§ pm
5300	642, 1440, 2250.	0	1.3	+4	159	84(3+)
—	750.	—	—	+3	126	80.(4+)
−269	2370., 5250.	<0	—	0	128	—
2300	581, 1139	<50	1.2	+3	177	89(3+)
−253	1310.	+73	2.2	−1, +1	78	154(1−)
2050	556, 1821	+29	1.8	+1, +3	167	72(3+)
184	1008, 1846	+295	2.7	−1, +1, +3, +5, +7	133	220.(1−)
4550	880.	+151	2.2	+3, +4	136	75(3+)
2760	759, 1561, 2957	+16	1.8	+2, +3	124	82(2+)
−153	1350., 2350.	<0	—	+2	—	169(1+)
3450	538, 1067, 1850.	+50.	1.1	+3	188	122(3+)
—	—	—	1.3	+3	—	88(3+)
1760	716, 1450.	+35	2.3	+2, +4	175	132(2+)
1347	519, 7298	+60.	1.0	+1	157	58(1+)
3400	524, 1340., 2022	<50	1.3	+3	172	85(3+)
1100	736, 1451	<0	1.3	+2	160.	72(2+)
2120	717, 1509	<0	1.6	+2, +3, +4, +7	137	91(2+)
—	840.	—	—	+2	—	83(2+)
—	635	—	1.3	+3	—	90.(3+)

(continued)

Element	Symbol	Atomic number	Molar mass, g/mol	Normal state*	Density, g/cm^3	Melting point, °C
mercury (the planet Mercury; Latin *hydrargyrum*, liquid silver)	Hg	80	200.59	l, m	13.55	−39
molybdenum (Greek *molybdos*, lead)	Mo	42	95.94	s, m	10.22	2620
neodymium (Greek *neos* + *didymos*, new twin)	Nd	60	144.24	s, m	7.00	1024
neon (Greek *neos*, new)	Ne	10	20.18	g, nm	1.21†	−249
neptunium (the planet Neptune)	Np	93	237.05	s, m	20.45	640
nickel (German *Nickel*, Old Nick, Satan)	Ni	28	58.71	s, m	8.91	1455
niobium (Niobe, daughter of Tantalos; see tantalum)	Nb	41	92.91	s, m	8.57	2425
nitrogen (Greek *nitron* + *genes*, soda-forming)	N	7	14.01	g, nm	1.04†	−210
nobelium (Alfred Nobel, the founder of the Nobel prizes)	No	102	255	s, m	—	—
osmium (Greek *osme*, a smell)	Os	76	190.2	s, m	22.58	3030
oxygen (Greek *oxys* + *genes*, acid-forming)	O	8	16.00	g, nm	1.14†	−218
palladium (the asteroid Pallas, discovered at about the same time)	Pd	46	106.4	s, m	12.00	1554
phosphorus (Greek *phosphoros*, light-bearing)	P	15	30.97	s, nm	1.82	44
platinum (Spanish *plata*, silver)	Pt	78	195.09	s, m	21.45	1772
plutonium (the planet Pluto)	Pu	94	239.05	s, m	19.81	640
polonium (Poland)	Po	84	210.	s, md	9.40	254
potassium (from potash; Latin *kalium* and Arabic *qali*, alkali)	K	19	39.10	s, m	0.86	64
praseodymium (Greek *prasios* + *didymos*, green twin)	Pr	59	140.91	s, m	6.78	935

(continued)

Boiling point, °C	Ionization energies, kJ/mol	Electron affinity, kJ/mol	Electronegativity	Principal oxidation states	Atomic radius, pm	Ionic radius,^{s} pm
357	1007, 1810.	−18	2.0	+1, +2	155	112(2+)
4830	685, 1558, 2621	+72	2.2	+4, +5, +6	140.	92(2+)
3100	530., 1035	<0	1.1	+3	182	104(3+)
−246	2080., 3952	0	—	0	—	—
—	597	—	1.4	+5	131	88(5+)
2150	737, 1753	+156	1.9	+2, +3	125	78(2+)
5000	664, 1382	+86	1.6	+5	147	69(5+)
−196	1400., 2856	−7	3.0	−3, +3, +5	74	171(3−)
—	642	—	1.3	+2	—	113(2+)
5000	840.	+106	2.2	+3, +4	135	81(3+)
−183	1310., 3388	+141, −844	3.4	−2	66	140.(2−)
3000	805, 1875	+54	2.2	+2, +4	137	86(2+)
280	1011, 1903, 2912	+72	2.2	−3, +3, +5	110.	212(3−)
3720	870., 1791	+205	2.3	+2, +4	139	85(2+)
3200	585	—	1.3	+3, +4	151	108(3+)
960	812	+174	2.0	+2, +4	167	65(4+)
774	418, 3051	+48	0.82	+1	235	138(1+)
3000	523, 1018	<50	1.1	+3	183	106(3+)

(continued)

Element	Symbol	Atomic number	Molar mass, g/mol	Normal state*	Density, g/cm³	Melting point, °C
promethium (Prometheus, the Greek god)	Pm	61	146.92	s, m	7.22	1168
protactinium (Greek *protos* + *aktis,* first ray)	Pa	91	231.04	s, m	15.37	1200
radium (Latin *radius,* ray)	Ra	88	226.03	s, m	5.00	700
radon (from radium)	Rn	86	222	g, nm	4.40†	−71
rhenium (Latin *Rhenus,* Rhine)	Re	75	186.2	s, m	21.02	3180
rhodium (Greek *rhodon,* rose; its aqueous solutions are often rose-colored)	Rh	45	102.91	s, m	12.42	1963
rubidium (Latin *rubidus,* deep red, "flushed")	Rb	37	85.47	s, m	1.53	39
ruthenium (Latin *Ruthenia,* Russia)	Ru	44	101.07	s, m	12.36	2310
rutherfordium (Ernest Rutherford)	Rf	104	261.11	—	—	—
samarium (from samarskite, a mineral)	Sm	62	150.35	s, m	7.54	1060
scandium (Latin *Scandia,* Scandinavia)	Sc	21	44.96	s, m	2.99	1540
seaborgium (Glenn Seaborg)	Sg	106	263.12	—	—	—
selenium (Greek *selēnē,* the moon)	Se	34	78.96	s, nm	4.81	220
silicon (Latin *silex,* flint)	Si	14	28.09	s, md	2.33	1410
silver (Anglo-Saxon *seolfor;* Latin *argentum*)	Ag	47	107.87	s, m	10.50	962
sodium (English soda; Latin *natrium*)	Na	11	22.99	s, m	0.97	98
strontium (Strontian, Scotland)	Sr	38	87.62	s, m	2.58	770
sulfur (Sanskrit *sulvere*)	S	16	32.06	s, nm	2.09	115
tantalum (Tantalos, Greek mythological figure)	Ta	73	180.95	s, m	16.67	3000

(continued)

Boiling point, °C	Ionization energies, kJ/mol	Electron affinity, kJ/mol	Electronegativity	Principal oxidation states	Atomic radius, pm	Ionic radius,§ pm
3300	536, 1052	<50	—	+3	181	106(3+)
4000	568	—	1.5	+5	161	89(5+)
1500	509, 979	—	0.9	+2	223	152(2+)
−62	1036, 1930.	<0	—	+2	—	—
5600	760., 1260.	+14	1.9	+4, +7	137	72(4+)
3700	720., 1744	+110	2.3	+3	134	75(3+)
688	402, 2632	+47	0.82	+1	250.	149(1+)
4100	711, 1617	+101	2.2	+2, +3, +4	134	77(3+)
—	490.	—	—	+4	150	67(4+)
1600	543, 1068	<50	1.2	+3	180.	100.(3+)
2800	631, 1235	+18	1.4	+3	164	83(3+)
—	730.	—	—	+6	132	86(5+)
685	941, 2044	+195	2.6	−2, +4, +6	117	198(2−)
2620	786, 1577	+134	1.9	+4	118	26(4+)
2212	731, 2073	+126	1.9	+1	144	113(1+)
883	494, 4562	+53	0.93	+1	191	102(1+)
1380	548, 1064	+5	0.95	+2	215	116(2+)
445	1000., 2251	+200., −532	2.6	−2, +4, +6	104	184(2−)
5400	761	+14	1.5	+5	147	72(3+)

(continued)

Element	Symbol	Atomic number	Molar mass, g/mol	Normal state*	Density, g/cm³	Melting point, °C
technetium (Greek *technētos*, artificial)	Tc	43	98.91	s, m	11.50	2200
tellurium (Latin *tellus*, earth)	Te	52	127.60	s, md	6.25	450
terbium (Ytterby, a town in Sweden)	Tb	65	158.92	s, m	8.27	1360
thallium (Greek *thallos*, a green shoot)	Tl	81	204.37	s, m	11.87	304
thorium (Thor, Norse god of thunder, weather, and crops)	Th	90	232.04	s, m	11.73	1700
thulium (*Thule*, early name for Scandinavia)	Tm	69	168.93	s, m	9.33	1550
tin (Anglo-Saxon *tin;* Latin *stannum*)	Sn	50	118.69	s, m	7.29	232
titanium (Titans, Greek mythological figures, sons of the Earth)	Ti	22	47.88	s, m	4.55	1660
tungsten (Swedish *tung* + *sten*, heavy stone; from wolframite)	W	74	183.85	s, m	19.30	3387
uranium (the planet Uranus)	U	92	238.03	s, m	18.95	1135
vanadium (Vanadis, Scandinavian mythological figure)	V	23	50.94	s, m	6.11	1920
xenon (Greek *xenos*, stranger)	Xe	54	131.30	g, nm	3.56†	−112
ytterbium (Ytterby, a town in Sweden)	Yb	70	173.04	s, m	6.97	824
yttrium (Ytterby, a town in Sweden)	Y	39	88.91	s, m	4.48	1510
zinc (Anglo-Saxon *zinc*)	Zn	30	65.37	s, m	7.14	420
zirconium (Arabic *zargun*, gold color)	Zr	40	91.22	s, m	6.51	1850

Boiling point, °C	Ionization energies, kJ/mol	Electron affinity, kJ/mol	Electronegativity	Principal oxidation states	Atomic radius, pm	Ionic radius,[§] pm
4600	702, 1472	+96	1.9	+4, +7	135	97(4+)
990	870., 1775	+190.	2.1	−2, +4	137	221(2−)
2500	565, 1112	<50	—	+3	178	93(3+)
1460	590., 1971	+19	2.0	+1, +3	171	88(3+)
4500	587, 1110.	—	1.3	+4	180.	99(4+)
2000	597, 1163	<50	1.2	+3	175	104(3+)
2720	707, 1412	+116	2.0	+2, +4	158	93(2+)
3300	658, 1310.	+7.6	1.5	+4	147	69(4+)
5420	770.	+79	2.4	+5, +6	141	62(6+)
4000	584, 1420.	—	1.4	+6	138	80.(6+)
3400	650., 1414	+51	1.6	+4, +5	135	61(4+)
−108	1170., 2046	<0	2.6	+2, +4, +6	218	190.(1+)
1500	603, 1176	<50	—	+3	194	113(3+)
3300	616, 1181	+30.	1.2	+3	182	106(3+)
907	906, 1733	+9	1.6	+2	137	83(2+)
4400	660., 1267	+41	1.3	+4	160.	87(4+)

*The normal state is the state of the element at normal temperature and pressure (25°C and 1 atm). s denotes
 solid; l, liquid; and g, gas; m denotes metal; nm, nonmetal; and md, metalloid.
†The density quoted is for the liquid.
‡The solid sublimes.
§Charge in parentheses.

2E The Top 25 Chemicals by Industrial Production in the United States in 1997

Production data are compiled annually by the American Chemical Society and published in *Chemical and Engineering News*. This table is based on the information about production in 1997 that was published in the June 29, 1998, issue, and will be updated annually on the web site for this book. Water, sodium chloride, and steel traditionally are not included and would outrank the rest if they were. Hydrogen is heavily used but almost always "on site" as soon as it has been prepared.

Rank	Name	Annual production, 10^9 kg	Comment on source
1	ethene (ethylene)	46.3	thermal cracking of petroleum
2	sulfuric acid	43.4	contact process
3	propene	25.0	thermal cracking
4	dichloroethane	23.9	chlorination of ethene
5	calcium hydroxide	19.3	decomposition of limestone
6	ammonia	17.4	Haber process
7	urea	14.1	ammonia + carbon dioxide
8	sulfur	13.5	Frasch process
9	phosphoric acid	12.2	from phosphate rocks
10	chlorine	11.8	electrolysis
11	ethylbenzene	11.5	Friedel-Crafts alkylation of benzene
12	sodium carbonate	10.4	mining
13	sodium hydroxide	10.3	electrolysis of brine
14	styrene	10.3	dehydration of ethylbenzene
15	nitric acid	8.2	Ostwald process
16	xylene	8.1	catalytic reforming
17	ammonium nitrate	7.5	ammonia + nitric acid
18	ethylene oxide	7.5	addition of O_2 to ethene
19	cumene (isopropylbenzene)	5.6	Friedel-Crafts alkylation
20	hydrogen chloride	3.8	by-product of hydrocarbon chlorination
21	1,3-butadiene	3.7	dehydrogenation of butane
22	acrylonitrile	3.0	HCN + ethyne
23	ammonium sulfate	2.5	ammonia + sulfuric acid
24	benzene	2.1	catalytic reforming
25	potash*	1.4	mined, electrolysis

*Potash refers collectively to K_2CO_3, KOH, K_2SO_4, KCl, and KNO_3, and is expressed in terms of the equivalent mass of K_2O.

3A The Nomenclature of Polyatomic Ions

Charge number	Chemical formula	Name	Oxidation number of central element	Charge number	Chemical formula	Name	Oxidation number of central element
2+	Hg_2^{2+}	mercury(I)	+1		O_3^-	ozonide	$-\frac{1}{3}$
	UO_2^{2+}	uranyl	+6		OH^-	hydroxide	$-2(O)$
	VO^{2+}	vanadyl	+4		SCN^-	thiocyanate	—
1+	NH_4^+	ammonium	−3	2−	C_2^{2-}	carbide	−1
	PH_4^+	phosphonium	−3		CO_3^{2-}	carbonate	+4
1−	$CH_3CO_2^-$	acetate, ethanoate	—		$C_2O_4^{2-}$	oxalate	+3
	HCO_2^-	formate,			CrO_4^{2-}	chromate	+6
		methanoate	—		$Cr_2O_7^{2-}$	dichromate	+6
	CN^-	cyanide	$+2(C)$		O_2^{2-}	peroxide	−1
	ClO_4^-	perchlorate*	+7		S_2^{2-}	disulfide	−1
	ClO_3^-	chlorate*	+5		SiO_3^{2-}	metasilicate	+4
	ClO_2^-	chlorite*	+3		SO_4^{2-}	sulfate	+6
	ClO^-	hypochlorite*	+1		SO_3^{2-}	sulfite	+4
	MnO_4^-	permanganate	+7		$S_2O_3^{2-}$	thiosulfate	+2
	NO_3^-	nitrate	+5	3−	AsO_4^{3-}	arsenate	+5
	NO_2^-	nitrite	+3		BO_3^{3-}	borate	+3
	N_3^-	azide	$-\frac{1}{3}$		PO_4^{3-}	phosphate	+5

*These names are representative of the halogen oxoanions.

When a hydrogen ion bonds to a −2 or −3 anion, add "hydrogen" before the name of the anion. For example, HSO_3^- is the hydrogen sulfite (or hydrogensulfite) ion. If two hydrogen ions bond to a −3 anion, add "dihydrogen" before the name of the anion. For example, $H_2PO_4^-$ is dihydrogen phosphate.

Oxoacids and Oxoanions

The names of oxoanions and their parent acids can be determined by noting the oxidation number of the central atom and then referring to the table on the right. For example, the nitrogen in $N_2O_2^{2-}$ has an oxidation number of +1; because nitrogen belongs to Group 15, the ion is a hyponitrite ion.

Group number					
14	15	16	17	Oxoanion	Oxoacid
—	—	—	+7	per . . . ate	per . . . ic acid
+4	+5	+6	+5	. . . ate	. . . ic acid
—	+3	+4	+3	. . . ite	. . . ous acid
—	+1	+2	+1	hypo . . . ite	hypo . . . ous acid

3B Common Names of Chemicals

Many chemicals have acquired common names, sometimes as a result of their use over hundreds of years and sometimes because they appear on the labels of consumer products, such as detergents, beverages, and antacids. The following substances are just a few that have found their way into the language of everyday life.

Common name	Formula	Chemical name
baking soda	$NaHCO_3$	sodium hydrogen carbonate (sodium bicarbonate)
bleach, laundry	$NaClO$	sodium hypochlorite
borax	$Na_2B_4O_7 \cdot 10H_2O$	sodium tetraborate decahydrate
brimstone	S_8	sulfur
calamine	$ZnCO_3$	zinc carbonate
chalk	$CaCO_3$	calcium carbonate
Epsom salts	$MgSO_4 \cdot 7H_2O$	magnesium sulfate heptahydrate
fool's gold	FeS_2	iron(II) disulfide
gypsum	$CaSO_4 \cdot 2H_2O$	calcium sulfate dihydrate
lime (quicklime)	CaO	calcium oxide
lime (slaked lime)	$Ca(OH)_2$	calcium hydroxide
limestone	$CaCO_3$	calcium carbonate
lye, caustic soda	$NaOH$	sodium hydroxide
marble	$CaCO_3$	calcium carbonate
milk of magnesia	$Mg(OH)_2$	magnesium hydroxide
plaster of Paris	$CaSO_4 \cdot \frac{1}{2}H_2O$	calcium sulfate hemihydrate
potash*	K_2CO_3	potassium carbonate
quartz	SiO_2	silicon dioxide
table salt	$NaCl$	sodium chloride
vinegar	CH_3COOH	acetic acid
washing soda	$Na_2CO_3 \cdot 10H_2O$	sodium carbonate decahydrate

*Potash also refers collectively to K_2CO_3, KOH, K_2SO_4, KCl, and KNO_3.

3C Names of Common Cations with Variable Charge Numbers

Modern nomenclature includes the oxidation number of elements with a variable oxidation state in the names of their compounds, as in cobalt(II) chloride. However, the traditional nomenclature, in which the suffixes -ous and -ic are used, is still encountered. The following table translates from one system to the other.

Element	Cation	Old-style name	Modern name
cobalt	Co^{2+}	cobaltous	cobalt(II)
	Co^{3+}	cobaltic	cobalt(III)
copper	Cu^+	cuprous	copper(I)
	Cu^{2+}	cupric	copper(II)
iron	Fe^{2+}	ferrous	iron(II)
	Fe^{3+}	ferric	iron(III)
lead	Pb^{2+}	plumbous	lead(II)
	Pb^{4+}	plumbic	lead(IV)
manganese	Mn^{2+}	manganous	manganese(II)
	Mn^{3+}	manganic	manganese(III)
mercury	Hg_2^{2+}	mercurous	mercury(I)
	Hg^{2+}	mercuric	mercury(II)
tin	Sn^{2+}	stannous	tin(II)
	Sn^{4+}	stannic	tin(IV)

3D   Naming *d*-Metal Complexes

1. When naming a *d*-metal complex, name the ligands first, then the metal atom or ion.

2. Neutral ligands have the same name as the molecule, such as en (ethylenediamine), except for H_2O (aqua), NH_3 (ammine), CO (carbonyl), and NO (nitrosyl).

3. Anionic ligands end in -o; for anions that end in -ide (like chloride), -ate (like sulfate), and -ite (like nitrite), change the endings as follows:

    -ide $\longrightarrow$ -o   -ate $\longrightarrow$ -ato   -ite $\longrightarrow$ -ito

    Examples: chloro, cyano, sulfato, carbonato, sulfito, and nitrito.

4. Greek prefixes are used to denote the number of each type of ligand in the complex ion:

2	3	4	5	6	...
di-	tri-	tetra-	penta-	hexa-	...

    If the ligand already contains a Greek prefix (as in ethylenediamine) or if it is polydentate (able to attach at more than one binding site), then the prefixes

2	3	4	...
dis-	tris-	tetrakis-	...

    are used instead.

5. Ligands are named in alphabetical order, regardless of the Greek prefix that indicates the number of each one present. (Notice that Cl_2 in the coordination sphere of the second complex in rule 7 (below) represents two chloride ligands, named as a dichloro, and not a Cl_2 molecular ligand.)

6. The chemical symbols of anionic ligands (such as Cl^-) precede those of neutral ligands (such as H_2O and NH_3) in the chemical formula of the complex (but not necessarily in its name).

7. The oxidation number of the central metal ion is given as a Roman numeral after its chemical name. *Examples:*

    $[FeCl(H_2O)_5]^+$ pentaaquachloroiron(II) ion

    $[CrCl_2(NH_3)_4]^+$ tetraamminedichloro-chromium(III) ion

    $[Co(en)_3]^{3+}$ tris(ethylenediamine) cobalt(III) ion

8. If the complex overall has a negative charge (an anionic complex), the suffix -ate is added to the stem of the metal's name. If the symbol of the metal originates from a Latin name (as listed in Appendix 2D), then the Latin stem is used. For example, the symbol for iron is Fe, from the Latin *ferrum*. Therefore, any anionic complex of iron ends with -ferrate followed by the oxidation number of the metal in Roman numerals:

    $[Fe(CN)_6]^{4-}$ hexacyanoferrate(II) ion

    $[Ni(CN)_4]^{2-}$ tetracyanonickelate(II) ion

9. The name of a coordination compound (as distinct from a complex cation or anion) is built in the same way as that of a simple compound, with the (possibly complex) cation named before the (possibly complex) anion. *Examples:*

    $NH_4[PtCl_3(NH_3)]$ ammonium amminetri-chloroplatinate(II)

    $[Cr(OH)_2(NH_3)_4]Br$ tetraamminedihydroxo-chromium(III) bromide

Summary

Identify the cation and anion, name each separately, and then combine them. For each complex, first note the name and oxidation number of the metal ion; then identify the ligands; and finally, string the names together in alphabetical order with the appropriate prefixes, ending with the name of the metal. For example, suppose we needed to name the coordination compound

$$[Co(NH_3)_3(H_2O)_3]_2(SO_4)_3$$

The charge on the complex cation must be +3 to ensure charge neutrality of the compound (there are three SO_4^{2-} ions for every two complex ions), so the complex cation is $[Co(NH_3)_3(H_2O)_3]^{3+}$. Because all the ligands are neutral, the cobalt must be present as cobalt(III). It follows that the name of the cation is triamminetriaquacobalt(III), so the compound is triamminetriaquacobalt(III) sulfate.

absolute zero ($T = 0$; that is, 0 on the *Kelvin scale*) The lowest possible temperature ($-273.15°C$).

absorb To accept one substance into and throughout the bulk of another substance. Compare with *adsorb*.

absorbance A A measure of the extent of absorption of radiation by a sample: $A = \log(I_0/I)$.

absorbed dose (of radiation) The energy deposited in a given mass of a sample when it is exposed to radiation (particularly but not exclusively nuclear radiation). Absorbed dose is measured in *gray* or *rad*.

absorption spectrum The wavelength dependence of the absorption of a sample, determined by measuring the extent to which the sample absorbs electromagnetic radiation as the wavelength is varied over a range.

abundance (of an isotope) The percentage (in terms of the numbers of atoms) of the isotope present in a sample of the element. See also *natural abundance*.

acceleration The rate of change of velocity (either its direction or magnitude).

acceleration of free fall g The acceleration experienced by a body due to the gravitational field at the surface of the Earth.

accuracy Freedom from systematic error. Compare with *precision*.

accurate measurements Measurements that have small systematic error and give a result close to the accepted value of the property.

achiral (molecule or complex) Not chiral: identical to its mirror image. See also *chiral*.

acid See *Arrhenius acid; Brønsted acid; Lewis acid*. Used alone, "acid" normally means a Brønsted acid.

acid anhydride A compound that forms an oxoacid when it reacts with water. A *formal anhydride* is a compound that has the formula of an acid minus the elements of water but does not react with water to produce the acid. *Examples:* SO_3, the anhydride of sulfuric acid; CO, the formal anhydride of formic acid, HCOOH.

acid-base indicator See *indicator*.

acid-base titration See *titration*.

acid buffer See *buffer*.

acid ionization (dissociation) constant K_a See *acidity constant*.

acidic hydrogen atom (1) A hydrogen atom (more exactly, the proton of that hydrogen atom) that can be donated to a base. (2) A hydrogen atom that can be released in water to form a hydronium ion.

acidic ion An ion that acts as a Brønsted acid. *Examples:* NH_4^+; $[Al(H_2O)_6]^{3+}$.

acidic oxide An oxide that reacts with water to give an acid; the oxides of nonmetallic elements generally are acidic oxides. *Examples:* CO_2; SO_3.

acidic solution A solution with pH < 7.

acidity constant K_a The equilibrium constant for proton transfer to water; for an acid HA, $K_a = [H_3O^+][A^-]/[HA]$ at equilibrium.

actinide A member of the second row of the *f* block (actinium through nobelium).

activated complex An unstable combination of reactant molecules that can either go on to form products or fall apart into the unchanged reactants.

activated complex theory A theory of reaction rates in which it is supposed that the reactants form an activated complex.

activation energy E_a (1) The minimum energy needed for reaction. (2) An empirical parameter that describes the temperature dependence of the rate constant.

active site (1) The region of an enzyme molecule where the substrate reacts. (2) The effective catalytic site on the surface of a heterogeneous catalyst.

activity In radioactivity, the number of nuclear disintegrations that occur per second.

addition polymerization The polymerization, usually of alkenes, by an addition reaction often propagated by radicals.

addition reaction A chemical reaction in which atoms or groups bond to two atoms initially joined by a multiple bond. The product of an addition reaction contains all the reactant atoms. *Example:*
$CH_3CH{=}CH_2 + HBr \rightarrow CH_3CH_2CH_2Br$.

adhesion Binding to a surface.

adhesive forces Forces that bind a substance to a surface.

adsorb To bind a substance to a surface; the surface *adsorbs* the substance. Distinguish from *absorb*.

alcohol An organic molecule containing an $-OH$ group attached to a carbon atom that is not part of a carbonyl group or an aromatic ring. Alcohols are classified as *primary, secondary,* and *tertiary* according to the number of carbon atoms attached to the C$-$OH carbon atom. *Examples:* CH_3CH_2OH (primary); $(CH_3)_2CHOH$ (secondary); $(CH_3)_3COH$ (tertiary).

aldehyde An organic compound containing a carbonyl group on a terminal carbon atom; the $-CHO$ group.

Examples: CH_3CHO, ethanal (acetaldehyde); C_6H_5CHO, benzaldehyde.

aliphatic hydrocarbon   A hydrocarbon that does not have benzene rings in its structure.

alkali   An aqueous solution of a strong base. *Example:* aqueous NaOH.

alkali metal   A member of Group 1 of the periodic table (the lithium family).

alkaline earth metal   Calcium, strontium, and barium; more informally, a member of Group 2 of the periodic table (the beryllium family).

alkaline solution   An aqueous solution with pH > 7.

alkane   (1) A hydrocarbon with no carbon-carbon multiple bonds. (2) A saturated hydrocarbon. (3) A member of a series of hydrocarbons with formulas derived from methane by the repetitive insertion of $-CH_2-$ groups; alkanes have molecular formula C_nH_{2n+2}. *Examples:* CH_4; CH_3CH_3; $CH_3(CH_2)_6CH_3$.

alkene   (1) A hydrocarbon with at least one carbon-carbon double bond. (2) A member of a series of hydrocarbons with formulas derived from that of ethene by the repetitive insertion of $-CH_2-$ groups; alkenes with one double bond have molecular formula C_nH_{2n}. *Examples:* $CH_2=CH_2$; $CH_3CH=CH_2$; $CH_3CH=CHCH_2CH_3$.

alkyl group R   A group of atoms that can be regarded as derived from an alkane by loss of a hydrogen atom. *Examples:* $-CH_3$, methyl; $-CH_2CH_3$, ethyl.

alkyne   (1) A hydrocarbon with at least one carbon-carbon triple bond. (2) A member of a series of hydrocarbons with formulas derived from ethyne by the repetitive insertion of $-CH_2-$ groups; alkynes with one triple bond have molecular formula C_nH_{2n-2}. *Examples:* $CH\equiv CH$; $CH_3C\equiv CCH_3$.

allotropes   Alternative forms of an element that differ in the way the atoms are linked. *Examples:* O_2 and O_3; white and gray tin.

alloy   A mixture of two or more metals formed by melting, mixing, and then cooling. A *substitutional alloy* is an alloy in which atoms of one metal are substituted for atoms of another metal. An *interstitial alloy* is an alloy in which atoms of one metal lie in the gaps in the lattice formed by atoms of another metal. A *homogeneous alloy* is an alloy in which the atoms of the elements are distributed uniformly. A *heterogeneous alloy* is an alloy that consists of (micro)crystalline phases with different compositions.

alpha (α) decay   Nuclear decay that involves the emission of an α particle.

alpha (α) helix   One type of secondary structure adopted by a polypeptide chain, in the form of a right-handed helix.

alpha (α) particle   Positively charged, subatomic particle emitted from some radioactive nuclei; nucleus of a helium atom ($^4_2He^{2+}$).

amide   An organic compound formed by the reaction of an amine and a carboxylic acid in which the acidic $-OH$ group has been replaced by an amino group or a substituted amino group. An amide contains the group $-CONR_2$. *Example:* CH_3CONH_2, acetamide.

amine   A compound derived from ammonia by replacing various numbers of H atoms with organic groups; the number of hydrogen atoms replaced determines the classification as *primary, secondary,* or *tertiary. Examples:* CH_3NH_2 (primary); $(CH_3)_2NH$ (secondary); $(CH_3)_3N$ (tertiary); $(CH_3)_4N^+$ (quaternary ammonium ion).

amino acid   A carboxylic acid that also contains an amino group. The *essential amino acids* are amino acids that must be ingested as a part of the diet. *Example:* NH_2CH_2COOH, glycine (nonessential).

amino group   The functional group $-NH_2$ characteristic of *amines.*

amorphous solid   A solid in which the atoms, ions, or molecules lie in a random jumble, with no long-range order. *Examples:* glass, butter. Compare with *crystalline solid.*

ampere A   The SI unit of electric current.

amphiprotic   Having the ability both to donate and to accept protons. *Examples:* H_2O; HCO_3^{2-}. Amphiprotic species are often called *amphoteric.*

amphoteric   The ability to react with both acids and bases. *Examples:* Al; Al_2O_3.

amplitude   The height of a wave above its center. On a graph depicting a wave, the height of the wave above the center line.

analysis   See *chemical analysis.*

analyte   The solution of unknown concentration in a titration. Normally, the analyte is in the flask, not the buret.

angular momentum quantum number   See *orbital angular momentum quantum number.*

angular wavefunction   The angular part of a wavefunction, particularly the angular component of the wavefunctions of the hydrogen atom.

anhydride   See *acid anhydride.*

anhydrous   Lacking water. *Example:* $CuSO_4$, the anhydrous form of copper(II) sulfate. Compare with *hydrated.*

anion   A negatively charged ion. *Examples:* F^-; SO_4^{2-}.

anisotropic   Depending on orientation.

anode   The electrode at which oxidation occurs.

antibonding orbital   A molecular orbital that, when occupied, results in an overall raising of the energy of a molecule.

antioxidant   A substance that reacts with radicals and so prevents the oxidation of another substance.

aqueous solution   A solution in which the solvent is water.

arene   An aromatic hydrocarbon.

aromatic hydrocarbon   An organic hydrocarbon that includes a benzene ring as part of its structure. *Examples:* C_6H_6, benzene; $C_6H_5CH_3$, toluene; $C_{10}H_8$, naphthalene.

Arrhenius acid   A compound that contains hydrogen and releases hydrogen ions (H^+) in water. *Examples:* HCl; CH_3COOH; but not CH_4.

Arrhenius base   A compound that produces hydroxide ions (OH^-) in water. *Examples:* NaOH; NH_3; but not Na.

Arrhenius behavior   A reaction shows Arrhenius behavior if a plot of ln k against $1/T$ is a straight line. See also *Arrhenius equation.*

Arrhenius equation   The equation ln k = ln A − E_a/RT for the commonly observed temperature dependence of a rate constant k. An *Arrhenius plot* is a graph of ln k against $1/T$; if the plot is a straight line, then the reaction is said to show *Arrhenius behavior.*

Arrhenius parameters   The *Arrhenius parameters* are the *pre-exponential factor A* (also called the *frequency factor*) and the *activation energy* E_a. See also *Arrhenius equation.*

aryl group   An aromatic group. *Example:* $-C_6H_5$, phenyl.

atactic polymer   See *polymer.*

atmosphere   (1) The layer of gases surrounding a planet (specifically, the air for the planet Earth). (2) A unit of pressure (1 atm = $1.013\ 25 \times 10^5$ Pa).

atom   (1) The smallest particle of an element that has the chemical properties of that element. (2) An electrically neutral species consisting of a nucleus and its surrounding electrons.

atomic emission spectrum   The spectrum of electromagnetic radiation emitted by atoms that have been heated to high temperatures or exposed to an electrical discharge.

atomic hypothesis   The proposal advanced by John Dalton that matter is composed of atoms.

atomic mass unit   u (formerly amu)   The mass of one atom of carbon-12.

atomic number Z The number of protons in the nucleus of an atom; this number determines the identity of the element and the number of electrons in the neutral atom.

atomic orbital   A region of space in which there is a high probability of finding an electron in an atom. An *s-orbital* is a spherical region; a *p-orbital* has two lobes, on opposite sides of the nucleus; a *d-orbital* typically has four lobes, with the nucleus at the center; an *f-orbital* has a more complicated arrangement of lobes.

atomic radius   Half the distance between the centers of neighboring atoms in an elemental solid or a homonuclear molecule.

atomic structure   The arrangement of subatomic particles in an atom, particularly the arrangement of electrons around the nucleus of an atom.

atomic weight   The ratio of the average mass of the atoms of an element to the *atomic mass unit*. See also *molar mass.*

Aufbau principle   See *building-up principle.*

autoionization   See *autoprotolysis.*

autoprotolysis   A reaction in which a proton is transferred between two molecules of the same substance. The products are the conjugate acid and conjugate base of the substance. *Example:* 2 $H_2O(l) \rightleftharpoons H_3O^+(aq) + OH^-(aq)$.

autoprotolysis constant   The equilibrium constant for an autoprotolysis reaction. *Example:* For water, K_w, with $K_w = [H_3O^+][OH^-]$.

average bond enthalpy ΔH_B(A—B)   The average of A—B bond enthalpies for a number of different molecules containing the A—B bond. See also *bond enthalpy.*

average reaction rate   The reaction rate calculated by measuring the change in concentration of a reactant or product over a finite time interval (and hence an average of the changing rate within that interval). See also *reaction rate.*

Avogadro constant   The number of objects per mole of objects ($N_A = 6.022\ 14 \times 10^{23}$/mol). *Avogadro's number* is the number of objects *in* one mole of objects (that is, the dimensionless number 6.02214×10^{23}).

Avogadro's principle   The volume of a sample of gas at a given temperature and pressure is proportional to the amount of gas molecules in the sample: $V \propto n$.

axial bond   A bond that is perpendicular to the molecular plane in a bipyramidal molecule.

axial lone pair   A lone pair lying on the axis of a bipyramidal molecule.

azimuthal quantum number l See *orbital angular momentum quantum number.*

background radiation   The average nuclear radiation to which the Earth's inhabitants are exposed daily.

balanced equation   See *chemical equation.*

ball-and-stick model   A depiction of a molecule in which atoms are represented by balls and bonds are represented by sticks.

Balmer series   A family of spectral lines (some of which lie in the visible region) in the spectrum of atomic hydrogen.

band gap   A range of energies for which there are no orbitals in a solid. The band gap lies between the valence band and the conduction band.

band of stability   A region in a plot of mass number against atomic number that corresponds to the existence of stable nuclei.

bar   A unit of pressure: 1 bar = 10^5 Pa.

barometer   An instrument for measuring atmospheric pressure.

base   See *Arrhenius base; Brønsted base; Lewis base.* Used alone, "base" normally means a Brønsted base.

base buffer   See *buffer.*

base ionization constant   See *basicity constant.*

base pair   Two specific nucleotides that link one complementary strand of a DNA molecule to the other by means of hydrogen bonding: adenine pairs with thymine and guanine pairs with cytosine.

base units   The units of measurement in the International System (SI) in terms of which all other units are defined. *Examples: kilogram* for mass; *meter* for length; *second* for time; *kelvin* for temperature; *ampere* for electric current.

basic ion   An ion that acts as a Brønsted base. *Example:* $CH_3CO_2^-$.

basic oxide   An oxide that is a Brønsted base. The oxides of metallic elements are generally basic. *Examples:* Na_2O; MgO.

basic oxygen process   The production of iron by forcing oxygen and powdered limestone through the molten metal.

basic solution   A solution with pH > 7.

basicity constant K_b The equilibrium constant for proton transfer from water to a base: for a base B, $K_b = [BH^+][OH^-]/[B]$ at equilibrium.

becquerel Bq   The SI unit of radioactivity (1 disintegration per second).

beta (β) decay   Nuclear decay that involves the emission of a β particle.

beta (β) particle   A fast electron emitted from a nucleus in a radioactive decay.

beta (β)-pleated sheet   One type of planar secondary structure adopted by a polypeptide, in the form of a pleated sheet.

bimolecular reaction   An elementary reaction in which two molecules, atoms, or ions come together and form a product. *Example:* $O + O_3 \rightarrow 2 O_2$.

binary   Consisting of two components, as in *binary mixture* and *binary (ionic or molecular) compound. Examples:* acetone and water (a binary mixture); HCl, $CaCl_2$, C_6H_6 (binary compounds; $CaCl_2$ is ionic, HCl and C_6H_6 are molecular).

binary hydride   See *hydride.*

bioenergetics   The deployment and utilization of energy in living cells.

biomass   The organic material of the planet produced annually, mostly by photosynthesis.

biradical   A species with two unpaired electrons. *Example:* $\cdot OCH_2CH_2O\cdot$.

block (*s* block, *p* block, *d* block, *f* block)   The region of the periodic table containing elements for which, according to the building-up principle, the corresponding subshell is currently being filled.

body-centered cubic structure bcc   A crystal structure with a unit cell in which an atom lies at the center of a cube formed by eight others.

boiling point b.p.   See *boiling temperature; normal boiling point.*

boiling-point constant k_b The constant of proportionality between the boiling-point elevation and the molality of a solvent.

boiling-point elevation   The increase in normal boiling point of a solvent caused by the presence of a solute (a colligative property).

boiling temperature   (1) The temperature at which a liquid boils. (2) The temperature at which a liquid is in equilibrium with its vapor at the pressure of the surroundings; vaporization then occurs throughout the liquid, not only at the liquid's surface.

Boltzmann constant k The value of R/N_A, where R is the gas constant and N_A is the Avogadro constant; $k = 1.38066 \times 10^{-23}$ J/K.

Boltzmann formula (for the entropy)   The formula $S = k \ln W$, where k is the Boltzmann constant and W is the number of atomic arrangements that correspond to the same energy.

bond   A link between atoms. See also *covalent bond; double bond; ionic bond; triple bond.*

bond angle   In an A—B—C molecule or part of a molecule, the angle between the B—A and B—C bonds.

bond enthalpy $\Delta H_B(X—Y)$ The enthalpy change accompanying the dissociation of a bond. *Example:* $H_2(g) \rightarrow 2 H(g)$, $\Delta H_B(H—H) = +436$ kJ/mol.

bond length   The distance between the nuclei of two atoms joined by a bond.

bond order   The number of electron-pair bonds that link a specific pair of atoms.

bonding molecular orbital   See *bonding orbital.*

bonding orbital   A molecular orbital that, when occupied, contributes to an overall lowering of the energy of a molecule.

Born-Haber cycle   A closed series of reactions used to express the enthalpy of formation of an ionic solid in terms of contributions that include the lattice enthalpy.

Born interpretation   The interpretation of the square of the wavefunction, ψ, of a particle as the probability of finding the particle in each region of space.

boundary surface   The surface showing the region of an orbital within which there is a high probability of finding an electron.

Boyle's law At constant temperature, and for a given sample of gas, the volume is inversely proportional to the pressure: $P \propto 1/V$.

Bragg equation An equation relating the angle of diffraction of x-rays to the spacing of layers of atoms in a crystal ($\lambda = 2d\sin\theta$).

branched alkane An alkane with hydrocarbon side chains.

branching Description of a step in a chain reaction in which more than one chain carrier is formed in a propagation step. *Example:* $\cdot O\cdot + H_2 \rightarrow \cdot OH + \cdot H$.

breeder reactor A reactor used to generate nuclear fuel by making use of neutrons that are not moderated.

Brønsted acid A proton donor (a source of hydrogen ions, H^+). *Examples:* HCl; CH_3COOH; HCO_3^-; NH_4^+.

Brønsted base A proton acceptor (a species to which hydrogen ions, H^+, can bond). *Examples:* OH^-; $CH_3CO_2^-$; HCO_3^-; NH_3.

Brønsted-Lowry theory A theory of acids and bases expressed in terms of proton transfer equilibria.

Brownian motion The ceaseless jittering motion of colloidal particles caused by the impact of solvent molecules.

buffer A solution that resists any change in pH when small amounts of acid or base are added. An *acid buffer* stabilizes solutions at pH < 7 and a *base buffer* stabilizes solutions at pH > 7. *Examples:* a solution containing CH_3COOH and $CH_3CO_2^-$ (acid buffer); a solution containing NH_3 and NH_4^+ (base buffer).

buffer capacity An indication of the amount of acid or base that can be added before a buffer loses its ability to resist the change in pH.

building-up principle The procedure for arriving at the ground-state electron configurations of atoms and molecules.

bulk properties Properties that depend on the collective behavior of large numbers of atoms. *Examples:* melting point; vapor pressure; internal energy.

buret A narrow, graduated tube fitted with a stopcock, used to measure the volume of liquid delivered into another vessel.

calibration Interpretation of an observation by comparison with known information.

calorie cal A unit of energy. The unit is now defined in terms of the joule by 1 cal $= 4.184$ J exactly. The dietary Calorie is 1 kcal.

calorimeter An apparatus used to determine the heat released or absorbed by measuring the temperature change.

calorimetry The use of a calorimeter to measure the thermochemical properties of reactions.

capillary action The rise of liquids up narrow tubes.

carbide A binary compound of a metal or metalloid and carbon: carbides may be *saline* (saltlike compounds formed with some metals), *covalent* (compounds formed with metalloids), and *interstitial,* in which the carbon atoms lie in the interstices of a transition metal crystal lattice. *Examples:* CaC_2 (saline); SiC (covalent); WC (interstitial).

carbohydrate A compound of general formula $C_m(H_2O)_n$, although small deviations from this general formula are often encountered. Carbohydrates include cellulose, starches, and sugars. *Examples:* $C_6H_{12}O_6$, glucose; $C_{12}H_{22}O_{11}$, sucrose.

carbon cycle The sequence of physical processes and chemical reactions by means of which carbon atoms circulate through the environment.

carbonyl group A $>C=O$ group in an inorganic or organic compound.

carboxyl group The functional group $-COOH$. See also *carboxylic acid.*

carboxylic acid An organic compound containing the carboxyl group $-COOH$. *Examples:* CH_3COOH, acetic acid; C_6H_5COOH, benzoic acid.

catalyst A substance that increases the rate of a reaction without being consumed in the reaction. A catalyst is *homogeneous* if it is present in the same phase as the reactants and *heterogeneous* if it is in a different phase from the reactants. *Examples:* homogeneous, Br^-(aq) for the decomposition of H_2O_2(aq); heterogeneous, Pt in the Ostwald process.

catenate To form chains or rings of atoms. *Examples:* O_3; S_8.

cathode The electrode at which reduction occurs.

cathodic protection Protection of a metal object by connecting it to a more strongly reducing metal, which acts as a sacrificial anode.

cation A positively charged ion. *Examples:* Na^+; NH_4^+; Al^{3+}.

cell diagram A statement of the arrangement of electrodes in an electrochemical cell. A cell is written with the anode on the left and the cathode on the right. *Example:* $Zn(s)|Zn^{2+}(aq)||Cu^{2+}(aq)|Cu(s)$.

cell potential E (1) The electron-pushing and -pulling power of a reaction in an electrochemical cell. (2) The potential difference between the electrodes of an electrochemical cell when it is producing no current. Cell potential, which is also called *electromotive force* (emf), is always positive. Compare with *standard potential.*

Celsius scale A temperature scale on which the freezing point of water is at 0 degrees and its normal boiling point is at 100 degrees. Units on this scale are degrees Celsius, °C.

ceramic An inorganic solid that has been subjected to heat treatment; usually containing oxides, borides, or carbides.

cesium-chloride structure   A crystal structure the same as that of solid cesium chloride.

chain branching   A step in a chain reaction in which more than one chain carrier is formed.

chain carrier   An intermediate in a chain reaction.

chain reaction   A reaction in which an intermediate reacts to produce another intermediate in a series of elementary reactions. *Example:* $Br\cdot + H_2 \rightarrow HBr + H\cdot$; $H\cdot + Br_2 \rightarrow HBr + Br\cdot$.

chalcogens   Oxygen, sulfur, selenium, and tellurium in Group 16 of the periodic table.

change of state   The change of a substance from one of its physical states to another of its physical states. *Examples:* melting, solid → liquid; gray tin → white tin.

charge   A measure of the strength with which a particle can interact electrostatically with another particle.

Charles's law   The volume of a given sample of gas at constant pressure is directly proportional to its absolute temperature: $V \propto T$.

chelate   A complex containing at least one polydentate ligand that forms a ring of atoms including the central metal atom. *Example:* $[Co(en)_3]^{3+}$.

chemical amount   (1) The formal name for the quantity for which the units are moles. (2) The number of particles of a substance, measured in moles.

chemical analysis   The determination of the chemical composition of a sample. See also *qualitative* and *quantitative*.

chemical bond   See *bond*.

chemical change   The conversion of one or more substances into different substances.

chemical element   See *element*.

chemical equation   A statement in terms of chemical formulas summarizing the qualitative information about the chemical changes taking place in a reaction and the quantitative information that atoms are neither created nor destroyed in a chemical reaction. In a *balanced chemical equation* (commonly called a "chemical equation"), the same number of atoms of each element appear on both sides of the equation.

chemical equilibrium   A dynamic equilibrium between reactants and products in a chemical reaction.

chemical formula   A collection of chemical symbols and subscripts that shows the composition of a substance. See also *empirical formula; molecular formula; structural formula*.

chemical kinetics   The study of the rates of reactions and the steps by which they occur.

chemical nomenclature   The systematic naming of compounds.

chemical plating   The deposition of a metal surface on an object by making use of a chemical reduction reaction.

chemical property   The ability of a substance to participate in a chemical reaction.

chemical reaction   A chemical change in which one substance responds to the presence of another, to a change of temperature, or to some other influence.

chemical symbol   One- or two-letter abbreviation of an element's name.

chemiluminescence   The emission of light by products formed in energetically excited states during a chemical reaction.

chemistry   The branch of science concerned with the study of matter and the changes that matter can undergo.

chiral   (molecule or complex)   Not able to be superimposed on its own mirror image. *Examples:* $CH_3CH(NH_2)COOH$; $CHBrClF$; $[Co(en)_3]^{3+}$.

chloralkali process   The production of chlorine and sodium hydroxide by the electrolysis of aqueous sodium chloride.

cholesteric   Having layers of parallel molecules twisted relative to one another in such a way that the orientations of the liquid-crystal molecules form a spiral structure.

chromatogram   The record of the signal from the detector, or the plate or paper record, obtained in a chromatographic analysis of a mixture.

chromatography   A separation technique that relies on the ability of different phases to adsorb substances to different extents.

cis isomer   A geometrical isomer in which two atoms or groups are attached on the same side of a double bond, ring, or complex. Compare with *trans isomer*.

classical mechanics   The laws of motion proposed by Isaac Newton in which particles travel in definite paths in response to forces.

clathrate   A compound in which a molecule of one component sits in a cage made up of molecules of another component, typically water. *Example:* SO_2 in water.

Claus process   A process for obtaining sulfur from the H_2S in oil wells by the oxidation of H_2S with SO_2; the latter is formed by the oxidation of H_2S with oxygen.

Clausius-Clapeyron equation   An equation that gives the quantitative dependence of the vapor pressure of a substance on the temperature.

close-packed structure   A crystal structure in which atoms occupy the smallest total volume with the least empty space. *Examples:* hexagonal close packing and cubic close packing of identical spheres.

closed shell (or subshell)   A shell (or subshell) containing the maximum number of electrons allowed by the exclusion principle. *Example:* the neonlike core $1s^2 2s^2 2p^6$.

closed system   See *system*.

coagulation   The formation of aggregates from colloidal particles.

cohesion The act or state in which the particles of a substance stick to one another.

cohesive forces The forces that bind the molecules of a substance together to form a bulk material and that are responsible for condensation.

coinage metals The elements copper, silver, and gold.

colligative property A property that depends only on the relative number of solute and solvent particles present in a solution, and not on the chemical identity of the solute. *Examples:* elevation of boiling point; depression of freezing point; osmosis.

collision theory The theory of elementary gas-phase bimolecular reactions in which it is assumed that molecules react only if they collide with a characteristic minimum kinetic energy.

colloid (or *colloidal* suspension) A suspension of tiny particles with lengths between 1 nm and 1 μm in a gas, liquid, or solid. *Example:* milk.

combined gas law A form of the *ideal gas law* used to predict the effect of changes in conditions; expressed as $P_1V_1/n_1T_1 = P_2V_2/n_2T_2$.

combustion A reaction in which an element or compound burns in oxygen. *Example:* $CH_4(g) + 2 O_2(g) \rightarrow CO_2(g) + 2 H_2O(g)$.

combustion analysis The determination of the composition of a sample by measuring the masses of the products of its combustion.

common-ion effect Reduction of the solubility of one salt by the presence of another salt with one ion in common. *Example:* the lower solubility of AgCl in NaCl(aq) than in pure water.

common name An informal name for a compound that may give little or no clue to the compound's composition. *Examples:* water; aspirin; acetic acid.

competing reaction A reaction occurring at the same time as the reaction of interest and using some of the same reactants.

complementary color The color that white light becomes when one of the colors present in it is removed.

complete ionic equation A balanced chemical equation expressed in terms of the cations and anions present in solution. *Example:* $Ag^+(aq) + NO_3^-(aq) + Na^+(aq) + Cl^-(aq) \rightarrow AgCl(s) + NO_3^-(aq) + Na^+(aq)$.

complex (1) The combination of a Lewis acid and a Lewis base linked by a coordinate covalent bond. (2) A species consisting of several ligands (the Lewis bases) that have an independent existence bonded to a single central metal atom or ion (the Lewis acid). *Examples:* (1) $H_3N{-}BF_3$; (2) $[Fe(H_2O)_6]^{3+}$; $[PtCl_4]^-$.

composite material A synthetic material composed of a polymer and one or more other substances that have been solidified together.

compound (1) A specific combination of elements that can be separated into its elements by using chemical techniques. (2) A substance consisting of atoms of two or more elements in a definite ratio.

compress To reduce the volume of a sample.

compressibility The ability to be confined into a smaller volume.

compression factor Z The quantity $Z = PV_m/RT$; the ratio of the actual molar volume of a gas to the molar volume of an ideal gas under the same conditions.

concentration The quantity of a substance in a given volume. See also *molar concentration.*

condensation The formation of a liquid or solid phase from a gas phase.

condensation polymer A polymer formed by a series of condensation reactions. *Examples:* polyesters; polyamides (nylon).

condensation reaction A reaction in which two molecules combine to form a larger one and a small molecule is eliminated. *Example:* $CH_3COOH + C_2H_5OH \rightarrow CH_3COOC_2H_5 + H_2O$.

condensed phase A solid or liquid phase; not a gas.

condensed structural formula A version of the molecular formula showing how the atoms are grouped together. *Example:* $CH_3CH(CH_3)CH_3$ for methylpropane.

conduction band An incompletely occupied band of energy levels in a solid.

configuration See *electron configuration.*

conformations Molecular shapes that can be interchanged by rotation about bonds, without bond breakage and reformation.

conjugate acid The Brønsted acid formed when a Brønsted base has accepted a proton. *Example:* NH_4^+ is the conjugate acid of NH_3.

conjugate acid-base pair A Brønsted acid and its conjugate base. *Examples:* HCl and Cl^-; NH_4^+ and NH_3.

conjugate base The Brønsted base formed when a Brønsted acid has donated a proton. *Example:* NH_3 is the conjugate base of NH_4^+.

constructive interference Interference that results in an increased amplitude of a wave. Compare with *destructive interference.*

contact process The production of sulfuric acid by the combustion of sulfur and the catalyzed oxidation of sulfur dioxide to sulfur trioxide.

convection The bulk motion of regions of a fluid, often as a result of differences in density brought about by differences in temperature.

conversion factor A factor that is used to convert a measurement from one unit to another.

cooling curve   A graph of the variation of the temperature of a sample as it is cooled at a constant rate.

coordinate   Use of a lone pair to form a coordinate covalent bond. *Examples:* $F_3B + :NH_3 \rightarrow F_3B{-}NH_3$; $Ni + 4\,CO \rightarrow Ni(CO)_4$.

coordinate covalent bond   A bond formed between a Lewis base and a Lewis acid by sharing an electron pair originally belonging to the Lewis base.

coordination compound   A neutral complex or an ionic compound in which at least one of the ions is a complex. *Examples:* $Ni(CO)_4$; $K_3[Fe(CN)_6]$.

coordination isomer   A type of ionization isomer in which one or more ligands are exchanged between a cationic complex and an anionic complex. *Example:* $[Cr(NH_3)_6][Fe(CN)_6]$ and $[Fe(NH_3)_6][Cr(CN)_6]$ are coordination isomers.

coordination number   (1) The number of nearest neighbors of an atom in a solid. (2) For ionic solids, the coordination number of an ion is the number of nearest neighbors of opposite charge. (3) For complexes, the number of ligands attached to the central metal ion.

coordination sphere   The ligands directly attached to the central ion in a complex.

core   The inner closed shells of an atom.

core electrons   The electrons that belong to an atom's core.

corrosion   The unwanted reaction of a material that results in the dissolution or consumption of the material. *Example:* the oxidation of a metal.

corrosive   (1) A reagent that can cause corrosion. (2) Having a high reactivity, such as the reactivity of a strong oxidizing agent or a concentrated acid or base.

Coulomb potential energy   The potential energy of an electric charge in the vicinity of another electric charge; the potential energy is inversely proportional to the separation of the charges.

couple   See *redox couple*.

covalent bond   A pair of electrons shared between two atoms.

covalent carbide   See *carbide*.

covalent radius   The contribution of an atom to the length of a covalent bond.

cracking   The process of converting petroleum fractions into smaller molecules with more double bonds. *Example:* $CH_3(CH_2)_6CH_3 \rightarrow CH_3(CH_2)_3CH_3 + CH_3CH{=}CH_2$.

critical mass   The mass of fissionable material above which so few neutrons escape from a sample of nuclear fuel that the fission chain reaction is sustained; a greater mass is *supercritical* and a smaller mass is *subcritical*.

critical pressure P_c The vapor pressure of a liquid extrapolated to its critical temperature.

critical temperature T_c The temperature at and above which a substance cannot exist as a liquid.

cryogenics   The study of matter at very low temperatures.

cryoscopy   The measurement of molar mass by using the depression of freezing point.

crystal face   A flat plane forming an edge of a crystal.

crystal field   The electrostatic influence of the ligands (modeled as point negative charges) on the central ion of a complex.

crystal field theory   A rationalization of the optical, magnetic, and thermodynamic properties of complexes in terms of the crystal field of their ligands.

crystalline solid   A solid in which the atoms, ions, or molecules lie in an orderly array. *Examples:* NaCl; diamond; graphite. Compare with *amorphous solid*.

crystallization   The process in which a solute comes out of solution as crystals.

cubic close-packed structure ccp   A close-packed structure with an ABCABC . . . pattern of layers.

curie Ci   The unit of activity (for radioactivity).

current I The rate of supply of charge; current is measured in amperes (A), with $1\,A = 1\,C/s$.

cycle   (1) In thermodynamics, a sequence of changes that begins and ends at the same state. (2) In spectroscopy, one complete reversal of the direction of the electromagnetic field and its return to the original direction.

cycloalkane   A saturated aliphatic hydrocarbon in which the carbon atoms form a ring. *Example:* C_6H_{12}, cyclohexane.

Dalton's law of partial pressures   See *law of partial pressures*.

Daniell cell   A galvanic cell in which the cathode consists of copper in copper(II) sulfate solution and the anode consists of zinc in zinc sulfate solution.

data   The information provided or collected from experiments.

daughter nucleus   A nucleus that is the product of a nuclear decay.

de Broglie relation   The proposal that every particle has wavelike properties and that its wavelength, λ, is related to its mass by $\lambda = h/(\text{mass} \times \text{velocity})$.

debye D   The unit used to report electric dipole moments: $1\,D = 3.336 \times 10^{-30}\,C{\cdot}m$.

decant   To pour off a liquid from another, denser liquid or from a solid.

decay constant k The rate constant for radioactive decay.

decomposition   A reaction in which a substance is broken down into simpler substances; *thermal decomposition* is decomposition brought about by heat. *Example:* $CaCO_3(s) \xrightarrow{\Delta} CaO(s) + CO_2(g)$.

dehydrating agent   A reagent that removes water or the elements of water from a compound. *Example:* H_2SO_4.

delocalized   Spread over a region. In particular, *delocalized* *electrons* are electrons that spread over several atoms in a molecule.

delta Δ (in a chemical equation) A symbol that signifies that the reaction occurs at elevated temperatures.

delta hazard $\triangle$ A symbol that indicates that a skeletal equation is not balanced.

delta X ΔX The difference between the final and initial values of a property, $\Delta X = X_{final} - X_{initial}$. *Examples:* ΔT; ΔE.

denaturation The loss of structure of a large molecule such as a protein.

density d The mass of a sample of a substance divided by its volume: $d = m/V$.

deprotonation Loss of a proton from a Brønsted acid. *Example:* $NH_4^+(aq) + H_2O(l) \rightarrow H_3O^+(aq) + NH_3(aq)$.

derived unit A combination of base units. *Examples:* centimeters cubed (cm^3); joules $(kg \cdot m^2/s^2)$.

destructive interference Interference that results in a reduced amplitude of a wave. Compare with *constructive interference*.

deuteron The nucleus of a deuterium atom, $^2H^+$, consisting of a proton and a neutron.

diagonal relationship A similarity in properties between diagonal neighbors in the periodic table, especially for elements in Periods 2 and 3 at the left of the table. *Examples:* Li and Mg; Be and Al.

diamagnetic (substance) Pushed out of a magnetic field; consisting of atoms, ions, or molecules with no unpaired electrons. *Example:* most common substances.

diatomic molecule A molecule that consists of two atoms. *Examples:* H_2; CO.

diffraction The deflection of waves and the resulting interference caused by an object in their path. See also *x-ray diffraction*.

diffraction pattern The pattern of bright spots against a dark background resulting from diffraction.

diffusion The spreading of one substance through another substance.

dilute To reduce the concentration of a solute by adding more solvent.

dimer The union of two identical molecules. *Example:* Al_2Cl_6 formed from two $AlCl_3$ molecules.

diol An organic compound with two —OH groups.

dipeptide An *oligopeptide* formed by the condensation of two amino acids.

dipole See *electric dipole; instantaneous dipole moment*.

dipole-dipole interaction The interaction between two electric dipoles: like partial charges repel and opposite partial charges attract.

diprotic An acid with two acidic hydrogen atoms. See also *polyprotic acid or base*.

dispersion See *suspension*.

dispersion force See *London force*.

disproportionation A redox reaction in which a single element is simultaneously oxidized and reduced. *Example:* $2 Cu^+(aq) \rightarrow Cu(s) + Cu^{2+}(aq)$.

dissociation (1) The breaking of a bond. (2) The separation of ions that occurs when an ionic solid dissolves.

dissociation constant See *acidity constant*.

distillate A liquid obtained by distillation.

distillation The separation of the components of a mixture by making use of their different volatilities.

distribution (of molecular speeds) The fraction of gas molecules moving at each speed at any instant.

disulfide link An —S—S— link that contributes to the secondary and tertiary structures of polypeptides.

domain A region of a metal in which the electron spins of the atoms are aligned, thereby resulting in *ferromagnetism*.

doping The addition of a known, small amount of a second substance to an otherwise pure solid substance.

d-orbital See *atomic orbital*.

dose equivalent The actual dose of radiation experienced by a sample modified to take into account the *relative biological effectiveness* of the radiation. The dose equivalent is measured in *sievert*. See also *roentgen equivalent man*.

double bond (1) Two electron pairs shared by neighboring atoms. (2) One σ-bond and one π-bond between neighboring atoms.

Downs process The production of sodium and chlorine by the electrolysis of molten sodium chloride.

Dow process The electrolytic production of magnesium from molten magnesium chloride.

dry cell A galvanic cell in which the electrolyte is a moist paste. Commercial dry cells contain a graphite cathode and a zinc anode.

drying agent A substance that absorbs water and thus maintains a dry atmosphere. *Example:* phosphorus(V) oxide.

ductile Having the ability to be drawn out into a wire (as for a metal).

duplet The $1s^2$ electron pair of the heliumlike electron configuration.

dynamic equilibrium The condition in which a forward process and its reverse are occurring simultaneously at equal rates. *Examples:* vaporizing and condensing; chemical reactions at equilibrium.

effective nuclear charge Z_{eff} The net nuclear charge after taking into account the shielding caused by other electrons in the atom.

effervesce To bubble out of solution as a gas.

effusion The escape of a substance (particularly a gas) through a small hole.

elasticity The ability to return to the original shape after distortion.

elastomer An elastic polymer. *Example:* rubber (polyisoprene).

electric current See *current*.

electric dipole A positive charge next to an equal but opposite negative charge.

electric dipole moment μ The magnitude of the electric dipole (in debye).

electric field A region of influence that affects charged particles. A component of an *electromagnetic field*.

electrical conduction The conduction of electric charge through matter. See also *electronic conduction; ionic conduction*.

electrochemical cell A system consisting of two electrodes in contact with an electrolyte. A *galvanic cell* is an electrochemical cell used to produce electricity, and an *electrolytic cell* is an electrochemical cell in which an electric current is used to cause chemical change.

electrochemical series Redox couples arranged in order of oxidizing and reducing strengths; usually arranged with strong oxidizing agents at the top of the list and strong reducing agents at the bottom.

electrochemistry The branch of chemistry that deals with the use of chemical reactions to produce electricity, the relative strengths of oxidizing and reducing agents, and the use of electricity to produce chemical change.

electrode A metallic conductor that makes contact with an electrolyte in an electrochemical cell.

electrolysis (1) A process in which a chemical change is produced by passing an electric current through a liquid. (2) The process of driving a reaction in a nonspontaneous direction by passing an electric current through a solution.

electrolyte A substance that dissolves to give an electrically conducting solution. A *strong electrolyte* is a substance that is fully ionized in solution. A *weak electrolyte* is a molecular substance that is only partially ionized in solution. A *nonelectrolyte* does not ionize in solution. *Examples:* NaCl is a strong electrolyte; CH_3COOH is a weak electrolyte; $C_6H_{12}O_6$ is a nonelectrolyte.

electrolyte solution An ionically conducting (usually aqueous) solution.

electrolytic cell See *electrochemical cell*.

electromagnetic field The region of influence generated by accelerated charged particles.

electromagnetic radiation A wave of oscillating electric and magnetic fields; includes light, x-rays, and γ rays.

electromotive force emf See *cell potential*.

electron e^- A negatively charged subatomic particle found outside the nucleus of an atom.

electron affinity E_{ea} The energy released when an electron is added to a gas-phase atom or monatomic ion.

electron arrangement (VSEPR model) The three-dimensional geometry of the arrangement of bonds and lone pairs about a central atom.

electron capture The capture by a nucleus of one of its own atom's *s*-electrons.

electron configuration The occupancy of orbitals in an atom or molecule. *Example:* N, $1s^22s^22p^3$.

electron-deficient compound A compound with too few valence electrons for it to be assigned a valid Lewis structure. *Example:* B_2H_6.

electronegative element An element with a high electronegativity. *Examples:* O; F.

electronegativity χ (chi) The ability of an atom to attract electrons to itself when it is part of a compound.

electronic conduction The conduction of electricity by the motion of electrons. See also *electrical conduction*.

electronic conductor A substance capable of supporting *electronic conduction*.

electronic structure The details of the distribution of the electrons that surround the nuclei in atoms and molecules.

electroplating The deposition of a thin film of metal on an object by electrolysis.

electropositive element An element that has a low electro-negativity and is likely to give up electrons to another element on compound formation. *Examples:* Cs; Mg.

electrostatic potential diagram A representation of molecular structure in which the electron distribution in a molecule is depicted by different colors.

element (1) A substance that cannot be separated into simpler components by using chemical techniques. (2) A substance consisting of atoms of the same atomic number. *Examples:* hydrogen; gold; uranium.

elementary reaction An individual reaction step in a mechanism.

elimination reaction A reaction in which two groups or atoms on neighboring carbon atoms are removed from a molecule, thereby leaving a multiple bond between the carbon atoms. *Example:* $CH_3CHBrCH_3 + OH^- \rightarrow CH_3CH{=}CH_2 + H_2O + Br^-$.

empirical formula A chemical formula that shows the relative numbers of atoms of each element in a compound by using the simplest whole number subscripts. *Examples:* P_2O_5; CH for benzene.

emulsion A suspension of droplets of one liquid dispersed throughout another liquid.

enantiomers A pair of optical isomers that are mirror images of, but not superimposable on, each other.

end point The stage in a titration at which enough titrant has been added to bring the indicator to a color halfway between its initial and final colors.

endothermic process   A process, particularly a chemical reaction, that absorbs heat ($\Delta H > 0$). *Example:* $N_2O_4(g) \rightarrow 2\,NO_2(g)$.

energy *E*   The capacity of a system to do work or supply heat. *Kinetic energy* is the energy of motion, and *potential energy* is the energy arising from position. The *total energy* is the sum of the potential and kinetic energies.

energy level   A permitted value of the energy in a quantized system such as an atom or a molecule.

enrich   In nuclear chemistry, to increase the abundance of a specific isotope.

enthalpy *H*   A state property that is equal to the quantity of heat transferred at constant pressure; $H = U + PV$.

enthalpy density (of a fuel)   The enthalpy of combustion per liter (without the negative sign).

enthalpy of freezing   The negative of the *enthalpy of fusion*.

enthalpy of fusion ΔH_{fus} The enthalpy change per mole accompanying fusion (melting).

enthalpy of hydration ΔH_{hyd}.   The enthalpy change accompanying the hydration of gas-phase ions.

enthalpy of melting ΔH_{melt} See *enthalpy of fusion*.

enthalpy of reaction   See *reaction enthalpy*.

enthalpy of solution ΔH_{sol} The change in enthalpy that occurs when a substance dissolves. The *limiting enthalpy of solution* is the enthalpy of solution for the formation of an infinitely dilute solution.

enthalpy of sublimation ΔH_{sub} The enthalpy change per mole accompanying sublimation (the direct conversion of a solid to a vapor).

enthalpy of vaporization ΔH_{vap} The enthalpy change per mole accompanying vaporization (the conversion of a substance from the liquid state to the vapor state).

entropy *S*   (1) A measure of the disorder of a system. (2) A change in entropy is equal to the heat supplied reversibly to a system divided by the temperature at which the transfer occurs.

entropy of fusion ΔS_{fus} The entropy change per mole accompanying fusion (the conversion of a substance from the solid state to the liquid state).

entropy of vaporization ΔS_{vap} The entropy change per mole accompanying vaporization (the conversion of a substance from the liquid state to the vapor state).

enzyme   A biological catalyst.

e-orbital   One of the orbitals d_{z^2} or $d_{x^2-y^2}$ in an octahedral or tetrahedral complex.

equatorial bond   A bond perpendicular to the axis of a molecule (particularly trigonal bipyramidal and octahedral molecules).

equatorial lone pair   A lone pair lying in the plane perpendicular to the molecular axis.

equilibrium   See *chemical equilibrium; dynamic equilibrium*.

equilibrium constant *K*   An expression characteristic of the equilibrium composition of the reaction mixture, with a form given by the law of mass action. *Example:* $N_2(g) + 3\,H_2(g) \rightleftharpoons 2\,NH_3(g)$, $K = P_{NH_3}{}^2/P_{N_2}P_{H_2}{}^3$.

equilibrium table   A table used to calculate the composition of a reaction mixture at equilibrium, given the initial composition. The columns are headed by the species and the rows are, successively, the initial composition, the change to reach equilibrium, and the equilibrium composition.

equivalence point   See *stoichiometric point*.

essential amino acid   An amino acid that is an essential component of the diet because it cannot be synthesized in the body.

essential oil   An oil that can be distilled from flowers and leaves (and conveys the "essence" of a plant).

ester   The product (other than water) of the reaction between a carboxylic acid and an alcohol and having the formula RCOOR'. *Example:* $CH_3COOC_2H_5$, ethyl acetate.

ether   An organic compound of the form R—O—R'. *Example:* $CH_3OC_2H_5$, ethyl methyl ether; $C_2H_5OC_2H_5$, diethyl ether.

evaporate   Vaporize completely.

excited state   A state other than the state of lowest energy.

exclusion principle   No more than two electrons can occupy any given orbital; and when two electrons do occupy one orbital, their spins must be paired.

exothermic process   A process, particularly a chemical reaction, that releases heat ($\Delta H < 0$). *Example:* $N_2(g) + 3\,H_2(g) \rightarrow 2\,NH_3(g)$.

expanded valence shell   A valence shell containing more than eight electrons. Also called an *expanded octet*. *Examples:* the valence shells of P and S in PCl_5 and SF_6.

expansion work   See *work*.

experiment   A test carried out under carefully controlled conditions.

exponential   The exponential of *x* is the natural antilogarithm of *x*, namely, e^x.

exponential decay   A variation with time of the form e^{-kt}. *Example:* $[A] = [A]_0 e^{-kt}$.

extensive property   A physical property of a substance that depends on the size of the sample. *Examples:* mass; internal energy; enthalpy; entropy.

extrapolate   To extend a graph outside the region covered by the data.

face   See *crystal face*.

face-centered cubic structure fcc   A crystal structure built from a cubic unit cell in which there is an atom at the center of each face and one at each corner.

Fahrenheit scale   A temperature scale on which the freezing point of water is at 32 degrees and the normal boiling point is at 212 degrees. Units on this scale are degrees Fahrenheit, °F.

fallout The fine dust that settles from clouds of airborne particles after a nuclear bomb test.

Faraday constant F The magnitude of the charge per mole of electrons; $F = 9.648\,53 \times 10^4$ C/mol.

Faraday's law of electrolysis The number of moles of product formed by an electric current is chemically equivalent to the number of moles of electrons supplied.

fat An ester of glycerol and carboxylic acids with long hydrocarbon chains; fats act as long-term energy storage.

fatty acid A carboxylic acid with a long hydrocarbon chain. *Example:* $CH_3(CH_2)_{16}COOH$, stearic acid.

ferroalloy An alloy of a metal with iron and, often, carbon. *Example:* ferrovanadium.

ferromagnetism The ability of some substances to be permanently magnetized. *Examples:* iron; magnetite, Fe_3O_4.

field An influence spreading over a region of space. *Examples:* an *electric field* from a charge; a *magnetic field* from a magnet.

filtration The separation of a heterogeneous mixture of a solid and liquid by passing the mixture through a fine mesh.

first ionization energy See *ionization energy.*

first law of thermodynamics The internal energy of an isolated system is constant.

first-order reaction A reaction in which the rate is proportional to the first power of the concentration of a substance.

fissile Having the ability to undergo fission induced by slow neutrons. *Example:* ^{235}U is fissile.

fission (nuclear) The breakup of a nucleus into two smaller nuclei of similar mass; fission may be *spontaneous* or *induced* (particularly by the impact of neutrons). *Examples:* $^{244}_{95}Am \rightarrow {}^{134}_{53}I + {}^{107}_{42}Mo + 3\,{}^{1}_{0}n$ (spontaneous); $^{235}_{92}U + {}^{1}_{0}n \rightarrow {}^{142}_{56}Ba + {}^{91}_{36}Kr + 3\,{}^{1}_{0}n$ (induced).

fissionable Having the ability to undergo induced fission.

fixation of nitrogen Conversion of elemental nitrogen to its compounds, particularly ammonia.

flocculation The joining of aggregated particles into larger masses that can be filtered.

f-orbital See *atomic orbital.*

force F An influence that changes the state of motion of an object. *Examples:* an electrostatic force from an electric charge; a mechanical force from an impact.

formal anhydride See *acid anhydride.*

formal charge (1) The electric charge of an atom assigned on the assumption that there is nonpolar covalent bonding. (2) Formal charge (FC) = number of valence electrons in the free atom − (number of lone-pair electrons + $\frac{1}{2} \times$ number of shared electrons).

formation constant K_f The equilibrium constant for complex formation. The *overall formation constant* is the product of *stepwise formation constants*. The inverse of the formation constant ($1/K_f$) is called the *stability constant.*

formula unit The group of ions that matches the formula of the smallest unit of an ionic compound. *Example:* NaCl, one Na^+ ion and one Cl^- ion.

formula weight See *molar mass.*

fossil fuels The partially decomposed remains of vegetable and marine life (mainly coal, oil, and natural gas).

fraction Samples of distillate obtained in different ranges of boiling temperatures.

fractional distillation Separation of the components of a liquid mixture by repeated distillation, making use of their differing volatilities.

free energy G The energy of a system that is free to do work at constant temperature and pressure: $G = H - TS$. The direction of spontaneous change at constant pressure and temperature is the direction of decreasing free energy.

freezing-point constant k_f The constant of proportionality between the freezing-point depression and the molality of a solute.

freezing-point depression The lowering of the freezing point of a solution caused by the presence of a solute (a colligative property).

freezing temperature The temperature at which a liquid freezes. The *normal freezing point* is the freezing temperature under a pressure of 1 atm.

frequency (of radiation) ν (nu) The number of cycles (repeats of the waveform) per second (unit: *hertz,* Hz).

froth flotation A process for separating the mineral from unwanted rock in an ore by blowing air through a mixture that contains oil, water, and detergents to generate a froth or foam.

fuel cell A primary electrochemical cell in which the reactants are supplied continuously from outside while the cell is in use.

functional group A group of atoms that brings a characteristic set of chemical properties to an organic molecule. *Examples:* −OH ; −Br ; −COOH.

fusion (1) Melting. (2) (nuclear) The merging of nuclei to form the nucleus of a heavier element.

galvanic cell See *electrochemical cell.*

galvanize To coat a metal with an unbroken film of zinc.

gamma (γ) radiation Very high frequency, short wavelength electromagnetic radiation emitted by nuclei.

gas A fluid form of matter that fills the container it occupies and can easily be compressed into a much smaller volume. (The distinction between a gas and a vapor is as follows: a *gas* is a substance at a temperature above its critical temperature; a *vapor* is a gaseous form of matter at a temperature below its critical temperature.)

gas constant R The constant of proportionality that appears in the ideal gas law. Values are listed inside the back cover of the book.

gas-liquid chromatography   A version of chromatography in which a gas carries the sample over a stationary liquid phase.

gauge pressure   The pressure inside a container less that outside the container.

Geiger counter   A device that is used to detect and measure radioactivity by relying on ionization caused by incident radiation.

gel   A soft, solid colloid.

geometrical isomers   Stereoisomers that differ in the spatial arrangement of the atoms. Geometrical isomers exhibit *cis-trans* isomerism.

Gibbs free energy   See *free energy*.

glass   An ionic solid with an amorphous structure resembling that of a liquid.

glass electrode   A thin-walled glass bulb containing an electrolyte solution and a metallic contact; used for measuring pH.

Graham's law of effusion   The rate of effusion of a gas is inversely proportional to the square root of its molar mass.

gravimetric analysis   The use of measurements of mass to determine the amount of substance present.

gray   Gy   The SI unit of *absorbed dose;* 1 Gy corresponds to an energy deposit of 1 J/kg. See also *rad*.

greenhouse effect   The blocking by some atmospheric gases (notably carbon dioxide) of the radiation of heat from the surface of the Earth back into space, leading to the possibility of a worldwide rise in temperature.

greenhouse gas   A gas that contributes to the greenhouse effect.

ground state   The state of lowest energy.

group   A vertical column in the periodic table.

Haber process (Haber-Bosch process)   The catalyzed synthesis of ammonia at high pressure and high temperature.

half-life $t_{1/2}$ (1) In chemical kinetics, the time needed for the concentration of a substance to fall to half its initial value. (2) In radioactivity, the time needed for half the initial number of radioactive nuclei to disintegrate.

half-reaction   A hypothetical oxidation or reduction reaction showing either electron loss or electron gain. *Examples:* $Na(s) \rightarrow Na^+(aq) + e^-$ (oxidation); $Cl_2(g) + 2\,e^- \rightarrow 2\,Cl^-(aq)$ (reduction).

halide ion   An anion formed from a halogen atom. *Examples:* F^-; I^-.

Hall process   The production of aluminum by the electrolysis of aluminum oxide dissolved in molten cryolite.

haloalkane   An alkane with a halogen substituent. *Example:* CH_3Cl, chloromethane.

halogenation   The incorporation of a halogen into a compound (particularly, into an organic compound).

halogens   The elements in Group 17.

hard water   Water that contains dissolved calcium and magnesium salts.

heat q A transfer of energy that occurs as the result of a temperature difference.

heat capacity C (1) The ratio of heat supplied to the temperature rise produced. (2) The heat required to raise the temperature of an object by 1°C. The *heat capacity at constant pressure,* C_P, and the *heat capacity at constant volume,* C_V, are normally distinguished. See also *specific heat capacity*.

heating   The act of transferring energy as heat.

heating curve   A graph of the variation of the temperature of a sample as it is heated at a constant rate.

Heisenberg uncertainty principle   The position and velocity of a particle cannot both be known simultaneously with arbitrary precision.

Henderson-Hasselbalch equation   An approximate equation for estimating the pH of a solution containing a conjugate acid and base. (See Section 16.9.)

Henry's law   The solubility of a gas in a liquid is proportional to its partial pressure above the liquid: solubility = k_H × partial pressure.

Henry's law constant   The constant k_H that appears in Henry's law.

hertz   Hz   The SI unit of frequency: 1 Hz is one complete cycle per second.

Hess's law   A reaction enthalpy is the sum of the enthalpies of any sequence of reactions (at the same temperature and pressure) into which the overall reaction can be divided.

heterogeneous alloy   See *alloy*.

heterogeneous catalyst   See *catalyst*.

heterogeneous equilibrium   An equilibrium in which at least one substance is in a different phase from that of the rest. *Example:* $AgCl(s) \rightleftharpoons Ag^+(aq) + Cl^-(aq)$.

heterogeneous mixture   A mixture in which the individual components, although mixed together, lie in distinct regions that can be distinguished on a microscopic scale. *Example:* a mixture of sand and sugar.

heteronuclear diatomic molecule   A molecule consisting of two atoms of different elements. *Examples:* HCl; CO.

hexagonal close-packed structure   hcp   A close-packed structure with an ABABA . . . pattern of layers.

high-spin complex A d^n complex with the maximum number of unpaired electron spins.

high-temperature superconductor A material that becomes superconducting at temperatures well above the transition temperature for the first generation of superconductors, typically, 100 K and above.

homogeneous alloy See *alloy*.

homogeneous catalyst See *catalyst*.

homogeneous equilibrium A chemical equilibrium in which all the substances taking part are in the same phase. *Example:* $H_2(g) + I_2(g) \rightleftharpoons 2 HI(g)$.

homogeneous mixture A mixture in which the individual components are uniformly mixed, even on a molecular scale. *Examples:* air; solutions.

homolytic dissociation Dissociation into radicals. *Example:* $CH_3I \rightarrow \cdot CH_3 + \cdot I$.

homonuclear diatomic molecule A molecule consisting of two atoms of the same element. *Examples:* H_2; N_2.

Hund's rule If more than one orbital in a subshell is available, electrons fill empty orbitals of that subshell before pairing in one of them.

hybrid orbital A mixed orbital formed by blending together atomic orbitals on the same atom. *Example:* an sp^3 hybrid orbital.

hybridization The formation of hybrid orbitals.

hydrate A solid compound containing H_2O molecules. *Example:* $CuSO_4 \cdot 5H_2O$.

hydrate isomers A type of ionization isomer resulting from the exchange of another ligand for an H_2O molecule in the coordination sphere.

hydrated Having water molecules attached.

hydration (1) (of ions) The attachment of water molecules to a central ion. (2) (of organic compounds) The addition of water across a multiple bond (H to one carbon atom, OH to the other). *Example:* $CH_2=CH_2 + H_2O \rightarrow CH_3CH_2OH$.

hydride A binary compound of a metal or metalloid with hydrogen; the term is often extended to include all binary compounds of hydrogen. A *saline* or *saltlike hydride* is a compound of hydrogen and a strongly electropositive metal; a *molecular hydride* is a compound of hydrogen and a nonmetal; a *metallic hydride* is a compound of certain *d*-block metals and hydrogen.

hydrocarbon A binary compound of carbon and hydrogen. *Examples:* CH_4; C_6H_6.

hydrogen bond A link formed by a hydrogen atom lying between two strongly electronegative atoms and covalently bonded to one of them (O, N, or F). The electronegative atoms may be located on different molecules or in different regions of the same molecule.

hydrogen economy The widespread use of hydrogen as a fuel.

hydrogen electrode An electrode consisting of platinum in contact with hydrogen gas and a solution containing hydrogen ions.

hydrogenation The addition of hydrogen to multiple bonds. *Example:* $CH_3CH=CH_2 + H_2 \rightarrow CH_3CH_2CH_3$.

hydrolysis reaction The reaction of water with a substance, resulting in the formation of a new element-oxygen bond. *Example:* $PCl_5(s) + 4 H_2O(l) \rightarrow H_3PO_4(aq) + 5 HCl(aq)$.

hydrometallurgical extraction The extraction of metals by reduction of their ions in aqueous solution. *Example:* $Cu^{2+}(aq) + Fe(s) \rightarrow Cu(s) + Fe^{2+}(aq)$.

hydronium ion The ion H_3O^+.

hydrophilic Water-attracting. *Example:* hydroxyl groups are hydrophilic.

hydrophobic Water-repelling. *Example:* hydrocarbon chains are hydrophobic.

hydrostatic pressure The pressure exerted by a column of water or liquid solution.

hydroxyl group An $-OH$ group in an organic compound.

hypertonic Having a higher osmotic pressure than a solution on the other side of a semipermeable membrane.

hypothesis A suggestion put forward to account for a series of observations. *Example:* Dalton's atomic hypothesis.

hypotonic Having a lower osmotic pressure than a solution on the other side of a semipermeable membrane.

ideal gas A gas that satisfies the ideal gas law and is described by the *kinetic model*.

ideal gas law $PV = nRT$ All gases obey the law more and more closely as the pressure is reduced to very low values.

ideal solution A solution that obeys Raoult's law at any concentration; all solutions behave ideally as the concentration approaches 0. *Example:* benzene and toluene.

***i* factor** A factor that takes into account the existence of ions in an electrolyte solution, particularly for the interpretation of colligative properties. It indicates the number of particles formed from one formula unit of the solute. *Example:* $i \approx 2$ for very dilute NaCl(aq).

incomplete octet A valence shell of an atom that has fewer than eight electrons. *Example:* the valence shell of B in BF_3.

indicator A substance that changes color when it goes from its acidic to its basic form (an *acid-base indicator*) or from its oxidized to its reduced form (a *redox indicator*).

induced-fit mechanism A model of the action of an enzyme in which the enzyme molecule adjusts its shape to accommodate the incoming substrate molecule. A modification of the *lock-and-key mechanism* of enzyme action.

induced nuclear fission See *fission*.

inert   (1) Unreactive. (2) Thermodynamically unstable but survives for long periods.

inert-pair effect   The observation that an element displays a valence lower than expected from its group number. An *inert pair* is a pair of valence shell *s*-electrons that are tightly bound to the atom and that might not participate in bond formation.

infrared radiation   Electromagnetic radiation with a lower frequency (longer wavelength) than that of red light but a higher frequency (shorter wavelength) than microwave radiation.

initial concentration (of a weak acid or base)   The concentration as prepared, as if no deprotonation or protonation had occurred.

initial rate   The rate at the start of a reaction when products are present in concentrations too low to affect the rate.

initiation   The formation of reactive intermediates that serve as chain carriers from a reactant at the start of a chain reaction. *Example:* $Br_2 \rightarrow 2\ Br\cdot$.

inner transition metal   A member of the *f* block of the periodic table (the *lanthanides* and *actinides*).

inorganic compound   A compound that is not organic. See also *organic compound.*

insoluble substance   A substance that is not soluble in a specified solvent. When the solvent is not specified, water is generally meant.

instantaneous dipole moment   A dipole moment that arises from a transient redistribution of charge and is responsible for the London force.

instantaneous rate   The slope of the tangent of a graph of concentration against time.

insulator (electrical)   A substance that does not conduct electricity. *Examples:* nonmetallic elements; molecular solids.

integrated rate law   An expression for the concentration of a reactant or product in terms of the time, obtained from the rate law of the reaction. *Example:* $[A] = [A]_0 e^{-kt}$ for a first-order reaction.

intensity   The brightness of radiation. The intensity of a wave is proportional to the square of its *amplitude.*

intensive property   A physical property of a substance that is independent of the size of the sample. *Examples:* density; molar volume; temperature.

intercept (of a graph)   The value at which the graph cuts the vertical axis. See also Appendix 1E.

interference   Interaction between waves, leading to a greater amplitude (*constructive interference*) or to a smaller one (*destructive interference*).

interhalogen   A binary compound of two halogens. *Example:* IF_3.

intermediate   See *reaction intermediate.*

intermolecular   Between molecules.

intermolecular forces   The forces of attraction and repulsion between molecules; *van der Waals forces.* *Examples:* hydrogen bonding; dipole-dipole interaction; London force.

internal energy *U*   (1) The measure of a system's capacity to do work and supply heat. (2) The total energy of a system.

International System SI   See *SI.*

internuclear axis   The straight line between the nuclei of two bonded atoms.

interpolate   Find a value between two measured values.

interstice   A hole or gap in a crystal lattice.

interstitial alloy   See *alloy.*

interstitial carbide   See *carbide.*

ion   An electrically charged atom or bonded group of atoms. *Examples:* Al^{3+}; SO_4^{2-}. See also *cation; anion.*

ion-dipole interaction   The attraction between an ion and the opposite partial charge of the electric dipole of a polar molecule.

ion exchange   The exchange of one type of ion in solution for another.

ionic bond   The attraction between the opposite charges of cations and anions.

ionic compound   A compound that consists of ions. *Examples:* NaCl; KNO_3.

ionic conduction   Electrical conduction in which the charge is carried by ions. See also *electrical conduction.*

ionic equation   See *complete ionic equation.*

ionic model   The description of bonding in terms of ions.

ionic radius   The contribution of an ion to the distance between neighboring ions in a solid ionic compound. In practice, the radius of an ion is defined as the distance between the centers of neighboring ions, with the radius of the O^{2-} ion set equal to 140. pm.

ionic solid   A solid built from cations and anions. *Examples:* NaCl; KNO_3.

ionization   (1) (of atoms and molecules) Conversion to cations by the removal of electrons. *Example:* $K(g) \rightarrow K^+(g) + e^-(g)$. (2) (of an acid) The donation of a proton from a neutral acid molecule to a base. (3) (of a base) The formation of the conjugate base (an anion, in this instance) of the acid. *Example:* $CH_3COOH(aq) + H_2O(l) \rightarrow H_3O^+(aq) + CH_3CO_2^-(aq)$.

ionization constant   See *acidity constant.*

ionization energy *I*   The minimum energy required to remove an electron from the ground state of a gaseous atom, molecule, or ion; also called the *first ionization energy.* The *second ionization energy* is the ionization energy for removal of a second electron, and so on.

ionization isomers   Isomers that differ by the exchange of a ligand with an anion or neutral molecule outside the coordination sphere.

ionizing radiation   High-energy radiation (typically but not necessarily nuclear radiation) that can cause ionization.

isoelectronic species   Species with the same number of atoms and the same number of valence electrons. *Examples:* F^- and Ne; SO_2 and O_3; CN^- and CO.

isolated system   See *system.*

isomer   One of two or more compounds that contain the same number of the same atoms in different arrangements. In *structural isomers,* the atoms have different partners or lie in a different order; in *stereoisomers,* the atoms have the same partners but are in different arrangements in space. *Examples:* $CH_3—O—CH_3$ and $CH_3CH_2—OH$ (structural isomers); *cis-* and *trans-*2-butene (stereoisomers). See also *geometrical isomers; optical isomers.*

isotactic polymer   See *polymer.*

isotherm   A line of constant temperature on a graph.

isothermal process   A change that occurs at constant temperature.

isotonic   Having an osmotic pressure the same as that of a solution on the other side of a semipermeable membrane.

isotope   One of two or more atoms that have the same atomic number but different atomic masses. *Example:* ^1H, ^2H, and ^3H are all isotopes of hydrogen.

isotopic abundance   See *abundance.*

isotopic dating   The determination of the age of objects by measuring the activity of the radioactive isotopes they contain, particularly ^{14}C.

isotopic label   See *tracer.*

isotropic   Not depending on orientation.

joule   J   The SI unit of energy (1 J = 1 kg·m²/s²).

Kekulé structures   Two Lewis structures of benzene, consisting of alternating single and double bonds.

kelvin   K   The SI unit of temperature.

Kelvin scale   A fundamental scale of temperature on which the triple point of water lies at 273.16 K and the lowest attainable temperature is at 0. The unit on the Kelvin scale is the kelvin, K.

ketone   An organic compound containing a carbonyl group between two carbon atoms, having the form $R—CO—R'$. *Example:* $CH_3—CO—CH_2CH_3$, butanone.

kilogram   kg   The SI unit of mass.

kinetic energy E_K The energy of a particle due to its motion. *Example:* the kinetic energy of a particle of mass m and speed v is $\frac{1}{2}mv^2$.

kinetic model   A model of the properties of an ideal gas in which pointlike molecules are in continuous random motion in straight lines until collisions occur between them.

kinetic molecular theory   The theory that discusses the *kinetic model* of gases.

labile   A species that survives only for short periods.

lachrymator   A substance that stimulates the production of tears. *Example:* PAN.

lanthanide   A member of the first row of the *f* block (lanthanum through ytterbium).

lanthanide contraction   The reduction of atomic radius of the elements following the lanthanides below the values that would be expected by extrapolation of the trend down a group (and arising from the poor shielding ability of *f*-electrons).

lattice   An orderly array of atoms, molecules, or ions in a crystal.

lattice enthalpy ΔH_L The standard enthalpy change for the conversion of an ionic solid to a gas of ions.

law   In science, a summary of experience.

law of combining volumes   At the same temperature and pressure, the volumes of gases react with one another in the ratios of small whole numbers.

law of conservation of energy   (1) Energy can be neither created nor destroyed. (2) The total energy of an isolated body is constant.

law of conservation of mass   Matter (and specifically atoms) is neither created nor destroyed in a chemical reaction.

law of constant composition   A compound has the same composition whatever its source.

law of mass action   For an equilibrium of the form $aA + bB \rightleftharpoons cC + dD$, the ratio $[C]^c[D]^d/[A]^a[B]^b$ evaluated at equilibrium is equal to a constant K_c, which has a specific value for a given reaction and temperature.

law of partial pressures   The total pressure of a mixture of gases is the sum of the partial pressures of its components.

law of radioactive decay   The rate of decay is proportional to the number of radioactive nuclides in the sample.

lead-acid cell   A secondary cell in which the electrodes are lead and the electrolyte is dilute sulfuric acid.

Le Chatelier's principle   When a stress is applied to a system in dynamic equilibrium, the equilibrium tends to adjust to minimize the effect of the stress. *Example:* a reaction at equilibrium tends to proceed in the endothermic reaction when the temperature is raised.

leveling   The observation that strong acids all have the same strength in water, and all behave as though they were solutions of H_3O^+ ions.

Lewis acid   An electron pair acceptor. *Examples:* H^+; Fe^{3+}; BF_3.

Lewis base An electron pair donor. *Examples:* OH^-; H_2O; NH_3.

Lewis formula (for an ionic compound) A representation of the structure of an ionic compound showing the formula unit of ions in terms of their Lewis symbols.

Lewis structure A diagram showing how electron pairs are shared between atoms in a molecule. *Examples:* $H-\ddot{C}\ddot{l}$; $\ddot{O}=C=\ddot{O}$.

Lewis symbol (for atoms and ions) The chemical symbol of an element with a dot for each valence electron.

ligand A group attached to the central metal ion in a complex; a *polydentate ligand* occupies more than one binding site.

ligand field splitting Δ The energy separation of the *e*- and *t*-orbitals induced by the presence of ligands in a complex.

light See *visible radiation.*

light-water reactor A nuclear reactor in which ordinary water is used as the moderator.

limiting enthalpy of solution See *enthalpy of solution.*

limiting law A law that is accurately obeyed only at the limit of a property, such as when a property (the pressure of a gas, for example) is made very small.

limiting reactant The reactant that governs the theoretical yield of product in a given reaction.

line structure A representation of the structure of an organic molecule in terms of lines showing the bonds; carbon atoms, and the hydrogen atoms attached to them, are not usually shown explicitly.

linear combination of atomic orbitals (LCAO) A molecular orbital formed by superimposing atomic orbitals.

linkage isomers Isomers that differ in the identity of the atom that a ligand uses to attach to the metal ion.

lipid A naturally occurring organic compound that dissolves in hydrocarbons but not in water. *Examples:* fats; steroids; terpenes; the molecules that form cell membranes.

liquid A fluid form of matter that has a well-defined surface and takes the shape of the part of the container it occupies.

liquid crystal A substance that flows like a liquid but has molecules that lie in a moderately orderly array. Liquid crystals may be *nematic, smectic,* or *cholesteric,* depending on the arrangement of the molecules.

lock-and-key mechanism A model of enzyme action in which the enzyme is thought of as a lock and its substrate as a matching key.

London force The force of attraction that arises from the interaction between instantaneous electric dipoles on neighboring polar or nonpolar molecules.

lone pair A pair of valence electrons that is not involved in bonding.

long period A period of the periodic table with more than eight members.

long-range order An orderly arrangement of atoms or molecules that is repeated over long distances.

low-spin complex A d^n complex with the minimum number of unpaired electron spins.

Lyman series A set of lines in the ultraviolet region of the spectrum of hydrogen.

lyotropic liquid crystal A liquid crystal that results from the action of a solvent on a solute.

macroscopic level The level at which visible objects can be observed directly.

magic numbers The numbers of protons or neutrons that correlate with enhanced nuclear stability. *Examples:* 2, 8, 20, 50, 82, and 126.

magnetic field A region of influence that affects the motion of moving charged particles.

magnetic quantum number m_l The quantum number that identifies the individual orbitals of a subshell of an atom and determines their orientation in space.

main group Any one of the groups forming the *s* and *p* blocks of the periodic table (Groups 1, 2, and 13 through 18).

malleable Deformable by striking with a hammer (as a metal).

many-electron atom An atom with more than one electron.

mass m The quantity of matter in a sample.

mass number A The total number of nucleons (protons plus neutrons) in the nucleus of an atom. *Example:* $^{14}_{6}C$, with mass number 14, has 14 nucleons (6 protons and 8 neutrons).

mass percentage composition The mass of a substance present in a sample, expressed as a percentage of the total mass of the sample.

mass spectrometer An instrument used in *mass spectrometry.*

mass spectrometry Technique for measuring the masses and abundances of atoms and molecules by passing a beam of ions through a magnetic field.

mass spectrum The display of the relative number of particles with each specified mass; the output generated by the detector of a mass spectrometer. See also *mass spectrometry.*

matter Anything that has mass and takes up space.

Maxwell distribution of molecular speeds The formula for calculating the percentage of molecules that move at any given speed in a gas at a specified temperature.

mean free path The average distance that a molecule travels between collisions.

mechanism See *reaction mechanism.*

melting temperature   The temperature at which a substance melts. The *normal melting point* is the melting point under a pressure of 1 atm.

meniscus   The curved surface that a liquid forms in a narrow tube.

metal   (1) A substance that conducts electricity, has a metallic luster, is malleable and ductile, forms cations, and has basic oxides. (2) A metal consists of cations held together by a sea of electrons. *Examples:* iron; copper; uranium.

metallic conductor   An electronic conductor with a resistance that increases as the temperature is raised.

metallic hydride   See *hydride.*

metallic solid   See *metal.*

metalloid   An element that has the physical appearance and properties of a metal but behaves chemically like a nonmetal. *Examples:* arsenic; polonium.

meter   m   The SI unit of length.

method of initial rates   A technique for determining the rate law for a reaction by measuring the initial rate of the reaction for different concentrations of reactants.

micelle   A compact, often nearly spherical, cluster of oriented detergent (surfactant) molecules.

microscopic level   A level of description that refers to the very small, such as atoms.

microwaves   Electromagnetic radiation with wavelengths close to 1 cm.

minerals   Substances that are mined; more generally, inorganic substances.

mixture   A type of matter that consists of more than one substance and may be separated into its components by making use of the different physical properties of the substances present.

model   A simplified description of nature.

moderator   A substance that slows neutrons. *Examples:* graphite; heavy water.

molality   The number of moles of solute per kilogram of solvent.

molar   Referring to the quantity per mole. *Examples: molar mass,* the mass per mole; *molar volume,* the volume per mole. (*Molar concentration* and some related quantities are exceptions.)

molar concentration M, [X]   The moles of solute per liter of solution, the *molarity.*

molar mass   (1) The mass per mole of atoms of an element (formerly, *atomic weight*). (2) The mass per mole of molecules of a compound (formerly, *molecular weight*). (3) The mass per mole of formula units of an ionic compound (formerly, *formula weight*).

molar solubility S The molar concentration of a saturated solution of a substance. The *dimensionless* molar solubility is denoted s.

molar volume   The volume of a sample divided by the number of moles of atoms, molecules, or formula units it contains.

molarity M Molar concentration.

mole   mol   The SI unit of chemical amount.

mole fraction x The number of moles of molecules (or ions) of a substance in a mixture expressed as a fraction of the total number of moles of ions and molecules in the mixture.

mole ratio   The stoichiometric relation between two species in a chemical reaction written as a conversion factor. *Example:* $(2 \text{ mol } H_2)/(1 \text{ mol } O_2)$ in the reaction $2 H_2(g) + O_2(g) \rightarrow 2 H_2O(l)$.

molecular compound   A compound that consists of molecules. *Examples:* water; sulfur hexafluoride; benzoic acid.

molecular formula   A combination of chemical symbols and subscripts showing the actual numbers of atoms of each element present in a molecule. *Examples:* H_2O; SF_6; C_6H_5COOH.

molecular hydride   See *hydride.*

molecular orbital   A one-electron wavefunction that spreads throughout the molecule and gives the probability (through its square) that an electron will be found at each location.

molecular orbital energy-level diagram   A portrayal of the relative energies of the molecular orbitals in a molecule.

molecular orbital theory   The description of molecular structure in which electrons occupy orbitals that spread throughout the molecule.

molecular solid   A solid consisting of a collection of individual molecules held together by intermolecular forces. *Examples:* glucose; aspirin; sulfur.

molecular weight   See *molar mass.*

molecularity   The number of reactant molecules (or free atoms or ions) taking part in an elementary reaction. See also *bimolecular reaction; termolecular reaction; unimolecular reaction.*

molecule   (1) The smallest particle of a compound that possesses the chemical properties of the compound. (2) A definite and distinct, electrically neutral group of bonded atoms. *Examples:* H_2; NH_3; CH_3COOH.

monatomic ion   An ion formed from a single atom. *Examples:* Na^+; Cl^-.

Mond process   The purification of nickel by the formation and decomposition of nickel carbonyl.

monomer   A small molecule from which a polymer is formed. *Examples:* $CH_2=CH_2$ for polyethylene; $NH_2(CH_2)_6NH_2$ for nylon.

monoprotic acid   A Brønsted acid with one acidic hydrogen atom. *Example:* CH_3COOH.

multiple bond   A double or triple bond between two atoms.

native (of an element)   Occurring in an uncombined state as the element itself.

natural abundance (of an isotope)   The abundance of an isotope in a sample of a naturally occurring material.

natural product   An organic substance that occurs naturally in the environment.

naturally occurring   Found in nature without needing to be synthesized.

nematic   Having rod-shaped molecules that form a liquid-crystal phase in which the long axes of the molecules are arranged parallel to one another, but in which the molecules are staggered with respect to one another in other directions.

Nernst equation   The equation expressing the cell potential in terms of the concentrations of the reagents taking part in the cell reaction; $E = E° - (RT/nF)\ln Q$.

net ionic equation   The equation showing the net change in a chemical reaction, obtained by canceling the spectator ions in a complete ionic equation. *Example:* $Ag^+(aq) + Cl^-(aq) \rightarrow AgCl(s)$.

network solid   A solid consisting of atoms linked together covalently throughout its extent. *Examples:* diamond; silica.

neutralization reaction   The reaction of an acid with a base to form salt and water or another molecular compound. *Example:* $HCl(aq) + NaOH(aq) \rightarrow NaCl(aq) + H_2O(l)$.

neutron n   An electrically neutral subatomic particle found in the nucleus of an atom; it has approximately the same mass as a proton.

neutron emission   A nuclear decay process in which a neutron is emitted. In neutron emission, the mass number decreases by 1, but the charge number remains the same.

neutron-induced transmutation   The conversion of one nucleus into another by the impact of a neutron. *Example:* $^{58}_{26}Fe + 2\,^1_0n \rightarrow\, ^{60}_{27}Co + \,^0_{-1}e$.

neutron-rich nucleus   Having such a high proportion of neutrons that the nucleus lies above the *band of stability*.

nickel-cadmium cell (nicad cell)   A rechargeable cell in which the reaction is the reduction of nickel(III) and the oxidation of cadmium.

NO$_x$ An oxide, or mixture of oxides, of nitrogen, typically in atmospheric chemistry.

noble gas   A member of Group 18 of the periodic table (the helium family).

nodal plane   A plane on which an electron will not be found.

node   A point or surface at which an electron occupying an orbital will not be found.

nomenclature   See *chemical nomenclature.*

nonaqueous solution   A solution in which the solvent is not water. *Example:* sulfur in carbon disulfide.

nonbonding orbital   A valence-shell atomic orbital that has not been used to form a bond to another atom.

nonelectrolyte   A substance that dissolves to give a solution that does not conduct electricity. *Example:* sucrose.

nonideal solution   A solution that does not obey Raoult's law. Compare with *ideal solution.*

nonmetal   A substance that does not conduct electricity and is neither malleable nor ductile. *Examples:* all gases; phosphorus; sodium chloride.

nonpolar bond   (1) A covalent bond between two atoms that have zero partial charges. (2) A covalent bond between two atoms with the same or nearly the same electronegativity.

nonpolar molecule   A molecule with zero electric dipole moment.

nonspontaneous change   A change that occurs only if it is driven. *Example:* an object can be heated to a higher temperature than its surroundings by forcing an electric current through it.

normal boiling point T_b (1) The boiling temperature when the pressure is 1 atm. (2) The temperature at which the vapor pressure of a liquid is 1 atm.

normal form   The form of a substance under typical everyday conditions (for instance, close to 1 atm, $25°C$).

normal freezing point T_f The temperature at which a liquid freezes at 1 atm.

normal melting point   The melting temperature of a substance at a pressure of 1 atm.

n-type semiconductor   See *semiconductor.*

nuclear atom   The structure of the atom proposed by Rutherford: a central small, very dense, positively charged nucleus surrounded by electrons.

nuclear binding energy E_{bind} The energy released when Z protons and $A - Z$ neutrons come together to form a nucleus. The greater the binding energy per nucleon, the lower the energy of the nucleus.

nuclear chemistry   The study of the chemical consequences of nuclear reactions.

nuclear decay   The spontaneous partial breakup of a nucleus (including its fission). Nuclear decay is also referred to as *nuclear disintegration.* *Example:* $^{226}_{88}Ra \rightarrow\, ^{222}_{86}Rn + \,^4_2\alpha$.

nuclear equation   A summary of the changes in a nuclear reaction written in a form resembling a chemical equation.

nuclear fission   See *fission.*

nuclear fusion   See *fusion.*

nuclear model   A model of the atom in which the electrons surround a minute central nucleus.

nuclear reaction   A change that a nucleus undergoes (such as a nuclear transmutation).

nuclear reactor   A device for achieving controlled self-sustaining nuclear fission.

nuclear transmutation   The conversion of one element into another. *Example:* $^{12}_{6}C + ^{4}_{2}\alpha \rightarrow ^{16}_{8}O + \gamma$.

nucleic acid   (1) The product of a condensation of nucleotides. (2) A molecule containing an organism's genetic information.

nucleon   A proton or a neutron; thus, either of the two principal components of a nucleus.

nucleoside   A combination of an organic base and a deoxyribose molecule.

nucleosynthesis   The formation of elements.

nucleotide   A nucleoside with a phosphate group attached to the carbohydrate ring; one of the units from which nucleic acids are made.

nucleus   The small, positively charged particle at the center of an atom that is responsible for most of its mass.

nuclide   A specific type of nucleus. *Examples:* $^{2}_{1}H$; $^{16}_{8}O$.

occupy   Have the characteristics of the wavefunction of a specified state; to be in a specified state.

octahedral complex   A complex in which six ligands are arranged at the corners of a regular octahedron, with the metal atom at the center. *Example:* $[Fe(CN)_6]^{4-}$.

octet   An s^2p^6 valence-electron configuration.

octet rule   When atoms form bonds, they proceed as far as possible toward completing their octets by sharing electron pairs.

oligopeptide   A short chain of amino acids connected by amide (peptide) bonds.

open system   See *system*.

optical activity   The ability of a substance to rotate the plane of polarized light passing through it.

optical isomers   Isomers that are related like an object and its mirror image. *Optical isomerism* is the existence of optical isomers and is a type of *stereoisomerism*.

orbital   See *atomic orbital; molecular orbital*.

orbital angular momentum quantum number l The quantum number that specifies the subshell of a given shell in an atom and determines the shapes of the orbitals in the subshell.; $l = 0, 1, 2, \ldots, n - 1$. *Examples:* $l = 0$ for the *s*-subshell; $l = 1$ for the *p*-subshell. (The quantum number l also specifies the magnitude of the angular momentum of the electron around the nucleus.)

order of reaction   The power to which the concentration of a single substance is raised in a rate law. *Example:* if rate $= k[SO_2][SO_3]^{-1/2}$, then the reaction is first order in SO_2 and of order $-\frac{1}{2}$ in SO_3.

ore   The natural mineral source of a metal. *Example:* Fe_2O_3, hematite, an iron ore.

organic chemistry   The branch of chemistry that deals with organic compounds.

organic compound   A compound containing the element carbon and usually hydrogen. (The carbonates are normally excluded.)

oscillating   Varying in a periodic manner with time.

osmometry   The measurement of the molar mass of a solute from observations of osmotic pressure.

osmosis   The tendency of a solvent to flow through a semipermeable membrane into a more concentrated solution (a colligative property).

osmotic pressure Π (pi)   The pressure needed to stop the flow of solvent through a semipermeable membrane. See also *osmosis*.

overall order   The sum of the powers to which individual concentrations are raised in the rate law of a reaction. *Example:* if the rate $= k[SO_2][SO_3]^{-1/2}$, then the overall order is $\frac{1}{2}$.

overall reaction   The net outcome of a sequence of reactions.

overlap   The merging of orbitals belonging to different atoms of a molecule.

overpotential   The additional potential difference that must be applied beyond the cell potential to cause appreciable electrolysis.

oxidation   (1) Combination with oxygen. (2) A reaction in which an atom, ion, or molecule loses an electron. (3) A reaction in which the oxidation number of an element is increased. *Examples:* (1, 2, 3) $2\,Mg(s) + O_2(g) \rightarrow 2\,MgO(s)$; (2, 3) $Mg(s) \rightarrow Mg^{2+}(s) + 2\,e^-$.

oxidation number   The effective charge on an atom in a compound, calculated according to a set of rules (Toolbox 3.3). An increase in oxidation number corresponds to oxidation and a decrease, to reduction.

oxidation-reduction reaction   See *redox reaction*.

oxidation state   The actual condition of a species with a specified oxidation number.

oxidizing agent   A species that removes electrons from a species being oxidized (and is itself reduced) in a redox reaction. *Examples:* O_2; O_3; MnO_4^-; Fe^{3+}.

oxoacid   An acid that contains oxygen. *Examples:* H_2CO_3; HNO_3; HNO_2; $HClO$.

oxoanion   An anion of an oxoacid. *Examples:* HCO_3^-; CO_3^{2-}.

paired electrons   Two electrons with opposite spins ($\uparrow\downarrow$).

parallel spins   Electrons with spin in the same direction ($\uparrow\uparrow$).

paramagnetic   The tendency to be pulled into a magnetic field. A paramagnetic substance is composed of atoms or molecules with unpaired electrons. *Examples:* O_2; $[Fe(CN)_6]^{3-}$.

parent nucleus   In a nuclear reaction, the nucleus that undergoes disintegration or transmutation.

partial charge A charge arising from small shifts in the distributions of electrons. A partial charge can be either *positive* ($\delta+$) or *negative* ($\delta-$).

partial pressure P_X The pressure a gas (X) in a mixture would exert if it alone occupied the container.

parts per million ppm (1) The ratio of the mass of a solute to the mass of the solvent, multiplied by 10^6. (2) The mass percentage composition multiplied by 10^4. (Parts per billion, ppb, the mass ratio multiplied by 10^9, may also be used.) Parts per million by mass is equal to the mass of the solute in milligrams divided by the mass of solvent in kilograms. Parts per million by volume is equal to the volume of solute in microliters divided by the volume of solvent in liters.

pascal Pa The SI unit of pressure (1 Pa = 1 $kg/m \cdot s^2$).

passivation Protection from further reaction by a surface film. *Example:* aluminum in air.

Pauli exclusion principle See *exclusion principle*.

p-electron An electron in a *p*-orbital.

penetration The possibility that an *s*-electron may be found inside the inner shells of an atom and hence close to the nucleus.

peptide A molecule formed by a condensation reaction between amino acids; often described in terms of the number of units, for example, dipeptide, oligopeptide, polypeptide.

peptide bond The $-$CONH$-$ group.

percentage deprotonation The fraction of a weak acid, expressed as a percentage, that is present as its conjugate base in a solution.

percentage protonated The fraction of a base, expressed as a percentage, that is present as its conjugate acid in a solution.

percentage yield The percentage of the theoretical yield of a product achieved in practice.

period A horizontal row in the periodic table; the number of the period is equal to the principal quantum number of the valence shell of the atoms.

periodic table A chart in which the elements are arranged in order of atomic number and divided into groups and periods in a manner that shows the relationships between the properties of the elements.

pH The negative logarithm of the hydronium ion molarity in a solution: pH = $-\log[H_3O^+]$. pH < 7 indicates an acidic solution; pH = 7, a neutral solution; and pH > 7, a basic solution.

pH curve The graph of the pH of a reaction mixture against volume of titrant added in an acid-base titration.

phase A particular physical state of matter that has a uniform composition. A substance may exist in solid, liquid, and gaseous phases and, in certain cases, in more than one solid or liquid phase. *Examples:* white and gray tin are two solid phases of tin; ice, liquid, and vapor are three phases of water.

phase boundary A line separating two areas in a phase diagram; the points on a phase boundary represent the conditions at which the two adjoining phases are in dynamic equilibrium.

phase diagram A summary in graphical form of the conditions of temperature and pressure at which the various solid, liquid, and gaseous phases of a substance exist. A *one-component phase diagram* is a phase diagram for a single substance.

phase transition The conversion of one phase of a substance to another phase. *Examples:* vaporization; white tin $\rightarrow$ gray tin.

phenol An organic compound in which a hydroxyl group is attached directly to an aromatic ring (Ar$-$OH). *Example:* C_6H_5OH, phenol.

phenolic resin A polymer resulting from the condensation reaction between phenol and formaldehyde.

phosphorescence Long-lasting luminescence.

photochemical reaction A reaction caused by light. *Example:* $H_2(g) + Cl_2(g) \xrightarrow{light} 2\,HCl(g)$.

photoelectric effect The emission of electrons from the surface of a metal when electromagnetic radiation strikes it.

photon A particlelike packet of electromagnetic radiation. The energy of a photon of frequency ν is $E = h\nu$.

physical property A characteristic that we observe or measure without changing the identity of the substance.

physical sciences The branches of science that make extensive use of quantitative measurements.

physical state The condition of being a solid, a liquid, or a gas at a particular temperature.

pi-bond (π-bond) A bond formed by the side-to-side overlap of two *p*-orbitals.

pi-orbital (π-orbital) A molecular orbital that has one nodal plane cutting through the internuclear axis.

piezoelectric A substance is piezoelectric if it becomes electrically charged when it is mechanically distorted. *Example:* $BaTiO_3$.

pipet A narrow tube, often with a central bulb, calibrated to deliver a specified volume.

pK_a and pK_b The negative common logarithms of the acidity and basicity constants: pK = $-\log K$. The larger the value of pK_a or pK_b, the weaker the acid or base, respectively.

Planck constant h A fundamental constant of nature with the value $6.626\,08 \times 10^{-34}$ J$\cdot$s.

plasma (1) An ionized gas. (2) In biology, the colorless component of blood in which the red and white blood cells are dispersed.

p-n junction A solid-state electronic device in which an n-type semiconductor is in contact with a p-type semiconductor. An electric current can flow across this junction in only one direction.

pOH The negative logarithm of the hydroxide ion molarity in a solution; $pOH = -\log[OH^-]$.

poison To inactivate a catalyst.

polar covalent bond A covalent bond between atoms that have partial electric charges. *Examples:* $H-Cl$; $O-S$.

polar molecule A molecule with a nonzero electric dipole moment. *Examples:* HCl; NH_3.

polarizable Easily polarized species.

polarize To distort the electron cloud of an atom or ion.

polarized light Plane-polarized light is light in which the wave motion occurs in a single plane.

polarizing power The ability of an ion to polarize a neighboring atom or ion.

polyamide A polymer in which the monomers are linked by amide bonds formed by condensation polymerization. Often, the monomers are a dicarboxylic acid and a diamine. *Example:* nylon.

polyatomic ion An ion consisting of two or more atoms linked by covalent bonds. *Examples:* NH_4^+; NO_3^-; SiF_6^{2-}.

polyatomic molecule A molecule that consists of more than two atoms. *Examples:* O_3; $C_{12}H_{22}O_{11}$.

polydentate ligand A ligand that can attach at several binding sites.

polyester A polymer in which the monomers are linked by ester groups formed by condensation polymerization. Often, the monomers are a dicarboxylic acid and a diol (a dialcohol).

polymer A substance with large molecules consisting of chains of covalently linked repeating units formed from small molecules known as *monomers*. *Examples:* polyethylene; nylon. An *atactic polymer* is a polymer in which substituents are attached to each side of the chain at random. An *isotactic polymer* is a polymer in which the substituents are all on the same side of the chain. A *syndiotactic polymer* is a polymer in which the substituents alternate on either side of the chain.

polynucleotide A polymer built from nucleotide units. *Examples:* DNA; RNA.

polypeptide A polymer formed by the condensation of amino acids.

polyprotic acid or base A Brønsted acid or base that can donate or accept more than one proton. (A polyprotic acid is sometimes called a polybasic acid.) *Examples:* H_3PO_4, triprotic acid; N_2H_4, diprotic base.

polysaccharide A chain of many saccharide units, such as glucose, linked together. *Examples:* cellulose; amylose.

p-orbital See *atomic orbital*.

positional disorder Disorder related to the locations of molecules.

positron An elementary particle with the same mass as an electron but with opposite charge.

positron emission A mode of radioactive decay in which a nucleus emits a positron.

potential energy E_p The energy arising from position. *Example:* the Coulomb potential energy of a charge is inversely proportional to its distance from another charge.

precipitate The solid formed in a precipitation reaction.

precipitation The process in which a solute comes out of solution rapidly as a finely divided powder, called a *precipitate.*

precipitation reaction A reaction in which a solid product is formed when two solutions are mixed. *Example:* $KBr(aq) + AgNO_3(aq) \rightarrow KNO_3(aq) + AgBr(s)$.

precise measurements (1) Measurements with a large number of significant figures. (2) A series of measurements with small random error and hence in close agreement.

precision Freedom from random error. Compare with *accuracy.*

pre-exponential factor A The constant obtained from the intercept in an Arrhenius plot.

pressure P Force divided by the area to which it is applied.

primary alcohol See *alcohol.*

primary cell A galvanic cell that produces electricity from chemicals sealed within it at the time of manufacture. It cannot be recharged.

primary pollutant A pollutant introduced directly into the environment. *Example:* SO_2.

primary structure The sequence of amino acids in the polypeptide chain of a protein.

primitive cubic structure A structure in which spheres (representing atoms or ions) lie at the corners of a cube.

principal quantum number n The quantum number that specifies the energy of an electron in a hydrogen atom and labels the shells of the atom.

product A species formed in a chemical reaction.

promotion (of an electron) The conceptual excitation of an electron to an orbital of higher energy in the description of bond formation.

propagation A series of steps in a chain reaction in which one chain carrier reacts with a reactant molecule to produce another carrier. *Examples:* $Br\cdot + H_2 \rightarrow HBr + H\cdot$; $H\cdot + Br_2 \rightarrow HBr + Br\cdot$.

properties The characteristics of matter. *Examples:* vapor pressure; color; density; temperature.

protective oxide An oxide that protects a metal from oxidation. *Example:* aluminum oxide.

proton p A positively charged subatomic particle found in the nucleus of an atom.

proton emission   A nuclear decay process in which a proton is emitted. In proton emission, the mass and charge numbers of the nucleus both decrease by 1.

proton-rich nucleus   Having such a low proportion of neutrons that the nucleus lies below the *band of stability.*

proton transfer equilibrium   The equilibrium involving the transfer of a hydrogen ion between an acid and a base.

protonation   Proton transfer to a Brønsted base. *Example:* $2\,H_3O^+(aq) + S^{2-}(s) \rightarrow H_2S(g) + 2\,H_2O(l).$

pseudofirst-order rate law   A rate law that is effectively first order because all but one substance have a virtually constant concentration.

p-type semiconductor   See *semiconductor.*

pX   The quantity $-\log X$. *Example:* $pOH = -\log[OH^-].$

pyrometallurgical process   The extraction of metals by using reactions at high temperatures. *Example:* $Fe_2O_3(s) + 3\,CO(g) \xrightarrow{\Delta} 2\,Fe(l) + 3\,CO_2(g).$

quadratic equation   An equation of the form $ax^2 + bx + c = 0$. See also Appendix 1D.

qualitative   A nonnumerical description of the properties of a substance, system, or process. *Example:* qualitative analysis, the identification of the substances present in a sample.

quanta   The plural of *quantum.*

quantitative   A numerical description of the properties of a substance, system, or process. *Example:* quantitative analysis, the determination of the amounts or concentrations of substances present in a sample.

quantization   The restriction of a property to certain values. *Examples:* the quantization of energy and angular momentum.

quantum   A packet of energy.

quantum mechanics   The description of matter that takes into account the wave-particle duality of matter and the fact that the energy of an object may be changed only in discrete steps.

quantum number   An integer (sometimes, a half-integer) that labels a wavefunction and specifies the value of a property. *Example:* principal quantum number *n.*

quaternary ammonium ion   An ion of the form R_4N^+, where R denotes hydrogen or an alkyl group (the four groups may be different).

quaternary structure   The manner in which neighboring polypeptide units stack together to form a protein molecule.

racemic mixture   A mixture containing equal concentrations of two enantiomers.

rad   A unit of *absorbed dose* of radiation; 1 rad corresponds to an energy deposit of 0.01 J/kg. See also *gray.*

radial wavefunction   The radial part of a wavefunction, particularly the radial component of the wavefunctions of the hydrogen atom; the probability amplitude of an electron as a function of distance from the nucleus.

radical   An atom, molecule, or ion with at least one unpaired electron. *Examples:* $\cdot NO$; $\cdot O\cdot$; $\cdot CH_3$.

radical chain reaction   A chain reaction propagated by radicals.

radical polymerization   A polymerization procedure that utilizes a radical chain reaction.

radioactive   The ability of a nuclide to emit radiation.

radioactive series   A stepwise nuclear decay path in which α and β particles are successively ejected and that terminates at a stable nuclide (often of lead).

radioactivity   The emission of radiation by nuclei. Such nuclei are radioactive.

radiocarbon dating   *Isotopic dating* based specifically on the use of carbon-14.

radioisotopes   Radioactive isotopes.

radius ratio   The ratio of the radius of the smaller ion in an ionic solid to the radius of the larger ion. The radius ratio controls which crystal structure is adopted by a simple ionic solid.

random error   An error that varies randomly from measurement to measurement, sometimes giving a high value and sometimes a low one.

Raoult's law   The vapor pressure of a solvent in the presence of a nonvolatile solute is directly proportional to the mole fraction of the solvent: $P = x_{solvent}P_{pure}$, where P_{pure} is the vapor pressure of the pure solvent.

rate   The change in a property divided by the time interval.

rate constant *k*   The constant of proportionality in a rate law.

rate-determining step   The slowest step in a multistep reaction sequence and therefore the step that governs the rate of the overall reaction. *Example:* the step $O + O_3 \rightarrow 2\,O_2$ in the decomposition of ozone.

rate law   An equation expressing the instantaneous reaction rate in terms of the concentrations, at that instant, of the substances taking part in the reaction. *Example:* rate $= k[NO_2]^2$.

rate of reaction   See *instantaneous rate; reaction rate.*

reactant   A species acting as a starting material in a chemical reaction; a reagent taking part in a specified reaction.

reaction enthalpy ΔH_r The change of enthalpy per mole of species indicated by the stoichiometric coefficients in the chemical equation for the reaction. *Example:* $CH_4(g) + 2\,O_2(g) \rightarrow CO_2(g) + 2\,H_2O(l)$, $\Delta H_r = -890.$ kJ/mol.

reaction free energy ΔG_r The free energy change associated with a chemical reaction. It is equal to the difference in the molar free energies of the products and the molar free energies of the reactants, each value multiplied by the stoichiometric coefficient of the species in the chemical equation.

reaction intermediate   A species that is produced and consumed during a reaction but does not occur in the overall chemical equation.

reaction mechanism   The pathway that is proposed for an overall reaction and accounts for the experimental rate law.

reaction order   See *order of reaction.*

reaction profile   The variation in potential energy that occurs as two reactants meet, form an activated complex, and separate as products.

reaction quotient Q The ratio of the molar concentrations or partial pressures of the products to those of the reactants, each raised to a power equal to the stoichiometric coefficient (as in the definition of the equilibrium constant, but at an arbitrary stage of a reaction). *Example:* for $N_2(g) + 3 H_2(g) \rightarrow 2 NH_3(g)$, $Q = P_{NH_3}^{2}/P_{N_2}P_{H_2}^{3}$.

reaction rate   The rate of a chemical reaction calculated by dividing the change in concentration of a substance by the interval during which the change occurs. See also *average reaction rate; instantaneous rate.*

reaction sequence   A series of reactions in which the product or products of one reaction take part as reactants in the next. *Example:* $2 C(s) + O_2(g) \rightarrow 2 CO(g)$, followed by $2 CO(g) + O_2(g) \rightarrow 2 CO_2(g)$.

reaction stoichiometry   The quantitative relation between the amounts of reactants consumed and products formed in chemical reactions as expressed by the balanced chemical equation for the reaction.

reagent   A substance or a solution that reacts with other substances.

real gas   An actual gas; a gas that differs from an ideal gas in its behavior.

redox couple   The oxidized and reduced forms of a substance taking part in a reduction or oxidation half-reaction. The notation is oxidized species/reduced species. *Example:* H^+/H_2.

redox indicator   See *indicator.*

redox reaction   A reaction in which oxidation and reduction occur. *Example:* $S(s) + 3 F_2(g) \rightarrow SF_6(g)$.

reducing agent   The species that supplies electrons to a substance being reduced (and is itself oxidized) in a redox reaction. *Examples:* $H_2; H_2S; SO_3^{2-}$.

reduction   (1) The removal of oxygen from, or the addition of hydrogen to, a compound. (2) A reaction in which an atom, ion, or molecule gains an electron. (3) A reaction in which the oxidation number of an element is decreased. *Example:* $Cl_2(g) + 2 e^- \rightarrow 2 Cl^-(aq)$.

re-forming reaction   A reaction in which a hydrocarbon is converted to carbon monoxide and hydrogen over a nickel catalyst.

refractory   Able to withstand high temperatures.

relative biological effectiveness Q A factor used when assessing the damage caused by a given dose of radiation.

rem   See *roentgen equivalent man.*

repeating unit   The combination of atoms in a polymer that repeats over and over again to produce the polymer chain.

representative elements   The elements in Periods 1, 2, and 13–18 of the periodic table.

residue   An amino acid in a polypeptide chain.

resistance (electrical)   A measure of the ability of matter to conduct electricity: the lower the resistance, the better it conducts.

resonance   A blending of Lewis structures into a single composite, hybrid structure. *Example:* $:\ddot{O}-\ddot{S}=\ddot{O} \leftrightarrow \ddot{O}=\ddot{S}-\ddot{O}:$.

resonance hybrid   The composite structure that results from resonance.

reverse osmosis   The passage of solvent out of a solution when a pressure greater than the osmotic pressure is applied on the solution side of a semipermeable membrane.

reversible process   A process that can be reversed by an infinitesimal change in a variable.

roast   To heat a metal ore in air. *Example:* $2 CuFeS_2(s) + 3 O_2(g) \rightarrow 2 CuS(s) + 2 FeO(s) + 2 SO_2(g)$ in the extraction of copper.

rock-salt structure   A crystal structure the same as that of a mineral form of sodium chloride.

roentgen equivalent man rem   The unit for reporting dose equivalent. See also *sievert.*

root mean square speed v The square root of the average value of the squares of the speeds of the molecules in a sample.

Rydberg constant R_H The constant that occurs in the formula for the frequencies of the lines in the spectrum of atomic hydrogen; $R_H = 3.289\ 84 \times 10^{15}$ Hz.

sacrificial anode   See *cathodic protection.*

saline carbide   See *carbide.*

saline hydride   See *hydride.*

salt   (1) An ionic compound. (2) The product (other than water) of the reaction between an acid and a base. *Examples:* $NaCl; K_2SO_4$.

salt bridge   A bridge-shaped tube containing a concentrated salt (potassium chloride or potassium nitrate in a jelly) that acts as an electrolyte and provides a conducting path between two compartments of an electrochemical cell.

sample   A representative part of a whole.

saturated   Unable to take up further material.

saturated hydrocarbon   A hydrocarbon with no carbon-carbon multiple bonds. *Example:* CH_3CH_3.

saturated solution   A solution in which the dissolved and undissolved solute are in dynamic equilibrium.

Schrödinger equation   An equation for calculating the wavefunction of a particle, especially of an electron in an atom or molecule. See also *wavefunction.*

scientific method   A set of procedures employed to develop a scientific understanding of nature.

scientific notation   The expression of numbers in the form $A \times 10^a$.

scintillation counter   A device for detecting and measuring radioactivity that makes use of the fact that certain substances give a flash of light when they are exposed to radiation.

sea of instability   A region in a graph of mass number against atomic number corresponding to unstable nuclei that decay with the emission of radiation. See also *band of stability.*

second s   The SI unit of time.

second law of thermodynamics   A spontaneous change is accompanied by an increase in the total entropy of the system and its surroundings.

second-order reaction   (1) A reaction with a rate law that is proportional to the square of the molar concentration of a reactant. (2) A reaction with an overall order of 2.

secondary alcohol   See *alcohol.*

secondary cell   A galvanic cell that must be charged (or recharged) by using an externally supplied current before it can be used.

secondary pollutant   A pollutant formed by the chemical reaction of another species in the environment. *Example:* SO_3 from the oxidation of SO_2 in air.

secondary structure   The manner in which a polypeptide chain is coiled. *Examples:* α helix; β-pleated sheet.

selective precipitation   The precipitation of one compound in the presence of other, more soluble compounds.

s-electron   An electron in an *s*-orbital.

semiconductor   An *electronic conductor* with a resistance that decreases as the temperature is raised. In an *n-type semiconductor,* the current is carried by electrons in a largely empty band; in a *p-type semiconductor,* the conduction is a result of electrons missing from otherwise filled bands.

semipermeable membrane   A membrane that allows only certain types of molecules or ions to pass.

shell   All the orbitals of a given principal quantum number. *Example:* the single 2*s*- and three 2*p*-orbitals of the shell with $n = 2$.

shielding   The repulsion that is experienced by an electron in an atom; it arises from the other electrons present and opposes the attraction exerted by the nucleus.

shift reaction   A reaction between carbon monoxide and water vapor: $CO(g) + H_2O(g) \rightarrow CO_2(g) + H_2(g)$; the reaction is used in the manufacture of hydrogen.

short-range order   Atoms or molecules lying in a regular arrangement that does not extend very far past their nearest neighbors.

SI (*Système International de Unités*)   The International System of units; a collection of definitions of units and their employment. It is an extension and rationalization of the metric system.

side chain   A hydrocarbon substituent on a hydrocarbon chain.

sievert Sv   The SI unit of *dose equivalent:* 1 Sv = 100 rem.

sigma-bond (σ-bond)   Two electrons in a cylindrically symmetrical cloud between two atoms.

sigma-orbital (σ-orbital)   A molecular orbital that has no nodal plane when viewed along the internuclear axis.

significant figures sf (in a measurement)   The digits in the measurement, up to and including the first uncertain digit in scientific notation. *Example:* 0.0260 mL (2.60×10^{-2} mL), a measurement with three significant figures (3 sf).

single bond   A shared electron pair.

skeletal equation   An unbalanced chemical equation that summarizes the qualitative information about the reaction. *Example:* $H_2 + O_2 \rightarrow H_2O$ △.

slope (of graph)   The gradient of a graph. See Appendix 1E.

smectic   Having molecules that lie parallel to one another and form layers in a liquid-crystal phase.

smelt   To melt a metal ore with a reducing agent. *Example:* $CuS(l) + O_2(g) \xrightarrow{\Delta} Cu(l) + SO_2(g)$.

sol   A colloidal dispersion of solid particles in a liquid.

solid   A rigid form of matter that maintains the same shape whatever the shape of its container.

solid emulsion   A colloidal dispersion of a liquid in a solid. *Example:* butter, an emulsion of water in butterfat.

solid solution   A solid homogeneous mixture of two or more substances.

solubility   The concentration of a saturated solution of a substance.

solubility constant   See *solubility product.*

solubility product K_{sp} The product of ionic molar concentrations in a saturated solution; the dissolution equilibrium constant. *Example:* $Hg_2Cl_2(s) \rightleftharpoons Hg_2^{2+}(aq) + 2 Cl^-(aq)$, $K_{sp} = [Hg_2^{2+}][Cl^-]^2$.

solubility rules   A summary of the solubility pattern of a range of common compounds in water. (See also Table 3.1.)

soluble substance   A substance that dissolves to a significant extent in a specified solvent; when the solvent is not specified, water is generally meant.

solute   A dissolved substance.

solution A homogeneous mixture. See also *solute; solvent.*

solvation The surrounding of a solute species by solvent molecules. (*Hydration* is a special case when the solvent is water.)

solvent (1) The most abundant component of a solution. (2) The component of a solution that determines its state of matter.

solvent extraction A process for separating a mixture of substances that makes use of their differing solubilities in various solvents.

s-orbital See *atomic orbital.*

space-filling model A depiction of a molecule in which atoms are represented by spheres that indicate the space they occupy.

species An atom, an ion, or a molecule.

specific enthalpy (of a fuel) The enthalpy of combustion per gram (without the negative sign).

specific heat capacity The heat capacity of a sample divided by its mass in grams.

spectator ion An ion that is present but remains unchanged during a reaction. *Examples:* Na^+ and NO_3^- in $NaCl(aq) + AgNO_3(aq) \rightarrow NaNO_3(aq) + AgCl(s)$.

spectral line Radiation of a single wavelength emitted or absorbed by an atom or a molecule.

spectrochemical series Ligands ordered according to the strength of the ligand field splitting they produce.

spectrometer An instrument for generating and recording the spectrum of a sample.

spectroscopy The analysis of the electromagnetic radiation emitted or absorbed by substances.

spectrum The frequencies or wavelengths of the electromagnetic radiation emitted or absorbed by substances.

sp^n hybrid A hybrid orbital constructed from an *s*-orbital and *n* *p*-orbitals. There are two *sp* hybrids, three sp^2 hybrids, and four sp^3 hybrids.

spin The intrinsic angular momentum of an electron; the spin cannot be eliminated and may occur in only two senses, denoted ↑ and ↓.

spin magnetic quantum number m_s The quantum number that distinguishes the two spin states of an electron: $m_s = +\frac{1}{2}$ (↑) and $m_s = -\frac{1}{2}$ (↓).

spontaneous change A natural change, one that has a tendency to occur without needing to be driven by an external influence. *Examples:* a gas expanding; a hot object cooling; methane burning.

spontaneous nuclear fission See *fission.*

square planar complex A complex in which four ligands lie at the corners of a square with the metal atom at the center.

stability constant See *formation constant.*

stable See *thermodynamically unstable compound.*

standard cell potential $E°$ The cell potential when the concentration of each type of solute taking part in the cell reaction is 1 mol/L and all the gases are at 1 atm. The *standard cell potential* is the difference of its two standard electrode potentials:
$E° = E°(\text{cathode}) - E°(\text{anode})$.

standard enthalpy of combustion $\Delta H_c°$ The change of enthalpy per mole of substance when it burns (reacts with oxygen) completely under standard conditions.

standard enthalpy of formation $\Delta H_f°$ The standard reaction enthalpy per mole of compound for the compound's synthesis from its elements in their most stable form at 1 atm and the specified temperature.

standard free energy of formation $\Delta G_f°$ The standard reaction free energy per mole for the formation of a compound from its elements in their most stable form.

standard free energy of reaction See *standard reaction free energy.*

standard hydrogen electrode SHE A hydrogen electrode that is in its standard state (hydrogen ions at 1 mol/L and hydrogen gas at 1 atm) and is defined as having $E° = 0$.

standard molar entropy $S_m°$ The entropy per mole of a pure substance at 1 atm pressure.

standard potential $E°$ (1) The contribution of an electrode to the standard cell potential. (2) The standard potential of a cell when a standard hydrogen electrode is on the left in the cell diagram and the electrode of interest is on the right.

standard reaction enthalpy $\Delta H_r°$ The difference between the molar enthalpies of the products of a reaction in their standard states and the molar enthalpies of the reactants in their standard states (in kilojoules per mole); in practice, $\Delta H_r° = \sum n \Delta H_f°(\text{products}) - \sum n \Delta H_f°(\text{reactants})$, where *n* represents the stoichiometric coefficients.

standard reaction entropy $\Delta S_r°$ The difference between the molar entropies of the products of a reaction in their standard states and the molar entropies of the reactants in their standard states (in joules per kelvin per mole). $\Delta S_r° = \sum n S_m°(\text{products}) - \sum n S_m°(\text{reactants})$, where *n* represents the stoichiometric coefficients.

standard reaction free energy $\Delta G_r°$ The difference between the molar free energies of the products of a reaction in their standard states and those of the reactants in their standard states (in kilojoules per mole); in practice, $\Delta G_r° = \sum n \Delta G_f°(\text{products}) - \sum n \Delta G_f°(\text{reactants})$, where *n* denotes the stoichiometric coefficients.

standard state The pure form of a substance at 1 atm; for a solute, the concentration 1 mol/L.

standard temperature and pressure STP 0°C (273.15 K) and 1 atm (101.325 kPa).

state function   See *state property*.

state of matter   The physical condition of a sample. The most common states of a pure substance are solid, liquid, and gas (vapor).

state property   A property of a substance that is independent of how the sample was prepared. *Examples:* pressure; enthalpy; entropy.

state symbol   A symbol denoting the state of a species. *Examples:* s (solid); l (liquid); g (gas); aq (aqueous solution).

stereoisomers   Isomers in which atoms have the same partners arranged differently in space. See also *isomers*.

stereoregular polymer   A polymer in which each unit or pair of repeating units has the same relative orientation.

steric requirement   A constraint on an elementary reaction in which the successful collision of two molecules depends on their relative orientation.

Stern-Gerlach experiment   The demonstration of the quantization of electron spin by passing a beam of atoms through a magnetic field.

Stock number   A Roman numeral equal to the number of electrons lost by an atom on formation of a compound (the oxidation number of the element) and sometimes added in parentheses to a name. *Example:* copper(II) in compounds containing Cu^{2+}.

stock solution   A solution stored in concentrated form.

stoichiometric coefficients   The numbers multiplying chemical formulas in a chemical equation. *Examples:* 1, 1, and 2 in $H_2 + Br_2 \rightarrow 2 HBr$.

stoichiometric point   The stage in a titration when exactly the right volume of solution needed to complete the reaction has been added.

stoichiometric relation   An expression that equates the relative amounts of reactants and products that participate in a reaction. *Example:* 1 mol $H_2 \simeq 2$ mol HBr.

stoichiometry   See *reaction stoichiometry*.

STP   See *standard temperature and pressure*.

strong acids and bases   Acids and bases that are fully ionized in solution. *Examples:* HCl, $HClO_4$ (strong acids); NaOH, $Ca(OH)_2$ (strong bases).

strong electrolyte   See *electrolyte solution*.

strong-field ligand   A ligand that produces a large *ligand field splitting* and that lies above H_2O in the *spectrochemical series*.

strong force   A short-range but very strong force that acts between nucleons and binds them together to form a nucleus.

structural formula   A chemical formula that shows how atoms in a compound are attached to one another.

structural isomers   Isomers in which the atoms have different partners. See also *isomers*.

subatomic particle   A particle smaller than an atom. *Examples:* electron; proton; neutron.

subcritical mass   See *critical mass*.

sublimation   The direct conversion of a solid to a vapor without first forming a liquid.

subshell   All the atomic orbitals of a given shell of an atom that have the same value of the quantum number *l*. *Example:* the five 3*d*-orbitals of an atom.

substance   A single, pure type of matter; either a compound or an element.

substitution reaction   (1) A reaction in which an atom (or a group of atoms) replaces an atom or group of atoms in the original molecule. (2) In complexes, a reaction in which one Lewis base expels another and takes its place. *Examples:* (1) $C_6H_5OH + Br_2 \rightarrow BrC_6H_4OH + HBr$; (2) $[Fe(H_2O)_6]^{3+}(aq) + 6 CN^-(aq) \rightarrow [Fe(CN)_6]^{3-}(aq) + 6 H_2O(l)$.

substitutional alloy   See *alloy*.

substrate   The chemical species on which an enzyme acts.

superconductor   An *electronic conductor* that conducts electricity with zero resistance. See also *high-temperature superconductor*.

supercooled   A liquid cooled to below its freezing point but not yet frozen.

supercritical fluid   A substance above its *critical temperature* and *critical pressure*.

supercritical mass   See *critical mass*.

superfluidity   The ability to flow without viscosity.

supersaturated solution   A solution in which the concentration of a solute is higher than indicated by its solubility.

surface-active agent   See *surfactant*.

surface tension γ The tendency of molecules at the surface of a liquid to be pulled inward, thereby resulting in a smooth surface.

surfactant   A substance that accumulates at the surface of a solution and lowers the surface tension of the solvent; a component of detergents. *Example:* the stearate ion of soaps.

surroundings   The region outside a system.

suspension   A mist of small particles.

symbolic language   The expression of chemical phenomena in terms of chemical symbols and chemical and mathematical equations.

syndiotactic polymer   See *polymer*.

synthesis   A reaction in which a substance is formed from simpler starting materials. *Example:* $N_2(g) + 3 H_2(g) \rightarrow 2 NH_3(g)$.

synthesis gas   A mixture of carbon monoxide and hydrogen produced by the catalyzed reaction of a hydrocarbon and water.

system   The object of study, usually a reaction vessel and its contents. An *open system* can exchange both matter and energy with the surroundings. A *closed system* has a fixed amount of matter but can exchange energy with the surroundings. An *isolated system* has no contact with its surroundings.

systematic error   An error that persists in a series of measurements and does not average out. See also *accuracy.*

systematic name   The name of a compound that reveals which elements are present (and, in its most complete form, how the atoms are arranged). *Example:* methylbenzene is the systematic name for toluene

Système International d'Unités   See *SI.*

temperature T (1) How hot or cold a sample is. (2) The intensive property that determines the direction in which heat will flow between two objects in contact.

termination   A step in a *chain reaction* in which chain carriers combine to form products. *Example:* $Br\cdot + Br\cdot \rightarrow Br_2$.

termolecular reaction   An *elementary reaction* involving the simultaneous collision of three species.

tertiary alcohol   See *alcohol.*

tertiary structure   The shape into which the α helix and β-pleated sheet sections of a polypeptide are twisted as a result of interactions between peptide groups lying in different parts of the primary structure.

tetrahedral complex   A complex in which four ligands lie at the corners of a regular tetrahedron with the metal atom at the center. *Example:* $[Cu(NH_3)_4]^{2+}$.

theoretical yield   The maximum quantity of product that can be obtained, according to the reaction stoichiometry, from a given quantity of a specified reactant.

theory   A collection of ideas and concepts used to account for a scientific law.

thermal decomposition   See *decomposition.*

thermal disorder   Disorder arising from the thermal motion of molecules.

thermal motion   The random, chaotic motion of atoms.

thermal pollution   The damage caused to the environment by the waste heat of an industrial process.

thermite reaction   (thermite process)   The reduction of a metal oxide by aluminum. *Example:* $2 Al(s) + Fe_2O_3(s) \rightarrow Al_2O_3(s) + 2 Fe(l)$.

thermochemical equation   An expression consisting of both the balanced chemical equation and the reaction enthalpy for the chemical reaction exactly as written.

thermochemistry   The study of the heat released or absorbed by chemical reactions; a branch of thermodynamics.

thermodynamically stable compound   (1) A compound with no thermodynamic tendency to decompose into its elements.

(2) A compound with a negative free energy of formation.

thermodynamically unstable compound   (1) A compound with a thermodynamic tendency to decompose into its elements. (2) A compound with a positive free energy of formation.

thermodynamics   The study of the transformations of energy from one form to another in assemblies of large numbers of particles. See also *first law of thermodynamics; second law of thermodynamics; third law of thermodynamics.*

thermonuclear explosion   An explosion resulting from uncontrolled nuclear fusion.

thermotropic liquid crystal   A liquid crystal prepared by melting the solid phase.

third law of thermodynamics   The entropies of all perfect crystals are the same at the absolute zero of temperature.

three-center bond   A chemical bond in which a hydrogen atom lies between two other atoms (typically boron atoms) and one electron pair binds all three atoms together.

titrant   The solution of known concentration added from a buret in a titration.

titration   The analysis of composition by measuring the volume of one solution needed to react with a given volume of another solution. In an *acid-base titration,* an acid is titrated with a base; in a *redox titration,* an oxidizing agent is titrated with a reducing agent.

***t*-orbital**   One of the orbitals d_{xy}, d_{yz}, and d_{zx} in an octahedral or tetrahedral complex.

total energy   See *energy.*

tracer   (in nuclear chemistry)   An isotope that can be tracked from compound to compound in the course of a sequence of reactions.

trans isomer   A geometrical isomer in which two atoms or groups are attached on opposite sides of a double bond, ring, or complex. Compare with *cis isomer.*

transition   A change of state. (1) In thermodynamics, a change of physical state. (2) In spectroscopy, a change of quantum state.

transition metal   An element that belongs to Groups 3 through 11. *Examples:* vanadium; iron; gold.

transmutation   See *nuclear transmutation.*

transuranium elements   The elements beyond uranium; those with $Z > 92$.

triple bond   (1) Three electron pairs shared by two neighboring atoms. (2) One σ-bond and two π-bonds between neighboring atoms.

triple point   The point where three phase boundaries meet in a phase diagram; under the conditions represented by the triple point, all three adjoining phases coexist in dynamic equilibrium.

triprotic   See *polyprotic acid or base.*

tube structure   A molecular representation in which the bonds between atoms are represented by colored tubes.

ultraviolet radiation   Electromagnetic radiation with a higher frequency (shorter wavelength) than that of violet light.

unbranched alkane   An alkane with no side chains, in which all the carbon atoms lie in a linear chain.

uncertainty principle   See *Heisenberg uncertainty principle*.

unimolecular reaction   An elementary reaction in which a single reactant molecule changes into products. *Example:* $O_3(g) \rightarrow O(g) + O_2(g)$.

unit   See *base units*.

unit cell   The smallest unit that, when stacked together repeatedly without any gaps, can reproduce an entire crystal.

unsaturated hydrocarbon   A hydrocarbon with at least one carbon-carbon multiple bond. *Examples:* $CH_2{=}CH_2$; C_6H_6.

valence   The number of bonds that an atom can form.

valence band   In the theory of solids, a band of energy levels fully occupied by electrons.

valence-bond theory   The description of bond formation in terms of the pairing of spins in the atomic orbitals of neighboring atoms.

valence electrons   The electrons that belong to the valence shell.

valence shell   The outermost shell of an atom. *Example:* The $n = 2$ shell of Period 2 atoms.

valence-shell electron-pair repulsion model   VSEPR model   A model for predicting the shapes of molecules, using the fact that electron pairs repel one another.

van der Waals coefficients   The experimentally determined coefficients that appear in the van der Waals equation and are unique for each real gas. The coefficient a is an indication of attractive intermolecular forces and the coefficient b is an indication of repulsive intermolecular forces. See also *van der Waals equation*.

van der Waals equation   An approximate equation of state for a real gas in which two parameters represent the effects of intermolecular forces.

van der Waals forces   See *intermolecular forces*.

van't Hoff equation   (1) The equation for the osmotic pressure in terms of the molarity, $\Pi = iRT \times$ molarity. (2) An equation that shows how the equilibrium constant varies with temperature.

van't Hoff *i* factor   See *i factor*.

vapor   The gaseous phase of a substance (specifically, of a substance that is a liquid or a solid at the temperature in question). See also *gas*.

vapor pressure   The pressure exerted by the vapor of a liquid (or a solid) when the vapor and the liquid (or solid) are in dynamic equilibrium.

vaporization   The formation of a gas or a vapor from a liquid.

variable covalence   The ability of an element to form different numbers of covalent bonds. *Example:* sulfur forms SO_2 and SO_3.

variable valence   The ability of an element to form ions with different charges. *Example:* In^+ and In^{3+}.

viscosity   The resistance of a fluid (a gas or a liquid) to flow: the higher the viscosity, the slower the flow.

visible light   See *visible radiation*.

visible radiation   Electromagnetic radiation that can be detected by the human eye, with wavelengths in the range 700 to 400 nm. Visible radiation is also called *visible light* or simply *light*.

volatility   The readiness with which a substance vaporizes. A substance is typically regarded as *volatile* if its boiling point is below 100°C.

voltaic cell   See *electrochemical cell*.

volume V The amount of space a sample occupies.

volumetric analysis   An analytical method using measurement of volume.

volumetric flask   A flask calibrated to contain a specified volume.

VSEPR formula   A "formula" used to identify the different combinations of atoms and lone pairs attached to the central atom. In this formula, A represents a central atom, X represents an atom bonded to the central atom, and E represents a lone pair of electrons on the central atom.

VSEPR model   See *valence-shell electron-pair repulsion theory*.

vulcanization   The heating of rubber with sulfur to increase its elasticity.

water autoprotolysis constant K_w The equilibrium constant for the autoprotolysis (autoionization) of water, $2\,H_2O(l) \rightleftharpoons H_3O^+(aq) + OH^-(aq)$, $K_w = [H_3O^+][OH^-]$.

water of hydration   See *hydration*.

wavefunction   A solution of the Schrödinger equation; the probability amplitude.

wavelength λ (lambda)   The peak-to-peak distance of a wave.

wave-particle duality   The combined wavelike and particlelike character of both radiation and matter.

weak acids and bases   Acids and bases that are only partially ionized in aqueous solutions at normal concentrations. *Examples:* HF, CH_3COOH (weak acids); NH_3, CH_3NH_2 (weak bases).

weak electrolyte   See *electrolyte*.

weak-field ligand   A ligand that produces a small *ligand field splitting* and that lies below NH_3 in the *spectrochemical series*.

weight   The gravitational force on a sample.

work *w* A transfer of energy that takes place when an object is moved against an opposing force. In *expansion work,* the system expands against an opposing pressure; *nonexpansion work* is work that does not involve a change in volume.

x-ray Electromagnetic radiation with wavelengths from about 10 pm up to about 1000 pm.

x-ray diffraction The analysis of crystal structures by studying the interference pattern in a beam of x-rays.

yield See *percentage yield; theoretical yield.*

zero-order reaction A reaction with a rate that is independent of the concentration of the reactant. *Example:* the catalyzed decomposition of ammonia.

Ziegler-Natta catalyst A stereospecific catalyst for polymerization reactions, consisting of titanium tetrachloride and triethylaluminum.

zinc-blende structure A crystal structure in which the cations occupy half the tetrahedral holes in a nearly close-packed cubic lattice of anions; also known as the sphalerite structure.

zone refining A method for purifying a solid by repeatedly passing a molten zone along the length of a sample.

Answers

SELF-TESTS B

Chapter 1

1.1B (a) Sn; (b) Na; (c) iodine; (d) yttrium

1.2B (a) 16; (b) 22

1.3B (a) 8, 8, 8; (b) 94, 145, 94

1.4B (a) ^{73}Ge; (b) ^{76}Ge

1.5B (a) Se, Group 16, Period 4, nonmetal; (b) Sn, Group 14, Period 5, metal; (c) N, Group 15, Period 2, nonmetal

1.6B S^{2-} and K^+

1.7B (a) Ca, P, and O atoms are present in the ratio 3:2:8; (b) ionic compound

1.8B (a) chemical; (b) chemical; (c) physical

1.9B (a) iron(III) ion; (b) fluoride ion; (c) nitrate ion; (d) silver ion

1.10B (a) gold(III) chloride; (b) calcium sulfide; (c) manganese(III) oxide

1.11B (a) phosphorus trichloride; (b) sulfur trioxide; (c) hydrogen bromide

1.12B (a) phosphoric acid; (b) $HClO_3$

1.13B (a) boron tribromide; (b) thallium(I) fluoride; (c) magnesium selenide

1.14B (a) $Cs_2S \cdot 4H_2O$; (b) Mn_2O_7; (c) H_2S; (d) S_2Cl_2

Chapter 2

2.1B (a) 1.05 g/cm^3; (b) yes (29.4 g)

2.2B 8.82 oz

2.3B 1.100 g/cm^3

2.4B 506–508 K

2.5B (a) 4 sf; (b) 3 sf; (c) 7 sf

2.6B (a) 8.1 g; (b) 1540.°C

2.7B (a) 6.9×10^{-4} L; (b) 1.38 g/L

2.8B Mass measurements had greater precision.

2.9B 1.89×10^{24} H_2O molecules

2.10B $(0.6917 \times 62.94) + (0.3083 \times 64.93)\text{g/mol} = 63.55$ g/mol

2.11B 62 g U

2.12B 1.260 mol O atoms

2.13B (a) 63.02 g/mol; (b) 342.14 g/mol

2.14B (a) 13.5 mol $Ca(OH)_2$; (b) 0.51 g

2.15B 63.50%

2.16B 88.1% C and 11.9% H

2.17B OSF_2

2.18B $C_2H_2O_4$

Chapter 3

3.1B $2 \text{ Co(NO}_3)_3\text{(aq)} + 3 \text{ (NH}_4)_2\text{S(aq)} \rightarrow Co_2S_3\text{(s)} + 6 \text{ NH}_4\text{NO}_3\text{(aq)}$

3.2B $C_3H_8\text{(g)} + 5 O_2\text{(g)} \rightarrow 3 CO_2\text{(g)} + 4 H_2O\text{(l)}$

3.3B $3 Hg_2^{2+}\text{(aq)} + 2 PO_4^{3-}\text{(aq)} \rightarrow (Hg_2)_3(PO_4)_2\text{(s)}$

3.4B Sodium sulfate and strontium nitrate solutions: $Sr^{2+}\text{(aq)} + SO_4^{2-}\text{(aq)} \rightarrow SrSO_4\text{(s)}$

3.5B (b) and (c) are acids; (d) is a base; (a) is neither

3.6B HCN is a weak acid; thus, only a small fraction of the acid molecules is ionized. The concentration of hydronium ions is much smaller than the concentration of nonionized HCN molecules.

3.7B (a) strong acid so strong electrolyte, conducts electricity; (b) weak base so weak electrolyte, conducts electricity to small extent

3.8B $3 \text{ Ca(OH)}_2\text{(aq)} + 2 H_3PO_4\text{(aq)} \rightarrow Ca_3(PO_4)_2\text{(s)} + 6 H_2O\text{(l)}$

3.9B $NH_3\text{(aq)} + HCN\text{(aq)} \rightarrow NH_4^+\text{(aq)} + CN^-\text{(aq)}$

3.10B $HCl\text{(aq)} + NaHCO_3\text{(aq)} \rightarrow NaCl\text{(aq)} + H_2O\text{(l)} + CO_2\text{(g)}$

3.11B $Cu^+\text{(aq)}$ is oxidized, $I_2\text{(s)}$ is reduced

3.12B (a) +4; (b) +3; (c) +5

3.13B Sulfuric acid is the oxidizing agent and sodium iodide is the reducing agent.

3.14B $2 Ce^{4+}\text{(aq)} + 2 I^-\text{(aq)} \rightarrow I_2\text{(s)} + 2 Ce^{3+}\text{(aq)}$

3.15B $2 NH_3\text{(g)} + 3 CuO\text{(s)} \rightarrow N_2\text{(g)} + 3 Cu\text{(s)} + 3 H_2O\text{(l)}$; redox

3.16B (a) $CO_3^{2-}\text{(aq)} + 2 H^+\text{(aq)} \rightarrow H_2O\text{(l)} + CO_2\text{(g)}$; alternatively, $CO_3^{2-}\text{(aq)} + 2 H_3O^+\text{(aq)} \rightarrow 2 H_2O\text{(l)} + CO_2\text{(g)}$; (b) gas formation reaction

Chapter 4

4.1B (a) 2 mol $N_2 \approx 3$ mol O_2; (b) 2 mol $N_2 \approx 2$ mol N_2O_3 (or 1 mol $N_2 \approx 1$ mol N_2O_3)

4.2B 50. mol Al

4.3B 396 g $CaSiO_3$

4.4B 41.47% barium

4.5B (15 kg Fe_2O_3)/159.69 g/mol) $\times$ (2 mol Fe)/(1 mol Fe_2O_3) $\times$ (55.85 g/mol) $=$ 10.5 kg Fe; 8.8 kg/10.5 kg $\times$ 100% $=$ 84%

4.6B There is 0.61 mol NO_2 and 1.0 mol H_2O. From the reaction stoichiometry, 1 mol H_2O requires 3 mol NO_2. There is not enough NO_2, so NO_2 is the limiting reactant.

4.7B the sample contains 0.0118 mol C (0.142 g C) and 0.0105 mol H (0.0106 g H). Mass of O = 0.236 − (0.142 + 0.0105)g = 0.0834 g O (0.00521 mol O). The C:H:O mole ratios are 0.0118:0.0105:0.00521, or 2.26:2.02:1. Multiplying these numbers by 4 gives 9:8:4 and the empirical formula $C_9H_8O_4$

4.8B [(1.368 g)/(180.16 g/mol)]/(0.050 L) = 0.1519 M $C_6H_{12}O_6$(aq)

4.9B 1.45×10^{-3} mol urea

4.10B 7.12 mL

4.11B (0.125 mol/L) × (0.05000 L) × (90.04 g $C_2H_2O_4$/mol) = 0.563 g $C_2H_2O_4$

4.12B $(1.59 \times 10^{-5}$ mol/L) × (0.02500 L)/(0.152 mol/L) = 2.62×10^{-6} L, or 2.62×10^{-3} mL

4.13B (0.255 × 0.01645) mol KOH × (1 mol H_2SO_4/2 mol KOH) × 1/(0.02500 L) = 0.0839 M H_2SO_4(aq)

4.14B (0.0100 × 0.02815) mol $KMnO_4$ × (5 mol As_4O_6/8 mol $KMnO_4$) × 395.68 g/mol = 6.962×10^{-2} g As_4O_6

Chapter 5

5.1B 8.40×10^4 Pa

5.2B l0. L

5.3B 5.33 atm. Under a pressure this high, the can would be expected to explode.

5.4B 0.10 L

5.5B 24.6 L/min

5.6B The pressure would be increased by a factor of four.

5.7B 12 L

5.8B [(1.00 × 10^3 × 0.791)/32.04] mol CH_3OH × [(3 mol O_2)/(2 mol CH_3OH)] × 22.41 L/mol = 830. L

5.9B (1.04 g/L) × [(0.08206 × 450.)/(200./760)] L/mol = 146 g/mol

5.10B 396 mol

5.11B Ozone effuses more slowly than nitrogen dioxide because its molar mass is greater (48 g/mol compared with 46 g/mol).

5.12B CH_4 molecules (molar mass 16.0 g/mol) have a greater rms speed than Cl_2 molecules (molar mass 71 g/mol) at the same temperature.

Chapter 6

6.1B (a) closed; (b) open

6.2B (a) −750. kJ; (b) −50. kJ; (c) −800. kJ

6.3B 3.10 kJ

6.4B (a) +50 kJ; (b) +70 kJ

6.5B 7.10 kJ

6.6B 0.26 J/°C·g

6.7B 4.16 kJ/3.24°C = 1.28 kJ/°C

6.8B 22 kJ / (23 g/46.07 g/mol) = 44 kJ/mol

6.9B C_2H_5OH(l) + 3 O_2(g) → 2 CO_2(g) + 3 H_2O(l), ΔH = −(2.16 kJ/°C × 6.333°C) / (0.461g/46.07 g/mol) = −1.37 × 10^3 kJ

6.10B −328.10 kJ

6.11B 350 kJ / (−1368 kJ/mol) × 46.07 g/mol = 11.8 g

6.12B methanol; its enthalpy density is higher, so less volume would be required for the same heat output

6.13B C(dia) + O_2(g) → CO_2(g); (−393.51 − 1.895) kJ/mol = −395.40 kJ/mol

Chapter 7

7.1B yellow

7.2B 4.5×10^{-19} J

7.3B $\lambda = \dfrac{6.64 \times 10^{-34}\ \text{J·s}}{5.00 \times 10^{-3}\ \text{kg·2·310 m/s}} = 2.14 \times 10^{-34}$ m

7.4B $\nu = \dfrac{1}{h}\Delta E = (3.29 \times 10^{-15}\ \text{Hz})\left(\dfrac{1}{2^2} - \dfrac{1}{5^2}\right)$
$= 6.91 \times 10^{14}$ Hz
$\lambda = (3.00 \times 10^8)/(6.91 \times 10^{14})$ m = 4.34×10^{-7} m
= 434 nm; violet line

7.5B n^2

7.6B $3p$

7.7B $1s^2 2s^2 2p_x^2 2p_y^1 2p_z^1$ or $1s^2 2s^2 2p^4$

7.8B $1s^2 2s^2 2p^6 3s^2$ or [Ne]$3s^2$

7.9B [Kr]$4d^2 5s^2$

7.10B [Ar]$3d^5$, [Ar]$3d^2$

7.11B I^-, [Kr]$4d^{10}5s^2 5p^6$

7.12B (a) 5; (b) ns^2np^3

7.13B (a) $r(Ca^{2+}) < r(K^+)$; (b) $r(Cl^-) < r(S^{2-})$

7.14B Typically, ionization energy decreases down a group. This trend results from the increasing average distance of the valence electrons from the nucleus and the lower effective nuclear charge they experience. However, gallium, unlike aluminum above it, has inner $3d$ electrons. These $3d$ electrons are less effective in shielding the outer $4s$- and $4p$-electrons from the nuclear charge than are inner s- and p-electrons. For this reason, the effective nuclear charge experienced by gallium's $4s$- and $4p$-electrons is larger than one might otherwise expect. Correspondingly, the outer $4p^1$-electron is more strongly bound. This effect counterbalances the usual periodic trend down the group.

7.15B In fluorine, a member of Group 17, an additional electron fills the single vacancy in the valence shell; the shell now has the noble-gas configuration of neon and is complete. In neon, an additional electron would have to enter a new shell, where it would be farther from the attraction of the nucleus.

7.16B (a) beryllium; elements in the main groups exhibit a similarity in properties between diagonal neighbors as one proceeds from upper left to lower right. This phenomenon can be explained by the general trends in atomic radius and ionization energy, both of which affect chemical properties. Radius decreases across the periodic table but increases down the table; ionization energy increases across but decreases down. Thus, values for diagonally related elements are often similar.

Chapter 8

8.1B (a) Al_2S_3; (b) Sr_3P_2

8.2B $[Pb:]^{2+} [:\ddot{O}:]^{2-}$ and $[:\ddot{O}:]^{2-} Pb^{4+} [:\ddot{O}:]^{2-}$

8.3B 147.70 (vaporize Mg) $+ 2 \times 111.88$ ($-\Delta H_f^\circ$ for 2 Br(g)) $+ 736$ (form Mg^+) $+ 1450$ ($Mg^+ \rightarrow Mg^{2+} + e^-$) $- (2 \times 325)$ (add electron to Br) $+ 524.3$ ($-\Delta H_f^\circ$) kJ $= 2463$ kJ $= 2432$ kJ/mol

8.4B $H-\ddot{B}r:$; no lone pairs on hydrogen and three on bromine

8.5B The C atom in CO ($:C\equiv O:$) has a nonbonding electron pair that can bind to hemoglobin, whereas the C atom in CO_2 ($\ddot{O}=C=\ddot{O}$) has no nonbonding pairs available.

8.6B
$$H-C=C-H$$
$$\;\;\;\;\;|\;\;\;\;|$$
$$\;\;\;\;H\;\;\;H$$

8.7B $H-\ddot{N}-H]^{-}$

8.8B $\ddot{O}=N-\ddot{O}:]^{-} \leftrightarrow :\ddot{O}-N=\ddot{O}]^{-}$
$\;\;\;\;0\;\;\;\;+1\;\;\;-1$

8.9B $\ddot{O}=\ddot{O}-\ddot{O}:$

8.10B $\ddot{O}=N-\ddot{O}:$

8.11B $:\ddot{I}-\ddot{I}-\ddot{I}:]^{-}$

8.12B
$$\begin{array}{c} \;\;\;\;\;\;\;0 \\ :\ddot{O}: \\ \;\;\;\;\parallel 0 \\ :\ddot{O}-As-\ddot{O}: \;\;\;]^{3-}\\ {}_{-1}\;\;\;|\;\;\;{}_{-1} \\ :\ddot{O}: \\ {}_{-1} \end{array}$$

8.13B See structure **43**. One Cl atom on a neighboring $AlCl_3$ molecule donates an electron pair to Al and vice versa.

8.14B (a) NH_3; (b) N in NH_3 and O in NO_2

8.15B (a)

Chapter 9

9.1B tetrahedral

9.2B linear

9.3B (a) AX_2E; (b) trigonal planar; (c) bent

9.4B square planar

9.5B

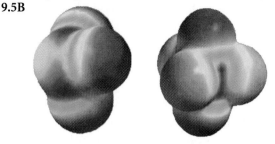

(a) polar; (b) nonpolar

9.6B $\Delta H = (2 \times 412) + (2 \times 158) - (2 \times 484) - (2 \times 565)$ kJ $= -958$ kJ

9.7B H—O, 111 pm; O—Cl, 172 pm

9.8B (a) three σ, no π; (b) three σ, one π

9.9B Two σ-bonds are formed from carbon sp^3 hybrids and hydrogen $1s$-orbitals, and two σ-bonds are formed from carbon sp^3 hybrids and chlorine $3p$-orbitals, in a tetrahedral arrangement.

9.10B octahedral; octahedral; sp^3d^2 hybridization

9.11B An atom of the CH_3 group is sp^3 hybridized and forms four σ-bonds with bond angles of about 109.5°. The other two C atoms are both sp^2 hybridized and form three σ-bonds and one π-bond (the π-bond is between these two C atoms); bond angles are about 120°.

9.12B lower energy

9.13B $\sigma_{2s}^2\sigma_{2s}^{*2}\pi_{2p}^4\sigma_{2p}^2\pi_{2p}^{*1}$, BO $= 2.5$

Chapter 10

10.1B 1,1-dichloroethene

10.2B CHF_3 is polar, whereas CF_4 is nonpolar. Dipole-dipole interactions would be expected to result in a higher boiling point for CHF_3. On the other hand, CF_4 has more electrons than CHF_3 and stronger London forces, which might lead us to predict a higher boiling point for CF_4. Because, experimentally, CHF_3 has a higher boiling point, dipole-dipole interactions must in this case be more important than London interactions.

10.3B (a) and (c)

10.4B (a) ethanol, because of dipole-dipole interactions and hydrogen bonding

10.5B 6

10.6B 4; $\frac{1}{8} \times 8$ (1 on each of eight corners), $\frac{1}{2} \times 2$ on opposite faces, and 2 inside

10.7B yes; the bcc structure has 2 atoms per unit cell; the length of a side $= 4r/\sqrt{3}$. Assuming a bcc structure, the density of a unit cell is

$$\frac{2 \times (55.85 / 6.022 \times 10^{23})\text{ g}}{[4 \times (1.24 \times 10^{-8})/\sqrt{3}]^3 \text{ cm}^3} = 7.89 \text{ g/cm}^3$$

This value is consistent with the experimentally determined density.

10.8B one; $\frac{1}{8} \times 1$ at each of eight corners. (Equivalently, each chloride ion can be thought of as being at the center of a bcc unit cell, with eight cesium cations at its corners.)

10.9B $CH_3CH_2CH_3$. Molecules of the two substances have the same molar mass, thus the same number of electrons and comparable London forces. In addition, CH_3CHO is polar and experiences dipole-dipole forces.

10.10B 66.5°C (actual: 64.7°C)

10.11B vapor

10.12B rhombic sulfur, because higher pressures favor its formation from solid monoclinic sulfur

10.13B The liquid CO_2 vaporizes.

10.14B Critical temperature appears to increase as capacity for hydrogen bonding increases. NH_3, which can form hydrogen bonds, has a higher critical temperature than CH_4, which cannot form hydrogen bonds. H_2O, in turn, has a capacity to form more hydrogen bonds than NH_3, and has an even higher critical temperature.

Chapter 11

11.1B

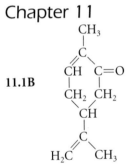

11.2B (a) because it has the greater molar mass and thus stronger London forces

11.3B 4-ethyl-2-methylhexane;
$CH_3CH_2C(CH_2CH_3)_2CH_2CH(CH_2CH_3)CH_2CH_3$

11.4B 1-ethyl-3,5-dimethyl-2-propylbenzene

11.5B (a) no reaction in the dark; (b) CH_3CH_2Cl

11.6B (a) $HCOOCH_2CH_3$, ethyl formate;
(b) $CH_3CH_2CH_2COOH$ and CH_3OH

11.7B C: sp^2; N: sp^3

11.8B (a) 3-pentanol; (b) 3-pentanone;
(c) ethylmethylamine

11.9B $CH_3CH_2CH_2CH_2Br$, $CH_3CH_2CHBrCH_3$,
$(CH_3)_2CHCH_2Br$, $(CH_3)_3CBr$

11.10B 41a is *cis*-1-phenyl-1-propene;
41b is *trans*-1-phenyl-1-propene

11.11B (c), because it has four different groups on one carbon atom.

11.12B (a) $CH_2{=}C(CH_3)(COOCH_3)$;
(b) $-NH-CH(CH_3)CONH-CH(CH_3)CO-$

Chapter 12

12.1B (c) formaldehyde

12.2B 0.021 mol

12.3B Rb^+ would be expected to have a less negative hydration enthalpy than Sr^{2+} because it is less highly charged.

12.4B moles NaCl = moles Na^+ = moles Cl^- = 5.00 g NaCl / 58.44 g NaCl/mol = 0.0856 mol;
moles H_2O = 5.55 mol;
total moles = 2(0.0856) + 5.55 mol = 5.72 mol;
x_{Na^+} = x_{Cl^-} = 0.0856 mol / 5.72 mol = 0.0150;
x_{H_2O} = 5.55 mol / 5.72 mol = 0.970

12.5B [7.36 g $KClO_3$ / 122.55 g $KClO_3$/mol] / 0.020 kg = 3.00 mol/kg

12.6B (0.255 mol/kg) × (0.250 kg) × (180.16 g/mol) = 11.5 g

12.7B Assume 0.250 mol methanol and 0.750 mol water.
0.250 mol methanol / 0.0135 kg water = 18.5 mol/kg

12.8B Assume 1.00 L solution.
mass solution = 1000. cm^3 × 1.07 g/cm^3 = 1070 g;
mass solute = 1.83 mol NaCl × 58.44 g NaCl/mol = 107 g NaCl;
mass water = 1070 g solution − 107 g NaCl = 963 g = 0.963 kg;
molality = 1.83 mol NaCl / 0.963 kg = 1.90 mol/kg

12.9B $x_{ethanol}$ = 0.986; P = 0.986 × 5.3 kPa = 5.2 kPa

12.10B 1.00 mol (0.25 mol Co^{3+} ions and 0.75 mol Cl^- ions); i = 4

12.11B molality = 0.0128 mol/kg
molar mass = [0.200 g/(0.0128 mol/kg)] / (0.100 kg)] = 1.6 × 10^2 g/mol (actual: 154.2 g/mol)

12.12B molarity = 8.516 × 10^{-4} mol/L;
moles = 8.516 × 10^{-4} mol/L × 0.175 L = 1.49 × 10^{-4} mol; molar mass = (1.50 g)/(1.49 × 10^{-4} mol) = 1.01 × 10^4 g/mol or 10.1 kg/mol

Chapter 13

13.1B 0.575 (mol N_2)/L·h

13.2B second order

13.3B (a) first order in CH_3Br and OH^-; second order overall; (b) L/mol·s

13.4B rate = $k[NO]^2[O_2]$, k = 3.5 × 10^4 L^2/mol^2·s

13.5B $[C_3H_6]$ = 0.100 mol/L × $e^{-(0.00067/s) \times (200. s)}$ = 0.087 mol/L

13.6B The slope of the plot of ln[penicillin] against time is −0.15/week. Therefore, k = 0.15/week.

13.7B (a) $t_{1/2}$ = (ln 2) /k = (ln 2) / 5.5 × 10^{-4} /s = 1.3 × 10^3 s; (b) concentration falls to 1/16 of its initial value after four half-lives; 4 × (1.3 × 10^3 s) = 5.0 × 10^3 s

13.8B The slope of a plot of $\ln k$ against $1/T$ is -2610 K. $E_a = -(8.3145\text{ J/Kmol}) \times (-2610\text{ K}) = 21700\text{ J/mol} = 21.7\text{ kJ/mol}$.

13.9B $\ln \{k'/(6.7 \times 10^{-4}/\text{s})\} =$

$$\frac{272\text{ kJ/mol}}{8.314 \times 10^{-3}\text{ kJ/K·mol}}\left(\frac{1}{773\text{ K}} - \frac{1}{573\text{ K}}\right) = -14.9$$

$$\frac{k'}{6.7 \times 10^{-4}/\text{s}} = e^{-14.9} = 3.84 \times 10^{-7}$$

$k' = 2.6 \times 10^{-10}/\text{s}$

13.10B (a) The rate law and rate constant will both be changed. (b) It provides an alternative path, typically with a lower activation energy.

13.11B (a) bimolecular; (b) unimolecular

13.12B rate $= (k_1 k_2/k_1')\,[\text{NO}]^2[\text{O}_2]$, or rate $= k[\text{NO}]^2[\text{O}_2]$, where $k = k_1 k_2/k_1'$

Chapter 14

14.1B $K_c = [\text{O}_2]^3/[\text{O}_3]^2$

14.2B 8.5×10^{-7}

14.3B $K_c = 1/[\text{O}_2]^5$

14.4B $K_p = P_{\text{SO}_3}^2/P_{\text{O}_2}^3$

14.5B $\Delta n = -2; K_p = K_c/(0.082\,06 \times 347)^2 = K_c/(9.88 \times 10^3)$

14.6B (a) $K = \dfrac{[\text{H}_2\text{CO}_3]}{P_{\text{CO}_2}}$; (b) $K_c = \dfrac{[\text{HCO}_3^-][\text{H}_3\text{O}^+]}{[\text{H}_2\text{CO}_3]}$

14.7B 31.9

14.8B $[\text{NO}] = \sqrt{K_c[\text{N}_2][\text{O}_2]} = 6.9 \times 10^{-19}$, about 10^{-15} times the concentration of N_2 and O_2

14.9B $P_{\text{CO}} = \dfrac{75}{\sqrt{(2.8 \times 10^{20})(1.4 \times 10^{-9})}} = 1.2 \times 10^{-4}$ atm

14.10B $Q_p = 0.60; Q_p < K_p$; NO_2 will tend to increase its partial pressure

14.11B

	N_2	H_2	NH_3
initial	0.40	0.90	0
change	$-x$	$-3x$	$+2x$
equilibrium	$0.40 - x$	$0.90 - 3x$	$2x$

$2x = 0.20; x = 0.10$
At equilibrium, $[\text{N}_2] = 0.30$ mol/L, $[\text{H}_2] = 0.60$ mol/L, $[\text{NH}_3] = 0.20$ mol/L
$K_c = 0.20^2/(0.60^3 \times 0.30) = 0.62$

14.12B $K_c = 20 = (2x)^2/(0.200 - x)(0.100 - x)$
$16x^2 - 6x - 0.40 = 0$
Using quadratic formula, $x = 0.29, 0.088$
Discard 0.29 because it would yield negative concentrations at equilibrium.
$[\text{ClF}] = 2x = 2 \times 0.088 = 0.18$ mol/L

14.13B $K_c = 3.5 \times 10^{-32} \approx (2x)^2 x/0.012^2$;
$x \approx 1.1 \times 10^{-12}$. The concentrations are 0.012 mol/L HCl;

2.2×10^{-12} mol/L HI; 1.1×10^{-12} mol/L Cl_2. Some I_2 remains as a pure solid.

14.14B The equilibrium tends to shift toward (a) products; (b) products; (c) reactants

14.15B The equilibrium tends to shift toward products.

14.16B CO_2 favored

Chapter 15

15.1B (a) NH_4^+ and HCO_3^-; (b) NH_2^- and NO_3^-

15.2B 0.010 mol/L H_3O^+; 1.0×10^{-12} mol/L OH^-

15.3B $[\text{H}_3\text{O}^+] = K_w/[\text{OH}^-] = 4.5 \times 10^{-12}$ mol/L; 2.2×10^{-3} mol/L OH^-

15.4B $[\text{H}_3\text{O}^+] = K_w/[\text{OH}^-] = 1.0 \times 10^{-14}/0.077 = 1.3 \times 10^{-13}$; pH $= 12.89$

15.5B $[\text{H}_3\text{O}^+] = 10^{-8.2} = 6 \times 10^{-9}$ mol/L

15.6B (a) Brønsted acid, NH_4^+; Brønsted base, HCO_3^-; (b) conjugate base, NH_3; conjugate acid, H_2CO_3

15.7B (a) weak base; (b) strong base; (c) weak base

15.8B stronger acids: (b) $\text{C}_5\text{H}_5\text{NH}^+$, (c) HIO_3; stronger bases: (a) NH_2NH_2, (d) ClO_2^-

15.9B (a) HClO_2, oxoacid, rule 1; (b) HI, binary acid, rule 2

15.10B $\text{CH}_3\text{COOH} < \text{CH}_2\text{ClCOOH} < \text{CHCl}_2\text{COOH}$

15.11B pH $= 1.77$ (must use exact solution)

15.12B $K_b = 1.0 \times 10^{-6} \approx x^2/0.012$; $[\text{OH}^-] = x \approx 1.1 \times 10^{-4}$; pOH $= 3.96$;
pH $= 14.00 - 3.96 = 10.04$;
$(1.1 \times 10^{-4})/0.012 \times 100\% = 0.92\%$

15.13B $0.012 = (0.10 + x)x/(0.10 - x)$;
$x^2 + 0.11x - 0.0012 = 0$; from the quadratic equation, $x = 0.010$; $[\text{H}_3\text{O}^+] = 0.10 + 0.010 = 0.11$; pH $= 0.96$

15.14B $K_{a1} = 1.5 \times 10^{-2} = x^2/(0.10 - x)$; from the quadratic equation, $x = 0.032$; pH $= 1.49$. Note that if $[\text{H}_3\text{O}^+]$ is not rounded, pH $= 1.50$.

Chapter 16

16.1B (a) basic; (b) acidic; (c) neutral

16.2B $K_a = 5.6 \times 10^{-10} = x^2/(0.10 - x)$;
$x = [\text{H}_3\text{O}^+] = 7.5 \times 10^{-6}$; pH ≈ 5.1

16.3B $K_b = 2.9 \times 10^{-11} = x^2/(0.020 - x) \approx x^2/0.020$;
$x = [\text{OH}^-] \approx 7.6 \times 10^{-7}$; pOH ≈ 6.12, pH ≈ 7.9

16.4B $K_a = 3.0 \times 10^{-8} \approx x(2.0 \times 10^{-4})/0.010$;
$x = [\text{H}_3\text{O}^+] = 1.5 \times 10^{-6}$; pH $= 5.82$

16.5B A total of 12.0 mL titrant has been added.
moles of $\text{OH}^- = 6.25 \times 10^{-3}$ mol
moles of $\text{H}_3\text{O}^+ = 4.08 \times 10^{-3}$ mol
moles of OH^- remaining after neutralization $= 2.17 \times 10^{-3}$ mol

total volume of solution $= (25.0 + 12.0)$ mL $=$
37.0 mL $= 0.037$ L
molarity of OH$^-$ $= 0.0586$ mol/L
pOH $= 1.23$; pH $= 12.77$
16.6B moles of H$_3$O$^+$ $= 8.50 \times 10^{-3}$ mol
moles of OH$^-$ $= 6.25 \times 10^{-3}$ mol
moles of H$_3$O$^+$ remaining after reaction $=$
2.25×10^{-3} mol
total volume of solution $= (25.0 + 25.0)$ mL $=$
50.0 mL $= 0.050$ L
molarity of H$_3$O$^+$ $= 0.045$ mol/L
pH $= 1.35$
16.7B salt present at stoichiometric point is KClO;
moles of KClO $=$ moles of HClO used $=$
$(0.010$ mol/L$) \times (0.025\,00$ L$) = 2.5 \times 10^{-4}$ mol;
volume (titrant) $= (2.5 \times 10^{-4}$ mol$)/(0.020$ mol/L$) =$
0.0125 L (the 5 is not a significant figure here)
total volume $= (0.025\,00 + 0.0125)$ L $= 0.0375$ L;
molarity of KClO $= 2.5 \times 10^{-4}$ mol $/$ 0.0375 L $=$
6.67×10^{-3} mol/L;
$K_b(ClO^-) = 3.3 \times 10^{-7} = x^2/[(6.67 \times 10^{-3}) - x]$
$x = [OH^-] = 4.7 \times 10^{-5}$; pOH ≈ 4.3; pH ≈ 9.7
16.8B total volume of titrant $= 15.00$ mL
total volume of solution $= 40.00$ mL;
moles of HCO$_2^-$ formed $=$ moles of titrant
added $= 2.25 \times 10^{-3}$;
moles of HCOOH remaining $= (2.50 \times 10^{-3}) -$
(2.25×10^{-3}) mol $= 2.5 \times 10^{-4}$ mol;
molarity of HCO$_2^-$ $= (2.25 \times 10^{-3}$ mol$)/(0.0400$ L$) =$
0.0562 mol/L;
molarity of HCOOH $= (2.5 \times 10^{-4}$ mol$)/(0.0400$ L$) =$
6.25×10^{-3} mol/L;
$1.8 \times 10^{-4} \approx x(0.0562)/(6.25 \times 10^{-3})$; $x = 2.00 \times 10^{-5}$;
pH $= 4.70$
16.9B (a) 15 mL; (b) 30. mL; (c) 45 mL
16.10B pH $= 3.37 + \log(0.20/0.15) = 3.49$
16.11B moles of CH$_3$CO$_2^-$ $= (0.040$ mol/L$) \times$
$(0.300$ L$) + 0.0200$ mol $= 0.032$ mol;
moles of CH$_3$COOH $= (0.080$ mol/L$) \times (0.300$ L$) -$
0.0200 mol $= 0.0040$ mol;
molarity of CH$_3$CO$_2^-$ $= (0.032$ mol$)/(0.300$ L$) =$
0.107 mol/L (the 7 is not a significant figure here);
molarity of CH$_3$COOH $= (0.0040$ mol$)/(0.300$ L$) =$
0.0133 mol/L (second 3 not significant);
pH $= 4.75 + \log(0.107 / 0.0133) = 5.65$ or about 5.6, an
increase of about 1.2
16.12B (CH$_3$)$_3$NH$^+$/(CH$_3$)$_3$N would be the best
candidate. NH$_4^+$/NH$_3$ would also be acceptable
16.13B $\log\{[C_6H_5CO_2^-]/[C_6H_5COOH]\} = pH - pK_a =$
$3.50 - 4.19 = -0.69$
$[C_6H_5CO_2^-]/[C_6H_5COOH] = 1{:}4.9$ or 0.20:1

16.14B $K_{sp} = 4s^3 = 4(3.8 \times 10^{-3})^3 = 2.2 \times 10^{-7}$
16.15B $K_{sp} = [Ag^+]^2[SO_4^{2-}] = s^3$; $S = 15$ mmol/L
16.16B $[Ag^+] = K_{sp}/[Br^-] = 7.7 \times 10^{-13}/0.20 =$
3.8×10^{-12} mol/L
16.17B no; $Q_{sp} = (3.3 \times 10^{-4})(6.7 \times 10^{-4})^2 =$
1.5×10^{-10}; $Q_{sp} < K_{sp}$
16.18B AgBr precipitates first, at 7.7×10^{-10} mol/L Br$^-$;
then PbBr$_2$ at 0.63 mol/L Br$^-$
16.19B $K = K_{sp} \times K_f = 1.7 \times 10^6$; solubility $=$
0.30 mol/L. Use the equilibrium solver in the calculator
tool on the CD to solve the cubic equation.

Chapter 17
17.1B liquid mercury at 0°C
17.2B -0.797 J/K
17.3B $+97.6$ J/K·mol
17.4B (a) Pb(l); (b) SbCl$_5$(g)
17.5B $\{229.60 - (219.56 + 130.68)\}$ J/K·mol $=$
-120.64 J/K·mol
17.6B $-(71,500$ J/150 K$) = -477$ J/K
17.7B The reaction is not spontaneous, because
$\Delta S_{tot}° < 0$.
$\Delta S_r° = +93.89$ J/K·mol; $\Delta H_r° = +132.72$ kJ/mol;
$\Delta S_{surr}° = -(132.72/298)$ kJ/K·mol $= -0.445$ kJ/K·mol;
$\Delta S_{tot}° = \Delta S_r° + \Delta S_{surr}° = -351$ J/K·mol
17.8B yes
17.9B $(35,300$ J/K$)/104.7$ K $= 337$ K
(experimental: 337 K)
17.10B 3 C(s, gr) + 3 H$_2$(g) $\rightarrow$ C$_3$H$_6$(g); 53 300 J $-$
298 K $[237.4 - (3 \times 5.740 + 3 \times 130.68)]$ J/K $=$
$+104\,500$ J/mol or $+104.5$ kJ/mol
17.11B no; for methylamine, $\Delta G_f° = +32.16$ kJ/mol
17.12B $\{-910 - [6 \times (-394.36) +$
$(-237.13)]\}$ kJ/mol $= +2879$ kJ/mol
17.13B $+8.96$ kJ/mol; toward reactants
17.14B $\Delta G_r° = \{2 \times (51.31) - 2 \times (86.55)\}$ kJ/mol $=$
-70.48 kJ/mol;
$\ln K_p = -(-70.48$ kJ/mol$)/(2.479$ kJ/mol$) = 28.43$;
$K_p = 2.2 \times 10^{12}$
17.15B MgCO$_3$(s) $\rightarrow$ MgO(s) + CO$_2$(g);
$\Delta H_r° = \{(-601.70 - 393.51) - (-1095.8)\}$ kJ/mol $=$
$+100.6$ kJ/mol;
$\Delta S_r° = \{26.94 + 213.74 - (+65.7)\}$ J/K·mol $=$
$+175.0$ J/K·mol;
$T = (100\,600$ J/mol$)/(175.0$ J/K·mol$) = 574.8$ K

Chapter 18
18.1B (a) Al(s) $\rightarrow$ Al^{3+}(aq) + 3 e$^-$;
(b) S$_8$(s) + 16 e$^-$ $\rightarrow$ 8 S^{2-}(aq)
18.2B 5 H$_2$SO$_3$(aq) + 2 MnO$_4^-$(aq) + H$^+$(aq) $\rightarrow$
5 HSO$_4^-$(aq) + 2 Mn^{2+}(aq) + 3 H$_2$O(l)

18.3B $3 IO_3^-(aq) + 9 H_2O(l) + 24 I^-(aq) \rightarrow$
$9 I_3^-(aq) + 18 OH^-(aq)$
18.4B $Mn(s)|Mn^{2+}(aq)||Cu^{2+}(aq),Cu^+(aq)|Pt(s)$
18.5B (a) $Cd^{2+}(aq) + 2 OH^-(aq) \rightarrow Cd(OH)_2(s)$;
(b) $Cd(s)|Cd(OH)_2(s)|OH^-(aq)||Cd^{2+}(aq)|Cd(s)$
18.6B -0.13 V
18.7B yes; the potential for reduction of chlorine lies above (is more positive than) the potential for the reduction of oxygen to water, so chlorine is the better oxidizing agent.
18.8B Ag^+; $+0.46$ V; $2 Ag^+(aq) + Cu(s) \rightarrow$
$2 Ag(s) + Cu^{2+}(aq)$
18.9B (a) 2; (b) $\Delta G° = -nFE° =$
$-\{2(96.485)(1.30)\}$ J $= -251$ kJ
18.10B $E° = -0.81 - (-0.40)$ V $= -0.41$ V; $n = 2$
$\ln K_{sp} = 2 \times (-0.41$ V$) / 0.025\,693$ V
$K_{sp} = 1.4 \times 10^{-14}$
18.11B $E° = 0$; $Q = [Ag^+]_{anode} / [Ag^+]_{cathode} =$
$0.0010/0.010 = 0.10$; $n = 1$;
$E = 0 - (0.025\,69$ V$)\ln 0.10 = 0.059$ V
18.12B $\Delta G_r° = -1405$ kJ; $n = 12$;
$E° = -\Delta G_r°/nF = 1.21$ V
18.13B (b) zinc
18.14B cathode, $H_2(g)$; anode, $O_2(g)$ and $Br_2(l)$, H^+. (On the basis of half-reaction potentials, it appears that H_2O, rather than Br^-, should be oxidized preferentially; thus, O_2 should be the product at the anode. However, because of the high overpotential for oxygen production, bromine may also be produced.)
18.15B $\dfrac{2.50 \text{ C/s} \times 36.0 \text{ h} \times 3600 \text{ s/h}}{96\,485 \text{ C/mol e}^-} \times$
$\dfrac{1 \text{ mol F}_2}{2 \text{ mol e}^-} \times \dfrac{38.0 \text{ g F}_2}{1 \text{ mol F}_2} = 63.8 \text{ g F}_2$
18.16B $12.00 \text{ g Cr} \times \dfrac{1 \text{ mol Cr}}{52.00 \text{ g}} \times \dfrac{6 \text{ mol e}^-}{1 \text{ mol Cr}} \times$
$\dfrac{96\,485 \text{ C}}{1 \text{ mol e}^-} \times \dfrac{1 \text{ s}}{6.20 \text{ C}} = 2.16 \times 10^4 \text{ s} = 5.99 \text{ h}$

Chapter 19
19.1B Hydrogen is a nonmetal. In its elemental state at (and well below) room temperature and pressure, it occurs as a diatomic molecule in the gas phase. It has a considerably higher ionization energy than the Group 1 ns^1 elements. It forms binary compounds with almost all the other main-group elements other than the noble gases; these compounds can have oxidation state $+1$ or -1. For example, when combined with strongly electropositive metallic elements such as potassium, hydrogen forms ionic compounds or hydrides and has oxidation number -1.

19.2 The high melting point and thermal stability of MgO are attributable to the small ionic radii and high charge of the Mg^{2+} and O^{2-} ions; these small, doubly charged ions interact with each other strongly by electrostatic forces. The high lattice enthalpy of MgO is consistent with such interaction.
19.3 Aluminum is resistant to corrosion because a stable oxide film passivates its surface.
19.4 Singly bonded silicon atoms can act as Lewis acids because they have empty d-orbitals in their valence shells. Thus, they can accept pairs of electrons from Lewis bases such as OH^-. In addition, Si atoms are larger than C atoms, which makes it easier for Lewis bases to attack them.

Chapter 20
20.1B Phosphorus, unlike nitrogen, has d-orbitals available for bonding. In a phosphorus atom, two $3d$-orbitals can hybridize with one $3s$-orbital and three $3p$-orbitals to form five sp^3d^2 hybrid orbitals; thus, five pairs of electrons can be accommodated around a central P atom. Because nitrogen does not have $2d$-orbitals, only four pairs of electrons can be accommodated around a central N atom. Furthermore, the small size of a nitrogen atom precludes the placement of five large chlorine atoms around it.
20.2B Oxidation half-reaction: $2 H_2O(l) \rightarrow$
$4 H^+(aq) + O_2(g) + 4 e^-$
Reduction half-reaction: $F_2(g) + 2 e^- \rightarrow 2 F^-(aq)$
Overall: $2 H_2O(l) + 2 F_2(g) \rightarrow 4 H^+(aq) + O_2(g) + 4 F^-(aq)$
$E° = 1.64$ V
20.3B Sulfuric acid forms strong hydrogen bonds with water, which would be consistent with the volume contraction.
20.4B $I^-(aq)$

20.5B square pyramid

Chapter 21
21.1B Densities increase from left to right across a row until the last few elements of the d block; at this point, densities level off and then decrease. Densities increase down a group.
21.2B Pig iron has a higher carbon content than steel. The relatively high carbon content of pig iron makes it hard and brittle and thus unsuitable as a building material.
21.3B (a) linkage; (b) hydrate

21.4B (a) chiral; (b) not chiral; (c) chiral;
(d) not chiral. There are no pairs of enantiomers shown.
21.5B 392 kJ/mol
21.6B (a); CN^- is a stronger field ligand than NH_3
21.7B (a) t^6e^1; (b) t^5e^2
21.8B Both are paramagnetic. Both have two unpaired electrons.

Chapter 22

22.1B (a) $^{235}_{92}U \rightarrow {}^{4}_{2}\alpha + {}^{231}_{90}Th$; (b) $^{11}_{6}C \rightarrow {}^{11}_{5}B + {}^{0}_{+1}e$
22.2B (a) $^{7}_{4}Be + {}^{0}_{-1}e \rightarrow {}^{7}_{3}Li$; (b) $^{228}_{88}Ra \rightarrow {}^{228}_{89}Ac + {}^{0}_{-1}e$
22.3B (c)
22.4B (a) $^{11}_{5}B$; (b) $^{246}_{96}Cm$
22.5B 3.6 μg
22.6B 1.0×10^2 y

22.7B $\dfrac{5.73 \times 10^3 \text{ y}}{\ln 2} \times \ln\left(\dfrac{14\,000}{18\,400}\right) = 2.3 \times 10^3$ y

22.8B
$\Delta m = [235.0439 - (92 \times 1.0078 + 143 \times 1.0087)] \times (1.6605 \times 10^{-27}) = -3.1845 \times 10^{-27}$ kg;
$E_{bind} = 1.73 \times 10^{14}$ J/mol $= 1.73 \times 10^{11}$ kJ/mol
22.9B For one atom:
$\Delta m = (235.92 - 236.05) \times (1.6605 \times 10^{-27})$ kg $= -2.16 \times 10^{-28}$ kg
$\Delta E = (-2.16 \times 10^{-28}$ kg$) \times (3.00 \times 10^8 \text{ m/s})^2 = -1.943 \times 10^{-11}$ J
Because there are 2.56×10^{21} atoms,
$\Delta E = (2.56 \times 10^{21}) \times (-1.943 \times 10^{-11}$ J$) = -4.97 \times 10^{10}$ J
Answer: 5.0×10^{10} J released

Odd-Numbered Exercises and Connection Questions

Chapter 1

1.1 (a) charge = −1, mass = 9.109×10^{-28} g; (b) J. J. Thomson

1.3 A law is a statement about a basic relationship in nature for which there are no known exceptions. Theories are tested explanations or formal explanations of a law.

1.5 (a) 2.00 g S/1.00 g S = 2:1; (b) 2.00 g S/0.667 g S = 3:1; (c) 1.00 g S/0.667 g S = 3:2

1.7 (a) As, 33; (b) S, 16; (c) Pd, 46; (d) Au, 79

1.9 (a) no error; (b) molybdenum, Mo; (c) yttrium, Y; (d) strontium, Sr

1.11 (a) p = 6, n = 7, e = 6; (b) p = 17; n = 20; e = 17; (c) p = 17, n = 18, e = 17; (d) p = 92, n = 143, e = 92

1.13 (a) ^{111}Cd; (b) ^{82}Kr; (c) ^{11}B

1.15 (a) number of protons and electrons; (b) number of neutrons

1.17 (a) lithium, Group 1, metal; (b) gallium, Group 13, metal; (c) xenon, Group 18, nonmetal; (d) potassium, Group 1, metal

1.19 (a) Cl, nonmetal; (b) Co, metal; (c) As, metalloid

1.21 (a) I, nonmetal; (b) Cr, metal; (c) Hg, metal; (d) Al, metal

1.23 lithium, Li, 3; sodium, Na, 11; potassium, K, 19; rubidium, Rb, 37; cesium, Cs, 55; francium, Fr, 87. The alkali metals react vigorously with water to form hydrogen gas.

1.25 $C_{12}H_{20}O_2$

1.27 (a) anion, S^{2-}; (b) cation, K^+; (c) cation, Sr^{2+}; (d) anion, Cl^-

1.29 (a) Group 16; (b) selenium, Se

1.31 (a) p = 1, n = 1, e = 0; (b) p = 4, n = 5, e = 2; (c) p = 35, n = 45, e = 36; (d) p = 16, n = 16, e = 18

1.33 (a) $^{19}F^-$; (b) $^{24}Mg^{2+}$; (c) $^{128}Te^{2-}$; (d) $^{86}Rb^+$

1.35 (a) element; (b) element (diatomic); (c) compound

1.37 (a) mixture; (b) element

1.39 (a) physical; (b) physical; (c) chemical

1.41 (a) physical; (b) physical; (c) chemical

1.43 (a) physical; (b) physical; (c) chemical; (d) physical

1.45 physical properties and changes: temperature, evaporation, and humidity;

1.47 (a) solubility differences; (b) abilities of substances to adsorb, or stick, to surfaces; (c) boiling-point differences

1.49 (a) homogeneous, distillation; (b) heterogeneous, filtration (salt dissolves in water, chalk does not and can be filtered); (c) homogeneous, evaporation or recrystallization

1.51 (a) chloride ion; (b) oxide ion; (c) carbide ion; (d) phosphide ion

1.53 (a) phosphate ion; (b) sulfate ion; (c) nitride ion; (d) sulfite ion; (e) iodite ion; (f) iodide ion

1.55 (a) ClO_3^-; (b) NO_3^-; (c) CO_3^{2-}; (d) ClO^-; (e) HSO_4^-

1.57 (a) plumbous ion, lead(II) ion; (b) ferrous ion, iron(II) ion; (c) cobaltic ion, cobalt(III) ion; (d) cuprous ion, copper(I) ion

1.59 (a) Cu^{2+}; (b) ClO_2^-; (c) P^{3-}; (d) H^-

1.61 (a) MgO; (b) $Ca_3(PO_4)_2$; (c) $Al_2(SO_4)_3$; (d) Ca_3N_2

1.63 (a) potassium phosphate; (b) ferrous iodide, iron(II) iodide; (c) niobium(V) oxide; (d) cupric sulfate, copper(II) sulfate

1.65 (a) copper(II) nitrate hexahydrate; (b) neodymium(III) chloride hexahydrate; (c) nickel(II) fluoride tetrahydrate

1.67 (a) ionic; (b) $AlBr_3$

1.69 (a) $Na_2CO_3 \cdot H_2O$; (b) $In(NO_3)_3 \cdot 5H_2O$; (c) $Cu(ClO_4)_2 \cdot 6H_2O$

1.71 (a) SeO_3; (b) CCl_4; (c) CS_2; (d) SF_6; (e) As_2S_3; (f) PCl_5; (g) N_2O; (h) ClF_3

1.73 (a) sulfur tetrafluoride; (b) dinitrogen pentoxide; (c) nitrogen triiodide; (d) xenon tetrafluoride; (e) arsenic tribromide; (f) chlorine dioxide; (g) diphosphorus pentoxide

1.75 (a) hydrochloric acid; (b) sulfuric acid; (c) nitric acid; (d) acetic acid; (e) sulfurous acid; (f) phosphoric acid

1.77 (a) Na_2O; (b) K_2SO_4; (c) AgF; (d) $Zn(NO_3)_2$; (e) Al_2S_3

1.79 (a) no; (b) a heterogeneous mixture

1.81

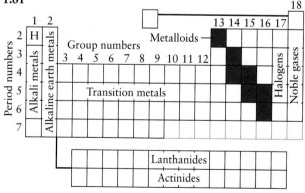

1.83 (a) Ionic compounds consist of an interlocking array of positive and negative ions held together by the attraction between their opposite charges. (b) Molecular

compounds consist of molecules, which are discrete units of electrically neutral atoms bonded together. (c) Ionic compounds are solids at room temperature and have high melting and boiling points. Molecular compounds have lower melting and boiling points; many of them exist as gases or liquids at room temperature.

1.85 (a) silver sulfide; (b) zinc chloride; (c) chlorine pentafluoride; (d) magnesium hydroxide; (e) nickel(II) sulfate hexahydrate; (f) phosphorus pentachloride; (g) chromium(III) dihydrogen phosphate; (h) diarsenic trioxide; (i) manganese(II) chloride

1.87 (a) $(NH_4)_2SO_3$; (b) Fe_2O_3; (c) $Cu(BrO_3)_2$; (d) PH_3; (e) $Ca(HCO_3)_2$; (f) HCN; (g) $LiHSO_4$; (h) SeF_4; (i) $FeSO_4 \cdot 7H_2O$

1.89 (a) 10 p, 8 n, 10 e; (b) total mass of protons: $1.672\,62 \times 10^{-23}$ g, total mass of neutrons: $1.339\,94 \times 10^{-23}$ g, total mass of electrons: $9.109\,39 \times 10^{-27}$ g, total mass of molecule: $3.013\,47 \times 10^{-23}$ g; (c) $0.4446 \times$ (your mass)

1.91 The ingot could be tested by pulsed fast-neutron analysis or spectrometry. X-ray fluorescence analysis may also be used.

1.93 (a) HCl; (b) 39, $^2H^{37}Cl$; 38, $^1H^{37}Cl$; 37, ^{37}Cl or $^2H^{35}Cl$; 36, $^1H^{35}Cl$; 35, ^{35}Cl; 1, ^1H

1.95 distillation (or boiling the water), followed by condensation

1.97 Initial observational data might include the frequency and severity of headaches, and environmental conditions like food eaten, noise levels, or odors. Possible hypotheses include (1) food caused the headaches; (2) the room caused the headaches; (3) homework caused the headaches. Three experiments are (1) eliminate the food (ideally, one food at a time, so several experiments are required); (2) eliminate the room (stay with a friend or do homework elsewhere); (3) eliminate the homework. Other experimental variations are possible. The data to be collected are (1) food eaten; (2) room used; (3) homework begun (yes or no); (4) headache (yes or no).

Chapter 2

2.1 (a) 5.556×10^{12}; (b) $1.169\,811 \times 10^6$; (c) 6×10^{-6} g; (d) 1×10^{-7} m

2.3 (a) 4.3×10^{-1}; (b) 1.492×10^1; (c) 5.1×10^{-9}; (d) 2.37×10^{14}

2.5 (a) 0.250 kg; (b) 2.54 cm; (c) 0.250 ms; (d) 0.149 dm; (e) 2.48×10^{-2} g; (f) 2.835×10^{-2} kg

2.7 (a) 1×10^{-6} m; (b) 5.50×10^{-4} mm; (c) 1.0×10^2 mg; (d) 1.05×10^{-4} μm

2.9 2.4 g/cm^3

2.11 4.5×10^3 g

2.13 1.71×10^{-2} cm^3

2.15 (a) 2.5×10^{-2} m^3; (b) 2.5×10^2 mg/dL^2; (c) 1.54×10^3 pm/μs; (d) 2.66×10^{-6} μg/μm^3; (e) 3.2×10^{-4} mL/s^2; (f) 2.47×10^2 peso/L

2.17 (a) 37.0°C; (b) −40.°F; (c) −273.15°C, −459.67°F; (d) 4 K

2.19 (a) 1×10^{-6} m^3; (b) 3.0×10^{-3} cm/μs; (c) 2.2×10^5 cm^2; (d) 25 mL

2.21 1.0×10^2 mm^2; 1.00×10^{-5} m^3; 1.00×10^{-1} L; 25.0 cm^3

2.23 (a) 3; (b) 3; (c) 3; (d) an integer, infinite; (e) 2; (f) an exact number, infinite

2.25 (a) 6.60 mL; (b) 26.0 mL

2.27 4.4 g

2.29 1.64 g

2.31 3

2.33 The precision is very good; there is very little variation between the measurements. The accuracy is not as good, because there is a difference between the experimental results (all low) and the accepted value.

2.35 10^{-2} mol

2.37 (a) 9.5×10^{-15} mol people; (b) 3.4×10^6 years

2.39 52.0 g/mol; chromium, Cr

2.41 12.01 g/mol

2.43 79.91 g/mol

2.45 (a) 8.01×10^{-2} mol ^{35}Cl; (b) 3.49×10^{-2} mol Cu; (c) 1.84 mol He; (d) 1.60×10^{-7} mol Fe

2.47 (a) 2.39×10^{24} atoms; (b) 2.45×10^{17} atoms; (c) 1.11×10^{12} atoms; (d) 2.28×10^{20} atoms

2.49 (a) 59 g; (b) 14 g

2.51 (a) 199.90 g/mol; (b) 114.22 g/mol; (c) 262.86 g/mol; (d) 44.01 g/mol; (e) 16.04 g/mol

2.53 (a) Two-thirds of the atoms in N_2O_4 are O atoms. (b) 5.0 mol O_2 was produced. (c) 1.000 mol Cl_2 contains 6.022×10^{23} chlorine molecules.

2.55 (a) 0.0650 mol, 3.92×10^{22} molecules CCl_4; (b) 1.29×10^{-5} mol, 7.77×10^{18} molecules HI; (c) 1.18×10^{-4} mol, 7.11×10^{19} molecules N_2H_4; (d) 1.46 mol, 8.77×10^{23} molecules sucrose; (e) 0.146 mol, 8.77×10^{22} O atoms, 4.38×10^{22} O_2 molecules

2.57 (a) 0.0140 mol Ag^+; (b) 2.10 mol UO_3; (c) 7.75×10^{-5} mol Cl^-; (d) 5.89×10^{-3} mol H_2O

2.59 (a) 4.03×10^{23} FU of $AgNO_3$; (b) 1.06×10^2 mg Rb_2SO_4; (c) 5.90×10^{25} FU of $NaHCO_2$ (FU = formula units)

2.61 (a) 3.5×10^{-6} mol testosterone; (b) 79.1% C, 9.8% H, 11.1% O

2.63 (a) 2.99×10^{-23} g H_2O; (b) 3.34×10^{22} H_2O molecules

2.65 (a) 0.0186 mol $CuBr_2 \cdot 4H_2O$; (b) 0.0372 mol Br^-; (c) 4.48×10^{22} H_2O molecules; (d) 0.215

2.67 (a) molecular: $C_4H_6Cl_2$; empirical: C_2H_3Cl;
(b) molecular: $C_3H_8O_3$; empirical: $C_3H_8O_3$
2.69 (a) Na_3AlF_6;   (b) $KClO_3$;   (c) $NH_4H_2PO_4$
2.71 PCl_5
2.73 $C_6H_6Cl_6$
2.75 $C_8H_{10}N_4O_2$
2.77 (a) kg;   (b) pm;   (c) g;   (d) μm
2.79 (a) 1 atom;   (b) $1 \text{ g}/1.6605 \times 10^{-24}$ g $=$
6.0223×10^{23}; the Avogadro constant
2.81 42.2 km
2.83 2×10^{-2} cm/s
2.85 400. K, 261°F; 90. K, -297°F
2.87 2.54 g Cu
2.89 $C_9H_{11}NO_4$
2.91 (a) new temperature/°X $=$
$[2 \times$ (Celsius temperature/°C)$] + 50$;   (b) 94°X
2.93 ^{10}B $=$ 26% abundance, ^{11}B $=$ 74% abundance
2.95 $C_{52}H_{81}N_6O_{14}$
2.97 (a) 0.6800 g C;   (b) 0.072 66 g H;   (c) 0.2473 g O;
68.00% C, 7.266% H, 24.73% O;   (d) $C_{11}H_{14}O_3$;
(e) $C_{22}H_{28}O_6$
2.99 (a) 63.4 g/mol;   (b) copper(I) oxide
2.101 (a) 12.010 g/mol;   (b) 35.48 g/mol
2.103 (a) $Cu(HSO_4)_2$;   (b) Cu^{2+} and $HSO_4{}^-$;
(c) 1 Cu^{2+} ion and 2 $HSO_4{}^-$ ions;   (d) 1 Cu atom, 2 H
atoms, 2 S atoms, and 8 O atoms;   (e) 4 mol Cu,
8 mol each H and S, 32 mol O;   (f) 257.68 g/mol;
(g) 8.06×10^{21} FU;   (h) 49.67%;   (i) 61.54%. The
answers in (h) and (i) differ because (h) asks about the
mass of oxygen; the masses of the other elements differ
from that of oxygen. Part (i) asks for the percentage of
atoms in the compound; all atoms count equally.

Chapter 3

3.1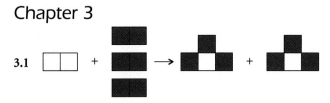

3.3 (a) $P_4O_{10}(s) + 6 H_2O(l) \rightarrow 4 H_3PO_4(l)$
(b) $Cd(NO_3)_2(aq) + Na_2S(aq) \rightarrow CdS(s) + 2 NaNO_3(aq)$
(c) $4 KClO_3(s) \xrightarrow{\Delta} 3 KClO_4(s) + KCl(s)$
(d) $2 HCl(aq) + Ca(OH)_2(aq) \rightarrow CaCl_2(aq) + 2 H_2O(l)$
3.5 (a) $2 Na(s) + 2 H_2O(l) \rightarrow H_2(g) + 2 NaOH(aq)$
(b) $Na_2O(s) + H_2O(l) \rightarrow 2 NaOH(aq)$
(c) $6 Li(s) + N_2(g) \rightarrow 2 Li_3N(s)$
(d) $Ca(s) + 2 H_2O(l) \rightarrow Ca(OH)_2(aq) + H_2(g)$
3.7 1st stage: $3 Fe_2O_3(l) + CO(g) \rightarrow 2 Fe_3O_4(l) + CO_2(g)$
2nd stage: $Fe_3O_4(l) + 4 CO(g) \rightarrow 3 Fe(l) + 4 CO_2(g)$
3.9 engine: $N_2(g) + O_2(g) \rightarrow 2 NO(g)$
atmosphere: $2 NO(g) + O_2(g) \rightarrow 2 NO_2(g)$

3.11 $4 HF(aq) + SiO_2(s) \rightarrow SiF_4(g) + 2 H_2O(l)$
3.13 First, identify all soluble ionic compounds. Then
write the complete ionic equation by writing all the soluble
ionic compounds in ionic form. Cancel all ions common to
both sides of the reaction. What remains is the net ionic
equation.
3.15 (a) soluble; all nitrates are soluble;
(b) slightly soluble;   (c) soluble; all nitrates are soluble;
(d) soluble; Group 1 compounds are soluble
3.17 (a) $Na^+(aq), I^-(aq)$;   (b) insoluble, but
very slight amounts of $Ag^+(aq)$ and $CO_3{}^{2-}(aq)$ form;
(c) $NH_4{}^+(aq), PO_4{}^{3-}(aq)$;   (d) $Fe^{2+}(aq), SO_4{}^{2-}(aq)$
3.19 (a) $Fe^{3+}(aq) + 3 OH^-(aq) \rightarrow Fe(OH)_3(s); Na^+, Cl^-$
(b) $Ag^+(aq) + I^-(aq) \rightarrow AgI(s); K^+, NO_3{}^-$
(c) $Pb^{2+}(aq) + SO_4{}^{2-}(aq) \rightarrow PbSO_4(s); K^+, NO_3{}^-$
(d) $Pb^{2+}(aq) + CrO_4{}^{2-}(aq) \rightarrow PbCrO_4(s); Na^+, NO_3{}^-$
(e) $Hg_2{}^{2+}(aq) + SO_4{}^{2-}(aq) \rightarrow Hg_2SO_4(s); K^+, NO_3{}^-$
3.21 (a) $(NH_4)_2CrO_4(aq) + BaCl_2(aq) \rightarrow$
$$BaCrO_4(s) + 2 NH_4Cl(aq)$$
$2 NH_4{}^+(aq) + CrO_4{}^{2-}(aq) + Ba^{2+}(aq) + 2 Cl^-(aq) \rightarrow$
$$BaCrO_4(s) + 2 NH_4{}^+(aq) + 2 Cl^-(aq)$$
$Ba^{2+}(aq) + CrO_4{}^{2-}(aq) \rightarrow BaCrO_4(s)$
$NH_4{}^+$ and Cl^- are spectator ions.
(b) $CuSO_4(aq) + Na_2S(aq) \rightarrow CuS(s) + Na_2SO_4(aq)$
$Cu^{2+}(aq) + SO_4{}^{2-}(aq) + 2 Na^+(aq) + S^{2-}(aq) \rightarrow$
$$CuS(s) + 2 Na^+(aq) + SO_4{}^{2-}(aq)$$
$Cu^{2+}(aq) + S^{2-}(aq) \rightarrow CuS(s)$
Na^+ and $SO_4{}^{2-}$ are spectator ions.
(c) $3 FeCl_2(aq) + 2 (NH_4)_3PO_4(aq) \rightarrow$
$$Fe_3(PO_4)_2(s) + 6 NH_4Cl(aq)$$
$3 Fe^{2+}(aq) + 6 Cl^-(aq) + 6 NH_4{}^+(aq) + 2 PO_4{}^{3-}(aq) \rightarrow$
$$Fe_3(PO_4)_2(s) + 6 NH_4{}^+(aq) + 6 Cl^-(aq)$$
$3 Fe^{2+}(aq) + 2 PO_4{}^{3-}(aq) \rightarrow Fe_3(PO_4)_2(s)$
$NH_4{}^+$ and Cl^- are spectator ions.
(d) $K_2C_2O_4(aq) + Ca(NO_3)_2(aq) \rightarrow$
$$2 KNO_3(aq) + CaC_2O_4(s)$$
$2 K^+(aq) + C_2O_4{}^{2-}(aq) + Ca^{2+}(aq) + 2 NO_3{}^-(aq) \rightarrow$
$$2 K^+(aq) + 2 NO_3{}^-(aq) + CaC_2O_4(s)$$
$Ca^{2+}(aq) + C_2O_4{}^{2-}(aq) \rightarrow CaC_2O_4(s)$
K^+ and $NO_3{}^-$ are spectator ions.
(e) $NiSO_4(aq) + Ba(NO_3)_2(aq) \rightarrow Ni(NO_3)_2(aq) + BaSO_4(s)$
$Ni^{2+}(aq) + SO_4{}^{2-}(aq) + Ba^{2+}(aq) + 2 NO_3{}^-(aq) \rightarrow$
$$Ni^{2+}(aq) + 2 NO_3{}^-(aq) + BaSO_4(s)$$
$Ba^{2+}(aq) + SO_4{}^{2-}(aq) \rightarrow BaSO_4(s)$
Ni^{2+} and $NO_3{}^-$ are spectator ions.
3.23 (a) $Pb^{2+}(aq) + 2 ClO_4{}^-(aq) + 2 Na^+(aq) +$
$2 Br^-(aq) \rightarrow PbBr_2(s) + 2 Na^+(aq) + 2 ClO_4{}^-(aq)$
$Pb^{2+}(aq) + 2 Br^-(aq) \rightarrow PbBr_2(s)$
(b) $Ag^+(aq) + NO_3{}^-(aq) + NH_4{}^+(aq) + Cl^-(aq) \rightarrow$
$$AgCl(s) + NH_4{}^+(aq) + NO_3{}^-(aq)$$
$Ag^+(aq) + Cl^-(aq) \rightarrow AgCl(s)$

(c) $2 Na^+(aq) + 2 OH^-(aq) + Cu^{2+}(aq) + 2 NO_3^-(aq) \rightarrow$
$\qquad Cu(OH)_2(s) + 2 Na^+(aq) + 2 NO_3^-(aq)$
$Cu^{2+}(aq) + 2 OH^-(aq) \rightarrow Cu(OH)_2(s)$

3.25   (a) $Pb^{2+}(aq) + SO_4^{2-}(aq) \rightarrow PbSO_4(s)$
(b) $Cu^{2+}(aq) + S^{2-}(aq) \rightarrow CuS(s)$
(c) $Co^{2+}(aq) + CO_3^{2-}(aq) \rightarrow CoCO_3(s)$
(d) for (a), $Pb(NO_3)_2$, Na_2SO_4 [spectators Na^+, NO_3^-]
for (b), $Cu(NO_3)_2$, Na_2S [spectators Na^+, NO_3^-]
for (c), $Co(NO_3)_2$, Na_2CO_3 [spectators Na^+, NO_3^-]

3.27 Ag^+, Zn^{2+}

3.29   Acids are molecules or ions that contain hydrogen and produce hydronium ions, H_3O^+, in water. Bases are molecules or ions that produce hydroxide ions, OH^-, in water. A base does not need to contain the hydroxide ion.

3.31   (a) base;   (b) acid;   (c) base;   (d) acid;
(e) base

3.33   (a) $HCl(aq) + NaOH(aq) \rightarrow H_2O(l) + NaCl(aq)$
$H_3O^+(aq) + Cl^-(aq) + Na^+(aq) + OH^-(aq) \rightarrow$
$\qquad 2 H_2O(l) + Na^+(aq) + Cl^-(aq)$
$H_3O^+(aq) + OH^-(aq) \rightarrow 2 H_2O(l)$
(b) $NH_3(aq) + HNO_3(aq) \rightarrow NH_4NO_3(aq)$
$NH_3(aq) + H_3O^+(aq) + NO_3^-(aq) \rightarrow$
$\qquad NH_4^+(aq) + NO_3^-(aq) + H_2O(l)$
$NH_3(aq) + H_3O^+(aq) \rightarrow NH_4^+(aq) + H_2O(l)$
(c) $CH_3NH_2(aq) + HI(aq) \rightarrow CH_3NH_3I(aq)$
$CH_3NH_2(aq) + H_3O^+(aq) + I^-(aq) \rightarrow$
$\qquad CH_3NH_3^+(aq) + I^-(aq) + H_2O(l)$
$CH_3NH_2(aq) + H_3O^+(aq) \rightarrow CH_3NH_3^+(aq) + H_2O(l)$

3.35   (a) HBr and KOH
$HBr(aq) + KOH(aq) \rightarrow KBr(aq) + 2 H_2O(l)$
$H_3O^+(aq) + OH^-(aq) \rightarrow 2 H_2O(l)$
(b) HNO_2 and $Ba(OH)_2$
$2 HNO_2(aq) + Ba(OH)_2(aq) \rightarrow Ba(NO_2)_2(aq) + 2 H_2O(l)$
$H_3O^+(aq) + OH^-(aq) \rightarrow 2 H_2O(l)$
(c) HCN and $Ca(OH)_2$
$2 HCN(aq) + Ca(OH)_2(aq) \rightarrow Ca(CN)_2(aq) + 2 H_2O(l)$
$HCN(aq) + OH^-(aq) \rightarrow CN^-(aq) + H_2O(l)$
(d) H_3PO_4 and KOH
$H_3PO_4(aq) + 3 KOH(aq) \rightarrow K_3PO_4(aq) + 3 H_2O(l)$
$H_3PO_4(aq) + 3 OH^-(aq) \rightarrow 3 H_2O(l) + PO_4^{3-}(aq)$

3.37   (a) $CH_3NH_2(aq)$ [base], $H_3O^+(aq)$ [acid];
(b) $C_2H_5NH_2(aq)$ [base], $HCl(aq)$ [acid];   (c) $HI(aq)$
[acid], $CaO(s)$ [base]

3.39   (a) basic;   (b) acidic;   (c) acidic;   (d) basic

3.41   Oxidation is electron loss. Reduction is electron gain.

3.43   (a) $2 P(s) + 3 Br_2(l) \rightarrow 2 PBr_3(s)$
(b) $2 Fe^{2+}(aq) + Sn^{4+}(aq) \rightarrow 2 Fe^{3+}(aq) + Sn^{2+}(aq)$
(c) $8 H_2(g) + S_8(s) \rightarrow 8 H_2S(g)$
(d) $2 NO(g) + O_2(g) \rightarrow 2 NO_2(g)$

3.45   (a) $Mg(s) + Cu^{2+}(aq) \rightarrow Mg^{2+}(aq) + Cu(s)$
(b) $2 Fe^{2+}(aq) + Pb^{4+}(aq) \rightarrow 2 Fe^{3+}(aq) + Pb^{2+}(aq)$

(c) $H_2(g) + Cl_2(g) \rightarrow 2 HCl(g)$
(d) $4 Fe(s) + 3 O_2(g) \rightarrow 2 Fe_2O_3(s)$

3.47   The oxidation number of an element is a number assigned on the basis of a set of rules and is used to monitor whether an element has been oxidized or reduced.

3.49   (a) +4;   (b) +1;   (c) +2;   (d) +5;   (e) +4;
(f) −2

3.51   (a) +7;   (b) +2;   (c) +6;   (d) +6;   (e) +6

3.53   (a) This is a substitution reaction; no oxidation or reduction occurs.   (b) BrO_3^- is reduced and Br^- is oxidized.   (c) F_2 is reduced and water is oxidized.

3.55   Oxidizing agents contain an element that gains electrons and decreases its oxidation number. Reducing agents lose electrons to the oxidizing agent. A loss of electrons means an increase in oxidation number for an element in the reducing agent.

3.57   (a) Cl_2 is stronger because it has a more positive oxidation number than Cl^-.   (b) N_2O_5 is stronger because the nitrogen has a more positive oxidation number than the nitrogen in N_2O.

3.59   (a) Zn is the reducing agent; HCl is the oxidizing agent.   (b) H_2S is the reducing agent; SO_2 is the oxidizing agent.   (c) Mg is the reducing agent; B_2O_3 is the oxidizing agent.

3.61   (a) oxidizing agent;   (b) oxidizing agent;
(c) reducing agent;   (d) reducing agent

3.63 $2 NaCl(l) \xrightarrow{\text{electrolysis at } 600°C} 2 Na(l) + Cl_2(g)$
Therefore, chlorine is produced by oxidation and sodium by reduction.

3.65   (a) strong acid;   (b) base;   (c) base;
(d) soluble ionic;   (e) weak acid;   (f) insoluble ionic;
(g) insoluble ionic;   (h) strong acid

3.67   (a) HNO_3, $Ba(OH)_2$;   (b) H_2SO_4, NaOH;
(c) $HClO_4$, KOH;   (d) HCl, CsOH

3.69   (a) nonelectrolyte;   (b) strong electrolyte;
(c) strong electrolyte

3.71   (a) acid-base neutralization; HCl, acid; $Mg(OH)_2$, base;   (b) acid-base neutralization; H_2SO_4, acid; $Ba(OH)_2$, base; or precipitation; $Ba^{2+}(aq) + SO_4^{2-}(aq) \rightarrow BaSO_4(s)$;
(c) redox; O_2 is the oxidizing agent, SO_2 is the reducing agent

3.73   (a) redox; I_2O_5 is the oxidizing agent, CO is the reducing agent;   (b) redox; I_2 is the oxidizing agent, $S_2O_3^{2-}(aq)$ is the reducing agent;   (c) precipitation; $Ag^+(aq) + Br^-(aq) \rightarrow AgBr(s)$;   (d) redox; UF_4 is the oxidizing agent, Mg is the reducing agent

3.75 $2 C_8H_{18}(l) + 25 O_2(g) \rightarrow 16 CO_2(g) + 18 H_2O(g)$

3.77 $4 C_{10}H_{15}N(s) + 55 O_2(g) \rightarrow$
$\qquad 40 CO_2(g) + 30 H_2O(l) + 2 N_2(g)$

3.79   (a) $-\frac{1}{2}$;   (b) −1;   (c) −1;   (d) −1;   (e) $-\frac{1}{3}$

3.81 $2 CO_2(g) + CaSiO_3(s) + H_2O(l) \rightarrow$
$\qquad SiO_2(s) + Ca(HCO_3)_2(aq)$

3.83 NaOH, lowest molar mass

3.85 $H_2S(g) + 2 NaOH(aq) \rightarrow Na_2S(aq) + 2 H_2O(l)$
$4 H_2S(g) + Na_2S(alc) \rightarrow Na_2S_5(alc) + 4 H_2(g)$
$10 H_2O(l) + 9 O_2(g) + 2 Na_2S_5(alc) \rightarrow$
$\qquad\qquad\qquad 2 Na_2S_2O_3 \cdot 5H_2O + 6 SO_2(g)$

3.87 (a) CHO_2; (b) $H_2C_2O_4$; (c) $H_2C_2O_4(aq) +$
$2 NaOH(aq) \rightarrow Na_2C_2O_4(aq) + 2 H_2O(l); H_2C_2O_4(aq) +$
$2 OH^-(aq) \rightarrow C_2O_4^{2-}(aq) + 2 H_2O(l)$; (d) oxalic acid;
water and sodium oxalate; (e) neutralization

3.89 (a) potassium ion, chloride ion; (b) copper(II) ion,
chloride ion; (c) silver ion, nitrate ion

3.91 (a)

Group	Maximum oxidation number	Minimum oxidation number
1	+1	0
2	+2	0
13	+3	0
14	+4	−4
15	+5	−3
16	+6	−2
17	+7	−1

(b) Going to the right across the periodic table, the
maximum oxidation number increases by one with each
group and the minimum oxidation number increases in
the same way from Group 14 to Group 17.

Chapter 4

4.1 (a) 11.0 mol NO_2; (b) 8.4×10^{-3} mol MnO_4^-

4.3 9.0 mol CO_2

4.5 5.50 mol H_2O

4.7 (a) 0.088 mol H_2O; (b) 329 g O_2

4.9 (a) 205 g CO_2; (b) 0.133 mol H_2O

4.11 (a) 18.9 g Al_2O_3; (b) 8.90 g O_2

4.13 (a) 5.7×10^3 kg Al; (b) 9.448×10^3 kg Al_2O_3

4.15 (a) 2.8×10^3 g H_2O; (b) 7.3 g O_2

4.17 1.1×10^3 g H_2O

4.19 86.7%

4.21

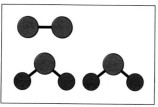

4.23 O_2 is the limiting reactant.

4.25 (a) H_2S is the limiting reactant. (b) 5.7 g SO_2;
(c) 8.6 g S, 3.2 g H_2O; (d) 11.4 g SO_2 + 6.08 g H_2S =
17.5 g of reactants and 5.7 g SO_2 + 8.6 g S + 3.2 g H_2O =
17.5 g of excess reactant and products; so everything
checks with the law of the conservation of mass.

4.27 (a) Cl_2 is the limiting reactant.
(b) 200. g $AlCl_3$

4.29 $C_7H_6O_2$

4.31 C_6H_7N

4.33 Empirical and molecular formulas are both
$C_{14}H_{18}N_2O_5$.

4.35 empirical formula: $C_4H_5N_2O$, molecular formula:
$C_8H_{10}N_4O_2$; the combustion equation is $2 C_8H_{10}N_4O_2(s) +$
$19 O_2(g) \rightarrow 16 CO_2(g) + 10 H_2O(g) + 4 N_2(g)$

4.37 (a) 6.268 mol/L; (b) 0.0241 mol/L

4.39 0.658 g $AgNO_3$

4.41 (a) 17 mL; (b) 29 mL; (c) 4.8 mL

4.43 (a) We would determine the mass of 0.010 mol of
$KMnO_4$ (1.58 g) and then dissolve the mass of $KMnO_4$ in
enough water to make 1.00 L of solution. (b) We dilute
the 0.050 M $KMnO_4$(aq) by a factor of 5; for example, we
could take 10 mL of 0.050 M $KMnO_4$(aq), put it into a
50-mL volumetric flask, and add water up to the
calibration mark.

4.45 (a) 4.51 mL; (b) 12.0 mL of NaOH is needed. A
60.0-mL volumetric flask is not normally available, but
a 100.0-mL flask can be found. So 20.0. mL of the initial
solution can be added to a 100.0-mL volumetric flask and
diluted with water to the mark; 60.0 mL of this solution
can then be used.

4.47 The stoichiometric point is the point at which the
exact amount of one reactant, the titrant, has been added
to complete the reaction with the other reactant, the
analyte, according to the balanced chemical equation. An
example is an acid-base reaction; $CH_3COOH(aq) +$
$NaOH(aq) \rightarrow NaCH_3CO_2(aq) + H_2O(l)$, when the same
number of moles of CH_3COOH and NaOH are present.

4.49 (a) 0.271 mol/L; (b) 0.163 g NaOH

4.51 (a) 0.2087 M HNO_3; (b) 0.3289 g HNO_3

4.53 63.0 g/mol

4.55 The theoretical yield is the maximum quantity of
product(s) that can be obtained, according to the reaction
stoichiometry, from a given quantity of a specified reactant
(the limiting reactant). The percentage yield of a product is
the percentage of its theoretical yield that is actually
achieved. It is calculated as

$$\text{percentage yield} = \frac{\text{actual yield}}{\text{theoretical yield}} \times 100\%$$

The percentage yield may be less than 100% for a specified
product, because alternate reactions may take place in
addition to the desired reaction, forming products other
than the desired product(s).

4.57 (a) 3.34 mol H_2O_2; (b) 5.69×10^{-2} mol HNO_3;
(c) 3.00×10^{-2} g H_2O

4.59 (a) 4.482 mg Mn; (b) 53.24%

4.61 (a) 5.56×10^{-3} L O_2; (b) 2.6×10^{-2} L air

4.63 (a) Two molecules of ammonia can form, two molecules of N_2 remain unreacted. (b) H_2 is the limiting reactant.

4.65 3.6×10^{15} kg
4.67 (a) 8.0 g $CuSO_4$; (b) 12 g $CuSO_4 \cdot 5H_2O$
4.69 Seven sandwiches can be made.
4.71 (a) 0.0123 mol $(NH)_3PO_4 \cdot 3H_2O$; 7.07×10^{-3} mol K_3PO_4; 27.7 mol H_2O; (b) 0.0194 mol PO_4^{3-}; (c) 1.84 g; (d) 501 g H_2O
4.73 (a) This error would have no effect.
(b) This error would show a concentration that is too high.
(c) This error would show a concentration that is too high.
4.75 18.4 mol/L
4.77 440. kg CO_2
4.79 (a) 85.63% C; (b) 59.96% C; (c) 83.90% C;
(d) $C_2H_4(g) + 3 O_2(g) \rightarrow 2 CO_2(g) + 2 H_2O(g)$
$2 C_3H_7OH(l) + 9 O_2(g) \rightarrow 6 CO_2(g) + 8 H_2O(g)$
$C_7H_{16}(l) + 11 O_2(g) \rightarrow 7 CO_2(g) + 8 H_2O(g)$
(e) The C_2H_4 produces the greatest amount of CO_2 per gram of fuel.
4.81 (a) 8.70×10^{-4} mol H_2SO_4; (b) 0.434%
4.83 (a) 0.1185 M I_3^-(aq); (b) 0.0412 M HCN(aq)
4.85 (a) H_2S is the limiting reactant. (b) The oxidizing agent is SO_2; the reducing agent is H_2S.
4.87 (a) This is a redox reaction. (b) Sb is the limiting reactant. (c) 1.2 mol O_2; (d) 2.5 mol Sb_2O_3;
(e) 80.%
4.89 (a) The reactant compound is $SnCl_2$ and the product compound is $SnCl_4$. (b) I_3^-(aq) + Sn^{2+}(aq) $\rightarrow$ 3 I^-(aq) + Sn^{4+}(aq); (c) Sn in $SnCl_2$ is +2; Sn in $SnCl_4$ is +4

Connection 1

1. 0.502 g HCl. The advertising claim is apparently not correct. However, it is possible that the active ingredient is only a small part of the tablet. To test whether the active ingredient itself can neutralize 40 times its mass of stomach acid, determine the mass of this ingredient.
3. (a) empirical: $C_{13}H_{17}O_7$ molecular: $C_{26}H_{34}O_{14}$;
(b) The baptisin has absorbed water as waters of hydration. The new formula is $C_{26}H_{34}O_{14} \cdot 9H_2O$.
5. Discard $Fe(CN)_2$, PbI_2, and CaC_2O_4—all have toxic components (CN^-, Pb^{2+}, $C_2O_4^{2-}$). For the remaining compounds, $FeSO_4$, 7 H_2O, KI, and $CaCO_3$ have the lowest cost per gram of nutrient. Assuming RDAs of 15 mg Fe, 150 μg I and 1200 mg Ca, your supplement needs 0.075 g $FeSO_4 \cdot 7H_2O$ at a cost of $0.0090, 2.0×10^{-4} g KI at a cost of 5.6×10^{-5}, and 3.0 g $CaCO_3$ at a cost of $0.60. Total price: $0.61. Supplement price: $2.44.

Chapter 5

5.1 Pressure is a force exerted divided by the area of surface. Air pressure exerted on the surface of the mercury in the dish (see Fig. 5.5) is transmitted through the liquid mercury and supports the mercury column. Equilibrium requires that the pressure at the base of the column of mercury is balanced by the pressure on the surface of the mercury in the dish. Thus, the height of the column is proportional to the pressure. The space above the mercury is a vacuum, so it adds no pressure. A column of mercury 760 mm high represents 1 atm. As in a thermometer, calibration makes it a true measuring device.
5.3 (a) 0.987 atm; (b) 1.33×10^2 Pa;
(c) 9.87×10^{-3} atm
5.5 (a) 8×10^1 kbar; (b) 8×10^9 Pa
5.7 9.51×10^2 cm
5.9 (a) 7.24×10^{-3} L; (b) 13.2 mL
5.11 (a) 1.0×10^2 Pa; (b) 3.6×10^3 Torr
5.13 (a) (1) 42.0 inHg, (2) 60.0 inHg; (b) 22.4 inches
5.15 194 mL
5.17 (a) 29°C; (b) 5°C
5.19 1.6 atm
5.21 46°C
5.23 The temperature must be doubled.
5.25 Boyle, Charles, and Avogadro demonstrated the following proportionalities:

Boyle: $V \propto \dfrac{1}{P}$ or $P \propto \dfrac{1}{V}$ at constant n and T

Charles: $V \propto T$ at constant n and P
Avogadro: $V \propto n$ at constant P and T
These proportions can be combined into one equation by introducing a constant of proportionality, R. This combination results in the ideal gas law, $PV = nRT$.
5.27 (a) 4.5×10^{-3} mol; (b) 1.3×10^{-2} Torr;
(c) 3.94×10^4 L
5.29 (a) 0.90 g; (b) 1.7×10^{14} atoms
5.31 6.0×10^{-3} g NH_3
5.33 (a) 0.361 g; (b) 0.349 g
5.35 632 Torr
5.37 0.377 L
5.39 (a) 24.4 L; (b) 9.31 L; (c) 73.4 L;
(d) 5.65 mL
5.41 (a) 3.21 g; (b) 6.35 mg; (c) 3.94 kg;
(d) 2.59 mg

5.43 0.299 L

5.45 (a) 127 mL; (b) 5.18×10^{-3} mol; (c) 85%

5.47 (a) 2.15×10^6 L; (b) 2.43×10^6 L

5.49 12.3 L CO_2, 24.7 L NH_3

5.51 Conditions (b) would produce the larger volume of CO_2 by combustion because the volume of CH_4 in system (b) is larger.

5.53 NO_2 (c) is the most dense.

5.55 (a) 4.88 g/L; (b) 3.90 g/L

5.57 (a) 3.18 g/L; (b) 77.9 g/mol

5.59 (a) 130. g/mol; (b) 5.30 g/L

5.61 (a) mole fraction of HCl: 0.9; mole fraction of benzene: 0.1; (b) pressure HCl: 0.7 atm, pressure benzene: 0.08 atm

5.63 (a) 0.90 atm N_2, 3.26 atm O_2, 4.16 atm total; (b) 0.67 atm H_2, 1.3 atm NH_3, 0.047 atm He, 2.0 atm total

5.65 1.56×10^{-2} g H_2O

5.67 (a) 2.3×10^2 kPa N_2, 33.7 kPa Ar; (b) 2.63×10^2 kPa

5.69 8.42×10^{-2} g $KClO_3$

5.71 (a) argon; (b) H_2

5.73 1.2×10^3 g/mol

5.75 Boyle's law states that pressure and volume are inversely proportional. As we decrease the volume, the distance that a gas molecule must travel prior to collision with the container wall decreases. It follows that more collisions will occur per unit time. Because pressure is the collective effect of these collisions, it should increase as the volume decreases.

5.77 (a) 1.84×10^3 m/s; (b) 238 m/s; (c) 1.36×10^3 m/s

5.79 (a) The rms speed increases by a factor of 1.12. (b) Air at room temperature has a rms speed that is 1.09 times that of the colder air.

5.81 Gases behave most ideally at high temperatures and low pressures. These conditions minimize the opportunity for interactions between molecules, that is, attractive and repulsive forces, which are the cause of deviations from ideality. The farther apart molecules are, the less interaction there will be.

5.83 Figure 5.30 shows this relationship. As two molecules approach, the strength of attraction increases until they are close enough to touch, then they repel each other strongly. An energy of interaction curve shows this effect as a decrease to a minimum value followed by a sharp increase in intermolecular energy.

5.85 (a) Intermolecular attractions in C_2H_4 result in its having a lower pressure than that of an ideal gas. Thus the ideal gas vessel has the greater pressure. (b) The free space is almost equal. Although the molecules of C_2H_4 do take up some space, this space is a small percentage of the total volume.

5.87 (a) 10.0 atm; (b) 399 atm; (c) 10.5 atm CO_2, 30.6 atm He; (d) 5% error for CO_2, 23% error for helium

5.89 Use the equation $V_2 = (T_2/T_1)(P_1/P_2)V_1$ to perform this conversion.

5.91 (a) number of atoms (Cl_2 is diatomic); (c) mass (Cl_2 has the higher molar mass); (d) average molecular speed (He atoms have higher average speeds); (e) density (Cl_2 has the greater density under the same conditions)

5.93 2.8 Torr

5.95 3.2×10^8 L

5.97 59 L

5.99 1.1×10^2 Torr

5.101 (a) 2.45 g CO; (b) 5.76 g/L; (c) Because the mass and volume are fixed quantities in this experiment, the density does not change.

5.103 (a) 154 s; (b) 123 s; (c) 33.0 s; (d) 186 s

5.105 4.38×10^2 m/s at $-50.°C$, 5.02×10^2 m/s at 20.°C.

5.107 (a) Hydrogen and helium molecules are very light. Consequently, they have speeds high enough to escape the Earth's gravitational pull. (b) The CO_2 of the Earth's atmosphere may have been consumed by growing plants, which use CO_2 in the synthesis of glucose.

5.109 $O_3 < NO_2 < O_2 < NO$

5.111 (a) 4 mol; (b) 4 μL; (c) 4 cm^3; (d) 3×10^{-4} Torr

5.113 (a) 2.95 kg CO_2; (b) 0.398 atm

5.115 0.481 M NaOH(aq)

5.117 C_3H_6

5.119 110. g/mol, C_8H_{12}

5.121 SO_2Cl_2

5.123 N_2H_4

5.125 (a) 146 g; (b) 0.732 M H_2SO_3(aq)

Chapter 6

6.1 (a) isolated; (b) closed; (c) closed

6.3 67 kJ

6.5 (a) Heat is absorbed by the system; q is positive; the system looks like it contracts, so work is done on the system; w is positive. (b) Heat is released by the system; q is negative; the system looks like it contracts, so work is done on the system; w is positive.

6.7 1.25×10^3 kJ

6.9 1.2×10^5 J

6.11 400. kJ

6.13 As the temperature of a system increases, the average velocity of the molecules in a system also increases.

6.15 -37 kJ

6.17 (a) endothermic ($\Delta H > 0$); (b) exothermic ($\Delta H < 0$)

6.19 -1.3 kJ, exothermic

6.21 (a) 625 J; (b) 5.4×10^2 °C

6.23　1.60×10^5 J; 88%

6.25　(a) 2.1×10^3 J;　(b) 35 g

6.27　0.53 J/g · °C

6.29　13.93 kJ/°C

6.31　-3.8 kJ

6.33　(a) -1.42×10^3 J;　(b) -57 kJ/mol

6.35　(a) 8.21 kJ/mol;　(b) 43.5 kJ/mol

6.37　(a) $+226$ kJ;　(b) $+199$ kJ

6.39　22 kJ

6.41　Heating curve for bromine:

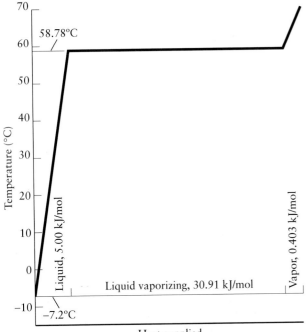

6.43　The standard state of a substance is its pure form at 1 atmosphere pressure. Standard state data are usually reported at 25°C, but they can be reported for any temperature.

6.45　(a) $+72$ kJ;　(b) $+149$ kJ;　(c) 184 g

6.47　(a) 2.5×10^2 g octane;　(b) -1.3×10^8 J

6.49　-197.78 kJ

6.51　-1775 kJ

6.53　-312 kJ

6.55　-184.7 kJ

6.57　$+1.90$ kJ/mol

6.59　48.44 kJ/g; 3.3×10^4 kJ/L

6.61　24.75 kJ/g for Mg, 31.05 kJ/g for Al; so Al is better.

6.63　(a) $K(s) + \frac{1}{2} Cl_2(g) + \frac{3}{2} O_2(g) \rightarrow KClO_3(s)$, $\Delta H_f^\circ = -397.73$ kJ/mol;　(b) $\frac{5}{2} H_2(g) + \frac{1}{2} N_2(g) + 2 C(graphite) + O_2(g) \rightarrow H_2NCH_2COOH(s)$, $\Delta H_f^\circ = -532.9$ kJ/mol;　(c) $2 Al(s) + \frac{3}{2} O_2(g) \rightarrow Al_2O_3(s)$, $\Delta H_f^\circ = -1675.7$ kJ/mol

6.65　(a) -15.4 kJ;　(b) -128.5 kJ

6.67　(a) $+8.77$ kJ;　(b) -233.57 kJ;　(c) -905.48 kJ

6.69　$+11.3$ kJ/mol

6.71　-444 kJ/mol

6.73　(a) The internal energy of an open system can be increased by adding matter, heating it, or by doing work on the system.　(b) None of these methods can be used to increase the internal energy of an isolated system because neither matter nor energy can be exchanged in an isolated system.

6.75　(a) ΔU, internal energy, is the sum of all kinetic and potential energies of all atoms and molecules in a sample. Internal energy is identified with the heat supplied at constant volume. ΔH, enthalpy, is identified with the heat supplied at constant pressure.　(b) ΔH equals ΔU when $P\Delta V$ work is 0; in other words, $\Delta H = \Delta U$ when there is no volume change at constant pressure.

6.77　(a) true only when $P\Delta V$ work is 0;　(b) always true;　(c) always false;　(d) true only when $P\Delta V$ work is 0; when this is true, however, $q = 0$ as well, so $\Delta U = 0$. (e) true, because $\Delta U = q + w$ and $q = 0$

6.79　(a) -6.81 kJ;　(b) -6271.0 kJ/mol; (c) -6277.8 kJ

6.81　1.4 g

6.83　31.1°C

6.85　-277 kJ/mol

6.87　-16.4 kJ

6.89　-175 kJ

6.91　-22.25 kJ

6.93　5.54°C

6.95　30 min

6.97　(a) 1.9×10^5 kJ;　(b) 7.6×10^6 kJ

6.99　(a) Brand X, 5.3×10^3 J; Brand ABC, 5.1×10^3 J; (b) Brand X, 5.3×10^2 kJ; Brand ABC, 5.1×10^2 kJ; (c) Brand X, 130. Cal; Brand ABC, 120. Cal

6.101　$+23.9 \times 10^3$ kJ/L (higher than the enthalpy density of hydrogen or methanol, but not as high as that of octane)

6.103　(a) 14.7 mol H_2;　(b) 4.20×10^3 kJ

6.105　(a) $Cu(s) + 2 AgNO_3(aq) \rightarrow Cu(NO_3)_2(aq) + 2 Ag(s)$, oxidation-reduction reaction;　(b) -1.28 kJ; (c) exothermic

6.107　(a) $CO_2(g) \rightarrow C(s) + O_2(g)$;　(b) 5.7×10^2 kJ; (c) heated;　(d) CO_2

Chapter 7

7.1　(a) 4×10^{14} Hz to 7×10^{14} Hz;　(b) 1.2×10^6 Hz

7.3　(a) 420 nm;　(b) 150 pm

7.5　(a) 3.38×10^{-19} J;　(b) 2.03×10^5 J

7.7　(a) 2.3×10^{-15} J;　(b) 2.5×10^5 J

7.9　(a) rubidium;　(b) The kinetic energy drops to 0 at different frequencies because each metal requires that light of some threshold frequency or higher be present for an electron to be ejected. The frequency at which electrons can be ejected is characteristic of the metal; different metals have different frequencies.   (c) rubidium

7.11   (a) 4.10×10^{-7} m;   (b) violet

7.13   (a) 3.29×10^{15} Hz;   (b) 9.12×10^{-8} m;   (c) x-ray or gamma ray

7.15   (a) 4.8×10^{-11} m;   (b) 2.6×10^{-14} m

7.17 $\lambda = h/mv$. If v = velocity, and each person runs at the same speed, $\lambda \propto 1/m$, and the larger the mass, the shorter the wavelength. Therefore, the person weighing 80 kg has the shorter wavelength.

7.19   A hard sphere, such as a billiard ball, has a uniform density and its mass is confined within a fixed and well-defined region of space. In atoms, the electron density spreads out from the central nucleus in a nonuniform manner, producing an electron cloud that is neither hard nor confined to a limited region. The electron density varies with distance in the manner shown in Fig. 7.17.

7.21   (a) 2;   (b) 3;   (c) The values for l can be 0, 1, and 2.

7.23   (a) 1;   (b) 5;   (c) 3;   (d) 7

7.25   (a) 2;   (b) 5;   (c) 4;   (d) 5

7.27   (a) $3d$, 5;   (b) $1s$, 1;   (c) $6f$, 7;   (d) $2p$, 3

7.29   (a) 6;   (b) 2;   (c) 8;   (d) 2

7.31   (a) not allowed;   (b) allowed;   (c) not allowed;   (d) allowed

7.33   (a) allowed;   (b) not allowed; if $l = 0$, then m_l can only be 0;   (c) not allowed; n cannot equal l

7.35   (a) $n = 2, l = 1, m_l = 0, m_s = -\frac{1}{2}$;   (b) $n = 5$, $l = 2, m_l = +1, m_s = +\frac{1}{2}$

7.37   The average distance from the nucleus for a $2s$-electron is much less than for a $3s$-electron; consequently, the electrostatic force of attraction between the nucleus and the $2s$-electron is much greater than that for a $3s$-electron. The electron cloud of the $2s$-electron is bunched more densely around the nucleus than is the $3s$-electron cloud.

7.39   (a) $n = 1, 2, 3, 4$; energy increases from 1 to 4;   (b) $l = 0, 1, 2, 3$; energy increases from 1 to 3

7.41 $1s, 2p, 3s, 3d, 5d$; energy increases left to right

7.43   (a) Ca, $[Ar]4s^2$;   (b) N, $1s^22s^22p^3$;   (c) Br, $[Ar]3d^{10}4s^24p^5$;   (d) U, $[Rn]5f^36d^17s^2$

7.45   (a) Ni, $[Ar]3d^84s^2$;   (b) Cd, $[Kr]4d^{10}5s^2$;   (c) Pb, $[Xe]4f^{14}5d^{10}6s^26p^2$;   (d) Ag, $[Kr]4d^{10}5s^1$

7.47   (a) Fe^{2+}, $[Ar]3d^6$;   (b) Cl^-, $[Ne]3s^23p^6$;   (c) Tl^+, $[Xe]4f^{14}5d^{10}6s^2$

7.49   (a) Si, $[Ne]3s^23p^2$;   (b) Ne, $[He]2s^22p^6$;   (c) Cs, $[Xe]6s^1$;   (d) S, $[Ne]3s^23p^4$

7.51   (a) Group 2: ns^2;   (b) Group 18: ns^2np^6

7.53   The outermost electrons of an atom determine the atomic radius. Proceeding down a group, the outermost electrons occupy shells that lie farther and farther from the nucleus, so the size (radius) increases down a group.

7.55   (a) Group 1 (alkali metals);   (b) The radius of the cation is less than the radius of the neutral atom.   (c) Cations with the largest number of protons (among a set of cations with the same number of electrons) will have the smallest radius.

7.57   (a) S (size decreases from left to right);   (b) S^{2-} (same number of electrons; Cl^- has more protons, so it is smaller);   (c) Na (size decreases from left to right);   (d) Mg^{2+} (same number of electrons; Al^{3+} has more protons, so it is smaller)

7.59   First ionization energies decrease down a group because the outermost electron occupies a shell that is farther from the nucleus and is therefore less tightly bound. Ionization energies increase across a period because the effective nuclear charge increases as we go from left to right across a given period. As a result, the outermost electron is gripped more tightly and the ionization energies increase.

7.61   (a) Mg;   (b) N;   (c) P

7.63   The ionization energies for Group 16 elements are less than those for Group 15 elements. The group configurations are Group 16, ns^2np^4, and Group 15, ns^2np^3. The half-filled subshell of Group 15 is more stable than simple theory suggests. This stability makes removal of the electron more difficult; therefore, the atom has a higher ionization energy. In Group 16, the fourth p-electron pairs with another electron, thereby producing stronger electron repulsion, which makes it easier to remove this electron. Although there is the competing effect of increasing effective nuclear charge, the electron repulsion effect predominates.

7.65   Both the first and second electrons lost from Mg are $3s$. The second electron for sodium must be removed from a $2p$ level that not only is filled but also is considerably lower in energy than the $3s$ level. Thus the second ionization energy of sodium is very high and sodium exists only as Na^+, not Na^{2+}.

7.67   (a) Group 17;   (b) Electron affinities generally increase (left to right) across a period.

7.69   (a) Cl;   (b) O;   (c) Cl

7.71   Atomic radii increase and ionization energies decrease from top to bottom within a group. Atomic radii decrease and ionization energies increase (left to right) across a period.

7.73   A diagonal relationship is a similarity between diagonal neighbors in the periodic table, especially for elements at the left of the table. Two examples are (1) Li and Mg, which both burn in nitrogen to form the nitride; (2) Be and Al, which are amphoteric, reacting with both acids and bases.

7.75 (a) The reactivity of metals decreases in going from left to right across a period.   (b) The reactivity of metals increases in going from top to bottom of a main group.

7.77 (a) metal;   (b) nonmetal;   (c) metalloid;   (d) metalloid

7.79 (a) do;   (b) do;   (c) do, but not strongly;   (d) do, but not strongly

7.81 1×10^{10} Hz

7.83 1.025×10^{-10} m

7.85 (a) 36;   (b) 7;   (c) 10;   (d) 3

7.87 (b) is a p_y orbital

7.89 (a) unacceptable;   (b) unacceptable;   (c) acceptable;   (d) acceptable

7.91 (a) Group 15;   (b) Group 12;   (c) Group 1;   (d) Group 18;   (e) Group 8;   (f) Group 2

7.93 (a) Zr; $[Kr]4d^25s^2$ (e) Sb; $[Kr]4d^{10}5s^25p^3$
(b) Se; $[Ar]3d^{10}4s^24p^4$ (f) Pu; $[Rn]5f^67s^2$
(c) Rb; $[Kr]5s^1$ (g) Si; $[Ne]3s^23p^2$
(d) Cl; $[Ne]3s^23p^5$ (h) Ar; $[Ne]3s^23p^6$

7.95 (a) $n = 3; l = 1; m_l = +1; m_s = +\frac{1}{2}$;
(b) $n = 3; l = 0; m_l = 0; m_s = -\frac{1}{2}$

7.97 (a) S   (e) Ca   (i) N   (m) $4g$
(b) F^- (f) Fe   (j) Cl   (n) 4
(c) Cs   (g) Ba^{2+} (k) I   (o) 7
(d) F   (h) S^{2-} (l) K

7.99 (a) Ag; $[Kr]4d^{10}5s^1$; Cu; $[Ar]3d^{10}4s^1$; Au; $[Xe]4f^{14}5d^{10}6s^1$. Silver, copper, and gold each have one electron in a high s-orbital. This electron can be removed easily, making these metals good conductors.   (b) Group 1 metals are very reactive, especially with water (see Chapter 19), forming salts that do not conduct as solids.

7.101 (a) d block;   (b) s block;   (c) p block;   (d) d block;   (e) p block;   (f) d block

7.103 (a) Zn, Cd, and Hg lose two outermost, highest energy s-electrons to form Zn^{2+}, Cd^{2+}, and Hg^{2+}, just as the alkaline earth metals do. For example, Mg loses two outermost, highest energy s-electrons to form Mg^{2+}. Zn, Cd, and Hg's next highest level, d, is completely filled and a significant energy difference exists between the highest energy s- and d-electrons.   (b) Differences between Groups IIA and IIB may be explained by the effect of the filled d-orbitals in IIB elements. As we go across a transition row, the d-orbitals fill, and the elements become more similar to the p-block elements. Group IIB is at the end of the row, closest to the p block. "A" metals are more reactive than "B" metals because their ionization energies are lower; their outermost electrons can be lost more easily and thus they are more reactive.

7.105 The molar volume roughly parallels atomic size (volume), which increases as the s-sublevel begins to fill and subsequently decreases as the p-sublevel fills (refer to textbook discussion of periodic variation of atomic radii).

In the plot below, this effect is most clearly seen in passing from Ne (10) to Na (11) and Mg (12), then to Al (13) and Si (14). Ne has a filled $2p$-sublevel; the $3s$-sublevel fills with Na and Mg; and the $3p$-sublevel begins to fill with Al.

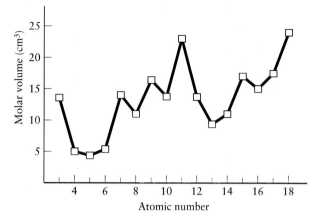

7.107 (a) Some of the energy of the incoming radiation, $h\nu$, is required to ionize the atom (free the electron); the remainder is the kinetic energy for the electron.
(b) 6.72×10^{-19} J/atom

7.109 -4.09×10^{-19} J/atom (-2.46×10^5 J/mol) at 487 nm;
-3.80×10^{-19} J/atom (-2.29×10^5 J/mol) at 524 nm;
-3.66×10^{-19} J/atom (-2.20×10^5 J/mol) at 543 nm;
-3.60×10^{-19} J/atom (-2.17×10^5 J/mol) at 553 nm;
-3.44×10^{-19} J/atom (-2.07×10^5 J/mol) at 578 nm

7.111 (a) 365 nm;   (b) Balmer series;   (c) longer

7.113 (a) Cesium has the lowest ionization energy of all the metals, so could be used with the lowest frequency ultraviolet radiation. In practice, semiconductor detectors are used.   (b) Cesium reacts very vigorously (explosively) with water and moist air.

7.115 $H_2 < He < Ne < N_2 < O_2 < F_2 < Ar < Cl_2 < Kr < Xe < Rn$

7.117 $Ba(s) + 2 H_2O(l) \rightarrow Ba(OH)_2(aq) + H_2(g)$
$Ca(s) + 2 H_2O(l) \rightarrow Ca(OH)_2(aq) + H_2(g)$

7.119 688.9 kJ

Chapter 8
8.1 (a) +1;   (b) −2;   (c) +2;   (d) +3

8.3 Na^+ has the configuration $1s^22s^22p^6$, which is a very stable electronic configuration, identical to that of Ne. These electrons are all core electrons, with high ionization energies. To lose one, as required to form Na^{2+}, would take a great deal of energy, because these core electrons are so tightly held.

8.5 (a) Ca:;   (b) ·S̈·;   (c) :Ö:$^{2-}$;   (d) :N̈:$^{3-}$

8.7 (a) $K^+[:F̈:]^-$
(b) $[:S̈:]^{2-} Al^{3+} [:S̈:]^{2-} Al^{3+} [:S̈:]^{2-}$

(c) $Ca^{2+}[:\ddot{\underset{..}{N}}:]^{3-} Ca^{2+}[:\ddot{\underset{..}{N}}:]^{3-} Ca^{2+}$

8.9 The inert-pair effect, or the tendency to form cations two units lower in charge than expected from the group number, can be observed in indium and thallium. Indium forms both In^{3+} and In^+ and thallium Tl^{3+} and Tl^+.

8.11 This difference is a result of the difference in ionic radii between Mg^{2+} and Ba^{2+}. See Fig. 7.36. The radius of Mg^{2+} (72 pm) is smaller than the radius of Ba^{2+} (136 pm), so the distance between Mg^{2+} and O^{2-} ions is less than that between Ba^{2+} and O^{2-} ions in the crystal lattice. Thus, the lattice enthalpy of MgO exceeds that of BaO.

8.13 $+1071$ kJ/mol

8.15 (a) endothermic; (b) exothermic; (c) endothermic; (d) exothermic; (e) exothermic

8.17 (a) $H-\ddot{\underset{..}{F}}:$; (b) $:\ddot{\underset{..}{Cl}}-\ddot{\underset{..}{O}}-\ddot{\underset{..}{Cl}}:$; (c) $:\ddot{\underset{..}{F}}-C-\ddot{\underset{..}{F}}:$ with $:\ddot{\underset{..}{F}}:$ above and $:\ddot{\underset{..}{F}}:$ below

8.19 (a) $\left[H-\overset{\displaystyle H}{\underset{\displaystyle H}{N}}-H\right]^{+}$; (b) $\left[:\ddot{\underset{..}{Cl}}-\ddot{\underset{..}{O}}:\right]^{-}$;

(c) $\left[:\ddot{\underset{..}{F}}-\overset{\displaystyle :\ddot{\underset{..}{F}}:}{\underset{\displaystyle :\ddot{\underset{..}{F}}:}{B}}-\ddot{\underset{..}{F}}:\right]^{-}$

8.21 (a) $\left[H-\overset{\displaystyle H}{\underset{\displaystyle H}{N}}-H\right]^{+}$ $\left[:\ddot{\underset{..}{Cl}}:\right]^{-}$; (b) K^+ $\left[:\ddot{\underset{..}{O}}-\overset{\displaystyle :O:}{\underset{\displaystyle :O:}{P}}-\ddot{\underset{..}{O}}:\right]^{3-}$ K^+ K^+ ;

(c) Na^+ $\left[:\ddot{\underset{..}{Cl}}-\ddot{\underset{..}{O}}:\right]^{-}$

8.23 (a) X is silicon, Y is sulfur; (b) X is carbon, Y is boron

8.25 (a) $H-C\overset{\displaystyle \ddot{O}.}{\underset{\displaystyle H}{\big\langle}}$; (b) $H-\overset{\displaystyle H}{\underset{\displaystyle H}{C}}-\ddot{\underset{..}{O}}-H$

8.27 An example is the benzene molecule, which can be represented as a resonance hybrid of two principal Lewis structures, although other structures contribute slightly.

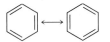

Two consequences of resonance are that (1) the blended structure, or resonance hybrid, has a significantly lower energy than any of the individual structures. Thus, resonance plays a significant role in the chemical properties of the compound. The compound is less reactive than

would have been predicted for a molecule existing in the form of just one of the structures. (2) Bond lengths are significantly different from what would be expected without resonance. In benzene, the C—C bond distance is intermediate between that expected for a C—C (single) bond and a C=C (double) bond, but the distance is not merely an average of the two.

8.29 (a) $\left[:\ddot{\underset{..}{O}}-\ddot{N}=\ddot{O}:\right]^{-}\longleftrightarrow\left[:\ddot{O}=\ddot{N}-\ddot{\underset{..}{O}}:\right]^{-}$

(b) $:\ddot{\underset{..}{O}}-\overset{\displaystyle :\ddot{Cl}:}{N}=\ddot{\underset{..}{O}} \longleftrightarrow \ddot{\underset{..}{O}}=\overset{\displaystyle :\ddot{Cl}:}{N}-\ddot{\underset{..}{O}}:$

8.31

(a) $\left[H-\ddot{\underset{..}{O}}-\overset{\displaystyle :\ddot{O}:}{\underset{\displaystyle :\ddot{O}:}{P}}-\ddot{\underset{..}{O}}-H\right]^{-} \longleftrightarrow \left[H-\ddot{\underset{..}{O}}-\overset{\displaystyle :O:}{\underset{\displaystyle :\ddot{O}:}{P}}-\ddot{\underset{..}{O}}-H\right]^{-} \longleftrightarrow \left[H-\ddot{\underset{..}{O}}-\overset{\displaystyle :O:}{\underset{\displaystyle :O:}{P}}-\ddot{\underset{..}{O}}-H\right]^{-}$ 2 ways

(b) $\left[:\ddot{\underset{..}{O}}-\overset{\displaystyle :\ddot{O}:}{\underset{\displaystyle :O:}{S}}:\right]^{2-} \longleftrightarrow \left[:\ddot{O}=\overset{\displaystyle :\ddot{O}:}{\underset{\displaystyle :O:}{S}}:\right]^{2-} \longleftrightarrow \left[:\ddot{O}=\overset{\displaystyle ·\ddot{O}·}{\underset{\displaystyle :O:}{S}}:\right]^{2-}$ 3 ways 3 ways

(c) $\left[:\ddot{\underset{..}{O}}-\overset{\displaystyle :\ddot{O}:}{\underset{\displaystyle :O:}{Cl}}:\right]^{2-} \longleftrightarrow \left[:\ddot{O}-\overset{\displaystyle ·\ddot{O}·}{\underset{\displaystyle :O:}{Cl}}:\right]^{2-} \longleftrightarrow \left[:\ddot{O}-\overset{\displaystyle ·\ddot{O}·}{\underset{\displaystyle ·\ddot{O}·}{Cl}}:\right]^{2-}$ 3 ways 3 ways

8.33 Formal charge is a good "bookkeeping" system for electrons and is based on the assumption that the molecule is "perfectly covalent." Formal charge is determined by dividing the electrons in a bond equally between the atoms connected by the bond. The concept is used primarily for deciding between otherwise equally plausible Lewis structures.

8.35 (a) Formal charge on N is 0 and on H is 0. (b) Formal charge on the central N is $+1$, that on each terminal N is -1.

8.37 (a) first structure: charge on S is $+1$, O= is 0, and O— is -1. second structure: charge on S is 0, and each O= is 0. The second structure is more plausible. (b) first structure: charge on S is 0 and each O= is also 0. second structure: charge on S is $+1$, O= is 0, and O— is -1. The first structure is more plausible.

8.39 Structure (a) has a charge of $+1$ on S and -1 on O—; structure (b) has a charge of 0 on both S and O=. Structure (b) is the more plausible structure.

8.41 (a) The first structure is dominant. (b) The first structure is dominant.

8.43 A radical is any species with an unpaired electron. Examples are

(1) ·CH₃ or
$$\cdot \overset{\displaystyle H}{\underset{\displaystyle H}{\overset{|}{\underset{|}{C}}}} - H$$

(2) NO or :N̈=Ö:

(3) NO₂ or Ö=N̈−Ö: ⟷ :Ö−N̈=Ö

8.45 (a)

$$\overset{:\ddot{C}l}{\underset{:\ddot{C}l\quad :\ddot{C}l:}{\overset{|}{S:}}}$$
1 lone pair on S

(b)
$$:\ddot{C}l - \overset{:\ddot{C}l:}{\underset{:\ddot{C}l:}{\overset{|}{\underset{|}{I}}}}$$
2 lone pairs on I

(c)
$$\left[\overset{:\ddot{F}\quad \ddot{F}:}{\underset{:\ddot{F}\quad \ddot{F}:}{\overset{\diagdown\diagup}{\underset{\diagup\diagdown}{:I:}}}} \right]^-$$
2 lone pairs on I

8.47 (a) 4 electron pairs; 2 bonding, 2 lone;
(b) 6 electron pairs; 4 bonding, 2 lone; (c) 5 electron pairs; 3 bonding, 2 lone; (d) 6 electron pairs, 5 bonding, 1 lone

8.49 (a) $[:\ddot{O}-\ddot{O}\cdot]^-$ 2 resonance forms; a radical

(b) :F̈−N̈−F̈: a radical

(c) :N=Ö−C̈l: one resonance form shown; not a radical

8.51 (a)
$$\overset{:\ddot{F}\; - \;\dot{X}\!e\; - \;\ddot{F}:}{\underset{\displaystyle :O:}{\overset{\displaystyle \|}{}}}$$
2 lone pairs on Xe

(b) :F̈=Ẍe−F̈: 3 lone pairs on Xe

(c)
$$\left[\overset{\displaystyle :O:}{\underset{\displaystyle :O:}{\overset{\diagup H}{\underset{|}{\overset{|}{:O=X\!e.=O:}}}}} \right]^-$$
1 lone pair on Xe

3 ways

8.53 A Lewis acid is an electron pair acceptor, therefore its electronic structure must allow an additional electron pair to become attached to it. A Lewis base is an electron pair donor, therefore it must contain a lone pair of electrons that it can donate. Lewis acids: H⁺, Al³⁺, BF₃; Lewis bases: OH⁻, NH₃, H₂O

8.55 The net ionic reaction for an acid-base neutralization is

$$H^+ \quad + \quad [:\ddot{O}-H]^- \quad \longrightarrow \quad \overset{\displaystyle \cdot\ddot{O}\cdot}{\underset{\displaystyle H\quad H}{\diagup\;\diagdown}}$$

Lewis acid + Lewis base ⟶ Lewis complex

8.57 (a) base; (b) acid; (c) acid; (d) base

8.59
$$\left[\overset{\displaystyle H}{\underset{\displaystyle H}{\overset{|}{\underset{|}{H-C-\ddot{O}:}}}} \right]^-$$
There are available electron pairs on the O; thus CH₃O⁻ would be expected to be a Lewis base.

8.61 (a)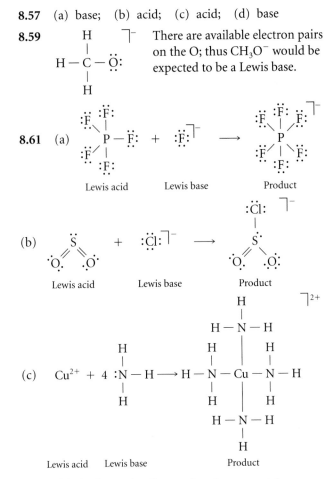

(b)
$$\overset{:\ddot{S}:}{\underset{\cdot\ddot{O}\quad\ddot{O}\cdot}{\diagup\;\diagdown}} \quad + \quad [:\ddot{C}l:]^- \quad \longrightarrow \quad \overset{[:\ddot{C}l:]^-}{\underset{\cdot\ddot{O}\quad\ddot{O}\cdot}{\overset{|}{\underset{\diagup\;\diagdown}{\dot{S}\cdot}}}}$$
Lewis acid   Lewis base   Product

(c) Cu²⁺ + 4 :N̈−H ⟶ Product

Lewis acid   Lewis base   Product

8.63 (a) ionic, made of a metal and a nonmetal;
(b) nonionic, made of a nonmetal and a nonmetal;
(c) ionic, made of a metal and a nonmetal
8.65 (a) ionic; (b) covalent; (c) covalent
8.67 The order is most likely Rb⁺ < Sr²⁺ < Be²⁺, which corresponds to the order of ionic size in reverse. Higher charge predominates over lower charge.
8.69 O²⁻ < N³⁻ < Cl⁻ < Br⁻. The order parallels the ionic size: Br⁻ is largest; O²⁻ is smallest.
8.71 (a) HCl; (b) CF₄; (c) CO₂
8.73 (a) ionic; (b) significantly covalent; (c) mainly ionic; (d) ionic
8.75 Boron is in Group 13 and Period 2 of the periodic table and has only 3 valence electrons. To have an octet of electrons, boron would have to add 5 electrons by sharing with other atoms. That is improbable, especially because the other atoms involved are likely to be highly electronegative nonmetals. Other elements in Period 2, or later periods, that form covalent bonds have to acquire no more than 4 additional electrons to complete their octet and thus are not likely to form electron-deficient compounds.
8.77 Br—Cl (least ionic) < C—Cl < Al—Cl < NaCl

8.79 (a) Li^+; $[:H]^-$; (b) $[:\ddot{\underset{..}{Cl}}:]^-$ Cu^{2+} $[:\ddot{\underset{..}{Cl}}:]^-$

(c) Ba^{2+} $[:\ddot{\underset{..}{N}}:]^{3-}$ Ba^{2+} $[:\ddot{\underset{..}{N}}:]^{3-}$ Ba^{2+}

(d) $[:\ddot{\underset{..}{O}}:]^{2-}$ Ga^{3+} $[:\ddot{\underset{..}{O}}:]^{2-}$ Ga^{3+} $[:\ddot{\underset{..}{O}}:]^{2-}$

8.81 (a) $+2564$ kJ/mol; (b) $+5485$ kJ/mol

8.83 (a) Lewis acid; (b) Lewis acid; (c) either Lewis acid (at Ga) or a Lewis base (at I)

8.85

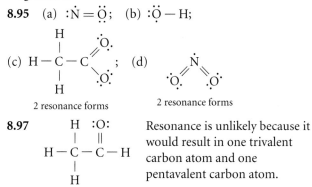

Lewis acid   Lewis base   Product

8.87 CH_3^-, because the Lewis structure of CH_3^- shows a lone pair on the C atom; there is no lone pair in CH_4.

8.89 (a) first structure: each atom has a formal charge of 0; second structure: each $O-$ has a formal charge of -1, $H-$ and $-O-$ are 0, and Cl is $+2$. The first structure has the lowest energy. (b) Each atom has a formal charge of 0. (c) Each atom has a formal charge of 0. (d) first structure: Each N has a formal charge of -1, and the C is 0; second structure: $N\equiv$ and $C=$ have charges of 0, $N-$ has a charge of -2. The first structure has the lower energy. (e) first structure: Each $O-$ has a charge of -1, and the As has a charge of $+1$; second structure: The $O=$ has a charge of 0, each $O-$ has a charge of -1, and the As has a charge of 0. The second structure has the lower energy.

8.91 Yes, because all bonds involve a pair of electrons influenced by two nuclear centers. It becomes a question of degree as to where the electron pair more closely resides. Nonpolar covalent bonds share the electron pair equally; polar covalent bonds less so, and ionic bonds hardly at all. The octet rule is useful in the understanding of all the bonding situations mentioned to the extent that it applies. Recall that there are exceptions. The coordinate covalent bond that results from complex formation is like any other covalent bond once it has formed.

8.93 The formation of MgCl releases 60. kJ/mol, whereas the formation of $MgCl_2$ releases 645 kJ/mol. The difference in ΔH_f° is 525 kJ. $MgCl_2$ is the far more likely chloride of magnesium.

8.95 (a) $:\dot{N}=\ddot{O}$; (b) $:\ddot{O}-H$;

(c) $H-\overset{\overset{\displaystyle H}{|}}{\underset{\underset{\displaystyle H}{|}}{C}}-C\overset{\ddot{O}}{\underset{\ddot{O}}{\diagup}}$; (d) $\ddot{\underset{..}{O}}=\overset{\dot{N}}{\diagdown}\ddot{\underset{..}{O}}$

2 resonance forms   2 resonance forms

8.97 $H-\overset{\overset{\displaystyle H}{|}}{\underset{\underset{\displaystyle H}{|}}{C}}-\overset{:O:}{\underset{}{C}}-H$ Resonance is unlikely because it would result in one trivalent carbon atom and one pentavalent carbon atom.

8.99 $CO_3^{2-}(s) \rightarrow CO_2(g) + O^{2-}(s)$

Complex   Lewis acid   Lewis base

$Ca^{2+}O^{2-}(s) + SiO_2(s) \rightarrow CaSiO_3(s)$

Lewis base   Lewis acid   Complex

8.101 Large anions, such as chromate and phosphate (which also have higher charges) are more polarizable than smaller anions like carbonate, chloride, and nitrate. In general, it appears that the more covalent character, and thus the more polarizable, the ion, the more insoluble the salts of the anion tend to be.

8.103 (a) The electron configurations of Mg and I are such that the octet rule may be easily satisfied by an electron transfer process. After the transfer, the small Mg^{2+} ion, with its highly concentrated positive charge, has a strong electrostatic attraction to the large iodide ions. The bonding energy is then primarily electrostatic or ionic. (b) In SiF_4, for the octet rule to be satisfied on Si by electron transfer to F, the Si^{4+} ion would have to generate such a highly charged ion that, given the bonding energy in SiF_4, one should conclude that the bond is primarily a result of electron sharing or covalent bonding. The radius of Si^{4+} is also quite small (26 pm), leading to significant covalent character.

8.105 (a) Cd, $[Kr]4d^{10}5s^2$; In^+, $[Kr]4d^{10}5s^2$; Sn^{2+}, $[Kr]4d^{10}5s^2$. All are the same, because the ions form by the loss of $5p$-electrons, and cadmium itself has no $5p$-electrons. (b) There are no unpaired electrons in any of the species. (c) In^{3+}, $[Kr]4d^{10}$, which is isoelectronic with palladium.

Chapter 9

9.1 (a) trigonal bipyramidal, $120°$ bond angles in equatorial plane, $90°$ between axial bonds and equatorial plane; (b) linear, $180°$; (c) trigonal pyramidal, slightly less than $109°$; (d) angular, slightly less than $109°$; (e) tetrahedral, $109.5°$

9.3 (a) must be; (b) may be

9.5 (a) AX_2E_2, angular; (b) AX_3E, trigonal pyramidal; (c) AX_2, linear; (d) AX_2E, angular

9.7 (a) AX_3E, trigonal planar; (b) AX_4, tetrahedral; (c) AX_4E, seesaw; (d) AX_3, trigonal planar

9.9 (a) AX_4E, seesaw; (b) AX_3E_2, T-shaped; (c) AX_4E_2, square planar; (d) AX_3E, trigonal pyramidal

9.11 (a) linear, $180°$; (b) T-shaped, slightly less than $90°$; (c) tetrahedral, $109.5°$; (d) octahedral, $90°$

9.13 (a) $:\ddot{\underset{..}{F}}-\overset{\overset{\displaystyle :\ddot{\underset{..}{Cl}}:}{|}}{\underset{\underset{\displaystyle :\ddot{\underset{..}{F}}:}{|}}{C}}-\ddot{\underset{..}{F}}:$ AX_4, tetrahedral, $109.5°$;

(b) :Ï—Ga with :Ï: and :Ï: AX₃, trigonal planar, 120°;

(c) [C̈ with H H H]⁻   AX₃E, trigonal pyramidal, slightly less than 109°;

(d) S with :F̈: positions around (F above, F, F, F, F below)   AX₆, octahedral, 90°;

(e) Xe with :F̈:, :F̈:, :F̈:, :F̈:, and .Ö.   AX₅E, square pyramidal, 90°

9.15

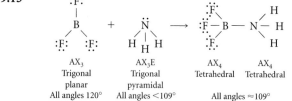

AX₃	AX₃E	AX₄	AX₄
Trigonal planar	Trigonal pyramidal	Tetrahedral	Tetrahedral
All angles 120°	All angles <109°	All angles ≈109°	

9.17 An electric dipole ("two poles") is a positive charge next to an equal but opposite negative charge. An electric dipole moment is a measure of the magnitude of the electric dipole in debye (D) units.

9.19 (a) toward O; (b) toward F; (c) toward F; (d) toward O

9.21 (a) one nonpolar bond; (b) four polar N—H bonds, one nonpolar N—N bond; (c) all C—H bonds are slightly polar; (d) both bonds are polar.

9.23 (a) trigonal planar, nonpolar; (b) square planar, nonpolar; (c) trigonal pyramidal, polar; (c) angular, polar

9.25

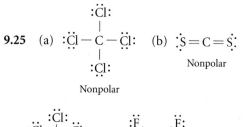

(a) :C̈l—C—C̈l: with :C̈l: top and :C̈l: bottom, Nonpolar

(b) :S̈=C=S̈: Nonpolar

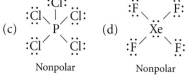

(c) P with :C̈l:, :C̈l:, :C̈l:, :C̈l: (top Cl, two side Cl, bottom Cl), Nonpolar

(d) Xe with :F̈:, :F̈:, :F̈:, :F̈:, Nonpolar

9.27 (a) (1) is polar, (2) is polar, (3) is nonpolar; (b) structure (1)

9.29 (a) +926 kJ/mol; (b) +1486 kJ/mol; (c) +3150 kJ/mol; (d) +2317 kJ/mol

9.31 (a) −92 kJ/mol; (b) −151 kJ/mol; (c) −868 kJ/mol; (d) −38 kJ/mol

9.33 (a) −232 kJ; (b) −55 kJ; (c) −287 kJ

9.35 (c) < (b) < (a)

9.37 (a) N—H, 112 pm; N—N, 150 pm; (b) C=O, 127 pm; (c) N—C, 152 pm; N—H, 112 pm; C=O, 127 pm; (d) N=N, 120 pm; N—H, 112 pm

9.39 The $3p_z$-$3p_z$ overlap forms a σ-bond.

9.41 (a) tetrahedral; (b) linear; (c) octahedral; (d) trigonal planar

9.43 (a) dsp^3; (b) sp^2; (c) sp^3; (d) sp

9.45 (a) sp^2 [↑][↑][↑] [ ]
    sp^2 sp^2 sp^2 p

(b) sp^3 [↑][↑][↑][ ]
    sp^3 sp^3 sp^3 sp^3

(c) sp^3 [↑↓][↑↓][↑↓][↑]
    sp^3 sp^3 sp^3 sp^3

(d) sp^2 [↑][↑][↑] [ ]
    sp^2 sp^2 sp^2 p

9.47 (a) sp^3; (b) sp; (c) dsp^3; (d) sp^3

9.49 (a) $\sigma_{2s}^2\sigma_{2s}^{*2}$, BO = 0; (b) $\sigma_{2s}^2\sigma_{2s}^{*2}\pi_{2p}^2$, BO = 1; (c) $\sigma_{2s}^2\sigma_{2s}^{*2}\sigma_{2p}^2\pi_{2p}^4\pi_{2p}^{*4}\sigma_{2p}^{*2}$, BO = 0

9.51 All three are paramagnetic.

9.53 (a) F₂, BO = 1; F₂²⁻, BO = 0; F₂ bond is stronger; (b) B₂, BO = 1; B₂⁺; BO = ½; B₂ bond is stronger

9.55 HeH⁻ has a bond order of 0; it probably doesn't exist. HeH⁺ has a bond order of 1. This species is more stable.

9.57 (a) S with .Ö, .C̈l:, .Ö, .C̈l: — distorted tetrahedron, polar

(b) As with :C̈l: :C̈l: :C̈l: top and :C̈l: :C̈l: bottom — trigonal bipyramidal, nonpolar

(c) :F̈—Si—F̈: with :F̈: top and :F̈: bottom — tetrahedral, nonpolar

(d) :S with :F̈: top, F̈:, :F̈: bottom — seesaw, polar

(e) H—C≡N: linear, polar

(f) :Cl̈—Ï—Cl̈:   T-shaped,
         | polar
        :Cl̈:

9.59 (b) < (c) < (a)

9.61 (a) :F̈—Të—F̈:   AX$_4$E,
                 ⟋   ⟍   seesaw
               :F̈.   .F̈:

(b)   ·N̈· ⌐   AX$_2$E$_2$,
       ⟋ ⟍   angular
      H   H

(c) S̈=C=S̈   AX$_2$, linear

(d)   H ⌐$^+$
           | AX$_4$,
           N   tetrahedral
        ⟋ | ⟍
      H   H   H

(e)   H⟍   ⟋H   AX$_4$,
         Ge   tetrahedral
      H⟋   ⟍H

(f) Ö=C=S̈   AX$_2$, about C, linear

9.63 (a) H⟍   ⟋H   ; (b) :Cl̈—C≡N̈: ;
              C=C
           H⟋   ⟍H   180°
              120°

(c) :Ö   C̈l: ; (d) H⟍   ⟋H
        ⟍ ⟋   N̈—N̈
         P   H⟋   ⟍H
      ⟋   ⟍
   :C̈l   C̈l:
      109°   109°

9.65 (a) 149 pm; (b) 183 pm; (c) 213 pm. The bond
length is proportional to the increasing size of the atom
attached to the fluorine atom; and size increases down a
group in the periodic table.

9.67 :Ö—F̈:   sp^3 hybridization (two sp^3—p bonds),
          | AX$_2$E$_2$, angular; the bond angle is
         :F̈:   somewhat less than 109°.

9.69 −228 kJ/mol

9.71

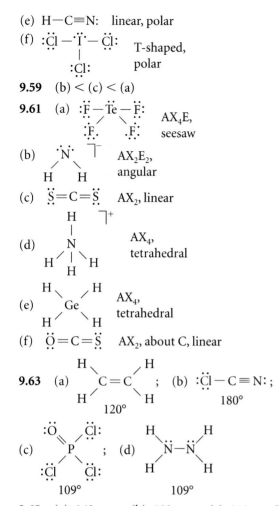

The hexagonal structure of the
molecule dictates that all bond
angles be 120°, thus we expect
that each B and N atom is sp^2
hybridized.

9.73 (a) $\sigma_{2s}^2\sigma_{2s}^{*2}\pi_{2p}^4\sigma_{2p}^2$, BO = 3;
(b) $\sigma_{2s}^2\sigma_{2s}^{*2}\pi_{2p}^4\sigma_{2p}^1$, BO = $\frac{5}{2}$;
(c) $\sigma_{2s}^2\sigma_{2s}^{*2}\pi_{2p}^4\sigma_{2p}^2\pi_{2p}^{*3}$, BO = $\frac{3}{5}$

9.75 (a)

Phenolphthalein (acid form)

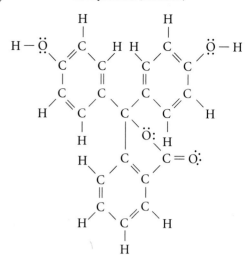

Phenolphthalein (base form)

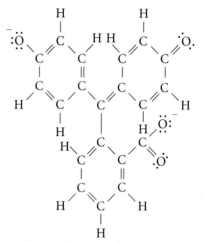

(b) The *C in the acid form is sp^3 hybridized, whereas the
same carbon in the base form is sp^2 hybridized.   (c) The
change of hybridization on the asterisk-labeled carbon
changes the electronic structure of phenolphthalein. The
base form has a more extensive conjugated system (see
Investigating Matter 9.2), which may also account for the
highly colored base form.

9.77 (a) $\sigma_{2s}^2\sigma_{2s}^{*2}\pi_{2p}^4\sigma_{2p}^2$;
(b) $\sigma_{2s}^2\sigma_{2s}^{*2}\pi_{2p}^4\sigma_{2p}^2\pi_{2p}^{*1}$;
(c) $\sigma_{2s}^2\sigma_{2s}^{*2}\pi_{2p}^4\sigma_{2p}^2$

9.79 (a) approximately 120°;   (b) Each carbon is
sp^2 hybridized, the O is sp^2 hybridized.   (c) There are
7 σ-bonds and 2 π-bonds in the molecule.

9.81 (a)   H   ; (b) C and O have sp^3
                     | ..   hybridization.
              H—C—Ö—H   (c) The molecule is polar.
                     | ..
                     H

9.83 (a), (b), and (c) are shown below:

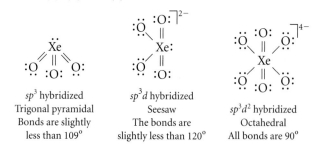

sp^3 hybridized	sp^3d hybridized	
Trigonal pyramidal	Seesaw	sp^3d^2 hybridized
Bonds are slightly	The bonds are	Octahedral
less than 109°	slightly less than 120°	All bonds are 90°

(d) Xe is +6 in XeO_3, +6 in XeO_4^{2-}, and +8 in XeO_6^{4-}

9.85 (a) $\Delta H_f^\circ = 153$ kJ/mol for N_2O as $\ddot{N}=N=\ddot{O}$;
$\Delta H_f^\circ = 128$ kJ/mol for NO_2 as $\ddot{O}-\dot{N}=\ddot{O}$;
(b) 71 kJ/mol for N_2O, 95 kJ/mol for NO_2; (c) NO_2 must have more resonance forms that contribute to its actual structure than N_2O does.

9.87 (a) $\sigma_{2s}^3\sigma_{2s}^{*3}\sigma_{2p}^3\pi_{2p}^1$, based on Fig. 9.57 and the assumption that the number of orbitals won't change. Oxygen atoms will then have 5 valence electrons.
(b) Yes, the valence electron configuration would be $\sigma_{3s}^3\sigma_{3s}^{*3}\pi_{3p}^6\sigma_{3p}^3\pi_{3p}^{*1}$, BO $=\frac{5}{3}$; (c) The "octet rule" would be the "dozen rule," because each atom would be most stable with 12 valence electrons. Single bonds would require 3 shared electrons, a "lone triplet" would have 3 electrons as well. Double bonds would contain 6 electrons, and triple bonds 9 electrons. BO $= \frac{1}{3} \times (B - A)$

Connection 2

1. Enthalpy densities: 2.3×10^4 kJ/L for ethanol; 2.7×10^4 kJ/L for ETBE. Tank sizes: 1.1×10^2 L for methanol; 70. L for ETBE; 1.5×10^4 L for H_2; 4.8×10^3 L for CH_4.

3. (a)

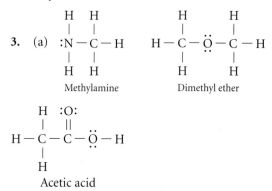

(b) 95.07 kJ/g for methylamine; 20.6 kJ/g for dimethyl ether; 9.61 kJ/g for acetic acid. (c) From the Merck Index, 12th ed. (1996), we find that methylamine irritates the eyes and respiratory system, may lead to coughing, skin and mucous membrane burns, and skin conditions. Its high specific enthalpy makes methylamine a decent fuel. Dimethyl ether, or methyl ether, has no listed hazards in

the Merck Index, though this compound can act as an anesthetic. The specific enthalpy is not particularly high, and is similar to that of methanol. Acetic acid (glacial) corrodes the mouth and gastrointestinal tract upon ingestion. Death may result if exposure is severe. Eye irritation, bronchitis, and erosion of dental enamel may also result. Its low enthalpy density makes acetic acid a poor choice of fuel.

For comparison, gasoline (octane) may irritate the eyes and nose, lead to dermatitis or chemical pneumonia. Methylamine is too toxic, but dimethyl ether would make a reasonable fuel.

5. (a) CO, (c) O_3, (d) SO_2, and (e) N_2O are likely greenhouse gases.

7. (a) 4.0×10^2 g; (b) 25 g.

Chapter 10

10.1 (a) London forces; (b) dipole-dipole, London forces; (c) London forces; (d) dipole-dipole, London forces

10.3 (a) $:C\equiv O:$ London forces; very weak dipole-dipole forces;

(b) hydrogen bonding; dipole-dipole and London forces also occur;

(c) London forces, very weak dipole-dipole forces;

(d) London forces

10.5 A hydrogen bond may be described as A—H···B with the dotted bond, ···, representing the hydrogen bond. A must be N, O, or F. They are the only elements sufficiently electronegative to produce the strongly polar A—H bond that is needed to attract the lone pair of electrons on the neighboring B atoms. Likewise, the B atoms should also be N, O, or F, because they are the smallest of the highly electronegative elements. Their small size results in the strongest interaction with the positive side of the polar A—H bond.

10.7 (a) HF, (c) NH_3, and (d) CH_3OH can form hydrogen bonds.

10.9 (a)

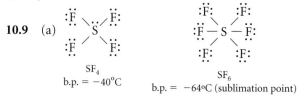

SF$_4$ has the seesaw shape, whereas SF$_6$ is octahedral; SF$_4$ has the higher boiling point.

(b)

$$:\ddot{F}:$$
$$|$$
$$B$$
$$:\ddot{F} \diagdown \ddot{F}:$$

BF$_3$
b.p. = −99.9°C

$$:\ddot{F}:$$
$$|$$
$$:\ddot{Cl}-\ddot{F}:$$
$$|$$
$$:\ddot{F}:$$

ClF$_3$
b.p. = 11.3°C

BF$_3$ is trigonal planar and nonpolar, whereas ClF$_3$ is T-shaped and polar. Therefore, ClF$_3$ has the higher boiling point.

(c)

$$:\ddot{F} \diagdown \diagup \ddot{F}:$$
$$S$$
$$:\ddot{F} \diagup \diagdown \ddot{F}:$$

SF$_4$
b.p. = −40°C

$$:\ddot{F}:$$
$$|$$
$$:\ddot{F}-C-\ddot{F}:$$
$$|$$
$$:\ddot{F}:$$

CF$_4$
b.p. = −128°C

CF$_4$ is tetrahedral; SF$_4$ is seesaw. SF$_4$ has the higher boiling point.

(d)

$$\begin{array}{cc} H \diagdown & \diagup H \\ & C=C \\ Cl \diagup & \diagdown Cl \end{array}$$

cis-CHCl=CHCl
b.p. = 60°C

$$\begin{array}{cc} H \diagdown & \diagup Cl \\ & C=C \\ Cl \diagup & \diagdown H \end{array}$$

trans-CHCl=CHCl
b.p. = 48°C

Both molecules have the same shape, but the atomic arrangement in the cis compound produces a dipole moment and, thus, a higher boiling point.

10.11 The intermolecular forces are stronger between water molecules than between water molecules and the hydrocarbon molecules in wax. Water molecules hydrogen bond to one another but not to hydrocarbon molecules. These intermolecular forces in water are displayed as high surface tension.

10.13 (a) ethanol; (b) propanone

10.15 (a) ionic; (b) molecular; (c) molecular; (d) metallic

10.17 A, ionic; B, metallic; C, molecular

10.19 (a) 2 atoms; (b) 8

10.21 (a) 3.54×10^{-8} cm; (b) 2.25×10^{22} unit cells

10.23 (a) 1 atom; (b) 6; (c) 334 pm

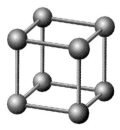

10.25 (a) 8.78 g/cm^3; (b) 1.48 g/cm^3

10.27 (a) 144 pm; (b) 131 pm

10.29 The fundamental difference is that the charge is

carried by electrons in electronic conduction and by ions in ionic conduction. This difference is apparent from their names.

10.31 (a) substitutional; (b) The melting point of the alloy will be lower than that of pure lead.

10.33 (a) n-type; (b) p-type; (c) n-type

10.35 In an electrical conductor, the many molecular orbitals are very close together, forming a nearly continuous band (Fig. 10.26). If the molecular orbitals are not filled completely and are closely spaced, then electrons can be easily excited to empty molecular orbitals. An electric current then moves through the solid. In an electrical insulator, a full band of orbitals, called a valence band, forms. There is a large energy gap between the full orbitals and the next available orbital band. Electrons are only excited to the higher-energy orbitals if a large amount of energy is available. The electrons are not easily mobile, and we call the material an electrical insulator.

10.37 Heating in a vacuum, where the partial pressure of oxygen is below its equilibrium value, causes partial loss of oxygen by way of the reaction

$$ZnO(s) \rightarrow ZnO_{1-x} + \tfrac{x}{2}O_2$$

Thus, the zinc oxide is left nonstoichiometric, for example, $ZnO_{0.95}$. For every oxygen atom formed, two conducting electrons are left in the zinc oxide lattice, and its conductivity increases. This trend is reversed when the ZnO is heated in oxygen.

10.39 (a) 4 Na$^+$ ions, 4 Cl$^-$ ions, 4 formula units; (b) 4 Ca^{2+} ions, 8 F$^-$ ions, 4 formula units; coordination numbers are 8 for Ca^{2+} and 4 for F$^-$

10.41 CaTiO$_3$

10.43 5.56×10^{18} unit cells

10.45 (a) 6; (b) 6; (c) 8

10.47 B, C, Si, Ge, P, As

10.49 356 kJ/mol

10.51 (a) London forces; (b) dipole-dipole, London forces, and hydrogen bonding; (c) dipole-dipole and London forces

10.53 A$_2$B$_2$C$_4$

10.55 Dynamic equilibrium is a condition in which a forward process and its reverse are occurring simultaneously at equal rates. Thus, in a dynamic equilibrium, processes (at the molecular level) are occurring; it is only the net effect that is unchanging. In a static equilibrium, no process is occurring at any level.

10.57 Hydrogen peroxide also undergoes hydrogen bonding, but its greater molar mass allows for greater London forces. These factors combine to produce lower vapor pressure and a higher boiling point.

10.59 0.017 g H$_2$O

10.61 (a) 99.2°C; (b) 99.7°C

10.63 350. K or 77°C

10.65 (a) vapor; (b) liquid; (c) vapor; (d) vapor
10.67 (a) The liquid water existing at 0°C and 2 atm would freeze as the pressure drops. (b) The liquid water existing at 50°C and 2 atm would vaporize as the pressure drops.
10.69 (a) ca. 2.4 K; (b) 10 atm (c) about 5 K; (d) no
10.71 (a) At the lower pressure triple point, liquid helium I and II are in equilibrium with helium gas; at the higher pressure triple point, liquid helium I and II are in equilibrium with solid helium. (b) The negative slope of the helium II/helium I phase boundary line suggests that— as in the case of the solid/liquid water phase boundary— helium I is the more dense, despite the fact that it is the higher temperature phase.
10.73 The intermolecular forces in isopropanol are not as strong as they are in water; isopropanol has a lower boiling point (82.3°C) than water (100°C) does and a higher vapor pressure than water at that temperature. Vaporization is an endothermic process, so heat is extracted from the skin; and because of its greater rate of evaporation than water, isopropanol has a more pronounced cooling effect.
10.75 (a) The attraction between atoms in both Xe and Ar is a result of London forces, but, because Xe has more electrons, it is bigger and therefore more polarizable; thus, it has stronger London forces and a higher melting point. (b) HI and HCl have both London forces and dipole-dipole attractions, but the dominant attraction is the London force, which is greater in HI. (c) There is strong hydrogen bonding in water, but none in $C_2H_5OC_2H_5$, so water has the lower vapor pressure at the same temperature.
10.77 (a) HCl has strong dipole-dipole interactions, but these forces are not nearly as strong as the ion-ion interactions in NaCl; thus NaCl has the higher normal boiling point. (b) These compounds have the same structure, but SiH_4 has the greater molar mass; thus, we expect that it has the higher normal boiling point. (c) HF, because of stronger hydrogen bonding; (d) H_2O, because of more opportunities for hydrogen bonding; there are two O—H bonds
10.79 (a) H_2O; it has two O—H bonds, both of which are capable of strong hydrogen bonding. (b) London forces increase with increasing molar mass, so C_3H_7OH will have the greatest surface tension. (c) CO_2; this is a nonpolar compound; SO_2 is polar and SiO_2 is a network solid. (d) Very likely HCl; it has the weakest London forces, although it has the strongest dipole-dipole forces. The London forces predominate (see Example 10.2). (e) H_2S; its hydrogen bonding forces are much less than in H_2O, and its London forces are much less than in H_2Te. (f) H_2; it has the weakest London forces. (g) NH_3, due to relatively strong hydrogen bonding forces. (h) Na_2O; ion-ion forces are stronger than all other forces. (i) All have the same shape, but the electronegativity difference is

greatest in SF_2; so it probably has the strongest dipole-dipole forces. (j) GeF_4; because it has the largest atoms with the most electrons.
10.81 (a) 78.6%; (b) At 25°C, the vapor pressure of water is 23.76 Torr; therefore, some of the water vapor in the air would condense as dew or fog.
10.83 The spacing between the carbon layers increases, resulting in decreased perpendicular conductivity. The parallel conductivity increases, although the effect may not be dramatic, because the band is already half-filled.
10.85 $$d\ (g/cm^3) = \frac{2.94 \times 10^{-25}\ M\ (g/mol)}{r^3\ (m^3)}$$
170 pm for Ne, 192 pm for Ar, 206 pm for Kr, 221 pm for Xe, 246 pm for Rn
10.87 21.0 g/cm^3
10.89 This material is a solid at room temperature and pressure, so it can't be nitrogen, oxygen, fluorine, or neon. The melting point is not high enough for carbon or boron. So the material must be beryllium or lithium. By consulting Appendix 2D, we find that the phase diagram is for lithium.
10.91 362 pm
10.93 (a) anisotropic; (b) isotropic; (c) anisotropic; (d) anisotropic; (e) anisotropic
10.95 Polar groups aid in the alignment of the molecules, the positive end of one molecule adhering to the negative end of another, and vice versa. That is, the dipoles align antiparallel to one another.
10.97 (a) Light of an appropriate wavelength may provide the energy needed for some electrons, or more electrons, to jump the band gap. Electrical conductivity would increase. (b) 685 nm; (c) Infrared radiation does not contain enough energy to excite electrons across the band gap in selenium. So amorphous selenium cannot be used in infrared burglar alarms.
10.99 (a) CH_3OH; (b) KI; (c) LiBr
10.101 65% Fe^{2+}, 35% Fe^{3+}
10.103 From Chapter 5, recall that a represents the effect of attractions. The more attractive forces, the larger a becomes. NH_3 molecules will hydrogen bond, thus ammonia has the largest value of a. CO_2 and O_2 have weak, intermolecular forces (London forces). CO_2 has somewhat stronger intermolecular forces because it has a larger molecular mass and thus a larger van der Waals a constant.

Chapter 11
11.1 The difference can be traced to the weaker London forces that exist between branched molecules. Atoms in neighboring branched molecules cannot lie as close together as they do in the unbranched isomers. As a result of the molecules' irregular shapes, the atoms in neighboring branched molecules are more effectively

shielded from one another than they are in neighboring unbranched molecules.

11.3 (a) four σ-type single bonds; (b) two σ-type single bonds and one double bond, consisting of one σ-bond and one π-bond; (c) one σ-type single bond and one triple bond, consisting of one σ-bond and two π-bonds

11.5 (a) $CH_4 + Cl_2 \xrightarrow{light} CH_3Cl + HCl$, substitution; (b) $CH_2{=}CH_2 + Br_2 \rightarrow CH_2Br{-}CH_2Br$, addition

11.7 (a) $CH_3CH_3 + 2\,Cl_2 \xrightarrow{light} CH_2ClCH_2Cl + 2\,HCl$, substitution ($CH_2ClCH_3$ may also be produced); (b) $CH_2{=}CH_2 + Cl_2 \rightarrow CH_2ClCH_2Cl$, addition; (c) $HC{\equiv}CH + 2\,Cl_2 \rightarrow CHCl_2CHCl_2$, addition

11.9 (a) ethane; (b) hexane; (c) octane; (d) heptane

11.11 (a) methyl; (b) pentyl; (c) propyl

11.13 (a) propane; (b) ethane; (c) pentane

11.15 (a) 4-methyl-2-pentene; (b) 2,3-dimethyl-2-phenylpentane

11.17 (a) propene (no geometrical isomers); (b) *cis*-2-hexene, *trans*-2-hexene; (c) 1-butyne (no geometrical isomers); (d) 2-butyne (no geometrical isomers)

11.19 (a) $CH_2CHCH(CH_3)CH_2CH_3$; (b) $CH_3CH_2C(CH_3)_2CH(CH_2CH_3)(CH_2)_2CH_3$; (c) $CHC(CH_2)_2C(CH_3)_3$; (d) $CH_3CH(CH_3)CH(CH_2CH_3)CH(CH_3)_2$

11.21

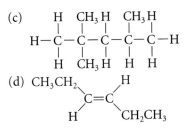

11.23 (a) 1-ethyl-3-methylbenzene; (b) 1,2,3,4,5-pentamethylbenzene (or, because there is only one possible structure, pentamethylbenzene)

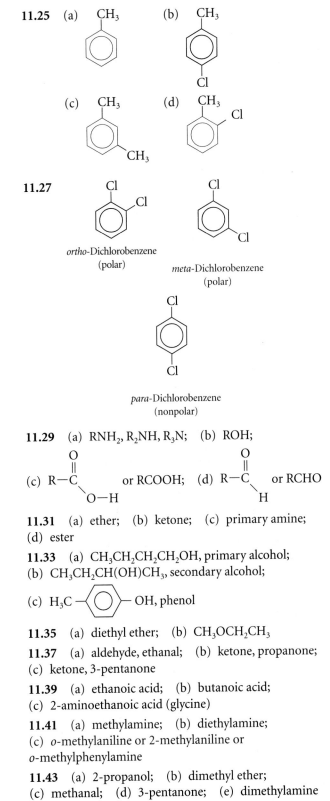

11.25 (a), (b), (c), (d)

11.27 *ortho*-Dichlorobenzene (polar); *meta*-Dichlorobenzene (polar); *para*-Dichlorobenzene (nonpolar)

11.29 (a) RNH_2, R_2NH, R_3N; (b) ROH; (c) $R{-}C(=O)O{-}H$ or $RCOOH$; (d) $R{-}C(=O)H$ or $RCHO$

11.31 (a) ether; (b) ketone; (c) primary amine; (d) ester

11.33 (a) $CH_3CH_2CH_2CH_2OH$, primary alcohol; (b) $CH_3CH_2CH(OH)CH_3$, secondary alcohol; (c) $H_3C{-}C_6H_4{-}OH$, phenol

11.35 (a) diethyl ether; (b) $CH_3OCH_2CH_3$

11.37 (a) aldehyde, ethanal; (b) ketone, propanone; (c) ketone, 3-pentanone

11.39 (a) ethanoic acid; (b) butanoic acid; (c) 2-aminoethanoic acid (glycine)

11.41 (a) methylamine; (b) diethylamine; (c) *o*-methylaniline or 2-methylaniline or *o*-methylphenylamine

11.43 (a) 2-propanol; (b) dimethyl ether; (c) methanal; (d) 3-pentanone; (e) dimethylamine

11.45 (a) alcohol, ether, aldehyde; (b) ketone; (c) tertiary amine, amide, ketone

11.47 (a)

H₂C=O (with H, H on left, C=O)

(a) structure: H\ /C=O with H below

(b) CH₃\ /C=O with CH₃ below

(c) CH₃\ /C=O with CH₃(CH₂)₃CH₂ below

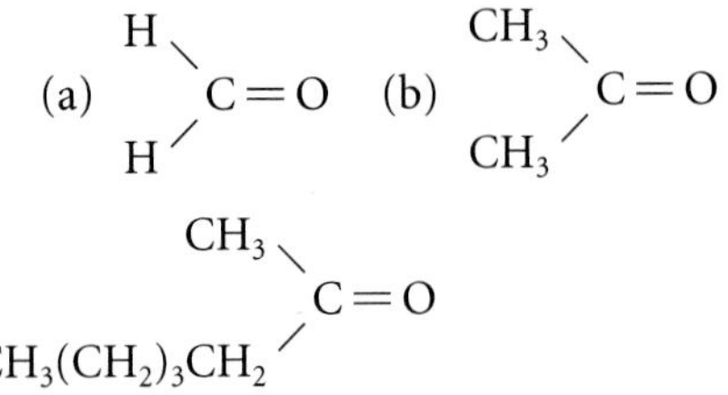

11.49 (a) ethanol; (b) 2-octanol; (c) 5-methyl-1-octanol. These reactions can be accomplished with an oxidizing agent such as acidified sodium dichromate, Na₂Cr₂O₇, although (a) and (c) may require a milder oxidizing agent, such as Ag and heat.

11.51 (a)

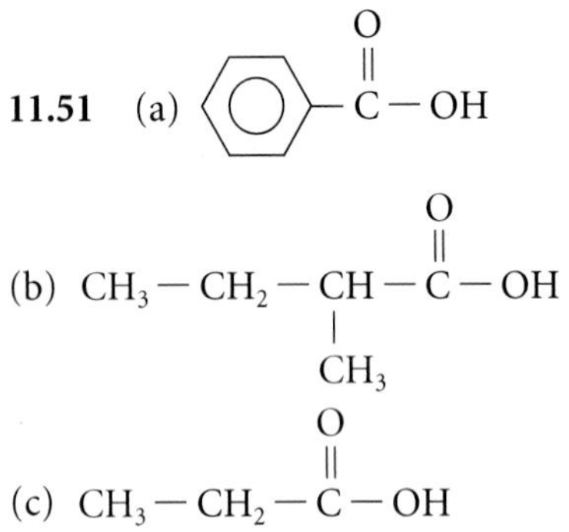

C₆H₅—C(=O)—OH

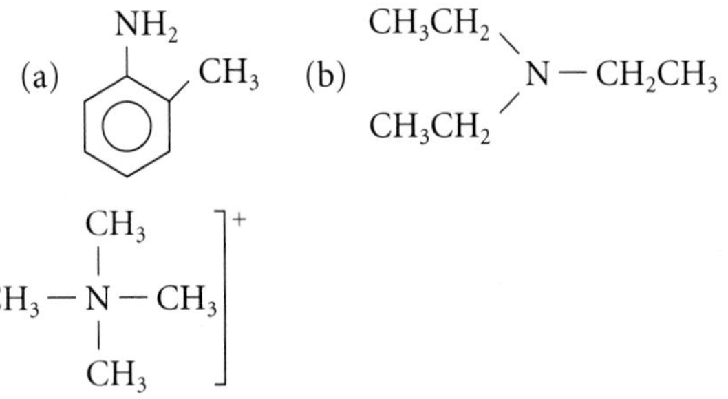

(b) CH₃—CH₂—CH—C(=O)—OH with CH₃ branch

(c) CH₃—CH₂—C(=O)—OH

11.53 (a)

aromatic ring with NH₂ and CH₃

(b) CH₃CH₂\ , CH₃CH₂/ N—CH₂CH₃

(c) [CH₃—N(CH₃)(CH₃)—CH₃]⁺ with CH₃ above and CH₃ below, charge +

11.55 (a) CH₃CH₂CH₂C(=O)O—CH(CH₃)CH₃

(b) CH₃C(=O)O—CH₂CH₂CH₂CH₂CH₃

(c) CH₃CH₂CH₂CH₂CH₂C(=O)N(CH₃)CH₂CH₃

(d) CH₃C(=O)NHCH₂CH₂CH₃

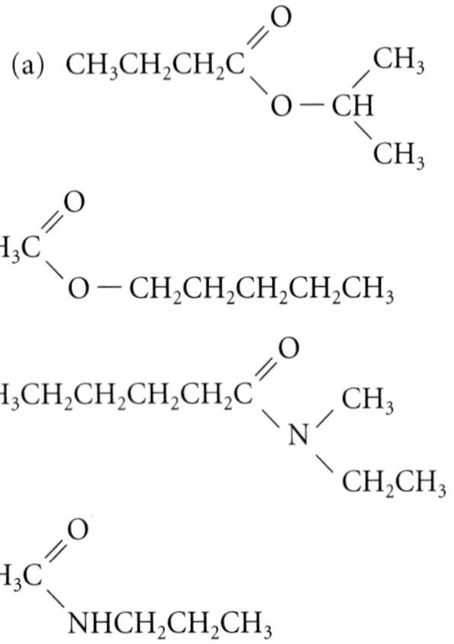

11.57 The following procedures may be used:
(1) Use an acid-base indicator and look for a color change.
(2) CH₃CH₂CHO $\xrightarrow{\text{Tollens reagent}}$ CH₃CH₂COOH + Ag(s)
(3) CH₃COCH₃ $\xrightarrow{\text{Tollens reagent}}$ no reaction
Procedure (1) distinguishes ethanoic acid from propanal and 2-propanone; (2) and (3) distinguish propanal from 2-propanone.

11.59

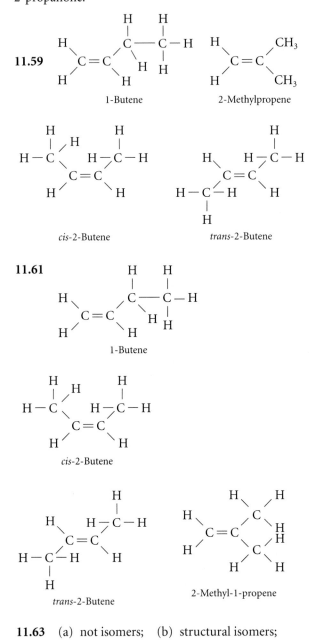

1-Butene

2-Methylpropene

cis-2-Butene

trans-2-Butene

11.61

1-Butene

cis-2-Butene

trans-2-Butene

2-Methyl-1-propene

11.63 (a) not isomers; (b) structural isomers; (c) structural isomers; (d) geometrical isomers; (e) not isomers

11.65 If only two isomeric products are formed, and they are both branched, then the only possibilities are

(a) $CH_3 - \underset{\underset{CH_3}{|}}{\overset{\overset{H}{|}}{C}} - CH_3$

(b) $CH_3 - \underset{\underset{CH_3}{|}}{\overset{\overset{Cl}{|}}{C}} - CH_3$ $Cl - \underset{\underset{H}{|}}{\overset{\overset{H}{|}}{C}} - \underset{\underset{CH_3}{|}}{\overset{\overset{H}{|}}{C}} - CH_3$

Note: All methyl groups are equivalent.

11.67 (a), (c), and (d) are optically active. C* indicates the chiral carbon atom.

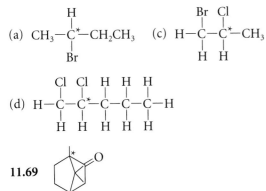

(a) $CH_3 - \underset{\underset{Br}{|}}{\overset{\overset{H}{|}}{C^*}} - CH_2CH_3$ (c) $H - \underset{\underset{H}{|}}{\overset{\overset{Br}{|}}{C}} - \underset{\underset{H}{|}}{\overset{\overset{Cl}{|}}{C^*}} - CH_3$

(d) $H - \underset{\underset{H}{|}}{\overset{\overset{Cl}{|}}{C}} - \underset{\underset{H}{|}}{\overset{\overset{Cl}{|}}{C^*}} - \underset{\underset{H}{|}}{\overset{\overset{H}{|}}{C}} - \underset{\underset{H}{|}}{\overset{\overset{H}{|}}{C}} - \underset{\underset{H}{|}}{\overset{\overset{H}{|}}{C}} - H$

11.69

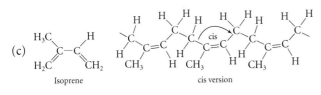

11.71

(a)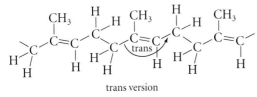

(b) $CH_2{=}CH{-}CN$ $-\underset{\underset{CN}{|}}{CH}-CH_2-\underset{\underset{CN}{|}}{CH}-CH_2-\underset{\underset{CN}{|}}{CH}-CH_2-$

Acrylonitrile

(c)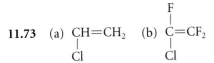

Isoprene   cis version

trans version

11.73 (a) $\underset{\underset{Cl}{|}}{CH}{=}CH_2$ (b) $\underset{\underset{Cl}{|}}{\overset{\overset{F}{|}}{C}}{=}CF_2$

11.75 (a) $-OC-CO-NH-(CH_2)_4-NHCO-CO-$
$NH-(CH_2)_4-NH-$

(b) $-OC-\underset{\underset{CH_3}{|}}{CH}-NH-OC-\underset{\underset{CH_3}{|}}{CH}-NH-$

11.77 An isotactic polymer is a polymer in which the substituents are all on the same side of the chain. A syndiotactic polymer is a polymer in which the substituent groups alternate, one side of the chain to the other. An atactic polymer is a polymer in which the groups are randomly attached, one side or the other, along the chain.

11.79 Longer chain length allows for greater intertwining of the chains, making them more difficult to pull apart. This intertwining results in (a) higher softening points, (b) greater viscosity, and (c) greater mechanical strength.

11.81 Highly linear, unbranched chains allow for maximum interaction between chains. The greater the intermolecular contact between chains, the stronger the forces between them, and the greater the strength of the material.

11.83 (a) 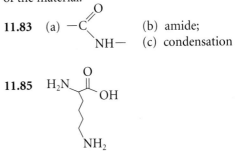   (b) amide;   (c) condensation

11.85

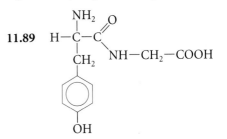

11.87 Serine, threonine, tyrosine, aspartic acid, glutamic acid, lysine, arginine, histidine, asparagine, and glutamine. Proline and tryptophan generally do not contribute to hydrogen bonding because they are typically found in hydrophobic regions of proteins. However, they are also capable of hydrogen bonding.

11.89

$H-\underset{\underset{\underset{\underset{OH}{\bigcirc}}{|}}{CH_2}}{\overset{\overset{NH_2}{|}}{C}}-\overset{\overset{O}{\|}}{C}\!\!\diagdown_{NH-CH_2-COOH}$

11.91 alcohol, aldehyde

11.93 (a) GTACTCAAT;   (b) ACTTAACGT

11.95 There are a large number and a wide variety of organic compounds because of the ability of carbon to form four bonds with as many as four different atoms or groups of atoms, the ability of carbon to bond directly to other carbon atoms to form long chains of carbon-carbon bonds (the basis for polymerization), and the ability of carbon-containing compounds to form a variety of isomers, compounds with the same molecular formula, but different structural formulas.

11.97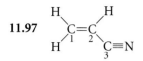

Carbon 1 has 3 σ-bonds and 1 π-bond; carbon 2 has 3 σ-bonds and one π-bond; carbon 3 has 2 σ-bonds and 2 π-bonds. The CCC bond angle is approximately 120°. Carbons 1 and 2 have sp^2 hybridization; carbon 3 has sp hybridization.

11.99 $CH_3-C\equiv C-CH_3 + 2\ HBr \rightarrow$
$CH_3-CHBr-CHBr-CH_3$

11.101 (a)

$$CH_3 \overset{1}{-} \overset{O}{\overset{||}{C}} \overset{2}{-} (CH_2)_5 \overset{3-7}{} \overset{H}{\underset{}{\diagdown}} C=C \overset{8\ 9}{} \overset{O}{\overset{||}{C}} \overset{10}{} \overset{}{\underset{H}{\diagup}} OH$$

(b) carbonyl, carbon-carbon double bond, carboxyl;
(c) 28 σ-bonds, 3 π-bonds; (d) carbons 1 and 3–7 are sp^3 hybridized; carbons 2 and 8–10 are sp^2 hybridized, as are both carbonyl oxygens.

11.103

$$H-\underset{\underset{H}{|}}{\overset{\overset{H}{|}}{C}}-\underset{\underset{H}{|}}{\overset{\overset{H}{|}}{C}}-\underset{\underset{Br}{|}}{\overset{\overset{H}{|}}{C}}-\underset{\underset{H}{|}}{\overset{\overset{H}{|}}{C}}-\underset{\underset{H}{|}}{\overset{\overset{H}{|}}{C}}-H \text{ and } H-\underset{\underset{H}{|}}{\overset{\overset{H}{|}}{C}}-\underset{\underset{Br}{|}}{\overset{\overset{H}{|}}{C}}-\underset{\underset{H}{|}}{\overset{\overset{H}{|}}{C}}-\underset{\underset{H}{|}}{\overset{\overset{H}{|}}{C}}-\underset{\underset{H}{|}}{\overset{\overset{H}{|}}{C}}-H$$

11.105 $HOCH_2$—⟨O⟩— CH_2OH

1,4-Di(hydroxymethyl)benzene

$HOOC$—⟨O⟩— $COOH$

Terephthalic acid

11.107 $CH_3(CH_2)_2CH=CH_2$
11.109 (a) The Lewis structure of ethyne is H—C≡C—H.

(b) $-C=C-C=\longleftrightarrow=C-C=C-$
$\quad\ \ |\ \ |\ \ |\qquad\qquad |\ \ |\ \ |$
$\quad\ \ H\ H\ H\qquad\quad\ H\ H\ H$

(c) Each carbon is sp^2 hybridized. (d) Because of resonance, every carbon-carbon bond in polyacetylene has considerable double-bond character. Free rotation cannot occur about double bonds; and even though these bonds are only partially double, we expect polyacetylenes to be fairly stiff.

11.111 (a) [benzene ring with NH₂] (b) sp^3 hybridized;
(c) sp^2 hybridized;
(d) Each nitrogen has a lone pair.

(e) The double and single bonds on the nitrogen atoms are linked to the alternating double bonds on the carbon atoms. Electrons are delocalized along this long chain, so yes, the nitrogens do help carry the current.

11.113 The empirical formula is C_2H_5. The molecular formula might be C_4H_{10}, which matches the general formula for alkanes, C_nH_{2n+2}. The compound cannot be an alkene or alkyne, because they all have mol H/mol C ratios less than 2.5.

11.115 (a)

$$H-\underset{\underset{H}{|}}{\overset{\overset{H}{|}}{C}}-\underset{\underset{H}{|}}{\overset{\overset{H}{|}}{C}}-O-\underset{\underset{H}{|}}{\overset{\overset{H}{|}}{C}}-\underset{\underset{H}{|}}{\overset{\overset{H}{|}}{C}}-H \qquad H-\underset{\underset{H}{|}}{\overset{\overset{H}{|}}{C}}-\underset{\underset{H}{|}}{\overset{\overset{H}{|}}{C}}-\underset{\underset{H}{|}}{\overset{\overset{H}{|}}{C}}-\underset{\underset{H}{|}}{\overset{\overset{H}{|}}{C}}-O-H$$

Dimethyl ether   1-Butanol

(b) 1-Butanol is capable of hydrogen bonding, but diethyl ether is not. The stronger interactions between 1-butanol molecules lead to its higher boiling point.

11.117 (a) 24.4 L of H_2; (b) 73.4 L of H_2;
(c) −124 kJ/mol for cyclohexene; −144 kJ/mol for benzene. (d) The calculations do not support the statement. The enthalpy of benzene is lower than its Lewis structures suggest, because of resonance. Without resonance the enthalpy of hydrogenation for benzene would be 3 × (−124 kJ/mol) = −376 kJ/mol.

11.119 (a) $H_2C=CHCH_3(g) + Br_2(l) \rightarrow$
$CH_2Br-CHBr-CH_3$
(b) Br_2 limits; (c) 22.2 g

Chapter 12

12.1 Both water and methanol are capable of hydrogen bonding, and they readily hydrogen bond to each other. Thus they intermingle at the molecular level. Toluene, on the other hand, cannot hydrogen bond to methanol. The much weaker London forces between methanol and toluene result in only limited solubility.

12.3 (a) water; (b) benzene; (c) water
12.5 (a) SF_4; (b) IF_5
12.7 (a) hydrophilic; (b) hydrophobic;
(c) hydrophilic
12.9 (a) Grease is composed principally of nonpolar hydrocarbons, which are expected to dissolve in gasoline (also a nonpolar hydrocarbon mixture) but not in the very polar water. (b) Soaps are long-chain molecules with both polar and nonpolar ends. The polar end of the soap dissolves in water, and the nonpolar end dissolves in the nonpolar grease. As such, the soap serves as a "bridge" to "connect" the grease to the water so that it can be washed away. See Fig. 12.8 and the similar figure here.

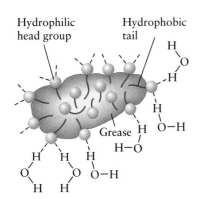

Hydrophilic head group

Hydrophobic tail

Grease

12.11 (a) sol and emulsion; (b) foam

12.13 (a) 6.4×10^{-4} mol/L; (b) 1.5×10^{-2} mol/L; (c) 2.3×10^{-3} mol/L

12.15 0.59 L

12.17 The equilibrium is $CO_2(aq) \rightleftharpoons CO_2(g)$. (a) If the partial pressure of CO_2 in the air above the solution is doubled; the equilibrium (above) will shift to the left, and the concentration of CO_2 in solution will double. (b) If the total pressure of the gas is increased by the addition of nitrogen, no change in the equilibrium (above) will occur; the partial pressure of CO_2 is unchanged, and the concentration is unchanged.

12.19 Ion hydration enthalpies for ions of the same charge within a group of elements become progressively less negative (or more positive) as we proceed down a group. This trend parallels increasing ion size. Because hydration energy is an ion-dipole interaction, it should be highest for small, high charge density ions.

12.21 The enthalpy of hydration is larger.

12.23 (a) $+6.7 \times 10^2$ J for NaCl; (b) -60 J for NaBr; (c) -2.47×10^4 J for $AlCl_3$; (d) $+3.21 \times 10^3$ J for NH_4NO_3

12.25 -51 kJ/mol

12.27 LiCl, because the hydration enthalpy is higher.

12.29 (a) The disorder in the system increases. This is true of any solution process, be it exothermic or endothermic. In dissolving exothermically, the energy (enthalpy) of the system decreases. (b) The disorder in the surroundings increases as energy is dispersed into it from the system. Thus, both the energy and disorder increase when a solute dissolves exothermically.

12.31 (a) 4.0% NaCl; (b) 3.8% NaCl; (c) 0.823% $C_{12}H_{22}O_{11}$

12.33 (a) $x_{H_2O} = 0.563$, $x_{C_2H_5OH} = 0.437$; (b) $x_{H_2O} = 0.471$, $x_{CH_3OH} = 0.529$

12.35 (a) 0.684 m; (b) 9.6 m; (c) 0.162 m

12.37 (a) 0.62 g $KClO_3$; (b) 7.4 g $KClO_3$

12.39 (a) 0.0613; (b) 3.62 m

12.41 (a) 1.8×10^{-3} for Na^+ and Cl^-, 0.996 ($\sim$1.0) for H_2O; (b) 7.1×10^{-3} for Na^+, 3.6×10^{-3} for CO_3^{2-}, and 0.99 for H_2O

12.43 (a) 0.36 m; (b) 0.0064

12.45 (a) 684 Torr; (b) 759 Torr

12.47 (a) 4.6 Torr; (b) 3.5×10^2 Torr; (c) 0.18 Torr

12.49

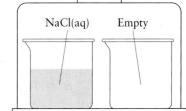

NaCl(aq)   Empty

The vapor pressure of the 0.010 m NaCl solution is lower than that of pure water. More water will evaporate from the pure water beaker until the system approaches equilibrium. The solute molecules in the 0.010 m NaCl(aq) prevent as much water from evaporating. Because the vapor pressure of the pure water is always higher than that of the water in the solution, at equilibrium it has all evaporated.

12.51 (a) 0.052; (b) 1.1×10^2 g/mol

12.53 (a) 0.051°C, 100.051°C; (b) 0.22°C, 100.22°C; (c) 0.090°C, 100.090°C

12.55 1.7×10^2 g/mol

12.57 (a) 0.19°C, -0.19°C; (b) 0.82°C, -0.82°C; (c) 0.172°C, -0.172°C

12.59 181 g/mol

12.61 (a) 1.7°C; (b) 1.63 m

12.63 (a) 1.84; (b) 84% dissociation

12.65 -0.20°C

12.67 (a) 0.24 atm; (b) 48 atm; (c) 0.72 atm

12.69 2.0×10^3 g/mol

12.71 3.3×10^3 g/mol

12.73 (a) 1.2 atm; (b) 0.048 atm; (c) 8.2×10^{-5} atm

12.75 The equilibrium between solid and aqueous $CuSO_4$ is a dynamic process. Cu^{2+} and SO_4^{2-} ions continually leave the solid and are replaced with ions returning to the solid from solution.

12.77 (a) 59.1 mL; (b) 1.13 m; (c) 32.1 g H_2SO_4

12.79 45 g of 70.% HNO_3

12.81 (a) greater; (b) less; (c) less; (d) greater

12.83 2.5×10^5 g/mol

12.85 The process of osmosis tries to equalize the concentration of water in the solution, both in the cells of the fish and the surrounding water; water from the aquarium passes through the cell membranes, causing them to expand and rupture.

12.87 (a) 4.76×10^3 g/mol; (b) -3.91×10^{-4}°C; (c) osmotic pressure

12.89 (a) panel (a): carboxylate and quaternary ammonium ions; panel (c): carboxylic acids and amides; panel (d): amides; (b) The interactions are strongest in panel (a). (c) The interactions are weakest in panel (b).

12.91 0.40 L

12.93 Amino acid (a) is more polar, because it has two carboxylic acid groups and an amine group. It should be absorbed by the stationary phase more strongly than (b). Amino acid (b) will thus have traveled more rapidly and further than amino acid (a).

12.95 (a) 6.90×10^3 g/mol; (b) 164; (c) 7.58×10^4 pm or 75.8 nm

12.97

$$\begin{array}{c}
\text{H} \\
|
\end{array}$$

H—O—C—C—H

H O   O H

H—C—C—O—H

H

Molecules of acetic acid can attract one another. The carboxyl groups form hydrogen bonds with each other, reducing the apparent number of solute species.

12.99 (a) 150. g/mol; (b) $C_6H_4Cl_2$; (c) 146.99 g/mol

Connection 3

1. (a) 0.15 mol/L for Na^+ and Cl^-, 0.30 mol/L overall; (b) 27 g

3. (a) There are three carboxylic acid groups and one alcohol group (—OH) in citric acid. (b) The three carbons in the carboxylic acid groups are all sp^2 hybridized. The other three carbons are sp^3 hybridized. (c) Citric acid can undergo hydrogen bonding at each carboxylic acid functional group and at the alcohol group. (d) Due to extensive intermolecular attractions (hydrogen bonding) and the molar mass of the material, citric acid is a solid at room temperature. Citric acid is also water soluble.

5. (a) sodium chloride, potassium citrate, glucose, citric acid, flavoring, and a color. Suitable masses are

360 mg NaCl (140 mg sodium)

250 mg $K_3C_6H_5O_7$ (70 mg potassium)

50 g glucose

Trace amounts of citric acid, flavoring, and color. (b) Your choice. There are many possibilities. (c) 0.29 m; an 8-oz. serving has 20 kcal (from the glucose).

Chapter 13

13.1 (a) $\frac{1}{3}$; (b) $\frac{2}{3}$; (c) 2

13.3 (a) 1.0 (mmol O_3)/L·s (b) 0.28 (mol $Cr_2O_4^{2-}$)/L·s

13.5 (a, c)

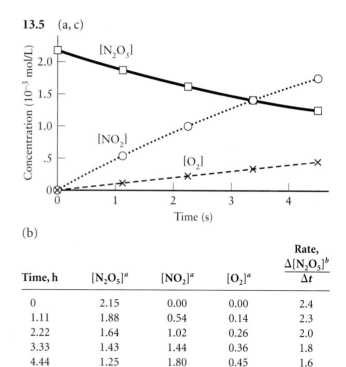

(b)

Time, h	$[N_2O_5]^a$	$[NO_2]^a$	$[O_2]^a$	Rate, $\dfrac{\Delta[N_2O_5]^b}{\Delta t}$
0	2.15	0.00	0.00	2.4
1.11	1.88	0.54	0.14	2.3
2.22	1.64	1.02	0.26	2.0
3.33	1.43	1.44	0.36	1.8
4.44	1.25	1.80	0.45	1.6

^{a}Units = 10^{-3} mol/L. ^{b}Units = 10^{-4} mol/L·h.

13.7 (a) rate = $k[X]^2[Y]^{1/2}$; (b) rate = $k[A][B][C]^{-1/2}$

13.9 (a) (mol A)/L·s; (b) /s; (c) L/(mol A)·s

13.11 third order

13.13 9.6×10^{-5} (mol N_2O_5)/L·s

13.15 (a) 2.4×10^{-5} mol/L·s; (b) by a factor of 2

13.17 (a) first order in H_2; first order in I_2; second order overall; (b) first order in SO_2; zero order in O_2; minus one-half order in SO_3; one-half order overall; (c) second order in A; zero order in B; minus second order in C; zero order overall

13.19 Because the rate increased in direct proportion to the concentrations of both reactants, the rate is first order in both reactants. rate = $k[CH_3Br][OH^-]$

13.21 (a) second order in A, first order in B, third order overall; (b) rate = $k[A]^2[B]$; (c) $k = 1.2 \times 10^2$ L²/mol²·s; (d) 0.84 mol/L·s

13.23 (a) rate = $k[ICl][H_2]$; (b) $k = 0.16$ L/mol·s (c) 2.0 m mol/L·s

13.25 (a) rate = $k[A][B]^2[C]^2$; (b) fifth order; (c) 2.85×10^{12} L⁴/mol⁴·s; (d) 1.13×10^{-2} mol/L·s

13.27 (a) 6.93×10^{-4} /s; (b) 1.8×10^{-2} /s; (c) 7.6×10^{-3} /s

13.29 (a) 200. s; (b) 800. s; (c) 634 s

13.31 (a) 5.2 h; (b) 1.78×10^{-2} mol/L; (c) 1.3×10^2 min

13.33 (a) 247 min; (b) 819 min; (c) 11 g

13.35 (a) 0.013 /s; (b) 1.7×10^2 s

13.37 The reaction is first order.

13.39 (a)

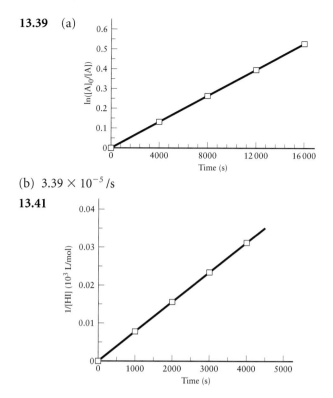

(b) 3.39×10^{-5} /s

13.41

(b) 8.0×10^{-3} L/mol·s

13.43 As temperature increases, the average speed of the reacting molecules also increases; therefore, the number of collisions between molecules increases as well. The rate of reaction is proportional to the number of collisions between molecules; thus the rate increases in proportion to this factor. This collision frequency factor is contained within the pre-exponential factor, A, in the Arrhenius equation. The fraction of molecules having the necessary energy to react increases with temperature in a manner governed by the Maxwell distribution of speeds. It turns out that this fraction is proportional to $e^{-E_a/RT}$. Therefore, we can write

$$k \propto \text{collision frequency} \times e^{-E_a/RT} \quad \text{or} \quad k = Ae^{-E_a/RT}$$

13.45

$E_a{}'$ is related to E_a as follows:
$E_a = E_a{}' + \Delta H \quad \text{or} \quad E_a{}' = E_a - \Delta H$

13.47 (a) 2.7×10^2 kJ/mol; (b) 8.0×10^{-2} /s

13.49 2.4×10^2 kJ/mol

13.51 2.6×10^9 L/mol·s

13.53 9.3×10^{-4} /s

13.55 by a factor of 3.4×10^6

13.57 81 kJ/mol

13.59 (a) rate = $k[NO]^2$, bimolecular; (b) rate = $k[Cl_2]$, unimolecular; (c) rate = $k[NO_2]^2$, bimolecular; (d) (b) and (c), because Cl and NO are radicals

13.61 (a) $O_3 + O \rightarrow 2\,O_2$;
(b) step 1: rate = $k[O_3][NO]$, bimolecular
step 2: rate = $k[NO_2][O]$, bimolecular
(c) NO is the catalyst. (d) NO_2 is the reaction intermediate.

13.63 $2\,HA + B^{2-} \rightarrow H_2B + 2\,A^-$; HB^- is a reaction intermediate.

13.65 rate = $k[NO][Cl_2]$

13.67 rate = $k[CO][Cl_2]^{3/2}$

13.69 Mechanism (a) gives a rate law of rate = $k[O_3]^2$. This does not agree with the expression given. Mechanism (b) gives a rate law of rate = $k[O_3]$, because the slow step governs the overall rate law. This mechanism does not agree with the expression given. Mechanism (c) remains, its rate law is

rate = $k_2[O_3][O]$ and
$k_1[O_3] = k_{-1}[O_2][O]$, solve for [O]
$$[O] = \frac{k_1}{k_{-1}} \frac{[O_3]}{[O_2]} \quad \text{and}$$
$$\text{rate} = k_2 \frac{k_1}{k_{-1}} [O_3] \frac{[O_3]}{[O_2]} \quad \text{or} \quad \text{rate} = k[O_3]^2[O_2]^{-1}$$
This agrees with the experimental rate law given.

13.71 (a) rate = $k[A][B]^{1/2}$; (b) $L^{1/2}/mol^{1/2}$·min

13.73 (a) 5.00 s; (b)

13.75 (a) exothermic; (b) the reverse reaction; the greater the activation energy, the more rapidly k increases with temperature

13.77 0.74 μg

13.79 Not every collision occurs with the correct geometrical arrangement or orientation of the atoms required for the products to form. Those collisions with incorrect placement of the atoms will not produce products.

13.81 (b), (c), (d), and (f) are linear

13.83 (a) If step 2 is the slow step, if step 1 is a rapid equilibrium, and if step 3 is fast also, then our proposed rate law will be rate = $k_2[N_2O_2][H_2]$. Consider the equilibrium of step 1:

$$k_1[NO]^2 = k_1'[N_2O_2] \quad [N_2O_2] = \frac{k}{k_1'}[NO]^2$$

Substituting in our proposed rate law, we have

$$\text{rate} = k_2\left(\frac{k_1}{k_1'}\right)[NO]^2[H_2] = k[NO]^2[H_2], \text{ where}$$

$$k = k_2\left(\frac{k_1}{k_1'}\right)$$

The assumptions made above reproduce the observed rate law; therefore, step 2 is the slow step.

(b)

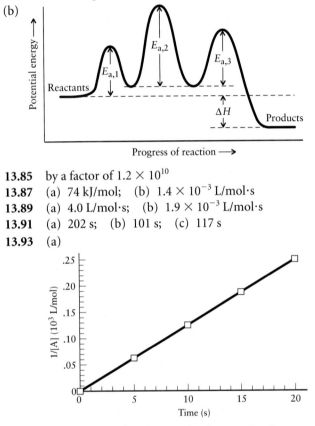

13.85 by a factor of 1.2×10^{10}

13.87 (a) 74 kJ/mol; (b) 1.4×10^{-3} L/mol·s

13.89 (a) 4.0 L/mol·s; (b) 1.9×10^{-3} L/mol·s

13.91 (a) 202 s; (b) 101 s; (c) 117 s

13.93 (a)

The linear plot shows that the reaction is second order.
(b) 12 L/mol·s

13.95 (a) $t = t_{1/2}$; (b) $t = 3t_{1/2}$; (c) $t = 15t_{1/2}$

13.97 (a) 2.5×10^2 min; (b) 4.9×10^2 min

13.99 At high concentrations of substrate, the plot of rate against substrate concentration levels off to a constant rate. Therefore,

$$\text{rate} \propto [S]^a = \text{constant}$$

implies that $a = 0$ and $[S]^0 = 1$ constant. So the reaction is zero order in substrate at high substrate concentrations.

13.101 There are many types of inhibitors. Suppose, for example, that the inhibitor functions by competing with the substrate to form an EI complex; then the slope of the curve will be reduced, and the rate of product formation for a given substrate concentration will also be less.

13.103 (a) Cl is the catalyst; (b) ClO (c) Cl, ClO, O;
(d) step 1 is initiating; step 2 is propagating;
(e) $Cl + Cl \rightarrow Cl_2$

13.105 (a) 8.0×10^{-6} mol/L·s; (b) H^+ limits;
(c) 0.20 g Se

13.107 (a) 0.0506 mg; (b) 0.106 atm

Chapter 14

14.1 (a) At the molecular level, chemical equilibrium is a dynamic process and interconversion of reactants and products continues to occur. However, at the macroscopic level, no change in concentrations of reactants and products is observable at equilibrium, because the rates of product and reactant formation are equal. (b) At a fixed temperature, the equilibrium constant is truly a constant, independent of concentrations. However, if the concentration of a reactant is increased, the concentrations of other chemical species will adjust because the equilibrium constant remains the same.

14.3 The concentrations of reactants and products stop changing when equilibrium has been attained. Equilibrium occurs when the forward and reverse rates of reaction are equal, a condition that will always be reached eventually, given enough time.

14.5 (a)–(d) The equilibrium composition in each system will be that which results in the same value of K. Thus, (d) and its inverse are identical in the two containers, but (a), (b), and (c) will be different, because the second container contains five times as much material. (e) The time required to reach equilibrium might be different in the two containers. There is no fixed simple relation between the time required to reach equilibrium and the concentration of reactant.

14.7 (a) $K_c = \dfrac{[COCl][Cl]}{[CO][Cl_2]}$; (b) $K_c = \dfrac{[HBr]^2}{[H_2][Br_2]}$;

(c) $K_c = \dfrac{[SO_2]^2[H_2O]^2}{[O_2]^3[H_2S]^2}$

14.9 (a) 0.024; (b) 6.4; (c) 1.7×10^3

14.11 (a) At equilibrium, the rates of the forward and reverse reactions are the same. The rate constants, k_1 and k_{-1}, do not change (unless temperature changes).
(b) If K_c is large for an elementary reaction, then the rate constant for the forward reaction is much larger than the rate constant of the reverse reaction. $K_c = k_1/k_{-1}$.

14.13 $K_c = 48.8$ for case 1, $K_c = 48.9$ for cases 2 and 3

14.15 (a) 4.4×10^{-4}; (b) 1.90

14.17 All the equilibria are heterogeneous, because in each case more than one phase is involved in the equilibrium. (a) $K_p = P_{H_2S}P_{NH_3}$; (b) $K_p = P_{NH_3}{}^2P_{CO_2}$;
(c) $K_p = P_{O_2}$

14.19 (a) $K_c = 1/[Cl_2]$; (b) $[N_2O][H_2O]^2$;
(c) $K_c = [CO_2]$

14.21 2.09×10^{-5} mol/L

14.23 19 atm

14.25 (a) $K_c = 160.$, $Q_c = 0.50$; (b) no, $Q_c \neq K_c$; (c) $Q_c < K_c$, tendency to form products

14.27 (a) 6.9; (b) $Q_c < K_c$, more $SO_3(g)$ will form

14.29 6.20×10^{-3}

14.31 1.58×10^{-8}

14.33 $[H_2O] = [CH_3COOC_2H_5]$; so, in experiment 1, we find 0.171 mol/L H_2O and $K_c = 3.9$; in experiment 2, 0.667 mol/L H_2O and $K_c = 4.0$; in experiment 3, 0.966 mol/L H_2O and $K_c = 3.9$

14.35 (a) 1.1×10^{-5} mol/L Cl, 1.0×10^{-3} mol/L Cl_2, 0.55%; (b) 3.18×10^{-4} mol/L F, 8.41×10^{-4} mol/L F_2, 19%; (c) Cl_2 is more stable because less decomposes.

14.37 1.7×10^{-3} %; equilibrium composition is 1.2×10^{-3} mol/L HBr, 1.1×10^{-8} mol/L H_2, and 1.1×10^{-8} mol/L Br_2

14.39 (a) 1.0×10^{-2} mol/L PCl_3, 1.0×10^{-2} mol/L Cl_2, 9.0×10^{-3} mol/L PCl_5; (b) 53%

14.41 0.20 mol/L NH_3, 8.0×10^{-4} mol/L H_2S

14.43 4.97×10^{-2} mol/L Cl_2, 2.00×10^{-1} mol/L PCl_3, 3.00×10^{-4} mol/L PCl_5

14.45 3.60×10^{-4} mol/L NO, 0.114 mol/L N_2, 0.114 mol/L O_2

14.47 1.06

14.49 (a) According to Le Chatelier's principle, an increase in the partial pressure of CO_2, a product, will favor a shift in the equilibrium to the left, thus decreasing the partial pressure of H_2. (b) According to Le Chatelier's principle, a decrease in the partial pressure of CO, a reactant, will favor a shift in the equilibrium to the left, thus decreasing the partial pressure of CO_2. (c) According to Le Chatelier's principle, an increase in the concentration of one of the reactants, CO, will favor the formation of products; thus the H_2 concentration will increase. (d) Nothing; the equilibrium constant is a constant independent of concentrations.

14.51

Change	Quantity	Effect
add NO	amount of H_2O	decrease
add NO	amount of O_2	increase
remove H_2O	amount of NO	increase
remove O_2	amount of NH_3	increase
add NH_3	K_c	no change
remove NO	amount of NH_3	decrease
add NH_3	amount of O_2	decrease

14.53 (a) reactants favored, smaller number of moles of gas; (b) reactants favored, smaller number of moles of gas; (c) reactants favored, smaller number of moles of gas; (d) no effect, same number of moles on each side; (e) reactants favored, smaller number of moles of gas

14.55 (a) The partial pressure of NH_3 increases when the partial pressure of NO increases. (b) The partial pressure of O_2 increases when the partial pressure of NH_3 decreases.

14.57 (a) products favored (endothermic reaction); (b) reactants favored (exothermic reaction); (c) reactants favored (exothermic reaction)

14.59 The amount of ammonia will be less at 700. K because K_p is smaller at 700. K than at 600. K.

14.61 More molecules of X_2 have decomposed to form X atoms. Because bond breaking is endothermic, then (a) will lead to more X atoms dissociating. Neither (b), (c), nor (d) will lead to the change shown.

14.63 (a) 1.7×10^6; (b) 7.6×10^{-4}

14.65 The reaction is not at equilibrium; there is a tendency to form product.

14.67 1.4 mol/L

14.69 3.91

14.71 0.0389 mol/L NO, 0.0489 mol/L SO_3, 0.0011 mol/L SO_2, 0.0211 mol/L NO_2

14.73 (a) 0.0100 mol/L N_2O, 0.0410 mol/L O_2; (b) 23.2

14.75 3.9×10^{-18} atm H_2, 3.9×10^{-18} atm Cl_2, 0.22 atm HCl

14.77 (a) 0.846 atm Cl_2, 1.69 atm NO, 92.7 atm NOCl; (b) 1.8%

14.79 $[CO_2] = 8.5 \times 10^{-5}$ mol/L; $[CO] = 4.9 \times 10^{-3}$ mol/L; $[O_2] = 4.6 \times 10^{-4}$ mol/L

14.81 $[NO_2] = 6.88 \times 10^{-3}$ mol/L; $[N_2O_4] = 1.02 \times 10^{-2}$ mol/L

14.83 $\alpha = \sqrt{\dfrac{K_p}{K_p + P}}$; (a) 0.95; (b) 0.91

14.85 180. kJ/mol

14.87 (a) 70. atm N_2, 10. atm H_2, 60. atm NH_3; (b) 67 atm N_2, 2 atm H_2, 7.8 atm NH_3

14.89 (c), (d), and (e) will shift the reaction toward the products; (a) and (f) will shift toward reactants; (b) will result in no change, because there are the same number of moles of gas molecules as products and as reactants.

14.91 (a) 0.32 mol/L; (b) $\Delta H_r^\circ = -2.94$ kJ/mol, so the reaction is slightly exothermic. Therefore, if the temperature is decreased, slightly more products will be present at equilibrium.

14.93 (a) $K_c = [Pb^{2+}][CrO_4^{2-}]$. Chromates are insoluble, so K_c is small. (b) $K_c = [Cs^+]^2[SO_4^{2-}]$. Cs_2SO_4 is soluble, so K_c is large. (c) $K_c = [Fe^{3+}][OH^-]^3$. $Fe(OH)_3$ is insoluble, so K_c is small.

Chapter 15

15.1 (a) amphiprotic; (b) base; (c) base; (d) acid

15.3 (a) $CH_3NH_3^+$; (b) $NH_2NH_3^+$; (c) H_2CO_3; (d) CO_3^{2-}; (e) $C_6H_5O^-$; (f) $CH_3CO_2^-$

15.5 (a) chloric acid; , where ⬤ = Cl, ⚫ = O

(b) nitrous acid; , where ⚫ = N, ⚪ = O

15.7 (a) H_2SO_4 (acid 1) + H_2O (base 2) →
H_3O^+ (acid 2) + HSO_4^- (base 1);
(b) $C_6H_5NH_3^+$ (acid 1) + H_2O (base 2) ⇌
H_3O^+ (acid 2) + $C_6H_5NH_2$ (base 1);
(c) $H_2PO_4^-$ (acid 1) + H_2O (base 2) ⇌ H_3O^+ (acid 2) +
HPO_4^{2-} (base 1); (d) HCOOH (acid 1) +
H_2O (base 2) ⇌ H_3O^+ (acid 2) + HCO_2^- (base 1);
(e) $NH_2NH_3^+$ (acid 1) + H_2O (base 2) ⇌ H_3O^+ (acid 2) +
NH_2NH_2 (base 1)

15.9 (a) HCO_3^- as an acid:
HCO_3^- (acid 1) + H_2O (base 2) ⇌ H_3O^+ (acid 2) +
CO_3^{2-} (base 1)
HCO_3^- as a base:
H_2O (acid 1) + HCO_3^- (base 2) ⇌ H_2CO_3 (acid 2) +
OH^- (base 1)
(b) HPO_4^{2-} as an acid:
HPO_4^{2-} (acid 1) + H_2O (base 2) ⇌ H_3O^+ (acid 2) +
PO_4^{3-} (base 1)
HPO_4^{2-} as a base:
H_2O (acid 1) + HPO_4^{2-} (base 2) ⇌ $H_2PO_4^-$ (acid 2) +
OH^- (base 1)

15.11 (a) Brønsted acid: HNO_3; Brønsted base: HPO_4^{2-};
(b) conjugate base to HNO_3: NO_3^-; conjugate acid to
HPO_4^{2-}: $H_2PO_4^-$

15.13 (a) 3.2×10^{-13} mol/L; (b) 1.0×10^{-10} mol/L;
(c) 5.0×10^{-14} mol/L

15.15 (a) 1.6×10^{-7} mol/L, 6.80; (b) 1.6×10^{-7}
mol/L, 6.80

15.17 $[HCl]_{initial} = [H_3O^+] = [Cl^-] = 0.96$ mol/L,
$[OH^-] = 1.0 \times 10^{-14}$ mol/L

15.19 $[Ba(OH_2)] = [Ba^{2+}] = 2.9 \times 10^{-3}$ mol/L,
$[OH^-] = 0.058$ mol/L, $[H_3O^+] = 1.7 \times 10^{-13}$ mol/L

15.21 (a) 5×10^{-4} mol/L; (b) 2×10^{-7} mol/L;
(c) 4×10^{-5} mol/L; (d) 5×10^{-6} mol/L;
(e) (b) < (d) < (c) < (a)

15.23 (a) pH = 4.70, pOH = 9.30; (b) pH = 0.00,
pOH = 14.00; (c) pH = 13.30, pOH = 0.70;
(d) pH = 4.300, pOH = 9.700

15.25 (a) pH = 2.00, pOH = 12.00; (b) pH = 0.66,
pOH = 13.34; (c) pH = 11.30, pOH = 2.70;
(d) pH = 11.146, pOH = 2.854

15.27 (a) formic acid, $K_a = 1.8 \times 10^{-4}$, $pK_a = 3.75$;
(b) acetic acid, $K_a = 1.8 \times 10^{-5}$, $pK_a = 4.75$;
(c) trichloroacetic acid, $K_a = 3.0 \times 10^{-1}$, $pK_a = 0.52$;

(d) benzoic acid, $K_a = 6.5 \times 10^{-5}$, $pK_a = 4.19$; acetic
acid < benzoic acid < formic acid < trichloroacetic
acid

15.29 (a) $K_a = 7.6 \times 10^{-3}$; (b) $K_a = 0.010$;
(c) $K_a = 5.5 \times 10^{-3}$; (d) $K_a = 5.13 \times 10^{-10}$;
$H_3AsO_3 < H_3PO_4 < H_3AsO_4 < H_3PO_3$

15.31 (a) strongest base, dihydrogen phosphite ion,
$H_2PO_3^-$; weakest base, dihydrogen arsenite ion, $H_2AsO_3^{2-}$;
(b) for $H_2PO_3^-$, $K_b = 1.0 \times 10^{-12}$; for $H_2AsO_3^{2-}$,
$K_b = 1.9 \times 10^{-5}$; (c) hydrogen arsenite ion

15.33 For oxoacids, the greater the number of highly
electronegative O atoms attached to the central atom, the
stronger the acid. This effect is related to the increased
oxidation number of the central atom as the number of O
atoms increases. Therefore, HIO_3 is the stronger acid, with
the lower pK_a.

15.35 (a) HCl is the stronger acid, because its bond
strength is much weaker than the bond in HF, and bond
strength is the dominant factor in determining the strength
of binary acids. (b) $HClO_2$ is stronger. There is one more
O atom attached to the Cl atom in $HClO_2$ than in HClO.
The additional O in $HClO_2$ helps to pull the electron of
the H atom out of the H—O bond. The oxidation state of
Cl is higher in $HClO_2$ than in HClO. (c) $HClO_2$ is
stronger. Cl has a greater electronegativity than Br, making
the H—O bond $HClO_2$ more polar than in $HBrO_2$.
(d) $HClO_4$ is stronger. Cl has a greater electronegativity
than P. (e) HNO_3 is stronger. The explanation is the
same as that for part (b). HNO_3 has one more O atom.
(f) H_2CO_3 is stronger. C has greater electronegativity than
Ge. See part (c).

15.37 (a) The $-CCl_3$ group that is bonded to the
carboxyl group, $-COOH$, in trichloroacetic acid, is more
electron withdrawing than the $-CH_3$ group in acetic acid.
Thus, trichloroacetic acid is the stronger acid. (b) The
$-CH_3$ group in acetic acid has electron-donating
properties, which means that it is less electron withdrawing
than the $-H$ attached to the carboxyl group in formic
acid, HCOOH. Thus, formic acid is a slightly stronger acid
than acetic acid. However, it is not nearly as strong as
trichloroacetic acid. The order is $CCl_3COOH \gg$
$HCOOH > CH_3COOH$.

15.39 The larger the K_a, the stronger the corresponding
acid. 2,4,6-Trichlorophenol is the stronger acid because
the chlorines have a greater electron-withdrawing power
than the hydrogens they replaced in the unsubstituted
phenol.

15.41 The larger the pK_a of an acid, the stronger the
corresponding conjugate base; hence, the order is
aniline < ammonia < methylamine < ethylamine.
Although one should not draw conclusions from such

a small data set, we might suggest the possibility that
(1) aryl amines < ammonia < alkyl amines, and
(2) methyl < ethyl < etc.

15.43 (a) $[H_3O^+] = 3.6 \times 10^{-3}$ mol/L, $[OH^-] = 2.8 \times 10^{-12}$ mol/L
(b) $[H_3O^+] = 1.7 \times 10^{-11}$ mol/L, $[OH^-] = 5.8 \times 10^{-4}$ mol/L
(c) $[H_3O^+] = 2.8 \times 10^{-12}$ mol/L, $[OH^-] = 3.6 \times 10^{-3}$ mol/L

15.45 (a) 2.22; (b) 10.65; (c) 2.51; (d) 9.76

15.47 (a) pH = 2.78, pOH = 11.22; (b) pH = 0.96, pOH = 13.04; (c) pH = 2.28, pOH = 11.72

15.49 (a) pH = 11.13, pOH = 2.87, 1.3%;
(b) pH = 9.14, pOH = 4.86, 0.080%; (c) pH = 11.56, pOH = 2.44, 1.8%; (d) pH = 10.25, pOH = 3.75, 0.90%

15.51 (a) $K_a = 0.1$, p$K_a = 1.0$; (b) $K_b = 5.7 \times 10^{-4}$, p$K_b = 3.25$

15.53 (a) 0.02 mol/L; (b) 0.02 mol/L

15.55 $K_a = 6.5 \times 10^{-5}$, pH = 2.58

15.57 pH = 11.83, $K_b = 5 \times 10^{-4}$

15.59 (a) $H_2SO_4 + H_2O \rightarrow H_3O^+ + HSO_4^-$
$HSO_4^- + H_2O \rightleftharpoons H_3O^+ + SO_4^{2-}$
(b) $H_3AsO_4 + H_2O \rightleftharpoons H_3O^+ + H_2AsO_4^-$
$H_2AsO_4^- + H_2O \rightleftharpoons H_3O^+ + HAsO_4^{2-}$
$HAsO_4^{2-} + H_2O \rightleftharpoons H_3O^+ + AsO_4^{3-}$
(c) $C_6H_4(COOH)_2 + H_2O \rightleftharpoons H_3O^+ + C_6H_4(COOH)CO_2^-$
$C_6H_4(COOH)CO_2^- + H_2O \rightleftharpoons H_3O^+ + C_6H_4(CO_2)_2^{2-}$

15.61 0.80

15.63 (a) 4.68; (b) 3.79

15.65 $[H_3O^+] = [OH^-]$

15.67 (a) C_2H_5COOH (acid 1) + H_2O (base 2) $\rightleftharpoons$ H_3O^+ (acid 2) + $C_2H_5CO_2^-$ (base 1); (b) $HClO_3$ (acid 1) + H_2O (base 2) $\rightleftharpoons H_3O^+$ (acid 2) + ClO_3^- (base 1); (c) $C_8H_7O_2COOH$ (acid 1) + H_2O (base 2) $\rightleftharpoons H_3O^+$ (acid 2) + $C_8H_7O_2COO^-$ (base 1); (d) H_2O (acid 1) + $C_8H_{10}N_4O_2$ (base 2) $\rightleftharpoons C_8H_{10}N_4O_2H^+$ (acid 2) + OH^- (base 1); (e) H_2O (acid 1) + C_5H_5N (base 2) $\rightleftharpoons$ $C_5H_5NH^+$ (acid 2) + OH^- (base 1)

15.69 (a) If a solution is a concentrated solution of an acid with $[H_3O^+] > 1$ mol/L, the pH will be negative. (b) If a solution is very basic, with $[OH^-] > 1$ M (which means $[H_3O^+] < 10^{-14}$ mol/L), the pH will be greater than 14.

15.71 (a) 4.7×10^{-10} mol/L; (b) 1.1×10^{-8} mol/L; (c) 0.98 mol/L; (d) 4.7×10^{-5} mol/L; (e) 0.010 mol/L; (f) 1.1×10^{-12} mol/L

15.73 (a) $C_6H_5OH + H_2O \rightleftharpoons H_3O^+ + C_6H_5O^-$
$NH_4^+ + H_2O \rightleftharpoons H_3O^+ + NH_3$
(b) p$K_a(C_6H_5OH) = 9.89$, p$K_a(NH_4^+) = 9.25$; (c) NH_4^+

15.75 (a) $HBrO_4$; because Br is slightly more electronegative than I. (b) HI; because H—I is a weaker bond than H—F. (c) HIO_3; because I has more oxygens attached to it in HIO_3 than in HIO_2. (d) H_2SeO_4; because Se is more electronegative than As.

15.77 0.12%, p$K_b = 6.89$, p$K_a = 7.11$

15.79 (a) $NH_3 + NH_3 \rightleftharpoons NH_4^+ + NH_2^-$;
(b) acid = NH_4^+, base = NH_2^-; (c) 33.0;
(d) 3×10^{-17} mol/L; (e) 17; (f) pNH_4 + pNH_2 = pK_{am}

15.81 (a) $D_2O + D_2O \rightleftharpoons D_3O^+ + OD^-$; (b) $K_w = 1.35 \times 10^{-15}$, p$K_w = 14.870$; (c) $[D_3O^+] = [OD^-] = 3.67 \times 10^{-8}$ mol/L; (d) pD = 7.435 = pOD; (e) pD + pOD = 14.870 = pK_w

15.83 (a) 50. kg SO_2; (b) pH = 4.82; (c) 4.52

15.85 pH = 5.82, which is close to 5.7

15.87 (a) pH = 10.94, pOH = 3.06; (b) pH = 4.06, pOH = 9.94

15.89 125 kJ/mol

Chapter 16

16.1 (a) pH < 7, acidic; $NH_4^+(aq) + H_2O(l) \rightleftharpoons H_3O^+(aq) + NH_3(aq)$;
(b) pH > 7, basic, $H_2O(l) + CO_3^{2-}(aq) \rightleftharpoons HCO_3^-(aq) + OH^-(aq)$;
(c) pH > 7, basic, $H_2O(l) + F^-(aq) \rightleftharpoons HF(aq) + OH^-(aq)$;
(d) pH = 7, neutral, K^+ is not an acid, Br^- is not a base;
(e) pH < 7, acidic; $Al(H_2O)_6^{3+}(aq) + H_2O(l) \rightleftharpoons H_3O^+(aq) + Al(H_2O)_5OH^{2+}(aq)$;
(f) pH < 7, acidic; $Cu(H_2O)_6^{2+}(aq) + H_2O(l) \rightleftharpoons H_3O^+(aq) + Cu(H_2O)_5OH^+(aq)$

16.3 (a) 5.6×10^{-10}; (b) 1.8×10^{-4}; (c) 2.8×10^{-11}; (d) 3.3×10^{-7}; (e) 2.3×10^{-8}; (f) 1.5×10^{-10}

16.5 (a) 9.00; (b) 5.13; (c) 2.92; (d) 11.23

16.7 (a) 9.17; (b) 4.74

16.9 (a) 6.7×10^{-6} mol/L; (b) 5.04

16.11 (a) The concentration of H_3O^+ decreases.
(b) The percentage of benzoic acid that is deprotonated decreases. (c) The pH decreases.

16.13 (a) p$K_a = 3.08$, $K_a = 8.3 \times 10^{-4}$; (b) 2.78

16.15 (a) 4.0×10^{-9} mol/L; (b) 2.8×10^{-10} mol/L; (c) 1.0×10^{-9} mol/L; (d) 2.8×10^{-11} mol/L

16.17 (a) pH = 2.22, pOH = 11.78; (b) pH = 1.62, pOH = 12.38; (c) pH = 1.92, pOH = 12.08

16.19 (a) 9.70; (b) 8.93; (c) 9.31

16.21 (a) 1.30; (b) 12.70; (c) 1.04

16.23 13.52

16.25 19.7 mL

16.27

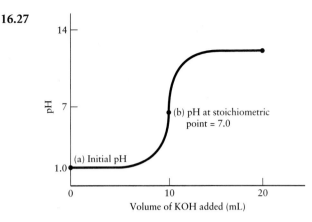

16.29 (a) 9.17×10^{-3} L; (b) 0.0183 L;
(c) 0.0635 mol/L; (d) 2.25

16.31 (a) 423 mL; (b) 0.142 mol/L

16.33 (a) 13.04; (b) 12.82; (c) 12.55; (d) 7.00;
(e) 1.81; (f) 1.55

16.35 (a) 2.87; (b) 4.57; (c) 12.5 mL; (d) 4.74;
(e) 25.0 mL; (f) 8.72; (g) thymol blue or
phenolphthalein

16.37 (a) 11.20; (b) 8.96; (c) 11.2 mL; (d) 9.25;
(e) 22.5 mL; (f) 5.24; (g) methyl red or bromocresol
green

16.39 (a) 2.02; (b) 2.6; (c) 3.15; (d) 7.9;
(e) 12.2; (f) 12.45; (g) phenol red

16.41 (a) 3.2–4.4; (b) 5.0–8.0; (c) 4.8–6.0;
(d) 8.2–10.0

16.43 The pH at the end point is about 8.9, so (c) and (d)
are suitable indicators.

16.45 (a) not a buffer; strong acid/salt of strong acid
and strong base; (b) weak acid/conjugate base; thus a
buffer:

$HClO(aq) + H_2O(l) \rightleftharpoons H_3O^+(aq) + ClO^-(aq)$

(c) weak base/conjugate acid; thus, a buffer:
$(CH_3)_3N(aq) + H_2O(l) \rightleftharpoons (CH_3)_3NH^+(aq) + OH^-(aq)$
(d) After partial neutralization, we have a weak acid
(CH_3COOH) and its salt $(NaCH_3CO_2)$ in solution. Thus,
weak acid/conjugate base and a buffer:

$CH_3COOH(aq) + H_2O(l) \rightleftharpoons H_3O^+(aq) + CH_3CO_2^-(aq)$

(e) not a buffer; complete neutralization results in a salt
and water

16.47 9.3×10^{-2}:1

16.49 (a) $pK_a = 3.08$; pH range, 2–4; (b) $pK_a = 4.19$;
pH range, 3–5; (c) $pK_{a3} = 12.68$; pH range, 11.5–13.5;
(d) $pK_{a2} = 7.21$; pH range, 6–8; (e) $pK_b = 7.97, pK_a = 6.03$; pH range, 5–7

16.51 (a) $HClO_2$ and $NaClO_2$, $pK_a = 2.00$;
(b) NaH_2PO_4 and Na_2HPO_4, $pK_{a2} = 7.21$;
(c) $CH_2ClCOOH$ and $NaCH_2ClCO_2$, $pK_a = 2.85$ or

$CH_3CH(OH)COOH$ and $NaCH_3CH(OH)CO_2$, $pK_a = 3.08$;
(d) Na_2HPO_4 and Na_3PO_4, $pK_a = 12.68$

16.53 (a) 5.6:1; (b) 77 g; (c) 1.8 g;
(d) 2.8×10^2 mL

16.55 (a) 4.75; (b) 5.02, $\Delta pH = 0.27$; (c) 4.15,
$\Delta pH = -0.60$

16.57 (a) 6.34, $\Delta pH = 1.59$; (b) 4.57, $\Delta pH = -0.18$

16.59 (a) $K_{sp} = [Ag^+][Br^-]$; (b) $K_{sp} = [Ag^+]^2[S^{2-}]$;
(c) $K_{sp} = [Ca^{2+}][OH^-]^2$; (d) $K_{sp} = [Ag^+]^2[CrO_4^{2-}]$

16.61 (a) 7.7×10^{-13}; (b) 1.7×10^{-14};
(c) 5.3×10^{-3}; (d) 6.9×10^{-9}

16.63 (a) 1.2×10^{-3} mol/L; (b) 1.0×10^{-5} mol/L;
(c) 0.23 mol/L

16.65 1.0×10^{-12}

16.67 (a) 8.0×10^{-10} mol/L; (b) 1.3×10^{-16} mol/L;
(c) 4.0×10^{-4} mol/L; (d) 6.3×10^{-6} mol/L

16.69 (a) 1.6×10^{-5} mol/L; (b) 2.7×10^2 µg

16.71 6.40

16.73 (a) A precipitate will form. (b) No precipitate
will form.

16.75 (a) A precipitate will form. (b) No precipitate
will form.

16.77 Because the sulfides of interest have the same
general formula (MS), each formula unit forms two ions in
solution. Therefore, $K_{sp} = [M^{2+}][S^{2-}]$ and the values of K_{sp}
can be compared directly. CuS precipitates first, followed
by FeS, then PbS.

16.79 (a) $Ni(OH)_2$ first, then $Mg(OH)_2$, and $Ca(OH)_2$
last. (b) $Ni(OH)_2$: 6.9, $Mg(OH)_2$: 10.0, $Ca(OH)_2$: 12.9

16.81 2.0×10^{-3} mol/L

16.83 (a) 1.0×10^{-12} mol/L; (b) 3.0×10^{-5} mol/L;
(c) 2.0×10^{-3} mol/L; (d) 0.20 mol/L

16.85 (a) $CaF_2(s) + 2 H_2O(l) \rightleftharpoons Ca^{2+}(aq) + 2 HF(aq) + 2 OH^-(aq)$, $K = 3.4 \times 10^{-32}$; (b) 9.5×10^{-7} mol/L;
(c) 2.0×10^{-5} mol/L

16.87 (a) KI is neutral, because neither K^+ nor I^- is
acidic or basic;
(b) CsF is basic, because F^- is basic;
(c) CrI_3 is acidic, because $Cr(H_2O)_6^{3+}$ is an acid;
(d) $C_6H_5NH_3Cl$ is acidic, because $C_6H_5NH_3^+$ is the
conjugate acid of the weak base $C_6H_5NH_2$;
(e) Na_2CO_3 is basic, because CO_3^{2-} is a base;
(f) $Cu(NO_3)_2$ is acidic, because $Cu(H_2O)_6^{2+}$ is an acid

16.89 12.81

16.91 The end point of an acid-base titration is the
pH at which the indicator color change is observed.
The stoichiometric point is the pH at which an exactly
equivalent amount of titrant (in moles) has been added
to the solution being titrated.

16.93 7.00

16.95 (a) 1.70; (b) 2.04; (c) 2.8; (d) 11.6;
(e) 11.88; (f) 11 mL

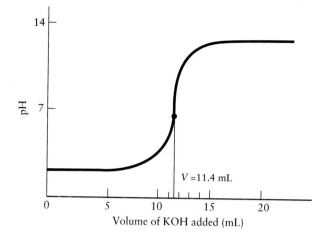

$V = 11.4$ mL

Volume of KOH added (mL)

16.97 (a) 1.58×10^{-2} L; (b) 10.54; (c) alizarin yellow R

16.99 2.8×10^{-2}:1

16.101 Equilibrium equation: $HCO_3^-(aq) + H_2O(l) \rightleftharpoons H_3O^+(aq) + CO_3^{2-}(aq)$

$$K_a = \frac{[H_3O^+][CO_3^{2-}]}{[HCO_3^-]}$$

$$pH = pK_a + \log\left(\frac{[CO_3^{2-}]}{[HCO_3^-]}\right)$$

$$10 = -\log(5.6 \times 10^{-11}) + \log\left(\frac{[CO_3^{2-}]}{[HCO_3^-]}\right)$$

$$\log\left(\frac{[CO_3^{2-}]}{[HCO_3^-]}\right) = 10 - 10.25 = -0.25$$

$$\frac{[CO_3^{2-}]}{[HCO_3^-]} = 0.56$$

Prepare a solution containing Na_2CO_3 and $NaHCO_3$ in a mole ratio of 0.56:1.

16.103 (a) 1.0×10^{-5} mol/L; (b) 5.5×10^{-7} mol/L; (c) 1.3 mg $BaSO_4$

16.105 No precipitate will form.

16.107 first: 1.54; second: 7.46

16.109 (a) 1.0×10^{-5} mol/L; (b) 1.3×10^{-4} mol/L; (c) 8.5×10^{-7} mol/L

16.111 (a) The equilibria involved are

$$CO_2(g) + H_2O(l) \rightleftharpoons H_2CO_3(aq)$$

$$H_2CO_3(aq) + H_2O(l) \rightleftharpoons H_3O^+(aq) + HCO_3^-(aq)$$

Hyperventilation decreases the amount of $CO_2(g)$, thus driving the first equilibrium to the left, which in turn drives the second equilibrium to the left. Therefore, both $[HCO_3^-]$ and $[H_2CO_3]$ are diminished, but it is not immediately apparent how their ratio is affected. We can analyze the effect on the ratio through the use of the Henderson-Hasselbalch equation:

$$pH = -\log[H_3O^+] = pK_{a1} + \log\left(\frac{[HCO_3^-]}{[H_2CO_3]}\right)$$

With loss of CO_2 during hyperventilation, there is a decrease in $[H_3O^+]$, which corresponds to an increased pH, which in turn corresponds to an increased $[HCO_3^-]/[H_2CO_3]$ ratio, or a decreased $[H_2CO_3]/[HCO_3^-]$ ratio. (b) Paramedics should anticipate alkalosis.

16.113 5.37 to 7.37

16.115 No precipitate will form.

16.117 (a) $H_2SO_4(aq) + Ba(OH)_2(aq) \rightarrow BaSO_4(s) + 2 H_2O(l)$; (b) 25 mL; (c) At the beginning of the titration, H_3O^+, $HSO_4^-(aq)$ and $SO_4^{2-}(aq)$ are all available to carry the current. As the titration proceeds, $BaSO_4(s)$ precipitates, leaving $H_3O^+(aq)$ and $OH^-(aq)$ to carry the current, but their concentrations are very low as they are in pure water (10^{-7} mol/L) at the stoichiometric point. As more $Ba(OH)_2$ is added, the concentrations of Ba^{2+} and OH^- ions increase and these ions carry the current.

16.119 (a) 0.19%; (b) 1.0×10^{-5}

16.121 (a) $SO_2(g) + 2 H_2O(l) \rightarrow H_3O^+(aq) + HSO_3^-(aq)$; (b) 4.7×10^{-7} mol; (c) 0.77 ppm

Connection 4

1. (a) $Pb^{2+}(aq) + Ca(OH)_2(aq) \rightarrow Pb(OH)_2(s) + Ca^{2+}(aq)$ and $Pb^{2+}(aq) + 2 OH^-(aq) \rightarrow Pb(OH)_2(s)$; $2 Fe^{3+} + 3 Ca(OH)_2(aq) \rightarrow 2 Fe(OH)_3(s) + 3 Ca^{2+}(aq)$ and $Fe^{3+} + 3 OH^-(aq) \rightarrow Fe(OH)_3(s)$; (b) $C_2H_5OH(aq) + 3 O_2(g) \rightarrow 2 CO_2(g) + 3 H_2O(l)$

3. The high surface areas of zeolites and activated charcoal are necessary because the larger surface areas provide many sites for organic compounds or metal ions to adhere to and be removed from the water. These materials (zeolites and activated charcoal) decolorize and deodorize water. Mg^{2+} and Ca^{2+} ions are also removed.

5. The low pH of the glacier may be due to acid precipitation. The pH of the water becomes less acidic as minerals dissolve in the water. The pH jumps to 7.2 as a result of municipal water treatment processes. Treated water is often made slightly basic to reduce corrosion of pipes; the pH increase to 8.4 is likely due to further treatment of the water to remove waste, disinfect, and reduce pipe corrosion. Hardness and total dissolved salts increase from glacier to city entry because salts from the limestone and granite canyon dissolve in the water as it flows through. Water leaving the city contains more dissolved salts and is harder as a result of ions added during use and not removed by waste treatment facilities; these facilities focus more on removing toxic metals and bacteria than on removing less harmful ions such as calcium and magnesium.

Chapter 17

17.1 (a) HBr(g); (b) NH_3(g); (c) I_2(l); (d) Ar(g) at 1.00 atm

17.3 $C(s) < H_2O(s) < H_2O(l) < H_2O(g)$. Ice has a more complex crystalline structure than carbon (either as diamond or graphite) and so has a higher entropy. $H_2O(l)$ has more disorder than $H_2O(s)$; in turn, $H_2O(g)$ has more disorder than $H_2O(l)$.

17.5 Entropy increases, because liquids are more disordered than solids.

17.7 (a) decreases; (b) increases; (c) decreases

17.9 (a) -22.0 J/K·mol; (b) 242 J/K

17.11 (a) -88.84 J/K·mol. The entropy charge is negative because the moles of gas decreased. (b) -173.00 J/K·mol. The entropy charge is negative because the moles of gas decreased. (c) $+160.6$ J/K·mol. The entropy charge is positive because the moles of gas increased. (d) -36.8 J/K·mol. The entropy charge is negative because four moles of solid products are more ordered than the four moles of solid reactants.

17.13 (a) ΔH is negative, because heat leaves the coffee as it cools. Therefore, $\Delta S_{surr} = -\Delta H/T$ is positive. Although ΔS of the system is negative, ΔS_{surr} is positive and of a greater magnitude, because the surroundings are at a lower temperature than the system (coffee). (b) No energy in the form of heat is transferred to or from the surroundings, so $\Delta S_{surr} = 0$. But ΔS of the system is positive as a result of the greater disorder generated by the mixing of the two substances. (c) When gasoline burns, both ΔS of the system and ΔS_{surr} are positive; thus ΔS_{tot} is positive. ΔS of the system is positive, because the number of moles of gas increases as CO_2(g) and H_2O(g) are formed by the combustion process. ΔS_{surr} is positive, because the reaction is exothermic and much heat is transferred to the surroundings. (d) The interdiffusion of two gases results in an increase in disorder: the gases are all mixed up. Hence, ΔS of the system is positive. $\Delta S_{surr} = 0$, because there is no energy transferred as heat; thus, $\Delta S_{tot} = \Delta S + \Delta S_{surr} = \Delta S$ is positive and the process is spontaneous.

17.15 A vapor is condensing into a liquid. ΔH is negative. The disorder of the surroundings increases, whereas the disorder of the system decreases. Because the process is spontaneous, ΔS_{tot} is positive. Thus, $\Delta S_{surr} > \Delta S$.

17.17 (a) 0.341 J/K·s; (b) 2.95×10^4 J/K·d; (c) Less, because, in the equation $\Delta S_{surr} = -\Delta H/T$, if T is higher, ΔS_{surr} is smaller.

17.19 (a) $+0.02$ J/K; (b) $+0.01$ J/K; (c) As seen in parts (a) and (b), the magnitude of the entropy change in the system is greater when the block is at 25°C. The same amount of heat has a greater effect on entropy changes at lower temperature. At high temperatures, matter is already more chaotic.

17.21 (a) $\Delta S_{surr} = -73$ J/K, $\Delta S = +73$ J/K; (b) $\Delta S_{surr} = -29$ J/K, $\Delta S = +29$ J/K; (c) $\Delta S_{surr} = +29$ J/K, $\Delta S = -29$ J/K

17.23 -116.16 J/K. $\Delta S_{tot}°$ is negative, so the process is not spontaneous.

17.25 Under constant temperature and pressure conditions, a negative value of ΔG_r corresponds to spontaneity. Frequently, the magnitude of ΔH_r is much greater than the magnitude of $T\Delta S_r$, so no matter what the sign of ΔS_r, ΔG_r will be negative if ΔH_r is negative, and the reaction will be spontaneous.

17.27 (a) negative; (b) cannot predict; depends on magnitude of ΔS, ΔH, and T; (c) A temperature change affects the sign in (b) by altering the magnitude of the $-T\Delta S$ term in ΔG.

17.29 -34°C

17.31 (a) 30. kJ/mol; (b) -11 J/K

17.33 (a) -457.2 kJ/mol. The reaction is spontaneous under standard conditions. (b) -514.4 kJ/mol. The reaction is spontaneous under standard conditions. (c) $+130.4$ kJ/mol. The reaction is not spontaneous under standard conditions, but may be at higher temperatures. (d) -133.1 kJ/mol. The reaction is spontaneous under standard conditions.

17.35 (a) $\frac{1}{2} N_2$(g) $+ \frac{3}{2} H_2$(g) $\rightarrow NH_3$(g), -16.5 kJ/mol; (b) H_2(g) $+ \frac{1}{2} O_2$(g) $\rightarrow H_2O$(g), -228.6 kJ/mol; (c) C (s, graphite) $+ \frac{1}{2} O_2$(g) $\rightarrow CO$(g), -137.2 kJ/mol; (d) $\frac{1}{2} N_2$(g) $+ O_2$(g) $\rightarrow NO_2$(g), 51.3 kJ/mol

17.37 (a) and (d) are stable, because they have negative values for $\Delta G_f°$ and thus positive values for $\Delta G°$ (decomposition).

17.39 When $\Delta S°$ (formation) is negative, the compound is more unstable at high temperatures, so (a).

17.41 (a) -141.74 kJ/mol, spontaneous; (b) $+130.4$ kJ/mol, not spontaneous; (c) $+33.1$ kJ/mol, not spontaneous; (d) $-10\,590.9$ kJ/mol, spontaneous

17.43 Free energy approaches a minimum for the total reacting system as equilibrium is approached. That is, G_{total} decreases until equilibrium is reached.

17.45 (a) 1.2×10^{80}; (b) 1.3×10^{90}; (c) 1.4×10^{-23}

17.47 (a) -12.3 kJ/mol; (b) -60.0 kJ/mol

17.49 There is a tendency to form reactants.

17.51 $+22$ kJ/mol, spontaneous toward reactants

17.53 (a) $+0.89$ kJ/mol; (b) Reactants will be formed.

17.55 (a) 4.8 kJ/mol; (b) -4.0 kJ/mol; (c) 52°C, $K_p = 1$

17.57 (a) There is a thermodynamic tendency for this reaction to occur at 25°C. (b) The tendency is greater at lower temperatures.

17.59 Reduction is not possible in either case.

17.61 Enthalpy is defined in terms of energy as $H = U + PV$. The change in enthalpy, ΔH, is related to the change in U under constant pressure conditions by the relation $\Delta H = \Delta U + P\Delta V = q_p$ (q_p is the heat effect at constant temperature.) That is, ΔH is a measure of the heat that may be obtained from the system when a change occurs at constant pressure. Entropy, on the other hand, is a measure of disorder and its total change in system and surroundings determines the spontaneity of a process. Both enthalpy and entropy changes are required: the first gives the heat of a process, the second determines whether a process occurs spontaneously.

17.63 The $Cl_2(g)$ molecule has twice the mass and not only has translational energy but also rotational and vibrational energy. The disordering due to this energy contributes to the overall energy disorder. However, matter is more organized in the molecule, so the molar entropy of the molecule is not twice that of the atom.

17.65 (a) 5×10^3 J/K; (b) 3×10^{-3} J/K

17.67 (a) $+125.8$ J/K·mol. There are more moles of gas molecules as products, so entropy increases. (b) -395.4 J/K·mol. The number of moles of gas molecules decreases, so entropy decreases. (c) $+214.62$ J/K·mol. There are more moles of gas molecules as products, so entropy increases. (d) $+446.43$ J/K·mol. There are more moles of gas as products, so entropy increases.

17.69 (a) $+91.4$ kJ/mol; (b) $708°C$, $K_p = 1$

17.71 1.6×10^2 kJ/mol

17.73 (a) 346 K; (b) 4×10^2 K; (c) 432 K; (d) 370.9 K

17.75 (a) 6.0 J/K; (b) 1.2 J/K; (c) The entropy change for vaporization is larger because gaseous water molecules are more disordered than liquid water molecules.

17.77 (a) The first part of this statement is true but the second part is false. Thermodynamics provides no information about the rates of reaction. (b) $\Delta G_r = 0$, not $\Delta G_r° = 0$, determines equilibrium. (c) Pure elements are assigned a free energy of formation equal to 0 only for their most stable forms. The most stable form corresponds to only one particular state of matter. (d) $\Delta G_r° = \Delta H_r° - T\Delta S_r°$. For the case described, $\Delta H_r°$ is negative and $\Delta S_r°$ is positive, thus $\Delta G_r° = (-) - (+)(+) = -$.

17.79 (a) $\Delta H_r° = -2808$ kJ/mol; $\Delta S_r° = 0.259$ kJ/K·mol; $\Delta G_r° = -2885$ kJ/mol; (b) As the temperature is raised, $\Delta G_r°$ becomes more negative, and more energy becomes available to do work, $\Delta G_r° = w_{max}$. Because $\Delta S_r°$ is positive, the reaction becomes more spontaneous as the temperature is raised.

17.81 94.6 mol ATP

17.83 (a) $\Delta G_r° = -326.4$ kJ/mol, $\Delta S_r° = +137.56$ J/K·mol; (b) 1.6×10^{57}; (c) The conversion of ozone to oxygen is spontaneous; if no matter or energy were added, the ozone would be used up. The rate at which this occurs, however, is another matter; thermodynamics does not give kinetic information.

17.85 (a) zero; (b) $+34$ J/mol; (c) The systems are different. The water is not pure in part (b). $\Delta G_{vap}° = 0$, only for pure water at 100.°C and 1 atm pressure. A nonvolatile solute depresses the vapor pressure.

17.87 (a) $\Delta G_r° = +91.73$ kJ/mol, $\Delta S_r° = +68.5$ J/K·mol, $K_{sp} = 8.33 \times 10^{-17}$; ΔS_{surr} must decrease more than ΔS increases (the dissolution is endothermic). (b) $\Delta G_r° = +46.41$ kJ/mol, $\Delta S_r° = -198.7$ J/K·mol, $K_{sp} = 7.32 \times 10^{-9}$; (c) $\Delta G_r° = +30.43$ kJ/mol, $\Delta S_r° = -158.0$ J/K·mol, $K_{sp} = 4.63 \times 10^{-9}$

17.89 (a) $+79.91$ kJ/mol. (b) The concentration decreases as temperature increases.

17.91 -32 kJ/mol. The reaction is spontaneous in the direction of I(g).

Chapter 18

18.1 (a) $VO^{2+}(aq) + 2 H^+(aq) + e^- \rightarrow V^{3+}(aq) + H_2O(l)$, gain of electron, reduction; (b) $PbSO_4(s) + 2 H_2O(l) \rightarrow PbO_2(s) + SO_4^{2-}(aq) + 4 H^+(aq) + 2 e^-$, loss of electrons, oxidation; (c) $H_2O_2(aq) \rightarrow O_2(g) + 2 H^+(aq) + 2 e^-$, loss of electrons, oxidation

18.3 (a) $ClO^-(aq) + H_2O(l) + 2 e^- \rightarrow Cl^-(aq) + 2 OH^-(aq)$ (b) $IO_3^-(aq) + 2 H_2O(l) + 4 e^- \rightarrow IO^-(aq) + 4 OH^-(aq)$ (c) $2 SO_3^{2-}(aq) + 2 H_2O(l) + 2 e^- \rightarrow S_2O_4^{2-}(aq) + 4 OH^-(aq)$. In each case, the reactants gain electrons, so all are reduction half-reactions.

18.5 (a) $4 Cl_2(g) + S_2O_3^{2-}(aq) + 5 H_2O(l) \rightarrow 8 Cl^-(aq) + 2 SO_4^{2-}(aq) + 10 H^+(aq)$. Cl_2 is the oxidizing agent and $S_2O_3^{2-}$ is the reducing agent. (b) $2 MnO_4^-(aq) + H^+(aq) + 5 H_2SO_3(aq) \rightarrow 2 Mn^{2+}(aq) + 3 H_2O(l) + 5 HSO_4^-(aq)$. MnO_4^- is the oxidizing agent and H_2SO_3 is the reducing agent. (c) $H_2S(aq) + Cl_2(aq) \rightarrow S(s) + 2 Cl^-(aq) + 2 H^+(aq)$. H_2S is the reducing agent and Cl_2 is the oxidizing agent. (d) $H_2O(l) + Cl_2(g) \rightarrow HOCl(aq) + H^+(aq) + Cl^-(aq)$. Cl_2 is both the oxidizing and the reducing agent.

18.7 (a) $3 O_3(g) + Br^-(aq) \rightarrow 3 O_2(g) + BrO_3^-(aq)$. O_3 is the oxidizing agent and Br^- is the reducing agent. (b) $3 Br_2(l) + 6 OH^-(aq) \rightarrow 5 Br^-(aq) + BrO_3^-(aq) + 3 H_2O(l)$. Br_2 is both the oxidizing agent and the reducing agent. (c) $2 Cr^{3+}(aq) + 4 OH^-(aq) + 3 MnO_2(s) \rightarrow 2 CrO_4^{2-}(aq) + 2 H_2O(l) + 3 Mn^{2+}(aq)$. Cr^{3+} is the reducing agent and MnO_2 is the oxidizing agent. (d) $P_4(s) + 3 H_2O(l) + 3 OH^-(aq) \rightarrow 3 H_2PO_2^-(aq) + PH_3(g)$. $P_4(s)$ is both the oxidizing agent and the reducing agent.

18.9 (a) $HSO_3^-(aq) + H_2O(l) \rightarrow HSO_4^-(aq) + 2 H^+(aq) + 2 e^-$ and $2 HSO_3^-(aq) \rightarrow S_2O_6^{2-}(aq) + 2 H^+(aq) + 2 e^-$; (b) $I_2(aq) + 2 e^- \rightarrow 2 I^-(aq)$ and $I_2(aq) + HSO_3^-(aq) + H_2O(l) \rightarrow 2 I^-(aq) + HSO_4^-(aq) + 2 H^+(aq)$

18.11 $MnO_4^-(aq) + 8 H^+(aq) + 5 e^- \rightarrow Mn^{2+}(aq) + 4 H_2O(l)$
$C_6H_{12}O_6(aq) + 6 H_2O(l) \rightarrow 6 CO_2(g) + 24 H^+(aq) + 24 e^-$;
$24 MnO_4^-(aq) + 72 H^+(aq) + 5 C_6H_{12}O_6(aq)$
$\rightarrow 24 Mn^{2+}(aq) + 66 H_2O(l) + 30 CO_2(g)$

18.13 (a) anode; (b) positive

18.15 (a) $Zn^{2+}(aq) + 2 e^- \rightarrow Zn(s)$; (b) $Fe^{3+}(aq) + e^- \rightarrow Fe^{2+}(aq)$. Platinum metal is an inert electrode and does not participate in the electrode reaction.
(c) $Cl_2(g) + 2 e^- \rightarrow 2 Cl^-(aq)$; (d) $Hg_2Cl_2(s) + 2 e^- \rightarrow 2 Hg(l) + 2 Cl^-(aq)$

18.17 (a) $Ni^{2+}(aq) + 2 e^- \rightarrow Ni(s)$ (cathode), $Zn(s) \rightarrow Zn^{2+}(aq) + 2 e^-$ (anode), $Ni^{2+}(aq) + Zn(s) \rightarrow Ni(s) + Zn^{2+}(aq)$ and $Zn(s)|Zn^{2+}(aq)||Ni^{2+}(aq)|Ni(s)$;
(b) $Ce^{4+}(aq) + e^- \rightarrow Ce^{3+}(aq)$ (cathode), $2 I^-(aq) \rightarrow 2 e^- + I_2(s)$ (anode), $2 I^-(aq) + 2 Ce^{4+}(aq) \rightarrow 2 Ce^{3+}(aq) + I_2(s)$ and $Pt(s)|I^-(aq)|I_2(s)||Ce^{4+}(aq), Ce^{3+}(aq)|Pt(s)$;
(c) $Cl_2(g) + 2 e^- \rightarrow 2 Cl^-(aq)$, (cathode), $H_2(g) \rightarrow 2 H^+(aq) + 2 e^-$ (anode), $H_2(g) + Cl_2(g) \rightarrow 2 HCl(aq)$ and $Pt(s)|H_2(g)|H^+(aq)||Cl^-(aq)|Cl_2(g)|Pt(s)$;
(d) $Au^+(aq) + e^- \rightarrow Au(s)$ (cathode), $Au(s) \rightarrow Au^{3+}(aq) + 3 e^-$ (anode), $3 Au^+(aq) \rightarrow 2 Au(s) + Au^{3+}(aq)$ and $Au(s)|Au^{3+}(aq)||Au^+(aq)|Au(s)$

18.19 (a) cathode: $Cl_2(g) + 2 e^- \rightarrow 2 Cl^-(aq)$; anode: $H_2(g) \rightarrow 2 H^+(aq) + 2 e^-$; $Cl_2(g) + H_2(g) \rightarrow 2 H^+(aq) + 2 Cl^-(aq)$; (b) cathode: $V^{2+}(aq) + 2 e^- \rightarrow V(s)$; anode: $U(s) \rightarrow U^{3+}(aq) + 3 e^-$; $3 V^{2+}(aq) + 2 U(s) \rightarrow 2 U^{3+}(aq) + 3 V(s)$; (c) cathode: $O_2(g) + 2 H_2O(l) + 4 e^- \rightarrow 4 OH^-(aq)$; anode: $2 H_2O(l) \rightarrow O_2(g) + 4 H^+(aq) + 4 e^-$; $H_2O(l) \rightarrow H^+(aq) + OH^-(aq)$; (d) cathode: $Hg_2Cl_2(s) + 2 e^- \rightarrow 2 Hg(l) + 2 Cl^-(aq)$; anode: $Sn^{2+}(aq) \rightarrow Sn^{4+}(aq) + 2 e^-$; $Sn^{2+}(aq) + Hg_2Cl_2(s) \rightarrow 2 Hg(l) + 2 Cl^-(aq) + Sn^{4+}(aq)$

18.21 (a) $MnO_4^-(aq) + 8 H^+(aq) + 5 e^- \rightarrow Mn^{2+}(aq) + 4 H_2O(l)$ (cathode half-reaction), $5 [Fe^{3+}(aq) + e^- \rightarrow Fe^{2+}(aq)]$ (anode half-reaction); (b) Reversing the anode reaction and adding the two equations yields $MnO_4^-(aq) + 5 Fe^{2+}(aq) + 8 H^+(aq) \rightarrow Mn^{2+}(aq) + 5 Fe^{3+}(aq) + 4 H_2O(l)$; (c) The cell diagram is $Pt(s)|Fe^{2+}(aq), Fe^{3+}(aq)||MnO_4^-(aq), Mn^{2+}(aq), H^+(aq)|Pt(s)$

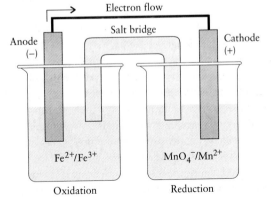

18.23 (a) cathode: $Ag^+(aq) + e^- \rightarrow Ag(s)$; anode: $Ag(s) + Br^-(aq) \rightarrow AgBr(s) + e^-$; $Ag(s)|Ag^+(aq)||Br^-(aq)|AgBr(s)|Ag(s)$; (b) cathode: $O_2(g) + 4 H^+(aq) + 4 e^- \rightarrow 2 H_2O(l)$; anode: $4 OH^-(aq) \rightarrow O_2(g) + 2 H_2O(l) + 4 e^-$; $Pt(s)|O_2(g)|OH^-(aq)||H^+(aq)|O_2(g)|Pt(s)$; (c) cathode: $Ni(OH)_3(s) + e^- \rightarrow Ni(OH)_2(s) + OH^-(aq)$; anode: $Cd(s) + 2 OH^- \rightarrow Cd(OH)_2(s) + 2 e^-$; $Cd(s)|Cd(OH)_2(s)|KOH(aq)||Ni(OH)_3(s)|Ni(OH)_2(s)|Ni(s)$

18.25 (a) 24; (b) 6; (c) 4

18.27 (a) $+0.75$ V; (b) $+0.37$ V; (c) $+0.52$ V

18.29 (a) -1.4×10^2 kJ/mol; (b) -36 kJ/mol; (c) -1.0×10^2 kJ/mol

18.31 (a) Cu < Fe < Zn < Cr (for the +2 ions); (b) Mg < Na < K < Li; (c) V < Ti < Al < U; (d) Au < Ag < Sn < Ni (for formation of Au^+, Ag^+, Sn^{2+}, and Ni^{2+})

18.33 -0.37 V

18.35 (a) Pt^{2+}/Pt, $E° = +1.20$ V, Pt^{2+} is oxidizing agent (cathode), AgF/Ag, F^-, $E° = +0.78$ V, Ag is reducing agent (anode), $Ag(s)|AgF(s)|F^-(aq)||Pt^{2+}(aq)|Pt(s)$, $+0.42$ V; (b) I_3^-/I^-, $E° = +0.53$ V, I_3^- is oxidizing agent (cathode), Cr^{3+}/Cr^{2+}, $E° = -0.41$ V, Cr^{2+} is reducing agent (anode), $Pt(s)|Cr^{2+}(aq), Cr^{3+}(aq)||I^-(aq), I_3^-(aq)|Pt(s)$, $+0.94$ V

18.37 (a) no; (b) yes; $3 Pb^{2+}(aq) + 2 Cr(s) \rightarrow 3 Pb(s) + 2 Cr^{3+}(aq)$, $+0.61$ V; (c) yes, $5 Cu(s) + 2 MnO_4^-(aq) + 16 H^+(aq) \rightarrow 5 Cu^{2+}(aq) + 2 Mn^{2+}(aq) + 8 H_2O(l)$, $+1.17$ V; (d) no

18.39 (a) reduction: $I_2 + 2 e^- \rightarrow 2 I^-$; oxidation: $H_2 \rightarrow 2 H^+ + 2 e^-$, $\Delta G° = -1.0 \times 10^2$ kJ/mol; (b) not spontaneous; (c) reduction: $Pb^{2+} + 2 e^- \rightarrow Pb$; oxidation: $Al \rightarrow Al^{3+} + 3 e^-$, $\Delta G° = -886$ kJ/mol

18.41 O_2 could be used because $E° (O_2, H^+/H_2O) > E° (Br_2/Br^-)$. It is not used because that reaction is so much slower than the one with Cl_2.

18.43 (a) $K_c = [H^+]^2[Cl^-]^2/[H_2]$; (b) $K_c = [NO][Fe^{3+}]^3/[Fe^{2+}]^3[H^+]^4[NO_3^-]$

18.45 (a) 6×10^{-16}, or about 10^{-15}; (b) 1×10^{31} or about 10^{31}; (c) 1×10^2

18.47 The aqueous sodium persulfate will work; $K = 2 \times 10^2$

18.49 (a) E_{cell} decreases; (b) E_{cell} increases; (c) E_{cell} increases; (d) no effect

18.51 (a) 3×10^6; (b) 1.0

18.53 (a) $+0.030$ V; (b) $+6 \times 10^{-2}$ V

18.55 (a) $+0.47$ V; (b) -1.44 V; (c) $+0.30$ V

18.57 (a) pH = 1.0; (b) $[Cl^-] = 9.9 \times 10^{-2}$ mol/L

18.59 A primary cell is the primary source of the electrical energy produced by its operation. A secondary cell is an energy storage device that stores electrical energy produced elsewhere and releases it upon operation (discharge). It is the secondary source of the electrical energy. Most primary cells produce electricity from chemicals that were sealed into them when they were made. They are not normally rechargeable. A secondary cell is one that must be charged from some other electrical supply before use. It is normally rechargeable.

18.61 (a) The electrolyte is $KOH(aq)/HgO(s)$. (b) The oxidizing agent is $HgO(s)$. (c) $HgO(s) + Zn(s) \rightarrow Hg(l) + ZnO(s)$

18.63 The anode reaction is $Zn(s) \rightarrow Zn^{2+}(aq) + 2 e^-$; this reaction supplies the electrons to the external circuit. The cathode reaction is $MnO_2(s) + H_2O(l) + e^- \rightarrow MnO(OH)_2(s) + OH^-(aq)$. The $OH^-(aq)$ produced reacts with $NH_4^+(aq)$ from the $NH_4Cl(aq)$ present: $NH_4^+(aq) + OH^-(aq) \rightarrow H_2O(l) + NH_3(g)$. The $NH_3(g)$ produced complexes with the $Zn^{2+}(aq)$ produced in the anode reaction: $Zn^{2+}(aq) + 4 NH_3(g) \rightarrow [Zn(NH_3)_4]^{2+}(aq)$.

18.65 (a) $KOH(aq)$; (b) $2 Ni(OH)_2(s) + 2 OH^-(aq) \rightarrow 2 Ni(OH)_3(s) + 2 e^-$

18.67 Comparison of the standard potentials shows that Cr is more easily oxidized than Fe, so the presence of Cr retards the rusting of Fe.

18.69 (a) $Fe_2O_3 \cdot H_2O$; (b) H_2O and O_2 jointly oxidize iron. (c) Water is more highly conducting if it contains dissolved ions, so the rate of rusting is increased.

18.71 (a) aluminum or magnesium; (b) cost, availability, toxicity of products to the environment; (c) Fe could act as the anode of an electrochemical cell if Cu^{2+} or Cu^+ are present; hence it could be oxidized at the point of contact. Water with dissolved ions acts as the electrolyte.

18.73 (a) anode; (b) positive

18.75 (a) $Co^{2+}(aq) + 2 e^- \rightarrow Co(s)$; (b) $2 H_2O(l) \rightarrow O_2(g) + 4 H^+(ag) + 4 e^-$; (c) $+1.10$ V

18.77 (a) $Cu^{2+}(aq) + 2 e^- \rightarrow Cu(s)$, reduction, cathode; (b) $Na^+(l) + e^- \rightarrow Na(l)$, reduction, cathode; (c) $2 Cl^-(l) \rightarrow Cl_2(g) + 2 e^-$, oxidation, anode; (d) $2 H_2O(l) + 2 e^- \rightarrow H_2(g) + 2 OH^-(aq)$, reduction, cathode

18.79 (a) Water is reduced. (b) Water is reduced. (c) Metal ion is reduced. (d) Metal ion is reduced.

18.81 (a) 0.161 mol e^- (b) 22.2 mol e^-; (c) 35.7 mol e^-

18.83 (a) 7.9 s; (b) 1.3 mg Cu

18.85 (a) 0.52 A; (b) 0.19 A

18.87 $+2$

18.89 0.92 g Zn

18.91 (a) $3 I^-(aq) \rightarrow I_3^-(aq) + 2 e^-$, oxidation; (b) $SeO_4^{2-}(aq) + H_2O(l) + 2 e^- \rightarrow SeO_3^{2-} + 2 OH^-(aq)$, reduction

18.93 (a) -0.27 V; (b) $+0.07$ V

18.95 The $E_{cell}°$ for the gold/permanganate cell is positive and thus spontaneous, so gold will be oxidized. The $E_{cell}°$

for the gold/dichromate cell is negative and not spontaneous, so gold will not be oxidized.

18.97 (a) $+0.17$ V; -98 kJ/mol; (b) not spontaneous; (c) $+0.36$ V; -2.1×10^2 kJ/mol

18.99 (a) 10^{-3} mol/L; (b) 10^{-6} mol/L; (c) 1×10^{-4} mol/L

18.101 (a) $+0.03$ V; (b) -0.27 V. The cell changes from spontaneous to nonspontaneous as a function of concentration.

18.103 (a) $+0.63$ V; (b) 9.8×10^{-3} mol/L

18.105 $+4$

18.107 105 s

18.109 (a) 52 g/mol; (b) chromium

18.111 Set up a cell at which one electrode is the silver–silver chloride electrode (anode, assume $[Cl^-] = 1.0$ mol/L) and the other electrode is the hydrogen electrode (cathode, assume $P_{H_2} = 1.00$ atm). The E of this cell will be sensitive to $[H^+]$ and can be used to obtain pH. For this system, (a) pH = $(E - 0.22$ V$)/0.0592$ V; (b) pOH = $14.00 -$ pH

18.113 (a) 7.46×10^{-2} mol; (b) 0.43

18.115 (a) $2 H_2O(l) + 3 H_2(g) + N_2(g) \rightarrow 2 NH_4^+(aq) + 2 OH^-(aq)$; (b) 1.16×10^3 kJ; (c) This fuel cell is not thermodynamically feasible.

18.117 93 A

18.119 0.08 V to 0.09 V

18.121 (a) closed; (b) open; (c) closed

18.123 4×10^{-6}

18.125 7.94×10^4 s

18.127 (a) 1.1×10^2 kg; (b) 1.0×10^5 L

Connection 5

1. Lithium is lightweight, has a low melting point, and has a very negative electrode potential. Sodium is also lightweight and has a negative electrode potential. Magnesium may also be appropriate. Potassium, rubidium, and cesium have negative electrode potentials, but they are rather dense materials and may not be as suitable.

3. (a) Lead-acid, nicad, and the zinc-air cell are labeled as least expensive. If cost were the only factor, the least expensive of these three would be the best choice. (b) The zinc-air cell appears to be the smallest. (c) The lead-acid cell, NiMH cell, and lithium-ion/lithium-polymer cell all have long shelf lives and long cycle lives. (d) Minimal mass suggests the zinc-air cell, because it has little mass. (e) Our preference is for a battery with a long cycle life that also is as harmless to the environment as possible. Our choice would be either the lithium-ion/lithium-polymer cell or the NiMH cell. Both are expensive, but worth it. If power concerns became problematic, we would choose the lithium-ion/lithium-polymer cell.

5. (a) Lead compounds are toxic. The acid in the battery may also damage surrounding foliage or damage storage containers. (b) Sodium reacts violently with moisture, forming hydrogen gas (explosive) and strong base. (c) Both nickel and cadmium are toxic metals (cadmium is more problematic).

Chapter 19

19.1 (a) saline; (b) molecular; (c) molecular; (d) metallic

19.3 (a) $H_2(g) + Cl_2(g) \xrightarrow{\text{light}} 2\,HCl(g)$

(b) $H_2(g) + 2\,Na(l) \xrightarrow{\Delta} 2\,NaH(s)$

(c) $P_4(s) + 6\,H_2(g) \rightarrow 4\,PH_3(g)$

19.5 (a) $Li^+[H\!:\!]^-$ (b) $H\!-\!\overset{\displaystyle H}{\underset{\displaystyle H}{Si}}\!-\!H$ (c) $H\!-\!\overset{\displaystyle\cdot\cdot}{Sb}\!-\!H$ with H below

19.7 (a) Binary hydrogen compounds increase in acidity from left to right across Period 2. As the element attached to hydrogen becomes more electronegative, the electrons are pulled more toward that element. The hydrogen can then be transferred to another molecule as H^+ (e.g., $H^+ + H_2O \rightarrow H_3O^+$), and thus the compound serves as an acid. (b) Oxides also become more acidic (and less basic) from left to right across Period 2.

19.9 In the majority of its reactions, hydrogen acts as a reducing agent. Examples are $2\,H_2(g) + O_2(g) \rightarrow 2\,H_2O(l)$ and various ore reduction processes, such as $NiO(s) + H_2(g) \xrightarrow{\Delta} Ni(s) + H_2O(g)$. With highly electropositive elements, such as the alkali metals, $H_2(g)$ acts as an oxidizing agent and forms metal hydrides, for example, $2\,K(s) + H_2(g) \rightarrow 2\,KH(s)$.

19.11 (a) $CO(g) + H_2O(g) \xrightarrow{400°C,\ Fe/Cu} CO_2(g) + H_2(g)$

(b) $2\,Li(s) + 2\,H_2O(l) \rightarrow 2\,LiOH(aq) + H_2(g)$

(c) $Mg(s) + 2\,H_2O(l) \rightarrow Mg(OH)_2(aq) + H_2(g)$

(d) $2\,K(s) + H_2(g) \rightarrow 2\,KH(s)$

19.13 (a) 206.10 kJ/mol; (b) −41.16 kJ/mol; (c) 164.94 kJ/mol

19.15 0.9

19.17 (a) red; (b) violet; (c) yellow; (d) violet

19.19 (a) $4\,Li(s) + O_2(g) \rightarrow 2\,Li_2O(s)$

(b) $6\,Li(s) + N_2(g) \xrightarrow{\Delta} 2\,Li_3N(s)$

(c) $2\,Na(s) + 2\,H_2O(l) \rightarrow 2\,NaOH(aq) + H_2(g)$

(d) $4\,KO_2(s) + 2\,H_2O(g) \rightarrow 4\,KOH(s) + 3\,O_2(g)$

19.21 (a) $Ca(s) + H_2(g) \xrightarrow{\Delta} CaH_2(s)$

(b) $2\,NaHCO_3(s) \rightarrow Na_2O(s) + 2\,CO_2(g) + H_2O(g)$

19.23 (a) sodium chloride, NaCl; (b) potassium chloride, KCl

19.25 7.15 g $Na_2CO_3 \cdot 10H_2O$

19.27 Be is the weakest reducing agent, Mg is stronger, but weaker than the remaining members of the group, all of which have approximately the same reducing strength. This effect is related to the very small radius of the Be^{2+} ion, 27 pm; its strong polarizing power introduces much covalent character into its compounds. Thus, Be attracts electrons more strongly and does not release them as readily as other members of the group. Mg^{2+} is also a small ion, 58 pm, so the same reasoning applies to it also, but to a lesser extent. The remaining ions of the group are considerably larger, release electrons more readily, and are better reducing agents.

19.29 (a) beryl; (b) limestone or dolomite; (c) dolomite

19.31 (a) magnesium sulfate heptahydrate, $MgSO_4 \cdot 7H_2O$; (b) calcium carbonate, $CaCO_3$; (c) magnesium hydroxide, $Mg(OH)_2$

19.33 (a) $2\,Al(s) + 2\,OH^-(aq) + 6\,H_2O(l) \rightarrow$
$$2\,[Al(OH)_4]^-(aq) + 3\,H_2(g)$$

(b) $Be(s) + 2\,OH^-(aq) + 2\,H_2O(l) \rightarrow$
$$[Be(OH)_4]^{2-}(aq) + H_2(g)$$

The similarity of Be and Al in chemical reactions is an example of the diagonal relationship in the periodic table; namely, the similar chemical behavior of elements that are diagonal neighbors of each other, such as Be and Al.

19.35 (a) $Mg(OH)_2(s) + 2\,HCl(aq) \rightarrow$
$$MgCl_2(aq) + 2\,H_2O(l)$$

(b) $Ca(s) + 2\,H_2O(l) \rightarrow Ca(OH)_2(aq) + H_2(g)$

(c) $BaCO_3(s) \xrightarrow{\Delta} BaO(s) + CO_2(g)$

19.37 (a) $:\!\overset{\cdot\cdot}{\underset{\cdot\cdot}{Cl}}\!-\!Be\!-\!\overset{\cdot\cdot}{\underset{\cdot\cdot}{Cl}}\!:$ (b) 180°; (c) *sp*

19.39 51.18%

19.41 The overall equation for the electrolytic reduction in the Hall process is $4\,Al^{3+}(\text{melt}) + 6\,O^{2-}(\text{melt}) + 3\,C(s,\,gr) \rightarrow 4\,Al(l) + 3\,CO_2(g)$

19.43 (a) boric acid, $B(OH)_3$; (b) alumina, Al_2O_3; (c) borax, $Na_2B_4O_7 \cdot 10H_2O$

19.45 (a) $B_2O_3(s) + 3\,Mg(l) \xrightarrow{\Delta} 2\,B(s) + 3\,MgO(s)$

(b) $2\,Al(l) + 3\,Cl_2(g) \rightarrow 2\,AlCl_3(s)$

(c) $4\,Al(s) + 3\,O_2(g) \rightarrow 2\,Al_2O_3(s)$

19.47 (a) The hydrate of $AlCl_3$, that is, $AlCl_3 \cdot 6H_2O$, functions as a deodorant and antiperspirant. (b) α-Alumina is corundum. It is used as an abrasive in sandpaper. (c) $B(OH)_3$ is an antiseptic and insecticide.

19.49 (a) $B_2H_6(g) + 6\,H_2O(l) \rightarrow 2\,B(OH)_3(aq) + 6\,H_2(g)$

(b) $B_2H_6(g) + 3\,O_2(g) \rightarrow B_2O_3(g) + 3\,H_2O(l)$

19.51 Silicon occurs widely in the Earth's crust in the form of silicates in rocks and as silicon dioxide in sand. It is obtained from quartzite, a form of quartz (SiO_2), by the following processes:
(1) reduction in an electric arc furnace

$$SiO_2(s) + 2\ C(s) \longrightarrow Si(s, crude) + 2\ CO(g)$$

(2) purification of the crude product in two steps

$$Si(s, crude) + 2\ Cl_2(g) \longrightarrow SiCl_4(l)$$

followed by reduction with hydrogen to the pure element

$$SiCl_4(l) + 2\ H_2(g) \longrightarrow Si(s, pure) + 4\ HCl(g)$$

19.53 In diamond, carbon is sp^3 hybridized and forms a tetrahedral, three-dimensional network structure, which is extremely rigid. Graphite carbon is sp^2 hybridized and planar, and its application as a lubricant results from the fact that the two-dimensional sheets can "slide" across one another, thereby reducing friction. In diamond, all electrons are localized in sp^3 hybridized C—C σ-bonds, so diamond is a poor conductor of electricity.

19.55 (a) carborundum, SiC; (b) silica, SiO_2; (c) zircon, $ZrSiO_4$

19.57 (a) $SiCl_4(l) + 2\ H_2(g) \rightarrow Si(s) + 4\ HCl(g)$;

(b) $SiO_2(s) + 3\ C(s) \xrightarrow{2000°C} SiC(s) + 2\ CO(g)$;
(c) $Ge(s) + 2\ F_2(g) \rightarrow GeF_4(s)$; (d) $CaC_2(s) + 2\ H_2O(l) \rightarrow Ca(OH)_2(s) + C_2H_2(g)$

19.59

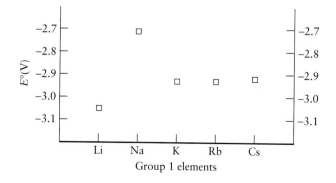

formal charges: Si = 0, O = −1
oxidation numbers: Si = +4, O = −2
This is an AX_4 VSEPR structure; therefore the shape is tetrahedral.

19.61 46.75%

19.63 4.00 mg HF

19.65 $2.4 \times 10^4\ m^2$

19.67 (a) The $Si_2O_7^{6-}$ ion is built from two SiO_4^{4-} tetrahedral ions in which the silicate tetrahedra share one O atom. See Figs. 19.43 and 19.44a. This is the only case in which one O is shared. (b) The pyroxenes, for example, jade, $NaAl(SiO_3)_2$, consist of chains of SiO_4 units in which two O atoms are shared by neighboring units. The repeating unit has the formula SiO_3^{2-}. See Fig. 19.46.

19.69 (a)

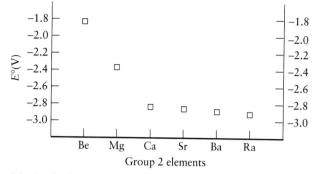

(b) For both groups, the trend in standard potentials with increasing atomic number is overall downward (they become more negative), but lithium is anomalous. This overall downward trend makes sense, because we expect that it is easier to remove electrons that are farther away from the nuclei. However, because there are several factors that influence ease of removal, the trend is not smooth. The potentials are a net composite of the free energies of sublimation of solids, dissociation of gaseous molecules, ionization enthalpies, and enthalpies of hydration of gaseous ions. The origin of the anomalously strong reducing power of Li is the strongly exothermic energy of hydration of the very small Li^+ ion, which favors the ionization of the element in aqueous solution.

19.71 (a, b) $H_2(g) + F_2(g) \rightarrow 2\ HF(g)$, explosive; $H_2(g) + Cl_2(g) \rightarrow 2\ HCl(g)$, explosive; $H_2(g) + Br_2(g) \rightarrow 2\ HBr(g)$, vigorous; $H_2(g) + I_2(g) \rightarrow 2\ HI(g)$, less vigorous (c) hydrofluoric acid, hydrochloric acid, hydrobromic acid, hydroidic acid

19.73 (a) $Ba^{2+}[:\ddot{O}—\ddot{O}:]^{2-}$ (d)
(b) H—Be—H
(c) $Na^+Na^+[:\ddot{O}—\ddot{O}:]^{2-}$

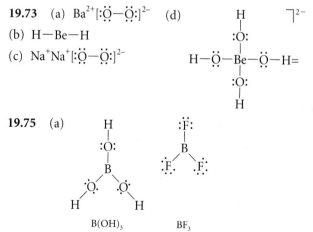

19.75 (a)

B(OH)₃   BF₃

(b) VSEPR predicts a trigonal planar structure with 120° bond angles for each. **(c)** sp^2 hybridization correlates with a trigonal planar arrangement.

19.77 (a) Ionization energies decrease (become less positive) down Groups 13 (B → Tl) and 14 (C → Pb). Atomic radii increase down Groups 13 (B → Tl) and 14 (C → Pb). (b) The ionization energies generally decrease down a group. As the atomic number of an element increases, atomic shells and subshells that are farther from the nucleus are filled. The outermost valence electrons are consequently easier to remove. The radii increase down a group for the same reason. The radii are primarily determined by the outer shell electrons, which are farther from the nucleus in the heavy elements.

(c) The trends correlate well with elemental properties; for example, the greater ease of outermost electron removal correlates with increased metallic character, that is, ability to form positive ions by losing one or more electrons.

19.79 (a) $2 AlCl_3(s) + 3 H_2O(l) \rightarrow 6 HCl(g) + Al_2O_3(s)$

(b) $B_2H_6(g) \xrightarrow{\text{high temp.}} 2 B(s) + 3 H_2(g)$

(c) $4 BF_3 + 3 BH_4^- \xrightarrow{\text{organic solvent}} 3 BF_4^- + 2 B_2H_6$

19.81 In the majority of its reactions, hydrogen acts as a reducing agent, like the alkali metals. However, it may also act as an oxidizing agent, like the halogens: $H_2(g) + 2 e^- \rightarrow 2 H^-(aq)$, $E° = -2.25$ V. Consequently, H_2 will oxidize elements with standard potentials more negative than -2.25 V, such as the alkali and alkaline earth metals (except Be). The compounds formed are hydrides and contain the H^- ion; the singly charged negative ion is reminiscent of the halide ions. Hydrogen also forms diatomic molecules and covalent bonds like the halogens. Consequently, hydrogen could be placed in either Group 1 or Group 17. But it is probably best to think of hydrogen as a unique element that has properties in common with both metals and nonmetals; therefore, it should probably be centered in the periodic table, as it is shown in the table in the text.

19.83 (a)

Ion	Radius, pm	Polarizing ability (×1000)	Ion	Radius pm	Polarizing ability (×1000)
Li$^+$	58	17	Be^{2+}	27	74
Na$^+$	102	9.8	Mg^{2+}	72	28
K$^+$	138	7.2	Ca^{2+}	100	20
Rb$^+$	149	6.7	Sr^{2+}	116	17
Cs$^+$	170	5.9	Ba^{2+}	136	15

(b) These data roughly support the diagonal relationship. Li$^+$ is more like Mg^{2+} than Be^{2+}, and Na^{2+} is more like Ca^{2+} than Mg^{2+}; but further down the group, the correlation fails. Charge divided by r^3 would be a better measure of polarizing ability.

19.85 Several reactions occur, depending on the silica to alkali ratio:

(1) $SiO_2(s) + OH^-(aq) \rightarrow HSiO_3^-(aq)$

(2) $SiO_2(s) + 2 OH^-(aq) \rightarrow H_2O(l) + SiO_3^{2-}(aq)$, metasilicate ion

(3) $SiO_2(s) + 4 OH^-(aq) \rightarrow 2 H_2O(l) + SiO_4^{4-}(aq)$, orthosilicate ion

(4) $2 SiO_2(s) + 6 OH^-(aq) \rightarrow 3 H_2O(l) + Si_2O_7^{6-}(aq)$, pyrosilicate ion

19.87 Glasses form from materials in which inhibited recrystallization occurs. Such materials are likely to be covalently bonded materials involving extensive network structures and materials with large complex molecules. On the basis of these principles, we predict that, of the substances listed, the following would probably solidify as a glass: (a) tar; (c) molten granite; (e) low-density polyethylene; and (f) a highly branched polymer.

19.89 (a) The "hardness" of water is due to the presence of calcium and magnesium salts (particularly their hydrogen carbonates). In laundering and bathing, Ca^{2+} and Mg^{2+} cations convert soluble Na$^+$ soaps to insoluble Ca^{2+} and Mg^{2+} soaps.

(b) $Ca(HCO_3)_2(aq) + Ca(OH)_2(aq) \rightarrow$
$$2 CaCO_3(s) + 2 H_2O(l)$$

$Mg(HCO_3)_2(aq) + Ca(OH)_2(aq) \rightarrow$
$$Mg(OH)_2(s) + Ca(HCO_3)_2(aq)$$

19.91 (a) 68 g Mg; (b) 9.21×10^5 L Cl$_2$

19.93 5.80 L H$_2$

19.95 for BH$_3$, ΔH(B—H) = 372 kJ/mol; for B$_2$H$_6$, ΔH(B—H—B) = 228 kJ/mol and 372 kJ/mol for (B—H) terminal bonds. Bond length and bond enthalpy are (very roughly) inversely related, so the stronger terminal B—H bonds are expected to be shorter.

19.97 $\Delta H° = -565.96$ kJ/mol; $\Delta S° = -173.00$ J/K · mol; $\Delta G° = -514.41$ kJ/mol; below 3271 K.

19.99 (a) $SnO_2(s) + C(s) \rightarrow Sn(s) + CO_2(g)$ (reaction 1) $SnO_2(s) + 2 CO(g) \rightarrow Sn(s) + 2 CO_2(g)$ (reaction 2)
(b) for reaction 1, $\Delta G° = 125.4$ kJ/mol; for reaction 2, $\Delta G° = 5.44$ kJ/mol; (c) In both cases, the reaction is spontaneous at high temperatures.

19.101 2.7×10^5 g Al

19.103 (a) -65.17 kJ/mol; (b) 62.3°C

Chapter 20

20.1 (a) He, $1s^2$; (b) O, [He]$2s^22p^4$; (c) F, [He]$2s^22p^5$; (d) As, [Ar]$3d^{10}4s^24p^3$

20.3 The S in SO$_3$ is in the +6 oxidation state and cannot be oxidized further; S in SO$_2$ is in the +4 oxidation state and can be further oxidized.

20.5 As $\approx$ P < S < O

20.7 The first step is the liquefaction of air, which is 76% by mass nitrogen. Air is cooled to below its boiling point by a series of expansion and compression steps in a kind of refrigerator. Nitrogen gas is then obtained by distillation of liquid air. The nitrogen boils off at -196°C, but gases with higher boiling points, principally O$_2$, remain as a liquid. The pure nitrogen gas is then liquefied by repeating the process.

20.9 (a) nitrous acid; (b) nitrogen monoxide or nitric oxide; (c) phosphoric acid; (d) dinitrogen trioxide

20.11 (a) NH$_4$NO$_3$; (b) Mg$_3$N$_2$; (c) Ca$_3$P$_2$; (d) H$_2$NNH$_2$

20.13 (a) +2; (b) +1; (c) +3; (d) $-\frac{1}{3}$

20.15 $CO(NH_2)_2 + 2 H_2O \rightarrow (NH_4)_2 CO_3$; 8.0×10^3 g

20.17 N_2O_3: HNO_2; $N_2O_3(g) + H_2O(l) \rightarrow 2 HNO_2(aq)$
N_2O_5: HNO_3; $N_2O_5(g) + H_2O(l) \rightarrow 2 HNO_3(aq)$

20.19

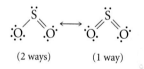

PCl$_4^+$, AX$_4$
Tetrahedral

PCl$_6^-$, AX$_6$
Octahedral

20.21 (a) 12.23%; (b) 26.46%

20.23 (a) H_2SO_4; (b) $CaSO_3$; (c) O_3; (d) BaO_2

20.25 (a) $4 Li(s) + O_2(g) \overset{\Delta}{\rightarrow} 2 Li_2O(s)$
(b) $2 Na(s) + 2 H_2O(l) \rightarrow 2 NaOH(aq) + H_2(g)$
(c) $2 F_2(g) + 2 H_2O(l) \rightarrow 4 HF(aq) + O_2(g)$
(d) $2 H_2O(l) \rightarrow O_2(g) + 4 H^+(aq) + 4 e^-$

20.27 (a) $2 H_2S(g) + 3 O_2(g) \rightarrow 2 SO_2(g) + 2 H_2O(g)$
(b) $PCl_5(s) + 4 H_2O(l) \rightarrow H_3PO_4(aq) + 5 HCl(g)$

20.29

Each O in H_2O_2 is an AX$_2$E$_2$ structure; the bond angle is predicted to be <109.5°. In actuality, it is 97°.

20.31 (a) When we consider formal charges and allow the possibility of expanded octets in S atoms, we can visualize two types of resonance structures that might contribute to the overall structure of SO_2:

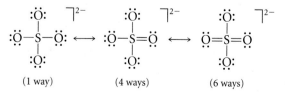

(2 ways)   (1 way)

The completely double-bonded structure has zero formal charges on all atoms and, as a result, may predominate in the resonance hybrid. Both structures have angular geometry.

(b) SF$_4$ is an AX$_4$E VSEPR structure. Therefore, its shape is seesaw.

(c) When we consider formal charges and the possibility of expanded octets in S atoms, we can visualize three types of resonance structures that might contribute to the overall structure of SO_4^{2-}:

(1 way)   (4 ways)   (6 ways)

The third type of structure, which has smaller average differences in formal charge between the S and O atoms, may be the predominant structure. All three structures have tetrahedral geometry.

20.33 Fluorine comes from the minerals fluorspar, CaF_2; cryolite, Na_3AlF_6; and the fluoraparites, $Ca_5F(PO_4)_3$. The free element is prepared from HF and KF by electrolysis, but the HF and the KF needed for the electrolysis are prepared in the laboratory. Chlorine primarily comes from the mineral rock salt, NaCl. The pure element is obtained by electrolysis of liquid NaCl.

20.35 fluorine: KF acts as an electrolyte for the electrolytic process, but the net reaction is $2 H^+ + 2 F^- \xrightarrow{\text{current}} H_2(g) + F_2(g)$
chlorine: $2 NaCl(l) \xrightarrow{\text{current}} 2 Na(l) + Cl_2(g)$

20.37 (a) hydrobromic acid; (b) iodine bromide; (c) chlorine dioxide; (d) sodium iodate

20.39 (a) $HClO_4(aq)$; (b) $NaClO_3$; (c) $HI(aq)$; (d) NaI_3

20.41 (a) +1; (b) +4; (c) +7; (d) +5

20.43 (a)

AX$_4$, tetrahedral electronic arrangement and shape. *Note:* This structure is the preferred structure based on formal charge considerations; alternative structures with 0, 1, 2, and 4 double bonds could be drawn. All structures have the same geometry, however.

(b)

AX$_3$E, tetrahedral electronic arrangement, trigonal pyramidal shape. *Note:* Lewis structures with three, one, or no iodine-oxygen double bonds are also possible. All structures have the same geometry. See the note in part (a).

(c)

AX$_3$E$_2$, trigonal bipyramidal electronic arrangement, T-shaped

20.45 (a) $4 KClO_3(l) \overset{\Delta}{\rightarrow} 3 KClO_4(s) + KCl(s)$
(b) $Br_2(l) + H_2O(l) \rightarrow HBrO(aq) + HBr(aq)$
(c) $NaCl(s) + H_2SO_4(aq) \rightarrow NaHSO_4(aq) + HCl(g)$
(d) (a) and (b) are redox reactions. In (a), Cl is both oxidized and reduced. In (b), Br is both oxidized and reduced. (c) is a Brønsted acid-base reaction; H_2SO_4 is the acid, and Cl^- the base.

20.47 (a) $HClO < HClO_2 < HClO_3 < HClO_4$, $HClO_4$ is strongest; HClO, weakest; (b) The oxidation number of Cl increases from HClO to $HClO_4$. In $HClO_4$, chlorine has its highest oxidation number of +7, so $HClO_4$ will be the strongest oxidizing agent.

20.49 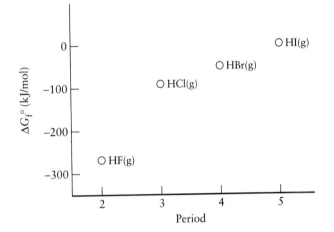 AX$_2$E$_2$, angular, slightly less than 109°

20.51 The thermodynamic stability of the hydrogen halides decreases down the group. The $\Delta G_f°$ values of HCl, HBr, and HI fit nicely on a straight line; whereas HF is anomalous. In other properties, HF is also the anomalous member of the group, in particular, its acidity.

20.53 Helium occurs as a component of natural gases found under rock formations in certain locations, especially some in Texas. Argon is obtained by distillation of liquid air.

20.55 (a) +2; (b) +6; (c) +4; (d) +6

20.57 XeF$_4$ + 4 H$^+$ + 4 e$^-$ → Xe + 4 HF

20.59 Because H$_4$XeO$_6$ has a greater number of highly electronegative O atoms bonded to Xe, we predict that H$_4$XeO$_6$ is more acidic than H$_2$XeO$_4$.

20.61 AX$_4$E$_2$, square planar, 90°

20.63 Ionization energies increase; electron affinities also increase in magnitude. Large ionization energies and electron affinities are characteristic of nonmetals. Electronegativities and standard potentials increase as well; large values of these properties are also characteristic of nonmetals.

20.65 oxidation: As$_2$S$_3$(s) + 8 H$_2$O(l) →
2 AsO$_4^{3-}$(aq) + 3 S^{2-}(aq) + 16 H$^+$(aq) + 4 e$^-$
reduction: H$_2$O$_2$(aq) + 2 H$^+$(aq) + 2 e$^-$ → 2 H$_2$O(l)
overall: As$_2$S$_3$(s) + 2 H$_2$O$_2$(aq) + 4 H$_2$O(l) →
2 AsO$_4^{3-}$(aq) + 3 S^{2-}(aq) + 12 H$^+$(aq)

20.67 8.16 × 10^2 L concentrated H$_2$SO$_4$

20.69 (a) SO$_2$(g) + H$_2$O(l) → H$_2$SO$_3$(l); this is a Lewis acid-base reaction. (b) 2 F$_2$(g) + 2 NaOH(aq) →
OF$_2$(g) + 2 NaF(aq) + H$_2$O(l); this is a redox reaction in basic solution. (c) S$_2$O$_3^{2-}$(aq) + 4 Cl$_2$(g) + 13 H$_2$O(l) →
2 HSO$_4^-$(aq) + 8 H$_3$O$^+$(aq) + 8 Cl$^-$(aq); this is a redox reaction. (d) 2 XeF$_6$(s) + 16 OH$^-$(aq) → XeO$_6^{4-}$(aq) +
Xe(g) + 12 F$^-$(aq) + 8 H$_2$O(l) + O$_2$(g); this is a redox reaction.

20.71 (a) I$_2$(s) + 3 F$_2$(g) → 2 IF$_3$(s); (b) I$_2$(aq) +
I$^-$(aq) → I$_3^-$(aq); (c) Cl$_2$(g) + H$_2$O(l) → HCl(aq) +
HOCl(aq); But there are competing reactions, such as
Cl$_2$(g) + H$_2$O(l) → 2 HCl(aq) + $\frac{1}{2}$O$_2$(g). The predominant reaction is determined by the temperature and pH.
(d) 2 F$_2$(g) + 2 H$_2$O(l) → 4 HF(aq) + O$_2$(g)

20.73 (a) CaCl$_2$(s) + 2 H$_2$SO$_4$(aq, conc) →
Ca(HSO$_4$)$_2$(aq) + 2 HCl(g)

(b) KBr(s) + H$_3$PO$_4$(aq) $\xrightarrow{\Delta}$ KH$_2$PO$_4$(aq) + HBr(g)

(c) KI(s) + H$_3$PO$_4$(aq) $\xrightarrow{\Delta}$ KH$_2$PO$_4$(aq) + HI(g)

20.75 (a) Some of the unusual properties of fluorine relative to the other halogens are (1) its much higher electronegativity; (2) its restriction to a negative oxidation number (−1) in all of its compounds; (3) the weak acidity of its hydrogen compound, HF; (4) the high lattice enthalpies of its ionic fluorides; (5) the lower solubility of the ionic fluorides; (6) the weakness of the F—F bond; (7) the high rate of its reactions; (8) the high volatility of most of its covalent compounds; (9) the low volatility of HF. Properties (1), (2), (4), (5), and (6) are related to the smaller size of the fluorine atom; property (3) is due to the greater strength of the H—F bond; property (7) is a result of property (6), the weakness of the F—F bond; property (8) is a result of weaker London forces, which in turn is due to the smaller size of the fluorine atom; and property (9) is related to stronger hydrogen bonding in HF, which in turn is due to its smaller size and greater electronegativity.
(b) This statement relates to its great strength as an oxidizing agent.

20.77 (a) Xe(g) + 2 F$_2$(g) $\xrightarrow{\Delta}$ XeF$_4$(s)
(b) Pt(s) + XeF$_4$(s) → Xe(g) + PtF$_4$(s)

20.79 physical: Melting points and boiling points increase down the group in a regular fashion consistent with the increase in the molar masses of the elements. Ionization energies decrease down the group in a regular manner consistent with the fact that further electron shells are being successively filled.
chemical: Chemical reactivity increases down the group. This is related to the decrease in ionization energy; it is easier for the heavier elements to share their electrons. He, Ne, and Ar form no stable compounds, but Kr, Xe, and Rn do, especially with fluorine.

20.81 Because this chapter deals with elements from Groups 15–18, only Period 2 and Period 3 elements from these four groups are included in the answer. Other examples are possible.

(a) N: NH_3, used in the production of fertilizers.
O: H_2O, found in all life forms, covers much of earth's surface.
F: Na_3AlF_6, used in the electrolytic refining of aluminum.
Ne: no known compounds, but the element is used in "neon" lights.

(b) P: found in phosphate groups that help hold DNA together (see Chapter 11 for the structure of this compound). Without it, the reproduction of higher life forms on this planet would not be possible. Phosphorus is also important in the conversion of ADP to ATP, the energy storage mechanism in living cells.
S: H_2SO_4, sulfuric acid, used in the production of fertilizers and detergents. One of the most heavily manufactured chemicals in the world.
Cl: $Ca(ClO)_2$, calcium hypochlorite, used as dry bleach and in swimming pools.
Ar: no known compounds, but the element is used in welding and as an inert gas in light bulbs.

20.83 (a) The method of organization is determined by whether one wishes to emphasize similarities or differences. It seems easier and more logical to emphasize similarities rather than differences. Trends within groups from top to bottom tend to be quantitative (small numerical differences) rather than qualitative (large numerical differences leading to distinctly different behavior), as are observed from left to right within a period. Trends from metallic to nonmetallic behavior across a period would be more apparent in the organization by period, as would the related trends in ionization energy and electron affinity. One could also more readily see changes in valence as, say, represented by the formulas of common compounds, such as the oxides and chlorides. Trends in melting points and boiling points of the elements and their compounds would also be m... apparent, as well as many other physical properties ...ods of there are advantages and disadvantages to both ...eems organization, but the organization by group s... ...ot permit the preferable, as organization by period would ...eful features of generalizations and summaries that are ... ...ase of the the present arrangement. (b) In th... ...in a period can be transition elements, similarities w... ...e, organization by period stronger than within groups; h... ...properties, and this is the allows a logical structuring c... ...s. usual choice among autho...

20.85 The ionic radii ...t the halide ions increase down the group. The small... the ion, the greater the lattice enthalpy of the co...pounds of the ion due to the greater concentration o...the charge in the ion. Increased lattice enthalpy resu...s in higher melting and boiling points; the

ions cannot as easily break free from each other. Thus, melting and boiling points decrease from fluoride to iodide for ionic halides. The predominant forces between covalent halogen compounds are London forces, which are greater for larger atoms. Thus, the melting and boiling points increase from fluoride to iodide for molecular halides.

20.87 Cl is reduced, Al and N are oxidized.
$$3\,Al + 3\,H_2O \rightarrow Al^{3+} + Al_2O_3 + 6\,H^+ + 9\,e^-$$
$$NH_4^+ + H_2O \rightarrow NO + 6\,H^+ + 5\,e^-$$
$$ClO_4^- + 8\,H^+ + 8\,e^- \rightarrow Cl^- + 4\,H_2O$$

20.89 -4738 kJ/mol

20.91 (a) $2\,NaCl(aq) + 2H_2O(l) \xrightarrow{current} 2\,NaOH(aq) + H_2(g) + Cl_2(g)$; (b) 3.0×10^{16} C; (c) 9.5×10^8 A

20.93 (a) $0.23 < N_2(g)$; (b) $Hg(N_3)_2$ would produce a larger volume, because its molar mass is less.

20.95 4.9 L O_2

20.97 9.79

20.99 PbO_2 and SO_2 formation results in a positive $\Delta G°$. $PbSO_4$ is the favored product.

20.101 2.9×10^{38}

20.103 0.014 mol/L

20.105 (a) 1×10^{10} kg S; (b) 2×10^{12} L $SO_3(g)$

20.107 (a) $4\,Zn(s) + NO_3^-(aq) + 7\,OH^-(aq) + 6\,H_2O(l) \rightarrow NH_3(aq) + 4\,Zn(OH)_4^{2-}(aq)$
$$NH_3(g) + HCl(aq) \rightarrow NH_4Cl(aq)$$
$$HCl(aq) + NaOH(aq) \rightarrow H_2O(l) + NaCl(aq)$$
(b) 0.33 mol/L

Chapter 1

21.1 (a) Mn, $[Ar]3d^54s^2$; (b) Cd, $[Kr]4d^{10}5s^2$; (c) Zr, $[Ar]3d^{10}4s^2$; (d) Zr, $[Kr]4d^25s^2$

21.3 (a) Sc, $[Ar]3d^14s^2$: one unpaired $3d$-electron; V, $[Ar]3d^34s^2$: three unpaired $3d$-electrons; (c) Cu, $[Ar]3d^{10}4s^1$: one unpaired $4s$-electron

21.5 iron, Fe; cobalt, Co; nickel, Ni

21.7 See Figs. 21.2 and 21.4. (a) Sc; (b) Au; (c) Nb

21.9 See Fig. 7.39 and Appendix 2D. (a) Ti; (b) Cu; (c) Zn

21.11 The lanthanide contraction accounts for the failure of the third-row (Period 6) metallic radii to increase as expected relative to the radii of the second row (Period 5). It results from the presence of the f-block orbitals. The f-electrons present in the lanthanides are even poorer as nuclear shields than d-electrons, and a marked decrease in metallic radius occurs along the f-block elements as a result of the increased effective nuclear charge, which pulls the electrons inward. When the d-block resumes (at lutetium), the metallic radius has "contracted" from 188 pm to 157 pm. Therefore, all the elements following lutetium have smaller-than-expected radii. Examples of this effect include the high density of the Period 6 elements and lack of reactivity of gold and platinum.

21.13 Hg is much more dense than Cd, because the shrinkage in atomic radius that occurs between $Z = 58$ and $Z = 71$ (the lanthanide contraction) causes the atoms following the rare earths to be smaller than might have been expected for their atomic masses and atomic numbers. Zn and Cd have densities that are not too dissimilar, because the radius of Cd is subject to only a smaller *d*-block contraction.

21.15 Proceeding down a group in the *d*-block (for example, from Cr to Mo to W), there is an increasing probability of finding the elements in higher oxidation states. That is, higher oxidation states become more stable on going down a group.

21.17 In MO_3, M has an oxidation number of +6. Of these three elements, the +6 oxidation state is most stable for Cr. See Fig. 21.7.

21.19 (a) $TiO_2(s) + 2 C(s) + Cl_2(g) \xrightarrow{1000°C}$
$$TiCl_4(g) + 2 CO(g),$$
followed by $TiCl_4(g) + 2 Mg(l) \xrightarrow{700°C} Ti(s) + 2 MgCl_2(s)$
(b) $V_2O_5(g) + 5 Ca(l) \rightarrow 5 CaO(s) + 2 V(s)$
or $VCl_2(s) + Mg(l) \xrightarrow{\Delta} V(s) + MgCl_2(s)$

21.21 (a) $Ti(s), MgCl_2(s)$
$TiCl_4(g) + 2 Mg(l) \rightarrow Ti(s) + 2 MgCl_2(s)$
(b) $Co^{2+}(aq), HCO_3^-(aq), NO_3^-(aq)$
$CoCO_3(s) + HNO_3(aq) \rightarrow$
$$Co^{2+}(aq) + HCO_3^-(aq) + NO_3^-(aq)$$
(c) $V(s), CaO(s)$
$V_2O_5(s) + 5 Ca(l) \xrightarrow{\Delta} 2 V(s) + 5 CaO(s)$

21.23 (a) titanium(IV) oxide, TiO_2; (b) iron(III) oxide, Fe_2O_3; (c) manganese(IV) oxide, MnO_2

21.25 (a) The products are V^{3+}, H_2, and Cl^-; (b) no reaction; (c) The products are Co^{2+}, H_2, and Cl^-.

21.27 All three elements in this group—Cu, Ag, and Au—are chemically rather inert, Ag more so than Cu, and Au more so than Ag. The standard potentials of their ions are all positive, in the order Au > Ag > Cu, so they are not readily oxidized. They have a common electron configuration, $(n - 1)d^{10}ns^1$.

21.29 (a) chalcopyrite, $CuFeS_2$, copper iron sulfide; (b) sphalerite, ZnS, zinc sulfide; (c) cinnabar, HgS, mercury(II) sulfide

21.31 (a) $2 ZnS(s) + 3 O_2(g) \xrightarrow{\Delta} 2 ZnO(s) + 2 SO_2(g)$, followed by $ZnO(s) + C(s) \xrightarrow{\Delta} Zn(l) + CO(g)$
(b) $HgS(s) + O_2(g) \xrightarrow{\Delta} Hg(g) + SO_2(g)$

21.33 (a) +2; (b) +3; (c) +3; (d) +3

21.35 (a) 4; (b) 2; (c) 6 (en is bidentate); (d) 6 (edta is hexadentate)

21.37 (a) hexacyanoferrate(II) ion;

(b) hexaamminecobalt(III) ion;
(c) aquapentacyanocobaltate(III) ion;
(d) pentaamminesulfatocobalt(III) ion

21.39 (a) $K_3[Cr(CN)_6]$; (b) $[Co(NH_3)_5(SO_4)]Cl$;
(c) $[Co(NH_3)_4(H_2O)_2]Br_3$; (d) $Na[Fe(H_2O)_2(C_2O_4)_2]$

21.41 (a) structural isomers, linkage isomers;
(b) structural isomers, ionization isomers;
(c) structural isomers, linkage isomers;
(d) structural isomers, ionization isomers

21.43 $[Co(H_2O)_6]Cl_3$, $[CoCl(H_2O)_5]Cl_2 \cdot H_2O$, $[CoCl_2(H_2O)_4]Cl \cdot 2H_2O$, and $[CoCl_3(H_2O)_3] \cdot 3H_2O$

21.45 $[CoCl(NO_2)(en)_2]Cl$ and $[CoCl(ONO)(en)_2]Cl$

21.47 (a) yes

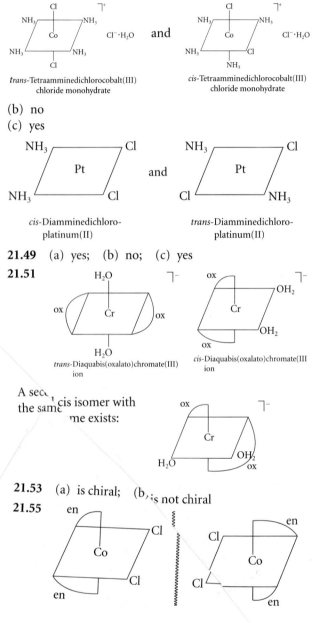

21.49 (a) yes; (b) no; (c) yes

21.51

21.53 (a) is chiral; (b) is not chiral

21.55

A_VERS
A

21.57 (a) d^7; (b) d^8; (c) d^5; (d) d^3

21.59 (a) octahedral: strong-field ligand, $6e^-$

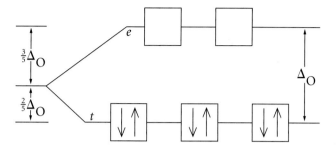

(b) tetrahedral: weak-field ligand, $8e^-$

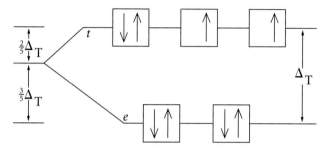

(c) octahedral: weak-field ligand, $5e^-$

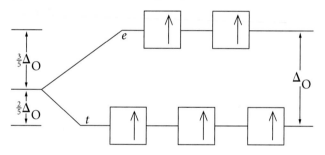

(d) octahedral: strong-field ligand, $5e^-$

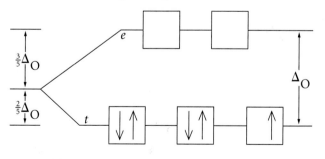

21.61 (a) zero; (b) two

21.63 Weak-field ligands do not interact strongly with the d-electrons in the metal ion, so they produce only a small crystal field splitting of the d-electron energy states. The opposite is true of strong-field ligands. With weak-field ligands, unpaired electrons remain unpaired if there are unfilled orbitals; hence a weak-field ligand is likely to lead to a high-spin complex. Strong-field ligands cause electrons in excess of three to pair up with electrons in lower energy orbitals. A strong-field ligand is likely to lead to a low-spin complex. Ligands arranged in the spectrochemical series help to distinguish strong-field and weak-field ligands. Measurement of magnetic susceptibility (specifically, paramagnetism) can be used to determine the number of unpaired electrons, which in turn establishes whether the associated ligand is weak-field or strong-field in nature.

21.65 Because F^- is a weak-field ligand and en a strong-field ligand, the splitting between levels is less in (a) than in (b). Therefore, (a) will absorb light of longer wavelength than will (b) and consequently will display a shorter wavelength color. Blue light is shorter in wavelength than yellow light, so (a) $[CoF_6]^{3-}$ is blue and (b) $[Co(en)_3]^{3+}$ is yellow.

21.67 (a) 410. nm, yellow; (b) 650. nm, green; (c) 480. nm, orange; (d) 590. nm, blue

21.69 In Zn^{2+}, the $3d$-orbitals are filled (d^{10}). Therefore, there can be no electronic transitions between the t and e levels; hence no visible light is absorbed and the aqueous ion is colorless. The d^{10} configuration has no unpaired electrons, so Zn compounds would not be paramagnetic.

21.71 (a) 162 kJ/mol; (b) 260. kJ/mol; (c) 208 kJ/mol; $Cl < H_2O < NH_3$

21.73 (a) A compound with unpaired electrons is paramagnetic and is pulled into a magnetic field. A diamagnetic substance has no unpaired electrons and is weakly pushed out of a magnetic field. (b) Paramagnetism is a property of any substance with unpaired electrons, whereas ferromagnetism is a property of certain substances that can become permanently magnetized. Ferromagnetism results when a large number of electrons in the metal have parallel spins. This parallel alignment can be retained even in the absence of a magnetic field. In a paramagnetic substance, the alignment is lost when the magnetic field is removed.

21.75 $[Sc(H_2O)_6]^{3+}(aq) + H_2O(l) \rightarrow$
$$[Sc(H_2O)_5OH]^{2+}(aq) + H_3O^+(aq)$$

21.77 (a) $CuSO_4(s) + 5\ H_2O(l) \rightarrow CuSO_4 \cdot 5H_2O(s)$
The pentahydrate has a complicated structure. It consists of the $Cu(H_2O)_4^{2+}$ ion, which is linked to the sulfate ion through the fifth water molecule. The four water molecules of the tetrahydrate ion form an approximate square planar structure around the copper atom.
(b) $CuSO_4(s) + 6\ H_2O(l) \rightarrow Cu(H_2O)_6^{2+}(aq) + SO_4^{2-}(aq)$
(c) $Cu(H_2O)_6^{2+}(aq) + 4\ NH_3(aq) \rightarrow$
$$Cu(NH_3)_4^{2+}(aq) + 6\ H_2O(l)$$

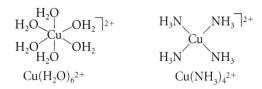

$Cu(H_2O)_6^{2+}$ $Cu(NH_3)_4^{2+}$

21.79 (a) Cr^{3+} ions in water form the complex $[Cr(H_2O)_6]^{3+}(aq)$, which behaves as a Brønsted acid: $[Cr(H_2O)_6]^{3+}(aq) + H_2O(l) \rightleftharpoons [Cr(H_2O)_5OH]^{2+}(aq) + H_3O^+(aq)$
(b) The gelatinous precipitate is the hydroxide $Cr(OH)_3$. The precipitate dissolves as the $Cr(OH)_4^-$ complex ion is formed:
$Cr^{3+}(aq) + 3\ OH^-(aq) \rightarrow Cr(OH)_3(s)$
$Cr(OH)_3(s) + OH^-(aq) \rightarrow Cr(OH)_4^-(aq)$
21.81 Oxidation of Fe^{2+} to Fe^{3+} readily occurs because a half-filled d-subshell is obtained, which is not the case with either Co^{2+} or Ni^{2+}. Half-filled subshells are low-energy electron arrangements; hence they have a strong tendency to form.
21.83 If the compound were tetrahedral, there would be only one compound, not two.
21.85 $[CrNH_3Cl(H_2O)_2]Cl_2$

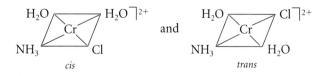

cis   and   *trans*

21.87 NH_3 is a strong-field ligand relative to H_2O; as such, it will yield a low-spin complex. Low-spin complexes exhibit diamagnetic properties.
21.89 The correct structure for $[Co(NH_3)_6]Cl_3$ consists of four ions, $Co(NH_3)_6^{3+}$ and 3 Cl^- in aqueous solution. The chloride ions can be easily precipitated as AgCl. This would not be possible if they were bonded to the other (NH_3) ligands. If the structure were $Co(NH_3—NH_3—Cl)_3$, VSEPR theory would predict that the Co^{3+} ion would have a trigonal planar ligand arrangement. The splitting of the d-orbital energies would not be the same as the octahedral arrangement and would lead to different spectroscopic and magnetic properties inconsistent with the experimental evidence. In addition, neither optical nor geometrical isomers would be observed.
21.91 (a) an alloy of copper and tin; (b) $Cu_2(OH)_2CO_3$; (c) pure gold; (d) osmium
21.93 "Fixing," that is, removing undeveloped AgI, was not a part of the process of producing a daguerreotype. Consequently, on further exposure to light, reduction of some of the remaining Ag^+ ions continued and caused a darkening or fading of the image.
    In photochromic sunglasses, the darkening of the glass

is a result of the reversible redox reaction $Ag^+(s) + Cu^+(s) \xrightarrow{\text{light}} Ag(s) + Cu^{2+}(s)$. This reaction is driven to the right in the presence of light, and the formation of $Ag(s)$ causes a darkening of the lens. When the light is removed, $Ag(s)$ is oxidized by $Cu^{2+}(s)$ back to Ag^+.
    The images on photographic film are permanent because further reduction of Ag^+ is prevented by "fixing," that is, removing undeveloped AgBr with sodium hyposulfite. The water-soluble ions are then washed away.
21.95 mercury, gold, platinum, and copper. However, mercury compounds are highly toxic and gold and platinum are very expensive. Thus, copper is the best choice.
21.97 (a) CO
(b) $Fe_2O_3(s) + 3\ CO(g) \rightarrow 2\ Fe(s) + 3\ CO_2(g)$ (Zone C)
$Fe_2O_3(s) + CO(g) \rightarrow 2\ FeO(s) + CO_2(g)$ (Zone D)
followed by $FeO(s) + CO(g) \rightarrow Fe(s) + CO_2(g)$ (Zone C)
(c) carbon
21.99 (a) yes; (b) no
21.101 10^6
21.103 0.691 kg $FeCr_2O_4$

Chapter 22

22.1 α particles: composed of $_2^4He$ nuclei, mass of 4.0 u, charge of $+2$, low penetrating power. β particles: composed of electrons, mass of 0.000 55 u, charge of -1, moderate penetrating power. γ particles: composed of photons, mass of 0 u, no charge, high penetrating power.
22.3 Let p = protons, n = neutrons, nu = nucleons.
(a) 1 p, 1 n, 2 nu; (b) 12 p, 12 n, 24 nu; (c) 104 p, 159 n, 263 nu; (d) 27 p, 33 n, 60 nu; (e) 94 p, 144 n, 238 nu; (f) 101 p, 157 n, 258 nu
22.5 Let p = protons, n = neutrons, nu = nucleons.
(a) $_{35}^{81}Br$, 35 p, 46 n, 81 nu; (b) $_{36}^{90}Kr$, 36 p, 54 n, 90 nu;
(c) $_{96}^{244}Cm$, 96 p, 148 n, 244 nu; (d) $_{53}^{128}I$, 53 p, 75 n, 128 nu;
(e) $_{16}^{32}S$, 16 p, 16 n, 32 nu; (f) $_{95}^{241}Am$, 95 p, 146 n, 241 nu
22.7 (a) $_1^3T \rightarrow _{-1}^0e + _2^3He$; (b) $_{39}^{83}Y \rightarrow _{+1}^0e + _{38}^{83}Sr$;
(c) $_{36}^{87}Kr \rightarrow _{-1}^0e + _{37}^{87}Rb$; (d) $_{91}^{225}Pa \rightarrow _2^4\alpha + _{89}^{221}Ac$
22.9 (a) $_5^8B \rightarrow _{+1}^0e + _4^8Be$; (b) $_{28}^{63}Ni \rightarrow _{-1}^0e + _{29}^{63}Cu$;
(c) $_{79}^{185}Au \rightarrow _2^4\alpha + _{77}^{181}Ir$; (d) $_4^7Be + _{-1}^0e \rightarrow _3^7Li$
22.11 (a) $_{11}^{24}Na \rightarrow _{12}^{24}Mg + _{-1}^0e$; a β particle is emitted.
(b) $_{50}^{128}Sn \rightarrow _{51}^{128}Sb + _{-1}^0e$; a β particle is emitted.
(c) $_{57}^{140}La \rightarrow _{56}^{140}Ba + _{+1}^0e$; a positron ($\beta^+$) is emitted.
(d) $_{90}^{228}Th \rightarrow _{88}^{224}Ra + _2^4\alpha$; an α particle is emitted.
22.13 (a) ^{40}Ca; (b) ^{208}Pb; both are more stable because nuclei with an even number of protons and neutrons are more stable than nuclei with other combinations.
22.15 (a) α particles are emitted in the decay of nuclei with $Z > 83$. A few other nuclei that have N/Z ratios lying near the band of stability and with $60 < Z < 70$ also emit α particles. (b) β particles are emitted in the decay of nuclei with N/Z ratios greater than the N/Z ratio of the band of

stability. The emission of a β particle is the equivalent of converting a neutron into a proton; thus, the N/Z ratio becomes smaller to fit the values required for stability.

22.17 (a) β decay, $^{68}_{30}Zn$; (b) $β^+$ decay, $^{103}_{47}Ag$; (c) α decay, $^{239}_{95}Am$; (d) α decay, $^{256}_{103}Lr$

22.19 $^{235}_{92}U \rightarrow ^4_2α + ^{231}_{90}Th$; $^{231}_{90}Th \rightarrow ^0_{-1}e + ^{231}_{91}Pa$; $^{231}_{91}Pa \rightarrow ^4_2α + ^{227}_{89}Ac$; $^{227}_{89}Ac \rightarrow ^0_{-1}e + ^{227}_{90}Th$; $^{227}_{90}Th \rightarrow ^4_2α + ^{223}_{88}Ra$; $^{223}_{88}Ra \rightarrow ^4_2α + ^{219}_{86}Rn$; $^{219}_{86}Rn \rightarrow ^4_2α + ^{215}_{84}Po$; $^{215}_{84}Po \rightarrow ^0_{-1}e + ^{215}_{85}At$; $^{215}_{85}At \rightarrow ^4_2α + ^{211}_{83}Bi$; $^{211}_{83}Bi \rightarrow ^0_{-1}e + ^{211}_{84}Po$; $^{211}_{84}Po \rightarrow ^4_2α + ^{207}_{82}Pb$

22.21 (a) $^{14}_7N + ^4_2α \rightarrow ^{17}_8O + ^1_1p$; (b) $^{248}_{96}Cm + ^1_0n \rightarrow ^{249}_{97}Bk + ^0_{-1}e$; (c) $^{243}_{95}Am + ^1_0n \rightarrow ^{244}_{96}Cm + ^0_{-1}e + γ$; (d) $^{13}_6C + ^1_0n \rightarrow ^{14}_6C + γ$

22.23 (a) $^{20}_{10}Ne + ^4_2α \rightarrow ^8_4Be + ^{16}_8O$; (b) $^{20}_{10}Ne + ^{20}_{10}Ne \rightarrow ^{24}_{12}Mg + ^{16}_8O$; (c) $^{44}_{20}Ca + ^4_2α \rightarrow γ + ^{48}_{22}Ti$; (d) $^{27}_{13}Al + ^2_1H \rightarrow ^1_1p + ^{28}_{13}Al$

22.25 (a) $^{14}_7N + ^4_2α \rightarrow ^{17}_8O + ^1_1p$; (b) $^{239}_{94}Pu + ^1_0n \rightarrow ^{240}_{95}Am + ^0_{-1}e$

22.27 (a) $^{244}_{95}Am \rightarrow ^{134}_{53}I + ^{107}_{42}Mo + 3\ ^1_0n$; (b) $^{235}_{92}U + ^1_0n \rightarrow ^{96}_{40}Zr + ^{138}_{52}Te + 2\ ^1_0n$; (c) $^{235}_{92}U + ^1_0n \rightarrow ^{101}_{42}Mo + ^{132}_{50}Sn + 3\ ^1_0n$;

22.29 1.0×10^{-4} Ci

22.31 (a) 3.7×10^{10} dps; (b) 3.0×10^9 dps; (c) 3.7×10^4 dps

22.33 1.0 Gy, 1.0 Sv

22.35 100 days

22.37 (a) 5.63×10^{-2}/year; (b) 0.82/s; (c) 0.0693/min

22.39 (a) 5.3×10^3 years; (b) 8.4×10^9 years

22.41 6.0×10^{-3} Ci

22.43 (a) 88.6%; (b) 32.4%

22.45 (a) 0.25; (b) 1.26×10^9 years

22.47 3.53×10^3 years

22.49 (a) 9.8×10^{-4} Ci; (b) 2.8×10^{-4} Ci; (c) 0.40 Ci

22.51 The term *critical mass* refers to self-sustaining nuclear chain reactions. For a chain reaction to be self-sustaining, each nucleus that splits must provide an average of at least one new neutron that results in the fission of another nucleus. If the mass of the fissionable material is too small, the neutrons will escape before they can produce fission. The critical mass is the smallest mass that can sustain a nuclear chain reaction.

22.53 (a) 9.0×10^{13} J; (b) 8.182×10^{-14} J

22.55 -4.3×10^9 kg/s

22.57 (a) -1.13×10^{-12} J/nucleon; (b) -1.213×10^{-12} J/nucleon; (c) -1.8×10^{-13} J/nucleon; (d) -1.410×10^{-12} J/nucleon; ^{56}Fe is the most stable because it has the largest binding energy per nucleon.

22.59 (a) -7.8×10^{10} J/g; (b) -3.52×10^{11} J/g; (c) -2.09×10^{11} J/g; (d) -3.52×10^{11} J/g

22.61 The order of penetrating power is $γ > β > α$.

Although α particles do not penetrate deeply, they are very damaging because of their large mass. Consequently, they can dislodge atoms from molecules, thereby altering the structure of the molecule, which in turn alters the ability of the molecule to function properly. If the molecule is DNA, a necessary enzyme, or another essential molecule in a living system, the result may be cancer.

22.63 The stabilities of nuclei vary and the greatest stabilities are associated with certain numbers of nucleons. These numbers are referred to as magic numbers; they are 2, 8, 20, 50, 82, 126. The existence of this series of numbers reminds us of the magic numbers of electrons in the electronic configurations of the noble gases: 2, 10, 18, 36, 54, 86. Because this series of numbers corresponds to a shell model for electrons in atoms, the analogous series of numbers for nucleons suggests a nuclear shell model.

22.65 (a) $^{11}_5B + ^4_2α \rightarrow ^{13}_7N + 2\ ^1_0n$; (b) $^{35}_{17}Cl + ^2_1D \rightarrow ^{36}_{18}Ar + ^1_0n$; (c) $^{96}_{42}Mo + ^2_1D \rightarrow ^{97}_{43}Tc + ^1_0n$; (d) $^{45}_{21}Sc + ^1_0n \rightarrow ^{42}_{19}K + ^4_2α$

22.67 (a) 9 dpm; (b) 7×10^5 decays

22.69 5.83 mg

22.71 (a) 14 d; (b) 0.23

22.73 1.0×10^7 dps, 2.7×10^{-4} Ci

22.75 -7.8×10^{-13} J

22.77 44.2 years

22.79 Positron emission results in a lowering of positive charge in the nucleus; that is, Z is decreased, and A/Z is increased. Consequently, isotopes that are below the band of stability are likely candidates for positron emission, because such emissions will move them in the direction of the band. (a) $^{18}_8O$, not suitable; (b) $^{13}_7N \rightarrow ^0_1e + ^{13}_6C$; (c) $^{11}_6C \rightarrow ^0_1e + ^{11}_5B$; (d) $^{20}_9F$, not suitable; (e) $^{15}_8O \rightarrow ^0_1e + ^{15}_7N$

22.81 $^{98}_{42}Mo + ^1_0n \rightarrow ^{99}_{42}Mo \rightarrow ^{99}_{43}Tc + ^0_{-1}e$

22.83 (a) 26 d; (b) Because it takes only 26 days for 99.0% of the radon to be removed by disintegration, it does not seem likely that radon formed deep in the Earth's crust could leak to the surface in such a relatively short time. So most of the radon observed must have been formed near the Earth's surface. (c) Radon enters homes from the soil into the basements. It is naturally given off by concrete, cinder block, and stone building materials. An absorbing material such as a polymer could be used to absorb the emitted radon gas before it entered the basement. Most of the radon would disintegrate before it freed itself from the polymer because its half-life is short.

22.85 The decay process can generate much heat, which would speed up the corrosion rate. The decay process can also result in the production of new, possibly corrosive, chemicals as a result of nuclear fission and nuclear transmutation. Chemical breakdown can occur as a result of ionizing radiation, resulting in new, highly corrosive gases and other substances.

22.87 Radioactive "fallout" contains the products of an atmospheric fission process (bomb explosion). It consists of many different nuclides, most of them radioactive. They are dispersed rapidly in an explosion and are carried many miles by air currents before they settle to the ground. These radioactive products are dangerous to the health of humans and animals when absorbed into a living system by breathing, drinking contaminated water, or eating contaminated plant and animal products.

22.89 (a) 3.2×10^{-12} m, 3.8×10^{10} J/mol; (b) 5.3×10^{-14}m, 2.3×10^{12} J/mol; (c) 8.1×10^{-13} m, 1.5×10^{11} J/mol; (d) 4.1×10^{-15}m, 2.9×10^{13} J/mol

22.91 (a) 3.21×10^{20} Hz, 9.34×10^{-13} m; (b) 3.97×10^{20} Hz, 7.56×10^{-13} m; (c) 2.65×10^{20} Hz, 1.13×10^{-12} m

22.93 (a) 2.3×10^{-10} g; (b) -1.86×10^{-10} g; (c) -8.33×10^{-9} g

22.95 To find out whether sodium and potassium mixtures can dissolve carbon from steel, we could use carbon-14 to make a sheet of steel and then cut it into pieces. We could expose one piece of the steel to a hot, liquid sodium-potassium mixture, stirring it to simulate the flowing of the liquid in steel pipes. After a certain period of time, the steel could be removed, weighed, and burned at very high temperatures in the presence of oxygen. Any carbon remaining in the steel would be converted to carbon dioxide. A piece of the steel that was not exposed should also be weighed and pyrolyzed. In each case, the gas given off by the molten steel would be passed through a gas chromatograph and past a scintillator. Carbon-14 is radioactive, so when the carbon dioxide passes out of the gas chromatograph and through the scintillator, the amount of carbon-14 that had been contained in the piece of steel will be counted. The percentage of carbon can be determined in this way for each piece of steel. If the percentage of carbon is less for the steel that was exposed to the alkali metal mixture, then it is possible that the metals had leached the carbon from it. If it is the same, then another study must be conducted.

An alternative approach would be to react the alkali metal mixture with water, evaporate the water, and use the scintillation counter to look for carbon-14 in the residue.

Such experiments should be repeated several times to ensure accuracy.

Connection 6

Your plan may take many forms. A very brief outline of one plan, touching on only a few options, follows:
(a) Raising the temperature will require balancing the release or synthesis of greenhouse gases with increasing the concentration of nitrogen and oxygen in the atmosphere. We might initiate the process by using geothermal energy to melt some CO_2(s) ice caps, or nuclear processes if geothermal processes were not sufficient. Addition of gases like CF_4 or $C_2Cl_2F_4$ will also help trap infrared radiation.
(b) Addition of CO_2 will increase atmospheric pressure. Decomposition of carbonate rocks may add CO_2 to the atmosphere; for example, $CaCO_3 \xrightarrow{\Delta} CaO + CO_2$. Careful study of bacteria and their by-products will be undertaken to try to add N_2 or CH_4. (c) Ultimately, the heat from geothermal vents, nuclear plants, and increasing greenhouse gases may melt any water present. Use of fuel cells to power electronic devices will also produce some water. Introduction of bacteria, perhaps some engineered to produce methane, nitrogen, or even oxygen, will help produce an environment more welcoming of life.
(d) As the atmospheric pressure increases and gases that filter out harmful radiation increase as well, radiation levels on the surface should drop.

Another good source is R. Zubrin, *The Case for Mars*, New York: The Free Press, 1996. Do check the suggested references and Web sites. Some day you may find yourself working on this problem "for real"—good luck!

Cover: John Maher/ Stock Market.

Chapter 1

0: Opener, Jean Froissart "Chroniques." Musee Conde, Chantilly, France. Courtesy of Giraudon/ Art Resource. **2:** Fig. 1.1, Rita E. Johnson, American Chemical Society; Fig. 1.2, W. H. Freeman photo by Ken Karp. **3:** Fig. 1.4, Randall M. Feenstra, IBM Thomas J. Watson Research Center, Yorktown Heights, New York **4:** Fig. 1.5, W. H. Freeman photo by Ken Karp. **6:** Box 1.1, Jeff Isaac Greenberg/Photo Researchers. **7:** Fig. 1.8, Donald Clegg, Champaign, Illinois. **12:** Fig. 1.13, W. H. Freeman photo by Ken Karp. **13:** Fig. 1.14, J. Hester and P. Scowen (Arizona State University)/NASA. **15:** Fig. 1.16, top, Alexander Boden; Fig. 1.17, bottom, Chip Clark. **16:** Fig. 1.18, Chip Clark. **18:** Fig. 1.20, Chip Clark. **20:** Fig. 1.24, Andrew Syred/Science Photo Library/ Photo Researchers. **22:** Fig. 1.27, Chip Clark. **26:** Fig. 1.31 left, center, right, W. H. Freeman photos by Ken Karp. **27:** Fig. 1.32, Peter Kresan. **28:** Fig. 1.33 a, b, W. H. Freeman photos by Ken Karp; Fig. 1.33 c, Kristen Brochmann/Fundamental Photographs; Fig. 1.34, Richard Megna/Fundamental Photographs. **29:** Fig. 1.35, Chip Clark. **30:** Fig. 1.38, W. H. Freeman photo by Ken Karp. **32:** Case Study 1, Randi Anglin/Image Works. **33:** Case Study 1, Ancore Corporation. **35:** Fig. 1.41, Paul Silverman/ Fundamental Photographs.

Chapter 2

48: Mark A. Johnson/Stock Market. **50:** Fig. 2.1, W. H. Freeman photo by Ken Karp. **51:** Fig. 2.2, Bureau International Des Poids et Mesures. **53:** Fig. 2.3, Houston Museum of Natural Science. **65:** Fig. 2.12, Burndy Library. **66:** Case Study 2, Kenneth L. Rinehart, University of Illinois at Urbana-Champaign. **67:** Case Study 2, Visuals Unlimited. **72:** Fig. 2.15, Chip Clark. **73:** Fig. 2.16, W. H. Freeman photo by Ken Karp. **74:** Fig. 2.17, Chip Clark. **77:** Fig. 2.18, Mark Gibson/Visuals Unlimited. **81:** Fig. 2.19, Michael Newmann/Photo Edit. **86:** Exercise 2.95, Fred Mcnnaughey/Photo Researchers.

Chapter 3

88: John Cancalosi/Stock Boston. **91:** Fig. 3.1 b, Chip Clark. **95:** Fig. 3.5 b, Pacific Gas and Electric. **96:** Fig. 3.6, Richard Megna/ Fundamental Photographs; Fig. 3.7 a, b, W. H. Freeman photos by Ken Karp. **97:** Fig. 3.9 a, b, c, Fundamental Photographs. **98:** Fig. 3.11, Richard Megna/Fundamental Photographs. **99:** Fig. 3.14, Chip Clark. **100:** Fig. 3.15, Chip Clark. **103:** Fig. 3.17, Chip Clark. **104:** Fig. 3.18, Ken Reid/FPG International. **105:** Fig. 3.19, W. H. Freeman photo by Ken Karp. **107:** Fig. 3.20, W. H. Freeman photo by Ken Karp. **108:** Fig. 3.21, Grant Heilman Photography. **110:** Fig. 3.22, Michael Dalton/Fundamental Photographs. **114:** Fig. 3.24 a, left, Chip Clark; b, W. H. Freeman photo by Ken Karp. **115:** Fig. 3.25, Chip Clark. **116:** Fig. 3.26, W. H. Freeman photo by Ken Karp. **117:** Fig. 3.27, W. H. Freeman photo by Ken Karp. **121:** Fig. 3.30 a, W. H. Freeman photo by Ken Karp; Fig. 3.31 a,b, W. H. Freeman photo by Ken Karp. **123:** Fig. 3.32, W. H. Freeman photo by Ken Karp. **124:** Case Study 3, NASA.

Chapter 4

136: Tony Stone Images. **138:** Fig. 4.1, Air Products and Chemicals, Inc. **143:** Fig. 4.5, Richard Megna/Fundamental Photographs. **147:** Case Study 4, left, right, Lonnie G. Thompson, Byrd Polar Resesarch Center, Ohio State University. **154:** Fig. 4.9, Air Products and Chemicals, Inc. **155:** Fig. 4.10 a,b,c, W. H. Freeman photo by Ken Karp. **160:** Fig. 4.14, W. H. Freeman photo by Ken Karp. **161:** Fig. 4.15, W. H. Freeman photo by Ken Karp. **164:** Fig. 4.18 a,b, Richard Megna/ Fundamental Photographs. **167:** Exercise 4.13, Hank Morgan/ Photo Researchers. **171:** Exercise 4.74, Westinghouse/Visuals Unlimited. **172:** Exercise 4.80, Rosenfeld Images LTD/Science Photo Library/Photo Researchers.

Connection 1

174: James Prince/ Photo Researchers; inset, Gary Retherford/Science Source/ Photo Researchers.

Chapter 5

176: Garry McMichael/Photo Researchers. **178:** Fig. 5.1, NASA. **181:** Fig. 5.6, Peter Ginter/Bilderberg. **191:** Fig. 5.15, W. H. Freeman photo by Ken Karp. **194:** Fig. 5.18, U.S. Department of the Interior, Bureau of Mines. **195:** Fig. 5.19, TRW Inc; Fig. 5.20, Craig Strong/Liaison. **203:** Case Study 5, top, left, NASA/Science Source/Photo Researchers; bottom, right Philippe Plailly/Science Photo Library/Photo Researchers **206:** Box 5.1, John Bazemore/ AP/Wide World Photos. **208:** Fig. 5.29, W. H. Freeman photo by Ken Karp. **212:** Exercise 5.45, Richard Megna/Fundamental Photographs. **216:** Exercise 5.114, Charles Thatcher/Tony Stone Images.

Chapter 6

218: Richard Thom/Visuals Unlimited. **222:** Fig. 6.5, NASA. **227:** Fig. 6.12, The Granger Collection, New York. **231:** Fig. 6.15, W. H. Freeman photo by Ken Karp. **232:** Fig. 6.18, W. H. Freeman photo by Ken Karp. **234:** Fig. 6.20, Kathy McLaughlin/ Image Works. **239:** Fig. 6.25, USGS, Flagstaff, Arizona. **242:** Fig. 6.28, Warren Gretz/NREL (National Renewable Energy Laboratory). **244:** Box 6.1, Chevron Corporation. **245:** Box 6.1, Martin Rogers/FPG International. **251:** Fig. 6.33, Imperial War Museum. **252:** Fig. 6.34, Catherine Pouedras/Eurelios/S P L/ Photo Researchers. **254:** Case Study 6, Arthur Tilley/Tony Stone Images. **264:** Exercise 6.105, Richard Megna/Fundamental Photographs.

Chapter 7

266: Anglo-Australian Observatory. **268:** Fig. 7.1b, David Olsen/ Tony Stone Images. **270:** Fig. 7.3b, Century Lubricants Specialists. **274:** Fig. 7.8, Chip Clark. **277:** Fig. 7.13, Science Museum, London. **278:** Fig. 7.15, UPI/Corbis-Bettmann. **287:** Fig. 7.26 a-d, W. H. Freeman photo by Ken Karp. **290:** Case Study 7, John Gillmoure/Stock Market. **298:** Box 7.2, Edgar Fahs Smith Collection, ACS Center for History of Chemistry, University of Pennsylvania. **300:** Fig. 7.32, W. H. Freeman photo by Ken Karp; Fig. 7.33, Hank Morgan/Science Source/Photo Researchers. **308:** Fig. 7.44, Chip Clark. **309:** Fig. 7.46, Chip

Chapter 17

754: Paul Harris/Tony Stone Images. 756: Fig. 17.1, top & bottom, W. H. Freeman photo by Ken Karp. 757: Fig. 17.2, W. H. Freeman photo by Ken Karp. 760: Box 17.1, Dieter Flamm. 769: Fig. 17.11, Yale University. 772: Case Study 17, Runk/Schoenberger/Grant Heilman Photography.

Chapter 18

790: Esbin/Andreson/Omni-Photo Communications. 798: Fig. 18.3 a,b, Chip Clark. 799: Fig. 18.6, , W. H. Freeman photo by Ken Karp. 800: Fig. 18.7, W. H. Freeman photo by Ken Karp. 807: Fig. 18.15, W. H. Freeman photo by Ken Karp. 816: Box 18.1, left, & right, W. H. Freeman photo by Ken Karp. 819: Fig. 18.20, Ken Lucas/Planet Earth Pictures. 820: Case Study 18, NASA 821: Case Study 18, Courtesy Ballard Power Systems, Inc., Burnaby, B.C., Canada. 822: Fig. 18.21, W. H. Freeman photo by Ken Karp. 823: Fig. 18.23, St. Joe Zinc Company/American Hot Dip Galvanizers Association. 827: Fig. 18.27, Corbis/Bettmann. 830: Fig. 18.31, Comstock. 839: Exercise 18.118, Warren Gretz/NREL(National Renewable Energy Laboratory).

Connection 5

840: General Motors Corporation. 841: Takeshi Takahara/Photo Researchers.

Chapter 19

842: Warren Salowe, Long Island City, New York. 847: Fig. 19.6, W. H. Freeman photo by Ken Karp. 848: Fig. 19.8, W. H. Freeman photo by Ken Karp. 850: Fig. 19.9 a-d, W. H. Freeman photo by Ken Karp. 852: Fig. 19.12 a-c, W. H. Freeman photo by Ken Karp; Fig. 19.13, P. P. Edwards, University of Birmingham. 853: Fig. 19.14, W. H. Freeman photo by Ken Karp; Fig. 19.15, Duracell International, Inc. 854: Fig. 19.16, Texasgulf; Fig. 19.18, Photo Researchers. 855: Fig. 19.19, Bill Tronca/Tom Stack & Associates. 856: Fig. 19.20 a-e, W. H. Freeman photo by Ken Karp. 857: Fig. 19.21, Houston Museum of Natural Science; Fig. 19.22, Dow Chemical. 858: Fig. 19.23, Alexander Boden. 860: Fig. 19.25, Travis Amos; Fig. 19.26, Field Museum of Natural History, Chicago. 861: Fig. 19.27, Aalborg Portland Betonforskningslaboratorium, Karlslunde. 862: Fig. 19.28, U.S. Borax. 864: Fig. 19.31 a, Houston Museum of Natural Science; b, Lee Boltin Picture Library; c, Houston Museum of Natural Science; Fig. 19.32, Chip Clark. 866: Fig. 19.34, W. H. Freeman photo by Ken Karp; Fig. 19.35, Viacom Consumer Products/Paramount Pictures Corporation. 868: Fig. 19.36, Chip Clark. 871: Fig. 19.38, Max Planck/Institute fur Kenphysik; Fig. 19.39, from "Carbon Nanotube Chemistry," by Jessica Cook, Jeremy Sloan, and Malcolm L. H. Green, August 10, 1996, *Chemistry & Industry,* and "Microscopy in the Study of Fullerene-Related Carbons," by Peter J. F. Harris, September 1994, *Microscopy and Analysis.* 872: Fig. 19.40 a-c, Field Museum of Natural History, Chicago. 873: Fig. 19.42, W. H. Freeman photo by Ken Karp. 875: Fig. 19.47, Chip Clark; Fig. 19.48, Chip Clark. 876: Case Study 19, left, Corning Glass; right, The Corning Museum of Glass; Fig. 19.49, Phillip Hayson/Photo Researchers. 877: Case Study 19, The Corning Museum of Glass. 878: Fig. 19.51, Chip Clark. 883: Exercise 19.98, Ken Kay/Fundamental Photographs.

Chapter 20

884: Hank Morgan/Photo Researchers. 887: Fig. 20.1, Chip Clark; Fig. 20.2, Photo Researchers. 888: Fig. 20.3, Chip Clark. 889: Fig. 20.5, Chip Clark; Fig. 20.4, W. H. Freeman photo by Ken Karp 890: Fig. 20.6, Chip Clark. 891: Fig. 20.7, W. H. Freeman photo by Ken Karp. 894: Fig. 20.8, Chip Clark; Fig. 20.9, Ross Chapple. 895: Fig. 20.10, Chip Clark; Fig. 20.11 a, Chip Clark; b, National Park Services. 896: Fig. 20.12, Chip Clark. 897: Fig. 20.15, W. H. Freeman photo by Ken Karp. 898: Fig. 20.16, Lee Boltin Picture Library. 899: Fig. 20.17, W. H. Freeman photo by Ken Karp; Fig. 20.18, W. H. Freeman photo by Ken Karp. 900: Fig. 20.19 a-c, Chip Clark. 902: Fig. 20.22, Martin Marietta Energy Systems; Fig. 20.23, Chip Clark. 903: Fig. 20.24, Chip Clark; Fig. 20.25, Chip Clark. 905: Fig. 20.26 a,b, W. H. Freeman photo by Ken Karp. 908: Case Study 20, Morton Thiokol. 909: Case Study 20, left & right, NASA. 910: Fig. 20.27, Greater Pittsburgh Neon/Tom Anthony. 911: Fig. 20.29, Argonne National Laboratory.

Chapter 21

916: Chris Baker/Tony Stone Images. 921: Fig. 21.8 a, Field Museum of Natural History, Chicago; b, Hagley Museum and Library; c, Oremet Titanium. 923: Fig. 21.9, Tom Carroll/FPG International. 924: Fig. 21.10, W. H. Freeman photo by Ken Karp; Fig. 21.11 , W. H. Freeman photo by Ken Karp. 925: Fig. 21.12, Institute of Oceanographic Sciences/NERC/SPL/Photo Researchers. 927: Fig. 21.14, U.S. Steel. 929: Fig. 21.15, Chip Clark; Fig. 21.17, Paul Chesley/Tony Stone Images. 930: Fig. 21.18, Lee Boltin Picture Library. 930: Fig. 21.19, Field Museum of Natural History, Chicago. 931: Fig. 21.20, Thierry Borredon/ Tony Stone Images. 932: Case Study 21, a, b, F. S. Judd, Research Laboratories, Kodak Ltd. 933: Case Study 21, Edward Keating. 934: Fig. 21.21, W. H. Freeman photo by Ken Karp. 935: Fig. 21.22, W. H. Freeman photo by Ken Karp. 937: Fig. 21.24, Carol Moralejo, Concordia University. 947: Fig. 21.42, Carol Moralejo, Concordia University. 956: Exercise 21.91, Andy Levin/Photo Researchers.

Chapter 22

958: Solar and Astrophysics Laboratory of the Lockheed-Martin Advanced Technology Center. 960: Fig. 22.1, The Granger Collection, New York; Fig. 22.2, The Granger Collection, New York. 969: Fig. 22.15, Fermilab. 971: Fig. 22.16, U.S. Department of Energy. 972: Case Study 22, Marcus E. Raichle and Washington University School of Medicine. 973: Case Study 22, adapted from "Positron Emission Tomographic Studies of the Cortical Anatomy of Single Word Processing," by S. E. Peterson, P. T. Fox, M. I. Posner, M. A. Mintum, M. E. Raichle, 1988, *Nature,* 331:585-589. 974: Fig. 22.17, Yoav Levy/Phototake NYC. 975: Fig. 22.18 a, Chip Clark; Fig. 22.19, Mitch Kezar/Phototake NYC. 979: Fig. 22.22, University of Arizona. 980: Box 22.1, left, CNRI/Phototake NYC; right, U.S. Department of Energy/SPL/ Photo Researchers. 987: Fig. 22.28, Tom Carroll/Phototake NYC. 988: Fig. 22.29, Princeton Plasma Physics Laboratory. 989: Fig. 22.30, Chip Clark; Fig. 22.31, Martin Marietta Energy Systems, Oak Ridge, Tennessee. 990: Fig. 22.32, Phil Schofield; Fig. 22.33, U.S. Department of Energy; Fig. 22.34, Photo Researchers. 993: Exercise 22.47, Georges Ollen/Sygma. 995: Exercise 22.81, David Parker/Science Photo Library/Photo Researchers.

Connection 6

996: Left, NASA; right, Michael Carroll, Littleton, Colorado. 997: Left & right, Michael Carroll, Littleton, Colorado.

Index

INDEX

THE ELEMENTS

Element	Symbol	Atomic number	Molar mass, g/mol	Element	Symbol	Atomic number	Molar mass, g/mol
Actinium	Ac	89	227.03	Mendelevium	Md	101	258.10
Aluminum	Al	13	26.98	Mercury	Hg	80	200.59
Americium	Am	95	241.06	Molybdenum	Mo	42	95.94
Antimony	Sb	51	121.75	Neodymium	Nd	60	144.24
Argon	Ar	18	39.95	Neon	Ne	10	20.18
Arsenic	As	33	74.92	Neptunium	Np	93	237.05
Astatine	At	85	210	Nickel	Ni	28	58.71
Barium	Ba	56	137.34	Niobium	Nb	41	92.91
Berkelium	Bk	97	249.08	Nitrogen	N	7	14.01
Beryllium	Be	4	9.01	Nobelium	No	102	255
Bismuth	Bi	83	208.98	Osmium	Os	76	190.2
Bohrium	Bh	107	262.12	Oxygen	O	8	16.00
Boron	B	5	10.81	Palladium	Pd	46	106.4
Bromine	Br	35	79.91	Phosphorus	P	15	30.97
Cadmium	Cd	48	112.40	Platinum	Pt	78	195.09
Calcium	Ca	20	40.08	Plutonium	Pu	94	239.05
Californium	Cf	98	251.08	Polonium	Po	84	210
Carbon	C	6	12.01	Potassium	K	19	39.10
Cerium	Ce	58	140.12	Praseodymium	Pr	59	140.91
Cesium	Cs	55	132.91	Promethium	Pm	61	146.92
Chlorine	Cl	17	35.45	Protactinium	Pa	91	231.04
Chromium	Cr	24	52.00	Radium	Ra	88	226.03
Cobalt	Co	27	58.93	Radon	Rn	86	222
Copper	Cu	29	63.54	Rhenium	Re	75	186.2
Curium	Cm	96	247.07	Rhodium	Rh	45	102.91
Dubnium	Db	105	262.11	Rubidium	Rb	37	85.47
Dysprosium	Dy	66	162.50	Ruthenium	Ru	44	101.07
Einsteinium	Es	99	254.09	Rutherfordium	Rf	104	261.11
Erbium	Er	68	167.26	Samarium	Sm	62	150.35
Europium	Eu	63	151.96	Scandium	Sc	21	44.96
Fermium	Fm	100	257.10	Seaborgium	Sg	106	263.12
Fluorine	F	9	19.00	Selenium	Se	34	78.96
Francium	Fr	87	223	Silicon	Si	14	28.09
Gadolinium	Gd	64	157.25	Silver	Ag	47	107.87
Gallium	Ga	31	69.72	Sodium	Na	11	22.99
Germanium	Ge	32	72.59	Strontium	Sr	38	87.62
Gold	Au	79	196.97	Sulfur	S	16	32.06
Hafnium	Hf	72	178.49	Tantalum	Ta	73	180.95
Hassium	Hs	108	265	Technetium	Tc	43	98.91
Helium	He	2	4.00	Tellurium	Te	52	127.60
Holmium	Ho	67	164.93	Terbium	Tb	65	158.92
Hydrogen	H	1	1.0079	Thallium	Tl	81	204.37
Indium	In	49	114.82	Thorium	Th	90	232.04
Iodine	I	53	126.90	Thulium	Tm	69	168.93
Iridium	Ir	77	192.2	Tin	Sn	50	118.69
Iron	Fe	26	55.85	Titanium	Ti	22	47.88
Krypton	Kr	36	83.80	Tungsten	W	74	183.85
Lanthanum	La	57	138.91	Uranium	U	92	238.03
Lawrencium	Lr	103	262.1	Vanadium	V	23	50.94
Lead	Pb	82	207.19	Xenon	Xe	54	131.30
Lithium	Li	3	6.94	Ytterbium	Yb	70	173.04
Lutetium	Lu	71	174.97	Yttrium	Y	39	88.91
Magnesium	Mg	12	24.31	Zinc	Zn	30	65.37
Manganese	Mn	25	54.94	Zirconium	Zr	40	91.22
Meitnerium	Mt	109	266				

FUNDAMENTAL CONSTANTS

Name	Symbol	Value
Atomic mass unit	u	1.66054×10^{-24} g
Avogadro constant	N_A	6.02214×10^{23}/mol
Boltzmann constant	k	1.38066×10^{-23} J/K
Elementary charge	e	1.60218×10^{-19} C
Faraday constant	$F = N_A e$	9.64853×10^4 C/mol
Gas constant	R	8.31451 J/K·mol
		8.31451 L·kPa/K·mol
		8.20578×10^{-2} L·atm/K·mol
		62.3639 L·Torr/K·mol
		8.31451×10^{-2} L·bar/K·mol
Mass of electron	m_e	9.10939×10^{-28} g
Mass of neutron	m_n	1.67493×10^{-24} g
Mass of proton	m_p	1.67262×10^{-24} g
Planck constant	h	6.62608×10^{-34} J·s
	$\hbar = h/2\pi$	1.05457×10^{-34} J·s
Rydberg constant	$\mathcal{R}$	3.28984×10^{15} Hz
Speed of light	c	2.99792×10^8 m/s
Standard acceleration of free fall	g	9.80665 m/s^2

SI PREFIXES

f
femto
10^{-15}

p
pico-
10^{-12}

n
nano-
10^{-9}

μ
micro-
10^{-6}

m
milli-
10^{-3}

c
centi-
10^{-2}

d
deci-
10^{-1}

da
deca-
10

h
hecto-
10^2

k
kilo-
10^3

M
mega-
10^6

G
giga-
10^9

T
tera-
10^{12}

RELATIONS BETWEEN UNITS*

Property	Common unit	SI unit
Mass	2.205 lb (lb = pound)	1.000 kg
	1.000 lb	453.6 g
	1.000 oz (oz = ounce)	28.35 g
	1.000 ton (= 2000 lb)	907.2 kg
	1 t (t = tonne, metric ton)	**10^3 kg**
Length	1.094 yd (yd = yard)	1.000 m
	0.3937 in. (in. = inch)	1.000 cm
	0.6214 mi (mi = mile)	1.000 km
	1 in.	**2.54 cm**
	1 ft (ft = foot)	**30.48 cm**
	1.000 yd	0.9144 m
	1 Å (Å = ångström)	**10^{-10} m**
Volume	**1 L (L = liter)**	**10^3 cm^3, 10^{-3} m^3**
	1.000 gal (gal = gallon)[†]	3.785×10^3 cm^3 (3.785 L)
	1.00 ft^3 (ft^3 = cubic foot)	2.83×10^{-2} m^3 (28.3 L)
	1.00 qt (qt = quart)[†]	9.46×10^2 cm^3 (0.946 L)
Time	**1 min (min = minute)**	**60 s**
	1 h (h = hour)	**3600 s**
	1 day	**86400 s**
Pressure	**1 atm (atm = atmosphere)**	**1.01325×10^5 Pa**
	1.000 Torr or 1.000 mmHg	133.3 Pa
	1.000 psi (psi = pounds per square inch)	6.895×10^3 Pa
	1 bar	**10^5 Pa**
Energy	**1 cal**	**4.184 J**
	1 eV	1.6022×10^{-19} J; 96.485 kJ/mol
	1 C·V	**1 J**
	1 kWh	**3.600×10^3 kJ**
	1 L·atm	**101.325 J**
Temperature conversions	(Fahrenheit temperature)/°F = $\frac{9}{5} \times$ (Celsius temperature)/°C **+ 32**	
	(Celsius temperature)/°C = $\frac{5}{9} \times$ {(Fahrenheit temperature)/°F **− 32**}	
	(Kelvin temperature)/K = (Celsius temperature)/°C **+ 273.15**	

*Entries in bold type are exact. All numbers in the temperature conversion formulas are exact.
[†]The European and Canadian Imperial quart and gallon are 1.201 times larger.